Properties of Real Numbers

$$a + b = b + a$$
$$ab = ba$$
$$a + (b + c) = (a + b) + c$$
$$a(bc) = (ab)c$$
$$a(b + c) = ab + ac$$
$$a(b - c) = ab - ac$$
$$(a + b)c = ac + bc$$
$$(a - b)c = ac - bc$$
$$a + 0 = a$$
$$a \cdot 0 = 0$$
$$a \cdot 1 = a$$
$$a + (-a) = 0$$
$$-(-a) = a$$
$$(-1)a = -a$$
$$a - b = a + (-b)$$
$$a - (-b) = a + b$$
$$a\left(\frac{1}{a}\right) = 1$$
$$\frac{a}{b} = a \cdot \frac{1}{b}$$
$$(-a)b = -(ab) = a(-b)$$
$$(-a)(-b) = ab$$
$$\frac{-a}{-b} = \frac{a}{b}$$
$$\frac{-a}{b} = -\frac{a}{b} = \frac{a}{-b}$$
$$\frac{a}{c} + \frac{b}{c} = \frac{a + b}{c}$$
$$\frac{a}{c} - \frac{b}{c} = \frac{a - b}{c}$$
$$\frac{a}{b} \cdot \frac{c}{d} = \frac{ac}{bd}$$
$$\frac{a/b}{c/d} = \frac{ad}{bc}$$
$$\frac{a}{b} = \frac{ac}{bc}$$

Exponents

$$a^0 = 1 \qquad (a \neq 0)$$
$$a^{-n} = \frac{1}{a^n} \qquad (a \neq 0)$$
$$a^m a^n = a^{m+n}$$
$$(a^m)^n = a^{mn}$$
$$(ab)^n = a^n b^n$$
$$\left(\frac{a}{b}\right)^n = \frac{a^n}{b^n}$$
$$\frac{a^m}{a^n} = a^{m-n}$$

Radicals

$$\sqrt[n]{a} = a^{1/n}$$
$$(\sqrt[n]{a})^n = a, \sqrt[n]{a^n} = a \qquad (a > 0)$$
$$\sqrt[n]{a^m} = (\sqrt[n]{a})^m = a^{m/n}$$
$$\sqrt[n]{ab} = \sqrt[n]{a}\sqrt[n]{b}$$
$$\sqrt[n]{\frac{a}{b}} = \frac{\sqrt[n]{a}}{\sqrt[n]{b}}$$
$$\sqrt[m]{\sqrt[n]{a}} = \sqrt[mn]{a}$$

Special Products

$$x(y + z) = xy + xz$$
$$(x + a)(x + b) = x^2 + (a + b)x + ab$$
$$(x + a)^2 = x^2 + 2ax + a^2$$
$$(x - a)^2 = x^2 - 2ax + a^2$$
$$(x + a)(x - a) = x^2 - a^2$$
$$(x + a)^3 = x^3 + 3ax^2 + 3a^2x + a^3$$
$$(x - a)^3 = x^3 - 3ax^2 + 3a^2x - a^3$$

Factoring Formulas

$$ab + ac = a(b + c)$$
$$a^2 - b^2 = (a + b)(a - b)$$
$$a^2 + 2ab + b^2 = (a + b)^2$$
$$a^2 - 2ab + b^2 = (a - b)^2$$
$$a^3 + b^3 = (a + b)(a^2 - ab + b^2)$$
$$a^3 - b^3 = (a - b)(a^2 + ab + b^2)$$

Quadratic Formula

If $ax^2 + bx + c = 0$, where $a \neq 0$, then

$$x = \frac{-b \pm \sqrt{b^2 - 4ac}}{2a}.$$

Straight Lines

$m = \dfrac{y_2 - y_1}{x_2 - x_1}$	(slope formula)
$y - y_1 = m(x - x_1)$	(point-slope form)
$y = mx + b$	(slope-intercept form)
$x = \text{constant}$	(vertical line)
$y = \text{constant}$	(horizontal line)

Inequalities

If $a < b$, then $a + c < b + c$.

If $a < b$ and $c > 0$, then $ac < bc$.

If $a < b$ and $c > 0$, then $a(-c) > b(-c)$.

Logarithms

$\log_b x = y$ where $x = b^y$

$\log_b(mn) = \log_b m + \log_b n$

$\log_b \dfrac{m}{n} = \log_b m - \log_b n$

$\log_b m^r = r \log_b m$

$\log_b 1 = 0$

$\log_b b = 1$

$\log_b b^r = r$

$b^{\log_b m} = m$

$\log_b m = \dfrac{\log_a m}{\log_a b}$

Counting

$${}_nP_r = \frac{n!}{(n-r)!}$$

$${}_nC_r = \frac{n!}{r!(n-r)!}$$

Greek Alphabet

alpha	A	α	nu	N	ν
beta	B	β	xi	Ξ	ξ
gamma	Γ	γ	omicron	O	o
delta	Δ	δ	pi	Π	π
epsilon	E	ε	rho	P	ρ
zeta	Z	ζ	sigma	Σ	σ
eta	H	η	tau	T	τ
theta	Θ	θ	upsilon	Υ	υ
iota	I	ι	phi	Φ	$\phi\varphi$
kappa	K	κ	chi	X	χ
lambda	Λ	λ	psi	Ψ	ψ
mu	M	μ	omega	Ω	ω

ELEVENTH EDITION

Introductory Mathematical Analysis

For Business, Economics, and the Life and Social Sciences

Ernest F. Haeussler, Jr.
The Pennsylvania State University

Richard S. Paul
The Pennsylvania State University

R. J. Wood
Dalhousie University

PEARSON
Prentice Hall

Pearson Education International

CRL
515 HAE
7 city

Executive Acquisitions Editor: *Petra Recter*
Editor in Chief: *Sally Yagan*
Project Manager: *Jacquelyn M. Riotto*
Production Editor: *Interactive Composition Corporation*
Senior Managing Editor: *Linda Mihatov Behrens*
Executive Managing Editor: *Kathleen Schiaparelli*
Vice President/Director of Production and Manufacturing: *David W. Riccardi*
Assistant Manufacturing Manager/Buyer: *Michael Bell*
Manufacturing Manager: *Trudy Pisciotti*
Marketing Manager: *Krista M. Bettino*
Marketing Assistant: *Annett Uebel*
Editorial Assistant/Print Supplements Editor: *Joanne Wendelken*
Art Director: *Maureen Eide*
Interior/Cover Designer: *Wanda Espana*
Art Editor: *Thomas Benfatti*
Creative Director: *Carole Anson*
Director of Creative Services: *Paul Belfanti*
Cover Photo: *AP Wide World Photos*
Photoresearch:
Director, Image Resource Center: *Melinda Reo*
Manager, Rights and Permissions: *Zina Arabia*
Manager, Visual Research: *Beth Brenzel*
Manager, Cover Visual Research & Permissions: *Karen Sanatar*
Image Permission Coordinator: *Nancy Seise*
Photo Researcher: *Diana Gongora*
Art Studio: *Laserwords*

PEARSON Prentice Hall

Pearson Prentice Hall
Pearson Education, Inc.
Upper Saddle River, NJ 07458

Printed in the United States of America

10 9 8 7 6 5 4 3 2 1

ISBN: 0-13-127629-8

Pearson Education Ltd.
Pearson Education Australia Pty. Ltd.
Pearson Education Singapore, Pte. Ltd.
Pearson Education North Asia Ltd.
Pearson Education Canada, Inc.
Pearson Educación de Mexico, S.A. de C.V.
Pearson Education—Japan
Pearson Education Malaysia, Pte. Ltd.
Pearson Education, Inc., Upper Saddle River, New Jersey

CONTENTS

For Allison

PREFACE

The eleventh edition of *Introductory Mathematical Analysis* continues to provide a mathematical foundation for students in business, economics, and the life and social sciences. It begins with noncalculus topics such as equations, functions, matrix algebra, linear programming, mathematics of finance, and probability. Then it progresses through both single-variable and multivariable calculus, including continuous random variables. Technical proofs, conditions, and the like are sufficiently described but are not overdone. Our guiding philosophy led us to include those proofs and general calculations that shed light on how the corresponding calculations are done in applied problems. At times, informal intuitive arguments are given to preserve clarity.

Applications

An abundance and variety of applications for the intended audience appear throughout the book; students continually see how the mathematics they are learning can be used. These applications cover such diverse areas as business, economics, biology, medicine, sociology, psychology, ecology, statistics, earth science, and archaeology. Many of these real-world situations are drawn from literature and are documented by references. In some, the background and context are given in order to stimulate interest. However, the text is virtually self-contained, in the sense that it assumes no prior exposure to the concepts on which the applications are based.

Organization Changes to the Eleventh Edition

The material from earlier editions has been rearranged to reflect what we understand to be the usage patterns of adoptees. Chapter 0 (Algebra Refresher) has been expanded and subsumes what was formerly Chapter 1 (Equations). It seems useful to us to teach the Mathematics of Finance, now Chapter 5, immediately after students have become aquainted (or reaquainted) with the Exponential and Logaithmic Functions in Chapter 4.

The chapters on differentiation have been rearranged to provide more unified themes. For example, the section on Elasticity of Demand has been moved to Chapter 12 (Additional Differentiation Topics) where it is placed immediately before Implicit Differentiation, which in turn is followed by Logarithmic Differentiation. All three of these topics have a similar flavour. Since Applications are stressed throughout it was decided to move the topics that formerly appeared in a Chapter titled Applications of Differentiation so as to reinforce the applicability of those topics to which they are mathematically related. In particular, Applied Maxima and Minima now provides the conclusion to Chapter 13 (Curve Sketching).

Chapter 14 (Integration) and Chapter 15 (Methods and Applications of Integration) have also been rearranged as a unit, beginning with the section on Differentials. Approximate Integration (14.9) now follows the definition of the Definite Integral as a limit of sums (14.7), sooner than in earlier editions. While the Fundamental Theorem of Calculus (14.8) provides the preferred method of evaluating a definite integral when the requisite antiderivative can be found, it is important for an applied text such as this to stress that the definite integral is a number that can be computed as accurately as one requires using only elementary arithmetic.

Inverse Functions

Earlier editions of the text emphasized solving equations that arise from mathematical modelling. Equations which are given in terms of invertible functions are particularly easy to solve if one has at hand the inverse of the given function. In the eleventh edition there is a new emphasis on inverse functions that is felt at a number of places in the text. Thus, the logarithmic function, base b, is defined as the inverse of the exponential function, base b, and matrix inverses are seen to provide inverses to functions given by multiplication by a fixed matrix. The derivative of the inverse of a function is derived following the motivation provided by the log and exponential pair. The idea is revisited in the work on differentials and, in the exercises, seen yet again with respect to integration.

Simplex Method Displays

In connection with the Simplex Method, many students become confused by what can be a bewildering array of kinds of variables. We now emphasize somewhat more the *'basic variables'*—in particular how they change and how we keep track of them. We have made some fine tuning to the display of the simplex tableaux which we believe will be of benefit to the student.

Combinatorial Foundations

In the combinatorial sections that precede the work on Probability in Chapters 8 and (to a lesser extent) 9 we have taken a somewhat more set-theoretical approach, surely justified by current school curricula. Thus it is stressed that a combination of n objects taken r at a time is precisely an r-element subset of an n-element set. This allows us to provide conceptual derivations for some of the clasical formulas for combinations and to juxtapose these with equational derivations.

Improved Pedagogy

Throughout the text we have adopted the popular "Now work Problem n" approach of other Prentice Hall books. The idea is that after a worked example, students are directed to an end of section problem that reinforces the ideas of the worked example. For the most part, problems led to are odd-numbered so that students can check their answers in the back of the book or find complete solutions in the Student Solutions Manual. In the same vein, we have expanded the use of cautionary warnings to the student. These are represented by *CAUTION* icons that point out commonly made errors. As before, **Definitions** are clearly stated and displayed. Key concepts, as well as important rules and formulas, are boxed to emphasize their importance.

Each chapter (except Chapter 0) has a review section that contains a list of important terms and symbols, a chapter summary, and numerous review problems.

Answers to odd-numbered problems appear at the end of the book. For many of the differentiation problems, the answers appear in both 'unsimplified' and 'simplified' forms. (Of course 'simplified' is in any event a subjective term when applied to mathematical expressions, that tends to presuppose the nature of subsequent calculations with such expressions.) This allows students to readily check their work.

Examples and Exercises

More than 850 examples are worked out in detail. Some include a **strategy** that is specifically designed to guide the student through the logistics of the solution before the solution is obtained. An abundant number of diagrams (almost 500) and exercises (more than 5000) are included. Of the exercises, more than 900 are new for the eleventh edition. In each exercise set, grouped problems are given in increasing order of difficulty. In many exercise sets the problems progress from the basic mechanical-drill type to more interesting thought-provoking problems. Many real-world type problems with real data are included. Considerable effort has been made to produce a proper balance between the drill-type exercises and the problems requiring the integration of the concepts learned.

Applications in Practice

The **Applications in Practice** element provides students with even more applications. Located in the margins of Chapters 1 through 17, these additional exercises give students real-world applications and more opportunities to see the chapter material put into practice.

Answers to Applications in Practice questions appear at the end of the text and complete solutions to these questions are found in the Solutions Manual.

Technology

In order that a student appreciate the value of current **technology,** optional graphing calculator material appears throughout the text both in the exposition and exercises. It appears for a variety of reasons: as a mathematical tool, to visualize a concept, as a computing aid, and to reinforce concepts. Although calculator displays for a TI-83 Plus accompany the corresponding technology discussion, our approach is general enough so that it can be applied to other fine graphing calculators.

In the exercise sets, graphing calculator problems are indicated by an icon. To provide flexibility for an instructor in planning assignments, these problems are usually placed at the end of an exercise set.

Course Planning

Because instructors plan a course outline to serve the individual needs of a particular class and curriculum, we shall not attempt to provide detailed sample outlines. However, depending on the background of the students, some instructors will choose to omit Chapter 0 (Algebra Refresher). A considerable number of courses can be served by the book.

A program that allows three quarters of Mathematics for well-prepared Business students can start a first course with with Chapter 1 and choose such topics as are of interest, up to and including Chapter 9. For example, if the students are concurrently taking a Finance course then it may be desirable to exclude Chapter 5 on the Mathematics of Finance (to avoid duplication of material for distinct credits). Others will find that Chapter 7 on Linear Programming includes more material than their students need. In this case, specific sections such as 7.3, 7.5, and 7.8 can be excluded without loss of continuity. On the other hand, Section 1.1 introduces some business terms, such as total revenue, fixed cost, variable cost and profit that recur throughout the book. Similarly, Section 3.2 introduces the notion of supply and demand equations, and Section 3.6 discusses the equilibrium point and the break-even point, all of fundamental importance for business applications.

A second, single quarter, course on Differential Calculus will use Chapter 10 on Limits and Continuity, followed by the the three differentiation chapters, 11 through 13 inclusive. Here, Section 12.6 on Newton's Method can be omitted without loss of continuity while some instructors may prefer to review Chapter 4 on Exponential and Logarithmic Functions prior to their study as differentiable functions.

Finally, Chapters 14 through 17 inclusive could define a third, single quarter, course on Integral Calculus with an introduction to Multivariable Calculus. In an applied course it is well to stress the use of tables to find integrals and thus the techniques of 'parts' and 'partial fractions', in 15.1 and 15.2 respectively, should be considered optional. Chapter 16 is certainly not needed for Chapter 17 and Section 15.7 on Improper Integrals can be safely omitted if Chapter 16 is not covered.

Schools which have two academic terms per year tend to give business students a term devoted to Finite Mathematics and a term devoted to Calculus. For these schools we recommend Chapters 1 through 9 for the first course, starting wherever the preparation of the students allows, and Chapters 10 through 17 for the second—deleting most optional material.

Supplements

For Instructors

Instructor's Solution Manual. Worked out solutions to all exercises and Applications in Practice problems.

Test Item File. Provides over 1700 test questions, keyed to chapter and section. Includes an editing feature which allows questions to be added or changed.

TestGen. An algorithmic test generator that allows for multiple tests to be created. It is fully editable.

For Students

Student Solutions Manual. Worked out solutions for every odd-numbered exercise and all Applications in Practice problems.

For Instructors and Students

PH Companion Website. Designed to complement and expand upon the text, the PH Companion Website offers a variety of interactive learning tools, including: links to related websites for students. `www.prenhall.com/haeussler`

Acknowledgments

We express our appreciation to the following colleagues who contributed comments and suggestions that were valuable to us in the evolution of this text:

E. Adibi (*Chapman University*); R. M. Alliston (*Pennsylvania State University*); R. A. Alo (*University of Houston*); K. T. Andrews (*Oakland University*); M. N. de Arce (*University of Puerto Rico*); E. Barbut (*University of Idaho*); G. R. Bates (*Western Illinois University*); D. E. Bennett (*Murray State University*); C. Bernett (*Harper College*); A. Bishop (*Western Illinois University*); S. A. Book (*California State University*); A. Brink (*St. Cloud State University*); R. Brown (*York University*); R. W. Brown (*University of Alaska*); S. D. Bulman-Fleming (*Wilfrid Laurier University*); D. Calvetti (*National College*); D. Cameron (*University of Akron*); K. S. Chung (*Kapiolani Community College*); D. N. Clark (*University of Georgia*); E. L. Cohen (*University of Ottawa*); J. Dawson (*Pennsylvania State University*); A. Dollins (*Pennsylvania State University*); G. A. Earles (*St. Cloud State University*); B. H. Edwards (*University of Florida*); J. R. Elliott (*Wilfrid Laurier University*); J. Fitzpatrick (*University of Texas at El Paso*); M. J. Flynn (*Rhode Island Junior College*); G. J. Fuentes (*University of Maine*); S. K. Goel (*Valdosta State University*); G. Goff (*Oklahoma State University*); J. Goldman (*DePaul University*); J. T. Gresser (*Bowling Green State University*); L. Griff (*Pennsylvania State University*); F. H. Hall (*Pennsylvania State University*); V. E. Hanks (*Western Kentucky University*); R. C. Heitmann (*The University of Texas at Austin*); J. N. Henry (*California State University*); W. U. Hodgson (*West Chester State College*); B. C. Horne, Jr. (*Virginia Polytechnic Institute and State University*); J. Hradnansky (*Pennsylvania State University*); C. Hurd (*Pennsylvania State University*); J. A. Jiminez (*Pennsylvania State University*); W. C. Jones (*Western Kentucky University*); R. M. King (*Gettysburg College*); M. M. Kostreva (*University of Maine*); G. A. Kraus (*Gannon University*); J. Kucera (*Washington State University*); M. R. Latina (*Rhode Island Junior College*); P. Lockwood-Cooke (*West Texas A&M University*); J. F. Longman (*Villanova University*); I. Marshak (*Loyola University of Chicago*); D. Mason (*Elmhurst College*); F. B. Mayer (*Mt. San Antonio College*); P. McDougle (*University of Miami*); F. Miles (*California State University*); E. Mohnike (*Mt. San Antonio College*); C. Monk (*University of Richmond*); R. A. Moreland (*Texas Tech University*); J. G. Morris (*University of Wisconsin-Madison*); J. C. Moss (*Paducah Community College*); D. Mullin (*Pennsylvania State University*); E. Nelson (*Pennsylvania State University*); S. A. Nett (*Western Illinois University*); R. H. Oehmke (*University of Iowa*); Y. Y. Oh (*Pennsylvania State University*); J. U. Overall (*University of La Verne*); A. Panayides (*William Patterson

University); D. Parker (*University of Pacific*); N. B. Patterson (*Pennsylvania State University*); V. Pedwaydon (*Lawrence Technical University*); E. Pemberton (*Wilfrid Laurier University*); M. Perkel (*Wright State University*); D. B. Priest (*Harding College*); J. R. Provencio (*University of Texas*); L. R. Pulsinelli (*Western Kentucky University*); M. Racine (*University of Ottawa*); N. M. Rice (*Queen's University*); A. Santiago (*University of Puerto Rico*); J. R. Schaefer (*University of Wisconsin–Milwaukee*); S. Sehgal (*The Ohio State University*); W. H. Seybold, Jr. (*West Chester State College*); G. Shilling (*The University of Texas at Arlington*); S. Singh (*Pennsylvania State University*); L. Small (*Los Angeles Pierce College*); E. Smet (*Huron College*); M. Stoll (*University of South Carolina*); A. Tierman (*Saginaw Valley State University*); B. Toole (*University of Maine*); J. W. Toole (*University of Maine*); D. H. Trahan (*Naval Postgraduate School*); J. P. Tull (*The Ohio State University*); L. O. Vaughan, Jr. (*University of Alabama in Birmingham*); L. A. Vercoe (*Pennsylvania State University*); M. Vuilleumier (*The Ohio State University*); B. K. Waits (*The Ohio State University*); A. Walton (*Virginia Polytechnic Institute and State University*); H. Walum (*The Ohio State University*); E. T. H. Wang (*Wilfrid Laurier University*); A. J. Weidner (*Pennsylvania State University*); L. Weiss (*Pennsylvania State University*); N. A. Weigmann (*California State University*); G. Woods (*The Ohio State University*); C. R. B. Wright (*University of Oregon*); C. Wu (*University of Wisconsin–Milwaukee*).

Some exercises are taken from problem supplements used by students at Wilfrid Laurier University. We wish to extend special thanks to the Department of Mathematics of Wilfrid Laurier University for granting Prentice Hall permission to use and publish this material, and also to thank Prentice Hall, who in turn allowed us to make use of this material.

Finally, we express our sincere gratitude to the faculty and course coordinators of The Ohio State University and Columbus State University who took a keen interest in the eleventh and tenth editions, offering a number of valuable suggestions.

Special thanks are due to Cindy Trimble of C Trimble & Associates for her careful checking of the manuscript solutions manuals, and answer pages. Her work was extraordinarily detailed and helpful to the authors.

Ernest F. Haeussler, Jr.
Richard S. Paul
RJ Wood

CHAPTER 0

Algebra Refresher

Anyone running a business needs to keep track of how well things are going. But how is this done? Financial professionals often measure company performance by calculating fractions called financial ratios. There are more than fifty different financial ratios in common use. Which one gets used depends on whether the analyst is trying to assess a company's growth, its profitability, its level of debt, or some other aspect of its performance.

One important ratio in retail sales is the *inventory turnover ratio*. For a given time period,

$$\text{inventory turnover ratio} = \frac{\text{net sales}}{\text{average inventory}}$$

where inventory is measured in total dollar value at point of sale. When we substitute appropriate expressions for net sales and average inventory, the formula becomes

$$\text{inventory turnover ratio} = \frac{\text{gross sales} - \text{returns and allowances}}{\dfrac{\text{opening inventory} + \text{closing inventory}}{2}}$$

The inventory turnover ratio measures how quickly the retailer's stock of goods is being sold and resupplied: the higher the ratio, the faster the turnover. Too low a ratio means a large inventory in which items sit on the shelf for long periods and are subject to spoilage. Too high a ratio means a small inventory and an associated risk, to the retailer, of either losing sales by running out of items or else having to pay high prices to resupply stock in small lots. The ideal inventory turnover ratio varies from industry to industry, but an ideal annual ratio of 6 is reasonable for a retailer selling durable goods, such as hardware or appliances. A greengrocer's ratio will of course need to be much higher.

The inventory turnover ratio is an example of an algebraic expression. Calculating its value involves substituting real numbers for the variable quantities (gross sales and so on) and performing arithmetic operations (addition, subtraction, and division). This chapter will review real numbers and algebraic expressions and the basic operations on them. It is designed to give you a brief review of some terms and methods of manipulative mathematics. No doubt you have been exposed to much of this material before. However, because these topics are important in handling the mathematics that comes later, an immediate second exposure to them can be beneficial. Devote whatever time is necessary to the sections in which you need review.

OBJECTIVE *To become familiar with sets, the classification of real numbers, and the real-number line.*

0.1 Sets of Real Numbers

In simplest terms, a *set* is a collection of objects. For example, we can speak of the set of even numbers between 5 and 11, namely, 6, 8, and 10. An object in a set is called an *element* or *member* of that set.

One way to specify a set is by listing its elements, in any order, inside braces. For example, the previous set is {6, 8, 10}, which we can denote by a letter such as A. Note that {8, 10, 6} also denotes the same set, as does {10, 10, 6, 8}. A set is determined by its elements, and repetitions in a listing do not affect the set. A set A is said to be a subset of a set B if and only if every element of A is also an element of B. For example, if $A = \{6, 8, 10\}$ and $B = \{6, 8, 10, 12\}$, then A is a subset of B.

Certain sets of numbers have special names. The numbers 1, 2, 3, and so on, form the set of **positive integers** (or **natural numbers**):

$$\textit{set of positive integers} = \{1, 2, 3, \ldots\}$$

The three dots mean that the listing of elements is unending, although we know what the elements are.

The positive integers, together with 0 and the **negative integers** $-1, -2, -3, \ldots,$ form the set of **integers:**

$$\textit{set of integers} = \{\ldots, -3, -2, -1, 0, 1, 2, 3, \ldots\}$$

CAUTION: The reason for $q \neq 0$ is that we cannot divide by zero.

Every integer is a rational number.

The set of **rational numbers** consists of numbers, such as $\frac{1}{2}$ and $\frac{5}{3}$, that can be written as a ratio (quotient) of two integers. That is, a rational number is a number that can be written as p/q, where p and q are integers and $q \neq 0$. (The symbol "$\neq$" is read "is not equal to.") For example, the numbers $\frac{19}{20}$, $\frac{-2}{7}$, and $\frac{-6}{-2}$, are rational. We remark that $\frac{2}{4}$, $\frac{1}{2}$, $\frac{3}{6}$, $\frac{-4}{-8}$, and 0.5 all represent the same rational number. The integer 2 is rational, since $2 = \frac{2}{1}$. In fact, every integer is rational.

All rational numbers can be represented by decimal numbers that *terminate*, such as $\frac{3}{4} = 0.75$ and $\frac{3}{2} = 1.5$, or by *nonterminating repeating decimal numbers* (composed of a group of digits that repeats without end), such as $\frac{2}{3} = 0.666\ldots$, $\frac{-4}{11} = -0.3636\ldots$, and $\frac{2}{15} = 0.1333\ldots$. Numbers represented by *nonterminating nonrepeating* decimals are called **irrational numbers.** An irrational number cannot be written as an integer divided by an integer. The numbers π (pi) and $\sqrt{2}$ are irrational.

The real numbers consist of all decimal numbers.

Together, the rational numbers and irrational numbers form the set of **real numbers.** Real numbers can be represented by points on a line. First we choose a point on the line to represent zero. This point is called the *origin.* (See Figure 0.1.) Then a standard measure of distance, called a "unit distance," is chosen and is successively marked off both to the right and to the left of the origin. With each point on the line we associate a directed distance, or *signed number,* which depends on the position of the point with respect to the origin. Positions to the right of the origin are considered positive (+) and positions to the left are negative (−). For example, with the point $\frac{1}{2}$ unit to the right of the origin there corresponds the signed number $\frac{1}{2}$, which is called the **coordinate** of that point. Similarly, the coordinate of the point 1.5 units to the left of the origin is -1.5. In Figure 0.1, the coordinates of some points are marked. The arrowhead indicates that the direction to the right along the line is considered the positive direction.

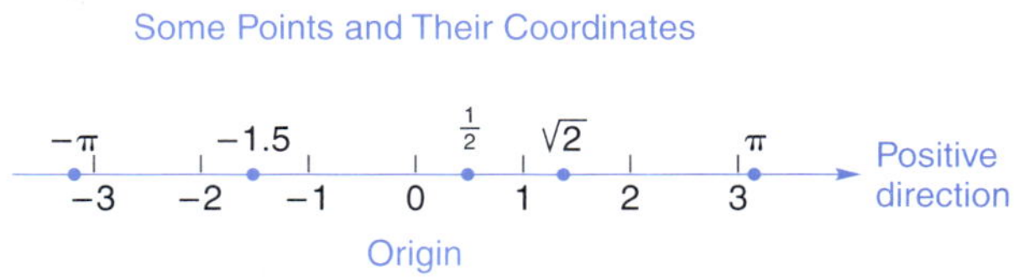

FIGURE 0.1 The real-number line.

To each point on the line there corresponds a unique real number, and to each real number there corresponds a unique point on the line. For this reason, we say that there is a *one-to-one correspondence* between points on the line and real numbers. We call this line a **coordinate line** or the **real-number line.** We feel free to treat real numbers as points on a real-number line and vice versa.

Exercise 0.1

In Problems 1–12, classify the statement as either true or false. If false, give a reason.

1. -13 is an integer.

2. $\frac{1}{6}$ is rational.

3. -3 is a natural number.

4. 0 is not rational.

5. 5 is rational.

6. $\frac{7}{0}$ is a rational number.

7. $\sqrt{25}$ is not a positive integer.

8. $\sqrt{2}$ is a real number.

9. $\frac{0}{6}$ is rational.

10. $\sqrt{3}$ is a natural number.

11. -3 is to the right of -4 on the real-number line.

12. Every integer is positive or negative.

OBJECTIVE *To name, illustrate, and relate properties of the basic operations on real numbers.*

0.2 Some Properties of Real Numbers

We now state a few important properties of the real numbers. Let a, b, and c be real numbers.

1. The Transitive Property of Equality

$$\text{If} \quad a = b \quad \text{and} \quad b = c \quad \text{then} \quad a = c$$

Thus, two numbers that are both equal to a third number are equal to each other. For example, if $x = y$ and $y = 7$, then $x = 7$.

2. The Commutative Properties of Addition and Multiplication

$$a + b = b + a \quad \text{and} \quad ab = ba$$

This means that two numbers can be added or multiplied in any order. For example, $3 + 4 = 4 + 3$ and $7(-4) = (-4)(7)$.

3. The Associative Properties of Addition and Multiplication

$$a+(b+c)=(a+b)+c \quad \text{and} \quad a(bc)=(ab)c$$

This means that in addition or multiplication, numbers can be grouped in any order. For example, $2+(3+4)=(2+3)+4$; in both cases, the sum is 9. Similarly, $2x+(x+y)=(2x+x)+y$ and $6(\frac{1}{3}\cdot 5)=(6\cdot\frac{1}{3})\cdot 5$.

4. The Inverse Properties

For each real number a, there is a unique real number denoted $-a$ such that

$$a+(-a)=0$$

The number $-a$ is called the **additive inverse,** or **negative,** of a.

For example, since $6+(-6)=0$, the additive inverse of 6 is -6. The additive inverse of a number is not necessarily a negative number. For example, the additive inverse of -6 is 6, since $(-6)+(6)=0$. That is, the negative of -6 is 6, so we can write $-(-6)=6$.

For each real number a, **except 0,** there is a unique real number denoted a^{-1} such that

$$a\cdot a^{-1}=1$$

The number a^{-1} is called the **multiplicative inverse** of a.

CAUTION Zero does not have a multiplicative inverse because there is no number that, when multiplied by 0, gives 1.

Thus, all numbers **except 0** have a multiplicative inverse. You may recall that a^{-1} can be written $\frac{1}{a}$ and is also called the *reciprocal* of a. For example, the multiplicative inverse of 3 is $\frac{1}{3}$, since $3(\frac{1}{3})=1$. Hence, $\frac{1}{3}$ is the reciprocal of 3. The reciprocal of $\frac{1}{3}$, is 3, since $(\frac{1}{3})(3)=1$. **The reciprocal of 0 is not defined.**

5. The Distributive Properties

$$a(b+c)=ab+ac \quad \text{and} \quad (b+c)a=ba+ca$$

For example, although $2(3+4)=2(7)=14$, we can write

$$2(3+4)=2(3)+2(4)=6+8=14$$

Similarly,

$$(2+3)(4)=2(4)+3(4)=8+12=20$$

and

$$x(z+4)=x(z)+x(4)=xz+4x$$

The distributive property can be extended to the form

$$a(b+c+d)=ab+ac+ad$$

as is to be shown in Problem 27. In fact, it can be extended to sums involving any number of terms.

Subtraction is defined in terms of addition:

$$a - b \text{ means } a + (-b)$$

where $-b$ is the additive inverse of b. Thus, $6 - 8$ means $6 + (-8)$.

In a similar way, we define **division** in terms of multiplication. If $b \neq 0$, then $a \div b$, or $\frac{a}{b}$, is defined by

$$\frac{a}{b} = a(b^{-1})$$

Since $b^{-1} = \frac{1}{b}$

$$\frac{a}{b} = a(b^{-1}) = a\left(\frac{1}{b}\right)$$

$\frac{a}{b}$ means a times the reciprocal of b.

Thus, $\frac{3}{5}$ means 3 times $\frac{1}{5}$, where $\frac{1}{5}$ is the multiplicative inverse of 5. Sometimes we refer to $a \div b$ or $\frac{a}{b}$ as the *ratio* of a to b. We remark that since 0 does not have a multiplicative inverse, **division by 0 is not defined.**

The following examples show some manipulations involving the preceding properties.

EXAMPLE 1 Applying Properties of Real Numbers

a. $x(y - 3z + 2w) = (y - 3z + 2w)x$, by the commutative property of multiplication.

b. By the associative property of multiplication, $3(4 \cdot 5) = (3 \cdot 4)5$. Thus, the result of multiplying 3 by the product of 4 and 5 is the same as the result of multiplying the product of 3 and 4 by 5. In either case, the result is 60.

▶ **Now work Problem 13.**

EXAMPLE 2 Applying Properties of Real Numbers

a. *Show that* $2 - \sqrt{2} = -\sqrt{2} + 2$.

Solution: By the definition of subtraction, $2 - \sqrt{2} = 2 + (-\sqrt{2})$. However, by the commutative property of addition, $2 + (-\sqrt{2}) = -\sqrt{2} + 2$. Hence, by the transitive property, $2 - \sqrt{2} = -\sqrt{2} + 2$. More concisely, we omit intermediate steps and directly write

$$2 - \sqrt{2} = -\sqrt{2} + 2$$

b. *Show that* $(8 + x) - y = 8 + (x - y)$.

Solution: Beginning with the left side, we have

$$\begin{aligned}(8 + x) - y &= (8 + x) + (-y) && \text{(definition of subtraction)}\\ &= 8 + [x + (-y)] && \text{(associative property)}\\ &= 8 + (x - y) && \text{(definition of subtraction)}\end{aligned}$$

Hence, by the transitive property,

$$(8 + x) - y = 8 + (x - y)$$

c. *Show that* $3(4x + 2y + 8) = 12x + 6y + 24$.

Solution: By the distributive property,

$$3(4x + 2y + 8) = 3(4x) + 3(2y) + 3(8)$$

But by the associative property of multiplication,

$$3(4x) = (3 \cdot 4)x = 12x \quad \text{and similarly} \quad 3(2y) = 6y$$

Thus, $3(4x + 2y + 8) = 12x + 6y + 24$

▶ **Now work Problem 21.**

EXAMPLE 3 Applying Properties of Real Numbers

a. *Show that* $\frac{ab}{c} = a\left(\frac{b}{c}\right)$ *for* $c \neq 0$.

Solution: By the definition of division,

$$\frac{ab}{c} = (ab) \cdot \frac{1}{c} \quad \text{for } c \neq 0$$

But by the associative property,

$$(ab) \cdot \frac{1}{c} = a\left(b \cdot \frac{1}{c}\right)$$

However, by the definition of division, $b \cdot \frac{1}{c} = \frac{b}{c}$. Thus,

$$\frac{ab}{c} = a\left(\frac{b}{c}\right)$$

We can also show that $\frac{ab}{c} = \left(\frac{a}{c}\right)b$

b. *Show that* $\frac{a+b}{c} = \frac{a}{c} + \frac{b}{c}$ *for* $c \neq 0$.

Solution: By the definition of division and the distributive property,

$$\frac{a+b}{c} = (a+b)\frac{1}{c} = a \cdot \frac{1}{c} + b \cdot \frac{1}{c}$$

However,

$$a \cdot \frac{1}{c} + b \cdot \frac{1}{c} = \frac{a}{c} + \frac{b}{c}$$

Hence,

$$\frac{a+b}{c} = \frac{a}{c} + \frac{b}{c}$$

▶ **Now work Problem 27.**

Finding the product of several numbers can be done only by considering products of numbers taken two at a time. For example, to find the product of x, y, and z, we could first multiply x by y and then multiply that product by z; that is, we find $(xy)z$. Or alternatively, we could multiply x by the product of y and z; that is, we find $x(yz)$. The associative property of multiplication guarantees that both results are identical, regardless of how the numbers are grouped. Thus, it is not ambiguous to write xyz. This concept can be extended to more than three numbers and applies equally well to addition.

Not only should you be aware of the manipulative aspects of the properties of the real numbers; you should also be aware of, and familiar with, the terminology involved.

The following list states important properties of real numbers that you should study thoroughly. Being able to manipulate real numbers is essential to your success in mathematics. A numerical example follows each property. All denominators are different from zero. A knowledge of addition and subtraction of real numbers is assumed.

Property	*Example(s)*
1. $a - b = a + (-b)$	$2 - 7 = 2 + (-7) = -5$
2. $a - (-b) = a + b$	$2 - (-7) = 2 + 7 = 9$
3. $-a = (-1)(a)$	$-7 = (-1)(7)$
4. $a(b + c) = ab + ac$	$6(7 + 2) = 6 \cdot 7 + 6 \cdot 2 = 54$
5. $a(b - c) = ab - ac$	$6(7 - 2) = 6 \cdot 7 - 6 \cdot 2 = 30$
6. $-(a + b) = -a - b$	$-(7 + 2) = -7 - 2 = -9$
7. $-(a - b) = -a + b$	$-(2 - 7) = -2 + 7 = 5$
8. $-(-a) = a$	$-(-2) = 2$
9. $a(0) = 0$	$2(0) = 0$
10. $(-a)(b) = -(ab) = a(-b)$	$(-2)(7) = -(2 \cdot 7) = 2(-7) = -14$
11. $(-a)(-b) = ab$	$(-2)(-7) = 2 \cdot 7 = 14$
12. $\frac{a}{1} = a$	$\frac{7}{1} = 7, \frac{-2}{1} = -2$
13. $\frac{a}{b} = a\left(\frac{1}{b}\right)$	$\frac{2}{7} = 2\left(\frac{1}{7}\right)$
14. $\frac{a}{-b} = -\frac{a}{b} = \frac{-a}{b}$	$\frac{2}{-7} = -\frac{2}{7} = \frac{-2}{7}$
15. $\frac{-a}{-b} = \frac{a}{b}$	$\frac{-2}{-7} = \frac{2}{7}$
16. $\frac{0}{a} = 0$ when $a \neq 0$	$\frac{0}{7} = 0$
17. $\frac{a}{a} = 1$ when $a \neq 0$	$\frac{2}{2} = 1, \frac{-5}{-5} = 1$
18. $a\left(\frac{b}{a}\right) = b$ when $a \neq 0$	$2\left(\frac{7}{2}\right) = 7$
19. $a \cdot \frac{1}{a} = 1$ when $a \neq 0$	$2 \cdot \frac{1}{2} = 1$
20. $\frac{a}{b} \cdot \frac{c}{d} = \frac{ac}{bd}$	$\frac{2}{3} \cdot \frac{4}{5} = \frac{2 \cdot 4}{3 \cdot 5} = \frac{8}{15}$
21. $\frac{ab}{c} = \left(\frac{a}{c}\right)b = a\left(\frac{b}{c}\right)$	$\frac{2 \cdot 7}{3} = \frac{2}{3} \cdot 7 = 2 \cdot \frac{7}{3}$
22. $\frac{a}{bc} = \left(\frac{a}{b}\right)\left(\frac{1}{c}\right) = \left(\frac{1}{b}\right)\left(\frac{a}{c}\right)$	$\frac{2}{3 \cdot 7} = \frac{2}{3} \cdot \frac{1}{7} = \frac{1}{3} \cdot \frac{2}{7}$

Property	*Example(s)*
23. $\frac{a}{b} = \left(\frac{a}{b}\right)\left(\frac{c}{c}\right) = \frac{ac}{bc}$ when $c \neq 0$	$\frac{2}{7} = \left(\frac{2}{7}\right)\left(\frac{5}{5}\right) = \frac{2 \cdot 5}{7 \cdot 5}$
24. $\frac{a}{b(-c)} = \frac{a}{(-b)(c)} = \frac{-a}{bc} = \frac{-a}{(-b)(-c)} = -\frac{a}{bc}$	$\frac{2}{3(-5)} = \frac{2}{(-3)(5)} = \frac{-2}{3(5)} = \frac{-2}{(-3)(-5)} = -\frac{2}{3(5)} = -\frac{2}{15}$
25. $\frac{a(-b)}{c} = \frac{(-a)b}{c} = \frac{ab}{-c} = \frac{(-a)(-b)}{-c} = -\frac{ab}{c}$	$\frac{2(-3)}{5} = \frac{(-2)(3)}{5} = \frac{2(3)}{-5} = \frac{(-2)(-3)}{-5} = -\frac{2(3)}{5} = -\frac{6}{5}$
26. $\frac{a}{c} + \frac{b}{c} = \frac{a+b}{c}$	$\frac{2}{9} + \frac{3}{9} = \frac{2+3}{9} = \frac{5}{9}$
27. $\frac{a}{c} - \frac{b}{c} = \frac{a-b}{c}$	$\frac{2}{9} - \frac{3}{9} = \frac{2-3}{9} = \frac{-1}{9}$
28. $\frac{a}{b} + \frac{c}{d} = \frac{ad+bc}{bd}$	$\frac{4}{5} + \frac{2}{3} = \frac{4 \cdot 3 + 5 \cdot 2}{5 \cdot 3} = \frac{22}{15}$
29. $\frac{a}{b} - \frac{c}{d} = \frac{ad-bc}{bd}$	$\frac{4}{5} - \frac{2}{3} = \frac{4 \cdot 3 - 5 \cdot 2}{5 \cdot 3} = \frac{2}{15}$
30. $\frac{\frac{a}{b}}{\frac{c}{d}} = \frac{a}{b} \div \frac{c}{d} = \frac{a}{b} \cdot \frac{d}{c} = \frac{ad}{bc}$	$\frac{\frac{2}{3}}{\frac{7}{5}} = \frac{2}{3} \div \frac{7}{5} = \frac{2}{3} \cdot \frac{5}{7} = \frac{2 \cdot 5}{3 \cdot 7} = \frac{10}{21}$
31. $\frac{a}{\frac{b}{c}} = a \div \frac{b}{c} = a \cdot \frac{c}{b} = \frac{ac}{b}$	$\frac{2}{\frac{3}{5}} = 2 \div \frac{3}{5} = 2 \cdot \frac{5}{3} = \frac{2 \cdot 5}{3} = \frac{10}{3}$
32. $\frac{\frac{a}{b}}{c} = \frac{a}{b} \div c = \frac{a}{b} \cdot \frac{1}{c} = \frac{a}{bc}$	$\frac{\frac{2}{3}}{5} = \frac{2}{3} \div 5 = \frac{2}{3} \cdot \frac{1}{5} = \frac{2}{3 \cdot 5} = \frac{2}{15}$

Property 23 is essentially the **fundamental principle of fractions,** which states that *multiplying or dividing both the numerator and the denominator of a fraction by the same nonzero number results in a fraction that is equivalent to (that is, has the same value as) the original fraction.* Thus,

$$\frac{7}{\frac{1}{8}} = \frac{7 \cdot 8}{\frac{1}{8} \cdot 8} = \frac{56}{1} = 56$$

By Properties 28 and 23, we have

$$\frac{2}{5} + \frac{4}{15} = \frac{2 \cdot 15 + 5 \cdot 4}{5 \cdot 15} = \frac{50}{75} = \frac{2 \cdot 25}{3 \cdot 25} = \frac{2}{3}$$

We can also do this problem by converting $\frac{2}{5}$ and $\frac{4}{15}$ into equivalent fractions that have the same denominators and then using Property 26. The fractions $\frac{2}{5}$ and $\frac{4}{15}$ can be written with a common denominator of $5 \cdot 15$:

$$\frac{2}{5} = \frac{2 \cdot 15}{5 \cdot 15} \quad \text{and} \quad \frac{4}{15} = \frac{4 \cdot 5}{15 \cdot 5}$$

However, 15 is the *least* such common denominator and is called the *least common denominator* (LCD) of $\frac{2}{5}$ and $\frac{4}{15}$. Thus,

$$\frac{2}{5}+\frac{4}{15}=\frac{2\cdot 3}{5\cdot 3}+\frac{4}{15}=\frac{6}{15}+\frac{4}{15}=\frac{6+4}{15}=\frac{10}{15}=\frac{2}{3}$$

Similarly,

$$\begin{aligned}\frac{3}{8}-\frac{5}{12}&=\frac{3\cdot 3}{8\cdot 3}-\frac{5\cdot 2}{12\cdot 2} \qquad \text{(LCD = 24)}\\ &=\frac{9}{24}-\frac{10}{24}=\frac{9-10}{24}\\ &=-\frac{1}{24}\end{aligned}$$

Exercise 0.2

In Problems 1–10, classify the statements as either true or false.

1. Every real number has a reciprocal.
2. The reciprocal of $\frac{7}{3}$ is $\frac{3}{7}$.
3. The additive inverse of 5 is $\frac{1}{5}$.
4. $2(3\cdot 4)=(2\cdot 3)(2\cdot 4)$
5. $-x+y=-y+x$
6. $(x+2)(4)=4x+8$
7. $\dfrac{x+2}{2}=\dfrac{x}{2}+1$
8. $3\left(\dfrac{x}{4}\right)=\dfrac{3x}{4}$
9. $x(5\cdot y)=(x5)\cdot(xy)$
10. $8(9x)=72x$

In Problems 11–20, state which properties of the real numbers are being used.

11. $2(x+y)=2x+2y$
12. $(x+5)+y=y+(x+5)$
13. *$2(3y)=(2\cdot 3)y$
14. $\dfrac{5}{11}=\dfrac{1}{11}\cdot 5$
15. $2(x-y)=(x-y)(2)$
16. $y+(x+y)=(y+x)+y$
17. $8-y=8+(-y)$
18. $5(4+7)=5(7+4)$
19. $(8+a)b=8b+ab$
20. $(-1)[-3+4]=(-1)(-3)+(-1)(4)$

In Problems 21–26, show that the statements are true by using properties of the real numbers.

21. *$2x(y-7)=2xy-14x$
22. $(2-x)+y=2+(y-x)$
23. $(x+y)(2)=2x+2y$
24. $2[27+(x+y)]=2[(y+27)+x]$
25. $x[(2y+1)+3]=2xy+4x$
26. $(1+a)(b+c)=b+c+ab+ac$
27. *Show that $a(b+c+d)=ab+ac+ad$. [*Hint:* $b+c+d=(b+c)+d$.]

Simplify each of the following if possible.

28. $-2+(-4)$
29. $-6+2$
30. $6+(-4)$
31. $7-2$
32. $7-(-4)$
33. $-5-(-13)$
34. $-8-(-6)$
35. $(-2)(9)$
36. $7(-9)$
37. $(-2)(-12)$
38. $19(-1)$
39. $-(-9)$
40. $-(-6+x)$
41. $-7(x)$
42. $-12(x-y)$
43. $-[-6+(-y)]$
44. $-3\div 15$
45. $-9\div(-27)$
46. $4\div(-2)$
47. $2(-6+2)$
48. $3[-2(3)+6(2)]$
49. $(-2)(-4)(-1)$
50. $(-12)(-12)$
51. $x(0)$
52. $3(x-4)$
53. $4(5+x)$
54. $-(x-2)$
55. $0(-x)$
56. $8\left(\frac{1}{11}\right)$
57. $\dfrac{5}{1}$
58. $\dfrac{-5x}{7y}$
59. $\dfrac{3}{-2x}$
60. $\dfrac{2}{3}\cdot\dfrac{1}{x}$
61. $\dfrac{a}{c}(3b)$
62. $(5a)\left(\dfrac{7}{5a}\right)$
63. $\dfrac{-18y}{-3x}$
64. $\dfrac{7}{y}\cdot\dfrac{1}{x}$
65. $\dfrac{2}{x}\cdot\dfrac{5}{y}$
66. $\dfrac{1}{2}+\dfrac{1}{3}$
67. $\dfrac{5}{12}+\dfrac{3}{4}$
68. $\dfrac{3}{10}-\dfrac{7}{15}$
69. $\dfrac{4}{5}+\dfrac{6}{5}$
70. $\dfrac{x}{9}-\dfrac{y}{9}$
71. $\dfrac{3}{2}-\dfrac{1}{4}+\dfrac{1}{6}$
72. $\dfrac{2}{5}-\dfrac{3}{8}$
73. $\dfrac{6}{\frac{x}{y}}$
74. $\dfrac{\frac{l}{3}}{m}$
75. $\dfrac{\frac{-5}{4}}{\frac{7}{10}}$
76. $\dfrac{7}{0}$
77. $\dfrac{0}{7}$
78. $\dfrac{0}{0}$
79. $0\cdot 0$

OBJECTIVE *To review positive integral exponents, the zero exponent, negative integral exponents, rational exponents, principal roots, radicals, and the procedure of rationalizing the denominator.*

0.3 Exponents and Radicals

The product $x \cdot x \cdot x$ is abbreviated x^3. In general, for n a positive integer, x^n is the abbreviation for the product of n x's. The letter n in x^n is called the *exponent,* and x is called the *base.* More specifically, if n is a positive integer, we have

1. $x^n = \underbrace{x \cdot x \cdot x \cdot \cdots \cdot x}_{n \text{ factors}}$

2. $x^{-n} = \dfrac{1}{x^n} = \dfrac{1}{\underbrace{x \cdot x \cdot x \cdot \cdots \cdot x}_{n \text{ factors}}}$

3. $\dfrac{1}{x^{-n}} = x^n$

4. $x^0 = 1$ if $x \neq 0$. 0^0 is not defined.

EXAMPLE 1 **Exponents**

a. $\left(\frac{1}{2}\right)^4 = \left(\frac{1}{2}\right)\left(\frac{1}{2}\right)\left(\frac{1}{2}\right)\left(\frac{1}{2}\right) = \frac{1}{16}$

b. $3^{-5} = \frac{1}{3^5} = \frac{1}{3 \cdot 3 \cdot 3 \cdot 3 \cdot 3} = \frac{1}{243}$

c. $\frac{1}{3^{-5}} = 3^5 = 243$

d. $2^0 = 1, \pi^0 = 1, (-5)^0 = 1$

e. $x^1 = x$

▸ Now work Problem 5.

If $r^n = x$, where n is a positive integer, then r is an nth *root* of x. For example, $3^2 = 9$, so 3 is a second root (usually called a *square root*) of 9. Since $(-3)^2 = 9$, -3 is also a square root of 9. Similarly, -2 is a *cube root* of -8, since $(-2)^3 = -8$.

Some numbers do not have an nth root that is a real number. For example, since the square of any real number is nonnegative, there is no real number that is a square root of -4.

The **principal *n*th root** of x is the nth root of x that is positive if x is positive and is negative if x is negative and n is odd. We denote the principal nth root of x by $\sqrt[n]{x}$. Thus,

$$\sqrt[n]{x} \text{ is } \begin{cases} \text{positive if } x \text{ is positive} \\ \text{negative if } x \text{ is negative and } n \text{ is odd} \end{cases}$$

CAUTION Although 2 and -2 are square roots of 4, the **principal** square root of 4 is 2, not -2. Hence, $\sqrt{4} = 2$.

For example, $\sqrt[2]{9} = 3$, $\sqrt[3]{-8} = -2$, and $\sqrt[3]{\frac{1}{27}} = \frac{1}{3}$. We define $\sqrt[n]{0} = 0$.

The symbol $\sqrt[n]{x}$ is called a **radical.** Here n is the *index,* x is the *radicand,* and $\sqrt{\ }$ is the *radical sign.* With principal square roots, we usually omit the index and write $\sqrt{x}$ instead of $\sqrt[2]{x}$. Thus, $\sqrt{9} = 3$.

If x is positive, the expression $x^{p/q}$, where p and q are integers and q is positive, is defined to be $\sqrt[q]{x^p}$. Hence,

$$x^{3/4} = \sqrt[4]{x^3}; \quad 8^{2/3} = \sqrt[3]{8^2} = \sqrt[3]{64} = 4$$

$$4^{-1/2} = \sqrt[2]{4^{-1}} = \sqrt{\frac{1}{4}} = \frac{1}{2}$$

Here are the basic laws of exponents and radicals:[1]

Law	*Example(s)*
1. $x^m \cdot x^n = x^{m+n}$	$2^3 \cdot 2^5 = 2^8 = 256;\ x^2 \cdot x^3 = x^5$
2. $x^0 = 1$ if $x \neq 0$	$2^0 = 1$
3. $x^{-n} = \dfrac{1}{x^n}$	$2^{-3} = \dfrac{1}{2^3} = \dfrac{1}{8}$
4. $\dfrac{1}{x^{-n}} = x^n$	$\dfrac{1}{2^{-3}} = 2^3 = 8;\ \dfrac{1}{x^{-5}} = x^5$
5. $\dfrac{x^m}{x^n} = x^{m-n} = \dfrac{1}{x^{n-m}}$	$\dfrac{2^{12}}{2^8} = 2^4 = 16;\ \dfrac{x^8}{x^{12}} = \dfrac{1}{x^4}$
6. $\dfrac{x^m}{x^m} = 1$	$\dfrac{2^4}{2^4} = 1$
7. $(x^m)^n = x^{mn}$	$(2^3)^5 = 2^{15};\ (x^2)^3 = x^6$
8. $(xy)^n = x^n y^n$	$(2 \cdot 4)^3 = 2^3 \cdot 4^3 = 8 \cdot 64 = 512$
9. $\left(\dfrac{x}{y}\right)^n = \dfrac{x^n}{y^n}$	$\left(\dfrac{2}{3}\right)^3 = \dfrac{2^3}{3^3};\ \left(\dfrac{1}{3}\right)^5 = \dfrac{1^5}{3^5} = \dfrac{1}{3^5} = 3^{-5}$
10. $\left(\dfrac{x}{y}\right)^{-n} = \left(\dfrac{y}{x}\right)^n$	$\left(\dfrac{3}{4}\right)^{-2} = \left(\dfrac{4}{3}\right)^2 = \dfrac{16}{9}$
11. $x^{1/n} = \sqrt[n]{x}$	$3^{1/5} = \sqrt[5]{3}$
12. $x^{-1/n} = \dfrac{1}{x^{1/n}} = \dfrac{1}{\sqrt[n]{x}}$	$4^{-1/2} = \dfrac{1}{4^{1/2}} = \dfrac{1}{\sqrt{4}} = \dfrac{1}{2}$
13. $\sqrt[n]{x}\sqrt[n]{y} = \sqrt[n]{xy}$	$\sqrt[3]{9}\sqrt[3]{2} = \sqrt[3]{18}$
14. $\dfrac{\sqrt[n]{x}}{\sqrt[n]{y}} = \sqrt[n]{\dfrac{x}{y}}$	$\dfrac{\sqrt[3]{90}}{\sqrt[3]{10}} = \sqrt[3]{\dfrac{90}{10}} = \sqrt[3]{9}$
15. $\sqrt[m]{\sqrt[n]{x}} = \sqrt[mn]{x}$	$\sqrt[3]{\sqrt[4]{2}} = \sqrt[12]{2}$
16. $x^{m/n} = \sqrt[n]{x^m} = (\sqrt[n]{x})^m$	$8^{2/3} = \sqrt[3]{8^2} = (\sqrt[3]{8})^2 = 2^2 = 4$
17. $(\sqrt[m]{x})^m = x$	$(\sqrt[8]{7})^8 = 7$

When computing $x^{m/n}$, it is often easier first to find $\sqrt[n]{x}$ and then raise the result to the mth power. Thus, $(-27)^{4/3} = (\sqrt[3]{-27})^4 = (-3)^4 = 81$.

EXAMPLE 2 Exponents and Radicals

a. By Law 1,

$$x^6x^8 = x^{6+8} = x^{14}$$

$$a^3b^2a^5b = a^3a^5b^2b^1 = a^8b^3$$

$$x^{11}x^{-5} = x^{11-5} = x^6$$

$$z^{2/5}z^{3/5} = z^1 = z$$

$$xx^{1/2} = x^1x^{1/2} = x^{3/2}$$

---

[1] Although some laws involve restrictions, they are not vital to our discussion.

b. By Law 16,

$$\left(\frac{1}{4}\right)^{3/2} = \left(\sqrt{\frac{1}{4}}\right)^3 = \left(\frac{1}{2}\right)^3 = \frac{1}{8}$$

c. $$\left(-\frac{8}{27}\right)^{4/3} = \left(\sqrt[3]{\frac{-8}{27}}\right)^4 = \left(\frac{\sqrt[3]{-8}}{\sqrt[3]{27}}\right)^4 \qquad \text{(Laws 16 and 14)}$$

$$= \left(\frac{-2}{3}\right)^4$$

$$= \frac{(-2)^4}{3^4} = \frac{16}{81} \qquad \text{(Law 9)}$$

d. $$(64a^3)^{2/3} = 64^{2/3}(a^3)^{2/3} \qquad \text{(Law 8)}$$

$$= (\sqrt[3]{64})^2a^2 \qquad \text{(Laws 16 and 7)}$$

$$= (4)^2a^2 = 16a^2$$

▶ **Now work Problem 39.**

Rationalizing the denominator of a fraction is a procedure in which a fraction having a radical in its denominator is expressed as an equivalent fraction without a radical in its denominator. We use the fundamental principle of fractions, as Example 3 shows.

EXAMPLE 3 **Rationalizing Denominators**

a. $$\frac{2}{\sqrt{5}} = \frac{2}{5^{1/2}} = \frac{2 \cdot 5^{1/2}}{5^{1/2} \cdot 5^{1/2}} = \frac{2 \cdot 5^{1/2}}{5^1} = \frac{2\sqrt{5}}{5}$$

b. $$\frac{2}{\sqrt[6]{3x^5}} = \frac{2}{\sqrt[6]{3} \cdot \sqrt[6]{x^5}} = \frac{2}{3^{1/6}x^{5/6}} = \frac{2 \cdot 3^{5/6}x^{1/6}}{3^{1/6}x^{5/6} \cdot 3^{5/6}x^{1/6}}$$

$$= \frac{2(3^5x)^{1/6}}{3x} = \frac{2\sqrt[6]{3^5x}}{3x}$$

▶ **Now work Problem 63.**

The following examples illustrate various applications of the laws of exponents and radicals.

EXAMPLE 4 **Exponents**

a. *Eliminate negative exponents in* $\dfrac{x^{-2}y^3}{z^{-2}}$.

Solution:

$$\frac{x^{-2}y^3}{z^{-2}} = x^{-2} \cdot y^3 \cdot \frac{1}{z^{-2}} = \frac{1}{x^2} \cdot y^3 \cdot z^2 = \frac{y^3z^2}{x^2}$$

By comparing our answer with the original expression, we conclude that we can bring a factor of the numerator down to the denominator, and vice versa, by changing the sign of the exponent.

b. *Simplify* $\dfrac{x^2y^7}{x^3y^5}$.

Solution:

$$\frac{x^2y^7}{x^3y^5} = \frac{y^{7-5}}{x^{3-2}} = \frac{y^2}{x}$$

c. *Simplify* $(x^5y^8)^5$.

Solution:

$$(x^5y^8)^5 = (x^5)^5(y^8)^5 = x^{25}y^{40}$$

d. *Simplify* $(x^{5/9}y^{4/3})^{18}$

Solution:

$$(x^{5/9}y^{4/3})^{18} = (x^{5/9})^{18}(y^{4/3})^{18} = x^{10}y^{24}$$

e. *Simplify* $\left(\dfrac{x^{1/5}y^{6/5}}{z^{2/5}}\right)^5$.

Solution:

$$\left(\frac{x^{1/5}y^{6/5}}{z^{2/5}}\right)^5 = \frac{(x^{1/5}y^{6/5})^5}{(z^{2/5})^5} = \frac{xy^6}{z^2}$$

f. *Simplify* $\dfrac{x^3}{y^2} \div \dfrac{x^6}{y^5}$.

Solution:

$$\frac{x^3}{y^2} \div \frac{x^6}{y^5} = \frac{x^3}{y^2} \cdot \frac{y^5}{x^6} = \frac{y^3}{x^3}$$

▶ **Now work Problem 51.**

EXAMPLE 5 **Exponents**

a. *Eliminate negative exponents in* $x^{-1} + y^{-1}$ *and simplify.*

Solution:

$$x^{-1} + y^{-1} = \frac{1}{x} + \frac{1}{y} = \frac{y + x}{xy}$$

b. *Simplify* $x^{3/2} - x^{1/2}$ *by using the distributive law.*

Solution:

$$x^{3/2} - x^{1/2} = x^{1/2}(x - 1)$$

c. *Eliminate negative exponents in* $7x^{-2} + (7x)^{-2}$.

Solution:

$$7x^{-2} + (7x)^{-2} = \frac{7}{x^2} + \frac{1}{(7x)^2} = \frac{7}{x^2} + \frac{1}{49x^2}$$

d. *Eliminate negative exponents in* $(x^{-1} - y^{-1})^{-2}$.

Solution:

$$(x^{-1} - y^{-1})^{-2} = \left(\frac{1}{x} - \frac{1}{y}\right)^{-2} = \left(\frac{y - x}{xy}\right)^{-2}$$

$$= \left(\frac{xy}{y - x}\right)^2 = \frac{x^2y^2}{(y - x)^2}$$

e. *Apply the distributive law to* $x^{2/5}(y^{1/2} + 2x^{6/5})$.

Solution:

$$x^{2/5}(y^{1/2} + 2x^{6/5}) = x^{2/5}y^{1/2} + 2x^{8/5}$$

▶ **Now work Problem 57.**

EXAMPLE 6 Radicals

a. *Simplify* $\sqrt[4]{48}$.

Solution:

$$\sqrt[4]{48} = \sqrt[4]{16 \cdot 3} = \sqrt[4]{16}\sqrt[4]{3} = 2\sqrt[4]{3}$$

b. *Rewrite* $\sqrt{2+5x}$ *without using a radical sign.*

Solution:

$$\sqrt{2+5x} = (2+5x)^{1/2}$$

c. *Rationalize the denominator of* $\dfrac{\sqrt[5]{2}}{\sqrt[3]{6}}$ *and simplify.*

Solution:

$$\frac{\sqrt[5]{2}}{\sqrt[3]{6}} = \frac{2^{1/5} \cdot 6^{2/3}}{6^{1/3} \cdot 6^{2/3}} = \frac{2^{3/15}6^{10/15}}{6} = \frac{(2^3 6^{10})^{1/15}}{6} = \frac{\sqrt[15]{2^3 6^{10}}}{6}$$

d. *Simplify* $\dfrac{\sqrt{20}}{\sqrt{5}}$.

Solution:

$$\frac{\sqrt{20}}{\sqrt{5}} = \sqrt{\frac{20}{5}} = \sqrt{4} = 2$$

▶ **Now work Problem 71.**

EXAMPLE 7 Radicals

a. *Simplify* $\sqrt[3]{x^6y^4}$.

Solution:

$$\sqrt[3]{x^6y^4} = \sqrt[3]{(x^2)^3y^3y} = \sqrt[3]{(x^2)^3} \cdot \sqrt[3]{y^3} \cdot \sqrt[3]{y}$$

$$= x^2y\sqrt[3]{y} \qquad \text{(Law 17)}$$

b. *Simplify* $\sqrt{\dfrac{2}{7}}$.

Solution:

$$\sqrt{\frac{2}{7}} = \sqrt{\frac{2 \cdot 7}{7 \cdot 7}} = \sqrt{\frac{14}{7^2}} = \frac{\sqrt{14}}{\sqrt{7^2}} = \frac{\sqrt{14}}{7}$$

c. *Simplify* $\sqrt{250} - \sqrt{50} + 15\sqrt{2}$.

Solution:

$$\begin{aligned}\sqrt{250} - \sqrt{50} + 15\sqrt{2} &= \sqrt{25 \cdot 10} - \sqrt{25 \cdot 2} + 15\sqrt{2}\\ &= 5\sqrt{10} - 5\sqrt{2} + 15\sqrt{2}\\ &= 5\sqrt{10} + 10\sqrt{2}\end{aligned}$$

d. *If x is any real number, simplify* $\sqrt{x^2}$.

Solution:

$$\sqrt{x^2} = \begin{cases} x & \text{if } x \geq 0 \\ -x & \text{if } x < 0 \end{cases}$$

Thus, $\sqrt{2^2} = 2$ and $\sqrt{(-3)^2} = -(-3) = 3$

▶ **Now work Problem 75.**

Exercise 0.3

In Problems 1–14, simplify and express all answers in terms of positive exponents.

1. $(2^3)(2^2)$ **2.** x^6x^9 **3.** w^4w^8

4. z^3zz^2 ***5.** $\dfrac{x^2x^6}{y^7y^{10}}$ **6.** $(x^{12})^4$

7. $\dfrac{(a^3)^7}{(b^4)^5}$ **8.** $\left(\dfrac{x^2}{y^3}\right)^5$ **9.** $(2x^2y^3)^3$

10. $\left(\dfrac{w^2s^3}{y^2}\right)^2$ **11.** $\dfrac{x^9}{x^5}$ **12.** $\left(\dfrac{3m^3}{9n^2}\right)^5$

13. $\dfrac{(x^3)^6}{x(x^3)}$ **14.** $\dfrac{(x^2)^3(x^3)^2}{(x^3)^4}$

In Problems 15–28, evaluate the expressions.

15. $\sqrt{25}$ **16.** $\sqrt[4]{81}$ **17.** $\sqrt[5]{-32}$

18. $\sqrt{0.04}$ **19.** $\sqrt[4]{\dfrac{1}{16}}$ **20.** $\sqrt[3]{-\dfrac{8}{27}}$

21. $(49)^{1/2}$ **22.** $(64)^{1/3}$ **23.** $9^{3/2}$

24. $(25)^{-3/2}$ **25.** $(32)^{-2/5}$ **26.** $(0.09)^{-1/2}$

27. $\left(\dfrac{1}{32}\right)^{4/5}$ **28.** $\left(-\dfrac{64}{27}\right)^{2/3}$

In Problems 29–40, simplify the expressions.

29. $\sqrt{32}$ **30.** $\sqrt[3]{54}$ **31.** $\sqrt[3]{2x^3}$

32. $\sqrt{4x}$ **33.** $\sqrt{16x^4}$ **34.** $\sqrt[4]{\dfrac{x}{16}}$

35. $2\sqrt{8} - 5\sqrt{27} + \sqrt[3]{128}$ **36.** $\sqrt{\dfrac{5}{11}}$

37. $(9z^4)^{1/2}$ **38.** $(16y^8)^{3/4}$

***39.** $\left(\dfrac{27t^3}{8}\right)^{2/3}$ **40.** $\left(\dfrac{256}{x^{12}}\right)^{-3/4}$

In Problems 41–52, write the expressions in terms of positive exponents only. Avoid all radicals in the final form. For example,

$$y^{-1}\sqrt{x} = \frac{x^{1/2}}{y}$$

41. $\dfrac{x^3y^{-2}}{z^2}$ **42.** $\sqrt[5]{x^2y^3z^{-10}}$ **43.** $5m^{-2}m^{-7}$

44. $x + y^{-1}$ **45.** $(3t)^{-2}$ **46.** $(3 - z)^{-4}$

47. $\sqrt[5]{5x^2}$ **48.** $(x^{-2}y^2)^{-2}$ **49.** $\sqrt{x} - \sqrt{y}$

50. $\dfrac{u^{-2}v^{-6}w^3}{vw^{-5}}$ ***51.** $x^2\sqrt[4]{xy^{-2}z^3}$ **52.** $\sqrt[4]{a^{-3}b^{-2}a^5b^{-4}}$

In Problems 53–58, write the exponential forms in equivalent forms involving radicals.

53. $(8x - y)^{4/5}$ **54.** $(ab^2c^3)^{3/4}$

55. $x^{-4/5}$ **56.** $2x^{1/2} - (2y)^{1/2}$

***57.** $3w^{-3/5} - (3w)^{-3/5}$ **58.** $[(x^{-4})^{1/5}]^{1/6}$

In Problems 59–68, rationalize the denominators.

59. $\dfrac{6}{\sqrt{5}}$ **60.** $\dfrac{5}{\sqrt[3]{9}}$ **61.** $\dfrac{4}{\sqrt{2x}}$

62. $\dfrac{y}{\sqrt{2y}}$ ***63.** $\dfrac{1}{\sqrt[3]{3x}}$ **64.** $\dfrac{2}{3\sqrt[3]{y^2}}$

65. $\dfrac{\sqrt{32}}{\sqrt{2}}$ **66.** $\dfrac{\sqrt{18}}{\sqrt{2}}$ **67.** $\dfrac{\sqrt[5]{2}}{\sqrt[4]{a^2b}}$

68. $\dfrac{\sqrt{2}}{\sqrt[3]{3}}$

In Problems 69–90, simplify the expressions. Express all answers in terms of positive exponents. Rationalize the denominator where necessary to avoid fractional exponents in the denominator.

69. $2x^2y^{-3}x^4$

70. $\dfrac{3}{u^{5/2}v^{1/2}}$

***71.** $\dfrac{\sqrt{243}}{\sqrt{3}}$

72. $\{[(2x^2)^3]^{-4}\}^{-1}$

73. $\dfrac{2^0}{(2^{-2}x^{1/2}y^{-2})^3}$

74. $\dfrac{\sqrt{s^5}}{\sqrt[3]{s^2}}$

***75.** $\sqrt[3]{x^2yz^3}\sqrt[3]{xy^2}$

76. $(\sqrt[4]{3})^8$

77. $3^2(81)^{-3/4}$

78. $(\sqrt[5]{x^2y})^{2/5}$

79. $(2x^{-1}y^2)^2$

80. $\dfrac{3}{\sqrt[3]{y}\sqrt[4]{x}}$

81. $\sqrt{x}\sqrt{x^2y^3}\sqrt{xy^2}$

82. $\sqrt{75k^4}$

83. $\dfrac{(ab^{-3}c)^8}{(a^{-1}c^2)^{-3}}$

84. $\sqrt[3]{5(25)}$

85. $\dfrac{(x^2)^3}{x^4} \div \left[\dfrac{x^3}{(x^3)^2}\right]^2$

86. $\sqrt{(-6)(-6)}$

87. $-\dfrac{8s^{-2}}{2s^3}$

88. $(a^5b^{-3}\sqrt{c})^3$

89. $(2x^2y \div 3y^3z^{-2})^2$

90. $\dfrac{1}{\left(\dfrac{\sqrt{2}x^{-2}}{\sqrt{16}x^3}\right)^2}$

OBJECTIVE *To add, subtract, multiply, and divide algebraic expressions. To define a polynomial, to use special products, and to use long division to divide polynomials.*

0.4 Operations with Algebraic Expressions

If numbers, represented by symbols, are combined by any or all of the operations of addition, subtraction, multiplication, division, exponentiation, and extraction of roots, then the resulting expression is called an *algebraic expression*.

EXAMPLE 1 Algebraic Expressions

a. $\sqrt[3]{\dfrac{3x^3 - 5x - 2}{10 - x}}$ is an algebraic expression in the variable x.

b. $10 - 3\sqrt{y} + \dfrac{5}{7 + y^2}$ is an algebraic expression in the variable y.

c. $\dfrac{(x + y)^3 - xy}{y} + 2$ is an algebraic expression in the variables x and y.

CAUTION The words *polynomial* and *multinomial* should not be used interchangeably. For example, $\sqrt{x} + 2$ is a multinomial, but not a polynomial. On the other hand, $x + 2$ is a multinomial and a polynomial.

The algebraic expression $5ax^3 - 2bx + 3$ consists of three *terms:* $+5ax^3$, $-2bx$, and $+3$. Some of the *factors* of the first term, $5ax^3$, are 5, a, x, x^2, x^3, $5ax$, and ax^2. Also, $5a$ is the coefficient of x^3, and 5 is the *numerical coefficient* of ax^3. If a and b represent fixed numbers throughout a discussion, then a and b are called *constants*.

Algebraic expressions with exactly one term are called *monomials*. Those having exactly two terms are *binomials*, and those with exactly three terms are *trinomials*. Algebraic expressions with more than one term are called *multinomials*. Thus, the multinomial $2x - 5$ is a binomial; the multinomial $3\sqrt{y} + 2y - 4y^2$ is a trinomial.

A *polynomial in x* is an algebraic expression of the form[2]

$$c_nx^n + c_{n-1}x^{n-1} + \cdots + c_1x + c_0$$

where n is a nonnegative integer and the coefficients $c_0, c_1, \ldots, c_n$ are constants with $c_n \neq 0$. We call n the *degree* of the polynomial. Hence, $4x^3 - 5x^2 + x - 2$ is a polynomial in x of degree 3, and $y^5 - 2$ is a polynomial in y of degree 5.

[2]The three dots indicate all other terms that are understood to be included in the sum.

A nonzero constant is a polynomial of degree zero; thus, 5 is a polynomial of degree zero. The constant 0 is considered to be a polynomial; however, no degree is assigned to it.

In the following examples, we illustrate operations with algebraic expressions.

EXAMPLE 2 Adding Algebraic Expressions

Simplify $(3x^2y - 2x + 1) + (4x^2y + 6x - 3)$.

Solution: We first remove the parentheses. Next, using the commutative property of addition, we gather all similar terms together. *Similar terms* are terms that differ only by their numerical coefficients. In this example, $3x^2y$ and $4x^2y$ are similar, as are the pairs $-2x$ and $6x$, and 1 and -3. Thus,

$$\begin{aligned}(3x^2y - 2x + 1) + (4x^2y + 6x - 3) &= 3x^2y - 2x + 1 + 4x^2y + 6x - 3\\ &= 3x^2y + 4x^2y - 2x + 6x + 1 - 3\end{aligned}$$

By the distributive property,

$$3x^2y + 4x^2y = (3 + 4)x^2y = 7x^2y$$

and

$$-2x + 6x = (-2 + 6)x = 4x$$

Hence, $(3x^2y - 2x + 1) + (4x^2y + 6x - 3) = 7x^2y + 4x - 2$

▶ **Now work Problem 7.**

EXAMPLE 3 Subtracting Algebraic Expressions

Simplify $(3x^2y - 2x + 1) - (4x^2y + 6x - 3)$.

Solution: Here we apply the definition of subtraction and the distributive property:

$$\begin{aligned}&(3x^2y - 2x + 1) - (4x^2y + 6x - 3)\\ &\quad= (3x^2y - 2x + 1) + (-1)(4x^2y + 6x - 3)\\ &\quad= (3x^2y - 2x + 1) + (-4x^2y - 6x + 3)\\ &\quad= 3x^2y - 2x + 1 - 4x^2y - 6x + 3\\ &\quad= 3x^2y - 4x^2y - 2x - 6x + 1 + 3\\ &\quad= (3 - 4)x^2y + (-2 - 6)x + 1 + 3\\ &\quad= -x^2y - 8x + 4\end{aligned}$$

▶ **Now work Problem 13.**

EXAMPLE 4 Removing Grouping Symbols

Simplify $3\{2x[2x + 3] + 5[4x^2 - (3 - 4x)]\}$.

Solution: We first eliminate the innermost grouping symbols (the parentheses). Then we repeat the process until all grouping symbols are removed—combining similar terms whenever possible. We have

$$\begin{aligned}3\{2x[2x+3]+5[4x^2-(3-4x)]\} &= 3\{2x[2x+3]+5[4x^2-3+4x]\}\\ &= 3\{4x^2+6x+20x^2-15+20x\}\\ &= 3\{24x^2+26x-15\}\\ &= 72x^2+78x-45\end{aligned}$$

▶ **Now work Problem 15.**

The distributive property is the key tool in multiplying expressions. For example, to multiply $ax+c$ by $bx+d$ we can consider $ax+c$ to be a single number and then use the distributive property:

$$(ax+c)(bx+d) = (ax+c)bx + (ax+c)d$$

Using the distributive property again, we have

$$\begin{aligned}(ax+c)bx+(ax+c)d &= abx^2+cbx+adx+cd\\ &= abx^2+(ad+cb)x+cd\end{aligned}$$

Thus, $(ax+c)(bx+d) = abx^2+(ad+cb)x+cd$. In particular, if $a=2$, $b=1$, $c=3$, and $d=-2$, then

$$\begin{aligned}(2x+3)(x-2) &= 2(1)x^2+[2(-2)+3(1)]x+3(-2)\\ &= 2x^2-x-6\end{aligned}$$

We now give a list of special products that may be obtained from the distributive property and are useful in multiplying algebraic expressions.

Special Products

1. $x(y+z) = xy+xz$ (distributive property)
2. $(x+a)(x+b) = x^2+(a+b)x+ab$
3. $(ax+c)(bx+d) = abx^2+(ad+cb)x+cd$
4. $(x+a)^2 = x^2+2ax+a^2$ (square of a binomial)
5. $(x-a)^2 = x^2-2ax+a^2$ (square of a binomial)
6. $(x+a)(x-a) = x^2-a^2$ (product of sum and difference)
7. $(x+a)^3 = x^3+3ax^2+3a^2x+a^3$ (cube of a binomial)
8. $(x-a)^3 = x^3-3ax^2+3a^2x-a^3$ (cube of a binomial)

EXAMPLE 5 Special Products

a. By Rule 2,

$$\begin{aligned}(x+2)(x-5) &= [x+2][x+(-5)]\\ &= x^2+(2-5)x+2(-5)\\ &= x^2-3x-10\end{aligned}$$

b. By Rule 3,

$$\begin{aligned}(3z+5)(7z+4) &= 3\cdot 7z^2 + (3\cdot 4 + 5\cdot 7)z + 5\cdot 4\\ &= 21z^2 + 47z + 20\end{aligned}$$

c. By Rule 5,

$$\begin{aligned}(x-4)^2 &= x^2 - 2(4)x + 4^2\\ &= x^2 - 8x + 16\end{aligned}$$

d. By Rule 6,

$$\begin{aligned}(\sqrt{y^2+1}+3)(\sqrt{y^2+1}-3) &= (\sqrt{y^2+1})^2 - 3^2\\ &= (y^2+1) - 9\\ &= y^2 - 8\end{aligned}$$

e. By Rule 7,

$$\begin{aligned}(3x+2)^3 &= (3x)^3 + 3(2)(3x)^2 + 3(2)^2(3x) + (2)^3\\ &= 27x^3 + 54x^2 + 36x + 8\end{aligned}$$

▶ **Now work Problem 19.**

EXAMPLE 6 Multiplying Multinomials

Find the product $(2t-3)(5t^2+3t-1)$.

Solution: We treat $2t-3$ as a single number and apply the distributive property twice:

$$\begin{aligned}(2t-3)(5t^2+3t-1) &= (2t-3)5t^2 + (2t-3)3t - (2t-3)1\\ &= 10t^3 - 15t^2 + 6t^2 - 9t - 2t + 3\\ &= 10t^3 - 9t^2 - 11t + 3\end{aligned}$$

▶ **Now work Problem 35.**

In Example 3(b) of Section 0.2, we showed that $\frac{a+b}{c} = \frac{a}{c} + \frac{b}{c}$. Similarly, $\frac{a-b}{c} = \frac{a}{c} - \frac{b}{c}$. Using these results, we can divide a multinomial by a monomial by dividing each term in the multinomial by the monomial.

EXAMPLE 7 Dividing a Multinomial by a Monomial

a. $\frac{x^3+3x}{x} = \frac{x^3}{x} + \frac{3x}{x} = x^2 + 3$

b. $$\begin{aligned}\frac{4z^3 - 8z^2 + 3z - 6}{2z} &= \frac{4z^3}{2z} - \frac{8z^2}{2z} + \frac{3z}{2z} - \frac{6}{2z}\\ &= 2z^2 - 4z + \frac{3}{2} - \frac{3}{z}\end{aligned}$$

▶ **Now work Problem 47.**

Long Division

To divide a polynomial by a polynomial, we use so-called long division when the degree of the divisor is less than or equal to the degree of the dividend, as the next example shows.

EXAMPLE 8 Long Division

Divide $2x^3 - 14x - 5$ *by* $x - 3$.

Solution: Here $2x^3 - 14x - 5$ is the *dividend* and $x - 3$ is the *divisor*. To avoid errors, it is best to write the dividend as $2x^3 + 0x^2 - 14x - 5$. Note that the powers of x are in decreasing order. We have

$$\begin{array}{r l}
 & 2x^2 + 6x + 4 \leftarrow \text{quotient} \\
x - 3 & \overline{)2x^3 + 0x^2 - 14x - 5} \\
 & \underline{2x^3 - 6x^2} \\
 & \qquad 6x^2 - 14x \\
 & \qquad \underline{6x^2 - 18x} \\
 & \qquad\qquad 4x - 5 \\
 & \qquad\qquad \underline{4x - 12} \\
 & \qquad\qquad\qquad 7 \leftarrow \text{remainder}
\end{array}$$

Note that we divided x (the first term of the divisor) into $2x^3$ and got $2x^2$. Then we multiplied $2x^2$ by $x - 3$, getting $2x^3 - 6x^2$. Subtracting $2x^3 - 6x^2$ from $2x^3 + 0x^2$, we obtained $6x^2$ and then "brought down" the term $-14x$. This process is continued until we arrive at 7, the *remainder*. We always stop when the remainder is 0 or is a polynomial whose degree is less than the degree of the divisor. Our answer may be written as

$$2x^2 + 6x + 4 + \frac{7}{x - 3}$$

That is, the answer has the form

$$\text{quotient} + \frac{\text{remainder}}{\text{divisor}}$$

A way of checking a division is to verify that

$$(\text{quotient})(\text{divisor}) + \text{remainder} = \text{dividend}$$

By using this equation, you should be able to verify the result of the example.

▸ **Now work Problem 51.**

Exercise 0.4

Perform the indicated operations and simplify.

1. $(8x - 4y + 2) + (3x + 2y - 5)$

2. $(6x^2 - 10xy + 2) + (2z - xy + 4)$

3. $(8t^2 - 6s^2) + (4s^2 - 2t^2 + 6)$

4. $(\sqrt{x} + 2\sqrt{x}) + (\sqrt{x} + 3\sqrt{x})$

5. $(\sqrt{a} + 2\sqrt{3b}) - (\sqrt{c} - 3\sqrt{3b})$

6. $(2x + 3y - 5) - (7x - 6y + 2)$

***7.** $(6x^2 - 10xy + \sqrt{2}) - (2z - xy + 4)$

8. $(\sqrt{x} + 2\sqrt{x}) - (\sqrt{x} + 3\sqrt{x})$

9. $(\sqrt{x} + \sqrt{2y}) - (\sqrt{x} + \sqrt{3z})$

10. $4(2z - w) - 3(w - 2z)$

11. $3(3x + 3y - 7) - 3(8x - 2y + 2)$

12. $(u-3v)+(-5u-4v)+(u-3)$

***13.** $3(x^2+y^2)-x(y+2x)+2y(x+3y)$

14. $2-[3+4(s-3)]$

***15.** $2\{3[3(x^2+2)-2(x^2-5)]\}$

16. $4\{3(t+5)-t[1-(t+1)]\}$

17. $-5(4x^2(2x+2)-2(x^2-(5-2x)))$

18. $-\{-2[2a+3b-1]+4[a-2b]-a[2(b-3)]\}$

***19.** $(x+4)(x+5)$

20. $(u+2)(u+5)$

21. $(w+2)(w-5)$

22. $(z-7)(z-3)$

23. $(2x+3)(5x+2)$

24. $(t-5)(2t+7)$

25. $(x+3)^2$

26. $(2x-1)^2$

27. $(x-5)^2$

28. $(\sqrt{x}-1)(2\sqrt{x}+5)$

29. $(\sqrt{3x}+5)^2$

30. $(y-4)(y+4)$

31. $(2s-1)(2s+1)$

32. $(z^2-3w)(z^2+3w)$

33. $(x^2-3)(x+4)$

34. $(x+1)(x^2+x+3)$

***35.** $(x^2-4)(3x^2+2x-1)$

36. $(3y-2)(4y^3+2y^2-3y)$

37. $x\{3(x-1)(x-2)+2[x(x+7)]\}$

38. $[(2z+1)(2z-1)](4z^2+1)$

39. $(x+y+2)(3x+2y-4)$

40. $(x^2+x+1)^2$

41. $(2a+3)^3$

42. $(x-2)^3$

43. $(2x-3)^3$

44. $(x+2y)^3$

45. $\dfrac{z^2-18z}{z}$

46. $\dfrac{2x^3-7x+4}{x}$

***47.** $\dfrac{6x^5+4x^3-1}{2x^2}$

48. $\dfrac{(3y-4)-(9y+5)}{3y}$

49. $(x^2+3x-1)\div(x+3)$

50. $(x^2-5x+4)\div(x-4)$

***51.** $(3x^3-2x^2+x-3)\div(x+2)$

52. $(x^4+2x^2+1)\div(x-1)$

53. $x^3\div(x+2)$

54. $(4x^2+6x+1)\div(2x-1)$

55. $(3x^2-4x+3)\div(3x+2)$

56. $(z^3+z^2+z)\div(z^2-z+1)$

OBJECTIVE *To state the basic rules for factoring and apply them to factor expressions.*

0.5 Factoring

If two or more expressions are multiplied together, the expressions are called *factors* of the product. Thus, if $c = ab$, then a and b are both factors of the product c. The process by which an expression is written as a product of its factors is called *factoring*.

Listed next are rules for factoring expressions, most of which arise from the special products discussed in Section 0.4. The right side of each identity is the factored form of the left side.

Rules for Factoring

1. $xy+xz=x(y+z)$ (common factor)
2. $x^2+(a+b)x+ab=(x+a)(x+b)$
3. $abx^2+(ad+cb)x+cd=(ax+c)(bx+d)$
4. $x^2+2ax+a^2=(x+a)^2$ (perfect-square trinomial)
5. $x^2-2ax+a^2=(x-a)^2$ (perfect-square trinomial)
6. $x^2-a^2=(x+a)(x-a)$ (difference of two squares)
7. $x^3+a^3=(x+a)(x^2-ax+a^2)$ (sum of two cubes)
8. $x^3-a^3=(x-a)(x^2+ax+a^2)$ (difference of two cubes)

When factoring a polynomial, we usually choose factors that themselves are polynomials. For example, $x^2-4=(x+2)(x-2)$. We shall not usually write $x-4$ as $(\sqrt{x}+2)(\sqrt{x}-2)$.

Always factor completely. For example,

$$2x^2 - 8 = 2(x^2 - 4) = 2(x + 2)(x - 2)$$

EXAMPLE 1 Common Factors

a. *Factor* $3k^2x^2 + 9k^3x$ *completely.*

Solution: Since $3k^2x^2 = (3k^2x)(x)$ and $9k^3x = (3k^2x)(3k)$, each term of the original expression contains the common factor $3k^2x$. Thus, by Rule 1,

$$3k^2x^2 + 9k^3x = 3k^2x(x + 3k)$$

Note that although $3k^2x^2 + 9k^3x = 3(k^2x^2 + 3k^3x)$, we do not say that the expression is completely factored, since $k^2x^2 + 3k^3x$ can yet be factored.

b. *Factor* $8a^5x^2y^3 - 6a^2b^3yz - 2a^4b^4xy^2z^2$ *completely.*

Solution:

$$8a^5x^2y^3 - 6a^2b^3yz - 2a^4b^4xy^2z^2 = 2a^2y(4a^3x^2y^2 - 3b^3z - a^2b^4xyz^2)$$

▶ **Now work Problem 5.**

EXAMPLE 2 Factoring Trinomials

a. *Factor* $3x^2 + 6x + 3$ *completely.*

Solution: First we remove a common factor. Then we factor the resulting expression completely. Thus, we have

$$\begin{aligned} 3x^2 + 6x + 3 &= 3(x^2 + 2x + 1) \\ &= 3(x + 1)^2 \qquad \text{(Rule 4)} \end{aligned}$$

b. *Factor* $x^2 - x - 6$ *completely.*

Solution: If this trinomial factors into the form $(x + a)(x + b)$, which is a product of two binomials, then we must determine the values of a and b. Since $(x + a)(x + b) = x^2 + (a + b)x + ab$, it follows that

$$x^2 + (-1)x + (-6) = x^2 + (a + b)x + ab$$

By equating corresponding coefficients, we want

$$a + b = -1 \quad \text{and} \quad ab = -6$$

If $a = -3$ and $b = 2$, then both conditions are met, and hence,

$$x^2 - x - 6 = (x - 3)(x + 2)$$

As a check, it is wise to multiply the right side to see if it agrees with the left side.

c. *Factor* $x^2 - 7x + 12$ *completely.*

Solution:

$$x^2 - 7x + 12 = (x - 3)(x - 4)$$

▶ **Now work Problem 9.**

EXAMPLE 3 **Factoring**

The following is an assortment of expressions that are completely factored. The numbers in parentheses refer to the rules used.

a. $x^2 + 8x + 16 = (x + 4)^2$ (4)

b. $9x^2 + 9x + 2 = (3x + 1)(3x + 2)$ (3)

c. $6y^3 + 3y^2 - 18y = 3y(2y^2 + y - 6)$ (1)

$= 3y(2y - 3)(y + 2)$ (3)

d. $x^2 - 6x + 9 = (x - 3)^2$ (5)

e. $z^{1/4} + z^{5/4} = z^{1/4}(1 + z)$ (1)

f. $x^4 - 1 = (x^2 + 1)(x^2 - 1)$ (6)

$= (x^2 + 1)(x + 1)(x - 1)$ (6)

g. $x^{2/3} - 5x^{1/3} + 4 = (x^{1/3} - 1)(x^{1/3} - 4)$ (2)

h. $ax^2 - ay^2 + bx^2 - by^2 = (ax^2 - ay^2) + (bx^2 - by^2)$

$= a(x^2 - y^2) + b(x^2 - y^2)$ (1)

$= (x^2 - y^2)(a + b)$ (1)

$= (x + y)(x - y)(a + b)$ (6)

i. $8 - x^3 = (2)^3 - (x)^3 = (2 - x)(4 + 2x + x^2)$ (8)

j. $x^6 - y^6 = (x^3)^2 - (y^3)^2 = (x^3 + y^3)(x^3 - y^3)$ (6)

$= (x + y)(x^2 - xy + y^2)(x - y)(x^2 + xy + y^2)$ (7), (8)

Note in Example 3(f) that $x^2 - 1$ is factorable, but $x^2 + 1$ is not. In Example 3(h), we factored by making use of grouping.

Exercise 0.5

Factor the following expressions completely.

1. $6x + 4$
2. $6y^2 - 4y$
3. $10xy + 5xz$
4. $3x^2y - 9x^3y^3$
***5.** $8a^3bc - 12ab^3cd + 4b^4c^2d^2$
6. $6u^3v^3 + 18u^2vw^4 - 12u^2v^3$
7. $z^2 - 49$
8. $x^2 + 3x - 4$
***9.** $p^2 + 4p + 3$
10. $s^2 - 6s + 8$
11. $16x^2 - 9$
12. $x^2 + 2x - 24$
13. $a^2 + 5a + 6$
14. $4t^2 - 9s^2$
15. $x^2 + 6x + 9$
16. $y^2 - 15y + 50$
17. $5x^2 + 25x + 30$
18. $3t^2 + 12t - 15$
19. $3x^2 - 3$
20. $4y^2 - 8y + 3$
21. $6y^2 + 13y + 2$
22. $4x^2 - x - 3$
23. $12s^3 + 10s^2 - 8s$
24. $9z^2 + 30z + 25$
25. $u^{8/3}v^2 - 4u^{2/3}v^3$
26. $9x^{4/7} - 1$
27. $2x^3 + 2x^2 - 12x$
28. $x^2y^2 - 4xy + 4$
29. $(4x + 2)^2$
30. $2x^2(2x - 4x^2)^2$
31. $x^3y^2 - 14x^2y + 49x$
32. $(3x^2 + x) + (6x + 2)$
33. $(x^3 - 4x) + (8 - 2x^2)$
34. $(x^2 - 1) + (x^2 - x - 2)$
35. $(y^4 + 8y^3 + 16y^2) - (y^2 + 8y + 16)$
36. $x^3y - 4xy + z^2x^2 - 4z^2$
37. $a^3 + 27$
38. $x^3 - 1$
39. $x^6 - 1$
40. $27 + 8x^3$
41. $(x + 3)^3(x - 1) + (x + 3)^2(x - 1)^2$
42. $(a + 5)^3(a + 1)^2 + (a + 5)^2(a + 1)^3$
43. $P(1 + r) + P(1 + r)r$
44. $(x - 3)(2x + 3) - (2x + 3)(x + 5)$
45. $x^4 - 16$
46. $81x^4 - y^4$
47. $y^8 - 1$
48. $t^4 - 4$
49. $X^4 + 2X^2 - 3$
50. $x^4 - 10x^2 + 9$
51. $x^4y - 2x^2y + y$
52. $4x^3 - 6x^2 - 4x$

OBJECTIVE *To simplify, add, subtract, multiply, and divide algebraic fractions. To rationalize the denominator of a fraction.*

0.6 Fractions

Simplifying Fractions

By using the fundamental principle of fractions (Section 0.2), we may be able to simplify algebraic expressions that are fractions. That principle allows us to multiply or divide both the numerator and the denominator of a fraction by the same nonzero quantity. The resulting fraction will be equivalent to the original one. The fractions that we shall consider are assumed to have nonzero denominators. Thus, all the factors of the denominators in our examples are assumed to be nonzero. This will often mean that certain values are excluded for the variables that occur in the denominators.

EXAMPLE 1 **Simplifying Fractions**

a. *Simplify* $\dfrac{x^2-x-6}{x^2-7x+12}$.

Solution: First, we completely factor the numerator and the denominator:

$$\frac{x^2-x-6}{x^2-7x+12}=\frac{(x-3)(x+2)}{(x-3)(x-4)}$$

Dividing both numerator and denominator by the common factor $x-3$, we have

$$\frac{(x-3)(x+2)}{(x-3)(x-4)}=\frac{1(x+2)}{1(x-4)}=\frac{x+2}{x-4}$$

Usually, we just write

$$\frac{x^2-x-6}{x^2-7x+12}=\frac{\overset{1}{\cancel{(x-3)}}(x+2)}{\underset{1}{\cancel{(x-3)}}(x-4)}=\frac{x+2}{x-4}$$

or

$$\frac{x^2-x-6}{x^2-7x+12}=\frac{(x-3)(x+2)}{(x-3)(x-4)}=\frac{x+2}{x-4}$$

The process of eliminating the common factor $x-3$ is commonly referred to as "cancellation."

b. *Simplify* $\dfrac{2x^2+6x-8}{8-4x-4x^2}$.

Solution:

$$\frac{2x^2+6x-8}{8-4x-4x^2}=\frac{2(x^2+3x-4)}{4(2-x-x^2)}=\frac{2(x-1)(x+4)}{4(1-x)(2+x)}$$

$$=\frac{2(x-1)(x+4)}{2(2)[(-1)(x-1)](2+x)}$$

$$=\frac{x+4}{-2(2+x)}=-\frac{x+4}{2(x+2)}$$

Note how $1-x$ is written as $(-1)(x-1)$ to allow cancellation.

▸ **Now work Problem 3.**

Multiplication and Division of Fractions

The rule for multiplying $\frac{a}{b}$ by $\frac{c}{d}$ is

$$\frac{a}{b} \cdot \frac{c}{d} = \frac{ac}{bd}$$

EXAMPLE 2 **Multiplying Fractions**

a. $\frac{x}{x+2} \cdot \frac{x+3}{x-5} = \frac{x(x+3)}{(x+2)(x-5)}$

b. $\frac{x^2-4x+4}{x^2+2x-3} \cdot \frac{6x^2-6}{x^2+2x-8} = \frac{[(x-2)^2][6(x+1)(x-1)]}{[(x+3)(x-1)][(x+4)(x-2)]}$

$$= \frac{6(x-2)(x+1)}{(x+3)(x+4)}$$

To divide $\frac{a}{b}$ by $\frac{c}{d}$, where $c \neq 0$, we have

In short, we invert the divisor and multiply.

$$\frac{a}{b} \div \frac{c}{d} = \frac{\frac{a}{b}}{\frac{c}{d}} = \frac{a}{b} \cdot \frac{d}{c}$$

EXAMPLE 3 **Dividing Fractions**

a. $\frac{x}{x+2} \div \frac{x+3}{x-5} = \frac{x}{x+2} \cdot \frac{x-5}{x+3} = \frac{x(x-5)}{(x+2)(x+3)}$

b. $\frac{\frac{x-5}{x-3}}{2x} = \frac{\frac{x-5}{x-3}}{\frac{2x}{1}} = \frac{x-5}{x-3} \cdot \frac{1}{2x} = \frac{x-5}{2x(x-3)}$

c. $\frac{\frac{4x}{x^2-1}}{\frac{2x^2+8x}{x-1}} = \frac{4x}{x^2-1} \cdot \frac{x-1}{2x^2+8x} = \frac{4x(x-1)}{[(x+1)(x-1)][2x(x+4)]}$

$$= \frac{2}{(x+1)(x+4)}$$

▶ **Now work Problem 11.**

Rationalizing the Denominator

Sometimes the denominator of a fraction has two terms and involves square roots, such as $2-\sqrt{3}$ or $\sqrt{5}+\sqrt{2}$. The denominator may then be rationalized by multiplying by an expression that makes the denominator a difference of two squares. For example,

$$\frac{4}{\sqrt{5}+\sqrt{2}} = \frac{4}{\sqrt{5}+\sqrt{2}} \cdot \frac{\sqrt{5}-\sqrt{2}}{\sqrt{5}-\sqrt{2}}$$

$$= \frac{4(\sqrt{5}-\sqrt{2})}{(\sqrt{5})^2-(\sqrt{2})^2} = \frac{4(\sqrt{5}-\sqrt{2})}{5-2}$$

$$= \frac{4(\sqrt{5}-\sqrt{2})}{3}$$

Rationalizing the *numerator* is a similar procedure.

EXAMPLE 4 Rationalizing Denominators

a. $$\frac{x}{\sqrt{2}-6} = \frac{x}{\sqrt{2}-6}\cdot\frac{\sqrt{2}+6}{\sqrt{2}+6} = \frac{x(\sqrt{2}+6)}{(\sqrt{2})^2-6^2}$$
$$= \frac{x(\sqrt{2}+6)}{2-36} = -\frac{x(\sqrt{2}+6)}{34}$$

b. $$\frac{\sqrt{5}-\sqrt{2}}{\sqrt{5}+\sqrt{2}} = \frac{\sqrt{5}-\sqrt{2}}{\sqrt{5}+\sqrt{2}}\cdot\frac{\sqrt{5}-\sqrt{2}}{\sqrt{5}-\sqrt{2}}$$
$$= \frac{(\sqrt{5}-\sqrt{2})^2}{5-2} = \frac{5-2\sqrt{5}\sqrt{2}+2}{3} = \frac{7-2\sqrt{10}}{3}$$

▸ **Now work Problem 53.**

Addition and Subtraction of Fractions

In Example 3(b) of Section 0.2, it was shown that $\frac{a}{c}+\frac{b}{c}=\frac{a+b}{c}$. That is, if we add two fractions having a common denominator, then the result is a fraction whose denominator is the common denominator. The numerator is the sum of the numerators of the original fractions. Similarly, $\frac{a}{c}-\frac{b}{c}=\frac{a-b}{c}$.

EXAMPLE 5 Adding and Subtracting Fractions

a. $$\frac{p^2-5}{p-2}+\frac{3p+2}{p-2} = \frac{(p^2-5)+(3p+2)}{p-2}$$
$$= \frac{p^2+3p-3}{p-2}$$

b. $$\frac{x^2-5x+4}{x^2+2x-3}-\frac{x^2+2x}{x^2+5x+6} = \frac{(x-1)(x-4)}{(x-1)(x+3)}-\frac{x(x+2)}{(x+2)(x+3)}$$
$$= \frac{x-4}{x+3}-\frac{x}{x+3} = \frac{(x-4)-x}{x+3} = -\frac{4}{x+3}$$

c. $$\frac{x^2+x-5}{x-7}-\frac{x^2-2}{x-7}+\frac{-4x+8}{x^2-9x+14}$$
$$= \frac{x^2+x-5}{x-7}-\frac{x^2-2}{x-7}+\frac{-4(x-2)}{(x-7)(x-2)}$$
$$= \frac{(x^2+x-5)-(x^2-2)+(-4)}{x-7}$$
$$= \frac{x-7}{x-7} = 1$$

▸ **Now work Problem 29.**

To add (or subtract) two fractions with different denominators, use the fundamental principle of fractions to rewrite the fractions as equivalent fractions that have the same denominator. Then proceed with the addition (or subtraction) by the method just described.

For example, to find

$$\frac{2}{x^3(x-3)}+\frac{3}{x(x-3)^2}$$

we can convert the first fraction into an equivalent fraction by multiplying the numerator and denominator by $x - 3$:

$$\frac{2(x-3)}{x^3(x-3)^2}$$

we can convert the second fraction by multiplying the numerator and denominator by x^2:

$$\frac{3x^2}{x^3(x-3)^2}$$

These fractions have the same denominator. Hence,

$$\begin{aligned}\frac{2}{x^3(x-3)} + \frac{3}{x(x-3)^2} &= \frac{2(x-3)}{x^3(x-3)^2} + \frac{3x^2}{x^3(x-3)^2} \\ &= \frac{3x^2 + 2x - 6}{x^3(x-3)^2}\end{aligned}$$

We could have converted the original fractions into equivalent fractions with *any* common denominator. However, we chose to convert them into fractions with the denominator $x^3(x-3)^2$. This is the **least common denominator (LCD)** of the fractions $2/[x^3(x-3)]$ and $3/[x(x-3)^2]$.

In general, to find the LCD of two or more fractions, first factor each denominator completely. *The LCD is the product of each of the distinct factors appearing in the denominators, each raised to the highest power to which it occurs in any single denominator.*

EXAMPLE 6 **Adding and Subtracting Fractions**

a. *Subtract:* $\dfrac{t}{3t+2} - \dfrac{4}{t-1}$.

Solution: The LCD is $(3t+2)(t-1)$. Thus, we have

$$\begin{aligned}\frac{t}{3t+2} - \frac{4}{t-1} &= \frac{t(t-1)}{(3t+2)(t-1)} - \frac{4(3t+2)}{(3t+2)(t-1)} \\ &= \frac{t(t-1) - 4(3t+2)}{(3t+2)(t-1)} \\ &= \frac{t^2 - t - 12t - 8}{(3t+2)(t-1)} = \frac{t^2 - 13t - 8}{(3t+2)(t-1)}\end{aligned}$$

b. *Add:* $\dfrac{4}{q-1} + 3$.

Solution: The LCD is $q - 1$.

$$\begin{aligned}\frac{4}{q-1} + 3 &= \frac{4}{q-1} + \frac{3(q-1)}{q-1} \\ &= \frac{4 + 3(q-1)}{q-1} = \frac{3q+1}{q-1}\end{aligned}$$

▸ **Now work Problem 33.**

EXAMPLE 7 **Subtracting Fractions**

$$\frac{x-2}{x^2+6x+9}-\frac{x+2}{2(x^2-9)}$$

$$=\frac{x-2}{(x+3)^2}-\frac{x+2}{2(x+3)(x-3)} \qquad [\text{LCD}=2(x+3)^2(x-3)]$$

$$=\frac{(x-2)(2)(x-3)}{(x+3)^2(2)(x-3)}-\frac{(x+2)(x+3)}{2(x+3)(x-3)(x+3)}$$

$$=\frac{(x-2)(2)(x-3)-(x+2)(x+3)}{2(x+3)^2(x-3)}$$

$$=\frac{2(x^2-5x+6)-(x^2+5x+6)}{2(x+3)^2(x-3)}$$

$$=\frac{2x^2-10x+12-x^2-5x-6}{2(x+3)^2(x-3)}$$

$$=\frac{x^2-15x+6}{2(x+3)^2(x-3)}$$

▶ **Now work Problem 39.**

Example 8 shows two methods of simplifying a "complex" fraction.

EXAMPLE 8 **Combined Operations with Fractions**

Simplify $\dfrac{\dfrac{1}{x+h}-\dfrac{1}{x}}{h}$.

Solution: First we combine the fractions in the numerator and obtain

$$\frac{\dfrac{1}{x+h}-\dfrac{1}{x}}{h}=\frac{\dfrac{x}{x(x+h)}-\dfrac{x+h}{x(x+h)}}{h}=\frac{\dfrac{x-(x+h)}{x(x+h)}}{h}$$

$$=\frac{\dfrac{-h}{x(x+h)}}{\dfrac{h}{1}}=\frac{-h}{x(x+h)h}=-\frac{1}{x(x+h)}$$

The original fraction can also be simplified by multiplying the numerator and denominator by the LCD of the fractions involved in the numerator (and denominator), namely, $x(x+h)$:

$$\frac{\dfrac{1}{x+h}-\dfrac{1}{x}}{h}=\frac{\left[\dfrac{1}{x+h}-\dfrac{1}{x}\right]x(x+h)}{h[x(x+h)]}$$

$$=\frac{x-(x+h)}{x(x+h)h}=\frac{-h}{x(x+h)h}=-\frac{1}{x(x+h)}$$

▶ **Now work Problem 47.**

Exercise 0.6

In Problems 1–6, simplify the expressions.

1. $\dfrac{a^2-9}{a^2-3a}$

2. $\dfrac{x^2-5x-6}{x^2-2x-3}$

*3. $\dfrac{x^2-9x+20}{x^2+x-20}$

4. $\dfrac{3x^2-27x+24}{2x^3-16x^2+14x}$

5. $\dfrac{6x^2+x-2}{2x^2+3x-2}$

6. $\dfrac{12x^2-19x+4}{6x^2-17x+12}$

In Problems 7–48, perform the operations and simplify as much as possible.

7. $\dfrac{y^2}{y-3}\cdot\dfrac{-1}{y+2}$

8. $\dfrac{t^2-9}{t^2+3t}\cdot\dfrac{t^2}{t^2-6t+9}$

9. $\dfrac{2x-3}{x-2}\cdot\dfrac{2-x}{2x+3}$

10. $\dfrac{x^2-y^2}{x+y}\cdot\dfrac{x^2+2xy+y^2}{y-x}$

*11. $\dfrac{2x-2}{x^2-2x-8}\div\dfrac{x^2-1}{x^2+5x+4}$

12. $\dfrac{x^2+2x}{3x^2-18x+24}\div\dfrac{x^2-x-6}{x^2-4x+4}$

13. $\dfrac{\frac{X^2}{8}}{\frac{X}{4}}$

14. $\dfrac{\frac{4x^3}{9x}}{\frac{x}{18}}$

15. $\dfrac{\frac{2m}{n^2}}{\frac{6m}{n^3}}$

16. $\dfrac{\frac{c+d}{c}}{\frac{c-d}{2c}}$

17. $\dfrac{\frac{4x}{3}}{2x}$

18. $\dfrac{4x}{\frac{3}{2x}}$

19. $\dfrac{-9x^3}{\frac{x}{3}}$

20. $\dfrac{\frac{-12Y^4}{Y}}{4}$

21. $\dfrac{\frac{x-5}{x^2-7x+10}}{x-2}$

22. $\dfrac{\frac{x^2+6x+9}{x}}{x+3}$

23. $\dfrac{\frac{10x^3}{x^2-1}}{\frac{5x}{x+1}}$

24. $\dfrac{\frac{x^2-x-6}{x^2-9}}{\frac{x^2-4}{x^2+2x-3}}$

25. $\dfrac{\frac{x^2+7x+10}{x^2+6x+5}}{\frac{x^2-2x-8}{x^2-3x-4}}$

26. $\dfrac{\frac{(x+2)^2}{3x-2}}{\frac{9x+18}{4-9x^2}}$

27. $\dfrac{\frac{4x^2-9}{x^2+3x-4}}{\frac{2x-3}{1-x^2}}$

28. $\dfrac{\frac{6x^2y+7xy-3y}{xy-x+5y-5}}{\frac{x^3y+4x^2y}{xy-x+4y-4}}$

*29. $\dfrac{x^2}{x+3}+\dfrac{5x+6}{x+3}$

30. $\dfrac{2}{x+2}+\dfrac{x}{x+2}$

31. $\dfrac{2}{t}+\dfrac{1}{3t}$

32. $\dfrac{9}{X^3}-\dfrac{1}{X^2}$

*33. $1-\dfrac{p^2}{p^2-1}$

34. $\dfrac{4}{s+4}+s$

35. $\dfrac{4}{2x-1}+\dfrac{x}{x+3}$

36. $\dfrac{x+1}{x-1}-\dfrac{x-1}{x+1}$

37. $\dfrac{1}{x^2-2x-3}+\dfrac{1}{x^2-9}$

38. $\dfrac{2}{3y^2-5y-2}-\dfrac{y}{3y^2-7y+2}$

*39. $\dfrac{4}{x-1}-3+\dfrac{-3x^2}{5-4x-x^2}$

40. $\dfrac{2x-3}{2x^2+11x-6}-\dfrac{3x+1}{3x^2+16x-12}+\dfrac{1}{3x-2}$

41. $(1+x^{-1})^2$

42. $(x^{-1}+y^{-1})^2$

43. $(x^{-1}-y)^{-1}$

44. $(a+b^{-1})^2$

45. $\dfrac{4+\frac{1}{x}}{3}$

46. $\dfrac{\frac{x+3}{x}}{x-\frac{9}{x}}$

*47. $\dfrac{3-\frac{1}{2x}}{x+\frac{x}{x+2}}$

48. $\dfrac{\frac{x-1}{x^2+5x+6}-\frac{1}{x+2}}{3+\frac{x-7}{3}}$

In Problems 49 and 50, perform the indicated operations, but do not rationalize the denominators.

49. $\dfrac{3}{\sqrt[3]{x+h}}-\dfrac{3}{\sqrt[3]{x}}$

50. $\dfrac{v\sqrt{v}}{\sqrt{2+v}}+\dfrac{1}{\sqrt{v}}$

In Problems 51–60, simplify and express your answer in a form that is free of radicals in the denominator.

51. $\dfrac{1}{2+\sqrt{3}}$

52. $\dfrac{1}{1-\sqrt{2}}$

*53. $\dfrac{\sqrt{2}}{\sqrt{3}-\sqrt{6}}$

54. $\dfrac{5}{\sqrt{6}+\sqrt{7}}$

55. $\dfrac{2\sqrt{2}}{\sqrt{2}-\sqrt{3}}$

56. $\dfrac{2\sqrt{5}}{\sqrt{3}-\sqrt{7}}$

57. $\dfrac{1}{x+\sqrt{5}}$

58. $\dfrac{x-3}{\sqrt{x}-1}+\dfrac{4}{\sqrt{x}-1}$

59. $\dfrac{5}{2+\sqrt{3}}-\dfrac{4}{1-\sqrt{2}}$

60. $\dfrac{4}{\sqrt{x}+2}\cdot\dfrac{x^2}{3}$

OBJECTIVE *To discuss equivalent equations and to develop techniques for solving linear equations, including literal equations and fractional and radical equations that lead to linear equations.*

0.7 Linear Equations

Equations

An **equation** is a statement that two expressions are equal. The two expressions that make up an equation are called its **sides.** They are separated by the **equality sign,** "=."

EXAMPLE 1 Examples of Equations

a. $x + 2 = 3$

b. $x^2 + 3x + 2 = 0$

c. $\dfrac{y}{y-4} = 6$

d. $w = 7 - z$

In Example 1, each equation contains at least one variable. A **variable** is a symbol that can be replaced by any one of a set of different numbers. The most popular symbols for variables are letters from the latter part of the alphabet, such as x, y, z, w, and t. Hence, Equations (a) and (c) are said to be in the variables x and y, respectively. Equation (d) is in the variables w and z. In the equation $x + 2 = 3$, the numbers 2 and 3 are called *constants*. They are fixed numbers.

Here we discuss restrictions on variables.

We *never* allow a variable in an equation to have a value for which any expression in that equation is undefined. For example, in

$$\frac{y}{y-4} = 6$$

y cannot be 4, because this would make the denominator zero. (We cannot divide by zero.) In some equations, the allowable values of a variable are restricted for physical reasons. For example, if the variable t represents time, negative values of t may not make sense. Hence, we should assume that $t \geq 0$.

To *solve* an equation means to find all values of its variables for which the equation is true. These values are called *solutions* of the equation and are said to *satisfy* the equation. When only one variable is involved, a solution is also called a **root.** The set of all solutions is called the **solution set** of the equation. Sometimes a letter representing an unknown quantity in an equation is simply called an *unknown*. Let us illustrate these terms.

EXAMPLE 2 Terminology for Equations

a. In the equation $x + 2 = 3$, the variable x is the unknown. The only value of x that satisfies the equation is obviously 1. Hence, 1 is a root and the solution set is $\{1\}$.

b. -2 is a root of $x^2 + 3x + 2 = 0$ because substituting -2 for x makes the equation true: $(-2)^2 + 3(-2) + 2 = 0$. Hence -2 is a member of the solution set, but in this case it is not the only one.

c. $w = 7 - z$ is an equation in two unknowns. One solution is the pair of values $w = 4$ and $z = 3$. However, there are infinitely many solutions. Can you think of another?

▸ **Now work Problem 1.**

Equivalent Equations

Solving an equation may involve performing operations on it. We prefer that any such operation result in another equation having exactly the same solutions as the given equation. When this occurs, the equations are said to be **equivalent.** There are three operations that guarantee equivalence:

1. Adding (subtracting) the same polynomial[3] to (from) both sides of an equation, where the polynomial is in the same variable as that occurring in the equation.

For example, if $-5x = 5 - 6x$, then adding $6x$ to both sides gives the equivalent equation $-5x + 6x = 5 - 6x + 6x$, or $x = 5$.

Equivalence is not guaranteed if both sides are multiplied or divided by an expression involving a variable.

2. Multiplying (dividing) both sides of an equation by the same nonzero constant.

For example, if $10x = 5$, then dividing both sides by 10 gives the equivalent equation $\frac{10x}{10} = \frac{5}{10}$, or $x = \frac{1}{2}$.

3. Replacing either side of an equation by an equal expression.

For example, if $x(x + 2) = 3$, then replacing the left side by the equal expression $x^2 + 2x$ gives the equivalent equation $x^2 + 2x = 3$.

We repeat: Operations 1–3 guarantee that the resulting equation is equivalent to the given one. However, sometimes in solving an equation we have to apply operations other than 1–3. These operations may *not* necessarily result in equivalent equations. They include the following.

Operations That May Not Produce Equivalent Equations

4. Multiplying both sides of an equation by an expression involving the variable;
5. Dividing both sides of an equation by an expression involving the variable;
6. Raising both sides of an equation to equal powers.

Operation 6 includes taking roots of both sides.

Let us illustrate the last three operations. For example, by inspection, the only root of $x - 1 = 0$ is 1. Multiplying each side by x (Operation 4) gives $x^2 - x = 0$, which is satisfied if x is 0 or 1. (Check this by substitution.) But 0 *does not* satisfy the *original* equation. Thus, the equations are not equivalent.

Continuing, you may check that the equation $(x - 4)(x - 3) = 0$ is satisfied when x is 4 or 3. Dividing both sides by $x - 4$ (Operation 5) gives $x - 3 = 0$, whose only root is 3. Again, we do not have equivalence, since in this case a root has been "lost." Note that when x is 4, division by $x - 4$ implies division by 0, an invalid operation.

Finally, squaring each side of the equation $x = 2$ (Operation 6) gives $x^2 = 4$, which is true if $x = 2$ or -2. But -2 is not a root of the given equation.

From our discussion, it is clear that when Operations 4–6 are performed, we must be careful about drawing conclusions concerning the roots of a given equation. Operations 4 and 6 *can* produce an equation with more roots. Thus, you should check whether or not each "solution" obtained by the these

[3]See Section 0.4 for the definition of a polynomial.

operations satisfies the *original* equation. Operation 5 *can* produce an equation with fewer roots. In this case, any "lost" root may never be determined. Thus, avoid Operation 5 whenever possible.

In summary, an equation may be thought of as a set of restrictions on any variable in the equation. Operations 4–6 may increase or decrease the number of restrictions, giving solutions different from those of the original equation. However, Operations 1–3 never affect the restrictions.

TECHNOLOGY

A graphing calculator can be used to test for a root. For example, suppose we want to determine whether 3/2 is a root of the equation

$$2x^3 + 7x^2 = 19x + 60$$

First, we rewrite the equation so that one side is 0. Subtracting $19x + 60$ from both sides gives the equivalent equation

$$2x^3 + 7x^2 - 19x - 60 = 0$$

Using a TI-83 Plus graphing calculator, we enter the expression $2x^3 + 7x^2 - 19x - 60$ as Y_1 and then evaluate Y_1 at $x = 3/2$. Figure 0.2 shows that the result is -66, which is not 0. Thus, 3/2 is not a root. However, Y_1 evaluated at $x = -5/2$ *is* 0. So $-5/2$ is a root of the original equation.

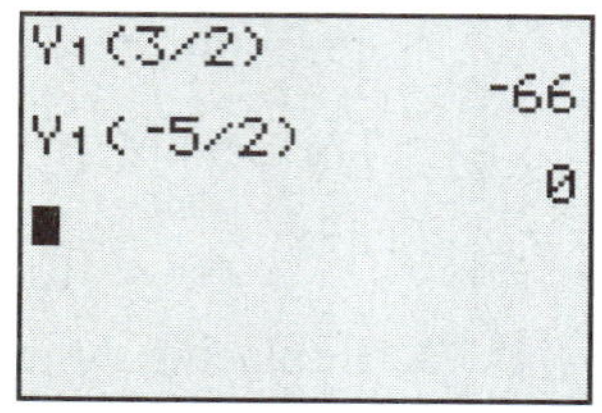

FIGURE 0.2 For $2x^3 + 7x^2 - 19x - 60 = 0$, $\frac{3}{2}$ is not a root, but $\frac{-5}{2}$ is a root.

It is worth noting that if the original equation had been in terms of the variable t, that is,

$$2t^3 + 7t^2 = 19t + 60$$

then we must replace t by x because the calculator evaluates Y_1 at a specified value of x, not t.

Linear Equations

The principles presented so far will now be demonstrated in the solution of a *linear equation*.

Definition

A ***linear equation*** in the variable x is an equation that can be written in the form

$$ax + b = 0 \tag{1}$$

where a and b are constants and $a \neq 0$.

A linear equation is also called a first-degree equation or an equation of degree one, since the highest power of the variable that occurs in Equation (1) is the first.

To solve a linear equation, we perform operations on it until we have an equivalent equation whose solutions are *obvious*. This means an equation in which the variable is isolated on one side, as the following examples show.

EXAMPLE 3 **Solving a Linear Equation**

Solve $5x - 6 = 3x$.

Solution: We begin by getting the terms involving x on one side and the constant on the other. Then we solve for x by the appropriate mathematical operation. We have

$$
\begin{aligned}
5x - 6 &= 3x \\
5x - 6 + (-3x) &= 3x + (-3x) && \text{(adding } -3x \text{ to both sides)} \\
2x - 6 &= 0 && \text{(simplifying, that is, Operation 3)} \\
2x - 6 + 6 &= 0 + 6 && \text{(adding 6 to both sides)} \\
2x &= 6 && \text{(simplifying)} \\
\frac{2x}{2} &= \frac{6}{2} && \text{(dividing both sides by 2)} \\
x &= 3
\end{aligned}
$$

Clearly, 3 is the only root of the last equation. Since each equation is equivalent to the one before it, we conclude that 3 must be the only root of $5x - 6 = 3x$. That is, the solution set is $\{3\}$. We can describe the first step in the solution as moving a term from one side of an equation to the other while changing its sign; this is commonly called *transposing*. Note that since the original equation can be put in the form $2x + (-6) = 0$, it is a linear equation.

▶ **Now work Problem 27.**

EXAMPLE 4 **Solving a Linear Equation**

Solve $2(p + 4) = 7p + 2$.

Solution: First, we remove parentheses. Then we collect similar terms and solve. We have

$$
\begin{aligned}
2(p + 4) &= 7p + 2 \\
2p + 8 &= 7p + 2 && \text{(distributive property)} \\
2p &= 7p - 6 && \text{(subtracting 8 from both sides)} \\
-5p &= -6 && \text{(subtracting } 7p \text{ from both sides)} \\
p &= \frac{-6}{-5} && \text{(dividing both sides by } -5\text{)} \\
p &= \frac{6}{5}
\end{aligned}
$$

▶ **Now work Problem 29.**

EXAMPLE 5 **Solving a Linear Equation**

Solve $\frac{7x+3}{2} - \frac{9x-8}{4} = 6.$

Solution: We first clear the equation of fractions by multiplying *both* sides by the least common denominator (LCD),[4] which is 4. Then we use various algebraic operations to obtain a solution. Thus,

The distributive property requires that *both* terms within the parentheses be multiplied by 4.

$$4\left(\frac{7x+3}{2} - \frac{9x-8}{4}\right) = 4(6)$$

$$4 \cdot \frac{7x+3}{2} - 4 \cdot \frac{9x-8}{4} = 24 \qquad \text{(distributive property)}$$

$$2(7x+3) - (9x-8) = 24 \qquad \text{(simplifying)}$$

$$14x + 6 - 9x + 8 = 24 \qquad \text{(distributive property)}$$

$$5x + 14 = 24 \qquad \text{(simplifying)}$$

$$5x = 10 \qquad \text{(subtracting 14 from both sides)}$$

$$x = 2 \qquad \text{(dividing both sides by 5)}$$

▶ **Now work Problem 31.**

Every linear equation has exactly one root.

Each equation in Examples 3–5 has one and only one root. This is true of every linear equation in one variable.

Literal Equations

Equations in which some of the constants are not specified but are represented by letters, such as a, b, c, or d, are called **literal equations,** and the letters are called **literal constants** or **arbitrary constants.** For example, in the literal equation $x + a = 4b$, we may consider a and b to be arbitrary constants. Formulas, such as $I = Prt$, that express a relationship between certain quantities may be regarded as literal equations. If we want to express a particular letter in a formula in terms of the others, this letter is considered the unknown.

EXAMPLE 6 **Solving Literal Equations**

a. *The equation $I = Prt$ is the formula for the simple interest I on a principal of P dollars at the annual interest rate of r for a period of t years. Express r in terms of I, P, and t.*

Solution: Here we consider r to be the unknown. To isolate r, we divide both sides by Pt. We have

$$I = Prt$$

$$\frac{I}{Pt} = \frac{Prt}{Pt}$$

$$\frac{I}{Pt} = r \quad \text{so} \quad r = \frac{I}{Pt}$$

[4]The least common denominator of two or more fractions is the smallest number with all the denominators as factors. That is, the LCD is the least common multiple of all the denominators.

When we divided both sides by Pt, we assumed that $Pt \neq 0$, since we cannot divide by 0. Similar assumptions will be made in solving other literal equations.

b. *The equation $S = P + Prt$ is the formula for the value S of an investment of a principal of P dollars at a simple annual interest rate of r for a period of t years. Solve for P.*

Solution:

$$S = P + Prt$$

$$S = P(1 + rt) \qquad \text{(factoring)}$$

$$\frac{S}{1 + rt} = P \qquad \text{(dividing both sides by } 1 + rt\text{)}$$

▶ **Now work Problem 87.**

EXAMPLE 7 Solving a Literal Equation

Solve $(a + c)x + x^2 = (x + a)^2$ for x.

Solution: We first simplify the equation and then get all terms involving x on one side:

$$(a + c)x + x^2 = (x + a)^2$$

$$ax + cx + x^2 = x^2 + 2ax + a^2$$

$$ax + cx = 2ax + a^2$$

$$cx - ax = a^2$$

$$x(c - a) = a^2$$

$$x = \frac{a^2}{c - a}$$

▶ **Now work Problem 89.**

Fractional Equations

A **fractional equation** is an equation in which an unknown is in a denominator. We illustrate that solving such a nonlinear equation may lead to a linear equation.

EXAMPLE 8 Solving a Fractional Equation

Solve $\dfrac{5}{x - 4} = \dfrac{6}{x - 3}$.

Solution:

Strategy: We first write the equation in a form that is free of fractions. Then we use standard algebraic techniques to solve the resulting linear equation.

An alternative solution that avoids multiplying both sides by the LCD is as follows:

$$\frac{5}{x-4} - \frac{6}{x-3} = 0$$

Assuming that x is neither 3 nor 4 and combining fractions gives

$$\frac{9-x}{(x-4)(x-3)} = 0$$

A fraction can be 0 only when its numerator is 0 and its denominator is not. Hence, $x = 9$.

Multiplying both sides by the LCD, $(x-4)(x-3)$, we have

$$(x-4)(x-3)\left(\frac{5}{x-4}\right) = (x-4)(x-3)\left(\frac{6}{x-3}\right)$$

$$5(x-3) = 6(x-4) \qquad \text{(linear equation)}$$

$$5x - 15 = 6x - 24$$

$$9 = x$$

In the first step, we multiplied each side by an expression involving the *variable* x. As we stated earlier, this means that we are not guaranteed that the last equation is equivalent to the *original* equation. Thus, we must check whether or not 9 satisfies the *original* equation. Since

$$\frac{5}{9-4} = \frac{5}{5} = 1 \quad \text{and} \quad \frac{6}{9-3} = \frac{6}{6} = 1$$

we see that 9 indeed satisfies the original equation.

▶ **Now work Problem 55.**

Some equations that are not linear do not have any solutions. In that case, we say that the solution set is the **empty set** or **null set,** which we denote by { } or ∅. Example 9 will illustrate.

EXAMPLE 9 Solving Fractional Equations

a. *Solve* $\frac{3x+4}{x+2} - \frac{3x-5}{x-4} = \frac{12}{x^2-2x-8}$.

Solution: Observing the denominators and noting that

$$x^2 - 2x - 8 = (x+2)(x-4)$$

we conclude that the LCD is $(x+2)(x-4)$. Multiplying both sides by the LCD, we have

$$(x+2)(x-4)\left(\frac{3x+4}{x+2} - \frac{3x-5}{x-4}\right) = (x+2)(x-4)\cdot\frac{12}{(x+2)(x-4)}$$

$$(x-4)(3x+4) - (x+2)(3x-5) = 12$$

$$3x^2 - 8x - 16 - (3x^2 + x - 10) = 12$$

$$3x^2 - 8x - 16 - 3x^2 - x + 10 = 12$$

$$-9x - 6 = 12$$

$$-9x = 18$$

$$x = -2 \qquad \textbf{(2)}$$

However, the *original* equation is not defined for $x = -2$ (we cannot divide by zero), so there are no roots. Thus, the solution set is ∅. Although -2 is a solution of Equation (2), it is not a solution of the *original* equation and is called an **extraneous solution** of that equation.

b. *Solve* $\frac{4}{x-5} = 0$.

Solution: The only way a fraction can equal zero is if the numerator is 0 but its denominator is not. Since the numerator, 4, is never 0, the solution set is $\emptyset$.

▶ **Now work Problem 49.**

EXAMPLE 10 **Literal Equation**

If $s = \dfrac{u}{au + v}$, *express u in terms of the remaining letters; that is, solve for u.*

Solution:

Strategy: Since the unknown, u, is in the denominator, we first clear fractions and then solve for u.

$$\begin{aligned}
s &= \frac{u}{au + v} \\
s(au + v) &= u \qquad \text{(multiplying both sides by } au + v\text{)} \\
sau + sv &= u \\
sau - u &= -sv \\
u(sa - 1) &= -sv \\
u &= \frac{-sv}{sa - 1} = \frac{sv}{1 - sa}
\end{aligned}$$

▶ **Now work Problem 91.**

Radical Equations

A **radical equation** is one in which an unknown occurs in a radicand. The next two examples illustrate the techniques employed to solve such equations.

EXAMPLE 11 **Solving a Radical Equation**

Solve $\sqrt{x^2 + 33} - x = 3$.

Solution: To solve this radical equation, we raise both sides to the same power to eliminate the radical. This operation does *not* guarantee equivalence, so we must check any resulting "solutions." We begin by isolating the radical on one side. Then we square both sides and solve using standard techniques. Thus,

$$\begin{aligned}
\sqrt{x^2 + 33} &= x + 3 \\
x^2 + 33 &= (x + 3)^2 \qquad \text{(squaring both sides)} \\
x^2 + 33 &= x^2 + 6x + 9 \\
24 &= 6x \\
4 &= x
\end{aligned}$$

You should show by substitution that 4 is indeed a root.

▶ **Now work Problem 79.**

With some radical equations, you may have to raise both sides to the same power more than once, as Example 12 shows.

EXAMPLE 12 **Solving a Radical Equation**

The reason why we desire one radical on each side is to eliminate squaring a binomial with two different radicals.

Solve $\sqrt{y-3} - \sqrt{y} = -3$.

Solution: When an equation has two terms involving radicals, first write the equation so that one radical is on each side, if possible. Then square and solve. We have

$$\sqrt{y-3} = \sqrt{y} - 3$$
$$y - 3 = y - 6\sqrt{y} + 9 \qquad \text{(squaring both sides)}$$
$$6\sqrt{y} = 12$$
$$\sqrt{y} = 2$$
$$y = 4 \qquad \text{(squaring both sides)}$$

Substituting 4 into the left side of the *original* equation gives $\sqrt{1} - \sqrt{4}$, which is -1. Since this does not equal the right side, -3, there is no solution. That is, the solution set is $\emptyset$. Here 4 is an extraneous solution.

▸ Now work Problem 77.

Exercise 0.7

In Problems 1–6, determine by substitution which of the given numbers, if any, satisfy the given equation.

*1. $9x - x^2 = 0$; 1, 0

2. $12 - 7x = -x^2$; 4, 3

3. $y + 2(y-3) = 4$; $\frac{10}{3}$, 1

4. $2x + x^2 - 8 = 0$; 2, -4

5. $x(6+x) - 2(x+1) - 5x = 4$; -2, 0

6. $x(x+1)^2(x+2) = 0$; 0, -1, 2

In Problems 7–16, determine what operations were applied to the first equation to obtain the second. State whether or not the operations guarantee that the equations are equivalent. Do not solve the equations.

7. $x - 5 = 4x + 10$; $x = 4x + 15$

8. $8x - 4 = 16$; $x - \frac{1}{2} = 2$

9. $x = 4$; $x^3 = 64$

10. $\frac{1}{2}x^2 + 3 = x - 9$; $x^2 + 6 = 2x - 18$

11. $x^2 - 2x = 0$; $x - 2 = 0$

12. $\dfrac{2}{x-2} + x = x^2$; $2 + x(x-2) = x^2(x-2)$

13. $\dfrac{x^2-1}{x-1} = 3$; $x^2 - 1 = 3(x-1)$

14. $(x+3)(x+11)(x+7) = (x+3)(x+2)$; $(x+11)(x+7) = x+2$

15. $\dfrac{x(x+1)}{x-5} = x(x+9)$; $x + 1 = (x+9)(x-5)$

16. $2x^2 - 9 = x$; $x^2 - \frac{1}{2}x = \frac{9}{2}$

In Problems 17–80, solve the equations.

17. $4x = 10$

18. $0.2x = 7$

19. $3y = 0$

20. $2x - 4x = -5$

21. $-8x = 12 - 20$

22. $3 - 2x = 4$

23. $5x - 3 = 9$

24. $\sqrt{2}x + 3 = 8$

25. $7x + 7 = 2(x+1)$

26. $4s + 3s - 1 = 41$

*27. $2(p-1) - 3(p-4) = 4p$

28. $t = 2 - 2[2t - 3(1-t)]$

*29. $\dfrac{x}{5} = 2x - 6$

30. $\dfrac{5y}{7} - \dfrac{6}{7} = 2 - 4y$

*31. $7 + \dfrac{4x}{9} = \dfrac{x}{2}$

32. $\dfrac{x}{3} - 4 = \dfrac{x}{5}$

33. $r = \dfrac{4}{3}r - 5$

34. $\dfrac{x}{2} + \dfrac{x}{3} = 7$

35. $3x + \dfrac{x}{5} - 5 = \dfrac{1}{5} + 5x$

36. $y - \dfrac{y}{2} + \dfrac{y}{3} - \dfrac{y}{4} = \dfrac{y}{5}$

37. $\frac{2y-3}{4} = \frac{6y+7}{3}$

38. $\frac{t}{4} + \frac{5}{3}t = \frac{7}{2}(t-1)$

39. $w + \frac{w}{2} - \frac{w}{3} + \frac{w}{4} = 5$

40. $\frac{7+2(x+1)}{3} = \frac{6x}{5}$

41. $\frac{x+2}{3} - \frac{2-x}{6} = x - 2$

42. $\frac{x}{5} + \frac{2(x-4)}{10} = 7$

43. $\frac{9}{5}(3-x) = \frac{3}{4}(x-3)$

44. $\frac{2y-7}{3} + \frac{8y-9}{14} = \frac{3y-5}{21}$

45. $\frac{4}{3}(5x-2) = 7[x-(5x-2)]$

46. $(3x-1)^2 - (5x-3)^2 = -(4x-2)^2$

47. $\frac{5}{x} = 25$

48. $\frac{4}{x-1} = 2$

***49.** $\frac{7}{3-x} = 0$

50. $\frac{3x-5}{x-3} = 0$

51. $\frac{4}{8-x} = \frac{3}{4}$

52. $\frac{x+3}{x} = \frac{2}{5}$

53. $\frac{q}{5q-4} = \frac{1}{3}$

54. $\frac{4p}{7-p} = 1$

***55.** $\frac{1}{p-1} = \frac{2}{p-2}$

56. $\frac{2x-3}{4x-5} = 6$

57. $\frac{1}{x} + \frac{1}{7} = \frac{3}{7}$

58. $\frac{4}{t-3} = \frac{3}{t-4}$

59. $\frac{3x-2}{2x+3} = \frac{3x-1}{2x+1}$

60. $\frac{x+2}{x-1} + \frac{x+1}{3-x} = 0$

61. $\frac{y-6}{y} - \frac{6}{y} = \frac{y+6}{y-6}$

62. $\frac{y-2}{y+2} = \frac{y-2}{y+3}$

63. $\frac{-4}{x-1} = \frac{7}{2-x} + \frac{3}{x+1}$

64. $\frac{1}{x-3} - \frac{3}{x-2} = \frac{4}{1-2x}$

65. $\frac{9}{x-3} = \frac{3x}{x-3}$

66. $\frac{x}{x+3} - \frac{x}{x-3} = \frac{3x-4}{x^2-9}$

67. $\sqrt{x+5} = 4$

68. $\sqrt{z-2} = 3$

69. $\sqrt{3x-4} - 8 = 0$

70. $6 - \sqrt{2x+5} = 0$

71. $\sqrt{\frac{x}{2}+1} = \frac{2}{3}$

72. $(x+6)^{1/2} = 7$

73. $\sqrt{4x-6} = \sqrt{x}$

74. $\sqrt{4+3x} = \sqrt{2x+5}$

75. $(x-3)^{3/2} = 8$

76. $\sqrt{y^2-9} = 9 - y$

***77.** $\sqrt{y} + \sqrt{y+2} = 3$

78. $\sqrt{x} - \sqrt{x+1} = 1$

***79.** $\sqrt{z^2+2z} = 3 + z$

80. $\sqrt{\frac{1}{w}} - \sqrt{\frac{2}{5w-2}} = 0$

In Problems 81–92, express the indicated symbol in terms of the remaining symbols.

81. $I = Prt; r$

82. $ax + b = 0; x$

83. $p = 8q - 1; q$

84. $p = -3q + 6; q$

85. $S = P(1+rt); r$

86. $r = \frac{2mI}{B(n+1)}; I$

***87.** $S = \frac{n}{2}(a_1 + a_n); a_1$

88. $S = \frac{R[(1+i)^n - 1]}{i}; R$

***89.** $r = \frac{d}{1-dt}; t$

90. $\frac{x-a}{b-x} = \frac{x-b}{a-x}; x$

***91.** $r = \frac{2mI}{B(n+1)}; n$

92. $\frac{1}{p} + \frac{1}{q} = \frac{1}{f}; q$

93. Geometry Use the formula $P = 2l + 2w$ to find the length l of a rectangle whose perimeter P is 660 m and whose width w is 160 m.

94. Geometry Use the formula $A = \frac{1}{2}bh$ to find the height h of a triangle whose area A is 75 cm^2 and whose base b is 15 cm.

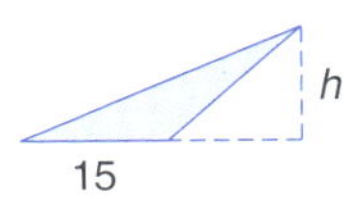

95. Sales Tax A salesperson needs to calculate the cost of an item with a sales tax of 8.25%. Write an equation that represents the total cost c of an item costing x dollars.

96. Revenue A day care center's total monthly revenue from the care of x toddlers is given by $r = 450x$, and its total monthly costs are given by $c = 380x + 3500$. How many toddlers need to be enrolled each month to break even? In other words, when will revenue equal costs?

97. Straight-Line Depreciation If you purchase an item for business use, in preparing your income tax you may be able to spread out its expense over the life of the item. This is called *depreciation*. One method of depreciation is *straight-line depreciation,* in which the annual depreciation is computed by dividing the cost of the item, less its estimated salvage value, by its useful life. Suppose the cost is C dollars, the useful life is N years, and there is no salvage value. Then the value V (in dollars) of the item at the end of n years is given by

$$V = C\left(1 - \frac{n}{N}\right)$$

If new office furniture is purchased for \$3200, has a useful life of 8 years, and has no salvage value, after how many years will it have a value of \$2000?

98. Radar Beam When radar is used on a highway to determine the speed of a car, a radar beam is sent out and reflected from the moving car. The difference F (in cycles per second) in frequency between the original and reflected beams is given by

$$F = \frac{vf}{334.8}$$

where v is the speed of the car in miles per hour and f is the frequency of the original beam (in megacycles per second).

Suppose you are driving along a highway with a speed limit of 65 mi/h. A police officer aims a radar beam with a frequency of 2500 megacycles per second at your car, and the officer observes the difference in frequency to be 495 cycles per second. Can the officer claim that you were speeding?

99. Savings Paula and Sam want to buy a house, so they have decided to save one-fourth of each of their salaries. Paula earns \$24.00 per hour and receives an extra \$8.00 a week because she declined company benefits, and Sam earns \$28.00 per hour plus benefits. They want to save at least \$405.00 each week. How many hours must they each work each week?

100. Predator–Prey Relation To study a predator–prey relationship, an experiment[5] was conducted in which a blindfolded subject, the "predator," stood in front of a 3-ft-square table on which uniform sandpaper discs, the "prey," were placed. For 1 minute the "predator" searched for the discs by tapping with a finger. Whenever a disc was found, it was removed and searching resumed. The experiment was repeated for various disc densities (number of discs per 9 ft^2). It was estimated that if y is the number of discs picked up in 1 minute when x discs are on the table, then

$$y = a(1 - by)x$$

where a and b are constants. Solve this equation for y.

[5]C. S. Holling, "Some Characteristics of Simple Types of Predation and Parasitism," *Canadian Entomologist* 91, no. 7 (1959), 385–98.

101. Prey Density In a certain area, the number y of moth larvae consumed by a single predatory beetle over a given period of time is given by

$$y = \frac{1.4x}{1 + 0.09x}$$

where x is the *prey density* (the number of larvae per unit of area). What prey density would allow a beetle to survive if it needs to consume 10 larvae over the given period?

102. Store Hours Suppose the ratio of the number of hours a video store is open to the number of daily customers is constant. When the store is open 8 hours, the number of customers is 92 less than the maximum number of customers. When the store is open 10 hours, the number of customers is 46 less than the maximum number of customers. Write an equation describing this situation, and find the maximum number of daily customers.

103. Travel Time The time it takes a boat to travel a given distance upstream (against the current) can be calculated by dividing the distance by the difference of the speed of the boat and the speed of the current. Write an equation that calculates the time t it takes a boat moving at a speed r against a current c to travel a distance d. Solve your equation for c.

104. Ramp Length The difference between the length of a ramp and the length of the horizontal distance it covers is 5 feet. The square of the vertical distance the ramp covers is 45 square feet. Write an equation for the difference and solve. What is the length of the ramp?

105. Automobile Skidding Police have used the formula $s = \sqrt{30fd}$ to estimate the speed s (in miles per hour) of a car if it skidded d feet when stopping. The literal number f is the coefficient of friction, determined by the kind of road (such as concrete, asphalt, gravel, or tar) and whether the road is wet or dry. At 45 mi/h, about how many feet will a car skid on a dry concrete road? The value of f for dry concrete is 0.8. Give your answer to the nearest foot.

In Problems 106–109, use a graphing calculator to determine which of the given numbers, if any, are roots of the given equation.

106. $112x^2 = 6x + 1; \frac{1}{8}, -\frac{2}{5}, -\frac{1}{14}$

107. $8x^3 + 11x + 21 = 58x^2; 5, -\frac{1}{2}, \frac{2}{3}$

108. $\frac{3.1t - 7}{4.8t - 2} = 7; \sqrt{6}, -\frac{47}{52}, \frac{14}{61}$

109. $\left(\frac{v}{v+3}\right)^2 = v; 0, \frac{27}{4}, \frac{13}{3}$

OBJECTIVE *To solve quadratic equations by factoring or by using the quadratic formula.*

0.8 Quadratic Equations

To learn how to solve more complicated problems, we turn to methods of solving *quadratic equations*.

Definition

A ***quadratic equation*** in the variable x is an equation that can be written in the form

$$ax^2 + bx + c = 0 \tag{1}$$

where a, b, and c are constants and $a \neq 0$.

A quadratic equation is also called a *second-degree equation* or an *equation of degree two,* since the highest power of the variable that occurs is the second. Whereas a linear equation has only one root, a quadratic equation may have two different roots.

Solution by Factoring

A useful method of solving quadratic equations is based on factoring, as the following example shows.

EXAMPLE 1 Solving a Quadratic Equation by Factoring

a. *Solve* $x^2 + x - 12 = 0$.

Solution: The left side factors easily:

$$(x - 3)(x + 4) = 0$$

Think of this as two quantities, $x - 3$ and $x + 4$, whose product is zero. **Whenever the product of two or more quantities is *zero,* at least one of the quantities *must* be zero.** This means that either

$$x - 3 = 0 \quad \text{or} \quad x + 4 = 0$$

Solving these gives $x = 3$ and $x = -4$. Thus, the roots of the original equation are 3 and -4, and the solution set is $\{3, -4\}$.

b. *Solve* $6w^2 = 5w$.

We do not divide both sides by w (a variable) since equivalence is not guaranteed and we may "lose" a root.

Solution: We write the equation as

$$6w^2 - 5w = 0$$

so that one side is 0. Factoring gives

$$w(6w - 5) = 0$$

Setting each factor equal to 0, we have

$$\begin{aligned} w = 0 \quad &\text{or} \quad 6w - 5 = 0 \\ w = 0 \quad &\text{or} \quad \phantom{6w-{}}6w = 5 \end{aligned}$$

Thus, the roots are $w = 0$ and $w = \frac{5}{6}$. Note that if we had divided both sides of $6w^2 = 5w$ by w and obtained $6w = 5$, our only solution would be $w = \frac{5}{6}$. That is, we would lose the root $w = 0$. This confirms our discussion of Operation 5 in Section 0.7.

▸ Now work Problem 3.

CAUTION You should approach a problem like this with caution. If the product of two quantities is equal to -2, it is not true that at least one of the quantities must be -2. Why? You should **not** set each factor equal to -2; doing so will not provide solutions to the given equation.

EXAMPLE 2 Solving a Quadratic Equation by Factoring

Solve $(3x - 4)(x + 1) = -2$.

Solution: We first multiply the factors on the left side:

$$3x^2 - x - 4 = -2$$

Rewriting this equation so that 0 appears on one side, we have

$$3x^2 - x - 2 = 0$$
$$(3x + 2)(x - 1) = 0$$
$$x = -\frac{2}{3}, 1$$

▶ **Now work Problem 7.**

Some equations that are not quadratic may be solved by factoring, as Example 3 shows.

EXAMPLE 3 Solving a Higher Degree Equation by Factoring

a. *Solve* $4x - 4x^3 = 0$.

Solution: This is called a *third-degree equation*. We proceed to solve it as follows:

$$4x - 4x^3 = 0$$
$$4x(1 - x^2) = 0 \quad \text{(factoring)}$$
$$4x(1 - x)(1 + x) = 0 \quad \text{(factoring)}$$

CAUTION Do not neglect the fact that the factor x gives rise to a root.

Setting each factor equal to 0 gives $4 = 0$ (impossible), $x = 0$, $1 - x = 0$, or $1 + x = 0$. Thus,

$$x = 0, 1, -1$$

which we can write as $x = 0, \pm 1$.

b. *Solve* $x(x + 2)^2(x + 5) + x(x + 2)^3 = 0$.

Solution: Factoring $x(x + 2)^2$ from both terms on the left side, we have

$$x(x + 2)^2[(x + 5) + (x + 2)] = 0$$
$$x(x + 2)^2(2x + 7) = 0$$

Hence, $x = 0$, $x + 2 = 0$, or $2x + 7 = 0$, from which it follows that $x = 0, -2, -\frac{7}{2}$.

▶ **Now work Problem 23.**

EXAMPLE 4 A Fractional Equation Leading to a Quadratic Equation

Solve

$$\frac{y + 1}{y + 3} + \frac{y + 5}{y - 2} = \frac{7(2y + 1)}{y^2 + y - 6} \quad \textbf{(2)}$$

Solution: Multiplying both sides by the LCD, $(y+3)(y-2)$, we get

$$(y-2)(y+1)+(y+3)(y+5)=7(2y+1) \qquad (3)$$

Since Equation (2) was multiplied by an expression involving the variable y, remember (from Section 0.7) that Equation (3) is not necessarily equivalent to Equation (2). After simplifying Equation (3), we have

$$2y^2-7y+6=0 \qquad \text{(quadratic equation)}$$
$$(2y-3)(y-2)=0 \qquad \text{(factoring)}$$

We have shown that *if* y satisfies the original equation *then* $y=\frac{3}{2}$ or $y=2$. Thus, $\frac{3}{2}$ and 2 are the only *possible* roots of the given equation. But 2 cannot be a root of Equation (2), since substitution leads to a denominator of 0. However, you should check that $\frac{3}{2}$ does indeed satisfy the *original* equation. Hence, its only root is $\frac{3}{2}$.

▶ **Now work Problem 53.**

EXAMPLE 5 **Solution by Factoring**

Solve $x^2=3$.

Solution:

CAUTION Do not hastily conclude that the solution of $x^2=3$ consists only of $x=\sqrt{3}$.

$$x^2=3$$
$$x^2-3=0$$

Factoring, we obtain

$$(x-\sqrt{3})(x+\sqrt{3})=0$$

Thus $x-\sqrt{3}=0$ or $x+\sqrt{3}=0$, so $x=\pm\sqrt{3}$.

A more general form of the equation $x^2=3$ is $u^2=k$. In the same manner as the preceding we can show that

$$\text{If } u^2=k \quad \text{then} \quad u=\pm\sqrt{k}.$$

Quadratic Formula

Solving quadratic equations by factoring can be quite difficult, as is evident by trying that method on $0.7x^2-\sqrt{2}x-8\sqrt{5}=0$. However, there is a formula called the *quadratic formula* that gives the roots of any quadratic equation.

Quadratic Formula

The roots of the quadratic equation $ax^2+bx+c=0$, where a, b, and c are constants and $a\neq 0$, are given by

$$x=\frac{-b\pm\sqrt{b^2-4ac}}{2a}$$

EXAMPLE 6 **A Quadratic Equation with Two Real Roots**

Solve $4x^2 - 17x + 15 = 0$ *by the quadratic formula.*

Solution: Here $a = 4$, $b = -17$, and $c = 15$. Thus,

$$x = \frac{-b \pm \sqrt{b^2 - 4ac}}{2a} = \frac{-(-17) \pm \sqrt{(-17)^2 - 4(4)(15)}}{2(4)}$$

$$= \frac{17 \pm \sqrt{49}}{8} = \frac{17 \pm 7}{8}$$

The roots are $\frac{17+7}{8} = \frac{24}{8} = 3$ and $\frac{17-7}{8} = \frac{10}{8} = \frac{5}{4}$.

▶ **Now work Problem 31.**

EXAMPLE 7 **A Quadratic Equation with One Real Root**

Solve $2 + 6\sqrt{2}y + 9y^2 = 0$ *by the quadratic formula.*

Solution: Look at the arrangement of the terms. Here $a = 9$, $b = 6\sqrt{2}$, and $c = 2$. Hence,

$$y = \frac{-b \pm \sqrt{b^2 - 4ac}}{2a} = \frac{-6\sqrt{2} \pm \sqrt{0}}{2(9)}$$

Thus,

$$y = \frac{-6\sqrt{2} + 0}{18} = -\frac{\sqrt{2}}{3} \quad \text{or} \quad y = \frac{-6\sqrt{2} - 0}{18} = -\frac{\sqrt{2}}{3}$$

Therefore, the only root is $-\frac{\sqrt{2}}{3}$.

▶ **Now work Problem 33.**

EXAMPLE 8 **A Quadratic Equation with No Real Solution**

Solve $z^2 + z + 1 = 0$ *by the quadratic formula.*

Solution: Here $a = 1$, $b = 1$, and $c = 1$. The roots are

$$z = \frac{-b \pm \sqrt{b^2 - 4ac}}{2a} = \frac{-1 \pm \sqrt{-3}}{2}$$

Now, $\sqrt{-3}$ denotes a number whose square is -3. However, no such real number exists, since the square of any real number is nonnegative. Thus, the equation has no real roots.[6]

▶ **Now work Problem 37.**

This describes the nature of the roots of a quadratic equation.

From Examples 6–8, you can see that a quadratic equation has either two different real roots, exactly one real root, or no real roots, depending on whether $b^2 - 4ac > 0$, $= 0$, or < 0, respectively.

[6] $\frac{-1 \pm \sqrt{-3}}{2}$ can be expressed as $\frac{-1 \pm i\sqrt{3}}{2}$, where $i = \sqrt{-1}$ is called the *imaginary unit*.

TECHNOLOGY

Using the program feature of a graphing calculator, one can create a program that gives the real roots of the quadratic equation $\mathrm{A}x^2+\mathrm{B}x+\mathrm{C}=0$. Figure 0.3 shows such a program for the TI-83 Plus graphing calculator. To execute it for

$$20x^2 - 33x + 10 = 0$$

```
PROGRAM:QUADROOT
:Prompt A,B,C
:If B²-4AC<0
:Then
:Disp "NOREALROO
T"
:Stop
:End
```

```
PROGRAM:QUADROOT
:Disp (-B+√(B²-4
AC))/(2A)
:Disp (-B-√(B²-4
AC))/(2A)
```

FIGURE 0.3 Program to find real roots of $\mathrm{A}x^2+\mathrm{B}x+\mathrm{C}=0$.

you are prompted to enter the values of A, B, and C. (See Figure 0.4.) The resulting roots are $x = 1.25$ and $x = 0.4$.

```
prgmQUADROOT
A=?20
B=?-33
C=?10
                1.25
                  .4
                Done
```

FIGURE 0.4 Roots of $20x^2 - 33x + 10 = 0$.

Quadratic-Form Equations

Sometimes an equation that is not quadratic can be transformed into a quadratic equation by an appropriate substitution. In this case, the given equation is said to be a **quadratic-form equation.** The next example will illustrate.

EXAMPLE 9 **Solving a Quadratic-Form Equation**

Solve $\dfrac{1}{x^6} + \dfrac{9}{x^3} + 8 = 0$.

Solution: This equation can be written as

$$\left(\frac{1}{x^3}\right)^2 + 9\left(\frac{1}{x^3}\right) + 8 = 0$$

so it is quadratic in $1/x^3$ and hence is a quadratic-form equation. Substituting the variable w for $1/x^3$ gives a quadratic equation in the variable w, which we can then solve:

$$w^2 + 9w + 8 = 0$$

$$(w+8)(w+1) = 0$$

$$w = -8 \quad \text{or} \quad w = -1$$

Do not assume that -8 and -1 are solutions of the *original* equation.

Returning to the variable x, we have

$$\frac{1}{x^3} = -8 \quad \text{or} \quad \frac{1}{x^3} = -1$$

Thus,

$$x^3 = -\frac{1}{8} \quad \text{or} \quad x^3 = -1$$

from which it follows that

$$x = -\frac{1}{2} \quad \text{or} \quad x = -1$$

Checking, we find that these values of x satisfy the original equation.

▶ **Now work Problem 49.**

Exercise 0.8

In Problems 1–30, solve by factoring.

1. $x^2 - 4x + 4 = 0$
2. $t^2 + 3t + 2 = 0$
*3. $t^2 - 8t + 15 = 0$
4. $x^2 + x - 12 = 0$
5. $x^2 - 2x - 3 = 0$
6. $x^2 - 16 = 0$
*7. $u^2 - 13u = -36$
8. $3w^2 - 12w + 12 = 0$
9. $x^2 - 4 = 0$
10. $3u^2 - 6u = 0$
11. $z^2 - 8z = 0$
12. $x^2 + 9x = -14$
13. $4x^2 + 1 = 4x$
14. $2z^2 + 9z = 5$
15. $v(3v - 5) = -2$
16. $8 + 2x - 3x^2 = 0$
17. $-x^2 + 3x + 10 = 0$
18. $\frac{1}{7}y^2 = \frac{3}{7}y$
19. $2p^2 = 3p$
20. $-r^2 - r + 12 = 0$
21. $x(x + 4)(x - 1) = 0$
22. $(w - 3)^2(w + 1)^2 = 0$
*23. $x^3 - 64x = 0$
24. $x^3 - 4x^2 - 5x = 0$
25. $6x^3 + 5x^2 - 4x = 0$
26. $(x + 1)^2 - 5x + 1 = 0$
27. $(x - 3)(x^2 - 4) = 0$
28. $3(x^2 + 3x - 10)(x - 8) = 0$
29. $p(p - 3)^2 - 4(p - 3)^3 = 0$
30. $x^4 - 3x^2 + 2 = 0$

In Problems 31–44, find all real roots by using the quadratic formula.

*31. $x^2 + 2x - 24 = 0$
32. $x^2 - 2x - 15 = 0$
*33. $4x^2 - 12x + 9 = 0$
34. $q^2 - 5q = 0$
35. $p^2 - 7p + 3 = 0$
36. $2 - 2x + x^2 = 0$
*37. $4 - 2n + n^2 = 0$
38. $2x^2 + x = 5$
39. $4x^2 + 5x - 2 = 0$
40. $w^2 - 2\sqrt{2}w + 2 = 0$
41. $0.02w^2 - 0.3w = 20$
42. $0.01x^2 + 0.2x - 0.6 = 0$
43. $2x^2 + 4x = 5$
44. $-2x^2 - 6x + 5 = 0$

In Problems 45–54, solve the given quadratic-form equation.

45. $x^4 - 5x^2 + 6 = 0$
46. $X^4 - 3X^2 - 10 = 0$
47. $\dfrac{2}{x^2} + \dfrac{3}{x} - 2 = 0$
48. $x^{-2} + x^{-1} - 12 = 0$
*49. $x^{-4} - 9x^{-2} + 20 = 0$
50. $\dfrac{1}{x^4} - \dfrac{9}{x^2} + 8 = 0$
51. $(X - 5)^2 + 7(X - 5) + 10 = 0$
52. $(x + 5)^2 - 8(x + 5) = 0$
*53. $\dfrac{1}{(x - 2)^2} - \dfrac{12}{x - 2} + 35 = 0$
54. $\dfrac{2}{(x + 4)^2} + \dfrac{7}{x + 4} + 3 = 0$

In Problems 55–76, solve by any method.

55. $x^2 = \dfrac{x + 3}{2}$
56. $\dfrac{x}{2} = \dfrac{7}{x} - \dfrac{5}{2}$
57. $\dfrac{3}{x - 4} + \dfrac{x - 3}{x} = 2$
58. $\dfrac{2}{2x + 1} - \dfrac{6}{x - 1} = 5$
59. $\dfrac{6x + 7}{2x + 1} - \dfrac{6x + 1}{2x} = 1$
60. $\dfrac{6(w + 1)}{2 - w} + \dfrac{w}{w - 1} = 3$
61. $\dfrac{2}{r - 2} - \dfrac{r + 1}{r + 4} = 0$
62. $\dfrac{2x - 3}{2x + 5} + \dfrac{2x}{3x + 1} = 1$
63. $\dfrac{t + 1}{t + 2} + \dfrac{t + 3}{t + 4} = \dfrac{t + 5}{t^2 + 6t + 8}$
64. $\dfrac{3}{t + 1} + \dfrac{4}{t} = \dfrac{12}{t + 2}$
65. $\dfrac{2}{x^2 - 1} - \dfrac{1}{x(x - 1)} = \dfrac{2}{x^2}$
66. $5 - \dfrac{3(x + 3)}{x^2 + 3x} = \dfrac{1 - x}{x}$
67. $\sqrt{2x - 3} = x - 3$
68. $3\sqrt{x + 4} = x - 6$
69. $q + 2 = 2\sqrt{4q - 7}$
70. $x + \sqrt{4x} - 5 = 0$
71. $\sqrt{x + 7} - \sqrt{2x} - 1 = 0$
72. $\sqrt{x} - \sqrt{2x - 8} - 2 = 0$
73. $\sqrt{x} - \sqrt{2x + 1} + 1 = 0$
74. $\sqrt{y - 2} + 2 = \sqrt{2y + 3}$
75. $\sqrt{x + 3} + 1 = 3\sqrt{x}$
76. $\sqrt{\sqrt{x} + 2} = \sqrt{2x - 4}$

In Problems 77 and 78, find the roots, rounded to two decimal places.

77. $0.04x^2 - 2.7x + 8.6 = 0$ **78.** $0.01x^2 + 0.2x - 0.6 = 0$

79. Geometry The area of a rectangular picture with a width 2 inches less than its length is 48 square inches. What are the dimensions of the picture?

80. Temperature The temperature has been rising X degrees per day for X days. X days ago it was 15 degrees. Today it is 51 degrees. How much has the temperature been rising each day? How many days has it been rising?

81. Economics One root of the economics equation

$$\overline{M} = \frac{Q(Q+10)}{44}$$

is $-5 + \sqrt{25 + 44\overline{M}}$. Verify this by using the quadratic formula to solve for Q in terms of $\overline{M}$. Here Q is real income and $\overline{M}$ is the level of money supply.

82. Diet for Rats A group of biologists studied the nutritional effects on rats that were fed a diet containing 10% protein.[7] The protein was made up of yeast and corn flour. By changing the percentage P (expressed as a decimal) of yeast in the protein mix, the group estimated that the average weight gain g (in grams) of a rat over a period of time was given by

$$g = -200P^2 + 200P + 20$$

What percentage of yeast gave an average weight gain of 60 grams?

83. Drug Dosage There are several rules for determining doses of medicine for children when the adult dose has been specified. Such rules may be based on weight, height, and so on. If A is the age of the child, d is the adult dose, and c is the child's dose, then here are two rules:

$$\text{Young's rule}: \quad c = \frac{A}{A+12}d$$

$$\text{Cowling's rule}: \quad c = \frac{A+1}{24}d$$

At what age are the children's doses the same under both rules? Round your answer to the nearest year.

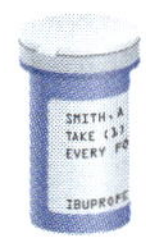

84. Delivered Price of a Good In a discussion of the delivered price of a good from a mill to a customer, DeCanio[8] arrives at and solves the two quadratic equations

$$(2n-1)v^2 - 2nv + 1 = 0$$

and

$$nv^2 - (2n+1)v + 1 = 0$$

where $n \geq 1$.

(a) Solve the first equation for v.
(b) Solve the second equation for v if $v < 1$.

85. Motion Suppose the height h of an object thrown straight upward from the ground is given by

$$h = 39.2t - 4.9t^2$$

where h is in meters and t is the elapsed time in seconds.

(a) After how many seconds does the object strike the ground?
(b) When is the object at a height of 68.2 m?

In Problems 86–91, use a program to determine any real roots of the equation. Round answers to three decimal places. For Problems 86 and 87, confirm your result algebraically.

86. $2x^2 - 3x - 27 = 0$

87. $8x^2 - 18x + 9 = 0$

88. $15x^2 + 7x - 3 = 0$

89. $27x^2 - \frac{11}{8}x + 5 = 0$

90. $\frac{9}{2}z^2 - 6.3 = \frac{z}{3}(1.1 - 7z)$

91. $(\pi t - 4)^2 = 4.1t - 3$

[7] Adapted from R. Bressani, "The Use of Yeast in Human Foods," in R. I. Mateles and S. R. Tannenbaum (eds.), *Single-Cell Protein* (Cambridge, MA: MIT Press, 1968).

[8] S. J. DeCanio, "Delivered Pricing and Multiple Basing Point Equilibria: A Revolution," *Quarterly Journal of Economics* 99, no. 2 (1984), 329–49.

Modeling Load Cell Behavior[9]

What do a bathroom scale and a crash test dummy have in common? They both contain load cells. A load cell is a device that measures force by translating it into an electrical signal. In a scale, one or more load cells measure the weight resting on the scale. In a crash test dummy, load cells distributed over the dummy's body measure impact forces as the dummy collides with the car's interior.

As measuring devices, load cells have to behave predictably and consistently. A common requirement is that the voltage output, V, be related to the force input, F, by an equation that is linear in form:

$$V = aF + b$$

Linear equations were discussed in Section 0.7. A linear response permits a simple translation of the voltage into a meter readout.

Lifting equipment often contains load cells that provide warning when the equipment is reaching its rated limit. Suppose a company that manufactures load cells for use in cranes puts a sample cell through a calibration test and obtains the following data (with force measured in thousands of pounds and voltage measured in volts).

Force	Voltage	Force	Voltage
150.000	0.11019	1650.000	1.20001
300.000	0.21956	1800.000	1.30822
450.000	0.32949	1950.000	1.41599
600.000	0.43899	2100.000	1.52399
750.000	0.54803	2250.000	1.63194
900.000	0.65694	2400.000	1.73947
1050.000	0.76562	2550.000	1.84646
1200.000	0.87487	2700.000	1.95392
1350.000	0.98292	2850.000	2.06128
1500.000	1.09146	3000.000	2.16844

If the load cell is behaving properly, a linear equation will be a good *model* of the data. In other words, when the data values are plotted as points on a graph and a line is superimposed, the points and the line should coincide.

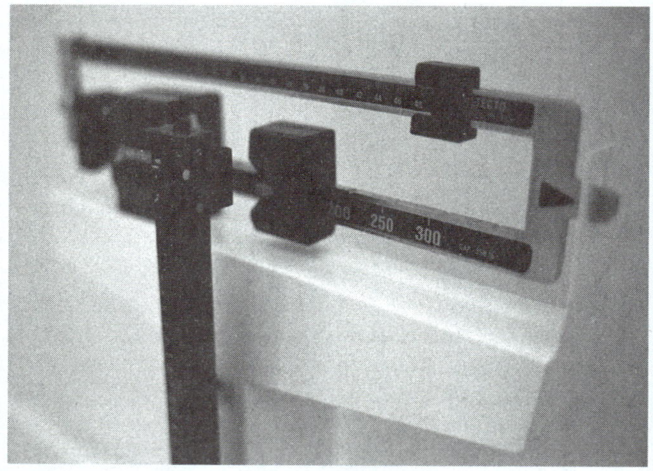

The mathematics for finding the line that best models the data is quite tedious. Fortunately, a graphing calculator can do it automatically. The result is

$$V = 0.0007221F + 0.006081368$$

Plotting the data and the equation together produces the result shown in Figure 0.5.

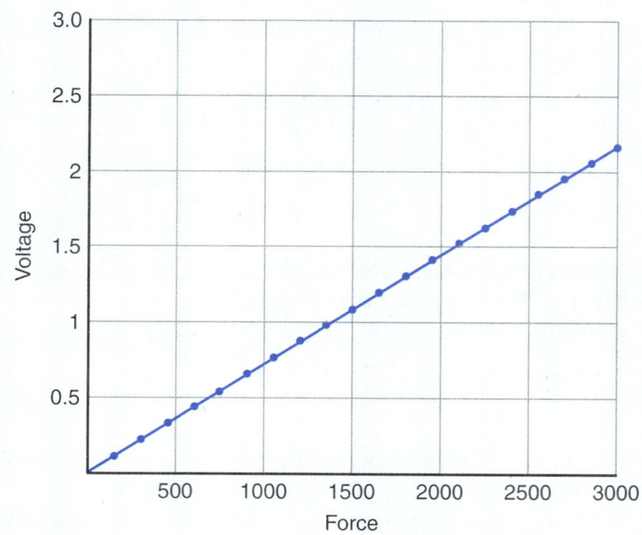

FIGURE 0.5 The linear model.

[9]Based on Section 4.6.1 of *Engineering Statistics Handbook*, National Institute of Standards and Technology/SEMATECH, www.nist.gov/itl/div898/handbook/pmd/section6/pmd61.htm.

It looks as if the linear model is a very good fit indeed. But is it as good as it can be? Let us look at the differences between the measured voltages and the respective values predicted by the linear model. For each force magnitude, we subtract the corresponding voltage along the line from the measured voltage at that force magnitude. The differences we calculate are called the residuals.

Once we have calculated the residuals, we can plot them just as we did the original data (Figure 0.6).

Apparently, the data points in the middle of Figure 0.5 actually lie slightly above the line (positive residuals), while those toward the ends of the line lie slightly below it (negative residuals). The pattern of data points, in other words, has a slight curvature to it, which becomes evident only when we plot the residuals and "zoom in" on the vertical scale.

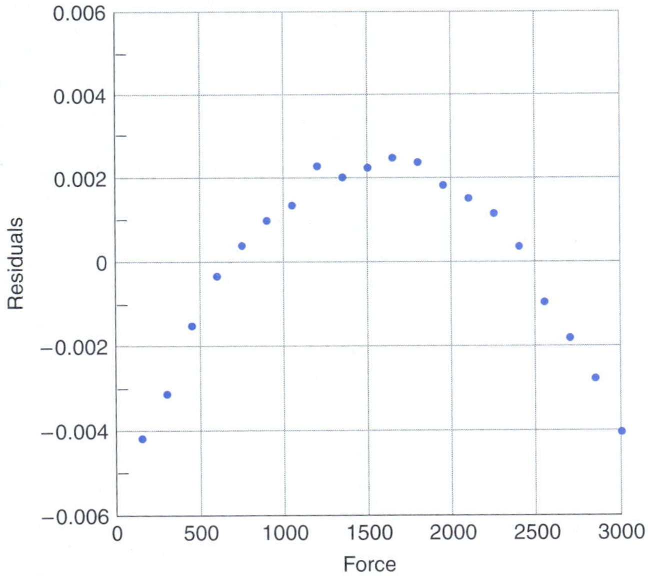

FIGURE 0.6 Plot of residuals.

The plot of the residuals looks like a parabola (see Chapter 3). Because a parabola has an equation containing a squared term, we can expect that a quadratic equation will be a better model of the data than a linear one. Using the quadratic regression function on a graphing calculator produces the equation

$$V = (-3.22693 \times 10^{-9})F^2 + (1.01652 \times 10^{-5})F - 0.00559066$$

The small coefficient on the F-squared term indicates a slight nonlinearity in the data.

For the manufacturer of the load cell, the slight nonlinearity will call for a decision. On the one hand, a nonlinear load cell response might lead to dangerously inaccurate measurements in some applications, especially if the cell is having to measure forces well outside the range of the test data. (Cranes mounted on cargo ships sometimes carry loads of up to 5000 tons, or 10 million pounds.) On the other hand, all manufacturing processes involve compromises between what is ideal and what is practically feasible.

Exercises

1. Enter the force and voltage values as two separate lists on a graphing calculator, and then use the Linear Regression function on the Statistics menu to generate a regression equation. Compare your result to the linear equation in the preceding discussion.
2. On most graphing calculators, if you multiply the list of forces by 0.0007221 and add 0.006081368, and then subtract the result from the list of voltages, you will have the list of residuals. Why does this work? Store the residuals as a new list; then plot them and compare your result with Figure 0.6.
3. Use the Quadratic Regression function on a graphing calculator to generate a new regression equation. Compare your result with the quadratic equation in the preceding discussion.
4. The quadratic model also has residuals, which when plotted look like this:

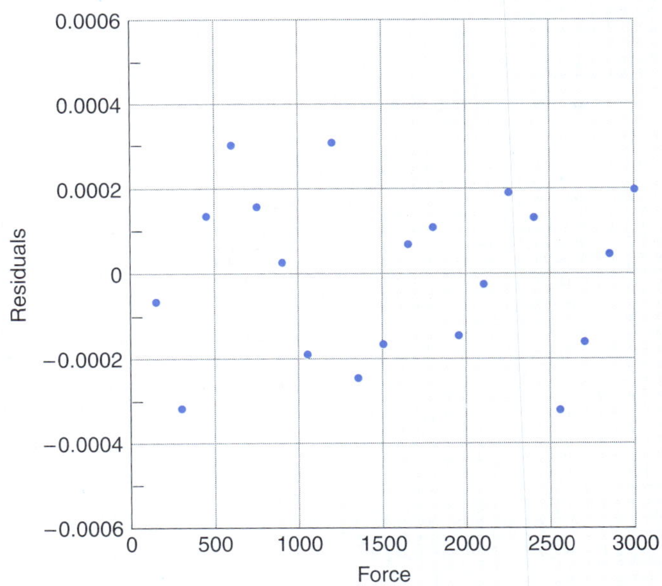

Compare the scale on the vertical axis with the one in Figure 0.6. What does the comparison suggest? What does the pattern of data points for the quadratic residuals suggest?

CHAPTER

1

Applications of Equations and Inequalities

In this chapter, we will apply equations to various real-life situations. We will then do the same with inequalities, which are statements that one quantity is greater than, less than, no greater than, or no less than some other quantity.

One application of inequalities is the regulation of sporting equipment. Some two dozen baseballs are used in a typical major league game, and it would be unrealistic to expect that every ball weigh exactly $5\frac{1}{8}$ ounces. But it is reasonable to require that each one weigh no less than 5 ounces and no more than $5\frac{1}{4}$ ounces, which is how the official rules read (www.majorleaguebaseball.com).

Another inequality applies to the sailboats used in the America's Cup race, which takes place every three to four years. The International America's Cup Class (IACC) for yachts is defined by the following rule:

$$\frac{L + 1.25\sqrt{S} - 9.8\sqrt[3]{DSP}}{0.679} \leq 24.000 \text{ m}$$

The "$\leq$" signifies that the expression on the left must come out as less than or equal to the 24 m on the right. The L, S, and DSP are themselves specified by complicated formulas, but roughly, L stands for length, S for sail area, and DSP for displacement (the hull volume below the waterline).

The IACC formula gives yacht designers some latitude. Suppose a yacht has $L = 20.2$ m, $S = 282$ m^2, and $DSP = 16.4$ m^3. Since the formula is an inequality, the designer could reduce the sail area while leaving the length and displacement unchanged. Typically, however, values of L, S, and DSP are used that make the expression on the left as close to 24 m as possible.

In addition to looking at applications of linear equations and inequalities, this chapter will review the concept of absolute value.

OBJECTIVE *To model situations described by linear or quadratic equations.*

1.1 Applications of Equations

In most cases, to solve practical problems you must translate the relationships stated in the problems into mathematical symbols. This is called *modeling*. The following examples illustrate basic techniques and concepts. Examine each of them carefully before going to the exercises.

EXAMPLE 1 **Mixture**

A chemist must prepare 350 ml *of a chemical solution made up of two parts alcohol and three parts acid. How much of each should be used?*

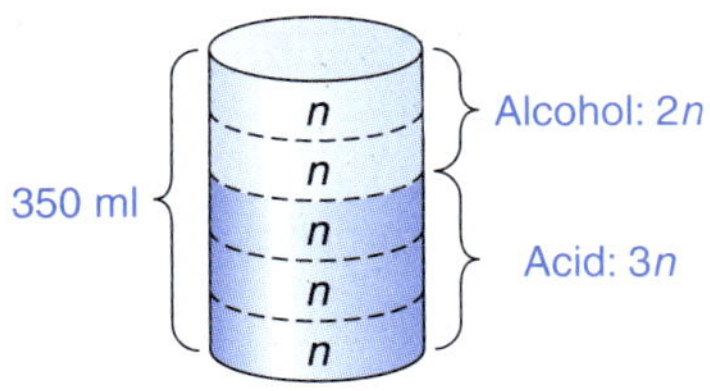

FIGURE 1.1 Chemical solution (Example 1).

Solution: Let n be the number of milliliters in each part. Figure 1.1 shows the situation. From the diagram, we have

$$2n + 3n = 350$$
$$5n = 350$$
$$n = \frac{350}{5} = 70$$

CAUTION Note that the solution to an equation is not necessarily the solution to the problem posed.

But $n = 70$ is *not* the answer to the original problem. Each *part* has 70 ml. The amount of alcohol is $2n = 2(70) = 140$, and the amount of acid is $3n = 3(70) = 210$. Thus, the chemist should use 140 ml of alcohol and 210 ml of acid. This example shows how helpful a diagram can be in setting up a word problem.

▸ **Now work Problem 5.**

EXAMPLE 2 **Observation Deck**

A rectangular observation deck overlooking a scenic valley is to be built. [See Figure 1.2(a).] The deck is to have dimensions 6 m *by* 12 m. *A rectangular shelter of area* 40 m^2 *is to be centered over the deck. The uncovered part of the deck is to be a walkaway of uniform width. How wide should this walkway be?*

Solution: A diagram of the deck is shown in Figure 1.2(b). Let w be the width (in meters) of the walkway. Then the part of the deck for the shelter has dimensions $12 - 2w$ by $6 - 2w$. Since its area must be 40 m^2, where

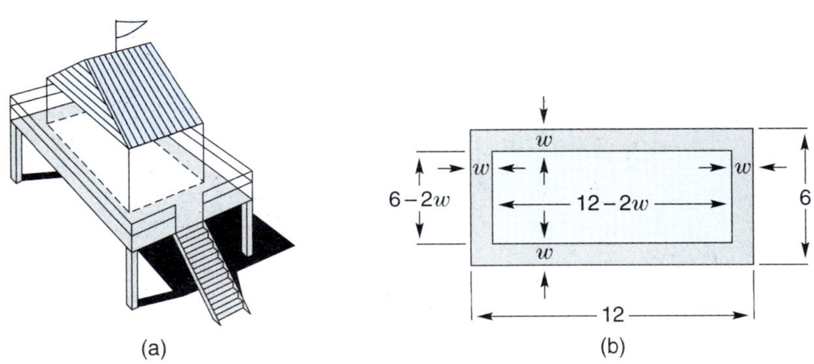

FIGURE 1.2 Deck walkway (Example 2).

area = (length)(width), we have

$$\begin{aligned}(12-2w)(6-2w) &= 40\\ 72-36w+4w^2 &= 40 \qquad \text{(multiplying)}\\ 4w^2-36w+32 &= 0\\ w^2-9w+8 &= 0 \qquad \text{(dividing both sides by 4)}\\ (w-8)(w-1) &= 0\\ w &= 8, 1\end{aligned}$$

Although 8 is a solution to the equation, it is *not* a solution to our problem, because one of the dimensions of the deck itself is only 6 m. Thus, the only possible solution is that the walkway be 1 m wide.

▶ **Now work Problem 7.**

The key words introduced here are *fixed cost, variable cost, total cost, total revenue,* and *profit.* This is the time for you to gain familiarity with these terms because they recur throughout the book.

In the next example, we shall use some business terms relative to a manufacturing firm. **Fixed cost** (or *overhead*) is the sum of all costs that are independent of the level of production, such as rent, insurance, and so on. This cost must be paid whether or not output is produced. **Variable cost** is the sum of all costs that are dependent on the level of output, such as labor and material. **Total cost** is the sum of variable cost and fixed cost:

$$\text{total cost} = \text{variable cost} + \text{fixed cost}$$

Total revenue is the money that the manufacturer receives for selling the output:

$$\text{total revenue} = (\text{price per unit})\ (\text{number of units sold})$$

Profit is total revenue minus total cost:

$$\text{profit} = \text{total revenue} - \text{total cost}$$

EXAMPLE 3 Profit

The Anderson Company produces a product for which the variable cost per unit is \$6 *and fixed cost is* \$80,000. *Each unit has a selling price of* \$10. *Determine the number of units that must be sold for the company to earn a profit of* \$60,000.

Solution: Let q be the number of units that must be sold. (In many business problems, q represents quantity.) Then the variable cost (in dollars) is $6q$. The *total* cost for the business is therefore $6q + 80,000$. The total revenue from the sale of q units is $10q$. Since

$$\text{profit} = \text{total revenue} - \text{total cost}$$

our model for this problem is

$$60{,}000 = 10q - (6q + 80{,}000)$$

Solving gives

$$\begin{aligned}60{,}000 &= 10q - 6q - 80{,}000\\ 140{,}000 &= 4q\\ 35{,}000 &= q\end{aligned}$$

Thus, 35,000 units must be sold to earn a profit of \$60,000.

▶ **Now work Problem 9.**

EXAMPLE 4 **Pricing**

Sportcraft manufactures women's sportswear and is planning to sell its new line of slacks to retail outlets. The cost to the retailer will be \$33 per pair of slacks. As a convenience to the retailer, Sportcraft will attach a price tag to each pair. What amount should be marked on the price tag so that the retailer may reduce this price by 20% during a sale and still make a profit of 15% on the cost?

Note that *price = cost + profit*.

Solution: Here we use the fact that

$$\text{selling price} = \text{cost per pair} + \text{profit per pair}$$

Let p be the tag price per pair, in dollars. During the sale, the retailer actually receives $p - 0.2p$. This must equal the cost, 33, plus the profit, (0.15)(33). Hence,

$$\begin{aligned} \text{selling price} &= \text{cost} + \text{profit} \\ p - 0.2p &= 33 + (0.15)(33) \\ 0.8p &= 37.95 \\ p &= 47.4375 \end{aligned}$$

From a practical point of view, Sportcraft should mark the price tag at \$47.44.

▶ **Now work Problem 13.**

EXAMPLE 5 **Investment**

A total of \$10,000 was invested in two business ventures, A and B. At the end of the first year, A and B yielded returns of 6% and $5\frac{3}{4}$%, respectively, on the original investments. How was the original amount allocated if the total amount earned was \$588.75?

Solution: Let x be the amount (in dollars) invested at 6%. Then $10{,}000 - x$ was invested at $5\frac{3}{4}$%. The interest earned from A was $(0.06)(x)$, and from B it was $(0.0575)(10{,}000 - x)$, which total 588.75. Hence,

$$\begin{aligned} (0.06)x + (0.0575)(10{,}000 - x) &= 588.75 \\ 0.06x + 575 - 0.0575x &= 588.75 \\ 0.0025x &= 13.75 \\ x &= 5500 \end{aligned}$$

Thus, \$5500 was invested at 6%, and \$10,000 − \$5500 = \$4500 was invested at $5\frac{3}{4}$%.

▶ **Now work Problem 11.**

EXAMPLE 6 **Bond Redemption**

The board of directors of Maven Corporation agrees to redeem some of its bonds in two years. At that time, \$1,102,500 will be required. Suppose the firm presently sets aside \$1,000,000. At what annual rate of interest, compounded annually, will this money have to be invested in order that its future value be sufficient to redeem the bonds?

Solution: Let r be the required annual rate of interest. At the end of the first year, the accumulated amount will be \$1,000,000 plus the interest, $1{,}000{,}000r$, for a total of

$$1{,}000{,}000 + 1{,}000{,}000r = 1{,}000{,}000(1 + r)$$

Under compound interest, at the end of the second year the accumulated amount will be $1{,}000{,}000(1 + r)$ plus the interest on this, which is $1{,}000{,}000(1 + r)r$. Thus, the total value at the end of the second year will be

$$1{,}0000{,}000(1 + r) + 1{,}000{,}000(1 + r)r$$

This must equal \$1,102,500:

$$1{,}000{,}000(1 + r) + 1{,}000{,}000(1 + r)r = 1{,}102{,}500 \qquad (1)$$

Since $1{,}000{,}000(1+r)$ is a common factor of both terms on the left side, we have

$$\begin{aligned}
1{,}000{,}000(1 + r)(1 + r) &= 1{,}102{,}500\\
1{,}000{,}000(1 + r)^2 &= 1{,}102{,}500\\
(1 + r)^2 &= \frac{1{,}102{,}500}{1{,}000{,}000} = \frac{11{,}025}{10{,}000} = \frac{441}{400}\\
1 + r &= \pm\sqrt{\frac{441}{400}} = \pm\frac{21}{20}\\
r &= -1 \pm \frac{21}{20}
\end{aligned}$$

Thus, $r = -1 + (21/20) = 0.05$, or $r = -1 - (21/20) = -2.05$. Although 0.05 and -2.05 are roots of Equation (1), we reject -2.05, since we want r to be positive. Hence, $r = 0.05$ so the desired rate is 5%.

▸ **Now work Problem 15.**

At times there may be more than one way to model a word problem, as Example 7 shows.

EXAMPLE 7 Apartment Rent

A real-estate firm owns the Parklane Garden Apartments, which consist of 96 apartments. At \$550 per month, every apartment can be rented. However, for each \$25 per month increase, there will be three vacancies with no possibility of filling them. The firm wants to receive \$54,600 per month from rents. What rent should be charged for each apartment?

Solution:

Method I. Suppose r is the rent (in dollars) to be charged per apartment. Then the increase over the \$550 level is $r - 550$. Thus, the number of \$25 increases is $\frac{r - 550}{25}$. Because each \$25 increase results in three vacancies, the total number of vacancies will be $3\left(\frac{r-550}{25}\right)$. Hence, the total number of apartments rented will be $96 - 3\left(\frac{r-550}{25}\right)$. Since

$$\text{total rent} = (\text{rent per apartment})(\text{number of apartments rented})$$

we have

$$54{,}600 = r\left[96 - \frac{3(r-550)}{25}\right]$$

$$54{,}600 = r\left[\frac{2400 - 3r + 1650}{25}\right]$$

$$54{,}600 = r\left[\frac{4050 - 3r}{25}\right]$$

$$1{,}365{,}000 = r(4050 - 3r)$$

Thus,

$$3r^2 - 4050r + 1{,}365{,}000 = 0$$

By the quadratic formula,

$$r = \frac{4050 \pm \sqrt{(-4050)^2 - 4(3)(1{,}365{,}000)}}{2(3)}$$

$$= \frac{4050 \pm \sqrt{22{,}500}}{6} = \frac{4050 \pm 150}{6} = 675 \pm 25$$

Hence, the rent for each apartment should be either \$650 or \$700.

Method II. Suppose n is the number of \$25 increases. Then the increase in rent per apartment will be $25n$ and there will be $3n$ vacancies. Since

total rent = (rent per apartment)(number of apartments rented)

we have

$$54{,}600 = (550 + 25n)(96 - 3n)$$

$$54{,}600 = 52{,}800 + 750n - 75n^2$$

$$75n^2 - 750n + 1800 = 0$$

$$n^2 - 10n + 24 = 0$$

$$(n-6)(n-4) = 0$$

Thus, $n = 6$ or $n = 4$. The rent charged should be either $550 + 25(6) = \$700$ or $550 + 25(4) = \$650$.

▶ **Now work Problem 29.**

Exercise 1.1

1. **Fencing** A fence is to be placed around a rectangular plot so that the enclosed area is 800 ft^3 and the length of the plot is twice the width. How many feet of fencing must be used?

2. **Geometry** The perimeter of a rectangle is 300 ft, and the length of the rectangle is twice the width. Find the dimensions of the rectangle.

3. **Gypsy Moth** One of the most troublesome defoliating insects is the gypsy moth caterpillar, which feeds on foliage of shade, forest, and fruit trees. A homeowner lives in an area in which the gypsy moth has become a problem. She wishes to spray the trees on her property before more defoliation occurs. She needs 128 oz of a solution made up of 3 parts of insecticide A and 5 parts of insecticide B. The solution is then mixed with water. How many ounces of each insecticide should be used?

4. **Concrete Mix** A builder makes a certain type of concrete by mixing together 1 part portland cement (made from lime and clay), 3 parts sand, and 5 parts crushed stone (by volume). If 765 ft^3 of concrete are

needed, how many cubic feet of each ingredient does he need?

*5. **Furniture Finish** According to *The Consumer's Handbook* [Paul Fargis, ed. (New York: Hawthorn, 1974)], a good oiled furniture finish contains two parts boiled linseed oil and one part turpentine. If you need a pint (16 fluid oz) of this furniture finish, how many fluid ounces of turpentine are needed?

6. **Forest Management** A lumber company owns a forest that is of rectangular shape, 1 mi by 2 mi. If the company cuts a uniform strip of trees along the outer edges of this forest, how wide should the strip be if $\frac{3}{4}$ sq mi of forest is to remain?

*7. **Garden Pavement** A rectangular plot, 4 m by 8 m, is to be used for a garden. It is decided to put a pavement inside the entire border so that 12 m^2 of the plot is left for flowers. How wide should the pavement be?

8. **Ventilating Duct** The diameter of a circular ventilating duct is 140 mm. This duct is joined to a square duct system as shown in Figure 1.3. To ensure smooth airflow, the areas of the circle and square sections must be equal. To the nearest millimeter, what should the length x of a side of the square section be?

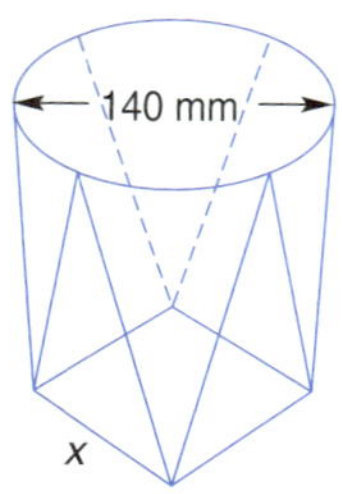

FIGURE 1.3 Ventilating duct (Problem 8).

*9. **Profit** A corn refining company produces corn gluten cattle feed at a variable cost of \$82 per ton. If fixed costs are \$120,000 per month and the feed sells for \$134 per ton, how many tons must be sold for the company to have a monthly profit of \$560,000?

10. **Sales** The Clark Company management would like to know the total sales units that are required for the company to earn a profit of \$100,000. The following data are available: unit selling price of \$20; variable cost per unit of \$15; total fixed cost of \$600,000. From these data, determine the required sales units.

*11. **Investment** A person wishes to invest \$20,000 in two enterprises so that the total income per year will be \$1440. One enterprise pays 6% annually; the other has more risk and pays $7\frac{1}{2}$% annually. How much must be invested in each?

12. **Investment** A person invested \$20,000, part at an interest rate of 6% annually and the remainder at 7% annually. The total interest at the end of 1 year was equivalent to an annual $6\frac{3}{4}$% rate on the entire \$20,000. How much was invested at each rate?

*13. **Pricing** The cost of a product to a retailer is \$3.40. If the retailer wishes to make a profit of 20% on the selling price, at what price should the product be sold?

14. **Bond Retirement** In three years, a company will require \$1,125,800 in order to retire some bonds. If the company now invests \$1,000,000 for this purpose, what annual rate of interest, compounded annually, must it receive on that amount in order to retire the bonds?

*15. **Expansion Program** In two years, a company will begin an expansion program. It has decided to invest \$2,000,000 now so that in two years the total value of the investment will be \$2,163,200, the amount required for the expansion. What is the annual rate of interest, compounded annually, that the company must receive to achieve its purpose?

16. **Business** A company finds that if it produces and sells q units of a product, its total sales revenue in dollars is $100\sqrt{q}$. If the variable cost per unit is \$2 and the fixed cost is \$1200, find the values of q for which

$$\text{total sales revenue} = \text{variable cost} + \text{fixed cost}.$$

(That is, profit is zero.)

17. **Dormitory Housing** A college dormitory houses 210 students. This fall, rooms are available for 76 freshmen. On the average, 95% of those freshmen who request room applications actually reserve a room. How many room applications should the college send out if it wants to receive 76 reservations?

18. **Poll** A group of people were polled, and 20%, or 700, of them favored a new product over the best-selling brand. How many people were polled?

19. **Prison Guard Salary** It was reported that in a certain women's jail, female prison guards, called matrons, received 30% (or \$200) a month less than their male counterparts, deputy sheriffs. Find the yearly salary of a deputy sheriff. Give your answer to the nearest dollar.

20. **Striking Drivers** A few years ago, cement drivers were on strike for 46 days. Before the strike, these drivers earned \$7.50 per hour and worked 260 eight-hour days a year. What percentage increase is needed in yearly income to make up for the lost time within 1 year?

21. Break Even A manufacturer of video-game cartridges sells each cartridge for \$21.95. The manufacturing cost of each cartridge is \$14.92. Monthly fixed costs are \$8500. During the first month of sales of a new game, how many cartridges must be sold in order for the manufacturer to break even (that is, in order that total revenue equal total cost)?

22. Investment Club An investment club bought a bond of an oil corporation for \$5000. The bond yields 8% per year. The club now wants to buy shares of stock in a hospital supply company. The stock sells at \$20 per share and earns a dividend of \$0.50 per share per year. How many shares should the club buy so that its total investment in stocks and bonds yields 5% per year?

23. Vision Care As a fringe benefit for its employees, a company established a vision-care plan. Under this plan, each year the company will pay the first \$35 of an employee's vision-care expenses and 80% of all additional vision-care expenses, up to a maximum *total* benefit payment of \$100. For an employee, find the total annual vision-care expenses covered by this program.

24. Quality Control Over a period of time, the manufacturer of a caramel-center candy bar found that 3.1% of the bars were rejected for imperfections.

(a) If c candy bars are made in a year, how many would the manufacturer expect to be rejected?

(b) This year, annual consumption of the candy is projected to be 600,000,000 bars. Approximately how many bars will have to be made if rejections are taken into consideration?

25. Business Suppose that consumers will purchase q units of a product when the price is $(80 - q)/4$ dollars *each*. How many units must be sold in order that sales revenue be \$400?

26. Investment How long would it take to double an investment at simple interest with a rate of 4% per year? (*Hint:* See Example 6(a) of Section 0.7, and express 4% as 0.04.)

27. Business Alternatives The inventor of a new toy offers the Kiddy Toy Company exclusive rights to manufacture and sell the toy for a lump-sum payment of \$25,000. After estimating that future sales possibilities beyond one year are nonexistent, the company management is reviewing an alternative proposal to give a lump-sum payment of \$2000 plus a royalty of \$0.50 for each unit sold. How many units must be sold the first year to make this alternative as economically attractive to the inventor as the original request? (*Hint:* Determine when the incomes under both proposals are the same.)

28. Parking Lot A company parking lot is 120 ft long and 80 ft wide. Due to an increase in personnel, it is decided to double the area of the lot by adding strips of equal width to one end and one side. Find the width of one such strip.

*29. **Rentals** You are the chief financial advisor to a corporation that owns an office complex consisting of 50 units. At \$400 per month, every unit can be rented. However, for each \$20 per month increase, there will be two vacancies with no possibility of filling them. The corporation wants to receive a total of \$20,240 per month from rents in the complex. You are asked to determine the rent that should be charged for each unit. What is your reply?

30. Investment Six months ago, an investment company had a \$3,100,000 portfolio consisting of blue-chip and glamour stocks. Since then, the value of the blue-chip investment increased by $\frac{1}{10}$, whereas the value of the glamour stocks decreased by $\frac{1}{10}$. The current value of the portfolio is \$3,240,000. What is the *current* value of the blue-chip investment?

31. Revenue The monthly revenue of a certain company is given by $R = 800p - 7p^2$, where p is the price in dollars of the product the company manufactures. At what price will the revenue be \$10,000 if the price must be greater than \$50?

32. Price–Earnings Ratio The *price–earnings ratio* (P/E) of a company is the ratio of the market value of one share of the company's outstanding common stock to the earnings per share. If P/E increases by 10% and the earnings per share increase by 20%, determine the percentage increase in the market value per share of the common stock.

33. Market Equilibrium When the price of a product is p dollars each, suppose that a manufacturer will supply $2p - 10$ units of the product to the market and that consumers will demand to buy $200 - 3p$ units. At the value of p for which supply equals demand, the market is said to be in equilibrium. Find this value of p.

34. Market Equilibrium Repeat Problem 33 for the following conditions: At a price of p dollars each, the supply is $3p^2 - 4p$ and the demand is $24 - p^2$.

35. Security Fence For security reasons, a company will enclose a rectangular area of 11,200 ft^2 in the rear of its plant. One side will be bounded by the building and the other three sides by fencing. (See Figure 1.4.) If 300 ft of fencing will be used, what will be the dimensions of the rectangular area?

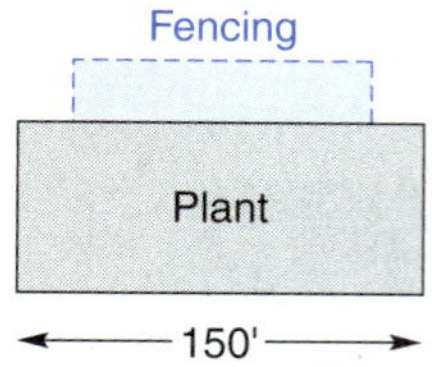

FIGURE 1.4 Security fencing (Problem 35).

36. Package Design A company is designing a package for its product. One part of the package is to be an open box made from a square piece of aluminum by cutting out a 2-in. square from each corner and folding up the sides. (See Figure 1.5.) The box is to contain 50 in^3. What are the dimensions of the square piece of aluminum that must be used?

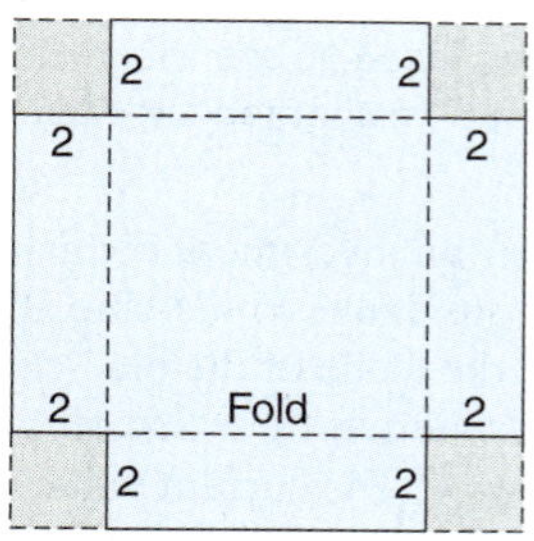

FIGURE 1.5 Box construction (Problem 36).

37. Product Design A candy company makes the popular Dandy Bar. The rectangular-shaped bar is 10 centimeters (cm) long, 5 cm wide, and 2 cm thick. (See Figure 1.6.) Because of increasing costs, the company has decided to cut the volume of the bar by a drastic 28%. The thickness will be the same, but the length and width will be reduced by equal amounts. What will be the length and width of the new bar?

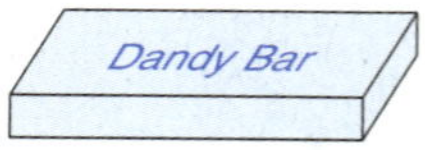

FIGURE 1.6 Candy bar (Problem 37).

38. Product Design A candy company makes a washer-shaped candy (a candy with a hole in it); see Figure 1.7. Because of increasing costs, the company will cut the volume of candy in each piece by 22%. To do this, the firm will keep the same thickness and outer radius, but will make the inner radius larger. At present the thickness is 2.1 millimeters (mm), the inner radius is 2 mm, and the outer radius is 7.1 mm. Find the inner radius of the new-style candy. (*Hint:* The volume V of a solid disk is $\pi r^2 h$, where r is the radius and h is the thickness of the disk.)

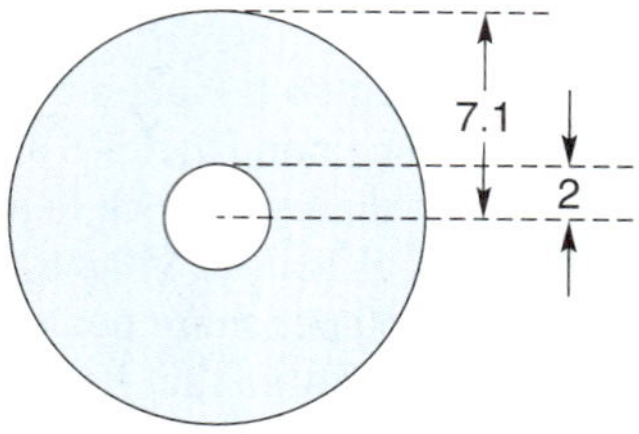

FIGURE 1.7 Washer-shaped candy (Problem 38).

39. Compensating Balance *Compensating balance* refers to that practice wherein a bank requires a borrower to maintain on deposit a certain portion of a loan during the term of the loan. For example, if a firm takes out a $100,000 loan that requires a compensating balance of 20%, it would have to leave $20,000 on deposit and would have the use of $80,000. To meet the expenses of retooling, the Victor Manufacturing Company must borrow $95,000. The Third National Bank, with whom the firm has had no prior association, requires a compensating balance of 15%. To the nearest thousand dollars, what amount of the loan is required to obtain the needed funds?

40. Incentive Plan A machine company has an incentive plan for its salespeople. For each machine that a salesperson sells, the commission is $40. The commission for *every* machine sold will increase by $0.04 for each machine sold over 600. For example, the commission on each of 602 machines sold is $40.08. How many machines must a salesperson sell in order to earn $30,800?

41. Real Estate A land investment company purchased a parcel of land for $7200. After having sold all but 20 acres at a profit of $30 per acre over the original cost per acre, the company regained the entire cost of the parcel. How many acres were sold?

42. Margin of Profit The *margin of profit* of a company is the net income divided by the total sales. A company's margin of profit increased by 0.02 from last year. Last year the company sold its product at $3.00 each and had a net income of $4500. This year it increased the price of its product by \$0.50 each, sold 2000 more, and had a net income of \$7140. The company never has had a margin of profit greater than 0.15. How many of its product were sold last year and how many were sold this year?

43. Business A company manufactures products A and B. The cost of producing each unit of A is \$2 more than that of B. The costs of production of A and B are \$1500 and \$1000, respectively, and 25 more units of A are produced than of B. How many of each are produced?

OBJECTIVE *To solve linear inequalities in one variable and to introduce interval notation.*

1.2 Linear Inequalities

Suppose a and b are two points on the real-number line. Then either a and b coincide, or a lies to the left of b, or a lies to the right of b. (See Figure 1.8.)

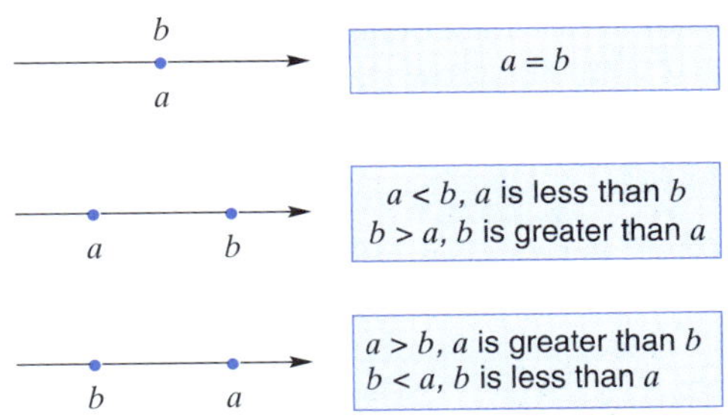

FIGURE 1.8 Relative positions of two points.

If a and b coincide, then $a = b$. If a lies to the left of b, we say that a is less than b and write $a < b$, where the *inequality symbol* "<" is read "is less than." On the other hand, if a lies to the right of b, we say that a is greater than b, written $a > b$. The statements $a > b$ and $b < a$ are equivalent. (If you have trouble keeping these symbols straight, it may help to notice that < looks somewhat like the letter "L" for "left" and that we have $a < b$ precisely when "a is left of b.")

Another inequality symbol "≤" is read "is less than or equal to" and is defined as follows: $a \leq b$ if and only if $a < b$ or $a = b$. Note that ≤ can also be read as "no greater than" or "no more than." These phrasings occur frequently in applications. Similarly, the symbol "≥" is defined as follows: $a \geq b$ if and only if $a > b$ or $a = b$. In this case, we say that a is greater than or equal to b. Again, ≥ is often used to translate "no less than" in applied problems.

We shall use the words *real numbers* and *points* interchangeably, since there is a one-to-one correspondence between real numbers and points on a line. Thus, we can speak of the points −5, −2, 0, 7, and 9, and can write $7 < 9$, $-2 > -5$, $7 \leq 7$, and $7 \geq 0$. (See Figure 1.9.) Clearly, if $a > 0$, then a is positive; if $a < 0$, then a is negative.

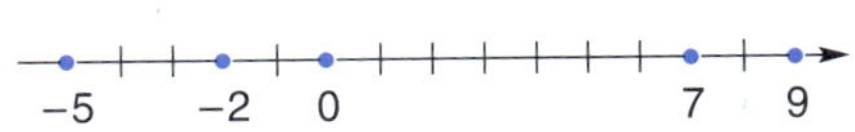

FIGURE 1.9 Points on number line.

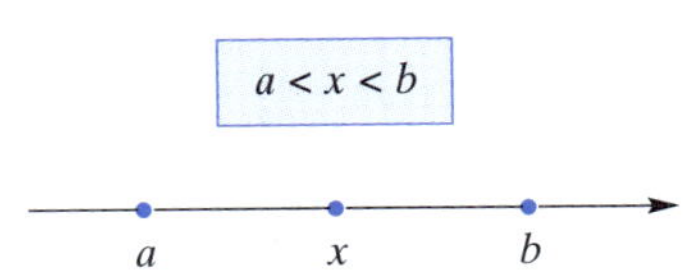

FIGURE 1.10 $a < x$ and $x < b$.

Suppose that $a < b$ and x is between a and b. (See Figure 1.10.) Then not only is $a < x$, but also, $x < b$. We indicate this by writing $a < x < b$, which can be thought of as a compound inequality. For example, $0 < 7 < 9$. (See Figure 1.9.)

The next Definition is stated in terms of the less-than relation ($<$), but it applies also to the other relations ($>$, $\leq$, $\geq$).

Definition

An ***inequality*** is a statement that one number is less than another number.

Of course, we represent inequalities by means of inequality symbols. If two inequalities have their inequality symbols pointing in the same direction, then the inequalities are said to have the *same sense*. If not, they are said to be *opposite in sense,* or one is said to have the *reverse sense* of the other. Hence, $a < b$ and $c < d$ have the same sense, but $a < b$ has the reverse sense of $c > d$.

Solving an inequality, such as $2(x - 3) < 4$, means finding all values of the variable for which the inequality is true. This involves the application of certain rules, which we now state.

Rules for Inequalities

1. If the same number is added to or subtracted from both sides of an inequality, the resulting inequality has the same sense as the original inequality. Symbolically,

$$\text{if} \quad a < b, \quad \text{then} \quad a + c < b + c \quad \text{and} \quad a - c < b - c$$

For example, $7 < 10$, so $7 + 3 < 10 + 3$.

Keep in mind that the rules also apply to $\leq$, $>$, and $\geq$.

2. If both sides of an inequality are multiplied or divided by the same ***positive*** number, the resulting inequality has the same sense as the original inequality. Symbolically,

$$\text{if} \quad a < b \quad \text{and} \quad c > 0, \quad \text{then} \quad ac < bc \quad \text{and} \quad \frac{a}{c} < \frac{b}{c}$$

For example, $3 < 7$ and $2 > 0$, so $3(2) < 7(2)$ and $\frac{3}{2} < \frac{7}{2}$.

CAUTION The sense of an inequality must be reversed when multiplying or dividing both sides by a negative number.

3. If both sides of an inequality are multiplied or divided by the same ***negative*** number, then the resulting inequality has the ***reverse*** sense of the original inequality. Symbolically,

$$\text{if} \quad a < b \quad \text{and} \quad c > 0, \quad \text{then} \quad a(-c) > b(-c) \quad \text{and} \quad \frac{a}{-c} > \frac{b}{-c}$$

For example, $4 < 7$ but $4(-2) > 7(-2)$ and $\frac{4}{-2} > \frac{7}{-2}$.

4. Any side of an inequality can be replaced by an expression equal to it. Symbolically,

$$\text{if} \quad a < b \quad \text{and} \quad a = c, \quad \text{then} \quad c < b$$

For example, if $x < 2$ and $x = y + 4$, then $y + 4 < 2$.

5. If the sides of an inequality are either both positive or both negative, then their respective reciprocals[1] are unequal in the ***reverse*** sense. For example, $2 < 4$, but $\frac{1}{2} > \frac{1}{4}$.

[1]The *reciprocal* of a nonzero number a is defined to be $\frac{1}{a}$.

6. If both sides of an inequality are positive and we raise each side to the same positive power, then the resulting inequality has the same sense as the original inequality. Thus, if $0 < a < b$ and $n > 0$, then

$$a^n < b^n \quad \text{and} \quad \sqrt[n]{a} < \sqrt[n]{b}$$

where we assume that n is a positive integer in the latter inequality. For example, $4 < 9$, so $4^2 < 9^2$ and $\sqrt{4} < \sqrt{9}$.

The result of applying Rules 1–4 to an inequality is called an *equivalent inequality*. This is an inequality whose solution is exactly the same as that of the original inequality. We shall apply these rules to a *linear inequality*.

Definition

A ***linear inequality*** in the variable x is an inequality that can be written in the form

$$ax + b < 0,$$

where a and b are constants and $a \neq 0$.

The definition also applies to $\leq$, $>$, and $\geq$.

APPLICATIONS IN PRACTICE 1

Solving a Linear Inequality

A salesman has a monthly income given by $I = 200 + 0.8S$, where S is the number of products sold in a month. How many products must he sell to make at least \$4500 a month?

EXAMPLE 1 Solving a Linear Inequality

Solve $2(x - 3) < 4$.

Solution:

Strategy: We shall replace the given inequality by equivalent inequalities until the solution is evident.

$$\begin{aligned} 2(x-3) &< 4 & \\ 2x - 6 &< 4 & \text{(Rule 4)} \\ 2x - 6 + 6 &< 4 + 6 & \text{(Rule 1)} \\ 2x &< 10 & \text{(Rule 4)} \\ \frac{2x}{2} &< \frac{10}{2} & \text{(Rule 2)} \\ x &< 5 & \text{(Rule 4)} \end{aligned}$$

All of the foregoing inequalities are equivalent. Thus, the original inequality is true for *all* real numbers x such that $x < 5$. For example, the inequality is true for $x = -10, -0.1, 0, \frac{1}{2}$, and 4.9. We may write our solution simply as $x < 5$ and can geometrically represent it by the color half-line in Figure 1.11. The parenthesis indicates that 5 is *not included* in the solution.

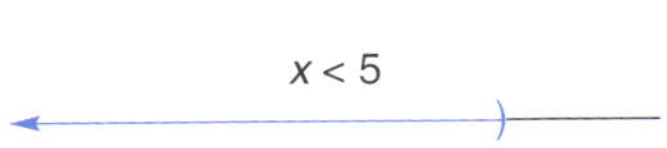

FIGURE 1.11 All real numbers less than 5.

▸ **Now work Problem 9.**

In Example 1 the solution consisted of a set of numbers, namely, all numbers less than 5. In general, it is common to use the term **interval** to describe such a set. In the case of Example 1, the *set of all* x such that $x < 5$ can be denoted by the *interval notation* $(-\infty, 5)$. The symbol $-\infty$ is not a number, but is merely a convenience for indicating that the interval extends indefinitely to the left.

There are other types of intervals. For example, the set of all numbers x for which $a \leq x \leq b$ is called a **closed interval** and includes the numbers a and b, which are called *endpoints* of the interval. This interval is denoted by $[a, b]$ and is shown in Figure 1.12(a). The square brackets indicated that a and b *are included* in the interval. On the other hand, the set of all x for which $a < x < b$ is called an **open interval** and is denoted by (a, b). The endpoints are *not* part of this set. (See Figure 1.12(b).) Extending these concepts, we have the intervals shown in Figure 1.13.

Closed interval $[a, b]$
(a)

Open interval (a, b)
(b)

FIGURE 1.12 Closed and open intervals.

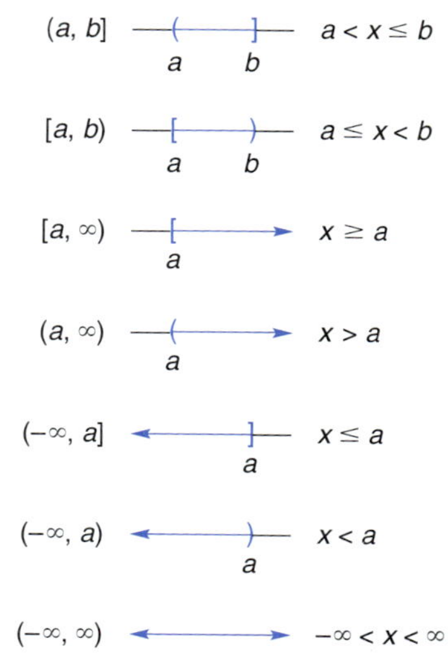

FIGURE 1.13 Intervals.

EXAMPLE 2 **Solving a Linear Inequality**

Solve $3 - 2x \leq 6$.

Solution:

$$3 - 2x \leq 6$$

$$-2x \leq 3 \qquad \text{(Rule 1)}$$

$$x \geq -\frac{3}{2} \qquad \text{(Rule 3)}$$

CAUTION Dividing both sides by -2 reverses the sense of the inequality.

The solution is $x \geq -\frac{3}{2}$, or, in interval notation, $[-\frac{3}{2}, \infty)$. This is represented geometrically in Figure 1.14.

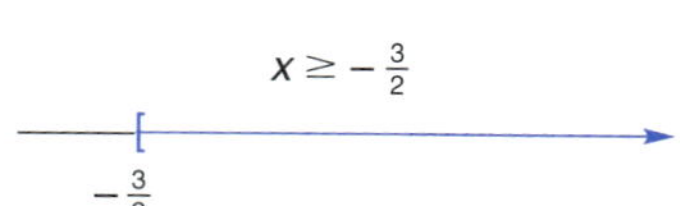

FIGURE 1.14 The interval $[-\frac{3}{2}, \infty)$.

▶ **Now work Problem 11.**

APPLICATIONS IN PRACTICE 2

Solving a Linear Inequality

A zoo veterinarian can purchase four different animal foods with various nutrient values for the zoo's grazing animals. Let x_1 represent the number of bags of food 1, x_2 represent the number of bags of food 2, and so on. The number of bags of each food needed can be described by the following equations:

$$x_1 = 150 - x_4$$

$$x_2 = 3x_4 - 210$$

$$x_3 = x_4 + 60$$

Develop four inequalities from these equations, assuming that each variable must be nonnegative.

EXAMPLE 3 Solving a Linear Inequality

Solve $\frac{3}{2}(s-2)+1 > -2(s-4)$.

Solution:

$$\frac{3}{2}(s-2)+1 > -2(s-4)$$

$$2\left[\frac{3}{2}(s-2)+1\right] > 2[-2(s-4)] \qquad \text{(Rule 2)}$$

$$3(s-2)+2 > -4(s-4) \qquad \text{(Rule 4)}$$

$$3s-4 > -4s+16 \qquad \text{(Rule 4)}$$

$$7s > 20 \qquad \text{(Rule 1)}$$

$$s > \frac{20}{7} \qquad \text{(Rule 2)}$$

The solution is $(\frac{20}{7}, \infty)$; see Figure 1.15.

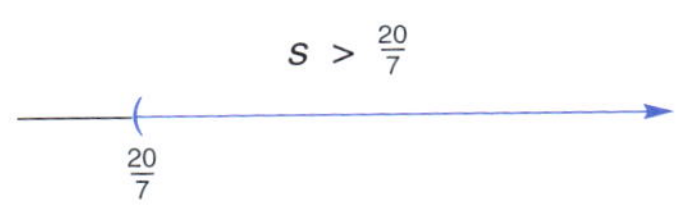

FIGURE 1.15 The interval $(\frac{20}{7}, \infty)$.

▶ **Now work Problem 19.**

EXAMPLE 4 Solving Linear Inequalities

a. *Solve* $2(x-4)-3 > 2x-1$.

Solution:

$$2(x-4)-3 > 2x-1$$

$$2x-8-3 > 2x-1$$

$$-11 > -1$$

Since it is never true that $-11 > -1$, there is no solution, and the solution set is $\emptyset$.

b. *Solve* $2(x-4)-3 < 2x-1$.

Solution: Proceeding as in part (a), we obtain $-11 < -1$. This is true for all real numbers x, so the solution is $(-\infty, \infty)$; see Figure 1.16.

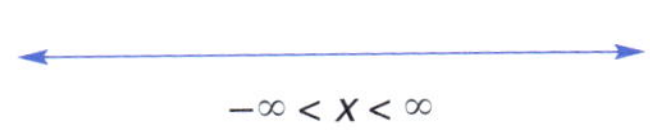

FIGURE 1.16 The interval $(-\infty, \infty)$.

▶ **Now work Problem 15.**

Exercise 1.2

In Problems 1–34, solve the inequalities. Give your answer in interval notation, and indicate the answer geometrically on the real-number line.

1. $3x > 12$
2. $4x < -2$
3. $5x - 11 \leq 9$
4. $3x \geq 0$
5. $-4x \geq 2$
6. $2y + 1 > 0$
7. $5 - 7s > 3$
8. $4s - 1 < -5$
*9. $3 < 2y + 3$
10. $4 \leq 3 - 2y$
*11. $2x - 3 \leq 4 + 7x$
12. $-3 \geq 8(2 - x)$
13. $3(2 - 3x) > 4(1 - 4x)$
14. $8(x + 1) + 1 < 3(2x) + 1$
*15. $2(4x - 2) > 4(2x + 1)$
16. $3 - 2(x - 1) \leq 2(4 + x)$
17. $x + 2 < \sqrt{3} - x$
18. $\sqrt{2}(x + 2) > \sqrt{8}(3 - x)$
*19. $\frac{5}{6}x < 40$
20. $-\frac{2}{3}x > 6$
21. $\frac{9y + 1}{4} \leq 2y - 1$
22. $\frac{3y - 2}{3} \geq \frac{1}{4}$
23. $4x - 1 \geq 4(x - 2) + 7$
24. $0x \leq 0$
25. $\frac{1 - t}{2} < \frac{3t - 7}{3}$
26. $\frac{3(2t - 2)}{2} > \frac{6t - 3}{5} + \frac{t}{10}$
27. $2x + 13 \geq \frac{1}{3}x - 7$
28. $4x - \frac{1}{2} \leq \frac{3}{2}x$
29. $\frac{2}{3}r < \frac{5}{6}r$
30. $\frac{7}{4}t > -\frac{8}{3}t$
31. $\frac{y}{2} + \frac{y}{3} > y + \frac{y}{5}$
32. $9 - 0.1x \leq \frac{2 - 0.01x}{0.2}$
33. $0.1(0.03x + 4) \geq 0.02x + 0.434$
34. $\frac{3y - 1}{-3} < \frac{5(y + 1)}{-3}$
35. **Earnings** Each month last year, a company had earnings that were greater than \$37,000, but less than \$53,000. If S represents the total earnings for the year, describe S by using inequalities.
36. Using inequalities, symbolize the following statement: The number of labor hours x to produce a product is not less than $2\frac{1}{2}$ nor more than 4.
37. **Geometry** In a right triangle, one of the acute angles x is less than 3 times the other acute angle plus 10 degrees. Solve for x.
38. **Spending** A student has \$360 to spend on a stereo system and some compact discs. If she buys a stereo that costs \$219 and the discs cost \$18.95 each, find the greatest number of discs she can buy.

OBJECTIVE *To model situations in terms of inequalities.*

1.3 Applications of Inequalities

Solving word problems may sometimes involve inequalities, as the following examples illustrate.

EXAMPLE 1 Profit

For a company that manufactures aquarium heaters, the combined cost for labor and material is \$21 per heater. Fixed costs (costs incurred in a given period, regardless of output) are \$70,000. If the selling price of a heater is \$35, how many must be sold for the company to earn a profit?

Solution:

Strategy: Recall that

$$\text{profit} = \text{total revenue} - \text{total cost}$$

We shall find total revenue and total cost and then determine when their difference is positive.

Let q be the number of heaters sold. Then their cost is $21q$. The total cost for the company is therefore $21q + 70{,}000$. The total revenue from the sale of q heaters will be $35q$. Now,

$$\text{profit} = \text{total revenue} - \text{total cost}$$

and we want profit > 0. Thus,

$$\text{total revenue} - \text{total cost} > 0$$
$$35q - (21q + 70{,}000) > 0$$
$$14q > 70{,}000$$
$$q > 5000$$

Therefore, at least 5001 heaters must be sold for the company to earn a profit.

▶ **Now work Problem 1.**

EXAMPLE 2 **Renting vs. Purchasing**

A builder must decide whether to rent or buy an excavating machine. If he were to rent the machine, the rental fee would be \$3000 per month (on a yearly basis), and the daily cost (gas, oil, and driver) would be \$180 for each day the machine is used. If he were to buy it, his fixed annual cost would be \$20,000, and daily operating and maintenance costs would be \$230 for each day the machine is used. What is the least number of days each year that the builder would have to use the machine to justify renting it rather than buying it?

Solution:

Strategy: We shall determine expressions for the annual cost of renting and the annual cost of purchasing. We then find when the cost of renting is less than that of purchasing.

Let d be the number of days each year that the machine is used. If the machine is rented, the total yearly cost consists of rental fees, which are (12)(3000), and daily charges of $180d$. If the machine is purchased, the cost per year is $20{,}000 + 230d$. We want

$$\text{cost}_{\text{rent}} < \text{cost}_{\text{purchase}}$$
$$12(3000) + 180d < 20{,}000 + 230d$$
$$36{,}000 + 180d < 20{,}000 + 230d$$
$$16{,}000 < 50d$$
$$320 < d$$

Thus, the builder must use the machine at least 321 days to justify renting it.

▶ **Now work Problem 3.**

EXAMPLE 3 **Current Ratio**

The current ratio *of a business is the ratio of its current assets (such as cash, merchandise inventory, and accounts receivable) to its current liabilities (such as short-term loans and taxes payable).*

After consulting with the comptroller, the president of the Ace Sports Equipment Company decides to take out a short-term loan to build up inventory. The company has current assets of \$350,000 and current liabilities of \$80,000. How much can the company borrow if the current ratio is to be no less than 2.5?

(Note: The funds received are considered as current assets and the loan as a current liability.)

Solution: Let x denote the amount the company can borrow. Then current assets will be $350{,}000 + x$, and current liabilities will be $80{,}000 + x$. Thus,

$$\text{current ratio} = \frac{\text{current assets}}{\text{current liabilities}} = \frac{350{,}000 + x}{80{,}000 + x}$$

We want

$$\frac{350{,}000 + x}{80{,}000 + x} \geq 2.5$$

Although the inequality that must be solved is not linear, it leads to a linear inequality.

Since x is positive, so is $80{,}000 + x$. Hence, we can multiply both sides of the inequality by $80{,}000 + x$ and the sense of the inequality will remain the same. We have

$$350{,}000 + x \geq 2.5(80{,}000 + x)$$
$$150{,}000 \geq 1.5x$$
$$100{,}000 \geq x$$

Consequently, the company may borrow as much as \$100,000 and yet maintain a current ratio of no less than 2.5.

EXAMPLE 4 **Publishing**

A publishing company finds that the cost of publishing each copy of a certain magazine is \$1.50. The revenue from dealers is \$1.40 per copy. The advertising revenue is 10% of the revenue received from dealers for all copies sold beyond 10,000. What is the least number of copies that must be sold so as to have a profit for the company?

Solution:

Strategy: We have profit = total revenue − total cost, so we find an expression for profit and then set it greater than 0.

Let q be the number of copies that are sold. The revenue from dealers is $1.40q$, and the revenue from advertising is $(0.10)[(1.40)(q - 10{,}000)]$. The total cost of publication is $1.50q$. Thus,

$$\text{total revenue} - \text{total cost} > 0$$
$$1.40q + (0.10)[(1.40)(q - 10{,}000)] - 1.50q > 0$$
$$1.4q + 0.14q - 1400 - 1.5q > 0$$
$$0.04q - 1400 > 0$$
$$0.04q > 1400$$
$$q > 35{,}000$$

Therefore, the total number of copies must be greater than 35,000. That is, at least 35,001 copies must be sold to guarantee a profit.

▸ **Now work Problem 5.**

Exercise 1.3

*1. **Profit** The Davis Company manufactures a product that has a unit selling price of \$20 and a unit cost of \$15. If fixed costs are \$600,000, determine the least number of units that must be sold for the company to have a profit.

2. **Profit** To produce 1 unit of a new product, a company determines that the cost for material is \$2.50 and the cost of labor is \$4. The constant overhead, regardless of sales volume, is \$5000. If the cost to a wholesaler is \$7.40 per unit, determine the least number of units that must be sold by the company to realize a profit.

*3. **Leasing vs. Purchasing** A businesswoman wants to determine the difference between the costs of owning and leasing an automobile. She can lease a car for \$420 per month (on an annual basis). Under this plan, the cost per mile (gas and oil) is \$0.06. If she were to purchase the car, the fixed annual expense would be \$4700, and other costs would amount to \$0.08 per mile. What is the least number of miles she would have to drive per year to make leasing no more expensive than purchasing?

4. **Shirt Manufacturer** A T-shirt manufacturer produces N shirts at a total labor cost (in dollars) of $1.3N$ and a total material cost of $0.4N$. The constant overhead for the plant is \$6500. If each shirt sells for \$3.50, how many must be sold by the company to realize a profit?

*5. **Publishing** The cost of publication of each copy of a magazine is \$0.65. It is sold to dealers for \$0.60 each, and the amount received for advertising is 10% of the amount received for all magazines issued beyond 10,000. Find the least number of magazines that can be published without loss—that is, such that profit ≥ 0. (Assume that all issues will be sold.)

6. **Production Allocation** A company produces alarm clocks. During the regular workweek, the labor cost for producing one clock is \$2.00. However, if a clock is produced on overtime, the labor cost is \$3.00. Management has decided to spend no more than a total of \$25,000 per week for labor. The company must produce 11,000 clocks this week. What is the minimum number of clocks that must be produced during the regular workweek?

7. **Investment** A company invests a total of \$30,000 of surplus funds at two annual rates of interest: 5% and $6\frac{3}{4}$%. It wishes an annual yield of no less than $6\frac{1}{2}$%. What is the least amount of money that the company must invest at the $6\frac{3}{4}$% rate?

8. **Current Ratio** The current ratio of Precision Machine Products is 3.8. If the firm's current assets are \$570,000, what are its current liabilities? To raise additional funds, what is the maximum amount the company can borrow on a short-term basis if the current ratio is to be no less than 2.6? (See Example 3 for an explanation of current ratio.)

9. **Sales Allocation** At present, a manufacturer has 2500 units of product in stock. The product is now selling at \$4 per unit. Next month the unit price will increase by \$0.50. The manufacturer wants the total revenue received from the sale of the 2500 units to be no less than \$10,750. What is the maximum number of units that can be sold this month?

10. **Revenue** Suppose consumers will purchase q units of a product at a price of $\frac{100}{q} + 1$ dollars per unit. What is the minimum number of units that must be sold in order that sales revenue be greater than \$5000?

11. **Hourly Rate** Painters are often paid either by the hour or on a per-job basis. The rate they receive can affect their working speed. For example, suppose they can work either for \$9.00 per hour or for \$320 plus \$3 for each hour less than 40 if they complete the job in less than 40 hours. Suppose the job will take t hours. If $t \geq 40$, clearly the hourly rate is better. If $t < 40$, for what values of t is the hourly rate the better pay scale?

12. **Compensation** Suppose a company offers you a sales position with your choice of two methods of determining your yearly salary. One method pays \$12,600 plus a bonus of 2% of your yearly sales. The other method pays a straight 8% commission on your sales. For what yearly sales amount is it better to choose the first method?

13. **Acid Test Ratio** The *acid test ratio* (or *quick ratio*) of a business is the ratio of its liquid assets—cash and securities plus accounts receivable—to its current liabilities. The minimum acid test ratio for a financially healthy company is around 1.0, but the standard varies somewhat from industry to industry. If a company has \$450,000 in cash and securities and has \$398,000 in current liabilities, how much does it need to be carrying as accounts receivable in order to keep its acid test ratio at or above 1.3?

OBJECTIVE *To solve equations and inequalities involving absolute values.*

1.4 Absolute Value

Absolute-Value Equations

On the real-number line, the distance of a number x from 0 is called the **absolute value** of x and is denoted by $|x|$. For example, $|5| = 5$ and $|-5| = 5$ because both 5 and -5 are 5 units from 0. (See Figure 1.17.) Similarly, $|0| = 0$. Notice that $|x|$ can never be negative; that is, $|x| \geq 0$.

Basically, the absolute value of a real number is its value when its sign is ignored.

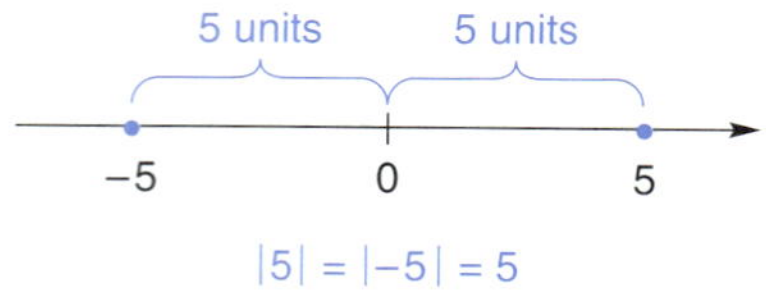

FIGURE 1.17 Absolute value.

If x is positive or zero, then $|x|$ is simply x itself, so we can omit the vertical bars and write $|x| = x$. On the other hand, consider the absolute value of a negative number, like $x = -5$.

$$|x| = |-5| = 5 = -(-5) = -x$$

Thus, if x is negative, then $|x|$ is the positive number $-x$. The minus sign indicates that we have changed the sign of x. The geometrical definition of absolute value is equivalent to the following:

Definition

The ***absolute value*** of a real number x, written $|x|$, is defined as

$$|x| = \begin{cases} x, & \text{if } x \geq 0 \\ -x, & \text{if } x < 0 \end{cases}$$

Observe that $|-x| = |x|$ follows from the definition.

Applying the definition, we have $|3| = 3$, $|-8| = -(-8) = 8$, and $|\frac{1}{2}| = \frac{1}{2}$. Also, $-|2| = -2$ and $-|-2| = -2$.

Also, $|-x|$ is not necessarily x and, thus, $|-x-1|$ is not necessarily $x+1$. For example, if we let $x = -3$, then $|-(-3)| \neq -3$, and

$$|-(-3)-1| \neq -3+1$$

CAUTION $\sqrt{x^2}$ is not necessarily x, but

$$\sqrt{x^2} = |x|$$

For example, $\sqrt{(-2)^2} = |-2| = 2$, not -2. This agrees with the fact that

$$\sqrt{(-2)^2} = \sqrt{4} = 2$$

EXAMPLE 1 Solving Absolute-Value Equations

a. *Solve* $|x-3| = 2$.

Solution: This equation states that $x-3$ is a number 2 units from 0. Thus, either

$$x-3 = 2 \quad \text{or} \quad x-3 = -2$$

Solving these equations gives $x = 5$ or $x = 1$.

b. *Solve* $|7-3x| = 5$.

Solution: The equation is true if $7-3x = 5$ or if $7-3x = -5$. Solving these equations gives $x = \frac{2}{3}$ or $x = 4$.

c. *Solve* $|x-4| = -3$.

Solution: The absolute value of a number is never negative, so the solution set is $\emptyset$.

▶ **Now work Problem 19.**

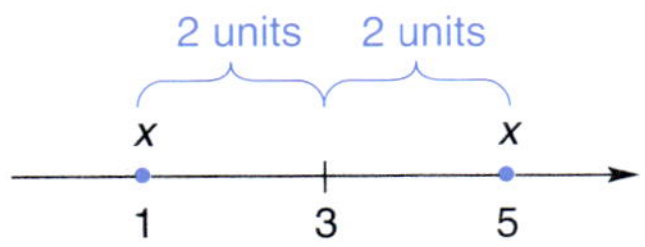

FIGURE 1.18 The solution of $|x - 3| = 2$ is 1 or 5.

We may interpret $|a - b| = |-(b - a)| = |b - a|$ as the distance between a and b. For example, the distance between 5 and 9 can be calculated by

$$\begin{aligned} &\text{either} \quad |9 - 5| = |4| = 4 \\ &\text{or} \quad |5 - 9| = |-4| = 4 \end{aligned}$$

Similarly, the equation $|x - 3| = 2$ states that the distance between x and 3 is 2 units. Thus, x can be 1 or 5, as shown in Example 1(a) and Figure 1.18.

Absolute-Value Inequalities

Let us turn now to inequalities involving absolute values. If $|x| < 3$, then x is less than 3 units from 0. Hence, x must lie between -3 and 3, that is, on the interval $-3 < x < 3$. (See Figure 1.19(a).) On the other hand, if $|x| > 3$, then x must be greater than 3 units from 0. Hence, there are two intervals in the solution: Either $x < -3$ or $x > 3$. (See Figure 1.19(b).) We can extend these ideas as follows: If $|x| \le 3$, then $-3 \le x \le 3$; if $|x| \ge 3$, then $x \le -3$ or $x \ge 3$. Table 1.1 gives a summary of the solutions to absolute-value inequalities.

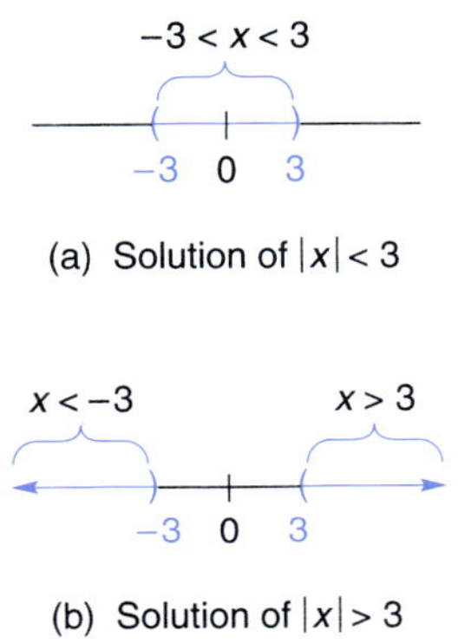

FIGURE 1.19 Solutions to $|x| < 3$ and $|x| > 3$.

TABLE 1.1

Inequality ($d > 0$)	Solution
$\lvert x\rvert < d$	$-d < x < d$
$\lvert x\rvert \le d$	$-d \le x \le d$
$\lvert x\rvert > d$	$x < -d$ or $x > d$
$\lvert x\rvert \ge d$	$x \le -d$ or $x \ge d$

EXAMPLE 2 Solving Absolute-Value Inequalities

a. *Solve* $|x - 2| < 4$.

Solution: The number $x - 2$ must be less than 4 units from 0. From the preceding discussion, this means that $-4 < x - 2 < 4$. We may set up the procedure for solving this inequality as follows:

$$\begin{aligned} -4 &< x - 2 < 4 \\ -4 + 2 &< x < 4 + 2 \quad \text{(adding 2 to each member)} \\ -2 &< x < 6 \end{aligned}$$

Thus, the solution is the open interval $(-2, 6)$. This means that all numbers between -2 and 6 satisfy the original inequality. (See Figure 1.20.)

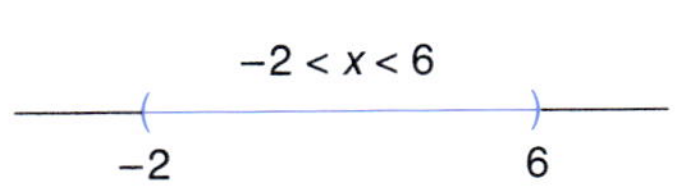

FIGURE 1.20 The solution of $|x - 2| < 4$ is the interval $(-2, 6)$.

b. *Solve* $|3 - 2x| \le 5$.

Solution:

$$\begin{aligned} -5 \le 3 - 2x &\le 5 \\ -5 - 3 \le -2x &\le 5 - 3 \quad \text{(subtracting 3 throughout)} \\ -8 \le -2x &\le 2 \\ 4 \ge x &\ge -1 \quad \text{(dividing throughout by } -2\text{)} \\ -1 \le x &\le 4 \quad \text{(rewriting)} \end{aligned}$$

Note that the sense of the original inequality was *reversed* when we divided by a negative number. The solution is the closed interval $[-1, 4]$.

▶ **Now work Problem 29.**

EXAMPLE 3 Solving Absolute-Value Inequalities

a. *Solve* $|x+5| \geq 7$.

Solution: Here $x + 5$ must be *at least* 7 units from 0. Thus, either $x+5 \leq -7$ *or* $x+5 \geq 7$. This means that either $x \leq -12$ *or* $x \geq 2$. Thus, the solution consists of two intervals: $(-\infty, -12]$ and $[2, \infty)$. We can abbreviate this collection of numbers by writing

$$(-\infty, -12] \cup [2, \infty)$$

where the connecting symbol $\cup$ is called the *union* symbol. (See Figure 1.21.) More formally, the **union** of sets A and B is the set consisting of all elements that are in either A or B (or in both A and B).

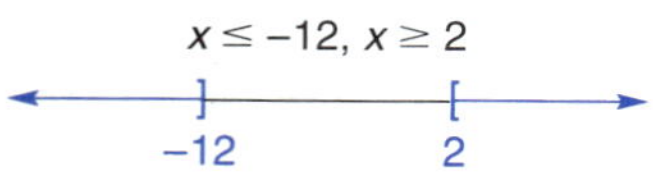

FIGURE 1.21 The union $(-\infty, -12] \cup [2, \infty)$.

b. *Solve* $|3x-4| > 1$.

Solution: Either $3x-4 < -1$ *or* $3x-4 > 1$. Thus, either $3x < 3$ *or* $3x > 5$. Therefore, $x < 1$ *or* $x > \frac{5}{3}$, so the solution consists of all numbers in the set $(-\infty, 1) \cup (\frac{5}{3}, \infty)$.

CAUTION The inequalities $x < 1$ *or* $x > \frac{5}{3}$ cannot be combined as we did with Examples 1 and 2.

▸ **Now work Problem 31.**

APPLICATIONS IN PRACTICE 1

Absolute-Value Notation

Express the following statement using absolute-value notation: The actual weight w of a box of cereal must be within 0.3 oz of the weight stated on the box, which is 22 oz.

EXAMPLE 4 Absolute-Value Notation

Using absolute-value notation, express the following statements:

a. x is less than 3 units from 5.

Solution:

$$|x-5| < 3$$

b. x differs from 6 by at least 7.

Solution:

$$|x-6| \geq 7$$

c. $x < 3$ and $x > -3$ simultaneously.

Solution:

$$|x| < 3$$

d. x is more than 1 unit from -2.

Solution:

$$|x-(-2)| > 1$$
$$|x+2| > 1$$

e. x is less than σ (a Greek letter read "sigma") units from μ (a Greek letter read "mu").

Solution:

$$|x-\mu| < \sigma$$

▸ **Now work Problem 11.**

Properties of Absolute Value

Four basic properties of absolute value are as follows:

1. $|ab| = |a| \cdot |b|$
2. $\left|\frac{a}{b}\right| = \frac{|a|}{|b|}$
3. $|a - b| = |b - a|$
4. $-|a| \leq a \leq |a|$

For example, Property 1 states that the absolute value of the product of two numbers is equal to the product of the absolute values of the numbers.

EXAMPLE 5 **Properties of Absolute Value**

a. $|(-7) \cdot 3| = |-7| \cdot |3| = 21$

b. $|4 - 2| = |2 - 4| = 2$

c. $|7 - x| = |x - 7|$

d. $\left|\frac{-7}{3}\right| = \frac{|-7|}{|3|} = \frac{7}{3}$; $\left|\frac{-7}{-3}\right| = \frac{|-7|}{|-3|} = \frac{7}{3}$

e. $\left|\frac{x-3}{-5}\right| = \frac{|x-3|}{|-5|} = \frac{|x-3|}{5}$

f. $-|2| \leq 2 \leq |2|$

Exercise 1.4

In Problems 1–10, write an equivalent form without the absolute-value symbol.

1. $|-13|$ **2.** $|2^{-1}|$ **3.** $|8 - 2|$

4. $|(-4 - 6)/2|$ **5.** $|2(-\frac{7}{2})|$ **6.** $|2 - 7| - |7 - 2|$

7. $|x| < 4$ **8.** $|x| < 10$ **9.** $|2 - \sqrt{5}|$

10. $|\sqrt{5} - 2|$

***11.** Using the absolute value symbol, express each fact.

(a) x is less than 3 units from 7.
(b) x differs from 2 by less than 3.
(c) x is no more than 5 units from 7.
(d) The distance between 7 and x is 4.
(e) $x + 4$ is less than 2 units from 0.
(f) x is between -3 and 3, but is not equal to 3 or -3.
(g) $x < -6$ or $x > 6$.
(h) $x - 6 > 4$ or $x - 6 < -4$.
(i) The number x of hours that a machine will operate efficiently differs from 105 by less than 3.
(j) The average monthly income x (in dollars) of a family differs from 850 by less than 100.

12. Use absolute-value notation to indicate that x and μ differ by no more than σ.

13. Use absolute-value notation to indicate that the prices p_1 and p_2 of two products may differ by no more than 9 (dollars).

14. Find all values of x such that $|x - \mu| \leq 2\sigma$.

In Problems 15–36, solve the given equation or inequality.

15. $|x| = 7$ **16.** $|-x| = 2$

17. $\left|\frac{x}{5}\right| = 7$ **18.** $\left|\frac{4}{x}\right| = 8$

***19.** $|x - 5| = 8$ **20.** $|4 + 3x| = 6$

21. $|5x - 2| = 0$ **22.** $|7x + 3| = x$

23. $|7 - 4x| = 5$ **24.** $|5 - 3x| = 2$

25. $|x| < 4$ **26.** $|-x| < 3$

27. $\left|\frac{x}{4}\right| > 2$ **28.** $\left|\frac{x}{3}\right| > \frac{1}{2}$

***29.** $|x + 9| < 5$ **30.** $|5x - 1| < -6$

***31.** $|x - \frac{1}{2}| > \frac{1}{2}$ **32.** $|1 - 3x| > 2$

33. $|5 - 8x| \leq 1$ **34.** $|4x - 1| \geq 0$

35. $\left|\frac{3x-8}{2}\right| \geq 4$ **36.** $\left|\frac{x-7}{3}\right| \leq 5$

In Problems 37–38, express the statement using absolute value notation.

37. In a science experiment, the measurement of a distance d is 17.2 m, and is accurate to ± 30 cm.

38. The difference in temperature between two chemicals that are to be mixed must be no less than 5 degrees and no more than 10 degrees.

39. Statistics In statistical analysis, the Chebyshev inequality asserts that if x is a random variable, μ is its mean, and σ is its standard deviation, then

$$(\text{probability that } |x - \mu| > h\sigma) \geq \frac{1}{h^2}$$

Find those values of x such that $|x - \mu| > h\sigma$.

40. Manufacturing Tolerance In the manufacture of widgets, the average dimension of a part is 0.01 cm. Using the absolute-value symbol, express the fact that an individual measurement x of a part does not differ from the average by more than 0.005 cm.

1.5 Review

Important Terms and Symbols

Section 1.1	fixed cost overhead variable cost total cost total revenue profit
Section 1.2	$a < b$ $a \leq b$ $a > b$ $a \geq b$ $a < x < b$ inequality sense of an inequality equivalent inequalities linear inequality $-\infty < x < \infty$ open interval closed interval endpoints interval notation
Section 1.4	absolute value, $\|x\|$ union, $\cup$

Summary

With a word problem, an equation is not handed to you. Instead, you must set it up by translating verbal statements into an equation (or inequality). This is *mathematical modeling*. It is important that you first read the problem more than once so that you clearly understand what facts are given and what you are asked to find. Then choose a letter to represent the unknown quantity that you want to find. Use the relationships and facts given in the problem and translate them into an equation involving the letter. Finally, solve the equation, and see if your solution answers what was asked. Sometimes the solution to the *equation* will not be the answer to the *problem,* but it may be useful in obtaining that answer.

Some basic relationships that are used in solving business problems are as follows:

$$\begin{aligned} \text{total cost} &= \text{variable cost} + \text{fixed cost} \\ \text{total revenue} &= (\text{price per unit})(\text{number of units sold}) \\ \text{profit} &= \text{total revenue} - \text{total cost} \end{aligned}$$

The inequality symbols $<$, $\leq$, $>$, and $\geq$ are used to represent an inequality, which is a statement that one number is, for example, less than another number. Three basic operations that, when applied to an inequality, guarantee an equivalent inequality are:

1. Adding (or subtracting) the same number to (or from) both sides.
2. Multiplying (or dividing) both sides by the same positive number.
3. Multiplying (or dividing) both sides by the same negative number and reversing the sense of the inequality.

These operations are useful in solving a linear inequality (an inequality that can be put in the form $ax + b < 0$ or $ax + b \leq 0$, where $a \neq 0$).

An algebraic definition of absolute value is

$$|x| = x \text{ if } x \geq 0 \quad \text{and} \quad |x| = -x \text{ if } x < 0$$

We interpret $|a - b|$ or $|b - a|$ as the distance between a and b. If $d > 0$, then the solution to the inequality $|x| < d$ is the interval $(-d, d)$. The solution to $|x| > d$ consists of two intervals and is given by $(-\infty, -d) \cup (d, \infty)$. Some basic properties of absolute value are:

1. $|ab| = |a| \cdot |b|$
2. $\left|\dfrac{a}{b}\right| = \dfrac{|a|}{|b|}$
3. $|a - b| = |b - a|$
4. $-|a| \leq a \leq |a|$

Review Problems

Problem numbers shown in color indicate problems suggested for use as a practice chapter test.

In Problems 1–15, solve the equation or inequality.

1. $3x - 8 \geq 4(x - 2)$
2. $2x - (7 + x) \leq x$
3. $-(5x + 2) < -(2x + 4)$
4. $-2(x + 6) > x + 4$
5. $3p(1 - p) > 3(2 + p) - 3p^2$
6. $3\left(5 - \frac{7}{3}q\right) < 9$
7. $\frac{x+5}{3} - \frac{1}{2} \leq 2$
8. $\frac{x}{2} + \frac{x}{3} > \frac{x}{4}$
9. $\frac{1}{4}s - 3 \leq \frac{1}{8}(3 + 2s)$
10. $\frac{1}{3}(t + 2) \geq \frac{1}{4}t + 4$
11. $|3 - 2x| = 7$
12. $\left|\frac{5x - 6}{13}\right| = 0$
13. $|5t - 2| < 3$
14. $4 < \left|\frac{2}{3}x + 5\right|$
15. $|3 - 2x| \geq 4$
16. **Profit** A profit of 40% on the selling price of a product is equivalent to what percent profit on the cost?
17. **Stock Exchange** On a certain day, there were 1132 different issues traded on the New York Stock Exchange. There were 48 more issues showing an increase than showing a decline, and no issues remained the same. How many issues suffered a decline?
18. **Sales Tax** The sales tax in a certain state is 6.5%. If a total of \$3039.29 in purchases, including tax, is made in the course of a year, how much of it is tax?
19. **Production Allocation** A company will manufacture a total of 10,000 units of its product at plants A and B. Available data are as follows:

	Plant A	Plant B
Unit cost for labor and material	\$5	\$5.50
Fixed cost	\$30,000	\$35,000

Between the two plants the company has decided to allot no more than \$117,000 for total costs. What is the minimum number of units that must be produced at plant A?

20. **Storage Tanks** A company is replacing two cylindrical oil-storage tanks with one new tank. The old tanks are each 16 ft high. One has a radius of 15 ft and the other a radius of 20 ft. The new tank will also be 16 ft high. Find its radius if it is to have the same volume as the old tanks combined. (*Hint:* The volume V of a cylindrical tank is $V = \pi r^2 h$, where r is the radius of the circular base and h is the height of the tank.)
21. **Operating Ratio** The *operating ratio* of a retail business is the ratio, expressed as a percentage, of operating costs (everything from advertising expenses to equipment depreciation) to net sales (i.e., gross sales minus returns and allowances). An operating ratio less than 100% indicates a profitable operation, while an operating ratio in the 80%–90% range is extremely good. If a company has net sales of \$236,460 in one period, write an inequality describing the operating costs that would keep the operating ratio below 90%.

Mathematical
SNAPSHOT

Variable-Quality Recording[2]

If you are like the millions of others who have a VCR, you have seen how convenient it is to record television shows for future viewing. With a VHS format, you may have your choice of standard play speed (SP), long play speed (LP), or extended play speed (EP). SP is the fastest speed and provides the best picture quality. LP, a slower speed, gives lower quality recording, and EP, which is the slowest speed, gives the lowest picture quality.

With the common T-120 videotape, the maximum recording time at SP is 2 hours. At LP it is 4 hours, and at EP it is 6 hours. In the following discussion you may assume that these recording times are exact and that the amount of tape used changes uniformly with recording time.

If you wish to record a movie that is no more than 2 hours long, obviously SP should be used to get the best picture quality. However, for recording a 3-hour movie on a single T-120 tape, using only SP speed would cause the tape to be filled 1 hour before the movie is over. You can overcome this difficulty by using SP in conjunction with another speed, making sure that the time at SP is maximized.

For example, you can begin at LP and then complete the recording at SP. Obviously your problem is to determine when the changeover to SP should be made. Let t be the time, in hours, that LP is used. Then $3 - t$ hours of the movie remain to be recorded at SP. Since the tape speed in the LP mode is $\frac{1}{4}$ tape per hour and the speed in the SP mode is $\frac{1}{2}$ tape per hour, the portion of the tape used at LP is $t/4$ and the portion at SP is $(3 - t)/2$. The sum of these portions must be 1 because the whole tape would be used up. Therefore, you need to solve a linear equation.

$$\frac{t}{4} + \frac{3-t}{2} = 1$$
$$t + 2(3-t) = 4$$
$$6 - t = 4$$
$$t = 2$$

[2]Adapted from Gregory N. Fiore, "An Application of Linear Equations to the VCR," *Mathematics Teacher* 81 (October 1988), 570–72. By permission of the National Council of Teachers of Mathematics.

Thus you should record at LP for 2 hours and then switch to SP for the remaining $3 - t = 3 - 2 = 1$ hour. This means that one-third of the movie will be recorded with the best picture quality.

Instead of restricting yourself to a 3-hour movie, you can generalize the preceding problem to handle a movie of length l hours, where $2 < l \le 4$. This situation gives

$$\frac{t}{4} + \frac{l-t}{2} = 1$$

whose solution is

$$t = 2l - 4$$

Alternatively, you might feel that there is not much difference between the picture qualities at LP and EP speeds. If you desire to begin at EP and end with SP, you could handle a movie of length l, where $2 < l \le 6$. Let t be the time, in hours, that EP is used. Then

$$\frac{t}{6} + \frac{l-t}{2} = 1$$
$$t + 3(l-t) = 6$$
$$-2t + 3l = 6$$
$$3l - 6 = 2t$$
$$t = \frac{3}{2}l - 3$$

For example, with a 3-hour movie you would record at EP for $t = \frac{3}{2}(3) - 3 = 1\frac{1}{2}$ hours and then at SP for $3 - 1\frac{1}{2} = 1\frac{1}{2}$ hours. This shows that using EP instead of

LP gives you $\frac{1}{2}$ hour more of quality recording at SP. As a second example, consider recording a 4-hour-and-20-minute movie. Here $l = 4\frac{1}{3}$ hours, so you would use EP for

$$t = \frac{3}{2}\left(\frac{13}{3}\right) - 3 = 3\frac{1}{2}\text{hours}$$

and use SP for the rest of the movie.

The same method can be used to maximize audio quality on recordable CDs. A standard CD can hold about 74 minutes of high-fidelity stereo sound. However, you can fit many hours of audio onto a CD by using software that sacrifices a little sound quality in order to shrink the amount of CD space the recording takes up. Depending on the method used, a recording can be compressed to one-twelfth or even one-twentieth of its original size. This is especially useful for archiving recordings in large volume.

Suppose you work for a radio station that uses compression in archiving its broadcasts. You have two compression schemes to choose from, one of which shrinks the space a recording takes up by a factor of 12 with very little loss in sound quality, and the other of which shrinks the recording by a factor of 20 with a noticeable loss in quality. You have 18 hours of audio to archive. How much of that 18 hours, or 1080 minutes, should be compressed at the higher quality level to maximize overall quality while still fitting the entire 18 hours onto a single CD?

To find the answer, let t equal the number of minutes compressed at a ratio of 12 to 1. This part of the recording will take up $t/12$ minutes' worth of space on the CD. The other $1080 - t$ minutes will be compressed at a ratio of 20 to 1 and will take up $(1080 - t)/20$ minutes' worth of space. Since there are 74 minutes' worth of room on the CD, the answer is found by solving the linear equation

$$\frac{t}{12} + \frac{1080 - t}{20} = 74$$

The solution looks like this:

$$\begin{aligned} 5t + 3(1080 - t) &= 4440 \\ 2t + 3240 &= 4440 \\ 2t &= 1200 \\ t &= 600 \end{aligned}$$

So, you should process 600 minutes, or 10 hours, at a compression of 12 to 1 and the remaining 8 hours at a compression of 20 to 1.

To learn more about audio compression schemes, visit www.webopedia.com and look up "data compression" and related terms.

Exercises

1. If LP and SP modes on a VCR are used to record a $2\frac{1}{2}$-hour movie, how long after the start of the movie should the switch from LP to SP be made?

2. If EP and SP modes are used to record a $2\frac{1}{2}$-hour show, how many minutes after the start of the show should the switch from EP to SP be made?

3. If EP and SP modes are used to record a movie of length 2 hours and 40 minutes, how long after the start of the movie should the switch from EP to SP be made?

4. EP and SP modes are used to record a 3-hour movie. How many minutes of the movie should be recorded at EP if the viewer eliminates 8 minutes of commercials each hour?

5. Use the Solver function on a graphing calculator to solve the equation

$$\frac{x}{12} + \frac{1080 - x}{20} = 74$$

Then, in a similar manner, solve the equation

$$\frac{x}{15} + \frac{1590 - x}{24} = 74$$

6. What, in the context of recording compressed audio on CDs, does the second equation in Problem 5 represent?

CHAPTER

2

Functions and Graphs

Mathematical SNAPSHOT

A Taxing Experience!

Suppose a 180-pound man drinks four beers in quick succession. We know that his blood alcohol concentration, or BAC, will first rise, then gradually fall back to zero. But what is the best way to describe how quickly the BAC rises, where it peaks, and how fast it falls again?

If we obtain measured BAC values for this particular drinker, we can display them in a table, as follows:

Time (h)	**1**	**2**	**3**	**4**	**5**	**6**
BAC (%)	0.0820	0.0668	0.0516	0.0364	0.0212	0.0060

However, a table can show only a limited number of values and so does not really give the overall picture.

We might instead relate the BAC to time t using a combination of linear and quadratic equations (recall Chapter 0):

$$\text{BAC} = -0.1025t^2 + 0.1844t \qquad \text{if } t \leq 0.97$$

$$\text{BAC} = -0.0152t + 0.0972 \qquad \text{if } t > 0.97$$

As with the table, however, it is hard to look at the equations and quickly understand what is happening with BAC over time.

Probably the best description of changes in the BAC over time is a graph, like the one at left. Here we easily see what happens. The blood alcohol concentration climbs rapidly, peaks at 0.083% after about an hour, and then gradually tapers off over the next five-and-a-half hours. Note that for most of three hours, this male drinker's BAC is above 0.05%, the point at which one's driving skills typically begin to decline. The curve will vary from one drinker to the next, but women are generally affected more severely than men, not only because of weight differences but also because of the different water contents in men's and women's bodies.

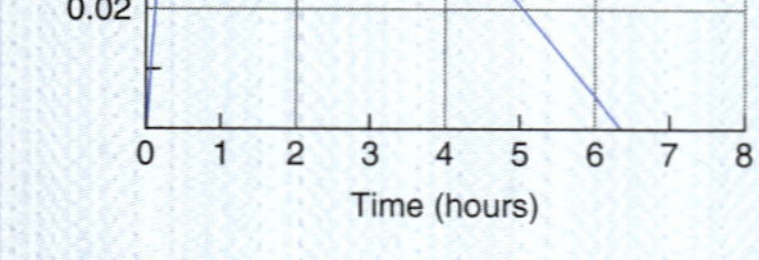

The relationship between time and blood alcohol content is an example of a function. This chapter deals in depth with functions and their graphs.

OBJECTIVE *To understand what a function is and to determine domains and function values.*

2.1 Functions

In the 17th century, Gottfried Wilhelm Leibniz, one of the inventors of calculus, introduced the term *function* into the mathematical vocabulary. The concept of a function is one of the most basic in all of mathematics, and it is essential to the study of calculus.

Briefly, a function is a special type of relation that expresses how one quantity (the *output*) depends on another quantity (the *input*). For example, when money is invested at some interest rate, the interest I (output) depends on the length of time t (input) that the money is invested. To express this dependence, we say that I is a "function of" t. Functional relations like this are usually specified by a formula that shows what must be done to the input to find the output.

To illustrate, suppose \$100 earns simple interest at an annual rate of 6%. Then it can be shown that interest and time are related by the formula

$$I = 100(0.06)t \tag{1}$$

where I is in dollars and t is in years. For example,

$$\text{if } t = \frac{1}{2} \quad \text{then} \quad I = 100(0.06)\left(\frac{1}{2}\right) = 3 \tag{2}$$

Thus, Formula (1) assigns the output 3 to the input $\frac{1}{2}$. We can think of Formula (1) as defining a *rule:* Multiply t by 100(0.06). The rule assigns to each input number t exactly one output number I, which we symbolize by the following arrow notation:

$$t \to I \quad \text{or} \quad t \to 100(0.06)t$$

This rule is an example of a *function* in the following sense:

Definition

A ***function*** is a rule that assigns to each input number exactly one output number. The set of all input numbers to which the rule applies is called the ***domain*** of the function. The set of all output numbers is called the ***range.***

For the interest function defined by Formula (1), the input number t cannot be negative, because negative time makes no sense. Thus, the domain consists of all nonnegative numbers—that is, all $t \geq 0$. From (2), we see that when the input is $\frac{1}{2}$, the output is 3. Thus, 3 is in the range.

We have been using the term *function* in a restricted sense because, in general, the inputs or outputs do not have to be numbers. For example, a list of states and their capitals assigns to each state its capital (exactly one output). Hence, a function is implied. However, for the time being, we shall consider only functions whose domains and ranges consist of real numbers.

A variable that represents input numbers for a function is called an **independent variable.** A variable that represents output numbers is called a **dependent variable** because its value *depends* on the value of *the independent variable*. We say that the dependent variable is a *function of the independent variable*. That is, output is a function of input. Thus, for the interest formula $I = 100(0.06)t$, the independent variable is t, the dependent variable is I, and I is a function of t.

As another example, the equation (or formula)

$$y = x + 2 \tag{3}$$

defines y as a function of x. The equation gives the rule, "Add 2 to x." This rule assigns to each input x exactly one output $x + 2$, which is y. If $x = 1$, then $y = 3$; if $x = -4$, then $y = -2$. The independent variable is x and the dependent variable is y.

CAUTION In $y^2 = x$, x and y are related, but the relationship is not a function of x.

Not all equations in x and y define y as a function of x. For example, let $y^2 = x$. If x is 9, then $y^2 = 9$, so $y = \pm 3$. Hence, to the input 9, there are assigned not one, but *two*, output numbers: 3 and -3. This violates the definition of a function, so y is **not** a function of x.

On the other hand, some equations in two variables define either variable as a function of the other variable. For example, if $y = 2x$, then for each input x, there is exactly one output, $2x$. Thus, y is a function of x. However, solving the equation for x gives $x = y/2$. For each input y, there is exactly one output, $y/2$. Consequently, x is a function of y.

Usually, the letters f, g, h, F, G, and so on are used to represent function rules. For example, Equation (2.3), $y = x+2$, defines y as a function of x, where the rule is "Add 2 to the input." Suppose we let f represent this rule. Then we say that f is the function. To indicate that f assigns the output 3 to the input 1, we write $f(1) = 3$, which is read "f of 1 equals 3." Similarly, $f(-4) = -2$. More generally, if x is any input, we have the following notation:

$f(x)$ is an output number.

$f(x)$, which is read "f of x," means the output number in the range of f that corresponds to the input number x in the domain of f.

$$\overset{\text{input}}{\downarrow}$$
$$\underbrace{f(x)}_{\substack{\uparrow \\ \text{output}}}$$

Thus, the output $f(x)$ is the same as y. But since $y = x + 2$, we may write $y = f(x) = x + 2$, or simply,

$$f(x) = x + 2$$

For example, to find $f(3)$, which is the output corresponding to the input 3, we replace each x in $f(x) = x + 2$ by 3:

$$f(3) = 3 + 2 = 5$$

Likewise,

$$f(8) = 8 + 2 = 10$$

$$f(-4) = -4 + 2 = -2$$

CAUTION $f(x)$ does **not** mean f times x. $f(x)$ is the output that corresponds to the input x.

Output numbers such as $f(-4)$ are called **function values.** Keep in mind that they are in the range of f.

Functional notation is used extensively in calculus.

Quite often, functions are defined by "functional notation." For example, the equation $g(x) = x^3 + x^2$ defines the function g that assigns the output number $x^3 + x^2$ to an input number x:

$$g: x \to x^3 + x^2$$

In other words, g adds the cube and the square of an input number. Some function values are

$$g(2) = 2^3 + 2^2 = 12$$

$$g(-1) = (-1)^3 + (-1)^2 = -1 + 1 = 0$$

$$g(t) = t^3 + t^2$$

$$g(x + 1) = (x + 1)^3 + (x + 1)^2$$

The idea of *replacement* is very important in determining function values.

Note that $g(x+1)$ was found by replacing each x in x^3+x^2 by the input $x+1$.

When we refer to the function g defined by $g(x)=x^3+x^2$, we shall feel free to call the equation itself a function. Thus, we speak of "the function $g(x)=x^3+x^2$," and, similarly, "the function $y=x+2$."

Let's be specific about the domain of a function. Unless otherwise stated, the domain consists of all real numbers for which the rule of the function makes sense; that is, the set of all real numbers for which the rule gives function values that are real numbers.

For example, suppose

$$h(x)=\frac{1}{x-6}$$

Here any real number can be used for x except 6, because the denominator is 0 when x is 6. So the domain of h is understood to be all real numbers except 6.

Equality of Functions

To say that two functions f and g are equal, denoted $f=g$, is to say that

1. The domain of f is equal to the domain of g;
2. For every x in the domain of f and g, $f(x)=g(x)$.

Requirement **1.** says that a number x is in the domain of f if and only if it is in the domain of g. Thus, if we have $f(x)=x^2$, with no explicit mention of domain, and $g(x)=x^2$ for $x\geq 0$, then $f\neq g$. For here the domain of f is the whole real line $(-\infty,\infty)$ and the domain of g is $[0,\infty)$. On the other hand, if we have $f(x)=(x+1)^2$ and $g(x)=x^2+2x+1$, then for both f and g the domain is understood to be $(-\infty,\infty)$ and the issue for deciding if $f=g$ is whether, for each real number x, we have $(x+1)^2=x^2+2x+1$. But this is true; it is a special case of 4. in the Special Products of Section 0.4. In fact, older textbooks refer to statements like $(x+1)^2=x^2+2x+1$ as 'identities', to indicate that they are true for any admissible value of the variable and to distinguish them from statements like $(x+1)^2=0$, which are true only for some values of x.

Given functions f and g, it follows that we have $f\neq g$ if *either* the domain of f is different from the domain of g *or* there is some x for which $f(x)\neq g(x)$.

EXAMPLE 1 Determining Equality of Functions

Determine which of the following functions are equal.

a. $f(x)=\dfrac{(x+2)(x-1)}{(x-1)}$ **b.** $g(x)=x+2$

c. $h(x)=\begin{cases} x+2 & \text{if } x\neq 1 \\ 0 & \text{if } x=1 \end{cases}$ **d.** $k(x)=\begin{cases} x+2 & \text{if } x\neq 1 \\ 3 & \text{if } x=1 \end{cases}$

Solution: The domain of f is the set of all real numbers other than 1, while that of g is the set of all real numbers. (For these we are following the convention that the domain is the set of all real numbers for which the rule make sense.) We will have more to say about functions like h and k that are defined by *cases* in Example 4 of Section 2.2. Here we observe that the domain of h and the domain of k are both $(-\infty, \infty)$, since for both we have a rule that makes sense for each real number. The domains of g, h, and k are equal to each other but that of f is different. So by requirement **1.** for equality of functions, $f \neq g$, $f \neq h$, and $f \neq k$. By definition, $g(x) = h(x) = k(x)$ for all $x \neq 1$, so the matter of equality of g, h, and k depends on their values at 1. Since $g(1) = 3$, $h(1) = 0$, and $k(1) = 3$, we conclude that $g = k$ and $g \neq h$ (and $h \neq k$). While this example might appear to be contrived, it is typical of an issue that arises frequently in calculus.

▶ **Now work Problem 3.**

APPLICATIONS IN PRACTICE 1

Finding Domains

The area of a circle depends on the length of the radius of the circle.

a. Write a function $a(r)$ for the area of a circle when the length of the radius is r.

b. What is the domain of this function out of context?

c. What is the domain of this function in the given context?

EXAMPLE 2 Finding Domains

Find the domain of each function.

a. $f(x) = \dfrac{x}{x^2 - x - 2}$

Solution: We cannot divide by zero, so we must find any values of x that make the denominator 0. These *cannot* be input numbers. Thus, we set the denominator equal to 0 and solve for x:

$$\begin{aligned} x^2 - x - 2 &= 0 && \text{(quadratic equation)} \\ (x-2)(x+1) &= 0 && \text{(factoring)} \\ x &= 2, -1 \end{aligned}$$

Therefore, the domain of f is all real numbers *except* 2 and -1.

b. $g(t) = \sqrt{2t - 1}$

Solution: $\sqrt{2t-1}$ is a real number if $2t - 1$ is greater than or equal to 0. If $2t - 1$ is negative, then $\sqrt{2t-1}$ is not a real number. (It is an *imaginary number.*) Since function values must be real numbers, we must assume that

$$\begin{aligned} 2t - 1 &\geq 0 \\ 2t &\geq 1 && \text{(adding 1 to both sides)} \\ t &\geq \frac{1}{2} && \text{(dividing both sides by 2)} \end{aligned}$$

Thus, the domain is the interval $[\frac{1}{2}, \infty)$.

▶ **Now work Problem 7.**

APPLICATIONS IN PRACTICE 2

Finding Domain and Function Values

The time it takes to go a given distance depends on the speed at which one is traveling.

a. Write a function $t(r)$ for the time it takes if the distance is 300 miles and the speed is r.

b. What is the domain of this function out of context?

c. What is the domain of this function in the given context?

d. Find $t(x)$, $t\left(\frac{x}{2}\right)$, and $t\left(\frac{x}{4}\right)$.

e. What happens to the time if the speed is reduced (divided) by a constant c? Describe this situation using an equation.

CAUTION Don't be confused by notation. In Example 3(c), we found $g(x+h)$ by replacing each x in $g(x) = 3x^2 - x + 5$ by the input $x + h$. Don't write the function and then add h. In this example, $g(x+h)$ is quite different from $g(x) + h$:

$$g(x+h) \neq 3x^2 - x + 5 + h$$

Also, $g(x+h)$ is quite different from $g(x) + g(h)$.

The difference quotient of a function is an important concept for calculus.

EXAMPLE 3 **Finding Domain and Function Values**

Let $g(x) = 3x^2 - x + 5$. Any real number can be used for x, so the domain of g is all real numbers.

a. *Find $g(z)$.*

Solution: Replacing each x in $g(x) = 3x^2 - x + 5$ by z gives

$$g(z) = 3(z)^2 - z + 5 = 3z^2 - z + 5$$

b. *Find $g(r^2)$.*

Solution: Replacing each x in $g(x) = 3x^2 - x + 5$ by r^2 gives

$$g(r^2) = 3(r^2)^2 - r^2 + 5 = 3r^4 - r^2 + 5$$

c. *Find $g(x+h)$.*

Solution:

$$\begin{aligned} g(x+h) &= 3(x+h)^2 - (x+h) + 5 \\ &= 3(x^2 + 2hx + h^2) - x - h + 5 \\ &= 3x^2 + 6hx + 3h^2 - x - h + 5 \end{aligned}$$

▶ **Now work Problem 31(a).**

EXAMPLE 4 **Finding a Difference Quotient**

If $f(x) = x^2$, find $\dfrac{f(x+h) - f(x)}{h}$.

Solution: The expression $\dfrac{f(x+h) - f(x)}{h}$ is referred to as a **difference quotient.** Here the numerator is a difference of function values. We have

$$\begin{aligned} \frac{f(x+h) - f(x)}{h} &= \frac{(x+h)^2 - x^2}{h} \\ &= \frac{x^2 + 2hx + h^2 - x^2}{h} \\ &= \frac{2hx + h^2}{h} \\ &= \frac{h(2x+h)}{h} = 2x + h \end{aligned}$$

▶ **Now work Problem 35.**

In some cases, the domain of a function is restricted for physical or economic reasons. For example, the previous interest function $I = 100(0.06)t$ has $t \geq 0$ because t represents time. Example 5 provides another illustration.

APPLICATIONS IN PRACTICE 3

Demand Function

Suppose the weekly demand function for pizzas at a local pizza parlor is $p = 26 - \frac{q}{40}$.

a. If the current price is \$18.50 per pizza, how many pizzas are sold each week?

b. If 200 pizzas are sold each week, what is the current price?

c. If the owner wants to double the number of pizzas sold each week (to 400), what should the price be?

EXAMPLE 5 **Demand Function**

Suppose that the equation $p = 100/q$ describes the relationship between the price per unit p of a certain product and the number of units q of the product that consumers will buy (that is, demand) per week at the stated price. This equation is called a *demand equation* for the product. If q is an input number, then to each value of q, there is assigned exactly one output number p:

$$q \to \frac{100}{q} = p$$

For example,

$$20 \to \frac{100}{20} = 5$$

that is, when q is 20, p is 5. Thus, price p is a function of quantity demanded, q. This function is called a **demand function.** The independent variable is q, and p is the dependent variable. Since q cannot be 0 (division by 0 is not defined) and cannot be negative (q represents quantity), the domain is all values of q such that $q > 0$.

We have seen that a function is essentially a *correspondence* whereby to each input number in the domain there is assigned exactly one output number in the range. For the correspondence given by $f(x) = x^2$, some sample assignments are shown by the arrows in Figure 2.1. The next example discusses a functional correspondence that is not given by an algebraic formula.

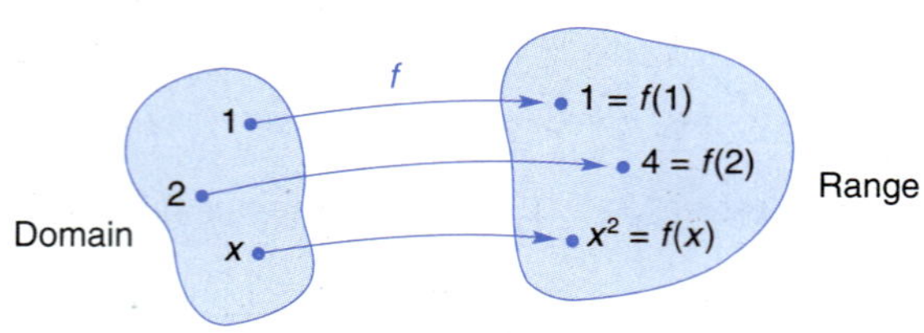

FIGURE 2.1 Functional correspondence for $f(x) = x^2$.

Supply Schedule

p Price per Unit in Dollars	q Quantity Supplied per Week
500	11
600	14
700	17
800	20

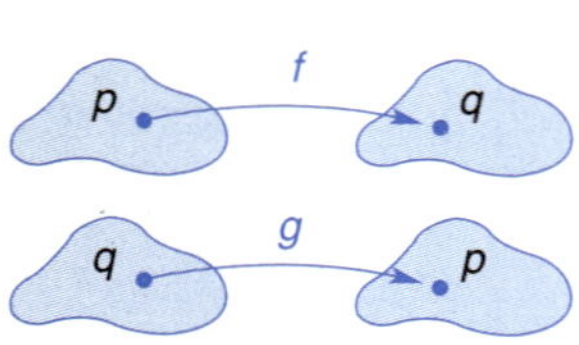

FIGURE 2.2 Supply schedule and supply functions.

EXAMPLE 6 **Supply Schedule**

The table in Figure 2.2 is a *supply schedule.* Such a table gives a correspondence between the price p of a certain product and the quantity q the producers will supply per week at that price. For each price, there corresponds exactly one quantity, and vice versa.

If p is the independent variable, then q is a function of p, say, $q = f(p)$, and

$$f(500) = 11 \quad f(600) = 14 \quad f(700) = 17 \quad \text{and} \quad f(800) = 20$$

Notice that as price per unit increases, the producers are willing to supply more units per week.

On the other hand, if q is the independent variable, then p is a function of q, say $p = g(q)$, and

$$g(11) = 500 \quad g(14) = 600 \quad g(17) = 700 \quad \text{and} \quad g(20) = 800$$

We speak of f and g as **supply functions.**

TECHNOLOGY

Function values are easily computed with a graphing calculator. For example, suppose

$$f(x) = 17x^4 - 13x^3 + 7$$

and we wish to find $f(0.7)$, $f(-2.31)$, and $f(10)$. With a TI-83 Plus, we first enter the function as Y_1:

$$Y_1 = 17X\hat{}4 - 13X\hat{}3 + 7$$

After pressing the TABLE key, we successively enter the x-values .7, −2.31, and 10. The results are shown in Figure 2.3. We remark that there are other methods of determining function values with the TI-83 Plus.

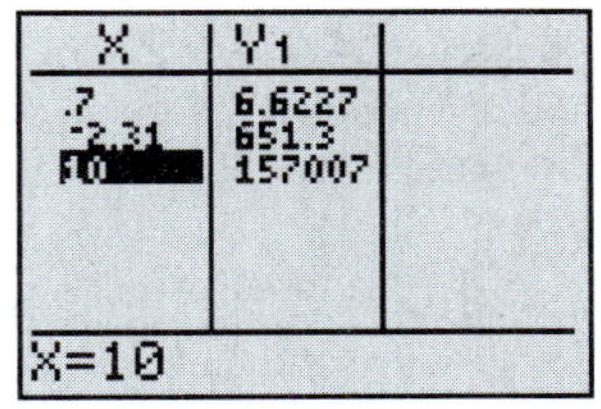

FIGURE 2.3 Table of function values for $f(x) = 17x^4 - 13x^3 + 7$.

Exercise 2.1

In Problems 1–4, determine whether the given functions are equal.

1. $f(x) = \sqrt{x^2}$; $g(x) = x$

2. $F(r) = (\sqrt{r})^2$; $G(r) = r$

*__3.__ $h(x) = \dfrac{|x|}{x}$; $k(x) = \begin{cases} 1 & \text{if } x \geq 0 \\ -1 & \text{if } x < 0 \end{cases}$

4. $f(x) = \begin{cases} \dfrac{x^2 - 4x + 3}{x - 3} & \text{if } x \neq 3 \\ 2 & \text{if } x = 3 \end{cases}$; $g(x) = x - 1$

In Problems 5–16, give the domain of each function.

5. $f(x) = \dfrac{8}{x}$

6. $g(x) = \dfrac{x}{5}$

*__7.__ $h(x) = \sqrt{x - 3}$

8. $K(z) = \dfrac{1}{\sqrt{z - 1}}$

9. $F(t) = 4t^2 - 6$

10. $H(x) = \dfrac{x}{x + 8}$

11. $f(x) = \dfrac{9x - 9}{2x + 7}$

12. $g(x) = \sqrt{4x + 3}$

13. $g(y) = \dfrac{4}{y^2 - 4y + 4}$

14. $f(x) = \dfrac{x + 1}{x^2 + 6x + 5}$

15. $h(s) = \dfrac{4 - s^2}{2s^2 - 7s - 4}$

16. $G(r) = \dfrac{2}{r^2 + 1}$

In Problems 17–28, find the function values for each function.

17. $f(x) = 2x + 1$; $f(0)$, $f(3)$, $f(-4)$

18. $H(s) = 5s^2 - 3$; $H(4)$, $H(\sqrt{2})$, $H\left(\frac{2}{3}\right)$

19. $G(x) = 2 - x^2$; $G(-8)$, $G(u)$, $G(u^2)$

20. $F(x) = -5x$; $F(s)$, $F(t + 1)$, $F(x + 3)$

21. $g(u) = u^2 + u$; $g(-2)$, $g(2v)$, $g(-x^2)$

22. $h(v) = \dfrac{1}{\sqrt{v}}$; $h(16)$, $h\left(\frac{1}{4}\right)$, $h(1 - x)$

23. $f(x) = x^2 + 2x + 1$; $f(1)$, $f(-1)$, $f(x + h)$

24. $H(x) = (x + 4)^2$; $H(0)$, $H(2)$, $H(t - 4)$

25. $k(x) = \dfrac{x - 7}{x^2 + 2}$; $k(5)$, $k(3x)$, $k(x + h)$

26. $H(x) = \sqrt{4 + x}$; $H(-4)$, $H(-3)$, $H(x + 1) - H(x)$

27. $f(x) = x^{4/3}$; $f(0)$, $f(64)$, $f\left(\frac{1}{8}\right)$

28. $g(x) = x^{2/5}$; $g(32)$, $g(-64)$, $g(t^{10})$

In Problems 29–36, find (a) $f(x + h)$ and (b) $\dfrac{f(x + h) - f(x)}{h}$; simplify your answers.

29. $f(x) = 4x - 5$

30. $f(x) = \dfrac{x}{2}$

*__31.__ $f(x) = x^2 + 2x$

32. $f(x) = 3x^2 - 2x - 1$

33. $f(x) = 2 - 4x - 3x^2$

34. $f(x) = x^3$

*__35.__ $f(x) = \dfrac{1}{x}$

36. $f(x) = \dfrac{x + 8}{x}$

37. If $f(x) = 5x + 3$, find $\dfrac{f(3 + h) - f(3)}{h}$

38. If $f(x) = x^2 - x$, find $\dfrac{f(x) - f(4)}{x - 4}$

In Problems 39–42, is y a function of x? Is x a function of y?

39. $9y - 3x - 4 = 0$

40. $x^2 + y = 0$

41. $y = 7x^2$

42. $x^2 + y^2 = 1$

43. The formula for the area of a circle of radius r is $A = \pi r^2$. Is the area a function of the radius?

44. Suppose $f(b) = a^2b^3 + a^3b^2$. (a) Find $f(a)$. (b) Find $f(ab)$.

45. Value of Business A business with an original capital of \$20,000 has income and expenses each week of \$4000 and \$3200, respectively. If all profits are retained in the business, express the value V of the business at the end of t weeks as a function of t.

46. Depreciation If a \$30,000 machine depreciates 2% of its original value each year, find a function f that expresses the machine's value V after t years have elapsed.

47. Profit Function If q units of a certain product are sold (q is nonnegative), the profit P is given by the equation $P = 1.25q$. Is P a function of q? What is the dependent variable? the independent variable?

48. Demand Function Suppose the yearly demand function for a particular actor to star in a film is $p = \dfrac{1{,}200{,}000}{q}$, where q is the number of films he stars in during the year. If the actor currently charges \$600,000 per film, how many films does he star in each year? If he wants to star in four films per year, what should his price be?

49. Supply Function Suppose the weekly supply function for a pound of house-blend coffee at a local coffee shop is $p = \dfrac{q}{48}$, where q is the number of pounds of coffee supplied per week. How many pounds of coffee per week will be supplied if the price is \$8.39 a pound? How many pounds of coffee per week will be supplied if the price is \$19.49 a pound? How does the amount supplied change as the price increases?

50. Hospital Discharges An insurance company examined the records of a group of individuals hospitalized for a particular illness. It was found that the total proportion discharged at the end of t days of hospitalization is given by

$$f(t) = 1 - \left(\frac{300}{300+t}\right)^3$$

Evaluate (a) $f(0)$, (b) $f(100)$, and (c) $f(300)$. (d) At the end of how many days was 99.9% (0.999) of the group discharged?

51. Psychology A psychophysical experiment was conducted to analyze human response to electrical shocks.[1] The subjects received a shock of a certain intensity. They were told to assign a magnitude of 10 to this particular shock, called the standard stimulus. Then other shocks (stimuli) of various intensities were given. For each one, the response R was to be a number that indicated the perceived magnitude of the shock relative to that of the standard stimulus. It was found that R was a function of the intensity I of the shock (I in microamperes) and was estimated by

$$R = f(I) = \frac{I^{4/3}}{2500} \qquad 500 \le I \le 3500$$

Evaluate (a) $f(1000)$ and (b) $f(2000)$. (c) Suppose that I_0 and $2I_0$ are in the domain of f. Express $f(2I_0)$ in terms of $f(I_0)$. What effect does the doubling of intensity have on response?

52. Psychology In a paired-associate learning experiment,[2] the probability of a correct response as a function of the number n of trials has the form

$$P(n) = 1 - \frac{1}{2}(1-c)^{n-1} \qquad n \ge 1$$

where the estimated value of c is 0.344. Find $P(1)$ and $P(2)$ using this value of c.

53. Demand Schedule The following table is called a *demand schedule*. It gives a correspondence between the price p of a product and the quantity q that consumers will demand (that is, purchase) at that price. (a) If $p = f(q)$, list the numbers in the domain of f. Find $f(2900)$ and $f(3000)$. (b) If $q = g(p)$, list the numbers in the domain of g. Find $g(10)$ and $g(17)$.

Price per Unit, p	Quantity Demanded per Week, q
\$10	3000
12	2900
17	2300
20	2000

In Problems 54–57, use your calculator to find the indicated values for the given function. Round answers to two decimal places.

54. $f(x) = 2.03x^3 - 5.27x^2 - 13.71$ **(a)** $f(1.73)$ **(b)** $f(-5.78)$ **(c)** $f(\sqrt{2})$

55. $f(x) = \dfrac{14.7x^2 - 3.95x - 15.76}{24.3 - x^3}$ **(a)** $f(4)$ **(b)** $f\left(-\frac{17}{4}\right)$ **(c)** $f(\pi)$

56. $f(x) = (20.3 - 3.2x)(2.25x^2 - 7.1x - 16)^4$ **(a)** $f(0.3)$ **(b)** $f(-0.02)$ **(c)** $f(1.9)$

57. $f(x) = \sqrt{\dfrac{\sqrt{2}x^2 + 47.62(x+1)}{9.07}}$ **(a)** $f(15.93)$ **(b)** $f(-146)$ **(c)** $f(0)$

[1] Adapted from H. Babkoff, "Magnitude Estimation of Short Electrocutaneous Pulses," *Psychological Research* 39, no. 1 (1976), 39–49.

[2] D. Laming, *Mathematical Psychology* (New York: Academic Press, 1983).

OBJECTIVE *To introduce constant functions, polynomial functions, rational functions, case-defined functions, the absolute-value function, and factorial notation.*

2.2 Special Functions

In this section, we shall look at functions having special forms and representations. We begin with perhaps the simplest type of function there is: a *constant function*.

APPLICATIONS IN PRACTICE 1

Constant Functions

Suppose the monthly health insurance premiums for an individual are \$125.00.

a. Write the monthly health insurance premiums as a function of the number of visits the individual makes to the doctor.

b. How do the health insurance premiums change as the number of visits to the doctor increases?

c. What kind of function is this?

EXAMPLE 1 Constant Functions

Let $h(x) = 2$. The domain of h is all real numbers. All function values are 2. For example,

$$h(10) = 2 \qquad h(-387) = 2 \qquad h(x+3) = 2$$

We call h a *constant function* because all the function values are the same. More generally, we have this definition:

A function of the form $h(x) = c$, where c is a *constant,* is called a **constant function.**

A constant function belongs to a broader class of functions, called *polynomial functions*.

A function of the form

$$f(x) = c_n x^n + c_{n-1}x^{n-1} + \cdots + c_1 x + c_0$$

where n is a nonnegative integer and $c_n, c_{n-1}, \cdots, c_0$ are constants with $c_n \neq 0$, is called a **polynomial function** (in x). The number n is called the **degree** of the polynomial, and c_n is the **leading coefficient.**

Each term in a polynomial function is either a constant or a constant times a positive integral power of x.

Thus,

$$f(x) = 3x^2 - 8x + 9$$

is a polynomial function of degree 2 with leading coefficient 3. Likewise, $g(x) = 4 - 2x$ has degree 1 and leading coefficient -2. Polynomial functions of degree 1 or 2 are called **linear** or **quadratic functions,** respectively. For example, $g(x) = 4 - 2x$ is linear and $f(x) = 3x^2 - 8x + 9$ is quadratic. Note that a nonzero constant function, such as $f(x) = 5$ [which can be written as $f(x) = 5x^0$], is a polynomial function of degree 0. The constant function $f(x) = 0$ is also considered a polynomial function, but has no degree assigned to it. The domain of any polynomial function is the set of all real numbers.

APPLICATIONS IN PRACTICE 2

Polynomial Functions

The function $d(t) = 3t^2$ represents the distance in meters a car will go in t seconds when it has a constant acceleration of 6 m per second.

a. What kind of function is this?

b. What is its degree?

c. What is its leading coefficient?

EXAMPLE 2 Polynomial Functions

a. $f(x) = x^3 - 6x^2 + 7$ is a polynomial (function) of degree 3 with leading coefficient 1.

b. $g(x) = \frac{2x}{3}$ is a linear function with leading coefficient $\frac{2}{3}$.

c. $f(x) = \dfrac{2}{x^3}$ is *not* a polynomial function. Because $f(x) = 2x^{-3}$ and the exponent for x is not a nonnegative integer, this function does not have the proper form for a polynomial. Similarly, $g(x) = \sqrt{x}$ is not a polynomial, because $g(x) = x^{1/2}$.

▶ **Now work Problem 3.**

Another type of function is a *rational function,* which involves polynomials.

A function that is a quotient of polynomial functions is called a **rational function.**

EXAMPLE 3 **Rational Functions**

a. $f(x) = \dfrac{x^2 - 6x}{x + 5}$ is a rational function, since the numerator and denominator are each polynomials. Note that this rational function is not defined for $x = -5$.

b. $g(x) = 2x + 3$ is a rational function, since $2x + 3 = \dfrac{2x+3}{1}$. In fact, every polynomial function is also a rational function.

Every polynomial function is a rational function.

▶ **Now work Problem 5.**

Sometimes more than one expression is needed to define a function, as Example 4 shows.

APPLICATIONS IN PRACTICE 3

Case-defined Functions

To reduce inventory, a department store charges three rates. If you buy 0–5 pairs of socks, the price is \$3.50 per pair. If you buy 6–10 pairs of socks, the price is \$3.00 per pair. If you buy more than 10 pairs, the price is \$2.75 per pair. Write a case-defined function to represent the cost of buying n pairs of socks.

EXAMPLE 4 **Case-defined Functions**

Let

$$F(s) = \begin{cases} 1 & \text{if } -1 \le s < 1 \\ 0 & \text{if } 1 \le s \le 2 \\ s - 3 & \text{if } 2 < s \le 8 \end{cases}$$

This is called a **case-defined function** because the rule for specifying it is given by rules for each of several disjoint cases. Here s is the independent variable, and the domain of F is all s such that $-1 \le s \le 8$. The value of s determines which expression to use.

Find $F(0)$: Since $-1 \le 0 < 1$, we have $F(0) = 1$.

Find $F(2)$: Since $1 \le 2 \le 2$, we have $F(2) = 0$.

Find $F(7)$: Since $2 < 7 \le 8$, we substitute 7 for s in $s - 3$.

$$F(7) = 7 - 3 = 4$$

▶ **Now work Problem 11.**

TECHNOLOGY

To illustrate how to enter a case-defined function with a TI-83 Plus, Figure 2.4 shows a key sequence for the function

$$f(x) = \begin{cases} 2x & \text{if } x < 0 \\ x^2 & \text{if } 0 \leq x < 10 \\ -x & \text{if } x \geq 10 \end{cases}$$

FIGURE 2.4 Entering a case-defined function.

The absolute-value function can be considered a case-defined function.

EXAMPLE 5 Absolute-Value Function

The function $f(x) = |x|$ is called the *absolute-value function*. Recall that the **absolute value,** of a real number x is denoted $|x|$ and is defined by

$$|x| = \begin{cases} x & \text{if } x \geq 0 \\ -x & \text{if } x < 0 \end{cases}$$

Thus, the domain of f is all real numbers. Some function values are

$$f(16) = |16| = 16$$
$$f\left(-\frac{4}{3}\right) = \left|-\frac{4}{3}\right| = -\left(-\frac{4}{3}\right) = \frac{4}{3}$$
$$f(0) = |0| = 0$$

In our next examples, we make use of *factorial notation*.

The symbol $r!$, with r a positive integer, is read "***r* factorial.**" It represents the product of the first r positive integers:

$$r! = 1 \cdot 2 \cdot 3 \cdots r$$

We define $0!$ to be 1.

Observe that $f(n) = n!$ defines a function whose domain is the set of nonnegative integers.

APPLICATIONS IN PRACTICE 4

Factorials

Seven different books are to be placed on a shelf. How many ways can they be arranged? Represent the question as a factorial problem and give the solution.

EXAMPLE 6 Factorials

a. $5! = 1 \cdot 2 \cdot 3 \cdot 4 \cdot 5 = 120$

b. $3!(6-5)! = 3! \cdot 1! = (3 \cdot 2 \cdot 1)(1) = (6)(1) = 6$

c. $\dfrac{4!}{0!} = \dfrac{1 \cdot 2 \cdot 3 \cdot 4}{1} = \dfrac{24}{1} = 24$

EXAMPLE 7 **Genetics**

Suppose two black guinea pigs are bred and produce exactly five offspring. Under certain conditions, it can be shown that the probability P that exactly r of the offspring will be brown and the others black is a function of r, say, $P = P(r)$, where

Factorials occur frequently in probability theory.

$$P(r) = \frac{5!\left(\frac{1}{4}\right)^r\left(\frac{3}{4}\right)^{5-r}}{r!(5-r)!} \qquad r = 0, 1, 2, \ldots, 5$$

The letter P in $P = P(r)$ is used in two ways. On the right side, P represents the function rule. On the left side, P represents the dependent variable. The domain of P is all integers from 0 to 5, inclusive. Find the probability that exactly three guinea pigs will be brown.

Solution: We want to find $P(3)$. We have

$$P(3) = \frac{5!\left(\frac{1}{4}\right)^3\left(\frac{3}{4}\right)^2}{3!2!} = \frac{120\left(\frac{1}{64}\right)\left(\frac{9}{16}\right)}{6(2)} = \frac{45}{512}$$

▶ **Now work Problem 35.**

Exercise 2.2

In Problems 1–4, determine whether the given function is a polynomial function.

1. $f(x) = x^2 - x^4 + 4$

2. $f(x) = \dfrac{x^3 + 7x - 3}{3}$

***3.** $g(x) = \dfrac{3}{x^2 + 7}$

4. $g(x) = 3^{-2}x^2$

In Problems 5–8, determine whether the given function is a rational function.

***5.** $f(x) = \dfrac{x^2 + x}{x^3 + 4}$

6. $f(x) = \dfrac{3}{2x + 1}$

7. $g(x) = \begin{cases} 1 & \text{if } x < 5 \\ 4 & \text{if } x \geq 5 \end{cases}$

8. $g(x) = 4x^{-4}$

In Problems 9–12, find the domain of each function.

9. $h(z) = 19$

10. $f(t) = \pi$

***11.** $f(x) = \begin{cases} 5x & \text{if } x > 1 \\ 4 & \text{if } x \leq 1 \end{cases}$

12. $f(x) = \begin{cases} 4 & \text{if } x = 3 \\ x^2 & \text{if } 1 \leq x < 3 \end{cases}$

In Problems 13–16, state (a) the degree and (b) the leading coefficient of the given polynomial function.

13. $F(x) = 7x^3 - 2x^2 + 6$

14. $g(x) = 7x$

15. $f(x) = 2 - 3x^4 + 2x$

16. $f(x) = 9$

In Problems 17–22, find the function values for each function.

17. $f(x) = 8$; $f(2)$, $f(t + 8)$, $f(-\sqrt{17})$

18. $g(x) = |x - 3|$; $g(10)$, $g(3)$, $g(-3)$

19. $F(t) = \begin{cases} 1 & \text{if } t > 0 \\ 0 & \text{if } t = 0 \\ -1 & \text{if } t < 0 \end{cases}$

$F(10)$, $F(-\sqrt{3})$, $F(0)$, $F\left(-\frac{18}{5}\right)$

20. $f(x) = \begin{cases} 4 & \text{if } x \geq 0 \\ 3 & \text{if } x < 0 \end{cases}$

$f(3)$, $f(-4)$, $f(0)$

21. $G(x) = \begin{cases} x - 1 & \text{if } x \geq 3 \\ 3 - x^2 & \text{if } x < 3 \end{cases}$

$G(8)$, $G(3)$, $G(-1)$, $G(1)$

22. $h(r) = \begin{cases} 3r - 1 & \text{if } r > 2 \\ r^2 - 4r + 7 & \text{if } r < 2 \end{cases}$

$h(3)$, $h(-3)$, $h(2)$

In Problems 23–28, determine the value of each expression.

23. $6!$

24. $0!$

25. $(4 - 2)!$

26. $6! \cdot 2!$

27. $\dfrac{5!}{4!}$

28. $\dfrac{8!}{5!(8-5)!}$

29. Train Trip A daily round-trip train ticket to the city costs \$4.50. Write the cost of a daily round-trip ticket as a function of a passenger's income. What kind of function is this?

30. Geometry A rectangular prism has length three more than its width and height one less than twice the width.

Write the volume of the rectangular prism as a function of the width. What kind of function is this?

31. Cost Function In manufacturing a component for a machine, the initial cost of a die is \$850 and all other additional costs are \$3 per unit produced. (a) Express the total cost C (in dollars) as a linear function of the number q of units produced. (b) How many units are produced if the total cost is \$1600?

32. Investment If a principal of P dollars is invested at a simple annual interest rate of r for t years, express the total accumulated amount of the principal and interest as a function of t. Is your result a linear function of t?

33. Sales To encourage large group sales, a theater charges two rates. If your group is fewer than 12, each ticket costs \$9.50. If your group is 12 or more, each ticket costs \$8.75. Write a case-defined function to represent the cost of buying n tickets.

34. Factorials At an amusement park, a group of friends wants to ride the log ride in every possible order. How many rides will a group of three have to take? How about a group of four? Five?

*35. **Genetics** Under certain conditions, if two brown-eyed parents have exactly three children, the probability that there will be exactly r blue-eyed children is given by the function $P = P(r)$, where

$$P(r) = \frac{3!\left(\frac{1}{4}\right)^r\left(\frac{3}{4}\right)^{3-r}}{r!(3-r)!}, \quad r = 0, 1, 2, 3.$$

Find the probability that exactly two of the children will be blue-eyed.

36. Genetics In Example 7, find the probability that all five offspring will be brown.

37. Bacteria Growth Bacteria are growing in a culture. The time t (in hours) for the bacteria to double in number (the generation time) is a function of the temperature T (in °C) of the culture. If this function is given by[3]

$$t = f(T) = \begin{cases} \frac{1}{24}T + \frac{11}{4} & \text{if } 30 \le T \le 36 \\ \frac{4}{3}T - \frac{175}{4} & \text{if } 36 < T \le 39 \end{cases}$$

(a) determine the domain of f and (b) find $f(30)$, $f(36)$, and $f(39)$.

In Problems 38–41, use your calculator to find the indicated function values for the given function. Round answers to two decimal places.

38. $f(x) = \begin{cases} 0.19x^4 - 27.99 & \text{if } x \ge 5.99 \\ 0.63x^5 - 57.42 & \text{if } x < 5.99 \end{cases}$

(a) $f(7.98)$ **(b)** $f(2.26)$ **(c)** $f(9)$

39. $f(x) = \begin{cases} 47.1x^5 + 30.4 & \text{if } x > 0 \\ 9.4x^3 - x & \text{if } x \le 0 \end{cases}$

(a) $f(5.5)$ **(b)** $f(-3.6)$ **(c)** $f\left(\frac{6}{7}\right)$

40. $f(x) = \begin{cases} 4.07x - 2.3 & \text{if } x < -8 \\ 19.12 & \text{if } -8 \le x < -2 \\ x^2 - 4x^{-2} & \text{if } x \ge -2 \end{cases}$

(a) $f(-5.8)$ **(b)** $f(-14.9)$ **(c)** $f(7.6)$

41. $f(x) = \begin{cases} x/(x+3) & \text{if } x < -5 \\ x(x-4)^2 & \text{if } -5 \le x < 0 \\ \sqrt{2.1x+3} & \text{if } x \ge 0 \end{cases}$

(a) $f(-\sqrt{30})$ **(b)** $f(46)$ **(c)** $f\left(-\frac{2}{3}\right)$

[3] Adapted from F. K. E. Imrie and A. J. Vlitos, "Production of Fungal Protein from Carob," in *Single-Cell Protein II*, ed. S. R. Tannenbaum and D. I. C. Wang (Cambridge, MA: MIT Press, 1975).

OBJECTIVE *To combine functions by means of addition, subtraction, multiplication, division, multiplication by a constant, and composition.*

2.3 Combinations of Functions

There are several ways of combining two functions to create a new function. Suppose f and g are the functions given by

$$f(x) = x^2 \quad \text{and} \quad g(x) = 3x$$

Adding $f(x)$ and $g(x)$ gives

$$f(x) + g(x) = x^2 + 3x$$

This operation defines a new function called the *sum* of f and g, denoted $f + g$.

Its function value at x is $f(x) + g(x)$. That is,

$$(f + g)(x) = f(x) + g(x) = x^2 + 3x$$

For example,

$$(f + g)(2) = 2^2 + 3(2) = 10$$

In general, for any functions f and g, we define the **sum** $f + g$, the **difference** $f - g$, the **product** fg, and the **quotient** $\frac{f}{g}$ as follows:[4]

$$\begin{aligned}
(f + g)(x) &= f(x) + g(x) \\
(f - g)(x) &= f(x) - g(x) \\
(fg)(x) &= f(x) \cdot g(x) \\
\frac{f}{g}(x) &= \frac{f(x)}{g(x)} \quad \text{for } g(x) \neq 0
\end{aligned}$$

A special case of fg deserves separate mention. For any real number c and any function f we define cf by

$$cf(x) = cf(x)$$

This restricted case of product is sometimes called **scalar product.** (In several branches of Mathematics the scalar product tends to share properties with sums that are not enjoyed by general products.)

Thus, for $f(x) = x^2$, and $g(x) = 3x$, we have

$$\begin{aligned}
(f + g)(x) &= f(x) + g(x) = x^2 + 3x \\
(f - g)(x) &= f(x) - g(x) = x^2 - 3x \\
(fg)(x) &= f(x) \cdot g(x) = x^2(3x) = 3x^3 \\
\frac{f}{g}(x) &= \frac{f(x)}{g(x)} = \frac{x^2}{3x} = \frac{x}{3} \quad \text{for } x \neq 0 \\
(\sqrt{2}f)(x) &= \sqrt{2}f(x) = \sqrt{2}x^2
\end{aligned}$$

EXAMPLE 1 **Combining Functions**

If $f(x) = 3x - 1$ and $g(x) = x^2 + 3x$, find

a. $(f + g)(x)$ **b.** $(f - g)(x)$ **c.** $(fg)(x)$

d. $\frac{f}{g}(x)$ **e.** $\left(\frac{1}{2}f\right)(x)$

Solution:

a. $(f + g)(x) = f(x) + g(x) = (3x - 1) + (x^2 + 3x) = x^2 + 6x - 1$

b. $(f - g)(x) = f(x) - g(x) = (3x - 1) - (x^2 + 3x) = -1 - x^2$

c. $(fg)(x) = f(x)g(x) = (3x - 1)(x^2 + 3x) = 3x^3 + 8x^2 - 3x$

---

[4]In each of the four combinations, we assume that x is in the domains of both f and g. In the quotient, we also do not allow any value of x for which $g(x)$ is 0.

d. $\dfrac{f}{g}(x) = \dfrac{f(x)}{g(x)} = \dfrac{3x-1}{x^2+3x}$

e. $\left(\dfrac{1}{2}f\right)(x) = \dfrac{1}{2}(f(x)) = \dfrac{1}{2}(3x-1) = \dfrac{3x-1}{2}$

▶ **Now work Problem 3(a)–(f).**

Composition

We may also combine two functions by first applying one function to a number and then applying the other function to the result. For example, suppose $g(x) = 3x$, $f(x) = x^2$, and $x = 2$. Then $g(2) = 3 \cdot 2 = 6$. Thus, g sends the input 2 to the output 6:

$$2 \xrightarrow{g} 6$$

Next, we let the output 6 become the input for f:

$$f(6) = 6^2 = 36.$$

So f sends 6 to 36:

$$6 \xrightarrow{f} 36$$

By first applying g and then f, we send 2 to 36:

$$2 \xrightarrow{g} 6 \xrightarrow{f} 36$$

To be more general, let's replace the 2 by x, where x is in the domain of g. (See Figure 2.5). Applying g to x, we get the number $g(x)$, which we shall assume is in the domain of f. By applying f to $g(x)$, we get $f(g(x))$, read "f of g of x," which is in the range of f. The operation of applying g and then applying f to the result is called *composition* and the resulting function, denoted $f \circ g$, is called the *composite* of f with g. This function assigns the output number $f(g(x))$ to the input number x. (See the bottom arrow in Figure 2.5.) Thus, $(f \circ g)(x) = f(g(x))$.

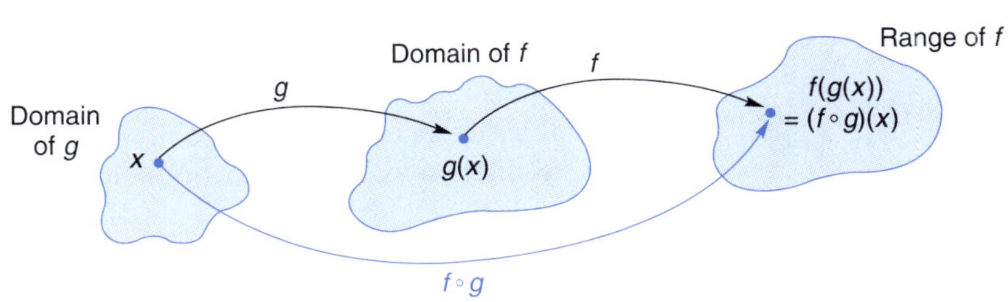

FIGURE 2.5 Composite of f with g.

Definition

If f and g are functions, the ***composite of f with g*** is the function $f \circ g$ defined by

$$(f \circ g)(x) = f(g(x))$$

where the domain of $f \circ g$ is the set of all those x in the domain of g such that $g(x)$ is in the domain of f.

For $f(x) = x^2$ and $g(x) = 3x$, we can get a simple form for $f \circ g$:

$$(f \circ g)(x) = f(g(x)) = f(3x) = (3x)^2 = 9x^2$$

For example, $(f \circ g)(2) = 9(2)^2 = 36$, as we saw before.

When dealing with real numbers and the operation of addition, 0 is special in that, for any real number a, we have

$$a + 0 = a = 0 + a$$

The number 1 enjoys a similar property with respect to multiplication. For any real number a, we have

$$a1 = a = 1a$$

With regard to composition of functions there is a function with properties similar to 0 for addition and 1 for multiplication. For reference in Section 2.4 we note that the function I defined by $I(x) = x$, satisfies, for any function f,

$$f \circ I = f = I \circ f$$

where here we mean equality of functions as defined in Section 2.1. Indeed, for any x,

$$(f \circ I)(x) = f(I(x)) = f(x) = I(f(x)) = (I \circ f)(x)$$

APPLICATIONS IN PRACTICE 1

Composition

A CD costs x dollars wholesale. The price the store pays is given by the function $s(x) = x + 3$. The price the customer pays is $c(x) = 2x$, where x is the price the store pays. Write a composite function to find the customer's price as a function of the wholesale price.

EXAMPLE 2 Composition

Let $f(x) = \sqrt{x}$ and $g(x) = x + 1$. Find

a. $(f \circ g)(x)$, **b.** $(g \circ f)(x)$.

Solution:

a. $(f \circ g)(x)$ is $f(g(x))$. Now g adds 1 to x, and f takes the square root of the result. Thus,

$$(f \circ g)(x) = f(g(x)) = f(x + 1) = \sqrt{x + 1}.$$

The domain of g is all real numbers x, and the domain of f is all nonnegative reals. Hence, the domain of the composite is all x for which $g(x) = x + 1$ is nonnegative. That is, the domain is all $x \geq -1$, or equivalently, the interval $[-1, \infty)$.

b. $(g \circ f)(x)$ is $g(f(x))$. Now f takes the square root of x, and g adds 1 to the result. Thus, g adds 1 to $\sqrt{x}$, and we have

$$(g \circ f)(x) = g(f(x)) = g(\sqrt{x}) = \sqrt{x} + 1.$$

The domain of f is all $x \geq 0$, and the domain of g is all reals. Hence, the domain of the composite is all $x \geq 0$ for which $f(x) = \sqrt{x}$ is real, namely, all $x \geq 0$.

▶ **Now work Problem 7.**

CAUTION

Generally, $f \circ g$ and $g \circ f$ are quite different. In Example 2,

$$(f \circ g)(x) = \sqrt{x+1}$$

but

$$(g \circ f)(x) = \sqrt{x} + 1$$

Observe that $(f \circ g)(1) = \sqrt{2}$, while $(g \circ f)(1) = 2$. Also, do not confuse $f(g(x))$ with $(fg)(x)$, which is the product $f(x)g(x)$. Here

$$f(g(x)) = \sqrt{x+1}$$

but

$$f(x)g(x) = \sqrt{x}(x+1)$$

Composition is *associative*, meaning that for any three functions f, g, and h,

$$(f \circ g) \circ h = f \circ (g \circ h)$$

EXAMPLE 3 **Composition**

If $F(p) = p^2 + 4p - 3$, $G(p) = 2p + 1$, *and* $H(p) = |p|$ *find*

a. $F(G(p))$ **b.** $F(G(H(p)))$ **c.** $G(F(1))$

Solution:

a. $F(G(p)) = F(2p+1) = (2p+1)^2 + 4(2p+1) - 3 = 4p^2 + 12p + 2 = (F \circ G)(p)x$

b. $F(G(H(p))) = (F \circ (G \circ H))(p) = ((F \circ G) \circ H)(p) = (F \circ G)(H(p)) = (F \circ G)(|p|) = 4|p|^2 + 12|p| + 2 = 4p^2 + 12|p| + 2$

c. $G(F(1)) = G(1^2 + 4 \cdot 1 - 3) = G(2) = 2 \cdot 2 + 1 = 5$

▶ **Now work Problem 9.**

In calculus, it is necessary at times to think of a particular function as a composite of two simpler functions, as the next example shows.

APPLICATIONS IN PRACTICE 2

Expressing a Function as a Composite

Suppose the area of a square garden is $g(x) = (x+3)^2$. Express g as a composite of two functions, and explain what each function represents.

EXAMPLE 4 **Expressing a Function as a Composite**

Express $h(x) = (2x-1)^3$ *as a composite.*

Solution:

We note that $h(x)$ is obtained by finding $2x - 1$ and cubing the result. Suppose we let $g(x) = 2x - 1$ and $f(x) = x^3$. Then

$$h(x) = (2x-1)^3 = [g(x)]^3 = f(g(x)) = (f \circ g)(x)$$

which gives h as a composite of two functions.

▶ **Now work Problem 13.**

TECHNOLOGY

Two functions can be combined by using a graphing calculator. Consider the functions

$$f(x) = 2x + 1 \quad \text{and} \quad g(x) = x^2$$

which we enter as Y_1 and Y_2, as shown in Figure 2.6. The sum of f and g is given by $Y_3 = Y_1 + Y_2$ and the composite $f \circ g$ by $Y_4 = Y_1(Y_2)$. For example, $f(g(3))$ is obtained by evaluating Y_4 at 3.

FIGURE 2.6 Y_3 and Y_4 are combinations of Y_1 and Y_2.

Exercise 2.3

1. If $f(x) = x + 3$ and $g(x) = x + 5$, find the following.
   (a) $(f + g)(x)$ **(b)** $(f + g)(0)$ **(c)** $(f - g)(x)$
   (d) $(fg)(x)$ **(e)** $(fg)(-2)$ **(f)** $\frac{f}{g}(x)$
   (g) $(f \circ g)(x)$ **(h)** $(f \circ g)(3)$ **(i)** $(g \circ f)(x)$
   (j) $(g \circ f)(3)$

2. If $f(x) = 2x$ and $g(x) = 6 + x$, find the following.
   (a) $(f + g)(x)$ **(b)** $(f - g)(x)$ **(c)** $(f - g)(4)$
   (d) $(fg)(x)$ **(e)** $\frac{f}{g}(x)$ **(f)** $\frac{f}{g}(2)$
   (g) $(f \circ g)(x)$ **(h)** $(g \circ f)(x)$ **(i)** $(g \circ f)(2)$

*3. If $f(x) = x^2 + 1$ and $g(x) = x^2 - x$, find the following.
   (a) $(f + g)(x)$ **(b)** $(f - g)(x)$ **(c)** $(f - g)\left(-\frac{1}{2}\right)$
   (d) $(fg)(x)$ **(e)** $\frac{f}{g}(x)$ **(f)** $\frac{f}{g}\left(-\frac{1}{2}\right)$
   (g) $(f \circ g)(x)$ **(h)** $(g \circ f)(x)$ **(i)** $(g \circ f)(-3)$

4. If $f(x) = x^2 - 1$ and $g(x) = 4$, find the following.
   (a) $(f + g)(x)$ **(b)** $(f + g)\left(\frac{1}{2}\right)$ **(c)** $(f - g)(x)$
   (d) $(fg)(x)$ **(e)** $(fg)(4)$ **(f)** $\frac{f}{g}(x)$
   (g) $(f \circ g)(x)$ **(h)** $(f \circ g)(100)$ **(i)** $(g \circ f)(x)$

5. If $f(x) = 3x^2 + 6$ and $g(x) = 4 - 2x$, find $f(g(2))$ and $g(f(2))$.

6. If $f(p) = \frac{4}{p}$ and $g(p) = \frac{p-2}{3}$, find $(f \circ g)(p)$ and $(g \circ f)(p)$.

*7. If $F(t) = t^2 + 7t + 1$ and $G(t) = \frac{2}{t-1}$, find $(F \circ G)(t)$ and $(G \circ F)(t)$.

8. If $F(t) = \sqrt{t}$ and $G(t) = 3t^2 + 4t + 2$, find $(F \circ G)(t)$ and $(G \circ F)(t)$.

*9. If $f(w) = \frac{1}{w^2 + 1}$ and $g(v) = \sqrt{v + 2}$, find $(f \circ g)(v)$ and $(g \circ f)(w)$.

10. If $f(x) = x^2 + 2x - 1$, find $(f \circ f)(x)$.

In Problems 11–16, find functions f and g such that $h(x) = f(g(x))$.

11. $h(x) = (4x - 3)^5$

12. $h(x) = \sqrt{x^2 - 2}$

*13. $h(x) = \frac{1}{x^2 - 2}$

14. $h(x) = (9x^3 - 5x)^3 - (9x^3 - 5x)^2 + 11$

15. $h(x) = \sqrt[4]{\frac{x^2 - 1}{x + 3}}$

16. $h(x) = \frac{x + 1}{(x + 1)^2 + 2}$

17. **Profit** A coffeehouse sells a pound of coffee for \$9.75. Expenses are \$4500 each month, plus \$4.25 for each pound of coffee sold.
   (a) Write a function $r(x)$ for the total monthly revenue as a function of the number of pounds of coffee sold.
   (b) Write a function $e(x)$ for the total monthly expenses as a function of the number of pounds of coffee sold.
   (c) Write a function $(r - e)(x)$ for the total monthly profit as a function of the number of pounds of coffee sold.

18. **Geometry** Suppose the volume of a cube is $v(x) = (4x - 2)^3$. Express v as a composite of two functions, and explain what each function represents.

19. **Business** A manufacturer determines that the total number of units of output per day, q, as a function of the number of employees, m, where

$$q = f(m) = \frac{40m - m^2}{4}$$

The total revenue r that is received for selling q units is given by the function g, where $r = g(q) = 40q$. Find $(g \circ f)(m)$. What does this composite function describe?

20. **Sociology** Studies have been conducted concerning the statistical relations between a person's status, education, and income.[5] Let S denote a numerical value of status based on annual income I. For a certain population, suppose

$$S = f(I) = 0.45(I - 1000)^{0.53}$$

Furthermore, suppose a person's income I is a function of the number of years of education E, where

$$I = g(E) = 7202 + 0.29E^{3.68}$$

Find $(f \circ g)(E)$. What does this function describe?

In Problems 21–24, for the given functions f and g, find the indicated function values. Round answers to two decimal places.

21. $f(x) = (4x - 13)^2$, $g(x) = 0.2x^2 - 4x + 3$
   (a) $(f + g)(4.5)$ **(b)** $(f \circ g)(-2)$

22. $f(x) = \sqrt{\frac{x - 3}{x + 1}}$, $g(x) = 11.2x + 5.39$
   (a) $\frac{f}{g}(-2)$ **(b)** $(g \circ f)(-10)$

23. $f(x) = x^{2/3}$, $g(x) = x^3 - 7$
   (a) $(fg)(5)$ **(b)** $(g \circ f)(2.25)$

24. $f(x) = \frac{5}{x + 3}$, $g(x) = \frac{2}{x^2}$
   (a) $(f - g)(7.3)$ **(b)** $(f \circ g)(-4.17)$

[5] R. K. Leik and B. F. Meeker, *Mathematical Sociology* (Englewood Cliffs, NJ: Prentice-Hall, 1975).

2.4 Inverse Functions

OBJECTIVE *To introduce inverse functions, their properties, and uses.*

Just as $-a$ is the number for which

$$a + (-a) = 0 = (-a) + a$$

and, for $a \neq 0$, a^{-1} is the number for which

$$aa^{-1} = 1 = a^{-1}a$$

so, given a function f, we can inquire about the existence of a function g satisfying

$$f \circ g = I = g \circ f \qquad \textbf{(1)}$$

where I is the function, introduced in Section 2.3 and given by $I(x) = x$. Suppose that we have g as above and a function h which also satisfies (1) so that

$$f \circ h = I = h \circ f$$

Then

$$h = h \circ I = h \circ (f \circ g) = (h \circ f) \circ g = I \circ g = g$$

shows that there is at most one function satisfying the requirements of g in (1). In mathematical jargon, g is uniquely determined by f and is therefore given a name, $g = f^{-1}$, that reflects its unique dependence on f. The function f^{-1} is read as f **inverse** and called the **inverse** of f.

CAUTION Do not confuse f^{-1}, the inverse of f, and $\frac{1}{f}$, the multiplicative reciprocal of f. Unfortunately, the nomenclature for inverse functions clashes with the numerical use of $(-)^{-1}$. Usually, $f^{-1}(x)$ is different from $\frac{1}{f}(x) = \frac{1}{f(x)}$. For example, $I^{-1} = I$ (since $I \circ I = I$) so $I^{-1}(x) = x$, but $\frac{1}{I}(x) = \frac{1}{I(x)} = \frac{1}{x}$.

The additive inverse $-a$ exists for any number a; the multiplicative inverse a^{-1} exists precisely if $a \neq 0$. The existence of f^{-1} places a strong requirement on a function f. It can be shown that f^{-1} exists if and only if, for all a and b, whenever $f(a) = f(b)$, then $a = b$. It may be helpful to think that such an f can be *canceled (on the left)*.

A function f that satisfies

for all a and b, if $f(a) = f(b)$, then $a = b$

is called a **one-to-one** function.

Thus, we can say that a function has an inverse precisely if it is one-to-one. An equivalent way to express the one-to-one condition is:

for all a and b, if $a \neq b$, then $f(a) \neq f(b)$

so that distinct inputs give rise to distinct outputs. Observe that this condition is not met for many simple functions. For example, if $f(x) = x^2$, then $f(-1) = (-1)^2 = 1 = (1)^2 = f(1)$ and $-1 \neq 1$ shows that the squaring function is not one-to-one. Similarly, $f(x) = |x|$ is not one-to-one.

In general, the domain of f^{-1} is the range of f and the range of f^{-1} is the domain of f.

Let us note here that (1) is equivalent to

$$f(f^{-1}(x)) = x = f^{-1}(f(x)) \qquad \textbf{(2)}$$

The first equation holds for all x in the domain of f^{-1} and the second equation holds for all x in the domain of f. In general the domain of f^{-1}, which is equal to the range of f, can be quite different from the domain of f.

EXAMPLE 1 **Inverses of Linear Functions**

According to Section 2.2, a function of the form $f(x) = ax + b$, where $a \neq 0$, is a linear function. *Show that a linear function is one-to-one. Find the inverse of $f(x) = ax + b$ and show that it is also linear.*

Solution: Assume that $f(u) = f(v)$, that is

$$au + b = av + b \tag{3}$$

To show that f is one-to-one, we must show that $u = v$ follows from this assumption. Subtracting b from both sides of (3) gives $au = av$ from which $u = v$ follows by dividing both sides by a. (We assumed that $a \neq 0$.) Since f is given by first multiplying by a and then adding b, we might expect that the effect of f can be undone by first subtracting b and then dividing by a. So consider $g(x) = \frac{x-b}{a}$. We have

$$(f \circ g)(x) = f(g(x)) = a\frac{x-b}{a} + b = (x - b) + b = x$$

and

$$(g \circ f)(x) = g(f(x)) = \frac{(ax+b)-b}{a} = \frac{ax}{a} = x$$

Since g satisfies the requirements of (1), it follows that g is the inverse of f. That is $f^{-1}(x) = \frac{x-b}{a} = \frac{1}{a}x + \frac{-b}{a}$ and the last equality shows that f^{-1} is also a linear function.

▶ **Now work Problem 1.**

EXAMPLE 2 **Identities for Inverses**

Show that

a. *If f and g are one-to-one functions, the composite $f \circ g$ is also one-to-one and $(f \circ g)^{-1} = g^{-1} \circ f^{-1}$.*

b. *If f is one-to-one $(f^{-1})^{-1} = f$.*

Solution:

a. Assume $(f \circ g)(a) = (f \circ g)(b)$, that is $f(g(a)) = f(g(b))$. Since f is one-to-one, $g(a) = g(b)$. Since g is one-to-one, $a = b$ and this shows that $f \circ g$ is one-to-one. The equations

$$(f \circ g) \circ (g^{-1} \circ f^{-1}) = f \circ (g \circ g^{-1}) \circ f^{-1} = f \circ I \circ f^{-1} = f \circ f^{-1} = I$$

and

$$(g^{-1} \circ f^{-1}) \circ (f \circ g) = g^{-1} \circ (f^{-1} \circ f) \circ g = g^{-1} \circ I \circ g = g^{-1} \circ g = I$$

show that $g^{-1} \circ f^{-1}$ is the inverse of $f \circ g$, which, in symbols, is the statement $g^{-1} \circ f^{-1} = (f \circ g)^{-1}$.

b. This follows immediately from (2).

EXAMPLE 3 **Inverses Used to Solve Equations**

Many equations take the form $f(x) = 0$, where f is a function. If f is a one-to-one function, then the equation has $x = f^{-1}(0)$ as its unique solution. Applying f^{-1} to both sides of $f(x) = 0$ gives $f^{-1}(f(x)) = f^{-1}(0)$ and $f^{-1}(f(x)) = x$ shows that $x = f^{-1}(0)$ is the only possible solution. Since $f(f^{-1}(0)) = 0$, $f^{-1}(0)$ is indeed a solution.

EXAMPLE 4 **Restricting the Domain of a Function**

It may happen that a function f whose domain is the natural one consisting of all numbers for which the defining rule makes sense is not one-to-one, and yet a one-to-one function g can be obtained by restricting the domain of f. For example, we have shown that the function $f(x) = x^2$ is not one-to-one but the function $g(x) = x^2$ *with domain explicitly given as* $[0, \infty)$ is one-to-one. Since $(\sqrt{x})^2 = x$ and $\sqrt{x^2} = x$, for $x \geq 0$, it follows that $\sqrt{}$ is the inverse of the restricted squaring function g.

Here is a more contrived example. Let $f(x) = |x|$ (with its natural domain). Let $g(x) = |x|$ *with domain explicitly given as* $(-\infty, -1) \cup [0, 1]$. The function g is one-to-one and hence has an inverse.

EXAMPLE 5 **Finding the Inverse of a Function**

To find the inverse of a one-to-one function f, solve the equation $y = f(x)$ for x in terms of y obtaining $x = g(y)$. Then $f^{-1}(x) = g(x)$. To illustrate, *find* $f^{-1}(x)$ *if* $f(x) = (x-1)^2$, *for* $x \geq 1$.

Solution: Let $y = (x-1)^2$, for $x \geq 1$. Then $x - 1 = \sqrt{y}$ and hence $x = \sqrt{y} + 1$. It follows that $f^{-1}(x) = \sqrt{x} + 1$.

▸ **Now work Problem 5.**

Exercise 2.4

In Problems 1–6, find the inverse of the given function.

*1. $f(x) = 3x + 7$

2. $g(x) = 2x + 1$

3. $F(x) = \frac{1}{2}x - 7$

4. $G(x) = (3x-2)^2$, for $x \geq \frac{2}{3}$

*5. $A(r) = \pi r^2$, for $r \geq 0$

6. $V(r) = \frac{4}{3}\pi r^3$

In Problems 7–10, determine whether or not the function is one-to-one.

7. $f(x) = 5x + 12$

8. $g(x) = (5x + 12)^2$

9. $h(x) = (5x+12)^2$, for $x \geq -\frac{5}{12}$

10. $F(x) = |x - 9|$

In Problems 11 and 12, solve each equation by finding an inverse function.

11. $(3x-2)^2 = 17$, for $x \geq \frac{2}{3}$

12. $\frac{4}{3}\pi r^3 = 100$

OBJECTIVE *To graph equations and functions in rectangular coordinates, to determine intercepts, to apply the vertical-line test, the horizontal-line test, and to determine the domain and range of a function from a graph.*

2.5 Graphs in Rectangular Coordinates

A **rectangular** (or **Cartesian**) **coordinate system** allows us to specify and locate points in a plane. It also provides a geometric way to represent equations in two variables as well as functions.

In a plane, two real-number lines, called *coordinate axes,* are constructed perpendicular to each other so that their origins coincide, as in Figure 2.7. Their point of intersection is called the *origin* of the coordinate system. For now, we shall call the horizontal line the *x-axis* and the vertical line the *y-axis*. The unit distance on the x-axis need not be the same as on the y-axis.

The plane on which the coordinate axes are placed is called a *rectangular coordinate plane* or, more simply, an *x,y-plane*. Every point in the x,y-plane can be labeled to indicate its position. To label point P in Figure 2.8(a), we draw perpendiculars from P to the x-axis and y-axis. They meet these axes at 4 and 2, respectively. Thus, P determines two numbers, 4 and 2. We say that the **rectangular coordinates** of P are given by the **ordered pair** (4, 2). The word *ordered* is important. In Figure 2.8(b), the point corresponding to (4, 2) is not the same as that for (2, 4):

$$(4, 2) \neq (2, 4)$$

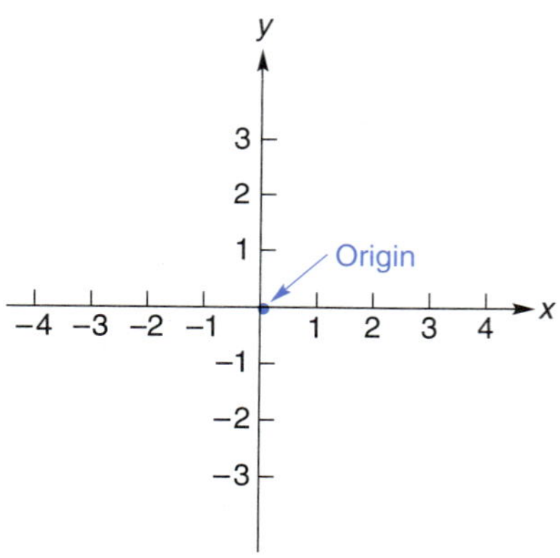

FIGURE 2.7 Coordinate axes.

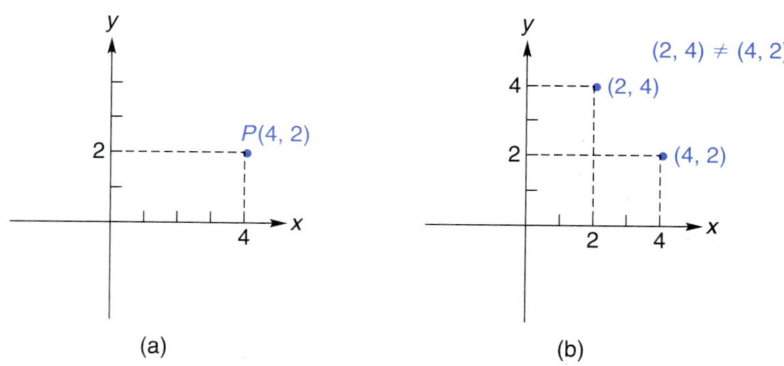

FIGURE 2.8 Rectangular coordinates.

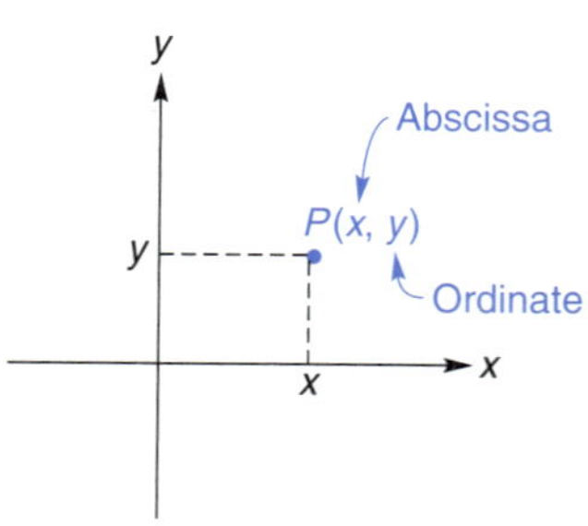

FIGURE 2.9 Coordinates of P.

In general, if P is any point, then its rectangular coordinates will be given by an ordered pair of the form (x, y). (See Figure 2.9.) We call x the *abscissa* or *x-coordinate* of P, and y the *ordinate* or *y-coordinate* of P.

Accordingly, with each point in a given coordinate plane, we can associate exactly one ordered pair (x, y) of real numbers. Also, it should be clear that with each ordered pair (x, y) of real numbers, we can associate exactly one point in that plane. Since there is a *one-to-one correspondence* between the points in the plane and all ordered pairs of real numbers, we shall refer to a point P with abscissa x and ordinate y simply as the point (x, y), or as $P(x, y)$. Moreover, we shall use the words *point* and *ordered pair* interchangeably.

In Figure 2.10, the coordinates of various points are indicated. For example, the point $(1, -4)$ is located one unit to the right of the y-axis and four units below the x-axis. The origin is $(0, 0)$. The x-coordinate of every point on the y-axis is 0, and the y-coordinate of every point on the x-axis is 0.

The coordinate axes divide the plane into four regions called *quadrants* (Figure 2.11). For example, quadrant I consists of all points (x_1, y_1) with $x_1 > 0$ and $y_1 > 0$. The points on the axes do not lie in any quadrant.

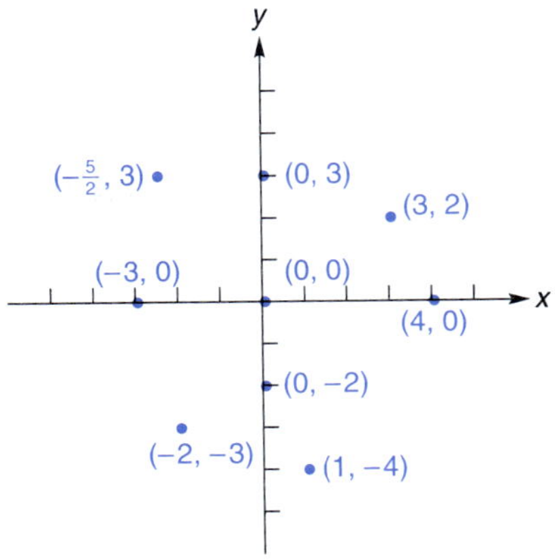

FIGURE 2.10 Coordinates of points.

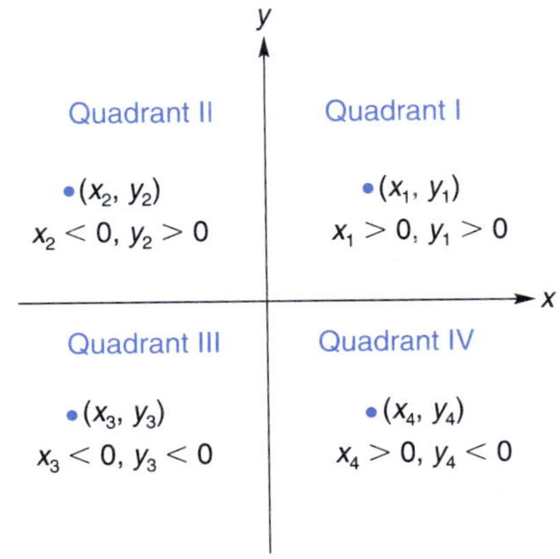

FIGURE 2.11 Quadrants.

Using a rectangular coordinate system, we can geometrically represent equations in two variables. For example, let us consider

$$y = x^2 + 2x - 3 \tag{1}$$

A solution of this equation is a value of x and a value of y that make the equation true. For example, if $x = 1$, substituting into Equation (1) gives

$$y = 1^2 + 2(1) - 3 = 0$$

Thus, $x = 1$, $y = 0$ is a solution. Similarly,

$$\text{if } x = -2 \quad \text{then} \quad y = (-2)^2 + 2(-2) - 3 = -3$$

and so $x = -2$, $y = -3$ is also a solution. By choosing other values for x, we can get more solutions. (See Figure 2.12(a).) It should be clear that there are infinitely many solutions of Equation (1).

x	y
−4	5
−3	0
−2	−3
−1	−4
0	−3
1	0
2	5

(a)

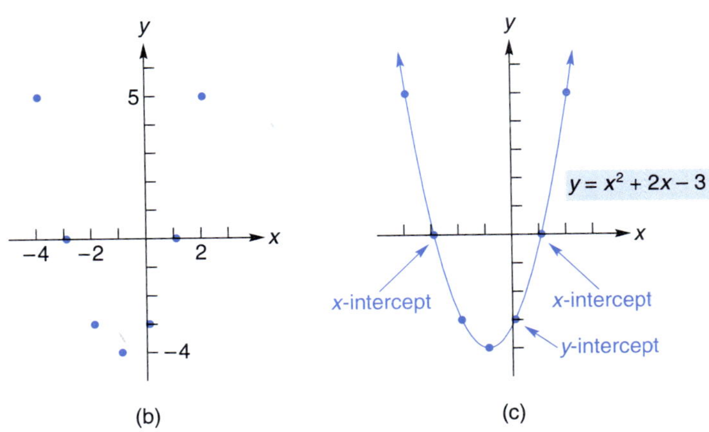

FIGURE 2.12 Graphing $y = x^2 + 2x - 3$.

Each solution gives rise to a point (x, y). For example, to $x = 1$ and $y = 0$ corresponds $(1, 0)$. The **graph** of $y = x^2 + 2x - 3$ is the geometric representation of all its solutions. In Figure 2.12(b), we have plotted the points corresponding to the solutions in the table.

Since the equation has infinitely many solutions, it seems impossible to determine its graph precisely. However, we are concerned only with the graph's general shape. For this reason, we plot enough points so that we may intelligently guess its proper shape. (The calculus techniques that we will study in Chapter 13 will make such "guesses" much more intelligent.) Then we join these points by a smooth curve wherever conditions permit. This gives the curve in Figure 2.12(c). Of course, the more points we plot, the better our graph is. Here we assume that the graph extends indefinitely upward, which is indicated by arrows.

Often, we simply say that the y-intercept is -3 and the x-intercepts are -3 and 1.

The point $(0, -3)$ where the curve intersects the y-axis is called the *y-intercept*. The points $(-3, 0)$ and $(1, 0)$ where the curve intersects the x-axis are called the *x-intercepts*. In general, we have the following definition.

Definition

An ***x-intercept*** of the graph of an equation in x and y is a point where the graph intersects the x-axis. A ***y-intercept*** is a point where the graph intersects the y-axis.

To find the x-intercepts of the graph of an equation in x and y, we first set $y = 0$ and solve the resulting equation for x. To find the y-intercepts, we first set $x = 0$ and solve for y. For example, let us find the x-intercepts for the graph of $y = x^2 + 2x - 3$. Setting $y = 0$ and solving for x gives

$$0 = x^2 + 2x - 3$$

$$0 = (x + 3)(x - 1)$$

$$x = -3, 1$$

Thus, the x-intercepts are $(-3, 0)$ and $(1, 0)$, as we saw before. If $x = 0$, then

$$y = 0^2 + 2(0) - 3 = -3$$

So $(0, -3)$ is the y-intercept. Keep in mind that an x-intercept has its y-coordinate 0, and a y-intercept has its x-coordinate 0. Intercepts are useful because they indicate precisely where the graph intersects the axes.

APPLICATIONS IN PRACTICE 1

Intercepts and Graph

Rachel has saved \$7250 for college expenses. She plans to spend \$600 a month from this account. Write an equation to represent the situation, and identify the intercepts.

EXAMPLE 1 Intercepts and Graph

Find the x- and y-intercepts of the graph of $y = 2x + 3$, and sketch the graph.

Solution: If $y = 0$, then

$$0 = 2x + 3 \quad \text{so that} \quad x = -\frac{3}{2}$$

Thus, the x-intercept is $(-\frac{3}{2}, 0)$. If $x = 0$, then

$$y = 2(0) + 3 = 3$$

So the y-intercept is $(0, 3)$. Figure 2.13 shows a table of other points on the graph and a sketch of the graph.

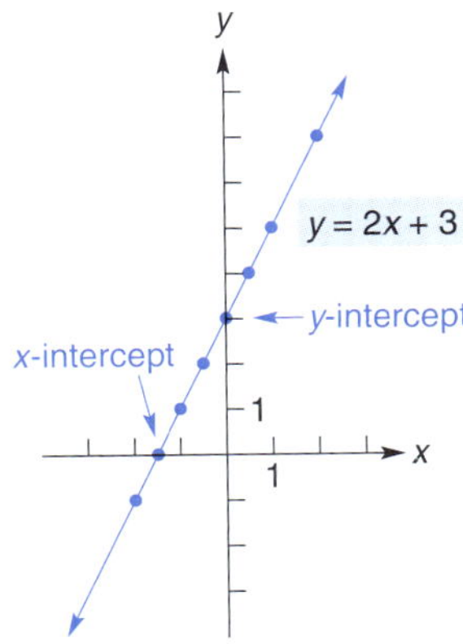

x	0	$-\frac{3}{2}$	$\frac{1}{2}$	$-\frac{1}{2}$	1	-1	2	-2
y	3	0	4	2	5	1	7	-1

FIGURE 2.13 Graph of $y = 2x + 3$.

▶ **Now work Problem 9.**

APPLICATIONS IN PRACTICE 2

Intercepts and Graph

The price of admission to an amusement park is \$24.95. This fee allows the customer to ride all the rides at the park as often as he or she likes. Write an equation that represents the relationship between the number of rides, x, that a customer takes and the cost per ride, y, to that customer. Describe the graph of this equation, and identify the intercepts. Assume $x > 0$.

EXAMPLE 2 **Intercepts and Graph**

Determine the intercepts, if any, of the graph of $s = \dfrac{100}{t}$, *and sketch the graph.*

Solution: For the graph, we shall label the horizontal axis t and the vertical axis s (Figure 2.14). Because t cannot equal 0 (division by 0 is not defined), there is no s-intercept. Thus, the graph has no point corresponding to $t = 0$. Moreover, there is no t-intercept, because if $s = 0$, then the equation

$$0 = \frac{100}{t}$$

has no solution. Remember, the only way that a fraction can be 0 is by having its numerator 0. Figure 2.14 shows the graph. In general, the graph of $s = k/t$, where k is a nonzero constant, is called a *hyperbola.*

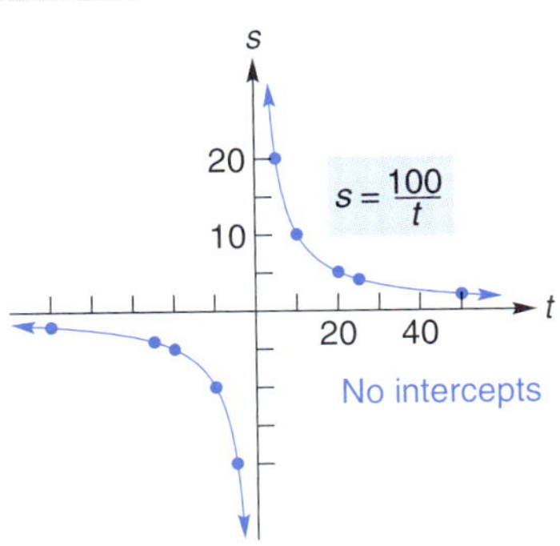

t	5	-5	10	-10	20	-20	25	-25	50	-50
s	20	-20	10	-10	5	-5	4	-4	2	-2

FIGURE 2.14 Graph of $s = \dfrac{100}{t}$.

▶ **Now work Problem 11.**

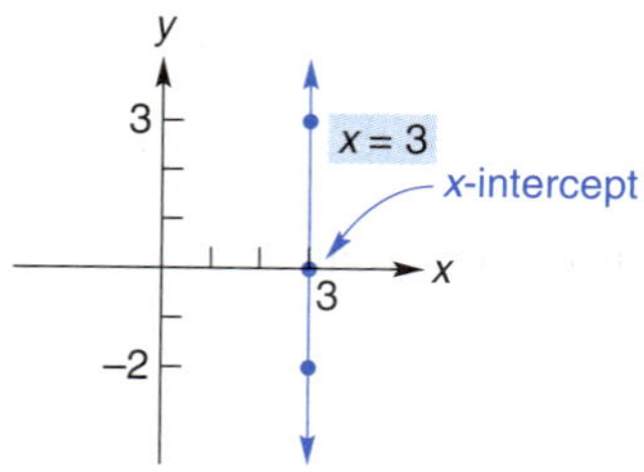

x	3	3	3
y	0	3	−2

FIGURE 2.15 Graph of $x = 3$.

EXAMPLE 3 Intercepts and Graph

Determine the intercepts of the graph of $x = 3$, and sketch the graph.

Solution: We can think of $x = 3$ as an equation in the variables x and y if we write it as $x = 3 + 0y$. Here y can be any value, but x must be 3. Because $x = 3$ when $y = 0$, the x-intercept is $(3, 0)$. There is no y-intercept, because x cannot be 0. (See Figure 2.15.) The graph is a vertical line.

▶ **Now work Problem 13.**

In addition to equations, we can also represent functions in a coordinate plane. If f is a function with independent variable x and dependent variable y, then the graph of f is simply the graph of the equation $y = f(x)$. This graph consists of all points (x, y) or $(x, f(x))$, where x is in the domain of f. The vertical axis can be labeled either y or $f(x)$ and is referred to as the **function-value axis.** *We always label the horizontal axis with the independent variable.*

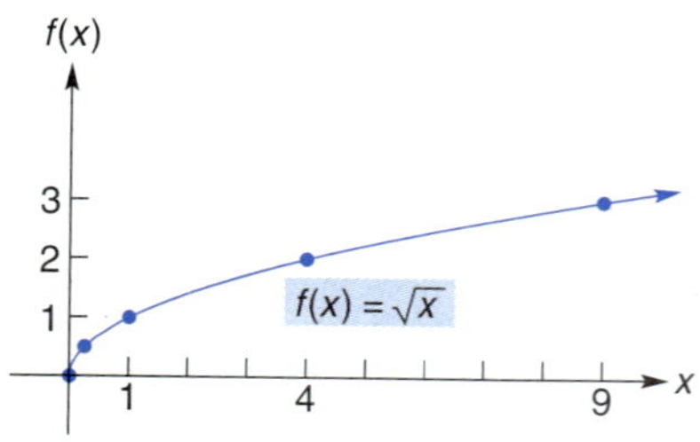

x	0	$\frac{1}{4}$	1	4	9
f(x)	0	$\frac{1}{2}$	1	2	3

FIGURE 2.16 Graph of $f(x) = \sqrt{x}$.

EXAMPLE 4 Graph of the Square-Root Function

Graph $f(x) = \sqrt{x}$.

Solution: The graph is shown in Figure 2.16. We label the vertical axis as $f(x)$. Recall that $\sqrt{x}$ denotes the *principal* square root of x. Thus, $f(9) = \sqrt{9} = 3$, not ± 3. Also, we cannot choose negative values for x, because we don't want imaginary numbers for $\sqrt{x}$. That is, we must have $x \geq 0$. Let us now consider intercepts. If $f(x) = 0$, then $\sqrt{x} = 0$, or $x = 0$. Also, if $x = 0$, then $f(x) = 0$. Thus, the x-intercept and the vertical-axis intercept are the same, namely, $(0, 0)$.

▶ **Now work Problem 29.**

APPLICATIONS IN PRACTICE 3

Graph of the Absolute-Value Function

Brett rented a bike from a rental shop, rode at a constant rate of 12 mi/h for 2.5 hours along a bike path, and then returned along the same path. Graph the absolute-value function to represent Brett's distance from the rental shop as a function of time over the appropriate domain.

EXAMPLE 5 Graph of the Absolute-Value Function

Graph $p = G(q) = |q|$.

Solution: We use the independent variable q to label the horizontal axis. The function-value axis can be labeled either $G(q)$ or p. (See Figure 2.17.) Notice that the q- and p-intercepts are the same point, $(0, 0)$.

q	0	1	−1	3	−3	5	−5
p	0	1	1	3	3	5	5

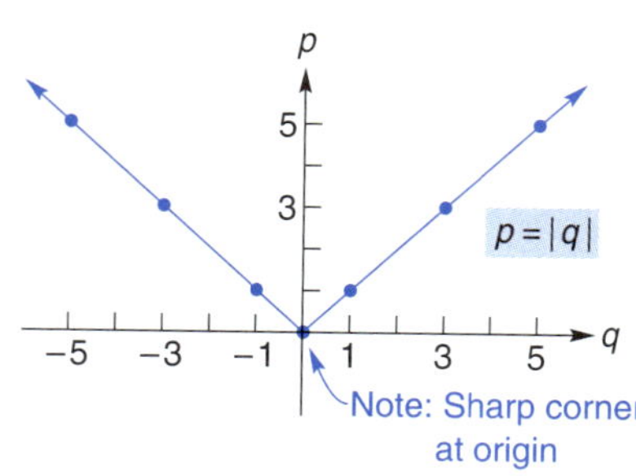

FIGURE 2.17 Graph of $p = |q|$.

▶ **Now work Problem 31.**

In general, the real solutions to an equation $f(x) = 0$ are the real zeros of f.

The notion of a *zero* is important in the study of functions.

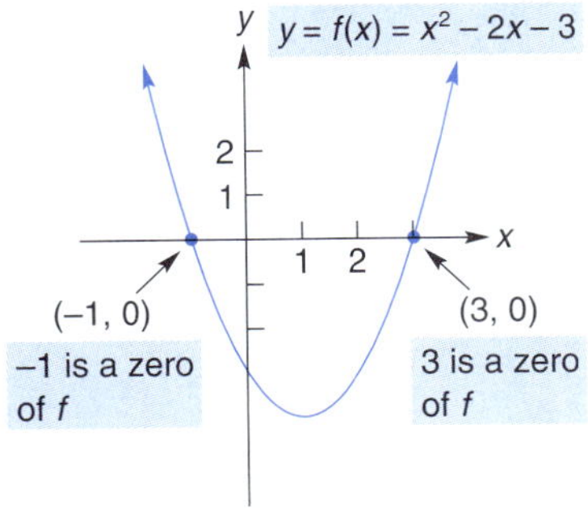

FIGURE 2.18 Zeros of a function.

Definition

A *zero* of a function f is any value of x for which $f(x) = 0$.

For example, a zero of the function $f(x) = 2x - 6$ is 3 because $f(3) = 2(3) - 6 = 0$. Here we call 3 a *real* zero because it is a real number. We note that zeros of f can be found by setting $f(x) = 0$ and solving for x. Thus, the real zeros of a function are precisely the x-intercepts of its graph, because it is at these points that $f(x) = 0$.

To further illustrate, Figure 2.18 shows the graph of the function $y = f(x) = x^2 - 2x - 3$. The x-intercepts of the graph are -1 and 3. Hence, -1 and 3 are zeros of f, or equivalently, -1 and 3 are solutions to the equation $x^2 - 2x - 3 = 0$.

TECHNOLOGY

To solve the equation $x^3 = 3x - 1$ with a graphing calculator, we first express the equation in the form $f(x) = 0$:

$$f(x) = x^3 - 3x + 1 = 0$$

Next we graph f and then estimate the x-intercepts, either by using zoom and trace or by using the root operation. (See Figure 2.19.) Note that we defined our window to be $-4 \le x \le 4$ and $-5 \le y \le 5$.

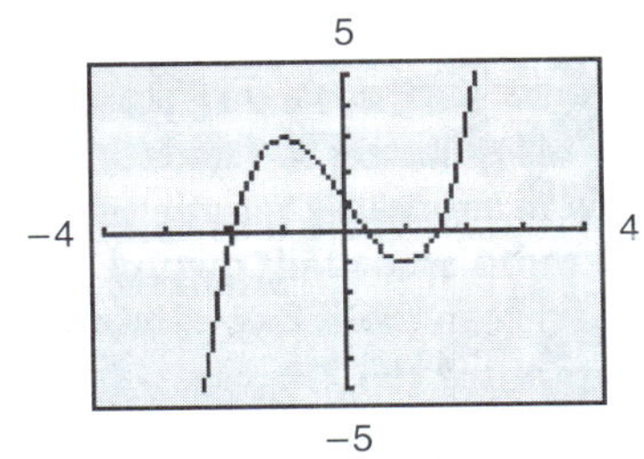

FIGURE 2.19 The roots of $x^3 - 3x + 1 = 0$ are approximately -1.88, 0.35, and 1.53.

Figure 2.20 shows the graph of a function $y = f(x)$. The point $(x, f(x))$ implies that corresponding to the input number x on the horizontal axis is the output number $f(x)$ on the vertical axis, as indicated by the arrow. For example, corresponding to the input 4 is the output 3, so $f(4) = 3$.

From the shape of the graph, it seems reasonable to assume that, for any value of x, there is an output number, so the domain of f is all real numbers. Notice that the set of all y-coordinates of points on the graph is the set of all nonnegative numbers. Thus, the range of f is all $y \ge 0$. This shows that we can

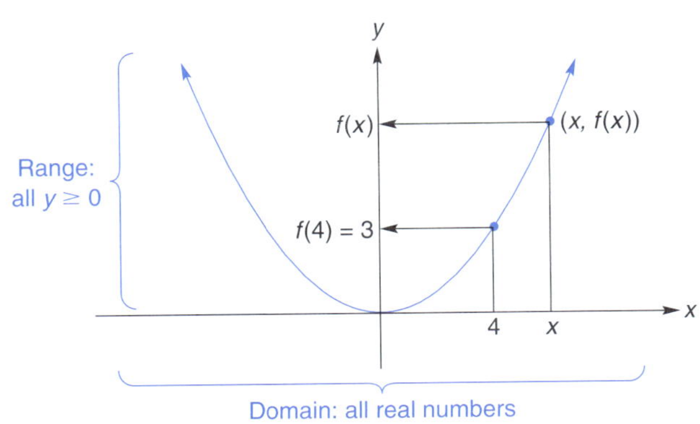

FIGURE 2.20 Domain, range, function values.

make an "educated" guess about the domain and range of a function by looking at its graph. *In general, the domain consists of all x-values that are included in the graph, and the range is all y-values that are included.* For example, Figure 2.16 implies that both the domain and range of $f(x) = \sqrt{x}$ are all nonnegative numbers. From Figure 2.17, it is clear that the domain of $p = G(q) = |q|$ is all real numbers and the range is all $p \geq 0$.

EXAMPLE 6 Domain, Range, and Function Values

Figure 2.21 shows the graph of a function F. To the right of 4, assume that the graph repeats itself indefinitely. Then the domain of F is all $t \geq 0$. The range is $-1 \leq s \leq 1$. Some function values are

$$F(0) = 0 \quad F(1) = 1 \quad F(2) = 0 \quad F(3) = -1$$

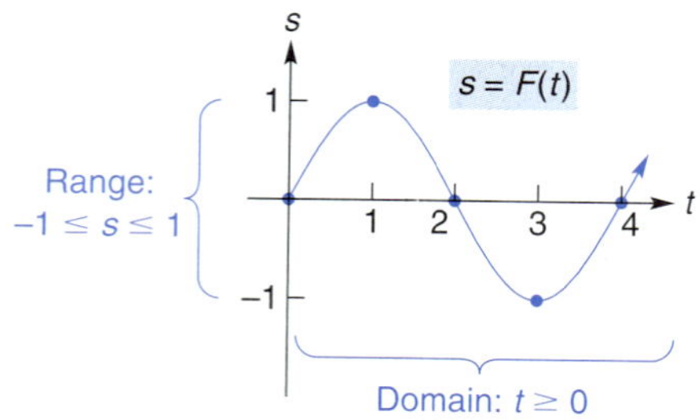

FIGURE 2.21 Domain, range, function values.

TECHNOLOGY

Using a graphing calculator, we can estimate the range of a function. The graph of

$$f(x) = 6x^4 - 8.1x^3 + 1$$

is shown in Figure 2.22. The lowest point on the graph corresponds to the minimum value of $f(x)$, and the range is all reals greater than or equal to this minimum. We can estimate this minimum y-value either by using trace and zoom or by selecting the "minimum" operation.

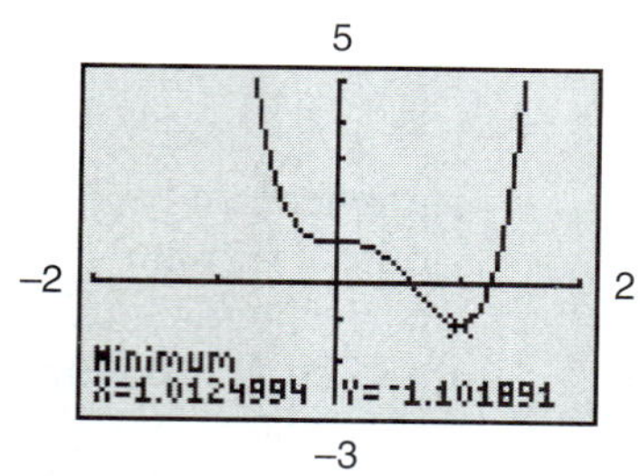

FIGURE 2.22 The range of $f(x) = 6x^4 - 8.1x^3 + 1$ is approximately $[-1.10, \infty)$.

APPLICATIONS IN PRACTICE 4

Graph of a Case-Defined Function

To encourage conservation, a gas company charges two rates. You pay \$0.53 per therm for 0–70 therms and \$0.74 for each therm over 70. Graph the case-defined function that represents the monthly cost of t therms of gas.

EXAMPLE 7 Graph of a Case-Defined Function

Graph the case-defined function

$$f(x) = \begin{cases} x & \text{if } 0 \leq x < 3 \\ x - 1 & \text{if } 3 \leq x \leq 5 \\ 4 & \text{if } 5 < x \leq 7 \end{cases}$$

Solution: The domain of f is $0 \leq x \leq 7$. The graph is given in Figure 2.23, where the *hollow dot* means that the point is *not* included in the graph. Notice that the range of f is all real numbers y such that $0 \leq y \leq 4$.

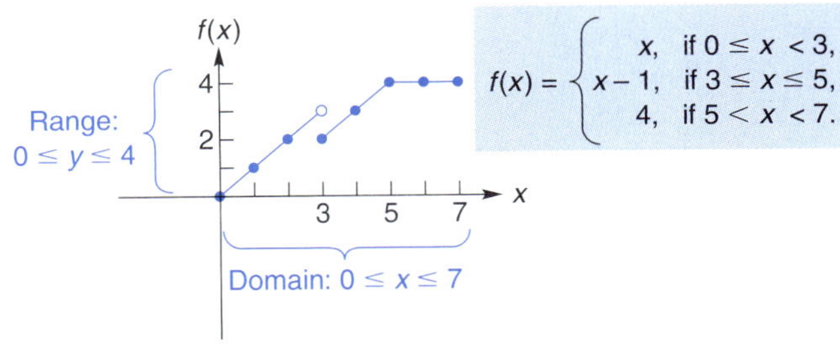

x	0	1	2	3	4	5	6	7
$f(x)$	0	1	2	2	3	4	4	4

FIGURE 2.23 Graph of case-defined function.

▶ **Now work Problem 37.**

There is an easy way to tell whether a curve is the graph of a function. In Figure 2.24(a), notice that with the given x there are associated *two* values of y: y_1 and y_2. Thus, the curve is *not* the graph of a function of x. Looking at it another way, we have the following general rule, called the **vertical-line test.** If a *vertical* line L can be drawn that intersects a curve in at least two points, then the curve is *not* the graph of a function of x. When no such vertical line can be drawn, the curve *is* the graph of a function of x. Consequently, the curves in Figure 2.24 do not represent functions of x, but those in Figure 2.25 do.

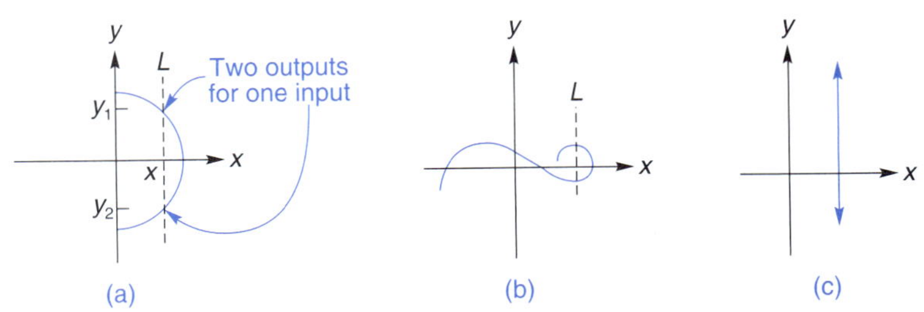

FIGURE 2.24 y is not a function of x.

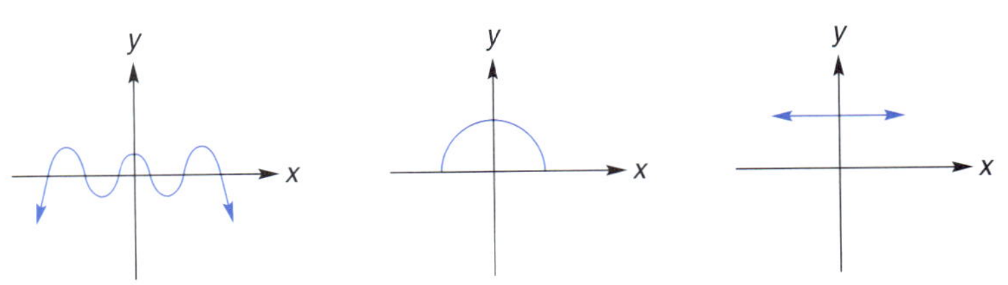

FIGURE 2.25 Functions of x.

EXAMPLE 8 A Graph That Does Not Represent a Function of x

Graph $x = 2y^2$.

Solution: Here it is easier to choose values of y and then find the corresponding values of x. Figure 2.26 shows the graph. By the vertical-line test, the equation $x = 2y^2$ does not define a function of x.

x	0	2	2	8	8	18	18
y	0	1	−1	2	−2	3	−3

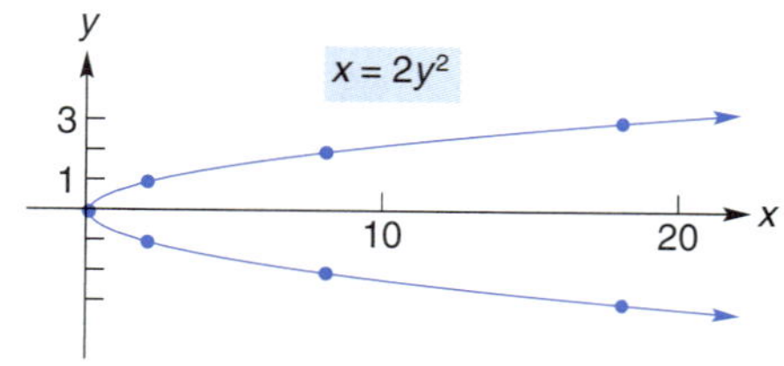

FIGURE 2.26 Graph of $x = 2y^2$.

▶ **Now work Problem 39.**

After we have determined whether a curve is the graph of a function, perhaps using the vertical-line test, there is an easy way to tell whether the function in question is one-to-one. In Figure 2.20 we see that $f(4) = 3$ and, apparently, also $f(-4) = 3$. Since the distinct input values -4 and 4 produce the same output, the function is not one-to-one. Looking at it another way, we have the following general rule, called the **horizontal-line test.** If a *horizontal* line L can be drawn that intersects the graph of a function in at least two points, then the function is *not* one-to-one. When no such horizontal line can be drawn, the function is one-to-one.

Exercise 2.5

In Problems 1 and 2, locate and label each of the points, and give the quadrant, if possible, in which each point lies.

1. $(2, 7)$, $(8, -3)$, $(-\frac{1}{2}, -2)$, $(0, 0)$.

2. $(-4, 5)$, $(3, 0)$, $(1, 1)$, $(0, -6)$.

3. Figure 2.27(a) shows the graph of $y = f(x)$.
- **(a)** Estimate $f(0)$, $f(2)$, $f(4)$, and $f(-2)$.
- **(b)** What is the domain of f?
- **(c)** What is the range of f?
- **(d)** What is a real zero of f?

4. Figure 2.27(b) shows the graph of $y = f(x)$.
- **(a)** Estimate $f(0)$ and $f(2)$.
- **(b)** What is the domain of f?
- **(c)** What is the range of f?
- **(d)** What is a real zero of f?

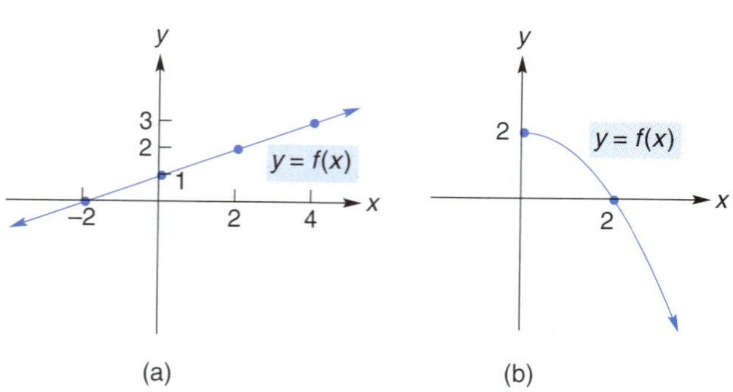

FIGURE 2.27 Diagram for Problems 3 and 4.

5. Figure 2.28(a) shows the graph of $y = f(x)$.
- **(a)** Estimate $f(0)$, $f(1)$, and $f(-1)$.
- **(b)** What is the domain of f?
- **(c)** What is the range of f?
- **(d)** What is a real zero of f?

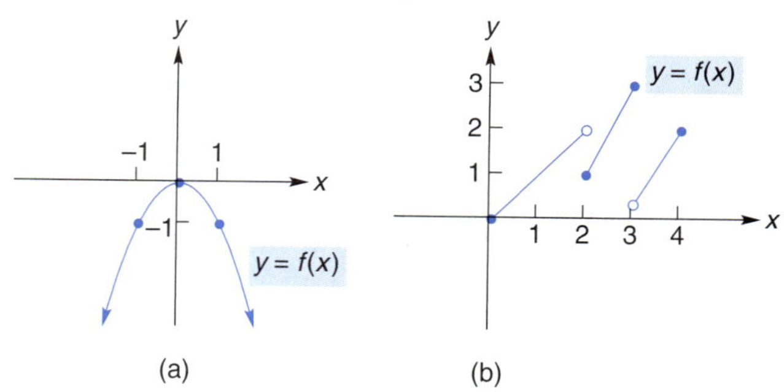

FIGURE 2.28 Diagram for Problems 5 and 6.

6. Figure 2.28(b) shows the graph of $y = f(x)$.
- **(a)** Estimate $f(0)$, $f(2)$, $f(3)$, and $f(4)$.
- **(b)** What is the domain of f?
- **(c)** What is the range of f?
- **(d)** What is a real zero of f?

In Problems 7–20, determine the intercepts of the graph of each equation, and sketch the graph. Based on your graph, is y a function of x, and if so, is it one-to-one and what are the domain and range?

7. $y = 2x$

8. $y = x + 1$

***9.** $y = 3x - 5$

10. $y = 3 - 2x$

*11. $y = x^4$

12. $y = \dfrac{3}{x}$

*13. $x = 0$

14. $y = 4x^2 - 16$

15. $y = x^3$

16. $x = -9$

17. $x = -3y^2$

18. $x^2 = y^2$

19. $2x + y - 2 = 0$

20. $x + y = 1$

In Problems 21–34, graph each function and give the domain and range. Also determine the intercepts.

21. $s = f(t) = 4 - t^2$

22. $f(x) = 5 - 2x^2$

23. $y = h(x) = 3$

24. $G(s) = -8$

25. $y = h(x) = x^2 - 4x + 1$

26. $y = f(x) = x^2 + 2x - 8$

27. $f(t) = -t^3$

28. $p = h(q) = 1 + 2q + q^2$

*29. $s = F(r) = \sqrt{r - 5}$

30. $F(r) = -\dfrac{1}{r}$

*31. $f(x) = |2x - 1|$

32. $v = H(u) = |u - 3|$

33. $F(t) = \dfrac{16}{t^2}$

34. $y = f(x) = \dfrac{2}{x - 4}$

In Problems 35–38, graph each case-defined function and give the domain and range.

35. $c = g(p) = \begin{cases} p + 1 & \text{if } 0 \le p < 7 \\ 5 & \text{if } p \ge 7 \end{cases}$

36. $f(x) = \begin{cases} 2x + 1 & \text{if } -1 \le x < 2 \\ 9 - x^2 & \text{if } x \ge 2 \end{cases}$

*37. $g(x) = \begin{cases} x + 6 & \text{if } x \ge 3 \\ x^2 & \text{if } x < 3 \end{cases}$

38. $f(x) = \begin{cases} x + 1 & \text{if } 0 < x \le 3 \\ 4 & \text{if } 3 < x \le 5 \\ x - 1, & \text{if } x > 5. \end{cases}$

*39. Which of the graphs in Figure 2.29 represent functions of x?

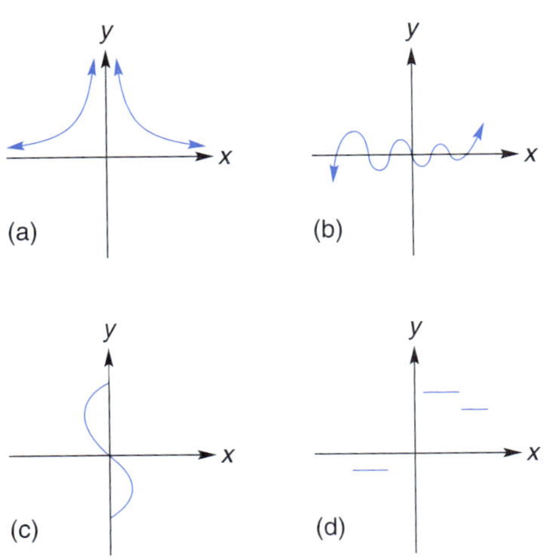

FIGURE 2.29 Diagram for Problem 39.

40. Which of the graphs in Figure 2.30 represent one-to-one functions of x?

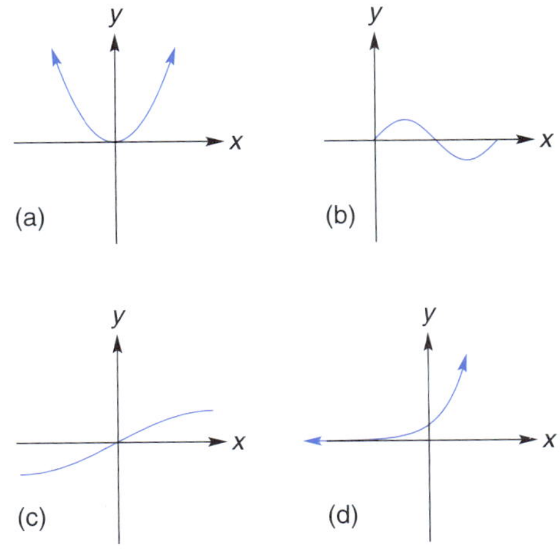

FIGURE 2.30 Diagram for Problem 40.

41. **Debt Payments** Janelle has charged \$1800 on her credit cards. She plans to pay them off at the rate of \$175 per month. Write an equation to represent the amount she owes, excluding any finance charges, and identify the intercepts.

42. **Pricing** To encourage an even flow of customers, a restaurant varies the price of an item throughout the day. From 6:00 P.M. to 8:00 P.M., customers pay full price. At lunch, from 10:30 A.M. until 2:30 P.M., customers pay half price. From 2:30 P.M. until 4:30 P.M., customers get a dollar off the lunch price. From 4:30 P.M. until 6:00 P.M., customers get \$5.00 off the dinner price. From 8:00 P.M. until closing time at 10:00 P.M., customers get \$5.00 off the dinner price. Graph the case-defined function that represents the cost of an item throughout the day for a dinner price of \$18.

43. **Supply Schedule** Given the following supply schedule (see Example 6 of Section 2.1), plot each quantity–price pair by choosing the horizontal axis for the possible quantities. Approximate the points in between the data by connecting the data points with a smooth curve. The result is a *supply curve*. From the graph, determine the relationship between price and supply. (That is, as price increases, what happens to the quantity supplied?) Is price per unit a function of quantity supplied?

Quantity Supplied per Week, q	Price per Unit, p
30	\$10
100	20
150	30
190	40
210	50

44. Demand Schedule The following table is called a *demand schedule.* It indicates the quantities of brand X that consumers will demand (that is, purchase) each week at certain prices per unit (in dollars). Plot each quantity–price pair by choosing the vertical axis for the possible prices. Connect the points with a smooth curve. In this way, we approximate points in between the given data. The result is called a *demand curve.* From the graph, determine the relationship between the price of brand X and the amount that will be demanded. (That is, as price decreases, what happens to the quantity demanded?) Is price per unit a function of quantity demanded?

Quantity Demanded, q	Price per Unit, p
5	\$20
10	10
20	5
25	4

45. Inventory Sketch the graph of

$$y = f(x) = \begin{cases} -100x + 1000 & \text{if } 0 \le x < 7 \\ -100x + 1700 & \text{if } 7 \le x < 14 \\ -100x + 2400 & \text{if } 14 \le x < 21 \end{cases}$$

A function such as this might describe the inventory y of a company at time x.

46. Psychology In a psychological experiment on visual information, a subject briefly viewed an array of letters and was then asked to recall as many letters as possible from the array. The procedure was repeated several times. Suppose that y is the average number of letters recalled from arrays with x letters. The results approximately fit the graph of

$$y = f(x) = \begin{cases} x & \text{if } 0 \le x \le 4 \\ \frac{1}{2}x + 2 & \text{if } 4 < x \le 5 \\ 4.5 & \text{if } 5 < x \le 12 \end{cases}$$

Plot this function.[6]

In Problems 47–50, use a graphing calculator to find all real roots of the given equation. Round answers to two decimal places.

47. $5x^3 + 7x = 3$

48. $x(x^2 - 3) = x^4 + 1$

49. $(9x + 3.1)^2 = 7.4 - 4x^2$

50. $(x - 2)^3 = x^2 - 3$

In Problems 51–54, use a graphing calculator to find all real zeros of the given function. Round answers to two decimal places.

51. $f(x) = x^3 + 5x + 7$

52. $f(x) = 2x^4 - 1.5x^3 + 2$

53. $g(x) = x^4 - 2.5x^3 + x$

54. $g(x) = \sqrt{3}x^5 - 4x^2 + 1$

In Problems 55–57, use a graphing calculator to find (a) the maximum value of $f(x)$ and (b) the minimum value of $f(x)$ for the indicated values of x. Round answers to two decimal places.

55. $f(x) = x^4 - 4.1x^3 + x^2 + 10 \qquad 1 \le x \le 4$

56. $f(x) = x(2.1x^2 - 3)^2 - x^3 + 1 \qquad -1 \le x \le 1$

57. $f(x) = \dfrac{x^2 - 4}{2x - 5} \qquad 3 \le x \le 5$

58. From the graph of $f(x) = \sqrt{2}x^3 + 1.1x^2 + 4$, find (a) the range and (b) the intercepts. Round values to two decimal places.

59. From the graph of $f(x) = 1 - 4x^3 - x^4$, find (a) the maximum value of $f(x)$, (b) the range of f, and (c) the real zeros of f. Round values to two decimal places.

60. From the graph of $f(x) = \dfrac{x^2 + 9.1}{3.8 + \sqrt{x}}$, find (a) the minimum value of $f(x)$, (b) the range of f, and (c) the intercepts. (d) Does f have any real zeros? Round values to two decimal places.

61. Graph $f(x) = \dfrac{4.1x^3 + \sqrt{2}}{x^2 - 3}$ for $2 \le x \le 5$. Determine (a) the maximum value of $f(x)$, (b) the minimum value of $f(x)$, (c) the range of f, and (d) all intercepts. Round values to two decimal places.

[6]Adapted from G. R. Loftus and E. F. Loftus, *Human Memory: The Processing of Information* (New York: Lawrence Erlbaum Associates, Inc., distributed by the Halsted Press, Division of John Wiley & Sons, Inc., 1976).

OBJECTIVE *To study symmetry about the x-axis, the y-axis, and the origin, and to apply symmetry to curve sketching.*

2.6 Symmetry

Examining the graphical behavior of equations is a basic part of mathematics. In this section, we shall examine equations to determine whether their graphs have *symmetry*. In a later chapter, you will see that calculus is a *great* aid in graphing because it helps determine the shape of a graph. It provides powerful techniques for determining whether or not a curve "wiggles" between points.

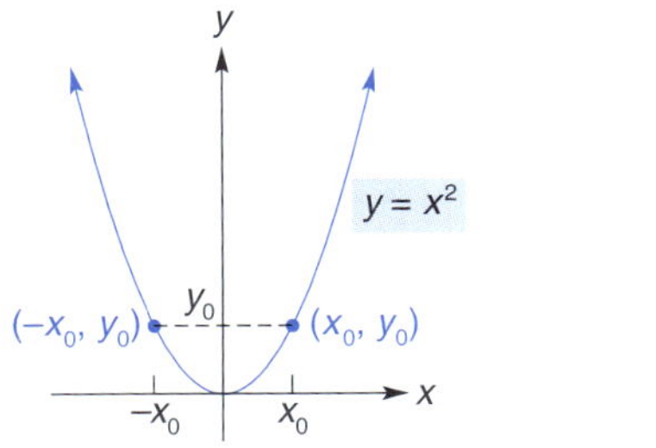

FIGURE 2.31 Symmetry about y-axis.

Consider the graph of $y = x^2$ in Figure 2.31. The portion to the left of the y-axis is the reflection (or mirror image) through the y-axis of that portion to the right of the y-axis, and vice versa. More precisely, if (x_0, y_0) is any point on this graph, then the point $(-x_0, y_0)$ must also lie on the graph. We say that this graph is *symmetric about the y-axis.*

Definition

A graph is ***symmetric about the y-axis*** if and only if $(-x_0, y_0)$ lies on the graph when (x_0, y_0) does.

EXAMPLE 1 ***y*-Axis Symmetry**

Use the preceding definition to show that the graph of $y = x^2$ is symmetric about the y-axis.

Solution: Suppose (x_0, y_0) is *any* point on the graph of $y = x^2$. Then

$$y_0 = x_0^2$$

We must show that the coordinates of $(-x_0, y_0)$ satisfy $y = x^2$:

$$y_0 = (-x_0)^2?$$
$$y_0 = x_0^2?$$

But we already know that $y_0 = x_0^2$. Thus, the graph *is* symmetric about the y-axis.

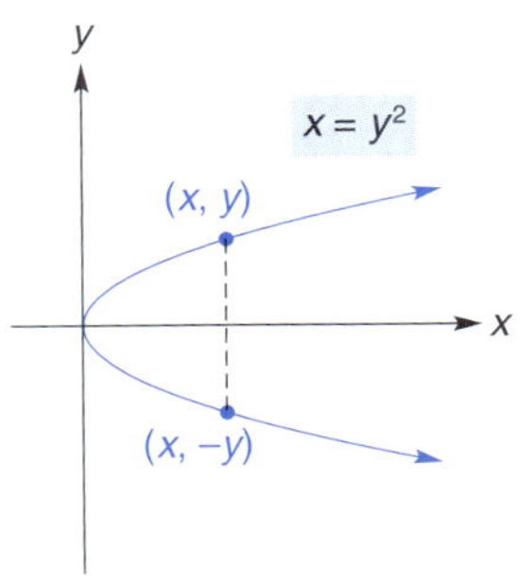

FIGURE 2.32 Symmetry about x-axis.

When one is testing for symmetry in Example 1, (x_0, y_0) can be any point on the graph. In the future, for convenience, we shall omit the subscripts. This means that a graph is symmetric about the y-axis if replacing x by $-x$ in its equation results in an equivalent equation.

Another type of symmetry is shown by the graph of $x = y^2$ in Figure 2.32. Here the portion below the x-axis is the reflection through the x-axis of that portion above the x-axis, and vice versa. If the point (x, y) lies on the graph, then $(x, -y)$ also lies on it. This graph is said to be *symmetric about the x-axis.*

Definition

A graph is ***symmetric about the x-axis*** if and only if $(x, -y)$ lies on the graph when (x, y) does.

Thus, the graph of an equation in x and y has x-axis symmetry if replacing y by $-y$ results in an equivalent equation. For example, applying this test to the graph of $x = y^2$ shown in Figure 2.32 gives

$$x = (-y)^2$$
$$x = y^2$$

which is equivalent to the original equation. Hence, the graph is symmetric about the x-axis.

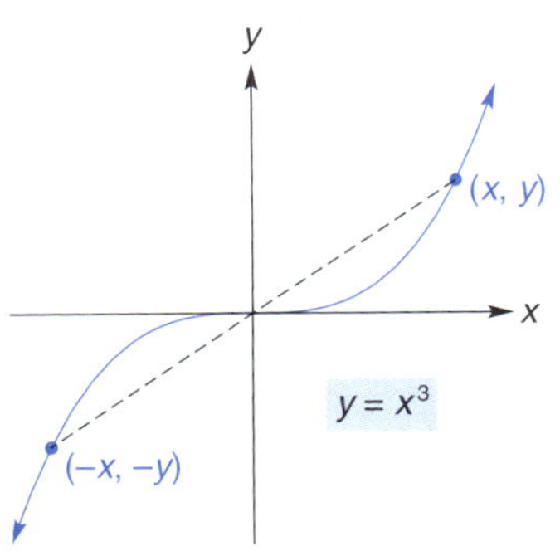

FIGURE 2.33 Symmetry about origin.

A third type of symmetry, *symmetry about the origin,* is illustrated by the graph of $y = x^3$ (Figure 2.33). Whenever the point (x, y) lies on the graph, $(-x, -y)$ also lies on it. As a result, the line segment joining points (x, y) and $(-x, -y)$ is bisected by the origin.

TABLE 2.1 Tests for Symmetry

Symmetry about x-axis	Replace y by $-y$ in given equation. Symmetric if equivalent equation is obtained.
Symmetry about y-axis	Replace x by $-x$ in given equation. Symmetric if equivalent equation is obtained.
Symmetry about origin	Replace x by $-x$ and y by $-y$ in given equation. Symmetric if equivalent equation is obtained.

Definition

A graph is ***symmetric about the origin*** if and only if $(-x, -y)$ lies on the graph when (x, y) does.

Thus, the graph of an equation in x and y has symmetry about the origin if replacing x by $-x$ and y by $-y$ results in an equivalent equation. For example, applying this test to the graph of $y = x^3$ shown in Figure 2.33 gives

$$-y = (-x)^3$$
$$-y = -x^3$$
$$y = x^3$$

which is the original equation. Accordingly, the graph is symmetric about the origin.

Table 2.1 summarizes the tests for symmetry. When we know that a graph has symmetry, we can sketch it by plotting fewer points than would otherwise be needed.

EXAMPLE 2 **Graphing with Intercepts and Symmetry**

Test $y = \dfrac{1}{x}$ for symmetry about the x-axis, the y-axis, and the origin. Then find the intercepts and sketch the graph.

Solution:

Symmetry *x-axis:* Replacing y by $-y$ in $y = 1/x$ gives

$$-y = \frac{1}{x} \quad \text{that is} \quad y = -\frac{1}{x}$$

which is not equivalent to the given equation. Thus, the graph is *not* symmetric about the x-axis.

y-axis: Replacing x by $-x$ in $y = 1/x$ gives

$$y = \frac{1}{-x} \quad \text{that is} \quad y = -\frac{1}{x}$$

which is not equivalent to the given equation. Hence, the graph is *not* symmetric about the y-axis.

Origin: Replacing x by $-x$ and y by $-y$ in $y = 1/x$ gives

$$-y = \frac{1}{-x} \quad \text{that is} \quad y = \frac{1}{x}$$

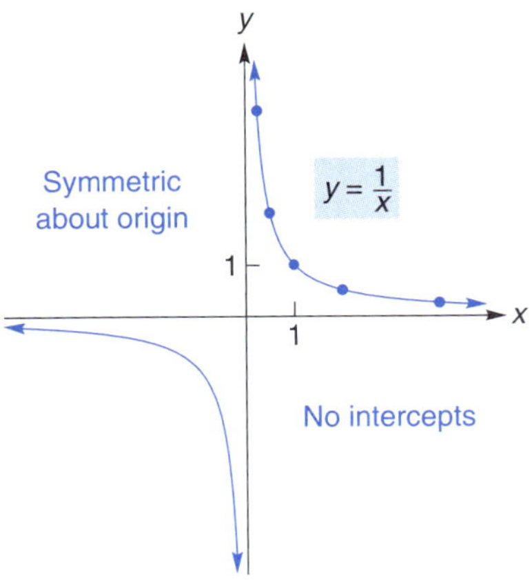

x	$\frac{1}{4}$	$\frac{1}{2}$	1	2	4
y	4	2	1	$\frac{1}{2}$	$\frac{1}{4}$

FIGURE 2.34 Graph of $y = \frac{1}{x}$.

which is equivalent to the given equation. Consequently, the graph *is* symmetric about the origin.

Intercepts Since x cannot be 0, the graph has no y-intercept. If y is 0, then $0 = 1/x$, and this equation has no solution. Thus, no x-intercept exists.

Discussion Because no intercepts exist, the graph cannot intersect either axis. If $x > 0$, we obtain points only in quadrant I. Figure 2.34 tabulates some points of the graph in quadrant I. By symmetry, we reflect that portion through the origin to obtain the entire graph.

▶ **Now work Problem 9.**

EXAMPLE 3 **Graphing with Intercepts and Symmetry**

Test $y = f(x) = 1 - x^4$ for symmetry about the x-axis, the y-axis, and the origin. Then find the intercepts and sketch the graph.

Solution:

Symmetry *x-axis:* Replacing y by $-y$ in $y = 1 - x^4$ gives

$$-y = 1 - x^4 \quad \text{that is} \quad y = -1 + x^4$$

which is not equivalent to the given equation. Thus, the graph is *not* symmetric about the x-axis.

y-axis: Replacing x by $-x$ in $y = 1 - x^4$ gives

$$y = 1 - (-x)^4 \quad \text{that is} \quad y = 1 - x^4$$

which is equivalent to the given equation. Hence, the graph *is* symmetric about the y-axis.

Origin: Replacing x by $-x$ and y by $-y$ in $y = 1 - x^4$ gives

$$-y = 1 - (-x)^4 \qquad -y = 1 - x^4 \qquad y = -1 + x^4$$

which is not equivalent to the given equation. Thus, the graph is *not* symmetric about the origin.

Intercepts Testing for x-intercepts, we set $y = 0$ in $y = 1 - x^4$. Then

$$1 - x^4 = 0$$
$$(1 - x^2)(1 + x^2) = 0$$
$$(1 - x)(1 + x)(1 + x^2) = 0$$
$$x = 1 \quad \text{or} \quad x = -1$$

The x-intercepts are therefore $(1, 0)$ and $(-1, 0)$. Testing for y-intercepts, we set $x = 0$. Then $y = 1$, so $(0, 1)$ is the only y-intercept.

Discussion If the intercepts and some points (x, y) to the right of the y-axis are plotted, we can sketch the *entire* graph by using symmetry about the y-axis (Figure 2.35).

x	y
0	1
$\frac{1}{2}$	$\frac{15}{16}$
$\frac{3}{4}$	$\frac{175}{256}$
1	0
$\frac{3}{2}$	$-\frac{65}{16}$

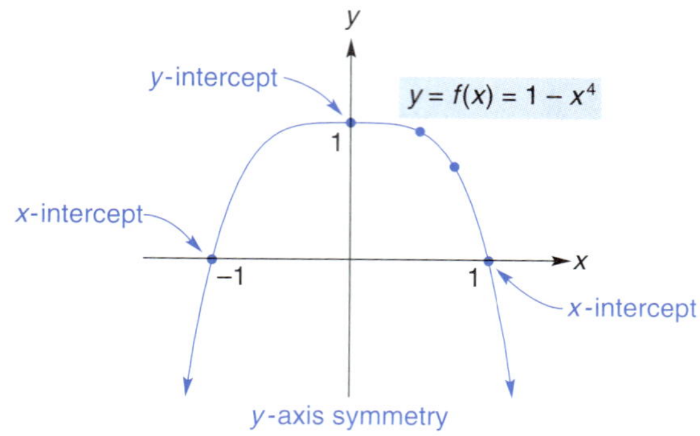

FIGURE 2.35 Graph of $y = 1 - x^4$.

▸ **Now work Problem 19.**

In Example 3, we showed that the graph of $y = f(x) = 1 - x^4$ does not have x-axis symmetry. With the exception of the constant function $f(x) = 0$, *the graph of any **function** $y = f(x)$ cannot be symmetric about the x-axis,* because such symmetry implies two y-values with the same x-value, which violates the definition of a function.

EXAMPLE 4 Graphing with Intercepts and Symmetry

Test the graph of $4x^2 + 9y^2 = 36$ *for intercepts and symmetry. Sketch the graph.*

Solution:

Intercepts If $y = 0$, then $4x^2 = 36$, so $x = \pm 3$. Thus, the x-intercepts are $(3, 0)$ and $(-3, 0)$. If $x = 0$, then $9y^2 = 36$, so $y = \pm 2$. Hence, the y-intercepts are $(0, 2)$ and $(0, -2)$.

Symmetry *x-axis:* Replacing y by $-y$ in $4x^2 + 9y^2 = 36$ gives

$$4x^2 + 9(-y)^2 = 36 \quad \text{that is} \quad 4x^2 + 9y^2 = 36$$

Since we obtain the original equation, there is symmetry about the x-axis.

y-axis: Replacing x by $-x$ in $4x^2 + 9y^2 = 36$ gives

$$4(-x)^2 + 9y^2 = 36 \quad \text{that is} \quad 4x^2 + 9y^2 = 36$$

Again we have the original equation, so there is also symmetry about the y-axis.

Origin: Replacing x by $-x$ and y by $-y$ in $4x^2 + 9y^2 = 36$ gives

$$4(-x)^2 + 9(-y)^2 = 36 \quad \text{that is} \quad 4x^2 + 9y^2 = 36$$

Since this is the original equation, the graph is also symmetric about the origin.

Discussion In Figure 2.36, the intercepts and some points in the first quadrant are plotted. The points in that quadrant are then connected by a smooth curve. By symmetry about the x-axis, the points in the fourth quadrant are obtained. Then, by symmetry about the y-axis, the complete graph is found. There are other ways of graphing the equation by using symmetry. For example, after plotting the intercepts and some points in the first quadrant, we can obtain the points in the third quadrant by symmetry about the origin. By symmetry about the x-axis (or y-axis), we can then obtain the entire graph.

x	y
± 3	0
0	± 2
1	$\frac{4\sqrt{2}}{3}$
2	$\frac{2\sqrt{5}}{3}$
$\frac{5}{2}$	$\frac{\sqrt{11}}{3}$

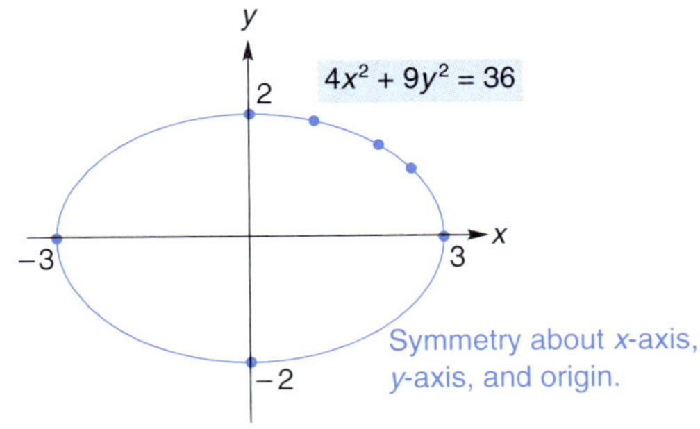

FIGURE 2.36 Graph of $4x^2 + 9y^2 = 36$.

▶ **Now work Problem 23.**

In Example 4, the graph is symmetric about the x-axis, the y-axis, and the origin. It can be shown that **for any graph, if any two of the three types of symmetry exist, then the remaining type must also exist.**

This fact can be a time-saving device in checking for symmetry.

EXAMPLE 5 Symmetry About the Line $y = x$

Definition

A graph is ***symmetric about the line*** $y = x$ if and only if (b, a) lies on the graph when (a, b) does.

Use the preceding definition to show that $x^2 + y^2 = 1$ *is symmetric about the line* $y = x$.

Solution: Suppose (a, b) is any point on the graph of $x^2 + y^2 = 1$. Then

$$a^2 + b^2 = 1$$

We must show that the coordinates of (b, a) satisfy $x^2 + y^2 = 1$:

$$b^2 + a^2 = 1?$$

But $b^2 + a^2 = a^2 + b^2$ so indeed $b^2 + a^2 = 1$ and thus the graph is symmetric about $y = x$.

The point with coordinates (b, a) is the mirror image in the line $y = x$ of the point (a, b). If f is a one-to-one function, $b = f(a)$ if and only if $a = f^{-1}(b)$. Thus the graph of f^{-1} is the mirror image in the line $y = x$ of the graph of f.

EXAMPLE 6 Symmetry and Inverse Functions

Sketch the graph of $g(x) = 2x + 1$ *and its inverse in the same plane.*

Solution: As we shall study in greater detail in Chapter 3, the graph of g is the straight line with slope 2 and y-intercept 1. This line, the line $y = x$, and the reflection of $y = 2x + 1$ in $y = x$ are shown in Figure 2.37

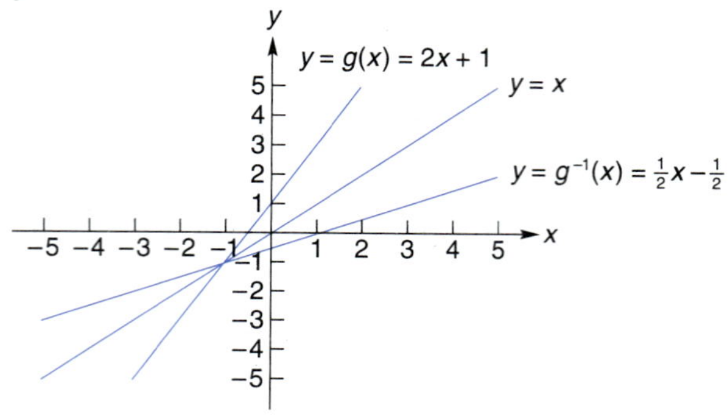

FIGURE 2.37 Graph of $y = g(x)$ and $y = g^{-1}(x)$.

▶ **Now work Problem 27.**

Exercise 2.6

In Problems 1–16, find the x- and y-intercepts of the graphs of the equations. Also test for symmetry about the x-axis, the y-axis, the origin, and the line y = x. Do not sketch the graphs.

1. $y = 5x$

2. $y = f(x) = x^2 - 4$

3. $2x^2 + y^2x^4 = 8 - y$

4. $x = y^3$

5. $16x^2 - 9y^2 = 25$

6. $y = 7$

7. $x = -2$

8. $y = |2x| - 2$

*__9.__ $x = -y^{-4}$

10. $y = \sqrt{x^2 - 25}$

11. $x - 4y - y^2 + 21 = 0$

12. $x^2 + xy + y^3 = 0$

13. $y = f(x) = \dfrac{x^3}{x^2 + 5}$

14. $x^2 + xy + y^2 = 0$

15. $y = \dfrac{3}{x^3 + 8}$

16. $y = \dfrac{x^4}{x + y}$

In Problems 17–24, find the x- and y-intercepts of the graphs of the equations. Also test for symmetry about the x-axis, the y-axis, the origin, and the line y = x. Then sketch the graphs.

17. $3x + y^2 = 9$

18. $x = y^4$

*__19.__ $y = f(x) = x^3 - 4x$

20. $3y = 5x - x^3$

21. $|x| - |y| = 0$

22. $x^2 + y^2 = 16$

*__23.__ $4x^2 + y^2 = 16$

24. $x^2 - y^2 = 4$

25. Prove that the graph of $y = f(x) = 2 - 0.03x^2 - x^4$ is symmetric about the y-axis, and then graph the function. (a) Make use of symmetry, where possible, to find all intercepts. Determine (b) the maximum value of $f(x)$ and (c) the range of f. Round all values to two decimal places.

26. Prove that the graph of $y = f(x) = 2x^4 - 7x^2 + 5$ is symmetric about the y-axis, and then graph the function. Find all real zeros of f. Round your answers to two decimal places.

*__27.__ Sketch the graph of $f(x) = -3x + 2$ and its inverse in the same plane.

●● **OBJECTIVE** *To be familiar with the shapes of the graphs of six basic functions and to consider translation, reflection, and vertical stretching or shrinking of the graph of a function.*

2.7 Translations and Reflections

Up to now, our approach to graphing has been based on plotting points and making use of any symmetry that exists. But this technique is not necessarily the preferred way. Later in this book, we shall analyze graphs by using other techniques. However, some functions and their associated graphs occur so frequently—for example, for illustrative purposes—that we find it worthwhile to memorize them. Figure 2.38 shows six such functions.

At times, by altering a function through an *algebraic* manipulation, the graph of the new function can be obtained from the graph of the original function by performing a *geometric* manipulation. For example, we can use the graph of $f(x) = x^2$ to graph $y = x^2 + 2$. Note that $y = f(x) + 2$. Thus, for each

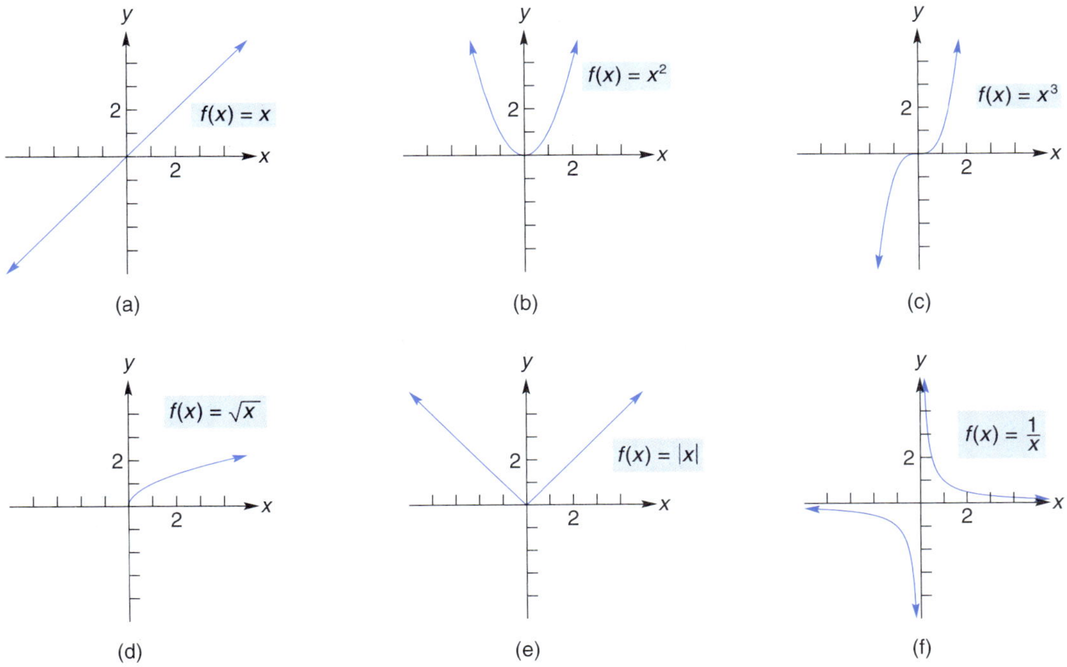

FIGURE 2.38 Functions frequently used.

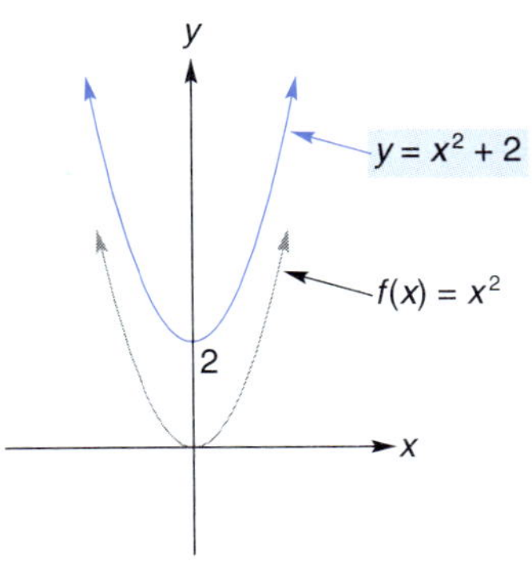

FIGURE 2.39 Graph of $y = x^2 + 2$.

x, the corresponding ordinate for the graph of $y = x^2 + 2$ is 2 more than the ordinate for the graph of $f(x) = x^2$. This means that the graph of $y = x^2 + 2$ is simply the graph of $f(x) = x^2$, shifted, or *translated,* 2 units upward. (See Figure 2.39.) We say that the graph of $y = x^2 + 2$ is a *transformation* of the graph of $f(x) = x^2$. Table 2.2 gives a list of basic types of transformations.

TABLE 2.2 Transformations, $c > 0$

Equation	How to transform graph of $y = f(x)$ to obtain graph of equation
1. $y = f(x) + c$	shift c units upward
2. $y = f(x) - c$	shift c units downward
3. $y = f(x - c)$	shift c units to right
4. $y = f(x + c)$	shift c units to left
5. $y = -f(x)$	reflect about x-axis
6. $y = f(-x)$	reflect about y-axis
7. $y = cf(x) \quad c > 1$	vertically stretch away from x-axis by a factor of c
8. $y = cf(x) \quad c < 1$	vertically shrink toward x-axis by a factor of c

EXAMPLE 1 **Horizontal Translation**

Sketch the graph of $y = (x - 1)^3$.

Solution: We observe that $(x - 1)^3$ is x^3 with x replaced by $x - 1$. Thus, if $f(x) = x^3$, then $y = (x - 1)^3 = f(x - 1)$, which has the form $f(x - c)$, where $c = 1$. From Table 2.2, the graph of $y = (x - 1)^3$ is the graph of $f(x) = x^3$, shifted 1 unit to the right. (See Figure 2.40.)

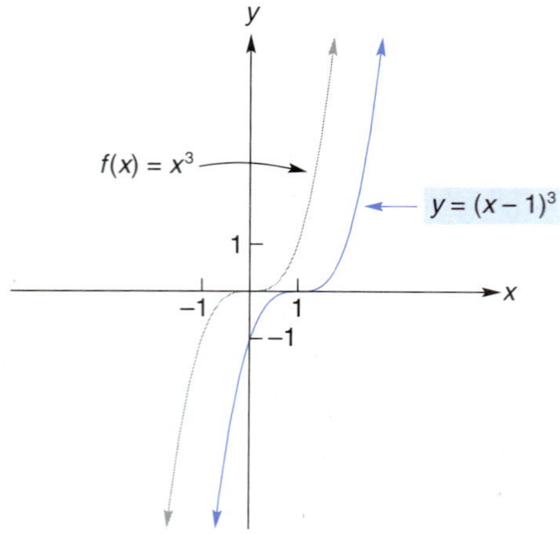

FIGURE 2.40 Graph of $y = (x - 1)^3$.

▶ Now work Problem 3.

EXAMPLE 2 **Shrinking and Reflection**

Sketch the graph of $y = -\frac{1}{2}\sqrt{x}$.

Solution: We can do this problem in two steps. First, observe that $\frac{1}{2}\sqrt{x}$ is $\sqrt{x}$ multiplied by $\frac{1}{2}$. Thus, if $f(x) = \sqrt{x}$, then $\frac{1}{2}\sqrt{x} = \frac{1}{2}f(x)$, which has the form $cf(x)$, where $c = \frac{1}{2}$. So the graph of $y = \frac{1}{2}\sqrt{x}$ is the graph of f shrunk vertically toward the x-axis by a factor of $\frac{1}{2}$ (transformation 8, Table 2.2; see Figure 2.41). Second, the minus sign in $y = -\frac{1}{2}\sqrt{x}$ causes a reflection in the graph of $y = \frac{1}{2}\sqrt{x}$ about the x-axis (transformation 5, Table 2.2; see Figure 2.41).

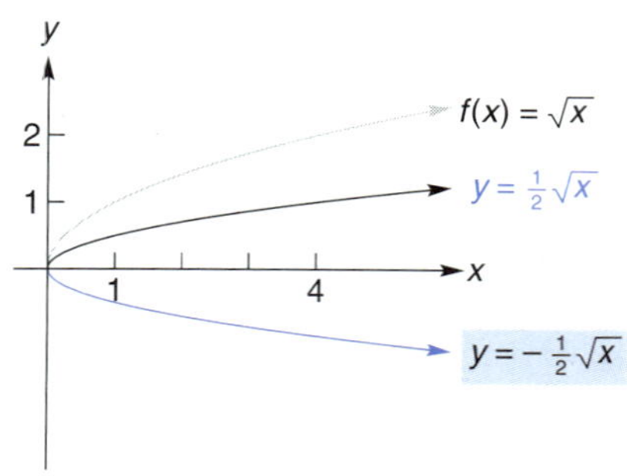

FIGURE 2.41 To graph $y = -\frac{1}{2}\sqrt{x}$, shrink $y = \sqrt{x}$ and reflect result about x-axis.

▶ Now work Problem 5.

Exercise 2.7

In Problems 1–12, use the graphs of the functions in Figure 2.38 and transformation techniques to plot the given functions.

1. $y = x^2 - 2$
2. $y = -x^2$
*3. $y = \dfrac{1}{x-2}$
4. $y = \sqrt{x+2}$
*5. $y = \dfrac{2}{3x}$
6. $y = |x| - 2$
7. $y = |x+1| - 2$
8. $y = -\frac{1}{2}x^3$
9. $y = 1 - (x-1)^2$
10. $y = (x-1)^2 + 1$
11. $y = \sqrt{-x}$
12. $y = \dfrac{5}{2-x}$

In Problems 13–16, describe what must be done to the graph of $y = f(x)$ to obtain the graph of the given equation.

13. $y = f(x-2)+1$

14. $y = f(x+3)-4$

15. $y = f(-x)-5$

16. $y = -f(x+3)$

17. Graph the function $y = \sqrt[3]{x} + k$ for $k = 0, 1, 2, 3, -1, -2,$ and -3. Observe the vertical translations compared to the first graph.

18. Graph the function $y = \sqrt[5]{x+k}$ for $k = 0, 1, 2, 3, -1, -2,$ and -3. Observe the horizontal translations compared to the first graph.

19. Graph the function $y = k\sqrt[3]{x}$ for $k = 1, 2, \frac{1}{2},$ and 3. Observe the vertical stretching and shrinking compared to the first graph. Graph the function for $k = -2$. Observe that the graph is the same as that obtained by stretching the reflection of $y = \sqrt[3]{x}$ about the x-axis by a factor of 2.

2.8 Review

Important Terms and Symbols

Section 2.1 function   domain   range   independent variable   dependent variable $f(x)$
function value   difference quotient, $\dfrac{f(x+h)-f(x)}{h}$ demand function   supply function

Section 2.2 constant function   polynomial function   linear function   quadratic function
case-defined function   absolute value, $|x|$ factorial, $r!$ rational function

Section 2.3 $f+g$ $f-g$ fg f/g $f \circ g$ composite function

Section 2.4 inverse function f^{-1}

Section 2.5 rectangular coordinate system   coordinate axes   origin xy-plane   ordered pair, (x, y)
coordinates of a point x-coordinate y-coordinate   abscissa   ordinate   quadrant
graph of equation y-intercept x-intercept   graph of function   function-value axis
zeros of function   vertical-line test   horizontal-line test

Section 2.6 x-axis symmetry y-axis symmetry   symmetry about origin   symmetry about $y = x$

Summary

A function f is a rule of correspondence that assigns exactly one output number $f(x)$ to each input number x. Usually, a function is specified by an equation that indicates what must be done to an input x to obtain $f(x)$. To obtain a particular function value $f(a)$, we replace each x in the equation by a.

The domain of a function consists of all input numbers, and the range consists of all output numbers. Unless otherwise specified, the domain of f consists of all real numbers x for which $f(x)$ is also a real number.

Some special types of functions are constant functions, polynomial functions, and rational functions. A function that is defined by more than one expression is called a case-defined function.

A function has an inverse if and only if it is one-to-one.

In economics, supply functions and demand functions give a correspondence between the price p of a product and the number of units q of the product that producers (or consumers) will supply (or buy) at that price.

Two functions f and g can be combined to form a sum, difference, product, quotient, or composite as follows:

$$(f+g)(x) = f(x)+g(x)$$
$$(f-g)(x) = f(x)-g(x)$$
$$(fg)(x) = f(x)g(x)$$
$$\frac{f}{g}(x) = \frac{f(x)}{g(x)}$$
$$(f \circ g)(x) = f(g(x))$$

A rectangular coordinate system allows us to geometrically represent equations in two variables, as well as functions. The graph of an equation in x and y consists of all points (x, y) that correspond to the solutions of the equation. We plot a sufficient number of points and connect them (where appropriate) so that the basic shape of the graph is apparent.

Points where the graph intersects the x- and y-axes are called x-intercepts and y-intercepts, respectively. An x-intercept is found by letting y be 0 and solving for x; a y-intercept is found by letting x be 0 and solving for y.

The graph of a function f is the graph of the equation $y = f(x)$ and consists of all points $(x, f(x))$ such that x is in the domain of f. The zeros of f are the values of x for which $f(x) = 0$. From the graph of a function, it is easy to determine the domain and range.

The fact that a graph represents a function can be determined by using the vertical-line test. A vertical line cannot cut the graph of a function at more than one point.

The fact that a function is one-to-one can be determined by using the horizontal-line test on its graph. A horizontal line cannot cut the graph of a one-to-one function at more than one point. When this is the case, the graph of the inverse can be obtained by reflecting the original graph in the line $y=x$.

When the graph of an equation has symmetry, the mirror-image effect allows us to sketch the graph by plotting fewer points than would otherwise be needed. The tests for symmetry are as follows:

Symmetry about x-axis	Replace y by $-y$ in given equation. Symmetric if equivalent equation is obtained.
Symmetry about y-axis	Replace x by $-x$ in given equation. Symmetric if equivalent equation is obtained.
Symmetry about origin	Replace x by $-x$ and y by $-y$ in given equation. Symmetric if equivalent equation is obtained.

Sometimes the graph of a function can be obtained from that of a familiar function by means of a vertical shift upward or downward, a horizontal shift to the right or left, a reflection about the x-axis or y-axis, or a vertical stretching or shrinking away from or toward the x-axis. Such transformations are indicated in Table 2.2 in Section 2.7.

Review Problems

Problem numbers shown in color indicate problems suggested for use as a practice chapter test.

In Problems 1–6, give the domain of each function.

1. $f(x) = \dfrac{x}{x^2 - 6x + 5}$

2. $g(x) = x^2 - 6|x|$

3. $F(t) = 7t + 4t^2$

4. $G(x) = 18$

5. $h(x) = \dfrac{\sqrt{x}}{x - 1}$

6. $H(s) = \dfrac{\sqrt{s-5}}{4}$

In Problems 7–14, find the function values for the given function.

7. $f(x) = 3x^2 - 4x + 7$; $f(0), f(-3), f(5), f(t)$

8. $h(x) = 7$; $h(4), h\left(\frac{1}{100}\right), h(-156), h(x+4)$

9. $G(x) = \sqrt{x-1}$; $G(1), G(5), G(t+1), G(x^3)$

10. $F(x) = \dfrac{x-3}{x+4}$; $F(-1), F(0), F(5), F(x+3)$

11. $h(u) = \dfrac{\sqrt{u+4}}{u}$; $h(5), h(-4), h(x), h(u-4)$

12. $H(s) = \dfrac{(s-4)^2}{3}$; $H(-2), H(7), H\left(\frac{1}{2}\right), H(x^2)$

13. $f(x) = \begin{cases} -3 & \text{if } x < 1 \\ 4 + x^2 & \text{if } x > 1 \end{cases}$

$f(4), f(-2), f(0), f(1)$

14. $h(q) = \begin{cases} q & \text{if } -1 \le q < 0 \\ 3 - q & \text{if } 0 \le q < 3 \\ 2q^2 & \text{if } 3 \le q \le 5 \end{cases}$

$h(0), h(4), h\left(-\frac{1}{2}\right), h\left(\frac{1}{2}\right)$

In Problems 15–18, find (a) $f(x+h)$ and (b) $\dfrac{f(x+h) - f(x)}{h}$, and simplify your answers.

15. $f(x) = 3 - 7x$

16. $f(x) = 11x^2 + 4$

17. $f(x) = 4x^2 + 2x - 5$

18. $f(x) = \dfrac{7}{x+1}$

19. If $f(x) = 3x - 1$ and $g(x) = 2x + 3$, find the following:

(a) $(f+g)(x)$ **(b)** $(f+g)(4)$ **(c)** $(f-g)(x)$
(d) $(fg)(x)$ **(e)** $(fg)(1)$ **(f)** $\dfrac{f}{g}(x)$
(g) $(f \circ g)(x)$ **(h)** $(f \circ g)(5)$ **(i)** $(g \circ f)(x)$

20. If $f(x) = -x^2$ and $g(x) = 3x - 2$, find the following:

(a) $(f+g)(x)$ **(b)** $(f-g)(x)$ **(c)** $(f-g)(-3)$
(d) $(fg)(x)$ **(e)** $\dfrac{f}{g}(x)$ **(f)** $\dfrac{f}{g}(2)$
(g) $(f \circ g)(x)$ **(h)** $(g \circ f)(x)$ **(i)** $(g \circ f)(-4)$

In Problems 21–24, find $(f \circ g)(x)$ and $(g \circ f)(x)$.

21. $f(x) = \dfrac{1}{x}$, $g(x) = x - 1$

22. $f(x) = \dfrac{x+1}{4}$, $g(x) = \sqrt{x}$

23. $f(x) = \sqrt{x+2}$, $g(x) = x^3$

24. $f(x) = 2$, $g(x) = 3$

In Problems 25 and 26, find the intercepts of the graph of each equation, and test for symmetry about the x-axis, the y-axis, the origin, and $y = x$. Do not sketch the graph.

25. $y = 3x - x^3$

26. $\dfrac{xy^2}{x^2+1} = 4$

In Problems 27 and 28, find the x- and y-intercepts of the graphs of the equations. Also test for symmetry about the x-axis, the y-axis, and the origin. Then sketch the graphs.

27. $y = 9 - x^2$

28. $y = 3x - 7$

In Problems 29–32, graph each function and give its domain and range. Also determine the intercepts.

29. $G(u) = \sqrt{u+4}$

30. $f(x) = |x| + 1$

31. $y = g(t) = \dfrac{2}{|t-4|}$

32. $h(u) = \sqrt{-5u}$

33. Graph the following case-defined function, and give its domain and range:

$$y = f(x) = \begin{cases} 1 - x & \text{if } x \le 0 \\ 1 & \text{if } x > 0 \end{cases}$$

34. Use the graph of $f(x) = \sqrt{x}$ to sketch the graph of $y = \sqrt{x-2} - 1$.

35. Use the graph of $f(x) = x^2$ to sketch the graph of $y = -\frac{1}{2}x^2 + 2$.

36. Trend Equation The projected annual sales (in dollars) of a new product is given by the equation $S = 150{,}000 + 3000t$, where t is the time in years from 2001. Such an equation is called a *trend equation*. Find the projected annual sales for 2006. Is S a function of t?

37. In Figure 2.42, which graphs represent functions of x?

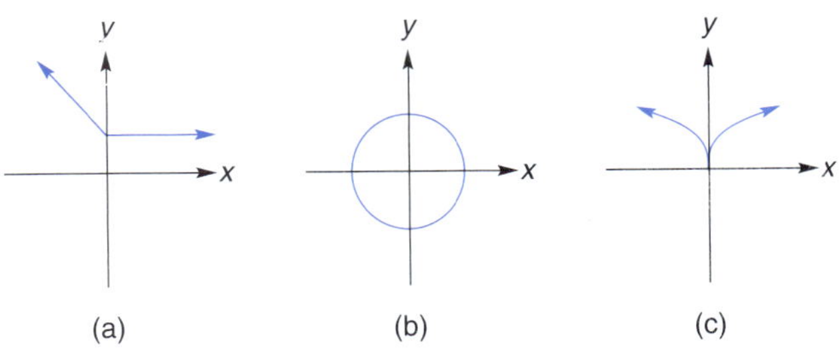

FIGURE 2.42 Diagram for Problem 37.

38. If $f(x) = (4x^{2.3} - 3x^3 + 7)^5$, find (a) $f(2)$ and (b) $f(2.3)$. Round your answers to two decimal places.

39. Find all real roots of the equation

$$5x^3 - 7x^2 = 4x - 2$$

Round your answers to two decimal places.

40. Find all real roots of the equation

$$x^4 - 4x^3 = (2x - 1)^2$$

Round your answers to two decimal places.

41. Find all real zeros of

$$f(x) = x(2.1x^2 - 3)^2 - x^3 + 1$$

Round your answers to two decimal places.

42. Determine the range of

$$f(x) = \begin{cases} -2.5x - 4 & \text{if } x < 0 \\ 6 + 4.1x - x^2 & \text{if } x \ge 0 \end{cases}$$

43. From the graph of $f(x) = -x^3 + 0.04x + 7$, find (a) the range and (b) the intercepts. Round values to two decimal places.

44. From the graph of $f(x) = \sqrt{x+5}(x^2 - 4)$, find (a) the minimum value of $f(x)$, (b) the range of f, and (c) all real zeros of f. Round values to two decimal places.

45. Graph $y = f(x) = x^3 + x^k$ for $k = 0$, 1, 2, 3, and 4. For which values of k does the graph have (a) symmetry about the y-axis? (b) symmetry about the origin?

Mathematical SNAPSHOT

A Taxing Experience!

You probably have heard the old saying: "There are only two things that are certain in life: death and taxes." Here you will see how functions can be applied to one of these "certainties," namely, taxes.

You will be using the 2000 federal tax rates for a married couple filing a joint return. Suppose that you want to determine a formula for the function f such that $f(x)$ is the tax in dollars on a taxable income of x dollars. The tax is based on various ranges of taxable income. According to Schedule Y-1 of the Internal Revenue Service (see Figure 2.43):

- If x is \$0 or less, the tax is \$0.
- If x is over \$0, but not over \$43,850, the tax is 15% of x.
- If x is over \$43,850, but not over \$105,950, the tax is \$6,577.50, plus 28% of the amount over \$43,850.
- If x is over \$105,950, but not over \$161,450, the tax is \$23,965.50, plus 31% of the amount over \$105,950.
- If x is over \$161,450, but not over \$288,350, the tax is \$41,170.50, plus 36% of the amount over \$161,450.
- If x is over \$288,350, the tax is \$86,854.50, plus 39.6% of the amount over \$288,350.

Schedule Y-1 — Use if your filing status is **Married filing jointly** or **Qualifying widow(er)**

If the amount on Form 1040 line 38, is *Over–*	*But not over–*	Enter on Form 1040, line 39		*of the amount over–*
\$0	\$43,850	-----------	**15%**	**\$0**
43,850	105,950	**\$6,577.50 +**	**28%**	**43,850**
105,950	161,450	**23,695.50 +**	**31%**	**105,950**
161,450	288,350	**41,170.50 +**	**36%**	**161,450**
288,350	-----------	**86,854.50 +**	**39.6%**	**288,350**

FIGURE 2.43 Internal Revenue Service 2000 Schedule Y-1.

Clearly, if $x \le 0$ then

$$f(x) = 0$$

If $0 < x \le 43{,}850$ then

$$f(x) = 0.15x$$

Notice that the tax on a taxable income of \$43,850 is

$$f(43{,}850) = 0.15(43{,}850) = 6{,}577.50$$

If $43{,}850 < x \le 105{,}950$, then the amount over 43,850 is $x - 43{,}850$, so

$$f(x) = 6{,}577.50 + 0.28(x - 43{,}850)$$

Because 6,577.50 is 15% of 43,850, for a taxable income between \$43,850 and \$105,950 you are essentially taxed at the rate of 15% for the first \$43,850 of income and at the rate of 28% for the remaining income. Notice that the tax on \$105,950 is

$$\begin{aligned} f(105{,}950) &= 6{,}577.50 + 0.28(105{,}950 - 43{,}850) \\ &= 6{,}577.50 + 0.28(62{,}100) \\ &= 6{,}577.50 + 17{,}388 = 23{,}965.50 \end{aligned}$$

If $105{,}950 < x \le 161{,}450$, then the amount over 105,950 is $x - 105{,}950$, so

$$f(x) = 23{,}965.50 + 0.31(x - 105{,}950)$$

Because \$23,965.50 is the tax on \$105,950, for a taxable income between \$105,950 and \$161,450 you are taxed at the rate of 15% for the first \$43,850 of income, at the rate of 28% for the next \$62,100 of income ($105{,}950 - 43{,}850 = 62{,}100$), and at the rate of 31% for the remaining income. Notice that the tax on $161,450 is

$$\begin{aligned} f(161{,}450) &= 23{,}965.50 + 0.31(161{,}450 - 105{,}950) \\ &= 23{,}965.50 + 0.31(55{,}500) \\ &= 23{,}965.50 + 17{,}205 \\ &= 41{,}170.50 \end{aligned}$$

If $161{,}450 < x \leq 288{,}350$, then the amount over 161,450 is $x - 161{,}450$, so

$$f(x) = 41{,}170.50 + 0.36(x - 161{,}450)$$

Because $41,170.50 is the tax on $161,450, for a taxable income between $161,450 and $288,350 you are taxed at the rate of 15% for the first $43,850 of income, at the rate of 28% for the next $62,100 of income, at the rate of 31% for the next $55,500 of income ($161{,}450 - 105{,}950 = 55{,}500$), and at the rate of 36% for the remaining income. Notice that the tax on $288,350 is

$$\begin{aligned} f(288{,}350) &= 41{,}170.50 + 0.36(288{,}350 - 161{,}450) \\ &= 41{,}170.50 + 0.36(126{,}900) \\ &= 41{,}170.50 + 45{,}684 = 86{,}854.50 \end{aligned}$$

If $x > 288{,}350$, then the amount over 288,350 is $x - 288{,}350$, so

$$f(x) = 86{,}854.50 + 0.396(x - 288{,}350).$$

Because $86,854.50 is the tax on $288,350, for a taxable income over $288,350 you are taxed at the rate of 15% for the first $43,850 of income, at the rate of 28% for the next $62,100 of income, at the rate of 31% for the next $55,500 of income, at the rate of 36% for the next $126,900 in income, and at the rate of 39.6% for the remaining income.

Summarizing all of these results gives the case-defined function

$$f(x) = \begin{cases} 0 & \text{if } x \leq 0 \\ 0.15x & \text{if } 0 < x \leq 43{,}850 \\ 6{,}577.50 + 0.28(x - 43{,}850) & \text{if } 43{,}850 < x \leq 105{,}950 \\ 23{,}965.50 + 0.31(x - 105{,}950) & \text{if } 105{,}950 < x \leq 161{,}450 \\ 41{,}170.50 + 0.36(x - 161{,}450) & \text{if } 161{,}450 < x \leq 288{,}350 \\ 86{,}854.50 + 0.396(x - 288{,}350) & \text{if } x > 288{,}350 \end{cases}$$

With these formulas, you can geometrically depict the income tax function, as in Figure 2.44.

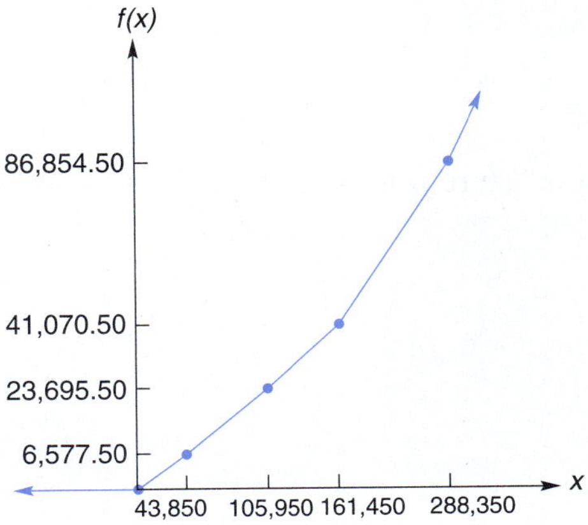

FIGURE 2.44 Income tax function.

Exercises

Use the preceding income tax function f to determine the tax on the given taxable income in the year 2000.

1. $120,000
2. $35,350
3. $290,000
4. $162,700
5. Look up the most recent Schedule Y-1 at www.irs.gov (Inst 1040 Tax Tables) and repeat Problems 1–4 using that information.
6. Why was it significant that $f(105{,}950) =$ $23,965.50, $f(161{,}450) =$ $41,170.50, and so on?

CHAPTER

3

Lines, Parabolas, and Systems

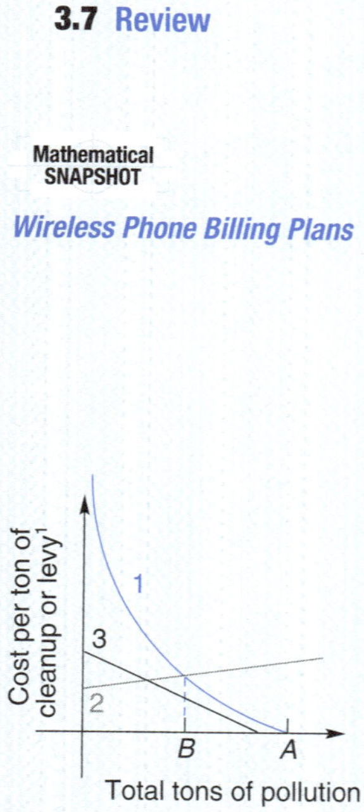

For the problem of industrial pollution, some people advocate a market-based solution: Let manufacturers pollute, but make them pay for the privilege. The more pollution, the greater the fee, or levy. The idea is to give manufacturers an incentive not to pollute more than necessary.

Does this approach work? In the figure to the left below, line 1 represents the cost per ton of cutting pollution. A company polluting indiscriminately can normally do some pollution reduction at minimal cost.[1] As the amount of pollution is reduced, however, the cost per ton rises and eventually goes through the roof. This is illustrated by line 1 rising indefinitely as the total tons of pollution produced approaches 0.

Line 2 is a levy scheme that goes easy on clean-running operations but charges an increasing per-ton fee as the total pollution amount goes up. Line 3, by contrast, is a scheme in which low-pollution manufacturers pay a high per-ton levy while gross polluters pay less per ton (but more overall). Questions of fairness aside, how well will each scheme work as a pollution control measure?

Faced with a pollution levy, a company tends to cut pollution *so long as it saves more in levy costs than it incurs in reduction costs*. The reduction efforts continue until the levy savings and the reduction costs start to balance each other out.

The latter half of this chapter deals with systems of equations. Here, lines 1 and 2 represent one system of equations, and lines 1 and 3 represent an alternative system. Once you have learned how to solve systems of equations, you can return to this page and verify that the line-2 scheme leads to a pollution reduction from amount A to amount B, while the line-3 scheme fails as a pollution control measure, leaving the pollution level at A.

---

[1] Technically, this is the *marginal* cost per ton (see Section 11.3).

OBJECTIVE *To develop the notion of slope and different forms of equations of lines.*

3.1 Lines

Slope of a Line

Many relationships between quantities can be represented conveniently by straight lines. One feature of a straight line is its "steepness." For example, in Figure 3.1, line L_1 rises faster as it goes from left to right than does line L_2. In this sense, L_1 is steeper.

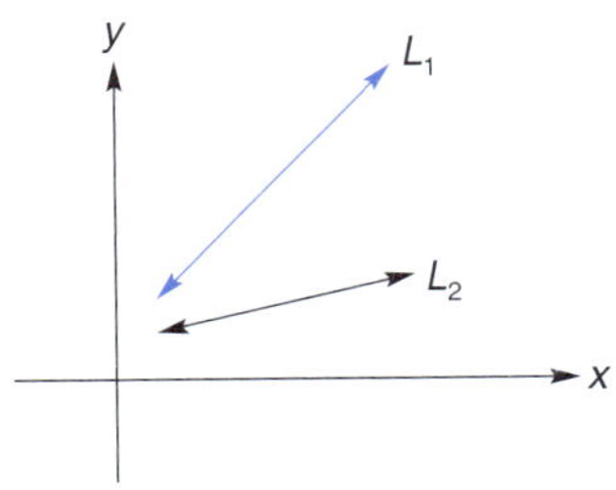

FIGURE 3.1 Line L_1 is "steeper" than L_2.

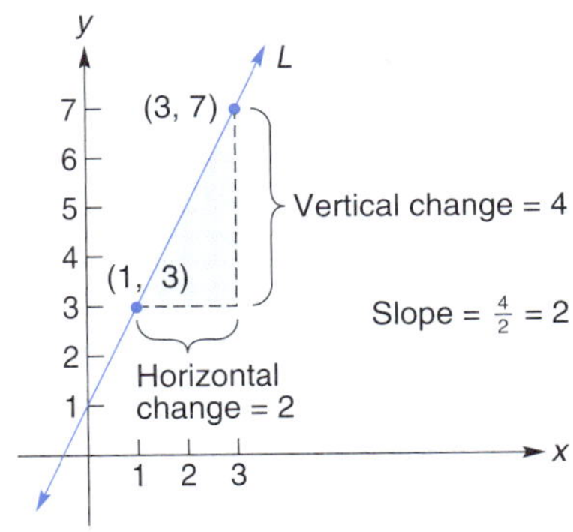

FIGURE 3.2 Slope of a line.

To measure the steepness of a line, we use the notion of *slope*. In Figure 3.2, as we move along line L from (1, 3) to (3, 7), the x-coordinate increases from 1 to 3, and the y-coordinate increases from 3 to 7. The average rate of change of y with respect to x is the ratio

$$\frac{\text{change in } y}{\text{change in } x} = \frac{\text{vertical change}}{\text{horizontal change}} = \frac{7-3}{3-1} = \frac{4}{2} = 2$$

The ratio of 2 means that for each 1-unit increase in x, there is a 2-unit *increase* in y. Because of the increase, the line *rises* from left to right. It can be shown that, regardless of which two points on L are chosen to compute the ratio of the change in y to the change in x, the result is always 2, which we call the *slope* of the line.

Definition

Let (x_1, y_1) and (x_2, y_2) be two different points on a nonvertical line. The slope of the line is

$$m = \frac{y_2 - y_1}{x_2 - x_1} \left(= \frac{\textit{vertical change}}{\textit{horizontal change}}\right) \qquad (1)$$

CAUTION Having no slope does not mean having a slope of zero

A vertical line does not have a slope, because any two points on it must have $x_1 = x_2$ (see Figure 3.3(a)), which gives a denominator of zero in Equation (1). For a horizontal line, any two points must have $y_1 = y_2$. (See Figure 3.3(b).) This gives a numerator of zero in Equation (1), and hence, the slope of the line is zero.

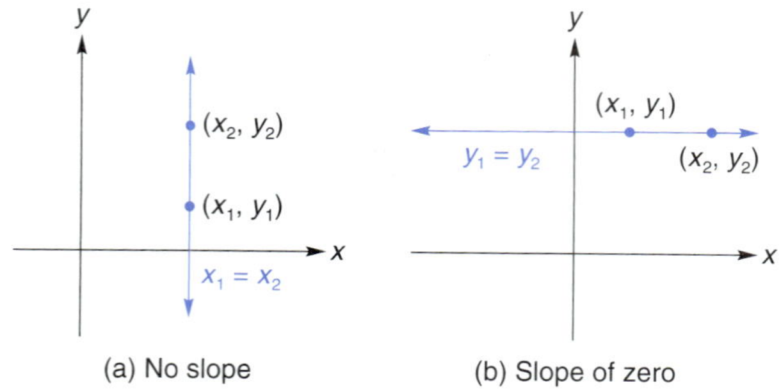

FIGURE 3.3 Vertical and horizontal lines.

This example shows how the slope can be interpreted.

APPLICATIONS IN PRACTICE 1

Price-Quantity Relationship

A doctor purchased a new car in 1991 for $32,000. In 1994, he sold it to a friend for $26,000. Draw a line showing the relationship between the selling price of the car and the year in which it was sold. Find and interpret the slope.

EXAMPLE 1 **Price-Quantity Relationship**

The line in Figure 3.4 shows the relationship between the price p of a widget (in dollars) and the quantity q of widgets (in thousands) that consumers will buy at that price. Find and interpret the slope.

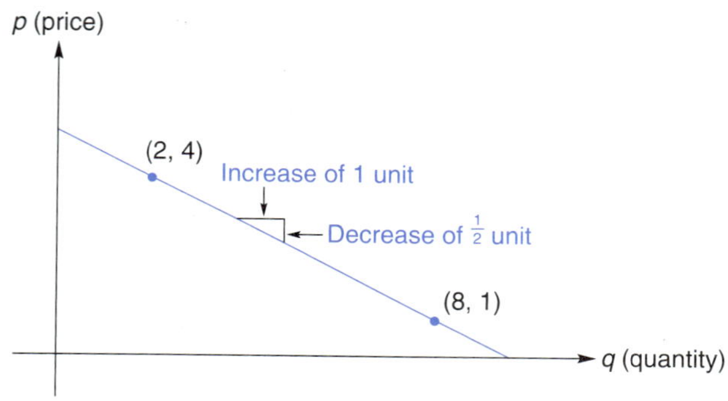

FIGURE 3.4 Price-quantity line.

Solution: In the slope formula (1), we replace the x's by q's and the y's by p's. Either point in Figure 3.4 may be chosen as (q_1, p_1). Letting $(2, 4) = (q_1, p_1)$ and $(8, 1) = (q_2, p_2)$, we have

$$m = \frac{p_2 - p_1}{q_2 - q_1} = \frac{1-4}{8-2} = \frac{-3}{6} = -\frac{1}{2}$$

The slope is negative, $-\frac{1}{2}$. This means that, for each 1-unit increase in quantity (one thousand widgets), there corresponds a **decrease** in price of $\frac{1}{2}$ (dollar per widget). Because of this decrease, the line **falls** from left to right.

▸ **Now work Problem 3.**

In summary, we can characterize the orientation of a line by its slope:

Zero slope: horizontal line
Undefined slope: vertical line
Positive slope: line rises from left to right
Negative slope: line falls from left to right

Lines with different slopes are shown in Figure 3.5. Notice that *the closer the slope is to 0, the more nearly horizontal is the line. The greater the absolute value of the slope, the more nearly vertical is the line.* We remark that two lines are parallel if and only if they have the same slope or are both vertical.

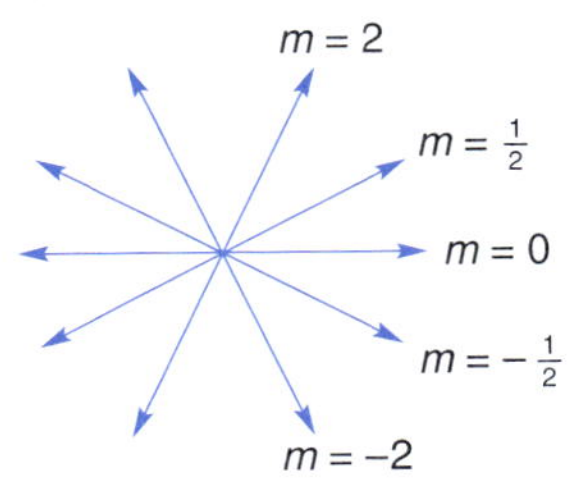

FIGURE 3.5 Slopes of lines.

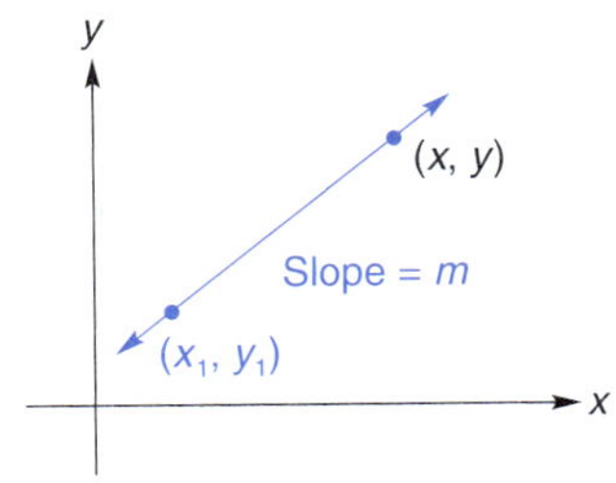

FIGURE 3.6 Line through (x_1, y_1) with slope m.

Equations of Lines

If we know a point on a line and the slope of the line, we can find an equation whose graph is that line. Suppose that line L has slope m and passes through the point (x_1, y_1). If (x, y) is *any* other point on L (see Figure 3.6), we can find an algebraic relationship between x and y. Using the slope formula on the points (x_1, y_1) and (x, y) gives

$$\frac{y - y_1}{x - x_1} = m$$

$$y - y_1 = m(x - x_1) \tag{2}$$

Every point on L satisfies Equation (2). It is also true that every point satisfying Equation (2) must lie on L. Thus, Equation (2) is an equation for L and is given a special name:

$$y - y_1 = m(x - x_1)$$

is a ***point–slope form*** *of an equation of the line through* (x_1, y_1) *with slope* m.

APPLICATIONS IN PRACTICE 2

Point–Slope Form

A new applied mathematics program at a university has grown in enrollment by 14 students per year for the last five years. If the program enrolled 50 students in its third year, what is an equation for the number of students S in the program as a function of the number of years T since its inception?

EXAMPLE 2 **Point–Slope Form**

Find an equation of the line that has slope 2 and passes through $(1, -3)$.

Solution: Using a point–slope form with $m = 2$ and $(x_1, y_1) = (1, -3)$ gives

$$y - y_1 = m(x - x_1)$$

$$y - (-3) = 2(x - 1)$$

$$y + 3 = 2x - 2$$

which can be rewritten as

$$2x - y - 5 = 0$$

▶ **Now work Problem 9.**

An equation of the line passing through two given points can be found easily, as Example 3 shows.

APPLICATIONS IN PRACTICE 3

Determining a Line from Two Points

Find an equation of the line passing through the given points. A temperature of 41°F is equivalent to 5°C, and a temperature of 77°F is equivalent to 25°C.

EXAMPLE 3 Determining a Line from Two Points

Find an equation of the line passing through $(-3, 8)$ *and* $(4, -2)$.

Solution:

Strategy: First we find the slope of the line from the given points. Then we substitute the slope and one of the points into a point–slope form.

The line has slope

$$m = \frac{-2-8}{4-(-3)} = -\frac{10}{7}$$

Using a point–slope form with $(-3, 8)$ as (x_1, y_1) gives

$$y - 8 = -\frac{10}{7}[x - (-3)]$$
$$y - 8 = -\frac{10}{7}(x + 3)$$
$$7y - 56 = -10x - 30$$
$$10x + 7y - 26 = 0$$

Choosing $(4, -2)$ as (x_1, y_1) would give the same result.

▸ **Now work Problem 13.**

Recall that a point $(0, b)$ where a graph intersects the y-axis is called a y-intercept (Figure 3.7). If the slope m and y-intercept b of a line are known, an equation for the line is [by using a point–slope form with $(x_1, y_1) = (0, b)$]

$$y - b = m(x - 0)$$

Solving for y gives $y = mx + b$, called the *slope–intercept form* of an equation of the line:

$$y = mx + b$$

*is the **slope–intercept form** of an equation of the line with slope m and y-intercept b.*

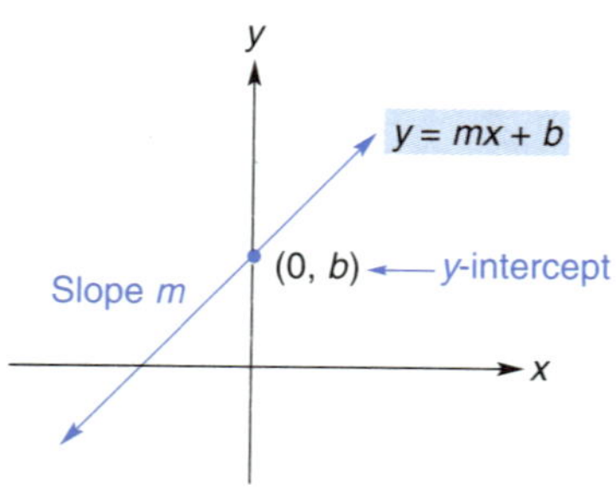

FIGURE 3.7 Line with slope m and y-intercept b.

EXAMPLE 4 Slope–Intercept Form

Find an equation of the line with slope 3 *and y-intercept* -4.

Solution: Using the slope–intercept form $y = mx + b$ with $m = 3$ and $b = -4$ gives

$$y = 3x + (-4)$$
$$y = 3x - 4$$

▸ **Now work Problem 17.**

APPLICATIONS IN PRACTICE 4

Find the Slope and y-Intercept of a Line

One formula for the recommended dosage (in milligrams) of medication for a child t years old is

$$y = \frac{1}{24}(t+1)a,$$

where a is the adult dosage. For an over-the-counter pain reliever, $a = 1000$. Find the slope and y-intercept of this equation.

EXAMPLE 5 **Find the Slope and y-Intercept of a Line**

Find the slope and y-intercept of the line with equation $y = 5(3 - 2x)$.

Solution:

Strategy: We shall rewrite the equation so it has the slope–intercept form $y = mx + b$. Then the slope is the coefficient of x and the y-intercept is the constant term.

We have

$$y = 5(3 - 2x)$$
$$y = 15 - 10x$$
$$y = -10x + 15$$

Thus, $m = -10$ and $b = 15$, so the slope is -10 and the y-intercept is 15.

▸ **Now work Problem 25.**

If a *vertical* line passes through (a, b) (see Figure 3.8), then any other point (x, y) lies on the line if and only if $x = a$. The y-coordinate can have any value. Hence, an equation of the line is $x = a$. Similarly, an equation of the *horizontal* line passing through (a, b) is $y = b$. (See Figure 3.9.) Here the x-coordinate can have any value.

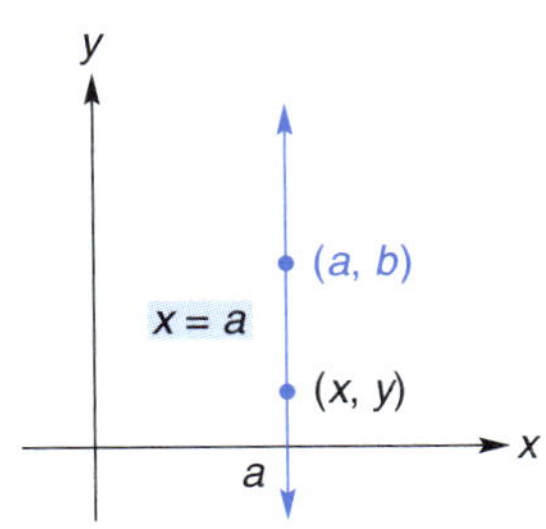

FIGURE 3.8 Vertical line through (a, b).

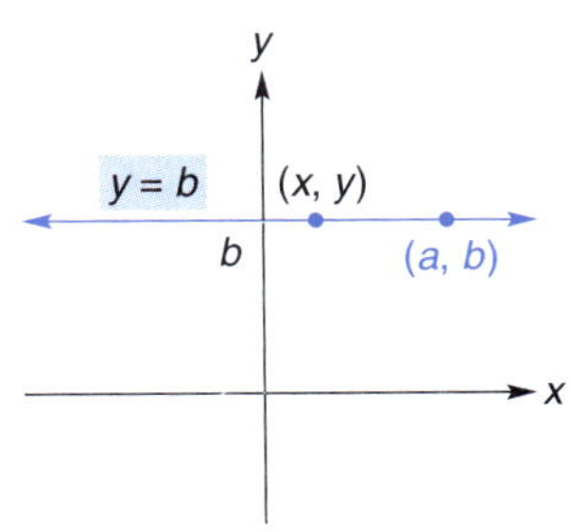

FIGURE 3.9 Horizontal line through (a, b).

EXAMPLE 6 **Equations of Horizontal and Vertical Lines**

a. An equation of the vertical line through $(-2, 3)$ is $x = -2$. An equation of the horizontal line through $(-2, 3)$ is $y = 3$.

b. The x- and y-axes are horizontal and vertical lines, respectively. Because $(0, 0)$ lies on both axes, an equation of the x-axis is $y = 0$, and an equation of the y-axis is $x = 0$.

▸ **Now work Problems 21 and 23.**

From our discussions, we can show that every straight line is the graph of an equation of the form $Ax + By + C = 0$, where A, B, and C are constants and A and B are not both zero. We call this a **general linear equation** (or an *equation of the first degree*) ***in the variables x and y***, and x and y are said to be **linearly related.** For example, a general linear equation for $y = 7x - 2$ is $(-7)x + (1)y + (2) = 0$. Conversely, the graph of a general linear equation is a straight line. Table 3.1 gives the various forms of equations of straight lines.

CAUTION Do not confuse the forms of equations of horizontal and vertical lines. Remember which one has the form $x =$ constant and which one has the form $y =$ constant.

TABLE 3.1 Forms of Equations of Straight Lines

Point–slope form	$y - y_1 = m(x - x_1)$
Slope–intercept form	$y = mx + b$
General linear form	$Ax + By + C = 0$
Vertical line	$x = a$
Horizontal line	$y = b$

Example 3 suggests that we could add another entry to the table. For if we know that points (x_1, y_1) and (x_2, y_2) are points on a line, then the slope of that line is $m = \dfrac{y_2 - y_1}{x_2 - x_1}$ and we could say that $y - y_1 = \dfrac{y_2 - y_1}{x_2 - x_1}(x - x_1)$ *is* ***two-point form*** *for an equation of a line passing through points* (x_1, y_1) *and* (x_2, y_2). Whether one chooses to remember many formulas or a few problem-solving principles is very much a matter of individual taste.

APPLICATIONS IN PRACTICE 5

Converting Forms of Equations of Lines

Find a general linear form of the Fahrenheit–Celsius conversion equation whose slope–intercept form is $F = \frac{9}{5}C + 32$.

EXAMPLE 7 Converting Forms of Equations of Lines

a. *Find a general linear form of the line whose slope–intercept form is*

$$y = -\frac{2}{3}x + 4$$

Solution: Getting one side to be 0, we obtain

$$\frac{2}{3}x + y - 4 = 0$$

which is a general linear form with $A = \frac{2}{3}$, $B = 1$, and $C = -4$. An alternative general form can be obtained by clearing fractions:

This illustrates that a general linear form of a line is not unique

$$2x + 3y - 12 = 0$$

b. *Find the slope–intercept form of the line having a general linear form* $3x + 4y - 2 = 0$.

Solution: We want the form $y = mx + b$, so we solve the given equation for y. We have

$$3x + 4y - 2 = 0$$
$$4y = -3x + 2$$
$$y = -\frac{3}{4}x + \frac{1}{2}$$

which is the slope–intercept form. Note that the line has slope $-\frac{3}{4}$ and y-intercept $\frac{1}{2}$.

▸ **Now work Problem 37.**

APPLICATIONS IN PRACTICE 6

Graphing a General Linear Equation

Sketch the graph of the Fahrenheit–Celsius conversion equation that you found in Applications in Practice 5. How could you use this graph to convert a Celsius temperature to Fahrenheit?

EXAMPLE 8 Graphing a General Linear Equation

Sketch the graph of $2x - 3y + 6 = 0$.

Solution:

Strategy: Since this is a general linear equation, its graph is a straight line. Thus, we need only determine two different points on the graph in order to sketch it. We shall find the intercepts.

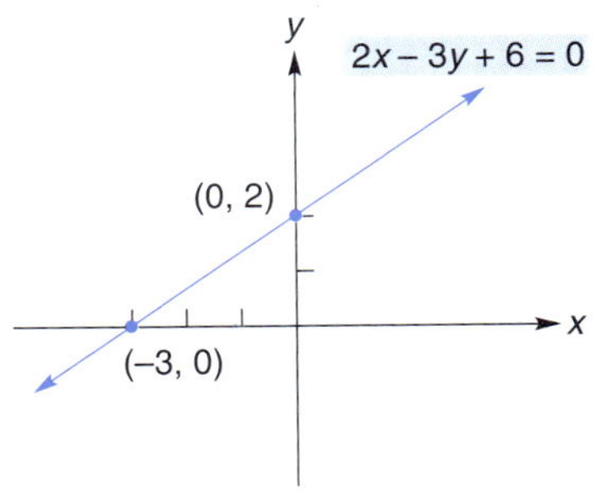

FIGURE 3.10 Graph of $2x - 3y + 6 = 0$.

If $x = 0$, then $-3y + 6 = 0$, so the y-intercept is 2. If $y = 0$, then $2x + 6 = 0$, so the x-intercept is -3. We now draw the line passing through (0, 2) and $(-3, 0)$. (See Figure 3.10.)

TECHNOLOGY

To graph the equation of Example 8 with a graphing calculator, we first express y in terms of x:

$$2x - 3y + 6 = 0$$
$$3y = 2x + 6$$
$$y = \frac{1}{3}(2x + 6)$$

Essentially, y is expressed as a function of x; the graph is shown in Figure 3.11.

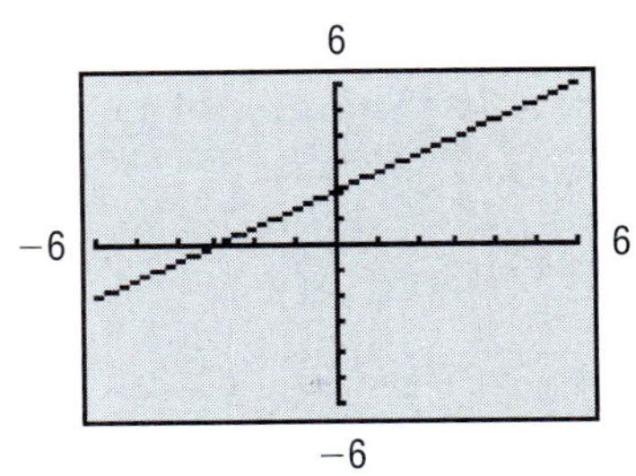

FIGURE 3.11 Calculator graph of $2x - 3y + 6 = 0$.

Parallel and Perpendicular Lines

As stated previously, there is a rule for parallel lines:

Parallel Lines: *Two lines are parallel if and only if they have the same slope or are both vertical.*

There is also a rule for perpendicular lines. Look back to Figure 3.5 and observe that the line with slope $-\frac{1}{2}$ is perpendicular to the line with slope 2. The fact that the slope of either of these lines is the negative reciprocal of the slope of the other line is not a coincidence, as the following rule states.

Perpendicular Lines: *Two lines with slopes m_1 and m_2 are perpendicular to each other if and only if*

$$m_1 = -\frac{1}{m_2}$$

Moreover, any horizontal line and any vertical line are perpendicular to each other.

APPLICATIONS IN PRACTICE 7

Parallel and Perpendicular Lines

Show that a triangle with vertices at $A(0, 0)$, $B(6, 0)$, and $C(7, 7)$ is not a right triangle.

EXAMPLE 9 Parallel and Perpendicular Lines

Figure 3.12 shows two lines passing through $(3, -2)$. *One is parallel to the line* $y = 3x + 1$, *and the other is perpendicular to it. Find equations of these lines.*

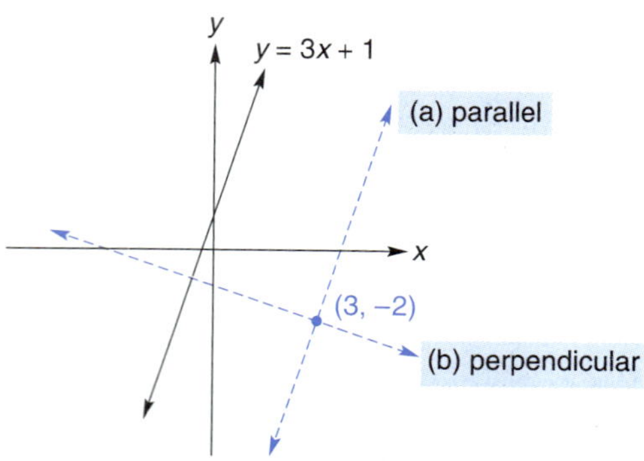

FIGURE 3.12 Lines parallel and perpendicular to $y = 3x + 1$.

Solution: The slope of $y = 3x + 1$ is 3. Thus, the line through $(3, -2)$ that is *parallel* to $y = 3x + 1$ also has slope 3. Using a point–slope form, we get

$$y - (-2) = 3(x - 3)$$
$$y + 2 = 3x - 9$$
$$y = 3x - 11$$

The slope of a line *perpendicular* to $y = 3x + 1$ must be $-\frac{1}{3}$ (the negative reciprocal of 3). Using a point–slope form, we get

$$y - (-2) = -\frac{1}{3}(x - 3)$$
$$y + 2 = -\frac{1}{3}x + 1$$
$$y = -\frac{1}{3}x - 1$$

▸ **Now work Problem 55.**

Exercise 3.1

In Problems 1–8, find the slope of the straight line that passes through the given points.

1. $(4, 1), (7, 10)$

2. $(-2, 10), (5, 3)$

***3.** $(4, -2), (-6, 3)$

4. $(2, -4), (3, -4)$

5. $(5, 3), (5, -8)$

6. $(0, -6), (3, 0)$

7. $(5, -2), (4, -2)$

8. $(1, -7), (9, 0)$

In Problems 9–24, find a general linear equation ($Ax + By + C = 0$) of the straight line that has the indicated properties, and sketch each line.

***9.** Passes through $(1, 9)$ and has slope 7.

10. Passes through the origin and has slope -5.

11. Passes through $(-2, 5)$ and has slope $-\frac{1}{4}$.

12. Passes through $(-\frac{5}{2}, 5)$ and has slope $\frac{1}{3}$.

***13.** Passes through $(-6, 1)$ and $(1, 4)$.

14. Passes through $(5, 2)$ and $(6, -4)$.

15. Passes through $(3, -1)$ and $(-2, -9)$.

16. Passes through $(0, 0)$ and $(2, 3)$.

***17.** Has slope 2 and y-intercept 4.

18. Has slope 5 and y-intercept -7.

19. Has slope $-\frac{1}{2}$ and y-intercept -3.

20. Has slope 0 and y-intercept $-\frac{1}{2}$.

***21.** Is horizontal and passes through $(-5, -3)$.

22. Is vertical and passes through $(-1, 4)$.

***23.** Passes through $(2, -3)$ and is vertical.

24. Passes through the origin and is horizontal.

In Problems 25–34, find, if possible, the slope and y-intercept of the straight line determined by the equation, and sketch the graph.

***25.** $y = 4x - 6$

26. $x - 2 = 6$

27. $x + 2y - 3 = 0$

28. $y + 4 = 7$

29. $x = -5$

30. $x - 9 = 5y + 3$

31. $y = 3x$

32. $y - 7 = 3(x - 4)$

33. $y = 3$

34. $2y - 3 = 0$

In Problems 35–40, find a general linear form and the slope–intercept form of the given equation.

35. $2x = 5 - 3y$

36. $3x + 2y = 6$

***37.** $4x + 9y - 5 = 0$

38. $3(x - 4) - 7(y + 1) = 2$

39. $\frac{x}{2} - \frac{y}{3} = -4$

40. $y = \frac{1}{300}x + 8$

In Problems 41–50, determine whether the lines are parallel, perpendicular, or neither.

41. $y = 7x + 2, y = 7x - 3$

42. $y = 4x + 3, y = 5 + 4x$

43. $y = 5x + 2, -5x + y - 3 = 0$

44. $y = x, y = -x$

45. $x + 3y + 5 = 0, y = -3x$

46. $x + 2y = 0, x + y - 4 = 0$

47. $y = 3, x = -\frac{1}{3}$

48. $x = 3, x = -3$

49. $3x + y = 4, x - 3y + 1 = 0$

50. $x - 2 = 3, y = 2$

In Problems 51–60, find an equation of the line satisfying the given conditions. Give the answer in slope–intercept form if possible.

51. Passing through $(-3, 2)$ and parallel to $y = 4x - 5$.

52. Passing through $(2, -8)$ and parallel to $x = -4$.

53. Passing through $(2, 1)$ and parallel to $y = 2$.

54. Passing through $(3, -4)$ and parallel to $y = 3 + 2x$.

***55.** Perpendicular to $y = 3x - 5$ and passing through $(3, 4)$.

56. Perpendicular to $y = -4$ and passing through $(1, 1)$.

57. Passing through $(5, 2)$ and perpendicular to $y = -3$.

58. Passing through $(-5, 4)$ and perpendicular to the line $2y = -x + 1$.

59. Passing through $(-7, -5)$ and parallel to the line $2x + 3y + 6 = 0$.

60. Passing through $(-4, 10)$ and parallel to the y-axis.

61. A straight line passes through $(1, 2)$ and $(-3, 8)$. Find the point on it that has an x-coordinate of 5.

62. A straight line has slope 3 and y-intercept $(0, 1)$. Does the point $(-1, -2)$ lie on the line?

63. Stock In 1988, the stock in a biotechnology company traded for \$30 per share. In 1998, the company started having trouble and the stock price dropped to \$10 per share. Draw a line showing the relationship between the price per share and the year in which it traded, with years on the x-axis and price on the y-axis. Find and interpret the slope.

In Problems 64–65, find an equation of the line describing the following information.

64. Home Runs In one season, a major league baseball player has hit 14 home runs by the end of the third month and 20 home runs by the end of the fifth month.

65. Business A delicatessen owner starts her business with debts of \$100,000. After operating for five years, she has accumulated a profit of \$40,000.

66. Due Date The length, L, of a human fetus more than 12 weeks old can be estimated by the formula $L = 1.53t - 6.7$, where L is in centimeters and t is in weeks from conception. An obstetrician uses the length of a fetus, measured by ultrasound, to determine the approximate age of the fetus and establish a due date for the mother. The formula must be rewritten to result in an age, t, given a fetal length, L. Find the slope and L-intercept of the equation.

67. **Discus Throw** A mathematical model can approximate the winning distance for the Olympic discus throw by the formula $d = 184 + t$, where d is in feet and $t = 0$ corresponds to the year 1948. Find a general linear form of this equation.

68. **Campus Map** A coordinate map of a college campus gives the coordinates (x, y) of three major buildings as follows: computer center, $(3.5, -1.5)$; engineering lab, $(0.5, 0.5)$; and library $(-1, -2.5)$. Find the equations (in slope–intercept form) of the straight-line paths connecting (a) the engineering lab with the computer center and (b) the engineering lab with the library. Are these two paths perpendicular to each other?

69. **Geometry** Show that the points $A(0, 0)$, $B(0, 4)$, $C(2, 3)$, and $D(2, 7)$ are the vertices of a parallelogram. (Opposite sides of a parallelogram are parallel.)

70. **Approach Angle** A small plane is landing at an airport with an approach angle of 45 degrees, or slope of -1. The plane begins its descent when it has an elevation of 3300 feet. Find the equation that describes the relationship between the craft's altitude and distance traveled, assuming that at distance 0 it starts the approach angle. Graph your equation on a graphing calculator. What does the graph tell you about the approach if the airport is 4000 feet from where the plane starts its landing.

71. **Cost Equation** The average daily cost, C, for a room at a city hospital has risen by \$59.82 per year for the years 1990 through 2000. If the average cost in 1996 was \$1128.50, what is an equation that describes the average cost during this decade, as a function of the number of years, T, since 1990?

72. **Revenue Equation** A small business predicts its revenue growth by a straight-line method with a slope of \$50,000 per year. In its fifth year, it had revenues of \$330,000. Find an equation that describes the relationship between revenues, R, and the number of years, T, since it opened for business.

73. Graph $y = -0.9x + 7.3$ and verify that the y-intercept is 7.3.

74. Graph the lines whose equations are

$$y = 1.5x + 1$$

$$y = 1.5x - 1$$

and

$$y = 1.5x + 2.5$$

What do you observe about the orientation of these lines? Why would you expect this result from the equations of the lines themselves?

75. Graph the line $y = 3.4x - 2.3$. Find the coordinates of any two points on the line, and use them to estimate the slope. What is the actual slope of the line?

76. Using the standard window, graph the lines with equations

$$0.1875x - 0.3y + 0.94 = 0$$

and

$$0.32x + 0.2y + 1.01 = 0$$

on the same viewing rectangle. Now, change the window to a square window (for example, on a TI-83 Plus, use ZOOM, ZSquare). Notice that the lines appear to be perpendicular to each other. Prove that this is indeed the case.

3.2 Applications and Linear Functions

OBJECTIVE *To develop the notion of demand and supply curves and to introduce linear functions.*

Many situations in economics can be described by using straight lines, as evidenced by Example 1.

APPLICATIONS IN PRACTICE 1

Production Levels

A sporting-goods manufacturer allocates 1000 units of time per day to make skis and ski boots. If it takes 8 units of time to make a ski, and 14 units of time to make a boot, find an equation to describe all possible production levels of the two products.

EXAMPLE 1 **Production Levels**

Suppose that a manufacturer uses 100 lb of material to produce products A and B, which require 4 lb and 2 lb of material per unit, respectively. If x and y denote the number of units produced of A and B, respectively, then all levels of production are given by the combinations of x and y that satisfy the equation

$$4x + 2y = 100 \qquad \text{where } x, \ y \geq 0$$

Thus, the levels of production of A and B are linearly related. Solving for y gives

$$y = -2x + 50 \qquad \text{(slope–intercept form),}$$

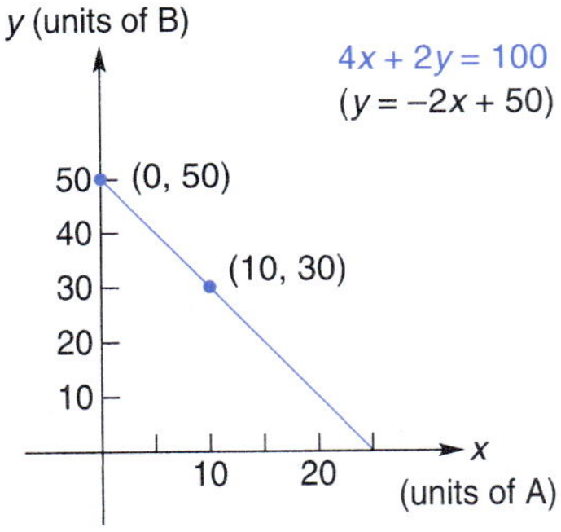

FIGURE 3.13 Linearly related production levels.

so the slope is -2. The slope reflects the rate of change of the level of production of B with respect to the level of production of A. For example, if 1 more unit of A is to be produced, it will require 4 more pounds of material, resulting in $\frac{4}{2} = 2$ *fewer* units of B. Accordingly, as x increases by 1 unit, the corresponding value of y decreases by 2 units. To sketch the graph of $y = -2x + 50$, we can use the y-intercept $(0, 50)$ and the fact that when $x = 10$, $y = 30$. (See Figure 3.13.)

Demand-and-Supply Curves

For each price level of a product, there is a corresponding quantity of that product that consumers will demand (that is, purchase) during some time period. Usually, the higher the price, the smaller is the quantity demanded; as the price falls, the quantity demanded increases. If the price per unit of the product is given by p and the corresponding quantity (in units) is given by q, then an equation relating p and q is called a **demand equation.** Its graph is called a **demand curve.** Figure 3.14(a) shows a demand curve. In keeping with the practice of most economists, the horizontal axis is the q-axis and the vertical axis is the p-axis. We shall assume that the price per unit is given in dollars and the period is one week. Thus, the point (a, b) in Figure 3.14(a) indicates that, at a price of b dollars per unit, consumers will demand a units per week. Since negative prices or quantities are not meaningful, both a and b must be nonnegative. For most products, an increase in the quantity demanded corresponds to a decrease in price. Thus, a demand curve typically falls from left to right, as in Figure 3.14(a).

CAUTION Typically, a demand curve falls from left to right and a supply curve rises from left to right. However, there are exceptions. For example, the demand for insulin could be represented by a vertical line, since this demand can remain constant regardless of price.

In response to various prices, there is a corresponding quantity of product that *producers* are willing to supply to the market during some time period. Usually, the higher the price per unit, the larger is the quantity that producers are willing to supply; as the price falls, so will the quantity supplied. If p denotes the price per unit and q denotes the corresponding quantity, then an equation relating p and q is called a **supply equation,** and its graph is called a **supply curve.** Figure 3.14(b) shows a supply curve. If p is in dollars and the period is one week, then the point (c, d) indicates that, at a price of d dollars each, producers will supply c units per week. As before, c and d are nonnegative. A supply curve usually rises from left to right, as in Figure 3.14(b). This indicates that a producer will supply more of a product at higher prices.

Observe that a function whose graph either falls from left to right or rises from left to right *throughout its entire domain* will pass the horizontal-line test of Section 2.5. Certainly, the demand curve and the supply curve in Figure 3.14 are each cut at most once by any horizontal line. Thus if the demand curve is

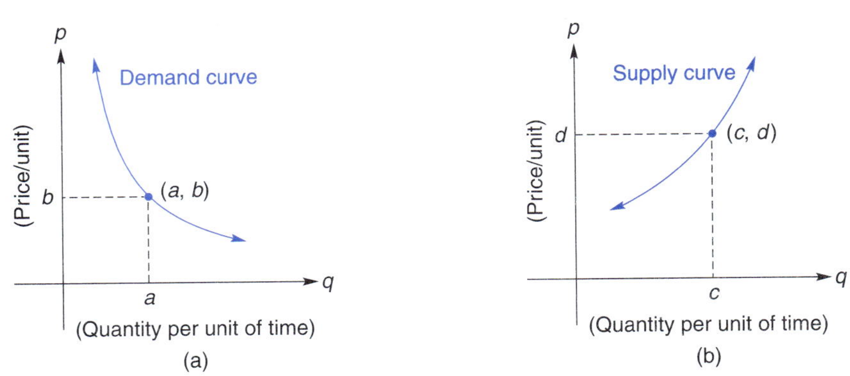

FIGURE 3.14 Demand-and-supply curves.

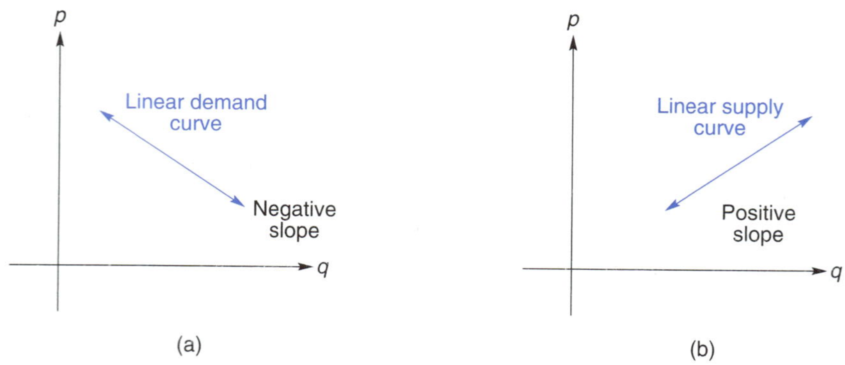

FIGURE 3.15 Linear demand and supply curves.

the graph of a function $p = D(q)$, then D will have an inverse and we can solve for q uniquely to get $q = D^{-1}(p)$. Similarly, if the supply curve is the graph of a function $p = S(q)$, then S is also one-to-one and has an inverse S^{-1} and we can write $q = S^{-1}(p)$.

We shall now focus on demand and supply curves that are straight lines (Figure 3.15). They are called *linear* demand and *linear* supply curves. Such curves have equations in which p and q are linearly related. Because a demand curve typically falls from left to right, a linear demand curve has a negative slope. (See Figure 3.15(a).) However, the slope of a linear supply curve is positive, because the curve rises from left to right. (See Figure 3.15(b).)

APPLICATIONS IN PRACTICE 2

Finding a Demand Equation

The demand per week for 26-inch television sets is 1200 units when the price is \$575 each and 800 units when the price is \$725 each. Find the demand equation for the sets, assuming that it is linear.

EXAMPLE 2 Finding a Demand Equation

Suppose the demand per week for a product is 100 units when the price is \$58 per unit and 200 units at \$51 each. Determine the demand equation, assuming that it is linear.

Solution:

Strategy: Since the demand equation is linear, the demand curve must be a straight line. We are given that quantity q and price p are linearly related such that $p = 58$ when $q = 100$ and $p = 51$ when $q = 200$. Thus, the given data can be represented in a q,p-coordinate plane (see Figure 3.15(a)) by points (100, 58) and (200, 51). With these points, we can find an equation of the line—that is, the demand equation.

The slope of the line passing through (100, 58) and (200, 51) is

$$m = \frac{51 - 58}{200 - 100} = -\frac{7}{100}$$

An equation of the line (point-slope form) is

$$p - p_1 = m(q - q_1)$$

$$p - 58 = -\frac{7}{100}(q - 100)$$

Simplifying gives the demand equation

$$p = -\frac{7}{100}q + 65 \qquad (1)$$

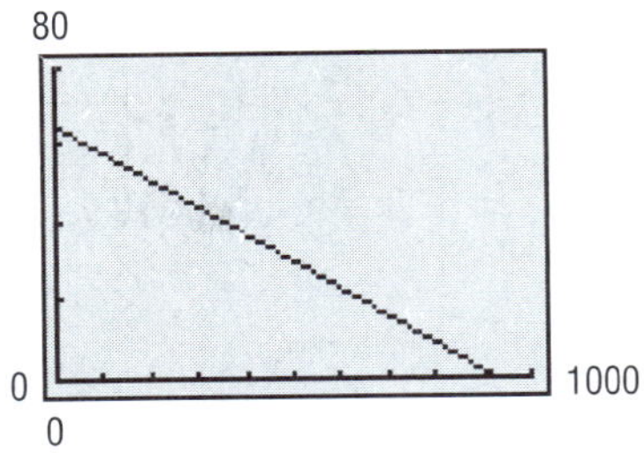

FIGURE 3.16 Graph of demand function $p = -\frac{7}{100}q + 65$.

Customarily, a demand equation (as well as a supply equation) expresses p in terms of q and actually defines a function of q. For example, Equation (1) defines p as a function of q and is called the *demand function* for the product. (See Figure 3.16.)

▶ **Now work Problem 15.**

Linear Functions

A *linear function* was described in Section 2.2. Here is another way to do so.

Definition

A function f is a *linear function* if and only if $f(x)$ can be written in the form $f(x) = ax + b$, where a and b are constants and $a \neq 0$.

Suppose that $f(x) = ax + b$ is a linear function, and let $y = f(x)$. Then $y = ax + b$, which is an equation of a straight line with slope a and y-intercept b. Thus, **the graph of a linear function is a straight line that is neither vertical nor horizontal.** We say that the function $f(x) = ax + b$ has slope a.

APPLICATIONS IN PRACTICE 3

Graphing Linear Functions

A computer repair company charges a fixed amount plus an hourly rate for a service call. If x is the number of hours needed for a service call, the total cost of a call is described by the function $f(x) = 40x + 60$. Graph the function by finding and plotting two points.

EXAMPLE 3 **Graphing Linear Functions**

a. *Graph* $f(x) = 2x - 1$.

Solution: Here f is a linear function (with slope 2), so its graph is a straight line. Since two points determine a straight line, we need only plot two points and then draw a line through them. (See Figure 3.17(a).) Note that one of the points plotted is the vertical-axis intercept, -1, which occurs when $x = 0$.

b. *Graph* $g(t) = \dfrac{15 - 2t}{3}$.

Solution: Notice that g is a linear function, because we can express it in the form $g(t) = at + b$.

$$g(t) = \frac{15 - 2t}{3} = \frac{15}{3} - \frac{2t}{3} = -\frac{2}{3}t + 5$$

The graph of g is shown in Figure 3.17(b). Since the slope is $-\frac{2}{3}$, observe that as t increases by 3 units, $g(t)$ *decreases* by 2.

x	$f(x)$
0	−1
2	3

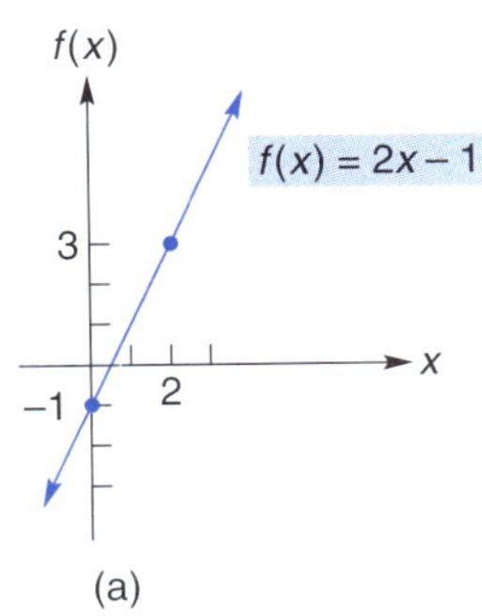

(a)

t	$g(t)$
0	5
6	1

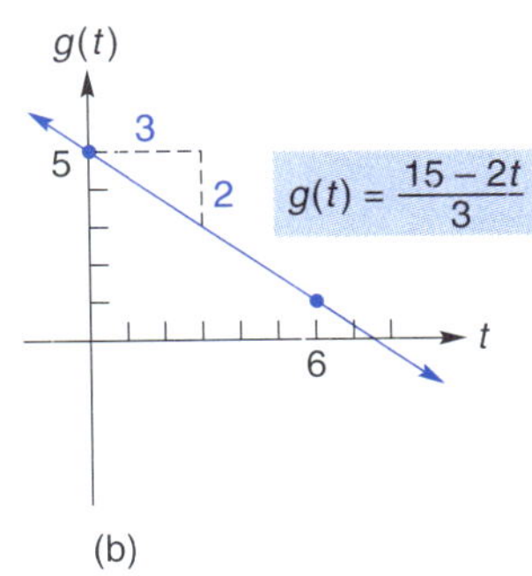

(b)

FIGURE 3.17 Graphs of linear functions.

▶ **Now work Problem 3.**

APPLICATIONS IN PRACTICE 4

Determining a Linear Function

The height of children between the ages of 6 years and 10 years can be modeled by a linear function of age t in years. The height of one child changes by 2.3 inches per year, and she is 50.6 inches tall at age 8. Find a function that describes the height of this child at age t.

EXAMPLE 4 Determining a Linear Function

Suppose f is a linear function with slope 2 and $f(4) = 8$. Find $f(x)$.

Solution: Since f is linear, it has the form $f(x) = ax + b$. The slope is 2, so $a = 2$, and we have

$$f(x) = 2x + b \qquad (2)$$

Now we determine b. Since $f(4) = 8$, we replace x by 4 in Equation (2) and solve for b:

$$\begin{aligned} f(4) &= 2(4) + b \\ 8 &= 8 + b \\ 0 &= b \end{aligned}$$

Hence, $f(x) = 2x$.

▸ **Now work Problem 7.**

APPLICATIONS IN PRACTICE 5

Determining a Linear Function

An antique necklace is expected to be worth \$360 after 3 years and \$640 after 7 years. Find a function that describes the value of the necklace after x years.

EXAMPLE 5 Determining a Linear Function

If $y = f(x)$ is a linear function such that $f(-2) = 6$ and $f(1) = -3$, find $f(x)$.

Solution:

Strategy: The function values correspond to points on the graph of f. With these points we can determine an equation of the line and hence the linear function.

The condition that $f(-2) = 6$ means that when $x = -2$, then $y = 6$. Thus, $(-2, 6)$ lies on the graph of f, which is a straight line. Similarly, $f(1) = -3$ implies that $(1, -3)$ also lies on the line. If we set $(x_1, y_1) = (-2, 6)$ and $(x_2, y_2) = (1, -3)$, the slope of the line is given by

$$m = \frac{y_2 - y_1}{x_2 - x_1} = \frac{-3 - 6}{1 - (-2)} = \frac{-9}{3} = -3$$

We can find an equation of the line by using a point–slope form:

$$\begin{aligned} y - y_1 &= m(x - x_1) \\ y - 6 &= -3[x - (-2)] \\ y - 6 &= -3x - 6 \\ y &= -3x \end{aligned}$$

Because $y = f(x)$, $f(x) = -3x$. Of course, the same result is obtained if we set $(x_1, y_1) = (1, -3)$.

▸ **Now work Problem 9.**

In many studies, data are collected and plotted on a coordinate system. An analysis of the results may indicate a functional relationship between the variables involved. For example, the data points may be approximated by points on a straight line. This would indicate a linear functional relationship, such as the one in the next example.

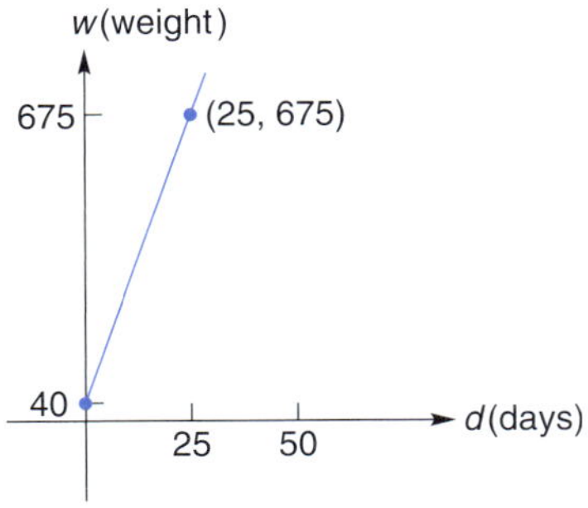

FIGURE 3.18 Linear function describing diet for hens.

EXAMPLE 6 **Diet for Hens**

In testing an experimental diet for hens, it was determined that the average live weight w (in grams) of a hen was statistically a linear function of the number of days d after the diet began, where $0 \leq d \leq 50$. Suppose the average weight of a hen beginning the diet was 40 grams and 25 days later it was 675 grams.

a. *Determine w as a linear function of d.*

Solution: Since w is a linear function of d, its graph is a straight line. When $d = 0$ (the beginning of the diet), $w = 40$. Thus, (0, 40) lies on the graph. (See Figure 3.18.) Similarly, (25, 675) lies on the graph. If we set $(d_1, w_1) = (0, 40)$ and $(d_2, w_2) = (25, 675)$, the slope of the line is

$$m = \frac{w_2 - w_1}{d_2 - d_1} = \frac{675 - 40}{25 - 0} = \frac{635}{25} = \frac{127}{5}$$

Using a point–slope form, we have

$$w - w_1 = m(d - d_1)$$
$$w - 40 = \frac{127}{5}(d - 0)$$
$$w - 40 = \frac{127}{5}d$$
$$w = \frac{127}{5}d + 40$$

which expresses w as a linear function of d.

b. *Find the average weight of a hen when $d = 10$.*

Solution: When $d = 10$, $w = \frac{127}{5}(10) + 40 = 254 + 40 = 294$. Thus, the average weight of a hen 10 days after the beginning of the diet is 294 grams.

▸ **Now work Problem 19.**

Exercise 3.2

In Problems 1–6 find the slope and vertical-axis intercept of the linear function, and sketch the graph.

1. $y = f(x) = -4x$

2. $y = f(x) = x + 1$

***3.** $h(t) = 5t - 7$

4. $g(t) = 2(4 - t)$

5. $h(q) = \dfrac{2 - q}{7}$

6. $h(q) = 0.5q + 0.25$

In Problems 7–14, find $f(x)$ if f is a linear function that has the given properties.

***7.** slope $= 4$, $f(2) = 8$

8. $f(0) = 3$, $f(4) = -5$

***9.** $f(1) = 2$, $f(-2) = 8$

10. slope $= -2$, $f(\frac{2}{5}) = -7$

11. slope $= -\frac{1}{2}$, $f(-\frac{1}{2}) = 4$

12. $f(1) = 1$, $f(2) = 2$

13. $f(-2) = -1$, $f(-4) = -3$

14. slope $= 0.01$, $f(0.1) = 0.01$

***15.** **Demand Equation** Suppose consumers will demand 40 units of a product when the price is \$12.75 per unit and 25 units when the price is \$18.75 each. Find the demand equation, assuming that it is linear. Find the price per unit when 37 units are demanded.

16. **Demand Equation** The demand per week for a best-selling book is 26,000 books when the price is \$16 each, and 10,000 books when the price is \$24 each. Find the demand equation for the book, assuming that it is linear.

17. **Supply Equation** A refrigerator manufacturer will produce 3000 units when the price is \$940, and 2200 units when the price is \$740. Assume that price, p, and quantity, q, produced are linearly related and find the supply equation.

18. Supply Equation Suppose a manufacturer of shoes will place on the market 50 (thousand pairs) when the price is 35 (dollars per pair) and 35 when the price is 30. Find the supply equation, assuming that price p and quantity q are linearly related.

*19. **Cost Equation** Suppose the cost to produce 10 units of a product is \$40 and the cost of 20 units is \$70. If cost, c, is linearly related to output, q, find a linear equation relating c and q. Find the cost to produce 35 units.

20. Cost Equation An advertiser goes to a printer and is charged \$79 for 100 copies of one flyer and \$88 for 400 copies of another flyer. This printer charges a fixed setup cost plus a charge for every copy of single-page flyers. Find a function that describes the cost of a printing job, if x is the number of copies made.

21. Electric Rates An electric utility company charges residential customers 12.5 cents per kilowatt hour plus a base charge each month. One customer's monthly bill comes to \$51.65 for 380 kilowatt hours. Find a linear function that describes the total monthly charges for electricity if x is the number of kilowatt hours used in a month.

22. Radiation Therapy A cancer patient is to receive drug and radiation therapies. Each cubic centimeter of the drug to be used contains 210 curative units, and each minute of radiation exposure gives 305 curative units. The patient requires 2410 curative units. If d cubic centimeters of the drug and r minutes of radiation are administered, determine an equation relating d and r. Graph the equation for $d \geq 0$ and $r \geq 0$; label the horizontal axis as d.

23. Depreciation Suppose the value of a piece of machinery decreases each year by 10% of its original value. If the original value is \$8000, find an equation that expresses the value v of the machinery t years after purchase, where $0 \leq t \leq 10$. Sketch the equation, choosing t as the horizontal axis and v as the vertical axis. What is the slope of the resulting line? This method of considering the value of equipment is called *straight-line depreciation*.

24. Depreciation A new television depreciates \$120 per year, and it is worth \$340 after four years. Find a function that describes the value of this television, if x is the age of the television in years.

25. Appreciation A new apartment building was sold for \$960,000 five years after it was purchased. The original owners calculated that the building appreciated \$45,000 per year while they owned it. Find a linear function that describes the appreciation of the building, if x is the number of years since the original purchase.

26. Appreciation A house purchased for \$198,000 is expected to double in value in 18 years. Find a linear equation that describes the house's value after x years.

27. Repair Charges A business-copier repair company charges a fixed amount plus an hourly rate for a service call. If a customer is billed \$159 for a one-hour service call and \$287 for a three-hour service call, find a linear function that describes the price of a service call, where x is the number of hours of service.

28. Sheep Wool Length For sheep maintained at high environmental temperatures, respiratory rate, r (per minute), increases as wool length, l (in centimeters), decreases.[2] Suppose sheep with a wool length of 2 cm have an (average) respiratory rate of 160, and those with a wool length of 4 cm have a respiratory rate of 125. Assume that r and l are linearly related. (a) Find an equation that gives r in terms of l. (b) Find the respiratory rate of sheep with a wool length of 1 cm.

29. Isocost Line In production analysis, an *isocost line* is a line whose points represent all combinations of two factors of production that can be purchased for the same amount. Suppose a farmer has allocated \$20,000 for the purchase of x tons of fertilizer (costing \$200 per ton) and y acres of land (costing \$2000 per acre). Find an equation of the isocost line that describes the various combinations that can be purchased for \$20,000. Observe that neither x nor y can be negative.

30. Isoprofit Line A manufacturer produces products X and Y for which the profits per unit are \$4 and \$6, respectively. If x units of X and y units of Y are sold, then the total profit P is given by $P = 4x + 6y$, where $x, y \geq 0$. (a) Sketch the graph of this equation for $P = 240$. The result is called an *isoprofit line,* and its points represent all combinations of sales that produce a profit of \$240. (b) Determine the slope for $P = 240$. (c) If $P = 600$, determine the slope. (d) Are isoprofit lines for products X and Y parallel?

31. Grade Scaling For reasons of comparison, a professor wants to rescale the scores on a set of test papers so that the maximum score is still 100, but the mean (average)

[2]Adapted from G. E. Folk, Jr., *Textbook of Environmental Physiology,* 2d ed. (Philadelphia: Lea & Febiger, 1974).

is 80 instead of 56. (a) Find a linear equation that will do this. [*Hint*: You want 56 to become an 80 and 100 to remain 100. Consider the points (56, 80) and (100, 100) and, more generally, (x, y), where x is the old score and y is the new score. Find the slope and use a point–slope form. Express y in terms of x.] (b) If 60 on the new scale is the lowest passing score, what was the lowest passing score on the original scale?

32. Psychology The result of Sternberg's psychological experiment[3] on information retrieval is that a person's reaction time R, in milliseconds, is statistically a linear function of memory set size N as follows:

$$R = 38N + 397.$$

Sketch the graph for $1 \leq N \leq 5$. What is the slope?

33. Psychology In a certain learning experiment involving repetition and memory,[4] the proportion, p, of items recalled was estimated to be linearly related to the effective study time, t (in seconds), where t is between 5 and 9. For an effective study time of 5 seconds, the proportion of items recalled was 0.32. For each 1-second increase in study time, the proportion recalled increased by 0.059. (a) Find an equation that gives p in terms of t. (b) What proportion of items was recalled with 9 seconds of effective study time?

34. Diet for Pigs In testing an experimental diet for pigs, it was determined that the (average) live weight, w (in kilograms), of a pig was statistically a linear function of the number of days, d, after the diet was initiated, where $0 \leq d \leq 100$. If the weight of a pig beginning the diet was 21 kg, and thereafter the pig gained 6.3 kg every 10 days, determine w as a function of d, and find the weight of a pig 55 days after the beginning of the diet.

35. Cricket Chirps Biologists have found that the number of chirps made per minute by crickets of a certain species is related to the temperature. The relationship is very close to being linear. At 68°F, the crickets chirp about 124 times a minute. At 80°F, they chirp about 172 times a minute. (a) Find an equation that gives Fahrenheit temperature, t, in terms of the number of chirps, c, per minute. (b) If you count chirps for only 15 seconds, how can you quickly estimate the temperature?

[3]G. R. Loftus and E. F. Loftus, *Human Memory: The Processing of Information* (New York: Lawrence Erlbaum Associates, Inc., distributed by the Halsted Press, Division of John Wiley & Sons, Inc., 1976).

[4]D. L. Hintzman, "Repetition and Learning," in *The Psychology of Learning*, vol. 10, ed. G. H. Bower (New York: Academic Press, Inc., 1976), p. 77.

OBJECTIVE *To sketch parabolas arising from quadratic functions.*

3.3 Quadratic Functions

In Section 2.2, a *quadratic function* was described as a polynomial function of degree 2. Here is another way of stating the definition.

Definition

A function f is a ***quadratic function*** if and only if $f(x)$ can be written in the form $f(x) = ax^2 + bx + c$, where a, b, and c are constants and $a \neq 0$.

For example, the functions $f(x) = x^2 - 3x + 2$ and $F(t) = -3t^2$ are quadratic. However, $g(x) = \frac{1}{x^2}$ is *not* quadratic, because it cannot be written in the form $g(x) = ax^2 + bx + c$.

The graph of the quadratic function $y = f(x) = ax^2 + bx + c$ is called a **parabola** and has a shape like the curves in Figure 3.19. If $a > 0$, the graph extends upward indefinitely, and we say that the parabola *opens upward* (Figure 3.19(a)). If $a < 0$, the parabola *opens downward* (Figure 3.19(b)).

Each parabola in Figure 3.19 is *symmetric* about a vertical line, called the **axis of symmetry** of the parabola. That is, if the page were folded on one of these lines, then the two halves of the corresponding parabola would coincide. The axis (of symmetry) is *not* part of the parabola, but is a useful aid in sketching the parabola.

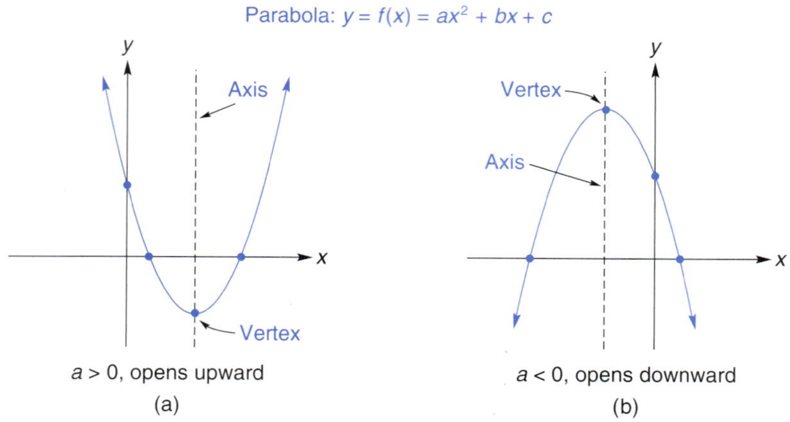

FIGURE 3.19 Parabolas.

Figure 3.19 also shows points labeled **vertex,** where the axis cuts the parabola. If $a > 0$, the vertex is the "lowest" point on the parabola. This means that $f(x)$ has a minimum value at this point. By performing algebraic manipulations on $ax^2 + bx + c$ (referred to as *completing the square*), we can determine not only this minimum value, but also where it occurs. We have

$$f(x) = ax^2 + bx + c = (ax^2 + bx) + c$$

Adding and subtracting $\dfrac{b^2}{4a}$ gives

$$\begin{aligned} f(x) &= \left(ax^2 + bx + \frac{b^2}{4a}\right) + c - \frac{b^2}{4a} \\ &= a\left(x^2 + \frac{b}{a}x + \frac{b^2}{4a^2}\right) + c - \frac{b^2}{4a} \end{aligned}$$

so that

$$f(x) = a\left(x + \frac{b}{2a}\right)^2 + c - \frac{b^2}{4a}$$

Since $\left(x + \dfrac{b}{2a}\right)^2 \geq 0$ and $a > 0$, it follows that $f(x)$ has a minimum value when $x + \dfrac{b}{2a} = 0$, that is, when $x = -\dfrac{b}{2a}$. The y-coordinate corresponding to this value of x is $f\left(-\dfrac{b}{2a}\right)$. Thus, the vertex is given by

$$\text{vertex} = \left(-\frac{b}{2a}, f\left(-\frac{b}{2a}\right)\right)$$

This is also the vertex of a parabola that opens downward ($a < 0$), but in this case $f\left(-\dfrac{b}{2a}\right)$ is the maximum value of $f(x)$. (See Figure 3.19(b).)

Observe that a function whose graph is a parabola is not one-to-one, in either the opening upward or opening downward case, since many horizontal lines will cut the graph twice. However, if we restrict the domain of a quadratic function to either $\left[-\dfrac{b}{2a}, \infty\right)$ or $\left(-\infty, -\dfrac{b}{2a}\right]$, then the restricted function will pass the horizontal line test and therefore be one-to-one. It follows that such restricted quadratic functions have inverse functions.

The point where the parabola $y = ax^2 + bx + c$ intersects the y-axis (that is, the y-intercept) occurs when $x = 0$. The y-coordinate of this point is c, so the y-intercept is $(0, c)$, or, more simply, c. In summary, we have the following.

Graph of Quadratic Function

The graph of the quadratic function $y = f(x) = ax^2 + bx + c$ is a parabola.

1. If $a > 0$, the parabola opens upward. If $a < 0$, it opens downward.
2. The vertex is $\left(-\frac{b}{2a}, f\left(-\frac{b}{2a}\right)\right)$.
3. The y-intercept is c.

We can quickly sketch the graph of a quadratic function by first locating the vertex, the y-intercept, and a few other points, such as those where the parabola intersects the x-axis. These *x-intercepts* are found by setting $y = 0$ and solving for x. Once the intercepts and vertex are found, it is then relatively easy to pass the appropriate parabola through these points. In the event that the x-intercepts are very close to the vertex or that no x-intercepts exist, we find a point on each side of the vertex, so that we can give a reasonable sketch of the parabola. Keep in mind that passing a (dashed) vertical line through the vertex gives the axis of symmetry. By plotting points to one side of the axis, we can use symmetry and obtain corresponding points on the other side.

APPLICATIONS IN PRACTICE 1

Graphing a Quadratic Function

The daily profit for a car dealership from the sale of a minivan is given by $P(x) = -x^2 + 2x + 399$, where x is the number of minivans sold. Find the function's vertex and intercepts, and graph the function.

EXAMPLE 1 Graphing a Quadratic Function

Graph the quadratic function $y = f(x) = -x^2 - 4x + 12$.

Solution: Here $a = -1$, $b = -4$, and $c = 12$. Since $a < 0$, the parabola opens downward and thus has a highest point. The x-coordinate of the vertex is

$$-\frac{b}{2a} = -\frac{-4}{2(-1)} = -2$$

The y-coordinate is $f(-2) = -(-2)^2 - 4(-2) + 12 = 16$. Thus, the vertex is $(-2, 16)$, so the maximum value of $f(x)$ is 16. Since $c = 12$, the y-intercept is 12. To find the x-intercepts, we let y be 0 in $y = -x^2 - 4x + 12$ and solve for x:

$$0 = -x^2 - 4x + 12$$
$$0 = -(x^2 + 4x - 12)$$
$$0 = -(x + 6)(x - 2)$$

Hence, $x = -6$ or $x = 2$, so the x-intercepts are -6 and 2. Now we plot the vertex, axis of symmetry, and intercepts. (See Figure 3.20(a).) Since $(0, 12)$ is *two*

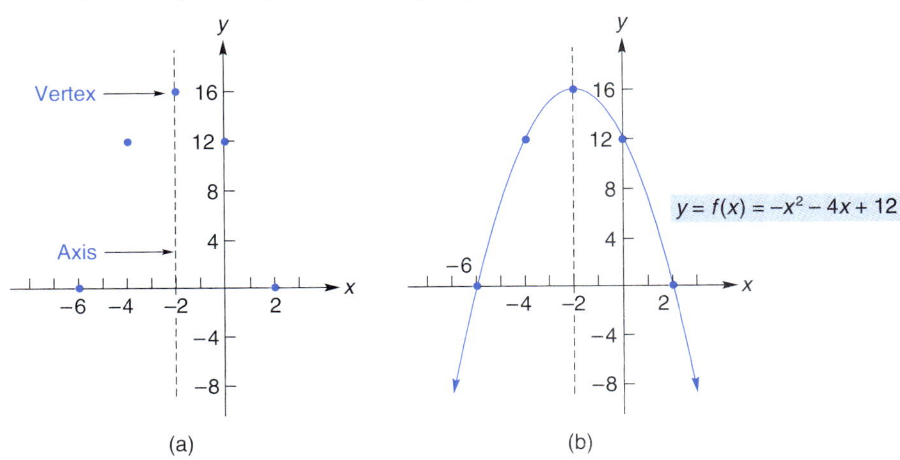

FIGURE 3.20 Graph of parabola $y = f(x) = -x^2 - 4x + 12$.

units to the *right* of the axis, there is a corresponding point *two* units to the *left* of the axis with the same y-coordinate. Thus, we get the point $(-4, 12)$. Through all points, we draw a parabola opening downward. (See Figure 3.20(b).)

▶ **Now work Problem 15.**

EXAMPLE 2 Graphing a Quadratic Function

Graph $p = 2q^2$.

Solution: Here p is a quadratic function of q, where $a = 2$, $b = 0$, and $c = 0$. Since $a > 0$, the parabola opens upward and thus has a lowest point. The q-coordinate of the vertex is

$$-\frac{b}{2a} = -\frac{0}{2(2)} = 0$$

and the p-coordinate is $2(0)^2 = 0$. Consequently, the *minimum* value of p is 0 and the vertex is $(0, 0)$. In this case, the p-axis is the axis of symmetry. A parabola opening upward with vertex at $(0, 0)$ cannot have any other intercepts. Hence, to draw a reasonable graph, we plot a point on each side of the vertex. If $q = 2$, then $p = 8$. This gives the point $(2, 8)$ and, by symmetry, the point $(-2, 8)$. (See Figure 3.21.)

q	p
2	8
−2	8

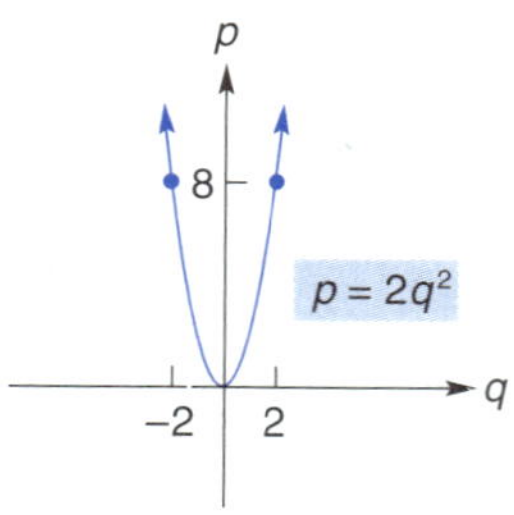

FIGURE 3.21 Graph of parabola $p = 2q^2$.

▶ **Now work Problem 13.**

Example 3 illustrates that finding intercepts may require use of the quadratic formula.

APPLICATIONS IN PRACTICE 2

Graphing a Quadratic Function

A man standing on a pitcher's mound throws a ball straight up with an initial velocity of 32 feet per second. The height h of the ball in feet t seconds after it was thrown is described by the function $h(t) = -16t^2 + 32t + 8$, for $t \geq 0$. Find the function's vertex and intercepts, and graph the function.

EXAMPLE 3 Graphing a Quadratic Function

Graph $g(x) = x^2 - 6x + 7$.

Solution: Here g is a quadratic function, where $a = 1$, $b = -6$, and $c = 7$. The parabola opens upward, because $a > 0$. The x-coordinate of the vertex (lowest point) is

$$-\frac{b}{2a} = -\frac{-6}{2(1)} = 3$$

and $g(3) = 3^2 - 6(3) + 7 = -2$, which is the minimum value of $g(x)$. Thus, the vertex is $(3, -2)$. Since $c = 7$, the vertical-axis intercept is 7. To find x-intercepts, we set $g(x) = 0$.

$$0 = x^2 - 6x + 7$$

The right side does not factor easily, so we shall use the quadratic formula to solve for x:

$$\begin{aligned} x &= \frac{-b \pm \sqrt{b^2 - 4ac}}{2a} = \frac{-(-6) \pm \sqrt{(-6)^2 - 4(1)(7)}}{2(1)} \\ &= \frac{6 \pm \sqrt{8}}{2} = \frac{6 \pm \sqrt{4 \cdot 2}}{2} = \frac{6 \pm 2\sqrt{2}}{2} \\ &= \frac{6}{2} \pm \frac{2\sqrt{2}}{2} = 3 \pm \sqrt{2} \end{aligned}$$

Therefore, the x-intercepts are $3+\sqrt{2}$ and $3-\sqrt{2}$. After plotting the vertex, intercepts, and (by symmetry) the point (6, 7), we draw a parabola opening upward in Figure 3.22.

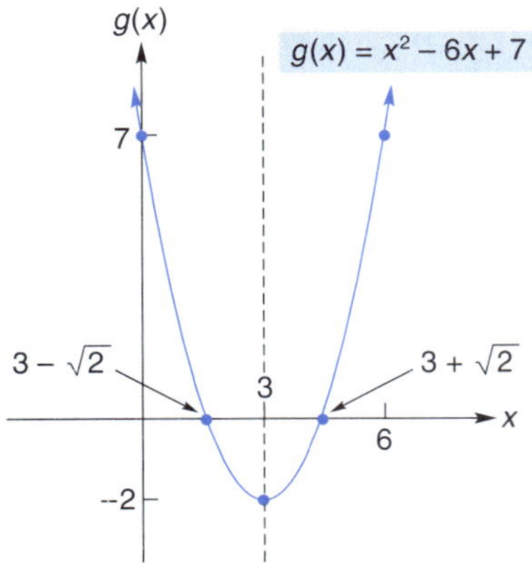

FIGURE 3.22 Graph of parabola $g(x)=x^2-6x+7$.

▶ **Now work Problem 17.**

EXAMPLE 4 **Graphing a Quadratic Function**

Graph $y=f(x)=2x^2+2x+3$ and find the range of f.

Solution: This function is quadratic with $a=2$, $b=2$, and $c=3$. Since $a>0$, the graph is a parabola opening upward. The x-coordinate of the vertex is

$$-\frac{b}{2a}=-\frac{2}{2(2)}=-\frac{1}{2}$$

and the y-coordinate is $2(-\frac{1}{2})^2+2(-\frac{1}{2})+3=\frac{5}{2}$. Thus, the vertex is $(-\frac{1}{2},\frac{5}{2})$. Since $c=3$, the y-intercept is 3. A parabola opening upward with its vertex above the x-axis has no x-intercepts. In Figure 3.23 we plotted the y-intercept, the vertex, and an additional point $(-2, 7)$ to the left of the vertex. By symmetry, we also get the point (1, 7). Passing a parabola through these points gives the desired graph. From the figure, we see that the range of f is all $y\geq\frac{5}{2}$, that is, the interval $[\frac{5}{2},\infty)$.

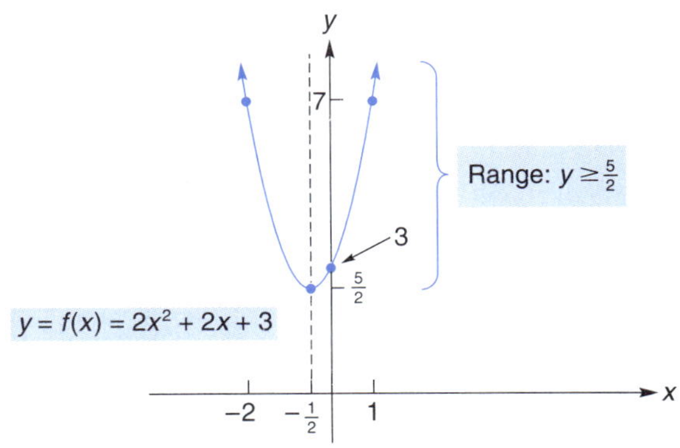

x	y
−2	7
1	7

FIGURE 3.23 Graph of $y=f(x)=2x^2+2x+3$.

▶ **Now work Problem 21.**

EXAMPLE 5 **Finding and Graphing an Inverse**

For the parabola given by the function

$$y=f(x)=ax^2+bx+c$$

determine the inverse of the restricted function given by $g(x)=ax^2+bx+c$, for $x\geq-\frac{b}{2a}$. Graph g and g^{-1} in the same plane, in the case where $a=2$, $b=2$ and $c=3$.

Solution: Following the procedure described in Example 5 of Section 2.4, we begin by solving $y=ax^2+bx+c$, where $x\geq-\frac{b}{2a}$, for x in terms of y. We do this by applying the quadratic formula to $ax^2+bx+c-y=0$, which gives $x=\frac{-b\pm\sqrt{b^2-4a(c-y)}}{2a}=\frac{-b}{2a}\pm\frac{\sqrt{b^2-4a(c-y)}}{2a}$. Whenever

$\sqrt{b^2 - 4a(c - y)}$ is defined (as a real number) it is nonnegative. Therefore, the sign of $\dfrac{\sqrt{b^2 - 4a(c - y)}}{2a}$ depends on a. It is nonnegative when a is positive, that is, when the parabola opens upward, and nonpositive when a is negative, that is, when the parabola opens downward. Thus, in order to satisfy $x \geq -\dfrac{b}{2a}$ we must take the + in ± when $a > 0$ and the parabola opens upward and the − in ± when $a < 0$ and the parabola opens downward. For definiteness now, let us deal with the case of $a > 0$. It follows, returning to the procedure of Example 5 of Section 2.4, that $g^{-1}(x) = \dfrac{-b + \sqrt{b^2 - 4a(c - x)}}{2a}$. The vertex of any parabola has y-coordinate given by $f\left(-\dfrac{b}{2a}\right) = a\left(-\dfrac{b}{2a}\right)^2 + b\left(-\dfrac{b}{2a}\right) + c = -\dfrac{b^2 - 4ac}{4a}$. The domain of g is by definition $\left[-\dfrac{b}{2a}, \infty\right)$. It is now apparent that in the upward-opening case, the range of g is $\left[-\dfrac{b^2 - 4ac}{4a}, \infty\right)$. As was stated in Section 2.4, it is a general fact that the domain of g^{-1} is the range of g. Let us verify that claim in this situation by considering the domain of $\dfrac{-b + \sqrt{b^2 - 4a(c - x)}}{2a}$ directly. The domain is the set of all x for which $b^2 - 4a(c - x) \geq 0$. Evidently, this inequality is equivalent to $b^2 - 4ac + 4ax \geq 0$, which in turn is equivalent to $4ax \geq -(b^2 - 4ac)$. In other words, $x \geq -\dfrac{b^2 - 4ac}{4a}$ as required.

To complete the exercise, observe that we have in Figure 3.23 provided the graph of $y = 2x^2 + 2x + 3$. For the task at hand, we redraw that part of the curve that lies to the right of the axis of symmetry. This provides the graph of g. Next we provide a dotted copy of the line $y = x$. Finally, we draw the mirror image of g in the line $y = x$ to obtain the graph of g^{-1} as in Figure 3.24.

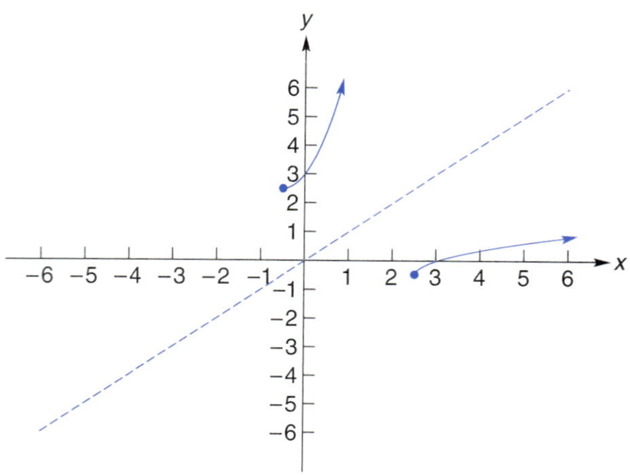

FIGURE 3.24 Graphs of g and g^{-1}.

APPLICATIONS IN PRACTICE 3

Maximum Revenue

The demand function for a publisher's line of cookbooks is $p = 6 - 0.003q$, where p is the price (in dollars) per unit when q units are demanded (per day) by consumers. Find the level of production that will maximize the manufacturer's total revenue, and determine this revenue.

EXAMPLE 6 **Maximum Revenue**

The demand function for a manufacturer's product is $p = 1000 - 2q$, where p is the price (in dollars) per unit when q units are demanded (per week) by consumers. Find the level of production that will maximize the manufacturer's total revenue, and determine this revenue.

Solution:

Strategy: To maximize revenue, we must determine the revenue function, $r = f(q)$. Using the relation

$$\textbf{total revenue} = \textbf{(price)(quantity)}$$

we have

$$r = pq$$

Using the demand equation, we can express p in terms of q, so r will be strictly a function of q.

The formula for total revenue should be added to your repertoire of relationships in business and economics.

We have

$$\begin{aligned} r &= pq \\ &= (1000 - 2q)q \\ r &= 1000q - 2q^2 \end{aligned}$$

Note that r is a quadratic function of q, with $a = -2$, $b = 1000$, and $c = 0$. Since $a < 0$ (the parabola opens downward), r is maximum at the vertex (q, r), where

$$q = -\frac{b}{2a} = -\frac{1000}{2(-2)} = 250$$

The maximum value of r is given by

$$\begin{aligned} r &= 1000(250) - 2(250)^2 \\ &= 250{,}000 - 125{,}000 = 125{,}000 \end{aligned}$$

Thus, the maximum revenue that the manufacturer can receive is \$125,000, which occurs at a production level of 250 units. Figure 3.25(a) shows the graph of the revenue function. Only that portion for which $q \geq 0$ and $r \geq 0$ is drawn, since quantity and revenue cannot be negative.

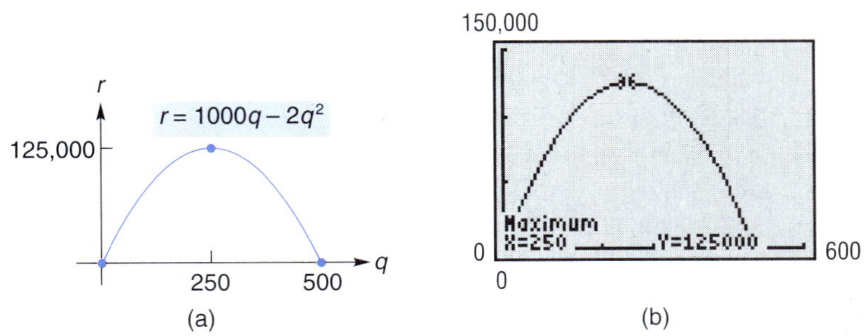

FIGURE 3.25 Graph of revenue function.

TECHNOLOGY

The maximum (or minimum) value of a function can be conveniently found with a graphing calculator either by using trace and zoom or by using the "maximum" (or "minimum") feature. Figure 3.25(b) shows the display for the revenue function of Example 6, namely, the graph of $y = 1000x - 2x^2$. Note that we replaced r by y and q by x.

Exercise 3.3

In Problems 1–8, state whether the function is quadratic.

1. $f(x) = 5x^2$

2. $g(x) = \dfrac{1}{2x^2 - 4}$

3. $g(x) = 7 - 6x$

4. $k(v) = 3v^2(v^2 + 2)$

5. $h(q) = (q + 4)^2$

6. $f(t) = 2t(3 - t) + 4t$

7. $f(s) = \dfrac{s^2 - 9}{2}$

8. $g(t) = (t^2 - 1)^2$

In Problems 9–12, do not include a graph.

9. (a) For the parabola $y = f(x) = -4x^2 + 8x + 7$, find the vertex. (b) Does the vertex correspond to the highest point or the lowest point on the graph?

10. Repeat Problem 9 if $y = f(x) = 8x^2 + 4x - 1$.

11. For the parabola $y = f(x) = x^2 + x - 6$, find (a) the y-intercept, (b) the x-intercepts, and (c) the vertex.

12. Repeat Problem 11 if $y = f(x) = 3 + x - 2x^2$.

In Problems 13–22, graph each function. Give the vertex and intercepts, and state the range.

*13. $y = f(x) = x^2 - 6x + 5$

14. $y = f(x) = -4x^2$

*15. $y = g(x) = -2x^2 - 6x$

16. $y = f(x) = x^2 - 4$

*17. $s = h(t) = t^2 + 2t + 1$

18. $s = h(t) = 2t^2 + 3t - 2$

19. $y = f(x) = -9 + 8x - 2x^2$

20. $y = H(x) = 1 - x - x^2$

*21. $t = f(s) = s^2 - 8s + 14$

22. $t = f(s) = s^2 + 6s + 11$

In Problems 23–26, state whether f(x) has a maximum value or a minimum value, and find that value.

23. $f(x) = 49x^2 - 10x + 17$

24. $f(x) = -2x^2 - 16x + 3$

25. $f(x) = 4x - 50 - 0.1x^2$

26. $f(x) = x(x + 3) - 12$

27. Revenue The demand function for a manufacturer's product is $p = f(q) = 1200 - 3q$, where p is the price (in dollars) per unit when q units are demanded (per week). Find the level of production that maximizes the manufacturer's total revenue and determine this revenue.

28. Revenue The demand function for an office supply company's line of plastic rulers is $p = 0.85 - 0.00045q$, where p is the price (in dollars) per unit when q units are demanded (per day) by consumers. Find the level of production that will maximize the manufacturer's total revenue, and determine this revenue.

29. Revenue The demand function for an electronics company's laptop computer line is $p = 2400 - 6q$, where p is the price (in dollars) per unit when q units are demanded (per week) by consumers. Find the level of production that will maximize the manufacturer's total revenue, and determine this revenue.

30. Marketing A marketing firm estimates that n months after the introduction of a client's new product, $f(n)$ thousand households will use it, where

$$f(n) = \frac{10}{9}n(12 - n), \quad 0 \le n \le 12.$$

Estimate the maximum number of households that will use the product.

31. Profit The daily profit for the garden department of a store from the sale of trees is given by $P(x) = -x^2 + 18x + 144$, where x is the number of trees sold. Find the function's vertex and intercepts, and graph the function.

32. Psychology A prediction made by early psychology relating the magnitude of a stimulus, x, to the magnitude of a response, y, is expressed by the equation $y = kx^2$, where k is a constant of the experiment. In an experiment on pattern recognition, $k = 2$. Find the function's vertex and graph the equation. (Assume no restriction on x.)

33. Biology Biologists studied the nutritional effects on rats that were fed a diet containing 10% protein.[5] The protein consisted of yeast and corn flour. By varying the percentage, P, of yeast in the protein mix, the group estimated that the average weight gain (in grams) of a rat over a period of time was

$$f(P) = -\frac{1}{50}P^2 + 2P + 20, \quad 0 \le P \le 100.$$

Find the maximum weight gain.

34. Height of Ball Suppose that the height, s, of a ball thrown vertically upward from the ground is given by

$$s = -4.9t^2 + 58.8t,$$

where s is in meters and t is elapsed time in seconds. (See Figure 3.26.) After how many seconds will the ball reach its maximum height? What is the maximum height?

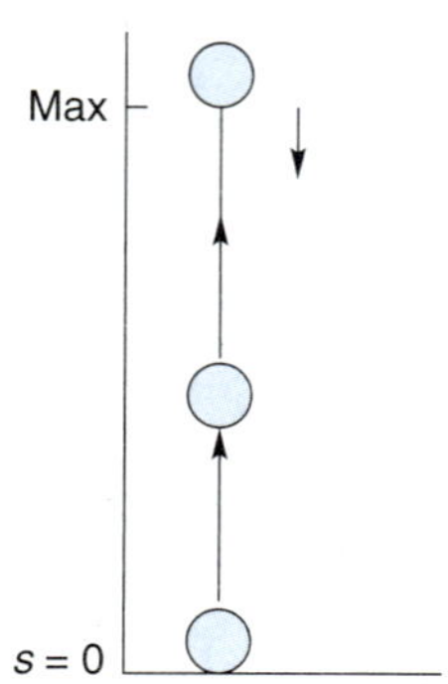

FIGURE 3.26 Ball thrown upward (Problem 34).

[5]Adapted from R. Bressani, "The Use of Yeast in Human Foods," in *Single-Cell Protein,* ed. R. I. Mateles and S. R. Tannenbaum (Cambridge, MA: MIT Press, 1968).

35. Archery A boy standing on a hill shoots an arrow straight up with an initial velocity of 85 feet per second. The height, h, of the arrow in feet, t seconds after it was released, is described by the function $h(t) = -16t^2 + 85t + 22$. What is the maximum height reached by the arrow? How many seconds after release does it take to reach this height?

36. Toy Toss A 6-year-old girl standing on a toy chest throws a doll straight up with an initial velocity of 16 feet per second. The height h of the doll in feet, t seconds after it was released, is described by the function $h(t) = -16t^2 + 16t + 4$. How long does it take the doll to reach its maximum height? What is the maximum height?

37. Rocket Launch A toy rocket is launched straight up from the roof of a garage with an initial velocity of 80 feet per second. The height, h, of the rocket in feet, t seconds after it was released, is described by the function $h(t) = -16t^2 + 80t + 16$. Find the function's vertex and intercepts, and graph the function.

38. Area Express the area of the rectangle shown in Figure 3.27 as a quadratic function of x. For what value of x will the area be a maximum?

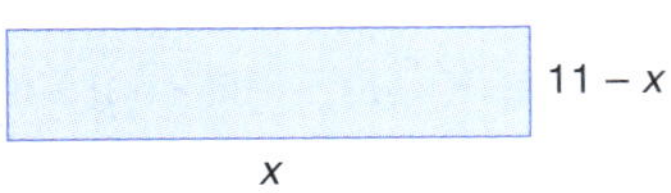

FIGURE 3.27 Diagram for Problem 38.

39. Enclosing Plot A building contractor wants to fence in a rectangular plot adjacent to a straight stream by using the stream for one side of the enclosed area. (See Figure 3.28.) If the contractor has 200 feet of fence, find the dimensions of the maximum enclosed area.

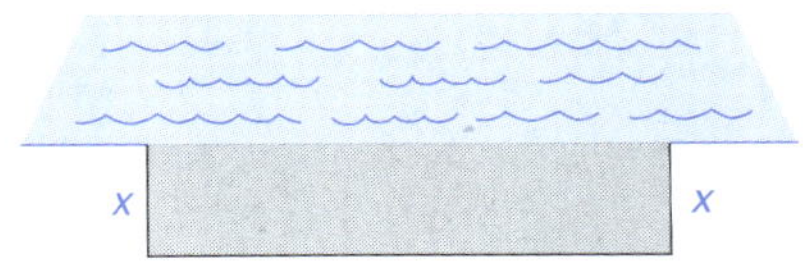

FIGURE 3.28 Diagram for Problem 39.

40. Find two numbers whose sum is 78 and whose product is a maximum.

41. From the graph of $y = 1.4x^2 - 3.1x + 4.6$, determine the coordinates of the vertex. Round values to two decimal places. Verify your answer by using the vertex formula.

42. Find the zeros of $f(x) = -\sqrt{2}x^2 + 3x + 8.5$ by examining the graph of f. Round values to two decimal places.

43. Determine the number of real zeros for each of the following quadratic functions:

(a) $f(x) = 4.2x^2 - 8.1x + 10.4$

(b) $f(x) = 5x^2 - 2\sqrt{35}x + 7$

(c) $f(x) = \dfrac{5.1 - 7.2x - x^2}{4.8}$

44. Find the maximum value (rounded to two decimal places) of the function $f(x) = 5.4 + 12x - 4.1x^2$ from its graph.

45. Find the minimum value (rounded to two decimal places) of the function $f(x) = 20x^2 - 13x + 7$ from its graph.

OBJECTIVE *To solve systems of linear equations in both two and three variables by using the technique of elimination by addition or by substitution. (In Chapter 6, other methods are shown.)*

3.4 Systems of Linear Equations

Two-Variable Systems

When a situation must be described mathematically, it is not unusual for a *set* of equations to arise. For example, suppose that the manager of a factory is setting up a production schedule for two models of a new product. Model A requires 4 widgets and 9 klunkers. Model B requires 5 widgets and 14 klunkers. From its suppliers, the factory gets 335 widgets and 850 klunkers each day. How many of each model should the manager plan to make each day so that all the widgets and klunkers are used?

It's a good idea to construct a table that summarizes the important information. Table 3.2 shows the number of widgets and klunkers required for each model, as well as the total number available.

Suppose we let x be the number of model A made each day and y be the number of model B. Then these require a total of $4x + 5y$ widgets and $9x + 14y$ klunkers. Since 335 widgets and 850 klunkers are available, we have

$$\begin{cases} 4x + 5y = 335 & (1) \\ 9x + 14y = 850 & (2) \end{cases}$$

TABLE 3.2

	Model A	Model B	Total Available
Widgets	4	5	335
Klunkers	9	14	850

We call this set of equations a **system** of two linear equations in the variables (or unknowns) x and y. The problem is to find values of x and y for which *both* equations are true *simultaneously*. These are called *solutions* of the system.

Since Equations (1) and (2) are linear, their graphs are straight lines; call these lines L_1 and L_2. Now, the coordinates of any point on a line satisfy the equation of that line; that is, they make the equation true. Thus, the coordinates of any point of intersection of L_1 and L_2 will satisfy both equations. This means that a point of intersection gives a solution of the system.

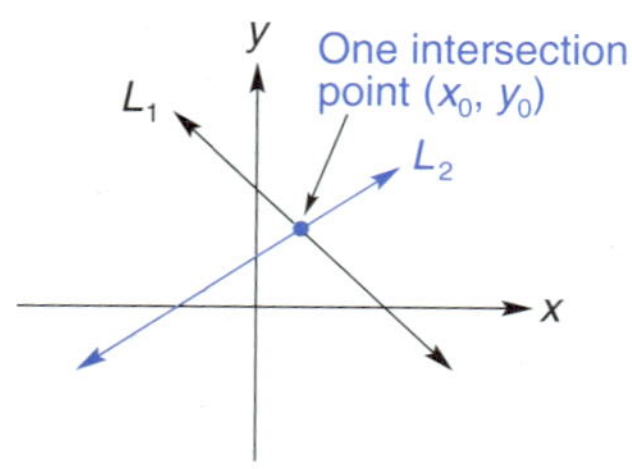

FIGURE 3.29 Linear system (one solution).

If L_1 and L_2 are drawn on the same plane, there are three situations that could occur:

1. L_1 and L_2 may intersect at exactly one point, say, (x_0, y_0). (See Figure 3.29.) Thus, the system has the solution $x = x_0$ and $y = y_0$.
2. L_1 and L_2 may be parallel and have no points in common. (See Figure 3.30.) In this case, there is no solution.
3. L_1 and L_2 may be the same line. (See Figure 3.31.) Here the coordinates of any point on the line are a solution of the system. Consequently, there are infinitely many solutions.

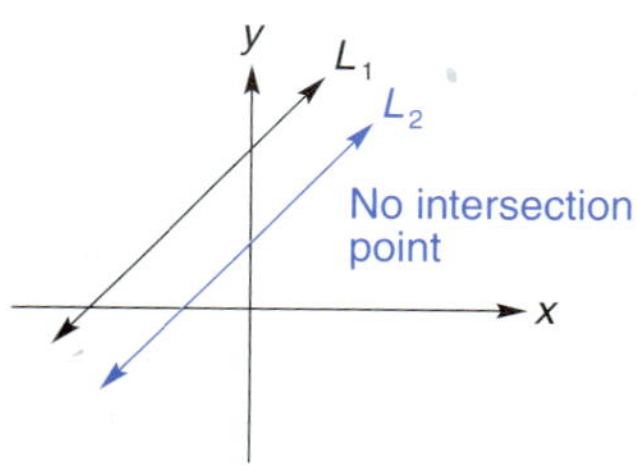

FIGURE 3.30 Linear system (no solution).

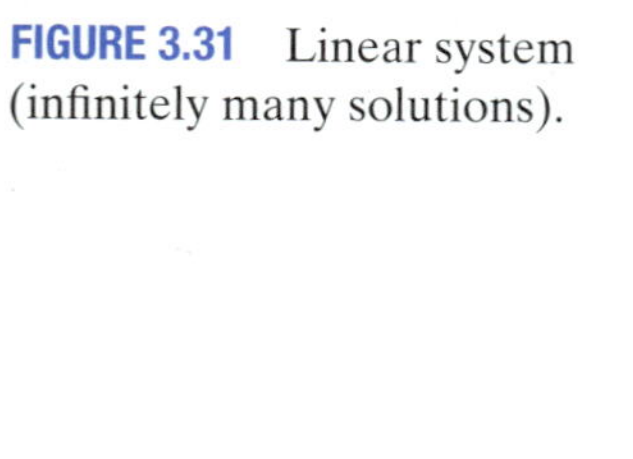

FIGURE 3.31 Linear system (infinitely many solutions).

Our main concern in this section is algebraic methods of solving a system of linear equations. Essentially, we successively replace the system by other systems that have the same solutions. Generalizing the terminology of Section 0.7 on equivalent equations, we say that two systems are *equivalent* if their sets of solutions are equal. The replacement systems have progressively more desirable forms for determining the solution. More precisely, we seek an equivalent system containing an equation in which one of the variables does not appear. (That is, one of the variables is eliminated.) In dealing with systems of *linear* equations, our passage from a system to an equivalent system will always be accomplished by one of the following procedures:

1. Interchanging two equations,
2. Multiplying one equation by a nonzero constant,
3. Replacing an equation by itself plus a multiple of another equation.

We will return to these procedures in more detail in Chapter 6. For the moment, since we will also consider nonlinear systems in this chapter, it is convenient to express our solutions in terms of the very general principles of Section 0.7 that guarantee equivalence of equations.

We shall illustrate the elimination procedure for the system in the problem originally posed:

$$\begin{cases} 4x + 5y = 335 & \textbf{(1)} \\ 9x + 14y = 850 & \textbf{(2)} \end{cases}$$

To begin, we shall obtain an equivalent system in which x does not appear in one equation. First we find an equivalent system in which the coefficients of the x-terms in each equation are the same except for their sign. Multiplying Equation (1) by 9 [that is, multiplying both sides of Equation (1) by 9] and multiplying Equation (2) by -4 gives

$$\begin{cases} 36x + 45y = 3015 & (3) \\ -36x - 56y = -3400 & (4) \end{cases}$$

The left and right sides of Equation (4) are equal, so each side can be *added* to the corresponding side of Equation (3). This results in

$$-11y = -385$$

which has only one variable, as planned. Solving gives

$$y = 35$$

so we obtain the equivalent system

$$\begin{cases} y = 35 & (5) \\ -36x - 56y = -3400 & (6) \end{cases}$$

Replacing y in Equation (6) by 35, we get

$$\begin{aligned} -36x - 56(35) &= -3400 \\ -36x - 1960 &= -3400 \\ -36x &= -1440 \\ x &= 40 \end{aligned}$$

Thus, the original system is equivalent to

$$\begin{cases} y = 35 & (7) \\ x = 40 & (8) \end{cases}$$

We can check our answer by substituting $x = 40$ and $y = 35$ into *both* of the original equations. In Equation (1), we get $4(40)+5(35) = 335$, or $335 = 335$. In Equation (2), we get $9(40) + 14(35) = 850$, or $850 = 850$. Hence, the solution is

$$x = 40 \quad \text{and} \quad y = 35$$

Each day the manager should plan to make 40 of model A and 35 of model B. Our procedure is referred to as **elimination by addition.** Although we chose to eliminate x first, we could have done the same for y by a similar procedure.

APPLICATIONS IN PRACTICE 1

Elimination-by-Addition Method

A computer consultant has \$200,000 invested for retirement, part at 9% and part at 8%. If the total yearly income from the investments is \$17,200, how much is invested at each rate?

EXAMPLE 1 Elimination-by-Addition Method

Use elimination by addition to solve the system.

$$\begin{cases} 3x - 4y = 13 \\ 3y + 2x = 3 \end{cases}$$

Solution: Aligning the x- and y-terms for convenience gives

$$\begin{cases} 3x - 4y = 13 & (9) \\ 2x + 3y = 3 & (10) \end{cases}$$

To eliminate y, we multiply Equation (9) by 3 and Equation (10) by 4:

$$\begin{cases} 9x - 12y = 39 & \textbf{(11)} \\ 8x + 12y = 12 & \textbf{(12)} \end{cases}$$

Adding Equation (11) to Equation (12) gives $17x = 51$, from which $x = 3$. We have the equivalent system

$$\begin{cases} 9x - 12y = 39 & \textbf{(13)} \\ x = 3 & \textbf{(14)} \end{cases}$$

Replacing x by 3 in Equation (13) results in

$$9(3) - 12y = 39$$
$$-12y = 12$$
$$y = -1$$

so the original system is equivalent to

$$\begin{cases} y = -1 \\ x = 3 \end{cases}$$

The solution is $x = 3$ and $y = -1$. Figure 3.32 shows a graph of the system.

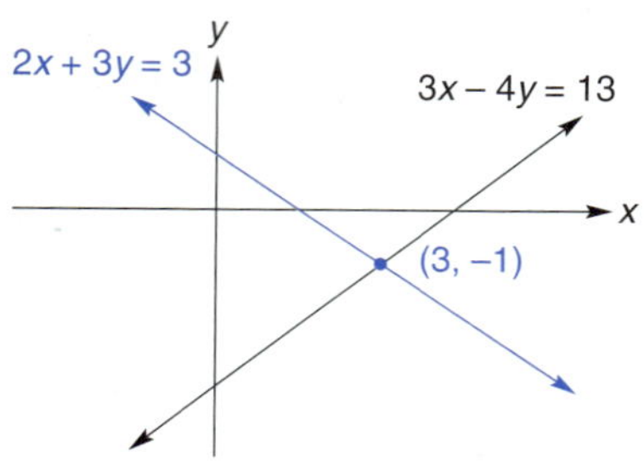

FIGURE 3.32 Linear system of Example 1: one solution.

▸ **Now work Problem 1.**

The system in Example 1,

$$\begin{cases} 3x - 4y = 13 & \textbf{(15)} \\ 2x + 3y = 3 & \textbf{(16)} \end{cases}$$

can be solved another way. We first choose one of the equations—for example, Equation (15)—and solve it for one unknown in terms of the other, say x in terms of y. Here, Equation (15) is equivalent to $3x = 4y + 13$, or

$$x = \frac{4}{3}y + \frac{13}{3}$$

and we obtain

$$\begin{cases} x = \frac{4}{3}y + \frac{13}{3} & \textbf{(17)} \\ 2x + 3y = 3 & \textbf{(18)} \end{cases}$$

Substituting the right side of Equation (17) for x in Equation (18) gives

$$2\left(\frac{4}{3}y + \frac{13}{3}\right) + 3y = 3 \qquad \textbf{(19)}$$

Thus, x has been eliminated. Solving Equation (19), we have

$$\frac{8}{3}y + \frac{26}{3} + 3y = 3$$
$$8y + 26 + 9y = 9 \qquad \text{(clearing fractions)}$$
$$17y = -17$$
$$y = -1$$

Replacing y in Equation (17) by -1 gives $x = 3$, and the original system is equivalent to

$$\begin{cases} x = 3 \\ y = -1 \end{cases}$$

as before. This method is called **elimination by substitution.**

APPLICATIONS IN PRACTICE 2

Method of Elimination by Substitution

Two species of deer, A and B, living in a wildlife refuge are given extra food in the winter. Each week, they receive 2 tons of food pellets and 4.75 tons of hay. Each deer of species A requires 4 pounds of the pellets and 5 pounds of hay. Each deer of species B requires 2 pounds of the pellets and 7 pounds of hay. How many of each species of deer will the food support so that all of the food is consumed each week?

EXAMPLE 2 Method of Elimination by Substitution

Use elimination by substitution to solve the system

$$\begin{cases} x + 2y - 8 = 0 \\ 2x + 4y + 4 = 0 \end{cases}$$

Solution: It is easy to solve the first equation for x. Doing so gives the equivalent system

$$\begin{cases} x = -2y + 8 & \textbf{(20)} \\ 2x + 4y + 4 = 0 & \textbf{(21)} \end{cases}$$

Substituting $-2y + 8$ for x in Equation (21) yields

$$2(-2y + 8) + 4y + 4 = 0$$
$$-4y + 16 + 4y + 4 = 0$$

The latter equation simplifies to $20 = 0$. Thus, we have the system

$$\begin{cases} x = -2y + 8 & \textbf{(22)} \\ 20 = 0 & \textbf{(23)} \end{cases}$$

Since Equation (23) is *never* true, there is **no solution** of the original system. The reason is clear if we observe that the original equations can be written in slope–intercept form as

$$y = -\frac{1}{2}x + 4$$

and

$$y = -\frac{1}{2}x - 1$$

These equations represent straight lines having slopes of $-\frac{1}{2}$, but different y-intercepts, namely, 4 and -1. That is, they determine different parallel lines. (See Figure 3.33.)

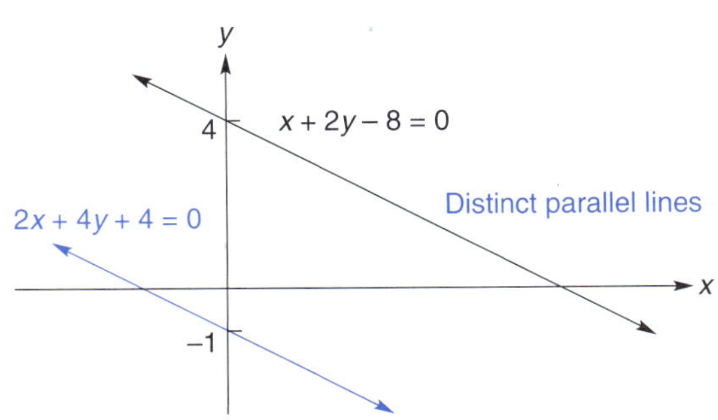

FIGURE 3.33 Linear system of Example 2: no solution.

▶ **Now work Problem 9.**

APPLICATIONS IN PRACTICE 3

A Linear System with Infinitely Many Solutions

Two species of fish, A and B, are raised in one pond at a fish farm where they are fed two vitamin supplements. Each day, they receive 100 grams of the first supplement and 200 grams of the second supplement. Each fish of species A requires 15 mg of the first supplement and 30 mg of the second supplement. Each fish of species B requires 20 mg of the first supplement and 40 mg of the second supplement. How many of each species of fish will the pond support so that all of the supplements are consumed each day?

EXAMPLE 3 A Linear System with Infinitely Many Solutions

Solve

$$\begin{cases} x + 5y = 2 & (24) \\ \frac{1}{2}x + \frac{5}{2}y = 1 & (25) \end{cases}$$

Solution: We begin by eliminating x from the second equation. Multiplying Equation (25) by -2, we have

$$\begin{cases} x + 5y = 2 & (26) \\ -x - 5y = -2 & (27) \end{cases}$$

Adding Equation (26) to Equation (27) gives

$$\begin{cases} x + 5y = 2 & (28) \\ 0 = 0 & (29) \end{cases}$$

Because Equation (29) is *always* true, any solution of Equation (28) is a solution of the system. Now let us see how we can express our answer. From Equation (28), we have $x = 2 - 5y$, where y can be any real number, say, r. Thus, we can write $x = 2 - 5r$. The complete solution is

$$x = 2 - 5r$$

$$y = r$$

where r is any real number. In this situation r is called a **parameter,** and we say that we have a one-parameter family of solutions. Each value of r determines a particular solution. For example, if $r = 0$, then $x = 2$ and $y = 0$ is a solution; if $r = 5$, then $x = -23$ and $y = 5$ is another solution. Clearly, the given system has infinitely many solutions.

It is worthwhile to note that by writing Equations (24) and (25) in their slope–intercept forms, we get the equivalent system

$$\begin{cases} y = -\frac{1}{5}x + \frac{2}{5} \\ y = -\frac{1}{5}x + \frac{2}{5} \end{cases}$$

in which both equations represent the same line. Hence, the lines coincide (Figure 3.34), and Equations (24) and (25) are equivalent. The solution of the system consists of the coordinate pairs of all points on the line $x + 5y = 2$, and these points are given by our parametric solution.

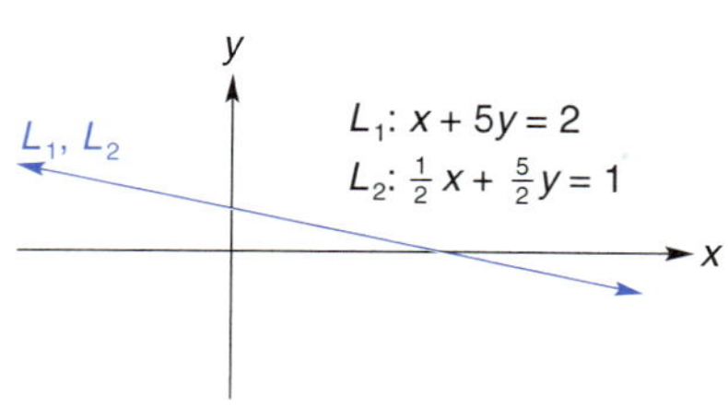

FIGURE 3.34 Linear system of Example 3: infinitely many solutions.

▶ **Now work Problem 13.**

TECHNOLOGY

Graphically solve the system

$$\begin{cases} 9x + 4.1y = 7 \\ 2.6x - 3y = 18 \end{cases}$$

Solution: First we solve each equation for y, so that each equation has the form $y = f(x)$:

$$y = \frac{1}{4.1}(7 - 9x)$$

$$y = -\frac{1}{3}(18 - 2.6x)$$

Next we enter these functions as Y_1 and Y_2 and display them on the same viewing rectangle. (See Figure 3.35.) Finally, either using trace and zoom or using the intersection feature, we estimate the solution to be $x = 2.52$, $y = -3.82$.

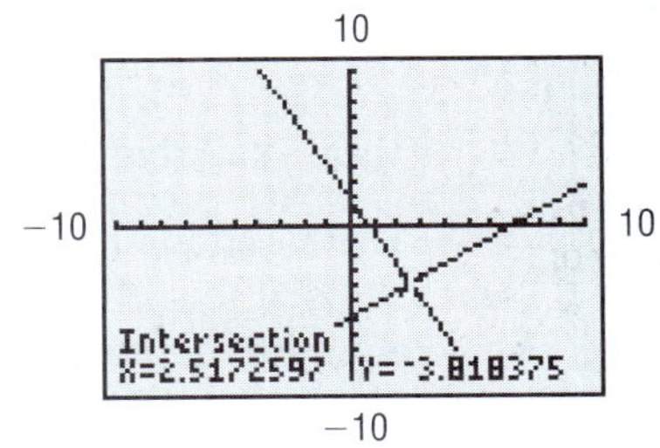

FIGURE 3.35 Graphical solution of system.

EXAMPLE 4 **Mixture**

A chemical manufacturer wishes to fill an order for 500 liters of a 25% acid solution. (Twenty-five percent by volume is acid.) If solutions of 30% and 18% are available in stock, how many liters of each must be mixed to fill the order?

Solution: Let x and y be the number of liters of the 30% and 18% solutions, respectively, that should be mixed. Then

$$x + y = 500$$

To help visualize the situation, we draw the diagram in Figure 3.36. In 500 liters of a 25% solution, there will be $0.25(500) = 125$ liters of acid. This acid comes from two sources: $0.30x$ liters of it come from the 30% solution, and $0.18y$ liters of it come from the 18% solution. Hence,

$$0.30x + 0.18y = 125$$

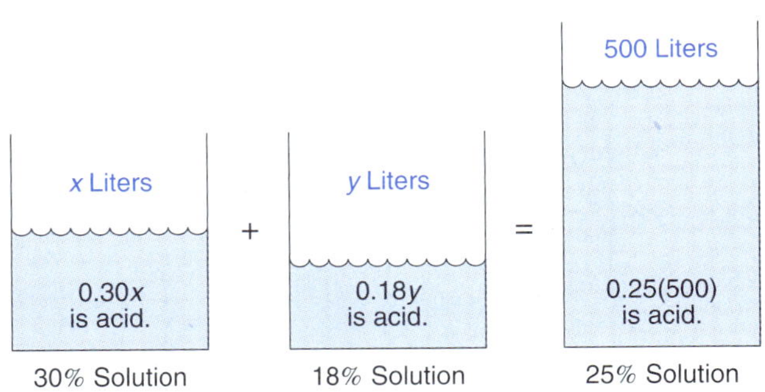

FIGURE 3.36 Mixture problem.

These two equations form a system of two linear equations in two unknowns.

Solving the first for x gives $x = 500 - y$. Substituting in the second gives

$$0.30(500 - y) + 0.18y = 125$$

Solving this equation for y, we find that $y = 208\frac{1}{3}$ liters. Thus, $x = 500 - 208\frac{1}{3} = 291\frac{2}{3}$ liters. (See Figure 3.37.)

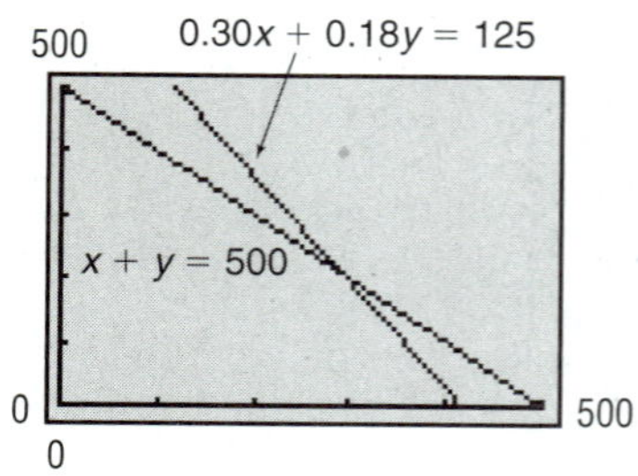

FIGURE 3.37 Graph for Example 4.

▶ **Now work Problem 25.**

Three-Variable Systems

The methods used in solving a two-variable system of linear equations can be used to solve a three-variable system of linear equations. A **general linear equation in the three variables x, y, and z** is an equation having the form

$$Ax + By + Cz = D$$

where A, B, C, and D are constants and A, B, and C are not all zero. For example, $2x - 4y + z = 2$ is such an equation. Geometrically, a general linear equation in three variables represents a *plane* in space, and a solution to a system of such equations is the intersection of planes. Example 5 shows how to solve a system of three linear equations in three variables.

APPLICATIONS IN PRACTICE 4

Solving a Three-Variable Linear System

A coffee shop specializes in blending gourmet coffees. From type A, type B, and type C coffees, the owner wants to prepare a blend that will sell for \$8.50 for a 1-pound bag. The cost per pound of these coffees is \$12, \$9, and \$7, respectively. The amount of type B is to be twice the amount of type A. How much of each type of coffee will be in the final blend?

EXAMPLE 5 **Solving a Three-Variable Linear System**

Solve

$$\begin{cases} 2x + y + z = 3 & (30) \\ -x + 2y + 2z = 1 & (31) \\ x - y - 3z = -6 & (32) \end{cases}$$

Solution: This system consists of three linear equations in three variables. From Equation (32), $x = y + 3z - 6$. By substituting for x in Equations (30) and (31), we obtain

$$\begin{cases} 2(y + 3z - 6) + y + z = 3 \\ -(y + 3z - 6) + 2y + 2z = 1 \\ x = y + 3z - 6 \end{cases}$$

Simplifying gives

$$\begin{cases} 3y + 7z = 15 & (33) \\ y - z = -5 & (34) \\ x = y + 3z - 6 \end{cases}$$

Note that x does not appear in Equations (33) and (34). Since any solution of the original system must satisfy Equations (33) and (34), we shall consider their solution first:

$$\begin{cases} 3y + 7z = 15 & (33) \\ y - z = -5 & (34) \end{cases}$$

From Equation (34), $y = z - 5$. This means that we can replace Equation (33) by

$$3(z - 5) + 7z = 15 \qquad \text{that is} \qquad z = 3$$

Since z is 3, we can replace Equation (34) with $y = -2$. Hence, the previous system is equivalent to

$$\begin{cases} z = 3 \\ y = -2 \end{cases}$$

The original system becomes

$$\begin{cases} z = 3 \\ y = -2 \\ x = y + 3z - 6 \end{cases}$$

from which $x = 1$. The solution is $x = 1$, $y = -2$, and $z = 3$, which you should verify.

▶ **Now work Problem 15.**

Just as a two-variable system may have a one-parameter family of solutions, a three-variable system may have a one- or two-parameter family of solutions.[6] The next two examples illustrate.

EXAMPLE 6 **One-Parameter Family of Solutions**

Solve

$$\begin{cases} x - 2y = 4 & (35) \\ 2x - 3y + 2z = -2 & (36) \\ 4x - 7y + 2z = 6 & (37) \end{cases}$$

Solution: Note that since Equation (35) can be written $x - 2y + 0z = 4$, we can view Equations (35) to (37) as a system of three linear equations in the variables x, y, and z. From Equation (35), we have $x = 2y + 4$. Using this equation and substitution, we can eliminate x from Equations (36) and (37):

$$\begin{cases} x = 2y + 4, \\ 2(2y + 4) - 3y + 2z = -2 \\ 4(2y + 4) - 7y + 2z = 6 \end{cases}$$

Or, more simply,

$$\begin{cases} x = 2y + 4 & (38) \\ y + 2z = -10 & (39) \\ y + 2z = -10 & (40) \end{cases}$$

Multiplying Equation (40) by -1 gives

$$\begin{cases} x = 2y + 4 \\ y + 2z = -10 \\ -y - 2z = 10 \end{cases}$$

Adding the second equation to the third yields

$$\begin{cases} x = 2y + 4 \\ y + 2z = -10 \\ 0 = 0 \end{cases}$$

[6]Note to instructor: Examples 6 and 7 may be omitted without loss of continuity.

Since the equation $0 = 0$ is always true, we can essentially deal with the system

$$\begin{cases} x = 2y + 4 & \textbf{(41)} \\ y + 2z = -10 & \textbf{(42)} \end{cases}$$

Solving Equation (42) for y, we have

$$y = -10 - 2z$$

which expresses y in terms of z. We can also express x in terms of z. From Equation (41),

$$\begin{aligned} x &= 2y + 4 \\ &= 2(-10 - 2z) + 4 \\ &= -16 - 4z \end{aligned}$$

Thus, we have

$$\begin{cases} x = -16 - 4z \\ y = -10 - 2z \end{cases}$$

Since no restriction is placed on z, this suggests a parametric family of solutions. Setting $z = r$, we have the following family of solutions of the given system:

$$\begin{aligned} x &= -16 - 4r \\ y &= -10 - 2r \\ z &= r \end{aligned}$$

Other parametric representations of the solution are possible.

where r can be any real number. We see, then, that the given system has infinitely many solutions. For example, setting $r = 1$ gives the particular solution $x = -20$, $y = -12$, and $z = 1$.

▸ **Now work Problem 19.**

EXAMPLE 7 Two-Parameter Family of Solutions

Solve the system

$$\begin{cases} x + 2y + z = 4 \\ 2x + 4y + 2z = 8 \end{cases}$$

Solution: This is a system of two linear equations in three variables. We shall eliminate x from the second equation by first multiplying that equation by $-\frac{1}{2}$:

$$\begin{cases} x + 2y + z = 4 \\ -x - 2y - z = -4 \end{cases}$$

Adding the first equation to the second gives

$$\begin{cases} x + 2y + z = 4 \\ 0 = 0 \end{cases}$$

From the first equation, we obtain

$$x = 4 - 2y - z$$

Since no restriction is placed on either y or z, they can be arbitrary real numbers, giving us a two-parameter family of solutions. Setting $y = r$ and $z = s$, we find that the solution of the given system is

$$x = 4 - 2r - s$$
$$y = r$$
$$z = s$$

where r and s can be any real numbers. Each assignment of values to r and s results in a solution of the given system, so there are infinitely many solutions. For example, letting $r = 1$ and $s = 2$ gives the particular solution $x = 0$, $y = 1$, and $z = 2$.

▶ **Now work Problem 23.**

Exercise 3.4

In Problems 1–24, solve the systems algebraically.

*1. $\begin{cases} x + 4y = 3 \\ 3x - 2y = -5 \end{cases}$

2. $\begin{cases} 4x + 2y = 9 \\ 5y - 4x = 5 \end{cases}$

3. $\begin{cases} 3x - 4y = 13 \\ 2x + 3y = 3 \end{cases}$

4. $\begin{cases} 2x - y = 1 \\ -x + 2y = 7 \end{cases}$

5. $\begin{cases} u + v = 5 \\ u - v = 7 \end{cases}$

6. $\begin{cases} -p - q = -3 \\ 3p + 2q = 19 \end{cases}$

7. $\begin{cases} x - 2y = -7 \\ 5x + 3y = -9 \end{cases}$

8. $\begin{cases} 3x + 5y = 7 \\ 5x + 9y = 7 \end{cases}$

*9. $\begin{cases} 4x - 3y - 2 = 3x - 7y \\ x + 5y - 2 = y + 4 \end{cases}$

10. $\begin{cases} 5x + 7y + 2 = 9y - 4x + 6 \\ \frac{21}{2}x - \frac{4}{3}y - \frac{11}{4} = \frac{3}{2}x + \frac{2}{3}y + \frac{5}{4} \end{cases}$

11. $\begin{cases} \frac{2}{3}x + \frac{1}{2}y = 2 \\ \frac{3}{8}x + \frac{5}{6}y = -\frac{11}{2} \end{cases}$

12. $\begin{cases} \frac{1}{2}z - \frac{1}{4}w = \frac{1}{6} \\ \frac{1}{2}z + \frac{1}{4}w = \frac{1}{6} \end{cases}$

*13. $\begin{cases} 4p + 12q = 6 \\ 2p + 6q = 3 \end{cases}$

14. $\begin{cases} 5x - 3y = 2 \\ -10x + 6y = 4 \end{cases}$

*15. $\begin{cases} 2x + y + 6z = 3 \\ x - y + 4z = 1 \\ 3x + 2y - 2z = 2 \end{cases}$

16. $\begin{cases} x + y + z = -1 \\ 3x + y + z = 1 \\ 4x - 2y + 2z = 0 \end{cases}$

17. $\begin{cases} x + 4y + 3z = 10 \\ 4x + 2y - 2z = -2 \\ 3x - y + z = 11 \end{cases}$

18. $\begin{cases} 3x - 2y + z = 0 \\ -2x + y - 3z = 15 \\ \frac{3}{2}x + \frac{4}{5}y + 4z = 10 \end{cases}$

*[7]19. $\begin{cases} x - 2z = 1 \\ y + z = 3 \end{cases}$

[7]20. $\begin{cases} 2y + 3z = 1 \\ 3x - 4z = 0 \end{cases}$

[7]21. $\begin{cases} x - y + 2z = 0 \\ 2x + y - z = 0 \\ x + 2y - 3z = 0 \end{cases}$

[7]22. $\begin{cases} x - 2y - z = 0 \\ 2x - 4y - 2z = 0 \\ -x + 2y + z = 0 \end{cases}$

*[7]23. $\begin{cases} 2x + 2y - z = 3 \\ 4x + 4y - 2z = 6 \end{cases}$

[7]24. $\begin{cases} 5x + y + z = 17 \\ 4x + y + z = 14 \end{cases}$

[7]Refer to concepts in Examples 6 and 7.

*25. **Mixture** A chemical manufacturer wishes to fill an order for 700 gallons of a 24% acid solution. Solutions of 20% and 30% are in stock. How many gallons of each solution must be mixed to fill the order?

26. **Mixture** A gardener has two fertilizers that contain different concentrations of nitrogen. One is 3% nitrogen and the other is 11% nitrogen. How many pounds of each should she mix to obtain 20 pounds of a 9% concentration?

27. **Fabric** A textile mill produces fabric made from different fibers. From cotton, polyester, and nylon, the owners want to produce a fabric blend that will cost $3.25 per pound to make. The cost per pound of these fibers is $4.00, $3.00, and $2.00, respectively. The amount of nylon is to be the same as the amount of polyester. How much of each fiber will be in the final fabric?

28. **Taxes** A company has taxable income of $312,000. The federal tax is 25% of that portion left after the state tax has been paid. The state tax is 10% of that portion left after the federal tax has been paid. Find the federal and state taxes.

29. **Airplane Speed** An airplane travels 900 mi in 2 h, 55 min, with the aid of a tail wind. It takes 3 h, 26 min, for the return trip, flying against the same wind. Find the speed of the airplane in still air and the speed of the wind.

30. **Speed of Raft** On a trip on a raft, it took $\frac{3}{4}$ h to travel 12 mi downstream. The return trip took $1\frac{1}{2}$ h. Find the

speed of the raft in still water and the speed of the current.

31. Furniture Sales A manufacturer of dining-room sets produces two styles: early American and contemporary. From past experience, management has determined that 20% more of the early American styles can be sold than the contemporary styles. A profit of \$250 is made on each early American set sold, whereas a profit of \$350 is made on each contemporary set. If, in the forthcoming year, management desires a total profit of \$130,000, how many units of each style must be sold?

32. Survey National Surveys was awarded a contract to perform a product-rating survey for Crispy Crackers. A total of 250 people were interviewed. National Surveys reported that 62.5% more people like Crispy Crackers than disliked them. However, the report did not indicate that 16% of those interviewed had no comment. How many of those surveyed liked Crispy Crackers? How many disliked them? How many had no comment?

33. Equalizing Cost United Products Co. manufactures calculators and has plants in the cities of Exton and Whyton. At the Exton plant, fixed costs are \$7000 per month, and the cost of producing each calculator is \$7.50. At the Whyton plant, fixed costs are \$8800 per month, and each calculator costs \$6.00 to produce. Next month, United Products must produce 1500 calculators. How many must be made at each plant if the total cost at each plant is to be the same?

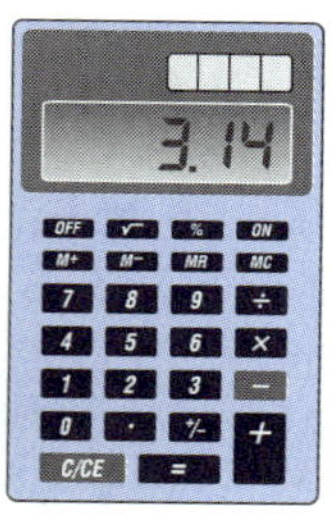

34. Coffee Blending A coffee wholesaler blends together three types of coffee that sell for \$2.20, \$2.30, and \$2.60 per pound, so as to obtain 100 lb of coffee worth \$2.40 per pound. If the wholesaler uses the same amount of the two higher priced coffees, how much of each type must be used in the blend?

35. Commissions A company pays its salespeople on a basis of a certain percentage of the first \$100,000 in sales, plus a certain percentage of any amount over \$100,000 in sales. If one salesperson earned \$8500 on sales of \$175,000 and another salesperson earned \$14,800 on sales of \$280,000, find the two percentages.

36. Yearly Profits In news reports, profits of a company this year (T) are often compared with those of last year (L), but actual values of T and L are not always given. This year, a company had profits of \$25 million more than last year. The profits were up 30%. Determine T and L from these data.

37. Production Universal Control Co. makes industrial control units. Their new models are the Argon I and the Argon II. To make each Argon I unit, they use 6 doodles and 3 skeeters. To make each Argon II unit, they use 10 doodles and 8 skeeters. The company receives a total of 760 doodles and 500 skeeters each day from its supplier. How many units of each model of the Argon can the company make each day? Assume that all the parts are used.

38. Investments A person made two investments, and the percentage return per year on each was the same. Of the total amount invested, $\frac{3}{10}$ of it plus \$600 was invested in one venture, and at the end of 1 year the person received a return of \$384 from that venture. If the total return after 1 year was \$1120, find the total amount invested.

39. Production Run A company makes three types of patio furniture: chairs, rockers, and chaise lounges. Each requires wood, plastic, and aluminum, in the amounts shown in the following table. The company has in stock 400 units of wood, 600 units of plastic, and 1500 units of aluminum. For its end-of-the-season production run, the company wants to use up all the stock. To do this, how many chairs, rockers, and chaise lounges should it make?

	Wood	Plastic	Aluminum
Chair	1 unit	1 unit	2 units
Rocker	1 unit	1 unit	3 units
Chaise lounge	1 unit	2 units	5 units

40. Investments A total of \$35,000 was invested at three interest rates: 7, 8, and 9%. The interest for the first year was \$2830, which was not reinvested. The second year the amount originally invested at 9% earned 10% instead, and the other rates remained the same. The total interest the second year was \$2960. How much was invested at each rate?

41. Hiring Workers A company pays skilled workers in its assembly department \$16 per hour. Semiskilled workers in that department are paid \$9.50 per hour. Shipping clerks are paid \$10 per hour. Because of an increase in orders, the company needs to hire a total of 70 workers in the assembly and shipping departments. It will pay a total of \$725 per hour to these employees. Because of a union contract, twice as many semiskilled workers as skilled workers must be employed. How many semiskilled workers, skilled workers, and shipping clerks should the company hire?

42. Solvent Storage A 10,000-gallon railroad tank car is to be filled with solvent from two storage tanks, A and B. Solvent from A is pumped at the rate of 20 gal/min. Solvent from B is pumped at 30 gal/min. Usually, both pumps operate at the same time. However, because of a blown fuse, the pump on A is delayed 10 minutes. How many gallons from each storage tank will be used to fill the car?

43. Verify your answer to Problem 1 by using your graphing calculator.

44. Verify your answer to Problem 11 by using your graphing calculator.

45. Graphically solve the system.

$$\begin{cases} 0.24x - 0.34y = 0.04 \\ 0.11x + 0.21y = 0.75 \end{cases}$$

46. Graphically solve the system

$$\begin{cases} x + y = 2 \\ \frac{1}{4}x + \frac{2}{5}y = \frac{3}{5} \end{cases}$$

Round the x- and y-values to two decimal places.

47. Graphically solve the system

$$\begin{cases} 0.5736x - 0.3420y = 0 \\ 0.8192x + 0.9397y = 20 \end{cases}$$

Round the x- and y-values to one decimal place.

3.5 Nonlinear Systems

OBJECTIVE *To use substitution to solve nonlinear systems.*

A system of equations in which at least one equation is not linear is called a **nonlinear system.** We can often solve a nonlinear system by substitution, as was done with linear systems. The following examples illustrate.

EXAMPLE 1 Solving a Nonlinear System

Solve

$$\begin{cases} x^2 - 2x + y - 7 = 0 & (1) \\ 3x - y + 1 = 0 & (2) \end{cases}$$

Solution:

Strategy: If a nonlinear system contains a linear equation, we usually solve the linear equation for one variable and substitute for that variable in the other equation.

Solving Equation (2) for y gives

$$y = 3x + 1 \qquad (3)$$

Substituting into Equation (1) and simplifying, we have

$$x^2 - 2x + (3x + 1) - 7 = 0$$
$$x^2 + x - 6 = 0$$
$$(x + 3)(x - 2) = 0$$
$$x = -3 \text{ or } x = 2$$

If $x = -3$, then Equation (3) implies that $y = -8$; if $x = 2$, then $y = 7$. You should verify that each pair of values satisfies the given system. Hence, the solutions are $x = -3,\ y = -8$ and $x = 2,\ y = 7$. These solutions can be seen geometrically in the graph of the system in Figure 3.38. Notice that the graph of Equation (1) is a parabola and the graph of Equation (2) is a line. The solutions correspond to the intersection points $(-3, -8)$ and $(2, 7)$.

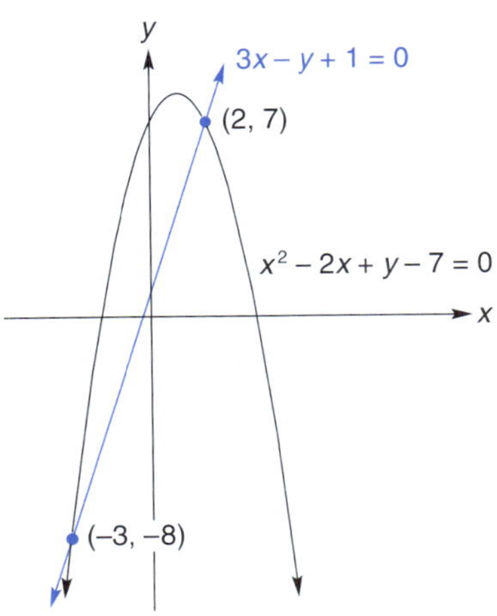

FIGURE 3.38 Nonlinear system of equations.

▸ **Now work Problem 1.**

EXAMPLE 2 **Solving a Nonlinear System**

Solve

$$\begin{cases} y = \sqrt{x+2} \\ x + y = 4 \end{cases}$$

CAUTION This example illustrates the need for checking all "solutions."

Solution: Solving the second equation, which is linear, for y gives

$$y = 4 - x \quad (4)$$

Substituting into the first equation yields

$$4 - x = \sqrt{x+2}$$
$$16 - 8x + x^2 = x + 2 \quad \text{(squaring both sides)}$$
$$x^2 - 9x + 14 = 0$$
$$(x-2)(x-7) = 0$$

Thus, $x = 2$ or $x = 7$. From Equation (4), if $x = 2$, then $y = 2$; if $x = 7$, then $y = -3$. Since we performed the operation of squaring both sides, we must check our results. Although the pair $x = 2$, $y = 2$ satisfies both of the original equations, this is not the case for $x = 7$ and $y = -3$. Thus, the solution is $x = 2$, $y = 2$. (See Figure 3.39.)

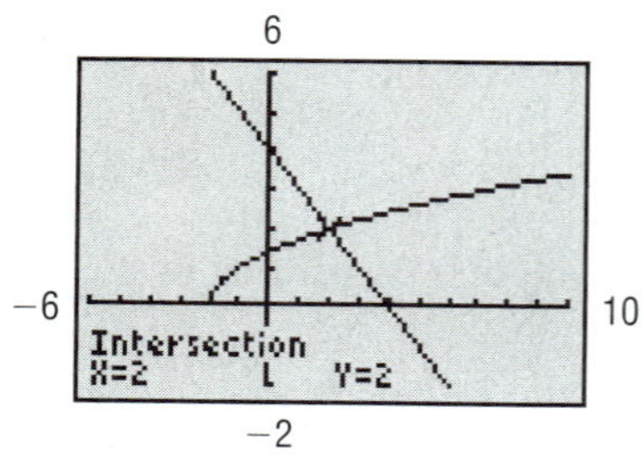

FIGURE 3.39 Nonlinear system of Example 2.

▶ **Now work Problem 13.**

TECHNOLOGY

Graphically solve the equation $0.5x^2 + x = 3$, *where* $x \geq 0$.

Solution: To solve the equation, we could find zeros of the function $f(x) = 0.5x^2 + x - 3$. Alternatively, we can think of this problem as solving the nonlinear system

$$y = 0.5x^2 + x$$
$$y = 3$$

In Figure 3.40, the intersection point is estimated to be $x = 1.65$, $y = 3$. Note that the graph of $y = 3$ is a horizontal line. The solution of the given equation is approximately $x = 1.65$.

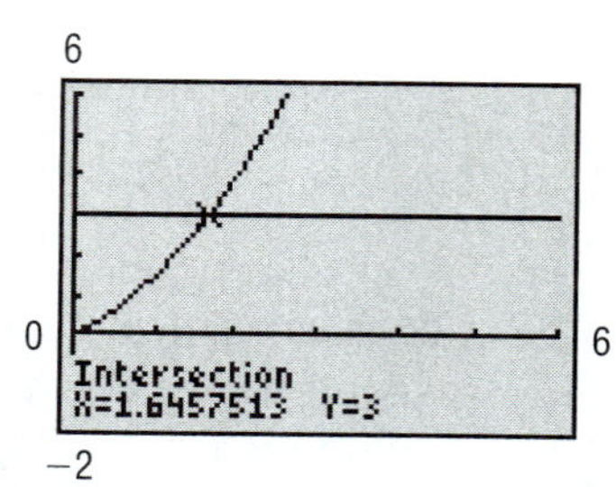

FIGURE 3.40 Solution of $0.5x^2 + x = 3$.

Exercise 3.5

In Problems 1–14, solve the given nonlinear system.

*1. $\begin{cases} y = 4 - x^2 \\ 3x + y = 0 \end{cases}$

2. $\begin{cases} y = x^3 \\ x - y = 0 \end{cases}$

3. $\begin{cases} p^2 = 5 - q \\ p = q + 1 \end{cases}$

4. $\begin{cases} y^2 - x^2 = 28 \\ x - y = 14 \end{cases}$

5. $\begin{cases} x = y^2 \\ y = x^2 \end{cases}$

6. $\begin{cases} p^2 - q + 1 = 0 \\ 5q - 3p - 2 = 0 \end{cases}$

7. $\begin{cases} y = 4x - x^2 + 8 \\ y = x^2 - 2x \end{cases}$

8. $\begin{cases} x^2 - y = 8 \\ y - x^2 = 0 \end{cases}$

9. $\begin{cases} p = \sqrt{q} \\ p = q^2 \end{cases}$

10. $\begin{cases} z = 4/w \\ 3z = 2w + 2 \end{cases}$

11. $\begin{cases} x^2 = y^2 + 13 \\ y = x^2 - 15 \end{cases}$

12. $\begin{cases} x^2 + y^2 - 2xy = 1 \\ 3x - y = 5 \end{cases}$

*13. $\begin{cases} 2x = y + 8 \\ y = 5\sqrt{x + 1} \end{cases}$

14. $\begin{cases} y = \dfrac{x^2}{x - 1} + 1 \\ y = \dfrac{1}{x - 1} \end{cases}$

15. **Decorations** The shape of a paper streamer suspended above a dance floor can be described by the function $y = 0.01x^2 + 0.01x + 7$, where y is the height of the streamer (in feet) above the floor and x is the horizontal distance (in feet) from the center of the room. A rope holding up other decorations touches the streamer and is described by the function $y = 0.01x + 8.0$. Where does the rope touch the paper streamer?

16. **Awning** The shape of a decorative awning over a storefront can be described by the function $y = 0.06x^2 + 0.012x + 8$, where y is the height of the edge of the awning (in feet) above the sidewalk and x is the distance (in feet) from the center of the store's doorway. A vandal pokes a stick through the awning, piercing it in two places. The position of the stick can be described by the function $y = 0.912x + 5$. Where are the holes in the awning caused by the vandal?

17. Graphically determine how many solutions there are to the system

$$\begin{cases} y = \dfrac{1}{x} \\ y = x^2 - 4 \end{cases}$$

18. Graphically solve the system

$$\begin{cases} 2y = x^3 \\ y = 8 - x^2 \end{cases}$$

and round to one-decimal-place.

19. Graphically solve the system

$$\begin{cases} y = x^2 - 2x + 1 \\ y = x^3 + x^2 - 2x + 3 \end{cases}$$

and round to one-decimal-place accuracy.

20. Graphically solve the system

$$\begin{cases} y = x^3 - x \\ y = 4x \end{cases}$$

and round to one-decimal-place accuracy.

In Problems 21–23, graphically solve the equation by treating it as a system. Round answers to two decimal places.

21. $0.8x^2 + 2x = 6$ where $x \geq 0$

22. $\sqrt{x + 2} = 5 - x$

23. $x^3 - 3x^2 = x - 8$

> **OBJECTIVE** *To solve systems describing equilibrium and break-even points.*

3.6 Applications of Systems of Equations

Equilibrium

Recall from Section 3.2 that an equation that relates price per unit and quantity demanded (supplied) is called a *demand equation (supply equation)*. Suppose that, for product Z, the demand equation is

$$p = -\frac{1}{180}q + 12$$

and the supply equation is

$$p = \frac{1}{300}q + 8$$

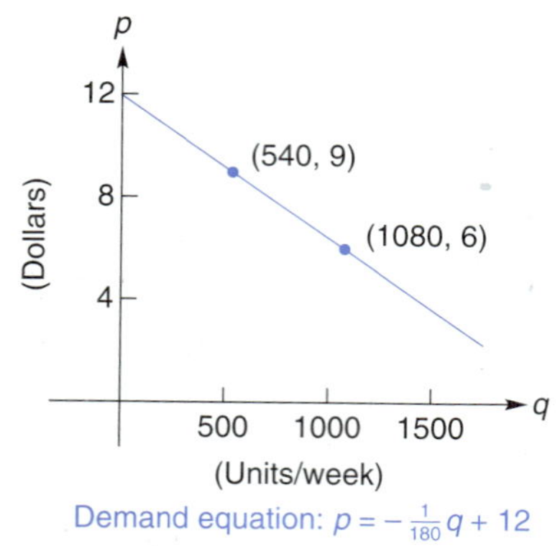

FIGURE 3.41 Demand curve.

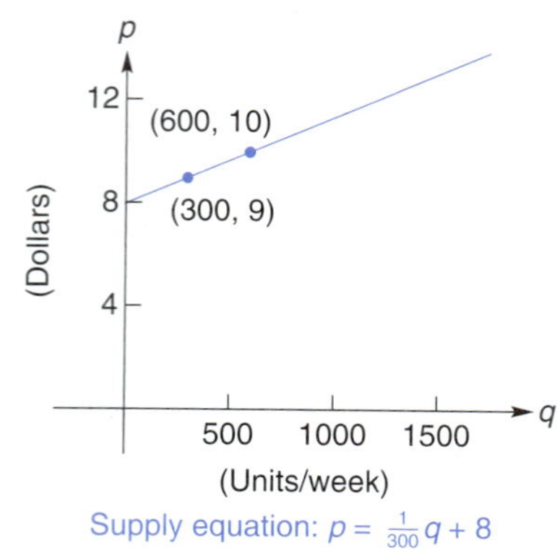

FIGURE 3.42 Supply curve.

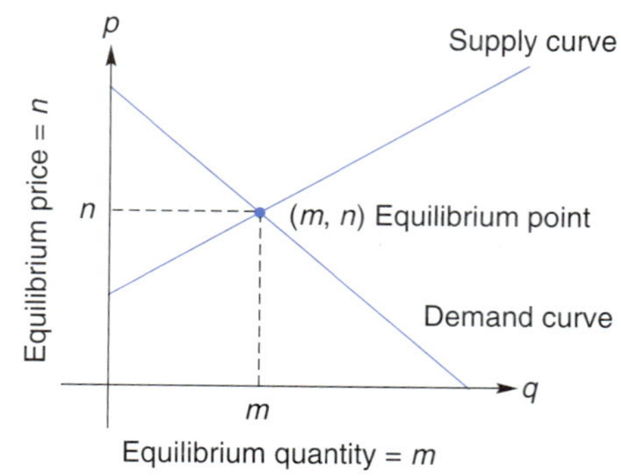

FIGURE 3.43 Equilibrium.

where $q, p \geq 0$. The corresponding demand and supply curves are the lines in Figures 3.41 and 3.42, respectively. In analyzing Figure 3.41, we see that consumers will purchase 540 units per week when the price is \$9 per unit, 1080 units when the price is \$6, and so on. Figure 3.42 shows that when the price is \$9 per unit producers will place 300 units per week on the market, at \$10 they will supply 600 units, and so on.

When the demand and supply curves of a product are represented on the same coordinate plane, the point (m, n) where the curves intersect is called the **point of equilibrium.** (See Figure 3.43.) The price n, called the **equilibrium price,** is the price at which consumers will purchase the same quantity of a product that producers wish to sell at that price. In short, n is the price at which stability in the producer-consumer relationship occurs. The quantity m is called the **equilibrium quantity.**

To determine precisely the equilibrium point, we solve the system formed by the supply and demand equations. Let us do this for our previous data, namely, the system

$$\begin{cases} p = -\dfrac{1}{180}q + 12 & \text{(demand equation)}, \\ p = \dfrac{1}{300}q + 8 & \text{(supply equation)}. \end{cases}$$

By substituting $\frac{1}{300}q + 8$ for p in the demand equation, we get

$$\frac{1}{300}q + 8 = -\frac{1}{180}q + 12$$

$$\left(\frac{1}{300} + \frac{1}{180}\right)q = 4$$

$$q = 450 \qquad \text{(equilibrium quantity)}$$

Thus,

$$p = \frac{1}{300}(450) + 8$$

$$= 9.50 \qquad \text{(equilibrium price)}$$

and the equilibrium point is (450, 9.50). Therefore, at the price of \$9.50 per unit, manufacturers will produce exactly the quantity (450) of units per week that consumers will purchase at that price. (See Figure 3.44.)

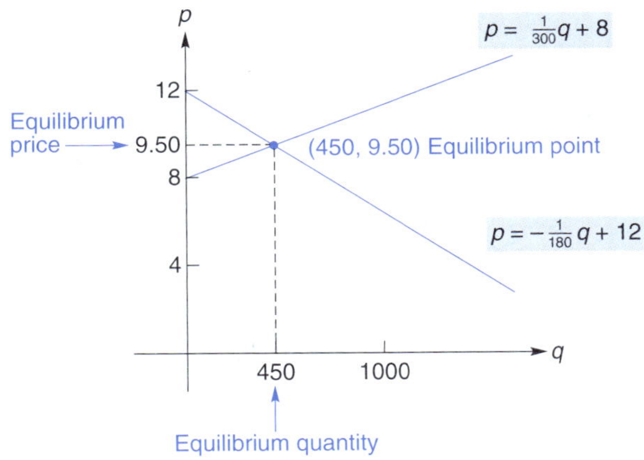

FIGURE 3.44 Equilibrium.

EXAMPLE 1 **Tax Effect on Equilibrium**

Let $p = \frac{8}{100}q + 50$ *be the supply equation for a manufacturer's product, and suppose the demand equation is* $p = -\frac{7}{100}q + 65$.

a. *If a tax of \$1.50 per unit is to be imposed on the manufacturer, how will the original equilibrium price be affected if the demand remains the same?*

Solution: Before the tax, the equilibrium price is obtained by solving the system

$$\begin{cases} p = \frac{8}{100}q + 50 \\ p = -\frac{7}{100}q + 65 \end{cases}$$

By substitution,

$$-\frac{7}{100}q + 65 = \frac{8}{100}q + 50$$
$$15 = \frac{15}{100}q$$
$$100 = q$$

and

$$p = \frac{8}{100}(100) + 50 = 58$$

Thus, \$58 is the original equilibrium price. Before the tax, the manufacturer supplies q units at a price of $p = \frac{8}{100}q + 50$ per unit. After the tax, he will sell the same q units for an additional \$1.50 per unit. The price per unit will be $\left(\frac{8}{100}q + 50\right) + 1.50$, so the new supply equation is

$$p = \frac{8}{100}q + 51.50$$

Solving the system

$$\begin{cases} p = \dfrac{8}{100}q + 51.50 \\ p = -\dfrac{7}{100}q + 65 \end{cases}$$

will give the new equilibrium price:

$$\begin{aligned} \frac{8}{100}q + 51.50 &= -\frac{7}{100}q + 65 \\ \frac{15}{100}q &= 13.50 \\ q &= 90 \\ p &= \frac{8}{100}(90) + 51.50 = 58.70 \end{aligned}$$

The tax of \$1.50 per unit increases the equilibrium price by \$0.70. (See Figure 3.45.) Note that there is also a decrease in the equilibrium quantity from $q = 100$ to $q = 90$, because of the change in the equilibrium price. (In the exercises, you are asked to find the effect of a subsidy given to the manufacturer, which will reduce the price of the product.)

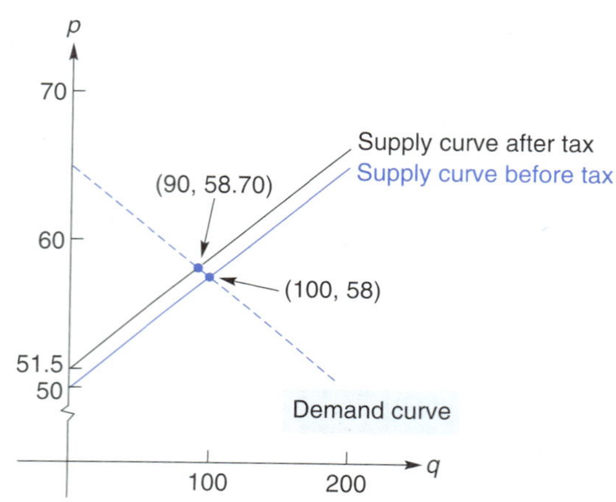

FIGURE 3.45 Equilibrium before and after tax.

▶ **Now work Problem 15.**

b. *Determine the total revenue obtained by the manufacturer at the equilibrium point both before and after the tax.*

Solution: If q units of a product are sold at a price of p dollars each, then the total revenue is given by

$$y_{\mathrm{TR}} = pq$$

Before the tax, the revenue at (100, 58) is (in dollars)

$$y_{\mathrm{TR}} = (58)(100) = 5800$$

After the tax, it is

$$y_{TR} = (58.70)(90) = 5283$$

which is a decrease.

EXAMPLE 2 Equilibrium with Nonlinear Demand

Find the equilibrium point if the supply and demand equations of a product are $p = \frac{q}{40} + 10$ *and* $p = \frac{8000}{q}$, *respectively.*

Solution: Here the demand equation is not linear. Solving the system

$$\begin{cases} p = \frac{q}{40} + 10 \\ p = \frac{8000}{q} \end{cases}$$

by substitution gives

$$\frac{8000}{q} = \frac{q}{40} + 10$$

$$320{,}000 = q^2 + 400q \qquad \text{(multiplying both sides by } 40q\text{)}$$

$$q^2 + 400q - 320{,}000 = 0$$

$$(q + 800)(q - 400) = 0$$

$$q = -800 \text{ or } q = 400$$

We disregard $q = -800$, since q represents quantity. Choosing $q = 400$, we have $p = 8000/400 = 20$, so the equilibrium point is (400, 20). (See Figure 3.46.)

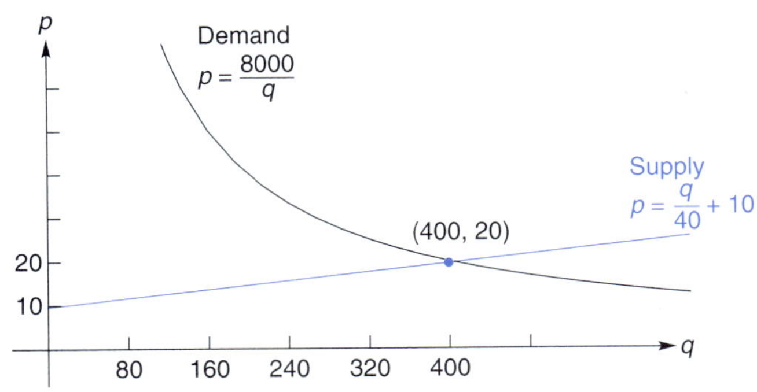

FIGURE 3.46 Equilibrium with nonlinear demand.

Break-Even Points

Suppose a manufacturer produces product A and sells it at \$8 per unit. Then the total revenue y_{TR} received (in dollars) from selling q units is

$$y_{TR} = 8q \qquad \text{(total revenue)}$$

The difference between the total revenue received for q units and the total cost of q units is the manufacturer's profit (or loss if the difference is negative):

$$\textbf{profit (or loss) = total revenue − total cost}$$

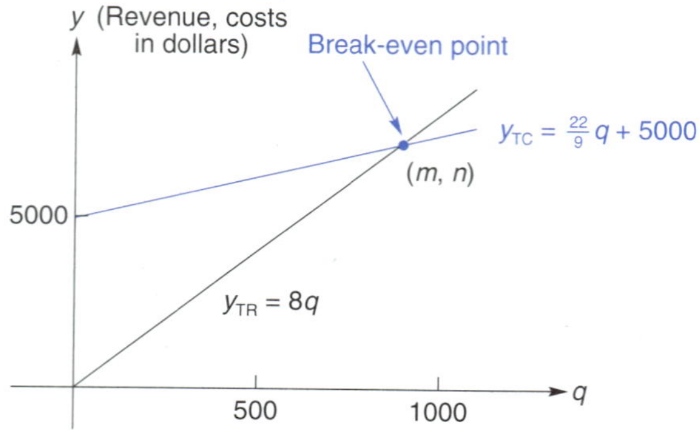

FIGURE 3.47 Break-even chart.

Total cost, y_{TC}, is the sum of total variable costs y_{VC} and total fixed costs y_{FC}:

$$y_{TC} = y_{VC} + y_{FC}$$

Fixed costs are those costs that, under normal conditions, do not depend on the level of production; that is, over some period of time they remain constant at all levels of output. (Examples are rent, officers' salaries, and normal maintenance.) **Variable costs** are those costs that vary with the level of production (such as the cost of materials, labor, maintenance due to wear and tear, etc.). For q units of product A, suppose that

$$y_{FC} = 5000 \qquad \text{(fixed cost)}$$

$$\text{and } y_{VC} = \frac{22}{9}q \qquad \text{(variable cost)}$$

Then

$$y_{TC} = \frac{22}{9}q + 5000 \qquad \text{(total cost)}$$

The graphs of total cost and total revenue appear in Figure 3.47. The horizontal axis represents the level of production, q, and the vertical axis represents the total dollar value, be it revenue or cost. The **break-even point** is the point at which total revenue equals total cost (TR = TC). It occurs when the levels of production and sales result in neither a profit nor a loss to the manufacturer. In the diagram, called a *break-even chart,* the break-even point is the point (m, n) at which the graphs of $y_{TR} = 8q$ and $y_{TC} = \frac{22}{9}q + 5000$ intersect. We call m the **break-even quantity** and n the **break-even revenue.** When total cost and revenue are linearly related to output, as in this case, for any production level greater than m, total revenue is greater than total cost, resulting in a profit. However, at any level less than m units, total revenue is less than total cost, resulting in a loss. At an output of m units, the profit is zero. In the following example, we shall examine our data in more detail.

EXAMPLE 3 **Break-Even Point, Profit, and Loss**

A manufacturer sells a product at \$8 per unit, selling all that is produced. Fixed cost is \$5000 and variable cost per unit is $\frac{22}{9}$ (dollars).

a. *Find the total output and revenue at the break-even point.*

Solution: At an output level of q units, the variable cost is $y_{VC} = \frac{22}{9}q$ and the total revenue is $y_{TR} = 8q$. Hence,

$$y_{TR} = 8q$$
$$y_{TC} = y_{VC} + y_{FC} = \frac{22}{9}q + 5000$$

At the break-even point, total revenue equals total cost. Thus, we solve the system formed by the foregoing equations. Since

$$y_{TR} = y_{TC}$$

we have

$$8q = \frac{22}{9}q + 5000$$
$$\frac{50}{9}q = 5000$$
$$q = 900$$

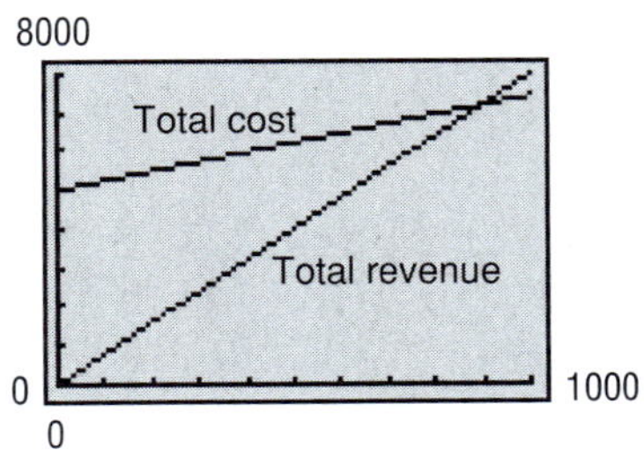

FIGURE 3.48 Equilibrium point (900, 7200).

Hence, the desired output is 900 units, resulting in a total revenue (in dollars) of

$$y_{TR} = 8(900) = 7200$$

(See Figure 3.48.)

b. *Find the profit when* 1800 *units are produced.*

Solution: Since profit = total revenue − total cost, when $q = 1800$ we have

$$y_{TR} - y_{TC} = 8(1800) - \left[\frac{22}{9}(1800) + 5000\right]$$
$$= 5000$$

The profit when 1800 units are produced and sold is \$5000.

c. *Find the loss when* 450 *units are produced.*

Solution: When $q = 450$,

$$y_{TR} - y_{TC} = 8(450) - \left[\frac{22}{9}(450) + 5000\right] = -2500$$

A loss of \$2500 occurs when the level of production is 450 units.

d. *Find the output required to obtain a profit of* \$10,000.

Solution: In order to obtain a profit of \$10,000, we have

$$\text{profit} = \text{total revenue} - \text{total cost}$$
$$10{,}000 = 8q - \left(\frac{22}{9}q + 5000\right)$$
$$15{,}000 = \frac{50}{9}q$$
$$q = 2700$$

Thus, 2700 units must be produced.

▶ **Now work Problem 9.**

EXAMPLE 4 **Break-Even Quantity**

Determine the break-even quantity of XYZ Manufacturing Co., given the following data: total fixed cost, \$1200; variable cost per unit, \$2; total revenue for selling q units, $y_{TR} = 100\sqrt{q}$.

Solution: For q units of output,

$$y_{TR} = 100\sqrt{q}$$
$$y_{TC} = 2q + 1200$$

Equating total revenue to total cost gives

$$100\sqrt{q} = 2q + 1200$$
$$50\sqrt{q} = q + 600 \qquad \text{(dividing both sides by 2).}$$

Squaring both sides, we have

$$2500q = q^2 + 1200q + (600)^2$$
$$0 = q^2 - 1300q + 360{,}000$$

By the quadratic formula,

$$q = \frac{1300 \pm \sqrt{250{,}000}}{2}$$
$$q = \frac{1300 \pm 500}{2}$$
$$q = 400 \quad \text{or} \quad q = 900$$

Although both $q = 400$ and $q = 900$ are break-even quantities, observe in Figure 3.49 that when $q > 900$, total cost is greater than total revenue, so there will always be a loss. This occurs because here total revenue is not linearly related to output. Thus, producing more than the break-even quantity does not necessarily guarantee a profit. The graph shows that, in this example, production quantity should satisfy $400 < q < 900$ in order to obtain a profit.

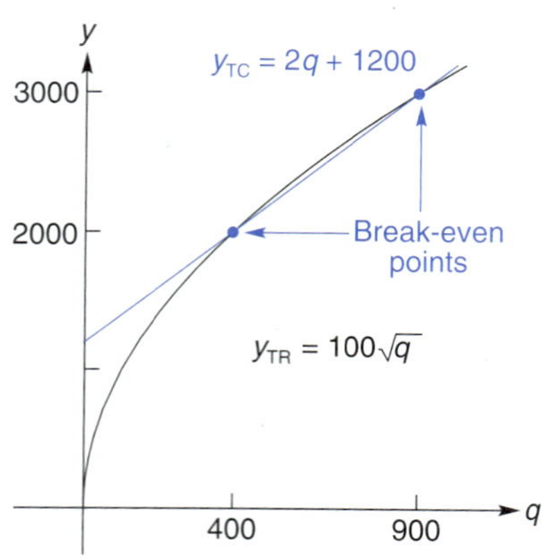

FIGURE 3.49 Two break-even points.

▸ **Now work Problem 13.**

Exercise 3.6

In Problems 1–8, you are given a supply equation and a demand equation for a product. If p represents price per unit in dollars and q represents the number of units per unit of time, find the equilibrium point. In Problems 1 and 2, sketch the system.

1. Supply: $p = \frac{4}{100}q + 3$, Demand: $p = -\frac{6}{100}q + 13$.

2. Supply: $p = \frac{1}{2000}q + 3$, Demand: $p = -\frac{1}{2500}q + \frac{42}{5}$.

3. Supply: $35q - 2p + 250 = 0$,
Demand: $65q + p - 537.5 = 0$.

4. Supply: $246p - 3.25q - 2460 = 0$,
Demand: $410p + 3q - 14{,}452.5 = 0$.

5. Supply: $p = 2q + 20$, Demand: $p = 200 - 2q^2$.

6. Supply: $p = (q + 10)^2$, Demand: $p = 388 - 16q - q^2$.

7. Supply: $p = \sqrt{q + 10}$, Demand: $p = 20 - q$.

8. Supply: $p = \frac{1}{4}q + 6$, Demand: $p = \dfrac{2240}{q + 12}$.

In Problems 9–14, y_{TR} represents total revenue in dollars and y_{TC} represents total cost in dollars for a manufacturer. If q represents both the number of units produced and the number of units sold, find the break-even quantity. Sketch a break-even chart in Problems 9 and 10.

***9.** $y_{TR} = 3q$
$y_{TC} = 2q + 4500$

10. $y_{TR} = 14q$
$y_{TC} = \frac{40}{3}q + 1200$

11. $y_{TR} = 0.05q$
$y_{TC} = 0.85q + 600$

12. $y_{TR} = 0.25q$
$y_{TC} = 0.16q + 360$

***13.** $y_{TR} = 90 - \dfrac{900}{q + 3}$
$y_{TC} = 1.1q + 37.3$

14. $y_{TR} = 0.1q^2 + 7q$
$y_{TC} = 2q + 500$

***15. Business** Supply and demand equations for a certain product are

$$3q - 200p + 1800 = 0$$

and

$$3q + 100p - 1800 = 0$$

respectively, where p represents the price per unit in dollars and q represents the number of units sold per time period.

(a) Find the equilibrium price algebraically, and sketch the break-even chart.

(b) Find the equilibrium price when a tax of 27 cents per unit is imposed on the supplier.

16. Business A manufacturer of a product sells all that is produced. The total revenue is given by $y_{TR} = 7q$, and the total cost is given by $y_{TC} = 6q + 800$, where q represents the number of units produced and sold.

(a) Find the level of production at the break-even point, and sketch the break-even chart.

(b) Find the level of production at the break-even point if the total cost increases by 5%.

17. Business A manufacturer sells a product at $8.35 per unit, selling all produced. The fixed cost is $2116 and the variable cost is $7.20 per unit. At what level of production will there be a profit of $4600? At what level of production will there be a loss of $1150? At what level of production will the break-even point occur?

18. Business The market equilibrium point for a product occurs when 13,500 units are produced at a price of $4.50 per unit. The producer will supply no units at $1, and the consumers will demand no units at $20. Find the supply and demand equations if they are both linear.

19. Business A manufacturer of a children's toy will break even at a sales volume of $200,000. Fixed costs are $40,000, and each unit of output sells for $5. Determine the variable cost per unit.

20. Business The Bigfoot Sandal Co. manufactures sandals for which the material cost is $0.85 per pair and the labor cost is $0.96 per pair. Additional variable costs amount to $0.32 per pair. Fixed costs are $70,500. If each pair sells for $2.63, how many pairs must be sold for the company to break even?

21. Business Find the break-even point for company Z, which sells all it produces, if the variable cost per unit is $2, fixed costs are $1050, and $y_{TR} = 50\sqrt{q}$, where q is the number of units of output produced.

22. Business A company has determined that the demand equation for its product is $p = 1000/q$, where p is the price per unit for q units produced and sold in some period. Determine the quantity demanded when the price per unit is (a) $4, (b) $2, and (c) $0.50. For each of these prices, determine the total revenue that the company will receive. What will be the revenue regardless of the price? (*Hint:* Find the revenue when the price is p dollars.)

23. Business Using the data in Example 1, determine how the original equilibrium price will be affected if the company is given a government subsidy of $1.50 per unit.

24. Business The Monroe Forging Company sells a corrugated steel product to the Standard Manufacturing Company and is in competition on such sales with other suppliers of the Standard Manufacturing Co. The vice president of sales of Monroe Forging Co. believes that by reducing the price of the product, a 40% increase in the volume of units sold to the Standard Manufacturing Co. could be secured. As the manager of the cost and analysis department, you have been asked to analyze the proposal of the vice president and submit your recommendations as to whether it is financially beneficial to the Monroe Forging Co. You are specifically requested to determine the following.

(a) Net profit or loss based on the pricing proposal.
(b) Unit sales volume under the proposed price that is required to make the same $40,000 profit, which is now earned at the current price and unit sales volume.

Use the following data in your analysis:

	Current Operations	**Proposal of Vice President of Sales**
Unit price	$2.50	$2.00
Unit sales volume	200,000 units	280,000 units
Variable cost		
Total	$350,000	$490,000
Per unit	$1.75	$1.75
Fixed cost	$110,000	$110,000
Profit	$40,000	?

25. Business Suppose products A and B have demand and supply equations that are related to each other. If q_A and q_B are the quantities produced and sold of A and B, respectively, and p_A and p_B are their respective prices, the demand equations are

$$q_A = 7 - p_A + p_B$$

and

$$q_B = 21 + p_A - p_B$$

and the supply equations are

$$q_A = -3 + 4p_A - 2p_B$$

and

$$q_B = -5 - 2p_A + 4p_B$$

Eliminate q_A and q_B to get the equilibrium prices.

26. Business The supply equation for a product is

$$p = 0.3q^2 + 14.6$$

and the demand equation is

$$p = \frac{35.2}{1 + 0.3q}$$

Here p represents price per unit in dollars and q represents number of units (in thousands) per unit time. Graph both equations, and from your graph, determine the equilibrium price and equilibrium quantity rounded to one decimal place.

27. Business For a manufacturer, the total-revenue equation is

$$y_{TR} = 20.5\sqrt{q + 4} - 41$$

and the total-cost equation is

$$y_{TC} = 0.02q^3 + 10.4,$$

where q represents (in thousands) both the number of units produced and the number of units sold. Graph a break-even chart and find the break-even quantity.

3.7 Review

Important Terms and Symbols

Section 3.1	slope of a line; linearly related; point–slope form; slope–intercept form; general linear equation in x and y
Section 3.2	demand equation; demand curve; supply equation; supply curve; linear function
Section 3.3	quadratic function; parabola; axis of symmetry; vertex
Section 3.4	system of equations; parameter; equivalent systems; general linear equation in x, y, and z; elimination by addition; elimination by substitution
Section 3.5	nonlinear system
Section 3.6	point of equilibrium; equilibrium price; equilibrium quantity; profit; total cost; fixed cost; variable cost; break-even point; break-even quantity; break-even revenue

Summary

The orientation of a nonvertical line is characterized by the slope of the line given by

$$m = \frac{y_2 - y_1}{x_2 - x_1}$$

where (x_1, y_1) and (x_2, y_2) are two different points on the line. The slope of a vertical line is not defined, and the slope of a horizontal line is zero. Rising lines have positive slopes; falling lines have negative slopes. Two lines are parallel if and only if they have the same slope or are vertical. Two lines with slopes m_1 and m_2 are perpendicular to each other if and only if $m_1 = -\frac{1}{m_2}$. A horizontal line and a vertical line are perpendicular to each other.

Basic forms of equations of lines are as follows:

$y - y_1 = m(x - x_1)$	(point–slope form)
$y = mx + b$	(slope–intercept form)
$x = a$	(vertical line)
$y = b$	(horizontal line)
$Ax + By + C = 0$	(general)

The linear function $f(x) = ax + b$ $(a \neq 0)$ has a straight line for its graph.

In economics, supply functions and demand functions have the form $p = f(q)$ and play an important role. Each gives a correspondence between the price p of a product and the number of units q of the product that manufacturers (or consumers) will supply (or purchase) at that price during some time period.

A quadratic function has the form

$$f(x) = ax^2 + bx + c \quad (a \neq 0)$$

The graph of f is a parabola that opens upward if $a > 0$ and downward if $a < 0$. The vertex is

$$\left(-\frac{b}{2a}, f\left(-\frac{b}{2a}\right)\right)$$

and the y-intercept is c. The axis of symmetry, as well as the x- and y-intercepts, are useful in sketching the graph.

A system of linear equations may be solved with the method of elimination by addition or elimination by substitution. A solution may involve one or more parameters. Substitution is also useful in solving nonlinear systems.

Solving a system formed by the supply and demand equations for a product gives the equilibrium point, which indicates the price at which consumers will purchase the same quantity of a product that producers wish to sell at that price.

Profit is total revenue minus total cost, where total cost is the sum of fixed costs and variable costs. The break-even point is the point where total revenue equals total cost.

Review Problems

Problem numbers shown in color indicate problems suggested for use as a practice chapter test.

1. The slope of the line through (2, 5) and (3, k) is 4. Find k.

2. The slope of the line through (5, 4) and (k, 4) is 0. Find k.

In Problems 3–9, determine the slope–intercept form and a general linear form of an equation of the straight line that has the indicated properties.

3. Passes through (3, −2) and has y-intercept 1.

4. Passes through (−1, −1) and is parallel to the line $y = 3x - 4$.

5. Passes through (10, 4) and has slope $\frac{1}{2}$.

6. Passes through (3, 5) and is vertical.

7. Passes through (−2, 4) and is horizontal.

8. Passes through (1, 2) and is perpendicular to the line $-3y + 5x = 7$.

9. Has y-intercept −3 and is perpendicular to $2y + 5x = 2$.

10. Determine whether the point (0, −7) lies on the line through (1, −3) and (4, 9).

In Problems 11–16, determine whether the lines are parallel, perpendicular, or neither.

11. $x + 4y + 2 = 0, \quad 8x - 2y - 2 = 0$

12. $y - 2 = 2(x - 1), \quad 2x + 4y - 3 = 0$

13. $x - 3 = 2(y + 4), \quad y = 4x + 2$

14. $2x + 7y - 4 = 0, \quad 6x + 21y = 90$

15. $y = \frac{1}{2}x + 5, \quad 2x = 4y - 3$

16. $y = 7x, \quad y = 7$

In Problems 17–20, write each line in slope–intercept form, and sketch. What is the slope of the line?

17. $3x - 2y = 4$

18. $x = -3y + 4$

19. $4 - 3y = 0$

20. $y = 2x$

In Problems 21–30, graph each function. For those that are linear, give the slope and the vertical-axis intercept. For those that are quadratic, give all intercepts and the vertex.

21. $y = f(x) = 17 - 5x$

22. $s = g(t) = 8 - 2t - t^2$

23. $y = f(x) = 9 - x^2$

24. $y = f(x) = 3x - 7$

25. $y = h(t) = t^2 - 4t - 5$

26. $y = k(t) = -3 - 3t$

27. $p = g(t) = 3t$

28. $y = F(x) = (2x - 1)^2$

29. $y = F(x) = -(x^2 + 2x + 3)$

30. $y = f(x) = \frac{x}{3} - 2$

In Problems 31–44, solve the given system.

31. $\begin{cases} 2x - y = 6 \\ 3x + 2y = 5 \end{cases}$

32. $\begin{cases} 8x - 4y = 7 \\ y = 2x - 4 \end{cases}$

33. $\begin{cases} 7x + 5y = 5 \\ 6x + 5y = 3 \end{cases}$

34. $\begin{cases} 3x + 6y = 9 \\ 4x + 8y = 12 \end{cases}$

35. $\begin{cases} \frac{1}{4}x - \frac{3}{2}y = -4 \\ \frac{3}{4}x + \frac{1}{2}y = 8 \end{cases}$

36. $\begin{cases} \frac{1}{3}x - \frac{1}{4}y = \frac{1}{12} \\ \frac{4}{3}x + 3y = \frac{5}{3} \end{cases}$

37. $\begin{cases} 3x - 2y + z = -2 \\ 2x + y + z = 1 \\ x + 3y - z = 3 \end{cases}$

38. $\begin{cases} 2x + \frac{3y + x}{3} = 9 \\ y + \frac{5x + 2y}{4} = 7 \end{cases}$

39. $\begin{cases} x^2 - y + 2x = 7 \\ x^2 + y = 5 \end{cases}$

40. $\begin{cases} y = \frac{18}{x + 4} \\ x - y + 7 = 0 \end{cases}$

[8]**41.** $\begin{cases} x + 2z = -2 \\ x + y + z = 5 \end{cases}$

[8]**42.** $\begin{cases} x + y + z = 0 \\ x - y + z = 0 \\ x + z = 0 \end{cases}$

[8]**43.** $\begin{cases} x - y - z = 0 \\ 2x - 2y + 3z = 0 \end{cases}$

[8]**44.** $\begin{cases} 2x - 5y + 6z = 1 \\ 4x - 10y + 12z = 2 \end{cases}$

45. Suppose a and b are linearly related so that $a = 1$ when $b = 2$ and $a = 5$ when $b = 3$. Find a general linear form of an equation that relates a and b. Also, find a when $b = 5$.

46. Temperature and Heart Rate When the temperature, T (in degrees Celsius), of a cat is reduced, the cat's heart rate, r (in beats per minute), decreases. Under laboratory conditions, a cat at a temperature of 37°C had a heart rate of 220, and at a temperature of 32°C its heart rate was 150. If r is linearly related to T, where T is between 26 and 38, (a) determine an equation for r in terms of T and (b) determine the cat's heart rate at a temperature of 28°C.

47. Suppose f is a linear function such that $f(1) = 5$ and $f(x)$ decreases by four units for every three-unit increase in x. Find $f(x)$.

[8]Refer to concepts in Examples 6 and 7 in Section 3.4.

48. If f is a linear function such that $f(-1) = 8$ and $f(2) = 5$, find $f(x)$.

49. Maximum Revenue The demand function for a manufacturer's product is $p = f(q) = 200 - 2q$, where p is the price (in dollars) per unit when q units are demanded. Find the level of production that maximizes the manufacturer's total revenue, and determine this revenue.

50. Sales Tax The difference in price of two items before a 5% sales tax is imposed is \$3.50. The difference in price after the sales tax is imposed is allegedly \$4.10. Show that this scenario is not possible.

51. Equilibrium Price If the supply and demand equations of a certain product are $125p - q - 250 = 0$ and $100p + q - 1100 = 0$, respectively, find the equilibrium price.

52. Psychology In psychology, the term *semantic memory* refers to our knowledge of the meaning and relationships of words, as well as the means by which we store and retrieve such information.[9] In a network model of semantic memory, there is a hierarchy of levels at which information is stored. In an experiment by Collins and Quillian based on a network model, data were obtained on the reaction time to respond to simple questions about nouns. The graph of the results shows that, on the average, the reaction time, R (in milliseconds), is a linear function of the level, L, at which a characterizing property of the noun is stored. At level 0, the reaction time is 1310; at level 2, the reaction time is 1460. (a) Find the linear function. (b) Find the reaction time at level 1. (c) Find the slope and determine its significance.

53. Break-Even Point A manufacturer of a certain product sells all that is produced. Determine the break-even point if the product is sold at \$16 per unit, fixed cost is \$10,000, and variable cost is given by $y_{VC} = 8q$, where q is the number of units produced (y_{VC} expressed in dollars).

54. Temperature Conversion Celsius temperature, C, is a linear function of Fahrenheit temperature, F. Use the facts that 32°F is the same as 0°C and 212°F is the same as 100°C to find this function. Also, find C when $F = 50$.

[9]G. R. Loftus and E. F. Loftus, *Human Memory: The Processing of Information* (New York: Lawrence Erlbaum Associates, Inc., distributed by the Halsted Press, Division of John Wiley & Sons, Inc., 1976).

55. Pollution In one province of a developing nation, water pollution is analyzed using a supply-and-demand model. The *environmental supply equation* $L = 0.0183 - \frac{0.0042}{p}$ describes the levy-per-ton, L (in dollars), as a function of total pollution, p (in tons per square kilometer), for $p \geq 0.2295$. The *environmental demand equation*, $L = 0.0005 + \frac{0.0378}{p}$, describes the per-ton abatement cost as a function of total pollution for $p > 0$. Find the expected equilibrium level of total pollution rounded to two decimal places.[10]

56. Graphically solve the linear system

$$\begin{cases} 3x + 4y = 20 \\ 7x + 5y = 64 \end{cases}$$

57. Graphically solve the linear system

$$\begin{cases} 0.3x - 0.4y = 2.5 \\ 0.5x + 0.7y = 3.1 \end{cases}$$

Round x and y to two decimal places.

[10]See Hua Wang and David Wheeler, "Pricing Industrial Pollution in China: An Economic Analysis of the Levy System," World Bank Policy Research Working Paper #1644, September 1996.

58. Graphically solve the nonlinear system

$$\begin{cases} y = \dfrac{2}{x} \quad \text{where } x > 0 \\ y = x^2 - 6 \end{cases}$$

Round x and y to two decimal places.

59. Graphically solve the nonlinear system

$$\begin{cases} y = x^3 + 1 \\ y = 2 - x^2 \end{cases}$$

Round x and y to two decimal places.

60. Graphically solve the equation

$$x^2 + 4 = x^3 - 3x$$

by treating it as a system. Round x to two decimal places.

Mathematical SNAPSHOT

Wireless Phone Billing Plans

Changes in technology and the law have transformed the communications industry in recent decades. Some of the changes have been a mixed blessing. Consider, for instance, the matter of choosing a cellular phone plan. In most urban areas, cell phone users have literally dozens of plans to choose from. The plans include monthly access fees, free minutes, charges for additional airtime, regional roaming charges, national roaming charges, peak and off-peak rates, and long-distance charges (not to mention activation fees, cancellation fees, and the like). How in the world, given all these factors, is a consumer to make an intelligent choice?

While finding the guaranteed best choice is a lot of work, making a reasonably good choice requires only a little math. Consider the plans offered by just one telecommunications company, called XY&Z Corporation, and assume that the bulk of calls are local, made (or received) in town during peak hours. In other words, we will ignore roaming charges, off-peak rates, and long-distance charges. In December 2000, this corporation offered the following plans:

Basic: \$19.99 per month buys 60 minutes. Additional time costs \$0.40 per minute.

Advantage I: \$29.99 per month buys 120 minutes. Additional time costs \$0.30 per minute.

Advantage II: \$39.99 per month buys 200 minutes. Additional time costs \$0.30 per minute.

Advantage III: \$49.99 per month buys 400 minutes. Additional time costs \$0.30 per minute.

Premier: \$59.99 per month buys 450 minutes. Additional time costs \$0.35 per minute.

To represent these plans mathematically, we have to write total monthly cost as a function of time for each plan. For the Basic plan, the monthly charge B will depend on the total number of calling minutes according to the function

$$B(t) = \begin{cases} 19.99 & \text{if } t \le 60 \\ 19.99 + 0.40(t - 60) & \text{if } t > 60 \end{cases}$$

Similarly, representing the three Advantage plans by $A1$, $A2$, and $A3$, respectively, and the Premier plan by

P, we have these functions:

$$A1(t) = \begin{cases} 29.99 & \text{if } t \le 120 \\ 29.99 + 0.30(t - 120) & \text{if } t > 120 \end{cases}$$

$$A2(t) = \begin{cases} 39.99 & \text{if } t \le 200 \\ 39.99 + 0.30(t - 200) & \text{if } t > 200 \end{cases}$$

$$A3(t) = \begin{cases} 49.99 & \text{if } t \le 400 \\ 49.99 + 0.30(t - 400) & \text{if } t > 400 \end{cases}$$

$$P(t) = \begin{cases} 59.99 & \text{if } t \le 450 \\ 59.99 + 0.35(t - 450) & \text{if } t > 450 \end{cases}$$

Faced with this much information, we are well advised to get the overview by constructing a graph. We could do this by hand, but here is a good opportunity to use a handy capability of a graphing calculator. We enter the function $B(t)$ as

$$\text{Y1} = 19.99 + 0.40(\text{X} - 60)(\text{X} > 60)$$

The $>$ comes from the TEST menu, and the expression $(\text{X} > 60)$ equals either 1 or 0, depending on whether x is, or is not, greater than 60. Entering the other four functions in similar fashion and graphing them all together, we get the display shown in Figure 3.50.

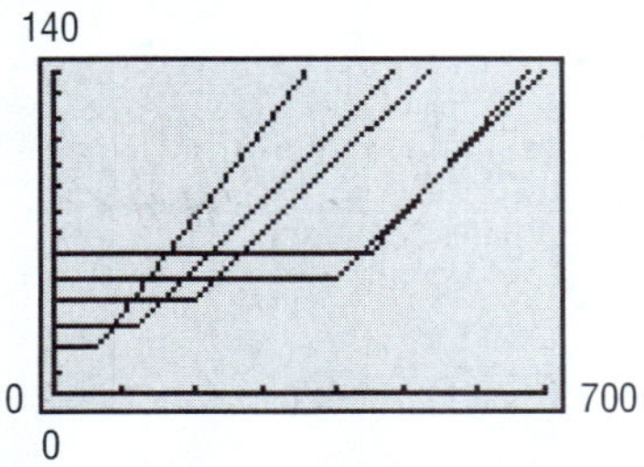

FIGURE 3.50 Costs under different plans.

Which plan is best depends on the amount of calling time: for any given monthly airtime, the best plan is the one whose graph is lowest at that point.

For very low calling times, the Basic plan is best, but at some point it starts to become more expensive than the Advantage I plan. Let us find out where this occurs—the t-value at which those two plans' graphs intersect. Notice that if we had not already graphed all the functions, we would not know which part of each function's definition to use; as it is, we can see that we use the second part of the definition of $B(t)$ (the part whose graph slants) and the first part of the definition of $A1(t)$ (the part whose graph is level). In other words, we solve the system of equations

$$\begin{cases} B(t) = 19.99 + 0.40(t - 60) \\ A1(t) = 29.99 \\ B(t) = A1(t) \end{cases}$$

Through substitution, this simplifies to a single equation that is readily solvable:

$$\begin{aligned} 19.99 + 0.40(t - 60) &= 29.99 \\ 0.40t - 24 &= 10 \\ 0.40t &= 34 \\ t &= 85 \end{aligned}$$

So the Advantage I plan becomes better than the Basic plan for monthly calling times over 85 minutes.

From the graph we can also see that at some point the Advantage II plan starts to be the best plan, and at a point beyond that, the Advantage III, in turn, becomes best. Notice, however, something interesting about the Advantage III and Premier plans: The Advantage III plan is better at first, and then the Premier plan is better for a while, but for very high monthly usage times, the Advantage III plan is again better.[11] Let us find the last switchover point. It is the t-value at which the two slanting portions of the graphs of $A3(t)$ and $P(t)$ intersect. Instead of solving algebraically, this time let us use the calculator to find the intersection point automatically. Our result is shown in Figure 3.51.

So the Advantage III plan is better than the Premier plan for calling times over 550 minutes.

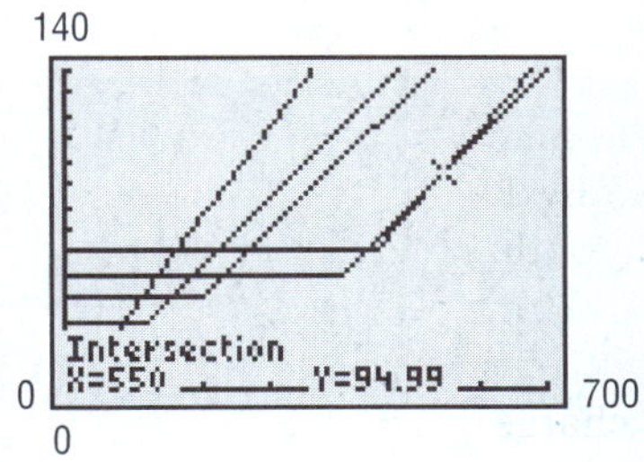

FIGURE 3.51 Advantage III and Premier

Given what we know so far, we can construct the following partial table.

Airtime (min)	Best Plan
0 to 85	Basic
	Advantage I
	Advantage II
	Advantage III
	Premier
550 and above	Advantage III

Finishing the table is left for the Exercises.

To research wireless phone services plans in your area, visit www.point.com.

Exercises

1. Copy the preceding table onto a separate piece of paper. Then use algebraic solution techniques to fill in the first two blanks in the airtime column.
2. Use the graphing calculator to fill in the remaining two blanks.
3. What happens when you try to use the calculator to find an intersection point but are not careful with your initial guess?
4. Why might it be that XY&Z offers five different plans, instead of just offering a single plan that gives the company a reasonable profit for any customer's airtime?

[11] The plans also differ significantly in regional roaming charges, but we are not considering those.

CHAPTER

4

Exponential and Logarithmic Functions

Just as biological viruses spread through contact between organisms, so computer viruses spread when computers interact via networks or e-mail. As scientists study how to fight computer viruses, which cause a lot of damage in the form of deleted and corrupted files, one thing the scientists do is devise mathematical models of how quickly viruses spread. For instance, on Friday, March 26, 1999, the first cases of the virus known as Melissa were reported; by Monday, March 29, Melissa had reached more than 100,000 computers.

Exponential functions, which this chapter discusses in detail, provide one plausible model. Consider a computer virus that hides in an e-mail attachment and, once the attachment is downloaded, automatically sends a message with a similar attachment to every address in the host computer's e-mail address book. If the typical address book contains 20 addresses, and if the typical computer user retrieves his or her e-mail once a day, then a virus on a single machine will have infected 20 machines after one day, $20^2 = 400$ machines after two days, $20^3 = 8000$ after three days, and in general, after t days, the number N of infected computers will be given by the exponential function $N(t) = 20^t$.

This model assumes that all the computers involved are linked together, via their address book lists, into a single, well-connected group. Exponential models are most accurate for small values of t; this model, in particular, ignores the slowdown that occurs when most e-mails start going to computers already infected, which happens as several days pass. For example, our model tells us that after seven days, the virus will infect $20^7 = 1.28$ billion computers—roughly every computer on Earth! But despite its limitations, the exponential model does explain why new viruses often infect many thousands of machines before antivirus experts have had time to react.

OBJECTIVE *To study exponential functions and their applications to such areas as compound interest, population growth, and radioactive decay.*

4.1 Exponential Functions

There are functions that have an important role not only in mathematics, but also in business, economics, and other areas of study. They involve a constant raised to a variable power, such as $f(x) = 2^x$. We call such functions *exponential functions*.

Definition

The function f defined by

$$f(x) = b^x$$

where $b > 0, b \neq 1$, and the exponent x is any real number, is called an ***exponential function*** with base b.[1]

CAUTION Do not confuse the exponential function $y = 2^x$ with the *power function* $y = x^2$, which has a variable base and a constant exponent.

Since the exponent in b^x can be any real number, you may wonder how we assign a value to something like $2^{\sqrt{2}}$, where the exponent is an irrational number. Stated simply, we use approximations. Because $\sqrt{2} = 1.41421\ldots$, $2^{\sqrt{2}}$ is approximately $2^{1.4} = 2^{7/5} = \sqrt[5]{2^7}$, which *is* defined. Better approximations are $2^{1.41} = 2^{141/100} = \sqrt[100]{2^{141}}$, and so on. In this way, the meaning of $2^{\sqrt{2}}$ becomes clear. A calculator value of $2^{\sqrt{2}}$ is (approximately) 2.66514.

When you work with exponential functions, it may be necessary to apply rules for exponents. These rules are as follows, where m and n are real numbers and a and b are positive.

POINTER To review exponents, refer to Section 0.3.

Rules for Exponents:

1. $a^m a^n = a^{m+n}$
2. $\dfrac{a^m}{a^n} = a^{m-n}$
3. $(a^m)^n = a^{mn}$
4. $(ab)^n = a^n b^n$
5. $\left(\dfrac{a}{b}\right)^n = \dfrac{a^n}{b^n}$
6. $a^1 = a$
7. $a^0 = 1$
8. $a^{-n} = \dfrac{1}{a^n}$

Some functions that do not appear to have the exponential form b^x can be put in that form by applying the preceding rules. For example, $2^{-x} = 1/(2^x) = \left(\frac{1}{2}\right)^x$ and $3^{2x} = (3^2)^x = 9^x$.

APPLICATIONS IN PRACTICE 1

Bacteria Growth

The number of bacteria in a culture that doubles every hour is given by $N(t) = A \cdot 2^t$, where A is the number originally present and t is the number of hours the bacteria have been doubling. Use a graphing calculator to plot this function for various values of $A > 1$. How are the graphs similar? How does the value of A alter the graph?

EXAMPLE 1 **Bacteria Growth**

The number of bacteria present in a culture after t minutes is given by

$$N(t) = 300\left(\frac{4}{3}\right)^t$$

Note that $N(t)$ is a constant multiple of the exponential function $\left(\dfrac{4}{3}\right)^t$.

a. *How many bacteria are present initially?*

Solution: Here we want to find $N(t)$ when $t = 0$. We have

$$N(0) = 300\left(\frac{4}{3}\right)^0 = 300(1) = 300$$

Thus, 300 bacteria are initially present.

[1] If $b = 1$, then $f(x) = 1^x = 1$. This function is already known to us as a constant function.

b. *Approximately how many bacteria are present after 3 minutes?*

Solution:

$$N(3) = 300\left(\frac{4}{3}\right)^3 = 300\left(\frac{64}{27}\right) = \frac{6400}{9} \approx 711$$

Hence, approximately 711 bacteria are present after 3 minutes.

▶ **Now work Problem 31.**

Graphs of Exponential Functions

EXAMPLE 2 Graphing Exponential Functions with $b > 1$

Graph the exponential functions $f(x) = 2^x$ and $f(x) = 5^x$.

Solution: By plotting points and connecting them, we obtain the graphs in Figure 4.1. For the graph of $f(x) = 5^x$, because of the unit distance chosen on the y-axis, the points $(-2, \frac{1}{25})$, $(2, 25)$, and $(3, 125)$ are not shown.

x	2^x
−2	$\frac{1}{4}$
−1	$\frac{1}{2}$
0	1
1	2
2	4
3	8

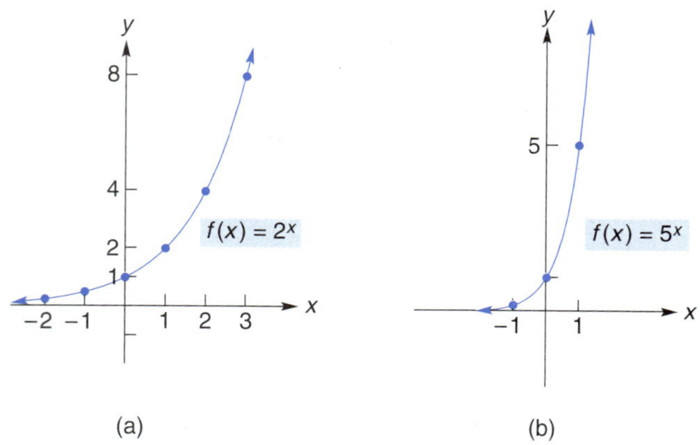

x	5^x
−2	$\frac{1}{25}$
−1	$\frac{1}{5}$
0	1
1	5
2	25
3	125

FIGURE 4.1 Graphs of $f(x) = 2^x$ and $f(x) = 5^x$.

APPLICATIONS IN PRACTICE 2

Graphing Exponential Functions with $b > 1$

Suppose an investment increases by 10% every year. Make a table of the factor by which the investment increases from the original amount for 0 to 4 years. For each year, write an expression for the increase as a power of some base. What base did you use? How does that base relate to the problem? Use your table to graph the multiplicative increase as a function of the number of years. Use your graph to determine when the investment will double.

We can make some observations about these graphs. The domain of each function consists of all real numbers, and the range consists of all positive real numbers. Each graph has y-intercept $(0, 1)$. Moreover, the graphs have the same general shape. Each *rises* from left to right. As x increases, $f(x)$ also increases. In fact, $f(x)$ increases without bound. However, in quadrant I, the graph of $f(x) = 5^x$ rises more quickly than that of $f(x) = 2^x$, because the base in 5^x is *greater* than the base in 2^x (that is, $5 > 2$). Looking at quadrant II, we see that as x becomes very negative, the graphs of both functions approach the x-axis.[2] This implies that the function values get very close to 0.

▶ **Now work Problem 1.**

[2]We say that the x-axis is an *asymptote* for each graph.

APPLICATIONS IN PRACTICE 3

Graphing Exponential Functions with $0 < b < 1$

Suppose the value of a car depreciates by 15% every year. Make a table of the factor by which the value decreases from the original amount for 0 to 3 years. For each year, write an expression for the decrease as a power of some base. What base did you use? How does that base relate to the problem? Use your table to graph the multiplicative decrease as a function of the number of years. Use your graph to determine when the car will be worth half as much as its original price.

The observations made in Example 2 are true for all exponential functions whose base b is greater than 1. Example 3 will examine the case for a base between 0 and 1 ($0 < b < 1$).

EXAMPLE 3 Graphing Exponential Functions with $0 < b < 1$

Graph the exponential function $f(x) = \left(\frac{1}{2}\right)^x$.

Solution: By plotting points and connecting them, we obtain the graph in Figure 4.2. Notice that the domain consists of all real numbers, and the range consists of all positive real numbers. The graph has y-intercept $(0, 1)$. Compared to the graphs in Example 2, the graph here *falls* from left to right. That is, as x increases, $f(x)$ decreases. Notice that as x becomes very positive, $f(x)$ takes on values close to 0 and the graph approaches the x-axis. However, as x becomes very negative, the function values are unbounded.

x	$\left(\frac{1}{2}\right)^x$
−3	8
−2	4
−1	2
0	1
1	$\frac{1}{2}$
2	$\frac{1}{4}$

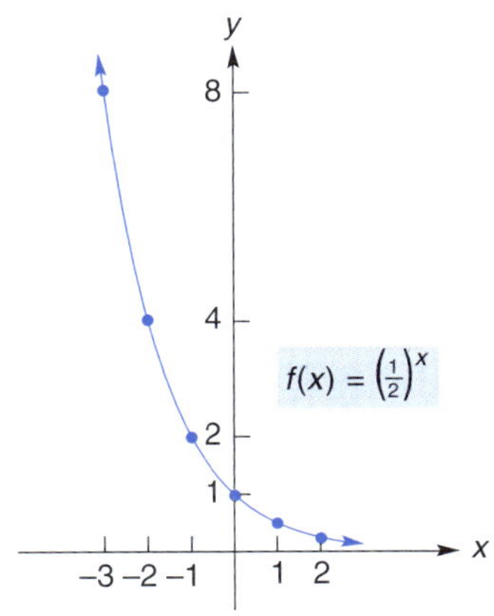

FIGURE 4.2 Graph of $f(x) = \left(\frac{1}{2}\right)^x$.

▶ **Now work Problem 3.**

There are two basic shapes for the graphs of exponential functions, and they depend on the base involved.

In general, the graph of an exponential function has one of two shapes, depending on the value of the base b. This is illustrated in Figure 4.3. It is important to observe that in either case the graph passes the horizontal line test. Thus all exponential functions are one-to-one. The basic properties of an exponential function and its graph are summarized in Table 4.1.

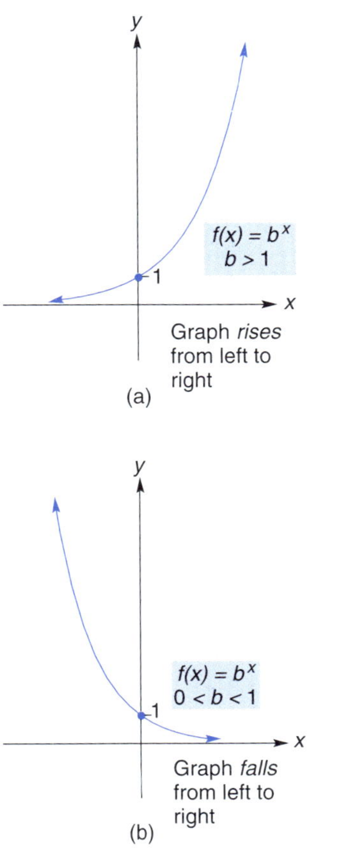

FIGURE 4.3 General shapes of $f(x) = b^x$.

TABLE 4.1 Properties of the Exponential Function $f(x) = b^x$

1. The domain of an exponential function consists of all real numbers. The range consists of all positive numbers.
2. The graph of $f(x) = b^x$ has y-intercept $(0, 1)$. There is no x-intercept.
3. If $b > 1$, the graph *rises* from left to right. If $0 < b < 1$, the graph *falls* from left to right.
4. If $b > 1$, the graph approaches the x-axis as x becomes more and more negative. If $0 < b < 1$, the graph approaches the x-axis as x becomes more and more positive.

Example 4 makes use of transformations from Table 2.2.

Recall from Section 2.7 that the graph of one function may be related to that of another by means of a certain transformation. Our next example has to do with this concept.

APPLICATIONS IN PRACTICE 4

Transformations of Exponential Functions

After watching his sister's money grow for three years in a plan with an 8% yearly return, George started a savings account with the same plan. If $y = 1.08^t$ represents the multiplicative increase in his sister's account, write an equation that will represent the multiplicative increase in George's account, using the same time reference. If George has a graph of the multiplicative increase in his sister's money at time t years since she started saving, how could he use the graph to project the increase in his money?

EXAMPLE 4 Transformations of Exponential Functions

a. *Use the graph of $y = 2^x$ to plot $y = 2^x - 3$.*

Solution: The function has the form $f(x) - c$, where $f(x) = 2^x$ and $c = 3$. Thus its graph is obtained by shifting the graph of $f(x) = 2^x$ three units downward. (See Figure 4.4.)

b. *Use the graph of $y = \left(\frac{1}{2}\right)^x$ to graph $y = \left(\frac{1}{2}\right)^{x-4}$.*

Solution: The function has the form $f(x-c)$, where $f(x) = \left(\frac{1}{2}\right)^x$ and $c = 4$. Hence, its graph is obtained by shifting the graph of $f(x) = \left(\frac{1}{2}\right)^x$ four units to the right. (See Figure 4.5.)

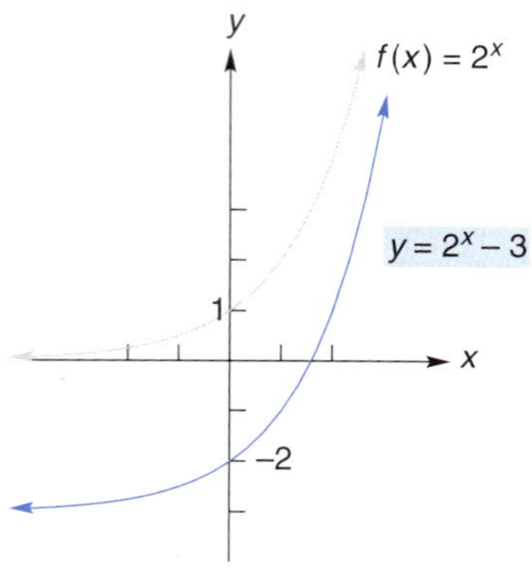

FIGURE 4.4 Graph of $y = 2^x - 3$.

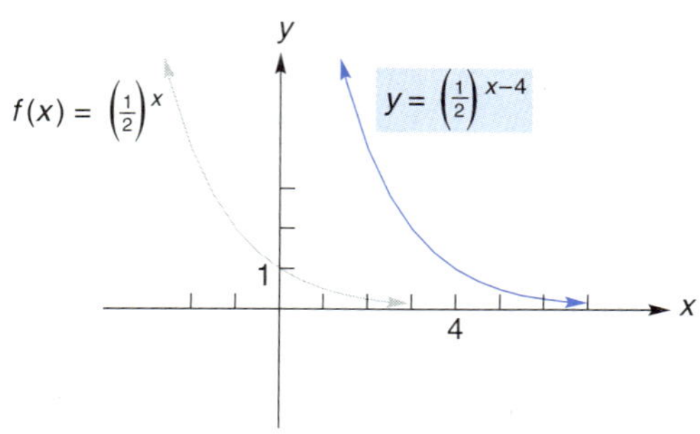

FIGURE 4.5 Graph of $y = \left(\frac{1}{2}\right)^{x-4}$.

▶ Now work Problem 7.

EXAMPLE 5 Graph of a Function with a Constant Base

Graph $y = 3^{x^2}$.

Solution: Although this is not an exponential function, it does have a constant base. We see that replacing x by $-x$ results in the same equation. Thus, the graph is symmetric about the y-axis. Plotting some points and using symmetry gives the graph in Figure 4.6.

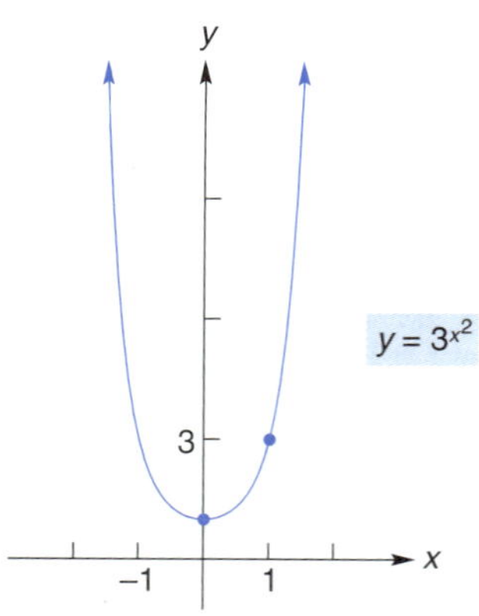

x	0	1	2
y	1	3	81

FIGURE 4.6 Graph of $y = 3^{x^2}$.

TECHNOLOGY

If $y = 4^x$, consider the problem of finding x when $y = 6$. One way to solve this equation is to find the intersection of the graphs of $y = 6$ and $y = 4^x$. Figure 4.7 shows that x is approximately 1.29.

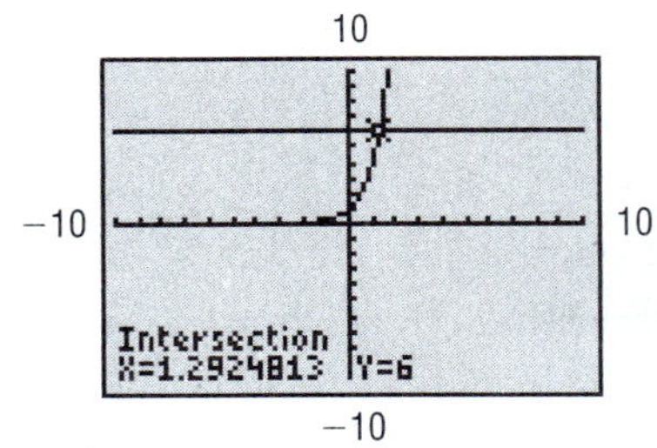

FIGURE 4.7 Solving the equation $6 = 4^x$.

Compound Interest

Exponential functions are involved in **compound interest,** whereby the interest earned by an invested amount of money (or **principal**) is reinvested so that it, too, earns interest. That is, the interest is converted (or *compounded*) into principal, and hence, there is "interest on interest."

For example, suppose that \$100 is invested at the rate of 5% compounded annually. At the end of the first year, the value of the investment is the original principal (\$100), plus the interest on the principal [100(0.05)]:

$$100 + 100(0.05) = \$105$$

This is the amount on which interest is earned for the second year. At the end of the second year, the value of the investment is the principal at the end of the first year (\$105), plus the interest on that sum [105(0.05)]:

$$105 + 105(0.05) = \$110.25$$

Thus, each year the principal increases by 5%. The \$110.25 represents the original principal, plus all accrued interest; it is called the **accumulated amount** or **compound amount.** The difference between the compound amount and the original principal is called the **compound interest.** Here the compound interest is 110.25 − 100 = 10.25.

More generally, if a principal of P dollars is invested at a rate of $100r$ percent compounded annually (for example, at 5%, r is 0.05), the compound amount after 1 year is $P + Pr$, or, by factoring, $P(1 + r)$. At the end of the second year, the compound amount is

$$\begin{aligned} &P(1+r) + [P(1+r)]r \\ &= P(1+r)[1+r] \qquad \text{(factoring)} \\ &= P(1+r)^2 \end{aligned}$$

This pattern continues. After three years, the compound amount is $P(1 + r)^3$. In general, **the compound amount S of the principal P at the end of n years at the rate of r compounded annually** is given by

$$S = P(1+r)^n \qquad (1)$$

Notice from Equation (1) that, for a given principal and rate, S is a function of n. In fact, S involves an exponential function with base $1 + r$.

APPLICATIONS IN PRACTICE 5

Compound Amount and Compound Interest

Suppose \$2000 is invested at 13% compounded annually. Find the value of the investment after five years. Find the interest earned over the first five years.

EXAMPLE 6 **Compound Amount and Compound Interest**

Suppose \$1000 is invested for 10 years at 6% compounded annually.

a. *Find the compound amount.*

Solution: We use Equation (1) with $P = 1000$, $r = 0.06$, and $n = 10$:

$$S = 1000(1 + 0.06)^{10} = 1000(1.06)^{10} \approx \$1790.85$$

Figure 4.8 shows the graph of $S = 1000(1.06)^n$. Notice that as time goes on, the compound amount grows dramatically.

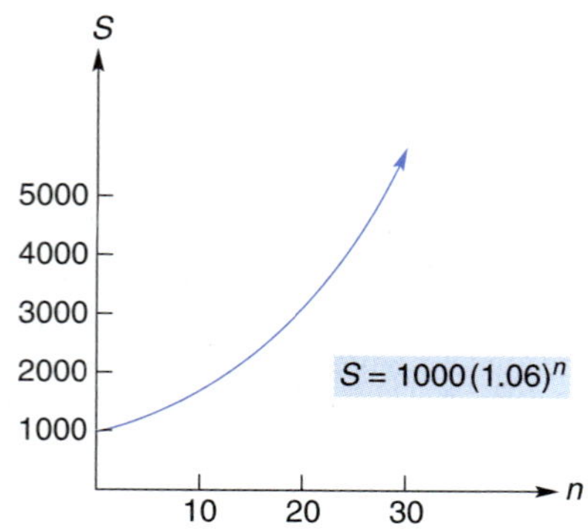

FIGURE 4.8 Graph of $S = 1000(1.06)^n$.

b. *Find the compound interest.*

Solution: Using the results from part (a), we have

$$\begin{aligned}\textbf{compound interest} &= \boldsymbol{S - P}\\ &= 1790.85 - 1000 = \$790.85\end{aligned}$$

▶ **Now work Problem 19.**

Suppose the principal of \$1000 in Example 6 is invested for 10 years as before, but this time the compounding takes place every three months (that is, *quarterly*) at the rate of $1\frac{1}{2}\%$ *per quarter*. Then there are four **interest periods** or **conversion periods** per year, and in 10 years there are $10(4) = 40$ interest periods. Thus, the compound amount with $r = 0.015$ is now

$$1000(1.015)^{40} \approx \$1814.02$$

and the compound interest is \$814.02. Usually, the interest rate per conversion period is stated as an annual rate. Here we would speak of an annual rate of 6% compounded quarterly, so that the rate per interest period, or the **periodic rate,** is $6\%/4 = 1.5\%$. This *quoted* annual rate of 6% is called the **nominal rate** or the **annual percentage rate (A.P.R.).** Unless otherwise stated, all interest rates will be assumed to be annual (nominal) rates. Thus a rate of 15% compounded monthly corresponds to a periodic rate of $15\%/12 = 1.25\%$.

The abbreviation A.P.R. is a common one and is found on credit card statements and in advertising.

On the basis of our discussion, we can generalize Equation (1). The formula

$$S = P(1 + r)^n \qquad (2)$$

gives **the compound amount S of a principal P at the end of n interest periods at the periodic rate of r.**

We have seen that for a principal of \$1000 at a nominal rate of 6% over a period of 10 years, annual compounding results in a compound interest of \$790.85, and with quarterly compounding the compound interest is \$814.02. It is typical that for a given nominal rate, the more frequent the compounding, the greater is the compound interest. However, as the number of interest periods increases, the effect tends to be less meaningful. For example, with weekly compounding the compound interest is

$$1000\left(1 + \frac{0.06}{52}\right)^{10(52)} - 1000 \approx \$821.49$$

and with daily compounding it is

$$1000\left(1 + \frac{0.06}{365}\right)^{10(365)} - 1000 \approx \$822.03$$

CAUTION A nominal rate of 6% does not necessarily mean that an investment increases in value by 6% in a year's time. The increase depends on the frequency of compounding.

The difference is not too significant here.

Sometimes the phrase "money is worth" is used to express an annual interest rate. Thus, saying that money is worth 6% compounded quarterly refers to an annual (nominal) rate of 6% compounded quarterly.

Population Growth

Equation (2) can be applied not only to the growth of money, but also to other types of growth, such as population growth. For example, suppose the population P of a town of 10,000 is increasing at the rate of 2% per year. Then P is a function of time t, in years. It is common to indicate this functional dependence by writing

$$P = P(t)$$

Here the letter P is used in two ways: On the right side, P represents the function; on the left side, P represents the dependent variable. From Equation (2), we have

$$P(t) = 10{,}000(1 + 0.02)^t = 10{,}000(1.02)^t$$

APPLICATIONS IN PRACTICE 6

Population Growth

A new company with five employees expects the number of employees to grow at the rate of 120% per year. Find the number of employees in four years.

EXAMPLE 7 **Population Growth**

The population of a town of 10,000 grows at the rate of 2% per year. Find the population three years from now.

Solution: From the preceding discussion,

$$P(t) = 10{,}000(1.02)^t$$

For $t = 3$, we have

$$P(3) = 10{,}000(1.02)^3 \approx 10{,}612$$

Thus, the population three years from now will be 10,612. (See Figure 4.9.)

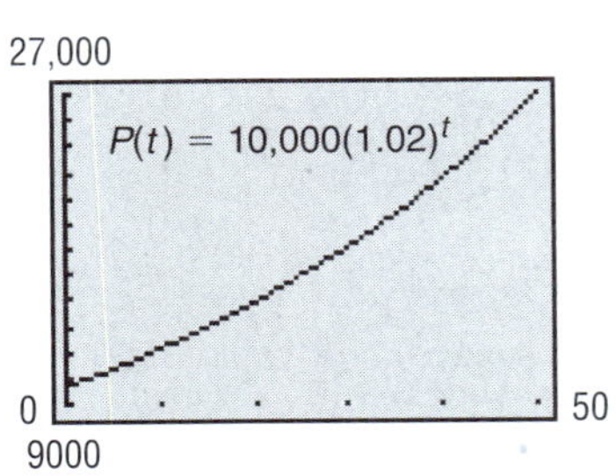

FIGURE 4.9 Graph of population function $P(t) = 10{,}000(1.02)^t$.

▸ **Now work Problem 15.**

The Number e

It is useful to conduct a "thought experiment", based on the considerations of Example 6, to introduce an important number. Suppose that a single dollar is invested for one year with an A.P.R. of 100% (remember, this is a thought experiment!) compounded annually. Then certainly the compound amount S at the end of the year is given by

$$S = 1(1+1)^1 = 2^1 = 2$$

Without changing any of the other data we now consider the effect of increasing the number of interest periods per year. If there are n interest periods per year, then the compound amount is given by

$$S = 1\left(1+\frac{1}{n}\right)^n = \left(\frac{n+1}{n}\right)^n$$

In the following table we give approximate values for $\left(\frac{n+1}{n}\right)^n$ for some values of n.

TABLE 4.2 Approximations of e

n	$\left(\frac{n+1}{n}\right)^n$
1	$\left(\frac{2}{1}\right)^1 = 2.00000$
2	$\left(\frac{3}{2}\right)^2 = 2.25000$
3	$\left(\frac{4}{3}\right)^3 \approx 2.37037$
4	$\left(\frac{5}{4}\right)^4 \approx 2.44141$
5	$\left(\frac{6}{5}\right)^5 = 2.48832$
10	$\left(\frac{11}{10}\right)^{10} \approx 2.59374$
100	$\left(\frac{101}{100}\right)^{100} \approx 2.70481$
1000	$\left(\frac{1,001}{1,000}\right)^{1,000} \approx 2.71692$
10,000	$\left(\frac{10,001}{10,000}\right)^{10,000} \approx 2.71815$
100,000	$\left(\frac{100,001}{100,000}\right)^{100,000} \approx 2.71827$
1,000,000	$\left(\frac{1,000,001}{1,000,000}\right)^{1,000,000} \approx 2.71828$

Apparently, the numbers $\left(\frac{n+1}{n}\right)^n$ increase as n does. However, they do not increase without bound. For example, it is possible to show that for any positive integer n, $\left(\frac{n+1}{n}\right)^n < 3$. In terms of our thought experiment this means that if you start with \$1.00 invested at 100%, then no matter how many interest periods there are per year, you will always have less than \$3.00 at the end of a year. There is a smallest real number that is greater than all of the numbers $\left(\frac{n+1}{n}\right)^n$. It is denoted by the letter e, in honor of the Swiss mathematician Leonhard Euler (1707–1783). The number e is irrational so that its decimal expansion is nonrepeating, like those of π and $\sqrt{2}$ that we mentioned in Section 0.1. However, each of the numerical values for $\left(\frac{n+1}{n}\right)^n$ can be considered to be a decimal approximation of e. The approximate value $\left(\frac{1{,}000{,}001}{1{,}000{,}000}\right)^{1{,}000{,}000} \approx 2.71828$ gives an approximation of e that is correct to 5 decimal places.

Exponential Function with Base e

The number

$$e \approx 2.718281828459$$

where the approximation is given correct to 12 decimal places, is most useful for a base in an exponential function. The exponential function with base e is called the **natural exponential function.**

Although e may seem to be a strange base, it arises quite naturally in calculus (as you will see in a later chapter). It also occurs in economic analysis and problems involving natural growth or decay, such as population studies, compound interest, and radioactive decay. Approximate values of e^x can be found with calculators. The graph of $y = e^x$ is shown in Figure 4.10. The accompanying table indicates y-values rounded to two decimal places. Of course, the graph has the general shape of an exponential function with base greater than 1.

APPLICATIONS IN PRACTICE 7

Graphs of Functions Involving e

The multiplicative decrease in purchasing power P after t years of inflation at 6% can be modeled by $P = e^{-0.06t}$. Graph the decrease in purchasing power as a function of t years.

You should gain familiarity with the graph of the natural exponential function in Figure 4.10.

EXAMPLE 8 **Graphs of Functions Involving e**

a. *Graph $y = e^{-x}$.*

Solution: Since $e^{-x} = \left(\frac{1}{e}\right)^x$ and $0 < \frac{1}{e} < 1$, the graph is that of an exponential function falling from left to right. (See Figure 4.11.) Alternatively,

x	y
−2	0.14
−1	0.37
0	1
1	2.72
2	7.39

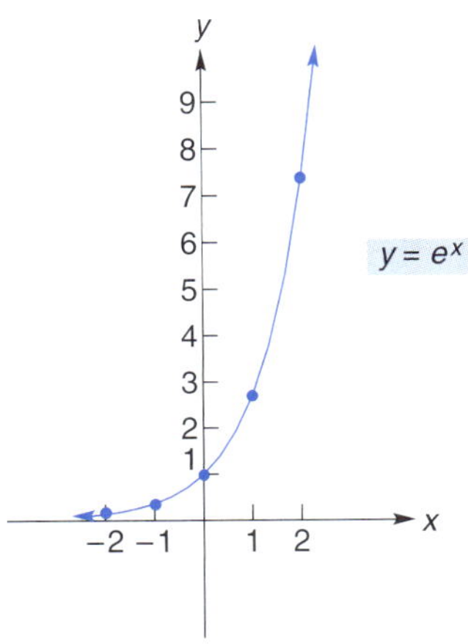

FIGURE 4.10 Graph of natural exponential function.

x	y
−2	7.39
−1	2.72
0	1
1	0.37
2	0.14

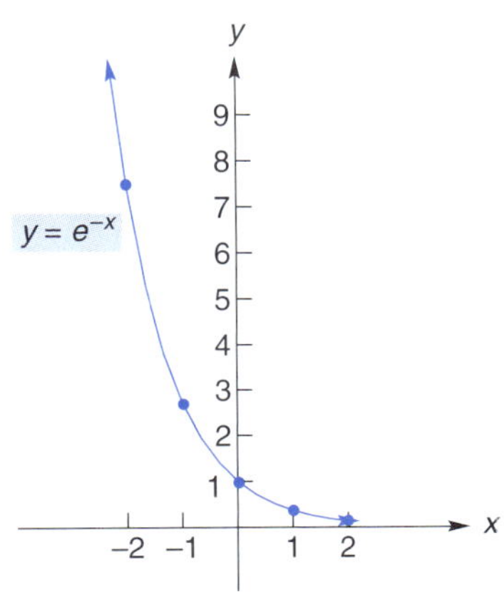

FIGURE 4.11 Graph of $y = e^{-x}$.

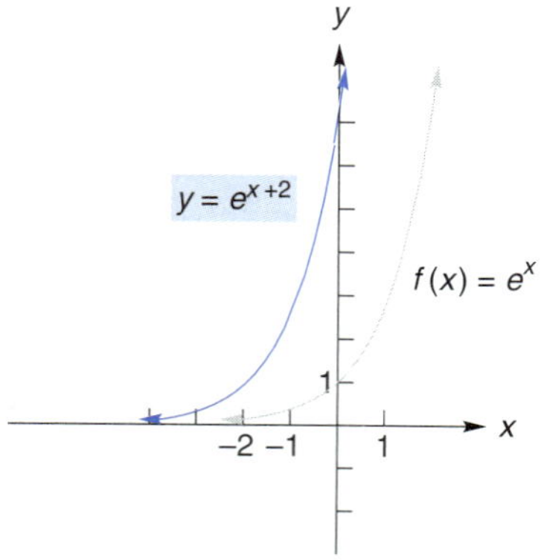

FIGURE 4.12 Graph of $y = e^{x+2}$.

we can consider the graph of $y = e^{-x}$ as a transformation of the graph of $f(x) = e^x$. Because $e^{-x} = f(-x)$, the graph of $y = e^{-x}$ is simply the reflection of the graph of f about the y-axis. (Compare the graphs in Figures 4.10 and 4.11.)

b. *Graph* $y = e^{x+2}$.

Solution: The graph of $y = e^{x+2}$ is related to that of $f(x) = e^x$. Since e^{x+2} is $f(x + 2)$, we can obtain the graph of $y = e^{x+2}$ by horizontally shifting the graph of $f(x) = e^x$ two units to the left. (See Figure 4.12.)

EXAMPLE 9 Population Growth

The projected population P of a city is given by

$$P = 100{,}000e^{0.05t}$$

where t is the number of years after 1990. *Predict the population for the year* 2010.

Solution: The number of years from 1990 to 2010 is 20, so let $t = 20$. Then

$$P = 100{,}000e^{0.05(20)} = 100{,}000e^1 = 100{,}000e \approx 271{,}828$$

Many forecasts are based on population studies.

In statistics, an important function used as a model in describing events occurring in nature is the **Poisson distribution function:**

$$f(n) = \frac{e^{-\mu}\mu^n}{n!} \qquad n = 0, 1, 2, \ldots.$$

The symbol μ (read "mu") is a Greek letter. In certain situations, $f(n)$ gives the probability that exactly n events will occur in an interval of time or space. The constant μ is the mean, or average, number of occurrences in the interval. The next example illustrates the Poisson distribution.

EXAMPLE 10 Hemocytometer and Cells

A hemocytometer is a counting chamber divided into squares and is used in studying the number of microscopic structures in a liquid. In a well-known experiment,[3] *yeast cells were diluted and thoroughly mixed in a liquid, and the mixture was placed in a hemocytometer. With a microscope, the number of yeast cells on each square was counted. The probability that there were exactly x yeast cells on a hemocytometer square was found to fit a Poisson distribution with* $\mu =$ 1.8. *Find the probability that there were exactly four cells on a particular square.*

Solution: We use the Poisson distribution function with $\mu = 1.8$ and $n = 4$:

$$f(n) = \frac{e^{-\mu}\mu^n}{n!}$$

$$f(4) = \frac{e^{-1.8}(1.8)^4}{4!} \approx 0.072$$

For example, this means that in 400 squares we would *expect* $400(0.072) \approx 29$ squares to contain exactly 4 cells. (In the experiment, the actual number observed in 400 squares was 30.)

[3]R. R. Sokal and F. J. Rohlf, *Introduction to Biostatistics* (San Francisco: W. H. Freeman and Company, Publishers, 1973).

Radioactive Decay

Radioactive elements are such that the amount of the element decreases with respect to time. We say that the element *decays*. It can be shown that, if N is the amount at time t, then

$$N = N_0 e^{-\lambda t} \tag{3}$$

where N_0 and λ (a Greek letter read "lambda") are positive constants. Notice that N involves an exponential function of t. We say that N follows an **exponential law of decay.** If $t = 0$, then $N = N_0 e^0 = N_0 \cdot 1 = N_0$. Thus, the constant N_0 represents the amount of the element present at time $t = 0$ and is called the **initial amount.** The constant λ depends on the particular element involved and is called the **decay constant.**

Because N decreases as time progresses, suppose we let T be the length of time it takes for the element to decrease to half of the initial amount. Then at time $t = T$, we have $N = N_0/2$. Equation (3) implies that

$$\frac{N_0}{2} = N_0 e^{-\lambda T}$$

We shall now use this fact to show that over *any* time interval of length T, half of the amount of the element decays. Consider the interval from time t to $t + T$, which has length T. At time t, the amount of the element is $N_0 e^{-\lambda t}$, and at time $t + T$ it is

$$\begin{aligned} N_0 e^{-\lambda(t+T)} &= N_0 e^{-\lambda t} e^{-\lambda T} = (N_0 e^{-\lambda T}) e^{-\lambda t} \\ &= \frac{N_0}{2} e^{-\lambda t} = \frac{1}{2}(N_0 e^{-\lambda t}) \end{aligned}$$

which is half of the amount at time t. This means that if the initial amount present, N_0, were 1 gram, then at time T, $\frac{1}{2}$ gram would remain; at time $2T$, $\frac{1}{4}$ gram would remain; and so on. The value of T is called the **half-life** of the radioactive element. Figure 4.13 shows a graph of radioactive decay.

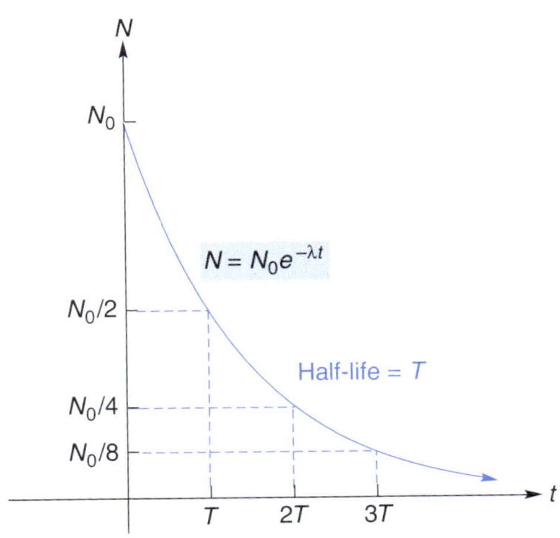

FIGURE 4.13 Radioactive decay.

EXAMPLE 11 **Radioactive Decay**

A radioactive element decays such that after t days the number of milligrams present is given by

$$N = 100e^{-0.062t}$$

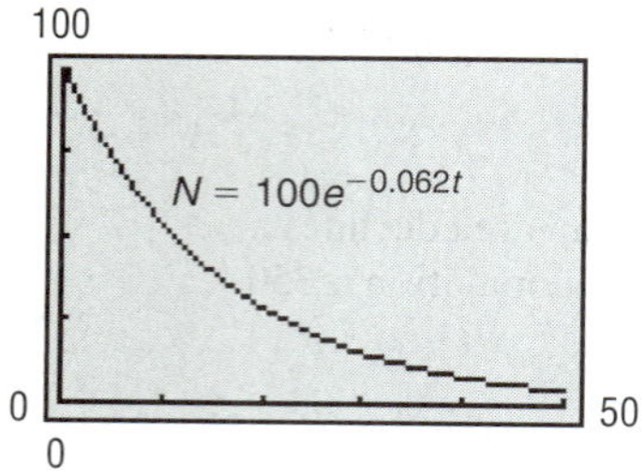

FIGURE 4.14 Graph of radioactive decay function $N = 100e^{-0.062t}$.

a. *How many milligrams are initially present?*

Solution: This equation has the form of Equation (3), $N = N_0e^{-\lambda t}$, where $N_0 = 100$ and $\lambda = 0.062$. N_0 is the initial amount and corresponds to $t = 0$. Thus, 100 milligrams are initially present. (See Figure 4.14.)

b. *How many milligrams are present after 10 days?*

Solution: When $t = 10$,

$$N = 100e^{-0.062(10)} = 100e^{-0.62} \approx 53.8$$

Therefore, approximately 53.8 milligrams are present after 10 days.

▶ **Now work Problem 47.**

Exercise 4.1

In Problems 1–12, graph each function.

*1. $y = f(x) = 4^x$

2. $y = f(x) = 3^x$

*3. $y = f(x) = \left(\frac{1}{3}\right)^x$

4. $y = f(x) = \left(\frac{1}{8}\right)^x$

5. $y = f(x) = 2\left(\frac{1}{4}\right)^x$

6. $y = f(x) = 3(2)^x$

*7. $y = f(x) = 3^{x+2}$

8. $y = f(x) = 2^{x-1}$

9. $y = f(x) = 2^x - 1$

10. $y = f(x) = 3^{x-1} - 1$

11. $y = f(x) = 3^{-x}$

12. $y = f(x) = \frac{1}{5}(3^{x/2})$

Problems 13 and 14 refer to Figure 4.15, which shows the graphs of $y = 0.4^x$, $y = 2^x$, and $y = 5^x$.

13. Of the curves A, B, and C, which is the graph of $y=5^x$?

14. Of the curves A, B, and C, which is the graph of $y=0.4^x$?

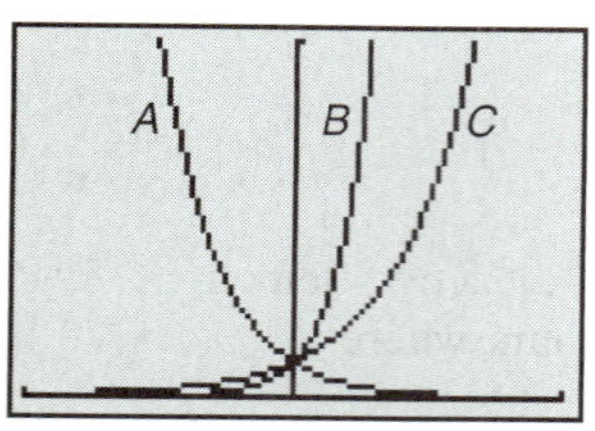

FIGURE 4.15 Diagram for Problems 13 and 14.

*15. **Population** The projected population of a city is given by $P = 125{,}000(1.11)^{t/20}$, where t is the number of years after 1995. What is the projected population in 2015?

16. **Population** For a certain city, the population P grows at the rate of 1.5% per year. The formula $P = 1{,}527{,}000(1.015)^t$ gives the population t years after 1998. Find the population in (a) 1999 and (b) 2000.

17. **Paired-Associate Learning** In a psychological experiment involving learning, subjects were asked to give particular responses after being shown certain stimuli.[4] Each stimulus was a pair of letters, and each response was either the digit 1 or 2. After each response, the subject was told the correct answer. In this so-called *paired-associate* learning experiment, the theoretical probability P that a subject makes a correct response on the nth trial is given by

$$P = 1 - \frac{1}{2}(1-c)^{n-1}, \quad n \geq 1, \quad 0 < c < 1.$$

Find P when $n = 1$.

18. Express $y = 2^{3x}$ as an exponential function in base 8.

In Problems 19–27, find (a) the compound amount and (b) the compound interest for the given investment and annual rate.

*19. \$4000 for 7 years at 6% compounded annually.

20. \$5000 for 20 years at 5% compounded annually.

21. \$700 for 15 years at 7% compounded semiannually.

22. \$4000 for 12 years at $7\frac{1}{2}$% compounded semiannually.

23. \$3000 for 16 years at $8\frac{3}{4}$% compounded quarterly.

24. \$900 for 11 years at 10% compounded quarterly.

25. \$5000 for $2\frac{1}{2}$ years at 9% compounded monthly.

26. \$500 for 5 years at 11% compounded semiannually.

27. \$8000 for 3 years at $6\frac{1}{4}$% compounded daily. (Assume that there are 365 days in a year.)

28. **Investment** Suppose \$900 is placed in a savings account that earns interest at the rate of 4.5% compounded semiannually. (a) What is the value of the account at the end of five years? (b) If the account had

[4]D. Laming, *Mathematical Psychology* (New York: Academic Press, Inc., 1973).

earned interest at the rate of 4.5% compounded annually, what would be the value after five years?

29. Investment A \$6000 certificate of deposit is purchased for \$6000 and is held seven years. If the certificate earns 8% compounded quarterly, what is it worth at the end of seven years?

30. Population Growth The population of a town of 5000 grows at the rate of 3% per year. (a) Determine an equation that gives the population t years from now. (b) Find the population three years from now. Give your answer to (b) to the nearest integer.

*__31. Bacteria Growth__ Bacteria are growing in a culture, and their number is increasing at the rate of 5% an hour. Initially, 400 bacteria are present. (a) Determine an equation that gives the number, N, of bacteria present after t hours. (b) How many bacteria are present after one hour? (c) After four hours? Give your answers to (b) and (c) to the nearest integer.

32. Bacteria Reduction A certain medicine reduces the bacteria present in a person by 10% each hour. Currently, 100,000 bacteria are present. Make a table of values for the number of bacteria present each hour for 0 to 4 hours. For each hour, write an expression for the number of bacteria as a product of 100,000 and a power of $\frac{9}{10}$. Use the expressions to make an entry in your table for the number of bacteria after t hours. Write a function N for the number of bacteria after t hours.

33. Recycling Suppose the amount of plastic being recycled increases by 30% every year. Make a table of the factor by which recycling increases over the original amount for 0 to 3 years. For each year, write an expression for the increase as a power of some base. What base did you use? How does that base relate to the problem? Use your table to graph the multiplicative increase as a function of years. Use your graph to determine when the recycling will triple.

34. Population Growth Cities A and B presently have populations of 70,000 and 60,000, respectively. City A grows at the rate of 4% per year, and B grows at the rate of 5% per year. Determine the difference in the populations of the cities at the end of five years. Give your answer to the nearest integer.

Problems 35 and 36 involve a declining population. If a population declines at the rate of r per time period, then the population after t time periods is given by

$$P = P_0(1 - r)^t$$

where P_0 is the initial population (the population when $t = 0$).

35. Population Because of an economic downturn, the population of a certain urban area declines at the rate of 1.5% per year. Initially, the population is 350,000. To the nearest person, what is the population after three years?

36. Workforce In a dramatic cost-cutting effort, a company will reduce its workforce at the rate of 2% per month for 12 months. If the company presently employs 500 workers how many workers, to the nearest worker, will it employ 12 months from now?

In Problems 37–40, use a calculator to find the value (rounded to four decimal places) of each expression.

37. $e^{1.5}$ **38.** $e^{3.4}$ **39.** $e^{-0.7}$ **40.** $e^{-2/3}$

In Problems 41 and 42, graph the functions.

41. $y = -e^x$ **42.** $y = 2e^x$

43. Telephone Calls The probability that a telephone operator will receive exactly x calls during a certain period is given by

$$P = \frac{e^{-3}3^x}{x!}$$

Find the probability that the operator will receive exactly three calls. Round your answer to four decimal places.

44. Normal Distribution An important function used in economic and business decisions is the *normal distribution density function,* which, in standard form, is

$$f(x) = \frac{1}{\sqrt{2\pi}}e^{-(\frac{1}{2})x^2}$$

Evaluate $f(0)$, $f(-1)$, and $f(1)$. Round your answers to three decimal places.

45. Express e^{kt} in the form b^t.

46. Express $\frac{1}{e^x}$ in the form b^x.

*__47. Radioactive Decay__ A radioactive element is such that N grams remain after t hours, where

$$N = 12e^{-0.031t}$$

(a) How many grams are initially present? To the nearest tenth of a gram, how many grams remain after (b) 10 hours? (c) 44 hours? (d) Based on your answer to part (c), what is your estimate of the half-life of this element?

48. Radioactive Decay At a certain time, there are 100 milligrams of a radioactive substance. The substance decays so that after t years the number of milligrams present, N, is given by

$$N = 100e^{-0.035t}$$

How many milligrams are present after 20 years? Give your answer to the nearest milligram.

49. Radioactive Decay If a radioactive substance has a half-life of 8 years, how long does it take for 1 gram of the substance to decay to $\frac{1}{16}$ gram?

50. Marketing A mail-order company advertises in a national magazine. The company finds that, of all small towns, the percentage (given as a decimal) in which exactly x people respond to an ad fits a Poisson distribution with $\mu = 0.5$. From what percentage of small towns can the company expect exactly two people to respond? Round your answer to four decimal places.

51. Emergency-Room Admissions Suppose the number of patients admitted into a hospital emergency room during a certain hour of the day has a Poisson distribution with $\mu = 4$. Find the probability that during that hour there will be exactly two emergency patients. Round your answer to four decimal places.

52. Graph $y = 17^x$ and $y = (\frac{1}{17})^x$ on the same screen. Determine the intersection point.

53. Graph $y = 2^x$ and $y = 4 \cdot 2^x$ on the same screen. It appears that the graph of $y = 4 \cdot 2^x$ is the graph of $y = 2^x$ shifted two units to the left. Prove algebraically that this is indeed true.

54. For $y = 7^x$, find x if $y = 4$. Round your answer to two decimal places.

55. For $y = 2^x$, find x if $y = 9$. Round your answer to two decimal places.

56. Cell Growth Cells are growing in a culture, and their number is increasing at the rate of 7% per hour. Initially, 1000 cells are present. After how many full hours will there be at least 3000 cells?

57. Bacteria Growth Refer to Example 1. How long will it take for 1000 bacteria to be present? Round your answer to the nearest tenth of a minute.

58. Demand Equation The demand equation for a new toy is

$$q = 10{,}000(0.95123)^p$$

(a) Evaluate q to the nearest integer when $p = 10$.
(b) Convert the demand equation to the form

$$q = 10{,}000e^{-xp}$$

(*Hint:* Find a number x such that $0.95123 \approx e^{-x}$.)
(c) Use the equation in part (b) to evaluate q to the nearest integer when $p = 10$. Your answers in parts (a) and (c) should be the same.

59. Investment If \$2500 is invested in a savings account that earns interest at the rate of 4.3% compounded annually, after how many full years will the amount at least double?

4.2 Logarithmic Functions

OBJECTIVE *To introduce logarithmic functions and their graphs. Properties of logarithms will be discussed in Section 4.3.*

Since all exponential functions pass the horizontal-line test, they are all one-to-one functions. It follows that each exponential function has an inverse. These functions, inverse to the exponential functions, are called the *logarithmic functions*.

POINTER To review inverse functions, refer to Section 2.4.

More precisely, if $f(x) = b^x$, the exponential function with base b (where $0 < b < 1$ or $1 < b$), then the inverse function $f^{-1}(x)$ is called the *logarithm function base b* and is denoted $\log_b x$. It follows from our general remarks about inverse functions in Section 2.4 that

$$y = \log_b x \quad \text{if and only if} \quad b^y = x$$

Said otherwise, given positive x, $\log_b x$ is the unique number with the property that $b^{\log_b x} = x$. The generalities about inverse functions also enable us to see immediately what the graph of a logarithmic function looks like.

In Figure 4.16 we have shown the graph of the particular exponential function $y = f(x) = 2^x$, whose general shape is typical of exponential functions $y = b^x$ for which the base b satisfies $1 < b$. We have added a (dashed) copy of the line $y = x$. The graph of $y = f^{-1}(x) = \log_2 x$ is obtained as the mirror image of $y = f(x) = 2^x$ in the line $y = x$.

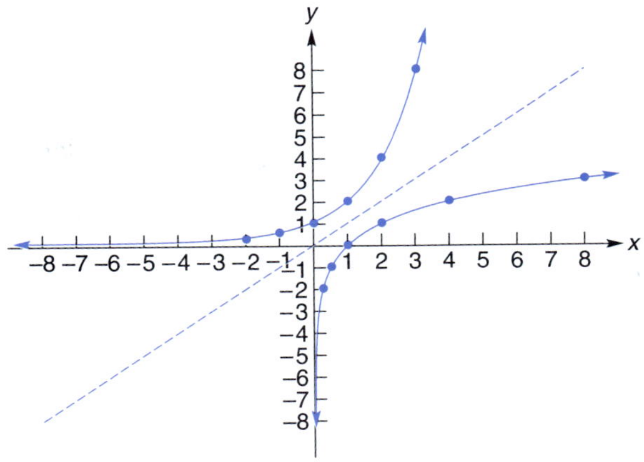

FIGURE 4.16 Graphs of $y = 2^x$ and $y = \log_2 x$.

TABLE 4.3 Selected function values

x	2^x	x	$\log_2 x$
−2	$\frac{1}{4}$	$\frac{1}{4}$	−2
−1	$\frac{1}{2}$	$\frac{1}{2}$	−1
0	1	1	0
1	2	2	1
2	4	4	2
3	8	8	3

In Table 4.3 we have tabulated the function values that appear as ordinates on the dots in Figure 4.16.

It is clear that the exponential function base 2 and the logarithm function base 2 "undo" the effects of each other. Thus, for all x in the domain of 2^x [which is $(-\infty, \infty)$] we have

$$\log_2 2^x = x$$

and, for all x in the domain of $\log_2 x$ [which is the range of 2^x which is $(0, \infty)$] we have

$$2^{\log_2 x} = x$$

It cannot be said too often that

$$y = \log_b x \quad \text{means} \quad b^y = x$$

and conversely

$$b^y = x \quad \text{means} \quad y = \log_b x$$

In this sense, *a logarithm of a number is an exponent:* $\log_b x$ is the power to which we must raise b to get x. For example,

$$\log_2 8 = 3 \quad \text{because} \quad 2^3 = 8$$

We say that $\log_2 8 = 3$ is the **logarithmic form** of the **exponential form** $2^3 = 8$. (See Figure 4.17.)

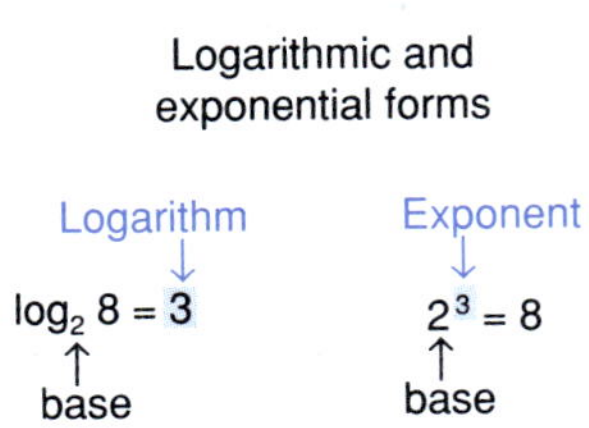

FIGURE 4.17 A logarithm can be considered an exponent.

APPLICATIONS IN PRACTICE 1

Converting from Exponential to Logarithmic Form

If bacteria have been doubling every hour and the current amount is 16 times the amount first measured, then the situation can be represented by $16 = 2^t$. Represent this equation in logarithmic form. What does t represent?

EXAMPLE 1 **Converting from Exponential to Logarithmic Form**

		Exponential Form		*Logarithmic Form*
a.	Since	$5^2 = 25$	it follows that	$\log_5 25 = 2$
b.	Since	$3^4 = 81$	it follows that	$\log_3 81 = 4$
c.	Since	$10^0 = 1$	it follows that	$\log_{10} 1 = 0$

APPLICATIONS IN PRACTICE 2

Converting from Logarithmic to Exponential Form

An earthquake measuring 8.3 on the Richter scale can be represented by $8.3 = \log_{10}\left(\frac{I}{I_0}\right)$, where I is the intensity of the earthquake and I_0 is the intensity of a zero-level earthquake. Represent this equation in exponential form.

APPLICATIONS IN PRACTICE 3

Graph of a Logarithmic Function with $b > 1$

Suppose a recycling plant has found that the amount of material being recycled has increased by 50% every year since the plant's first year of operation. Graph each year as a function of the multiplicative increase in recycling since the first year. Label the graph with the name of the function.

APPLICATIONS IN PRACTICE 4

Graph of a Logarithmic Function with $0 < b < 1$

Suppose a boat depreciates 20% every year. Graph the number of years the boat is owned as a function of the multiplicative decrease in its original value. Label the graph with the name of the function.

EXAMPLE 2 Converting from Logarithmic to Exponential Form

	Logarithmic Form		*Exponential Form*
a.	$\log_{10} 1000 = 3$	means	$10^3 = 1000$
b.	$\log_{64} 8 = \frac{1}{2}$	means	$64^{1/2} = 8$
c.	$\log_2 \frac{1}{16} = -4$	means	$2^{-4} = \frac{1}{16}$

EXAMPLE 3 Graph of a Logarithmic Function with $b > 1$

Examine again the graph of $y = \log_2 x$ in Figure 4.16. This graph is typical for a logarithmic function with $0 < b < 1$.

EXAMPLE 4 Graph of a Logarithmic Function with $0 < b < 1$

Graph $y = \log_{1/2} x$.

Solution: To plot points, we plot the inverse function $y = \left(\frac{1}{2}\right)^x$ and reflect the graph in the line $y = x$.

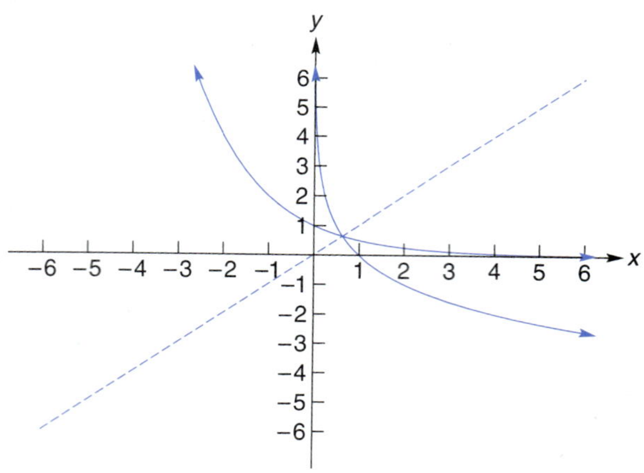

FIGURE 4.18 Graphs of $y = \left(\frac{1}{2}\right)^x$ and $y = \log_{1/2} x$.

From the graph, we can see that the domain of $y = \log_{1/2} x$ is the set of all positive reals, for that is the range of $y = \left(\frac{1}{2}\right)^x$ and the range of $y = \log_{1/2} x$ is the set of all real numbers, which is the domain of $y = \left(\frac{1}{2}\right)^x$. The graph falls from left to right. Numbers between 0 and 1 have positive base $\frac{1}{2}$ logarithms and the closer a number is to 0, the larger is its base $\frac{1}{2}$ logarithm. Numbers greater than 1 have negative base $\frac{1}{2}$ logarithms. The logarithm of 1 is 0, *regardless of the base b* and corresponds to the x-intercept $(1, 0)$. This graph is typical for a logarithmic function with $0 < b < 1$.

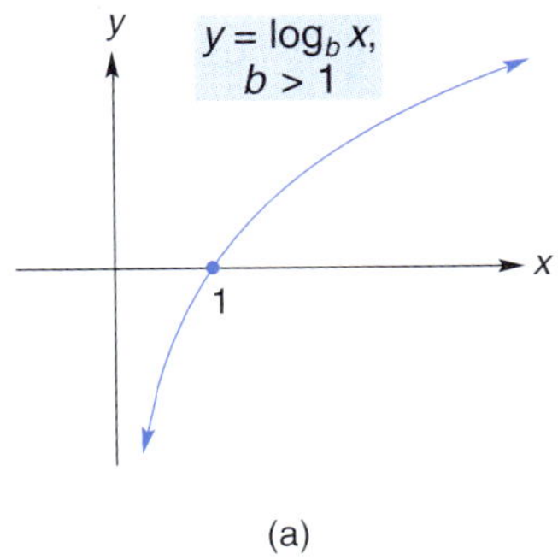

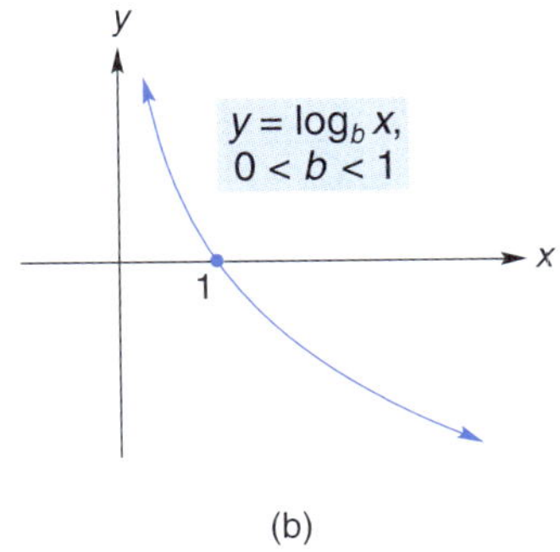

FIGURE 4.19 General shapes of $y = \log_b x$.

Summarizing the results of Examples 3 and 4, we can say that the graph of a logarithmic function has one of two general shapes, depending on whether $b > 1$ or $0 < b < 1$. (See Figure 4.19.) For $b > 1$, the graph rises from left to right; as x gets closer and closer to 0, the function values decrease without bound, and the graph gets closer and closer to the y-axis. For $0 < b < 1$, the graph falls from left to right; as x gets closer and closer to 0, the function values increase without bound, and the graph gets closer and closer to the y-axis. In each case, note that

1. The domain of a logarithmic function is the interval $(0, \infty)$. Thus, the logarithm of either a negative number or 0 does not exist.
2. The range is the interval $(-\infty, \infty)$.
3. The logarithm of 1 is 0, which corresponds to the x-intercept $(1, 0)$.

You should gain familiarity with the graph of the natural logarithmic function in Figure 4.20.

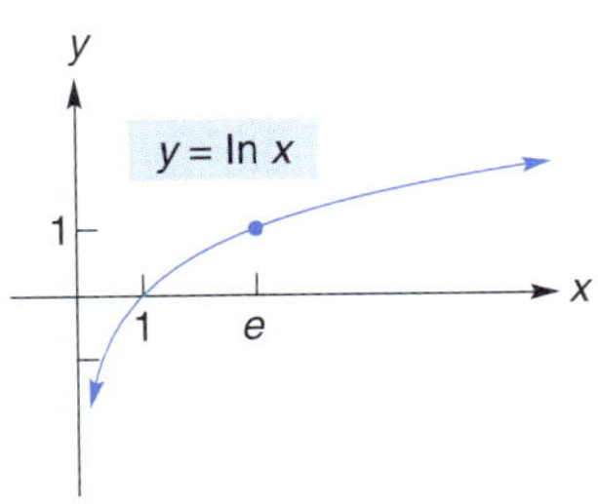

FIGURE 4.20 Graph of natural logarithmic function.

Logarithms to the base 10 are called **common logarithms.** They were frequently used for computational purposes before the calculator age. The subscript 10 is usually omitted from the notation:

$$\log x \quad \text{means} \quad \log_{10} x$$

Important in calculus are logarithms to the base e, called **natural logarithms.** We use the notation "ln" for such logarithms:

$$\ln x \quad \text{means} \quad \log_e x$$

The symbol $\ln x$ may be read "natural log of x." Your calculator gives approximate values for natural and common logarithms. For example, verify that $\ln 2 \approx 0.69315$. This means that $e^{0.69315} \approx 2$. Figure 4.20 shows the graph of $y = \ln x$. Because $e > 1$, the graph has the general shape of that of a logarithmic function with $b > 1$ (see Figure 4.19(a)) and rises from left to right.

APPLICATIONS IN PRACTICE 5

Finding Logarithms

The number of years it takes for an amount invested at an annual rate of r and compounded continuously to quadruple is a function of the annual rate r given by $t(r) = \dfrac{\ln 4}{r}$. Use a calculator to find the rate needed to quadruple an investment in 10 years.

EXAMPLE 5 Finding Logarithms

a. *Find* log 100.

Solution: Here the base is 10. Thus, log 100 is the power to which we must raise 10 to get 100. Since $10^2 = 100$, $\log 100 = 2$.

b. *Find* ln 1.

Solution: Here the base is e. Because $e^0 = 1$, $\ln 1 = 0$.

c. *Find* log 0.1.

Solution: Since $0.1 = \frac{1}{10} = 10^{-1}$, $\log 0.1 = -1$.

d. *Find* $\ln e^{-1}$.

Remember: A logarithm (in a sense) is an exponent.

Solution: Since $\ln e^{-1}$ is the power to which e must be raised to obtain e^{-1}, clearly $\ln e^{-1} = -1$.

e. *Find* $\log_{36} 6$.

Solution: Because $36^{1/2}$ (or $\sqrt{36}$) is 6, $\log_{36} 6 = \frac{1}{2}$.

▶ **Now work Problem 3.**

Many equations involving logarithmic or exponential forms can be solved for an unknown quantity by first transforming from logarithmic form to exponential form or vice versa. Example 6 will illustrate.

APPLICATIONS IN PRACTICE 6

Solving Logarithmic and Exponential Equations

The multiplicative increase m of an amount invested at an annual rate of r compounded continuously for a time t is given by $m = e^{rt}$. What annual percentage rate is needed to triple the investment in 12 years?

EXAMPLE 6 **Solving Logarithmic and Exponential Equations**

a. *Solve* $\log_2 x = 4$.

Solution: We can get an explicit expression for x by writing the equation in exponential form. This gives

$$2^4 = x$$

so $x = 16$.

b. *Solve* $\ln(x + 1) = 7$.

Solution: The exponential form yields $e^7 = x + 1$. Thus, $x = e^7 - 1$.

c. *Solve* $\log_x 49 = 2$.

Solution: In exponential form, $x^2 = 49$, so $x = 7$. We reject $x = -7$ because a negative number cannot be the base of a logarithmic function.

d. *Solve* $e^{5x} = 4$.

Solution: We can get an explicit expression for x by writing the equation in logarithmic form. We have

$$\ln 4 = 5x$$
$$x = \frac{\ln 4}{5}$$

▶ **Now work Problem 49.**

Radioactive Decay and Half-Life

From our discussion of the decay of a radioactive element in Section 4.1, we know that the amount of the element present at time t is given by

$$N = N_0 e^{-\lambda t} \tag{1}$$

where N_0 is the initial amount (the amount at time $t = 0$) and λ is the decay constant. Let us now determine the half-life T of the element. At time T, half of the initial amount is present. That is, when $t = T$, $N = N_0/2$. Thus, from

Equation (1), we have

$$\frac{N_0}{2} = N_0 e^{-\lambda T}$$

Solving for T gives

$$\frac{1}{2} = e^{-\lambda T}$$

$$2 = e^{\lambda T} \qquad \text{(taking reciprocals of both sides)}$$

To get an explicit expression for T, we shall convert to logarithmic form. This results in

$$\lambda T = \ln 2$$

$$T = \frac{\ln 2}{\lambda}$$

Summarizing, we have the following:

If a radioactive element has decay constant λ, then the half-life of the elements is given by

$$T = \frac{\ln 2}{\lambda} \qquad (2)$$

EXAMPLE 7 Finding Half-Life

A 10-milligram sample of radioactive polonium 210 (^{210}Po) *decays according to the equation*

$$N = 10e^{-0.00501t}$$

where N is the number of milligrams present after t days. (See Figure 4.21.) Determine the half-life of ^{210}Po.

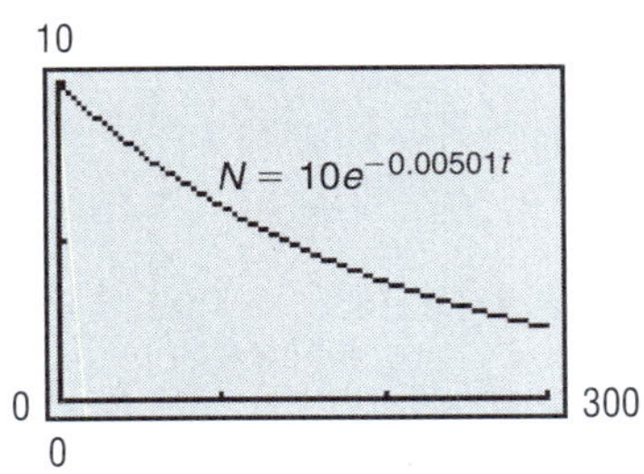

FIGURE 4.21 Radioactive decay function $N = 10e^{-0.00501t}$.

Solution: Here the decay constant λ is 0.00501. By Equation (2), the half-life is given by

$$T = \frac{\ln 2}{\lambda} = \frac{\ln 2}{0.00501} \approx 138.4 \text{ days}$$

▸ **Now work Problem 63.**

Exercise 4.2

In Problems 1–8, express each logarithmic form exponentially and each exponential form logarithmically.

1. $10^4 = 10{,}000$ **2.** $2 = \log_{12} 144$

*__3.__ $\log_2 64 = 6$ **4.** $8^{2/3} = 4$

5. $e^3 = 20.0855$ **6.** $e^{0.33647} = 1.4$

7. $\ln 3 = 1.09861$ **8.** $\log 5 = 0.6990$

In Problems 9–16, graph the functions.

9. $y = f(x) = \log_3 x$ **10.** $y = f(x) = \log_4 2x$

11. $y = f(x) = \log_{1/4} x$ **12.** $y = f(x) = \log_{1/5} x$

13. $y = f(x) = \log_2(x - 4)$ **14.** $y = f(x) = \log_2(-x)$

15. $y = f(x) = -2 \ln x$ **16.** $y = f(x) = \ln(x + 2)$

In Problems 17–28, evaluate the expression.

17. $\log_6 36$ **18.** $\log_2 64$

19. $\log_3 27$ **20.** $\log_{16} 4$

21. $\log_7 7$ **22.** $\log 10{,}000$

23. $\log 0.01$ **24.** $\log_2 \sqrt[3]{2}$

25. $\log_5 1$ **26.** $\log_5 \frac{1}{25}$

27. $\log_2 \frac{1}{8}$ **28.** $\log_4 \sqrt[5]{4}$

In Problems 29–48, find x.

29. $\log_3 x = 4$ **30.** $\log_2 x = 8$

31. $\log_5 x = 3$

32. $\log_4 x = 0$

33. $\log x = -1$

34. $\ln x = 1$

35. $\ln x = -3$

36. $\log_x 25 = 2$

37. $\log_x 8 = 3$

38. $\log_x 3 = \frac{1}{2}$

39. $\log_x \frac{1}{6} = -1$

40. $\log_x y = 1$

41. $\log_3 x = -3$

42. $\log_x(2x-3) = 1$

43. $\log_x(6-x) = 2$

44. $\log_8 64 = x - 1$

45. $2 + \log_2 4 = 3x - 1$

46. $\log_3(x+2) = -2$

47. $\log_x(2x+8) = 2$

48. $\log_x(6+4x-x^2) = 2$

In Problems 49–52, find x and express your answer in terms of natural logarithms.

***49.** $e^{3x} = 2$

50. $0.1e^{0.1x} = 0.5$

51. $e^{2x-5} + 1 = 4$

52. $6e^{2x} - 1 = \frac{1}{2}$

In Problems 53–56, use your calculator to find the approximate value of each expression. Round your answer to five decimal places.

53. $\ln 5$

54. $\ln 4.27$

55. $\ln 7.39$

56. $\ln 9.98$

57. Appreciation Suppose an antique gains 10% in value every year. Graph the number of years it is owned as a function of the multiplicative increase in its original value. Label the graph with the name of the function.

58. Cost Equation The cost for a firm producing q units of a product is given by the cost equation

$$c = (3q \ln q) + 12$$

Evaluate the cost when $q = 6$. (Round your answer to two decimal places.)

59. Supply Equation A manufacturer's supply equation is

$$p = \log\left(10 + \frac{q}{2}\right)$$

where q is the number of units supplied at a price p per unit. At what price will the manufacturer supply 1980 units?

60. Earthquake The magnitude, M, of an earthquake and its energy, E, are related by the equation[5]

$$1.5M = \log\left(\frac{E}{2.5 \times 10^{11}}\right)$$

where M is given in terms of Richter's preferred scale of 1958 and E is in ergs. Solve the equation for E.

61. Biology For a certain population of cells, the number of cells at time t is given by $N = N_0(2^{t/k})$, where N_0 is the number of cells at $t = 0$ and k is a positive constant. (a) Find N when $t = k$. (b) What is the significance of k? (c) Show that the time it takes to have population N_1 can be written

$$t = k \log_2 \frac{N_1}{N_0}$$

62. Inferior Good In a discussion of an inferior good, Persky[6] solves an equation of the form

$$u_0 = A\ln(x_1) + \frac{x_2^2}{2}$$

for x_1, where x_1 and x_2 are quantities of two products, u_0 is a measure of utility, and A is a positive constant. Determine x_1.

***63. Radioactive Decay** A 1-gram sample of radioactive lead 211 (^{211}Pb) decays according to the equation $N = e^{-0.01920t}$, where N is the number of grams present after t minutes. Find the half-life of ^{211}Pb to the nearest tenth of a minute.

64. Radioactive Decay A 100-milligram sample of radioactive actinium 227 (^{227}Ac) decays according to the equation

$$N = 100e^{-0.03194t}$$

where N is the number of milligrams present after t years. Find the half-life of ^{227}Ac to the nearest tenth of a year.

65. If $\log_y x = 3$ and $\log_z x = 2$, find a formula for z as an explicit function of y only.

66. Solve for y as an explicit function of x if

$$x + 3e^{2y} - 8 = 0$$

67. Suppose $y = f(x) = x \ln x$. (a) For what values of x is $y < 0$? [*Hint:* Determine when the graph is below the x-axis.] (b) Determine the range of f.

68. Find the x-intercept of $y = x^2 \ln x$.

69. Use the graph of $y = e^x$ to estimate $\ln 3$. Round your answer to two decimal places.

70. Use the graph of $y = \ln x$ to estimate e^2. Round your answer to two decimal places.

71. Determine the x-values of points of intersection of the graphs of $y = (x-2)^2$ and $y = \ln x$. Round your answers to two decimal places.

[5] K. E. Bullen, *An Introduction to the Theory of Seismology* (Cambridge, U.K.: Cambridge at the University Press, 1963).

[6] A. L. Persky, "An Inferior Good and a Novel Indifference Map," *American Economist* 29, no. 1 (spring 1985).

OBJECTIVE *To study basic properties of logarithmic functions.*

4.3 Properties of Logarithms

The logarithmic function has many important properties. For example, the logarithm of a product of two numbers is the sum of the logarithms of the numbers. Symbolically, we have $\log_b(mn) = \log_b m + \log_b n$. To prove this, we let $x = \log_b m$ and $y = \log_b n$. Then $b^x = m$, $b^y = n$, and

$$mn = b^x b^y = b^{x+y}$$

Thus, $mn = b^{x+y}$. In logarithmic form, this means that $\log_b(mn) = x + y$. Therefore, $\log_b(mn) = \log_b m + \log_b n$.

1. $\log_b(mn) = \log_b m + \log_b n$

That is, the logarithm of a product is the sum of the logarithms of the factors.

We shall not prove the next two properties, since their proofs are similar to that of Property 1.

2. $\log_b \dfrac{m}{n} = \log_b m - \log_b n$

That is, the logarithm of a quotient is the difference of the logarithm of the numerator and the logarithm of the denominator.

CAUTION Make sure that you clearly understand Properties 1–3. They do not apply to the log of a sum $[\log_b(m+n)]$, to the log of a difference $[\log_b(m-n)]$, to the product of two logs $[(\log_b m)(\log_b n)]$, or to the quotient of two logs $\left[\dfrac{\log_b m}{\log_b n}\right]$.

3. $\log_b m^r = r \log_b m$

That is, the logarithm of a power of a number is the exponent times the logarithm of the number.

Table 4.4 gives the values of a few common logarithms. Most entries are approximate. For example, $\log 4 \approx 0.6021$, which means $10^{0.6021} \approx 4$. To illustrate the use of properties of logarithms, we shall use this table in some of the examples that follow.

TABLE 4.4 Common Logarithms

x	$\log x$	x	$\log x$
2	0.3010	7	0.8451
3	0.4771	8	0.9031
4	0.6021	9	0.9542
5	0.6990	10	1.0000
6	0.7782	e	0.4343

EXAMPLE 1 **Finding Logarithms by Using Table 4.4**

Although the logarithms in Example 1 can be found with a calculator, we shall make use of properties of logarithms.

a. *Find* log 56.

Solution: Log 56 is not in the table. But we can write 56 as the product $8 \cdot 7$. Thus, by Property 1,

$$\log 56 = \log(8 \cdot 7) = \log 8 + \log 7 \approx 0.9031 + 0.8451 = 1.7482$$

b. *Find* $\log \frac{9}{2}$.

Solution: By Property 2,

$$\log \frac{9}{2} = \log 9 - \log 2 \approx 0.9542 - 0.3010 = 0.6532$$

c. *Find* log 64.

Solution: Since $64 = 8^2$, by Property 3,

$$\log 64 = \log 8^2 = 2\ \log 8 \approx 2(0.9031) = 1.8062$$

d. *Find* $\log \sqrt{5}$.

Solution: By Property 3, we have

$$\log \sqrt{5} = \log 5^{1/2} = \frac{1}{2} \log 5 \approx \frac{1}{2}(0.6990) = 0.3495$$

e. *Find* $\log \frac{16}{21}$.

Solution:

$$\begin{aligned} \log \frac{16}{21} &= \log 16 - \log 21 = \log(4^2) - \log(3 \cdot 7) \\ &= 2\ \log 4 - [\log 3 + \log 7] \\ &\approx 2(0.6021) - [0.4771 + 0.8451] = -0.1180 \end{aligned}$$

▶ **Now work Problem 3.**

EXAMPLE 2 **Rewriting Logarithmic Expressions**

a. *Express* $\log \frac{1}{x^2}$ *in terms of* $\log x$.

Solution:

$$\log \frac{1}{x^2} = \log x^{-2} = -2\ \log x \qquad \text{(Property 3)}$$

Here we have assumed that $x > 0$. Although $\log(1/x^2)$ is defined for $x \neq 0$, the expression $-2 \log x$ is defined only if $x > 0$.

b. *Express* $\log \frac{1}{x}$ *in terms of* $\log x$.

Solution: By Property 3,

$$\log \frac{1}{x} = \log x^{-1} = -1\ \log x = -\log x$$

From Example 2(b), we see that $\log(1/x) = -\log x$. Generalizing gives the following property:

4. $\log_b \dfrac{1}{m} = -\log_b m$

That is, the logarithm of the reciprocal of a number is the negative of the logarithm of the number.

For example, $\log \dfrac{2}{3} = -\log \dfrac{3}{2}$.

▶ **Now work Problem 19.**

EXAMPLE 3 **Writing Logarithms in Terms of Simpler Logarithms**

Manipulations such as those in Example 3 are frequently used in calculus.

a. *Write* $\ln \dfrac{x}{zw}$ *in terms of* $\ln x$, $\ln z$, *and* $\ln w$.

Solution:

$$\begin{aligned} \ln \frac{x}{zw} &= \ln x - \ln(zw) && \text{(Property 2)} \\ &= \ln x - (\ln z + \ln w) && \text{(Property 1)} \\ &= \ln x - \ln z - \ln w \end{aligned}$$

b. *Write* $\ln \sqrt[3]{\dfrac{x^5(x-2)^8}{x-3}}$ *in terms of* $\ln x$, $\ln(x-2)$, *and* $\ln(x-3)$.

Solution:

$$\begin{aligned} \ln \sqrt[3]{\frac{x^5(x-2)^8}{x-3}} &= \ln \left[\frac{x^5(x-2)^8}{x-3}\right]^{1/3} = \frac{1}{3} \ln \frac{x^5(x-2)^8}{x-3} \\ &= \frac{1}{3}\{\ln[x^5(x-2)^8] - \ln(x-3)\} \\ &= \frac{1}{3}[\ln x^5 + \ln(x-2)^8 - \ln(x-3)] \\ &= \frac{1}{3}[5 \ln x + 8 \ln(x-2) - \ln(x-3)] \end{aligned}$$

▶ **Now work Problem 29.**

APPLICATIONS IN PRACTICE 1

Combining Logarithms

The Richter scale measure of an earthquake is given by $R = \log\left(\dfrac{I}{I_0}\right)$, where I is the intensity of the earthquake and I_0 is the intensity of a zero-level earthquake. How much more on the Richter scale is an earthquake with intensity 900,000 times the intensity of a zero-level earthquake than an earthquake with intensity 9000 times the intensity of a zero-level earthquake? Write the answer as an expression involving logarithms. Simplify the expression by combining logarithms, and then evaluate the resulting expression.

EXAMPLE 4 **Combining Logarithms**

a. *Write* $\ln x - \ln(x+3)$ *as a single logarithm.*

Solution:

$$\ln x - \ln(x+3) = \ln \frac{x}{x+3} \qquad \text{(Property 2)}$$

b. *Write* $\ln 3 + \ln 7 - \ln 2 - 2\ln 4$ *as a single logarithm.*

Solution:

$$\begin{aligned}
&\ln 3 + \ln 7 - \ln 2 - 2\ln 4 \\
&\quad = \ln 3 + \ln 7 - \ln 2 - \ln(4^2) && \text{(Property 3)} \\
&\quad = \ln 3 + \ln 7 - [\ln 2 + \ln(4^2)] \\
&\quad = \ln(3\cdot 7) - \ln(2\cdot 4^2) && \text{(Property 1)} \\
&\quad = \ln 21 - \ln 32 \\
&\quad = \ln\frac{21}{23} && \text{(Property 2)}
\end{aligned}$$

▶ **Now work Problem 37.**

Since $b^0 = 1$ and $b^1 = b$, by converting to logarithmic forms we have the following properties:

5. $\log_b 1 = 0$

6. $\log_b b = 1$

By Property 3, $\log_b b^r = r\log_b b$. But by Property 6, $\log_b b = 1$. Thus, we have the next property:

7. $\log_b b^r = r$

If we write $f(x) = b^x$, then $f^{-1}(x) = \log_b x$ and Property 7 is then just $f^{-1}(f(r)) = r$, as in Equation (2) of Section 2.4.

APPLICATIONS IN PRACTICE 2

Simplifying Logarithmic Expressions

If an earthquake is 10,000 times as intense as a zero-level earthquake, what is its measurement on the Richter scale? Write the answer as a logarithmic expression and simplify it. (See previous Application in Practice for the formula.)

EXAMPLE 5 **Simplifying Logarithmic Expressions**

a. *Find* $\ln e^{3x}$.

Solution: By Property 7 with $b = e$, we have $\ln e^{3x} = 3x$. Alternatively, by Properties 3 and 6,

$$\ln e^{3x} = 3x\ln e = 3x(1) = 3x$$

b. *Find* $\log 1 + \log 1000$.

Solution: By Property 5, $\log 1 = 0$. Thus,

$$\begin{aligned}
\log 1 + \log 1000 &= 0 + \log 10^3 \\
&= 0 + 3 && \text{(Property 7 with } b = 10) \\
&= 3
\end{aligned}$$

c. *Find* $\log_7 \sqrt[9]{7^8}$.

Solution:

$$\log_7 \sqrt[9]{7^8} = \log_7 7^{8/9} = \frac{8}{9}$$

d. *Find* $\log_3\left(\frac{27}{81}\right)$.

Solution:

$$\log_3\left(\frac{27}{81}\right) = \log_3\left(\frac{3^3}{3^4}\right) = \log_3(3^{-1}) = -1$$

e. *Find* $\ln e + \log\frac{1}{10}$.

Solution:

$$\begin{aligned}\ln e + \log\frac{1}{10} &= \ln e + \log 10^{-1} \\ &= 1 + (-1) = 0\end{aligned}$$

▶ **Now work Problem 41.**

Do not confuse $\ln x^2$ with $(\ln x)^2$. We have

$$\ln x^2 = \ln(x \cdot x)$$

but

$$(\ln x)^2 = (\ln x)(\ln x)$$

Sometimes $(\ln x)^2$ is written as $\ln^2 x$. This is not a new formula but merely a notation. More generally, some people write $f^2(x)$ for $(f(x))^2$. We recommend avoiding the notation $f^2(x)$.

Our next property is the following:

8. $b^{\log_b m} = m$

and, in particular,

$$10^{\log x} = x \quad \text{and} \quad e^{\ln x} = x$$

Property 8 is true because it states, in logarithmic form, that $\log_b m = \log_b m$.

If we again write $f(x) = b^x$, then $f^{-1}(x) = \log_b x$ and Property 8 is then just $f(f^{-1}(m)) = m$, as in Equation (2) of Section 2.4.

EXAMPLE 6 **Use of Property 8**

a. *Find* $e^{\ln x^2}$.

Solution: By Property 8, $e^{\ln x^2} = x^2$.

b. *Solve* $10^{\log x^2} = 25$ *for* x.

Solution:

$$\begin{aligned}10^{\log x^2} &= 25 \\ x^2 &= 25 \qquad \text{(Property 8)} \\ x &= \pm 5\end{aligned}$$

▶ **Now work Problem 45.**

EXAMPLE 7 **Evaluating a Logarithm Base 5**

Use a calculator to find $\log_5 2$.

Solution: Calculators typically have keys for logarithms in base 10 and base e, but not for base 5. However, we can convert logarithms in one base to logarithms in another base. Let us convert from base 5 to base 10. First, let $x = \log_5 2$. Then $5^x = 2$. Taking the common logarithms of both sides of $5^x = 2$ gives

$$\begin{aligned} \log 5^x &= \log 2 \\ x \log 5 &= \log 2 \\ x &= \frac{\log 2}{\log 5} \approx 0.4307 \end{aligned}$$

If we had taken natural logarithms of both sides, the result would be $x = (\ln 2)/(\ln 5) \approx 0.4307$, the same as before.

Generalizing the method used in Example 7, we obtain the so-called *change-of-base* formula:

Change-of-Base Formula

9. $\log_b m = \dfrac{\log_a m}{\log_a b}$

Some students find the change of base formula more memorable when it is expressed in the form:

$$(\log_a b)(\log_b m) = \log_a m$$

in which the two instances of b apparently cancel. Let us see how to prove this identity, for ability to see the truth of such statements greatly enhances one's ability to use them in practical applications. Since $\log_a m = y$ precisely if $a^y = m$ our task is equivalently to show that

$$a^{(\log_a b)(\log_b m)} = m$$

and we have

$$\begin{aligned} a^{(\log_a b)(\log_b m)} &= \left(a^{\log_a b}\right)^{\log_b m} \\ &= b^{\log_b m} \\ &= m \end{aligned}$$

as required.

The change-of-base formula allows logarithms to be converted from one base (a) to another (b).

EXAMPLE 8 **Change-of-Base Formula**

Express $\log x$ *in terms of natural logarithms.*

Solution: We must transform from base 10 into base e. Thus, we use the change-of-base formula (Property 9) with $b = 10$, $m = x$, and $a = e$:

$$\log x = \log_{10} x = \frac{\log_e x}{\log_e 10} = \frac{\ln x}{\ln 10}$$

▸ **Now work Problem 49.**

TECHNOLOGY

Problem: *Display the graph of* $y = \log_2 x$.

Solution: To enter the function, we must first convert it to base e or base 10. We shall choose base e. By Property 9,

$$y = \log_2 x = \frac{\log_e x}{\log_e 2} = \frac{\ln x}{\ln 2}$$

Now we graph $y = (\ln x)/(\ln 2)$, which is shown in Figure 1.22.

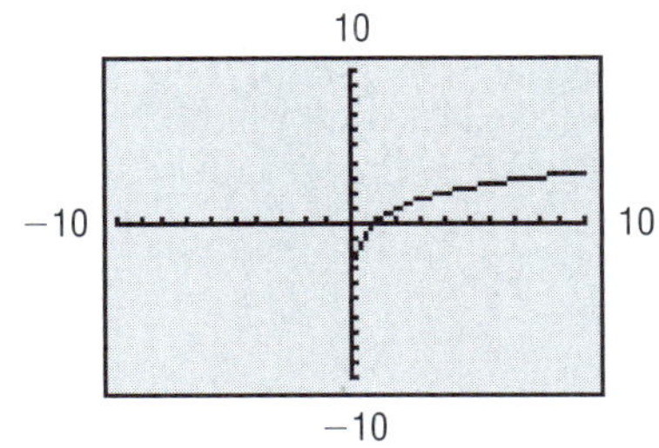

FIGURE 4.22 Graph of $y = \log_2 x$.

Exercise 4.3

In Problems 1–10, let $\log 2 = a$, $\log 3 = b$, *and* $\log 5 = c$. *Express the indicated logarithm in terms of a, b, or c.*

1. $\log 15$ **2.** $\log 16$ *__3.__ $\log \frac{2}{3}$

4. $\log \frac{5}{2}$ **5.** $\log \frac{8}{3}$ **6.** $\log \frac{6}{25}$

7. $\log 36$ **8.** $\log 0.0002$ **9.** $\log_2 3$

10. $\log_3 5$

In Problems 11–20, determine the value of the expression without the use of a calculator.

11. $\log_7 7^{48}$ **12.** $\log_5(5\sqrt{5})^5$ **13.** $\log 0.00001$

14. $10^{\log 3.4}$ **15.** $\ln e^{5.01}$ **16.** $\ln e$

17. $\ln \dfrac{1}{e^2}$ **18.** $\log_3 81$ *__19.__ $\log \frac{1}{10} + \ln e^3$

20. $e^{\ln 6}$

In Problems 21–32, write the expression in terms of $\ln x$, $\ln(x+1)$, *and/or* $\ln(x+2)$.

21. $\ln[x(x+1)^2]$ **22.** $\ln \dfrac{\sqrt{x}}{x+1}$

23. $\ln \dfrac{x^2}{(x+1)^3}$ **24.** $\ln[x(x+1)]^3$

25. $\ln\left(\dfrac{x}{x+1}\right)^3$ **26.** $\ln\sqrt{x(x+1)(x+2)}$

27. $\ln \dfrac{x}{(x+1)(x+2)}$ **28.** $\ln \dfrac{x^2(x+1)}{x+2}$

*__29.__ $\ln \dfrac{\sqrt{x}}{(x+1)^2(x+2)^3}$ **30.** $\ln \dfrac{x^2}{(x-1)(x-2)}$

31. $\ln\left[\dfrac{1}{x+2}\sqrt[5]{\dfrac{x^2}{x+1}}\right]$ **32.** $\ln\sqrt{\dfrac{x^4(x+1)^3}{x+2}}$

In Problems 33–40, express each of the given forms as a single logarithm.

33. $\log 6 + \log 4$ **34.** $\log_3 10 - \log_3 5$

35. $\log_2(2x) - \log_2(x+1)$ **36.** $2\log x - \frac{1}{2}\log(x-2)$

*__37.__ $9\log 7 + 5\log 23$

38. $5(2\log x + 3\log y - 2\log z)$

39. $2 + 10\log 1.05$

40. $\dfrac{1}{2}(\log 215 + 8\log 6 - 3\log 169)$

In Problems 41–44, determine the values of the expressions without using a calculator.

*__41.__ $e^{4\ln 3 - 3\ln 4}$

42. $\log_2[\ln(\sqrt{5+e^2}+\sqrt{5}) + \ln(\sqrt{5+e^2}-\sqrt{5})]$

43. $\log_6 54 - \log_6 9$

44. $\log_2\sqrt{2} + \log_3\sqrt[3]{3} - \log_4\sqrt[4]{4}$

In Problems 45–48, find x.

*__45.__ $e^{\ln(2x)} = 5$ **46.** $4^{\log_4 x + \log_4 2} = 3$

47. $10^{\log x^2} = 4$ **48.** $e^{3\ln x} = 8$

In Problems 49–52, write each expression in terms of natural logarithms.

*__49.__ $\log(x+6)$ **50.** $\log_3(x^2+2x+2)$

51. $\log_3(x^2+1)$ **52.** $\log_5(9-x^2)$

53. If $e^{\ln z} = 7e^y$, solve for y in terms of z.

54. **Statistics** In statistics, the sample regression equation $y = ab^x$ is reduced to a linear form by taking logarithms of both sides. Express $\log y$ in terms of x, $\log a$, and log b and explain what is meant by saying that the resulting expression is linear.

55. Military Compensation In a study of military enlistments, Brown[7] considers total military compensation C as the sum of basic military compensation B (which includes the value of allowances, tax advantages, and base pay) and educational benefits E. Thus, $C = B + E$. Brown states that

$$\ln C = \ln B + \ln\left(1 + \frac{E}{B}\right)$$

Verify this.

56. Earthquake According to Richter,[8] the magnitude M of an earthquake occurring 100 km from a certain type of seismometer is given by $M = \log(A) + 3$, where A is the recorded trace amplitude (in millimeters) of the quake. (a) Find the magnitude of an earthquake that records a trace amplitude of 1 mm. (b) If a particular earthquake has amplitude A_1 and magnitude M_1, determine the magnitude of a quake with amplitude $100A_1$. Express your answer to part (b) in terms of M_1.

57. Display the graph of $y = \log_6 x$.

58. Display the graph of $y = \log_4(x + 2)$.

59. Display the graphs of $y = \log x$ and $y = \dfrac{\ln x}{\ln 10}$ on the same screen. The graphs appear to be identical. Why?

60. On the same screen, display the graphs of $y = \ln x$ and $y = \ln(4x)$. It appears that the graph of $y = \ln(4x)$ is the graph of $y = \ln x$ shifted upward. Determine algebraically the value of this shift.

61. On the same screen, display the graphs of $y = \ln x$ and $y = \ln(x/\pi)$. It appears that the graph of $y = \ln(x/\pi)$ is the graph of $y = \ln x$ shifted downward. Determine algebraically the value of this shift and calculate its approximate value using your calculator.

[7]C. Brown, "Military Enlistments: What Can We Learn from Geographic Variation?" *American Economic Review* 75, no. 1 (1985), 228–34.

[8]C. F. Richter, *Elementary Seismology* (San Francisco: W. H. Freeman and Company, Publisher, 1958).

OBJECTIVE *To develop techniques for solving logarithmic and exponential equations.*

4.4 Logarithmic and Exponential Equations

Here we shall solve *logarithmic* and *exponential equations*. A **logarithmic equation** is an equation that involves the logarithm of an expression containing an unknown. For example, $2\ln(x + 4) = 5$ is a logarithmic equation. On the other hand, an **exponential equation** has the unknown appearing in an exponent, as in $2^{3x} = 7$.

To solve some logarithmic equations, it is convenient to use the fact that, for any base b, the function $y = \log_b x$ is one-to-one. This means of course that

$$\text{if} \quad \log_b m = \log_b n \quad \text{then} \quad m = n$$

This is visually apparent by inspecting the two possible shapes of $y = \log_b x$ given in Figure 4.19. In either event it is clear that the function passes the horizontal line test of Section 2.5. On the other hand, we had already observed that the exponential functions $y = b^x$ are one-to-one, meaning that

$$\text{if} \quad b^m = b^n \quad \text{then} \quad m = n$$

so each has an inverse, namely $y = \log_b x$, and since $(f^{-1})^{-1} = f$, each function $y = \log_b x$ has an inverse and is thus one-to-one. Also useful for solving logarithmic and exponential equations are Properties 7 and 8 in Section 4.3.

EXAMPLE 1 **Oxygen Composition**

An experiment was conducted with a particular type of small animal.[9] The logarithm of the amount of oxygen consumed per hour was determined for a number

[9]R. W. Poole, *An Introduction to Quantitative Ecology* (New York: McGraw-Hill Book Company, 1974).

of the animals and was plotted against the logarithms of the weights of the animals. It was found that

$$\log y = \log 5.934 + 0.885 \log x$$

where y is the number of microliters of oxygen consumed per hour and x is the weight of the animal (in grams). Solve for y.

Solution: We first combine the terms on the right side into a single logarithm:

$$\begin{aligned}\log y &= \log 5.934 + 0.885 \log x \\ &= \log 5.934 + \log x^{0.885} && \text{(Property 3 of Section 4.3)} \\ &= \log(5.934x^{0.885}) && \text{(Property 1 of Section 4.3)}\end{aligned}$$

Since log is one-to-one we have

$$y = 5.934x^{0.885}$$

▶ **Now work Problem 1.**

APPLICATIONS IN PRACTICE 1

Solving an Exponential Equation

Greg took a number and multiplied it by a power of 32. Jean started with the same number and got the same result when she multiplied it by 4 raised to a number that was nine less than three times the exponent that Greg used. What power of 32 did Greg use?

EXAMPLE 2 **Solving an Exponential Equation**

Find x if $(25)^{x+2} = 5^{3x-4}$.

Solution: Since $25 = 5^2$, we can express both sides of the equation as powers of 5:

$$\begin{aligned}(25)^{x+2} &= 5^{3x-4} \\ (5^2)^{x+2} &= 5^{3x-4} \\ 5^{2x+4} &= 5^{3x-4}\end{aligned}$$

Since 5^x is a one-to-one function

$$\begin{aligned}2x + 4 &= 3x - 4 \\ x &= 8\end{aligned}$$

▶ **Now work Problem 7.**

APPLICATIONS IN PRACTICE 2

Using Logarithms to Solve an Exponential Equation

The sales manager at a fast-food chain finds that breakfast sales begin to fall after the end of a promotional campaign. The sales in dollars as a function of the number of days d after the campaign's end is given by $S = 800\left(\frac{4}{3}\right)^{-0.1d}$. If the manager does not want sales to drop below 450 per day before starting a new campaign, when should he start such a campaign?

Some exponential equations can be solved by taking the logarithm of both sides after the equation is put in a desirable form. The following example illustrates.

EXAMPLE 3 **Using Logarithms to Solve an Exponential Equation**

Solve $5 + (3)4^{x-1} = 12$.

Solution: We first isolate the exponential expression 4^{x-1} on one side of the equation:

$$\begin{aligned}5 + (3)4^{x-1} &= 12 \\ (3)4^{x-1} &= 7 \\ 4^{x-1} &= \frac{7}{3}\end{aligned}$$

Now we take the natural logarithm of both sides:

$$\ln 4^{x-1} = \ln \frac{7}{3}$$

Simplifying gives

$$(x-1)\ln 4 = \ln \frac{7}{3}$$

$$x - 1 = \frac{\ln \frac{7}{3}}{\ln 4}$$

$$x = \frac{\ln \frac{7}{3}}{\ln 4} + 1 \approx 1.61120$$

▶ **Now work Problem 13.**

In Example 3, we used natural logarithms to solve the given equation. However, logarithms in any base can be employed. Generally, natural or common logarithms are used if a decimal form of the solution is desired. If we used common logarithms, we would obtain

$$x = \frac{\log \frac{7}{3}}{\log 4} + 1 \approx 1.61120$$

TECHNOLOGY

Figure 4.23 shows a graphical solution of the equation $5 + (3)4^{x-1} = 12$ of Example 3. This solution occurs at the intersection of the graphs of $y = 5 + (3)4^{x-1}$ and $y = 12$.

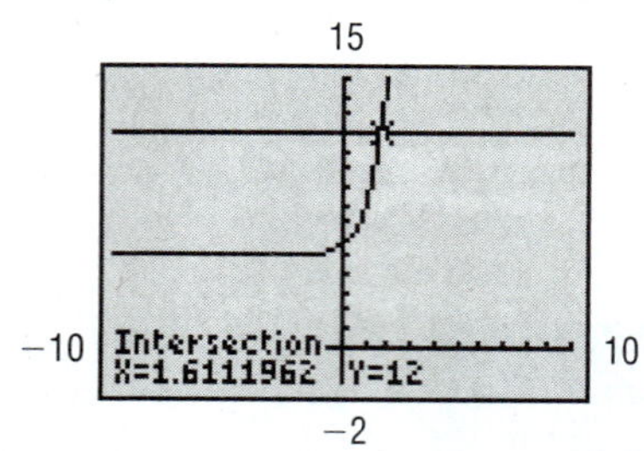

FIGURE 4.23 The solution of $5 + (3)4^{x-1} = 12$ is approximately 1.61120.

EXAMPLE 4 **Demand Equation**

The demand equation for a product is $p = 12^{1-0.1q}$. *Use common logarithms to express* q *in terms of* p.

Solution: Figure 4.24 shows the graph of this demand equation for $q \geq 0$. As is typical of a demand equation, the graph falls from left to right. We want to solve the equation for q. Taking the common logarithms of both sides of $p = 12^{1-0.1q}$

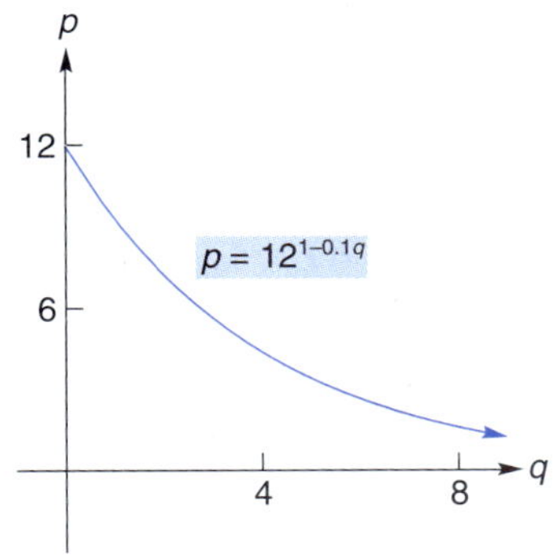

FIGURE 4.24 Graph of the demand equation $p = 12^{1-0.1q}$.

gives

$$\log p = \log(12^{1-0.1q})$$
$$\log p = (1 - 0.1q)\log 12$$
$$\frac{\log p}{\log 12} = 1 - 0.1q$$
$$0.1q = 1 - \frac{\log p}{\log 12}$$
$$q = 10\left(1 - \frac{\log p}{\log 12}\right)$$

▶ **Now work Problem 43.**

To solve some exponential equations involving base e or base 10, such as $10^{2x} = 3$, the process of taking logarithms of both sides may be combined with the identity $\log_b b^r = r$ to transform the equation into an equivalent logarithmic form. In this case, we have

$$10^{2x} = 3$$
$$2x = \log 3 \qquad \text{(logarithmic form)}$$
$$x = \frac{\log 3}{2} \approx 0.2386$$

EXAMPLE 5 **Predator-Prey Relation**

In an article concerning predators and prey, Holling[10] *refers to an equation of the form*

$$y = K(1 - e^{-ax})$$

where x is the prey density, y is the number of prey attacked, and K and a are constants. Verify his claim that

$$\ln \frac{K}{K - y} = ax$$

[10]C. S. Holling, "Some Characteristics of Simple Types of Predation and Parasitism," *Canadian Entomologist* 91, no. 7 (1959), 385–98.

Solution: To find ax, we shall first solve the given equation for e^{-ax}:

$$y = K(1 - e^{-ax})$$
$$\frac{y}{K} = 1 - e^{-ax}$$
$$e^{-ax} = 1 - \frac{y}{K}$$
$$e^{-ax} = \frac{K - y}{K}$$

Now we convert to logarithmic form:

$$\ln \frac{K-y}{K} = -ax$$
$$-\ln \frac{K-y}{K} = ax$$
$$\ln \frac{K}{K-y} = ax \qquad \text{(Property 4 of Section 4.3)}$$

as was to be shown.

▶ **Now work Problem 39.**

Some logarithmic equations can be solved by rewriting them in exponential forms.

APPLICATIONS IN PRACTICE 3

Solving a Logarithmic Equation

The Richter scale measure of an earthquake is given by $R = \log\left(\frac{I}{I_0}\right)$, where I is the intensity of the earthquake, and I_0 is the intensity of a zero-level earthquake. An earthquake that is 675,000 times as intense as a zero-level earthquake has a magnitude on the Richter scale that is 4 more than another earthquake. What is the intensity of the other earthquake?

EXAMPLE 6 Solving a Logarithmic Equation

Solve $\log_2 x = 5 - \log_2(x+4)$.

Solution: Here we must assume that both x and $x + 4$ are positive, so that their logarithms are defined. Both conditions are satisfied if $x > 0$. To solve the equation, we first place all logarithms on one side so that we can combine them:

$$\log_2 x + \log_2(x+4) = 5$$
$$\log_2[x(x+4)] = 5$$

In exponential form, we have

$$x(x+4) = 2^5$$
$$x^2 + 4x = 32$$
$$x^2 + 4x - 32 = 0 \qquad \text{(quadratic equation)}$$
$$(x-4)(x+8) = 0$$
$$x = 4 \quad \text{or} \quad x = -8$$

Because we must have $x > 0$, the only solution is 4, as can be verified by substituting into the original equation. Indeed, replacing x by 4 in $\log_2 x$ gives $\log_2 4 = \log_2 2^2 = 2$ while replacing x by 4 in $5 - \log_2(x+4)$ gives $5 - \log_2(4+4) = 5 - \log_2(8) = 5 - \log_2 2^3 = 5 - 3 = 2$. Since the results are the same, 4 is a solution of the equation.

In solving a logarithmic equation, it is a good idea to check for extraneous solutions.

▶ **Now work Problem 5.**

Exercise 4.4

In Problems 1–36, find x. Round any approximate answers to three decimal places.

*1. $\log(3x+2) = \log(2x+5)$

2. $\log x + \log 3 = \log 5$

3. $\log 7 - \log(x-1) = \log 4$

4. $\log_2 x + 3\log_2 2 = \log_2 \dfrac{2}{x}$

*5. $\ln(-x) = \ln(x^2 - 6)$

6. $\ln(4-x) + \ln 2 = 2\ln x$

*7. $e^{2x} \cdot e^{5x} = e^{14}$

8. $(e^{3x-2})^3 = e^3$

9. $(16)^{3x} = 2$

10. $(27)^{2x+1} = \frac{1}{3}$

11. $e^{2x} = 9$

12. $e^{4x} = \frac{3}{4}$

*13. $2e^{5x+2} = 17$

14. $6e^{1-x} + 1 = 25$

15. $10^{4/x} = 6$

16. $\dfrac{4(10)^{0.2x}}{5} = 3$

17. $\dfrac{5}{10^{2x}} = 7$

18. $2(10)^x + (10)^{x+1} = 4$

19. $2^x = 5$

20. $7^{2x+3} = 9$

21. $5^{2x-5} = 9$

22. $4^{x/2} = 20$

23. $2^{-2x/3} = \dfrac{4}{5}$

24. $5(3^x - 6) = 10$

25. $(4)5^{3-x} - 7 = 2$

26. $\dfrac{9}{2^x} = 5$

27. $\log(x-3) = 3$

28. $\log_2(x+1) = 4$

29. $\log_4(9x-4) = 2$

30. $\log_4(2x+4) - 3 = \log_4 3$

31. $\log(3x-1) - \log(x-3) = 2$

32. $\log(x-3) + \log(x-5) = 1$

33. $\log_3(2x+3) = 4 - \log_3(x+6)$

34. $\log(x+2)^2 = 2$, where $x > 0$.

35. $\log_2\left(\dfrac{2}{x}\right) = 3 + \log_2 x$

36. $\ln(x-2) = \ln(2x-1) + 3$

37. **Rooted Plants** In a study of rooted plants in a certain geographic region,[11] it was determined that on plots of size A (in square meters), the average number of species that occurred was S. When $\log S$ was graphed as a function of $\log A$, the result was a straight line given by

$$\log S = \log 12.4 + 0.26 \log A$$

Solve for S.

38. **Radioactivity** The number of milligrams of a radioactive substance present after t years is given by

$$Q = 100e^{-0.035t}$$

(a) How many milligrams are present after 0 years?
(b) After how many years will there be 20 milligrams present?

Give your answer to the nearest year.

*39. **Gross National Product** In an article, Taagepera and Hayes refer to an equation of the form

$$\log T = 1.7 + 0.2068 \log P - 0.1334 \log^2 P$$

Here T is the percentage of a country's gross national product (GNP) that corresponds to foreign trade (exports plus imports), and P is the country's population (in units of 100,000).[12] Verify the claim that

$$T = 50P^{(0.2068-0.1334\ \log P)}$$

You may assume that $\log 50 = 1.7$.

40. **Blood Sample** On the surface of a glass slide is a grid that divides the surface into 225 equal squares. Suppose a blood sample containing N red cells is spread on the slide and the cells are randomly distributed. Then the number of squares containing no cells is (approximately) given by $225e^{-N/225}$. If 100 of the squares contain no cells, estimate the number of cells the blood sample contained.

41. **Population** In one city, the population P grows at the rate of 2% per year. The equation $P = 1{,}000{,}000(1.02)^t$ gives the population t years after 1998. Find the value of t for which the population is 1,500,000. Give your answer to the nearest tenth.

42. **Market Penetration** In a discussion of market penetration by new products, Hurter and Rubenstein[13] refer to the function

$$F(t) = \frac{q - pe^{-(t+C)(p+q)}}{q[1 + e^{(t+C)(p+q)}]}$$

[11] R. W. Poole, *An Introduction to Quantitative Ecology* (New York: McGraw-Hill Book Company, 1974).

[12] R. Taagepera and J. P. Hayes, "How Trade/GNP Ratio Decreases with Country Size," *Social Science Research* 6 (1977), 108–32.

[13] A. P. Hurter, Jr., A. H. Rubenstein, et al., "Market Penetration by New Innovations: The Technological Literature," *Technological Forecasting and Social Change* 11 (1978), 197–221.

where p, q, and C are constants. They claim that if $F(0) = 0$, then

$$C = -\frac{1}{p+q} \ln \frac{q}{p}$$

Show that their claim is true.

*43. **Demand Equation** The demand equation for a consumer product is $q = 80 - 2^p$. Solve for p and express your answer in terms of common logarithms, as in Example 4. Evaluate p to two decimal places when $q = 60$.

44. **Investment** The equation $A = P(1.105)^t$ gives the value A at the end of t years of an investment of P dollars compounded annually at an annual interest rate of 10.5%. How many years will it take for an investment to double? Give your answer to the nearest year.

45. **Sales** After t years the number of units of a product sold per year is given by $q = 1000\left(\frac{1}{2}\right)^{0.8^t}$. Such an equation is called a *Gompertz equation* and describes natural growth in many areas of study. Solve this equation for t in the same manner as in Example 4, and show that

$$t = \frac{\log \dfrac{3 - \log q}{\log 2}}{(3 \log 2) - 1}$$

46. **Learning Equation** Suppose that the daily output of units of a new product on the tth day of a production run is given by

$$q = 500(1 - e^{-0.2t})$$

Such an equation is called a *learning equation* and indicates that as time progresses, output per day will increase. This may be due to a gain in a worker's proficiency at his or her job. Determine, to the nearest complete unit, the output on (a) the first day and (b) the tenth day after the start of a production run. (c) After how many days will a daily production run of 400 units be reached? Give your answer to the nearest day.

47. Verify that 4 is the only solution to the logarithmic equation in Example 6 by graphing the function

$$y = 5 - \log_2(x + 4) - \log_2 x$$

and observing when $y = 0$.

48. Solve $2^{3x+0.5} = 17$. Round your answer to two decimal places.

49. Solve $\ln(x + 2) = 5 - x$. Round your answer to two decimal places.

50. Graph the equation $4x + (3)4^y = 1$. [*Hint:* Solve for y as an explicit function of x.]

4.5 Review

Important Terms and Symbols

Section 4.1 exponential function, b^x compound interest principal compound amount interest period periodic rate nominal rate e natural exponential function, e^x exponential law of decay initial amount decay constant half-life

Section 4.2 logarithmic function, $\log_b x$ common logarithm, $\log x$ natural logarithm, $\ln x$

Section 4.3 change-of-base formula

Section 4.4 logarithmic equation exponential equation

Summary

An exponential function has the form $f(x) = b^x$. The graph of $f(x) = b^x$ has one of two general shapes, depending on the value of the base b. (See Figure 4.3.) An exponential function is involved in the compound interest formula

$$S = P(1 + r)^n$$

where S is the compound amount of a principal of P at the end of n interest periods at the periodic rate r.

A frequently used base in an exponential function is the irrational number $e \approx 2.71828$. This base occurs in economic analysis and many situations involving growth or decay, such as population studies and radioactive decay. Radioactive elements follow the exponential law of decay,

$$N = N_0 e^{-\lambda t}$$

where N is the amount of an element present at time t, N_0 is the initial amount, and λ is the decay constant. The time required for half of the amount of the element to decay is called the half-life.

The logarithmic function is the inverse function of the exponential function, and vice versa. The logarithmic function with base b is denoted $\log_b$, and $y = \log_b x$ if and only if $b^y = x$. The graph of $y = \log_b x$ has one of two general shapes, depending on the value of the base b. (See Figure 4.19.) Logarithms with base e are called natural logarithms and are denoted ln; those with base 10 are called common logarithms and are denoted log. The half-life T of a radioactive element can be given in terms of a natural logarithm and the decay constant: $T = (\ln 2)/\lambda$.

Some important properties of logarithms are the following:

$$\log_b(mn) = \log_b m + \log_b n$$

$$\log_b \frac{m}{n} = \log_b m - \log_b n$$

$$\log_b m^r = r \log_b m$$

$$\log_b \frac{1}{m} = -\log_b m$$

$$\log_b 1 = 0$$

$$\log_b b = 1$$

$$\log_b b^r = r$$

$$b^{\log_b m} = m$$

$$\log_b m = \frac{\log_a m}{\log_a b}$$

Moreover, if $\log_b m = \log_b n$, then $m = n$. Similarly, if $b^m = b^n$, then $m = n$. Many of these properties are used in solving logarithmic and exponential equations.

Review Problems

Problem numbers shown in color indicate problems suggested for use as a practice chapter test.

In Problems 1–6, write each exponential form logarithmically and each logarithmic form exponentially.

1. $3^5 = 243$ **2.** $\log_5 625 = 4$ **3.** $\log_{16} 2 = \frac{1}{4}$

4. $10^5 = 100{,}000$ **5.** $e^4 = 54.598$ **6.** $\log_9 9 = 1$

In Problems 7–12, find the value of the expression without using a calculator.

7. $\log_5 125$ **8.** $\log_4 16$ **9.** $\log_3 \frac{1}{81}$

10. $\log_{1/3} \frac{1}{9}$ **11.** $\log_{1/3} 9$ **12.** $\log_4 2$

In Problems 13–18, find x without using a calculator.

13. $\log_5 625 = x$ **14.** $\log_x \frac{1}{81} = -4$ **15.** $\log x = -2$

16. $\ln \frac{1}{e} = x$ **17.** $\ln(2x + 3) = 0$ **18.** $e^{\ln(x+4)} = 7$

In Problems 19 and 20, let $\log 2 = a$ and $\log 3 = b$. Express the given logarithm in terms of a and b.

19. $\log 8000$ **20.** $\log \frac{9}{\sqrt{2}}$

In Problems 21–26, write each expression as a single logarithm.

21. $3 \log 7 - 2 \log 5$ **22.** $6 \ln x + 4 \ln y$

23. $2 \ln x + \ln y - 3 \ln z$ **24.** $\log_6 2 - \log_6 4 - 9 \log_6 3$

25. $\frac{1}{2} \log_2 x + 2 \log_2(x^2) - 3 \log_2(x + 1) - 4 \log_2(x + 2)$

26. $4 \log x + 2 \log y - 3(\log z + \log w)$

In Problems 27–32, write the expression in terms of ln x, ln y, and ln z.

27. $\ln \frac{x^2 y}{z^3}$ **28.** $\ln \frac{\sqrt{x}}{(yz)^2}$

29. $\ln \sqrt[3]{xyz}$ **30.** $\ln \left[\frac{xy^3}{z^2}\right]^4$

31. $\ln \left[\frac{1}{x}\sqrt{\frac{y}{z}}\right]$ **32.** $\ln \left[\left(\frac{x}{y}\right)^2 \left(\frac{x}{z}\right)^3\right]$

33. Write $\log_3(x + 5)$ in terms of natural logarithms.

34. Write $\log_3(5x^2 + 2)$ in terms of common logarithms.

35. Suppose that $\log_2 19 = 4.2479$ and $\log_2 5 = 2.3219$. Find $\log_5 19$.

36. Use natural logarithms to determine the value of $\log_4 5$.

37. If $\ln 3 = x$ and $\ln 4 = y$, express $\ln(16\sqrt{3})$ in terms of x and y.

38. Express $\log \frac{x^2 \sqrt[3]{x + 1}}{\sqrt[5]{x^2 + 2}}$ in terms of $\log x$, $\log(x + 1)$, and $\log(x^2 + 2)$.

39. Simplify $e^{\ln x} + \ln e^x + \ln 1$.

40. Simplify $\log 10^2 + \log 1000 - 5$.

41. If $\ln y = x^2 + 2$, find y.

42. Sketch the graphs of $y = 3^x$ and $y = \log_3 x$.

43. Sketch the graph of $y = 2^{x+3}$.

44. Sketch the graph of $y = -2 \log_2 x$.

In Problems 45–52, find x.

45. $\log(5x + 1) = \log(4x + 6)$

46. $\log x + \log 2 = 1$

47. $3^{4x} = 9^{x+1}$

48. $4^{3-x} = \frac{1}{16}$

49. $\log x + \log(10x) = 3$

50. $\log_2(x+4) = \log_2(x-2) + 3$

51. $\ln(\log_x 2) = -1$

52. $\log_2 x + \log_4 x = 3$

In Problems 53–58, find x. Round your answers to three decimal places.

53. $e^{3x} = 14$

54. $10^{3x/2} = 5$

55. $3(10^{x+4} - 3) = 9$

56. $7e^{3x-1} - 2 = 1$

57. $4^{x+3} = 7$

58. $4^{3/x} = 5$

59. Investment If \$2600 is invested for $6\frac{1}{2}$ years at 6% compounded quarterly, find (a) the compound amount and (b) the compound interest.

60. Investment Find the compound amount of an investment of \$4000 for five years at the rate of 11% compounded monthly.

61. Interest Find the nominal rate that corresponds to a periodic rate of $1\frac{1}{6}\%$ per month.

62. Bacteria Growth Bacteria are growing in a culture, and their number is increasing at the rate of 5% an hour. Initially, 600 bacteria are present. (a) Determine an equation that gives the number, N, of bacteria present after t hours. (b) How many bacteria are present after one hour? (c) After five hours? Give your answer to (c) to the nearest integer.

63. Population Growth The population of a town of 8000 grows at the rate of 2% per year. (a) Determine an equation that gives the population, P, after t years from now. (b) Find the population two years from now. Give your answer to part (b) to the nearest integer.

64. Revenue Owing to ineffective advertising, the Kleer-Kut Razor Company finds that its annual revenues have been cut sharply. Moreover, the annual revenue, R, at the end of t years of business satisfies the equation $R = 200{,}000e^{-0.2t}$. Find the annual revenue at the end of two years and at the end of three years.

65. Radioactivity A radioactive substance decays according to the formula

$$N = 10e^{-0.41t}$$

where N is the number of milligrams present after t hours. (a) Determine the initial amount of the substance present. (b) To the nearest tenth of a milligram, determine the amount present after 2 hours. (c) After 10 hours. (d) To the nearest tenth of an hour, determine the half-life of the substance, and (e) determine the number of hours for 1 milligram to remain.

66. Radioactivity If a radioactive substance has a half-life of 10 days, in how many days will $\frac{1}{8}$ of the initial amount be present?

67. Marketing A marketing-research company needs to determine how people adapt to the taste of a new cough drop. In one experiment, a person was given a cough drop and was asked periodically to assign a number, on a scale from 0 to 10, to the perceived taste. This number was called the *response magnitude*. The number 10 was assigned to the initial taste. After conducting the experiment several times, the company estimated that the response magnitude is given by

$$R = 10e^{-t/40},$$

where t is the number of seconds after the person is given the cough drop. (a) Find the response magnitude after 20 seconds. Give your answer to the nearest integer. (b) After how many seconds does a person have a response magnitude of 5? Give your answer to the nearest second.

68. Sediment in Water The water in a midwestern lake contains sediment, and the presence of the sediment reduces the transmission of light through the water. Experiments indicate that the intensity of light is reduced by 10% by passage through 20 cm of water. Suppose that the lake is uniform with respect to the amount of sediment contained by the water. A measuring instrument can detect light at the intensity of 0.17% of full sunlight. This measuring instrument is lowered into the lake. At what depth will it first cease to record the presence of light? Give your answer to the nearest 10 cm.

69. Body Cooling In a discussion of the rate of cooling of isolated portions of the body when they are exposed to low temperatures, there occurs the equation[14]

$$T_t - T_e = (T_t - T_e)_o e^{-at}$$

where T_t is the temperature of the portion at time t, T_e is the environmental temperature, the subscript o refers to the initial temperature difference, and a is a constant. Show that

$$a = \frac{1}{t} \ln \frac{(T_t - T_e)_o}{T_t - T_e}$$

70. Depreciation An alternative to straight-line depreciation is *declining balance* depreciation. This

[14]R. W. Stacy et al., *Essentials of Biological and Medical Physics* (New York: McGraw-Hill Book Company, 1955).

method assumes that an item loses value more steeply at the beginning of its life than later on. A fixed percentage of the value is subtracted each year. Suppose an item's initial cost is C and its useful life is N years. Then the value, V (in dollars), of the item at the end of n years is given by

$$V = C\left(1 - \frac{1}{N}\right)^n$$

where each year brings a depreciation of $\frac{100}{N}$ percent. (This is called single declining balance depreciation; if the annual depreciation were $\frac{200}{N}$ percent, that would be double declining balance.) If a new photocopier is purchased for \$2145 and has a useful life of 6 years, after how many years does its value drop below \$900? Give the answer rounded up to the nearest integer.

71. If $y = f(x) = \frac{\ln x}{x}$, determine the range of f. Round values to two decimal places.

72. Determine the points of intersection of the graphs of $y = \ln(x + 2)$ and $y = x^2 - 7$. Round your answers to two decimal places.

73. Solve $\ln x = 6 - 2x$. Round your answer to two decimal places.

74. Solve $6^{3-4x} = 15$. Round your answer to two decimal places.

75. Display the graph of $y = \log_3(x^2 + 1)$.

76. Display the graph of the equation $(6)5^y + x = 2$. [*Hint:* Solve for y as an explicit function of x.]

77. Graph $y = 3^x$ and $y = \frac{3^x}{9}$ on the same screen. It appears that the graph of $y = \frac{3^x}{9}$ is the graph of $y = 3^x$ shifted two units to the right. Prove algebraically that this is indeed true.

Mathematical SNAPSHOT

Drug Dosages[15]

Determining and prescribing drug dosages are extremely important aspects of the medical profession. Quite often, caution must be taken because of possible adverse side or toxic effects of drugs.

Many drugs are used up by the human body in such a way that the amount present follows an exponential law of decay. That is, if N is the amount of the drug present in the body at time t, then

$$N = N_0 e^{-kt} \qquad (1)$$

where k is a positive constant and N_0 is the amount present at time $t = 0$. If H is the half-life of such a drug, then from Section 4.2, $H = (\ln 2)/k$ or, equivalently, $k = (\ln 2)/H$.

Suppose that you want to analyze the situation whereby equal doses of such a drug are introduced into a patient's system every I units of time until a therapeutic level is attained, and then the dosage is reduced to maintain the therapeutic level. The reason for *reduced* maintenance doses is frequently related to the toxic effects of drugs.

In particular, assume that there are d doses of P units each, a dose is given at times $t = 0, I, 2I, \ldots,$ and $(d-1)/I$, and that the therapeutic level, T, is attained at $t = dI$, which occurs one time interval after the last dose is administered. You will now see how to determine a formula that gives the therapeutic level.

At time $t = 0$ the patient receives the first P units, so the amount of drug in the body is P. At time $t = I$ the amount present from the first dose is [from Equation (1)] Pe^{-kI}. In addition, at $t = I$ the second P units are given. Thus the total amount of the drug present is

$$P + Pe^{-kI}$$

At time $t = 2I$, the amount remaining from the first dose is Pe^{-2kI}; from the second dose, which has been in the system for only one time interval, the amount present is Pe^{-kI}. Also, at time $t = 2I$ the third dose of P units is given, so the total amount of the drug present is

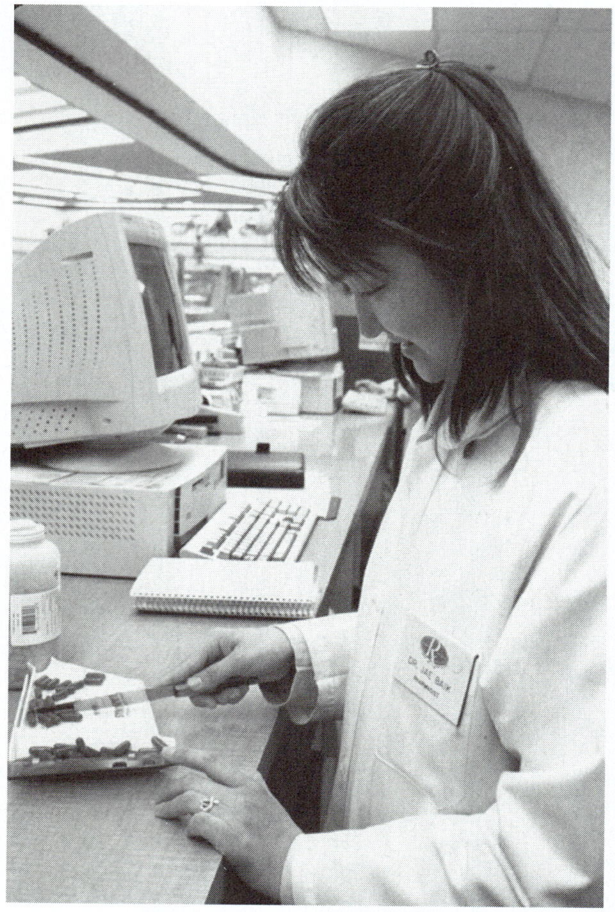

$$P + Pe^{-kI} + Pe^{-2kI}$$

Continuing in this way, the amount T of drug present in the system at time dI, one time interval after the last dose, is given by

$$T = Pe^{-kI} + Pe^{-2kI} + \cdots + Pe^{-dkI} \qquad (2)$$

You can express the right side of Equation (2) in a different form. First, multiply both sides of Equation (2) by e^{-kI}.

$$e^{-kI}T = e^{-kI}(Pe^{-kI} + Pe^{-2kI} + \cdots + Pe^{-dkI})$$

$$e^{-kI}T = Pe^{-2kI} + Pe^{-3kI} + \cdots + Pe^{-(d+1)kI} \qquad (3)$$

[15]This discussion is adapted from Gerald M. Armstrong and Calvin P. Midgley, "The Exponential-Decay Law Applied to Medical Dosages," *Mathematics Teacher* 80, no. 3 (February 1987), 110–13. By permission of the National Council of Teachers of Mathematics.

Subtracting the sides of Equation (3) from the corresponding sides of Equation (2), you have

$$T - e^{-kI}T = Pe^{-kI} - Pe^{-(d+1)kI}$$

Simplifying and solving for T gives

$$(1 - e^{-kI})T = Pe^{-kI}(1 - e^{-dkI})$$

$$T = \frac{Pe^{-kI}(1 - e^{-dkI})}{1 - e^{-kI}} \tag{4}$$

$$T = \frac{P(1 - e^{-dkI})}{e^{kI}(1 - e^{-kI})}$$

$$T = \frac{P(1 - e^{-dkI})}{e^{kI} - 1} \tag{5}$$

Equation (5) allows you to determine the therapeutic level, T, in terms of the dose, P, the length of the time intervals, I, the number of doses, d, and the half-life, H, of the drug [since $k = (\ln 2)/H$]. Among other possibilities, you can determine the dose P if T, H, I, and d are known.

The objective now is to maintain the therapeutic level in the patient's system. To do this, a reduced dose, R, is given at times $t = dI$, $(d+1)I$, $(d+2)I$, and so on. You can determine a formula for R in the following way.

At time $t = (d + 1)I$, but before the second reduced dose is given, the amount of drug in the system from the first reduced dose is Re^{-kI} and the amount that remains from the therapeutic level is Te^{-kI}. Suppose you require that the sum of these amounts be the therapeutic level, T; that is,

$$T = Re^{-kI} + Te^{-kI}$$

Solving for R gives

$$Re^{-kI} = T - Te^{-kI}$$

$$R = T(1 - e^{-kI})e^{kI}$$

Replacing T by the right side of Equation (4) gives

$$R = \frac{Pe^{-kI}(1 - e^{-dkI})}{1 - e^{-kI}}(1 - e^{-kI})e^{kI}$$

or, more simply,

$$R = P(1 - e^{-dkI}) \tag{6}$$

By continuing the reduced doses at time intervals of length I, you are assured that the drug level in the system never falls below T. Furthermore, note that because $-dkI < 0$, then $0 < e^{-dkI} < 1$. Consequently, the factor $1 - e^{-dkI}$ in Equation (6) is between 0 and 1. This ensures that R is less than P; hence R is indeed a *reduced* dose.

It is interesting to note that Armstrong and Midgley state that "the therapeutic amount T must be chosen from a range of empirically determined values. Medical discretion and experience are needed to select proper intervals and durations of time to administer the drug. Even the half-life of a drug can vary somewhat among different patients." More on medical drugs and their safe use can be found at www.fda.gov/cder.

Exercises

1. Solve Equation (5) for (a) P and (b) d.
2. Show that if I is equal to the half-life of the drug, Equation (5) can be written as

$$T = \left(1 - \frac{1}{2^d}\right)P$$

Note that $0 < 1 - (1/2^d) < 1$ for $d > 0$. Hence, the foregoing equation implies that when doses of P units are administered at time intervals equal to the half-life of the drug, it follows that at one time interval after any dose is given, but before the next dose is given, the total level of the drug in a patient's system is less than P.

3. Theophylline is a drug used to treat bronchial asthma and has a half-life of 8 hours in the system of a relatively healthy nonsmoking patient. Suppose that such a patient achieves the desired therapeutic level of this drug in 12 hours when 100 milligrams is administered every 4 hours. Here $d = 3$. Because of toxicity, the dose must be reduced thereafter. To the nearest milligram, determine (a) the therapeutic level and (b) the reduced dose.
4. Use a graphing calculator to generate a drug concentration graph and verify that Equation (6) correctly gives the maintenance dose. On the calculator, enter 0.5 → K, 3 → D, 1 → I and 1 → P. Then enter Y1 = P(1 − e^(−D*K*I)) to represent R. Finally, enter Y2 = P e^(−K X) + P e^(−K(X−I))*(X ≥ I)+P e^(−K(X−2I))*(X ≥ 2I)+Y1 e^(−K(X−3I))*(X ≥ 3I)+Y1 e^(−K(X−4I))*(X ≥ 4I). Then select only Y2 to be graphed and graph the function. Experiment with different values for k, d, I, and P. What adjustment is necessary in the expression for Y2 as you change d?

CHAPTER

5

Mathematics of Finance

For people who like cars and can afford to buy a nice one, a trip to an auto dealership can be a lot of fun. However, buying a car also has a side that many people find unpleasant: the negotiating. The verbal tug-of-war with the salesperson is especially difficult if the buyer is planning to pay on an installment plan and does not understand the numbers being quoted.

How, for instance, does the fact that the salesperson is offering the car for \$12,800 translate into a monthly payment of \$281.54? The answer is amortization. The term comes via French from the Latin root *mort-*, meaning "dead"; from this we also get *mortal* and *mortified*. A debt that is gradually paid off is eventually "killed," and the payment plan for doing this is called an amortization schedule. The schedule is determined by a formula you will learn in Section 5.3 and apply in Section 5.4.

Using the formula, we can calculate the monthly payment for the car. If one makes a \$900 down payment on a \$12,800 car and pays the rest off over four years at 4.8% A.P.R. compounded monthly, the monthly payment for principal and interest should be only \$272.97. If the payment is higher than that, it may contain additional charges such as sales tax, registration fees, or insurance premiums that the buyer should ask about since some of them may be optional. Understanding the mathematics of finance can clearly help consumers make more informed decisions about purchases and investments.

● ● **OBJECTIVE** *To extend the notion of compound interest to include effective rates and to solve interest problems whose solutions require logarithms.*

5.1 Compound Interest

In this chapter we shall model selected topics in finance that deal with the time value of money, such as investments, loans, and so on. In later chapters, when more mathematics is at our disposal, certain topics will be revisited and expanded.

Let us first review some facts from Section 4.1 where the notion of compound interest was introduced. Under compound interest, at the end of each interest period the interest earned for that period is added to the *principal* (the invested amount) so that it, too, earns interest over the next interest period. The basic formula for the value (or *compound amount*) of an investment after n interest periods under compound interest is as follows:

Compound Interest Formula

For an original principal of P, the formula

$$S = P(1+r)^n \tag{1}$$

gives the **compound amount** S at the end of n *interest periods at the periodic rate* of r.

The compound amount is also called the *accumulated amount*, and the difference between the compound amount and the original principal, $S - P$, is called the *compound interest*.

Recall that an interest rate is usually quoted as an *annual* rate, called the *nominal rate* or the *annual percentage rate* (A.P.R.). The periodic rate (or rate per interest period) is obtained by dividing the nominal rate by the number of interest periods per year.

For example, let us compute the compound amount when \$1000 is invested for 5 years at the nominal rate of 8% compounded quarterly. The rate *per period* is 0.08/4, and the number of interest periods is 5×4. From Equation (1), we have

Keep a calculator handy as you read this chapter.

$$\begin{aligned} S &= 1000\left(1+\frac{0.08}{4}\right)^{5\times 4} \\ &= 1000(1+0.02)^{20} \approx \$1485.95 \end{aligned}$$

APPLICATIONS IN PRACTICE 1

Compound Interest

Suppose you leave an initial amount of \$518 in a savings account for three years. If interest is compounded daily (365 times per year), use a graphing calculator to graph the compound amount S as a function of the nominal rate of interest. From the graph, estimate the nominal rate of interest so that there is \$600 after three years.

EXAMPLE 1 **Compound Interest**

Suppose that \$500 amounted to \$588.38 in a savings account after three years. If interest was compounded semiannually, find the nominal rate of interest, compounded semiannually, that was earned by the money.

Solution: Let r be the semiannual rate. There are $2 \times 3 = 6$ interest periods. From Equation (1),

$$\begin{aligned} 500(1+r)^6 &= 588.38 \\ (1+r)^6 &= \frac{588.38}{500} \\ 1+r &= \sqrt[6]{\frac{588.38}{500}} \\ r &= \sqrt[6]{\frac{588.38}{500}} - 1 \approx 0.0275 \end{aligned}$$

Thus, the semiannual rate was 2.75%, so the nominal rate was $5\frac{1}{2}$% compounded semiannually.

▸ **Now work Problem 7.**

EXAMPLE 2 Doubling Money

At what nominal rate of interest, compounded yearly, will money double in eight years?

Solution: Let r be the rate at which a principal of P doubles in eight years. Then the compound amount is $2P$. From Equation (1),

$$P(1+r)^8 = 2P$$
$$(1+r)^8 = 2$$
$$1+r = \sqrt[8]{2}$$
$$r = \sqrt[8]{2} - 1 \approx 0.0905$$

Note that the doubling rate is independent of the principal P.

Hence, the desired rate is 9.05%.

▸ **Now work Problem 11.**

We can determine how long it takes for a given principal to accumulate to a particular amount by using logarithms, as Example 3 shows.

APPLICATIONS IN PRACTICE 2

Compound Interest

Suppose you leave an initial amount of \$520 in a savings account at an annual rate of 5.2% compounded daily (365 days per year). Use a graphing calculator to graph the compound amount S as a function of the interest periods. From the graph, estimate how long it takes for the amount to accumulate to \$750.

EXAMPLE 3 Compound Interest

How long will it take for \$600 to amount to \$900 at an annual rate of 6% compounded quarterly?

Solution: The periodic rate is $r = 0.06/4 = 0.015$. Let n be the number of interest periods it takes for a principal of $P = 600$ to amount to $S = 900$. Then, from Equation (1),

$$900 = 600(1.015)^n \qquad (2)$$
$$(1.015)^n = \frac{900}{600}$$
$$(1.015)^n = 1.5$$

To solve for n, we first take the natural logarithms of both sides:

$$\ln(1.015)^n = \ln 1.5$$
$$n \ln 1.015 = \ln 1.5 \qquad \text{since } \ln m^r = r \ln m$$
$$n = \frac{\ln 1.5}{\ln 1.015} \approx 27.233$$

The number of years that corresponds to 27.233 quarterly interest periods is $27.233/4 \approx 6.8083$, which is about 6 years, $9\frac{1}{2}$ months. Actually, the principal doesn't amount to \$900 until 7 years pass, because interest is compounded quarterly.

▸ **Now work Problem 20.**

TECHNOLOGY

We can solve Equation (2) in Example 3 by graphing

$$Y_1 = 900$$

$$Y_2 = 600(1.015)\text{^}X$$

and finding the intersection. (See Figure 5.1.)

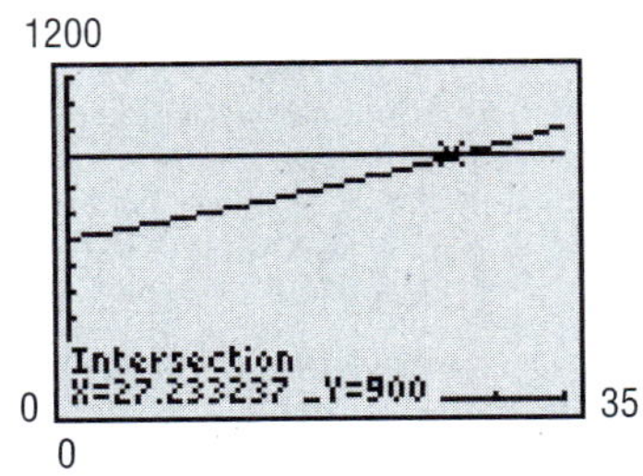

FIGURE 5.1 Solution of Example 3.

Effective Rate

If P dollars are invested at a nominal rate of 10% compounded quarterly for one year, the principal will earn more than 10% that year. In fact, the compound interest is

$$S - P = P\left(1 + \frac{0.10}{4}\right)^4 - P = [(1.025)^4 - 1]P$$
$$\approx 0.103813P$$

which is about 10.38% of P. That is, 10.38% is the approximate rate of interest compounded *annually* that is actually earned, and that rate is called the **effective rate** of interest, or the **yield.** The effective rate is independent of P. In general, the effective interest rate is just the rate of *simple* interest earned over a period of one year. Thus, we have shown that the nominal rate of 10% compounded quarterly is equivalent to an effective rate of 10.38%. Following the preceding procedure, we can generalize our result:

Effective Rate

The **effective rate** r_e that is equivalent to a nominal rate of r compounded n times a year is given by

$$r_e = \left(1 + \frac{r}{n}\right)^n - 1 \qquad (3)$$

APPLICATIONS IN PRACTICE 3

Effective Rate

An investment is compounded monthly. Use a graphing calculator to graph the effective rate r_e as a function of the nominal rate r. Then use the graph to find the nominal rate that is equivalent to an effective rate of 8%.

EXAMPLE 4 Effective Rate

What effective rate is equivalent to a nominal rate of 6% compounded (a) semiannually and (b) quarterly?

Solution:

a. From Equation (3), the effective rate is

$$r_e = \left(1 + \frac{0.06}{2}\right)^2 - 1 = (1.03)^2 - 1 = 0.0609 \qquad \text{that is} \qquad 6.09\%$$

b. The effective rate is

$$r_e = \left(1 + \frac{0.06}{4}\right)^4 - 1 = (1.015)^4 - 1 \approx 0.061364 \qquad \text{that is} \qquad 6.14\%$$

▶ **Now work Problem 9.**

Example 4 illustrates that, for a given nominal rate r, the effective rate increases as the number of interest periods per year (n) increases. However, in Section 10.3 it is shown that, regardless of how large n is, the maximum effective rate that can be obtained is $e^r - 1$.

EXAMPLE 5 **Effective Rate**

To what amount will \$12,000 *accumulate in* 15 *years if it is invested at an effective rate of* 5%?

Solution: Since an effective rate is the rate that is compounded annually, we have

$$S = 12{,}000(1.05)^{15} \approx \$24{,}947.14$$

▶ **Now work Problem 15.**

EXAMPLE 6 **Doubling Money**

How many years will it take for money to double at the effective rate of r?

Solution: Let n be the number of years it takes for a principal of P to double. Then the compound amount is $2P$. Thus,

$$2P = P(1 + r)^n$$

$$2 = (1 + r)^n$$

$$\ln 2 = n \ln(1 + r) \qquad \text{(taking logs of both sides)}$$

Hence,

$$n = \frac{\ln 2}{\ln(1 + r)}$$

For example, if $r = 0.06$, the number of years it takes to double a principal is

$$\frac{\ln 2}{\ln 1.06} \approx 11.9 \text{ years}$$

We remark that when alternative interest rates are available to an investor, effective rates are used to compare them—that is, to determine which of them is the "best." The next example illustrates.

APPLICATIONS IN PRACTICE 4

Comparing Interest Rates

Suppose you have two investment opportunities. You can invest \$10,000 at 11% compounded monthly, or you can invest \$9700 at 11.25% compounded quarterly. Which has the better effective rate of interest? Which is the better investment over 20 years?

EXAMPLE 7 Comparing Interest Rates

If an investor has a choice of investing money at 6% compounded daily or $6\frac{1}{8}$% compounded quarterly, which is the better choice?

Solution:

Strategy: We determine the equivalent effective rate of interest for each nominal rate and then compare our results.

The respective effective rates of interest are

$$r_e = \left(1 + \frac{0.06}{365}\right)^{365} - 1 \approx 6.18\%$$

and

$$r_e = \left(1 + \frac{0.06125}{4}\right)^{4} - 1 \approx 6.27\%$$

Since the second choice gives the higher effective rate, it is the better choice (in spite of the fact that daily compounding may be psychologically more appealing).

▶ **Now work Problem 21.**

Exercise 5.1

In Problems 1 and 2, find (a) the compound amount and (b) the compound interest for the given investment and rate.

1. \$6000 for eight years at an effective rate of 8%

2. \$750 for 12 months at an effective rate of 7%

In Problems 3–6, find the effective rate that corresponds to the given nominal rate. Round answers to three decimal places.

3. 3% compounded semiannually

4. 6% compounded monthly

5. 4% compounded daily

6. 6% compounded daily

***7.** Find the effective rate of interest (rounded to three decimal places) that is equivalent to a nominal rate of 10% compounded

(a) yearly, **(b)** semiannually,
(c) quarterly, **(d)** monthly,
(e) daily.

8. Find (i) the compound interest (rounded to two decimal places) and (ii) the effective rate (to three decimal places) if \$1000 is invested for five years at an annual rate of 7% compounded

(a) quarterly, **(b)** monthly,
(c) weekly, **(d)** daily.

***9.** Over a five-year period, an original principal of \$2000 accumulated to \$2950 in an account in which interest was compounded quarterly. Determine the effective rate of interest, rounded to two decimal places.

10. Suppose that over a seven-year period, \$1000 accumulated to \$1835 in an investment certificate in which interest was compounded monthly. Find the nominal rate of interest, compounded monthly, that was earned. Round your answer to two decimal places.

In Problems 11 and 12, find how many years it would take to double a principal at the given effective rate. Give your answer to one decimal place.

***11.** 8% **12.** 5%

13. A \$6000 certificate of deposit is purchased for \$6000 and is held seven years. If the certificate earns an effective rate of 8%, what is it worth at the end of that period?

14. How many years will it take for money to triple at the effective rate of r?

***15.** **College Costs** Suppose attending a certain college cost \$21,500 in the 2000–2001 school year. This price includes tuition, room, board, books, and other expenses. Assuming an effective 6% inflation rate for these costs, determine what the college costs will be in the 2010–2011 school year.

16. **College Costs** Repeat Problem 15 for an inflation rate of 6% compounded quarterly.

17. **Finance Charge** A major credit-card company has a finance charge of $1\frac{1}{2}\%$ per month on the outstanding indebtedness. (a) What is the nominal rate compounded monthly? (b) What is the effective rate?

18. How long would it take for a principal of P to double if money is worth 12% compounded monthly? Give your answer to the nearest month.

19. To what sum will \$2000 amount in eight years if invested at a 6% effective rate for the first four years and at 6% compounded semiannually thereafter?

***20.** How long will it take for \$500 to amount to \$700 if invested at 8% compounded quarterly?

***21.** An investor has a choice of investing a sum of money at 8% compounded annually or at 7.8% compounded semiannually. Which is the better of the two rates?

22. What nominal rate of interest, compounded monthly, corresponds to an effective rate of 4.5%?

23. **Savings Account** A bank advertises that it pays interest on savings accounts at the rate of $5\frac{1}{4}\%$ compounded daily. Find the effective rate if the bank assumes that a year consists of (a) 360 days or (b) 365 days in determining the *daily rate*. Assume that compounding occurs 365 times a year, and round your answer to two decimal places.

24. **Savings Account** Suppose that \$700 amounted to \$801.06 in a savings account after two years. If interest was compounded quarterly, find the nominal rate of interest, compounded quarterly, that was earned by the money.

25. **Inflation** As a hedge against inflation, an investor purchased a painting in 1990 for \$100,000. It was sold in 2000 for \$300,000. At what effective rate did the painting appreciate in value?

26. **Inflation** If the rate of inflation for certain goods is $7\frac{1}{4}\%$ compounded daily, how many years will it take for the average price of such a good to double?

27. **Zero-Coupon Bond** A *zero-coupon bond* is a bond that is sold for less than its face value (that is, it is *discounted*) and has no periodic interest payments. Instead, the bond is redeemed for its face value at maturity. Thus, in this sense, interest is paid at maturity. Suppose that a zero-coupon bond sells for \$420 and can be redeemed in 14 years for its face value of \$1000. The bond earns interest at what nominal rate, compounded semiannually?

28. **Inflation** Suppose \$1000 is hidden under a mattress for safekeeping. Each year, because of inflation, the purchasing power of the money is 96.5% of what it was the previous year. After six years, what is the purchasing power of the \$1000? (*Hint:* Consider Equation (1) with $r = -0.035$.)

OBJECTIVE To study present value and to solve problems involving the time value of money by using equations of value. To introduce the net present value of cash flows.

5.2 Present Value

Suppose that \$100 is deposited in a savings account that pays 6% compounded annually. Then at the end of two years, the account is worth

$$100(1.06)^2 = 112.36$$

To describe this relationship, we say that the compound amount of \$112.36 is the *future value* of the \$100, and \$100 is the *present value* of the \$112.36. In general, there are times when we may know the future value of an investment and wish to find the present value. To obtain a formula for doing this, we solve the equation $S = P(1+r)^n$ for P. The result is $P = S/(1+r)^n = S(1+r)^{-n}$.

Present Value

The principal P that must be invested at the periodic rate of r for n interest periods so that the compound amount is S is given by

$$P = S(1+r)^{-n} \tag{1}$$

and is called the **present value** of S.

EXAMPLE 1 Present Value

Find the present value of \$1000 due after three years if the interest rate is 9% compounded monthly.

Solution: We use Equation (1) with $S=1000$, $r=0.09/12=0.0075$, and $n=3(12)=36$:

$$P = 1000(1.0075)^{-36} \approx \$764.15$$

This means that \$764.15 must be invested at 9% compounded monthly to have \$1000 in three years.

▶ **Now work Problem 1.**

If the interest rate in Example 1 were 10% compounded monthly, the present value would be

$$P = 1000\left(1+\frac{0.1}{12}\right)^{-36} \approx \$741.74$$

which is less than before. It is typical that the present value for a given future value decreases as the interest rate per interest period increases.

EXAMPLE 2 Single-Payment Trust Fund

A trust fund for a child's education is being set up by a single payment so that at the end of 15 years there will be \$50,000. If the fund earns interest at the rate of 7% compounded semiannually, how much money should be paid into the fund?

Solution: We want the present value of \$50,000, due in 15 years. From Equation (1) with $S = 50{,}000$, $r = 0.07/2 = 0.035$, and $n = 15(2) = 30$, we have

$$P = 50{,}000(1.035)^{-30} \approx \$17,813.92$$

▶ **Now work Problem 13.**

Equations of Value

Suppose that Mr. Smith owes Mr. Jones two sums of money: \$1000, due in two years, and \$600, due in five years. If Mr. Smith wishes to pay off the total debt now by a single payment, how much should the payment be? Assume an interest rate of 8% compounded quarterly.

The single payment x due now must be such that it would grow and eventually pay off the debts when they are due. That is, it must equal the sum of the present values of the future payments. As shown in Figure 5.2, we have

$$x = 1000(1.02)^{-8} + 600(1.02)^{-20} \tag{2}$$

Figure 5.2 is a useful tool for visualizing the time value of money. It is a great aid in setting up an equation of value.

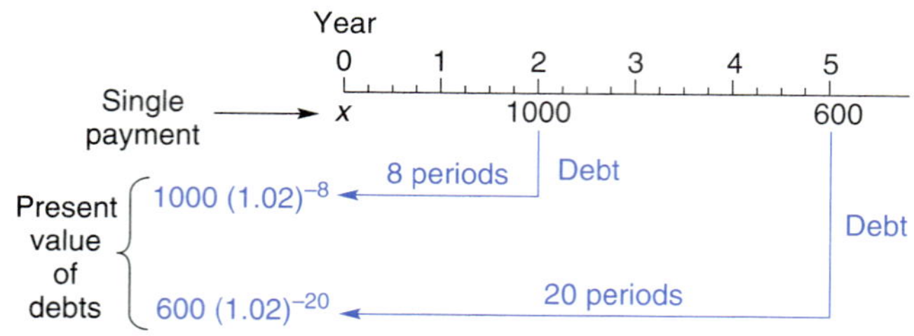

FIGURE 5.2 Replacing two future payments by a single payment now.

This equation is called an *equation of value*. We find that

$$x \approx \$1257.27$$

Thus, the single payment now due is \$1257.27. Let us analyze the situation in more detail. There are two methods of payment of the debt: a single payment now or two payments in the future. Notice that Equation (2) indicates that the value *now* of all payments under one method must equal the value *now* of all payments under the other method. In general, this is true not just *now*, but *at any time*. For example, if we multiply both sides of Equation (2) by $(1.02)^{20}$, we get the equation of value

$$x(1.02)^{20} = 1000(1.02)^{12} + 600 \tag{3}$$

The left side of Equation (3) gives the value five years from now of the single payment (see Figure 5.3), while the right side gives the value five years from now of all payments under the other method. Solving Equation (3) for x gives the same result, $x \approx 1257.27$. In general, an **equation of value** illustrates that when one is considering two methods of paying a debt (or of making some other transaction), *at any time* the value of all payments under one method must equal the value of all payments under the other method.

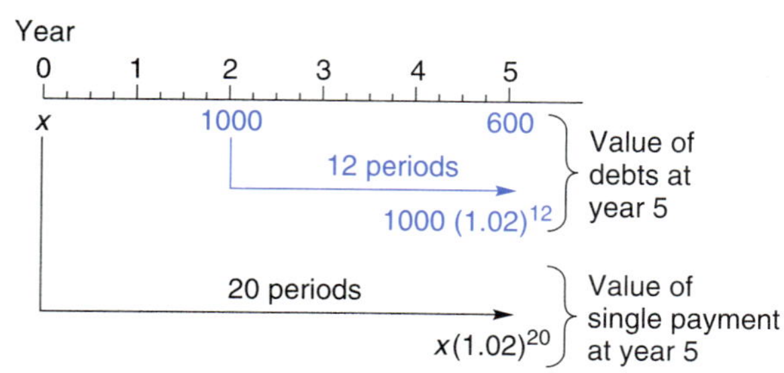

FIGURE 5.3 Diagram for equation of value.

In certain situations, one equation of value may be more convenient to use than another, as Example 3 illustrates.

EXAMPLE 3 **Equation of Value**

A debt of \$3000 due six years from now is instead to be paid off by three payments: \$500 now, \$1500 in three years, and a final payment at the end of five years. What would this payment be if an interest rate of 6% compounded annually is assumed?

Solution: Let x be the final payment due in five years. For computational convenience, we shall set up an equation of value to represent the situation at the end of that time, for in that way the coefficient of x will be 1, as seen in Figure 5.4. Notice that at year 5 we compute the future values of \$500 and \$1500, and the present value of \$3000. The equation of value is

$$500(1.06)^5 + 1500(1.06)^2 + x = 3000(1.06)^{-1}$$

so

$$x = 3000(1.06)^{-1} - 500(1.06)^5 - 1500(1.06)^2$$
$$\approx \$475.68$$

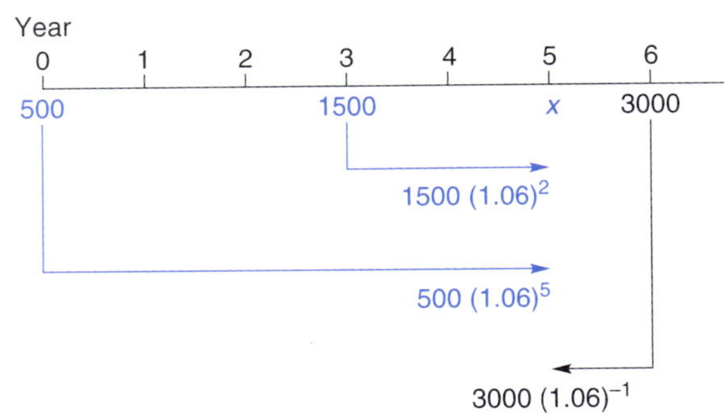

FIGURE 5.4 Time values of payments for Example 3.

▶ **Now work Problem 15.**

When one is considering a choice of two investments, a comparison should be made of the value of each investment at a certain time, as Example 4 shows.

EXAMPLE 4 **Comparing Investments**

Suppose that you had the opportunity of investing \$5000 in a business such that the value of the investment after five years would be \$6300. On the other hand, you could instead put the \$5000 in a savings account that pays 6% compounded semiannually. Which investment is better?

Solution: Let us consider the value of each investment at the end of five years. At that time the business investment would have a value of \$6300, while the savings account would have a value of $5000(1.03)^{10} \approx \$6719.58$. Clearly, the better choice is putting the money in the savings account.

Net Present Value

If an initial investment will bring in payments at future times, the payments are called **cash flows.** The **net present value,** denoted NPV, of the cash flows is defined to be the sum of the present values of the cash flows, minus the initial investment. If NPV > 0, then the investment is profitable; if NPV < 0, the investment is not profitable.

EXAMPLE 5 **Net Present Value**

Suppose that you can invest \$20,000 in a business that guarantees you cash flows at the end of years, 2, 3, and 5, as indicated in the table below. Assume an interest rate of 7% compounded annually, and find the net present value of the cash flows.

Year	Cash Flow
2	\$10,000
3	8000
5	6000

Solution: Subtracting the initial investment from the sum of the present values of the cash flows gives

$$\begin{aligned}\text{NPV} &= 10{,}000(1.07)^{-2} + 8000(1.07)^{-3} + 6000(1.07)^{-5} - 20{,}000 \\ &\approx -\$457.31\end{aligned}$$

Since NPV < 0, the business venture is not profitable if one considers the time value of money. It would be better to invest the \$20,000 in a bank paying 7%, since the venture is equivalent to investing only \$20,000 − \$457.31 = \$19,542.69.

▶ **Now work Problem 19.**

Exercise 5.2

In Problems 1–10, find the present value of the given future payment at the specified interest rate.

*1. \$6000 due in 20 years at 5% compounded annually

2. \$3500 due in eight years at 6% effective

3. \$4000 due in 12 years at 7% compounded semiannually

4. \$1740 due in two years at 18% compounded monthly

5. \$8000 due in $7\frac{1}{2}$ years at 6% compounded quarterly

6. \$6000 due in $6\frac{1}{2}$ years at 10% compounded semiannually

7. \$8000 due in five years at 10% compounded monthly

8. \$500 due in three years at $8\frac{3}{4}$% compounded quarterly

9. \$10,000 due in four years at $9\frac{1}{2}$% compounded daily

10. \$1250 due in $1\frac{1}{2}$ years at $13\frac{1}{2}$% compounded weekly

11. A bank account pays 5.3% annual interest, compounded monthly. How much must be deposited now so that the account contains exactly \$12,000 at the end of one year?

12. Repeat Problem 11 for the nominal rate of 3% compounded quarterly.

*13. **Trust Fund** A trust fund for a 10-year-old child is being set up by a single payment so that at age 21 the child will receive \$27,000. Find how much the payment is if an interest rate of 6% compounded semiannually is assumed.

14. A debt of \$550 due in four years and \$550 due in five years is to be repaid by a single payment now. Find how much the payment is if an interest rate of 10% compounded quarterly is assumed.

*15. A debt of \$600 due in three years and \$800 due in four years is to be repaid by a single payment two years from now. If the interest rate is 8% compounded semiannually, how much is the payment?

16. A debt of \$7000 due in five years is to be repaid by a payment of \$3000 now and a second payment at the end

of five years. How much should the second payment be if the interest rate is 8% compounded monthly?

17. A debt of \$5000 due five years from now and \$5000 due ten years from now is to be repaid by a payment of \$2000 in two years, a payment of \$4000 in four years, and a final payment at the end of six years. If the interest rate is 2.5% compounded annually, how much is the final payment?

18. A debt of \$2000 due in three years and \$3000 due in seven years is to be repaid by a single payment of \$1000 now and two equal payments that are due one year from now and four years from now. If the interest rate is 6% compounded annually, how much are each of the equal payments?

*19. **Cash Flows** An initial investment of \$25,000 in a business guarantees the following cash flows:

Year	Cash Flow
3	\$8,000
4	10,000
6	14,000

Assume an interest rate of 5% compounded semiannually.

(a) Find the net present value of the cash flows.
(b) Is the investment profitable?

20. **Cash Flows** Repeat Problem 19 for the interest rate of 6% compounded semiannually.

21. **Decision Making** Suppose that a person has the following choices of investing \$10,000:

(a) placing the money in a savings account paying 6% compounded semiannually;

(b) investing in a business such that the value of the investment after 8 years is \$16,000.

Which is the better choice?

22. A owes B two sums of money: \$1000 plus interest at 7% compounded annually, which is due in five years, and \$2000 plus interest at 8% compounded semiannually, which is due in seven years. If both debts are to be paid off by a single payment at the end of six years, find the amount of the payment if money is worth 6% compounded quarterly.

23. **Purchase Incentive** A jewelry store advertises that for every \$1000 spent on diamond jewelry, the purchaser receives a \$1000 bond at absolutely no cost. In reality, the \$1000 is the full maturity value of a zero-coupon bond (see Problem 27 of Exercise 5.1), which the store purchases at a heavily reduced price. If the bond earns interest at the rate of 7.5% compounded quarterly and matures after 20 years, how much does the bond cost the store?

24. Find the present value of \$4500 due in three years at a bank rate of 5.2% compounded daily. Assume that the bank uses 360 days in determining the daily rate and that there are 365 days in a year; that is, compounding occurs 365 times in a year.

25. **Promissory Note** A *(promissory) note* is a written statement agreeing to pay a sum of money either on demand or at a definite future time. When a note is purchased for its present value at a given interest rate, the note is said to be *discounted,* and the interest rate is called the *discount rate*. Suppose a \$10,000 note due eight years from now is sold to a financial institution for \$4700. What is the nominal discount rate with quarterly compounding?

OBJECTIVE *To introduce the notions of ordinary annuities and annuities due. To use geometric series to model the present value and future value of an annuity. To determine payments to be placed in a sinking fund.*

5.3 Annuities

Sequences and Geometric Series

In mathematics we use the word **sequence** to describe a list of numbers, called *terms,* that are arranged in a definite order. For example, the list

$$2, 4, 6, 8$$

is a (finite) sequence. The first term is 2, the second is 4, and so on.

In the sequence

$$3, 6, 12, 24, 48$$

each term, after the first, can be obtained by multiplying the preceding term by 2:

$$6 = 3(2), \qquad 12 = 6(2), \qquad \text{and so on.}$$

This means that the *ratio* of every two consecutive terms is 2:

$$\frac{6}{3} = 2, \qquad \frac{12}{6} = 2, \qquad \text{and so on.}$$

We call the sequence a *geometric sequence* with *common ratio* 2. Note that it can be written as

$$3,\ 3(2),\ 3(2)(2),\ 3(2)(2)(2),\ 3(2)(2)(2)(2)$$

or in the form

$$3,\ 3(2),\ 3(2^2),\ 3(2^3),\ 3(2^4)$$

More generally, if a geometric sequence has n terms such that the first term is a and the common ratio is the constant r, then the sequence has the form

$$a, ar, ar^2, ar^3, \ldots, ar^{n-1}$$

Note that the nth term in the sequence is ar^{n-1}.

Definition

The sequence of n numbers

$$a, ar, ar^2, \ldots, ar^{n-1}, \quad \text{where } a \neq 0^1$$

is called a ***geometric sequence*** with ***first term a*** and ***common ratio r.***

APPLICATIONS IN PRACTICE 1

Geometric Sequences

A rubber ball always bounces back $\frac{3}{4}$ of its previous height. If the ball is dropped from a height of 64 feet, what are the next five heights of the ball?

EXAMPLE 1 Geometric Sequences

a. The geometric sequence with $a = 3$, common ratio $\frac{1}{2}$, and $n = 5$ is

$$3,\ 3\left(\frac{1}{2}\right),\ 3\left(\frac{1}{2}\right)^2,\ 3\left(\frac{1}{2}\right)^3,\ 3\left(\frac{1}{2}\right)^4$$

that is

$$3,\ \frac{3}{2},\ \frac{3}{4},\ \frac{3}{8},\ \frac{3}{16}$$

b. The numbers

$$1,\ 0.1,\ 0.01,\ 0.001$$

form a geometric sequence with $a = 1$, $r = 0.1$, and $n = 4$.

▶ **Now work Problem 1.**

[1] If $a = 0$, the sequence is $0, 0, 0, \ldots$. We shall not consider this uninteresting case.

APPLICATIONS IN PRACTICE 2

Geometric Sequence

Suppose bacteria increases at a rate of 50% each minute for six minutes. If the starting population is 500, list the population at the end of each minute as a geometric sequence.

EXAMPLE 2 **Geometric Sequence**

If \$100 is invested at the rate of 6% compounded annually, then the list of compound amounts at the end of each year for eight years is

$$100(1.06), 100(1.06)^2, 100(1.06)^3, \ldots, 100(1.06)^8$$

This is a geometric sequence with common ratio 1.06.

The indicated sum of the terms of the geometric sequence $a, ar, ar^2, \ldots, ar^{n-1}$ is called a **geometric series:**

$$a + ar + ar^2 + \cdots + ar^{n-1} \quad \textbf{(1)}$$

For example,

$$1 + \frac{1}{2} + \left(\frac{1}{2}\right)^2 + \cdots + \left(\frac{1}{2}\right)^6$$

is a geometric series with $a = 1$, common ratio $r = \frac{1}{2}$, and $n = 7$.

Let us compute the sum s of the geometric series in Equation (1):

$$s = a + ar + ar^2 + \cdots + ar^{n-1} \quad \textbf{(2)}$$

We can express s in a more compact form. Multiplying both sides by r gives

$$rs = ar + ar^2 + ar^3 + \cdots + ar^n \quad \textbf{(3)}$$

Subtracting corresponding sides of Equation (3) from Equation (2) yields

$$s - rs = a - ar^n$$

$$s(1 - r) = a(1 - r^n) \qquad \text{(factoring)}$$

Dividing both sides by $1 - r$, we have $s = a(1 - r^n)/(1 - r)$.

Sum of Geometric Series

The **sum** s of a geometric series[2] of n terms whose first term is a and common ratio is r is given by

$$s = \frac{a(1 - r^n)}{1 - r} \quad \textbf{(4)}$$

APPLICATIONS IN PRACTICE 3

Sum of Geometric Series

A ball rebounds $\frac{2}{3}$ of its previous height after each bounce. If the ball is tossed up to a height of 6 meters, how far has it traveled in the air when it hits the ground for the 12th time?

EXAMPLE 3 **Sum of Geometric Series**

Find the sum of the geometric series

$$1 + \frac{1}{2} + \left(\frac{1}{2}\right)^2 + \cdots + \left(\frac{1}{2}\right)^6$$

[2]This formula assumes that $r \neq 1$. However, if $r = 1$, then $s = a + \cdots + a = na$.

Solution: Here $a = 1, r = \frac{1}{2}$, and $n = 7$ (not 6). From Equation (4), we have

$$s = \frac{a(1-r^n)}{1-r} = \frac{1\left[1-\left(\frac{1}{2}\right)^7\right]}{1-\frac{1}{2}} = \frac{\frac{127}{128}}{\frac{1}{2}} = \frac{127}{64}$$

▶ **Now work Problem 5.**

APPLICATIONS IN PRACTICE 4

Sum of Geometric Series

A company earns a profit of \$2000 in its first month. Suppose the profit increases by 10% each month for two years. Find the amount of profit the company earns in its first two years.

EXAMPLE 4 **Sum of Geometric Series**

Find the sum of the geometric series

$$3^5 + 3^6 + 3^7 + \cdots + 3^{11}$$

Solution: Here $a = 3^5, r = 3$, and $n = 7$. From Equation (4),

$$s = \frac{3^5(1-3^7)}{1-3} = \frac{243(1-2187)}{-2} = 265{,}599$$

▶ **Now work Problem 7.**

Present Value of an Annuity

The notion of a geometric series is the basis of the mathematical model of an *annuity*. Essentially, an **annuity** is a sequence of payments made at fixed periods of time over a given interval. The fixed period is called the **payment period,** and the given interval is the **term** of the annuity. An example of an annuity is the depositing of \$100 in a savings account every three months for a year.

The **present value of an annuity** is the sum of the *present values* of all the payments. It represents the amount that must be invested now to purchase the payments due in the future. Unless otherwise specified, we assume that each payment is made at the *end* of a payment period; such an annuity is called an **ordinary annuity.** We also assume that interest is computed at the end of each payment period.

Let us consider an annuity of n payments of R (dollars) each, where the interest rate *per period* is r (see Figure 5.5) and the first payment is due one period from now. The present value of the annuity is given by

$$A = R(1+r)^{-1} + R(1+r)^{-2} + \cdots + R(1+r)^{-n}$$

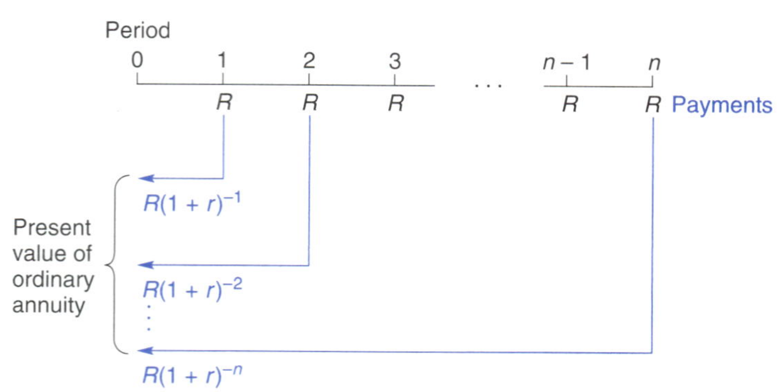

FIGURE 5.5 Present value of ordinary annuity.

This is a geometric series of n terms with first term $R(1+r)^{-1}$ and common ratio $(1+r)^{-1}$. Hence, from Equation (4), we obtain the formula

$$\begin{aligned} A &= \frac{R(1+r)^{-1}[1-(1+r)^{-n}]}{1-(1+r)^{-1}} \\ &= \frac{R[1-(1+r)^{-n}]}{(1+r)[1-(1+r)^{-1}]} = \frac{R[1-(1+r)^{-n}]}{(1+r)-1} \\ &= R \cdot \frac{1-(1+r)^{-n}}{r} \end{aligned}$$

Present Value of Annuity

The formula

$$A = R \cdot \frac{1-(1+r)^{-n}}{r} \tag{5}$$

gives the **present value** A of an ordinary annuity of R dollars per payment period for n periods at the interest rate of r per period.

In Equation (5), the expression $[1-(1+r)^{-n}]/r$ is denoted $a_{\overline{n}\rceil r}$ and (letting $R = 1$) represents the present value of an annuity of \$1 per period. The symbol $a_{\overline{n}\rceil r}$ is read "a angle n at r." Thus, Equation (5) can be written as

$$A = Ra_{\overline{n}\rceil r} \tag{6}$$

Whenever a desired value of $a_{\overline{n}\rceil r}$ is not in Appendix B, we shall use a calculator to compute it.

Selected values of $a_{\overline{n}\rceil r}$ are given in Appendix B. (Most are approximate.)

APPLICATIONS IN PRACTICE 5

Present Value of Annuity

Given a payment of \$500 per month for six years, use a graphing calculator to graph the present value A as a function of the interest rate per month, r. Determine the nominal rate if the present value of the annuity is \$30,000.

EXAMPLE 5 **Present Value of Annuity**

Find the present value of an annuity of \$100 per month for $3\frac{1}{2}$ years at an interest rate of 6% compounded monthly.

Solution: Substituting in Equation (6), we set $R = 100$, $r = 0.06/12 = 0.005$, and $n = (3\frac{1}{2})(12) = 42$. Thus,

$$A = 100a_{\overline{42}\rceil 0.005}$$

From Appendix B, $a_{\overline{42}\rceil 0.005} \approx 37.798300$. Hence,

$$A \approx 100(37.798300) = \$3779.83$$

▶ **Now work Problem 13.**

APPLICATIONS IN PRACTICE 6

Present Value of Annuity

Suppose a man purchases a house with an initial down payment of \$20,000 and then makes quarterly payments: \$2000 at the end of each quarter for six years and \$3500 at the end of each quarter for eight more years. Given an interest rate of 6% compounded quarterly, find the present value of the payments and the list price of the house.

EXAMPLE 6 **Present Value of Annuity**

Given an interest rate of 5% compounded annually, find the present value of an annuity of \$2000 due at the end of each year for three years and \$5000 due thereafter at the end of each year for four years. (See Figure 5.6.)

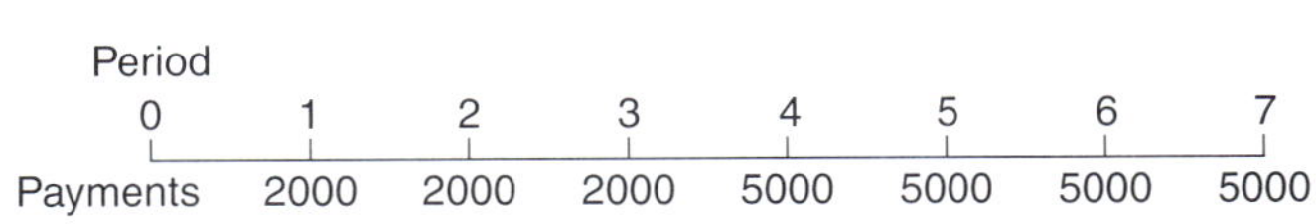

FIGURE 5.6 Annuity of Example 6.

Solution: The present value is obtained by summing the present values of all payments:

$$2000(1.05)^{-1} + 2000(1.05)^{-2} + 2000(1.05)^{-3} + 5000(1.05)^{-4} + 5000(1.05)^{-5}$$
$$+ 5000(1.05)^{-6} + 5000(1.05)^{-7}$$

Rather than evaluating this expression, we can simplify our work by considering the payments to be an annuity of \$5000 for seven years, minus an annuity of \$3000 for three years, so that the first three payments are \$2000 each. Thus, the present value is

$$5000a_{\overline{7}|0.05} - 3000a_{\overline{3}|0.05}$$
$$\approx 5000(5.786373) - 3000(2.723248)$$
$$\approx \$20{,}762.12$$

▶ **Now work Problem 25.**

APPLICATIONS IN PRACTICE 7

Periodic Payment of Annuity

Given an annuity with equal payments at the end of each quarter for six years and an interest rate of 4.8% compounded quarterly, use a graphing calculator to graph the present value A as a function of the monthly payment R. Determine the monthly payment if the present value of the annuity is \$15,000.

EXAMPLE 7 Periodic Payment of Annuity

If \$10,000 is used to purchase an annuity consisting of equal payments at the end of each year for the next four years and the interest rate is 6% compounded annually, find the amount of each payment.

Solution: Here $A = \$10{,}000$, $n = 4$, $r = 0.06$, and we want to find R. From Equation (6),

$$10{,}000 = Ra_{\overline{4}|0.06}$$

Solving for R gives

$$R = \frac{10{,}000}{a_{\overline{4}|0.06}} \approx \frac{10{,}000}{3.465106} \approx \$2885.91$$

In general, the formula

$$R = \frac{A}{a_{\overline{n}|r}}$$

gives the periodic payment R of an ordinary annuity whose present value is A.

▶ **Now work Problem 27.**

APPLICATIONS IN PRACTICE 8

Annuity Due

A man makes house payments of \$1200 at the beginning of every month. If the man wishes to pay one year's worth of payments in advance, how much should he pay, provided that the interest rate is 6.8% compounded monthly?

EXAMPLE 8 Annuity Due

The premiums on an insurance policy are \$50 per quarter, payable at the beginning of each quarter. If the policyholder wishes to pay one year's premiums in advance, how much should be paid, provided that the interest rate is 4% compounded quarterly?

Solution: We want the present value of an annuity of \$50 per period for four periods at a rate of 1% per period. However, each payment is due at the *beginning* of the payment period. Such an annuity is called an **annuity due.** The

An example of a situation involving an annuity due is an apartment lease for which the first payment is made immediately.

given annuity can be thought of as an initial payment of \$50, followed by an ordinary annuity of \$50 for three periods. (See Figure 5.7.) Thus, the present value is

$$50 + 50a_{\overline{3}|0.01} \approx 50 + 50(2.940985) \approx \$197.05$$

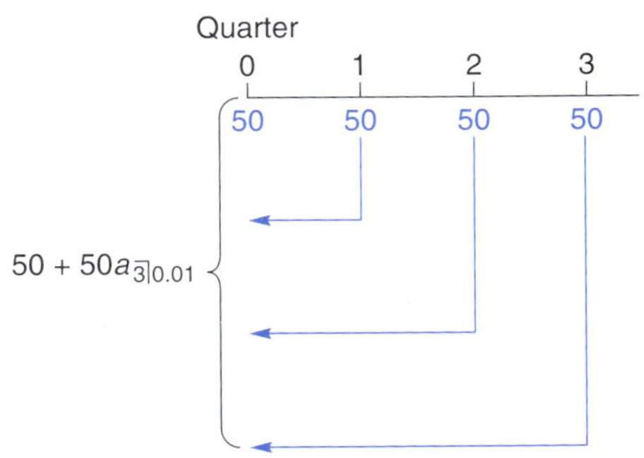

FIGURE 5.7 Annuity due (present value).

We remark that the general formula for the **present value of an annuity due** is $A = R + Ra_{\overline{n-1}|r}$, that is

$$A = R(1 + a_{\overline{n-1}|r})$$

▸ **Now work Problem 17.**

Amount of an Annuity

The **amount** (or **future value**) **of an annuity** is the value, at the end of the term, of all payments. That is, it is the sum of the compound amounts of all payments. Let us consider an ordinary annuity of n payments of R (dollars) each, where the interest rate per period is r. The compound amount of the last payment is R, since it occurs at the end of the last interest period and hence does not accrue interest. (See Figure 5.8.) The $(n-1)$th payment earns interest for one period, and so on, and the first payment earns interest for $n-1$ periods. Thus, the future value of the annuity is

$$R + R(1+r) + R(1+r)^2 + \cdots + R(1+r)^{n-1}$$

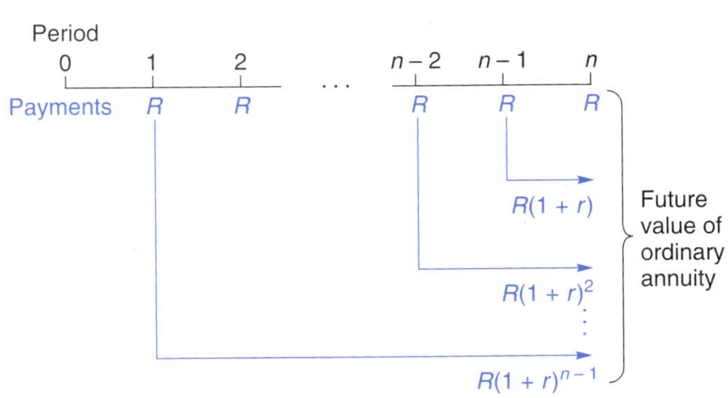

FIGURE 5.8 Future value of ordinary annuity.

This is a geometric series of n terms with first term R and common ratio $1+r$. Consequently, its sum S is (by using Equation (4))

$$S = \frac{R[1-(1+r)^n]}{1-(1+r)} = R \cdot \frac{1-(1+r)^n}{-r} = R \cdot \frac{(1+r)^n - 1}{r}$$

Amount of an Annuity

The formula

$$S = R \cdot \frac{(1+r)^n - 1}{r} \tag{7}$$

gives the **amount** S of an ordinary annuity of R (dollars) per payment period for n periods at the interest rate of r per period.

The expression $[(1+r)^n - 1]/r$ is abbreviated $s_{\overline{n}|r}$, and some approximate values of $s_{\overline{n}|r}$ are given in Appendix B. Thus,

$$S = Rs_{\overline{n}|r} \tag{8}$$

APPLICATIONS IN PRACTICE 9

Amount of Annuity

Suppose you invest in an IRA by depositing $2000 at the end of every tax year for the next 15 years. If the interest rate is 5.7% compounded annually, how much will you have at the end of 15 years?

EXAMPLE 9 Amount of Annuity

Find the amount of an annuity consisting of payments of \$50 at the end of every three months for three years at the rate of 6% compounded quarterly. Also, find the compound interest.

Solution: To find the amount of the annuity, we use Equation (8) with $R = 50$, $n = 4(3) = 12$, and $r = 0.06/4 = 0.015$

$$S = 50s_{\overline{12}|0.015} \approx 50(13.041211) \approx \$652.06$$

The compound interest is the difference between the amount of the annuity and the sum of the payments, namely,

$$652.06 - 12(50) = 652.06 - 600 = \$52.06$$

▸ **Now work Problem 19.**

APPLICATIONS IN PRACTICE 10

Amount of Annuity Due

Suppose you invest in an IRA by depositing $2000 at the beginning of every tax year for the next 15 years. If the interest rate is 5.7% compounded annually, how much will you have at the end of 15 years?

EXAMPLE 10 Amount of Annuity Due

At the beginning of each quarter, \$50 is deposited into a savings account that pays 6% compounded quarterly. Find the balance in the account at the end of three years.

Solution: Since the deposits are made at the beginning of a payment period, we want the amount of an *annuity due*, as defined in Example 8. (See Figure 5.9.) The given annuity can be thought of as an ordinary annuity of $50 for 13 periods, minus the final payment of $50. Thus, the amount is

$$50s_{\overline{13}|0.015} - 50 \approx 50(14.236830) - 50 \approx \$661.84$$

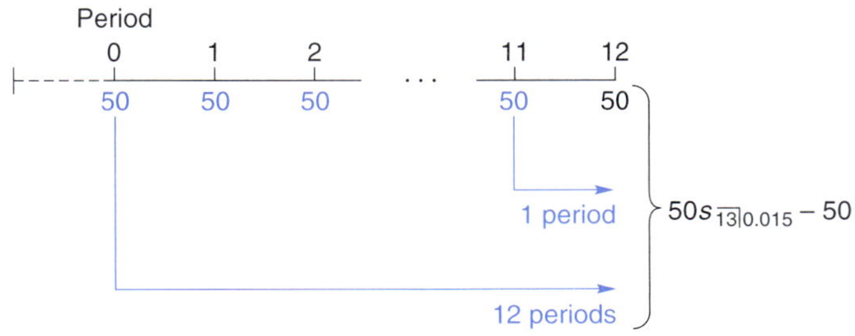

FIGURE 5.9 Future value of annuity due.

The formula for the **future value of an annuity due** is $S = Rs_{\overline{n+1}|r} - R$, or

$$S = R(s_{\overline{n+1}|r} - 1)$$

▶ **Now work Problem 23.**

Sinking Fund

Our final examples involve the notion of a *sinking fund.*

EXAMPLE 11 **Sinking Fund**

*A **sinking fund** is a fund into which periodic payments are made in order to satisfy a future obligation. Suppose a machine costing \$7000 is to be replaced at the end of eight years, at which time it will have a salvage value of \$700. In order to provide money at that time for a new machine costing the same amount, a sinking fund is set up. The amount in the fund at the end of eight years is to be the difference between the replacement cost and the salvage value. If equal payments are placed in the fund at the end of each quarter and the fund earns 8% compounded quarterly, what should each payment be?*

Solution: The amount needed after eight years is $7000 - 700 = \$6300$. Let R be the quarterly payment. The payments into the sinking fund form an annuity with $n = 4(8) = 32$, $r = 0.08/4 = 0.02$, and $S = 6300$. Thus, from Equation (8), we have

$$6300 = Rs_{\overline{32}|0.02}$$

$$R = \frac{6300}{s_{\overline{32}|0.02}} \approx \frac{6300}{44.227030} \approx \$142.45$$

In general, the formula

$$R = \frac{S}{s_{\overline{n}|r}}$$

gives the periodic payment R of an annuity that is to amount to S.

EXAMPLE 12 Sinking Fund

A rental firm estimates that, if purchased, a machine will yield an annual net return of \$1000 for six years, after which the machine would be worthless. How much should the firm pay for the machine if it wants to earn 7% annually on its investment and also set up a sinking fund to replace the purchase price? For the fund, assume annual payments and a rate of 5% compounded annually.

Solution: Let x be the purchase price. Each year, the return on the investment is $0.07x$. Since the machine gives a return of \$1000 a year, the amount left to be placed into the fund each year is $1000 - 0.07x$. These payments must accumulate to x. Hence,

$$
\begin{aligned}
(1000 - 0.07x)s_{\overline{6}|0.05} &= x \\
1000s_{\overline{6}|0.05} - 0.07xs_{\overline{6}|0.05} &= x \\
1000s_{\overline{6}|0.05} &= x(1 + 0.07s_{\overline{6}|0.05}) \\
\frac{1000s_{\overline{6}|0.05}}{1 + 0.07s_{\overline{6}|0.05}} &= x \\
x &\approx \frac{1000(6.801913)}{1 + 0.07(6.801913)} \\
&\approx \$4607.92
\end{aligned}
$$

Another way to look at the problem is as follows: Each year, the \$1000 must account for a return of $0.07x$ and also a payment of $\frac{x}{s_{\overline{6}|0.05}}$, into the sinking fund. Thus, we have $1000 = 0.07x + \frac{x}{s_{\overline{6}|0.05}}$, which, when solved, gives the same result.

▶ **Now work Problem 33.**

Exercise 5.3

In Problems 1–4, write the geometric sequence satisfying the given conditions. Simplify the terms.

*1. $a = 64, r = \frac{1}{2}, n = 5$

2. $a = 2, r = -3, n = 4$

3. $a = 100, r = 1.02, n = 3$

4. $a = 81, r = 3^{-1}, n = 4$

In Problems 5–8, find the sum of the given geometric series by using Equation (4) of this section.

*5. $\frac{4}{7} + \left(\frac{4}{7}\right)^2 + \cdots + \left(\frac{4}{7}\right)^5$

6. $1 + \frac{1}{4} + \left(\frac{1}{4}\right)^2 + \cdots + \left(\frac{1}{4}\right)^5$

*7. $1 + 0.1 + (0.1)^2 + \cdots + (0.1)^5$

8. $(1.1)^{-1} + (1.1)^{-2} + \cdots + (1.1)^{-6}$

In Problems 9–12, use Appendix B and find the value of the given expression.

9. $a_{\overline{35}|0.04}$

10. $a_{\overline{15}|0.07}$

11. $s_{\overline{8}|0.0075}$

12. $s_{\overline{11}|0.0125}$

In Problems 13–16, find the present value of the given (ordinary) annuity.

*13. \$500 per year for five years at the rate of 7% compounded annually

14. \$1000 every six months for four years at the rate of 10% compounded semiannually

15. \$2000 per quarter for $4\frac{1}{2}$ years at the rate of 8% compounded quarterly

16. \$1500 per month for 15 months at the rate of 9% compounded monthly

In Problems 17 and 18, find the present value of the given annuity due.

*17. \$800 paid at the beginning of each six-month period for six years at the rate of 7% compounded semiannually

18. \$100 paid at the beginning of each month for four years at the rate of 6% compounded monthly

In Problems 19–22, find the future value of the given (ordinary) annuity.

*19. \$2000 per month for three years at the rate of 15% compounded monthly

20. \$600 per quarter for four years at the rate of 8% compounded quarterly

21. \$5000 per year for 20 years at the rate of 7% compounded annually

22. \$2000 every six months for 10 years at the rate of 6% compounded semiannually

In Problems 23 and 24, find the future value of the given annuity due.

*23. \$1200 each year for 12 years at the rate of 8% compounded annually

24. \$600 every quarter for $7\frac{1}{2}$ years at the rate of 10% compounded quarterly

*25. For an interest rate of 6% compounded monthly, find the present value of an annuity of \$50 at the end of each month for six months and \$75 thereafter at the end of each month for two years

26. **Leasing Office Space** A company wishes to lease temporary office space for a period of six months. The rental fee is \$1500 a month, payable in advance. Suppose that the company wants to make a lump-sum payment, at the beginning of the rental period, to cover all rental fees due over the six-month period. If money is worth 9% compounded monthly, how much should the payment be?

*27. An annuity consisting of equal payments at the end of each quarter for three years is to be purchased for \$5000. If the interest rate is 6% compounded quarterly, how much is each payment?

28. **Equipment Purchase** A machine is purchased for \$3000 down and payments of \$250 at the end of every six months for six years. If interest is at 8% compounded semiannually, find the corresponding cash price of the machine.

29. Suppose \$50 is placed in a savings account at the end of each month for four years. If no further deposits are made, (a) how much is in the account after six years, and (b) how much of this amount is compound interest? Assume that the savings account pays 6% compounded monthly.

30. **Insurance Settlement Options** The beneficiary of an insurance policy has the option of receiving a lump-sum payment of \$157,000 or 10 equal yearly payments, where the first payment is due at once. If interest is at 3% compounded annually, find the yearly payment.

31. **Sinking Fund** In 10 years, a \$40,000 machine will have a salvage value of \$4000. A new machine at that time is expected to sell for \$52,000. In order to provide funds for the difference between the replacement cost and the salvage value, a sinking fund is set up into which equal payments are placed at the end of each year. If the fund earns 7% compounded annually, how much should each payment be?

32. **Sinking Fund** A paper company is considering the purchase of a forest that is estimated to yield an annual return of \$50,000 for 10 years, after which the forest will have no value. The company wants to earn 8% on its investment and also set up a sinking fund to replace the purchase price. If money is placed in the fund at the end of each year and earns 6% compounded annually, find the price the company should pay for the forest. Round your answer to the nearest hundred dollars.

*33. **Sinking Fund** In order to replace a machine in the future, a company is placing equal payments into a sinking fund at the end of each year so that after 10 years the amount in the fund is \$25,000. The fund earns 6% compounded annually. After 6 years, the interest rate increases and the fund pays 7% compounded annually. Because of the higher interest rate, the company decreases the amount of the remaining payments. Find the amount of the new payment. Round your answer to the nearest dollar.

34. A owes B the sum of \$5000 and agrees to pay B the sum of \$1000 at the end of each year for five years and a final payment at the end of the sixth year. How much should the final payment be if interest is at 8% compounded annually?

In Problems 35–43, use the following formulas:

$$a_{\overline{n}|r} = \frac{1-(1+r)^{-n}}{r}$$

$$s_{\overline{n}|r} = \frac{(1+r)^{-n}-1}{r}$$

$$R = \frac{A}{a_{\overline{n}|r}} = \frac{Ar}{1-(1+r)^{-n}} = \frac{Ar(1+r)^n}{(1+r)^n-1}$$

$$R = \frac{S}{s_{\overline{n}|r}} = \frac{Sr}{(1+r)^n-1}$$

35. Find $s_{\overline{60}|0.017}$ to five decimal places.

36. Find $a_{\overline{9}|0.052}$ to five decimal places.

37. Find $700a_{\overline{360}|0.0125}$ to two decimal places.

38. Find $1000s_{\overline{120}|0.01}$ to two decimal places.

39. Equal payments are to be deposited in a savings account at the end of each quarter for five years so that at the end of that time there will be \$3000. If interest is at $5\frac{1}{2}$%

compounded quarterly, find the quarterly payment.

40. Insurance Proceeds Suppose that insurance proceeds of \$25,000 are used to purchase an annuity of equal payments at the end of each month for five years. If interest is at the rate of 10% compounded monthly, find the amount of each payment.

41. Lottery Mary Jones won a state \$4,000,000 lottery and will receive a check for \$200,000 now and a similar one each year for the next 19 years. To provide these 20 payments, the State Lottery Commission purchased an annuity due at the interest rate of 10% compounded annually. How much did the annuity cost the commission?

42. Pension Plan Options Suppose an employee of a company is retiring and has the choice of two benefit options under the company pension plan. Option A consists of a guaranteed payment of \$540 at the end of each month for 15 years. Alternatively, under option B, the employee receives a lump-sum payment equal to the present value of the payments described under option A.

(a) Find the sum of the payments under option A.

(b) Find the lump-sum payment under option B if it is determined by using an interest rate of 6% compounded monthly. Round your answer to the nearest dollar.

43. An Early Start to Investing An insurance agent offers services to clients who are concerned about their personal financial planning for retirement. To emphasize the advantages of an early start to investing, she points out that a 25-year-old person who saves \$2000 a year for 10 years (and makes no more contributions after age 34) will earn more than by waiting 10 years and then saving \$2000 a year from age 35 until retirement at age 65 (a total of 30 contributions). Find the net earnings (compound amount minus total contributions) at age 65 for both situations. Assume an effective annual rate of 7%, and suppose that deposits are made at the beginning of each year. Round answers to the nearest dollar.

OBJECTIVE *To learn how to amortize a loan and set up an amortization schedule.*

5.4 Amortization of Loans

Suppose that a bank lends a borrower \$1500 and charges interest at the nominal rate of 12% compounded monthly. The \$1500 plus interest is to be repaid by equal payments of R dollars at the end of each month for three months. Essentially, by paying the borrower \$1500, the bank is purchasing an annuity of three payments of R each. Using the formula from Example 7 of the preceding section, we find that the monthly payment is given by

$$R = \frac{A}{a_{\overline{n}|r}} = \frac{1500}{a_{\overline{3}|0.01}} \approx \frac{1500}{2.940985} \approx \$510.0332$$

We shall round the payment to \$510.03, which may result in a slightly higher final payment. However, it is not unusual for a bank to round *up* to the nearest cent, in which case the final payment may be less than the other payments.

The bank can consider each payment as consisting of two parts: (1) interest on the outstanding loan and (2) repayment of part of the loan. This is called **amortizing.** A loan is **amortized** when part of each payment is used to pay interest and the remaining part is used to reduce the outstanding principal. Since each payment reduces the outstanding principal, the interest portion of a payment decreases as time goes on. Let us analyze the loan just described.

At the end of the first month, the borrower pays \$510.03. The interest on the outstanding principal is $0.01(1500) = \$15$. The balance of the payment, $510.03 - 15 = \$495.03$, is then applied to reduce the principal. Hence, the principal outstanding is now $1500 - 495.03 = \$1004.97$. At the end of the second month, the interest is $0.01(1004.97) \approx \$10.05$. Thus, the amount of the loan repaid is $510.03 - 10.05 = \$499.98$, and the outstanding balance is $1004.97 - 499.98 = \$504.99$. The interest due at the end of the third and final month is $0.01(504.99) \approx \$5.05$, so the amount of the loan repaid is $510.03 - 5.05 = \$504.98$. This would leave an outstanding balance of

Many end-of-year mortgage statements are issued in the form of an amortization schedule.

504.99 − 504.98 = \$0.01, so we take the final payment to be \$510.04, and the debt is paid off. As we said earlier, the final payment is adjusted to offset rounding errors. An analysis of how each payment in the loan is handled can be given in a table called an **amortization schedule.** (See Table 5.1.) The total interest paid is \$30.10, which is often called the **finance charge.**

TABLE 5.1 Amortization Schedule

Period	Principal Outstanding at Beginning of Period	Interest for Period	Payment at End of Period	Principal Repaid at End of Period
1	\$1500	\$15	\$510.03	\$495.03
2	1004.97	10.05	510.03	499.98
3	504.99	5.05	510.04	504.99
Total		30.10	1530.10	1500.00

When one is amortizing a loan, at the beginning of any period the principal outstanding is the present value of the remaining payments. Using this fact together with our previous development, we obtain the formulas listed in Table 5.2, which describe the amortization of an interest-bearing loan of A dollars, at a rate r per period, by n equal payments of R dollars each and such that a payment is made at the end of each period. In particular, notice that Formula 1 for the periodic payment R involves $a_{\overline{n}|r}$, which, as you recall, is defined as $[1-(1+r)^{-n}]/r$.

TABLE 5.2 Amortization Formulas

1. Periodic payment: $R = \dfrac{A}{a_{\overline{n}|r}} = A \cdot \dfrac{r}{1-(1+r)^{-n}}$
2. Principal outstanding at beginning of kth period:
$$Ra_{\overline{n-k+1}|r} = R \cdot \frac{1-(1+r)^{-n+k-1}}{r}$$
3. Interest in kth payment: $Rra_{\overline{n-k+1}|r}$
4. Principal contained in kth payment: $R[1 - ra_{\overline{n-k+1}|r}]$
5. Total interest paid: $R(n - a_{\overline{n}|r})$, $(= nR - A)$

EXAMPLE 1 **Amortizing a Loan**

A person amortizes a loan of \$170,000 *for a new home by obtaining a* 20-*year mortgage at the rate of* 7.5% *compounded monthly. Find (a) the monthly payment, (b) the total interest charges, and (c) the principal remaining after* 5 *years.*

Solution:

a. The number of payment periods is $n = 12(20) = 240$, the interest rate per period is $r = 0.075/12 = 0.00625$, and $A = 170{,}000$. From Formula 1 in Table 5.2, the monthly payment R is $170{,}000/a_{\overline{240}|0.00625}$. Since $a_{\overline{240}|0.00625}$

is not in Appendix B, we use the following equivalent formula and a calculator:

$$R = 170{,}000\left[\frac{0.00625}{1-(1.00625)^{-240}}\right]$$
$$\approx \$1369.51$$

b. From Formula 5, the total interest charges are

$$240(1369.51) - 170{,}000 = 328{,}682.40 - 170{,}000$$
$$= \$158{,}682.40$$

This is almost as much as the loan itself.

c. After 5 years, we are at the beginning of the 61st period. Using Formula 2 with $n-k+1 = 240-61+1 = 180$, we find that the principal remaining is

$$1369.51\left[\frac{1-(1.00625)^{-180}}{0.00625}\right] \approx \$147{,}733.74$$

▶ **Now work Problem 1.**

At one time, a very common type of installment loan involved the "add-on method" of determining the finance charge. With that method, the finance charge is found by applying a quoted annual interest rate [under simple (that is, noncompounded) interest] to the borrowed amount of the loan. The charge is then added to the principal, and the total is divided by the number of *months* of the loan to determine the monthly installment payment. In loans of this type, the borrower may not immediately realize that the true annual rate is significantly higher than the quoted rate, as the following technology example shows.

TECHNOLOGY

Problem: A \$1000 loan is taken for one year at 9% interest under the add-on method. Estimate the true annual interest rate if monthly compounding is assumed.

Solution: Since the add-on method is used, payments will be made monthly. The finance charge for \$1000 at 9% simple interest for one year is $0.09(1000) = \$90$. Adding this to the loan amount gives $1000 + 90 = \$1090$. Thus, the monthly installment payment is $1090/12 \approx \$90.83$. Hence, we have a loan of \$1000 with 12 equal payments of \$90.83. From Formula 1 in Table 5.2,

$$R = \frac{A}{a_{\overline{n}|r}}$$
$$\frac{1090}{12} = \frac{1000}{a_{\overline{12}|r}}$$
$$a_{\overline{12}|r} = \frac{1000(12)}{1090} \approx 11.009174$$

We now solve $a_{\overline{12}|r} = 11.009174$ for the monthly rate r. We have

$$\frac{1-(1+r)^{-12}}{r} = 11.009174$$

Graphing

$$Y_1 = (1-(1+X) \wedge -12)/X$$

$$Y_2 = 11.009174$$

and finding the intersection (see Figure 5.10) gives

$$r \approx 0.01351374$$

which corresponds to an annual rate of

$$12(0.01351374) \approx 0.1622, \quad \text{or} \quad 16.22\%.$$

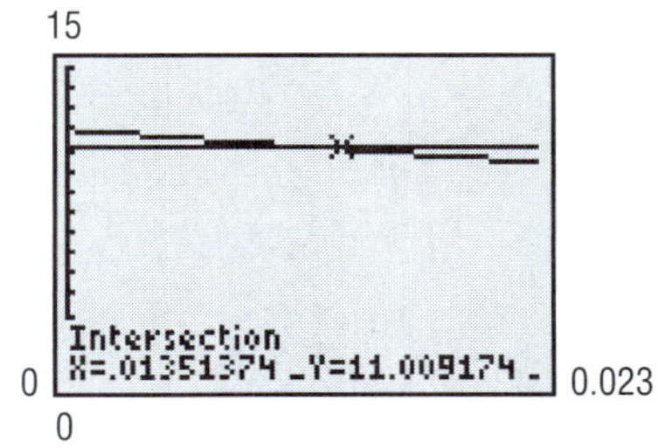

FIGURE 5.10 Solution of $a_{\overline{12}|r} = 11.009174$.

Thus, the true annual rate is 16.22%. Federal regulations concerning truth-in-lending laws have made add-on loans virtually obsolete.

The annuity formula

$$A = R \cdot \frac{1-(1+r)^{-n}}{r}$$

can be solved for n to give the number of periods of a loan. Multiplying both sides by $\frac{r}{R}$ gives

$$\frac{Ar}{R} = 1-(1+r)^{-n}$$

$$(1+r)^{-n} = 1 - \frac{Ar}{R} = \frac{R-Ar}{R}$$

$$-n\ln(1+r) = \ln\left(\frac{R-Ar}{R}\right) \qquad \text{(taking logs of both sides)}$$

$$n = -\frac{\ln\left(\frac{R-Ar}{R}\right)}{\ln(1+r)}$$

Using properties of logarithms, we eliminate the minus sign by inverting the quotient in the numerator:

$$n = \frac{\ln\left(\frac{R}{R-Ar}\right)}{\ln(1+r)} \tag{1}$$

EXAMPLE 2 **Periods of a Loan**

Muhammar Smith recently purchased a computer for \$1500 *and agreed to pay it off by making monthly payments of* \$75. *If the store charges interest at the rate of* 12% *compounded monthly, how many months will it take to pay off the debt?*

Solution: From Equation (1),

$$n = \frac{\ln\left[\dfrac{75}{75 - 1500(0.01)}\right]}{\ln(1.01)} \approx 22.4 \text{ months}$$

In reality, there are 23 payments; however, the final payment will be less than \$75.

▶ **Now work Problem 11.**

Exercise 5.4

*1. A person borrows \$2000 from a bank and agrees to pay it off by equal payments at the end of each month for three years. If interest is at 15% compounded monthly, how much is each payment?

2. A person wishes to make a three-year loan and can afford payments of \$50 at the end of each month. If interest is at 12% compounded monthly, how much can the person afford to borrow?

3. **Finance Charge** Determine the finance charge on a 36-month \$8000 auto loan with monthly payments if interest is at the rate of 4% compounded monthly.

4. For a one-year loan of \$500 at the rate of 15% compounded monthly, find (a) the monthly installment payment and (b) the finance charge.

5. **Car Loan** A person is amortizing a 36-month car loan of \$7500 with interest at the rate of 4% compounded monthly. Find (a) the monthly payment, (b) the interest in the first month, and (c) the principal repaid in the first payment.

6. **Real-Estate Loan** A person is amortizing a 48-month loan of \$35,000 for a house lot. If interest is at the rate of 7.8% compounded monthly, find (a) the monthly payment, (b) the interest in the first payment, and (c) the principal repaid in the first payment.

In Problems 7–10, construct amortization schedules for the indicated debts. Adjust the final payments if necessary.

7. \$5000 repaid by four equal yearly payments with interest at 7% compounded annually.

8. \$8000 repaid by six equal semiannual payments with interest at 8% compounded semiannually.

9. \$900 repaid by five equal quarterly payments with interest at 10% compounded quarterly.

10. \$10,000 repaid by five equal monthly payments with interest at 9% compounded monthly.

*11. A loan of \$1000 is being paid off by quarterly payments of \$100. If interest is at the rate of 8% compounded quarterly, how many *full* payments will be made?

12. A loan of \$2000 is being amortized over 48 months at an interest rate of 12% compounded monthly. Find:
 (a) the monthly payment;
 (b) the principal outstanding at the beginning of the 36th month;
 (c) the interest in the 36th payment;
 (d) the principal in the 36th payment;
 (e) the total interest paid.

13. A debt of \$16,000 is being repaid by 15 equal semiannual payments, with the first payment to be made six months from now. Interest is at the rate of 6% compounded semiannually. However, after two years, the interest rate increases to 8% compounded semiannually. If the debt must be paid off on the original date agreed upon, find the new semiannual payment. Give your answer to the nearest dollar.

14. A person borrows \$2000 and will pay off the loan by equal payments at the end of each month for five years. If interest is at the rate of 16.8% compounded monthly, how much is each payment?

15. **Mortgage** A \$245,000 mortgage for 25 years for a new home is obtained at the rate of 9.2% compounded monthly. Find (a) the monthly payment, (b) the interest in the first payment, (c) the principal repaid in the first payment, and (d) the finance charge.

16. **Auto Loan** An automobile loan of \$8500 is to be amortized over 48 months at an interest rate of 13.2% compounded monthly. Find (a) the monthly payment and (b) the finance charge.

17. **Furniture Loan** A person purchases furniture for \$2000 and agrees to pay off this amount by monthly payments of \$100. If interest is charged at the rate of 18% compounded monthly, how many *full* payments will there be?

18. Find the monthly payment of a five-year loan for \$9500 if interest is at 9.24% compounded monthly.

19. **Mortgage** Bob and Mary Rodgers want to purchase a new house and feel that they can afford a mortgage

payment of \$600 a month. They are able to obtain a 30-year 7.6% mortgage (compounded monthly), but must put down 25% of the cost of the house. Assuming that they have enough savings for the down payment, how expensive a house can they afford? Give your answer to the nearest dollar.

20. **Mortgage** Suppose you have the choice of taking out a \$180,000 mortgage at 8% compounded monthly for either 15 years or 30 years. How much savings is there in the finance charge if you were to choose the 15-year mortgage?

21. On a \$25,000 five-year loan, how much less is the monthly payment if the loan were at the rate of 12% compounded monthly rather than at 15% compounded monthly?

22. **Home Loan** The federal government has a program to aid low-income homeowners in urban areas. This program allows certain qualified homeowners to obtain low-interest home improvement loans. Each loan is processed through a commercial bank. The bank makes home improvement loans at an annual rate of $9\frac{1}{4}\%$ compounded monthly. However, the government subsidizes the bank so that the loan to the homeowner is at the annual rate of 4% compounded monthly. If the monthly payment at the 4% rate is x dollars (x dollars is the homeowner's monthly payment) and the monthly payment at the $9\frac{1}{4}\%$ rate is y dollars (y dollars is the monthly payment the bank must receive), then the government makes up the difference $y - x$ to the bank each month. From a practical point of view, the government does not want to bother with *monthly* payments. Instead, at the beginning of the loan, the government pays the present value of all such monthly differences, at an annual rate of $9\frac{1}{4}\%$ compounded monthly.

    If a qualified homeowner takes out a loan for \$5000 for five years, determine the government's payment to the bank at the beginning of the loan.

5.5 Review

Important Terms and Symbols

Section 5.1	effective rate yield
Section 5.2	present value future value equation of value cash flows net present value
Section 5.3	geometric sequence geometric series common ratio annuity ordinary annuity annuity due present value of annuity, $a_{\overline{n}|r}$ amount of annuity, $s_{\overline{n}|r}$
Section 5.4	amortizing amortization schedule finance charge

Summary

The concept of compound interest lies at the heart of any discussion dealing with the time value of money—that is, the present value of money due in the future or the future value of money presently invested. Under compound interest, interest is converted into principal and earns interest itself. The basic compound-interest formulas are

$$S = P(1+r)^n \qquad \text{(future value)}$$
$$P = S(1+r)^{-n} \qquad \text{(present value)}$$

where S = compound amount (future value)
P = principal (present value)
r = periodic rate
n = number of interest periods

Interest rates are usually quoted as an annual rate called the nominal rate. The periodic rate is obtained by dividing the nominal rate by the number of interest periods each year. The effective rate is the annual simple-interest rate that is equivalent to the nominal rate of r compounded n times a year and is given by

$$r_e = \left(1 + \frac{r}{n}\right)^n - 1 \qquad \text{(effective rate)}$$

Effective rates are used to compare different interest rates.

An annuity is a sequence of payments made at fixed periods of time over some interval. The mathematical basis for formulas dealing with annuities is the notion of the sum of a geometric series—that is,

$$s = \frac{a(1-r^n)}{1-r} \qquad \text{(sum of geometric series)}$$

where s = sum
a = first term
r = common ratio
n = number of terms

An ordinary annuity is an annuity in which each payment is made at the *end* of a payment period, whereas an annuity due is an annuity in which each payment is made at the *beginning* of a payment period. The basic formulas dealing with ordinary annuities are

$$A = R \cdot \frac{1-(1+r)^{-n}}{r} = Ra_{\overline{n}|r} \qquad \text{(present value)}$$

$$S = R \cdot \frac{(1+r)^{n}-1}{r} = Rs_{\overline{n}|r} \qquad \text{(future value)}$$

where A = present value of annuity
S = amount (future value) of annuity
R = amount of each payment
n = number of payment periods
r = periodic rate

For an annuity due, the corresponding formulas are

$$A = R(1 + a_{\overline{n-1}|r}) \qquad \text{(present value)}$$
$$S = R(s_{\overline{n+1}|r} - 1) \qquad \text{(future value)}$$

A loan, such as a mortgage, is amortized when part of each installment payment is used to pay interest and the remaining part is used to reduce the principal. A complete analysis of each payment is given in an amortization schedule. The following formulas deal with amortizing a loan of A dollars, at the periodic rate of r, by n equal payments of R dollars each and such that a payment is made at the end of each period:

Periodic payment:

$$R = \frac{A}{a_{\overline{n}|r}} = A \cdot \frac{r}{1-(1+r)^{-n}}$$

Principal outstanding at beginning of kth period:

$$Ra_{\overline{n-k+1}|r} = R \cdot \frac{1-(1+r)^{-n+k-1}}{r}$$

Interest in kth payment: $Rra_{\overline{n-k+1}|r}$

Principal contained in kth payment:

$$R[1 - ra_{\overline{n-k+1}|r}]$$

Total interest paid: $R(n - a_{\overline{n}|r}) = nR - A$

Review Problems

Problem numbers shown in color indicate problems suggested for use as a practice chapter test.

1. Find the sum of the geometric series

$$3 + 2 + 2 \cdot \frac{2}{3} + \cdots + 3\left(\frac{2}{3}\right)^5$$

2. Find the effective rate that corresponds to a nominal rate of 6% compounded quarterly.

3. An investor has a choice of investing a sum of money at either 8.5% compounded annually or 8.2% compounded semiannually. Which is the better choice?

4. **Cash Flows** Find the net present value of the following cash flows, which can be purchased by an initial investment of $7000:

Year	Cash Flow
2	$3400
4	3500

Assume that interest is at 7% compounded semiannually.

5. A debt of $1200 due in four years and $1000 due in six years is to be repaid by a payment of $1000 now and a second payment at the end of two years. How much should the second payment be if interest is at 8% compounded semiannually?

6. Find the present value of an annuity of $250 at the end of each month for four years if interest is at 6% compounded monthly.

7. For an annuity of $200 at the end of every six months for $6\frac{1}{2}$ years, find (a) the present value and (b) the future value at an interest rate of 8% compounded semiannually.

8. Find the amount of an annuity due that consists of 13 yearly payments of $150, provided that the interest rate is 4% compounded annually.

9. Suppose $100 is initially placed in a savings account and $100 is deposited at the end of every six months for the next four years. If interest is at 7% compounded semiannually, how much is in the account at the end of four years?

10. A savings account pays interest at the rate of 5% compounded semiannually. What amount must be deposited now so that $250 can be withdrawn at the end of every six months for the next 10 years?

11. **Sinking Fund** A company borrows $5000 on which it will pay interest at the end of each year at the annual rate of 11%. In addition, a sinking fund is set up so that the $5000 can be repaid at the end of five years. Equal payments are placed in the fund at the end of each year, and the fund earns interest at the effective rate of 6%. Find the annual payment in the *sinking fund*.

12. **Car Loan** A debtor is to amortize a \$7000 car loan by making equal payments at the end of each month for 36 months. If interest is at 4% compounded monthly, find (a) the amount of each payment and (b) the finance charge.

13. A person has debts of \$500 due in three years with interest at 5% compounded annually and \$500 due in four years with interest at 6% compounded semiannually. The debtor wants to pay off these debts by making two payments: the first payment now, and the second, which is double the first payment, at the end of the third year. If money is worth 7% compounded annually, how much is the first payment?

14. Construct an amortization schedule for a loan of \$2700 repaid by three monthly payments with interest at 13.5% compounded monthly.

15. Construct an amortization schedule for a loan of \$15,000 repaid by five monthly payments with interest at 9% compounded monthly.

16. Find the present value of an ordinary annuity of \$540 every month for seven years at the rate of 10% compounded monthly.

17. **Auto Loan** Determine the finance charge for a 48-month auto loan of \$11,000 with monthly payments at the rate of 5.5% compounded monthly.

Mathematical SNAPSHOT

Treasury Securities

The safest single type of investment is in securities issued by the U.S. Treasury. These pay fixed returns on a predetermined schedule, which can span as little as three months or as much as thirty years. The finish date is called the date of maturity.

Although Treasury securities are initially sold by the government, they trade on the open market. Because the prices are free to float up and down, the securities' rates of return can change over time. Consider, for example, a six-month Treasury bill, or T-bill, bearing a \$10,000 face value and purchased on the date of issue for \$9832.84. T-bills pay no interest before maturity but upon maturity are redeemed by the government at face value. This T-bill, if held the full six months, will pay back $\frac{10{,}000}{9832.84} \approx 101.7\%$ of the original investment, for an annualized yield (effective annual rate of return) of $1.017^2 - 1 \approx 3.429\%$. If the same T-bill, however, is sold at midterm for \$9913.75, the new owner has a prospective annualized yield of $\left(\frac{10{,}000}{9913.75}\right)^4 - 1 \approx 3.526\%$ over the remaining three months.

Like T-bills, Treasury notes and bonds are redeemed at face value upon maturity. In addition, however, notes and bonds pay interest twice a year according to a fixed nominal rate.[3] A 6.5%, \$20,000 seven-year note pays $0.065(20{,}000) = \$1300$ every six months. At the end of seven years, the holder receives the final interest payment plus the face value, for a total of \$21,300.

Mathematically, it is easier to calculate the present value of a note or bond from an assumed yield than to find the yield given an assumed present value (or price). Notes and bonds differ only in times to maturity: one to ten years for notes, ten to thirty years for bonds. Each note or bond is a guarantee of a lump sum at a future date plus an annuity until then. The present value of the note or bond, then, is the sum of the present value of the future lump sum and the present value of the annuity. We will assume that notes and bonds are evaluated at times when the next interest payment is exactly six months away; that way we can use the formula for the present value of an annuity in Section 5.3.

[3] In this context, *nominal rate* does *not* refer to the annual percentage rate. The former is constant, while the latter changes in tandem with the yield.

With semiannual compounding, an annual yield of r corresponds to an interest payment of $\sqrt{1+r} - 1$ every six months. Making the appropriate substitution in the formulas from Sections 5.2 and 5.3, we obtain the following general formula for the present value of a Treasury note or bond

$$P = S(1 + \sqrt{1+r} - 1)^{-2n} + R \cdot \frac{1 - (1 + \sqrt{1+r} - 1)^{-2n}}{\sqrt{1+r} - 1}$$

which simplifies to

$$P = S(1+r)^{-n} + R \cdot \frac{1 - (1+r)^{-n}}{\sqrt{1+r} - 1}$$

where S is the face value, r is the assumed annual yield, and n is a multiple of $\frac{1}{2}$ representing the number of years to maturity (so that $2n$ is the number of six-month periods). R is the amount of the semiannual interest payment, that is, S times the half the bond's nominal rate ($0.03S$ for a 6% bond, and so on).

Since we can treat a T-bill as a shorter-term note with a nominal rate of 0%, this formula covers T-bills as well, for which (since there is no annuity component) n need not be a multiple of $\frac{1}{2}$.

To illustrate, if we are seeking a 7.4% yield on a new-issue, one-year, \$30,000 T-bill (for which $R = 0$), we should be willing to pay

$$30{,}000(1.074)^{-1} \approx \$27{,}932.96$$

But if we are seeking a 7.4% yield on a 5.5%, \$30,000 bond with 17 years left to maturity ($R = 0.0275 \cdot 30{,}000 = 825$), we should be willing to pay only

$$30{,}000(1.074)^{-17} + 825 \cdot \frac{1 - (1.074)^{-17}}{\sqrt{1.074} - 1} \approx 24{,}870.66$$

Of course, it may happen that our yield expectations are unrealistic and that no bonds are for sale at the price we calculate. In that case, we may need to look at market prices and consider whether we can accept the corresponding yields. But how do we find a security's yield r from its market price? For T-bills, the second term on the right side of the present value formula drops out, and we can solve the simplified formula for r to obtain

$$r = \left(\frac{S}{P}\right)^{1/n} - 1$$

Calculations for three- and six-month T-bills use $n = \frac{1}{4}$ and $n = \frac{1}{2}$ (as, for example, in the calculations near the beginning of this Snapshot).

Calculating the yield on a note or bill, on the other hand, involves solving the complete present value equations—and this cannot be done algebraically. However, it can be with a graphing calculator. We set Y_1 equal to the left side of the equation, Y_2 equal to the right side, and find where Y_1 and Y_2 are equal. Suppose, for example, that a 6.8%, \$26,000 bond is selling at \$26,617.50 eleven years from maturity. Each of the 22 interest payments will amount to 0.034(26,000) = \$884. To find the yield, set

$$Y_1 = 26{,}617.50$$

and

$$Y_2 = 26{,}000(1 + X)^\wedge - 11$$
$$+ 884(1 - (1 + X)^\wedge - 11)/(\sqrt{}(1 + X) - 1)$$

Then graph Y_1 and Y_2 and find where the two graphs intersect (Figure 5.11).

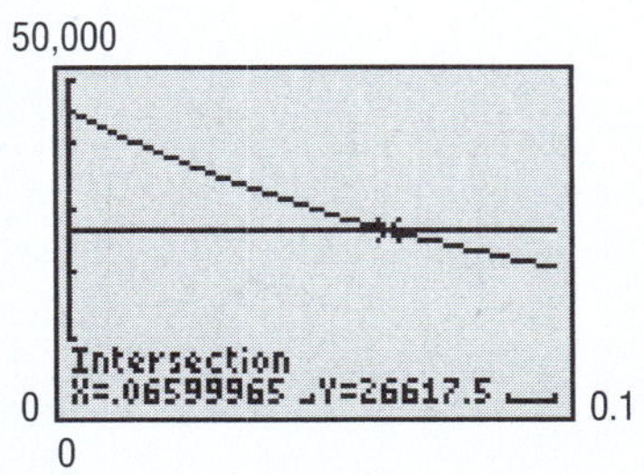

FIGURE 5.11 Finding yield.

The graphs intersect for $X \approx 0.0660$, which means that the yield is 6.6%.

The graph describing the current yields of Treasury securities as a function of time to maturity is called the yield curve. Economists keep a daily watch on this curve; you can monitor it yourself at www.bloomberg.com/usatoday. Most typically, the yield curve looks something like that shown in Figure 5.12.

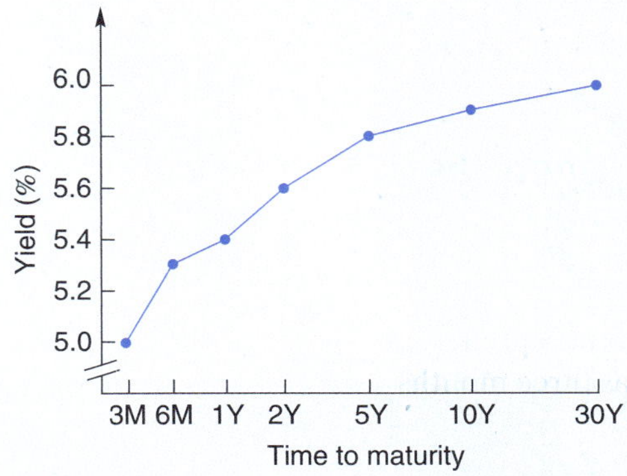

FIGURE 5.12 A normal yield curve.

You can see that the longer the time to maturity, the greater the yield. The usual explanation for this pattern is that having money tied up in a long-term investment means a loss of short-term flexibility—of liquidity, as it is called. To attract buyers, long-term securities must generally be priced for slightly higher yields than short-term securities.

Exercises

1. Find the present value of a 7.5%, 20-year, \$15,000 bond, assuming an annual yield of 7.25%.
2. Find the yield on a 6.5%, \$10,000 note that is selling at \$10,389 with seven years left to maturity.
3. In late December of 2000, the yield curve for Treasury securities had the atypical shape shown in Figure 5.13.

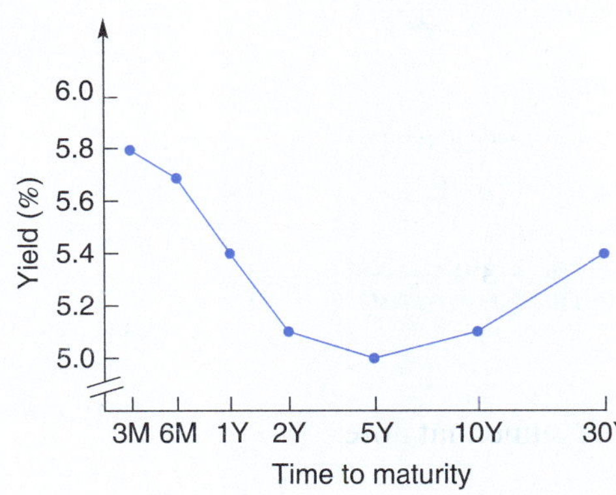

FIGURE 5.13

T-bills were earning higher yields than five-year notes, contrary to what one would expect. How might investor expectations about future earnings possibilities explain the yield curve?

CHAPTER 6

Matrix Algebra

Matrices, the subject of this chapter, are simply arrays of numbers. Matrices and matrix algebra have potential application whenever numerical information can be meaningfully arranged into rectangular blocks.

One area of application for matrix algebra is computer graphics. An object in a coordinate system can be represented by a matrix that contains the coordinates of each corner. For example, we might set up a connect-the-dots scheme in which the lightning bolt shown is represented by the matrix to its right.

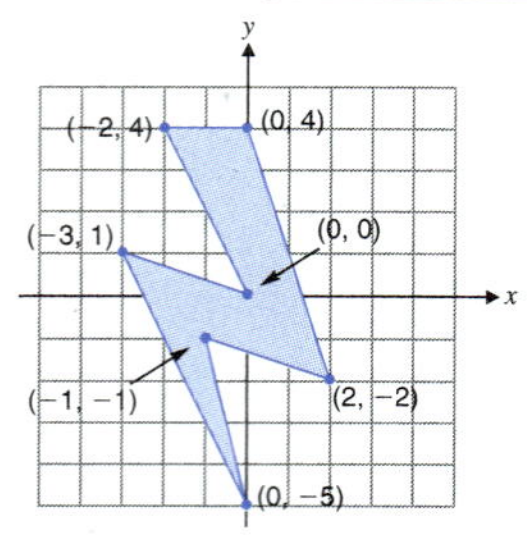

$$\begin{array}{cc} x & y \\ \downarrow & \downarrow \end{array}$$
$$\begin{bmatrix} 0 & 0 \\ -2 & 4 \\ 0 & 4 \\ 2 & -2 \\ -1 & -1 \\ 0 & -5 \\ -3 & 1 \end{bmatrix}$$

Computer graphics often show objects rotating in space. Computationally, rotation is done by matrix multiplication. The lightning bolt is rotated clockwise 52 degrees about the origin by matrix multiplication, involving a matrix whose entries are trigonometric functions of the rotation angle:

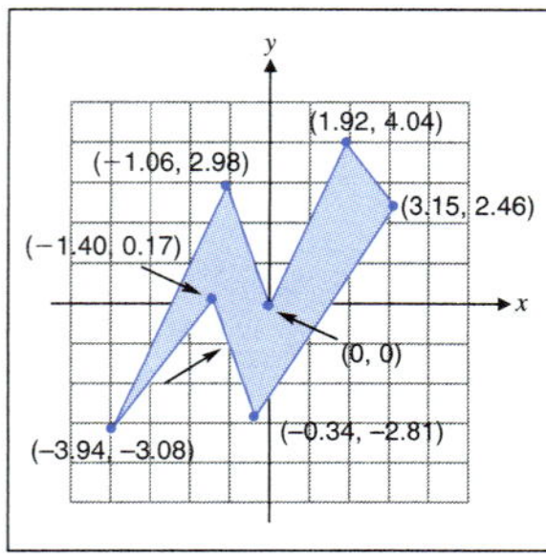

$$\begin{bmatrix} 0 & 0 \\ -2 & 4 \\ 0 & 4 \\ 2 & -2 \\ -1 & -1 \\ 0 & -5 \\ -3 & 1 \end{bmatrix} \begin{bmatrix} \cos 52^\circ & -\sin 52^\circ \\ \sin 52^\circ & \cos 52^\circ \end{bmatrix} = \begin{bmatrix} 0 & 0 \\ 1.92 & 4.04 \\ 3.15 & 2.46 \\ -0.34 & -2.81 \\ -1.40 & 0.17 \\ -3.94 & -3.08 \\ -1.06 & 2.98 \end{bmatrix}$$

OBJECTIVE *To introduce the concept of a matrix and to consider special types of matrices.*

6.1 Matrices

Finding ways to describe many situations in mathematics and economics leads to the study of rectangular arrays of numbers. Consider, for example, the system of linear equations

$$\begin{cases} 3x + 4y + 3z = 0 \\ 2x + y - z = 0 \\ 9x - 6y + 2z = 0 \end{cases}$$

The features that characterize this system are the numerical coefficients in the equations, together with their relative positions. For this reason, the system can be described by the rectangular array

$$\begin{bmatrix} 3 & 4 & 3 \\ 2 & 1 & -1 \\ 9 & -6 & 2 \end{bmatrix}$$

CAUTION Do not use vertical bars, ||, instead of brackets or parentheses, for they have a different meaning.

which is called a *matrix* (plural: *matrices*, pronounced may′ tri sees). We shall consider such rectangular arrays to be objects in themselves, and our custom, as just shown, will be to enclose them in brackets. Parentheses are also commonly used. In symbolically representing matrices, we shall use bold capital letters such as **A**, **B**, **C**, and so on.

In economics it is often convenient to use matrices in formulating problems and displaying data. For example, a manufacturer who produces products A, B, and C could represent the units of labor and material involved in one week's production of these items as in Table 6.1. More simply, the data can be represented by the matrix

TABLE 6.1

	Product		
	A	B	C
Labor	10	12	16
Material	5	9	7

$$\mathbf{A} = \begin{bmatrix} 10 & 12 & 16 \\ 5 & 9 & 7 \end{bmatrix}$$

The horizontal rows of a matrix are numbered consecutively from top to bottom, and the vertical columns are numbered from left to right. For the foregoing matrix **A**, we have

$$\begin{array}{r} \\ \text{row 1} \\ \text{row 2} \end{array} \begin{array}{c} \begin{array}{ccc} \text{column 1} & \text{column 2} & \text{column 3} \end{array} \\ \begin{bmatrix} \quad 10 \quad & \quad 12 \quad & \quad 16 \quad \\ 5 & 9 & 7 \end{bmatrix} \end{array} = \mathbf{A}$$

Since **A** has two rows and three columns, we say that **A** has *size*, 2 × 3 (read "2 by 3"), where the number of rows is specified first. Similarly, the matrices

$$\mathbf{B} = \begin{bmatrix} 1 & 6 & -2 \\ 5 & 1 & -4 \\ -3 & 5 & 0 \end{bmatrix} \quad \text{and} \quad \mathbf{C} = \begin{bmatrix} 1 & 2 \\ -3 & 4 \\ 5 & 6 \\ 7 & -8 \end{bmatrix}$$

have sizes 3 × 3 and 4 × 2, respectively.

The numbers in a matrix are called its **entries** or **elements.** To denote arbitrary entries in a matrix of size 2 × 3, say, there are two common methods. First, we may use different letters:

$$\begin{bmatrix} a & b & c \\ d & e & f \end{bmatrix}$$

Second, a single letter may be used, say, a, along with appropriate *double* subscripts to indicate position:

$$\begin{bmatrix} a_{11} & a_{12} & a_{13} \\ a_{21} & a_{22} & a_{23} \end{bmatrix}$$

CAUTION The row subscript appears to the left of the column subscript. In general, a_{ij} and a_{ji} are quite different.

For the entry a_{12} (read "a sub one-two" or just "a one-two"), the first subscript, 1, specifies the row and the second subscript, 2, the column in which the entry appears. Similarly, the entry a_{23} (read "a two-three") is the entry in the second row and the third column. Generalizing, we say that the symbol a_{ij} denotes the entry in the ith row and jth column.

Our concern in this chapter is the manipulation and application of various types of matrices. For completeness, we now give a formal definition of a matrix.

Definition

A rectangular array of numbers consisting of m horizontal rows and n vertical columns,

$$\begin{bmatrix} a_{11} & a_{12} & \cdots & a_{1n} \\ a_{21} & a_{22} & \cdots & a_{2n} \\ \cdot & \cdot & \cdots & \cdot \\ \cdot & \cdot & \cdots & \cdot \\ \cdot & \cdot & \cdots & \cdot \\ a_{m1} & a_{m2} & \cdots & a_{mn} \end{bmatrix}$$

is called an ***m* × *n* matrix** or a ***matrix of size m* × *n***. For the entry a_{ij}, we call i the row subscript and j the column subscript.

CAUTION Do not confuse the general entry a_{ij} with the matrix $[a_{ij}]$.

The number of entries in an $m \times n$ matrix is mn. For brevity, an $m \times n$ matrix can be denoted by the symbol $[a_{ij}]_{m\times n}$ or, more simply, $[a_{ij}]$, where the size is understood to be that which is appropriate for the given context. This notation merely indicates what types of symbols we are using to denote the general entry.

A matrix that has exactly one row, such as the 1×4 matrix

$$\mathbf{A} = [1 \quad 7 \quad 12 \quad 3]$$

is called a **row matrix** or a **row vector.** A matrix consisting of a single column, such as the 5×1 matrix

$$\begin{bmatrix} 1 \\ -2 \\ 15 \\ 9 \\ 16 \end{bmatrix}$$

is called a **column matrix** or a **column vector.**

APPLICATIONS IN PRACTICE 1

Size of a Matrix

A manufacturer who uses raw materials A and B is interested in tracking the costs of these materials from three different sources. What is the size of the matrix she would use?

EXAMPLE 1 **Size of a Matrix**

a. The matrix $[1 \quad 2 \quad 0]$ has size 1×3.

b. The matrix $\begin{bmatrix} 1 & -6 \\ 5 & 1 \\ 9 & 4 \end{bmatrix}$ has size 3×2.

c. The matrix [7] has size 1×1.

d. The matrix $\begin{bmatrix} 1 & 3 & 7 & -2 & 4 \\ 9 & 11 & 5 & 6 & 8 \\ 6 & -2 & -1 & 1 & 1 \end{bmatrix}$ has size 3×5 and $3(5) = 15$ entries.

▶ **Now work Problem 1a.**

APPLICATIONS IN PRACTICE 2

Constructing Matrices

An analysis of a workplace uses a 3×5 matrix to describe the time spent on each of three phases of five different projects. Project 1 requires 1 hour for each phase, project 2 requires twice as much time as project 1, project 3 requires twice as much time as project 2, ..., and so on. Construct this time-analysis matrix.

EXAMPLE 2 **Constructing Matrices**

a. *Construct a three-entry column matrix such that $a_{21} = 6$ and $a_{i1} = 0$ otherwise.*

Solution: Since $a_{11} = a_{31} = 0$, the matrix is

$$\begin{bmatrix} 0 \\ 6 \\ 0 \end{bmatrix}$$

b. *If $\mathbf{A} = [a_{ij}]$ has size 3×4 and $a_{ij} = i + j$, find* **A**.

Solution: Here $i = 1, 2, 3$ and $j = 1, 2, 3, 4$, and **A** has $(3)(4) = 12$ entries. Since $a_{ij} = i + j$, the entry in row i and column j is obtained by adding the numbers i and j. Hence, $a_{11} = 1 + 1 = 2$, $a_{12} = 1 + 2 = 3$, $a_{13} = 1 + 3 = 4$, and so on. Thus,

$$\mathbf{A} = \begin{bmatrix} 1+1 & 1+2 & 1+3 & 1+4 \\ 2+1 & 2+2 & 2+3 & 2+4 \\ 3+1 & 3+2 & 3+3 & 3+4 \end{bmatrix} = \begin{bmatrix} 2 & 3 & 4 & 5 \\ 3 & 4 & 5 & 6 \\ 4 & 5 & 6 & 7 \end{bmatrix}$$

c. *Construct the 3×3 matrix* **I**, *given that $a_{11} = a_{22} = a_{33} = 1$ and $a_{ij} = 0$ otherwise.*

Solution: The matrix is given by

$$\mathbf{I} = \begin{bmatrix} 1 & 0 & 0 \\ 0 & 1 & 0 \\ 0 & 0 & 1 \end{bmatrix}$$

▶ **Now work Problem 11.**

Equality of Matrices

We now define what is meant by saying that two matrices are *equal*.

Definition

Matrices $\mathbf{A} = [a_{ij}]$ and $\mathbf{B} = [b_{ij}]$ are ***equal*** if and only if they have the same size and $a_{ij} = b_{ij}$ for each i and j (that is, corresponding entries are equal).

Thus,

$$\begin{bmatrix} 1+1 & \frac{2}{2} \\ 2 \cdot 3 & 0 \end{bmatrix} = \begin{bmatrix} 2 & 1 \\ 6 & 0 \end{bmatrix}$$

but

$$[1 \quad 1] \neq \begin{bmatrix} 1 \\ 1 \end{bmatrix} \quad \text{and} \quad [1 \quad 1] \neq [1 \quad 1 \quad 1] \qquad \text{(different sizes)}$$

A matrix equation can define a system of equations. For example, suppose that

$$\begin{bmatrix} x & y+1 \\ 2z & 5w \end{bmatrix} = \begin{bmatrix} 2 & 7 \\ 4 & 2 \end{bmatrix}$$

By equating corresponding entries, we must have

$$\begin{cases} x = 2 \\ y + 1 = 7 \\ 2z = 4 \\ 5w = 2 \end{cases}$$

Solving gives $x = 2$, $y = 6$, $z = 2$, and $w = \frac{2}{5}$.

Transpose of a Matrix

If $\mathbf{A}$ is a matrix, the matrix formed from $\mathbf{A}$ by interchanging its rows with its columns is called the *transpose* of $\mathbf{A}$.

Definition

The ***transpose*** of an $m \times n$ matrix $\mathbf{A}$, denoted $\mathbf{A}^{\mathrm{T}}$, is the $n \times m$ matrix whose ith row is the ith column of $\mathbf{A}$.

EXAMPLE 3 Transpose of a Matrix

If $\mathbf{A} = \begin{bmatrix} 1 & 2 & 3 \\ 4 & 5 & 6 \end{bmatrix}$, *find* $\mathbf{A}^{\mathrm{T}}$.

Solution: Matrix $\mathbf{A}$ is 2×3, so $\mathbf{A}^{\mathrm{T}}$ is 3×2. Column 1 of $\mathbf{A}$ becomes row 1 of $\mathbf{A}^{\mathrm{T}}$, column 2 becomes row 2, and column 3 becomes row 3. Thus,

$$\mathbf{A}^{\mathrm{T}} = \begin{bmatrix} 1 & 4 \\ 2 & 5 \\ 3 & 6 \end{bmatrix}.$$

▸ **Now work Problem 19.**

Observe that the columns of $\mathbf{A}^{\mathrm{T}}$ are the rows of $\mathbf{A}$. Also, if we take the transpose of our answer, the original matrix $\mathbf{A}$ is obtained. That is, the transpose operation has the property that

$$(\mathbf{A}^{\mathrm{T}})^{\mathrm{T}} = \mathbf{A}$$

TECHNOLOGY

Graphing calculators have the ability to manipulate matrices. For example, Figure 6.1 shows the result of applying the transpose operation to matrix **A**.

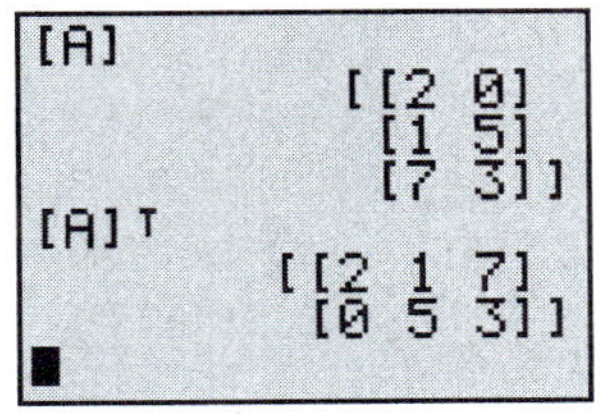

FIGURE 6.1 **A** and $\mathbf{A}^T$.

Special Matrices

Certain types of matrices play important roles in matrix theory. We now consider some of these special types.

An $m \times n$ matrix whose entries are all 0 is called the $m \times n$ **zero matrix** and is denoted by $\mathbf{O}_{m \times n}$ or, more simply, by **O** if its size is understood. Thus, the 2×3 zero matrix is

$$\mathbf{O} = \begin{bmatrix} 0 & 0 & 0 \\ 0 & 0 & 0 \end{bmatrix}$$

CAUTION: Do not confuse the matrix **O** with the real number 0.

and in general we have

$$\mathbf{O} = \begin{bmatrix} 0 & 0 & \cdots & 0 \\ 0 & 0 & \cdots & 0 \\ \vdots & \vdots & \ddots & \vdots \\ 0 & 0 & \cdots & 0 \end{bmatrix}$$

A matrix having the same number of columns as rows—for example, n rows and n columns—is called a **square matrix** of order n. That is, an $m \times n$ matrix is square if and only if $m = n$. For example, matrices

$$\begin{bmatrix} 2 & 7 & 4 \\ 6 & 2 & 0 \\ 4 & 6 & 1 \end{bmatrix} \quad \text{and} \quad [3]$$

are square with orders 3 and 1, respectively.

In a square matrix of order n, the entries $a_{11}, a_{22}, a_{33}, \ldots, a_{nn}$ that lie on the "main" diagonal extending from the upper left corner to the lower right corner are called the *main-diagonal* entries, or more simply the **main diagonal.** Thus, in the matrix

$$\begin{bmatrix} 1 & 2 & 3 \\ 4 & 5 & 6 \\ 7 & 8 & 9 \end{bmatrix}$$

the main diagonal (see the shaded region) consists of $a_{11} = 1, a_{22} = 5$, and $a_{33} = 9$.

A square matrix $\mathbf{A}$ is called a **diagonal matrix** if all the entries that are off the main diagonal are zero—that is, if $a_{ij} = 0$ for $i \neq j$. Examples of diagonal matrices are

$$\begin{bmatrix} 1 & 0 \\ 0 & 1 \end{bmatrix} \quad \text{and} \quad \begin{bmatrix} 3 & 0 & 0 \\ 0 & 6 & 0 \\ 0 & 0 & 9 \end{bmatrix}$$

It follows that a diagonal matrix is both upper triangular and lower triangular.

A square matrix $\mathbf{A}$ is said to be an **upper triangular matrix** if all entries *below* the main diagonal are zero—that is, if $a_{ij} = 0$ for $i > j$. Similarly, a matrix $\mathbf{A}$ is said to be a **lower triangular matrix** if all entries *above* the main diagonal are zero—that is, if $a_{ij} = 0$ for $i < j$. When a matrix is either upper triangular or lower triangular, it is called a **triangular matrix.** Thus, the matrices

$$\begin{bmatrix} 5 & 1 & 1 \\ 0 & -3 & 7 \\ 0 & 0 & 4 \end{bmatrix} \quad \text{and} \quad \begin{bmatrix} 7 & 0 & 0 & 0 \\ 3 & 2 & 0 & 0 \\ 6 & 5 & -4 & 0 \\ 1 & 6 & 0 & 1 \end{bmatrix}$$

are upper and lower triangular matrices, respectively, and are therefore triangular matrices.

Exercise 6.1

*1. Let

$$\mathbf{A} = \begin{bmatrix} 1 & -6 & 2 \\ -4 & 2 & 1 \end{bmatrix} \quad \mathbf{B} = \begin{bmatrix} 1 & 2 & 3 \\ 4 & 5 & 6 \\ 7 & 8 & 9 \end{bmatrix} \quad \mathbf{C} = \begin{bmatrix} 1 & 1 \\ 2 & 2 \\ 3 & 3 \end{bmatrix}$$

$$\mathbf{D} = \begin{bmatrix} 1 & 0 \\ 2 & 3 \end{bmatrix} \quad \mathbf{E} = \begin{bmatrix} 1 & 2 & 3 & 4 \\ 0 & 1 & 6 & 0 \\ 0 & 0 & 2 & 0 \\ 0 & 0 & 6 & 1 \end{bmatrix} \quad \mathbf{F} = [6 \quad 2]$$

$$\mathbf{G} = \begin{bmatrix} 5 \\ 6 \\ 1 \end{bmatrix} \quad \mathbf{H} = \begin{bmatrix} 1 & 6 & 2 \\ 0 & 0 & 0 \\ 0 & 0 & 0 \end{bmatrix} \quad \mathbf{J} = [4]$$

(a) State the size of each matrix.
(b) Which matrices are square?
(c) Which matrices are upper triangular? lower triangular?
(d) Which are row vectors?
(e) Which are column vectors?

In Problems 2–9, let

$$\mathbf{A} = [a_{ij}] = \begin{bmatrix} 7 & -2 & 14 & 6 \\ 6 & 2 & 3 & -2 \\ 5 & 4 & 1 & 0 \\ 8 & 0 & 2 & 0 \end{bmatrix}$$

2. What is the order of $\mathbf{A}$?

Find the following entries.

3. a_{21} **4.** a_{12} **5.** a_{32}
6. a_{34} **7.** a_{44} **8.** a_{55}

9. What are the main-diagonal entries?

10. Write the upper triangular matrix of order 4, given that all entries that are not required to be 0 are equal to the sum of their subscripts. (For example, $a_{23} = 2 + 3 = 5$.)

*11. Construct the matrix $\mathbf{A} = [a_{ij}]$ if $\mathbf{A}$ is 3×4 and $a_{ij} = 4i + 2j$.

12. Construct the matrix $\mathbf{B} = [b_{ij}]$ if $\mathbf{B}$ is 2×2 and $b_{ij} = (-1)^{i+j}(i^2 + j^2)$.

13. If $\mathbf{A} = [a_{ij}]$ is 12×10, how many entries does $\mathbf{A}$ have? If $a_{ij} = 1$ for $i = j$ and $a_{ij} = 0$ for $i \neq j$, find a_{33}, a_{52}, $a_{10,10}$, and $a_{12,10}$.

14. List the main diagonal of

(a) $\begin{bmatrix} 1 & 4 & -2 & 0 \\ 7 & 0 & 4 & -1 \\ -6 & 6 & -5 & 1 \\ 2 & 1 & 7 & 2 \end{bmatrix}$ **(b)** $\begin{bmatrix} x & 1 & y \\ 9 & y & 7 \\ y & 0 & z \end{bmatrix}$

15. Write the zero matrix of order (a) 4 and (b) 6.

16. If $\mathbf{A}$ is a 5×7 matrix, what is the size of $\mathbf{A}^{\mathrm{T}}$?

In Problems 17–20, find $\mathbf{A}^{\mathrm{T}}$.

17. $\mathbf{A} = \begin{bmatrix} 6 & -3 \\ 2 & 4 \end{bmatrix}$ **18.** $\mathbf{A} = [2 \quad 4 \quad 6 \quad 8]$

*19. $\mathbf{A} = \begin{bmatrix} 1 & 3 & 7 & 3 \\ 3 & 2 & -2 & 0 \\ -4 & 5 & 0 & 1 \end{bmatrix}$

20. $\mathbf{A} = \begin{bmatrix} 2 & -1 & 0 \\ -1 & 5 & 1 \\ 0 & 1 & 3 \end{bmatrix}$

21. Let

$$\mathbf{A} = \begin{bmatrix} 7 & 0 \\ 0 & 6 \end{bmatrix} \quad \mathbf{B} = \begin{bmatrix} 1 & 0 & 0 \\ 0 & 2 & 0 \\ 0 & 10 & -3 \end{bmatrix}$$

$$\mathbf{C} = \begin{bmatrix} 0 & 0 & 0 \\ 0 & 0 & 0 \\ 0 & 0 & 0 \end{bmatrix} \quad \mathbf{D} = \begin{bmatrix} 2 & 0 & -1 \\ 0 & 4 & 0 \\ 0 & 0 & 6 \end{bmatrix}$$

(a) Which are diagonal matrices?
(b) Which are triangular matrices?

22. A matrix is *symmetric* if $\mathbf{A}^T = \mathbf{A}$. Is the matrix of Problem 20 symmetric?

23. If

$$\mathbf{A} = \begin{bmatrix} 1 & 2 & 3 \\ 4 & 5 & 6 \\ 7 & 8 & 9 \end{bmatrix}$$

verify the general property that $(\mathbf{A}^T)^T = \mathbf{A}$ by finding $\mathbf{A}^T$ and then $(\mathbf{A}^T)^T$.

In Problems 24–27, solve the matrix equation.

24. $\begin{bmatrix} 2x & y \\ z & 3w \end{bmatrix} = \begin{bmatrix} 4 & 6 \\ 0 & 7 \end{bmatrix}$

25. $\begin{bmatrix} 6 & 2 \\ x & 7 \\ 3y & 2z \end{bmatrix} = \begin{bmatrix} 6 & 2 \\ 6 & 7 \\ 2 & 7 \end{bmatrix}$

26. $\begin{bmatrix} 4 & 2 & 1 \\ 3x & y & 3z \\ 0 & w & 7 \end{bmatrix} = \begin{bmatrix} 4 & 2 & 1 \\ 6 & 7 & 9 \\ 0 & 9 & 8 \end{bmatrix}$

27. $\begin{bmatrix} 2x & 7 \\ 7 & 2y \end{bmatrix} = \begin{bmatrix} y & 7 \\ 7 & y \end{bmatrix}$

28. Inventory A grocer sold 250 cans of tomato soup, 375 cans of beans, and 220 cans of tuna. Write a row vector that gives the number of each item sold. If the items sell for \$0.75, \$0.83, and \$1.05 each, respectively, write this information as a column vector.

29. Sales Analysis The Widget Company has its monthly sales reports given by means of matrices whose rows, in order, represent the number of regular, deluxe, and super-duper models sold, and the columns, in order, give the number of red, white, blue, and purple units sold. The matrices for January (**J**) and February (**F**) are

$$\mathbf{J} = \begin{bmatrix} 2 & 6 & 1 & 2 \\ 0 & 1 & 3 & 5 \\ 2 & 7 & 9 & 0 \end{bmatrix} \quad \mathbf{F} = \begin{bmatrix} 0 & 2 & 8 & 4 \\ 2 & 3 & 3 & 2 \\ 4 & 0 & 2 & 6 \end{bmatrix}$$

(a) How many white super-duper models were sold in January? (b) How many blue deluxe models were sold in February? (c) In which month were more purple regular models sold? (d) Which model and color sold the same number of units in both months? (e) In which month were more deluxe models sold? (f) In which month were more red widgets sold? (g) How many widgets were sold in January?

30. Input-Output Matrix Input-output matrices, which were developed by W. W. Leontief, indicate the interrelationships that exist among the various sectors of an economy during some period of time. A hypothetical example for a simplified economy is given by matrix **M** at the end of this problem. The consuming sectors are the same as the producing sectors and can be thought of as manufacturers, government, steel, agriculture, households, and so on. Each row shows how the output of a given sector is consumed by the four sectors. For example, of the total output of industry A, 50 went to industry A itself, 70 to B, 200 to C, and 360 to all others. The sum of the entries in row 1, namely, 680, gives the total output of A for a given period. Each column gives the output of each sector that is consumed by a given sector. For example, in producing 680 units, industry A consumed 50 units of A, 90 of B, 120 of C, and 420 from all other producers. For each column, find the sum of the entries. Do the same for each row. What do you observe in comparing these totals? Suppose sector A increases its output by 20%, namely, by 136 units. Assuming that this results in a uniform 20% increase of all its inputs, by how many units will sector B have to increase its output? Answer the same question for C and for all other producers.

M =

	CONSUMERS			
PRODUCERS	Industry A	Industry B	Industry C	All Other Consumers
Industry A	50	70	200	360
Industry B	90	30	270	320
Industry C	120	240	100	1050
All Other Producers	420	370	940	4960

31. Find all the values of x for which

$$\begin{bmatrix} x^2 + 2000x & \sqrt{x^2} \\ x^2 & \ln(e^x) \end{bmatrix} = \begin{bmatrix} 2001 & -x \\ 2001 - 2000x & x \end{bmatrix}$$

In Problems 32 and 33, find $\mathbf{A}^T$.

32. $\mathbf{A} = \begin{bmatrix} 3 & -4 & 5 \\ -2 & 1 & 6 \end{bmatrix}$

33. $\mathbf{A} = \begin{bmatrix} 3 & 1 & 4 & 2 \\ 1 & 7 & 3 & 6 \\ 1 & 4 & 1 & 2 \end{bmatrix}$

OBJECTIVE *To define matrix addition and scalar multiplication and to consider properties related to these operations.*

6.2 Matrix Addition and Scalar Multiplication

Matrix Addition

Consider a snowmobile dealer who sells two models, Deluxe and Super. Each is available in one of two colors, red and blue. Suppose that the sales for January and February are represented by the matrices

$$\mathbf{J} = \begin{array}{c} \\ \text{red} \\ \text{blue} \end{array}\begin{array}{c} \text{Deluxe} \quad \text{Super} \\ \begin{bmatrix} 1 & & & 2 \\ 3 & & & 5 \end{bmatrix} \end{array} \qquad \mathbf{F} = \begin{bmatrix} 3 & 1 \\ 4 & 2 \end{bmatrix}$$

Each row of **J** and **F** gives the number of each model sold for a given color. Each column gives the number of each color sold for a given model. A matrix representing total sales for each model and color over the two months can be obtained by adding the corresponding entries in **J** and **F**:

$$\begin{bmatrix} 4 & 3 \\ 7 & 7 \end{bmatrix}$$

This situation provides some motivation for introducing the operation of matrix addition for two matrices of the same size.

Definition

If $\mathbf{A} = [a_{ij}]$ and $\mathbf{B} = [b_{ij}]$ are both $m \times n$ matrices, then the ***sum*** $\mathbf{A} + \mathbf{B}$ is the $m \times n$ matrix obtained by adding corresponding entries of **A** and **B**; that is, $\mathbf{A} + \mathbf{B} = [a_{ij} + b_{ij}]$.

For example, let

$$\mathbf{A} = \begin{bmatrix} 3 & 0 & -2 \\ 2 & -1 & 4 \end{bmatrix} \quad \text{and} \quad \mathbf{B} = \begin{bmatrix} 5 & -3 & 6 \\ 1 & 2 & -5 \end{bmatrix}$$

Since **A** and **B** are the same size (2×3), their sum is defined. We have

$$\mathbf{A} + \mathbf{B} = \begin{bmatrix} 3+5 & 0+(-3) & -2+6 \\ 2+1 & -1+2 & 4+(-5) \end{bmatrix} = \begin{bmatrix} 8 & -3 & 4 \\ 3 & 1 & -1 \end{bmatrix}$$

APPLICATIONS IN PRACTICE 1

Matrix Addition

An office furniture company manufactures desks and tables at two plants, A and B. Matrix **J** represents the production of the two plants in January, and matrix **F** represents the production of the two plants in February. Write a matrix that represents the total production at the two plants for the two months. **J** and **F** are as follows:

$$\mathbf{J} = \begin{array}{c} \\ \text{desks} \\ \text{tables} \end{array}\begin{array}{c} \text{A} \quad \text{B} \\ \begin{bmatrix} 120 & 80 \\ 105 & 130 \end{bmatrix} \end{array}$$

$$\mathbf{F} = \begin{array}{c} \text{desks} \\ \text{tables} \end{array}\begin{bmatrix} 110 & 140 \\ 85 & 125 \end{bmatrix}$$

EXAMPLE 1 **Matrix Addition**

a. $$\begin{bmatrix} 1 & 2 \\ 3 & 4 \\ 5 & 6 \end{bmatrix} + \begin{bmatrix} 7 & -2 \\ -6 & 4 \\ 3 & 0 \end{bmatrix} = \begin{bmatrix} 1+7 & 2-2 \\ 3-6 & 4+4 \\ 5+3 & 6+0 \end{bmatrix} = \begin{bmatrix} 8 & 0 \\ -3 & 8 \\ 8 & 6 \end{bmatrix}$$

b. $\begin{bmatrix} 1 & 2 \\ 3 & 4 \end{bmatrix} + \begin{bmatrix} 2 \\ 1 \end{bmatrix}$ is not defined, since the matrices are not the same size.

▶ **Now work Problem 1.**

If **A**, **B**, **C**, and **O** have the same size, then the following properties hold for matrix addition:

These properties of matrix addition are similar to the corresponding properties of real numbers.

Properties of Matrix Addition

1. $\mathbf{A} + \mathbf{B} = \mathbf{B} + \mathbf{A}$ (commutative property)
2. $\mathbf{A} + (\mathbf{B} + \mathbf{C}) = (\mathbf{A} + \mathbf{B}) + \mathbf{C}$ (associative property)
3. $\mathbf{A} + \mathbf{O} = \mathbf{O} + \mathbf{A} = \mathbf{A}$ (identity property)

Property 1 states that matrices can be added in any order, and Property 2 allows matrices to be grouped for the addition operation. Property 3 states that the zero matrix plays the same role in matrix addition as does the number 0 in the addition of real numbers. These properties are illustrated in the next example.

EXAMPLE 2 **Properties of Matrix Addition**

Let

$$\mathbf{A} = \begin{bmatrix} 1 & 2 & 1 \\ -2 & 0 & 1 \end{bmatrix} \qquad \mathbf{B} = \begin{bmatrix} 0 & 1 & 2 \\ 1 & -3 & 1 \end{bmatrix}$$

$$\mathbf{C} = \begin{bmatrix} -2 & 1 & -1 \\ 0 & -2 & 1 \end{bmatrix} \qquad \mathbf{O} = \begin{bmatrix} 0 & 0 & 0 \\ 0 & 0 & 0 \end{bmatrix}$$

a. *Show that* $\mathbf{A} + \mathbf{B} = \mathbf{B} + \mathbf{A}$.

Solution:

$$\mathbf{A} + \mathbf{B} = \begin{bmatrix} 1 & 3 & 3 \\ -1 & -3 & 2 \end{bmatrix} \qquad \mathbf{B} + \mathbf{A} = \begin{bmatrix} 1 & 3 & 3 \\ -1 & -3 & 2 \end{bmatrix}$$

Thus, $\mathbf{A} + \mathbf{B} = \mathbf{B} + \mathbf{A}$.

b. *Show that* $\mathbf{A} + (\mathbf{B} + \mathbf{C}) = (\mathbf{A} + \mathbf{B}) + \mathbf{C}$.

Solution:

$$\mathbf{A} + (\mathbf{B} + \mathbf{C}) = \mathbf{A} + \begin{bmatrix} -2 & 2 & 1 \\ 1 & -5 & 2 \end{bmatrix} = \begin{bmatrix} -1 & 4 & 2 \\ -1 & -5 & 3 \end{bmatrix}$$

$$(\mathbf{A} + \mathbf{B}) + \mathbf{C} = \begin{bmatrix} 1 & 3 & 3 \\ -1 & -3 & 2 \end{bmatrix} + \mathbf{C} = \begin{bmatrix} -1 & 4 & 2 \\ -1 & -5 & 3 \end{bmatrix}$$

c. *Show that* $\mathbf{A} + \mathbf{O} = \mathbf{A}$.

Solution:

$$\mathbf{A} + \mathbf{O} = \begin{bmatrix} 1 & 2 & 1 \\ -2 & 0 & 1 \end{bmatrix} + \begin{bmatrix} 0 & 0 & 0 \\ 0 & 0 & 0 \end{bmatrix} = \begin{bmatrix} 1 & 2 & 1 \\ -2 & 0 & 1 \end{bmatrix} = \mathbf{A}$$

▸ **Now work Problem 7.**

EXAMPLE 3 **Demand Vectors for an Economy**

Consider a simplified hypothetical economy having three industries, say, coal, electricity, and steel, and three consumers, 1, 2, and 3. Suppose that each consumer may use some of the output of each industry and also that each industry uses some of the output of each other industry. Then the needs of each consumer and industry can be represented by a (row) demand vector whose entries, in order, give the amount of coal, electricity, and steel needed by the consumer or industry in some convenient units. For example, the demand vectors for the consumers might be

$$\mathbf{D}_1 = [3 \quad 2 \quad 5] \qquad \mathbf{D}_2 = [0 \quad 17 \quad 1] \qquad \mathbf{D}_3 = [4 \quad 6 \quad 12]$$

and for the industries they might be

$$\mathbf{D}_{\mathrm{C}} = [0 \quad 1 \quad 4] \qquad \mathbf{D}_{\mathrm{E}} = [20 \quad 0 \quad 8] \qquad \mathbf{D}_{\mathrm{S}} = [30 \quad 5 \quad 0]$$

where the subscripts C, E, and S stand for coal, electricity, and steel, respectively. The total demand for these goods by the consumers is given by the sum

$$\mathbf{D}_1 + \mathbf{D}_2 + \mathbf{D}_3 = [3 \quad 2 \quad 5] + [0 \quad 17 \quad 1] + [4 \quad 6 \quad 12] = [7 \quad 25 \quad 18]$$

The total industrial demand is given by the sum

$$\mathbf{D}_{\mathrm{C}} + \mathbf{D}_{\mathrm{E}} + \mathbf{D}_{\mathrm{S}} = [0 \quad 1 \quad 4] + [20 \quad 0 \quad 8] + [30 \quad 5 \quad 0] = [50 \quad 6 \quad 12]$$

Therefore, the total overall demand is given by

$$[7 \quad 25 \quad 18] + [50 \quad 6 \quad 12] = [57 \quad 31 \quad 30]$$

Thus, the coal industry sells a total of 57 units, the total units of electricity sold is 31, and the total units of steel that are sold is 30.[1]

Scalar Multiplication

Returning to the snowmobile dealer, recall that February sales were given by the matrix

$$\mathbf{F} = \begin{bmatrix} 3 & 1 \\ 4 & 2 \end{bmatrix}$$

If, in March, the dealer doubles February's sales of each model and color of snowmobile, the sales matrix for March could be obtained by multiplying each entry in **F** by 2, yielding

$$\mathbf{M} = \begin{bmatrix} 2(3) & 2(1) \\ 2(4) & 2(2) \end{bmatrix}$$

It seems reasonable to write this operation as

$$\mathbf{M} = 2\mathbf{F} = 2\begin{bmatrix} 3 & 1 \\ 4 & 2 \end{bmatrix} = \begin{bmatrix} 2 \cdot 3 & 2 \cdot 1 \\ 2 \cdot 4 & 2 \cdot 2 \end{bmatrix} = \begin{bmatrix} 6 & 2 \\ 8 & 4 \end{bmatrix}$$

which is thought of as multiplying a matrix by a real number. Indeed, we have the following definition.

Definition

If **A** is an $m \times n$ matrix and k is a real number (also called a scalar), then, by $k\mathbf{A}$, we denote the $m \times n$ matrix obtained by multiplying each entry in **A** by k. This operation is called ***scalar multiplication,*** and $k\mathbf{A}$ is called a ***scalar multiple*** of **A**.

For example,

$$-3\begin{bmatrix} 1 & 0 & -2 \\ 2 & -1 & 4 \end{bmatrix} = \begin{bmatrix} -3(1) & -3(0) & -3(-2) \\ -3(2) & -3(-1) & -3(4) \end{bmatrix} = \begin{bmatrix} -3 & 0 & 6 \\ -6 & 3 & -12 \end{bmatrix}$$

[1]This example and some others in this chapter are from John G. Kemeny, J. Laurie Snell, and Gerald L. Thompson, *Introduction to Finite Mathematics,* 3d ed. © 1974. Reprinted by permission of Prentice-Hall, Inc., Englewood Cliffs, New Jersey.

EXAMPLE 4 **Scalar Multiplication**

Let

$$\mathbf{A} = \begin{bmatrix} 1 & 2 \\ 4 & -2 \end{bmatrix} \qquad \mathbf{B} = \begin{bmatrix} 3 & -4 \\ 7 & 1 \end{bmatrix} \qquad \mathbf{O} = \begin{bmatrix} 0 & 0 \\ 0 & 0 \end{bmatrix}$$

Compute the following:

a. $5\mathbf{A}$

Solution:

$$5\mathbf{A} = 5\begin{bmatrix} 1 & 2 \\ 4 & -2 \end{bmatrix} = \begin{bmatrix} 5(1) & 5(2) \\ 5(4) & 5(-2) \end{bmatrix} = \begin{bmatrix} 5 & 10 \\ 20 & -10 \end{bmatrix}$$

b. $-\frac{2}{3}\mathbf{B}$

Solution:

$$-\frac{2}{3}\mathbf{B} = \begin{bmatrix} -\frac{2}{3}(3) & -\frac{2}{3}(-4) \\ -\frac{2}{3}(7) & -\frac{2}{3}(1) \end{bmatrix} = \begin{bmatrix} -2 & \frac{8}{3} \\ -\frac{14}{3} & -\frac{2}{3} \end{bmatrix}$$

c. $\frac{1}{2}\mathbf{A} + 3\mathbf{B}$

Solution:

$$\frac{1}{2}\mathbf{A} + 3\mathbf{B} = \frac{1}{2}\begin{bmatrix} 1 & 2 \\ 4 & -2 \end{bmatrix} + 3\begin{bmatrix} 3 & -4 \\ 7 & 1 \end{bmatrix}$$

$$= \begin{bmatrix} \frac{1}{2} & 1 \\ 2 & -1 \end{bmatrix} + \begin{bmatrix} 9 & -12 \\ 21 & 3 \end{bmatrix} = \begin{bmatrix} \frac{19}{2} & -11 \\ 23 & 2 \end{bmatrix}$$

d. $0\mathbf{A}$

Solution:

$$0\mathbf{A} = 0\begin{bmatrix} 1 & 2 \\ 4 & -2 \end{bmatrix} = \begin{bmatrix} 0 & 0 \\ 0 & 0 \end{bmatrix} = \mathbf{O}$$

e. $k\mathbf{O}$

Solution:

$$k\mathbf{O} = k\begin{bmatrix} 0 & 0 \\ 0 & 0 \end{bmatrix} = \begin{bmatrix} 0 & 0 \\ 0 & 0 \end{bmatrix} = \mathbf{O}$$

▶ **Now work Problem 5.**

If $\mathbf{A}$, $\mathbf{B}$, and $\mathbf{O}$ are the same size, then, for any scalars k, k_1, and k_2, we have the following properties of scalar multiplication:

Properties of Scalar Multiplication

1. $k(\mathbf{A} + \mathbf{B}) = k\mathbf{A} + k\mathbf{B}$

2. $(k_1 + k_2)\mathbf{A} = k_1\mathbf{A} + k_2\mathbf{A}$

3. $k_1(k_2\mathbf{A}) = (k_1k_2)\mathbf{A}$

4. $0\mathbf{A} = \mathbf{O}$

5. $k\mathbf{O} = \mathbf{O}$

Remember that $\mathbf{O} \neq 0$, for 0 is a *scalar* and $\mathbf{O}$ is a zero *matrix*.

Properties 4 and 5 were illustrated in Examples 4(d) and (e); the others will be illustrated in the exercises.

We also have the following properties of the transpose operation, where **A** and **B** are of the same size and k is any scalar:

$$(\mathbf{A}+\mathbf{B})^{\mathrm{T}} = \mathbf{A}^{\mathrm{T}} + \mathbf{B}^{\mathrm{T}}$$
$$(k\mathbf{A})^{\mathrm{T}} = k\mathbf{A}^{\mathrm{T}}$$

The first property states that *the transpose of a sum is the sum of the transposes.*

Subtraction of Matrices

If **A** is any matrix, then the scalar multiple $(-1)\mathbf{A}$ is simply written as $-\mathbf{A}$ and is called the **negative of A**:

$$-\mathbf{A} = (-1)\mathbf{A}$$

Thus, if

$$\mathbf{A} = \begin{bmatrix} 3 & 1 \\ -4 & 5 \end{bmatrix}$$

then

$$-\mathbf{A} = (-1)\begin{bmatrix} 3 & 1 \\ -4 & 5 \end{bmatrix} = \begin{bmatrix} -3 & -1 \\ 4 & -5 \end{bmatrix}$$

Note that $-\mathbf{A}$ is the matrix obtained by multiplying each entry of **A** by -1.

Subtraction of matrices is defined in terms of matrix addition:

Definition

If **A** and **B** are the same size, then, by $\mathbf{A}-\mathbf{B}$, we mean $\mathbf{A}+(-\mathbf{B})$.

More simply, to find $\mathbf{A}-\mathbf{B}$, we can subtract each entry in **B** from the corresponding entry in **A**.

EXAMPLE 5 **Matrix Subtraction**

a. $$\begin{bmatrix} 2 & 6 \\ -4 & 1 \\ 3 & 2 \end{bmatrix} - \begin{bmatrix} 6 & -2 \\ 4 & 1 \\ 0 & 3 \end{bmatrix} = \begin{bmatrix} 2 & 6 \\ -4 & 1 \\ 3 & 2 \end{bmatrix} + (-1)\begin{bmatrix} 6 & -2 \\ 4 & 1 \\ 0 & 3 \end{bmatrix}$$
$$= \begin{bmatrix} 2 & 6 \\ -4 & 1 \\ 3 & 2 \end{bmatrix} + \begin{bmatrix} -6 & 2 \\ -4 & -1 \\ 0 & -3 \end{bmatrix}$$
$$= \begin{bmatrix} 2-6 & 6+2 \\ -4-4 & 1-1 \\ 3+0 & 2-3 \end{bmatrix} = \begin{bmatrix} -4 & 8 \\ -8 & 0 \\ 3 & -1 \end{bmatrix}$$

b. If $\mathbf{A} = \begin{bmatrix} 6 & 0 \\ 2 & -1 \end{bmatrix}$ and $\mathbf{B} = \begin{bmatrix} 3 & -3 \\ 1 & 2 \end{bmatrix}$ then

$$\mathbf{A}^{\mathrm{T}} - 2\mathbf{B} = \begin{bmatrix} 6 & 2 \\ 0 & -1 \end{bmatrix} - \begin{bmatrix} 6 & -6 \\ 2 & 4 \end{bmatrix} = \begin{bmatrix} 0 & 8 \\ -2 & -5 \end{bmatrix}$$

▸ **Now work Problem 17.**

APPLICATIONS IN PRACTICE 2

Matrix Equation

A manufacturer of doors, windows, and cabinets writes her yearly profit (in thousands of dollars) for each category in a vector as $\mathbf{P} = \begin{bmatrix} 248 \\ 319 \\ 532 \end{bmatrix}$. Her fixed costs of production can be described by the vector $\mathbf{C} = \begin{bmatrix} 40 \\ 30 \\ 60 \end{bmatrix}$. She calculates that, with a new pricing structure that generates an income that is 80% of her competitor's income, she can double her profit, assuming that her fixed costs remain the same. This calculation can be represented by

$$0.8\begin{bmatrix} x_1 \\ x_2 \\ x_3 \end{bmatrix} - \begin{bmatrix} 40 \\ 30 \\ 60 \end{bmatrix} = 2\begin{bmatrix} 248 \\ 319 \\ 532 \end{bmatrix}.$$

Solve for x_1, x_2, and x_3, which represent her competitor's income from each category.

EXAMPLE 6 Matrix Equation

Solve the equation $2\begin{bmatrix} x_1 \\ x_2 \end{bmatrix} - \begin{bmatrix} 3 \\ 4 \end{bmatrix} = 5\begin{bmatrix} 5 \\ -4 \end{bmatrix}$.

Solution:

Strategy: We first simplify each side into one matrix. Then, by equality of matrices, we equate corresponding entries.

We have

$$2\begin{bmatrix} x_1 \\ x_2 \end{bmatrix} - \begin{bmatrix} 3 \\ 4 \end{bmatrix} = 5\begin{bmatrix} 5 \\ -4 \end{bmatrix}$$

$$\begin{bmatrix} 2x_1 \\ 2x_2 \end{bmatrix} - \begin{bmatrix} 3 \\ 4 \end{bmatrix} = \begin{bmatrix} 25 \\ -20 \end{bmatrix}$$

$$\begin{bmatrix} 2x_1 - 3 \\ 2x_2 - 4 \end{bmatrix} = \begin{bmatrix} 25 \\ -20 \end{bmatrix}$$

By equality of matrices, we must have $2x_1 - 3 = 25$, which gives $x_1 = 14$; from $2x_2 - 4 = -20$, we get $x_2 = -8$.

▶ **Now work Problem 35.**

TECHNOLOGY

The matrix operations of addition, subtraction, and scalar multiplication can be performed on a graphing calculator. For example, Figure 6.2 shows $2\mathbf{A} - 3\mathbf{B}$, where

$$\mathbf{A} = \begin{bmatrix} -2 & 0 \\ 1 & 3 \end{bmatrix} \quad \text{and} \quad \mathbf{B} = \begin{bmatrix} 1 & 2 \\ 4 & 1 \end{bmatrix}$$

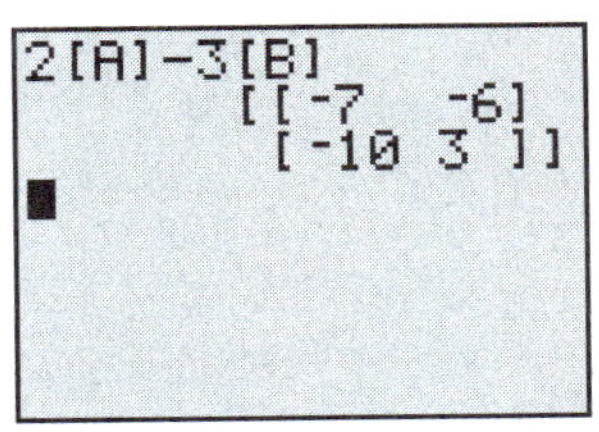

FIGURE 6.2 Matrix operations with graphing calculator.

Exercise 6.2

In Problems 1–12, perform the indicated operations.

***1.** $\begin{bmatrix} 2 & 0 & -3 \\ -1 & 4 & 0 \\ 1 & -6 & 5 \end{bmatrix} + \begin{bmatrix} 2 & -3 & 4 \\ -1 & 6 & 5 \\ 9 & 11 & -2 \end{bmatrix}$

2. $\begin{bmatrix} 2 & -7 \\ -6 & 4 \end{bmatrix} + \begin{bmatrix} 7 & -4 \\ -2 & 1 \end{bmatrix} + \begin{bmatrix} 2 & 7 \\ 7 & 2 \end{bmatrix}$

3. $\begin{bmatrix} 1 & 4 \\ -2 & 7 \\ 6 & 9 \end{bmatrix} - \begin{bmatrix} 6 & -1 \\ 7 & 2 \\ 1 & 0 \end{bmatrix}$

4. $\frac{1}{2}\begin{bmatrix} 4 & -2 & 6 \\ 2 & 10 & -12 \\ 0 & 0 & 8 \end{bmatrix}$

***5.** $3[1 \ \ -3 \ \ 1] + 2[-6 \ \ 1 \ \ 4] - 0[-2 \ \ 7 \ \ 4]$

6. $[7 \ \ 7] + 66$

***7.** $\begin{bmatrix} 1 & 2 \\ 3 & 4 \end{bmatrix} + \begin{bmatrix} 7 \\ 2 \end{bmatrix}$

8. $\begin{bmatrix} 2 & -1 \\ 7 & 4 \end{bmatrix} + 3\begin{bmatrix} 0 & 0 \\ 0 & 0 \end{bmatrix}$

9. $-6\begin{bmatrix} 2 & -6 & 7 & 1 \\ 7 & 1 & 6 & -2 \end{bmatrix}$

10. $\begin{bmatrix} 1 & -1 \\ 2 & 0 \\ 3 & -6 \\ 4 & 9 \end{bmatrix} - 3\begin{bmatrix} -6 & 9 \\ 2 & 6 \\ 1 & -2 \\ 4 & 5 \end{bmatrix}$

11. $\begin{bmatrix} 1 & -5 & 0 \\ -2 & 7 & 0 \\ 4 & 6 & 10 \end{bmatrix} + \frac{1}{5}\begin{bmatrix} 10 & 0 & 30 \\ 0 & 5 & 0 \\ 5 & 20 & 25 \end{bmatrix}$

12. $2\begin{bmatrix} 1 & 0 & 0 \\ 0 & 1 & 0 \\ 0 & 0 & 1 \end{bmatrix} - 3\left(\begin{bmatrix} 2 & 1 & 0 \\ 1 & -2 & 3 \\ 1 & 0 & 0 \end{bmatrix} - \begin{bmatrix} 6 & -2 & 1 \\ -5 & 1 & -2 \\ 0 & 1 & 3 \end{bmatrix}\right)$

In Problems 13–24, compute the required matrices if

$$\mathbf{A} = \begin{bmatrix} 2 & 1 \\ 3 & -3 \end{bmatrix} \qquad \mathbf{B} = \begin{bmatrix} -6 & -5 \\ 2 & -3 \end{bmatrix}$$

$$\mathbf{C} = \begin{bmatrix} -2 & -1 \\ -3 & 3 \end{bmatrix} \qquad \mathbf{O} = \begin{bmatrix} 0 & 0 \\ 0 & 0 \end{bmatrix}$$

13. $-\mathbf{B}$

14. $-(\mathbf{A} - \mathbf{B})$

15. $2\mathbf{O}$

16. $\mathbf{A} - \mathbf{B} + \mathbf{C}$

***17.** $2(\mathbf{A} - 2\mathbf{B})$

18. $0(\mathbf{A} + \mathbf{B})$

19. $3(\mathbf{A} - \mathbf{C}) + 6$

20. $\mathbf{A} + (\mathbf{C} + \mathbf{B})$

21. $2\mathbf{B} - 3\mathbf{A} + 2\mathbf{C}$

22. $3\mathbf{C} - 2\mathbf{B}$

23. $\frac{1}{2}\mathbf{A} - 2(\mathbf{B} + 2\mathbf{C})$

24. $3\mathbf{A} - \frac{1}{3}(\mathbf{B} + \mathbf{C})$

In Problems 25–28, verify the equations for the preceding matrices **A**, **B**, *and* **C**.

25. $3(\mathbf{A} + \mathbf{B}) = 3\mathbf{A} + 3\mathbf{B}$

26. $(2 + 3)\mathbf{A} = 2\mathbf{A} + 3\mathbf{A}$

27. $k_1(k_2\mathbf{A}) = (k_1k_2)\mathbf{A}$

28. $k(\mathbf{A} - 2\mathbf{B} + \mathbf{C}) = k\mathbf{A} - 2k\mathbf{B} + k\mathbf{C}$

In Problems 29–34, let

$$\mathbf{A} = \begin{bmatrix} 1 & 2 \\ 0 & -1 \\ 7 & 0 \end{bmatrix} \qquad \mathbf{B} = \begin{bmatrix} 1 & 3 \\ 4 & -1 \end{bmatrix}$$

$$\mathbf{C} = \begin{bmatrix} 1 & 0 \\ 1 & 2 \end{bmatrix} \qquad \mathbf{D} = \begin{bmatrix} 1 & 2 & -1 \\ 1 & 0 & 2 \end{bmatrix}$$

Compute the indicated matrices if possible.

29. $3\mathbf{A}^T + \mathbf{D}$

30. $(\mathbf{B} - \mathbf{C})^T$

31. $2\mathbf{B}^T - 3\mathbf{C}^T$

32. $2\mathbf{B} + \mathbf{B}^T$

33. $\mathbf{C}^T - \mathbf{D}$

34. $(\mathbf{D} - 2\mathbf{A}^T)^T$

***35.** Express the matrix equation

$$x\begin{bmatrix} 3 \\ 2 \end{bmatrix} - y\begin{bmatrix} -4 \\ 7 \end{bmatrix} = 3\begin{bmatrix} 2 \\ 4 \end{bmatrix}$$

as a system of linear equations and solve.

36. In the reverse of the manner used in Problem 35, write the system

$$\begin{cases} 3x + 5y = 16 \\ 2x - 6y = -4 \end{cases}$$

as a matrix equation.

In Problems 37–40, solve the matrix equations.

37. $3\begin{bmatrix} x \\ y \end{bmatrix} - 3\begin{bmatrix} -2 \\ 4 \end{bmatrix} = 4\begin{bmatrix} 6 \\ -2 \end{bmatrix}$

38. $3\begin{bmatrix} x \\ 2 \end{bmatrix} - 4\begin{bmatrix} 7 \\ -y \end{bmatrix} = \begin{bmatrix} -x \\ 2y \end{bmatrix}$

39. $\begin{bmatrix} 2 \\ 4 \\ 6 \end{bmatrix} + 2\begin{bmatrix} x \\ y \\ 4z \end{bmatrix} = \begin{bmatrix} -10 \\ -24 \\ 14 \end{bmatrix}$

40. $x\begin{bmatrix} 2 \\ 0 \\ 2 \end{bmatrix} + 2\begin{bmatrix} -1 \\ 0 \\ 6 \end{bmatrix} + y\begin{bmatrix} 0 \\ 2 \\ -5 \end{bmatrix} = \begin{bmatrix} 10 \\ 6 \\ 2x + 12 - 5y \end{bmatrix}$

41. Production An electronics company manufactures televisions, VCRs, and CD players at two plants, A and B. Matrix **X** represents the production of the two plants for retailer X, and matrix **Y** represents the production of the two plants for retailer Y. Write a matrix that represents the total production at the two plants for both retailers. Matrices **X** and **Y** are as follows:

$$\mathbf{X} = \begin{matrix} \text{TV} \\ \text{VCR} \\ \text{CD} \end{matrix}\overset{\begin{matrix} \text{A} & \text{B} \end{matrix}}{\begin{bmatrix} 20 & 40 \\ 45 & 30 \\ 15 & 10 \end{bmatrix}} \qquad \mathbf{Y} = \begin{matrix} \text{TV} \\ \text{VCR} \\ \text{CD} \end{matrix}\overset{\begin{matrix} \text{A} & \text{B} \end{matrix}}{\begin{bmatrix} 15 & 25 \\ 30 & 25 \\ 10 & 5 \end{bmatrix}}$$

42. Sales Let matrix **A** represent the sales (in thousands of dollars) of a toy company in 1998 in three cities, and let **B** represent the sales in the same cities in 2000, where **A** and **B** are given by

$$\mathbf{A} = \begin{matrix} \text{Action} \\ \text{Educational} \end{matrix}\begin{bmatrix} 400 & 350 & 150 \\ 450 & 280 & 850 \end{bmatrix}$$

$$\mathbf{B} = \begin{matrix} \text{Action} \\ \text{Educational} \end{matrix}\begin{bmatrix} 380 & 330 & 220 \\ 460 & 320 & 750 \end{bmatrix}$$

If the company bought a competitor and doubled its 2000 sales in 2001, what was the change in sales between 1998 and 2001?

43. Suppose the prices of products A, B, and C are given, in that order, by the price vector

$$\mathbf{P} = [p_1 \quad p_2 \quad p_3]$$

If the prices are to be increased by 10%, the vector for the new prices can be obtained by multiplying **P** by what scalar?

44. Prove that $(\mathbf{A} - \mathbf{B})^T = \mathbf{A}^T - \mathbf{B}^T$. [*Hint:* Use the definition of subtraction and properties of the transpose operation.]

In Problems 45–47, compute the given matrices if

$$\mathbf{A} = \begin{bmatrix} 3 & -4 & 5 \\ -2 & 1 & 6 \end{bmatrix} \qquad \mathbf{B} = \begin{bmatrix} 1 & 4 & 2 \\ 4 & 1 & 2 \end{bmatrix} \qquad \mathbf{C} = \begin{bmatrix} -1 & 1 & 3 \\ 2 & 6 & -6 \end{bmatrix}$$

45. $4\mathbf{A} + 3\mathbf{B}$

46. $-3(\mathbf{A} + 2\mathbf{B}) + \mathbf{C}$

47. $2(3\mathbf{C} - \mathbf{A}) + 2\mathbf{B}$

> **OBJECTIVE** *To define multiplication of matrices and to consider associated properties. To express a system as a single matrix equation by using matrix multiplication.*

6.3 Matrix Multiplication

Besides the operations of matrix addition and scalar multiplication, the product **AB** of matrices **A** and **B** can be defined under certain conditions, namely, that the number of columns of **A** is equal to the number of rows of **B**. Although the following definition of matrix multiplication may not appear to you to be a natural one (it would seem more natural to simply multiply corresponding entries), a more thorough study of matrices would convince you that the definition is appropriate and extremely practical for applications.

Definition

Let **A** be an $m \times n$ matrix and **B** be an $n \times p$ matrix. Then the product **AB** is the $m \times p$ matrix **C** whose entry c_{ij} in row i and column j is obtained as follows: Sum the products formed by multiplying, in order, each entry (that is, first, second, etc.) in row i of **A** by the "corresponding" entry (that is, first, second, etc.) in column j of **B**.

Three points must be completely understood concerning this definition of **AB**. First, the condition that **A** be $m \times n$ and **B** be $n \times p$ is equivalent to saying that the number of columns of **A** must be equal to the number of rows of **B**. Second, the product will be a matrix of size $m \times p$; it will have as many rows as **A** and as many columns as **B**.

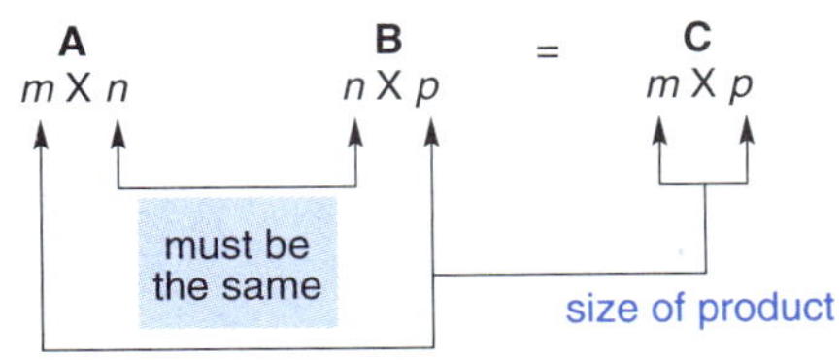

Third, the definition refers to the product **AB**, *in that order;* **A** is the left factor and **B** is the right factor. For **AB**, we say that **B** is *premultiplied* by **A** or **A** is *postmultiplied* by **B**.

To apply the definition, let us find the product

$$\mathbf{AB} = \begin{bmatrix} 2 & 1 & -6 \\ 1 & -3 & 2 \end{bmatrix} \begin{bmatrix} 1 & 0 & -3 \\ 0 & 4 & 2 \\ -2 & 1 & 1 \end{bmatrix}$$

Matrix **A** has size $2 \times 3 (m \times n)$ and matrix **B** has size $3 \times 3 (n \times p)$. The number of columns of **A** is equal to the number of rows of $\mathbf{B} (n = 3)$, so the product **C** is defined and will be a $2 \times 3 (m \times p)$ matrix; that is,

$$\mathbf{C} = \begin{bmatrix} c_{11} & c_{12} & c_{13} \\ c_{21} & c_{22} & c_{23} \end{bmatrix}$$

The entry c_{11} is obtained by summing the products of each entry in row 1 of **A** by the "corresponding" entry in column 1 of **B**. Thus,

row 1 entries of **A**

$$c_{11} = (2)(1) + (1)(0) + (-6)(-2) = 14.$$

column 1 entries of **B**

At this stage, we have

$$\begin{bmatrix} 2 & 1 & -6 \\ 1 & -3 & 2 \end{bmatrix} \begin{bmatrix} 1 & 0 & -3 \\ 0 & 4 & 2 \\ -2 & 1 & 1 \end{bmatrix} = \begin{bmatrix} 14 & c_{12} & c_{13} \\ c_{21} & c_{22} & c_{23} \end{bmatrix}$$

Similarly, for c_{12}, we use the entries in row 1 of **A** and those in column 2 of **B**:

row 1 entries of **A**

$$c_{12} = (2)(0) + (1)(4) + (-6)(1) = -2.$$

column 2 entries of **B**

We now have

$$\begin{bmatrix} 2 & 1 & -6 \\ 1 & -3 & 2 \end{bmatrix} \begin{bmatrix} 1 & 0 & -3 \\ 0 & 4 & 2 \\ -2 & 1 & 1 \end{bmatrix} = \begin{bmatrix} 14 & -2 & c_{13} \\ c_{21} & c_{22} & c_{23} \end{bmatrix}$$

For the remaining entries of **AB**, we obtain

$$c_{13} = (2)(-3) + (1)(2) + (-6)(1) = -10$$

$$c_{21} = (1)(1) + (-3)(0) + (2)(-2) = -3$$

$$c_{22} = (1)(0) + (-3)(4) + (2)(1) = -10$$

$$c_{23} = (1)(-3) + (-3)(2) + (2)(1) = -7$$

Thus,

$$\mathbf{AB} = \begin{bmatrix} 2 & 1 & -6 \\ 1 & -3 & 2 \end{bmatrix} \begin{bmatrix} 1 & 0 & -3 \\ 0 & 4 & 2 \\ -2 & 1 & 1 \end{bmatrix} = \begin{bmatrix} 14 & -2 & -10 \\ -3 & -10 & -7 \end{bmatrix}$$

Note that if we reverse the order of the factors, then the product

$$\mathbf{BA} = \begin{bmatrix} 1 & 0 & -3 \\ 0 & 4 & 2 \\ -2 & 1 & 1 \end{bmatrix} \begin{bmatrix} 2 & 1 & -6 \\ 1 & -3 & 2 \end{bmatrix}$$

is *not* defined, because the number of columns of **B** does *not* equal the number of rows of **A**. This shows that matrix multiplication is not commutative. That is, for any matrices **A** and **B**, it is usually the case that **AB** and **BA** are different, even when both products are defined, so the order in which the matrices in a product are written is extremely important.

CAUTION

Matrix multiplication is not commutative.

EXAMPLE 1 Sizes of Matrices and Their Product

Let **A** be a 3×5 matrix and **B** be a 5×3 matrix. Then **AB** is defined and is a 3×3 matrix. Moreover, **BA** is also defined and is a 5×5 matrix.

If **C** is a 3×5 matrix and **D** is a 7×3 matrix, then **CD** is undefined, but **DC** is defined and is a 7×5 matrix.

▸ **Now work Problem 7.**

EXAMPLE 2 **Matrix Product**

Compute the matrix product

$$\mathbf{AB} = \begin{bmatrix} 2 & -4 & 2 \\ 0 & 1 & -3 \end{bmatrix} \begin{bmatrix} 2 & 1 \\ 0 & 4 \\ 2 & 2 \end{bmatrix}$$

Solution: Since $\mathbf{A}$ is 2×3 and $\mathbf{B}$ is 3×2, the product $\mathbf{AB}$ is defined and will have order 2×2. By simultaneously moving the index finger of your left hand along the rows of $\mathbf{A}$ and the index finger of your right hand along the columns of $\mathbf{B}$, it should not be difficult for you to mentally find the entries of the product. We obtain

$$\begin{bmatrix} 2 & -4 & 2 \\ 0 & 1 & -3 \end{bmatrix} \begin{bmatrix} 2 & 1 \\ 0 & 4 \\ 2 & 2 \end{bmatrix} = \begin{bmatrix} 8 & -10 \\ -6 & -2 \end{bmatrix}$$

▶ **Now work Problem 19.**

APPLICATIONS IN PRACTICE 1

Matrix Products

A bookstore has 100 dictionaries, 70 cookbooks, and 90 thesauruses in stock. If the value of each dictionary is \$28, each cookbook is \$22, and each thesaurus is \$16, use a matrix product to find the total value of the bookstore's inventory.

EXAMPLE 3 **Matrix Products**

a. *Compute* $[1 \quad 2 \quad 3] \begin{bmatrix} 4 \\ 5 \\ 6 \end{bmatrix}$.

Solution: The product has order 1×1:

$$[1 \quad 2 \quad 3] \begin{bmatrix} 4 \\ 5 \\ 6 \end{bmatrix} = [32]$$

b. *Compute* $\begin{bmatrix} 1 \\ 2 \\ 3 \end{bmatrix} [1 \quad 6]$.

Solution: The product has order 3×2:

$$\begin{bmatrix} 1 \\ 2 \\ 3 \end{bmatrix} [1 \quad 6] = \begin{bmatrix} 1 & 6 \\ 2 & 12 \\ 3 & 18 \end{bmatrix}$$

c. $$\begin{bmatrix} 1 & 3 & 0 \\ -2 & 2 & 1 \\ 1 & 0 & -4 \end{bmatrix} \begin{bmatrix} 1 & 0 & 2 \\ 5 & -1 & 3 \\ 2 & 1 & -2 \end{bmatrix} = \begin{bmatrix} 16 & -3 & 11 \\ 10 & -1 & 0 \\ -7 & -4 & 10 \end{bmatrix}$$

d. $$\begin{bmatrix} a_{11} & a_{12} \\ a_{21} & a_{22} \end{bmatrix} \begin{bmatrix} b_{11} & b_{12} \\ b_{21} & b_{22} \end{bmatrix} = \begin{bmatrix} a_{11}b_{11} + a_{12}b_{21} & a_{11}b_{12} + a_{12}b_{22} \\ a_{21}b_{11} + a_{22}b_{21} & a_{21}b_{12} + a_{22}b_{22} \end{bmatrix}$$

▶ **Now work Problem 25.**

Example 4 shows that even when the matrix products **AB** and **BA** are both defined and the same size, they are not necessarily equal.

EXAMPLE 4 **Matrix Products**

Compute **AB** *and* **BA** *if*

$$\mathbf{A} = \begin{bmatrix} 2 & -1 \\ 3 & 1 \end{bmatrix} \quad \textit{and} \quad \mathbf{B} = \begin{bmatrix} -2 & 1 \\ 1 & 4 \end{bmatrix}.$$

Solution: We have

$$\mathbf{AB} = \begin{bmatrix} 2 & -1 \\ 3 & 1 \end{bmatrix}\begin{bmatrix} -2 & 1 \\ 1 & 4 \end{bmatrix} = \begin{bmatrix} -5 & -2 \\ -5 & 7 \end{bmatrix}$$

$$\mathbf{BA} = \begin{bmatrix} -2 & 1 \\ 1 & 4 \end{bmatrix}\begin{bmatrix} 2 & -1 \\ 3 & 1 \end{bmatrix} = \begin{bmatrix} -1 & 3 \\ 14 & 3 \end{bmatrix}$$

Note that although both **AB** and **BA** are defined, and the same size, **AB** and **BA** are not equal.

▸ **Now work Problem 37.**

TECHNOLOGY

Figure 6.3 shows the result of using a graphing calculator to find the product **AB** in Example 4.

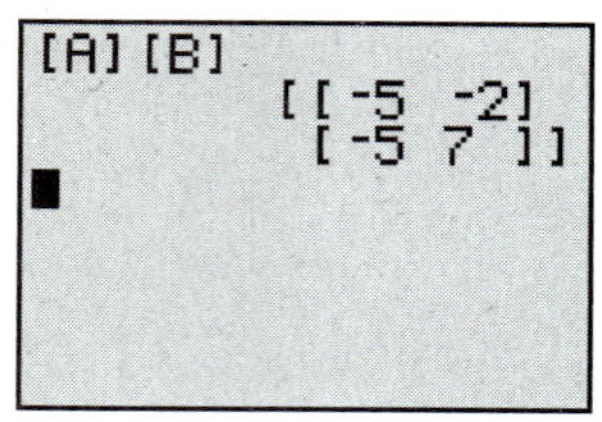

FIGURE 6.3 Calculator solution of matrix product of Example 4.

APPLICATIONS IN PRACTICE 2

Cost Vector

The prices (in dollars per unit) for three textbooks are represented by the price vector $\mathbf{P} = [26.25 \quad 34.75 \quad 28.50]$. A university bookstore orders these books in the quantities given by the column vector $\mathbf{Q} = \begin{bmatrix} 250 \\ 325 \\ 175 \end{bmatrix}$. Find the total cost (in dollars) of the purchase.

EXAMPLE 5 **Cost Vector**

Suppose that the prices (in dollars per unit) for products A, B, and C are represented by the price vector

$$\begin{matrix} & \text{Price of} & \\ \text{A} & \text{B} & \text{C} \end{matrix}$$
$$\mathbf{P} = [2 \quad 3 \quad 4]$$

If the quantities (in units) of A, B, and C that are purchased are given by the column vector

$$\mathbf{Q} = \begin{bmatrix} 7 \\ 5 \\ 11 \end{bmatrix} \begin{matrix} \text{units of A} \\ \text{units of B} \\ \text{units of C} \end{matrix}$$

then the total cost (in dollars) of the purchases is given by the entry in the cost vector

$$\mathbf{PQ} = [2 \quad 3 \quad 4]\begin{bmatrix} 7 \\ 5 \\ 11 \end{bmatrix} = [(2 \cdot 7) + (3 \cdot 5) + (4 \cdot 11)] = [73]$$

▶ **Now work Problem 27.**

EXAMPLE 6 Profit for an Economy

In Example 3 of Section 6.2, suppose that in the hypothetical economy the price of coal is \$10,000 per unit, the price of electricity is \$20,000 per unit, and the price of steel is \$40,000 per unit. These prices can be represented by the (column) price vector

$$\mathbf{P} = \begin{bmatrix} 10{,}000 \\ 20{,}000 \\ 40{,}000 \end{bmatrix}$$

Consider the steel industry. It sells a total of 30 units of steel at \$40,000 per unit, and its total income is therefore \$1,200,000. Its costs for the various goods are given by the matrix product

$$\mathbf{D}_S\mathbf{P} = [30 \quad 5 \quad 0]\begin{bmatrix} 10{,}000 \\ 20{,}000 \\ 40{,}000 \end{bmatrix} = [400{,}000]$$

Hence, the profit for the steel industry is \$1,200,000 − \$400,000 = \$800,000.

▶ **Now work Problem 67.**

Matrix multiplication satisfies the following properties, provided that all sums and products are defined:

Properties of Matrix Multiplication

1. $\mathbf{A}(\mathbf{BC}) = (\mathbf{AB})\mathbf{C}$ (associative property)
2. $\mathbf{A}(\mathbf{B}+\mathbf{C}) = \mathbf{AB}+\mathbf{AC}$, $(\mathbf{A}+\mathbf{B})\mathbf{C} = \mathbf{AC}+\mathbf{BC}$ (distributive properties)

EXAMPLE 7 Associative Property

If

$$\mathbf{A} = \begin{bmatrix} 1 & -2 \\ -3 & 4 \end{bmatrix} \qquad \mathbf{B} = \begin{bmatrix} 3 & 0 & -1 \\ 1 & 1 & 2 \end{bmatrix} \qquad \mathbf{C} = \begin{bmatrix} 1 & 0 \\ 0 & 2 \\ 1 & 1 \end{bmatrix}$$

compute **ABC** *in two ways.*

Solution: Grouping **BC** gives

$$\mathbf{A}(\mathbf{BC}) = \begin{bmatrix} 1 & -2 \\ -3 & 4 \end{bmatrix}\left(\begin{bmatrix} 3 & 0 & -1 \\ 1 & 1 & 2 \end{bmatrix}\begin{bmatrix} 1 & 0 \\ 0 & 2 \\ 1 & 1 \end{bmatrix}\right)$$

$$= \begin{bmatrix} 1 & -2 \\ -3 & 4 \end{bmatrix}\begin{bmatrix} 2 & -1 \\ 3 & 4 \end{bmatrix} = \begin{bmatrix} -4 & -9 \\ 6 & 19 \end{bmatrix}$$

Alternatively, grouping **AB** gives

$$\begin{aligned}(\mathbf{AB})\mathbf{C} &= \left(\begin{bmatrix} 1 & -2 \\ -3 & 4 \end{bmatrix}\begin{bmatrix} 3 & 0 & -1 \\ 1 & 1 & 2 \end{bmatrix}\right)\begin{bmatrix} 1 & 0 \\ 0 & 2 \\ 1 & 1 \end{bmatrix} \\ &= \begin{bmatrix} 1 & -2 & -5 \\ -5 & 4 & 11 \end{bmatrix}\begin{bmatrix} 1 & 0 \\ 0 & 2 \\ 1 & 1 \end{bmatrix} \\ &= \begin{bmatrix} -4 & -9 \\ 6 & 19 \end{bmatrix}\end{aligned}$$

Note that $\mathbf{A}(\mathbf{BC}) = (\mathbf{AB})\mathbf{C}$.

EXAMPLE 8 **Distributive Property**

Verify that $\mathbf{A}(\mathbf{B} + \mathbf{C}) = \mathbf{AB} + \mathbf{AC}$ *if*

$$\mathbf{A} = \begin{bmatrix} 1 & 0 \\ 2 & 3 \end{bmatrix} \qquad \mathbf{B} = \begin{bmatrix} -2 & 0 \\ 1 & 3 \end{bmatrix} \qquad \mathbf{C} = \begin{bmatrix} -2 & 1 \\ 0 & 2 \end{bmatrix}$$

Solution: On the left side, we have

$$\begin{aligned}\mathbf{A}(\mathbf{B} + \mathbf{C}) &= \begin{bmatrix} 1 & 0 \\ 2 & 3 \end{bmatrix}\left(\begin{bmatrix} -2 & 0 \\ 1 & 3 \end{bmatrix} + \begin{bmatrix} -2 & 1 \\ 0 & 2 \end{bmatrix}\right) \\ &= \begin{bmatrix} 1 & 0 \\ 2 & 3 \end{bmatrix}\begin{bmatrix} -4 & 1 \\ 1 & 5 \end{bmatrix} = \begin{bmatrix} -4 & 1 \\ -5 & 17 \end{bmatrix}\end{aligned}$$

On the right side,

$$\begin{aligned}\mathbf{AB} + \mathbf{AC} &= \begin{bmatrix} 1 & 0 \\ 2 & 3 \end{bmatrix}\begin{bmatrix} -2 & 0 \\ 1 & 3 \end{bmatrix} + \begin{bmatrix} 1 & 0 \\ 2 & 3 \end{bmatrix}\begin{bmatrix} -2 & 1 \\ 0 & 2 \end{bmatrix} \\ &= \begin{bmatrix} -2 & 0 \\ -1 & 9 \end{bmatrix} + \begin{bmatrix} -2 & 1 \\ -4 & 8 \end{bmatrix} = \begin{bmatrix} -4 & 1 \\ -5 & 17 \end{bmatrix}\end{aligned}$$

Thus, $\mathbf{A}(\mathbf{B} + \mathbf{C}) = \mathbf{AB} + \mathbf{AC}$.

▸ **Now work Problem 69.**

EXAMPLE 9 **Raw Materials and Cost**

Suppose that a building contractor has accepted orders for 5 ranch-style houses, 7 Cape Cod–style houses, and 12 colonial-style houses. Then his orders can be represented by the row vector

$$\mathbf{Q} = [5 \quad 7 \quad 12]$$

Furthermore, suppose that the "raw materials" that go into each type of house are steel, wood, glass, paint, and labor. The entries in the following matrix **R** give the number of units of each raw material going into each type of house (the entries are not necessarily realistic, but are chosen for convenience):

$$\begin{array}{l} \\ \text{Ranch} \\ \text{Cape Cod} \\ \text{Colonial} \end{array} \begin{array}{c} \begin{array}{ccccc} \text{Steel} & \text{Wood} & \text{Glass} & \text{Paint} & \text{Labor} \end{array} \\ \begin{bmatrix} 5 & 20 & 16 & 7 & 17 \\ 7 & 18 & 12 & 9 & 21 \\ 6 & 25 & 8 & 5 & 13 \end{bmatrix} \end{array} = \mathbf{R}$$

Each row indicates the amount of each raw material needed for a given type of house; each column indicates the amount of a given raw material needed for each type of house. Suppose now that the contractor wishes to compute the amount of each raw material needed to fulfill his orders. Then such information is given by the matrix

$$\mathbf{QR} = \begin{bmatrix} 5 & 7 & 12 \end{bmatrix} \begin{bmatrix} 5 & 20 & 16 & 7 & 17 \\ 7 & 18 & 12 & 9 & 21 \\ 6 & 25 & 8 & 5 & 13 \end{bmatrix}$$

$$= \begin{bmatrix} 146 & 526 & 260 & 158 & 388 \end{bmatrix}$$

Thus, the contractor should order 146 units of steel, 526 units of wood, 260 units of glass, and so on.

The contractor is also interested in the costs he will have to pay for these materials. Suppose steel costs \$2500 per unit, wood costs \$1200 per unit, and glass, paint, and labor cost \$800, \$150, and \$1500 per unit, respectively. These data can be written as the column cost vector

$$\mathbf{C} = \begin{bmatrix} 2500 \\ 1200 \\ 800 \\ 150 \\ 1500 \end{bmatrix}$$

Then the cost of each type of house is given by the matrix

$$\mathbf{RC} = \begin{bmatrix} 5 & 20 & 16 & 7 & 17 \\ 7 & 18 & 12 & 9 & 21 \\ 6 & 25 & 8 & 5 & 13 \end{bmatrix} \begin{bmatrix} 2500 \\ 1200 \\ 800 \\ 150 \\ 1500 \end{bmatrix} = \begin{bmatrix} 75{,}850 \\ 81{,}550 \\ 71{,}650 \end{bmatrix}$$

Consequently, the cost of materials for the ranch-style house is \$75,850, for the Cape Code house \$81,550, and for the colonial house \$71,650.

The total cost of raw materials for all the houses is given by

$$\mathbf{QRC} = \mathbf{Q}(\mathbf{RC}) = \begin{bmatrix} 5 & 7 & 12 \end{bmatrix} \begin{bmatrix} 75{,}850 \\ 81{,}550 \\ 71{,}650 \end{bmatrix} = [1{,}809{,}900]$$

The total cost is \$1,809,900.

▶ **Now work Problem 65.**

Another property of matrices involves scalar and matrix multiplications. If k is a scalar and the product $\mathbf{AB}$ is defined, then

$$k(\mathbf{AB}) = (k\mathbf{A})\mathbf{B} = \mathbf{A}(k\mathbf{B})$$

The product $k(\mathbf{AB})$ may be written simply as $k\mathbf{AB}$. Thus,

$$k\mathbf{AB} = k(\mathbf{AB}) = (k\mathbf{A})\mathbf{B} = \mathbf{A}(k\mathbf{B})$$

For example,

$$3\begin{bmatrix}2 & 1\\0 & -1\end{bmatrix}\begin{bmatrix}1 & 3\\2 & 0\end{bmatrix} = \left(3\begin{bmatrix}2 & 1\\0 & -1\end{bmatrix}\right)\begin{bmatrix}1 & 3\\2 & 0\end{bmatrix}$$

$$= \begin{bmatrix}6 & 3\\0 & -3\end{bmatrix}\begin{bmatrix}1 & 3\\2 & 0\end{bmatrix}$$

$$= \begin{bmatrix}12 & 18\\-6 & 0\end{bmatrix}$$

There is an interesting property concerning the transpose of a matrix product:

$$(\mathbf{AB})^{\mathrm{T}} = \mathbf{B}^{\mathrm{T}}\mathbf{A}^{\mathrm{T}}$$

In words, the transpose of a product of matrices is equal to the product of their transposes in the *reverse* order.

This property can be extended to the case of more than two factors. For example,

$$(\mathbf{A}^{\mathrm{T}}\mathbf{BC})^{\mathrm{T}} = \mathbf{C}^{\mathrm{T}}\mathbf{B}^{\mathrm{T}}(\mathbf{A}^{\mathrm{T}})^{\mathrm{T}} = \mathbf{C}^{\mathrm{T}}\mathbf{B}^{\mathrm{T}}\mathbf{A}$$

Here we used the fact that $(\mathbf{A}^{\mathrm{T}})^{\mathrm{T}} = \mathbf{A}$.

EXAMPLE 10 Transpose of a Product

Let

$$\mathbf{A} = \begin{bmatrix}1 & 0\\1 & 2\end{bmatrix} \quad \textit{and} \quad \mathbf{B} = \begin{bmatrix}1 & 2\\1 & 0\end{bmatrix}$$

Show that $(\mathbf{AB})^{\mathrm{T}} = \mathbf{B}^{\mathrm{T}}\mathbf{A}^{\mathrm{T}}$.

Solution: We have

$$\mathbf{AB} = \begin{bmatrix}1 & 2\\3 & 2\end{bmatrix} \quad \text{so} \quad (\mathbf{AB})^{\mathrm{T}} = \begin{bmatrix}1 & 3\\2 & 2\end{bmatrix}$$

Now,

$$\mathbf{A}^{\mathrm{T}} = \begin{bmatrix}1 & 1\\0 & 2\end{bmatrix} \quad \text{and} \quad \mathbf{B}^{\mathrm{T}} = \begin{bmatrix}1 & 1\\2 & 0\end{bmatrix}$$

Thus,

$$\mathbf{B}^{\mathrm{T}}\mathbf{A}^{\mathrm{T}} = \begin{bmatrix}1 & 1\\2 & 0\end{bmatrix}\begin{bmatrix}1 & 1\\0 & 2\end{bmatrix} = \begin{bmatrix}1 & 3\\2 & 2\end{bmatrix} = (\mathbf{AB})^{\mathrm{T}}$$

so $(\mathbf{AB})^{\mathrm{T}} = \mathbf{B}^{\mathrm{T}}\mathbf{A}^{\mathrm{T}}$.

Just as the zero matrix plays an important role as the identity in matrix addition, there is a special matrix, called the *identity matrix,* that plays a corresponding role in matrix multiplication:

The $n \times n$ **identity matrix,** denoted $\mathbf{I}_n$, is the diagonal matrix whose main diagonal entries are 1's.

For example, the identity matrices $\mathbf{I}_3$ and $\mathbf{I}_4$ are

$$\mathbf{I}_3 = \begin{bmatrix}1 & 0 & 0\\0 & 1 & 0\\0 & 0 & 1\end{bmatrix} \quad \text{and} \quad \mathbf{I}_4 = \begin{bmatrix}1 & 0 & 0 & 0\\0 & 1 & 0 & 0\\0 & 0 & 1 & 0\\0 & 0 & 0 & 1\end{bmatrix}$$

When the size of an identity matrix is understood to be appropriate for an operation to be defined, we shall omit the subscript and simply denote the matrix by $\mathbf{I}$. It should be clear that

$$\mathbf{I}^{\mathrm{T}} = \mathbf{I}$$

The identity matrix plays the same role in matrix multiplication as does the number 1 in the multiplication of real numbers. That is, just as the product of a real number and 1 is the number itself, the product of a matrix and the identity matrix is the matrix itself. For example,

$$\begin{bmatrix} 2 & 4 \\ 1 & 5 \end{bmatrix} \mathbf{I} = \begin{bmatrix} 2 & 4 \\ 1 & 5 \end{bmatrix} \begin{bmatrix} 1 & 0 \\ 0 & 1 \end{bmatrix} = \begin{bmatrix} 2 & 4 \\ 1 & 5 \end{bmatrix}$$

and

$$\mathbf{I} \begin{bmatrix} 2 & 4 \\ 1 & 5 \end{bmatrix} = \begin{bmatrix} 1 & 0 \\ 0 & 1 \end{bmatrix} \begin{bmatrix} 2 & 4 \\ 1 & 5 \end{bmatrix} = \begin{bmatrix} 2 & 4 \\ 1 & 5 \end{bmatrix}$$

In general, if $\mathbf{I}$ is $n \times n$ and $\mathbf{A}$ has n columns, then $\mathbf{AI} = \mathbf{A}$. If $\mathbf{B}$ has n rows, then $\mathbf{IB} = \mathbf{B}$. Moreover, if $\mathbf{A}$ is $n \times n$, then

$$\mathbf{AI} = \mathbf{IA} = \mathbf{A}$$

EXAMPLE 11 **Matrix Operations Involving I and O**

If

$$\mathbf{A} = \begin{bmatrix} 3 & 2 \\ 1 & 4 \end{bmatrix} \qquad \mathbf{B} = \begin{bmatrix} \frac{2}{5} & -\frac{1}{5} \\ -\frac{1}{10} & \frac{3}{10} \end{bmatrix}$$

$$\mathbf{I} = \begin{bmatrix} 1 & 0 \\ 0 & 1 \end{bmatrix} \qquad \mathbf{O} = \begin{bmatrix} 0 & 0 \\ 0 & 0 \end{bmatrix}$$

compute each of the following:

a. $\mathbf{I} - \mathbf{A}$

Solution:

$$\mathbf{I} - \mathbf{A} = \begin{bmatrix} 1 & 0 \\ 0 & 1 \end{bmatrix} - \begin{bmatrix} 3 & 2 \\ 1 & 4 \end{bmatrix} = \begin{bmatrix} -2 & -2 \\ -1 & -3 \end{bmatrix}$$

b. $3(\mathbf{A} - 2\mathbf{I})$

Solution:

$$\begin{aligned} 3(\mathbf{A} - 2\mathbf{I}) &= 3\left(\begin{bmatrix} 3 & 2 \\ 1 & 4 \end{bmatrix} - 2\begin{bmatrix} 1 & 0 \\ 0 & 1 \end{bmatrix}\right) \\ &= 3\left(\begin{bmatrix} 3 & 2 \\ 1 & 4 \end{bmatrix} - \begin{bmatrix} 2 & 0 \\ 0 & 2 \end{bmatrix}\right) \\ &= 3\begin{bmatrix} 1 & 2 \\ 1 & 2 \end{bmatrix} = \begin{bmatrix} 3 & 6 \\ 3 & 6 \end{bmatrix} \end{aligned}$$

c. $\mathbf{AO}$

Solution:

$$\mathbf{AO} = \begin{bmatrix} 3 & 2 \\ 1 & 4 \end{bmatrix} \begin{bmatrix} 0 & 0 \\ 0 & 0 \end{bmatrix} = \begin{bmatrix} 0 & 0 \\ 0 & 0 \end{bmatrix} = \mathbf{O}$$

In general, if **AO** and **OA** are defined, then

$$\mathbf{AO} = \mathbf{OA} = \mathbf{O}$$

d. **AB**

Solution:

$$\mathbf{AB} = \begin{bmatrix} 3 & 2 \\ 1 & 4 \end{bmatrix} \begin{bmatrix} \frac{2}{5} & -\frac{1}{5} \\ -\frac{1}{10} & \frac{3}{10} \end{bmatrix} = \begin{bmatrix} 1 & 0 \\ 0 & 1 \end{bmatrix} = \mathbf{I}$$

If **A** is a square matrix, we can speak of a *power* of **A**:

If **A** is a square matrix and p is a positive integer, then the ***p*th power** of **A**, written $\mathbf{A}^p$, is the product of p factors of **A**:

$$\mathbf{A}^p = \underbrace{\mathbf{A} \cdot \mathbf{A} \cdots \mathbf{A}}_{p \text{ factors}}$$

If **A** is $n \times n$, we define $\mathbf{A}^0 = \mathbf{I}_n$.

We remark that $\mathbf{I}^p = \mathbf{I}$.

EXAMPLE 12 **Power of a Matrix**

If $\mathbf{A} = \begin{bmatrix} 1 & 0 \\ 1 & 2 \end{bmatrix}$, *compute* $\mathbf{A}^3$.

Solution: Since $\mathbf{A}^3 = (\mathbf{A}^2)\mathbf{A}$ and

$$\mathbf{A}^2 = \begin{bmatrix} 1 & 0 \\ 1 & 2 \end{bmatrix} \begin{bmatrix} 1 & 0 \\ 1 & 2 \end{bmatrix} = \begin{bmatrix} 1 & 0 \\ 3 & 4 \end{bmatrix}$$

we have

$$\mathbf{A}^3 = \mathbf{A}^2\mathbf{A} = \begin{bmatrix} 1 & 0 \\ 3 & 4 \end{bmatrix} \begin{bmatrix} 1 & 0 \\ 1 & 2 \end{bmatrix} = \begin{bmatrix} 1 & 0 \\ 7 & 8 \end{bmatrix}$$

▶ **Now work Problem 45.**

TECHNOLOGY

Use of the graphing calculator to compute $\mathbf{A}^4$, where $\mathbf{A} = \begin{bmatrix} 2 & -3 \\ 1 & 4 \end{bmatrix}$, is shown in Figure 6.4.

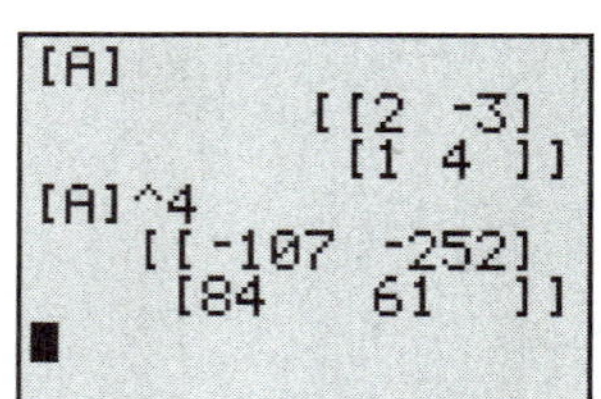

FIGURE 6.4 Power of a matrix.

Matrix Equations

Systems of linear equations can be represented by using matrix multiplication. For example, consider the matrix equation

$$\begin{bmatrix} 1 & 4 & -2 \\ 2 & -3 & 1 \end{bmatrix} \begin{bmatrix} x_1 \\ x_2 \\ x_3 \end{bmatrix} = \begin{bmatrix} 4 \\ -3 \end{bmatrix} \qquad (1)$$

The product on the left side has size 2×1 and hence is a column matrix. Thus,

$$\begin{bmatrix} x_1 + 4x_2 - 2x_3 \\ 2x_1 - 3x_2 + x_3 \end{bmatrix} = \begin{bmatrix} 4 \\ -3 \end{bmatrix}$$

By equality of matrices, corresponding entries must be equal, so we obtain the system

$$\begin{cases} x_1 + 4x_2 - 2x_3 = 4 \\ 2x_1 - 3x_2 + x_3 = -3 \end{cases}$$

Hence, this system of linear equations can be defined by matrix Equation (1). We usually describe Equation (1) by saying that it has the form

$$\mathbf{AX} = \mathbf{B}$$

where $\mathbf{A}$ is the matrix obtained from the coefficients of the variables, $\mathbf{X}$ is a column matrix obtained from the variables, and $\mathbf{B}$ is a column matrix obtained from the constants. Matrix $\mathbf{A}$ is called the *coefficient matrix* for the system.

APPLICATIONS IN PRACTICE 3

Matrix Form of a System Using Matrix Multiplication

Write the following pair of lines in matrix form, using matrix multiplication.

$$y = -\frac{8}{5}x + \frac{8}{5}, \quad y = -\frac{1}{3}x + \frac{5}{3}.$$

EXAMPLE 13 Matrix Form of a System Using Matrix Multiplication

Write the system

$$\begin{cases} 2x_1 + 5x_2 = 4 \\ 8x_1 + 3x_2 = 7 \end{cases}$$

in matrix form by using matrix multiplication.

Solution: If

$$\mathbf{A} = \begin{bmatrix} 2 & 5 \\ 8 & 3 \end{bmatrix} \qquad \mathbf{X} = \begin{bmatrix} x_1 \\ x_2 \end{bmatrix} \qquad \mathbf{B} = \begin{bmatrix} 4 \\ 7 \end{bmatrix}$$

then the given system is equivalent to the single matrix equation

$$\mathbf{AX} = \mathbf{B}$$

that is

$$\begin{bmatrix} 2 & 5 \\ 8 & 3 \end{bmatrix} \begin{bmatrix} x_1 \\ x_2 \end{bmatrix} = \begin{bmatrix} 4 \\ 7 \end{bmatrix}$$

▶ Now work Problem 59.

Exercise 6.3

If $\mathbf{A} = \begin{bmatrix} 1 & 3 & -2 \\ -2 & 1 & -1 \\ 0 & 4 & 3 \end{bmatrix}$, $\mathbf{B} = \begin{bmatrix} 0 & -2 & 3 \\ -2 & 4 & -2 \\ 3 & 1 & -1 \end{bmatrix}$, *and* $\mathbf{AB} = \mathbf{C}$, *find each of the following:*

1. c_{11} **2.** c_{23} **3.** c_{32}

4. c_{33} **5.** c_{31} **6.** c_{13}

If $\mathbf{A}$ *is* 2×3, $\mathbf{B}$ *is* 3×1, $\mathbf{C}$ *is* 2×5, $\mathbf{D}$ *is* 4×3, $\mathbf{E}$ *is* 3×2, *and* $\mathbf{F}$ *is* 2×3, *find the size and number of entries of each of the following:*

***7.** $\mathbf{AE}$ **8.** $\mathbf{DE}$ **9.** $\mathbf{EC}$

10. $\mathbf{DB}$ **11.** $\mathbf{FB}$ **12.** $\mathbf{BC}$

13. $\mathbf{EA}$ **14.** $\mathbf{E(AE)}$ **15.** $\mathbf{E(FB)}$

16. $\mathbf{(F + A)B}$

Write the identity matrix that has the following order:

17. 4 **18.** 5

In Problems 19–36, perform the indicated operations.

***19.** $\begin{bmatrix} 2 & -4 \\ 3 & 2 \end{bmatrix} \begin{bmatrix} 4 & 0 \\ -1 & 3 \end{bmatrix}$

20. $\begin{bmatrix} -1 & 1 \\ 0 & 4 \\ 2 & 1 \end{bmatrix} \begin{bmatrix} 1 & -2 \\ 3 & 4 \end{bmatrix}$

21. $\begin{bmatrix} 2 & 0 & 3 \\ -1 & 4 & 5 \end{bmatrix} \begin{bmatrix} 1 \\ 4 \\ 7 \end{bmatrix}$

22. $\begin{bmatrix} 1 & 0 & 6 & 2 \end{bmatrix} \begin{bmatrix} 0 \\ 1 \\ 2 \\ 3 \end{bmatrix}$

23. $\begin{bmatrix} 1 & 4 & -1 \\ 0 & 0 & 2 \\ -2 & 1 & 1 \end{bmatrix} \begin{bmatrix} 2 & 1 & 0 \\ 0 & -1 & 1 \\ 1 & 1 & 2 \end{bmatrix}$

24. $\begin{bmatrix} 4 & 2 & -2 \\ 3 & 10 & 0 \\ 1 & 0 & 2 \end{bmatrix} \begin{bmatrix} 3 & 1 & 1 & 0 \\ 0 & 0 & 0 & 0 \\ 0 & 1 & 0 & 1 \end{bmatrix}$

***25.** $\begin{bmatrix} -1 & 2 & 3 \end{bmatrix} \begin{bmatrix} 3 & 1 & -1 & 2 \\ 0 & 4 & 3 & 1 \\ -1 & 3 & 1 & -2 \end{bmatrix}$

26. $\begin{bmatrix} 1 & -4 \end{bmatrix} \begin{bmatrix} -2 & 1 \\ 0 & 1 \\ 5 & 0 \end{bmatrix}$

***27.** $\begin{bmatrix} 2 \\ 3 \\ -4 \\ 1 \end{bmatrix} \begin{bmatrix} 2 & 3 & -2 & 3 \end{bmatrix}$

28. $\begin{bmatrix} 0 & 1 \\ 2 & 3 \end{bmatrix} \left(\begin{bmatrix} 1 & 0 & 1 \\ 1 & 1 & 0 \end{bmatrix} + \begin{bmatrix} 0 & 1 & 0 \\ 0 & 0 & 1 \end{bmatrix} \right)$

29. $3\left(\begin{bmatrix} -2 & 0 & 2 \\ 3 & -1 & 1 \end{bmatrix} + 2\begin{bmatrix} -1 & 0 & 2 \\ 1 & 1 & -2 \end{bmatrix} \right) \begin{bmatrix} 1 & 2 \\ 3 & 4 \\ 5 & 6 \end{bmatrix}$

30. $\begin{bmatrix} -1 & 2 \\ -2 & 0 \end{bmatrix} \begin{bmatrix} -1 & 0 & 2 & 7 & -1 \\ 2 & 1 & 4 & -3 & -2 \end{bmatrix}$

31. $\begin{bmatrix} 1 & 2 \\ 3 & 4 \end{bmatrix} \left(\begin{bmatrix} 2 & 0 & 1 \\ 1 & 0 & -2 \end{bmatrix} \begin{bmatrix} 1 & -2 \\ 2 & 1 \\ 3 & 0 \end{bmatrix} \right)$

32. $3\begin{bmatrix} 1 & 2 \\ -1 & 4 \end{bmatrix} - 4\left(\begin{bmatrix} 1 & 0 \\ 0 & 1 \end{bmatrix} \begin{bmatrix} -2 & 4 \\ 6 & 1 \end{bmatrix} \right)$

33. $\begin{bmatrix} 0 & 0 & 1 \\ 0 & 1 & 0 \\ 1 & 0 & 0 \end{bmatrix} \begin{bmatrix} x \\ y \\ z \end{bmatrix}$

34. $\begin{bmatrix} a_{11} & a_{12} \\ a_{21} & a_{22} \end{bmatrix} \begin{bmatrix} x_1 \\ x_2 \end{bmatrix}$

35. $\begin{bmatrix} 2 & 1 & 3 \\ 4 & 9 & 7 \end{bmatrix} \begin{bmatrix} x_1 \\ x_2 \\ x_3 \end{bmatrix}$

36. $\begin{bmatrix} 2 & -3 \\ 0 & 1 \\ 2 & 1 \end{bmatrix} \begin{bmatrix} x_1 \\ x_2 \end{bmatrix}$

In Problems 37–44, compute the required matrices if

$$\mathbf{A} = \begin{bmatrix} 1 & -2 \\ 0 & 3 \end{bmatrix} \quad \mathbf{B} = \begin{bmatrix} -2 & 3 & 0 \\ 1 & -4 & 1 \end{bmatrix} \quad \mathbf{C} = \begin{bmatrix} -1 & 1 \\ 0 & 3 \\ 2 & 4 \end{bmatrix}$$

$$\mathbf{D} = \begin{bmatrix} 1 & 0 & 0 \\ 0 & 1 & 1 \\ 1 & 2 & 1 \end{bmatrix} \quad \mathbf{E} = \begin{bmatrix} 3 & 0 & 0 \\ 0 & 6 & 0 \\ 0 & 0 & 3 \end{bmatrix} \quad \mathbf{F} = \begin{bmatrix} \frac{1}{3} & 0 & 0 \\ 0 & \frac{1}{6} & 0 \\ 0 & 0 & \frac{1}{3} \end{bmatrix}$$

$$\mathbf{I} = \begin{bmatrix} 1 & 0 & 0 \\ 0 & 1 & 0 \\ 0 & 0 & 1 \end{bmatrix}$$

***37.** $\mathbf{DI} - \frac{1}{3}\mathbf{E}$ **38.** $\mathbf{DD}$ **39.** $3\mathbf{A} - 2\mathbf{BC}$

40. $\mathbf{B}(\mathbf{D} + \mathbf{E})$ **41.** $3\mathbf{I} - \frac{2}{3}\mathbf{FE}$ **42.** $\mathbf{E}(2\mathbf{D} - 3\mathbf{I})$

43. $(\mathbf{DC})\mathbf{A}$ **44.** $\mathbf{A}(\mathbf{BC})$

In Problems 45–58, compute the required matrix, if it exists, given that

$$\mathbf{A} = \begin{bmatrix} 1 & -1 & 0 \\ 0 & 1 & 1 \end{bmatrix} \quad \mathbf{B} = \begin{bmatrix} 0 & 0 & -1 \\ 2 & -1 & 0 \\ 0 & 0 & 2 \end{bmatrix} \quad \mathbf{C} = \begin{bmatrix} 1 & 0 \\ 2 & -1 \\ 0 & 1 \end{bmatrix}$$

$$\mathbf{I} = \begin{bmatrix} 1 & 0 & 0 \\ 0 & 1 & 0 \\ 0 & 0 & 1 \end{bmatrix} \quad \mathbf{O} = \begin{bmatrix} 0 & 0 & 0 \\ 0 & 0 & 0 \\ 0 & 0 & 0 \end{bmatrix}$$

***45.** $\mathbf{A}^2$ **46.** $\mathbf{A}^{\mathrm{T}}\mathbf{A}$ **47.** $\mathbf{B}^3$

48. $\mathbf{A}(\mathbf{B}^{\mathrm{T}})^2\mathbf{C}$ **49.** $(\mathbf{AC})^2$ **50.** $\mathbf{A}^{\mathrm{T}}(2\mathbf{C}^{\mathrm{T}})$

51. $(\mathbf{BA}^{\mathrm{T}})^{\mathrm{T}}$ **52.** $(2\mathbf{B})^{\mathrm{T}}$ **53.** $(2\mathbf{I})^2 - 2\mathbf{I}^2$

54. $\mathbf{C}^0\mathbf{I}$ **55.** $\mathbf{A}(\mathbf{I} - \mathbf{O})$ **56.** $\mathbf{I}^{\mathrm{T}}\mathbf{O}$

57. $(\mathbf{AB})(\mathbf{AB})^{\mathrm{T}}$ **58.** $\mathbf{B}^2 - 3\mathbf{B} + 2\mathbf{I}$

In Problems 59–61, represent the given system by using matrix multiplication.

***59.** $\begin{cases} 3x + y = 6 \\ 2x - 9y = 5 \end{cases}$

60. $\begin{cases} 3x + y + z = 2 \\ x - y + z = 4 \\ 5x - y + 2z = 12 \end{cases}$

61. $\begin{cases} 4r - s + 3t = 9 \\ 3r - t = 7 \\ 3s + 2t = 15 \end{cases}$

62. Secret Messages Secret messages can be encoded by using a code and an encoding matrix. Suppose we have the following code:

a	*b*	*c*	*d*	*e*	*f*	*g*	*h*	*i*	*j*	*k*	*l*	*m*
1	2	3	4	5	6	7	8	9	10	11	12	13
n	*o*	*p*	*q*	*r*	*s*	*t*	*u*	*v*	*w*	*x*	*y*	*z*
14	15	16	17	18	19	20	21	22	23	24	25	26

Let the encoding matrix be $\mathbf{E} = \begin{bmatrix} 1 & 3 \\ 2 & 4 \end{bmatrix}$. Then we can encode a message by taking every two letters of the message, converting them to their corresponding numbers, creating a 2×1 matrix, and then multiplying each matrix by **E**. Use this code and matrix to encode the message "the/falcon/has/landed," leaving the slashes to separate words.

63. Inventory A pet store has 6 kittens, 10 puppies, and 7 parrots in stock. If the value of each kitten is \$55, each puppy is \$150, and each parrot is \$35, using matrix multiplication, find the total value of the pet store's inventory.

64. Stocks A stockbroker sold a customer 200 shares of stock A, 300 shares of stock B, 500 shares of stock C, and 250 shares of stock D. The prices per share of A, B,

C, and D are \$100, \$150, \$200, and \$300, respectively. Write a row vector representing the number of shares of each stock bought. Write a column vector representing the price per share of each stock. Using matrix multiplication, find the total cost of the stocks.

*65. **Construction Cost** In Example 9, assume that the contractor is to build 5 ranch-style, 2 Cape Cod–style, and 4 colonial-style houses. Using matrix multiplication, compute the total cost of raw materials.

66. Costs In Example 9, assume that the contractor wishes to take into account the cost of transporting raw materials to the building site, as well as the purchasing cost. Suppose the costs are given in the following matrix:

$$\mathbf{C} = \begin{array}{c} \text{Purchase} \quad \text{Transport} \\ \begin{bmatrix} 2500 & 45 \\ 1200 & 20 \\ 800 & 30 \\ 150 & 5 \\ 1500 & 0 \end{bmatrix} \end{array} \begin{array}{l} \\ \text{Steel} \\ \text{Wood} \\ \text{Glass} \\ \text{Paint} \\ \text{Labor} \end{array}$$

(a) By computing **RC**, find a matrix whose entries give the purchase and transportation costs of the materials for each type of house.

(b) Find the matrix **QRC** whose first entry gives the total purchase price and whose second entry gives the total transportation cost.

(c) Let $\mathbf{Z} = \begin{bmatrix} 1 \\ 1 \end{bmatrix}$, and then compute **QRCZ**, which gives the total cost of materials and transportation for all houses being built.

*67. Perform the following calculations for Example 6.

(a) Compute the amount that each industry and each consumer have to pay for the goods they receive.

(b) Compute the profit earned by each industry.

(c) Find the total amount of money that is paid out by all the industries and consumers.

(d) Find the proportion of the total amount of money found in part (c) paid out by the industries. Find the proportion of the total amount of money found in part (c) that is paid out by the consumers.

68. Prove that if $\mathbf{AB} = \mathbf{BA}$, then $(\mathbf{A} + \mathbf{B})(\mathbf{A} - \mathbf{B}) = \mathbf{A}^2 - \mathbf{B}^2$.

*69. Show that if

$$\mathbf{A} = \begin{bmatrix} 1 & 2 \\ 1 & 2 \end{bmatrix} \quad \text{and} \quad \mathbf{B} = \begin{bmatrix} 2 & -3 \\ -1 & \frac{3}{2} \end{bmatrix}$$

then $\mathbf{AB} = \mathbf{O}$. Observe that since neither **A** nor **B** is the zero matrix, the algebraic rule for real numbers, "if $ab = 0$, then either a or b is zero," does not hold for matrices. It can also be shown that the cancellation law is not true for matrices; that is, if $\mathbf{AB} = \mathbf{AC}$, then it is not necessarily true that $\mathbf{B} = \mathbf{C}$.

70. Let $\mathbf{D}_1$ and $\mathbf{D}_2$ be two arbitrary 3×3 diagonal matrices. By computing $\mathbf{D}_1\mathbf{D}_2$ and $\mathbf{D}_2\mathbf{D}_1$, show that

(a) Both $\mathbf{D}_1\mathbf{D}_2$ and $\mathbf{D}_2\mathbf{D}_1$ are diagonal matrices.

(b) $\mathbf{D}_1$ and $\mathbf{D}_2$ commute.

In Problems 71–74, compute the required matrices, given that

$$\mathbf{A} = \begin{bmatrix} 3.2 & -4.1 & 5.1 \\ -2.6 & 1.2 & 6.8 \end{bmatrix} \qquad \mathbf{B} = \begin{bmatrix} 1.1 & 4.8 \\ -2.3 & 3.2 \\ 4.6 & -1.4 \end{bmatrix}$$

$$\mathbf{C} = \begin{bmatrix} -1.2 & 1.5 \\ 2.4 & 6.2 \end{bmatrix}$$

71. $\mathbf{A}(2\mathbf{B})$

72. $-3.1(\mathbf{CA})$

73. $(-\mathbf{C})(3\mathbf{A})\mathbf{B}$

74. $\mathbf{C}^3$

OBJECTIVE *To show how to reduce a matrix and to use matrix reduction to solve a linear system.*

6.4 Solving Systems by Reducing Matrices

In this section, we shall illustrate a method by which matrices can be used to solve a system of linear equations. In developing this *method of reduction,* we shall first solve a system by the usual method of elimination. Then we shall obtain the same solution by using matrices.

Let us consider the system

$$\begin{cases} 3x - y = 1 & (1) \\ x + 2y = 5 & (2) \end{cases}$$

consisting of two linear equations in two unknowns, x and y. Although this system can be solved by various algebraic methods, we shall solve it by a method that is readily adapted to matrices.

For reasons that will be obvious later, we begin by replacing Equation (1) by Equation (2), and Equation (2) by Equation (1), thus obtaining the equivalent system,[2]

$$\begin{cases} x + 2y = 5 & \textbf{(3)} \\ 3x - y = 1 & \textbf{(4)} \end{cases}$$

Multiplying both sides of Equation (3) by -3 gives $-3x - 6y = -15$. Adding the left and right sides of this equation to the corresponding sides of Equation (4) produces an equivalent system in which x is eliminated from the second equation:

$$\begin{cases} x + 2y = 5 & \textbf{(5)} \\ 0x - 7y = -14 & \textbf{(6)} \end{cases}$$

Now we shall eliminate y from the first equation. Multiplying both sides of Equation (6) by $-\frac{1}{7}$ gives the equivalent system,

$$\begin{cases} x + 2y = 5 & \textbf{(7)} \\ 0x + y = 2 & \textbf{(8)} \end{cases}$$

From Equation (8), $y = 2$, and hence, $-2y = -4$. Adding the sides of $-2y = -4$ to the corresponding sides of Equation (7), we get the equivalent system,

$$\begin{cases} x + 0y = 1 \\ 0x + y = 2 \end{cases}$$

Therefore, $x = 1$ and $y = 2$, so the original system is solved.

Note that in solving the original system, we successively replaced it by an equivalent system that was obtained by performing one of the following three operations (called *elementary operations*), which leave the solution unchanged:

1. Interchanging two equations.
2. Adding a constant multiple of the sides of one equation to the corresponding sides of another equation.
3. Multiplying one equation by a nonzero constant.

Before showing a matrix method of solving the original system,

$$\begin{cases} 3x - y = 1 \\ x + 2y = 5 \end{cases}$$

we first need to define some terms. Recall from Section 6.3 that the matrix

$$\begin{bmatrix} 3 & -1 \\ 1 & 2 \end{bmatrix}$$

is the **coefficient matrix** of this system. The entries in the first column correspond to the coefficients of the x's in the equations. For example, the entry in the first row and first column corresponds to the coefficient of x in the first equation; and the entry in the second row and first column corresponds to the coefficient of x in the second equation. Similarly, the entries in the second column correspond to the coefficients of the y's.

---

[2]Recall from Section 3.4 that two or more systems are equivalent if they have the same solution.

Another matrix associated with this system is called the **augmented coefficient matrix** and is given by

$$\left[\begin{array}{rr|r} 3 & -1 & 1 \\ 1 & 2 & 5 \end{array}\right]$$

The first and second columns are the first and second columns, respectively, of the coefficient matrix. The entries in the third column correspond to the constant terms in the system: the entry in the first row of this column is the constant term of the first equation, whereas the entry in the second row is the constant term of the second equation. Although it is not necessary to include the vertical line in the augmented coefficient matrix, it serves to remind us that the 1 and the 5 are the constant terms that appear on the right sides of the equations. The augmented coefficient matrix itself completely describes the system of equations.

The procedure that was used to solve the original system involved a number of equivalent systems. With each of these systems, we can associate its augmented coefficient matrix. Following are the systems that were involved, together with their corresponding augmented coefficient matrices, which we have labeled **A**, **B**, **C**, **D**, and **E**:

$$\begin{cases} 3x - y = 1 \\ x + 2y = 5 \end{cases} \qquad \left[\begin{array}{rr|r} 3 & -1 & 1 \\ 1 & 2 & 5 \end{array}\right] = \mathbf{A}$$

$$\begin{cases} x + 2y = 5 \\ 3x - y = 1 \end{cases} \qquad \left[\begin{array}{rr|r} 1 & 2 & 5 \\ 3 & -1 & 1 \end{array}\right] = \mathbf{B}$$

$$\begin{cases} x + 2y = 5 \\ 0x - 7y = -14 \end{cases} \qquad \left[\begin{array}{rr|r} 1 & 2 & 5 \\ 0 & -7 & -14 \end{array}\right] = \mathbf{C}$$

$$\begin{cases} x + 2y = 5 \\ 0x + y = 2 \end{cases} \qquad \left[\begin{array}{rr|r} 1 & 2 & 5 \\ 0 & 1 & 2 \end{array}\right] = \mathbf{D}$$

$$\begin{cases} x + 0y = 1 \\ 0x + y = 2 \end{cases} \qquad \left[\begin{array}{rr|r} 1 & 0 & 1 \\ 0 & 1 & 2 \end{array}\right] = \mathbf{E}$$

Let us see how these matrices are related.

B can be obtained from **A** by interchanging the first and second rows of **A**. This operation corresponds to interchanging the two equations in the original system.

C can be obtained from **B** by adding to each entry in the second row of **B** -3 times the corresponding entry in the first row of **B**:

$$\begin{aligned} \mathbf{C} &= \left[\begin{array}{cc|c} 1 & 2 & 5 \\ 3 + (-3)(1) & -1 + (-3)(2) & 1 + (-3)(5) \end{array}\right] \\ &= \left[\begin{array}{rr|r} 1 & 2 & 5 \\ 0 & -7 & -14 \end{array}\right] \end{aligned}$$

This operation is described as the addition of -3 times the first row of **B** to the second row of **B**.

D can be obtained from **C** by multiplying each entry in the second row of **C** by $-\frac{1}{7}$. This operation is referred to as multiplying the second row of **C** by $-\frac{1}{7}$.

E can be obtained from **D** by adding -2 times the second row of **D** to the first row of **D**.

Observe that **E**, which essentially gives the solution, was obtained from **A** by successively performing one of three matrix operations, called **elementary row operations**:

Elementary Row Operation

1. Interchanging two rows of a matrix;
2. Adding a multiple of one row of a matrix to a different row of that matrix;
3. Multiplying a row of a matrix by a nonzero scalar.

These elementary row operations correspond to the three elementary operations used in the algebraic method of elimination. Whenever a matrix can be obtained from another by one or more elementary row operations, we say that the matrices are **equivalent.** Thus, **A** and **E** are equivalent. (We could also obtain **A** from **E** by performing similar row operations in the reverse order, so the term *equivalent* is appropriate.) When describing particular elementary row operations, we will use the following notation for convenience:

Notation	Corresponding Row Operation
$R_i \leftrightarrow R_j$	Interchange rows R_i and R_j
kR_i	Multiply row R_i by the constant k
$kR_i + R_j$	Add k times row R_i to row R_j (but row R_i remains the same)

For example, writing

$$\begin{bmatrix} 1 & 0 & -2 \\ 4 & -2 & 1 \\ 5 & 0 & 3 \end{bmatrix} \xrightarrow{-4R_1 + R_2} \begin{bmatrix} 1 & 0 & -2 \\ 0 & -2 & 9 \\ 5 & 0 & 3 \end{bmatrix}$$

means that the second matrix was obtained from the first by adding -4 times row 1 to row 2. Note that we may write $(-k)R_i$ as $-kR_i$.

We are now ready to describe a matrix procedure for solving a system of linear equations. First, we form the augmented coefficient matrix of the system; then, by means of elementary row operations, we determine an equivalent matrix that clearly indicates the solution. Let us be quite specific as to what we mean by a matrix that clearly indicates the solution. This is a matrix, called a *reduced matrix,* defined as follows:

Reduced Matrix

A matrix is said to be a **reduced matrix,**[3] provided that all of the following are true:

1. If a row does not consist entirely of zeros, then the first nonzero entry in the row, called the **leading entry,** is 1, whereas all other entries in the column in which the 1 appears are zeros.
2. The first nonzero entry in each row is to the right of the first nonzero entry in each row above it.
3. Any rows that consist entirely of zeros are at the bottom of the matrix.

---

[3]Or in *reduced row echelon form.*

In other words, to solve the system, we must find the reduced matrix such that the augmented coefficient matrix is equivalent to it. In our previous discussion of elementary row operations, the matrix

$$\mathbf{E} = \left[\begin{array}{cc|c} 1 & 0 & 1 \\ 0 & 1 & 2 \end{array}\right]$$

is a reduced matrix.

EXAMPLE 1 **Reduced Matrices**

For each of the following matrices, determine whether it is reduced or not reduced.

1. $\begin{bmatrix} 1 & 0 \\ 0 & 3 \end{bmatrix}$ **2.** $\begin{bmatrix} 1 & 0 & 0 \\ 0 & 1 & 0 \end{bmatrix}$ **3.** $\begin{bmatrix} 0 & 1 \\ 1 & 0 \end{bmatrix}$

4. $\begin{bmatrix} 0 & 0 & 0 \\ 0 & 0 & 0 \end{bmatrix}$ **5.** $\begin{bmatrix} 1 & 0 & 0 \\ 0 & 0 & 0 \\ 0 & 1 & 0 \end{bmatrix}$ **6.** $\begin{bmatrix} 0 & 1 & 0 & 3 \\ 0 & 0 & 1 & 2 \\ 0 & 0 & 0 & 0 \end{bmatrix}$

Solution:

1. Not a reduced matrix, because the leading entry in the second row is not 1.
2. Reduced matrix.
3. Not a reduced matrix, because the leading entry in the second row is not to the right of the first nonzero entry in the first row.
4. Reduced matrix.
5. Not a reduced matrix, because the second row, consisting entirely of zeros, is not at the bottom of the matrix.
6. Reduced matrix.

▶ **Now work Problem 1.**

EXAMPLE 2 **Reducing a Matrix**

Reduce the matrix

$$\begin{bmatrix} 0 & 0 & 1 & 2 \\ 3 & -6 & -3 & 0 \\ 6 & -12 & 2 & 11 \end{bmatrix}$$

Strategy: To reduce the matrix, we must get the leading entry to be a 1 in the first row, a 1 in the second row, and so on, until we arrive at a zero row, if there is any. Moreover, we must work from left to right, because the leading 1 in each row must be to the *left* of all other leading 1's in the rows below it.

Solution: Since there are no zero rows to move to the bottom, we proceed to find the first column that contains a nonzero entry; this turns out to be column 1. Accordingly, in the reduced matrix, the leading 1 in the first row must be in column 1. To accomplish this, we shall begin by interchanging the first two rows so that a nonzero entry is in row 1 of column 1:

$$\begin{bmatrix} 0 & 0 & 1 & 2 \\ 3 & -6 & -3 & 0 \\ 6 & -12 & 2 & 11 \end{bmatrix} \xrightarrow{R_1 \leftrightarrow R_2} \begin{bmatrix} 3 & -6 & -3 & 0 \\ 0 & 0 & 1 & 2 \\ 6 & -12 & 2 & 11 \end{bmatrix}$$

Next, we multiply row 1 by $\frac{1}{3}$ so that the leading entry is a 1:

$$\xrightarrow{\frac{1}{3}R_1} \begin{bmatrix} 1 & -2 & -1 & 0 \\ 0 & 0 & 1 & 2 \\ 6 & -12 & 2 & 11 \end{bmatrix}$$

Now, because we must have zeros below (and above) each leading 1, we add -6 times row 1 to row 3:

$$\xrightarrow{-6R_1 + R_3} \begin{bmatrix} 1 & -2 & -1 & 0 \\ 0 & 0 & 1 & 2 \\ 0 & 0 & 8 & 11 \end{bmatrix}$$

Next, we move to the right of column 1 to find the first column that has a nonzero entry in row 2 or below; this is column 3. Consequently, in the reduced matrix, the leading 1 in the second row must be in column 3. The foregoing matrix already does have a leading 1 there. Thus, all we need do to get zeros below and above the leading 1 is add 1 times row 2 to row 1 and add -8 times row 2 to row 3:

$$\xrightarrow[-8R_2 + R_3]{(1)R_2 + R_1} \begin{bmatrix} 1 & -2 & 0 & 2 \\ 0 & 0 & 1 & 2 \\ 0 & 0 & 0 & -5 \end{bmatrix}$$

Again, we move to the right to find the first column that has a nonzero entry in row 3; namely, column 4. To make the leading entry a 1, we multiply row 3 by $-\frac{1}{5}$:

$$\xrightarrow{-\frac{1}{5}R_3} \begin{bmatrix} 1 & -2 & 0 & 2 \\ 0 & 0 & 1 & 2 \\ 0 & 0 & 0 & 1 \end{bmatrix}$$

CAUTION The sequence of steps that is used to reduce a matrix is not unique; however, the reduced form *is* unique.

Finally, to get all other entries in column 4 to be zeros, we add -2 times row 3 to both row 1 and row 2:

$$\xrightarrow[-2R_3 + R_2]{-2R_3 + R_1} \begin{bmatrix} 1 & -2 & 0 & 0 \\ 0 & 0 & 1 & 0 \\ 0 & 0 & 0 & 1 \end{bmatrix}$$

The last matrix is in reduced form.

▸ **Now work Problem 9.**

TECHNOLOGY

Although elementary row operations can be performed on a graphing calculator, the procedure is rather awkward.

The method of reduction described for solving our original system can be generalized to systems consisting of m linear equations in n unknowns. To solve such a system as

$$\begin{cases} a_{11}x_1 + a_{12}x_2 + \cdots + a_{1n}x_n = c_1 \\ a_{21}x_1 + a_{22}x_2 + \cdots + a_{2n}x_n = c_2 \\ \quad \vdots \qquad\quad \vdots \qquad\qquad\quad \vdots \qquad\ \vdots \\ a_{m1}x_1 + a_{m2}x_2 + \cdots + a_{mn}x_n = c_m \end{cases}$$

involves

1. determining the augmented coefficient matrix of the system, which is

$$\left[\begin{array}{cccc|c} a_{11} & a_{12} & \cdots & a_{1n} & c_1 \\ a_{21} & a_{22} & \cdots & a_{2n} & c_2 \\ \vdots & \vdots & & \vdots & \vdots \\ a_{m1} & a_{m2} & \cdots & a_{mn} & c_m \end{array}\right]$$

   and

2. determining a reduced matrix such that the augmented coefficient matrix is equivalent to it.

Frequently, step 2 is called *reducing the augmented coefficient matrix*.

APPLICATIONS IN PRACTICE 1

Solving a System by Reduction

An investment firm offers three stock portfolios: A, B, and C. The number of blocks of each type of stock in each of these portfolios is summarized in the following table:

		Portfolio		
		A	B	C
	High	6	1	3
Risk:	Moderate	3	2	3
	Low	1	5	3

A client wants 35 blocks of high-risk stock, 22 blocks of moderate-risk stock, and 18 blocks of low-risk stock. How many of each portfolio should be suggested?

EXAMPLE 3 **Solving a System by Reduction**

By using matrix reduction, solve the system

$$\begin{cases} 2x + 3y = -1 \\ 2x + \ \ y = \ \ 5 \\ \ \ x + \ \ y = \ \ 1 \end{cases}$$

Solution: Reducing the augmented coefficient matrix of the system, we have

$$\left[\begin{array}{cc|c} 2 & 3 & -1 \\ 2 & 1 & 5 \\ 1 & 1 & 1 \end{array}\right] \xrightarrow{R_1 \leftrightarrow R_3} \left[\begin{array}{cc|c} 1 & 1 & 1 \\ 2 & 1 & 5 \\ 2 & 3 & -1 \end{array}\right]$$

$$\xrightarrow{-2R_1 + R_2} \left[\begin{array}{cc|c} 1 & 1 & 1 \\ 0 & -1 & 3 \\ 2 & 3 & -1 \end{array}\right]$$

$$\xrightarrow{-2R_1 + R_3} \left[\begin{array}{cc|c} 1 & 1 & 1 \\ 0 & -1 & 3 \\ 0 & 1 & -3 \end{array}\right]$$

$$\xrightarrow{(-1)R_2} \left[\begin{array}{cc|c} 1 & 1 & 1 \\ 0 & 1 & -3 \\ 0 & 1 & -3 \end{array}\right]$$

$$\xrightarrow{-R_2 + R_1} \left[\begin{array}{cc|c} 1 & 0 & 4 \\ 0 & 1 & -3 \\ 0 & 1 & -3 \end{array}\right]$$

$$\xrightarrow{-R_2 + R_3} \left[\begin{array}{cc|c} 1 & 0 & 4 \\ 0 & 1 & -3 \\ 0 & 0 & 0 \end{array}\right]$$

The last matrix is reduced and corresponds to the system

$$\begin{cases} x + 0y = 4 \\ 0x + y = -3 \\ 0x + 0y = 0 \end{cases}$$

Since the original system is equivalent to this system, it has a unique solution, namely,

$$x = 4$$
$$y = -3$$

▶ **Now work Problem 13.**

APPLICATIONS IN PRACTICE 2

Solving a System by Reduction

A health spa customizes the diet and vitamin supplements of each of its clients. The spa offers three different vitamin supplements, each containing different percentages of the recommended daily allowance (RDA) of vitamins A, C, and D. One tablet of supplement X provides 40% of the RDA of A, 20% of the RDA of C, and 10% of the RDA of D. One tablet of supplement Y provides 10% of the RDA of A, 10% of the RDA of C, and 30% of the RDA of D. One tablet of supplement Z provides 10% of the RDA of A, 50% of the RDA of C, and 20% of the RDA of D. The spa staff determines that one client should take 180% of the RDA of vitamin A, 200% of the RDA of vitamin C, and 190% of the RDA of vitamin D each day. How many tablets of each supplement should she take each day?

EXAMPLE 4 Solving a System by Reduction

Using matrix reduction, solve

$$\begin{cases} x + 2y + 4z - 6 = 0 \\ 2z + y - 3 = 0 \\ x + y + 2z - 1 = 0 \end{cases}$$

Solution: Rewriting the system so that the variables are aligned and the constant terms appear on the right sides of the equations, we have

$$\begin{cases} x + 2y + 4z = 6 \\ y + 2z = 3 \\ x + y + 2z = 1 \end{cases}$$

Reducing the augmented coefficient matrix, we obtain

$$\left[\begin{array}{ccc|c} 1 & 2 & 4 & 6 \\ 0 & 1 & 2 & 3 \\ 1 & 1 & 2 & 1 \end{array}\right] \xrightarrow{-R_1 + R_3} \left[\begin{array}{ccc|c} 1 & 2 & 4 & 6 \\ 0 & 1 & 2 & 3 \\ 0 & -1 & -2 & -5 \end{array}\right]$$

$$\xrightarrow[(1)R_2 + R_3]{-2R_2 + R_1} \left[\begin{array}{ccc|c} 1 & 0 & 0 & 0 \\ 0 & 1 & 2 & 3 \\ 0 & 0 & 0 & -2 \end{array}\right]$$

$$\xrightarrow{-\frac{1}{2}R_3} \left[\begin{array}{ccc|c} 1 & 0 & 0 & 0 \\ 0 & 1 & 2 & 3 \\ 0 & 0 & 0 & 1 \end{array}\right]$$

$$\xrightarrow{-3R_3 + R_2} \left[\begin{array}{ccc|c} 1 & 0 & 0 & 0 \\ 0 & 1 & 2 & 0 \\ 0 & 0 & 0 & 1 \end{array}\right]$$

The last matrix is reduced and corresponds to

$$\begin{cases} x = 0 \\ y + 2z = 0 \\ 0 = 1 \end{cases}$$

Whenever we get a row with all 0's to the left side of the vertical rule and a nonzero entry to the right, no solution exists.

Since $0 \neq 1$, there are no values of x, y, and z for which all equations are satisfied simultaneously. Thus, the original system has no solution.

▶ **Now work Problem 15.**

APPLICATIONS IN PRACTICE 3

Parametric Form of a Solution

A zoo veterinarian can purchase animal food of four different types: A, B, C, and D. Each food comes in the same size bag, and the number of grams of each of three nutrients in each bag is summarized in the following table:

		Food			
		A	B	C	D
	N_1	5	5	10	5
Nutrient	N_2	10	5	30	10
	N_3	5	15	10	25

For one animal, the veterinarian determines that she needs to combine the bags to get 10,000 g of N_1, 20,000 g of N_2, and 20,000 g of N_3. How many bags of each type of food should she order?

EXAMPLE 5 Parametric Form of a Solution

Using matrix reduction, solve

$$\begin{cases} 2x_1 + 3x_2 + 2x_3 + 6x_4 = 10 \\ \phantom{2x_1 + {}} x_2 + 2x_3 + x_4 = 2 \\ 3x_1 \phantom{{}+3x_2} - 3x_3 + 6x_4 = 9 \end{cases}$$

Solution: Reducing the augmented coefficient matrix, we have

$$\left[\begin{array}{cccc|c} 2 & 3 & 2 & 6 & 10 \\ 0 & 1 & 2 & 1 & 2 \\ 3 & 0 & -3 & 6 & 9 \end{array}\right] \xrightarrow{\frac{1}{2}R_1} \left[\begin{array}{cccc|c} 1 & \frac{3}{2} & 1 & 3 & 5 \\ 0 & 1 & 2 & 1 & 2 \\ 3 & 0 & -3 & 6 & 9 \end{array}\right]$$

$$\xrightarrow{-3R_1 + R_3} \left[\begin{array}{cccc|c} 1 & \frac{3}{2} & 1 & 3 & 5 \\ 0 & 1 & 2 & 1 & 2 \\ 0 & -\frac{9}{2} & -6 & -3 & -6 \end{array}\right]$$

$$\xrightarrow[\frac{9}{2}R_2 + R_3]{-\frac{3}{2}R_2 + R_1} \left[\begin{array}{cccc|c} 1 & 0 & -2 & \frac{3}{2} & 2 \\ 0 & 1 & 2 & 1 & 2 \\ 0 & 0 & 3 & \frac{3}{2} & 3 \end{array}\right]$$

$$\xrightarrow{\frac{1}{3}R_3} \left[\begin{array}{cccc|c} 1 & 0 & -2 & \frac{3}{2} & 2 \\ 0 & 1 & 2 & 1 & 2 \\ 0 & 0 & 1 & \frac{1}{2} & 1 \end{array}\right]$$

$$\xrightarrow[-2R_3 + R_2]{2R_3 + R_1} \left[\begin{array}{cccc|c} 1 & 0 & 0 & \frac{5}{2} & 4 \\ 0 & 1 & 0 & 0 & 0 \\ 0 & 0 & 1 & \frac{1}{2} & 1 \end{array}\right]$$

This matrix is reduced and corresponds to the system

$$\begin{cases} x_1 + \frac{5}{2}x_4 = 4 \\ x_2 = 0 \\ x_3 + \frac{1}{2}x_4 = 1 \end{cases}$$

Thus,

$$x_1 = -\tfrac{5}{2}x_4 + 4 \quad (9)$$

$$x_2 = 0 \quad (10)$$

$$x_3 = -\tfrac{1}{2}x_4 + 1 \quad (11)$$

$$x_4 = x_4 \quad (12)$$

If x_4 is any real number, r, then Equations (9)–(12) determine a particular solution of the original system. For example, if $r = 0$ (that is, $x_4 = 0$), then a *particular* solution is

$$x_1 = 4 \qquad x_2 = 0 \qquad x_3 = 1 \qquad x_4 = 0$$

If $r = 2$, then

$$x_1 = -1 \qquad x_2 = 0 \qquad x_3 = 0 \qquad x_4 = 2$$

Recall (see Examples 3 and 6 of Section 3.4) that the variable r, on which x_1, x_3, and x_4 depend, is called a **parameter.** There are infinitely many solutions of the system—one corresponding to each value of the parameter. We say that the *general* solution of the original system is given by

$$x_1 = -\tfrac{5}{2}r + 4$$

$$x_2 = 0$$

$$x_3 = -\tfrac{1}{2}r + 1$$

$$x_4 = r$$

where r is any real number, and we speak of having a *one-parameter family* of solutions. Now, with matrix addition and scalar multiplication at hand, we can say a little more about such families. Observe that

$$\begin{bmatrix} x_1 \\ x_2 \\ x_3 \\ x_4 \end{bmatrix} = \begin{bmatrix} 4 \\ 0 \\ 1 \\ 0 \end{bmatrix} + r\begin{bmatrix} -\frac{5}{2} \\ 0 \\ -\frac{1}{2} \\ 1 \end{bmatrix}$$

Readers familiar with analytic geometry will see that the solutions form a *line* in $x_1x_2x_3x_4$-space, passing through the *point* $\begin{bmatrix} 4 \\ 0 \\ 1 \\ 0 \end{bmatrix}$ and in the *direction* of the line segment joining $\begin{bmatrix} 0 \\ 0 \\ 0 \\ 0 \end{bmatrix}$ and $\begin{bmatrix} -\frac{5}{2} \\ 0 \\ -\frac{1}{2} \\ 1 \end{bmatrix}$.

▸ **Now work Problem 17.**

Examples 3–5 illustrate the fact that a system of linear equations may have a unique solution, no solution, or infinitely many solutions.

Exercise 6.4

In Problems 1–6, determine whether the matrix is reduced or not reduced.

*1. $\begin{bmatrix} 1 & 2 \\ 3 & 0 \end{bmatrix}$

2. $\begin{bmatrix} 1 & 0 & 0 & 3 \\ 0 & 0 & 1 & 2 \end{bmatrix}$

3. $\begin{bmatrix} 1 & 0 & 0 \\ 0 & 1 & 0 \\ 0 & 0 & 1 \end{bmatrix}$

4. $\begin{bmatrix} 1 & 1 \\ 0 & 1 \\ 0 & 0 \\ 0 & 0 \end{bmatrix}$

5. $\begin{bmatrix} 0 & 0 & 0 & 0 \\ 0 & 1 & 0 & 0 \\ 0 & 0 & 1 & 0 \\ 0 & 0 & 0 & 0 \end{bmatrix}$

6. $\begin{bmatrix} 0 & 0 & 1 \\ 1 & 0 & 3 \\ 0 & 1 & 5 \\ 0 & 0 & 0 \end{bmatrix}$

In Problems 7–12, reduce the given matrix.

7. $\begin{bmatrix} 1 & 3 \\ 4 & 0 \end{bmatrix}$

8. $\begin{bmatrix} 0 & -2 & 0 & 1 \\ 1 & 2 & 0 & 4 \end{bmatrix}$

*9. $\begin{bmatrix} 2 & 4 & 6 \\ 1 & 2 & 3 \\ 1 & 2 & 3 \end{bmatrix}$

10. $\begin{bmatrix} 2 & 3 \\ 1 & -6 \\ 4 & 8 \\ 1 & 7 \end{bmatrix}$

11. $\begin{bmatrix} 2 & 0 & 3 & 1 \\ 1 & 4 & 2 & 2 \\ -1 & 3 & 1 & 4 \\ 0 & 2 & 1 & 0 \end{bmatrix}$

12. $\begin{bmatrix} 0 & 0 & 2 \\ 2 & 0 & 3 \\ 0 & -1 & 0 \\ 0 & 4 & 1 \end{bmatrix}$

Solve the systems in Problems 13–26 by the method of reduction.

*13. $\begin{cases} 3x + 5y = 25 \\ x - 2y = 1 \end{cases}$

14. $\begin{cases} x - 3y = -11 \\ 4x + 3y = 9 \end{cases}$

*15. $\begin{cases} 3x + y = 4 \\ 12x + 4y = 2 \end{cases}$

16. $\begin{cases} x + 2y - 3z = 0 \\ -2x - 4y + 6z = 1 \end{cases}$

*17. $\begin{cases} x + 2y + z - 4 = 0 \\ 3x + 2z - 5 = 0 \end{cases}$

18. $\begin{cases} x + 3y + 2z - 1 = 0 \\ x + y + 5z - 10 = 0 \end{cases}$

19. $\begin{cases} x_1 - 3x_2 = 0 \\ 2x_1 + 2x_2 = 3 \\ 5x_1 - x_2 = 1 \end{cases}$

20. $\begin{cases} x_1 + 3x_2 = 6 \\ 2x_1 + x_2 = 7 \\ x_1 + x_2 = 4 \end{cases}$

21. $\begin{cases} x - y - 3z = -5 \\ 2x - y - 4z = -8 \\ x + y - z = -1 \end{cases}$

22. $\begin{cases} x + y - z = 7 \\ 2x - 3y - 2z = 4 \\ x - y - 5z = 23 \end{cases}$

23. $\begin{cases} 2x - 4z = 8 \\ x - 2y - 2z = 14 \\ x + y - 2z = -1 \\ 3x + y + z = 0 \end{cases}$

24. $\begin{cases} x + 3z = -1 \\ 3x + 2y + 11z = 1 \\ x + y + 4z = 1 \\ 2x - 3y + 3z = -8 \end{cases}$

25. $\begin{cases} x_1 + x_2 - x_3 + x_4 + x_5 = 0 \\ x_1 + x_2 + x_3 - x_4 + x_5 = 0 \\ x_1 - x_2 - x_3 + x_4 - x_5 = 0 \\ x_1 + x_2 - x_3 - x_4 - x_5 = 0 \end{cases}$

26. $\begin{cases} x_1 + x_2 + x_3 - x_4 = 0 \\ x_1 - x_2 - x_3 + x_4 = 0 \\ x_1 + x_2 - x_3 - x_4 = 0 \\ x_1 + x_2 - x_3 + x_4 = 0 \end{cases}$

Solve Problems 27–33 by using matrix reduction.

27. **Taxes** A company has taxable income of \$312,000. The federal tax is 25% of that portion that is left after the state tax has been paid. The state tax is 10% of that portion that is left after the federal tax has been paid. Find the company's federal and state taxes.

28. **Decision Making** A manufacturer produces two products, A and B. For each unit of A sold, the profit is \$8, and for each unit of B sold, the profit is \$11. From past experience, it has been found that 25% more of A can be sold than of B. Next year the manufacturer desires a total profit of \$42,000. How many units of each product must be sold?

29. **Production Scheduling** A manufacturer produces three products: A, B, and C. The profits for each unit sold of A, B, and C are \$1, \$2, and \$3, respectively. Fixed costs are \$17,000 per year, and the costs of producing each unit of A, B, and C are \$4, \$5, and \$7, respectively. Next year, a total of 11,000 units of all three products is to be produced and sold, and a total profit of \$25,000 is to be realized. If total cost is to be \$80,000, how many units of each of the products should be produced next year?

30. **Production Allocation** National Desk Co. has plants for producing desks on both the East and West Coasts. At the East Coast plant, fixed costs are \$20,000 per year and the cost of producing each desk is \$90. At the West Coast plant, fixed costs are \$18,000 per year and the cost of producing each desk is \$95. Next year the company wants to produce a total of 800 desks. Determine the production order for each plant for the forthcoming year if the total cost for each plant is to be the same.

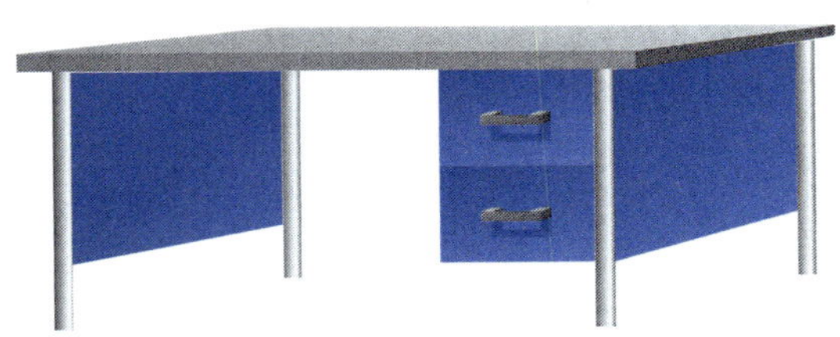

31. **Vitamins** A person is ordered by a doctor to take 10 units of vitamin A, 9 units of vitamin D, and 19 units of vitamin E each day. The person can choose from three brands of vitamin pills. Brand X contains 2 units of vitamin A, 3 units of vitamin D, and 5 units of vitamin E; brand Y has 1, 3, and 4 units, respectively; and brand Z has 1 unit of vitamin A, none of vitamin D, and 1 of vitamin E.

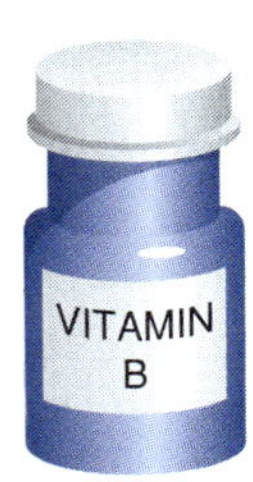

(a) Find all possible combinations of pills that will provide exactly the required amounts of vitamins.
(b) If brand X costs 1 cent a pill, brand Y 6 cents, and brand Z 3 cents, are there any combinations in part (a) costing exactly 15 cents a day?
(c) What is the least expensive combination in part (a)? The most expensive?

32. Production A firm produces three products, A, B, and C, that require processing by three machines, I, II, and III. The time in hours required for processing one unit of each product by the three machines is given by the following table:

	I	II	III
A	3	1	2
B	1	2	1
C	2	4	1

Machine I is available for 850 hours, machine II for 1200 hours, and machine III for 550 hours. Find how many units of each product should be produced to make use of all the available time on the machines.

33. Investments An investment company sells three types of pooled funds, Standard (S), Deluxe (D), and Gold Star (G).

Each unit of S contains 12 shares of stock A, 16 of stock B, and 8 of stock C.

Each unit of D contains 20 shares of stock A, 12 of stock B, and 28 of stock C.

Each unit of G contains 32 shares of stock A, 28 of stock B, and 36 of stock C.

Suppose an investor wishes to purchase exactly 220 shares of stock A, 176 shares of stock B, and 264 shares of stock C by buying units of the three funds.

(a) Determine those combinations of units of S, D, and G that will meet the investor's requirements exactly.
(b) Suppose each unit of S (respectively, D, G) costs the investor \$300 (respectively, \$400, \$600). Which of the combinations from part (a) will minimize the total cost to the investor?

OBJECTIVE *To focus our attention on nonhomogeneous systems that involve more than one parameter in their general solution and to solve, and consider the theory of, homogeneous systems.*

6.5 Solving Systems by Reducing Matrices (continued)[4]

As we saw in Section 6.4, a system of linear equations may have a unique solution, no solution, or infinitely many solutions. When there are infinitely many, the general solution is expressed in terms of at least one parameter. For example, the general solution in Example 5 was given in terms of the parameter r:

$$\begin{aligned} x_1 &= -\tfrac{5}{2}r + 4 \\ x_2 &= 0 \\ x_3 &= -\tfrac{1}{2}r + 1 \\ x_4 &= r \end{aligned}$$

At times, more than one parameter is necessary,[5] as the next example shows.

EXAMPLE 1 Two-Parameter Family of Solutions

Using matrix reduction, solve

$$\begin{cases} x_1 + 2x_2 + 5x_3 + 5x_4 = -3 \\ x_1 + x_2 + 3x_3 + 4x_4 = -1 \\ x_1 - x_2 - x_3 + 2x_4 = 3 \end{cases}$$

[4]This section may be omitted.

[5]See Example 7 of Section 3.4.

Solution: The augmented coefficient matrix is

$$\left[\begin{array}{rrrr|r} 1 & 2 & 5 & 5 & -3 \\ 1 & 1 & 3 & 4 & -1 \\ 1 & -1 & -1 & 2 & 3 \end{array}\right]$$

whose reduced form is

$$\left[\begin{array}{rrrr|r} 1 & 0 & 1 & 3 & 1 \\ 0 & 1 & 2 & 1 & -2 \\ 0 & 0 & 0 & 0 & 0 \end{array}\right]$$

Hence,

$$\begin{cases} x_1 + x_3 + 3x_4 = 1 \\ x_2 + 2x_3 + x_4 = -2 \end{cases}$$

from which it follows that

$$\begin{aligned} x_1 &= 1 - x_3 - 3x_4 \\ x_2 &= -2 - 2x_3 - x_4 \end{aligned}$$

Since no restriction is placed on either x_3 or x_4, they can be arbitrary real numbers, giving us a parametric family of solutions. Setting $x_3 = r$ and $x_4 = s$, we can give the solution of the given system as

$$\begin{aligned} x_1 &= 1 - r - 3s \\ x_2 &= -2 - 2r - s \\ x_3 &= r \\ x_4 &= s \end{aligned}$$

where the parameters r and s can be any real numbers. By assigning specific values to r and s, we get particular solutions. For example, if $r = 1$ and $s = 2$, then the corresponding particular solution is $x_1 = -6$, $x_2 = -6$, $x_3 = 1$, and $x_4 = 2$. As in the one-parameter case we can now go further and write

$$\begin{bmatrix} x_1 \\ x_2 \\ x_3 \\ x_4 \end{bmatrix} = \begin{bmatrix} 1 \\ -2 \\ 0 \\ 0 \end{bmatrix} + r\begin{bmatrix} -1 \\ -2 \\ 1 \\ 0 \end{bmatrix} + s\begin{bmatrix} -3 \\ -1 \\ 0 \\ 1 \end{bmatrix}$$

which can be shown to exhibit the family of solutions as a *plane* through $\begin{bmatrix} 1 \\ -2 \\ 0 \\ 0 \end{bmatrix}$ in $x_1x_2x_3x_4$-space.

▶ **Now work Problem 1.**

It is customary to classify a system of linear equations as being either *homogeneous* or *nonhomogeneous*, depending on whether the constant terms are all zero.

Definition

The system

$$\begin{cases} a_{11}x_1 + a_{12}x_2 + \cdots + a_{1n}x_n = c_1 \\ a_{21}x_1 + a_{22}x_2 + \cdots + a_{2n}x_n = c_2 \\ \vdots \qquad \vdots \qquad \vdots \qquad \vdots \\ a_{m1}x_1 + a_{m2}x_2 + \cdots + a_{mn}x_n = c_m \end{cases}$$

is called a ***homogeneous system*** if $c_1 = c_2 = \cdots = c_m = 0$. The system is a ***nonhomogeneous system*** if at least one of the c's is not equal to 0.

EXAMPLE 2 **Nonhomogeneous and Homogeneous Systems**

The system

$$\begin{cases} 2x + 3y = 4 \\ 3x - 4y = 0 \end{cases}$$

is nonhomogeneous because of the 4 in the top equation. The system

$$\begin{cases} 2x + 3y = 0 \\ 3x - 4y = 0 \end{cases}$$

is homogeneous.

If the homogeneous system

$$\begin{cases} 2x + 3y = 0 \\ 3x - 4y = 0 \end{cases}$$

were solved by the method of reduction, first the augmented coefficient matrix would be written

$$\left[\begin{array}{cc|c} 2 & 3 & 0 \\ 3 & -4 & 0 \end{array}\right]$$

Observe that the last column consists entirely of zeros. This is typical of the augmented coefficient matrix of any homogeneous system. We would then reduce this matrix by using elementary row operations:

$$\left[\begin{array}{cc|c} 2 & 3 & 0 \\ 3 & -4 & 0 \end{array}\right] \to \cdots \to \left[\begin{array}{cc|c} 1 & 0 & 0 \\ 0 & 1 & 0 \end{array}\right]$$

The last column of the reduced matrix also consists only of zeros. This does not occur by chance. When any elementary row operation is performed on a matrix that has a column consisting entirely of zeros, the corresponding column of the resulting matrix will also be all zeros. For convenience, it will be our custom when solving a homogeneous system by matrix reduction to delete the last column of the matrices involved. That is, we shall reduce only the *coefficient matrix* of the system. For the preceding system, we would have

$$\begin{bmatrix} 2 & 3 \\ 3 & -4 \end{bmatrix} \to \cdots \to \begin{bmatrix} 1 & 0 \\ 0 & 1 \end{bmatrix}$$

Here the reduced matrix, called the *reduced coefficient matrix,* corresponds to the system

$$\begin{cases} x + 0y = 0 \\ 0x + y = 0 \end{cases}$$

so the solution is $x = 0$ and $y = 0$.

Let us now consider the number of solutions of the homogeneous system

$$\begin{cases} a_{11}x_1 + a_{12}x_2 + \cdots + a_{1n}x_n = 0 \\ a_{21}x_1 + a_{22}x_2 + \cdots + a_{2n}x_n = 0 \\ \vdots \qquad \vdots \qquad \qquad \vdots \qquad \vdots \\ a_{m1}x_1 + a_{m2}x_2 + \cdots + a_{mn}x_n = 0 \end{cases}$$

One solution always occurs when $x_1 = 0, x_2 = 0, \ldots,$ and $x_n = 0$, since each equation is satisfied for these values. This solution, called the **trivial solution,** is a solution of *every* homogeneous system.

There is a theorem that allows us to determine whether a homogeneous system has a unique solution (the trivial solution only) or infinitely many solutions. The theorem is based on the number of nonzero rows that appear in the reduced coefficient matrix of the system. A *nonzero row* is a row that does not consist entirely of zeros.

Theorem

Let $\mathbf{A}$ be the *reduced* coefficient matrix of a homogeneous system of m linear equations in n unknowns. If $\mathbf{A}$ has exactly k nonzero rows, then $k \leq n$. Moreover,

1. if $k < n$, the system has infinitely many solutions, and
2. if $k = n$, the system has a unique solution (the trivial solution).

If a homogeneous system consists of m equations in n unknowns, then the coefficient matrix of the system has size $m \times n$. Thus, if $m < n$ and k is the number of nonzero rows in the reduced coefficient matrix, then $k \leq m$, and hence, $k < n$. By the foregoing theorem, the system must have infinitely many solutions. Consequently, we have the following corollary.

Corollary

A homogeneous system of linear equations with fewer equations than unknowns has infinitely many solutions.

CAUTION The preceding theorem and corollary apply only to **homogeneous** systems of linear equations. For example, consider the system

$$\begin{cases} x + y - 2z = 3, \\ 2x + 2y - 4z = 4, \end{cases}$$

which consists of two linear equations in three unknowns. We **cannot** conclude that this system has infinitely many solutions, since it is not homogeneous. Indeed, you should verify that it has no solution.

EXAMPLE 3 Number of Solutions of a Homogeneous System

Determine whether the system

$$\begin{cases} x + y - 2z = 0 \\ 2x + 2y - 4z = 0 \end{cases}$$

has a unique solution or infinitely many solutions.

Solution: There are two equations in this homogeneous system, and this number is less than the number of unknowns (three). Thus, by the previous corollary, the system has infinitely many solutions.

▸ **Now work Problem 9.**

APPLICATIONS IN PRACTICE 1

Solving Homogeneous Systems

A plane in three-dimensional space can be written as $ax + by + cz = d$. We can find the possible intersections of planes in this form by writing them as systems of linear equations and using reduction to solve them. If $d = 0$ in each equation, then we have a homogeneous system with either a unique solution or infinitely many solutions. Determine whether the intersection of the planes

$$5x + 3y + 4z = 0$$
$$6x + 8y + 7z = 0$$
$$3x + 1y + 2z = 0$$

has a unique solution or infinitely many solutions; then solve the system.

EXAMPLE 4 **Solving Homogeneous Systems**

Determine whether the following homogeneous systems have a unique solution or infinitely many solutions; then solve the systems.

a. $$\begin{cases} x - 2y + z = 0 \\ 2x - y + 5z = 0 \\ x + y + 4z = 0 \end{cases}$$

Solution: Reducing the coefficient matrix, we have

$$\begin{bmatrix} 1 & -2 & 1 \\ 2 & -1 & 5 \\ 1 & 1 & 4 \end{bmatrix} \rightarrow \cdots \rightarrow \begin{bmatrix} 1 & 0 & 3 \\ 0 & 1 & 1 \\ 0 & 0 & 0 \end{bmatrix}$$

The number of nonzero rows, 2, in the reduced coefficient matrix is less than the number of unknowns, 3, in the system. By the previous theorem, there are infinitely many solutions.

▶ **Now work Problem 13.**

Since the reduced coefficient matrix corresponds to

$$\begin{cases} x + 3z = 0 \\ y + z = 0 \end{cases}$$

the solution may be given in parametric form by

$$x = -3r$$
$$y = -r$$
$$z = r$$

where r is any real number.

b. $$\begin{cases} 3x + 4y = 0 \\ x - 2y = 0 \\ 2x + y = 0 \\ 2x + 3y = 0 \end{cases}$$

Solution: Reducing the coefficient matrix, we have

$$\begin{bmatrix} 3 & 4 \\ 1 & -2 \\ 2 & 1 \\ 2 & 3 \end{bmatrix} \rightarrow \cdots \rightarrow \begin{bmatrix} 1 & 0 \\ 0 & 1 \\ 0 & 0 \\ 0 & 0 \end{bmatrix}$$

The number of nonzero rows (2) in the reduced coefficient matrix equals the number of unknowns in the system. By the theorem, the system must have a unique solution, namely, the trivial solution $x = 0$, $y = 0$.

Exercise 6.5

In Problems 1–8, solve the systems by using matrix reduction.

*__1.__ $$\begin{cases} w + x - y - 9z = -3 \\ 2w + 3x + 2y + 15z = 12 \\ 2w + x + 2y + 5z = 8 \end{cases}$$

2. $$\begin{cases} 3w - x + 12y + 18z = -4 \\ w - 2x + 4y + 11z = -13 \\ w + x + 4y + 2z = 8 \end{cases}$$

3. $$\begin{cases} 3w - x - 3y - z = -2 \\ 2w - 2x - 6y - 6z = -4 \\ 2w - x - 3y - 2z = -2 \\ 3w + x + 3y + 7z = 2 \end{cases}$$

4. $$\begin{cases} w + x + 5z = 1 \\ w \quad y + 2z = 1 \\ w - 3x + 4y - 7z = 1 \\ x - y + 3z = 0 \end{cases}$$

5. $\begin{cases} w + x + 3y - z = 2 \\ 2w + x + 5y - 2z = 0 \\ 2w - x + 3y - 2z = -8 \\ 3w + 2x + 8y - 3z = 2 \\ w + 2y - z = -2 \end{cases}$

6. $\begin{cases} w + x + y + 2z = 4 \\ 2w + x + 2y + 2z = 7 \\ w + 2x + y + 4z = 5 \\ 3w - 2x + 3y - 4z = 7 \\ 4w - 3x + 4y - 6z = 9 \end{cases}$

7. $\begin{cases} 4x_1 - 3x_2 + 5x_3 - 10x_4 + 11x_5 = -8 \\ 2x_1 + x_2 + 5x_3 + 3x_5 = 6 \end{cases}$

8. $\begin{cases} x_1 + 3x_3 + x_4 + 4x_5 = 1 \\ x_2 + x_3 - 2x_4 = 0 \\ 2x_1 - 2x_2 + 3x_3 + 10x_4 + 15x_5 = 10 \\ x_1 + 2x_2 + 3x_3 - 2x_4 + 2x_5 = -2 \end{cases}$

For each of Problems 9–14, determine whether the system has infinitely many solutions or only the trivial solution. Do not solve the systems.

*9. $\begin{cases} 0.07x + 0.3y + 0.02z = 0 \\ 0.053x - 0.4y + 0.08z = 0 \end{cases}$

10. $\begin{cases} 3w + 5x - 4y + 2z = 0 \\ 7w - 2x + 9y + 3z = 0 \end{cases}$

11. $\begin{cases} 3x - 4y = 0 \\ x + 5y = 0 \\ 4x - y = 0 \end{cases}$

12. $\begin{cases} 2x + 3y + 12z = 0 \\ 3x - 2y + 5z = 0 \\ 4x + y + 14z = 0 \end{cases}$

*13. $\begin{cases} x + y + z = 0 \\ x - z = 0 \\ x - 2y - 5z = 0 \end{cases}$

14. $\begin{cases} 3x + 2y - 2z = 0 \\ 2x + 2y - 2z = 0 \\ 3x - 2y + 3z = 0 \end{cases}$

Solve each of the following systems.

15. $\begin{cases} x + y = 0 \\ 3x - 4y = 0 \end{cases}$

16. $\begin{cases} 2x - 5y = 0 \\ 8x - 20y = 0 \end{cases}$

17. $\begin{cases} x + 6y - 2z = 0 \\ 2x - 3y + 4z = 0 \end{cases}$

18. $\begin{cases} 4x + 7y = 0 \\ 2x + 3y = 0 \end{cases}$

19. $\begin{cases} x + y = 0 \\ 3x - 4y = 0 \\ 5x - 8y = 0 \end{cases}$

20. $\begin{cases} 2x - 3y + z = 0 \\ x + 2y - z = 0 \\ x + y + z = 0 \end{cases}$

21. $\begin{cases} x + y + z = 0 \\ 5x - 2y - 9z = 0 \\ 3x + y - z = 0 \\ 3x - 2y - 7z = 0 \end{cases}$

22. $\begin{cases} x + y + 7z = 0 \\ x - y - z = 0 \\ 2x - 3y - 6z = 0 \\ 3x + y + 13z = 0 \end{cases}$

23. $\begin{cases} w + x + y + 4z = 0 \\ w + x + 5z = 0 \\ 2w + x + 3y + 4z = 0 \\ w - 3x + 2y - 9z = 0 \end{cases}$

24. $\begin{cases} w + x + 2y + 7z = 0 \\ w - 2x - y + z = 0 \\ w + 2x + 3y + 9z = 0 \\ 2w - 3x - y + 4z = 0 \end{cases}$

OBJECTIVE *To determine the inverse of an invertible matrix and to use inverses to solve systems.*

6.6 Inverses

We have seen how useful the method of reduction is for solving systems of linear equations. But it is by no means the only method that uses matrices. In this section, we shall discuss a different method that applies to certain systems of n linear equations in n unknowns.

In Section 6.3, we showed how a system of linear equations can be written in matrix form as the single matrix equation $\mathbf{AX} = \mathbf{B}$, where $\mathbf{A}$ is the coefficient matrix. For example, the system

$$\begin{cases} x_1 + 2x_2 = 3 \\ x_1 - x_2 = 1 \end{cases}$$

can be written in the matrix form $\mathbf{AX} = \mathbf{B}$, where

$$\mathbf{A} = \begin{bmatrix} 1 & 2 \\ 1 & -1 \end{bmatrix} \qquad \mathbf{X} = \begin{bmatrix} x_1 \\ x_2 \end{bmatrix} \qquad \mathbf{B} = \begin{bmatrix} 3 \\ 1 \end{bmatrix}$$

If we can determine the values of the entries in the unknown matrix $\mathbf{X}$, we have a solution of the system. Thus, we would like to find a method to solve the matrix equation $\mathbf{AX} = \mathbf{B}$ for $\mathbf{X}$. Some motivation is provided by looking at the procedure for solving the algebraic equation $ax = b$. The latter equation is solved by simply multiplying both sides by the multiplicative inverse of a. [Recall that the multiplicative inverse of a nonzero number a is denoted a^{-1} (which is $1/a$) and has the property that $a^{-1}a = 1$.] For example, if $3x = 11$, then

$$3^{-1}(3x) = 3^{-1}(11) \quad \text{so} \quad x = \frac{11}{3}$$

If we can apply a similar procedure to the *matrix* equation

$$\mathbf{AX} = \mathbf{B} \tag{1}$$

then we need a multiplicative inverse of **A**—that is, a matrix **C** such that $\mathbf{CA} = \mathbf{I}$. Then we can simply multiply both sides of Equation (1) by **C**:

$$\mathbf{C}(\mathbf{AX}) = \mathbf{CB}$$
$$(\mathbf{CA})\mathbf{X} = \mathbf{CB}$$
$$\mathbf{IX} = \mathbf{CB}$$
$$\mathbf{X} = \mathbf{CB} \tag{2}$$

Thus, the solution is $\mathbf{X} = \mathbf{CB}$. Of course, this method is based on the existence of a matrix **C** such that $\mathbf{CA} = \mathbf{I}$. When such a matrix does exist, we say that it is an *inverse matrix* (or simply an *inverse*) of **A**.

Definition

If **A** is a square matrix and there exists a matrix **C** such that $\mathbf{CA} = \mathbf{I}$, then **C** is called an inverse of **A**, and **A** is said to be ***invertible.***

APPLICATIONS IN PRACTICE 1

Inverse of a Matrix

Secret messages can be encoded by using a code and an encoding matrix. Suppose we have the following code:

a	*b*	*c*	*d*	*e*	*f*	*g*	*h*	*i*	*j*	*k*	*l*	*m*
1	2	3	4	5	6	7	8	9	10	11	12	13

n	*o*	*p*	*q*	*r*	*s*	*t*	*u*	*v*	*w*	*x*	*y*	*z*
14	15	16	17	18	19	20	21	22	23	24	25	26

Let the encoding matrix be **E**. Then we can encode a message by taking every two letters of the message, converting them to their corresponding numbers, creating a 2×1 matrix, and then multiply each matrix by **E**. The message may be unscrambled with a decoding matrix that is the inverse of the coding matrix—that is, $\mathbf{E}^{-1}$. Determine whether the encoding matrices

$$\begin{bmatrix} 1 & 3 \\ 2 & 4 \end{bmatrix} \quad \text{and} \quad \begin{bmatrix} -2 & 1.5 \\ 1 & -0.5 \end{bmatrix}$$

are inverses of each other.

EXAMPLE 1 Inverse of a Matrix

Let $\mathbf{A} = \begin{bmatrix} 1 & 2 \\ 3 & 7 \end{bmatrix}$ and $\mathbf{C} = \begin{bmatrix} 7 & -2 \\ -3 & 1 \end{bmatrix}$. Since

$$\mathbf{CA} = \begin{bmatrix} 7 & -2 \\ -3 & 1 \end{bmatrix}\begin{bmatrix} 1 & 2 \\ 3 & 7 \end{bmatrix} = \begin{bmatrix} 1 & 0 \\ 0 & 1 \end{bmatrix} = \mathbf{I}$$

matrix **C** is an inverse of **A**.

It can be shown that an invertible matrix has one and only one inverse; that is, an inverse is unique. Thus, in Example 1, matrix **C** is the *only* matrix such that $\mathbf{CA} = \mathbf{I}$. For this reason, we can speak of *the* inverse of an invertible matrix **A**, which we denote by the symbol $\mathbf{A}^{-1}$. Accordingly, $\mathbf{A}^{-1}\mathbf{A} = \mathbf{I}$. Moreover, although matrix multiplication is not generally commutative, it is a fact that $\mathbf{A}^{-1}$ commutes with **A**:

$$\mathbf{A}^{-1}\mathbf{A} = \mathbf{A}\mathbf{A}^{-1} = \mathbf{I}.$$

Returning to the matrix equation $\mathbf{AX} = \mathbf{B}$, from Equation (2) we can now state the following:

If **A** is an invertible matrix, then the matrix equation $\mathbf{AX} = \mathbf{B}$ has the unique solution $\mathbf{X} = \mathbf{A}^{-1}\mathbf{B}$.

Probably the idea of an inverse matrix has an air of déjà vu about it. In Section 2.4 we discussed inverse functions and these can be used to further understand inverse matrices. First, we note that the idea of a function is not limited to the world of numbers. For example, if we have a rule that applies to matrices and gives, for each of certain matrices $\mathbf{X}$ a unique, well-defined matrix $f(\mathbf{X})$, then we would say that f is a function that takes matrices to matrices. To be a little more specific, suppose that $\mathbf{A}$ is an $m \times n$ matrix and define $f(\mathbf{X}) = \mathbf{AX}$, for $\mathbf{X}$ any $n \times 1$ matrix. Then this rule f provides a function from the set of $n \times 1$ matrices to the set of $m \times 1$ matrices. If $m = n$, it can be shown that the function given by $f(\mathbf{X}) = \mathbf{AX}$ has an inverse in the sense of Section 2.4 if and only if $\mathbf{A}$ has an inverse matrix $\mathbf{A}^{-1}$, in which case $f^{-1}(\mathbf{X}) = \mathbf{A}^{-1}\mathbf{X}$. If f is a function, in the most general possible sense, that has an inverse then any equation of the form $f(x) = b$ has a unique solution, namely $x = f^{-1}(b)$.

APPLICATIONS IN PRACTICE 2

Using the Inverse to Solve a System

Suppose the encoding matrix $\mathbf{E} = \begin{bmatrix} 1 & 3 \\ 2 & 4 \end{bmatrix}$ was used to encode a message. Use the code from Applications in Practice 1 and the inverse $\mathbf{E}^{-1} = \begin{bmatrix} -2 & 1.5 \\ 1 & -0.5 \end{bmatrix}$ to decode the message, broken into the following pieces:

28, 46, 65, 90

61, 82

59, 88, 57, 86

60, 84, 21, 34, 76, 102

EXAMPLE 2 **Using the Inverse to Solve a System**

Solve the system

$$\begin{cases} x_1 + 2x_2 = 5 \\ 3x_1 + 7x_2 = 18 \end{cases}$$

Solution: In matrix form, we have $\mathbf{AX} = \mathbf{B}$, where

$$\mathbf{A} = \begin{bmatrix} 1 & 2 \\ 3 & 7 \end{bmatrix} \qquad \mathbf{X} = \begin{bmatrix} x_1 \\ x_2 \end{bmatrix} \qquad \mathbf{B} = \begin{bmatrix} 5 \\ 18 \end{bmatrix}$$

In Example 1, we showed that

$$\mathbf{A}^{-1} = \begin{bmatrix} 7 & -2 \\ -3 & 1 \end{bmatrix}$$

Therefore,

$$\mathbf{X} = \mathbf{A}^{-1}\mathbf{B} = \begin{bmatrix} 7 & -2 \\ -3 & 1 \end{bmatrix} \begin{bmatrix} 5 \\ 18 \end{bmatrix} = \begin{bmatrix} -1 \\ 3 \end{bmatrix}$$

so $x_1 = -1$ and $x_2 = 3$.

In order that the method of Example 2 apply to a system, two conditions must be met:

1. The system must have the same number of equations as there are unknowns.
2. The coefficient matrix must be invertible.

As far as condition 2 is concerned, we caution that not all square matrices are invertible. For example, if

$$\mathbf{A} = \begin{bmatrix} 0 & 1 \\ 0 & 1 \end{bmatrix}$$

then

$$\begin{bmatrix} a & b \\ c & d \end{bmatrix} \begin{bmatrix} 0 & 1 \\ 0 & 1 \end{bmatrix} = \begin{bmatrix} 0 & a+b \\ 0 & c+d \end{bmatrix} \neq \begin{bmatrix} 1 & 0 \\ 0 & 1 \end{bmatrix}$$

Hence, there is no matrix that, when postmultiplied by **A**, yields the identity matrix. Thus, **A** is not invertible.

Before discussing a procedure for finding the inverse of an invertible matrix, we introduce the concept of *elementary matrices*. An $n \times n$ **elementary matrix** is a matrix obtained from the $n \times n$ identity matrix **I** by an elementary row operation. There are three basic types of elementary matrices:

Elementary Matrices

1. One obtained by interchanging two rows of **I**
2. One obtained by multiplying any row of **I** by a nonzero scalar
3. One obtained by adding a constant multiple of one row of **I** to another row

EXAMPLE 3 **Elementary Matrices**

The matrices

$$\mathbf{E}_1 = \begin{bmatrix} 1 & 0 & 0 \\ 0 & 0 & 1 \\ 0 & 1 & 0 \end{bmatrix} \qquad \mathbf{E}_2 = \begin{bmatrix} -4 & 0 \\ 0 & 1 \end{bmatrix} \qquad \mathbf{E}_3 = \begin{bmatrix} 1 & 0 \\ 3 & 1 \end{bmatrix}$$

are elementary matrices. $\mathbf{E}_1$ is obtained from the 3×3 identity matrix by interchanging the second and third rows. $\mathbf{E}_2$ is obtained from the 2×2 identity matrix by multiplying the first row by -4. $\mathbf{E}_3$ is obtained from the 2×2 identity matrix by adding 3 times the first row to the second.

Suppose that **E** is an $n \times n$ elementary matrix obtained from **I** by a certain elementary row operation and **A** is an $n \times n$ matrix. Then it can be shown that the product **EA** is equal to the matrix that is obtained from **A** by applying the same elementary row operation to **A**. For example, let

$$\mathbf{A} = \begin{bmatrix} 1 & 2 \\ 3 & 4 \end{bmatrix} \qquad \mathbf{E}_1 = \begin{bmatrix} 0 & 1 \\ 1 & 0 \end{bmatrix}$$

$$\mathbf{E}_2 = \begin{bmatrix} 1 & 0 \\ 0 & 2 \end{bmatrix} \qquad \mathbf{E}_3 = \begin{bmatrix} 1 & -2 \\ 0 & 1 \end{bmatrix}$$

Observe that $\mathbf{E}_1$, $\mathbf{E}_2$, and $\mathbf{E}_3$ are elementary matrices. $\mathbf{E}_1$ is obtained by interchanging the first and second rows of **I**. Likewise, the product

$$\mathbf{E}_1\mathbf{A} = \begin{bmatrix} 0 & 1 \\ 1 & 0 \end{bmatrix} \begin{bmatrix} 1 & 2 \\ 3 & 4 \end{bmatrix} = \begin{bmatrix} 3 & 4 \\ 1 & 2 \end{bmatrix}$$

is the matrix obtained from **A** by interchanging the first and second rows of **A**. Matrix $\mathbf{E}_2$ is obtained by multiplying the second row of **I** by 2. Accordingly, the product

$$\mathbf{E}_2\mathbf{A} = \begin{bmatrix} 1 & 0 \\ 0 & 2 \end{bmatrix} \begin{bmatrix} 1 & 2 \\ 3 & 4 \end{bmatrix} = \begin{bmatrix} 1 & 2 \\ 6 & 8 \end{bmatrix}$$

is the matrix obtained by multiplying the second row of **A** by 2. Matrix $\mathbf{E}_3$ is obtained by adding -2 times the second row of **I** to the first row. The product

$$\mathbf{E}_3\mathbf{A} = \begin{bmatrix} 1 & -2 \\ 0 & 1 \end{bmatrix} \begin{bmatrix} 1 & 2 \\ 3 & 4 \end{bmatrix} = \begin{bmatrix} -5 & -6 \\ 3 & 4 \end{bmatrix}$$

is the matrix obtained from **A** by the same elementary row operation.

If we wanted to reduce the matrix

$$\mathbf{A} = \begin{bmatrix} 1 & 0 \\ 2 & 2 \end{bmatrix}$$

we might proceed through a sequence of steps as follows:

$$\begin{bmatrix} 1 & 0 \\ 2 & 2 \end{bmatrix} \xrightarrow{-2R_1 + R_2} \begin{bmatrix} 1 & 0 \\ 0 & 2 \end{bmatrix}$$

$$\xrightarrow{\frac{1}{2}R_2} \begin{bmatrix} 1 & 0 \\ 0 & 1 \end{bmatrix}$$

Note that **A** reduces to **I**. Since our reduction procedure involves elementary row operations, it seems natural that elementary matrices can be used to reduce **A**. If **A** is premultiplied by the elementary matrix $\mathbf{E}_1 = \begin{bmatrix} 1 & 0 \\ -2 & 1 \end{bmatrix}$, then $\mathbf{E}_1\mathbf{A}$ is the matrix obtained from **A** by adding -2 times the first row to the second row:

$$\mathbf{E}_1\mathbf{A} = \begin{bmatrix} 1 & 0 \\ -2 & 1 \end{bmatrix} \begin{bmatrix} 1 & 0 \\ 2 & 2 \end{bmatrix} = \begin{bmatrix} 1 & 0 \\ 0 & 2 \end{bmatrix}$$

Premultiplying $\mathbf{E}_1\mathbf{A}$ by the elementary matrix $\mathbf{E}_2 = \begin{bmatrix} 1 & 0 \\ 0 & \frac{1}{2} \end{bmatrix}$ gives the matrix obtained by multiplying the second row of $\mathbf{E}_1\mathbf{A}$ by $\frac{1}{2}$:

$$\mathbf{E}_2(\mathbf{E}_1\mathbf{A}) = \begin{bmatrix} 1 & 0 \\ 0 & \frac{1}{2} \end{bmatrix} \begin{bmatrix} 1 & 0 \\ 0 & 2 \end{bmatrix} = \begin{bmatrix} 1 & 0 \\ 0 & 1 \end{bmatrix} = \mathbf{I}$$

Thus, we have reduced **A** by multiplying it by a product of elementary matrices.

Since $(\mathbf{E}_2\mathbf{E}_1)\mathbf{A} = \mathbf{E}_2(\mathbf{E}_1\mathbf{A}) = \mathbf{I}$, the product $\mathbf{E}_2\mathbf{E}_1$ is $\mathbf{A}^{-1}$. However,

$$\mathbf{A}^{-1} = \mathbf{E}_2\mathbf{E}_1 = (\mathbf{E}_2\mathbf{E}_1)\mathbf{I} = \mathbf{E}_2(\mathbf{E}_1\mathbf{I})$$

Consequently, $\mathbf{A}^{-1}$ can be obtained by applying the same elementary row operations, beginning with **I**, that were used to reduce **A** to **I**:

$$\begin{bmatrix} 1 & 0 \\ 0 & 1 \end{bmatrix} \xrightarrow{-2R_1 + R_2} \begin{bmatrix} 1 & 0 \\ -2 & 1 \end{bmatrix}$$

$$\xrightarrow{\frac{1}{2}R_2} \begin{bmatrix} 1 & 0 \\ -1 & \frac{1}{2} \end{bmatrix}$$

Therefore,

$$\mathbf{A}^{-1} = \begin{bmatrix} 1 & 0 \\ -1 & \frac{1}{2} \end{bmatrix}$$

Our result can be verified by showing that $\mathbf{A}^{-1}\mathbf{A} = \mathbf{I}$:

$$\mathbf{A}^{-1}\mathbf{A} = \begin{bmatrix} 1 & 0 \\ -1 & \frac{1}{2} \end{bmatrix} \begin{bmatrix} 1 & 0 \\ 2 & 2 \end{bmatrix} = \begin{bmatrix} 1 & 0 \\ 0 & 1 \end{bmatrix} = \mathbf{I}$$

In summary, to find $\mathbf{A}^{-1}$, we apply the identical elementary row operations, beginning with **I** and proceeding in the same order, as those that were used to reduce **A** to **I**. Finding $\mathbf{A}^{-1}$ by this technique can be done conveniently by using the following format. First, we write the matrix

$$[\mathbf{A} \mid \mathbf{I}] = \left[\begin{array}{cc|cc} 1 & 0 & 1 & 0 \\ 2 & 2 & 0 & 1 \end{array}\right]$$

Then we apply elementary row operations until $[\mathbf{A} \mid \mathbf{I}]$ is equivalent to a matrix that has **I** as its first two columns. The last two columns of this matrix will be $\mathbf{A}^{-1}$. Thus,

$$[\mathbf{A} \mid \mathbf{I}] = \left[\begin{array}{cc|cc} 1 & 0 & 1 & 0 \\ 2 & 2 & 0 & 1 \end{array}\right] \to \left[\begin{array}{cc|cc} 1 & 0 & 1 & 0 \\ 0 & 2 & -2 & 1 \end{array}\right]$$

$$\to \left[\begin{array}{cc|cc} 1 & 0 & 1 & 0 \\ 0 & 1 & -1 & \frac{1}{2} \end{array}\right] = [\mathbf{I} \mid \mathbf{A}^{-1}]$$

Note that the first two columns of $[\mathbf{I} \mid \mathbf{A}^{-1}]$ form a reduced matrix.

This procedure can be extended to find the inverse of *any* invertible matrix:

A matrix is invertible if and only if it is equivalent to the identity matrix.

Method to Find the Inverse of a Matrix

If **M** *is an invertible* $n \times n$ *matrix, form the* $n \times (2n)$ *matrix* $[\mathbf{M} \mid \mathbf{I}]$. *Then perform elementary row operations until the first n columns form a reduced matrix equal to* **I**. *The last n columns will be* $\mathbf{M}^{-1}$. *Symbolically,*

$$[\mathbf{M} \mid \mathbf{I}] \to \cdots \to [\mathbf{I} \mid \mathbf{M}^{-1}]$$

If a matrix **M** *does not reduce to* **I**, *then* $\mathbf{M}^{-1}$ *does not exist.*

APPLICATIONS IN PRACTICE 3

Finding the Inverse of a Matrix

We could extend the encoding scheme used in Applications in Practice 1 to a 3×3 matrix, encoding three letters of a message at a time. Find the inverses of the following 3×3 encoding matrices:

$$\mathbf{E} = \begin{bmatrix} 3 & 1 & 2 \\ 2 & 2 & 2 \\ 2 & 1 & 3 \end{bmatrix} \quad \mathbf{F} = \begin{bmatrix} 2 & 1 & 2 \\ 3 & 2 & 3 \\ 4 & 3 & 4 \end{bmatrix}$$

EXAMPLE 4 **Finding the Inverse of a Matrix**

Determine $\mathbf{A}^{-1}$ *if* **A** *is invertible.*

a. $\mathbf{A} = \begin{bmatrix} 1 & 0 & -2 \\ 4 & -2 & 1 \\ 1 & 2 & -10 \end{bmatrix}$.

Solution: Following the foregoing procedure, we have

$$[\mathbf{A} \mid \mathbf{I}] = \left[\begin{array}{ccc|ccc} 1 & 0 & -2 & 1 & 0 & 0 \\ 4 & -2 & 1 & 0 & 1 & 0 \\ 1 & 2 & -10 & 0 & 0 & 1 \end{array}\right]$$

$$\xrightarrow[-1R_1 + R_3]{-4R_1 + R_2} \left[\begin{array}{ccc|ccc} 1 & 0 & -2 & 1 & 0 & 0 \\ 0 & -2 & 9 & -4 & 1 & 0 \\ 0 & 2 & -8 & -1 & 0 & 1 \end{array}\right]$$

$$\xrightarrow{-\frac{1}{2}R_2}\left[\begin{array}{ccc|ccc}1 & 0 & -2 & 1 & 0 & 0\\ 0 & 1 & -\frac{9}{2} & 2 & -\frac{1}{2} & 0\\ 0 & 2 & -8 & -1 & 0 & 1\end{array}\right]$$

$$\xrightarrow{-2R_2+R_3}\left[\begin{array}{ccc|ccc}1 & 0 & -2 & 1 & 0 & 0\\ 0 & 1 & -\frac{9}{2} & 2 & -\frac{1}{2} & 0\\ 0 & 0 & 1 & -5 & 1 & 1\end{array}\right]$$

$$\xrightarrow[\frac{9}{2}R_3+R_2]{2R_3+R_1}\left[\begin{array}{ccc|ccc}1 & 0 & 0 & -9 & 2 & 2\\ 0 & 1 & 0 & -\frac{41}{2} & 4 & \frac{9}{2}\\ 0 & 0 & 1 & -5 & 1 & 1\end{array}\right]$$

The first three columns of the last matrix form **I**. Thus, **A** is invertible and

$$\mathbf{A}^{-1} = \begin{bmatrix}-9 & 2 & 2\\ -\frac{41}{2} & 4 & \frac{9}{2}\\ -5 & 1 & 1\end{bmatrix}$$

b. $\mathbf{A} = \begin{bmatrix}3 & 2\\ 6 & 4\end{bmatrix}$.

Solution: We have

$$[\mathbf{A}\,|\,\mathbf{I}] = \left[\begin{array}{cc|cc}3 & 2 & 1 & 0\\ 6 & 4 & 0 & 1\end{array}\right]\xrightarrow{-2R_1+R_2}\left[\begin{array}{cc|cc}3 & 2 & 1 & 0\\ 0 & 0 & -2 & 1\end{array}\right]$$

$$\xrightarrow{\frac{1}{3}R_1}\left[\begin{array}{cc|cc}1 & \frac{2}{3} & \frac{1}{3} & 0\\ 0 & 0 & -2 & 1\end{array}\right]$$

The first two columns of the last matrix form a reduced matrix different from **I**. Thus, **A** is not invertible.

▸ **Now work Problem 1.**

TECHNOLOGY

Finding the inverse of an invertible matrix with a graphing calculator can be a real timesaver. Figure 6.5 shows the inverse of

$$\mathbf{A} = \begin{bmatrix}3 & 2\\ 1 & 4\end{bmatrix}$$

Moreover, on a TI-83 Plus, we can display our answer with fractional entries.

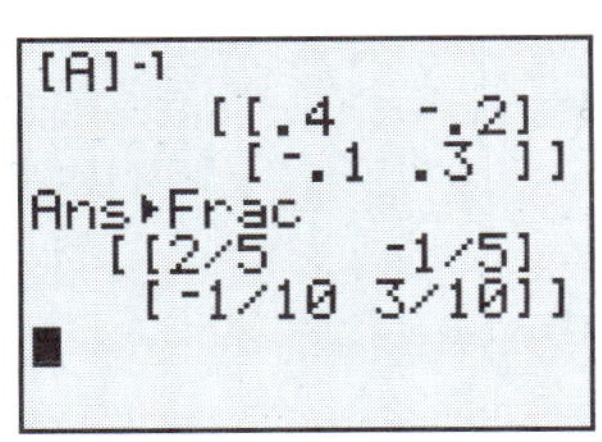

FIGURE 6.5 Inverse of **A** with decimal entries and with fractional entries.

Now we shall solve a system by using the inverse.

APPLICATIONS IN PRACTICE 4

Using the Inverse to Solve a System

A group of investors has \$500,000 to invest in the stocks of three companies. Company A sells for \$50 a share and has an expected growth of 13% per year. Company B sells for \$20 per share and has an expected growth of 15% per year. Company C sells for \$80 a share and has an expected growth of 10% per year. The group plans to buy twice as many shares of Company A as of Company C. If the group's goal is 12% growth per year, how many shares of each stock should the investors buy?

EXAMPLE 5 Using the Inverse to Solve a System

Solve the system

$$\begin{cases} x_1 \phantom{{}-2x_2} - 2x_3 = 1 \\ 4x_1 - 2x_2 + x_3 = 2 \\ x_1 + 2x_2 - 10x_3 = -1 \end{cases}$$

by finding the inverse of the coefficient matrix.

Solution: In matrix form the system is $\mathbf{AX} = \mathbf{B}$, where

$$\mathbf{A} = \begin{bmatrix} 1 & 0 & -2 \\ 4 & -2 & 1 \\ 1 & 2 & -10 \end{bmatrix}$$

is the coefficient matrix. From Example 4(a),

$$\mathbf{A}^{-1} = \begin{bmatrix} -9 & 2 & 2 \\ -\frac{41}{2} & 4 & \frac{9}{2} \\ -5 & 1 & 1 \end{bmatrix}$$

The solution is given by $\mathbf{X} = \mathbf{A}^{-1}\mathbf{B}$:

$$\begin{bmatrix} x_1 \\ x_2 \\ x_3 \end{bmatrix} = \begin{bmatrix} -9 & 2 & 2 \\ -\frac{41}{2} & 4 & \frac{9}{2} \\ -5 & 1 & 1 \end{bmatrix} \begin{bmatrix} 1 \\ 2 \\ -1 \end{bmatrix} = \begin{bmatrix} -7 \\ -17 \\ -4 \end{bmatrix}$$

so $x_1 = -7$, $x_2 = -17$, and $x_3 = -4$.

It can be shown that a system of n linear equations in n unknowns has a unique solution if and only if the coefficient matrix is invertible. Indeed, in the previous example the coefficient matrix is invertible, and a unique solution does in fact exist. When the coefficient matrix is not invertible, the system will have either no solution or infinitely many solutions.

EXAMPLE 6 A Coefficient Matrix That Is Not Invertible

Solve the system

$$\begin{cases} x - 2y + z = 0 \\ 2x - y + 5z = 0 \\ x + y + 4z = 0 \end{cases}$$

Solution: The coefficient matrix is

$$\begin{bmatrix} 1 & -2 & 1 \\ 2 & -1 & 5 \\ 1 & 1 & 4 \end{bmatrix}$$

Since

$$\left[\begin{array}{ccc|ccc} 1 & -2 & 1 & 1 & 0 & 0 \\ 2 & -1 & 5 & 0 & 1 & 0 \\ 1 & 1 & 4 & 0 & 0 & 1 \end{array}\right] \to \cdots \to \left[\begin{array}{ccc|ccc} 1 & 0 & 3 & -\frac{1}{3} & \frac{2}{3} & 0 \\ 0 & 1 & 1 & -\frac{2}{3} & \frac{1}{3} & 0 \\ 0 & 0 & 0 & 1 & -1 & 1 \end{array}\right]$$

the coefficient matrix is not invertible. Hence, the system *cannot* be solved by inverses. Instead, another method must be used. In Example 4(a) of Section 6.5, the solution was found to be $x = -3r$, $y = -r$, and $z = r$, where r is any real number (thus providing infinitely many solutions).

▶ **Now work Problem 7.**

TECHNOLOGY

To solve the system

$$\begin{cases} 3x + 2y = 6 \\ x + 4y = -8 \end{cases}$$

with a graphing calculator, we enter the coefficient matrix as [A] and the column matrix of constants as [B]. The product $[A]^{-1}$ [B] in Figure 6.6 gives the solution $x = 4$, $y = -3$.

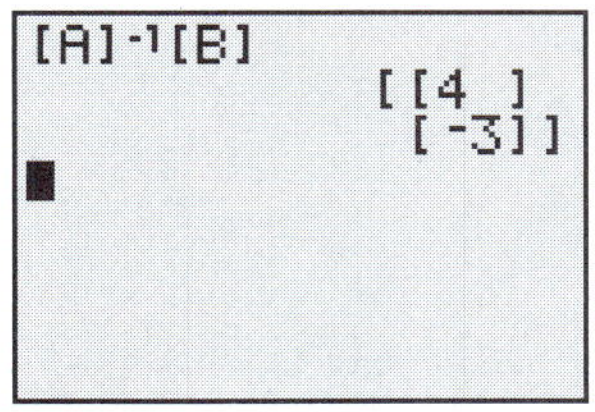

FIGURE 6.6 $[A]^{-1}$ [B] gives solution $x = 4$, $y = -3$ to system of equations.

Exercise 6.6

In Problems 1–18, if the given matrix is invertible, find its inverse.

*1. $\begin{bmatrix} 6 & 1 \\ 7 & 1 \end{bmatrix}$

2. $\begin{bmatrix} 2 & 4 \\ 3 & 6 \end{bmatrix}$

3. $\begin{bmatrix} 1 & 1 \\ 1 & 1 \end{bmatrix}$

4. $\begin{bmatrix} \frac{1}{4} & \frac{3}{8} \\ 0 & -\frac{1}{6} \end{bmatrix}$

5. $\begin{bmatrix} 1 & 0 & 0 \\ 0 & -3 & 0 \\ 0 & 0 & 4 \end{bmatrix}$

6. $\begin{bmatrix} 2 & 0 & 8 \\ -1 & 4 & 0 \\ 2 & 1 & 0 \end{bmatrix}$

*7. $\begin{bmatrix} 1 & 2 & 3 \\ 0 & 0 & 4 \\ 0 & 0 & 5 \end{bmatrix}$

8. $\begin{bmatrix} 2 & 0 & 0 \\ 0 & 0 & 0 \\ 0 & 0 & -4 \end{bmatrix}$

9. $\begin{bmatrix} 1 & 2 \\ 2 & 3 \\ 2 & 5 \end{bmatrix}$

10. $\begin{bmatrix} 0 & 0 & 0 \\ 0 & 0 & 0 \\ 0 & 0 & 0 \end{bmatrix}$

11. $\begin{bmatrix} 1 & 1 & 1 \\ 0 & 1 & 1 \\ 0 & 0 & 1 \end{bmatrix}$

12. $\begin{bmatrix} 1 & 2 & -1 \\ 0 & 1 & 4 \\ 1 & -1 & 2 \end{bmatrix}$

13. $\begin{bmatrix} 7 & 0 & -2 \\ 0 & 1 & 0 \\ -3 & 0 & 1 \end{bmatrix}$

14. $\begin{bmatrix} 2 & 3 & -1 \\ 1 & 2 & 1 \\ -1 & -1 & 3 \end{bmatrix}$

15. $\begin{bmatrix} 2 & 1 & 0 \\ 4 & -1 & 5 \\ 1 & -1 & 2 \end{bmatrix}$

16. $\begin{bmatrix} -5 & 4 & -3 \\ 10 & -7 & 6 \\ 8 & -6 & 5 \end{bmatrix}$

17. $\begin{bmatrix} 1 & 2 & 3 \\ 1 & 3 & 5 \\ 1 & 5 & 12 \end{bmatrix}$

18. $\begin{bmatrix} 2 & -1 & 3 \\ 0 & 2 & 0 \\ 2 & 1 & 1 \end{bmatrix}$

19. Solve $\mathbf{AX} = \mathbf{B}$ if

$$\mathbf{A}^{-1} = \begin{bmatrix} 1 & 2 \\ 8 & 1 \end{bmatrix} \quad \text{and} \quad \mathbf{B} = \begin{bmatrix} 2 \\ 4 \end{bmatrix}.$$

20. Solve $\mathbf{AX} = \mathbf{B}$ if

$$\mathbf{A}^{-1} = \begin{bmatrix} 1 & 0 & 1 \\ 0 & 3 & 0 \\ 2 & 0 & 4 \end{bmatrix} \quad \text{and} \quad \mathbf{B} = \begin{bmatrix} 10 \\ 2 \\ -1 \end{bmatrix}.$$

For Problems 21–34, if the coefficient matrix of the system is invertible, solve the system by using the inverse. If not, solve the system by the method of reduction.

21. $\begin{cases} 6x + 5y = 2 \\ x + y = -3 \end{cases}$

22. $\begin{cases} 2x + 4y = 5 \\ -x + 3y = -2 \end{cases}$

23. $\begin{cases} 2x + y = 5 \\ 3x - y = 0 \end{cases}$

24. $\begin{cases} 3x + 2y = 26 \\ 4x + 3y = 37 \end{cases}$

25. $\begin{cases} 2x + 6y = 2 \\ 3x + 9y = 3 \end{cases}$

26. $\begin{cases} 2x + 6y = 8 \\ 3x + 9y = 7 \end{cases}$

27. $\begin{cases} x + 2y + z = 4 \\ 3x + \quad z = 2 \\ x - y + z = 1 \end{cases}$

28. $\begin{cases} x + y + z = 2 \\ x - y + z = -2 \\ x - y - z = 0 \end{cases}$

29. $\begin{cases} x + y + z = 2 \\ x - y + z = 1 \\ x - y - z = 0 \end{cases}$

30. $\begin{cases} 2x + 8z = 8 \\ -x + 4y = 36 \\ 2x + y = 9 \end{cases}$

31. $\begin{cases} x + 3y + 3z = 7 \\ 2x + y + z = 4 \\ x + y + z = 4 \end{cases}$

32. $\begin{cases} x + 3y + 3z = 7 \\ 2x + y + z = 4 \\ x + y + z = 3 \end{cases}$

33. $\begin{cases} w + 2y + z = 4 \\ w - x + 2z = 12 \\ 2w + x + z = 12 \\ w + 2x + y + z = 12 \end{cases}$

34. $\begin{cases} w + x - y = 1 \\ x + y + z = 0 \\ -w + x + y = 1 \\ w - x - y + 2z = 1 \end{cases}$

For Problems 35 and 36, find $(\mathbf{I} - \mathbf{A})^{-1}$ *for the given matrix* **A**.

35. $\mathbf{A} = \begin{bmatrix} 2 & -1 \\ 1 & 3 \end{bmatrix}$

36. $\mathbf{A} = \begin{bmatrix} -3 & 2 \\ 4 & 3 \end{bmatrix}$

37. Auto Production Solve the following problems by using the inverse of the matrix involved.

(a) An automobile factory produces two models, A and B. Model A requires 1 labor hour to paint and $\frac{1}{2}$ labor hour to polish; model B requires 1 labor hour for each process. During each hour that the assembly line is operating, there are 100 labor hours available for painting and 80 labor hours for polishing. How many of each model can be produced each hour if all the labor hours available are to be utilized?

(b) Suppose each model A requires 10 widgets and 14 shims and each model B requires 7 widgets and 10 shims. The factory can obtain 800 widgets and 1130 shims each hour. How many cars of each model can it produce while using all the parts available?

38. If $\mathbf{A} = \begin{bmatrix} a & 0 & 0 \\ 0 & b & 0 \\ 0 & 0 & c \end{bmatrix}$, where $a, b, c \neq 0$, show that

$$\mathbf{A}^{-1} = \begin{bmatrix} 1/a & 0 & 0 \\ 0 & 1/b & 0 \\ 0 & 0 & 1/c \end{bmatrix}$$

39. **(a)** If **A** and **B** are invertible matrices with the same order, show that $(\mathbf{AB})^{-1} = \mathbf{B}^{-1}\mathbf{A}^{-1}$. [*Hint:* Show that

$$(\mathbf{B}^{-1}\mathbf{A}^{-1})(\mathbf{AB}) = \mathbf{I}$$

and use the fact that the inverse is unique.]

(b) If

$$\mathbf{A}^{-1} = \begin{bmatrix} 1 & 2 \\ 3 & 4 \end{bmatrix} \quad \text{and} \quad \mathbf{B}^{-1} = \begin{bmatrix} 1 & 1 \\ 1 & 2 \end{bmatrix}$$

find $(\mathbf{AB})^{-1}$.

40. If **A** is invertible, it can be shown that $(\mathbf{A}^{\mathrm{T}})^{-1} = (\mathbf{A}^{-1})^{\mathrm{T}}$. Verify this relationship if

$$\mathbf{A} = \begin{bmatrix} 1 & 3 \\ 2 & 4 \end{bmatrix}$$

41. A matrix **P** is said to be *orthogonal* if $\mathbf{P}^{-1} = \mathbf{P}^{\mathrm{T}}$. Is the matrix $\mathbf{P} = \frac{1}{5}\begin{bmatrix} 3 & -4 \\ 4 & 3 \end{bmatrix}$ orthogonal?

42. Secret Message A friend has sent you a secret message that consists of three row matrices of numbers as follows:

$$\mathbf{R}_1 = [33 \quad 87 \quad 70] \qquad \mathbf{R}_2 = [57 \quad 133 \quad 20]$$
$$\mathbf{R}_3 = [38 \quad 90 \quad 33]$$

Both you and your friend have committed the following matrix to memory (your friend used it to code the message):

$$\mathbf{A} = \begin{bmatrix} 1 & 2 & -1 \\ 2 & 5 & 2 \\ -1 & -2 & 2 \end{bmatrix}$$

Decipher the message by proceeding as follows:

(a) Calculate the three matrix products $\mathbf{R}_1\mathbf{A}^{-1}$, $\mathbf{R}_2\mathbf{A}^{-1}$, and $\mathbf{R}_3\mathbf{A}^{-1}$.

(b) Assume that the letters of the alphabet correspond to the numbers 1 through 26, replace the numbers in the preceding three matrices by letters, and determine the message.

43. Investing A group of investors decides to invest \$500,000 in the stocks of three companies. Company D sells for \$60 a share and has an expected growth of 16% per year. Company E sells for \$80 per share and has an expected growth of 12% per year. Company F sells for \$30 a share and has an expected growth of 9% per year. The group plans to buy four times as many shares of company F as of company E. If the group's goal is 13.68% growth per year, how many shares of each stock should the investors buy?

44. Investing The investors in Problem 43 decide to try a new investment strategy with the same companies. They wish to buy twice as many shares of company F as of company E, and they have a goal of 14.52% growth per year. How many shares of each stock should they buy?

In Problems 45 and 46 use a graphing calculator to (a) find $\mathbf{A}^{-1}$*, and express its entries in decimal form rounded to two decimal places. (b) Express the entries of* $\mathbf{A}^{-1}$ *in fractional form if your calculator has such capability. (Caution: For part (b), use the calculator matrix* $\mathbf{A}^{-1}$ *to convert to fractional entries; do not use the matrix of rounded values from part (a).)*

45. $\mathbf{A} = \begin{bmatrix} \frac{2}{3} & -\frac{1}{2} \\ -\frac{2}{7} & \frac{4}{5} \end{bmatrix}$

46. $\mathbf{A} = \begin{bmatrix} 2 & 6 & -3 \\ 4 & 8 & 9 \\ -7 & 2 & 5 \end{bmatrix}$

47. If $\mathbf{A} = \begin{bmatrix} 0.4 & 0.6 & -0.3 \\ 0.2 & 0.1 & -0.1 \\ 0.3 & 0.2 & -0.4 \end{bmatrix}$, find $(\mathbf{I} - \mathbf{A})^{-1}$, where $\mathbf{I}$ is the identity matrix of order 3. Round entries to two decimal places.

In Problems 48 and 49, use a graphing calculator to solve the system by using the inverse of the coefficient matrix.

48. $\begin{cases} 0.9x + 3y - 4.7z = 13 \\ 2x - 0.4y + 2z = 4.7 \\ x - 0.8y - 0.5z = 7.2 \end{cases}$

49. $\begin{cases} \frac{2}{5}w + 4x + \frac{1}{2}y - \frac{3}{7}z = \frac{14}{13} \\ \frac{5}{9}w - \frac{2}{3}x - 4y - z = \frac{7}{8} \\ x - \frac{4}{9}y + \frac{5}{6}z = 9 \\ \frac{1}{2}w + 4y - \frac{1}{3}z = \frac{4}{7} \end{cases}$

OBJECTIVE *To use the methods of this chapter to analyze the production of industrial sectors of an economy.*

6.7 Leontief's Input-Output Analysis

Input-output matrices, which were developed by Wassily W. Leontief,[6] indicate the supply and demand interrelationships that exist between the various sectors of an economy during some time period. The phrase "input-output" is used because the matrices show the values of outputs of each industry that are sold as inputs to each industry and for final use by consumers.

A hypothetical example for an oversimplified two-industry economy is given by the input-output matrix to be presented next. Before we present and explain the matrix, however, let us say that the *industrial* sectors can be thought of as manufacturing, steel, agriculture, coal, and so on. The *other production factors* sector consists of costs to the respective industries, such as labor, profits, and so on. The *final-demand* sector could be consumption by households, government, and so on. The matrix is as follows:

	Consumers (input)			
	Industry A	**Industry B**	**Final Demand**	
Producers (output):				***Totals***
Industry A	240	500	460	1200
Industry B	360	200	940	1500
Other Production Factors	600	800	—	
Totals	1200	1500		

Each industry appears in a row and column. The row shows the purchases of an industry's output by the industrial sectors and by consumers for final

[6] Leontief won the 1973 Nobel prize in economic science for the development of the "input-output" method and its applications to economic problems.

use (hence the term "final demand"). The entries represent the value of the products and might be in units of millions of dollars of product. For example, of the total output of industry A, 240 went as input to industry A itself (for internal use), 500 went to industry B, and 460 went directly to the final-demand sector. The total output of A is the sum of industrial demand and final demand $(240 + 500 + 460 = 1200)$.

Each industry column gives the value of what the industry purchased for input from each industry, as well as what it spent for other costs. For example, in order to produce its 1200 units, A purchased 240 units of output from itself, bought 360 of B's output, and had labor and other costs of 600 units.

Note that for each industry, the sum of the entries in its row is equal to the sum of the entries in its column. That is, the value of the total output of A is equal to the value of the total input to A.

Input-output analysis allows us to estimate the total production of each *industrial* sector if there is a change in final demand, *as long as the basic structure of the economy remains the same*. This important assumption means that for each industry, the amount spent on each input for each dollar's worth of output must remain fixed.

For example, in producing 1200 units' worth of product, industry A purchases 240 units' worth from industry A, purchases 360 units' worth from B, and spends 600 units on other costs. Thus, for each dollar's worth of output, industry A spends $\frac{240}{1200} = \frac{1}{5}(=\$0.20)$ on A, $\frac{360}{1200} = \frac{3}{10}(=\$0.30)$ on B, and $\frac{600}{1200} = \frac{1}{2}(=\$0.50)$ on other costs. Combining these fixed ratios of industry A with those of industry B, we can give the input requirements per dollar of output for each industry:

$$\begin{array}{c} \\ \textbf{A} \\ \textbf{B} \\ \textbf{Other} \end{array} \begin{array}{c} \quad\textbf{A} \qquad \textbf{B} \\ \begin{bmatrix} \frac{240}{1200} & \frac{500}{1500} \\ \frac{360}{1200} & \frac{200}{1500} \\ \frac{600}{1200} & \frac{800}{1500} \end{bmatrix} \end{array} = \begin{array}{c} \quad\textbf{A} \quad \textbf{B} \\ \begin{bmatrix} \frac{1}{5} & \frac{1}{3} \\ \frac{3}{10} & \frac{2}{15} \\ \frac{1}{2} & \frac{8}{15} \end{bmatrix} \end{array} \begin{array}{c} \\ \textbf{A} \\ \textbf{B} \\ \textbf{Other} \end{array}$$

The entries in the matrix are called **input-output coefficients.** The sum of each column is 1.

Now, suppose the value of final demand changes from 460 to 500 for industry A and from 940 to 1200 for industry B. We would like to estimate the value of *total* output that A and B must produce for both industry and final demand to meet this goal, provided that the structure in the preceding matrix remains the same.

Let the new values of total outputs for industries A and B be X_A and X_B, respectively. Now, for A,

$$\begin{array}{c}\text{total value of}\\ \text{output of } A\end{array} = \begin{array}{c}\text{value consumed}\\ \text{by } A\end{array} + \begin{array}{c}\text{value consumed}\\ \text{by } B\end{array} + \begin{array}{c}\text{value consumed}\\ \text{by final demand}\end{array}$$

so we have

$$X_A = \frac{1}{5}X_A + \frac{1}{3}X_B + 500$$

Similarly, for B

$$X_B = \frac{3}{10}X_A + \frac{2}{15}X_B + 1200$$

Using matrix notation, we can write

$$\begin{bmatrix} X_A \\ X_B \end{bmatrix} = \begin{bmatrix} \frac{1}{5} & \frac{1}{3} \\ \frac{3}{10} & \frac{2}{15} \end{bmatrix} \begin{bmatrix} X_A \\ X_B \end{bmatrix} + \begin{bmatrix} 500 \\ 1200 \end{bmatrix} \tag{1}$$

In this matrix equation, let

$$\mathbf{X} = \begin{bmatrix} X_A \\ X_B \end{bmatrix} \quad \mathbf{A} = \begin{bmatrix} \frac{1}{5} & \frac{1}{3} \\ \frac{3}{10} & \frac{2}{15} \end{bmatrix} \quad \mathbf{C} = \begin{bmatrix} 500 \\ 1200 \end{bmatrix}$$

We call **X** the **output matrix, A** the **coefficient matrix,** and **C** the **final-demand matrix.** From Equation (1),

$$\mathbf{X} = \mathbf{AX} + \mathbf{C}$$

$$\mathbf{X} - \mathbf{AX} = \mathbf{C}$$

Given output **X**, we can show that **AX** is the amount of the output that is required to self-sustain the economy. If we refer to this as the **internal-demand** then we arrive at the conceptual equation

$$\textbf{output} = \textbf{internal-demand} + \textbf{final-demand}$$

as a further justification for $\mathbf{X} = \mathbf{AX} + \mathbf{C}$. If **I** is the 2×2 identity matrix, then we can derive from the last equation

$$\mathbf{IX} - \mathbf{AX} = \mathbf{C}$$

$$(\mathbf{I} - \mathbf{A})\mathbf{X} = \mathbf{C}$$

FIGURE 6.7 Evaluating an output matrix.

If $(\mathbf{I} - \mathbf{A})^{-1}$ exists, then

$$\mathbf{X} = (\mathbf{I} - \mathbf{A})^{-1}\mathbf{C}$$

The matrix $\mathbf{I} - \mathbf{A}$ is called the **Leontief matrix.** We enter the matrices **A** and **C** into a graphing calculator. With a TI-83 Plus, the identity matrix of order 2 is obtained with the command "identity 2." Evaluating $(\mathbf{I} - \mathbf{A})^{-1}\mathbf{C}$ as shown in Figure 6.7 results in the output matrix

$$\mathbf{X} = (\mathbf{I} - \mathbf{A})^{-1}\mathbf{C} = \begin{bmatrix} 1404.49 \\ 1870.79 \end{bmatrix}$$

Here we rounded the entries in **X** to two decimal places. Thus, to meet the goal, industry A must produce 1404.49 units of value, and industry B must produce 1870.79. If we were interested in the value of other production factors for A, say, P_A, then

$$P_A = \frac{1}{2}X_A = 702.25$$

EXAMPLE 1 **Input-Output Analysis**

Given the input-output matrix

	Industry A	Industry B	Industry C	Final Demand
Industry: A	240	180	144	36
B	120	36	48	156
C	120	72	48	240
Other	120	72	240	—

suppose final demand changes to 77 *for A,* 154 *for B, and* 231 *for C. Find the output matrix for the economy. (The entries are in millions of dollars.)*

Solution: We separately add the entries in the first three rows. The total values of output for industries A, B, and C are 600, 360, and 480, respectively. To get the coefficient matrix **A**, we divide the industry entries in each industry column by the total value of output for that industry:

$$\mathbf{A} = \begin{bmatrix} \frac{240}{600} & \frac{180}{360} & \frac{144}{480} \\ \frac{120}{600} & \frac{36}{360} & \frac{48}{480} \\ \frac{120}{600} & \frac{72}{360} & \frac{48}{480} \end{bmatrix}$$

The final-demand matrix is

$$\mathbf{C} = \begin{bmatrix} 77 \\ 154 \\ 231 \end{bmatrix}$$

FIGURE 6.8 Evaluating the output matrix of Example 1.

Figure 6.8 shows the result of evaluating $(\mathbf{I}-\mathbf{A})^{-1}\mathbf{C}$. Thus, the output matrix is

$$\mathbf{X} = (\mathbf{I}-\mathbf{A})^{-1}\mathbf{C} = \begin{bmatrix} 692.5 \\ 380 \\ 495 \end{bmatrix}$$

▶ **Now work Problem 1.**

Exercise 6.7

*1. Given the input-output matrix

	Industry: Steel	Industry: Coal	Final Demand
Industry: Steel	200	500	500
Coal	400	200	900
Other	600	800	—

find the output matrix if final demand changes to 600 for Steel and 805 for Coal. Find the total value of the other production costs that this involves.

2. Given the input-output matrix

	Industry: Education	Industry: Government	Final Demand
Industry: Education	40	120	40
Government	120	90	90
Other	40	90	—

find the output matrix if final demand changes to (a) 200 for Education and 300 for Government; (b) 64 for Education and 64 for Government.

3. Given the input-output matrix

	Industry: Grain	Industry: Fertilizer	Industry: Cattle	Final Demand
Industry: Grain	15	30	45	10
Fertilizer	25	30	60	5
Cattle	50	40	60	30
Other	10	20	15	—

find the output matrix (with entries rounded to two decimal places) if final demand changes to (a) 15 for Grain, 10 for Fertilizer, and 35 for Cattle; (b) 10 for Grain, 10 for Fertilizer, and 10 for Cattle.

4. Given the input-output matrix

	Industry: Water	Electric Power	Agriculture	Final Demand
Industry: Water	100	400	240	260
Electric Power	100	80	480	140
Agriculture	300	160	240	500
Other	500	160	240	—

find the output matrix if final demand changes to 300 for Water, 200 for Electric Power, and 400 for Agriculture. Round your entries to two decimal places.

5. Given the input-output matrix

	Industry: Government	Agriculture	Manufacturing	Final Demand
Industry: Government	400	200	200	200
Agriculture	200	400	100	300
Manufacturing	200	100	300	400
Other	200	300	400	—

with entries in billions of dollars, find the output matrix for the economy if the final demand changes to 300 for Government, 350 for Agriculture, and 450 for Manufacturing. Round your entries to the nearest billion dollars.

6. Given the input-output matrix in Problem 5, find the output matrix for the economy if the final demand changes to 150 for Government, 200 for Agriculture, and 300 for Manufacturing. Round your entries to the nearest billion dollars.

7. Given the input-output matrix in Problem 5, find the output matrix for the economy if the final demand changes to 250 for Government, 300 for Agriculture, and 350 for Manufacturing. Round your entries to the nearest billion dollars.

8. Given the input-output matrix in Problem 5, find the output matrix for the economy if the final demand changes to 300 for Government, 400 for Agriculture, and 500 for Manufacturing. Round your entries to the nearest billion dollars.

6.8 Review

Important Terms and Symbols

Section 6.1 matrix size entry a_{ij} $[a_{ij}]$ row matrix (or vector) column matrix (or vector) equality of matrices transpose of matrix, $\mathbf{A}^T$ zero matrix, $\mathbf{O}$ square matrix main diagonal diagonal matrix upper (lower) triangular matrix

Section 6.2 scalar multiplication addition and subtraction of matrices

Section 6.3 matrix multiplication identity matrix, $\mathbf{I}$ power of a matrix matrix equation, $\mathbf{AX} = \mathbf{B}$

Section 6.4 coefficient matrix augmented coefficient matrix elementary row operation equivalent matrices reduced matrix leading entry parameter

Section 6.5 homogeneous system nonhomogeneous system trivial solution

Section 6.6 inverse matrix invertible matrix elementary matrix

Section 6.7 input-output matrix Leontief matrix

Summary

A matrix is a rectangular array of numbers enclosed within brackets. Three special types are the zero matrix $\mathbf{O}$, a square matrix, and the identity matrix $\mathbf{I}$. Besides the basic operation of scalar multiplication, there are the operations of matrix addition and subtraction, which apply to matrices of the same size. The product $\mathbf{AB}$ is defined when the number of columns of $\mathbf{A}$ is equal to the number of rows of $\mathbf{B}$. Although matrix addition is commutative, matrix multiplication is not. By using matrix multiplication, we can express a system of linear equations as the matrix equation $\mathbf{AX} = \mathbf{B}$.

A system of linear equations may have a unique solution, no solution, or infinitely many solutions. Two methods of solving a system of linear equations with matrices are (1) by using the three elementary row operations and (2) by using an inverse matrix. The first method, which is always available, involves applying elementary row operations to the augmented coefficient matrix of the system until an equivalent reduced matrix is obtained. The reduced matrix makes the solution(s) to the system obvious (assuming that any exist). If there are infinitely many solutions, the general solution involves at least one parameter.

The second method, which is available only if the number of equations is equal to the number of unknowns, involves finding the inverse of the coefficient matrix, if it exists. The inverse (if it exists) of a square matrix **A** is found by augmenting **A** with **I** and applying elementary row operations until **A** is reduced to **I**. (If the reduced form of **A** is not **I**, then $\mathbf{A}^{-1}$ does not exist.) The result of applying the same elementary row operations to **I** is $\mathbf{A}^{-1}$. If the inverse of an $n \times n$ matrix **A** exists, then the unique solution to $\mathbf{AX} = \mathbf{B}$ is given by $\mathbf{X} = \mathbf{A}^{-1}\mathbf{B}$. If **A** is not invertible, the system has either no solution or infinitely many solutions.

Our final application of matrices dealt with the interrelationships that exist between the various sectors of an economy and is known as input-output analysis.

Review Problems

Problem numbers shown in color indicate problems suggested for use as a practice chapter test.

In Problems 1–8, simplify.

1. $2\begin{bmatrix} 3 & 4 \\ -5 & 1 \end{bmatrix} - 3\begin{bmatrix} 1 & 0 \\ 2 & 4 \end{bmatrix}$

2. $8\begin{bmatrix} 1 & 2 \\ 7 & 0 \end{bmatrix} - 2\begin{bmatrix} 1 & 0 \\ 0 & 1 \end{bmatrix}$

3. $\begin{bmatrix} 1 & 7 \\ 2 & -3 \\ 1 & 0 \end{bmatrix}\begin{bmatrix} 1 & 0 & -2 \\ 0 & 6 & 1 \end{bmatrix}$

4. $[2 \;\; 3 \;\; 7]\begin{bmatrix} 2 & 3 \\ 0 & -1 \\ 5 & 2 \end{bmatrix}$

5. $\begin{bmatrix} 1 & 0 \\ -1 & 4 \end{bmatrix}\left(\begin{bmatrix} 1 & 4 \\ 6 & 5 \end{bmatrix} - \begin{bmatrix} 2 & 6 \\ 5 & 0 \end{bmatrix}\right)$

6. $-\left(\begin{bmatrix} 2 & 0 \\ 7 & 8 \end{bmatrix} + 2\begin{bmatrix} 0 & -5 \\ 6 & -4 \end{bmatrix}\right)$

7. $2\begin{bmatrix} 1 & -2 \\ 3 & 1 \end{bmatrix}^2 [1 \;\; -2]^{\mathrm{T}}$

8. $\frac{1}{3}\begin{bmatrix} 3 & 0 \\ 3 & 6 \end{bmatrix}\left(\begin{bmatrix} 1 & 0 \\ 1 & 3 \end{bmatrix}^{\mathrm{T}}\right)^2$

In Problems 9–12, compute the required matrix if

$$\mathbf{A} = \begin{bmatrix} 1 & 1 \\ -1 & 2 \end{bmatrix} \qquad \mathbf{B} = \begin{bmatrix} 1 & 0 \\ 0 & 2 \end{bmatrix}$$

9. $(2\mathbf{A})^{\mathrm{T}} - 3\mathbf{I}^2$

10. $\mathbf{A}(2\mathbf{I}) - \mathbf{AO}^{\mathrm{T}}$

11. $\mathbf{B}^3 + \mathbf{I}^5$

12. $(\mathbf{AB})^{\mathrm{T}} - \mathbf{B}^{\mathrm{T}}\mathbf{A}^{\mathrm{T}}$

In Problems 13 and 14, solve for x and y.

13. $\begin{bmatrix} 5 \\ 7 \end{bmatrix}[x] = \begin{bmatrix} 15 \\ y \end{bmatrix}$

14. $\begin{bmatrix} 1 & x \\ 2 & y \end{bmatrix}\begin{bmatrix} 2 & 1 \\ x & 3 \end{bmatrix} = \begin{bmatrix} 3 & 4 \\ 3 & y \end{bmatrix}$

In Problems 15–18, reduce the given matrices.

15. $\begin{bmatrix} 1 & 4 \\ 5 & 8 \end{bmatrix}$

16. $\begin{bmatrix} 0 & 0 & 7 \\ 0 & 5 & 9 \end{bmatrix}$

17. $\begin{bmatrix} 2 & 4 & 3 \\ 1 & 2 & 3 \\ 4 & 8 & 6 \end{bmatrix}$

18. $\begin{bmatrix} 0 & 0 & 0 & 1 \\ 0 & 0 & 0 & 0 \\ 1 & 0 & 0 & 0 \end{bmatrix}$

In Problems 19–22, solve each of the systems by the method of reduction.

19. $\begin{cases} 2x - 5y = 0 \\ 4x + 3y = 0 \end{cases}$

20. $\begin{cases} x - y + 2z = 3 \\ 3x + y + z = 5 \end{cases}$

21. $\begin{cases} x + y + 2z = 1 \\ 3x - 2y - 4z = -7 \\ 2x - y - 2z = 2 \end{cases}$

22. $\begin{cases} x - y - z - 2 = 0 \\ x + y + 2z + 5 = 0 \\ 2x + z + 3 = 0 \end{cases}$

In Problems 23–26, find the inverses of the matrices.

23. $\begin{bmatrix} 1 & 5 \\ 3 & 9 \end{bmatrix}$

24. $\begin{bmatrix} 0 & 1 \\ 1 & 0 \end{bmatrix}$

25. $\begin{bmatrix} 1 & 3 & -2 \\ 4 & 1 & 0 \\ 3 & -2 & 2 \end{bmatrix}$

26. $\begin{bmatrix} 5 & 0 & 0 \\ -5 & 2 & 1 \\ -5 & 1 & 3 \end{bmatrix}$

In Problems 27 and 28, solve the given system by using the inverse of the coefficient matrix.

27. $\begin{cases} 3x + y + 4z = 1 \\ x + z = 0 \\ 2y + z = 2 \end{cases}$

28. $\begin{cases} 5x = 3 \\ -5x + 2y + z = 0 \\ -5x + y + 3z = 2 \end{cases}$

29. Let $\mathbf{A} = \begin{bmatrix} 0 & 0 & 1 \\ 0 & 1 & 0 \\ 1 & 0 & 0 \end{bmatrix}$. Find the matrices $\mathbf{A}^2$, $\mathbf{A}^{-1}$, and $\mathbf{A}^{2000}$.

30. $\mathbf{A} = \begin{bmatrix} 2 & 0 \\ 0 & 4 \end{bmatrix}$, show that $(\mathbf{A}^{\mathrm{T}})^{-1} = (\mathbf{A}^{-1})^{\mathrm{T}}$.

31. A consumer wishes to supplement his vitamin intake by *exactly* 13 units of vitamin A, 22 units of vitamin B, and 31 units of vitamin C per week. There are three brands of vitamin capsules available. Brand I contains 1 unit each of vitamins A, B, and C per capsule; brand II contains 1 unit of vitamin A, 2 of B, and 3 of C; and brand III contains 4 units of A, 7 of B, and 10 of C.

(a) What combinations of capsules of brands I, II, and III will produce *exactly* the desired amounts?

(b) If brand I capsules cost 5 cents each, brand II 7 cents each, and brand III 20 cents each, what combination will minimize the consumer's weekly cost?

32. Suppose that **A** is an invertible $n \times n$ matrix.

(a) Prove that $\mathbf{A}^3$ is invertible.

(b) Prove that if **B** and **C** are $n \times n$ matrices such that $\mathbf{AB} = \mathbf{AC}$, then $\mathbf{B} = \mathbf{C}$.

(c) If $\mathbf{A}^2 = \mathbf{A}$ (we say that **A** is *idempotent*), find **A**.

33. If $\mathbf{A} = \begin{bmatrix} 10 & -3 \\ 4 & 7 \end{bmatrix}$ and $\mathbf{B} = \begin{bmatrix} 8 & 6 \\ -7 & -3 \end{bmatrix}$, find $3\mathbf{AB} - 4\mathbf{B}^2$.

34. Solve the system

$$\begin{cases} 9.7x - 3.4y + 7.2z = 19.1 \\ 4.3x + 8.5y - 6.7z = 20.8 \\ 5.4x - 2.6y - 4.7z = 30.9 \end{cases}$$

by using the inverse of the coefficient matrix. Round your answers to two decimal places.

35. Given the input-output matrix

	Industry A	Industry B	Final Demand
Industry: A	10	20	4
B	15	14	10
Other	9	5	—

find the output matrix if final demand changes to 10 for A and 5 for B. (Data are in tens of billions of dollars.)

Mathematical
SNAPSHOT

Insulin Requirements as a Linear Process[7]

A vacation lodge in the mountains of Washington State has a well-deserved reputation for attending to the special health needs of its guests. Next week the manager of the lodge is expecting four guests, each of whom has insulin-dependent diabetes. These guests plan to stay at the lodge for 7, 14, 21, and 28 days, respectively.

The lodge is quite a distance from the nearest drugstore, so before the arrival of the guests the manager plans to obtain the total amount of insulin that will be needed. Three different types of insulin are required: lente, semi-lente, and ultra-lente. The manager will store the insulin, and then lodge personnel will administer the daily dose of the three different types of insulin to each of the guests.

The daily requirements of the four guests are as follows:

Guest 1 20 insulin units of semi-lente, 30 units of lente, 10 units of ultra-lente;

Guest 2 40 insulin units of semi-lente, 0 units of lente, 0 units of ultra-lente;

Guest 3 30 insulin units of semi-lente, 10 units of lente, 30 units of ultra-lente;

Guest 4 10 insulin units of semi-lente, 10 units of lente, 50 units of ultra-lente.

This information will be represented by the following "requirement" matrix **A**:

$\mathbf{A} = [a_{ij}]_{3\times 4}$ where **A** is given by

$$\begin{array}{l} \\ \\ \textbf{semi-lente insulin} \\ \textbf{lente insulin} \\ \textbf{ultra-lente insulin} \end{array} \begin{array}{cccc} \textbf{Guest} & \textbf{Guest} & \textbf{Guest} & \textbf{Guest} \\ \textbf{1} & \textbf{2} & \textbf{3} & \textbf{4} \\ 20 & 40 & 30 & 10 \\ 30 & 0 & 10 & 10 \\ 10 & 0 & 30 & 50 \end{array}$$

Recall that Guest 1 will stay for 7 days, Guest 2 for 14 days, Guest 3 for 21 days, and Guest 4 for 28 days. You can let the following column vector **T** represent the time, in days, that each guest is staying at the lodge:

$$\mathbf{T} = \begin{bmatrix} 7 \\ 14 \\ 21 \\ 28 \end{bmatrix}$$

To determine the total amounts of the different types of insulin needed by the four guests, you compute the matrix product **AT**.

$$\mathbf{AT} = \begin{bmatrix} 20 & 40 & 30 & 10 \\ 30 & 0 & 10 & 10 \\ 10 & 0 & 30 & 50 \end{bmatrix} \begin{bmatrix} 7 \\ 14 \\ 21 \\ 28 \end{bmatrix}$$

$$= 10(7) \begin{bmatrix} 2 & 4 & 3 & 1 \\ 3 & 0 & 1 & 1 \\ 1 & 0 & 3 & 5 \end{bmatrix} \begin{bmatrix} 1 \\ 2 \\ 3 \\ 4 \end{bmatrix}$$

$$= 70 \begin{bmatrix} 23 \\ 10 \\ 30 \end{bmatrix} = \begin{bmatrix} 1610 \\ 700 \\ 2100 \end{bmatrix} = \mathbf{B}$$

Vector **B** (**AT**) indicates that a total of 1610 insulin units of semi-lente, 700 insulin units of lente, and 2100 insulin units of ultra-lente are required by the four guests.

Now, change the problem a bit. Suppose that each guest decided to double the original length of stay. The resulting vector that gives the total amount needed of semi-lente, lente, and ultra-lente insulin is

$$\mathbf{A}(2\mathbf{T}) = 2(\mathbf{AT}) = 2\mathbf{B} = \begin{bmatrix} 3220 \\ 1400 \\ 4200 \end{bmatrix}$$

In fact, if each guest planned to spend a factor k ($k \geq 0$) of the original time at the lodge (that is, Guest 1 planned to stay for $k \cdot 7$ days, Guest 2 for $k \cdot 14$ days, and so on), then the insulin requirements would be

$$\mathbf{A}(k\mathbf{T}) = k(\mathbf{AT}) = k\mathbf{B} = \begin{bmatrix} k \cdot 1610 \\ k \cdot 700 \\ k \cdot 2100 \end{bmatrix}$$

[7]Adapted from Richard F. Baum, "Insulin Requirements as a Linear Process," in R. M. Thrall, J. A. Mortimer, K. R. Rebman, and R. F. Baum (eds.), *Some Mathematical Models in Biology,* rev. ed. Report 40241-R-7. Prepared at the University of Michigan, 1967.

Similarly, if the guests decided to add 1, 3, 4, and 6 days, respectively, to the times they originally intended to stay, then the amounts of insulin required would be

$$\mathbf{A}(\mathbf{T}+\mathbf{T}_1) = \mathbf{AT} + \mathbf{AT}_1 \qquad \text{where } \mathbf{T}_1 = \begin{bmatrix} 1 \\ 3 \\ 4 \\ 6 \end{bmatrix}$$

Based on the results thus far, it is obvious that the following matrix equation generalizes the situation.

$$\mathbf{AX} = \mathbf{B}$$

or

$$\begin{bmatrix} 20 & 40 & 30 & 10 \\ 30 & 0 & 10 & 10 \\ 10 & 0 & 30 & 50 \end{bmatrix} \begin{bmatrix} x_1 \\ x_2 \\ x_3 \\ x_4 \end{bmatrix} = \begin{bmatrix} b_1 \\ b_2 \\ b_3 \end{bmatrix}$$

which represents the linear system

$$\begin{cases} 20x_1 + 40x_2 + 30x_3 + 10x_4 = b_1 \\ 30x_1 + 10x_3 + 10x_4 = b_2 \\ 10x_1 + 30x_3 + 50x_4 = b_3 \end{cases}$$

where x_i is the number of days that Guest i stays at the lodge, and b_1, b_2, b_3 give, respectively, the total number of units of semi-lente, lente, and ultra-lente insulin needed by the four guests for their entire stay at the lodge.

Finally, suppose once again that vector **T** represents the number of days that each guest originally planned to stay at the lodge. Furthermore, suppose vector **C** gives the cost (in cents) per insulin unit of the three types of insulin, where

$$\mathbf{C} = \begin{bmatrix} 9 \\ 8 \\ 10 \end{bmatrix} = \text{cost matrix}$$

That is, one unit of semi-lente costs 9, one unit of lente costs 8, and one unit of ultra-lente costs 10. Then the total amount paid by the lodge for all the insulin required by the four guests is

$$\mathbf{C}^{\mathrm{T}}(\mathbf{AT}) = \mathbf{C}^{\mathrm{T}}\mathbf{B} = \begin{bmatrix} 9 & 8 & 10 \end{bmatrix} \begin{bmatrix} 1610 \\ 700 \\ 2100 \end{bmatrix} = [41{,}090]$$

that is 41,090 cents which is \$410.90.

Exercises

1. Suppose that Guest 1 will stay at the lodge for 7 days, Guest 2 for 10 days, Guest 3 for 7 days, and Guest 4 for 5 days. Assume that the daily requirements of the four guests and the cost matrix are the same as given in the discussion. Find the total amount (in dollars) that the lodge must pay for all the insulin required by the guests.
2. Suppose that the insulin requirements of the four guests add up to 1180 insulin units of semi-lente, 580 of lente, and 1500 of ultra-lente. Assume the same daily requirements for the four guests are as given in the discussion. Using the inverse matrix method on a graphing calculator, find the length of stay for each guest if the total number of days for all four guests is 52.
3. Assume that the daily requirements of the four guests and the cost matrix are the same as given in the discussion. Is it possible, given only the total amount (in dollars), that the lodge must pay for all the required insulin, to determine the length of each guest's stay? Why or why not?

CHAPTER

7

Linear Programming

Linear programming sounds like something involving the writing of computer code. But while linear programming is often done on computers, the "programming" part of the name actually comes from World War II–era military terminology, in which training, supply, and unit-deployment plans were called programs. Each program was, formally speaking, a solution to a problem in resource allocation.

For example, suppose that military units in a combat theater need diesel fuel. Each unit has a certain number of tanks, trucks, and other vehicles; each unit uses its vehicles to accomplish an assigned mission; and each unit's mission has some relation to the overall goal of winning the campaign. What fuel distribution program will best contribute to overall victory?

Solving this problem requires quantifying its various elements. Counting gallons of fuel and numbers of each type of vehicle is easy, as is translating gallons of fuel into miles a vehicle can travel. Quantifying the relation between vehicle miles and unit mission accomplishment includes identifying constraints: the maximum gallons per load a tanker truck can carry, the minimum number of miles each unit must travel to reach its combat objective, and so on. Additional quantitative factors include probabilities, such as a unit's chances of winning a key engagement if it maneuvers along one route of travel rather than another.

Quantifying complicated real-world problems in this way is the province of operations research, as it is called. Linear programming, one of the oldest and still one of the most important tools of operations research, is used when a problem can be described using equations and inequalities that are all linear.

OBJECTIVE *To geometrically represent the solution of a linear inequality in two variables and to extend this representation to a system of linear inequalities.*

7.1 Linear Inequalities in Two Variables

Suppose a consumer receives a fixed income of \$60 per week and uses it all to purchase products A and B. If x kilograms of A cost \$2 per kilogram and y kilograms of B cost \$3 per kilogram, then

$$2x + 3y = 60 \quad \text{where} \quad x, y \geq 0$$

The solutions of this equation, called a *budget equation,* give the possible combinations of A and B that can be purchased for \$60. The graph of the equation is the *budget line* in Figure 7.1. Note that (15, 10) lies on the line. This means that if 15 kg of A are purchased, then 10 kg of B must be bought, for a total cost of \$60.

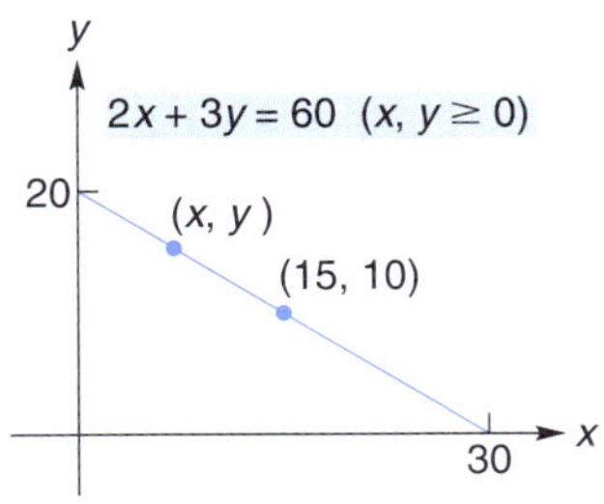

FIGURE 7.1 Budget line.

On the other hand, suppose the consumer does not necessarily wish to spend all of the \$60. In this case, the possible combinations are described by the inequality

$$2x + 3y \leq 60 \quad \text{where} \quad x, y \geq 0 \qquad (1)$$

When inequalities in one variable were discussed in Chapter 1, their solutions were represented geometrically by *intervals* on the real number line. However, for an inequality in two variables, like the inequality (1), the solution is usually represented by a *region* in the coordinate plane. We shall find the region corresponding to (1) after considering such inequalities in general.

Definition

A ***linear inequality*** in the variables x and y is an inequality that can be written in the form

$$ax + by + c < 0 \qquad (\text{or} \leq 0, \geq 0, > 0)$$

where a, b, and c are constants and not both a and b are zero.

Geometrically, the solution (or graph) of a linear inequality in x and y consists of all points (x, y) in the plane whose coordinates satisfy the inequality. For example, a solution of $x + 3y < 20$ is the point $(-2, 4)$, because substitution gives

$$-2 + 3(4) < 20,$$

$$10 < 20, \quad \text{which is true.}$$

Clearly, there are infinitely many solutions, which is typical of every linear inequality.

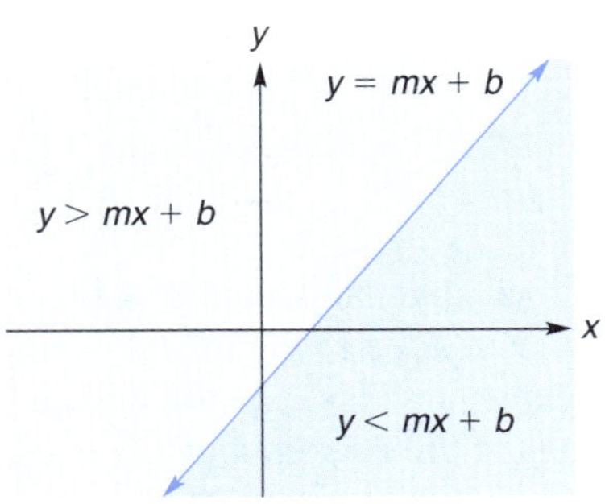

FIGURE 7.2 A nonvertical line determines two half planes.

To consider linear inequalities in general, we first note that the graph of a nonvertical line $y = mx + b$ separates the plane into three distinct parts (see Figure 7.2):

1. the line itself, consisting of all points (x, y) whose coordinates satisfy the equation $y = mx + b$;
2. the region *above* the line, consisting of all points (x, y) whose coordinates satisfy the inequality $y > mx + b$ (this region is called an *open* **half plane**);
3. the open half plane *below* the line, consisting of all points (x, y) whose coordinates satisfy the inequality $y < mx + b$.

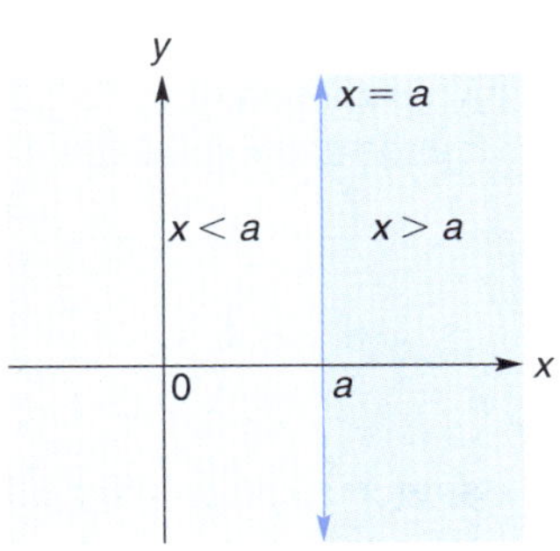

FIGURE 7.3 A vertical line determines two half planes.

In the situation where the strict inequality "<" is replaced by "≤", the solution of $y \leq mx + b$ consists of the line $y = mx + b$, as well as the half plane below it. In this case, we say that the solution is a *closed* half plane. A similar statement can be made when ">" is replaced by "≥". For a vertical line $x = a$ (see Figure 7.3), we speak of a half plane to the right $(x > a)$ of the line or to its left $(x < a)$. Since any linear inequality (in two variables) can be put into one of the forms we have discussed, we can say that *the solution of a linear inequality must be a half plane.*

To apply these facts, we shall solve the linear inequality

$$2x + y < 5$$

From our previous discussion, we know that the solution is a half plane. To find it, we begin by replacing the inequality symbol by an equality sign and then graphing the resulting *line*, $2x + y = 5$. This is easily done by choosing two points on the line—for instance, the intercepts $(\frac{5}{2}, 0)$ and $(0, 5)$. (See Figure 7.4.) Because points on the line do not satisfy the "<" inequality, we used a *dashed* line to indicate that the line is not part of the solution. We must now determine whether the solution is the half plane *above* the line or the one *below* it. This can be done by solving the inequality for y. Once y is isolated, the appropriate half plane will be apparent. We have

$$y < 5 - 2x$$

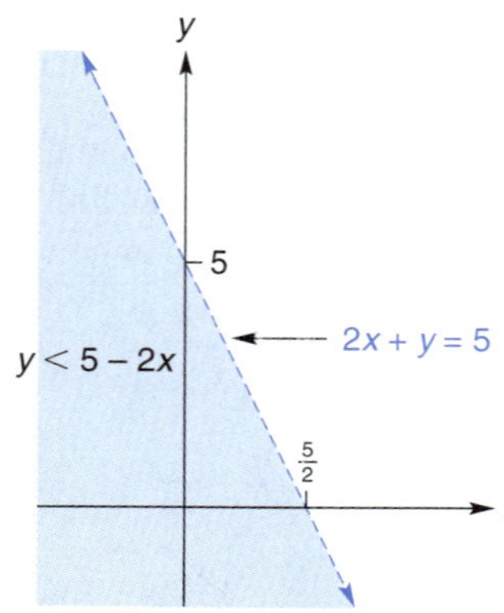

FIGURE 7.4 Graph of $2x + y < 5$.

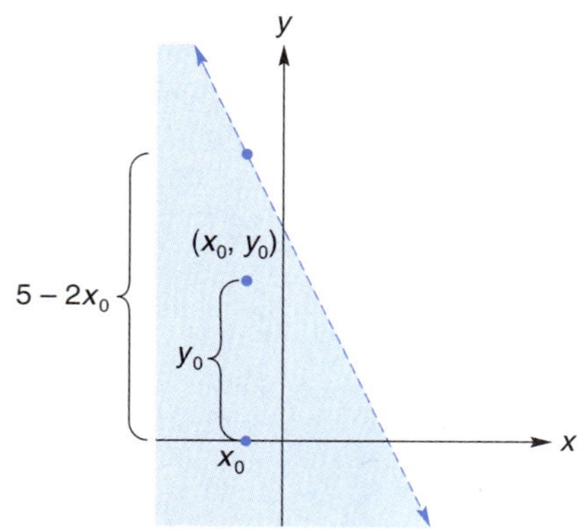

FIGURE 7.5 Analysis of point satisfying $y < 5 - 2x$.

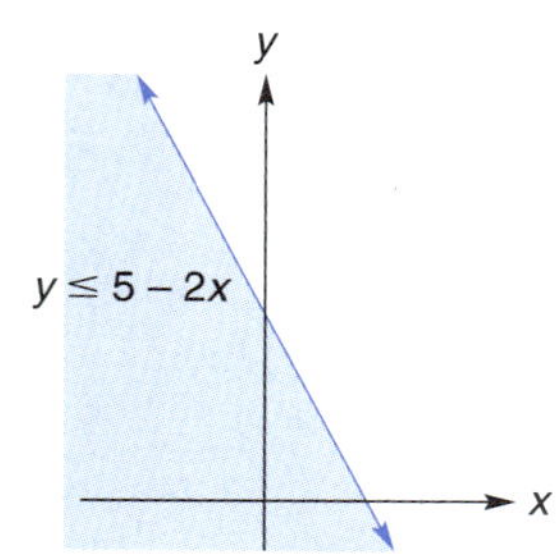

FIGURE 7.6 Graph of $y \le 5 - 2x$.

Geometrically, the solution of a linear inequality in one variable is an interval on the line, but the solution of a linear inequality in *two* variables is a *region* in the plane.

From the aforementioned statement 3, we conclude that the solution consists of the half plane *below* the line. Part of this region is shaded in Figure 7.4. Thus, if (x_0, y_0) is *any* point in this region, then its ordinate y_0 is less than the number $5 - 2x_0$. (See Figure 7.5.) For example, $(-2, -1)$ is in the region, and

$$-1 < 5 - 2(-2)$$
$$-1 < 9$$

If, instead, the original inequality had been $y \le 5 - 2x$, then the line $y = 5 - 2x$ would have been included in the solution. We would indicate its inclusion by using a solid line rather than a dashed line. This solution, which is a closed half plane, is shown in Figure 7.6. Keep in mind that **a solid line *is* included in the solution, and a dashed line *is not*.**

EXAMPLE 1 Solving a Linear Inequality

Find the region defined by the inequality $y \le 5$.

Solution: Since x does not appear, the inequality is assumed to be true for all values of x. The region consists of the line $y = 5$, together with the half plane below it. (See Figure 7.7.)

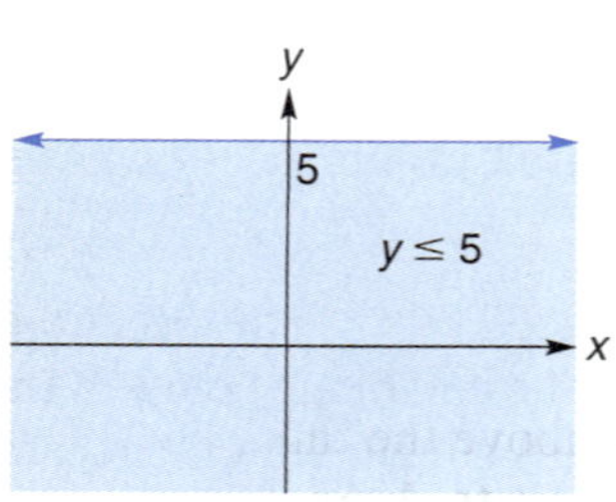

FIGURE 7.7 Graph of $y \le 5$.

APPLICATIONS IN PRACTICE 1

Solving a Linear Inequality

To earn some extra money, you make two types of refrigerator magnets, type A and type B, to sell. You have an initial start-up expense of \$50. The production cost for type A is \$0.90 per magnet, and the production cost for type B is \$0.70 per magnet. The price for type A is \$2.00 per magnet, and the price for type B is \$1.50 per magnet. Let x be the number of type A and y be the number of type B produced and sold. Write an inequality describing revenue greater than cost. Solve the inequality and describe the region. Also, describe what this result means in terms of magnets.

EXAMPLE 2 **Solving a Linear Inequality**

Solve the inequality $2(2x - y) < 2(x + y) - 4$.

Solution: We first solve the inequality for y, so that the appropriate half plane is obvious. The inequality is equivalent to

$$4x - 2y < 2x + 2y - 4$$
$$4x - 4y < 2x - 4$$
$$-4y < -2x - 4$$
$$y > \frac{x}{2} + 1 \qquad \text{(dividing both sides by } -4 \text{ and reversing the sense of the inequality)}$$

Using a dashed line, we now sketch $y = (x/2) + 1$ by noting that its intercepts are $(0, 1)$ and $(-2, 0)$. Because the inequality symbol is $>$, we shade the half plane above the line. (See Figure 7.8.) Each point in this region is a solution.

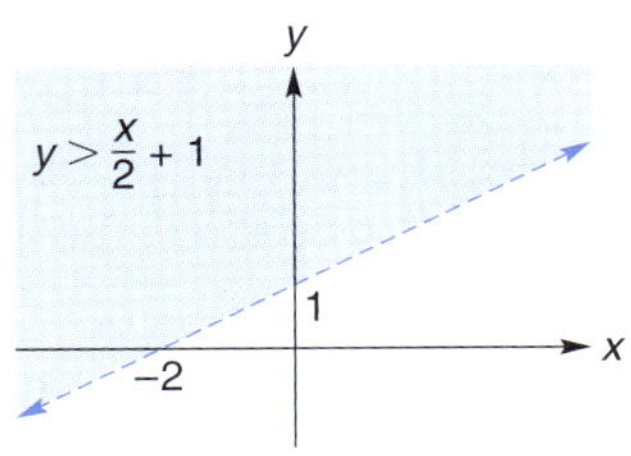

FIGURE 7.8 Graph of $y > \frac{x}{2} + 1$.

▸ **Now work Problem 1.**

Systems of Inequalities

The solution of a *system* of inequalities consists of all points whose coordinates simultaneously satisfy all of the given inequalities. Geometrically, it is the region that is common to all the regions determined by the given inequalities. For example, let us solve the system

$$\begin{cases} 2x + y > 3 \\ x \geq y \\ 2y - 1 > 0 \end{cases}$$

We first rewrite each inequality so that y is isolated. This gives the equivalent system

$$\begin{cases} y > -2x + 3 \\ y \leq x \\ y > \frac{1}{2} \end{cases}$$

Now we sketch the corresponding lines $y = -2x + 3$, $y = x$, and $y = \frac{1}{2}$. Noting that the first and third inequalities define *open* half planes, but the second one defines a *closed* half plane, we then shade the region that is simultaneously above the first line, on or below the second line, and above the third line. (See Figure 7.9.) When you are sketching the lines, **it is best to draw dashed lines everywhere until it is clear which portions are to be included in the solution.**

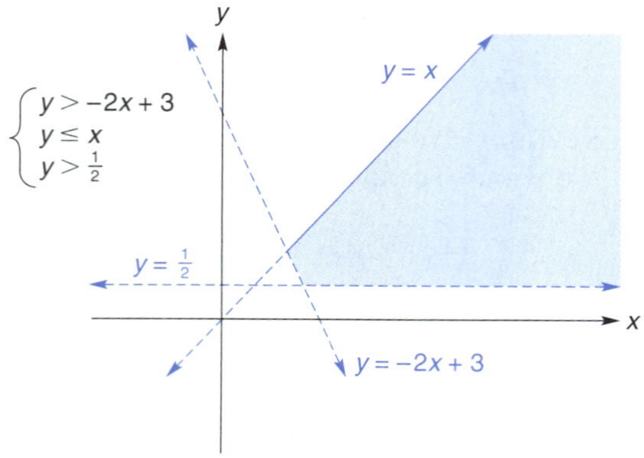

FIGURE 7.9 Solution of system of linear inequalities.

CAUTION The point where the graphs of $y = x$ and $y = -2x + 3$ intersect is *not* included in the solution.

APPLICATIONS IN PRACTICE 2

Solving a System of Linear Inequalities

A store sells two types of cameras. In order to cover overhead, it must sell at least 50 cameras per week, and in order to satisfy distribution requirements, it must sell at least twice as many of type I as type II. Write a system of inequalities to describe the situation. Let x be the number of type I that the store sells in a week and y be the number of type II that it sells in a week. Find the region described by the system of linear inequalities.

EXAMPLE 3 **Solving a System of Linear Inequalities**

Solve the system

$$\begin{cases} y \ge -2x + 10 \\ y \ge x - 2 \end{cases}$$

Solution: The solution consists of all points that are simultaneously on or above the line $y = -2x + 10$ and on or above the line $y = x - 2$. It is the shaded region in Figure 7.10.

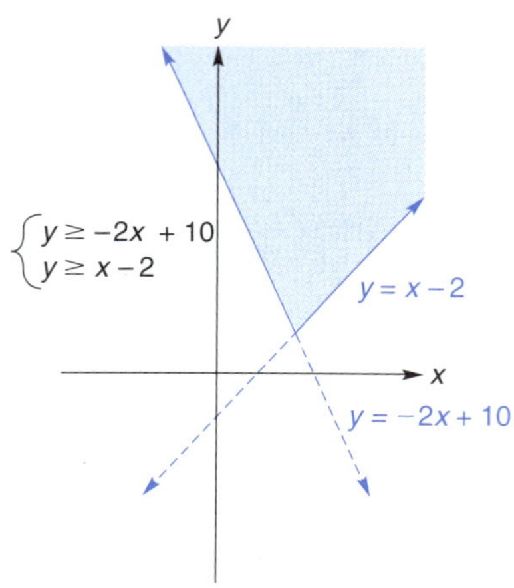

FIGURE 7.10 Solution of system of linear inequalities.

▶ **Now work Problem 9.**

EXAMPLE 4 **Solving a System of Linear Inequalities**

Find the region described by

$$\begin{cases} 2x + 3y \le 60 \\ x \ge 0 \\ y \ge 0 \end{cases}$$

Solution: This system relates to inequality (1) at the beginning of the section. The first inequality is equivalent to $y \le -\frac{2}{3}x + 20$. The last two inequalities restrict the solution to points that are both on or to the right of the y-axis *and* also on or above the x-axis. The desired region is shaded in Figure 7.11.

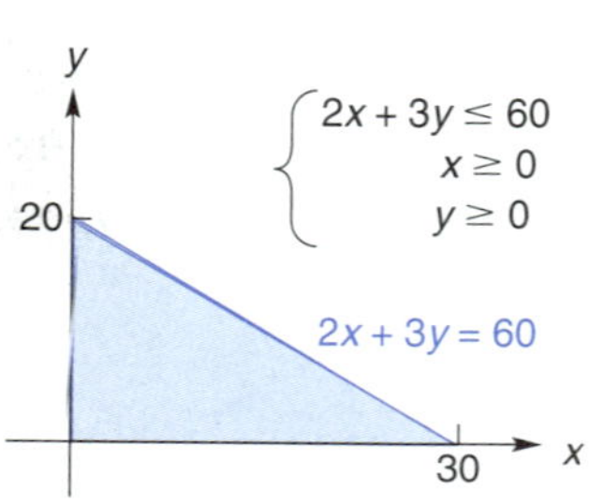

FIGURE 7.11 Solution of system of linear inequalities.

▶ **Now work Problem 17.**

Exercise 7.1

In Problems 1–24, solve the inequalities.

*1. $2x + 3y > 6$

2. $3x - 2y \geq 12$

3. $x + 2y \leq 7$

4. $y > 6 - 2x$

5. $-x \leq 2y - 4$

6. $3x + 5y \geq 12$

7. $3x + y < 0$

8. $x + 5y < -5$

*9. $\begin{cases} 3x - 2y < 6 \\ x - 3y > 9 \end{cases}$

10. $\begin{cases} 2x + 3y > -6 \\ 3x - y < 6 \end{cases}$

11. $\begin{cases} 2x + 3y \leq 6 \\ x \geq 0 \end{cases}$

12. $\begin{cases} 2y - 3x < 6 \\ x < 0 \end{cases}$

13. $\begin{cases} 2y - 4x < 7 \\ 3x - 4y > -5 \end{cases}$

14. $\begin{cases} x - y < 1 \\ y - x < 1 \end{cases}$

15. $\begin{cases} 2x - 2 \geq y \\ 2x \leq 3 - 2y \end{cases}$

16. $\begin{cases} 2y < 4x + 2 \\ y < 2x + 1 \end{cases}$

*17. $\begin{cases} x - y > 4 \\ x < 2 \\ y > -5 \end{cases}$

18. $\begin{cases} 5x + 2y < -3 \\ y > -x \\ 3x + 6 < 0 \end{cases}$

19. $\begin{cases} y < 2x + 4 \\ x \geq -2 \\ y < 1 \end{cases}$

20. $\begin{cases} 4x + 3y \geq 12 \\ y \geq x \\ 2y \leq 3x + 6 \end{cases}$

21. $\begin{cases} x + y > 1 \\ 3x - 5 \leq y \\ y < 2x \end{cases}$

22. $\begin{cases} 2x - 3y > -12 \\ 3x + y > -6 \\ y > x \end{cases}$

23. $\begin{cases} 3x + y > -6 \\ x - y > -5 \\ x \geq 0 \end{cases}$

24. $\begin{cases} 5y - 2x \leq 10 \\ 4x - 6y \leq 12 \\ y \geq 0 \end{cases}$

If a consumer wants to spend no more than P dollars to purchase quantities x and y of two products having prices of p_1 and p_2 dollars per unit, respectively, then $p_1x + p_2y \leq P$, where $x, y \geq 0$. In Problems 25 and 26, find geometrically the possible combinations of purchases by determining the solution of this system for the given values of p_1, p_2, and P.

25. $p_1 = 5, p_2 = 3, P = 15$

26. $p_1 = 7, p_2 = 3, P = 25$

27. If a manufacturer wishes to purchase a *total* of no more than 100 lb of product Z from suppliers A and B, set up a system of inequalities that describes the possible combinations of quantities that can be purchased from each supplier. Sketch the solution in the plane.

28. **Manufacturing** The XYZ Corporation produces two models of home computers: the Alpha model and the Beta model. Let x be the number of Alpha models and y the number of Beta models produced at the San Antonio factory per week. If the factory can produce at most 650 Alpha and Beta models, combined, in a week, write inequalities to describe this situation.

29. **Manufacturing** A chair company produces two models of chairs. The Sequoia model takes 3 worker-hours to assemble and $\frac{1}{2}$ worker-hour to paint. The Saratoga model takes 2 worker-hours to assemble and 1 worker-hour to paint. The maximum number of worker-hours available to assemble chairs is 240 per day, and the maximum number of worker-hours available to paint chairs is 80 per day. Write a system of linear inequalities to describe the situation. Let x represent the number of Sequoia models produced in a day and y represent the number of Saratoga models produced in a day. Find the region described by this system of linear inequalities.

7.2 Linear Programming

OBJECTIVE *To state the nature of a linear programming problem, to introduce terminology associated with it, and to solve it geometrically.*

Sometimes it is desired to maximize or minimize a function, subject to certain restrictions (or *constraints*). For example, a manufacturer may want to maximize a profit function, subject to production restrictions imposed by limitations on the use of machinery and labor.

We shall now consider how to solve such problems when the function to be maximized or minimized is *linear*. A **linear function in x and y** has the form

$$Z = ax + by$$

where a and b are constants. We shall also require that the corresponding constraints be represented by a system of linear inequalities (involving "$\leq$" or "$\geq$") or linear equations in x and y, and that all variables be nonnegative. A problem involving all of these conditions is called a *linear programming problem*.

Linear programming was developed by George B. Dantzig in the late 1940s and was first used by the U.S. Air Force as an aid in decision making. Today it has wide application in industrial and economic analysis.

CAUTION A great deal of terminology is used in discussing linear programming, and you are advised to learn this terminology as it is introduced.

In a linear programming problem, the function to be maximized or minimized is called the **objective function.** Although there are usually infinitely many solutions to the system of constraints (these are called **feasible solutions** or **feasible points**), the aim is to find one such solution that is an **optimum solution** (that is, one that gives the maximum or minimum value of the objective function).

We shall now give a geometrical approach to linear programming. Section 7.4 will discuss a matrix approach that will enable us to work with more than two variables and, hence, a wider range of problems.

We consider the following problem. A company produces two types of widgets: manual and electric. Each requires in its manufacture the use of three machines: A, B, and C. Table 7.1 gives data relating to the manufacture of these widgets. Each manual widget requires the use of machine A for 2 hours, machine B for 1 hour, and machine C for 1 hour. An electric widget requires 1 hour on A, 2 hours on B, and 1 hour on C. Furthermore, suppose the maximum numbers of hours available per month for the use of machines A, B, and C are 180, 160, and 100, respectively. The profit on a manual widget is \$4, and on an electric widget it is \$6. If the company can sell all the widgets it can produce, how many of each type should it make in order to maximize the monthly profit?

To solve the problem, let x and y denote the number of manual and electric widgets, respectively, that are made in a month. Since the number of widgets made is not negative,

$$x \geq 0 \quad \text{and} \quad y \geq 0$$

TABLE 7.1

	Manual	Electric	Hours Available
A	2 hr	1 hr	180
B	1 hr	2 hr	160
C	1 hr	1 hr	100
Profit/Unit	\$4	\$6	

For machine A, the time needed for working on x manual widgets is $2x$ hours, and the time needed for working on y electric widgets is $1y$ hours. The sum of these times cannot be greater than 180, so

$$2x + y \leq 180$$

Similarly, the restrictions for machines B and C give

$$x + 2y \leq 160 \quad \text{and} \quad x + y \leq 100$$

The profit is a function of x and y and is given by the *profit function*

$$P = 4x + 6y$$

Summarizing, we want to maximize the *objective function*

$$P = 4x + 6y \qquad (1)$$

subject to the condition that x and y must be a solution of the system of constraints

$$\begin{cases} 2x + y \leq 180 & (2) \\ x + 2y \leq 160 & (3) \\ x + y \leq 100 & (4) \\ x \geq 0 & (5) \\ y \geq 0 & (6) \end{cases}$$

Thus, we have a linear programming problem. Constraints (5) and (6) are called **nonnegativity conditions.** The region simultaneously satisfying constraints (2)–(6) is shaded in Figure 7.12. Each point in this region represents a feasible solution, and the region is called the **feasible region.** Although there are infinitely many feasible solutions, we must find one that maximizes the profit function.

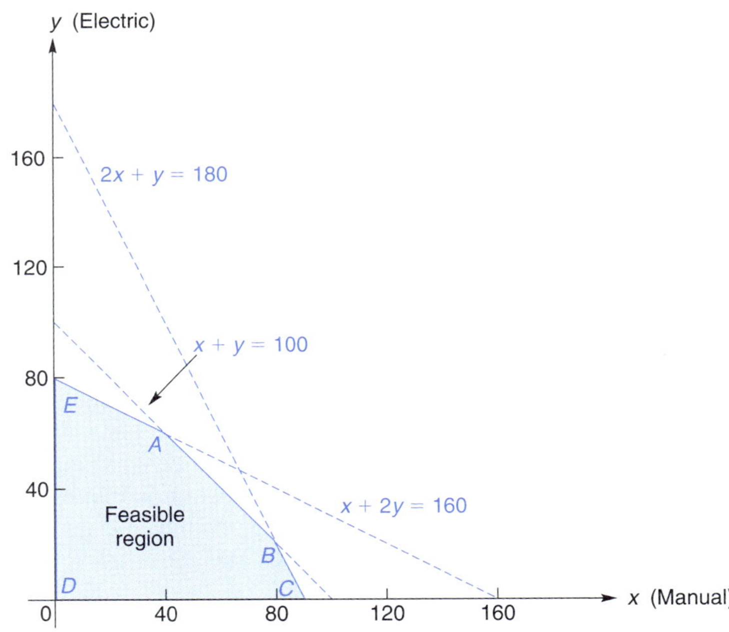

FIGURE 7.12 Feasible region.

Since the objective function, $P = 4x + 6y$, is equivalent to

$$y = -\frac{2}{3}x + \frac{P}{6}$$

it defines a so-called family of parallel lines, one for each possible value of P, each having a slope of $-2/3$ and y-intercept $(0, P/6)$. For example, if $P = 600$, then we obtain the line

$$y = -\frac{2}{3}x + 100$$

shown in Figure 7.13. This line, called an **isoprofit line,** gives all possible combinations of x and y that yield the same profit, \$600. Note that this isoprofit line has no point in common with the feasible region, whereas the isoprofit line for $P = 300$ has infinitely many such points. Let us look for the member of the family that contains a feasible point and whose P-value is maximum. *This*

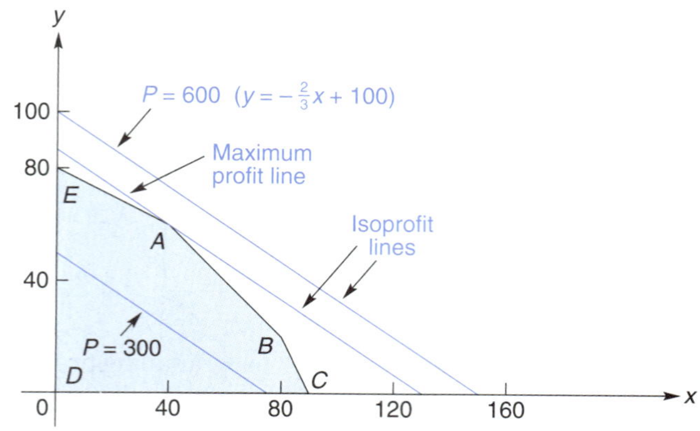

FIGURE 7.13 Isoprofit lines and feasible region.

will be the line whose y-intercept is farthest from the origin (giving a maximum value of P) and that has at least one point in common with the feasible region. It is not difficult to observe that such a line will contain the *corner point A*. Any isoprofit line with a greater profit will contain no points of the feasible region.

From Figure 7.12, we see that A lies on both the line $x + y = 100$ and the line $x + 2y = 160$. Thus, its coordinates may be found by solving the system

$$\begin{cases} x + y = 100 \\ x + 2y = 160 \end{cases}$$

This gives $x = 40$ and $y = 60$. Substituting these values into the equation $P = 4x + 6y$, we find that the maximum profit subject to the constraints is \$520, which is obtained by producing 40 manual widgets and 60 electric widgets per month.

If a feasible region can be contained within a circle, such as the region in Figure 7.13, it is called a **bounded feasible region.** Otherwise, it is **unbounded.** When a feasible region contains at least one point, it is said to be **nonempty.** Otherwise, it is **empty.** The region in Figure 7.13 is a nonempty bounded feasible region.

It can be shown that:

A linear function defined on a nonempty bounded feasible region has a maximum (minimum) value, and this value can be found at a corner point.

This statement gives us a way of finding an optimum solution without drawing isoprofit lines, as we did previously: We simply evaluate the objective function at each of the corner points of the feasible region and then choose a corner point at which the function is optimum.

For example, in Figure 7.13 the corner points are A, B, C, D, and E. We found A before to be $(40, 60)$. To find B, we see from Figure 7.12 that we must solve $2x + y = 180$ and $x + y = 100$ simultaneously. This gives the point $B = (80, 20)$. In a similar way, we obtain all the corner points:

$$A = (40, 60) \qquad B = (80, 20) \qquad C = (90, 0)$$
$$D = (0, 0) \qquad E = (0, 80)$$

We now evaluate the objective function $P = 4x + 6y$ at each point:

$$P(A) = 4(40) + 6(60) = 520$$

$$P(B) = 4(80) + 6(20) = 440$$

$$P(C) = 4(90) + 6(0) = 360$$

$$P(D) = 4(0) + 6(0) = 0$$

$$P(E) = 4(0) + 6(80) = 480$$

Thus, P has a maximum value of 520 at A, where $x = 40$ and $y = 60$.

The optimum solution to a linear programming problem is given by the optimum value of the objective function and the point where the optimum value of the objective function occurs.

EXAMPLE 1 **Solving a Linear Programming Problem**

Maximize the objective function $Z = 3x + y$ subject to the constraints

$$2x + y \leq 8$$

$$2x + 3y \leq 12$$

$$x \geq 0$$

$$y \geq 0$$

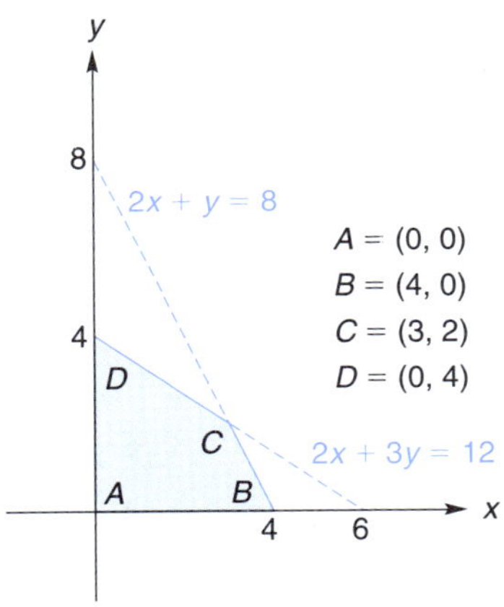

FIGURE 7.14 A, B, C, and D are corner points of feasible region.

Solution: In Figure 7.14, the feasible region is nonempty and bounded. Thus, Z is maximum at one of the four corner points. The coordinates of A, B, and D are obvious on inspection. To find the coordinates of C, we solve the equations $2x + y = 8$ and $2x + 3y = 12$ simultaneously, which gives $x = 3$, $y = 2$. Thus,

$$A = (0, 0) \qquad B = (4, 0) \qquad C = (3, 2) \qquad D = (0, 4)$$

Evaluating Z at these points, we obtain

$$Z(A) = 3(0) + 0 = 0$$

$$Z(B) = 3(4) + 0 = 12$$

$$Z(C) = 3(3) + 2 = 11$$

$$Z(D) = 3(0) + 4 = 4$$

Hence, the maximum value of Z, subject to the constraints, is 12, and it occurs when $x = 4$ and $y = 0$.

▶ **Now work Problem 1.**

Empty Feasible Region

The next example illustrates a situation where no optimum solution exists.

EXAMPLE 2 **Empty Feasible Region**

Minimize the objective function $Z = 8x - 3y$, subject to the constraints

$$-x + 3y = 21$$

$$x + y \leq 5$$

$$x \geq 0$$

$$y \geq 0$$

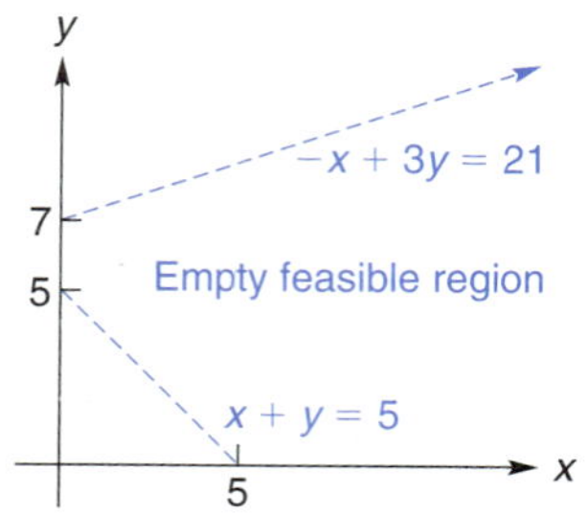

FIGURE 7.15 Empty feasible region.

Solution: Notice that the first constraint, $-x + 3y = 21$, is an *equality*. The portions of the lines $-x + 3y = 21$ and $x + y = 5$ for which $x \geq 0$ and $y \geq 0$ are shown in Figure 7.15. They will remain dashed lines until we determine whether or not they are to be included in the feasible region. A feasible point (x, y) must have $x \geq 0$ and $y \geq 0$, and must lie both on the top dashed line and on or below the bottom dashed line (since $y \leq 5 - x$). However, no such point exists. Hence, the feasible region is *empty,* and the problem has *no* optimum solution.

▶ **Now work Problem 5.**

The situation in Example 2 can be made more general:

Whenever the feasible region of a linear programming problem is empty, no optimum solution exists.

Unbounded Feasible Region

Suppose a feasible region is defined by

$$y = 2$$
$$x \geq 0$$
$$y \geq 0$$

This region is the portion of the horizontal line $y = 2$ indicated in Figure 7.16. Since the region cannot be contained within a circle, it is *unbounded*. Let us consider maximizing

$$Z = x + y$$

subject to the foregoing constraints. Since $y = 2$, $Z = x + 2$. Clearly, as x increases without bound, so does Z. Thus, no feasible point maximizes Z, so no optimum solution exists. In this case, we say that the solution is "unbounded." On the other hand, suppose we want to *minimize* $Z = x + y$ over the same region. Since $Z = x + 2$, then Z is minimum when x is as small as possible, namely, when $x = 0$. This gives a minimum value of $Z = x + y = 0 + 2 = 2$, and the optimum solution is the corner point $(0, 2)$.

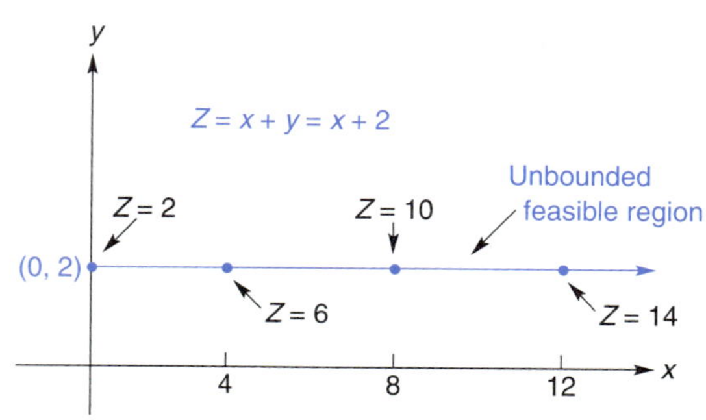

FIGURE 7.16 Unbounded feasible region on which Z has no maximum.

In general, it can be shown that:

If a feasible region is unbounded, and if the objective function has a maximum (or minimum) value, then that value occurs at a corner point.

EXAMPLE 3 **Unbounded Feasible Region**

A produce grower is purchasing fertilizer containing three nutrients, A, B, *and* C. *The minimum needs are* 160 *units of* A, 200 *units of* B, *and* 80 *units of* C. *There are two popular brands of fertilizer on the market. Fast Grow, costing* \$8 *a bag, contains* 3 *units of* A, 5 *units of* B, *and* 1 *unit of* C. *Easy Grow, costing* \$6 *a bag, contains* 2 *units of each nutrient. If the grower wishes to minimize cost while still maintaining the nutrients required, how many bags of each brand should be bought? The information is summarized as follows:*

	Fast Grow	Easy Grow	Units Required
A	3 units	2 units	160
B	5 units	2 units	200
C	1 unit	2 units	80
Cost/Bag	\$8	\$6	

Solution: Let x be the number of bags of Fast Grow that are bought and y the number of bags of Easy Grow that are bought. Then we wish to *minimize* the cost function

$$C = 8x + 6y \tag{7}$$

subject to the constraints

$$3x + 2y \geq 160 \tag{8}$$

$$5x + 2y \geq 200 \tag{9}$$

$$x + 2y \geq 80 \tag{10}$$

$$x \geq 0 \tag{11}$$

$$y \geq 0 \tag{12}$$

The feasible region satisfying constraints (8)–(12) is shaded in Figure 7.17, along with *isocost lines* for $C = 400$ and $C = 600$. The feasible region is unbounded. The member of the family of lines $C = 8x + 6y$ that gives a minimum cost, subject to the constraints, intersects the feasible region at the corner point B. Here we chose the isocost line whose y-intercept was *closest* to the origin and that had at least one point in common with the feasible region. The coordinates

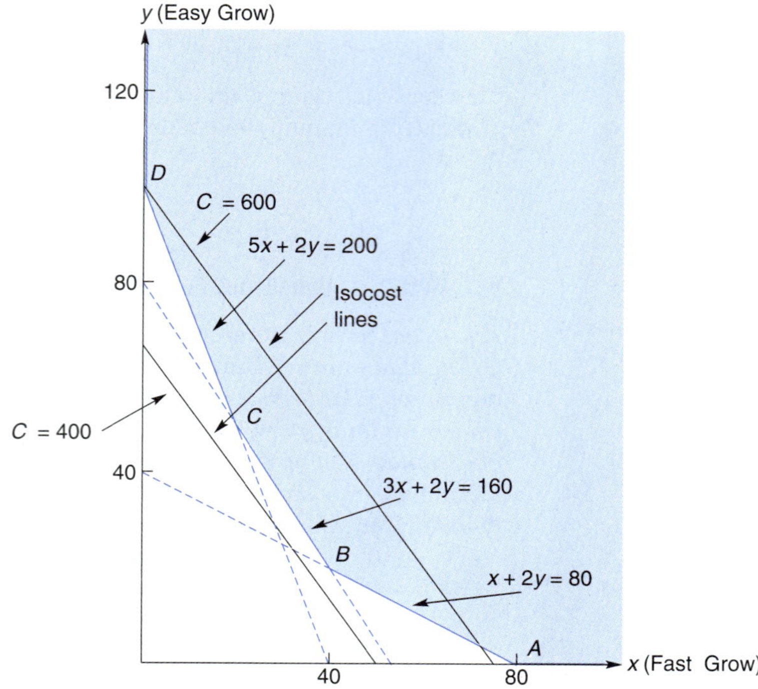

FIGURE 7.17 Minimum cost at corner point B of unbounded feasible region.

of B are found by solving the system

$$\begin{cases} 3x + 2y = 160 \\ x + 2y = 80 \end{cases}$$

Thus, $x = 40$ and $y = 20$, which gives a minimum cost of \$440. The produce grower should buy 40 bags of Fast Grow and 20 bags of Easy Grow.

▶ **Now work Problem 15.**

In Example 3, we found that the function $C = 8x + 6y$ has a minimum value at a corner point of the unbounded feasible region. On the other hand, suppose we want to *maximize* C over that region and take the approach of evaluating C at all corner points. These points are

$$A = (80, 0) \qquad B = (40, 20) \qquad C = (20, 50) \qquad D = (0, 100)$$

from which we obtain

$$C(A) = 8(80) + 6(0) = 640$$
$$C(B) = 8(40) + 6(20) = 440$$
$$C(C) = 8(20) + 6(50) = 460$$
$$C(D) = 8(0) + 6(100) = 600$$

CAUTION When working with an unbounded feasible region, do not simply conclude that an optimum solution exists at a corner point, since there may not be an optimum solution.

A hasty conclusion is that the maximum value of C is 640. This is *false!* There is *no* maximum value, since isocost lines with arbitrarily large values of C intersect the feasible region.

TECHNOLOGY

Problem: Maximize $Z = 4.1x - 3.2y$ subject to the constraints

$$y \leq 8 - x \tag{13}$$

$$y \leq 6 - 0.2x \tag{14}$$

$$y \geq 2 + 0.3x \tag{15}$$

and

$$x \geq 0 \qquad y \geq 0$$

Solution: As shown in Figure 7.18, we enter the objective function as Y_1, where y is entered as "alpha Y." Next, the equations corresponding to constraints (13)–(15) are entered as Y_2, Y_3, and Y_4. To begin, the Y_1 function is "turned off" (that is, the symbol "=" is not highlighted), and we obtain the graphs of Y_2, Y_3, and Y_4. (See Figure 7.19.) It is a good idea to make a pencil sketch of the graphs and label the lines. From the sketch, we determine the feasible region and label any corner points. In Figure 7.19, the feasible region is shaded and the corner points are A, B, C, and D. Because the feasible region is nonempty and bounded, the maximum value of Z will occur at one of the corner points.

FIGURE 7.18 Entering objective function and equations corresponding to constraints and "turning off" objective function.

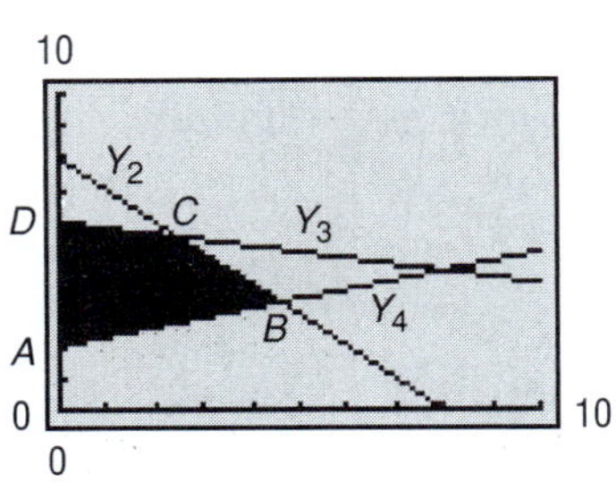

FIGURE 7.19 Determining feasible region and labeling corner points.

Point A is the y-intercept of Y_4 and is easily found to be (0, 2). The value of Z at A is found by storing 0 in X, and 2 in Y and then evaluating Y_1. (See Figure 7.20.) Thus, $Z = -6.4$ at this corner point.

Point B is the intersection of Y_2 and Y_4. To find the coordinates of B, we first turn off the Y_3 function and highlight only Y_2 and Y_4. After displaying the graphs of Y_2 and Y_4, we find their intersection point. On the TI-83 it is convenient to use the "intersection" feature. (See Figure 7.21.) The values of X and Y at the intersection are automatically stored in the X and Y registers. Returning to the home screen, we evaluate Y_1 and obtain 8.09 (rounded to two decimal places). Thus, $Z = 8.09$ at corner point B.

Continuing in this fashion, we find the coordinates of C and D and evaluate Y_1 (or Z) there:

$$C = (2.5, 5.5) \qquad Z = -7.35$$

$$D = (0, 6) \qquad Z = -19.2$$

Hence, the maximum value of Z is 8.09 and occurs at the corner point B, where $x \approx 4.62$ and $y \approx 3.38$.

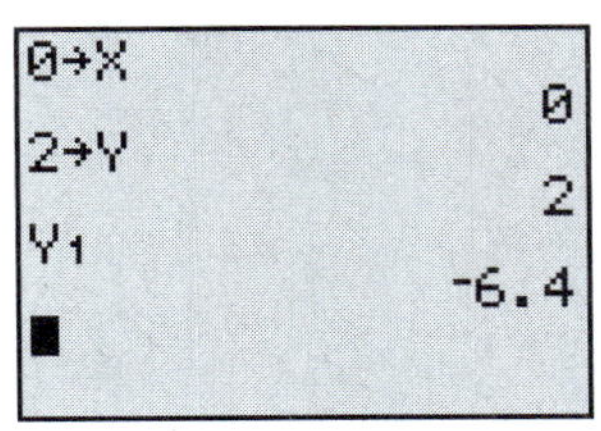

FIGURE 7.20 Evaluating objective function at corner point $A = (0, 2)$.

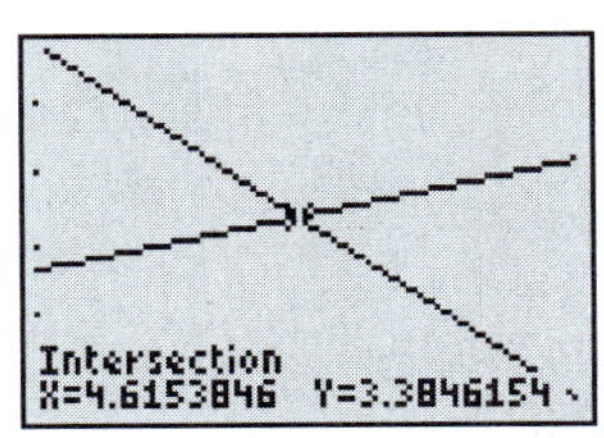

FIGURE 7.21 Determining corner point B.

Exercise 7.2

*1. Maximize

$$P = 5x + 7y$$

subject to

$$\begin{aligned} 2x + 3y &\le 45 \\ 4x - 21y &\ge 6 \\ x, y &\ge 0 \end{aligned}$$

2. Maximize

$$P = 5x + 6y$$

subject to

$$\begin{aligned} x + y &\le 80 \\ 3x + 2y &\le 220 \\ 2x + 3y &\le 210 \\ x, y &\ge 0 \end{aligned}$$

3. Maximize

$$Z = 4x - 6y$$

subject to

$$\begin{aligned} y &\le 7 \\ 3x - y &\le 3 \\ x + y &\ge 5 \\ x, y &\ge 0 \end{aligned}$$

4. Minimize

$$Z = x + y$$

subject to

$$\begin{aligned} x - y &\ge 0 \\ 4x + 3y &\ge 12 \\ 9x + 11y &\le 99 \\ x &\le 8 \\ x, y &\ge 0 \end{aligned}$$

*5. Maximize

$$Z = 4x - 10y$$

subject to

$$\begin{aligned} x - 4y &\ge 4 \\ 2x - y &\le 2 \\ x, y &\ge 0 \end{aligned}$$

6. Minimize

$$Z = 20x + 30y$$

subject to

$$\begin{aligned} 2x + y &\le 10 \\ 3x + 4y &\le 24 \\ 8x + 7y &\ge 56 \\ x, y &\ge 0 \end{aligned}$$

7. Minimize

$$Z = 7x + 3y$$

subject to

$$\begin{aligned} 3x - y &\ge -2 \\ x + y &\le 9 \\ x - y &= -1 \\ x, y &\ge 0 \end{aligned}$$

8. Maximize

$$Z = 0.4x - 0.2y$$

subject to

$$\begin{aligned} 2x - 5y &\ge -3 \\ 2x - y &\le 5 \\ 3x + y &= 6 \\ x, y &\ge 0 \end{aligned}$$

9. Minimize

$$C = 2x + y$$

subject to

$$\begin{aligned} 3x + y &\ge 3 \\ 4x + 3y &\ge 6 \\ x + 2y &\ge 2 \\ x, y &\ge 0 \end{aligned}$$

10. Minimize

$$C = 2x + 2y$$

subject to

$$\begin{aligned} x + 2y &\ge 80 \\ 3x + 2y &\ge 160 \\ 5x + 2y &\ge 200 \\ x, y &\ge 0 \end{aligned}$$

11. Maximize

$$Z = 10x + 2y$$

subject to

$$\begin{aligned} x + 2y &\ge 4 \\ x - 2y &\ge 0 \\ x, y &\ge 0 \end{aligned}$$

12. Minimize

$$Z = y - x$$

subject to

$$\begin{aligned} x &\ge 3 \\ x + 3y &\ge 6 \\ x - 3y &\ge -6 \\ x, y &\ge 0 \end{aligned}$$

13. Production for Maximum Profit A toy manufacturer preparing a production schedule for two new toys, trucks and spinning tops, must use the information concerning their construction times given in the following table:

	Machine A	Machine B	Finishing
Truck	2 hr	3 hr	5 hr
Spinning Top	1 hr	1 hr	1 hr

For example, each truck requires 2 hours on machine A. The available employee hours per week are as follows: for operating machine A, 80 hours; for B, 50 hours; for finishing, 70 hours. If the profits on each truck and spinning top are \$7 and \$2, respectively, how many of each toy should be made per week in order to maximize profit? What would the maximum profit be?

14. Production for Maximum Profit A manufacturer produces two types of barbecue grills: Old Smokey and Blaze Away. During production, the grills require the use of two machines, A and B. The number of hours needed on both machines are indicated in the following table:

	Machine A	Machine B
Old Smokey	2 hr	4 hr
Blaze Away	4 hr	2 hr

If each machine can be used 24 hours a day, and the profits on the Old Smokey and Blaze Away models are \$4 and \$6, respectively, how many of each type of grill should be made per day to obtain maximum profit? What is the maximum profit?

*15. **Diet Formulation** A diet is to contain at least 16 units of carbohydrates and 20 units of protein. Food A contains 2 units of carbohydrates and 4 of protein; food B contains 2 units of carbohydrates and 1 of protein. If food A costs \$1.20 per unit and food B costs \$0.80 per unit, how many units of each food should be purchased in order to minimize cost? What is the minimum cost?

16. Fertilizer Nutrients A produce grower is purchasing fertilizer containing three nutrients: A, B, and C. The minimum weekly requirements are 80 units of A, 120 of B, and 240 of C. There are two popular blends of fertilizer on the market. Blend I, costing \$8 a bag, contains 2 units of A, 6 of B, and 4 of C. Blend II, costing \$10 a bag, contains 2 units of A, 2 of B, and 12 of C. How many bags of each blend should the grower buy each week to minimize the cost of meeting the nutrient requirements?

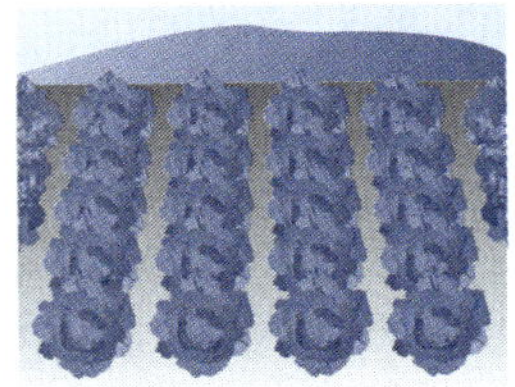

17. Mineral Extraction A company extracts minerals from ore. The numbers of pounds of minerals A and B that can be extracted from each ton of ores I and II are given in the following table, together with the costs per ton of the ores:

	Ore I	Ore II
Mineral A	100 lb	200 lb
Mineral B	200 lb	50 lb
Cost per ton	\$50	\$60

If the company must produce at least 3000 lb of A and 2500 lb of B, how many tons of each ore should be processed in order to minimize cost? What is the minimum cost?

18. Production Scheduling An oil company that has two refineries needs at least 8000, 14,000, and 5000 barrels of low-, medium-, and high-grade oil, respectively. Each day, Refinery I produces 2000 barrels of low-, 3000 barrels of medium-, and 1000 barrels of high-grade oil, whereas Refinery II produces 1000 barrels each of low- and high- and 2000 barrels of medium-grade oil. If it costs \$25,000 per day to operate Refinery I and \$20,000 per day to operate Refinery II, how many days should each refinery be operated to satisfy the production requirements at minimum cost? What is the minimum cost? (Assume that a minimum cost exists.)

19. Construction Cost A chemical company is designing a plant for producing two types of polymers, P_1 and P_2. The plant must be capable of producing at least 100 units of P_1 and 420 units of P_2 each day. There are two possible designs for the basic reaction chambers that are to be included in the plant. Each chamber of type A costs \$600,000 and is capable of producing 10 units of P_1 and 20 units of P_2 per day; type B is of cheaper design, costing \$300,000, and is capable of producing 4 units of P_1 and 30 units of P_2 per day. Because of operating costs, it is necessary to have at least four chambers of each type in the plant. How many chambers of each type should be included to minimize the cost of construction and still meet the required production schedule? (Assume that a minimum cost exists.)

20. Pollution Control Because of new federal regulations on pollution, a chemical company has introduced into its plant a new, more expensive process to supplement or replace an older process in the production of a particular chemical. The older process discharges 25 grams of carbon dioxide and 50 grams of particulate matter into the atmosphere for each liter of chemical produced. The new process discharges 15 grams of carbon dioxide and 40 grams of particulate matter into the atmosphere for each liter produced. The company makes a profit of 40 cents per liter and 15 cents per liter on the old and new processes, respectively. If the government allows the plant to discharge no more than 12,525 grams of carbon dioxide and no more than 20,000 grams of particulate matter into the atmosphere each day, how many liters of chemical should be produced daily, by each process, to maximize daily profit? What is the maximum daily profit?

21. Construction Discount The highway department has decided to add exactly 200 km of highway and exactly 100 km of expressway to its road system this year. The standard price for road construction is \$1 million per kilometer of highway and \$5 million per kilometer of expressway. Only two contractors, company A and company B, can do this kind of construction, so the entire 300 km of road must be built by these two companies. However, company A can construct at most 200 km of roadway (highway and expressway), and company B can construct at most 150 km. For political reasons, each company must be awarded a contract with a standard price of at least \$250 million (before discounts). Company A offers a discount of \$1000 per kilometer of highway and \$6000 per kilometer of expressway; company B offers a discount of \$2000 for each kilometer of highway and \$5000 for each kilometer of expressway.

(a) Let x and y represent the number of kilometers of highway and expressway, respectively, awarded to company A. Show that the total discount received from both companies is given by

$$D = 900 - x + y$$

where D is in thousands of dollars.

(b) The highway department wishes to maximize the total discount D. Show that this problem is equivalent to the following linear programming problem, by showing exactly how the first four constraints arise:

$$\text{Maximize } D = 900 - x + y$$

subject to

$$\begin{aligned} x + y &\le 200 \\ x + y &\ge 150 \\ x + 5y &\ge 250 \\ x + 5y &\le 450 \\ x, y &\ge 0 \end{aligned}$$

(c) Find the values of x and y that maximize D.

In Problems 22–25, round answers to two decimal places.

22. Maximize

$$Z = 2x + 0.3y$$

subject to

$$\begin{aligned} y &\le 6 - 4x \\ y &\ge 2 - 0.5x \\ x, y &\ge 0 \end{aligned}$$

23. Maximize

$$Z = 14x - 3y$$

subject to

$$\begin{aligned} y &\ge 12.5 - 4x \\ y &\le 9.3 - x \\ y &\ge 4.7 + 0.8x \\ x, y &\ge 0 \end{aligned}$$

24. Minimize

$$Z = 5.1y - 3.5x$$

subject to

$$\begin{aligned} 7.5x + 2y &\ge 35 \\ 2.5x + y &\le 7.4 \\ 0.6x - y &\ge -0.8 \\ x, y &\ge 0 \end{aligned}$$

25. Minimize

$$Z = 17.3x - 14.4y$$

subject to

$$\begin{aligned} 0.73x - y &\le -2.4 \\ 1.22x - y &\ge -5.1 \\ 0.45x - y &\ge -12.4 \\ x, y &\ge 0 \end{aligned}$$

OBJECTIVE *To consider situations in which a linear programming problem has more than one optimum solution.*

APPLICATIONS IN PRACTICE 1

Multiple Optimum Solutions

Suppose a TV dealer has stores A and B and warehouses C and D. The cost of shipping a TV is \$18 from C to A, \$9 from C to B, \$24 from D to A, and \$15 from D to B. Suppose that store A orders 25 TV sets and store B orders 30 TV sets. Suppose also that warehouse C has 45 TV sets and warehouse D has 40 TV sets available. Find the best way to minimize cost, and find the minimum cost. (*Hint:* Let x be the number of TV sets shipped from C to A and y be the number of TV sets shipped from C to B. Then $25 - x$ is the number of TV sets shipped from D to A and $30 - y$ the number of TV sets shipped from D to B.)

7.3 Multiple Optimum Solutions[1]

Sometimes an objective function attains its optimum value at more than one feasible point, in which case **multiple optimum solutions** are said to exist. Example 1 will illustrate.

EXAMPLE 1 **Multiple Optimum Solutions**

Maximize $Z = 2x + 4y$ *subject to the constraints*

$$x - 4y \leq -8$$
$$x + 2y \leq 16$$
$$x, y \geq 0$$

Solution: The feasible region appears in Figure 7.22. Since the region is nonempty and bounded, Z has a maximum value at a corner point. The corner points are $A = (0, 2)$, $B = (8, 4)$, and $C = (0, 8)$. Evaluating the objective function at

$$A = (0, 2) \qquad B = (8, 4) \qquad C = (0, 8)$$

gives

$$Z(A) = 2(0) + 4(2) = 8$$
$$Z(B) = 2(8) + 4(4) = 32$$
$$Z(C) = 2(0) + 4(8) = 32$$

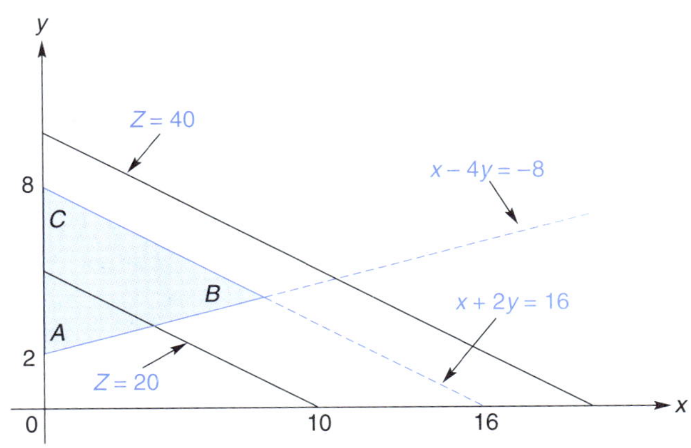

FIGURE 7.22 $Z = 2x + 4y$ is maximized at each point on the line segment $\overline{BC}$.

Thus, the maximum value of Z over the region is 32, and it occurs at *two* corner points, B and C. In fact, this maximum value also occurs at *all* points on the line segment *joining* B and C, for the following reason. Each member of the family of lines $Z = 2x + 4y$ has slope $-\frac{1}{2}$. Moreover, the constraint line $x + 2y = 16$, which contains B and C, also has slope $-\frac{1}{2}$ and hence is parallel to each member of $Z = 2x + 4y$. Figure 7.22 shows lines for $Z = 20$ and $Z = 40$. Note that the member of the family that maximizes Z contains not only B and C, but also all points on the line segment $\overline{BC}$. It thus has infinitely many points in common

[1] This section can be omitted.

with the feasible region. Hence, this linear programming problem has infinitely many optimum solutions. In fact, it can be shown that

If (x_1, y_1) and (x_2, y_2) are two corner points at which an objective function is optimum, then the function will also be optimum at all points (x, y) where

$$x = (1 - t)x_1 + tx_2$$
$$y = (1 - t)y_1 + ty_2$$

and

$$0 \leq t \leq 1$$

In our case, if $(x_1, y_1) = B = (8, 4)$ and $(x_2, y_2) = C = (0, 8)$, then Z is maximum at any point (x, y) where

$$x = (1 - t)8 + t \cdot 0 = 8(1 - t)$$
$$y = (1 - t)4 + t \cdot 8 = 4(1 + t)$$
$$\text{for} \quad 0 \leq t \leq 1$$

These equations give the coordinates of any point on the line segment $\overline{BC}$. In particular, if $t = 0$, then $x = 8$ and $y = 4$, which gives the corner point $B = (8, 4)$. If $t = 1$, we get the corner point $C = (0, 8)$. The value $t = \frac{1}{2}$ gives the point (4, 6). Notice that at (4, 6), $Z = 2(4) + 4(6) = 32$, which is the maximum value of Z.

▶ **Now work Problem 1.**

Exercise 7.3

*1. Minimize

$$Z = 3x + 9y$$

subject to

$$y \geq -\tfrac{3}{2}x + 6$$
$$y \geq -\tfrac{1}{3}x + \tfrac{11}{3}$$
$$y \geq x - 3$$
$$x, y \geq 0$$

2. Maximize

$$Z = 2x + 2y$$

subject to

$$2x - y \geq -4$$
$$x - 2y \leq 4$$
$$x + y = 6$$
$$x, y \geq 0$$

3. Maximize

$$Z = 18x + 9y$$

subject to

$$2x + 3y \leq 12$$
$$2x + y \leq 8$$
$$x, y \geq 0$$

4. Minimize Cost Suppose a car dealer has showrooms in Atherton and Berkeley and warehouses in Concord and Dublin. The cost of delivering a car is \$60 from Concord to Atherton, \$45 from Concord to Berkeley, \$50 from Dublin to Atherton, and \$35 from Dublin to Berkeley. Suppose that the showroom in Atherton orders seven cars and the showroom in Berkeley orders four cars. Suppose also that the warehouse in Concord has six cars and the warehouse in Dublin has eight cars available. Find the best way to minimize cost, and find the minimum cost. (*Hint:* Let x be the number of cars delivered from Concord to Atherton and y be the number of cars delivered from Concord to Berkeley. Then $7 - x$ is the number of cars delivered from Dublin to Atherton and $4 - y$ the number of cars shipped from Dublin to Berkeley.)

OBJECTIVE *To show how the simplex method is used to solve a standard linear programing problem. This method will allow you to solve problems that cannot be solved geometrically.*

7.4 The Simplex Method

Up to now, we have solved linear programming problems by a geometric method. This method is not practical when the number of variables increases to three, and is not possible beyond that. Now we shall look at a different technique—the **simplex method,** whose name is linked in more advanced discussions to a geometrical object called a simplex.

The simplex method begins with a feasible solution and tests whether it is optimal. If it is not, the method proceeds to a better solution. We say "better" in the sense that the new solution usually brings you closer to optimization of the objective function.[2] Should this new solution not be optimal, we repeat the procedure. Eventually, the simplex method leads to an optimal solution, if one exists.

Besides being efficient, the simplex method has other advantages. For one, it is completely mechanical. (We use matrices, elementary row operations, and basic arithmetic.) Moreover, no geometry is involved; this allows us to solve linear programming problems having any number of constraints and variables.

In this section, we shall consider only so-called **standard linear programming problems.** These can be put in the following form.

Standard Linear Programming Problem

Maximize the linear function $Z = c_1x_1 + c_2x_2 + \cdots + c_nx_n$ subject to the constraints

$$\left.\begin{aligned} a_{11}x_1 + a_{12}x_2 + \cdots + a_{1n}x_n &\le b_1 \\ a_{21}x_1 + a_{22}x_2 + \cdots + a_{2n}x_n &\le b_2 \\ \vdots \qquad \vdots \qquad\quad \vdots \quad &\quad \vdots \\ a_{m1}x_1 + a_{m2}x_2 + \cdots + a_{mn}x_n &\le b_m \end{aligned}\right\} \qquad (1)$$

where $x_1, x_2, \ldots, x_n$ and $b_1, b_2, \ldots, b_m$ are nonnegative.

It is helpful to formulate the problem in matrix notation so as to make its structure more memorable. Let

$$\mathbf{C} = \begin{bmatrix} c_1 & c_2 & \cdots & c_n \end{bmatrix} \quad \text{and} \quad \mathbf{X} = \begin{bmatrix} x_1 \\ x_2 \\ \vdots \\ x_n \end{bmatrix}$$

Then the objective function can be written as

$$W = \mathbf{CX}$$

Now if we write

$$\mathbf{A} = \begin{bmatrix} a_{11} & a_{12} & \cdots & a_{1n} \\ a_{21} & a_{22} & \cdots & a_{2n} \\ \vdots & \vdots & & \vdots \\ a_{m1} & a_{m2} & \cdots & a_{mn} \end{bmatrix} \quad \text{and} \quad \mathbf{B} = \begin{bmatrix} b_1 \\ b_2 \\ \vdots \\ b_m \end{bmatrix}$$

[2] In most cases this is true. In some situations, however, the new solution may be only "just as good" as the previous one. Example 2 will illustrate this.

then we can say that a Standard Linear Programming Problem is one that can be put in the form:

Note that $B \geq 0$ is a condition on the data and is not considered to be a "constraint".

$$\text{Maximixe} \qquad W = \mathbf{CX}$$

$$\text{subject to} \qquad \begin{cases} \mathbf{AX} \leq \mathbf{B} \\ \mathbf{X} \geq \mathbf{0} \end{cases}$$

$$\text{where} \qquad \mathbf{B} \geq \mathbf{0}$$

(Matrix inequalities are to be understood like matrix equality. The comparisons refer to matrices of the same size and the inequality is required to hold for all corresponding entries.)

Other types of linear programming problems will be discussed in Sections 7.6 and 7.7.

Note that one feasible solution to a *standard* linear programming problem is always $x_1 = 0, x_2 = 0, \ldots, x_n = 0$ and that at this feasible solution the value of the objective function Z is 0.

The procedure that we follow here will be outlined later in this section.

We shall now apply the simplex method to the problem in Example 1 of Section 7.2, which has the form

$$\text{maximize } Z = 3x_1 + x_2$$

subject to the constraints

$$2x_1 + x_2 \leq 8 \tag{2}$$

and

$$2x_1 + 3x_2 \leq 12 \tag{3}$$

where $x_1 \geq 0$ and $x_2 \geq 0$. This problem is of standard form. We begin by expressing constraints (2) and (3) as equations. In (2), $2x_1 + x_2$ will *equal* 8 if we add some nonnegative number s_1 to $2x_1 + x_2$, so that

$$2x_1 + x_2 + s_1 = 8 \quad \text{for some } s_1 \geq 0$$

We call s_1 a **slack variable,** since it makes up for the "slack" on the left side of (2) to give us equality. Similarly, inequality (3) can be written as an equation by using the slack variable s_2; we have

$$2x_1 + 3x_2 + s_2 = 12 \quad \text{for some } s_2 \geq 0$$

The variables x_1 and x_2 are called **decision variables.**

Now we can restate the problem in terms of equations:

$$\text{Maximize } Z = 3x_1 + x_2 \tag{4}$$

subject to

$$2x_1 + x_2 + s_1 = 8 \tag{5}$$

and

$$2x_1 + 3x_2 + s_2 = 12 \qquad (6)$$

where x_1, x_2, s_1, and s_2 are nonnegative.

From Section 7.2, we know that the optimum solution occurs at a corner point of the feasible region in Figure 7.23. At each of these points, at least *two* of the variables x_1, x_2, s_1, and s_2 are 0, as the following listing indicates:

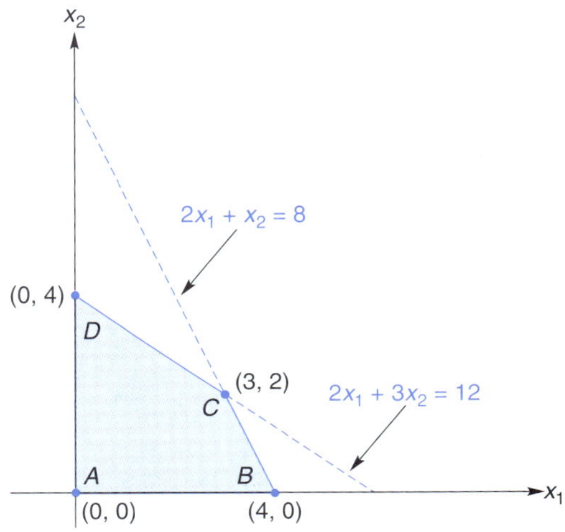

FIGURE 7.23 Optimum solution must occur at corner point of feasible region.

1. At *A*, we have $x_1 = 0$ and $x_2 = 0$.
2. At *B*, $x_1 = 4$ and $x_2 = 0$. But from Equation (5), $2(4) + 0 + s_1 = 8$. Thus, $s_1 = 0$.
3. At *C*, $x_1 = 3$ and $x_2 = 2$. But from Equation (5), $2(3) + 2 + s_1 = 8$. Hence, $s_1 = 0$. From Equation (6), $2(3) + 3(2) + s_2 = 12$. Therefore, $s_2 = 0$.
4. At *D*, $x_1 = 0$ and $x_2 = 4$. From Equation (6), $2(0) + 3(4) + s_2 = 12$. Thus, $s_2 = 0$.

CAUTION We remarked earlier that there is a great deal of terminology used in the discussion of linear programming. In particular there are many types of *variables*. It is important to understand that the variables called decision variables $x_1, x_2, \ldots, x_n$ remain decision variables throughout the solution of a problem and the same remark applies to the slack variables $s_1, s_2, \ldots, s_m$. In the process of examining the corner points of the feasible region we find solutions to the system in which at least n of the $n + m$ variables are 0. Precisely n of these are called nonbasic variables and the remaining m are called basic variables. Which m of the $n + m$ variables are *basic* depends on the corner point under consideration. Among other things, the procedure that we are describing provides a mechanical way of keeping track of which variables, at any time, are basic.

It can also be shown that any solution to Equations (5) and (6), such that at least *two* of the four variables x_1, x_2, s_1, and s_2 are zero, corresponds to a corner point. Any such solution where at least two of these variables are zero is called a **basic feasible solution** (abbreviated BFS). This number, 2, is determined by the number n of decision variables, 2 in the present example. For any particular BFS, the variables held at 0 are called **nonbasic variables,** and all the others are called **basic variables** for that BFS. Since there is a total of $n + m$ variables, the number of basic variables in the general system that arises from (1) is m, the number of constraints (other than those expressing nonnegativity). Thus, for the BFS corresponding to item 3 in the preceding list, s_1 and s_2 are the nonbasic variables x_1 and x_2 are the basic variables, but for the BFS corresponding to item 4, the nonbasic variables are x_1 and s_2 and the basic variables are x_2 and s_1.

We shall first find an initial BFS, and hence an initial corner point, and then determine whether the corresponding value of Z can be increased by a different BFS. Since $x_1 = 0$ and $x_2 = 0$ is a feasible solution to this standard linear programming problem, let us initially find the BFS such that the decision variables x_1 and x_2 are nonbasic and hence the slack variables s_1 and s_2 are basic. That is, we choose $x_1 = 0$ and $x_2 = 0$ and find the corresponding values of s_1, s_2, and Z. This can be done most conveniently by matrix techniques, based on the methods developed in Chapter 6.

If we write Equation (4) as $-3x_1 - x_2 + Z = 0$ then Equations (5), (6), and (4) form the linear system

$$\begin{cases} 2x_1 + x_2 + s_1 & = 8 \\ 2x_1 + 3x_2 + s_2 & = 12 \\ -3x_1 - x_2 + Z & = 0 \end{cases}$$

in the variables $x_1, x_2, s_1, s_2,$ *and* Z. Thus, in general when we add the objective function to the system that provides the constraints we have $m + 1$ equations in $n+m+1$ unknowns. In terms of an augmented coefficient matrix, called the **initial simplex tableau,** we have

$$\begin{array}{c} \text{B} \\ s_1 \\ s_2 \\ Z \end{array} \begin{array}{c} \begin{array}{cccccc} x_1 & x_2 & s_1 & s_2 & Z & \text{R} \end{array} \\ \left[\begin{array}{rrrrr|r} 2 & 1 & 1 & 0 & 0 & 8 \\ 2 & 3 & 0 & 1 & 0 & 12 \\ \hline -3 & -1 & 0 & 0 & 1 & 0 \end{array}\right] \end{array}$$

It is convenient to be generous with labels for matrices that are being used as simplex tableaux. Thus, the columns in the matrix to the left of the vertical bar are labeled, naturally enough, by the variables to which they correspond. We have chosen R as a label for the column that provides the R)ight sides of the system of equations. We have chosen B to label the list of row labels. The first two rows correspond to the constraints, and the last row, called the **objective row,** corresponds to the objective equation—thus the horizontal separating line. Notice that if $x_1 = 0$ and $x_2 = 0$, then, from rows 1, 2, and 3, we can directly read off the values of s_1, s_2, and Z: $s_1 = 8$, $s_2 = 12$, and $Z = 0$. Thus the rows of this *initial* simplex tableau are labeled to the left by s_1, s_2, and Z. We remind you that s_1 and s_2 are the basic variables. So the column heading B can be understood to stand for B)asic variables. Our initial basic feasible solution is

$$x_1 = 0 \qquad x_2 = 0 \qquad s_1 = 8 \qquad s_2 = 12$$

at which $Z = 0$.

Let us see if we can find a BFS that gives a larger value of Z. The variables x_1 and x_2 are nonbasic in the preceding BFS. We shall now look for a BFS in which one of these variables is basic while the other remains nonbasic. Which one should we choose as the basic variable? Let us examine the possibilities. From the Z-row of the matrix above, $Z = 3x_1 + x_2$. If x_1 is allowed to become basic, then x_2 remains at 0 and $Z = 3x_1$; thus, for each one-unit increase in x_1, Z increases by three units. On the other hand, if x_2 is allowed to become basic, then x_1 remains at 0 and $Z = x_2$; hence, for each one-unit increase in x_2, Z increases by one unit. Consequently, we get a *greater* increase in the value of Z if x_1, rather than x_2, enters the basic-variable category. In this case, we call x_1 an **entering variable.** Thus, in terms of the simplex tableau that follows (which is the same as the matrix presented earlier, except for some additional labeling), the entering variable can be found by looking at the "most negative" of the numbers enclosed by the brace in the Z-row. (By *most negative*, we mean the negative indicator having the greatest magnitude.) Since that number is -3 and appears in the x_1-column, x_1 is the entering variable. The numbers in the brace are called **indicators.**

entering variable ↓

B	x_1	x_2	s_1	s_2	Z	R
s_1	2	1	1	0	0	8
s_2	2	3	0	1	0	12
Z	-3	-1	0	0	1	0

indicators (the entries -3, -1, 0, 0 of the Z-row)

Let us summarize the information that can be obtained from this tableau. It gives a BFS where s_1 and s_2 are the basic variables and x_1 and x_2 are nonbasic. The BFS is $s_1 = 8$ (the right-hand side of the s_1-row), $s_2 = 12$ (the right-hand side of the s_2-row), $x_1 = 0$, and $x_2 = 0$. The -3 in the x_1-column of the Z-row indicates that if x_2 remains 0, then Z increases three units for each one-unit increase in x_1. The -1 in the x_2-column of the Z-row indicates that if x_1 remains 0, then Z increases one unit for each one-unit increase in x_2. The column in which the most negative indicator, -3, lies gives the entering variable x_1—that is, the variable that should become basic in the next BFS.

In our new BFS, the larger the increase in x_1 (from $x_1 = 0$), the larger is the increase in Z. Now, by how much can we increase x_1? Since x_2 is still held at 0, from rows 1 and 2 of the simplex tableau, it follows that

$$s_1 = 8 - 2x_1$$

and

$$s_2 = 12 - 2x_1$$

Since s_1 and s_2 are nonnegative, we have

$$8 - 2x_1 \geq 0$$

and

$$12 - 2x_1 \geq 0$$

From the first inequality, $x_1 \leq \frac{8}{2} = 4$; from the second, $x_1 \leq \frac{12}{2} = 6$. Thus, x_1 must be less than or equal to the smaller of the quotients $\frac{8}{2}$ and $\frac{12}{2}$, which is $\frac{8}{2}$. Hence, x_1 can increase at most by 4. However, in a BFS, two variables must be 0. We already have $x_2 = 0$. Since $s_1 = 8 - 2x_1$, s_1 must be 0 for $x_1 = 4$. Therefore, we have a new BFS with x_1 replacing s_1 as a basic variable. That is, s_1 will *depart* from the list of basic variables in the previous BFS and will be nonbasic in the new BFS. We say that s_1 is the **departing variable** for the previous BFS. In summary, for our new BFS, we want x_1 and s_2 as basic variables with $x_1 = 4$, $s_2 = 12$ (as before) and x_2 and s_1 as nonbasic variables ($x_2 = 0, s_1 = 0$).

Before proceeding, let us update our tableau. To the right of the following tableau, the quotients $\frac{8}{2}$ and $\frac{12}{2}$ are indicated:

entering variable
(most negative indicator)
↓

$$\begin{array}{c} \text{departing} \leftarrow \\ \text{variable} \end{array}\quad \begin{array}{c} \text{B} \\ s_1 \\ s_2 \\ Z \end{array} \begin{array}{c} \begin{array}{cccccc} x_1 & x_2 & s_1 & s_2 & Z & \text{R} \end{array} \\ \left[\begin{array}{ccccc|c} 2 & 1 & 1 & 0 & 0 & 8 \\ 2 & 3 & 0 & 1 & 0 & 12 \\ \hline -3 & -1 & 0 & 0 & 1 & 0 \end{array}\right] \end{array} \quad \begin{array}{l} \textit{Quotients} \\ 8 \div 2 = 4 \text{ (smaller)} \\ 12 \div 2 = 6 \\ \end{array}$$

These quotients are obtained by dividing each entry in the first two rows of the r-column by the entry in the corresponding row of the entering-variable column, that is, the x_1-column. Notice that the departing variable is in the same row as the *smaller* quotient, $8 \div 2$.

Since x_1 and s_2 will be basic variables in our new BFS, it would be convenient to change our previous tableau by elementary row operations into a form in which the values of x_1, s_2, and Z can be read off with ease (just as we were able to do with the solution corresponding to $x_1 = 0$ and $x_2 = 0$). To do this, we want to find a matrix that is equivalent to the preceding tableau, but that has the form

$$\begin{array}{c} \text{B} \\ x_1 \\ s_2 \\ Z \end{array} \begin{array}{c} \begin{array}{cccccc} x_1 & x_2 & s_1 & s_2 & Z & \text{R} \end{array} \\ \left[\begin{array}{ccccc|c} 1 & ? & ? & 0 & 0 & ? \\ 0 & ? & ? & 1 & 0 & ? \\ \hline 0 & ? & ? & 0 & 1 & ? \end{array}\right] \end{array}$$

where the question marks represent numbers to be determined. Notice here that if $x_2 = 0$ and $s_1 = 0$, then x_1 equals the number in row x_1 of column R, s_2 equals the number in row s_2 of column R and Z is the number in row Z of column R. Thus, we must transform the tableau

entering
variable
↓

$$\begin{array}{c} \text{departing} \leftarrow \\ \text{variable} \end{array}\quad \begin{array}{c} \text{B} \\ s_1 \\ s_2 \\ Z \end{array} \begin{array}{c} \begin{array}{cccccc} x_1 & x_2 & s_1 & s_2 & Z & \text{R} \end{array} \\ \left[\begin{array}{ccccc|c} \boxed{2} & 1 & 1 & 0 & 0 & 8 \\ 2 & 3 & 0 & 1 & 0 & 12 \\ \hline -3 & -1 & 0 & 0 & 1 & 0 \end{array}\right] \end{array} \tag{7}$$

into an equivalent matrix that has a 1 where the "shaded" entry appears and 0's elsewhere in the x_1-column. The shaded entry is called the **pivot entry**—it is in the column of the entering variable (called the *pivot column*) and the row of the departing variable (called the *pivot row*). By elementary row operations, we have

$$\begin{array}{c} \begin{array}{ccccc} x_1 & x_2 & s_1 & s_2 & Z \end{array} \\ \left[\begin{array}{ccccc|c} \boxed{2} & 1 & 1 & 0 & 0 & 8 \\ 2 & 3 & 0 & 1 & 0 & 12 \\ \hline -3 & -1 & 0 & 0 & 1 & 0 \end{array}\right] \end{array}$$

$$\xrightarrow{\frac{1}{2}\text{R}_1} \left[\begin{array}{ccccc|c} 1 & \frac{1}{2} & \frac{1}{2} & 0 & 0 & 4 \\ 2 & 3 & 0 & 1 & 0 & 12 \\ \hline -3 & -1 & 0 & 0 & 1 & 0 \end{array}\right]$$

$$\xrightarrow[3R_1+R_3]{-2R_1+R_2}\left[\begin{array}{ccccc|c} 1 & \frac{1}{2} & \frac{1}{2} & 0 & 0 & 4 \\ 0 & 2 & -1 & 1 & 0 & 4 \\ \hline 0 & \frac{1}{2} & \frac{3}{2} & 0 & 1 & 12 \end{array}\right]$$

Thus, we have a new simplex tableau:

$$\begin{array}{c} \\ x_1 \\ s_2 \\ Z \end{array}\begin{array}{c} \begin{array}{cccccc} x_1 & x_2 & s_1 & s_2 & Z & R \end{array} \\ \left[\begin{array}{ccccc|c} 1 & \frac{1}{2} & \frac{1}{2} & 0 & 0 & 4 \\ 0 & 2 & -1 & 1 & 0 & 4 \\ \hline 0 & \underbrace{\frac{1}{2} \quad \frac{3}{2}}_{\text{indicators}} & & 0 & 1 & 12 \end{array}\right] \end{array} \tag{8}$$

For $x_2 = 0$ and $s_1 = 0$, from the first row, we have $x_1 = 4$; from the second, we obtain $s_2 = 4$. These values give us the new BFS. Note that we replaced the s_1 located to the left of the initial tableau (7) by x_1 in our new tableau (8), so that s_1 *departed* and x_1 *entered*. From row 3, for $x_2 = 0$ and $s_1 = 0$, we get $Z = 12$, which is a larger value than we had before ($Z = 0$).

In our present BFS, x_2 and s_1 are nonbasic variables ($x_2 = 0, s_1 = 0$). Suppose we look for another BFS that gives a larger value of Z and is such that one of x_2 or s_1 is basic. The equation corresponding to the Z-row is given by $\frac{1}{2}x_2 + \frac{3}{2}s_1 + Z = 12$, or

$$Z = 12 - \frac{1}{2}x_2 - \frac{3}{2}s_1. \tag{9}$$

If x_2 becomes basic and therefore s_1 remains nonbasic, then

$$Z = 12 - \frac{1}{2}x_2 \quad (\text{since } s_1 = 0).$$

Here, each one-unit increase in x_2 *decreases* Z by $\frac{1}{2}$ unit. Thus, any increase in x_2 would make Z smaller than before. On the other hand, if s_1 becomes basic and x_2 remains nonbasic, then, from Equation (9),

$$Z = 12 - \frac{3}{2}s_1 \quad (\text{since } x_2 = 0).$$

Here each one-unit increase in s_1 *decreases* Z by $\frac{3}{2}$ units. Hence, any increase in s_1 would make Z smaller than before. Consequently, we cannot move to a better BFS. In short, no BFS gives a larger value of Z than the BFS $x_1 = 4$, $s_2 = 4$, $x_2 = 0$, and $s_1 = 0$ (which gives $Z = 12$).

In fact, since $x_2 \geq 0$ and $s_1 \geq 0$, and since the coefficients of x_2 and s_1 in Equation (9) are negative, Z is maximum when $x_2 = 0$ and $s_1 = 0$. That is, in tableau (8), *having all nonnegative indicators means that we have an optimum solution*.

In terms of our original problem, if

$$Z = 3x_1 + x_2$$

subject to

$$2x_1 + x_2 \leq 8 \qquad 2x_1 + 3x_2 \leq 12 \qquad x_1, x_2 \geq 0$$

then Z is maximized when $x_1 = 4$ and $x_2 = 0$, and the maximum value of Z is 12. (This confirms the result of Example 1 of Section 7.2.) Note that the values of s_1 and s_2 do not have to appear here.

Let us outline the simplex method for a standard linear programming problem with three decision variables and four constraints, not counting nonnegativity conditions. The outline suggests how the simplex method works for any number of decision variables and any number of constraints.

Simplex Method

Problem:

$$\text{Maximize } Z = c_1x_1 + c_2x_2 + c_3x_3$$

subject to

$$a_{11}x_1 + a_{12}x_2 + a_{13}x_3 \leq b_1$$
$$a_{21}x_1 + a_{22}x_2 + a_{23}x_3 \leq b_2$$
$$a_{31}x_1 + a_{32}x_2 + a_{33}x_3 \leq b_3$$
$$a_{41}x_1 + a_{42}x_2 + a_{43}x_3 \leq b_4$$

where x_1, x_2, x_3 and b_1, b_2, b_3, b_4 are nonnegative.

Method:

1. Set up the initial simplex tableau:

$$\begin{array}{c} \text{B} \\ s_1 \\ s_2 \\ s_3 \\ s_4 \\ \\ Z \end{array} \begin{array}{c} \begin{array}{cccccccc|c} x_1 & x_2 & x_3 & s_1 & s_2 & s_3 & s_4 & Z & \text{R} \end{array} \\ \left[\begin{array}{cccccccc|c} a_{11} & a_{12} & a_{13} & 1 & 0 & 0 & 0 & 0 & b_1 \\ a_{21} & a_{22} & a_{23} & 0 & 1 & 0 & 0 & 0 & b_2 \\ a_{31} & a_{32} & a_{33} & 0 & 0 & 1 & 0 & 0 & b_3 \\ a_{41} & a_{42} & a_{43} & 0 & 0 & 0 & 1 & 0 & b_4 \\ \hline -c_1 & -c_2 & -c_3 & 0 & 0 & 0 & 0 & 1 & 0 \end{array}\right] \\ \underbrace{\hphantom{-c_1 \quad -c_2 \quad -c_3 \quad 0 \quad 0 \quad 0 \quad 0 \quad 1}}_{\text{indicators}} \end{array}$$

There are four slack variables: s_1, s_2, s_3, and s_4—one for each constraint.

2. If all the indicators in the last row are nonnegative, then Z has a maximum when $x_1 = 0$, $x_2 = 0$, and $x_3 = 0$. The maximum value is 0. If there are any negative indicators, locate the column in which the most negative indicator appears. This pivot column gives the entering variable. (If more than one column contains the most negative indicator, the choice of pivot column is arbitrary.)

3. Divide each *positive*[3] entry above the objective row in the entering-variable column *into* the corresponding value of column R.

4. Mark the entry in the pivot column that corresponds to the smallest quotient in step 3. This is the pivot entry and the row in which it is located is the pivot row. The departing variable is the one that labels the pivot row.

5. Use elementary row operations to transform the tableau into a new equivalent tableau that has a 1 where the pivot entry was and 0's elsewhere in that column.

---

[3]This will be discussed after Example 1.

6. In the labels column B of this tableau, the entering variable replaces the departing variable.
7. If the indicators of the new tableau are all nonnegative, we have an optimum solution. The maximum value of Z is the entry in the last row and last column. It occurs when the basic variables as found in the label column, B, are equal to the corresponding entries in column R. All other variables are 0. If at least one of the indicators is negative, repeat the process, beginning with step 2 applied to the new tableau.

As an aid in understanding the simplex method, we should be able to interpret certain entries in a tableau. Suppose that we obtain a tableau in which the last row is as shown in the following array:

$$\begin{array}{c} \\ \\ \\ \\ Z \end{array}\begin{array}{c} \begin{array}{cccccccccc} \text{B} & x_1 & x_2 & x_3 & s_1 & s_2 & s_3 & s_4 & Z & \text{R} \end{array} \\ \left[\begin{array}{ccccccccc|c} \cdot & \cdot & \cdot & \cdot & \cdot & \cdot & \cdot & \cdot & & \cdot \\ \cdot & \cdot & \cdot & \cdot & \cdot & \cdot & \cdot & \cdot & & \cdot \\ \cdot & \cdot & \cdot & \cdot & \cdot & \cdot & \cdot & \cdot & & \cdot \\ \hline a & b & c & d & e & f & g & 1 & & h \end{array}\right] \end{array}$$

We can interpret the entry b, for example, as follows: If x_2 is nonbasic and were to become basic, then, for each one-unit increase in x_2,

$$\begin{aligned} &\text{if } b < 0, && Z \textit{ increases} \text{ by } |b| \text{ units} \\ &\text{if } b > 0, && Z \textit{ decreases} \text{ by } |b| \text{ units} \\ &\text{if } b = 0, && \text{there is no change in } Z \end{aligned}$$

EXAMPLE 1 **The Simplex Method**

Maximize $Z = 5x_1 + 4x_2$ *subject to*

$$\begin{aligned} x_1 + x_2 &\leq 20 \\ 2x_1 + x_2 &\leq 35 \\ -3x_1 + x_2 &\leq 12 \end{aligned}$$

and $x_1, x_2 \geq 0$.

Solution: This linear programming problem fits the standard form. The initial simplex tableau is

entering variable ↓ (column x_1)

B	x_1	x_2	s_1	s_2	s_3	Z	R	Quotient
s_1	1	1	1	0	0	0	20	$20 \div 1 = 20$
departing variable ← s_2	2	1	0	1	0	0	35	$35 \div 2 = \frac{35}{2}$
s_3	−3	1	0	0	1	0	12	no quotient, $-3 \not> 0$
Z	−5	−4	0	0	0	1	0	

indicators (the entries −5, −4, 0, 0, 0 of the Z-row)

The most negative indicator, −5, occurs in the x_1-column. Thus, x_1 is the entering variable. The smaller quotient is $\frac{35}{2}$, so s_2 is the departing variable. The pivot entry is 2. Using elementary row operations to get a 1 in the pivot position and 0's elsewhere in its column, we have

$$\begin{array}{c} \begin{array}{cccccc|c} x_1 & x_2 & s_1 & s_2 & s_3 & Z & \\ \end{array} \\ \left[\begin{array}{cccccc|c} 1 & 1 & 1 & 0 & 0 & 0 & 20 \\ 2 & 1 & 0 & 1 & 0 & 0 & 35 \\ -3 & 1 & 0 & 0 & 1 & 0 & 12 \\ \hline -5 & -4 & 0 & 0 & 0 & 1 & 0 \end{array}\right] \end{array}$$

$$\xrightarrow{\frac{1}{2}R_2} \left[\begin{array}{cccccc|c} 1 & 1 & 1 & 0 & 0 & 0 & 20 \\ 1 & \frac{1}{2} & 0 & \frac{1}{2} & 0 & 0 & \frac{35}{2} \\ -3 & 1 & 0 & 0 & 1 & 0 & 12 \\ \hline -5 & -4 & 0 & 0 & 0 & 1 & 0 \end{array}\right]$$

$$\xrightarrow[\substack{3R_2+R_3 \\ 5R_2+R_4}]{-1R_2+R_1} \left[\begin{array}{cccccc|c} 0 & \frac{1}{2} & 1 & -\frac{1}{2} & 0 & 0 & \frac{5}{2} \\ 1 & \frac{1}{2} & 0 & \frac{1}{2} & 0 & 0 & \frac{35}{2} \\ 0 & \frac{5}{2} & 0 & \frac{3}{2} & 1 & 0 & \frac{129}{2} \\ \hline 0 & -\frac{3}{2} & 0 & \frac{5}{2} & 0 & 1 & \frac{175}{2} \end{array}\right]$$

Our new tableau is

entering variable ↓ (x_2)

departing variable ← (s_1)

$$\begin{array}{c|cccccc|c} B & x_1 & x_2 & s_1 & s_2 & s_3 & Z & R \\ \hline s_1 & 0 & \frac{1}{2} & 1 & -\frac{1}{2} & 0 & 0 & \frac{5}{2} \\ x_1 & 1 & \frac{1}{2} & 0 & \frac{1}{2} & 0 & 0 & \frac{35}{2} \\ s_3 & 0 & \frac{5}{2} & 0 & \frac{3}{2} & 1 & 0 & \frac{129}{2} \\ \hline Z & 0 & -\frac{3}{2} & 0 & \frac{5}{2} & 0 & 1 & \frac{175}{2} \end{array}$$

indicators (bottom row, under x_1 through s_3)

Quotients

$\frac{5}{2} \div \frac{1}{2} = 5$

$\frac{35}{2} \div \frac{1}{2} = 35$

$\frac{129}{2} \div \frac{5}{2} = 25\frac{4}{5}$

Note that in column B, which keeps track of which variables are basic, x_1 has replaced s_2. Since we still have a negative indicator, $-\frac{3}{2}$, we must continue our process. Evidently, $-\frac{3}{2}$ is the most negative indicator and the entering variable is now x_2. The smallest quotient is 5. Hence, s_1 is the departing variable and $\frac{1}{2}$ is the pivot entry. Using elementary row operations, we have

$$\begin{array}{c} \begin{array}{cccccc|c} x_1 & x_2 & s_1 & s_2 & s_3 & Z & R \\ \end{array} \\ \left[\begin{array}{cccccc|c} 0 & \frac{1}{2} & 1 & -\frac{1}{2} & 0 & 0 & \frac{5}{2} \\ 1 & \frac{1}{2} & 0 & \frac{1}{2} & 0 & 0 & \frac{35}{2} \\ 0 & \frac{5}{2} & 0 & \frac{3}{2} & 1 & 0 & \frac{129}{2} \\ \hline 0 & -\frac{3}{2} & 0 & \frac{5}{2} & 0 & 1 & \frac{175}{2} \end{array}\right] \end{array}$$

$$\xrightarrow[\substack{-5R_1+R_3 \\ 3R_1+R_4}]{-1R_1+R_2} \left[\begin{array}{cccccc|c} 0 & \frac{1}{2} & 1 & -\frac{1}{2} & 0 & 0 & \frac{5}{2} \\ 1 & 0 & -1 & 1 & 0 & 0 & 15 \\ 0 & 0 & -5 & 4 & 1 & 0 & 52 \\ \hline 0 & 0 & 3 & 1 & 0 & 1 & 95 \end{array}\right]$$

$$\xrightarrow{2R_1}\left[\begin{array}{cccccc|c} 0 & 1 & 2 & -1 & 0 & 0 & 5 \\ 1 & 0 & -1 & 1 & 0 & 0 & 15 \\ 0 & 0 & -5 & 4 & 1 & 0 & 52 \\ \hline 0 & 0 & 3 & 1 & 0 & 1 & 95 \end{array}\right]$$

Our new tableau is

$$\begin{array}{c} \\ x_2 \\ x_1 \\ s_3 \\ Z \end{array} \begin{array}{c} \begin{array}{cccccccc} \text{B} & x_1 & x_2 & s_1 & s_2 & s_3 & Z & \text{R} \end{array} \\ \left[\begin{array}{cccccc|c} 0 & 1 & 2 & -1 & 0 & 0 & 5 \\ 1 & 0 & -1 & 1 & 0 & 0 & 15 \\ 0 & 0 & -5 & 4 & 1 & 0 & 52 \\ \hline 0 & 0 & 3 & 1 & 0 & 1 & 95 \end{array}\right] \end{array}$$

indicators

where x_2 replaced s_1 in column B. Since all indicators are nonnegative, the maximum value of Z is 95 and occurs when $x_2 = 5$ and $x_1 = 15$ (and $s_3 = 52$, $s_1 = 0$, and $s_2 = 0$).

▶ **Now work Problem 1.**

It is interesting to see how the values of Z got progressively "better" in successive tableaux in Example 1. These are the entries in the last row and column of each tableau. In the initial tableau, we had $Z = 0$. From then on, we obtained $Z = \frac{175}{2} = 87\frac{1}{2}$ and then $Z = 95$, the maximum.

In Example 1, you may wonder why no quotient is considered in the third row of the initial tableau. The BFS for this tableau is

$$s_1 = 20, \quad s_2 = 35, \quad s_3 = 12, \quad x_1 = 0, \quad x_2 = 0$$

where x_1 is the entering variable. The quotients 20 and $\frac{35}{2}$ reflect that, for the next BFS, we have $x_1 \leq 20$ and $x_1 \leq \frac{35}{2}$. Since the third row represents the equation $s_3 = 12 + 3x_1 - x_2$, and $x_2 = 0$, it follows that $s_3 = 12 + 3x_1$. But $s_3 \geq 0$, so $12 + 3x_1 \geq 0$, which implies that $x_1 \geq -\frac{12}{3} = -4$. Thus, we have

$$x_1 \leq 20, \quad x_1 \leq \frac{35}{2}, \quad \text{and} \quad x_1 \geq -4$$

Hence, x_1 can increase at most by $\frac{35}{2}$. The condition $x_1 \geq -4$ has no influence in determining the maximum increase in x_1. That is why the quotient $12/(-3) = -4$ is not considered in row 3. In general, *no quotient is considered for a row if the entry in the entering-variable column is negative* (*or, of course,* 0).

Although the simplex procedure that has been developed in this section applies only to linear programming problems of standard form, other forms may be adapted to fit this form. Suppose that a constraint has the form

$$a_1x_1 + a_2x_2 + \cdots + a_nx_n \geq -b$$

where $b > 0$. Here the inequality symbol is "$\geq$," and the constant on the right side is *negative*. Thus, the constraint is not in standard form. However, multiplying both sides by -1 gives

$$-a_1x_1 - a_2x_2 - \cdots - a_nx_n \leq b,$$

which *does* have the proper form. Accordingly, it may be necessary to rewrite a constraint before proceeding with the simplex method.

In a simplex tableau, several indicators may "tie" for being most negative. In this case, we choose any one of these indicators to give the column for the entering variable. Likewise, there may be several quotients that "tie" for being the smallest. We may then choose any one of these quotients to determine the departing variable and pivot entry. Example 2 will illustrate this situation. When a tie for the smallest quotient exists, then, along with the nonbasic variables, a BFS will have a basic variable that is 0. In this case we say that the BFS is *degenerate* or that the linear programming problem has a *degeneracy*. More will be said about degeneracies in Section 7.5.

APPLICATIONS IN PRACTICE 1

The Simplex Method

The What If Company has \$30,000 for the purchase of materials to make three types of gadgets. The company has allocated a total of 1200 hours of assembly time and 180 hours of packaging time for the gadgets. The following table gives the cost per gadget, the number of hours per gadget, and the profit per gadget for each type:

	Type 1	Type 2	Type 3
Cost/gadget	\$300	\$300	\$400
Assembly Hours/gadget	15	15	10
Packaging Hours/gadget	2	2	3
Profit	\$150	\$250	\$200

Find the number of gadgets of each type the company should produce to maximize profit.

EXAMPLE 2 **The Simplex Method**

Maximize $Z = 3x_1 + 4x_2 + \frac{3}{2}x_3$ *subject to*

$$-x_1 - 2x_2 \geq -10 \qquad (10)$$

$$2x_1 + 2x_2 + x_3 \leq 10$$

$$x_1, x_2, x_3 \geq 0$$

Solution: Constraint (10) does not fit the standard form. However, multiplying both sides of inequality (10) by -1 gives

$$x_1 + 2x_2 \leq 10$$

which *does* have the proper form. Thus, our initial simplex tableau is tableau I:

SIMPLEX TABLEAU I

entering variable ↓ (x_2)

	B	x_1	x_2	x_3	s_1	s_2	Z	R	*Quotients*
departing variable ←	s_1	1	**2**	0	1	0	0	10	$10 \div 2 = 5$
	s_2	2	2	1	0	1	0	10	$10 \div 2 = 5$
	Z	-3	-4	$-\frac{3}{2}$	0	0	1	0	

indicators (the Z-row entries under x_1 through s_2)

The entering variable is x_2. Since there is a tie for the smallest quotient, we can choose either s_1 or s_2 as the departing variable. Let us choose s_1. The pivot entry is shaded. Using elementary row operations, we get tableau II:

SIMPLEX TABLEAU II

entering variable ↓ (x_3)

	B	x_1	x_2	x_3	s_1	s_2	Z	R	*Quotients*
	x_2	$\frac{1}{2}$	1	0	$\frac{1}{2}$	0	0	5	no quotient $0 \not> 0$
departing variable ←	s_2	1	0	**1**	-1	1	0	0	$0 \div 1 = 0$
	Z	-1	0	$-\frac{3}{2}$	2	0	1	20	

indicators (the Z-row entries under x_1 through s_2)

Tableau II corresponds to a BFS in which a basic variable, s_2, is zero. Thus, the BFS is degenerate. Since there are negative indicators, we continue. The entering variable is now x_3, the departing variable is s_2, and the pivot is shaded. Using elementary row operations, we get tableau III:

SIMPLEX TABLEAU III

B	x_1	x_2	x_3	s_1	s_2	Z	R
x_2	$\frac{1}{2}$	1	0	$\frac{1}{2}$	0	0	5
x_3	1	0	1	-1	1	0	0
Z	$\frac{1}{2}$	0	0	$\frac{1}{2}$	$\frac{3}{2}$	1	20

indicators

Since all indicators are nonnegative, Z is maximized when $x_2 = 5$, $x_3 = 0$, and $x_1 = s_1 = s_2 = 0$. The maximum value is $Z = 20$. Note that this value is the same as the value of Z in tableau II. In degenerate problems, it is possible to arrive at the same value of Z at various stages of the simplex process. In Problem 7 you are asked to solve this example by using s_2 as the departing variable in the initial tableau.

Because of its mechanical nature, the simplex procedure is readily adaptable to computers to solve linear programming problems involving many variables and constraints.

Exercise 7.4

Use the simplex method to solve the following problems.

*1. Maximize

$$Z = x_1 + 2x_2$$

subject to

$$\begin{aligned} 2x_1 + x_2 &\leq 8 \\ 2x_1 + 3x_2 &\leq 12 \\ x_1, x_2 &\geq 0 \end{aligned}$$

2. Maximize

$$Z = 2x_1 + x_2$$

subject to

$$\begin{aligned} -x_1 + x_2 &\leq 4 \\ x_1 + x_2 &\leq 6 \\ x_1, x_2 &\geq 0 \end{aligned}$$

3. Maximize

$$Z = -x_1 + 2x_2$$

subject to

$$\begin{aligned} 3_1 + 2x_2 &\leq 5 \\ -x_1 + 3x_2 &\leq 3 \\ x_1, x_2 &\geq 0 \end{aligned}$$

4. Maximize

$$Z = 3x_1 + 8x_2$$

subject to

$$\begin{aligned} x_1 + 2x_2 &\leq 8 \\ x_1 + 6x_2 &\leq 12 \\ x_1, x_2 &\geq 0 \end{aligned}$$

5. Maximize

$$Z = 8x_1 + 2x_2$$

subject to

$$\begin{aligned} x_1 - x_2 &\leq 1 \\ x_1 + 2x_2 &\leq 8 \\ x_1 + x_2 &\leq 5 \\ x_1, x_2 &\geq 0 \end{aligned}$$

6. Maximize

$$Z = 2x_1 - 6x_2$$

subject to

$$\begin{aligned} x_1 - x_2 &\leq 4 \\ -x_1 + x_2 &\leq 4 \\ x_1 + x_2 &\leq 6 \\ x_1, x_2 &\geq 0 \end{aligned}$$

7. Solve the problem in Example 2 by choosing s_2 as the departing variable in tableau I.

8. Maximize

$$Z = 2x_1 - x_2 + x_3$$

subject to

$$\begin{aligned} 2x_1 + x_2 - x_3 &\le 4 \\ x_1 + x_2 + x_3 &\le 2 \\ x_1, x_2, x_3 &\ge 0 \end{aligned}$$

9. Maximize

$$Z = 2x_1 + x_2 - x_3$$

subject to

$$\begin{aligned} x_1 + x_2 &\le 1 \\ x_1 - 2x_2 - x_3 &\ge -2 \\ x_1, x_2, x_3 &\ge 0 \end{aligned}$$

10. Maximize

$$Z = -2x_1 + 3x_2$$

subject to

$$\begin{aligned} x_1 + x_2 &\le 1 \\ x_1 - x_2 &\le 2 \\ x_1 - x_2 &\ge -3 \\ x_1 &\le 5 \\ x_1, x_2 &\ge 0 \end{aligned}$$

11. Maximize

$$Z = x_1 + x_2$$

subject to

$$\begin{aligned} x_1 - x_2 &\le 4 \\ -x_1 + x_2 &\le 4 \\ 8x_1 + 5x_2 &\le 40 \\ 2x_1 + x_2 &\le 6 \\ x_1, x_2 &\ge 0 \end{aligned}$$

12. Maximize

$$W = 2x_1 + x_2 - 2x_3$$

subject to

$$\begin{aligned} -2x_1 + x_2 + x_3 &\ge -2 \\ x_1 - x_2 + x_3 &\le 4 \\ x_1 + x_2 + 2x_3 &\le 6 \\ x_1, x_2, x_3 &\ge 0 \end{aligned}$$

13. Maximize

$$W = x_1 - 12x_2 + 4x_3$$

subject to

$$\begin{aligned} 4x_1 + 3x_2 - x_3 &\le 1 \\ x_1 + x_2 - x_3 &\ge -2 \\ -x_1 + x_2 + x_3 &\ge -1 \\ x_1, x_2, x_3 &\ge 0 \end{aligned}$$

14. Maximize

$$W = 4x_1 + 0x_2 - x_3$$

subject to

$$\begin{aligned} x_1 + x_2 + x_3 &\le 6 \\ x_1 - x_2 + x_3 &\le 10 \\ x_1 - x_2 - x_3 &\le 4 \\ x_1, x_2, x_3 &\ge 0 \end{aligned}$$

15. Maximize

$$Z = 60x_1 + 0x_2 + 90x_3 + 0x_4$$

subject to

$$\begin{aligned} x_1 - 2x_2 &\le 2 \\ x_1 + x_2 &\le 5 \\ x_3 + x_4 &\le 4 \\ x_3 - 2x_4 &\le 7 \\ x_1, x_2, x_3, x_4 &\ge 0 \end{aligned}$$

16. Maximize

$$Z = 4x_1 + 5x_2 - 3x_3 - x_4$$

subject to

$$\begin{aligned} x_1 + x_3 - x_4 &\le 2 \\ x_1 + x_2 + x_4 &\le 5 \\ x_1 + x_2 - x_3 + x_4 &\le 3 \\ x_1, x_2, x_3, x_4 &\ge 0 \end{aligned}$$

17. Freight Shipments A freight company handles shipments by two corporations, A and B, that are located in the same city. Corporation A ships boxes that weigh 3 lb each and have a volume of 2 ft^3; B ships 1 ft^3 boxes that weigh 5 lb each. Both A and B ship to the same destination. The transportation cost for each box from A is \$0.75, and from B it is \$0.50. The freight company has a truck with 2400 ft^3 of cargo space and a maximum capacity of 36,800 lb. In one haul, how many boxes from each corporation should be transported by this truck so that the freight company receives maximum revenue? What is the maximum revenue?

18. Production A company manufactures three products: X, Y, and Z. Each product requires machine time and finishing time as shown in the following table:

	Machine Time	Finishing Time
X	1 hr	4 hr
Y	2 hr	4 hr
Z	3 hr	8 hr

The numbers of hours of machine time and finishing time available per month are 900 and 5000, respectively. The unit profit on X, Y, and Z is \$6, \$8, and \$12, respectively. What is the maximum profit per month that can be obtained?

19. Production A company manufactures three types of patio furniture: chairs, rockers, and chaise lounges. Each requires wood, plastic, and aluminum as shown in the following table:

	Wood	Plastic	Aluminum
Chair	1 unit	1 unit	2 units
Rocker	1 unit	1 unit	3 units
Chaise lounge	1 unit	2 units	5 units

The company has available 400 units of wood, 500 units of plastic, and 1450 units of aluminum. Each chair, rocker, and chaise lounge sells at \$21, \$24, and \$36, respectively. Assuming that all furniture can be sold, determine a production order so that total revenue will be maximum. What is the maximum revenue?

OBJECTIVE *To consider the simplex method in relation to degeneracy, unbounded solutions, and multiple optimum solutions.*

7.5 Degeneracy, Unbounded Solutions, and Multiple Optimum Solutions[4]

Degeneracy

In the preceding section, we stated that a BFS is **degenerate** if, along with the nonbasic variables, one of the basic variables is 0. Suppose x_1, x_2, x_3, and x_4 are the variables in a degenerate BFS, where x_1 and x_2 are basic with $x_1 = 0$, x_3 and x_4 are nonbasic, and x_3 is the entering variable. The corresponding simplex tableau has the form

$$\begin{array}{c} \\ \text{departing} \leftarrow \\ \text{variable} \\ \\ \end{array}
\begin{array}{c} \text{B} \\ x_1 \\ x_2 \\ Z \end{array}
\begin{array}{c} \begin{array}{cccccc} x_1 & x_2 & \overset{\text{entering variable}\ \downarrow}{x_3} & x_4 & Z & \text{R} \end{array} \\
\left[\begin{array}{ccccc|c} 1 & 0 & a_{13} & a_{14} & 0 & 0 \\ 0 & 1 & a_{23} & a_{24} & 0 & a \\ \hline 0 & 0 & d_1 & d_2 & 1 & d_3 \end{array}\right] \\
\underbrace{\hphantom{0 \quad 0 \quad d_1 \quad d_2}}_{\text{indicators}} \end{array}
\begin{array}{c} \\ 0 \div a_{13} = 0 \\ \\ \\ \end{array}$$

Thus, the BFS is

$$x_1 = 0 \quad x_2 = a \quad x_3 = 0 \quad x_4 = 0$$

Suppose $a_{13} > 0$. Then the smaller quotient is 0, and we can choose a_{13} as the pivot entry. Therefore, x_1 is the departing variable. Elementary row operations give the following tableau, where the question marks represent numbers to be determined:

$$\begin{array}{c} \text{B} \\ x_3 \\ x_2 \\ Z \end{array}
\begin{array}{c} \begin{array}{cccccc} x_1 & x_2 & x_3 & x_4 & Z & \text{R} \end{array} \\
\left[\begin{array}{ccccc|c} ? & 0 & 1 & ? & 0 & 0 \\ ? & 1 & 0 & ? & 0 & a \\ \hline ? & 0 & 0 & ? & 1 & d_3 \end{array}\right] \end{array}$$

For the BFS corresponding to this tableau, x_3 and x_2 are basic variables, and x_1 and x_4 are nonbasic. The BFS is

$$x_3 = 0 \quad x_2 = a \quad x_1 = 0 \quad x_4 = 0$$

[4]This section may be omitted.

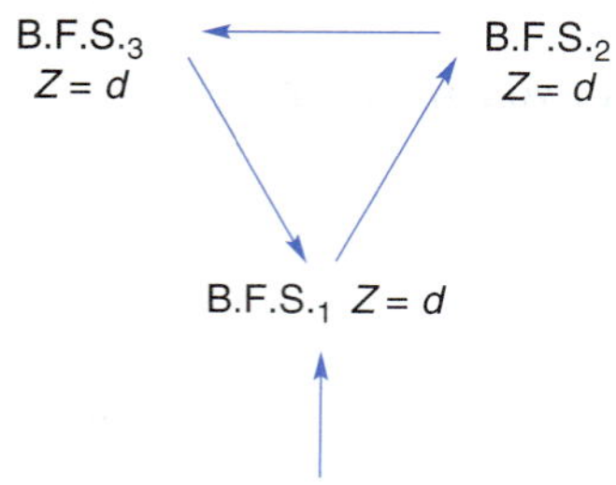

FIGURE 7.24 Cycling.

which is the same BFS as before. Actually, these are usually considered different BFS's, with the only distinction being that x_1 is basic in the first BFS, whereas in the second it is nonbasic. The value of Z for both BFS's is the same: d_3. Thus, no "improvement" in Z is obtained.

In a degenerate situation, some problems may develop in the simplex procedure. It is possible to obtain a sequence of tableaux that correspond to BFS's that give the same Z-value. Moreover, we may eventually return to the first tableau in the sequence. In Figure 7.24, we arrive at BFS_1, proceed to BFS_2, go on to BFS_3, and finally return to BFS_1. This is called *cycling*. When cycling occurs, we may never obtain the optimum value of Z. This phenomenon is encountered rarely in practical linear programming problems; however, there are techniques (which will not be considered in this text) for resolving such difficulties.

A degenerate BFS will occur when two quotients in a simplex tableau tie for being the smallest. For example, consider the following (partial) tableau:

$$\begin{array}{c} \\ x_1 \\ x_2 \end{array}\!\!\begin{array}{c} \text{B} \quad x_3 \qquad\quad \text{R} \\ \left[\begin{array}{c|c} q_1 & p_1 \\ q_2 & p_2 \end{array}\right] \end{array} \quad \begin{array}{c} \textit{Quotients} \\ p_1/q_1 \\ p_2/q_2 \end{array}$$

Here x_1 and x_2 are basic variables. Suppose x_3 is nonbasic and entering, and $p_1/q_1 = p_2/q_2$ are the smallest quotients. Choosing q_1 as the pivot entry, we obtain, by elementary row operations,

$$\begin{array}{c} \\ x_3 \\ x_2 \end{array}\!\!\begin{array}{c} \text{B} \quad x_3 \qquad\qquad \text{R} \\ \left[\begin{array}{c|c} 1 & p_1/q_1 \\ 0 & p_2 - q_2\frac{p_1}{q_1} \end{array}\right] \end{array}$$

Since $p_1/q_1 = p_2/q_2$, we have $p_2 - q_2(p_1/q_1) = 0$. Thus, the BFS corresponding to this tableau has $x_2 = 0$, which gives a *degenerate* BFS. Although such a BFS may produce cycling, we shall not encounter such situations in this book.

Unbounded Solutions

We now turn our attention to "unbounded problems." In Section 7.2, we saw that a linear programming problem may have no maximum value because the feasible region is such that the objective function may become arbitrarily large therein. In this case, the problem is said to have an **unbounded solution.** This is a way of saying specifically that no optimum solution exists. Such a situation occurs when no quotients are possible in a simplex tableau for an entering variable. For example, consider the following tableau:

entering variable
↓

$$\begin{array}{c} \\ x_1 \\ x_3 \\ Z \end{array}\!\!\begin{array}{c} \text{B} \quad x_1 \quad x_2 \quad x_3 \quad x_4 \quad Z \qquad \text{R} \\ \left[\begin{array}{rrrrr|r} 1 & -3 & 0 & 2 & 0 & 5 \\ 0 & 0 & 1 & 4 & 0 & 1 \\ \hline 0 & -5 & 0 & -2 & 1 & 10 \end{array}\right] \end{array} \quad \begin{array}{l} \\ \text{no quotient} \\ \text{no quotient} \\ \\ \end{array}$$

$\underbrace{\hspace{6em}}_{\text{indicators}}$

Here x_2 is the entering variable, and for each one-unit increase in x_2, Z increases by 5. Since there are no positive entries in the first two rows of the x_2 column, no quotients exist. From rows 1 and 2 we get

$$x_1 = 5 + 3x_2 - 2x_4$$

and

$$x_3 = 1 - 4x_4$$

If we try to proceed to the next BFS, what is an upper bound on x_2? In that BFS, x_4 will remain nonbasic ($x_4 = 0$). Thus, $x_1 = 5 + 3x_2$ and $x_3 = 1$. Since $x_1 \geq 0$, $x_2 \geq -\frac{5}{3}$. Therefore, there is no upper bound on x_2. Hence, Z can be arbitrarily large, and we have an unbounded solution. In general:

If no quotients exist in a simplex tableau, then the linear programming problem has an unbounded solution.

EXAMPLE 1 **Unbounded Solution**

Maximize $Z = x_1 + 4x_2 - x_3$, *subject to*

$$-5x_1 + 6x_2 - 2x_3 \leq 30$$

$$-x_1 + 3x_2 + 6x_3 \leq 12$$

$$x_1, x_2, x_3 \geq 0$$

Solution: The initial simplex tableau is

$$
\begin{array}{rc|cccccc|c|l}
 & \text{B} & x_1 & \overset{\substack{\text{entering}\\ \text{variable}\\ \downarrow}}{x_2} & x_3 & s_1 & s_2 & Z & \text{R} & \textit{Quotients} \\
\text{departing} \leftarrow & s_1 & -5 & 6 & -2 & 1 & 0 & 0 & 30 & 30 \div 6 = 5 \\
\text{variable} & s_2 & -1 & 3 & 6 & 0 & 1 & 0 & 12 & 12 \div 3 = 4 \\
\cline{3-9}
 & Z & -1 & -4 & 1 & 0 & 0 & 1 & 0 & \\
 & & \multicolumn{5}{c}{\underbrace{\hphantom{-1 \quad -4 \quad 1 \quad 0 \quad 0}}_{\text{indicators}}} & & &
\end{array}
$$

The second tableau is

$$
\begin{array}{c|cccccc|c|l}
\text{B} & \overset{\substack{\text{entering}\\ \text{variable}\\ \downarrow}}{x_1} & x_2 & x_3 & s_1 & s_2 & Z & \text{R} & \\
s_1 & -3 & 0 & -14 & 1 & -2 & 0 & 6 & \text{no quotient} \\
x_2 & -\frac{1}{3} & 1 & 2 & 0 & \frac{1}{3} & 0 & 4 & \text{no quotient} \\
\cline{2-8}
Z & -\frac{7}{3} & 0 & 9 & 0 & \frac{4}{3} & 1 & 16 & \\
 & \multicolumn{5}{c}{\underbrace{\hphantom{-\tfrac{7}{3} \quad 0 \quad 9 \quad 0 \quad \tfrac{4}{3}}}_{\text{indicators}}} & & &
\end{array}
$$

Here the entering variable is x_1. Since the entries in the first two rows of the x_1-column are negative, no quotients exist. Hence, the problem has an unbounded solution.

▸ **Now work Problem 3.**

Multiple Optimum Solutions

We conclude this section with a discussion of "multiple optimum solutions." Suppose that

$$x_1 = a_1 \quad x_2 = a_2 \quad \ldots \quad x_n = a_n$$

and

$$x_1 = b_1 \quad x_2 = b_2 \quad \ldots \quad x_n = b_n$$

are two *different* BFS's for which a linear programming problem is optimum. By "different BFS's," we mean that $a_i \neq b_i$ for some i, where $1 \leq i \leq n$. It can be shown that the values

$$\begin{aligned} x_1 &= (1-t)a_1 + tb_1 \\ x_2 &= (1-t)a_2 + tb_2 \\ &\vdots \\ x_n &= (1-t)a_n + tb_n \end{aligned} \tag{1}$$

$$\text{for any } t \text{ such that } 0 \leq t \leq 1$$

also give an optimum solution (although it may not necessarily be a BFS). Thus, there are *multiple (optimum) solutions* to the problem.

We can determine the possibility of multiple optimum solutions from a simplex tableau that gives an optimum solution, such as the following (partial) tableau:

$$\begin{array}{c} \text{B} \\ x_1 \\ x_2 \\ Z \end{array} \begin{array}{c} \begin{array}{cccccc} x_1 & x_2 & x_3 & x_4 & Z & \text{R} \end{array} \\ \left[\begin{array}{ccccc|c} & & & & & p_1 \\ & & & & & q_1 \\ \hline \underbrace{0 \quad 0 \quad a \quad 0}_{\text{indicators}} & & & & 1 & r \end{array}\right] \end{array}$$

Here a must be nonnegative. The corresponding BFS is

$$x_1 = p_1 \qquad x_2 = q_1 \qquad x_3 = 0 \qquad x_4 = 0$$

and the maximum value of Z is r. If x_4 were to become basic, the indicator 0 in the x_4-column means that for each one-unit increase in x_4, Z does not change. Thus, we can find a BFS in which x_4 is basic and the corresponding Z-value is the same as before. This is done by treating x_4 as an entering variable in the preceding tableau. If, for instance, x_1 is the departing variable, then the pivot element will be the one in row x_1 and column x_4. The new BFS will have the form

$$x_1 = 0 \quad x_2 = q_2 \quad x_3 = 0 \quad x_4 = p_2$$

where q_2 and p_2 are numbers that result from the pivoting process. If this BFS is different from the previous one, multiple solutions exist. In fact, from Equation (1), an optimum solution is given by any values of x_1, x_2, x_3, and x_4, such that

$$\begin{aligned} x_1 &= (1-t)p_1 + t \cdot 0 = (1-t)p_1 \\ x_2 &= (1-t)q_1 + tq_2 \\ x_3 &= (1-t) \cdot 0 + t \cdot 0 = 0 \\ x_4 &= (1-t) \cdot 0 + tp_2 = tp_2 \end{aligned}$$

$$\text{where } 0 \leq t \leq 1$$

Note that when $t = 0$, we get the first optimum BFS; when $t = 1$, we get the second. Of course, it may be possible to repeat the procedure by using the tableau corresponding to the last BFS and obtain more optimum solutions by using Equation (1).

In general,

In a tableau that gives an optimum solution, a zero indicator for a nonbasic variable suggests the possibility of multiple optimum solutions.

APPLICATIONS IN PRACTICE 1

Multiple Solutions

A company produces three kinds of devices requiring three different production procedures. The company has allocated a total of 190 hours for procedure 1, 180 hours for procedure 2, and 165 hours for procedure 3. The following table gives the number of hours per device for each procedure.

	Device 1	Device 2	Device 3
Procedure 1	5.5	5.5	6.5
Procedure 2	3.5	6.5	7.5
Procedure 3	4.5	6.0	6.5

If the profit is \$50 per device 1, \$50 per device 2, and \$50 per device 3, find the number of devices of each kind the company should produce to maximize profit. Enter the initial tableau into a matrix on your graphing calculator, and perform the necessary row operations to find the answer. Round answers to the nearest whole number.

EXAMPLE 2 **Multiple Solutions**

Maximize $Z = -x_1 + 4x_2 + 6x_3$ *subject to*

$$x_1 + 2x_2 + 3x_3 \leq 6$$
$$-2x_1 - 5x_2 + x_3 \leq 10$$
$$x_1, x_2, x_3 \geq 0$$

Solution: Our initial simplex tableau is

$$\begin{array}{c} \\ \text{departing} \leftarrow s_1 \\ \text{variable} \quad s_2 \\ Z \end{array}
\begin{array}{c} \\ \\ \\ \\ \end{array}
\overset{\begin{array}{c}\text{entering variable} \\ \downarrow\end{array}}{
\begin{array}{c}
\begin{array}{ccccccc|c} x_1 & x_2 & x_3 & s_1 & s_2 & Z & & R \end{array} \\
\left[\begin{array}{cccccc|c}
1 & 2 & 3 & 1 & 0 & 0 & 6 \\
-2 & -5 & 1 & 0 & 1 & 0 & 10 \\
\hline
1 & -4 & -6 & 0 & 0 & 1 & 0
\end{array}\right]
\end{array}}
\begin{array}{l} \textit{Quotients} \\ 6 \div 3 = 2 \\ 10 \div 1 = 10 \\ \\ \end{array}$$

(indicators: $1, -4, -6, 0, 0$)

Since there is a negative indicator, we continue and obtain

$$\begin{array}{c} \\ \text{departing} \leftarrow x_3 \\ \text{variable} \quad s_2 \\ Z \end{array}
\overset{\begin{array}{c}\text{entering variable} \\ \downarrow\end{array}}{
\begin{array}{c}
\begin{array}{ccccccc|c} x_1 & x_2 & x_3 & s_1 & s_2 & Z & & R \end{array} \\
\left[\begin{array}{cccccc|c}
\frac{1}{3} & \frac{2}{3} & 1 & \frac{1}{3} & 0 & 0 & 2 \\
-\frac{7}{3} & -\frac{17}{3} & 0 & -\frac{1}{3} & 1 & 0 & 8 \\
\hline
3 & 0 & 0 & 2 & 0 & 1 & 12
\end{array}\right]
\end{array}}
\begin{array}{l} \textit{Quotients} \\ 2 \div \frac{2}{3} = 3 \\ \text{no quotient} \\ \\ \end{array}$$

(indicators: $3, 0, 0, 2, 0$)

All indicators are nonnegative: hence, an optimum solution occurs for the BFS

$$x_3 = 2 \quad s_2 = 8 \quad x_1 = 0 \quad x_2 = 0 \quad s_1 = 0$$

and the maximum value of Z is 12. However, since x_2 is a nonbasic variable and its indicator is 0, we shall check for multiple solutions. Treating x_2 as an entering variable, we obtain the following tableau:

$$\begin{array}{c} \\ x_2 \\ s_2 \\ Z \end{array}
\begin{array}{c}
\begin{array}{ccccccc|c} x_1 & x_2 & x_3 & s_1 & s_2 & Z & & R \end{array} \\
\left[\begin{array}{cccccc|c}
\frac{1}{2} & 1 & \frac{3}{2} & \frac{1}{2} & 0 & 0 & 3 \\
\frac{1}{2} & 0 & \frac{17}{2} & \frac{5}{2} & 1 & 0 & 25 \\
\hline
3 & 0 & 0 & 2 & 0 & 1 & 12
\end{array}\right]
\end{array}$$

The BFS here is

$$x_2 = 3 \qquad s_2 = 25 \qquad x_1 = 0 \qquad x_3 = 0 \qquad s_1 = 0$$

(for which $Z = 12$, as before) and is different from the previous one. Thus, multiple solutions exist. Since we are concerned only with values of the decision variables, we have an optimum solution,

$$x_1 = (1-t)\cdot 0 + t\cdot 0 = 0$$
$$x_2 = (1-t)\cdot 0 + t\cdot 3 = 3t$$
$$x_3 = (1-t)\cdot 2 + t\cdot 0 = 2(1-t)$$

for each value of t such that $0 \le t \le 1$. (For example, if $t = \frac{1}{2}$, then $x_1 = 0$, $x_2 = \frac{3}{2}$, and $x_3 = 1$ is an optimum solution.)

In the last BFS, x_3 is nonbasic and its indicator is 0. However, if we repeated the process for determining other optimum solutions, we would return to the second tableau. Therefore, our procedure gives no other optimum solutions.

▶ **Now work Problem 5.**

Exercise 7.5

In each of Problems 1 and 2, does the linear programming problem associated with the given tableau have a degeneracy? If so, why?

1.

B	x_1	x_2	s_1	s_2	Z	R
x_1	1	2	4	0	0	6
s_2	0	1	1	1	0	3
Z	0	−3	−2	0	1	10

indicators

2.

B	x_1	x_2	x_3	s_1	s_2	Z	R
s_1	2	0	2	1	1	0	4
x_2	3	1	1	0	1	0	0
Z	−5	0	1	0	−3	1	2

indicators

In Problems 3–11, use the simplex method.

***3.** Maximize

$$Z = 2x_1 + 7x_2$$

subject to

$$4x_1 - 3x_2 \le 4$$
$$3x_1 - x_2 \le 6$$
$$5x_1 \le 8$$
$$x_1, x_2 \ge 0$$

4. Maximize

$$Z = 2x_1 + x_2$$

subject to

$$x_1 - x_2 \le 7$$
$$-x_1 + x_2 \le 5$$
$$8x_1 + 5x_2 \le 40$$
$$2x_1 + x_2 \le 6$$
$$x_1, x_2 \ge 0$$

***5.** Maximize

$$Z = 3x_1 - 3x_2$$

subject to

$$x_1 - x_2 \le 4$$
$$-x_1 + x_2 \le 4$$
$$x_1 + x_2 \le 6$$
$$x_1, x_2 \ge 0$$

6. Maximize

$$Z = 8x_1 + 2x_2 + 4x_3$$

subject to

$$x_1 - x_2 + 4x_3 \le 6$$
$$x_1 - x_2 - x_3 \ge -4$$
$$x_1 - 6x_2 + x_3 \le 8$$
$$x_1, x_2, x_3 \ge 0$$

7. Maximize

$$Z = 5x_1 + 6x_2 + x_3$$

subject to

$$\begin{aligned} 9x_1 + 3x_2 - 2x_3 &\leq 5 \\ 4x_1 + 2x_2 - x_3 &\leq 2 \\ x_1 - 4x_2 + x_3 &\leq 3 \\ x_1, x_2, x_3 &\geq 0 \end{aligned}$$

8. Maximize

$$Z = 2x_1 + x_2 - 4x_3$$

subject to

$$\begin{aligned} 6x_1 + 3x_2 - 3x_3 &\leq 10 \\ x_1 - x_2 + x_3 &\leq 1 \\ 2x_1 - x_2 + 2x_3 &\leq 12 \\ x_1, x_2, x_3 &\geq 0 \end{aligned}$$

9. Maximize

$$Z = 6x_1 + 2x_2 + x_3$$

subject to

$$\begin{aligned} 2x_1 + x_2 + x_3 &\leq 7 \\ -4x_1 - x_2 &\geq -6 \\ x_1, x_2, x_3 &\geq 0 \end{aligned}$$

10. Maximize

$$P = x_1 + 2x_2 + x_3 + 2x_4$$

subject to

$$\begin{aligned} x_1 - x_2 &\leq 2 \\ x_2 - x_3 &\leq 3 \\ x_2 - 3x_3 + x_4 &\leq 4 \\ x_1, x_2, x_3, x_4 &\geq 0 \end{aligned}$$

11. Production A company manufactures three types of patio furniture: chairs, rockers, and chaise lounges. Each requires wood, plastic, and aluminum as given in the following table:

	Wood	Plastic	Aluminum
Chair	1 unit	1 unit	2 units
Rocker	1 unit	1 unit	3 units
Chaise lounge	1 unit	2 units	5 units

The company has available 400 units of wood, 600 units of plastic, and 1500 units of aluminum. Each chair, rocker, and chaise lounge sells at \$24, \$32, and \$48, respectively. Assuming that all furniture can be sold, what is the maximum total revenue that can be obtained? Determine the possible production orders that will generate this revenue.

OBJECTIVE *To handle maximization problems that are not of standard form by introducing artificial variables.*

7.6 Artificial Variables

To initiate the simplex method, a BFS is required. For a standard linear programming problem, we begin with the BFS in which all decision variables are zero. However, for a maximization problem that is not of standard form, such a BFS may not exist. In this section, we will learn how the simplex method is used in such situations.

Let us consider the following problem:

$$\text{Maximize } Z = x_1 + 2x_2$$

subject to

$$x_1 + x_2 \leq 9 \quad (1)$$

$$x_1 - x_2 \geq 1 \quad (2)$$

$$x_1, x_2 \geq 0$$

Since constraint (2) cannot be written as $a_1x_1 + a_2x_2 \leq b$, where b is nonnegative, this problem cannot be put into standard form. Note that (0, 0) is not a feasible point. To solve the problem, we begin by writing constraints (1) and (2) as equations. Constraint (1) becomes

$$x_1 + x_2 + s_1 = 9 \quad (3)$$

where $s_1 \geq 0$ is a slack variable. For constraint (2), $x_1 - x_2$ will equal 1 if we *subtract* a nonnegative slack variable s_2 from $x_1 - x_2$. That is, by subtracting s_2, we are making up for the "surplus" on the left side of (2), so that we have equality. Thus,

$$x_1 - x_2 - s_2 = 1 \tag{4}$$

where $s_2 \geq 0$. We can now restate the problem:

$$\text{Maximize } Z = x_1 + 2x_2 \tag{5}$$

subject to

$$x_1 + x_2 + s_1 = 9 \tag{6}$$

$$x_1 - x_2 - s_2 = 1 \tag{7}$$

$$x_1, x_2, s_1, s_2 \geq 0$$

Since $(0, 0)$ is not in the feasible region, we do not have a BFS in which $x_1 = x_2 = 0$. In fact, if $x_1 = 0$ and $x_2 = 0$ are substituted into Equation (7), then $0 - 0 - s_2 = 1$, which gives $s_2 = -1$. But this contradicts the condition that $s_2 \geq 0$.

To get the simplex method started, we need an initial BFS. Although none is obvious, there is an ingenious method to arrive at one *artificially*. It requires that we consider a related linear programming problem called the *artificial problem*. First, a new equation is formed by adding a nonnegative variable t to the left side of the equation in which the coefficient of the slack variable is -1. The variable t is called an **artificial variable.** In our case, we replace Equation (7) by $x_1 - x_2 - s_2 + t = 1$. Thus, Equations (6) and (7) become

$$x_1 + x_2 + s_1 = 9 \tag{8}$$

$$x_1 - x_2 - s_2 + t = 1 \tag{9}$$

$$x_1, x_2, s_1, s_2, t \geq 0$$

An obvious solution to Equations (8) and (9) is found by setting x_1, x_2, and s_2 equal to 0. This gives

$$x_1 = x_2 = s_2 = 0 \quad s_1 = 9 \quad t = 1$$

Note that these values do not satisfy Equation (7). However, it is clear that any solution of Equations (8) and (9) for which $t = 0$ will give a solution to Equations (6) and (7), and conversely.

We can eventually force t to be 0 if we alter the original objective function. We define the **artificial objective function** to be

$$W = Z - Mt = x_1 + 2x_2 - Mt \tag{10}$$

where the constant M is a large positive number. We shall not worry about the particular value of M and shall proceed to maximize W by the simplex method. Since there are $m = 2$ constraints (excluding the nonnegativity conditions) and $n = 5$ variables in Equations (8) and (9), any BFS must have at least $n - m = 3$ variables equal to zero. We start with the following BFS:

$$x_1 = x_2 = s_2 = 0 \qquad s_1 = 9 \qquad t = 1 \tag{11}$$

In this initial BFS, the nonbasic variables are the decision variables and the "surplus" variable s_2. The corresponding value of W is $W = x_1 + 2x_2 - Mt = -M$, which is "extremely" negative. A significant improvement in W will occur if we can find another BFS for which $t = 0$. Since the simplex method seeks better values of W at each stage, we shall apply it until we reach such a BFS, if possible. That solution will be an initial BFS for the original problem.

To apply the simplex method to the artificial problem, we first write Equation (10) as

$$-x_1 - 2x_2 + Mt + W = 0 \tag{12}$$

The augmented coefficient matrix of Equations (8), (9), and (12) is

$$\begin{array}{c} \\ s_1 \\ t \\ \\ \end{array}\left[\begin{array}{cccccc|c} x_1 & x_2 & s_1 & s_2 & t & W & \\ 1 & 1 & 1 & 0 & 0 & 0 & 9 \\ 1 & -1 & 0 & -1 & 1 & 0 & 1 \\ \hline -1 & -2 & 0 & 0 & M & 1 & 0 \end{array}\right] \tag{13}$$

An initial BFS is given by (11). Notice that, from row s_1, when $x_1 = x_2 = s_2 = 0$, we can directly read the value of s_1, namely, $s_1 = 9$. From row 2, we get $t = 1$. From row 3, $Mt + W = 0$. Since $t = 1$, $W = -M$. But in a simplex tableau we want the value of W to appear in the last row and last column. This is not so in (13); thus, we modify that matrix.

To do this, we transform (13) into an equivalent matrix whose last row has the form

$$\begin{array}{cccccc|c} x_1 & x_2 & s_1 & s_2 & t & W & \\ ? & ? & 0 & ? & 0 & 1 & ? \end{array}$$

That is, the M in the t-column is replaced by 0. As a result, if $x_1 = x_2 = s_2 = 0$, then W equals the last entry. Proceeding to obtain such a matrix, we have, by pivoting at the shaded element in column t:

$$\left[\begin{array}{cccccc|c} x_1 & x_2 & s_1 & s_2 & t & W & R \\ 1 & 1 & 1 & 0 & 0 & 0 & 9 \\ 1 & -1 & 0 & -1 & \boxed{1} & 0 & 1 \\ \hline -1 & -2 & 0 & 0 & M & 1 & 0 \end{array}\right]$$

$$\xrightarrow{-MR_2 + R_3} \left[\begin{array}{cccccc|c} x_1 & x_2 & s_1 & s_2 & t & W & R \\ 1 & 1 & 1 & 0 & 0 & 0 & 9 \\ 1 & -1 & 0 & -1 & 1 & 0 & 1 \\ \hline -1-M & -2+M & 0 & M & 0 & 1 & -M \end{array}\right]$$

Let us now check things out. If $x_1 = 0$, $x_2 = 0$, and $s_2 = 0$, then from row 1 we get $s_1 = 9$, from row 2, $t = 1$, and from row 3, $W = -M$. Thus, we now have initial simplex tableau I:

SIMPLEX TABLEAU I

entering variable ↓ (x_1)

$$\begin{array}{c} B \\ s_1 \\ t \\ W \end{array}\left[\begin{array}{cccccc|c} x_1 & x_2 & s_1 & s_2 & t & W & R \\ 1 & 1 & 1 & 0 & 0 & 0 & 9 \\ \boxed{1} & -1 & 0 & -1 & 1 & 0 & 1 \\ \hline -1-M & -2+M & 0 & M & 0 & 1 & -M \end{array}\right] \begin{array}{c} \textit{Quotients} \\ 9 \div 1 = 9 \\ 1 \div 1 = 1 \\ \\ \end{array}$$

departing variable ← t

indicators (bottom row entries under x_1 through t)

From this point, we can use the procedures of Section 7.4. Since M is a large positive number, the most negative indicator is $-1 - M$. Thus, the entering variable is x_1. From the quotients, we get t as the departing variable. The pivot entry is shaded. Using elementary row operations to get 1 in the pivot position and 0's elsewhere in that column, we get tableau II:

SIMPLEX TABLEAU II

entering variable ↓ (x_2)

B	x_1	x_2	s_1	s_2	t	W	R	Quotients
departing variable ← s_1	0	2	1	1	−1	0	8	$8 \div 2 = 4$
x_1	1	−1	0	−1	1	0	1	no quotient
W	0	−3	0	−1	$1+M$	1	1	

indicators

From tableau II, we have the following BFS:

$$s_1 = 8, \quad x_1 = 1, \quad x_2 = 0, \quad s_2 = 0, \quad t = 0$$

Since $t = 0$, the values $s_1 = 8$, $x_1 = 1$, $x_2 = 0$, and $s_2 = 0$ form an initial BFS for the *original* problem! The artificial variable has served its purpose. For succeeding tableaux, we shall delete the t-column (since we want to solve the original problem) and change the W's to Z's (since $W = Z$ for $t = 0$). From tableau II, the entering variable is x_2, the departing variable is s_1, and the pivot entry is shaded. Using elementary row operations (omitting the t-column), we get tableau III:

SIMPLEX TABLEAU III

B	x_1	x_2	s_1	s_2	Z	R
x_2	0	1	$\frac{1}{2}$	$\frac{1}{2}$	0	4
x_1	1	0	$\frac{1}{2}$	$-\frac{1}{2}$	0	5
Z	0	0	$\frac{3}{2}$	$\frac{1}{2}$	1	13

indicators

Since all the indicators are nonnegative, the maximum value of Z is 13. It occurs when $x_1 = 5$ and $x_2 = 4$.

Here is a summary of the procedure involving artificial variables.

It is worthwhile to review the steps we performed to solve our problem:

$$\text{Maximize } Z = x_1 + 2x_2$$

subject to

$$x_1 + x_2 \leq 9 \tag{14}$$

$$x_1 - x_2 \geq 1 \tag{15}$$

$$x_1, x_2 \geq 0$$

We write Equation (14) as

$$x_1 + x_2 + s_1 = 9 \tag{16}$$

Since Equation (15) involves the symbol $\geq$, and the constant on the right side is nonnegative, we write Equation (15) in a form having both a surplus variable and an artificial variable:

$$x_1 - x_2 - s_2 + t = 1 \tag{17}$$

The artificial objective equation to consider is $W = x_1 + 2x_2 - Mt$, or equivalently,

$$-x_1 - 2x_2 + Mt + W = 0 \tag{18}$$

The augmented coefficient matrix of the system formed by Equations (16)–(18) is

$$\begin{array}{c} \text{B} \\ s_1 \\ t \\ W \end{array} \begin{array}{c} \begin{array}{cccccc|c} x_1 & x_2 & s_1 & s_2 & t & W & \text{R} \end{array} \\ \left[\begin{array}{cccccc|c} 1 & 1 & 1 & 0 & 0 & 0 & 9 \\ 1 & -1 & 0 & -1 & 1 & 0 & 1 \\ \hline -1 & -2 & 0 & 0 & M & 1 & 0 \end{array}\right] \end{array}$$

Next, we remove the M from the artificial variable column and replace it by 0 by using elementary row operations. The resulting simplex tableau I corresponds to the initial BFS of the artificial problem in which the decision variables, x_1 and x_2, and the surplus variable s_2 are each 0:

SIMPLEX TABLEAU I

$$\begin{array}{c} \text{B} \\ s_1 \\ t \\ W \end{array} \begin{array}{c} \begin{array}{cccccc|c} x_1 & x_2 & s_1 & s_2 & t & W & \text{R} \end{array} \\ \left[\begin{array}{cccccc|c} 1 & 1 & 1 & 0 & 0 & 0 & 9 \\ 1 & -1 & 0 & -1 & 1 & 0 & 1 \\ \hline -1-M & -2+M & 0 & M & 0 & 1 & -M \end{array}\right] \end{array}$$

The basic variables s_1 and t in column B of the tableau correspond to the nondecision variables in Equations (16) and (17) that have positive coefficients. We now apply the simplex method until we obtain a BFS in which the artificial variable t equals 0. Then we can delete the artificial variable column, change the W's to Z's, and continue the procedure until the maximum value of Z is obtained.

EXAMPLE 1 Artificial Variables

Use the simplex method to maximize $Z = 2x_1 + x_2$ subject to

$$x_1 + x_2 \leq 12 \tag{19}$$

$$x_1 + 2x_2 \leq 20 \tag{20}$$

$$-x_1 + x_2 \geq 2 \tag{21}$$

$$x_1, x_2 \geq 0$$

Solution: Equations (19)–(21) will involve two slack variables, s_1 and s_2, for the two $\leq$ constraints, and a surplus variable s_3 and an artificial variable t for the $\geq$ constraint. We thus have

$$x_1 + x_2 + s_1 = 12 \tag{22}$$

$$x_1 + 2x_2 + s_2 = 20 \tag{23}$$

$$-x_1 + x_2 - s_3 + t = 2 \tag{24}$$

We consider $W = Z - Mt = 2x_1 + x_2 - Mt$ as the artificial objective equation, or equivalently,

$$-2x_1 - x_2 + Mt + W = 0 \tag{25}$$

where M is a large positive number. Now we construct the augmented

APPLICATIONS IN PRACTICE 1

Artificial Variables

The GHI Company manufactures two models of snowboards, standard and deluxe, at two different manufacturing plants. The maximum output at plant I is 1200 per month, while the maximum output at plant II is 1000 per month. Owing to contractual obligations, the number of deluxe models produced at plant I cannot exceed the number of standard models produced at plant I by more than 200. The profit per standard and deluxe snowboard manufactured at plant I is \$40 and \$60, respectively, while the profit per standard and deluxe snowboard manufactured at plant II is \$45 and \$50, respectively. This month, GHI received an order for 1000 standard and 800 deluxe models. Find how many of each model should be produced at each plant to satisfy the order and maximize the profit. (*Hint:* Let x_1 represent the number of standard models and x_2 represent the number of deluxe models manufactured at plant I.)

coefficient matrix of Equations (22)–(25):

$$\begin{array}{c} \begin{matrix} x_1 & x_2 & s_1 & s_2 & s_3 & t & W \end{matrix} \\ \left[\begin{array}{rrrrrrr|r} 1 & 1 & 1 & 0 & 0 & 0 & 0 & 12 \\ 1 & 2 & 0 & 1 & 0 & 0 & 0 & 20 \\ -1 & 1 & 0 & 0 & -1 & 1 & 0 & 2 \\ \hline -2 & -1 & 0 & 0 & 0 & M & 1 & 0 \end{array}\right] \end{array}$$

To get simplex tableau I, we replace the M in the artificial variable column by zero by adding $-M$ times row 3 to row 4:

SIMPLEX TABLEAU I

entering variable ↓ (x_2)

$$\begin{array}{c} \\ s_1 \\ s_2 \\ \text{departing variable} \leftarrow t \\ W \end{array} \begin{array}{c} \begin{matrix} x_1 & x_2 & s_1 & s_2 & s_3 & t & W & R \end{matrix} \\ \left[\begin{array}{cccccccc|c} 1 & 1 & 1 & 0 & 0 & 0 & 0 & & 12 \\ 1 & 2 & 0 & 1 & 0 & 0 & 0 & & 20 \\ -1 & \boxed{1} & 0 & 0 & -1 & 1 & 0 & & 2 \\ \hline -2+M & -1-M & 0 & 0 & M & 0 & 1 & & -2M \end{array}\right] \end{array} \begin{array}{r} \text{Quotients} \\ 12 \div 1 = 12 \\ 20 \div 2 = 10 \\ 2 \div 1 = 2 \\ \\ \end{array}$$

indicators (under x_1 through t in the W row)

The variables s_1, s_2, and t in column B, the basic variables, are the nondecision variables with positive coefficients in Equations (22)–(24). Since M is a large positive number, $-1-M$ is the most negative indicator. The entering variable is x_2, the departing variable is t, and the pivot entry is shaded. Proceeding, we get tableau II:

SIMPLEX TABLEAU II

entering variable ↓ (x_1)

$$\begin{array}{c} \\ \text{departing variable} \leftarrow s_1 \\ s_2 \\ x_2 \\ W \end{array} \begin{array}{c} \begin{matrix} x_1 & x_2 & s_1 & s_2 & s_3 & t & W & R \end{matrix} \\ \left[\begin{array}{ccccccc|c} \boxed{2} & 0 & 1 & 0 & 1 & -1 & 0 & 10 \\ 3 & 0 & 0 & 1 & 2 & -2 & 0 & 16 \\ -1 & 1 & 0 & 0 & -1 & 1 & 0 & 2 \\ \hline -3 & 0 & 0 & 0 & -1 & 1+M & 1 & 2 \end{array}\right] \end{array} \begin{array}{r} \text{Quotients} \\ 10 \div 2 = 5 \\ 16 \div 3 = 5\frac{1}{3} \\ \\ \\ \end{array}$$

indicators (under x_1 through t in the W row)

The BFS corresponding to tableau II has $t = 0$. Thus, we shall delete the t-column and change W's to Z's in succeeding tableaux. Continuing, we obtain tableau III:

SIMPLEX TABLEAU III

$$\begin{array}{c} \\ x_1 \\ s_2 \\ x_2 \\ Z \end{array} \begin{array}{c} \begin{matrix} x_1 & x_2 & s_1 & s_2 & s_3 & Z & R \end{matrix} \\ \left[\begin{array}{cccccc|c} 1 & 0 & \frac{1}{2} & 0 & \frac{1}{2} & 0 & 5 \\ 0 & 0 & -\frac{3}{2} & 1 & \frac{1}{2} & 0 & 1 \\ 0 & 1 & \frac{1}{2} & 0 & -\frac{1}{2} & 0 & 7 \\ \hline 0 & 0 & \frac{3}{2} & 0 & \frac{1}{2} & 1 & 17 \end{array}\right] \end{array}$$

indicators (under x_1 through s_3 in the Z row)

All indicators are nonnegative. Hence, the maximum value of Z is 17. It occurs when $x_1 = 5$ and $x_2 = 7$.

▶ **Now work Problem 1.**

Equality Constraints

When an *equality* constraint of the form

$$a_1x_1 + a_2x_2 + \cdots + a_nx_n = b \quad \text{where } b \geq 0$$

occurs in a linear programming problem, artificial variables are used in the simplex method. To illustrate, consider the following problem:

$$\text{Maximize } Z = x_1 + 3x_2 - 2x_3$$

subject to

$$x_1 + x_2 - x_3 = 6 \tag{26}$$

$$x_1, x_2, x_3 \geq 0$$

Constraint (26) is already expressed as an equation, so no slack variable is necessary. Since $x_1 = x_2 = x_3 = 0$ is not a feasible solution, we do not have an obvious starting point for the simplex procedure. Thus, we create an artificial problem by first adding an artificial variable t to the left side of Equation (26):

$$x_1 + x_2 - x_3 + t = 6$$

Here an obvious BFS is $x_1 = x_2 = x_3 = 0, t = 6$. The artificial objective function is

$$W = Z - Mt = x_1 + 3x_2 - 2x_3 - Mt$$

where M is a large positive number. The simplex procedure is applied to this artificial problem until we obtain a BFS in which $t = 0$. This solution will give an initial BFS for the original problem, and we then proceed as before.

In general, the simplex method may be used to

$$\text{maximize } Z = c_1x_1 + c_2x_2 + \cdots + c_nx_n$$

subject to

$$\left.\begin{array}{l} a_{11}x_1 + a_{12}x_2 + \cdots + a_{1n}x_n\{\leq, \geq, =\}\ b_1 \\ a_{21}x_1 + a_{22}x_2 + \cdots + a_{2n}x_n\{\leq, \geq, =\}\ b_2 \\ \quad\vdots \qquad\quad \vdots \qquad\qquad\quad \vdots \qquad\qquad \vdots \\ a_{m1}x_1 + a_{m2}x_2 + \cdots + a_{mn}x_n\{\leq, \geq, =\}\ b_m \end{array}\right\} \tag{27}$$

where $x_1, x_2, \ldots, x_n$ and $b_1, b_2, \ldots, b_m$ are nonnegative. The symbolism $\{\leq, \geq, =\}$ means that one of the relations "$\leq$," "$\geq$," or "$=$" exists for a constraint. If all constraints involve "$\leq$," the problem is of standard form and the simplex techniques of the previous sections apply. If any constraint involves "$\geq$" or "$=$," we begin with an artificial problem, which is obtained as follows.

Each constraint that contains "$\leq$" is written as an equation involving a slack variable s_i (with coefficient +1):

$$a_{i1}x_1 + a_{i2}x_2 + \cdots + a_{in}x_n + s_i = b_i$$

Each constraint that contains "$\geq$" is written as an equation involving a surplus variable s_j (with coefficient -1) and an artificial variable t_j:

$$a_{j1}x_1 + a_{j2}x_2 + \cdots + a_{jn}x_n - s_j + t_j = b_j$$

A nonnegative artificial variable t_k is inserted into each equality constraint:

$$a_{k1}x_1 + a_{k2}x_2 + \cdots + a_{kn}x_n + t_k = b_k$$

Should the artificial variables involved in this problem be, for example, t_1, t_2, and t_3, then the artificial objective function is

$$W = Z - Mt_1 - Mt_2 - Mt_3$$

where M is a large positive number. An initial BFS occurs when $x_1 = x_2 = \cdots = x_n = 0$ and each surplus variable equals 0. After obtaining an initial simplex tableau, we apply the simplex procedure until we arrive at a tableau that corresponds to a BFS in which *all* artificial variables are 0. We then delete the artificial variable columns, change W's to Z's, and continue by using the procedures of the previous sections.

EXAMPLE 2 An Equality Constraint

Use the simplex method to maximize $Z = x_1 + 3x_2 - 2x_3$ subject to

$$-x_1 - 2x_2 - 2x_3 = -6 \tag{28}$$

$$-x_1 - x_2 + x_3 \leq -2 \tag{29}$$

$$x_1, x_2, x_3 \geq 0 \tag{30}$$

Solution: Constraints (28) and (29) will have the forms indicated in (27) (that is, b's positive) if we multiply both sides of each constraint by -1:

$$x_1 + 2x_2 + 2x_3 = 6 \tag{31}$$

$$x_1 + x_2 - x_3 \geq 2 \tag{32}$$

Since constraints (31) and (32) involve "=" and "≥," two artificial variables, t_1 and t_2, will occur. The equations for the artificial problem are

$$x_1 + 2x_2 + 2x_3 \quad + t_1 \quad = 6 \tag{33}$$

and

$$x_1 + x_2 - x_3 - s_2 \quad + t_2 = 2 \tag{34}$$

Here the subscript 2 on s_2 reflects the order of the equations. The artificial objective function is $W = Z - Mt_1 - Mt_2$, or equivalently,

$$-x_1 - 3x_2 + 2x_3 + Mt_1 + Mt_2 + W = 0 \tag{35}$$

where M is a large positive number. The augmented coefficient matrix of Equations (33)–(35) is

$$\begin{array}{c} \begin{matrix} x_1 & x_2 & x_3 & s_2 & t_1 & t_2 & W & \end{matrix} \\ \left[\begin{array}{rrrrrrr|r} 1 & 2 & 2 & 0 & 1 & 0 & 0 & 6 \\ 1 & 1 & -1 & -1 & 0 & 1 & 0 & 2 \\ \hline -1 & -3 & 2 & 0 & M & M & 1 & 0 \end{array}\right] \end{array}$$

We now use elementary row operations to remove the bottom-row M's from *all* the artificial variable columns. By adding $-M$ times row 1 to row 3 and adding

$-M$ times row 2 to row 3, we get initial simplex tableau I:

SIMPLEX TABLEAU I

entering variable ↓ (x_2)

	B	x_1	x_2	x_3	s_2	t_1	t_2	W	R	*Quotients*
	t_1	1	2	2	0	1	0	0	6	$6 \div 2 = 3$
departing ← variable	t_2	1	1	-1	-1	0	1	0	2	$2 \div 1 = 2$
	W	$-1-2M$	$-3-3M$	$2-M$	M	0	0	1	$-8M$	

indicators

Proceeding, we obtain simplex tableaux II and III:

SIMPLEX TABLEAU II

entering variable ↓ (x_3)

	B	x_1	x_2	x_3	s_2	t_1	t_2	W	R	*Quotients*
departing ← variable	t_1	-1	0	4	2	1	-2	0	2	$2 \div 4 = \frac{1}{2}$
	x_2	1	1	-1	-1	0	1	0	2	
	W	$2+M$	0	$-1-4M$	$-3-2M$	0	$3+3M$	1	$6-2M$	

indicators

SIMPLEX TABLEAU III

entering variable ↓ (s_2)

	B	x_1	x_2	x_3	s_2	t_1	t_2	W	R	*Quotients*
departing ← variable	x_3	$-\frac{1}{4}$	0	1	$\frac{1}{2}$	$\frac{1}{4}$	$-\frac{1}{2}$	0	$\frac{1}{2}$	$\frac{1}{2} \div \frac{1}{2} = 1$
	x_2	$\frac{3}{4}$	1	0	$-\frac{1}{2}$	$\frac{1}{4}$	$\frac{1}{2}$	0	$\frac{5}{2}$	
	W	$\frac{7}{4}$	0	0	$-\frac{5}{2}$	$\frac{1}{4}+M$	$\frac{5}{2}+M$	1	$\frac{13}{2}$	

indicators

For the BFS corresponding to tableau III, the artificial variables t_1 and t_2 are both 0. We now can delete the t_1- and t_2-columns and change W's to Z's. Continuing, we obtain simplex tableau IV:

SIMPLEX TABLEAU IV

B	x_1	x_2	x_3	s_2	Z	R
s_2	$-\frac{1}{2}$	0	2	1	0	1
x_2	$\frac{1}{2}$	1	1	0	0	3
Z	$\frac{1}{2}$	0	5	0	1	9

indicators

Since all indicators are nonnegative, we have reached the final tableau. The maximum value of Z is 9, and it occurs when $x_1 = 0$, $x_2 = 3$, and $x_3 = 0$.

▸ **Now work Problem 5.**

Empty Feasible Regions

It is possible that the simplex procedure terminates and not all artificial variables are 0. It can be shown that in this situation *the feasible region of the original problem is empty,* and hence, there is *no optimum solution*. The following example will illustrate.

EXAMPLE 3 **An Empty Feasible Region**

Use the simplex method to maximize $Z = 2x_1 + x_2$ *subject to*

$$-x_1 + x_2 \geq 2 \tag{36}$$

$$x_1 + x_2 \leq 1 \tag{37}$$

$$x_1, x_2 \geq 0$$

and $x_1, x_2 \geq 0$.

Solution: Since constraint (36) is of the form $a_{11}x_1 + a_{12}x_2 \geq b_1$, where $b_1 \geq 0$, an artificial variable will occur. The equations to consider are

$$-x_1 + x_2 - s_1 \quad + t_1 = 2 \tag{38}$$

and

$$x_1 + x_2 \quad + s_2 = 1 \tag{39}$$

where s_1 and s_2 are slack variables and t_1 is artificial. The artificial objective function is $W = Z - Mt_1$, or equivalently,

$$-2x_1 - x_2 + Mt_1 + W = 0. \tag{40}$$

The augmented coefficient matrix of Equations (38)–(40) is

$$\begin{array}{c} \begin{matrix} x_1 & x_2 & s_1 & s_2 & t_1 & W & \end{matrix} \\ \left[\begin{array}{cccccc|c} -1 & 1 & -1 & 0 & 1 & 0 & 2 \\ 1 & 1 & 0 & 1 & 0 & 0 & 1 \\ \hline -2 & -1 & 0 & 0 & M & 1 & 0 \end{array}\right] \end{array}$$

The simplex tableaux are as follows:

SIMPLEX TABLEAU I

entering variable ↓ (x_2)

B	x_1	x_2	s_1	s_2	t_1	W	R	Quotients
t_1	-1	1	-1	0	1	0	2	$2 \div 1 = 2$
departing variable ← s_2	1	1	0	1	0	0	1	$1 \div 1 = 1$
W	$-2+M$	$-1-M$	M	0	0	1	$-2M$	

indicators

SIMPLEX TABLEAU II

B	x_1	x_2	s_1	s_2	t_1	W	R
t_1	-2	0	-1	-1	1	0	1
x_2	1	1	0	1	0	0	1
W	$-1+2M$	0	M	$1+M$	0	1	$1-M$

indicators

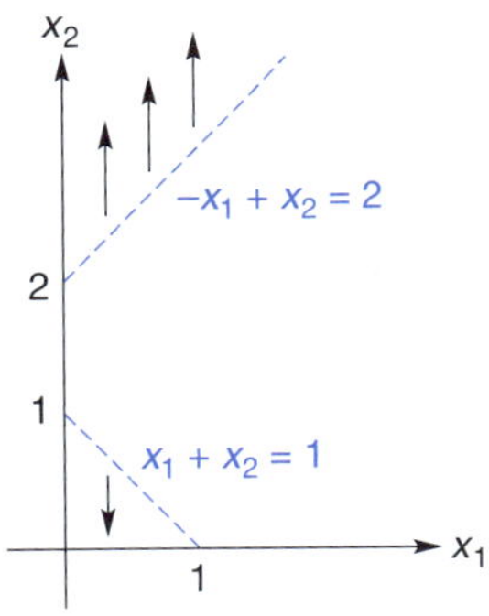

FIGURE 7.25 Empty feasible region (no solution exists).

Since M is a large positive number, the indicators in simplex tableau II are nonnegative, so the simplex procedure terminates. The value of the artificial variable t_1 is 1. Therefore, as previously stated, the feasible region of the original problem is empty, and hence, no solution exists. This result can be obtained geometrically. Figure 7.25 shows the graphs of $-x_1 + x_2 = 2$ and $x_1 + x_2 = 1$ for $x_1, x_2 \geq 0$. Since there is no point (x_1, x_2) that simultaneously lies above $-x_1 + x_2 = 2$ and below $x_1 + x_2 = 1$ such that $x_1, x_2 \geq 0$, the feasible region is empty, and thus, no solution exists.

▶ **Now work Problem 9.**

In the next section we shall use the simplex method on minimization problems.

Exercise 7.6

Use the simplex method to solve the following problems.

*1. Maximize

$$Z = 2x_1 + x_2$$

subject to

$$\begin{aligned} x_1 + x_2 &\leq 6 \\ -x_1 + x_2 &\geq 4 \\ x_1, x_2 &\geq 0 \end{aligned}$$

2. Maximize

$$Z = 3x_1 + 4x_2$$

subject to

$$\begin{aligned} x_1 + 2x_2 &\leq 8 \\ x_1 + 6x_2 &\geq 12 \\ x_1, x_2 &\geq 0 \end{aligned}$$

3. Maximize

$$Z = 2x_1 + x_2 - x_3$$

subject to

$$\begin{aligned} x_1 + 2x_2 + x_3 &\leq 5 \\ -x_1 + x_2 + x_3 &\geq 1 \\ x_1, x_2, x_3 &\geq 0 \end{aligned}$$

4. Maximize

$$Z = x_1 - x_2 + 4x_3$$

subject to

$$\begin{aligned} x_1 + x_2 + x_3 &\leq 9 \\ x_1 - 2x_2 + x_3 &\geq 6 \\ x_1, x_2, x_3 &\geq 0 \end{aligned}$$

*5. Maximize

$$Z = 3x_1 + 2x_2 + x_3$$

subject to

$$\begin{aligned} x_1 + x_2 + x_3 &\leq 10 \\ x_1 - x_2 - x_3 &= 6 \\ x_1, x_2, x_3 &\geq 0 \end{aligned}$$

6. Maximize

$$Z = x_1 + 2x_2 + 3x_3$$

subject to

$$\begin{aligned} x_2 - 2x_3 &\geq 5 \\ x_1 + x_2 + x_3 &= 8 \\ x_1, x_2, x_3 &\geq 0 \end{aligned}$$

7. Maximize

$$Z = x_1 - 10x_2$$

subject to

$$\begin{aligned} x_1 - x_2 &\leq 1 \\ x_1 + 2x_2 &\leq 8 \\ x_1 + x_2 &\geq 5 \\ x_1, x_2 &\geq 0 \end{aligned}$$

8. Maximize

$$Z = x_1 + 4x_2 - x_3$$

subject to

$$\begin{aligned} x_1 + x_2 - x_3 &\geq 5 \\ x_1 + x_2 + x_3 &\leq 3 \\ x_1 - x_2 + x_3 &= 7 \\ x_1, x_2, x_3 &\geq 0 \end{aligned}$$

*9. Maximize

$$Z = 3x_1 - 2x_2 + x_3$$

subject to

$$\begin{aligned} x_1 + x_2 + x_3 &\leq 1 \\ x_1 - x_2 + x_3 &\geq 2 \\ x_1 - x_2 - x_3 &\leq -6 \\ x_1, x_2, x_3 &\geq 0 \end{aligned}$$

10. Maximize

$$Z = x_1 + 4x_2$$

subject to

$$\begin{aligned} x_1 + 2x_2 &\leq 8 \\ x_1 + 6x_2 &\geq 12 \\ x_2 &\geq 2 \\ x_1, x_2 &\geq 0 \end{aligned}$$

11. Maximize

$$Z = -3x_1 + 2x_2$$

subject to

$$\begin{aligned} x_1 - x_2 &\leq 4 \\ -x_1 + x_2 &= 4 \\ x_1 &\geq 6 \\ x_1, x_2 &\geq 0 \end{aligned}$$

12. Maximize

$$Z = 2x_1 - 8x_2$$

subject to

$$\begin{aligned} x_1 - 2x_2 &\geq -12 \\ -x_1 + x_2 &\geq 2 \\ x_1 + x_2 &\geq 10 \\ x_1, x_2 &\geq 0 \end{aligned}$$

13. Production A company manufactures two types of bookcases: Standard and Executive. Each type requires assembly and finishing times as given in the following table:

	Assembly Time	Finishing Time	Profit per Unit
Standard	1 hr	2 hr	$30
Executive	2 hr	3 hr	36

The profit on each unit is also indicated. The number of hours available per week in the assembly department is 400, and in the finishing department it is 510. Because of a union contract, the finishing department is guaranteed at least 240 hours of work per week. How many units of each type should the company produce each week to maximize profit?

14. Production A company manufactures three products: X, Y, and Z. Each product requires the use of time on machines A and B as given in the following table:

	Machine A	Machine B
Product X	1 hr	1 hr
Product Y	2 hr	1 hr
Product Z	2 hr	2 hr

The numbers of hours per week that A and B are available for production are 40 and 30, respectively. The profit per unit on X, Y, and Z is $50, $60, and $75, respectively. At least five units of Z must be produced next week. What should be the production order for that period if maximum profit is to be achieved? What is the maximum profit?

15. Investments The prospectus of an investment fund states that all money is invested in bonds that are rated A, AA, and AAA; no more than 30% of the total investment is in A and AA bonds, and at least 50% is in AA and AAA bonds. The A, AA, and AAA bonds respectively yield 8%, 7%, and 6% annually. Determine the percentages of the total investment that should be committed to each type of bond so that the fund maximizes its annual yield. What is this yield?

OBJECTIVE *To show how to solve a minimization problem by altering the objective function so that a maximization problem results.*

7.7 Minimization

So far we have used the simplex method to *maximize* objective functions. In general, to *minimize* a function it suffices to maximize the negative of the function. To understand why, consider the function $f(x) = x^2 - 4$. In Figure 7.26(a), observe that the minimum value of f is -4, and it occurs when $x = 0$. Figure 7.26(b) shows the graph of $g(x) = -f(x) = -(x^2 - 4)$. This graph is the reflection through the x-axis of the graph of f. Notice that the maximum value of g is 4 and occurs when $x = 0$. Thus the minimum value of $x^2 - 4$ is

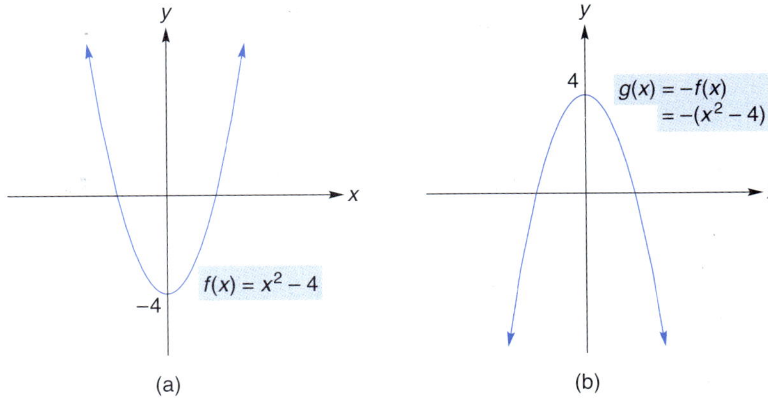

FIGURE 7.26 Minimum value of $f(x)$ is equal to the negative of the maximum value of $-f(x)$.

the negative of the maximum value of $-(x^2 - 4)$. That is,

$$\min f = -\max(-f)$$

Alternatively, think of a point C on the positive half of the number line moving to the left. As it does so, the point $-C$ moves to the right. It is clear that if for some reason C stops, then it stops at the minimum value that C encounters. If C stops, then so does $-C$, at the maximum value encountered by $-C$. Since this value of $-C$ is still the negative of the value of C, we see that

$$\min C = -\max(-C)$$

The problem in Example 1 will be solved more efficiently in Example 4 of Section 7.8.

EXAMPLE 1 **Minimization**

Use the simplex method to minimize $Z = x_1 + 2x_2$ *subject to*

$$-2x_1 + x_2 \geq 1 \quad (1)$$

$$-x_1 + x_2 \geq 2 \quad (2)$$

$$x_1, x_2 \geq 0 \quad (3)$$

and $x_1, x_2 \geq 0$.

Solution: To minimize Z, we can maximize $-Z = -x_1 - 2x_2$. Note that constraints (1) and (2) each have the form $a_1x_1 + a_2x_2 \geq b$, where $b \geq 0$. Thus, their equations involve two slack variables s_1 and s_2, each with coefficient -1, and two artificial variables t_1 and t_2:

$$-2x_1 + x_2 - s_1 + t_1 = 1 \quad (4)$$

$$-x_1 + x_2 - s_2 + t_2 = 2 \quad (5)$$

Since there are *two* artificial variables, we maximize the objective function

$$W = (-Z) - Mt_1 - Mt_2$$

where M is a large positive number. Equivalently,

$$x_1 + 2x_2 + Mt_1 + Mt_2 + W = 0 \quad (6)$$

The augmented coefficient matrix of Equations (4)–(6) is

$$\begin{array}{c} \begin{matrix} x_1 & x_2 & s_1 & s_2 & t_1 & t_2 & W & \end{matrix} \\ \left[\begin{array}{ccccccc|c} -2 & 1 & -1 & 0 & 1 & 0 & 0 & 1 \\ -1 & 1 & 0 & -1 & 0 & 1 & 0 & 2 \\ \hline 1 & 2 & 0 & 0 & M & M & 1 & 0 \end{array}\right] \end{array}$$

Proceeding, we obtain tableaux I, II, III:

SIMPLEX TABLEAU I

entering variable ↓ (x_2)

$$\begin{array}{c|ccccccc|c|c} B & x_1 & x_2 & s_1 & s_2 & t_1 & t_2 & W & R & \textit{Quotients} \\ \hline \text{departing variable} \leftarrow t_1 & -2 & 1 & -1 & 0 & 1 & 0 & 0 & 1 & 1 \div 1 = 1 \\ t_2 & -1 & 1 & 0 & -1 & 0 & 1 & 0 & 2 & 2 \div 1 = 2 \\ \hline W & 1+3M & 2-2M & M & M & 0 & 0 & 1 & -3M & \end{array}$$

indicators

SIMPLEX TABLEAU II

entering variable ↓ (s_1)

$$\begin{array}{c|ccccccc|c|c} B & x_1 & x_2 & s_1 & s_2 & t_1 & t_2 & W & R & \textit{Quotients} \\ \hline x_2 & -2 & 1 & -1 & 0 & 1 & 0 & 0 & 1 & \\ \text{departing variable} \leftarrow t_2 & 1 & 0 & 1 & -1 & -1 & 1 & 0 & 1 & 1 \div 1 = 1 \\ \hline W & 5-M & 0 & 2-M & M & -2+2M & 0 & 1 & -2-M & \end{array}$$

indicators

SIMPLEX TABLEAU III

$$\begin{array}{c|ccccccc|c} B & x_1 & x_2 & s_1 & s_2 & t_1 & t_2 & W & R \\ \hline x_2 & -1 & 1 & 0 & -1 & 0 & 1 & 0 & 2 \\ s_1 & 1 & 0 & 1 & -1 & -1 & 1 & 0 & 1 \\ \hline W & 3 & 0 & 0 & 2 & M & -2+M & 1 & -4 \end{array}$$

indicators

The BFS corresponding to tableau III has both artificial variables equal to 0. Thus, the t_1- and t_2-columns are no longer needed. However, the indicators in the x_1-, x_2-, s_1-, and s_2-columns are nonnegative, and hence an optimum solution has been reached. Since $W = -Z$ when $t_1 = t_2 = 0$, the maximum value of $-Z$ is -4. Consequently, the *minimum* value of Z is $-(-4)$, or 4. It occurs when $x_1 = 0$ and $x_2 = 2$.

▶ **Now work Problem 1.**

Here is an interesting example dealing with environmental controls.

EXAMPLE 2 **Reducing Dust Emissions**

A cement plant produces 2,500,000 *barrels of cement per year. The kilns emit* 2 *lb of dust for each barrel produced. A governmental agency dealing with environmental protection requires that the plant reduce its dust emissions to no more than* 800,000 *lb per year. There are two emission control devices available,* A *and* B. *Device* A *reduces emissions to* $\frac{1}{2}$ *lb per barrel, and its cost is* \$0.20 *per barrel of cement produced. For device* B, *emissions are reduced to* $\frac{1}{5}$ *lb per barrel, and the cost is* \$0.25 *per barrel of cement produced. Determine the most economical course of action that the plant should take so that it complies with the agency's requirement and also maintains its annual production of* 2,500,000 *barrels of cement.*[5]

Solution: We must minimize the annual cost of emission control. Let x_1, x_2, and x_3 be the annual numbers of barrels of cement produced in kilns that use device A, device B, and no device, respectively. Then $x_1, x_2, x_3 \geq 0$, and the annual emission control cost (in dollars) is

$$C = \frac{1}{5}x_1 + \frac{1}{4}x_2 + 0x_3 \tag{7}$$

Since 2,500,000 barrels of cement are produced each year,

$$x_1 + x_2 + x_3 = 2{,}500{,}000 \tag{8}$$

The numbers of pounds of dust emitted annually by the kilns that use device A, device B, and no device are $\frac{1}{2}x_1$, $\frac{1}{5}x_2$, and $2x_3$, respectively. Since the total number of pounds of dust emission is to be no more than 800,000,

$$\frac{1}{2}x_1 + \frac{1}{5}x_2 + 2x_3 \leq 800{,}000 \tag{9}$$

To minimize C subject to constraints (8) and (9), where $x_1, x_2, x_3 \geq 0$, we shall first maximize $-C$ by using the simplex method. The equations to consider are

$$x_1 + x_2 + x_3 + t_1 = 2{,}500{,}000 \tag{10}$$

and

$$\frac{1}{2}x_1 + \frac{1}{5}x_2 + 2x_3 + s_2 = 800{,}000 \tag{11}$$

where t_1 and s_2 are artificial and slack variables, respectively. The artificial objective equation is $W = (-C) - Mt_1$, or equivalently,

$$\frac{1}{5}x_1 + \frac{1}{4}x_2 + 0x_3 + Mt_1 + W = 0 \tag{12}$$

where M is a large positive number. The augmented coefficient matrix of Equations (10)–(12) is

$$\begin{array}{c} \\ \\ \\ \end{array}\begin{matrix} x_1 & x_2 & x_3 & s_2 & t_1 & W & \\ \end{matrix}$$
$$\left[\begin{array}{cccccc|c} 1 & 1 & 1 & 0 & 1 & 0 & 2{,}500{,}000 \\ \frac{1}{2} & \frac{1}{5} & 2 & 1 & 0 & 0 & 800{,}000 \\ \hline \frac{1}{5} & \frac{1}{4} & 0 & 0 & M & 1 & 0 \end{array}\right]$$

[5]This example is adapted from Robert E. Kohn, "A Mathematical Model for Air Pollution Control," *School Science and Mathematics* 69 (1969), 487–94.

Once we determine our initial simplex tableau, we proceed and obtain (after three additional tableaux) our final tableau:

$$\begin{array}{c} \\ x_2 \\ x_1 \\ -C \end{array}\begin{array}{c}\begin{array}{cccccc} \text{B}\ x_1 & x_2 & x_3 & s_2 & -C & \text{R} \end{array}\\ \left[\begin{array}{ccccc|c} 0 & 1 & -5 & -\frac{10}{3} & 0 & 1{,}500{,}000 \\ 1 & 0 & 6 & \frac{10}{3} & 0 & 1{,}000{,}000 \\ \hline 0 & 0 & \frac{1}{20} & \frac{1}{6} & 1 & -575{,}000 \end{array}\right]\end{array}$$

indicators

Notice that W is replaced by $-C$ when $t_1 = 0$. The maximum value of $-C$ is $-575{,}000$ and occurs when $x_1 = 1{,}000{,}000$, $x_2 = 1{,}500{,}000$, and $x_3 = 0$. Thus, the *minimum* annual cost of the emission control must be $-(-575{,}000) =$ \$575,000. Device A should be installed on kilns producing 1,000,000 barrels of cement annually, and device B should be installed on kilns producing 1,500,000 barrels annually.

▸ **Now work Problem 11.**

Exercise 7.7

Use the simplex method to solve the following problems.

*__1.__ Minimize

$$Z = 3x_1 + 6x_2$$

subject to

$$\begin{aligned} -x_1 + x_2 &\geq 6 \\ x_1 + x_2 &\geq 10 \\ x_1, x_2 &\geq 0 \end{aligned}$$

2. Minimize

$$Z = 8x_1 + 12x_2$$

subject to

$$\begin{aligned} 2x_1 + 2x_2 &\geq 1 \\ x_1 + 3x_2 &\geq 2 \\ x_1, x_2 &\geq 0 \end{aligned}$$

3. Minimize

$$Z = 12x_1 + 6x_2 + 3x_3$$

subject to

$$\begin{aligned} x_1 - x_2 - x_3 &\geq 18 \\ x_1, x_2, x_3 &\geq 0 \end{aligned}$$

4. Minimize

$$Z = x_1 + x_2 + 2x_3$$

subject to

$$\begin{aligned} x_1 + 2x_2 - x_3 &\geq 4 \\ x_1, x_2, x_3 &\geq 0 \end{aligned}$$

5. Minimize

$$Z = 2x_1 + 3x_2 + x_3$$

subject to

$$\begin{aligned} x_1 + x_2 + x_3 &\leq 6 \\ x_1 - x_3 &\leq -4 \\ x_2 + x_3 &\leq 5 \\ x_1, x_2, x_3 &\geq 0 \end{aligned}$$

6. Minimize

$$Z = 5x_1 + x_2 + 3x_3$$

subject to

$$\begin{aligned} 3x_1 + x_2 - x_3 &\leq 4 \\ 2x_1 + 2x_3 &\leq 5 \\ x_1 + x_2 + x_3 &\geq 2 \\ x_1, x_2, x_3 &\geq 0 \end{aligned}$$

7. Minimize

$$Z = x_1 - x_2 - 3x_3$$

subject to

$$\begin{aligned} x_1 + 2x_2 + x_3 &= 4 \\ x_2 + x_3 &= 1 \\ x_1 + x_2 &\leq 6 \\ x_1, x_2, x_3 &\geq 0 \end{aligned}$$

8. Minimize

$$Z = x_1 + x_2 - 2x_3$$

subject to

$$x_1 - x_2 + x_3 \le 4$$
$$2x_1 + x_2 - 3x_3 \ge 6$$
$$x_1 - x_2 - 2x_3 = 2$$
$$x_1, x_2, x_3 \ge 0$$

9. Minimize

$$Z = x_1 + 8x_2 + 5x_3$$

subject to

$$x_1 + x_2 + x_3 \ge 8$$
$$-x_1 + 2x_2 + x_3 \ge 2$$
$$x_1, x_2, x_3 \ge 0$$

10. Minimize

$$Z = 4x_1 + 4x_2 + 6x_3$$

subject to

$$x_1 - x_2 - x_3 \le 3$$
$$x_1 - x_2 + x_3 \ge 3$$
$$x_1, x_2, x_3 \ge 0$$

*__11.__ **Emission Control** A cement plant produces 3,300,000 barrels of cement per year. The kilns emit 2 lb of dust for each barrel produced. The plant must reduce its dust emissions to no more than 1,000,000 lb per year. There are two devices available, A and B, that will control emissions. Device A will reduce emissions to $\frac{1}{2}$ lb per barrel, and the cost is $0.25 per barrel of cement produced. For device B, emissions are reduced to $\frac{1}{4}$ lb per barrel, and the cost is $0.40 per barrel of cement produced. Determine the most economical course of action the plant should take so that it maintains an annual production of exactly 3,300,000 barrels of cement.

12. Delivery Truck Scheduling Because of increased business, a catering service finds that it must rent additional delivery trucks. The minimum needs are 12 units each of refrigerated and nonrefrigerated space. Two standard types of trucks are available in the rental market. Type A has 2 units of refrigerated space and 1 unit of nonrefrigerated space. Type B has 2 units of refrigerated space and 3 units of nonrefrigerated space. The costs per mile are $0.40 for A and $0.60 for B. How many of each type of truck should be rented so as to minimize total cost per mile? What is the minimum total cost per mile?

13. Transportation Costs A retailer has stores in Exton and Whyton and has warehouses A and B in two other cities. Each store requires delivery of exactly 15 refrigerators. In warehouse A there are 25 refrigerators, and in B there are 10.

The transportation costs to ship refrigerators from the warehouses to the stores are given in the following table:

	Exton	Whyton
Warehouse A	$15	$13
Warehouse B	11	12

For example, the cost to ship a refrigerator from A to the Exton store is $15. How should the retailer order the refrigerators so that the requirements of the stores are met and the total transportation costs are minimized? What is the minimum transportation cost?

14. Parts Purchasing An auto manufacturer purchases alternators from two suppliers, X and Y. The manufacturer has two plants, A and B, and requires delivery of exactly 7000 alternators to plant A and exactly 5000 to plant B. Supplier X charges $300 and $320 per alternator (including transportation cost) to A and B, respectively. For these prices, X requires that the auto manufacturer order at least a total of 3000 alternators. However, X can supply no more than 5000 alternators. Supplier Y charges $340 and $280 per alternator to A and B, respectively, and requires a minimum order of 7000 alternators. Determine how the auto manufacturer should order the necessary alternators so that their total cost is a minimum. What is this minimum cost?

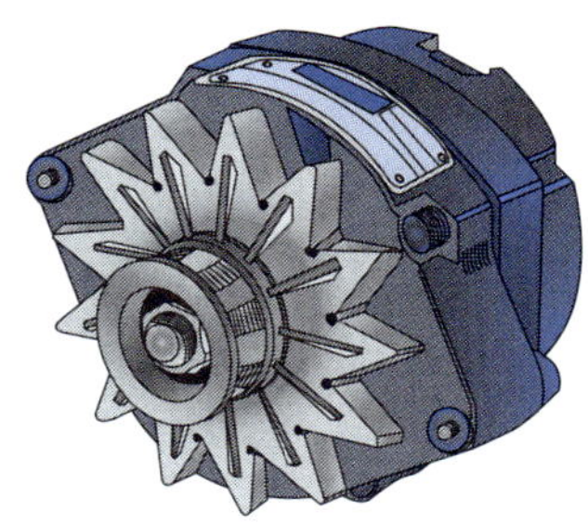

15. Producing Wrapping Paper A paper company stocks its holiday wrapping paper in 48-in.-wide rolls, called stock rolls, and cuts such rolls into smaller widths, depending on customers' orders. Suppose that an order for 50 rolls of 15-in.-wide paper and 60 rolls of 10-in.-wide paper is received. From a stock roll, the company can cut three 15-in.-wide rolls and one

3-in.-wide roll. (See Figure 7.27.) Since the 3-in.-wide roll cannot be used in the order, 3 in. is called the trim loss for this roll.

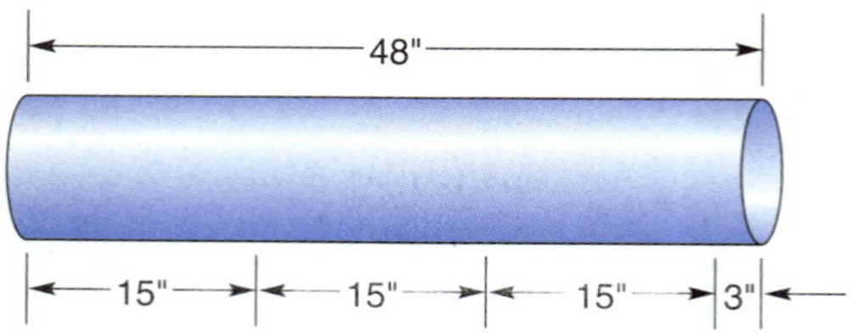

FIGURE 7.27 Diagram for Problem 15.

Similarly, from a stock roll, two 15-in.-wide rolls, one 10-in. wide roll, and one 8-in.-wide roll could be cut. Here the trim loss would be 8 in. The following table indicates the number of 15-in. and 10-in. rolls, together with trim loss, that can be cut from a stock roll:

Roll width	15 in.	3	2	1	—
	10 in.	0	1	—	—
Trim loss		3	8	—	—

(a) Complete the last two columns of the table.
(b) Assume that the company has a sufficient number of stock rolls to fill the order and that *at least* 50 rolls of 15-in.-wide and *at least* 60 rolls of 10-in.-wide wrapping paper will be cut. If x_1, x_2, x_3, and x_4 are the numbers of stock rolls that are cut in a manner described by columns 1–4 of the table, respectively, determine the values of the x's so that the total trim loss is minimized.
(c) What is the minimum amount of total trim loss?

OBJECTIVE *To first motivate and then formally define the dual of a linear programming problem.*

7.8 The Dual

There is a fundamental principle, called *duality,* that allows us to solve a maximization problem by solving a related minimization problem. Let us illustrate.

Suppose that a company produces two types of widgets, manual and electric, and each requires the use of machines A and B in its production. Table 7.2 indicates that a manual widget requires the use of A for 1 hour and B for 1 hour. An electric widget requires A for 2 hours and B for 4 hours. The maximum numbers of hours available per month for machines A and B are 120 and 180, respectively. The profit on a manual widget is \$10, and on an electric widget it is \$24. Assuming that the company can sell all the widgets it can produce, we shall determine the maximum monthly profit. If x_1 and x_2 are the numbers of manual and electric widgets produced per month, respectively, then we want to maximize the monthly profit function

$$P = 10x_1 + 24x_2$$

subject to

$$x_1 + 2x_2 \leq 120 \quad (1)$$

$$x_1 + 4x_2 \leq 180 \quad (2)$$

$$x_1, x_2 \geq 0$$

TABLE 7.2

	Machine A	Machine B	Profit/Unit
Manual	1 hr	1 hr	\$10
Electric	2 hr	4 hr	\$24
Hours available	120	180	

Writing constraints (1) and (2) as equations, we have

$$x_1 + 2x_2 + s_1 = 120$$

and **(3)**

$$x_1 + 4x_2 + s_2 = 180$$

where s_1 and s_2 are slack variables. In Equations (3), $x_1 + 2x_2$ is the number of hours that machine A is used. Since 120 hours on A are available, s_1 is the number of available hours that are *not* used for production. That is, s_1 represents unused capacity (in hours) for A. Similarly, s_2 represents unused capacity for B. Solving this problem by the simplex method, we find that the final tableau is

$$\begin{array}{c} \\ x_1 \\ x_2 \\ \\ P \end{array}\begin{array}{c} \begin{array}{cccccccc} B & x_1 & x_2 & s_1 & s_2 & P & & R \end{array} \\ \left[\begin{array}{cccccc|c} 1 & 0 & 2 & -1 & 0 & & 60 \\ 0 & 1 & -\frac{1}{2} & \frac{1}{2} & 0 & & 30 \\ \hline 0 & 0 & 8 & 2 & 1 & & 1320 \end{array}\right] \end{array} \quad (4)$$

indicators (under the P-row entries for x_1, x_2, s_1, s_2)

Thus, the maximum profit per month is \$1320, which occurs when $x_1 = 60$ and $x_2 = 30$.

Now let us look at the situation from a different point of view. Suppose that the company wishes to rent out machines A and B. What is the minimum monthly rental fee they should charge? Certainly, if the charge is too high, no one would rent the machines. On the other hand, if the charge is too low, it may not pay the company to rent them at all. Obviously, the minimum rent should be \$1320. That is, the minimum the company should charge is the profit it could make by using the machines itself. We can arrive at this minimum rental fee directly by solving a linear programming problem.

Let F be the total monthly rental fee. To determine F, suppose the company assigns values or "worths" to each hour of capacity on machines A and B. Let these worths be y_1 and y_2 dollars, respectively, where $y_1, y_2 \geq 0$. Then the monthly worth of machine A is $120y_1$, and for B it is $180y_2$. Thus,

$$F = 120y_1 + 180y_2$$

The total worth of machine time to produce a manual widget is $1y_1 + 1y_2$. This should be at least equal to the \$10 profit the company can earn by producing that widget. If not, the company would make more money by using the machine time to produce a manual widget. Accordingly,

$$1y_1 + 1y_2 \geq 10$$

Similarly, the total worth of machine time to produce an electric widget should be at least \$24:

$$2y_1 + 4y_2 \geq 24$$

Therefore, the company wants to

$$\text{minimize } F = 120y_1 + 180y_2$$

subject to

$$y_1 + y_2 \geq 10 \quad (5)$$

$$2y_1 + 4y_2 \geq 24 \quad (6)$$

$$y_1, y_2 \geq 0$$

To minimize F, we shall maximize $-F$. Since constraints (5) and (6) have the form $a_1y_1 + a_2y_2 \geq b$, where $b \geq 0$, we consider an artificial problem. If r_1 and r_2 are slack variables and t_1 and t_2 are artificial variables, then we want to maximize

$$W = (-F) - Mt_1 - Mt_2$$

where M is a large positive number, such that

$$y_1 + y_2 - r_1 + t_1 = 10$$

$$2y_1 + 4y_2 - r_2 + t_2 = 24$$

and the y's, r's, and t's are nonnegative. The final simplex tableau for this problem (with the artificial variable columns deleted and W changed to $-F$) is

$$\begin{array}{c} \text{B} \\ y_1 \\ y_2 \\ -R \end{array} \begin{array}{c} \begin{matrix} y_1 & y_2 & r_1 & r_2 & -F & & \text{R} \end{matrix} \\ \left[\begin{array}{ccccc|c} 1 & 0 & -2 & \frac{1}{2} & 0 & 8 \\ 0 & 1 & 1 & -\frac{1}{2} & 0 & 2 \\ \hline 0 & 0 & 60 & 30 & 1 & -1320 \end{array}\right] \end{array}$$

indicators

Since the maximum value of $-F$ is -1320, the *minimum* value of F is $-(-1320) = \$1320$ (as anticipated). It occurs when $y_1 = 8$ and $y_2 = 2$. We have therefore determined the optimum value of one linear programming problem (maximizing profit) by finding the optimum value of another problem (minimizing rental fee).

The values $y_1 = 8$ and $y_2 = 2$ could have been anticipated from the final tableau of the maximization problem. In (4), the indicator 8 in the s_1-column means that at the optimum level of production, if s_1 increases by one unit, then the profit P *decreases* by 8. That is, 1 unused hour of capacity on A decreases the maximum profit by \$8. Thus, 1 hour of capacity on A is worth \$8. We say that the **shadow price** or **accounting price** of 1 hour of capacity on A is \$8. Now, recall that y_1 in the rental problem is the worth of 1 hour of capacity on A. Therefore, y_1 must equal 8 in the optimum solution for that problem. Similarly, since the indicator in the s_2-column is 2, the shadow price of 1 hour of capacity on B is \$2, which is the value of y_2 in the optimum solution of the rental problem.

Let us now analyze the structure of our two linear programming problems:

Maximize	Minimize
$P = 10x_1 + 24x_2$	$F = 120y_1 + 180y_2$
subject to	subject to
$\left.\begin{aligned} x_1 + 2x_2 &\leq 120 \\ x_1 + 4x_2 &\leq 180 \end{aligned}\right\}$ **(7)**	$\left.\begin{aligned} y_1 + y_2 &\geq 10 \\ 2y_1 + 4y_2 &\geq 24 \end{aligned}\right\}$ **(8)**
and $x_1, x_2 \geq 0$.	and $y_1, y_2 \geq 0$.

Note that in (7) the inequalities are all $\leq$, but in (8) they are all $\geq$. The coefficients of the objective function in the minimization problem are the constant terms in (7). The constant terms in (8) are the coefficients of the objective function of the maximization problem. The coefficients of the y_1's in (8) are the coefficients of x_1 and x_2 in the first constraint of (7); the coefficients of the y_2's in (8) are the coefficients of x_1 and x_2 in the second constraint of (7). The minimization problem is called the *dual* of the maximization problem, and vice versa.

In general, with any given linear programming problem, we can associate another linear programming problem called its **dual.** The given problem is called **primal.** If the primal is a maximization problem, then its dual is a minimization problem. Similarly, if the primal involves minimization, then the dual involves maximization.

Any primal maximization problem can be written in the form indicated in Table 7.3. Note that there are no restrictions on the b's.[6] The corresponding dual minimization problem can be written in the form indicated in Table 7.4. Similarly, any primal minimization problem can be put in the form of Table 7.4, and its dual is the maximization problem in Table 7.3.

TABLE 7.3 Primal (Dual)

Maximize $Z = c_1x_1 + c_2x_2 + \cdots + c_nx_n$
subject to

$$\left.\begin{aligned} a_{11}x_1 + a_{12}x_2 + \cdots + a_{1n}x_n &\le b_1 \\ a_{21}x_1 + a_{22}x_2 + \cdots + a_{2n}x_n &\le b_2 \\ \vdots \qquad \vdots \qquad\quad \vdots \quad &\;\; \vdots \\ a_{m1}x_1 + a_{m2}x_2 + \cdots + a_{mn}x_n &\le b_m \end{aligned}\right\} \qquad (9)$$

and $x_1, x_2, \ldots, x_n \ge 0$

TABLE 7.4 Dual (Primal)

Minimize $W = b_1y_1 + b_2y_2 + \cdots + b_my_m$
subject to

$$\left.\begin{aligned} a_{11}y_1 + a_{21}y_2 + \cdots + a_{m1}y_m &\ge c_1 \\ a_{12}y_1 + a_{22}y_2 + \cdots + a_{m2}y_m &\ge c_2 \\ \vdots \qquad \vdots \qquad\quad \vdots \quad &\;\; \vdots \\ a_{1n}y_1 + a_{2n}y_2 + \cdots + a_{mn}y_m &\ge c_n \end{aligned}\right\} \qquad (10)$$

and $y_1, y_2, \ldots, y_m \ge 0$

Let us compare the primal and its dual in Tables 7.3 and 7.4. For convenience, when we refer to constraints, we shall mean those in (9) or (10); we shall not include the nonnegativity conditions. Observe that if all the constraints in the primal involve $\le$ ($\ge$), then all the constraints in its dual involve $\ge$ ($\le$). The coefficients in the dual's objective function are the constant terms in the primal's constraints. Similarly, the constant terms in the dual's constraints are

[6] If an inequality constraint involves $\ge$, multiplying both sides by -1 yields an inequality involving $\le$. If a constraint is an equality, it can be written in terms of two inequalities, one involving $\le$ and one involving $\ge$.

the coefficients of the primal's objective function. The coefficient matrix of the left sides of the dual's constraints is the *transpose* of the coefficient matrix of the left sides of the primal's constraints. That is,

$$\begin{bmatrix} a_{11} & a_{12} & \cdots & a_{1n} \\ a_{21} & a_{22} & \cdots & a_{2n} \\ \vdots & \vdots & & \vdots \\ a_{m1} & a_{m2} & \cdots & a_{mn} \end{bmatrix}^{\mathrm{T}} = \begin{bmatrix} a_{11} & a_{21} & \cdots & a_{m1} \\ a_{12} & a_{22} & \cdots & a_{m2} \\ \vdots & \vdots & & \vdots \\ a_{1n} & a_{2n} & \cdots & a_{mn} \end{bmatrix}$$

If the primal involves n decision variables and m slack variables, then the dual involves m decision variables and n slack variables. It should be noted that the dual of the *dual* is the primal.

There is an important relationship between the primal and its dual:

If the primal has an optimum solution, then so does the dual, and the optimum value of the primal's objective function is the *same* as that of its dual.

Moreover, suppose that the primal's objective function is

$$Z = c_1x_1 + c_2x_2 + \cdots + c_nx_n$$

Then

If s_i is the slack variable associated with the ith constraint in the dual, then the indicator in the s_i-column of the final simplex tableau of the dual is the value of x_i in the optimum solution of the primal.

Thus, we can solve the primal by merely solving its dual. At times this is more convenient than solving the primal directly. The link between the primal and the dual can be expressed very succinctly using matrix notation. Let

$$\mathbf{C} = [c_1 \quad c_2 \quad \cdots \quad c_n] \quad \text{and} \quad \mathbf{X} = \begin{bmatrix} x_1 \\ x_2 \\ \vdots \\ x_n \end{bmatrix}$$

Then the objective function of the primal problem can be written as

$$Z = \mathbf{CX}$$

Furthermore, if we write

$$\mathbf{A} = \begin{bmatrix} a_{11} & a_{12} & \cdots & a_{1n} \\ a_{21} & a_{22} & \cdots & a_{2n} \\ \vdots & \vdots & & \vdots \\ a_{m1} & a_{m2} & \cdots & a_{mn} \end{bmatrix} \quad \text{and} \quad \mathbf{B} = \begin{bmatrix} x_1 \\ x_2 \\ \vdots \\ x_n \end{bmatrix}$$

then the system of constraints for the primal problem becomes

$$\mathbf{AX} \leq \mathbf{B} \quad \text{and} \quad \mathbf{X} \geq \mathbf{0}$$

where we understand $\leq$ ($\geq$) between matrices of the same size to mean that the inequality holds for each pair of corresponding entries. Now let

$$\mathbf{Y} = \begin{bmatrix} y_1 \\ y_2 \\ \cdot \\ \cdot \\ \cdot \\ y_m \end{bmatrix}$$

The dual problem has objective function given by

$$W = \mathbf{B}^T\mathbf{Y}$$

and its system of constraints is

$$\mathbf{A}^T\mathbf{Y} \geq \mathbf{C}^T \quad \text{and} \quad \mathbf{Y} \geq \mathbf{0}$$

APPLICATIONS IN PRACTICE 1

Finding the Dual of a Maximization Problem

Find the dual of the following problem:
Suppose that the What If Company has \$60,000 for the purchase of materials to make three types of gadgets. The company has allocated a total of 2000 hours of assembly time and 120 hours of packaging time for the gadgets. The following table gives the cost per gadget, the number of hours per gadget, and the profit per gadget for each type:

	Type 1	Type 2	Type 3
Cost/ Gadget	\$300	\$220	\$180
Assembly Hours/ Gadget	20	40	20
Packaging Hours/ Gadget	3	1	2
Profit	\$300	\$200	\$200

EXAMPLE 1 **Finding the Dual of a Maximization Problem**

Find the dual of the following:

$$\textit{Maximize } Z = 3x_1 + 4x_2 + 2x_3$$

subject to

$$x_1 + 2x_2 + 0x_3 \leq 10$$
$$2x_1 + 2x_2 + x_3 \leq 10$$

and $x_1, x_2, x_3 \geq 0$

Solution: The primal is of the form of Table 7.3. Thus, the dual is

$$\text{minimize } W = 10y_1 + 10y_2$$

subject to

$$y_1 + 2y_2 \geq 3$$
$$2y_1 + 2y_2 \geq 4$$
$$0y_1 + y_2 \geq 2$$

and $y_1, y_2 \geq 0$

▸ **Now work Problem 1.**

APPLICATIONS IN PRACTICE 2

Finding the Dual of a Minimization Problem

Find the dual of the following problem:
A person decides to take two different dietary supplements. Each supplement contains two essential ingredients, A and B, for which there are minimum daily requirements, and each contains a third ingredient, C, that needs to be minimized.

	Supplement 1	Supplement 2	Daily Requirement
A	20 mg/oz	6 mg/oz	98 mg
B	8 mg/oz	16 mg/oz	80 mg
C	6 mg/oz	2 mg/oz	

APPLICATIONS IN PRACTICE 3

Applying the Simplex Method to the Dual

A company produces three kinds of devices requiring three different production procedures. The company has allocated a total of 300 hours for procedure 1, 400 hours for procedure 2, and 600 hours for procedure 3. The following table gives the number of hours per device for each procedure:

	Device 1	Device 2	Device 3
Procedure 1	30	15	10
Procedure 2	20	30	20
Procedure 3	40	30	25

If the profit is $30 per device 1, $20 per device 2, and $20 per device 3, then, using the dual and the simplex method, find the number of devices of each kind the company should produce to maximize profit.

EXAMPLE 2 **Finding the Dual of a Minimization Problem**

Find the dual of the following:

$$\textit{Minimize } Z = 4x_1 + 3x_2$$

subject to

$$3x_1 - x_2 \geq 2 \qquad \textbf{(11)}$$

$$x_1 + x_2 \leq 1 \qquad \textbf{(12)}$$

$$-4x_1 + x_2 \leq 3 \qquad \textbf{(13)}$$

and $x_1, x_2 \geq 0$

Solution: Since the primal is a minimization problem, we want constraints (12) and (13) to involve $\geq$. (See Table 7.4.) Multiplying both sides of (12) and (13) by -1, we get $-x_1 - x_2 \geq -1$ and $4x_1 - x_2 \geq -3$. Thus, constraints (11)–(13) become

$$3x_1 - x_2 \geq 2$$

$$-x_1 - x_2 \geq -1$$

$$4x_1 - x_2 \geq -3$$

The dual is

$$\text{maximize } W = 2y_1 - y_2 - 3y_3$$

subject to

$$3y_1 - y_2 + 4y_3 \leq 4$$

$$-y_1 - y_2 - y_3 \leq 3$$

and $y_1, y_2, y_3 \geq 0$

▸ **Now work Problem 3.**

EXAMPLE 3 **Applying the Simplex Method to the Dual**

Use the dual and the simplex method to

$$\textit{maximize } Z = 4x_1 - x_2 - x_3$$

subject to

$$3x_1 + x_2 - x_3 \leq 4$$

$$x_1 + x_2 + x_3 \leq 2$$

and $x_1, x_2, x_3 \geq 0$

Solution: The dual is

$$\text{minimize } W = 4y_1 + 2y_2$$

subject to

$$3y_1 + y_2 \geq 4 \qquad \textbf{(14)}$$

$$y_1 + y_2 \geq -1 \qquad \textbf{(15)}$$

$$-y_1 + y_2 \geq -1 \qquad \textbf{(16)}$$

and $y_1, y_2 \geq 0$. To use the simplex method, we must get nonnegative constants in (15) and (16). Multiplying both sides of these equations by -1 gives

$$-y_1 - y_2 \leq 1 \tag{17}$$

$$y_1 - y_2 \leq 1 \tag{18}$$

Since (14) involves $\geq$, an artificial variable is required. The corresponding equations of (14), (17), and (18) are, respectively,

$$3y_1 + y_2 - s_1 + t_1 = 4$$

$$-y_1 - y_2 + s_2 \qquad = 1$$

and

$$y_1 - y_2 + s_3 \qquad = 1$$

where t_1 is an artificial variable and s_1, s_2, and s_3 are slack variables. To minimize W, we maximize $-W$. The artificial objective function is $U = (-W) - Mt_1$, where M is a large positive number. After computations, we find that the final simplex tableau is

$$\begin{array}{c} \\ y_2 \\ s_2 \\ y_1 \\ -W \end{array}
\begin{array}{c}
\begin{array}{cccccccc} y_1 & y_2 & s_1 & s_2 & s_3 & -W & & R \end{array} \\
\left[\begin{array}{cccccc|c}
0 & 1 & -\frac{1}{4} & 0 & -\frac{3}{4} & 0 & \frac{1}{4} \\
0 & 0 & -\frac{1}{2} & 1 & -\frac{1}{2} & 0 & \frac{5}{2} \\
1 & 0 & -\frac{1}{4} & 0 & \frac{1}{4} & 0 & \frac{5}{4} \\
\hline
0 & 0 & \frac{3}{2} & 0 & \frac{1}{2} & 1 & -\frac{11}{2}
\end{array}\right]
\end{array}$$

(B = basic variables; the bottom-row entries under y_1 through s_3 are labeled "indicators".)

indicators

The maximum value of $-W$ is $-\frac{11}{2}$, so the *minimum* value of W is $\frac{11}{2}$. Hence, the maximum value of Z is also $\frac{11}{2}$. Note that the indicators in the s_1-, s_2-, and s_3-columns are $\frac{3}{2}$, 0, and $\frac{1}{2}$, respectively. Thus, the maximum value of Z occurs when $x_1 = \frac{3}{2}$, $x_2 = 0$, and $x_3 = \frac{1}{2}$.

▶ **Now work Problem 11.**

In Example 1 of Section 7.7 we used the simplex method to

$$\textit{minimize } Z = x_1 + 2x_2$$

subject to

$$-2x_1 + x_2 \geq 1$$

$$-x_1 + x_2 \geq 2$$

and $x_1, x_2 \geq 0$. The initial simplex tableau had 24 entries and involved two artificial variables. The tableau of the dual has only 18 entries and *no artificial variables* and is easier to handle, as Example 4 will show. Thus, there may be a distinct advantage in solving the dual to determine the solution of the primal.

This discussion shows the advantage of solving the dual problem.

EXAMPLE 4 **Using the Dual and the Simplex Method**

Use the dual and the simplex method to

$$\text{minimize } Z = x_1 + 2x_2$$

subject to

$$-2x_1 + x_2 \geq 1$$
$$-x_1 + x_2 \geq 2$$

and $x_1, x_2 \geq 0$.

Solution: The dual is

$$\text{maximize } W = y_1 + 2y_2$$

subject to

$$-2y_1 - y_2 \leq 1$$
$$y_1 + y_2 \leq 2$$

and $y_1, y_2 \geq 0$. The initial simplex tableau is tableau I:

SIMPLEX TABLEAU I

entering variable ↓ (y_2 column)

B	y_1	y_2	s_1	s_2	W	R	Quotients
s_1	−2	−1	1	0	0	1	
s_2 (departing ← variable)	1	1	0	1	0	2	$2 \div 1 = 2$
W	−1	−2	0	0	1	0	

indicators (y_1 through s_2 in the W row)

Continuing, we get tableau II.

SIMPLEX TABLEAU II

B	y_1	y_2	s_1	s_2	W	R
s_1	−1	0	1	1	0	3
y_2	1	1	0	1	0	2
W	1	0	0	2	1	4

indicators (y_1 through s_2 in the W row)

Since all indicators are nonnegative in tableau II, the maximum value of W is 4. Hence, the minimum value of Z is also 4. The indicators 0 and 2 in the s_1- and s_2-columns of tableau II mean that the minimum value of Z occurs when $x_1 = 0$ and $x_2 = 2$.

▸ **Now work Problem 9.**

Exercise 7.8

In Problems 1–8, find the duals. Do not solve.

*1. Maximize

$$Z = x_1 + 2x_2$$

subject to

$$x_1 + x_2 \le 5$$
$$-x_1 + x_2 \le 3$$
$$x_1, x_2 \ge 0$$

2. Maximize

$$Z = 2x_1 + x_2 - x_3$$

subject to

$$x_1 + x_2 \le 1$$
$$-x_1 + 2x_2 + x_3 \le 2$$
$$x_1, x_2, x_3 \ge 0$$

*3. Minimize

$$Z = x_1 + 8x_2 + 5x_3$$

subject to

$$x_1 + x_2 + x_3 \ge 8$$
$$-x_1 + 2x_2 + x_3 \ge 2$$
$$x_1, x_2, x_3 \ge 0$$

4. Minimize

$$Z = 8x_1 + 12x_2$$

subject to

$$2x_1 + 2x_2 \ge 1$$
$$x_1 + 3x_2 \ge 2$$
$$x_1, x_2 \ge 0$$

5. Maximize

$$Z = x_1 - x_2$$

subject to

$$-x_1 + 2x_2 \le 13$$
$$-x_1 + x_2 \ge 3$$
$$x_1 + x_2 \ge 11$$
$$x_1, x_2 \ge 0$$

6. Maximize

$$Z = x_1 - x_2 + 4x_3$$

subject to

$$x_1 + x_2 + x_3 \le 9$$
$$x_1 - 2x_2 + x_3 \ge 6$$
$$x_1, x_2, x_3 \ge 0$$

7. Minimize

$$Z = 4x_1 + 4x_2 + 6x_3$$

subject to

$$x_1 - x_2 - x_3 \le 3$$
$$x_1 - x_2 + x_3 \ge 3$$
$$x_1, x_2, x_3 \ge 0$$

8. Minimize

$$Z = 5x_1 + 4x_2$$

subject to

$$-4x_1 + 3x_1 \ge -10$$
$$8x_1 - 10x_2 \le 80$$
$$x_1, x_2 \ge 0$$

In Problems 9–14, solve by using the dual and the simplex method.

*9. Minimize

$$Z = 4x_1 + 4x_2 + 6x_3$$

subject to

$$x_1 - x_2 + x_3 \ge 1$$
$$-x_1 + x_2 + x_3 \ge 2$$
$$x_1, x_2, x_3 \ge 0$$

10. Minimize

$$Z = 2x_1 + 2x_2$$

subject to

$$x_1 + 4x_2 \ge 28$$
$$2x_1 - x_2 \ge 2$$
$$-3x_1 + 8x_2 \ge 16$$
$$x_1, x_2 \ge 0$$

*11. Maximize

$$Z = 3x_1 + 8x_2$$

subject to

$$x_1 + 2x_2 \le 8$$
$$x_1 + 6x_2 \le 12$$
$$x_1, x_2 \ge 0$$

12. Maximize

$$Z = 2x_1 + 6x_2$$

subject to

$$3x_1 + x_2 \le 12$$
$$x_1 + x_2 \le 8$$
$$x_1, x_2 \ge 0$$

13. Minimize

$$Z = 6x_1 + 4x_2$$

subject to

$$-x_1 + x_2 \leq 1$$
$$x_1 + x_2 \geq 3$$
$$x_1, x_2 \geq 0$$

14. Minimize

$$Z = x_1 + x_2 + 2x_3$$

subject to

$$x_1 - x_2 - x_3 \leq 1$$
$$-x_1 - x_2 + x_3 \geq 3$$
$$x_1, x_2, x_3 \geq 0$$

15. Advertising A firm is comparing the costs of advertising in two media: newspaper and radio. For every dollar's worth of advertising, the following table gives the number of people by income group, reached by these media:

	Under $40,000	Over $40,000
Newspaper	40	100
Radio	50	25

The firm wants to reach at least 80,000 persons earning under $40,000 and at least 60,000 earning over $40,000. Use the dual and the simplex method to find the amounts that the firm should spend on newspaper and radio advertising so as to reach these numbers of people at a minimum total advertising cost. What is the minimum total advertising cost?

16. Use the dual and the simplex method to find the minimum total cost per mile in Problem 12 of Exercise 7.7.

17. Labor Costs A company pays skilled and semiskilled workers in its assembly department $14 and $8 per hour, respectively. In the shipping department, shipping clerks are paid $9 per hour and shipping clerk apprentices are paid $6 per hour. The company requires at least 90 workers in the assembly department and at least 60 in the shipping department. Because of union agreements, at least twice as many semiskilled workers must be employed as skilled workers. Also, at least twice as many shipping clerks must be employed as shipping clerk apprentices. Use the dual and the simplex method to find the number of each type of worker that the company must employ so that the total hourly wage paid to these employees is a minimum. What is the minimum total hourly wage?

7.9 Review

Important Terms and Symbols

Section 7.1 linear inequality   half plane (open, closed)   system of inequalities

Section 7.2 constraints   linear function in x and y linear programming problem   objective function   feasible solution   nonnegativity conditions   feasible region   isoprofit line   corner point   bounded feasible region   unbounded feasible region   nonempty feasible region   empty feasible region   isocost line   unbounded solution

Section 7.3 multiple optimum solutions

Section 7.4 standard linear programming problem   slack variable   decision variable   basic feasible solution   nonbasic variable   basic variable   simplex tableau   objective row   entering variable   indicator   departing variable   pivot entry   simplex method   degeneracy

Section 7.6 artificial problem   artificial variable   artificial objective function

Section 7.8 shadow price   dual   primal

Summary

The solution of a system of linear inequalities consists of all points whose coordinates simultaneously satisfy all of the inequalities. Geometrically, for two variables, it is the region that is common to all of the regions determined by the inequalities.

Linear programming involves maximizing or minimizing a linear function (the objective function), subject to a system of constraints, which are linear inequalities or linear equations. One method for finding an optimum solution for a nonempty feasible region is the corner-point method. The objective function is evaluated at each of the corner points of the feasible region, and we choose a corner point at which the objective function is optimum.

For a problem involving more than two variables, the corner-point method is either impractical or impossible. Instead, we use a matrix method called the simplex method, which is efficient and completely mechanical.

Review Problems

Problem numbers shown in color indicate problems suggested for use as a practice chapter test.

In Problems 1–10, solve the given inequality or system of inequalities.

1. $-3x + 2y > -6$

2. $2x - 3y + 8 \geq 0$

3. $2y \leq -3$

4. $-x < 2$

5. $\begin{cases} y - 3x < 6 \\ x - y > -3 \end{cases}$

6. $\begin{cases} x - 2y > 4 \\ x + y > 1 \end{cases}$

7. $\begin{cases} x - y < 4 \\ y - x < 4 \end{cases}$

8. $\begin{cases} x > y \\ x + y < 0 \end{cases}$

9. $\begin{cases} 4x + 2y > -6 \\ 3x - 2y > -7 \\ x \geq 0 \end{cases}$

10. $\begin{cases} x - y > 4 \\ x < 2 \\ y < -4 \end{cases}$

In Problems 11–18, do not use the simplex method.

11. Maximize

$$Z = x - 2y$$

subject to

$$\begin{aligned} y - x &\leq 2 \\ x + y &\leq 4 \\ x &\leq 3 \\ x, y &\geq 0 \end{aligned}$$

12. Maximize

$$Z = 4x + 2y$$

subject to

$$\begin{aligned} x + 2y &\leq 10 \\ x &\leq 4 \\ y &\geq 1 \\ x, y &\geq 0 \end{aligned}$$

13. Minimize

$$Z = 2x - y$$

subject to

$$\begin{aligned} x - y &\geq -2 \\ x + y &\geq 1 \\ x - 2y &\leq 2 \\ x, y &\geq 0 \end{aligned}$$

14. Minimize

$$Z = x + y$$

subject to

$$\begin{aligned} x + 2y &\leq 12 \\ 4x + 3y &\leq 15 \\ x - 6y &\leq 0 \\ x, y &\geq 0 \end{aligned}$$

15. Minimize

$$Z = 4x - 3y$$

subject to

$$\begin{aligned} x + y &\leq 3 \\ 2x + 3y &\leq 12 \\ 5x + 8y &\geq 40 \\ x, y &\geq 0 \end{aligned}$$

[7]**16.** Minimize

$$Z = 2x + 2y$$

subject to

$$\begin{aligned} x + y &\geq 4 \\ -x + 3y &\leq 18 \\ x &\leq 6 \\ x, y &\geq 0 \end{aligned}$$

[7]**17.** Maximize

$$Z = 9x + 6y$$

subject to

$$\begin{aligned} x + 2y &\leq 8 \\ 3x + 2y &\leq 12 \\ x, y &\geq 0 \end{aligned}$$

18. Maximize

$$Z = 4x + y$$

subject to

$$\begin{aligned} x + 2y &\geq 16 \\ 3x + 2y &\geq 24 \\ x, y &\geq 0 \end{aligned}$$

[7]Problems 16 and 17 refer to Section 7.3.

In Problems 19–28, use the simplex method.

19. Maximize

$$Z = 4x_1 + 5x_2$$

subject to

$$x_1 + 6x_2 \leq 12$$
$$x_1 + 2x_2 \leq 8$$
$$x_1, x_2 \geq 0$$

20. Maximize

$$Z = 18x_1 + 20x_2$$

subject to

$$2x_1 + 3x_2 \leq 18$$
$$4x_1 + 3x_2 \leq 24$$
$$x_2 \leq 5$$
$$x_1, x_2 \geq 0$$

21. Minimize

$$Z = 3x_1 + 2x_2 + x_3$$

subject to

$$x_1 + 2x_2 + 3x_3 \geq 5$$
$$x_1, x_2, x_3 \geq 0$$

22. Minimize

$$Z = x_1 + x_2$$

subject to

$$3x_1 + 4x_2 \geq 24$$
$$x_2 \geq 3$$
$$x_1, x_2 \geq 0$$

23. Maximize

$$Z = x_1 + 2x_2$$

subject to

$$x_1 + x_2 \leq 12$$
$$x_1 + x_2 \geq 5$$
$$x_1 \leq 10$$
$$x_1, x_2 \geq 0$$

24. Minimize

$$Z = 2x_1 + x_2$$

subject to

$$x_1 + 2x_2 \leq 6$$
$$x_1 + x_2 \geq 1$$
$$x_1, x_2 \geq 0$$

25. Minimize

$$Z = x_1 + 2x_2 + x_3$$

subject to

$$x_1 - x_2 - x_3 \leq -1$$
$$6x_1 + 3x_2 + 2x_3 = 12$$
$$x_1, x_2, x_3 \geq 0$$

26. Maximize

$$Z = 2x_1 + 3x_2 + 5x_3$$

subject to

$$x_1 + x_2 + 3x_3 \geq 5$$
$$2x_1 + x_2 + 4x_3 \leq 5$$
$$x_1, x_2, x_3 \geq 0$$

[8]**27.** Maximize

$$Z = x_1 + 4x_2 + 2x_3$$

subject to

$$4x_1 - x_2 \leq 2$$
$$-10x_1 + x_2 + 3x_3 \leq 1$$
$$x_1, x_2, x_3 \leq 0$$

[8]**28.** Minimize

$$Z = x_1 + x_2$$

subject to

$$x_1 + x_2 + 2x_3 \leq 4$$
$$x_3 \geq 1$$
$$x_1, x_2, x_3 \geq 0$$

In Problems 29 and 30, solve by using the dual and the simplex method.

29. Minimize

$$Z = 2x_1 + 7x_2 + 8x_3$$

subject to

$$x_1 + 2x_2 + 3x_3 \geq 35$$
$$x_1 + x_2 + x_3 \geq 25$$
$$x_1, x_2, x_3 \geq 0$$

30. Maximize

$$Z = x_1 - 2x_2$$

subject to

$$x_1 - x_2 \leq 3$$
$$x_1 + 2x_2 \leq 4$$
$$4x_1 + x_2 \geq 2$$
$$x_1, x_2 \geq 0$$

[8]Problems 27 and 28 refer to Section 7.5.

31. Production Order A company manufactures three products: X, Y, and Z. Each product requires the use of time on machines A and B as given in the following table:

	Machine A	Machine B
Product X	1 hr	1 hr
Product Y	2 hr	1 hr
Product Z	2 hr	2 hr

The numbers of hours per week that A and B are available for production are 40 and 34, respectively. The profit per unit on X, Y, and Z is \$10, \$15, and \$22, respectively. What should be the weekly production order if maximum profit is to be obtained? What is the maximum profit?

32. Repeat Problem 31 if the company must produce at least a total of 24 units per week.

33. Oil Transportation An oil company has storage facilities for heating fuel in cities A, B, C, and D. Cities C and D are each in need of exactly 500,000 gal of fuel. The company determines that A and B can each sacrifice at most 600,000 gal to satisfy the needs of C and D. The following table gives the costs per gallon to transport fuel between the cities:

	To	
From	C	D
A	\$0.01	\$0.02
B	0.02	0.04

How should the company distribute the fuel in order to minimize the total transportation cost? What is the minimum transportation cost?

34. Profit J. Smith, who operates from a pay-phone booth, sells two electronic devices: the "Alpha" and the "Omega." These devices are built for Smith by three friends, named A, B, and C, each of whom must do some of the work on each device. The time that each must spend in the manufacture of each device is given in the following table:

	Friend A	Friend B	Friend C
Alpha	3 hr	2 hr	1 hr
Omega	1 hr	2 hr	3 hr

Smith's friends have other work to do, but they find that each month they can spend up to 50, 40, and 70 hours, respectively, working on Smith's products. Smith makes a profit of \$6 on each Alpha and \$8 on each Omega device. How many of each device should Smith have built each month to maximize profit, and what would be this maximum profit?

35. Diet Formulation A technician in a zoo must formulate a diet from two commercial products, food A and food B, for a certain group of animals. Two hundred grams of food A contains 16 grams of fat, 32 grams of carbohydrate, and 4 grams of protein. Two hundred grams of food B contains 8 grams of fat, 64 grams of carbohydrate, and 10 grams of protein. The minimum daily requirements are 176 grams of fat, 1024 grams of carbohydrate, and 200 grams of protein. If food A costs 8 cents per 100 grams and food B costs 22 cents per 100 grams, how many grams of each food should be used to meet the minimum daily requirements at the least cost? (Assume that a minimum cost exists.)

In Problems 36 and 37, do not use the simplex method. Round your answers to two decimal places.

36. Minimize

$$Z = 4.2x - 2.1y$$

subject to

$$\begin{aligned} y &\le 3.4 + 1.2x \\ y &\le -7.6 + 3.5x \\ y &\le 18.7 - 0.6x \\ x, y &\ge 0 \end{aligned}$$

37. Maximize

$$Z = 12.4x + 8.3y$$

subject to

$$\begin{aligned} 1.4x + 1.7y &\le 15.9 \\ -3.6x + 2.6y &\le -10.7 \\ -1.3x + 4.3y &\ge -5.2 \\ x, y &\ge 0 \end{aligned}$$

Mathematical SNAPSHOT

Drug and Radiation Therapies[9]

Frequently there are alternative forms of treatment available for a patient diagnosed to have a particular disease complex. With each treatment there may be not only positive effects on the patient but also negative effects, such as toxicity or discomfort. A physician must make the best choice of these treatments or combination of treatments. This choice will depend not only on curative effects but also on the effects of toxicity and discomfort.

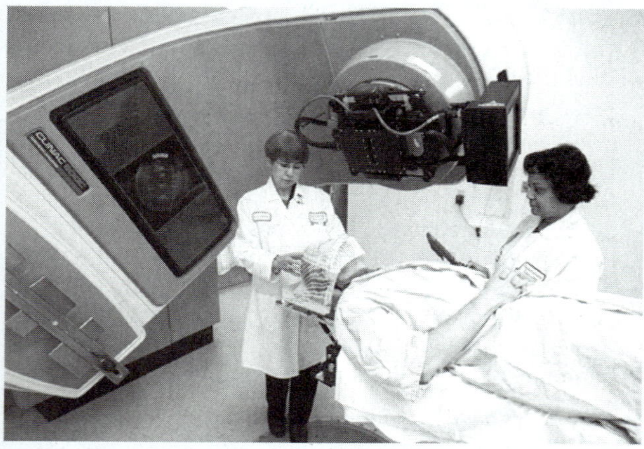

Suppose that you are a physician with a cancer patient under your care and there are two possible treatments available: drug administration and radiation therapy. Let us assume that the efficacies of the treatments are expressed in common units, say, curative units. The drug contains 1000 curative units per ounce, and the radiation gives 1000 curative units per minute. Your analysis indicates that the patient must receive at least 3000 curative units.

However, a degree of toxicity is involved with each treatment. Assume that the toxic effects of each treatment are measured in a common unit of toxicity, say, a toxic unit. The drug contains 400 toxic units per ounce, and the radiation induces 1000 toxic units per minute. Based on your studies, you believe that the patient must receive not more than 2000 toxic units.

In addition, each treatment involves a degree of discomfort to the patient. The drug is three times as noxious per ounce as the radiation per minute.

Table 7.5 summarizes the data. The problem posed to you is to determine the doses of the drug and radiation that will satisfy the curative and toxicity requirements and, at the same time, minimize the discomfort to the patient.

TABLE 7.5

	Curative Units	Toxic Units	Relative Discomfort
Drug (per ounce)	1000	400	3
Radiation (per minute)	1000	1000	1
Requirement	≥ 3000	≤ 2000	

Let x_1 be the number of ounces of the drug and x_2 be the number of minutes of radiation to be administered. Then you want to minimize the discomfort D given by

$$D = 3x_1 + x_2$$

subject to the curative condition

$$1000x_1 + 1000x_2 \geq 3000$$

and the toxic condition

$$400x_1 + 1000x_2 \leq 2000$$

where $x_1 \geq 0$ and $x_2 \geq 0$. You should recognize that this is a linear programming problem. By graphing, the feasible region shown in Figure 7.28 is obtained. The corner points are (3, 0), (5, 0), and $(\frac{5}{3}, \frac{4}{3})$.

Evaluating D at each corner point gives the following:

at (3, 0) $\quad D = 3(3) + 0 = 9$

at (5, 0) $\quad D = 3(5) + 0 = 15$

[9]Adapted from R. S. Ledley and L. B. Lusted, "Medical Diagnosis and Modern Decision Making," *Proceedings of Symposia in Applied Mathematics*, Vol. XIV; *Mathematical Problems in the Biological Sciences* (American Mathematical Society, 1962).

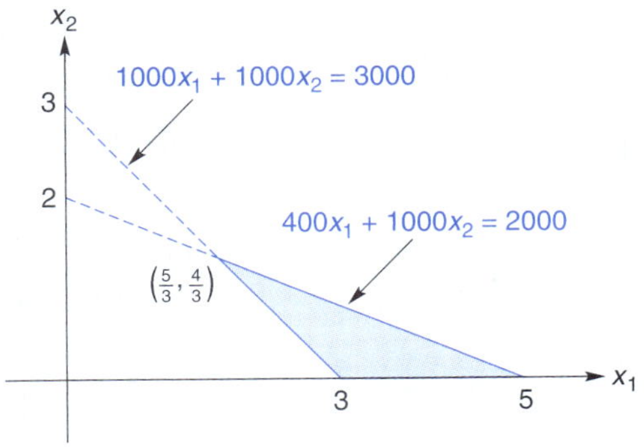

FIGURE 7.28 Feasible region for drug and radiation problem.

and

$$\text{at } \left(\frac{5}{3}, \frac{4}{3}\right) \qquad D = 3\left(\frac{5}{3}\right) + \frac{4}{3} = \frac{19}{3} \approx 6.3.$$

Because D is minimum at $(\frac{5}{3}, \frac{4}{3})$, you should prescribe a treatment of $\frac{5}{3}$ ounces of the drug and $\frac{4}{3}$ minutes of radiation. Thus, by solving a linear programming problem, you have determined the "best" treatment for the patient.

The National Institutes of Health maintain a Web site, www.nih.gov/health, that contains current information in a variety of health-related areas.

You may also want to search the Web for online sites that use applets to demonstrate the simplex method. Simply enter "simplex method" and "applet" in any Internet search engine.

Exercises

1. Suppose that drug and radiation treatments are available to a patient. Each ounce of the drug contains 500 curative units and 400 toxic units. Each minute of radiation gives 1000 curative units and 600 toxic units. The patient requires at least 2000 curative units and can tolerate no more than 1400 toxic units. If each ounce of the drug is as noxious as each minute of radiation, determine the doses of the drug and radiation so that the discomfort to the patient is minimized. Use the geometric method on a graphing calculator if one is available.
2. Suppose that drug A, drug B, and radiation therapy are treatments available to a patient. Each ounce of drug A contains 600 curative units and 500 toxic units. Each ounce of drug B contains 500 curative units and 100 toxic units. Each minute of radiation gives 1000 curative units and 1000 toxic units. The patient requires at least 3000 curative units and can tolerate no more than 2000 toxic units. If each ounce of drug A and each minute of radiation are equally as noxious, and each ounce of drug B is twice as noxious as each ounce of drug A, determine the doses of the drugs and radiation so that the discomfort to the patient is minimized. Use the simplex method.
3. Which method do you think is easier for doing linear programming, the simplex method or a technology-assisted geometric method? Give reasons for your answer.

CHAPTER

8

Introduction to Probability and Statistics

Modern probability theory began with a problem more practical than respectable. If a game between two gamblers is interrupted, the player who is ahead has, it would seem, a right to more than half the pot of money being contested—but not to all of it. How should the pot be divided? This problem was unsolved in 1654, when a French count shared it with Blaise Pascal.

The solution involves probability. If a player has a 30% chance of winning \$150, the current dollar value of the player's situation is $0.30(\$150) = \$45$. Why? Because a player who repeatedly finds himself with a 30% chance of winning \$150 will in the long run win just as much money by staying in the game as by allowing himself to be bought out for \$45 every time. Each player, then, should get a share proportional to his or her probability, just before interruption, of winning if the game had been played to completion.

But how is that probability calculated? Working together, Pascal and Pierre de Fermat established the following result. Suppose a game consists of a series of rounds and the overall winner is the one who first wins a certain number of rounds. If, at interruption, Player 1 is r rounds short of winning while player 2 is s rounds short of winning, then Player 1's probability of winning is

$$\sum_{k=0}^{s-1} \frac{\binom{n}{k}}{2^n}$$

where $n = r + s - 1$ is the maximum number of rounds the game could still have lasted. Player 2's probability of winning is 1 minus Player 1's. The notation $\binom{n}{k}$ may be unfamiliar to you; you will learn about it in this chapter. Then you can verify that if Player 1 is 1 round away from winning and Player 2 is 3 rounds away from winning, Player 1 should get $\frac{7}{8}$ of the pot to Player 2's $\frac{1}{8}$.

The term *probability* is familiar to most of us. It is not uncommon to hear such phrases as "the probability of precipitation," "the probability of flooding," and "the probability of receiving an A in a course." Loosely speaking, probability refers to a number that indicates the degree of likelihood that some future event will have a particular outcome. For example, before tossing a well-balanced coin, you do not know with certainty whether the outcome will be a head or a tail. However, no doubt you consider these outcomes as being equally likely to occur. This means that if the coin were tossed a large number of times, you would expect that approximately half of the tosses would give heads. Thus, we say that the probability of a head occurring on any toss is $\frac{1}{2}$, or 50%.

The field of probability forms the basis of the study of statistics. In statistics, we are concerned about making an inference—that is, a prediction or decision—about a population (a large set of objects under consideration) by using a sample of data drawn from that population. In other words, in statistics, we make an inference about a population based on a known sample. For example, by drawing a sample of units from an assembly line, we can statistically make an inference about *all* the units in a production run. However, in the study of probability, we work with a known population and consider the likelihood (or probability) of drawing a particular sample from it. For example, if we select a card from a deck, we may be interested in the probability that it will be the ace of hearts.

OBJECTIVE *To develop and apply the Basic Counting Principle and to extend it to permutations.*

8.1 Basic Counting Principle and Permutations

Basic Counting Principle

Later on, you will find that computing a probability may require you to calculate the number of elements in a set. Because counting the elements individually may be extremely tedious (or even prohibitive), we shall spend some time developing efficient counting techniques. We begin by motivating the *Basic Counting Principle,* which is useful in solving a wide variety of problems.

Suppose a manufacturer wants to produce coffee brewers in 2-, 8-, and 10-cup capacities, with each capacity available in colors of white, beige, red, and green. How many types of brewers must the manufacturer produce? To answer the question, it is not necessary that we count the capacity-color pairs one by one (such as 2-white and 8-beige). Since there are three capacities, and for each capacity there are four colors, the number of types is the product $3 \cdot 4$, or 12. We can systematically list the different types by using the **tree diagram** of Figure 8.1. From the starting point, there are three branches that indicate the possible capacities. From each of these branches are four more branches that indicate the possible colors. This tree determines 12 paths, each beginning at the starting point and ending at a tip. Each path determines a different type of coffee brewer. We refer to the diagram as being a *two-level* tree: There is a level for capacity and a level for color.

We can consider the listing of the types of coffee brewers as a two-stage procedure. In the first stage we indicate a capacity and in the second a color. The number of types of coffee brewers is the number of ways the first stage can occur (3), times the number of ways the second stage can occur (4), which yields $3 \cdot 4 = 12$. This multiplication procedure can be generalized into the

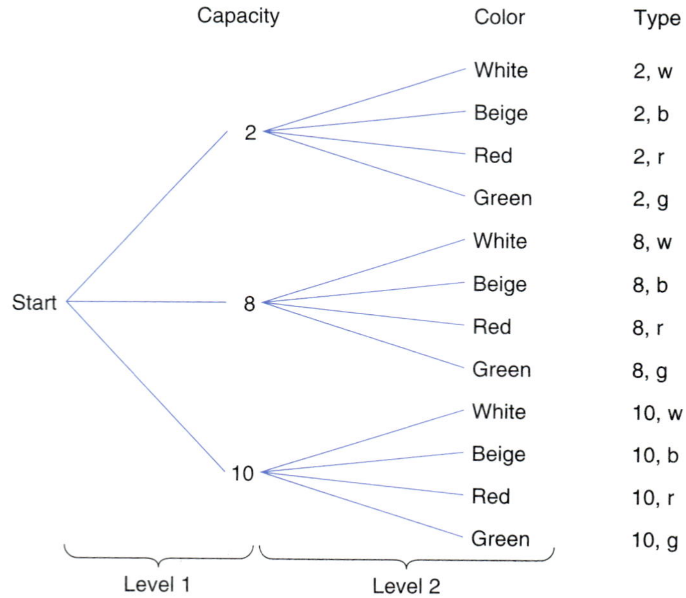

FIGURE 8.1 Two-level tree diagram for types of coffee brewers.

Basic Counting Principle:

> **Basic Counting Principle**
>
> Suppose that a procedure involves a sequence of k stages. Let n_1 be the number of ways the first can occur and n_2 be the number of ways the second can occur after the first stage has occurred. Continuing in this way, let n_k be the number of ways the kth stage can occur after the first $k-1$ stages have occurred. Then the total number of different ways the procedure can occur is
>
> $$n_1 \cdot n_2 \cdot \cdots \cdot n_k$$

EXAMPLE 1 **Travel Routes**

Two roads connect cities A *and* B, *four connect* B *and* C, *and five connect* C *and* D. *(See Figure 8.2.) To drive from* A, *to* B, *to* C, *and then to city* D, *how many different routes are possible?*

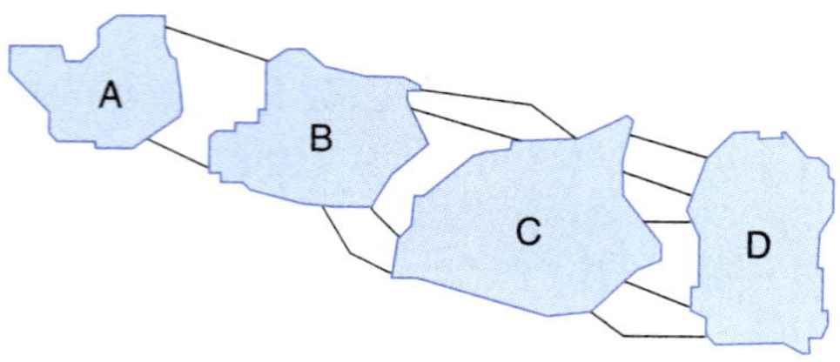

FIGURE 8.2 Roads connecting cities A, B, C, D.

Solution: Here we have a three-stage procedure. The first $(A \to B)$ has two possibilities, the second $(B \to C)$ has four, and the third $(C \to D)$ has five. By the Basic Counting Principle, the total number of routes is $2 \cdot 4 \cdot 5 = 40$.

▶ **Now work Problem 1.**

EXAMPLE 2 Coin Tosses and Roll of a Die

When a coin is tossed, a head (H) *or a tail* (T) *may show. If a die is rolled, a* 1, 2, 3, 4, 5, *or* 6 *may show. Suppose a coin is tossed twice and then a die is rolled, and the result is noted (such as* H *on first toss,* T *on second, and* 4 *on roll of die). How many different results can occur?*

Solution: Tossing a coin twice and then rolling a die can be considered a three-stage procedure. Each of the first two stages (the coin toss) has two possible outcomes. The third stage (rolling the die) has six possible outcomes. By the Basic Counting Principle, the number of different results for the procedure is

$$2 \cdot 2 \cdot 6 = 24$$

▶ **Now work Problem 3.**

EXAMPLE 3 Answering a Quiz

In how many different ways can a quiz be answered under each of the following conditions?

a. *The quiz consists of three multiple-choice questions with four choices for each.*

Solution: Successively answering the three questions is a three-stage procedure. The first question can be answered in any of four ways. Likewise, each of the other two questions can be answered in four ways. By the Basic Counting Principle, the number of ways to answer the quiz is

$$4 \cdot 4 \cdot 4 = 4^3 = 64$$

b. *The quiz consists of three multiple-choice questions (with four choices for each) and five true-false questions.*

Solution: Answering the quiz can be considered a two-stage procedure. First we can answer the multiple-choice questions (the first stage), and then we can answer the true-false questions (the second stage). From part (a), the three multiple-choice questions can be answered in $4 \cdot 4 \cdot 4$ ways. Each of the true-false questions has two choices ("true" or "false"), so the total number of ways of answering all five of them is $2 \cdot 2 \cdot 2 \cdot 2 \cdot 2$. By the Basic Counting Principle, the number of ways the entire quiz can be answered is

$$\underbrace{(4 \cdot 4 \cdot 4)}_{\text{multiple-choice}} \; \underbrace{(2 \cdot 2 \cdot 2 \cdot 2 \cdot 2)}_{\text{true-false}} = 4^3 \cdot 2^5 = 2048$$

▶ **Now work Problem 5.**

EXAMPLE 4 Letter Arrangements

From the five letters A, B, C, D, *and* E, *how many three-letter horizontal arrangements (called "words") are possible if no letter may be repeated? (A "word" need not make sense.) For example,* BDE *and* DEB *are two acceptable words, but* CAC *is not.*

Solution: To form a word, we must successively fill the positions __ __ __ with different letters. Thus, we have a three-stage procedure. For the first position, we can choose any of the five letters. After filling that position with some letter, we can fill the second position with any of the remaining four letters. After that position is filled, the third position can be filled with any of the three letters that have not yet been used. By the Basic Counting Principle, the total number of three-letter words is

$$5 \cdot 4 \cdot 3 = 60$$

If repetitions are allowed, the number of words is $5 \cdot 5 \cdot 5 = 125$.

▶ **Now work Problem 7.**

Permutations

In Example 4, we selected three different letters from five letters and arranged them in an *order*. Each result is called a *permutation of five letters taken three at a time*. More generally, we have the following definition.

Definition

An ordered arrangement of r objects, without repetition, selected from n distinct objects is called a *permutation of n objects taken r at a time*. The number of such permutations is denoted ${}_nP_r$.

Thus, in Example 4, we found that

$${}_5P_3 = 5 \cdot 4 \cdot 3 = 60$$

By a similar analysis, we shall now find a general formula for ${}_nP_r$. In making an ordered arrangement of r objects from n objects, for the first position we may choose any one of the n objects. (See Figure 8.3.) After the first position is filled, there remain $n-1$ objects that may be chosen for the second position. After that position is filled, there are $n-2$ objects that may be chosen for the third position. Continuing in this way and using the Basic Counting Principle, we arrive at the following formula:

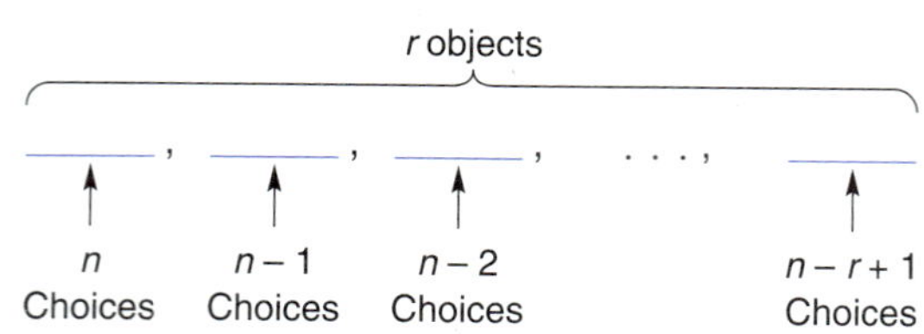

FIGURE 8.3 An ordered arrangement of r objects selected from n objects.

The number of permutations of n objects taken r at a time is given by

$${}_nP_r = \underbrace{n(n-1)(n-2)\cdots(n-r+1)}_{r \text{ factors}} \qquad \textbf{(1)}$$

The formula for ${}_nP_r$ can be expressed in terms of factorials.[1] Multiplying the right side of Equation (1) by

$$\frac{(n-r)(n-r-1)\cdots(2)(1)}{(n-r)(n-r-1)\cdots(2)(1)}$$

[1] Factorials are discussed in Section 2.2.

gives

$$_nP_r = \frac{n(n-1)(n-2)\cdots(n-r+1)\cdot(n-r)(n-r-1)\cdots(2)(1)}{(n-r)(n-r-1)\cdots(2)(1)}$$

The numerator is simply $n!$, and the denominator is $(n-r)!$. Thus, we have the following result:

The number of permutations of n objects taken r at a time is given by

$$_nP_r = \frac{n!}{(n-r)!} \tag{2}$$

For example, from Equation (2), we have

$$_7P_3 = \frac{7!}{(7-3)!} = \frac{7!}{4!} = \frac{7\cdot6\cdot5\cdot4\cdot3\cdot2\cdot1}{4\cdot3\cdot2\cdot1} = 210$$

Many calculators can directly calculate $_nP_r$.

This calculation can be obtained easily with a calculator by using the factorial key. Alternatively, we can conveniently write

$$\frac{7!}{4!} = \frac{7\cdot6\cdot5\cdot4!}{4!} = 7\cdot6\cdot5 = 210$$

Notice how 7! was written so that the 4!'s would cancel.

EXAMPLE 5 **Club Officers**

A club has 20 members. The offices of president, vice president, secretary, and treasurer are to be filled, and no member may serve in more than one office. How many different slates of candidates are possible?

Solution: We shall consider a slate in the order of president, vice president, secretary, and treasurer. Each ordering of four members constitutes a slate, so the number of possible slates is $_{20}P_4$. By Equation (1),

$$_{20}P_4 = 20\cdot19\cdot18\cdot17 = 116{,}280$$

Alternatively, using Equation (2) gives

$$_{20}P_4 = \frac{20!}{(20-4)!} = \frac{20!}{16!} = \frac{20\cdot19\cdot18\cdot17\cdot16!}{16!}$$
$$= 20\cdot19\cdot18\cdot17 = 116{,}280$$

Note the large number of slates that are possible!

▶ **Now work Problem 11.**

EXAMPLE 6 **Political Questionnaire**

A politician sends a questionnaire to her constituents to determine their concerns about six important national issues: unemployment, the environment, taxes, interest rates, national defense, and social security. A respondent is to select four issues of personal concern and rank them by placing the number 1, 2, 3, or 4 after each issue to indicate the degree of concern, with 1 indicating the greatest concern and 4 the least. In how many ways can a respondent reply to the questionnaire?

Solution: A respondent is to rank four of the six issues. Thus, we can consider a reply as an ordered arrangement of six items taken four at a time, where the first item is the issue with rank 1, the second is the issue with rank 2, and so on.

Hence, we have a permutation problem, and the number of possible replies is

$$_6P_4 = \frac{6!}{(6-4)!} = \frac{6!}{2!} = \frac{6\cdot5\cdot4\cdot3\cdot2!}{2!} = 6\cdot5\cdot4\cdot3 = 360$$

▶ **Now work Problem 21.**

In case you want to find the number of permutations of n objects taken all at a time, setting $r = n$ in Equation (2) gives

$$_nP_n = \frac{n!}{(n-n)!} = \frac{n!}{0!} = \frac{n!}{1} = n!$$

Each of these permutations is simply called a **permutation of n objects.**

The number of permutations of n objects is $n!$

For example, the number of permutations of the letters in the word SET is 3!, or 6. These permutations are

SET STE EST ETS TES TSE

EXAMPLE 7 Name of Legal Firm

Lawyers Smith, Jones, Jacobs, and Bell want to form a legal firm and will name it by using all four of their last names. How many possible names are there?

Solution: Since order is important, we must find the number of permutations of four names, which is

$$4! = 4\cdot3\cdot2\cdot1 = 24$$

Thus, there are 24 possible names for the firm.

▶ **Now work Problem 19.**

Exercise 8.1

*1. **Production Process** In a production process, a product goes through one of the assembly lines A, B, or C and then goes through one of the finishing lines D or E. Draw a tree diagram that indicates the possible production routes for a unit of the product. How many production routes are possible?

2. **Air Conditioner Models** A manufacturer produces air conditioners having 6000-, 8000-, and 10,000-BTU capacities. Each capacity is available with one- or two-speed fans. Draw a tree diagram that represents all types of models. How many types are there?

*3. **Dice Rolls** A red die is rolled and then a green die is rolled. Draw a tree diagram to indicate the possible results. How many results are possible?

4. **Coin Toss** A coin is tossed three times. Draw a tree diagram to indicate the possible results. How many results are possible?

In Problems 5–10, use the Basic Counting Principle.

*5. **Course Selection** A student must take a science course and a humanities course. The available science courses are biology, chemistry, physics, computer science, and mathematics. In humanities, the available courses are English, history, speech communications, and classics. How many two-course selections can the student make?

6. **Auto Routes** A person lives in city A and commutes by automobile to city B. There are five roads connecting A and B. (a) How many routes are possible for a round-trip? (b) How many round-trip routes are

possible if a different road is to be used for the return trip?

*7. **Dinner Choices** At a restaurant, a complete dinner consists of an appetizer, an entrée, a dessert, and, a beverage. The choices for the appetizer are soup and salad; for the entrée, the choices are chicken, fish, steak, and lamb; for the dessert, the choices are cherries jubilee, fresh peach cobbler, chocolate truffle cake, and blueberry roly-poly; for the beverage, the choices are coffee, tea, and milk. How many complete dinners are possible?

8. **Multiple-Choice Exam** In how many ways is it possible to answer a six-question multiple-choice examination if each question has four choices (and one choice is selected for each question)?

9. **True-False Exam** In how many ways is it possible to answer a 10-question true-false examination?

10. **Canadian Postal Codes** A Canadian postal code consists of a string of six characters, of which three are letters and three are digits, which begins with a letter, and for which each letter is followed by a (single) digit. (For readability, the string is broken into strings of three. For example, B3H 3J5 is a valid postal code.) How many Canadian postal codes are possible?

In Problems 11–16, determine the values.

*11. ${}_5P_2$ 12. ${}_{95}P_1$ 13. ${}_6P_6$

14. ${}_9P_4$ 15. ${}_4P_2 \cdot {}_5P_3$ 16. $\dfrac{{}_7P_4}{{}_7P_3}$

17. Compute 1000!/999! without using a calculator. Now try it with your calculator, using the factorial feature.

18. Determine $\dfrac{{}_nP_r}{n!}$.

In Problems 19–42, use any appropriate counting method.

*19. **Name of Firm** Flynn, Peters, and Walters are forming an advertising firm and agree to name it by their three last names. How many names for the firm are possible?

20. **Softball** If a softball league has six teams, how many different end-of-the-season rankings are possible? Assume that there are no ties.

*21. **Contest** In how many ways can a judge award first, second, and third prizes in a contest having eight contestants?

22. **Matching-Type Exam** On a history exam, each of six items in one column is to be matched with exactly one of eight items in another column. No item in the second column can be selected more than once. In how many ways can the matching be done?

23. **Die Roll** A die is rolled three times and the outcome of each roll is noted. How many results are possible?

24. **Coin Toss** A coin is tossed eight times. How many results are possible if the order of the tosses is considered?

25. **Problem Assignment** In a mathematics class with 12 students, the instructor wants homework problems 1, 3, and 5 put on the board by three different students. In how many ways can the instructor assign the problems?

26. **Combination Lock** A combination lock has 26 different letters, and a sequence of three different letters must be selected for the lock to open. How many combinations are possible?

27. **Student Questionnaire** A university issues a questionnaire whereby each student must rank the four items with which he or she is most dissatisfied. The items are

tuition fees	professors
parking fees	cafeteria food
dormitory rooms	class sizes

The ranking is to be indicated by the numbers 1, 2, 3, and 4, where 1 indicates the item involving the greatest dissatisfaction and 4 the least. In how many ways can a student answer the questionnaire?

28. **Die Roll** A die is rolled four times. How many results are possible if the order of the rolls is considered and the second roll is even?

29. **Letter Arrangements** How many six-letter words from the letters in the word MEADOW are possible if no letter is repeated?

30. **Letter Arrangements** Using the letters in the word DISC, how many four-letter words are possible if no letter is repeated?

31. **Book Arrangements** In how many ways can five of seven books be arranged on a bookshelf? In how many ways can all seven books be arranged on the shelf?

32. **Lecture Hall** A lecture hall has five doors.

(a) In how many ways can a student enter the hall by one door and exit by a different door?

(b) Exit by any door?

33. **Poker Hand** A poker hand consists of five cards drawn from a deck of 52 playing cards. The hand is said to be "four of a kind" if 4 of the cards have the same face value. For example, hands with four 10's or four jacks or four 2's are four-of-a-kind hands. How many such hands are possible?

34. Merchandise Choice In a merchandise catalog, a CD rack is available in the colors of black, red, yellow, gray, and blue. When placing an order for one, customers must indicate their first and second color choices. In how many ways can this be done?

35. Diner Order Four students go to a diner and order a hamburger, a cheeseburger, a fish sandwich, and a roast beef sandwich (one sandwich for each). When the waiter returns with the food, she forgets which student ordered which item and simply places a sandwich before each student. In how many ways can the waiter do this?

36. Group Photograph In how many ways can three men and two women line up for a group picture? In how many ways can they line up if a woman is to be at each end?

37. Club Officers A club has 12 members.

(a) In how many ways can the offices of president, vice president, secretary, and treasurer be filled if no member can serve in more than one office?

(b) In how many ways can the four offices be filled if the president and vice president must be different members?

38. Fraternity Names Suppose a fraternity is named by three Greek letters. (There are 24 letters in the Greek alphabet.)

(a) How many names are possible?

(b) How many names are possible if no letter can be used more than once?

39. Basketball In how many ways can a basketball coach assign positions to her five-member team if two of the members are qualified for the center position and all five are qualified for all the other positions?

40. Call Letters Suppose the call letters of a radio station consist of four letters, of which the first must be a K or a W. How many such identifications not ending in O are possible?

41. Baseball A baseball manager determines that, of his nine team members, four are strong hitters and five are weak. If the manager wants the strong hitters to be the first four batters in a batting order, how many batting orders are possible?

42. Signal Flags When at least one of four flags colored red, green, yellow, and blue are arranged vertically on a flagpole, the result indicates a signal (or message). Different arrangements give different signals.

(a) How many different signals are possible if all four flags are used?

(b) How many different signals are possible if at least one flag is used?

OBJECTIVE *To discuss combinations, permutations with repeated objects, and assignments to cells.*

8.2 Combinations and Other Counting Principles

Combinations

We continue our discussion of counting methods by considering the following: In a 20-member club the offices of president, vice president, secretary, and treasurer are to be filled, and no member may serve in more than one office. If these offices, in the order given, are filled by members A, B, C, and D, respectively, then we can represent this slate by

ABCD

A different slate is

BACD

These two slates represent different permutations of 20 members taken four at a time. Now, as a different situation, let us consider four-person *committees* that may be formed from the 20 members. In that case, the two arrangements

ABCD and BACD

represent the *same* committee. Here *the order of listing the members is of no concern*. These two arrangements are considered to give the same *combination* of A, B, C, and D.

The important phrase here is "without regard to order," since order implies a permutation rather than a combination.

Definition

An arrangement of r objects, without regard to order and without repetition, selected from n distinct objects is called a *combination of n objects taken r at a time*. The number of such combinations is denoted ${}_nC_r$.

EXAMPLE 1 **Comparing Combinations and Permutations**

List all combinations and all permutations of the four letters

A, B, C, *and* D

when they are taken three at a time.

Solution: The combinations are

ABC ABD ACD BCD

There are four combinations, so ${}_4C_3 = 4$. The permutations are

ABC	ABD	ACD	BCD
ACB	ADB	ADC	BDC
BAC	BAD	CAD	CBD
BCA	BDA	CDA	CDB
CAB	DAB	DAC	DBC
CBA	DBA	DCA	DCB

There are 24 permutations.

▶ **Now work Problem 1.**

In Example 1, notice that each column consists of all the permutations for the same combination of letters. With this observation, we can determine a formula for ${}_nC_r$—the number of combinations of n objects taken r at a time. Suppose one such combination is

$$x_1x_2 \cdots x_r$$

The number of permutations of these r objects is $r!$. If we listed all other such combinations and then listed all permutations of these combinations, we would obtain a complete list of permutations of the n objects taken r at a time. Thus,

$${}_nC_r \cdot r! = {}_nP_r$$

Solving for ${}_nC_r$ gives

$${}_nC_r = \frac{{}_nP_r}{r!} = \frac{\dfrac{n!}{(n-r)!}}{r!} = \frac{n!}{r!(n-r)!}$$

Many calculators can directly compute ${}_nC_r$.

The number of combinations of n objects taken r at a time is given by

$${}_nC_r = \frac{n!}{r!(n-r)!}$$

EXAMPLE 2 **Committee Selection**

If a club has 20 members, how many different four-member committees are possible?

Solution: Order is not important because, no matter how the members of a committee are arranged, we have the same committee. Thus, we simply have to compute the number of combinations of 20 objects taken four at a time, ${}_{20}C_4$:

$$\begin{aligned} {}_{20}C_4 &= \frac{20!}{4!(20-4)!} = \frac{20!}{4!16!} \\ &= \frac{20 \cdot 19 \cdot 18 \cdot 17 \cdot 16!}{4 \cdot 3 \cdot 2 \cdot 1 \cdot 16!} = 4845 \end{aligned}$$

Observe how 20! was written so that the 16!'s would cancel.

There are 4845 possible committees.

▶ **Now work Problem 9.**

It is important to remember that if a selection of objects is made and *order is important,* then *permutations* should be considered. If *order is not important,* consider *combinations.* A key aid to memory is that ${}_nP_r$ is the number of executive slates with r ranks that can be chosen from n people while ${}_nC_r$ is the number of committees with r members that can be chosen from n people. An executive slate can be thought of as a committee in which every individual has been ranked. There are $r!$ ways to rank the members of a committee with r members. Thus if we think of forming an executive slate as a two-stage procedure, then using the Basic Counting Principle of Section 8.1 we again get

$${}_nP_r = {}_nC_r \cdot r!$$

EXAMPLE 3 Poker Hand

*A **poker hand** consists of five cards dealt from an ordinary deck of 52 cards. How many different poker hands are there?*

Solution: One possible hand is

2 of hearts, 3 of diamonds, 6 of clubs,
4 of spades, king of hearts,

which we can abbreviate as

2H 3D 6C 4S KH

The order in which the cards are dealt does not matter, so this hand is the same as

KH 4S 6C 3D 2H

Thus, the number of possible hands is the number of ways that five objects can be selected from 52, without regard to order. This is a combination problem. We have

$$\begin{aligned} {}_{52}C_5 &= \frac{52!}{5!(52-5)!} = \frac{52!}{5!47!} \\ &= \frac{52 \cdot 51 \cdot 50 \cdot 49 \cdot 48 \cdot 47!}{5 \cdot 4 \cdot 3 \cdot 2 \cdot 1 \cdot 47!} \\ &= \frac{52 \cdot 51 \cdot 50 \cdot 49 \cdot 48}{5 \cdot 4 \cdot 3 \cdot 2} = 2{,}598{,}960 \end{aligned}$$

▶ **Now work Problem 11.**

EXAMPLE 4 **Majority Decision and Sum of Combinations**

A college promotion committee consists of five members. In how many ways can the committee reach a majority decision in favor of a promotion?

Strategy: A favorable majority decision is reached if, and only if,

exactly three members vote favorably,
or exactly four members vote favorably,
or all five members vote favorably.

To determine the total number of ways to reach a favorable majority decision, we *add* the number of ways that each of the preceding votes can occur.

Solution: Suppose exactly three members vote favorably. The order of the members is of no concern, and thus, we can think of these members as forming a combination. Hence, the number of ways three of the five members can vote favorably is ${}_5C_3$. Similarly, the number of ways exactly four can vote favorably is ${}_5C_4$, and the number of ways all five can vote favorably is ${}_5C_5$ (which, of course, is 1). Thus, the number of ways to reach a majority decision in favor of a promotion is

$$\begin{aligned}{}_5C_3 + {}_5C_4 + {}_5C_5 &= \frac{5!}{3!(5-3)!} + \frac{5!}{4!(5-4)!} + \frac{5!}{5!(5-5)!}\\ &= \frac{5!}{3!2!} + \frac{5!}{4!1!} + \frac{5!}{5!0!}\\ &= \frac{5\cdot 4\cdot 3!}{3!\cdot 2\cdot 1} + \frac{5\cdot 4!}{4!\cdot 1} + 1\\ &= 10 + 5 + 1 = 16\end{aligned}$$

▶ **Now work Problem 15.**

Combinations and Sets

The previous example leads rather naturally to some properties of combinations that are quite useful in the study of probability. For example, we will show that

$${}_5C_0 + {}_5C_1 + {}_5C_2 + {}_5C_3 + {}_5C_4 + {}_5C_5 = 2^5$$

and, for any nonnegative integer n,

$${}_nC_0 + {}_nC_1 + \cdots + {}_nC_{n-1} + {}_nC_n = 2^n \qquad \textbf{(1)}$$

We can build on the last equation of Example 4 to verify the first of these equations:

$$\begin{aligned}{}_5C_0 + {}_5C_1 + {}_5C_2 + {}_5C_3 + {}_5C_4 + {}_5C_5 &= {}_5C_0 + {}_5C_1 + {}_5C_2 + 16\\ &= \frac{5!}{0!(5-0)!} + \frac{5!}{1!(5-1)!} + \frac{5!}{2!(5-2)!} + 16\\ &= \frac{5!}{0!5!} + \frac{5!}{1!4!} + \frac{5!}{2!3!} + 16\\ &= 1 + \frac{5\cdot 4!}{4!} + \frac{5\cdot 4\cdot 3!}{2!3!} + 16\\ &= 1 + 5 + 10 + 16\\ &= 32\\ &= 2^5\end{aligned}$$

However, this calculation is not illuminating and would be impractical if we were to adapt it for values of n much larger than 5.

Thus far we have primarily looked at *sets* in the context of sets of numbers. In examples in the study of probability, we often look at things like sets of playing cards, sets of die rolls, sets of ordered pairs of dice rolls, and the like. Typically these sets are finite. If a set S has n elements, we can in principle list its elements. For example, we might write:

$$S = \{s_1, s_2, \ldots, s_n\}$$

A *subset* E of S is a set with the property that *every element of E is also an element of S*. When this is the case we write $E \subseteq S$. Formally:

$E \subseteq S$ if and only if (for all x)(if x is an element of E then x is an element of S)

For any set S, we always have $\emptyset \subseteq S$ and $S \subseteq S$. If a set S has n elements, then any subset of S has r elements, where $0 \leq r \leq n$. The empty set, $\emptyset$, is the only subset of S that has 0 elements. The whole set, S, is the only subset of S that has n elements. What is a general subset of S, containing r elements, where $0 \leq r \leq n$? According to our previous definition of a *combination,* such a subset is exactly a combination of n objects taken r at a time and the number of such combinations is denoted ${}_nC_r$. Thus we may also think of ${}_nC_r$ as *the number of r-element subsets of an n-element set.*

For any set S, we can form the set of *all* subsets of S. It is called the *power set* of S and sometimes denoted 2^S. We claim that if S has n elements then 2^S has 2^n elements. This is quite easy to see. If

$$S = \{s_1, s_2, \ldots, s_n\}$$

then specification of a subset E of S can be thought of as a procedure involving n stages. The first stage is to ask the question "Is s_1 an element of E?", the second stage is to ask the question "Is s_2 an element of E?". We continue to ask such questions until we come to the nth stage—the last stage—"Is s_n an element of E?" Observe that each of these questions can be answered in exactly two ways: namely yes or no. According to the Basic Counting Principle of Section 8.1 the total number of ways that specification of a subset of S can occur is

$$\underbrace{2 \cdot 2 \cdot \cdots \cdot 2}_{n \text{ factors}} = 2^n$$

It follows that there are 2^n subsets of an n-element set. It is convenient to write $\#(S)$ for the number of elements of set S. Thus we have

$$\#(2^S) = 2^{\#(S)} \tag{2}$$

If $\#(S) = n$, then for each E in 2^S, we have $\#(E) = r$, for some r satisfying $0 \leq r \leq n$. For each such r, let us write $\mathcal{S}_r$ for the subset of 2^S consisting of all those elements E with $\#(E) = r$. Thus $\mathcal{S}_r$ is the set of all r-element subsets of the n-element set S. From our observations in the last paragraph it follows that

$$\#(\mathcal{S}_r) = {}_nC_r \tag{3}$$

Now we claim that

$$\#(\mathcal{S}_0) + \#(\mathcal{S}_1) + \cdots + \#(\mathcal{S}_{n-1}) + \#(\mathcal{S}_n) = \#(2^S) \tag{4}$$

since every element E of 2^S is in *exactly* one of the sets $\mathcal{S}_r$. Substituting Equation (3), for each $0 \leq r \leq n$, and Equation 2 in Equation (4), we have Equation (1).

EXAMPLE 5 **A Basic Combinatorial Identity**

Establish the identity

$$_nC_r + {_nC_{r+1}} = {_{n+1}C_{r+1}}$$

Solution 1: We can calculate using $_nC_r = \dfrac{n!}{r!(n-r)!}$:

$$\begin{aligned}
{_nC_r} + {_nC_{r+1}} &= \frac{n!}{r!(n-r)!} + \frac{n!}{(r+1)!(n-r-1)!} \\
&= \frac{n!}{r!(n-r)(n-r-1)!} + \frac{n!}{(r+1)r!(n-r-1)!} \\
&= \frac{(r+1)n! + (n-r)n!}{(r+1)!(n-r)!} \\
&= \frac{((r+1)+(n-r))n!}{(r+1)!(n-r)!} \\
&= \frac{(n+1)n!}{(r+1)!((n+1)-(r+1))!} \\
&= \frac{(n+1)!}{(r+1)!((n+1)-(r+1))!} \\
&= {_{n+1}C_{r+1}}
\end{aligned}$$

Solution 2: We can reason using the idea that $_nC_r$ is the number of r-element subsets of an n-element set.

Let S be an n-element set that does not contain s_* as an element. Then $S \cup \{s_*\}$ is an $(n+1)$-element set. Now the $(r+1)$-element subsets of $S \cup \{s_*\}$ are disjointly of two kinds:

1. those that contain s_* as an element;
2. those that do not contain s_* as an element.

Let us write $\mathcal{S}_*$ for the $(r+1)$-element subsets of $S \cup \{s_*\}$ that contain s_* and $\mathcal{S}$ for the $(r+1)$-element subsets of $S \cup \{s_*\}$ that do not contain s_*. Then

$$_{n+1}C_{r+1} = \#(\mathcal{S}_*) + \#(\mathcal{S})$$

because every $r+1$-element subset of $S \cup \{s_*\}$ is in exactly one of $\mathcal{S}_*$ and $\mathcal{S}$. Now the $(r+1)$-element subsets of $S \cup \{s_*\}$ that contain s_* are in one-to-one correspondence with the r-element subsets of S so we have:

$$\#(\mathcal{S}_*) = {_nC_r}$$

On the other hand, $(r+1)$-element subsets of $S \cup \{s_*\}$ that do not contain s_* are in one-to-one correspondence with the $(r+1)$-element subsets of S so:

$$\#(\mathcal{S}) = {_nC_{r+1}}$$

Assembling the last three displayed equations gives:

$$_{n+1}C_{r+1} = {_nC_r} + {_nC_{r+1}}$$

as required.

The first solution is good computational practice but the second solution is illustrative of ideas and arguments that are often useful in the study of probability. The identity we have just established together with

$$_nC_0 = 1 = {_nC_n}$$

for all n, allows us to generate *Pascal's Triangle:*

$$\begin{array}{ccccccccccc} & & & & & 1 & & & & & \\ & & & & 1 & & 1 & & & & \\ & & & 1 & & 2 & & 1 & & & \\ & & 1 & & 3 & & 3 & & 1 & & \\ & 1 & & 4 & & 6 & & 4 & & 1 & \\ 1 & & 5 & & 10 & & 10 & & 5 & & 1 \\ & & & & & \vdots & & & & & \end{array}$$

You should convince yourself that the $(r + 1)$th entry in the $(n + 1)$th row of Pascal's Triangle is $_nC_r$.

Permutations with Repeated Objects

In Section 8.1, we discussed permutations of objects that were all different. Now we examine the case where some of the objects are alike (or *repeated*). For example, consider determining the number of different permutations of the seven letters in the word

SUCCESS

Here the letters C and S are repeated. If the two C's were interchanged, the resulting permutation would be indistinguishable from SUCCESS. Thus, the number of distinct permutations is not 7!, as it would be with 7 different objects. To determine the number of distinct permutations, we use an approach that involves combinations.

Figure 8.4(a) shows boxes representing the different letters in the word SUCCESS. In these boxes we place the integers from 1 through 7. We place three integers in the S's box (because there are three S's), one in the U box, two in the C's box, and one in the E box. A typical placement is indicated in Figure 8.4(b). That placement can be thought of as indicating a permutation of the seven letters in SUCCESS, namely, the permutation in which (going from left to right) the S's are in the second, third, and sixth positions, the U is in the first position, and so on. Thus, Figure 8.4(b) corresponds to the permutation

USSECSC

To count the number of distinct permutations, it suffices to determine the number of ways the integers from 1 to 7 can be placed in the boxes. Since the order in which they are placed into a box is not important, the S's box can

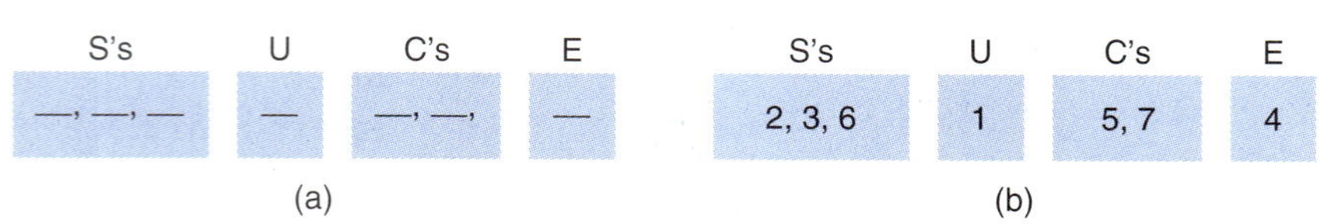

FIGURE 8.4 Permutations with repeated objects.

be filled in ${}_7C_3$ ways. Then the U box can be filled with one of the remaining four integers in ${}_4C_1$ ways. Finally, the C's and E boxes can be successively filled in ${}_3C_2$ and ${}_1C_1$ ways, respectively. Since we have a four-stage procedure, by the Basic Counting Principle the total number of ways to fill the boxes or, equivalently, the number of distinguishable permutations of the letters in SUCCESS is

$$\begin{aligned} {}_7C_3 \cdot {}_4C_1 \cdot {}_3C_2 \cdot {}_1C_1 &= \frac{7!}{3!4!} \cdot \frac{4!}{1!3!} \cdot \frac{3!}{2!1!} \cdot \frac{1!}{1!0!} \\ &= \frac{7!}{3!1!2!1!} \\ &= 420 \end{aligned}$$

In summary, the word SUCCESS has four types of letters: S, U, C, and E. There are three S's, one U, two C's, and one E, and the number of distinguishable permutations of the 7 letters is

$$\frac{7!}{3!1!2!1!}$$

Observing the forms of the numerator and denominator, we can make the following generalization:

Permutations with Repeated Objects

The number of distinguishable permutations of n objects such that n_1 are of one type, n_2 are of a second type, ..., and n_k are of a kth type, where $n_1 + n_2 + \cdots + n_k = n$, is

$$\frac{n!}{n_1!n_2!\cdots n_k!} \tag{5}$$

In problems of this kind there are often a number of quite different solutions to the same problem. A solution that is found straightforward by one person may seem complicated to another. Accordingly, we present another solution to the problem of counting the number, N, of different permutations of the letters of

SUCCESS

We will begin by tagging the letters so that they become distinguishable, thus obtaining

$$S_1U_1C_1C_2E_1S_2S_3$$

Giving a permutation of these seven "different" letters can be described as a multistage procedure. We can begin by permuting as if we can't see the subscripts and by definition there are N ways to accomplish this task. For each of these ways, there are 3! ways to permute the three Ss, 1! ways to permute the one U, 2! ways to permute the two Cs, and 1! ways to permute the one E. According to the Basic Counting Principle of Section 8.1, there are

$$N \cdot 3! \cdot 1! \cdot 2! \cdot 1!$$

ways to permute the seven "different" letters of $S_1U_1C_1C_2E_1S_2S_3$. On the other hand, we already know that there are

$$7!$$

permutations of seven different letters so we must have:

$$N \cdot 3! \cdot 1! \cdot 2! \cdot 1! = 7!$$

From this we find

$$N = \frac{7!}{3!1!2!1!}$$

in agreement with our earlier finding.

EXAMPLE 6 **Letter Arrangements with and without Repetition**

For each of the following words, how many distinguishable permutations of the letters are possible?

a. APOLLO

Solution: The word APOLLO has six letters with repetition. We have one A, one P, two O's, and two L's. Using Equation (5), we find that the number of permutations is

$$\frac{6!}{1!1!2!2!} = 180$$

b. GERM

Solution: None of the four letters in GERM is repeated, so the number of permutations is

$${}_4P_4 = 4! = 24$$

▶ **Now work Problem 17.**

EXAMPLE 7 **Name of Legal Firm**

A group of four lawyers, Smith, Jones, Smith, and Bell (the Smiths are cousins), want to form a legal firm and will name it by using all of their last names. How many possible names exist?

Solution: Each different permutation of the last four names is a name for the firm. There are two Smiths, one Jones, and one Bell. From Equation (5), the number of distinguishable names is

$$\frac{4!}{2!1!1!} = 12$$

▶ **Now work Problem 19.**

Cells

At times, we want to find the number of ways in which objects can be placed into "compartments," or *cells*. For example, suppose that from a group of five people, three are to be assigned to room A and two to room B. In how many ways can this be done? Figure 8.5 shows one such assignment, where the numbers 1, 2, ..., 5 represent the people. Obviously, the order in which people are placed into the rooms is of no concern. The boxes (or cells) remind us of those in Figure 8.4(b), and, by an analysis similar to the discussion of permutations with

FIGURE 8.5 Assignment of people to rooms.

repeated objects, the number of ways to assign the people is

$$\frac{5!}{3!2!} = \frac{5 \cdot 4 \cdot 3!}{3!2!} = 10$$

In general, we have the following principle:

Assignment to Cells

Suppose n distinct objects are assigned to k ordered cells with n_i objects in cell $i\,(i = 1, 2, \ldots, k)$ and the order in which the objects are assigned to cell i is of no concern. The number of all such assignment is

$$\frac{n!}{n_1!n_2!\cdots n_k!} \tag{6}$$

where $n_1 + n_2 + \cdots + n_k = n$.

To say it again, slightly differently, there are ${}_{n_1+n_2+\cdots+n_k}C_{n_1}$ ways to choose n_1 objects to put in the first cell, and for each of these ways there are ${}_{n_2+n_3+\cdots+n_k}C_{n_2}$ ways to choose n_2 objects to put in the second cell, and so on, giving, by the Basic Counting Principle of Section 8.1,

$$\begin{aligned}
&({}_{n_1+n_2+\cdots+n_k}C_{n_1})({}_{n_2+n_3+\cdots+n_k}C_{n_2})\cdots({}_{n_{k-1}+n_k}C_{n_{k-1}})({}_{n_k}C_{n_k}) \\
&= \frac{(n_1+n_2+\cdots+n_k)!}{n_1!(n_2+n_3+\cdots+n_k)!} \cdot \frac{(n_2+n_3+\cdots+n_k)!}{n_2!(n_3+n_4+\cdots+n_k)!} \cdots \frac{(n_{k-1}+n_k)!}{n_{k-1}!n_k!} \cdot \frac{n_k!}{n_k!0!} \\
&= \frac{(n_1+n_2+\cdots+n_k)!}{n_1!n_2!\cdots n_k!}
\end{aligned}$$

which is the number in (6).

EXAMPLE 8 **Assigning Mourners to Limousines**

A funeral director must assign 15 *mourners to three limousines:* 6 *in the first limousine,* 5 *in the second, and* 4 *in the third. In how many ways can this be done?*

Solution: Here 15 people are placed into three cells (limousines): 6 in cell 1, 5 in cell 2, and 4 in cell 3. By Equation (6), the number of ways this can be done is

$$\frac{15!}{6!5!4!} = 630{,}630$$

▶ **Now work Problem 23.**

Example 9 will show three different approaches to a counting problem. As we have said, many counting problems have alternative methods of solution.

EXAMPLE 9 **Art Exhibit**

An artist has created 20 *original paintings, and she will exhibit some of them in three galleries. Four paintings will be sent to gallery* A, 4 *to gallery* B, *and* 3 *to gallery* C. *In how many ways can this be done?*

Solution:

Method 1 The artist must send $4 + 4 + 3 = 11$ paintings to the galleries, and the 9 that are not sent can be thought of as staying in her studio. Thus, we can think of this situation as placing 20 paintings into four cells:

4 in gallery A
4 in gallery B
3 in gallery C
9 in the artist's studio

From Equation (6), the number of ways this can be done is

$$\frac{20!}{4!4!3!9!} = 1{,}939{,}938{,}000$$

Method 2 We can handle the problem in terms of a two-stage procedure and use the Basic Counting Principle. First, 11 paintings are selected for exhibit. Then, these are split into three groups (cells) corresponding to the three galleries. We proceed as follows.

Selecting 11 of the 20 paintings for exhibit (order is of no concern) can be done in ${}_{20}C_{11}$ ways. Once a selection is made, 4 of the paintings go into one cell (gallery A), 4 go to a second cell (gallery B), and 3 go to a third cell (gallery C). By Equation (6), this can be done in $\frac{11!}{4!4!3!}$ ways. Applying the Basic Counting Principle gives the number of ways the artist can send the paintings to the galleries:

$${}_{20}C_{11} \cdot \frac{11!}{4!4!3!} = \frac{20!}{11!9!} \cdot \frac{11!}{4!4!3!} = 1{,}939{,}938{,}000$$

Method 3 Another approach to this problem is in terms of a three-stage procedure. First, 4 of the 20 paintings are selected for shipment to gallery A. This can be done in ${}_{20}C_4$ ways. Then, from the remaining 16 paintings, the number of ways that 4 can be selected for gallery B is ${}_{16}C_4$. Finally, the number of ways that 3 can be sent to gallery C from the 12 paintings that have not yet been selected is ${}_{12}C_3$. By the Basic Counting Principle, the entire procedure can be done in

$${}_{20}C_4 \cdot {}_{16}C_4 \cdot {}_{12}C_3$$

ways, which gives the previous answer, as expected!

▸ **Now work Problem 27.**

Exercise 8.2

In Problems 1–6, determine the values.

*1. ${}_6C_4$ 2. ${}_6C_2$ 3. ${}_{100}C_{100}$

4. ${}_{1001}C_1$ 5. ${}_3P_2 \cdot {}_3C_2$ 6. ${}_4P_2 \cdot {}_5C_3$

7. Verify that ${}_nC_r = {}_nC_{n-r}$. 8. Determine ${}_nC_n$.

*9. **Committee** In how many ways can a four-member committee be formed from a group of 17 people?

10. **Horse Race** In a horse race, a horse is said to *finish in the money* if it finishes in first, second, or third place. For an eight-horse race, in how many ways can the horses finish in the money? Assume no ties.

*11. **Math Exam** On a 13-question mathematics examination, a student must answer any 9 questions. In how many ways can the 9 questions be chosen (without regard to order)?

12. **Cards** From a deck of 52 playing cards, how many 4-card hands are there?

13. **Quality Control** A quality-control technician must select a sample of 10 dresses from a production lot of 74 couture dresses. How many different samples are possible? Express your answer in terms of factorials.

14. **Packaging** A jelly producer makes seven types of jelly. The producer packages gift boxes containing four jars of jelly, no two of which are of the same type. To reflect the three national chains through which the jelly is distributed, the producer uses three types of boxes. How many different gift boxes are possible?

*15. **Scoring on Exam** In a 10-question examination, each question is worth 10 points and is graded right or wrong. Considering the individual questions, in how many ways can a student score 80 or better?

16. **Team Results** A sports team plays eleven games. In how many ways can the outcomes of the games result in four wins, five losses, and two ties?

*17. **Letter Arrangements** How many distinguishable permutations of the letters in the word REMEMBER are possible?

18. **Letter Arrangements** How many distinguishable permutations of the letters in the word ALABAMA are possible?

*19. **Coin Toss** If a coin is tossed seven times and the outcome of each toss is noted, in how many ways can four heads and three tails occur?

20. **Die Roll** A die is rolled six times and the order of the rolls is considered. In how many ways can two 2's, three 3's, and one 4 occur?

21. **Repair Scheduling** An appliance repairman must go out on six service calls. In how many ways can he arrange his schedule?

22. **Baseball** A Little League baseball team has 12 members and must play an away game. Three cars will be used for transportation. In how many ways can the manager assign the members to specific cars if each car can accommodate four members?

*23. **Project Assignment** The director of research and development for a company has nine scientists who are equally qualified to work on projects A, B, and C. In how many ways can the director assign three scientists to each project?

24. **Holly Bushes** A landscaper plants a row hedge of seven young holly plants. The nursery guarantees that exactly four of them are female (and will thus have the characteristic red berries); however, it is not clear to the landscaper which of the young plants are the females. How many different arrangements of male and female plants may become evident when the females have berries?

25. **True-False Exam** A biology instructor includes several true-false questions on quizzes. From past experience, a student believes that half of the questions are true and half are false. If there are ten true-false questions on the next quiz, in how many ways can the student answer half of them "true" and the other half "false"?

26. **Food Order** A waiter takes the following order from a table with seven people: three hamburgers, two cheese-burgers, and two steak sandwiches. Upon returning with the food, he forgets who ordered what item and simply places an item in front of each person. In how many ways can the waiter do this?

*27. **Caseworker Assignment** A social services office has 15 new clients. The supervisor wants to assign 5 clients to each of three specific caseworkers. In how many ways can this be done?

28. **Hockey** There are 11 members on a hockey team, and all but one, the goalie, are qualified for the other five positions. In how many ways can the coach form a starting lineup?

29. **Flag Signals** Colored flags arranged vertically on a flagpole indicate a signal (or message). How many different signals are possible if

(a) two red, two green, and two yellow flags are used?
(b) two red, two green, and three yellow flags are available, and all the red and green flags and at least one yellow flag are used?

30. **Hiring** A company personnel director must hire six people: four for the assembly department and two for the shipping department. There are 10 applicants who are equally qualified to work in each department. In how many ways can the personnel director fill the positions?

31. **Financial Portfolio** A financial advisor wants to create a portfolio consisting of eight stocks and four bonds. If twelve stocks and seven bonds are acceptable for the portfolio, in how many ways can the portfolio be created?

32. **World Series** A baseball team wins the World Series if it is the first team in the series to win four games. Thus, a series could range from four to seven games. For example, a team winning the first four games would be the champion. Likewise, a team losing the first three games and winning the last four would be champion. In how many ways can a team win the World Series?

33. **Subcommittee** A committee has seven members, three of whom are male and four female. In how many ways can a subcommittee be selected if it is to consist exactly of

(a) three males? **(b)** four females?
(c) two males and two females?

34. **Subcommittee** A committee has four male and four female members. In how many ways can a subcommittee of four be selected if at least two females are to serve on it?

35. **Poker Hand** A poker hand consists of 5 cards from a deck of 52 playing cards. The hand is a "full house" if there are 3 cards of one denomination and 2 cards of another. For example, three 10's and two jacks form a full house. How many full-house hands are possible?

36. **Poker Hand** In poker, 2 cards of the same denomination form a "pair." For example, two 8's form a pair. A poker hand (5 cards from a 52-card deck) is

said to be a "two-pair" hand if it contains two pairs and there are three different face values involved in the five cards. For example, a pair of 3's, a pair of 8's, and a 10 constitute a two-pair hand. How many two-pair hands, which contain a pair of 8's, are possible?

37. **Tram Loading** At a tourist attraction, two trams carry sightseers up a picturesque mountain. One tram can accommodate seven people and the other eight. A busload of 18 tourists arrives, and both trams are at the bottom of the mountain. Obviously, only 15 tourists can initially go up the mountain. In how many ways can the attendant load 15 tourists onto the two trams?

38. **Discussion Groups** A history instructor wants to split a class of 10 students into three discussion groups. One group will consist of 4 students and discuss topic A. The second and third groups will discuss topics B and C, respectively, and consist of 3 students each.
 (a) In how many ways can the instructor form the groups?
 (b) If the instructor designates a group leader and a secretary (different students) for each group, in how many ways can the class be split?

OBJECTIVE *To determine a sample space and to consider events associated with it. (These notions involve sets and subsets.) To represent a sample space and events by means of a Venn diagram. To introduce the notions of complement, union, and intersection.*

8.3 Sample Spaces and Events

Sample Spaces

Inherent in any discussion of probability is the performance of an experiment (a procedure) in which a particular result, or *outcome,* involves chance. For example, consider the experiment of tossing a coin. There are only two ways the coin can fall, a head (H) or a tail (T), but the actual outcome is determined by chance. (We assume that the coin does not land on its edge.) The set of possible outcomes,

$$\{H, T\}$$

is called a *sample space* for the experiment, and H and T are called *sample points*.

Definition

A *sample space S* for an experiment is the set of all possible outcomes of the experiment such that each outcome corresponds to exactly one element in S. The elements of S are called *sample points*. If there is a finite number of sample points, that number is denoted $n(S)$, and S is said to be a *finite sample space*.

When determining "possible outcomes" of an experiment, we must be sure that they reflect the situation about which we are concerned. For example, consider the experiment of rolling a die and observing the top face. We could say that a sample space is

The order in which sample points are listed in a sample space is of no concern.

$$S_1 = \{1, 2, 3, 4, 5, 6\}$$

where the possible outcomes are the number of dots on the top face. However, other possible outcomes are

odd number of dots appear (odd)

and even number of dots appear (even)

Thus, the set

$$S_2 = \{\text{odd, even}\}$$

is also a sample space for the experiment, so you can see that an experiment may have more than one sample space.

If an outcome in S_1 occurred, then we know which outcome in S_2 occurred, but the reverse is not true. To describe this asymmetry, we say that S_1 is a **more**

primitive sample space than S_2. Usually, the more primitive a sample space is, the more questions pertinent to the experiment it allows us to answer. For example, with S_1, we can answer such questions as

"Did a three occur?"
"Did a number greater than two occur?"
"Did a number less than four occur?"

But with S_2, we cannot answer these questions. As a rule of thumb, the more primitive a sample space is, the more elements it has and the more detail it indicates. Unless otherwise stated, when an experiment has more than one sample space, it will be our practice to consider only a sample space that gives sufficient detail to answer all pertinent questions relative to the experiment. For example, for the experiment of rolling a die and observing the top face, it will be tacitly understood that we are observing the number of dots. Thus, we shall consider the sample space to be

$$S_1 = \{1, 2, 3, 4, 5, 6\}$$

and shall refer to it as the *usual* sample space for the experiment.

EXAMPLE 1 **Sample Space: Toss of Two Coins**

Two different coins are tossed, and the result (H *or* T) *for each coin is observed. Determine a sample space.*

Solution: One possible outcome is a head on the first coin and a head on the second, which we can indicate by the ordered pair (H, H) or, more simply, HH. Similarly, we indicate a head on the first coin and a tail on the second by HT, and so on. A sample space is

$$S = \{\text{HH, HT, TH, TT}\}$$

A tree diagram is given in Figure 8.6 and, in a sense, indicates the sample space. We remark that S is also a sample space for the experiment of tossing a single coin twice in succession. In fact, these two experiments can be considered one and the same. Although other sample spaces can be contemplated, we take S to be the *usual* sample space for these experiments.

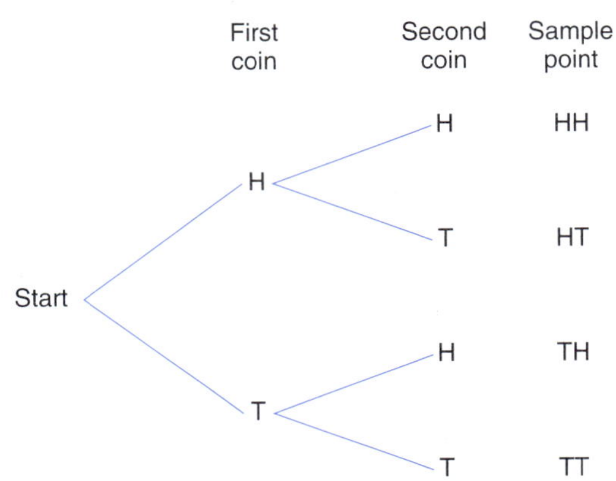

FIGURE 8.6 Tree diagram for toss of two coins.

▶ **Now work Problem 3.**

APPLICATIONS IN PRACTICE 1

Sample Space: Three Tosses of Coin

A video store has 400 different movies to rent. A customer wants to rent 3 movies. If she chooses the videos at random, how many 3-movie choices (sample points) does she have?

EXAMPLE 2 **Sample Space: Three Tosses of Coin**

A coin is tossed three times, and the result of each toss is observed. Describe a sample space and determine the number of sample points.

Solution: Because there are three tosses, we shall choose a sample point to be an ordered *triple*, such as HHT, where each component is either H or T. By the Basic Counting Principle, the total number of sample points is $2 \cdot 2 \cdot 2$, or 8. A sample space (the *usual* one) is

$$S = \{\text{HHH, HHT, HTH, HTT, THH, THT, TTH, TTT}\}$$

and a tree diagram appears in Figure 8.7. Note that it is not necessary to list the entire sample space to determine the number of sample points in it.

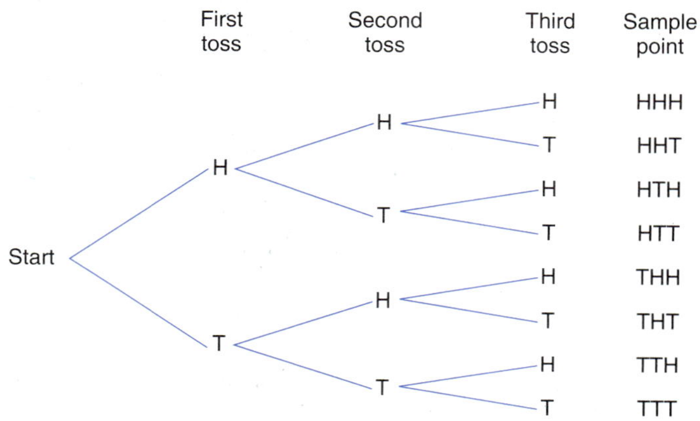

FIGURE 8.7 Tree diagram for three tosses of a coin.

▶ **Now work Problem 9.**

EXAMPLE 3 **Sample Space: Marbles in Urn**

An urn contains four marbles: one red, one pink, one black, and one white. (See Figure 8.8.)

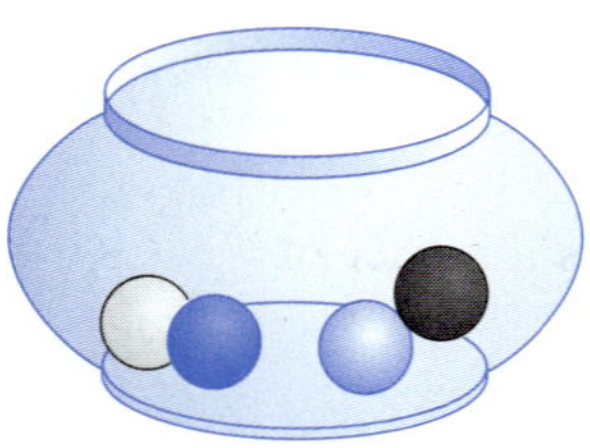

FIGURE 8.8 Four colored marbles in urn.

a. *A marble is drawn at random, its color is noted, and it is placed back into the urn. Then a marble is again randomly drawn and its color noted. Describe a sample space and determine the number of sample points.*

Solution: In this experiment we say that the two marbles are drawn **with replacement.** Let R, P, B, and W denote drawing a red, pink, black, and white marble, respectively. Then our sample space consists of the sample points RW, PB, RB, WW, and so on, where (for example) RW represents the outcome that the first marble drawn is red and the second is white. There are four possibilities for the first draw and, since that marble is placed back, four possibilities for the second draw. By the Basic Counting Principle, the number of sample points is $4 \cdot 4$, or 16.

b. *Determine the number of sample points in the sample space if two marbles are selected in succession* **without replacement** *and the colors are noted.*

Solution: The first marble drawn can have any of four colors. Since it is *not* returned to the urn, the second marble drawn can have any of the *three* remaining colors. Thus, the number of sample points is $4 \cdot 3$, or 12. Alternatively, there are ${}_4P_2 = 12$ sample points.

▶ **Now work Problem 7.**

EXAMPLE 4 **Sample Space: Poker Hand**

From an ordinary deck of 52 playing cards, a poker hand is dealt. Describe a sample space and determine the number of sample points.

Solution: A sample space consists of all combinations of 52 cards taken 5 at a time. From Example 3 of Section 8.2, the number of sample points is ${}_{52}C_5 = 2{,}598{,}960$.

▶ **Now work Problem 13.**

EXAMPLE 5 **Sample Space: Roll of Two Dice**

A pair of dice is rolled once, and for each die, the number that turns up is observed. Describe a sample space.

Solution: Think of the dice as being distinguishable, as if one were red and the other green. Each die can turn up in six ways, so we can take a sample point to be an ordered pair in which each component is an integer between 1 and 6, inclusive. For example, $(4, 6)$, $(3, 2)$, and $(2, 3)$ are three different sample points. By the Basic Counting Principle, the number of sample points is $6 \cdot 6$, or 36.

▶ **Now work Problem 11.**

Events

At times, we are concerned with the outcomes of an experiment that satisfy a particular relationship. For example, we may be interested in whether the outcome of rolling a single die is an even number. This relationship can be considered as the set of outcomes $\{2, 4, 6\}$, which is a subset of the sample space

$$S = \{1, 2, 3, 4, 5, 6\}$$

In general, any subset of a sample space is called an *event* for the experiment. Thus,

$$\{2, 4, 6\}$$

is the event that an even number turns up, which can also be described by

$$\{\text{an even number}\}$$

Note that although an event is a set, it can be described verbally. Usually, an event is denoted by E. When several events are involved in a discussion, they may be denoted by E, F, G, H, and so on, or by E_1, E_2, E_3, and so on.

Definition

Any subset E of a sample space for an experiment is called an *event* for the experiment. If the outcome of the experiment is a sample point in E, then event E is said to *occur*.

In the previous experiment of rolling a die, we saw that {2, 4, 6} is an event. Thus, if the outcome is a 2, that event occurs. Some other events are

$$E = \{1, 3, 5\} = \{\text{an odd number}\}$$

$$F = \{3, 4, 5, 6\} = \{\text{a number} \geq 3\}$$

$$G = \{1\}$$

A sample space is a subset of itself, so it, too, is an event, called a **certain event;** it must occur no matter what the outcome. An event, such as {1}, that consists of a single sample point is called a **simple event.** We can also consider an event such as "7 occurs." This event contains no sample points, so it is the empty set ∅ (the set with no elements in it). In fact, ∅ is called an **impossible event,** because it can never occur.

EXAMPLE 6 Events

A coin is tossed three times, and the result of each toss is noted. The usual sample space (from Example 2) is

$$\{\text{HHH, HHT, HTH, HTT, THH, THT, TTH, TTT}\}$$

Determine the following events.

a. $E = \{\text{one head and two tails}\}$

Solution:

$$E = \{\text{HTT, THT, TTH}\}$$

b. $F = \{\text{at least two heads}\}$

Solution:

$$F = \{\text{HHH, HHT, HTH, THH}\}$$

c. $G = \{\text{all heads}\}$

Solution:

$$G = \{\text{HHH}\}$$

d. $I = \{\text{head on first toss}\}$

Solution:

$$I = \{\text{HHH, HHT, HTH, HTT}\}$$

▶ **Now work Problem 15.**

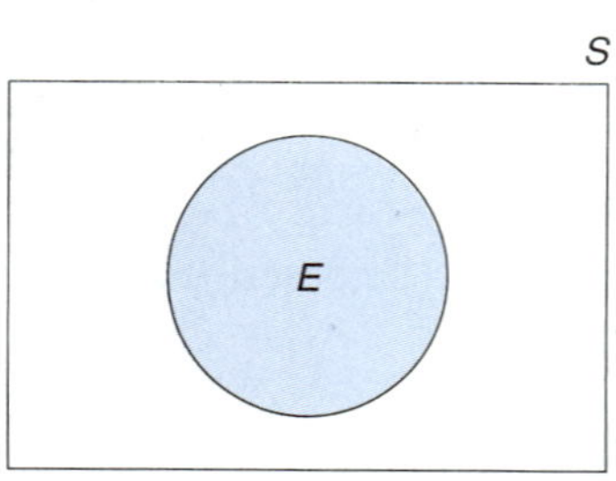

FIGURE 8.9 Venn diagram for sample space S and event E.

Sometimes it is convenient to represent a sample space S and an event E by a *Venn diagram,* as in Figure 8.9. The region inside the rectangle represents the sample points in S. (The sample points are not specifically shown.) The sample points in E are represented by the points inside the circle. Because E is a subset of S, the circular region cannot extend outside the rectangle.

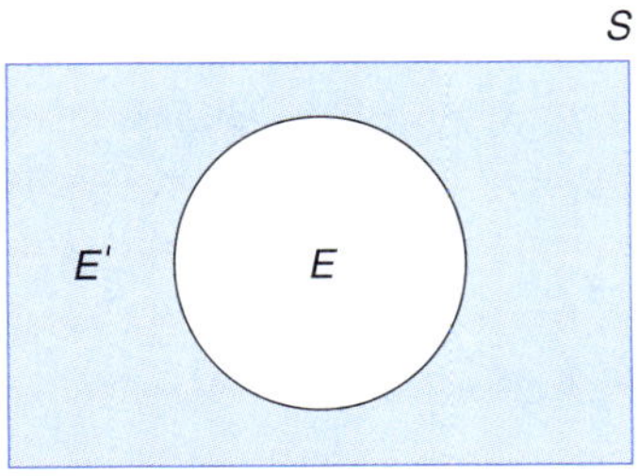

E' is the shaded region

FIGURE 8.10 Venn diagram for the complement of E.

With Venn diagrams, it is easy to see how events for an experiment can be used to form other events. Figure 8.10 shows sample space S and event E. The shaded region inside the rectangle, but outside the circle, represents the set of all sample points in S that are not in E. This set is an event called the *complement of* E and is denoted by E'. Figure 8.11(a) shows two events, E and F. The shaded region represents the set of all sample points either in E, or in F, or in both E and F. This set is an event called the *union* of E and F and is denoted by $E \cup F$. The shaded region in Figure 8.11(b) represents the event consisting of all sample points that are common to both E and F. This event is called the *intersection* of E and F and is denoted by $E \cap F$. In summary, we have the following definitions.

Definition

Let S be a sample space for an experiment with events E and F. The *complement* of E, denoted by E', is the event consisting of all sample points in S that are not in E. The *union* of E and F, denoted by $E \cup F$, is the event consisting of all sample points that are either in E, or in F, or in both E and F. The **intersection** of E and F, denoted by $E \cap F$, is the event consisting of all sample points that are common to both E and F.

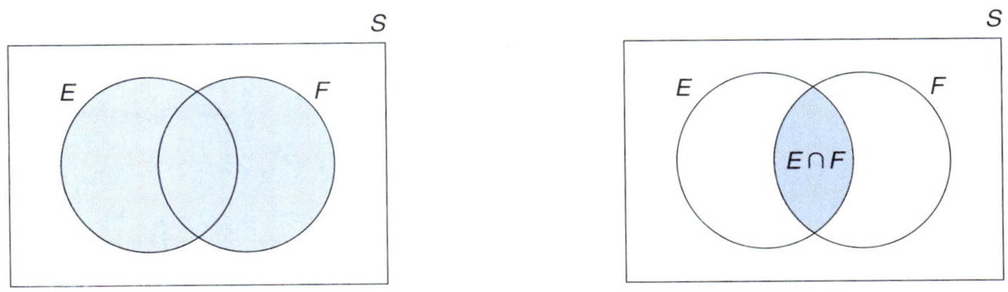

$E \cup F$, union of E and F

(a)

$E \cap F$, intersection of E and F

(b)

FIGURE 8.11 Representation of $E \cup F$ and $E \cap F$.

Note that if a sample point is in the event $E \cup F$, then the point is in at least one of the sets E and F. Thus, for the event $E \cup F$ to occur, *at least one* of the events E and F must occur, and conversely. On the other hand, if event $E \cap F$ occurs, then *both* E and F must occur, and conversely. If event E' occurs, then E *does not* occur, and conversely.

EXAMPLE 7 Complement, Union, Intersection

Given the usual sample space

$$S = \{1, 2, 3, 4, 5, 6\}$$

for the rolling of a die, let E, F, and G be the events

$$E = \{1, 3, 5\} \quad F = \{3, 4, 5, 6\} \quad G = \{1\}$$

Determine each of the following events.

a. E'

Solution: Event E' consists of those sample points in S that are not in E, so

$$E' = \{2, 4, 6\}$$

We note that E' is the event that an even number appears.

b. $E \cup F$

Solution: We want the sample points in E, or F, or both. Thus,

$$E \cup F = \{1, 3, 4, 5, 6\}$$

c. $E \cap F$

Solution: The sample points common to both E and F are 3 and 5, so

$$E \cap F = \{3, 5\}$$

d. $F \cap G$

Solution: Since F and G have no sample point in common,

$$F \cap G = \emptyset$$

e. $E \cup E'$

Solution: Using the result of part (a), we have

$$E \cup E' = \{1, 3, 5\} \cup \{2, 4, 6\} = \{1, 2, 3, 4, 5, 6\} = S$$

f. $E \cap E'$

Solution:

$$E \cap E' = \{1, 3, 5\} \cap \{2, 4, 6\} = \emptyset$$

▶ **Now work Problem 17.**

The results of Examples 7(e) and 7(f) can be generalized as follows:

If E is any event for an experiment with sample space S, then

$$E \cup E' = S \quad \text{and} \quad E \cap E' = \emptyset$$

Thus, the union of an event and its complement is the sample space; the intersection of an event and its complement is the empty set. These and other properties of events are listed in Table 8.1.

When two events E and F have no sample point in common, that is

$$E \cap F = \emptyset$$

they are called *mutually exclusive* or *disjoint* events. For example, in the rolling of a die, the events

$$E = \{2, 4, 6\} \quad \text{and} \quad F = \{1\}$$

are mutually exclusive (see Figure 8.12).

Definition

Events E and F are said to be *mutually exclusive* events if and only if $E \cap F = \emptyset$.

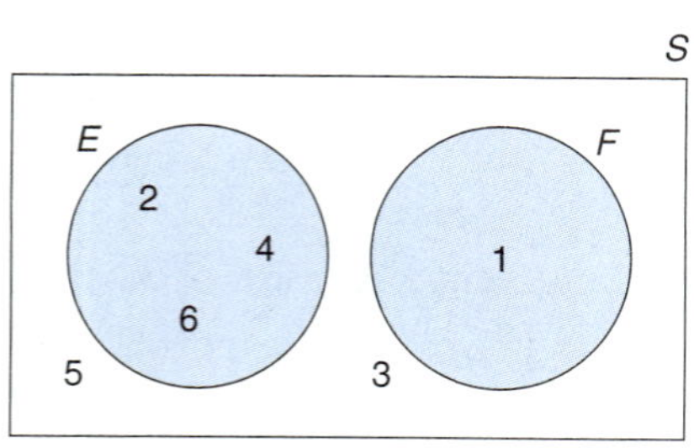

FIGURE 8.12 Mutually exclusive events.

TABLE 8.1 Properties of Events

If E, F, and G are any events for an experiment with sample space S, then

1. $E \cup E = E$	
2. $E \cap E = E$	
3. $(E')' = E$	(the complement of the complement of an event is the event)
4. $E \cup E' = S$	
5. $E \cap E' = \emptyset$	
6. $E \cup S = S$	
7. $E \cap S = E$	
8. $E \cup \emptyset = E$	
9. $E \cap \emptyset = \emptyset$	
10. $E \cup F = F \cup E$	(commutative property of union)
11. $E \cap F = F \cap E$	(commutative property of intersection)
12. $(E \cup F)' = E' \cap F'$	(the complement of a union is the intersection of complements)
13. $(E \cap F)' = E' \cup F'$	(the complement of an intersection is the union of complements)
14. $E \cup (F \cup G) = (E \cup F) \cup G$	(associative property of union)
15. $E \cap (F \cap G) = (E \cap F) \cap G$	(associative property of intersection)
16. $E \cap (F \cup G) = (E \cap F) \cup (E \cap G)$	(distributive property of intersection through union)
17. $E \cup (F \cap G) = (E \cup F) \cap (E \cup G)$	(distributive property of union through intersection)

When two events are mutually exclusive, the occurrence of one event means that the other event cannot occur; that is, the two events cannot occur simultaneously. An event and its complement are mutually exclusive, since $E \cap E' = \emptyset$.

EXAMPLE 8 **Mutually Exclusive Events**

If E, F, and G are events for an experiment and F and G are mutually exclusive, show that events $E \cap F$ and $E \cap G$ are also mutually exclusive.

Solution: Given that $F \cap G = \emptyset$, we must show that the intersection of $E \cap F$ and $E \cap G$ is the empty set. Using the properties in Table 8.1, we have

$$\begin{aligned}(E \cap F) \cap (E \cap G) &= (E \cap F \cap E) \cap G && \text{(property 15)}\\ &= (E \cap E \cap F) \cap G && \text{(property 11)}\\ &= (E \cap F) \cap G && \text{(property 2)}\\ &= E \cap (F \cap G) && \text{(property 15)}\\ &= E \cap \emptyset && \text{(given)}\\ &= \emptyset && \text{(property 9)}\end{aligned}$$

▸ **Now work Problem 31.**

Exercise 8.3

In Problems 1–6, determine a sample space for the given experiment.

1. **Card Selection** A card is drawn from a four-card deck consisting of the 9 of diamonds, 9 of hearts, 9 of clubs, and 9 of spades.

2. **Coin Toss** A coin is tossed four times in succession, and the faces showing are observed.

*3. **Die Roll and Coin Toss** A die is rolled and then a coin is tossed.

4. **Dice Roll** Two dice are rolled, and the sum of the numbers that turn up is observed.

5. **Digit Selection** Two different digits are selected, in succession, from those in the number "64901."

6. **Genders of Children** The genders of the first, second, and third children of a three-child family are noted. (Let, for example, BGB denote that the first, second, and third children are boy, girl, boy, respectively.)

*7. **Marble Selection** An urn contains three colored marbles: one red, one white, and one blue. Determine a sample space if (a) two marbles are selected with replacement and (b) two marbles are selected without replacement.

8. **Manufacturing Process** A company makes a product that goes through three processes during its manufacture. The first is an assembly line, the second is a finishing line, and the third is an inspection line. There are three assembly lines (A, B, and C); two finishing lines (D and E), and two inspection lines (F and G). For each process, the company chooses a line at random. Determine a sample space.

In Problems 9–14, describe the nature of a sample space for the given experiment, and determine the number of sample points.

*9. **Coin Toss** A coin is tossed six times in succession, and the faces showing are observed.

10. **Dice Roll** Five dice are rolled, and the numbers that turn up are observed.

*11. **Card and Die** A card is drawn from an ordinary deck of 52 cards, and then a die is rolled.

12. **Rabbit Selection** From a hat containing nine rabbits, four rabbits are pulled successively without replacement.

*13. **Card Deal** A 13-card hand is dealt from a deck of 52 cards. Do not simplify your answer.

14. **Letter Selection** A four-letter "word" is formed by successively choosing any four letters from the alphabet with replacement.

Suppose that $S = \{1, 2, 3, 4, 5, 6, 7, 8, 9, 10\}$ is the sample space for an experiment with events

$$E = \{1, 3, 5\} \quad F = \{3, 5, 7, 9\} \quad G = \{2, 4, 6, 8\}$$

In Problems 15–22, determine the indicated events.

*15. $E \cup F$ 16. G' *17. $E' \cap F$

18. $F \cap G'$ 19. F' 20. $(E \cup F)'$

21. $(F \cap G)'$ 22. $(E \cup G) \cap F'$

23. Of the following events, which pairs are mutually exclusive?

$$E_1 = \{1, 2, 3\} \quad E_2 = \{3, 4, 5\}$$
$$E_3 = \{1, 2\} \quad E_4 = \{5, 6, 7\}$$

24. **Card Selection** From a standard deck of 52 playing cards, 2 cards are drawn without replacement. Suppose E_J is the event that both cards are jacks, E_C is the event that both cards are clubs, and E_3 is the event that both cards are 3's. Which pairs of these events are mutually exclusive?

25. **Card Selection** From a standard deck of 52 playing cards, 1 card is selected. Which pairs of the following events are mutually exclusive?

$$E = \{\text{diamond}\}$$
$$F = \{\text{ace}\}$$
$$G = \{\text{red}\}$$
$$H = \{\text{club}\}$$
$$I = \{\text{ace of diamonds}\}$$

26. **Dice** A red die and a green die are thrown, and the numbers on each are noted. Which pairs of the following events are mutually exclusive?

$$E = \{\text{both are even}\}$$
$$F = \{\text{both are odd}\}$$
$$G = \{\text{sum is 2}\}$$
$$H = \{\text{sum is 4}\}$$
$$I = \{\text{sum is greater than 10}\}$$

27. **Coin Toss** A coin is tossed three times in succession, and the results are observed. Determine each of the following:

(a) The usual sample space S
(b) The event E_1 that at least one head occurs
(c) The event E_2 that at least one tail occurs
(d) $E_1 \cup E_2$ **(e)** $E_1 \cap E_2$
(f) $(E_1 \cup E_2)'$ **(g)** $(E_1 \cap E_2)'$

28. **Genders of Children** A husband and wife have two children. The outcome of the first child being a boy and the second a girl can be represented by BG. Determine

each of the following:

(a) Sample space that describes all the orders of the possible genders of the children.
(b) The event that at least one child is a girl.
(c) The event that at least one child is a boy.
(d) Is the event in part (c) the complement of the event in part (b)?

29. Arrivals Persons A, B, and C enter a building at different times. The outcome of A arriving first, B second, and C third can be indicated by ABC. Determine each of the following:

(a) The sample space involved for the arrivals.
(b) The event that A arrives first.
(c) The event that A does not arrive first.

30. Supplier Selection A manufacturer can order electronic components from suppliers U, V, W, and X and mechanical components from suppliers U, V, Y, and Z. The manufacturer selects one supplier for each type of component. The outcome of U being selected for electronic components and V for mechanical components can be represented by UV.

(a) Determine a sample space.
(b) Determine the event E that the suppliers are the same.
(c) Determine E' and give a verbal description of this event.

*__31.__ If E and F are events for an experiment, prove that events $E \cap F$ and $E \cap F'$ are mutually exclusive.

32. If E and F are events for an experiment, show that

$$(E \cap F) \cup (E \cap F') = E$$

Note that from Problem 31, $E \cap F$ and $E \cap F'$ are mutually exclusive events. Thus, the foregoing equation expresses E as a union of mutually exclusive events. (*Hint:* Make use of a distributive property.)

OBJECTIVE *To define what is meant by the probability of an event. To develop formulas that are used in computing probabilities. Emphasis is placed on equiprobable spaces.*

8.4 Probability

Equiprobable Spaces

We now introduce the basic concepts underlying the study of probability. Consider tossing a well-balanced die and observing the number that turns up. The usual sample space for the experiment is

$$S = \{1, 2, 3, 4, 5, 6\}$$

Before the experiment is performed, we cannot predict with certainty which of the six possible outcomes (sample points) will occur. But it does seem reasonable that each outcome has the same chance of occurring; that is, the outcomes are *equally likely*. This does not mean that in six tosses each number must turn up once. Rather, it means that if the experiment were performed a large number of times, each outcome would occur about $\frac{1}{6}$ of the time.

To be more specific, let the experiment be performed n times. Each performance of an experiment is called a **trial.** Suppose that we are interested in the event of obtaining a 1 (that is, the simple event consisting of the sample point 1). If a 1 occurs in k of these n trials, then the proportion of times that 1 occurs is k/n. This ratio is called the **relative frequency** of the event. Because getting a 1 is just one of six possible equally likely outcomes, we expect that in the long run a 1 will occur $\frac{1}{6}$ of the time. That is, as n becomes very large, we expect the relative frequency k/n to approach $\frac{1}{6}$. The number $\frac{1}{6}$ is taken to be the probability of getting a 1 on the toss of a well-balanced die, which is denoted $P(1)$. Thus, $P(1) = \frac{1}{6}$. Similarly, $P(2) = \frac{1}{6}$, $P(3) = \frac{1}{6}$, and so on.

In this experiment, all of the simple events in the sample space were understood to be equally likely to occur. To describe this equal likelihood, we say that S is an *equiprobable space.*

Definition

A sample space S is called an *equiprobable space* if and only if all the simple events are equally likely to occur.

We remark that besides the phrase "equally likely," other words and phrases used in the context of an equiprobable space are "well balanced," "fair," "unbiased," and "at random." For example, we may have a *well-balanced* die (as above), a *fair* coin, or *unbiased* dice, or we may select a marble *at random* from an urn.

We now generalize our discussion of the die experiment to other (finite) equiprobable spaces.

Definition

If S is an equiprobable sample space with N sample points (or outcomes) $s_1, s_2, \ldots, s_N$, then the *probability of the simple event* $\{s_i\}$ is given by

$$P(s_i) = \frac{1}{N}$$

for $i = 1, 2, \ldots, N$.

We remark that $P(s_i)$ can be interpreted as the relative frequency of $\{s_i\}$ occurring in the long run.

We can also assign probabilities to events that are not simple. For example, in the die experiment, consider the event E of a 1 or a 2 turning up:

$$E = \{1, 2\}$$

Because the die is well balanced, in n trials (where n is large) we expect that a 1 should turn up approximately $\frac{1}{6}$ of the time and a 2 should turn up approximately $\frac{1}{6}$ of the time. Thus, a 1 or 2 should turn up approximately $\frac{1}{6} + \frac{1}{6}$ of the time, or $\frac{2}{6}$ of the time. Hence, it is reasonable to assume that the long-run relative frequency of E is $\frac{2}{6}$. For this reason, we define $\frac{2}{6}$ to be the probability of E and denote it $P(E)$.

$$P(E) = \frac{1}{6} + \frac{1}{6} = \frac{2}{6}$$

Note that $P(E)$ is simply the sum of the probabilities of the simple events that form E. Equivalently, $P(E)$ is the ratio of the number of sample points in E (two) to the number of sample points in the sample space (six).

Definition

If S is a finite equiprobable space for an experiment and $E = \{s_1, s_2, \ldots, s_j\}$ is an event, then the *probability of* E is given by

$$P(E) = P(s_1) + P(s_2) + \cdots + P(s_j)$$

Equivalently,

$$P(E) = \frac{\#(E)}{\#(S)}$$

where $\#(E)$ is the number of outcomes in E and $\#(S)$ is the number of outcomes in S.

Note that we can think of P as a function that associates with each event E the probability of E, namely, $P(E)$. The probability of E can be interpreted as the relative frequency of E occurring in the long run. Thus, in k trials, we would expect E to occur approximately $k \cdot P(E)$ times, provided that k is large.

EXAMPLE 1 **Coin Tossing**

Two fair coins are tossed. Determine the probability that

a. *two heads occur*

b. *at least one head occurs*

Solution: The usual sample space is

$$S = \{\text{HH, HT, TH, TT}\}$$

Since the four outcomes are equally likely, S is equiprobable and $n(S) = 4$.

a. If $E = \{\text{HH}\}$, then E is a simple event, so

$$P(E) = \frac{\#(E)}{\#(S)} = \frac{1}{4}$$

b. Let $F = \{\text{at least one head}\}$. Then

$$F = \{\text{HH, HT, TH}\}$$

which has three outcomes. Thus,

$$P(F) = \frac{\#(F)}{\#(S)} = \frac{3}{4}$$

Alternatively,

$$\begin{aligned} P(F) &= P(\text{HH}) + P(\text{HT}) + P(\text{TH}) \\ &= \frac{1}{4} + \frac{1}{4} + \frac{1}{4} = \frac{3}{4} \end{aligned}$$

Consequently, in 1000 trials of this experiment, we would expect F to occur approximately $1000 \cdot \frac{3}{4}$, or 750, times.

▶ **Now work Problem 1.**

EXAMPLE 2 **Cards**

From an ordinary deck of 52 playing cards, 2 cards are randomly drawn without replacement. If E is the event that one card is a 2 and the other a 3, find $P(E)$.

Solution: We can disregard the order in which the 2 cards are drawn. As our sample space S, we choose the set of all combinations of the 52 cards taken 2 at a time. Thus, S is equiprobable and $\#(S) = {}_{52}C_2$. To find $\#(E)$, we note that since there are four suits, a 2 can be drawn in four ways and a 3 in four ways. Hence, a 2 and a 3 can be drawn in $4 \cdot 4$ ways, so

$$P(E) = \frac{\#(E)}{\#(S)} = \frac{4 \cdot 4}{{}_{52}C_2} = \frac{16}{1326} = \frac{8}{663}$$

▶ **Now work Problem 7.**

EXAMPLE 3 **Four-of-a-Kind Poker Hand**

Find the probability of drawing four of a kind in a poker hand (for example, four 10's and one 4).

Solution: The set of all combinations of 52 cards taken 5 at a time is an equiprobable sample space. (The order in which the cards are dealt is of no concern.) Thus, $\#(S) = {}_{52}C_5$. We now must find $\#(E)$, where E is the event of drawing four of a kind. Each of the four suits has 13 denominations, so four cards of one denomination can be drawn in 13 ways. There are $52 - 4 = 48$ possible selections for the fifth card. Hence, four of a kind can be drawn in $13 \cdot 48$ ways, and we have

$$P(\text{four of a kind}) = \frac{\#(E)}{\#(S)} = \frac{13 \cdot 48}{{}_{52}C_5} = \frac{13 \cdot 48}{2{,}598{,}960} \approx 0.00024$$

▶ **Now work Problem 13.**

EXAMPLE 4 **Selecting a Subcommittee**

From a committee of three males and four females, a subcommittee of four is to be randomly selected. Find the probability that it consists of two males and two females.

Solution: Since order of selection is not important, the number of subcommittees of four that can be selected from the seven members is ${}_7C_4$. The two males can be selected in ${}_3C_2$ ways and the two females in ${}_4C_2$ ways. By the Basic Counting Principle, the number of subcommittees of two males and two females is ${}_3C_2 \cdot {}_4C_2$. Thus,

$$P(\text{two males and two females}) = \frac{{}_3C_2 \cdot {}_4C_2}{{}_7C_4} = \frac{\dfrac{3!}{2!1!} \cdot \dfrac{4!}{2!2!}}{\dfrac{7!}{4!3!}} = \frac{18}{35}$$

▶ **Now work Problem 21.**

Properties of Probability

We now develop some properties of probability. Let S be an equiprobable sample space with N outcomes; that is, $\#(S) = N$. (We assume a finite sample space throughout this section.) If E is an event, then $0 \leq \#(E) \leq N$. Dividing each member by $\#(S)$, or N, gives

$$0 \leq \frac{\#(E)}{\#(S)} \leq \frac{N}{N}$$

But $\dfrac{\#(E)}{\#(S)} = P(E)$, so we have the following property:

$$0 \leq P(E) \leq 1$$

That is, the probability of an event is a number between 0 and 1, inclusive.

Moreover, $P(\emptyset) = \dfrac{\#(\emptyset)}{\#(S)} = \dfrac{0}{N} = 0$. Thus,

$$P(\emptyset) = 0$$

Also, $P(S) = \dfrac{\#(S)}{\#(S)} = \dfrac{N}{N} = 1$, so

$$P(S) = 1$$

Accordingly, the probability of an impossible event is 0 and the probability of a certain event is 1.

Since $P(\mathrm{S})$ is the sum of the probabilities of the outcomes in the sample space, we conclude that the sum of the probabilities of all the simple events for a sample space is 1.

Now let us focus on the probability of the union of two events E and F. The event $E \cup F$ occurs if and only if at least one of the events (E or F) occurs. Thus, $P(E \cup F)$ is the probability that *at least one* of the events E and F occurs. We know that

$$P(E \cup F) = \frac{\#(E \cup F)}{\#(S)}$$

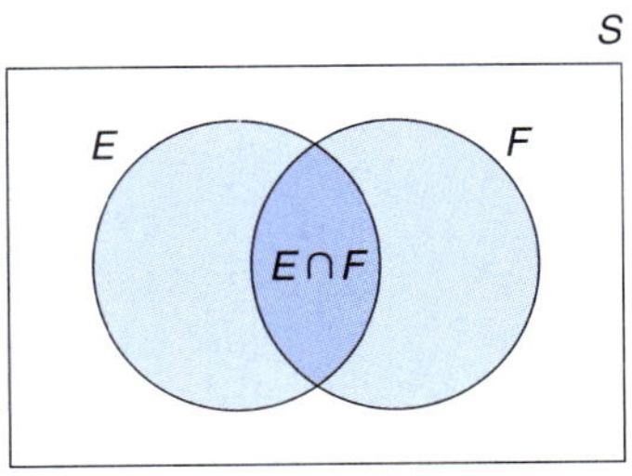

FIGURE 8.13 $E \cap F$ is contained in both E and F.

Although you might think that $\#(E \cup F) = \#(E) + \#(F)$, this is not necessarily true. Since event $E \cap F$ is contained in both E and F (see Figure 8.13), the sum $\#(E) + \#(F)$ includes $\#(E \cap F)$ twice. Thus, we must subtract $\#(E \cap F)$ from the sum to obtain $\#(E \cup F)$:

$$\#(E \cup F) = \#(E) + \#(F) - \#(E \cap F)$$

Dividing both sides by $n(\mathrm{S})$ gives the following result:

The probability of a union is not, in general, the sum of the probabilities. Note that while we derived Equation (1) for an equiprobable sample space, the result is in fact a general one.

Probability of a Union of Events

If E and F are events, then

$$P(E \cup F) = P(E) + P(F) - P(E \cap F) \tag{1}$$

For example, let a fair die be rolled, and let $E = \{1, 3, 5\}$ and $F = \{1, 2, 3\}$. Then $E \cap F = \{1, 3\}$, so

$$\begin{aligned} P(E \cup F) &= P(E) + P(F) - P(E \cap F) \\ &= \frac{3}{6} + \frac{3}{6} - \frac{2}{6} = \frac{2}{3} \end{aligned}$$

Alternatively, $E \cup F = \{1, 2, 3, 5\}$, so $P(E \cup F) = \frac{4}{6} = \frac{2}{3}$.

If E and F are mutually exclusive events, then $E \cap F = \emptyset$ so $P(E \cap F) = P(\emptyset) = 0$. Hence, from Equation (1), we obtain the following law:

Addition Law for Mutually Exclusive Events

If E and F are *mutually exclusive* events, then

$$P(E \cup F) = P(E) + P(F)$$

For example, let a fair die be rolled, and let $E = \{2, 3\}$ and $F = \{1, 5\}$. Then $E \cap F = \emptyset$, so

$$P(E \cup F) = P(E) + P(F) = \frac{2}{6} + \frac{2}{6} = \frac{2}{3}$$

The addition law can be extended to more than two mutually exclusive events.[2] For example, if events E, F, and G are mutually exclusive, then

$$P(E \cup F \cup G) = P(E) + P(F) + P(G)$$

An event and its complement are mutually exclusive, so by the addition law,

$$P(E \cup E') = P(E) + P(E')$$

But $P(E \cup E') = P(S) = 1$. Thus,

$$1 = P(E) + P(E')$$

$$P(E') = 1 - P(E)$$

In order to find the probability of an event, sometimes it may be more convenient to first find the probability of its complement and then subtract the result from 1. See, especially, Example 6(c).

or, equivalently,

$$P(E) = 1 - P(E')$$

Accordingly, if we know the probability of an event, then we can easily find the probability of its complement, and vice versa. For example, if $P(E) = \frac{1}{4}$, then $P(E') = 1 - \frac{1}{4} = \frac{3}{4}$. $P(E')$ is the probability that E does not occur.

EXAMPLE 5 Quality Control

From a production run of 5000 light bulbs, 2% of which are defective, 1 bulb is selected at random. What is the probability that the bulb is defective? What is the probability that it is not defective?

Solution: The sample space S consists of the 5000 bulbs. Since a bulb is selected at random, the possible outcomes are equally likely. Let E be the event of selecting a defective bulb. The number of outcomes in E is $0.02 \cdot 5000 = 100$. Thus,

$$P(E) = \frac{\#(E)}{\#(S)} = \frac{100}{5000} = \frac{1}{50} = 0.02$$

Alternatively, since the probability of selecting a particular bulb is $\frac{1}{5000}$ and E contains 100 sample points, by summing probabilities we have

$$P(E) = 100 \cdot \frac{1}{5000} = 0.02$$

The event that the bulb selected is *not* defective is E'. Hence,

$$P(E') = 1 - P(E) = 1 - 0.02 = 0.98$$

▶ **Now work Problem 17.**

EXAMPLE 6 Dice

A pair of well-balanced dice is rolled, and the number on each die is noted. Determine the probability that the sum of the numbers that turn up is (a) 7, (b) 7 or 11, and (c) greater than 3.

---

[2]Two or more events are **mutually exclusive** if and only if no two of them can occur at the same time. That is, given any two of them, their intersection must be empty. For example, to say the events E, F, and G are mutually exclusive means that

$$E \cap F = E \cap G = F \cap G = \emptyset$$

Solution: Since each die can turn up in any of six ways, by the Basic Counting Principle the number of possible outcomes is $6 \cdot 6$, or 36. Our sample space consists of the following ordered pairs:

$$\begin{array}{cccccc}
(1,1) & (1,2) & (1,3) & (1,4) & (1,5) & (1,6) \\
(2,1) & (2,2) & (2,3) & (2,4) & (2,5) & (2,6) \\
(3,1) & (3,2) & (3,3) & (3,4) & (3,5) & (3,6) \\
(4,1) & (4,2) & (4,3) & (4,4) & (4,5) & (4,6) \\
(5,1) & (5,2) & (5,3) & (5,4) & (5,5) & (5,6) \\
(6,1) & (6,2) & (6,3) & (6,4) & (6,5) & (6,6)
\end{array}$$

The outcomes are equally likely, so the probability of each outcome is $\frac{1}{36}$.

a. Let E_7 be the event that the sum of the numbers appearing is 7. Then

$$E_7 = \{(1,6), (2,5), (3,4), (4,3), (5,2), (6,1)\}$$

which has six outcomes. Thus,

$$P(E_7) = \frac{6}{36} = \frac{1}{6}$$

b. Let $E_{7 \text{ or } 11}$ be the event that the sum is 7 or 11. If E_{11} is the event that the sum is 11, then

$$E_{11} = \{(5,6), (6,5)\}$$

which has two outcomes. Since $E_{7 \text{ or } 11} = E_7 \cup E_{11}$ and E_7 and E_{11} are mutually exclusive, we have

$$P(E_{7 \text{ or } 11}) = P(E_7) + P(E_{11}) = \frac{6}{36} + \frac{2}{36} = \frac{8}{36} = \frac{2}{9}$$

Alternatively, we can determine $P(E_{7 \text{ or } 11})$ by counting the number of outcomes in $E_{7 \text{ or } 11}$. We obtain

$$E_{7 \text{ or } 11} = \{(1,6), (2,5), (3,4), (4,3), (5,2), (6,1), (5,6), (6,5)\}$$

which has eight outcomes. Thus,

$$P(E_{7 \text{ or } 11}) = \frac{8}{36} = \frac{2}{9}$$

c. Let E be the event that the sum is greater than 3. The number of outcomes in E is relatively large. Thus, to determine $P(E)$, it is easier to find E', rather than E, and then use the formula $P(E) = 1 - P(E')$. Here E' is the event that the sum is 2 or 3. We have

$$E' = \{(1,1), (1,2), (2,1)\}$$

which has three outcomes. Hence,

$$P(E) = 1 - P(E') = 1 - \frac{3}{36} = \frac{11}{12}$$

▶ **Now work Problem 27.**

Probability Functions in General

Many of the properties of equiprobable spaces carry over to sample spaces that are not equiprobable. To illustrate, consider the experiment of tossing two fair coins and observing the number of heads. The coins can fall in one of four ways, namely,

$$\text{HH} \quad \text{HT} \quad \text{TH} \quad \text{TT}$$

which correspond to two heads, one head, one head, and zero heads, respectively. Because we are interested in the number of heads, we can choose a sample space to be

$$S = \{0, 1, 2\}$$

However, the simple events in S are *not* equally likely to occur, because of the four possible ways in which the coins can fall: Two of these ways correspond to the one-head outcome, whereas only one corresponds to the two-head outcome and similarly for the zero-head outcome. In the long run, it is reasonable to expect repeated trials to result in one head about $\frac{2}{4}$ of the time, zero heads about $\frac{1}{4}$ of the time, and two heads about $\frac{1}{4}$ of the time. If we were to assign probabilities to these simple events, it is natural to have

$$P(0) = \frac{1}{4} \qquad P(1) = \frac{2}{4} = \frac{1}{2} \qquad P(2) = \frac{1}{4}$$

Although S is not equiprobable, these probabilities lie between 0 and 1, inclusive, and their sum is 1. This is consistent with what was stated for an equiprobable space.

Based on our discussion, we can consider a *probability function* that relates to sample spaces in general.

Definition

Let $S = \{s_1, s_2, \ldots, s_N\}$ be a sample space for an experiment. The function P is called a *probability function* if both of the following are true:

1. $0 \leq P(s_i) \leq 1 \quad \text{for } i = 1 \text{ to } N$
2. $P(s_1) + P(s_2) + \cdots + P(s_N) = 1$

If E is an event, then $P(E)$ is the sum of the probabilities of the sample points in E. We define $P(\emptyset)$ to be 0.

From a mathematical point of view, any function P that satisfies conditions 1 and 2 is a probability function for a sample space. For example, consider the sample space for the previous experiment of tossing two fair coins and observing the number of heads:

$$S = \{0, 1, 2\}.$$

We could assign probabilities as follows:

$$P(0) = 0.1 \qquad P(1) = 0.2 \qquad P(2) = 0.7$$

Here P satisfies both conditions 1 and 2 and thus is a legitimate probability function. However, this assignment does not reflect the long-run interpretation of probability and, consequently, would not be acceptable from a practical point of view.

In general, for any probability function defined on a sample space (finite or infinite), the following properties hold:

$$P(E') = 1 - P(E)$$

$$P(S) = 1$$

$$P(E_1 \cup E_2) = P(E_1) + P(E_2) \quad \text{if } E_1 \cap E_2 = \emptyset$$

Empirical Probability

You have seen how easy it is to assign probabilities to simple events when we have an equiprobable sample space. For example, when a fair coin is tossed, we have $S = \{H, T\}$ and $P(H) = P(T) = \frac{1}{2}$. These probabilities are determined, in essence, by the intrinsic nature of the experiment—namely, that there are two possible outcomes that should have the same probability because the outcomes are equally likely. Such probabilities are called *theoretical* probabilities. However, suppose the coin is not fair. How can probabilities then be assigned? By tossing the coin a number of times, we can determine the relative frequencies of heads and tails occurring. For example, suppose that in 1000 tosses, heads occurs 517 times and tails occurs 483 times. Then the relative frequencies of heads and tails occurring are $\frac{517}{1000}$ and $\frac{483}{1000}$, respectively. In this situation, the assignment $P(H) = 0.517$ and $P(T) = 0.483$ would be quite reasonable. Probabilities assigned in this way are called *empirical probabilities*. In general, probabilities based on sample or historical data are empirical. Now suppose that the coin were tossed 2000 times, and the relative frequencies of heads and tails occurring were $\frac{1023}{2000} = 0.5115$ and $\frac{977}{2000} = 0.4885$, respectively. Then in this case, the assignment $P(H) = 0.5115$ and $P(T) = 0.4885$ would be acceptable. In a certain sense, the latter probabilities may be more indicative of the true nature of the coin than would be the probabilities associated with 1000 tosses.

In the next example, probabilities (empirical) are assigned on the basis of sample data.

EXAMPLE 7 **Opinion Survey**

An opinion survey of a sample of 150 adult residents of a town was conducted. Each person was asked his or her opinion about floating a bond issue to build a community swimming pool. The results are summarized in Table 8.2.

Suppose an adult resident from the town is randomly selected. Let M be the event "male selected" and F be the event "selected person favors the bond issue." Find each of the following:

a. $P(M)$ **b.** $P(F)$

c. $P(M \cap F)$ **d.** $P(M \cup F)$

TABLE 8.2

	Favor	Oppose	Total
Male	60	20	80
Female	40	30	70
Total	100	50	150

Strategy: We shall assume that proportions that apply to the sample also apply to the adult population of the town.

Solution:

a. Of the 150 persons in the sample, 80 are males. Thus, for the adult population of the town (the sample space), we assume that $\frac{80}{150}$ are male. Hence, the (empirical) probability of selecting a male is

$$P(M) = \frac{80}{150} = \frac{8}{15}$$

b. Of the 150 persons in the sample, 100 favor the bond issue. Therefore,

$$P(F) = \frac{100}{150} = \frac{2}{3}$$

c. Table 8.2 indicates that 60 males favor the bond issue. Hence,

$$P(M \cap F) = \frac{60}{150} = \frac{2}{5}$$

d. To find $P(M \cup F)$, we use Equation (1):

$$P(M \cup F) = P(M) + P(F) - P(M \cap F)$$
$$= \frac{80}{150} + \frac{100}{150} - \frac{60}{150} = \frac{120}{150} = \frac{4}{5}$$

▸ **Now work Problem 31.**

Odds

The probability of an event is sometimes expressed in terms of *odds*, especially in gaming situations.

Definition

The *odds* in favor of event E occurring is the ratio

$$\frac{P(E)}{P(E')}$$

provided that $P(E') \neq 0$. Odds are usually expressed as the ratio $\frac{p}{q}$ (or $p : q$) of two positive integers, which is read "p to q."

EXAMPLE 8 Odds for an "A" in an Exam

A student believes that the probability of getting an A on the next mathematics exam is 0.2. What are the odds (in favor) of this occurring?

Solution: If E = "gets an A," then $P(E) = 0.2$ and $P(E') = 1 - 0.2 = 0.8$. Hence, the odds of getting an A are

$$\frac{P(E)}{P(E')} = \frac{0.2}{0.8} = \frac{2}{8} = \frac{1}{4} = 1:4$$

That is, the odds are 1 to 4. (We remark that the odds *against* getting an A are 4 to 1.)

If the odds that event E occurs are $a : b$, then the probability of E can be easily determined. We are given that

$$\frac{P(E)}{1 - P(E)} = \frac{a}{b}$$

Solving for $P(E)$ gives

$$bP(E) = [1 - P(E)]a \qquad \text{(clearing fractions)}$$
$$aP(E) + bP(E) = a$$
$$(a + b)P(E) = a$$
$$P(E) = \frac{a}{a + b}$$

Finding Probability from Odds

If the odds that event E occurs are $a : b$, then

$$P(E) = \frac{a}{a + b}$$

Over the long run, if the odds that E occurs are $a : b$, then, on the average, E should occur a times in every $a + b$ trials of the experiment.

EXAMPLE 9 **Probability of Winning a Prize**

A \$1000 savings bond is one of the prizes listed on a contest brochure received in the mail. The odds in favor of winning the bond are stated to be 1:10,000. What is the probability of winning this prize?

Solution: Here $a = 1$ and $b = 10{,}000$. From the preceding rule,

$$P(\text{winning prize}) = \frac{a}{a+b}$$

$$= \frac{1}{1+10{,}000} = \frac{1}{10{,}001}$$

Exercise 8.4

*1. In 2000 trials of an experiment, how many times would you expect event E to occur if $P(E) = 0.3$?

2. In 3000 trials of an experiment, how many times would you expect event E to occur if $P(E') = 0.45$?

3. If $P(E) = 0.2$, $P(F) = 0.3$, and $P(E \cap F) = 0.1$, find (a) $P(E')$ and (b) $P(E \cup F)$.

4. If $P(E) = \frac{1}{4}$, $P(F) = \frac{1}{2}$, and $P(E \cap F) = \frac{1}{8}$, find (a) $P(E')$ and (b) $P(E \cup F)$.

5. If $P(E \cap F) = 0.831$, are E and F mutually exclusive?

6. If $P(E) = \frac{1}{2}$, $P(E \cup F) = \frac{13}{20}$, and $P(E \cap F) = \frac{1}{10}$, find $P(F)$.

*7. **Dice** A pair of well-balanced dice is tossed. Find the probability that the sum of the numbers is (a) 8, (b) 2 or 3, (c) 3, 4, or 5, (d) 12 or 13, (e) even, (f) odd, and (g) less than 10.

8. **Dice** A pair of fair dice is tossed. Determine the probability that at least one die shows a 2.

9. **Card Selection** A card is randomly selected from a standard deck of 52 playing cards. Determine the probability that the card is (a) the king of hearts, (b) a diamond, (c) a jack, (d) red, (e) a heart or a club, (f) a club and a 4, (g) a club or a 4, (h) red and a king, and (i) a spade and a heart.

10. **Coin and Die** A fair coin and a fair die are tossed. Find the probability that (a) a head and a 5 show, (b) a head shows, (c) a 3 shows, and (d) a head and an even number show.

11. **Coin, Die, and Card** A fair coin and a fair die are tossed, and a card is randomly selected from a standard deck of 52 playing cards. Determine the probability that the coin, die, and card respectively show (a) a tail, a 3, and the queen of hearts, (b) a tail, a 3, and a queen, (c) a head, a 2 or 3, and a queen, and (d) a head, an even number, and a diamond.

12. **Coins** Three fair coins are tossed. Find the probability that (a) three heads show, (b) exactly one tail shows, (c) no more than two heads show, and (d) no more than one tail shows.

*13. **Card Selection** Three cards from a standard deck of 52 playing cards are successively drawn at random without replacement. Find the probability that (a) all 3 cards are aces and (b) 2 cards are diamonds and the other is a heart.

14. **Card Selection** Two cards from a standard deck of 52 playing cards are successively drawn at random with replacement. Find the probability that (a) both cards are kings and (b) 1 card is a king and the other is a heart.

15. **Genders of Children** Assuming that the gender of a person is determined at random, determine the probability that a family with three children has (a) three girls, (b) exactly one boy, (c) no girls, and (d) at least one girl.

16. **Marble Selection** A marble is randomly drawn from an urn that contains seven red, three white, and eight blue marbles. Find the probability that the marble is (a) blue, (b) not red, (c) red or white, (d) neither red nor blue, (e) yellow, and (f) red or yellow.

*17. **Stock Selection** A stock is selected at random from a list of 60 utility stocks, 48 of which have an annual dividend yield of 6% or more. Find the probability that the stock pays an annual dividend that yields (a) 6% or more and (b) less than 6%.

18. **Inventory** A clothing store maintains its inventory of ties so that 40% are 100% pure silk. If a tie is selected at random, what is the probability that it is (a) 100% pure silk and (b) not 100% pure silk?

19. **Examination Grades** On an examination given to 40 students, 10% received an A, 25% a B, 35% a C, 25% a D, and 5% an F. If a student is selected at random, what is the probability that the student (a) received an A, (b) received an A or a B, (c) received neither a D nor an F, and (d) did not receive an F? (e) Answer questions (a)–(d) if the number of students that were given the examination is unknown.

20. Marble Selection Two urns contain colored marbles. Urn 1 contains three red and two green marbles, and urn 2 contains four red and five green marbles. A marble is selected at random from each urn. Find the probability that (a) both marbles are red and (b) one marble is red and the other is green.

*21. **Committee Selection** From a group of two women and three men, two persons are selected at random to form a committee. Find the probability that the committee consists of women only.

22. Committee Selection For the committee selection in Problem 21, find the probability that the committee consists of a man and a woman.

23. Examination Score A student answers each question on a 10-question true-false examination in a random fashion. If each question is worth 10 points, what is the probability that the student scores (a) 100 points and (b) 90 or more points?

24. Multiple-Choice Examination On an eight-question, multiple-choice examination there are four choices for each question, only one of which is correct. If a student answers each question in a random fashion, find the probability that the student answers (a) each question correctly and (b) exactly four questions correctly.

25. Poker Hand Find the probability of being dealt a full house in a poker game. A full house is three of one kind and two of another, such as three queens and two 10's. Express your answer using the symbol ${}_nC_r$.

26. Suppose $P(E) = \frac{1}{4}$, $P(E \cup F) = \frac{5}{14}$, and $P(E \cap F) = \frac{1}{7}$.

(a) Find $P(F)$
(b) Find $P(E' \cup F)$

(*Hint:*

$$F = (E \cap F) \cup (E' \cap F)$$

where $E \cap F$ and $E' \cap F$ are mutually exclusive.)

*27. **Faculty Committee** The classification of faculty at a college is indicated in Table 8.3. If a committee of three faculty members is selected at random, what is the probability that it consists of (a) all females; (b) a professor and two associate professors?

TABLE 8.3 Faculty Classification

	Male	Female	Total
Professor	12	3	15
Associate Professor	15	9	24
Assistant Professor	18	8	26
Instructor	20	15	35
Total	65	35	100

28. Biased Die A die is biased such that $P(1) = \frac{3}{10}$, $P(2) = P(5) = \frac{2}{10}$, and $P(3) = P(4) = P(6) = \frac{1}{10}$. If the die is tossed, find $P(\text{even number})$.

29. Biased Die When a biased die is tossed, the probabilities of 1, 3, and 5 showing are the same. The probabilities of 2, 4, and 6 showing are also the same, but are twice those of 1, 3, and 5. Determine $P(1)$.

30. For the sample space $\{a, b, c, d, e, f\}$, suppose that the probabilities of a, b, c, and d are the same and that the probabilities of e and f are the same. Is it possible to determine $P(f)$? If it is also known that $P(\{a, e\}) = \frac{1}{3}$ what more can be said?

*31. **Tax Increase** A legislative body is considering a tax increase to support education. An opinion survey of 100 registered voters was conducted, and the results are indicated in Table 8.4. Assume that the survey reflects the opinions of the voting population. If a person from that population is selected at random, determine each of the following (empirical) probabilities.

(a) P (favors tax increase)
(b) P (opposes tax increase)
(c) P (is a Republican with no opinion)

TABLE 8.4 Tax Increase Survey

	Favor	Oppose	No Opinion	Total
Democrat	32	26	2	60
Republican	15	17	3	35
Other	4	1	0	5
Total	51	44	5	100

32. Camcorder Sales A department store chain has stores in the cities of Exton and Whyton. Each store sells three brands of camcorders, A, B, and C. Over the past year, the average monthly unit sales of the camcorders was determined, and the results are indicated in Table 8.5. Assume that future sales follow the pattern indicated in the table.

(a) Determine the probability that a sale of a camcorder next month is for brand B.
(b) Next month, if a sale occurs at the Exton store, find the probability that it is for brand C.

TABLE 8.5 Unit Sales Per Month

	A	B	C
Exton	25	40	30
Whyton	20	25	30

In Problems 33–36, for the given probability, find the odds that E will occur.

33. $P(E) = \frac{4}{5}$

34. $P(E) = \frac{1}{6}$

35. $P(E) = 0.3$

36. $P(E) = 0.001$

In Problems 37–40, the odds that E will occur are given. Find P(E).

37. $7:5$

38. $100:1$

39. $4:10$

40. $a:a$

41. Weather Forecast A television weather forecaster reported a 75% chance of rain tomorrow. What are the odds that it will rain tomorrow?

42. If the odds of event E *not* occurring are $a:b$, what are the odds that E does occur?

OBJECTIVE *To discuss conditional probability via a reduced sample space as well as the original space. To analyze a stochastic process with the aid of a probability tree. To develop the general multiplication law for $P(E \cap F)$.*

8.5 Conditional Probability and Stochastic Processes

Conditional Probability

The probability of an event could be affected when additional related information about the experiment is known. For example, if you guess at the answer to a multiple-choice question having five choices, the probability of getting the correct answer is $\frac{1}{5}$. However, if you know that answers A and B are wrong and thus can be ignored, the probability of guessing the correct answer increases to $\frac{1}{3}$. In this section, we consider similar situations in which we want the probability of an event E when it is known that some other event F has occurred. This is called a **conditional probability** and is denoted by $P(E|F)$, which is read "the conditional probability of E, given F." For instance, in the situation involving the multiple-choice question, we have

$$P(\text{guessing correct answer}|\text{A and B eliminated}) = \frac{1}{3}$$

To investigate the notion of conditional probability, we consider the following situation. A fair die is rolled, and we are interested in the probability of the event

$$E = \{\text{even number shows}\}$$

The usual equiprobable sample space for this experiment is

$$S = \{1, 2, 3, 4, 5, 6\}$$

so

$$E = \{2, 4, 6\}$$

Thus,

$$P(E) = \frac{\#(E)}{\#(S)} = \frac{3}{6} = \frac{1}{2}$$

Now we change the situation a bit. Suppose the die is rolled out of our sight, and then we are told that a number greater than 3 occurred. In light of this additional information, what now is the probability of an even number? To answer that question we reason as follows. The event F of a number greater than 3 is

$$F = \{4, 5, 6\}$$

Since F already occurred, the set of possible outcomes is no longer S; it is F. That is, F becomes our new sample space, called a **reduced sample space** or a *subspace* of S. The outcomes in F are equally likely, and, of these, only 4 and 6 are favorable to E; that is,

$$E \cap F = \{4, 6\}$$

Since two of the three outcomes in the reduced sample space are favorable to an even number occurring, we say that $\frac{2}{3}$ is *the conditional probability of an even number, given that a number greater than 3 occurred:*

$$P(E|F) = \frac{\#(E \cap F)}{\#(F)} = \frac{2}{3} \tag{1}$$

The Venn diagram in Figure 8.14 illustrates the situation.

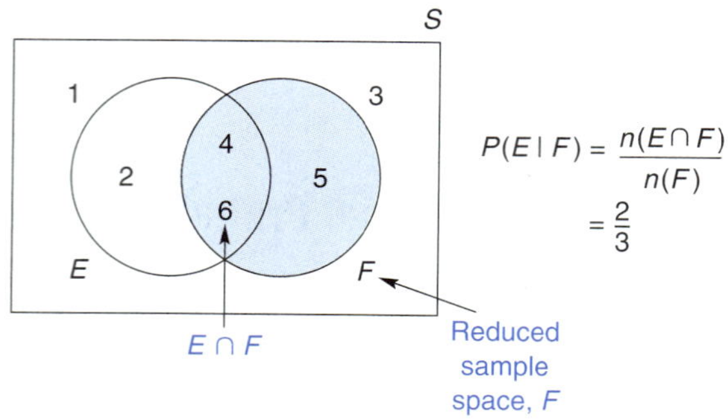

FIGURE 8.14 Venn diagram for conditional probability.

If we compare the conditional probability $P(E|F) = \frac{2}{3}$ with the "unconditional" probability $P(E) = \frac{1}{2}$, we see that $P(E|F) > P(E)$. This means that knowing that a number greater than 3 occurred *increases* the likelihood that an even number occurred. There are situations, however, in which conditional and unconditional probabilities are the same. These are discussed in the next section.

In summary, we have the following generalization of Equation (1):

Formula for a Conditional Probability

If E and F are events associated with an equiprobable sample space and $F \neq \emptyset$, then

$$P(E|F) = \frac{\#(E \cap F)}{\#(F)} \tag{2}$$

It can be shown that

$$P(E'|F) = 1 - P(E|F)$$

EXAMPLE 1 **Marbles in Urn**

An urn contains two blue marbles (say, B_1 and B_2) and two white marbles (W_1 and W_2). If two marbles are randomly drawn without replacement, find the probability that the second marble drawn is white, given that the first marble drawn is blue. (See Figure 8.15.)

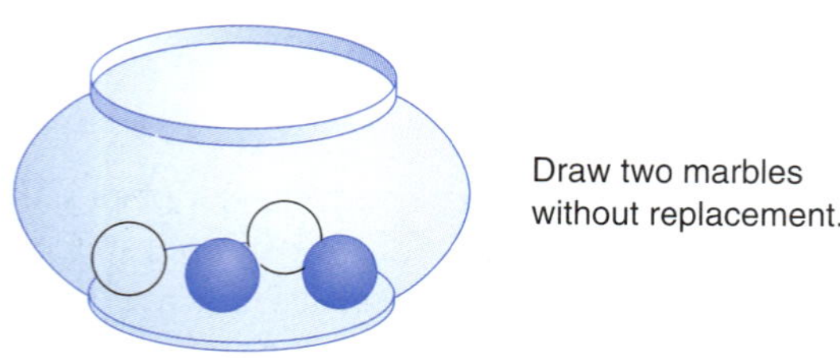

FIGURE 8.15 Two white and two blue marbles in urn.

Solution: For our equiprobable sample space, we take all ordered pairs, such as B_1W_2 and W_2W_1, whose components indicate the marbles selected on the first and on the second draw. Let B and W be the events

$$B = \{\text{blue on first draw}\}$$
$$W = \{\text{white on second draw}\}$$

We are interested in

$$P(W|B) = \frac{\#(W \cap B)}{\#(B)}$$

The reduced sample space B consists of all outcomes in which a blue marble is drawn first:

$$B = \{B_1B_2, B_1W_1, B_1W_2, B_2B_1, B_2W_1, B_2W_2\}$$

Event $W \cap B$ consists of the outcomes in B for which the second marble is white:

$$W \cap B = \{B_1W_1, B_1W_2, B_2W_1, B_2W_2\}$$

Since $\#(B) = 6$ and $\#(W \cap B) = 4$, we have

$$P(W|B) = \frac{4}{6} = \frac{2}{3}$$

▶ **Now work Problem 1.**

Example 1 showed how efficient the use of a reduced sample space can be. Note that it was not necessary to list all the outcomes either in the original sample space or in event W. Although we listed the outcomes in B, we could have found $\#(B)$ by using counting methods:

There are two ways in which the first marble can be blue, and three possibilities for the second marble, which can be either the remaining blue marble or one of the two white marbles. Thus, $\#(B) = 2 \cdot 3 = 6$.

The number $\#(W \cap B)$ could also be found by means of counting methods.

EXAMPLE 2 **Survey**

In a survey of 150 *people, each person was asked his or her marital status and opinion about floating a bond issue to build a community swimming pool. The results are summarized in Table 8.6. If one of these persons is randomly selected, find each of the following conditional probabilities.*

TABLE 8.6

	Favor (F)	Oppose (F')	Total
Married (M)	60	20	80
Single (M')	40	30	70
Total	100	50	150

a. *The probability that the person favors the bond issue, given that the person is married.*

Solution: We are interested in $P(F|M)$. The reduced sample space (M) contains 80 married persons, of which 60 favor the bond issue. Thus,

$$P(F|M) = \frac{\#(F \cap M)}{\#(M)} = \frac{60}{80} = \frac{3}{4}$$

b. *The probability that the person is married, given that the person favors the bond issue.*

Solution: We want to find $P(M|F)$. The reduced sample space (F) contains 100 persons who favor the bond issue. Of these, 60 are married. Hence,

$$P(M|F) = \frac{\#(M \cap F)}{\#(F)} = \frac{60}{100} = \frac{3}{5}$$

Note *here* that $P(M|F) \neq P(F|M)$ but this is not a general fact.

Another method of computing a conditional probability is by means of a formula involving *probabilities* with respect to the *original* sample space itself. Before stating the formula, we shall provide some motivation so that it seems reasonable to you. (The discussion that follows is oversimplified in the sense that certain assumptions are tacitly made.)

To consider $P(E|F)$, we shall assume that event F has probability $P(F)$ and event $E \cap F$ has probability $P(E \cap F)$. Let the experiment associated with this problem be repeated n times, where n is very large. Then the number of trials in which F occurs is approximately $n \cdot P(F)$. Of these, the number in which event E *also* occurs is approximately $n \cdot P(E \cap F)$. For large n, we estimate $P(E|F)$ by the relative frequency of the number of occurrences of $E \cap F$ with respect to the number of occurrences of F, which is approximately

$$\frac{n \cdot P(E \cap F)}{n \cdot P(F)} = \frac{P(E \cap F)}{P(F)}$$

This result strongly suggests the formula that appears in the following formal definition of conditional probability. (The definition applies to any sample space, whether or not it is equiprobable.)

Definition

The *conditional probability* of an event E, given that event F has occurred, is denoted $P(E|F)$ and is defined by

$$P(E|F) = \frac{P(E \cap F)}{P(F)} \quad \text{if } P(F) \neq 0 \tag{3}$$

Similarly,

$$P(F|E) = \frac{P(F \cap E)}{P(E)} \quad \text{if } P(E) \neq 0 \tag{4}$$

We emphasize that **the probabilities in Equations (3) and (4) are with respect to the original sample space.** Here we do *not* deal directly with a reduced sample space.

EXAMPLE 3 **Quality Control**

After the initial production run of a new style of steel desk, a quality control technician found that 40% *of the desks had an alignment problem and* 10% *had both a defective paint job and an alignment problem. If a desk is randomly selected from this run, and it has an alignment problem, what is the probability that it also has a defective paint job?*

Solution: Let A and D be the events

$$A = \{\text{alignment problem}\}$$

$$D = \{\text{defective paint job}\}$$

We are interested in $P(D|A)$, the probability of a defective paint job, given an alignment problem. From the given data, we have $P(A) = 0.4$ and $P(D \cap A) = 0.1$. Substituting into Equation (3) gives

$$P(D|A) = \frac{P(D \cap A)}{P(A)} = \frac{0.1}{0.4} = \frac{1}{4}$$

Note that we cannot use Equation (2) to solve this problem, because we are given probabilities rather than information about the sample space.

▶ **Now work Problem 7.**

EXAMPLE 4 **Genders of Offspring**

If a family has two children, find the probability that both are boys, given that one of the children is a boy. Assume that a child of either gender is equally likely and that, for example, having a girl first and a boy second is just as likely as having a boy first and a girl second.

Solution: Let E and F be the events

$$E = \{\text{both children are boys}\}$$

$$F = \{\text{at least one of the children is a boy}\}$$

We are interested in $P(E|F)$. Letting the letter B denote "boy" and G denote "girl," we use the equiprobable sample space

$$S = \{\text{BB, BG, GG, GB}\}$$

where, in each outcome, the order of the letters indicates the order in which the children are born. Thus,

$$E = \{\text{BB}\} \quad F = \{\text{BB, BG, GB}\} \quad \text{and} \quad E \cap F = \{\text{BB}\}$$

From Equation (3),

$$P(E|F) = \frac{P(E \cap F)}{P(F)} = \frac{\frac{1}{4}}{\frac{3}{4}} = \frac{1}{3}$$

Alternatively, this problem can be solved by using the reduced sample space F:

$$P(E|F) = \frac{\#(E \cap F)}{\#(F)} = \frac{1}{3}$$

▶ **Now work Problem 9.**

Equations (3) and (4) can be rewritten in terms of products by clearing fractions. This gives

$$P(E \cap F) = P(F)P(E|F)$$

and

$$P(F \cap E) = P(E)P(F|E)$$

By the commutative law, $P(E \cap F) = P(F \cap E)$, so we can combine the preceding equations to get an important law:

General Multiplication Law

$$P(E \cap F) = P(E)P(F|E) \qquad (5)$$
$$= P(F)P(E|F)$$

The general multiplication law states that the probability that two events *both* occur is equal to the probability that one of them occurs, times the conditional probability that the other one occurs, given that the first has occurred.

EXAMPLE 5 Advertising

A computer hardware company placed an ad for its new modem in a popular computer magazine. The company believes that the ad will be read by 32% of the magazine's readers and that 2% of those who read the ad will buy the modem. Assume that this is true, and find the probability that a reader of the magazine will read the ad and buy the modem.

Solution: Letting R denote the event "read ad" and B denote "buy modem," we are interested in $P(R \cap B)$. We are given that $P(R) = 0.32$. The fact that 2% of the readers of the ad will buy the modem can be written $P(B|R) = 0.02$. By the general multiplication law, Equation (5),

$$P(R \cap B) = P(R)P(B|R) = (0.32)(0.02) = 0.0064$$

▶ **Now work Problem 11.**

Stochastic Processes

The general multiplication law is also called the **law of compound probability.** The reason is that it is extremely useful when applied to an experiment that can be expressed as a *sequence* (or a compounding) of two or more other experiments, called **trials** or **stages.** The original experiment is called a **compound experiment,** and the sequence of trials is called a **stochastic process.** The probabilities of the events associated with each trial (beyond the first) could depend on what events occurred in the preceding trials, so they are conditional probabilities.

When we analyze a compound experiment, a tree diagram is extremely useful in keeping track of the possible outcomes at each stage. A complete path from the start to a tip of the tree gives an outcome of the experiment.

The notion of a compound experiment is discussed in detail in the next example. Read it carefully! Although the discussion is lengthy for the sake of developing a new idea, the actual computation takes little time.

EXAMPLE 6 Cards and Probability Tree

Two cards are drawn without replacement from a standard deck of cards. Find the probability that the second card is red.

Solution: The experiment of drawing two cards without replacement can be thought of as a compound experiment consisting of a sequence of two trials: The first is drawing a card, and the second is drawing a card after the first card has been drawn. The first trial has two possible outcomes:

$$R_1 = \{\text{red card}\} \quad \text{or} \quad B_1 = \{\text{black card}\}$$

(Here the subscript "1" refers to the first trial.) In Figure 8.16, these outcomes are represented by the two branches in the first level of the tree. Keep in mind that these outcomes are mutually exclusive, and they are also *exhaustive* in the sense that there are no other possibilities. Since there are 26 cards of each color, we have

$$P(R_1) = \frac{26}{52} \quad \text{and} \quad P(B_1) = \frac{26}{52}$$

These *unconditional* probabilities are written along the corresponding branches. We appropriately call Figure 8.16 a **probability tree.**

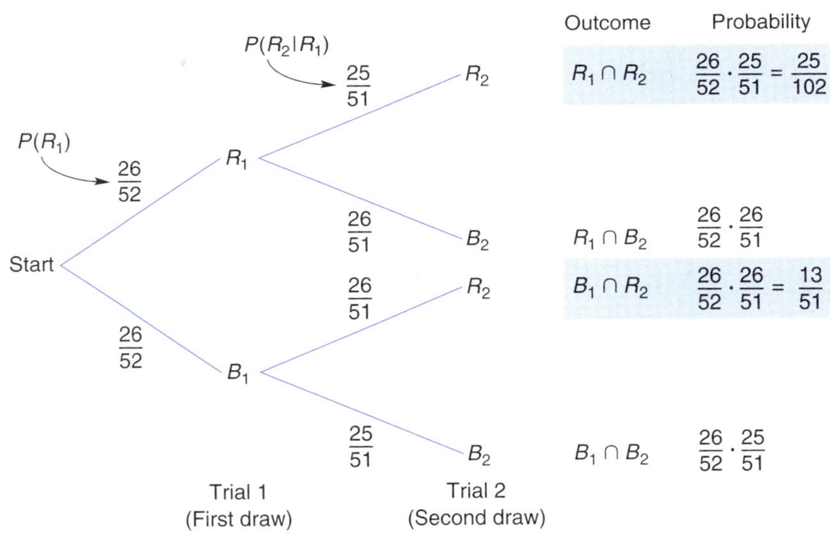

FIGURE 8.16 Probability tree for compound experiment.

Now, if a red card is obtained in the first trial, then, of the remaining 51 cards, 25 are red and 26 are black. The card drawn in the second trial can be red (R_2) or black (B_2). Thus, in the tree, the fork at R_1 has two branches: red and black. The *conditional* probabilities $P(R_2|R_1) = \frac{25}{51}$ and $P(B_2|R_1) = \frac{26}{51}$ are placed along these branches. Similarly, if a black card is obtained in the first trial, then, of the remaining 51 cards, 26 are red and 25 are black. Hence, $P(R_2|B_1) = \frac{26}{51}$ and $P(B_2|B_1) = \frac{25}{51}$, as indicated alongside the two branches emanating from B_1. The complete tree has two levels (one for each trial) and four paths (one for each of the four mutually exclusive and exhaustive events of the compound experiment).

Note that the sum of the probabilities along the branches from the vertex "Start" to R_1 and B_1 is 1:

$$\frac{26}{52} + \frac{26}{52} = 1$$

In general, the sum of the probabilities along all the branches emanating from a single vertex to an outcome of that trial must be 1. Thus, for the vertex at R_1,

$$\frac{25}{51} + \frac{26}{51} = 1$$

and for the vertex at B_1,

$$\frac{26}{51} + \frac{25}{51} = 1$$

Now, consider the topmost path. It represents the event "red on first draw and red on second draw." By the general multiplication law,

$$P(R_1 \cap R_2) = P(R_1)P(R_2|R_1) = \frac{26}{52} \cdot \frac{25}{51} = \frac{25}{102}$$

That is, *the probability of an event is obtained by multiplying the probabilities in the branches of the path for that event.* The probabilities for the other three paths are also indicated in the tree.

Returning to the original question, we see that two paths give a red card on the second draw, namely, the paths for $R_1 \cap R_2$ and $B_1 \cap R_2$. Therefore, the event "second card red" is the union of two mutually exclusive events. By the addition law, the probability of the event is the sum of the probabilities for the two paths:

$$P(R_2) = \frac{26}{52} \cdot \frac{25}{51} + \frac{26}{52} \cdot \frac{26}{51} = \frac{25}{102} + \frac{13}{51} = \frac{1}{2}$$

Note how easy it was to find $P(R_2)$ by using a probability tree.

Here is a summary of what we have done:

$$\begin{aligned} R_2 &= (R_1 \cap R_2) \cup (B_1 \cap R_2) \\ P(R_2) &= P(R_1 \cap R_2) + P(B_1 \cap R_2) \\ &= P(R_1)P(R_2|R_1) + P(B_1)P(R_2|B_1) \\ &= \frac{26}{52} \cdot \frac{25}{51} + \frac{26}{52} \cdot \frac{26}{51} = \frac{25}{102} + \frac{13}{51} = \frac{1}{2} \end{aligned}$$

▸ **Now work Problem 29.**

EXAMPLE 7 **Cards**

Two cards are drawn without replacement from a standard deck of cards. Find the probability that both cards are red.

Solution: Refer back to the probability tree in Figure 8.16. Only one path gives a red card on both draws, namely, that for $R_1 \cap R_2$. Thus, multiplying the probabilities along this path gives the desired probability:

$$P(R_1 \cap R_2) = P(R_1)P(R_2|R_1) = \frac{26}{52} \cdot \frac{25}{51} = \frac{25}{102}$$

▸ **Now work Problem 33.**

EXAMPLE 8 **Defective Computer Chips**

A company uses one computer chip in assembling each unit of a product. The chips are purchased from suppliers A, B, and C and are randomly picked for assembling a unit. Twenty percent come from A, 30% come from B, and the remainder come from C. The company believes that the probability that a chip from A will prove to be defective in the first 24 hours of use is 0.03, and the corresponding probabilities for B and C are 0.04 and 0.01, respectively. If an assembled unit is chosen at random and tested for 24 continuous hours, what is the probability that the chip in it is defective?

Solution: In this problem, there is a sequence of two trials: selecting a chip (A, B, or C) and then testing the selected chip [defective (D) or nondefective (D')]. We are given the unconditional probabilities

$$P(A) = 0.2 \quad \text{and} \quad P(B) = 0.3$$

Since A, B, and C are mutually exclusive and exhaustive,

$$P(C) = 1 - (0.2 + 0.3) = 0.5$$

From the statement of the problem, we also have the conditional probabilities

$$P(D|A) = 0.03 \quad P(D|B) = 0.04 \quad P(D|C) = 0.01$$

We want to find $P(D)$. To begin, we construct the two-level probability tree shown in Figure 8.17. We see that the paths that give a defective chip are those for the events

$$A \cap D \quad B \cap D \quad C \cap D$$

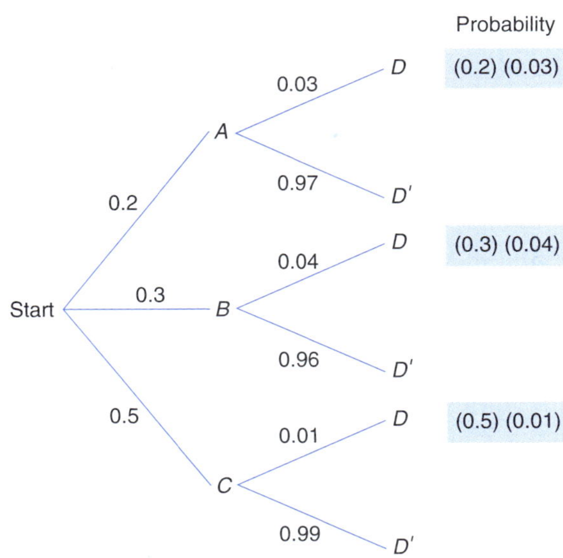

FIGURE 8.17 Probability tree for Example 8.

Since these events are mutually exclusive,

$$\begin{aligned} P(D) &= P(A \cap D) + P(B \cap D) + P(C \cap D) \\ &= P(A)P(D|A) + P(B)P(D|B) + P(C)P(D|C) \\ &= (0.2)(0.03) + (0.3)(0.04) + (0.5)(0.01) = 0.023 \end{aligned}$$

▶ **Now work Problem 47.**

The general multiplication law can be extended so that it applies to more than two events. For n events, we have

$$P(E_1 \cap E_2 \cap \cdots \cap E_n)$$
$$= P(E_1)P(E_2|E_1)P(E_3|E_1 \cap E_2) \cdots P(E_n|E_1 \cap E_2 \cap \cdots \cap E_{n-1})$$

(We assume that all conditional probabilities are defined.) In words, the probability that two or more events all occur is equal to the probability that one of them occurs, times the conditional probability that a second one occurs given that the first occurred, times the conditional probability that a third occurs given that the first two occurred, and so on. For example, in the manner of Example 7, the probability of drawing three red cards from a deck without replacement is

$$P(R_1 \cap R_2 \cap R_3) = P(R_1)P(R_2|R_1)P(R_3|R_1 \cap R_2) = \frac{26}{52} \cdot \frac{25}{51} \cdot \frac{24}{50}$$

EXAMPLE 9 **Marbles in Urn**

Urn I contains one black and two red marbles, and Urn II contains one pink marble. (See Figure 8.18.) An urn is selected at random. Then a marble is randomly drawn from it and placed in the other urn. A marble is then randomly drawn from that urn. Find the probability this marble is pink.

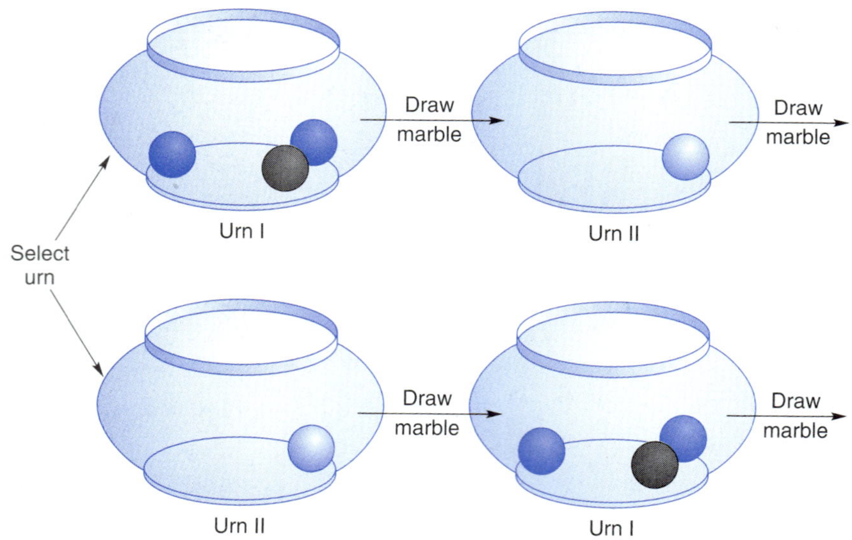

FIGURE 8.18 Marble selections from urns.

Solution: This is a compound experiment with three trials:

a. Selecting an urn.

b. Drawing a marble from the urn.

c. Drawing a marble from the other urn after the marble drawn in trial 2 is placed in it.

We want to find P(pink marble on second draw).

We analyze the situation by constructing a three-level probability tree. (See Figure 8.19.) The first trial has two equally likely possible outcomes, "Urn I" or "Urn II," so each has probability of $\frac{1}{2}$.

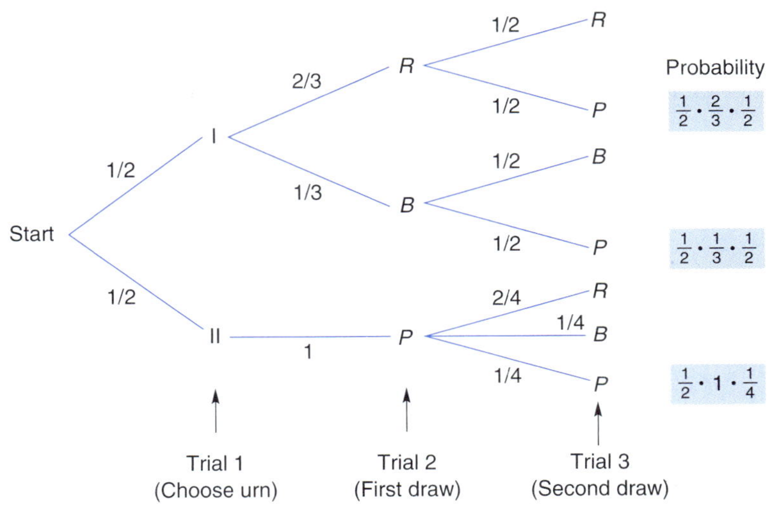

FIGURE 8.19 Three-level probability tree.

If Urn I was selected, the second trial has two possible outcomes, "red" (R) or "black" (B), with conditional probabilities $P(R|\text{I}) = \frac{2}{3}$ and $P(B|\text{I}) = \frac{1}{3}$. If Urn II was selected, there is one possible outcome, "pink" (P), so $P(P|\text{II}) = 1$. Thus, the second level of the tree has three branches.

Now we turn to the third trial. If Urn I was selected and a red marble drawn from it and placed in Urn II, then Urn II contains one red and one pink marble. Hence, at the end of the second trial, the fork at vertex R has two branches, R and P, with conditional probabilities

$$P(R|\text{I} \cap R) = \frac{1}{2} \quad \text{and} \quad P(P|\text{I} \cap R) = \frac{1}{2}$$

Similarly, the tree shows the two possibilities if Urn I was initially selected and a black marble was placed into Urn II. Now, if Urn II was selected in the first trial, then the pink marble in it was drawn and placed into Urn I, so Urn I contains two red, one black, and one pink marble. Thus, the fork at P has *three* branches, one with probability $\frac{2}{4}$ and two with probability $\frac{1}{4}$.

We see that three paths give a pink marble on the third trial, so for each, we multiply the probabilities along its branches. For example, the second path from the top represents $\text{I} \to R \to P$; the probability of this event is

$$\begin{aligned} P(\text{I} \cap R \cap P) &= P(\text{I})P(R|\text{I})P(P|\text{I} \cap R) \\ &= \frac{1}{2} \cdot \frac{2}{3} \cdot \frac{1}{2} \end{aligned}$$

Adding the probabilities for the three paths gives

$$\begin{aligned} P(\text{pink marble on second draw}) &= \frac{1}{2} \cdot \frac{2}{3} \cdot \frac{1}{2} + \frac{1}{2} \cdot \frac{1}{3} \cdot \frac{1}{2} + \frac{1}{2} \cdot 1 \cdot \frac{1}{4} \\ &= \frac{1}{6} + \frac{1}{12} + \frac{1}{8} = \frac{3}{8} \end{aligned}$$

▸ **Now work Problem 43.**

Exercise 8.5

*1. Given the equiprobable sample space

$$S = \{1, 2, 3, 4, 5, 6, 7, 8, 9\}$$

and events

$$E = \{1, 3\}$$
$$F = \{1, 2, 4, 5, 6\}$$
$$G = \{5, 6, 7, 8, 9\}$$

find each of the following:

(a) $P(E|F)$ **(b)** $P(E'|F)$ **(c)** $P(E|F')$
(d) $P(F|E)$ **(e)** $P(E|F \cap G)$

2. Given the equiprobable sample space

$$S = \{1, 2, 3, 4\}$$

and events

$$E = \{1\}$$
$$F = \{2\}$$
$$G = \{1, 2\}$$

find each of the following:

(a) $P(E)$ **(b)** $P(E|F)$ **(c)** $P(E|G)$
(d) $P(G|E)$ **(e)** $P(G|F')$ **(f)** $P(E'|F')$

3. If $P(E) > 0$, find $P(E|E)$.

4. If $P(E) > 0$, find $P(\emptyset|E)$.

5. If $P(E|F) = 0.57$, find $P(E'|F)$.

6. If F and G are mutually exclusive events with positive probabilities, find $P(F|G)$.

*7. If $P(E) = \frac{1}{4}$, $P(F) = \frac{1}{3}$, and $P(E \cap F) = \frac{1}{6}$, find each of the following:

(a) $P(E|F)$ **(b)** $P(F|E)$

8. If $P(E) = \frac{1}{4}$, $P(F) = \frac{1}{3}$, and $P(E|F) = \frac{3}{4}$, find $P(E \cup F)$. (*Hint:* Use the addition law to find $P(E \cup F)$.)

*9. If $P(E) = \frac{1}{3}$, $P(E \cup F) = \frac{8}{15}$, and $P(E \cap F) = \frac{1}{5}$, find each of the following:

(a) $P(F|E)$
(b) $P(F)$ (*Hint:* Use the addition law.)
(c) $P(E|F)$
(d) $P(E|F')$ (*Hint:* Find $P(E \cap F')$ by using the identity $P(E) = P(E \cap F) + P(E \cap F')$.)

10. If $P(E) = \frac{3}{5}$, $P(F) = \frac{3}{10}$, and $P(E \cup F) = \frac{7}{10}$, find $P(E|F)$.

*11. **Gypsy Moth** Because of gypsy moth infestation of three large areas that are densely populated with trees, consideration is being given to aerial spraying to destroy larvae. A survey was made of the 200 residents of these areas to determine whether or not they favor the spraying. The resulting data are shown in Table 8.7. Suppose that a resident is randomly selected. Let I be the event "the resident is from Area I" and so on. Find each of the following:

TABLE 8.7

	Area I	Area II	Area II	Total
Favor (F)	46	35	44	125
Opposed (O)	22	15	10	47
No opinion (N)	10	8	10	28
Total	78	58	64	200

(a) $P(F)$ **(b)** $P(F|\text{II})$ **(c)** $P(O|\text{I})$
(d) $P(\text{III})$ **(e)** $P(\text{III}|O)$ **(f)** $P(\text{II}|N')$

12. **College Selection and Family Income** A survey of 175 students resulted in the data shown in Table 8.8. It shows the type of college the student attends and the income level of the student's family. Suppose a student in the survey is randomly selected.

TABLE 8.8

	College		
Income	**Private**	**Public**	**Total**
High	14	11	25
Middle	25	55	80
Low	10	60	70
Total	49	126	175

(a) Find the probability that the student attends a public college, given that the student comes from a middle-income family.
(b) Find the probability that the student is from a high-income family, given that the student attends a private college.
(c) If the student comes from a high-income family, find the probability that the student attends a private college.
(d) Find the probability that the student attends a public college or comes from a low-income family.

13. **Cola Preference** A survey was taken among cola drinkers to see which of two popular brands people preferred. It was found that 45% liked brand A, 40% liked brand B, and 20% liked both brands. Suppose that a person in the survey is randomly selected.

(a) Find the probability that the person liked brand A, given that he or she liked brand B.
(b) Find the probability that the person liked brand B, given that he or she liked brand A.

14. Quality Control Of the cars produced by an automobile manufacturer, 18% have defective assembly and 11% have both defective assembly and poor paintwork. If a car is randomly selected from a lot and it has defective assembly, what is the probability that it has poor paintwork?

In Problems 15 and 16, assume that a child of either gender is equally likely and that, for example, having a girl first and a boy second is just as likely as having a boy first and a girl second.

15. Genders of Offspring If a family has two children, what is the probability that one child is a girl, given that at least one child is a boy?

16. Genders of Offspring If a family has three children, find each of the following:

(a) The probability that it has two girls, given that at least one child is a boy.
(b) The probability that it has at least two girls, given that the oldest child is a girl.

17. Coin Toss If a fair coin is tossed three times in succession, find each of the following:

(a) The probability of getting exactly two tails, given that the second toss is a tail.
(b) The probability of getting exactly two tails, given that the second toss is a head.

18. Coin Toss If a fair coin is tossed four times in succession, find the odds of getting four tails, given that the first toss is a tail.

19. Die Roll If a fair die is rolled, find the probability of getting a number less than 4, given that the number is odd.

20. Cards If a card is drawn randomly from a standard deck, find the probability of getting a face card, given that the card is red.

21. Dice Roll If two fair dice are rolled, find the probability that two 6's occur, given that at least one die shows a 6.

22. Dice Roll If a fair red die and a fair green die are rolled, find the probability that the sum is greater than 9, given that a 5 shows on the red die.

23. Dice Roll If a fair red die and a fair green die are rolled, find the probability of getting a total of 7, given that the green die shows an even number.

24. Dice Roll A fair die is rolled two times in succession.

(a) Find the probability that the sum is 6, given that the second roll is neither a 2 nor a 4.
(b) Find the probability that the sum is 6 and that the second roll is neither a 2 nor a 4.

25. Die Roll If a fair die is rolled two times in succession, find the probability of getting a total greater than 7, given that the first roll is greater than 3.

26. Coin and Die If a fair coin and a fair die are thrown, find the probability that the coin shows heads, given that the number on the die is even.

27. Cards If a card is randomly drawn from a deck of 52 cards, find the probability that the card is a king, given that it is a heart.

28. Cards If a card is randomly drawn from a deck of 52 cards, find the probability that the card is a heart, given that it is a face card (a jack, queen, or king).

*29. **Cards** If two cards are randomly drawn without replacement from a standard deck, find the probability that the second card is not a face card, given that the first card is a face card (a jack, queen, or king).

In Problems 30–35, consider the experiment to be a compound experiment.

30. Cards If two cards are randomly drawn from a standard deck, find the probability that both cards are face cards if

(a) the cards are drawn without replacement.
(b) the cards are drawn with replacement.

31. Cards If three cards are randomly drawn without replacement from a standard deck, find the probability of getting a king, a queen, and a jack, in that order.

32. Cards If three cards are randomly drawn without replacement from a standard deck, find the probability of getting the ace of spades, the ace of hearts, and the ace of diamonds, in that order.

*33. **Cards** If three cards are randomly drawn without replacement from a standard deck, find the probability that all three cards are diamonds.

34. Cards If two cards are randomly drawn without replacement from a standard deck of cards, find the probability that the second card is a heart.

35. Cards If two cards are randomly drawn without replacement from a standard deck, find the probability of getting two diamonds, given that the first card is red.

36. Wake-Up Call Barbara Smith, a sales representative, is staying overnight at a hotel and has a breakfast meeting with an important client the following morning. She asked room service to give her a 7 A.M. wake-up call that morning in order that she be prompt for the meeting. The probability that she will get the call is 0.9. If she gets the call, the probability that she will be on time is 0.9. If the call is not given, the probability that she will be on time is 0.4. Find the probability that she will be on time for the meeting.

37. Taxpayer Survey In a certain school district, a questionnaire was sent to all property-tax payers concerning whether or not a new high school should be built. Of those that responded, 60% favored its construction, 30% opposed it, and 10% had no opinion. Further analysis of the data concerning the area in which the respondents lived gave the results in Table 8.9.

TABLE 8.9

	Urban	Suburban
Favor	45%	55%
Oppose	55%	45%
No opinion	35%	65%

(a) If one of the respondents is selected at random, what is the probability that he or she lives in an urban area?

(b) If a respondent is selected at random, use the result of part (a) to find the probability that he or she favors the construction of the school, given that the person lives in an urban area.

38. Marketing A travel agency has a computerized telephone that randomly selects telephone numbers for advertising group cruiseboat sailings. The telephone automatically dials the selected number and plays a prerecorded message to the recipient of the call. Experience has shown that 3% of those called show interest and contact the agency. However, of these, 14% actually agree to purchase a berth on a cruise.

(a) Find the probability that a person called will contact the agency and purchase a berth on a cruise.

(b) If 10,000 people are called, how many can be expected to contact the agency and purchase a berth on a cruise?

39. Marbles in Urn An urn contains three yellow and two red marbles.

(a) If two marbles are randomly drawn without replacement, find the probability that the second marble drawn is yellow, given that the first marble drawn is red.

(b) Repeat part (a), but assume that the first marble is replaced before the second marble is drawn.

40. Marbles in Urn Urn 1 contains four green and three red marbles, and Urn 2 contains three green, one white, and two red marbles. A marble is randomly drawn from Urn 1 and placed into Urn 2. If a marble is then randomly drawn from Urn 2, find the probability that the marble is green.

41. Balls in Urn Urn 1 contains three red and two white balls. Urn 2 contains two red and two white balls. An urn is chosen at random and then a ball is chosen at random from it. What is the probability that the ball is white?

42. Balls in Urn Urn 1 contains two red and three white balls. Urn 2 contains three red and four white balls. Urn 3 contains two red, two white, and two green balls. An urn is chosen at random, and then a ball is chosen at random from it.

(a) Find the probability that the ball is white.

(b) Find the probability that the ball is red.

(c) Find the probability that the ball is green.

***43. Marbles in Urn** Urn 1 contains one green and one red marble, and Urn 2 contains one white and one red marble. An urn is selected at random. A marble is randomly drawn from it and placed in the other urn. A marble is then randomly drawn from that urn. Find the probability that the marble is white.

44. Dead Batteries During Hurricane Juan's passage through Halifax, Ms. Wood's lights went out and in the dark she reached in the kitchen drawer for 4 batteries for her flashlight. There were 10 batteries in the drawer, but 5 of them were dead ones (that Mr. Wood should have thrown out). Find the probability that the 4 batteries Ms. Wood took from the drawer were all dead.

45. Quality Control A manufacturing process requires the use of a robotic welder on each of two assembly lines. Assembly line A produces 200 units of product per day, and assembly line B produces 400 units per day. Over a period of time, it has been found that the welder on A produces 2% defective units, whereas the welder on B produces 5% defective units. At the end of a day, a unit was selected at random from the total production. Find the probability that the unit has defective welding.

46. Game Show A TV game show host presents the following situation to a contestant. On a table are three identical boxes. One of them contains two identical envelopes. In one is a check for \$5000, and in the other is a check for \$1. Another box contains two envelopes with a check for \$5000 in each and six envelopes with a check for \$1 in each. The remaining box contains one envelope with a check for \$5000 inside and five envelopes with a check for \$1 inside each. If the contestant must select a box at random and then randomly draw an envelope, find the probability that a check for \$5000 is inside.

***47. Quality Control** A company uses one computer chip in assembling each unit of a product. The chips are purchased from suppliers A, B, and C and are randomly picked for assembling a unit. Ten percent come from A, 20% come from B, and the remainder come from C. The probability that a chip from A will prove to be defective in the first 24 hours of use is 0.06, and the corresponding probabilities for B and C are 0.04 and 0.05, respectively. If an assembled unit is chosen at random and tested for 24 continuous hours, what is the probability that the chip in it will prove to be defective?

48. Quality Control A manufacturer of widgets has four assembly lines: A, B, C, and D. The percentages of output produced by the lines are 30%, 20%, 35%, and 15%, respectively, and the percentages of defective units they produce are 6%, 3%, 2%, and 5%. If a widget is randomly selected from stock, what is the probability that it is defective?

49. Voting In a certain town, 40% of eligible voters are registered Democrats, 35% are Republicans, and the remainder are Independents. In the last primary election, 15% of the Democrats, 20% of the Republicans, and 10% of the Independents voted.

(a) If an eligible voter is chosen at random, what is the probability that he or she is a Democrat who voted?

(b) If an eligible voter is chosen at random, what is the probability that he or she voted?

50. Job Applicants A garage has four openings for mechanics. Suppose Frank, Joan, Sam, Barbara, Tom, Richard, and Doris are the only applicants for these jobs, and all are equally qualified. If four are hired at random, find the probability that Frank, Joan, Tom, and Doris were chosen, given that Richard was not hired.

51. Committee Selection Suppose six female and five male students wish to fill three openings on a campus committee on cultural diversity. If three of the students are chosen at random for the committee, find the probability that all three are female, given that at least one is female.

OBJECTIVE *To develop the notion of independent events and apply the special multiplication law.*

8.6 Independent Events

In our discussion of conditional probability, you saw that the probability of an event may be affected by the knowledge that another event has occurred. In this section, we consider the situation where the additional information has no effect. That is, the conditional probability $P(E|F)$ and the unconditional probability $P(E)$ are the same.

When $P(E|F) = P(E)$, we say that E is *independent* of F. If E is independent of F, it follows that F is independent of E (and vice versa). To prove this, assume that $P(E|F) = P(E)$ and $P(E) \neq 0$. Then

$$P(F|E) = \frac{P(E \cap F)}{P(E)} = \frac{P(F)P(E|F)}{P(E)} = \frac{P(F)P(E)}{P(E)} = P(F)$$

which means that F is independent of E. Thus, to prove independence, it suffices to show that either $P(E|F) = P(E)$ or $P(F|E) = P(F)$, and when one of these is true, we simply say that E and F are *independent events*.

Independence of two events is defined by probabilities, not by a causal relationship.

Definition

Let E and F be events with positive probabilities. Then E and F are said to be *independent events* if either

$$P(E|F) = P(E) \quad \textbf{(1)}$$

or

$$P(F|E) = P(F) \quad \textbf{(2)}$$

If E and F are not independent, they are said to be *dependent events*.

Thus, with dependent events, the occurrence of one of the events *does* affect the probability of the other. If E and F are independent events, it can be shown that the events in each of the following pairs are also independent:

$$E \text{ and } F' \qquad E' \text{ and } F \qquad E' \text{ and } F'$$

EXAMPLE 1 **Showing That Two Events Are Independent**

A fair coin is tossed twice. Let E and F be the events

$$E = \{head\ on\ first\ toss\}$$

$$F = \{head\ on\ second\ toss\}$$

Determine whether or not E and F are independent events.

Solution: We suspect that they are independent, because one coin toss should not influence the outcome of another toss. To confirm our suspicion, we shall compare $P(E)$ with $P(E|F)$. For the equiprobable sample space $S = \{\text{HH, HT, TH, TT}\}$, we have $E = \{\text{HH, HT}\}$ and $F = \{\text{HH, TH}\}$. Thus,

$$P(E) = \frac{\#(E)}{\#(S)} = \frac{2}{4} = \frac{1}{2}$$

$$P(E|F) = \frac{\#(E \cap F)}{\#(F)} = \frac{n(\{\text{HH}\})}{n(F)} = \frac{1}{2}$$

Since $P(E|F) = P(E)$, events E and F are independent.

▸ **Now work Problem 7.**

In Example 1 we suspected the result, and certainly there are other situations where we have an intuitive feeling as to whether or not two events are independent. For example, if a red die and green die are tossed, we expect (and it is indeed true) that the events "3 on red die" and "6 on green die" are independent, because the outcome on one die should not be influenced by the outcome on the other die. Similarly, if two cards are drawn *with replacement* from a deck of cards, we would assume that the events "first card is a jack" and "second card is a jack" are independent. However, suppose the cards are drawn *without replacement*. Because the first card drawn is not put back in the deck, it should have an effect on the outcome of the second draw, so we expect the events to be dependent. In many problems, your intuitive notion of independence or the context of the problem may make it clear whether or not independence can be assumed. In spite of one's intuition (which can prove to be wrong!), the only sure way to determine whether events E and F are independent (or dependent) is by showing that Equation (1) or Equation (2) is true (or is not true).

EXAMPLE 2 Smoking and Sinusitis

In a study of smoking and sinusitis, 4000 people were studied, with the results as given in Table 8.10. Suppose a person from the study is selected at random. On the basis of the data, determine whether or not the events "having sinusitis" (L) and "smoking" (S) are independent events.

TABLE 8.10 Smoking and Sinusitis

	Smoker	**Nonsmoker**	**Total**
Sinusitis	432	1018	1450
No sinusitis	528	2022	2550
Total	960	3040	4000

Solution: We shall compare $P(L)$ with $P(L|S)$. The number $P(L)$ is the proportion of the people studied that have sinusitis:

$$P(L) = \frac{1450}{4000} = \frac{29}{80} = 0.3625$$

For $P(L|S)$, the sample space is reduced to 960 smokers, of which 432 have sinusitis:

$$P(L|S) = \frac{432}{960} = \frac{9}{20} = 0.45$$

Since $P(L|S) \neq P(L)$, having sinusitis and smoking are dependent.

▶ **Now work Problem 9.**

The general multiplication law takes on an extremely important form for independent events. Recall that law:

$$\begin{aligned} P(E \cap F) &= P(E)P(F|E) \\ &= P(F)P(E|F) \end{aligned}$$

If events E and F are independent, then $P(F|E) = P(F)$, so substitution in the first equation gives

$$P(E \cap F) = P(E)P(F)$$

The same result is obtained from the second equation. Thus, we have the following law:

Special Multiplication Law

If E and F are *independent events,* then

$$P(E \cap F) = P(E)P(F) \tag{3}$$

Equation (3) states that if E and F are independent events, then the probability that E and F both occur is the probability that E occurs times the probability that F occurs. Keep in mind that Equation (3) is *not* valid when E and F are dependent.

EXAMPLE 3 Survival Rates

Suppose the probability of the event "Bob lives 20 more years" (B) is 0.8 and the probability of the event "Doris lives 20 more years" (D) is 0.85. Assume that B and D are independent events.

a. *Find the probability that both Bob and Doris live 20 more years.*

Solution: We are interested in $P(B \cap D)$. Since B and D are independent events, the special multiplication law applies:

$$P(B \cap D) = P(B)P(D) = (0.8)(0.85) = 0.68$$

b. *Find the probability that at least one of them lives 20 more years.*

Solution: Here we want $P(B \cup D)$. By the addition law,

$$P(B \cup D) = P(B) + P(D) - P(B \cap D)$$

From part (a), $P(B \cap D) = 0.68$, so

$$P(B \cup D) = 0.8 + 0.85 - 0.68 = 0.97$$

c. *Find the probability that exactly one of them lives 20 more years.*

Solution: We first express the event

$$E = \{\text{exactly one of them lives 20 more years}\}$$

in terms of the given events, B and D. Now, event E can occur in one of two *mutually exclusive* ways: Bob lives 20 more years but Doris does not $(B \cap D')$, or Doris lives 20 more years but Bob does not $(B' \cap D)$. Thus,

$$E = (B \cap D') \cup (B' \cap D)$$

By the addition law (for mutually exclusive events),

$$P(E) = P(B \cap D') + P(B' \cap D) \quad (4)$$

To compute $P(B \cap D')$, we note that, since B and D are independent, so are B and D' (from the statement preceding Example 1). Accordingly, we can use the multiplication law and the rule for complements:

$$\begin{aligned} P(B \cap D') &= P(B)P(D') \\ &= P(B)[1 - P(D)] = (0.8)(0.15) = 0.12 \end{aligned}$$

Similarly,

$$P(B' \cap D) = P(B')P(D) = (0.2)(0.85) = 0.17$$

Substituting into Equation (4) gives

$$P(E) = 0.12 + 0.17 = 0.29$$

▸ **Now work Problem 25.**

In Example 3, it was assumed that events B and D are independent. However, if Bob and Doris are related in some way, it is quite possible that the survival of one of them has a bearing on the survival of the other. In that case the assumption of independence is not justified, and we could not use the special multiplication law, Equation (3).

EXAMPLE 4 **Cards**

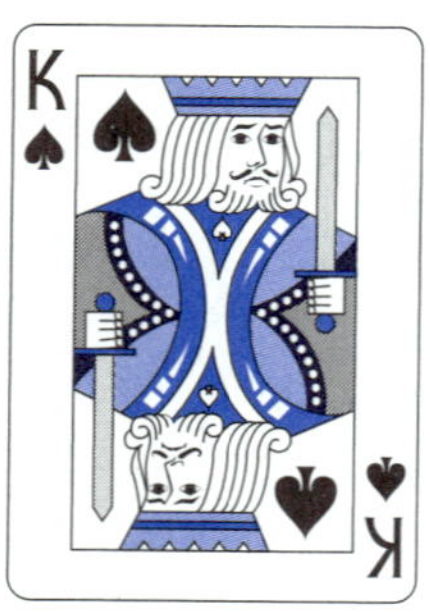

In a math exam, a student was given the following two-part problem. A card is randomly drawn from a deck of 52 cards. Let H, K, and R be the events

$$H = \{heart\ drawn\}$$

$$K = \{king\ drawn\}$$

$$R = \{red\ card\ drawn\}$$

Find $P(H \cap K)$ and $P(H \cap R)$.

For the first part, the student wrote

$$P(H \cap K) = P(H)P(K) = \frac{13}{52} \cdot \frac{4}{52} = \frac{1}{52}$$

and for the second part, she wrote

$$P(H \cap R) = P(H)P(R) = \frac{13}{52} \cdot \frac{26}{52} = \frac{1}{8}$$

The answer was correct for $P(H \cap K)$, but not for $P(H \cap R)$. Why?

Solution: The reason is that the student assumed independence in *both* parts by using the special multiplication law to multiply unconditional probabilities when, in fact, that assumption should *not* have been made. Let us examine the first part of the exam problem for independence. We shall see whether $P(H)$ and $P(H|K)$ are the same. We have

$$P(H) = \frac{13}{52} = \frac{1}{4}$$

and

$$P(H|K) = \frac{1}{4} \qquad \text{(1 heart out of four kings)}$$

Since $P(H) = P(H|K)$, events H and K are independent, so the student's procedure is valid. (The student was lucky!) For the second part, again we have $P(H) = \frac{1}{4}$, but

$$P(H|R) = \frac{13}{26} = \frac{1}{2} \qquad \text{(13 hearts out of 26 red cards)}$$

Since $P(H|R) \neq P(H)$, events H and R are dependent, so the student should not have multiplied the unconditional probabilities. (The student was unlucky.) However, the student would have been safe by using the *general* multiplication law, that is,

$$P(H \cap R) = P(H)P(R|H) = \frac{13}{52} \cdot 1 = \frac{1}{4}$$

or

$$P(H \cap R) = P(R)P(H|R) = \frac{26}{52} \cdot \frac{13}{26} = \frac{1}{4}$$

More simply, observe that $H \cap R = H$, so

$$P(H \cap R) = P(H) = \frac{13}{52} = \frac{1}{4}$$

Equation (3) is often used as an alternative means of defining independent events, and we shall consider it as such:

Events E and F are independent if and only if

$$P(E \cap F) = P(E)P(F) \qquad \textbf{(3)}$$

▶ **Now work Problem 33.**

Putting everything together, we can say that to prove events E and F, with nonzero probabilities, are independent, only one of the following relationships has to be shown:

$$P(E|F) = P(E) \qquad \textbf{(1)}$$

or

$$P(F|E) = P(F) \qquad \textbf{(2)}$$

or

$$P(E \cap F) = P(E)P(F) \qquad \textbf{(3)}$$

In other words, if any one of these equations is true, then all of them are true; if any is false, then all of them are false, and E and F are dependent.

EXAMPLE 5 **Dice**

Two fair dice, one red and the other green, are rolled, and the numbers on the top faces are noted. Let E and F be the events

$$E = \{number\ on\ red\ die\ is\ even\}$$
$$F = \{sum\ is\ 7\}$$

Test whether $P(E \cap F) = P(E)P(F)$ to determine whether E and F are independent.

Solution: Our usual sample space for the roll of two dice has $6 \cdot 6 = 36$ equally likely outcomes. For event E, the red die can fall in any of three ways and the green die any of six ways, so E consists of $3 \cdot 6 = 18$ outcomes. Thus, $P(E) = \frac{18}{36} = \frac{1}{2}$. Event F has six outcomes:

$$F = \{(1, 6), (2, 5), (3, 4), (4, 3), (5, 2), (6, 1)\} \qquad (5)$$

where, for example, we take (1, 6) to mean "1" on the red die and "6" on the green die. Therefore, $P(F) = \frac{6}{36} = \frac{1}{6}$, so

$$P(E)P(F) = \frac{1}{2} \cdot \frac{1}{6} = \frac{1}{12}$$

Now, event $E \cap F$ consists of all outcomes in which the red die is even and the sum is 7. Using (5) as an aid, we see that

$$E \cap F = \{(2, 5), (4, 3), (6, 1)\}$$

Thus,

$$P(E \cap F) = \frac{3}{36} = \frac{1}{12}$$

Since $P(E \cap F) = P(E)P(F)$, events E and F are independent. This fact may not have been obvious before the problem was solved.

▶ **Now work Problem 17.**

EXAMPLE 6 **Genders of Offspring**

For a family with at least two children, let E and F be the events

$$E = \{at\ most\ one\ boy\},$$
$$F = \{at\ least\ one\ child\ of\ each\ gender\}.$$

Assume that a child of either gender is equally likely and that, for example, having a girl first and a boy second is just as likely as having a boy first and a girl second. Determine whether E and F are independent in each of the following situations:

a. *The family has exactly two children.*

Solution: We will use the equiprobable sample space

$$S = \{\text{BB, BG, GG, GB}\}$$

and test whether $P(E \cap F) = P(E)P(F)$. We have

$$E = \{\text{BG, GB, GG}\} \quad F = \{\text{BG, GB}\} \quad E \cap F = \{\text{BG, GB}\}$$

Thus, $P(E) = \frac{3}{4}$, $P(F) = \frac{2}{4} = \frac{1}{2}$, and $P(E \cap F) = \frac{2}{4} = \frac{1}{2}$. We ask whether

$$P(E \cap F) \stackrel{?}{=} P(E)P(F)$$

and see that

$$\frac{1}{2} \neq \frac{3}{4} \cdot \frac{1}{2} = \frac{3}{8}$$

so E and F are dependent events.

b. *The family has exactly three children.*

Solution: Based on the result of part (a), you may have an intuitive feeling that E and F are dependent. Nevertheless, we must test this conjecture. For three children, we use the equiprobable sample space

$$S = \{\text{BBB, BBG, BGB, BGG, GBB, GBG, GGB, GGG}\}$$

Again we test whether $P(E \cap F) = P(E)P(F)$. We have

$$E = \{\text{BGG, GBG, GGB, GGG}\}$$
$$F = \{\text{BBG, BGB, BGG, GBB, GBG, GGB}\}$$
$$E \cap F = \{\text{BGG, GBG, GGB}\}$$

Hence, $P(E) = \frac{4}{8} = \frac{1}{2}$, $P(F) = \frac{6}{8} = \frac{3}{4}$, and $P(E \cap F) = \frac{3}{8}$, so

$$P(E)P(F) = \frac{1}{2} \cdot \frac{3}{4} = \frac{3}{8} = P(E \cap F)$$

Therefore, we have the *unexpected* result that events E and F are independent. (*Moral:* You cannot always trust your intuition.)

▶ **Now work Problem 27.**

We now generalize our discussion of independence to the case of more than two events.

Definition

The events $E_1, E_2, \ldots, E_n$ are said to be *independent* if and only if for each set of two or more of the events, the probability of the intersection of the events in the set is equal to the product of the probabilities of the events in that set.

For instance, let us apply the definition to the case of three events ($n = 3$). We say that E, F, and G are independent events if the special multiplication law is true for these events, taken two at a time and three at a time. That is, each of the following equations must be true:

$$\left.\begin{aligned} P(E \cap F) &= P(E)P(F) \\ P(E \cap G) &= P(E)P(G) \\ P(F \cap G) &= P(F)P(G) \end{aligned}\right\} \text{Two at a time}$$

$$P(E \cap F \cap G) = P(E)P(F)P(G)\} \text{ Three at a time}$$

As another example, if events E, F, G, and H are independent, then we can assert such things as

$$P(E \cap F \cap G \cap H) = P(E)P(F)P(G)P(H)$$

$$P(E \cap G \cap H) = P(E)P(G)P(H)$$

and

$$P(F \cap H) = P(F)P(H)$$

Similar conclusions can be made if any of the events are replaced by their complements.

EXAMPLE 7 **Cards**

Four cards are randomly drawn, with replacement, from a deck of 52 cards. Find the probability that the cards chosen, in order, are a king (K), a queen (Q), a jack (J), and a heart (H).

Solution: Since there is replacement, what happens on any draw does not affect the outcome on any other draw, so we can assume independence and multiply the unconditional probabilities. We obtain

$$\begin{aligned} P(K \cap Q \cap J \cap H) &= P(K)P(Q)P(J)P(H) \\ &= \frac{4}{52} \cdot \frac{4}{52} \cdot \frac{4}{52} \cdot \frac{13}{52} = \frac{1}{8788} \end{aligned}$$

▶ **Now work Problem 35.**

EXAMPLE 8 **Aptitude Test**

Personnel Temps, a temporary-employment agency, requires that each job applicant take the company's aptitude test, which has 80% accuracy.

a. *Find the probability that the test will be accurate for the next three applicants who are tested.*

Solution: Let A, B, and C be the events that the test will be accurate for applicants A, B, and C, respectively. We are interested in

$$P(A \cap B \cap C)$$

Since the accuracy of the test for one applicant should not affect the accuracy for any of the others, it seems reasonable to assume that A, B, and C are independent. Thus, we can multiply probabilities:

$$\begin{aligned} P(A \cap B \cap C) &= P(A)P(B)P(C) \\ &= (0.8)(0.8)(0.8) = (0.8)^3 = 0.512 \end{aligned}$$

b. *Find the probability that the test will be accurate for at least two of the next three applicants who are tested.*

Solution: Here, "at least two" means exactly two or three. In the first case, the possible ways of choosing the two tests that are accurate are

$$A \text{ and } B \qquad A \text{ and } C \qquad B \text{ and } C$$

In each of these three possibilities, the test for the remaining applicant is not accurate. For example, choosing A and B gives the event $A \cap B \cap C'$, whose probability is

$$P(A)P(B)P(C') = (0.8)(0.8)(0.2) = (0.8)^2(0.2)$$

You should verify that the probability for each of the other two possibilities is also $(0.8)^2(0.2)$. Summing the three probabilities gives

$$P(\text{exactly two accurate}) = 3[(0.8)^2(0.2)] = 0.384$$

Using this result and that of part (a), we obtain

$$P(\text{at least two accurate}) = P(\text{exactly two accurate}) + P(\text{three accurate})$$
$$= 0.384 + 0.512 = 0.896$$

Alternatively, the problem could be solved by computing

$$1 - [P(\text{none accurate}) + P(\text{exactly one accurate})]$$

Why?

▶ **Now work Problem 21.**

We conclude with a note of caution: **Do not confuse independent events with mutually exclusive events.** The concept of independence is defined in terms of probability, whereas mutual exclusiveness is not. When two events are independent, the occurrence of one of them does not affect the probability of the other. However, when two events are mutually exclusive, they cannot occur simultaneously. Although these two concepts are not the same, we can draw some conclusions about their relationship. If E and F are mutually exclusive events with positive probabilities, the occurrence of F means that E cannot occur. Hence,

$$P(E|F) = 0 \neq P(E) \quad \text{since } P(E) > 0$$

Thus, E and F must be dependent. In short, *mutually exclusive events with positive probabilities must be dependent.* Another way of saying this is that *independent events with positive probabilities are not mutually exclusive.*

Exercise 8.6

1. If events E and F are independent with $P(E) = \frac{1}{3}$ and $P(F) = \frac{3}{4}$, find each of the following:

(a) $P(E \cap F)$ **(b)** $P(E \cup F)$ **(c)** $P(E|F)$

(d) $P(E'|F)$ **(e)** $P(E \cap F')$ **(f)** $P(E \cup F')$

(g) $P(E|F')$

2. If events E, F, and G are independent with $P(E) = 0.1$, $P(F) = 0.3$, and $P(G) = 0.6$, find each of the following:

(a) $P(E \cap F)$ **(b)** $P(F \cap G)$

(c) $P(E \cap F \cap G)$ **(d)** $P(E|(F \cap G))$

(e) $P(E' \cap F \cap G')$

3. If events E and F are independent with $P(E) = \frac{2}{5}$ and $P(E \cap F) = \frac{1}{3}$, find $P(F)$.

4. If events E and F are independent with $P(E|F) = \frac{1}{3}$, find $P(E')$.

In Problems 5 and 6, events E and F satisfy the given conditions. Determine whether E and F are independent or dependent.

5. $P(E) = \frac{3}{4}$, $P(F) = \frac{8}{9}$, $P(E \cap F) = \frac{2}{3}$

6. $P(E) = 0.28$, $P(F) = 0.15$, $P(E \cap F) = 0.038$

***7. Stockbrokers** Six hundred investors were surveyed to determine whether a person who uses a full-service stockbroker has better performance in his or her investment portfolio than one who uses a discount broker. In general, discount brokers usually offer no investment advice to their clients, whereas full-service brokers usually offer help in selecting stocks, but charge larger fees. The data, based on the last 12 months, are given in Table 8.11. Determine whether the event of having a full-service broker and the event of having an increase in portfolio value are independent or dependent.

TABLE 8.11 Portfolio Value

	Increase	Decrease	Total
Full service	320	80	400
Discount	160	40	200
Total	480	120	600

8. Cinema Offenses An observation of 175 patrons in a theater resulted in the data shown in Table 8.12. The table shows three types of cinema offenses committed by male and female patrons. Crunchers include noisy eaters of popcorn and other morsels, as well as cold-drink slurpers. Determine whether the event of being a male and the event of being a cruncher are independent or dependent. (See page 5D of the July 21, 1991, issue of *USA TODAY* for the article "Pests Now Appearing at a Theater Near You.")

TABLE 8.12 Theater Patrons

	Male	Female	Total
Talkers	60	10	70
Crunchers	55	25	80
Seat kickers	15	10	25
Total	130	45	175

*9. **Dice** Two fair dice are rolled, one red and one green, and the numbers on the top faces are noted. Let event E be "number on red die is neither 2 nor 3" and event F be "sum is 8." Determine whether E and F are independent or dependent.

10. Cards A card is randomly drawn from an ordinary deck of 52 cards. Let R and F be the events "red card drawn" and "a jack, queen, or king drawn," respectively. Determine whether R and F are independent or dependent.

11. Coins If two fair coins are tossed, let E be the event "at most one head" and F be the event "exactly one head." Determine whether E and F are independent or dependent.

12. Coins If three fair coins are tossed, let E be the event "at most one head" and F be the event "at least one head and one tail." Determine whether E and F are independent or dependent.

13. Chips in Urn An urn contains seven chips numbered from 1 to 7. Two chips are randomly drawn with replacement. Let E, F, and G be the events

$$E = 3 \text{ on first draw}$$
$$F = 3 \text{ on second draw}$$
$$G = \text{sum is odd}$$

(a) Determine whether E and F are independent or dependent.
(b) Determine whether E and G are independent or dependent.
(c) Determine whether F and G are independent or dependent.
(d) Are E, F, and G independent?

14. Chips in Urn An urn contains six chips numbered from 1 to 6. A chip is randomly drawn. Let E be the event of drawing a 3 and F be the event of drawing a 5.

(a) Are E and F mutually exclusive?
(b) Are E and F independent?

In Problems 15 and 16, events E and F satisfy the given conditions. Determine whether E and F are independent or dependent.

15. $P(E|F) = 0.6$, $P(E \cap F) = 0.28$, $P(F|E) = 0.4$

16. $P(E|F) = \frac{2}{3}$, $P(E \cup F) = \frac{17}{18}$, $P(E \cap F) = \frac{5}{9}$

In Problems 17–37, you may make use of your intuition concerning independent events if nothing to that effect is specified.

*17. **Dice** Two fair dice are rolled, one red and one green. Find the probability that the red die is a 4 and the green die is a number greater than 4.

18. Die If a fair die is rolled three times, find the probability that a 2 or 3 comes up each time.

19. Fitness Classes At a certain fitness center, the probability that a member regularly attends an aerobics class is $\frac{1}{5}$. If two members are randomly selected, find the probability that both attend the class regularly. Assume independence.

20. Monopoly In the game of Monopoly, a player rolls two fair dice. One special situation that can arise is that the numbers on the top faces of the dice are the same (such as two 3's). This result is called a "double," and when it occurs, the player continues his or her turn and rolls the dice again. The pattern continues, unless the player is unfortunate enough to throw doubles three consecutive times. In that case, the player goes to jail. Find the probability that a player goes to jail in this way.

*21. **Cards** Three cards are randomly drawn, with replacement, from an ordinary deck of 52 cards. Find the probability that the cards drawn, in order, are an ace, a face card (a jack, queen, or king), and a spade.

22. Die If a fair die is rolled seven times, find each of the following.

(a) The probability of getting a 1 each time.
(b) The probability of getting a number less than 5 each time.

23. Exam Grades In a sociology course, the probability that Bill gets an A on the final exam is $\frac{3}{4}$, and for Jim and Linda, the probabilities are $\frac{1}{2}$ and $\frac{4}{5}$, respectively. Assume independence and find each of the following.

(a) The probability that all three of them get an A on the exam.
(b) The probability that none of them get an A on the exam.
(c) The probability that, of the three, only Linda gets an A.

24. Die If a fair die is rolled three times, find the probability of getting at least one 6.

*__25. Survival Rates__ The probability that person A survives 15 more years is $\frac{2}{3}$, and the probability that person B survives 15 more years is $\frac{3}{5}$. Find the probability of each of the following. Assume independence.

(a) A and B both survive 15 years.
(b) B survives 15 years, but A does not.
(c) Exactly one of A and B survives 15 years.
(d) At least one of A and B survives 15 years.
(e) Neither A nor B survives 15 years.

26. Matching In his desk, a secretary has a drawer containing a mixture of two sizes of paper (A and B) and another drawer containing a mixture of envelopes of two corresponding sizes. The percentages of each size of paper and envelope in the drawers are given in Table 8.13. If a piece of paper and an envelope are randomly drawn, find the probability that they are of the same size.

TABLE 8.13 Paper and Envelopes

	Drawers	
Size	**Paper**	**Envelopes**
A	63%	57%
B	37%	43%

*__27. Marbles in Urn__ An urn contains four red, six white, and five green marbles. If two marbles are randomly drawn with replacement, find each of the following.

(a) The probability that the first marble is white and the second is green.
(b) The probability that one marble is white and the other one is green.

28. Dice Suppose two fair dice are rolled twice. Find the probability of getting a total of 7 on one of the rolls and a total of 12 on the other one.

29. Marbles in Urn An urn contains three red, seven white, and nine green marbles. If two marbles are randomly drawn with replacement, find the probability that they have the same color.

30. Die Find the probability of rolling the same number in three throws of a fair die.

31. Tickets in Hat Twenty tickets numbered from 1 to 20 are placed in a hat. If two tickets are randomly drawn with replacement, find the probability that the sum is 35.

32. Coins and Dice Suppose two fair coins are tossed and then two fair dice are rolled. Find each of the following:

(a) The probability that two tails and two 3's occur.
(b) The probability that two heads, one 4, and one 6 occur.

*__33. Carnival Game__ In a carnival game, a well-balanced roulette-type wheel has 12 equally spaced slots that are numbered from 1 to 12. The wheel is spun, and a ball travels along the rim of the wheel. When the wheel stops, the number of the slot in which the ball finally rests is considered the result of the spin. If the wheel is spun three times, find each of the following:

(a) The probability that the first number will be 4 and the second and third numbers will be 5.
(b) The probability that there will be one even number and two odd numbers.

34. Cards Three cards are randomly drawn, with replacement, from an ordinary deck of 52 cards. Find each of the following:

(a) The probability of drawing, in order, a jack, a heart, and a red ace.
(b) The probability of drawing exactly two jacks.
(c) The probability that one jack, one heart, and one red ace are drawn.
(d) The probability of drawing exactly one jack.

*__35. Multiple-Choice Exam__ A quiz contains five multiple-choice problems. Each problem has four choices for the answer, but only one of them is correct. Suppose a student randomly guesses the answer to each problem. Find each of the following by assuming that the guesses are independent.

(a) The probability that the student gets exactly four correct answers.
(b) The probability that the student gets at least four correct answers.
(c) The probability that the student gets three or more correct answers.

36. Shooting Gallery At a shooting gallery, suppose Bill, Jim, and Linda each take one shot at a moving target. The probability that Bill hits the target is 0.5, and for Jim and Linda, the probabilities are 0.4 and 0.7, respectively. Assume independence and find each of the following.

(a) The probability that none of them hit the target.
(b) The probability that Linda is the only one of them that hits the target.
(c) The probability that exactly one of them hits the target.
(d) The probability that exactly two of them hit the target.
(e) The probability that all of them hit the target.

37. Decision Making[3] The president of Zeta Construction Company must decide which of two actions to take, namely, to rent or to buy expensive excavating equipment. The probability that the vice president makes a faulty analysis and thus recommends the wrong decision to the president is 0.04. To be thorough, the president hires two consultants, who study the problem independently and make their recommendations. After having observed them at work, the president estimates that the first consultant is likely to recommend the wrong decision with probability 0.05, the other with probability 0.1. He decides to take the action recommended by a majority of the three recommendations he receives. What is the probability that he will make the wrong decision?

[3] Samuel Goldberg, *Probability: An Introduction* (Prentice-Hall, Inc., 1960, Dover Publications, Inc., 1986), p. 113. Adapted by permission of the author.

8.7 Bayes's Formula

OBJECTIVE *To solve a Bayes's problem, which basically involves the reverse of the usual conditional probability pattern. To develop Bayes's formula.*

In this section, we shall be dealing with a two-stage experiment in which we know the outcome of the second stage and are interested in the probability that a particular outcome has occurred in the first stage.

To illustrate, suppose it is believed that of the total population (our sample space), 8% have a particular disease. Imagine also that there is a new blood test for detecting the disease and that researchers have evaluated its effectiveness. Data from extensive testing show that the blood test is not perfect: Not only is it positive for only 95% of those who have the disease, but it is also positive for 3% of those who do not. Suppose a person from the population is selected at random and given the blood test. If the result is positive, what is the probability that the person has the disease?

To analyze this problem, we consider the following events:

$$D_1 = \{\text{having the disease}\}$$

$$D_2 = \{\text{not having the disease}\}$$

$$T_1 = \{\text{testing positive}\}$$

$$T_2 = \{\text{testing negative}\}$$

We are given that

$$P(D_1) = 0.08$$

so

$$P(D_2) = 1 - 0.08 = 0.92$$

because D_1 and D_2 are complements. It is reasonable to assume that T_1 and T_2 are also complements; in that case, we have the conditional probabilities

$$P(T_1|D_1) = 0.95 \qquad P(T_2|D_1) = 1 - 0.95 = 0.05$$

$$P(T_1|D_2) = 0.03 \qquad P(T_2|D_2) = 1 - 0.03 = 0.97$$

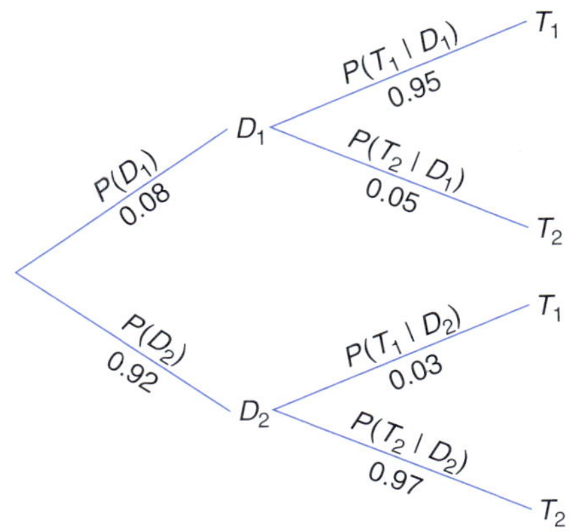

FIGURE 8.20 Two-stage probability tree.

Figure 8.20 shows a two-stage probability tree that reflects this information. The first stage takes into account either having or not having the disease, and the second stage shows possible test results.

We are interested in the probability that a person who tests positive has the disease. That is, we want to find the conditional probability that D_1 occurred in the first stage, given that T_1 occurred in the second stage:

$$P(D_1 | T_1)$$

It is important that you understand the difference between the conditional probabilities $P(D_1|T_1)$ and $P(T_1|D_1)$. The probability $P(T_1|D_1)$, which is *given* to us, is a "typical" conditional probability, in that it deals with the probability of an outcome in the second stage *after* an outcome in the first stage has occurred. However, with $P(D_1|T_1)$, we have a "reverse" situation. Here we must find the probability of an outcome in the *first* stage, given that an outcome in the second stage occurred. In a sense, we have the "cart before the horse" in that this probability does not fit the usual (and more natural) pattern of a typical conditional probability. Fortunately, we have all the tools needed to find $P(D_1|T_1)$. We proceed as follows.

From the definition of conditional probability,

$$P(D_1|T_1) = \frac{P(D_1 \cap T_1)}{P(T_1)}. \qquad \textbf{(1)}$$

Consider the numerator. Applying the general multiplication law gives

$$\begin{aligned} P(D_1 \cap T_1) &= P(D_1)P(T_1|D_1) \\ &= (0.08)(0.95) = 0.076 \end{aligned}$$

which is indicated in the path through D_1 and T_1 in Figure 8.21. The denominator, $P(T_1)$, is the sum of the probabilities for all paths of the tree ending in T_1. Thus,

$$\begin{aligned} P(T_1) &= P(D_1 \cap T_1) + P(D_2 \cap T_1) \\ &= P(D_1)P(T_1|D_1) + P(D_2)P(T_1|D_2) \\ &= (0.08)(0.95) + (0.92)(0.03) = 0.1036 \end{aligned}$$

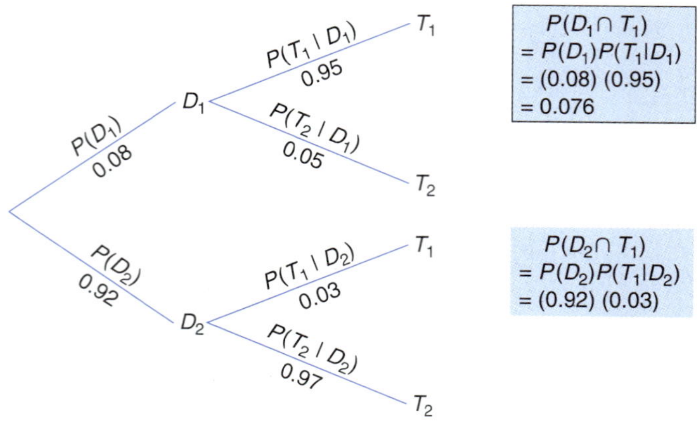

FIGURE 8.21 Probability tree to determine $P(D_1|T_1)$.

Hence,

$$\begin{aligned} P(D_1|T_1) &= \frac{P(D_1 \cap T_1)}{P(T_1)} \\ &= \frac{\text{probability of path through } D_1 \text{ and } T_1}{\text{sum of probabilities of all paths to } T_1} \\ &= \frac{0.076}{0.1036} = \frac{760}{1036} = \frac{190}{259} \approx 0.734 \end{aligned}$$

So the probability that the person has the disease, given that the test is positive, is approximately 0.734. In other words, about 73.4% of people who test positive actually have the disease. You would probably agree that this probability was relatively easy to find by using basic principles (Equation [1]) and a probability tree (Figure 8.21).

At this point, some terminology should be introduced. The *unconditional* probabilities $P(D_1)$ and $P(D_2)$ are called **prior** (or *a priori*) **probabilities,** because they are given *before* we have any knowledge about the outcome of a blood test. The conditional probability $P(D_1|T_1)$ is called a **posterior** (or *a posteriori*) **probability,** because it is found after the outcome (T_1) of the test is known.

From our answer for $P(D_1|T_1)$, we can easily find the posterior probability of not having the disease given a positive test result:

$$P(D_2|T_1) = 1 - P(D_1|T_1) = 1 - \frac{190}{259} = \frac{69}{259} \approx 0.266$$

Of course, this can also be found by using the probability tree:

$$\begin{aligned} P(D_2|T_1) &= \frac{\text{probability of path through } D_2 \text{ and } T_1}{\text{sum of probabilities of all paths to } T_1} \\ &= \frac{(0.92)(0.03)}{0.1036} = \frac{0.0276}{0.1036} = \frac{276}{1036} = \frac{69}{259} \approx 0.266 \end{aligned}$$

It is not really necessary to use a probability tree to find $P(D_1|T_1)$. Instead, a formula can be developed. We know that

$$P(D_1|T_1) = \frac{P(D_1 \cap T_1)}{P(T_1)} = \frac{P(D_1)P(T_1|D_1)}{P(T_1)} \tag{2}$$

Although we used a probability tree to express $P(T_1)$ conveniently as a sum of probabilities, the sum can be found another way. Take note that events D_1 and D_2 have two properties: They are mutually exclusive and their union is the sample space S. Such events are collectively called a **partition** of S. Using this partition, we can break up event T_1 into mutually exclusive "pieces":

$$T_1 = T_1 \cap S = T_1 \cap (D_1 \cup D_2)$$

Then, by the distributive and commutative laws,

$$T_1 = (D_1 \cap T_1) \cup (D_2 \cap T_1) \tag{3}$$

Since D_1 and D_2 are mutually exclusive, so are events $D_1 \cap T_1$ and $D_2 \cap T_1$.[4] Thus, T_1 has been expressed as a union of mutually exclusive events. In this form, we can find $P(T_1)$ by adding probabilities. Applying the addition law for mutually exclusive events to Equation (3) gives

$$\begin{aligned} P(T_1) &= P(D_1 \cap T_1) + P(D_2 \cap T_1) \\ &= P(D_1)P(T_1|D_1) + P(D_2)P(T_1|D_2) \end{aligned}$$

Substituting into Equation (2), we obtain

$$P(D_1|T_1) = \frac{P(D_1)P(T_1|D_1)}{P(D_1)P(T_1|D_1) + P(D_2)P(T_1|D_2)} \tag{4}$$

which is a formula for computing $P(D_1|T_1)$.

Equation (4) is a special case (namely, for a partition of S into two events) of the following general formula, called **Bayes's formula,**[5] which has had wide application in decision making:

Bayes's Formula

Suppose $F_1, F_2 \ldots, F_n$ are n events that partition a sample space S. That is, the F_i's are mutually exclusive and their union is S. Furthermore, suppose that E is any event in S, where $P(E) > 0$. Then the conditional probability of F_i given that event E has occurred is expressed by

$$P(F_i|E) = \frac{P(F_i)P(E|F_i)}{P(F_1)P(E|F_1) + P(F_2)P(E|F_2) + \cdots + P(F_n)P(E|F_n)}$$

for each value of i, where $i = 1, 2, \ldots, n$.

Rather than memorize the formula, a probability tree can be used to obtain $P(F_i|E)$. Using the tree in Figure 8.22, we have

$$P(F_i|E) = \frac{\text{probability for path through } F_i \text{ and } E}{\text{sum of all probabilities for paths to } E}$$

EXAMPLE 1 Quality Control

A camcorder manufacturer uses one microchip in assembling each camcorder it produces. The microchips are purchased from suppliers A, B, *and* C *and are randomly picked for assembling each camcorder. Twenty percent of the microchips come from* A, 35% *come from* B, *and the remainder come from* C. *Based on past experience, the manufacturer believes that the probability that a microchip from* A *is defective is* 0.03, *and the corresponding probabilities for* B *and* C *are* 0.02 *and* 0.01, *respectively. A camcorder is selected at random from a day's production, and its microchip is found to be defective. Find the probability that it was supplied (a) from* A, *(b) from* B, *and (c) from* C. *(d) From what supplier was the microchip most likely supplied?*

---

[4]See Example 8 of Section 8.3.

[5]After Thomas Bayes (1702–1761), the 18th-century English minister who discovered the formula.

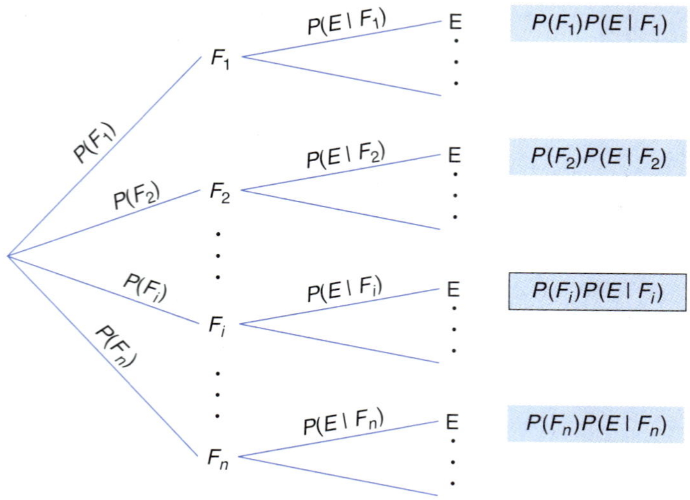

FIGURE 8.22 Probability tree for $P(F_i|E)$.

Solution: We define the following events:

$$S_1 = \{\text{supplier A}\}$$
$$S_2 = \{\text{supplier B}\}$$
$$S_3 = \{\text{supplier C}\}$$
$$D = \{\text{defective microchip}\}$$

We have

$$P(S_1) = 0.2 \quad P(S_2) = 0.35 \quad P(S_3) = 0.45$$

and the conditional probabilities

$$P(D|S_1) = 0.03 \quad P(D|S_2) = 0.02 \quad P(D|S_3) = 0.01$$

which are reflected in the probability tree in Figure 8.23. Note that the figure shows only the portion of the complete probability tree that relates to event D. This is all that actually needs to be drawn, and this abbreviated form is often called a *Bayes's probability tree.*

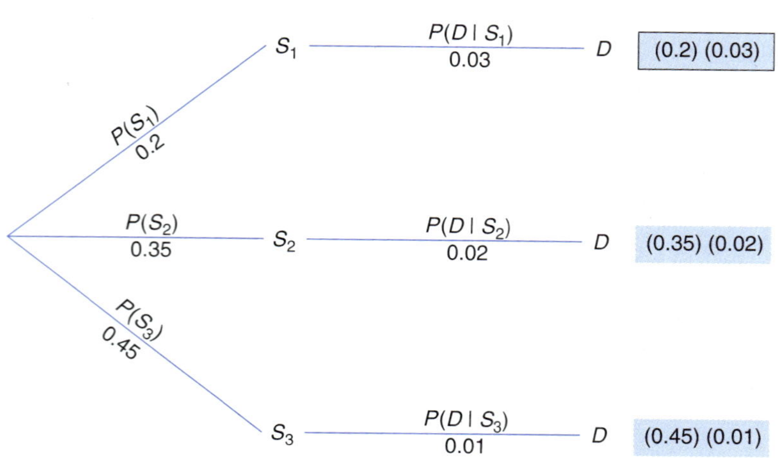

FIGURE 8.23 Bayes's probability tree for Example 1.

For part (a), we want to find the probability of S_1 given that D has occurred. That is,

$$P(S_1|D) = \frac{\text{probability of path through } S_1 \text{ and } D}{\text{sum of probabilities of all paths to } D}$$

$$= \frac{(0.2)(0.03)}{(0.2)(0.03) + (0.35)(0.02) + (0.45)(0.01)}$$

$$= \frac{0.006}{0.006 + 0.007 + 0.0045}$$

$$= \frac{0.006}{0.0175} = \frac{60}{175} = \frac{12}{35} \approx 0.343$$

This means that approximately 34.3% of the defective microchips come from supplier A.

For part (b), we have

$$P(S_2|D) = \frac{\text{probability of path through } S_2 \text{ and } D}{\text{sum of probabilities of all paths to } D}$$

$$= \frac{(0.35)(0.02)}{0.0175} = \frac{0.007}{0.0175} = \frac{70}{175} = \frac{14}{35}$$

For part (c),

$$P(S_3|D) = \frac{\text{probability of path through } S_3 \text{ and } D}{\text{sum of probabilities of all paths to } D}$$

$$= \frac{(0.45)(0.01)}{0.0175} = \frac{0.0045}{0.0175} = \frac{45}{175} = \frac{9}{35}$$

For part (d), the greatest of $P(S_1|D)$, $P(S_2|D)$, and $P(S_3|D)$ is $P(S_2|D)$. Thus, the defective microchip was most likely supplied by B.

▶ **Now work Problem 9.**

EXAMPLE 2 **Marbles in Urn**

Two identical urns, Urn I and Urn II, are on a table. Urn I contains one red and one black marble; Urn II contains two red marbles. (See Figure 8.24.) An urn is selected at random, and then a marble is randomly drawn from it. The marble is red. What is the probability that the other marble in the selected urn is red?

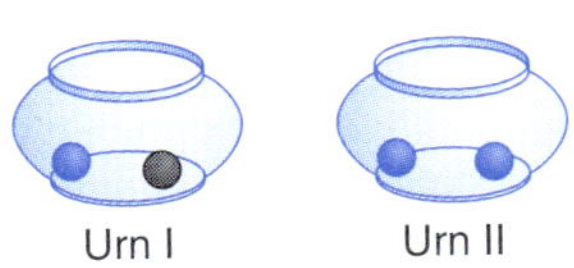

FIGURE 8.24 Diagram for Example 2.

Solution: Because the other marble could be red or black, you might hastily conclude that the answer is $\frac{1}{2}$. Sorry, your intuition fails! The question can be restated as follows: Find the probability that the marble came from Urn II, given that the marble is red. We define the events

$$U_1 = \{\text{Urn I selected}\}$$

$$U_2 = \{\text{Urn II selected}\}$$

$$R = \{\text{red marble selected}\}$$

We want to find $P(U_2|R)$. Since an urn is selected at random, $P(U_1) = P(U_2) = \frac{1}{2}$. From Figure 8.24, we conclude that

$$P(R|U_1) = \frac{1}{2} \quad \text{and} \quad P(R|U_2) = 1$$

We will show two methods of solving this problem, the first with a probability tree and the second with Bayes's formula.

Method 1: Probability Tree Figure 8.25 shows a Bayes's probability tree for our problem. Since all paths end at R,

$$P(U_2|R) = \frac{\text{probability for path through } U_2 \text{ and } R}{\text{sum of probabilities of all paths}}$$

$$= \frac{\left(\frac{1}{2}\right)(1)}{\left(\frac{1}{2}\right)\left(\frac{1}{2}\right) + \left(\frac{1}{2}\right)(1)} = \frac{\frac{1}{2}}{\frac{3}{4}} = \frac{2}{3}$$

Note that the unconditional probability of choosing Urn II, namely, $P(U_2) = \frac{1}{2}$, increases to $\frac{2}{3}$, given that a red marble was drawn. An increase is reasonable: Since there are only red marbles in Urn II, choosing a red marble should make it more likely that it came from Urn II.

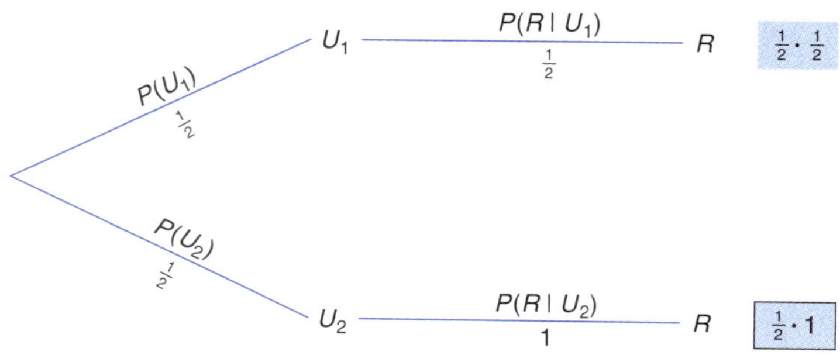

FIGURE 8.25 Bayes's probability tree for Example 2.

Method 2: Bayes's Formula Because U_1 and U_2 partition the sample space, by Bayes's formula we have

$$P(U_2|R) = \frac{P(U_2)P(R|U_2)}{P(U_1)P(R|U_1) + P(U_2)P(R|U_2)}$$

$$= \frac{\left(\frac{1}{2}\right)(1)}{\left(\frac{1}{2}\right)\left(\frac{1}{2}\right) + \left(\frac{1}{2}\right)(1)} = \frac{\frac{1}{2}}{\frac{3}{4}} = \frac{2}{3}$$

▶ **Now work Problem 7.**

Exercise 8.7

1. Suppose events E and F partition a sample space S, where E and F have probabilities

$$P(E) = \frac{2}{5} \quad P(F) = \frac{3}{5}$$

If D is an event such that

$$P(D|E) = \frac{1}{10} \quad P(D|F) = \frac{1}{5}$$

find the probabilities $P(E|D)$ and $P(F|D')$.

2. A sample space is partitioned by events E_1, E_2, and E_3, whose probabilities are $\frac{1}{5}$, $\frac{3}{10}$, and $\frac{1}{2}$, respectively. Suppose S is an event such that the following conditional probabilities hold:

$$P(S|E_1) = \frac{2}{5} \quad P(S|E_2) = \frac{7}{10} \quad P(S|E_3) = \frac{1}{2}$$

Find the probabilities $P(E_1|S)$ and $P(E_3|S')$.

3. Voting In a certain precinct, 42% of the eligible voters are registered Democrats, 33% are Republicans, and the remainder are Independents. During the last primary election, 25% of the Democrats, 27% of the Republicans, and 15% of the Independents voted. Find the probability that a person who voted is a Democrat.

4. Imported vs. Domestic Tires Out of 3000 tires in the warehouse of a tire distributor, 2000 tires are domestic and 1000 are imported. Among the domestic tires, 40% are whitewalls; of the imported tires, 10% are whitewalls. If a tire is selected at random and it is a whitewall, what is the probability that it is imported?

5. Disease Testing A new test was developed for detecting Gamma's disease, which is believed to affect 3% of the population. Results of extensive testing indicate that 86% of persons who have this disease will have a positive reaction to the test, whereas 7% of those who do not have the disease will also have a positive reaction.

(a) What is the probability that a randomly selected person who has a positive reaction will actually have Gamma's disease?

(b) What is the probability that a randomly selected person who has a negative reaction will actually have Gamma's disease?

6. Earnings and Dividends Of the companies in a particular sector of the economy, it is believed that one-third will have an increase in quarterly earnings. Of those which do, the percentage that declare a dividend is 60%. Of those that do not have an increase, the percentage that declare a dividend is 10%. What percentage of companies that declare a dividend will have an increase in quarterly earnings?

***7. Marbles in Urn** An urn contains four red and two green marbles, and a second urn contains two red and three green marbles. An urn is selected at random and a marble is randomly drawn from it. The marble is red. What is the probability that it came from the first urn?

8. Balls in Urn Urn I contains two red and three white balls. Urn II contains four red and three white balls. Urn III contains two red, two white, and two green balls. An urn is chosen at random, and then a ball is chosen at random from it. The ball is white. Find the probability that it came from Urn I.

***9. Quality Control** A manufacturing process requires the use of a robotic welder on each of two assembly lines, A and B, which produce 300 and 500 units of product per day, respectively. Based on past experience, it is believed that the welder on A produces 2% defective units, whereas the welder on B produces 5% defective units. At the end of a day, a unit was selected at random from the total production and was found to be defective. What is the probability that it came from line A?

10. Quality Control An automobile manufacturer has four plants: A, B, C, and D. The percentages of total daily output that are produced by the four plants are 35%, 20%, 30%, and 15%, respectively. The percentages of defective units produced by the plants are estimated to be 2%, 5%, 3%, and 4%, respectively. Suppose that a car on a dealer's lot is randomly selected and found to be defective. What is the probability that it came from assembly line (a) A? (b) B? (c) C? (d) D?

11. Wake-Up Call Barbara Smith, a sales representative, is staying overnight at a hotel and has a breakfast meeting with an important client the following morning. She asked room service to give her a 7 A.M. wake-up call that morning in order that she be prompt for the meeting. The probability that room service makes the call is 0.9. If the call is made, the probability that she will be on time is 0.9, but if the call is not made, the probability that she will be on time is 0.8. If she is on time for the meeting, what is the probability that the call was made?

12. Candy Snatcher On a high shelf are two identical opaque candy jars containing 50 raisin clusters each. The clusters in one of the jars are made with dark chocolate. In the other jar, 20 are made with dark chocolate and 30 are made with milk chocolate. (They are mixed well, however.) Bob Jones, who has a sudden craving for chocolate, reaches up and randomly takes out a raisin cluster from one of the jars. If it is made with dark chocolate, what is the probability that it was drawn from the jar containing only dark chocolate?

13. Physical Fitness Activity The week of National Employee Health and Fitness Day, the employees of a large company were asked to exercise a minimum of three times that week for at least 20 minutes per session. The purpose was to generate "exercise miles." All participants who completed this requirement received a certificate acknowledging their contribution. The activities reported were walking, bicycling, and running. Of all who participated, $\frac{1}{2}$ reported walking, $\frac{1}{4}$ reported bicycling, and $\frac{1}{4}$ reported running. Suppose that the probability that a participant who walks will complete the requirement is $\frac{9}{10}$, and for bicycling and running it is $\frac{4}{5}$ and $\frac{2}{3}$, respectively. What percentage of persons who completed the requirement do you expect reported walking? (Assume that each participant got his or her exercise from one activity only.)

14. Battery Reliability When the weather is extremely frigid, a motorist must charge his car battery during the night in order to improve the likelihood that the car will start early the following morning. If he does not charge it, the probability that the car will not start is $\frac{4}{5}$. If he does charge it, the probability that the car will not start is $\frac{1}{8}$. Past experience shows that the probability that he remembers to charge the battery is $\frac{9}{10}$. One morning, during a cold spell, he cannot start his car. What is the probability that he forgot to charge the battery?

15. Automobile Satisfaction Survey In a customer satisfaction survey, $\frac{3}{5}$ of those surveyed had a Japanese-made car, $\frac{1}{10}$ a European-made car, and $\frac{3}{10}$ an American-made car. Of the first group, 85% said they would buy the same make of car again, and for the other two groups the corresponding percentages are 50% and 40%. What is the probability that a person who said she would buy the same make again had a Japanese-made car?

16. Mineral Test Borings A geologist believes that the probability that the rare earth mineral junko occurs in a particular region of the country is 0.01. If junko is present in that region, the geologist's test borings will have a positive result 78% of the time. However, if junko is not present, a negative result will occur 86% of the time.

(a) If a test is positive on a site in the region, find the probability that junko is there.

(b) If a test is negative on such a site, find the probability that junko is there.

17. Physics Exam After a physics exam was given, it turned out that only 75% of the class answered every question. Of those who did, 80% passed, but of those who did not, only 50% passed. If a student passed the exam, what is the probability that the student answered every question? (P.S.: The instructor eventually reached the conclusion that the test was too long and curved the exam grades, to be fair and merciful.)

18. Giving Up Smoking In a 1990 survey of smokers, 75% predicted that they would still be smoking five years later. Five years later, 70% of those who predicted that they would be smoking did not smoke, and of those who predicted that they would not be smoking, 90% did not smoke. What percentage of those who do not smoke predicted that they would be smoking?

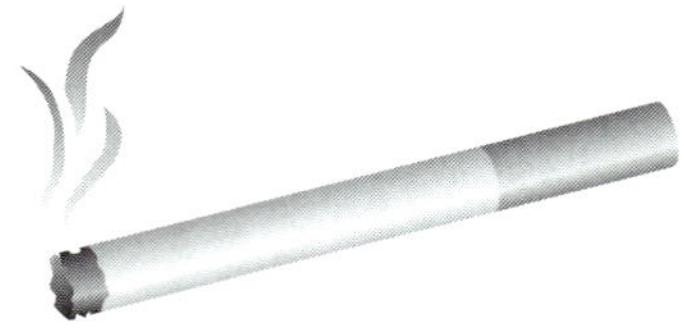

19. Alien Communication B. G. Cosmos, a scientist, believes that the probability is $\frac{2}{5}$ that aliens from an advanced civilization on Planet X are trying to communicate with us by sending high-frequency signals to Earth. By using sophisticated equipment, Cosmos hopes to pick up these signals. The manufacturer of the equipment, Trekee, Inc., claims that if aliens are indeed sending signals, the probability that the equipment will detect them is $\frac{3}{5}$. However, if aliens are not sending signals, the probability that the equipment will seem to detect such signals is $\frac{1}{10}$. If the equipment detects signals, what is the probability that aliens are actually sending them?

20. Calculus Grades In an honors Calculus I class, 60% of students had an A average at midterm. Of these, 70% ended up with a course grade of A, and of those who did not have an A average at midterm, 60% ended up with a course grade of A. If one of the students is selected at random and is found to have received an A for the course, what is the probability that the student did not have an A average at midterm?

21. Movie Critique A well-known pair of highly influential movie critics have a popular TV show on which they review new movie releases. Over the past 10 years, they gave a "Two Thumbs Up" to 70% of movies that turned out to be box-office successes; they gave a "Two Thumbs Down" to 80% of movies that proved to be unsuccessful. A new movie, *Math Wizard*, whose release is imminent, is considered favorably by "experts" in the industry who have previewed it; in fact, they give it a prior probability of success of $\frac{8}{10}$. Find the probability that it will be a success, given that the pair of TV critics give it a "Two Thumbs Up" after seeing it. Assume that all films are given either "Two Thumbs Up" or "Two Thumbs Down" and that box-office success coincides precisely with favorable consideration by the "experts."

22. Balls in Urn Urn 1 contains five green and four red balls, and Urn 2 contains three green, one white, and three red balls. A ball is randomly drawn from Urn 1 and placed into Urn 2. A ball is then randomly drawn from Urn 2. If the ball is green, find the probability that a green ball was drawn from Urn 1.

23. Risky Loan In the loan department of Third National Bank, past experience indicates that 15% of loan requests are considered by bank examiners to fall into the "substandard" class and should not be approved. However, the bank's loan reviewer, I. M. Risky, is lax at times and concludes that a request is not in the substandard class when it is, and vice versa. Suppose that 20% of requests that are actually substandard are not considered substandard by Risky and that 10% of requests that are not substandard are considered by Risky to be substandard and, hence, not approved.

(a) Find the probability that a request is considered to be substandard by Risky.

(b) Find the probability that a request is substandard, given that it is considered to be substandard by Risky.

(c) Find the probability that Risky makes an error in considering a request. (An error occurs when the request is not substandard, but is considered substandard, or when the request is substandard, but considered to be not substandard.)

24. Coins in Chests Each of three identical chests has two drawers. The first chest contains a gold coin in each drawer. The second chest contains a silver coin in each drawer, and the third contains a silver coin in one drawer and a gold coin in the other. A chest is chosen at random and a drawer is opened. There is a gold coin in it. What is the probability that the coin in the other drawer of that chest is silver?

25. Product Identification after Flood[6] After a severe flood, a distribution warehouse is stocked with waterproofed boxes of fireworks from which the identification labels have been washed off. There are three types of fireworks: low quality, medium quality, and high quality, each packed in units of 100 in identical boxes. None of the individual fireworks have markings on them, but it is believed that in the entire warehouse, the proportions of boxes with low-, medium-, and high-quality fireworks are 0.25, 0.25, and 0.5, respectively. Because detonating a firework destroys it, extensive testing of the fireworks is impractical. Instead, the distributor decides that two fireworks from each box will be tested. The quality of the fireworks will then be decided on the basis of how many of the two are defective. The manufacturer, on the basis of past experience, estimates the conditional probabilities given in Table 8.14. Suppose two fireworks are selected from a box, and tested, and both are found to detonate satisfactorily. Let the events L, M, and H be that the box contains low-, medium-, and high-quality fireworks, respectively. Furthermore, let E be the observed event that neither of the fireworks was defective.

(a) Find $P(L|E)$, the probability that the box contains low-quality fireworks, given E.

(b) Find the probability that the box contains medium-quality fireworks, given E.

(c) Find the probability that the box contains high-quality fireworks, given E.

TABLE 8.14 Conditional Probabilities of Finding x Defectives, Given That Two Fireworks Were Tested from a Box of Known Quality

Number of Defectives x	Quality of Box: Low	Medium	High
0	0.49	0.64	0.81
1	0.44	0.32	0.18
2	0.07	0.04	0.01

(d) What is the most likely quality of the fireworks in the box, given E?

26. Product Identification after Flood

(a) Repeat Problem 25 if E is the observed event that exactly one of the tested fireworks is defective.

(b) Repeat Problem 25 if E is the observed event that both of the tested fireworks are defective.

27. Weather Forecasting[7] J. B. Smith, who has lived in the same city many years, assigns a prior probability of 0.4 that today's weather will be inclement. (He thinks that today will be fair with probability 0.6.) Smith listens to an early morning weather forecast to get information on the day's weather. The forecaster makes one of three predictions: fair weather, inclement weather, or uncertain weather. Smith has made estimates of conditional probabilities of the different predictions, given the day's weather, as shown in Table 8.15. For example, Smith believes that, of the fair days, 70% are correctly forecast, 20% are forecast as inclement, and 10% are forecast as uncertain. Suppose that Smith hears the forecaster predict fair weather. What is the posterior probability of fair weather?

TABLE 8.15 Weather and Forecast

Day's Weather	Forecast: Fair	Inclement	Uncertain
Fair	0.7	0.2	0.1
Inclement	0.3	0.6	0.1

[6] Samuel Goldberg, *Probability: An Introduction* (Prentice-Hall, Inc., 1960, Dover Publications, Inc., 1986), pp. 97–98. Adapted by permission of the author.

[7] Samuel Goldberg, *Probability: An Introduction* (Prentice-Hall, Inc., 1960, Dover Publications, Inc., 1986), pp. 99–100. Adapted by permission of the author.

8.8 Review

Important Terms and Symbols

Section 8.1 tree diagram Basic Counting Principle permutation, ${}_nP_r$

Section 8.2 combination, ${}_nC_r$ permutation with repeated objects cells

Section 8.3 sample space sample point finite sample space event certain event impossible event simple event Venn diagram complement, E' union, $\cup$ intersection, $\cap$ mutually exclusive events

Section 8.4 equally likely outcomes trial relative frequency equiprobable space probability of event, $P(E)$ addition law for mutually exclusive events empirical probability odds

Section 8.5 conditional probability, $P(E|F)$ reduced sample space general multiplication law (law of compound probability) trial (stage) compound experiment probability tree

Section 8.6 independent events dependent events special multiplication law

Section 8.7 partition prior probability posterior probability Bayes's formula Bayes's probability tree

Summary

It is important to know the number of ways a procedure can occur. Suppose a procedure involves a sequence of k stages. Let n_1 be the number of ways the first stage can occur, and let n_i be the number of ways the ith stage can occur after the first $i-1$ stages have occurred, for $i = 2, 3, \ldots, k$. Then the number of ways the procedure can occur is

$$n_1 \cdot n_2 \cdots n_k$$

This result is called the Basic Counting Principle.

An ordered arrangement of r objects, without repetition, selected from n distinct objects is called a permutation of the n objects taken r at a time. The number of such permutations is denoted ${}_nP_r$ and is given by

$$ {}_nP_r = \underbrace{n(n-1)(n-2)\cdots(n-r+1)}_{r\text{ factors}} = \frac{n!}{(n-r)!}$$

If the arrangement is made without regard to order, then it is simply an r-element subset of an n-element set and is called a combination of n objects taken r at a time. The number of such combinations is denoted ${}_nC_r$ and is given by

$${}_nC_r = \frac{n!}{r!(n-r)!}$$

When some of the objects are repeated, the number of distinguishable permutations of n objects, such that n_1 are of one type, n_2 are of a second type, ..., and n_k are of the kth type, is

$$\frac{n!}{n_1!n_2!\cdots n_k!} \tag{1}$$

where $n_1 + n_2 + \cdots + n_k = n$.

The expression in Equation (1) can also be used to determine the number of assignments of objects to cells. If n distinct objects are placed into k ordered cells, with n_i objects in cell i ($i = 1, 2, \ldots, k$), then the number of such assignments is

$$\frac{n!}{n_1!n_2!\cdots n_k!}$$

where $n_1 + n_2 + \cdots + n_k = n$.

A sample space for an experiment is a set S of all possible outcomes of the experiment. These outcomes are called sample points. A subset E of S is called an event. Two special events are the sample space itself, which is a certain event, and the empty set, which is an impossible event. An event consisting of a single sample point is called a simple event. Two events are said to be mutually exclusive when they have no sample point in common.

A sample space whose outcomes are equally likely is called an equiprobable space. If E is an event for a finite equiprobable space S, then the probability that E occurs is given by

$$P(E) = \frac{\#(E)}{\#(S)}$$

If F is also an event in S, we have

$$P(E \cup F) = P(E) + P(F) - P(E \cap F)$$

$$P(E \cup F) = P(E) + P(F) \quad \text{if } E \text{ and } F \text{ are mutually exclusive}$$

$$P(E') = 1 - P(E)$$

$$P(S) = 1$$

$$P(\emptyset) = 0$$

For an event E, the ratio

$$\frac{P(E)}{P(E')} \left(\text{or, equivalently, } \frac{P(E)}{1 - P(E)} \right)$$

gives the odds that E occurs. Conversely, if the odds that E occurs are $a : b$, then

$$P(E) = \frac{a}{a+b}$$

The probability that an event E occurs, given that event F has occurred, is called a conditional probability. It is denoted by $P(E|F)$ and can be computed either by considering a reduced equiprobable sample space and using the formula

$$P(E|F) = \frac{n(E \cap F)}{n(F)}$$

or from the formula

$$P(E|F) = \frac{P(E \cap F)}{P(F)}$$

which involves probabilities with respect to the original sample space.

To find the probability that two events both occur, we may use the general multiplication law:

$$P(E \cap F) = P(E)P(F|E) = P(F)P(E|F)$$

Here we multiply the probability that one of the events occurs by the conditional probability that the other one occurs, given that the first has occurred. For more than two events, the corresponding law is

$$P(E_1 \cap E_2 \cap \cdots \cap E_n) = P(E_1)P(E_2|E_1)P(E_3|E_1 \cap E_2) \cdots P(E_n|E_1 \cap E_2 \cap \cdots \cap E_{n-1})$$

The general multiplication law is also called the law of compound probability, because it is useful when applied to a compound experiment—one that can be expressed as a sequence of two or more other experiments, called trials or stages.

When we analyze a compound experiment, a probability tree is extremely useful in keeping track of the possible outcomes for each trial of the experiment. A path is a complete sequence of branches from the start to a tip of the tree. Each path represents an outcome of the compound experiment, and the probability of that path is the product of the probabilities for the branches of the path.

Events E and F are independent when the occurrence of one of them does not affect the probability of the other; that is,

$$P(E|F) = P(E) \quad \text{or} \quad P(F|E) = P(F)$$

Events that are not independent are dependent.

If E and F are independent, the general multiplication law simplifies into the special multiplication law:

$$P(E \cap F) = P(E)P(F)$$

Here the probability that E and F both occur is the probability of E times the probability of F. The preceding equation forms the basis of an alternative definition of independence: Events E and F are independent if and only if

$$P(E \cap F) = P(E)P(F)$$

Three or more events are independent if and only if for each set of two or more of the events, the probability of the intersection of the events in that set is equal to the product of the probabilities of those events.

A partition divides a sample space into mutually exclusive events. If E is an event and $F_1, F_2, \ldots, F_n$ is a partition, then, to find the conditional probability of event F_i, given E, when prior and conditional probabilities are known, we may use Bayes's formula:

$$P(F_i|E) = \frac{P(F_i)P(E|F_i)}{P(F_1)P(E|F_1) + P(F_2)P(E|F_2) + \cdots + P(F_n)P(E|F_n)}$$

A Bayes's-type problem may also be solved with the aid of a Bayes's probability tree.

Review Problems

Problem numbers shown in color indicate problems suggested for use as a practice chapter test.

In Problems 1–4, determine the values.

1. ${}_8P_3$ **2.** ${}_{20}P_1$ **3.** ${}_9C_7$ **4.** ${}_{12}C_5$

5. License Plate A five-symbol license plate consists of two letters followed by three numbers, the first of which is not zero. How many different license plates are possible?

6. Dinner In a restaurant, a complete dinner consists of one appetizer, one entrée and one dessert. The choices for the appetizer are soup and salad; for the entrée, chicken, steak, lobster, and veal; and for the dessert, ice cream, pie, and pudding. How many complete dinners are possible?

7. Garage-Door Opener The transmitter for an electric garage-door opener transmits a coded signal to a receiver. The code is determined by five switches, each of which is either in an "on" or "off" position. Determine the number of different codes that may be transmitted.

8. Baseball A baseball manager must determine a batting order for his nine-member team. How many batting orders are possible?

9. Softball A softball league has seven teams. In terms of first, second, and third place, in how many ways can the season end? Assume that there are no ties.

10. Trophies In a trophy case, nine different trophies are to be placed—two on the top shelf, three on the middle, and four on the bottom. Considering the order of arrangement on each shelf, in how many ways can the trophies be placed in the case?

11. Groups Eleven stranded wait-listed passengers surge to the counter for boarding passes. But there are only six boarding passes available. How many different groups of passengers can board?

12. Cards From a 52-card deck of playing cards, a three-card hand is dealt. In how many ways can two of the cards be of the same face value and the other of a different face value?

13. Lightbulbs A carton contains 24 lightbulbs, one of which is defective. (a) In how many ways can three bulbs be selected? (b) In how many ways can three bulbs be selected if one is defective?

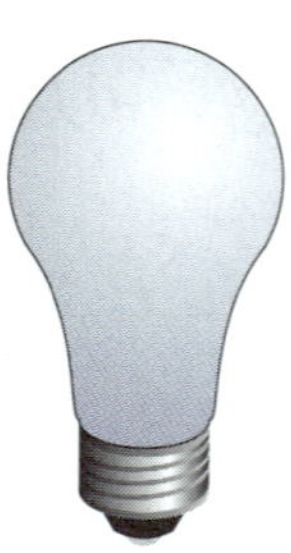

14. Multiple-Choice Exam Each question of a 10-question multiple-choice examination has four choices, only one of which is correct, and is worth 10 points. By guessing, in how many ways is it possible to receive a score of 90 or better?

15. Letter Arrangement How many distinguishable permutations of the letters in MISSISSIPPI are possible?

16. Flag Signals Colored flags arranged vertically on a flagpole indicate a signal (or message). How many different signals are possible if two red, three green, and four white flags are all used?

17. Personnel Agency A nursing personnel agency provides nurses on a temporary basis to hospitals that are short of staff. The manager has a pool of eight nurses and must send three to hospital A and two to hospital B. In how many ways can the manager make assignments?

18. Tour Operator A tour operator has three vans, and each can accommodate seven tourists. Suppose 14 people arrive for a city sightseeing tour and the operator will use only two vans. In how many ways can the operator assign the people to the vans?

19. Suppose $S = \{1, 2, 3, 4, 5, 6, 7, 8\}$ is the sample space and $E_1 = \{1, 2, 3, 4, 5, 6\}$ and $E_2 = \{4, 5, 6, 7\}$ are events for an experiment. Find (a) $E_1 \cup E_2$, (b) $E_1 \cap E_2$, (c) $E_1' \cup E_2$, (d) $E_1 \cap E_1'$, and (e) $(E_1 \cap E_2')'$. (f) Are E_1 and E_2 mutually exclusive?

20. Die and Coin A die is rolled and then a coin is tossed. (a) Determine a sample space for this experiment. Determine the events that (b) a 2 shows and (c) a head and an even number show.

21. Marbles in Urn Three urns, labeled 1, 2, and 3, each contain two marbles, one red and the other green. A marble is selected at random from each urn. (a) Determine a sample space for this experiment. Determine the events that (b) exactly two marbles are red and (c) the marbles are the same color.

22. Suppose that E_1 and E_2 are events for an experiment with a finite number of sample points. If $P(E_1) = 0.6$, $P(E_1 \cup E_2) = 0.7$, and $P(E_1 \cap E_2) = 0.2$, find $P(E_2)$.

23. Quality Control A manufacturer of computer chips packages 10 chips to a box. For quality control, two chips are selected at random from each box and tested. If any one of the chips is defective, the entire box of chips is rejected for sale. For a box that contains exactly one defective chip, what is the probability that the box is rejected?

24. Drugs Each of 100 white rats was injected with one of four drugs, A, B, C, or D. Drug A was given to 38%, B to 23%, and C to 17%. If a rat is chosen at random, determine the probability that it was injected with either C or D. If the experiment is repeated on a larger group of 200 rats but with the drugs given in the same proportion, what is the effect on the previous probability?

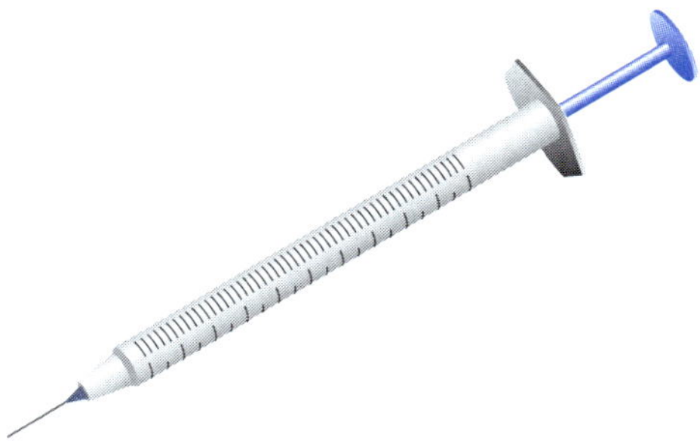

25. Multiple-Choice Exam Each question on a five-question multiple-choice examination has four choices, only one of which is correct. If a student answers each question in a random fashion, what is the probability that the student answers exactly two questions incorrectly?

26. Cola Preference To determine the national preference of cola drinkers, an advertising agency conducted a survey of 200 of them. Two cola brands, A and B, were involved. The results of the survey are indicated in Table 8.16. If a cola drinker is selected at random, determine the (empirical) probability that the person

(a) Likes both A and B.

(b) Likes A, but not B.

TABLE 8.16 Cola Preference

Like A only	70
Like B only	80
Like both A and B	35
Like neither A nor B	15
Total	200

27. Marbles in Urn An urn contains four red and six green marbles.

(a) If two marbles are randomly selected in succession with replacement, determine the probability that both are red.

(b) If the selection is made without replacement, determine the probability that both are red.

28. Dice A pair of fair dice is rolled. Determine the probability that the sum of the numbers is (a) 2 or 7, (b) a multiple of 3, and (c) no less than 7.

29. Cards Two cards from a standard deck of 52 playing cards are randomly drawn in succession with replacement. Determine the probability that (a) both cards are red and (b) one card is red and the other is a club.

30. Cards Two cards from a standard deck of 52 playing cards are randomly drawn in succession without replacement. Determine the probability that (a) both are hearts and (b) one is an ace and the other is a red king.

In Problems 31 and 32, for the given value of $P(E)$, find the odds that E will occur.

31. $P(E) = \frac{3}{8}$

32. $P(E) = 0.92$

In Problems 33 and 34, the odds that E will occur are given. Find $P(E)$.

33. $6:1$

34. $3:4$

35. Cards If a card is randomly drawn from a fair deck of 52 cards, find the probability that it is not a face card (a jack, queen, or king), given that it is a heart.

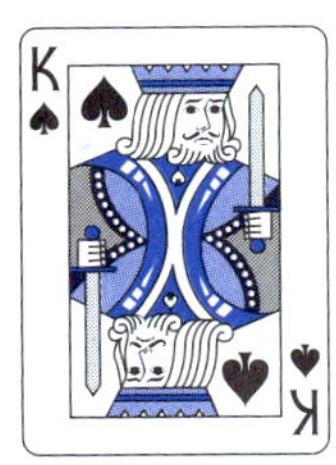

36. Dice If two fair dice are rolled, find the probability that the sum is greater than 9, given that a 4 shows on at least one of the dice.

37. Novel and TV Movie The probability that a particular novel will be successful is 0.6, and if it is successful, the probability that the rights will be purchased for a made-for-TV movie is 0.7. Find the probability that the novel will be successful and made into a TV movie.

38. Cards Three cards are drawn from a standard deck of cards. Find the probability that the cards are, in order, a queen, a heart, and the ace of clubs if the cards are drawn with replacement.

39. Dice If two dice are thrown, find each of the following.

(a) The probability of getting a total of 7, given that a 4 occurred on at least one die.

(b) The probability of getting a total of 7 and that a 4 occurred on at least one die.

40. Die A fair die is tossed two times in succession. Find the probability that the first toss is less than 4, given that the total is greater than 8.

41. Die If a fair die is tossed two times in succession, find the probability that the first number is greater than or equal to the second number, given that the second number is greater than 4.

42. Cards Three cards are drawn without replacement from a standard deck of cards. Find the probability that the third card is a heart.

43. Seasoning Survey A survey of 600 adults was made to determine whether or not they liked the taste of a new seasoning. The results are summarized in Table 8.17.

TABLE 8.17 Seasoning Survey

	Like	Dislike	Total
Male	80	40	120
Female	320	160	480
Total	400	200	600

(a) If a person in the survey is selected at random, find the probability that the person dislikes the seasoning (L'), given that the person is a female (F).

(b) Determine whether the events $L = \{\text{liking the seasoning}\}$ and $M = \{\text{being a male}\}$ are independent or dependent.

44. Chips An urn contains six chips numbered from 1 to 6. Two chips are randomly drawn with replacement. Let E be the event of getting a 4 on the first draw and F be the event of getting a 4 on the second draw.

(a) Are E and F mutually exclusive?

(b) Are E and F independent?

45. College and Family Income A survey of 175 students resulted in the data shown in Table 8.18. The table shows the type of college the student attends and the income level of the student's family. If a student is selected at random, determine whether the event of attending a public college and the event of coming from a middle-class family are independent or dependent.

TABLE 8.18 Student Survey

	College		
Income	**Private**	**Public**	**Total**
High	15	10	25
Middle	25	55	80
Low	10	60	70
Total	50	125	175

46. If $P(E) = \frac{1}{4}$, $P(F) = \frac{1}{3}$, and $P(E|F) = \frac{1}{6}$, find $P(E \cup F)$.

47. Shrubs When a certain type of shrub is planted, the probability that it will take root is 0.7. If four shrubs are planted, find each of the following. Assume independence.

(a) The probability that none of them take root.

(b) The probability that exactly two of them take root.

(c) The probability that at most two of them take root.

48. Antibiotic A certain antibiotic is effective for 70% of the people who take it. Suppose three persons take this drug. What is the probability that it will be effective for at least two of them? Assume independence.

49. Urns and Marbles Urn I contains 3 green and 2 red marbles, and Urn II contains 4 red, 2 green, and 2 white marbles. A marble is randomly drawn from Urn I and placed into Urn II. If a marble is then randomly drawn from Urn II, find the probability that the marble is red.

50. Urns and Marbles Urn I contains 4 red and 2 white marbles. Urn II contains 2 red and 3 white marbles. An urn is chosen at random, and then a marble is randomly drawn from it.

(a) What is the probability that the marble is white?

(b) If the marble is white, what is the probability that it was drawn from Urn II?

51. Grade Distribution Last semester, the grade distribution for a certain class taking an upper-level college course was analyzed. It was found that the proportion of students receiving a grade of A was 0.4 and the proportion getting an A and being a graduate student was 0.1. If a student is randomly selected from this class and is found to have received an A, find the probability that the student is a graduate student.

52. Alumni Reunion At the most recent alumni day at Omega College, 507 persons attended. Of these, 409 lived within the state, and 40% of them were attending for the first time. Among the alumni who lived out of the state, 73% were attending for the first time. That day a raffle was held, and the person who won had also won it the year before. Find the probability that the winner was from out of state.

53. Quality Control A manufacturer produces widgits on two shifts. The first shift produces 300 units of product per day, and the second produces 200. From past experience, it is believed that, of the output produced by the first and second shifts, 1 and 2% are defective, respectively. At the end of a day, a unit was selected at random from the total production.

(a) Find the probability that the unit is defective.

(b) If the unit is defective, find the probability that it came from the second shift.

54. Aptitude Test In the past, a company has hired only experienced personnel for its word-processing department. Because of a shortage in this field, the company has decided to hire inexperienced persons and will provide on-the-job training. It has supplied an employment agency with a new aptitude test that has been designed for applicants who desire such a training position. Of those who recently took the test, 35% passed. In order to gauge the effectiveness of the test, everyone who took the test was put in the training program. Of those who passed, 80% performed satisfactorily, whereas of those who failed, only 30% did satisfactorily. If one of the new trainees is selected at random and is found to be satisfactory, what is the probability that the person passed the exam?

Mathematical SNAPSHOT

Probability and Cellular Automata[8]

Systems of one, two, or three equations are good for modeling simple processes, like the path of a thrown object or the accrual of interest in a bank account. But how do we model something complicated and irregular, like a bolt of lightning or the spread of a rumor? For this, instead of trying to write and solve a system of equations, we can use a different modeling technique: cellular automata.

Cellular automata represent large, complex phenomena using collections of many small entities each following a few simple rules. The best-known system of cellular automata is the game called LIFE developed by John Conway in the late 1970s. It can be played by hand on a checkerboard, but using a computer program is faster and easier. Downloadable freeware can be found on the Internet. (Go to any search engine and search for "LIFE" and "Conway.")

Intriguing though LIFE is, it is not particularly good for modeling real-life processes. Better for such a task are cellular automata whose rules contain an element of randomness. Here is an example. Let us model the seepage of an oil spill into the soil below. We will model the soil as a pattern of cells layered like bricks (Figure 8.26).

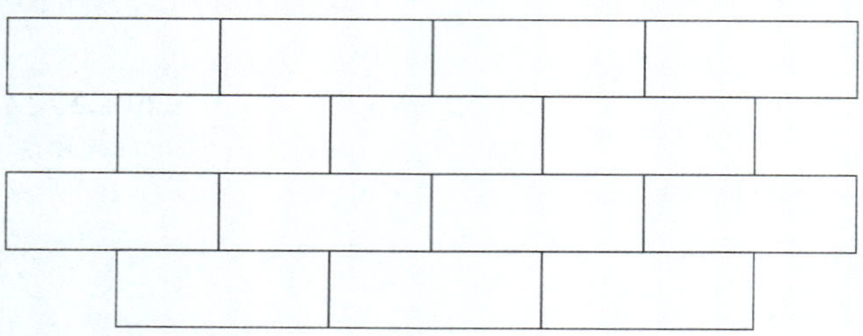

FIGURE 8.26 Cells for an oil spill model.

Each cell represents a pore in the soil, a pocket of space between dirt particles. All cells start in the "empty" state. To simulate an oil spill, we switch the entire top layer of cells (the soil surface) from empty to "filled." Depending on the microstructure of the pore arrangement, oil might flow from a pore either to both pores below, or just to the left one, or just to the right one, or to neither. We will model this by supposing that at every junction between a filled cell and an empty one below it, the filled cell has a probability P of switching the empty cell to filled.

For a TI-83 Plus graphing calculator, the following program models the process:

```
PROGRAM:OIL
:Input "P?", P
:ClrDraw
:AxesOff
:0 → X : 0 → N
:For(Y,0,46)
:Pxl–On(X,2Y + N)
:Pxl–On(X,2Y + N + 1)
:End
:For(X,1,62)
:X–2iPart(X/2) → N
:For(Y,0,46)
:If ((pxl–Test(X–1,2Y + N)and rand < P) or
    (pxl–Test(X–1,2Y + N + 1)and rand < P))
:Then
:Pxl–On(X,2Y + N)
:Pxl–On(X,2Y + N + 1)
:End:End:End
```

After you have entered this program, set the standard viewing window and run the program. At the P? prompt, enter some value between 0 and 1. Then watch as the screen fills with a simulated oil spill percolating downward through the soil. You should find that for $P < 0.55$, sooner or later the oil stops seeping downward; that for $P > 0.75$, the oil shows no sign of slowdown; and that for $0.55 < P < 0.75$, the seepage exhibits a highly irregular pattern something like the one in Figure 8.27.

FIGURE 8.27 Oil seepage with $P = 0.6$.

[8]Adapted from L. Charles Biehl, "Forest Fires, Oil Spills, and Fractal Geometry, Part 1: Cellular Automata and Modeling Natural Phenomena," *The Mathematics Teacher* 91 (November 1998), 682–87. By permission of the National Council of Teachers of Mathematics.

The oil spill model is one dimensional in the sense that at any instant, only one row of cells is in transition. Other systems of cellular automata run in two dimensions. We can use a scheme somewhat like the LIFE game to model the spread of a fad—be it mood rings (1975–1977) or collapsible metal scooters (1999–2000). A fad is, by nature, a transient thing: People who pick up a fad soon tire of it and from then on are "immune." The spread of a fad is somewhat like the spread of a forest fire, where trees catch fire and in turn pass the fire on to others before burning out.

We can model the fad-spreading process on a grid where each square represents a person in one of three states: prefad, midfad, and postfad. A prefad square that shares a side with a midfad square has a probability P of being "infected" and being turned into a midfad cell. The midfad state only lasts one cycle and is followed by a permanent postfad state.

This model would be hard to implement on a graphing calculator but can readily be done using a computer algebra system, like *Maple* or *Mathematica*. The results are analogous to the ones for the oil spill: For $P < 0.4$, the fad dies out; for $P > 0.6$, the fad spreads almost uniformly throughout the population; and for $0.4 < P < 0.6$, the fad spreads in an irregular and unpredictable pattern (Figure 8.28).

The interesting result here is that the difference between a fad that spreads widely and one that quickly peters out can be quite small. A bit of fashion that passes from friend to friend with a probability of 0.45 may never really catch on, while a fashion transmitted with a probability of 0.55 has a good chance of becoming a major craze. Of course the model we used is quite simple. A more sophisticated model might allow for the fact that some people have more friends than others, that some friendships are more fad-transmission prone than others, and so on.

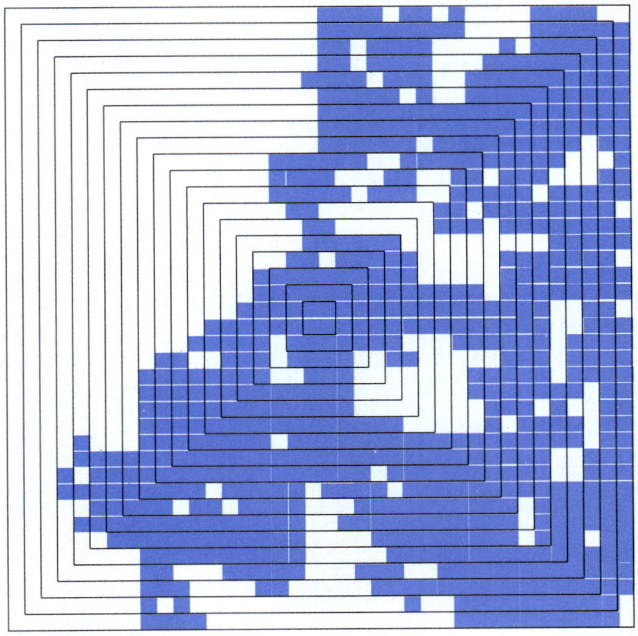

FIGURE 8.28 Fad spreading with $P = 0.5$.

Exercises

1. By repeated runs on a graphing calculator, estimate the critical value of P at which the oil spill in the simulation starts to penetrate deeply into the soil. That is, what is the smallest value of P for which the spill does not diminish significantly between the middle and the bottom of the screen?

2. What other phenomena might be modeled using cellular automata? What would the rules look like?

CHAPTER

9

Additional Topics in Probability

As we saw at the beginning of the last chapter, probability can be used to solve the problem of dividing up the pot of money between two gamblers when their game is interrupted. Now we might ask a follow-up question: What are the chances that a game will be interrupted in the first place?

Our answer depends on the details, of course. If the gamblers know in advance that they will play a fixed number of rounds—the "interruption" being scheduled in advance, as it were—then it might be fairly easy to calculate the probability that time will run out before they finish. Or, if the amount of time available is unknown, we might calculate the *expected duration* of a complete game and an *expected time* before the next interruption. Then, if the expected game length came out well under the expected time to the next interruption, we could say that the probability of having to break the game off in midplay was low. But if we wanted to give a more exact, numerical answer we would have to do a more complicated calculation.

The kind of problem encountered here is not unique to gambling. In industry, manufacturers need to know how likely they are to have to interrupt a production cycle due to equipment breakdown. One way they keep this probability low is by logging the usage hours on each machine and replacing it as the hours approach "mean time to failure"—the expected value of the number of usage hours the machine provides in its lifetime. Medical researchers face a related problem when they consider the possibility of having to break off an experiment because too many test subjects drop out. To keep this probability low, researchers often calculate an expected number of drop-outs in advance and include this number, plus a cushion, in the number of people recruited for a study.

The idea of the expected value for a number—the length of time until something happens, or the number of people who drop out of a study—is one of the key concepts of this chapter.

OBJECTIVE *To develop the probability distribution of a random variable and to represent that distribution geometrically by a graph or a histogram. To compute the mean, variance, and standard deviation of a random variable.*

9.1 Discrete Random Variables and Expected Value

With some experiments, we are interested in events associated with numbers. For example, if two coins are tossed, our interest may be in the *number* of heads that occur. Thus, we consider the events

$$\{0\} \quad \{1\} \quad \{2\}$$

If we let X be a variable that represents the number of heads that occur, then the only values that X can assume are 0, 1, and 2. The value of X is determined by the outcome of the experiment, and hence by chance. In general, a variable whose values depend on the outcome of a random process is called a **random variable.** Usually, random variables are denoted by capital letters such as X, Y, or Z, and the values that these variables assume may be denoted by corresponding lowercase letters (x, y, or z). Thus, for the number of heads (X) that occur in the tossing of two coins, we may indicate the possible values by writing

$$X = x \qquad \text{where } x = 0, 1, 2$$

or, more simply,

$$X = 0, 1, 2$$

EXAMPLE 1 Random Variables

a. Suppose a die is rolled and X is the number that turns up. Then X is a random variable and $X = 1, 2, 3, 4, 5, 6$.

b. Suppose a coin is successively tossed until a head appears. If Y is the number of such tosses, then Y is a random variable and

$$Y = y \qquad \text{where } y = 1, 2, 3, 4, \ldots.$$

Note that Y may assume infinitely many values.

c. A student is taking an exam with a one-hour time limit. If X is the number of minutes it takes to complete the exam, then X is a random variable. The values that X may assume form the interval (0, 60]. That is, $0 < X \leq 60$.

A random variable is called a **discrete random variable** if it may assume only a finite number of values or if its values can be placed in one-to-one correspondence with the positive integers. In Examples 1(a) and 1(b), X and Y are discrete. A random variable is called a **continuous random variable** if it may assume any value in some interval or intervals, such as X does in Example 1(c). In this chapter, we shall be concerned with discrete random variables; Chapter 16 deals with continuous random variables.

If X is a random variable, the probability of the event that X assumes the value x is denoted $P(X = x)$. Similarly, we can consider the probabilities of events such as $X \leq x$ and $X > x$. If X is discrete, then the function f that assigns the number $P(X = x)$ to each possible value of X is called the **probability function,** the **probability distribution,** or—more simply—the **distribution** of the random variable X. Thus,

$$f(x) = P(X = x)$$

EXAMPLE 2 Distribution of a Random Variable

Suppose that X is the number of heads that appear on the toss of two well-balanced coins. Determine the distribution of X.

Solution: We must find the probabilities of the events $X = 0$, $X = 1$, and $X = 2$. The usual equiprobable sample space is

$$S = \{HH, HT, TH, TT\}$$

Hence,

$$\text{the event } X = 0 \text{ is } \{TT\}$$

$$\text{the event } X = 1 \text{ is } \{HT, TH\}$$

$$\text{the event } X = 2 \text{ is } \{HH\}$$

Probability Table

x	$P(X = x)$
0	$\frac{1}{4}$
1	$\frac{2}{4} = \frac{1}{2}$
2	$\frac{1}{4}$

The probability for each of these events is given in the **probability table** in the margin. If f is the distribution for X, that is, $f(x) = P(X = x)$, then

$$f(0) = \frac{1}{4} \qquad f(1) = \frac{1}{2} \qquad f(2) = \frac{1}{4}$$

In Example 2, the distribution f was indicated by the listing

$$f(0) = \frac{1}{4} \qquad f(1) = \frac{1}{2} \qquad f(2) = \frac{1}{4}$$

However, the probability table for X gives the same information and is an acceptable way of expressing the probability distribution. Another way is by the graph of the distribution, as shown in Figure 9.1. The vertical lines from the x-axis to the points on the graph merely emphasize the heights of the points. Another representation of the distribution of X is the rectangle diagram in Figure 9.2, called the **probability histogram** for X. Here a rectangle is centered over each value of X. The rectangle above x has width 1 and height $P(X = x)$. Thus, its *area* is the probability $1 \cdot P(X = x) = P(X = x)$. This interpretation of probability as an area is important in Chapter 16.

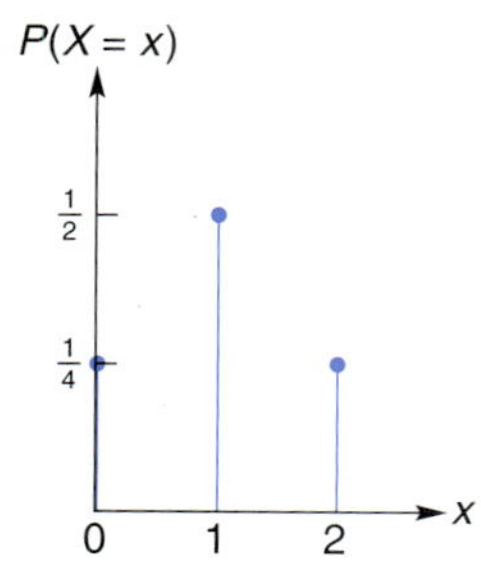

FIGURE 9.1 Graph of the distribution of X.

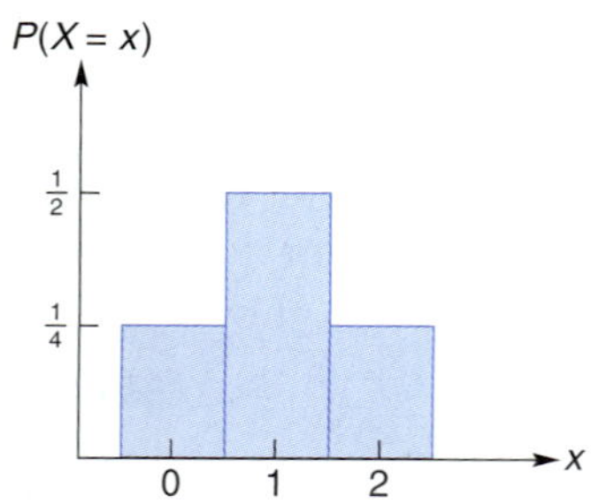

FIGURE 9.2 Probability histogram for X.

Note in Example 2 that the sum of $f(0)$, $f(1)$, and $f(2)$ is 1:

$$f(0) + f(1) + f(2) = \frac{1}{4} + \frac{1}{2} + \frac{1}{4} = 1$$

This must be the case, because the events $X = 0$, $X = 1$, and $X = 2$ are mutually exclusive and the union of all three is the sample space [and $P(S) = 1$]. We can conveniently indicate the sum $f(0) + f(1) + f(2)$ by the notation

$$\sum_x f(x)$$

which is called *sigma notation*[1] because the Greek letter Σ (sigma) is used. Here $\sum_x f(x)$ means that we are to sum all terms of the form $f(x)$ for all values of x, which in this case are 0, 1, and 2. Thus,

$$\sum_x f(x) = f(0) + f(1) + f(2)$$

[1]More thorough discussions of sigma notation occur in Section 14.6 and Appendix A.

In general, for any probability distribution f, we have $0 \le f(x) \le 1$ for all x, and the sum of all function values is 1. Therefore,

$$\sum_x f(x) = 1$$

This means that in any probability histogram, the sum of the areas of the rectangles is 1.

The probability distribution for a random variable X gives the relative frequencies of the values of X in the long run. However, it is often useful to determine the "average" value of X in the long run. In Example 2, for instance, suppose that the two coins were tossed n times, which resulted in $X = 0$ occurring k_0 times, $X = 1$ occurring k_1 times, and $X = 2$ occurring k_2 times. Then the average value of X for these n tosses is

$$\frac{0 \cdot k_0 + 1 \cdot k_1 + 2 \cdot k_2}{n}$$

or, equivalently,

$$0 \cdot \frac{k_0}{n} + 1 \cdot \frac{k_1}{n} + 2 \cdot \frac{k_2}{n}$$

But the fractions k_0/n, k_1/n, and k_2/n are the relative frequencies of the events $X = 0$, $X = 1$, and $X = 2$, respectively, that occur in the n tosses. If n is very large, then these relative frequencies approach the probabilities of the events $X = 0$, $X = 1$, and $X = 2$. Thus, it seems reasonable that the average value of X in the long run is

$$0 \cdot f(0) + 1 \cdot f(1) + 2 \cdot f(2) = 0 \cdot \frac{1}{4} + 1 \cdot \frac{1}{2} + 2 \cdot \frac{1}{4} = 1 \tag{1}$$

This means that if we tossed the coins many times, the average number of heads appearing per toss is very close to 1. We define the sum in Equation (1) to be the *mean, expected value,* or *expectation of* X and denote it by μ (the Greek letter "mu") or $E(X)$. Note that from Equation (1), μ has the form $\sum_x xf(x)$. In general, we have the following definition.

Definition

If X is a discrete random variable with probability distribution f, then the ***mean*** (or ***expected value*** or ***expectation***) **of X**, denoted by μ, or $E(X)$, is given by

$$\mu = E(X) = \sum_x xf(x)$$

The mean of X can be interpreted as the average value of X in the long run. The mean does not necessarily have to be an outcome of the experiment.

EXAMPLE 3 Expected Gain

An insurance company offers a \$180,000 *catastrophic fire insurance policy to homeowners of a certain type of house. The policy provides protection in the event that such a house is totally destroyed by fire in a one-year period. The company has determined that the probability of such an event is* 0.002. *If the annual policy premium is* \$379, *find the expected gain per policy for the company.*

Strategy: If an insured house does not suffer a catastrophic fire, the company gains \$379. However, if there is such a fire, the company loses \$180,000 – \$379 (insured value of house minus premium), or \$179,621. If X is the gain (in dollars) to the company, then X is a random variable that may assume the values 379 and $-179{,}621$. (A loss is considered a negative gain.) The expected gain per policy for the company is the expected value of X.

Solution: If f is the probability function for X, then

$$f(-179{,}621) = P(X = -179{,}621) = 0.002$$

and

$$f(379) = P(X = 379) = 1 - 0.002 = 0.998$$

The expected value of X is given by

$$E(X) = \sum_x xf(x) = -179{,}621\,f(-179{,}621) + 379\,f(379)$$

$$= -179{,}621(0.002) + 379(0.998) = 19$$

Thus, if the company sold many policies, it could expect to gain approximately \$19 per policy, which could be applied to such expenses as advertising, overhead, and profit.

▶ **Now work Problem 19.**

Since $E(X)$ is the average value of X in the long run, it is a measure of the central tendency of X. However, $E(X)$ does not indicate the dispersion or spread of X from the mean in the long run. For example, Figure 9.3 shows the graphs of two probability distributions, f and g, for the random variables X and Y. It can easily be demonstrated that both X and Y have the same mean: $E(X) = 2$ and $E(Y) = 2$. You should verify these results. But from Figure 9.3, X is more likely to assume the values 1 or 3 than is Y, because $f(1)$ and $f(3)$ are $\frac{2}{5}$, whereas $g(1)$ and $g(3)$ are $\frac{1}{5}$. Thus, X has more likelihood of assuming values away from the mean than does Y, so there is more dispersion for X in the long run.

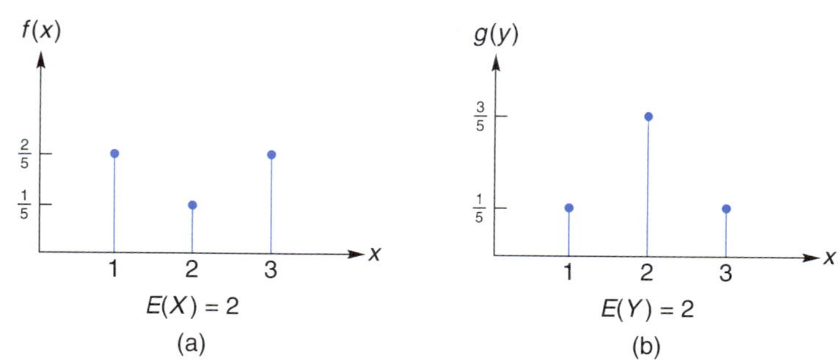

FIGURE 9.3 Probability distributions.

There are various ways to measure dispersion for a random variable X. One way is to determine the long-run average of the absolute values of the deviations from the mean μ—that is, $E(|X-\mu|)$:[2]

$$E(|X-\mu|) = \sum_x |x-\mu| f(x)$$

However, since absolute value is involved here, $E(|X-\mu|)$ is a mathematically awkward expression.

Although many other measures of dispersion can be considered, two are most widely accepted. One is the *variance,* and the other (which is related to the variance) is the *standard deviation*. The **variance of X,** denoted by $\mathrm{Var}(X)$, is the long-run average of the *squares* of the deviations of X from μ:[3]

Variance of X

$$\mathrm{Var}(X) = E[(X-\mu)^2] = \sum_x (x-\mu)^2 f(x) \qquad (2)$$

Since $(X-\mu)^2$ is involved in $\mathrm{Var}(X)$, and both X and μ have the same units of measurement, the units for $\mathrm{Var}(X)$ are those of X^2. For instance, in Example 3, X is in dollars; thus, $\mathrm{Var}(X)$ has units of dollars squared. It is convenient to have a measure of dispersion in the same units as X. Such a measure is $\sqrt{\mathrm{Var}(X)}$, which is called the **standard deviation of X** and is denoted by σ (the lowercase Greek letter sigma):

Standard Deviation of X

$$\sigma = \sqrt{\mathrm{Var}(X)}$$

Note that σ has the property that

$$\sigma^2 = \mathrm{Var}(X)$$

Both $\mathrm{Var}(X)$ [or σ^2] and σ are measures of the dispersion of X. The greater the value of $\mathrm{Var}(X)$, or σ, the greater is the dispersion. One result of a famous theorem, *Chebyshev's inequality,* is that the probability of X falling within two standard deviations of the mean is at least $\frac{3}{4}$.

We can write the formula for variance in Equation (2) in a different way. Suppose the possible values of X are x_1 and x_2, and f is the probability function for X. Then by Equation (2),

$$\begin{aligned}
\mathrm{Var}(X) &= \sum_x (x-\mu)^2 f(x) \\
&= (x_1-\mu)^2 f(x_1) + (x_2-\mu)^2 f(x_2) \\
&= \left(x_1^2 - 2x_1\mu + \mu^2\right) f(x_1) + \left(x_2^2 - 2x_2\mu + \mu^2\right) f(x_2) \\
&= x_1^2 f(x_1) - 2x_1\mu f(x_1) + \mu^2 f(x_1) + x_2^2 f(x_2) - 2x_2\mu f(x_2) + \mu^2 f(x_2)
\end{aligned}$$

---

[2] It can be shown that if $Y = g(X)$, then $E(Y) = \sum_x g(x) f(x)$, where f is the probability function for X. For example, if $Y = |X-\mu|$, then $E(|X-\mu|) = \sum_x |x-\mu| f(x)$.

[3] By the preceding footnote, if $Y = (X-\mu)^2$, then $E[(X-\mu)^2] = \sum_x (x-\mu)^2 f(x)$.

Regrouping and factoring, we have

$$\begin{aligned}\operatorname{Var}(X) &= \left[x_1^2 f(x_1) + x_2^2 f(x_2)\right] - 2\mu[x_1 f(x_1) + x_2 f(x_2)] + \mu^2[f(x_1) + f(x_2)] \\ &= \sum_x x^2 f(x) - 2\mu \sum_x x f(x) + \mu^2 \sum_x f(x)\end{aligned}$$

But $\sum_x x^2 f(x) = E(X^2)$,[4] $\sum_x x f(x) = \mu$ and $\sum_x f(x) = 1$. Therefore,

$$\operatorname{Var}(X) = \sigma^2 = E(X^2) - 2\mu^2 + \mu^2$$

That is

$$\begin{aligned}\operatorname{Var}(X) = \sigma^2 &= E(X^2) - \mu^2 \\ &= \sum_x x^2 f(x) - \mu^2 \end{aligned} \tag{3}$$

This formula for variance is quite useful, since it usually simplifies computations.

EXAMPLE 4 **Mean, Variance, and Standard Deviation**

An urn contains 10 *marbles, each of which shows a number. Five marbles show* 1, *two show* 2, *and three show* 3. *A marble is drawn at random. If X is the number that shows, determine μ,* $\operatorname{Var}(X)$, *and σ.*

Solution: The sample space consists of 10 equally likely outcomes (the marbles). The values that X can assume are 1, 2, and 3. The events $X = 1$, $X = 2$, and $X = 3$ contain 5, 2, and 3 sample points, respectively. Thus, if f is the probability function for X,

$$\begin{aligned} f(1) &= P(X=1) = \tfrac{5}{10} = \tfrac{1}{2} \\ f(2) &= P(X=2) = \tfrac{2}{10} = \tfrac{1}{5} \\ f(3) &= P(X=3) = \tfrac{3}{10} \end{aligned}$$

Calculating the mean gives

$$\begin{aligned}\mu &= \sum_x x f(x) = 1 \cdot f(1) + 2 \cdot f(2) + 3 \cdot f(3) \\ &= 1 \cdot \frac{5}{10} + 2 \cdot \frac{2}{10} + 3 \cdot \frac{3}{10} = \frac{18}{10} = \frac{9}{5}\end{aligned}$$

To find $\operatorname{Var}(X)$, either Equation (2) or Equation (3) can be used. Both will be used here, so that we can compare the arithmetical computations involved. By Equation (2),

$$\begin{aligned}\operatorname{Var}(X) &= \sum_x (x-\mu)^2 f(x) \\ &= \left(1 - \frac{9}{5}\right)^2 f(1) + \left(2 - \frac{9}{5}\right)^2 f(2) + \left(3 - \frac{9}{5}\right)^2 f(3) \\ &= \left(-\frac{4}{5}\right)^2 \cdot \frac{5}{10} + \left(\frac{1}{5}\right)^2 \cdot \frac{2}{10} + \left(\frac{6}{5}\right)^2 \cdot \frac{3}{10} \\ &= \frac{16}{25} \cdot \frac{5}{10} + \frac{1}{25} \cdot \frac{2}{10} + \frac{36}{25} \cdot \frac{3}{10} \\ &= \frac{80 + 2 + 108}{250} = \frac{190}{250} = \frac{19}{25}\end{aligned}$$

[4]This is a consequence of footnote 2.

By Equation (3),

$$\begin{aligned}\text{Var}(X) &= \sum_x x^2 f(x) - \mu^2 \\ &= 1^2 \cdot f(1) + 2^2 \cdot f(2) + 3^2 \cdot f(3) - \left(\frac{9}{5}\right)^2 \\ &= 1 \cdot \frac{5}{10} + 4 \cdot \frac{2}{10} + 9 \cdot \frac{3}{10} - \frac{81}{25} \\ &= \frac{5+8+27}{10} - \frac{81}{25} = \frac{40}{10} - \frac{81}{25} \\ &= 4 - \frac{81}{25} = \frac{19}{25}\end{aligned}$$

Notice that Equation (2) involves $(x - \mu)^2$, but Equation (3) involves x^2. Because of this, it is often easier to compute variances by Equation (3) than by Equation (2).

Since $\sigma^2 = \text{Var}(X) = \frac{19}{25}$, the standard deviation is

$$\sigma = \sqrt{\text{Var}(X)} = \sqrt{\frac{19}{25}} = \frac{\sqrt{19}}{5}$$

▶ **Now work Problem 1.**

Exercise 9.1

In Problems 1–4, the distribution of the random variable X is given. Determine μ, Var(X), and σ. In Problem 1, construct the probability histogram. In Problem 2, graph the distribution.

*1. $f(0) = 0.1,\ f(1) = 0.4,\ f(2) = 0.2,\ f(3) = 0.3$

2. $f(4) = 0.4,\ f(5) = 0.6$

3. See Figure 9.4.

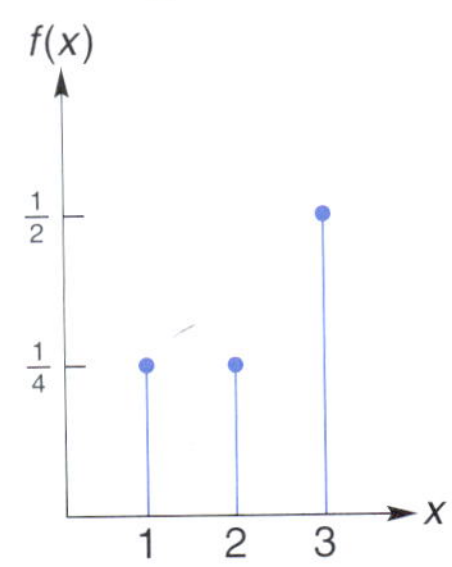

FIGURE 9.4 Distribution for Problem 3.

4. See Figure 9.5.

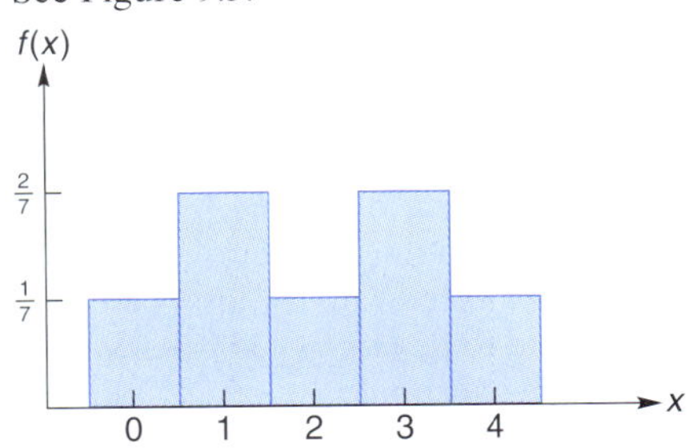

FIGURE 9.5 Distribution for Problem 4.

5. The random variable X has the following distribution:

x	$P(X = x)$
3	
5	0.3
6	0.2
7	0.4

(a) Find $P(X = 3)$
(b) Find μ
(c) Find σ^2

6. The random variable X has the following distribution:

x	$P(X = x)$
2	$2n$
4	n
6	0.4

(a) Find $P(X = 2)$ and $P(X = 4)$
(b) Find μ

In Problems 7–10, determine $E(X)$, σ^2, and σ for the random variable X.

7. **Coin Toss** Three fair coins are tossed. Let X be the number of heads that occur.

8. **Marbles in Urn** An urn contains six marbles, each of which shows a number. Four marbles show a 1 and two show a 2. A marble is randomly selected and the number that shows, X, is observed.

9. **Committee** From a group of two women and three men, two persons are selected at random to form a committee. Let X be the number of men on the committee.

10. **Marbles in Urn** An urn contains two red and three green marbles. Two marbles are randomly drawn in succession with replacement, and the number of red marbles, X, is observed.

11. **Marbles in Urn** An urn contains three red and two white marbles. Two marbles are randomly drawn in succession without replacement. Let $X =$ the number of red marbles drawn. Find the distribution f for X.

12. **Subcommittee** From a state government committee consisting of four Whigs and six Tories, a subcommittee of three is to be randomly selected. Let X be the number of Whigs in the subcommittee. Find a general formula, in terms of combinations, that gives $P(X = x)$, where $x = 0, 1, 2, 3$.

13. **Raffle** A charitable organization is having a raffle for a single prize of \$8500. Each raffle ticket costs \$1 and 10,000 tickets have been sold.
    (a) Find the expected gain for the purchaser of a single ticket.
    (b) Find the expected gain for the purchaser of two tickets.

14. **Coin Game** Consider the following game. You are to toss three fair coins. If three heads or three tails turn up, your friend pays you \$10. If either one or two heads turn up, you must pay your friend \$6. What are your expected winnings or losses per game?

15. **Earnings** A landscaper earns \$200 per day when working and loses \$30 per day when not working. If the probability of working on any day is $\frac{4}{7}$, find the landscaper's expected daily earnings.

16. **Fast-Food Restaurant** A fast-food chain estimates that if it opens a restaurant in a shopping center, the probability that the restaurant is successful is 0.65. A successful restaurant earns an annual profit of \$75,000; a restaurant that is not successful loses \$20,000. What is the expected gain to the chain if it opens a restaurant in a shopping center?

17. **Insurance** An insurance company offers a hospitalization policy to individuals in a certain group. For a one-year period, the company will pay \$100 per day, up to a maximum of five days, for each day the policyholder is hospitalized. The company estimates that the probability that any person in this group is hospitalized for exactly one day is 0.001; for exactly two days, 0.002; for exactly three days, 0.003; for exactly four days, 0.004; and for five or more days, 0.008. Find the expected gain per policy to the company if the annual premium is \$10.

18. **Demand** The following table for a small car rental company gives the probability that x cars are rented daily:

x	0	1	2	3	4	5	6
$P(X = x)$	0.05	0.10	0.15	0.25	0.20	0.15	0.1

Determine the expected daily demand for their cars.

*19. **Insurance Premium** In Example 3, if the company wants an expected gain of \$50 per policy, determine the annual premium.

20. **Roulette** In the game of roulette, there is a wheel with 37 slots numbered with the integers from 0 to 36, inclusive. A player bets \$1 (for example) and chooses a number. The wheel is spun and a ball rolls on the wheel. If the ball lands in the slot showing the chosen number, the player receives the \$1 bet plus \$35. Otherwise, the player loses the \$1 bet. Assume that all numbers are equally likely, and determine the expected gain or loss per play.

21. **Coin Game** Suppose that you pay \$1.25 to play a game in which two fair coins are tossed. You receive the number of dollars equal to the number of heads that occur. What is your expected gain (or loss) on each play? The game is said to be *fair* to you when your expected gain is \$0. What should you pay to play if this is to be a fair game?

OBJECTIVE *To develop the binomial distribution and relate it to the binomial theorem.*

9.2 The Binomial Distribution

Binomial Theorem

Later in this section you will see that the terms in the expansion of a power of a binomial are useful in describing the probability distributions of certain random variables. It is worthwhile, therefore, to first discuss the *binomial theorem,* which is a formula for expanding $(a+b)^n$, where n is a positive integer.

Regardless of n, there are patterns in the expansion of $(a+b)^n$. To illustrate, we shall consider the *cube* of the binomial $a+b$. By successively applying the distributive law, we have

$$\begin{aligned}(a+b)^3 &= [(a+b)(a+b)](a+b)\\ &= [a(a+b)+b(a+b)](a+b)\\ &= [aa+ab+ba+bb](a+b)\\ &= aa(a+b)+ab(a+b)+ba(a+b)+bb(a+b)\\ &= aaa+aab+aba+abb+baa+bab+bba+bbb \qquad \textbf{(1)}\end{aligned}$$

so that

$$(a+b)^3 = a^3+3a^2b+3ab^2+b^3 \qquad \textbf{(2)}$$

Three observations can be made about the right side of Equation (2). First, notice that the number of terms is four, which is one more than the power to which $a+b$ is raised (3). Second, the first and last terms are the *cubes* of a and b; the powers of a *decrease* from left to right (from 3 to 0), and the powers of b *increase* (from 0 to 3). Third, for each term, the sum of the exponents of a and b is 3, which is the power to which $a+b$ is raised.

Let us now focus on the coefficients of the terms in Equation (2). Consider the coefficient of the ab^2-term. It is the number of terms in Equation (1) that involve exactly two b's, namely, 3. But let us see *why* there are three terms that involve two b's. Notice in Equation (1) that each term is the product of three numbers, each of which is either a or b. Because of the distributive law, each of the three $a+b$ factors in $(a+b)^3$ contributes either an a or b to the term. Thus, the number of terms involving one a and two b's is equal to the number of ways of choosing two of the three factors to supply a b, namely, ${}_3C_2 = \frac{3!}{2!1!} = 3$. Similarly,

$$\text{the coefficient of the } a^3\text{-term is } {}_3C_0$$

$$\text{the coefficient of the } a^2b\text{-term is } {}_3C_1$$

and

$$\text{the coefficient of the } b^3\text{-term is } {}_3C_3$$

Generalizing our results, we obtain a formula for expanding $(a+b)^n$, called the *binomial theorem.*

Binomial Theorem

If n is a positive integer, then

$$\begin{aligned}(a+b)^n &= {}_nC_0a^n + {}_nC_1a^{n-1}b + {}_nC_2a^{n-2}b^2\\ &\quad + \cdots + {}_nC_{n-1}ab^{n-1} + {}_nC_nb^n\end{aligned}$$

The numbers ${}_nC_r$ are also called **binomial coefficients** for this reason.

EXAMPLE 1 **Binomial Theorem**

Use the binomial theorem to expand $(q + p)^4$.

Solution: Here $n = 4$, $a = q$, and $b = p$. Thus,

$$(q + p)^4 = {}_4C_0 q^4 + {}_4C_1 q^3 p + {}_4C_2 q^2 p^2 + {}_4C_3 q p^3 + {}_4C_4 p^4$$
$$= \frac{4!}{0!4!} q^4 + \frac{4!}{1!3!} q^3 p + \frac{4!}{2!2!} q^2 p^2 + \frac{4!}{3!1!} q p^3 + \frac{4!}{4!0!} p^4$$

Recalling that $0! = 1$, we have

$$(q + p)^4 = q^4 + 4q^3 p + 6q^2 p^2 + 4qp^3 + p^4$$

Binomial Distribution

We now turn our attention to repeated trials of an experiment in which the outcome of any trial does not affect the outcome of any other trial. These are referred to as **independent trials.** For example, when a fair die is rolled five times, the outcome on one roll does not affect the outcome on any other roll. Here we have five independent trials of rolling a die. Together, these five trials can be considered as a five-stage compound experiment involving independent events, so we can use the special multiplication law of Section 8.6 to determine the probability of obtaining specific outcomes of the trials.

To illustrate, let us find the probability of getting exactly two 4's in the five rolls of the die. We shall consider getting a 4 as a *success* (S) and getting any of the other five numbers as a *failure* (F). For example, the sequence

SSFFF

denotes getting

4, 4, followed by three other numbers.

This sequence can be considered as the intersection of five independent events: success on the first trial, success on the second, failure on the third, and so on. Since the probability of success on any trial is $\frac{1}{6}$, and the probability of failure is $1 - \frac{1}{6} = \frac{5}{6}$, by the special multiplication law for the intersection of independent events, the probability of the sequence SSFFF occurring is

$$\frac{1}{6} \cdot \frac{1}{6} \cdot \frac{5}{6} \cdot \frac{5}{6} \cdot \frac{5}{6} = \left(\frac{1}{6}\right)^2 \left(\frac{5}{6}\right)^3$$

In fact, this is the probability for *any* particular order of the two S's and three F's. Let us determine how many ways a sequence of two S's and three F's can be formed. Out of five trials, the number of ways of choosing the two trials for success is ${}_5C_2$. Another way to look at this problem is that we are counting "Permutations with Repeated Objects" as in Section 8.2 of the "word" SSFFF. There are $\frac{5!}{2! \cdot 3!} = {}_5C_2$ of these. So the probability of getting exactly two 4's in the five rolls is

$${}_5C_2 \left(\frac{1}{6}\right)^2 \left(\frac{5}{6}\right)^3 \qquad (3)$$

If we denote the probability of success by p and the probability of failure by $q(= 1 - p)$, then (3) takes the form

$${}_5C_2 p^2 q^3$$

which is the term involving p^2 in the expansion of $(q + p)^5$.

More generally, consider the probability of getting exactly x 4's in n rolls of the die. Then $n - x$ of the rolls must be some other number. For a particular order, the probability is

$$p^x q^{n-x}$$

The number of possible orders is ${}_nC_x$, which again we can see as the question of finding the number of permutations of n symbols, where x of them are S (success) and the remaining $n - x$ are F (failure). According to the result in Section 8.2 on "Permutations with Repeated Objects," there are

$$\frac{n!}{x! \cdot (n-x)!} = {}_nC_x$$

of these and therefore,

$$P(X = x) = {}_nC_x p^x q^{n-x}$$

which is a general expression for the terms in $(q + p)^n$. In summary, the distribution for X (the number of 4's that occur in n rolls) is given by the terms in $(q + p)^n$.

Whenever we have n independent trials of an experiment in which each trial has only two possible outcomes (success and failure) and the probability of success in each trial remains the same, the trials are called **Bernoulli trials.** Because the distribution of the number of successes corresponds to the expansion of a power of a binomial, the experiment is called a **binomial experiment,** and the distribution of the number of successes is called a **binomial distribution.**

Binomial Distribution

If X is the number of successes in n independent trials of a binomial experiment with probability p of success and q of failure on any trial, then the distribution f for X is given by

$$f(x) = P(X = x) = {}_nC_x p^x q^{n-x}$$

where x is an integer such that $0 \leq x \leq n$ and $q = 1 - p$. Any random variable with this distribution is called a **binomial random variable** and is said to have a **binomial distribution.** The mean and standard deviation of X are given, respectively, by

$$\mu = np \qquad \sigma = \sqrt{npq}$$

APPLICATIONS IN PRACTICE 1

Binomial Distribution

Let X be the number of persons out of four job applicants who are hired. If the probability of any one applicant being hired is 0.3, find the distribution of X.

EXAMPLE 2 Binomial Distribution

Suppose X is a binomial random variable with $n = 4$ and $p = \frac{1}{3}$. Find the distribution for X.

Solution: Here $q = 1 - p = 1 - \frac{1}{3} = \frac{2}{3}$. So we have

$$P(X = x) = {}_nC_x p^x q^{n-x} \qquad x = 0, 1, 2, 3, 4$$

Thus,

$$P(X=0) = {}_4C_0\left(\frac{1}{3}\right)^0\left(\frac{2}{3}\right)^4 = \frac{4!}{0!4!}\cdot 1\cdot\frac{16}{81} = 1\cdot 1\cdot\frac{16}{81} = \frac{16}{81}$$

$$P(X=1) = {}_4C_1\left(\frac{1}{3}\right)^1\left(\frac{2}{3}\right)^3 = \frac{4!}{1!3!}\cdot\frac{1}{3}\cdot\frac{8}{27} = 4\cdot\frac{1}{3}\cdot\frac{8}{27} = \frac{32}{81}$$

$$P(X=2) = {}_4C_2\left(\frac{1}{3}\right)^2\left(\frac{2}{3}\right)^2 = \frac{4!}{2!2!}\cdot\frac{1}{9}\cdot\frac{4}{9} = 6\cdot\frac{1}{9}\cdot\frac{4}{9} = \frac{8}{27}$$

$$P(X=3) = {}_4C_3\left(\frac{1}{3}\right)^3\left(\frac{2}{3}\right)^1 = \frac{4!}{3!1!}\cdot\frac{1}{27}\cdot\frac{2}{3} = 4\cdot\frac{1}{27}\cdot\frac{2}{3} = \frac{8}{81}$$

$$P(X=4) = {}_4C_4\left(\frac{1}{3}\right)^4\left(\frac{2}{3}\right)^0 = \frac{4!}{4!0!}\cdot\frac{1}{81}\cdot 1 = 1\cdot\frac{1}{81}\cdot 1 = \frac{1}{81}$$

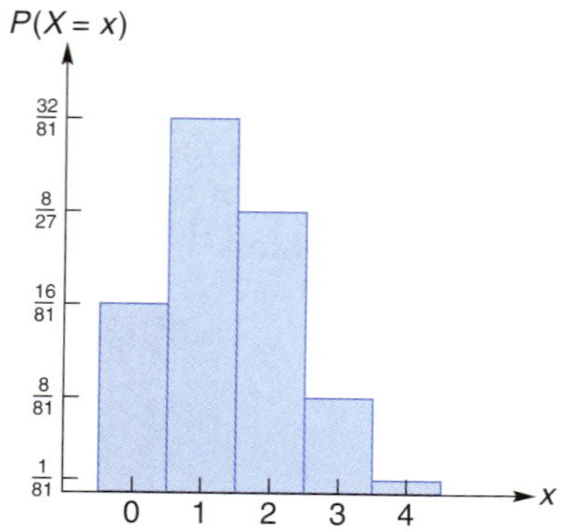

FIGURE 9.6 Binomial distribution, $n = 4$, $p = \frac{1}{3}$.

The probability histogram for X is given in Figure 9.6. Note that the mean μ for X is $np = 4(\frac{1}{3}) = \frac{4}{3}$, and the standard deviation is

$$\sigma = \sqrt{npq} = \sqrt{4\cdot\frac{1}{3}\cdot\frac{2}{3}} = \sqrt{\frac{8}{9}} = \frac{2\sqrt{2}}{3}$$

▶ **Now work Problem 1.**

EXAMPLE 3 **At Least Two Heads in Eight Coin Tosses**

A fair coin is tossed eight times. Find the probability of getting at least two heads.

Solution: If X is the number of heads that occur, then X has a binomial distribution with $n = 8$, $p = \frac{1}{2}$, and $q = \frac{1}{2}$. To simplify our work, we use the fact that

$$P(X \geq 2) = 1 - P(X < 2)$$

Now,

$$\begin{aligned} P(X<2) &= P(X=0) + P(X=1) \\ &= {}_8C_0\left(\frac{1}{2}\right)^0\left(\frac{1}{2}\right)^8 + {}_8C_1\left(\frac{1}{2}\right)^1\left(\frac{1}{2}\right)^7 \\ &= \frac{8!}{0!8!}\cdot 1\cdot\frac{1}{256} + \frac{8!}{1!7!}\cdot\frac{1}{2}\cdot\frac{1}{128} \\ &= 1\cdot 1\cdot\frac{1}{256} + 8\cdot\frac{1}{2}\cdot\frac{1}{128} = \frac{9}{256} \end{aligned}$$

Thus,

$$P(X \geq 2) = 1 - \frac{9}{256} = \frac{247}{256}$$

A probability histogram for X is given in Figure 9.7.

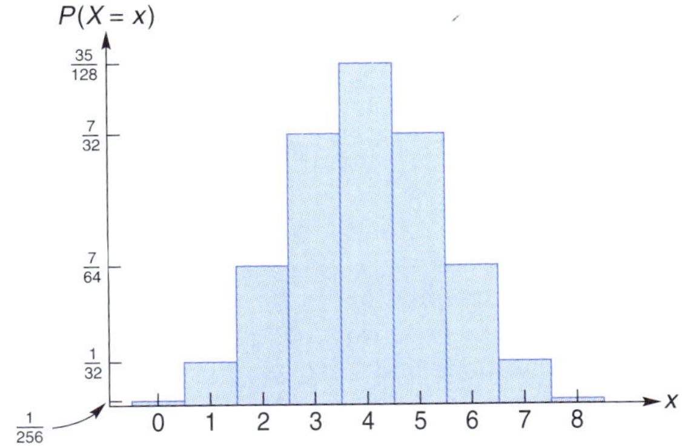

FIGURE 9.7 Binomial distribution, $n = 8$, $p = \frac{1}{2}$.

▶ **Now work Problem 17.**

EXAMPLE 4 **Income Tax Audit**

For a particular group of individuals, 20% of their income tax returns are audited each year. Of five randomly chosen individuals, what is the probability that exactly two will have their returns audited?

Solution: We shall consider this to be a binomial experiment with five trials (selecting an individual). Actually, the experiment is not truly binomial, because selecting an individual from this group affects the probability that another individual's return will be audited. For example, if there are 5000 individuals, then 20%, or 1000, will be audited. The probability that the first individual selected will be audited is $\frac{1000}{5000}$. If that event occurs, the probability that the second individual selected will be audited is $\frac{999}{4999}$. Thus, the trials are not independent. However, we assume that the number of individuals is large, so that for all practical purposes, the probability of auditing an individual remains constant from trial to trial.

For each trial, the two outcomes are *being audited* and *not being audited.* Here we shall define a success as being audited. Letting X be the number of returns audited, $p = 0.2$, and $q = 1 - 0.2 = 0.8$, we have

$$P(X = 2) = {}_5C_2(0.2)^2(0.8)^3 = \frac{5!}{2!3!}(0.04)(0.512)$$
$$= 10(0.04)(0.512) = 0.2048$$

▶ **Now work Problem 15.**

Exercise 9.2

In Problems 1–4, determine the distribution f for the binomial random variable X if the number of trials is n and the probability of success on any trial is p. Also, find μ and σ.

*1. $n = 2$, $p = \frac{1}{4}$

2. $n = 3$, $p = \frac{1}{2}$

3. $n = 3$, $p = \frac{2}{3}$

4. $n = 4$, $p = 0.4$

In Problems 4–10, determine the given probability if X is a binomial random variable, n is the number of trials, and p is the probability of success on any trial.

5. $P(X = 5)$; $n = 6$, $p = 0.2$

6. $P(X = 2)$; $n = 5$, $p = \frac{1}{3}$

7. $P(X = 2)$; $n = 4, p = \frac{4}{5}$

8. $P(X = 6)$; $n = 8, p = 0.1$

9. $P(X < 2)$; $n = 5, p = \frac{1}{2}$

10. $P(X \geq 2)$; $n = 6, p = \frac{2}{3}$

11. Coin A fair coin is tossed 11 times. What is the probability that exactly eight heads occur?

12. Multiple-Choice Quiz Each question in a six-question multiple-choice quiz has four choices, only one of which is correct. If a student guesses at all six questions, find the probability that exactly three will be correct.

13. Marbles An urn contains five red and seven green marbles. Four marbles are randomly drawn in succession with replacement. Determine the probability that exactly two marbles are green.

14. Cards From a deck of 52 playing cards, 3 cards are randomly selected in succession with replacement. Determine the probability that exactly two cards are aces.

***15. Quality Control** A manufacturer produces electrical switches, of which 2% are defective. From a production run of 50,000 switches, four are randomly selected and each is tested. Determine the probability that the sample contains exactly two defective switches. Round your answer to three decimal places. Assume that the four trials are independent and that the number of defective switches in the sample has a binomial distribution.

16. Coin A coin is biased so that $P(H) = 0.2$ and $P(T) = 0.8$. If X is the number of heads in three tosses, determine a formula for $P(X = x)$.

***17. Coin** A biased coin is tossed three times in succession. The probability of heads on any toss is $\frac{1}{4}$. Find the probability that (a) exactly two heads occur and (b) two or three heads occur.

18. Cards From an ordinary deck of 52 playing cards, 7 cards are randomly drawn in succession with replacement. Find the probability that there are (a) exactly four hearts and (b) at least four hearts.

19. Quality Control In a large production lot of electronic devices, it is believed that one-fifth are defective. If a sample of six is randomly selected, find the probability that no more than one will be defective.

20. Computer For a certain large population, the probability that a randomly selected person has a computer is 0.6. If four persons are selected at random, find the probability that at least three have a computer.

21. Baseball The probability that a certain baseball player gets a hit is 0.300. Find the probability that if he goes to bat four times, he will get at least one hit.

22. Stocks A financial advisor claims that 60% of the stocks that he recommends for purchase increase in value. From a list of 200 recommended stocks, a client selects 4 at random. Determine the probability, rounded to two decimal places, that at least 2 of the chosen stocks increase in value. Assume that the selections of the stocks are independent trials and that the number of stocks that increase in value has a binomial distribution.

23. Genders of Children If a family has five children, find the probability that at least two are girls. (Assume that the probability that a child is a girl is $\frac{1}{2}$.)

24. If X is a binomially distributed random variable with $n = 50$ and $p = \frac{2}{5}$, find σ^2.

25. Suppose X is a binomially distributed random variable such that $\mu = 2$ and $\sigma^2 = \frac{3}{2}$. Find $P(X = 1)$.

26. Quality Control In a production process, the probability of a defective unit is 0.06. Suppose a sample of 15 units is selected at random. Let X be the number of defectives.

(a) Find the expected number of defective units.

(b) Find $\text{Var}(X)$.

(c) Find $P(X \leq 1)$. Round your answer to two decimal places.

OBJECTIVE *To develop the notions of a Markov chain and the associated transition matrix. To find state vectors and the steady-state vector.*

9.3 Markov Chains

We conclude this chapter with a discussion of a special type of stochastic process called a *Markov chain*.[5]

Markov Chain

A **Markov chain** is a sequence of trials of an experiment in which the possible outcomes of each trial remain the same from trial to trial, are finite in number, and have probabilities that depend only upon the outcome of the previous trial.

[5] After the Russian mathematician Andrei Markov (1856–1922).

To illustrate a Markov chain, we consider the following situation. Imagine that a small town has only two service stations—say, stations 1 and 2—that handle the servicing needs of the town's automobile owners. (These customers form the population under consideration.) Each time a customer needs car servicing, he or she must make a *choice* of which station to use.

Thus, each customer can be placed into a category according to which of the two stations he or she most recently chose. We can view a customer and the service stations as a *system*. If a customer most recently chose station 1, we shall refer to this as *state* 1 of the system. Similarly, if a customer most recently chose station 2, we say that the system is currently in state 2. Hence, at any given time, the system is in one of its two states. Of course, over a period of time, the system may move from one state to the other. For example, the sequence 1, 2, 2, 1 indicates that in four successive car servicings, the system changed from state 1 to state 2, remained at state 2, and then changed to state 1.

This situation can be thought of as a sequence of trials of an experiment (choosing a service station) in which the possible outcomes for each trial are the two states (station 1 and station 2). Each trial involves observing the state of the system at that time.

If we know the current state of the system, we realize that we cannot be sure of its state at the next observation. However, we may know the *likelihood* of its being in a particular state. For example, suppose that if a customer most recently used station 1, then the probability that the customer uses station 1 the next time is 0.7. (This means that, of those customers who used station 1 most recently, 70% continued to use station 1 the next time and 30% changed to station 2.) Assume also that if a customer used station 2 most recently, the probability is 0.8 that the customer also uses station 2 the next time. You may recognize these probabilities as being *conditional* probabilities. That is,

$$P(\text{remaining in state 1}|\text{currently in state 1}) = 0.7$$

$$P(\text{changing to state 2}|\text{currently in state 1}) = 0.3$$

$$P(\text{remaining in state 2}|\text{currently in state 2}) = 0.8$$

$$P(\text{changing to state 1}|\text{currently in state 2}) = 0.2$$

These four probabilities can be organized in a square matrix $\mathbf{T} = [t_{ij}]$ by letting entry t_{ij} be the probability that a customer currently in state i will be in state j at the next observation:

$$\begin{array}{cc} & \begin{array}{cc} & \textit{Next State} \\ \textit{Current State} & \text{State 1} \quad \text{State 2} \end{array} \\ \mathbf{T} = & \begin{array}{c} \text{State 1} \\ \text{State 2} \end{array} \begin{bmatrix} 0.7 & 0.3 \\ 0.2 & 0.8 \end{bmatrix} \end{array}$$

Matrix $\mathbf{T}$ is called a *transition matrix* because it indicates the probabilities of transition from one state to another in *one step*—that is, as we go from one observation period to the next. The entries are called *transition probabilities*. We emphasize that *the transition matrix remains the same at every stage of the sequence of observations*. Note that all entries of the matrix are nonnegative, because they are probabilities. Moreover, the sum of the entries in each row must be 1, because, for each current state, the probabilities account for all possible transitions.

For example, the sum of the entries in row 1 of $\mathbf{T}$ is $0.7 + 0.3 = 1$.

Let us summarize our service station situation up to now. We have a sequence of trials in which the possible outcomes (or states) are the same from trial to trial and are finite in number (two). The probability that the system is in a particular state for a given trial depends only on the state for the preceding trial. Thus, we have a so-called *two-state Markov chain*. A Markov chain determines a square matrix $\mathbf{T}$, called a transition matrix.

Transition Matrix

A **transition matrix** for a k-state Markov chain is a $k \times k$ matrix $\mathbf{T} = [t_{ij}]$ in which the entry t_{ij} is the probability of moving from state i to state j from one trial to the next. All entries are nonnegative, and the sum of the entries in each row is 1.

A subscript of 0 is used for the initial state vector.

Suppose that when observations are initially made, 60% of all customers used station 1 most recently and 40% used station 2. This means that, before any additional trials (car servicings) are considered, the probabilities that a customer is in state 1 or 2 are 0.6 and 0.4, respectively. These probabilities are called *initial state probabilities* and are collectively referred to as being the *initial distribution*. They can be represented by a row vector, called an **initial state vector,** which is denoted by $\mathbf{X_0}$. In this case,

$$\mathbf{X_0} = [0.6 \quad 0.4]$$

We would like to find the vector that gives the state probabilities for a customer's *next* visit to a service station. This state vector is denoted by $\mathbf{X_1}$. More generally, a state vector is defined as follows:

State Vector

The **state vector $\mathbf{X_n}$** for a k-state Markov chain is a k-entry row vector in which the entry x_j is the probability of being in state j after the nth trial.

We can find the entries for $\mathbf{X_1}$ from the probability tree in Figure 9.8. We see that the probability of being in state 1 after the next visit is the sum

$$(0.6)(0.7) + (0.4)(0.2) = 0.5 \qquad \textbf{(1)}$$

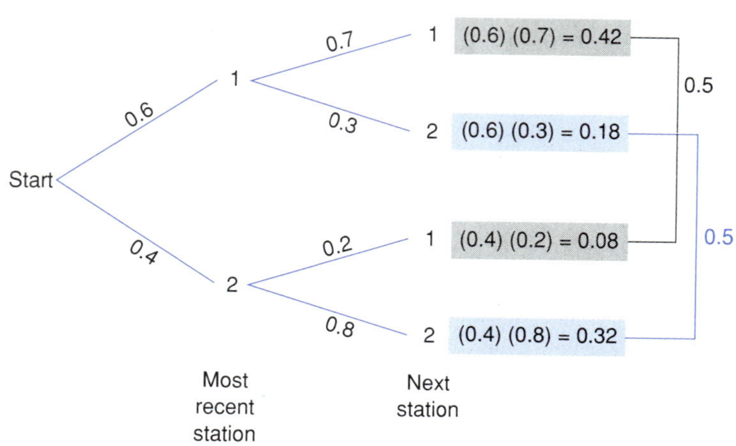

FIGURE 9.8 Probability tree for two-state Markov chain.

and the probability of being in state 2 is

$$(0.6)(0.3) + (0.4)(0.8) = 0.5 \tag{2}$$

Thus,

$$\mathbf{X_1} = [0.5 \quad 0.5]$$

The sums of products on the left sides of Equations (1) and (2) remind us of matrix multiplication. In fact, they are the entries in the matrix $\mathbf{X_0T}$ obtained by multiplying the initial state vector by the transition matrix:

$$\mathbf{X_1} = \mathbf{X_0T} = [0.6 \quad 0.4]\begin{bmatrix} 0.7 & 0.3 \\ 0.2 & 0.8 \end{bmatrix} = [0.5 \quad 0.5]$$

This pattern of taking the product of a state vector and the transition matrix to get the next state vector continues, allowing us to find state probabilities for future observations. For example, to find $\mathbf{X_2}$, the state vector that gives the probabilities for each state after two trials (following the initial observation), we have

$$\mathbf{X_2} = \mathbf{X_1T} = [0.5 \quad 0.5]\begin{bmatrix} 0.7 & 0.3 \\ 0.2 & 0.8 \end{bmatrix} = [0.45 \quad 0.55]$$

Thus, the probability of being in state 1 after two car servicings is 0.45. Note that, since $\mathbf{X_1} = \mathbf{X_0T}$, we can write

$$\mathbf{X_2} = (\mathbf{X_0T})\mathbf{T}$$

so that

$$\mathbf{X_2} = \mathbf{X_0T}^2$$

In general, the nth state vector $\mathbf{X_n}$ can be found by multiplying the previous state vector $\mathbf{X_{n-1}}$ by $\mathbf{T}$.

If $\mathbf{T}$ is the transition matrix for a Markov chain, then the state vector $\mathbf{X_n}$ for the nth trial is given by

$$\mathbf{X_n} = \mathbf{X_{n-1}T}$$

Equivalently, we can find $\mathbf{X_n}$ by using only the initial state row-vector $\mathbf{X_0}$ and the transition matrix $\mathbf{T}$:

Here we find $\mathbf{X_n}$ by using powers of $\mathbf{T}$.

$$\mathbf{X_n} = \mathbf{X_0T}^n \tag{3}$$

Let us now consider the situation in which we know the initial state of the system. For example, take the case of observing initially that a customer has most recently chosen station 1. This means the probability that the system is in state 1 is 1, so the initial state vector must be

$$\mathbf{X_0} = [1 \quad 0]$$

Suppose we determine $\mathbf{X_2}$, the state vector that gives the state probabilities after the next two visits. This is given by

$$\mathbf{X_2} = \mathbf{X_0T}^2 = [1 \quad 0]\begin{bmatrix} 0.7 & 0.3 \\ 0.2 & 0.8 \end{bmatrix}^2$$

$$= [1 \quad 0]\begin{bmatrix} 0.55 & 0.45 \\ 0.30 & 0.70 \end{bmatrix} = [0.55 \quad 0.45]$$

Thus, for this customer, the probabilities of using station 1 or station 2 after two steps are 0.55 and 0.45, respectively. Observe that these probabilities form the *first row* of $\mathbf{T}^2$. On the other hand, if the system were initially in state 2, then the state vector after two steps would be

$$[0 \quad 1]\mathbf{T}^2 = [0 \quad 1]\begin{bmatrix} 0.55 & 0.45 \\ 0.30 & 0.70 \end{bmatrix} = [0.30 \quad 0.70]$$

Hence, for this customer, the probabilities of using station 1 or station 2 after two steps are 0.30 and 0.70, respectively. Observe that these probabilities form the *second row* of $\mathbf{T}^2$. Based on our observations, we now have a way of interpreting $\mathbf{T}^2$. The entries in

$$\mathbf{T}^2 = \begin{array}{c} \\ 1 \\ 2 \end{array}\begin{array}{c} \begin{array}{cc} 1 & \quad 2 \end{array} \\ \begin{bmatrix} 0.55 & 0.45 \\ 0.30 & 0.70 \end{bmatrix} \end{array}$$

give the probabilities of moving from one state to another in *two* steps. In general, we have the following:

This gives the significance of the entries in $\mathbf{T}^n$.

If $\mathbf{T}$ is a transition matrix, then for $\mathbf{T}^n$ the entry in row i and column j gives the probability of moving from state i to state j in n steps.

EXAMPLE 1 **Demography**

A certain county is divided into three demographic regions. Research indicates that each year 20% of the residents in region 1 move to region 2 and 10% move to region 3. (The others remain in region 1.) Of the residents in region 2, 10% move to region 1 and 10% move to region 3. Of the residents in region 3, 20% move to region 1 and 10% move to region 2.

a. *Find the transition matrix* $\mathbf{T}$ *for this situation.*

Solution: We have

$$\begin{array}{c} \textit{From} \\ \textit{Region} \\ 1 \\ \mathbf{T} = 2 \\ 3 \end{array} \begin{array}{c} \textit{To Region} \\ \begin{array}{ccc} 1 & 2 & 3 \end{array} \\ \begin{bmatrix} 0.7 & 0.2 & 0.1 \\ 0.1 & 0.8 & 0.1 \\ 0.2 & 0.1 & 0.7 \end{bmatrix} \end{array}$$

Note that to find t_{11}, we subtracted the sum of the other two entries in the first row from 1. The entries t_{22} and t_{33} are found similarly.

b. *Find the probability that a resident of region 1 this year is a resident of region 1 next year; in two years.*

Solution: From entry t_{11} in transition matrix $\mathbf{T}$, the probability that a resident of region 1 remains in region 1 after one year is 0.7. The probabilities of moving from one region to another in two steps are given by $\mathbf{T}^2$:

$$\mathbf{T}^2 = \begin{array}{c} \\ 1 \\ 2 \\ 3 \end{array}\begin{array}{c} \begin{array}{ccc} 1 & \quad 2 & \quad 3 \end{array} \\ \begin{bmatrix} 0.53 & 0.31 & 0.16 \\ 0.17 & 0.67 & 0.16 \\ 0.29 & 0.19 & 0.52 \end{bmatrix} \end{array}$$

Powers of $\mathbf{T}$ can be conveniently found with a graphing calculator.

Thus, the probability that a resident of region 1 is in region 1 after two years is 0.53.

c. *This year, suppose 40% of county residents live in region 1, 30% live in region 2, and 30% live in region 3. Find the probability that a resident of the county lives in region 2 after three years.*

Solution: The initial state vector is

$$\mathbf{X}_0 = [0.40 \quad 0.30 \quad 0.30]$$

The distribution of the population after three years is given by state vector $\mathbf{X}_3$. From Equation (3) with $n = 3$, we have

$$\begin{aligned}\mathbf{X}_3 &= \mathbf{X}_0\mathbf{T}^3 = \mathbf{X}_0\mathbf{T}^2\mathbf{T}\\ &= [0.40 \quad 0.30 \quad 0.30]\begin{bmatrix}0.53 & 0.31 & 0.16\\ 0.17 & 0.67 & 0.16\\ 0.29 & 0.19 & 0.52\end{bmatrix}\begin{bmatrix}0.7 & 0.2 & 0.1\\ 0.1 & 0.8 & 0.1\\ 0.2 & 0.1 & 0.7\end{bmatrix}\\ &= [0.3368 \quad 0.4024 \quad 0.2608]\end{aligned}$$

Of course, $\mathbf{X}_3$ can be easily obtained with a graphing calculator: Enter $\mathbf{X}_0$ and $\mathbf{T}$, and then evaluate $\mathbf{X}_0\mathbf{T}^3$ directly.

This result means that in three years, 33.68% of the county residents live in region 1, 40.24% live in region 2, and 26.08% live in region 3. Thus, the probability that a resident lives in region 2 in three years is 0.4024.

▶ **Now work Problem 33.**

Steady-State Vectors

Let us now return to our service station problem. Recall that if the initial state vector is

$$\mathbf{X}_0 = [0.6 \quad 0.4],$$

then

$$\begin{aligned}\mathbf{X}_1 &= [0.5 \quad 0.5]\\ \mathbf{X}_2 &= [0.45 \quad 0.55]\end{aligned}$$

Some state vectors beyond the second are

$$\begin{aligned}\mathbf{X}_3 &= \mathbf{X}_2\mathbf{T} = [0.45 \quad 0.55]\begin{bmatrix}0.7 & 0.3\\ 0.2 & 0.8\end{bmatrix} = [0.425 \quad 0.575]\\ \mathbf{X}_4 &= \mathbf{X}_3\mathbf{T} = [0.425 \quad 0.575]\begin{bmatrix}0.7 & 0.3\\ 0.2 & 0.8\end{bmatrix} = [0.4125 \quad 0.5875]\\ \mathbf{X}_5 &= \mathbf{X}_4\mathbf{T} = [0.4125 \quad 0.5875]\begin{bmatrix}0.7 & 0.3\\ 0.2 & 0.8\end{bmatrix} = [0.40625 \quad 0.59375]\\ &\vdots\\ \mathbf{X}_{10} &= \mathbf{X}_9\mathbf{T} \approx [0.40020 \quad 0.59980]\end{aligned}$$

These results strongly suggest, and it is indeed the case, that, as the number of trials increases, the entries in the state vectors tend to get closer and closer to the corresponding entries in the vector

$$\mathbf{Q} = [0.40 \quad 0.60]$$

(Equivalently, it can be shown that the entries in each row of $\mathbf{T}^n$ approach the corresponding entries in those of $\mathbf{Q}$ as n increases.) Vector $\mathbf{Q}$ has a special property. Observe the result of multiplying $\mathbf{Q}$ by the transition matrix $\mathbf{T}$:

$$\mathbf{QT} = [0.40 \quad 0.60]\begin{bmatrix} 0.7 & 0.3 \\ 0.2 & 0.8 \end{bmatrix} = [0.40 \quad 0.60] = \mathbf{Q}$$

We thus have

$$\mathbf{QT} = \mathbf{Q},$$

which indicates that $\mathbf{Q}$ *remains unchanged from trial to trial.*

In summary, as the number of trials increases, the state vectors get closer and closer to $\mathbf{Q}$, which remains unchanged from trial to trial. Essentially, the distribution of the population between the service stations is stabilizing. That is, in the long run, approximately 40% of the population will have their cars serviced at station 1 and 60% at station 2. To describe this, we say that $\mathbf{Q}$ is the **steady-state vector** of this process. It can be shown that the steady-state vector is unique. (There is only one such vector.) Moreover, $\mathbf{Q}$ does not depend on the initial state vector $\mathbf{X}_0$ but depends only on the transition matrix $\mathbf{T}$. For this reason, we say that $\mathbf{Q}$ is the *steady-state vector for* $\mathbf{T}$.

The steady-state vector is unique and does not depend on the initial distribution.

What we need now is a procedure for finding the steady-state vector $\mathbf{Q}$ without having to compute state vectors for large values of n. Fortunately, the previously stated property that $\mathbf{QT} = \mathbf{Q}$ can be used to find $\mathbf{Q}$. If we let $\mathbf{Q} = [q_1 \quad q_2]$, we have

$$\mathbf{QT} = \mathbf{Q}$$

$$[q_1 \quad q_2]\begin{bmatrix} 0.7 & 0.3 \\ 0.2 & 0.8 \end{bmatrix} = [q_1 \quad q_2]$$

$$[0.7q_1 + 0.2q_2 \quad 0.3q_1 + 0.8q_2] = [q_1 \quad q_2]$$

By matrix equality, this implies that

$$\begin{cases} 0.7q_1 + 0.2q_2 = q_1 \\ 0.3q_1 + 0.8q_2 = q_2 \end{cases}$$

Simplifying gives

$$\begin{cases} -0.3q_1 + 0.2q_2 = 0 \\ 0.3q_1 - 0.2q_2 = 0 \end{cases}$$

These equations are equivalent, so there are infinitely many solutions. However, it can be shown that $q_1 + q_2$ must be 1. So we now have the system

$$\begin{cases} -0.3q_1 + 0.2q_2 = 0 \\ 0.3q_1 - 0.2q_2 = 0 \\ q_1 + q_2 = 1 \end{cases}$$

Solving this system by elimination or matrix techniques gives $q_1 = 0.40$ and $q_2 = 0.60$. Thus, the steady-state vector is

$$\mathbf{Q} = [0.40 \quad 0.60]$$

which confirms our previous assumption.

We must point out that for Markov chains in general, the state vectors do not always approach a steady-state vector. However, it can be shown that a steady-state vector for $\mathbf{T}$ does exist, provided that $\mathbf{T}$ is *regular:*

A transition matrix $\mathbf{T}$ is **regular** if there exists an integer power of $\mathbf{T}$ for which all entries are positive.

Only regular transition matrices will be considered in this section. A Markov chain whose transition matrix is regular is called a **regular Markov chain.**

In summary, we have the following:

Suppose $\mathbf{T}$ is the transition matrix for a regular Markov chain. Then the steady-state row vector

$$\mathbf{Q} = [q_1 \quad q_2 \quad \cdots \quad q_k]$$

is the solution to the matrix equation

$$\mathbf{QT} = \mathbf{Q} \tag{4}$$

subject to the condition that

$$q_1 + q_2 + \cdots + q_k = 1 \tag{5}$$

Equations (4) and (5) are at the heart of finding the steady-state vector.

EXAMPLE 2 Steady-State Vector

For the demography problem of Example 1, *in the long run what percentage of county residents will live in each region?*

Solution: The population distribution in the long run is given by the steady-state vector $\mathbf{Q}$, which we now proceed to find. From Equation (4), $\mathbf{QT} = \mathbf{Q}$; that is,

$$[q_1 \quad q_2 \quad q_3]\begin{bmatrix} 0.7 & 0.2 & 0.1 \\ 0.1 & 0.8 & 0.1 \\ 0.2 & 0.1 & 0.7 \end{bmatrix} = [q_1 \quad q_2 \quad q_3].$$

Multiplying the matrices on the left side gives a row matrix. By equating the entries of this matrix with the corresponding entries of the row matrix on the right side, we obtain the system

$$\begin{cases} 0.7q_1 + 0.1q_2 + 0.2q_3 = q_1 \\ 0.2q_1 + 0.8q_2 + 0.1q_3 = q_2 \\ 0.1q_1 + 0.1q_2 + 0.7q_3 = q_3 \end{cases}$$

Combining terms gives

$$\begin{cases} -0.3q_1 + 0.1q_2 + 0.2q_3 = 0 \\ 0.2q_1 - 0.2q_2 + 0.1q_3 = 0 \\ 0.1q_1 + 0.1q_2 - 0.3q_3 = 0 \end{cases}$$

Solving this system, together with the condition that

$$q_1 + q_2 + q_3 = 1$$

gives

$$\mathbf{Q} = [0.3125 \quad 0.4375 \quad 0.2500].$$

Thus, in the long run, the percentages of county residents living in regions 1, 2, and 3 are 31.25%, 43.75%, and 25%, respectively.

▶ **Now work Problem 37.**

We conclude this section by developing a matrix procedure for finding the steady-state vector

$$\mathbf{Q} = [q_1 \quad q_2 \quad \cdots \quad q_k]$$

for the transition matrix $\mathbf{T}$ of a k-state regular Markov chain. Since

$$\mathbf{QT} = \mathbf{Q}$$

we have

$$\mathbf{QT} - \mathbf{Q} = \mathbf{O}$$
$$\mathbf{QT} - \mathbf{QI} = \mathbf{O}$$
$$\mathbf{Q}(\mathbf{T} - \mathbf{I}) = \mathbf{O}$$

Here we make use of a number of properties of the transpose of a matrix. These properties were discussed in Chapter 6.

Taking transposes gives

$$[\mathbf{Q}(\mathbf{T} - \mathbf{I})]^{\mathrm{T}} = \mathbf{O}^{\mathrm{T}}$$
$$(\mathbf{T} - \mathbf{I})^{\mathrm{T}}\mathbf{Q}^{\mathrm{T}} = \mathbf{O}$$
$$(\mathbf{T}^{\mathrm{T}} - \mathbf{I}^{\mathrm{T}})\mathbf{Q}^{\mathrm{T}} = \mathbf{O}$$
$$(\mathbf{T}^{\mathrm{T}} - \mathbf{I})\mathbf{Q}^{\mathrm{T}} = \mathbf{O}$$

The last equation corresponds to a homogeneous system of linear equations. This system, along with the condition

$$q_1 + q_2 + \cdots + q_k = 1$$

form a nonhomogeneous system that can be solved by reducing a matrix of the form

$$\left[\begin{array}{c|c} \mathbf{T}^{\mathrm{T}} - \mathbf{I} & \mathbf{O} \\ \\ 1 \ldots 1 & 1 \end{array}\right]$$

To illustrate, for the situation in Example 2 with

$$\mathbf{T} = \begin{bmatrix} 0.7 & 0.2 & 0.1 \\ 0.1 & 0.8 & 0.1 \\ 0.2 & 0.1 & 0.7 \end{bmatrix}$$

we have

$$\mathbf{T}^{\mathrm{T}} - \mathbf{I} = \begin{bmatrix} 0.7 & 0.1 & 0.2 \\ 0.2 & 0.8 & 0.1 \\ 0.1 & 0.1 & 0.7 \end{bmatrix} - \begin{bmatrix} 1 & 0 & 0 \\ 0 & 1 & 0 \\ 0 & 0 & 1 \end{bmatrix}$$
$$= \begin{bmatrix} -0.3 & 0.1 & 0.2 \\ 0.2 & -0.2 & 0.1 \\ 0.1 & 0.1 & -0.3 \end{bmatrix}$$

We now reduce the augmented matrix

$$\left[\begin{array}{ccc|c} -0.3 & 0.1 & 0.2 & 0 \\ 0.2 & -0.2 & 0.1 & 0 \\ 0.1 & 0.1 & -0.3 & 0 \\ 1 & 1 & 1 & 1 \end{array}\right]$$

which gives

$$\left[\begin{array}{ccc|c} 1 & 0 & 0 & 0.3125 \\ 0 & 1 & 0 & 0.4375 \\ 0 & 0 & 1 & 0.2500 \\ 0 & 0 & 0 & 0 \end{array}\right]$$

Thus, $q_1 = 0.3125$, $q_2 = 0.4375$, and $q_3 = 0.2500$, as before.

Exercise 9.3

In Problems 1–6, can the given matrix be a transition matrix for a Markov chain?

1. $\begin{bmatrix} \frac{1}{2} & -\frac{3}{2} \\ \frac{2}{3} & \frac{1}{3} \end{bmatrix}$ **2.** $\begin{bmatrix} 0.1 & 0.9 \\ 1 & 0 \end{bmatrix}$

3. $\begin{bmatrix} \frac{1}{2} & -\frac{1}{4} & \frac{3}{4} \\ \frac{1}{8} & \frac{5}{8} & \frac{1}{4} \\ \frac{1}{3} & \frac{1}{3} & \frac{1}{3} \end{bmatrix}$ **4.** $\begin{bmatrix} 0.2 & 0.7 & 0.1 \\ 0.6 & 0.2 & 0.2 \\ 0 & 0 & 0 \end{bmatrix}$

5. $\begin{bmatrix} 0.4 & 0.2 & 0.4 \\ 0 & 0.1 & 0.9 \\ 0.5 & 0.3 & 0.2 \end{bmatrix}$ **6.** $\begin{bmatrix} 0.5 & 0.4 & 0.6 \\ 0.1 & 0.3 & 0.6 \\ 0.3 & 0.3 & 0.4 \end{bmatrix}$

In Problems 7–10, a transition matrix for a Markov chain is given. Determine the values of the letter entries.

7. $\begin{bmatrix} \frac{2}{3} & a \\ b & \frac{1}{4} \end{bmatrix}$ **8.** $\begin{bmatrix} a & \frac{5}{12} \\ b & a \end{bmatrix}$

9. $\begin{bmatrix} 0.4 & a & 0.3 \\ a & 0.1 & b \\ a & b & c \end{bmatrix}$ **10.** $\begin{bmatrix} a & a & a \\ a & b & \frac{1}{4} \\ a & b & c \end{bmatrix}$

In Problems 11–14, determine whether the given vector could be a state vector for a Markov chain.

11. $[0.4 \quad 0.6]$ **12.** $[1 \quad 0]$

13. $[0.2 \quad 0.7 \quad 0.5]$ **14.** $[0.9 \quad -0.1 \quad 0.2]$

In Problems 15–20, a transition matrix **T** *and an initial state vector* $\mathbf{X}_0$ *are given. Compute the state vectors* $\mathbf{X}_1$, $\mathbf{X}_2$, *and* $\mathbf{X}_3$.

15. $\mathbf{T} = \begin{bmatrix} \frac{2}{3} & \frac{1}{3} \\ 1 & 0 \end{bmatrix}$, $\mathbf{X}_0 = \begin{bmatrix} \frac{1}{4} & \frac{3}{4} \end{bmatrix}$

16. $\mathbf{T} = \begin{bmatrix} \frac{1}{2} & \frac{1}{2} \\ \frac{1}{4} & \frac{3}{4} \end{bmatrix}$, $\mathbf{X}_0 = \begin{bmatrix} \frac{1}{2} & \frac{1}{2} \end{bmatrix}$

17. $\mathbf{T} = \begin{bmatrix} 0.3 & 0.7 \\ 0.5 & 0.5 \end{bmatrix}$, $\mathbf{X}_0 = [0.4 \quad 0.6]$

18. $\mathbf{T} = \begin{bmatrix} 0.1 & 0.9 \\ 0.9 & 0.1 \end{bmatrix}$, $\mathbf{X}_0 = [0.2 \quad 0.8]$

19. $\mathbf{T} = \begin{bmatrix} 0.1 & 0.2 & 0.7 \\ 0 & 0.4 & 0.6 \\ 0.3 & 0.3 & 0.4 \end{bmatrix}$, $\mathbf{X}_0 = [0.2 \quad 0 \quad 0.8]$

20. $\mathbf{T} = \begin{bmatrix} 0.4 & 0 & 0.4 & 0.2 \\ 0.1 & 0.1 & 0.7 & 0.1 \\ 0.2 & 0.3 & 0.4 & 0.1 \\ 0.1 & 0.3 & 0.4 & 0.2 \end{bmatrix}$, $\mathbf{X}_0 = [0.1 \quad 0.3 \quad 0.4 \quad 0.2]$

In Problems 21–24, a transition matrix **T** *is given.*

(a) Compute $\mathbf{T}^2$ and $\mathbf{T}^3$.
(b) What is the probability of going from state 1 to state 2 after two steps?
(c) What is the probability of going from state 2 to state 1 after three steps?

21. $\begin{bmatrix} \frac{1}{4} & \frac{3}{4} \\ \frac{3}{4} & \frac{1}{4} \end{bmatrix}$ **22.** $\begin{bmatrix} \frac{1}{3} & \frac{2}{3} \\ \frac{1}{2} & \frac{1}{2} \end{bmatrix}$

23. $\begin{bmatrix} 0 & 1 & 0 \\ 0.5 & 0.4 & 0.1 \\ 0.3 & 0.3 & 0.4 \end{bmatrix}$ **24.** $\begin{bmatrix} 0.1 & 0.2 & 0.7 \\ 0.1 & 0.1 & 0.8 \\ 0.1 & 0.1 & 0.8 \end{bmatrix}$

In Problems 25–30, find the steady-state vector for the given transition matrix.

25. $\begin{bmatrix} \frac{1}{2} & \frac{1}{2} \\ \frac{2}{3} & \frac{1}{3} \end{bmatrix}$ **26.** $\begin{bmatrix} \frac{1}{2} & \frac{1}{2} \\ \frac{1}{4} & \frac{3}{4} \end{bmatrix}$

27. $\begin{bmatrix} \frac{1}{5} & \frac{4}{5} \\ \frac{3}{5} & \frac{2}{5} \end{bmatrix}$ **28.** $\begin{bmatrix} \frac{1}{4} & \frac{3}{4} \\ \frac{1}{3} & \frac{2}{3} \end{bmatrix}$

29. $\begin{bmatrix} 0.4 & 0.3 & 0.3 \\ 0.6 & 0.3 & 0.1 \\ 0.6 & 0.1 & 0.3 \end{bmatrix}$ **30.** $\begin{bmatrix} 0.1 & 0.2 & 0.7 \\ 0.4 & 0.2 & 0.4 \\ 0.3 & 0.3 & 0.4 \end{bmatrix}$

31. Spread of Flu A flu has attacked a college dorm that has 200 students. Suppose the probability that a student having the flu will still have it 4 days later is 0.1. However, for a student who does not have the flu, the probability of having the flu 4 days later is 0.2.

(a) Find a transition matrix for this situation.
(b) If 120 students now have the flu, how many students (to the nearest integer) can be expected to have the flu 8 days from now? 12 days from now?

32. Physical Fitness A physical-fitness center has found that, of those members who perform high-impact exercising on one visit, 55% will do the same on the next visit and 45% will do low-impact exercising. Of those who perform low-impact exercising on one visit, 75% will do the same on the next visit and 25% will do high-impact exercising. On the last visit, suppose that 65% of members did high-impact exercising and 35% did low-impact exercising. After two more visits, what percentage of members will be performing high-impact exercising?

***33. Newspapers** In a certain area, two daily newspapers are available. It has been found that if a customer buys newspaper A on one day, then the probability is 0.1 that he or she will change to the other newspaper the next day. If a customer buys newspaper B on one day, then the probability is 0.8 that he or she will buy the same newspaper the next day.

(a) Find the transition matrix for this situation.
(b) Find the probability that a person who buys A on Monday will buy A on Thursday.

34. Video Rentals A video rental store has three locations in a city. A video can be rented from any of the three locations and returned to any of them. Studies show that videos are rented from one location and returned to a location according to the probabilities given by the following matrix:

$$\begin{array}{cc} \text{Rented} & \text{Returned to} \\ \text{from} & \begin{array}{ccc} 1 & 2 & 3 \end{array} \\ \begin{array}{c} 1 \\ 2 \\ 3 \end{array} & \begin{bmatrix} 0.7 & 0.1 & 0.2 \\ 0.2 & 0.8 & 0 \\ 0.2 & 0.2 & 0.6 \end{bmatrix} \end{array}$$

Suppose that 20% of the videos are initially rented from location 1, 50% from 2, and 30% from 3. Find the percentages of videos that can be expected to be returned to each location:

(a) After this rental.
(b) After the next rental.

35. Voting In a certain region, voter registration was analyzed according to party affiliation: Democratic, Republican, and other. It was found that on a year-to-year basis, the probability that a voter switches registration from Democratic to Republican is 0.1; from Democratic to other, 0.1; from Republican to Democratic, 0.1; from Republican to other, 0.1; from other to Democratic, 0.3; and from other to Republican, 0.2.

(a) Find a transition matrix for this situation.
(b) What is the probability that a presently registered Republican voter will be registered Democratic two years from now?
(c) If 40% of the present voters are Democratic and 40% are Republican, what percentage can be expected to be Republican one year from now?

36. Demography The residents of a certain region are classified as urban (U), suburban (S), or rural (R). A marketing firm has found that over successive 5-year periods, residents shift from one classification to another according to the probabilities given by the following matrix:

$$\begin{array}{cc} & \begin{array}{ccc} U & S & R \end{array} \\ \begin{array}{c} U \\ S \\ R \end{array} & \begin{bmatrix} 0.7 & 0.1 & 0.2 \\ 0.1 & 0.8 & 0.1 \\ 0.1 & 0.1 & 0.8 \end{bmatrix} \end{array}$$

(a) Find the probability that a suburban resident will be a rural resident in 15 years.
(b) Suppose the initial population of the region is 50% urban, 25% suburban, and 25% rural. Determine the expected population distribution in 15 years.

***37. Long-Distance Telephone Service** A major long-distance telephone company (company A) has studied the tendency of telephone users to switch from one carrier to another. The company believes that over successive six-month periods, the probability that a customer who uses A's service will switch to a competing service is 0.2 and the probability that a customer of any competing service will switch to A is 0.3.

(a) Find a transition matrix for this situation.
(b) If A presently controls 70% of the market, what percentage can it expect to control six months from now?
(c) What percentage of the market can A expect to control in the long run?

38. Automobile Purchases In a certain region, a study of car ownership was made. It was determined that if a person presently owns a Ford, then the probability that the next car the person buys is also a Ford is 0.65. If a person does not presently own a Ford, then the probability that the person will buy a Ford on the next car purchase is 0.35.

(a) Find the transition matrix for this situation.
(b) In the long run, what proportion of car purchases in the region can be expected to be Fords?

39. Laboratory Mice Suppose 100 mice are in a two-compartment cage and are free to move between the compartments. At regular time intervals, the number of mice in each compartment is observed. It has been found that if a mouse is in compartment 1 at one observation, then the probability that the mouse will be in compartment 1 at the next observation is $\frac{5}{7}$. If a mouse is in compartment 2 at one observation, then the probability that the mouse will be in compartment 2 at the next observation is $\frac{4}{7}$. Initially, suppose that 50 mice are placed into each compartment.

(a) Find the transition matrix for this situation.
(b) After two observations, what percentage of mice (rounded to two decimal places) can be expected to be in each compartment?
(c) In the long run, what percentage of mice can be expected in each compartment?

40. Vending Machines A typical cry of college students is "Don't put your money into that soda machine; I tried it and the machine didn't work!" Suppose that if a vending machine is working properly one time, then the probability that it will work properly the next time is 0.8. On the other hand, suppose that if the machine is not working properly one time, then the probability that it will not work properly the next time is 0.9.

(a) Find a transition matrix for this situation.
(b) Suppose that four people line up at such a soda machine. If the first person receives a soda, what is the probability that the fourth person receives a soda? (Assume nobody makes more than one attempt.)
(c) If there are 42 such vending machines on a college campus, in the long run how many machines do you expect to work properly?

41. Advertising A supermarket chain sells bread from bakeries A and B. Presently, A accounts for 50% of the chain's daily bread sales. To increase sales, A launches a new advertising campaign. The bakery believes that the change in bread sales at the chain will be based on the following transition matrix:

$$\begin{array}{cc} & \begin{array}{cc} A & B \end{array} \\ \begin{array}{c} A \\ B \end{array} & \begin{bmatrix} \frac{3}{4} & \frac{1}{4} \\ \frac{1}{2} & \frac{1}{2} \end{bmatrix} \end{array}$$

(a) Find the steady-state vector.

(b) In the long run, by what percentage can A expect to increase present sales at the chain? Assume that the total daily sales of bread at the chain remain the same.

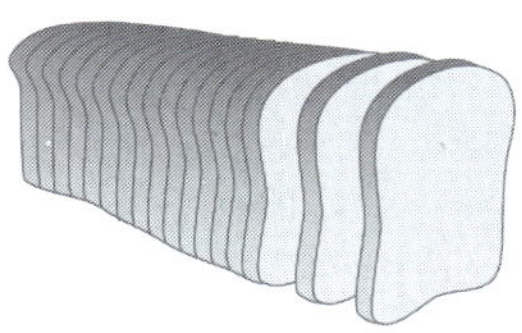

42. Bank Branches A bank with three branches, A, B, and C, finds that customers usually return to the same branch for their banking needs. However, at times a customer may go to a different branch because of a changed circumstance. For example, a person who usually goes to branch A may sometimes deviate and go to branch B because the person has business to conduct in the vicinity of branch B. For customers of branch A, suppose that 80% return to A on their next visit, 10% go to B, and 10% go to C. For customers of branch B, suppose that 70% return to B on their next visit, 20% go to A, and 10% go to C. For customers of branch C, suppose that 70% return to C on their next visit, 20% go to A, and 10% go to B.

(a) Find a transition matrix for this situation.

(b) If a customer most recently went to branch B, what is the probability that the customer returns to B on the second bank visit from now?

(c) Initially, suppose 200 customers go to A, 200 go to B, and 100 go to C. On their next visit, how many can be expected to go to A? To B? To C?

(d) Of the initial 500 customers, in the long run how many can be expected to go to A? To B? To C?

43. Show that the transition matrix $\mathbf{T} = \begin{bmatrix} \frac{1}{2} & \frac{1}{2} \\ 1 & 0 \end{bmatrix}$ is regular.

(*Hint:* Examine the entries in $\mathbf{T}^2$.)

44. Show that the transition matrix $\begin{bmatrix} 0 & 1 \\ 1 & 0 \end{bmatrix}$ is not regular.

9.4 Review

Important Terms and Symbols

Section 9.1 discrete random variable continuous random variable probability function probability distribution probability histogram sigma notation, $\sum_x f(x)$ mean, μ expected value, $E(X)$ variance, $\text{Var}(X)$, σ^2 standard deviation, σ

Section 9.2 binomial theorem binomial coefficients independent trials Bernoulli trials binomial experiment binomial distribution

Section 9.3 Markov chain states of system transition matrix, $\mathbf{T}$ state vector, $\mathbf{X_n}$ regular transition matrix steady-state vector, $\mathbf{Q}$

Summary

If X is a discrete random variable and f is the function such that $f(x) = P(X = x)$, then f is called the probability function, or probability distribution, of X. In general,

$$\sum_x f(x) = 1$$

The mean, or expected value, of X is the long-run average of X and is denoted μ or $E(X)$:

$$\mu = E(X) = \sum_x x f(x)$$

The mean can be interpreted as a measure of the central tendency of X in the long run. A measure of the dispersion of X is the variance, denoted $\text{Var}(X)$ and is given by

$$\text{Var}(X) = \sum_x (x - \mu)^2 f(x)$$

or, equivalently, by

$$\text{Var}(X) = \sum_x x^2 f(x) - \mu^2$$

Another measure of dispersion of X is the standard deviation σ:

$$\sigma = \sqrt{\text{Var}(X)}$$

If an experiment is repeated several times, then each performance of the experiment is called a trial. The trials are independent when the outcome of any single trial does not affect the outcome of any other. If there are only two possible outcomes (success and failure) for each independent trial, and the probabilities of success and failure do not change from trial to trial, then the experiment is called a binomial experiment. For such an experiment, if X is the number of successes in n trials, then the distribution f of X is called a binomial distribution, and

$$f(x) = P(X = x) = {}_nC_x p^x q^{n-x}$$

where p is the probability of success on any trial and $q = 1 - p$ is the probability of failure. The mean μ and standard deviation σ of X are given by

$$\mu = np \quad \text{and} \quad \sigma = \sqrt{npq}$$

A binomial distribution is intimately connected with the binomial theorem, which is a formula for expanding the nth power of a binomial, namely,

$$(a+b)^n = {}_nC_0a^n + {}_nC_1a^{n-1}b + \cdots + {}_nC_{n-1}ab^{n-1} + {}_nC_nb^n$$

where n is a positive integer.

A Markov chain is a sequence of trials of an experiment in which the possible outcomes of each trial, which are called states, remain the same from trial to trial, are finite in number, and have probabilities that depend only upon the outcome of the previous trial. For a k-state Markov chain, if the probability of moving from state i to state j from one trial to the next is represented by t_{ij}, then the $k \times k$ matrix $\mathbf{T} = [t_{ij}]$ is called the transition matrix for the chain. The entries in the nth power of $\mathbf{T}$ also represent probabilities; the entry in the ith row and jth column of $\mathbf{T}^n$ gives the probability of moving from state i to state j in n steps. A k-entry row vector in which the entry x_j is the probability of being in state j after the nth trial is called a state vector and is denoted $\mathbf{X_n}$. The initial state probabilities are represented by the initial state vector $\mathbf{X_0}$. The state vector $\mathbf{X_n}$ can be found by multiplying the previous state vector $\mathbf{X_{n-1}}$ by the transition matrix $\mathbf{T}$:

$$\mathbf{X_n} = \mathbf{X_{n-1}T}$$

Alternatively, $\mathbf{X_n}$ can be found by multiplying the initial state vector $\mathbf{X_0}$ by $\mathbf{T}^n$:

$$\mathbf{X_n} = \mathbf{X_0T}^n$$

If the transition matrix $\mathbf{T}$ is regular (that is, if there is a positive integer power of $\mathbf{T}$ for which all entries are positive), then, as the number n of trials increases, $\mathbf{X_n}$ gets closer and closer to the vector $\mathbf{Q}$, called the steady-state vector of $\mathbf{T}$. If

$$\mathbf{Q} = [q_1 \quad q_2 \quad \cdots \quad q_k]$$

then the entries of $\mathbf{Q}$ indicate the long-run probability distribution of the states. The vector $\mathbf{Q}$ can be found by solving the matrix equation

$$\mathbf{QT} = \mathbf{Q}$$

subject to the condition that

$$q_1 + q_2 + \cdots + q_k = 1$$

From a procedural point of view, the entries of $\mathbf{Q}$ can be found by reducing a matrix of the form

$$\left[\begin{array}{c|c} \mathbf{T}^{\text{T}} - \mathbf{I} & \mathbf{O} \\ \\ 1 \cdots 1 & 1 \end{array}\right]$$

Review Problems

Problem numbers shown in color indicate problems suggested for use as a practice chapter test.

In Problems 1 and 2, the distribution for the random variable X is given. Construct the probability histogram and determine μ, $Var(X)$, and σ.

1. $f(1) = 0.7$, $f(2) = 0.1$, $f(3) = 0.2$
2. $f(0) = \frac{1}{6}$, $f(1) = \frac{1}{2}$, $f(2) = \frac{1}{3}$
3. **Coin and Die** A fair coin and a fair die are tossed. Let X be the sum of the number of heads and the number of dots that show. Determine (a) the distribution f for X and (b) $E(X)$.
4. **Cards** Two cards from a standard deck of 52 playing cards are randomly drawn in succession without replacement, and the number of aces, X, is observed. Determine (a) the distribution f for X and (b) $E(X)$.
5. **Card Game** In a game, a player pays \$0.25 to randomly draw 2 cards, with replacement, from a standard deck of 52 playing cards. For each 10 that appears, the player receives \$1. What is the player's expected gain or loss? Give your answer to the nearest cent.
6. **Gas Station Profits** An oil company determines that the probability that a gas station located along an interstate highway is successful is 0.45. A successful station earns an annual profit of \$40,000; a station that is not successful loses \$10,000 annually. What is the expected gain to the company if it locates a station along an interstate highway?
7. **Mail-Order Computers** A mail-order computer company offers a 30-day money-back guarantee to any customer who is not completely satisfied with its

product. The company realizes a profit of \$200 for each computer sold, but assumes a loss of \$100 for shipping and handling for each unit returned. The probability that a unit is returned is 0.08.

(a) What is the expected gain for each unit shipped?
(b) If the distributor ships 4000 units per year, what is the expected annual profit?

8. Lottery In a lottery, you pay \$1.00 to choose one of 41 million number combinations. If that combination is drawn, you win \$15 million. What is your expected gain (or loss) per play?

In Problems 9 and 10, determine the distribution f for the binomial random variable X if the number of trials is n and the probability of success on any trial is p. Also, find μ and σ.

9. $n = 4,\ p = 0.15$ **10.** $n = 4,\ p = \frac{1}{4}$

In Problems 11 and 12, determine the given probability if X is a binomial random variable, n is the number of trials, and p is the probability of success on any trial.

11. $P(X \leq 1);\ n = 5,\ p = \frac{3}{4}$ **12.** $P(X > 2);\ n = 6,\ p = \frac{2}{3}$

13. Die A fair die is rolled four times. Find the probability that exactly three of the rolls result in a 2 or a 3.

14. Planting Success The probability that a certain type of bush survives planting is 0.9. If four bushes are planted, what is the probability that all of them die?

15. Coin A biased coin is tossed four times. The probability that a head occurs on any toss is $\frac{1}{3}$. Find the probability that at least two heads occur.

16. Marbles An urn contains two red, three green, and five black marbles. Five marbles are randomly drawn in succession with replacement. Find the probability that at most two of the marbles are red.

In Problems 17 and 18, a transition matrix for a Markov chain is given. Determine the values of a, b, and c.

17. $\begin{bmatrix} 0.1 & a & 0.6 \\ 2a & b & b \\ a & b & c \end{bmatrix}$ **18.** $\begin{bmatrix} a & a & 0.2 \\ a & b & c \\ b & b & a \end{bmatrix}$

In Problems 19 and 20, a transition matrix **T** *and an initial state vector* $\mathbf{X}_0$ *for a Markov chain are given. Compute the state vectors* $\mathbf{X}_1$, $\mathbf{X}_2$, *and* $\mathbf{X}_3$.

19. $\mathbf{T} = \begin{bmatrix} 0.1 & 0.2 & 0.7 \\ 0.3 & 0.4 & 0.3 \\ 0.1 & 0.1 & 0.8 \end{bmatrix}$ $\mathbf{X}_0 = [0.5\ \ 0\ \ 0.5]$

20. $\mathbf{T} = \begin{bmatrix} 0.4 & 0.2 & 0.4 \\ 0.1 & 0.6 & 0.3 \\ 0.1 & 0.5 & 0.4 \end{bmatrix}$ $\mathbf{X}_0 = [0.1\ \ 0.3\ \ 0.6]$

In Problems 21 and 22, a transition matrix **T** *for a Markov chain is given.*

(a) Compute $\mathbf{T}^2$ and $\mathbf{T}^3$.
(b) What is the probability of going from state 1 to state 2 after two steps?
(c) What is the probability of going from state 2 to state 1 after three steps?

21. $\begin{bmatrix} \frac{1}{7} & \frac{6}{7} \\ \frac{3}{7} & \frac{4}{7} \end{bmatrix}$ **22.** $\begin{bmatrix} 0 & 0 & 1 \\ 0.4 & 0.3 & 0.3 \\ 0.3 & 0.5 & 0.2 \end{bmatrix}$

In Problems 23 and 24, find the steady-state vector for the given transition matrix for a Markov chain.

23. $\begin{bmatrix} \frac{1}{3} & \frac{2}{3} \\ \frac{2}{3} & \frac{1}{3} \end{bmatrix}$. **24.** $\begin{bmatrix} 0.4 & 0.3 & 0.3 \\ 0.4 & 0.2 & 0.4 \\ 0.3 & 0.3 & 0.4 \end{bmatrix}$.

25. Automobile Market For a particular segment of the automobile market, the results of a survey indicate that 80% of Japanese-car owners would buy a Japanese car the next time and 20% would buy a non-Japanese car. Of non-Japanese-car owners, 40% would buy a non-Japanese car the next time and 60% would buy a Japanese car.

(a) Of those who currently own a Japanese car, what percentage will buy a Japanese car two cars later?
(b) If 60% of this segment currently own a Japanese car and 40% own a non-Japanese car, what will be the distribution for this segment of the market two cars from now?
(c) How will this segment be distributed in the long run?

26. Voting Suppose that the probabilities of voting for particular parties in a future election depend on the voting patterns in the previous election. For a certain region where there is a three-party political system, assume that these probabilities are contained in the matrix

$$\mathbf{T} = [t_{ij}] = \begin{bmatrix} 0.7 & 0.2 & 0.1 \\ 0.4 & 0.5 & 0.1 \\ 0.1 & 0.1 & 0.8 \end{bmatrix}$$

where t_{ij} represents the probability that a voter who voted for party i in the last election will vote for party j in the next election.

(a) At the last election, 50% of the electorate voted for party 1, 30% for party 2, and 20% for party 3. What is the expected percentage distribution of votes for the next election?
(b) In the long run, what is the percentage distribution of votes? Give your answers to the nearest percent.

Mathematical SNAPSHOT

Markov Chains in Game Theory

Game theory is the mathematical study of how people behave in competitive and cooperative situations. Each situation is represented by a table showing the players' options and their payoffs for various outcomes. Figure 9.9 shows a famous game, called the Prisoner's Dilemma. Each player chooses between "Cooperate" and "Defect." The players' subsequent payoffs are in the cell to the right of Player 1's choice and below Player 2's. Player 1 gets the payoff in the lower left corner of the cell, Player 2 the payoff in the upper right corner.

		Player 2: Cooperate	Player 2: Defect
Player 1	Cooperate	Player 1: 4, Player 2: 4	Player 1: 1, Player 2: 5
	Defect	Player 1: 5, Player 2: 1	Player 1: 2, Player 2: 2

FIGURE 9.9 The Prisoner's Dilemma.

Most people who have studied the game think that the rational choice for a player trying to maximize his or her payoff is to defect—a move that, if played by both players, leads to a payoff of 2 for each. The paradox is that both players would be better off (payoff of 4) with the supposedly irrational choice, Cooperate. You should study the game for a few minutes and see what you think.

The game gets its name from a fable about two prisoners, but the same puzzle arises whenever two people agree to do each other a favor but have no way of checking up on each other. Both benefit if both cooperate, but each is tempted to cheat. On a larger scale, voting and voluntary garbage recycling are cooperative behaviors in a many-person Prisoner's Dilemma where the players are all the members of a community.

Game theorists have long been troubled by the seeming irrationality of Cooperate. Trying to understand how it might make sense after all, the theorists have studied scenarios in which the Prisoner's Dilemma is played not just once, but repeatedly. In this Iterated Prisoner's Dilemma, more than just two strategies are possible, because a player's play in a given round can be based on what happened in the previous round.

One much-discussed strategy, called Tit-for-Tat, is to cooperate initially and from then on to do whatever the other player did in the preceding round. If the other player defected in round n, Tit-for-Tat defects in round $n + 1$; if the other player cooperated in round n, Tit-for-Tat cooperates in round $n + 1$. Of course, "Always Cooperate" and "Always Defect" are also options. "Tit-for-Tat," however, models a more sophisticated way of thinking: It is prepared to cooperate but does not allow itself to be repeatedly exploited by the other's defections.

The succession of rounds in the Iterated Prisoner's Dilemma is a Markov chain with four states, where each state represents a combination of players' choices. Let us number the states as follows: Both cooperate = State 1; Player 1 cooperates/Player 2 defects = State 2; Player 1 defects/Player 2 cooperates = State 3; Both defect = State 4. The transitions between states are fully determined by the players' choice of strategies. If both are playing Tit-for-Tat, then State 1 leads to State 1, State 2 leads to State 3, State 3 leads to State 2, and State 4 leads to State 4. You should take a moment to convince yourself of this. The transition matrix looks like this:

$$\mathbf{T} = \begin{array}{c} \\ 1 \\ 2 \\ 3 \\ 4 \end{array} \begin{array}{c} \begin{array}{cccc} 1 & 2 & 3 & 4 \end{array} \\ \begin{bmatrix} 1 & 0 & 0 & 0 \\ 0 & 0 & 1 & 0 \\ 0 & 1 & 0 & 0 \\ 0 & 0 & 0 & 1 \end{bmatrix} \end{array}$$

With this matrix, the course of play crucially depends on what choice the two players open with. If Player 1 cooperates and Player 2 defects, initially, then the successive state vectors look like this:

$$\mathbf{X_0} = [0 \quad 1 \quad 0 \quad 0]$$
$$\mathbf{X_0T} = [0 \quad 0 \quad 1 \quad 0]$$
$$\mathbf{X_0T}^2 = [0 \quad 1 \quad 0 \quad 0]$$
$$\mathbf{X_0T}^3 = [0 \quad 0 \quad 1 \quad 0]$$
$$\vdots$$

The players oscillate back and forth between States 2 and 3. On the other hand, States 1 and 4 are both "traps": If two Tit-for-Tat players start off both cooperating, they will cooperate all down the line, but if they both start with Defect, they will continue in that pattern.

It is regrettable that two Tit-for-Tat players, both ready to cooperate, may get locked into the cycle of defection instead. What if we modify Tit-for-Tat slightly, so that it occasionally forgives defection and cooperates in the next round, as a sort of peace offering? We can model this by giving each side, let us say, a 10% chance of cooperating after the other has defected. The transition matrix then ends up looking like this:

$$\mathbf{T} = \begin{array}{c} \\ 1 \\ 2 \\ 3 \\ 4 \end{array} \begin{array}{c} \begin{array}{cccc} 1 & 2 & 3 & 4 \end{array} \\ \begin{bmatrix} 1 & 0 & 0 & 0 \\ 0.1 & 0 & 0.9 & 0 \\ 0.1 & 0.9 & 0 & 0 \\ 0.01 & 0.09 & 0.09 & 0.81 \end{bmatrix} \end{array}$$

If both defect (State 4), there is a $0.9 \times 0.9 = 0.81$ chance that they will both defect in the next round (State 4 again), a $0.1 \times 0.9 = 0.09$ that Player 1 will cooperate and Player 2 will defect (State 2), and so on. Make sure you understand how the probabilities in the matrix are obtained.

Now two Tit-for-Tat-ers can escape State 4 and settle permanently into State 1. How long will that take? In principle the game could stay in State 4 for a long time. But sooner or later, it will find its way to State 1. How many iterations are necessary before there is at least a 0.50 probability of the players having made it to mutual cooperation? Formally, if $\mathbf{X_0} = [0 \quad 0 \quad 0 \quad 1]$, what value of n puts the first entry of the state vector $\mathbf{X_n} = \mathbf{X_0T}^n$ at greater than or equal to 0.5? The answer is 12:

$$\mathbf{X_0T}^{12} = [0.5149 \quad 0.2027 \quad 0.2027 \quad 0.0798]$$

So if a game between modified Tit-for-Tat-ers runs to 100 rounds, it is highly likely that most of those rounds will have cooperative outcomes, even if the game started out with mutual distrust. In computer simulations involving interaction between many different strategies, Tit-for-Tat generally emerges as quite profitable—more profitable, it should be noted, than Always Defect. Cooperation, it is comforting to discover, need not be irrational after all.

To learn more on the Internet, visit a search engine and type in "Iterated Prisoner's Dilemma."

Exercises

1. Using a graphing calculator, model play between two modified Tit-for-Tat-ers where one starts with Cooperate and the other with Defect. How many rounds are needed to reach at least 0.50 probability of mutual cooperation?
2. Prove that $[1 \quad 0 \quad 0 \quad 0]$ is the steady-state vector for a game between two modified Tit-for-Tat-ers.
3. Assume that all states are initially possible, but beginning with the second round, Player 2 adopts the strategy of Always Defect, Always Cooperate, or regular Tit-for-Tat. What is the transition matrix for each of these three games if Player 1 is using modified Tit-for-Tat in each case?
4. Use the results from question 3 to describe what will happen in the long run with each game.

CHAPTER

10

Limits and Continuity

The philosopher Zeno of Elea was fond of paradoxes about motion. His most famous one goes something like this. The warrior Achilles agrees to run a race against a tortoise. Achilles can run 10 meters per second and the tortoise only 1 meter per second, so the tortoise gets a 10-meter head start. Since Achilles is so much faster, he should still win. But by the time he has covered his first 10 meters and reached the place where the tortoise started, the tortoise has advanced 1 meter and is still that far ahead. And after Achilles has covered that 1 meter, the tortoise has advanced another 0.1 meter and is still ahead. And so on. Therefore, Achilles gets closer and closer to the tortoise but can never catch up.

Zeno's audience knew, of course, that the argument was fishy. We can write an algebraic equation with Achilles' total progress on the left, the tortoise's on the right, and t representing the time in seconds at which Achilles pulls even with the tortoise:

$$(10 \text{ m/s})t = (1 \text{ m/s})t + 10 \text{ m}$$

The solution is $t = 1\frac{1}{9}$ seconds, at which time Achilles will have run $\left(1\frac{1}{9}\text{s}\right)(10 \text{ m/s}) = 11\frac{1}{9}$ meters.

What puzzled Zeno and his listeners is how it could be that

$$10 + 1 + \frac{1}{10} + \frac{1}{100} + \cdots = 11\frac{1}{9}$$

where the left side represents an *infinite sum* and the right side is a finite result. The modern solution to this problem is the concept of a limit, which is the key topic of this chapter. The left side of the equation is an infinite geometric series. Using limit notation and the formula from Section 5.3 for the sum of a geometric series, we write

$$\lim_{k\to\infty} \sum_{n=0}^{k} 10^{1-n} = \lim_{k\to\infty} \frac{10\left(1 - \left(\frac{1}{10}\right)^{k+1}\right)}{1 - \frac{1}{10}} = \frac{100}{9} = 11\frac{1}{9}$$

OBJECTIVE *To study limits and their basic properties.*

10.1 Limits

Perhaps you have been in a parking-lot situation in which you must "inch up" to the car in front, but yet you do not want to bump or touch it. This notion of getting closer and closer to something, but yet not touching it, is very important in mathematics and is involved in the concept of *limits,* which lies at the foundation of calculus. Basically, we will let a variable "inch up" to a particular value and examine the effect it has on the values of a function.

For example, consider the function

$$f(x) = \frac{x^3 - 1}{x - 1}$$

TABLE 10.1

$x < 1$		$x > 1$	
x	$f(x)$	x	$f(x)$
0.8	2.44	1.2	3.64
0.9	2.71	1.1	3.31
0.95	2.8525	1.05	3.1525
0.99	2.9701	1.01	3.0301
0.995	2.985025	1.005	3.015025
0.999	2.997001	1.001	3.003001

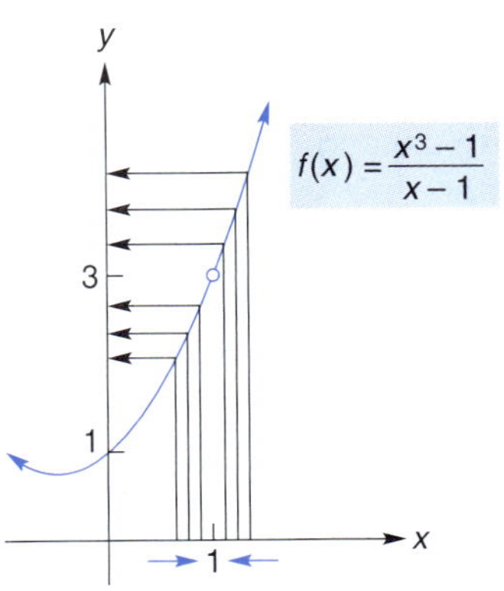

FIGURE 10.1 $\lim_{x\to 1} \frac{x^3 - 1}{x - 1} = 3.$

Although this function is not defined at $x = 1$, we may be curious about the behavior of the function values as x gets very close to 1. Table 10.1 gives some values of x that are slightly less than 1 and some that are slightly greater, as well as their corresponding function values. Notice that as x takes on values closer and closer to 1, regardless of whether x approaches it *from the left* ($x < 1$) or *from the right* ($x > 1$), the corresponding values of $f(x)$ get closer and closer to one and only one number, 3. This is also clear from the graph of f in Figure 10.1. Notice there that even though the function is not defined at $x = 1$ (as indicated by the hollow dot), the function values get closer and closer to 3 as x gets closer and closer to 1. To express this, we say that the **limit** of $f(x)$ as x approaches 1 is 3 and write

$$\lim_{x\to 1} \frac{x^3 - 1}{x - 1} = 3$$

We can make $f(x)$ as close to 3 as we wish, and keep it that close, by taking x sufficiently close to, but not equal to, 1. The limit exists at 1, even though 1 is not in the domain of f.

We can also consider the limit of a function as x approaches a number that is in the domain. Let us examine the limit of $f(x) = x + 3$ as x approaches 2:

$$\lim_{x\to 2}(x + 3)$$

Obviously, if x is close to 2 (but not equal to 2), then $x+3$ is close to 5. This is also apparent from the table and graph in Figure 10.2. Thus,

$$\lim_{x \to 2}(x+3) = 5$$

$x < 2$

x	$f(x)$
1.5	4.5
1.9	4.9
1.95	4.95
1.99	4.99
1.999	4.999

$x > 2$

x	$f(x)$
2.5	5.5
2.1	5.1
2.05	5.05
2.01	5.01
2.001	5.001

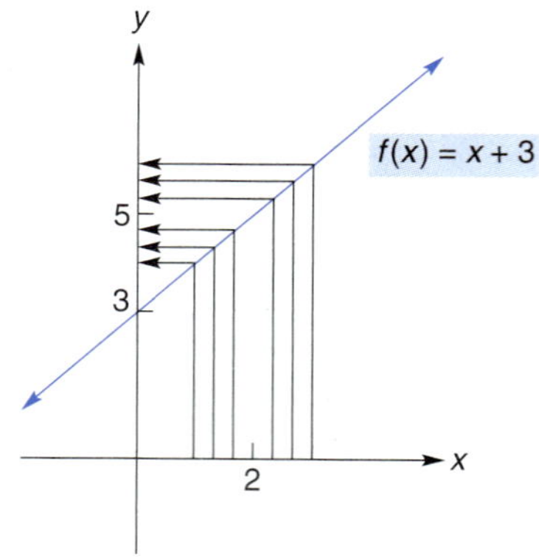

FIGURE 10.2 $\lim_{x \to 2}(x+3) = 5$.

Although $x+3$ is 5 when $x = 2$, this has no bearing on the existence of a limit.

In general, for any function f, we have the following definition of a limit.

Definition

The ***limit*** of $f(x)$ as x approaches a is the number L, written

$$\lim_{x \to a} f(x) = L$$

provided that $f(x)$ is arbitrarily close to L for all x sufficiently close to, but not equal to, a.

We emphasize that, when finding a limit, we are concerned not with what happens to $f(x)$ when x equals a, but only with what happens to $f(x)$ when x is close to a. Moreover, a limit must be independent of the way in which x *approaches* a, meaning the way in which x gets close to a. That is, the limit must be the same whether x approaches a from the left or from the right (for $x < a$ or $x > a$, respectively).

EXAMPLE 1 Estimating a Limit from a Graph

a. *Estimate* $\lim_{x \to 1} f(x)$, *where the graph of f is given in Figure 10.3(a).*

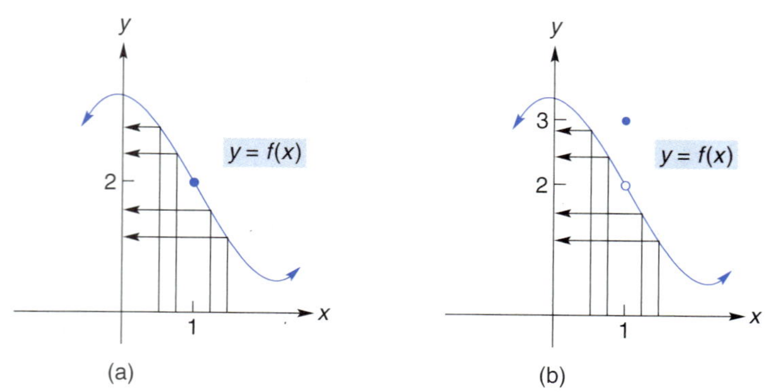

FIGURE 10.3 Investigation of $\lim_{x \to 1} f(x)$.

Solution: If we look at the graph for values of x near 1, we see that $f(x)$ is near 2. Moreover, as x gets closer and closer to 1, $f(x)$ appears to get closer and closer to 2. Thus, we estimate that

$$\lim_{x\to 1} f(x) \text{ is } 2$$

b. *Estimate* $\lim_{x\to 1} f(x)$*, where the graph of f is given in Figure 10.3(b).*

Solution: Although $f(1) = 3$, this fact has no bearing whatsoever on the limit of $f(x)$ as x approaches 1. We see that as x gets closer and closer to 1, $f(x)$ appears to get closer and closer to 2. Thus, we estimate that

$$\lim_{x\to 1} f(x) \text{ is } 2$$

▶ **Now work Problem 1.**

Up to now, all of the limits that we have considered did indeed exist. Next we shall look at some situations in which a limit does not exist.

APPLICATIONS IN PRACTICE 1

Limits That Do Not Exist

The greatest integer function in mathematics, denoted $f(x) = [x]$, is used every day by cashiers making change for customers. This function tells the amount of paper money for each amount of change owed. (For example, if a customer is owed \$1.25 in change, he or she would get \$1 in paper money; thus, $[1.25] = 1$.) Formally, $[x]$ is defined as the greatest integer less than or equal to x. Graph f, sometimes called a step function, on your graphing calculator in the standard viewing rectangle. (You will find it in the numbers menu; it's called "integer part.") Explore this graph using TRACE. Determine whether $\lim_{x\to a} f(x)$ exists.

EXAMPLE 2 Limits That Do Not Exist

a. *Estimate* $\lim_{x\to -2} f(x)$ *if it exists, where the graph of f is given in Figure 10.4.*

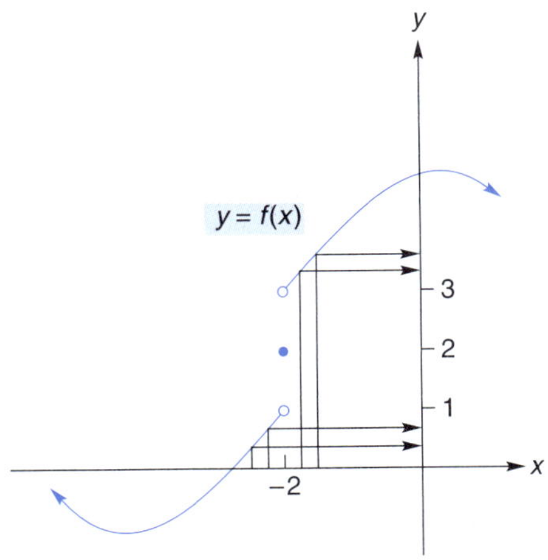

FIGURE 10.4 $\lim_{x\to -2} f(x)$ does not exist.

Solution: As x approaches -2 from the left ($x < -2$), the values of $f(x)$ appear to get closer to 1. But as x approaches -2 from the right ($x > -2$), $f(x)$ appears to get closer to 3. Hence, as x approaches -2, the function values do not settle down to one and only one number. We conclude that

$$\lim_{x\to -2} f(x) \text{ does not exist}$$

Note that the limit does not exist even though the function is defined at $x = -2$.

b. *Estimate* $\lim_{x\to 0} \dfrac{1}{x^2}$ *if it exists.*

Solution: Let $f(x) = \dfrac{1}{x^2}$. The table in Figure 10.5 gives values of $f(x)$ for some values of x near 0. As x gets closer and closer to 0, the values of $f(x)$

x	$f(x)$
±1	1
±0.5	4
±0.1	100
±0.01	10,000
±0.001	1,000,000

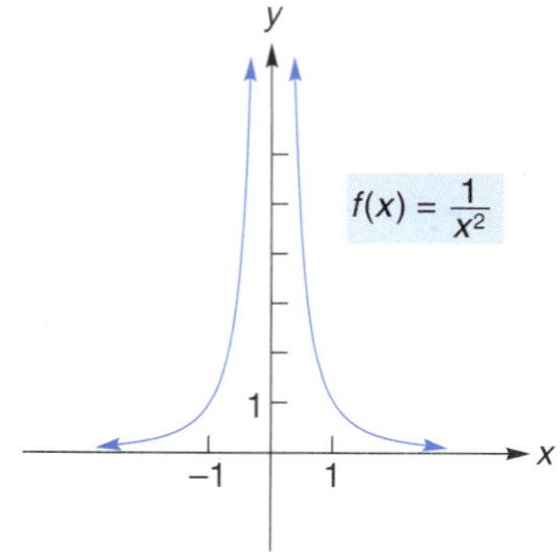

FIGURE 10.5 $\lim_{x \to 0} \frac{1}{x^2}$ does not exist.

get larger and larger without bound. This is also clear from the graph. Since the values of $f(x)$ do not approach a number as x approaches 0,

$$\lim_{x \to 0} \frac{1}{x^2} \text{ does not exist}$$

▶ **Now work Problem 3.**

TECHNOLOGY

Problem: Estimate $\lim_{x \to 2} f(x)$ if

$$f(x) = \frac{x^3 + 2.1x^2 - 10.2x + 4}{x^2 + 2.5x - 9}$$

Solution: One method of finding the limit is by constructing a table of function values $f(x)$ when x is close to 2. From Figure 10.6, we estimate the limit to be 1.57. Alternatively, we can estimate the limit from the graph of f. Figure 10.7 shows the graph of f in the standard window of $[-10, 10] \times [-10, 10]$. First we zoom in several times around $x = 2$ and obtain Figure 10.8. After tracing around $x = 2$, we estimate the limit to be 1.57.

X	Y1
1.9	1.4688
1.99	1.5592
1.999	1.5682
1.9999	1.5691
2.01	1.5793
2.001	1.5702
2.0001	1.5693

X=2.0001

FIGURE 10.6 $\lim_{x \to 2} f(x) \approx 1.57$.

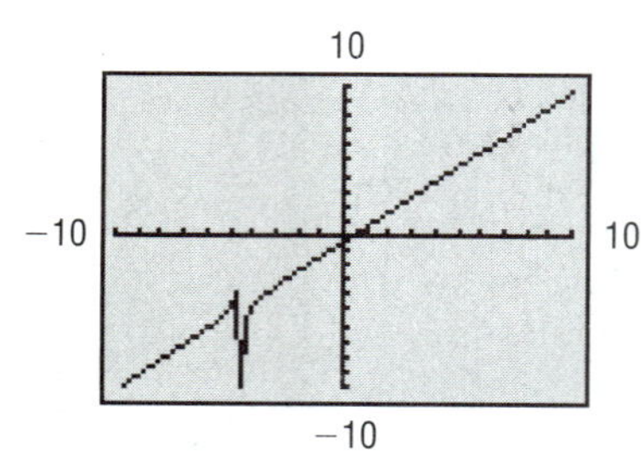

FIGURE 10.7 Graph of $f(x)$ in standard window.

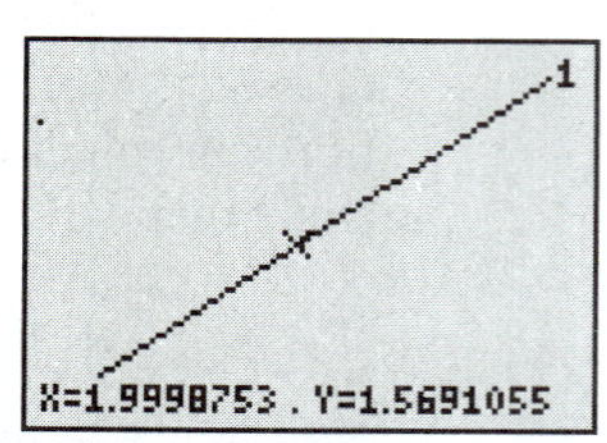

FIGURE 10.8 Zooming and tracing around $x = 2$ gives $\lim_{x \to 2} f(x) \approx 1.57$.

Properties of Limits

To determine limits, we do not always want to compute function values or sketch a graph. Alternatively, there are several properties of limits that we may be able to employ. The following properties may seem reasonable to you:

1. If $f(x) = c$ is a constant function, then

$$\lim_{x \to a} f(x) = \lim_{x \to a} c = c$$

2. $\lim_{x \to a} x^n = a^n$, for any positive integer n

EXAMPLE 3 **Applying Limit Properties 1 and 2**

a. $\lim_{x \to 2} 7 = 7$; $\lim_{x \to -5} 7 = 7$

b. $\lim_{x \to 6} x^2 = 6^2 = 36$

c. $\lim_{t \to -2} t^4 = (-2)^4 = 16$

▸ **Now work Problem 9.**

Some other properties of limits are as follows:

If $\lim_{x \to a} f(x)$ and $\lim_{x \to a} g(x)$ exist, then

3. $\lim_{x \to a}[f(x) \pm g(x)] = \lim_{x \to a} f(x) \pm \lim_{x \to a} g(x)$

That is, the limit of a sum or difference is the sum or difference, respectively, of the limits.

4. $\lim_{x \to a}[f(x) \cdot g(x)] = \lim_{x \to a} f(x) \cdot \lim_{x \to a} g(x)$

That is, the limit of a product is the product of the limits.

5. $\lim_{x \to a}[cf(x)] = c \cdot \lim_{x \to a} f(x)$, where c is a constant

That is, the limit of a constant times a function is the constant times the limit of the function.

APPLICATIONS IN PRACTICE 2

Applying Limit Properties

The volume of helium in a spherical balloon (in cubic centimeters), as a function of the radius r in centimeters, is given by $V(r) = \frac{4}{3}\pi r^3$. Find $\lim_{r \to 1} V(r)$.

EXAMPLE 4 **Applying Limit Properties**

a.

$$\begin{aligned} \lim_{x \to 2}(x^2 + x) &= \lim_{x \to 2} x^2 + \lim_{x \to 2} x && \text{(Property 3)} \\ &= 2^2 + 2 = 6 && \text{(Property 2)} \end{aligned}$$

b. Property 3 can be extended to the limit of a finite number of sums and differences. For example,

$$\begin{aligned} \lim_{q \to -1}(q^3 - q + 1) &= \lim_{q \to -1} q^3 - \lim_{q \to -1} q + \lim_{q \to -1} 1 \\ &= (-1)^3 - (-1) + 1 = 1 \end{aligned}$$

c. $$\lim_{x\to 2}[(x+1)(x-3)] = \lim_{x\to 2}(x+1)\cdot \lim_{x\to 2}(x-3) \quad \text{(Property 4)}$$
$$= [\lim_{x\to 2} x + \lim_{x\to 2} 1]\cdot[\lim_{x\to 2} x - \lim_{x\to 2} 3]$$
$$= (2+1)\cdot(2-3) = 3(-1) = -3$$

d. $$\lim_{x\to -2} 3x^3 = 3\cdot \lim_{x\to -2} x^3 \quad \text{(Property 5)}$$
$$= 3(-2)^3 = -24$$

▶ **Now work Problem 11.**

APPLICATIONS IN PRACTICE 3

Limit of a Polynomial Function

The revenue function for a certain product is given by $R(x) = 500x - 6x^2$. Find $\lim_{x\to 8} R(x)$.

EXAMPLE 5 Limit of a Polynomial Function

Let $f(x) = c_n x^n + c_{n-1}x^{n-1} + \cdots + c_1 x + c_0$ define a polynomial function. Then

$$\lim_{x\to a} f(x) = \lim_{x\to a}(c_n x^n + c_{n-1}x^{n-1} + \cdots + c_1 x + c_0)$$
$$= c_n \cdot \lim_{x\to a} x^n + c_{n-1}\cdot \lim_{x\to a} x^{n-1} + \cdots + c_1 \cdot \lim_{x\to a} x + \lim_{x\to a} c_0$$
$$= c_n a^n + c_{n-1}a^{n-1} + \cdots + c_1 a + c_0 = f(a)$$

Thus, we have the following property:

If f is a polynomial function, then

$$\lim_{x\to a} f(x) = f(a)$$

That is, the limit of a polynomial function as x approaches a is merely the value of the function at a.

▶ **Now work Problem 13.**

The result of Example 5 allows us to find many limits as $x \to a$ by just substituting a for x. For example, we can find

$$\lim_{x\to -3}(x^3 + 4x^2 - 7)$$

by substituting -3 for x because $x^3 + 4x^2 - 7$ is a polynomial function:

$$\lim_{x\to -3}(x^3 + 4x^2 - 7) = (-3)^3 + 4(-3)^2 - 7 = 2$$

Similarly,

$$\lim_{h\to 3}[2(h-1)] = 2(3-1) = 4$$

We want to stress that one does not evaluate limits simply by "plugging in," unless one is relying on a rule that covers the situation. We were able to find the previous two limits by direct substitution because we have a rule that applies to limits of polynomial functions. However, indiscriminate use of substitution can lead to erroneous results. To illustrate, in Example 1(b) we have $f(1) = 3$, which is not the limit as $x \to 1$; in Example 2(a), $f(-2) = 2$, which is not the limit as $x \to -2$.

The next two limit properties concern quotients and roots.

If $\lim_{x\to a} f(x)$ and $\lim_{x\to a} g(x)$ exist, then

6.

$$\lim_{x\to a}\frac{f(x)}{g(x)}=\frac{\lim_{x\to a} f(x)}{\lim_{x\to a} g(x)} \quad \text{if} \quad \lim_{x\to a} g(x)\neq 0$$

That is, the limit of a quotient is the quotient of limits, provided that the denominator does not have a limit of 0.

7.

$$\lim_{x\to a}\sqrt[n]{f(x)}=\sqrt[n]{\lim_{x\to a} f(x)}$$ [1]

Note that in Example 6(a) the numerator and denominator of the function are polynomials. In general, we can determine the limit of a rational function as $x \to a$ by direct substitution, provided that the denominator is not 0 at a.

APPLICATIONS IN PRACTICE 4

Applying Limit Property

The rate of change of productivity p (in number of units produced per hour) increases with time on the job by the function

$$p=\frac{50(t^2+4t)}{t^2+3t+20}. \text{ Find } \lim_{t\to 2} p.$$

EXAMPLE 6 **Applying Limit Properties 6 and 7**

a. $\lim_{x\to 1}\frac{2x^2+x-3}{x^3+4}=\frac{\lim_{x\to 1}(2x^2+x-3)}{\lim_{x\to 1}(x^3+4)}=\frac{2+1-3}{1+4}=\frac{0}{5}=0$

b. $\lim_{t\to 4}\sqrt{t^2+1}=\sqrt{\lim_{t\to 4}(t^2+1)}=\sqrt{17}$

c. $\lim_{x\to 3}\sqrt[3]{x^2+7}=\sqrt[3]{\lim_{x\to 3}(x^2+7)}=\sqrt[3]{16}=\sqrt[3]{8\cdot 2}=2\sqrt[3]{2}$

▶ **Now work Problem 15.**

Limits and Algebraic Manipulation

We now consider limits to which our limit properties do not apply and which cannot be evaluated by direct substitution. Our technique will be to algebraically manipulate $f(x)$ so as to obtain a form to which our limit properties will apply.

EXAMPLE 7 **Finding a Limit by Factoring and Cancellation**

Find $\lim_{x\to -1}\frac{x^2-1}{x+1}$.

Solution: As $x \to -1$, both numerator and denominator approach zero. Because the limit of the denominator is 0, we *cannot* use Property 6. However, since what happens to the quotient when x equals -1 is of no concern, we can assume that $x \neq -1$ and simplify the fraction:

$$\frac{x^2-1}{x+1}=\frac{(x+1)(x-1)}{x+1}=x-1 \qquad \text{for } x\neq -1$$

This algebraic manipulation (factoring and cancellation) on the original function $\frac{x^2-1}{x+1}$ yields a new function $x-1$, which is the same as the original function for $x \neq -1$. Thus,

$$\lim_{x\to -1}\frac{x^2-1}{x+1}=\lim_{x\to -1}\frac{(x+1)(x-1)}{x+1}=\lim_{x\to -1}(x-1)=-1-1=-2$$

Notice that, although the original function is not defined at -1, it *does* have a limit as $x \to -1$.

▶ **Now work Problem 21.**

---

[1] If n is even, we require that $\lim_{x\to a} f(x)$ be positive.

When both $f(x)$ and $g(x)$ approach 0 as $x \to a$, then the limit

$$\lim_{x \to a} \frac{f(x)}{g(x)}$$

is said to have the *form* $0/0$.

In Example 7, the method of finding a limit by direct substitution does not work. Replacing x by -1 gives $0/0$, which has no meaning. When the meaningless form $0/0$ arises, algebraic manipulation (as in Example 7) may result in a form for which the limit *can* be determined.

In the beginning of this section, we found

$$\lim_{x \to 1} \frac{x^3 - 1}{x - 1}$$

by examining a table of function values of $f(x) = (x^3 - 1)/(x - 1)$ and also by considering the graph of f. This limit has the form $0/0$. Now we shall determine the limit by using the technique described in Example 7 (the technique of factoring and cancellation).

CAUTION There is frequently confusion about what principle is being used in this example and in Example 7. It is this:

$$\text{If } f(x) = g(x) \quad \text{for } x \neq a$$

$$\text{then } \lim_{x \to a} f(x) = \lim_{x \to a} g(x)$$

EXAMPLE 8 Form 0/0

Find $\lim_{x \to 1} \frac{x^3 - 1}{x - 1}$.

Solution: As $x \to 1$, both the numerator and denominator approach 0. Thus, we shall try to express the quotient in a different form for $x \neq 1$. By factoring, we have

$$\frac{x^3 - 1}{x - 1} = \frac{(x-1)(x^2 + x + 1)}{(x-1)} = x^2 + x + 1 \qquad \text{for } x \neq 1$$

(Alternatively, long division would give the same result.) Therefore,

$$\lim_{x \to 1} \frac{x^3 - 1}{x - 1} = \lim_{x \to 1}(x^2 + x + 1) = 1^2 + 1 + 1 = 3$$

as we showed before.

▶ **Now work Problem 23.**

APPLICATIONS IN PRACTICE 5

Form 0/0

The length of a material increases as it is heated up according to the equation $l = 125 + 2x$. The rate at which the length is increasing is given by

$$\lim_{h \to 0} \frac{125 + 2(x + h) - (125 + 2x)}{h}$$

Calculate this limit.

EXAMPLE 9 Form 0/0

If $f(x) = x^2 + 1$, *find* $\lim_{h \to 0} \frac{f(x+h) - f(x)}{h}$.

Solution:

$$\lim_{h \to 0} \frac{f(x+h) - f(x)}{h} = \lim_{h \to 0} \frac{[(x+h)^2 + 1] - (x^2 + 1)}{h}$$

Here we treat x as a constant because h, not x, is changing. As $h \to 0$, both the numerator and denominator approach 0. Therefore, we shall try to express the quotient in a different form, for $h \neq 0$. We have

$$\begin{aligned}
\lim_{h \to 0} \frac{[(x+h)^2 + 1] - (x^2 + 1)}{h} &= \lim_{h \to 0} \frac{[x^2 + 2xh + h^2 + 1] - x^2 - 1}{h} \\
&= \lim_{h \to 0} \frac{2xh + h^2}{h} \\
&= \lim_{h \to 0} \frac{h(2x + h)}{h} \\
&= \lim_{h \to 0}(2x + h) \\
&= 2x
\end{aligned}$$

The expression

$$\frac{f(x+h) - f(x)}{h}$$

is called a *difference quotient*. The limit of the difference quotient lies at the heart of differential calculus. You will encounter such limits in Chapter 11.

CAUTION When $\frac{h(2x+h)}{h}$ and $2x+h$ are considered as *functions of* h, they are seen to be equal, for all $h \neq 0$. It follows that their limits as h approaches 0 are equal.

Note well. It is the fourth equality above, $\lim_{h\to 0} \frac{h(2x+h)}{h} = \lim_{h\to 0}(2x+h)$, that uses the crucial idea mentioned in the CAUTION to the left.

▸ **Now work Problem 35.**

A Special Limit

We conclude this section with a note concerning a most important limit, namely,

$$\lim_{x\to 0}(1+x)^{1/x}$$

Figure 10.9 shows the graph of $f(x) = (1+x)^{1/x}$. Although $f(0)$ does not exist, as $x \to 0$ it is clear that the limit of $(1+x)^{1/x}$ exists. It is approximately 2.71828 and is denoted by the letter e. This, you may recall, is the base of the system of natural logarithms. The limit

This limit will be used in Chapter 12, Section 1.

$$\lim_{x\to 0}(1+x)^{1/x} = e$$

can actually be considered the definition of e. In Section 10.3 we show this to be the same as the definition of e that we gave in Section 4.1.

x	$(1+x)^{1/x}$
0.5	2.2500
0.1	2.5937
0.01	2.7048
0.001	2.7169

x	$(1+x)^{1/x}$
−0.5	4.0000
−0.1	2.8680
−0.01	2.7320
−0.001	2.7196

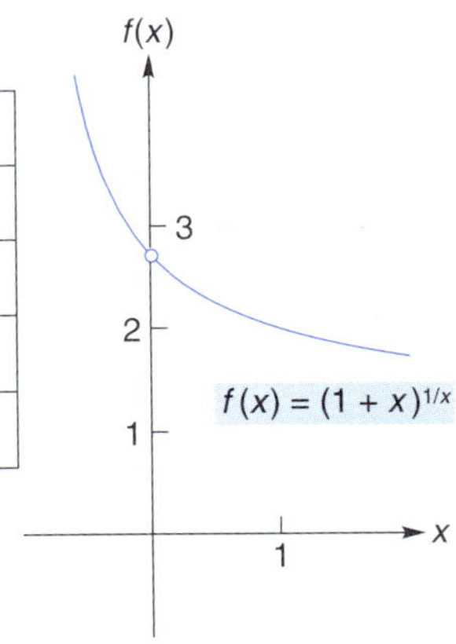

FIGURE 10.9 $\lim_{x\to 0}(1+x)^{1/x} = e$.

Exercise 10.1

In Problems 1–4, use the graph of f to estimate each limit if it exists.

*1. Graph of f appears in Figure 10.10.

(a) $\lim_{x\to 0} f(x)$ **(b)** $\lim_{x\to 1} f(x)$ **(c)** $\lim_{x\to 2} f(x)$

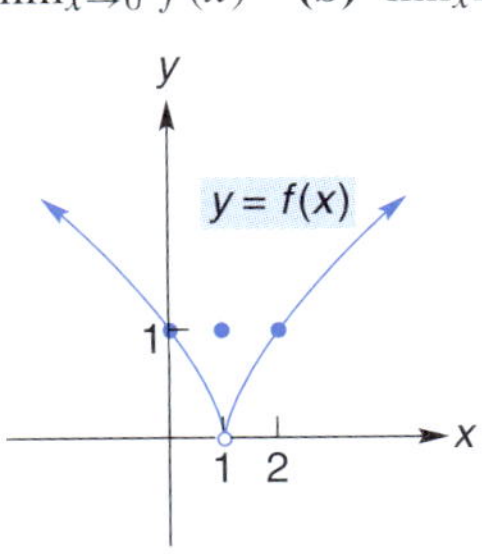

FIGURE 10.10 Diagram for Problem 1.

2. Graph of f appears in Figure 10.11.

(a) $\lim_{x\to -1} f(x)$ **(b)** $\lim_{x\to 0} f(x)$ **(c)** $\lim_{x\to 1} f(x)$

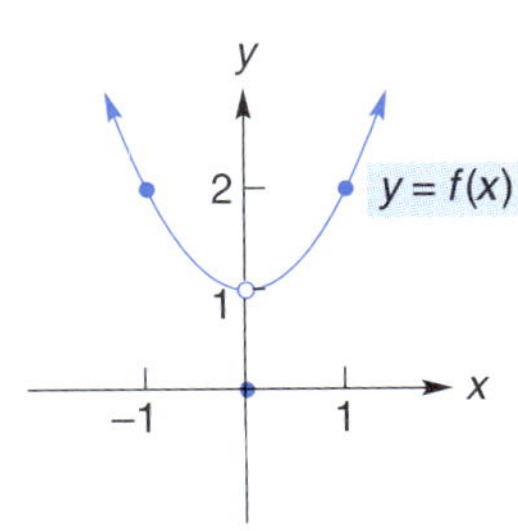

FIGURE 10.11 Diagram for Problem 2.

*3. Graph of f appears in Figure 10.12.
(a) $\lim_{x\to-1} f(x)$ **(b)** $\lim_{x\to1} f(x)$ **(c)** $\lim_{x\to2} f(x)$

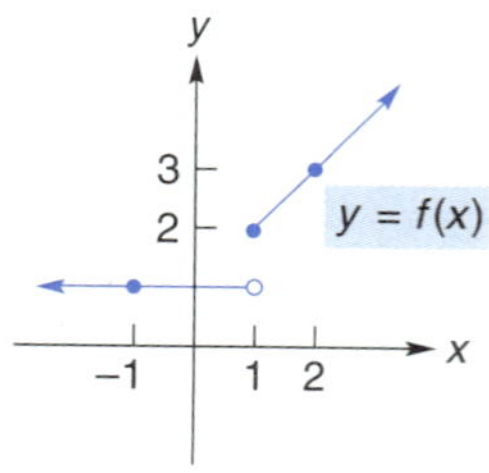

FIGURE 10.12 Diagram for Problem 3.

4. Graph of f appears in Figure 10.13.
(a) $\lim_{x\to-1} f(x)$ **(b)** $\lim_{x\to0} f(x)$ **(c)** $\lim_{x\to1} f(x)$

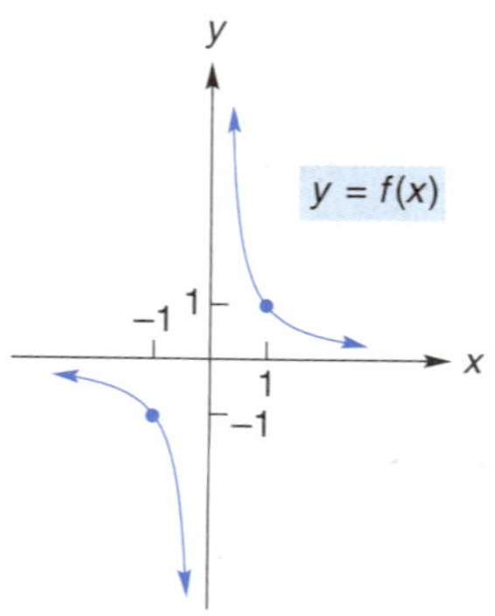

FIGURE 10.13 Diagram for Problem 4.

In Problems 5–8, use your calculator to complete the table, and use your results to estimate the given limit.

5. $\lim\limits_{x\to-1} \dfrac{3x^2+2x-1}{x+1}$

x	−0.9	−0.99	−0.999	−1.001	−1.01	−1.1
$f(x)$						

6. $\lim\limits_{x\to-2} \dfrac{x^2-4}{x+2}$

x	−2.1	−2.01	−2.001	−1.999	−1.99	−1.9
$f(x)$						

7. $\lim\limits_{x\to0} \dfrac{e^x-1}{x}$

x	−0.1	−0.01	−0.001	0.001	0.01	0.1
$f(x)$						

8. $\lim\limits_{h\to0} \dfrac{\sqrt{1+h}-1}{h}$

h	−0.1	−0.01	−0.001	0.001	0.01	0.1
$f(x)$						

In Problems 9–34, find the limits.

*9. $\lim\limits_{x\to2} 16$

10. $\lim\limits_{x\to3} 2x$

*11. $\lim\limits_{t\to-5} (t^2-5)$

12. $\lim\limits_{t\to1/3} (5t-7)$

*13. $\lim\limits_{x\to-1} (x^3-3x^2-2x+1)$

14. $\lim\limits_{r\to9} \dfrac{4r-3}{11}$

*15. $\lim\limits_{t\to-3} \dfrac{t-2}{t+5}$

16. $\lim\limits_{x\to-6} \dfrac{x^2+6}{x-6}$

17. $\lim\limits_{h\to0} \dfrac{h}{h^2-7h+1}$

18. $\lim\limits_{h\to0} \dfrac{h^3-3h^2-4}{h^2+1}$

19. $\lim_{p\to4} \sqrt{p^2+p+5}$

20. $\lim_{y\to15} \sqrt{y+3}$

*21. $\lim\limits_{x\to-2} \dfrac{x^2+2x}{x+2}$

22. $\lim\limits_{x\to-1} \dfrac{x+1}{x+1}$

*23. $\lim\limits_{x\to2} \dfrac{x^2-x-2}{x-2}$

24. $\lim\limits_{t\to0} \dfrac{t^3+3t^2}{t^3-4t^2}$

25. $\lim\limits_{x\to-1} \dfrac{x^2+2x+1}{x+1}$

26. $\lim\limits_{t\to2} \dfrac{t^2-4}{t-2}$

27. $\lim\limits_{x\to3} \dfrac{x-3}{x^2-9}$

28. $\lim\limits_{x\to0} \dfrac{x^2-2x}{x}$

29. $\lim\limits_{x\to4} \dfrac{x^2-9x+20}{x^2-3x-4}$

30. $\lim\limits_{x\to-2} \dfrac{x^4-16}{x^2-x-6}$

31. $\lim\limits_{x\to2} \dfrac{3x^2-x-10}{x^2+5x-14}$

32. $\lim\limits_{x\to-4} \dfrac{x^2+2x-8}{x^2+5x+4}$

33. $\lim\limits_{h\to0} \dfrac{(2+h)^2-2^2}{h}$

34. $\lim\limits_{x\to0} \dfrac{(x+2)^2-4}{x}$

*35. Find $\lim\limits_{h\to0} \dfrac{(x+h)^2-x^2}{h}$ by treating x as a constant.

36. Find $\lim\limits_{h\to0} \dfrac{3(x+h)^2+7(x+h)-3x^2-7x}{h}$ by treating x as a constant.

In Problems 37–42, find $\lim\limits_{h\to0} \dfrac{f(x+h)-f(x)}{h}$.

37. $f(x)=4-x$

38. $f(x)=2x+3$

39. $f(x)=x^2-3$

40. $f(x)=x^2+x+1$

41. $f(x)=x^3-4x^2$

42. $f(x)=5-2x-3x^2$

43. Find $\lim\limits_{x\to6} \dfrac{\sqrt{x-2}-2}{x-6}$ (*Hint:* First rationalize the numerator by multiplying both the numerator and denominator by $\sqrt{x-2}+2$.)

44. Find the constant c so that $\lim\limits_{x\to3} \dfrac{x^2+x+c}{x^2-5x+6}$ exists. For that value of c, determine the limit. (*Hint:* Find the value of c for which $x-3$ is a factor of the numerator.)

45. Power Plant The maximum theoretical efficiency of a power plant is given by

$$E=\frac{T_h-T_c}{T_h}$$

where T_h and T_c are the respective absolute temperatures of the hotter and colder reservoirs. Find (a) $\lim_{T_c \to 0} E$ and (b) $\lim_{T_c \to T_h} E$.

46. Satellite When a 3200-lb satellite revolves about the earth in a circular orbit of radius r ft, the total mechanical energy E of the earth-satellite system is given by

$$E = -\frac{7.0 \times 10^{17}}{r} \text{ ft-lb}$$

Find the limit of E as $r \to 7.5 \times 10^7$ ft.

In Problems 47–50, use a graphing calculator to graph the functions, and then estimate the limits. Round your answers to two decimal places.

47. $\lim_{x \to 2} \frac{x^4 + x^3 - 24}{x^2 - 4}$

48. $\lim_{x \to 0} x^x$

49. $\lim_{x \to 16} \frac{x - \sqrt{x} - 12}{4 - \sqrt{x}}$

50. $\lim_{x \to 1} \frac{x^3 + x^2 - 5x + 3}{x^3 + 2x^2 - 7x + 4}$

51. Water Purification The cost of purifying water is given by $C = \frac{50{,}000}{p} - 6500$, where p is the percent of impurities remaining after purification. Graph this function on your graphing calculator, and determine $\lim_{p \to 0} C$. Discuss what this means.

52. Profit Function The profit function for a certain business is given by $P(x) = 224x - 3.1x^2 - 800$. Graph this function on your graphing calculator, and use the evaluation function to determine $\lim_{x \to 53.2} P(x)$, utilizing the rule about the limit of a polynomial function.

OBJECTIVE *To study one-sided limits, infinite limits, and limits at infinity.*

10.2 Limits (Continued)

One-Sided Limits

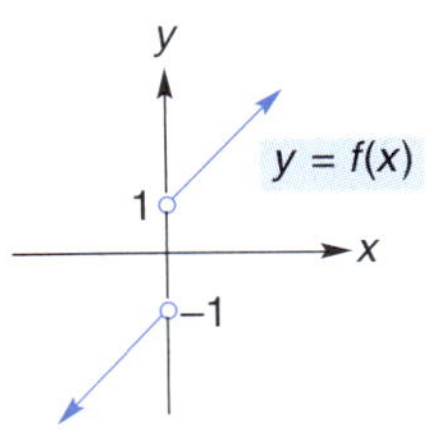

FIGURE 10.14 $\lim_{x \to 0} f(x)$ does not exist.

Figure 10.14 shows the graph of a function f. Notice that $f(x)$ is not defined when $x = 0$. As x approaches 0 *from the right*, $f(x)$ approaches 1. We write this as

$$\lim_{x \to 0^+} f(x) = 1$$

On the other hand, as x approaches 0 *from the left*, $f(x)$ approaches -1, and we write

$$\lim_{x \to 0^-} f(x) = -1$$

Limits like these are called **one-sided limits.** From the preceding section, we know that the limit of a function as $x \to a$ is independent of the way x approaches a. Thus, the limit will exist if and only if both one-sided limits exist and are equal. We therefore conclude that

$$\lim_{x \to 0} f(x) \text{ does not exist}$$

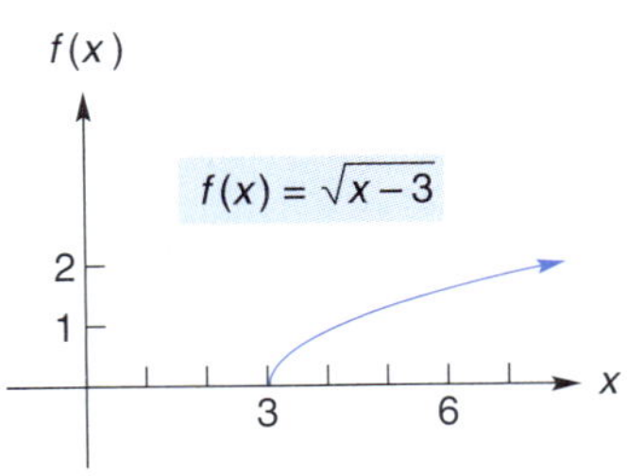

FIGURE 10.15 $\lim_{x \to 3^+} \sqrt{x - 3} = 0$.

As another example of a one-sided limit, consider $f(x) = \sqrt{x - 3}$ as x approaches 3. Since f is defined only when $x \geq 3$, we may speak of the limit as x approaches 3 from the right. If x is slightly greater than 3, then $x - 3$ is a positive number that is close to 0, so $\sqrt{x - 3}$ is close to 0. We conclude that

$$\lim_{x \to 3^+} \sqrt{x - 3} = 0$$

This limit is also evident from Figure 10.15.

Infinite Limits

In the previous section, we considered limits of the form 0/0—that is, limits where both the numerator and denominator approach 0. Now we shall examine limits where the denominator approaches 0, but the numerator approaches a number different from 0. For example, consider

$$\lim_{x \to 0} \frac{1}{x^2}$$

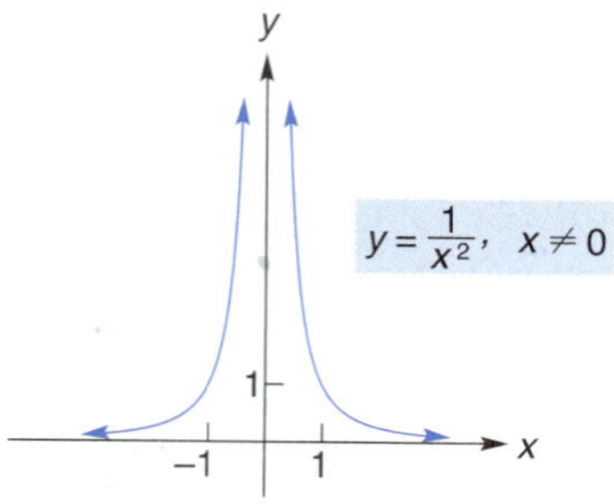

x	f(x)
±1	1
±0.5	4
±0.1	100
±0.01	10,000
±0.001	1,000,000

FIGURE 10.16 $\lim_{x\to 0} \frac{1}{x^2} = \infty$.

CAUTION The use of the "equality" sign in this situation does not mean that the limit exists. On the contrary, the symbolism here (∞) is a way of saying specifically that there is no limit, and it indicates why there is no limit.

This CAUTION is extremely important.

Here, as x approaches 0, the denominator approaches 0 and the numerator approaches 1. Let us investigate the behavior of $f(x) = 1/x^2$ when x is close to 0. The number x^2 is positive and also close to 0. Thus, dividing 1 by such a number results in a very large number. In fact, the closer x is to 0, the larger the value of $f(x)$. For example, see the table of values in Figure 10.16, which also shows the graph of f. Clearly, as $x \to 0$ both from the left and from the right, $f(x)$ increases without bound. Hence, no limit exists at 0. We say that as $x \to 0$, $f(x)$ becomes positively infinite, and symbolically, we express this "infinite limit" by writing

$$\lim_{x\to 0} \frac{1}{x^2} = \infty$$

If $\lim_{x\to a} f(x)$ does not exist, it may be for a reason other than that the values $f(x)$ become arbitrarily large as x gets close to a. For example, look at the situation in Example 2 of Section 10.1. Here we have

$$\lim_{x\to -2} f(x) \text{ does not exist, but } \lim_{x\to -2} f(x) \neq \infty$$

Consider now the graph of $y = f(x) = 1/x$ for $x \neq 0$. (See Figure 10.17.) As x approaches 0 from the right, $1/x$ becomes positively infinite; as x approaches 0 from the left, $1/x$ becomes negatively infinite. Symbolically, these infinite limits are written

$$\lim_{x\to 0^+} \frac{1}{x} = \infty \quad \text{and} \quad \lim_{x\to 0^-} \frac{1}{x} = -\infty$$

x	f(x)
0.01	100
0.001	1000
0.0001	10,000
−0.01	−100
−0.001	−1000
−0.0001	−10,000

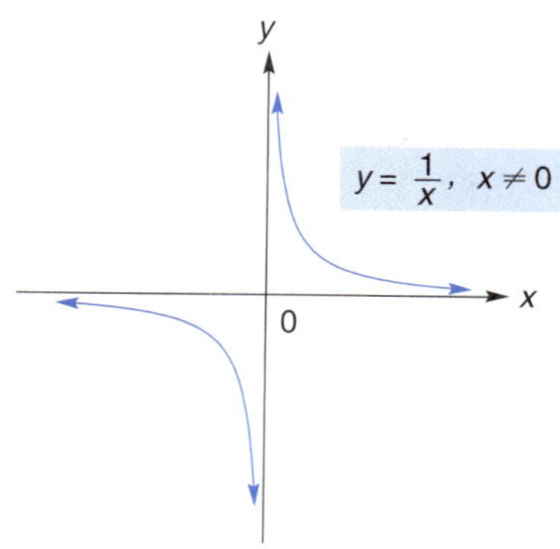

FIGURE 10.17 $\lim_{x\to 0} \frac{1}{x}$ does not exist.

Either one of these facts implies that

$$\lim_{x\to 0} \frac{1}{x} \text{ does not exist}$$

EXAMPLE 1 **Infinite Limits**

Find the limit (if it exists).

a. $\lim_{x\to -1^+} \frac{2}{x+1}$

Solution: As x approaches −1 from the right (think of values of x such as −0.9, −0.99, and so on, as shown in Figure 10.18), $x + 1$ approaches 0, but is always positive. Since we are dividing 2 by positive numbers approaching

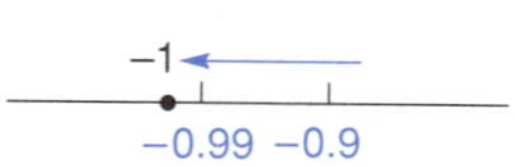

FIGURE 10.18 $x \to -1^+$.

0, the results, $2/(x+1)$, are positive numbers that are becoming arbitrarily large. Thus,

$$\lim_{x \to -1^+} \frac{2}{x+1} = \infty$$

and the limit does not exist. By a similar analysis, you should be able to show that

$$\lim_{x \to -1^-} \frac{2}{x+1} = -\infty$$

b. $\lim_{x \to 2} \frac{x+2}{x^2-4}$

Solution: As $x \to 2$, the numerator approaches 4 and the denominator approaches 0. Hence, we are dividing numbers near 4 by numbers near 0. The results are numbers that become arbitrarily large in magnitude. At this stage, we can write

$$\lim_{x \to 2} \frac{x+2}{x^2-4} \text{ does not exist}$$

However, let us see if we can use the symbol ∞ or $-\infty$ to be more specific about "does not exist." Notice that

$$\lim_{x \to 2} \frac{x+2}{x^2-4} = \lim_{x \to 2} \frac{x+2}{(x+2)(x-2)} = \lim_{x \to 2} \frac{1}{x-2}$$

Since

$$\lim_{x \to 2^+} \frac{1}{x-2} = \infty \qquad \text{and} \qquad \lim_{x \to 2^-} \frac{1}{x-2} = -\infty$$

$\lim_{x \to 2} \frac{x+2}{x^2-4}$ is neither ∞ nor $-\infty$.

▶ **Now work Problem 31.**

Example 1 considered limits of the form $k/0$, where $k \neq 0$. It is important that you distinguish the form $k/0$ from the form $0/0$, which was discussed in Section 10.1. These two forms are handled quite differently.

EXAMPLE 2 Finding a Limit

Find $\lim_{t \to 2} \frac{t-2}{t^2-4}$.

Solution: As $t \to 2$, *both* numerator and denominator approach 0 (form $0/0$). Thus, we first simplify the fraction, as we did in Section 10.1, and then take the limit:

$$\lim_{t \to 2} \frac{t-2}{t^2-4} = \lim_{t \to 2} \frac{t-2}{(t+2)(t-2)} = \lim_{t \to 2} \frac{1}{t+2} = \frac{1}{4}$$

▶ **Now work Problem 37.**

Limits at Infinity

Now let us examine the function

$$f(x) = \frac{1}{x}$$

as x becomes infinite, first in a positive sense and then in a negative sense. From Table 10.2, you can see that as x increases without bound through positive

TABLE 10.2 Behavior of $f(x)$ as $x \to \pm\infty$

x	$f(x)$	x	$f(x)$
1000	0.001	−1000	−0.001
10,000	0.0001	−10,000	−0.0001
100,000	0.00001	−100,000	−0.00001
1,000,000	0.000001	−1,000,000	−0.000001

You should be able to obtain

$$\lim_{x\to\infty} \frac{1}{x} \quad \text{and} \quad \lim_{x\to-\infty} \frac{1}{x}$$

without the benefit of a graph or a table. Conceptually, for $x \to \infty$, increasing the value of the denominator in $1/x$ means that the fraction is getting smaller and smaller in magnitude—that is, closer to 0. Alternatively, dividing 1 by a large positive number results in a number close to 0. A similar argument can be made for the limit as $x \to -\infty$.

values, the values of $f(x)$ approach 0. Likewise, as x decreases without bound through negative values, the values of $f(x)$ also approach 0. These observations are also apparent from the graph in Figure 10.17. There, as you move to the right along the curve through positive x-values, the corresponding y-values approach 0. Similarly, as you move to the left along the curve through negative x-values, the corresponding y-values approach 0. Symbolically, we write

$$\lim_{x\to\infty} \frac{1}{x} = 0 \qquad \text{and} \qquad \lim_{x\to-\infty} \frac{1}{x} = 0$$

Both of these limits are called *limits at infinity*.

APPLICATIONS IN PRACTICE 1

Limits at Infinity

The demand function for a certain product is given by $p(x) = \frac{10{,}000}{(x+1)^2}$, where p is the price in dollars and x is the quantity sold. Graph this function on your graphing calculator in the window $[0, 10] \times [0, 10{,}000]$. Use the TRACE function to find $\lim_{x\to\infty} p(x)$. Determine what is happening to the graph and what this means about the demand function.

EXAMPLE 3 Limits at Infinity

Find the limit (if it exists).

a. $\lim_{x\to\infty} \dfrac{4}{(x-5)^3}$

Solution: As x becomes very large, so does $x - 5$. Since the cube of a large number is also large, $(x-5)^3 \to \infty$. Dividing 4 by very large numbers results in numbers near 0. Thus,

$$\lim_{x\to\infty} \frac{4}{(x-5)^3} = 0$$

b. $\lim_{x\to-\infty} \sqrt{4-x}$

Solution: As x gets negatively infinite, $4 - x$ becomes positively infinite. Because square roots of large numbers are large numbers, we conclude that

$$\lim_{x\to-\infty} \sqrt{4-x} = \infty$$

In our next discussion we shall need a certain limit, namely, $\lim_{x\to\infty} 1/x^p$, where $p > 0$. As x becomes very large, so does x^p. Dividing 1 by very large numbers results in numbers near 0. Thus, $\lim_{x\to\infty} 1/x^p = 0$. In general,

$$\lim_{x\to\infty} \frac{1}{x^p} = 0 \qquad \text{and} \qquad \lim_{x\to-\infty} \frac{1}{x^p} = 0$$

where $p > 0$.[2] For example,

$$\lim_{x\to\infty} \frac{1}{\sqrt[3]{x}} = \lim_{x\to\infty} \frac{1}{x^{1/3}} = 0$$

[2]For $\lim_{x\to-\infty} 1/x^p$, we assume that p is such that $1/x^p$ is defined for $x < 0$.

Let us now find the limit of the rational function

$$f(x) = \frac{4x^2+5}{2x^2+1}$$

as $x \to \infty$. (Recall from Section 2.2 that a rational function is a quotient of polynomials.) As x gets larger and larger, *both* the numerator and denominator of $f(x)$ become infinite. However, the form of the quotient can be changed, so that we can draw a conclusion as to whether or not it has a limit. To do this, we divide both the numerator and denominator by the greatest power of x that occurs in the denominator. Here it is x^2. This gives

$$\lim_{x\to\infty} \frac{4x^2+5}{2x^2+1} = \lim_{x\to\infty} \frac{\frac{4x^2+5}{x^2}}{\frac{2x^2+1}{x^2}} = \lim_{x\to\infty} \frac{\frac{4x^2}{x^2}+\frac{5}{x^2}}{\frac{2x^2}{x^2}+\frac{1}{x^2}}$$

$$= \lim_{x\to\infty} \frac{4+\frac{5}{x^2}}{2+\frac{1}{x^2}} = \frac{\lim\limits_{x\to\infty} 4 + 5\cdot \lim\limits_{x\to\infty} \frac{1}{x^2}}{\lim\limits_{x\to\infty} 2 + \lim\limits_{x\to\infty} \frac{1}{x^2}}$$

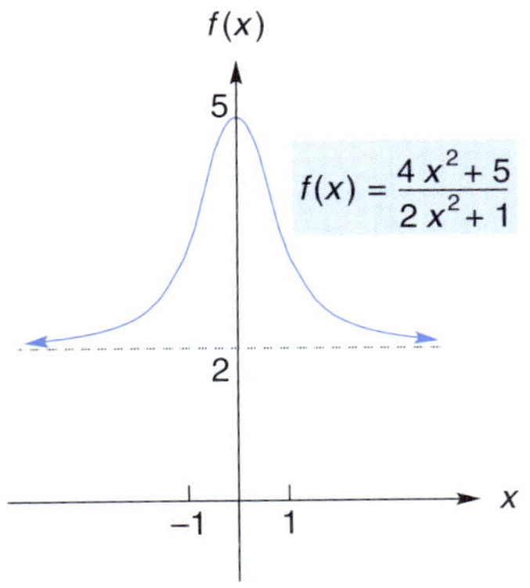

FIGURE 10.19 $\lim_{x\to\infty} f(x) = 2$ and $\lim_{x\to-\infty} f(x) = 2$.

Since $\lim_{x\to\infty} 1/x^p = 0$ for $p > 0$,

$$\lim_{x\to\infty} \frac{4x^2+5}{2x^2+1} = \frac{4+5(0)}{2+0} = \frac{4}{2} = 2$$

Similarly, the limit as $x \to -\infty$ is 2. These limits are clear from the graph of f in Figure 10.19.

For the preceding function, there is an easier way to find $\lim_{x\to\infty} f(x)$. For *large* values of x, in the numerator the term involving the greatest power of x, namely, $4x^2$, dominates the sum $4x^2+5$, and the dominant term in the denominator, $2x^2+1$, is $2x^2$. Thus, as $x \to \infty$, $f(x)$ can be approximated by $(4x^2)/(2x^2)$. As a result, to determine the limit of $f(x)$, it suffices to determine the limit of $(4x^2)/(2x^2)$. That is,

$$\lim_{x\to\infty} \frac{4x^2+5}{2x^2+1} = \lim_{x\to\infty} \frac{4x^2}{2x^2} = \lim_{x\to\infty} 2 = 2,$$

as we saw before. In general, we have the following rule:

Limits at Infinity for Rational Functions

If $f(x)$ is a *rational function* and a_nx^n and b_mx^m are the terms in the numerator and denominator, respectively, with the greatest powers of x, then

$$\lim_{x\to\infty} f(x) = \lim_{x\to\infty} \frac{a_nx^n}{b_mx^m}$$

and

$$\lim_{x\to-\infty} f(x) = \lim_{x\to-\infty} \frac{a_nx^n}{b_mx^m}$$

Let us apply this rule to the situation where the degree of the numerator is greater than the degree of the denominator. For example,

$$\lim_{x\to-\infty} \frac{x^4-3x}{5-2x} = \lim_{x\to-\infty} \frac{x^4}{-2x} = \lim_{x\to-\infty} \left(-\frac{1}{2}x^3\right) = \infty$$

(Note that in the next-to-last step, as x becomes very negative, so does x^3;

moreover, $-\frac{1}{2}$ times a very negative number is very positive.) Similarly,

$$\lim_{x\to\infty}\frac{x^4-3x}{5-2x}=\lim_{x\to\infty}\left(-\frac{1}{2}x^3\right)=-\infty$$

From this illustration, we make the following conclusion: If the degree of the numerator of a *rational function* is greater than the degree of the denominator, then the function has no limit as $x\to\infty$ or as $x\to-\infty$.

APPLICATIONS IN PRACTICE 2

Limits at Infinity for Rational Functions

The yearly amount of sales y of a certain company (in thousands of dollars) is related to the amount the company spends on advertising, x (in thousands of dollars), according to the equation $y(x)=\frac{500x}{x+20}$. Graph this function on your graphing calculator in the window $[0, 1000]\times[0, 550]$. Use TRACE to explore $\lim_{x\to\infty} y(x)$, and determine what this means to the company.

EXAMPLE 4 Limits at Infinity for Rational Functions

Find the limit (if it exists).

a. $\displaystyle\lim_{x\to\infty}\frac{x^2-1}{7-2x+8x^2}$

Solution:

$$\lim_{x\to\infty}\frac{x^2-1}{7-2x+8x^2}=\lim_{x\to\infty}\frac{x^2}{8x^2}=\lim_{x\to\infty}\frac{1}{8}=\frac{1}{8}$$

b. $\displaystyle\lim_{x\to-\infty}\frac{x}{(3x-1)^2}$

Solution:

$$\lim_{x\to-\infty}\frac{x}{(3x-1)^2}=\lim_{x\to-\infty}\frac{x}{9x^2-6x+1}=\lim_{x\to-\infty}\frac{x}{9x^2}$$

$$=\lim_{x\to-\infty}\frac{1}{9x}=\frac{1}{9}\cdot\lim_{x\to-\infty}\frac{1}{x}=\frac{1}{9}(0)=0$$

c. $\displaystyle\lim_{x\to\infty}\frac{x^5-x^4}{x^4-x^3+2}$

Solution: Since the degree of the numerator is greater than that of the denominator, there is no limit. More precisely,

$$\lim_{x\to\infty}\frac{x^5-x^4}{x^4-x^3+2}=\lim_{x\to\infty}\frac{x^5}{x^4}=\lim_{x\to\infty}x=\infty$$

▸ **Now work Problem 21.**

CAUTION: Our preceding technique applies only to limits of rational functions at *infinity*.

To find $\displaystyle\lim_{x\to0}\frac{x^2-1}{7-2x+8x^2}$, we do not determine the limit of $\frac{x^2}{8x^2}$, because x does not approach ∞ or $-\infty$. Instead, we have

$$\lim_{x\to0}\frac{x^2-1}{7-2x+8x^2}=\frac{0-1}{7-0+0}=-\frac{1}{7}$$

Let us now consider the limit of the polynomial function $f(x)=8x^2-2x$ as $x\to\infty$:

$$\lim_{x\to\infty}(8x^2-2x)$$

Because a polynomial is a rational function with denominator 1, we have

$$\lim_{x\to\infty}(8x^2-2x)=\lim_{x\to\infty}\frac{8x^2-2x}{1}=\lim_{x\to\infty}\frac{8x^2}{1}=\lim_{x\to\infty}8x^2$$

That is, the limit of $8x^2-2x$ as $x\to\infty$ is the same as the limit of the term involving the greatest power of x, namely, $8x^2$. As x becomes very large, so does $8x^2$. Thus,

$$\lim_{x\to\infty}(8x^2-2x)=\lim_{x\to\infty}8x^2=\infty$$

In general, we have the following:

As $x \to \infty$ (or $x \to -\infty$), the limit of a *polynomial function* is the same as the limit of its term that involves the greatest power of x.

APPLICATIONS IN PRACTICE 3

Limits at Infinity for Polynomial Functions

The cost C of producing x units of a certain product is given by $C(x) = 50{,}000 + 200x + 0.3x^2$. Use your graphing calculator to explore $\lim_{x\to\infty} C(x)$ and determine what this means.

EXAMPLE 5 Limits at Infinity for Polynomial Functions

a. $\lim_{x\to-\infty}(x^3 - x^2 + x - 2) = \lim_{x\to-\infty} x^3$. As x becomes very negative, so does x^3. Thus,

$$\lim_{x\to-\infty}(x^3 - x^2 + x - 2) = \lim_{x\to-\infty} x^3 = -\infty$$

b. $\lim_{x\to-\infty}(-2x^3 + 9x) = \lim_{x\to-\infty} -2x^3 = \infty$, because -2 times a very negative number is very positive.

▶ **Now work Problem 9.**

CAUTION

Do not use dominant terms when a function is not rational.

The technique of focusing on dominant terms to find limits as $x \to \infty$ or $x \to -\infty$ is valid for *rational functions*, but it is not necessarily valid for other types of functions. For example, consider

$$\lim_{x\to\infty}(\sqrt{x^2 + x} - x) \qquad (1)$$

Notice that $\sqrt{x^2 + x} - x$ is not a rational function. It is *incorrect* to infer that because x^2 dominates in $x^2 + x$, the limit in (1) is the same as

$$\lim_{x\to\infty}(\sqrt{x^2} - x) = \lim_{x\to\infty}(x - x) = \lim_{x\to\infty} 0 = 0$$

It can be shown (see Problem 62) that the limit in (1) is not 0, but is $\frac{1}{2}$.

The ideas discussed in this section will now be applied to a case-defined function.

APPLICATIONS IN PRACTICE 4

Limits for a Case-Defined Function

A plumber charges \$100 for the first hour of work at your house and \$75 for every hour (or fraction thereof) afterwards. The function for what an x-hour visit will cost you is

$$f(x) = \begin{cases} \$100 & \text{if } 0 < x \le 1 \\ \$175 & \text{if } 1 < x \le 2 \\ \$250 & \text{if } 2 < x \le 3 \\ \$325 & \text{if } 3 < x \le 4 \end{cases}$$

Find $\lim_{x\to1} f(x)$ and $\lim_{x\to2.5} f(x)$.

EXAMPLE 6 Limits for a Case-Defined Function

If $f(x) = \begin{cases} x^2 + 1 & \text{if } x \ge 1 \\ 3 & \text{if } x < 1 \end{cases}$, *find the limit (if it exists).*

a. $\lim_{x\to1^+} f(x)$

Solution: Here x gets close to 1 from the right. For $x > 1$, we have $f(x) = x^2 + 1$. Thus,

$$\lim_{x\to1^+} f(x) = \lim_{x\to1^+}(x^2 + 1)$$

If x is greater than 1, but close to 1, then $x^2 + 1$ is close to 2. Therefore,

$$\lim_{x\to1^+} f(x) = \lim_{x\to1^+}(x^2 + 1) = 2$$

b. $\lim_{x\to1^-} f(x)$

Solution: Here x gets close to 1 from the left. For $x < 1$, $f(x) = 3$. Hence,

$$\lim_{x\to1^-} f(x) = \lim_{x\to1^-} 3 = 3$$

c. $\lim_{x\to 1} f(x)$

Solution: We want the limit as x approaches 1. However, the rule of the function depends on whether $x \geq 1$ or $x < 1$. Thus, we must consider one-sided limits. The limit as x approaches 1 will exist if and only if both one-sided limits exist and are the same. From parts (a) and (b),

$$\lim_{x\to 1^+} f(x) \neq \lim_{x\to 1^-} f(x), \qquad \text{since } 2 \neq 3$$

Therefore,

$$\lim_{x\to 1} f(x) \qquad \text{does not exist}$$

d. $\lim_{x\to\infty} f(x)$

Solution: For very large values of x, we have $x \geq 1$, so $f(x) = x^2 + 1$. Thus,

$$\lim_{x\to\infty} f(x) = \lim_{x\to\infty} (x^2+1) = \lim_{x\to\infty} x^2 = \infty$$

e. $\lim_{x\to-\infty} f(x)$

Solution: For very negative values of x, we have $x < 1$, so $f(x) = 3$. Hence,

$$\lim_{x\to-\infty} f(x) = \lim_{x\to-\infty} 3 = 3$$

All the limits in parts (a) through (c) should be obvious from the graph of f in Figure 10.20.

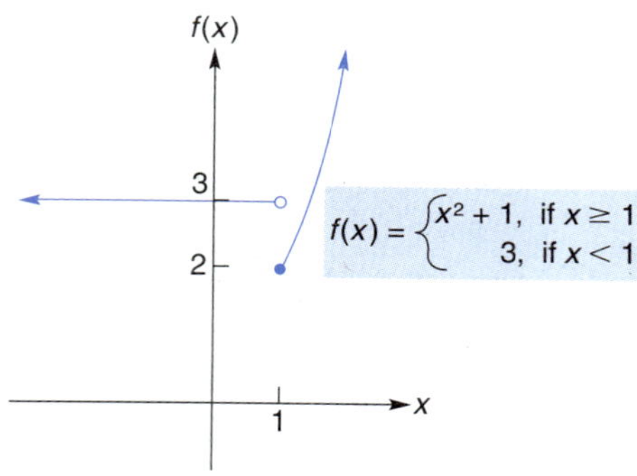

FIGURE 10.20 Graph of case-defined function.

Exercise 10.2

1. For the function f given in Figure 10.21, find the following limits. If the limit does not exist, so state that, or use the symbol ∞ or $-\infty$ where appropriate.

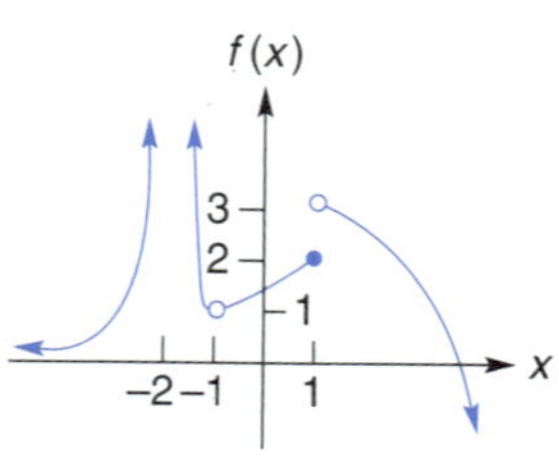

FIGURE 10.21 Diagram for Problem 1.

(a) $\lim_{x\to 1^-} f(x)$ **(b)** $\lim_{x\to 1^+} f(x)$
(c) $\lim_{x\to 1} f(x)$ **(d)** $\lim_{x\to\infty} f(x)$
(e) $\lim_{x\to -2^-} f(x)$ **(f)** $\lim_{x\to -2^+} f(x)$
(g) $\lim_{x\to -2} f(x)$ **(h)** $\lim_{x\to -\infty} f(x)$
(i) $\lim_{x\to -1^-} f(x)$ **(j)** $\lim_{x\to -1^+} f(x)$
(k) $\lim_{x\to -1} f(x)$

2. For the function f given in Figure 10.22, find the following limits. If the limit does not exist, so state that, or use the symbol ∞ or $-\infty$ where appropriate.

(a) $\lim_{x\to 0^-} f(x)$ **(b)** $\lim_{x\to 0^+} f(x)$
(c) $\lim_{x\to 0} f(x)$ **(d)** $\lim_{x\to -\infty} f(x)$

(e) $\lim_{x\to 1} f(x)$

(f) $\lim_{x\to\infty} f(x)$

(g) $\lim_{x\to 2^+} f(x)$

(h) $\lim_{x\to\infty} f(x)$

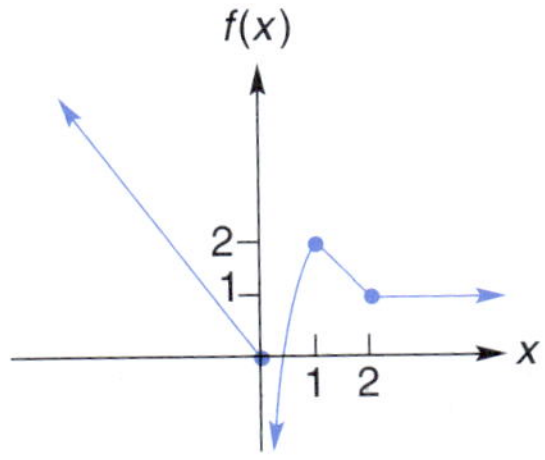

FIGURE 10.22 Diagram for Problem 2.

In Problems 3–54, find the limit. If the limit does not exist, so state that, or use the symbol ∞ or −∞ where appropriate.

3. $\lim_{x\to 3^+}(x-2)$

4. $\lim_{x\to -1^+}(1-x^2)$

5. $\lim_{x\to -\infty} 5x$

6. $\lim_{x\to -\infty} 19$

7. $\lim_{x\to 0^-}\dfrac{6x}{x^4}$

8. $\lim_{x\to 0}\dfrac{5}{x-1}$

***9.** $\lim_{x\to -\infty} x^2$

10. $\lim_{t\to\infty}(t-1)^3$

11. $\lim_{h\to 0^+}\sqrt{h}$

12. $\lim_{h\to 5^-}\sqrt{5-h}$

13. $\lim_{x\to -1}\dfrac{7}{x+1}$

14. $\lim_{x\to 0^-} 2^{1/2}$

15. $\lim_{x\to 1^+}(4\sqrt{x-1})$

16. $\lim_{x\to 2^+}(x\sqrt{x^2-4})$

17. $\lim_{x\to\infty}\sqrt{x+10}$

18. $\lim_{x\to -\infty} -\sqrt{1-10x}$

19. $\lim_{x\to\infty}\dfrac{3}{\sqrt{x}}$

20. $\lim_{x\to\infty}\dfrac{7}{2x\sqrt{x}}$

***21.** $\lim_{x\to\infty}\dfrac{x+8}{x-3}$

22. $\lim_{x\to\infty}\dfrac{2x-4}{3-2x}$

23. $\lim_{x\to -\infty}\dfrac{x^2-1}{x^3+4x-3}$

24. $\lim_{r\to\infty}\dfrac{r^3}{r^2+1}$

25. $\lim_{t\to\infty}\dfrac{5t^2+2t+1}{4t+7}$

26. $\lim_{x\to\infty}\dfrac{5x}{3x^7-x^3+4}$

27. $\lim_{x\to\infty}\dfrac{7}{2x+1}$

28. $\lim_{x\to -\infty}\dfrac{2}{(4x-1)^3}$

29. $\lim_{x\to\infty}\dfrac{3-4x-2x^3}{5x^3-8x+1}$

30. $\lim_{x\to -\infty}\dfrac{3-2x-2x^3}{7-5x^3+2x^2}$

***31.** $\lim_{x\to 3^-}\dfrac{x+3}{x^2-9}$

32. $\lim_{x\to -2^+}\dfrac{5x}{4-x^2}$

33. $\lim_{w\to\infty}\dfrac{2w^2-3w+4}{5w^2+7w-1}$

34. $\lim_{x\to\infty}\dfrac{4-3x^3}{x^3-1}$

35. $\lim_{x\to\infty}\dfrac{6-4x^2+x^3}{4+5x-7x^2}$

36. $\lim_{x\to -\infty}\dfrac{3x-x^3}{x^3+x+1}$

***37.** $\lim_{x\to -5}\dfrac{2x^2+9x-5}{x^2+5x}$

38. $\lim_{t\to 3}\dfrac{t^2-4t+3}{t^2-2t-3}$

39. $\lim_{x\to 1}\dfrac{x^2-3x+1}{x^2+1}$

40. $\lim_{x\to -1}\dfrac{3x^3-x^2}{2x+1}$

41. $\lim_{x\to 1^+}\left[1+\dfrac{1}{x-1}\right]$

42. $\lim_{x\to -\infty} -\dfrac{x^5+2x^3-1}{x^5-4x^2}$

43. $\lim_{x\to -7^-}\dfrac{x^2+1}{\sqrt{x^2-49}}$

44. $\lim_{x\to -3^+}\dfrac{x}{\sqrt{9-x^2}}$

45. $\lim_{x\to 0^+}\dfrac{5}{x+x^2}$

46. $\lim_{x\to\infty}\left(x+\dfrac{1}{x}\right)$

47. $\lim_{x\to 1} x(x-1)^{-1}$

48. $\lim_{x\to 1/2}\dfrac{1}{2x-1}$

49. $\lim_{x\to 0^+}\left(-\dfrac{3}{x}\right)$

50. $\lim_{x\to 3}\left(-\dfrac{7}{x-3}\right)$

51. $\lim_{x\to 0}|x|$

52. $\lim_{x\to 0}\left|\dfrac{1}{x}\right|$

53. $\lim_{x\to -\infty}\dfrac{x+1}{x}$

54. $\lim_{x\to\infty}\left[\dfrac{3}{x}-\dfrac{2x^2}{x^2+1}\right]$

In Problems 55–58, find the indicated limits. If the limit does not exist, so state that, or use the symbol ∞ or −∞ where appropriate.

55. $f(x)=\begin{cases}2 & \text{if } x\le 2\\ 1 & \text{if } x>2\end{cases}$

(a) $\lim_{x\to 2^+} f(x)$ **(b)** $\lim_{x\to 2^-} f(x)$ **(c)** $\lim_{x\to 2} f(x)$ **(d)** $\lim_{x\to\infty} f(x)$ **(e)** $\lim_{x\to -\infty} f(x)$

56. $f(x)=\begin{cases}x & \text{if } x\le 1\\ 2 & \text{if } x>1\end{cases}$

(a) $\lim_{x\to 1^+} f(x)$ **(b)** $\lim_{x\to 1^-} f(x)$ **(c)** $\lim_{x\to 1} f(x)$ **(d)** $\lim_{x\to\infty} f(x)$ **(e)** $\lim_{x\to -\infty} f(x)$

57. $g(x)=\begin{cases}x & \text{if } x<0\\ -x & \text{if } x>0\end{cases}$

(a) $\lim_{x\to 0^+} g(x)$ **(b)** $\lim_{x\to 0^-} g(x)$ **(c)** $\lim_{x\to 0} g(x)$ **(d)** $\lim_{x\to\infty} g(x)$ **(e)** $\lim_{x\to -\infty} g(x)$

58. $g(x)=\begin{cases}x^2 & \text{if } x<0\\ -x & \text{if } x>0\end{cases}$

(a) $\lim_{x\to 0^+} g(x)$ **(b)** $\lim_{x\to 0^-} g(x)$ **(c)** $\lim_{x\to 0} g(x)$ **(d)** $\lim_{x\to\infty} g(x)$ **(e)** $\lim_{x\to -\infty} g(x)$

59. Average Cost If c is the total cost in dollars to produce q units of a product, then the average cost per unit for an output of q units is given by $\bar{c}=c/q$. Thus, if the total cost equation is $c=5000+6q$, then

$$\bar{c}=\frac{5000}{q}+6$$

For example, the total cost of an output of 5 units is \$5030, and the average cost per unit at this level of production is \$1006. By finding $\lim_{q\to\infty}\bar{c}$, show that the average cost approaches a level of stability if the producer continually increases output. What is the limiting value of the average cost? Draw the graph of the average-cost function.

60. Average Cost Repeat Problem 59, given that the fixed cost is \$12,000 and the variable cost is given by the function $c_v = 7q$.

61. Population The population of a certain small city t years from now is predicted to be

$$N = 20{,}000 - \frac{5{,}000}{(t+1)^2}$$

Find the population in the long run; that is, find $\lim_{t\to\infty} N$.

62. Show that

$$\lim_{x\to\infty} (\sqrt{x^2+x} - x) = \frac{1}{2}$$

(*Hint:* Rationalize the numerator by multiplying the expression $\sqrt{x^2+x} - x$ by

$$\frac{\sqrt{x^2+x}+x}{\sqrt{x^2+x}+x}$$

Then express the denominator in a form such that x is a factor.)

63. Host-Parasite Relationship For a particular host-parasite relationship, it was determined that when the host density (number of hosts per unit of area) is x, the number of hosts parasitized over a period of time is

$$y = \frac{900x}{10+45x}$$

If the host density were to increase without bound, what value would y approach?

64. If $f(x) = \begin{cases} \sqrt{2-x} & \text{if } x < 2 \\ x^3 + k(x+1) & \text{if } x \geq 2 \end{cases}$, determine the value of the constant k for which $\lim_{x\to 2} f(x)$ exists.

In Problems 65 and 66, use a calculator to evaluate the given function when $x = 1, 0.5, 0.2, 0.1, 0.01, 0.001,$ and 0.0001. From your results, speculate about $\lim_{x\to 0^+} f(x)$.

65. $f(x) = x^{2x}$

66. $f(x) = e^{-1/x}$

67. Graph $f(x) = \sqrt{4x^2-1}$. Use the graph to estimate $\lim_{x\to 1/2^+} f(x)$.

68. Graph $f(x) = \dfrac{\sqrt{x^2-4}}{2-x}$. Use the graph to estimate $\lim_{x\to 2^+} f(x)$ if it exists. Use the symbol ∞ or $-\infty$ if appropriate.

69. Graph $f(x) = \begin{cases} 2x^2+3 & \text{if } x < 2 \\ 2x+5 & \text{if } x \geq 2 \end{cases}$. Use the graph to estimate each of the following limits if it exists:

(a) $\lim_{x\to 2^-} f(x)$ **(b)** $\lim_{x\to 2^+} f(x)$

(c) $\lim_{x\to 2} f(x)$

OBJECTIVE *To extend the notion of compound interest discussed in Chapters 4 and 5 to the situation where interest is compounded continuously. To develop formulas for compound amount and present value.*

10.3 Interest Compounded Continuously

When money is invested at a given annual rate, the interest earned each year depends on how frequently interest is compounded. For example, more interest is earned if it is compounded monthly rather than semiannually. We can successively get more interest by compounding it weekly, daily, per hour, and so on. However, there is a maximum interest that can be earned, which we now examine.

Suppose a principal of P dollars is invested for t years at an annual rate of r. If interest is compounded k times a year, then the rate per conversion period is r/k, and there are kt periods. From Section 4.1, the compound amount is given by

$$S = P\left(1+\frac{r}{k}\right)^{kt}$$

If $k \to \infty$, the number of conversion periods increases indefinitely, and the length of each period approaches 0. In this case, we say that interest is **compounded continuously**—that is, at every instant of time. The compound amount is

$$\lim_{k\to\infty} P\left(1+\frac{r}{k}\right)^{kt}$$

which may be written

$$P\left[\lim_{k\to\infty}\left(1+\frac{r}{k}\right)^{k/r}\right]^{rt}$$

By letting $x = r/k$, then as $k \to \infty$, we have $x \to 0$. Thus, the limit inside the brackets has the form $\lim_{x\to 0}(1+x)^{1/x}$ which, as we saw in Section 10.1, is e. Similarly, by taking $x = 1/n$, where n is a positive integer, we see that

By making a substitution, we get the limit to have a familiar form.

$$\lim_{x\to 0}(1+x)^{1/x} = \lim_{n\to\infty}\left(1+\frac{1}{n}\right)^n = \lim_{n\to\infty}\left(\frac{n+1}{n}\right)^n$$

In Section 4.1 we defined e as the least real number that is greater than all the values $\left(\frac{n+1}{n}\right)^n$, where n is a positive integer. It can be shown that this means precisely $\lim_{n\to\infty}\left(\frac{n+1}{n}\right)^n$, thus showing that our two definitions of e are in fact the same.

In any event, we have the following formula:

Compound Amount under Continuous Interest

The formula

$$S = Pe^{rt}$$

gives the compound amount S of a principal of P dollars after t years at an annual interest rate r compounded continuously.

EXAMPLE 1 **Compound Amount**

If \$100 is invested at an annual rate of 5% compounded continuously, find the compound amount at the end of

a. *1 year*

b. *5 years*

Solution:

a. Here $P = 100$, $r = 0.05$, and $t = 1$, so

The interest of \$5.13 is the maximum amount of compound interest that can be earned at an annual rate of 5%.

$$S = Pe^{rt} = 100e^{(0.05)(1)} \approx \$105.13$$

We can compare this value with the value after one year of a \$100 investment at an annual rate of 5% compounded semiannually—namely, $100(1.025)^2 \approx 105.06$. The difference is not significant.

b. Here $P = 100$, $r = 0.05$, and $t = 5$, so

$$S = 100e^{(0.05)(5)} = 100e^{0.25} \approx \$128.40$$

▸ **Now work Problem 1.**

We can find an expression that gives the effective rate that corresponds to an annual rate of r compounded continuously. (From Chapter 5, the effective rate is the equivalent rate compounded annually.) If i is the corresponding effective rate, then after one year, a principal P accumulates to $P(1+i)$. This must equal the accumulated amount under continuous interest, Pe^r. Thus, $P(1+i) = Pe^r$, from which it follows that $1+i = e^r$, so $i = e^r - 1$.

Effective Rate under Continuous Interest

The effective rate corresponding to an annual rate of r compounded continuously is

$$e^r - 1$$

EXAMPLE 2 Effective Rate

Find the effective rate that corresponds to an annual rate of 5% compounded continuously.

Solution: The effective rate is

$$e^r - 1 = e^{0.05} - 1 \approx 0.0513$$

which is 5.13%.

▶ **Now work Problem 5.**

If we solve $S = Pe^{rt}$ for P, we get $P = S/e^{rt}$. In this formula, P is the principal that must be invested now at an annual rate of r compounded continuously so that at the end of t years the compound amount is S. We call P the **present value** of S.

Present Value under Continuous Interest

The formula

$$P = Se^{-rt}$$

gives the present value P of S dollars due at the end of t years at an annual rate of r compounded continuously.

EXAMPLE 3 Trust Fund

A trust fund is being set up by a single payment so that at the end of 20 years there will be \$25,000 in the fund. If interest is compounded continuously at an annual rate of 7%, how much money (to the nearest dollar) should be paid into the fund initially?

Solution: We want the present value of \$25,000 due in 20 years. Therefore,

$$P = Se^{-rt} = 25{,}000e^{-(0.07)(20)}$$
$$= 25{,}000e^{-1.4} \approx 6165$$

Thus, \$6165 should be paid initially.

▶ **Now work Problem 13.**

Exercise 10.3

In Problems 1 and 2, find the compound amount and compound interest if \$4000 is invested for six years and interest is compounded continuously at the given annual rate.

*1. $6\frac{1}{4}\%$

2. 9%

In Problems 3 and 4, find the present value of \$2500 due eight years from now if interest is compounded continuously at the given annual rate.

3. $6\frac{3}{4}\%$

4. 8%

In Problems 5–8, find the effective rate of interest that corresponds to the given annual rate compounded continuously.

*5. 4%

6. 7%

7. 3%

8. 11%

9. **Investment** If \$100 is deposited in a savings account that earns interest at an annual rate of $4\frac{1}{2}\%$ compounded continuously, what is the value of the account at the end of two years?

10. **Investment** If \$1000 is invested at an annual rate of 3% compounded continuously, find the compound amount at the end of eight years.

11. **Stock Redemption** The board of directors of a corporation agrees to redeem some of its callable preferred stock in five years. At that time, \$1,000,000 will be required. If the corporation can invest money at an annual interest rate of 5% compounded continuously, how much should it presently invest so that the future value is sufficient to redeem the shares?

12. **Trust Fund** A trust fund is being set up by a single payment so that at the end of 30 years there will be \$50,000 in the fund. If interest is compounded continuously at an annual rate of 6%, how much money should be paid into the fund initially?

*13. **Trust Fund** As a gift for their newly born daughter's 21st birthday, the Smiths want to give her at that time a sum of money that has the same buying power as does \$21,000 on the date of her birth. To accomplish this, they will make a single initial payment into a trust fund set up specifically for the purpose.

(a) Assume that the annual effective rate of inflation is 3%. In 21 years, what sum will have the same buying power as does \$21,000 at the date of the Smith's daughter's birth? Round your answer to the nearest dollar.

(b) What should be the amount of the single initial payment into the fund if interest is compounded continuously at an annual rate of 5%? Round your answer to the nearest dollar.

14. **Investment** Presently, the Smiths have \$50,000 to invest for 18 months. They have two options open to them:

(a) Invest the money in a certificate paying interest at the nominal rate of 5% compounded quarterly;

(b) Invest the money in a savings account earning interest at the annual rate of 4.5% compounded continuously.

How much money will they have in 18 months with each option?

15. What annual rate compounded continuously is equivalent to an effective rate of 5%?

16. What annual rate r compounded continuously is equivalent to a nominal rate of 6% compounded semiannually? (*Hint:* First show that $r = 2\ln[1.03]$.)

17. **Continuous Annuity** An annuity in which R dollars is paid each year by uniform payments that are payable continuously is called a *continuous annuity* or a *continuous income stream*. The present value of a continuous annuity for t years is

$$R \cdot \frac{1 - e^{-rt}}{r}$$

where r is the annual rate of interest compounded continuously. Find the present value of a continuous annuity of \$100 a year for 20 years at 5% compounded continuously. Give your answer to the nearest dollar.

18. **Profit** Suppose a business has an annual profit of \$40,000 for the next five years and the profits are earned continuously throughout each year. Then the profits can be thought of as a continuous annuity. (See Problem 17.) If money is worth 4% compounded continuously, find the present value of the profits.

19. If interest is compounded continuously at an annual rate of 0.07, how many years would it take for a principal P to triple? Give your answer to the nearest year.

20. If interest is compounded continuously, at what annual rate will a principal of P triplele in 20 years? Give your answer to the nearest percent.

21. **Savings Options** On July 1, 2001, Mr. Green had \$1000 in a savings account at the First National Bank. This account earns interest at an annual rate of 3.5% compounded continuously. A competing bank was attempting to attract new customers by offering to add \$20 immediately to any new account opened with a minimum \$1000 deposit, and the new account would earn interest at the annual rate of 3.5% compounded semiannually. Mr. Green decided to choose one of the following three options on July 1, 2001:

(a) Leave the money at the First National Bank;

(b) Move the money to the competing bank;

(c) Leave half the money at the First National Bank and move the other half to the competing bank.

For each of these three options, find Mr. Green's accumulated amount on July 1, 2003.

22. Investment

(a) On November 1, 1990, Ms. Rodgers invested \$10,000 in a 10-year certificate of deposit that paid interest at the annual rate of 4% compounded continuously. When the certificate matured on November 1, 2000, she reinvested the entire accumulated amount in corporate bonds, which earn interest at the rate of 5% compounded annually. To the nearest dollar, what will be Ms. Rodgers's accumulated amount on November 1, 2005?

(b) If Ms. Rodgers had made a single investment of \$10,000 in 1990 that matured in 2005 and had an effective rate of interest of 4.5%, would her accumulated amount be more or less than that in part (a) and by how much (to the nearest dollar)?

23. Investment Strategy Suppose that you have \$9000 to invest.

(a) If you invest it with the First National Bank at the nominal rate of 5% compounded quarterly, find the accumulated amount at the end of one year.

(b) The First National Bank also offers certificates on which it pays 5.5% compounded continuously. However, a minimum investment of \$10,000 is required. Because you have only \$9000, the bank is willing to give you a 1-year loan for the extra \$1000 that you need. Interest for this loan is at an effective rate of 8%, and both principal and interest are payable at the end of the year. Determine whether or not this strategy of investment is preferable to the strategy in part (a).

OBJECTIVE *To study continuity in the context of limits and to find points of discontinuity for a function.*

10.4 Continuity

Many functions have the property that there is no "break" in their graphs. For example, compare the functions

$$f(x) = x \qquad \text{and} \qquad g(x) = \begin{cases} x & \text{if } x \neq 1 \\ 2 & \text{if } x = 1 \end{cases}$$

whose graphs appear in Figures 10.23 and 10.24, respectively. When $x = 1$, the graph of f is unbroken, but the graph of g has a break. Stated another way, if you were to trace both graphs with a pencil, you would have to lift the pencil off the graph of g when $x = 1$, but you would not have to lift it off the graph of f. These situations can be expressed by limits. As x approaches 1, compare the limit of each function with the value of the function at $x = 1$:

$$\lim_{x\to 1} f(x) = 1 = f(1)$$

whereas

$$\lim_{x\to 1} g(x) = 1 \neq 2 = g(1)$$

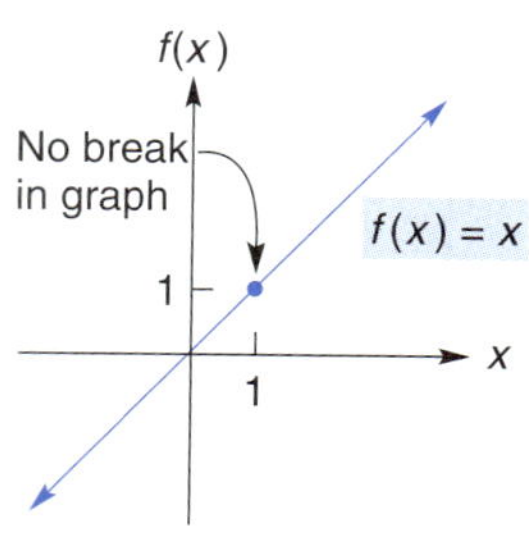

FIGURE 10.23 Continuous at 1.

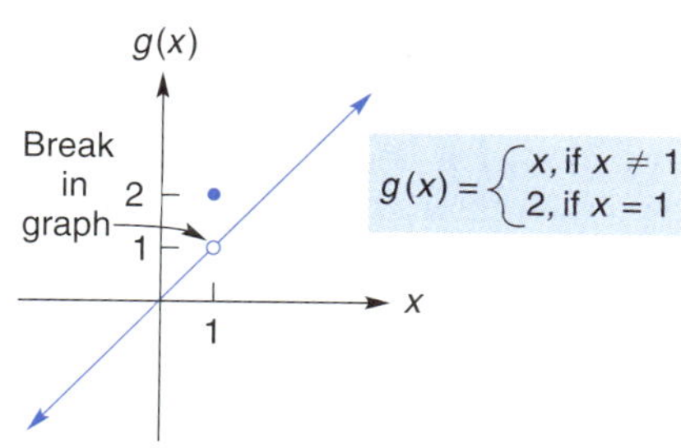

FIGURE 10.24 Discontinuous at 1.

The limit of f as $x \to 1$ is the same as $f(1)$, but the limit of g as $x \to 1$ is *not* the same as $g(1)$. For these reasons, we say that f is *continuous* at 1 and g is *discontinuous* at 1.

Definition

A function f is ***continuous*** at a if and only if the following three conditions are met:

1. $f(x)$ is defined at $x = a$; that is, a is in the domain of f
2. $\lim_{x \to a} f(x)$ exists
3. $\lim_{x \to a} f(x) = f(a)$

If f is defined on an open interval containing a, except perhaps at a itself, and f is not continuous at a, then f is said to be ***discontinuous*** at a, and a is called a ***point of discontinuity.***

EXAMPLE 1 Applying the Definition of Continuity

a. *Show that $f(x) = 5$ is continuous at 7.*

Solution: We must verify that the preceding three conditions are met. First, $f(7) = 5$, so f is defined at $x = 7$. Second,

$$\lim_{x \to 7} f(x) = \lim_{x \to 7} 5 = 5$$

Thus, f has a limit as $x \to 7$. Third,

$$\lim_{x \to 7} f(x) = 5 = f(7)$$

Therefore, f is continuous at 7. (See Figure 10.25.)

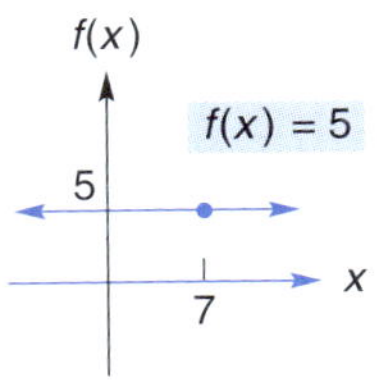

FIGURE 10.25 f is continuous at 7.

b. *Show that $g(x) = x^2 - 3$ is continuous at -4.*

Solution: The function g is defined at $x = -4: g(-4) = 13$. Also,

$$\lim_{x \to -4} g(x) = \lim_{x \to -4} (x^2 - 3) = 13 = g(-4)$$

Therefore, g is continuous at -4. (See Figure 10.26.)

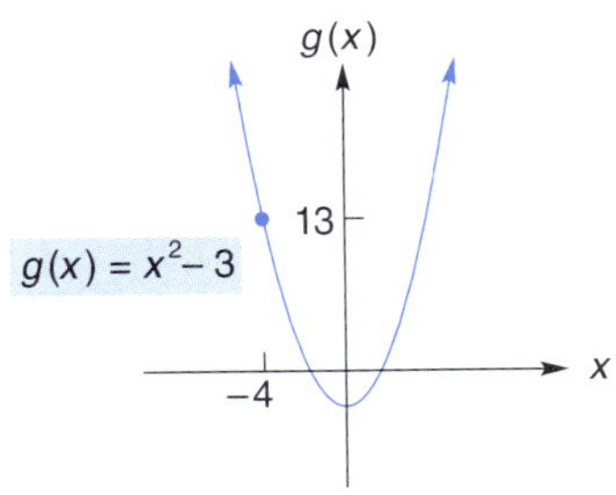

FIGURE 10.26 g is continuous at -4.

▶ **Now work Problem 1.**

We say that a function is *continuous on an interval* if it is continuous at each point there. In this situation, the graph of the function is connected over the interval. For example, $f(x) = x^2$ is continuous on the interval [2, 5]. In fact, in Example 5 of Section 10.1, we showed that, for *any* polynomial function f, $\lim_{x \to a} f(x) = f(a)$. This means that

A polynomial function is continuous at every point.

It follows that such a function is continuous on every interval. We say that polynomial functions are **continuous everywhere** or, more simply, that they are continuous.

EXAMPLE 2 Continuity of Polynomial Functions

The functions $f(x) = 7$ and $g(x) = x^2 - 9x + 3$ are polynomial functions. Therefore, they are continuous. For example, they are continuous at 3.

▶ **Now work Problem 13.**

When is a function discontinuous? We can say that a function f defined on an open interval containing a is discontinuous at a if

1. f has no limit as $x \to a$

or

2. as $x \to a$, f has a limit that is different from $f(a)$

Obviously, if f is not defined at a, it is not continuous there. However, if f is not defined at a, but *is* defined for all nearby values, then not only is f not continuous at a, but also, it is discontinuous there. In Figure 10.27, we can find points of discontinuity by inspection.

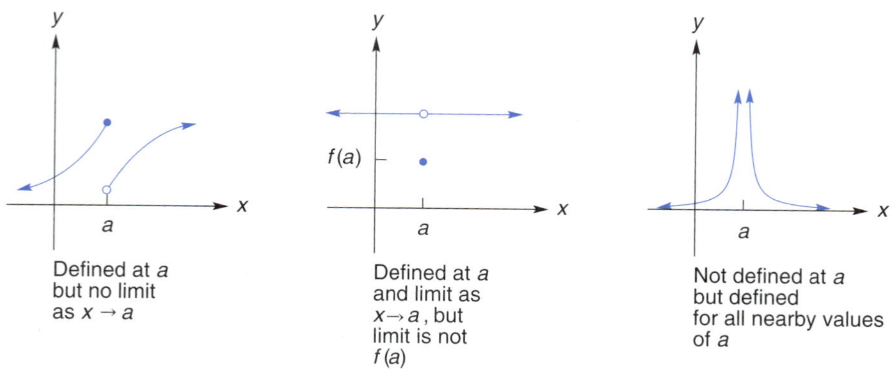

FIGURE 10.27 Discontinuities at a.

EXAMPLE 3 **Discontinuities**

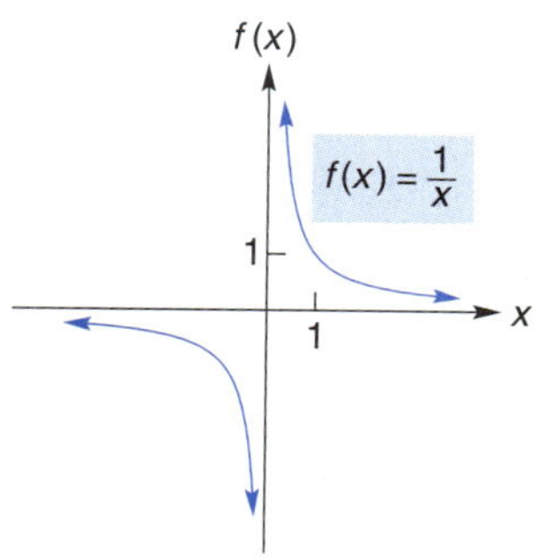

FIGURE 10.28 Infinite discontinuity at 0.

a. Let $f(x) = \dfrac{1}{x}$. (See Figure 10.28.) Note that f is not defined at $x = 0$, but it is defined for all other x nearby. Thus, f is discontinuous at 0. Moreover, $\lim_{x\to 0^+} f(x) = \infty$ and $\lim_{x\to 0^-} f(x) = -\infty$. A function is said to have an **infinite discontinuity** at a when at least one of the one-sided limits is either ∞ or $-\infty$ as $x \to a$. Hence, f has an *infinite discontinuity* at $x = 0$.

b. Let $f(x) = \begin{cases} 1 & \text{if } x > 0 \\ 0 & \text{if } x = 0 \\ -1 & \text{if } x < 0 \end{cases}$

(See Figure 10.29.) Although f is defined at $x = 0$, $\lim_{x\to 0} f(x)$ does not exist. Thus, f is discontinuous at 0.

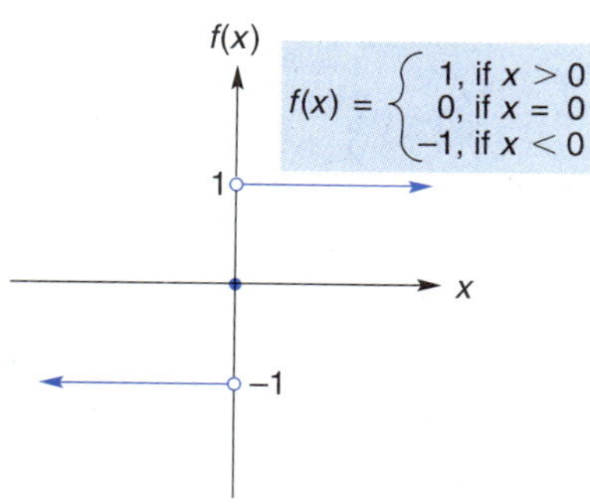

FIGURE 10.29 Discontinuous case-defined function.

▸ **Now work Problem 29.**

The following property indicates where the discontinuities of a rational function occur:

Discontinuities of a Rational Function

A rational function is discontinuous at points where the denominator is 0 and is continuous otherwise.

CAUTION Is the function $f(x) = \frac{x+1}{x+1}$ continuous everywhere? No, it is discontinuous at -1. Some students think that $f(x)$ is identically 1, but it is not. Actually,

$$f(x) = 1 \text{ for } x \neq -1$$

and f is not defined for $x = -1$.

EXAMPLE 4 **Locating Discontinuities in Rational Functions**

For each of the following functions, find all points of discontinuity.

a. $f(x) = \dfrac{x^2 - 3}{x^2 + 2x - 8}$

Solution: This rational function has denominator

$$x^2 + 2x - 8 = (x+4)(x-2)$$

which is 0 when $x = -4$ or $x = 2$. Thus, f is discontinuous only at -4 and 2.

b. $h(x) = \dfrac{x+4}{x^2+4}$

Solution: For this rational function, the denominator is never 0. (It is always positive.) Therefore, h has no discontinuity.

▶ **Now work Problem 19.**

EXAMPLE 5 **Locating Discontinuities in Case-Defined Functions**

For each of the following functions, find all points of discontinuity.

a. $f(x) = \begin{cases} x+6 & \text{if } x \geq 3 \\ x^2 & \text{if } x < 3 \end{cases}$

Solution: The only possible trouble may occur when $x = 3$, because this is the only place at which the graph of f could be disconnected. We know that $f(3) = 3 + 6 = 9$. So because

$$\lim_{x\to 3^+} f(x) = \lim_{x\to 3^+} (x+6) = 9$$

and

$$\lim_{x\to 3^-} f(x) = \lim_{x\to 3^-} x^2 = 9$$

we can conclude that $\lim_{x\to 3} f(x) = 9 = f(3)$. Thus, the function is continuous at 3, as well as at all other x. We can reach the same conclusion by inspecting the graph of f in Figure 10.30.

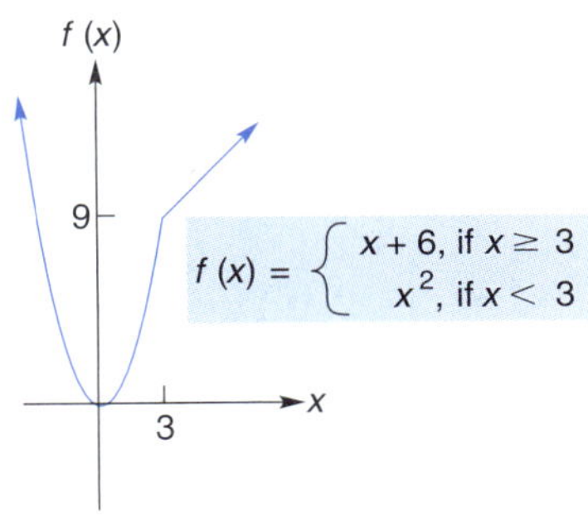

FIGURE 10.30 Continuous case-defined function.

b. $f(x) = \begin{cases} x+2 & \text{if } x > 2 \\ x^2 & \text{if } x < 2 \end{cases}$

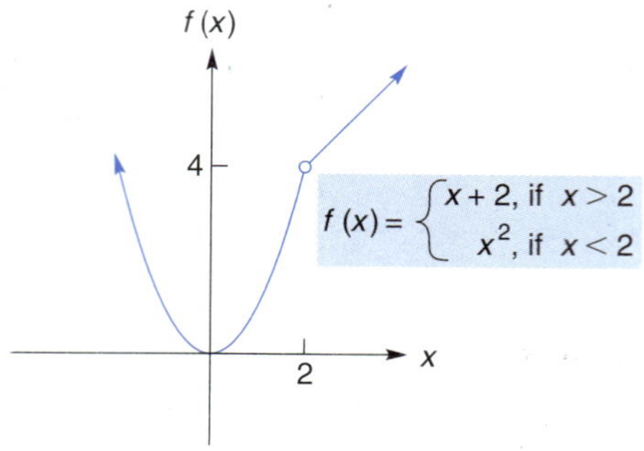

FIGURE 10.31 Discontinuous at 2.

Solution: Since f is not defined at $x = 2$, but it is defined for all other x nearby, f is discontinuous at 2. It is continuous for all other x. (See Figure 10.31.)

▶ **Now work Problem 31.**

EXAMPLE 6 "Post-Office" Function

The "post-office" function

$$c = f(x) = \begin{cases} 37 & \text{if } 0 < x \le 1 \\ 60 & \text{if } 1 < x \le 2 \\ 83 & \text{if } 2 < x \le 3 \\ 106 & \text{if } 3 < x \le 4 \end{cases}$$

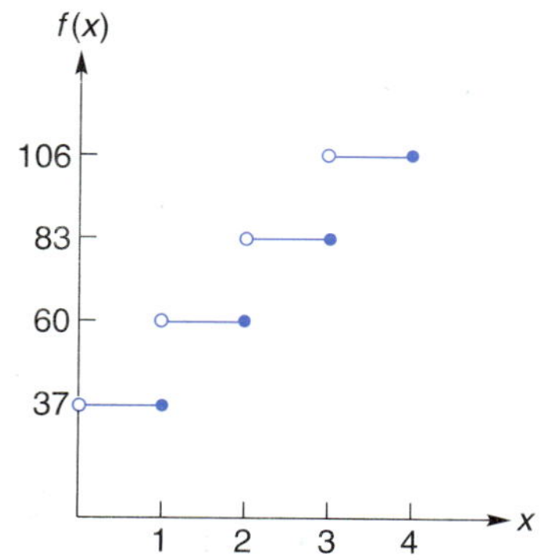

FIGURE 10.32 Post-office function.

gives the cost c (in cents) of mailing a parcel of weight x (ounces), for $0 < x \le 4$, in June 2002. It is clear from its graph in Figure 10.32 that f has discontinuities at 1, 2, and 3 and is constant for values of x between successive discontinuities. Such a function is called a *step function* because of the appearance of its graph.

▶ **Now work Problem 35.**

There is another way to express continuity besides that given in the definition. If we take the statement

$$\lim_{x \to a} f(x) = f(a)$$

This method of expressing continuity at a is used frequently in mathematical proofs.

and replace x by $a + h$, then as $x \to a$, we have $h \to 0$ (Figure 10.33). Thus, the statement

$$\lim_{h \to 0} f(a + h) = f(a)$$

defines continuity at a.

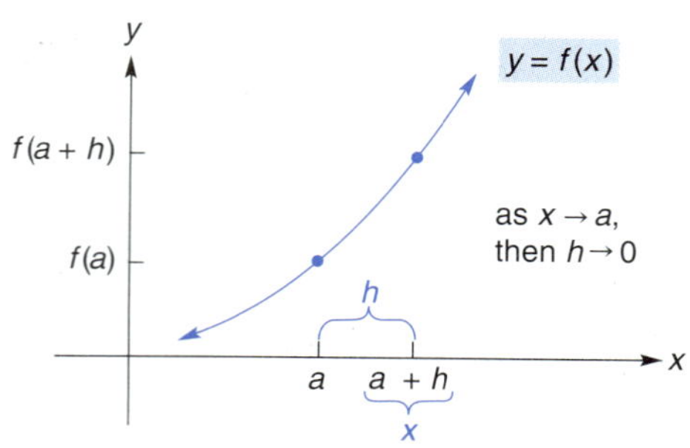

FIGURE 10.33 Diagram for continuity at a.

TECHNOLOGY

By observing the graph of a function, we may be able to determine where a discontinuity occurs. However, we can be fooled. For example, the function

$$f(x) = \frac{x-1}{x^2-1}$$

is discontinuous at ± 1, but the discontinuity at 1 is not obvious from the graph of f in Figure 10.34. On the other hand, the discontinuity at -1 is obvious.

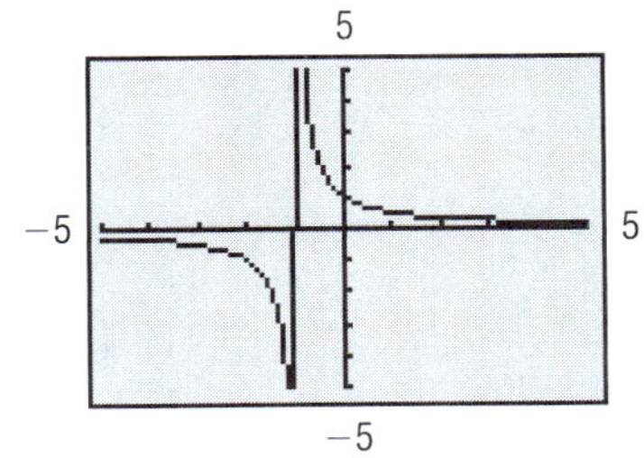

FIGURE 10.34 Discontinuity at 1 is not apparent from graph of $f(x) = \frac{x-1}{x^2-1}$.

TABLE 10.3 Demand Schedule

Price/Unit p	Quantity/Week q
\$20	0
10	5
5	15
4	20
2	45
1	95

Often, it is helpful to describe a situation by a continuous function. For example, the demand schedule in Table 10.3 indicates the number of units of a particular product that consumers will demand per week at various prices. This information can be given graphically, as in Figure 10.35(a), by plotting each quantity-price pair as a point. Clearly, the graph does not represent a continuous function. Furthermore, it gives us no information as to the price at which, say, 35 units would be demanded. However, if we connect the points in Figure 10.35(a) by a smooth curve (see Figure 10.35(b)), we get a so-called demand curve. From it, we could guess that at about \$2.50 per unit, 35 units would be demanded.

Frequently, it is possible and useful to describe a graph, as in Figure 10.35(b), by means of an equation that defines a continuous function f. Such a function not only gives us a demand equation, $p = f(q)$, which allows us to anticipate corresponding prices and quantities demanded, but also permits a convenient mathematical analysis of the nature and basic properties of demand. Of course, some care must be used in working with equations such as $p = f(q)$.

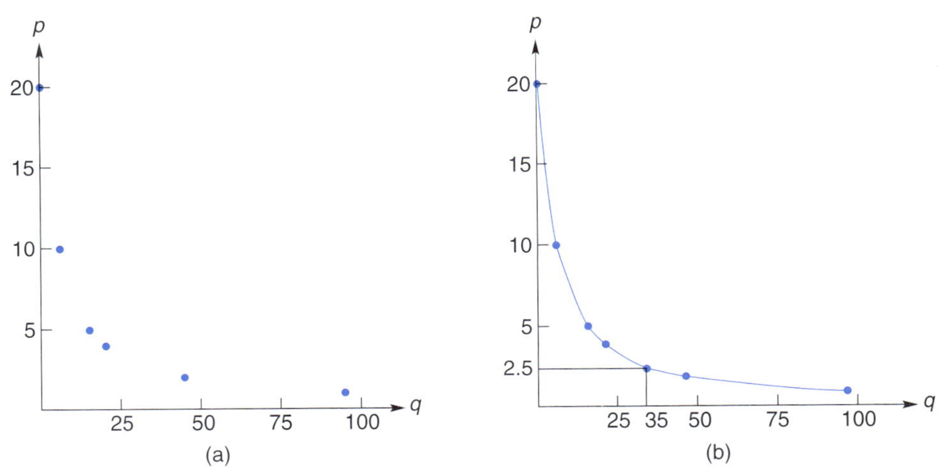

FIGURE 10.35 Viewing data via a continuous function.

Mathematically, f may be defined when $q = \sqrt{37}$, but from a practical standpoint, a demand of $\sqrt{37}$ units could be meaningless to our particular situation. For example, if a unit is an egg, then a demand of $\sqrt{37}$ eggs make no sense.

In general, it will be our desire to view practical situations in terms of continuous functions whenever possible, so that we may be better able to analyze their nature.

Exercise 10.4

In Problems 1–6, use the definition of continuity to show that the given function is continuous at the indicated point.

*1. $f(x) = x^3 - 5x;\ x = 2$

2. $f(x) = \dfrac{x-2}{4x};\ x = -2$

3. $g(x) = \sqrt{2-3x};\ x = 0$

4. $f(x) = \dfrac{x}{8};\ x = 2$

5. $h(x) = \dfrac{x-4}{x+4};\ x = 4$

6. $f(x) = \sqrt[3]{x};\ x = -1$

In Problems 7–12, determine whether the function is continuous at the given points.

7. $f(x) = \dfrac{x+4}{x-2};\ -2,\ 0$

8. $f(x) = \dfrac{x^2-4x+4}{6};\ 2,\ -2$

9. $g(x) = \dfrac{x-4}{x^2-16};\ 4,\ -4$

10. $h(x) = \dfrac{3}{x^2+4};\ 2,\ -2$

11. $f(x) = \begin{cases} x+2 & \text{if } x \geq 2 \\ x^2 & \text{if } x < 2 \end{cases};\ 2, 0$

12. $f(x) = \begin{cases} \dfrac{1}{x} & \text{if } x \neq 0 \\ 0 & \text{if } x \neq 0 \end{cases};\ 0, -1$

In Problems 13–16, give a reason why the function is continuous everywhere.

*13. $f(x) = 2x^2 - 3$

14. $f(x) = \dfrac{3+x}{7}$

15. $f(x) = \dfrac{x-1}{x^2+4}$

16. $f(x) = x(1-x)$

In Problems 17–34, find all points of discontinuity.

17. $f(x) = 3x^2 - 3$

18. $h(x) = x - 2$

*19. $f(x) = \dfrac{3}{x+4}$

20. $f(x) = \dfrac{x^2+3x-4}{x^2-4}$

21. $g(x) = \dfrac{(x^2-1)^2}{5}$

22. $f(x) = -1$

23. $f(x) = \dfrac{x^2+6x+9}{x^2+2x-15}$

24. $g(x) = \dfrac{x-3}{x^2+x}$

25. $h(x) = \dfrac{x-7}{x^3-x}$

26. $f(x) = \dfrac{x-1}{1-x}$

27. $p(x) = \dfrac{x}{x^2+1}$

28. $f(x) = \dfrac{x^4}{x^4-1}$

*29. $f(x) = \begin{cases} 1 & \text{if } x \geq 0 \\ -1 & \text{if } x < 0 \end{cases}$

30. $f(x) = \begin{cases} 2x+1 & \text{if } x \geq -1 \\ 1 & \text{if } x < -1 \end{cases}$

*31. $f(x) = \begin{cases} 0 & \text{if } x \leq 1 \\ x-1 & \text{if } x > 1 \end{cases}$

32. $f(x) = \begin{cases} x-3 & \text{if } x > 2 \\ 3-2x & \text{if } x < 2 \end{cases}$

33. $f(x) = \begin{cases} x^2 & \text{if } x > 2 \\ x-1 & \text{if } x < 2 \end{cases}$

34. $f(x) = \begin{cases} \dfrac{16}{x^2} & \text{if } x \geq 2 \\ 3x-2 & \text{if } x < 2 \end{cases}$

*35. **Telephone Rates** Suppose the long-distance rate for a telephone call from Hazleton, Pennsylvania, to Los Angeles, California, is \$0.10 for the first minute and \$0.06 for each additional minute or fraction thereof. If $y = f(t)$ is a function that indicates the total charge y for a call of t minutes' duration, sketch the graph of f for $0 < t \leq 4\frac{1}{2}$. Use your graph to determine the values of t, where $0 < t \leq 4\frac{1}{2}$, at which discontinuities occur.

36. The *greatest integer function*, $f(x) = [x]$, is defined to be the greatest integer less than or equal to x, where x is any real number. For example, $[3] = 3$, $[1.999] = 1$, $\left[\frac{1}{4}\right] = 0$ and $[-4.5] = -5$. Sketch the graph of this function for $-3.5 \leq x \leq 3.5$. Use your sketch to determine the values of x at which discontinuities occur.

37. **Inventory** Sketch the graph of

$$y = f(x) = \begin{cases} -100x + 600 & \text{if } 0 \leq x < 5 \\ -100x + 1100 & \text{if } 5 \leq x < 10 \\ -100x + 1600 & \text{if } 10 \leq x < 15 \end{cases}$$

A function such as this might describe the inventory y of a company at time x. Is f continuous at 2? At 5? At 10?

38. Graph $g(x) = e^{-1/|x|}$. Because g is not defined at $x = 0$, but is defined for all other nearby x, g is discontinuous at 0. Based on the graph of g, is

$$f(x) = \begin{cases} e^{-1/|x|} & \text{if } x \neq 0 \\ 0 & \text{if } x = 0 \end{cases}$$

continuous at 0?

10.5 Continuity Applied to Inequalities

OBJECTIVE *To develop techniques for solving nonlinear inequalities.*

In Section 1.2, we solved linear inequalities. We now turn our attention to showing how the notion of continuity can be applied to solving a nonlinear inequality such as $x^2 + 3x - 4 < 0$.

Recall (from Section 2.5) that there is a relationship between the x-intercepts of the graph of a function g (that is, the points where the graph meets the x-axis) and the roots of the equation $g(x) = 0$, which are called zeros of g. If the graph of g has an x-intercept $(r, 0)$, then $g(r) = 0$, so r is a root of the equation $g(x) = 0$, and thus, r is a zero of g. Hence, from the graph of $y = g(x)$ in Figure 10.36, we conclude that r_1, r_2, and r_3 are zeros of g. On the other hand, if r is any real root of the equation $g(x) = 0$, then $g(r) = 0$, and hence, $(r, 0)$ lies on the graph of g. This means that all real zeros of g can be represented by the points where the graph of g meets the x-axis. Note also in Figure 10.36 that the three zeros determine four open intervals on the x-axis:

$$(-\infty, r_1) \quad (r_1, r_2) \quad (r_2, r_3) \quad (r_3, \infty)$$

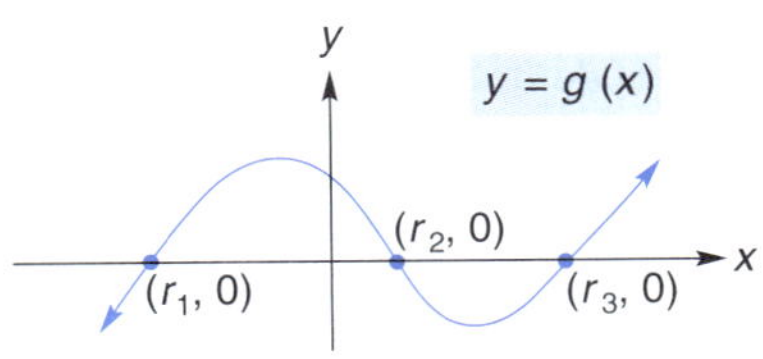

FIGURE 10.36 r_1, r_2, and r_3 are zeros of g.

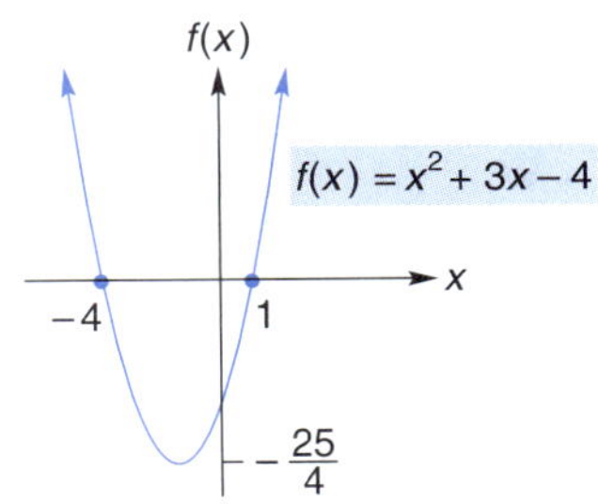

FIGURE 10.37 -4 and 1 are zeros of f.

To solve $x^2 + 3x - 4 > 0$, we shall let

$$f(x) = x^2 + 3x - 4 = (x + 4)(x - 1)$$

Because f is a polynomial function, it is continuous everywhere. The zeros of f are -4 and 1; hence, the graph of f has x-intercepts $(-4, 0)$ and $(1, 0)$. (See Figure 10.37.) The zeros—or, to be more precise, the x-intercepts—determine three intervals on the x-axis:

$$(-\infty, -4) \quad (-4, 1) \quad (1, \infty)$$

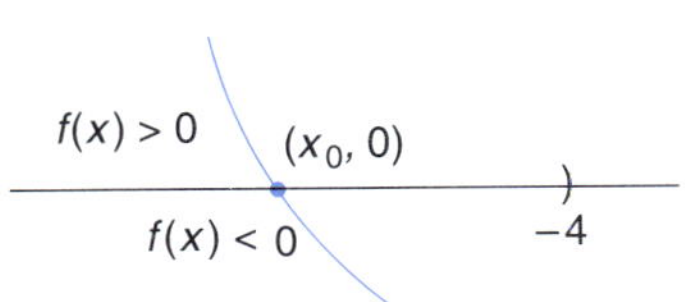

FIGURE 10.38 Change of sign for continuous function.

Consider the interval $(-\infty, -4)$. Since f is continuous on this interval, we claim that either $f(x) > 0$ or $f(x) < 0$ *throughout* the interval. If this were not the case, then $f(x)$ would indeed change sign on the interval. By the continuity of f, there would be a point where the graph intersects the x-axis—for example, at $(x_0, 0)$. (See Figure 10.38.) But then x_0 would be a zero of f. However, this cannot be, because there is no zero of f that is less than -4. Hence, $f(x)$ must be strictly positive or strictly negative on $(-\infty, -4)$. A similar argument can be made for each of the other intervals.

To determine the sign of $f(x)$ on any one of the three intervals, it suffices to determine its sign at any point in the interval. For instance, -5 is in $(-\infty, -4)$ and

$$f(-5) = 6 > 0. \qquad \text{Thus, } f(x) > 0 \text{ on } (-\infty, -4)$$

Similarly, 0 is in $(-4, 1)$, and

$$f(0) = -4 < 0. \qquad \text{Thus, } f(x) < 0 \text{ on } (-4, 1)$$

Finally, 3 is in $(1, \infty)$, and

$$f(3) = 14 > 0. \qquad \text{Thus, } f(x) > 0 \text{ on } (1, \infty)$$

(See the "sign chart" in Figure 10.39.) Therefore,

$$x^2 + 3x - 4 > 0 \quad \text{on} \quad (-\infty, -4) \quad \text{and} \quad (1, \infty)$$

Notice the importance of continuity in the solution of inequalities.

so we have solved the inequality. These results are obvious from the graph in Figure 10.37. The graph lies above the x-axis (that is, $f(x) > 0$) on $(-\infty, -4)$ and $(1, \infty)$.

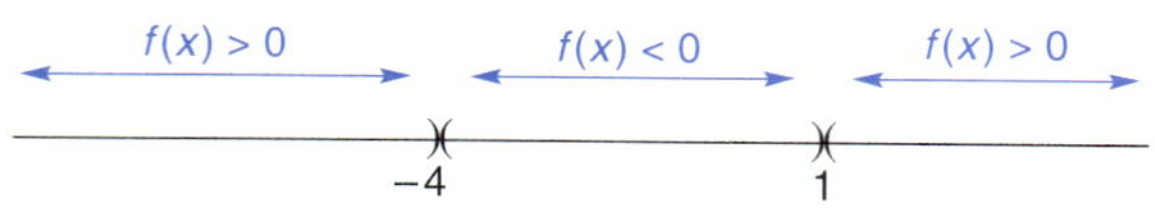

FIGURE 10.39 Sign chart for $x^2 + 3x - 4$.

EXAMPLE 1 Solving a Quadratic Inequality

Solve $x^2 - 3x - 10 < 0$

Solution: If $f(x) = x^2 - 3x - 10$, then f is a polynomial (quadratic) function and thus is continuous everywhere. To find the (real) zeros of f, we have

$$x^2 - 3x - 10 = 0$$

$$(x + 2)(x - 5) = 0$$

$$x = -2, 5$$

FIGURE 10.40 Zeros of $x^2 - 3x - 10$.

The zeros -2 and 5 are shown in Figure 10.40. These zeros determine three intervals:

$$(-\infty, -2) \quad (-2, 5) \quad (5, \infty)$$

To determine the sign of $f(x)$ on $(-\infty, -2)$, we choose a point in that interval, say, -3. The sign of $f(x)$ throughout $(-\infty, -2)$ is the same as that of $f(-3)$. Because

$$f(x) = (x + 2)(x - 5) \qquad \text{(factored form of } f(x))$$

we have

$$f(-3) = (-3 + 2)(-3 - 5) = (-)(-) = +$$

(Note how we conveniently found the sign of $f(-3)$ by using the signs of the factors of $f(x)$.) Thus, $f(x) > 0$ on $(-\infty, -2)$. To test the other two intervals, we shall choose the points 0 and 6. We find that

$$f(0) = (0 + 2)(0 - 5) = (+)(-) = -$$

so $f(x) < 0$ on $(-2, 5)$, and

$$f(6) = (6 + 2)(6 - 5) = (+)(+) = +$$

so $f(x) > 0$ on $(5, \infty)$. A summary of our results is in the sign chart in Figure 10.41. Therefore, $x^2 - 3x - 10 < 0$ on $(-2, 5)$.

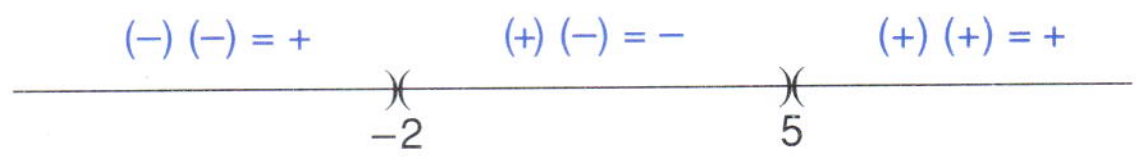

FIGURE 10.41 Sign chart for $(x+2)(x-5)$.

▶ **Now work Problem 1.**

APPLICATIONS IN PRACTICE 1

Solving a Polynomial Inequality

An open box is formed by cutting a square piece out of each corner of an 8-inch-by-10-inch piece of metal. If each side of the cut out squares is x inches long, the volume of the box is given by $V(x) = x(8-2x)(10-2x)$. This problem makes sense only when this volume is positive. Find the values of x for which the volume is positive.

FIGURE 10.42 Zeros of $x(x-1)(x+4)$.

EXAMPLE 2 **Solving a Polynomial Inequality**

Solve $x(x-1)(x+4) \le 0$

Solution: If $f(x) = x(x-1)(x+4)$, then f is a polynomial function and is continuous everywhere. The zeros of f are 0, 1, and −4, which are shown in Figure 10.42. These zeros determine four intervals:

$$(-\infty, -4) \quad (-4, 0) \quad (0, 1) \quad (1, \infty)$$

Now, at a test point in each interval, we find the sign of $f(x)$:

$$f(-5) = (-)(-)(-) = - \quad \text{so } f(x) < 0 \text{ on } (-\infty, -4)$$

$$f(-2) = (-)(-)(+) = + \quad \text{so } f(x) > 0 \text{ on } (-4, 0)$$

$$f\left(\frac{1}{2}\right) = (+)(-)(+) = - \quad \text{so } f(x) < 0 \text{ on } (0, 1)$$

$$f(2) = (+)(+)(+) = + \quad \text{so } f(x) > 0 \text{ on } (1, \infty)$$

Figure 10.43 shows the sign chart for $f(x)$. Thus, $x(x-1)(x+4) \le 0$ on $(-\infty, -4]$ and $[0, 1]$. Note that −4, 0, and 1 are included in the solution because these values satisfy the equality (=) part of the inequality (≤).

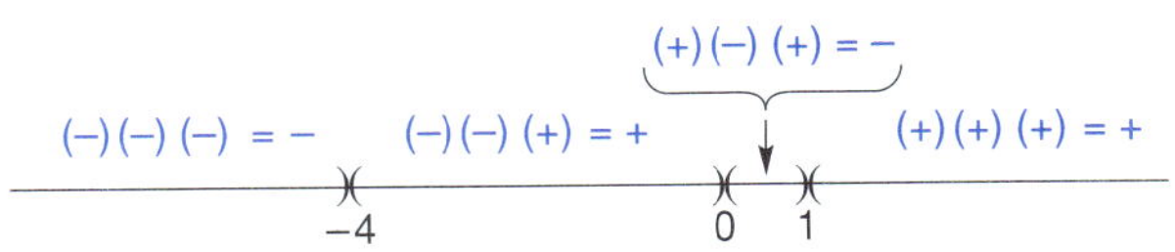

FIGURE 10.43 Sign chart for $x(x-1)(x+4)$.

▶ **Now work Problem 11.**

EXAMPLE 3 **Solving a Rational Function Inequality**

Solve $\dfrac{x^2 - 6x + 5}{x} \ge 0$

FIGURE 10.44 Zeros and points of discontinuity for $\dfrac{(x-1)(x-5)}{x}$.

Solution: Let $f(x) = \dfrac{x^2 - 6x + 5}{x} = \dfrac{(x-1)(x-5)}{x}$. For a rational function f, we solve the inequality by considering the intervals determined by the zeros of f and the points where f is discontinuous, for it is around such points that the sign of $f(x)$ may change. Here the zeros are 1 and 5. The function is discontinuous at 0 and continuous otherwise. In Figure 10.44, we have placed a hollow dot at 0 to indicate that f is not defined there. We thus consider the intervals

$$(-\infty, 0) \quad (0, 1) \quad (1, 5) \quad (5, \infty)$$

Determining the sign of $f(x)$ at a test point in each interval, we find that

$$f(-1) = \frac{(-)(-)}{(-)} = - \quad \text{so } f(x) < 0 \text{ on } (-\infty, 0)$$

$$f\left(\frac{1}{2}\right) = \frac{(-)(-)}{(+)} = + \quad \text{so } f(x) > 0 \text{ on } (0, 1)$$

$$f(2) = \frac{(+)(-)}{(+)} = - \quad \text{so } f(x) < 0 \text{ on } (1, 5)$$

$$f(6) = \frac{(+)(+)}{(+)} = + \quad \text{so } f(x) > 0 \text{ on } (5, \infty)$$

The sign chart for $f(x)$ is shown in Figure 10.45. Therefore, $f(x) \geq 0$ on $(0, 1]$ and $[5, \infty)$. Why are 1 and 5 included, but 0 excluded? Figure 10.46 shows the graph of f. Notice that the solution of $f(x) \geq 0$ consists of all x values for which the graph lies on or above the x-axis.

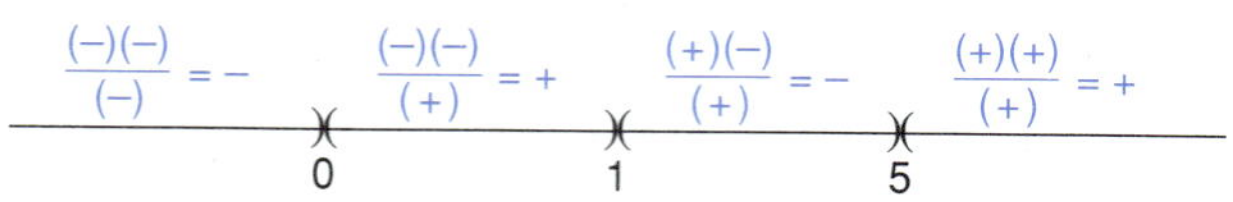

FIGURE 10.45 Sign chart for $\dfrac{(x-1)(x-5)}{x}$.

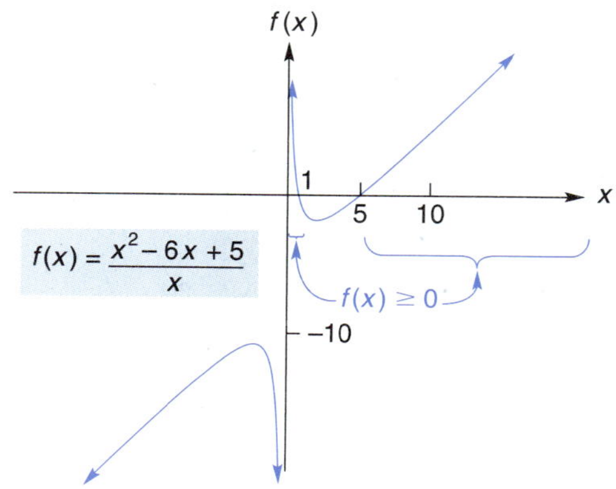

FIGURE 10.46 Graph of $f(x) = \dfrac{x^2 - 6x + 5}{x}$.

▶ Now work Problem 17.

EXAMPLE 4 Solving Nonlinear Inequalities

a. *Solve* $x^2 + 1 > 0$.

Solution: The equation $x^2 + 1 = 0$ has no real roots. Thus, $f(x) = x^2 + 1$ has no real zero. Also, f is continuous everywhere. Therefore, $f(x)$ is always positive or is always negative. But x^2 is always positive or zero, so $x^2 + 1$ is always positive. Hence, the solution of $x^2 + 1 > 0$ is $(-\infty, \infty)$.

b. *Solve* $x^2 + 1 < 0$.

Solution: From part (a), $x^2 + 1$ is always positive, so $x^2 + 1 < 0$ has no solution.

▶ Now work Problem 7.

Exercise 10.5

In Problems 1–26, solve the inequalities by the technique discussed in this section.

*1. $x^2 - 3x - 4 > 0$

2. $x^2 - 8x + 15 > 0$

3. $x^2 - 8x + 15 \leq 0$

4. $14 - 5x - x^2 \leq 0$

5. $2x^2 + 11x + 14 < 0$

6. $x^2 - 4 < 0$

*7. $x^2 + 4 < 0$

8. $2x^2 - x - 2 \leq 0$

9. $(x + 2)(x - 3)(x + 6) \leq 0$

10. $(x + 3)(x + 2)(x - 5) \geq 0$

*11. $-x(x - 5)(x + 4) > 0$

12. $(x + 2)^2 > 0$

13. $x^3+4x \geq 0$

14. $(x+2)^2(x^2-1)<0$

15. $x^3+2x^2-3x>0$

16. $x^3+6x^2+9x<0$

*17. $\dfrac{x}{x^2-9}<0$

18. $\dfrac{x^2-1}{x}<0$

19. $\dfrac{4}{x-1} \geq 0$

20. $\dfrac{3}{x^2-5x+6}>0$

21. $\dfrac{x^2-x-6}{x^2+4x-5} \geq 0$

22. $\dfrac{x^2+2x-3}{x^2-x-2} \geq 0$

23. $\dfrac{3}{x^2+6x+5} \leq 0$

24. $\dfrac{2x+1}{x^2} \leq 0$

25. $x^2+2x \geq 2$

26. $x^4-16 \geq 0$

27. **Revenue** Suppose that consumers will purchase q units of a product when the price of *each* unit is $23-0.2q$ dollars. How many units must be sold in order that sales revenue will be no less than \$650?

28. **Forest Management** A lumber company owns a forest that is of rectangular shape, 1 mi × 2 mi. The company wants to cut a uniform strip of trees along the outer edges of the forest. At most, how wide can the strip be if the company wants at least $1\frac{5}{16}$ mi^2 of forest to remain?

29. **Container Design** A container manufacturer wishes to make an open box by cutting a 4-in. square from each corner of a square sheet of aluminum and then turning up the sides. The box is to contain at least 324 in^3. Find the dimensions of the smallest sheet of aluminum that can be used.

30. **Workshop Participation** Imperial Education Services (I.E.S.) is offering a workshop in data processing to key personnel at Zeta Corporation. The price per person is \$50, and Zeta Corporation guarantees that at least 50 people will attend. Suppose I.E.S. offers to reduce the charge for *everybody* by \$0.50 for each person over the 50 who attends. How should I.E.S. limit the size of the group so that the total revenue it receives will never be less than that received for 50 persons?

31. Graph $f(x)=x^3+7x^2-5x+4$. Use the graph to determine the solution of

$$x^3+7x^2-5x+4 \leq 0$$

32. Graph $f(x)=\dfrac{3x^2-0.5x+2}{6.2-4.1x}$. Use the graph to determine the solution of

$$\frac{3x^2-0.5x+2}{6.2-4.1x}>0$$

A novel way of solving a nonlinear inequality like $f(x)>0$ is by examining the graph of $g(x)=f(x)/|f(x)|$, whose range consists only of 1 and -1:

$$g(x)=\frac{f(x)}{|f(x)|}=\begin{cases} 1 & \text{if } f(x)>0 \\ -1 & \text{if } f(x)<0 \end{cases}$$

The solution of $f(x)>0$ consists of all intervals for which $g(x)=1$. Using this technique, solve the inequalities in Problems 33 and 34.

33. $6x^2-x-2>0$

34. $\dfrac{x^2+x-3}{x^2-x-2}<0$

10.6 Review

Important Terms and Symbols

Section 10.1 $\lim_{x\to a} f(x)=L$

Section 10.2 one-sided limits $\lim_{x\to a^-} f(x)=L$ $\lim_{x\to a^+} f(x)=L$ $\lim_{x\to a} f(x)=\infty$ $\lim_{x\to -\infty} f(x)=L$ $\lim_{x\to \infty} f(x)=L$

Section 10.3 compounding continuously

Section 10.4 continuous discontinuous continuous on an interval continuous everywhere

Summary

The notion of a limit lies at the foundation of calculus. To say that $\lim_{x\to a} f(x)=L$ means that the values of $f(x)$ can be made as close to the number L as we choose by taking x sufficiently close to, but different from, a. If $\lim_{x\to a} f(x)$ and $\lim_{x\to a} g(x)$ exist and c is a constant, then:

1. $\lim_{x\to a} c=c$
2. $\lim_{x\to a} x^n=a^n$
3. $\lim_{x\to a}[f(x)\pm g(x)]=\lim_{x\to a} f(x)\pm\lim_{x\to a} g(x)$
4. $\lim_{x\to a}[f(x)\cdot g(x)]=\lim_{x\to a} f(x)\cdot\lim_{x\to a} g(x)$
5. $\lim_{x\to a}[cf(x)]=c\cdot\lim_{x\to a} f(x)$
6. $\lim_{x\to a}\dfrac{f(x)}{g(x)}=\dfrac{\lim_{x\to a} f(x)}{\lim_{x\to a} g(x)}$ if $\lim_{x\to a} g(x)\neq 0$,
7. $\lim_{x\to a}\sqrt[n]{f(x)}=\sqrt[n]{\lim_{x\to a} f(x)}$
8. If f is a polynomial function, then $\lim_{x\to a} f(x)=f(a)$

Property 8 implies that the limit of a polynomial function as $x\to a$ can be found by simply substituting a for x. However, with other functions, substitution may lead to the meaningless form 0/0. In such cases, algebraic manipulation such as factoring and cancellation may give a form from which the limit can be determined.

If $f(x)$ approaches L as x approaches a from the right, then we write $\lim_{x\to a^+} f(x) = L$. If $f(x)$ approaches L as x approaches a from the left, we write $\lim_{x\to a^-} f(x) = L$. These limits are called one-sided limits.

The infinity symbol ∞, which does not represent a number, is used in describing limits. The statement

$$\lim_{x\to\infty} f(x) = L$$

means that as x increases without bound, the values of $f(x)$ approach the number L. A similar statement applies for the situation when $x \to -\infty$, which means that x is decreasing without bound. In general, if $p > 0$, then

$$\lim_{x\to\infty} \frac{1}{x^p} = 0 \quad \text{and} \quad \lim_{x\to-\infty} \frac{1}{x^p} = 0$$

If $f(x)$ increases without bound as $x \to a$, then we write $\lim_{x\to a} f(x) = \infty$. Similarly, if $f(x)$ decreases without bound, we have $\lim_{x\to a} f(x) = -\infty$. To say that the limit of a function is ∞ (or $-\infty$) does not mean that the limit exists. Rather, it is a way of saying that the limit does not exist and tells *why* there is no limit.

There is a rule for evaluating the limit of a rational function (quotient of polynomials) as $x \to \infty$ or $-\infty$. If $f(x)$ is a rational function and $a_n x^n$ and $b_m x^m$ are the terms in the numerator and denominator, respectively, with the greatest powers of x, then

$$\lim_{x\to\infty} f(x) = \lim_{x\to\infty} \frac{a_n x^n}{b_m x^m} \quad \text{and} \quad \lim_{x\to-\infty} f(x) = \lim_{x\to-\infty} \frac{a_n x^n}{b_m x^m}$$

In particular, as $x \to \infty$ or $-\infty$, the limit of a polynomial is the same as the limit of the term that involves the greatest power of x. This means that, for a nonconstant polynomial, the limit as $x \to \infty$ or $-\infty$ is either ∞ or $-\infty$.

When interest is compounded at every instant of time, we say that it is compounded continuously. Under continuous compounding at an annual rate r for t years, the formula $S = Pe^{rt}$ gives the compound amount S of a principal of P dollars. The formula $P = Se^{-rt}$ gives the present value P of S dollars. The effective rate corresponding to an annual rate r compounded continuously is $e^r - 1$.

A function f is continuous at a if and only if

1. $f(x)$ is defined at $x = a$
2. $\lim_{x\to a} f(x)$ exists
3. $\lim_{x\to a} f(x) = f(a)$

Geometrically this means that the graph of f has no break when $x = a$. If a function is not continuous at a and is defined on an open interval containing a, except perhaps at a itself, then the function is said to be discontinuous at a. Polynomial functions are continuous everywhere, and rational functions are discontinuous only at points where the denominator is zero.

To solve the inequality $f(x) > 0$ [or $f(x) < 0$], we first find the real zeros of f and the values of x for which f is discontinuous. These values determine intervals, and on each interval, $f(x)$ is either always positive or always negative. To find the sign on any one of these intervals, it suffices to find the sign of $f(x)$ at any point there. After the signs are determined for all intervals, it is easy to give the solution of $f(x) > 0$ [or $f(x) < 0$].

Review Problems

Problem numbers shown in color indicate problems suggested for use as a practice chapter test.

In Problems 1–28, find the limits if they exist. If the limit does not exist, so state that, or use the symbol ∞ or $-\infty$ where appropriate.

1. $\lim_{x\to-1} (2x^2 + 6x - 1)$

2. $\lim_{x\to 0} \frac{2x^2 - 3x + 1}{2x^2 - 2}$

3. $\lim_{x\to 3} \frac{x^2 - 9}{x^2 - 3x}$

4. $\lim_{x\to -3} \frac{x + 1}{x^2 - 3}$

5. $\lim_{h\to 0} (x + h)$

6. $\lim_{x\to 2} \frac{x^2 - 4}{x^2 - 3x + 2}$

7. $\lim_{x\to -4} \frac{x^3 + 4x^2}{x^2 + 2x - 8}$

8. $\lim_{x\to 1} \frac{x^2 + x - 2}{x^2 + 4x - 5}$

9. $\lim_{x\to\infty} \frac{2}{x + 1}$

10. $\lim_{x\to\infty} \frac{x^2 + 1}{2x^2}$

11. $\lim_{x\to\infty} \frac{5x + 2}{9x - 3}$

12. $\lim_{x\to-\infty} \frac{1}{x^4}$

13. $\lim_{t\to 3} \frac{2t - 3}{t - 3}$

14. $\lim_{x\to-\infty} \frac{x^6}{x^5}$

15. $\lim_{x\to-\infty} \frac{x + 3}{1 - x}$

16. $\lim_{x\to 9} \sqrt{16}$

17. $\lim_{x\to\infty} \frac{x^2 - 1}{(3x + 2)^2}$

18. $\lim_{x\to 1} \frac{x^2 + x - 2}{x - 1}$

19. $\lim_{x\to 3^-} \frac{x + 3}{x^2 - 9}$

20. $\lim_{x\to 2} \frac{2 - x}{x - 2}$

21. $\lim_{x\to\infty} \sqrt{3x}$

22. $\lim_{y\to 5^+} \sqrt{y - 5}$

23. $\lim_{x\to\infty} \frac{x^{100} + (1/x^2)}{e - x^{98}}$

24. $\lim_{x\to-\infty} \frac{ex^2 - x^4}{31x - 2x^3}$

25. $\lim_{x\to 1} f(x)$ if $f(x) = \begin{cases} x^2 & \text{if } 0 \le x < 1 \\ x & \text{if } x > 1 \end{cases}$

26. $\lim_{x\to 3} f(x)$ if $f(x) = \begin{cases} x + 5 & \text{if } x < 3 \\ 6 & \text{if } x \ge 3 \end{cases}$

27. $\lim_{x\to 4^+} \frac{\sqrt{x^2-16}}{4-x}$ (*Hint:* For $x > 4$, $\sqrt{x^2-16} = \sqrt{x-4}\sqrt{x+4}$)

28. $\lim_{x\to 3^+} \frac{x^2-x-6}{\sqrt{x-3}}$ (*Hint:* For $x > 3$, $\frac{x-3}{\sqrt{x-3}} = \sqrt{x-3}$]

29. If $f(x) = 8x - 2$, find $\lim_{h\to 0} \frac{f(x+h)-f(x)}{h}$

30. If $f(x) = 2x^2 - 3$, find $\lim_{h\to 0} \frac{f(x+h)-f(x)}{h}$

31. Host-Parasite Relationship For a particular host-parasite relationship, it was determined that when the host density (number of hosts per unit of area) is x, then the number of hosts parasitized over a certain period of time is

$$y = 23\left(1 - \frac{1}{1+2x}\right)$$

If the host density were to increase without bound, what value would y approach?

32. Predator-Prey Relationship For a particular predator-prey relationship, it was determined that the number y of prey consumed by an individual predator over a period of time was a function of the prey density x (the number of prey per unit of area). Suppose

$$y = f(x) = \frac{10x}{1+0.1x}$$

If the prey density were to increase without bound, what value would y approach?

33. For an annual interest of 5% compounded continuously, find

(a) the compound amount of $2500 after 14 years.

(b) the present value of $2500 due in 14 years.

34. For an annual interest rate of 6% compounded continuously, find

(a) the compound amount of $800 after 11 years.

(b) the present value of $800 due in 11 years.

35. Find the effective rate equivalent to an annual rate of 6% compounded continuously.

36. Find the effective rate equivalent to an annual rate of 1% compounded continuously.

37. If interest is earned at the annual rate of 6.5% compounded continuously, determine the exact number of years required for a principal of P to double.

38. Investment The Smiths have just made two investments: $400 in account *A*, which earns interest at the rate of 6% compounded semiannually, and $400 in account *B*, which earns interest at the rate of 5.5% compounded continuously.

(a) Calculate the effective interest rate for each investment.

(b) Which investment is worth more at the end of 5 years and by how much?

39. Using the definition of continuity, show that the function $f(x) = x + 5$ is continuous at $x = 7$.

40. Using the definition of continuity, show that the function $f(x) = \frac{x-5}{x^2+2}$ is continuous at $x = 5$.

41. State whether $f(x) = \frac{x}{4}$ is continuous everywhere. Give a reason for your answer.

42. State whether $f(x) = x^2 - 2$ is continuous everywhere. Give a reason for your answer.

In Problems 43–50, find the points of discontinuity (if any) for each function.

43. $f(x) = \frac{x^2}{x+3}$

44. $f(x) = \frac{0}{x^3}$

45. $f(x) = \frac{x-1}{2x^2+3}$

46. $f(x) = (5-7x)^5$

47. $f(x) = \frac{4-x^2}{x^2+3x-4}$

48. $f(x) = \frac{2x+6}{x^3+x}$

49. $f(x) = \begin{cases} x+4 & \text{if } x > -2 \\ 3x+6 & \text{if } x \le -2 \end{cases}$

50. $f(x) = \begin{cases} \frac{1}{x} & \text{if } x < 1 \\ 1 & \text{if } x \ge 1 \end{cases}$

In Problems 51–58, solve the given inequalities.

51. $x^2 + 4x - 12 > 0$

52. $3x^2 - 3x - 6 \le 0$

53. $x^3 \ge 2x^2$

54. $x^3 + 8x^2 + 15x \ge 0$

55. $\frac{x+5}{x^2-1} < 0$

56. $\frac{x(x+5)(x+8)}{3} < 0$

57. $\frac{x^2+3x}{x^2+2x-8} \ge 0$

58. $\frac{x^2-4}{x^2+2x+1} \le 0$

59. Graph $f(x) = \frac{x^3+3x^2-19x+18}{x^3-2x^2+x-2}$. Use the graph to estimate $\lim_{x\to 2} f(x)$.

60. Graph $f(x) = \frac{\sqrt{x+3}-2}{x-1}$. From the graph, estimate $\lim_{x\to 1} f(x)$.

61. Graph $f(x) = x \ln x$. From the graph, estimate the one-sided limit $\lim_{x\to 0^+} f(x)$.

62. Graph $f(x) = \frac{e^x - 1}{(e^x+1)(e^{2x}-e^x)}$. Use the graph to estimate $\lim_{x\to 0} f(x)$.

63. Graph $f(x) = x^3 - x^2 + x - 6$. Use the graph to determine the solution of

$$x^3 - x^2 + x - 6 \ge 0$$

64. Graph $f(x) = \frac{x^5-4}{x^3+1}$. Use the graph to determine the solution of

$$\frac{x^5-4}{x^3+1} \le 0$$

Mathematical SNAPSHOT

National Debt

The size of the U.S. national debt is of great concern to many people and is frequently a topic in the news. The magnitude of the debt affects the confidence in the U.S. economy of both domestic and foreign investors, corporate officials, and political leaders. There are those who believe that to reduce the debt there must be cuts in government spending, which could affect government programs, or there must be an increase in revenues, possibly through tax increases.

Suppose that it is possible for the debt to be reduced continuously at an annual fixed rate. This is similar to compounding interest continuously, except that instead of adding interest to an amount at each instant of time, you would be subtracting from the debt at each instant. Let us see how you could model this situation.

Suppose the debt D_0 at time $t = 0$ is reduced at an annual rate r. Furthermore, assume that there are k time periods of equal length in a year. At the end of the first period, the original debt is reduced by $D_0\left(\frac{r}{k}\right)$, so the new debt is

$$D_0 - D_0\left(\frac{r}{k}\right) = D_0\left(1 - \frac{r}{k}\right)$$

At the end of the second period, this debt is reduced by $D_0\left(1 - \frac{r}{k}\right)\frac{r}{k}$, so the new debt is

$$\begin{aligned} D_0\left(1 - \frac{r}{k}\right) - D_0\left(1 - \frac{r}{k}\right)\frac{r}{k} &\\ = D_0\left(1 - \frac{r}{k}\right)\left(1 - \frac{r}{k}\right) &\\ = D_0\left(1 - \frac{r}{k}\right)^2 & \end{aligned}$$

The pattern continues. At the end of the third period the debt is $D_0\left(1 - \frac{r}{k}\right)^3$, and so on. At the end of t years the number of periods is kt and the debt is $D_0\left(1 - \frac{r}{k}\right)^{kt}$. If the debt is to be reduced at each instant of time, then $k \to \infty$. Thus you want to find

$$\lim_{k\to\infty} D_0\left(1 - \frac{r}{k}\right)^{kt}$$

which may be rewritten as

$$D_0\left[\lim_{k\to\infty}\left(1 - \frac{r}{k}\right)^{-k/r}\right]^{-r}$$

If you let $x = -\frac{r}{k}$, then the condition $k \to \infty$ implies that $x \to 0$. Hence the limit inside the brackets has the form $\lim_{x\to 0}(1 + x)^{1/x}$, which is well known to be e. Therefore, if the debt D_0 at time $t = 0$ is reduced continuously at an annual rate r, then t years later the debt D is given by

$$D = D_0 e^{-rt}$$

For example, assume a current debt of \$5345 billion and a continuous reduction rate of 6% annually. Then the debt t years from now is given by

$$D = 5345e^{-0.06t}$$

where D is in billions of dollars. This means that in 10 years, the debt will be $5345e^{-0.6} \approx \$2933$ billion. Figure 10.47 shows the graph of $D = 5345e^{-rt}$ for various rates r. Of course the greater the value of r, the faster the debt reduction. Notice that for $r = 0.06$, the debt at the end of 30 years is still considerable (approximately \$884 billion).

It is interesting to note that decaying radioactive elements also follow the model of continuous debt reduction, $D = D_0 e^{-rt}$.

To find out where the U.S. national debt currently stands, visit one of the national debt clocks on the Internet. You can find them by looking for "national debt clock" using any search engine.

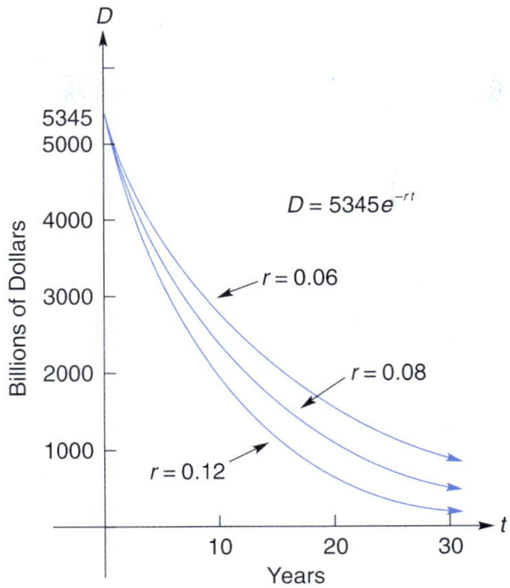

FIGURE 10.47 Budget debt is reduced continuously.

Exercises

In the following problems, assume a current national debt of \$5345 *billion.*

1. If the debt were reduced to \$4500 billion a year from now, what annual rate of continuous debt reduction would be involved? Give your answer to the nearest percent.
2. For a continuous debt reduction at an annual rate of 6%, determine the number of years from now required for the debt to be reduced by one-half. Give your answer to the nearest year.
3. What assumptions underlie a model of debt reduction that uses an exponential function? What are the limitations of this approach?

CHAPTER

11

Differentiation

Mathematical SNAPSHOT

Marginal Propensity to Consume

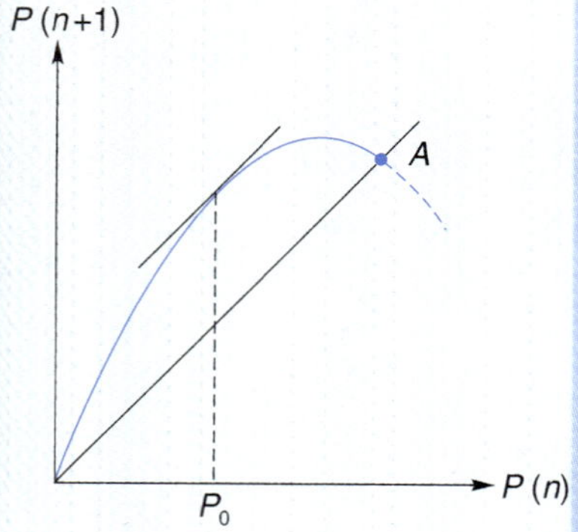

Government regulations generally limit the number of fish taken from a given fishing ground by commercial fishing boats in a season. This prevents overfishing, which depletes the fish population and leaves, in the long run, fewer fish to catch.

From a strictly commercial perspective, the ideal regulations would maximize the number of fish available for the year-to-year fish harvest. The key to finding those ideal regulations is a mathematical function called the reproduction curve. For a given fish habitat, this function estimates the fish population a year from now, $P(n+1)$, based on the population now, $P(n)$, assuming no external intervention (i.e., no fishing, no influx of predators, and so on).

The figure to the left shows a typical reproduction curve. Also graphed is the line $y = x$, the line along which the populations $P(n)$ and $P(n+1)$ would be equal. Notice the intersection of the curve and the straight line at point A. This is where, because of habitat crowding, the population has reached its maximum sustainable size. A population that is this size one year will be the same size the next year.

For any point on the horizontal axis, the distance between the reproduction curve and the line $y = x$ represents the sustainable harvest: the number of fish that could be caught, after the spawn have grown to maturity, so that in the end the population is back at the same size it was a year ago.

Commercially speaking, the optimal population size is the one where the distance between the reproduction curve and the line $y = x$ is the greatest. This condition is met where the slopes of the reproduction curve and the line $y = x$ are equal. Thus, for a maximum fish harvest year after year, regulations should aim to keep the fish population fairly close to P_0.

A central idea here is that of the slope of a curve at a given point. That idea is the cornerstone concept of this chapter.

Now we begin our study of calculus. The ideas involved in calculus are completely different from those of algebra and geometry. The power and importance of these ideas and their applications will be clear to you later in the book. In this chapter we introduce the *derivative* of a function, and you will learn important rules for finding derivatives. You will also see how the derivative is used to analyze the rate of change of a quantity, such as the rate at which the position of a body is changing.

11.1 The Derivative

OBJECTIVE *To develop the idea of a tangent line to a curve, to define the slope of a curve, and to define a derivative and give it a geometric interpretation. To compute derivatives by using the limit definition.*

One of the main problems with which calculus deals is finding the slope of the *tangent line* at a point on a curve. In geometry you probably thought of a tangent line, or *tangent,* to a circle as a line that meets the circle at exactly one point (Figure 11.1). However, this idea of a tangent is not very useful for other kinds of curves. For example, in Figure 11.2(a), the lines L_1 and L_2 intersect the curve at exactly one point P. Although we would not think of L_2 as the tangent at this point, it seems natural that L_1 is. In Figure 11.2(b) we intuitively would consider L_3 to be the tangent at point P, even though L_3 intersects the curve at other points.

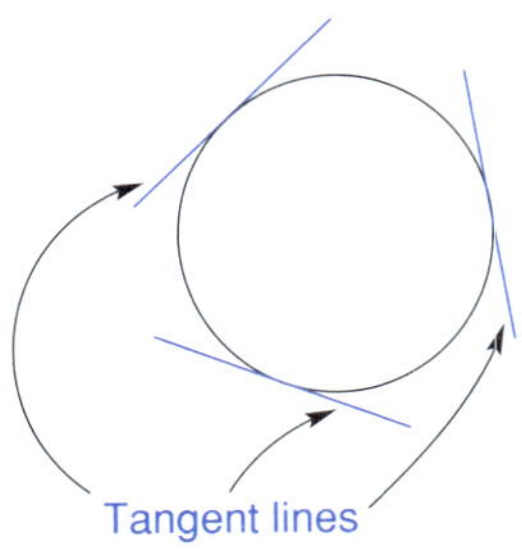

FIGURE 11.1 Tangent lines to a circle.

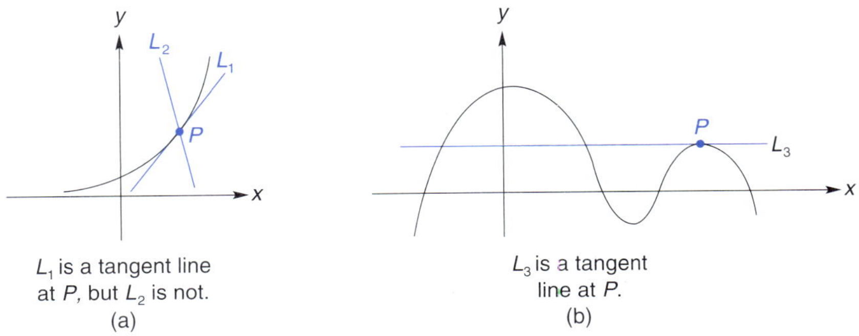

FIGURE 11.2 Tangent line at a point.

From the previous examples, you can see that we must drop the idea that a tangent is simply a line that intersects a curve at only one point. To obtain a suitable definition of tangent line, we use the limit concept and the geometrical notion of a *secant line.* A **secant line** is a line that intersects a curve at two or more points.

Look at the graph of the function $y = f(x)$ in Figure 11.3. We wish to define the tangent line at point P. If Q is a different point on the curve, the line PQ is a secant line. If Q moves along the curve and approaches P from the right (see Figure 11.4), typical secant lines are PQ', PQ'', and so on. As Q approaches P from the left, they are PQ_1, PQ_2, and so on. *In both cases, the secant lines approach the same limiting position.* This common limiting position of the secant lines is defined to be the **tangent line** to the curve at P. This definition seems reasonable and applies to curves in general, not just circles.

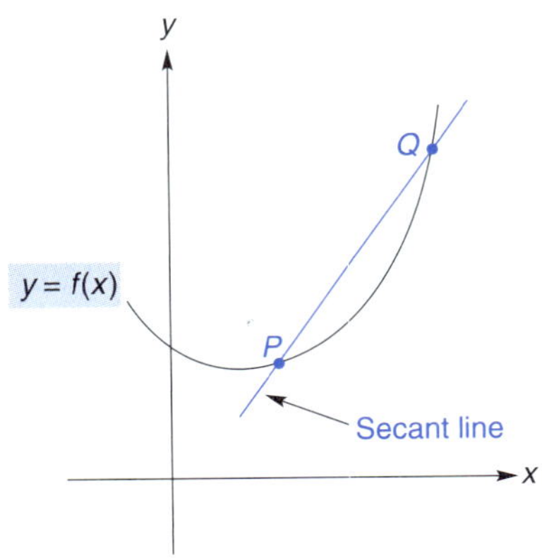

FIGURE 11.3 Secant line PQ.

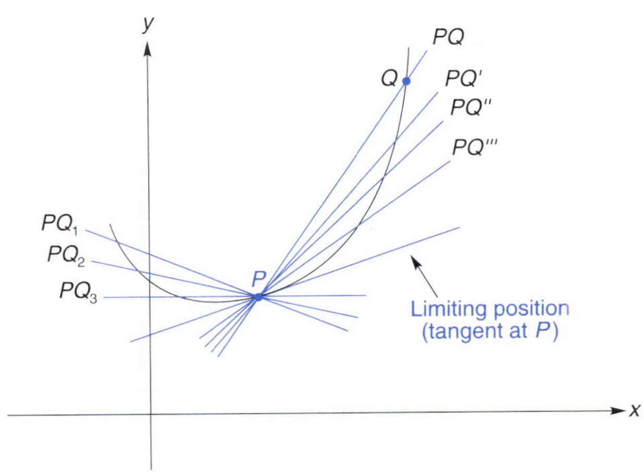

FIGURE 11.4 The tangent line is a limiting position of secant lines.

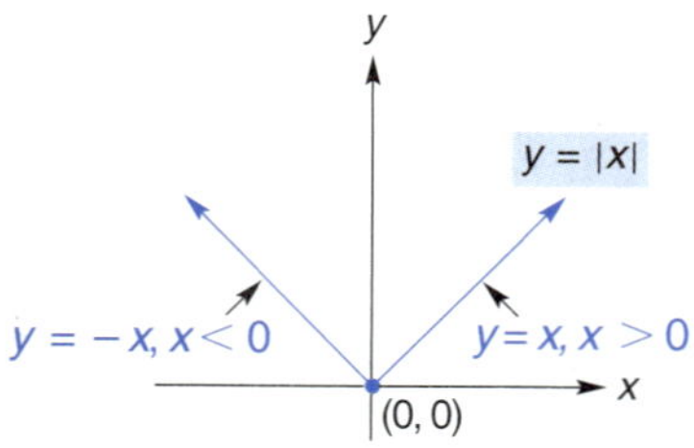

FIGURE 11.5 No tangent line to graph of $y = |x|$ at $(0, 0)$.

A curve does not necessarily have a tangent line at each of its points. For example, the curve $y = |x|$ does not have a tangent at $(0, 0)$. As you can see in Figure 11.5, a secant line through $(0, 0)$ and a nearby point to its right on the curve must always be the line $y = x$. Thus the limiting position of such secant lines is also the line $y = x$. However, a secant line through $(0, 0)$ and a nearby point to its left on the curve must always be the line $y = -x$. Hence, the limiting position of such secant lines is also the line $y = -x$. Since there is no common limiting position, there is no tangent line at $(0, 0)$.

Now that we have a suitable definition of a tangent to a curve at a point, we can define the *slope of a curve* at a point.

Definition

The ***slope of a curve*** at a point P is the slope, if it exists, of the tangent line at P.

Since the tangent at P is a limiting position of secant lines PQ, we consider the slope of the tangent to be the limiting value of the slopes of the secant lines as Q approaches P. For example, let us consider the curve $f(x) = x^2$ and the slopes of some secant lines PQ, where $P = (1, 1)$. For the point $Q = (2.5, 6.25)$, the slope of PQ (see Figure 11.6) is

$$m_{PQ} = \frac{y_2 - y_1}{x_2 - x_1} = \frac{6.25 - 1}{2.5 - 1} = 3.5$$

Table 11.1 includes other points Q on the curve, as well as the corresponding slopes of PQ. Notice that as Q approaches P, the slopes of the secant lines seem to approach 2. Thus, we expect the slope of the indicated tangent line at $(1, 1)$ to be 2. This will be confirmed later, in Example 1. But first, we wish to generalize our procedure.

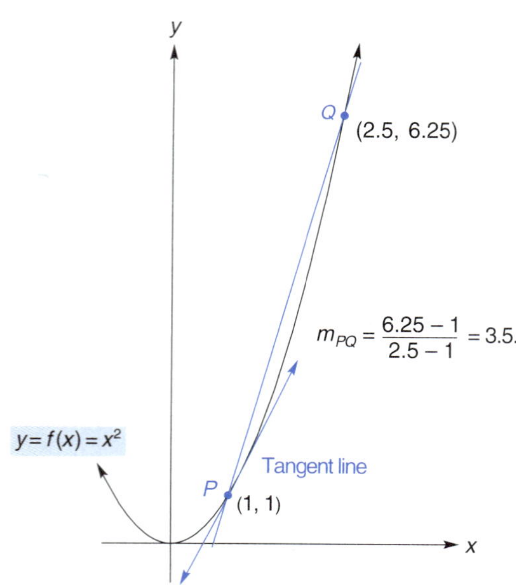

FIGURE 11.6 Secant line to $f(x) = x^2$ through $(1, 1)$ and $(2.5, 6.25)$.

TABLE 11.1 Slopes of secant lines to the curve $f(x) = x^2$ at $P = (1, 1)$

Q	Slope of PQ
(2.5, 6.25)	$(6.25 - 1)/(2.5 - 1) = 3.5$
(2, 4)	$(4 - 1)/(2 - 1) = 3$
(1.5, 2.25)	$(2.25 - 1)/(1.5 - 1) = 2.5$
(1.25, 1.5625)	$(1.5625 - 1)/(1.25 - 1) = 2.25$
(1.1, 1.21)	$(1.21 - 1)/(1.1 - 1) = 2.1$
(1.01, 1.0201)	$(1.021 - 1)/(1.01 - 1) = 2.01$

For the curve $y = f(x)$ in Figure 11.7, we shall find an expression for the slope at the point $P = (x_1, f(x_1))$. If $Q = (x_2, f(x_2))$, the slope of the secant line PQ is

$$m_{PQ} = \frac{f(x_2) - f(x_1)}{x_2 - x_1}$$

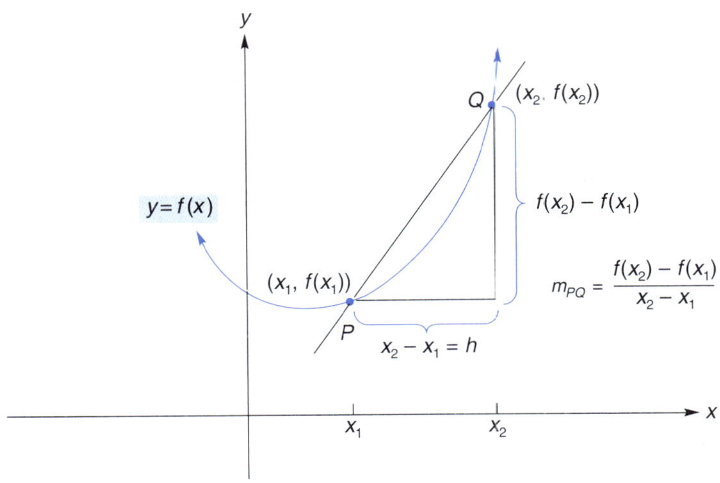

FIGURE 11.7 Secant line through P and Q.

If the difference $x_2 - x_1$ is called h, then we can write x_2 as $x_1 + h$. Here we must have $h \neq 0$, for if $h = 0$, then $x_2 = x_1$, and no secant line exists. Accordingly,

$$m_{PQ} = \frac{f(x_1 + h) - f(x_1)}{(x_1 + h) - x_1} = \frac{f(x_1 + h) - f(x_1)}{h}$$

As Q moves along the curve toward P, x_2 approaches x_1. This means that h approaches zero. The limiting value of the slopes of the secant lines—which is the slope of the tangent line at $(x_1, f(x_1))$—is

$$m_{\tan} = \lim_{h \to 0} \frac{f(x_1 + h) - f(x_1)}{h} \qquad \textbf{(1)}$$

In Example 1, we use this limit to confirm our previous conclusion that the slope of the tangent line to the curve $f(x) = x^2$ at $(1, 1)$ is 2.

EXAMPLE 1 **Finding the Slope of a Tangent Line**

Find the slope of the tangent line to the curve $y = f(x) = x^2$ at the point (1, 1).

Solution: The slope is the limit in Equation (1) with $f(x) = x^2$ and $x_1 = 1$:

$$\begin{aligned}\lim_{h\to 0}\frac{f(1+h)-f(1)}{h} &= \lim_{h\to 0}\frac{(1+h)^2-(1)^2}{h}\\ &= \lim_{h\to 0}\frac{1+2h+h^2-1}{h} = \lim_{h\to 0}\frac{2h+h^2}{h}\\ &= \lim_{h\to 0}\frac{h(2+h)}{h} = \lim_{h\to 0}(2+h) = 2\end{aligned}$$

Therefore, the tangent line to $y = x^2$ at (1, 1) has slope 2. (Refer back to Figure 11.6.)

▶ **Now work Problem 1.**

We can generalize Equation (1) so that it applies to any point $(x, f(x))$ on a curve. Replacing x_1 by x gives a function, called the *derivative* of f, whose input is x and whose output is the slope of the tangent line to the curve at $(x, f(x))$, provided that the tangent line *has* a slope (that is, provided that the tangent line is nonvertical). We thus have the following definition, which forms the basis of differential calculus:

Definition

The ***derivative*** of a function f is the function denoted f' (read "f prime") and defined by

$$f'(x) = \lim_{h\to 0}\frac{f(x+h)-f(x)}{h}$$

provided that this limit exists. If $f'(x_1)$ can be found, f is said to be ***differentiable*** at x_1, and $f'(x_1)$ is called the derivative of f at x_1 or the derivative of f with respect to x at x_1. The process of finding the derivative is called ***differentiation.***

In the definition of the derivative, the expression

$$\frac{f(x+h)-f(x)}{h}$$

is called a **difference quotient.** Thus $f'(x)$ is the limit of a difference quotient as $h \to 0$.

EXAMPLE 2 **Using the Definition to Find the Derivative**

If $f(x) = x^2$, find the derivative of f.

Solution: Applying the definition of a derivative gives

$$\begin{aligned}f'(x) &= \lim_{h\to 0}\frac{f(x+h)-f(x)}{h}\\ &= \lim_{h\to 0}\frac{(x+h)^2-x^2}{h} = \lim_{h\to 0}\frac{x^2+2xh+h^2-x^2}{h}\\ &= \lim_{h\to 0}\frac{2xh+h^2}{h} = \lim_{h\to 0}\frac{h(2x+h)}{h} = \lim_{h\to 0}(2x+h) = 2x\end{aligned}$$

CAUTION: Don't be sloppy when applying the limit definition of a derivative. Write $\lim_{h\to 0}$ at each step before the limit is actually taken. Unfortunately, some students neglect to take the final limit, and h appears in their answer. This is a quick way to lose points on an examination.

Observe that, in taking the limit, we treated x as a constant, because it was h, not x, that was changing. Also, note that $f'(x) = 2x$ defines a function of x, which we can interpret as giving the slope of the tangent line to the graph of f at $(x, f(x))$. For example, if $x = 1$, then the slope is $f'(1) = 2(1) = 2$, which confirms the result in Example 1.

▶ **Now work Problem 3.**

Besides the notation $f'(x)$, other common ways to denote the derivative of $y = f(x)$ at x are

CAUTION The notation $\frac{dy}{dx}$, which is called *Leibniz notation*, should **not** be thought of as a fraction, although it looks like one. It is a single symbol for a derivative. We have not yet attached any meaning to individual symbols such as dy and dx.

$\frac{dy}{dx}$	(pronounced "dee y, dee x" or "dee y by dee x")
$\frac{d}{dx}[f(x)]$	("dee $f(x)$, dee x")
y'	("y prime")
$D_x y$	("dee x of y")
$D_x[f(x)]$	("dee x of $f(x)$")

Because the derivative gives the slope of the tangent line, $f'(x_1)$ is the slope of the line tangent to the graph of $y = f(x)$ at $(x_1, f(x_1))$.

Two other notations for the derivative of f at x_1 are

$$\left.\frac{dy}{dx}\right|_{x=x_1} \quad \text{and} \quad y'(x_1)$$

EXAMPLE 3 **Finding an Equation of a Tangent Line**

If $f(x) = 2x^2 + 2x + 3$, find an equation of the tangent line to the graph of f at $(1, 7)$.

Solution:

Strategy: We shall first determine the slope of the tangent line by computing the derivative and evaluating it at $x = 1$. Using this result and the point $(1, 7)$ in a point–slope form gives an equation of the tangent line.

We have

$$\begin{aligned} f'(x) &= \lim_{h \to 0} \frac{f(x+h) - f(x)}{h} \\ &= \lim_{h \to 0} \frac{[2(x+h)^2 + 2(x+h) + 3] - (2x^2 + 2x + 3)}{h} \\ &= \lim_{h \to 0} \frac{2x^2 + 4xh + 2h^2 + 2x + 2h + 3 - 2x^2 - 2x - 3}{h} \\ &= \lim_{h \to 0} \frac{4xh + 2h^2 + 2h}{h} = \lim_{h \to 0}(4x + 2h + 2) \end{aligned}$$

CAUTION

In Example 3 it is ***not*** correct to say that, since the derivative is $4x+2$, the tangent line at $(1, 7)$ is $y-7=(4x+2)(x-1)$. The derivative must be **evaluated** at the point of tangency to determine the slope of the tangent line.

So

$$f'(x) = 4x + 2$$

and

$$f'(1) = 4(1) + 2 = 6$$

Thus, the tangent line to the graph at $(1, 7)$ has slope 6. A point–slope form of this tangent is

$$y - 7 = 6(x - 1)$$

It will be our custom to express equations of tangent lines in slope–intercept form:

$$y - 7 = 6x - 6$$

$$y = 6x + 1$$

▶ **Now work Problem 25.**

EXAMPLE 4 Finding the Slope of a Curve at a Point

Find the slope of the curve $y = 2x + 3$ at the point where $x = 6$.

Solution: The slope of the curve is the slope of the tangent line. Letting $y = f(x) = 2x + 3$, we have

$$\frac{dy}{dx} = \lim_{h \to 0} \frac{f(x+h) - f(x)}{h} = \lim_{h \to 0} \frac{[2(x+h)+3] - (2x+3)}{h}$$

$$= \lim_{h \to 0} \frac{2h}{h} = \lim_{h \to 0} 2 = 2$$

Since $dy/dx = 2$, the slope when $x = 6$, or in fact at any point, is 2. Note that the curve is a straight line and thus has the same slope at each point.

▶ **Now work Problem 19.**

EXAMPLE 5 A Function with a Vertical Tangent Line

Find $\frac{d}{dx}(\sqrt{x})$.

Solution: Letting $f(x) = \sqrt{x}$, we have

$$\frac{d}{dx}(\sqrt{x}) = \lim_{h \to 0} \frac{f(x+h) - f(x)}{h} = \lim_{h \to 0} \frac{\sqrt{x+h} - \sqrt{x}}{h}.$$

You should become familiar with the procedure of rationalizing the *numerator.*

As $h \to 0$, both the numerator and denominator approach zero. This can be avoided by rationalizing the *numerator:*

$$\frac{\sqrt{x+h} - \sqrt{x}}{h} = \frac{\sqrt{x+h} - \sqrt{x}}{h} \cdot \frac{\sqrt{x+h} + \sqrt{x}}{\sqrt{x+h} + \sqrt{x}}$$

$$= \frac{(x+h) - x}{h(\sqrt{x+h} + \sqrt{x})} = \frac{h}{h(\sqrt{x+h} + \sqrt{x})}$$

$$= \frac{1}{\sqrt{x+h} + \sqrt{x}}$$

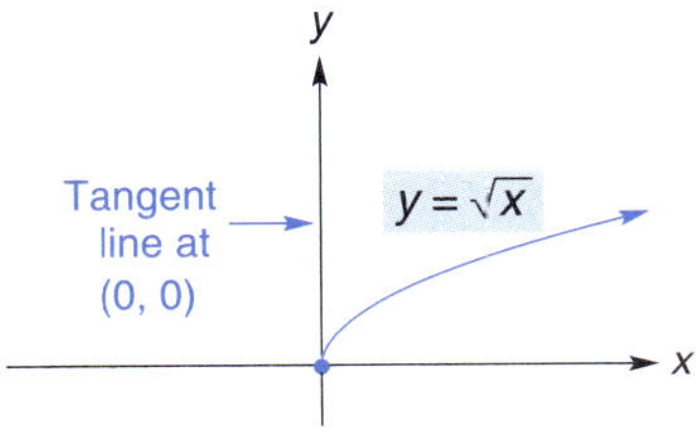

FIGURE 11.8 Vertical tangent line at (0, 0).

Therefore,

$$\frac{d}{dx}(\sqrt{x}) = \lim_{h\to 0}\frac{1}{\sqrt{x+h}+\sqrt{x}} = \frac{1}{\sqrt{x}+\sqrt{x}} = \frac{1}{2\sqrt{x}}$$

Note that the original function, $\sqrt{x}$, is defined for $x \geq 0$, but its derivative, $1/(2\sqrt{x})$, is defined only when $x > 0$. The reason for this is clear from the graph of $y = \sqrt{x}$ in Figure 11.8. When $x = 0$, the tangent is a vertical line, so its slope is not defined.

▶ **Now work Problem 17.**

In Example 5 we saw that the function $y = \sqrt{x}$ is not differentiable when $x = 0$, because the tangent line is vertical at that point. It is worthwhile mentioning that $y = |x|$ also is not differentiable when $x = 0$, but for a different reason: There is *no* tangent line at all at that point. (Refer back to Figure 11.5.)

It is wise for you to see variables other than x and y involved in a problem. Example 6 illustrates the use of other variables.

To indicate a derivative, Leibniz notation is often useful because it makes it convenient to emphasize the independent and dependent variables involved. For example, if the variable p is a function of the variable q, we speak of the derivative of p with respect to q, written dp/dq.

APPLICATIONS IN PRACTICE 1

Finding the Derivative of *H* with Respect to *t*

If a ball is thrown upward at a speed of 40 ft/s from a height of 6 feet, its height H in feet after t seconds is given by $H = 6 + 40t - 16t^2$. Find $\frac{dH}{dt}$.

EXAMPLE 6 **Finding the Derivative of *p* with Respect to *q***

If $p = f(q) = \frac{1}{2q}$, *find* $\frac{dp}{dq}$.

Solution:

$$\frac{dp}{dq} = \frac{d}{dq}\left(\frac{1}{2q}\right) = \lim_{h\to 0}\frac{f(q+h)-f(q)}{h}$$

$$= \lim_{h\to 0}\frac{\frac{1}{2(q+h)} - \frac{1}{2q}}{h} = \lim_{h\to 0}\frac{\frac{q-(q+h)}{2q(q+h)}}{h}$$

$$= \lim_{h\to 0}\frac{q-(q+h)}{h[2q(q+h)]} = \lim_{h\to 0}\frac{-h}{h[2q(q+h)]}$$

$$= \lim_{h\to 0}\frac{-1}{2q(q+h)} = -\frac{1}{2q^2}$$

Note that when $q = 0$, neither the function nor its derivative exist.

▶ **Now work Problem 15.**

Keep in mind that the derivative of $y = f(x)$ at x is nothing more than a limit, namely

$$\lim_{h\to 0}\frac{f(x+h)-f(x)}{h}$$

Although we can interpret the derivative as a function that gives the slope of the tangent line to the curve $y = f(x)$ at the point $(x, f(x))$, this interpretation is simply a geometric convenience that assists our understanding. The preceding limit may exist, aside from any geometric considerations at all. As you will see later, there are other useful interpretations of the derivative.

TECHNOLOGY

Many graphing calculators have a numerical derivative feature that estimates the derivative of a function at a point. With the TI-83 Plus we would use the "nDeriv" command, in which we must enter the function, the variable, and the value of the variable (separated by commas) in the format

nDeriv(function, variable, value of variable)

For example, the derivative of $f(x)=\sqrt{x^3+2}$ at $x=1$ is estimated in Figure 11.9. Thus, $f'(1)\approx 0.866$.

FIGURE 11.9 Numerical derivative.

On the other hand, we can take the "limit of a difference quotient" approach to estimate this derivative. To make use of the table feature of a graphing calculator, we can enter $f(x)$ as Y_1. Then, for Y_2, we enter the following form of the difference quotient:

$$(Y_1(1+X)-Y_1(1))/X$$

(Here, X plays the role of h.) Figure 11.10 shows a table for Y_2 as x approaches 0 from both the left and right. This table strongly suggests that $f'(1)\approx 0.866$.

FIGURE 11.10 Limit of a difference quotient as $x\to 0$.

Exercise 11.1

In Problems 1 and 2, a function f and a point P on its graph are given.

(a) *Find the slope of the secant line PQ for each point $Q=(x, f(x))$ whose x-value is given in the table. Round your answers to four decimal places.*

(b) *Use your results from part (a) to estimate the slope of the tangent line at P.*

*__1.__ $f(x)=x^3+3$, $P=(2, 11)$

x-value of Q	3	2.5	2.2	2.1	2.01	2.001
m_{PQ}						

2. $f(x)=e^{2x}$, $P=(0, 1)$

x-value of Q	1	0.5	0.2	0.1	0.01	0.001
m_{PQ}						

In Problems 3–18, use the definition of the derivative to find each of the following.

*__3.__ $f'(x)$ if $f(x)=x$

4. $f'(x)$ if $f(x)=4x-1$

5. $\frac{dy}{dx}$ if $y=3x+5$

6. $\frac{dy}{dx}$ if $y=-5x$

7. $\frac{d}{dx}(5-4x)$

8. $\frac{d}{dx}\left(2-\frac{x}{4}\right)$

9. $f'(x)$ if $f(x)=3$

10. $f'(x)$ if $f(x)=7.01$

11. $\frac{d}{dx}(x^2+4x-8)$

12. y' if $y=x^2+5x+1$

13. $\frac{dp}{dq}$ if $p=2q^2+5q-1$

14. $\frac{d}{dx}(x^2-x-3)$

*__15.__ y' if $y=\frac{6}{x}$

16. $\frac{dC}{dq}$ if $C=7+2q-3q^2$

*__17.__ $f'(x)$ if $f(x)=\sqrt{x+2}$

18. $H'(x)$ if $H(x)=\frac{3}{x-2}$

*__19.__ Find the slope of the curve $y=x^2+4$ at the point $(-2, 8)$.

20. Find the slope of the curve $y=2-3x^2$ at the point $(1, -1)$.

21. Find the slope of the curve $y=4x^2-5$ when $x=0$.

22. Find the slope of the curve $y=\sqrt{x}$ when $x=1$.

In Problems 23–28, find an equation of the tangent line to the curve at the given point.

23. $y=x+4$; $(3, 7)$

24. $y=3x^2-4$; $(1, -1)$

*25. $y = 3x^2 + 3x - 4$; $(-1, -4)$

26. $y = (x - 7)^2$; $(6, 1)$

27. $y = \dfrac{3}{x - 1}$; $(2, 3)$

28. $y = \dfrac{5}{1 - 3x}$; $(2, -1)$

29. Banking Equations may involve derivatives of functions. In an article on interest rate deregulation, Christofi and Agapos[1] solve the equation

$$r = \left(\frac{\eta}{1 + \eta}\right)\left(r_L - \frac{dC}{dD}\right)$$

for η (the Greek letter "eta"). Here r is the deposit rate paid by commercial banks, r_L is the rate earned by commercial banks, C is the administrative cost of transforming deposits into return-earning assets, D is the savings deposits level, and η is the deposit elasticity with respect to the deposit rate. Find η.

In Problems 30 and 31, use the numerical derivative feature of your graphing calculator to estimate the derivatives of the functions at the indicated values. Round your answers to three decimal places.

30. $f(x) = \sqrt{2x^2 + 3x}$; $x = 1$, $x = 2$

31. $f(x) = e^x(4x - 7)$; $x = 0$, $x = 1.5$

In Problems 32 and 33, use the "limit of a difference quotient" approach to estimate $f'(x)$ at the indicated values of x. Round your answers to three decimal places.

32. $f(x) = \dfrac{\ln x}{x + 3}$; $x = 0.5$, $x = 10$

33. $f(x) = \dfrac{x^2 + 4x + 2}{x^3 - 3}$; $x = 2$, $x = -4$

34. Find an equation of the tangent line to the curve $f(x) = x^2 + x$ at the point $(-2, 2)$. Graph both the curve and the tangent line. Notice that the tangent line is a good approximation to the curve near the point of tangency.

35. The derivative of $f(x) = x^3 - x + 2$ is $f'(x) = 3x^2 - 1$. Graph both the function f and its derivative f'. Observe that there are two points on the graph of f where the tangent line is horizontal. For the x-values of these points, what are the corresponding values of $f'(x)$? Why are these results expected? Observe the intervals where $f'(x)$ is positive. Notice that tangent lines to the graph of f have positive slopes over these intervals. Observe the interval where $f'(x)$ is negative. Notice that tangent lines to the graph of f have negative slopes over this interval.

[1] A. Christofi and A. Agapos, "Interest Rate Deregulation: An Empirical Justification," *Review of Business and Economic Research* 20, no. 1 (1984), 39–49.

11.2 Rules for Differentiation

OBJECTIVE *To develop basic differentiation rules, namely, formulas for the derivative of a constant, of x^n, of a constant times a function, and of sums and differences of functions.*

You would probably agree that differentiating a function by direct use of the definition of a derivative can be tedious. Fortunately, there are rules that give us completely mechanical and efficient procedures for differentiation. They also avoid the direct use of limits. We shall look at some of these rules in this section.

We begin by showing that the derivative of a constant function is zero. Recall that the graph of the constant function $f(x) = c$ is a horizontal line (see Figure 11.11), which has a slope of zero at each point. This means that $f'(x) = 0$ regardless of x. As a formal proof of this result, we apply the definition of the derivative to $f(x) = c$:

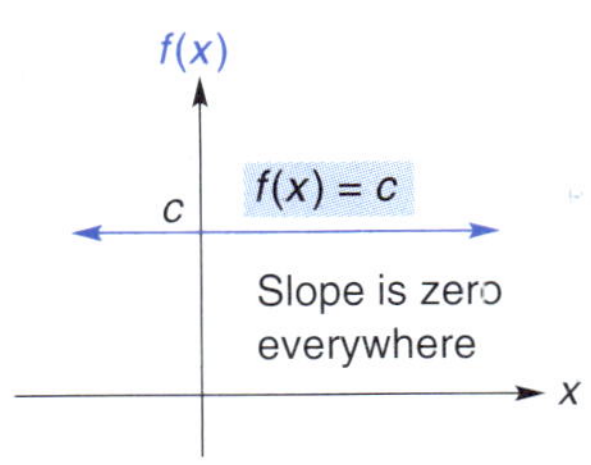

FIGURE 11.11 The slope of a constant function is 0.

$$f'(x) = \lim_{h\to 0} \frac{f(x + h) - f(x)}{h} = \lim_{h\to 0} \frac{c - c}{h}$$

$$= \lim_{h\to 0} \frac{0}{h} = \lim_{h\to 0} 0 = 0$$

Thus, we have our first rule:

Rule 1 Derivative of a Constant

If c is a constant, then

$$\frac{d}{dx}(c) = 0$$

That is, the derivative of a constant function is zero.

EXAMPLE 1 Derivatives of Constant Functions

a. $\frac{d}{dx}(3) = 0$ because 3 is a constant function.

b. If $g(x) = \sqrt{5}$, then $g'(x) = 0$ because g is a constant function. For example, the derivative of g when $x = 4$ is $g'(4) = 0$.

c. If $s(t) = (1{,}938{,}623)^{807.4}$, then $ds/dt = 0$.

▶ **Now work Problem 1.**

The next rule gives a formula for the derivative of x raised to a constant power—that is, the derivative of $f(x) = x^n$. A function of this form is called a **power function.** For example, $f(x) = x^2$ is a power function. To prove the rule, we must expand $(x+h)^n$. Recall that

$$(x+h)^2 = x^2 + 2xh + h^2$$

and

$$(x+h)^3 = x^3 + 3x^2h + 3xh^2 + h^3$$

In both expansions, the exponents of x decrease from left to right while those of h increase. This is true for the general case $(x+h)^n$, where n is a positive integer. It can be shown that

$$(x+h)^n = x^n + nx^{n-1}h + (\ )x^{n-2}h^2 + \cdots + (\ )xh^{n-1} + h^n$$

where the missing numbers inside the parentheses are certain constants. This formula is used to prove the following rule:

Rule 2 Derivative of x^n

If n is any real number, then

$$\frac{d}{dx}(x^n) = nx^{n-1}$$

provided that x^{n-1} is defined. That is, the derivative of constant power of x is the exponent times x raised to a power one less the given power.

Although the rule states that n can be any real number, our proof is for a positive integer only. The general proof is not given in this text.

Proof. We shall give a proof for the case where n is a positive integer. If $f(x) = x^n$, applying the definition of the derivative gives

$$f'(x) = \lim_{h\to 0} \frac{f(x+h) - f(x)}{h} = \lim_{h\to 0} \frac{(x+h)^n - x^n}{h}$$

By our previous discussion on expanding $(x+h)^n$,

$$f'(x) = \lim_{h\to 0} \frac{x^n + nx^{n-1}h + (\ )x^{n-2}h^2 + \cdots + h^n - x^n}{h}$$

In the numerator, the sum of the first and last terms is 0. Dividing each of the remaining terms by h gives

$$f'(x) = \lim_{h\to 0}[nx^{n-1} + (\ )x^{n-2}h + \cdots + h^{n-1}]$$

Each term after the first has h as a factor and must approach 0 as $h \to 0$. Hence, $f'(x) = nx^{n-1}$.

EXAMPLE 2 **Derivatives of Powers of x**

a. By Rule 2, $\frac{d}{dx}(x^2) = 2x^{2-1} = 2x$.

b. If $F(x) = x = x^1$, then $F'(x) = 1 \cdot x^{1-1} = 1 \cdot x^0 = 1$. Thus, the derivative of x with respect to x is 1.

c. If $f(x) = x^{-10}$, then $f'(x) = -10x^{-10-1} = -10x^{-11}$.

▶ **Now work Problem 3.**

When we apply a differentiation rule to a function, sometimes the function must first be rewritten so that it has the proper form for that rule. For example, to differentiate $f(x) = \frac{1}{x^{10}}$ we would first rewrite f as $f(x) = x^{-10}$ and then proceed as in Example 2(c).

EXAMPLE 3 **Rewriting Functions in the Form x^n**

a. To differentiate $y = \sqrt{x}$, we rewrite $\sqrt{x}$ as $x^{1/2}$ so that it has the form x^n. Thus,

$$\frac{dy}{dx} = \frac{1}{2}x^{(1/2)-1} = \frac{1}{2}x^{-1/2} = \frac{1}{2\sqrt{x}}$$

CAUTION In Example 3(b), do not rewrite $\frac{1}{x\sqrt{x}}$ as $\frac{1}{x^{3/2}}$ and then merely differentiate the denominator.

b. Let $h(x) = \frac{1}{x\sqrt{x}}$. To apply Rule 2, we *must* rewrite $h(x)$ as $h(x) = x^{-3/2}$ so that it has the form x^n. We have

$$h'(x) = \frac{d}{dx}(x^{-3/2}) = -\frac{3}{2}x^{(-3/2)-1} = -\frac{3}{2}x^{-5/2}$$

▶ **Now work Problem 39.**

Now that we can say immediately that the derivative of x^3 is $3x^2$, the question arises as to what we could say about the derivative of a *multiple* of x^3, such as $5x^3$. Our next rule will handle this situation of differentiating a constant times a function.

Rule 3 Constant Factor Rule

If f is a differentiable function and c is a constant, then $cf(x)$ is differentiable, and

$$\frac{d}{dx}[cf(x)] = cf'(x)$$

That is, the derivative of a constant times a function is the constant times the derivative of the function.

Proof. If $g(x) = cf(x)$, applying the definition of the derivative of g gives

$$g'(x) = \lim_{h\to 0}\frac{g(x+h)-g(x)}{h} = \lim_{h\to 0}\frac{cf(x+h)-cf(x)}{h}$$

$$= \lim_{h\to 0}\left[c \cdot \frac{f(x+h)-f(x)}{h}\right] = c \cdot \lim_{h\to 0}\frac{f(x+h)-f(x)}{h}$$

But $\lim_{h\to 0}\frac{f(x+h)-f(x)}{h}$ is $f'(x)$; so $g'(x) = cf'(x)$.

EXAMPLE 4 Differentiating a Constant Times a Function

Differentiate the following functions.

a. $g(x) = 5x^3$

Solution: Here g is a constant (5) times a function (x^3). So

$$\frac{d}{dx}(5x^3) = 5\frac{d}{dx}(x^3) \qquad \text{(Rule 3)}$$

$$= 5(3x^{3-1}) = 15x^2 \qquad \text{(Rule 2)}$$

b. $f(q) = \dfrac{13q}{5}$

Solution:

Strategy: We first rewrite f as a constant times a power function and then apply Rules 2 and 3.

Because $\dfrac{13q}{5} = \frac{13}{5}q$, f is the constant $\frac{13}{5}$ times the function q. Thus,

$$f'(q) = \frac{13}{5}\frac{d}{dq}(q) \qquad \text{(Rule 3)}$$

$$= \frac{13}{5} \cdot 1 = \frac{13}{5} \qquad \text{(Rule 2)}$$

c. $y = \dfrac{0.25}{\sqrt[5]{x^2}}$

Solution: We can express y as a constant times a function:

$$y = 0.25 \cdot \frac{1}{\sqrt[5]{x^2}} = 0.25x^{-2/5}$$

Hence,

$$y' = 0.25\frac{d}{dx}(x^{-2/5}) \qquad \text{(Rule 3)}$$

$$= 0.25\left(-\frac{2}{5}x^{-7/5}\right) = -0.1x^{-7/5} \qquad \text{(Rule 2)}$$

▸ **Now work Problem 7.**

CAUTION To differentiate $f(x) = (4x)^3$, you may be tempted to write $f'(x) = 3(4x)^2$. **This is incorrect!** Do you see why? The reason is that Rule 2 applies to a power of the variable x, **not** a power of an expression involving x, such as $4x$. To apply our rules, we must get a suitable form for $f(x)$. We can rewrite $(4x)^3$ as 4^3x^3 or $64x^3$. Thus,

$$f'(x) = 64\frac{d}{dx}(x^3) = 64(3x^2) = 192x^2$$

The next rule involves derivatives of sums and differences of functions.

Rule 4 Derivative of a Sum or Difference

If f and g are differentiable functions, then $f+g$ and $f-g$ are differentiable, and

$$\frac{d}{dx}[f(x)+g(x)] = f'(x)+g'(x)$$

and

$$\frac{d}{dx}[f(x) - g(x)] = f'(x) - g'(x)$$

That is, the derivative of the sum (difference) of two functions is the sum (difference) of their derivatives.

Proof. For the case of a sum, if $F(x) = f(x) + g(x)$, applying the definition of the derivative of F gives

$$\begin{aligned} F'(x) &= \lim_{h\to 0} \frac{F(x+h) - F(x)}{h} \\ &= \lim_{h\to 0} \frac{[f(x+h) + g(x+h)] - [f(x) + g(x)]}{h} \\ &= \lim_{h\to 0} \frac{[f(x+h) - f(x)] + [g(x+h) - g(x)]}{h} && \text{(regrouping)} \\ &= \lim_{h\to 0} \left[\frac{f(x+h) - f(x)}{h} + \frac{g(x+h) - g(x)}{h}\right] \end{aligned}$$

Because the limit of a sum is the sum of the limits,

$$F'(x) = \lim_{h\to 0} \frac{f(x+h) - f(x)}{h} + \lim_{h\to 0} \frac{g(x+h) - g(x)}{h}$$

But these two limits are $f'(x)$ and $g'(x)$. Thus,

$$F'(x) = f'(x) + g'(x)$$

The proof for the derivative of a difference of two functions is similar.

Rule 4 can be extended to the derivative of any number of sums and differences of functions. For example,

$$\frac{d}{dx}[f(x) - g(x) + h(x) + k(x)] = f'(x) - g'(x) + h'(x) + k'(x)$$

APPLICATIONS IN PRACTICE 1

Differentiating Sums and Differences of Functions

If the revenue function for a certain product is $r(q) = 50q - 0.3q^2$, find the derivative of this function, also known as the marginal revenue.

EXAMPLE 5 Differentiating Sums and Differences of Functions

Differentiate the following functions.

a. $F(x) = 3x^5 + \sqrt{x}$

Solution: Here F is the sum of two functions, $3x^5$ and $\sqrt{x}$. Therefore,

$$\begin{aligned} F'(x) &= \frac{d}{dx}(3x^5) + \frac{d}{dx}(x^{1/2}) && \text{(Rule 4)} \\ &= 3\frac{d}{dx}(x^5) + \frac{d}{dx}(x^{1/2}) && \text{(Rule 3)} \\ &= 3(5x^4) + \tfrac{1}{2}x^{-1/2} = 15x^4 + \frac{1}{2\sqrt{x}} && \text{(Rule 2)} \end{aligned}$$

b. $f(z) = \dfrac{z^4}{4} - \dfrac{5}{z^{1/3}}$

Solution: To apply our rules, we shall rewrite $f(z)$ in the form $f(z) = \frac{1}{4}z^4 - 5z^{-1/3}$. Since f is the difference of two functions,

$$\begin{aligned} f'(z) &= \frac{d}{dz}\left(\frac{1}{4}z^4\right) - \frac{d}{dz}(5z^{-1/3}) && \text{(Rule 4)} \\ &= \frac{1}{4}\frac{d}{dz}(z^4) - 5\frac{d}{dz}(z^{-1/3}) && \text{(Rule 3)} \\ &= \frac{1}{4}(4z^3) - 5\left(-\frac{1}{3}z^{-4/3}\right) && \text{(Rule 2)} \\ &= z^3 + \frac{5}{3}z^{-4/3} \end{aligned}$$

c. $y = 6x^3 - 2x^2 + 7x - 8$

Solution:

$$\begin{aligned} \frac{dy}{dx} &= \frac{d}{dx}(6x^3) - \frac{d}{dx}(2x^2) + \frac{d}{dx}(7x) - \frac{d}{dx}(8) \\ &= 6\frac{d}{dx}(x^3) - 2\frac{d}{dx}(x^2) + 7\frac{d}{dx}(x) - \frac{d}{dx}(8) \\ &= 6(3x^2) - 2(2x) + 7(1) - 0 \\ &= 18x^2 - 4x + 7 \end{aligned}$$

▶ **Now work Problem 17.**

In Examples 6 and 7, we need to rewrite the given function in a form to which our rules apply.

EXAMPLE 6 Evaluating a Derivative

Find the derivative of $f(x) = 2x(x^2 - 5x + 2)$ *when* $x = 2$.

Solution: We multiply and then differentiate each term:

$$\begin{aligned} f(x) &= 2x^3 - 10x^2 + 4x \\ f'(x) &= 2(3x^2) - 10(2x) + 4(1) \\ &= 6x^2 - 20x + 4 \\ f'(2) &= 6(2)^2 - 20(2) + 4 = -12 \end{aligned}$$

▶ **Now work Problem 75.**

EXAMPLE 7 Finding an Equation of a Tangent Line

Find an equation of the tangent line to the curve

$$y = \frac{3x^2 - 2}{x}$$

when $x = 1$.

Solution:

Strategy: First we find $\frac{dy}{dx}$, which gives the slope of the tangent line at any point. Evaluating $\frac{dy}{dx}$ when $x = 1$ gives the slope of the required tangent line. We then determine the y-coordinate of the point on the curve when $x = 1$. Finally, we substitute the slope and both of the coordinates of the point in a point–slope form to obtain an equation of the tangent line.

Rewriting y as a difference of two functions, we have

$$y = \frac{3x^2}{x} - \frac{2}{x} = 3x - 2x^{-1}$$

Thus,

$$\frac{dy}{dx} = 3(1) - 2[(-1)x^{-2}] = 3 + \frac{2}{x^2}$$

The slope of the tangent line to the curve when $x = 1$ is

$$\left.\frac{dy}{dx}\right|_{x=1} = 3 + \frac{2}{1^2} = 5$$

To find the y-coordinate of the point on the curve where $x = 1$, we substitute this value of x into the equation of the *curve*. This gives

CAUTION To obtain the y-value of the point on the curve when $x = 1$, substitute into the equation of the *curve*, not into the formula for the derivative!

$$y = \frac{3(1)^2 - 2}{1} = 1$$

Hence, the point (1, 1) lies on both the curve and the tangent line. Therefore, an equation of the tangent line is

$$y - 1 = 5(x - 1)$$

In slope–intercept form, we have

$$y = 5x - 4$$

▸ **Now work Problem 79.**

Exercise 11.2

In Problems 1–74, differentiate the functions.

*1. $f(x) = 5$

2. $f(x) = \left(\frac{11}{13}\right)^{4/5}$

*3. $y = x^6$

4. $f(x) = x^{21}$

5. $y = x^{80}$

6. $y = x^{5.3}$

*7. $f(x) = 9x^2$

8. $y = 4x^3$

9. $g(w) = 4w^5$

10. $v(x) = x^e$

11. $y = \frac{2}{3}x^4$

12. $f(p) = \sqrt{3}p^4$

13. $f(t) = \frac{t^7}{25}$

14. $y = \frac{x^3}{3}$

15. $f(x) = x + 3$

16. $f(x) = 3x - 2$

*17. $f(x) = 4x^2 - 2x + 3$

18. $F(x) = 5x^2 - 9x$

19. $g(p) = p^4 - 3p^3 - 1$

20. $f(t) = -13t^2 + 14t + 1$

21. $y = -x^8 + x^5$

22. $y = -8x^4 + \ln 2$

23. $y = -13x^3 + 14x^2 - 2x + 3$

24. $V(r) = r^8 - 7r^6 + 3r^2 + 1$

25. $f(x) = 2(13 - x^4)$

26. $\phi(s) = 3(s^6 - 2)$

27. $g(x) = \frac{13 - x^4}{3}$

28. $f(x) = \frac{5(x^4 - 6)}{2}$

29. $h(x) = 4x^4 + x^3 - \frac{9x^2}{2} + 8x$

30. $k(x) = -2x^2 + \frac{5}{3}x + 11$

31. $f(x) = \frac{3x^4}{10} + \frac{7}{3}x^3$

32. $p(x) = \frac{x^7}{7} + \frac{2x}{3}$

33. $f(x) = x^{7/2}$ **34.** $f(x) = 2x^{-14/5}$

35. $y = x^{3/4} + 2x^{5/3}$ **36.** $y = 5x^3 - x^{-2/5}$

37. $y = 11\sqrt{x}$ **38.** $y = \sqrt[5]{x^3}$

***39.** $f(r) = 6\sqrt[3]{r}$ **40.** $y = 4\sqrt[8]{x^2}$

41. $f(x) = x^{-4}$ **42.** $f(s) = 2s^{-3}$

43. $f(x) = x^{-3} + x^{-5} - 2x^{-6}$

44. $f(x) = 100x^{-3} + 10x^{1/2}$

45. $y = \frac{1}{x}$ **46.** $f(x) = \frac{2}{x^3}$

47. $y = \frac{8}{x^5}$ **48.** $y = \frac{1}{4x^5}$

49. $g(x) = \frac{4}{3x^3}$ **50.** $y = \frac{6}{3x^5}$

51. $f(t) = \frac{1}{2t}$ **52.** $g(x) = \frac{7}{9x}$

53. $f(x) = \frac{x}{7} + \frac{7}{x}$ **54.** $\Phi(x) = \frac{x^3}{3} - \frac{3}{x^3}$

55. $f(x) = -9x^{1/3} + 5x^{-2/5}$

56. $f(z) = 3z^{1/4} - 12^2 - 8z^{-3/4}$

57. $q(x) = \frac{1}{\sqrt[5]{x}}$ **58.** $f(x) = \frac{3}{\sqrt[4]{x^3}}$

59. $y = \frac{2}{\sqrt{x}}$ **60.** $y = \frac{1}{2\sqrt{x}}$

61. $y = x^2\sqrt{x}$ **62.** $f(x) = x^2(4x^3)$

63. $f(x) = x(3x^2 - 10x + 7)$

64. $f(x) = x^3(3x^6 - 5x^2 + 4)$

65. $f(x) = x^3(3x)^2$

66. $s(x) = \sqrt[3]{x}(\sqrt[4]{x} - 6x + 3)$ **67.** $v(x) = x^{-2/3}(x + 5)$

68. $f(x) = x^{3/5}(x^2 + 7x + 11)$

69. $f(q) = \frac{4q^3 + 7q - 4}{q}$

70. $f(w) = \frac{w - 5}{w^5}$ **71.** $f(x) = (x + 1)(x + 3)$

72. $f(x) = x^2(x - 2)(x + 4)$

73. $w(x) = \frac{x^2 + x^3}{x^2}$ **74.** $f(x) = \frac{5x^4 + x^2}{12\sqrt{x}}$

For each curve in Problems 75–78, find the slopes at the indicated points.

***75.** $y = 3x^2 + 4x - 8$; $(0, -8)$, $(2, 12)$, $(-3, 7)$

76. $y = 5 - 6x - 2x^3$; $(0, 5)$, $(\frac{3}{2}, -\frac{43}{4})$, $(-3, 77)$

77. $y = 4$; when $x = -4$, $x = 7$, $x = 22$

78. $y = 3x - 4\sqrt{x}$; when $x = 4$, $x = 9$, $x = 25$

In Problems 79–82, find an equation of the tangent line to the curve at the indicated point.

***79.** $y = 4x^2 + 5x + 6$; $(1, 15)$

80. $y = \frac{1 - x^2}{5}$; $(4, -3)$ **81.** $y = \frac{2}{x^2}$; $(1, 2)$

82. $y = -\sqrt[3]{x}$; $(8, -2)$

83. Find an equation of the tangent line to the curve

$$y = 3 + x - 5x^2 + x^4$$

when $x = 0$.

84. Repeat Problem 83 for the curve

$$y = \frac{\sqrt{x}(2 - x^2)}{x}$$

when $x = 4$.

85. Find all points on the curve

$$y = \frac{5}{2}x^2 - x^3$$

where the tangent line is horizontal.

86. Repeat Problem 85 for the curve

$$y = \frac{x^5}{5} - x + 1$$

87. Find all points on the curve

$$y = x^2 - 5x + 3$$

where the slope is 1.

88. Repeat Problem 87 for the curve

$$y = \frac{x^3}{3} - 3x + 4$$

89. If $f(x) = \sqrt{x} + \frac{1}{\sqrt{x}}$, evaluate the expression

$$\frac{x - 1}{2x\sqrt{x}} - f'(x)$$

90. Economics Eswaran and Kotwal[2] consider agrarian economies in which there are two types of workers, permanent and casual. Permanent workers are employed on long-term contracts and may receive benefits such as holiday gifts and emergency aid. Casual workers are hired on a daily basis and perform routine and menial tasks such as weeding, harvesting, and threshing. The difference z in the present-value cost of hiring a permanent worker over that of hiring a casual worker is given by

$$z = (1 + b)w_p - bw_c$$

where w_p and w_c are wage rates for permanent labor and casual labor, respectively, b is a constant, and w_p is a function of w_c. Eswaran and Kotwal claim that

$$\frac{dz}{dw_c} = (1 + b)\left[\frac{dw_p}{dw_c} - \frac{b}{1 + b}\right]$$

Verify this.

---

[2]M. Eswaran and A. Kotwal, "A Theory of Two-Tier Labor Markets in Agrarian Economies," *American Economic Review* 75, no. 1 (1985), 162–77.

91. Find an equation of the tangent line to the graph of $y = x^3 - 3x$ at the point (2, 2). Graph both the function and the tangent line on the same screen. Notice that the line passes through (2, 2) and the line appears to be tangent to the curve.

92. Find an equation of the tangent line to the graph of $y = \sqrt[3]{x}$, at the point (−8, −2). Graph both the function and the tangent line on the same screen. Notice that the line passes through (−8, −2) and the line appears to be tangent to the curve.

OBJECTIVE *To motivate the instantaneous rate of change of a function by means of velocity and to interpret the derivative as an instantaneous rate of change. To develop the "marginal" concept, which is frequently used in business and economics.*

11.3 The Derivative as a Rate of Change

We have given a geometrical interpretation of the derivative as being the slope of the tangent line to a curve at a point. Historically, an important application of the derivative involves the motion of an object traveling in a straight line. This gives us a convenient way to interpret the derivative as a *rate of change*.

To denote the change in a variable such as x, the symbol Δx (read "delta x") is commonly used. For example, if x changes from 1 to 3, then the change in x is $\Delta x = 3 - 1 = 2$. The new value of $x(= 3)$ is the old value plus the change, or $1 + \Delta x$. Similarly, if t increases by Δt, the new value is $t + \Delta t$. We shall use Δ-notation in the discussion that follows.

Suppose an object moves along the number line in Figure 11.12 according to the equation

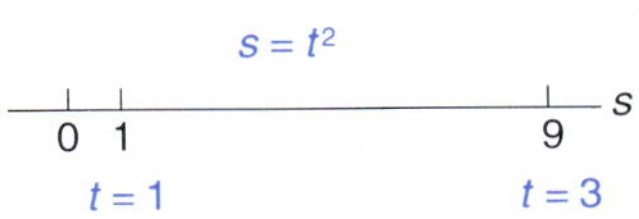

FIGURE 11.12 Motion along a number line.

$$s = f(t) = t^2$$

where s is the position of the object at time t. This equation is called an *equation of motion*, and f is called a ***position function.*** Assume that t is in seconds and s is in meters. At $t = 1$ the position is $s = f(1) = 1^2 = 1$, and at $t = 3$ the position is $s = f(3) = 3^2 = 9$. Over this two-second time interval, the object has a change in position, or a *displacement*, of $9 - 1 = 8$ m, and the *average velocity* of the object is defined as

$$v_{\text{ave}} = \frac{\text{displacement}}{\text{length of time interval}} \tag{1}$$

$$= \frac{8}{2} = 4 \text{ m/s}$$

To say that the average velocity is 4 m/s from $t = 1$ to $t = 3$ means that, *on the average*, the position of the object changed by 4 m to the right each second during that time interval. Let us denote the changes in s-values and t-values by Δs and Δt, respectively. Then the average velocity is given by

$$v_{\text{ave}} = \frac{\Delta s}{\Delta t} = 4 \text{ m/s} \qquad (\text{for the interval } t = 1 \text{ to } t = 3)$$

The ratio $\Delta s/\Delta t$ is also called the **average rate of change of s with respect to t** over the interval from $t = 1$ to $t = 3$.

Now, let the time interval be only 1 second long (that is, $\Delta t = 1$). Then, for the *shorter* interval from $t = 1$ to $t = 1 + \Delta t = 2$, we have $f(2) = 2^2 = 4$, so

$$v_{\text{ave}} = \frac{\Delta s}{\Delta t} = \frac{f(2) - f(1)}{\Delta t} = \frac{4 - 1}{1} = 3 \text{ m/s}$$

More generally, over the time interval from $t = 1$ to $t = 1 + \Delta t$, the object moves from position $f(1)$ to position $f(1 + \Delta t)$. Thus, its displacement is

$$\Delta s = f(1 + \Delta t) - f(1).$$

Since the time interval has length Δt, the object's average velocity is given by

$$v_{\text{ave}} = \frac{\Delta s}{\Delta t} = \frac{f(1+\Delta t) - f(1)}{\Delta t}$$

If Δt were to become smaller and smaller, the average velocity over the interval from $t = 1$ to $t = 1 + \Delta t$ would be close to what we might call the *instantaneous velocity* at time $t = 1$, that is, the velocity at a *point* in time ($t = 1$), as opposed to the velocity over an *interval* of time. For some typical values of Δt between 0.1 and 0.001, we get the average velocities in Table 11.2, which you may verify.

TABLE 11.2

Length of Time Interval Δt	Time Interval $t = 1$ to $t = 1 + \Delta t$	Average Velocity, $\frac{\Delta s}{\Delta t} = \frac{f(1+\Delta t) - f(1)}{\Delta t}$
0.1	$t = 1$ to $t = 1.1$	2.1 m/s
0.07	$t = 1$ to $t = 1.07$	2.07 m/s
0.05	$t = 1$ to $t = 1.05$	2.05 m/s
0.03	$t = 1$ to $t = 1.03$	2.03 m/s
0.01	$t = 1$ to $t = 1.01$	2.01 m/s
0.001	$t = 1$ to $t = 1.001$	2.001 m/s

The table suggests that as the length of the time interval approaches zero, the average velocity approaches the value 2 m/s. In other words, as Δt approaches 0, $\Delta s/\Delta t$ approaches 2 m/s. We define the limit of the average velocity as $\Delta t \to 0$ to be the **instantaneous velocity** (or simply the **velocity**), v, at time $t = 1$. This limit is also called the **instantaneous rate of change** of s with respect to t at $t = 1$:

$$v = \lim_{\Delta t \to 0} v_{\text{ave}} = \lim_{\Delta t \to 0} \frac{\Delta s}{\Delta t} = \lim_{\Delta t \to 0} \frac{f(1+\Delta t) - f(1)}{\Delta t}$$

If we think of Δt as h, then the limit on the right is simply the derivative of s with respect to t at $t = 1$. Thus, the instantaneous velocity of the object at $t = 1$ is just ds/dt at $t = 1$. Because $s = t^2$ and

$$\frac{ds}{dt} = 2t$$

the velocity at $t = 1$ is

$$v = \left.\frac{ds}{dt}\right|_{t=1} = 2(1) = 2 \text{ m/s}$$

which confirms our previous conclusion.

In summary, if $s = f(t)$ is the position function of an object moving in a straight line, then the average velocity of the object over the time interval $[t, t + \Delta t]$ is given by

$$v_{\text{ave}} = \frac{\Delta s}{\Delta t} = \frac{f(t+\Delta t) - f(t)}{\Delta t}$$

and the velocity at time t is given by

$$v = \lim_{\Delta t \to 0} \frac{f(t+\Delta t) - f(t)}{\Delta t} = \frac{ds}{dt}$$

EXAMPLE 1 **Finding Average Velocity and Velocity**

Suppose the position function of an object moving along a number line is given by $s = f(t) = 3t^2 + 5$, where t is in seconds and s is in meters.

a. Find the average velocity over the interval [10, 10.1].

b. Find the velocity when $t = 10$.

Solution:

a. Here $t = 10$ and $\Delta t = 10.1 - 10 = 0.1$. So we have

$$\begin{aligned} v_{\text{ave}} &= \frac{\Delta s}{\Delta t} = \frac{f(t+\Delta t) - f(t)}{\Delta t} \\ &= \frac{f(10+0.1) - f(10)}{0.1} \\ &= \frac{f(10.1) - f(10)}{0.1} \\ &= \frac{311.03 - 305}{0.1} = \frac{6.03}{0.1} = 60.3 \text{ m/s} \end{aligned}$$

b. The velocity at time t is given by

$$v = \frac{ds}{dt} = 6t$$

When $t = 10$, the velocity is

$$\left.\frac{ds}{dt}\right|_{t=10} = 6(10) = 60 \text{ m/s}$$

Notice that the average velocity over the interval [10, 10.1] is close to the velocity at $t = 10$. This is to be expected because the length of the interval is small.

▸ **Now work Problem 1.**

Our discussion of the rate of change of s with respect to t applies equally well to *any* function $y = f(x)$. This means that we have the following:

If $y = f(x)$, then

$$\frac{\Delta y}{\Delta x} = \frac{f(x+\Delta x) - f(x)}{\Delta x} = \begin{cases} \text{average rate of change} \\ \text{of } y \text{ with respect to } x \\ \text{over the interval from} \\ x \text{ to } x + \Delta x \end{cases}$$

and

$$\frac{dy}{dx} = \lim_{\Delta x \to 0} \frac{\Delta y}{\Delta x} = \begin{cases} \text{instantaneous rate of change} \\ \text{of } y \text{ with respect to } x \end{cases} \quad \textbf{(2)}$$

Because the instantaneous rate of change of $y = f(x)$ at a point is a derivative, it is also the *slope of the tangent line* to the graph of $y = f(x)$ at that point. For convenience, we usually refer to the instantaneous rate of change simply as the **rate of change.** The interpretation of a derivative as a rate of change is extremely important.

Let us now consider the significance of the rate of change of y with respect to x. From Equation (2), if Δx (a change in x) is close to 0, then $\Delta y/\Delta x$ is close to dy/dx. That is,

$$\frac{\Delta y}{\Delta x} \approx \frac{dy}{dx}$$

Therefore,

$$\Delta y \approx \frac{dy}{dx}\Delta x \tag{3}$$

That is, if x changes by Δx, then the change in y, Δy, is approximately dy/dx times the change in x. In particular,

if x changes by 1, an estimate of the change in y is $\dfrac{dy}{dx}$

APPLICATIONS IN PRACTICE 1

Estimating ΔP by Using dP/dp

Suppose that the profit P made by selling a certain product at a price p per unit is given by $P = f(p)$ and the rate of change of that profit with respect to change in price is $dP/dp = 5$ at $p = 25$. Estimate the change in the profit P if the price changes from 25 to 25.5.

EXAMPLE 2 Estimating Δy by Using dy/dx

Suppose that $y = f(x)$ and $dy/dx = 8$ when $x = 3$. Estimate the change in y if x changes from 3 to 3.5.

Solution: We have $dy/dx = 8$ and $\Delta x = 3.5 - 3 = 0.5$. The change in y is given by Δy, and, from (3),

$$\Delta y \approx \frac{dy}{dx}\Delta x = 8(0.5) = 4$$

We remark that, since $\Delta y = f(3.5) - f(3)$, we have $f(3.5) = f(3) + \Delta y$. For example, if $f(3) = 5$, then $f(3.5)$ can be estimated by $5 + 4 = 9$.

APPLICATIONS IN PRACTICE 2

Finding a Rate of Change

The position of an object thrown upward at a speed of 16 feet/s from a height of 0 feet is given by $y(t) = 16t - 16t^2$. Find the rate of change of y with respect to t, and evaluate it when $t = 0.5$. Use your graphing calculator to graph $y(t)$. Use the graph to interpret the behavior of the object when $t = 0.5$.

EXAMPLE 3 Finding a Rate of Change

Find the rate of change of $y = x^4$ with respect to x, and evaluate it when $x = 2$ and when $x = -1$. Interpret your results.

Solution: The rate of change is

$$\frac{dy}{dx} = 4x^3$$

When $x = 2$, $dy/dx = 4(2)^3 = 32$. This means that if x increases by a small amount, then y increases approximately 32 times as much. More simply, we say that y is increasing 32 times as fast as x does. When $x = -1$, $dy/dx = 4(-1)^3 = -4$. The significance of the minus sign on -4 is that y is *decreasing* 4 times as fast as x increases.

▶ Now work Problem 11.

EXAMPLE 4 **Rate of Change of Price with Respect to Quantity**

Let $p = 100 - q^2$ be the demand function for a manufacturer's product. Find the rate of change of price p per unit with respect to quantity q. How fast is the price changing with respect to q when $q = 5$? Assume that p is in dollars.

Solution: The rate of change of p with respect to q is

$$\frac{dp}{dq} = \frac{d}{dq}(100 - q^2) = -2q$$

Thus,

$$\left.\frac{dp}{dq}\right|_{q=5} = -2(5) = -10$$

This means that when five units are demanded, an *increase* of one extra unit demanded corresponds to a decrease of approximately \$10 in the price per unit that consumers are willing to pay.

EXAMPLE 5 **Rate of Change of Volume**

A spherical balloon is being filled with air. Find the rate of change of the volume of the balloon with respect to its radius. Evaluate this rate of change when the radius is 2 ft.

Solution: The formula for the volume V of a sphere of radius r is $V = \frac{4}{3}\pi r^3$. The rate of change of V with respect to r is

$$\frac{dV}{dr} = \frac{4}{3}\pi(3r^2) = 4\pi r^2$$

When $r = 2$ ft, the rate of change is

$$\left.\frac{dV}{dr}\right|_{r=2} = 4\pi(2)^2 = 16\pi\,\frac{\text{ft}^3}{\text{ft}}$$

This means that when the radius is 2 ft, changing the radius by 1 ft will change the volume by approximately 16π ft^3.

EXAMPLE 6 **Rate of Change of Enrollment**

A sociologist is studying various suggested programs that can aid in the education of preschool-age children in a certain city. The sociologist believes that x years after the beginning of a particular program, f(x) thousand preschoolers will be enrolled, where

$$f(x) = \frac{10}{9}(12x - x^2) \qquad 0 \le x \le 12$$

At what rate would enrollment change (a) after three years from the start of this program and (b) after nine years?

Solution: The rate of change of $f(x)$ is

$$f'(x) = \frac{10}{9}(12 - 2x)$$

a. After three years, the rate of change is

$$f'(3) = \frac{10}{9}[12 - 2(3)] = \frac{10}{9} \cdot 6 = \frac{20}{3} = 6\frac{2}{3}$$

Thus, enrollment would be increasing at the rate of $6\frac{2}{3}$ thousand preschoolers per year.

b. After nine years, the rate is

$$f'(9) = \frac{10}{9}[12 - 2(9)] = \frac{10}{9}[-6] = -\frac{20}{3} = -6\frac{2}{3}$$

Thus, enrollment would be *decreasing* at the rate of $6\frac{2}{3}$ thousand preschoolers per year.

▶ **Now work Problem 9.**

Applications of Rate of Change to Economics

A manufacturer's **total-cost function,** $c = f(q)$, gives the total cost c of producing and marketing q units of a product. The rate of change of c with respect to q is called the **marginal cost.** Thus,

$$\text{marginal cost} = \frac{dc}{dq}$$

For example, suppose $c = f(q) = 0.1q^2 + 3$ is a cost function, where c is in dollars and q is in pounds. Then

$$\frac{dc}{dq} = 0.2q$$

The marginal cost when 4 lb are produced is dc/dq, evaluated when $q = 4$:

$$\left.\frac{dc}{dq}\right|_{q=4} = 0.2(4) = 0.80$$

This means that if production is increased by 1 lb, from 4 lb to 5 lb, then the change in cost is approximately \$0.80. That is, the additional pound costs about \$0.80. In general, *we interpret marginal cost as the approximate cost of one additional unit of output.* [The actual cost of producing one more pound beyond 4 lb is $f(5) - f(4) = 5.5 - 4.6 = \$0.90$.]

If c is the total cost of producing q units of a product, then the **average cost per unit,** $\bar{c}$, is

$$\bar{c} = \frac{c}{q} \tag{4}$$

For example, if the total cost of 20 units is \$100, then the average cost per unit is $\bar{c} = 100/20 = \$5$. By multiplying both sides of Equation (4) by q, we have

$$c = q\bar{c}$$

That is, total cost is the product of the number of units produced and the average cost per unit.

▶ **Now work Problem 19.**

EXAMPLE 7 **Marginal Cost**

If a manufacturer's average-cost equation is

$$\bar{c} = 0.0001q^2 - 0.02q + 5 + \frac{5000}{q}$$

find the marginal-cost function. What is the marginal cost when 50 *units are produced?*

Solution:

Strategy: The marginal-cost function is the derivative of the total-cost function c. Thus, we first find c by multiplying $\bar{c}$ by q. We have

$$\begin{aligned} c &= q\bar{c} \\ &= q\left[0.0001q^2 - 0.02q + 5 + \frac{5000}{q}\right] \\ c &= 0.0001q^3 - 0.02q^2 + 5q + 5000 \end{aligned}$$

Differentiating c, we have the marginal-cost function:

$$\begin{aligned} \frac{dc}{dq} &= 0.0001(3q^2) - 0.02(2q) + 5(1) + 0 \\ &= 0.0003q^2 - 0.04q + 5 \end{aligned}$$

The marginal cost when 50 units are produced is

$$\left.\frac{dc}{dq}\right|_{q=50} = 0.0003(50)^2 - 0.04(50) + 5 = 3.75$$

If c is in dollars and production is increased by one unit, from $q = 50$ to $q = 51$, then the cost of the additional unit is approximately \$3.75. If production is increased by $\frac{1}{3}$ unit, from $q = 50$, then the cost of the additional output is approximately $\left(\frac{1}{3}\right)(3.75) = \$1.25$.

▸ **Now work Problem 21.**

Suppose $r = f(q)$ is the **total-revenue function** for a manufacturer. The equation $r = f(q)$ states that the total dollar value received for selling q units of a product is r. The **marginal revenue** is defined as the rate of change of the total dollar value received with respect to the total number of units sold. Hence, marginal revenue is merely the derivative of r with respect to q:

$$\text{marginal revenue} = \frac{dr}{dq}$$

Marginal revenue indicates the rate at which revenue changes with respect to units sold. We interpret it as *the approximate revenue received from selling one additional unit of output.*

EXAMPLE 8 **Marginal Revenue**

Suppose a manufacturer sells a product at \$2 per unit. If q units are sold, the total revenue is given by

$$r = 2q$$

The marginal-revenue function is

$$\frac{dr}{dq} = \frac{d}{dq}(2q) = 2$$

which is a constant function. Thus, the marginal revenue is 2 regardless of the number of units sold. This is what we would expect, because the manufacturer receives \$2 for each unit sold.

▶ **Now work Problem 23.**

Relative and Percentage Rates of Change

For the total-revenue function in Example 8, namely, $r = f(q) = 2q$, we have

$$\frac{dr}{dq} = 2$$

This means that revenue is changing at the rate of \$2 per unit, regardless of the number of units sold. Although this is valuable information, it may be more significant when compared with r itself. For example, if $q = 50$, then $r = 2(50) = 100$. Thus, the rate of change of revenue is $2/100 = 0.02$ of r. On the other hand, if $q = 5000$, then $r = 2(5000) = \$10{,}000$, so the rate of change of r is $2/10{,}000 = 0.0002$ of r. Although r changes at the same rate at each level, compared with r itself, this rate is relatively smaller when $r = 10{,}000$ than when $r = 100$. By considering the ratio

$$\frac{dr/dq}{r}$$

we have a means of comparing the rate of change of r with r itself. This ratio is called the *relative rate of change* of r. We have shown that the relative rate of change when $q = 50$ is

$$\frac{dr/dq}{r} = \frac{2}{100} = 0.02$$

and when $q = 5000$, it is

$$\frac{dr/dq}{r} = \frac{2}{10{,}000} = 0.0002$$

By multiplying relative rates by 100, we obtain so-called *percentage rates of change*. The percentage rate of change when $q = 50$ is $(0.02)(100) = 2\%$; when $q = 5000$, it is $(0.0002)(100) = 0.02\%$. For example, if an additional unit beyond 50 is sold, then revenue increases by approximately 2%.

In general, for any function f, we have the following definition:

Definition

The ***relative rate of change*** of $f(x)$ is

$$\frac{f'(x)}{f(x)}$$

The ***percentage rate of change*** of $f(x)$ is

$$\frac{f'(x)}{f(x)} \cdot 100$$

APPLICATIONS IN PRACTICE 3

Relative and Percentage Rates of Change

The volume V of a capsule-shaped container with a cylindrical height of 4 feet and radius r is given by

$$V(r) = \frac{4}{3}\pi r^3 + 4\pi r^2$$

Determine the relative and percentage rates of change of volume with respect to the radius when the radius is 2 feet.

EXAMPLE 9 Relative and Percentage Rates of Change

Determine the relative and percentage rates of change of

$$y = f(x) = 3x^2 - 5x + 25$$

when $x = 5$.

Solution: Here

$$f'(x) = 6x - 5$$

Since $f'(5) = 6(5) - 5 = 25$ and $f(5) = 3(5)^2 - 5(5) + 25 = 75$, the relative rate of change of y when $x = 5$ is

$$\frac{f'(5)}{f(5)} = \frac{25}{75} \approx 0.333$$

Multiplying 0.333 by 100 gives the percentage rate of change: $(0.333)(100) = 33.3\%$.

▶ **Now work Problem 35.**

Exercise 11.3

*1. Suppose that the position function of an object moving along a straight line is $s = f(t) = 2t^2 + 3t$, where t is in seconds and s is in meters. Find the average velocity $\Delta s/\Delta t$ over the interval $[1, 1 + \Delta t]$, where Δt is given in the following table:

Δt	1	0.5	0.2	0.1	0.01	0.001
$\Delta s/\Delta t$						

From your results, estimate the velocity when $t = 1$. Verify your estimate by using differentiation.

2. If $y = f(x) = \sqrt{2x + 5}$, find the average rate of change of y with respect to x over the interval $[3, 3 + \Delta x]$, where Δx is given in the following table:

Δx	1	0.5	0.2	0.1	0.01	0.001
$\Delta y/\Delta x$						

From your result, estimate the rate of change of y with respect to x when $x = 3$.

In each of Problems 3–8, a position function is given, where t is in seconds and s is in meters.

(a) *Find the position at the given t-value.*
(b) *Find the average velocity over the given interval.*
(c) *Find the velocity at the given t-value.*

3. $s = t^2 - 3t$; $[4, 4.5]$; $t = 4$

4. $s = \frac{1}{2}t + 1$; $[2, 2.1]$; $t = 2$

5. $s = 2t^3 + 6$; $[1, 1.02]$; $t = 1$

6. $s = -3t^2 + 2t + 1$; $[1, 1.25]$; $t = 1$

7. $s = t^4 - 2t^3 + t$; $[2, 2.1]$; $t = 2$

8. $s = 3t^4 - t^{7/2}$; $[0, \frac{1}{4}]$; $t = 0$

*9. **Income-Education** Sociologists studied the relation between income and number of years of education for members of a particular urban group. They found that a person with x years of education before seeking regular employment can expect to receive an average yearly income of y dollars per year, where

$$y = 5x^{5/2} + 5900 \qquad 4 \le x \le 16$$

Find the rate of change of income with respect to number of years of education. Evaluation the expression when $x = 9$.

10. Find the rate of change of the area A of a circle with respect to its radius r if

$$A = \pi r^2$$

Evaluate the expression when $r = 7$ in.

*11. **Skin Temperature** The approximate temperature T of the skin in terms of the temperature T_e of the environment is given by

$$T = 32.8 + 0.27(T_e - 20)$$

where T and T_e are in degrees Celsius.[3] Find the rate of change of T with respect to T_e.

12. **Biology** The volume V of a spherical cell is given by $V = \frac{4}{3}\pi r^3$, where r is the radius. Find the rate of change of volume with respect to the radius when $r = 6.3 \times 10^{-4}$ cm.

In Problems 13–18, cost functions are given, where c is the cost of producing q units of a product. In each case, find the marginal-cost function. What is the marginal cost at the given value(s) of q?

13. $c = 500 + 10q$; $q = 100$

14. $c = 5000 + 6q$; $q = 36$

15. $c = 0.3q^2 + 2q + 850$; $q = 3$

16. $c = 0.1q^2 + 3q + 2$; $q = 3$

17. $c = q^2 + 50q + 1000$; $q = 15, q = 16, q = 17$

18. $c = 0.04q^3 - 0.5q^2 + 4.4q + 7500$; $q = 5, q = 25, q = 1000$

In Problems 19–22, $\bar{c}$ represents average cost per unit, which is a function of the number q of units produced. Find the marginal-cost function and the marginal cost for the indicated values of q.

*19. $\bar{c} = 0.01q + 5 + \dfrac{500}{q}$; $q = 50, q = 100$

20. $\bar{c} = 2 + \dfrac{1000}{q}$; $q = 25, q = 235$

*21. $\bar{c} = 0.00002q^2 - 0.01q + 6 + \dfrac{20{,}000}{q}$; $q = 100, q = 500$

22. $\bar{c} = 0.001q^2 - 0.3q + 40 + \dfrac{7000}{q}$; $q = 10, q = 20$

In Problems 23–26, r represents total revenue and is a function of the number q of units sold. Find the marginal-revenue function and the marginal revenue for the indicated values of q.

*23. $r = 0.8q$; $q = 9, q = 300, q = 500$

24. $r = q(15 - \frac{1}{30}q)$; $q = 5, q = 15, q = 150$

25. $r = 250q + 45q^2 - q^3$; $q = 5, q = 10, q = 25$

26. $r = 2q(30 - 0.1q)$; $q = 10, q = 20$

27. **Hosiery Mill** The total-cost function for a hosiery mill is estimated by Dean[4] to be

$$c = -10{,}484.69 + 6.750q - 0.000328q^2$$

where q is output in dozens of pairs and c is total cost in dollars. Find the marginal-cost function and evaluate it when $q = 5000$.

28. **Light and Power Plant** The total-cost function for an electric light and power plant is estimated by Nordin[5] to be

$$c = 32.07 - 0.79q + 0.02142q^2 - 0.0001q^3 \quad 20 \le q \le 90$$

where q is the eight-hour total output (as a percentage of capacity) and c is the total fuel cost in dollars. Find the marginal-cost function and evaluate it when $q = 70$.

29. **Urban Concentration** Suppose the 100 largest cities in the United States in 1920 are ranked according to magnitude (areas of cities). From Lotka,[6] the following relation holds approximately:

$$PR^{0.93} = 5{,}000{,}000$$

Here, P is the population of the city having respective rank R. This relation is called the *law of urban concentration* for 1920. Solve for P in terms of R, and then find how fast the population is changing with respect to rank.

30. **Depreciation** Under the straight-line method of depreciation, the value v of a certain machine after t years have elapsed is given by

$$v = 85{,}000 - 10{,}500t$$

where $0 \le t \le 9$. How fast is v changing with respect to t when $t = 2$? $t = 3$? at any time?

31. **Winter Moth** A study of the winter moth was made in Nova Scotia (adapted from Embree).[7] The prepupae of the moth fall onto the ground from host trees. At a distance of x ft from the base of a host tree, the prepupal density (number of prepupae per square foot of soil) was y, where

$$y = 59.3 - 1.5x - 0.5x^2 \qquad 1 \le x \le 9$$

(a) At what rate is the prepupal density changing with respect to distance from the base of the tree when $x = 6$?

(b) For what value of x is the prepupal density decreasing at the rate of 6 prepupae per square foot per foot?

32. **Cost Function** For the cost function

$$c = 0.4q^2 + 4q + 5,$$

find the rate of change of c with respect to q when $q = 2$. Also, what is $\Delta c/\Delta q$ over the interval [2, 3]?

[3] R. W. Stacy et al., *Essentials of Biological and Medical Physics* (New York: McGraw-Hill Book Company, 1955).

[4] J. Dean, "Statistical Cost Functions of a Hosiery Mill," *Studies in Business Administration* 11, no. 4 (Chicago: University of Chicago Press, 1941).

[5] J. A. Nordin, "Note on a Light Plant's Cost Curves," *Econometrica* 15 (1947), 231–35.

[6] A. J. Lotka, *Elements of Mathematical Biology* (New York: Dover Publications, 1956).

[7] D. G. Embree, "The Population Dynamics of the Winter Moth in Nova Scotia, 1954–1962," *Memoirs of the Entomological Society of Canada*, no. 46 (1965).

In Problems 33–38, find (a) the rate of change of y with respect to x and (b) the relative rate of change of y. At the given value of x, find (c) the rate of change of y, (d) the relative rate of change of y, and (e) the percentage rate of change of y.

33. $y = f(x) = x + 4; x = 5$

34. $y = f(x) = 5 - 2x; x = 3$

***35.** $y = 3x^2 + 7; x = 2$

36. $y = 5 - 3x^3; x = 1$

37. $y = 8 - x^3; x = 1$

38. $y = x^2 + 3x - 4; x = -1$

39. Cost Function For the cost function

$$c = 0.2q^2 + 1.2q + 4$$

how fast does c change with respect to q when $q = 5$? Determine the percentage rate of change of c with respect to q when $q = 5$.

40. Organic Matters/Species Diversity In a discussion of contemporary waters of shallow seas, Odum[8] claims that in such waters the total organic matter y (in milligrams per liter) is a function of species diversity x (in number of species per thousand individuals). If $y = 100/x$, at what rate is the total organic matter changing with respect to species diversity when $x = 10$? What is the percentage rate of change when $x = 10$?

41. Revenue For a certain manufacturer, the revenue obtained from the sale of q units of a product is given by

$$r = 30q - 0.3q^2$$

(a) How fast does r change with respect to q? When $q = 10$, **(b)** find the relative rate of change of r, and **(c)** to the nearest percent, find the percentage rate of change of r.

42. Revenue Repeat Problem 41 for the revenue function given by $r = 10q - 0.2q^2$ and $q = 25$.

43. Weight of Limb The weight of a limb of a tree is given by $W = 2t^{0.432}$, where t is time. Find the relative rate of change of W with respect to t.

44. Response to Shock A psychological experiment[9] was conducted to analyze human responses to electrical shocks (stimuli). The subjects received shocks of various intensities. The response R to a shock of intensity I (in microamperes) was to be a number that indicated the perceived magnitude relative to that of a "standard" shock. The standard shock was assigned a magnitude of 10. Two groups of subjects were tested under slightly different conditions. The responses R_1 and R_2 of the first and second groups to a shock of intensity I were given by

$$R_1 = \frac{I^{1.3}}{1855.24} \qquad 800 \leq I \leq 3500$$

and

$$R_2 = \frac{I^{1.3}}{1101.29} \qquad 800 \leq I \leq 3500$$

(a) For each group, determine the relative rate of change of response with respect to intensity.
(b) How do these changes compare with each other?
(c) In general, if $f(x) = C_1x^n$ and $g(x) = C_2x^n$, where C_1 and C_2 are constants, how do the relative rates of change of f and g compare?

45. Cost A manufacturer of mountain bikes has found that when 20 bikes are produced per day, the average cost is \$150 and the marginal cost is \$125. Based on that information, approximate the total cost of producing 21 bikes per day.

46. Marginal and Average Costs Suppose that the cost function for a certain product is $c = f(q)$. If the relative rate of change of c (with respect to q) is $\frac{1}{q}$, prove that the marginal-cost function and the average-cost function are equal.

In Problems 47 and 48, use the numerical derivative feature of your graphing calculator.

47. If the total-cost function for a manufacturer is given by

$$c = \frac{5q^2}{\sqrt{q^2 + 3}} + 5000$$

where c is in dollars, find the marginal cost when 10 units are produced. Round your answer to the nearest cent.

48. The population of a city t years from now is given by

$$P = 250{,}000e^{0.04t}$$

Find the rate of change of population with respect to time t three years from now. Round your answer to the nearest integer.

[8] H. T. Odum, "Biological Circuits and the Marine Systems of Texas," in *Pollution and Marine Biology*, ed. T. A. Olsen and F. J. Burgess (New York: Interscience Publishers, 1967).

[9] H. Babkoff, "Magnitude Estimation of Short Electrocutaneous Pulses," *Psychological Research* 39, no. 1 (1976), 39–49.

OBJECTIVE *To relate differentiability to continuity.*

11.4 Differentiability and Continuity

In the next section, we shall make use of the following important relationship between differentiability and continuity:

If f is differentiable at a then f is continuous at a.

To establish this result, we shall assume that f is differentiable at a. Then $f'(a)$ exists, and

$$\lim_{h\to 0} \frac{f(a+h)-f(a)}{h} = f'(a).$$

Consider the numerator $f(a+h)-f(a)$ as $h \to 0$. We have

$$\begin{aligned}\lim_{h\to 0}[f(a+h)-f(a)] &= \lim_{h\to 0}\left[\frac{f(a+h)-f(a)}{h}\cdot h\right]\\ &= \lim_{h\to 0}\frac{f(a+h)-f(a)}{h}\cdot\lim_{h\to 0} h\\ &= f'(a)\cdot 0 = 0.\end{aligned}$$

Thus, $\lim_{h\to 0}[f(a+h)-f(a)] = 0$. This means that $f(a+h)-f(a)$ approaches 0 as $h \to 0$. Consequently,

$$\lim_{h\to 0} f(a+h) = f(a).$$

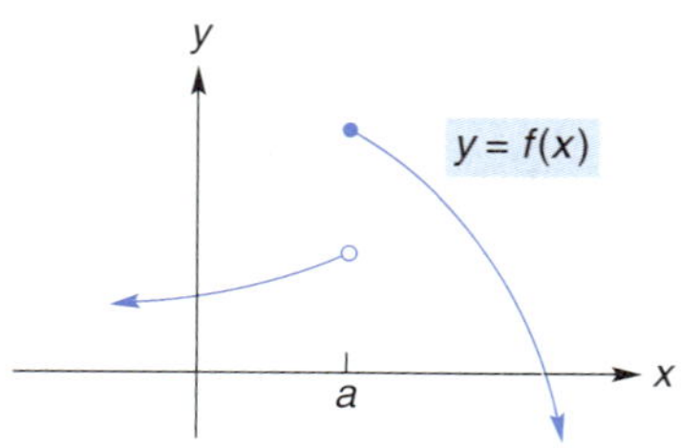

FIGURE 11.13 f is not continuous at a, so f is not differentiable at a.

As stated in Section 10.4, this condition means that f is continuous at a. The foregoing, then, proves that f is continuous at a when f is differentiable there. More simply, we say that **differentiability at a point implies continuity at that point.**

If a function is not continuous at a point, then it cannot have a derivative there. For example, the function in Figure 11.13 is discontinuous at a. The curve has no tangent at that point, so the function is not differentiable there.

EXAMPLE 1 **Continuity and Differentiability**

a. Let $f(x) = x^2$. The derivative, $2x$, is defined for all values of x, so $f(x) = x^2$ must be continuous for all values of x.

b. The function $f(p) = \frac{1}{2p}$ is not continuous at $p = 0$ because f is not defined there. Thus, the derivative does not exist at $p = 0$.

The converse of the statement that differentiability implies continuity is *false*. That is, it is false that continuity implies differentiability. In Example 2, you will see a function that is continuous at a point, but not differentiable there.

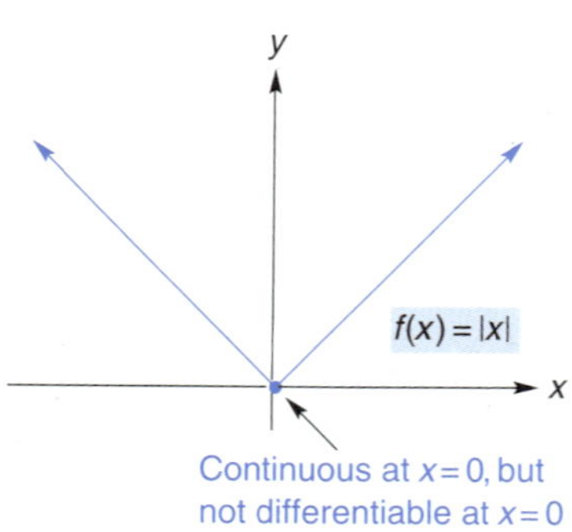

FIGURE 11.14 Continuity does not imply differentiability.

EXAMPLE 2 **Continuity Does Not Imply Differentiability**

The function $y = f(x) = |x|$ is continuous at $x = 0$. (See Figure 11.14.) As we mentioned in Section 11.1, there is no tangent line at $x = 0$. Thus, the derivative does not exist there. This shows that continuity does *not* imply differentiability.

OBJECTIVE *To find derivatives by applying the product and quotient rules, and to develop the concepts of marginal propensity to consume and marginal propensity to save.*

11.5 Product and Quotient Rules

The equation $F(x) = (x^2 + 3x)(4x + 5)$ expresses $F(x)$ as a product of two functions: $x^2 + 3x$ and $4x + 5$. To find $F'(x)$ by using only our previous rules, we first multiply the functions. Then we differentiate the result, term by term:

$$F(x) = (x^2 + 3x)(4x + 5) = 4x^3 + 17x^2 + 15x$$

$$F'(x) = 12x^2 + 34x + 15 \qquad (1)$$

However, in many problems that involve differentiating a product of functions, the multiplication is not as simple as it is here. At times, it is not even practical to attempt it. Fortunately, there is a rule for differentiating a product, and the rule avoids such multiplications. Since the derivative of a sum of functions is the sum of their derivatives, you might expect a similar rule for products. However, the situation is rather subtle.

Rule 1 Product Rule

If f and g are differentiable functions, then the product fg is differentiable, and

$$\frac{d}{dx}[f(x)g(x)] = f'(x)g(x) + f(x)g'(x)$$

That is, the derivative of the product of two functions is the derivative of the first function times the second, plus the first function times the derivative of the second.

$$\frac{d}{dx}(\text{product}) = \begin{pmatrix}\text{derivative}\\ \text{of first}\end{pmatrix}(\text{second}) + (\text{first})\begin{pmatrix}\text{derivative}\\ \text{of second}\end{pmatrix}$$

Proof. If $F(x) = f(x)g(x)$, then, by the definition of the derivative of F,

$$\begin{aligned} F'(x) &= \lim_{h\to 0} \frac{F(x+h) - F(x)}{h} \\ &= \lim_{h\to 0} \frac{f(x+h)g(x+h) - f(x)g(x)}{h} \end{aligned}$$

Now we use a "trick." Adding and subtracting $f(x)g(x+h)$ in the numerator, we have

$$F'(x) = \lim_{h\to 0} \frac{f(x+h)g(x+h) - f(x)g(x) + f(x)g(x+h) - f(x)g(x+h)}{h}$$

Regrouping gives

$$\begin{aligned} F'(x) &= \lim_{h\to 0} \frac{[f(x+h)g(x+h) - f(x)g(x+h)] + [f(x)g(x+h) - f(x)g(x)]}{h} \\ &= \lim_{h\to 0} \frac{[f(x+h) - f(x)]g(x+h) + f(x)[g(x+h) - g(x)]}{h} \\ &= \lim_{h\to 0} \frac{[f(x+h) - f(x)]g(x+h)}{h} + \lim_{h\to 0} \frac{f(x)[g(x+h) - g(x)]}{h} \\ &= \lim_{h\to 0} \frac{f(x+h) - f(x)}{h} \cdot \lim_{h\to 0} g(x+h) + \lim_{h\to 0} f(x) \cdot \lim_{h\to 0} \frac{g(x+h) - g(x)}{h} \end{aligned}$$

Since we assumed that f and g are differentiable, then

$$\lim_{h\to 0}\frac{f(x+h)-f(x)}{h}=f'(x)$$

and

$$\lim_{h\to 0}\frac{g(x+h)-g(x)}{h}=g'(x).$$

The differentiability of g implies that g is continuous, and from Section 10.4,

$$\lim_{h\to 0}g(x+h)=g(x)$$

Thus,

$$F'(x)=f'(x)g(x)+f(x)g'(x)$$

EXAMPLE 1 **Applying the Product Rule**

If $F(x)=(x^2+3x)(4x+5)$, find $F'(x)$.

Solution: We shall consider F as a product of two functions:

$$F(x)=\underbrace{(x^2+3x)}_{f(x)}\underbrace{(4x+5)}_{g(x)}$$

Therefore, we can apply the product rule:

$$\begin{aligned}F'(x)&=f'(x)g(x)+f(x)g'(x)\\&=\underbrace{\frac{d}{dx}(x^2+3x)}_{\text{Derivative of first}}\underbrace{(4x+5)}_{\text{Second}}+\underbrace{(x^2+3x)}_{\text{First}}\underbrace{\frac{d}{dx}(4x+5)}_{\text{Derivative of second}}\\&=(2x+3)(4x+5)+(x^2+3x)(4)\\&=12x^2+34x+15 \qquad \text{(simplifying)}\end{aligned}$$

CAUTION It is worthwhile to repeat that the derivative of the product of two functions is somewhat subtle. Do not be tempted to make up a simpler rule.

This agrees with our previous result. [See Equation (1).] Although there doesn't seem to be much advantage to using the product rule here, you will see that there are times when it is practical to use.

▸ **Now work Problem 1.**

APPLICATIONS IN PRACTICE 1

Applying the Product Rule

A taco stand usually sells 225 tacos per day at \$2 each. A business student's research tells him that for every \$0.15 decrease in the price, the stand will sell 20 more tacos per day. The revenue function for the taco stand is $R(x)=(2-0.15x)(225+20x)$, where x is the number of \$0.15 reductions in price. Find $\frac{dR}{dx}$.

EXAMPLE 2 **Applying the Product Rule**

If $y=(x^{2/3}+3)(x^{-1/3}+5x)$, find dy/dx.

Solution: Applying the product rule gives

$$\begin{aligned}\frac{dy}{dx}&=\frac{d}{dx}(x^{2/3}+3)(x^{-1/3}+5x)+(x^{2/3}+3)\frac{d}{dx}(x^{-1/3}+5x)\\&=\left(\frac{2}{3}x^{-1/3}\right)(x^{-1/3}+5x)+(x^{2/3}+3)\left(\frac{-1}{3}x^{-4/3}+5\right)\\&=\frac{25}{3}x^{2/3}+\frac{1}{3}x^{-2/3}-x^{-4/3}+15\end{aligned}$$

Alternatively, we could have found the derivative without the product rule by first finding the product $(x^{2/3}+3)(x^{-1/3}+5x)$ and then differentiating the result, term by term.

▸ **Now work Problem 15.**

EXAMPLE 3 **Differentiating a Product of Three Factors**

If $y = (x+2)(x+3)(x+4)$, *find* y'.

Solution:

Strategy: We would like to use the product rule, but it applies only to *two* factors. By treating the first two factors as a single factor, we can consider y to be a product of two functions:

$$y = [(x+2)(x+3)](x+4)$$

The product rule gives

$$\begin{aligned} y' &= \frac{d}{dx}[(x+2)(x+3)](x+4) + [(x+2)(x+3)]\frac{d}{dx}(x+4) \\ &= \frac{d}{dx}[(x+2)(x+3)](x+4) + [(x+2)(x+3)](1) \end{aligned}$$

Applying the product rule again, we have

$$\begin{aligned} y' &= \left[\frac{d}{dx}(x+2)(x+3) + (x+2)\frac{d}{dx}(x+3)\right](x+4) + [(x+2)(x+3)] \\ &= [(1)(x+3) + (x+2)(1)](x+4) + [(x+2)(x+3)] \end{aligned}$$

After simplifying, we obtain

$$y' = 3x^2 + 18x + 26$$

Two other ways of finding the derivative are as follows:

1. Multiply the first two factors of y to obtain

$$y = (x^2 + 5x + 6)(x+4)$$

and then apply the product rule.

2. Multiply all three factors to obtain

$$y = x^3 + 9x^2 + 26x + 24$$

and then differentiate, term by term.

▸ **Now work Problem 19.**

It is sometimes helpful to remember differentiation rules in more streamlined notation. For example,

$$(fg)' = f'g + fg'$$

is a correct equality of functions that expresses the product rule. We can then calculate:

$$\begin{aligned} (fgh)' &= [(fg)h]' \\ &= (fg)'h + (fg)h' \\ &= (f'g + fg')h + (fg)h' \\ &= f'gh + fg'h + fgh' \end{aligned}$$

It is not suggested that you try to commit to memory derived rules like:

$$(fgh)' = f'gh + fg'h + fgh'$$

Because $(fg)' = f'g + fg' = fg' + gf'$, using commutativity of both sum and product of functions, we can express the product rule with the derivatives as second factors. In applications this can make the scope of $\frac{d}{dx}$ a little clearer.

APPLICATIONS IN PRACTICE 2

Derivative of a Product without the Product Rule

One hour after x milligrams of a particular drug are given to a person, the change in body temperature $T(x)$, in degrees Fahrenheit, is given approximately by $T(x) = x^2\left(1 - \frac{x}{3}\right)$. The rate at which T changes with respect to the size of the dosage x, $T'(x)$, is called the *sensitivity* of the body to the dosage. Find the sensitivity when the dosage is 1 milligram. Do not use the product rule. The product rule (and quotient rule that follows) should not be applied when a more direct and efficient method is available.

EXAMPLE 4 Using the Product Rule to Find Slope

Find the slope of the graph of $f(x) = (7x^3 - 5x + 2)(2x^4 + 7)$ *when* $x = 1$.

Solution:

Strategy: We find the slope by evaluating the derivative when $x = 1$. Because f is a product of two functions, we can find the derivative by using the product rule.

We have

$$\begin{aligned} f'(x) &= (7x^3 - 5x + 2)\frac{d}{dx}(2x^4 + 7) + (2x^4 + 7)\frac{d}{dx}(7x^3 - 5x + 2) \\ &= (7x^3 - 5x + 2)(8x^3) + (2x^4 + 7)(21x^2 - 5) \end{aligned}$$

Since we must compute $f'(x)$ when $x = 1$, *there is no need to simplify* $f'(x)$ *before evaluating it*. Substituting into $f'(x)$, we obtain

$$f'(1) = 4(8) + 9(16) = 176$$

▶ **Now work Problem 49.**

Usually, we do not use the product rule when simpler ways are obvious. For example, if $f(x) = 2x(x + 3)$, then it is quicker to write $f(x) = 2x^2 + 6x$, from which $f'(x) = 4x + 6$. Similarly, we do not usually use the product rule to differentiate $y = 4(x^2 - 3)$. Since the 4 is a constant factor, by the constant-factor rule we have $y' = 4(2x) = 8x$.

The next rule is used for differentiating a *quotient* of two functions.

Rule 2 Quotient Rule

If f and g are differentiable functions and $g(x) \neq 0$, then the quotient f/g is also differentiable, and

$$\frac{d}{dx}\left[\frac{f(x)}{g(x)}\right] = \frac{g(x)f'(x) - f(x)g'(x)}{[g(x)]^2}$$

With the understanding about the denominator not being zero, we can write

$$\left(\frac{f}{g}\right)' = \frac{gf' - fg'}{g^2}$$

That is, the derivative of the quotient of two functions is the denominator times the derivative of the numerator, minus the numerator times the derivative of the denominator, all divided by the square of the denominator.

$$\frac{d}{dx}(\text{quotient}) = \frac{(\text{denominator})\left(\substack{\text{derivative}\\ \text{of numerator}}\right) - (\text{numerator})\left(\substack{\text{derivative}\\ \text{of denominator}}\right)}{(\text{denominator})^2}$$

Proof. If $F(x) = \dfrac{f(x)}{g(x)}$, then

$$F(x)g(x) = f(x)$$

By the product rule,

$$F(x)g'(x) + g(x)F'(x) = f'(x)$$

Solving for $F'(x)$, we have

$$F'(x) = \frac{f'(x) - F(x)g'(x)}{g(x)}$$

But $F(x) = f(x)/g(x)$. Thus,

$$F'(x) = \frac{f'(x) - \dfrac{f(x)g'(x)}{g(x)}}{g(x)}$$

Simplifying gives[10]

$$F'(x) = \frac{g(x)f'(x) - f(x)g'(x)}{[g(x)]^2}$$

CAUTION The derivative of the quotient of two functions is trickier still than the product rule. One must remember where the minus sign goes!

EXAMPLE 5 Applying the Quotient Rule

If $F(x) = \dfrac{4x^2+3}{2x-1}$, *find* $F'(x)$.

Solution:

Strategy: We recognize F as a quotient, so we can apply the quotient rule.

Let $f(x) = 4x^2 + 3$ and $g(x) = 2x - 1$. Then

$$F'(x) = \frac{g(x)f'(x) - f(x)g'(x)}{[g(x)]^2}$$

$$= \frac{\overbrace{(2x-1)}^{\text{Denominator}}\ \overbrace{\frac{d}{dx}(4x^2+3)}^{\text{Derivative of numerator}} - \overbrace{(4x^2+3)}^{\text{Numerator}}\ \overbrace{\frac{d}{dx}(2x-1)}^{\text{Derivative of numerator}}}{\underbrace{(2x-1)^2}_{\text{Square of denominator}}}$$

$$= \frac{(2x-1)(8x) - (4x^2+3)(2)}{(2x-1)^2}$$

$$= \frac{8x^2 - 8x - 6}{(2x-1)^2} = \frac{2(2x+1)(2x-3)}{(2x-1)^2}$$

▶ **Now work Problem 21.**

[10]You may have observed that this proof assumes the existence of $F'(x)$. However, the rule can be proven without this assumption.

EXAMPLE 6 **Rewriting before Differentiating**

Differentiate $y = \dfrac{1}{x + \dfrac{1}{x+1}}$.

Solution:

Strategy: To simplify the differentiation, we shall rewrite the function so that no fraction appears in the denominator.

We have

$$y = \frac{1}{x + \dfrac{1}{x+1}} = \frac{1}{\dfrac{x(x+1)+1}{x+1}} = \frac{x+1}{x^2+x+1}$$

$$\frac{dy}{dx} = \frac{(x^2+x+1)(1) - (x+1)(2x+1)}{(x^2+x+1)^2} \qquad \text{(quotient rule)}$$

$$= \frac{(x^2+x+1) - (2x^2+3x+1)}{(x^2+x+1)^2}$$

$$= \frac{-x^2-2x}{(x^2+x+1)^2} = -\frac{x^2+2x}{(x^2+x+1)^2}$$

▶ **Now work Problem 45.**

Although a function may have the form of a quotient, this does not necessarily mean that the quotient rule must be used to find the derivative. The next example illustrates some typical situations in which, although the quotient rule can be used, a simpler and more efficient method is available.

EXAMPLE 7 **Differentiating Quotients without Using the Quotient Rule**

Differentiate the following functions.

a. $f(x) = \dfrac{2x^3}{5}$

Solution: Rewriting, we have $f(x) = \frac{2}{5}x^3$. By the constant-factor rule,

$$f'(x) = \frac{2}{5}(3x^2)$$

$$= \frac{6x^2}{5}$$

b. $f(x) = \dfrac{4}{7x^3}$

Solution: Rewriting, we have $f(x) = \frac{4}{7}(x^{-3})$. Thus,

$$f'(x) = \frac{4}{7}(-3x^{-4})$$

$$= -\frac{12}{7x^4}$$

c. $f(x) = \dfrac{5x^2 - 3x}{4x}$

CAUTION To differentiate $f(x) = \dfrac{1}{x^2 - 2}$, you might be tempted to first rewrite the quotient as $(x^2 - 2)^{-1}$. It would be a mistake to do this because we presently have no rule for differentiating that form. In short, we have no choice but to use the quotient rule. However, in the next section we shall develop a rule that allows us to differentiate $(x^2 - 2)^{-1}$ in a direct and efficient way.

Solution: Rewriting, we have $f(x) = \dfrac{1}{4}\left(\dfrac{5x^2 - 3x}{x}\right) = \frac{1}{4}(5x - 3)$ for $x \neq 0$. Thus,

$$f'(x) = \frac{1}{4}(5) = \frac{5}{4} \qquad \text{for } x \neq 0$$

Since the function f is not defined for $x = 0$, f' is not defined for $x = 0$ either.

▶ **Now work Problem 17.**

EXAMPLE 8 **Marginal Revenue**

If the demand equation for a manufacturer's product is

$$p = \frac{1000}{q + 5}$$

where p is in dollars, find the marginal-revenue function and evaluate it when $q = 45$.

Solution:

Strategy: First we must find the revenue function. The revenue r received for selling q units when the price per unit is p is given by

$$\textbf{revenue} = \textbf{(price)(quantity)}, \qquad \text{that is} \qquad r = pq$$

Using the demand equation we shall express r in terms of q only. Then we shall differentiate to find the marginal-revenue function, dr/dq.

The revenue function is

$$r = \left(\frac{1000}{q + 5}\right)q = \frac{1000q}{q + 5}$$

Thus, the marginal-revenue function is given by

$$\begin{aligned}\frac{dr}{dq} &= \frac{(q+5)\dfrac{d}{dq}(1000q) - (1000q)\dfrac{d}{dq}(q+5)}{(q+5)^2} \\ &= \frac{(q+5)(1000) - (1000q)(1)}{(q+5)^2} = \frac{5000}{(q+5)^2}\end{aligned}$$

and

$$\left.\frac{dr}{dq}\right|_{q=45} = \frac{5000}{(45+5)^2} = \frac{5000}{2500} = 2$$

This means that selling one additional unit beyond 45 results in approximately \$2 more in revenue.

▶ **Now work Problem 59.**

Consumption Function

A function that plays an important role in economic analysis is the **consumption function.** The consumption function $C = f(I)$ expresses a relationship between the total national income I and the total national consumption C. Usually, both I and C are expressed in billions of dollars and I is restricted to some interval. The *marginal propensity to consume* is defined as the rate of change of consumption with respect to income. It is merely the derivative of C with respect to I:

$$\text{Marginal propensity to consume} = \frac{dC}{dI}$$

If we assume that the difference between income I and consumption C is savings S, then

$$S = I - C$$

Differentiating both sides with respect to I gives

$$\frac{dS}{dI} = \frac{d}{dI}(I) - \frac{d}{dI}(C) = 1 - \frac{dC}{dI}$$

We define dS/dI as the **marginal propensity to save.** Thus, the marginal propensity to save indicates how fast savings change with respect to income, and

$$\text{Marginal propensity to save} = 1 - \text{Marginal propensity to consume}$$

EXAMPLE 9 Finding Marginal Propensities to Consume and to Save

If the consumption function is given by

$$C = \frac{5(2\sqrt{I^3}+3)}{I+10}$$

determine the marginal propensity to consume and the marginal propensity to save when $I = 100$.

Solution:

$$\frac{dC}{dI} = 5\left[\frac{(I+10)\frac{d}{dI}(2I^{3/2}+3) - (2\sqrt{I^3}+3)\frac{d}{dI}(I+10)}{(I+10)^2}\right]$$

$$= 5\left[\frac{(I+10)(3I^{1/2}) - (2\sqrt{I^3}+3)(1)}{(I+10)^2}\right]$$

When $I = 100$, the marginal propensity to consume is

$$\left.\frac{dC}{dI}\right|_{I=100} = 5\left[\frac{1297}{12{,}100}\right] \approx 0.536$$

The marginal propensity to save when $I = 100$ is $1 - 0.536 = 0.464$. This means that if a current income of \$100 billion increases by \$1 billion, the nation consumes approximately 53.6% (536/1000) and saves 46.4% (464/1000) of that increase.

▶ Now work Problem 69.

Exercise 11.5

In Problems 1–48, differentiate the functions.

*1. $f(x) = (4x+1)(6x+3)$

2. $f(x) = (3x-1)(7x+2)$

3. $s(t) = (5-3t)(t^3-2t^2)$

4. $Q(x) = (5-2x)(x^2+1)$

5. $f(r) = (3r^2-4)(r^2-5r+1)$

6. $C(I) = (2I^2-3)(3I^2-4I+1)$

7. $f(x) = x^2(2x^2-5)$

8. $f(x) = 3x^3(x^2-2x+2)$

9. $y = (x^2+3x-2)(2x^2-x-3)$

10. $\phi(x) = (3-5x+2x^2)(2+x-4x^2)$

11. $f(w) = (8w^2+2w-3)(5w^3+2)$

12. $f(x) = (3x-x^2)(3-x-x^2)$

13. $y = (x^2-1)(3x^3-6x+5)-4(4x^2+2x+1)$

14. $h(x) = 4(x^5-3)+3(8x^2-5)(2x+2)$

*15. $F(p) = \frac{3}{2}(5\sqrt{p}-2)(3p-1)$

16. $g(x) = (\sqrt{x}-3x+1)(\sqrt[4]{x}-2\sqrt{x})$

*17. $y = 7 \cdot \frac{2}{3}$

18. $y = (x-1)(x-2)(x-3)$

*19. $y = (2x-1)(3x+4)(x+7)$

20. $y = \dfrac{2x-3}{4x+1}$

*21. $f(x) = \dfrac{5x}{x-1}$

22. $H(x) = \dfrac{-5x}{5-x}$

23. $f(x) = \dfrac{3}{2x^6}$

24. $f(x) = \dfrac{5(x^2-2)}{7}$

25. $y = \dfrac{x+2}{x-1}$

26. $h(w) = \dfrac{3w^2+5w-1}{w-3}$

*27. $h(z) = \dfrac{6-2z}{z^2-4}$

28. $z = \dfrac{3x^2-x+1}{x^2+2x+5}$

29. $y = \dfrac{8x^2-2x+1}{x^2-5x}$

30. $f(x) = \dfrac{x^3-x^2+1}{x^2+1}$

31. $y = \dfrac{x^2-4x+3}{2x^2-3x+2}$

32. $F(z) = \dfrac{z^4+4}{3z}$

33. $g(x) = \dfrac{1}{x^{100}+7}$

34. $y = \dfrac{-9}{2x^5}$

35. $u(v) = \dfrac{v^5-8}{v}$

36. $y = \dfrac{x-5}{8\sqrt{x}}$

37. $y = \dfrac{3x^2-x-1}{\sqrt[3]{x}}$

38. $y = \dfrac{x^{0.3}-2}{2x^{2.1}+1}$

39. $y = 7 - \dfrac{4}{x-8} + \dfrac{2x}{3x+1}$

40. $q(x) = 3x^2 + \dfrac{3x-2}{x+2} - \dfrac{4}{x^2}$

41. $y = \dfrac{x-5}{(x+2)(x-4)}$

42. $y = \dfrac{(9x-1)(3x+2)}{4-5x}$

43. $s(t) = \dfrac{t^2+3t}{(t^2-1)(t^3+7)}$

44. $f(s) = \dfrac{17}{s(5s^2-10s+4)}$

*45. $y = 3x - \dfrac{\frac{2}{x}-\frac{3}{x-1}}{x-2}$

46. $y = 3 - 12x^3 + \dfrac{1-\frac{5}{x^2+2}}{x^2+5}$

47. $f(x) = \dfrac{a-x}{a+x}$, where a is a constant

48. $f(x) = \dfrac{x^{-1}+a^{-1}}{x^{-1}-a^{-1}}$, where a is a constant, $a \neq 0$

*49. Find the slope of the curve $y = (4x^2+2x-5)(x^3+7x+4)$ at $(-1, 12)$.

50. Find the slope of the curve $y = \dfrac{x^3}{x^4+1}$ at $(-1, -\frac{1}{2})$.

In Problems 51–54, find an equation of the tangent line to the curve at the given point.

51. $y = \dfrac{6}{x-1}$; $(3, 3)$

52. $y = \dfrac{2x+3}{x^2}$; $(1, 5)$

53. $y = (2x+3)[2(x^4-5x^2+4)]$; $(0, 24)$

54. $y = \dfrac{x+1}{x^2(x-4)}$; $(2, -\frac{3}{8})$

In Problems 55 and 56, determine the relative rate of change of y with respect to x for the given value of x.

55. $y = \dfrac{x}{2x-6}$; $x = 1$

56. $y = \dfrac{1-x}{1+x}$; $x = 5$

57. **Motion** The position function for an object moving in a straight line is

$$s = \frac{2}{t^3+1}$$

where t is in seconds and s is in meters. Find the position and velocity of the object at $t = 1$.

58. **Motion** The position function for an object moving in a straight-line path is

$$s = \frac{t+3}{t^2+7}$$

where t is in seconds and s is in meters. Find the positive value(s) of t for which the velocity of the object is 0.

In Problems 59–62, each equation represents a demand function for a certain product, where p denotes the price per unit for q units. Find the marginal-revenue function in each case. Recall that revenue = pq.

*59. $p = 25 - 0.02q$

60. $p = 500/q$

61. $p = \dfrac{108}{q+2} - 3$

62. $p = \dfrac{q+750}{q+50}$

63. Consumption Function For the United States (1922–1942), the consumption function is estimated by[11]

$$C = 0.672I + 113.1$$

Find the marginal propensity to consume.

64. Consumption Function Repeat Problem 63 if $C = 0.712I + 95.05$ for the United States for 1929–1941.[12]

In Problems 65–68, each equation represents a consumption function. Find the marginal propensity to consume and the marginal propensity to save for the given value of I.

65. $C = 3 + \sqrt{I} + 2\sqrt[3]{I};\ I = 1$

66. $C = 6 + \dfrac{3I}{4} - \dfrac{\sqrt{I}}{3};\ I = 25$

67. $C = \dfrac{16\sqrt{I} + 0.8\sqrt{I^3} - 0.2I}{\sqrt{I}+4};\ I = 36$

68. $C = \dfrac{20\sqrt{I} + 0.5\sqrt{I^3} - 0.4I}{\sqrt{I}+5};\ I = 100$

***69. Consumption Function** Suppose that a country's consumption function is given by

$$C = \frac{10\sqrt{I} + 0.7\sqrt{I^3} - 0.2I}{\sqrt{I}}$$

where C and I are expressed in billions of dollars.

(a) Find the marginal propensity to save when income is 25 billion dollars.

(b) Determine the relative rate of change of C with respect to I when income is 25 billion dollars.

70. Marginal Propensities to Consume and to Save Suppose that the savings function of a country is

$$S = \frac{I - 2\sqrt{I} - 8}{\sqrt{I}+2}$$

where the national income (I) and the national savings (S) are measured in billions of dollars. Find the country's marginal propensity to consume and its marginal propensity to save when the national income is \$150 billion. [*Hint:* It may be helpful to first factor the numerator.]

71. Marginal Cost If the total-cost function for a manufacturer is given by

$$c = \frac{5q^2}{q+3} + 5000$$

find the marginal-cost function.

[11]T. Haavelmo, "Methods of Measuring the Marginal Propensity to Consume," *Journal of the American Statistical Association* 42 (1947), 105–22.

[12]Ibid.

72. Marginal and Average Costs Given the cost function $c = f(q)$, show that if $\frac{d}{dq}(\bar{c}) = 0$, then the marginal-cost function and average-cost function are equal.

73. Host-Parasite Relation For a particular host-parasite relationship, it is determined that when the host density (number of hosts per unit of area) is x, the number of hosts that are parasitized is y, where

$$y = \frac{900x}{10+45x}$$

At what rate is the number of hosts parasitized changing with respect to host density when $x = 2$?

74. Acoustics The persistence of sound in a room after the source of the sound is turned off is called *reverberation*. The *reverberation time* RT of the room is the time it takes for the intensity level of the sound to fall 60 decibels. In the acoustical design of an auditorium, the following formula may be used to compute the RT of the room:[13]

$$\text{RT} = \frac{0.05V}{A + xV}$$

Here V is the room volume, A is the total room absorption, and x is the air absorption coefficient. Assuming that A and x are positive constants, show that the rate of change of RT with respect to V is always positive. If the total room volume increases by one unit, does the reverberation time increase or decrease?

75. Predator-Prey In a predator-prey experiment,[14] it was statistically determined that the number of prey consumed, y, by an individual predator was a function of the prey density x (the number of prey per unit of area), where

$$y = \frac{0.7355x}{1 + 0.02744x}$$

Determine the rate of change of prey consumed with respect to prey density.

76. Social Security Benefits In a discussion of social security benefits, Feldstein[15] differentiates a function of the form

$$f(x) = \frac{a(1+x) - b(2+n)x}{a(2+n)(1+x) - b(2+n)x}$$

where a, b, and n are constants. He determines that

$$f'(x) = \frac{-1(1+n)ab}{[a(1+x) - bx]^2(2+n)}$$

Verify this. [*Hint:* For convenience, let $2 + n = c$.]

[13]L. L. Doelle, *Environmental Acoustics* (New York: McGraw-Hill Book Company, 1972).

[14]C. S. Holling, "Some Characteristics of Simple Types of Predation and Parasitism," *Canadian Entomologist* 91, no. 7 (1959), 385–98.

[15]M. Feldstein, "The Optimal Level of Social Security Benefits," *Quarterly Journal of Economics* 100, no. 2 (1985), 303–20.

77. Business The manufacturer of a product has found that when 20 units are produced per day, the average cost is \$150 and the marginal cost is \$125. What is the relative rate of change of average cost with respect to quantity when $q = 20$?

78. Use the result $(fgh)' = f'gh + fg'h + fgh'$ to find dy/dx if

$$y = (3x+1)(2x-1)(x-4)$$

OBJECTIVE *To introduce and apply the chain rule, to derive the power rule as a special case of the chain rule, and to develop the concept of the marginal-revenue product as an application of the chain rule.*

11.6 The Chain Rule and the Power Rule

Our next rule, the *chain rule,* is one of the most important rules for finding derivatives. It involves a situation in which y is a function of the variable u, but u is a function of x, and we want to find the derivative of y with respect to x. For example, the equations

$$y = u^2 \quad \text{and} \quad u = 2x+1$$

define y as a function of u and u as a function of x. If we substitute $2x+1$ for u in the first equation, we can consider y to be a function of x:

$$y = (2x+1)^2.$$

To find dy/dx, we first expand $(2x+1)^2$:

$$y = 4x^2 + 4x + 1$$

Then

$$\frac{dy}{dx} = 8x + 4$$

From this example, you can see that finding dy/dx by first performing a substitution could be quite involved. For instance, if originally we had been given $y = u^{100}$ instead of $y = u^2$, we wouldn't even want to try substituting. Fortunately, the chain rule will allow us to handle such situations with ease.

Rule 1 Chain Rule

If y is a differentiable function of u and u is a differentiable function of x, then y is a differentiable function of x and

$$\frac{dy}{dx} = \frac{dy}{du} \cdot \frac{du}{dx}$$

We can show you why the chain rule is reasonable by considering rates of change. Suppose

$$y = 8u + 5 \quad \text{and} \quad u = 2x - 3$$

Let x change by one unit. How does u change? To answer this question, we differentiate and find $du/dx = 2$. But for *each* one-unit change in u, there is a change in y of $dy/du = 8$. Therefore, what is the change in y if x changes by one unit; that is, what is dy/dx? The answer is $8 \cdot 2$, which is $\frac{dy}{du} \cdot \frac{du}{dx}$. Thus, $\frac{dy}{dx} = \frac{dy}{du} \cdot \frac{du}{dx}$.

We shall now use the chain rule to redo the problem at the beginning of this section. If

$$y = u^2 \quad \text{and} \quad u = 2x+1$$

then

$$\frac{dy}{dx} = \frac{dy}{du} \cdot \frac{du}{dx} = \frac{d}{du}(u^2) \cdot \frac{d}{dx}(2x+1)$$
$$= (2u)2 = 4u$$

Replacing u by $2x+1$ gives

$$\frac{dy}{dx} = 4(2x+1) = 8x+4$$

which agrees with our previous result.

APPLICATIONS IN PRACTICE 1

Using the Chain Rule

If an object moves horizontally according to $x = 6t$, where t is in seconds, and vertically according to $y = 4x^2$, find its vertical velocity $\frac{dy}{dt}$.

EXAMPLE 1 **Using the Chain Rule**

a. *If* $y = 2u^2 - 3u - 2$ *and* $u = x^2 + 4$, find dy/dx.

Solution: By the chain rule,

$$\frac{dy}{dx} = \frac{dy}{du} \cdot \frac{du}{dx} = \frac{d}{du}(2u^2 - 3u - 2) \cdot \frac{d}{dx}(x^2+4)$$
$$= (4u-3)(2x).$$

We can write our answer in terms of x alone by replacing u by x^2+4.

$$\frac{dy}{dx} = [4(x^2+4) - 3](2x) = [4x^2+13](2x) = 8x^3 + 26x$$

b. *If* $y = \sqrt{w}$ and $w = 7 - t^3$, *find* dy/dt.

Solution: Here, y is a function of w and w is a function of t, so we can view y as a function of t. By the chain rule,

$$\frac{dy}{dt} = \frac{dy}{dw} \cdot \frac{dw}{dt} = \frac{d}{dw}(\sqrt{w}) \cdot \frac{d}{dt}(7-t^3)$$
$$= \left(\frac{1}{2}w^{-1/2}\right)(-3t^2) = \frac{1}{2\sqrt{w}}(-3t^2)$$
$$= -\frac{3t^2}{2\sqrt{w}} = -\frac{3t^2}{2\sqrt{7-t^3}}$$

▸ **Now work Problem 1.**

EXAMPLE 2 **Using the Chain Rule**

If $y = 4u^3 + 10u^2 - 3u - 7$ *and* $u = 4/(3x-5)$, *find* dy/dx *when* $x = 1$.

Solution: By the chain rule,

$$\frac{dy}{dx} = \frac{dy}{du} \cdot \frac{du}{dx} = \frac{d}{du}(4u^3 + 10u^2 - 3u - 7) \cdot \frac{d}{dx}\left(\frac{4}{3x-5}\right)$$
$$= (12u^2 + 20u - 3) \cdot \frac{(3x-5)\frac{d}{dx}(4) - 4\frac{d}{dx}(3x-5)}{(3x-5)^2}$$
$$= (12u^2 + 20u - 3) \cdot \frac{-12}{(3x-5)^2}$$

Do not simply replace x by 1 and leave your answer in terms of u.

Even though dy/dx is in terms of x's and u's, we can evaluate it when $x = 1$ if we determine the corresponding value of u. When $x = 1$,

$$u = \frac{4}{3(1) - 5} = -2$$

Thus,

$$\left.\frac{dy}{dx}\right|_{x=1} = [12(-2)^2 + 20(-2) - 3] \cdot \frac{-12}{[3(1) - 5]^2}$$
$$= 5 \cdot (-3) = -15$$

▶ **Now work Problem 5.**

The chain rules states that if $y = f(u)$ and $u = g(x)$, then

$$\frac{dy}{dx} = \frac{dy}{du} \cdot \frac{du}{dx}$$

Actually, the chain rule applies to a composite function, because

$$y = f(u) = f(g(x)) = (f \circ g)(x)$$

Thus y, as a function of x, is $f \circ g$. This means that we can use the chain rule to differentiate a function when we recognize the function as a composition. However, we must first break down the function into composite parts.

For example, to differentiate

$$y = (x^3 - x^2 + 6)^{100}$$

we think of the function as a composition. Let

$$y = f(u) = u^{100} \quad \text{and} \quad u = g(x) = x^3 - x^2 + 6$$

Then $y = (x^3 - x^2 + 6)^{100} = [g(x)]^{100} = f(g(x))$. Now that we have a composite, we differentiate. Since $y = u^{100}$ and $u = x^3 - x^2 + 6$, by the chain rule we have

$$\frac{dy}{dx} = \frac{dy}{du} \cdot \frac{du}{dx}$$
$$= (100u^{99})(3x^2 - 2x)$$
$$= 100(x^3 - x^2 + 6)^{99}(3x^2 - 2x)$$

We have just used the chain rule to differentiate $y = (x^3 - x^2 + 6)^{100}$, which is a power of a *function* of x, not simply a power of x. The following rule, called the *power rule*, generalizes our result and is a special case of the chain rule:

Rule 2 Power Rule

If u is a differentiable function of x and n is any real number, then

$$\frac{d}{dx}(u^n) = nu^{n-1}\frac{du}{dx}$$

Proof. Let $y = u^n$. Since y is a differentiable function of u and u is a differentiable function of x, the chain rule gives

$$\frac{dy}{dx} = \frac{dy}{du} \cdot \frac{du}{dx}$$

But $dy/du = nu^{n-1}$. Thus,

$$\frac{dy}{dx} = nu^{n-1}\frac{du}{dx}$$

which is the power rule.

Another way of writing the power-rule formula is

$$\frac{d}{dx}([u(x)]^n) = n[u(x)]^{n-1}u'(x)$$

EXAMPLE 3 Using the Power Rule

If $y = (x^3 - 1)^7$, find y'.

Solution: Since y is a power of a *function* of x, the power rule applies. Letting $u(x) = x^3 - 1$ and $n = 7$, we have

$$\begin{aligned} y' &= n[u(x)]^{n-1}u'(x) \\ &= 7(x^3-1)^{7-1}\frac{d}{dx}(x^3-1) \\ &= 7(x^3-1)^6(3x^2) = 21x^2(x^3-1)^6 \end{aligned}$$

▸ **Now work Problem 9.**

EXAMPLE 4 Using the Power Rule

If $y = \sqrt[3]{(4x^2+3x-2)^2}$, find dy/dx when $x = -2$.

Solution: Since $y = (4x^2+3x-2)^{2/3}$, we use the power rule with

$$u = 4x^2 + 3x - 2$$

and $n = \frac{2}{3}$. We have

$$\begin{aligned} \frac{dy}{dx} &= \frac{2}{3}(4x^2+3x-2)^{(2/3)-1}\frac{d}{dx}(4x^2+3x-2) \\ &= \frac{2}{3}(4x^2+3x-2)^{-1/3}(8x+3) \\ &= \frac{2(8x+3)}{3\sqrt[3]{4x^2+3x-2}} \end{aligned}$$

Thus,

$$\left.\frac{dy}{dx}\right|_{x=-2} = \frac{2(-13)}{3\sqrt[3]{8}} = -\frac{13}{3}$$

▸ **Now work Problem 19.**

EXAMPLE 5 Using the Power Rule

If $y = \dfrac{1}{x^2-2}$, find $\dfrac{dy}{dx}$.

The technique used in Example 5 is frequently used when the numerator of a quotient is a constant and the denominator is not.

Solution: Although the quotient rule can be used here, a more efficient approach is to treat the right side as the power $(x^2-2)^{-1}$ and use the power rule. Let $u = x^2 - 2$. Then $y = u^{-1}$, and

$$\begin{aligned} \frac{dy}{dx} &= nu^{n-1}\frac{du}{dx} \\ &= (-1)(x^2-2)^{-1-1}\frac{d}{dx}(x^2-2) \\ &= (-1)(x^2-2)^{-2}(2x) \\ &= -\frac{2x}{(x^2-2)^2} \end{aligned}$$

▸ **Now work Problem 27.**

In Example 5, the technique of moving the denominator to the numerator is not usually used when the numerator and denominator of a quotient involve variables. For example, we would not rewrite

$$\frac{x^2+1}{x^2-2} \quad \text{as} \quad (x^2+1)(x^2-2)^{-1}$$

for the following reason: Using the quotient rule on the first form results in an expression that is easily simplified; but using the product rule on the second form results in a sum of terms involving negative exponents, which is not as easily simplified.

EXAMPLE 6 Differentiating a Power of a Quotient

If $z = \left(\frac{2s+5}{s^2+1}\right)^4$, *find* $\frac{dz}{ds}$.

The problem here is to recognize the basic form of the function to be differentiated. In this case it is a power, not a quotient.

Solution: Since z is a power of a function, we first use the power rule:

$$\frac{dz}{ds} = 4\left(\frac{2s+5}{s^2+1}\right)^{4-1} \frac{d}{ds}\left(\frac{2s+5}{s^2+1}\right)$$

Now we use the quotient rule:

$$\frac{dz}{ds} = 4\left(\frac{2s+5}{s^2+1}\right)^3 \left[\frac{(s^2+1)(2)-(2s+5)(2s)}{(s^2+1)^2}\right]$$

Simplifying, we have

$$\begin{aligned}\frac{dz}{ds} &= 4 \cdot \frac{(2s+5)^3}{(s^2+1)^3}\left[\frac{-2s^2-10s+2}{(s^2+1)^2}\right] \\ &= -\frac{8(s^2+5s-1)(2s+5)^3}{(s^2+1)^5}\end{aligned}$$

▶ **Now work Problem 41.**

EXAMPLE 7 Differentiating a Product of Powers

If $y = (x^2-4)^5(3x+5)^4$, *find* y'.

Solution: Since y is a product, we first apply the product rule:

$$y' = (x^2-4)^5\frac{d}{dx}[(3x+5)^4] + (3x+5)^4\frac{d}{dx}[(x^2-4)^5]$$

Now we use the power rule:

$$\begin{aligned}y' &= (x^2-4)^5[4(3x+5)^3(3)] + (3x+5)^4[5(x^2-4)^4(2x)] \\ &= 12(x^2-4)^5(3x+5)^3 + 10x(3x+5)^4(x^2-4)^4\end{aligned}$$

In differentiating a product in which at least one factor is a power, simplifying the derivative usually involves factoring.

To simplify, we first remove common factors:

$$\begin{aligned}y' &= 2(x^2-4)^4(3x+5)^3[6(x^2-4)+5x(3x+5)] \\ &= 2(x^2-4)^4(3x+5)^3(21x^2+25x-24)\end{aligned}$$

▶ **Now work Problem 39.**

Usually, the power rule should be used to differentiate $y = [u(x)]^n$. Although a function such as $y = (x^2+2)^2$ may be written $y = x^4 + 4x^2 + 4$ and differentiated easily, this method is impractical for a function such as $y = (x^2+2)^{1000}$. Since $y = (x^2+2)^{1000}$ is of the form $y = [u(x)]^n$, we have

$$y' = 1000(x^2+2)^{999}(2x)$$

Marginal-Revenue Product

Let us now use our knowledge of calculus to develop a concept relevant to economic studies. Suppose a manufacturer hires m employees who produce a total of q units of a product per day. We can think of q as a function of m. If r is the total revenue the manufacturer receives for selling these units, then r can also be considered a function of m. Thus, we can look at dr/dm, the rate of change of revenue with respect to the number of employees. The derivative dr/dm is called the **marginal-revenue product.** It is approximately the change in revenue that results when a manufacturer hires an extra employee.

EXAMPLE 8 Marginal-Revenue Product

A manufacturer determines that m employees will produce a total of q units of a product per day, where

$$q = \frac{10m^2}{\sqrt{m^2+19}} \tag{1}$$

If the demand equation for the product is $p = 900/(q+9)$, determine the marginal-revenue product when $m = 9$.

Solution: We must find dr/dm, where r is revenue. Note that, by the chain rule,

$$\frac{dr}{dm} = \frac{dr}{dq} \cdot \frac{dq}{dm}$$

Thus, we must find both dr/dq and dq/dm when $m = 9$. We begin with dr/dq. The revenue function is given by

$$r = pq = \left(\frac{900}{q+9}\right)q = \frac{900q}{q+9} \tag{2}$$

so, by the quotient rule,

$$\frac{dr}{dq} = \frac{(q+9)(900) - 900q(1)}{(q+9)^2} = \frac{8100}{(q+9)^2}$$

In order to evaluate this expression when $m = 9$, we first use the given equation $q = 10m^2/\sqrt{m^2+19}$ to find the corresponding value of q:

$$q = \frac{10(9)^2}{\sqrt{9^2+19}} = 81$$

Hence,

$$\left.\frac{dr}{dq}\right|_{m=9} = \left.\frac{dr}{dq}\right|_{q=81} = \frac{8100}{(81+9)^2} = 1$$

Now we turn to dq/dm. From the quotient and power rules, we have

$$\frac{dq}{dm} = \frac{d}{dm}\left(\frac{10m^2}{\sqrt{m^2+19}}\right)$$

$$= \frac{(m^2+19)^{1/2}\frac{d}{dm}(10m^2) - (10m^2)\frac{d}{dm}[(m^2+19)^{1/2}]}{[(m^2+19)^{1/2}]^2}$$

$$= \frac{(m^2+19)^{1/2}(20m) - (10m^2)\left[\frac{1}{2}(m^2+19)^{-1/2}(2m)\right]}{m^2+19}$$

so

$$\left.\frac{dq}{dm}\right|_{m=9} = \frac{(81+19)^{1/2}(20\cdot 9) - (10\cdot 81)\left[\frac{1}{2}(81+19)^{-1/2}(2\cdot 9)\right]}{81+19}$$

$$= 10.71$$

A direct formula for the marginal-revenue product is

$$\frac{dr}{dm} = \frac{dq}{dm}\left(p + q\frac{dp}{dq}\right)$$

Therefore, from the chain rule,

$$\left.\frac{dr}{dm}\right|_{m=9} = (1)(10.71) = 10.71$$

This means that if a tenth employee is hired, revenue will increase by approximately \$10.71 per day.

▶ **Now work Problem 80.**

TECHNOLOGY

In Example 8 the marginal-revenue product, dr/dm, was found by using the chain rule. Another method, which is well suited to a graphing calculator, is to use substitution to express r as a function of m and then differentiate directly. First we take Equation (1) and substitute for q in the revenue function, Equation (2); this gives r as a function of m. Here are the details: In our function menu, we enter

$$Y_1 = 10X\text{^}2/\sqrt{}(X\text{^}2+19)$$

$$Y_2 = 900Y_1/(Y_1+9)$$

Y_2 expresses revenue as a function of the number of employees. Finally, to find the marginal-revenue product when $m = 9$, we compute nDeriv(Y_2, X, 9). You should verify that this method gives (approximately) 10.71.

Exercise 11.6

In Problems 1–8, use the chain rule.

*1. If $y = u^2 - 2u$ and $u = x^2 - x$, find dy/dx.

2. If $y = 2u^3 - 8u$ and $u = 7x - x^3$, find dy/dx.

3. If $y = \frac{1}{w^2}$ and $w = 2 - x$, find dy/dx.

4. If $y = \sqrt[4]{z}$ and $z = x^5 - x^4 + 3$, find dy/dx.

*5. If $w = u^2$ and $u = \frac{t+1}{t-1}$, find dw/dt when $t = 3$.

6. If $z = u^2 + \sqrt{u} + 9$ and $u = 2s^2 - 1$, find dz/ds when $s = -1$.

7. If $y = 3w^2 - 8w + 4$ and $w = 2x^2 + 1$, find dy/dx when $x = 0$.

8. If $y = 3u^3 - u^2 + 7u - 2$ and $u = 5x - 2$, find dy/dx when $x = 1$.

In Problems 9–52, find y'.

*__9.__ $y = (3x+2)^6$

10. $y = (x^2 - 4)^4$

11. $y = (3 + 2x^3)^5$

12. $y = (x^2 - x)^3$

13. $y = 2(x^3 - 8x^2 + x)^{100}$

14. $y = \dfrac{(2x^2+1)^4}{2}$

15. $y = (x^2 - 2)^{-3}$

16. $y = (2x^3 - 8x)^{-12}$

17. $y = 3(2x^2 - 3x - 1)^{-10/3}$

18. $y = 4(7x - x^4)^{-3/2}$

*__19.__ $y = \sqrt{5x^2 - x}$

20. $y = \sqrt{3x^2 - 7}$

21. $y = \sqrt[4]{2x - 1}$

22. $y = \sqrt[3]{8x^2 - 1}$

23. $y = 2\sqrt[5]{(x^3+1)^2}$

24. $y = 3\sqrt[5]{(x^3-2)^4}$

25. $y = \dfrac{6}{2x^2 - x + 1}$

26. $y = \dfrac{3}{x^4 + 2}$

*__27.__ $y = \dfrac{1}{(x^2 - 3x)^2}$

28. $y = \dfrac{1}{(2+x)^4}$

29. $y = \dfrac{2}{\sqrt{8x - 1}}$

30. $y = \dfrac{3}{(3x^2 - x)^{2/3}}$

31. $y = \sqrt[3]{7x} + \sqrt[3]{7}x$

32. $y = \sqrt{2x} + \dfrac{1}{\sqrt{2x}}$

33. $y = x^2(x-4)^5$

34. $y = x(x+4)^4$

35. $y = 4x^2\sqrt{5x+1}$

36. $y = 2x\sqrt{1-x}$

37. $y = (x^2 + 2x - 1)^3(5x)$

38. $y = x^2(x^3 - 1)^4$

*__39.__ $y = (8x-1)^3(2x+1)^4$

40. $y = (3x+2)^5(4x-5)^2$

*__41.__ $y = \left(\dfrac{x-7}{x+4}\right)^{10}$

42. $y = \left(\dfrac{2x}{x+2}\right)^4$

43. $y = \sqrt{\dfrac{x-2}{x+3}}$

44. $y = \sqrt[3]{\dfrac{8x^2-3}{x^2+2}}$

45. $y = \dfrac{2x-5}{(x^2+4)^3}$

46. $y = \dfrac{(4x-2)^4}{3x^2+7}$

47. $y = \dfrac{(8x-1)^5}{(3x-1)^3}$

48. $y = \sqrt{(x-1)(x+2)^3}$

49. $y = 6(5x^2+2)\sqrt{x^4+5}$

50. $y = 6 + 3x - 4x(7x+1)^2$

51. $y = 8t + \dfrac{t-1}{t+4} - \left(\dfrac{8t-7}{4}\right)^2$

52. $y = \dfrac{(2x^3+6)(7x-5)}{(2x+4)^2}$

In Problems 53 and 54, use the quotient rule and power rule to find y'. Do not simplify your answer.

53. $y = \dfrac{(2x+1)(3x-5)^2}{(x^2-7)^4}$

54. $y = \dfrac{\sqrt{x+2}(4x^2-1)^2}{9x-3}$

55. If $y = (5u+6)^3$ and $u = (x^2+1)^4$, find dy/dx when $x = 0$.

56. If $z = 2y^2 - 4y + 5$, $y = 6x - 5$, and $x = 2t$, find dz/dt when $t = 1$.

57. Find the slope of the curve $y = (x^2 - 7x - 8)^3$ at the point $(8, 0)$.

58. Find the slope of the curve $y = \sqrt{x+1}$ at the point $(8, 3)$.

In Problems 59–62, find an equation of the tangent line to the curve at the given point.

59. $y = \sqrt[3]{(x^2-8)^2}$; $(3, 1)$

60. $y = (3x+1)^2$; $(-1, 4)$

61. $y = \dfrac{\sqrt{7x+2}}{x+1}$; $\left(1, \frac{3}{2}\right)$

62. $y = \dfrac{-3}{(3x^2+1)^3}$; $(0, -3)$

In Problems 63 and 64, determine the percentage rate of change of y with respect to x for the given value of x.

63. $y = (x^2+9)^3$; $x = 4$

64. $y = \dfrac{1}{(x^2-1)^3}$; $x = 2$

In Problems 65–68, q is the total number of units produced per day by m employees of a manufacturer, and p is the price per unit at which the q units are sold. In each case, find the marginal-revenue product for the given value of m.

65. $q = 2m$, $p = -0.5q + 20$; $m = 5$

66. $q = (200m - m^2)/20$, $p = -0.1q + 70$; $m = 40$

67. $q = 10m^2/\sqrt{m^2+9}$, $p = 525/(q+3)$; $m = 4$

68. $q = 100m/\sqrt{m^2+19}$, $p = 4500/(q+10)$; $m = 9$

69. Demand Equation Suppose $p = 100 - \sqrt{q^2 + 20}$ is a demand equation for a manufacturer's product.

(a) Find the rate of change of p with respect to q.

(b) Find the relative rate of change of p with respect to q.

(c) Find the marginal-revenue function.

70. Marginal-Revenue Product If $p = k/q$, where k is a constant, is the demand equation for a manufacturer's product and $q = f(m)$ defines a function that gives the total number of units produced per day by m employees, show that the marginal-revenue product is always zero.

71. Cost Function The cost c of producing q units of a product is given by

$$c = 5500 + 12q + 0.2q^2$$

If the price per unit p is given by the equation

$$q = 900 - 1.5p$$

use the chain rule to find the rate of change of cost with respect to price per unit when $p = 85$.

72. Hospital Discharges A governmental health agency examined the records of a group of individuals who were hospitalized with a particular illness. It was found that the total proportion that had been discharged at the end of t days of hospitalization was given by

$$f(t) = 1 - \left(\frac{300}{300+t}\right)^3$$

Find $f'(300)$ and interpret your answer.

73. Marginal Cost If the total-cost function for a manufacturer is given by

$$c = \frac{5q^2}{\sqrt{q^2+3}} + 5000$$

find the marginal-cost function.

74. Salary/Education For a certain population, if E is the number of years of a person's education and S represents average annual salary in dollars, then for $E \geq 7$,

$$S = 340E^2 - 4360E + 42{,}800$$

(a) How fast is salary changing with respect to education when $E = 16$?
(b) At what level of education does the rate of change of salary equal \$5000 per year of education?

75. Biology The volume of a spherical cell is given by $V = \frac{4}{3}\pi r^3$, where r is the radius. At time t seconds, the radius (in centimeters) is given by

$$r = 10^{-8}t^2 + 10^{-7}t$$

Use the chain rule to find dV/dt when $t = 10$.

76. Pressure in Body Tissue Under certain conditions, the pressure p developed in body tissue by ultrasonic beams is given as a function of the beam's intensity via the equation[16]

$$p = (2\rho VI)^{1/2}$$

where ρ (a Greek letter read "rho") is density of the affected tissue and V is the velocity of propagation of the beam. Here ρ and V are constants. (a) Find the rate of change of p with respect to I. (b) Find the relative rate of change of p with respect to I.

77. Demography Suppose that, for a certain group of 20,000 births, the number of people surviving to age x years is

$$l_x = -0.000354x^4 + 0.00452x^3 + 0.848x^2 - 34.9x + 20{,}000 \qquad 0 \leq x \leq 95.2$$

(a) Find the rate of change of l_x with respect to x, and evaluate your answer for $x = 55$.
(b) Find the relative rate of change of l_x when $x = 55$. Round your answers to three decimal places.

78. Muscle Contraction A muscle has the ability to shorten when a load, such as a weight, is imposed on it. The equation

$$(P+a)(v+b) = k$$

is called the "fundamental equation of muscle contraction."[17] Here P is the load imposed on the muscle, v is the velocity of the shortening of the muscle fibers, and a, b, and k are positive constants. Express v as a function of P. Use your result to find dv/dP.

79. Economics Suppose $pq = 100$ is the demand equation for a manufacturer's product. Let c be the total cost, and assume that the marginal cost is 0.01 when $q = 200$. Use the chain rule to find dc/dp when $q = 200$.

***80. Marginal-Revenue Product** A monopolist who employs m workers finds that they produce

$$q = 2m(2m+1)^{3/2}$$

units of product per day. The total revenue r (in dollars) is given by

$$r = \frac{50q}{\sqrt{1000+3q}}$$

(a) What is the price per unit (to the nearest cent) when there are 12 workers?
(b) Determine the marginal revenue when there are 12 workers.
(c) Determine the marginal-revenue product when $m = 12$.

81. Suppose $y = f(x)$, where $x = g(t)$. Given that $g(2) = 3$, $g'(2) = 4$, $f(2) = 5$, $f'(2) = 6$, $g(3) = 7$, $g'(3) = 8$, $f(3) = 9$, and $f'(3) = 10$, determine the value of $\left.\frac{dy}{dt}\right|_{t=2}$.

82. Business A manufacturer has determined that, for his product, the daily average cost (in hundreds of dollars) is given by

$$\bar{c} = \frac{324}{\sqrt{q^2+35}} + \frac{5}{q} + \frac{19}{18}$$

(a) As daily production increases, the average cost approaches a constant dollar amount. What is this amount?
(b) Determine the manufacturer's marginal cost when 17 units are produced per day.
(c) The manufacturer determines that if production (and sales) were increased to 18 units per day, revenue would increase by \$275. Should this move be made? Why?

---

[16] R. W. Stacy et al., *Essentials of Biological and Medical Physics* (New York: McGraw-Hill Book Company, 1955).

[17] Ibid.

83. If

$$y = (u+1)\sqrt{u+5}$$

and

$$u = x(x^2+5)^5$$

find dy/dx when $x = 0.1$. Round your answer to two decimal places.

84. If

$$y = \frac{4u+7}{2u^3+1}$$

and

$$u = \frac{2x-1}{(3x+2)^4}$$

find dy/dx when $x = -1$. Round your answer to two decimal places.

11.7 Review

Important Terms and Symbols

Section 11.1 secant line tangent line slope of a curve derivative $\lim_{h\to 0} \frac{f(x+h)-f(x)}{h}$ difference quotient $f'(x)$ y' $\frac{dy}{dx}$ $\frac{d}{dx}[f(x)]$

Section 11.3 Δx position function velocity rate of change total-cost function marginal cost average cost total-revenue function marginal revenue relative rate of change percentage rate of change

Section 11.5 product rule quotient rule consumption function marginal propensity to consume marginal propensity to save

Section 11.6 chain rule power rule marginal-revenue product

Summary

The tangent line (or tangent) to a curve at point P is the limiting position of secant lines PQ as Q approaches P along the curve. The slope of the tangent at P is called the slope of the curve at P.

If $y = f(x)$, the derivative of f at x is the function $f'(x)$ defined by the limit in the equation

$$f'(x) = \lim_{h\to 0} \frac{f(x+h)-f(x)}{h}$$

Geometrically, the derivative gives the slope of the curve $y = f(x)$ at the point $(x, f(x))$. An equation of the tangent at a particular point (x_1, y_1) is obtained by evaluating $f'(x_1)$, which is the slope m of the tangent, and substituting into the point–slope form of a line, $y - y_1 = m(x - x_1)$. Any function that is differentiable at a point must also be continuous there.

The basic rules for finding derivatives are as follows, where we assume that all functions are differentiable:

$$\frac{d}{dx}(c) = 0, \text{ where } c \text{ is any constant}$$

$$\frac{d}{dx}(x^n) = nx^{n-1}, \text{ where } n \text{ is any real number}$$

$$\frac{d}{dx}[cf(x)] = cf'(x), \text{ where } c \text{ is a constant}$$

$$\frac{d}{dx}[f(x)+g(x)] = f'(x)+g'(x)$$

$$\frac{d}{dx}[f(x)-g(x)] = f'(x)-g'(x)$$

$$\frac{d}{dx}[f(x)g(x)] = f(x)g'(x)+g(x)f'(x)$$

$$\frac{d}{dx}\left[\frac{f(x)}{g(x)}\right] = \frac{g(x)f'(x)-f(x)g'(x)}{[g(x)]^2}$$

$$\frac{dy}{dx} = \frac{dy}{du}\cdot\frac{du}{dx}, \text{ where } y \text{ is a function of } u \text{ and } u \text{ is a function of } x$$

$$\frac{d}{dx}(u^n) = nu^{n-1}\frac{du}{dx}, \text{ where } u \text{ is a function of } x \text{ and } n \text{ is any real number}$$

The derivative dy/dx can also be interpreted as giving the (instantaneous) rate of change of y with respect to x:

$$\frac{dy}{dx} = \lim_{\Delta x\to 0}\frac{\Delta y}{\Delta x} = \lim_{\Delta x\to 0}\frac{\text{change in } y}{\text{change in } x}$$

In particular, if $s = f(t)$ is a position function, where s is position at time t, then

$$\frac{ds}{dt} = \text{velocity at time } t$$

In economics, the term *marginal* is used to describe derivatives of specific types of functions. If $c = f(q)$ is a total-cost function (c is the total cost of q units of a product), then the rate of change

$$\frac{dc}{dq} \text{ is called marginal cost}$$

We interpret marginal cost as the approximate cost of one additional unit of output. (Average cost per unit, $\bar{c}$, is related to total cost c by $\bar{c} = c/q$, or $c = \bar{c}q$.)

A total-revenue function $r = f(q)$ gives a manufacturer's revenue r for selling q units of product. (Revenue r and price p are related by $r = pq$.) The rate of change

$$\frac{dr}{dq} \text{ is called marginal revenue}$$

which is interpreted as the approximate revenue obtained from selling one additional unit of output.

If r is the revenue that a manufacturer receives when the total output of m employees is sold, then the derivative dr/dm is called the marginal-revenue product and gives the approximate change in revenue that results when the manufacturer hires an extra employee.

If $C = f(I)$ is a consumption function, where I is national income and C is national consumption, then

$$\frac{dC}{dI} \text{ is marginal propensity to consume}$$

and

$$1 - \frac{dC}{dI} \text{ is marginal prospensity to save}$$

For any function, the relative rate of change of $f(x)$ is

$$\frac{f'(x)}{f(x)}$$

which compares the rate of change of $f(x)$ with $f(x)$ itself. The percentage rate of change is

$$\frac{f'(x)}{f(x)} \cdot 100$$

Review Problems

Problem numbers shown in color indicate problems suggested for use as a practice chapter test.

In Problems 1–4, use the definition of the derivative to find $f'(x)$.

1. $f(x) = 2 - x^2$

2. $f(x) = 2x^2 - 3x + 1$

3. $f(x) = \sqrt{3x}$

4. $f(x) = \dfrac{2}{1+4x}$

In Problems 5–38, differentiate.

5. $y = 7^4$

6. $y = x$

7. $y = 7x^4 - 6x^3 + 5x^2 + 1$

8. $y = 4(x^2 + 5) - 7x$

9. $f(s) = s^2(s^2 + 2)$

10. $y = \sqrt{x+3}$

11. $y = \dfrac{x^2+1}{5}$

12. $y = -\dfrac{2}{2x^2}$

13. $y = (x^2 + 6x)(x^3 - 6x^2 + 4)$

14. $y = (x^2 + 1)^{100}(x - 6)$

15. $f(x) = (2x^2 + 4x)^{100}$

16. $f(w) = w\sqrt{w} + w^2$

17. $y = \dfrac{3}{2x+1}$

18. $y = \dfrac{4x^3 - 8x^2}{5x}$

19. $y = (8 + 2x)(x^2 + 1)^4$

20. $g(z) = (2z)^{3/5} + 5$

21. $f(z) = \dfrac{z^2-1}{z^2+4}$

22. $y = \dfrac{x-5}{(x+2)^2}$

23. $y = \sqrt[3]{4x-1}$

24. $f(x) = (1 + 2^3)^{12}$

25. $y = \dfrac{1}{\sqrt{1-x}}$

26. $y = \dfrac{x(x+1)}{2x^2+3}$

27. $h(x) = (x - 6)^4(x + 5)^3$

28. $y = \dfrac{(x+3)^5}{x}$

29. $y = \dfrac{5x-4}{x+6}$

30. $f(x) = 8x^2\sqrt{1+2x^3}$

31. $y = 2x^{-3/8} + (2x)^{-3/8}$

32. $y = \sqrt{\dfrac{x}{2}} + \sqrt{\dfrac{2}{x}}$

33. $y = \dfrac{x^2+6}{\sqrt{x^2+5}}$

34. $y = \sqrt[3]{(7-3x^2)^2}$

35. $y = (x^3 + 6x^2 + 9)^{3/5}$

36. $z = 0.4x^2(x+1)^{-3} + 0.5$

37. $g(z) = \dfrac{-7z}{(z-1)^{-1}}$

38. $g(z) = \dfrac{-3}{4(z^5 + 2z - 5)^4}$

In Problems 39–42, find an equation of the tangent line to the curve at the point corresponding to the given value of x.

39. $y = x^2 - 6x + 4$, $x = 1$

40. $y = -2x^3 + 6x + 1$, $x = 2$

41. $y = \sqrt[3]{x}$, $x = 8$

42. $y = -\dfrac{x}{x-9}$, $x = 10$

43. If $f(x) = 4x^2 + 2x + 8$, find the relative and percentage rates of change of $f(x)$ when $x = 1$.

44. If $f(x) = x/(x + 4)$, find the relative and percentage rates of change of $f(x)$ when $x = 1$.

45. Marginal Revenue If $r = q(20 - 0.1q)$ is a total-revenue function, find the marginal-revenue function.

46. Marginal Cost If

$$c = 0.0001q^3 - 0.02q^2 + 3q + 6000$$

is a total-cost function, find the marginal cost when $q = 100$.

47. Consumption Function If

$$C = 7 + 0.6I - 0.25\sqrt{I}$$

is a consumption function, find the marginal propensity to consume and the marginal propensity to save when $I = 16$.

48. Demand Equation If $p = \dfrac{q+12}{q+5}$ is a demand equation, find the rate of change of price p with respect to quantity q.

49. Demand Equation If $p = -0.5q + 450$ is a demand equation, find the marginal-revenue function.

50. Average Cost If $\bar{c} = 0.03q + 1.2 + \dfrac{3}{q}$ is an average-cost function, find the marginal cost when $q = 100$.

51. Power-Plant Cost Function The total-cost function of an electric light and power plant is estimated by[18]

$$c = 16.68 + 0.125q + 0.00439q^2 \quad 20 \le q \le 90$$

where q is the eight-hour total output (as a percentage of capacity) and c is the total fuel cost in dollars. Find the marginal-cost function and evaluate it when $q = 70$.

52. Marginal-Revenue Product A manufacturer has determined that m employees will produce a total of q units of product per day, where

$$q = m(50 - m)$$

If the demand function is given by

$$p = -0.01q + 9$$

find the marginal-revenue product when $m = 10$.

53. Winter Moth In a study of the winter moth in Nova Scotia,[19] it was determined that the average number of eggs, y, in a female moth was a function of the female's abdominal width x (in millimeters), where

$$y = f(x) = 14x^3 - 17x^2 - 16x + 34$$

and $1.5 \le x \le 3.5$. At what rate does the number of eggs change with respect to abdominal width when $x = 2$?

54. Host-Parasite Relation For a particular host-parasite relationship, it is found that when the host density (number of hosts per unit of area) is x, the number of hosts that are parasitized is

$$y = 12\left(1 - \frac{1}{1+3x}\right) \quad x \ge 0$$

For what value of x does dy/dx equal $\frac{1}{4}$?

55. Bacteria Growth Bacteria are growing in a culture. The time t (in hours) for the number of bacteria to double in number (the generation time) is a function of the temperature T (in degrees Celsius) of the culture and is given by

$$t = f(T) = \begin{cases} \frac{1}{24}T + \frac{11}{4} & \text{if } 30 \le T \le 36 \\ \frac{4}{3}T - \frac{175}{4} & \text{if } 36 < T \le 39 \end{cases}$$

Find dt/dT when (a) $T = 38$ and (b) $T = 35$.

56. Motion The position function of a particle moving in a straight line is

$$s = \frac{9}{2t^2 + 3}$$

where t is in seconds and s is in meters. Find the velocity of the particle at $t = 1$.

57. Rate of Change The volume of a sphere is given by $V = \frac{1}{6}\pi d^3$, where d is the diameter. Find the rate of change of V with respect to d when $d = 4$ ft.

58. Motion The position function for a ball thrown vertically upward from the ground is

$$s = 218t - 16t^2$$

where s is the height in feet above the ground after t seconds. For what value(s) of t is the velocity 64 ft/s?

59. Find the marginal-cost function if the average-cost function is

$$\bar{c} = 2q + \frac{10{,}000}{q^2}$$

60. Find an equation of the tangent line to the curve

$$y = \frac{(x^3+2)\sqrt{x+1}}{x^4 + 2x}$$

at the point on the curve where $x = 1$.

61. A manufacturer has found that when m employees are working, the number of units of product produced per day is

$$q = 10\sqrt{m^2 + 3600} - 600$$

The demand equation for the product is

$$9q + p^2 - 7200 = 0$$

where p is the selling price when the demand for the product is q units per day.

(a) Determine the manufacturer's marginal-revenue product when $m = 80$.

(b) Find the relative rate of change of revenue with respect to the number of employees when $m = 80$.

(c) Suppose it would cost the manufacturer \$300 more per day to hire an additional employee. Would you advise the manufacturer to hire the 81st employee? Why?

---

[18] J. A. Nordin, "Note on a Light Plant's Cost Curves," *Econometrica* 15 (1947), 231–55.

[19] D. G. Embree, "The Population Dynamics of the Winter Moth in Nova Scotia, 1954–1962," *Memoirs of the Entomological Society of Canada* no. 46 (1965).

62. If $f(x) = x^2 \ln x$, use the "limit of a difference quotient" approach to estimate $f'(5)$. Round your answer to three decimal places.

63. If $f(x) = \sqrt[3]{x^2 + 3x - 4}$, use the numerical derivative feature of your graphing calculator to estimate the derivative when $x = 10$. Round your answer to three decimal places.

64. The total-cost function for a manufacturer is given by

$$c = \frac{5q^2 + 4}{\sqrt{q^2 + 6}} + 2500$$

where c is in dollars. Use the numerical derivative feature of your graphing calculator to estimate the marginal cost when 15 units are produced. Round your answer to the nearest cent.

65. If

$$y = (u + 3)\sqrt{u + 6}$$

and

$$u = \frac{x + 4}{x + 3}$$

find dy/dx when $x = 0.3$. Round your answer to two decimal places.

Marginal Propensity to Consume

A consumption function can be defined either for a nation, as in Section 11.5, or for an individual family. In either case, the function relates total consumption to total income. A savings function, similarly, relates total savings to total income, either at the national or at the family level.

Data about income, consumption, and savings for the United States as a whole can be found in the National Income and Product Accounts (NIPA) tables compiled by the Bureau of Economic Analysis, a division of the U.S. Department of Commerce. The tables are downloadable at www.bea.doc.gov. For the years 1959–1999, the national consumption function is indicated by the scatterplot in Figure 11.15.

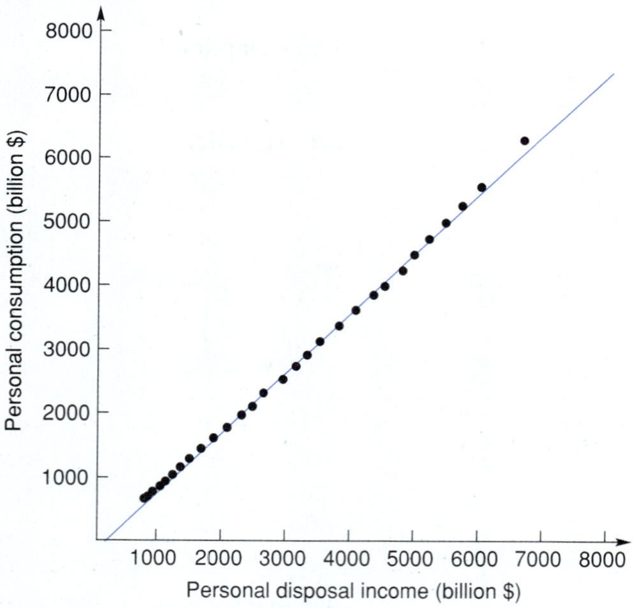

FIGURE 11.15 U.S. national consumption function.

Notice that the points lie more or less along a straight line. A linear regression gives the equation for this as $y = 0.9314x - 99.1936$.

The marginal propensity to consume derived from this graph is simply the slope of the line, that is, about 0.931 or 93.1%. At the national level, then, an increase of \$1 billion in total disposable income produces an increase of \$931 million in consumption. And if we assume that the rest is saved, there is an increase of \$69 million in total savings.[20]

Perhaps somewhat easier to relate to, because of the smaller numbers involved, is the consumption function for an individual household. This function is documented in Consumer Expenditure Surveys conducted by the Bureau of Labor Statistics, which is part of the U.S. Department of Labor. The survey results for each year can be downloaded at www.bls.gov/csxhome.htm.

Each year's survey gives information for five quintiles, as they are called, where a quintile represents one-fifth of American households. The quintiles are ordered by income, so that the bottom quintile represents the poorest 20% of Americans and the top quintile represents the richest 20%.

For the year 1999, income and consumption are as shown in Table 11.3. The numbers are average values within each quintile. If these data values are plotted using a graphing calculator, the points lie in a pattern that could be reasonably well approximated by a straight line but could be even better approximated by a curve—a curve shaped, qualitatively, like a square root function (Figure 11.16).

Most graphing calculators do not have a regression function for a square root–type function. They do,

[20]In reality, we must also account for interest payments and other outlays not counted as consumption. But we will ignore this complication from now on.

TABLE 11.3 U.S. family income and expenses, 1999

After-Tax Income	Total Expenses
$7101	$16,766
$17,576	$24,850
$30,186	$33,078
$48,607	$46,015
$98,214	$75,080

however, have a quadratic regression function—and the inverse of a quadratic function is a square root-type function. (Inverse functions were defined in Section 2.4.) So, we proceed as follows. First, using the statistics capabilities of a graphics calculator, enter the numbers in *second* column in Table 11.3 as x-values and those in the *first* column as y-values. Second, perform a quadratic regression. The function obtained is given by

$$y = (4.4627 \times 10^{-6})x^2 + 1.1517x - 13{,}461$$

Third, swap the lists of x- and y-values in preparation for plotting. Fourth, replace y with x and x with y in the quadratic regression equation and solve the result for y (using the quadratic formula) to obtain the equation

$$y = \frac{-1.1517 \pm \sqrt{1.1517^2 - 4(4.4627 \times 10^{-6})(-13{,}461 - x)}}{2(4.4627 \times 10^{-6})}$$

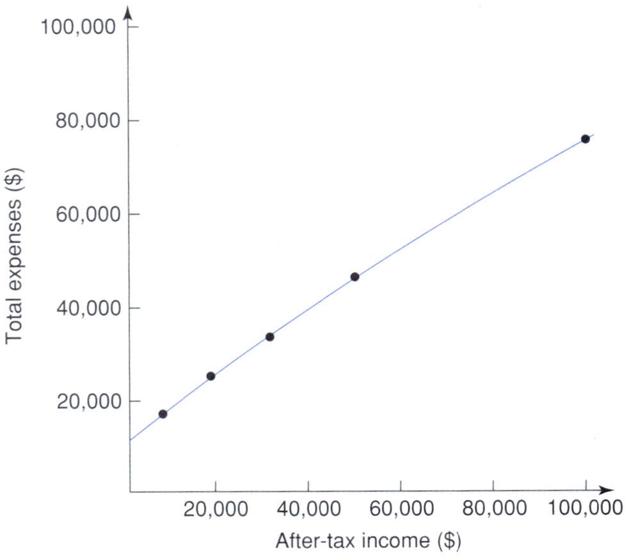

FIGURE 11.16 U.S. family consumption function.

or, more simply,

$$y = -129{,}036 \pm \sqrt{1.9667 \times 10^{10} + 224{,}080x}$$

Finally, enter the upper half of the curve (corresponding to the + part of the ± sign) as a function to be graphed; then display it together with a plot of the data. The result looks as shown in Figure 11.17.

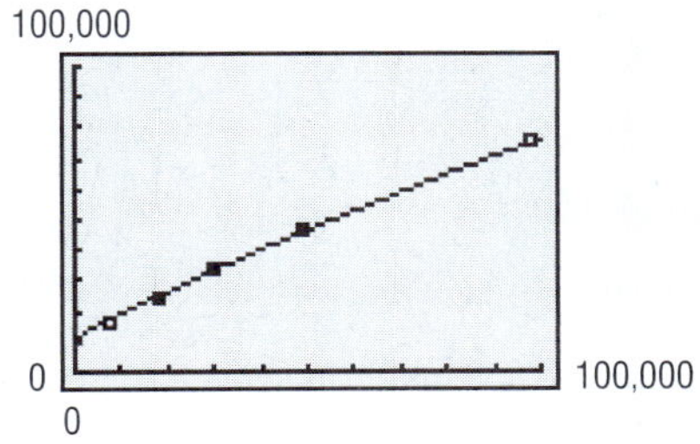

FIGURE 11.17 Graph of regression line.

To find the marginal consumption for a given income level, we now use the dy/dx function. To find the marginal consumption at $50,000, for instance, we select dy/dx, then enter 50000. The calculator returns the value 0.637675, which represents a marginal consumption of about 63.8%. In other words, a family earning $50,000 per year will, if given an extra $1000, spend $638 of it and save the rest.

Exercises

1. Compare the consumption function for Figure 11.15 with the consumption functions in Problems 63 and 64 of Section 11.5. How do these consumption functions differ significantly?
2. The first row in Table 11.3 has $7101 in the first column and $16,766 in the second column. What does this mean?
3. Suppose a family earning $25,000 per year in 1999 received an unexpected bonus check for $1000. How much of that check would you expect the family to spend? How much to save?
4. Suppose a family earning $90,000 per year in 1999 received an unexpected $1000 bonus check. How much would it spend?
5. What are the likely real-life reasons for the different answers in Problems 3 and 4?

CHAPTER

12

Additional Differentiation Topics

After an uncomfortable trip in a vehicle, passengers sometimes describe the ride as "jerky." But what is jerkiness, exactly? What does it mean for, say, an engineer designing a new transportation system?

Travel in a straight line at a constant speed is called *uniform motion,* and there is nothing jerky about it. But if either the path or the speed changes, the ride may become jerky. Change in speed over time is, formally, the derivative of velocity. Called acceleration, the change in velocity is the *second derivative* of position with respect to time—the derivative of the derivative of position. One of the important concepts covered in this chapter is that of a higher-order derivative, of which acceleration is an example.

But is acceleration responsible for jerkiness? The feeling of being jerked back and forth on a roller coaster is certainly related to acceleration. On the other hand, automotive magazines often praise a car for having *smooth* acceleration. So apparently acceleration has something to do with jerkiness but is not itself the cause.

The derivative of acceleration is the *third* derivative of position with respect to time. When this third derivative is large, the acceleration is changing rapidly. A roller coaster in a steady turn to the left is undergoing steady leftward acceleration. But when the coaster changes abruptly from a hard left turn to a hard right turn, the acceleration changes directions—and the riders experience a jerk. The third derivative of position is, in fact, so apt a measure of jerkiness that it is customarily called the jerk, just as the second derivative is called the acceleration.

Jerk has implications not only for passenger comfort in vehicles but also for equipment reliability. Engineers designing equipment for spacecraft, for instance, follow guidelines about the maximum jerk the equipment must be able to survive without damage to its internal components.

OBJECTIVE *To develop a differentiation formula for $y = \ln u$, to apply the formula, and to use it to differentiate a logarithmic function to a base other than e.*

12.1 Derivatives of Logarithmic Functions

In this section, we develop formulas for differentiating logarithmic functions. We begin with the derivative of $f(x) = \ln x$, where $x > 0$. By the definition of the derivative,

$$\frac{d}{dx}(\ln x) = \lim_{h\to 0} \frac{f(x+h) - f(x)}{h} = \lim_{h\to 0} \frac{\ln(x+h) - \ln x}{h}$$

Using the property of logarithms that $\ln m - \ln n = \ln(m/n)$, we have

$$\begin{aligned}\frac{d}{dx}(\ln x) &= \lim_{h\to 0} \frac{\ln\left(\dfrac{x+h}{x}\right)}{h} \\ &= \lim_{h\to 0}\left[\frac{1}{h}\ln\left(\frac{x+h}{x}\right)\right] = \lim_{h\to 0}\left[\frac{1}{h}\ln\left(1+\frac{h}{x}\right)\right]\end{aligned}$$

Writing $\dfrac{1}{h}$ as $\dfrac{1}{x}\cdot\dfrac{x}{h}$ gives

$$\begin{aligned}\frac{d}{dx}(\ln x) &= \lim_{h\to 0}\left[\frac{1}{x}\cdot\frac{x}{h}\ln\left(1+\frac{h}{x}\right)\right] \\ &= \lim_{h\to 0}\left[\frac{1}{x}\ln\left(1+\frac{h}{x}\right)^{x/h}\right] \qquad \text{(since } r\ln m = \ln m^r\text{)} \\ &= \frac{1}{x}\cdot\lim_{h\to 0}\left[\ln\left(1+\frac{h}{x}\right)^{x/h}\right]\end{aligned}$$

Because it can be shown that the limit of a logarithm is the logarithm of the limit ($\lim \ln u = \ln \lim u$), we have

$$\frac{d}{dx}(\ln x) = \frac{1}{x}\ln\left[\lim_{h\to 0}\left(1+\frac{h}{x}\right)^{x/h}\right] \tag{1}$$

To evaluate $\lim_{h\to 0}\left(1+\dfrac{h}{x}\right)^{x/h}$, first note that as $h \to 0$, then $\dfrac{h}{x} \to 0$. Thus, if we replace $\dfrac{h}{x}$ by k, the limit has the form

$$\lim_{k\to 0}(1+k)^{1/k}$$

As stated in Section 10.1, this limit is e. Hence, Equation (1) becomes

$$\frac{d}{dx}(\ln x) = \frac{1}{x}\ln e = \frac{1}{x}(1) = \frac{1}{x}$$

Therefore,

$$\frac{d}{dx}(\ln x) = \frac{1}{x} \tag{2}$$

EXAMPLE 1 **Differentiating Functions Involving ln *x***

a. *Differentiate* $f(x) = 5\ln x$.

Solution: Here f is a constant (5) times a function ($\ln x$), so by Equation (2), we have

$$f'(x) = 5\frac{d}{dx}(\ln x) = 5\cdot\frac{1}{x} = \frac{5}{x}$$

b. *Differentiate* $y = \frac{\ln x}{x^2}$.

Solution: By the quotient rule and Equation (2),

$$y' = \frac{x^2 \frac{d}{dx}(\ln x) - (\ln x)\frac{d}{dx}(x^2)}{(x^2)^2}$$

$$= \frac{x^2\left(\frac{1}{x}\right) - (\ln x)(2x)}{x^4} = \frac{x - 2x\ln x}{x^4} = \frac{1 - 2\ln x}{x^3}$$

▶ **Now work Problem 1.**

The chain rule is invaluable in developing the differentiation formula for ln u.

We shall now extend Equation (2) to cover a broader class of functions. Let $y = \ln u$, where u is a positive differentiable function of x. By the chain rule,

$$\frac{d}{dx}(\ln u) = \frac{dy}{du} \cdot \frac{du}{dx} = \frac{d}{du}(\ln u) \cdot \frac{du}{dx} = \frac{1}{u} \cdot \frac{du}{dx}$$

Thus,

$$\frac{d}{du}(\ln u) = \frac{1}{u} \cdot \frac{du}{dx} \qquad (3)$$

APPLICATIONS IN PRACTICE 1

Differentiating Functions Involving ln *u*

The supply of q units of a product at a price of p dollars per unit is given by $q(p) = 25 + 2\ln(3p^2 + 4)$. Find the rate of change of supply with respect to price, $\frac{dq}{dp}$.

EXAMPLE 2 Differentiating Functions Involving ln *u*

a. *Differentiate* $y = \ln(x^2 + 1)$.

Solution: This function has the form ln u with $u = x^2 + 1$. Using Equation (3) gives

$$\frac{dy}{dx} = \frac{1}{x^2+1}\frac{d}{dx}(x^2+1) = \frac{1}{x^2+1}(2x) = \frac{2x}{x^2+1}$$

b. *Differentiate* $y = x^2 \ln(4x + 2)$.

Solution: Using the product rule gives

$$\frac{dy}{dx} = x^2\frac{d}{dx}[\ln(4x+2)] + [\ln(4x+2)]\frac{d}{dx}(x^2)$$

By Equation (3) with $u = 4x + 2$,

$$\frac{dy}{dx} = x^2\left(\frac{1}{4x+2}\right)(4) + [\ln(4x+2)](2x)$$

$$= \frac{2x^2}{2x+1} + 2x\ln(4x+2)$$

c. *Differentiate* $y = \ln(\ln x)$.

Solution: This has the form $y = \ln u$ with $u = \ln x$. Using Equation (3), we obtain

$$y' = \frac{1}{\ln x}\frac{d}{dx}(\ln x) = \frac{1}{\ln x}\left(\frac{1}{x}\right) = \frac{1}{x\ln x}$$

▶ **Now work Problem 9.**

Frequently, we can reduce the work involved in differentiating the logarithm of a product, quotient, or power by using properties of logarithms to rewrite the logarithm *before* differentiating. The next example will illustrate.

EXAMPLE 3 Rewriting Logarithmic Functions before Differentiating

a. *Find* $\frac{dy}{dx}$ *if* $y = \ln(2x+5)^3$.

Solution: Here we have the logarithm of a power. First we simplify the right side by using properties of logarithms. Then we differentiate. We have

$$y = \ln(2x+5)^3 = 3\ln(2x+5)$$
$$\frac{dy}{dx} = 3\left(\frac{1}{2x+5}\right)(2) = \frac{6}{2x+5}$$

Comparing both methods, we note that the easier one is to simplify first and then differentiate.

Alternatively, if the simplification were not performed first, we would write

$$\frac{dy}{dx} = \frac{1}{(2x+5)^3}\frac{d}{dx}[(2x+5)^3]$$
$$= \frac{1}{(2x+5)^3}(3)(2x+5)^2(2) = \frac{6}{2x+5}$$

b. *Find* $f'(p)$ *if* $f(p) = \ln[(p+1)^2(p+2)^3(p+3)^4]$.

Solution: We simplify the right side and then differentiate:

$$f(p) = 2\ln(p+1) + 3\ln(p+2) + 4\ln(p+3)$$
$$f'(p) = 2\left(\frac{1}{p+1}\right)(1) + 3\left(\frac{1}{p+2}\right)(1) + 4\left(\frac{1}{p+3}\right)(1)$$
$$= \frac{2}{p+1} + \frac{3}{p+2} + \frac{4}{p+3}$$

▶ **Now work Problem 5.**

EXAMPLE 4 Differentiating Functions Involving Logarithms

a. *Find* $f'(w)$ *if* $f(w) = \ln\sqrt{\frac{1+w^2}{w^2-1}}$.

Solution: We simplify by using properties of logarithms and then differentiate:

$$f(w) = \frac{1}{2}[\ln(1+w^2) - \ln(w^2-1)]$$
$$f'(w) = \frac{1}{2}\left[\frac{1}{1+w^2}(2w) - \frac{1}{w^2-1}(2w)\right]$$
$$= \frac{w}{1+w^2} - \frac{w}{w^2-1} = -\frac{2w}{w^4-1}$$

b. *Find* $f'(x)$ *if* $f(x) = \ln^3(2x+5)$.

Solution: The exponent 3 refers to the cubing of $\ln(2x+5)$. That is,

$$f(x) = \ln^3(2x+5) = [\ln(2x+5)]^3$$

CAUTION Do not confuse $\ln^3(2x+5)$ with $\ln(2x+5)^3$, which occurred in Example 3(a). It is advisable to write $\ln^3(2x+5)$ explicitly as $[\ln(2x+5)]^3$ and avoid $\ln^3(2x+5)$ altogether.

By the power rule,

$$\begin{aligned} f'(x) &= 3[\ln(2x+5)]^2 \frac{d}{dx}[\ln(2x+5)] \\ &= 3[\ln(2x+5)]^2 \left[\frac{1}{2x+5}(2)\right] \\ &= \frac{6}{2x+5}[\ln(2x+5)]^2 \end{aligned}$$

▶ Now work Problem 39.

Derivatives of Logarithmic Functions to the Base *b*

To differentiate a logarithmic function to a base different from e, we can first convert the logarithm to natural logarithms via the change-of-base formula and then differentiate the resulting expression. For example, consider $y = \log_b u$, where u is a differentiable function of x. By the change-of-base formula,

$$y = \log_b u = \frac{\ln u}{\ln b}$$

Differentiating, we have

$$\frac{d}{dx}(\log_b u) = \frac{d}{dx}\left(\frac{\ln u}{\ln b}\right) = \frac{1}{\ln b}\frac{d}{dx}(\ln u) = \frac{1}{\ln b}\cdot\frac{1}{u}\frac{du}{dx}$$

Thus,

$$\frac{d}{dx}(\log_b u) = \frac{1}{(\ln b)u}\cdot\frac{du}{dx}$$

Rather than memorize this rule, we suggest that you remember the procedure used to obtain it.

Procedure to Differentiate $\log_b u$

Convert $\log_b u$ to natural logarithms to obtain $\frac{\ln u}{\ln b}$, and then differentiate.

EXAMPLE 5 Differentiating a Logarithmic Function to the Base 2

Differentiate $y = \log_2 x$.

Solution: Following the foregoing procedure, we have

$$\frac{d}{dx}(\log_2 x) = \frac{d}{dx}\left(\frac{\ln x}{\ln 2}\right) = \frac{1}{\ln 2}\frac{d}{dx}(\ln x) = \frac{1}{x\ln 2}$$

It is worth mentioning that we can write our answer in terms of the original base. Because

$$\frac{1}{\ln b} = \frac{1}{\frac{\log_b b}{\log_b e}} = \frac{\log_b e}{1} = \log_b e$$

we can express $\frac{1}{x\ln 2}$ as $\frac{1}{x}\log_2 e$.

▶ Now work Problem 15.

APPLICATIONS IN PRACTICE 2

Differentiating a Logarithmic Function to the Base 10

The intensity of an earthquake is measured on the Richter scale. The reading is given by $R = \log \frac{I}{I_0}$, where I is the intensity and I_0 is a standard minimum intensity. If $I_0 = 1$, find $\frac{dR}{dI}$, the rate of change of the Richter-scale reading with respect to the intensity.

EXAMPLE 6 Differentiating a Logarithmic Function to the Base 10

If $y = \log(2x+1)$, find the rate of change of y with respect to x.

Solution: The rate of change is dy/dx, and the base involved is 10. Therefore, we have

$$\frac{dy}{dx} = \frac{d}{dx}[\log(2x+1)] = \frac{d}{dx}\left[\frac{\ln(2x+1)}{\ln 10}\right]$$

$$= \frac{1}{\ln 10} \cdot \frac{1}{2x+1}(2) = \frac{2}{(2x+1)\ln 10}$$

Exercise 12.1

In Problems 1–44, differentiate the functions. If possible, first use properties of logarithms to simplify the given function.

*1. $y = 4\ln x$

2. $y = \frac{\ln x}{14}$

3. $y = \ln(3x-7)$

4. $y = \ln(5x-6)$

*5. $y = \ln x^2$

6. $y = \ln(3x^2+2x+1)$

7. $y = \ln(1-x^2)$

8. $y = \ln(-x^2+6x)$

*9. $f(p) = \ln(2p^3+3p)$

10. $f(r) = \ln(2r^4-3r^2+2r+1)$

11. $f(t) = t\ln t$

12. $y = x^2\ln x$

13. $y = x^3\ln(2x+5)$

14. $y = (2x+5)^2\ln(2x+5)$

*15. $y = \log_3(8x-1)$

16. $f(w) = \log(w^2+w)$

17. $y = x^2 + \log_2(x^2+4)$

18. $y = x^2\log_2 x$

19. $f(z) = \frac{\ln z}{z}$

20. $y = \frac{x^2}{\ln x}$

21. $y = \frac{x^2-1}{\ln x}$

22. $y = \ln x^{100}$

23. $y = \ln(x^2+4x+5)^3$

24. $y = 6\ln\sqrt[3]{x}$

25. $y = 9\ln\sqrt{1+x^2}$

26. $f(t) = \ln\left(\frac{t^4}{1+t+t^2}\right)$

27. $f(l) = \ln\left(\frac{1+l}{1-l}\right)$

28. $y = \ln\left(\frac{2x+3}{3x-4}\right)$

29. $y = \ln\sqrt[4]{\frac{1+x^2}{1-x^2}}$

30. $y = \ln\sqrt[3]{\frac{x^3-1}{x^3+1}}$

31. $y = \ln[(x^2+2)^2(x^3+x-1)]$

32. $y = \ln[(5x+2)^4(8x-3)^6]$

33. $y = 5\ln(x\sqrt{2x+1})$

34. $y = 6\ln\frac{x}{\sqrt{2x+1}}$

35. $y = (x^2+1)\ln(2x+1)$

36. $y = (ax+b)\ln(ax)$

37. $y = \ln x^3 + \ln^3 x$

38. $y = x^{\ln 2} + (\ln 2)^x$

*39. $y = \ln^4(ax)$

40. $y = \ln^2(2x+11)$

41. $y = x\ln\sqrt{x-1}$

42. $y = \ln(x^3\sqrt[4]{2x+1})$

43. $y = \sqrt{4+3\ln x}$

44. $y = \ln(x+\sqrt{1+x^2})$

45. Find an equation of the tangent line to the curve
$$y = \ln(x^2-2x-2)$$
when $x = 3$.

46. Find an equation of the tangent line to the curve
$$y = x[\ln(x)-1]$$
at the point where $x = e$.

47. Find the slope of the curve $y = \frac{x}{\ln x}$ when $x = 3$.

48. **Marginal Revenue** Find the marginal-revenue function if the demand function is $p = 25/\ln(q+2)$.

49. **Marginal Cost** A total-cost function is given by
$$c = 25\ln(q+1) + 12$$
Find the marginal cost when $q = 6$.

50. **Marginal Cost** A manufacturer's average-cost function, in dollars, is given by
$$\bar{c} = \frac{350}{\ln(q+2)}$$
Find the marginal cost (rounded to two decimal places) when $q = 40$.

51. **Supply Change** The supply of q units of a product at a price of p dollars per unit is given by $q(p) = 25 + 10\ln(2p+1)$. Find the rate of change of supply with respect to price, $\frac{dq}{dp}$.

52. **Sound Perception** The loudness of sound (L, measured in decibels) perceived by the human ear depends upon intensity levels (I) according to $L = 10\log\frac{I}{I_0}$, where I_0 is the standard threshold of audibility. If $I_0 = 17$, find $\frac{dL}{dI}$, the rate of change of the loudness with respect to the intensity.

53. **Biology** In a certain experiment with bacteria, it is observed that the relative activeness of a given bacteria

colony is described by

$$A = 6\ln\left(\frac{T}{a-T} - a\right)$$

where a is a constant and T is the surrounding temperature. Find the rate of change of A with respect to T.

54. Show that the relative rate of change of $y = f(x)$ with respect to x is equal to the derivative of $y = \ln f(x)$.

55. Show that $\frac{d}{dx}(\log_b u) = \frac{1}{u}(\log_b e)\frac{du}{dx}$.

In Problems 56 and 57, use differentiation rules to find $f'(x)$. Then use your graphing calculator to find all zeros of $f'(x)$. Round your answers to two decimal places.

56. $f(x) = x^3 \ln x$

57. $f(x) = \frac{\ln(2x)}{x}$

12.2 Derivatives of Exponential Functions

OBJECTIVE *To develop a differentiation formula for $y = e^u$, to apply the formula, and to use it to differentiate an exponential function with a base other than e.*

We now obtain a formula for the derivative of the exponential function

$$y = e^u$$

where u is a differentiable function of x. In logarithmic form, we have

$$u = \ln y$$

Differentiating both sides with respect to x gives

$$\frac{d}{dx}(u) = \frac{d}{dx}(\ln y)$$

$$\frac{du}{dx} = \frac{1}{y}\frac{dy}{dx}$$

Solving for dy/dx and then replacing y by e^u gives

$$\frac{dy}{dx} = y\frac{du}{dx} = e^u\frac{du}{dx}$$

Thus,

$$\frac{d}{dx}(e^u) = e^u\frac{du}{dx} \quad (1)$$

As a special case, let $u = x$. Then $du/dx = 1$, and

$$\frac{d}{dx}(e^x) = e^x \quad (2)$$

Note that the function and its derivative are the same.

CAUTION The power rule does not apply to e^x and other exponential functions, a^x. The power rule applies to power functions, x^a. Note the location of the variable.

EXAMPLE 1 Differentiating Functions Involving e^x

a. *Find* $\frac{d}{dx}(3e^x)$.

Solution: Since 3 is a constant factor,

$$\frac{d}{dx}(3e^x) = 3\frac{d}{dx}(e^x)$$
$$= 3e^x \quad \text{[by Equation (2)]}$$

b. *If* $y = \dfrac{x}{e^x}$, *find* $\dfrac{dy}{dx}$.

Solution: We *first* use the quotient rule and then use Equation (2):

$$\frac{dy}{dx} = \frac{e^x \frac{d}{dx}(x) - x\frac{d}{dx}(e^x)}{(e^x)^2} = \frac{e^x(1) - x(e^x)}{(e^x)^2} = \frac{e^x(1-x)}{e^{2x}} = \frac{1-x}{e^x}$$

c. *If* $y = e^2 + e^x + \ln 3$, *find* y'.

Solution: Since e^2 and $\ln 3$ are constants, $y' = 0 + e^x + 0 = e^x$.

▶ **Now work Problem 1.**

APPLICATIONS IN PRACTICE 1

Differentiating Functions Involving e^u

When an object is moved from one environment to another, the change in temperature of the object is given by $T = Ce^{kt}$, where C is the temperature difference between the two environments, t is the time in the new environment, and k is a constant. Find the rate of change of temperature with respect to time.

EXAMPLE 2 Differentiating Functions Involving e^u

a. *Find* $\dfrac{d}{dx}(e^{x^3+3x})$.

Solution: The function has the form e^u with $u = x^3 + 3x$. From Equation (1),

$$\frac{d}{dx}\left(e^{x^3+3x}\right) = e^{x^3+3x}\frac{d}{dx}(x^3+3x) = e^{x^3+3x}(3x^2+3)$$
$$= 3(x^2+1)e^{x^3+3x}$$

b. *Find* $\dfrac{d}{dx}[e^{x+1}\ln(x^2+1)]$.

Solution: By the product rule,

$$\frac{d}{dx}[e^{x+1}\ln(x^2+1)] = e^{x+1}\frac{d}{dx}[\ln(x^2+1)] + [\ln(x^2+1)]\frac{d}{dx}(e^{x+1})$$
$$= e^{x+1}\left(\frac{1}{x^2+1}\right)(2x) + [\ln(x^2+1)]e^{x+1}(1)$$
$$= e^{x+1}\left[\frac{2x}{x^2+1} + \ln(x^2+1)\right]$$

$\dfrac{d}{dx}(e^u) = e^u\dfrac{du}{dx}$. Don't forget the $\dfrac{du}{dx}$.

▶ **Now work Problem 3.**

EXAMPLE 3 The Normal-Distribution Density Function

An important function used in the social sciences is the ***normal-distribution density function***

$$y = f(x) = \frac{1}{\sigma\sqrt{2\pi}}e^{-(1/2)[(x-\mu)/\sigma]^2}$$

where σ (a Greek letter read "sigma") and μ (a Greek letter read "mu") are constants. The graph of this function, called the normal curve, is "bell shaped." (See Figure 12.1.) Determine the rate of change of y with respect to x when $x = \mu$.

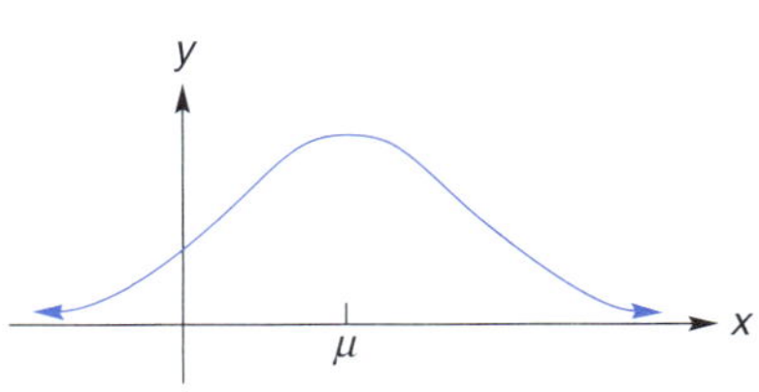

FIGURE 12.1 The normal-distribution density function.

Solution: The rate of change of y with respect to x is dy/dx. We note that the factor $\frac{1}{\sigma\sqrt{2\pi}}$ is a constant and the second factor has the form e^u, where

$$u = -\frac{1}{2}\left(\frac{x-\mu}{\sigma}\right)^2$$

Thus,

$$\frac{dy}{dx} = \frac{1}{\sigma\sqrt{2\pi}}\left[e^{-(1/2)[(x-\mu)/\sigma]^2}\right]\left[-\frac{1}{2}(2)\left(\frac{x-\mu}{\sigma}\right)\left(\frac{1}{\sigma}\right)\right]$$

Evaluating dy/dx when $x = \mu$, we obtain

$$\left.\frac{dy}{dx}\right|_{x=\mu} = 0$$

Geometrically, this means that the slope of the tangent line to the graph of $y = f(x)$ at $x = \mu$ is horizontal. (Refer back to Figure 12.1.)

Differentiating Exponential Functions to the Base *a*

Now that we are familiar with the derivative of e^u, we consider the derivative of the more general exponential function a^u. By replacing a with the equivalent form $e^{\ln a}$, we can express a^u as an exponential function with the base e, a form we can differentiate. We have

$$\begin{aligned}\frac{d}{dx}(a^u) &= \frac{d}{dx}[(e^{\ln a})^u] = \frac{d}{dx}(e^{u\ln a}) \\ &= e^{u\ln a}\frac{d}{dx}(u\ln a) \\ &= e^{u\ln a}\left(\frac{du}{dx}\right)\ln a \\ &= a^u(\ln a)\frac{du}{dx} \qquad \text{(since } e^{u\ln a} = a^u\text{)}\end{aligned}$$

Therefore,

$$\frac{d}{dx}(a^u) = a^u(\ln a)\frac{du}{dx} \tag{3}$$

Note that if $a = e$, then the factor $\ln a$ in Equation (3) is 1. Thus, if exponential functions to the base e are used, we have a simpler differentiation formula with which to work. This is the reason natural exponential functions are used extensively in calculus. Rather than memorizing Equation (3), we suggest that you remember the procedure for obtaining it.

Procedure to Differentiate a^u

Convert a^u to a natural exponential function by using the property that $a = e^{\ln a}$, and then differentiate.

The next example will illustrate this procedure.

EXAMPLE 4 **Differentiating an Exponential Function to the Base 4**

Verify the result by using Equation (3) directly.

Find $\frac{d}{dx}(4^x)$.

Solution: Using the preceding procedure, we have

$$\begin{aligned}\frac{d}{dx}(4^x) &= \frac{d}{dx}[(e^{\ln 4})^x] \\ &= \frac{d}{dx}[e^{(\ln 4)x}] && \left[\text{form: } \frac{d}{dx}(e^u)\right] \\ &= e^{(\ln 4)x}(\ln 4) && [\text{by Equation (1)}] \\ &= 4^x(\ln 4)\end{aligned}$$

▶ **Now work Problem 15.**

EXAMPLE 5 **Differentiating Different Forms**

Find $\frac{d}{dx}(e^2 + x^e + 2^{\sqrt{x}})$.

Solution: Here we must differentiate three different forms; do not confuse them! The first (e^2) is a constant base to a constant power, so it is a constant itself. Thus, its derivative is zero. The second (x^e) is a variable base to a constant power, so the power rule applies. The third ($2^{\sqrt{x}}$) is a constant base to a variable power, so we must differentiate an exponential function. Taken all together, we have

$$\begin{aligned}\frac{d}{dx}(e^2 + x^e + 2^{\sqrt{x}}) &= 0 + ex^{e-1} + \frac{d}{dx}\left[e^{(\ln 2)\sqrt{x}}\right] \\ &= ex^{e-1} + \left[e^{(\ln 2)\sqrt{x}}\right](\ln 2)\left(\frac{1}{2\sqrt{x}}\right) \\ &= ex^{e-1} + \frac{2^{\sqrt{x}}\ln 2}{2\sqrt{x}}\end{aligned}$$

▶ **Now work Problem 17.**

EXAMPLE 6 **Differentiating Inverse Functions**

We can apply the technique of this section to differentiate any inverse function f^{-1} once we know the derivative of f. Suppose that

$$y = f^{-1}(u)$$

where, as usual, u is a differentiable function of x and we wish to find dy/dx. In terms of f we have

$$f(y) = u$$

and differentiating both sides with respect to x we obtain

$$f'(y)\frac{dy}{dx} = \frac{du}{dx}$$

using the chain rule for the left side. Solving for dy/dx and replacing y by $f^{-1}(u)$ gives

$$\frac{dy}{dx} = \frac{1}{f'(f^{-1}(u))}\frac{du}{dx}$$

so that

$$\frac{d}{dx}(f^{-1}(u)) = \frac{1}{f'(f^{-1}(u))}\frac{du}{dx} \tag{4}$$

and in particular

$$\frac{d}{dx}(f^{-1}(x)) = \frac{1}{f'(f^{-1}(x))} \tag{5}$$

The reader may have noticed here, and in our treatment of the derivative of the exponential function, that we have assumed that f^{-1} is differentiable. The assumption can be avoided as it follows from the differentiability of f. A geometric plausibility argument can be given as follows. At each point $(x, f(x))$ on the graph of f there is a tangent line whose slope is $f'(x)$. Since the graph of f^{-1} is obtained from the graph of f by reflection in the line $y = x$, it seems clear that at each point $(x, f^{-1}(x)$ on the graph of f^{-1} there will also be a tangent line—whose slope is $(f^{-1})'(x)$.

As with the Chain Rule, Leibniz notation is well suited for inverse functions. Indeed, if $y = f^{-1}(x)$, then $\frac{d}{dx}(f^{-1}(x)) = \frac{dy}{dx}$. Now since $f(y) = x$, $\frac{dx}{dy} = f'(y)$ and we see that Equation 5 can be rewritten as:

$$\frac{dy}{dx} = \frac{1}{\frac{dx}{dy}} \tag{6}$$

Exercise 12.2

In Problems 1–28, differentiate the functions.

*1. $y = 5e^x$

2. $y = \dfrac{2e^x}{5}$

*3. $y = e^{x^2+4}$

4. $y = e^{2x^2+5}$

5. $y = e^{9-5x}$

6. $f(q) = e^{-q^3+6q-1}$

7. $f(r) = e^{3r^2+4r+4}$

8. $y = e^{x^2+6x^3+1}$

9. $y = xe^x$

10. $y = x^2e^{-x}$

11. $y = x^2e^{-x^2}$

12. $y = xe^{3x}$

13. $y = \dfrac{e^x + e^{-x}}{3}$

14. $y = \dfrac{e^x - e^{-x}}{e^x + e^{-x}}$

*15. $y = 4^{3x^2}$

16. $y = 2^x x^2$

*17. $f(w) = \dfrac{e^{2w}}{w^2}$

18. $y = e^{x-\sqrt{x}}$

19. $y = e^{1+\sqrt{x}}$

20. $y = (e^{2x} + 1)^3$

21. $y = x^5 - 5^x$

22. $f(z) = e^{1/z}$

23. $y = \dfrac{e^x - 1}{e^x + 1}$

24. $y = e^{2x}(x + 6)$

25. $y = \ln e^x$

26. $y = e^{-x}\ln x$

27. $y = e^{x\ln x}$

28. $y = \ln e^{4x+1}$

29. If $f(x) = ee^xe^{x^2}$, find $f'(-1)$.

30. If $f(x) = 5^{x^2\ln x}$, find $f'(1)$.

31. Find an equation of the tangent line to the curve $y = e^x$ when $x = -2$.

32. Find an equation of the tangent line to the curve $y = e^x$ at the point $(1, e)$.

For each of the demand equations in Problems 33 and 34, find the rate of change of price p with respect to quantity q. What is the rate of change for the indicated value of q?

33. $p = 15e^{-0.001q}$; $q = 500$

34. $p = 8e^{-3q/800}$; $q = 400$

In Problems 35 and 36, $\bar{c}$ is the average cost of producing q units of a product. Find the marginal-cost function and the marginal cost for the given values of q.

35. $\bar{c} = \dfrac{7000e^{q/700}}{q}$; $q = 350, q = 700$

36. $\bar{c} = \dfrac{850}{q} + 4000\dfrac{e^{(2q+6)/800}}{q}$; $q = 97, q = 197$

37. If $w = e^{x^3-4x} + x\ln(x-1)$ and $x = \frac{t+1}{t-1}$, find $\frac{dw}{dt}$ when $t = 3$.

38. If $f'(x) = x^3$ and $u = e^x$, show that

$$\frac{d}{dx}[f(u)] = e^{4x}$$

39. Determine the value of the positive constant c if

$$\frac{d}{dx}(c^x - x^c) = 0$$

when $x = 1$.

40. Calculate the relative rate of change of

$$f(x) = 10^{-x} + \ln(8+x) + 0.01e^{x-2}$$

when $x = 2$. Round your answer to four decimal places.

41. Production Run For a firm, the daily output on the tth day of a production run is given by

$$q = 500(1 - e^{-0.2t})$$

Find the rate of change of output q with respect to t on the tenth day.

42. Normal-Density Function For the normal-density function

$$f(x) = \frac{1}{\sqrt{2\pi}}e^{-x^2/2}$$

find $f'(1)$.

43. Population The population, in millions, of the greater Seattle area t years from 1970 is estimated by $P = 1.92e^{0.0176t}$. Show that $dP/dt = kP$, where k is a constant. This means that the rate of change of population at any time is proportional to the population at that time.

44. Market Penetration In a discussion of diffusion of a new process into a market, Hurter and Rubenstein[1] refer to an equation of the form

$$Y = k\alpha^{\beta^t}$$

where Y is the cumulative level of diffusion of the new process at time t and k, α, and β are positive constants. Verify their claim that

$$\frac{dY}{dt} = k\alpha^{\beta^t}(\beta^t \ln\alpha)\ln\beta$$

45. Finance After t years, the value S of a principal of P dollars invested at the annual rate of r compounded continuously is given by $S = Pe^{rt}$. Show that the relative rate of change of S with respect to t is r.

[1] A. P. Hurter, Jr., A. H. Rubenstein, et al., "Market Penetration by New Innovations: The Technological Literature," *Technological Forecasting and Social Change* 11 (1978), 197–221.

46. Predator-Prey Relationship In an article concerning predators and prey, Holling[2] refers to an equation of the form

$$y = K(1 - e^{-ax})$$

where x is the prey density, y is the number of prey attacked, and K and a are constants. Verify his statement that

$$\frac{dy}{dx} = a(K - y)$$

47. Earthquakes According to Richter,[3] the number of earthquakes of magnitude M or greater per unit of time is given by $N = 10^A 10^{-bM}$, where A and b are constants. Find dN/dM.

48. Psychology Short-term retention was studied by Peterson and Peterson.[4] The two researchers analyzed a procedure in which an experimenter verbally gave a subject a three-letter consonant syllable, such as CHJ, followed by a three-digit number, such as 309. The subject then repeated the number and counted backward by 3's, such as 309, 306, 303, ... After a period of time, the subject was signaled by a light to recite the three-letter consonant syllable. The time between the experimenter's completion of the last consonant to the onset of the light was called the *recall interval*. The time between the onset of the light and the completion of a response was referred to as *latency*. After many trials, it was determined that, for a recall interval of t seconds, the approximate proportion of correct recalls with latency below 2.83 seconds was

$$p = 0.89[0.01 + 0.99(0.85)^t]$$

(a) Find dp/dt and interpret your result.

(b) Evaluate dp/dt when $t = 2$. Round your answer to two decimal places.

49. Medicine Suppose a tracer, such as a colored dye, is injected instantly into the heart at time $t = 0$ and mixes uniformly with blood inside the heart. Let the initial concentration of the tracer in the heart be C_0, and assume that the heart has constant volume V. Also assume that, as fresh blood flows into the heart, the

[2] C. S. Holling, "Some Characteristics of Simple Types of Predation and Parasitism," *Canadian Entomologist* 91, no. 7 (1959), 385–98.

[3] C. F. Richter, *Elementary Seismology* (San Francisco: W. H. Freeman and Company, Publishers, 1958).

[4] L. R. Peterson and M. J. Peterson, "Short-Term Retention of Individual Verbal Items," *Journal of Experimental Psychology* 58 (1959), 193–98.

diluted mixture of blood and tracer flows out at the constant positive rate r. Then the concentration of the tracer in the heart at time t is given by

$$C(t) = C_0 e^{-(r/V)t}$$

Show that $dC/dt = (-r/V)C(t)$.

50. Medicine In Problem 49, suppose the tracer is injected at a constant rate R. Then the concentration at time t is

$$C(t) = \frac{R}{r}\left[1 - e^{-(r/V)t}\right]$$

(a) Find $C(0)$.

(b) Show that $\frac{dC}{dt} = \frac{R}{V} - \frac{r}{V}C(t)$.

51. Schizophrenia Several models have been used to analyze the length of stay in a hospital. For a particular group of schizophrenics, one such model is[5]

$$f(t) = 1 - e^{-0.008t}$$

where $f(t)$ is the proportion of the group that was discharged at the end of t days of hospitalization. Find the rate of discharge (the proportion discharged per day) at the end of 100 days. Round your answer to four decimal places.

52. Savings and Consumption A country's savings S (in billions of dollars) is related to its national income I (in billions of dollars) by the equation

$$S = \ln \frac{5}{3 + e^{-I}}$$

(a) Find the marginal propensity to consume as a function of income.

(b) To the nearest million dollars, what is the national income when the marginal propensity to save is $\frac{1}{8}$? [*Hint:* 1 billion = 1000 million.]

In Problems 53 and 54, use differentiation rules to find $f'(x)$. Then use your graphing calculator to find all real zeros of $f'(x)$. Round your answers to two decimal places.

53. $f(x) = e^{x^4 + 2x^2 - 4x}$

54. $f(x) = x - e^{-x}$

[5] W. W. Eaton and G. A. Whitmore, "Length of Stay as a Stochastic Process: A General Approach and Application to Hospitalization for Schizophrenia," *Journal of Mathematical Sociology* 5 (1977), 273–92.

12.3 Elasticity of Demand

OBJECTIVE *To give a mathematical analysis of the economic concept of elasticity.*

Elasticity of demand is a means by which economists measure how a change in the price of a product will affect the quantity demanded. That is, it refers to consumer response to price changes. Loosely speaking, elasticity of demand is the ratio of the resulting percentage change in quantity demanded to a given percentage change in price:

$$\frac{\text{percentage change in quantity}}{\text{percentage change in price}}$$

For example, if, for a price increase of 5%, quantity demanded were to decrease by 2%, we would loosely say that elasticity of demand is $-\frac{2}{5}$.

To be more general, suppose $p = f(q)$ is the demand function for a product. Consumers will demand q units at a price of $f(q)$ per unit and will demand $q + h$ units at a price of $f(q+h)$ per unit (Figure 12.2). The *percentage* change in quantity demanded from q to $q + h$ is

$$\frac{(q+h) - q}{q} \cdot 100 = \frac{h}{q} \cdot 100$$

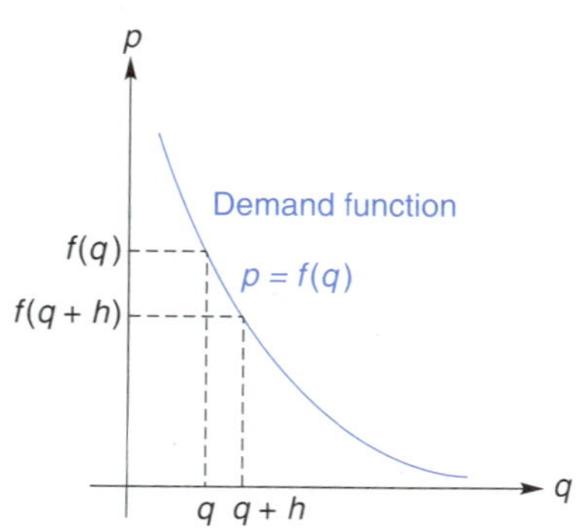

FIGURE 12.2 Change in demand.

The corresponding percentage change in price per unit is

$$\frac{f(q+h) - f(q)}{f(q)} \cdot 100$$

The ratio of these percentage changes is

$$\frac{\frac{h}{q}\cdot 100}{\frac{f(q+h)-f(q)}{f(q)}\cdot 100} = \frac{h}{q}\cdot\frac{f(q)}{f(q+h)-f(q)}$$

$$= \frac{f(q)}{q}\cdot\frac{h}{f(q+h)-f(q)}$$

$$= \frac{\frac{f(q)}{q}}{\frac{f(q+h)-f(q)}{h}} \tag{1}$$

If f is differentiable, then as $h \to 0$, the limit of $[f(q+h)-f(q)]/h$ is $f'(q) = dp/dq$. Thus, the limit of (1) is

$$\frac{\frac{f(q)}{q}}{\frac{dp}{dq}} = \frac{\frac{p}{q}}{\frac{dp}{dq}} \qquad \text{(since } p = f(q)\text{)}$$

which is called the *point elasticity of demand*.

Definition

If $p = f(q)$ is a differentiable demand function, the ***point elasticity of demand,*** denoted by the Greek letter η (eta), at (q, p) is given by

$$\eta = \eta(q) = \frac{\frac{p}{q}}{\frac{dp}{dq}}$$

Since p is a function of q, dp/dq is a function of q and thus the ratio that defines η is a function of q. That is why we write $\eta = \eta(q)$.

To illustrate, let us find the point elasticity of demand for the demand function $p = 1200 - q^2$. We have

$$\eta = \frac{\frac{p}{q}}{\frac{dp}{dq}} = \frac{\frac{1200-q^2}{q}}{-2q} = -\frac{1200-q^2}{2q^2} = -\left[\frac{600}{q^2} - \frac{1}{2}\right] \tag{2}$$

For example, if $q = 10$, then $\eta = -[(600/10^2) - \frac{1}{2}] = -5\frac{1}{2}$. Since

$$\eta \approx \frac{\text{\% change in demand}}{\text{\% change in price}}$$

we have

$$(\text{\% change in price})(\eta) \approx \text{\% change in demand}$$

Thus, if price were increased by 1% when $q = 10$, then quantity demanded would change by approximately

$$(1\%)\left(-5\frac{1}{2}\right) = -5\frac{1}{2}\%$$

That is, demand would decrease $5\frac{1}{2}$%. Similarly, decreasing price by $\frac{1}{2}$% when $q = 10$ results in a change in demand of approximately

$$\left(-\frac{1}{2}\%\right)\left(-5\frac{1}{2}\right) = 2\frac{3}{4}\%$$

Hence, demand increases by $2\frac{3}{4}$%.

Note that when elasticity is evaluated, no units are attached to it—it is nothing more than a real number. In fact, the 100's arising from the word percentage cancel, so that elasticity is really an approximation of the ratio

$$\frac{\text{relative change in quantity}}{\text{relative change in price}}$$

and each of the relative changes is no more than a real number. For usual behavior of demand, an increase (decrease) in price corresponds to a decrease (increase) in quantity. Thus, dp/dq will always be negative or 0, and η (where defined) will always be negative or 0. Some economists disregard the minus sign; in the preceding situation, they would consider the elasticity to be $5\frac{1}{2}$. We shall not adopt this practice.

There are three categories of elasticity:

1. When $|\eta| > 1$, demand is *elastic*.
2. When $|\eta| = 1$, demand has *unit elasticity*.
3. When $|\eta| < 1$, demand is *inelastic*.

For example, in Equation (2), since $|\eta| = 5\frac{1}{2}$ when $q = 10$, demand is elastic. If $q = 20$, then $|\eta| = |-[(600/20^2) - \frac{1}{2}]| = 1$ so demand has unit elasticity. If $q = 25$, then $|\eta| = |-\frac{23}{50}|$, and demand is inelastic.

Loosely speaking, for a given percentage change in price, there is a greater percentage change in quantity demanded if demand is elastic, a smaller percentage change if demand is inelastic, and an equal percentage change if demand has unit elasticity. To better understand elasticity, it is helpful to think of typical examples. Demand for an essential utility such as electricity tends to be inelastic through a wide range of prices. If electricity prices are increased by 10%, consumers can be expected to reduce their consumption somewhat but a full 10% decrease may not be possible if most of their electricity usage is for essentials of life such as heating and food preparation. On the other hand, demand for luxury goods tends to be quite elastic. A 10% increase in the price of jewelry, for example, may result in a 50% decrease in demand.

EXAMPLE 1 **Finding Point Elasticity of Demand**

Determine the point elasticity of the demand equation

$$p = \frac{k}{q} \qquad \textit{where } k > 0 \textit{ and } q > 0$$

Solution: From the definition, we have

$$\eta = \frac{\frac{p}{q}}{\frac{dp}{dq}} = \frac{\frac{k}{q^2}}{\frac{-k}{q^2}} = -1.$$

Thus, the demand has unit elasticity for all $q > 0$. The graph of $p = k/q$ is called an *equilateral hyperbola* and is often found in economics texts in discussions of elasticity. (See Figure 2.14 for a graph of such a curve.)

▸ **Now work Problem 1.**

If we are given $p = f(q)$ for our demand equation, as in our discussion thus far, then it is usually straightforward to calculate $dp/dq = f'(q)$. However, if instead we are given q as a function of p, then we will have $q = f^{-1}(p)$ and from Example 6 in Section 12.2,

$$\frac{dp}{dq} = \frac{1}{\frac{dq}{dp}}$$

It follows that

$$\eta = \frac{\frac{p}{q}}{\frac{dp}{dq}} = \frac{p}{q} \cdot \frac{dq}{dp} \tag{3}$$

provides another useful expression for η. Notice too that if $q = g(p)$, then

$$\eta = \eta(p) = \frac{p}{q} \cdot \frac{dq}{dp} = \frac{p}{g(p)} \cdot g'(p) = p \cdot \frac{g'(p)}{g(p)}$$

and thus

$$\text{elasticity} = \text{price} \cdot (\text{relative rate of change of quantity as a function of price}) \tag{4}$$

EXAMPLE 2 **Finding Point Elasticity of Demand**

Determine the point elasticity of the demand equation

$$q = p^2 - 40p + 400 \qquad \textit{where } q > 0$$

Solution: Here we have q given as a function of p and it is easy to see that $dq/dp = 2p - 40$. Thus,

$$\eta(p) = \frac{p}{q} \cdot \frac{dq}{dp} = \frac{p}{q(p)}(2p - 40)$$

For example, if $p = 15$, then $q = q(15) = 25$; hence, $\eta(15) = [15(-10)]/25 = -6$, so demand is elastic for $p = 15$.

▶ **Now work Problem 13.**

Here we analyze elasticity for linear demand.

Point elasticity for a *linear* demand equation is quite interesting. Suppose the equation has the form

$$p = mq + b \qquad \text{where } m < 0 \text{ and } b > 0$$

(See Figure 12.3.) We assume that $q > 0$; thus, $p < b$. The point elasticity of demand is

$$\eta = \frac{\frac{p}{q}}{\frac{dp}{dq}} = \frac{\frac{p}{q}}{m} = \frac{p}{mq} = \frac{p}{p - b}$$

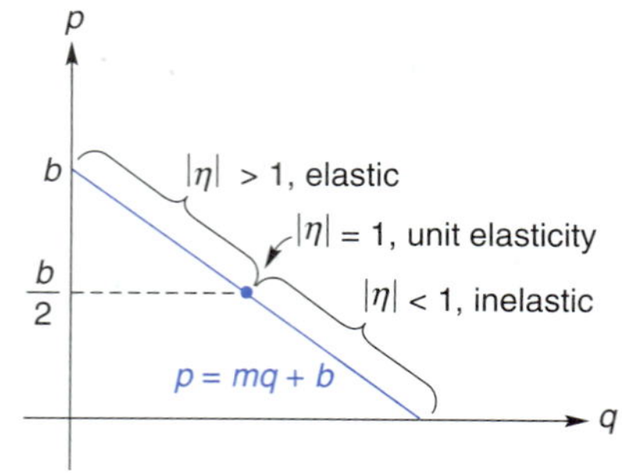

FIGURE 12.3 Elasticity for linear demand.

By considering $d\eta/dp$, we shall show that η is a decreasing function of p. By the quotient rule,

$$\frac{d\eta}{dp} = \frac{(p-b)-p}{(p-b)^2} = -\frac{b}{(p-b)^2}$$

Since $b > 0$ and $(p-b)^2 > 0$, it follows that $d\eta/dp < 0$, so η is a decreasing function of p: As p increases, η must decrease. However, p ranges between 0 and b, and at the midpoint of this range, $b/2$,

$$\eta = \frac{\frac{b}{2}}{\frac{b}{2}-b} = \frac{\frac{b}{2}}{-\frac{b}{2}} = -1$$

Therefore, if $p < \frac{b}{2}$, then $\eta > -1$; if $p > \frac{b}{2}$, then $\eta < -1$. Because we must have $\eta \le 0$, we can state these facts another way: When $p < \frac{b}{2}$, $|\eta| < 1$, and demand is inelastic; when $p = \frac{b}{2}$, $|\eta| = 1$, and demand has unit elasticity; when $p > \frac{b}{2}$, $|\eta| > 1$ and demand is elastic. This shows that the slope of a demand curve is not a measure of elasticity. The slope of the line in Figure 12.3 is m everywhere, but elasticity varies with the point on the line. Of course this is in accord with Equation (4).

Elasticity and Revenue

Here we analyze the relationship between elasticity and the rate of change of revenue.

Turning to a different situation, we can relate how elasticity of demand affects changes in revenue (marginal revenue). If $p = f(q)$ is a manufacturer's demand function, the total revenue is given by

$$r = pq$$

To find the marginal revenue, dr/dq, we differentiate r by using the product rule:

$$\frac{dr}{dq} = p + q\frac{dp}{dq}. \tag{5}$$

Factoring the right side of Equation (5), we have

$$\frac{dr}{dq} = p\left(1 + \frac{q}{p}\frac{dp}{dq}\right)$$

But

$$\frac{q}{p}\frac{dp}{dq} = \frac{\frac{dp}{dq}}{\frac{p}{q}} = \frac{1}{\eta}$$

Thus,

$$\frac{dr}{dq} = p\left(1 + \frac{1}{\eta}\right) \tag{6}$$

If demand is elastic, then $\eta < -1$, so $1 + \frac{1}{\eta} > 0$. If demand is inelastic, then $\eta > -1$, so $1 + \frac{1}{\eta} < 0$. We can assume that $p > 0$. From Equation (6) we can conclude that $dr/dq > 0$ on intervals for which demand is elastic. As we shall soon see, a function is increasing on intervals for which its derivative is positive and a function is decreasing on intervals for which its derivative is negative. Hence, total revenue r is increasing on intervals for which demand is elastic and total revenue is decreasing on intervals for which demand is inelastic.

Thus, we conclude from the preceding argument that as more units are sold, a manufacturer's total revenue increases if demand is elastic, but decreases if demand is inelastic. That is, if demand is elastic, a lower price will increase revenue. This means that a lower price will cause a large enough increase in demand to actually increase revenue. If demand is inelastic, a lower price will decrease revenue. For unit elasticity, a lower price leaves total revenue unchanged.

Exercise 12.3

In Problems 1–14, find the point elasticity of the demand equations for the indicated values of q or p, and determine whether demand is elastic, is inelastic, or has unit elasticity.

***1.** $p = 40 - 2q;\ q = 5$

2. $p = 10 - 0.04q;\ q = 100$

3. $p = \frac{3500}{q};\ q = 288$

4. $p = \frac{1000}{q^2};\ q = 156$

5. $p = \frac{500}{q+2};\ q = 104$

6. $p = \frac{800}{2q+1};\ q = 24$

7. $p = 150 - e^{q/100};\ q = 100$

8. $p = 100e^{-q/200};\ q = 200$

9. $q = 1200 - 150p;\ p = 4$

10. $q = 100 - p;\ p = 50$

11. $q = \sqrt{2500 - p};\ p = 900$

12. $q = \sqrt{2500 - p^2};\ p = 20$

***13.** $q = \frac{(p-100)^2}{2};\ p = 20$

14. $q = p^2 - 50p + 850;\ p = 20$

15. For the linear demand equation $p = 13 - 0.05q$, verify that demand is elastic when $p = 10$, is inelastic when $p = 3$, and has unit elasticity when $p = 6.50$.

16. For what value (or values) of q do the following demand equations have unit elasticity?

(a) $p = 26 - 0.10q$
(b) $p = 1200 - q^2$

17. The demand equation for a product is

$$q = 500 - 40p + p^2$$

where p is the price per unit (in dollars) and q is the quantity of units demanded (in thousands). Find the point elasticity of demand when $p = 15$. If this price of 15 is increased by $\frac{1}{2}$%, what is the approximate change in demand?

18. The demand equation for a certain product is

$$q = \sqrt{2500 - p^2}$$

where p is in dollars. Find the point elasticity of demand when $p = 30$, and use this value to compute the approximate percentage change in demand if the price of \$30 is decreased to \$28.50.

19. For the demand equation $p = 500 - 2q$, verify that demand is elastic and total revenue is increasing for $0 < q < 125$. Verify that demand is inelastic and total revenue is decreasing for $125 < q < 250$.

20. Verify that $\frac{dr}{dq} = p\left(1 + \frac{1}{\eta}\right)$ if $p = 50 - 3q$.

21. Repeat Problem 20 for $p = \frac{1000}{q^2}$.

22. Suppose $p = mq + b$ is a linear demand equation, where $m \neq 0$ and $b > 0$.

(a) Show that $\lim_{p \to b^-} \eta = -\infty$.
(b) Show that $\eta = 0$ when $p = 0$.

23. The demand equation for a manufacturer's product is

$$p = \frac{200}{\sqrt{6000 + 10q^2}}$$

(a) Verify that $q = 20$ when $p = 2$.
(b) Determine the point elasticity of demand when $p = 2$. Is demand elastic, is it inelastic, or does it have unit elasticity at this point?

(c) If the price when $p = 2$ is decreased by 2%, what is the approximate number of units by which demand changes?

(d) If the price when $p = 2$ is decreased by 2%, will the total revenue increase, decrease, or remain constant? Give a reason for your answer.

24. Given the demand equation $q^2(1 + p)^2 = p$, determine the point elasticity of demand when $p = 9$.

25. The demand equation for a product is

$$q = \frac{60}{p} + \ln(65 - p^3)$$

(a) Determine the point elasticity of demand when $p = 4$, and classify the demand as elastic, inelastic, or of unit elasticity at this price level.

(b) If the price is lowered by 2% (from \$4.00 to \$3.92), use the answer to part (a) to estimate the corresponding percentage change in quantity sold.

(c) Will the changes in part (b) result in an increase or decrease in revenue? Explain.

26. The demand equation for a manufacturer's product is

$$p = 50(151 - q)^{0.02\sqrt{q+19}}$$

(a) Find the value of dp/dq when 150 units are demanded.

(b) Using the result in part (a), determine the point elasticity of demand when 150 units are demanded. At this level, is demand elastic, inelastic, or of unit elasticity?

(c) Use the result in part (b) to approximate the price per unit if demand decreases from 150 to 140 units.

(d) If the current demand is 150 units, should the manufacturer increase or decrease price in order to increase revenue? (Justify your answer.)

27. A manufacturer of aluminum doors currently is able to sell 500 doors per week at a price of \$80 each. If the price were lowered to \$75 each, an additional 50 doors per week could be sold. Estimate the current elasticity of demand for the doors, and also estimate the current value of the manufacturer's marginal-revenue function.

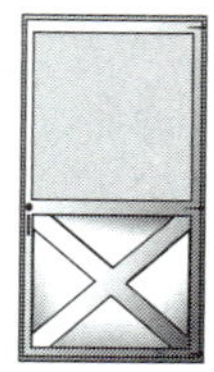

28. Given the demand equation

$$p = 1000 - q^2$$

where $5 \le q \le 30$, for what value of q is $|\eta|$ a maximum? For what value is it a minimum?

29. Repeat Problem 28 for

$$p = \frac{200}{q + 5}$$

such that $5 \le q \le 95$.

OBJECTIVE *To discuss the notion of a function defined implicitly and to determine derivatives by means of implicit differentiation.*

12.4 Implicit Differentiation

Implicit differentiation is a technique for differentiating functions that are not given in the usual form $y = f(x)$ (nor in the form $x = g(y)$). To introduce this technique, we shall find the slope of a tangent line to a circle. Let us take the circle of radius 2 whose center is at the origin (Figure 12.4). Its equation is

$$x^2 + y^2 = 4$$

$$x^2 + y^2 - 4 = 0 \qquad (1)$$

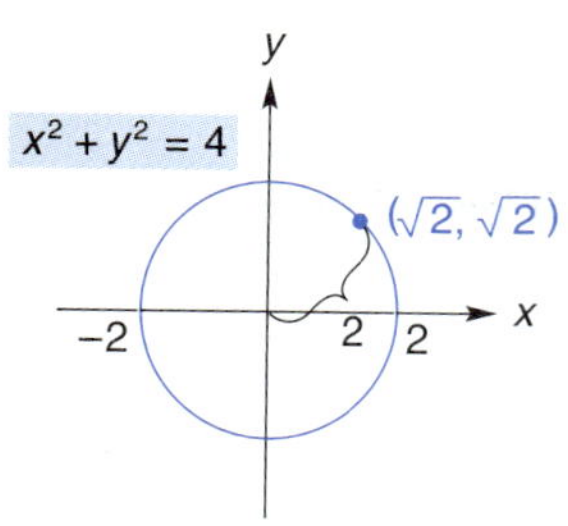

FIGURE 12.4 The circle $x^2 + y^2 = 4$.

The point $(\sqrt{2}, \sqrt{2})$ lies on the circle. To find the slope at this point, we need to find dy/dx there. Until now, we have always had y given explicitly (directly) in terms of x before determining y'; that is, our equation was always in the form $y = f(x)$. In Equation (1), this is not so. We say that Equation (1) has the form $F(x, y) = 0$, where $F(x, y)$ denotes a function of two variables. The obvious thing to do is solve Equation (1) for y in terms of x:

$$x^2 + y^2 - 4 = 0$$

$$y^2 = 4 - x^2$$

$$y = \pm\sqrt{4 - x^2} \qquad (2)$$

A problem now occurs: Equation (2) may give two values of y for a value of x. It does not define y explicitly as a function of x. We can, however, suppose that Equation (1) defines y as one of two different functions of x,

$$y = +\sqrt{4 - x^2} \quad \text{and} \quad y = -\sqrt{4 - x^2},$$

whose graphs are given in Figure 12.5. Since the point $(\sqrt{2}, \sqrt{2})$ lies on the graph of $y = \sqrt{4 - x^2}$, we should differentiate that function:

$$\begin{aligned} y &= \sqrt{4 - x^2} \\ \frac{dy}{dx} &= \frac{1}{2}(4 - x^2)^{-1/2}(-2x) \\ &= -\frac{x}{\sqrt{4 - x^2}} \end{aligned}$$

So
$$\left.\frac{dy}{dx}\right|_{x=\sqrt{2}} = -\frac{\sqrt{2}}{\sqrt{4-2}} = -1$$

Thus, the slope of the circle $x^2 + y^2 - 4 = 0$ at the point $(\sqrt{2}, \sqrt{2})$ is -1.

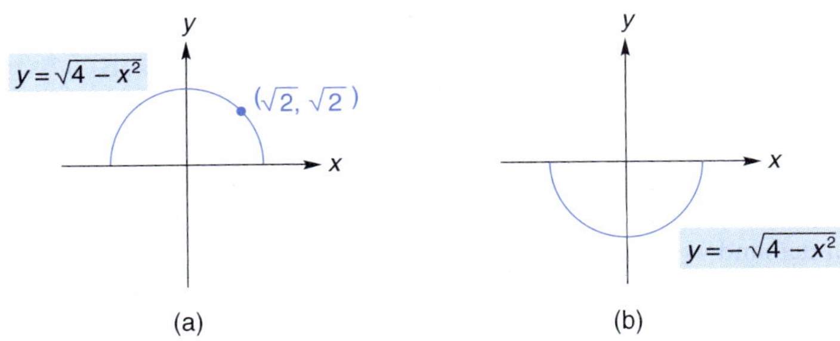

FIGURE 12.5 $x^2 + y^2 = 4$ gives rise to two different functions of x.

Let us summarize the difficulties we had. First, y was not originally given explicitly in terms of x. Second, after we tried to find such a relation, we ended up with more than one function of x. In fact, depending on the equation given, it may be very complicated or even impossible to find an explicit expression for y. For example, it would be difficult to solve $ye^x + \ln(x + y) = 0$ for y. We shall now consider a method that avoids such difficulties.

An equation of the form $F(x, y) = 0$, such as we had originally, is said to express y *implicitly* as a function of x. The word "implicitly" is used, since y is not given explicitly as a function of x. However, it is assumed or *implied* that the equation defines y as at least one differentiable function of x. Thus, we assume that Equation (1), $x^2 + y^2 - 4 = 0$, defines some differentiable function of x, say, $y = f(x)$. Next, we treat y as a function of x and differentiate both sides of Equation (1) with respect to x. Finally, we solve the result for dy/dx. Applying this procedure, we obtain

$$\frac{d}{dx}(x^2 + y^2 - 4) = \frac{d}{dx}(0)$$

$$\frac{d}{dx}(x^2) + \frac{d}{dx}(y^2) - \frac{d}{dx}(4) = \frac{d}{dx}(0) \quad (3)$$

We know that $\frac{d}{dx}(x^2) = 2x$ and that both $\frac{d}{dx}(4)$ and $\frac{d}{dx}(0)$ are 0. But $\frac{d}{dx}(y^2)$ is **not** $2y$, because we are differentiating with respect to x, not y. That is, y is not the independent variable. Since y is assumed to be a function of x, y^2

has the form u^n, where y plays the role of u. Just as the power rule states that $\frac{d}{dx}(u^2) = 2u\frac{du}{dx}$, we have $\frac{d}{dx}(y^2) = 2y\frac{dy}{dx}$. Hence, Equation (3) becomes

$$2x + 2y\frac{dy}{dx} = 0$$

Solving for dy/dx gives

$$2y\frac{dy}{dx} = -2x$$

$$\frac{dy}{dx} = -\frac{x}{y} \tag{4}$$

Notice that the expression for dy/dx involves the variable y as well as x. This means that to find dy/dx at a point, both coordinates of the point must be substituted into dy/dx. Thus,

$$\left.\frac{dy}{dx}\right|_{(\sqrt{2},\sqrt{2})} = -\frac{\sqrt{2}}{\sqrt{2}} = -1$$

as before. This method of finding dy/dx is called **implicit differentiation.** We note that Equation (4) is not defined when $y = 0$. Geometrically, this is clear, since the tangent line to the circle at either $(2, 0)$ or $(-2, 0)$ is vertical, and the slope is not defined.

Here are the steps to follow when differentiating implicitly:

Implicit Differentiation Procedure

For an equation that we assume defines y implicitly as a differentiable function of x, the derivative $\frac{dy}{dx}$ can be found as follows:

1. Differentiate both sides of the equation with respect to x.
2. Collect all terms involving $\frac{dy}{dx}$ on one side of the equation, and collect all other terms on the other side.
3. Factor $\frac{dy}{dx}$ from the side involving the $\frac{dy}{dx}$ terms.
4. Solve for $\frac{dy}{dx}$.

EXAMPLE 1 Implicit Differentiation

Find $\frac{dy}{dx}$ *by implicit differentiation if* $y + y^3 - x = 7$.

Solution: Here y is not given as an explicit function of x [that is, not in the form $y = f(x)$]. Thus, we assume that y is an implicit (differentiable) function of x and apply the preceding four-step procedure:

1. Differentiating both sides with respect to x, we have

$$\frac{d}{dx}(y + y^3 - x) = \frac{d}{dx}(7)$$

$$\frac{d}{dx}(y) + \frac{d}{dx}(y^3) - \frac{d}{dx}(x) = \frac{d}{dx}(7)$$

The derivative of y^3 with respect to x is $3y^2\frac{dy}{dx}$, not $3y^2$.

Now, $\frac{d}{dx}(y)$ can be written $\frac{dy}{dx}$, and $\frac{d}{dx}(x) = 1$. By the power rule,

$$\frac{d}{dx}(y^3) = 3y^2\frac{dy}{dx}$$

Hence, we obtain

$$\frac{dy}{dx} + 3y^2\frac{dy}{dx} - 1 = 0$$

2. Collecting all $\frac{dy}{dx}$ terms on the left side and all other terms on the right side gives

$$\frac{dy}{dx} + 3y^2\frac{dy}{dx} = 1$$

3. Factoring $\frac{dy}{dx}$ from the left side, we have

$$\frac{dy}{dx}(1 + 3y^2) = 1$$

In an implicit-differentiation problem, we are able to find the derivative of a function without knowing the function.

4. We solve for $\frac{dy}{dx}$ by dividing both sides by $1 + 3y^2$:

$$\frac{dy}{dx} = \frac{1}{1 + 3y^2}$$

Because step 4 of the process often involves division by an expression involving the variables, the answer obtained must often be restricted to exclude those values of the variables that would make the denominator zero. Here the denominator is always greater than or equal to 1, so there is no restriction.

▸ **Now work Problem 3.**

APPLICATIONS IN PRACTICE 1

Implicit Differentiation

Suppose that P, the proportion of people affected by a certain disease, is described by $\ln\left(\frac{P}{1-P}\right) = 0.5t$, where t is the time in months. Find $\frac{dP}{dt}$, the rate at which P grows with respect to time.

EXAMPLE 2 **Implicit Differentiation**

Find $\frac{dy}{dx}$ *if* $x^3 + 4xy^2 - 27 = y^4$.

Solution: Since y is not given explicitly in terms of x, we shall use the method of implicit differentiation:

1. Assuming that y is a function of x and differentiating both sides with respect to x, we get

$$\frac{d}{dx}(x^3 + 4xy^2 - 27) = \frac{d}{dx}(y^4)$$

$$\frac{d}{dx}(x^3) + 4\frac{d}{dx}(xy^2) - \frac{d}{dx}(27) = \frac{d}{dx}(y^4)$$

To find $\frac{d}{dx}(xy^2)$, we use the product rule:

$$3x^2 + 4\left[x\frac{d}{dx}(y^2) + y^2\frac{d}{dx}(x)\right] - 0 = 4y^3\frac{dy}{dx}$$

$$3x^2 + 4\left[x\left(2y\frac{dy}{dx}\right) + y^2(1)\right] = 4y^3\frac{dy}{dx}$$

$$3x^2 + 8xy\frac{dy}{dx} + 4y^2 = 4y^3\frac{dy}{dx}$$

2. Collecting $\frac{dy}{dx}$ terms on the left side and other terms on the right gives

$$8xy\frac{dy}{dx} - 4y^3\frac{dy}{dx} = -3x^2 - 4y^2$$

3. Factoring $\frac{dy}{dx}$ from the left side yields

$$\frac{dy}{dx}(8xy - 4y^3) = -3x^2 - 4y^2$$

4. Solving for $\frac{dy}{dx}$, we have

$$\frac{dy}{dx} = \frac{-3x^2 - 4y^2}{8xy - 4y^3} = \frac{3x^2 + 4y^2}{4y^3 - 8xy}$$

which gives the value of dy/dx at points (x, y) for which $4y^3 - 8xy \neq 0$.

▶ **Now work Problem 11.**

APPLICATIONS IN PRACTICE 2

Implicit Differentiation

The volume V of a spherical balloon is given by the equation $V = \frac{4}{3}\pi r^3$, where r is the radius of the balloon. If the radius is increasing at a rate of 5 inches/minute (that is, $\frac{dr}{dt} = 5$), then find $\frac{dV}{dt}$, the rate of increase of the volume of the balloon, when the radius is 12 inches.

EXAMPLE 3 Implicit Differentiation

Find the slope of the curve $x^3 = (y - x^2)^2$ *at* $(1, 2)$.

Solution: The slope at $(1, 2)$ is the value of dy/dx at that point. Finding dy/dx by implicit differentiation, we have

$$\frac{d}{dx}(x^3) = \frac{d}{dx}[(y - x^2)^2]$$

$$3x^2 = 2(y - x^2)\left(\frac{dy}{dx} - 2x\right)$$

$$3x^2 = 2\left(y\frac{dy}{dx} - 2xy - x^2\frac{dy}{dx} + 2x^3\right)$$

$$3x^2 = 2y\frac{dy}{dx} - 4xy - 2x^2\frac{dy}{dx} + 4x^3$$

$$3x^2 + 4xy - 4x^3 = 2y\frac{dy}{dx} - 2x^2\frac{dy}{dx}$$

$$3x^2 + 4xy - 4x^3 = 2\frac{dy}{dx}(y - x^2)$$

$$\frac{dy}{dx} = \frac{3x^2 + 4xy - 4x^3}{2(y - x^2)} \qquad \text{for } y - x^2 \neq 0$$

For the point $(1, 2)$, $y - x^2 = 2 - 1^2 = 1 \neq 0$. Thus, the slope of the curve at $(1, 2)$ is

$$\left.\frac{dy}{dx}\right|_{(1,2)} = \frac{3(1)^2 + 4(1)(2) - 4(1)^3}{2[2 - (1)^2]} = \frac{7}{2}$$

▶ **Now work Problem 25.**

APPLICATIONS IN PRACTICE 3

Implicit Differentiation

A 10-foot ladder is placed against a vertical wall. Suppose the bottom of the ladder slides away from the wall at a constant rate of 3 ft/s. (That is, $\frac{dx}{dt} = 3$.) How fast is the top of the ladder sliding down the wall when the top of the ladder is 8 feet (that is, when $y = 8$) from the ground? (That is, what is $\frac{dy}{dt}$?) (Use the Pythagorean theorem for right triangles, $x^2 + y^2 = z^2$, where x and y are the legs of the triangle and z is the hypotenuse.)

EXAMPLE 4 Implicit Differentiation

If $q - p = \ln q + \ln p$, find dq/dp.

Solution: We assume that q is a function of p and differentiate both sides with respect to p:

$$\frac{d}{dp}(q) - \frac{d}{dp}(p) = \frac{d}{dp}(\ln q) + \frac{d}{dp}(\ln p)$$

$$\frac{dq}{dp} - 1 = \frac{1}{q}\frac{dq}{dp} + \frac{1}{p}$$

$$\frac{dq}{dp} - \frac{1}{q}\frac{dq}{dp} = \frac{1}{p} + 1$$

$$\frac{dq}{dp}\left(1 - \frac{1}{q}\right) = \frac{1}{p} + 1$$

$$\frac{dq}{dp}\left(\frac{q-1}{q}\right) = \frac{1+p}{p}$$

$$\frac{dq}{dp} = \frac{(1+p)q}{p(q-1)} \qquad \text{for } p(q-1) \neq 0$$

▶ **Now work Problem 19.**

Exercise 12.4

In Problems 1–24, find dy/dx by implicit differentiation.

1. $x^2 + 4y^2 = 4$

2. $3x^2 + 6y^2 = 1$

*__3.__ $2y^3 - 7x^2 = 5$

4. $2x^2 - 3y^2 = 4$

5. $\sqrt{x} + \sqrt{y} = 3$

6. $x^{1/5} + y^{1/5} = 4$

7. $x^{3/4} + y^{3/4} = 5$

8. $y^3 = 4x$

9. $xy = 4$

10. $x^2 + xy - 2y^2 = 0$

*__11.__ $xy - y - 11x = 5$

12. $x^2 + y^2 = 2xy + 3$

13. $2x^3 + y^3 - 12xy = 0$

14. $2x^3 + 3xy + y^3 = 0$

15. $x = \sqrt{y} + \sqrt[4]{y}$

16. $x^3y^3 + x = 9$

17. $3x^2y^3 - x + y = 25$

18. $y^2 + y = \ln x$

*__19.__ $y \ln x = xe^y$

20. $\ln(xy) + x = 4$

21. $xe^y + y = 13$

22. $4x^2 + 9y^2 = 16$

23. $(1 + e^{3x})^2 = 3 + \ln(x + y)$

24. $y^2 = \ln(x + y)$

*__25.__ If $x + xy + y^2 = 7$, find dy/dx at (1, 2).

26. If $x\sqrt{y+1} = y\sqrt{x+1}$, find dy/dx at (3, 3).

27. Find the slope of the curve $4x^2 + 9y^2 = 1$ at the point $(0, \frac{1}{3})$; at the point (x_0, y_0).

28. Find the slope of the curve $(x^2 + y^2)^2 = 4y^2$ at the point (0, 2).

29. Find an equation of the tangent line to the curve of

$$x^3 + y^2 = 3$$

at the point (−1, 2).

30. Repeat Problem 29 for the curve

$$y^2 + xy - x^2 = 5$$

at the point (4, 3).

For the demand equations in Problems 31–34, find the rate of change of q with respect to p.

31. $p = 100 - q^2$

32. $p = 400 - \sqrt{q}$

33. $p = \dfrac{20}{(q+5)^2}$

34. $p = \dfrac{10}{q^2 + 3}$

35. Radioactivity The relative activity I/I_0 of a radioactive element varies with elapsed time according to the equation

$$\ln\left(\frac{I}{I_0}\right) = -\lambda t$$

where λ (a Greek letter read "lambda") is the disintegration *constant* and I_0 is the initial intensity (a constant). Find the rate of change of the intensity I with respect to the elapsed time t.

36. Earthquakes The magnitude M of an earthquake and its energy E are related by the equation[6]

$$1.5M = \log\left(\frac{E}{2.5 \times 10^{11}}\right)$$

Here M is given in terms of Richter's preferred scale of 1958 and E is in ergs. Determine the rate of change of energy with respect to magnitude.

37. Physical Scale The relationship between the speed (v), frequency (f), and wavelength (λ) of any wave is given by

$$v = f\lambda$$

Find $df/d\lambda$ by differentiating implicitly. (Treat v as a constant.) Then show that the same result is obtained if you first solve the equation for f and then differentiate with respect to λ.

38. Biology The equation $(P + a)(v + b) = k$ is called the "fundamental equation of muscle contraction."[7] Here P is the load imposed on the muscle, v is the velocity of the shortening of the muscle fibers, and a, b, and k are positive constants. Use implicit differentiation to show that, in terms of P,

$$\frac{dv}{dP} = -\frac{k}{(P + a)^2}$$

[6]K. E. Bullen, *An Introduction to the Theory of Seismology* (Cambridge, U.K.: Cambridge at the University Press, 1963).

[7]R. W. Stacy et al., *Essentials of Biological and Medical Physics* (New York: McGraw-Hill Book Company, 1955).

39. Marginal Propensity to Consume A country's savings S is defined implicitly in terms of its national income I by the equation

$$S^2 + \frac{1}{4}I^2 = SI + I$$

where both S and I are in billions of dollars. Find the marginal propensity to *consume* when $I = 16$ and $S = 12$.

40. Technological Substitution New products or technologies often tend to replace old ones. For example, today most commercial airlines use jet engines rather than prop engines. In discussing the forecasting of technological substitution, Hurter and Rubenstein[8] refer to the equation

$$\ln\frac{f(t)}{1 - f(t)} + \sigma\frac{1}{1 - f(t)} = C_1 + C_2 t$$

where $f(t)$ is the market share of the substitute over time t and C_1, C_2, and σ (a Greek letter read "sigma") are constants. Verify their claim that the rate of substitution is

$$f'(t) = \frac{C_2 f(t)[1 - f(t)]^2}{\sigma f(t) + [1 - f(t)]}$$

[8]A. P. Hurter, Jr., A. H. Rubenstein, et al., "Market Penetration by New Innovations: The Technological Literature," *Technological Forecasting and Social Change* 11 (1978), 197–221.

OBJECTIVE *To describe the method of logarithmic differentiation and to show how to differentiate a function of the form u^v.*

12.5 Logarithmic Differentiation

A technique called **logarithmic differentiation** often simplifies the differentiation of $y = f(x)$ when $f(x)$ involves products, quotients, or powers. The procedure is as follows:

Logarithmic Differentiation

To differentiate $y = f(x)$,

1. Take the natural logarithm of both sides. This results in
$$\ln y = \ln[f(x)].$$
2. Simplify $\ln[f(x)]$ by using properties of logarithms.
3. Differentiate both sides with respect to x.
4. Solve for $\frac{dy}{dx}$.
5. Express the answer in terms of x only. This requires substituting $f(x)$ for y.

The next example illustrates this procedure.

EXAMPLE 1 **Logarithmic Differentiation**

Find y' *if* $y = \dfrac{(2x-5)^3}{x^2\sqrt[4]{x^2+1}}$.

Solution: Differentiating this function in the usual way is messy because it involves the quotient, power, and product rules. Logarithmic differentiation makes the work less of a chore.

1. We take the natural logarithm of both sides:

$$\ln y = \ln \frac{(2x-5)^3}{x^2\sqrt[4]{x^2+1}}$$

2. Simplifying by using properties of logarithms, we have

$$\begin{aligned} \ln y &= \ln(2x-5)^3 - \ln[x^2\sqrt[4]{x^2+1}] \\ &= \ln(2x-5)^3 - [\ln x^2 + \ln(x^2+1)^{1/4}] \\ \ln y &= 3\ln(2x-5) - 2\ln x - \frac{1}{4}\ln(x^2+1) \end{aligned}$$

3. Differentiating with respect to x gives

Since y is a function of x, differentiating $\ln y$ with respect to x gives $\frac{1}{y}y'$.

$$\begin{aligned} \frac{1}{y}y' &= 3\left(\frac{1}{2x-5}\right)(2) - 2\left(\frac{1}{x}\right) - \frac{1}{4}\left(\frac{1}{x^2+1}\right)(2x) \\ \frac{y'}{y} &= \frac{6}{2x-5} - \frac{2}{x} - \frac{x}{2(x^2+1)} \end{aligned}$$

4. Solving for y' yields

$$y' = y\left[\frac{6}{2x-5} - \frac{2}{x} - \frac{x}{2(x^2+1)}\right]$$

5. Substituting the original expression for y gives y' in terms of x only:

$$y' = \frac{(2x-5)^3}{x^2\sqrt[4]{x^2+1}}\left[\frac{6}{2x-5} - \frac{2}{x} - \frac{x}{2(x^2+1)}\right]$$

▶ **Now work Problem 1.**

Logarithmic differentiation can also be used to differentiate a function of the form $y = u^v$, where both u and v are differentiable functions of x. Because neither the base nor the exponent is necessarily a constant, the differentiation techniques for u^n and a^u do not apply here.

EXAMPLE 2 **Differentiating the Form u^v**

Differentiate $y = x^x$ *by using logarithmic differentiation.*

Solution: This equation has the form $y = u^v$, where u and v are functions of x. Taking the natural logarithm of both sides gives $\ln y = \ln x^x$ or

$$\ln y = x\ln x$$

Differentiating both sides with respect to x yields

$$\frac{1}{y}y' = x\left(\frac{1}{x}\right) + (\ln x)(1),$$

$$\frac{y'}{y} = 1 + \ln x$$

Solving for y' and substituting x^x for y, we obtain

$$y' = y(1 + \ln x) = x^x(1 + \ln x)$$

It is worthwhile mentioning that an alternative technique for differentiating a function of the form $y = u^v$ is to convert it to an exponential function to the base e. To illustrate, for the function in Example 2, we have

$$y = x^x = (e^{\ln x})^x = e^{x \ln x}$$

$$y' = e^{x \ln x}\left(x \cdot \frac{1}{x} + \ln x\right)$$

$$= x^x(1 + \ln x)$$

▶ **Now work Problem 15.**

EXAMPLE 3 **Differentiating the Form u^v**

Find the derivative of $y = (1 + e^x)^{\ln x}$.

Solution: This has the form $y = u^v$, where $u = 1 + e^x$ and $v = \ln x$. Using logarithmic differentiation, we have

$$\ln y = \ln[(1 + e^x)^{\ln x}]$$

$$\ln y = (\ln x)\ln(1 + e^x)$$

$$\frac{1}{y}y' = (\ln x)\left[\frac{1}{1 + e^x} \cdot e^x\right] + [\ln(1 + e^x)]\left(\frac{1}{x}\right)$$

$$\frac{1}{y}y' = \frac{e^x \ln x}{1 + e^x} + \frac{\ln(1 + e^x)}{x}$$

$$y' = y\left[\frac{e^x \ln x}{1 + e^x} + \frac{\ln(1 + e^x)}{x}\right]$$

$$y' = (1 + e^x)^{\ln x}\left[\frac{e^x \ln x}{1 + e^x} + \frac{\ln(1 + e^x)}{x}\right]$$

▶ **Now work Problem 17.**

Alternatively, we can differentiate even a general function of the form $y = u(x)^{v(x)}$ with $u(x) > 0$ by using the equation

$$u^v = e^{v \ln u}$$

Indeed if $y = u(x)^{v(x)} = e^{v(x) \ln u(x)}$ for $u(x) > 0$, then

$$\frac{dy}{dx} = \frac{d}{dx}\left(e^{v(x) \ln u(x)}\right) = e^{v(x) \ln u(x)} \cdot \frac{d}{dx}(v(x) \ln u(x))$$

$$= u^v \cdot \frac{d}{dx}\left(v'(x) \ln u(x) + v(x)\frac{u'(x)}{u(x)}\right)$$

which could be summarized as

$$(u^v)' = u^v\left(v'\ln u + v\frac{u'}{u}\right)$$

As is often the case, there is no suggestion that you should memorize the last formula. The point here is that we have shown *any* function of the form u^v can be differentiated using the equation $u^v = e^{v\ln u}$. The same result will be obtained from logarithmic differentiation:

$$\ln y = \ln(u^v)$$

$$\ln y = v\ln u$$

$$\frac{y'}{y} = v'\ln u + v\frac{u'}{u}$$

$$y' = y\left(v'\ln u + v\frac{u'}{u}\right)$$

$$(u^v)' = u^v\left(v'\ln u + v\frac{u'}{u}\right)$$

After completing this section, you should understand how to differentiate each of the following forms:

$$y = \begin{cases} [f(x)]^n & \text{(a)} \\ a^{f(x)} & \text{(b)} \\ [f(x)]^{g(x)} & \text{(c)} \end{cases}$$

In form (a), you may utilize the power rule. In form (b), use the differentiation formula for exponential functions. [If $a \neq e$, you can first convert $a^{f(x)}$ to an e^u function.] In form (c), use logarithmic differentiation or first convert to an e^u function. Do not apply a rule in a situation where the rule does not apply. For example, the power rule does not apply to x^x.

Exercise 12.5

In Problems 1–12, find y' by using logarithmic differentiation.

*1. $y = (x+1)^2(x-2)(x^2+3)$

2. $y = (3x+4)(8x-1)^2(3x^2+1)^4$

3. $y = (3x^3-1)^2(2x+5)^3$

4. $y = (2x^2+1)\sqrt{8x^2-1}$

5. $y = \sqrt{x+1}\sqrt{x^2-2}\sqrt{x+4}$

6. $y = (x+2)\sqrt{x^2+9}\sqrt[3]{6x+1}$

7. $y = \dfrac{\sqrt{1-x^2}}{1-2x}$

8. $y = \sqrt{\dfrac{x^2+5}{x+9}}$

9. $y = \dfrac{(2x^2+2)^2}{(x+1)^2(3x+2)}$

10. $y = \dfrac{x(1+x^2)^2}{\sqrt{2+x^2}}$

11. $y = \sqrt{\dfrac{(x+3)(x-2)}{2x-1}}$

12. $y = \sqrt[3]{\dfrac{6(x^3+1)^2}{x^6e^{-4x}}}$

In Problems 13–20, find y'.

13. $y = x^{2x+1}$

14. $y = (2x)^{\sqrt{x}}$

*15. $y = x^{1/x}$

16. $y = \left(\dfrac{3}{x^2}\right)^x$

*17. $y = (3x+1)^{2x}$

18. $y = 4x^{x^2}$

19. $y = 4e^x x^{3x}$

20. $y = (\ln x)^{e^x}$

21. If $y = (4x-3)^{2x+1}$, find dy/dx when $x = 1$.

22. If $y = (\ln x)^{\ln x}$, find dy/dx when $x = e$.

23. Find an equation of the tangent line to

$$y = (x+1)(x+2)^2(x+3)^2$$

at the point where $x = 0$.

24. Find an equation of the tangent line to the graph of
$$y = x^x$$
at the point where $x = 1$.

25. Find an equation of the tangent line to the graph of
$$y = 3e^x(x^2 - x + 1)^x$$
at the point where $x = 1$.

26. If $y = x^x$, find the relative rate of change of y with respect to x when $x = 1$.

27. If $y = (3x)^{-2x}$, find the value of x for which the *percentage* rate of change of y with respect to x is 60.

28. Suppose $f(x)$ is a positive differentiable function and g is a differentiable function and $y = [f(x)]^{g(x)}$. Use logarithmic differentiation to show that
$$\frac{dy}{dx} = [f(x)]^{g(x)}\left(f'(x)\frac{g(x)}{f(x)} + g'(x)\ln[f(x)]\right)$$

OBJECTIVE *To approximate real roots of an equation by using calculus. The method shown is suitable for calculators.*

12.6 Newton's Method

It is quite easy to solve equations of the form $f(x) = 0$ when f is a linear or quadratic function. For example, we can solve $x^2 + 3x - 2 = 0$ by the quadratic formula. However, if $f(x)$ has degree greater than two or is not a polynomial, it may be difficult, or even impossible, to find solutions (or roots) of $f(x) = 0$ by the methods to which you are accustomed. For this reason, we may settle for approximate solutions, which can be obtained in a variety of efficient ways. For example, a graphing calculator may be used to estimate the real roots of $f(x) = 0$. In this section, you will learn how the derivative may be so used (provided that f is differentiable). The procedure we shall develop, called *Newton's method,* is well suited to a calculator or computer.

Newton's method requires an initial estimate for a root of $f(x) = 0$. One way of obtaining this estimate is by making a rough sketch of the graph of $y = f(x)$ and estimating the root from the graph. A point on the graph where $y = 0$ is an x-intercept, and the x-value of this point is a root of $f(x) = 0$. Another way of locating a root is based on the following fact:

If f is continuous on the interval $[a, b]$ and $f(a)$ and $f(b)$ have opposite signs, then the equation $f(x) = 0$ has at least one real root between a and b.

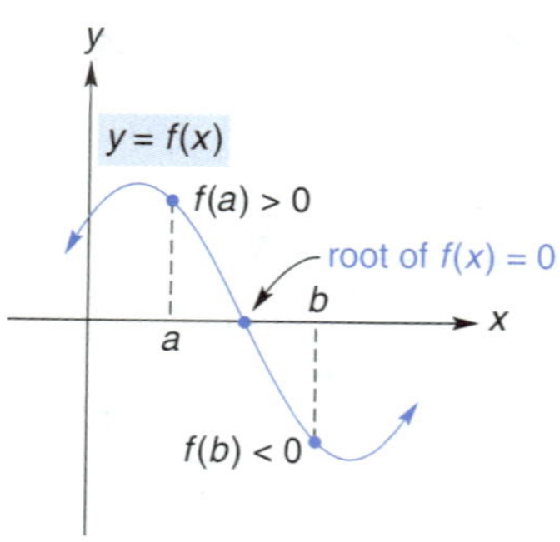

FIGURE 12.6 Root of $f(x) = 0$ between a and b, where $f(a)$ and $f(b)$ have opposite signs.

Figure 12.6 depicts this situation. The x-intercept between a and b corresponds to a root of $f(x) = 0$, and we can use either a or b to approximate this root.

Assuming that we have an estimated (but incorrect) value for a root, we turn to a way of getting a better approximation. In Figure 12.7, you can see that $f(r) = 0$, so r is a root of the equation $f(x) = 0$. Suppose x_1 is an initial approximation to r (and one that is close to r). Observe that the tangent line to the curve at $(x_1, f(x_1))$ intersects the x-axis at the point $(x_2, 0)$, and x_2 is a better approximation to r than is x_1.

We can find x_2 from the equation of the tangent line. The slope of the tangent line is $f'(x_1)$, so a point–slope form for this line is
$$y - f(x_1) = f'(x_1)(x - x_1) \qquad (1)$$

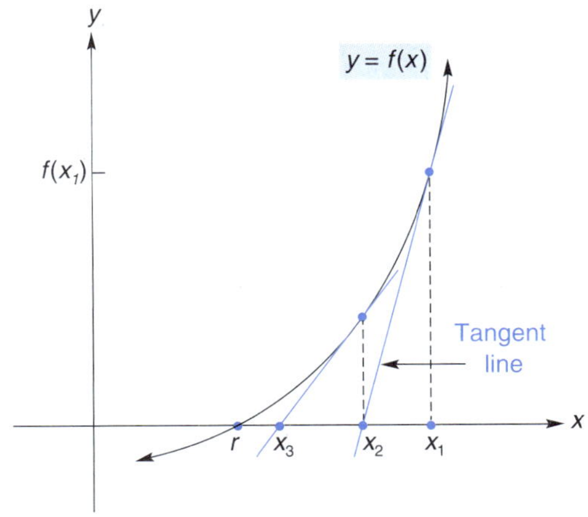

FIGURE 12.7 Improving approximation of root via tangent line.

Since $(x_2, 0)$ is on the tangent line, its coordinates must satisfy Equation (1). This gives

$$0 - f(x_1) = f'(x_1)(x_2 - x_1)$$

$$-\frac{f(x_1)}{f'(x_1)} = x_2 - x_1 \qquad (\text{if } f'(x_1) \neq 0)$$

Thus,

$$x_2 = x_1 - \frac{f(x_1)}{f'(x_1)} \tag{2}$$

To get a better approximation to r, we again perform the procedure described, but this time we use x_2 as our starting point. This gives the approximation

$$x_3 = x_2 - \frac{f(x_2)}{f'(x_2)} \tag{3}$$

Repeating (or *iterating*) this computation over and over, we hope to obtain better approximations, in the sense that the sequence of values

$$x_1, x_2, x_3, \ldots$$

will approach r. In practice, we terminate the process when we have reached a desired degree of accuracy.

If you analyze Equations (2) and (3), you can see how x_2 is obtained from x_1 and how x_3 is obtained from x_2. In general, x_{n+1} is obtained from x_n by means of the following general formula, called **Newton's method:**

Newton's Method

$$x_{n+1} = x_n - \frac{f(x_n)}{f'(x_n)} \qquad n = 1, 2, 3, \ldots \tag{4}$$

A formula, like Equation (4), that indicates how one number in a sequence is obtained from the preceding one is called a **recursion formula,** or an *iteration equation*.

APPLICATIONS IN PRACTICE 1

Approximating a Root by Newton's Method

If the total profit (in dollars) from the sale of x televisions is $P(x) = 20x - 0.01x^2 - 850 + 3\ln(x)$, use Newton's method to approximate the break-even quantities. (*Note:* There are two break-even quantities; one is between 10 and 50, and the other is between 1900 and 2000.) Give the x-value to the nearest integer.

In the event that a root lies between a and b, and $f(a)$ and $f(b)$ are equally close to 0, choose either a or b as the first approximation.

EXAMPLE 1 Approximating a Root by Newton's Method

Approximate the root of $x^4 - 4x + 1 = 0$ *that lies between* 0 *and* 1. *Continue the approximation procedure until two successive approximations differ by less than* 0.0001.

Solution: Letting $f(x) = x^4 - 4x + 1$, we have

$$f(0) = 0 - 0 + 1 = 1$$

and

$$f(1) = 1 - 4 + 1 = -2$$

(Note the change in sign.) Since $f(0)$ is closer to 0 than is $f(1)$, we choose 0 to be our first approximation, x_1. Now,

$$f'(x) = 4x^3 - 4$$

so

$$f(x_n) = x_n^4 - 4x_n + 1 \quad \text{and} \quad f'(x_n) = 4x_n^3 - 4$$

Substituting into Equation (4) gives the recursion formula

$$x_{n+1} = x_n - \frac{f(x_n)}{f'(x_n)} = x_n - \frac{x_n^4 - 4x_n + 1}{4x_n^3 - 4}$$

$$= \frac{4x_n^4 - 4x_n - x_n^4 + 4x_n - 1}{4x_n^3 - 4}$$

so

$$x_{n+1} = \frac{3x_n^4 - 1}{4x_n^3 - 4} \qquad (5)$$

Since $x_1 = 0$, letting $n = 1$ in Equation (5) gives

$$x_2 = \frac{3x_1^4 - 1}{4x_1^3 - 4} = \frac{3(0)^4 - 1}{4(0)^3 - 4} = 0.25$$

Letting $n = 2$ in Equation (5) gives

$$x_3 = \frac{3x_2^4 - 1}{4x_2^3 - 4} = \frac{3(0.25)^4 - 1}{4(0.25)^3 - 4} \approx 0.25099$$

Letting $n = 3$ in Equation (5) gives

$$x_4 = \frac{3x_3^4 - 1}{4x_3^3 - 4} = \frac{3(0.25099)^4 - 1}{4(0.25099)^3 - 4} \approx 0.25099$$

The data obtained thus far are displayed in Table 12.1. Since the values of x_3 and x_4 differ by less than 0.0001, we take the root to be 0.25099 (that is, x_4).

TABLE 12.1

n	x_n	x_{n+1}
1	0.00000	0.25000
2	0.25000	0.25099
3	0.25099	0.25099

▸ **Now work Problem 1.**

EXAMPLE 2 **Approximating a Root by Newton's Method**

Approximate the root of $x^3 = 3x - 1$ that lies between -1 and -2. Continue the approximation procedure until two successive approximations differ by less than 0.0001.

Solution: Letting $f(x) = x^3 - 3x + 1$ [we need the form $f(x) = 0$], we find that

$$f(-1) = (-1)^3 - 3(-1) + 1 = 3$$

and

$$f(-2) = (-2)^3 - 3(-2) + 1 = -1$$

(Note the change in sign.) Since $f(-2)$ is closer to 0 than is $f(-1)$, we choose -2 to be our first approximation, x_1. Now,

$$f'(x) = 3x^2 - 3$$

so

$$f(x_n) = x_n^3 - 3x_n + 1 \quad \text{and} \quad f'(x_n) = 3x_n^2 - 3$$

Substituting into Equation (4) gives the recursion formula

$$x_{n+1} = x_n - \frac{f(x_n)}{f'(x_n)} = x_n - \frac{x_n^3 - 3x_n + 1}{3x_n^2 - 3}$$

so

$$x_{n+1} = \frac{2x_n^3 - 1}{3x_n^2 - 3} \tag{6}$$

Since $x_1 = -2$, letting $n = 1$ in Equation (6) gives

$$x_2 = \frac{2x_1^3 - 1}{3x_1^2 - 3} = \frac{2(-2)^3 - 1}{3(-2)^2 - 3} \approx -1.88889$$

Continuing in this way, we obtain Table 12.2. Because the values of x_3 and x_4 differ by 0.00006, which is less than 0.0001, we take the root to be -1.87939 (that is, x_4).

TABLE 12.2

n	x_n	$x_n + 1$
1	−2.00000	−1.88889
2	−1.88889	−1.87945
3	−1.87945	−1.87939

▸ **Now work Problem 3.**

The situation where x_1 leads to a derivative of 0 occurs in Problems 2 and 8.

In case your choice for the initial approximation, x_1, gives the derivative a value of zero, choose a different number that is close to the desired root. A graph of f could be helpful in this situation. Finally, we remark that there are times when the sequence of approximations does not approach the root. A discussion of such situations is beyond the scope of this book.

TECHNOLOGY

Figure 12.8 gives a short TI-83 Plus program for Newton's method. Before the program is executed, the first approximation to the root of $f(x) = 0$ is stored as X, and $f(x)$ and $f'(x)$ are stored as Y_1 and Y_2, respectively.

When executed, the program computes the first iteration and pauses. Successive iterations are obtained by successively pressing ENTER. Figure 12.9 shows the iterations for the problem in Example 2. (Press ON to break out of the program.)

```
PROGRAM:NEWTON
:Lbl A
:X-Y1(X)/Y2(X)
:Ans→X
:Disp X
:Pause
:Goto A
```

FIGURE 12.8 Calculator program for Newton's method.

```
-2→X
                    -2
prgmNEWTON
          -1.888888889
          -1.879451567
          -1.879385245
```

FIGURE 12.9 Iterations for problem in Example 2.

Exercise 12.6

In Problems 1–10, use Newton's method to approximate the indicated root of the given equation. Continue the approximation procedure until the difference of two successive approximations is less than 0.0001.

***1.** $x^3 - 4x + 1 = 0$; root between 0 and 1

2. $x^3 + 2x^2 - 1 = 0$; root between 0 and 1

***3.** $x^3 - x - 1 = 0$; root between 1 and 2

4. $x^3 - 9x + 6 = 0$; root between 2 and 3

5. $x^3 + x + 1 = 0$; root between -1 and 0

6. $x^3 = 2x + 5$; root between 2 and 3

7. $x^4 = 3x - 1$; root between 0 and 1

8. $x^4 + 4x - 1 = 0$; root between -2 and -1

9. $x^4 - 2x^3 + x^2 - 3 = 0$; root between 1 and 2

10. $x^4 - x^3 + x - 2 = 0$; root between 1 and 2

11. Estimate, to three-decimal-place accuracy, the cube root of 71. [*Hint:* Show that the problem is equivalent to finding a root of $f(x) = x^3 - 71 = 0$. Choose 4 as the initial estimate. Continue the iteration until two successive approximations, rounded to three decimal places, are the same.]

12. Estimate $\sqrt[4]{19}$, to two-decimal-place accuracy. Use 2 as your initial estimate.

13. Find, to two-decimal-place accuracy, all real solutions of the equation $e^x = x + 5$. [*Hint:* A rough sketch of the graphs of $y = e^x$ and $y = x + 5$ makes it clear how many solutions there are. Use nearby integer values for your initial estimates.]

14. Find, to three-decimal-place accuracy, all real solutions of the equation $\ln x = 5 - x$.

15. Break-Even Quantity The cost of manufacturing q tons of a certain product is given by

$$c = 250 + 2q - 0.1q^3$$

and the revenue obtained by selling the q tons is given by

$$r = 3q$$

Approximate, to two-decimal-place accuracy, the break-even quantity. [*Hint:* Approximate a root of $r - c = 0$ by choosing 13 as your initial estimate.]

16. Break-Even Quantity The total cost of manufacturing q hundred pencils is c dollars, where

$$c = 40 + 3q + \frac{q^2}{1000} + \frac{1}{q}$$

Pencils are sold for \$7 per hundred.

(a) Show that the break-even quantity is a solution of the equation

$$f(q) = \frac{q^3}{1000} - 4q^2 + 40q + 1 = 0$$

(b) Use Newton's method to approximate the solution of $f(q) = 0$, where $f(q)$ is given in part (a). Use 10 as your initial approximation, and give your answer to two-decimal-place accuracy.

17. Equilibrium Given the supply equation $p = 2q + 5$ and the demand equation $p = \frac{100}{q^2+1}$, use Newton's method to estimate the market equilibrium quantity. Give your answer to three-decimal-place accuracy.

18. Equilibrium Given the supply equation

$$p = 0.2q^3 + 0.5q + 2$$

and the demand equation $p = 10 - q$, use Newton's method to estimate the market equilibrium quantity, and find the corresponding equilibrium price. Use 5 as an initial estimate for the required value of q, and give your answer to two-decimal-place accuracy.

19. Use Newton's method to approximate (to two-decimal-place accuracy) a value of x such that $f'(x) = 0$ for

$$f(x) = \frac{x^3}{3} - x^2 - 5x + 1$$

on the interval [3, 4].

OBJECTIVE *To find higher-order derivatives both directly and implicitly.*

12.7 Higher-Order Derivatives

We know that the derivative of a function $y = f(x)$ is itself a function, $f'(x)$. If we differentiate $f'(x)$, the resulting function is called the **second derivative** of f at x. It is denoted $f''(x)$, which is read "f double prime of x." Similarly, the derivative of the second derivative is called the **third derivative,** written $f'''(x)$. Continuing in this way, we get *higher-order derivatives*. Some notations for higher-order derivatives are given in Table 12.3. To avoid clumsy notation, primes are not used beyond the third derivative.

CAUTION The symbol d^2y/dx^2 represents the second derivative of y. It is not the same as $(dy/dx)^2$, the square of the first derivative of y.

TABLE 12.3

First derivative:	y'	$f'(x)$	$\frac{dy}{dx}$	$\frac{d}{dx}[f(x)]$	$D_x y$
Second derivative:	y''	$f''(x)$	$\frac{d^2y}{dx^2}$	$\frac{d^2}{dx^2}[f(x)]$	$D_x^2 y$
Third derivative:	y'''	$f'''(x)$	$\frac{d^3y}{dx^3}$	$\frac{d^3}{dx^3}[f(x)]$	$D_x^3 y$
Fourth derivative:	$y^{(4)}$	$f^{(4)}(x)$	$\frac{d^4y}{dx^4}$	$\frac{d^4}{dx^4}[f(x)]$	$D_x^4 y$

EXAMPLE 1 **Finding Higher-Order Derivatives**

a. *If* $f(x) = 6x^3 - 12x^2 + 6x - 2$, *find* all *higher-order derivatives.*

Solution: Differentiating $f(x)$ gives

$$f'(x) = 18x^2 - 24x + 6$$

Differentiating $f'(x)$ yields

$$f''(x) = 36x - 24$$

Similarly,

$$f'''(x) = 36$$
$$f^{(4)}(x) = 0$$

All successive derivatives are also 0: $f^{(5)}(x) = 0$, and so on.

b. *If $f(x) = 7$, find $f''(x)$.*

Solution:

$$f'(x) = 0$$
$$f''(x) = 0$$

▶ **Now work Problem 1.**

APPLICATIONS IN PRACTICE 1

Finding a Second-Order Derivative

The height $h(t)$ of a rock dropped off of a 200-foot building is given by $h(t) = 200 - 16t^2$, where t is the time measured in seconds.

Find $\frac{d^2h}{dt^2}$, the acceleration of the rock at time t.

EXAMPLE 2 Finding a Second-Order Derivative

If $y = e^{x^2}$, find $\frac{d^2y}{dx^2}$.

Solution:

$$\frac{dy}{dx} = e^{x^2}(2x) = 2xe^{x^2}$$

By the product rule,

$$\frac{d^2y}{dx^2} = 2\left[x\left(e^{x^2}\right)(2x) + e^{x^2}(1)\right] = 2e^{x^2}(2x^2 + 1)$$

▶ **Now work Problem 5.**

APPLICATIONS IN PRACTICE 2

Evaluating a Second-Order Derivative

If the cost to produce q units of a product is

$$c(q) = 7q^2 + 11q + 19$$

and the marginal cost function is $c'(q)$, find the rate of change of the marginal cost function with respect to q when $q = 3$.

EXAMPLE 3 Evaluating a Second-Order Derivative

If $y = f(x) = \frac{16}{x+4}$, find $\frac{d^2y}{dx^2}$ and evaluate it when $x = 4$.

Solution: Since $y = 16(x+4)^{-1}$, the power rule gives

$$\frac{dy}{dx} = -16(x+4)^{-2}$$
$$\frac{d^2y}{dx^2} = 32(x+4)^{-3} = \frac{32}{(x+4)^3}$$

Evaluating when $x = 4$, we obtain

$$\left.\frac{d^2y}{dx^2}\right|_{x=4} = \frac{32}{8^3} = \frac{1}{16}$$

The second derivative evaluated at $x = 4$ is also denoted $f''(4)$ or $y''(4)$.

▶ **Now work Problem 21.**

EXAMPLE 4 Finding the Rate of Change of $f''(x)$

If $f(x) = x \ln x$, find the rate of change of $f''(x)$.

The rate of change of $f''(x)$ is $f'''(x)$.

Solution: To find the rate of change of any function, we must find its derivative. Thus, we want the derivative of $f''(x)$, which is $f'''(x)$. Accordingly,

$$f'(x) = x\left(\frac{1}{x}\right) + (\ln x)(1) = 1 + \ln x$$

$$f''(x) = 0 + \frac{1}{x} = \frac{1}{x}$$

$$f'''(x) = \frac{d}{dx}(x^{-1}) = (-1)x^{-2} = -\frac{1}{x^2}$$

▸ **Now work Problem 17.**

Higher-Order Implicit Differentiation

We shall now find a higher-order derivative by means of implicit differentiation. Keep in mind that we shall assume y to be a function of x.

EXAMPLE 5 Higher-Order Implicit Differentiation

Find $\frac{d^2y}{dx^2}$ *if* $x^2 + 4y^2 = 4$.

Solution: Differentiating both sides with respect to x, we obtain

$$2x + 8y\frac{dy}{dx} = 0$$

$$\frac{dy}{dx} = \frac{-x}{4y} \qquad (1)$$

$$\frac{d^2y}{dx^2} = \frac{4y\frac{d}{dx}(-x) - (-x)\frac{d}{dx}(4y)}{(4y)^2}$$

$$= \frac{4y(-1) - (-x)\left(4\frac{dy}{dx}\right)}{16y^2}$$

$$= \frac{-4y + 4x\frac{dy}{dx}}{16y^2}$$

$$\frac{d^2y}{dx^2} = \frac{-y + x\frac{dy}{dx}}{4y^2} \qquad (2)$$

Although we have found an expression for d^2y/dx^2, our answer involves the derivative dy/dx. It is customary to express the answer without the derivative—that is, in terms of x and y only. This is easy to do. From Equation (1), $\frac{dy}{dx} = \frac{-x}{4y}$, so by substituting into Equation (2), we have

$$\frac{d^2y}{dx^2} = \frac{-y + x\left(\frac{-x}{4y}\right)}{4y^2} = \frac{-4y^2 - x^2}{16y^3} = -\frac{4y^2 + x^2}{16y^3}$$

In Example 5, the simplification of d^2y/dx^2 by making use of the original equation is not unusual.

We can further simplify the answer. Since $x^2 + 4y^2 = 4$ (the original equation),

$$\frac{d^2y}{dx^2} = -\frac{4}{16y^3} = -\frac{1}{4y^3}$$

▶ Now work Problem 23.

EXAMPLE 6 **Higher-Order Implicit Differentiation**

Find $\dfrac{d^2y}{dx^2}$ *if* $y^2 = e^{x+y}$.

Solution: Differentiating both sides with respect to x gives

$$2y\frac{dy}{dx} = e^{x+y}\left(1 + \frac{dy}{dx}\right)$$

Solving for dy/dx, we obtain

$$2y\frac{dy}{dx} = e^{x+y} + e^{x+y}\frac{dy}{dx}$$

$$2y\frac{dy}{dx} - e^{x+y}\frac{dy}{dx} = e^{x+y}$$

$$(2y - e^{x+y})\frac{dy}{dx} = e^{x+y}$$

$$\frac{dy}{dx} = \frac{e^{x+y}}{2y - e^{x+y}}$$

Since $y^2 = e^{x+y}$ (the original equation),

$$\frac{dy}{dx} = \frac{y^2}{2y - y^2} = \frac{y}{2 - y}$$

$$\frac{d^2y}{dx^2} = \frac{(2-y)\dfrac{dy}{dx} - y\left(-\dfrac{dy}{dx}\right)}{(2-y)^2} = \frac{2\dfrac{dy}{dx}}{(2-y)^2}$$

Now we express our answer without dy/dx. Since $\dfrac{dy}{dx} = \dfrac{y}{2-y}$,

$$\frac{d^2y}{dx^2} = \frac{2\left(\dfrac{y}{2-y}\right)}{(2-y)^2} = \frac{2y}{(2-y)^3}$$

▶ Now work Problem 31.

Exercise 12.7

In Problems 1–20, find the indicated derivatives.

*1. $y = 4x^3 - 12x^2 + 6x + 2$, y'''

2. $y = 2x^4 - 6x^2 + 7x - 2$, y'''

3. $y = 8 - x$, $\dfrac{d^2y}{dx^2}$

4. $y = -x - x^2$, $\dfrac{d^2y}{dx^2}$

*5. $y = x^3 + e^x$, $y^{(4)}$

6. $F(q) = \ln(q + 1)$, $\dfrac{d^3F}{dq^3}$

7. $f(x) = x^2 \ln x$, $f''(x)$

8. $y = \dfrac{1}{x}$, y'''

9. $f(p) = \dfrac{1}{6p^3}$, $f'''(p)$

10. $f(x) = \sqrt{x}$, $f''(x)$

11. $f(r) = \sqrt{9 - r}$, $f''(r)$

12. $y = e^{-4x^2}$, y''

13. $y = \dfrac{1}{2x + 3}$, $\dfrac{d^2y}{dx^2}$

14. $y = (2x + 1)^4$, y''

15. $y = \dfrac{x+1}{x-1}$, y''

16. $y = 2x^{1/2} + (2x)^{1/2}$, y''

*__17.__ $y = \ln[x(x+6)]$, y''

18. $y = \ln \dfrac{(2x+5)(5x-2)}{x+1}$, y''

19. $f(z) = z^2e^z$, $f''(z)$

20. $y = \dfrac{x}{e^x}$, $\dfrac{d^2y}{dx^2}$

*__21.__ If $y = e^{2x}$, find $\left.\dfrac{d^5y}{dx^5}\right|_{x=0}$

22. If $y = e^{2\ln(x^3+1)}$, find y'' when $x = 1$

In Problems 23–32, find y''.

*__23.__ $x^2 + 4y^2 - 16 = 0$

24. $x^2 - y^2 = 16$

25. $y^2 = 4x$

26. $4x^2 + 9y^2 = 36$

27. $\sqrt{x} + 4\sqrt{y} = 4$

28. $y^2 - 6xy = 4$

29. $xy + y - x = 4$

30. $x^2 + 2xy + y^2 = 1$

*__31.__ $y = e^{x+y}$

32. $e^x - e^y = x^2 + y^2$

33. If $x^2 + 8y = y^2$, find d^2y/dx^2 when $x = 3$ and $y = -1$.

34. Show that the equation

$$f''(x) + 4f'(x) + 4f(x) = 0$$

is satisfied if $f(x) = (3x-5)e^{-2x}$.

35. Find the rate of change of $f'(x)$ if $f(x) = (5x-3)^4$.

36. Find the rate of change of $f''(x)$ if

$$f(x) = 6\sqrt{x} + \frac{1}{6\sqrt{x}}$$

37. Marginal Cost If $c = 0.3q^2 + 2q + 850$ is a cost function, how fast is marginal cost changing when $q = 100$?

38. Marginal Revenue If $p = 500 - 50q - q^2$ is a demand equation, how fast is marginal revenue changing when $q = 5$?

39. If $f(x) = x^4 - 6x^2 + 5x - 6$, determine the values of x for which $f''(x) = 0$.

40. Suppose that $e^y = y^2e^x$. (a) Determine dy/dx, and express your answer in terms of y only. (b) Determine d^2y/dx^2, and express your answer in terms of y only.

In Problems 41 and 42, determine $f''(x)$. Then use your graphing calculator to find all real zeros of $f''(x)$. Round your answers to two decimal places.

41. $f(x) = 6e^x - x^3 - 15x^2$

42. $f(x) = \dfrac{x^5}{20} + \dfrac{x^4}{12} + \dfrac{5x^3}{6} + \dfrac{x^2}{2}$

12.8 Review

Important Terms and Symbols

Section 12.3	point elasticity of demand elastic inelastic unit elasticity
Section 12.4	implicit differentiation
Section 12.5	logarithmic differentiation
Section 12.6	Newton's method
Section 12.7	higher-order derivatives, $f''(x)$, $\dfrac{d^3y}{dx^3}$, $\dfrac{d^4}{dx^4}[f(x)]$, and so on

Summary

Point elasticity of demand is a function that measures how consumer demand is affected by a change in price. It is given by

$$\eta = \frac{p/q}{dp/dq}$$

where p is the price per unit at which q units are demanded. The three categories of elasticity are as follows:

$|\eta| > 1$ demand is elastic

$|\eta| = 1$ unit elasticity

$|\eta| < 1$ demand is inelastic

For a given percentage change in price, if there is a greater (respectively lesser) change in quantity demanded then demand is elastic (respectively inelastic) and conversely.

The relationship between elasticity and the rate of change of revenue is given by

$$\frac{dr}{dq} = p\left(1 + \frac{1}{\eta}\right)$$

If an equation implicitly defines y as a function of x [rather than defining it explicitly in the form $y = f(x)$], then dy/dx can be found by implicit differentiation. With this method, we treat y as a function of x and differentiate both sides of the equation with respect to x. When doing this, remember that

$$\frac{d}{dx}(y^n) = ny^{n-1}\frac{dy}{dx}$$

Finally, we solve the resulting equation for dy/dx.

The derivative formulas for natural logarithmic and exponential functions are

$$\frac{d}{dx}(\ln u) = \frac{1}{u}\frac{du}{dx}$$

and

$$\frac{d}{dx}(e^u) = e^u\frac{du}{dx}$$

To differentiate logarithmic and exponential functions in bases other than e, you can first transform the function to base e and then differentiate the result. Alternatively, differentiation formulas can be applied:

$$\frac{d}{dx}(\log_b u) = \frac{1}{(\ln b)u}\cdot\frac{du}{dx}$$

$$\frac{d}{dx}(a^u) = a^u(\ln a)\frac{du}{dx}$$

Suppose that $f(x)$ consists of products, quotients, or powers. To differentiate $y = \log_b[f(x)]$, it may be helpful to use properties of logarithms to rewrite $\log_b[f(x)]$ in terms of simpler logarithms and then differentiate that form. To differentiate $y = f(x)$, where $f(x)$ consists of products, quotients, or powers, the method of logarithmic differentiation may be used. In that method, we take the natural logarithm of both sides of $y = f(x)$ to obtain $\ln y = \ln[f(x)]$. After simplifying $\ln[f(x)]$ by using properties of logarithms, we differentiate both sides of $\ln y = \ln[f(x)]$ with respect to x and then solve for y'. Logarithmic differentiation can also be used to differentiate $y = u^v$, where both u and v are functions of x.

Newton's method is the name given to the following formula which is used to provide a sequence of approximations to a roots of the equation $f(x) = 0$, for f a differentiable function:

$$x_{n+1} = x_n - \frac{f(x_n)}{f'(x_n)} \qquad n = 1, 2, 3, \ldots$$

In most cases you might encounter, the approximations improve as n increases.

Because the derivative $f'(x)$ of a function $y = f(x)$ is itself a function, it can be successively differentiated to obtain the second derivative $f''(x)$, the third derivative $f'''(x)$, and other higher-order derivatives.

Review Problems

Problem numbers shown in color indicate problems suggested for use as a practice chapter test.

In Problems 1–30, differentiate.

1. $y = 3e^x + e^2 + e^{x^2} + x^{e^2}$

2. $f(w) = we^w + w^2$

3. $f(r) = \ln(r^2 + 5r)$

4. $y = e^{\ln x}$

5. $y = e^{x^2+4x+5}$

6. $f(t) = \log_6 \sqrt{t^2 + 1}$

7. $y = e^x(x^2 + 2)$

8. $y = 3^{5x^3}$

9. $y = \sqrt{(x-6)(x+5)(9-x)}$

10. $f(t) = e^{\sqrt{t}}$

11. $y = \dfrac{\ln x}{e^x}$

12. $y = \dfrac{e^x + e^{-x}}{x^2}$

13. $f(q) = \ln[(q+1)^2(q+2)^3]$

14. $y = (x+2)^3(x+1)^4(x-2)^2$

15. $y = 10^{2-7x}$

16. $y = (e + e^2)^0$

17. $y = \dfrac{4e^{3x}}{xe^{x-1}}$

18. $y = \dfrac{e^x}{\ln x}$

19. $y = \log_2(8x + 5)^2$

20. $y = \ln\left(\dfrac{5}{x^2}\right)$

21. $f(l) = \ln(1 + l + l^2 + l^3)$

22. $y = x^{x^3}$

23. $y = (x+1)^{x+1}$

24. $y = \dfrac{1+e^x}{1-e^x}$

25. $\phi(t) = \ln(t\sqrt{4-t^2})$

26. $y = (x+3)^{\ln x}$

27. $y = \dfrac{(x^2+2)^{3/2}(x^2+9)^{4/9}}{(x^3+6x)^{4/11}}$

28. $y = \dfrac{\ln x}{\sqrt{x}}$

29. $y = (x^x)^x$

30. $y = x^{(x^x)}$

In Problems 31–34, evaluate y' at the given value of x.

31. $y = (x+1)\ln x^2$, $x = 1$

32. $y = \dfrac{e^{x^2+1}}{\sqrt{x^2+1}}$, $x = 1$

33. $y = e^{e+x\ln(1/x)}$, $x = e$

34. $y = \left[\dfrac{4^{3x}(x^3 - x + 1)^{1/5}}{(x^2+x+1)^4}\right]^{-2}$, $x = 0$

In Problems 35 and 36, find an equation of the tangent line to the curve at the point corresponding to the given value of x.

35. $y = 3e^x$, $x = \ln 2$

36. $y = x + x^2 \ln x$, $x = 1$

37. Find the y-intercept of the tangent line to the graph of $y = x(2^{2-x^2})$ at the point where $x = 1$.

38. If $w = 2^{x+1} + \ln(1 + x^2)$ and

$$x = \log_2(t^2 + 1) - e^{(t-1)^2}$$

find w and dw/dt when $t = 1$.

In Problems 39–42, find the indicated derivative at the given point.

39. $y = e^{x^2-4}$, y'', $(2, 1)$

40. $y = x^2e^x$, y''', $(1, e)$

41. $y = \ln(2x)$, y''', $(1, \ln 2)$

42. $y = x\ln x$, y'', $(1, 0)$

In Problems 43–46, find dy/dx.

43. $2xy + y^2 = 10$

44. $x^3y^3 = 3$

45. $\ln(xy^2) = xy$

46. $(\ln y)e^{y \ln x} = e^2$

In Problems 47 and 48, find d^2y/dx^2 at the given point.

47. $x + xy + y = 5$, $(2, 1)$

48. $xy + y^2 = 2$, $(-1, 2)$

49. If y is defined implicitly by $e^y = (y+1)e^x$, determine both dy/dx and d^2y/dx^2 as explicit functions of y only.

50. If $\sqrt{x} + \sqrt{y} = 1$, find $\dfrac{d^2y}{dx^2}$

51. Schizophrenia Several models have been used to analyze the length of stay in a hospital. For a particular group of schizophrenics, one such model is[9]

$$f(t) = 1 - (0.8e^{-0.01t} + 0.2e^{-0.0002t})$$

where $f(t)$ is the proportion of the group that was discharged at the end of t days of hospitalization. Determine the discharge rate (proportion discharged per day) at the end of t days.

52. Earthquakes According to Richter,[10] the number N of earthquakes of magnitude M or greater per unit of time is given by $\log N = A - bM$, where A and b are constants. He claims that

$$\log\left(-\frac{dN}{dM}\right) = A + \log\left(\frac{b}{q}\right) - bM$$

where $q = \log e$. Verify this statement.

53. If $f(x) = e^{9x^4+4x^3-36x}$, find all real zeros of $f'(x)$. Round your answers to two decimal places.

[9]Adapted from W. W. Eaton and G. A. Whitmore, "Length of Stay as a Stochastic Process: A General Approach and Application to Hospitalization for Schizophrenia," *Journal of Mathematical Sociology* 5 (1977), 273–92.

[10]C. F. Richter, *Elementary Seismology* (San Francisco: W. H. Freeman and Company, Publishers, 1958).

54. If $f(x) = \dfrac{x^5}{10} + \dfrac{x^4}{6} + \dfrac{2x^3}{3} + x^2 + 1$, find all zeros of $f''(x)$. Round your answers to two decimal places.

For the demand equations in Problems 55–57, determine whether demand is elastic, is inelastic, or has unit elasticity for the indicated value of q.

55. $p = \dfrac{500}{q}$; $q = 200$.

56. $p = 900 - q^2$; $q = 10$.

57. $p = 18 - 0.02q$; $q = 600$.

58. The demand equation for a product is

$$p = 30 - \sqrt{q}.$$

(a) Find the point elasticity of demand when $p = 10$.
(b) Verify that demand is inelastic if $0 < p < 10$.

59. The demand equation of a product is

$$q = \sqrt{2500 - p^2}.$$

Find the point elasticity of demand when $p = 30$. If the price of 30 decreases $\frac{2}{3}\%$, what is the approximate change in demand?

60. The demand equation for a product is

$$q = \sqrt{100 - p}, \quad \text{where } 0 < p < 100.$$

(a) Find all prices that correspond to elastic demand.
(b) Compute the point elasticity of demand when $p = 40$. Use your answer to estimate the percentage increase or decrease in demand when price is increased by 5% to $p = 42$.

61. The equation $x^3 - 2x - 2 = 0$ has a root between 1 and 2. Use Newton's method to estimate the root. Continue the approximation procedure until the difference of two successive approximations is less than 0.0001. Round your answer to four decimal places.

62. Find, to an accuracy of three decimal places, all real solutions of the equation $e^x = 3x$.

Mathematical SNAPSHOT

Population Change over Time

Now that we know how to find the derivative of a function, we might ask whether there is a way to run the process in reverse: to find a function, given its derivative. Ultimately, this is what integration (Chapters 14–15) is all about. Meanwhile, however, we can use the derivative of a function to find the function *approximately* even without knowing how to do integration.

To illustrate, suppose we wish to describe the population, over time, of a small town in a frontier area. Let use imagine that the things we know about the town are all facts about how its population, P, changes over time, t, with population measured in people and time in years:

1. Births exceed deaths, so that over the course of a year there is a 25% increase in the population before other factors are accounted for. Thus, the annual change due to the birth/death difference is $0.25P$.
2. Every year, of the travelers passing through, ten decide to stop and settle down. This contributes a constant 10 to the annual change.
3. Loneliness causes some people to leave when the town is too small for them. At the extreme, 99% of people will leave over the course of a year if they are all alone (population = 1). When the population is 100, 10% of the residents leave per year due to loneliness.

   Assuming an exponential relationship, we write the likelihood that a given person leaves within a year due to loneliness as Ae^{-kP}, where A and k are positive constants. The numbers tell us that $Ae^{-k\cdot 1} = 0.99$ and $Ae^{-k\cdot 100} = 0.10$. Solving this pair of equations for A and k yields

   $$k = \frac{\ln 9.9}{99} \approx 0.02316$$

   and

   $$A = 0.99e^{(\ln 9.9)/99} \approx 1.01319$$

   And if Ae^{-kP} is the likelihood of a single person's leaving, the population change per year due to loneliness is $-P$ times that, namely $-1.01319Pe^{-0.02316P}$. (The negative sign is due to the fact that the change is downward.)

4. Crowding causes some people to leave when the town is too large for them. Nobody has a crowding problem when they are all alone (population = 1), but when the population is 100, 10% of the residents leave per year due to crowding.

Again assuming an exponential relationship, we write the likelihood that a given person leaves within a year due to crowding as $1 - Ae^{-kP}$. This time, the numbers tell us that $1 - Ae^{-k\cdot 1} = 0$ and $1 - Ae^{-k\cdot 100} = 0.10$. Solving this pair of equations for A and k yields

$$k = -\frac{\ln 0.9}{99} \approx 0.001064$$

and

$$A = e^{-(\ln 0.9)/99} \approx 1.001065$$

If $1 - Ae^{-kP}$ is the likelihood of a single person's leaving, the population change per year due to crowding is $-P$ times that, namely $-P(1 - 1.001065e^{-0.001064P})$.

The overall rate of change in the population, now, is the net effect of all these factors added together. In equation form,

$$\frac{dP}{dt} = 0.25P + 10 - 1.01319Pe^{-0.02316P}$$
$$-P(1 - 1.001065e^{-0.001064P})$$

Before we try to reconstruct the function $P(t)$, let us graph the derivative. On a graphing calculator, it looks as shown in Figure 12.10. Note that $\frac{dP}{dt}$ is depicted as a function of P. This is a different graph from what we would get if we knew P as a function of t, found its derivative, and graphed that in the standard manner, namely as function of t. Nonetheless, this graph reveals some significant facts. First, the derivative is positive from $P = 0$ to $P = 311$; this means that the population will have positive growth in that entire range and thus can be expected to grow from nothing into a substantial community.

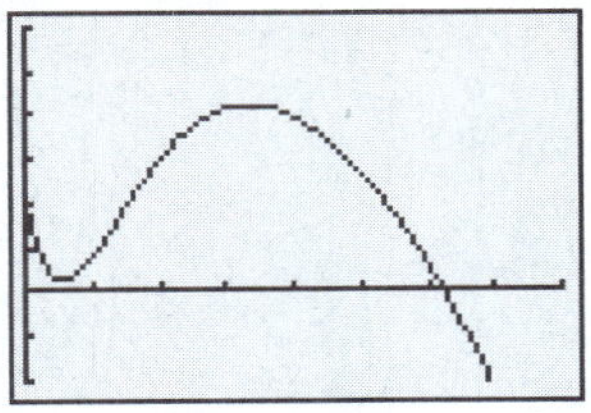

FIGURE 12.10 $\frac{dP}{dt}$ as a function of P.

Growth does fall to near zero at around $P = 30$. Apparently departures due to loneliness nearly bring growth to a halt when the population is still small. But once the town has grown through that phase, its size increases steadily, at one point (around $P = 170$) adding 21 people per year.

Eventually, departures due to crowding start to take a toll. Above 312 the derivative is negative. This means that if the population ever fluctuated above 312, population losses would shrink it back down to that level. In short, the population of this town stabilizes at 311 or 312—not exactly a city, but this is, after all, a frontier environment.

If we now wish to graph the town's population as a function of time, here is how we do this: We approximate the graph with a string of line segments, each of which has a slope given by the expression we obtained for dP/dt. We begin with a known time and population and calculate the initial slope. Let us grow the town from nothing, setting $t = 0$ at $P = 0$. Then $\frac{dP}{dt} = 10$. Now we advance the clock forward by a convenient time interval—let us choose 1 year—and, since the slope at (0, 10) equals 10, we increase the population from 0 to 10. The new values for t and P are 1 and 10, respectively, so we draw a line segment from (0, 0) to (1, 10). Now, with $t = 1$ and $P = 10$, we recalculate the slope and go through the same steps again, and we repeat this process until we have drawn as much of the curve as we want to see.

Obviously, this would be extremely tedious to do by hand. On a graphing calculator, however, we can use the programming and line-drawing features. For a TI-83 Plus, the following program does the job nicely, after the expression for $\frac{dP}{dt}$ is entered as Y_1 (keeping P as the variable):

```
PROGRAM:POPLTN
:Input "P?",P
:Input "T?",T
:ClrDraw
:T → S
:For(I, S + 1, S + 55)
:Line(T,P,I,P + Y1)
:I → T : (P + Y1) → P
:End
```

Deselect the function Y_1. Set the graphing window to display the coordinate plane from 0 to 55 horizontally and from 0 to 350 vertically. Then run the program and, at the prompt, give initial values for P and t. The program will draw 55 line segments, enough to take the population to its final size from $P = 0$, $t = 0$. The result is shown in Figure 12.11.

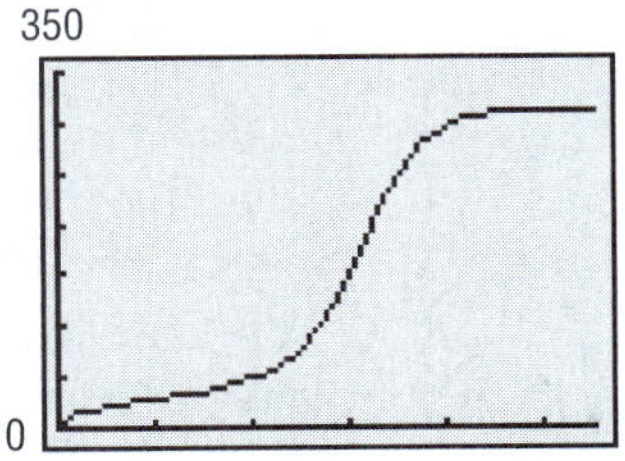

FIGURE 12.11 P as a function of t.

Exercises

1. What information does Figure 12.11 give that is not evident from Figure 12.10?
2. What happens when an initial value of 450 is selected for P? (The display should be adjusted to run from 0 to 500 vertically.) Does this seem right?
3. Why is this procedure for obtaining a graph of $P(t)$ only approximate? How could the approximation be improved?

CHAPTER

13

Curve Sketching

Mathematical SNAPSHOT

Sketching the Phillips Curve

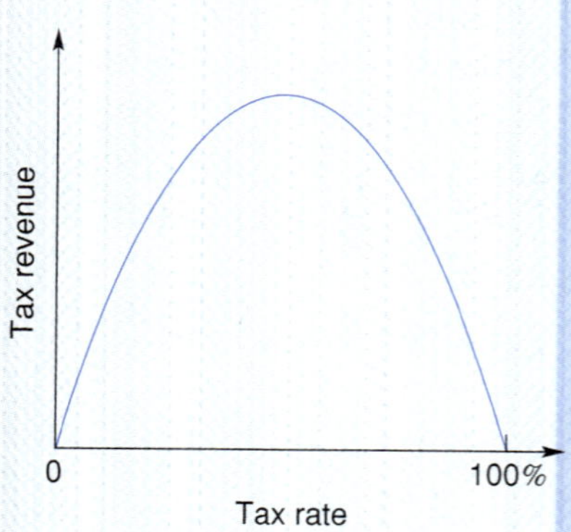

In the mid-1970s, economist Arthur Laffer was explaining his views on taxes to a politician—as the story goes, it was either presidential aspirant Ronald Reagan or Ford administration staff member Richard Cheney (later vice president under George W. Bush). To illustrate his argument, Laffer grabbed a paper napkin and sketched the graph that now bears his name: the Laffer curve.[1]

The Laffer curve describes total government tax revenue as a function of the tax rate. Obviously, if the tax rate is zero, the government gets nothing. But if the tax rate is 100%, revenue would again equal zero, because there is no incentive to earn money if it will all be taken away. Since tax rates between 0% and 100% do generate revenue, Laffer reasoned, the curve relating revenue to tax rate must look, qualitatively, more or less as shown in the figure to the left.

Laffer's argument was not meant to show that the optimal tax rate was 50%. What it was meant to show is that under some circumstances, namely when the tax rate is to the right of the peak of the curve, it is possible to *raise government revenue by lowering taxes*. This was a key argument made for the tax cuts passed by Congress during the first term of the Reagan presidency.

Because the Laffer curve is only a qualitative picture, it does not actually give an optimal tax rate. Revenue-based arguments for tax cuts involve the claim that the point of peak revenue lies to the left of the current taxation scheme on the horizontal axis. By the same token, those who urge raising taxes to raise government income are assuming either a different relationship between rates and revenues or a different location of the curve's peak.

By itself, then, the Laffer curve is too abstract to be of much help in determining the optimal tax rate. But even very simple sketched curves, like supply and demand curves and the Laffer curve, can help economists describe the causal factors that drive an economy. In this chapter, we will discuss techniques for sketching and interpreting curves.

[1] For one version of the story, see Jude Wanniski, *The Way the World Works*, 3rd ed. (Morristown, NJ: Polyconomics, 1989), p. 299.

OBJECTIVE *To find when a function is increasing or decreasing, to find critical values, to locate relative maxima and relative minima, and to state the first-derivative test. Also, to sketch the graph of a function by using the information obtained from the first derivative.*

13.1 Relative Extrema

Increasing or Decreasing Nature of a Function

Examining the graphical behavior of equations is a basic part of mathematics and has applications to many areas of study. When we sketch a curve, just plotting points may not give enough information about its shape. For example, the points $(-1, 0)$, $(0, -1)$, and $(1, 0)$ satisfy the equation given by $y = (x + 1)^3(x - 1)$. On the basis of these points, you might hastily conclude that the graph should appear as in Figure 13.1(a), but in fact the true shape is given in Figure 13.1(b). In this chapter we shall explore the powerful role that differentiation plays in analyzing a function, so that we may determine the true shape and behavior of its graph.

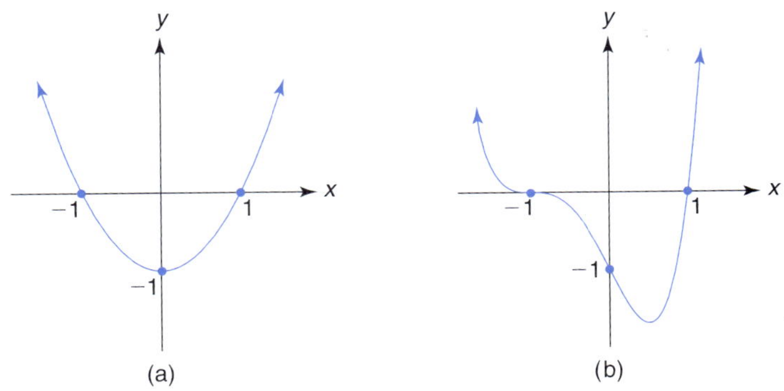

FIGURE 13.1 Curves passing through $(-1, 0)$, $(0, -1)$, and $(1, 0)$.

We begin by analyzing the graph of the function $y = f(x)$ in Figure 13.2. Notice that as x increases (goes from left to right) on the interval I_1, between a and b, the values of $f(x)$ increase and the curve is rising. Mathematically, this observation means that if x_1 and x_2 are any two points in I_1 such that $x_1 < x_2$, then $f(x_1) < f(x_2)$. Here f is said to be an *increasing function* on I_1. On the other hand, as x increases on the interval I_2 between c and d, the curve is falling. On this interval, $x_3 < x_4$ implies that $f(x_3) > f(x_4)$, and f is said to be a *decreasing function* on I_2. We summarize these observations in the following definition.

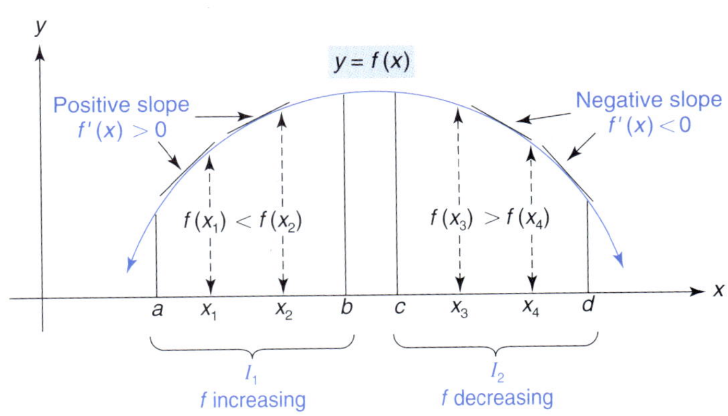

FIGURE 13.2 Increasing or decreasing nature of function.

Definition

A function f is said to be ***increasing*** on the interval I if, for any two numbers x_1, x_2 in I, where $x_1 < x_2$, then $f(x_1) < f(x_2)$. A function f is ***decreasing*** on the interval I if, for any two numbers x_1, x_2 in I, where $x_1 < x_2$, then $f(x_1) > f(x_2)$.

Turning again to Figure 13.2, we note that over the interval I_1, tangent lines to the curve have positive slopes, so $f'(x)$ must be positive for all x in I_1. Basically, a positive derivative implies that the curve is rising. Over the interval I_2, the tangent lines have negative slopes, so $f'(x) < 0$ for all x in I_2. Basically, the curve is falling where the derivative is negative. We thus have the following rule, which allows us to use the derivative to determine when a function is increasing or decreasing:

Rule 1 Criteria for Increasing or Decreasing Function

Let f be differentiable on the interval (a, b). If $f'(x) > 0$ for all x in (a, b), then f is increasing on (a, b). If $f'(x) < 0$ for all x in (a, b), then f is decreasing on (a, b).

To illustrate these ideas, we shall use Rule 1 to find the intervals on which $y = 18x - \frac{2}{3}x^3$ is increasing and the intervals on which y is decreasing. Letting $y = f(x)$, we must determine when $f'(x)$ is positive and when $f'(x)$ is negative. We have

$$f'(x) = 18 - 2x^2 = 2(9 - x^2) = 2(3 + x)(3 - x).$$

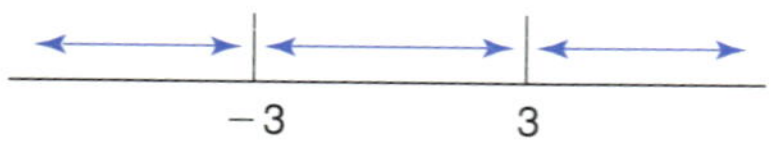

FIGURE 13.3 Intervals determined by roots of $f'(x) = 0$.

Using the technique of Section 10.5, we can find the sign of $f'(x)$ by testing the intervals determined by the roots of $2(3 + x)(3 - x) = 0$, namely, 3 and -3. (See Figure 13.3.) In each interval, the sign of $f'(x)$ is determined by the signs of its factors:

$$\begin{array}{lll} \text{If } x < -3, & \text{then } f'(x) = 2(-)(+) = -, & \text{so } f \text{ is } \textit{decreasing} \\ \text{If } -3 < x < 3, & \text{then } f'(x) = 2(+)(+) = +, & \text{so } f \text{ is } \textit{increasing} \\ \text{If } x > 3, & \text{then } f'(x) = 2(+)(-) = -, & \text{so } f \text{ is } \textit{decreasing} \end{array}$$

These results are indicated in Figure 13.4(a). Thus, f is decreasing on $(-\infty, -3)$ and $(3, \infty)$, and is increasing on $(-3, 3)$. This corresponds to the rising and falling nature of the graph of f shown in Figure 13.4(b). These results could be sharpened by noting that, by definition, f is decreasing on $(-\infty, -3]$ and $[3, \infty)$,

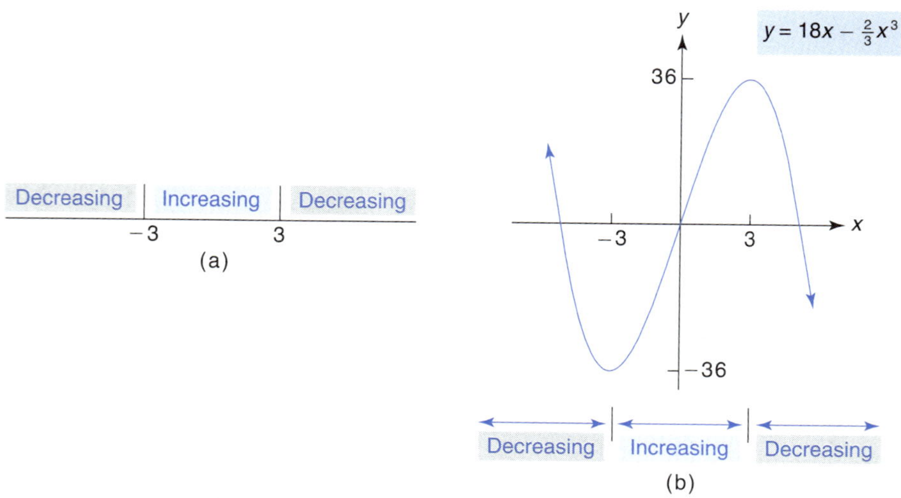

FIGURE 13.4 Increasing/decreasing for $y = 18x - \frac{2}{3}x^3$.

and increasing on $[-3, 3]$. However, for our purposes, open intervals are sufficient. *It will be our practice to determine* ***open*** *intervals on which a function is increasing or decreasing.*

Extrema

Look now at the graph of $y = f(x)$ in Figure 13.5. Some observations can be made. First, there is something special about the points P_1, P_2, and P_3. Notice that P_1 is *higher* than any other "nearby" point on the curve—and likewise for P_3. The point P_2 is *lower* than any other "nearby" point on the curve. Since P_1, P_2, and P_3 may not necessarily be the highest or lowest points on the *entire* curve, we simply say that the graph of f has a *relative maximum (point)* when $x = x_1$ (or at x_1) and when $x = x_3$, and has a *relative minimum (point)* when $x = x_2$. On the function level, when $x = x_1$ and $x = x_3$, f has *relative maximum values* of $f(x_1)$ and $f(x_3)$. Similarly, when $x = x_2$, f has a *relative minimum value* of $f(x_2)$. When we refer to a relative maximum or minimum, it is understood that we are referring to a point or value, depending on the context.

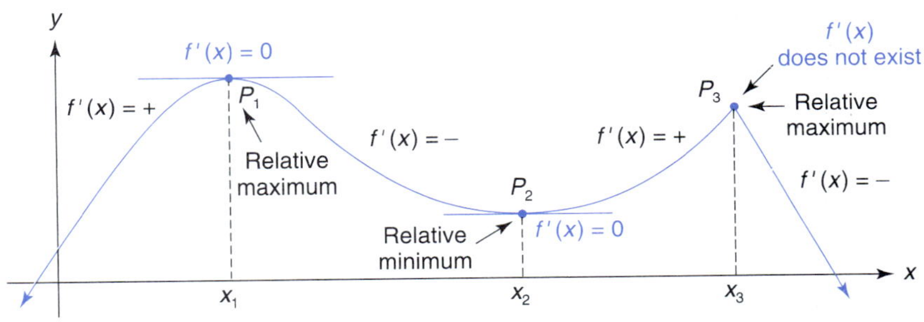

FIGURE 13.5 Relative maxima and relative minima.

Turning back to the graph, we see that there is an *absolute maximum* (highest point on the entire curve) when $x = x_1$, but there is no *absolute minimum* (lowest point on the entire curve) because the curve is assumed to extend downward indefinitely. More precisely, we define these new terms as follows:

Definition

A function f has a ***relative maximum*** when $x = x_0$ if there is an open interval containing x_0 on which $f(x_0) \geq f(x)$ for all x in the interval. The relative maximum is $f(x_0)$. A function f has a ***relative minimum*** when $x = x_0$ if there is an open interval containing x_0 on which $f(x_0) \leq f(x)$ for all x in the interval. The relative minimum is $f(x_0)$.

If it exists, an absolute maximum is unique; however, it may occur for more than one value of x. A similar statement is true for an absolute minimum.

Definition

A function f has an ***absolute maximum*** when $x = x_0$ if $f(x_0) \geq f(x)$ for all x in the domain of f. The absolute maximum is $f(x_0)$. A function f has an ***absolute minimum*** when $x = x_0$ if $f(x_0) \leq f(x)$ for all x in the domain of f. The absolute minimum is $f(x_0)$.

We refer to either a relative maximum or a relative minimum as a **relative extremum** (plural: *relative extrema*). Similarly, we speak of **absolute extrema.**

When dealing with relative extrema, we compare the function value at a point with values of nearby points; however, when dealing with absolute extrema, we compare the function value at a point with all other values determined by the domain. Thus, relative extrema are "local" in nature, whereas absolute extrema are "global" in nature.

Referring back to Figure 13.5, we notice that at a relative extremum the derivative may not be defined (as when $x = x_3$). But whenever it is defined at a relative extremum, it is 0 (as when $x = x_1$ and $x = x_2$), and hence, the tangent line is horizontal. We may state the following:

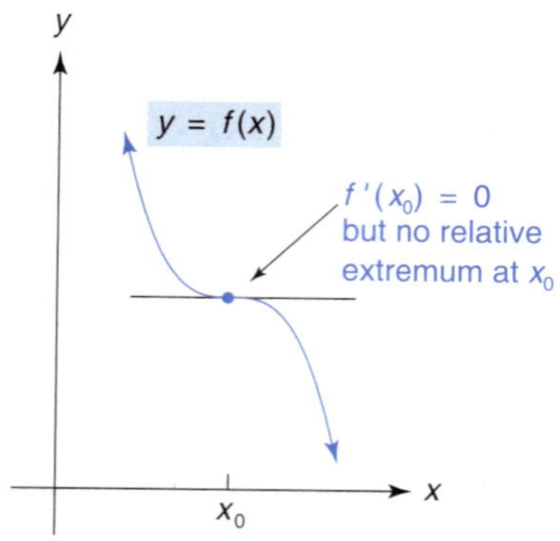

FIGURE 13.6 No relative extremum at x_0.

Rule 2 A Necessary Condition for Relative Extrema

If f has a relative extremum when $x = x_0$, then $f'(x_0) = 0$ or $f'(x_0)$ is not defined.

The implication in Rule 2 goes in only one direction:

$$\left.\begin{array}{c}\text{relative extremum}\\ \text{at } x_0\end{array}\right\} \Longrightarrow \left\{\begin{array}{c} f'(x_0) = 0 \\ \text{or} \\ f'(x_0) \text{ not defined}\end{array}\right.$$

Rule 2 does *not* say that if $f'(x_0)$ is 0 or is not defined, then there must be a relative extremum at x_0. In fact, there may not be one at all. For example, in Figure 13.6, $f'(x_0)$ is 0 because the tangent line is horizontal at x_0, but yet there is no relative extremum there.

In general, the best we can say is that relative extrema *may* occur at points on the graph of f where $f'(x) = 0$ or where $f'(x)$ is not defined. Because these points are so important for locating relative extrema, they are called *critical points*, and their x-coordinates are called *critical values*. Thus, in Figure 13.5, the numbers x_1, x_2, and x_3 are critical values, and P_1, P_2, and P_3 are critical points.

CAUTION We remind you again that not every critical value corresponds to a relative extremum. For example, if $y = f(x) = x^3$, then $f'(x) = 3x^2$. Since $f'(0) = 0$ and $f(0)$ is defined, 0 is a critical value. Now if $x < 0$, then $3x^2 > 0$. If $x > 0$, then $3x^2 > 0$. Since $f'(x)$ does not change sign, no relative maximum or minimum exists. Indeed, since $f'(x) \geq 0$ for all x, the graph of f never falls, and f is said to be *nondecreasing*. (See Figure 13.7.)

Definition

If x_0 is in the domain of f and either $f'(x_0) = 0$ or $f'(x_0)$ is not defined, then x_0 is called a ***critical value*** of f. If x_0 is a critical value, then the point $(x_0, f(x_0))$ is called a ***critical point.***

At a critical point, there may be a relative maximum, a relative minimum, or neither. Moreover, from Figure 13.5, we observe that each relative extremum occurs at a point around which the sign of $f'(x)$ is changing. For the relative maximum when $x = x_1$, $f'(x)$ goes from + for $x < x_1$ to − for $x > x_1$, *as long as x is near* x_1. At the relative minimum when $x = x_2$, $f'(x)$ goes from − to +, and at the relative maximum when $x = x_3$, it again goes from + to −. Thus, *around relative maxima, f is increasing and then decreasing, and the reverse holds for relative minima.* More precisely, we have the following rule:

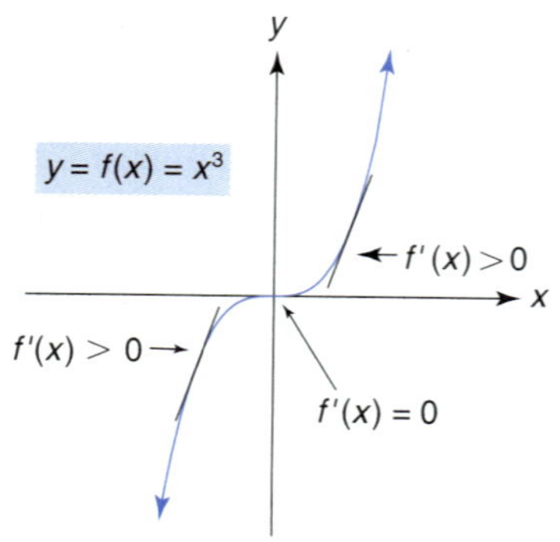

FIGURE 13.7 Zero is a critical value, but does not give a relative extremum.

Rule 3 Criteria for Relative Extrema

Suppose f is continuous on an open interval I that contains the critical value x_0 and f is differentiable on I, except possibly at x_0.

1. If $f'(x)$ changes from positive to negative as x increases through x_0, then f has a relative maximum when $x = x_0$.
2. If $f'(x)$ changes from negative to positive as x increases through x_0, then f has a relative minimum when $x = x_0$.

From our discussions and the preceding CAUTION, you should realize that a critical value is only a "candidate" for a relative extremum. It may correspond to a relative maximum, a relative minimum, or neither.

It is important to understand that not every value of x where $f'(x)$ does not exist is a critical value. For example, if

$$y = f(x) = \frac{1}{x^2} \quad \text{then} \quad f'(x) = -\frac{2}{x^3}$$

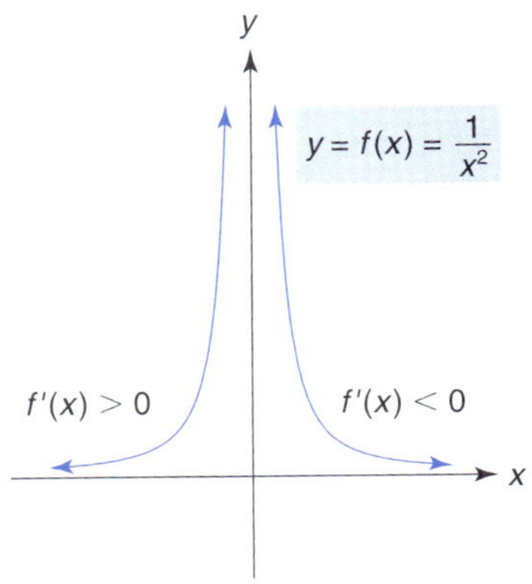

FIGURE 13.8 $f'(0)$ is not defined, but 0 is not a critical value because 0 is not in the domain of f.

Although $f'(x)$ is not defined when $x = 0$, 0 is not a critical value, because 0 is not in the domain of f. That is, no y-value corresponds to $x = 0$. Thus, a relative extremum cannot occur when $x = 0$. Nevertheless, the derivative may change sign around any x-value where $f'(x)$ is not defined, so such values are important in determining intervals over which f is increasing or decreasing.

$$\text{If } x < 0, \quad \text{then} \quad f'(x) = -\frac{2}{x^3} > 0$$

$$\text{If } x > 0, \quad \text{then} \quad f'(x) = -\frac{2}{x^3} < 0$$

Hence, f is increasing on $(-\infty, 0)$ and decreasing on $(0, \infty)$. (See Figure 13.8.)

In Rule 3 the hypotheses must be satisfied, or the conclusion need not hold. For example, consider the case-defined function

$$f(x) = \begin{cases} \dfrac{1}{x^2} & \text{if } x \neq 0 \\ 0 & \text{if } x = 0 \end{cases}$$

As you can see in Figure 13.9, 0 is in the domain and $f'(0)$ does not exist, so 0 is a critical value. Although $f'(x)$ is positive for $x < 0$ and $f'(x)$ is negative for $x > 0$, f does not have a relative maximum at 0. Rule 3 does not apply because f is not continuous at 0. Actually, 0 is an absolute minimum, according to the definition. This shows that if $f'(x_0)$ does not exist and f is not continuous at x_0, you need to examine carefully what happens around x_0.

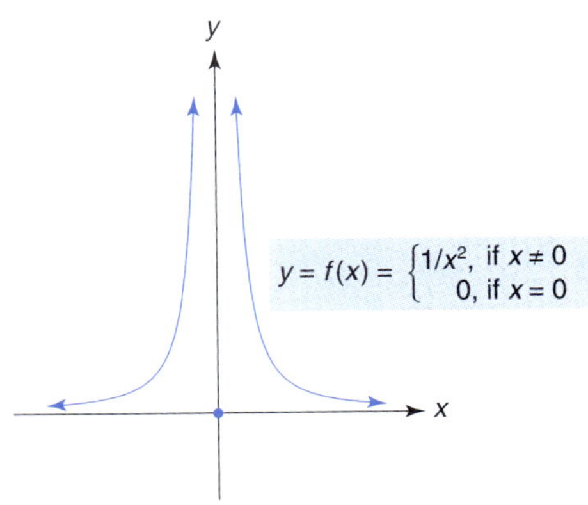

FIGURE 13.9 Zero is a critical value, but not a relative extremum.

Summarizing the results of this section, we have the *first-derivative test* for the relative extrema of $y = f(x)$:

First-Derivative Test for Relative Extrema

Step 1. Find $f'(x)$.

Step 2. Determine all values of x where $f'(x) = 0$ or $f'(x)$ is not defined. (These values include critical values and points of discontinuity.)

Step 3. On the intervals suggested by the values in step 2, determine whether f is increasing $[f'(x) > 0]$ or decreasing $[f'(x) < 0]$.

Step 4. For each critical value x_0 at which f is continuous, determine whether $f'(x)$ changes sign as x increases through x_0. There is a relative maximum when $x = x_0$ if $f'(x)$ changes from + to − going from left to right and a relative minimum if $f'(x)$ changes from − to + going from left to right. If $f'(x)$ does not change sign, there is no relative extremum when $x = x_0$.

APPLICATIONS IN PRACTICE 1

First-Derivative Test

The cost equation for a hot dog stand is given by $c(q) = 2q^3 - 21q^2 + 60q + 500$, where q is the number of hot dogs sold, and $c(q)$ is the cost in dollars. Use the first-derivative test to find when relative extrema occur.

EXAMPLE 1 **First-Derivative Test**

If $y = f(x) = x + \dfrac{4}{x+1}$, *use the first-derivative test to find when relative extrema occur.*

Solution:

Step 1. $f(x) = x + 4(x+1)^{-1}$, so

$$f'(x) = 1 + 4(-1)(x+1)^{-2} = 1 - \frac{4}{(x+1)^2}$$

$$= \frac{(x+1)^2 - 4}{(x+1)^2} = \frac{x^2 + 2x - 3}{(x+1)^2} = \frac{(x+3)(x-1)}{(x+1)^2}$$

Note that we expressed $f'(x)$ as a quotient with numerator and denominator factored. This enables us in step 2 to easily determine when $f'(x)$ is 0 or not defined.

Step 2. Setting $f'(x) = 0$ gives $x = -3, 1$. The denominator of $f'(x)$ is 0 when x is -1, so $f'(-1)$ does not exist. The values -3 and 1 are critical values, but -1 is not because $f(-1)$ is not defined (f is discontinuous at $x = -1$).

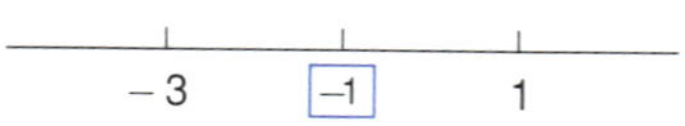

FIGURE 13.10 Four intervals to test for increasing/decreasing.

Step 3. The three values in step 2 lead us to test four intervals (Figure 13.10). (On each of these intervals, f is differentiable and is not zero.) Note that we have boxed the value -1 to indicate that it cannot correspond to a relative extremum. However, it is essential that -1 be considered in our analysis of increasing/decreasing.

If $x < -3$, then $f'(x) = \dfrac{(-)(-)}{(+)} = +$, so f is increasing

If $-3 < x < -1$, then $f'(x) = \dfrac{(+)(-)}{(+)} = -$, so f is decreasing

If $-1 < x < 1$, then $f'(x) = \dfrac{(+)(-)}{(+)} = -$, so f is decreasing

If $x > 1$, then $f'(x) = \dfrac{(+)(+)}{(+)} = +$, so f is increasing

(Figure 13.11)

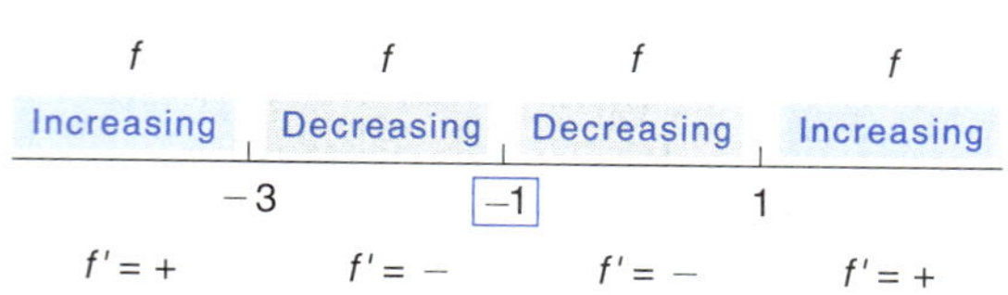

FIGURE 13.11 Sign chart for $f'(x) = \dfrac{(x+3)(x-1)}{(x+1)^2}$.

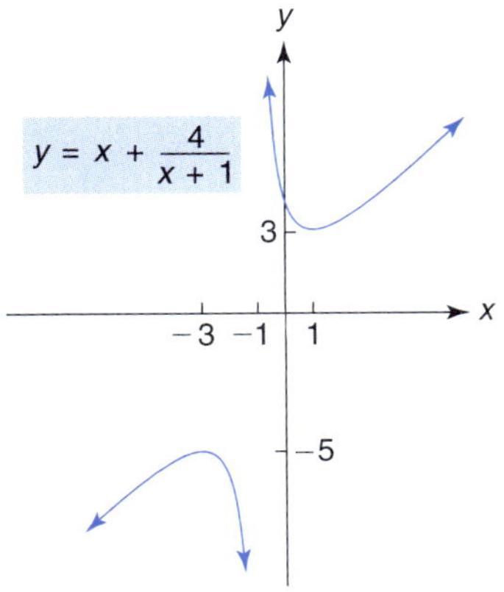

FIGURE 13.12 Graph of $y = x + \dfrac{4}{x+1}$.

Thus, f is increasing on the intervals $(-\infty, -3)$ and $(1, \infty)$ and is decreasing on $(-3, -1)$ and $(-1, 1)$.

Step 4. From Figure 13.11, we conclude that when $x = -3$, there is a relative maximum, since $f'(x)$ changes from $+$ to $-$. [This relative maximum value is $f(-3) = -3 + (4/-2) = -5$.] When $x = 1$, there is a relative minimum, because $f'(x)$ changes from $-$ to $+$. We ignore $x = -1$ since -1 is not a critical value. The graph is shown in Figure 13.12.

▶ **Now work Problem 37.**

EXAMPLE 2 **A Relative Extremum Where $f'(x)$ Does Not Exist**

Test $y = f(x) = x^{2/3}$ for relative extrema.

Solution: We have

$$f'(x) = \frac{2}{3}x^{-1/3} = \frac{2}{3\sqrt[3]{x}}.$$

When $x = 0$, $f'(x)$ is not defined, but $f(x)$ is defined. Thus, 0 is a critical value and there are no others. If $x < 0$, then $f'(x) < 0$. If $x > 0$, then $f'(x) > 0$. Therefore, there is a relative (as well as an absolute) minimum when $x = 0$. (See Figure 13.13.) Note that when $x = 0$, the tangent line exists and is vertical.

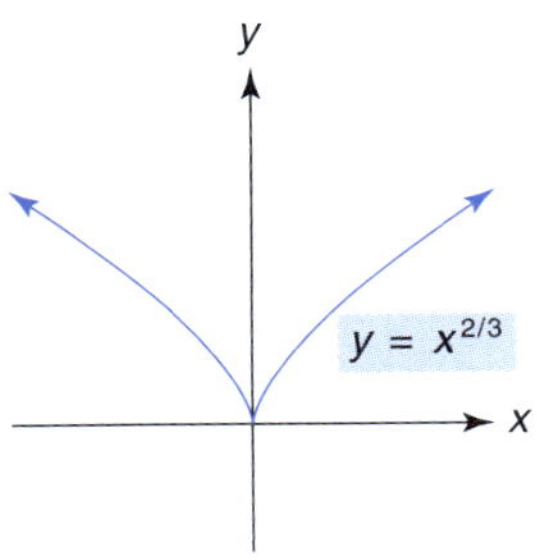

FIGURE 13.13 Derivative does not exist at 0 and there is a minimum at 0.

APPLICATIONS IN PRACTICE 2

Finding Relative Extrema

A drug is injected into a patient's bloodstream. The concentration of the drug in the bloodstream t hours after the injection is approximated by $C(t) = \dfrac{0.14t}{t^2 + 4t + 4}$. Find the relative extrema for $t > 0$, and use them to determine when the drug is at its greatest concentration.

EXAMPLE 3 **Finding Relative Extrema**

Test $y = f(x) = x^2e^x$ for relative extrema.

Solution: By the product rule,

$$f'(x) = x^2e^x + e^x(2x) = xe^x(x + 2)$$

Noting that e^x is always positive, we obtain the critical values 0 and -2. From the signs of $f'(x)$ given in Figure 13.14, we conclude that there is a relative maximum when $x = -2$ and a relative minimum when $x = 0$.

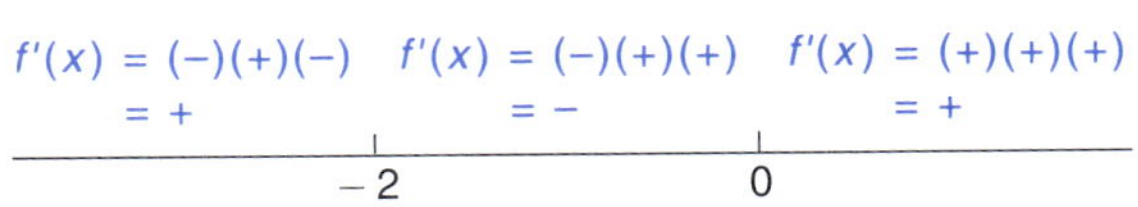

FIGURE 13.14 Sign chart for $f'(x) = xe^x(x + 2)$.

▶ **Now work Problem 49.**

Curve Sketching

In the next example we show how the first-derivative test, in conjunction with the notions of intercepts and symmetry, can be used as an aid in sketching the graph of a function.

EXAMPLE 4 Curve Sketching

Sketch the graph of $y = f(x) = 2x^2 - x^4$ with the aid of intercepts, symmetry, and the first-derivative test.

Solution:

Intercepts If $x = 0$, then $y = 0$. If $y = 0$, then

$$0 = 2x^2 - x^4 = x^2(2 - x^2) = x^2(\sqrt{2} + x)(\sqrt{2} - x),$$

so $x = 0, \pm\sqrt{2}$. Thus, the intercepts are $(0, 0)$, $(\sqrt{2}, 0)$, and $(-\sqrt{2}, 0)$.

Symmetry Testing for y-axis symmetry, we have

$$y = 2(-x)^2 - (-x)^4 \quad \text{that is} \quad y = 2x^2 - x^4$$

Since this is the original equation, there is y-axis symmetry. Because y is a function (and not the zero function), there is no x-axis symmetry and, hence, no symmetry about the origin.

First-Derivative Test

Step 1. $y' = 4x - 4x^3 = 4x(1 - x^2) = 4x(1 + x)(1 - x)$.

Step 2. Setting $y' = 0$ gives the critical values $x = 0, \pm 1$. Since we are interested in a graph, the critical *points* are important to us. By substituting the critical values into the *original* equation, $y = 2x^2 - x^4$, we obtain the y-coordinates of these points. We find the critical points to be $(-1, 1)$, $(0, 0)$, and $(1, 1)$.

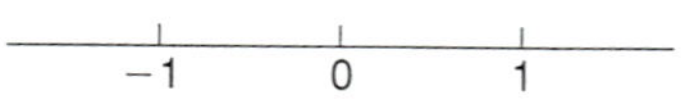

FIGURE 13.15 Intervals for sign chart of $y' = 4x(1 + x)(1 - x)$.

Step 3. There are four intervals to consider in Figure 13.15:

If $x < -1$, then $y' = 4(-)(-)(+) = +$, so f is increasing
If $-1 < x < 0$, then $y' = 4(-)(+)(+) = -$, so f is decreasing
If $0 < x < 1$, then $y' = 4(+)(+)(+) = +$, so f is increasing
If $x > 1$, then $y' = 4(+)(+)(-) = -$, so f is decreasing (Figure 13.16)

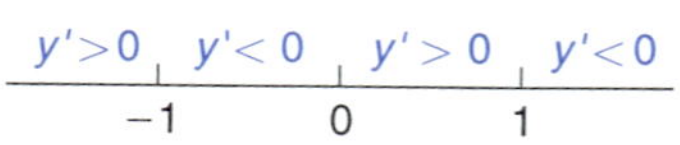

FIGURE 13.16 Sign chart of $y' = 4x(1 + x)(1 - x)$.

Step 4. From Figure 13.16, we conclude that relative maxima occur at $(-1, 1)$ and $(1, 1)$; a relative minimum occurs at $(0, 0)$.

Discussion In Figure 13.17(a), we have indicated the horizontal tangents at the relative maximum and minimum points. We know the curve rises from the left, has a relative maximum, then falls, has a relative minimum, then rises to a relative maximum, and falls thereafter. A sketch is shown in Figure 13.17(b).

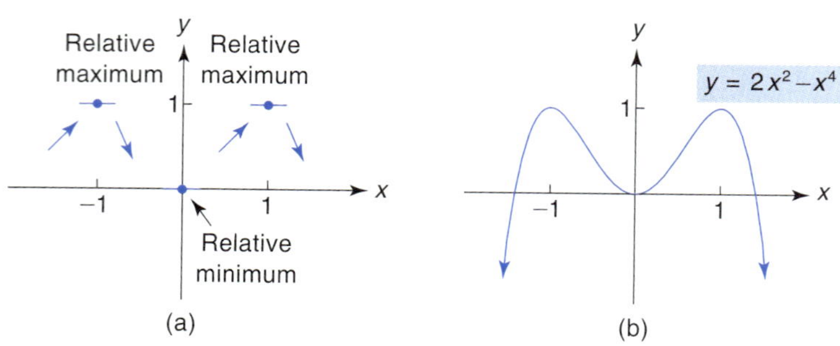

FIGURE 13.17 Putting together the graph of $y = 2x^2 - x^4$.

▸ **Now work Problem 59.**

As a passing comment, we note that in Example 4 *absolute* maxima occur at $x = \pm 1$. [See Figure 13.17(b).] There is no absolute minimum.

TECHNOLOGY

A graphing calculator is a powerful tool for investigating relative extrema. For example, consider the function

$$f(x) = 3x^4 - 4x^3 + 4$$

whose graph is shown in Figure 13.18. It appears that there is a relative minimum near $x = 1$. We can locate this minimum by either using "trace and zoom" or (on a TI-83 Plus) using the "minimum" feature. Figure 13.19 shows the latter approach. The relative minimum point is estimated to be (1.00, 3).

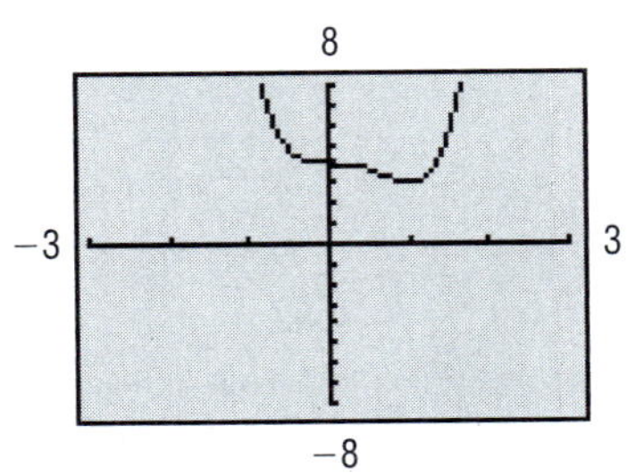

FIGURE 13.18 Graph of $f(x) = 3x^4 - 4x^3 + 4$.

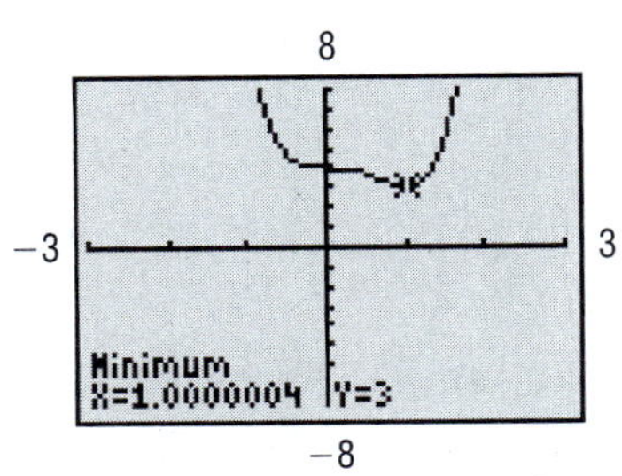

FIGURE 13.19 Relative minimum at (1.00, 3).

Now let us see how the graph of f' indicates when extrema occur. We have

$$f'(x) = 12x^3 - 12x^2$$

whose graph is shown in Figure 13.20. It appears that $f'(x)$ is 0 at two points. Using "trace and zoom" or the "zero" feature, we estimate the zeros of f' (the critical values of f) to be 1 and 0. Around $x = 1$, we see that $f'(x)$ goes from negative values to positive values. (That is, the graph of f' goes from below the x-axis to above it.) Thus, we conclude that f has a relative minimum at $x = 1$, which confirms our previous result.

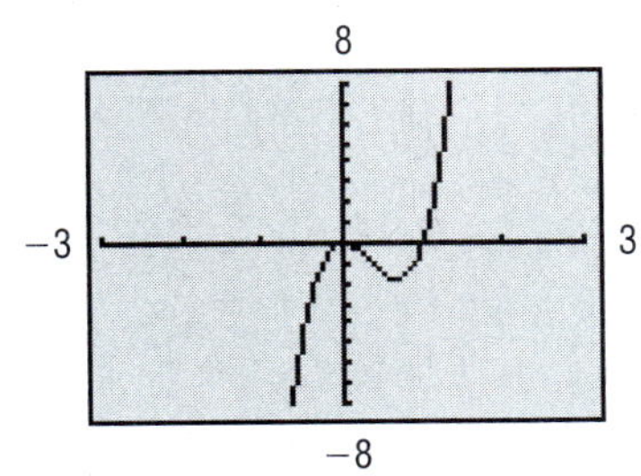

FIGURE 13.20 Graph of $f'(x) = 12x^3 - 12x^2$.

Around the critical value $x = 0$, the values of $f'(x)$ are negative. Since $f'(x)$ does not change sign, we conclude that there is no relative extremum at $x = 0$. This is also apparent from the graph in Figure 13.18.

It is worthwhile to note that we can approximate the graph of f' without determining $f'(x)$ itself. We make use of the "nDeriv" feature. First we enter the function f as Y_1. Then we set

$$Y_2 = \text{nDeriv}(Y_1, X, X)$$

The graph of Y_2 approximates the graph of $f'(x)$.

Exercise 13.1

In Problems 1–4, the graph of a function is given. Find the open intervals on which the function is increasing or decreasing, and find the coordinates of all relative extrema.

1.

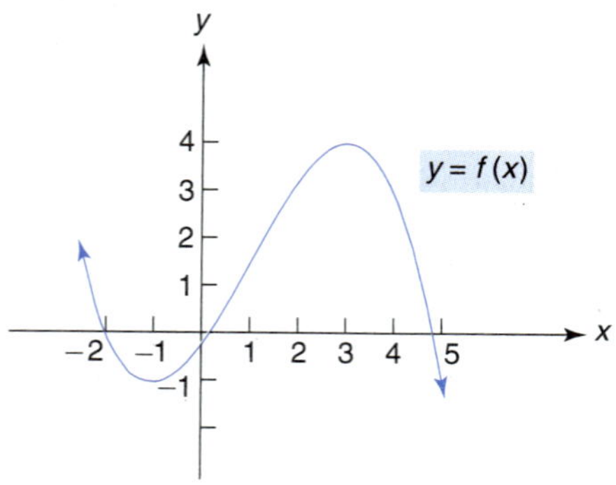

FIGURE 13.21 Graph for Problem 1.

2.

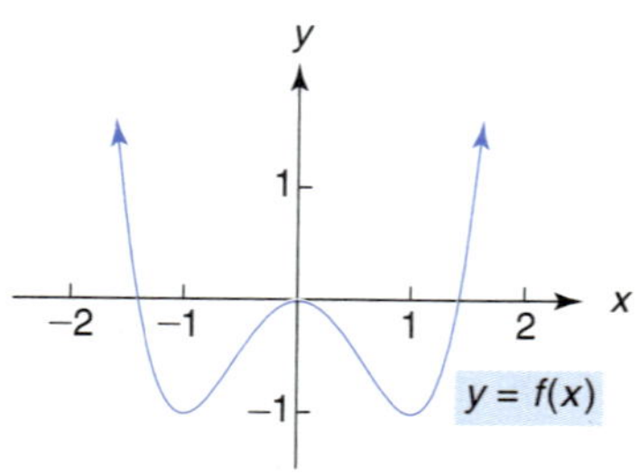

FIGURE 13.22 Graph for Problem 2.

3.

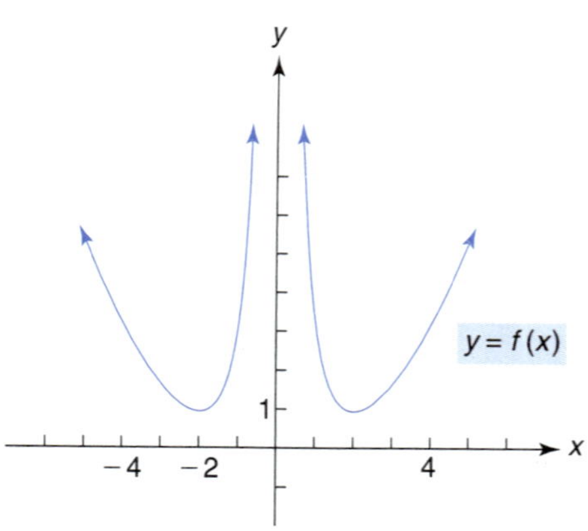

FIGURE 13.23 Graph for Problem 3.

4.

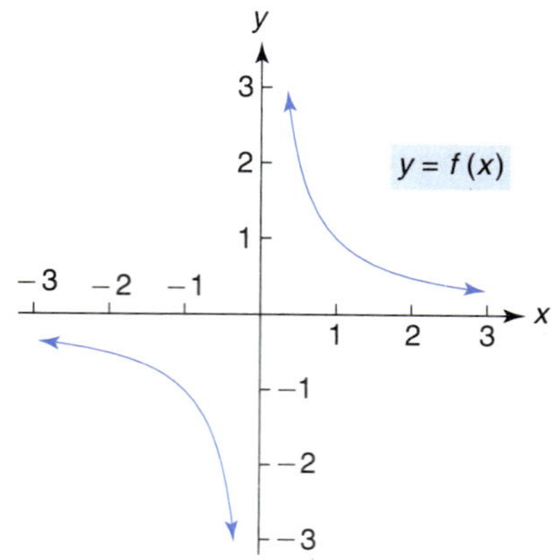

FIGURE 13.24 Graph for Problem 4.

In Problems 5–8, the derivative of the continuous function f is given. Find open intervals on which f is increasing or decreasing, and find the x-values of all relative extrema.

5. $f'(x) = (x+2)(x-1)$

6. $f'(x) = 2x(x-1)^3$

7. $f'(x) = (x+1)(x-3)^2$

8. $f'(x) = \dfrac{x(x+2)}{x^2+1}$

In Problems 9–52, determine when the function is increasing or decreasing, and determine when relative maxima and minima occur. Do not sketch the graph.

9. $y = 2x^3 + 1$

10. $y = x^2 + 4x + 3$

11. $y = x - x^2 + 2$

12. $y = 4x - x^2$

13. $y = -\dfrac{x^3}{3} - 2x^2 + 5x - 2$

14. $y = \dfrac{x^4}{4} + x^3$

15. $y = x^4 - 2x^2$

16. $y = -3 + 12x - x^3$

17. $y = x^3 - 6x^2 + 9x$

18. $y = x^3 - 6x^2 + 12x - 6$

19. $y = 2x^3 - \dfrac{11}{2}x^2 - 10x + 2$

20. $y = -5x^3 + x^2 + x - 7$

21. $y = \dfrac{x^3}{3} - 5x^2 + 22x + 1$

22. $y = \dfrac{9}{5}x^5 - \dfrac{47}{3}x^3 + 10x$

23. $y = 3x^5 - 5x^3$

24. $y = 6x - x^6$

25. $y = -x^5 - 5x^4 + 200$

26. $y = \dfrac{3x^4}{2} - 4x^3 + 17$

27. $y = 8x^4 - x^8$

28. $y = \dfrac{4}{5}x^5 - \dfrac{13}{3}x^3 + 3x + 4$

29. $y = (x^3 + 1)^3$

30. $y = \sqrt[3]{x}(x-2)$

31. $y = \dfrac{5}{x-1}$

32. $y = \dfrac{3}{x}$

33. $y = \dfrac{10}{\sqrt{x}}$

34. $y = \dfrac{3x}{2x+5}$

35. $y = \dfrac{x^2}{2-x}$

36. $y = x + \dfrac{4}{x}$

*37. $y = \frac{x^2 - 3}{x + 2}$

38. $y = \frac{2x^2}{4x^2 - 25}$

39. $y = \frac{5x + 2}{x^2 + 1}$

40. $y = \sqrt[3]{x^3 - 9x}$

41. $y = (x + 2)^3(x - 5)^2$

42. $y = x^2(x + 3)^4$

43. $y = x^3(x - 6)^4$

44. $y = x(1 - x)^{2/5}$

45. $y = e^{-\pi x} + \pi$

46. $y = x \ln x$

47. $y = x^2 - 9 \ln x$

48. $y = xe^x$

*49. $y = e^x + e^{-x}$

50. $y = e^{-x^2/2}$

51. $y = x \ln x - x$

52. $y = (x^2 + 1)e^{-x}$

In Problems 53–64, determine intervals on which the function is increasing or decreasing, relative maxima and minima, symmetry, and those intercepts that can be obtained conveniently. Then sketch the graph.

53. $y = x^2 - 6x - 7$

54. $y = 2x^2 - 5x - 12$

55. $y = 3x - x^3$

56. $y = x^4 - 16$

57. $y = 2x^3 - 9x^2 + 12x$

58. $y = 2x^3 - x^2 - 4x + 4$

*59. $y = x^4 + 4x^3 + 4x^2$

60. $y = x^5 - \frac{5}{4}x^4$

61. $y = (x - 1)^2(x + 2)^2$

62. $y = \sqrt{x}(x^2 - x - 2)$

63. $y = 2\sqrt{x} - x$

64. $y = x^{5/3} + 5x^{2/3}$

65. Sketch the graph of a continuous function f such that $f(1) = 2$, $f(3) = 1$, $f'(1) = f'(3) = 0$, $f'(x) > 0$ for $x < 1$, $f'(x) < 0$ for $1 < x < 3$, and f has a relative minimum when $x = 3$.

66. Sketch the graph of a continuous function f such that $f(1) = 2$, $f(4) = 5$, $f'(1) = 0$, $f'(x) \geq 0$ for $x < 4$, f has a relative maximum when $x = 4$, and there is a vertical tangent line when $x = 4$.

67. **Average Cost** If $c_f = 25{,}000$ is a fixed-cost function, show that the average fixed-cost function $\bar{c}_f = c_f/q$ is a decreasing function for $q > 0$. Thus, as output q increases, each unit's portion of fixed cost declines.

68. **Marginal Cost** If $c = 3q - 3q^2 + q^3$ is a cost function, when is marginal cost increasing?

69. **Marginal Revenue** Given the demand function

$$p = 400 - 2q$$

find when marginal revenue is increasing.

70. **Cost Function** For the cost function $c = \sqrt{q}$, show that marginal and average costs are always decreasing for $q > 0$.

71. **Revenue** For a manufacturer's product, the revenue function is given by $r = 240q + 57q^2 - q^3$. Determine the output for maximum revenue.

72. **Labor Markets** Eswaran and Kotwal[2] consider agrarian economies in which there are two types of workers, permanent and casual. Permanent workers are employed on long-term contracts and may receive benefits such as holiday gifts and emergency aid. Casual workers are hired on a daily basis and perform routine and menial tasks such as weeding, harvesting, and threshing. The difference z in the present-value cost of hiring a permanent worker over that of hiring a casual worker is given by

$$z = (1 + b)w_p - bw_c,$$

where w_p and w_c are wage rates for permanent labor and casual labor, respectively, b is a positive constant, and w_p is a function of w_c.

(a) Show that

$$\frac{dz}{dw_c} = (1 + b)\left[\frac{dw_p}{dw_c} - \frac{b}{1 + b}\right]$$

(b) If $dw_p/dw_c < b/(1 + b)$, show that z is a decreasing function of w_c.

73. **Thermal Pollution** In Shonle's discussion of thermal pollution,[3] the efficiency of a power plant is given by

$$E = 0.71\left(1 - \frac{T_c}{T_h}\right)$$

where T_h and T_c are the respective absolute temperatures of the hotter and colder reservoirs. Assume that T_c is a positive constant and that T_h is positive. Using calculus, show that as T_h increases, the efficiency increases.

74. **Telephone Service** In a discussion of the pricing of local telephone service, Renshaw[4] determines that total revenue r is given by

$$r = 2F + \left(1 - \frac{a}{b}\right)p - p^2 + \frac{a^2}{b}$$

where p is an indexed price per call, and a, b, and F are constants. Determine the value of p that maximizes revenue.

75. **Storage and Shipping Costs** In his model for storage and shipping costs of materials for a manufacturing process, Lancaster[5] derives the cost function

$$C(k) = 100\left(100 + 9k + \frac{144}{k}\right) \qquad 1 \leq k \leq 100$$

where $C(k)$ is the total cost (in dollars) of storage and transportation for 100 days of operation if a load of k tons of material is moved every k days.

(a) Find $C(1)$.

[2] M. Eswaran and A. Kotwal, "A Theory of Two-Tier Labor Markets in Agrarian Economics," *American Economic Review* 75, no. 1 (1985), 162–77.

[3] J. I. Shonle, *Environmental Applications of General Physics* (Reading, MA: Addison-Wesley Publishing Company, 1975).

[4] E. Renshaw, "A Note of Equity and Efficiency in the Pricing of Local Telephone Services," *American Economic Review* 75, no. 3 (1985), 515–18.

[5] P. Lancaster, *Mathematics: Models of the Real World* (Englewood Cliffs, NJ: Prentice-Hall, 1976).

(b) For what value of k does $C(k)$ have a minimum?
(c) What is the minimum value?

76. Physiology—The Bends When a deep-sea diver undergoes decompression or a pilot climbs to a high altitude, nitrogen may bubble out of the blood, causing what is commonly called *the bends*. Suppose the percentage P of people who suffer effects of the bends at an altitude of h thousand feet is given by[6]

$$P = \frac{100}{1 + 100{,}000e^{-0.36h}}$$

Is P an increasing function of h?

In Problems 77–80, from the graph of the function, find the coordinates of all relative extrema. Round your answers to two decimal places.

77. $y = 0.5x^2 + 4.1x + 6.2$

78. $y = 3x^4 - 4x^3 - 5x + 1$

79. $y = \dfrac{8.2x}{0.4x^2 + 3}$

80. $y = \dfrac{e^x(3 - x)}{7x^2 + 1}$

[6] Adapted from G. E. Folk, Jr., *Textbook of Environmental Physiology*, 2nd ed. (Philadelphia: Lea & Febiger, 1974).

81. Graph the function

$$f(x) = [x(x-2)(2x-3)]^2$$

in the window $-1 \le x \le 3$, $-1 \le y \le 3$. Upon first glance, it may appear that this function has two relative minimum points and one relative maximum point. However, in reality, it has three relative minimum points and two relative maximum points. Determine the x-values of all these points. Round answers to two decimal places.

82. If $f(x) = 3x^3 - 7x^2 + 4x + 2$, display the graphs of f and f' on the same screen. Notice that $f'(x) = 0$ where relative extrema of f occur.

83. Let $f(x) = 6 + 4x - 3x^2 - x^3$. (a) Find $f'(x)$. (b) Graph $f'(x)$. (c) Observe where $f'(x)$ is positive and where it is negative. Give the intervals (rounded to two decimal places) where f is increasing and where f is decreasing. (d) Graph f and f' on the same screen, and verify your results to part (c).

84. If $f(x) = 2x^4 - x^2 - (3x + 2)^2$, find $f'(x)$. Determine the critical values of f. Round your answers to two decimal places.

OBJECTIVE *To find extreme values on a closed interval.*

13.2 Absolute Extrema on a Closed Interval

If a function f is *continuous* on a *closed* interval $[a, b]$, it can be shown that of *all* the function values $f(x)$ for x in $[a, b]$, there must be an (absolute) maximum value and an (absolute) minimum value. These two values are called **extreme values** of f on that interval. This important property of continuous functions is called the *extreme-value theorem*.

Extreme-Value Theorem

If a function is continuous on a closed interval, then the function has *both* a maximum value and a minimum value on that interval.

For example, each function in Figure 13.25 is continuous on the closed interval $[1, 3]$. Geometrically, the extreme-value theorem assures us that over this interval each graph has a highest point and a lowest point.

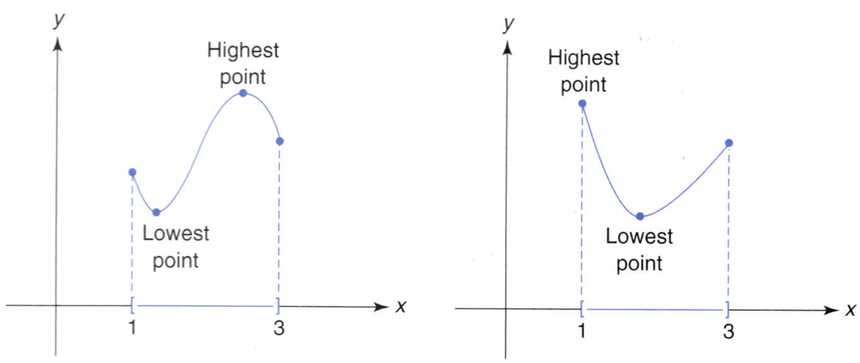

FIGURE 13.25 Illustrating the extreme-value theorem.

In the extreme-value theorem, you must realize that we are dealing with

1. a closed interval and
2. a function continuous on that interval.

If either condition (1) or condition (2) is not met, then extreme values are not guaranteed. For example, Figure 13.26(a) shows the graph of the continuous function $f(x) = x^2$ on the *open* interval $(-1, 1)$. You can see that f has no maximum value on the interval (although f has a minimum value there). Now consider the function $f(x) = 1/x^2$ on the closed interval $[-1, 1]$. Here f is *not continuous* at 0. From the graph of f in Figure 13.26(b), you can see that f has no maximum value (although there is a minimum value).

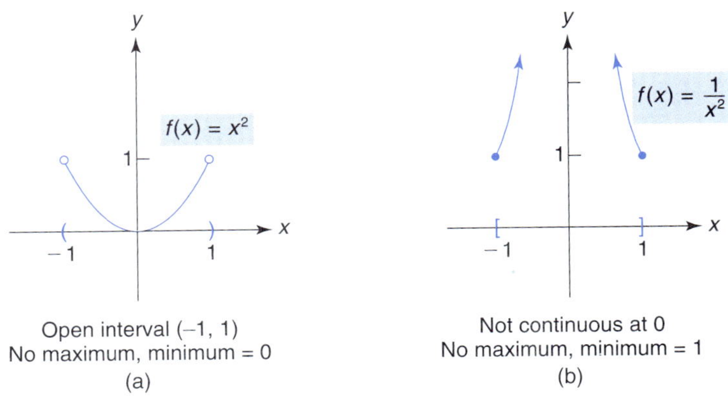

FIGURE 13.26 Extreme-value theorem does not apply.

In the previous section, our emphasis was on relative extrema. Now we shall focus our attention on absolute extrema and make use of the extreme-value theorem where possible. If the domain of a function is a closed interval, to determine *absolute* extrema we must examine the function not only at critical values, but also at the endpoints. For example, Figure 13.27 shows the graph of the continuous function $y = f(x)$ over $[a, b]$. The extreme-value theorem guarantees absolute extrema over the interval. Clearly, the important points on the graph occur at $x = a, b, c$, and d, which correspond to endpoints or

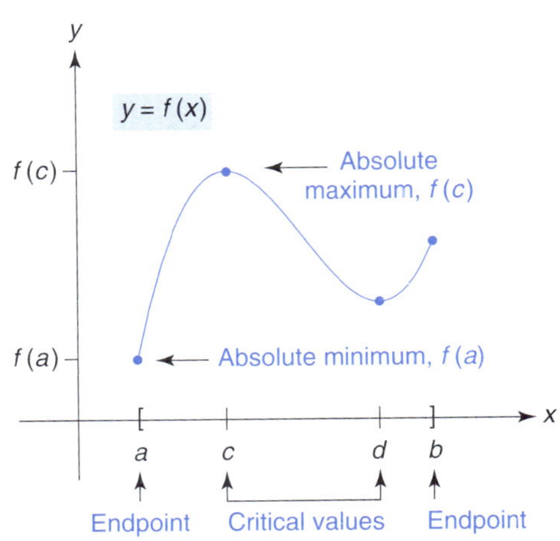

FIGURE 13.27 Absolute extrema.

critical values. Notice that the absolute maximum occurs at the critical value c and the absolute minimum occurs at the endpoint a. These results suggest the following procedure:

Procedure to Find Absolute Extrema for a Function f That Is Continuous on $[a, b]$

Step 1. Find the critical values of f.

Step 2. Evaluate $f(x)$ at the endpoints a and b and at the critical values on (a, b).

Step 3. The maximum value of f is the greatest of the values found in step 2. The minimum value of f is the least of the values found in step 2.

EXAMPLE 1 Finding Extreme Values on a Closed Interval

Find absolute extrema for $f(x) = x^2 - 4x + 5$ *over the closed interval* $[1, 4]$.

Solution: Since f is continuous on $[1, 4]$, the foregoing procedure applies.

Step 1. To find the critical values of f, we first find f':

$$f'(x) = 2x - 4 = 2(x - 2)$$

This gives the critical value $x = 2$.

Step 2. Evaluating $f(x)$ at the endpoints 1 and 4 and at the critical value 2, we have

$$\left.\begin{aligned} f(1) &= 2 \\ f(4) &= 5 \end{aligned}\right\} \text{ values of } f \text{ at endpoints}$$

and

$$f(2) = 1 \quad \text{value of } f \text{ at critical value 2 in } (1, 4)$$

Step 3. From the function values in step 2, we conclude that the maximum is $f(4) = 5$ and the minimum is $f(2) = 1$. (See Figure 13.28.)

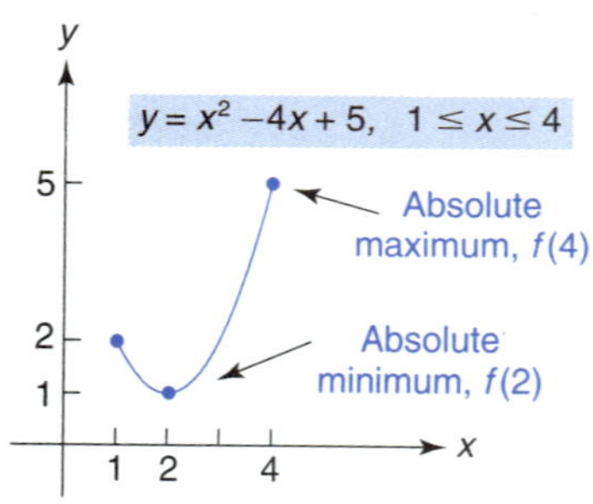

FIGURE 13.28 Extreme values for Example 1.

▸ **Now work Problem 1.**

Exercise 13.2

In Problems 1–14, find the absolute extrema of the given function on the given interval.

*1. $f(x) = x^2 - 2x + 3$, $[0, 3]$

2. $f(x) = -2x^2 - 6x + 5$, $[-3, 2]$

3. $f(x) = \frac{1}{3}x^3 + \frac{1}{2}x^2 - 2x + 1$, $[-1, 0]$

4. $f(x) = \frac{1}{4}x^4 - \frac{3}{2}x^2$, $[0, 1]$

5. $f(x) = 4x^3 + 3x^2 - 18x + 3$, $[\frac{1}{2}, 3]$

6. $f(x) = x^{4/3}$, $[-8, 8]$

7. $f(x) = -3x^5 + 5x^3$, $[-2, 0]$

8. $f(x) = \frac{7}{3}x^3 + 2x^2 - 3x + 1$, $[0, 3]$

9. $f(x) = 3x^4 - x^6$, $[-1, 2]$

10. $f(x) = \frac{1}{4}x^4 - \frac{1}{2}x^2 + 3$, $[-2, 3]$

11. $f(x) = x^4 - 9x^2 + 2$, $[-1, 3]$

12. $f(x) = \frac{x}{x^2+1}$, $[0, 2]$

13. $f(x) = x^{2/3}$, $[-2, 3]$

14. $f(x) = 0.2x^3 - 3.6x^2 + 2x + 1$, $[-1, 2]$

15. Consider the function

$$f(x) = x^4 + 8x^3 + 21x^2 + 20x + 9$$

over the interval $[-4, 9]$.

(a) Determine the value(s) (rounded to two decimal places) of x at which f attains a minimum value.
(b) What is the minimum value (rounded to two decimal places) of f?
(c) Determine the value(s) of x at which f attains a maximum value.
(d) What is the maximum value of f?

OBJECTIVE *To test a function for concavity and inflection points. Also, to sketch curves with the aid of the information obtained from the first and second derivatives.*

13.3 Concavity

You have seen that the first derivative provides much information for sketching curves. It is used to determine when a function is increasing or decreasing and to locate relative maxima and minima. However, to be sure we know the true shape of a curve, we may need more information. For example, consider the curve $y = f(x) = x^2$. Since $f'(x) = 2x$, $x = 0$ is a critical value. If $x < 0$, then $f'(x) < 0$, and f is decreasing; if $x > 0$, then $f'(x) > 0$, and f is increasing. Thus, there is a relative minimum when $x = 0$. In Figure 13.29, both curves meet the preceding conditions. But which one truly describes the curve $y = x^2$? This question will easily be settled by using the second derivative and the notion of *concavity*.

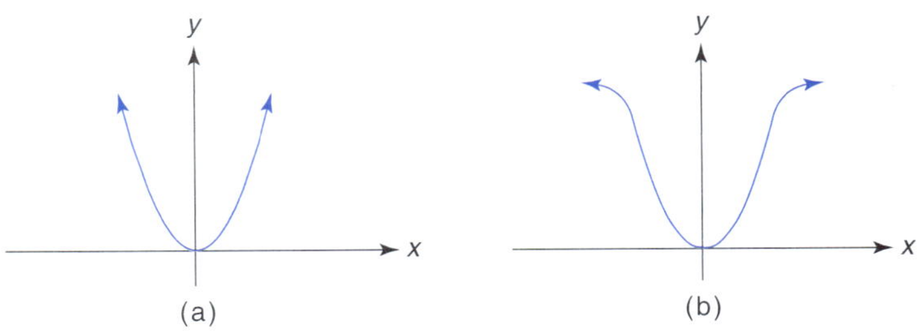

FIGURE 13.29 Two functions with $f'(x) < 0$ for $x < 0$ and $f'(x) > 0$ for $x > 0$.

In Figure 13.30, note that each curve $y = f(x)$ "bends" (or opens) upward. This means that if tangent lines are drawn to each curve, the curves lie *above* them. Moreover, the slopes of the tangent lines *increase* in value as x increases: In part (a), the slopes go from small positive values to larger values; in part (b), they are negative and approaching zero (and thus increasing); in part (c), they pass from negative values to positive values. Since $f'(x)$ gives the slope at a point, an increasing slope means that f' must be an increasing function. To describe this property, each curve (or function f) is said to be *concave up*.

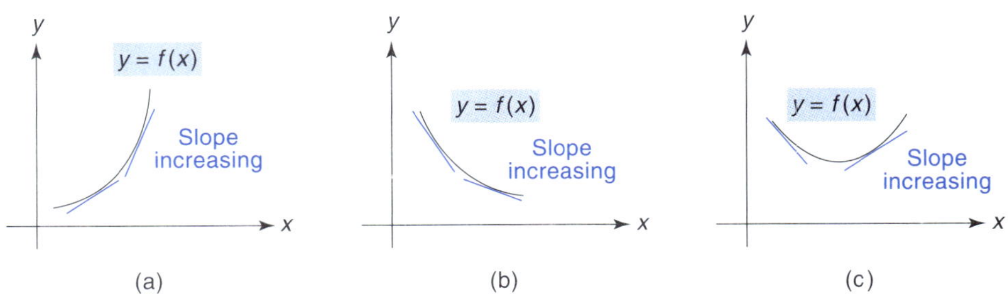

FIGURE 13.30 Each curve is concave up.

CAUTION Concavity relates to whether f', not f, is increasing or decreasing. In Figure 13.30(b), note that f is concave up and decreasing; however, in Figure 13.31(a), f is concave down and decreasing.

In Figure 13.31, it can be seen that each curve lies *below* the tangent lines and the curves are bending downward. As x increases, the slopes of the tangent lines are *decreasing*. Thus, f' must be a decreasing function here, and we say that f is *concave down*.

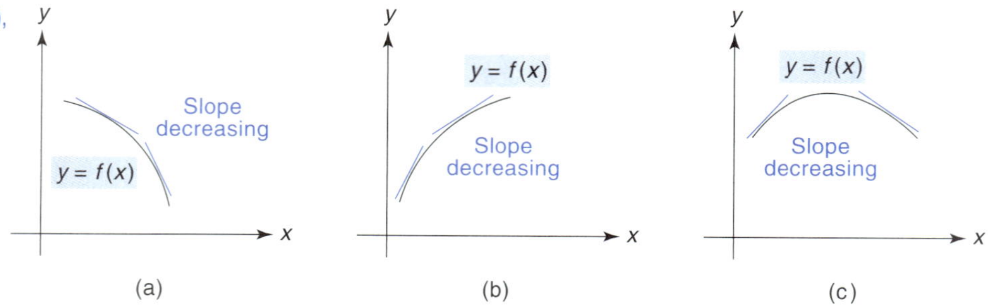

FIGURE 13.31 Each curve is concave down.

Definition

Let f be differentiable on the interval (a, b). Then f is said to be ***concave up*** [***concave down***] on (a, b) if f' is increasing [decreasing] on (a, b).

Remember: If f is concave up on an interval, then, geometrically, its graph is bending upward there. If f is concave down, then its graph is bending downward.

Since f' is increasing when its derivative $f''(x)$ is positive, and f' is decreasing when $f''(x)$ is negative, we can state the following rule:

Rule 1 Criteria for Concavity

Let f' be differentiable on the interval (a, b). If $f''(x) > 0$ for all x in (a, b), then f is concave up on (a, b). If $f''(x) < 0$ for all x in (a, b), then f is concave down on (a, b).

A function f is also said to be concave up at a point x_0 if there exists an open interval around x_0 on which f is concave up. In fact, for the functions that we shall consider, if $f''(x_0) > 0$, then f is concave up at x_0. Similarly, f is concave down at x_0 if $f''(x_0) < 0$.

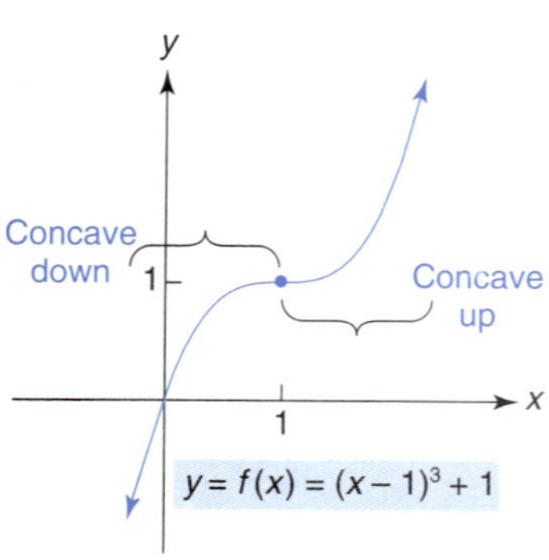

FIGURE 13.32 Concavity for $f(x) = (x-1)^3 + 1$.

EXAMPLE 1 Testing for Concavity

Determine where the given function is concave up and where it is concave down.

a. $y = f(x) = (x-1)^3 + 1$.

Solution: To apply Rule 1, we must examine the signs of y''. Now, $y' = 3(x-1)^2$, so

$$y'' = 6(x-1)$$

Thus, f is concave up when $6(x-1) > 0$, that is, when $x > 1$. And f is concave down when $6(x-1) < 0$, that is, when $x < 1$. (See Figure 13.32.)

b. $y = x^2$.

Solution: We have $y' = 2x$ and $y'' = 2$. Because y'' is always positive, the graph of $y = x^2$ must always be concave up, as in Figure 13.29(a). The graph cannot appear as in Figure 13.29(b), for that curve is sometimes concave down.

A point on a graph where concavity changes from concave down to concave up, or vice versa, such as (1, 1) in Figure 13.32, is called an *inflection point* or a *point of inflection*. Around such a point, the sign of $f''(x)$ must go from $-$ to $+$ or from $+$ to $-$. More precisely, we have the following definition:

The definition of an inflection point implies that x_0 is in the domain of f.

Definition

A function f has an ***inflection point*** when $x = x_0$ if and only if f is continuous at x_0 and f changes concavity at x_0.

▶ **Now work Problem 1.**

To test a function for concavity and inflection points, first find the values of x where $f''(x)$ is 0 or not defined. These values of x determine intervals. On each interval, determine whether $f''(x) > 0$ (f is concave up) or $f''(x) < 0$ (f is concave down). If concavity changes around one of these x-values and f is continuous there, then f has an inflection point at this x-value. The continuity requirement implies that the x-value must be in the domain of the function. In brief, a *candidate* for an inflection point must satisfy two conditions:

1. f'' must be 0 or not defined at that point.
2. f must be continuous at that point.

The candidate *will be* an inflection point if concavity changes around it. For example, if $f(x) = x^{1/3}$, then $f'(x) = \frac{1}{3}x^{-2/3}$ and

$$f''(x) = -\frac{2}{9}x^{-5/3} = -\frac{2}{9x^{5/3}}$$

Because f'' is not defined at 0, but f is continuous at 0, there is a candidate for an inflection point when $x = 0$. If $x > 0$, then $f''(x) < 0$, so f is concave down for $x > 0$; if $x < 0$, then $f''(x) > 0$, so f is concave up for $x < 0$. Because concavity changes at $x = 0$, there is an inflection point there. (See Figure 13.33.)

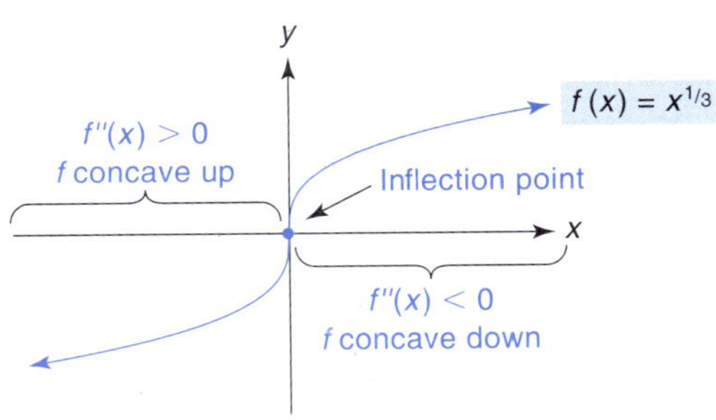

FIGURE 13.33 Inflection point for $f(x) = x^{1/3}$.

EXAMPLE 2 **Concavity and Inflection Points**

Test $y = 6x^4 - 8x^3 + 1$ *for concavity and inflection points.*

Solution: We have

$$y' = 24x^3 - 24x^2$$

$$y'' = 72x^2 - 48x = 24x(3x - 2)$$

FIGURE 13.34 Intervals to consider for concavity of $y = 6x^4 - 8x^3 + 1$.

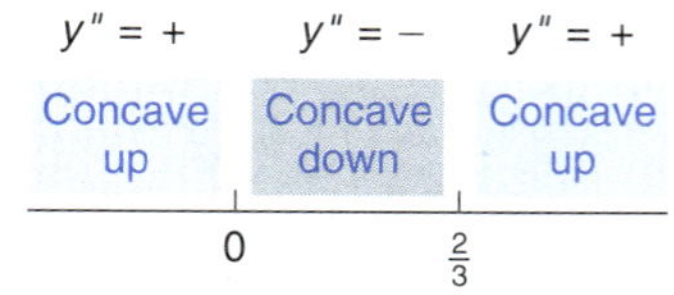

FIGURE 13.35 Sign chart of $y'' = 24x(3x - 2)$.

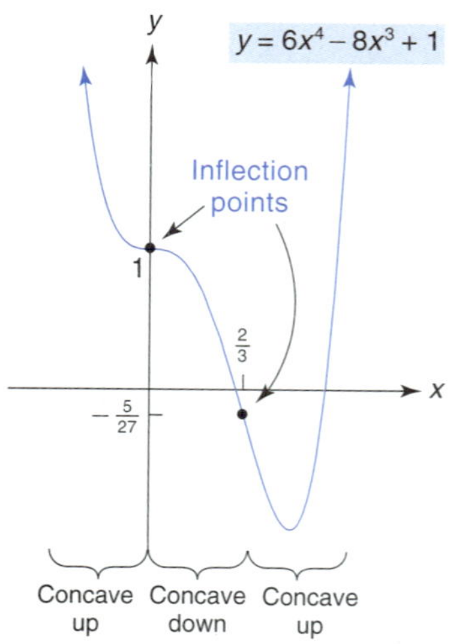

FIGURE 13.36 Graph of $y = 6x^4 - 8x^3 + 1$.

To find where $y'' = 0$, we set each factor in y'' equal to 0. This gives $x = 0, \frac{2}{3}$. We also note that y'' is never undefined. Thus, there are three intervals to consider. (See Figure 13.34.) Since y is continuous at 0 and $\frac{2}{3}$, these points are candidates for inflection points.

If $x < 0$, then $y'' = 24(-)(-) = +$, so the curve is concave up;

If $0 < x < \frac{2}{3}$, then $y'' = 24(+)(-) = -$, so the curve is concave down;

If $x > \frac{2}{3}$, then $y'' = 24(+)(+) = +$, so the curve is concave up. (See Figure 13.35.)

Since concavity changes at the points where $x = 0$ and $\frac{2}{3}$, these candidates are indeed inflection points. (See Figure 13.36.) In summary, the curve is concave up on $(-\infty, 0)$ and $(\frac{2}{3}, \infty)$, and is concave down on $(0, \frac{2}{3})$. Inflection points occur when $x = 0$ or $x = \frac{2}{3}$. These points are $(0, y(0)) = (0, 1)$ and $(\frac{2}{3}, y(\frac{2}{3})) = (\frac{2}{3}, -\frac{5}{27})$.

▶ **Now work Problem 13.**

EXAMPLE 3 **A Change in Concavity with No Inflection Point**

Discuss concavity and find all inflection points for $f(x) = \frac{1}{x}$.

Solution: Since $f(x) = x^{-1}$,

$$f'(x) = -x^{-2}$$

$$f''(x) = 2x^{-3} = \frac{2}{x^3}$$

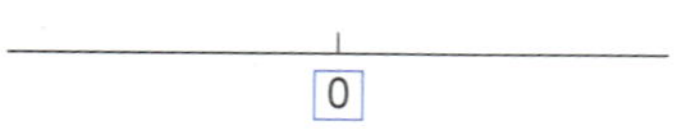

FIGURE 13.37 Intervals to consider for concavity analysis.

We see that $f''(x)$ is never 0, but it is not defined when $x = 0$. Since f is not continuous at 0, we conclude that 0 is not a candidate for an inflection point. Thus, the given function has no inflection point. However, 0 must be considered in an analysis of concavity. (See the number line in Figure 13.37; note that we have boxed the value 0 to indicate that it cannot correspond to an inflection point.) If $x > 0$, then $f''(x) > 0$; if $x < 0$, then $f''(x) < 0$. Hence, f is concave up on $(0, \infty)$ and concave down on $(-\infty, 0)$. (See Figure 13.38.) Although concavity changes around $x = 0$, there is no inflection point there because f is not continuous at 0 (nor is it even defined there).

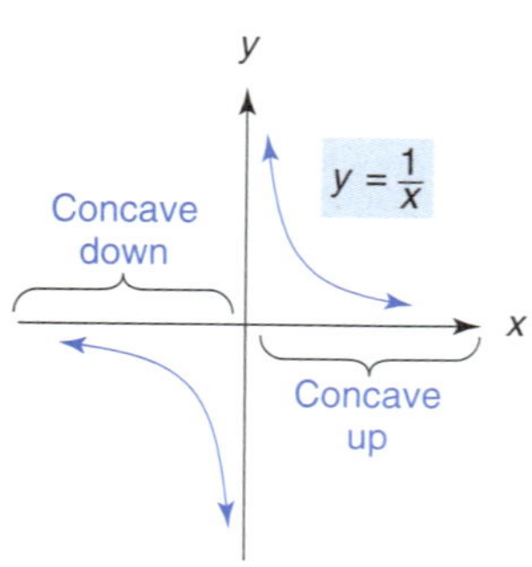

FIGURE 13.38 Graph of $y = \frac{1}{x}$.

▶ **Now work Problem 23.**

CAUTION A candidate for an inflection point may not necessarily be an inflection point. For example, if $f(x) = x^4$, then $f''(x) = 12x^2$ and $f''(0) = 0$. But $x < 0$ implies that $f''(x) > 0$, and $x > 0$ also implies that $f''(x) > 0$. Thus, concavity does not change, and there are no inflection points. (See Figure 13.39.)

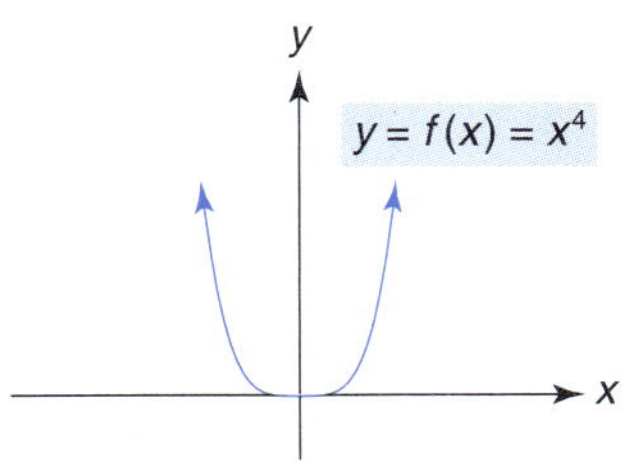

FIGURE 13.39 Graph of $f(x) = x^4$.

Curve Sketching

EXAMPLE 4 **Curve Sketching**

Sketch the graph of $y = 2x^3 - 9x^2 + 12x$.

Solution:

Intercepts If $x = 0$, then $y = 0$. Setting $y = 0$ gives $0 = x(2x^2 - 9x + 12)$. Clearly, $x = 0$, and using the quadratic formula on $2x^2 - 9x + 12 = 0$ gives no real roots. Thus, the only intercept is (0, 0).

Symmetry None.

Maxima and Minima Letting $y = f(x)$, we have

$$f'(x) = 6x^2 - 18x + 12 = 6(x^2 - 3x + 2) = 6(x - 1)(x - 2)$$

The critical values are $x = 1, 2$. (See Figure 13.40.)

If $x < 1$, then $f'(x) = 6(-)(-) = +$, so f is *increasing;*

If $1 < x < 2$, then $f'(x) = 6(+)(-) = -$, so f is *decreasing;*

If $x > 2$, then $f'(x) = 6(+)(+) = +$, so f is *increasing.* (See Figure 13.41.)

There is a relative maximum when $x = 1$ and a relative minimum when $x = 2$.

FIGURE 13.40 Critical values of $y = 2x^3 - 9x^2 + 12x$.

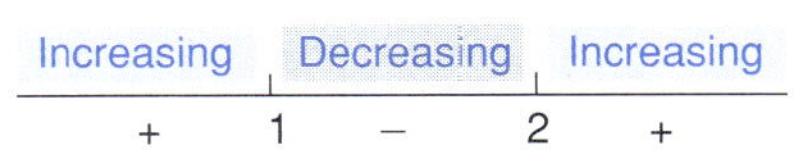

FIGURE 13.41 Sign chart of $f'(x)$.

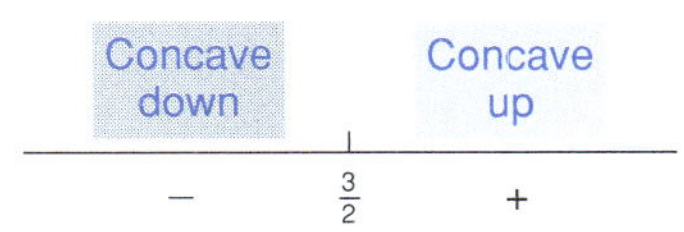

FIGURE 13.42 Sign chart of $f''(x)$.

Concavity

$$f''(x) = 12x - 18 = 6(2x - 3)$$

Setting $f''(x) = 0$ gives a possible inflection point at $x = \frac{3}{2}$.

If $x < \frac{3}{2}$, then $f''(x) < 0$, so f is concave down;

If $x > \frac{3}{2}$, then $f''(x) > 0$, so f is concave up. (See Figure 13.42.)

Since concavity changes around $x = \frac{3}{2}$ and f is continuous there, f has an inflection point when $x = \frac{3}{2}$.

Discussion We now find the coordinates of the important points on the graph (and any other points if there is doubt as to the behavior of the curve). We have the following table:

x	0	1	$\frac{3}{2}$	2
y	0	5	$\frac{9}{2}$	4

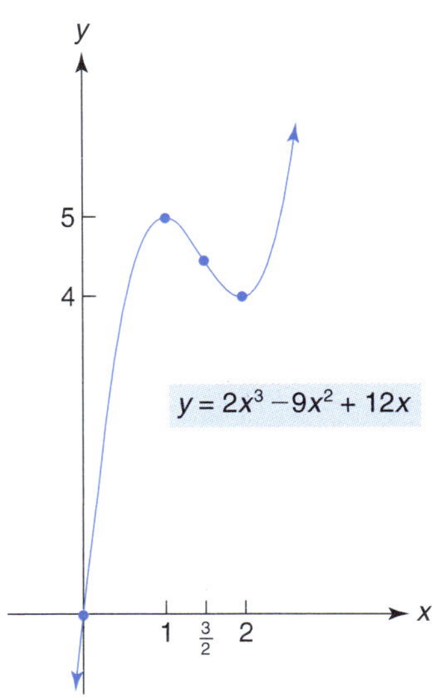

FIGURE 13.43 Graph of $y = 2x^3 - 9x^2 + 12x$.

As x increases, the function is first concave down and increases to a relative maximum at (1, 5); it then decreases to $(\frac{3}{2}, \frac{9}{2})$, after which it becomes concave up, but continues to decrease until it reaches a relative minimum at (2, 4); thereafter, it increases and is still concave up. (See Figure 13.43.)

▶ **Now work Problem 39.**

TECHNOLOGY

Suppose that you need to find the inflection points for

$$f(x) = \frac{1}{20}x^5 - \frac{17}{16}x^4 + \frac{273}{32}x^3 - \frac{4225}{128}x^2 + \frac{750}{4}$$

The second derivative of f is given by

$$f''(x) = x^3 - \frac{51}{4}x^2 + \frac{819}{16}x - \frac{4225}{64}$$

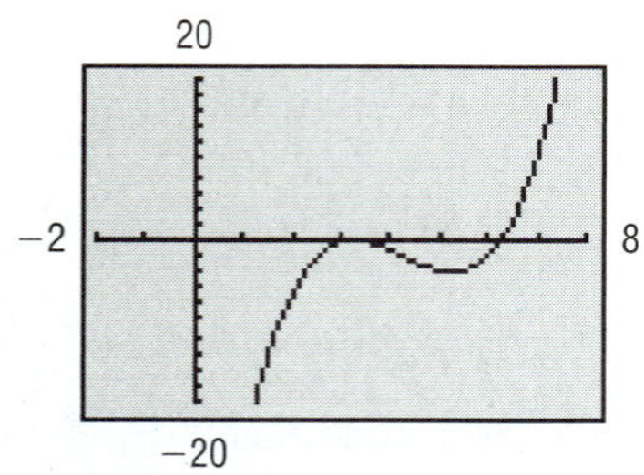

FIGURE 13.44 Graph of f''; zeros of f'' are 3.25 and 6.25.

Here the zeros of f'' are not obvious. Thus, we shall graph f'' using a graphing calculator. (See Figure 13.44.) We find that the zeros of f'' are approximately 3.25 and 6.25. Around $x = 6.25$, $f''(x)$ goes from negative to positive values. Therefore, at $x = 6.25$, there is an inflection point. Around $x = 3.25$, $f''(x)$ does not change sign, so no inflection point exists at $x = 3.25$. Comparing our results with the graph of f in Figure 13.45, we see that everything checks out.

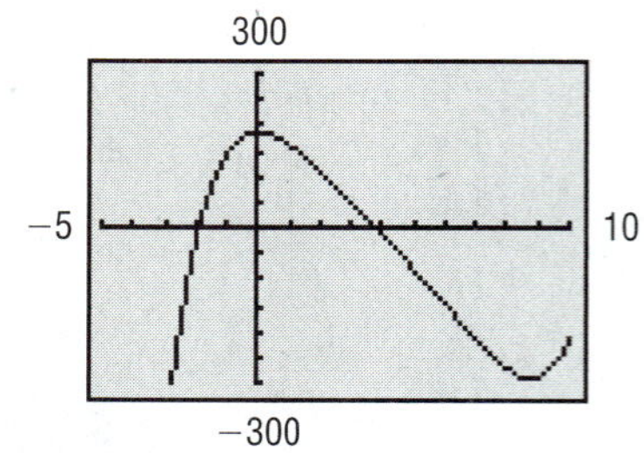

FIGURE 13.45 Graph of f; inflection point at $x = 6.25$, but not at $x = 3.25$.

Exercise 13.3

In Problems 1–6, a function and its second derivative are given. Determine the concavity of f and x-values where points of inflection occur.

*1. $f(x) = x^4 - 3x^3 + 7x - 5$; $f''(x) = 6x(2x - 3)$

2. $f(x) = \frac{x^5}{20} + \frac{x^4}{4} - 2x^2$; $f''(x) = (x - 1)(x + 2)^2$

3. $f(x) = \frac{2 + x - x^2}{x^2 - 2x + 1}$; $f''(x) = \frac{2(7 - x)}{(x - 1)^4}$

4. $f(x) = \frac{x^2}{(x - 1)^2}$; $f''(x) = \frac{2(2x + 1)}{(x - 1)^4}$

5. $f(x) = \frac{x^2 + 1}{x^2 - 2}$; $f''(x) = \frac{6(3x^2 + 2)}{(x^2 - 2)^3}$

6. $f(x) = x\sqrt{4 - x^2}$; $f''(x) = \frac{2x(x^2 - 6)}{(4 - x^2)^{3/2}}$

In Problems 7–34, determine concavity and the x-values where points of inflection occur. Do not sketch the graphs.

7. $y = -2x^2 + 4x$

8. $y = 3x^2 - 6x + 5$

9. $y = 4x^3 + 12x^2 - 12x$

10. $y = x^3 - 6x^2 + 9x + 1$

11. $y = 2x^3 - 5x^2 + 5x - 2$

12. $y = x^4 - 8x^2 - 6$

*13. $y = x^4 - 6x^2 + 5x - 6$

14. $y = -\frac{x^4}{4} + \frac{9x^2}{2} + 2x$

15. $y = 2x^{1/5}$

16. $y = \frac{7}{x^3}$

17. $y = \frac{x^4}{2} + \frac{19x^3}{6} - \frac{7x^2}{2} + x + 5$

18. $y = -\frac{5}{2}x^4 - \frac{1}{6}x^3 + \frac{1}{2}x^2 + \frac{1}{3}x - \frac{2}{5}$

19. $y = \frac{1}{20}x^5 - \frac{1}{4}x^4 + \frac{1}{6}x^3 - \frac{1}{2}x - \frac{2}{3}$

20. $y = \frac{9}{5}x^5 - \frac{32}{3}x^3 + 10x - 2$

21. $y = \frac{1}{30}x^6 - \frac{7}{12}x^4 + 5x^2 + 2x - 1$

22. $y = x^6 - 3x^4$

*23. $y = \frac{x + 1}{x - 1}$

24. $y = 1 - \frac{1}{x^2}$

25. $y = \frac{4x^2}{x^2 + 1}$

26. $y = \frac{4x^2}{x + 3}$

27. $y = \frac{21x + 40}{6(x + 3)^2}$

28. $y = 3(x^2 - 2)^2$

29. $y = 5e^x$

30. $y = e^x - e^{-x}$

31. $y = 3xe^x$

32. $y = 2xe^{-x}$

33. $y = \dfrac{\ln x}{2x}$

34. $y = \dfrac{x^2 + 1}{3e^x}$

In Problems 35–62, determine intervals on which the function is increasing, decreasing, concave up, and concave down; relative maxima and minima; inflection points; symmetry; and those intercepts that can be obtained conveniently. Then sketch each curve.

35. $y = x^2 - x - 6$

36. $y = x^2 + 2$

37. $y = 4x - x^2$

38. $y = x - x^2 + 2$

***39.** $y = x^3 - 9x^2 + 24x - 19$

40. $y = x^3 - 25x^2$

41. $y = \dfrac{x^3}{3} - 4x$

42. $y = x^3 - 6x^2 + 9x$

43. $y = x^3 - 3x^2 + 3x - 3$

44. $y = 2x^3 - 9x^2 + 12x$

45. $y = 4x^3 - 3x^4$

46. $y = -x^3 + 2x^2 - x + 4$

47. $y = -2 + 12x - x^3$

48. $y = (3 + 2x)^3$

49. $y = x^3 - 6x^2 + 12x - 6$

50. $y = \dfrac{x^5}{100} - \dfrac{x^4}{20}$

51. $y = 5x - x^5$

52. $y = x^2(x - 1)^2$

53. $y = 3x^4 - 4x^3 + 1$

54. $y = 3x^5 - 5x^3$

55. $y = 4x^2 - x^4$

56. $y = x^4 - 2x^2$

57. $y = x^{1/3}(x - 8)$

58. $y = (x - 1)^2(x + 2)^2$

59. $y = 4x^{1/3} + x^{4/3}$

60. $y = (x + 1)\sqrt{x + 4}$

61. $y = 2x + 3x^{2/3}$

62. $y = 5x^{2/3} - x^{5/3}$

63. Sketch the graph of a continuous function f such that $f(2) = 4$, $f'(2) = 0$, $f'(x) < 0$ if $x < 2$, and $f''(x) > 0$ if $x > 2$.

64. Sketch the graph of a continuous function f such that $f(4) = 4$, $f'(4) = 0$, $f''(x) < 0$ for $x < 4$, and $f''(x) > 0$ for $x > 4$.

65. Sketch the graph of a continuous function f such that $f(1) = 1$, $f'(1) = 0$, and $f''(x) < 0$ for all x.

66. Sketch the graph of a continuous function f such that $f(3) = 4$, both $f'(x) > 0$ and $f''(x) > 0$ for $x < 3$, and both $f'(x) < 0$ and $f''(x) > 0$ for $x > 3$.

67. Demand Equation Show that the graph of the demand equation $p = \dfrac{100}{q + 2}$ is decreasing and concave up for $q > 0$.

68. Average Cost For the cost function

$$c = 3q^2 + 5q + 6$$

show that the graph of the average-cost function $\bar{c}$ is always concave up for $q > 0$.

69. Species of Plants The number of species of plants on a plot may depend on the size of the plot. For example, in Figure 13.46, we see that on 1-m² plots there are three species (A, B, and C on the left plot, A, B, and D on the right plot), and on a 2-m² plot there are four species (A, B, C, and D).

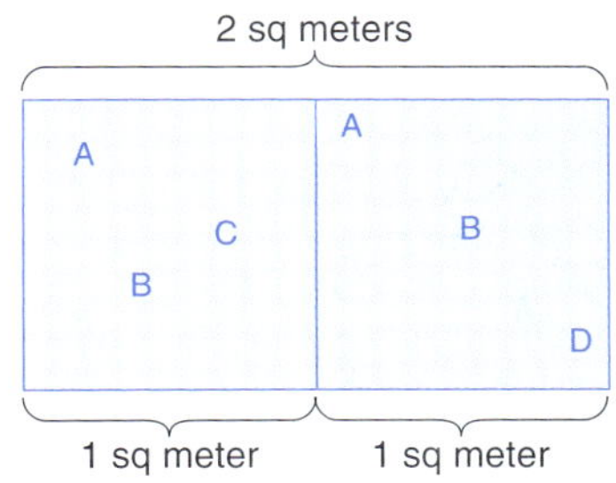

FIGURE 13.46 Species of plants.

In a study of rooted plants in a certain geographic region,[7] it was determined that the average number of species, S, occurring on plots of size A (in square meters) is given by

$$S = f(A) = 12\sqrt[4]{A} \qquad 0 \le A \le 625$$

Sketch the graph of f. (*Note:* Your graph should be rising and concave down. Thus, the number of species is increasing with respect to area, but at a decreasing rate.)

70. Inferior Good In a discussion of an inferior good, Persky[8] considers a function of the form

$$g(x) = e^{(U_0/A)}e^{-x^2/(2A)}$$

where x is a quantity of a good, U_0 is a constant that represents utility, and A is a positive constant. Persky claims that the graph of g is concave down for $x < \sqrt{A}$ and concave up for $x > \sqrt{A}$. Verify this.

71. Psychology In a psychological experiment involving conditioned response,[9] subjects listened to four tones, denoted 0, 1, 2, and 3. Initially, the subjects were conditioned to tone 0 by receiving a shock whenever this tone was heard. Later, when each of the four tones (stimuli) were heard without shocks, the subjects' responses were recorded by means of a tracking device that measures galvanic skin reaction. The average response to each stimulus (without shock) was determined, and the results were plotted on a coordinate plane where the x- and y-axes represent the stimuli (0, 1, 2, 3) and the average galvanic responses, respectively. It was determined that the points fit a curve that is approximated by the graph of

$$y = 12.5 + 5.8(0.42)^x$$

Show that this function is decreasing and concave up.

---

[7] Adapted from R. W. Poole, *An Introduction to Quantitative Ecology* (New York: McGraw-Hill Book Company, 1974).

[8] A. L. Persky, "An Inferior Good and a Novel Indifference Map," *American Economist* 29, no. 1 (1985), 67–69.

[9] Adapted from C. I. Hovland, "The Generalization of Conditioned Responses: I. The Sensory Generalization of Conditioned Responses with Varying Frequencies of Tone," *Journal of General Psychology* 17 (1937), 125–48.

72. Entomology In a study of the effects of food deprivation on hunger,[10] an insect was fed until its appetite was completely satisfied. Then it was deprived of food for t hours (the deprivation period). At the end of this period, the insect was refed until its appetite was again completely satisfied. The weight H (in grams) of the food that was consumed at this time was statistically found to be a function of t, where

$$H = 1.00\left[1 - e^{-(0.0464t+0.0670)}\right]$$

Here H is a measure of hunger. Show that H is increasing with respect to t and is concave down.

73. Insect Dispersal In an experiment on the dispersal of a particular insect,[11] a large number of insects are placed at a release point in an open field. Surrounding this point are traps that are placed in a concentric circular arrangement at a distance of 1 m, 2 m, 3 m, and so on, from the release point. Twenty-four hours after the insects are released, the number of insects in each trap is counted. It is determined that a distance of r meters from the release point, the average number of insects contained in a trap is

$$n = f(r) = 0.1\ln(r) + \frac{7}{r} - 0.8 \quad 1 \leq r \leq 10$$

(a) Show that the graph of f is always falling and concave up. (b) Sketch the graph of f. (c) When $r = 5$, at what rate is the average number of insects in a trap decreasing with respect to distance?

74. Graph $y = -0.35x^3 + 4.1x^2 + 8.3x - 7.4$, and from the graph, determine the number of (a) relative maximum points, (b) relative minimum points, and (c) inflection points.

75. Graph $y = x^5(x - 2.3)$, and from the graph, determine the number of inflection points.

76. Graph $y = 1 - 2^{-x^2}$, and from the graph, determine the number of inflection points.

77. Graph the curve $y = x^3 - 2x^2 + x + 3$, and also graph the tangent line to the curve at $x = 2$. Around $x = 2$, does the curve lie above or below the tangent line? From your observation determine the concavity at $x = 2$.

78. If $f(x) = 2x^3 + 3x^2 - 6x + 1$, find $f'(x)$ and $f''(x)$. Note that where f' has a relative minimum, f changes its direction of bending. Why?

79. If $f(x) = x^6 + 3x^5 - 4x^4 + 2x^2 + 1$, find the x-values (rounded to two decimal places) of the inflection points of f.

80. If $f(x) = \dfrac{x+3}{x^2+1}$, find the x-values (rounded to two decimal places) of the inflection points of f.

[10]C. S. Holling, "The Functional Response of Invertebrate Predators to Prey Density," *Memoirs of the Entomological Society of Canada*, no. 48 (1966).

[11]Adapted from Poole, *An Introduction to Quantitative Ecology*.

OBJECTIVE *To locate relative extrema by applying the second-derivative test.*

13.4 The Second-Derivative Test

The second derivative may be used to test certain critical values for relative extrema. Observe in Figure 13.47 that when $x = x_0$, there is a horizontal tangent; that is, $f'(x_0) = 0$. Furthermore, around x_0, the function is concave up. [That is, $f''(x_0) > 0$.] This leads us to conclude that there is a relative minimum at x_0. On the other hand, around x_1, the function is concave down. [That is, $f''(x_1) < 0$.] Because the tangent line is horizontal at x_1, we conclude that a relative maximum exists there. This technique of examining the second derivative at points where the first derivative is 0 is called the *second-derivative test* for relative extrema.

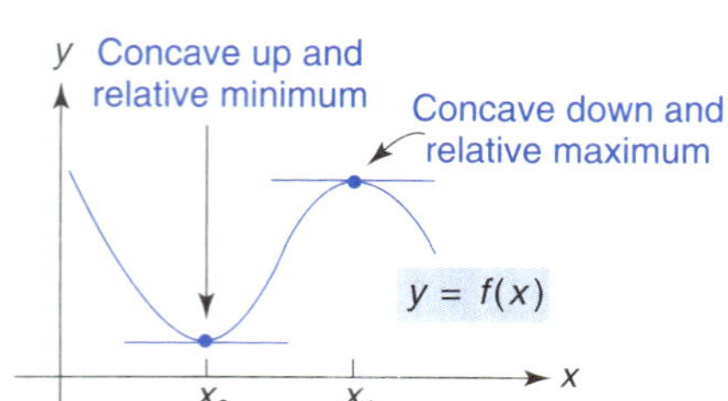

FIGURE 13.47 Relating concavity to relative extrema.

Second-Derivative Test for Relative Extrema

Suppose $f'(x_0) = 0$.

If $f''(x_0) < 0$, then f has a relative maximum at x_0.
If $f''(x_0) > 0$, then f has a relative minimum at x_0.

We want to emphasize that **the second-derivative test does *not* apply when $f'(x_0) = 0$ and $f''(x_0) = 0$.** Under these conditions, there may be a relative maximum, a relative minimum, or neither at x_0. In such cases, the first-derivative test should be used to analyze what is happening at x_0. Also, the second-derivative test does not apply when $f'(x_0)$ is not defined.

EXAMPLE 1 Second-Derivative Test

Test the following for relative maxima and minima. Use the second-derivative test if possible.

a. $y = 18x - \frac{2}{3}x^3$.

Solution:

$$y' = 18 - 2x^2 = 2(9 - x^2) = 2(3 + x)(3 - x)$$

$$y'' = -4x \qquad \left(\text{taking } \frac{d}{dx} \text{ of } 18 - 2x^2\right)$$

Solving $y' = 0$ gives the critical values $x = \pm 3$.

$$\text{If } x = 3, \quad \text{then } y'' = -4(3) = -12 < 0$$

so there is a relative maximum when $x = 3$.

$$\text{If } x = -3, \quad \text{then } y'' = -4(-3) = 12 > 0$$

so there is a relative minimum when $x = -3$. (Refer back to Figure 13.4.)

Although the second-derivative test is very useful, do not depend entirely on it. Not only may the test not apply, but at times it may be awkward to find the second derivative.

b. $y = 6x^4 - 8x^3 + 1$.

Solution:

$$y' = 24x^3 - 24x^2 = 24x^2(x - 1)$$

$$y'' = 72x^2 - 48x$$

Solving $y' = 0$ gives the critical values $x = 0, 1$. We see that

$$\text{if } x = 0, \quad \text{then } y'' = 0$$

and

$$\text{if } x = 1, \quad \text{then } y'' > 0$$

By the second-derivative test, there is a relative minimum when $x = 1$. We cannot apply the test when $x = 0$ because $y'' = 0$ there. To analyze what is happening at 0, we turn to the first-derivative test:

$$\text{If } x < 0, \quad \text{then } y' < 0$$

$$\text{If } 0 < x < 1, \quad \text{then } y' < 0$$

Thus, no maximum or minimum exists when $x = 0$. (Refer back to Figure 13.36.)

▶ **Now work Problem 5.**

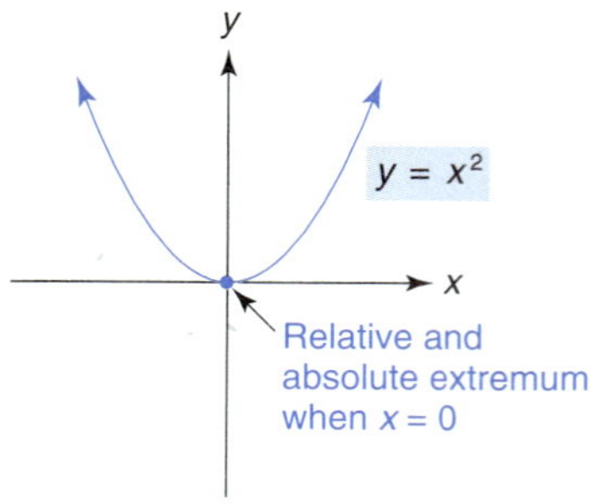

FIGURE 13.48 Exactly one relative extremum implies an absolute extremum.

If a continuous function has *exactly one* relative extremum on an interval, it can be shown that the relative extremum must also be an *absolute* extremum on the interval. To illustrate, in Figure 13.48 the function $y = x^2$ has a relative minimum when $x = 0$, and there are no other relative extrema. Since $y = x^2$ is continuous, this relative minimum is also an absolute minimum for the function.

EXAMPLE 2 **Absolute Extrema**

If $y = f(x) = x^3 - 3x^2 - 9x + 5$, *determine when absolute extrema occur on the interval* $(0, \infty)$.

Solution: We have

$$f'(x) = 3x^2 - 6x - 9 = 3(x^2 - 2x - 3)$$
$$= 3(x+1)(x-3)$$

The only critical value on the interval $(0, \infty)$ is 3. Applying the second-derivative test at this point gives

$$f''(x) = 6x - 6$$
$$f''(3) = 6(3) - 6 = 12 > 0$$

Thus, there is a relative minimum when $x = 3$. Since this is the only relative extremum on $(0, \infty)$ and f is continuous there, we conclude by our previous discussion that there is an *absolute* minimum value when $x = 3$; this value is $f(3) = -22$. (See Figure 13.49.)

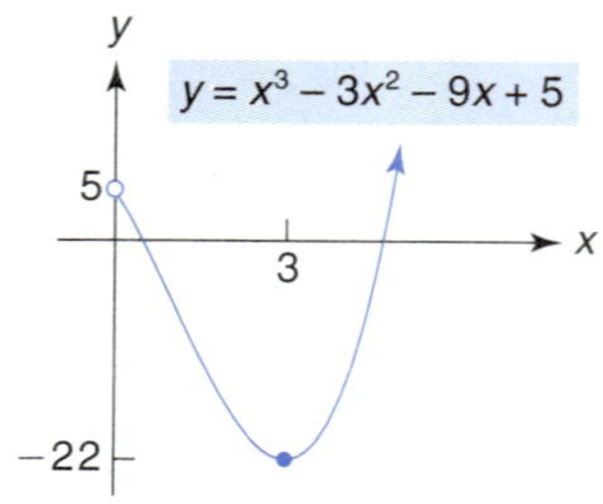

FIGURE 13.49 On $(0, \infty)$, there is an absolute minimum when $x = 3$.

▶ **Now work Problem 3.**

Observe that the definition of relative extrema in Section 13.1 does allow for the possibility of relative extrema at the the end points of a closed interval. Consider the function $f(x) = x^2$ for $1 \le x$. It is easy to see that f has exactly one relative extremum, namely a relative minimum at $x = 1$. As is also clear directly, it follows from the present discussion that f has an absolute minimum value of 1 at $x = 1$.

Exercise 13.4

In Problems 1–14, test for relative maxima and minima. Use the second-derivative test if possible. In Problems 1–4, state whether the relative extrema are also absolute extrema.

1. $y = x^2 - 5x + 6$

2. $y = -2x^2 + 6x + 12$

***3.** $y = -4x^2 + 2x - 8$

4. $y = 3x^2 - 5x + 6$

***5.** $y = \frac{1}{3}x^3 + 2x^2 - 5x + 1$

6. $y = x^3 - 12x + 1$

7. $y = -x^3 + 3x^2 + 1$

8. $y = x^4 - 2x^2 + 4$

9. $y = 3x^4 + 3$

10. $y = -2x^7$

11. $y = 81x^5 - 5x$

12. $y = \frac{55}{3}x^3 - x^2 - 21x - 3$

13. $y = (x^2 + 7x + 10)^2$

14. $y = \frac{x^3}{3} + 2x^2 + x - 5$

OBJECTIVE *To determine horizontal and vertical asymptotes for a curve and to sketch the graphs of functions having asymptotes.*

13.5 Asymptotes

Vertical Asymptotes

In this section, we conclude our discussion of curve-sketching techniques by investigating functions having *asymptotes*. Basically, an asymptote is a line that a curve approaches arbitrarily closely. For example, in each part of Figure 13.50,

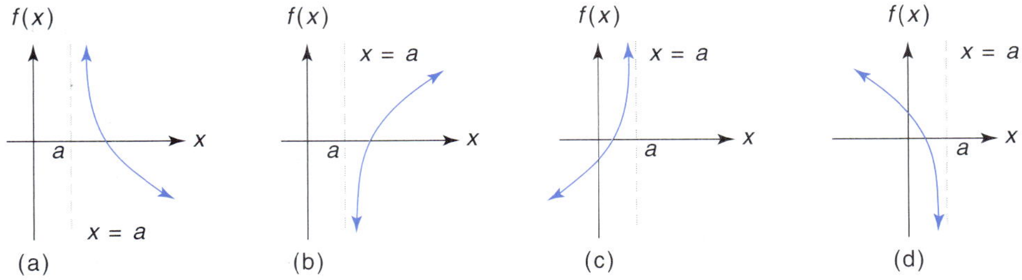

FIGURE 13.50 Vertical asymptotes $x = a$.

the dashed line $x = a$ is an asymptote. But to be precise about it, we need to make use of infinite limits. In Figure 13.50(a), notice that as $x \to a^+$, $f(x)$ becomes positively infinite:

$$\lim_{x \to a^+} f(x) = \infty$$

In Figure 13.50(b), as $x \to a^+$, $f(x)$ becomes negatively infinite:

$$\lim_{x \to a^+} f(x) = -\infty$$

In Figures 13.50(c) and (d), we have

$$\lim_{x \to a^-} f(x) = \infty \quad \text{and} \quad \lim_{x \to a^-} f(x) = -\infty$$

respectively.

Loosely speaking, we can say that each graph in Figure 13.50 has an "explosion" around the dashed vertical line $x = a$, in the sense that a one-sided limit of $f(x)$ at a is either ∞ or $-\infty$. The line $x = a$ is called a *vertical asymptote* for the graph. A vertical asymptote is not part of the graph, but is a useful aid in sketching it because part of the graph approaches the asymptote. Because of the explosion around $x = a$, the function is *not* continuous at a.

Definition

The line $x = a$ is a ***vertical asymptote*** for the graph of the function f if and only if at least one of the following is true:

$$\lim_{x \to a^+} f(x) = \pm\infty$$

or

$$\lim_{x \to a^-} f(x) = \pm\infty$$

To determine vertical asymptotes, we must find values of x around which $f(x)$ increases or decreases without bound. For a rational function (a quotient of two polynomials), these x-values are precisely those for which the denominator is zero but the numerator is not zero. For example, consider the rational function

$$f(x) = \frac{3x - 5}{x - 2}$$

When x is 2, the denominator is 0, but the numerator is not. If x is slightly larger than 2, then $x - 2$ is both close to 0 and positive, and $3x - 5$ is close to 1. Thus, $(3x - 5)/(x - 2)$ is very large, so

$$\lim_{x \to 2^+} \frac{3x - 5}{x - 2} = \infty$$

This limit is sufficient to conclude that the line $x = 2$ is a vertical asymptote. Because we are ultimately interested in the behavior of a function around a

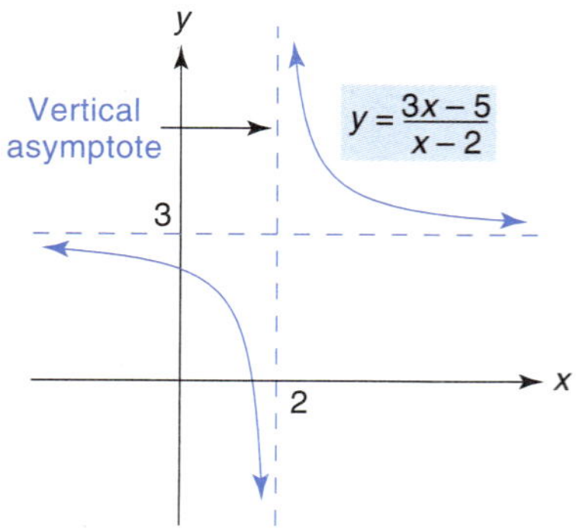

FIGURE 13.51 Graph of $y = \dfrac{3x-5}{x-2}$.

vertical asymptote, it is worthwhile to examine what happens to this function as x approaches 2 from the left. If x is slightly less than 2, then $x - 2$ is very close to 0 but negative, and $3x - 5$ is close to 1. Hence, $(3x-5)/(x-2)$ is "very negative," so

$$\lim_{x \to 2^-} \frac{3x-5}{x-2} = -\infty$$

We conclude that the function increases without bound as $x \to 2^+$ and decreases without bound as $x \to 2^-$. The graph appears in Figure 13.51.

In summary, we have a rule for vertical asymptotes.

Vertical-Asymptote Rule for Rational Functions

Suppose that

$$f(x) = \frac{P(x)}{Q(x)}$$

where P and Q are polynomial functions. The line $x = a$ is a vertical asymptote for the graph of f if and only if $Q(a) = 0$ and $P(a) \neq 0$.

EXAMPLE 1 **Finding Vertical Asymptotes**

Determine vertical asymptotes for the graph of

$$f(x) = \frac{x^2 - 4x}{x^2 - 4x + 3}$$

Solution: Since f is a rational function, the vertical-asymptote rule applies. Writing

$$f(x) = \frac{x(x-4)}{(x-3)(x-1)} \qquad \text{(factoring)}$$

makes it clear that the denominator is 0 when x is 3 or 1. Neither of these values makes the numerator 0. Thus, the lines $x = 3$ and $x = 1$ are vertical asymptotes. (See Figure 13.52.)

Although the vertical-asymptote rule guarantees that the lines $x = 3$ and $x = 1$ are vertical asymptotes, it does not indicate the precise nature of the "explosion" around these lines. A precise analysis requires the use of one-sided limits.

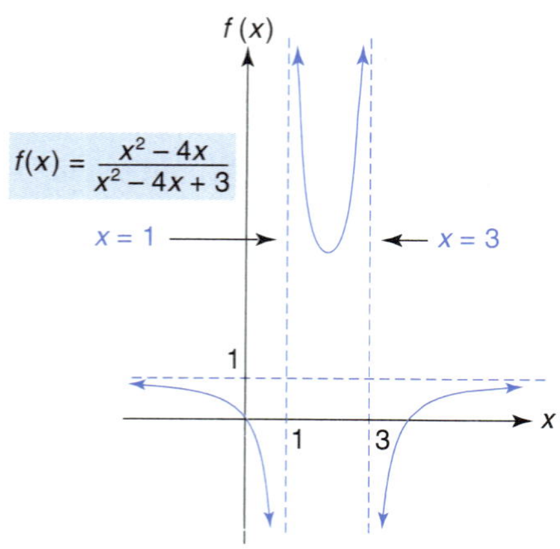

FIGURE 13.52 Graph of $f(x) = \dfrac{x^2 - 4x}{x^2 - 4x + 3}$.

Horizontal Asymptotes

A curve $y = f(x)$ may have another kind of asymptote. In Figure 13.53(a), as x increases without bound ($x \to \infty$), the graph approaches the horizontal line $y = b$. That is,

$$\lim_{x \to \infty} f(x) = b$$

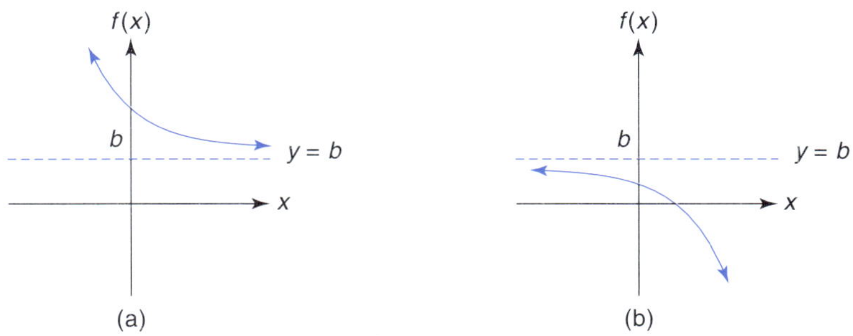

FIGURE 13.53 Horizontal asymptotes $y = b$.

In Figure 13.53(b), as x becomes negatively infinite, the graph approaches the horizontal line $y = b$. That is,

$$\lim_{x \to -\infty} f(x) = b$$

In each case, the dashed line $y = b$ is called a *horizontal asymptote* for the graph. It is a horizontal line around which the graph "settles" either as $x \to \infty$ or as $x \to -\infty$.

Although the graph of a horizontal line settles around itself as $x \to \infty$ or as $x \to -\infty$, a line is not considered to have asymptotes. In summary, we have the following definition:

Definition

Let f be a nonlinear function. The line $y = b$ is a ***horizontal asymptote*** for the graph of f if and only if at least one of the following is true:

$$\lim_{x \to \infty} f(x) = b \quad \text{or} \quad \lim_{x \to -\infty} f(x) = b$$

To test for horizontal asymptotes, we must find the limits of $f(x)$ as $x \to \infty$ and as $x \to -\infty$. To illustrate, we again consider

$$f(x) = \frac{3x - 5}{x - 2}$$

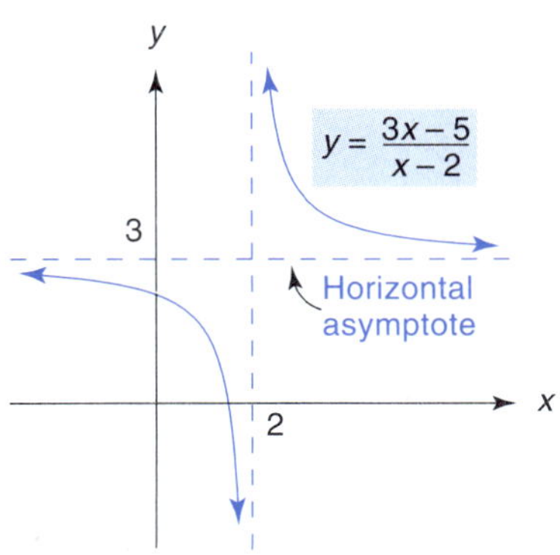

FIGURE 13.54 Graph of $f(x) = \dfrac{3x - 5}{x - 2}$.

Since this is a rational function, we can use the procedures of Section 10.2 to find the limits. Because the dominant term in the numerator is $3x$ and the dominant term in the denominator is x, we have

$$\lim_{x \to \infty} \frac{3x - 5}{x - 2} = \lim_{x \to \infty} \frac{3x}{x} = \lim_{x \to \infty} 3 = 3$$

Thus, the line $y = 3$ is a horizontal asymptote. (See Figure 13.54.) Also,

$$\lim_{x \to -\infty} \frac{3x - 5}{x - 2} = \lim_{x \to -\infty} \frac{3x}{x} = \lim_{x \to -\infty} 3 = 3$$

Hence, the graph settles down near the horizontal line $y = 3$ both as $x \to \infty$ and as $x \to -\infty$.

EXAMPLE 2 **Finding Horizontal Asymptotes**

Find horizontal asymptotes for the graph of

$$f(x) = \frac{x^2 - 4x}{x^2 - 4x + 3}.$$

Solution: We have

$$\lim_{x\to\infty} \frac{x^2 - 4x}{x^2 - 4x + 3} = \lim_{x\to\infty} \frac{x^2}{x^2} = \lim_{x\to\infty} 1 = 1$$

Therefore, the line $y = 1$ is a horizontal asymptote. The same result is obtained as $x \to -\infty$. (Refer back to Figure 13.52.)

▶ **Now work Problem 11.**

A few remarks about asymptotes are appropriate now. With vertical asymptotes, we are examining the behavior of a graph around specific x-values. However, with horizontal asymptotes we are examining the graph as x becomes unbounded. Although a graph may have numerous vertical asymptotes, it can have at most two different horizontal asymptotes.

From Section 10.2, when the numerator of a rational function has degree greater than that of the denominator, no limit exists as $x \to \infty$ or $x \to -\infty$. From this observation, we conclude that *whenever the degree of the numerator of a rational function is greater than the degree of the denominator, the graph of the function cannot have a horizontal asymptote.*

EXAMPLE 3 **Finding Horizontal and Vertical Asymptotes**

Find vertical and horizontal asymptotes for the graph of the polynomial function

$$y = f(x) = x^3 + 2x$$

Solution: We begin with vertical asymptotes. This is a rational function with denominator 1, which is never zero. By the vertical-asymptote rule, there are no vertical asymptotes. Because the degree of the numerator (3) is greater than the degree of the denominator (0), there are no horizontal asymptotes. However, let us examine the behavior of the graph of f as $x \to \infty$ and $x \to -\infty$. We have

$$\lim_{x\to\infty} (x^3 + 2x) = \lim_{x\to\infty} x^3 = \infty$$

and

$$\lim_{x\to-\infty} (x^3 + 2x) = \lim_{x\to-\infty} x^3 = -\infty$$

Thus, as $x \to \infty$, the graph must extend indefinitely upward, and as $x \to -\infty$, the graph must extend indefinitely downward. (See Figure 13.55.)

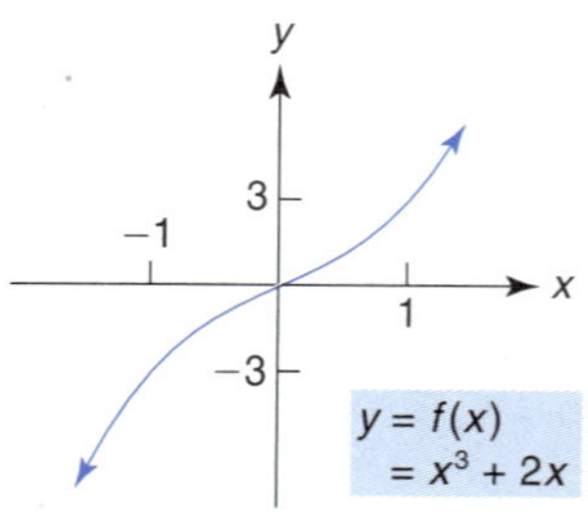

FIGURE 13.55 Graph of $y = x^3 + 2x$ has neither horizontal nor vertical asymptotes.

▶ **Now work Problem 9.**

The results in Example 3 can be generalized to any polynomial function:

A polynomial function has neither a horizontal nor a vertical asymptote.

EXAMPLE 4 **Finding Horizontal and Vertical Asymptotes**

Find horizontal and vertical asymptotes for the graph of $y = e^x - 1$.

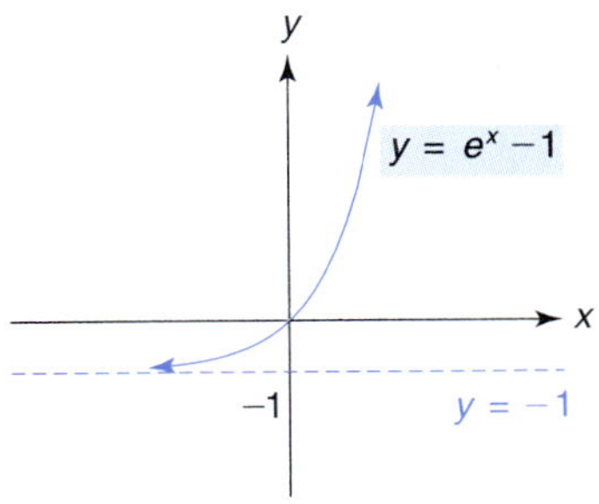

FIGURE 13.56 Graph of $y = e^x - 1$ has a horizontal asymptote.

Solution: Testing for horizontal asymptotes, we let $x \to \infty$. Then e^x increases without bound, so

$$\lim_{x\to\infty} (e^x - 1) = \infty$$

Thus, the graph does not settle down as $x \to \infty$. However, as $x \to -\infty$, we have $e^x \to 0$, so

$$\lim_{x\to-\infty} (e^x - 1) = \lim_{x\to-\infty} e^x - \lim_{x\to-\infty} 1 = 0 - 1 = -1$$

Therefore, the line $y = -1$ is a horizontal asymptote. The graph has no vertical asymptotes because $e^x - 1$ neither increases nor decreases without bound around any fixed value of x. (See Figure 13.56.)

▶ **Now work Problem 23.**

Curve Sketching

We conclude this section by showing you how to graph a function by making use of all the curve-sketching tools that we have developed.

EXAMPLE 5 **Curve Sketching**

Sketch the graph of $y = \dfrac{1}{4 - x^2}$.

Solution:

Intercepts When $x = 0$, $y = \frac{1}{4}$. If $y = 0$, then $0 = 1/(4 - x^2)$, which has no solution. Thus $(0, \frac{1}{4})$ is the only intercept.

Symmetry There is symmetry about the y-axis: Replacing x by $-x$ gives

$$y = \frac{1}{4 - (-x)^2} = \frac{1}{4 - x^2}$$

which is the original equation. It can be shown that there is no other symmetry.

Asymptotes Testing for horizontal asymptotes, we have

$$\lim_{x\to\infty} \frac{1}{4 - x^2} = \lim_{x\to\infty} \frac{1}{-x^2} = -\lim_{x\to\infty} \frac{1}{x^2} = 0$$

Similarly,

$$\lim_{x\to-\infty} \frac{1}{4 - x^2} = 0$$

Thus, $y = 0$ (the x-axis) is a horizontal asymptote. Because the denominator of $1/(4 - x^2)$ is 0 when $x = \pm 2$, and the numerator is not 0 for these values of x, the lines $x = 2$ and $x = -2$ are vertical asymptotes.

Maxima and Minima Since $y = (4 - x^2)^{-1}$,

$$y' = -1(4 - x^2)^{-2}(-2x) = \frac{2x}{(4 - x^2)^2}$$

We see that y' is 0 when $x = 0$ and y' is undefined when $x = \pm 2$. However, only 0 is a critical value, because y is not defined at ± 2. There are four intervals to test for increasing/decreasing:

If $x < -2$, then $y' < 0$

If $-2 < x < 0$, then $y' < 0$

If $0 < x < 2$, then $y' > 0$

If $x > 2$, then $y' > 0$

The function is decreasing on $(-\infty, -2)$ and $(-2, 0)$ and increasing on $(0, 2)$ and $(2, \infty)$. (See Figure 13.57.) There is a relative minimum when $x = 0$.

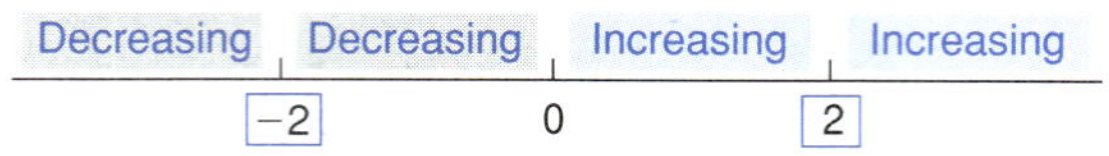

FIGURE 13.57 Increasing/decreasing analysis.

Concavity

$$y'' = \frac{(4-x^2)^2(2) - (2x)2(4-x^2)(-2x)}{(4-x^2)^4}$$

$$= \frac{2(4-x^2)[(4-x^2)-(2x)(-2x)]}{(4-x^2)^4} = \frac{2(4+3x^2)}{(4-x^2)^3}$$

Setting $y'' = 0$, we get no real roots. However, y'' is undefined when $x = \pm 2$. Although concavity may change around these values of x, the values cannot correspond to inflection points because they are not in the domain of the function. There are three intervals to test for concavity:

$$\begin{aligned} &\text{If } x < -2, &&\text{then } y'' < 0 \\ &\text{If } -2 < x < 2, &&\text{then } y'' > 0 \\ &\text{If } x > 2, &&\text{then } y'' < 0 \end{aligned}$$

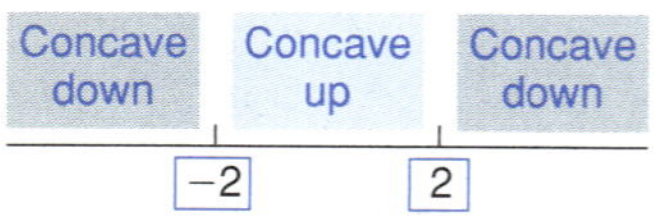

FIGURE 13.58 Concavity analysis.

The graph is concave up on $(-2, 2)$ and concave down on $(-\infty, -2)$ and $(2, \infty)$. (See Figure 13.58.) Although concavity changes around $x = \pm 2$, as we said before, these values cannot correspond to inflection points.

Discussion Plotting the points in the table in Figure 13.59, some arbitrarily chosen, and using the preceding information, we get the indicated graph. Owing to symmetry about the y-axis, our table has only $x \geq 0$.

x	0	1	3
y	$\frac{1}{4}$	$\frac{1}{3}$	$-\frac{1}{5}$

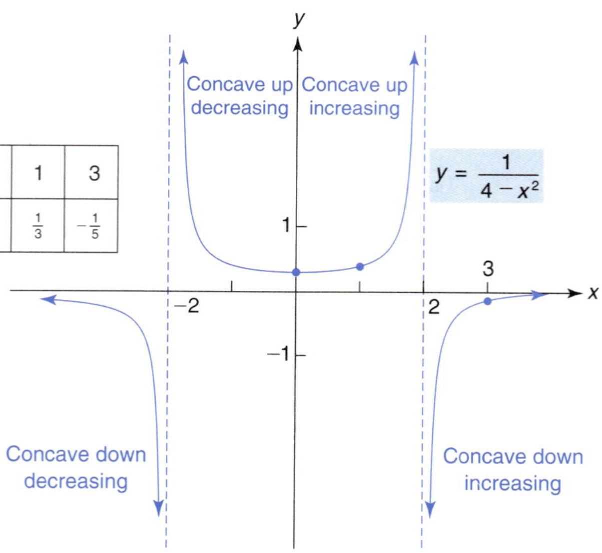

FIGURE 13.59 Graph of $y = \dfrac{1}{4-x^2}$.

▶ **Now work Problem 31.**

EXAMPLE 6 Curve Sketching

Sketch the graph of $y = \dfrac{4x}{x^2+1}$.

Solution:

Intercepts When $x = 0$, $y = 0$; when $y = 0$, $x = 0$. Thus, $(0, 0)$ is the only intercept.

Symmetry There is symmetry about the origin: Replacing x by $-x$ and y by $-y$ gives

$$-y = \frac{4(-x)}{(-x)^2 + 1} \quad \text{that is} \quad y = \frac{4x}{x^2 + 1}$$

which is the original equation. No other symmetry exists.

Asymptotes Testing for horizontal asymptotes, we have

$$\lim_{x \to \infty} \frac{4x}{x^2 + 1} = \lim_{x \to \infty} \frac{4x}{x^2} = \lim_{x \to \infty} \frac{4}{x} = 0$$

and similarly,

$$\lim_{x \to -\infty} \frac{4x}{x^2 + 1} = 0$$

Thus, $y = 0$ (the x-axis) is a horizontal asymptote. Because the denominator of $4x/(x^2 + 1)$ is never 0, there is no vertical asymptote.

Maxima and Minima Letting $y = f(x)$, we have

$$f'(x) = \frac{(x^2 + 1)(4) - 4x(2x)}{(x^2 + 1)^2} = \frac{4 - 4x^2}{(x^2 + 1)^2} = \frac{4(1 + x)(1 - x)}{(x^2 + 1)^2}$$

The critical values are $x = \pm 1$, so there are three intervals to consider:

If $x < -1$, then $f'(x) = \frac{4(-)(+)}{(+)} = -$, so f is *decreasing*.

If $-1 < x < 1$, then $f'(x) = \frac{4(+)(+)}{(+)} = +$, so f is *increasing*.

If $x > 1$, then $f'(x) = \frac{4(+)(-)}{(+)} = -$, so f is *decreasing*.

(See Figure 13.60.)

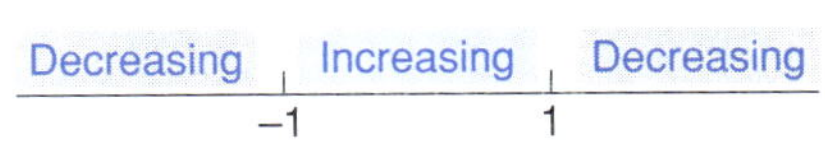

FIGURE 13.60 Increasing/decreasing analysis.

There is a relative minimum when $x = -1$ and a relative maximum when $x = 1$.

Concavity Since $f'(x) = \frac{4 - 4x^2}{(x^2 + 1)^2}$,

$$f''(x) = \frac{(x^2 + 1)^2(-8x) - (4 - 4x^2)(2)(x^2 + 1)(2x)}{(x^2 + 1)^4}$$

$$= \frac{8x(x^2 + 1)(x^2 - 3)}{(x^2 + 1)^4} = \frac{8x(x + \sqrt{3})(x - \sqrt{3})}{(x^2 + 1)^3}$$

Setting $f''(x) = 0$, we conclude that the possible points of inflection are when $x = \pm\sqrt{3}, 0$. There are four intervals to consider:

If $x < -\sqrt{3}$, then $f''(x) = \frac{8(-)(-)(-)}{(+)} = -$, so f is *concave down*.

If $-\sqrt{3} < x < 0$, then $f''(x) = \frac{8(-)(+)(-)}{(+)} = +$, so f is *concave up*.

If $0 < x < \sqrt{3}$, then $f''(x) = \frac{8(+)(+)(-)}{(+)} = -$, so f is *concave down*.

If $x > \sqrt{3}$, then $f''(x) = \frac{8(+)(+)(+)}{(+)} = +$, so f is *concave up*.

(See Figure 13.61.)

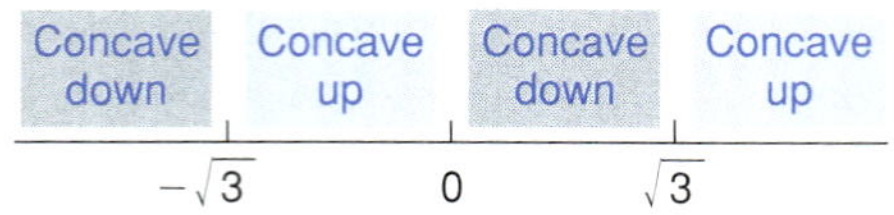

FIGURE 13.61 Concavity analysis.

Inflection points occur when $x = 0, \pm\sqrt{3}$.

Discussion After consideration of all of the preceding information, the graph of $y = 4x/(x^2 + 1)$ is given in Figure 13.62, together with a table of important points.

x	0	1	$\sqrt{3}$	-1	$-\sqrt{3}$
y	0	2	$\sqrt{3}$	-2	$-\sqrt{3}$

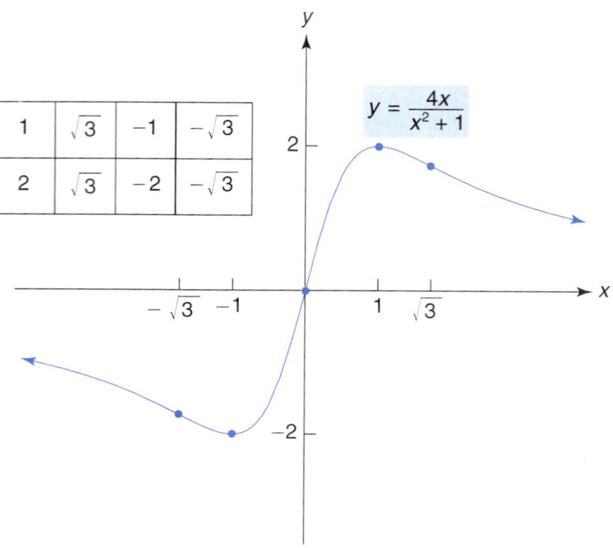

FIGURE 13.62 Graph of $y = \dfrac{4x}{x^2+1}$.

▶ **Now work Problem 39.**

Exercise 13.5

In Problems 1–24, find the horizontal and vertical asymptotes for the graphs of the functions. Do not sketch the graphs.

1. $y = \dfrac{x}{x-1}$

2. $y = \dfrac{x+1}{x}$

3. $f(x) = \dfrac{x-1}{2x+3}$

4. $y = \dfrac{2x+1}{2x+1}$

5. $y = \dfrac{4}{x}$

6. $y = 1 - \dfrac{2}{x^2}$

7. $y = \dfrac{1}{x^2-1}$

8. $y = \dfrac{x}{x^2-4}$

***9.** $y = x^2 - 5x + 5$

10. $y = \dfrac{x^3}{x^2-9}$

***11.** $f(x) = \dfrac{2x^2}{x^2+x-6}$

12. $f(x) = \dfrac{x^3}{5}$

13. $y = \dfrac{2x^2+3x+1}{x^2-5}$

14. $y = \dfrac{2x^3+1}{3x(2x-1)(4x-3)}$

15. $y = \dfrac{4}{x-6} + 7$

16. $f(x) = \dfrac{x^2-1}{2x^2-9x+4}$

17. $f(x) = \dfrac{3-x^4}{x^3+x^2}$

18. $y = \dfrac{x^2+4x^3+6x^4}{3x^2}$

19. $y = \dfrac{x^2-3x-4}{1+4x+4x^2}$

20. $y = \dfrac{x^4+1}{1-x^4}$

21. $y = \dfrac{9x^2-16}{2(3x+4)^2}$

22. $y = \dfrac{2}{9} + \dfrac{3x}{14x^2+x-3}$

***23.** $y = 2e^{x+2} + 4$

24. $f(x) = 12e^{-x}$

In Problems 25–46, determine intervals on which the function is increasing, decreasing, concave up, and concave down; relative maxima and minima; inflection points; symmetry; horizontal and vertical asymptotes; and those intercepts that can be obtained conveniently and then sketch the curve.

25. $y = \dfrac{3}{x}$

26. $y = \dfrac{2}{2x - 3}$

27. $y = \dfrac{x}{x+1}$

28. $y = \dfrac{10}{\sqrt{x}}$

29. $y = x^2 + \dfrac{1}{x^2}$

30. $y = \dfrac{3x^2 - 5x - 1}{x - 2}$

***31.** $y = \dfrac{1}{x^2 - 1}$

32. $y = \dfrac{1}{x^2 + 1}$

33. $y = \dfrac{1 + x}{1 - x}$

34. $y = \dfrac{1 - x}{x^2}$

35. $y = \dfrac{x^2}{7x + 4}$

36. $y = \dfrac{x^3 + 1}{x}$

37. $y = \dfrac{9}{9x^2 - 6x - 8}$

38. $y = \dfrac{8x^2 + 3x + 1}{2x^2}$

***39.** $y = \dfrac{2x - 3}{(2x - 9)^2}$

40. $y = \dfrac{3x + 1}{(6x + 5)^2}$

41. $y = \dfrac{x^2 - 1}{x^3}$

42. $y = \dfrac{3x}{(x - 2)^2}$

43. $y = x + \dfrac{1}{x + 1}$

44. $y = \dfrac{3x^4 + 1}{x^3}$

45. $y = \dfrac{-3x^2 + 2x - 5}{3x^2 - 2x - 1}$

46. $y = 2x + 1 + \dfrac{4}{2x + 1}$

47. Sketch the graph of a function f such that $f(0) = 0$, there is a horizontal asymptote $y = 1$ for $x \to \pm\infty$, there is a vertical asymptote $x = 2$, both $f'(x) < 0$ and $f''(x) < 0$ for $x < 2$, and both $f'(x) < 0$ and $f''(x) > 0$ for $x > 2$.

48. Sketch the graph of a function f such that $f(0) = 0$, there is a horizontal asymptote $y = 2$ for $x \to \pm\infty$, there is a vertical asymptote $x = -1$, both $f'(x) > 0$ and $f''(x) > 0$ for $x < -1$, and both $f'(x) > 0$ and $f''(x) < 0$ for $x > -1$.

49. Sketch the graph of a function f such that $f(0) = 0$, there is a horizontal asymptote $y = 0$ for $x \to \pm\infty$, there are vertical asymptotes $x = -1$ and $x = 2$, $f'(x) < 0$ for $x < -1$ and $-1 < x < 2$, and $f''(x) < 0$ for $x > 2$.

50. Sketch the graph of a function f such that $f(-2) = 2$, $f(0) = 0$, $f(2) = 0$, there is a horizontal asymptote $y = 1$ for $x \to \pm\infty$, there are vertical asymptotes $x = -1$ and $x = 1$, $f''(x) > 0$ for $x < -1$, and $f'(x) < 0$ for $-1 < x < 1$ and $f''(x) < 0$ for $1 < x$.

51. Purchasing Power In discussing the time pattern of purchasing, Mantell and Sing[12] use the curve

$$y = \frac{x}{a + bx}$$

as a mathematical model. They claim that $y = 1/b$ is an asymptote. Verify this.

52. Sketch the graphs of $y = 6 - 3e^{-x}$ and $y = 6 + 3e^{-x}$. Show that they are asymptotic to the same line. What is the equation of this line?

53. Market for Product For a new product, the yearly number of thousand packages sold, y, after t years from its introduction is predicted to be given by

$$y = f(t) = 150 - 76e^{-t}$$

Show that $y = 150$ is a horizontal asymptote for the graph. This reveals that after the product is established with consumers, the market tends to be constant.

54. Graph $y = \dfrac{x^2 - 2}{x^3 + \frac{7}{2}x^2 + 12x + 1}$. From the graph, locate any horizontal or vertical asymptotes.

55. Graph $y = \dfrac{6x^3 - 2x^2 + 6x - 1}{3x^3 - 2x^2 - 18x + 12}$. From the graph, locate any horizontal or vertical asymptotes.

56. Graph $y = \dfrac{\ln(x + 4)}{x^2 - 8x + 5}$, in the standard window. The graph suggests that there are two vertical asymptotes of the form $x = k$, where $k > 0$. Also, it appears that the graph "begins" near $x = -4$. As $x \to -4^+$, $\ln(x + 4) \to -\infty$ and $x^2 - 8x + 5 \to 53$. Thus, $\lim_{x \to -4^+} y = -\infty$. This gives the vertical asymptote $x = -4$. So in reality, there are *three* vertical asymptotes. Use the zoom feature to make the asymptote $x = -4$ apparent from the display.

57. Graph $y = \dfrac{0.34e^{0.7x}}{4.2 + 0.71e^{0.7x}}$, where $x > 0$. From the graph, determine an equation of the horizontal asymptote by examining the y-values as $x \to \infty$. To confirm this equation algebraically, find $\lim_{x \to \infty} y$ by first dividing both the numerator and denominator by $e^{0.7x}$.

[12] L. H. Mantell and F. P. Sing, *Economics for Business Decisions* (New York: McGraw-Hill Book Company, 1972), p. 107.

OBJECTIVE *To model situations involving maximizing or minimizing a quantity.*

13.6 Applied Maxima and Minima

By using techniques from this chapter, we can solve problems that involve maximizing or minimizing a quantity. For example, we might want to maximize profit or minimize cost. The crucial part is expressing the quantity to be maximized or minimized as a function of some variable in the problem. Then we differentiate and test the resulting critical values. For this, the first-derivative test or the second-derivative test may be used, although it may be obvious from the nature of the problem whether or not a critical value represents an appropriate answer. Because our interest is in *absolute* maxima and minima, sometimes we must examine endpoints of the domain of the function.

The aim of this example is to set up a cost function from which cost is minimized.

EXAMPLE 1 **Minimizing the Cost of a Fence**

For insurance purposes, a manufacturer plans to fence in a 10,800-ft^2 rectangular storage area adjacent to a building by using the building as one side of the enclosed area. The fencing parallel to the building faces a highway and will cost \$3 per foot installed, whereas the fencing for the other two sides costs \$2 per foot installed. Find the amount of each type of fence so that the total cost of the fence will be a minimum. What is the minimum cost?

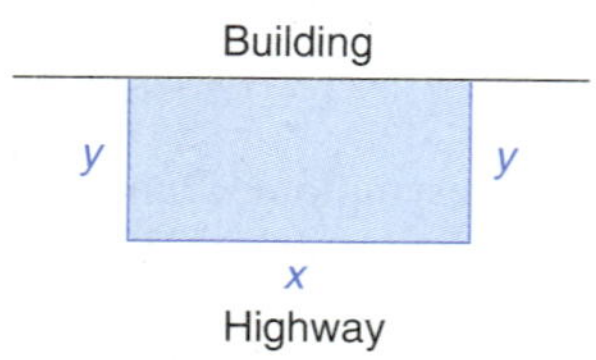

FIGURE 13.63 Fencing problem of Example 1.

Solution: As a first step in a problem like this, it is a good idea to draw a diagram that reflects the situation. In Figure 13.63, we have labeled the length of the side parallel to the building as x and the lengths of the other two sides as y, where x and y are in feet.

Since we want to minimize cost, our next step is to determine a function that gives cost. The cost obviously depends on how much fencing is along the highway and how much is along the other two sides. Along the highway the cost per foot is 3 (dollars), so the total cost of that fencing is $3x$. Similarly, along *each* of the other two sides, the cost is $2y$. Thus, the total cost of the fencing is given by the cost function

$$C = 3x + 2y + 2y$$

that is

$$C = 3x + 4y \tag{1}$$

We need to find the absolute minimum value of C. To do this, we use the techniques discussed in this chapter; that is, we examine C at critical values (and any endpoints) in the domain. But in order to differentiate, we need to first express C as a function of one variable only. [Equation (1) gives C as a function of *two* variables, x and y.] We can accomplish this by first finding a relationship between x and y. From the statement of the problem, we are told that the storage area, which is xy, must be 10,800:

$$xy = 10{,}800 \tag{2}$$

With this equation, we can express one variable (say, y) in terms of the other (x). Then, substitution into Equation (1) will give C as a function of one variable only. Solving Equation (2) for y gives

$$y = \frac{10{,}800}{x} \tag{3}$$

Substituting into Equation (1), we have

$$C = 3x + 4\left(\frac{10{,}800}{x}\right)$$

$$C = 3x + \frac{43{,}200}{x} \tag{4}$$

From the physical nature of the problem, the domain of C is $x > 0$.

We now find dC/dx, set it equal to 0, and solve for x. We have

$$\frac{dC}{dx} = 3 - \frac{43{,}200}{x^2} \qquad \left[\frac{d}{dx}(43{,}200x^{-1}) = -43{,}200x^{-2}\right]$$

$$3 - \frac{43{,}200}{x^2} = 0$$

$$3 = \frac{43{,}200}{x^2}$$

from which it follows that

$$x^2 = \frac{43{,}200}{3} = 14{,}400$$

$$x = 120 \qquad \text{(since } x > 0\text{)}$$

Thus, 120 is the *only* critical value, and there are no endpoints to consider. To test this value, we shall use the second-derivative test.

$$\frac{d^2C}{dx^2} = \frac{86{,}400}{x^3}$$

When $x = 120$, $d^2C/dx^2 > 0$, so we conclude that $x = 120$ gives a relative minimum. However, since 120 is the only critical value on the open interval $(0, \infty)$ and C is continuous on that interval, this relative minimum must also be an absolute minimum.

We are not done yet! The questions posed in the problem must be answered. For minimum cost, the number of feet of fencing along the highway is 120. When $x = 120$, we have, from Equation (3), $y = 10{,}800/120 = 90$. Therefore, the number of feet of fencing for the other two sides is $2y = 180$. It follows that 120 ft of the \$3 fencing and 180 ft of the \$2 fencing are needed. The minimum cost can be obtained from the cost function, Equation (4), and is

$$C = 3x + \frac{43{,}200}{x} = 3(120) + \frac{43{,}200}{120} = \$720$$

▶ Now work Problem 3.

Based on Example 1, the following guide may be helpful in solving an applied maximum or minimum problem:

Guide for Solving Applied Max–Min Problems

Step 1. When appropriate, draw a diagram that reflects the information in the problem.

Step 2. Set up a function for the quantity that you want to maximize or minimize.

Step 3. Express the function in step 2 as a function of one variable only, and note the domain of this function. The domain may be implied by the nature of the problem itself.

Step 4. Find the critical values of the function. After testing each critical value, determine which one gives the absolute extreme value you are seeking. If the domain of the function includes endpoints, be sure to also examine function values at these endpoints.

Step 5. Based on the results of step 4, answer the question(s) posed in the problem.

EXAMPLE 2 Maximizing Revenue

This example involves maximizing revenue when a demand equation is known.

The demand equation for a manufacturer's product is

$$p = \frac{80 - q}{4} \qquad 0 \le q \le 80$$

where q is the number of units and p is the price per unit. At what value of q will there be maximum revenue? What is the maximum revenue?

Solution: Let r represent total revenue, which is the quantity to be maximized. Since

$$\text{revenue} = (\text{price})(\text{quantity})$$

we have

$$r = pq = \frac{80 - q}{4} \cdot q = \frac{80q - q^2}{4}$$

where $0 \le q \le 80$. Setting $dr/dq = 0$, we obtain

$$\frac{dr}{dq} = \frac{80 - 2q}{4} = 0$$

$$80 - 2q = 0$$

$$q = 40$$

Thus, 40 is the only critical value. Now we see whether this gives a maximum. Examining the first derivative for $0 \le q < 40$, we have $dr/dq > 0$, so r is increasing. If $q > 40$, then $dr/dq < 0$, so r is decreasing. Because to the left of 40 we have r increasing, and to the right r is decreasing, we conclude that $q = 40$ gives the *absolute* maximum revenue, namely, $[80(40) - (40)^2]/4 = 400$.

▸ **Now work Problem 7.**

EXAMPLE 3 Minimizing Average Cost

This example involves minimizing average cost when the cost function is known.

A manufacturer's total-cost function is given by

$$c = \frac{q^2}{4} + 3q + 400$$

where c is the total cost of producing q units. At what level of output will average cost per unit be a minimum? What is this minimum?

Solution: The quantity to be minimized is the average cost $\bar{c}$. The average-cost function is

$$\bar{c} = \frac{c}{q} = \frac{\frac{q^2}{4} + 3q + 400}{q} = \frac{q}{4} + 3 + \frac{400}{q} \qquad (5)$$

Here q must be positive. To minimize $\bar{c}$, we differentiate:

$$\frac{d\bar{c}}{dq} = \frac{1}{4} - \frac{400}{q^2} = \frac{q^2 - 1600}{4q^2}$$

To get the critical values, we solve $d\bar{c}/dq = 0$:

$$\begin{aligned} q^2 - 1600 &= 0 \\ (q - 40)(q + 40) &= 0 \\ q &= 40 \qquad \text{(since } q > 0\text{)} \end{aligned}$$

To determine whether this level of output gives a relative minimum, we shall use the second-derivative test. We have

$$\frac{d^2\bar{c}}{dq^2} = \frac{800}{q^3}$$

which is positive for $q = 40$. Thus, $\bar{c}$ has a relative minimum when $q = 40$. We note that $\bar{c}$ is continuous for $q > 0$. Since $q = 40$ is the only relative extremum, we conclude that this relative minimum is indeed an absolute minimum. Substituting $q = 40$ in Equation (5) gives the minimum average cost $\bar{c} = 23$.

▶ **Now work Problem 5.**

This example is a biological application involving maximizing the rate at which an enzyme is formed. The equation involved is a literal equation.

EXAMPLE 4 Maximization Applied to Enzymes

An enzyme is a protein that acts as a catalyst for increasing the rate of a chemical reaction that occurs in cells. In a certain reaction, an enzyme is converted to another enzyme called the product. The product acts as a catalyst for its own formation. The rate R at which the product is formed (with respect to time) is given by

$$R = kp(l - p)$$

where l is the total initial amount of both enzymes, p is the amount of the product enzyme, and k is a positive constant. For what value of p will R be a maximum?

Solution: We can write $R = k(pl - p^2)$. Setting $dR/dp = 0$ and solving for p gives

$$\begin{aligned} \frac{dR}{dp} &= k(l - 2p) = 0 \\ p &= \frac{l}{2} \end{aligned}$$

Now, $d^2R/dp^2 = -2k$. Since $k > 0$, the second derivative is always negative. Hence, $p = l/2$ gives a relative maximum. Moreover, since R is a continuous function of p, we conclude that we indeed have an absolute maximum when $p = l/2$.

Calculus can be applied to inventory decisions, as the following example shows.

This example involves determining the number of units in a production run in order to minimize certain costs.

EXAMPLE 5 **Economic Lot Size**

A company annually produces and sells 10,000 *units of a product. Sales are uniformly distributed throughout the year. The company wishes to determine the number of units to be manufactured in each production run in order to minimize total annual setup costs and carrying costs. The same number of units is produced in each run. This number is referred to as the* ***economic lot size or economic order quantity.*** *The production cost of each unit is* \$20, *and carrying costs (insurance, interest, storage, etc.) are estimated to be* 10% *of the value of the average inventory. Setup costs per production run are* \$40. *Find the economic lot size.*

Solution: Let q be the number of units in a production run. Since sales are distributed at a uniform rate, we shall assume that inventory varies uniformly from q to 0 between production runs. Thus, we take the average inventory to be $q/2$ units. The production costs are \$20 per unit, so the value of the average inventory is $20(q/2)$. Carrying costs are 10% of this value:

$$0.10(20)\left(\frac{q}{2}\right)$$

The number of production runs per year is $10{,}000/q$. Hence, the total setup costs are

$$40\left(\frac{10{,}000}{q}\right)$$

Therefore, the total of the annual carrying costs and setup costs is given by

$$\begin{aligned} C &= 0.10(20)\left(\frac{q}{2}\right) + 40\left(\frac{10{,}000}{q}\right) \\ &= q + \frac{400{,}000}{q} \qquad (q > 0) \\ \frac{dC}{dq} &= 1 - \frac{400{,}000}{q^2} = \frac{q^2 - 400{,}000}{q^2} \end{aligned}$$

Setting $dC/dq = 0$, we get

$$q^2 = 400{,}000$$

Since $q > 0$, we choose

$$q = \sqrt{400{,}000} = 200\sqrt{10} \approx 632.5$$

To determine whether this value of q minimizes C, we shall examine the first derivative. If $0 < q < \sqrt{400{,}000}$, then $dC/dq < 0$. If $q > \sqrt{400{,}000}$, then $dC/dq > 0$. We conclude that there is an *absolute* minimum at $q = 632.5$. The number of production runs is $10{,}000/632.5 \approx 15.8$. For practical purposes, there would be 16 lots, each having an economic lot size of 625 units.

▸ **Now work Problem 29.**

The aim of this example is to set up a revenue function from which revenue is maximized over a closed interval.

EXAMPLE 6 **Maximizing TV Cable Company Revenue**

The Vista TV Cable Co. currently has 100,000 *subscribers who are each paying a monthly rate of* \$40. *A survey reveals that there will be* 1000 *more subscribers for each* \$0.25 *decrease in the rate. At what rate will maximum revenue be obtained, and how many subscribers will there be at this rate?*

Solution: Let x be the number of \$0.25 decreases. The monthly rate is then $40 - 0.25x$, where $0 \leq x \leq 160$ (the rate cannot be negative), and the number of *new* subscribers is $1000x$. Thus, the total number of subscribers is $100{,}000 + 1000x$. We want to maximize the revenue, which is given by

$$\begin{aligned} r &= (\text{number of subscribers})(\text{rate per subscriber}) \\ &= (100{,}000 + 1000x)(40 - 0.25x) \\ &= 1000(100 + x)(40 - 0.25x) \\ &= 1000(4000 + 15x - 0.25x^2) \end{aligned}$$

Setting $r' = 0$ and solving for x, we have

$$\begin{aligned} r' = 1000(15 - 0.5x) &= 0 \\ x &= 30 \end{aligned}$$

Since the domain of r is the closed interval $[0, 160]$, the absolute maximum value of r must occur at $x = 30$ or at one of the endpoints of the interval. We now compute r at these three points:

$$\begin{aligned} &\text{If } x = 0, && \text{then } r = 4{,}000{,}000 \\ &\text{If } x = 30, && \text{then } r = 4{,}225{,}000 \\ &\text{If } x = 160, && \text{then } r = 0 \end{aligned}$$

Accordingly, the maximum revenue occurs when $x = 30$. This corresponds to thirty \$0.25 decreases, for a total decrease of \$7.50; that is, the monthly rate is \$32.50. The number of subscribers at that rate is $100{,}000 + 30(1000) = 130{,}000$.

▶ **Now work Problem 19.**

Here we maximize a function over a closed interval.

EXAMPLE 7 Maximizing the Number of Recipients of Health-Care Benefits

An article in a sociology journal stated that if a particular health-care program for the elderly were initiated, then t years after its start, n thousand elderly people would receive direct benefits, where

$$n = \frac{t^3}{3} - 6t^2 + 32t \qquad 0 \leq t \leq 12$$

For what value of t does the maximum number receive benefits?

Solution: Setting $dn/dt = 0$, we have

$$\begin{aligned} \frac{dn}{dt} = t^2 - 12t + 32 &= 0 \\ (t - 4)(t - 8) &= 0 \\ t = 4 \quad \text{or} \quad t &= 8 \end{aligned}$$

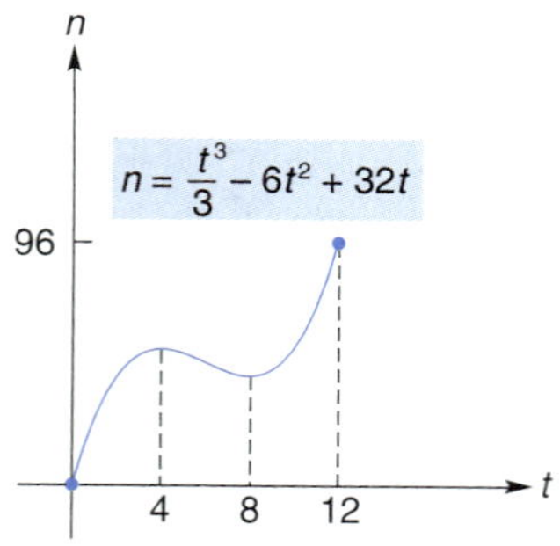

FIGURE 13.64 Graph of $n = \frac{t^3}{3} - 6t^2 + 32t$ on $[0, 12]$.

CAUTION: The preceding example illustrates that you should not ignore endpoints when finding absolute extrema on a closed interval.

Since the domain of n is the closed interval $[0, 12]$, the absolute maximum value of n must occur at $t = 0, 4, 8$, or 12:

$$\text{If } t = 0, \quad \text{then } n = 0$$

$$\text{If } t = 4, \quad \text{then } n = \frac{160}{3} = 53\frac{1}{3}$$

$$\text{If } t = 8, \quad \text{then } n = \frac{128}{3} = 42\frac{2}{3}$$

$$\text{If } t = 12, \quad \text{then } n = 96$$

Thus, an absolute maximum occurs when $t = 12$. A graph of the function is given in Figure 13.64.

▶ **Now work Problem 15.**

In the next example, we use the word *monopolist*. Under a situation of monopoly, there is only one seller of a product for which there are no similar substitutes, and the seller—that is, the monopolist—controls the market. By considering the demand equation for the product, the monopolist may set the price (or volume of output) so that maximum profit will be obtained.

This example involves maximizing profit when the demand and average-cost functions are known. In part 4, a tax is imposed on the monopolist, and a new profit function is analyzed.

EXAMPLE 8 Profit Maximization

Suppose that the demand equation for a monopolist's product is $p = 400 - 2q$ and the average-cost function is $\bar{c} = 0.2q + 4 + (400/q)$, where q is number of units, and both p and $\bar{c}$ are expressed in dollars per unit.

a. *Determine the level of output at which profit is maximized.*

b. *Determine the price at which maximum profit occurs.*

c. *Determine the maximum profit.*

d. *If, as a regulatory device, the government imposes a tax of \$22 per unit on the monopolist, what is the new price for profit maximization?*

Solution: We know that

$$\text{profit} = \text{total revenue} - \text{total cost}$$

Since total revenue r and total cost c are given by

$$r = pq = 400q - 2q^2$$

and

$$c = q\bar{c} = 0.2q^2 + 4q + 400$$

the profit is

$$P = r - c = 400q - 2q^2 - (0.2q^2 + 4q + 400)$$

so that

$$P = 396q - 2.2q^2 - 400 \qquad \text{for } q > 0 \tag{6}$$

a. To maximize profit, we set $dP/dq = 0$:

$$\frac{dp}{dq} = 396 - 4.4q = 0$$

$$q = 90$$

Now, $d^2P/dq^2 = -4.4$ is always negative, so it is negative at the critical value $q = 90$. By the second-derivative test then, there is a relative maximum there. Since $q = 90$ is the only critical value on $(0, \infty)$, we must have an absolute maximum there.

b. The price at which maximum profit occurs is obtained by setting $q = 90$ in the demand equation:

$$p = 400 - 2(90) = \$220$$

c. The maximum profit is obtained by replacing q by 90 in Equation (6), which gives

$$P = \$17{,}420$$

d. The tax of \$22 per unit means that for q units, the total cost increases by $22q$. The new cost function is $c_1 = 0.2q^2 + 4q + 400 + 22q$, and the new profit is given by

$$\begin{aligned} P_1 &= 400q - 2q^2 - (0.2q^2 + 4q + 400 + 22q) \\ &= 374q - 2.2q^2 - 400 \end{aligned}$$

Setting $dP_1/dq = 0$ gives

$$\frac{dP_1}{dq} = 374 - 4.4q = 0$$

$$q = 85$$

Since $d^2P_1/dq^2 = -4.4 < 0$, we conclude that, to maximize profit, the monopolist must restrict output to 85 units at a higher price of $p_1 = 400 - 2(85) = \$230$. Since this price is only \$10 more than before, only part of the tax has been shifted to the consumer, and the monopolist must bear the cost of the balance. The profit now is \$15,495, which is less than the former profit.

▶ **Now work Problem 13.**

This discussion leads to the economic principle that when profit is maximum, marginal revenue is equal to marginal cost.

We conclude this section by using calculus to develop an important principle in economics. Suppose $p = f(q)$ is the demand function for a firm's product, where p is price per unit and q is the number of units produced and sold. Then the total revenue is given by $r = qp = qf(q)$, which is a function of q. Let the total cost of producing q units be given by the cost function $c = g(q)$. Thus, the total profit, which is total revenue minus total cost, is also a function of q, namely,

$$P = r - c = qf(q) - g(q)$$

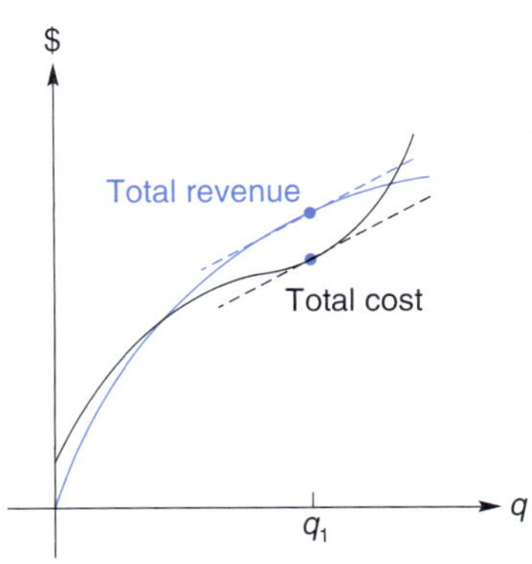

FIGURE 13.65 At maximum profit, marginal revenue equals marginal cost.

Let us consider the most profitable output for the firm. Ignoring special cases, we know that profit is maximized when $dP/dq = 0$ and $d^2P/dq^2 < 0$. We have

$$\frac{dP}{dq} = \frac{d}{dq}(r - c) = \frac{dr}{dq} - \frac{dc}{dq}$$

Consequently, $dP/dq = 0$ when

$$\frac{dr}{dq} = \frac{dc}{dq}$$

That is, at the level of maximum profit, the slope of the tangent to the total-revenue curve must equal the slope of the tangent to the total-cost curve (Figure 13.65). But dr/dq is the marginal revenue MR, and dc/dq is the marginal cost MC. Thus, under typical conditions, to maximize profit, it is necessary that

$$\text{MR} = \text{MC}$$

For this to indeed correspond to a maximum, it is necessary that $d^2P/dq^2 < 0$:

$$\frac{d^2P}{dq^2} = \frac{d^2}{dq^2}(r - c) = \frac{d^2r}{dq^2} - \frac{d^2c}{dq^2} < 0 \quad \text{equivalently} \quad \frac{d^2r}{dq^2} < \frac{d^2c}{dq^2}$$

That is, when MR = MC, in order to ensure maximum profit, the slope of the marginal-revenue curve must be less than the slope of the marginal-cost curve.

The condition that $d^2P/dq^2 < 0$ when $dP/dq = 0$ can be viewed another way. Equivalently, to have MR = MC correspond to a maximum, dP/dq must go from + to −; that is, it must go from $dr/dq - dc/dq > 0$ to $dr/dq - dc/dq < 0$. Hence, as output increases, we must have MR > MC and then MR < MC. This means that at the point q_1 of maximum profit, *the marginal-cost curve must cut the marginal-revenue curve from below* (Figure 13.66). For production up to q_1, the revenue from additional output would be greater than the cost of such output, and the total profit would increase. For output beyond q_1, MC > MR, and each unit of output would add more to total costs than to total revenue. Hence, total profits would decline.

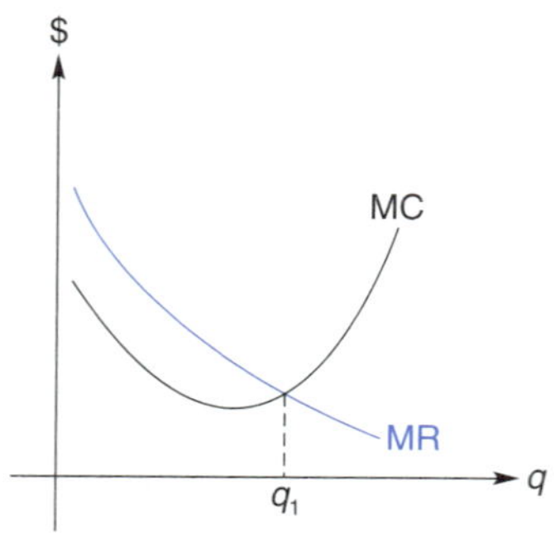

FIGURE 13.66 At maximum profit, the marginal-cost curve cuts the marginal-revenue curve from below.

Exercise 13.6

In this set of problems, unless otherwise specified, p is price per unit (in dollars) and q is output per unit of time. Fixed costs refer to costs that remain constant at all levels of production during a given time period. (An example is rent.)

1. Find two numbers whose sum is 82 and whose product is as big as possible.

2. Find two nonnegative numbers whose sum is 20 and which are such that the product of twice one number and the square of the other number will be a maximum.

*3. **Fencing** A company has set aside \$9000 to fence in a rectangular portion of land adjacent to a stream by using the stream for one side of the enclosed area. The cost of the fencing parallel to the stream is \$15 per foot installed, and the fencing for the remaining two sides costs \$9 per foot installed. Find the dimensions of the maximum enclosed area.

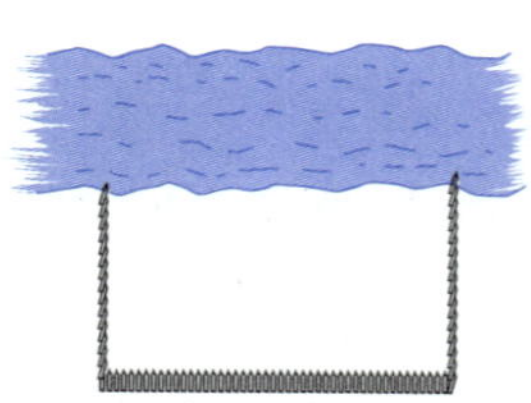

4. **Fencing** The owner of the Laurel Nursery Garden Center wants to fence in 1000 ft^2 of land in a rectangular plot to be used for different types of shrubs. The plot is to be divided into four equal plots with three fences parallel to the same pair of sides, as shown in Figure 13.67. What is the least number of feet of fence needed?

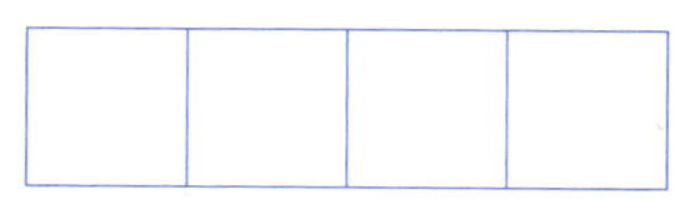

FIGURE 13.67 Diagram for Problem 4.

*5. **Average Cost** A manufacturer finds that the total cost c of producing a product is given by the cost function

$$c = 0.05q^2 + 5q + 500$$

At what level of output will average cost per unit be a minimum?

6. **Automobile Expense** The cost per hour (in dollars) of operating an automobile is given by

$$C = 0.12s - 0.0012s^2 + 0.08 \qquad 0 \le s \le 60$$

where s is the speed in miles per hour. At what speed is the cost per hour a minimum?

*7. **Revenue** The demand equation for a monopolist's product is

$$p = -5q + 30$$

At what price will revenue be maximized?

8. **Revenue** Suppose that the demand function for a monopolist's product is of the form

$$q = Ae^{-Bp}$$

for positive constants A and B. In terms of A and B, find the value of p for which maximum revenue is obtained. Can you explain why your answer does not depend on A?

9. Weight Gain A group of biologists studied the nutritional effects on rats that were fed a diet containing 10% protein.[13] The protein consisted of yeast and cottonseed flour. By varying the percent p of yeast in the protein mix, the group found that the (average) weight gain (in grams) of a rat over a period of time was

$$f(p) = 160 - p - \frac{900}{p+10} \qquad 0 \le p \le 100$$

Find (a) the maximum weight gain and (b) the minimum weight gain.

10. Drug Dose The severity of the reaction of the human body to an initial dose D of a drug is given by[14]

$$R = f(D) = D^2\left(\frac{C}{2} - \frac{D}{3}\right)$$

where the constant C denotes the maximum amount of the drug that may be given. Show that R has a maximum *rate of change* when $D = C/2$.

11. Profit For a monopolist's product, the demand function is

$$p = 72 - 0.04q$$

and the cost function is

$$c = 500 + 30q$$

At what level of output will profit be maximized? At what price does this occur, and what is the profit?

12. Profit For a monopolist, the cost per unit of producing a product is \$3, and the demand equation is

$$p = \frac{10}{\sqrt{q}}$$

What price will give the greatest profit?

***13. Profit** For a monopolist's product, the demand equation is

$$p = 42 - 4q$$

and the average-cost function is

$$\bar{c} = 2 + \frac{80}{q}$$

Find the profit-maximizing price.

14. Profit For a monopolist's product, the demand function is

$$p = \frac{40}{\sqrt{q}}$$

and the average-cost function is

$$\bar{c} = \frac{1}{3} + \frac{2000}{q}$$

Find the profit-maximizing price and output. At this level, show that marginal revenue is equal to marginal cost.

***15. Profit** A manufacturer can produce at most 120 units of a certain product each year. The demand equation for the product is

$$p = q^2 - 100q + 3200$$

and the manufacturer's average-cost function is

$$\bar{c} = \frac{2}{3}q^2 - 40q + \frac{10{,}000}{q}$$

Determine the profit-maximizing output q and the corresponding maximum profit.

16. Cost A manufacturer has determined that, for a certain product, the average cost (in dollars per unit) is given by

$$\bar{c} = 2q^2 - 36q + 210 - \frac{200}{q}$$

where $2 \le q \le 10$.

(a) At what level within the interval [2, 10] should production be fixed in order to minimize total cost? What is the minimum total cost?

(b) If production were required to lie within the interval [5, 10], what value of q would minimize total cost?

17. Profit For XYZ Manufacturing Co., total fixed costs are \$1200, material and labor costs combined are \$2 per unit, and the demand equation is

$$p = \frac{100}{\sqrt{q}}$$

What level of output will maximize profit? Show that this occurs when marginal revenue is equal to marginal cost. What is the price at profit maximization?

18. Revenue A real-estate firm owns 100 garden-type apartments. At \$400 per month, each apartment can be rented. However, for each \$10-per-month increase, there will be two vacancies with no possibility of filling

[13]Adapted from R. Bressani, "The Use of Yeast in Human Foods," in *Single-Cell Protein*, ed. R. I. Mateles and S. R. Tannenbaum (Cambridge, MA: MIT Press, 1968).

[14]R. M. Thrall, J. A. Mortimer, K. R. Rebman, and R. F. Baum, eds., *Some Mathematical Models in Biology*, rev. ed., Report No. 40241-R-7. Prepared at University of Michigan, 1967.

them. What rent per apartment will maximize monthly revenue?

*19. **Revenue** A TV cable company has 4800 subscribers who are each paying \$18 per month. It can get 150 more subscribers for each \$0.50 decrease in the monthly fee. What rate will yield maximum revenue, and what will this revenue be?

20. **Profit** A manufacturer of a product finds that, for the first 600 units that are produced and sold, the profit is \$40 per unit. The profit on each of the units beyond 600 is decreased by \$0.05 times the number of additional units produced. For example, the total profit when 602 units are produced and sold is $600(40) + 2(39.90)$. What level of output will maximize profit?

21. **Container Design** A container manufacturer is designing a rectangular box, open at the top and with a square base, that is to have a volume of 32 ft^3. If the box is to require the least amount of material, what must be its dimensions? (See Figure 13.68.)

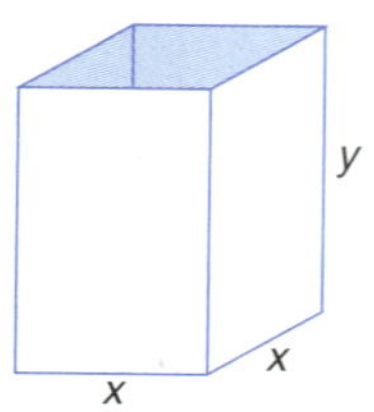

FIGURE 13.68 Open-top box for Problems 21 and 22.

22. **Container Design** An open-top box with a square base is to be constructed from 192 ft^2 of material. What should be the dimensions of the box if the volume is to be a maximum? What is the maximum volume? (See Figure 13.68.)

23. **Container Design** An open box is to be made by cutting equal squares from each corner of a 12-inch-square piece of cardboard and then folding up the sides. Find the length of the side of the square that must be cut out if the volume of the box is to be maximized. What is the maximum volume? (See Figure 13.69.)

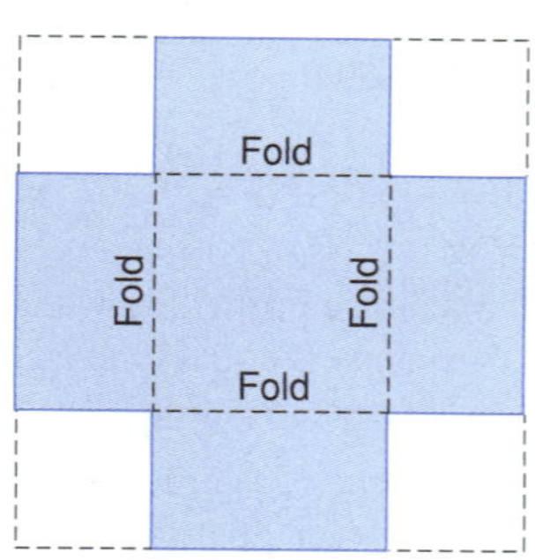

FIGURE 13.69 Box for Problem 23.

24. **Poster Design** A rectangular cardboard poster is to have 240 in^2 for printed matter. It is to have a 5-in. margin at the top and bottom and a 3-in. margin on each side. Find the dimensions of the poster so that the amount of cardboard used is minimized. (See Figure 13.70.)

[*Hint:* First find the values of x and y in Figure 13.70 that minimize the amount of cardboard.]

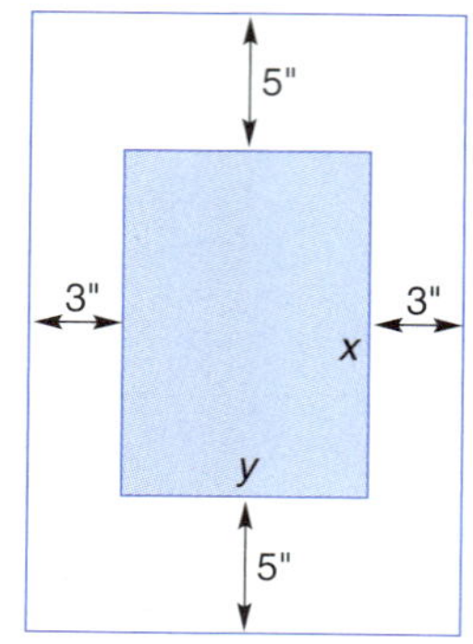

FIGURE 13.70 Poster for Problem 24.

25. **Container Design** A cylindrical can, open at the top, is to have a fixed volume of K. Show that if the least amount of material is to be used, then both the radius and height are equal to $\sqrt[3]{K/\pi}$. (See Figure 13.71.)

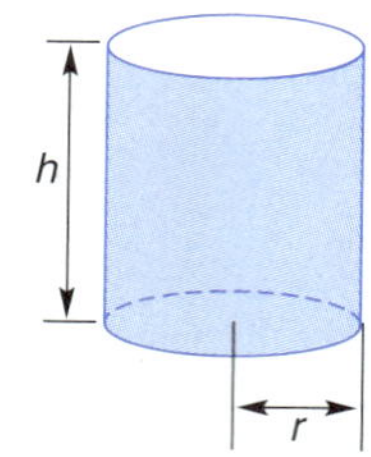

FIGURE 13.71 Can for Problems 25 and 26.

26. **Container Design** A cylindrical can, open at the top, is to be made from a fixed amount of material, K. If the volume is to be a maximum, show that both the radius and height are equal to $\sqrt{K/(3\pi)}$. (See Figure 13.71.)

27. **Profit** The demand equation for a monopolist's product is

$$p = 600 - 2q$$

and the total-cost function is

$$c = 0.2q^2 + 28q + 200$$

Find the profit-maximizing output and price, and determine the corresponding profit. If the government were to impose a tax of \$22 per unit on the manufacturer, what would be the new profit-maximizing output and price? What is the profit now?

28. Profit Use the *original* data in Problem 27, and assume that the government imposes a license fee of \$100 on the manufacturer. This is a lump-sum amount without regard to output. Show that the profit-maximizing price and output remain the same. Show, however, that there will be less profit.

*__29.__ **Economic Lot Size** A manufacturer has to produce annually 1000 units of a product that is sold at a uniform rate during the year. The production cost of each unit is \$10, and carrying costs (insurance, interest, storage, etc.) are estimated to be 12.8% of the value of average inventory. Setup costs per production run are \$40. Find the economic lot size.

30. Profit For a monopolist's product, the cost function is

$$c = 0.004q^3 + 20q + 5000$$

and the demand function is

$$p = 450 - 4q$$

Find the profit-maximizing output.

31. Workshop Attendance Imperial Educational Services (IES) is considering offering a workshop in resource allocation to key personnel at Acme Corp. To make the offering economically feasible, IES feels that at least 30 persons must attend at a cost of \$50 each. Moreover, IES will agree to reduce the charge for *everybody* by \$1.25 for each person over the 30 who attends. How many people should be in the group for IES to maximize revenue? Assume that the maximum allowable number in the group is 40.

32. Cost of Leasing Motor The Kiddie Toy Company plans to lease an electric motor that will be used 80,000 horsepower-hours per year in manufacturing. One horsepower-hour is the work done in 1 hour by a 1-horsepower motor. The annual cost to lease a suitable motor is \$200, plus \$0.40 per horsepower. The cost per horsepower-hour of operating the motor is $\$0.008/N$, where N is the horsepower. What size motor, in horsepower, should be leased in order to minimize cost?

33. Transportation Cost The cost of operating a truck on a thruway (excluding the salary of the driver) is

$$0.165 + \frac{s}{200}$$

dollars per mile, where s is the (steady) speed of the truck in miles per hour. The truck driver's salary is \$18 per hour. At what speed should the truck driver operate the truck to make a 700-mile trip most economical?

34. Cost For a manufacturer, the cost of making a part is \$30 per unit for labor and \$10 per unit for materials; overhead is fixed at \$20,000 per week. If more than 5000 units are made each week, labor is \$45 per unit for those units in excess of 5000. At what level of production will average cost per unit be a minimum?

35. Profit Ms. Jones owns a small insurance agency that sells policies for a large insurance company. For each policy sold, Ms. Jones, who does not sell policies herself, is paid a commission of \$50 by the insurance company. From previous experience, Ms. Jones has determined that, when she employs m salespeople,

$$q = m^3 - 12m^2 + 60m$$

policies can be sold per week. She pays each of the m salespeople a salary of \$750 per week, and her weekly fixed cost is \$2500. Current office facilities can accommodate at most seven salespeople. Determine the number of salespeople that Ms. Jones should hire to maximize her weekly profit. What is the corresponding maximum profit?

36. Profit A manufacturing company sells high-quality jackets through a chain of specialty shops. The demand equation for these jackets is

$$p = 400 - 50q$$

where p is the selling price (in dollars per jacket) and q is the demand (in thousands of jackets). If this company's marginal-cost function is given by

$$\frac{dc}{dq} = \frac{800}{q+5}$$

show that there is a maximum profit, and determine the number of jackets that must be sold to obtain this maximum profit.

37. Chemical Production Each day, a firm makes x tons of chemical A ($x \geq 4$) and

$$y = \frac{24 - 6x}{5 - x}$$

tons of chemical B. The profit on chemical A is \$2000 per ton, and on B it is \$1000 per ton. How much of chemical A should be produced per day to maximize profit? Answer the same question if the profit on A is P per ton and that on B is $P/2$ per ton.

38. Rate of Return To erect an office building, fixed costs are \$1.44 million and include land, architect's fees, a basement, a foundation, etc. If x floors are constructed, the cost (excluding fixed costs) is

$$c = 10x[120{,}000 + 3000(x - 1)]$$

The revenue per month is \$60,000 per floor. How many floors will yield a maximum rate of return on investment? (Rate of return = total revenue/total cost.)

39. Gait and Power Output of an Animal In a model by Smith,[15] the power output of an animal at a given speed as a function of its movement or *gait* j is found to be

$$P(j) = Aj\frac{L^4}{V} + B\frac{V^3L^2}{1 + j}$$

where A and B are constants, j is a measure of the "jumpiness" of the gait, L is a constant representing linear dimension, and V is a constant forward speed.

Assume that P is a minimum when $dP/dj = 0$. Show that when this occurs,

$$(1 + j)^2 = \frac{BV^4}{AL^2}$$

As a passing comment, Smith indicates that "at top speed, j is zero for an elephant, 0.3 for a horse, and 1 for a greyhound, approximately."

40. Traffic Flow In a model of traffic flow on a lane of a freeway, the number of cars the lane can carry per unit time is given by[16]

$$N = \frac{-2a}{-2at_r + v - \dfrac{2al}{v}}$$

where a is the acceleration of a car when stopping ($a < 0$), t_r is the reaction time to begin braking, v is the average speed of the cars, and l is the length of a car. Assume that a, t_r, and l are constant. To find how many cars a lane can carry at most, we want to find the speed v that maximizes N. To maximize N, it suffices to minimize the denominator

$$-2at_r + v - \frac{2al}{v}$$

(a) Find the value of v that minimizes the denominator.
(b) Evaluate your answer in part (a) when $a = -19.6$ (ft/sec^2), $l = 20$ (ft), and $t_r = 0.5$ (sec). Your answer will be in feet per second.
(c) Find the corresponding value of N to one decimal place. Your answer will be in cars per second. Convert your answer to cars per hour.

41. Average Cost During the Christmas season, a promotional company purchases cheap red felt stockings, glues fake white fur and sequins onto them, and packages them for distribution. The total cost of producing q cases of stockings is given by

$$c = 3q^2 + 50q - 18q \ln q + 120$$

Find the number of cases that should be processed in order to minimize the average cost per case. Determine (to two decimal places) this minimum average cost.

42. Profit A monopolist's demand equation is given by

$$p = q^2 - 20q + 160$$

where p is the selling price (in thousands of dollars) per ton when q tons of product are sold. Suppose that fixed cost is \$50 thousand and that each ton costs \$30 thousand to produce. If current equipment has a maximum production capacity of 12 tons, use the graph of the profit function to determine at what production level the maximum profit occurs. Find the corresponding maximum profit and selling price per ton.

15 J. M. Smith, *Mathematical Ideas in Biology* (London: Cambridge University Press, 1968).

16 J. I. Shonle, *Environmental Applications of General Physics* (Reading, MA: Addison-Wesley Publishing Co., 1975).

13.7 Review

Important Terms

Section 13.1	increasing function decreasing function relative maximum relative minimum relative extrema absolute extrema critical value critical point first-derivative test
Section 13.2	extreme-value theorem
Section 13.3	concave up concave down inflection point
Section 13.4	second-derivative test
Section 13.5	vertical asymptote horizontal asymptote vertical-asymptote rule for rational functions
Section 13.6	economic lot size

Summary

Calculus is a great aid in sketching the graph of a function. The first derivative is used to determine when a function is increasing or decreasing and to locate relative maxima and minima. If $f'(x)$ is positive throughout an interval, then over that interval, f is increasing and its graph rises (from left to right). If $f'(x)$ is negative throughout an interval, then over that interval, f is decreasing and its graph is falling.

A point (x_0, y_0) on the graph at which $f'(x)$ is 0 or is not defined is a candidate for a relative extremum, and x_0 is called a critical value. For a relative extremum to occur at x_0, the first derivative must change sign around x_0. The following procedure is the first-derivative test for the relative extrema of $y = f(x)$:

First-Derivative Test for Relative Extrema

Step 1. Find $f'(x)$.

Step 2. Determine all values of x where $f'(x) = 0$ or $f'(x)$ is not defined.

Step 3. On the intervals suggested by the values in step 2, determine whether f is increasing $[f'(x) > 0]$ or decreasing $[f'(x) < 0]$.

Step 4. For each critical value x_0 at which f is continuous, determine whether $f'(x)$ changes sign as x increases through x_0. There is a relative maximum when $x = x_0$ if $f'(x)$ changes from + to −, and a relative minimum if $f'(x)$ changes from − to +. If $f'(x)$ does not change sign, there is no relative extremum when $x = x_0$.

Under certain conditions, a function is guaranteed to have absolute extrema. The extreme-value theorem states that if f is continuous on a closed interval, then f has an absolute maximum value and an absolute minimum value over the interval. To locate absolute extrema, the following procedure may be used:

Procedure to Find Absolute Extrema for a Function f Continuous on $[a, b]$

Step 1. Find the critical values of f.

Step 2. Evaluate $f(x)$ at the endpoints a and b and at the critical values on (a, b).

Step 3. The maximum value of f is the greatest of the values found in step 2. The minimum value of f is the least of the values found in step 2.

The second derivative is used to determine concavity and points of inflection. If $f''(x) > 0$ throughout an interval, then f is concave up over that interval, and its graph bends upward. If $f''(x) < 0$ over an interval, then f is concave down throughout that interval, and its graph bends downward. A point on the graph where f is continuous and its concavity changes is an inflection point. The point (x_0, y_0) on the graph is a possible point of inflection if $f''(x_0)$ is 0 or is not defined and f is continuous at x_0.

The second derivative also provides a means for testing certain critical values for relative extrema:

Second-Derivative Test for Relative Extrema

Suppose $f'(x_0) = 0$. Then

if $f''(x_0) < 0$, f has a relative maximum at x_0

if $f''(x_0) > 0$, f has a relative minimum at x_0

Asymptotes are also aids in curve sketching. Graphs "explode" near vertical asymptotes, and they "settle" near horizontal asymptotes. The line $x = a$ is a vertical asymptote for the graph of a function f if $\lim f(x) = \infty$ or $-\infty$ as x approaches a from the right ($x \to a^+$) or the left ($x \to a^-$). For the case of a rational function, $f(x) = P(x)/Q(x)$, we can find vertical asymptotes without evaluating limits. If $Q(a) = 0$, but $P(a) \neq 0$, then the line $x = a$ is a vertical asymptote.

The line $y = b$ is a horizontal asymptote for the graph of a nonlinear function f if at least one of the following is true:

$$\lim_{x\to\infty} f(x) = b \quad \text{or} \quad \lim_{x\to-\infty} f(x) = b$$

In particular, a polynomial function has neither a horizontal nor a vertical asymptote. Moreover, a rational function whose numerator has degree greater than that of the denominator does not have a horizontal asymptote.

In applied work the importance of calculus in maximization and minimization problems can hardly be overstated. For example, in the area of economics, we can maximize profit or minimize cost. Some important relationships that are used in economics problems are the following:

$\bar{c} = \dfrac{c}{q}$ average cost per unit $= \dfrac{\text{total cost}}{\text{quantity}}$

$r = pq$ revenue = (price)(quantity)

$P = r - c$ profit = total revenue − total cost

Review Problems

Problem numbers shown in color indicate problems suggested for use as a practice chapter test.

In Problems 1–4, find horizontal and vertical asymptotes.

1. $y = \dfrac{3x^2}{x^2 - 16}$
2. $y = \dfrac{x + 3}{9x - 3x^2}$
3. $y = \dfrac{5x^2 - 3}{(3x + 2)^2}$
4. $y = \dfrac{4x + 1}{3x - 5} - \dfrac{3x + 1}{2x - 11}$

In Problems 5–8, find critical values.

5. $f(x) = \dfrac{7x^2}{2 - x^2}$
6. $f(x) = 8(x - 1)^2(x + 6)^4$
7. $f(x) = \dfrac{\sqrt[3]{x + 1}}{3 - 4x}$
8. $f(x) = \dfrac{13xe^{-5x/6}}{6x + 5}$

In Problems 9–12, find intervals on which the function is increasing or decreasing.

9. $f(x) = -\frac{5}{3}x^3 + 15x^2 + 35x + 10$
10. $f(x) = \dfrac{2x^2}{(x + 1)^2}$
11. $f(x) = \dfrac{6x^4}{x^2 - 3}$
12. $f(x) = 9\sqrt[3]{3x^3 - 4x}$

In Problems 13–18, find intervals on which the function is concave up or concave down.

13. $f(x) = x^4 - x^3 - 14$
14. $f(x) = \dfrac{x - 2}{x + 2}$
15. $f(x) = \dfrac{1}{2x - 1}$
16. $f(x) = x^3 + 2x^2 - 5x + 2$
17. $f(x) = (4x + 1)^3(4x + 9)$
18. $f(x) = (x^2 - x - 1)^2$

In Problems 19–24, test for relative extrema.

19. $f(x) = 2x^3 - 9x^2 + 12x + 7$
20. $f(x) = \dfrac{2x + 1}{x^2}$
21. $f(x) = \dfrac{x^{10}}{10} + \dfrac{x^5}{5}$
22. $f(x) = \dfrac{x^2}{x^2 - 4}$
23. $f(x) = x^{2/3}(x + 1)$
24. $f(x) = x^3(2x - 1)^4$

In Problems 25–30, find the x-values where inflection points occur.

25. $y = x^5 - 5x^4 + 3x$
26. $y = \dfrac{x^2 + 2}{5x}$
27. $y = 4(3x - 5)(x^4 + 2)$
28. $y = x^2 + 2\ln(-x)$
29. $y = \dfrac{x^2}{5e^x}$
30. $y = 6(x^2 - 4)^3$

In Problems 31–34, test for absolute extrema on the given interval.

31. $f(x) = 3x^4 - 4x^3$, $[0, 2]$
32. $f(x) = 2x^3 - 15x^2 + 36x$, $[0, 3]$
33. $f(x) = \dfrac{x}{(5x - 6)^2}$, $[-2, 0]$
34. $f(x) = (x + 1)^2(x - 1)^{2/3}$, $[2, 3]$
35. Let $f(x) = (x^2 + 1)e^{-x}$
    - **(a)** Determine the values of x at which relative maxima and relative minima, if any, occur.
    - **(b)** Determine the interval(s) on which the graph of f is concave down, and find the coordinates of all points of inflection.
36. Let $f(x) = \frac{x}{x^2 + 1}$. It can be shown that

$$f'(x) = \frac{1 - x^2}{(x^2 + 1)^2}$$

- **(a)** Determine whether the graph of f is symmetric about the x-axis, y-axis, or origin.
- **(b)** Find the interval(s) on which f is increasing.
- **(c)** Find the coordinates of all relative extrema of f.
- **(d)** Determine $\lim_{x\to-\infty} f(x)$ and $\lim_{x\to\infty} f(x)$.
- **(e)** Sketch the graph of f.
- **(f)** State the absolute minimum and absolute maximum values of $f(x)$ (if they exist).

In Problems 37–48, indicate intervals on which the function is increasing, decreasing, concave up, or concave down; indicate relative maximum points, relative minimum points, points of inflection, horizontal asymptotes, vertical asymptotes, symmetry, and those intercepts that can be obtained conveniently and sketch the graph.

37. $y = x^2 - 2x - 24$

38. $y = 2x^3 + 15x^2 + 36x + 9$

39. $y = x^3 - 12x + 20$

40. $y = x^4 - 4x^3 - 20x^2 + 150$

41. $y = x^3 + x$

42. $y = \dfrac{x+2}{x-3}$

43. $f(x) = \dfrac{100(x+5)}{x^2}$

44. $y = \dfrac{x^2-4}{x^2-1}$

45. $y = \dfrac{x}{(2x-1)^3}$

46. $y = 6x^{1/3}(2x-1)$

47. $f(x) = \dfrac{e^x + e^{-x}}{2}$

48. $f(x) = 1 + \ln(x^2)$

49. Are the following statements true or false?

(a) If $f'(x_0) = 0$, then f must have a relative extremum at x_0.

(b) Since the function $f(x) = 1/x$ is decreasing on the intervals $(-\infty, 0)$ and $(0, \infty)$, it is impossible to find x_1 and x_2 in the domain of f such that $x_1 < x_2$ and $f(x_1) < f(x_2)$.

(c) On the interval $(-1, 1]$, the function $f(x) = x^4$ has an absolute maximum and an absolute minimum.

(d) If $f''(x_0) = 0$, then $(x_0, f(x_0))$ must be a point of inflection.

(e) A function f defined on the interval $(-2, 2)$ with exactly one relative maximum must have an absolute maximum.

50. An important function in probability theory is the standard normal-density function

$$f(x) = \frac{1}{\sqrt{2\pi}} e^{-x^2/2}$$

(a) Determine whether the graph of f is symmetric about the x-axis, y-axis, or origin.

(b) Find the intervals on which f is increasing and those on which it is decreasing.

(c) Find the coordinates of all relative extrema of f.

(d) Find $\lim_{x\to-\infty} f(x)$ and $\lim_{x\to\infty} f(x)$.

(e) Find the intervals on which the graph of f is concave up and those on which it is concave down.

(f) Find the coordinates of all points of inflection.

(g) Sketch the graph of f.

(h) Find all absolute extrema.

51. Marginal Cost If $c = q^3 - 6q^2 + 12q + 18$ is a total-cost function, for what values of q is marginal cost increasing?

52. Marginal Revenue If $r = 320q^{3/2} - 2q^2$ is the revenue function for a manufacturer's product, determine the intervals on which the marginal revenue function is increasing.

53. Revenue Function The demand equation for a manufacturer's product is

$$p = 150 - \frac{\sqrt{q}}{10} \quad \text{where } q > 0$$

Show that the graph of the revenue function is concave down wherever it is defined.

54. Contraception In a model of the effect of contraception on birthrate,[17] the equation,

$$R = f(x) = \frac{x}{4.4 - 3.4x} \qquad 0 \le x \le 1$$

gives the proportional reduction R in the birthrate as a function of the efficiency x of a contraception method. An efficiency of 0.2 (or 20%) means that the probability of becoming pregnant is 80% of the probability of becoming pregnant without the contraceptive. Find the reduction (as a percentage) when efficiency is (a) 0, (b) 0.5, and (c) 1. Find dR/dx and d^2R/dx^2, and sketch the graph of the equation.

55. Learning and Memory If you were to recite members of a category, such as four-legged animals, the words that you utter would probably occur in "chunks," with distinct pauses between such chunks. For example, you might say the following for the category of four-legged animals:

dog, cat, mouse, rat,
(pause)
horse, donkey, mule,
(pause)
cow, pig, goat, lamb,
etc.

The pauses may occur because one has to mentally search for subcategories (animals around the house, beasts of burden, farm animals, etc.).

The elapsed time between onsets of successive words is called *interresponse time*. A function has been used to analyze the length of time for pauses and the chunk size (number of words in a chunk).[18] This function f is such that

$$f(t) = \begin{cases} \text{the average number of words} \\ \text{that occur in succession with} \\ \text{interresponse times less than } t \end{cases}$$

The graph of f has a shape similar to that in Figure 13.72 and is best fit by a third-degree polynomial, such as

$$f(t) = At^3 + Bt^2 + Ct + D$$

---

[17] R. K. Leik and B. F. Meeker, *Mathematical Sociology* (Englewood Cliffs, NJ: Prentice-Hall, 1975).

[18] A. Graesser and G. Mandler, "Limited Processing Capacity Constrains the Storage of Unrelated Sets of Words and Retrieval from Natural Categories," *Human Learning and Memory* 4, no. 1 (1978), 86–100.

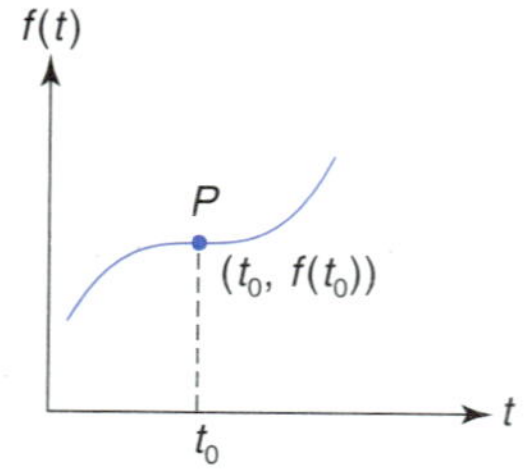

FIGURE 13.72 Diagram for Problem 55.

The point P has special meaning. It is such that the value t_0 separates interresponse times *within* chunks from those *between* chunks. Mathematically, P is a critical point that is also a point of inflection. Assume these two conditions, and show that (a) $t_0 = -B/(3A)$ and (b) $B^2 = 3AC$.

56. Market Penetration In a model for the market penetration of a new product, sales S of the product at time t are given by[19]

$$S = g(t) = \frac{m(p+q)^2}{p}\left[\frac{e^{-(p+q)t}}{\left(\frac{q}{p}e^{-(p+q)t}+1\right)^2}\right]$$

where p, q, and m are nonzero constants.

(a) Show that

$$\frac{dS}{dt} = \frac{\frac{m}{p}(p+q)^3e^{-(p+q)t}\left[\frac{q}{p}e^{-(p+q)t}-1\right]}{\left(\frac{q}{p}e^{-(p+q)t}+1\right)^3}$$

(b) Determine the value of t for which maximum sales occur. You may assume that S attains a maximum when $dS/dt = 0$.

In Problems 57–60, where appropriate, round your answers to two decimal places.

57. From the graph of $y = 4x^3 + 5.3x^2 - 7x + 3$, find the coordinates of all relative extrema.

58. From the graph of $f(x) = x^4 - 2x^3 + 3x - 1$, determine the absolute extrema of f over the interval $[-1, 1]$.

59. The graph of a function f has exactly one inflection point. If

$$f''(x) = \frac{x^3+3x+2}{5x^2-2x+4}$$

use the graph of f'' to determine the x-value of the inflection point of f.

[19] A. P. Hurter, Jr., A. H. Rubenstein, et al., "Market Penetration by New Innovations: The Technological Literature," *Technological Forecasting and Social Change* 11 (1978), 197–221.

60. Graph $y = \frac{4x-2x^2}{x^3+4x+4}$. From the graph, locate any horizontal or vertical asymptotes.

61. Maximization of Production A manufacturer determined that m employees on a certain production line will produce q units per month, where

$$q = 80m^2 - 0.1m^4$$

To obtain maximum monthly production, how many employees should be assigned to the production line?

62. Revenue The demand function for a manufacturer's product is given by $p = 100e^{-0.1q}$. For what value of q does the manufacturer maximize total revenue?

63. Revenue The demand function for a monopolist's product is

$$p = \sqrt{500-q}$$

If the monopolist wants to produce at least 100 units, but not more than 200 units, how many units should be produced to maximize total revenue?

64. Average Cost If $c = 0.01q^2 + 5q + 100$ is a cost function, find the average-cost function. At what level of production q is there a minimum average cost?

65. Profit The demand function for a monopolist's product is

$$p = 400 - 2q$$

and the average cost per unit for producing q units is

$$\bar{c} = q + 160 + \frac{2000}{q}$$

where p and $\bar{c}$ are in dollars per unit. Find the maximum profit that the monopolist can achieve.

66. Container Design A rectangular box is to be made by cutting out equal squares from each corner of a piece of cardboard 10 in. by 16 in. and then folding up the sides. What must be the length of the side of the square cut out if the volume of the box is to be maximum?

67. Fencing A rectangular portion of a field is to be enclosed by a fence and divided equally into three parts by two fences parallel to one pair of the sides. If a total of 800 ft of fencing is to be used, find the dimensions that will maximize the fenced area.

68. Poster Design A rectangular poster having an area of 500 in^2 is to have a 4-in. margin at each side and at the bottom and a 6-in. margin at the top. The remainder of the poster is for printed matter. Find the dimensions of the poster so that the area for the printed matter is maximized.

69. Cost A furniture company makes personal-computer stands. For a certain model, the total cost (in thousands of dollars) when q *hundred* stands are produced is given by

$$c = 2q^3 - 9q^2 + 12q + 20$$

(a) The company is currently capable of manufacturing between 75 and 600 stands (inclusive) per week. Determine the number of stands that should be produced per week to minimize the total cost, and find the corresponding average cost per stand.

(b) Suppose that between 300 and 600 stands must be produced. How many should the company now produce in order to minimize total cost?

70. Bacteria In a laboratory, an experimental antibacterial agent is applied to a population of 100 bacteria. Data indicate that the number of bacteria t hours after the agent is introduced is given by

$$N = \frac{12{,}100 + 110t + 100t^2}{121 + t^2}$$

For what value of t does the maximum number of bacteria in the population occur? What is this maximum number?

Mathematical SNAPSHOT

Sketching the Phillips Curve

Before the Laffer curve's rise to fame, economists debated another curve, also named after its originator: the Phillips curve. This curve is meant to describe the relationship between unemployment and inflation. Sketching the Phillips curve will give us a sense of how curve sketching is sometimes an uncertain process, involving not only a fit to known data values but also, necessarily, some conception of the process underlying the data.

The reasoning behind the Phillips curve is that when unemployment is high, the average worker is spending less money; and at the same time, employers do not have to give generous raises to retain workers. Employer thrift and limited worker spending keep prices and wages from rising rapidly and produce a low inflation rate. Conversely, when the country is at near-full employment, workers are earning more and spending more, and employers have to offer larger salary incentives to keep their workers. These factors drive wages and prices up.

There appears to be, then, some sort of broadly inverse relationship between the unemployment rate and the inflation rate. The graph of this relationship is the Phillips curve.

To the preceding reasoning, let us now add some empirical data. One source of data is the annual Economic Report of the President (downloadable at w3.access.gpo.gov/eop). Table 13.1 is based on the 2001 report. Inflation can be measured in several ways; here the measure used is the annual percentage change in the Gross Domestic Product (GDP) Price Index.

TABLE 13.1 U.S. Unemployment and Inflation, 1959–2000

Year	Unemployment Rate (%)	Inflation Rate (%)	Year	Unemployment Rate (%)	Inflation Rate (%)
1959	5.5	1.1	1980	7.1	9.2
1960	5.5	1.4	1981	7.6	9.3
1961	6.7	1.1	1982	9.7	6.2
1962	5.5	1.4	1983	9.6	3.9
1963	5.7	1.1	1984	7.5	3.7
1964	5.2	1.5	1985	7.2	3.2
1965	4.5	1.9	1986	7	2.2
1966	3.8	2.8	1987	6.2	3
1967	3.8	3.1	1988	5.5	3.4
1968	3.6	4.3	1989	5.3	3.8
1969	3.5	4.9	1990	5.6	3.9
1970	4.9	5.3	1991	6.8	3.6
1971	5.9	5	1992	7.5	2.4
1972	5.6	4.2	1993	6.9	2.4
1973	4.9	5.6	1994	6.1	2.1
1974	5.6	9	1995	5.6	2.2
1975	8.5	9.4	1996	5.4	1.9
1976	7.7	5.7	1997	4.9	1.9
1977	7.1	6.4	1998	4.5	1.3
1978	6.1	7.1	1999	4.2	1.5
1979	5.8	8.3	2000	4	2.1

Consider first the data values for the years 1959–1969, where the unemployment rate was between 3.5% and 6.7%. If these values are plotted, they suggest what the middle of the curve should look like, but they do not determine the end behavior. To fix the ends of the curve, we may reason that since negative unemployment is impossible and even perfect zero unemployment is an unrealistic scenario, the curve should not intersect the vertical axis but should, instead, have it as an asymptote.

At the right end of the curve, by contrast, we know that negative inflation is theoretically possible and, historically, a real (though unusual) occurrence. So there should probably be an intercept on the horizontal axis. With this information in hand, we can plot the data values with a sketched curve superimposed, as in Figure 13.73.

FIGURE 13.73 The years 1959–1969.

As things stood at the beginning of the 1970s, economists were generally confident that the Phillips curve represented an important, useful insight into the workings of the U.S. economy. Data from earlier decades, though not lining up exactly with the 1959–1969 data, fit the same overall pattern.

In 1970, however, this picture began to come apart. Figure 13.74 shows a scatterplot of the inflation and unemployment data for the years 1970–2000. This time it is hard to know what sort of curve to draw! The points do not suggest anything like a Phillips curve.

Looking at data from the 1970s, some economists suggested that perhaps there are really two Phillips curves, one for the short term that looks more or less as shown in Figure 13.73, and another, "truer" curve for the long run—simply a vertical line, corresponding to a natural level of unemployment, at around 5 or 6%. These economists figured that although the unemployment rate can fluctuate, ultimately there is an unemployment set-point, so to speak. If the rate of unemployment drops below that, it sets off a chain of events that eventually forces unemployment back up. This is not a universally accepted view, but it is one possible interpretation of the data.

Exercises

1. In Figure 13.74, examine the data points for 1998–2000. How well does this part of the data fit the Phillips curve sketched for 1959–1969?
2. In the data in Table 13.1, there is a pattern that does not correspond to any particular sketched curve. It appears in the scatterplot only when time is taken into account: a tendency toward a sort of clockwise spiral. To see this pattern, follow the year-to-year succession of points in Figure 13.74. In light of this pattern, and in light of claims about a vertical long-run Phillips curve, what might one expect inflation and unemployment to do in the years 2001–2005? How reliable are these expectations?

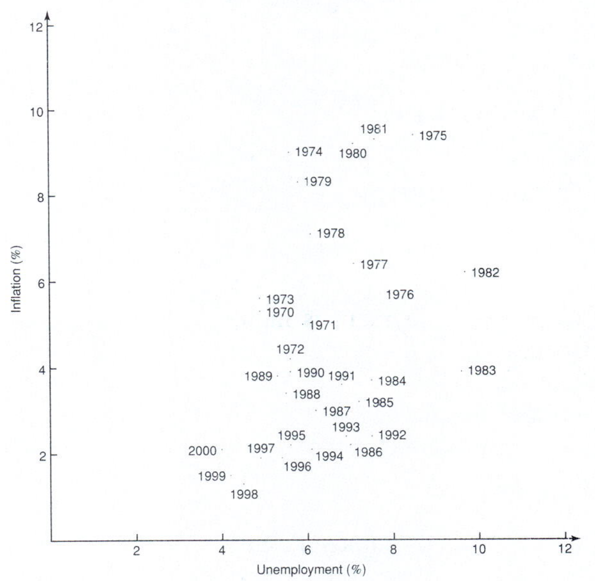

FIGURE 13.74 The years 1970–2000.

CHAPTER

14

Integration

Anyone who runs a business knows the need for accurate cost estimates. When jobs are individually contracted, determining how much a job will cost is generally the first step in deciding how much to bid.

For example, a painter must determine how much paint a job will take. Since a gallon of paint will cover a certain number of square feet, the key is to determine the area of the surfaces to be painted. Normally, even this requires only simple arithmetic—walls and ceilings are rectangular, and so total area is a sum of products of base and height.

But not all area calculations are as simple. Suppose, for instance, that the bridge shown below must be sandblasted to remove accumulated soot. How would the contractor who charges for sandblasting by the square foot calculate the area of the vertical face on each side of the bridge?

The area could be estimated as perhaps three-quarters of the area of the trapezoid formed by points A, B, C, and D. But a more accurate calculation—which might be desirable if the bid were for dozens of bridges of the same dimensions (as along a stretch of railroad); this would require a more refined approach.

If the shape of the bridge's arch can be described mathematically by a function, the contractor could use the method introduced in this chapter: integration. Integration has many applications, the simplest of which is finding areas of regions bounded by curves. Other applications include calculating the total deflection of a beam due to bending stress, calculating the distance traveled underwater by a submarine, and calculating the electricity bill for a company that consumes power at differing rates over the course of a month. Chapters 11–13 dealt with differential calculus. We differentiated a function and obtained another function, its derivative. *Integral calculus* is concerned with the reverse process: We are given the derivative of a function and must find the original function. The need for doing this arises in a natural way. For example, we may have a marginal-revenue function and want to find the revenue function from it. Integral calculus also involves a concept that allows us to take the limit of a special kind of sum as the number of terms in the sum becomes infinite. This is the real power of integral calculus! With such a notion, we may find the area of a region that cannot be found by any other convenient method.

OBJECTIVE *To define the differential, interpret it geometrically, and use it in approximations. Also, to restate the reciprocal relationship between dx/dy and dy/dx.*

14.1 Differentials

We shall soon give you a reason for using the symbol dy/dx to denote the derivative of y with respect to x. To do this, we introduce the notion of the *differential* of a function.

Definition

Let $y = f(x)$ be a differentiable function of x, and let Δx denote a change in x, where Δx can be any real number. Then the ***differential*** of y, denoted dy or $d[f(x)]$, is given by

$$dy = f'(x)\Delta x$$

Note that dy is a function of two variables, namely, x and Δx.

EXAMPLE 1 **Computing a Differential**

Find the differential of $y = x^3 - 2x^2 + 3x - 4$, and evaluate it when $x = 1$ and $\Delta x = 0.04$.

Solution: The differential is

$$\begin{aligned} dy &= \frac{d}{dx}(x^3 - 2x^2 + 3x - 4)\Delta x \\ &= (3x^2 - 4x + 3)\Delta x \end{aligned}$$

When $x = 1$ and $\Delta x = 0.04$,

$$dy = [3(1)^2 - 4(1) + 3](0.04) = 0.08$$

If $y = x$, then $dy = d(x) = 1\Delta x = \Delta x$. Hence, the differential of x is Δx. We abbreviate $d(x)$ by dx. Thus, $dx = \Delta x$. From now on, it will be our practice to write dx for Δx when finding a differential. For example,

$$d(x^2 + 5) = \frac{d}{dx}(x^2 + 5)\,dx = 2x\,dx$$

▶ **Now work Problem 1.**

Summarizing, we say that if $y = f(x)$ defines a differentiable function of x, then

$$dy = f'(x)\,dx$$

where dx is any real number. Provided that $dx \neq 0$, we can divide both sides by dx:

$$\frac{dy}{dx} = f'(x)$$

That is, dy/dx can be viewed either as the quotient of two differentials, namely,

dy divided by dx, or as one symbol for the derivative of f at x. It is for this reason that we introduced the symbol dy/dx to denote the derivative.

EXAMPLE 2 Finding a Differential in Terms of dx

a. If $f(x) = \sqrt{x}$, then

$$d(\sqrt{x}) = \frac{d}{dx}(\sqrt{x})\,dx = \frac{1}{2}x^{-1/2}dx = \frac{1}{2\sqrt{x}}\,dx$$

b. If $u = (x^2+3)^5$, then $du = 5(x^2+3)^4(2x)\,dx = 10x(x^2+3)^4\,dx$

▶ **Now work Problem 3.**

The differential can be interpreted geometrically. In Figure 14.1, the point $P(x, f(x))$ is on the curve $y = f(x)$. Suppose x changes by dx, a real number, to the new value $x + dx$. Then the new function value is $f(x + dx)$, and the corresponding point on the curve is $Q(x + dx,\ f(x + dx))$.

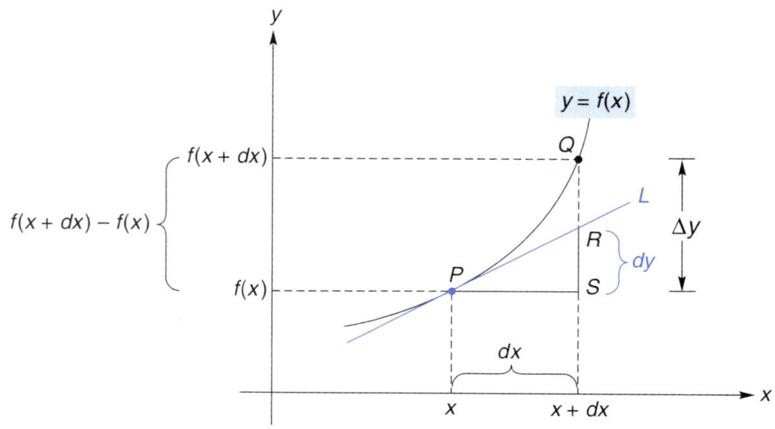

FIGURE 14.1 Geometric interpretation of dy and Δx.

Passing through P and Q are horizontal and vertical lines, respectively, that intersect at S. A line L tangent to the curve at P intersects segment QS at R, forming the right triangle PRS. Observe that the graph of f near P is approximated by the tangent line at P. The slope of L is $f'(x)$ but it is also given by $\overline{SR}/\overline{PS}$ so that:

$$f'(x) = \frac{\overline{SR}}{\overline{PS}}$$

Since $dy = f'(x)\,dx$ and $dx = \overline{PS}$,

$$dy = f'(x)\,dx = \frac{\overline{SR}}{\overline{PS}} \cdot \overline{PS} = \overline{SR}$$

Thus, if dx is a change in x at P, then dy is the corresponding vertical change along the **tangent line** at P. Note that for the same dx, the vertical change along the **curve** is $\Delta y = \overline{SQ} = f(x + dx) - f(x)$. Do not confuse Δy with dy. However, from Figure 14.1, the following is apparent:

When dx is close to 0, dy is an approximation to Δy. Therefore,

$$\Delta y \approx dy$$

This fact is useful in estimating Δy, a change in y, as Example 3 shows.

EXAMPLE 3 **Using the Differential to Estimate a Change in a Quantity**

A governmental health agency examined the records of a group of individuals who were hospitalized with a particular illness. It was found that the total proportion P that are discharged at the end of t days of hospitalization is given by

$$P = P(t) = 1 - \left(\frac{300}{300+t}\right)^3$$

Use differentials to approximate the change in the proportion discharged if t changes from 300 to 305.

Solution: The change in t from 300 to 305 is $\Delta t = dt = 305 - 300 = 5$. The change in P is $\Delta P = P(305) - P(300)$. We approximate ΔP by dP:

$$\Delta P \approx dP = P'dt = -3\left(\frac{300}{300+t}\right)^2\left[-\frac{300}{(300+t)^2}\right]dt$$

When $t = 300$ and $dt = 5$,

$$\begin{aligned} dP &= -3\left(\frac{300}{600}\right)^2\left[-\frac{300}{(600)^2}\right]5 \\ &= -3\left(\frac{1}{2}\right)^2\left[-\frac{1}{2(600)}\right]5 = \frac{1}{320} \approx 0.0031 \end{aligned}$$

For a comparison, the true value of ΔP is

$$P(305) - P(300) \approx 0.87807 - 0.87500 = 0.00307$$

(rounded to five decimal places)

▶ **Now work Problem 11.**

Equation (1) is used to approximate a function value, whereas the formula $\Delta y \approx dy$ is used to approximate a change in function values.

We said that if $y = f(x)$, then $\Delta y \approx dy$ if dx is close to zero. Thus,

$$\Delta y = f(x+dx) - f(x) \approx dy$$

so that

$$f(x+dx) \approx f(x) + dy \tag{1}$$

This formula gives us a way of estimating a function value $f(x + dx)$. For example, suppose we estimate ln(1.06). Letting $y = f(x) = \ln x$, we need to estimate $f(1.06)$. Since $d(\ln x) = (1/x)\,dx$, we have, from Equation (1).

$$\begin{aligned} f(x+dx) &\approx f(x) + dy \\ \ln(x+dx) &\approx \ln x + \frac{1}{x}\,dx \end{aligned}$$

We know the exact value of ln 1, so we shall let $x = 1$ and $dx = 0.06$. Then $x + dx = 1.06$, and dx is close to zero. Therefore,

$$\begin{aligned} \ln(1+0.06) &\approx \ln(1) + \frac{1}{1}(0.06) \\ \ln(1.06) &\approx 0 + 0.06 = 0.06 \end{aligned}$$

The true value of ln(1.06) rounded to five decimal places is 0.05827.

EXAMPLE 4 **Using the Differential to Estimate a Function Value**

The demand function for a product is given by

$$p = f(q) = 20 - \sqrt{q}$$

where p is the price per unit in dollars for q units. By using differentials, approximate the price when 99 units are demanded.

Solution: We want to approximate $f(99)$. By Equation (1),

$$f(q + dq) \approx f(q) + dp$$

where

$$dp = -\frac{1}{2\sqrt{q}}\,dq \qquad \left(\frac{dp}{dq} = -\frac{1}{2}q^{-1/2}\right)$$

We choose $q = 100$ and $dq = -1$ because $q + dq = 99$, dq is small, and it is easy to compute $f(100) = 20 - \sqrt{100} = 10$. We thus have

$$f(99) = f[100 + (-1)] \approx f(100) - \frac{1}{2\sqrt{100}}(-1)$$

$$f(99) \approx 10 + 0.05 = 10.05$$

Hence, the price per unit when 99 units are demanded is approximately \$10.05.

▶ **Now work Problem 17.**

The equation $y = x^3 + 4x + 5$ defines y as a function of x. We could write $f(x) = x^3 + 4x + 5$. However, the equations also defines x implicitly as a function of y. In fact, if we restrict the domain of f to some set of real numbers x so that $y = f(x)$ is a one-to-one function, then in principle we could solve for x in terms of y and get $x = f^{-1}(y)$. As we did in Example 6 of Section 12.2 we can look at the derivative of x with respect to y, dx/dy and we have seen that it is given by

$$\frac{dx}{dy} = \frac{1}{\frac{dy}{dx}} \qquad \text{provided that } dy/dx \neq 0$$

Since dx/dy can be considered a quotient of differentials, we now see that it is the reciprocal of the quotient of differentials dy/dx. Thus

$$\frac{dx}{dy} = \frac{1}{3x^2 + 4}$$

It is important to understand that it is not necessary to be able to solve $y = x^3 + 4x + 5$ for x in terms of y and the equation $\frac{dx}{dy} = \frac{1}{3x^2 + 4}$ holds for all x, not just for sets of numbers where f restricts to a one-to-one function.

EXAMPLE 5 **Finding dp/dq from dq/dp**

Find $\frac{dp}{dq}$ if $q = \sqrt{2500 - p^2}$.

Solution:

Strategy: There are a number of ways to find dp/dq. One approach is to solve the given equation for p explicitly in terms of q and then differentiate directly. Another approach to find dp/dq is to use implicit differentiation. However, since q is given explicitly as a function of p, we can easily find dq/dp and then use the preceding reciprocal relation to find dp/dq. We shall take this approach.

We have

$$\frac{dq}{dp} = \frac{1}{2}(2500 - p^2)^{-1/2}(-2p) = -\frac{p}{\sqrt{2500 - p^2}}$$

Hence,

$$\frac{dp}{dq} = \frac{1}{\frac{dq}{dp}} = -\frac{\sqrt{2500 - p^2}}{p}$$

▸ **Now work Problem 27.**

Exercise 14.1

In Problems 1–10, find the differential of the function in terms of x and dx.

*1. $y = 3x - 4$

2. $y = 2$

*3. $f(x) = \sqrt{x^4 - 9}$

4. $f(x) = (4x^2 - 5x + 2)^3$

5. $u = \frac{1}{x^2}$

6. $u = \frac{1}{\sqrt{x}}$

7. $p = \ln(x^2 + 7)$

8. $p = e^{x^3+5}$

9. $y = (9x + 3)e^{2x^2+3}$

10. $y = \ln \sqrt{x^2 + 12}$

In Problems 11–16, find Δy and dy for the given values of x and dx.

*11. $y = 4 - 7x$; $x = 3$, $dx = 0.02$

12. $y = 5x^2$; $x = -1$, $dx = -0.02$

13. $y = 4x^2 - 3x + 10$; $x = -1$, $dx = 0.25$

14. $y = (3x + 2)^2$; $x = -1$, $dx = -0.03$

15. $y = \sqrt{32 - x^2}$; $x = 4$, $dx = -0.05$ Round your answer to three decimal places.

16. $y = \ln(-x)$; $x = -5$, $dx = 0.1$

*17. Let $f(x) = \frac{x+5}{x+1}$

(a) Evaluate $f'(1)$.
(b) Use differentials to estimate the value of $f(1.1)$.

18. Let $f(x) = x^{3x}$

(a) Evaluate $f'(1)$.
(b) Use differentials to estimate the value of $f(0.98)$.

In Problems 19–26, approximate each expression by using differentials.

19. $\sqrt{99}$

20. $\sqrt{122}$

21. $\sqrt[3]{65.5}$

22. $\sqrt[4]{16.3}$

23. $\ln 0.97$

24. $\ln 1.01$

25. $e^{0.01}$

26. $e^{-0.01}$

In Problems 27–32, find dx/dy or dp/dq.

*27. $y = 2x - 1$

28. $y = 5x^2 + 3x + 2$

29. $q = (p^2 + 5)^3$

30. $q = \sqrt{p + 5}$

31. $q = \frac{1}{p}$

32. $q = e^{5-p}$

33. If $y = 7x^2 - 6x + 3$, find the value of dx/dy when $x = 3$.

34. If $y = \ln x^2$, find the value of dx/dy when $x = 3$.

In Problems 35 and 36, find the rate of change of q with respect to p for the indicated value of q.

35. $p = \frac{500}{q+2}$; $q = 18$

36. $p = 50 - \sqrt{q}$; $q = 100$

37. **Profit** Suppose that the profit (in dollars) of producing q units of a product is

$$P = 396q - 2.2q^2 - 400$$

Using differentials, find the approximate change in profit if the level of production changes from $q = 80$ to $q = 81$. Find the true change.

38. Revenue Given the revenue function

$$r = 250q + 45q^2 - q^3$$

use differentials to find the approximate change in revenue if the number of units increases from $q = 40$ to $q = 41$. Find the true change.

39. Demand The demand equation for a product is

$$p = \frac{10}{\sqrt{q}}$$

Using differentials, approximate the price when 24 units are demanded.

40. Demand Given the demand function

$$p = \frac{200}{\sqrt{q+8}}$$

use differentials to estimate the price per unit when 40 units are demanded.

41. If $y = f(x)$, then the *proportional change in y* is defined to be $\Delta y/y$, which can be approximated with differentials by dy/y. Use this last form to approximate the proportional change in the cost function

$$c = f(q) = \frac{q^4}{2} + 3q + 400$$

when $q = 10$ and $dq = 2$. Round your answer to one decimal place.

42. Status/Income Suppose that S is a numerical value of status based on a person's annual income I (in thousands of dollars). For a certain population, suppose $S = 20\sqrt{I}$. Use differentials to approximate the change in S if annual income decreases from \$45,000 to \$44,500.

43. Biology The volume of a spherical cell is given by $V = \frac{4}{3}\pi r^3$, where r is the radius. Estimate the change in volume when the radius changes from 6.5×10^{-4} cm to 6.6×10^{-4} cm.

44. Muscle Contraction The equation

$$(P + a)(v + b) = k$$

is called the "fundamental equation of muscle contraction."[1] Here P is the load imposed on the muscle, v is the velocity of the shortening of the muscle fibers, and a, b, and k are positive constants. Find v in terms of P, and then use the differential to approximate the change in v due to a small change in P.

45. Demand The demand, q, for a monopolist's product is related to the price per unit, p, according to the equation

$$2 + \frac{q^2}{200} = \frac{4000}{p^2}$$

(a) Verify that 40 units will be demanded when the price per unit is \$20.

(b) Show that $\frac{dq}{dp} = -2.5$ when the price per unit is \$20.

(c) Use differentials and the results of parts (a) and (b) to approximate the number of units that will be demanded if the price per unit is reduced to \$19.20.

46. Profit The demand equation for a monopolist's product is

$$p = \frac{1}{2}q^2 - 66q + 7000$$

and the average-cost function is

$$\bar{c} = 500 - q + \frac{80{,}000}{2q}$$

(a) Find the profit when 100 units are demanded.

(b) Use differentials and the result of part (a) to estimate the profit when 98 units are demanded.

[1] R. W. Stacy et al., *Essentials of Biological and Medical Physics* (New York: McGraw-Hill Book Company, 1955).

OBJECTIVE *To define the antiderivative and the indefinite integral and to apply basic integration formulas.*

14.2 The Indefinite Integral

Given a function f, if F is a function such that

$$F'(x) = f(x) \quad (1)$$

then F is called an **antiderivative** of f. Thus,

an antiderivative of f is simply a function whose derivative is f.

Multiplying both sides of Equation (1) by the differential dx gives $F'(x)\,dx = f(x)\,dx$. However, because $F'(x)\,dx$ is the differential of F, we have $dF = f(x)\,dx$. Hence, we can think of an antiderivative of f as a function whose differential is $f(x)\,dx$.

Definition

An ***antiderivative*** of a function f is a function F such that

$$F'(x) = f(x)$$

or equivalently, in differential notation,

$$dF = f(x)\,dx$$

For example, because the derivative of x^2 is $2x$, x^2 is an antiderivative of $2x$. However, it is not the only antiderivative of $2x$: Since

$$\frac{d}{dx}(x^2+1) = 2x \quad \text{and} \quad \frac{d}{dx}(x^2-5) = 2x$$

both x^2+1 and x^2-5 are also antiderivatives of $2x$. In fact, it is obvious that because the derivative of a constant is zero, x^2+C is also an antiderivative of $2x$ for *any* constant C. Thus, $2x$ has infinitely many antiderivatives. More important, *all* antiderivatives of $2x$ must be functions of the form x^2+C, because of the following fact:

Any two antiderivatives of a function differ only by a constant.

Since x^2+C describes all antiderivatives of $2x$, we can refer to it as being the *most general antiderivative* of $2x$, denoted by $\int 2x\,dx$, which is read "the *indefinite integral* of $2x$ with respect to x." Thus, we write

$$\int 2x\,dx = x^2 + C$$

The symbol $\int$ is called the **integral sign,** $2x$ is the **integrand,** and C is the **constant of integration.** The dx is part of the integral notation and indicates the variable involved. Here x is the **variable of integration.**

More generally, the **indefinite integral** of any function f with respect to x is written $\int f(x)\,dx$ and denotes the most general antiderivative of f. Since all antiderivatives of f differ only by a constant, if F is any antiderivative of f then

$$\int f(x)\,dx = F(x) + C \qquad \text{where } C \text{ is a constant}$$

To *integrate* f means to find $\int f(x)\,dx$. In summary,

$$\int f(x)\,dx = F(x) + C \quad \text{if and only if} \quad F'(x) = f(x)$$

APPLICATIONS IN PRACTICE 1

Finding an Indefinite Integral

If the marginal cost for a company is $f(q) = 28.3$, find $\int 28.3\,dq$, which gives the form of the cost function.

EXAMPLE 1 Finding an Indefinite Integral

Find $\int 5\,dx$.

Solution:

Strategy: First we must find (perhaps better words are "guess at") a function whose derivative is 5. Then we add the constant of integration.

Since we know that the derivative of $5x$ is 5, $5x$ is an antiderivative of 5. Therefore,

$$\int 5\,dx = 5x + C$$

CAUTION

A common mistake is to omit C, the constant of integration.

▸ **Now work Problem 1.**

Using differentiation formulas from Chapters 11 and 12, we have compiled a list of basic integration formulas in Table 14.1. These formulas are easily verified. For example, Formula 2 is true because the derivative of $x^{n+1}/(n+1)$ is x^n for $n \neq -1$. (We must have $n \neq -1$ because the denominator is 0 when $n = -1$.) Formula 2 states that the indefinite integral of a power of x (except x^{-1}) is obtained by increasing the exponent of x by one, dividing by the new exponent, and adding on the constant of integration. The indefinite integral of x^{-1} will be discussed in Section 14.4.

TABLE 14.1 Basic Integration Formulas

1. $\int k\,dx = kx + C$ k is a constant
2. $\int x^n\,dx = \dfrac{x^{n+1}}{n+1} + C$ $n \neq -1$
3. $\int e^x\,dx = e^x + C$
4. $\int kf(x)\,dx = k\int f(x)\,dx$ k is a constant.
5. $\int [f(x) \pm g(x)]\,dx = \int f(x)\,dx \pm \int g(x)\,dx$

To verify Formula 4, we must show that the derivative of $k\int f(x)\,dx$ is $kf(x)$. Since the derivative of $k\int f(x)\,dx$ is simply k times the derivative of $\int f(x)\,dx$, which is $f(x)$, Formula 4 is verified. You should verify the other formulas. Formula 5 can be extended to any number of sums or differences.

EXAMPLE 2 **Indefinite Integrals of a Constant and of a Power of *x***

a. *Find* $\int 1\,dx$.

Solution: By Formula 1 with $k = 1$

$$\int 1\,dx = 1x + C = x + C$$

Usually, we write $\int 1\,dx$ as $\int dx$. Thus, $\int dx = x + C$.

b. *Find* $\int x^5\,dx$.

Solution: By Formula 2 with $n = 5$,

$$\int x^5\,dx = \frac{x^{5+1}}{5+1} + C = \frac{x^6}{6} + C$$

▸ **Now work Problem 3.**

APPLICATIONS IN PRACTICE 2

Indefinite Integral of a Constant Times a Function of *t*

If the rate of change of a company's revenues can be modeled by $\frac{dR}{dt} = 0.12t^2$ then find $\int 0.12t^2\,dt$, which gives the form of the company's revenue function.

CAUTION Only a *constant* factor of the integrand can "pass through" an integral sign. Because x is not a constant, one cannot find $\int 7x\,dx$ as $7x$ times $\int dx$.

EXAMPLE 3 Indefinite Integral of a Constant Times a Function of *x*

Find $\int 7x\,dx$.

Solution: By Formula 4 with $k = 7$ and $f(x) = x$,

$$\int 7x\,dx = 7\int x\,dx$$

Since x is x^1, by Formula 2 we have

$$\int x^1\,dx = \frac{x^{1+1}}{1+1} + C_1 = \frac{x^2}{2} + C_1$$

where C_1 is the constant of integration. Therefore,

$$\int 7x\,dx = 7\int x\,dx = 7\left[\frac{x^2}{2} + C_1\right] = \frac{7}{2}x^2 + 7C_1$$

Since $7C_1$ is just an arbitrary constant, we shall replace it by C for simplicity. Thus,

$$\int 7x\,dx = \frac{7}{2}x^2 + C$$

It is not necessary to write all intermediate steps when integrating. More simply, we write

$$\int 7x\,dx = (7)\frac{x^2}{2} + C = \frac{7}{2}x^2 + C$$

▸ **Now work Problem 5.**

EXAMPLE 4 Indefinite Integral of a Constant Times a Function of *x*

Find $\int -\frac{3}{5}e^x\,dx$.

Solution:

$$\int -\frac{3}{5}e^x\,dx = -\frac{3}{5}\int e^x\,dx \qquad \text{(Formula 4)}$$

$$= -\frac{3}{5}e^x + C \qquad \text{(Formula 3)}$$

▸ **Now work Problem 7.**

APPLICATIONS IN PRACTICE 3

Finding Indefinite Integrals

Owing to new competition, the number of subscriptions to a certain magazine is declining at a rate of $\frac{dS}{dt} = -\frac{480}{t^3}$ subscriptions per month, where t is the number of months since the competition entered the market. Find the form of the equation for the number of subscribers to the magazine.

EXAMPLE 5 **Finding Indefinite Integrals**

a. *Find* $\int \frac{1}{\sqrt{t}}\, dt$.

Solution: Here t is the variable of integration. We rewrite the integrand so that a basic formula can be used. Since $1/\sqrt{t} = t^{-1/2}$, applying Formula 2 gives

$$\int \frac{1}{\sqrt{t}}\, dt = \int t^{-1/2}\, dt = \frac{t^{(-1/2)+1}}{-\frac{1}{2}+1} + C = \frac{t^{1/2}}{\frac{1}{2}} + C = 2\sqrt{t} + C$$

b. *Find* $\int \frac{1}{6x^3}\, dx$.

Solution:

$$\int \frac{1}{6x^3}\, dx = \frac{1}{6}\int x^{-3}\, dx = \left(\frac{1}{6}\right)\frac{x^{-3+1}}{-3+1} + C$$

$$= -\frac{x^{-2}}{12} + C = -\frac{1}{12x^2} + C$$

▶ **Now work Problem 9.**

APPLICATIONS IN PRACTICE 4

Indefinite Integral of a Sum

The rate of growth of the population of a new city is estimated by $\frac{dN}{dt} = 500 + 300\sqrt{t}$, where t is in years. Find

$$\int (500 + 300\sqrt{t})\, dt$$

EXAMPLE 6 **Indefinite Integral of a Sum**

Find $\int (x^2 + 2x)\, dx$.

Solution: By Formula 5,

$$\int (x^2 + 2x)\, dx = \int x^2\, dx + \int 2x\, dx$$

Now,

$$\int x^2\, dx = \frac{x^{2+1}}{2+1} + C_1 = \frac{x^3}{3} + C_1$$

and

$$\int 2x\, dx = 2\int x\, dx = (2)\frac{x^{1+1}}{1+1} + C_2 = x^2 + C_2$$

Thus,

$$\int (x^2 + 2x)\, dx = \frac{x^3}{3} + x^2 + C_1 + C_2$$

When integrating an expression involving more than one term, only one constant of integration is needed.

For convenience, we shall replace the constant $C_1 + C_2$ by C. We then have

$$\int (x^2 + 2x)\, dx = \frac{x^3}{3} + x^2 + C$$

Omitting intermediate steps, we simply integrate term by term and write

$$\int (x^2 + 2x)\, dx = \frac{x^3}{3} + (2)\frac{x^2}{2} + C = \frac{x^3}{3} + x^2 + C$$

▶ **Now work Problem 11.**

APPLICATIONS IN PRACTICE 5

Indefinite Integral of a Sum and Difference

Suppose the rate of savings in the United States is given by $\frac{dS}{dt} = 2.1t^2 - 65.4t + 491.6$, where t is the time in years and S is the amount of money saved in billions of dollars. Find the form of the equation for the amount of money saved.

EXAMPLE 7 **Indefinite Integral of a Sum and Difference**

Find $\int (2\sqrt[5]{x^4} - 7x^3 + 10e^x - 1)\,dx$.

Solution:

$$\int (2\sqrt[5]{x^4} - 7x^3 + 10e^x - 1)\,dx$$

$$= 2\int x^{4/5}\,dx - 7\int x^3\,dx + 10\int e^x\,dx - \int 1\,dx \qquad \text{(Formulas 5 \& 4)}$$

$$= (2)\frac{x^{9/5}}{\frac{9}{5}} - (7)\frac{x^4}{4} + 10e^x - x + C \qquad \text{(Formulas 1, 2, \& 3)}$$

$$= \frac{10}{9}x^{9/5} - \frac{7}{4}x^4 + 10e^x - x + C$$

▶ **Now work Problem 15.**

Sometimes, in order to apply the basic integration formulas, it is necessary to first perform algebraic manipulations on the integrand, as Example 8 shows.

EXAMPLE 8 **Using Algebraic Manipulation to Find an Indefinite Integral**

Find $\int y^2\left(y + \frac{2}{3}\right)dy$.

Solution: The integrand does not fit a familiar integration form. However, by multiplying the integrand we get

$$\int y^2\left(y + \frac{2}{3}\right)dy = \int \left(y^3 + \frac{2}{3}y^2\right)dy$$

$$= \frac{y^4}{4} + \left(\frac{2}{3}\right)\frac{y^3}{3} + C = \frac{y^4}{4} + \frac{2y^3}{9} + C$$

CAUTION In Example 8, we first multiplied the factors in the integrand. The answer could not have been found simply in terms of $\int y^2\,dy$ and $\int (y + \frac{2}{3})\,dy$. The integral of a product is *not* the product of integrals.

▶ **Now work Problem 41.**

EXAMPLE 9 **Using Algebraic Manipulation to Find an Indefinite Integral**

a. *Find* $\int \frac{(2x-1)(x+3)}{6}\,dx$.

Solution: By factoring out the constant $\frac{1}{6}$ and multiplying the binomials, we get

$$\int \frac{(2x-1)(x+3)}{6}\,dx = \frac{1}{6}\int (2x^2 + 5x - 3)\,dx$$

$$= \frac{1}{6}\left[(2)\frac{x^3}{3} + (5)\frac{x^2}{2} - 3x\right] + C$$

$$= \frac{x^3}{9} + \frac{5x^2}{12} - \frac{x}{2} + C$$

Another algebraic approach to b, is

$$\int \frac{x^3-1}{x^2}\,dx = \int (x^3-1)x^{-2}\,dx$$
$$= \int (x - x^{-2})\,dx$$

and so on.

b. *Find* $\int \frac{x^3-1}{x^2}\,dx$.

Solution: We can break up the integrand into fractions by dividing each term in the numerator by the denominator:

$$\int \frac{x^3-1}{x^2}\,dx = \int \left(\frac{x^3}{x^2} - \frac{1}{x^2}\right)dx = \int (x - x^{-2})\,dx$$
$$= \frac{x^2}{2} - \frac{x^{-1}}{-1} + C = \frac{x^2}{2} + \frac{1}{x} + C$$

▶ **Now work Problem 49.**

Exercise 14.2

In Problems 1–52, find the indefinite integrals.

*1. $\int 7\,dx$

2. $\int \frac{1}{2}\,dx$

*3. $\int x^8\,dx$

4. $\int 5x^{24}\,dx$

*5. $\int 5x^{-7}\,dx$

6. $\int \frac{z^{-3}}{3}\,dz$

*7. $\int \frac{2}{x^{10}}\,dx$

8. $\int \frac{7}{x^4\,dx}$

*9. $\int \frac{1}{y^{11/5}}\,dy$

10. $\int \frac{7}{2x^{9/4}}\,dx$

*11. $\int (4+t)\,dt$

12. $\int (r^3+2r)\,dr$

13. $\int (y^5-5y)\,dy$

14. $\int (7-3w-9w^2)\,dw$

*15. $\int (3t^2-4t+5)\,dt$

16. $\int (1+t^2+t^4+t^6)\,dt$

17. $\int (7+e)\,dx$

18. $\int (5-2^{-1})\,dx$

19. $\int \left(\frac{x}{7}-\frac{3}{4}x^4\right)dx$

20. $\int \left(\frac{2x^2}{7}-\frac{8}{3}x^4\right)dx$

21. $\int 6e^x\,dx$

22. $\int \left(\frac{e^x}{3}+2x\right)dx$

23. $\int (x^{8.3}-9x^6+3x^{-4}+x^{-3})\,dx$

24. $\int (0.7y^3+10+2y^{-3})\,dy$

25. $\int \frac{-2\sqrt{x}}{3}\,dx$

26. $\int dw$

27. $\int \frac{1}{4\sqrt[8]{x^2}}\,dx$

28. $\int \frac{-4}{(3x)^3}\,dx$

29. $\int \left(\frac{x^3}{3}-\frac{3}{x^3}\right)dx$

30. $\int \left(\frac{1}{2x^3}-\frac{1}{x^4}\right)dx$

31. $\int \left(\frac{3w^2}{2}-\frac{2}{3w^2}\right)dw$

32. $\int \frac{4}{e^{-s}}\,ds$

33. $\int \frac{2z-5}{7}\,dz$

34. $\int \frac{1}{12}\left(\frac{1}{3}e^x\right)dx$

35. $\int (u^e+e^u)\,du$

36. $\int \left(3y^3-2y^2+\frac{e^y}{6}\right)dy$

37. $\int (2\sqrt{x}-3\sqrt[4]{x})\,dx$

38. $\int 0\,dx$

39. $\int \left(-\frac{\sqrt[3]{x^2}}{5}-\frac{7}{2\sqrt{x}}+6x\right)dx$

40. $\int \left(\sqrt[3]{u}+\frac{1}{\sqrt{u}}\right)du$

*41. $\int (x^2+5)(x-3)\,dx$

42. $\int x^4(x^3+8x^2+7)\,dx$

43. $\int \sqrt{x}(x+3)\,dx$

44. $\int (z+2)^2\,dz$

45. $\int (2u+1)^2\,du$

46. $\int \left(\frac{2}{\sqrt[5]{x}}-1\right)^2 dx$

47. $\int v^{-2}(2v^4+3v^2-2v^{-3})\,dv$

48. $\int [6e^u - u^3(\sqrt{u} + 1)]\, du$

*__49.__ $\int \frac{z^4 + 10z^3}{2z^2}\, dz$

50. $\int \frac{x^4 - 3x^2 + 4x}{5x}\, dx$

51. $\int \frac{e^x + e^{2x}}{e^x}\, dx$

52. $\int \frac{(x^3 + 1)^2}{x^2}\, dx$

53. If $F(x)$ and $G(x)$ are such that $F'(x) = G'(x)$, is it true that $F(x) - G(x)$ must be zero?

54. **(a)** Find a function F such that $\int F(x)\, dx = xe^x + C$.
(b) Is there only one function F satisfying the equation given in part (a), or are there many such functions?

55. Find $\int \frac{d}{dx}\left(\frac{1}{\sqrt{x^2 + 1}}\right) dx$

● ● **OBJECTIVE** *To find a particular antiderivative of a function that satisfies certain conditions. This involves evaluating a constant of integration.*

14.3 Integration with Initial Conditions

If we know the rate of change, f', of the function f, then the function f itself is an antiderivative of f' (since the derivative of f is f'). Of course, there are many antiderivatives of f', and the most general one is denoted by the indefinite integral. For example, if

$$f'(x) = 2x$$

then

$$f(x) = \int f'(x)\, dx = \int 2x\, dx = x^2 + C \qquad (1)$$

That is, *any* function of the form $f(x) = x^2 + C$ has its derivative equal to $2x$. Because of the constant of integration, notice that we do not know $f(x)$ specifically. However, if f must assume a certain function value for a particular value of x, then we can determine the value of C and thus determine $f(x)$ specifically. For instance, if $f(1) = 4$, then from Equation (1),

$$f(1) = 1^2 + C$$

$$4 = 1 + C$$

$$C = 3$$

Thus,

$$f(x) = x^2 + 3$$

That is, we now know the particular function $f(x)$ for which $f'(x) = 2x$ and $f(1) = 4$. The condition $f(1) = 4$, which gives a function value of f for a specific value of x, is called an *initial condition* (or *boundary value*).

APPLICATIONS IN PRACTICE 1

Initial-Condition Problem

The rate of growth of a species of bacteria is estimated by $\frac{dN}{dt} = 800 + 200e^t$ where N is the number of bacteria (in thousands) after t hours. If $N(5) = 40{,}000$ find $N(t)$.

EXAMPLE 1 **Initial-Condition Problem**

If y is a function of x such that $y' = 8x - 4$ and $y(2) = 5$, find y. [Note: $y(2) = 5$ means that $y = 5$ when $x = 2$.] Also, find $y(4)$.

Solution: Here $y(2) = 5$ is the initial condition. Since $y' = 8x - 4$, y is an antiderivative of $8x - 4$:

$$y = \int (8x - 4)\, dx = 8 \cdot \frac{x^2}{2} - 4x + C = 4x^2 - 4x + C \qquad (2)$$

We can determine the value of C by using the initial condition. Because $y = 5$ when $x = 2$, from Equation (2), we have,

$$5 = 4(2)^2 - 4(2) + C$$

$$5 = 16 - 8 + C$$

$$C = -3$$

Replacing C by -3 in Equation (2) gives the function that we seek:

$$y = 4x^2 - 4x - 3 \tag{3}$$

To find $y(4)$, we let $x = 4$ in Equation (3):

$$y(4) = 4(4)^2 - 4(4) - 3 = 64 - 16 - 3 = 45$$

▶ **Now work Problem 1.**

APPLICATIONS IN PRACTICE 2

Initial-Condition Problem Involving y''

The acceleration of an object after t seconds is given by $y'' = 84t + 24$, the velocity at 8 seconds is given by $y'(8) = 2891$ ft/s, and the position at 2 seconds is given by $y(2) = 185$ ft. Find $y(t)$.

EXAMPLE 2 Initial-Condition Problem Involving y''

Given that $y'' = x^2 - 6$, $y'(0) = 2$, and $y(1) = -1$, find y.

Solution:

Strategy: To go from y'' to y, two integrations are needed: the first to take us from y'' to y' and the other to take us from y' to y. Hence, there will be two constants of integration, which we shall denote by C_1 and C_2.

Since $y'' = \frac{d}{dx}(y') = x^2 - 6$, y' is an antiderivative of $x^2 - 6$. Thus,

$$y' = \int (x^2 - 6)\, dx = \frac{x^3}{3} - 6x + C_1 \tag{4}$$

Now, $y'(0) = 2$ means that $y' = 2$ when $x = 0$; therefore, from Equation (4), we have

$$2 = \frac{0^3}{3} - 6(0) + C_1$$

Hence, $C_1 = 2$, so

$$y' = \frac{x^3}{3} - 6x + 2$$

By integration, we can find y:

$$y = \int \left(\frac{x^3}{3} - 6x + 2\right) dx$$

$$= \left(\frac{1}{3}\right)\frac{x^4}{4} - (6)\frac{x^2}{2} + 2x + C_2$$

so

$$y = \frac{x^4}{12} - 3x^2 + 2x + C_2 \tag{5}$$

Now, since $y = -1$ when $x = 1$, we have, from Equation (5),

$$-1 = \frac{1^4}{12} - 3(1)^2 + 2(1) + C_2$$

Thus, $C_2 = -\frac{1}{12}$, so

$$y = \frac{x^4}{12} - 3x^2 + 2x - \frac{1}{12}$$

▶ **Now work Problem 5.**

Integration with initial conditions is applicable to many applied situations, as the next three examples illustrate.

EXAMPLE 3 Income and Education

For a particular urban group, sociologists studied the current average yearly income y (in dollars) that a person can expect to receive with x years of education before seeking regular employment. They estimated that the rate at which income changes with respect to education is given by

$$\frac{dy}{dx} = 100x^{3/2} \qquad 4 \leq x \leq 16$$

where $y = 28{,}720$ *when* $x = 9$. *Find* y.

Solution: Here y is an antiderivative of $100x^{3/2}$. Thus,

$$y = \int 100x^{3/2}\,dx = 100\int x^{3/2}\,dx$$

$$= (100)\frac{x^{5/2}}{\frac{5}{2}} + C$$

$$y = 40x^{5/2} + C \tag{6}$$

The initial condition is that $y = 28{,}720$ when $x = 9$. By substituting these values into Equation (6), we can determine the value of C:

$$28{,}720 = 40(9)^{5/2} + C$$

$$= 40(243) + C$$

$$28{,}720 = 9720 + C$$

Therefore, $C = 19{,}000$, and

$$y = 40x^{5/2} + 19{,}000$$

▶ **Now work Problem 17.**

EXAMPLE 4 Finding the Demand Function from Marginal Revenue

If the marginal-revenue function for a manufacturer's product is

$$\frac{dr}{dq} = 2000 - 20q - 3q^2$$

find the demand function.

Solution:

Strategy: By integrating dr/dq and using an initial condition, we can find the revenue function r. But revenue is also given by the general relationship $r = pq$, where p is the price per unit. Thus, $p = r/q$. Replacing r in this equation by the revenue function yields the demand function.

Since dr/dq is the derivative of total revenue r,

$$r = \int (2000 - 20q - 3q^2)\,dq$$

$$= 2000q - (20)\frac{q^2}{2} - (3)\frac{q^3}{3} + C$$

so that

$$r = 2000q - 10q^2 - q^3 + C \tag{7}$$

Revenue is 0 when q is 0.

We assume that **when no units are sold, total revenue is 0;** that is, $r = 0$ when $q = 0$. This is our initial condition. Substituting these values into Equation (7) gives

$$0 = 2000(0) - 10(0)^2 - 0^3 + C$$

Although $q = 0$ gives $C = 0$, this is not true in general. It occurs in this section because the revenue functions are polynomials. In later sections, evaluating when $q = 0$ may produce a nonzero value for C.

Hence, $C = 0$, and

$$r = 2000q - 10q^2 - q^3$$

To find the demand function, we use the fact that $p = r/q$ and substitute for r:

$$p = \frac{r}{q} = \frac{2000q - 10q^2 - q^3}{q}$$

$$p = 2000 - 10q - q^2$$

▶ **Now work Problem 11.**

EXAMPLE 5 Finding Cost from Marginal Cost

In the manufacture of a product, fixed costs per week are \$4000. *Fixed costs are costs, such as rent and insurance, that remain constant at all levels of production during a given time period. If the marginal-cost function is*

$$\frac{dc}{dq} = 0.000001(0.002q^2 - 25q) + 0.2$$

where c is the total cost (in dollars) of producing q pounds of product per week, find the cost of producing 10,000 *lb in* 1 *week.*

Solution: Since dc/dq is the derivative of the total cost c,

$$c = \int [0.000001(0.002q^2 - 25q) + 0.2]\,dq$$

$$= 0.000001 \int (0.002q^2 - 25q)\,dq + \int 0.2\,dq$$

$$c = 0.000001\left(\frac{0.002q^3}{3} - \frac{25q^2}{2}\right) + 0.2q + C$$

When q is 0, total cost is equal to fixed cost.

Although $q = 0$ gives C a value equal to fixed costs, this is not true in general. It occurs in this section because the cost functions are polynomials. In later sections, evaluating when $q = 0$ may produce a value for C that is different from fixed cost.

Fixed costs are constant regardless of output. Therefore, when $q = 0$, $c = 4000$, which is our initial condition. By substitution, we find that $C = 4000$, so

$$c = 0.000001\left(\frac{0.002q^3}{3} - \frac{25q^2}{2}\right) + 0.2q + 4000 \tag{8}$$

From Equation (8), when $q = 10{,}000$, $c = 5416\frac{2}{3}$. Thus, the total cost for producing 10,000 pounds of product in 1 week is \$5416.67.

▶ **Now work Problem 15.**

Exercise 14.3

In Problems 1 and 2, find y, subject to the given conditions.

*1. $dy/dx = 3x - 4$; $y(-1) = \frac{13}{2}$

2. $dy/dx = x^2 - x$; $y(3) = \frac{19}{2}$

In Problems 3 and 4, if y satisfies the given conditions, find y(x) for the given value of x.

3. $y' = 4/\sqrt{x}$, $y(4) = 10$; $x = 9$

4. $y' = -x^2 + 2x$, $y(2) = 1$; $x = 1$

In Problems 5–8, find y, subject to the given conditions.

*5. $y'' = -3x^2 + 4x$; $y'(1) = 2$, $y(1) = 3$

6. $y'' = x + 1$; $y'(0) = 0$, $y(0) = 5$

7. $y''' = 2x$; $y''(-1) = 3$, $y'(3) = 10$, $y(0) = 13$

8. $y''' = e^x + 1$; $y''(0) = 1$, $y'(0) = 2$, $y(0) = 3$

In Problems 9–12, dr/dq is a marginal-revenue function. Find the demand function.

9. $dr/dq = 0.7$

10. $dr/dq = 15 - \frac{1}{15}q$

*11. $dr/dq = 275 - q - 0.3q^2$

12. $dr/dq = 5{,}000 - 3(2q + 2q^3)$

In Problems 13–16, dc/dq is a marginal-cost function, and fixed costs are indicated in braces. For Problems 13 and 14, find the total-cost function. For Problems 15 and 16, find the total cost for the indicated value of q.

13. $dc/dq = 1.35$; $\{200\}$

14. $dc/dq = 2q + 75$; $\{2000\}$

*15. $dc/dq = 0.09q^2 - 1.2q + 4.5$; $\{7700\}$; $q = 10$

16. $dc/dq = 0.000204q^2 - 0.046q + 6$; $\{15{,}000\}$; $q = 200$

*17. **Diet for Rats** A group of biologists studied the nutritional effects on rats that were fed a diet containing 10% protein.[2] The protein consisted of yeast and corn flour.

Over a period of time, the group found that the (approximate) rate of change of the average weight gain G (in grams) of a rat with respect to the percentage P of yeast in the protein mix was

$$\frac{dG}{dP} = -\frac{P}{25} + 2 \qquad 0 \le P \le 100$$

If $G = 38$ when $P = 10$, find G.

18. **Winter Moth** A study of the winter moth was made in Nova Scotia.[3] The prepupae of the moth fall onto the ground from host trees. It was found that the (approximate) rate at which prepupal density y (the number of prepupae per square foot of soil) changes with respect to distance x (in feet) from the base of a host tree is

$$\frac{dy}{dx} = -1.5 - x \qquad 1 \le x \le 9$$

If $y = 57.3$ when $x = 1$, find y.

19. **Fluid Flow** In the study of the flow of fluid in a tube of constant radius R, such as blood flow in portions of the body, one can think of the tube as consisting of concentric tubes of radius r, where $0 \le r \le R$. The velocity v of the fluid is a function of r and is given by[4]

$$v = \int -\frac{(P_1 - P_2)r}{2l\eta}\,dr$$

where P_1 and P_2 are pressures at the ends of the tube, η (a Greek letter read "eta") is fluid viscosity, and l is the length of the tube. If $v = 0$ when $r = R$, show that

$$v = \frac{(P_1 - P_2)(R^2 - r^2)}{4l\eta}$$

20. **Elasticity of Demand** The sole producer of a product has determined that the marginal-revenue function is

$$\frac{dr}{dq} = 100 - 3q^2$$

Determine the point elasticity of demand for the product when $q = 5$. [*Hint:* First find the demand function.]

21. **Average Cost** A manufacturer has determined that the marginal-cost function is

$$\frac{dc}{dq} = 0.003q^2 - 0.4q + 40$$

where q is the number of units produced. If marginal cost is \$27.50 when $q = 50$, and fixed costs are \$5000, what is the *average* cost of producing 100 units?

22. If $f''(x) = 12x^2 + 6x$ and $f'(1) = 7$, evaluate

$$f(97.335) - f(-97.335)$$

---

[2] Adapted from R. Bressani, "The Use of Yeast in Human Foods," in *Single-Cell Protein*, ed. R. I. Mateles and S. R. Tannenbaum (Cambridge, MA: MIT Press, 1968).

[3] Adapted from D. G. Embree, "The Population Dynamics of the Winter Moth in Nova Scotia, 1954–1962," *Memoirs of the Entomological Society of Canada*, no. 46 (1965).

[4] R. W. Stacy et al., *Essentials of Biological and Medical Physics* (New York: McGraw-Hill, 1955).

●●● **OBJECTIVE** *To use the formulas for $\int u^n\,du$, $\int e^u\,du$, and $\int \frac{1}{u}\,du$.*

14.4 More Integration Formulas

Power Rule for Integration

The formula

$$\int x^n\,dx = \frac{x^{n+1}}{n+1} + C \qquad \text{if } n \neq -1$$

which applies to a power of x, can be generalized to handle a power of a *function* of x. Let u be a differentiable function of x. By the power rule for differentiation, if $n \neq -1$, then

$$\frac{d}{dx}\left(\frac{[u(x)]^{n+1}}{n+1}\right) = \frac{(n+1)[u(x)]^n \cdot u'(x)}{n+1} = [u(x)]^n \cdot u'(x)$$

Thus,

$$\int [u(x)]^n \cdot u'(x)\,dx = \frac{[u(x)]^{n+1}}{n+1} + C \qquad n \neq -1$$

We call this the *power rule for integration*. Note that $u'(x)dx$ is the differential of u, namely du. In mathematical shorthand, we can replace $u(x)$ by u and $u'(x)\,dx$ by du:

Power Rule for Integration

If u is differentiable, then

$$\int u^n\,du = \frac{u^{n+1}}{n+1} + C \qquad \text{if } n \neq -1$$

It is essential that you realize the difference between the power rule for integration and the formula for $\int x^n\,dx$. In the power rule, u represents a function, whereas in $\int x^n\,dx$, x is a variable.

EXAMPLE 1 Applying the Power Rule for Integration

a. *Find* $\int (x+1)^{20}\,dx$.

Solution: Since the integrand is a power of the function $x+1$, we shall set $u = x+1$. Then $du = dx$, and $\int (x+1)^{20}\,dx$ has the form $\int u^{20}\,du$. By the power rule for integration,

$$\int (x+1)^{20}\,dx = \int u^{20}\,du = \frac{u^{21}}{21} + C = \frac{(x+1)^{21}}{21} + C$$

Note that we give our answer not in terms of u, but explicitly in terms of x.

b. *Find* $\int 3x^2(x^3+7)^3\,dx$.

Solution: We observe that the integrand contains a power of the function x^3+7. Let $u = x^3+7$. Then $du = 3x^2\,dx$. Fortunately, $3x^2$ appears as a factor in the integrand and can be used as part of du. We thus have

After integrating, you may wonder what happened to $3x^2$; $3x^2$, together with dx, forms the differential of u in the power rule.

$$\int 3x^2(x^3+7)^3\,dx = \int (x^3+7)^3[3x^2\,dx] = \int u^3\,du$$

$$= \frac{u^4}{4} + C = \frac{(x^3+7)^4}{4} + C$$

▸ **Now work Problem 3.**

In order to apply the power rule for integration, sometimes an adjustment must be made to obtain du in the integrand, as Example 2 illustrates.

EXAMPLE 2 **Adjusting for *du***

Find $\int x\sqrt{x^2+5}\,dx$.

Solution: We can write this as $\int x(x^2+5)^{1/2}\,dx$. Notice that the integrand contains a power of the function x^2+5. If $u = x^2+5$, then $du = 2x\,dx$. Since the *constant* factor 2 in du does *not* appear in the integrand, this integral does not have the form $\int u^n\,du$. However, we can put the given integral in this form by first multiplying and dividing the integrand by 2. This does not change its value. Thus,

$$\int x(x^2+5)^{1/2}\,dx = \int \frac{2}{2}x(x^2+5)^{1/2}\,dx = \int \frac{1}{2}(x^2+5)^{1/2}[2x\,dx]$$

Moving the *constant* factor $\frac{1}{2}$ in front of the integral sign, we have

$$\int x(x^2+5)^{1/2}\,dx = \frac{1}{2}\int (x^2+5)^{1/2}[2x\,dx] \qquad \textbf{(1)}$$

$$= \frac{1}{2}\int u^{1/2}\,du = \frac{1}{2}\left[\frac{u^{3/2}}{\frac{3}{2}}\right] + C$$

Going back to x gives

$$\int x\sqrt{x^2+5}\,dx = \frac{(x^2+5)^{3/2}}{3} + C$$

▶ **Now work Problem 15.**

In Example 2, we needed the factor 2 in the integrand. In Equation (1), it was inserted, and the integral was simultaneously multiplied by $\frac{1}{2}$. More generally, if k is a nonzero constant, then

$$\int f(x)\,dx = \int \frac{k}{k}f(x)\,dx = \frac{1}{k}\int kf(x)\,dx$$

CAUTION We can adjust for constant factors, but not variable factors.

In effect, we can multiply the integrand by a nonzero constant k, as long as we compensate for this by multiplying the entire integral by $1/k$. Such a manipulation **cannot** be done with *variable* factors.

When using the form $\int u^n\,du$, do not neglect du. For example,

$$\int (4x+1)^2\,dx \neq \frac{(4x+1)^3}{3} + C$$

The correct way to do this problem is as follows. Letting $u = 4x+1$, we have $du = 4\,dx$. Thus,

$$\int (4x+1)^2\,dx = \frac{1}{4}\int (4x+1)^2[4\,dx] = \frac{1}{4}\int u^2\,du$$

$$= \frac{1}{4}\cdot\frac{u^3}{3} + C = \frac{(4x+1)^3}{12} + C$$

EXAMPLE 3 **Applying the Power Rule for Integration**

a. *Find* $\int \sqrt[3]{6y}\,dy$.

Solution: The integrand is $(6y)^{1/3}$, a power of a function. Let us try to use the power rule for integration. If we set $u = 6y$, then $du = 6\,dy$. Since the factor 6 does not appear in the integrand, we insert a factor of 6 and adjust for it with a factor of $\frac{1}{6}$ in front of the integral. We then have

Example 3(a) can be handled without the use of the power rule by rewriting the integrand as $\sqrt[3]{6}y^{1/3}$. This gives the equivalent answer $\frac{3\sqrt[3]{6}}{4}y^{4/3} + C$.

$$\int \sqrt[3]{6y}\,dy = \int (6y)^{1/3}\,dy = \frac{1}{6}\int (6y)^{1/3}[6\,dy] = \frac{1}{6}\int u^{1/3}\,du$$

$$= \left(\frac{1}{6}\right)\frac{u^{4/3}}{\frac{4}{3}} + C = \frac{(6y)^{4/3}}{8} + C$$

b. *Find* $\int \frac{2x^3 + 3x}{(x^4 + 3x^2 + 7)^4}\,dx$.

Solution: We can write this as $\int (x^4 + 3x^2 + 7)^{-4}(2x^3 + 3x)\,dx$. Let us try to use the power rule for integration. If $u = x^4 + 3x^2 + 7$, then $du = (4x^3 + 6x)\,dx$, which is two times the quantity $(2x^3 + 3x)\,dx$ in the integral. Thus, we insert a factor of 2 and adjust for it with a factor of $\frac{1}{2}$ in front of the integral, as follows:

$$\int (x^4 + 3x^2 + 7)^{-4}(2x^3 + 3x)\,dx$$

$$= \frac{1}{2}\int (x^4 + 3x^2 + 7)^{-4}[2(2x^3 + 3x)\,dx] \qquad \text{(Factors inserted)}$$

$$= \frac{1}{2}\int (x^4 + 3x^2 + 7)^{-4}[(4x^3 + 6x)\,dx]$$

$$= \frac{1}{2}\int u^{-4}\,du = \frac{1}{2}\cdot\frac{u^{-3}}{-3} + C = -\frac{1}{6u^3} + C$$

$$= -\frac{1}{6(x^4 + 3x^2 + 7)^3} + C \qquad \text{(Rewrite in terms of } x\text{)}$$

▸ **Now work Problem 5.**

In using the power rule for integration, take care when making your choice for u. In Example 3(b), you would *not* be able to proceed very far if, for instance, you let $u = 2x^3 + 3x$. At times you may find it necessary to try many different choices. So don't just sit and look at the integral. Try something even if it is wrong, because it may give you a hint as to what might work. **Skill at integration comes only after many hours of practice and conscientious study.**

EXAMPLE 4 **An Integral to Which the Power Rule Does Not Apply**

Find $\int 4x^2(x^4 + 1)^2\,dx$.

Solution: If we set $u = x^4 + 1$, then $du = 4x^3\,dx$. To get du in the integral, we need an additional factor of the *variable* x. However, we can adjust only for **constant** factors. Thus, we cannot use the power rule. Instead, to find the integral, we shall first expand $(x^4+1)^2$:

$$\int 4x^2(x^4+1)^2\,dx = 4\int x^2(x^8+2x^4+1)\,dx$$
$$= 4\int (x^{10}+2x^6+x^2)\,dx$$
$$= 4\left(\frac{x^{11}}{11}+\frac{2x^7}{7}+\frac{x^3}{3}\right)+C$$

▶ **Now work Problem 67.**

Integrating Natural Exponential Functions

We now turn our attention to integrating exponential functions. If u is a differentiable function of x, then

$$\frac{d}{dx}(e^u) = e^u\frac{du}{dx}$$

Corresponding to this differentiation formula is the integration formula

$$\int e^u \frac{du}{dx}\,dx = e^u + C$$

But $\frac{du}{dx}\,dx$ is the differential of u, namely, du. Thus,

$$\int e^u\,du = e^u + C \qquad (2)$$

CAUTION Do not apply the power-rule formula for $\int u^n\,du$ to $\int e^u\,du$

APPLICATIONS IN PRACTICE 1

Integrals Involving Exponential Functions

When an object is moved from one environment to another, its temperature T changes at a rate given by $\frac{dT}{dt} = kCe^{kt}$, where t is the time (in hours) after changing environments, C is the temperature difference (original minus new) between the environments, and k is a constant. If the original environment is 70°, the new environment is 60°, and $k = -0.5$, find the general form of $T(t)$.

EXAMPLE 5 Integrals Involving Exponential Functions

a. *Find* $\int 2xe^{x^2}\,dx$.

Solution: Let $u = x^2$. Then $du = 2x\,dx$, and by Equation (2),

$$\int 2xe^{x^2}\,dx = \int e^{x^2}[2x\,dx] = \int e^u\,du$$
$$= e^u + C = e^{x^2} + C$$

b. *Find* $\int (x^2+1)e^{x^3+3x}\,dx$.

Solution: If $u = x^3+3x$, then $du = (3x^2+3)\,dx = 3(x^2+1)\,dx$. If the integrand contained a factor of 3, the integral would have the form $\int e^u\,du$. Thus, we write

$$\int (x^2+1)e^{x^3+3x}\,dx = \frac{1}{3}\int e^{x^3+3x}[3(x^2+1)\,dx]$$
$$= \frac{1}{3}\int e^u\,du = \frac{1}{3}e^u + C$$
$$= \frac{1}{3}e^{x^3+3x} + C$$

▶ **Now work Problem 41.**

Integrals Involving Logarithmic Functions

As you know, the power-rule formula $\int u^n\,du = u^{n+1}/(n+1) + C$ does not apply when $n = -1$. To handle that situation, namely, $\int u^{-1}\,du = \int \frac{1}{u}\,du$, we first recall that

$$\frac{d}{dx}(\ln u) = \frac{1}{u}\frac{du}{dx}$$

It would seem that $\int \frac{1}{u}\frac{du}{dx}\,dx = \int \frac{1}{u}\,du = \ln u + C$. However, the logarithm of u is defined if and only if u is positive. If $u < 0$, then $\ln u$ is not defined. Thus,

$$\int \frac{1}{u}\,du = \ln u + C \qquad \text{provided that } u > 0$$

On the other hand, if $u < 0$, then $-u > 0$, and $\ln(-u)$ is defined. Moreover,

$$\frac{d}{dx}[\ln(-u)] = \frac{1}{-u}(-1)\frac{du}{dx} = \frac{1}{u}\frac{du}{dx}$$

So in the case $u < 0$,

$$\int \frac{1}{u}\frac{du}{dx}\,dx = \int \frac{1}{u}\,du = \ln(-u) + C$$

In summary, if $u > 0$, then $\int \frac{1}{u}\,du = \ln u + C$; if $u < 0$, then $\int \frac{1}{u}\,du = \ln(-u) + C$. Combining the $u > 0$ and $u < 0$ cases, we have

$$\int \frac{1}{u}\,du = \ln|u| + C \tag{3}$$

In particular, if $u = x$, then $du = dx$, and

$$\int \frac{1}{x}\,dx = \ln|x| + C \tag{4}$$

APPLICATIONS IN PRACTICE 2

Integrals Involving $\frac{1}{u}\,du$

If the rate of vocabulary memorization of the average student in a foreign language is given by $\frac{dv}{dt} = \frac{35}{t+1}$, where v is the number of vocabulary words memorized in t hours of study, find the general form of $v(t)$.

EXAMPLE 6 Integrals Involving $\frac{1}{u}\,du$

a. *Find* $\int \frac{7}{x}\,dx$.

Solution: From Equation (4),

$$\int \frac{7}{x}\,dx = 7\int \frac{1}{x}\,dx = 7\ln|x| + C$$

Using properties of logarithms, we can write this answer another way:

$$\int \frac{7}{x}\,dx = \ln|x^7| + C$$

b. *Find* $\int \frac{2x}{x^2+5}\,dx$.

Solution: Let $u = x^2 + 5$. Then $du = 2x\,dx$. From Equation (3),

$$\int \frac{2x}{x^2+5}\,dx = \int \frac{1}{x^2+5}[2x\,dx] = \int \frac{1}{u}\,du$$
$$= \ln|u| + C = \ln|x^2+5| + C$$

Since $x^2 + 5$ is always positive, we can omit the absolute-value bars:

$$\int \frac{2x}{x^2+5}\,dx = \ln(x^2+5) + C$$

▶ **Now work Problem 31.**

EXAMPLE 7 An Integral Involving $\frac{1}{u}\,du$

Find $\displaystyle\int \frac{(2x^3+3x)\,dx}{x^4+3x^2+7}$.

Solution: If $u = x^4 + 3x^2 + 7$, then $du = (4x^3 + 6x)\,dx$, which is two times the numerator. To apply Equation (3), we insert a factor of 2 and adjust for it with a factor of $\frac{1}{2}$, as follows:

$$\begin{aligned}\int \frac{2x^3+3x}{x^4+3x^2+7}\,dx &= \frac{1}{2}\int \frac{2(2x^3+3x)}{x^4+3x^2+7}\,dx\\ &= \frac{1}{2}\int \frac{1}{x^4+3x^2+7}[(4x^3+6x)\,dx]\\ &= \frac{1}{2}\int \frac{1}{u}\,du = \frac{1}{2}\ln|u| + C\\ &= \frac{1}{2}\ln|x^4+3x^2+7| + C && \text{(Rewrite } u \text{ in terms of } x\text{)}\\ &= \ln\sqrt{|x^4+3x^2+7|} + C && \left(\frac{1}{2}\ln|u| = \ln|u|^{1/2}\right)\\ &= \ln\sqrt{x^4+3x^2+7} + C && (x^4+3x^2+7>0 \quad \text{for all } x)\end{aligned}$$

▶ **Now work Problem 51.**

EXAMPLE 8 An Integral Involving Two Forms

Find $\displaystyle\int \left[\frac{1}{(1-w)^2} + \frac{1}{w-1}\right] dw$.

Solution:

$$\begin{aligned}\int \left[\frac{1}{(1-w)^2} + \frac{1}{w-1}\right] dw &= \int (1-w)^{-2}dw + \int \frac{1}{w-1}dw\\ &= -1\int (1-w)^{-2}[-dw] + \int \frac{1}{w-1}dw\end{aligned}$$

The first integral has the form $\int u^{-2}\,du$, and the second has the form $\int \frac{1}{v}\,dv$. Thus,

$$\begin{aligned}\int \left[\frac{1}{(1-w)^2} + \frac{1}{w-1}\right] dw &= -\frac{(1-w)^{-1}}{-1} + \ln|w-1| + C\\ &= \frac{1}{1-w} + \ln|w-1| + C\end{aligned}$$

▶ **Now work Problem 69.**

For your convenience, we list in Table 14.2 the basic integration formulas so far discussed. We assume that u is a function of x.

TABLE 14.2 Basic Integration Formulas

1. $\displaystyle\int k\,du = ku + C \qquad k \text{ a constant}$
2. $\displaystyle\int u^n\,du = \frac{u^{n+1}}{n+1} + C \qquad n \neq -1$
3. $\displaystyle\int e^u\,du = e^u + C$
4. $\displaystyle\int \frac{1}{u}\,du = \ln|u| + C \qquad u \neq 0$
5. $\displaystyle\int kf(x)\,dx = k\int f(x)\,dx \qquad k \text{ a constant}$
6. $\displaystyle\int [f(x) \pm g(x)]\,dx = \int f(x)\,dx \pm \int g(x)\,dx$

Exercise 14.4

In Problems 1–80, find the indefinite integrals.

1. $\displaystyle\int (x+5)^7\,dx$
2. $\displaystyle\int 15(x+2)^4\,dx$
*3. $\displaystyle\int 2x(x^2+3)^5\,dx$
4. $\displaystyle\int (3x^2+14x)(x^3+7x^2+1)\,dx$
*5. $\displaystyle\int (3y^2+6y)(y^3+3y^2+1)^{2/3}\,dy$
6. $\displaystyle\int (15t^2-6t+1)(5t^3-3t^2+t)^{17}\,dt$
7. $\displaystyle\int \frac{5}{(3x-1)^3}\,dx$
8. $\displaystyle\int \frac{4x}{(2x^2-7)^{10}}\,dx$
9. $\displaystyle\int \sqrt{2x-1}\,dx$
10. $\displaystyle\int \frac{1}{\sqrt{x-2}}\,dx$
11. $\displaystyle\int (7x-6)^4\,dx$
12. $\displaystyle\int x^2(3x^3+7)^3\,dx$
13. $\displaystyle\int u(5u^2-9)^{14}\,du$
14. $\displaystyle\int 9x\sqrt{1+2x^2}\,dx$
*15. $\displaystyle\int 4x^4(27+x^5)^{1/3}\,dx$
16. $\displaystyle\int (3-2x)^{10}\,dx$
17. $\displaystyle\int 3e^{3x}\,dx$
18. $\displaystyle\int 5e^{3t+7}\,dt$
19. $\displaystyle\int (2t+1)e^{t^2+t}\,dt$
20. $\displaystyle\int -3w^2e^{-w^3}\,dw$
21. $\displaystyle\int xe^{7x^2}\,dx$
22. $\displaystyle\int x^3e^{4x^4}\,dx$
23. $\displaystyle\int 6e^{-2x}\,dx$
24. $\displaystyle\int x^4e^{-6x^5}\,dx$
25. $\displaystyle\int \frac{1}{x+5}\,dx$
26. $\displaystyle\int \frac{12x^2+4x+2}{x+x^2+2x^3}\,dx$
27. $\displaystyle\int \frac{3x^2+4x^3}{x^3+x^4}\,dx$
28. $\displaystyle\int \frac{9x^2-2x}{1-x^2+3x^3}\,dx$
29. $\displaystyle\int \frac{6z}{(z^2-6)^5}\,dz$
30. $\displaystyle\int \frac{3}{(5v-1)^4}\,dv$
*31. $\displaystyle\int \frac{4}{x}\,dx$
32. $\displaystyle\int \frac{3}{1+2y}\,dy$
33. $\displaystyle\int \frac{s^2}{s^3+5}\,ds$
34. $\displaystyle\int \frac{2x^2}{3-4x^3}\,dx$
35. $\displaystyle\int \frac{8}{5-3x}\,dx$
36. $\displaystyle\int \frac{7t}{5t^2-6}\,dt$
37. $\displaystyle\int \sqrt{5x}\,dx$
38. $\displaystyle\int \frac{1}{(3x)^6}\,dx$
39. $\displaystyle\int \frac{x}{\sqrt{x^2-4}}\,dx$
40. $\displaystyle\int \frac{11}{3-2x}\,dx$
*41. $\displaystyle\int 2y^3e^{y^4+1}\,dy$
42. $\displaystyle\int 2\sqrt{2x-1}\,dx$
43. $\displaystyle\int v^2e^{-2v^3+1}\,dv$
44. $\displaystyle\int \frac{x^2}{\sqrt[3]{2x^3+9}}\,dx$

45. $\int (e^{-5x} + 2e^x)\,dx$

46. $\int 4\sqrt[3]{y+1}\,dy$

47. $\int (x+1)(3-3x^2-6x)^3\,dx$

48. $\int 2ye^{3y^2}\,dy$

49. $\int \frac{x^2+2}{x^3+6x}\,dx$

50. $\int (e^x + 2e^{-3x} - e^{5x})\,dx$

***51.** $\int \frac{16s-4}{3-2s+4s^2}\,ds$

52. $\int (t^2+4t)(t^3+6t^2)^6\,dt$

53. $\int x(2x^2+1)^{-1}\,dx$

54. $\int (8w^5+w^2-2)(6w-w^3-4w^6)^{-4}\,dw$

55. $\int -(x^2-2x^5)(x^3-x^6)^{-10}\,dx$

56. $\int \frac{3}{5}(v-2)e^{2-4v+v^2}\,dv$

57. $\int (2x^3+x)(x^4+x^2)\,dx$

58. $\int (e^{3.1})^2\,dx$

59. $\int \frac{18+12x}{(4-9x-3x^2)^5}\,dx$

60. $\int (e^x-e^{-x})^2\,dx$

61. $\int x(2x+1)e^{4x^3+3x^2-4}\,dx$

62. $\int (u^3-ue^{6-3u^2})\,du$

63. $\int x\sqrt{(8-5x^2)^3}\,dx$

64. $\int e^{-x/4}\,dx$

65. $\int \left(\sqrt{2x}-\frac{1}{\sqrt{2x}}\right)dx$

66. $\int 3\frac{x^4}{e^{x^5}}\,dx$

***67.** $\int (x^2+1)^2\,dx$

68. $\int \left[x(x^2-16)^2-\frac{1}{2x+5}\right]dx$

***69.** $\int \left[\frac{x}{x^2+1}+\frac{x^5}{(x^6+1)^2}\right]dx$

70. $\int \left[\frac{3}{x-1}+\frac{1}{(x-1)^2}\right]dx$

71. $\int \left[\frac{1}{3x-5}-(x^2-2x^5)(x^3-x^6)^{-10}\right]dx$

72. $\int (r^3+5)^2\,dr$

73. $\int \left[\sqrt{3x+1}-\frac{x}{x^2+3}\right]dx$

74. $\int \left[\frac{x}{3x^2+5}-\frac{x^2}{(x^3+1)^3}\right]dx$

75. $\int \frac{e^{\sqrt{x}}}{\sqrt{x}}\,dx$

76. $\int (e^4-2^e)\,dx$

77. $\int \frac{1+e^{2x}}{4e^x}\,dx$

78. $\int \frac{2}{t^2}\sqrt{\frac{1}{t}+9}\,dt$

79. $\int \frac{x+1}{x^2+2x}\ln(x^2+2x)\,dx$

80. $\int \sqrt[3]{x}e^{\sqrt[3]{8x^4}}\,dx$

In Problems 81–84, find y, subject to the given conditions.

81. $y' = (3-2x)^2;\ y(0)=1$

82. $y' = \frac{x}{x^2+6};\ y(1)=0$

83. $y'' = \frac{1}{x^2};\ y'(-1)=1,\ y(1)=0$

84. $y'' = (x+1)^{3/2};\ y'(3)=0,\ y(3)=0$

85. Real Estate The rate of change of the value of a house that cost $350,000 to build can be modeled by $\frac{dV}{dt} = 8e^{0.05t}$, where t is the time in years since the house was built and V is the value (in thousands of dollars) of the house. Find $V(t)$.

86. Life Span If the rate of change of the expected life span l at birth of people born in the United States can be modeled by $\frac{dl}{dt} = \frac{12}{2t+50}$, where t is the number of years after 1940 and the expected life span was 63 years in 1940, find the expected life span for people born in 1998.

87. Oxygen in Capillary In a discussion of the diffusion of oxygen from capillaries,[5] concentric cylinders of radius r are used as a model for a capillary. The concentration C of oxygen in the capillary is given by

$$C = \int \left(\frac{Rr}{2K}+\frac{B_1}{r}\right)dr$$

where R is the constant rate at which oxygen diffuses from the capillary, and K and B_1 are constants. Find C. (Write the constant of integration as B_2.)

88. Find $f(2)$ if $f\left(\frac{1}{2}\right)=1$ and $f'(x)=e^{2x-1}-6x$.

[5] W. Simon, *Mathematical Techniques for Physiology and Medicine* (New York: Academic Press, 1972).

OBJECTIVE *To discuss techniques of handling more challenging integration problems, namely, by algebraic manipulation and by fitting the integrand to a familiar form. To integrate an exponential function with a base different from e and to find the consumption function, given the marginal propensity to consume.*

14.5 Techniques of Integration

Now that you have had some practice in determining indefinite integrals, suppose we consider some problems of a greater degree of difficulty.

When you are integrating fractions, sometimes a preliminary division is needed to get familiar integration forms, as the next example shows.

EXAMPLE 1 Preliminary Division before Integration

a. Find $\int \frac{x^3+x}{x^2}\,dx$.

Solution: A familiar integration form is not apparent. However, we can break up the integrand into two fractions by dividing each term in the numerator by the denominator. We then have

Here we split up the integrand.

$$\int \frac{x^3+x}{x^2}\,dx = \int \left[\frac{x^3}{x^2}+\frac{x}{x^2}\right]dx = \int \left[x+\frac{1}{x}\right]dx = \frac{x^2}{2}+\ln|x|+C$$

b. Find $\int \frac{2x^3+3x^2+x+1}{2x+1}\,dx$.

Solution: Here the integrand is a quotient of polynomials in which the degree of the numerator is greater than or equal to that of the denominator and the denominator has more than one term. In such a situation, in order to integrate, we first use long division until the degree of the remainder is less than that of the divisor. We obtain

Here we use long division to rewrite the integrand.

$$\begin{aligned}\int \frac{2x^3+3x^2+x+1}{2x+1}\,dx &= \int \left(x^2+x+\frac{1}{2x+1}\right)dx\\ &= \frac{x^3}{3}+\frac{x^2}{2}+\int \frac{1}{2x+1}\,dx\\ &= \frac{x^3}{3}+\frac{x^2}{2}+\frac{1}{2}\int \frac{1}{2x+1}[2\,dx]\\ &= \frac{x^3}{3}+\frac{x^2}{2}+\frac{1}{2}\ln|2x+1|+C\end{aligned}$$

▸ **Now work Problem 1.**

EXAMPLE 2 Indefinite Integrals

a. Find $\int \frac{1}{\sqrt{x}(\sqrt{x}-2)^3}\,dx$.

Solution: We can write this integral as $\int \frac{(\sqrt{x}-2)^{-3}}{\sqrt{x}}\,dx$. Let us try the power rule for integration with $u=\sqrt{x}-2$. Then $du=\frac{1}{2\sqrt{x}}\,dx$, and

Here the integral is fit to the form to which the power rule for integration applies.

$$\begin{aligned}\int \frac{(\sqrt{x}-2)^{-3}}{\sqrt{x}}\,dx &= 2\int (\sqrt{x}-2)^{-3}\left[\frac{1}{2\sqrt{x}}\,dx\right]\\ &= 2\int u^{-3}\,du = 2\left(\frac{u^{-2}}{-2}\right)+C\\ &= -\frac{1}{u^2}+C = -\frac{1}{(\sqrt{x}-2)^2}+C\end{aligned}$$

b. *Find* $\int \frac{1}{x \ln x}\, dx$.

Here the integral fits the familiar form $\int \frac{1}{u}\, du$.

Solution: If $u = \ln x$, then $du = \frac{1}{x}\, dx$, and

$$\int \frac{1}{x \ln x}\, dx = \int \frac{1}{\ln x}\left(\frac{1}{x}\, dx\right) = \int \frac{1}{u}\, du$$

$$= \ln |u| + C = \ln |\ln x| + C$$

c. *Find* $\int \frac{5}{w(\ln w)^{3/2}}\, dw$.

Here the integral is fit to the form to which the power rule for integration applies.

Solution: If $u = \ln w$, then $du = \frac{1}{w} dw$. Applying the power rule for integration, we have

$$\int \frac{5}{w(\ln w)^{3/2}}\, dw = 5\int (\ln w)^{-3/2}\left(\frac{1}{w}\, dw\right)$$

$$= 5\int u^{-3/2}\, du = 5 \cdot \frac{u^{-1/2}}{-\frac{1}{2}} + C$$

$$= \frac{-10}{u^{1/2}} + C = -\frac{10}{(\ln w)^{1/2}} + C$$

▶ **Now work Problem 23.**

Integrating a^u

In Section 14.4, we integrated an exponential function to the base e:

$$\int e^u\, du = e^u + C$$

Now let us consider the integral of an exponential function to a base other than e:

$$\int a^u\, du$$

To find this integral, we first convert a^u to an exponential function to the base e by using Property 8 of Section 4.3:

$$a = e^{\ln a} \qquad \textbf{(1)}$$

Example 3 will illustrate.

EXAMPLE 3 **An Integral Involving $a^u\, du$**

Find $\int 2^{3-x}\, dx$.

Solution:

Strategy: We want to integrate an exponential function to the base 2. To do this, we shall first convert from base 2 to base e by using Equation (1) to write 2 in terms of e.

Since $2 = e^{\ln 2}$, we have

$$\int 2^{3-x}\, dx = \int (e^{\ln 2})^{3-x}\, dx = \int e^{(\ln 2)(3-x)}\, dx$$

The integrand of the last integral is of the form e^u, where $u = (\ln 2)(3 - x)$. Since $du = -\ln 2\, dx$, we have

$$\int e^{(\ln 2)(3-x)}\, dx = -\frac{1}{\ln 2}\int e^{(\ln 2)(3-x)}[(-\ln 2)\, dx] \qquad \text{(form: } \textstyle\int e^u\, du)$$

$$= -\frac{1}{\ln 2}e^{(\ln 2)(3-x)} + C = -\frac{1}{\ln 2}2^{3-x} + C$$

Thus,

$$\int 2^{3-x}\, dx = -\frac{1}{\ln 2}2^{3-x} + C$$

Notice that we expressed our answer in terms of an exponential function to the base 2, the base of the original integrand.

▸ **Now work Problem 27.**

Generalizing the procedure described in Example 3, we can obtain a formula for integrating a^u:

$$\int a^u du = \int (e^{\ln a})^u du = \int e^{(\ln a)u} du$$

$$= \frac{1}{\ln a}\int e^{(\ln a)u}[(\ln a)\, du] \qquad (\ln a \text{ is constant})$$

$$= \frac{1}{\ln a}e^{(\ln a)u} + C = \frac{1}{\ln a}(e^{\ln a})^u + C$$

$$= \frac{1}{\ln a}a^u + C \qquad (e^{\ln a} = a)$$

Hence, we have

$$\int a^u\, du = \frac{1}{\ln a}a^u + C$$

Applying this formula to the integral in Example 3 gives

$$\int 2^{3-x}\, dx \qquad (a = 2, u = 3 - x)$$

$$= -\int 2^{3-x}(-dx) \qquad (du = -dx)$$

$$= -\frac{1}{\ln 2}2^{3-x} + C$$

which is the same result that we obtained before.

Application of Integration

We shall now consider an application of integration that relates a consumption function to the marginal propensity to consume.

EXAMPLE 4 **Find Consumption Function from Marginal Propensity to Consume**

For a certain country, the marginal propensity to consume is given by

$$\frac{dC}{dI} = \frac{3}{4} - \frac{1}{2\sqrt{3I}}$$

where consumption C is a function of national income I. Here I is expressed in *billions of slugs (50 slugs = \$0.01). Determine the consumption function for the country if it is known that consumption is 10 billion slugs ($C = 10$) when $I = 12$.*

Solution: Since the marginal propensity to consume is the derivative of C, we have

$$C = \int \left(\frac{3}{4} - \frac{1}{2\sqrt{3I}}\right) dI = \int \frac{3}{4} dI - \frac{1}{2}\int (3I)^{-1/2}\, dI$$

$$= \frac{3}{4}I - \frac{1}{2}\int (3I)^{-1/2}\, dI$$

If we let $u = 3I$, then $du = 3\,dI$, and

$$C = \frac{3}{4}I - \left(\frac{1}{2}\right)\frac{1}{3}\int (3I)^{-1/2}[3\,dI]$$

$$= \frac{3}{4}I - \frac{1}{6}\frac{(3I)^{1/2}}{\frac{1}{2}} + C_1$$

$$C = \frac{3}{4}I - \frac{\sqrt{3I}}{3} + C_1$$

This is an example of an initial-value problem.

When $I = 12$, then $C = 10$, so

$$10 = \frac{3}{4}(12) - \frac{\sqrt{3(12)}}{3} + C_1$$

$$10 = 9 - 2 + C_1$$

Thus, $C_1 = 3$, and the consumption function is

$$C = \frac{3}{4}I - \frac{\sqrt{3I}}{3} + 3$$

▶ **Now work Problem 61.**

Exercise 14.5

In Problems 1–56, determine the indefinite integrals.

*1. $\int \frac{2x^6 + 8x^4 - 4x^2}{2x^2}\, dx$

2. $\int \frac{9x^2 + 5}{3x}\, dx$

3. $\int (3x^2 + 2)\sqrt{2x^3 + 4x + 1}\, dx$

4. $\int \frac{x}{\sqrt[4]{x^2 + 1}}\, dx$

5. $\int \frac{15}{\sqrt{4 - 5x}}\, dx$

6. $\int \frac{2xe^{x^2}\, dx}{e^{x^2} - 2}$

7. $\int 4^{7x}\, dx$

8. $\int 5^t\, dt$

9. $\int 2x(7 - e^{x^2/4})\, dx$

10. $\int \left(e^x + x^e + ex + \frac{e}{x}\right) dx$

11. $\int \frac{6x^2 - 11x + 5}{3x - 1}\, dx$

12. $\int \frac{(2x - 1)(x + 3)}{x - 5}\, dx$

13. $\int \frac{5e^{2x}}{7e^{2x} + 4}\, dx$

14. $\int 6(e^{4-3x})^2\, dx$

15. $\int \frac{e^{7/x}}{x^2}\, dx$

16. $\int \frac{2x^4 - 6x^3 + x - 2}{x - 2}\, dx$

17. $\int \frac{2x^3}{x^2 - 4}\, dx$

18. $\int \frac{5 - 4x^2}{3 + 2x}\, dx$

19. $\int \frac{(\sqrt{x} + 2)^2}{3\sqrt{x}}\, dx$

20. $\int \frac{5e^s}{1 + 3e^s}\, ds$

21. $\int \frac{5(x^{1/3} + 2)^4}{\sqrt[3]{x^2}}\, dx$

22. $\int \frac{\sqrt{1 + \sqrt{x}}}{\sqrt{x}}\, dx$

*23. $\int \frac{\ln x}{x}\, dx$

24. $\int \sqrt{t}(5 - t\sqrt{t})^{0.4}\, dt$

25. $\int \frac{\ln^2(r + 1)}{r + 1}\, dr$

26. $\int \frac{9x^5 - 6x^4 - ex^3}{7x^2}\, dx$

*27. $\int \frac{3^{\ln x}}{x}\,dx$

28. $\int \frac{4}{x\ln(2x^2)}\,dx$

29. $\int x\sqrt{e^{x^2+3}}\,dx$

30. $\int \frac{x+3}{x+6}\,dx$

31. $\int \frac{8}{(x+3)\ln(x+3)}\,dx$

32. $\int (e^{e^2} + x^e - 2x)\,dx$

33. $\int \frac{x^3+x^2-x-3}{x^2-3}\,dx$

34. $\int \frac{4x\ln\sqrt{1+x^2}}{1+x^2}\,dx$

35. $\int \frac{6x\sqrt{\ln(x^2+1)^2}}{x^2+1}\,dx$

36. $\int 3(x^2+2)^{-1/2}xe^{\sqrt{x^2+2}}\,dx$

37. $\int \left(\frac{x^3-1}{\sqrt{x^4-4x}} - \ln 7\right)dx$

38. $\int \frac{x-x^{-2}}{x^2+2x^{-1}}\,dx$

39. $\int \frac{2x^4-8x^3-6x^2+4}{x^3}\,dx$

40. $\int \frac{e^x+e^{-x}}{e^x-e^{-x}}\,dx$

41. $\int \frac{x}{x-1}\,dx$

42. $\int \frac{2x}{(x^2+1)\ln(x^2+1)}\,dx$

43. $\int \frac{xe^{x^2}}{\sqrt{e^{x^2}+2}}\,dx$

44. $\int \frac{5}{(3x+1)[1+\ln(3x+1)]^2}\,dx$

45. $\int \frac{(e^{-x}+6)^2}{e^x}\,dx$

46. $\int \left[\frac{1}{8x+1} - \frac{1}{e^x(8+e^{-x})^2}\right]dx$

47. $\int (x^3+ex)\sqrt{x^2+e}\,dx$

48. $\int 2^{x\ln x}(1+\ln x)\,dx$

49. $\int \sqrt{x}\sqrt{(8x)^{3/2}+3}\,dx$

50. $\int \frac{2}{x(\ln x)^{2/3}}\,dx$

51. $\int \frac{\sqrt{s}}{e^{\sqrt{s^3}}}\,ds$

52. $\int \frac{\ln^3 x}{3x}\,dx$

53. $\int e^{\ln(x+2)}\,dx$

54. $\int dx$

55. $\int \frac{\ln(xe^x)}{x}\,dx$

56. $\int e^{f(x)+\ln(f'(x))}\,dx$ assuming $f' > 0$

In Problems 57 and 58, dr/dq is a marginal-revenue function. Find the demand function.

57. $\frac{dr}{dq} = \frac{200}{(q+2)^2}$

58. $\frac{dr}{dq} = \frac{900}{(2q+3)^3}$

In Problems 59 and 60, dc/dq is a marginal-cost function. Find the total-cost function if fixed costs in each case are 2000.

59. $\frac{dc}{dq} = \frac{20}{q+5}$

60. $\frac{dc}{dq} = 2e^{0.001q}$

In Problems 61–63, dC/dI represents the marginal propensity to consume. Find the consumption function, subject to the given condition.

*61. $\frac{dC}{dI} = \frac{1}{\sqrt{I}}$; $C(9) = 8$

62. $\frac{dC}{dI} = \frac{1}{2} - \frac{1}{2\sqrt{2I}}$; $C(2) = \frac{3}{4}$

63. $\frac{dC}{dI} = \frac{3}{4} - \frac{1}{6\sqrt{I}}$; $C(25) = 23$.

64. **Cost Function** The marginal-cost function for a manufacturer's product is given by

$$\frac{dc}{dq} = 10 - \frac{100}{q+10}$$

where c is the total cost in dollars when q units are produced. When 100 units are produced, the average cost is \$50 per unit. To the nearest dollar, determine the manufacturer's fixed cost.

65. **Cost Function** Suppose the marginal-cost function for a manufacturer's product is given by

$$\frac{dc}{dq} = \frac{100q^2 - 4998q + 50}{q^2 - 50q + 1}$$

where c is the total cost in dollars when q units are produced.

(a) Determine the marginal cost when 50 units are produced.
(b) If fixed costs are \$10,000, find the total cost of producing 50 units.
(c) Use the results of parts (a) and (b) and differentials to approximate the total cost of producing 52 units.

66. **Cost Function** The marginal-cost function for a manufacturer's product is given by

$$\frac{dc}{dq} = \frac{9}{10}\sqrt{q}\sqrt{0.04q^{3/4}+4}$$

where c is the total cost in dollars when q units are produced. Fixed costs are \$360.

(a) Determine the marginal cost when 25 units are produced.
(b) Find the total cost of producing 25 units.
(c) Use the results of parts (a) and (b) and differentials to approximate the total cost of producing 23 units.

67. **Value of Land** It is estimated that t years from now the value V (in dollars) of an acre of land near the ghost town of Cherokee, California, will be increasing at the rate of $\frac{8t^3}{\sqrt{0.2t^4+8000}}$ dollars per year. If the land is currently worth \$500 per acre, how much will it be worth in 10 years? Express your answer to the nearest dollar.

68. **Revenue Function** The marginal-revenue function for a manufacturer's product is of the form

$$\frac{dr}{dq} = \frac{a}{e^q + b}$$

for constants a and b, where r is the total revenue received (in dollars) when q units are produced and sold. Find the demand function, and express it in the form $p = f(q)$. [*Hint:* Rewrite dr/dq by multiplying both numerator and denominator by e^{-q}.]

69. **Savings** A certain country's marginal propensity to save is given by

$$\frac{dS}{dI} = \frac{5}{(I+2)^2}$$

where S and I represent total national savings and income, respectively, and are measured in billions of dollars. If total national consumption is \$7.5 billion when total national income is \$8 billion, for what value(s) of I is total national savings equal to zero?

70. **Consumption Function** A certain country's marginal propensity to save is given by

$$\frac{dS}{dI} = \frac{1}{2} - \frac{1.8}{\sqrt[3]{3I^2}}$$

where S and I represent total national savings and income, respectively, and are measured in billions of dollars.

(a) Determine the marginal propensity to consume when total national income is \$81 billion.

(b) Determine the consumption function, given that savings are \$3 billion when total national income is \$24 billion.

(c) Use the result in part (b) to show that consumption is \$54.9 billion when total national income is \$81 billion.

(d) Use differentials and the results in parts (a) and (c) to approximate consumption when total national income is \$78 billion.

14.6 Summation

OBJECTIVE *To introduce sigma notation and to give summation formulas that will be used in the next section.*

To prepare you for further applications of integration, we need to discuss certain sums.

Consider finding the sum S of the first n positive integers:

$$S = 1 + 2 + \cdots + (n-1) + n \tag{1}$$

Writing the terms on the right side of Equation (1) in reverse order, we have

$$S = n + (n-1) + \cdots + 2 + 1 \tag{2}$$

Adding the corresponding sides of Equations (1) and (2) gives

$$\begin{array}{rcccccccc} S = & 1 & + & 2 & + \cdots + & (n-1) & + & n \\ S = & n & + & (n-1) & + \cdots + & 2 & + & 1 \\ \hline 2S = & (n+1) & + & (n+1) & + \cdots + & (n+1) & + & (n+1) \end{array}$$

On the right side of the last equation, the term $(n+1)$ occurs n times. Thus, $2S = n(n+1)$, so

$$S = \frac{n(n+1)}{2} \quad \text{(the sum of the first } n \text{ positive integers)} \tag{3}$$

For example, the sum of the first 100 positive integers corresponds to $n = 100$ and is $100(100+1)/2$, or 5050.

For convenience, to indicate a sum we shall introduce *sigma notation,* so named because the Greek letter Σ (sigma) is used. For example, the notation

$$\sum_{k=1}^{3}(2k+5)$$

denotes the sum of those numbers obtained from the expression $2k + 5$ by first replacing k by 1, then by 2, and, finally, by 3. Thus,

$$\begin{aligned}\sum_{k=1}^{3}(2k+5) &= [2(1)+5]+[2(2)+5]+[2(3)+5] \\ &= 7+9+11 = 27\end{aligned}$$

The letter k is called the *index of summation;* the numbers 1 and 3 are the *limits of summation* (1 is the *lower limit* and 3 is the *upper limit*). The values of the index begin at the lower limit and progress through integer values to the upper limit. The symbol used for the index is a "dummy" symbol in the sense that it does not affect the sum of the terms. Any other letter can be used. For example,

$$\sum_{j=1}^{3}(2j+5) = 7+9+11 = \sum_{k=1}^{3}(2k+5)$$

EXAMPLE 1 **Sigma Notation**

a. *Evaluate* $\displaystyle\sum_{k=4}^{7}\frac{k^2+3}{2}$.

Solution: Here the sum begins with $k = 4$. So we have

$$\begin{aligned}\sum_{k=4}^{7}\frac{k^2+3}{2} &= \frac{4^2+3}{2}+\frac{5^2+3}{2}+\frac{6^2+3}{2}+\frac{7^2+3}{2} \\ &= \frac{19}{2}+\frac{28}{2}+\frac{39}{2}+\frac{52}{2} = 69\end{aligned}$$

b. *Evaluate* $\displaystyle\sum_{j=0}^{2}(-1)^{j+1}(j-1)^2$.

Solution:

$$\begin{aligned}&\sum_{j=0}^{2}(-1)^{j+1}(j-1)^2 \\ &= (-1)^{0+1}(0-1)^2+(-1)^{1+1}(1-1)^2+(-1)^{2+1}(2-1)^2 \\ &= (-1)+0+(-1) = -2\end{aligned}$$

▶ **Now work Problem 3.**

To express the sum of the first n positive integers in sigma notation, we can write

$$\sum_{k=1}^{n} k = 1+2+\cdots+n$$

By Equation (3),

$$\sum_{k=1}^{n} k = \frac{n(n+1)}{2} \tag{4}$$

Note in Equation (4) that $\sum_{k=1}^{n} k$ is a function of n alone, not of k.

EXAMPLE 2 **Applying Formula 4**

a. *Evaluate* $\sum_{k=1}^{60} k$.

Solution: Here we must find the sum of the first 60 positive integers. By Equation (4) with $n = 60$,

$$\sum_{k=1}^{60} k = \frac{60(60+1)}{2} = 1830$$

b. *Evaluate* $\sum_{k=1}^{n-1} k$.

Solution: Here we must add the first $n - 1$ positive integers. Replacing n by $n - 1$ in Equation (4), we obtain

$$\sum_{k=1}^{n-1} k = \frac{(n-1)[(n-1)+1]}{2} = \frac{(n-1)n}{2}$$

▶ **Now work Problem 9.**

Another useful formula is that for the sum of the *squares* of the first n positive integers:

$$\sum_{k=1}^{n} k^2 = \frac{n(n+1)(2n+1)}{6} \qquad (5)$$

EXAMPLE 3 **Applying Formula 5**

Evaluate $1 + 4 + 9 + 16 + 25 + 36$.

Solution: This sum can be written as $\sum_{k=1}^{6} k^2$. By Equation (5) with $n = 6$,

$$\sum_{k=1}^{6} k^2 = \frac{6(6+1)[2(6)+1]}{6} = 91$$

We conclude with a property of sigma. If $x_1, x_2, \ldots, x_n$, are real numbers and c is a constant, then

$$\sum_{i=1}^{n} cx_i = cx_1 + cx_2 + \cdots + cx_n$$

$$= c(x_1 + x_2 + \cdots + x_n) = c\sum_{i=1}^{n} x_i$$

Thus,

$$\sum_{i=1}^{n} cx_i = c\sum_{i=1}^{n} x_i$$

This means that a constant factor can "pass through" a sigma. For example,

$$\sum_{i=1}^{5} 3i^2 = 3\sum_{i=1}^{5} i^2$$

CAUTION Although constant factors can "pass through" a sigma, nothing else can. The situation is the same as for integration, where we have seen that constants, but only constants, can "pass through" an integral sign.

By Equation (5), we have

$$\sum_{i=1}^{5} 3i^2 = 3\sum_{i=1}^{5} i^2 = 3\left[\frac{5(6)(11)}{6}\right] = 165.$$

▸ **Now work Problem 21.**

Exercise 14.6

In Problems 1–10, evaluate the given sum.

1. $\sum_{k=1}^{5}(k+4)$ **2.** $\sum_{k=4}^{10}(5-3k)$ ***3.** $\sum_{j=1}^{10}(-1)^j$

4. $\sum_{j=0}^{5}2^j$ **5.** $\sum_{n=2}^{3}(3n^2-7)$ **6.** $\sum_{n=2}^{4}\frac{n+1}{n-1}$

7. $\sum_{k=3}^{4}\frac{(-1)^k(k+1)}{2^k}$ **8.** $\sum_{n=1}^{5}4$

***9.** $\sum_{k=2}^{5}\frac{(-1)^{k+1}(1+k^2)}{k-1}$ **10.** $\sum_{n=1}^{4}(n^2+n)$

In Problems 11–16, express the given sums in sigma notation.

11. $1+2+3+\cdots+19$ **12.** $7+8+9+10$

13. $1+3+5+7$ **14.** $3+6+9+\cdots+36$

15. $1^2+2^2+3^2+\cdots+10^2$ **16.** $3+6+9+12$

In Problems 17–22, evaluate the sums by using Equations (4) and (5).

17. $\sum_{k=1}^{450}k$ **18.** $\sum_{k=1}^{10}k^2$ **19.** $\sum_{j=1}^{6}4j$

20. $\sum_{i=1}^{40}\frac{i}{2}$ ***21.** $\sum_{i=1}^{6}3i^2$ **22.** $\sum_{n=1}^{8}\left(\frac{2n}{3}\right)^2$

23. A company has an asset whose original value is \$3200. The asset has no salvage value. The maintenance cost each year is \$100 and increases by \$100 each year. Show that the average annual total cost over a period of n years is

$$C = \frac{3200}{n} + 50(n+1)$$

Find the value of n that minimizes C. What is the average annual cost at this value of n?

OBJECTIVE *To motivate, by means of the concept of area, the definite integral as a limit of a special sum; to evaluate simple definite integrals by using a limiting process.*

14.7 The Definite Integral

Figure 14.2 shows the region R bounded by the lines $y = f(x) = 2x$, $y = 0$ (the x-axis), and $x = 1$. The region is simply a right triangle. If b and h are the lengths of the base and the height, respectively, then, from geometry, the area of the triangle is $A = \frac{1}{2}bh = \frac{1}{2}(1)(2) = 1$ square unit. (Henceforth, we will treat areas as pure numbers and write "square unit" only if it seems necessary for emphasis. We shall now find this area by another method, which, as you will see later, applies to more complex regions. This method involves the summation of areas of rectangles.

Let us divide the interval $[0, 1]$ on the x-axis into four subintervals of equal length by means of the equally spaced points $x_0 = 0$, $x_1 = \frac{1}{4}$, $x_2 = \frac{2}{4}$, $x_3 = \frac{3}{4}$, and $x_4 = \frac{4}{4} = 1$. (See Figure 14.3.) Each subinterval has length $\Delta x = \frac{1}{4}$. These subintervals determine four subregions of R: R_1, R_2, R_3, and R_4, as indicated.

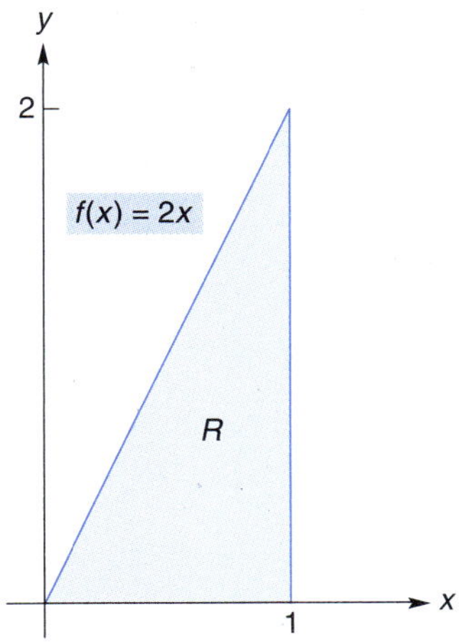

FIGURE 14.2 Region bounded by $f(x) = 2x$, $y = 0$, and $x = 1$.

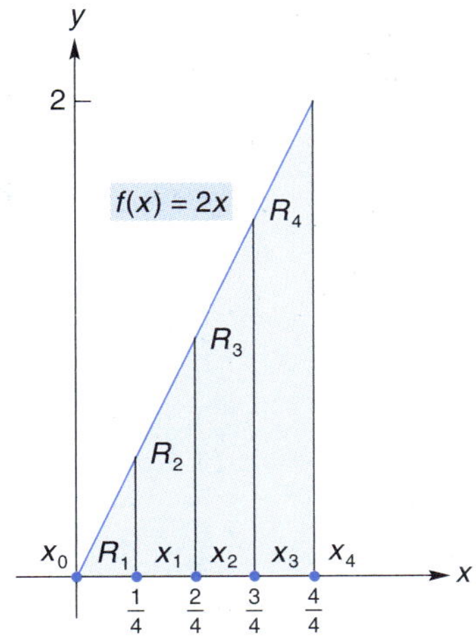

FIGURE 14.3 Four subregions of R.

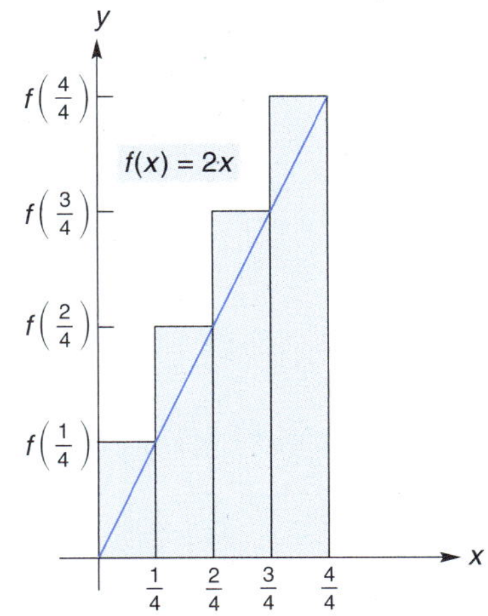

FIGURE 14.4 Four circumscribed rectangles.

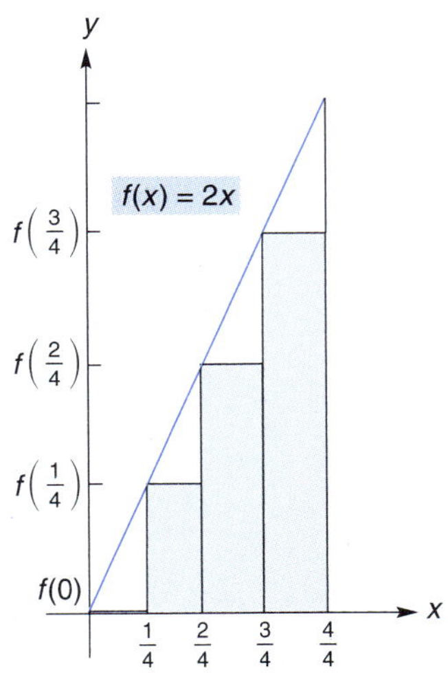

FIGURE 14.5 Four inscribed rectangles.

With each subregion, we can associate a *circumscribed* rectangle (Figure 14.4)—that is, a rectangle whose base is the corresponding subinterval and whose height is the *maximum* value of $f(x)$ on that subinterval. Since f is an increasing function, the maximum value of $f(x)$ on each subinterval occurs when x is the right-hand endpoint. Thus, the areas of the circumscribed rectangles associated with regions R_1, R_2, R_3, and R_4 are $\frac{1}{4}f(\frac{1}{4})$, $\frac{1}{4}f(\frac{2}{4})$, $\frac{1}{4}f(\frac{3}{4})$, and $\frac{1}{4}f(\frac{4}{4})$, respectively. The area of each rectangle is an approximation to the area of its corresponding subregion. Hence, the sum of the areas of these rectangles, denoted by $\overline{S}_4$ (read "S sub 4 upper bar" or "the fourth upper sum"), approximates the area A of the triangle. We have

$$\begin{aligned}\overline{S}_4 &= \tfrac{1}{4}f\left(\tfrac{1}{4}\right) + \tfrac{1}{4}f\left(\tfrac{2}{4}\right) + \tfrac{1}{4}f\left(\tfrac{3}{4}\right) + \tfrac{1}{4}f\left(\tfrac{4}{4}\right)\\ &= \tfrac{1}{4}\left[2\left(\tfrac{1}{4}\right) + 2\left(\tfrac{2}{4}\right) + 2\left(\tfrac{3}{4}\right) + 2\left(\tfrac{4}{4}\right)\right] = \tfrac{5}{4}\end{aligned}$$

You may verify that we can write $\overline{S}_4$ as $\overline{S}_4 = \sum_{i=1}^{4} f(x_i)\Delta x$. The fact that $\overline{S}_4$ is greater than the actual area of the triangle might have been expected, since $\overline{S}_4$ includes areas of shaded regions that are not in the triangle. (See Figure 14.4.)

On the other hand, with each subregion we can also associate an *inscribed* rectangle (Figure 14.5)—that is, a rectangle whose base is the corresponding subinterval, but whose height is the *minimum* value of $f(x)$ on that subinterval. Since f is an increasing function, the minimum value of $f(x)$ on each subinterval will occur when x is the left-hand endpoint. Thus, the areas of the four inscribed rectangles associated with R_1, R_2, R_3, and R_4 are $\frac{1}{4}f(0)$, $\frac{1}{4}f(\frac{1}{4})$, $\frac{1}{4}f(\frac{2}{4})$, and $\frac{1}{4}f(\frac{3}{4})$, respectively. Their sum, denoted $\underline{S}_4$ (read "S sub 4 lower bar" or "the fourth lower sum"), is also an approximation to the area A of the triangle. We have

$$\begin{aligned}\underline{S}_4 &= \tfrac{1}{4}f(0) + \tfrac{1}{4}f\left(\tfrac{1}{4}\right) + \tfrac{1}{4}f\left(\tfrac{2}{4}\right) + \tfrac{1}{4}f\left(\tfrac{3}{4}\right)\\ &= \tfrac{1}{4}\left[2(0) + 2\left(\tfrac{1}{4}\right) + 2\left(\tfrac{2}{4}\right) + 2\left(\tfrac{3}{4}\right)\right] = \tfrac{3}{4}\end{aligned}$$

Using sigma notation, we can write $\underline{S}_4 = \sum_{i=0}^{3} f(x_i)\Delta x$. Note that $\underline{S}_4$ is less than the area of the triangle, because the rectangles do not account for that portion of the triangle that is not shaded in Figure 14.5.

Since

$$\tfrac{3}{4} = \underline{S}_4 \leq A \leq \overline{S}_4 = \tfrac{5}{4}$$

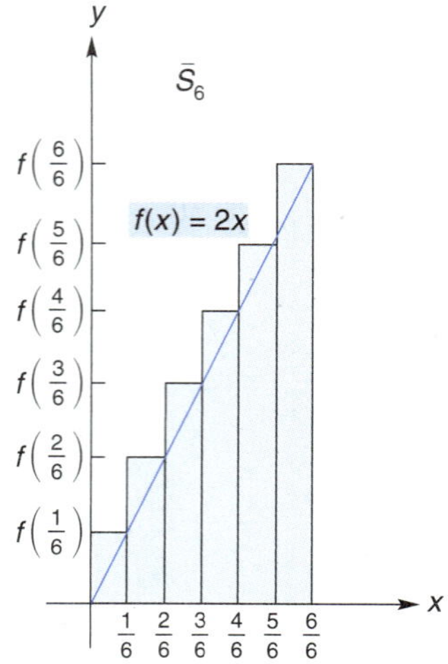

FIGURE 14.6 Six circumscribed rectangles.

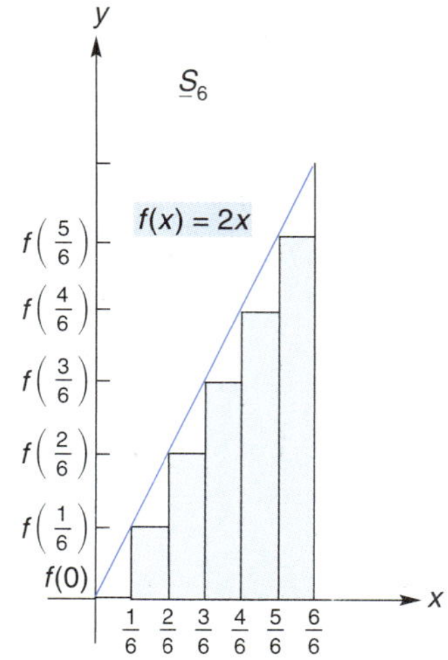

FIGURE 14.7 Six inscribed rectangles.

we say that $\underline{S}_4$ is an approximation to A from *below* and $\overline{S}_4$ is an approximation to A from *above*.

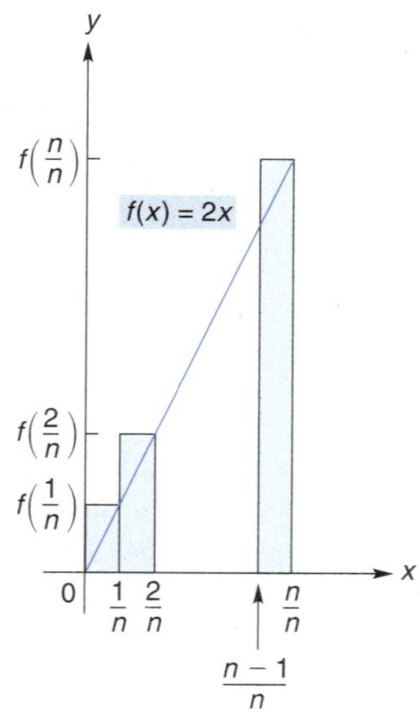

FIGURE 14.8 n circumscribed rectangles.

If [0, 1] is divided into more subintervals, we expect that better approximations to A will occur. To test this out, let us use six subintervals of equal length $\Delta x = \frac{1}{6}$. Then $\overline{S}_6$, the total area of six circumscribed rectangles (see Figure 14.6), and $\underline{S}_6$, the total area of six inscribed rectangles (see Figure 14.7), are

$$\begin{aligned}\overline{S}_6 &= \tfrac{1}{6}f\left(\tfrac{1}{6}\right) + \tfrac{1}{6}f\left(\tfrac{2}{6}\right) + \tfrac{1}{6}f\left(\tfrac{3}{6}\right) + \tfrac{1}{6}f\left(\tfrac{4}{6}\right) + \tfrac{1}{6}f\left(\tfrac{5}{6}\right) + \tfrac{1}{6}f\left(\tfrac{6}{6}\right)\\ &= \tfrac{1}{6}\left[2\left(\tfrac{1}{6}\right) + 2\left(\tfrac{2}{6}\right) + 2\left(\tfrac{3}{6}\right) + 2\left(\tfrac{4}{6}\right) + 2\left(\tfrac{5}{6}\right) + 2\left(\tfrac{6}{6}\right)\right] = \tfrac{7}{6}\end{aligned}$$

and

$$\begin{aligned}\underline{S}_6 &= \tfrac{1}{6}f(0) + \tfrac{1}{6}f\left(\tfrac{1}{6}\right) + \tfrac{1}{6}f\left(\tfrac{2}{6}\right) + \tfrac{1}{6}f\left(\tfrac{3}{6}\right) + \tfrac{1}{6}f\left(\tfrac{4}{6}\right) + \tfrac{1}{6}f\left(\tfrac{5}{6}\right)\\ &= \tfrac{1}{6}\left[2(0) + 2\left(\tfrac{1}{6}\right) + 2\left(\tfrac{2}{6}\right) + 2\left(\tfrac{3}{6}\right) + 2\left(\tfrac{4}{6}\right) + 2\left(\tfrac{5}{6}\right)\right] = \tfrac{5}{6}\end{aligned}$$

Note that $\underline{S}_6 \leq A \leq \overline{S}_6$, and, with appropriate labeling, both $\overline{S}_6$ and $\underline{S}_6$ will be of the form $\sum f(x)\Delta x$. Clearly, using six subintervals gives better approximations to the area than does four subintervals, as expected.

More generally, if we divide [0, 1] into n subintervals of equal length Δx, then $\Delta x = 1/n$, and the endpoints of the subintervals are $x = 0, 1/n, 2/n, \ldots, (n-1)/n$, and $n/n = 1$. (See Figure 14.8.) The total area of n *circumscribed* rectangles is

$$\begin{aligned}\overline{S}_n &= \frac{1}{n}f\left(\frac{1}{n}\right) + \frac{1}{n}f\left(\frac{2}{n}\right) + \cdots + \frac{1}{n}f\left(\frac{n}{n}\right) \qquad (1)\\ &= \frac{1}{n}\left[2\left(\frac{1}{n}\right) + 2\left(\frac{2}{n}\right) + \cdots + 2\left(\frac{n}{n}\right)\right]\\ &= \frac{2}{n^2}[1 + 2 + \cdots + n] \qquad \left(\textit{by factoring } \frac{2}{n} \textit{ from each term}\right)\end{aligned}$$

From Section 14.6, the sum of the first n positive integers is $\dfrac{n(n+1)}{2}$. Thus,

$$\overline{S}_n = \left(\frac{2}{n^2}\right)\frac{n(n+1)}{2} = \frac{n+1}{n}$$

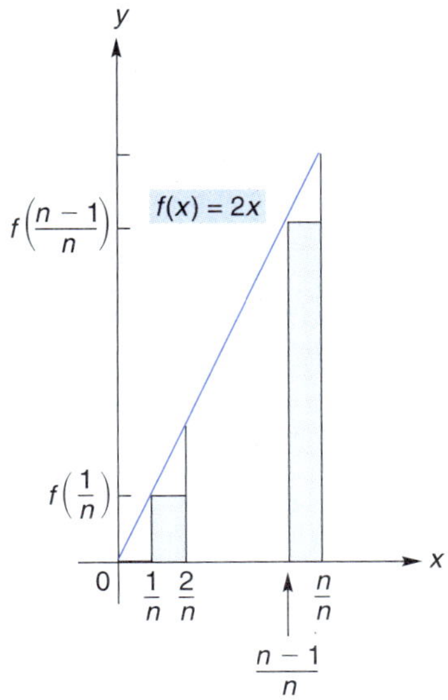

FIGURE 14.9 n inscribed rectangles.

For n *inscribed* rectangles, the total area determined by the subintervals (see Figure 14.9) is

$$\underline{S}_n = \frac{1}{n}f(0) + \frac{1}{n}f\left(\frac{1}{n}\right) + \cdots + \frac{1}{n}f\left(\frac{n-1}{n}\right) \tag{2}$$

$$= \frac{1}{n}\left[2(0) + 2\left(\frac{1}{n}\right) + \cdots + 2\left(\frac{n-1}{n}\right)\right]$$

$$= \frac{2}{n^2}[1 + \cdots + (n-1)]$$

Summing the first $n-1$ positive integers as we did in Example 2(b) of Section 14.6, we obtain

$$\underline{S}_n = \left(\frac{2}{n^2}\right)\frac{(n-1)n}{2} = \frac{n-1}{n}$$

From Equations (1) and (2), we again see that both $\overline{S}_n$ and $\underline{S}_n$ are sums of the form $\sum f(x)\Delta x$, namely, $\overline{S}_n = \sum_{k=1}^{n} f\left(\frac{k}{n}\right)\Delta x$ and $\underline{S}_n = \sum_{k=0}^{n-1} f\left(\frac{k}{n}\right)\Delta x$.

From the nature of $\overline{S}_n$ and $\underline{S}_n$, it seems reasonable—and it is indeed true—that

$$\underline{S}_n \leq A \leq \overline{S}_n$$

As n becomes larger, $\underline{S}_n$ and $\overline{S}_n$ become better approximations to A. In fact, let us take the limits of $\underline{S}_n$ and $\overline{S}_n$ as n approaches ∞ through positive integral values:

$$\lim_{n\to\infty} \underline{S}_n = \lim_{n\to\infty} \frac{n-1}{n} = \lim_{n\to\infty}\left(1 - \frac{1}{n}\right) = 1$$

$$\lim_{n\to\infty} \overline{S}_n = \lim_{n\to\infty} \frac{n+1}{n} = \lim_{n\to\infty}\left(1 + \frac{1}{n}\right) = 1$$

Since $\overline{S}_n$ and $\underline{S}_n$ have the same common limit, namely,

$$\lim_{n\to\infty} \overline{S}_n = \lim_{n\to\infty} \underline{S}_n = 1 \tag{3}$$

and since

$$\underline{S}_n \leq A \leq \overline{S}_n$$

we shall take this limit to be the area of the triangle. Thus, $A = 1$, which agrees with our prior finding.

We call the common limit of $\overline{S}_n$ and $\underline{S}_n$, namely, 1, the *definite integral* of $f(x) = 2x$ on the interval from $x = 0$ to $x = 1$, and we denote this quantity by writing

$$\int_0^1 2x\,dx = 1 \tag{4}$$

The reason for using the term *definite integral* and the symbolism in Equation (4) will become apparent in the next section. The numbers 0 and 1 appearing with the integral sign $\int$ in Equation (4) are called the *limits of integration;* 0 is the *lower limit* and 1 is the *upper limit.*

In general, for a function f defined on the interval from $x = a$ to $x = b$, where $a < b$, we can form the sums $\overline{S}_n$ and $\underline{S}_n$, which are obtained by considering the maximum and minimum values, respectively, on each of n subintervals of equal length Δx.[6] We can now state the following:

The common limit of $\overline{S}_n$ and $\underline{S}_n$ as $n \to \infty$, if it exists, is called the **definite integral** of f over $[a, b]$ and is written

$$\int_a^b f(x)\,dx$$

The numbers a and b are called **limits of integration;** a is the **lower limit** and b is the **upper limit.** The symbol x is called the **variable of integration** and $f(x)$ is the **integrand.**

In terms of a limiting process, we have

$$\sum f(x)\Delta x \to \int_a^b f(x)\,dx$$

The definite integral is the limit of the form $\sum f(x)\,\Delta x$. This interpretation will be useful in later sections.

Two points must be made about the definite integral. First, the definite integral is the limit of a sum of the form $\sum f(x)\,\Delta x$. In fact, we can think of the integral sign as an elongated "S," the first letter of "Summation." Second, for an arbitrary function f defined on an interval, we may be able to calculate the sums $\overline{S}_n$ and $\underline{S}_n$, and determine their common limit if it exists. However, some terms in the sums may be negative if $f(x)$ is negative at points in the interval. These terms are not areas of rectangles (an area is never negative), so the common limit may not represent an area. Thus, **the definite integral is nothing more than a real number; it may or may not represent an area.**

As you saw in Equation (3), $\lim_{n\to\infty} \underline{S}_n$ is equal to $\lim_{n\to\infty} \overline{S}_n$. For an arbitrary function, this is not always true. However, for the functions that we shall consider, these limits will be equal, and the definite integral will always exist. To save time, we shall just use the **right-hand endpoint** of each subinterval in computing a sum. For the functions in this section, this sum will be denoted S_n and will correspond to either $\underline{S}_n$ or $\overline{S}_n$.

APPLICATIONS IN PRACTICE 1

Computing an Area by Using Right-Hand Endpoints

A company has determined that its marginal-revenue function is given by $R'(x) = 600 - 0.5x$, where R is the revenue (in dollars) received when x units are sold. Find the total revenue received for selling 10 units by finding the area in the first quadrant bounded by $y = R'(x) = 600 - 0.5x$ and the lines $y = 0$, $x = 0$, and $x = 10$.

EXAMPLE 1 **Computing an Area by Using Right-Hand Endpoints**

Find the area of the region in the first quadrant bounded by $f(x) = 4 - x^2$ and the lines $x = 0$ and $y = 0$.

Solution: A sketch of the region appears in Figure 14.10. The interval over which x varies in this region is seen to be $[0, 2]$, which we divide into n subintervals of equal length Δx. Since the length of $[0, 2]$ is 2, we take $\Delta x = 2/n$. The endpoints of the subintervals are $x = 0, 2/n, 2(2/n), \ldots, (n-1)(2/n)$, and $n(2/n) = 2$, which are shown in Figure 14.11. The diagram also shows the corresponding rectangles obtained by using the right-hand endpoint of each subinterval. The area of each rectangle is the product of its width $(2/n)$ and its height, which is the function value at the right-hand endpoint of its base. Summing these areas, we get

In general, over $[a, b]$, we have

$$\Delta x = \frac{b-a}{n}$$

$$\begin{aligned}S_n &= \frac{2}{n} f\left(\frac{2}{n}\right) + \frac{2}{n} f\left[2\left(\frac{2}{n}\right)\right] + \cdots + \frac{2}{n} f\left[n\left(\frac{2}{n}\right)\right]\\ &= \frac{2}{n}\left[f\left(\frac{2}{n}\right) + f\left[2\left(\frac{2}{n}\right)\right] + \cdots + f\left[n\left(\frac{2}{n}\right)\right]\right]\\ &= \frac{2}{n}\left[\left\{4 - \left(\frac{2}{n}\right)^2\right\} + \left\{4 - \left[2\left(\frac{2}{n}\right)\right]^2\right\} + \cdots + \left\{4 - \left[n\left(\frac{2}{n}\right)\right]^2\right\}\right]\end{aligned}$$

[6]Here we assume that the maximum and minimum values exist.

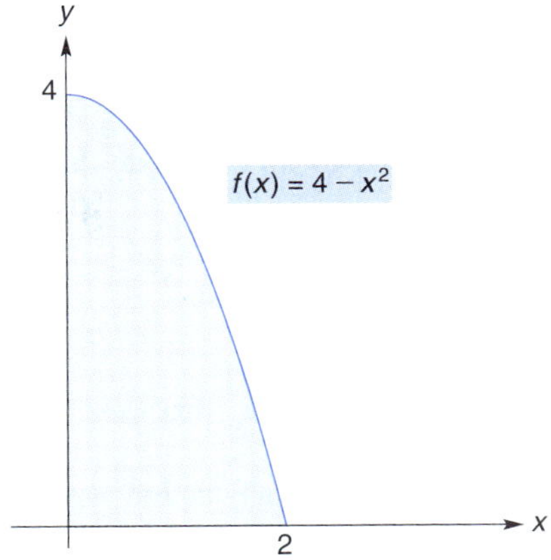

FIGURE 14.10 Region of Example 1.

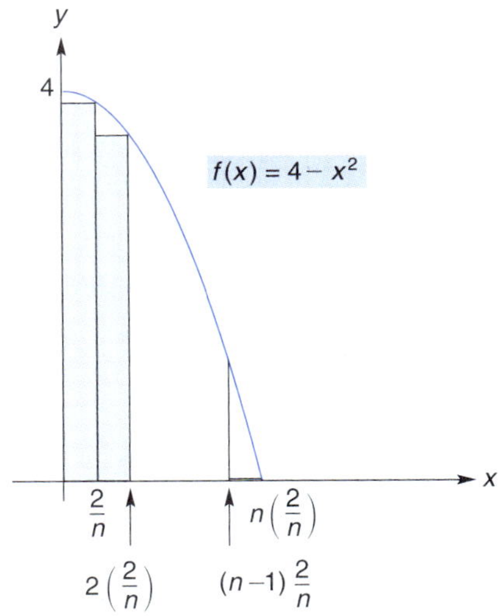

FIGURE 14.11 n subintervals and corresponding rectangles for Example 1.

Since the number 4 occurs n times in the sum, we can simplify S_n. We get

$$S_n = \frac{2}{n}\left[4n - \left(\frac{2}{n}\right)^2 - 2^2\left(\frac{2}{n}\right)^2 - \cdots - n^2\left(\frac{2}{n}\right)^2\right]$$

$$= \frac{2}{n}\left[4n - \left(\frac{2}{n}\right)^2\{1^2 + 2^2 + \cdots + n^2\}\right]$$

From Section 14.6, $\sum_{k=1}^{n} k^2 = \frac{n(n+1)(2n+1)}{6}$, so

$$S_n = \frac{2}{n}\left[4n - \left(\frac{2}{n}\right)^2 \frac{n(n+1)(2n+1)}{6}\right]$$

$$= 8 - \frac{4(n+1)(2n+1)}{3n^2} \qquad \text{(Distribute)}$$

$$= 8 - \frac{4}{3}\left(\frac{2n^2+3n+1}{n^2}\right) \qquad \text{(Expand)}$$

Finally, we take the limit of S_n as $n \to \infty$:

$$\lim_{n\to\infty} S_n = \lim_{n\to\infty}\left[8 - \frac{4}{3}\left(\frac{2n^2+3n+1}{n^2}\right)\right]$$

$$= \lim_{n\to\infty}\left[8 - \frac{4}{3}\left(2 + \frac{3}{n} + \frac{1}{n^2}\right)\right]$$

$$= 8 - \frac{8}{3} = \frac{16}{3}$$

Thus,

$$\lim_{n\to\infty} S_n = \frac{16}{3}$$

Hence, the area of the region is $\frac{16}{3}$.

▸ **Now work Problem 7.**

EXAMPLE 2 **Evaluating a Definite Integral**

Evaluate $\int_0^2 (4 - x^2)\,dx$.

Solution: We want to find the definite integral of $f(x) = 4 - x^2$ over the interval $[0, 2]$. Thus, we must compute $\lim_{n\to\infty} S_n$. But this limit is precisely the limit $\frac{16}{3}$ found in Example 1, so we conclude that

No units are attached to the answer, since a definite integral is simply a unitless number.

$$\int_0^2 (4 - x^2)\,dx = \frac{16}{3}$$

▶ **Now work Problem 19.**

EXAMPLE 3 **Integrating a Function over an Interval**

Integrate $f(x) = x - 5$ *from* $x = 0$ *to* $x = 3$; *that is, evaluate* $\int_0^3 (x - 5)\,dx$.

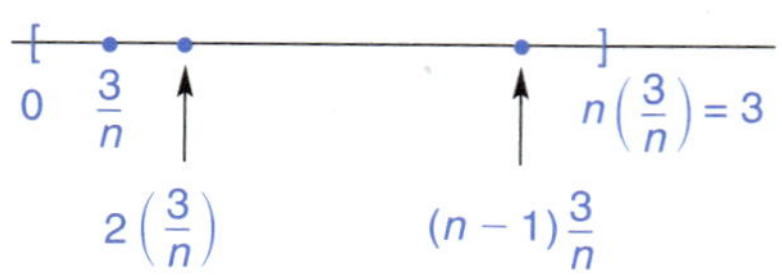

FIGURE 14.12 Dividing [0, 3] into n subintervals.

Solution: We first divide $[0, 3]$ into n subintervals of equal length $\Delta x = 3/n$. The endpoints are $0, 3/n, 2(3/n), \ldots, (n-1)(3/n), n(3/n) = 3$. (See Figure 14.12.) Using right-hand endpoints, we form the sum

$$S_n = \frac{3}{n} f\left(\frac{3}{n}\right) + \frac{3}{n} f\left[2\left(\frac{3}{n}\right)\right] + \cdots + \frac{3}{n} f\left[n\left(\frac{3}{n}\right)\right]$$

Simplifying, we have

$$\begin{aligned}
S_n &= \frac{3}{n}\left[\left\{\frac{3}{n} - 5\right\} + \left\{2\left(\frac{3}{n}\right) - 5\right\} + \cdots + \left\{n\left(\frac{3}{n}\right) - 5\right\}\right] \\
&= \frac{3}{n}\left[-5n + \frac{3}{n}\{1 + 2 + \cdots + n\}\right] \\
&= \frac{3}{n}\left[-5n + \left(\frac{3}{n}\right)\frac{n(n+1)}{2}\right] \qquad \left(\sum_{k=1}^{n} k = \frac{n(n+1)}{2}\right) \\
&= -15 + \frac{9}{2} \cdot \frac{n+1}{n} \\
&= -15 + \frac{9}{2}\left(1 + \frac{1}{n}\right)
\end{aligned}$$

Taking the limit, we obtain

$$\lim_{n\to\infty} S_n = \lim_{n\to\infty}\left[-15 + \frac{9}{2}\left(1 + \frac{1}{n}\right)\right] = -15 + \frac{9}{2} = -\frac{21}{2}$$

Thus,

$$\int_0^3 (x - 5)\,dx = -\frac{21}{2}$$

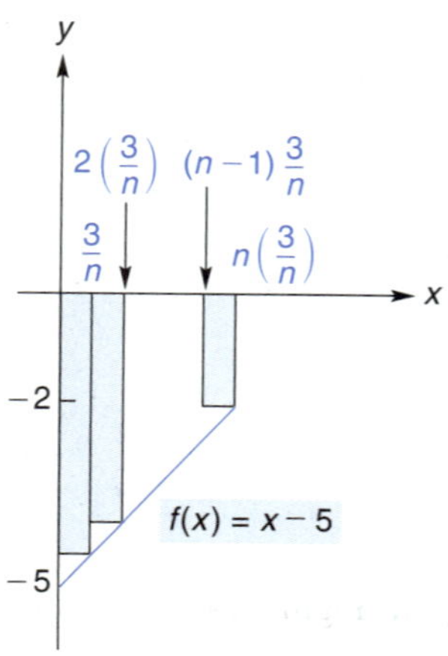

FIGURE 14.13 $f(x)$ is negative at each right-hand endpoint.

Note that the definite integral here is a *negative* number. The reason is clear from the graph of $f(x) = x - 5$ over the interval $[0, 3]$. (See Figure 14.13.) Since the value of $f(x)$ is negative at each right-hand endpoint, each term in S_n must also be negative. Hence, $\lim_{n\to\infty} S_n$, which is the definite integral, is negative.

Geometrically, each term in S_n is the negative of the area of a rectangle. (Refer again to Figure 14.13.) Although the definite integral is simply a number, here we can interpret it as representing the negative of the area of the region bounded by $f(x) = x - 5$, $x = 0$, $x = 3$, and the x-axis ($y = 0$).

▶ **Now work Problem 17.**

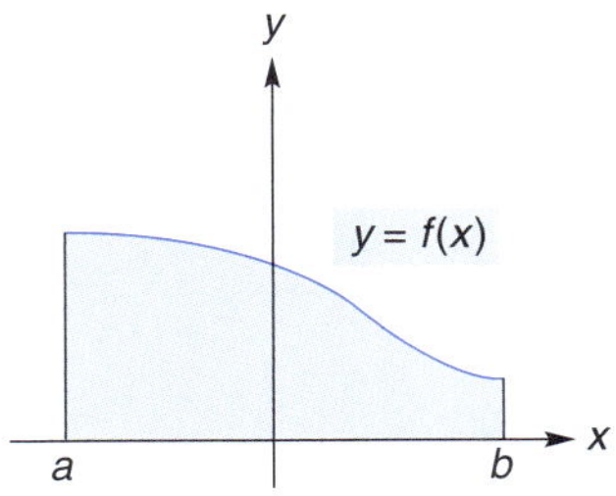

FIGURE 14.14 If f is continuous and $f(x) \geq 0$ on $[a, b]$, then $\int_a^b f(x)\,dx$ represents area.

In Example 3, it was shown that *the definite integral does not have to represent an area.* In fact, there the definite integral was negative. However, if f is continuous and $f(x) \geq 0$ on $[a, b]$, then $S_n \geq 0$ for all values of n. Therefore, $\lim_{n\to\infty} S_n \geq 0$, so $\int_a^b f(x)\,dx \geq 0$. Furthermore, this definite integral gives the area of the region bounded by $y = f(x)$, $y = 0$, $x = a$, and $x = b$. (See Figure 14.14.)

Although the approach that we took to discuss the definite integral is sufficient for our purposes, it is by no means rigorous. **The important thing to remember about the definite integral is that it is the limit of a special sum.**

TECHNOLOGY

Here is a program for the TI-83 Plus graphing calculator that will estimate the limit of S_n as $n \to \infty$ for a function f defined on $[a, b]$.

```
PROGRAM:RIGHTSUM
Lbl 1
Input "SUBINTV",N
(B − A)/N → H
0 → S
A + H → X
1 → I
Lbl 2
Y1 + S → S
X + H → X
I + 1 → I
If I ≤ N
Goto 2
H*S → S
Disp S
Pause
Goto 1
```

RIGHTSUM will compute S_n for a given number n of subintervals. Before executing the program, store $f(x)$, a, and b as Y_1, A, and B, respectively. Upon execution of the program, you will be prompted to enter the number of subintervals. Then the program proceeds to display the value of S_n. Each time ENTER

FIGURE 14.15 Values of S_n for $f(x) = x - 5$ on $[0, 3]$.

is pressed, the program repeats. In this way, a display of values of S_n for various numbers of subintervals may be obtained. Figure 14.15 shows values of S_n ($n = 100$, 1000, and 2000) for the function $f(x) = x - 5$ on the interval $[0, 3]$. As $n \to \infty$, it appears that $S_n \to -10.5$. Thus, we estimate that

$$\lim_{n\to\infty} S_n \approx -10.5$$

equivalently,

$$\int_0^3 (x - 5)\,dx \approx -10.5$$

which agrees with our result in Example 3.

It is interesting to note that the time required for a calculator to compute S_{2000} in Figure 14.15 was in excess of 1.5 minutes.

Exercise 14.7

In Problems 1–4, sketch the region in the first quadrant that is bounded by the given curves. Approximate the area of the region by the indicated sum. Use the right-hand endpoint of each subinterval.

1. $f(x) = x, y = 0, x = 1;\quad S_3$

2. $f(x) = 3x, y = 0, x = 1;\quad S_5$

3. $f(x) = x^2, y = 0, x = 1;\quad S_4$

4. $f(x) = x^2 + 1, y = 0, x = 0, x = 1;\quad S_2$

In Problems 5 and 6, by dividing the indicated interval into n subintervals of equal length, find S_n for the given function. Use the right-hand endpoint of each subinterval. Do not find $\lim_{n\to\infty} S_n$.

5. $f(x) = 4x;\quad [0, 1]$

6. $f(x) = 2x + 1;\quad [0, 2]$

In Problems 7 and 8, (a) simplify S_n and (b) find $\lim_{n\to\infty} S_n$.

***7.** $S_n = \frac{1}{n}\left[\left(\frac{1}{n}+1\right) + \left(\frac{2}{n}+1\right) + \cdots + \left(\frac{n}{n}+1\right)\right]$

8. $S_n = \frac{2}{n}\left[\left(\frac{2}{n}\right)^2 + \left(2\cdot\frac{2}{n}\right)^2 + \cdots + \left(n\cdot\frac{2}{n}\right)^2\right]$

In Problems 9–14, sketch the region in the first quadrant that is bounded by the given curves. Determine the exact area of the region by considering the limit of S_n as $n \to \infty$. Use the right-hand endpoint of each subinterval.

9. Region as described in Problem 1.

10. Region as described in Problem 2.

11. Region as described in Problem 3.

12. $y = x^2, y = 0, x = 1, x = 2$

13. $f(x) = 2x^2, y = 0, x = 2$

14. $f(x) = 9 - x^2, y = 0, x = 0$

In Problems 15–20, evaluate the given definite integral by taking the limit of S_n. Use the right-hand endpoint of each subinterval. Sketch the graph, over the given interval, of the function to be integrated.

15. $\int_1^3 5x\,dx$

16. $\int_0^4 9\,dx$

***17.** $\int_0^3 -4x\,dx$

18. $\int_0^3 (2x - 9)\,dx$

***19.** $\int_0^1 (x^2 + x)\,dx$

20. $\int_1^2 (x + 2)\,dx$

21. Find $D_x\left[\int_2^3 \sqrt{x^2+1}\,dx\right]$ without the use of limits.

22. Find $\int_0^3 f(x)\,dx$ without the use of limits, where

$$f(x) = \begin{cases} 2 & \text{if } 0 \le x < 1 \\ 4 - 2x & \text{if } 1 \le x < 2 \\ 5x - 10 & \text{if } 2 \le x \le 3 \end{cases}$$

23. Find $\int_{-1}^3 f(x)\,dx$ without the use of limits, where

$$f(x) = \begin{cases} 1 & \text{if } x \le 1 \\ 2 - x & \text{if } 1 \le x \le 2 \\ -1 + \frac{x}{2} & \text{if } x > 2 \end{cases}$$

In each of Problems 24–26, use a program, such as ***RIGHTSUM,*** *to estimate the area of the region in the first quadrant bounded by the given curves. Round your answer to one decimal place.*

24. $f(x) = x^3 + 1, y = 0, x = 2, x = 3.7$

25. $f(x) = \sqrt{x}, y = 0, x = 1.3, x = 4$

26. $f(x) = e^x, y = 0, x = 0, x = 1$

In each of Problems 27–30, use a program, such as ***RIGHTSUM,*** *to estimate the value of the definite integral. Round your answer to one decimal place.*

27. $\int_2^5 \frac{x+1}{x+2}\,dx$

28. $\int_{-3}^{-1} \frac{1}{x^2}\,dx$

29. $\int_{-1}^2 (4x^2 + x - 13)\,dx$

30. $\int_{0.1}^{0.2} \ln x\,dx$

14.8 The Fundamental Theorem of Integral Calculus

OBJECTIVE *To informally develop the Fundamental Theorem of Integral Calculus and to use it to compute definite integrals. To obtain a change in function values when the rate of change in the function is known.*

The Fundamental Theorem

Thus far, the limiting processes of both the derivative and definite integral have been considered separately. We shall now bring these fundamental ideas together and develop the important relationship that exists between them. As a result, we may evaluate definite integrals more efficiently.

The graph of a function f is given in Figure 14.16. Assume that f is continuous on the interval $[a, b]$ and that its graph does not fall below the x-axis. That is, $f(x) \geq 0$. From the preceding section, the area of the region below the graph and above the x-axis from $x = a$ to $x = b$ is given by $\int_a^b f(x)\,dx$. We shall now consider another way to determine this area.

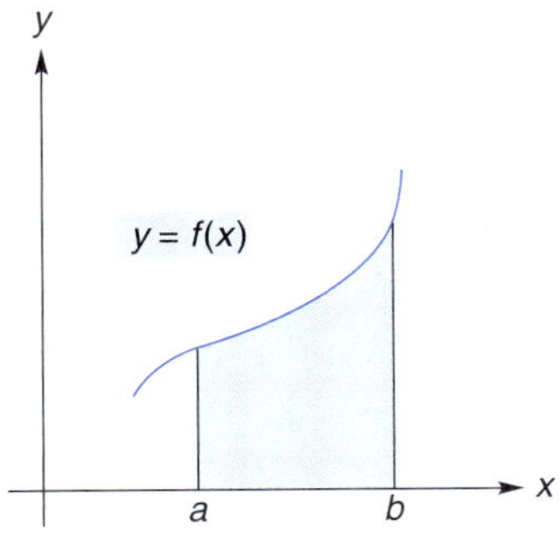

FIGURE 14.16 On $[a, b]$, f is continuous and $f(x) \geq 0$.

Suppose that there is a function $A = A(x)$, which we shall refer to as an area function, that gives the area of the region below the graph of f and above the x-axis from a to x, where $a \leq x \leq b$. This region is shaded in Figure 14.17. Do not confuse $A(x)$, which is an area, with $f(x)$, which is the height of the graph at x.

From its definition, we can state two properties of A immediately:

1. $A(a) = 0$, since there is "no area" from a to a
2. $A(b)$ is the area from a to b; that is

$$A(b) = \int_a^b f(x)\,dx$$

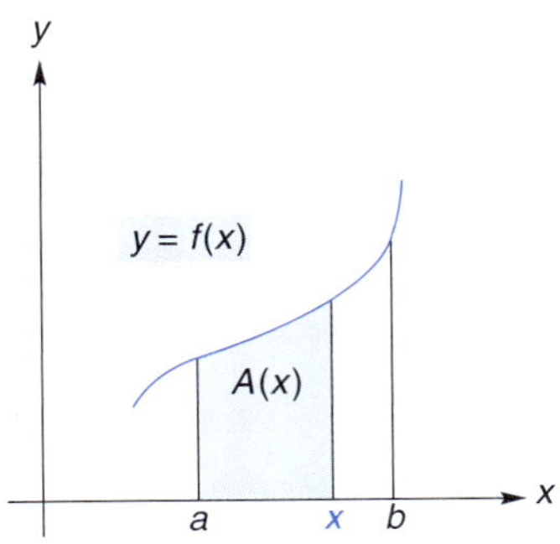

FIGURE 14.17 $A(x)$ is an area function.

If x is increased by h units, then $A(x+h)$ is the area of the shaded region in Figure 14.18. Hence, $A(x+h) - A(x)$ is the difference of the areas in Figures 14.18 and 14.17, namely, the area of the shaded region in Figure 14.19. For h sufficiently close to zero, the area of this region is the same as the area of a rectangle (Figure 14.20) whose base is h and whose height is some value $\overline{y}$ between $f(x)$ and $f(x+h)$. Here $\overline{y}$ is a function of h. Thus on the one hand, the area of the rectangle is $A(x+h) - A(x)$, and on the other hand it is $h\overline{y}$, so

$$A(x+h) - A(x) = h\overline{y}$$

or

$$\frac{A(x+h) - A(x)}{h} = \overline{y} \qquad \text{(dividing by } h\text{)}$$

As $h \to 0$, then $\overline{y}$ approaches the number $f(x)$, so

$$\lim_{h \to 0} \frac{A(x+h) - A(x)}{h} = f(x) \tag{1}$$

But the left side is merely the derivative of A. Thus, Equation (1) becomes

$$A'(x) = f(x)$$

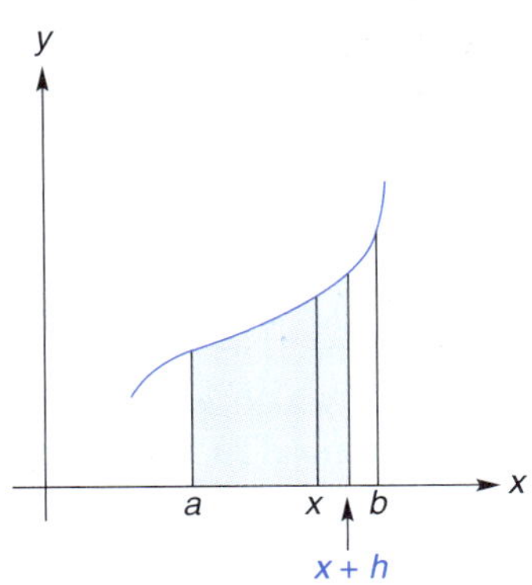

FIGURE 14.18 $A(x+h)$ gives the area of the shaded region.

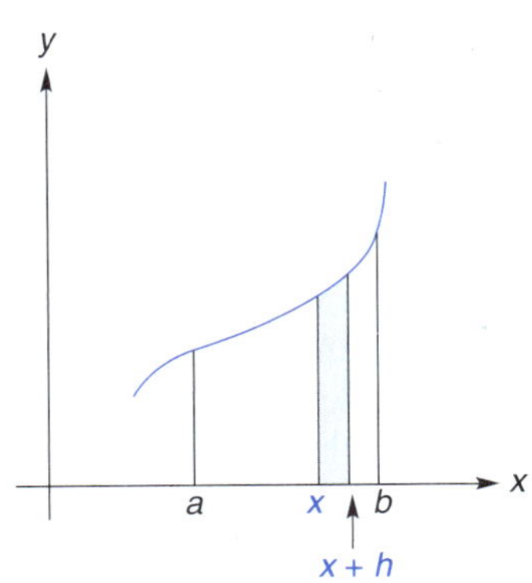

FIGURE 14.19 Area of shaded region is $A(x+h) - A(x)$.

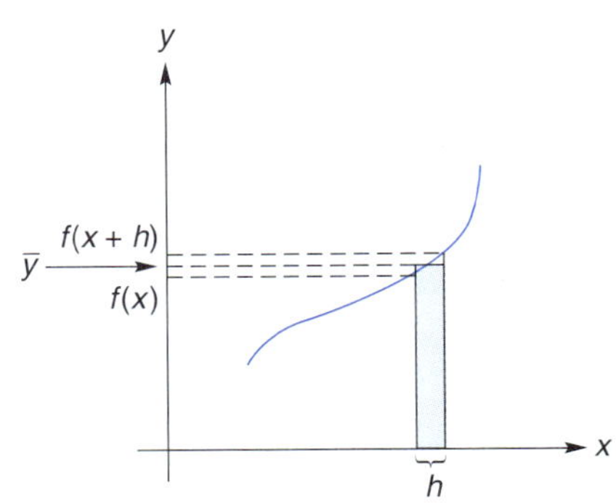

FIGURE 14.20 Area of rectangle is the same as area of shaded region in Figure 14.19.

We conclude that the area function A has the additional property that its derivative A' is f. That is, A is an antiderivative of f. Now, suppose that F is *any* antiderivative of f. Then, since both A and F are antiderivatives of the same function, they differ at most by a constant C:

$$A(x) = F(x) + C \tag{2}$$

Recall that $A(a) = 0$. So, evaluating both sides of Equation (2) when $x = a$ gives

$$0 = F(a) + C$$

so that

$$C = -F(a)$$

Thus, Equation (2) becomes

$$A(x) = F(x) - F(a) \tag{3}$$

If $x = b$, then, from Equation (3),

$$A(b) = F(b) - F(a) \tag{4}$$

But recall that

$$A(b) = \int_a^b f(x)\,dx \tag{5}$$

From Equations (4) and (5), we get

$$\int_a^b f(x)\,dx = F(b) - F(a)$$

A relationship between a definite integral and antidifferentiation has now become clear. To find $\int_a^b f(x)\,dx$, it suffices to find an antiderivative of f, say, F, and subtract the value of F at the lower limit a from its value at the upper limit b. We assumed here that f was continuous and $f(x) \geq 0$ so that we could appeal to the concept of an area. However, our result is true for any continuous function[7] and is known as the *Fundamental Theorem of Integral Calculus.*

Fundamental Theorem of Integral Calculus

If f is continuous on the interval $[a, b]$ and F is any antiderivative of f on that interval, then

$$\int_a^b f(x)\,dx = F(b) - F(a)$$

The definite integral is a number, and an indefinite integral is a function.

It is important that you understand the difference between a definite integral and an indefinite integral. The **definite integral** $\int_a^b f(x)\,dx$ is a **number** defined to be the limit of a sum. The Fundamental Theorem states that the **indefinite integral** $\int f(x)\,dx$ (an antiderivative of f), which is a **function** of x and is related to the differentiation process, can be used to determine this limit.

Suppose we apply the Fundamental Theorem to evaluate $\int_0^2 (4 - x^2)\,dx$. Here $f(x) = 4 - x^2$, $a = 0$, and $b = 2$. Since an antiderivative of $4 - x^2$ is

[7]If f is continuous on $[a, b]$, it can be shown that $\int_a^b f(x)\,dx$ does indeed exist.

$F(x) = 4x - (x^3/3)$, it follows that

$$\int_0^2 (4 - x^2)\, dx = F(2) - F(0) = \left(8 - \frac{8}{3}\right) - (0) = \frac{16}{3}$$

This confirms our result in Example 2 of Section 14.7. If we had chosen $F(x)$ to be $4x - (x^3/3) + C$, then we would have

$$F(2) - F(0) = \left[\left(8 - \frac{8}{3}\right) + C\right] - [0 + C] = \frac{16}{3}$$

as before. Since the choice of the value of C is immaterial, for convenience we shall always choose it to be 0, as originally done. Usually, $F(b) - F(a)$ is abbreviated by writing

$$F(x)\Big|_a^b$$

Since F in the Fundamental Theorem of Calculus is *any* antiderivative of f and $\int f(x)\, dx$ is the most general antiderivative of f, it showcases the notation to write

$$\int_a^b f(x)\, dx = \left(\int f(x)\, dx\right)\Bigg|_a^b$$

Using the $|_a^b$ notation, we have

$$\int_0^2 (4 - x^2)\, dx = \left(4x - \frac{x^3}{3}\right)\Bigg|_0^2 = \left(8 - \frac{8}{3}\right) - 0 = \frac{16}{3}$$

APPLICATIONS IN PRACTICE 1

Applying the Fundamental Theorem

The income (in dollars) from a fast-food chain is increasing at a rate of $f(t) = 10{,}000e^{0.02t}$, where t is in years. Find $\int_3^6 10{,}000e^{0.02t}\, dt$, which gives the total income for the chain between the third and sixth years.

EXAMPLE 1 Applying the Fundamental Theorem

Find $\int_{-1}^3 (3x^2 - x + 6)\, dx$.

Solution: An antiderivative of $3x^2 - x + 6$ is

$$x^3 - \frac{x^2}{2} + 6x$$

Thus,

$$\begin{aligned}\int_{-1}^3 (3x^2 - x + 6)\, dx &= \left(x^3 - \frac{x^2}{2} + 6x\right)\Bigg|_{-1}^3 \\ &= \left[3^3 - \frac{3^2}{2} + 6(3)\right] - \left[(-1)^3 - \frac{(-1)^2}{2} + 6(-1)\right] \\ &= \left(\frac{81}{2}\right) - \left(-\frac{15}{2}\right) = 48\end{aligned}$$

▶ **Now work Problem 1.**

Properties of the Definite Integral

For $\int_a^b f(x)\, dx$, we have assumed that $a < b$. We now define the cases in which $a > b$ or $a = b$. First,

$$\text{if } a > b \text{ then } \int_a^b f(x)\, dx = -\int_b^a f(x)\, dx$$

That is, interchanging the limits of integration changes the integral's sign. For example,

$$\int_2^0 (4 - x^2)\,dx = -\int_0^2 (4 - x^2)\,dx$$

If the limits of integration are equal, we have

$$\int_a^a f(x)\,dx = 0$$

Some properties of the definite integral deserve mention. The first of the properties that follow restates more formally our comment from the preceding section concerning area.

Properties of the Definite Integral

1. If f is continuous and $f(x) \geq 0$ on $[a, b]$, then $\int_a^b f(x)\,dx$ can be interpreted as the area of the region bounded by the curve $y = f(x)$, the x-axis, and the lines $x = a$ and $x = b$.

2. $\int_a^b kf(x)\,dx = k\int_a^b f(x)\,dx$ where k is a constant

3. $\int_a^b [f(x) \pm g(x)]\,dx = \int_a^b f(x)\,dx \pm \int_a^b g(x)\,dx$

Properties 2 and 3 are similar to rules for indefinite integrals because a definite integral may be evaluated by the Fundamental Theorem in terms of an antiderivative. Two more properties of definite integrals are as follows.

4. $\int_a^b f(x)\,dx = \int_a^b f(t)\,dt$

The variable of integration is a "dummy variable" in the sense that any other variable produces the same result—that is, the same number.

To illustrate property 4, you may verify, for example, that

$$\int_0^2 x^2\,dx = \int_0^2 t^2\,dt$$

5. If f is continuous on an interval I and a, b, and c are in I, then

$$\int_a^c f(x)\,dx = \int_a^b f(x)\,dx + \int_b^c f(x)\,dx$$

Property 5 means that the definite integral over an interval can be expressed in terms of definite integrals over subintervals. Thus,

$$\int_0^2 (4 - x^2)\,dx = \int_0^1 (4 - x^2)\,dx + \int_1^2 (4 - x^2)\,dx$$

We shall look at some examples of definite integration now and compute some areas in the next section.

EXAMPLE 2 **Using the Fundamental Theorem**

Find $\int_0^1 \frac{x^3}{\sqrt{1+x^4}}\,dx$.

Solution: To find an antiderivative of the integrand, we shall apply the power rule for integration:

$$\int_0^1 \frac{x^3}{\sqrt{1+x^4}}\,dx = \int_0^1 x^3(1+x^4)^{-1/2}\,dx$$

$$= \frac{1}{4}\int_0^1 (1+x^4)^{-1/2}[4x^3\,dx] = \left(\frac{1}{4}\right)\frac{(1+x^4)^{1/2}}{\frac{1}{2}}\Bigg|_0^1$$

$$= \frac{1}{2}(1+x^4)^{1/2}\Big|_0^1 = \frac{1}{2}(2)^{1/2} - \frac{1}{2}(1)^{1/2} = \frac{1}{2}(\sqrt{2}-1)$$

CAUTION In Example 2, the value of the antiderivative $\frac{1}{2}(1+x^4)^{1/2}$ at the lower limit 0 is $\frac{1}{2}(1)^{1/2}$. **Do not** assume that an evaluation at the limit zero will yield 0.

▸ **Now work Problem 13.**

EXAMPLE 3 **Evaluating Definite Integrals**

a. *Find* $\int_1^2 [4t^{1/3} + t(t^2+1)^3]\,dt$.

Solution:

$$\int_1^2 [4t^{1/3} + t(t^2+1)^3]\,dt = 4\int_1^2 t^{1/3}\,dt + \frac{1}{2}\int_1^2 (t^2+1)^3[2t\,dt]$$

$$= (4)\frac{t^{4/3}}{\frac{4}{3}}\Bigg|_1^2 + \left(\frac{1}{2}\right)\frac{(t^2+1)^4}{4}\Bigg|_1^2$$

$$= 3(2^{4/3}-1) + \frac{1}{8}(5^4 - 2^4)$$

$$= 3\cdot 2^{4/3} - 3 + \frac{609}{8}$$

$$= 6\sqrt[3]{2} + \frac{585}{8}$$

b. *Find* $\int_0^1 e^{3t}\,dt$.

Solution:

$$\int_0^1 e^{3t}\,dt = \frac{1}{3}\int_0^1 e^{3t}[3\,dt]$$

$$= \left(\frac{1}{3}\right)e^{3t}\Big|_0^1 = \frac{1}{3}(e^3 - e^0) = \frac{1}{3}(e^3-1)$$

▸ **Now work Problem 15.**

EXAMPLE 4 **Finding and Interpreting a Definite Integral**

Evaluate $\int_{-2}^1 x^3\,dx$.

Solution:

$$\int_{-2}^1 x^3\,dx = \frac{x^4}{4}\Big|_{-2}^1 = \frac{1^4}{4} - \frac{(-2)^4}{4} = \frac{1}{4} - \frac{16}{4} = -\frac{15}{4}$$

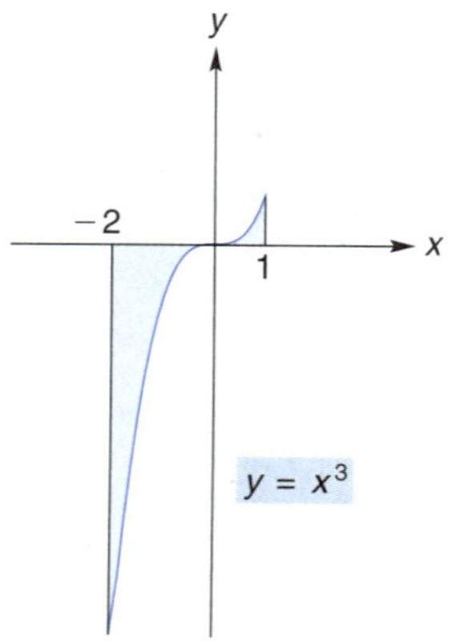

FIGURE 14.21 Graph of $y = x^3$ on the interval $[-2, 1]$.

The reason the result is negative is clear from the graph of $y = x^3$ on the interval $[-2, 1]$. (See Figure 14.21.) For $-2 \leq x < 0$, $f(x)$ is negative. Since a definite integral is a limit of a sum of the form $\Sigma f(x)\Delta x$, it follows that $\int_{-2}^{0} x^3\,dx$ is not only a negative number, but also the negative of the area of the shaded region in the third quadrant. On the other hand, $\int_0^1 x^3\,dx$ is the area of the shaded region in the first quadrant, since $f(x) \geq 0$ on $[0, 1]$. The definite integral over the entire interval $[-2, 1]$ is the *algebraic* sum of these numbers, because, from property 5,

$$\int_{-2}^{1} x^3\,dx = \int_{-2}^{0} x^3\,dx + \int_{0}^{1} x^3\,dx$$

Thus, $\int_{-2}^{1} x^3\,dx$ does not represent the area between the curve and the x-axis. However, if area is desired, it can be given by

$$\left|\int_{-2}^{0} x^3\,dx\right| + \int_{0}^{1} x^3\,dx$$

▸ **Now work Problem 25.**

CAUTION Remember that $\int_a^b f(x)\,dx$ is a limit of a sum. In some cases this limit represents an area. In others it does not. When $f(x) \geq 0$ on $[a, b]$, then the integral represents the area between the graph of f and the x-axis from $x = a$ to $x = b$.

The Definite Integral of a Derivative

Since a function f is an antiderivative of f', by the Fundamental Theorem we have

$$\int_a^b f'(x)\,dx = f(b) - f(a) \qquad (6)$$

But $f'(x)$ is the rate of change of f with respect to x. Hence, if we know the rate of change of f and want to find the difference in function values $f(b) - f(a)$, it suffices to evaluate $\int_a^b f'(x)\,dx$.

APPLICATIONS IN PRACTICE 2

Change in Function Values

A managerial service determines that the rate of increase in maintenance costs (in dollars per year) for a particular apartment complex is given by $M'(x) = 90x^2 + 5000$, where x is the age of the apartment complex in years and $M(x)$ is the total (accumulated) cost of maintenance for x years. Find the total cost for the first five years.

EXAMPLE 5 Finding a Change in Function Values by Definite Integration

A manufacturer's marginal-cost function is

$$\frac{dc}{dq} = 0.6q + 2$$

If production is presently set at $q = 80$ units per week, how much more would it cost to increase production to 100 units per week?

Solution: The total-cost function is $c = c(q)$, and we want to find the difference $c(100) - c(80)$. The rate of change of c is dc/dq, so by Equation (6),

$$c(100) - c(80) = \int_{80}^{100} \frac{dc}{dq}\,dq = \int_{80}^{100} (0.6q + 2)\,dq$$

$$= \left[\frac{0.6q^2}{2} + 2q\right]\Bigg|_{80}^{100} = [0.3q^2 + 2q]\Big|_{80}^{100}$$

$$= [0.3(100)^2 + 2(100)] - [0.3(80)^2 + 2(80)]$$

$$= 3200 - 2080 = 1120$$

If c is in dollars, then the cost of increasing production from 80 units to 100 units is \$1120.

▸ **Now work Problem 59.**

TECHNOLOGY

Many graphing calculators have the capability to estimate the value of a definite integral. On a TI-83 Plus, to estimate

$$\int_{80}^{100} (0.6q + 2)\, dq$$

we use the "fnInt(" command, as indicated in Figure 14.22. The four parameters that must be entered with this command are

function to be integrated	variable of integration	lower limit	upper limit

We see that the value of the definite integral is approximately 1120, which agrees with the result in Example 5.

Similarly, to estimate

$$\int_{-2}^{1} x^3\, dx$$

we enter

$$\text{fnInt}(X^3, X, -2, 1)$$

or alternatively, if we first store x^3 as Y_1, we can enter

$$\text{fnInt}(Y_1, X, -2, 1)$$

In each case we obtain −3.75, which agrees with the result in Example 4.

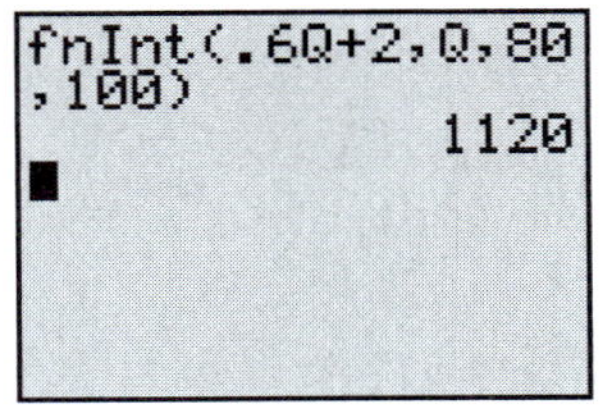

FIGURE 14.22 Estimating $\int_{80}^{100}(0.6q + 2)\, dq$.

Exercise 14.8

In Problems 1–43, evaluate the definite integral.

*1. $\int_0^2 7\, dx$

2. $\int_2^4 (1 - e)\, dx$

3. $\int_1^2 5x\, dx$

4. $\int_2^8 -5x\, dx$

5. $\int_{-3}^1 (2x - 3)\, dx$

6. $\int_{-1}^1 (4 - 9y)\, dy$

7. $\int_2^3 (y^2 - 2y + 1)\, dy$

8. $\int_3^2 (2t - t^2)\, dt$

9. $\int_{-2}^{-1} (3w^2 - w - 1)\, dw$

10. $\int_8^9 dt$

11. $\int_1^3 3t^{-3}\, dt$

12. $\int_1^2 \frac{x^{-2}}{2}\, dx$

*13. $\int_{-1}^1 \sqrt[3]{x^5}\, dx$

14. $\int_{1/2}^{3/2} (x^2 + x + 1)\, dx$

*15. $\int_{1/2}^3 \frac{1}{x^2}\, dx$

16. $\int_9^{36} (\sqrt{x} - 2)\, dx$

17. $\int_{-1}^1 (z + 1)^5\, dz$

18. $\int_1^8 (x^{1/3} - x^{-1/3})\, dx$

19. $\int_0^1 2x^2(x^3 - 1)^3\, dx$

20. $\int_1^3 (x + 3)^3\, dx$

21. $\int_1^8 \frac{4}{y}\, dy$

22. $\int_{-(e^e)}^{-1} \frac{6}{x}\, dx$

23. $\int_0^1 e^5\, dx$

24. $\int_2^{e+1} \frac{1}{x - 1}\, dx$

*25. $\int_0^2 x^2 e^{x^3}\, dx$

26. $\int_0^1 (3x^2 + 4x)(x^3 + 2x^2)^4\, dx$

27. $\int_4^5 \frac{2}{(x - 3)^3}\, dx$

28. $\int_{-1/3}^{20/3} \sqrt{3x + 5}\, dx$

29. $\int_{1/3}^2 \sqrt{10 - 3p}\, dp$

30. $\int_{-1}^1 q\sqrt{q^2 + 3}\, dq$

31. $\int_0^1 x^2 \sqrt[3]{7x^3 + 1}\, dx$

32. $\int_0^{\sqrt{7}} \left[2x - \frac{x}{(x^2 + 1)^{5/3}}\right] dx$

33. $\int_0^1 \frac{2x^3 + x}{x^2 + x^4 + 1}\, dx$

34. $\int_a^b (m + ny)\, dy$

35. $\int_0^1 \frac{e^x - e^{-x}}{2}\,dx$ **36.** $\int_{-2}^1 8|x|\,dx$

37. $\int_1^e 2(x^{-1} + x^{-2} - x^{-3})\,dx$

38. $\int_1^2 \left(6\sqrt{x} - \frac{1}{\sqrt{2x}}\right) dx$ **39.** $\int_1^3 (x+1)e^{x^2+2x}\,dx$

40. $\int_1^{95} \frac{x}{\ln e^x}\,dx$

41. $\int_0^2 \frac{x^6 + 6x^4 + x^3 + 8x^2 + x + 5}{x^3 + 5x + 1}\,dx$

42. $\int_{-1}^1 \frac{2}{1+e^x}\,dx$ [*Hint:* Multiply the integrand by $\frac{e^{-x}}{e^{-x}}$.]

43. $\int_0^2 f(x)\,dx$ where $f(x) = \begin{cases} 4x^2 & \text{if } 0 \le x \le \frac{1}{2} \\ 2x & \text{if } \frac{1}{2} \le x \le 2 \end{cases}$

44. Evaluate $\left(\int_2^3 x\,dx\right)^2 - \int_2^3 x^2\,dx$.

45. Suppose $f(x) = \int_1^x 3\frac{1}{t^2}\,dt$. Evaluate $\int_e^1 f(x)\,dx$.

46. Evaluate $\int_7^7 e^{x^2}\,dx + \int_0^{\sqrt{2}} \frac{1}{3\sqrt{2}}\,dx$.

47. If $\int_1^3 f(x)\,dx = 4$ and $\int_3^2 f(x)\,dx = 3$, find $\int_1^2 f(x)\,dx$.

48. If $\int_1^4 f(x)\,dx = 6$, $\int_2^4 f(x)\,dx = 5$, and $\int_1^3 f(x)\,dx = 2$, find $\int_2^3 f(x)\,dx$.

49. Evaluate $\int_1^2 \left(\frac{d}{dx}\int_1^2 e^{x^2}\,dx\right) dx$. [*Hint:* It is not necessary to find $\int_1^2 e^{x^2}\,dx$.]

50. Suppose that $f(x) = \int_e^x \frac{e^t - e^{-t}}{e^t + e^{-t}}\,dt$ where $x > e$. Find $f'(x)$.

51. Severity Index In discussing traffic safety, Shonle[8] considers how much acceleration a person can tolerate in a crash so that there is no major injury. The *severity index* is defined as

$$\text{S.I.} = \int_0^T \alpha^{5/2}\,dt$$

where α (a Greek letter read "alpha") is considered a constant involved with a weighted average acceleration, and T is the duration of the crash. Find the severity index.

52. Statistics In statistics, the mean μ (a Greek letter read "mu") of the continuous probability density function f defined on the interval $[a, b]$ is given by

$$\mu = \int_a^b [x \cdot f(x)]\,dx$$

and the variance σ^2 (σ is a Greek letter read "sigma") is given by

$$\sigma^2 = \int_a^b (x-\mu)^2 f(x)\,dx$$

Compute μ and then σ^2 if $a = 0$, $b = 1$, and $f(x) = 1$.

53. Distribution of Incomes The economist Pareto[9] has stated an empirical law of distribution of higher incomes that gives the number N of persons receiving x or more dollars. If

$$\frac{dN}{dx} = -Ax^{-B}$$

where A and B are constants, set up a definite integral that gives the total number of persons with incomes between a and b, where $a < b$.

54. Biology In a discussion of gene mutation,[10] the following integral occurs:

$$\int_0^{10^{-4}} x^{-1/2}\,dx$$

Evaluate this integral.

55. Continuous Income Flow The present value (in dollars) of a continuous flow of income of \$2000 a year for five years at 6% compounded continuously is given by

$$\int_0^5 2000e^{-0.06t}\,dt$$

Evaluate the present value to the nearest dollar.

56. Biology In biology, problems frequently arise involving the transfer of a substance between compartments. An example is a transfer from the bloodstream to tissue. Evaluate the following integral, which occurs in a two-compartment diffusion

[8] J. I. Shonle, *Environmental Applications of General Physics* (Reading, MA: Addison-Wesley Publishing Company, 1975).

[9] G. Tintner, *Methodology of Mathematical Economics and Econometrics* (Chicago: University of Chicago Press, 1967), p. 16.

[10] W. J. Ewens, *Population Genetics* (London: Methuen & Company, 1969).

problem:[11]

$$\int_0^t (e^{-a\tau} - e^{-b\tau})\, d\tau$$

Here, τ (read "tau") is a Greek letter; a and b are constants.

57. Demography For a certain population, suppose l is a function such that $l(x)$ is the number of persons who reach the age of x in any year of time. This function is called a *life table function*. Under appropriate conditions, the integral

$$\int_x^{x+n} l(t)\, dt$$

gives the expected number of people in the population between the exact ages of x and $x+n$, inclusive. If

$$l(x) = 10{,}000\sqrt{100 - x}$$

determine the number of people between the exact ages of 36 and 64, inclusive. Give your answer to the nearest integer, since fractional answers make no sense.

58. Mineral Consumption If c_0 is the yearly consumption of a mineral at time $t = 0$, then, under continuous consumption, the total amount of the mineral used in the interval $[0, t_1]$ is

$$\int_0^{t_1} c_0 e^{kt}\, dt$$

where k is the rate of consumption. For a rare-earth mineral, it has been determined that $c_0 = 3000$ units and $k = 0.05$. Evaluate the integral for these data.

***59. Marginal Cost** A manufacturer's marginal-cost function is

$$\frac{dc}{dq} = 0.2q + 8$$

If c is in dollars, determine the cost involved to increase production from 65 to 75 units.

60. Marginal Cost Repeat Problem 59 if

$$\frac{dc}{dq} = 0.004q^2 - 0.5q + 50$$

and production increases from 90 to 180 units.

61. Marginal Revenue A manufacturer's marginal-revenue function is

$$\frac{dr}{dq} = \frac{1000}{\sqrt{100q}}$$

If r is in dollars, find the change in the manufacturer's total revenue if production is increased from 400 to 900 units.

62. Marginal Revenue Repeat Problem 61 if

$$\frac{dr}{dq} = 250 + 90q - 3q^2$$

and production is increased from 10 to 20 units.

63. Crime Rate A sociologist is studying the crime rate in a certain city. She estimates that t months after the beginning of next year, the total number of crimes committed will increase at the rate of $8t + 10$ crimes per month. Determine the total number of crimes that can be expected to be committed next year. How many crimes can be expected to be committed during the last six months of that year?

64. Hospital Discharges For a group of hospitalized individuals, suppose the discharge rate is given by

$$f(t) = \frac{81 \times 10^6}{(300 + t)^4}$$

where $f(t)$ is the proportion of the group discharged per day at the end of t days. What proportion has been discharged by the end of 700 days?

65. Production Imagine a one-dimensional country of length $2R$. (See Figure 14.23.[12]) Suppose the production of goods for this country is continuously distributed from border to border. If the amount produced each year per unit of distance is $f(x)$, then the country's total yearly production is given by

$$G = \int_{-R}^{R} f(x)\, dx$$

Evaluate G if $f(x) = i$, where i is constant.

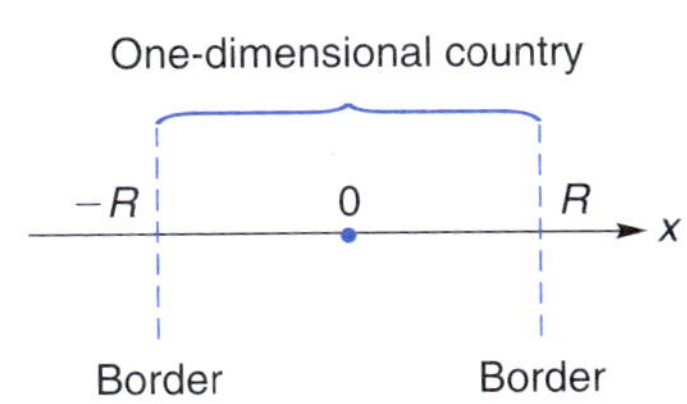

FIGURE 14.23 Diagram for Problem 65.

66. Exports For the one-dimensional country of Problem 65, under certain conditions the amount of the country's exports is given by

$$E = \int_{-R}^{R} \frac{i}{2}\left[e^{-k(R-x)} + e^{-k(R+x)}\right] dx$$

where i and k are constants ($k \neq 0$). Evaluate E.

[11] W. Simon, *Mathematical Techniques for Physiology and Medicine* (New York: Academic Press, 1972).

[12] R. Taagepera, "Why the Trade/GNP Ratio Decreases with Country Size," *Social Science Research* 5 (1976), 385–404.

67. Average Delivered Price In a discussion of a delivered price of a good from a mill to a customer, DeCanio[13] claims that the average delivered price paid by consumers is given by

$$A = \frac{\int_0^R (m+x)[1-(m+x)]\,dx}{\int_0^R [1-(m+x)]\,dx}$$

where m is mill price, and x is the maximum distance to the point of sale. DeCanio determines that

$$A = \frac{m + \frac{R}{2} - m^2 - mR - \frac{R^2}{3}}{1 - m - \frac{R}{2}}$$

Verify this.

[13] S. J. DeCanio, "Delivered Pricing and Multiple Basing Point Equilibria: A Reevaluation," *Quarterly Journal of Economics* 99, no. 2 (1984), 329–49.

In each of Problems 68–70, use the Fundamental Theorem of Integral Calculus to determine the value of the definite integral. Confirm your result with your calculator.

68. $\int_3^5 (1 + 3x + 2x^2)\,dx$ **69.** $\int_0^4 \frac{1}{(4x+4)^2}\,dx$

70. $\int_0^1 e^{3t}\,dt$ Round your answer to two decimal places.

In each of Problems 71–74, estimate the value of the definite integral. Round your answer to two decimal places.

71. $\int_{-1}^5 \frac{x^2+1}{x^2+4}\,dx$ **72.** $\int_3^4 \frac{1}{x \ln x}\,dx$

73. $\int_0^4 5\sqrt{t^2+3}\,dt$ **74.** $\int_{-1}^1 \frac{6\sqrt{q+1}}{q+3}\,dq$

OBJECTIVE *To estimate the value of a definite integral by using either the trapezoidal rule or Simpson's rule.*

14.9 Approximate Integration

Trapezoidal Rule

Any function f constructed from polynomials, exponentials, and logs using algebraic operations and composition can be differentiated and the resulting function f' is again of the same kind—one that can be constructed from polynomials, exponentials, and logs using algebraic operations and composition. Let us call such functions *elementary* (although the term usually has a slightly different meaning). In this terminology, the derivative of an elementary function is also elementary. Integration is more complicated. If an elementary function f has F as an antiderivative, then F may fail to be elementary. Said otherwise, even for a fairly simple-looking function f, it is sometimes impossible to find $\int f(x)\,dx$ in terms of the functions that we consider in this book. For example, there is no elementary function whose derivative is e^{x^2} so that you cannot expect to "do" the integral $\int e^{x^2}dx$.

On the other hand, consider a function f that is continuous on a closed interval $[a, b]$ with $f(x) \geq 0$ for all x in $[a, b]$. Then $\int_a^b f(x)\,dx$ is simply the *number,* which gives the area of the region bounded by the curves $y = f(x)$, $y = 0$, $x = a$, and $x = b$. It is unsatisfying, and perhaps impractical, to not say anything about the number $\int_a^b f(x)\,dx$ because of an inability to "do" the integral $\int f(x)\,dx$. This also applies when the integral $\int f(x)\,dx$ is merely too difficult for the person who needs to find the number $\int_a^b f(x)\,dx$.

Since $\int_a^b f(x)\,dx$ is defined as a limit of sums of the form $\sum f(x)\Delta x$, any particular well-formed sum of the form $\sum f(x)\Delta x$ can be regarded as an approximation to $\int_a^b f(x)\,dx$. At least for nonnegative f such sums can be regarded as sums of areas of thin rectagles. Consider for example Figure 14.11 in Section 14.7 in which two rectangles are explicitly shown. It is clear that the error that arises from such rectangles is associated with the small region at the top. The error would be reduced if we replaced the rectangles by shapes that

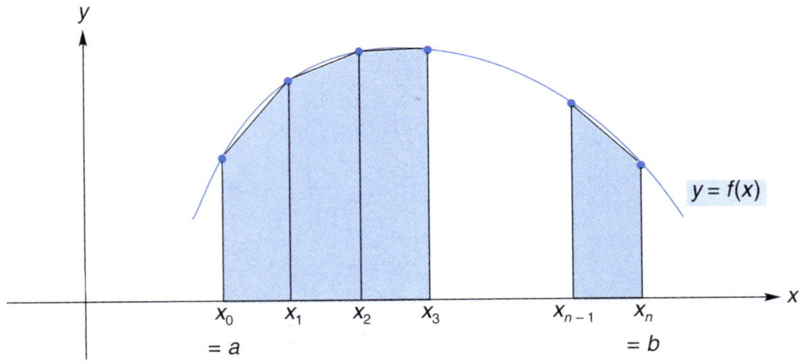

FIGURE 14.24 Approximating an area by using trapezoids.

have a top side that is closer to the shape of the curve. We shall consider two possibilities: using thin trapezoids rather than thin rectangles, the *trapezoidal rule,* and using thin regions surmounted by parabolic arcs, *Simpson's rule*. In each case only a finite number of numerical values of $f(x)$ need be known and the calculations involved are especially suitable for computers or calculators. In both cases, we assume that f is continuous on $[a, b]$.

In developing the trapezoidal rule, for convenience we shall also assume that $f(x) \geq 0$ on $[a, b]$, so that we can think in terms of area. Basically, this rule involves approximating the graph of f by straight-line segments.

In Figure 14.24, the interval $[a, b]$ is divided into n subintervals of equal length by the points $a = x_0, x_1, x_2, \ldots,$ and $x_n = b$. Since the length of $[a, b]$ is $b - a$, the length of each subinterval is $(b - a)/n$, which we shall call h.

Clearly,

$$x_1 = a + h, x_2 = a + 2h, \ldots, x_n = a + nh = b$$

With each subinterval, we can associate a trapezoid (a four-sided figure with two parallel sides). The area A of the region bounded by the curve, the x-axis, and the lines $x = a$ and $x = b$ is $\int_a^b f(x)\,dx$ and can be approximated by the sum of the areas of the trapezoids determined by the subintervals.

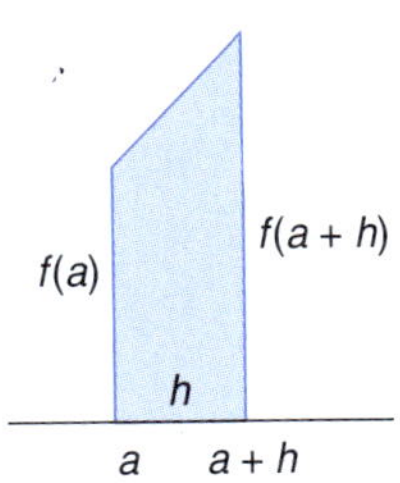

FIGURE 14.25 First trapezoid.

Consider the first trapezoid, which is redrawn in Figure 14.25. Since the area of a trapezoid is equal to one-half the base times the sum of the lengths of the parallel sides, this trapezoid has area

$$\tfrac{1}{2}h[f(a) + f(a + h)]$$

Similarly, the second trapezoid has area

$$\tfrac{1}{2}h[f(a + h) + f(a + 2h)]$$

The area A under the curve is approximated by the sum of the areas of n trapezoids:

$$\begin{aligned} A \approx\ & \tfrac{1}{2}h[f(a) + f(a + h)] + \tfrac{1}{2}h[f(a + h) + f(a + 2h)] \\ & + \tfrac{1}{2}h[f(a + 2h) + f(a + 3h)] + \cdots + \tfrac{1}{2}h[f(a + (n - 1)h) + f(b)] \end{aligned}$$

Since $A = \int_a^b f(x)\,dx$, by simplifying the preceding formula we have the trapezoidal rule:

The Trapezoidal Rule

$$\int_a^b f(x)\,dx \approx \frac{h}{2}[f(a) + 2f(a+h) + 2f(a+2h) + \cdots + 2f(a+(n-1)h) + f(b)]$$

where $h = (b-a)/n$.

The pattern of the coefficients inside the braces is 1, 2, 2, . . . , 2, 1. Usually, the more subintervals, the better is the approximation. In our development, we assumed for convenience that $f(x) \geq 0$ on $[a, b]$. However, the trapezoidal rule is valid without this restriction.

APPLICATIONS IN PRACTICE 1

Trapezoidal Rule

An oil tanker is losing oil at a rate of $R'(t) = \frac{60}{\sqrt{t^2+9}}$, where t is the time in minutes and $R(t)$ is the radius of the oil slick in feet. Use the trapezoidal rule with $n=5$ to approximate $\int_0^5 \frac{60}{\sqrt{t^2+9}}\,dt$, the size of the radius after five seconds.

EXAMPLE 1 Trapezoidal Rule

Use the trapezoidal rule to estimate the value of

$$\int_0^1 \frac{1}{1+x^2}\,dx$$

for $n = 5$. *Compute each term to four decimal places, and round off the answer to three decimal places.*

Solution: Here $f(x) = 1/(1+x^2)$, $n = 5$, $a = 0$, and $b = 1$. Thus,

$$h = \frac{b-a}{n} = \frac{1-0}{5} = \frac{1}{5} = 0.2$$

The terms to be added are:

$$\begin{aligned} f(a) &= f(0) &&= 1.0000 \\ 2f(a+h) &= 2f(0.2) &&= 1.9231 \\ 2f(a+2h) &= 2f(0.4) &&= 1.7241 \\ 2f(a+3h) &= 2f(0.6) &&= 1.4706 \\ 2f(a+4h) &= 2f(0.8) &&= 1.2195 \\ f(b) &= f(1) &&= \underline{0.5000} \qquad (a+nh=b) \\ & && 7.8373 = \text{sum} \end{aligned}$$

Hence, our estimate for the integral is

$$\int_0^1 \frac{1}{1+x^2}\,dx \approx \frac{0.2}{2}(7.8373) \approx 0.784$$

The actual value of the integral is approximately 0.785.

▸ **Now work Problem 1.**

Simpson's Rule

Another method for estimating $\int_a^b f(x)\,dx$ is given by Simpson's rule, which involves approximating the graph of f by parabolic segments. We shall omit the derivation.

Simpson's Rule

$$\int_a^b f(x)\,dx \approx \frac{h}{3}[f(a) + 4f(a+h) + 2f(a+2h) + \cdots + 4f(a+(n-1)h) + f(b)]$$

where $h = (b-a)/n$ and n is even.

The pattern of coefficients inside the braces is 1, 4, 2, 4, 2, ..., 2, 4, 1, which requires that ***n* be even.** Let us use this rule for the integral in Example 1.

APPLICATIONS IN PRACTICE 2

Simpson's Rule

A yeast culture is growing at the rate of $A'(t) = 0.3e^{0.2t^2}$, where t is the time in hours and $A(t)$ is the amount in grams. Use Simpson's rule with $n=8$ to approximate $\int_0^4 0.3e^{0.2t^2}\,dt$, the amount the culture grew over the first four hours.

EXAMPLE 2 Simpson's Rule

Use Simpson's rule to estimate the value of $\int_0^1 \frac{1}{1+x^2}\,dx$ *for* $n = 4$. *Compute each term to four decimal places, and round off the answer to three decimal places.*

Solution: Here $f(x) = 1/(1+x^2)$, $n = 4$, $a = 0$, and $b = 1$. Thus, $h = (b-a)/n = 1/4 = 0.25$. The terms to be added are

$$\begin{aligned} f(a) &= f(0) &&= 1.0000 \\ 4f(a+h) &= 4f(0.25) &&= 3.7647 \\ 2f(a+2h) &= 2f(0.5) &&= 1.6000 \\ 4f(a+3h) &= 4f(0.75) &&= 2.5600 \\ f(b) &= f(1) &&= \underline{0.5000} \\ & && \quad 9.4247 = \text{sum} \end{aligned}$$

Therefore, by Simpson's rule,

$$\int_0^1 \frac{1}{1+x^2}\,dx \approx \frac{0.25}{3}(9.4247) \approx 0.785$$

This is a better approximation than that which we obtained in Example 1 by using the trapezoidal rule.

▸ **Now work Problem 5.**

Both Simpson's rule and the trapezoidal rule may be used if we know only $f(a)$, $f(a+h)$, and so on; we need not know f itself. Example 3 will illustrate.

In Example 3, a definite integral is estimated from data points; the function itself is not known.

EXAMPLE 3 Demography

A function often used in demography (the study of births, marriages, mortality, etc., in a population) is the ***life-table function,*** *denoted* l. *In a population having 100,000 births in any year of time,* $l(x)$ *represents the number of persons who reach the age of* x *in any year of time. For example, if* $l(20) = 98{,}857$, *then the number of persons who attain age 20 in any year of time is 98,857. Suppose that the function* l *applies to all people born over an extended period of time. It can be shown that, at any time, the expected number of persons in the population between the exact ages of* x *and* $x + m$, *inclusive, is given by*

$$\int_x^{x+m} l(t)\,dt$$

The following table gives values of $l(x)$ for males and females in the U.S.[14] *Approximate the number of women in the 20–35 age group by using the trapezoidal rule with $n = 3$.*

Life Table

Age $= x$	$l(x)$ Males	$l(x)$ Females	Age $= x$	$l(x)$ Males	$l(x)$ Females
0	100,000	100,000	45	93,717	96,582
5	99,066	99,220	50	91,616	95,392
10	98,967	99,144	55	88,646	93,562
15	98,834	99,059	60	84,188	90,700
20	98,346	98,857	65	77,547	86,288
25	97,648	98,627	70	68,375	79,926
30	96,970	98,350	75	56,288	70,761
35	96,184	97,964	80	42,127	58,573
40	95,163	97,398			

Solution: We want to estimate

$$\int_{20}^{35} l(t)\, dt$$

We have $h = \dfrac{b-a}{n} = \dfrac{35-20}{3} = 5$. The terms to be added under the trapezoidal rule are

$$\begin{aligned} l(20) &= 98{,}857 \\ 2l(25) = 2(98{,}627) &= 197{,}254 \\ 2l(30) = 2(98{,}350) &= 196{,}700 \\ l(35) &= \underline{97{,}964} \\ & \ 590{,}775 = \text{sum} \end{aligned}$$

By the trapezoidal rule,

$$\int_{20}^{35} l(t)\, dt \approx \frac{5}{2}(590{,}775) = 1{,}476{,}937.5$$

▶ **Now work Problem 17.**

Formulas used to determine the accuracy of answers obtained with the trapezoidal or Simpson's rule may be found in standard texts on numerical analysis.

[14]*National Vital Statistics Report* 48, no. 18, February 7, 2001.

Exercise 14.9

In Problems 1 and 2, use the trapezoidal rule or Simpson's rule (as indicated) and the given value of n to estimate the integral.

*1. $\int_{-2}^{4} \frac{170}{1+x^2}\,dx$; trapezoidal rule, $n = 6$

2. $\int_{-1}^{5} \frac{170}{1+x^2}\,dx$; Simpson's rule, $n = 6$

In Problems 3–8, use the trapezoidal rule or Simpson's rule (as indicated) and the given value of n to estimate the integral. Compute each term to four decimal places, and round the answer to three decimal places. In Problems 3–6, also evaluate the integral by antidifferentiation (the Fundamental Theorem of Integral Calculus).

3. $\int_{0}^{1} x^2\,dx$; trapezoidal rule, $n = 5$

4. $\int_{0}^{1} x^2\,dx$; Simpson's rule, $n = 4$

*5. $\int_{1}^{4} \frac{dx}{x^2}$; Simpson's rule, $n = 4$

6. $\int_{1}^{4} \frac{dx}{x}$; trapezoidal rule, $n = 6$

7. $\int_{0}^{2} \frac{x\,dx}{x+1}$; trapezoidal rule, $n = 4$

8. $\int_{2}^{4} \frac{dx}{x+x^2}$; Simpson's rule, $n = 4$

In Problems 9 and 10, use the life table in Example 3 to estimate the given integrals by the trapezoidal rule.

9. $\int_{15}^{40} l(t)\,dt$, males, $n = 5$

10. $\int_{35}^{55} l(t)\,dt$, females, $n = 4$

In Problems 11 and 12, suppose the graph of a continuous function f, where $f(x) \geq 0$, contains the given points. Use Simpson's rule and all of the points to approximate the area between the graph and the x-axis on the given interval. Round the answer to one decimal place.

11. $(1, 0.4), (2, 0.6), (3, 1.2), (4, 0.8), (5, 0.5)$; $[1, 5]$

12. $(2, 0), (2.5, 6), (3, 10), (3.5, 11), (4, 14), (4.5, 15), (5, 16)$; $[2, 5]$

13. Using all the information given in Figure 14.26, estimate $\int_1^3 f(x)\,dx$ by using Simpson's rule. Give your answer in fractional form.

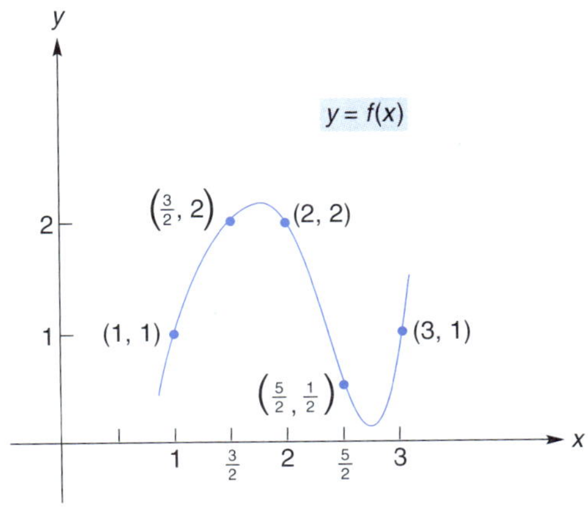

FIGURE 14.26 Graph of f for Problem 13.

In Problems 14 and 15, use Simpson's rule and the given value of n to estimate the integral. Compute each term to four decimal places, and round the answer to three decimal places.

14. $\int_{4}^{6} \frac{1}{\sqrt{1+x}}\,dx$; $n = 4$ Also, evaluate the integral by the Fundamental Theorem of Integral Calculus.

15. $\int_0^1 \sqrt{1-x^2}\,dx$; $n = 4$

16. **Revenue** Use Simpson's rule to approximate the total revenue received from the production and sale of 80 units of a product if the values of the marginal-revenue function dr/dq are as follows:

q (units)	0	10	20	30	40	50	60	70	80
$\frac{dr}{dq}$ (\$ per unit)	10	9	8.5	8	8.5	7.5	7	6.5	7

*17. **Area of Lake** A straight stretch of highway runs alongside a lake. A surveyor who wishes to know the approximate area of the lake measures the distance from various points along the road to the near and far shores of the lake according to the following table:

Distance along highway (km)	0.0	0.5	1.0	1.5	2.0	2.5	3.0	3.5	4.0
Distance to near shore (km)	0.5	0.3	0.7	1.0	0.5	0.2	0.5	0.8	1.0
Distance to far shore (km)	0.5	2.3	2.2	3.0	2.5	2.2	1.5	1.3	1.0

Draw a rough sketch of the geographical situation. Then use Simpson's rule to give the best estimate of the lake's area. Give your answer in fractional form.

18. Manufacturing A manufacturer estimated both marginal cost (MC) and marginal revenue (MR) at various levels of output (q). These estimates are given in the following table:

q **(units)**	0	20	40	60	80	100
MC ($ per unit)	260	250	240	200	240	250
MR ($ per unit)	410	350	300	250	270	250

(a) Using the trapezoidal rule, estimate the total variable costs of production for 100 units.

(b) Using the trapezoidal rule, estimate the total revenue from the sale of 100 units.

(c) If we assume that maximum profit occurs when MR = MC (that is, when $q = 100$), estimate the maximum profit if fixed costs are $2000.

OBJECTIVE *To use vertical strips and the definite integral to find the area of the region between a curve and the x-axis.*

14.10 Area

In Section 14.7 we saw that the area of a region could be found by evaluating the limit of a sum of the form $\sum f(x)\Delta x$, where $f(x)\ \Delta x$ represents the area of a rectangle. This limit is a special case of a definite integral, so it can easily be found by using the Fundamental Theorem.

When using the definite integral to determine area, you should make a rough sketch of the region involved. Let us consider the area of the region bounded by $y = f(x)$ and the x-axis from $x = a$ to $x = b$, where $f(x) \geq 0$ on $[a, b]$. (See Figure 14.27.) To set up the integral, a sample rectangle should be included in the sketch because the area of the region is a limit of sums of areas of rectangles. This will not only help you understand the integration process, but it will also help you find areas of more complicated regions. Such a rectangle (see Figure 14.27) is called a **vertical element of area** (or a **vertical strip**). In the diagram, the width of the vertical element is Δx. The height is the y-value of the curve. Hence, the rectangle has area $y\,\Delta x$ or $f(x)\Delta x$. The area of the entire region is found by summing the areas of all such elements between $x = a$ and $x = b$ and finding the limit of this sum, which is the definite integral. Symbolically, we have

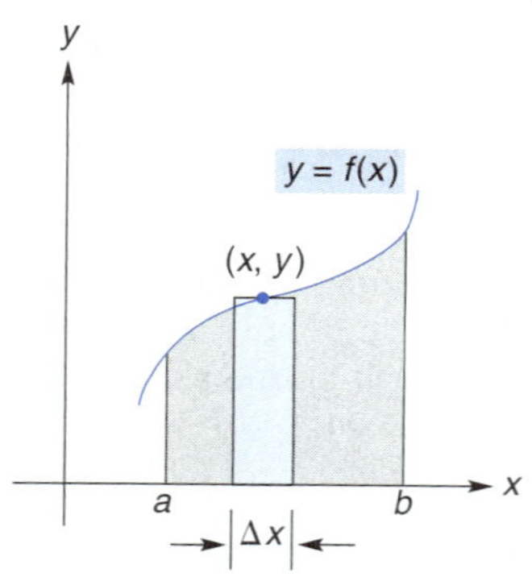

FIGURE 14.27 Region with vertical element.

$$\sum f(x)\,\Delta x \to \int_a^b f(x)\,dx = \text{area}$$

Example 1 will illustrate.

EXAMPLE 1 **Using the Definite Integral to Find Area**

Find the area of the region bounded by the curve

$$y = 6 - x - x^2$$

and the x-axis.

Solution: First we must sketch the curve so that we can visualize the region. Since

$$y = -(x^2 + x - 6) = -(x - 2)(x + 3)$$

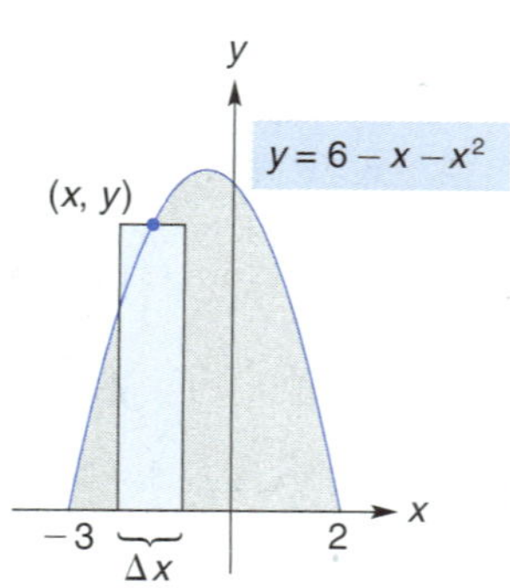

FIGURE 14.28 Region of Example 1 with vertical element.

the x-intercepts are $(2, 0)$ and $(-3, 0)$. Using techniques of graphing that were previously discussed, we obtain the graph and region shown in Figure 14.28. *With this region, it is crucial that the x-intercepts of the curve be found, because they determine the interval over which the areas of the elements must be summed.* That is, *these x-values are the limits of integration.* The vertical element shown has width Δx and height y. Hence, the area of the element is $y\,\Delta x$. Summing the areas of all such elements from $x = -3$ to $x = 2$ and taking the limit via the definite integral gives the area:

$$\sum y\,\Delta x \to \int_{-3}^{2} y\,dx = \text{area}$$

To evaluate the integral, we must express the integrand in terms of the variable of integration, x. Since $y = 6 - x - x^2$,

$$\begin{aligned}\text{area} &= \int_{-3}^{2} (6 - x - x^2)\,dx = \left(6x - \frac{x^2}{2} - \frac{x^3}{3}\right)\Bigg|_{-3}^{2} \\ &= \left(12 - \frac{4}{2} - \frac{8}{3}\right) - \left(-18 - \frac{9}{2} - \frac{-27}{3}\right) = \frac{125}{6}\end{aligned}$$

▶ **Now work Problem 1.**

EXAMPLE 2 **Finding the Area of a Region**

Find the area of the region bounded by the curve $y = x^2 + 2x + 2$, the x-axis, and the lines $x = -2$ and $x = 1$.

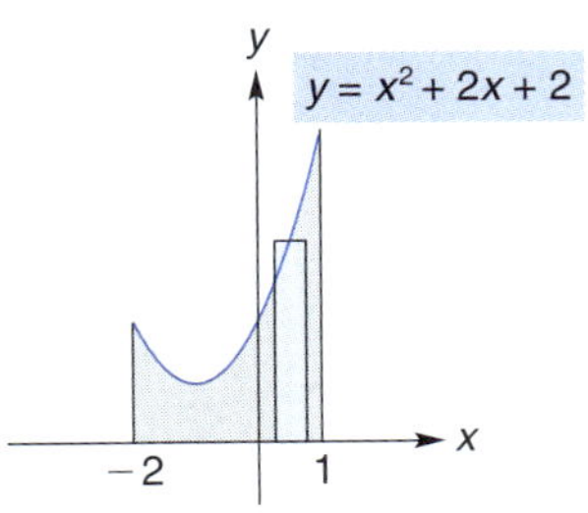

FIGURE 14.29 Diagram for Example 2.

Solution: A sketch of the region is given in Figure 14.29. We have

$$\begin{aligned}\text{area} &= \int_{-2}^{1} y\,dx = \int_{-2}^{1} (x^2 + 2x + 2)\,dx \\ &= \left(\frac{x^3}{3} + x^2 + 2x\right)\Bigg|_{-2}^{1} = \left(\frac{1}{3} + 1 + 2\right) - \left(-\frac{8}{3} + 4 - 4\right) \\ &= 6\end{aligned}$$

▶ **Now work Problem 7.**

EXAMPLE 3 **Finding the Area of a Region**

Find the area of the region between the curve $y = e^x$ and the x-axis from $x = 1$ to $x = 2$.

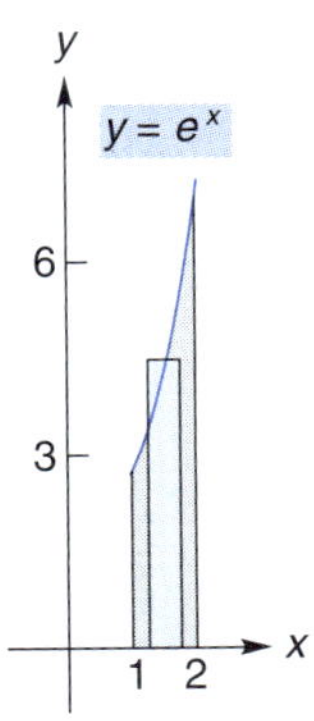

FIGURE 14.30 Diagram for Example 3.

Solution: A sketch of the region is given in Figure 14.30. We write

$$\begin{aligned}\text{area} &= \int_{1}^{2} y\,dx = \int_{1}^{2} e^x\,dx = e^x\Big|_{1}^{2} \\ &= e^2 - e = e(e - 1)\end{aligned}$$

▶ **Now work Problem 27.**

EXAMPLE 4 **An Area Requiring Two Definite Integrals**

Find the area of the region bounded by the curve

$$y = x^2 - x - 2$$

and the line $y = 0$ (the x-axis) from $x = -2$ to $x = 2$.

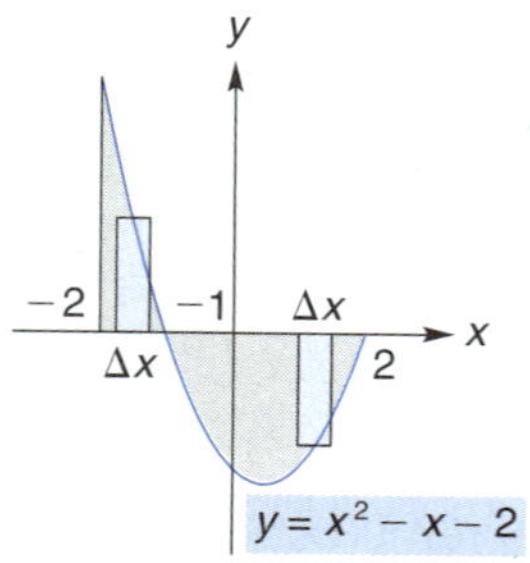

FIGURE 14.31 Diagram for Example 4.

CAUTION It is wrong to write hastily that the area is $\int_{-2}^{2} y\,dx$, for the following reason: For the left rectangle, the height is y. However, for the rectangle on the right, y is negative, so its height is the positive number $-y$. This points out the importance of sketching the region.

Solution: A sketch of the region is given in Figure 14.31. Notice that the x-intercepts are $(-1, 0)$ and $(2, 0)$.

On the interval $[-2, -1]$, the area of the element is

$$y\Delta x = (x^2 - x - 2)\Delta x$$

On $[-1, 2]$ the area is

$$-y\Delta x = -(x^2 - x - 2)\Delta x$$

Thus,

$$\begin{aligned}
\text{area} &= \int_{-2}^{-1} (x^2 - x - 2)\,dx + \int_{-1}^{2} -(x^2 - x - 2)\,dx \\
&= \left(\frac{x^3}{3} - \frac{x^2}{2} - 2x\right)\Bigg|_{-2}^{-1} - \left(\frac{x^3}{3} - \frac{x^2}{2} - 2x\right)\Bigg|_{-1}^{2} \\
&= \left[\left(-\frac{1}{3} - \frac{1}{2} + 2\right) - \left(-\frac{8}{3} - \frac{4}{2} + 4\right)\right] \\
&\quad - \left[\left(\frac{8}{3} - \frac{4}{2} - 4\right) - \left(-\frac{1}{3} - \frac{1}{2} + 2\right)\right] \\
&= \frac{19}{3}
\end{aligned}$$

▶ **Now work Problem 31.**

The next example shows the use of area as a probability in statistics.

EXAMPLE 5 Statistics Application

In statistics, a (probability) ***density function*** *f of a variable x, where x assumes all values in the interval $[a, b]$, has the following properties:*

a. $f(x) \geq 0$

b. $\int_a^b f(x)\,dx = 1$

c. *The probability that x assumes a value between c and d, which is written $P(c \leq x \leq d)$, where $a \leq c \leq d \leq b$, is represented by the area of the region bounded by the graph of f and the x-axis between $x = c$ and $x = d$. Hence (see Figure 14.32),*

$$P(c \leq x \leq d) = \int_c^d f(x)\,dx$$

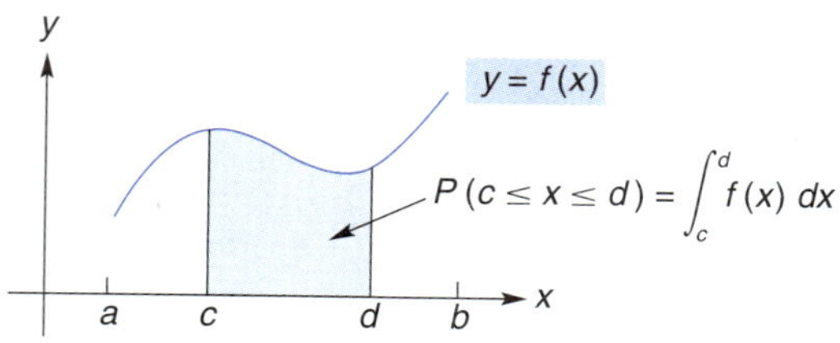

FIGURE 14.32 Probability as an area.

For the density function $f(x) = 6(x - x^2)$, where $0 \leq x \leq 1$, find each of the following probabilities.

a. $P(0 \leq x \leq \frac{1}{4})$

Solution: Here $[a, b]$ is $[0, 1]$, c is 0, and d is $\frac{1}{4}$. By property c, we have

$$P(0 \leq x \leq \tfrac{1}{4}) = \int_0^{1/4} 6(x - x^2)\,dx = 6\int_0^{1/4} (x - x^2)\,dx$$

$$= 6\left(\frac{x^2}{2} - \frac{x^3}{3}\right)\Bigg|_0^{1/4} = (3x^2 - 2x^3)\Bigg|_0^{1/4}$$

$$= \left[3\left(\frac{1}{4}\right)^2 - 2\left(\frac{1}{4}\right)^3\right] - 0 = \frac{5}{32}$$

b. $P\left(x \geq \frac{1}{2}\right)$

Solution: Since the domain of f is $0 \leq x \leq 1$, to say that $x \geq \frac{1}{2}$ means that $\frac{1}{2} \leq x \leq 1$. Thus,

$$P\left(x \geq \frac{1}{2}\right) = \int_{1/2}^{1} 6(x - x^2)\,dx = 6\int_{1/2}^{1} (x - x^2)\,dx$$

$$= 6\left(\frac{x^2}{2} - \frac{x^3}{3}\right)\Bigg|_{1/2}^{1} = (3x^2 - 2x^3)\Bigg|_{1/2}^{1} = \frac{1}{2}$$

▶ **Now work Problem 37.**

Exercise 14.10

In Problems 1–34, use a definite integral to find the area of the region bounded by the given curve, the x-axis, and the given lines. In each case, first sketch the region. Watch out for areas of regions that are below the x-axis.

*__1.__ $y = 4x, \quad x = 2$

2. $y = \frac{3}{4}x + 1, \quad x = 0, \quad x = 16$

3. $y = 3x + 2, \quad x = 2, \quad x = 3$

4. $y = x + 5, \quad x = 2, \quad x = 4$

5. $y = x - 1, \quad x = 5$

6. $y = 3x^2, \quad x = 1, \quad x = 3$

*__7.__ $y = x^2, \quad x = 2, \quad x = 3$

8. $y = 2x^2 - x, \quad x = -2, \quad x = -1$

9. $y = x^2 + 2, \quad x = -1, \quad x = 2$

10. $y = 2x + x^3, \quad x = 1$

11. $y = x^2 - 2x, \quad x = -3, \quad x = -1$

12. $y = 3x^2 - 4x, \quad x = -2, \quad x = -1$

13. $y = 6 - x - x^2$

14. $y = \frac{4}{x}, \quad x = 1, \quad x = 2$

15. $y = 1 - x - x^3, \quad x = -2, \quad x = 0$

16. $y = e^x, \quad x = 1, \quad x = 3$

17. $y = 3 + 2x - x^2$

18. $y = \frac{1}{(x-1)^2}, \quad x = 2, \quad x = 3$

19. $y = \frac{1}{x}, \quad x = 1, \quad x = e$

20. $y = \frac{1}{x}, \quad x = 1, \quad x = e^2$

21. $y = \sqrt{x + 9}, \quad x = -9, \quad x = 0$

22. $y = x^2 - 2x, \quad x = 1, \quad x = 3$

23. $y = \sqrt{2x - 1}, \quad x = 1, \quad x = 5$

24. $y = x^3 + 3x^2, \quad x = -2, \quad x = 2$

25. $y = \sqrt[3]{x}, \quad x = 2$

26. $y = x^2 + 4x - 5, \quad x = -5, \quad x = 1$

*__27.__ $y = e^x, \quad x = 0, \quad x = 2$

28. $y = |x|, \quad x = -2, \quad x = 2$

29. $y = x + \frac{2}{x}, \quad x = 1, \quad x = 2$

30. $y = 4 + 3x - x^2$

*__31.__ $y = x^3, \quad x = -2, \quad x = 4$

32. $y = \sqrt{x - 2}, \quad x = 2, \quad x = 6$

33. $y = 2x - x^2, \quad x = 1, \quad x = 3$

34. $y = x^2 - x + 1, \quad x = 0, \quad x = 1$

35. Given that

$$f(x) = \begin{cases} 3x^2 & \text{if } 0 \le x \le 2 \\ 16 - 2x & \text{if } x \ge 2 \end{cases}$$

determine the area of the region bounded by the graph of $y = f(x)$, the x-axis, and the line $x = 3$. Include a sketch of the region.

36. Under conditions of a continuous uniform distribution, a topic in statistics, the proportion of persons with incomes between a and t, where $a \le t \le b$, is the area of the region between the curve $y = 1/(b - a)$ and the x-axis from $x = a$ to $x = t$. Sketch the graph of the curve and determine the area of the given region.

***37.** Suppose $f(x) = x/8$, where $0 \le x \le 4$. If f is a density function (refer to Example 5), find each of the following:

(a) $P(0 \le x \le 1)$
(b) $P(2 \le x \le 4)$
(c) $P(x \ge 3)$

38. Suppose $f(x) = 3(1 - x)^2$, where $0 \le x \le 1$. If f is a density function (refer to Example 5), find each of the following:

(a) $P\left(\frac{1}{3} \le x \le \frac{2}{3}\right)$
(b) $P\left(\frac{1}{3} \le x \le \frac{3}{4}\right)$
(c) $P\left(x \le \frac{1}{3}\right)$
(d) $P\left(x \ge \frac{1}{3}\right)$ using your result from part (c)

39. Suppose $f(x) = 1/x$, where $e \le x \le e^2$. If f is a density function (refer to Example 5), find each of the following:

(a) $P(3 \le x \le 5)$
(b) $P(x \le 4)$
(c) $P(x \ge 3)$
(d) Verify that $P(e \le x \le e^2) = 1$.

40. (a) Let r be a real number, where $r > 1$. Evaluate

$$\int_1^r \frac{1}{x^2}\,dx$$

(b) Your answer to part (a) can be interpreted as the area of a certain region of the plane. Sketch this region.

(c) Evaluate $\lim\limits_{r \to \infty} \left(\int_1^r \frac{1}{x^2}\,dx \right)$.

(d) Your answer to part (c) can be interpreted as the area of a certain region of the plane. Sketch this region.

In each of Problems 41–44, use definite integration to estimate the area of the region bounded by the given curve, the x-axis, and the given lines. Round your answer to two decimal places.

41. $y = \dfrac{1}{x^2 + 1}, \quad x = -2, \quad x = 1$

42. $y = \dfrac{x}{\sqrt{x + 5}}, \quad x = 2, \quad x = 7$

43. $y = x^4 - 2x^3 - 2, \quad x = 1, \quad x = 3$

44. $y = 1 + 3x - x^4$

OBJECTIVE *To find the area of a region bounded by two or more curves by using either vertical or horizontal strips.*

14.11 Area between Curves

Vertical Elements

We shall now find the area of a region enclosed by several curves. As before, our procedure will be to draw a sample element of area and use the definite integral to "add together" the areas of all such elements.

For example, consider the area of the region in Figure 14.33 that is bounded on the top and bottom by the curves $y = f(x)$ and $y = g(x)$ and on the sides

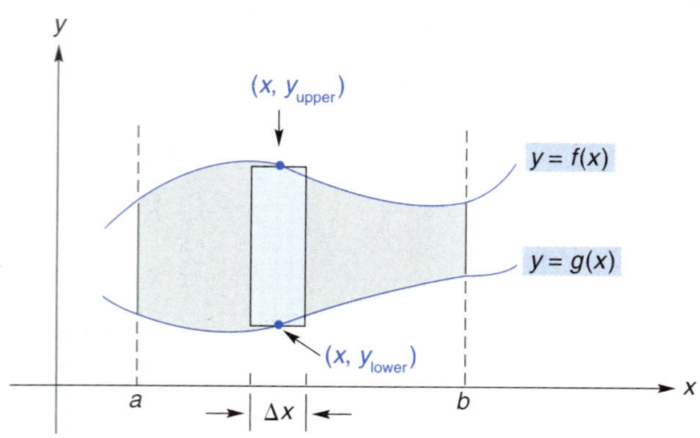

FIGURE 14.33 Region between curves.

by the lines $x = a$ and $x = b$. The width of the indicated vertical element is Δx, and the height is the y-value of the upper curve minus the y-value of the lower curve, which we shall write as $y_{\text{upper}} - y_{\text{lower}}$. Thus, the area of the element is

$$[y_{\text{upper}} - y_{\text{lower}}]\Delta x$$

or

$$[f(x) - g(x)]\Delta x$$

Summing the areas of all such elements from $x = a$ to $x = b$ by the definite integral gives the area of the region:

$$\sum [f(x) - g(x)]\Delta x \to \int_a^b [f(x) - g(x)]\,dx = \text{area}$$

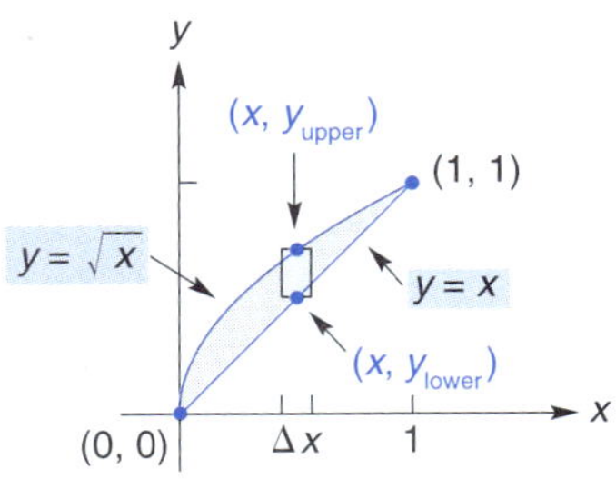

FIGURE 14.34 Diagram for Example 1.

EXAMPLE 1 **Finding an Area between Two Curves**

Find the area of the region bounded by the curves $y = \sqrt{x}$ and $y = x$.

Solution: A sketch of the region appears in Figure 14.34. To determine where the curves intersect, we solve the system formed by the equations $y = \sqrt{x}$ and $y = x$. Eliminating y by substitution, we obtain

$$\sqrt{x} = x$$

$$x = x^2 \qquad \text{(squaring both sides)}$$

$$0 = x^2 - x = x(x - 1)$$

$$x = 0 \quad \text{or} \quad x = 1$$

Since we "squared both sides," we must check the solutions found with respect to the *original* equation. It is easily determined that both $x = 0$ and $x = 1$ are solutions of $\sqrt{x} = x$. If $x = 0$, then $y = 0$; if $x = 1$, then $y = 1$. Thus, the curves intersect at (0, 0) and (1, 1). The width of the indicated element of area is Δx. The height is the y-value on the upper curve minus the y-value on the lower curve:

$$y_{\text{upper}} - y_{\text{lower}} = \sqrt{x} - x$$

It should be obvious to you that knowing the points of intersection is important in determining the limits of integration.

Hence, the area of the element is $(\sqrt{x} - x)\Delta x$. Summing the areas of all such elements from $x = 0$ to $x = 1$ by the definite integral, we get the area of the entire region:

$$\text{area} = \int_0^1 (\sqrt{x} - x)\,dx$$

$$= \int_0^1 (x^{1/2} - x)\,dx = \left(\frac{x^{3/2}}{\frac{3}{2}} - \frac{x^2}{2}\right)\Bigg|_0^1$$

$$= \left(\frac{2}{3} - \frac{1}{2}\right) - (0 - 0) = \frac{1}{6}$$

▸ **Now work Problem 9.**

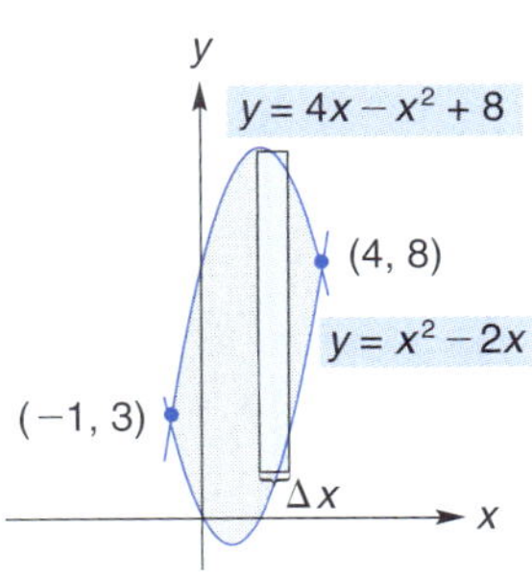

FIGURE 14.35 Diagram for Example 2.

EXAMPLE 2 **Finding an Area between Two Curves**

Find the area of the region bounded by the curves $y = 4x - x^2 + 8$ and $y = x^2 - 2x$.

Solution: A sketch of the region appears in Figure 14.35. To find where the curves intersect, we solve the system of equations $y = 4x - x^2 + 8$ and

$y = x^2 - 2x$:

$$\begin{aligned} 4x - x^2 + 8 &= x^2 - 2x, \\ -2x^2 + 6x + 8 &= 0, \\ x^2 - 3x - 4 &= 0, \\ (x+1)(x-4) &= 0, \qquad \text{(factoring)} \\ x = -1 \quad &\text{or} \quad x = 4 \end{aligned}$$

When $x = -1$, then $y = 3$; when $x = 4$, then $y = 8$. Thus, the curves intersect at $(-1, 3)$ and $(4, 8)$. The width of the indicated element is Δx. The height is the y-value on the upper curve minus the y-value on the lower curve:

$$y_{\text{upper}} - y_{\text{lower}} = (4x - x^2 + 8) - (x^2 - 2x)$$

Therefore, the area of the element is

$$[(4x - x^2 + 8) - (x^2 - 2x)]\Delta x = (-2x^2 + 6x + 8)\Delta x$$

Summing all such areas from $x = -1$ to $x = 4$, we have

$$\text{area} = \int_{-1}^{4} (-2x^2 + 6x + 8)\, dx = 41\tfrac{2}{3}$$

▶ **Now work Problem 25.**

EXAMPLE 3 **Area of a Region Having Two Different Upper Curves**

Find the area of the region between the curves $y = 9 - x^2$ and $y = x^2 + 1$ from $x = 0$ to $x = 3$.

Solution: The region is sketched in Figure 14.36. The curves intersect when

$$\begin{aligned} 9 - x^2 &= x^2 + 1 \\ 8 &= 2x^2 \\ 4 &= x^2 \\ x &= \pm 2 \qquad \text{(two solutions)} \end{aligned}$$

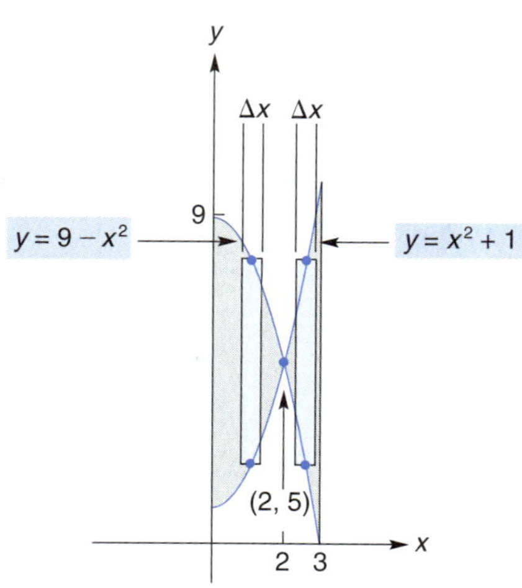

FIGURE 14.36 y_{upper} is $9 - x^2$ on $[0, 2]$ and is $x^2 + 1$ on $[2, 3]$.

When $x = \pm 2$, then $y = 5$, so the points of intersection are $(\pm 2, 5)$. Because we are interested in the region from $x = 0$ to $x = 3$, the intersection point that is of concern to us is $(2, 5)$. Notice in Figure 14.36 that in the region to the *left* of the intersection point $(2, 5)$, an element has

$$y_{\text{upper}} = 9 - x^2 \quad \text{and} \quad y_{\text{lower}} = x^2 + 1$$

but for an element to the *right* of $(2, 5)$ the reverse is true, namely,

$$y_{\text{upper}} = x^2 + 1 \quad \text{and} \quad y_{\text{lower}} = 9 - x^2$$

Thus, from $x = 0$ to $x = 2$, the area of an element is

$$\begin{aligned}(y_{\text{upper}} - y_{\text{lower}})\,\Delta x &= [(9 - x^2) - (x^2 + 1)]\,\Delta x\\ &= (8 - 2x^2)\,\Delta x,\end{aligned}$$

but from $x = 2$ to $x = 3$, it is

$$\begin{aligned}(y_{\text{upper}} - y_{\text{lower}})\,\Delta x &= [(x^2 + 1) - (9 - x^2)]\,\Delta x\\ &= (2x^2 - 8)\,\Delta x.\end{aligned}$$

Therefore, to find the area of the entire region, we need *two* integrals:

$$\begin{aligned}\text{area} &= \int_0^2 (8 - 2x^2)\,dx + \int_2^3 (2x^2 - 8)\,dx\\ &= \left(8x - \frac{2x^3}{3}\right)\Bigg|_0^2 + \left(\frac{2x^3}{3} - 8x\right)\Bigg|_2^3\\ &= \left[\left(16 - \frac{16}{3}\right) - 0\right] + \left[(18 - 24) - \left(\frac{16}{3} - 16\right)\right]\\ &= \frac{46}{3}\end{aligned}$$

▶ **Now work Problem 33.**

Horizontal Elements

Sometimes area can more easily be determined by summing areas of horizontal elements rather than vertical elements. In the following example, an area will be found by both methods. In each case, the element of area determines the form of the integral.

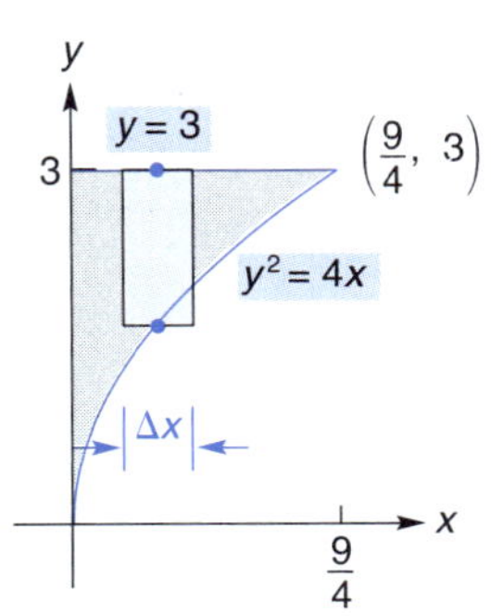

FIGURE 14.37 Vertical element of area.

EXAMPLE 4 The Methods of Vertical Elements and Horizontal Elements

Find the area of the region bounded by the curve $y^2 = 4x$ and the lines $y = 3$ and $x = 0$ (the y-axis).

Solution: The region is sketched in Figure 14.37. When the curves $y = 3$ and $y^2 = 4x$ intersect, $9 = 4x$, so $x = \frac{9}{4}$. Thus, the intersection point is $(\frac{9}{4}, 3)$. Since the width of the vertical strip is Δx, we integrate with respect to the variable x. Accordingly, y_{upper} and y_{lower} must be expressed as functions of x. For the lower curve, $y^2 = 4x$, we have $y = \pm 2\sqrt{x}$. But $y \geq 0$ for the portion of this curve that bounds the region, so we use $y = 2\sqrt{x}$. The upper curve is $y = 3$. Hence, the height of the strip is

$$y_{\text{upper}} - y_{\text{lower}} = 3 - 2\sqrt{x}$$

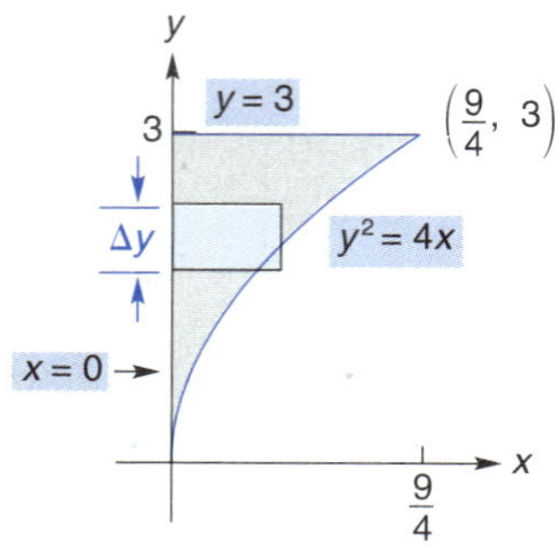

FIGURE 14.38 Horizontal element of area.

With horizontal elements, the width is Δy, not Δx.

Therefore, the strip has an area of $(3 - 2\sqrt{x})\Delta x$, and we wish to sum all such areas from $x = 0$ to $x = \frac{9}{4}$. We have

$$\text{area} = \int_0^{9/4} (3 - 2\sqrt{x})\, dx = \left(3x - \frac{4x^{3/2}}{3}\right)\Bigg|_0^{9/4}$$
$$= \left[3\left(\frac{9}{4}\right) - \frac{4}{3}\left(\frac{9}{4}\right)^{3/2}\right] - (0)$$
$$= \frac{27}{4} - \frac{4}{3}\left[\left(\frac{9}{4}\right)^{1/2}\right]^3 = \frac{27}{4} - \frac{4}{3}\left(\frac{3}{2}\right)^3 = \frac{9}{4}$$

Let us now approach this problem from the point of view of a **horizontal element of area** (or **horizontal strip**) as shown in Figure 14.38. The width of the element is Δy. The length of the element is the *x-value on the rightmost curve minus the x-value on the leftmost curve*. Thus, the area of the element is

$$(x_{\text{right}} - x_{\text{left}})\, \Delta y$$

We wish to sum all such areas from $y = 0$ to $y = 3$:

$$\sum (x_{\text{right}} - x_{\text{left}})\, \Delta y \rightarrow \int_0^3 (x_{\text{right}} - x_{\text{left}})\, dy$$

Since the variable of integration is y, we must express x_{right} and x_{left} as functions of y. The rightmost curve is $y^2 = 4x$ or, equivalently, $x = y^2/4$. The left curve is $x = 0$. Thus,

$$\text{area} = \int_0^3 (x_{\text{right}} - x_{\text{left}})\, dy = \int_0^3 \left(\frac{y^2}{4} - 0\right) dy = \frac{y^3}{12}\Bigg|_0^3 = \frac{9}{4}$$

Note that for this region, horizontal strips make the definite integral easier to evaluate (and set up) than an integral with vertical strips. In any case, remember that **the limits of integration are those limits for the variable of integration.**

▸ **Now work Problem 23.**

EXAMPLE 5 Advantage of Horizontal Elements

Find the area of the region bounded by the graphs of $y^2 = x$ and $x - y = 2$.

Solution: The region is sketched in Figure 14.39. The curves intersect when $y^2 - y = 2$. Thus, $y^2 - y - 2 = 0$; equivalently, $(y + 1)(y - 2) = 0$, from which it follows that $y = -1$ or $y = 2$. This gives the intersection points $(1, -1)$ and $(4, 2)$. Let us try vertical elements of area. [See Figure 14.39(a).] Solving $y^2 = x$ for y

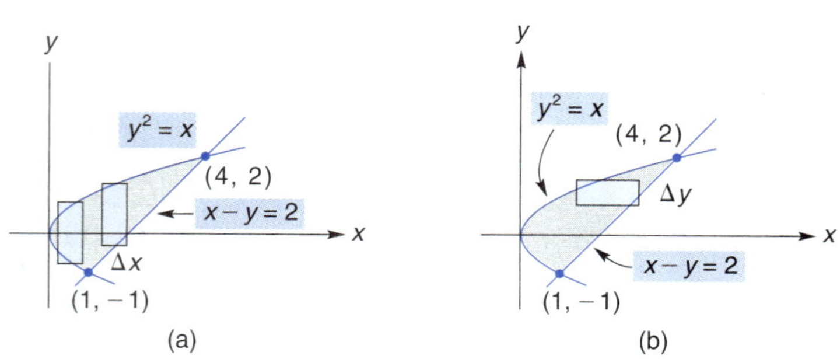

FIGURE 14.39 Region of Example 5 with vertical and horizontal elements.

gives $y = \pm\sqrt{x}$. As seen in Figure 14.39(a), to the *left* of $x = 1$, the upper end of the element lies on $y = \sqrt{x}$ and the lower end lies on $y = -\sqrt{x}$. To the *right* of $x = 1$, the upper curve is $y = \sqrt{x}$ and the lower curve is $x - y = 2$ (or $y = x - 2$). Thus, with vertical strips, *two* integrals are needed to evaluate the area:

$$\text{area} = \int_0^1 [\sqrt{x} - (-\sqrt{x})]\,dx + \int_1^4 [\sqrt{x} - (x - 2)]\,dx$$

Perhaps the use of horizontal strips can simplify our work. In Figure 14.39(b), the width of the strip is Δy. The rightmost curve is *always* $x - y = 2$ (or $x = y + 2$), and the leftmost curve is always $y^2 = x$ (or $x = y^2$). Therefore, the area of the horizontal strip is $[(y + 2) - y^2]\,\Delta y$, so the total area is

$$\text{area} = \int_{-1}^{2} (y + 2 - y^2)\,dy = \frac{9}{2}$$

Clearly, the use of horizontal strips is the most desirable approach to solving the problem. Only a single integral is needed, and it is much simpler to compute.

▶ **Now work Problem 19.**

TECHNOLOGY

Problem: Estimate the area of the region bounded by the graphs of

$$y = x^4 - 2x^3 - 2 \quad \text{and} \quad y = 1 + 2x - 2x^2$$

Solution: With a TI-83 Plus, we enter $x^4 - 2x^3 - 2$ as Y_1 and $1 + 2x - 2x^2$ as Y_2 and display their graphs. The region in question is shaded in Figure 14.40; y_{upper} corresponds to Y_2 and y_{lower} corresponds to Y_1. Using vertical strips, we have

$$\text{area} = \int_A^B (Y_2 - Y_1)\,dx$$

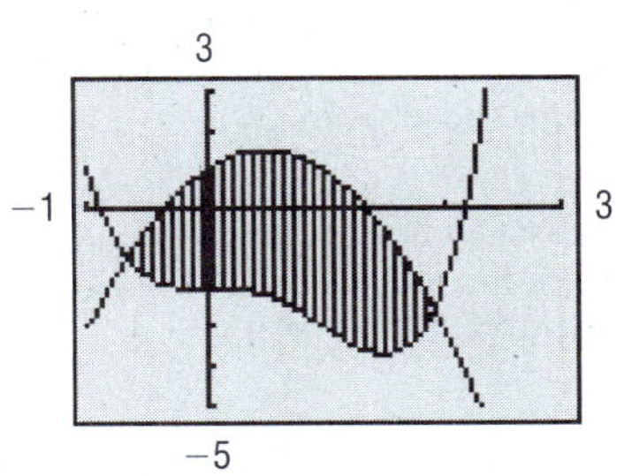

FIGURE 14.40 Graphs of $Y_1(y_{\text{lower}})$ and $Y_2(y_{\text{upper}})$.

where A and B are the x-values of the intersection points in Quadrants III and IV, respectively. With the intersection feature we find A, as indicated in Figure 14.41. This value of x is then stored from the home screen as A. (See Figure 14.42.) Similarly, we find the x-value of the intersection point in Quadrant IV, which we store as B. Using the "fnInt(" command (Figure 14.42), we estimate the area of the region to be 7.54 square units.

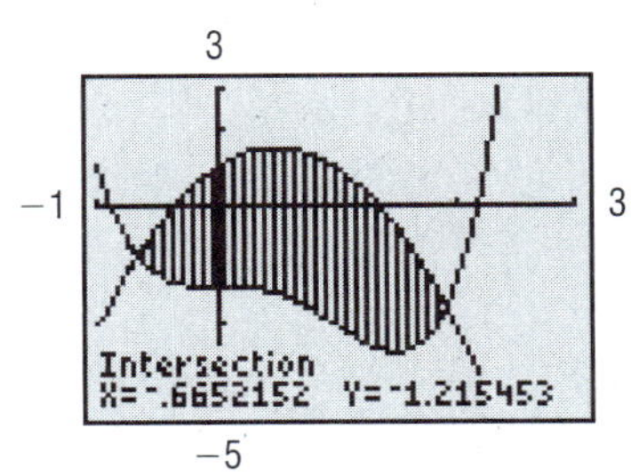

FIGURE 14.41 Intersection point in Quadrant III.

FIGURE 14.42 Storing x-values of intersection points and estimating area.

Exercise 14.11

In Problems 1–6, express the area of the shaded region in terms of an integral (or integrals). Do not evaluate your expression.

1. See Figure 14.43.

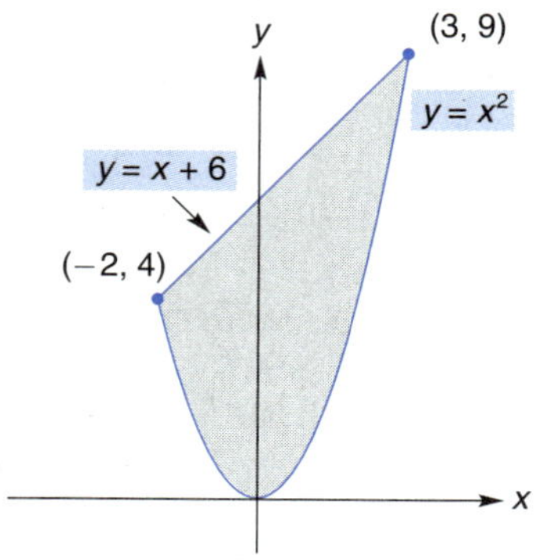

FIGURE 14.43 Region for Problem 1.

2. See Figure 14.44.

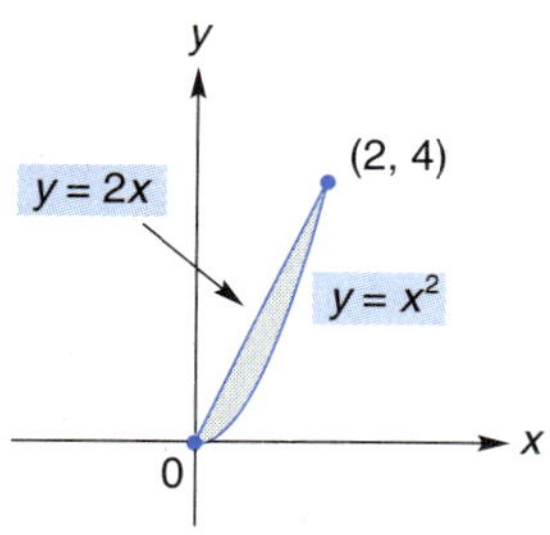

FIGURE 14.44 Region for Problem 2.

3. See Figure 14.45.

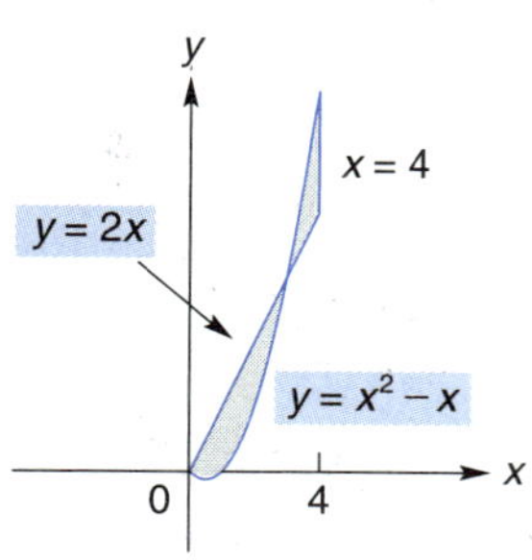

FIGURE 14.45 Region for Problem 3.

4. See Figure 14.46.

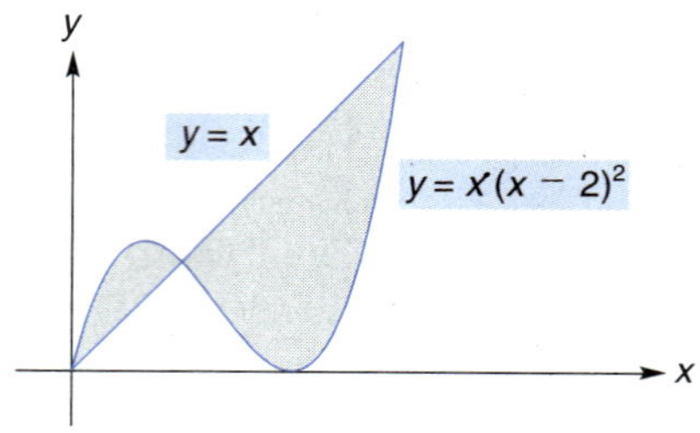

FIGURE 14.46 Region for Problem 4.

5. See Figure 14.47.

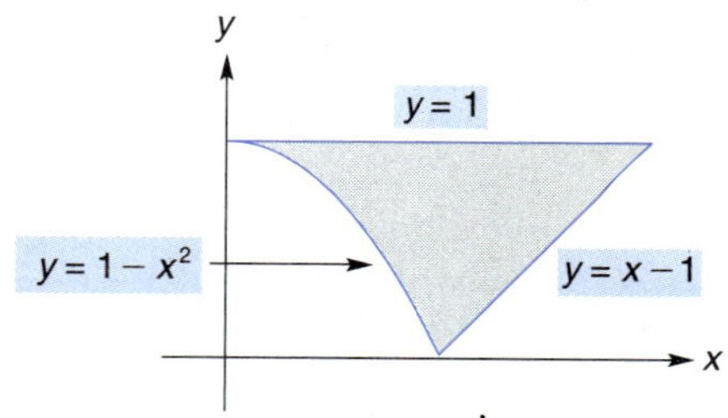

FIGURE 14.47 Region for Problem 5.

6. See Figure 14.48.

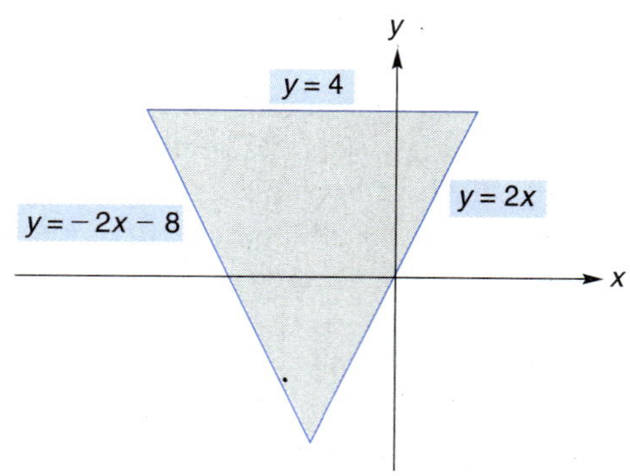

FIGURE 14.48 Region for Problem 6.

7. Express, in terms of a single integral, the total area of the region to the right of the line $x = 1$ that is between the curves $y = x^2 - 5$ and $y = 7 - 2x^2$. Do *not* evaluate the integral.

8. Express, in terms of a single integral, the total area of the region in the first quadrant bounded by the x-axis and the graphs of $y^2 = x$ and $2y = 3 - x$. Do *not* evaluate the integral.

In Problems 9–32, find the area of the region bounded by the graphs of the given equations. Be sure to find any needed points of intersection. Consider whether the use of horizontal strips makes the integral simpler than when vertical strips are used.

*9. $y = x^2, \quad y = 2x$

10. $y = x, \quad y = -x + 3, \quad y = 0$

11. $y = x^2, \quad x = 0, \quad y = 4(x \geq 0)$

12. $y = x^2 + 1, \quad y = x + 3$

13. $y = 10 - x^2, \quad y = 4$

14. $y^2 = x + 1, \quad x = 1$

15. $x = 8 + 2y, \quad x = 0, \quad y = -1, \quad y = 3$

16. $y = x - 4, \quad y^2 = 2x$

17. $y = 4 - x^2, \quad y = -3x$

18. $x = y^2 + 2, \quad x = 6$

*19. $y^2 = x, \quad y = x - 2$

20. $y = x^3, \quad y = x + 6, \quad x = 0.$
[*Hint:* The only real root of $x^3 - x - 6 = 0$ is 2.]

21. $2y = 4x - x^2, \quad 2y = x - 4$

22. $y = \sqrt{x}, \quad y = x^2$

*23. $y^2 = x, \quad 3x - 2y = 1$

24. $y = 2 - x^2, \quad y = x$

*25. $y = 8 - x^2, \quad y = x^2, \quad x = -1, \quad x = 1$

26. $y^2 = 6 - x, \quad 3y = x + 12$

27. $y = x^2, \quad y = 2, \quad y = 5$

28. $y = x^3 - x, \quad x$-axis

29. $y = x^3 - 1, \quad y = x - 1$

30. $y = x^3, \quad y = \sqrt{x}$

31. $4x + 4y + 17 = 0, \quad y = \dfrac{1}{x}$

32. $y^2 = -x - 2, \quad x - y = 5, \quad y = -1, \quad y = 1$

*33. Find the area of the region that is between the curves

$$y = x - 1 \quad \text{and} \quad y = 5 - 2x$$

from $x = 0$ to $x = 4$.

34. Find the area of the region that is between the curves

$$y = x^2 - 4x + 4 \quad \text{and} \quad y = 10 - x^2$$

from $x = 2$ to $x = 4$.

35. Lorentz Curve A *Lorentz curve* is used in studying income distributions. If x is the cumulative percentage of income recipients, ranked from poorest to richest, and y is the cumulative percentage of income, then equality of income distribution is given by the line $y = x$ in Figure 14.49, where x and y are expressed as decimals. For example, 10% of the people receive 10% of total income, 20% of the people receive 20% of the income, and so on. Suppose the actual distribution is given by the Lorentz curve defined by

$$y = \frac{20}{21}x^2 + \frac{1}{21}x$$

Note, for example, that 30% of the people receive only 10% of total income. The degree of deviation from

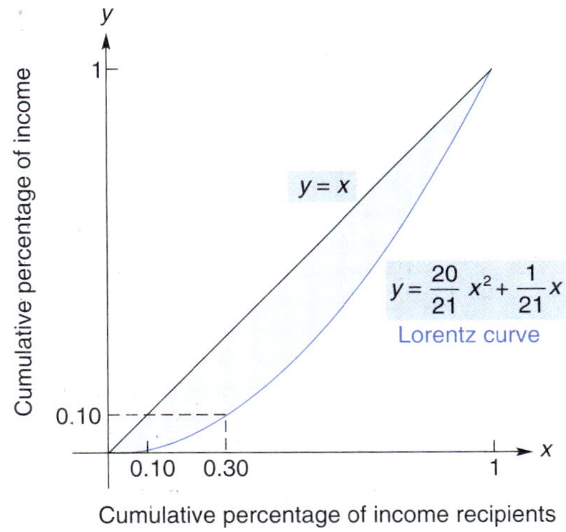

FIGURE 14.49 Diagram for Problem 35.

equality is measured by the *coefficient of inequality*[15] for a Lorentz curve. This coefficient is defined to be the area between the curve and the diagonal, divided by the area under the diagonal:

$$\frac{\text{area between curve and diagonal}}{\text{area under diagonal}}$$

For example, when all incomes are equal, the coefficient of inequality is zero. Find the coefficient of inequality for the Lorentz curve just defined.

36. Lorentz curve Find the coefficient of inequality as in Problem 35 for the Lorentz curve defined by $y = \frac{11}{12}x^2 + \frac{1}{12}x$.

37. Find the area of the region bounded by the graphs of the equations $y^2 = 3x$ and $y = mx$, where m is a positive constant.

38. **(a)** Find the area of the region bounded by the graphs of $y = x^2 - 1$ and $y = 2x + 2$.
(b) What percentage of the area in part (a) lies above the x-axis?

39. The region bounded by the curve $y = x^2$ and the line $y = 4$ is divided into two parts of equal area by the line $y = k$, where k is a constant. Find the value of k.

In each of Problems 40–44, estimate the area of the region bounded by the graphs of the given equations. Round your answer to two decimal places.

40. $y = x^2 - 2x - 1, \quad y = -\dfrac{2}{x}$

41. $y = \sqrt{25 - x^2}, \quad y = 7 - 2x - x^4$

42. $y = x^3 - 8x + 1, \quad y = x^2 - 5$

43. $y = x^5 - 3x^3 + 2x, \quad y = 3x^2 - 4$

44. $y = x^4 - 3x^3 - 15x^2 + 19x + 30, \quad y = x^3 + x^2 - 20x$

---

[15] G. Stigler, *The Theory of Price*, 3rd ed. (New York: Macmillan Company, 1966), pp. 293–94.

OBJECTIVE *To develop the economic concepts of consumers' surplus and producers' surplus, which are represented by areas.*

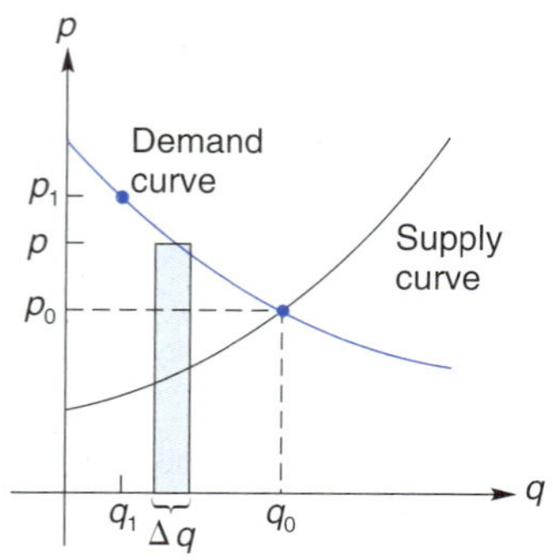

FIGURE 14.50 Supply and demand curves.

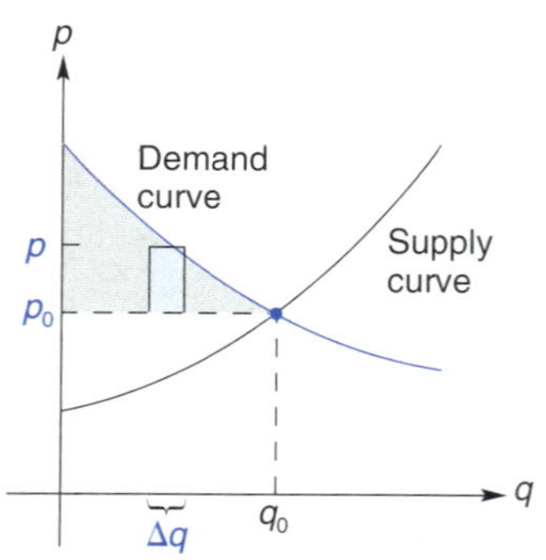

FIGURE 14.51 Benefit to consumers for Δq units.

14.12 Consumers' and Producers' Surplus

Determining the area of a region has applications in economics. Figure 14.50 shows a supply curve for a product. The curve indicates the price p per unit at which the manufacturer will sell (or supply) q units. The diagram also shows a demand curve for the product. This curve indicates the price p per unit at which consumers will purchase (or demand) q units. The point (q_0, p_0) where the two curves intersect is called the *point of equilibrium*. Here p_0 is the price per unit at which consumers will purchase the same quantity q_0 of a product that producers wish to sell at that price. In short, p_0 is the price at which stability in the producer-consumer relationship occurs.

Let us assume that the market is at equilibrium and the price per unit of the product is p_0. According to the demand curve, there are consumers who would be willing to pay *more* than p_0. For example, at the price per unit of p_1, consumers would buy q_1 units. These consumers are benefiting from the lower equilibrium price p_0.

The vertical strip in Figure 14.50 has area $p\Delta q$. This expression can also be thought of as the total amount of money that consumers would spend by buying Δq units of the product if the price per unit were p. Since the price is actually p_0, these consumers spend only $p_0\Delta q$ for the Δq units and thus benefit by the amount $p\Delta q - p_0\Delta q$. This expression can be written $(p - p_0)\Delta q$, which is the area of a rectangle of width Δq and height $p - p_0$. (See Figure 14.51.) Summing the areas of all such rectangles from $q = 0$ to $q = q_0$ by definite integration, we have

$$\int_0^{q_0} (p - p_0)\, dq$$

This integral, under certain conditions, represents the total gain to consumers who are willing to pay more than the equilibrium price. This total gain is called **consumers' surplus,** abbreviated CS. If the demand function is given by $p = f(q)$, then

$$\text{CS} = \int_0^{q_0} [f(q) - p_0]\, dq$$

Geometrically (see Figure 14.52), consumers' surplus is represented by the area between the line $p = p_0$ and the demand curve $p = f(q)$ from $q = 0$ to $q = q_0$.

Some of the producers also benefit from the equilibrium price, since they are willing to supply the product at prices *less* than p_0. Under certain conditions, the total gain to the producers is represented geometrically in Figure 14.53 by

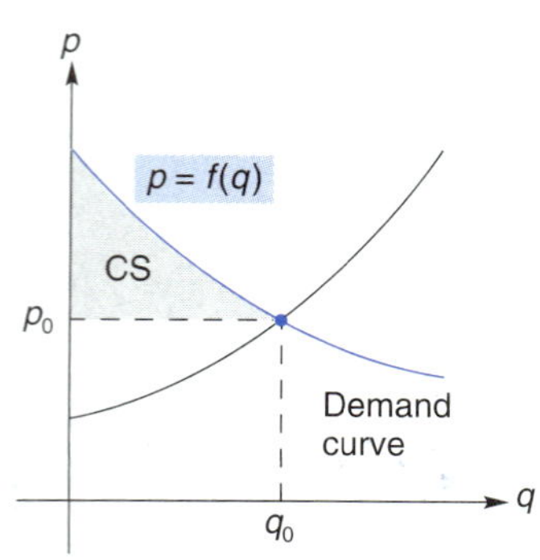

FIGURE 14.52 Consumers' surplus.

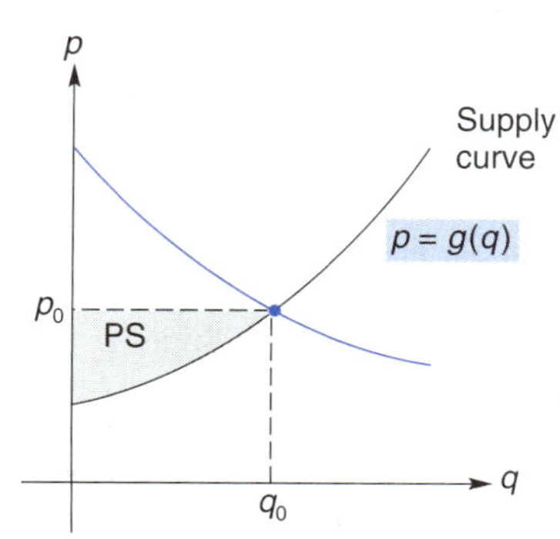

FIGURE 14.53 Producers' surplus.

the area between the line $p = p_0$ and the supply curve $p = g(q)$ from $q = 0$ to $q = q_0$. This gain, called **producers' surplus** and abbreviated PS, is given by

$$\text{PS} = \int_0^{q_0} [p_0 - g(q)]\,dq$$

EXAMPLE 1 **Finding Consumers' Surplus and Producers' Surplus**

The demand function for a product is

$$p = f(q) = 100 - 0.05q$$

where p is the price per unit (in dollars) for q units. The supply function is

$$p = g(q) = 10 + 0.1q$$

Determine consumers' surplus and producers' surplus under market equilibrium.

Solution: First we must find the equilibrium point (p_0, q_0) by solving the system formed by the functions $p = 100 - 0.05q$ and $p = 10 + 0.1q$. We thus equate the two expressions for p and solve:

$$\begin{aligned} 10 + 0.1q &= 100 - 0.05q \\ 0.15q &= 90 \\ q &= 600 \end{aligned}$$

When $q = 600$, then $p = 10 + 0.1(600) = 70$. Hence, $q_0 = 600$ and $p_0 = 70$. Consumers' surplus is

$$\begin{aligned} \text{CS} &= \int_0^{q_0} [f(q) - p_0]\,dq = \int_0^{600} (100 - 0.05q - 70)\,dq \\ &= \left(30q - 0.05\frac{q^2}{2}\right)\Bigg|_0^{600} = 9000 \end{aligned}$$

Producers' surplus is

$$\begin{aligned} PS &= \int_0^{q_0} [p_0 - g(q)]\,dq = \int_0^{600} [70 - (10 + 0.1q)]\,dq \\ &= \left(60q - 0.1\frac{q^2}{2}\right)\Bigg|_0^{600} = 18{,}000 \end{aligned}$$

Therefore, consumers' surplus is \$9000 and producers' surplus is \$18,000.

▸ **Now work Problem 1.**

EXAMPLE 2 **Using Horizontal Strips to Find Consumers' Surplus and Producers' Surplus**

The demand equation for a product is

$$q = f(p) = \frac{90}{p} - 2$$

and the supply equation is $q = g(p) = p - 1$. Determine consumers' surplus and producers' surplus when market equilibrium has been established.

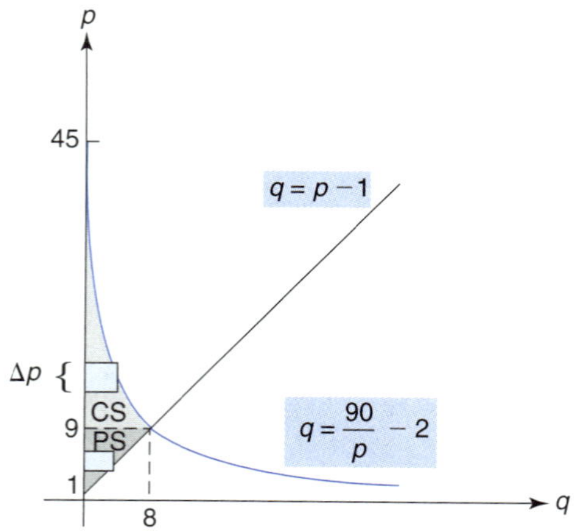

FIGURE 14.54 Diagram for Example 2.

Solution: Determining the equilibrium point, we have

$$p - 1 = \frac{90}{p} - 2$$
$$p^2 + p - 90 = 0$$
$$(p + 10)(p - 9) = 0$$

Thus, $p_0 = 9$, so $q_0 = 9 - 1 = 8$. (See Figure 14.54.) Note that the demand equation expresses q as a function of p. Since consumers' surplus can be considered an area, this area can be determined by means of horizontal strips of width Δp and length $q = f(p)$. The areas of these strips are summed from $p = 9$ to $p = 45$ by integrating with respect to p:

$$\text{CS} = \int_9^{45} \left(\frac{90}{p} - 2\right) dp = (90 \ln |p| - 2p)\Big|_9^{45}$$
$$= 90 \ln 5 - 72 \approx 72.85$$

Using horizontal strips for producers' surplus, we have

$$\text{PS} = \int_1^9 (p - 1)\, dp = \frac{(p-1)^2}{2}\Big|_1^9 = 32$$

▸ **Now work Problem 5.**

Exercise 14.12

In Problems 1–6, the first equation is a demand equation and the second is a supply equation of a product. In each case, determine consumers' surplus and producers' surplus under market equilibrium.

*__1.__ $p = 22 - 0.8q$
$p = 6 + 1.2q$

2. $p = 2200 - q^2$
$p = 400 + q^2$

3. $p = \dfrac{50}{q + 5}$
$p = \dfrac{q}{10} + 4.5$

4. $p = 400 - q^2$
$p = 20q + 100$

*__5.__ $q = 100(10 - p)$
$q = 80(p - 1)$

6. $q = \sqrt{100 - p}$
$q = \dfrac{p}{2} - 10$

7. The demand equation for a product is

$$q = 10\sqrt{100 - p}$$

Calculate consumers' surplus under market equilibrium that occurs at a price of \$84.

8. The demand equation for a product is

$$q = 400 - p^2$$

and the supply equation is

$$p = \frac{q}{60} + 5$$

Find producers' surplus and consumers' surplus under market equilibrium.

9. The demand equation for a product is $p = 2^{11-q}$, and the supply equation is $p = 2^{q+1}$, where p is the price per unit (in hundreds of dollars) when q units are demanded or supplied. Determine, to the nearest thousand dollars, consumers' surplus under market equilibrium.

10. The demand equation for a product is

$$(p + 10)(q + 20) = 1000$$

and the supply equation is

$$q - 4p + 10 = 0$$

(a) Verify, by substitution, that market equilibrium occurs when $p = 10$ and $q = 30$.
(b) Determine consumers' surplus under market equilibrium.

11. The demand equation for a product is

$$p = 60 - \frac{50q}{\sqrt{q^2 + 3600}}$$

and the supply equation is

$$p = 10\ln(q + 20) - 26$$

Determine consumers' surplus and producers' surplus under market equilibrium. Round your answers to the nearest integer.

12. **Producers' Surplus** The supply function for a product is given by the following table, where p is the price per unit (in dollars) at which q units are supplied to the market:

q	0	10	20	30	40	50
p	38	59	69	75	80	84

Use the trapezoidal rule to estimate the producers' surplus if the selling price is \$80.

14.13 Review

Important Terms and Symbols

Section 14.1	differential
Section 14.2	antiderivative indefinite integral $\int f(x)\,dx$ integral sign integrand variable of integration constant of integration
Section 14.3	initial condition
Section 14.4	power rule for integration
Section 14.6	$\sum$ index of summation limits of summation
Section 14.7	definite integral $\int_a^b f(x)\,dx$ lower limit of integration upper limit of integration
Section 14.8	Fundamental Theorem of Integral Calculus $F(x)\vert_a^b$
Section 14.9	trapezoidal rule Simpson's rule
Section 14.10	vertical element of area (vertical strip)
Section 14.11	horizontal element of area (horizontal strip)
Section 14.12	consumers' surplus producers' surplus

Summary

An antiderivative of a function f is a function F such that $F'(x) = f(x)$. Any two antiderivatives of f differ at most by a constant. The most general antiderivative of f is called the indefinite integral of f and is denoted $\int f(x)\,dx$. Thus,

$$\int f(x)\,dx = F(x) + C$$

where C is called the constant of integration.

Some basic integration formulas are as follows:

$$\int k\,dx = kx + C \qquad k \text{ a constant}$$

$$\int x^n\,dx = \frac{x^{n+1}}{n+1} + C \qquad n \neq -1$$

$$\int e^x\,dx = e^x + C$$

$$\int kf(x)\,dx = k\int f(x)\,dx \qquad k \text{ a constant}$$

and

$$\int [f(x) \pm g(x)]\,dx = \int f(x)\,dx \pm \int g(x)\,dx$$

Another formula is the power rule for integration:

$$\int u^n\,du = \frac{u^{n+1}}{n+1} + C, \quad \text{if } n \neq -1$$

Here u represents a differentiable function of x, and du is its differential. In applying the power rule to a given integral, it is important that the integral be written in a form that precisely matches the power rule. Other integration formulas are

$$\int e^u\,du = e^u + C$$

and

$$\int \frac{1}{u}\,du = \ln|u| + C \qquad u \neq 0$$

If the rate of change of a function f is known—that is, if f' is known—then f is an antiderivative of f'. In addition, if we know that f satisfies an initial condition, then we can find the particular antiderivative. For example, if a marginal-cost function dc/dq is given to us, then by integration, we can find the most general form of c. That form involves a constant of integration. However, if we are also given fixed costs (that is, costs involved when $q = 0$), then we can determine the value of the constant of integration and thus find the particular cost function c. Similarly, if we are given a marginal-revenue function dr/dq, then by integration and by using the fact that $r = 0$ when $q = 0$, we can determine the particular revenue function r. Once r is known, the corresponding demand equation can be found by using the equation $p = r/q$.

Sigma notation is convenient for representing sums. This notation is especially useful in determining areas. To find the area of the region bounded by $y = f(x)$ [where $f(x) \geq 0$ and f is continuous] and the x-axis from $x = a$ to $x = b$, we divide the interval $[a, b]$ into n subintervals of equal length Δx. If x_i is the right-hand endpoint of an arbitrary subinterval, then the product $f(x_i)\Delta x$ is the area of a rectangle. Denoting the sum of all such areas of rectangles for the n subintervals by S_n, we define the limit of S_n as $n \to \infty$ as the area of the entire region:

$$\lim_{n\to\infty} S_n = \lim_{n\to\infty} \sum_{i=1}^{n} f(x_i)\Delta x = \text{area}$$

If the restriction that $f(x) \geq 0$ is omitted, this limit is defined as the definite integral of f over $[a, b]$:

$$\lim_{n\to\infty} \sum_{i=1}^{n} f(x_i)\Delta x = \int_a^b f(x)\,dx$$

Instead of evaluating definite integrals by using limits, we may be able to employ the Fundamental Theorem of Integral Calculus. Mathematically,

$$\int_a^b f(x)\,dx = F(x)\Big|_a^b = F(b) - F(a)$$

where F is any antiderivative of f.

Some properties of the definite integral are

$$\int_a^b kf(x)\,dx = k\int_a^b f(x)\,dx \qquad k \text{ a constant}$$

$$\int_a^b [f(x) \pm g(x)]\,dx = \int_a^b f(x)\,dx \pm \int_a^b g(x)\,dx$$

and

$$\int_a^c f(x)\,dx = \int_a^b f(x)\,dx + \int_b^c f(x)\,dx$$

If the rate of change of a function f is known, then a change in function values of f can easily be found by the formula

$$\int_a^b f'(x)\,dx = f(b) - f(a)$$

If $f(x) \geq 0$ and is continuous on $[a, b]$, then the definite integral may be used to find the area of the region bounded by $y = f(x)$ and the x-axis from $x = a$ to $x = b$. The definite integral may also be used to find areas of more complicated regions. In these situations, an element of area should be drawn in the region. This will allow you to set up the proper definite integral. In some situations vertical elements should be considered, whereas in others horizontal elements are more advantageous.

One application of finding areas involves consumers' surplus and producers' surplus. Suppose the market for a product is at equilibrium and (q_0, p_0) is the equilibrium point (the point of intersection of the supply and demand curves for the product). Then consumers' surplus, CS, corresponds to the area from $q = 0$ to $q = q_0$, bounded above by the demand curve and below by the line $p = p_0$. Thus,

$$\text{CS} = \int_0^{q_0} [f(q) - p_0]\, dq$$

where f is the demand function. Producers' surplus, PS, corresponds to the area from $q = 0$ to $q = q_0$, bounded above by the line $p = p_0$ and below by the supply curve. Therefore,

$$\text{PS} = \int_0^{q_0} [p_0 - g(q)]\, dq$$

where g is the supply function.

Review Problems

Problem numbers shown in color indicate problems suggested for use as a practice chapter test.

In Problems 1–40, determine the integrals.

1. $\int (x^3 + 2x - 7)\, dx$

2. $\int dx$

3. $\int_0^8 (\sqrt{2x} + 2x)\, dx$

4. $\int \frac{4}{5 - 3x}\, dx$

5. $\int \frac{6}{(x+5)^3}\, dx$

6. $\int_4^{12} (y - 8)^{501}\, dy$

7. $\int \frac{6x^2 - 12}{x^3 - 6x + 1}\, dx$

8. $\int_0^2 xe^{4-x^2}\, dx$

9. $\int_0^1 \sqrt[3]{3t + 8}\, dt$

10. $\int \frac{4 - 2x}{7}\, dx$

11. $\int y(y+1)^2\, dy$

12. $\int_0^1 10^{-8}\, dx$

13. $\int \frac{\sqrt[4]{z} - \sqrt[3]{z}}{\sqrt{z}}\, dz$

14. $\int \frac{(0.5x - 0.1)^4}{0.4}\, dx$

15. $\int_1^3 \frac{2t^2}{3 + 2t^3}\, dt$

16. $\int \frac{4x^2 - x}{x}\, dx$

17. $\int x^2\sqrt{3x^3 + 2}\, dx$

18. $\int (2x^3 + x)(x^4 + x^2)^{3/4}\, dx$

19. $\int (e^{2y} - e^{-2y})\, dy$

20. $\int \frac{8x}{3\sqrt[3]{7 - 2x^2}}\, dx$

21. $\int \left(\frac{1}{x} + \frac{2}{x^2}\right) dx$

22. $\int_0^2 \frac{3e^{3x}}{1 + e^{3x}}\, dx$

23. $\int_{-2}^1 10(y^4 - y + 1)\, dy$

24. $\int_7^{70} dx$

25. $\int_{\sqrt{3}}^2 7x\sqrt{4 - x^2}\, dx$

26. $\int_0^1 (2x+1)(x^2+x)^4\, dx$

27. $\int_0^1 \left[2x - \frac{1}{(x+1)^{2/3}}\right] dx$

28. $\int_3^{27} 3(\sqrt{3x} - 2x + 1)\, dx$

29. $\int \frac{\sqrt{t} - 3}{t^2}\, dt$

30. $\int \frac{2z^2}{z - 1}\, dz$

31. $\int_{-1}^0 \frac{x^2 + 4x - 1}{x + 2}\, dx$

32. $\int \frac{(x^2 + 4)^2}{x^2}\, dx$

33. $\int 9\sqrt{x}\sqrt{x^{3/2} + 1}\, dx$

34. $\int \frac{e^{\sqrt{5x}}}{\sqrt{3x}}\, dx$

35. $\int_1^e \frac{e^{\ln x}}{x^2}\, dx$

36. $\int \frac{6x^2 + 4}{e^{x^3 + 2x}}\, dx$

37. $\int \frac{(1 + e^{3x})^2}{e^{-3x}}\, dx$

38. $\int \frac{3}{e^{3x}(6 + e^{-3x})^2}\, dx$

39. $\int 3\sqrt{10^{3x}}\, dx$

40. $\int \frac{5x^3 + 15x^2 + 37x + 3}{x^2 + 3x + 7}\, dx$

In Problems 41 and 42, find y, subject to the given condition.

41. $y' = e^{2x} + 3, \quad y(0) = -\frac{1}{2}$

42. $y' = \frac{x + 3}{x}, \quad y(1) = 5$

In Problems 43–50, determine the area of the region bounded by the given curve, the x-axis, and the given lines.

43. $y = x^2 - 1, \quad x = 2$

44. $y = 4e^x, \quad x = 0, \quad x = 3$

45. $y = \sqrt{x + 4}, \quad x = 0$

46. $y = x^2 - x - 6, \quad x = -4, \quad x = 3$

47. $y = 5x - x^2$

48. $y = \sqrt[4]{x}, \quad x = 1, \quad x = 16$

49. $y = \frac{1}{x} + 3, \quad x = 1, \quad x = 3$

50. $y = x^3 - 1, \quad x = -1$

In Problems 51–58, find the area of the region bounded by the given curves.

51. $y^2 = 4x, \quad x = 0, \quad y = 2$

52. $y = 3x^2 - 5, \quad x = 0, \quad y = 4$

53. $y = x^2 + 4x - 5, \quad y = 0$

54. $y = 2x^2, \quad y = x^2 + 9$

55. $y = x^2 - 2x, \quad y = 12 - x^2$

56. $y = \sqrt{x}, \quad x = 0, \quad y = 3$

57. $y = \ln x, \quad x = 0, \quad y = 0, \quad y = 1$

58. $y = 2 - x, \quad y = x - 3, \quad y = 0, \quad y = 2$

59. Marginal Revenue If marginal revenue is given by

$$\frac{dr}{dq} = 100 - \frac{3}{2}\sqrt{2q}$$

determine the corresponding demand equation.

60. Marginal Cost If marginal cost is given by

$$\frac{dc}{dq} = q^2 + 7q + 6$$

and fixed costs are 2500, determine the total cost of producing six units. Assume that costs are in dollars.

61. Marginal Revenue A manufacturer's marginal-revenue function is

$$\frac{dr}{dq} = 275 - q - 0.3q^2$$

If r is in dollars, find the increase in the manufacturer's total revenue if production is increased from 10 to 20 units.

62. Marginal Cost A manufacturer's marginal-cost function is

$$\frac{dc}{dq} = \frac{1000}{\sqrt{3q + 70}}$$

If c is in dollars, determine the cost involved to increase production from 10 to 33 units.

63. Hospital Discharges For a group of hospitalized individuals, suppose the discharge rate is given by

$$f(t) = 0.008e^{-0.008t}$$

where $f(t)$ is the proportion discharged per day at the end of t days of hospitalization. What proportion of the group is discharged at the end of 100 days?

64. Business Expenses The total expenditures (in dollars) of a business over the next five years is given by

$$\int_0^5 4000e^{0.05t}\, dt$$

Evaluate the expenditures.

65. Find the area of the region between the curves $y = 9 - 2x$ and $y = x$ from $x = 0$ to $x = 4$.

66. Find the area of the region between the curves $y = x^2$ and $y = 4 - 3x$ from $x = -1$ to $x = 2$.

67. Consumers' and Producers' Surplus The demand equation for a product is

$$p = 0.01q^2 - 1.1q + 30$$

and the supply equation is

$$p = 0.01q^2 + 8$$

Determine consumers' surplus and producers' surplus when market equilibrium has been established.

68. Consumers' Surplus The demand equation for a product is

$$p = (q - 5)^2$$

and the supply equation is

$$p = q^2 + q + 3$$

where p (in thousands of dollars) is the price per 100 units when q hundred units are demanded or supplied. Determine consumers' surplus under market equilibrium.

69. Biology In a discussion of gene mutation,[16] the equation

$$\int_{q_0}^{q_n} \frac{dq}{q - \hat{q}} = -(u + v) \int_0^n dt$$

occurs, where u and v are gene mutation rates, the q's are gene frequencies, and n is the number of generations. Assume that all letters represent constants, except q and t. Integrate both sides and then use your result to show that

$$n = \frac{1}{u + v} \ln \left| \frac{q_0 - \hat{q}}{q_n - \hat{q}} \right|$$

70. Fluid Flow In studying the flow of a fluid in a tube of constant radius R, such as blood flow in portions of the body, one can think of the tube as consisting of concentric tubes of radius r, where $0 \leq r \leq R$. The velocity v of the fluid is a function of r and is given by[17]

$$v = \frac{(P_1 - P_2)(R^2 - r^2)}{4\eta l}$$

---

[16] W. B. Mather, *Principles of Quantitative Genetics* (Minneapolis: Burgess Publishing Company, 1964).

[17] R. W. Stacy et al., *Essentials of Biological and Medical Physics* (New York: McGraw-Hill Book Company, 1955).

where P_1 and P_2 are pressures at the ends of the tube, η (a Greek letter read "eta") is the fluid viscosity, and l is the length of the tube. The volume rate of flow through the tube, Q, is given by

$$Q = \int_0^R 2\pi r v \, dr$$

Show that $Q = \dfrac{\pi R^4(P_1 - P_2)}{8\eta l}$. Note that R occurs as a factor to the fourth power. Thus, doubling the radius of the tube has the effect of increasing the flow by a factor of 16. The formula that you derived for the volume rate of flow is called *Poiseuille's law,* after the French physiologist Jean Poiseuille.

71. Inventory In a discussion of inventory, Barbosa and Friedman[18] refer to the function

$$g(x) = \frac{1}{k}\int_1^{1/x} ku^r \, du$$

where k and r are constants, $k > 0$ and $r > -2$, and $x > 0$. Verify the claim that

$$g'(x) = -\frac{1}{x^{r+2}}$$

[*Hint:* Consider two cases: when $r \neq -1$ and when $r = -1$.]

In Problems 72–74, estimate the area of the region bounded by the given curves. Round your answer to two decimal places.

72. $y = x^3 + 9x^2 + 14x - 24, \quad y = 0$

73. $y = x^3 - 2x - 3, \quad y = 4 + 2x - 3x^2$

74. $y = x^3 + x^2 - 5x - 3, \quad y = x^2 + 2x + 3$

75. The demand equation for a product is

$$p = \frac{200}{\sqrt{q + 20}}$$

and the supply equation is

$$p = 2\ln(q + 10) + 5$$

Determine consumers' surplus and producers' surplus under market equilibrium. Round your answers to the nearest integer.

[18] L. C. Barbosa and M. Friedman, "Deterministic Inventory Lot Size Models—a General Root Law," *Management Science* 24, no. 8 (1978), 819–26.

Mathematical SNAPSHOT

Delivered Price

Suppose that you are a manufacturer of a product whose sales occur within R miles of your mill. Assume that you charge customers for shipping at the rate s, in dollars per mile, for each unit of product sold. If m is the unit price (in dollars) at the mill, then the delivered unit price p to a customer x miles from the mill is the mill price plus the shipping charge sx:

$$p = m + sx \qquad 0 \le x \le R \tag{1}$$

The problem is to determine the average delivered price of the units sold.

Suppose that there is a function f such that $f(t) \ge 0$ on the interval $[0, R]$ and such that the area under the graph of f and above the t-axis from $t = 0$ to $t = x$ represents the total number of units Q sold to customers within x miles of the mill. [See Figure 14.55(a).] You can refer to f as the distribution of demand. Because Q is a function of x and is represented by area,

$$Q(x) = \int_0^x f(t)\, dt$$

In particular, the total number of units sold within the market area is

$$Q(R) = \int_0^R f(t)\, dt$$

[see Figure 14.55(b)]. For example, if $f(t) = 10$ and $R = 100$, then the total number of units sold within the

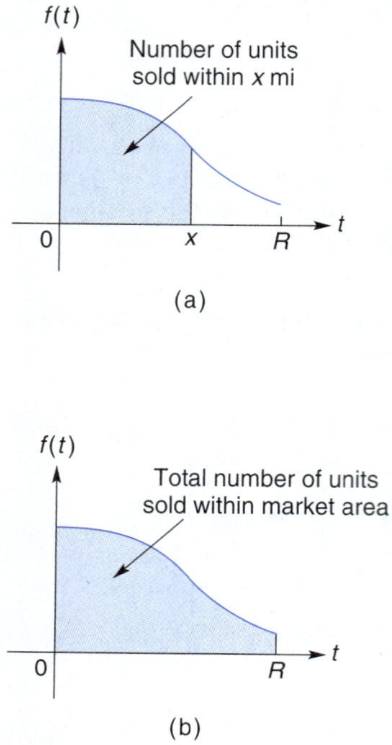

FIGURE 14.55 Number of units sold as an area.

market area is

$$Q(100) = \int_0^{100} 10\, dt = 10t \Big|_0^{100} = 1000 - 0 = 1000$$

The average delivered price A is given by

$$A = \frac{\text{total revenue}}{\text{total number of units sold}}$$

Because the denominator is $Q(R)$, A can be determined once the total revenue is found.

To find the total revenue, first consider the number of units sold over an interval. If $t_1 < t_2$ [see Figure 14.56(a)], then the area under the graph of f and above the t-axis from $t = 0$ to $t = t_1$ represents the number of units sold within t_1 miles of the mill. Similarly, the area under the graph of f and above the t-axis from $t = 0$ to $t = t_2$ represents the number of units sold within t_2 miles of the mill. Thus the difference in these areas is geometrically the area of the shaded region in Figure 14.56(a) and represents the number of units sold between t_1 and t_2 miles of the mill, which

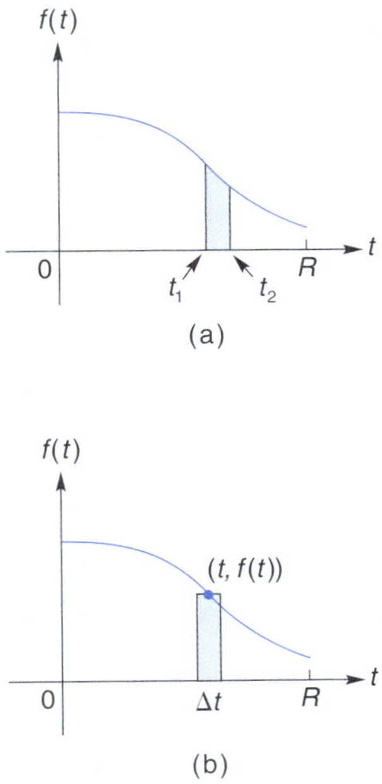

FIGURE 14.56 Number of units sold over an interval.

is $Q(t_2) - Q(t_1)$. Thus

$$Q(t_2) - Q(t_1) = \int_{t_1}^{t_2} f(t)\,dt$$

For example, if $f(t) = 10$, then the number of units sold to customers located between 4 and 6 miles of the mill is

$$Q(6) - Q(4) = \int_4^6 10\,dt = 10t\Big|_4^6 = 60 - 40 = 20$$

The area of the shaded region in Figure 14.56(a) can be approximated by the area of a rectangle [see Figure 14.56(b)] whose height is $f(t)$ and whose width is Δt, where $\Delta t = t_2 - t_1$. Thus the number of units sold over the interval of length Δt is approximately $f(t)\,\Delta t$. Because the price of each of these units is [from Equation (1)] approximately $m+st$, the revenue received is approximately

$$(m+st)\,f(t)\Delta t$$

The sum of all such products from $t = 0$ to $t = R$ approximates the total revenue. Definite integration gives

$$\sum(m+st)\,f(t)\Delta t \to \int_0^R (m+st)\,f(t)\,dt$$

Thus,

$$\text{total revenue} = \int_0^R (m+st)\,f(t)\,dt$$

Consequently, the average delivered price A is given by

$$A = \frac{\int_0^R (m+st)\,f(t)\,dt}{Q(R)}$$

equivalently,

$$A = \frac{\int_0^R (m+st)\,f(t)dt}{\int_0^R f(t)\,dt}$$

For example, if $f(t)=10$, $m=200$, $s=0.25$, and $R=100$, then

$$\begin{aligned}\int_0^R (m+st)\,f(t)\,dt &= \int_0^{100} (200+0.25t)\cdot 10\,dt\\ &= 10\int_0^{100}(200+0.25t)\,dt\\ &= 10\left(200t + \frac{t^2}{8}\right)\Bigg|_0^{100}\\ &= 10\left[\left(20{,}000 + \frac{10{,}000}{8}\right) - 0\right]\\ &= 212{,}500\end{aligned}$$

From before,

$$\int_0^R f(t)\,dt = \int_0^{100} 10\,dt = 1000$$

Thus, the average delivered price is $212{,}500/1000 = \$212.50$.

Exercises

1. If $f(t) = 50 - 2t$, determine the number of units sold to customers located (a) within 5 miles of the mill, and (b) between 10 and 15 miles.

2. If $f(t) = 40 - 0.5t$, $m = 50$, $s = 0.20$, and $R = 80$, determine (a) the total revenue, (b) the total number of units sold, and (c) the average delivered price.

3. If $f(t) = 900 - t^2$, $m = 100$, $s = 1$, and $R = 30$, determine (a) the total revenue, (b) the total number of units sold, and (c) the average delivered price. Use a graphing calculator if you like.

4. How do real-world sellers of such things as books and clothing generally determine shipping charges for an order? (Visit an online retailer to find out.) How would you calculate average delivered price for their products? Is the procedure fundamentally different from the one in the Snapshot?

CHAPTER

15

Methods and Applications of Integration

We now know how to find the derivative of a function, and in some cases, we know how to find a function from its derivative through integration. However, the integration process is not always straightforward.

Suppose we model the gradual disappearance of a chemical substance using the equations $M' = -0.004t$ and $M(0) = 3000$, where the amount M, in grams, is a function of time t in days. This initial-condition problem is easily solved by integration with respect to t: $M = -0.002t^2 + 3000$. But what if, instead, the disappearance of the substance were modeled by the equations $M' = -0.004M$ and $M(0) = 3000$? The simple replacement of t in the first equation with M changes the character of the problem. We have not yet learned how to find a function when its derivative is described in terms of the function itself.

If you worked through the Mathematical Snapshot in Chapter 12, you may remember a similar situation, involving an equation with P on one side and the derivative of P on the other. There, we used an approximation to solve the problem. In this chapter, we will learn a method that yields an exact solution in many problems of this type.

Equations of the form $y' = ky$, where k is a constant, are especially common. When y represents the amount of a radioactive substance, $y' = ky$ can represent the rate of its disappearance through radioactive decay. And if y is the temperature of a chicken just taken out of the oven or just put into a freezer, then a related formula, called Newton's law of cooling, can be used to describe the change in the chicken's internal temperature over time. Newton's law, which is discussed in this chapter, might be used to write procedures for a restaurant kitchen, so that food prone to contamination through bacterial growth does not spend too much time in the temperature danger zone (40°F to 140°F). (Bacterial growth, for that matter, also follows a $y' = ky$ law!)

OBJECTIVE *To develop and apply the formula for integration by parts.*

15.1 Integration by Parts[1]

Many integrals cannot be found by our previous methods. However, there are ways of changing certain integrals to forms that are easier to integrate. Of these methods, we shall discuss two: *integration by parts* and (in Section 15.2) *integration using partial fractions*.

If u and v are differentiable functions of x, we have, by the product rule,

$$(uv)' = uv' + vu'$$

Rearranging gives

$$uv' = (uv)' - vu'$$

Integrating both sides with respect to x, we get

$$\int uv'\,dx = \int (uv)'\,dx - \int vu'\,dx \tag{1}$$

For $\int (uv)'\,dx$, we must find a function whose derivative with respect to x is $(uv)'$. Clearly, uv is such a function. Hence $\int (uv)'\,dx = uv + C_1$, and Equation (1) becomes

$$\int uv'\,dx = uv + C_1 - \int vu'\,dx$$

Absorbing C_1 into the constant of integration for $\int vu'\,dx$ and replacing $v'\,dx$ by dv and $u'\,dx$ by du, we have the *formula for integration by parts:*

Formula for Integration by Parts

$$\int u\,dv = uv - \int v\,du \tag{2}$$

This formula expresses an integral, $\int u\,dv$, in terms of another integral, $\int v\,du$, that may be easier to find.

To apply the formula to a given integral $\int f(x)\,dx$, we must write $f(x)\,dx$ as the product of two factors (or *parts*) by choosing a function u and a differential dv such that $f(x)\,dx = u\,dv$. However, for the formula to be useful, we must be able to integrate the part chosen for dv. To illustrate, consider

$$\int xe^x\,dx$$

This integral cannot be determined by previous integration formulas. One way to write $xe^x\,dx$ in the form $u\,dv$ is by letting

$$u = x \quad \text{and} \quad dv = e^x\,dx$$

To apply the formula for integration by parts, we must find du and v:

$$du = dx \quad \text{and} \quad v = \int e^x\,dx = e^x + C_1$$

[1]This section may be omitted without loss of continuity.

Thus,

$$\begin{aligned}\int xe^x\,dx &= \int u\,dv \\ &= uv - \int v\,du \\ &= x(e^x + C_1) - \int (e^x + C_1)\,dx \\ &= xe^x + C_1x - e^x - C_1x + C \\ &= xe^x - e^x + C \\ &= e^x(x-1) + C\end{aligned}$$

The first constant, C_1, does not appear in the final answer. It is easy to prove that the constant involved in finding v from dv will always drop out so from now on we will not write it when we find v.

When you are using the formula for integration by parts, sometimes the "best choice" for u and dv may not be obvious. In some cases, one choice may be as good as another; in other cases, only one choice may be suitable. Insight into making a good choice (if any exists) will come only with practice and, of course, trial and error.

APPLICATIONS IN PRACTICE 1

Integration by Parts

The monthly sales of a computer keyboard are estimated to decline at the rate of $S'(t) = -4te^{0.1t}$ keyboards per month, where t is time in months and $S(t)$ is the number of keyboards sold each month. If 5000 keyboards are sold now $[S(0) = 5000]$, find $S(t)$.

EXAMPLE 1 Integration by Parts

Find $\int \frac{\ln x}{\sqrt{x}}\,dx$ *by integration by parts.*

Solution: We try

$$u = \ln x \quad \text{and} \quad dv = \frac{1}{\sqrt{x}}\,dx$$

Then

$$du = \frac{1}{x}\,dx \quad \text{and} \quad v = \int x^{-1/2}\,dx = 2x^{1/2}$$

Thus,

$$\begin{aligned}\int \ln x\left(\frac{1}{\sqrt{x}}\,dx\right) &= \int u\,dv \\ &= uv - \int v\,du \\ &= (\ln x)(2\sqrt{x}) - \int (2x^{1/2})\left(\frac{1}{x}\,dx\right) \\ &= 2\sqrt{x}\ln x - 2\int x^{-1/2}\,dx \\ &= 2\sqrt{x}\ln x - 2(2\sqrt{x}) + C \qquad (x^{1/2} = \sqrt{x}) \\ &= 2\sqrt{x}[\ln(x) - 2] + C\end{aligned}$$

▶ **Now work Problem 3.**

Example 2 shows how a poor choice for u and dv can be made. If you make a choice that does not work, do not become frustrated. Rather, make other choices until one that works is found (if such a choice exists).

EXAMPLE 2 Integration by Parts

Evaluate $\int_1^2 x\ln x\,dx$.

Solution: Since the integral does not fit a familiar form, we shall try integration by parts. Let $u = x$ and $dv = \ln x\,dx$. Then $du = dx$, but $v = \int \ln x\,dx$ is not apparent by inspection. So we shall make a different choice for u and dv. Let

$$u = \ln x \quad \text{and} \quad dv = x\,dx$$

Then

$$du = \frac{1}{x}\,dx \quad \text{and} \quad v = \int x\,dx = \frac{x^2}{2}$$

Therefore,

$$\begin{aligned}
\int_1^2 x \ln x\,dx &= (\ln x)\left(\frac{x^2}{2}\right)\Bigg|_1^2 - \int_1^2 \left(\frac{x^2}{2}\right)\frac{1}{x}\,dx \\
&= (\ln x)\left(\frac{x^2}{2}\right)\Bigg|_1^2 - \frac{1}{2}\int_1^2 x\,dx \\
&= \frac{x^2 \ln x}{2}\Bigg|_1^2 - \frac{1}{2}\left(\frac{x^2}{2}\right)\Bigg|_1^2 \\
&= (2\ln 2 - 0) - \left(1 - \frac{1}{4}\right) = 2\ln 2 - \frac{3}{4}
\end{aligned}$$

EXAMPLE 3 **Integration by Parts Where *u* Is the Entire Integrand**

Determine $\int \ln y\,dy$.

Solution: We cannot integrate $\ln y$ by previous methods, so we shall try integration by parts. Let $u = \ln y$ and $dv = dy$. Then $du = (1/y)\,dy$ and $v = y$. So we have

$$\begin{aligned}
\int \ln y\,dy &= (\ln y)(y) - \int y\left(\frac{1}{y}\,dy\right) \\
&= y \ln y - \int dy = y \ln y - y + C \\
&= y[\ln(y) - 1] + C
\end{aligned}$$

▸ **Now work Problem 37.**

Before trying integration by parts, see whether the technique is really needed. Sometimes the integral can be handled by a basic technique, as Example 4 shows.

EXAMPLE 4 **Basic Integration Form**

Determine $\int xe^{x^2}\,dx$.

Solution: This integral can be fit to the form $\int e^u\,du$.

CAUTION: Do not forget about basic integration forms. Integration by parts is not needed here!

$$\begin{aligned}
\int xe^{x^2}\,dx &= \frac{1}{2}\int e^{x^2}(2x\,dx) \\
&= \frac{1}{2}\int e^u\,du \qquad \text{(where } u = x^2\text{)} \\
&= \frac{1}{2}e^u + C = \frac{1}{2}e^{x^2} + C
\end{aligned}$$

▸ **Now work Problem 17.**

Sometimes integration by parts must be used more than once, as shown in the following example.

APPLICATIONS IN PRACTICE 2

Applying Integration by Parts Twice

Suppose a population of bacteria grows at a rate of

$$P'(t) = 0.1t(\ln t)^2$$

Find the general form of $P(t)$.

EXAMPLE 5 Applying Integration by Parts Twice

Determine $\int x^2e^{2x+1}\,dx$.

Solution: Let $u = x^2$ and $dv = e^{2x+1}\,dx$. Then $du = 2x\,dx$ and $v = e^{2x+1}/2$.

$$\begin{aligned}\int x^2e^{2x+1}\,dx &= \frac{x^2e^{2x+1}}{2} - \int \frac{e^{2x+1}}{2}(2x\,dx)\\ &= \frac{x^2e^{2x+1}}{2} - \int xe^{2x+1}\,dx\end{aligned}$$

To find $\int xe^{2x+1}\,dx$, we shall again use integration by parts. Here, let $u = x$ and $dv = e^{2x+1}\,dx$. Then $du = dx$ and $v = e^{2x+1}/2$, and we have

$$\begin{aligned}\int xe^{2x+1}\,dx &= \frac{xe^{2x+1}}{2} - \int \frac{e^{2x+1}}{2}\,dx\\ &= \frac{xe^{2x+1}}{2} - \frac{e^{2x+1}}{4} + C_1\end{aligned}$$

Thus,

$$\begin{aligned}\int x^2e^{2x+1}\,dx &= \frac{x^2e^{2x+1}}{2} - \frac{xe^{2x+1}}{2} + \frac{e^{2x+1}}{4} + C \qquad \text{(where } C = -C_1\text{)}\\ &= \frac{e^{2x+1}}{2}\left(x^2 - x + \frac{1}{2}\right) + C\end{aligned}$$

▸ **Now work Problem 23.**

Exercise 15.1

1. In applying integration by parts to

$$\int f(x)\,dx$$

a student found that $u = x$, $du = dx$, $dv = (x+5)^{1/2}$, and $v = \frac{2}{3}(x+5)^{3/2}$. Use this information to find $\int f(x)\,dx$.

2. Use integration by parts to find

$$\int xe^{5x+2}\,dx$$

by choosing $u = x$ and $dv = e^{5x+2}\,dx$.

In Problems 3–29, find the integrals.

***3.** $\int xe^{-x}\,dx$ **4.** $\int xe^{-5x}\,dx$ **5.** $\int y^3\ln y\,dy$

6. $\int x^2\ln x\,dx$ **7.** $\int \ln(4x)\,dx$ **8.** $\int \frac{t}{e^t}\,dt$

9. $\int 15x\sqrt{x+1}\,dx$ **10.** $\int \frac{12x}{\sqrt{1+4x}}\,dx$

11. $\int \frac{x}{(5x+2)^3}\,dx$ **12.** $\int \frac{\ln(x+1)}{2(x+1)}\,dx$

13. $\int \frac{\ln x}{x^2}\,dx$ **14.** $\int \frac{x+1}{e^x}\,dx$

15. $\int_1^2 4xe^{2x}\,dx$ **16.** $\int_1^2 2xe^{-3x}\,dx$

***17.** $\int_0^1 xe^{-x^2}\,dx$ **18.** $\int \frac{3x^3}{\sqrt{4-x^2}}\,dx$

19. $\int_1^2 \frac{3x}{\sqrt{4-x}}\,dx$ **20.** $\int (\ln x)^2\,dx$

21. $\int 2(2x-1)\ln(x-1)\,dx$ **22.** $\int \frac{xe^x}{(x+1)^2}\,dx$

***23.** $\int x^2e^x\,dx$ **24.** $\int_e^3 \sqrt[3]{x}\,\ln(x^5)\,dx$

25. $\int (x-e^{-x})^2\,dx$ **26.** $\int x^2e^{-2x}\,dx$

27. $\int x^3 e^{x^2}\,dx$

28. $\int x^5 e^{x^2}\,dx$

29. $\int (2^x + x)^2\,dx$

30. Find $\int \ln(x + \sqrt{x^2+1})\,dx$. *Hint:* Show that

$$\frac{d}{dx}[\ln(x + \sqrt{x^2+1})] = \frac{1}{\sqrt{x^2+1}}$$

31. Find the area of the region bounded by the x-axis, the curve $y = \ln x$, and the line $x = e^3$.

32. Find the area of the region bounded by the x-axis and the curve $y = x^2 e^x$ between $x = 0$ and $x = 1$.

33. Find the area of the region bounded by the x-axis and the curve $y = x\sqrt{2x+1}$ between $x = 0$ and $x = 4$.

34. Consumers' Surplus Suppose the demand equation for a manufacturer's product is given by

$$p = 10(q + 10)e^{-(0.1q+1)}$$

where p is the price per unit (in dollars) when q units are demanded. Assume that market equilibrium occurs when $q = 20$. Determine consumers' surplus at market equilibrium.

35. Revenue Suppose total revenue r and price per unit p are differentiable functions of output q.

(a) Use integration by parts to show that

$$\int p\,dq = r - \int q\frac{dp}{dq}\,dq$$

(b) Using part (a), show that

$$r = \int \left(p + q\frac{dp}{dq}\right)dq$$

(c) Using part (b), prove that

$$r(q_0) = \int_0^{q_0} \left(p + q\frac{dp}{dq}\right)dq$$

[*Hint:* Refer to Section 14.8.]

36. Suppose f is a differentiable function. Apply integration by parts to $\int f(x)e^x\,dx$ to prove that

$$\int f(x)e^x\,dx + \int f'(x)e^x\,dx = f(x)e^x + C$$

$$\left(\text{Hence, } \int [f(x) + f'(x)]e^x\,dx = f(x)e^x + C\right)$$

***37.** Suppose that f has an inverse and that $F' = f$. Use integration by parts to develop a useful formula for $\int f^{-1}(x)\,dx$ in terms of F and f^{-1}. *Hint:* Review Example 3. It used the idea required here, for the special case of $f(x) = e^x$.

OBJECTIVE *To show how to integrate a proper rational function by first expressing it as a sum of its partial fractions.*

15.2 Integration by Partial Fractions[2]

Recall that a *rational function* is a quotient of polynomials $N(x)/D(x)$ and that it is *proper* if N and D have no common polynomial factor and the degree of the numerator N is less than the degree of the denominator D. If N/D is not proper, then we can use long division to divide $N(x)$ by $D(x)$:

$$\begin{array}{r} Q(x) \\ D(x)\,\overline{)\,N(x)} \\ \vdots \\ \overline{R(x)} \end{array} \qquad \text{thus} \qquad \frac{N(x)}{D(x)} = Q(x) + \frac{R(x)}{D(x)}$$

Here the quotient $Q(x)$ and the remainder $R(x)$ are also polynomials and either $R(x)$ is the constant 0-polynomial or the degree of $R(x)$ is less than that of $D(x)$. Thus R/D is a simple rational function. Since

$$\int \frac{N(x)}{D(x)}\,dx = \int \left(Q(x) + \frac{R(x)}{D(x)}\right)dx = \int Q(x)\,dx + \int \frac{R(x)}{D(x)}\,dx$$

and we already know how to integrate a polynomial, it follows that the task of integrating rational functions reduces to that of integrating *proper* rational functions. We emphasize that the technique we are about to explain requires that a rational function be proper so that the long divison step is not optional.

[2]This section may be omitted without loss of continuity.

For example,

$$\int \frac{2x^4 - 3x^3 - 4x^2 - 17x - 6}{x^3 - 2x^2 - 3x}\,dx = \int \left(2x + 1 + \frac{4x^2 - 14x - 6}{x^3 - 2x^2 - 3x}\right) dx$$

$$= x^2 + x + \int \frac{4x^2 - 14x - 6}{x^3 - 2x^2 - 3x}\,dx$$

Distinct Linear Factors

Therefore, we shall consider

$$\int \frac{4x^2 - 14x - 6}{x^3 - 2x^2 - 3x}\,dx$$

It is essential that the denominator be expressed in factored form:

$$\int \frac{4x^2 - 14x - 6}{x(x+1)(x-3)}\,dx$$

Observe that in this example the denominator consists only of **linear factors** and that each factor occurs exactly once. It can be shown that, to each such factor $x - a$, there corresponds a *partial fraction* of the form

$$\frac{A}{x-a} \qquad (A \text{ a constant})$$

such that the integrand is the sum of the partial fractions. If there are n such *distinct* linear factors, there will be n such partial fractions, each of which is easily integrated. Applying these facts, we can write

$$\frac{4x^2 - 14x - 6}{x(x+1)(x-3)} = \frac{A}{x} + \frac{B}{x+1} + \frac{C}{x-3} \tag{1}$$

To determine the constants A, B, and C, we first combine the terms on the right side:

$$\frac{4x^2 - 14x - 6}{x(x+1)(x-3)} = \frac{A(x+1)(x-3) + Bx(x-3) + Cx(x+1)}{x(x+1)(x-3)}$$

Since the denominators of both sides are equal, we can equate their numerators:

$$4x^2 - 14x - 6 = A(x+1)(x-3) + Bx(x-3) + Cx(x+1) \tag{2}$$

Although Equation (1) is not defined for $x = 0$, $x = -1$, and $x = 3$, we want to find values for A, B, and C that will make Equation (2) true for all values of x, so that the two sides of the equality provide equal functions. By successively setting x in Equation (2) equal to any three different numbers, we can obtain a system of equations that can be solved for A, B, and C. In particular, the work can be simplified by letting x be the roots of $D(x) = 0$; in our case, $x = 0$, $x = -1$, and $x = 3$. Using Equation (2), we have, for $x = 0$,

$$-6 = A(1)(-3) + B(0) + C(0) = -3A, \quad \text{so } A = 2$$

If $x = -1$,

$$12 = A(0) + B(-1)(-4) + C(0) = 4B, \quad \text{so } B = 3$$

If $x = 3$,

$$-12 = A(0) + B(0) + C(3)(4) = 12C, \quad \text{so } C = -1$$

Thus Equation (1) becomes

$$\frac{4x^2 - 14x - 6}{x(x+1)(x-3)} = \frac{2}{x} + \frac{3}{x+1} - \frac{1}{x-3}$$

Hence,

$$\begin{aligned}\int \frac{4x^2 - 14x - 6}{x(x+1)(x-3)}\,dx &= \int \left(\frac{2}{x} + \frac{3}{x+1} - \frac{1}{x-3}\right) dx \\ &= 2\int \frac{dx}{x} + 3\int \frac{dx}{x+1} - \int \frac{dx}{x-3} \\ &= 2\ln|x| + 3\ln|x+1| - \ln|x-3| + C \\ &= \ln\left|\frac{x^2(x+1)^3}{x-3}\right| + C\end{aligned}$$

(using properties of logarithms)

For the *original* integral, we can now state that

$$\int \frac{2x^4 - 3x^3 - 4x^2 - 17x - 6}{x^3 - 2x^2 - 3x}\,dx = x^2 + x + \ln\left|\frac{x^2(x+1)^3}{x-3}\right| + C$$

An alternative method of determining A, B, and C involves expanding the right side of Equation (2) and combining similar terms:

$$\begin{aligned}4x^2 - 14x - 6 &= A(x^2 - 2x - 3) + B(x^2 - 3x) + C(x^2 + x) \\ &= Ax^2 - 2Ax - 3A + Bx^2 - 3Bx + Cx^2 + Cx \\ 4x^2 - 14x - 6 &= (A + B + C)x^2 + (-2A - 3B + C)x + (-3A)\end{aligned}$$

For this last equality to express an equality of functions the coefficients of corresponding powers of x on the left and right sides must be equal:

$$\begin{cases} 4 = A + B + C \\ -14 = -2A - 3B + C \\ -6 = -3A \end{cases}$$

Solving gives $A = 2$, $B = 3$, and $C = -1$ as before.

APPLICATIONS IN PRACTICE 1

Distinct Linear Factors

The marginal revenue for a company manufacturing q radios per week is given by $r'(q) = \dfrac{5(q+4)}{q^2 + 4q + 3}$, where $r(q)$ is the revenue in thousands of dollars. Find the equation for $r(q)$.

EXAMPLE 1 Distinct Linear Factors

Determine $\int \frac{2x+1}{3x^2 - 27}\,dx$ *by using partial fractions.*

Solution: Since the degree of the numerator is less than the degree of the denominator, no long division is necessary. The integral can be written as

$$\frac{1}{3}\int \frac{2x+1}{x^2 - 9}\,dx$$

Expressing $(2x+1)/(x^2 - 9)$ as a sum of partial fractions, we have

$$\frac{2x+1}{x^2 - 9} = \frac{2x+1}{(x+3)(x-3)} = \frac{A}{x+3} + \frac{B}{x-3}.$$

Combining terms and equating numerators gives

$$2x + 1 = A(x - 3) + B(x + 3)$$

If $x = 3$, then

$$7 = 6B, \quad \text{so } B = \frac{7}{6}$$

If $x = -3$, then

$$-5 = -6A, \quad \text{so } A = \frac{5}{6}$$

Thus,

$$\begin{aligned}\int \frac{2x+1}{3x^2-27}\,dx &= \frac{1}{3}\left[\int \frac{\frac{5}{6}\,dx}{x+3} + \int \frac{\frac{7}{6}\,dx}{x-3}\right]\\ &= \frac{1}{3}\left[\frac{5}{6}\ln|x+3| + \frac{7}{6}\ln|x-3|\right] + C\\ &= \frac{1}{18}\ln|(x+3)^5(x-3)^7| + C\end{aligned}$$

▶ **Now work Problem 1.**

Repeated Linear Factors

If the denominator of $N(x)/D(x)$ contains only linear factors, some of which are repeated, then, for each factor $(x-a)^k$, where k is the maximum number of times $x-a$ occurs as a factor, there will correspond the sum of k partial fractions:

$$\frac{A}{x-a} + \frac{B}{(x-a)^2} + \cdots + \frac{K}{(x-a)^k}$$

EXAMPLE 2 **Repeated Linear Factors**

Determine $\int \frac{6x^2+13x+6}{(x+2)(x+1)^2}\,dx$ *by using partial fractions.*

Solution: Since the degree of the numerator, namely 2, is less than that of the denominator, namely 3, no long division is necessary. In the denominator, the linear factor $x+2$ occurs once and the linear factor $x+1$ occurs twice. There will thus be three partial fractions and three constants to determine, and we have

$$\frac{6x^2+13x+6}{(x+2)(x+1)^2} = \frac{A}{x+2} + \frac{B}{x+1} + \frac{C}{(x+1)^2}$$

$$6x^2+13x+6 = A(x+1)^2 + B(x+2)(x+1) + C(x+2)$$

Let us choose $x=-2$, $x=-1$, and, for convenience, $x=0$. For $x=-2$, we have

$$4 = A$$

If $x = -1$, then

$$-1 = C$$

If $x = 0$, then

$$\begin{aligned}6 &= A + 2B + 2C = 4 + 2B - 2 = 2 + 2B\\ 4 &= 2B\\ 2 &= B\end{aligned}$$

Therefore,

$$\int \frac{6x^2 + 13x + 6}{(x+2)(x+1)^2}\,dx = 4\int \frac{dx}{x+2} + 2\int \frac{dx}{x+1} - \int \frac{dx}{(x+1)^2}$$

$$= 4 \ln|x+2| + 2 \ln|x+1| + \frac{1}{x+1} + C$$

$$= \ln\left[(x+2)^4(x+1)^2\right] + \frac{1}{x+1} + C$$

▶ **Now work Problem 5.**

Distinct Irreducible Quadratic Factors

Suppose a quadratic factor $x^2 + bx + c$ occurs in $D(x)$ and it cannot be expressed as a product of two linear factors with real coefficients. Such a factor is said to be an *irreducible quadratic factor over the real numbers*. To each distinct irreducible quadratic factor that occurs exactly once in $D(x)$, there will correspond a partial fraction of the form

$$\frac{Ax + B}{x^2 + bx + c}$$

Note that even after you have expressed a rational function in terms of partial fractions, you may still find it impossible to integrate using only the calculus you have been taught so far. For example, a very simple irreducible quadratic factor is $x^2 + 1$ and yet

$$\int \frac{1}{x^2+1}\,dx = \int \frac{dx}{x^2+1} = \tan^{-1} x + C$$

where $\tan^{-1}$ is the inverse of the trigonometric function $\tan(x)$ when $\tan(x)$ is restricted to $(-\pi/2, \pi/2)$. We will not discuss trigonometric functions in this book.

EXAMPLE 3 **An Integral with a Distinct Irreducible Quadratic Factor**

Determine $\int \frac{-2x-4}{x^3+x^2+x}\,dx$ *by using partial fractions.*

Solution: Since $x^3 + x^2 + x = x(x^2 + x + 1)$, we have the linear factor x and the quadratic factor $x^2 + x + 1$, which does not seem factorable on inspection. If it were factorable into $(x - r_1)(x - r_2)$, with r_1 and r_2 real, then r_1 and r_2 would be roots of the equation $x^2 + x + 1 = 0$. By the quadratic formula, the roots are

$$x = \frac{-1 \pm \sqrt{1-4}}{2}$$

Since there are no real roots, we conclude that $x^2 + x + 1$ is irreducible. Thus there will be two partial fractions and *three* constants to determine. We have

$$\frac{-2x-4}{x(x^2+x+1)} = \frac{A}{x} + \frac{Bx+C}{x^2+x+1}$$

$$-2x - 4 = A(x^2 + x + 1) + (Bx + C)x$$

$$= Ax^2 + Ax + A + Bx^2 + Cx$$

$$0x^2 - 2x - 4 = (A + B)x^2 + (A + C)x + A$$

Equating coefficients of like powers of x, we obtain

$$\begin{cases} 0 = A + B \\ -2 = A + C \\ -4 = A \end{cases}$$

Solving gives $A = -4$, $B = 4$ and $C = 2$. Hence,

$$\begin{aligned} \int \frac{-2x-4}{x(x^2+x+1)}\,dx &= \int \left(\frac{-4}{x} + \frac{4x+2}{x^2+x+1}\right) dx \\ &= -4\int \frac{dx}{x} + 2\int \frac{2x+1}{x^2+x+1}\,dx \end{aligned}$$

Both integrals have the form $\int \frac{du}{u}$, so

$$\begin{aligned} \int \frac{-2x-4}{x(x^2+x+1)}\,dx &= -4\ln|x| + 2\ln|x^2+x+1| + C \\ &= \ln\left[\frac{(x^2+x+1)^2}{x^4}\right] + C \end{aligned}$$

▶ **Now work Problem 7.**

Repeated Irreducible Quadratic Factors

Suppose $D(x)$ contains factors of the form $(x^2+bx+c)^k$, where k is the maximum number of times the irreducible factor x^2+bx+c occurs. Then, to each such factor, there will correspond a sum of k partial fractions of the form

$$\frac{A+Bx}{x^2+bx+c} + \frac{C+Dx}{(x^2+bx+c)^2} + \cdots + \frac{M+Nx}{(x^2+bx+c)^k}$$

EXAMPLE 4 **Repeated Irreducible Quadratic Factors**

Determine $\int \frac{x^5}{(x^2+4)^2}\,dx$ *by using partial fractions.*

Solution: Since the numerator has degree 5 and the denominator has degree 4, we first use long division, which gives

$$\frac{x^5}{x^4+8x^2+16} = x - \frac{8x^3+16x}{(x^2+4)^2}$$

The quadratic factor x^2+4 in the denominator of $(8x^3+16x)/(x^2+4)^2$ is irreducible and occurs as a factor twice. Thus, to $(x^2+4)^2$ there correspond two partial fractions and *four* coefficients to be determined. Accordingly we set

$$\frac{8x^3+16x}{(x^2+4)^2} = \frac{Ax+B}{x^2+4} + \frac{Cx+D}{(x^2+4)^2}$$

and obtain

$$\begin{aligned} 8x^3+16x &= (Ax+B)(x^2+4) + Cx + D \\ 8x^3+0x^2+16x+0 &= Ax^3 + Bx^2 + (4A+C)x + 4B + D \end{aligned}$$

Equating like powers of x yields

$$\begin{cases} 8 = A \\ 0 = B \\ 16 = 4A + C \\ 0 = 4B + D \end{cases}$$

Solving gives $A = 8$, $B = 0$, $C = -16$, and $D = 0$. Therefore,

$$\int \frac{x^5}{(x^2+4)^2}\,dx = \int \left(x - \left[\frac{8x}{x^2+4} - \frac{16x}{(x^2+4)^2} \right] \right) dx$$

$$= \int x\,dx - 4\int \frac{2x}{x^2+4}\,dx + 8\int \frac{2x}{(x^2+4)^2}\,dx$$

The second integral on the preceding line has the form $\int \frac{du}{u}$, and the third integral has the form $\int \frac{du}{u^2}$. So

$$\int \frac{x^5}{(x^2+4)^2} = \frac{x^2}{2} - 4\ln(x^2+4) - \frac{8}{x^2+4} + C$$

▶ **Now work Problem 27.**

CAUTION Do not forget about basic integration forms.

From our examples, you may have deduced that the number of constants needed to express $N(x)/D(x)$ by partial fractions is equal to the degree of $D(x)$, if it is assumed that $N(x)/D(x)$ defines a proper rational function. This is indeed the case. Note also that the representation of a proper rational function by partial fractions is unique; that is, there is only one choice of constants that can be made. Furthermore, regardless of the complexity of the polynomial $D(x)$, it can always (theoretically) be expressed as a product of linear and irreducible quadratic factors with real coefficients.

APPLICATIONS IN PRACTICE 2

An Integral Not Requiring Partial Fractions

The rate of change of the voting population of a city with respect to time t (in years) is estimated to be $V'(t) = \frac{300t^3}{t^2+6}$. Find the general form of $V(t)$.

EXAMPLE 5 An Integral Not Requiring Partial Fractions

Find $\int \frac{2x+3}{x^2+3x+1}\,dx$.

Solution: This integral has the form $\int \frac{1}{u}\,du$. Thus,

$$\int \frac{2x+3}{x^2+3x+1}\,dx = \ln|x^2+3x+1| + C$$

▶ **Now work Problem 12.**

Exercise 15.2

In Problems 1–8, express the given rational function in terms of partial fractions. Watch out for any preliminary divisions.

*1. $f(x) = \frac{10x}{x^2+7x+6}$

2. $f(x) = \frac{x+5}{x^2-1}$

3. $f(x) = \frac{x^2}{x^2+6x+8}$

4. $f(x) = \frac{2x^2-15}{x^2+5x}$

*5. $f(x) = \frac{x+4}{x^2+4x+4}$

6. $f(x) = \frac{2x+3}{x^2(x-1)}$

*7. $f(x) = \frac{x^2+3}{x^3+x}$

8. $f(x) = \frac{3x^2+5}{(x^2+4)^2}$

In Problems 9–30, determine the integrals.

9. $\int \frac{5x-2}{x^2-x}\,dx$

10. $\int \frac{3x+8}{x^2+2x}\,dx$

11. $\int \frac{x+10}{x^2-x-2}\,dx$

*__12.__ $\int \frac{2x-1}{x^2-x-12}\,dx$

13. $\int \frac{3x^3-3x+4}{4x^2-4}\,dx$

14. $\int \frac{7(4-x^2)}{(x-4)(x-2)(x+3)}\,dx$

15. $\int \frac{17x-12}{x^3-x^2-12x}\,dx$

16. $\int \frac{4-x}{x^4-x^2}\,dx$

17. $\int \frac{2(3x^5+4x^3-x)}{x^6+2x^4-x^2-2}\,dx$

18. $\int \frac{x^4-2x^3+6x^2-11x+2}{x^3-3x^2+2x}\,dx$

19. $\int \frac{2x^2-5x-2}{(x-2)^2(x-1)}\,dx$

20. $\int \frac{-3x^3+2x-3}{x^2(x^2-1)}\,dx$

21. $\int \frac{2(x^2+8)}{x^3+4x}\,dx$

22. $\int \frac{2x^3-6x^2-10x-6}{x^4-1}\,dx$

23. $\int \frac{-x^3+8x^2-9x+2}{(x^2+1)(x-3)^2}\,dx$

24. $\int \frac{5x^4+9x^2+3}{x(x^2+1)^2}\,dx$

25. $\int \frac{14x^3+24x}{(x^2+1)(x^2+2)}\,dx$

26. $\int \frac{12x^3+20x^2+28x+4}{3(x^2+2x+3)(x^2+1)}\,dx$

*__27.__ $\int \frac{3x^3+x}{(x^2+1)^2}\,dx$

28. $\int \frac{3x^2-8x+4}{x^3-4x^2+4x-6}\,dx$

29. $\int_0^1 \frac{2-2x}{x^2+7x+12}\,dx$

30. $\int_1^2 \frac{3x^2+15x+13}{x^2+4x+3}\,dx$

31. Find the area of the region bounded by the graph of

$$y=\frac{6(x^2+1)}{(x+2)^2}$$

and the x-axis from $x=0$ to $x=1$.

32. Consumers' Surplus Suppose the demand equation for a manufacturer's product is given by

$$p=\frac{200(q+3)}{q^2+7q+6}$$

where p is the price per unit (in dollars) when q units are demanded. Assume that market equilibrium occurs at the point $(q, p) = (10, 325/22)$. Determine consumers' surplus at market equilibrium.

OBJECTIVE *To illustrate the use of the table of integrals in Appendix B.*

15.3 Integration by Tables

Certain forms of integrals that occur frequently may be found in standard tables of integration formulas.[3] A short table appears in Appendix C, and its use will be illustrated in this section.

A given integral may have to be replaced by an equivalent form before it will fit a formula in the table. The equivalent form must match the formula *exactly*. Consequently, the steps that you perform should *not* be done mentally. *Write them down!* Failure to do this can easily lead to incorrect results. Before proceeding with the exercises, be sure you understand the illustrative examples *thoroughly*.

In the following examples, the formula numbers refer to the Table of Selected Integrals given in Appendix C.

EXAMPLE 1 **Integration by Tables**

Find $\int \frac{x\,dx}{(2+3x)^2}$.

[3]See, for example, W. H. Beyer (ed.), *CRC Standard Mathematical Tables and Formulae,* 30th ed. (Boca Raton, FL: CRC Press, 1996).

Solution: Scanning the table, we identify the integrand with Formula 7:

$$\int \frac{u\,du}{(a+bu)^2} = \frac{1}{b^2}\left(\ln|a+bu| + \frac{a}{a+bu}\right) + C$$

Now we see if we can exactly match the given integrand with that in the formula. If we replace x by u, 2 by a, and 3 by b, then $du = dx$, and by substitution, we have

$$\int \frac{x\,dx}{(2+3x)^2} = \int \frac{u\,du}{(a+bu)^2} = \frac{1}{b^2}\left(\ln|a+bu| + \frac{a}{a+bu}\right) + C$$

Returning to the variable x and replacing a by 2 and b by 3, we obtain

$$\int \frac{x\,dx}{(2+3x)^2} = \frac{1}{9}\left(\ln|2+3x| + \frac{2}{2+3x}\right) + C$$

Note that the answer must be given in terms of x, the *original* variable of integration.

▶ **Now work Problem 5.**

EXAMPLE 2 **Integration by Tables**

Find $\int x^2\sqrt{x^2-1}\,dx$.

Solution: This integral is identified with Formula 24:

$$\int u^2\sqrt{u^2 \pm a^2}\,du = \frac{u}{8}(2u^2 \pm a^2)\sqrt{u^2 \pm a^2} - \frac{a^4}{8}\ln|u + \sqrt{u^2 \pm a^2}| + C$$

In the preceding formula, if the bottommost sign in the dual symbol "±" on the left side is used, then the bottommost sign in the dual symbols on the right side must also be used. In the original integral, we let $u = x$ and $a = 1$. Then $du = dx$, and by substitution, the integral becomes

$$\int x^2\sqrt{x^2-1}\,dx = \int u^2\sqrt{u^2-a^2}\,du$$

$$= \frac{u}{8}(2u^2 - a^2)\sqrt{u^2-a^2} - \frac{a^4}{8}\ln|u + \sqrt{u^2-a^2}| + C$$

Since $u = x$ and $a = 1$,

$$\int x^2\sqrt{x^2-1}\,dx = \frac{x}{8}(2x^2-1)\sqrt{x^2-1} - \frac{1}{8}\ln|x + \sqrt{x^2-1}| + C$$

▶ **Now work Problem 17.**

This example, as well as Examples 4, 5, and 7, shows how to adjust an integral so that it conforms to one in the table.

EXAMPLE 3 **Integration by Tables**

Find $\int \frac{dx}{x\sqrt{16x^2+3}}$.

Solution: The integrand can be identified with Formula 28:

$$\int \frac{du}{u\sqrt{u^2+a^2}} = \frac{1}{a}\ln\left|\frac{\sqrt{u^2+a^2}-a}{u}\right| + C$$

If we let $u = 4x$ and $a = \sqrt{3}$, then $du = 4\,dx$. Watch closely how, by inserting 4's in the numerator and denominator, we transform the given integral into an equivalent form that matches Formula 28:

$$\int \frac{dx}{x\sqrt{16x^2+3}} = \int \frac{(4\,dx)}{(4x)\sqrt{(4x)^2+(\sqrt{3})^2}} = \int \frac{du}{u\sqrt{u^2+a^2}}$$

$$= \frac{1}{a}\ln\left|\frac{\sqrt{u^2+a^2}-a}{u}\right| + C$$

$$= \frac{1}{\sqrt{3}}\ln\left|\frac{\sqrt{16x^2+3}-\sqrt{3}}{4x}\right| + C$$

▸ **Now work Problem 7.**

EXAMPLE 4 **Integration by Tables**

Find $\int \frac{dx}{x^2(2-3x^2)^{1/2}}$.

Solution: The integrand is identified with Formula 21:

$$\int \frac{du}{u^2\sqrt{a^2-u^2}} = -\frac{\sqrt{a^2-u^2}}{a^2u} + C$$

Letting $u = \sqrt{3}x$ and $a^2 = 2$, we have $du = \sqrt{3}\,dx$. Hence, by inserting two factors of $\sqrt{3}$ in both the numerator and denominator of the original integral, we have

$$\int \frac{dx}{x^2(2-3x^2)^{1/2}} = \sqrt{3}\int \frac{(\sqrt{3}\,dx)}{(\sqrt{3}x)^2[2-(\sqrt{3}x)^2]^{1/2}} = \sqrt{3}\int \frac{du}{u^2(a^2-u^2)^{1/2}}$$

$$= \sqrt{3}\left[-\frac{\sqrt{a^2-u^2}}{a^2u}\right] + C = \sqrt{3}\left[-\frac{\sqrt{2-3x^2}}{2(\sqrt{3}x)}\right] + C$$

$$= -\frac{\sqrt{2-3x^2}}{2x} + C$$

▸ **Now work Problem 35.**

EXAMPLE 5 **Integration by Tables**

Find $\int 7x^2\ln(4x)\,dx$.

Solution: This is similar to Formula 42 with $n = 2$:

$$\int u^n \ln u\,du = \frac{u^{n+1}\ln u}{n+1} - \frac{u^{n+1}}{(n+1)^2} + C$$

If we let $u = 4x$, then $du = 4\,dx$. Hence,

$$\begin{aligned}\int 7x^2 \ln(4x)\,dx &= \frac{7}{4^3}\int (4x)^2 \ln(4x)(4\,dx)\\ &= \frac{7}{64}\int u^2 \ln u\,du = \frac{7}{64}\left(\frac{u^3 \ln u}{3} - \frac{u^3}{9}\right) + C\\ &= \frac{7}{64}\left[\frac{(4x)^3 \ln(4x)}{3} - \frac{(4x)^3}{9}\right] + C\\ &= 7x^3\left[\frac{\ln(4x)}{3} - \frac{1}{9}\right] + C\\ &= \frac{7x^3}{9}[3\ln(4x) - 1] + C\end{aligned}$$

▶ **Now work Problem 45.**

EXAMPLE 6 **Integral Table Not Needed**

Find $\int \frac{e^{2x}\,dx}{7+e^{2x}}$.

Solution: At first glance, we do not identify the integrand with any form in the table. Perhaps rewriting the integral will help. Let $u = 7 + e^{2x}$, then $du = 2e^{2x}\,dx$. So

$$\begin{aligned}\int \frac{e^{2x}\,dx}{7+e^{2x}} &= \frac{1}{2}\int \frac{(2e^{2x}\,dx)}{7+e^{2x}} = \frac{1}{2}\int \frac{du}{u} = \frac{1}{2}\ln|u| + C\\ &= \frac{1}{2}\ln|7+e^{2x}| + C = \frac{1}{2}\ln(7+e^{2x}) + C\end{aligned}$$

Thus, we had only to use our knowledge of basic integration forms. Actually, this form appears as Formula 2 in the table.

▶ **Now work Problem 11.**

EXAMPLE 7 **Finding a Definite Integral by Using Tables**

Evaluate $\int_1^4 \frac{dx}{(4x^2+2)^{3/2}}$.

Solution: We shall use Formula 32 to get the indefinite integral first:

$$\int \frac{du}{(u^2 \pm a^2)^{3/2}} = \frac{\pm u}{a^2\sqrt{u^2 \pm a^2}} + C$$

Letting $u = 2x$ and $a^2 = 2$, we have $du = 2\,dx$. Thus,

$$\begin{aligned}\int \frac{dx}{(4x^2+2)^{3/2}} &= \frac{1}{2}\int \frac{(2\,dx)}{[(2x)^2+2]^{3/2}} = \frac{1}{2}\int \frac{du}{(u^2+2)^{3/2}}\\ &= \frac{1}{2}\left[\frac{u}{2\sqrt{u^2+2}}\right] + C\end{aligned}$$

Here we determine the limits of integration with respect to u.

Instead of substituting back to x and evaluating from $x = 1$ to $x = 4$, we can determine the corresponding limits of integration with respect to u and then evaluate the last expression between those limits. Since $u = 2x$, when $x = 1$ we have $u = 2$; when $x = 4$ we have $u = 8$. Hence,

$$\int_1^4 \frac{dx}{(4x^2+2)^{3/2}} = \frac{1}{2}\int_2^8 \frac{du}{(u^2+2)^{3/2}}$$

$$= \frac{1}{2}\left(\frac{u}{2\sqrt{u^2+2}}\right)\Bigg|_2^8 = \frac{2}{\sqrt{66}} - \frac{1}{2\sqrt{6}}$$

CAUTION When changing the variable of integration x to the variable of integration u, be sure to change the limits of integration so that they agree with u.

▶ **Now work Problem 15.**

Integration Applied to Annuities

Tables of integrals are useful when one deals with integrals associated with annuities. Suppose that you must pay out \$100 at the end of each year for the next two years. Recall from Chapter 5 that a series of payments over a period of time, such as this, is called an *annuity*. If you were to pay off the debt now instead, you would pay the present value of the \$100 that is due at the end of the first year, plus the present value of the \$100 that is due at the end of the second year. The sum of these present values is the present value of the annuity. (The present value of an annuity is discussed in Section 5.3.) We shall now consider the present value of payments made continuously over the time interval from $t = 0$ to $t = T$, with t in years, when interest is compounded continuously at an annual rate of r.

Suppose a payment is made at time t such that on an annual basis this payment is $f(t)$. If we divide the interval $[0, T]$ into subintervals $[t_{i-1}, t_i]$ of length Δt (where Δt is small), then the total amount of all payments over such a subinterval is approximately $f(t_i)\Delta t$. [For example, if $f(t) = 2000$ and Δt were one day, the total amount of the payments would be $2000\left(\frac{1}{365}\right)$.] The present value of these payments is approximately $e^{-rt_i} f(t_i)\Delta t$. (See Section 10.3.) Over the interval $[0, T]$, the total of all such present values is

$$\sum e^{-rt_i} f(t_i)\Delta t$$

This sum approximates the present value A of the annuity. The smaller Δt is, the better the approximation. That is, as $\Delta t \to 0$, the limit of the sum is the present value. However, this limit is also a definite integral. That is,

$$A = \int_0^T f(t)e^{-rt}\,dt \qquad \textbf{(1)}$$

where A is the **present value of a continuous annuity** at an annual rate r (compounded continuously) for T years if a payment at time t is at the rate of $f(t)$ per year.

We say that Equation (1) gives the **present value of a continuous income stream.** Equation (1) may also be used to find the present value of future profits of a business. In this situation, $f(t)$ is the annual rate of profit at time t.

We can also consider the *future* value of an annuity rather than its present value. If a payment is made at time t, then it has a certain value at the *end* of the period of the annuity—that is, $T - t$ years later. This value is

$$\begin{pmatrix}\text{amount of}\\ \text{payment}\end{pmatrix} + \begin{pmatrix}\text{interest on this}\\ \text{payment for } T - t \text{ years}\end{pmatrix}$$

If S is the total of such values for all payments, then S is called the *accumulated amount of a continuous annuity* and is given by the formula

$$S = \int_0^T f(t)e^{r(T-t)}\,dt$$

where S is the **accumulated amount of a continuous annuity** at the end of T years at an annual rate r (compounded continuously) when a payment at time t is at the rate of $f(t)$ per year.

EXAMPLE 8 **Present Value of a Continuous Annuity**

Find the present value (to the nearest dollar) of a continuous annuity at an annual rate of 8% for 10 years if the payment at time t is at the rate of t^2 dollars per year.

Solution: The present value is given by

$$A = \int_0^T f(t)e^{-rt}\,dt = \int_0^{10} t^2 e^{-0.08t}\,dt$$

We shall use Formula 39,

$$\int u^n e^{au}\,du = \frac{u^n e^{au}}{a} - \frac{n}{a}\int u^{n-1}e^{au}\,du$$

This is called a *reduction formula,* since it reduces one integral to an expression that involves another integral that is easier to determine. If $u = t$, $n = 2$, and $a = -0.08$, then $du = dt$, and we have

$$A = \left.\frac{t^2e^{-0.08t}}{-0.08}\right|_0^{10} - \frac{2}{-0.08}\int_0^{10} te^{-0.08t}\,dt$$

In the new integral, the exponent of t has been reduced to 1. We can match this integral with Formula 38,

$$\int ue^{au}\,du = \frac{e^{au}}{a^2}(au - 1) + C$$

by letting $u = t$ and $a = -0.08$. Then $du = dt$, and

$$A = \int_0^{10} t^2e^{-0.08t}\,dt = \left.\frac{t^2e^{-0.08t}}{-0.08}\right|_0^{10} - \frac{2}{-0.08}\left[\left.\frac{e^{-0.08t}}{(-0.08)^2}(-0.08t - 1)\right]\right|_0^{10}$$

$$= \frac{100e^{-0.8}}{-0.08} - \frac{2}{-0.08}\left[\frac{e^{-0.8}}{(-0.08)^2}(-0.8 - 1) - \frac{1}{(-0.08)^2}(-1)\right]$$

$$\approx 185$$

The present value is \$185.

▶ Now work Problem 59.

Exercise 15.3

In Problems 1 and 2, use Formula 19 in Appendix C to determine the integrals.

1. $\displaystyle\int \frac{dx}{(9-x^2)^{3/2}}$

2. $\displaystyle\int \frac{dx}{(25-4x^2)^{3/2}}$

In Problems 3 and 4, use Formula 30 in Appendix C to determine the integrals.

3. $\displaystyle\int \frac{dx}{x^2\sqrt{16x^2+3}}$

4. $\displaystyle\int \frac{dx}{x^3\sqrt{x^4-4}}$

In Problems 5–38, find the integrals by using the table in Appendix C.

***5.** $\displaystyle\int \frac{dx}{x(6+7x)}$

6. $\displaystyle\int \frac{3x^2\,dx}{(2+5x)^2}$

***7.** $\displaystyle\int \frac{dx}{x\sqrt{x^2+9}}$

8. $\displaystyle\int \frac{dx}{(x^2+7)^{3/2}}$

9. $\displaystyle\int \frac{x\,dx}{(2+3x)(4+5x)}$

10. $\displaystyle\int 2^{5x}\,dx$

***11.** $\displaystyle\int \frac{dx}{4+3e^{2x}}$

12. $\displaystyle\int x^2\sqrt{1+x}\,dx$

13. $\displaystyle\int \frac{7\,dx}{x(5+2x)^2}$

14. $\displaystyle\int \frac{dx}{x\sqrt{5-11x^2}}$

***15.** $\displaystyle\int_0^1 \frac{x\,dx}{2+x}$

16. $\displaystyle\int \frac{5x^2\,dx}{2+5x}$

***17.** $\displaystyle\int \sqrt{x^2-3}\,dx$

18. $\displaystyle\int \frac{dx}{(1+5x)(2x+3)}$

19. $\displaystyle\int_0^{1/12} xe^{12x}\,dx$

20. $\displaystyle\int \sqrt{\frac{2+3x}{5+3x}}\,dx$

21. $\displaystyle\int x^2e^x\,dx$

22. $\displaystyle\int_1^2 \frac{4\,dx}{x^2(1+x)}$

23. $\displaystyle\int \frac{\sqrt{4x^2+1}}{x^2}\,dx$

24. $\displaystyle\int \frac{dx}{x\sqrt{2-x}}$

25. $\displaystyle\int \frac{x\,dx}{(1+3x)^2}$

26. $\displaystyle\int \frac{3\,dx}{\sqrt{(5+3x)(6+3x)}}$

27. $\displaystyle\int \frac{dx}{7-5x^2}$

28. $\displaystyle\int x^2\sqrt{2x^2-9}\,dx$

29. $\displaystyle\int 36x^5\ln(3x)\,dx$

30. $\displaystyle\int \frac{5\,dx}{x^2(3+2x)^2}$

31. $\displaystyle\int 270x\sqrt{1+3x}\,dx$

32. $\displaystyle\int 9x^2\ln x\,dx$

33. $\displaystyle\int \frac{dx}{\sqrt{4x^2-13}}$

34. $\displaystyle\int \frac{dx}{x\ln(2x)}$

***35.** $\displaystyle\int \frac{dx}{x^2\sqrt{9-4x^2}}$

36. $\displaystyle\int \frac{\sqrt{2-3x^2}}{x}\,dx$

37. $\displaystyle\int \frac{dx}{\sqrt{x}(\pi+7e^{4\sqrt{x}})}$

38. $\displaystyle\int_0^1 \frac{3x^2\,dx}{1+2x^3}$

In Problems 39–56, find the integrals by any method.

39. $\displaystyle\int \frac{x\,dx}{x^2+1}$

40. $\displaystyle\int \sqrt{x}e^{x^{3/2}}\,dx$

41. $\displaystyle\int 6x\sqrt{2x^2+1}\,dx$

42. $\displaystyle\int \frac{5x^3-\sqrt{x}}{2x}\,dx$

43. $\displaystyle\int \frac{dx}{x^2-5x+6}$

44. $\displaystyle\int \frac{e^{2x}}{\sqrt{e^{2x}+3}}\,dx$

***45.** $\displaystyle\int x^3\ln x\,dx$

46. $\displaystyle\int_0^3 xe^{-x}\,dx$

47. $\displaystyle\int 4xe^{2x}\,dx$

48. $\displaystyle\int_1^2 35x^2\sqrt{3+2x}\,dx$

49. $\displaystyle\int \ln^2 x\,dx$

50. $\displaystyle\int_1^e 3x\ln x^2\,dx$

51. $\displaystyle\int_1^2 \frac{x\,dx}{\sqrt{4-x}}$

52. $\displaystyle\int_1^2 x\sqrt{1+2x}\,dx$

53. $\displaystyle\int_0^1 \frac{2x\,dx}{\sqrt{8-x^2}}$

54. $\displaystyle\int_0^{\ln 2} x^2e^{3x}\,dx$

55. $\displaystyle\int_1^2 x\ln(2x)\,dx$

56. $\displaystyle\int_1^2 dx$

57. Biology In a discussion about gene frequency,[4] the integral

$$\int_{q_0}^{q_n} \frac{dq}{q(1-q)}$$

occurs, where the q's represent gene frequencies. Evaluate this integral.

58. Biology Under certain conditions, the number n of generations required to change the frequency of a gene from 0.3 to 0.1 is given by[5]

$$n = -\frac{1}{0.4}\int_{0.3}^{0.1} \frac{dq}{q^2(1-q)}$$

Find n (to the nearest integer).

***59. Continuous Annuity** Find the present value, to the nearest dollar, of a continuous annuity at an annual rate of r for T years if the payment at time t is at the annual rate of $f(t)$ dollars, given that

(a) $r = 0.06$ $\quad T = 10$ $\quad f(t) = 5000$

(b) $r = 0.05$ $\quad T = 8$ $\quad f(t) = 200t$

---

[4] W. B. Mather, *Principles of Quantitative Genetics* (Minneapolis: Burgess Publishing Company, 1964).

[5] E. O. Wilson and W. H. Bossert, *A Primer of Population Biology* (Stamford, CT: Sinauer Associates, 1971).

60. If $f(t) = k$, where k is a positive constant, show that the value of the integral in Equation (1) of this section is

$$k\left(\frac{1 - e^{-rT}}{r}\right)$$

61. Continuous Annuity Find the accumulated amount, to the nearest dollar, of a continuous annuity at an annual rate of r for T years if the payment at time t is at an annual rate of $f(t)$ dollars, given that

(a) $r = 0.06 \quad T = 10 \quad f(t) = 400$
(b) $r = 0.04 \quad T = 5 \quad f(t) = 40t$

62. Value of Business Over the next five years, the profits of a business at time t are estimated to be $50,000t$ dollars per year. The business is to be sold at a price equal to the present value of these future profits. To the nearest 10 dollars, at what price should the business be sold if interest is compounded continuously at the annual rate of 7%?

OBJECTIVE *To develop the concept of the average value of a function.*

15.4 Average Value of a Function

If we are given the three numbers 1, 2, and 9, then their average value, or *mean,* is their sum divided by 3. Denoting this average by $\overline{y}$, we have

$$\overline{y} = \frac{1 + 2 + 9}{3} = 4$$

Similarly, suppose we are given a function f defined on the interval $[a, b]$, and the points $x_1, x_2, \ldots, x_n$ are in the interval. Then the average value of the n corresponding function values $f(x_1), f(x_2), \ldots, f(x_n)$ is

$$\overline{y} = \frac{f(x_1) + f(x_2) + \cdots + f(x_n)}{n} = \frac{\sum_{i=1}^{n} f(x_i)}{n} \tag{1}$$

We can go a step further. Let us divide the interval $[a, b]$ into n subintervals of equal length. We shall choose x_1 to be in the first subinterval, x_2 to be in the second, etc. Because $[a, b]$ has length $b - a$, each subinterval has length $\frac{b-a}{n}$, which we shall call Δx. Thus, Equation (1) can be written

$$\overline{y} = \frac{\sum_{i=1}^{n} f(x_i)\left(\frac{\Delta x}{\Delta x}\right)}{n} = \frac{\frac{1}{\Delta x}\sum_{i=1}^{n} f(x_i)\Delta x}{n} = \frac{1}{n\,\Delta x}\sum_{i=1}^{n} f(x_i)\Delta x \tag{2}$$

Since $\Delta x = \frac{b-a}{n}$, it follows that $n\,\Delta x = b - a$. So the expression $\frac{1}{n\,\Delta x}$ in Equation (2) can be replaced by $\frac{1}{b-a}$. Moreover, as $n \to \infty$, the number of function values used in computing $\overline{y}$ increases, and we get the so-called *average value of the function f*, denoted by $\overline{f}$:

$$\overline{f} = \lim_{n\to\infty}\left[\frac{1}{b-a}\sum_{i=1}^{n} f(x_i)\Delta x\right] = \frac{1}{b-a}\lim_{n\to\infty}\sum_{i=1}^{n} f(x_i)\Delta x$$

But the limit on the right is just the definite integral $\int_a^b f(x)\,dx$. Thus, we have the following definition.

Definition

The ***average*** (or ***mean***) ***value of a function*** $y = f(x)$ over the interval $[a, b]$ is denoted $\overline{f}$ (or $\overline{y}$) and is given by

$$\overline{f} = \frac{1}{b-a}\int_a^b f(x)\,dx$$

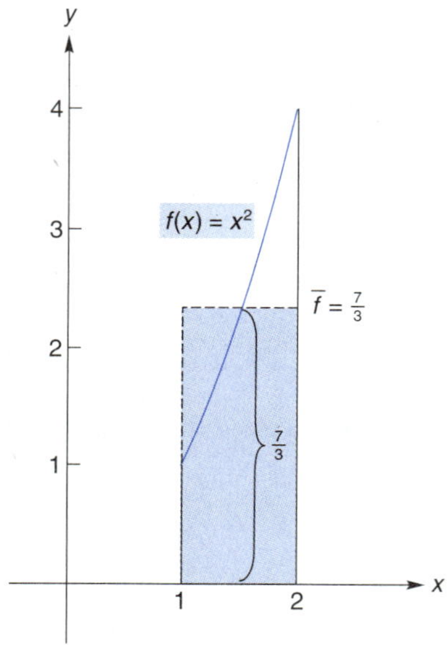

FIGURE 15.1 Geometric interpretation of the average value of a function.

EXAMPLE 1 Average Value of a Function

Find the average value of the function $f(x) = x^2$ over the interval $[1, 2]$.

Solution:

$$\overline{f} = \frac{1}{b-a}\int_a^b f(x)\,dx$$

$$= \frac{1}{2-1}\int_1^2 x^2\,dx = \left.\frac{x^3}{3}\right|_1^2 = \frac{7}{3}$$

In Example 1, we found that the average value of $y = f(x) = x^2$ over the interval $[1, 2]$ is $\frac{7}{3}$. We can interpret this value geometrically. Since

$$\frac{1}{2-1}\int_1^2 x^2\,dx = \frac{7}{3}$$

by solving for the integral we have

$$\int_1^2 x^2\,dx = \frac{7}{3}(2-1)$$

However, this integral gives the area of the region bounded by $f(x) = x^2$ and the x-axis from $x = 1$ to $x = 2$. (See Figure 15.1.) From the preceding equation, this area is $\left(\frac{7}{3}\right)(2-1)$, which is the area of a rectangle whose height is the average value $\overline{f} = \frac{7}{3}$ and whose width is $b - a = 2 - 1 = 1$.

▶ **Now work Problem 1.**

EXAMPLE 2 Average Flow of Blood

Suppose the flow of blood at time t in a system is given by

$$F(t) = \frac{F_1}{(1+\alpha t)^2} \qquad 0 \le t \le T$$

where F_1 and α (a Greek letter read "alpha") are constants.[6] *Find the average flow $\overline{F}$ on the interval $[0, T]$.*

Solution:

$$\overline{F} = \frac{1}{T-0}\int_0^T F(t)\,dt$$

$$= \frac{1}{T}\int_0^T \frac{F_1}{(1+\alpha t)^2}\,dt = \frac{F_1}{\alpha T}\int_0^T (1+\alpha t)^{-2}(\alpha\,dt)$$

$$= \frac{F_1}{\alpha T}\left[\left.\frac{(1+\alpha t)^{-1}}{-1}\right]\right|_0^T = \frac{F_1}{\alpha T}\left[-\frac{1}{1+\alpha T}+1\right]$$

$$= \frac{F_1}{\alpha T}\left[\frac{-1+1+\alpha T}{1+\alpha T}\right] = \frac{F_1}{\alpha T}\left[\frac{\alpha T}{1+\alpha T}\right] = \frac{F_1}{1+\alpha T}$$

▶ **Now work Problem 11.**

[6] W. Simon, *Mathematical Techniques for Physiology and Medicine* (New York: Academic Press, 1972).

Exercise 15.4

In Problems 1–8, find the average value of the function over the given interval.

*1. $f(x) = x^2$; $[-1, 3]$

2. $f(x) = 3x - 1$; $[1, 2]$

3. $f(x) = 2 - 3x^2$; $[-1, 2]$

4. $f(x) = x^2 + x + 1$; $[1, 3]$

5. $f(t) = 4t^3$; $[-2, 2]$

6. $f(t) = t\sqrt{t^2 + 9}$; $[0, 4]$

7. $f(x) = 6\sqrt{x}$; $[1, 9]$

8. $f(x) = 5/x^2$; $[1, 3]$

9. **Profit** The profit (in dollars) of a business is given by

$$P = P(q) = 369q - 2.1q^2 - 400$$

where q is the number of units of the product sold. Find the average profit on the interval from $q = 0$ to $q = 100$.

10. **Cost** Suppose the cost (in dollars) of producing q units of a product is given by

$$c = 4000 + 10q + 0.1q^2$$

Find the average cost on the interval from $q = 100$ to $q = 500$.

*11. **Investment** An investment of \$3000 earns interest at an annual rate of 5% compounded continuously. After t years, its value S (in dollars) is given by $S = 3000e^{0.05t}$. Find the average value of a two-year investment.

12. **Medicine** Suppose that colored dye is injected into the bloodstream at a constant rate R. At time t, let

$$C(t) = \frac{R}{F(t)}$$

be the concentration of dye at a location distant (distal) from the point of injection, where $F(t)$ is as given in Example 2. Show that the average concentration on $[0, T]$ is

$$\overline{C} = \frac{R(1 + \alpha T + \frac{1}{3}\alpha^2 T^2)}{F_1}$$

13. **Revenue** Suppose a manufacturer receives revenue r from the sale of q units of a product. Show that the average value of the marginal-revenue function over the interval $[0, q_0]$ is the price per unit when q_0 units are sold.

14. Find the average value of $f(x) = \dfrac{3}{x^2 + 2}$ over the interval $[0, 1]$. Round your answer to two decimal places.

OBJECTIVE *To solve a differential equation by using the method of separation of variables. To discuss particular solutions and general solutions. To develop interest compounded continuously in terms of a differential equation. To discuss exponential growth and decay.*

15.5 Differential Equations

Occasionally, you may have to solve an equation that involves the derivative of an unknown function. Such an equation is called a **differential equation.** An example is

$$y' = xy^2 \tag{1}$$

More precisely, Equation (1) is called a **first-order differential equation**, since it involves a derivative of the first order and none of higher order. A solution of Equation (1) is any function $y = f(x)$ that is defined on an interval and satisfies the equation for all x in the interval.

To solve $y' = xy^2$, equivalently,

$$\frac{dy}{dx} = xy^2 \tag{2}$$

we think of dy/dx as a quotient of differentials and algebraically "separate variables" by rewriting the equation so that each side contains only one variable and a differential is not in a denominator:

$$\frac{dy}{y^2} = x\,dx$$

Integrating both sides and combining the constants of integration, we obtain

$$\int \frac{1}{y^2}\,dy = \int x\,dx$$

$$-\frac{1}{y} = \frac{x^2}{2} + C_1$$

$$-\frac{1}{y} = \frac{x^2 + 2C_1}{2}$$

Since $2C_1$ is an arbitrary constant, we can replace it by C.

$$-\frac{1}{y} = \frac{x^2 + C}{2} \tag{3}$$

Solving Equation (3) for y, we have

$$y = -\frac{2}{x^2 + C} \tag{4}$$

We can verify by substitution that y is a solution to the differential Equation (2):

$$\frac{dy}{dx} = xy^2$$

$$\frac{4x}{(x^2 + C)^2} \stackrel{?}{=} x\left[-\frac{2}{x^2 + C}\right]^2$$

$$\frac{4x}{(x^2 + C)^2} = \frac{4x}{(x^2 + C)^2}$$

Note in Equation (4) that, for *each* value of C, a different solution is obtained. We call Equation (4) the **general solution** of the differential equation. The method that we used to find it is called **separation of variables.**

In the foregoing example, suppose we are given the condition that $y = -\frac{2}{3}$ when $x = 1$, that is, $y(1) = -\frac{2}{3}$. Then the *particular* function that satisfies both Equation (2) and this condition can be found by substituting the values $x = 1$ and $y = -\frac{2}{3}$ into Equation (4) and solving for C:

$$-\frac{2}{3} = -\frac{2}{1^2 + C}$$

$$C = 2$$

Therefore, the solution of $dy/dx = xy^2$ such that $y(1) = -\frac{2}{3}$ is

$$y = -\frac{2}{x^2 + 2} \tag{5}$$

We call Equation (5) a **particular solution** of the differential equation.

APPLICATIONS IN PRACTICE 1

Separation of Variables

For a clear liquid, light intensity diminishes at a rate of $\frac{dI}{dx} = -kI$, where I is the intensity of the light and x is the number of feet below the surface of the liquid. If $k = 0.0085$ and $I = I_0$ when $x = 0$, find I as a function of x.

EXAMPLE 1 Separation of Variables

Solve $y' = -\frac{y}{x}$ if $x, y > 0$.

Solution: Writing y' as dy/dx, separating variables, and integrating, we have

$$\frac{dy}{dx} = -\frac{y}{x}$$

$$\frac{dy}{y} = -\frac{dx}{x}$$

$$\int \frac{1}{y}\,dy = -\int \frac{1}{x}\,dx$$

$$\ln|y| = C_1 - \ln|x|$$

Since $x, y > 0$, we can omit the absolute-value bars:

$$\ln y = C_1 - \ln x \tag{6}$$

To solve for y, we convert Equation (6) to exponential form:

$$y = e^{C_1 - \ln x}$$

So

$$y = e^{C_1} e^{-\ln x} = \frac{e^{C_1}}{e^{\ln x}}$$

Replacing e^{C_1} by C, where $C > 0$, and rewriting $e^{\ln x}$ as x gives

$$y = \frac{C}{x} \qquad C, x > 0$$

▶ **Now work Problem 1.**

In Example 1, note that Equation (6) expresses the solution implicitly, whereas the final equation ($y = C/x$) states the solution y explicitly in terms of x. You will find that the solutions of certain differential equations are often expressed in implicit form for convenience (or necessity because of the difficulty involved in obtaining an explicit form).

Exponential Growth and Decay

In Section 10.3, the notion of interest compounded continuously was developed. Let us now take a different approach to this topic that involves a differential equation. Suppose P dollars are invested at an annual rate r compounded n times a year. Let the function $S = S(t)$ give the compound amount S (or total amount present) after t years from the date of the initial investment. Then the initial principal is $S(0) = P$. Furthermore, since there are n interest periods per year, each period has length $1/n$ years, which we shall denote by Δt. At the end of the first period, the accrued interest for that period is added to the principal, and the sum acts as the principal for the second period, and so on. Hence, if the beginning of an interest period occurs at time t, then the increase in the amount present at the end of a period Δt is $S(t + \Delta t) - S(t)$, which we write as ΔS. This increase, ΔS, is also the interest earned for the period. Equivalently, the interest earned is principal times rate times time:

$$\Delta S = S \cdot r \cdot \Delta t$$

Dividing both sides by Δt, we obtain

$$\frac{\Delta S}{\Delta t} = rS \tag{7}$$

As $\Delta t \to 0$, then $n = \dfrac{1}{\Delta t} \to \infty$, and consequently, interest is being *compounded continuously;* that is, the principal is subject to continuous growth at every instant. However, as $\Delta t \to 0$, then $\Delta S/\Delta t \to dS/dt$, and Equation (7) takes the form

$$\frac{dS}{dt} = rS \tag{8}$$

This differential equation means that *when interest is compounded continuously, the rate of change of the amount of money present at time t is proportional to the amount present at time t.* The constant of proportionality is r.

To determine the actual function S, we solve the differential Equation (8) by the method of separation of variables:

$$\frac{dS}{dt} = rS$$

$$\frac{dS}{S} = r\,dt$$

$$\int \frac{1}{S}\,dS = \int r\,dt$$

$$\ln|S| = rt + C_1$$

We assume that $S > 0$, so $\ln|S| = \ln S$. Thus,

$$\ln S = rt + C_1$$

To get an explicit form, we can solve for S by converting to exponential form.

$$S = e^{rt+C_1} = e^{C_1}e^{rt}$$

For simplicity, e^{C_1} can be replaced by C to obtain the general solution

$$S = Ce^{rt}$$

The condition $S(0) = P$ allows us to find the value of C:

$$P = Ce^{r(0)} = C \cdot 1$$

Hence $C = P$, so

$$S = Pe^{rt} \tag{9}$$

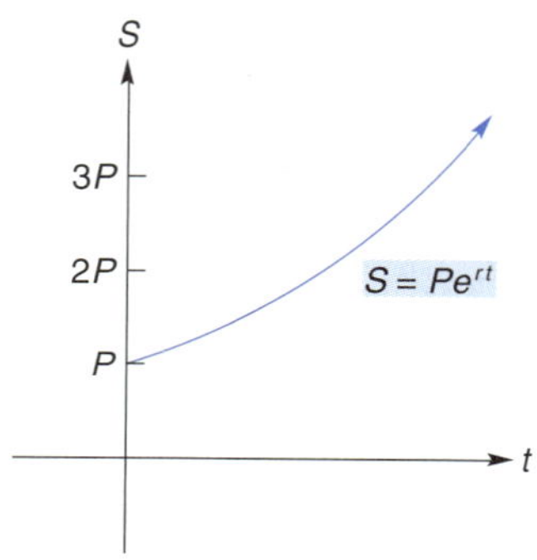

FIGURE 15.2 Compounding continuously.

Equation (9) gives the total value after t years of an initial investment of P dollars compounded continuously at an annual rate r. (See Figure 15.2.)

In our discussion of compound interest, we saw from Equation (8) that the rate of change in the amount present was proportional to the amount present. There are many natural quantities, such as population, whose rate of growth or decay at any time is considered proportional to the amount of that quantity present. If N denotes the amount of such a quantity at time t, then this rate of growth means that

$$\frac{dN}{dt} = kN$$

where k is a constant. If we separate variables and solve for N as we did for Equation (8), we get

$$N = N_0e^{kt} \tag{10}$$

where N_0 is a constant. In particular, if $t = 0$, then $N = N_0e^0 = N_0 \cdot 1 = N_0$. Thus, the constant N_0 is simply $N(0)$. Owing to the form of Equation (10), we say that the quantity follows an **exponential law of growth** if k is positive and an **exponential law of decay** if k is negative.

EXAMPLE 2 **Population Growth**

In a certain city, the rate at which the population grows at any time is proportional to the size of the population. If the population was 125,000 *in* 1970 *and* 140,000 *in* 1990, *what is the expected population in* 2010?

Solution: Let N be the size of the population at time t. Since the exponential law of growth applies,

$$N = N_0 e^{kt}$$

To find the population in 2010, we must first find the particular law of growth involved by determining the values of N_0 and k. Let the year 1970 correspond to $t = 0$. Then $t = 20$ in 1990 and $t = 40$ in 2010. We have

$$N_0 = N(0) = 125{,}000$$

Thus,

$$N = 125{,}000e^{kt}$$

To find k, we use the fact that $N = 140{,}000$ when $t = 20$:

$$140{,}000 = 125{,}000e^{20k}$$

Hence,

$$\begin{aligned} e^{20k} &= \frac{140{,}000}{125{,}000} = 1.12 \\ 20k &= \ln(1.12) && \text{(logarithmic form)} \\ k &= \tfrac{1}{20}\ln(1.12) \end{aligned}$$

Therefore, the law of growth is

$$\begin{aligned} N &= 125{,}000e^{(t/20)\ln 1.12} \\ &= 125{,}000[e^{\ln 1.12}]^{t/20} \end{aligned} \qquad \textbf{(11)}$$

so that

$$N = 125{,}000(1.12)^{t/20} \qquad \textbf{(12)}$$

Setting $t = 40$ gives the expected population in 2010:

$$N = N(40) = 125{,}000(1.12)^2 = 156{,}800$$

We remark that we can write Equation (11) in a form different from Equation (12). Because $\ln(1.12)/20 \approx 0.0057$, we have

$$N \approx 125{,}000e^{0.0057t}$$

▶ **Now work Problem 23.**

In Chapter 4, radioactive decay was discussed. Here we shall consider this topic from the perspective of a differential equation. The rate at which a radioactive element decays at any time is found to be proportional to the amount of that element present. If N is the amount of a radioactive substance at time t, then the rate of decay is given by

$$\frac{dN}{dt} = -\lambda N \qquad \textbf{(13)}$$

The positive constant λ (a Greek letter read "lambda") is called the **decay constant,** and the minus sign indicates that N is decreasing as t increases. Thus, we have exponential decay. From Equation (10), the solution of this differential equation is

$$N = N_0 e^{-\lambda t} \qquad \textbf{(14)}$$

If $t = 0$, then $N = N_0 \cdot 1 = N_0$, so N_0 represents the amount of the radioactive substance present when $t = 0$.

The time for one-half of the substance to decay is called the **half-life** of the substance. In Section 4.2, it was shown that the half-life is given by

$$\text{half-life} = \frac{\ln 2}{\lambda} \approx \frac{0.69315}{\lambda} \tag{15}$$

Note that the half-life depends on λ. In Chapter 4, Figure 4.13 shows the graph of radioactive decay.

EXAMPLE 3 **Finding the Decay Constant and Half-life**

If 60% of a radioactive substance remains after 50 days, find the decay constant and the half-life of the element.

Solution: From Equation (14),

$$N = N_0 e^{-\lambda t}$$

where N_0 is the amount of the element present at $t = 0$. When $t = 50$, then $N = 0.6N_0$, and we have

$$\begin{aligned}
0.6N_0 &= N_0 e^{-50\lambda} \\
0.6 &= e^{-50\lambda} \\
-50\lambda &= \ln(0.6) && \text{(logarithmic form)} \\
\lambda &= -\frac{\ln(0.6)}{50} \approx 0.01022
\end{aligned}$$

Thus, $N \approx N_0 e^{-0.01022t}$. The half-life, from Equation (15), is

$$\frac{\ln 2}{\lambda} \approx 67.85 \text{ days}$$

▸ **Now work Problem 27.**

Radioactivity is useful in dating such things as fossil plant remains and archaeological remains made from organic material. Plants and other living organisms contain a small amount of radioactive carbon 14 (^{14}C) in addition to ordinary carbon (^{12}C). The ^{12}C atoms are stable, but the ^{14}C atoms are decaying exponentially. However, ^{14}C is formed in the atmosphere due to the effect of cosmic rays. Eventually, this ^{14}C is taken up by plants during photosynthesis and replaces what has decayed. As a result, the ratio of ^{14}C atoms to ^{12}C atoms is considered constant over a long period of time. When a plant dies, it stops absorbing ^{14}C, and the remaining ^{14}C atoms decay. By comparing the proportion of ^{14}C to ^{12}C in a fossil plant to that of plants found today, we can estimate the age of the fossil. The half-life of ^{14}C is approximately 5600 years. Thus, if a fossil is found to have a ^{14}C-to-^{12}C ratio that is half that of a similar substance found today, we would estimate the fossil to be 5600 years old.

EXAMPLE 4 **Estimating the Age of an Ancient Tool**

A wood tool found in a Middle East excavation site is found to have a ^{14}C-to-^{12}C ratio that is 0.6 of the corresponding ratio in a present-day tree. Estimate the age of the tool to the nearest hundred years.

Solution: Let N be the amount of ^{14}C present in the wood t years after the tool was made. Then $N = N_0 e^{-\lambda t}$ where N_0 is the amount of ^{14}C when $t = 0$. Since the ratio of ^{14}C to ^{12}C is 0.6 of the corresponding ratio in a present-day tree, this means that we want to find the value of t for which $N = 0.6N_0$. Thus, we have

$$\begin{aligned} 0.6N_0 &= N_0 e^{-\lambda t} \\ 0.6 &= e^{-\lambda t} \\ -\lambda t &= \ln(0.6) \qquad \text{(logarithmic form)} \\ t &= -\frac{1}{\lambda}\ln(0.6) \end{aligned}$$

From Equation (15), the half-life is $(\ln 2)/\lambda$, which equals 5600, so $\lambda = (\ln 2)/5600$. Consequently,

$$\begin{aligned} t &= -\frac{1}{(\ln 2)/5600}\ln(0.6) \\ &= -\frac{5600\ln(0.6)}{\ln 2} \\ &\approx 4100 \text{ years} \end{aligned}$$

▶ **Now work Problem 29.**

Exercise 15.5

In Problems 1–8, solve the differential equations.

*1. $y' = 2xy^2$

2. $y' = x^2y^2$

3. $\dfrac{dy}{dx} - 3x\sqrt{x^2+1} = 0$

4. $\dfrac{dy}{dx} = \dfrac{x}{y}$

5. $\dfrac{dy}{dx} = y,\ y > 0$

6. $y' = e^x y^2$

7. $y' = \dfrac{y}{x},\ x, y > 0$

8. $\dfrac{dy}{dx} + xe^x = 0$

In Problems 9–18, solve each of the differential equations, subject to the given conditions.

9. $y' = \dfrac{1}{y^2};\ y(1) = 1$

10. $y' = e^{x-y};\ y(0) = 0$ [*Hint:* $e^{x-y} = e^x/e^y$]

11. $e^y y' - x^2 = 0;\quad y = 0$ when $x = 0$

12. $x^2y' + \dfrac{1}{y^2} = 0;\quad y(1) = 2$

13. $(4x^2+3)^2y' - 4xy^2 = 0;\quad y(0) = \frac{3}{2}$

14. $y' + x^3y = 0;\quad y = e$ when $x = 0$

15. $\dfrac{dy}{dx} = \dfrac{3x\sqrt{1+y^2}}{y};\quad y > 0,\ y(1) = \sqrt{8}$

16. $2y(x^3+2x+1)\dfrac{dy}{dx} = \dfrac{3x^2+2}{\sqrt{y^2+9}};\quad y(0) = 0$

17. $2\dfrac{dy}{dx} = \dfrac{xe^{-y}}{\sqrt{x^2+3}};\quad y(1) = 0$

18. $x(y^3+4)^{3/2}\,dx = 3e^{x^2}y^2\,dy;\quad y(0) = 0$

19. **Cost** Find the manufacturer's cost function $c = f(q)$ given that

$$(q+1)^2\frac{dc}{dq} = cq$$

and fixed cost is e.

20. Find $f(2)$, given that $f(1) = 0$ and that $y = f(x)$ satisfies the differential equation

$$\frac{dy}{dx} = xe^{x-y}$$

21. **Circulation of Money** A country has 1.00 billion dollars of paper money in circulation. Each week 25 million dollars is brought into the banks for deposit, and the same amount is paid out. The government decides to issue new paper money; whenever the old money comes into the banks, it is destroyed and replaced by new money. Let y be the amount of old money (in millions of dollars) in circulation at time t (in weeks). Then y satisfies the differential equation

$$\frac{dy}{dt} = -0.025y$$

How long will it take for 95% of the paper money in circulation to be new? Round your answer to the nearest week. [*Hint:* If money is 95% new, then y is 5% of 1000.]

22. Marginal Revenue and Demand Suppose that a monopolist's marginal-revenue function is given by the differential equation

$$\frac{dr}{dq} = (50 - 4q)e^{-r/5}$$

Find the demand equation for the monopolist's product.

*23. **Population Growth** In a certain town, the population at any time changes at a rate proportional to the population. If the population in 1985 was 40,000 and in 1995 was 48,000, find an equation for the population at time t, where t is the number of years past 1985. Write your answer in two forms, one involving e. You may assume that $\ln 1.2 = 0.18$. What is the expected population in 2005?

24. Population Growth The population of a town increases by natural growth at a rate proportional to the number N of persons present. If the population at time $t = 0$ is 50,000, find two expressions for the population N, t years later, if the population doubles in 50 years. Assume that $\ln 2 = 0.69$. Also, find N for $t = 100$.

25. Population Growth Suppose that the population of the world in 1930 was 2 billion and in 1960 was 3 billion. If the exponential law of growth is assumed, what is the expected population in 2010? Give your answer in terms of e.

26. Population Growth If exponential growth is assumed, in approximately how many years will a population double if it triples in 100 years? [*Hint:* Let the population at $t = 0$ be N_0.]

*27. **Radioactivity** If 30% of the initial amount of a radioactive sample remains after 100 seconds, find the decay constant and the half-life of the element.

28. Radioactivity If 30% of the initial amount of a radioactive sample has *decayed* after 100 seconds, find the decay constant and the half-life of the element.

*29. **Carbon Dating** An Egyptian scroll was found to have a ^{14}C-to-^{12}C ratio 0.7 of the corresponding ratio in similar present-day material. Estimate the age of the scroll, to the nearest hundred years.

30. Carbon Dating A recently discovered archaeological specimen has a ^{14}C-to-^{12}C ratio 0.2 of the corresponding ratio found in present-day organic material. Estimate the age of the specimen, to the nearest hundred years.

31. Population Growth Suppose that a population follows exponential growth given by $dN/dt = kN$ for $t \geq t_0$. Suppose also that $N = N_0$ when $t = t_0$. Find N, the population size at time t.

32. Radioactivity Polonium-210 has a half-life of about 140 days. (a) Find the decay constant in terms of ln 2. (b) What fraction of the original amount of a sample of polonium-210 remains after one year?

33. Radioactivity Radioactive isotopes are used in medical diagnoses as tracers to determine abnormalities that may exist in an organ. For example, if radioactive iodine is swallowed, after some time it is taken up by the thyroid gland. With the use of a detector, the rate at which it is taken up can be measured, and a determination can be made as to whether the uptake is normal. Suppose radioactive technetium-99m, which has a half-life of six hours, is to be used in a brain scan two hours from now. What should be its activity now if the activity when it is used is to be 10 units? Give your answer to one decimal place. [*Hint:* In Equation (14), let N = activity t hours from now and N_0 = activity now.]

34. Radioactivity A radioactive substance that has a half-life of eight days is to be temporarily implanted in a hospital patient until three-fifths of the amount originally present remains. How long should the implant remain in the patient?

35. Ecology In a forest, natural litter occurs, such as fallen leaves and branches, dead animals, etc.[7] Let $A = A(t)$ denote the amount of litter present at time t, where $A(t)$ is expressed in grams per square meter and t is in years. Suppose that there is no litter at $t = 0$. Thus, $A(0) = 0$. Assume that

(a) Litter falls to the ground continuously at a constant rate of 200 grams per square meter per year.
(b) The accumulated litter decomposes continuously at the rate of 50% of the amount present per year (which is $0.50A$).

The difference of the two rates is the rate of change of the amount of litter present with respect to time:

$$\begin{pmatrix}\text{rate of change}\\ \text{of litter present}\end{pmatrix} = \begin{pmatrix}\text{rate of falling}\\ \text{to ground}\end{pmatrix} - \begin{pmatrix}\text{rate of}\\ \text{decomposition}\end{pmatrix}$$

Therefore,

$$\frac{dA}{dt} = 200 - 0.50A$$

Solve for A. To the nearest gram, determine the amount of litter per square meter after one year.

36. Profit and Advertising A certain company determines that the rate of change of monthly net profit P, as a function of monthly advertising expenditure x, is proportional to the difference between a fixed amount, \$150,000, and $2P$; that is, dP/dx is proportional to \$150,000 $- 2P$. Furthermore, if no money is spent on monthly advertising, the monthly net profit is \$15,000; if \$1000 is spent on monthly advertising, the monthly net profit is \$70,000. What would the monthly net profit be if \$2000 were spent on advertising each month?

[7] R. W. Poole, *An Introduction to Quantitative Ecology* (New York: McGraw-Hill Book Company, 1974).

37. Value of Auto The value of a certain model of automobile depreciates 30% in the first year after the car is purchased. The rate at which the automobile subsequently depreciates is proportional to its value. Suppose that such an automobile was purchased new on July 1, 1999, for \$30,000 and was valued at \$18,900 on January 1, 2001.

(a) Determine a formula that expresses the value V of the automobile in terms of t, the number of years after July 1, 2000.

(b) Use the formula in part (a) to determine the year and month in which the car has a value of exactly \$14,000.

OBJECTIVE *To develop the logistic function as a solution of a differential equation. To model the spread of a rumor. To discuss and apply Newton's law of cooling.*

15.6 More Applications of Differential Equations

Logistic Growth

In the previous section, we found that if the number N of individuals in a population at time t follows an exponential law of growth, then $N = N_0 e^{kt}$, where $k > 0$ and N_0 is the population when $t = 0$. This law assumes that at time t the rate of growth, dN/dt, of the population is proportional to the number of individuals in the population. That is, $dN/dt = kN$.

Under exponential growth, a population would get infinitely large as time goes on. In reality, however, when the population gets large enough, environmental factors slow down the rate of growth. Examples are food supply, predators, overcrowding, and so on. These factors cause dN/dt to eventually decrease. It is reasonable to assume that the size of a population is limited to some maximum number M, where $0 < N < M$, and as $N \to M$, $dN/dt \to 0$, and the population size tends to be stable.

In summary, we want a population model that has exponential growth initially, but that also includes the effects of environmental resistance to large population growth. Such a model is obtained by multiplying the right side of $dN/dt = kN$ by the factor $(M - N)/M$:

$$\frac{dN}{dt} = kN\left(\frac{M - N}{M}\right)$$

Notice that if N is small, then $(M - N)/M$ is close to 1, and we have growth that is approximately exponential. As $N \to M$, then $M - N \to 0$ and $dN/dt \to 0$, as we wanted in our model. Because k/M is a constant, we can replace it by K. Thus,

$$\frac{dN}{dt} = KN(M - N) \tag{1}$$

This states that the rate of growth is proportional to the product of the size of the population and the difference between the maximum size and the actual size of the population. We can solve for N in the differential Equation (1) by the method of separation of variables:

$$\frac{dN}{N(M - N)} = K\,dt$$

$$\int \frac{1}{N(M - N)}\,dN = \int K\,dt \tag{2}$$

The integral on the left side can be found by using Formula 5 in the table of integrals in Appendix C. Thus, Equation (2) becomes

$$\frac{1}{M}\ln\left|\frac{N}{M - N}\right| = Kt + C$$

so

$$\ln\left|\frac{N}{M-N}\right| = MKt + MC$$

Since $N > 0$ and $M - N > 0$, we can write

$$\ln\frac{N}{M-N} = MKt + MC$$

In exponential form, we have

$$\frac{N}{M-N} = e^{MKt+MC} = e^{MKt}e^{MC}$$

Replacing the positive constant e^{MC} by A and solving for N gives

$$\begin{aligned}
\frac{N}{M-N} &= Ae^{MKt}\\
N &= (M-N)Ae^{MKt}\\
N &= MAe^{MKt} - NAe^{MKt}\\
NAe^{MKt} + N &= MAe^{MKt}\\
N(Ae^{MKt}+1) &= MAe^{MKt}\\
N &= \frac{MAe^{MKt}}{Ae^{MKt}+1}
\end{aligned}$$

Dividing numerator and denominator by Ae^{MKt}, we have

$$N = \frac{M}{1+\dfrac{1}{Ae^{MKt}}} = \frac{M}{1+\dfrac{1}{A}e^{-MKt}}$$

Replacing $1/A$ by b and MK by c yields the so-called *logistic function:*

Logistic Function

The function defined by

$$N = \frac{M}{1+be^{-ct}} \qquad (3)$$

is called the **logistic function** or the **Verhulst–Pearl logistic function.**

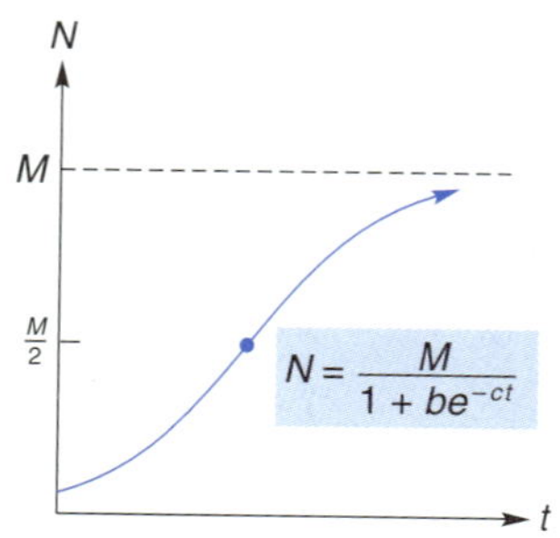

FIGURE 15.3 Logistic curve.

The graph of Equation (3), called a *logistic curve,* is S-shaped and appears in Figure 15.3. Notice that the line $N = M$ is a horizontal asymptote; that is,

$$\lim_{t\to\infty}\frac{M}{1+be^{-ct}} = \frac{M}{1+b(0)} = M$$

Moreover, from Equation (1), the rate of growth is

$$KN(M-N)$$

which can be considered a function of N. To find when the maximum rate of growth occurs, we solve $\frac{d}{dN}[KN(M-N)] = 0$ for N:

$$\begin{aligned}
\frac{d}{dN}[KN(M-N)] &= \frac{d}{dN}[K(MN-N^2)]\\
&= K[M-2N] = 0
\end{aligned}$$

Thus, $N = M/2$. In other words, the rate of growth increases until the population size is $M/2$ and decreases thereafter. The maximum rate of growth occurs when

$N = M/2$ and corresponds to a point of inflection in the graph of N. To find the value of t for which this occurs, we substitute $M/2$ for N in Equation (3) and solve for t:

$$\begin{aligned} \frac{M}{2} &= \frac{M}{1+be^{-ct}} \\ 1+be^{-ct} &= 2 \\ e^{-ct} &= \frac{1}{b} \\ e^{ct} &= b \\ ct &= \ln b \qquad \text{(logarithmic form)} \\ t &= \frac{\ln b}{c} \end{aligned}$$

Therefore, the maximum rate of growth occurs at the point $([\ln b]/c, M/2)$.

We remark that in Equation (3) we may replace e^{-c} by C, and then the logistic function has the following form:

Alternative Form of Logistic Function

$$N = \frac{M}{1+bC^t}$$

EXAMPLE 1 Logistic Growth of Club Membership

Suppose the membership in a new country club is to be a maximum of 800 persons, due to limitations of the physical plant. One year ago the initial membership was 50 persons, and now there are 200. Provided that enrollment follows a logistic function, how many members will there be three years from now?

Solution: Let N be the number of members enrolled t years after the formation of the club. Then, from Equation (3),

$$N = \frac{M}{1+be^{-ct}}$$

Here $M = 800$, and when $t = 0$, we have $N = 50$. So

$$\begin{aligned} 50 &= \frac{800}{1+b} \\ 1+b &= \frac{800}{50} = 16 \\ b &= 15 \end{aligned}$$

Thus,

$$N = \frac{800}{1+15e^{-ct}} \tag{4}$$

When $t = 1$, then $N = 200$, so we have

$$\begin{aligned} 200 &= \frac{800}{1+15e^{-c}} \\ 1+15e^{-c} &= \frac{800}{200} = 4 \\ e^{-c} &= \frac{3}{15} = \frac{1}{5} \end{aligned}$$

Hence, $c = -\ln\frac{1}{5} = \ln 5$. Rather than substituting this value of c into Equation (4), it is more convenient to substitute the value of e^{-c} into the same equation:

$$N = \frac{800}{1 + 15\left(\frac{1}{5}\right)^t}$$

Three years from now, t will be 4. Therefore,

$$N = \frac{800}{1 + 15\left(\frac{1}{5}\right)^4} \approx 781$$

▶ **Now work Problem 5.**

Modeling the Spread of a Rumor

Let us now consider a simplified model[8] of how a rumor spreads in a population of size M. A similar situation would be the spread of an epidemic or new fad.

Let $N = N(t)$ be the number of persons who know the rumor at time t. We shall assume that those who know the rumor spread it randomly in the population and that those who are told the rumor become spreaders of the rumor. Furthermore, we shall assume that each knower tells the rumor to k individuals per unit of time. (Some of these k individuals may already know the rumor.) We want an expression for the rate of increase of the knowers of the rumor. Over a unit of time, each of approximately N persons will tell the rumor to k persons. Thus, the total number of persons who are told the rumor over the unit of time is (approximately) Nk. However, we are interested only in *new* knowers. The proportion of the population that does not know the rumor is $(M - N)/M$. Hence, the total number of new knowers of the rumor is

$$Nk\left(\frac{M-N}{M}\right)$$

which can be written $(k/M)N(M - N)$. Therefore,

$$\begin{aligned}\frac{dN}{dt} &= \frac{k}{M}N(M-N)\\ &= KN(M-N) \qquad \text{where } K = \frac{k}{M}\end{aligned}$$

This differential equation has the form of Equation (1), so its solution, from Equation (3), is a *logistic function:*

$$N = \frac{M}{1 + be^{-ct}}$$

EXAMPLE 2 Campus Rumor

In a large university of 45,000 students, a sociology major is researching the spread of a new campus rumor. When she begins her research, she determines that 300 students know the rumor. After one week, she finds that 900 know it. Estimate the number of students who know it four weeks after the research begins by assuming logistic growth. Give the answer to the nearest thousand.

---

[8]More simplified, that is, than the model described in the Mathematical Snapshot for Chapter 8.

Solution: Let N be the number of students who know the rumor t weeks after the research begins. Then

$$N = \frac{M}{1 + be^{-ct}}$$

Here M, the size of the population, is 45,000, and when $t = 0$, $N = 300$. So we have

$$300 = \frac{45{,}000}{1 + b}$$
$$1 + b = \frac{45{,}000}{300} = 150$$
$$b = 149$$

Thus,

$$N = \frac{45{,}000}{1 + 149e^{-ct}}$$

When $t = 1$, then $N = 900$. Hence,

$$900 = \frac{45{,}000}{1 + 149e^{-c}}$$
$$1 + 149e^{-c} = \frac{45{,}000}{900} = 50$$

Therefore, $e^{-c} = \frac{49}{149}$, so

$$N = \frac{45{,}000}{1 + 149\left(\frac{49}{149}\right)^t}$$

When $t = 4$,

$$N = \frac{45{,}000}{1 + 149\left(\frac{49}{149}\right)^4} \approx 16{,}000$$

After four weeks, approximately 16,000 students know the rumor.

▶ **Now work Problem 3.**

Newton's Law of Cooling

We conclude this section with an interesting application of a differential equation. If a homicide is committed, the temperature of the victim's body will gradually decrease from 37°C (normal body temperature) to the temperature of the surroundings (ambient temperature). In general, the temperature of the cooling body changes at a rate proportional to the difference between the temperature of the body and the ambient temperature. This statement is known as **Newton's law of cooling.** Thus, if $T(t)$ is the temperature of the body at time t and the ambient temperature is a, then

$$\frac{dT}{dt} = k(T - a)$$

where k is the constant of proportionality. Therefore, Newton's law of cooling is a differential equation. It can be applied to determine the time at which a homicide was committed, as the next example illustrates.

EXAMPLE 3 **Time of Murder**

A wealthy industrialist was found murdered in his home. Police arrived on the scene at 11:00 P.M. The temperature of the body at that time was 31°C, and one hour later it was 30°C. The temperature of the room in which the body was found was 22°C. Estimate the time at which the murder occurred.

Solution: Let t be the number of hours after the body was discovered and $T(t)$ be the temperature (in degrees Celsius) of the body at time t. We want to find the value of t for which $T = 37$ (normal body temperature). This value of t will, of course, be negative. By Newton's law of cooling,

$$\frac{dT}{dt} = k(T - a)$$

where k is a constant and a (the ambient temperature) is 22. Thus,

$$\frac{dT}{dt} = k(T - 22)$$

Separating variables, we have

$$\frac{dT}{T - 22} = k\,dt$$

$$\int \frac{dT}{T - 22} = \int k\,dt$$

$$\ln|T - 22| = kt + C$$

Because $T - 22 > 0$,

$$\ln(T - 22) = kt + C$$

When $t = 0$, then $T = 31$. Therefore,

$$\ln(31 - 22) = k \cdot 0 + C$$

$$C = \ln 9$$

Hence,

$$\ln(T - 22) = kt + \ln 9$$

$$\ln(T - 22) - \ln 9 = kt$$

$$\ln\frac{T - 22}{9} = kt \qquad \left(\ln a - \ln b = \ln\frac{a}{b}\right)$$

When $t = 1$, then $T = 30$, so

$$\ln\frac{30 - 22}{9} = k \cdot 1$$

$$k = \ln\frac{8}{9}$$

Thus,

$$\ln\frac{T - 22}{9} = t\ln\frac{8}{9}$$

Now we find t when $T = 37$:

$$\ln\frac{37 - 22}{9} = t\ln\frac{8}{9}$$

$$t = \frac{\ln(15/9)}{\ln(8/9)} \approx -4.34$$

Accordingly, the murder occurred about 4.34 hours *before* the time of discovery of the body (11:00 P.M.). Since 4.34 hours is (approximately) 4 hours and 20 minutes, the industrialist was murdered about 6:40 P.M.

▶ **Now work Problem 9.**

Exercise 15.6

1. **Population** The population of a city follows logistic growth and is limited to 80,000. If the population in 1995 was 40,000 and in 2000 was 50,000, what will be the population in 2005? Give your answer to the nearest hundred.

2. **Production** A company believes that the production of its product in present facilities will follow logistic growth. Presently, 200 units per day are produced, and production will increase to 300 units per day in one year. If production is limited to 500 units per day, what is the anticipated daily production in two years? Give your answer to the nearest unit.

*3. **Spread of Rumor** In a university of 40,000 students, the administration holds meetings to discuss the idea of bringing in a major rock band for homecoming weekend. Before the plans are officially announced, students representatives on the administrative council spread information about the event as a rumor. At the end of one week, 100 people know the rumor. Assuming logistic growth, how many people know the rumor after two weeks? Give your answer to the nearest hundred.

4. **Spread of Fad** A new fad is sweeping a college campus of 30,000 students. The college newspaper feels that its readers would be interested in a series on the fad. It assigns a reporter when the number of faddists is 400. One week later, there are 1200 faddists. Assuming logistic growth, find a formula for the number N of faddists t weeks after the assignment of the reporter.

*5. **Flu Outbreak** In a city whose population is 100,000, an outbreak of flu occurs. When the city health department begins its record keeping, there are 500 infected persons. One week later, there are 1000 infected persons. Assuming logistic growth, estimate the number of infected persons two weeks after record keeping begins.

6. **Population** The logistic curve for the U.S. population from 1790 to 1910 is estimated to be[9]

$$N = \frac{197.30}{1 + 35.60e^{-0.031186t}}$$

where N is the population in millions and t is in years counted from 1800. If this logistic function were valid for years after 1910, for what year would the point of inflection occur? Round your answer to one decimal place.

7. **Biology** In an experiment,[10] five *Paramecia* were placed in a test tube containing a nutritive medium. The number N of *Paramecia* in the tube at the end of t days is given approximately by

$$N = \frac{375}{1 + e^{5.2-2.3t}}$$

(a) Show that this equation can be written as

$$N = \frac{375}{1 + 181.27e^{-2.3t}}$$

and hence is a logistic function.

(b) Find $\lim_{t\to\infty} N$.

8. **Biology** In a study of the growth of a colony of unicellular organisms,[11] the equation

$$N = \frac{0.2524}{e^{-2.128x} + 0.005125} \qquad 0 \le x \le 5$$

was obtained, where N is the estimated area of the growth in square centimeters and x is the age of the colony in days after being first observed.

(a) Put this equation in the form of a logistic function.

(b) Find the area when the age of the colony is 0.

*9. **Time of Murder** A waterfront murder was committed, and the victim's body was discovered at 4:15 A.M. by police. At that time, the temperature of the body was 28°C. One hour later, the body temperature was 20°C. After checking with the weather bureau, it was determined that the temperature at the waterfront was −10°C from 11:00 P.M. to 6:00 A.M. About what time did the murder occur?

10. **Enzyme Formation** An enzyme is a protein that acts as a catalyst for increasing the rate of a chemical reaction that occurs in cells. In a certain reaction, an enzyme A is converted to another enzyme B. Enzyme B acts as a catalyst for its own formation. Let p be the

---

[9] N. Keyfitz, *Introduction to the Mathematics of Population* (Reading, MA: Addison-Wesley Publishing Company, 1968).

[10] G. F. Gause, *The Struggle for Existence* (New York: Hafner Publishing Co., 1964).

[11] A. J. Lotka, *Elements of Mathematical Biology* (New York: Dover Publications, 1956).

amount of enzyme B at time t and I be the total amount of both enzymes when $t = 0$. Suppose the rate of formation of B is proportional to $p(I - p)$. Without directly using calculus, find the value of p for which the rate of formation will be a maximum.

11. Fund-raising A small town decides to conduct a fund-raising drive for a fire engine whose cost is \$70,000. The initial amount in the fund is \$10,000. On the basis of past drives, it is determined that at t months after the beginning of the drive, the rate dx/dt at which money is contributed to such a fund is proportional to the difference between the desired goal of \$70,000 and the total amount x in the fund at that time. After one month, a total of \$40,000 is in the fund. How much will be in the fund after three months?

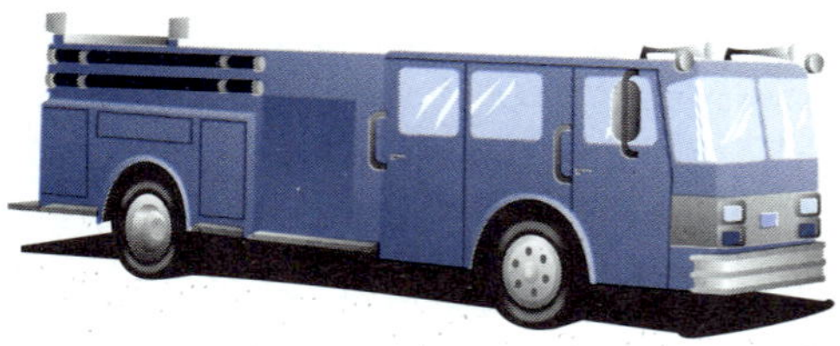

12. Birthrate In a discussion of unexpected properties of mathematical models of population, Bailey[12] considers the case in which the birthrate per *individual* is proportional to the population size N at time t. Since the birth rate per individual is $\frac{1}{N}\frac{dN}{dt}$, this means that

$$\frac{1}{N}\frac{dN}{dt} = kN$$

so that

$$\frac{dN}{dt} = kN^2 \qquad \text{(subject to } N = N_0 \text{ at } t = 0)$$

where $k > 0$. Show that

$$N = \frac{N_0}{1 - kN_0 t}$$

Use this result to show that

$$\lim N = \infty \quad \text{as} \quad t \to \left(\frac{1}{kN_0}\right)^-$$

This means that over a finite interval of time, there is an infinite amount of growth. Such a model might be useful only for rapid growth over a short interval of time.

13. Population Suppose that the rate of growth of a population is proportional to the difference between some maximum size M and the number N of individuals in the population at time t. Suppose that when $t = 0$, the size of the population is N_0. Find a formula for N.

[12]N. T. J. Bailey, *The Mathematical Approach to Biology and Medicine* (New York: John Wiley & Sons, 1967).

OBJECTIVE *To define and evaluate improper integrals.*

15.7 Improper Integrals[13]

Suppose $f(x)$ is continuous and nonnegative for $a \le x < \infty$. (See Figure 15.4.) We know that the integral $\int_a^r f(x)\,dx$ is the area of the region between the curve $y = f(x)$ and the x-axis from $x = a$ to $x = r$. As $r \to \infty$, we may think of

$$\lim_{r\to\infty} \int_a^r f(x)\,dx$$

as the area of the unbounded region that is shaded in Figure 15.5. This limit is

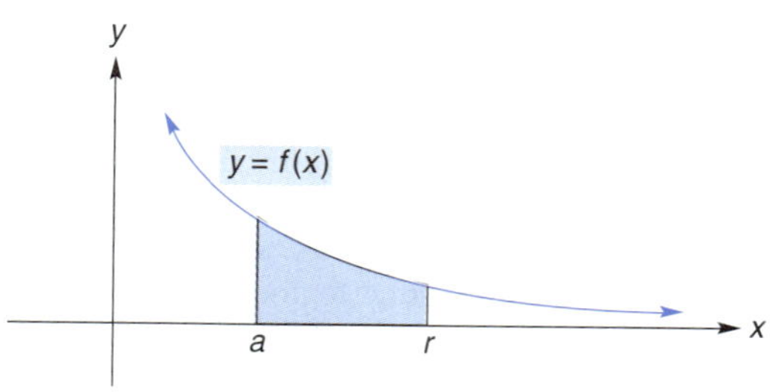

FIGURE 15.4 Area from a to r.

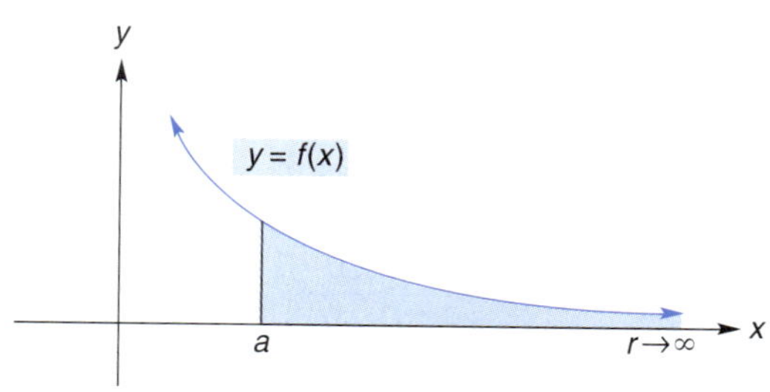

FIGURE 15.5 Area from a to r as $r \to \infty$.

[13]This section may be omitted if Chapter 16 is not covered.

abbreviated by

$$\int_a^{\infty} f(x)\,dx \tag{1}$$

called an **improper integral.** If the limit exists, $\int_a^{\infty} f(x)\,dx$ is said to be **convergent** or to *converge* to that limit. In this case the unbounded region is considered to have a finite area, and this area is represented by $\int_a^{\infty} f(x)\,dx$. If the limit does not exist, the improper integral is said to be **divergent,** and the region does not have a finite area.

We can remove the restriction that $f(x) \geq 0$. In general, the improper integral $\int_a^{\infty} f(x)\,dx$ is defined by

$$\int_a^{\infty} f(x)\,dx = \lim_{r\to\infty} \int_a^r f(x)\,dx$$

Other types of improper integrals are

$$\int_{-\infty}^{b} f(x)\,dx \tag{2}$$

and

$$\int_{-\infty}^{\infty} f(x)\,dx \tag{3}$$

In each of the three types of improper integrals [(1), (2), and (3)], the interval over which the integral is evaluated has infinite length. The improper integral in (2) is defined by

$$\int_{-\infty}^{b} f(x)\,dx = \lim_{r\to-\infty} \int_r^b f(x)\,dx$$

If this limit exists, $\int_{-\infty}^{b} f(x)\,dx$ is said to be convergent. Otherwise, it is divergent. We shall define the improper integral in (3) after the following example.

APPLICATIONS IN PRACTICE 1

Improper Integrals of the Form $\int_a^{\infty} f(x)dx$ and $\int_{-\infty}^{b} f(x)dx$

The rate at which the human body eliminates a certain drug from its system may be approximated by $R(t) = 3e^{-0.1t} - 3e^{-0.3t}$, where $R(t)$ is in milliliters per minute and t is the time in minutes since the drug was taken. Find $\int_0^{\infty}(3e^{-0.1t} - 3e^{-0.3t})\,dt$, the total amount of the drug that is eliminated.

EXAMPLE 1 Improper Integrals of the Form $\int_a^{\infty} f(x)dx$ and $\int_{-\infty}^{b} f(x)\,dx$

Determine whether the following improper integrals are convergent or divergent. For any convergent integral, determine its value.

a. $\displaystyle\int_1^{\infty} \frac{1}{x^3}\,dx.$

Solution:

$$\int_1^{\infty} \frac{1}{x^3}\,dx = \lim_{r\to\infty} \int_1^r x^{-3}\,dx = \lim_{r\to\infty} -\frac{x^{-2}}{2}\Bigg|_1^r$$

$$= \lim_{r\to\infty} \left[-\frac{1}{2r^2} + \frac{1}{2}\right] = -0 + \frac{1}{2} = \frac{1}{2}$$

Therefore, $\displaystyle\int_1^{\infty} \frac{1}{x^3}\,dx$ converges to $\frac{1}{2}$.

b. $\int_{-\infty}^{0} e^x\,dx$.

Solution:

$$\int_{-\infty}^{0} e^x\,dx = \lim_{r\to-\infty}\int_{r}^{0} e^x\,dx = \lim_{r\to-\infty} e^x\Big|_r^0$$
$$= \lim_{r\to-\infty}(1-e^r) = 1-0 = 1 \qquad (e^0 = 1)$$

(Here we used the fact that as $r \to -\infty$, the graph of $y = e^r$ approaches the r-axis, so $e^r \to 0$.) Therefore, $\int_{-\infty}^{0} e^x\,dx$ converges to 1.

c. $\int_{1}^{\infty} \frac{1}{\sqrt{x}}\,dx$.

Solution:

$$\int_{1}^{\infty} \frac{1}{\sqrt{x}}\,dx = \lim_{r\to\infty}\int_{1}^{r} x^{-1/2}\,dx = \lim_{r\to\infty} 2x^{1/2}\Big|_1^r$$
$$= \lim_{r\to\infty} 2(\sqrt{r}-1) = \infty$$

Therefore, the improper integral diverges.

▸ **Now work Problem 3.**

The improper integral $\int_{-\infty}^{\infty} f(x)\,dx$ is defined in terms of improper integrals of the forms (1) and (2):

$$\int_{-\infty}^{\infty} f(x)\,dx = \int_{-\infty}^{0} f(x)\,dx + \int_{0}^{\infty} f(x)\,dx \qquad \textbf{(4)}$$

If *both* integrals on the right side of Equation (4) are convergent, then $\int_{-\infty}^{\infty} f(x)dx$ is said to be convergent; otherwise, it is divergent.

EXAMPLE 2 **An Improper Integral of the Form $\int_{-\infty}^{\infty} f(x)\,dx$**

Determine whether $\int_{-\infty}^{\infty} e^x\,dx$ is convergent or divergent.

Solution:

$$\int_{-\infty}^{\infty} e^x\,dx = \int_{-\infty}^{0} e^x\,dx + \int_{0}^{\infty} e^x\,dx$$

By Example 1(b), $\int_{-\infty}^{0} e^x\,dx = 1$. On the other hand,

$$\int_{0}^{\infty} e^x\,dx = \lim_{r\to\infty}\int_{0}^{r} e^x\,dx = \lim_{r\to\infty} e^x\Big|_0^r = \lim_{r\to\infty}(e^r-1) = \infty$$

Since $\int_{0}^{\infty} e^x\,dx$ is divergent, $\int_{-\infty}^{\infty} e^x\,dx$ is also divergent.

▸ **Now work Problem 11.**

EXAMPLE 3 **Density Function**

In statistics, a function f is called a density function if $f(x) \geq 0$ *and*

$$\int_{-\infty}^{\infty} f(x)\,dx = 1$$

Suppose

$$f(x) = \begin{cases} ke^{-x} & \text{for } x \geq 0 \\ 0 & \text{elsewhere} \end{cases}$$

is a density function. Find k.

Solution: We write the equation $\int_{-\infty}^{\infty} f(x)\,dx = 1$ as

$$\int_{-\infty}^{0} f(x)\,dx + \int_{0}^{\infty} f(x)\,dx = 1$$

Since $f(x) = 0$ for $x < 0$, $\int_{-\infty}^{0} f(x)\,dx = 0$. Thus,

$$\int_{0}^{\infty} ke^{-x}\,dx = 1$$

$$\lim_{r\to\infty} \int_{0}^{r} ke^{-x}\,dx = 1$$

$$\lim_{r\to\infty} -ke^{-x}\Big|_{0}^{r} = 1$$

$$\lim_{r\to\infty} (-ke^{-r} + k) = 1$$

$$0 + k = 1 \qquad \left(\lim_{r\to\infty} e^{-r} = 0\right)$$

$$k = 1$$

▸ **Now work Problem 13.**

Exercise 15.7

In Problems 1–12, determine the integrals if they exist. Indicate those that are divergent.

1. $\int_{3}^{\infty} \frac{1}{x^2}\,dx$

2. $\int_{2}^{\infty} \frac{1}{(2x-1)^3}\,dx$

*3. $\int_{1}^{\infty} \frac{1}{x}\,dx$

4. $\int_{2}^{\infty} \frac{1}{\sqrt[3]{(x+2)^2}}\,dx$

5. $\int_{1}^{\infty} e^{-x}\,dx$

6. $\int_{0}^{\infty} (5+e^{-x})\,dx$

7. $\int_{1}^{\infty} \frac{1}{\sqrt{x}}\,dx$

8. $\int_{4}^{\infty} \frac{x\,dx}{\sqrt{(x^2+9)^3}}$

9. $\int_{-\infty}^{-2} \frac{1}{(x+1)^3}\,dx$

10. $\int_{-\infty}^{3} \frac{1}{\sqrt{7-x}}\,dx$

*11. $\int_{-\infty}^{\infty} 2xe^{-x^2}\,dx$

12. $\int_{-\infty}^{\infty} (5-3x)\,dx$

*13. **Density Function** The density function for the life x, in hours, of an electronic component in a radiation meter is given by

$$f(x) = \begin{cases} \dfrac{k}{x^2} & \text{for } x \geq 800 \\ 0 & \text{for } x < 800 \end{cases}$$

(a) If k satisfies the condition that $\int_{800}^{\infty} f(x)\,dx = 1$, find k.

(b) The probability that the component will last at least 1200 hours is given by $\int_{1200}^{\infty} f(x)\,dx$. Evaluate this integral.

14. Density Function Given the density function

$$f(x) = \begin{cases} ke^{-4x} & \text{for } x \geq 0 \\ 0 & \text{elsewhere} \end{cases}$$

find k. [*Hint:* See Example 3.]

15. Future Profits For a business, the present value of all future profits at an annual interest rate r compounded continuously is given by

$$\int_0^\infty p(t)e^{-rt}\,dt$$

where $p(t)$ is the profit per year in dollars at time t. If $p(t) = 240{,}000$ and $r = 0.06$, evaluate this integral.

16. Psychology In a psychological model for signal detection,[14] the probability α (a Greek letter read "alpha") of reporting a signal when no signal is present is given by

$$\alpha = \int_{x_c}^\infty e^{-x}\,dx \qquad x \geq 0$$

The probability β (a Greek letter read "beta") of detecting a signal when it is present is

$$\beta = \int_{x_c}^\infty ke^{-kx}\,dx \qquad x \geq 0$$

In both integrals, x_c is a constant (called a criterion value in this model). Find α and β if $k = \frac{1}{8}$.

17. Find the area of the region in the third quadrant bounded by the curve $y = e^{3x}$ and the x-axis.

18. Economics In discussing entrance of a firm into an industry, Stigler[15] uses the equation

$$V = \pi_0 \int_0^\infty e^{\theta t}e^{-\rho t}\,dt$$

where π_0, θ (a Greek letter read "theta"), and ρ (a Greek letter read "rho") are constants. Show that $V = \pi_0/(\rho - \theta)$ if $\theta < \rho$.

19. Population The predicted rate of growth per year of the population of a certain small city is given by

$$\frac{40{,}000}{(t+2)^2}$$

where t is the number of years from now. In the long run (that is, as $t \to \infty$), what is the expected change in population from today's level?

[14] D. Laming, *Mathematical Psychology* (New York: Academic Press, 1973).

[15] G. Stigler, *The Theory of Price*, 3d ed. (New York: Macmillan Publishing Company, 1966), p. 344.

15.8 Review

Important Terms and Symbols

Section 15.1	integration by parts
Section 15.2	proper rational function partial fractions
Section 15.3	present value of continuous annuity accumulated amount of continuous annuity
Section 15.4	average value of function
Section 15.5	first-order differential equation separation of variables exponential growth exponential decay decay constant half-life
Section 15.6	logistic function Newton's law of cooling
Section 15.7	improper integral, $\int_a^\infty f(x)\,dx$, $\int_{-\infty}^b f(x)\,dx$, $\int_{-\infty}^\infty f(x)\,dx$

Summary

Sometimes we can easily determine an integral whose form is $u\,dv$, where u and v are functions of the same variable, by applying the formula for integration by parts:

$$\int u\,dv = uv - \int v\,du$$

A proper rational function may be integrated by applying the technique of partial fractions. Here we express the rational function as a sum of fractions, each of which is easier to integrate than the original function.

To determine an integral that does not have a familiar form, you may be able to match it with a formula in a table of integrals. However, it may be necessary to transform the given integral into an equivalent form before the matching can occur.

An annuity is a series of payments over a period of time. Suppose payments are made continuously for T years such that a payment at time t is at the rate of $f(t)$ per year. If the annual rate of interest is r, compounded continuously, then the present value of the continuous annuity is given by

$$A = \int_0^T f(t)e^{-rt}\,dt$$

and the accumulated amount is given by

$$S = \int_0^T f(t)e^{r(T-t)}\,dt$$

The average value $\overline{f}$ of a function f over the interval $[a, b]$ is given by

$$\overline{f} = \frac{1}{b-a}\int_a^b f(x)\,dx$$

An equation that involves the derivative of an unknown function is called a differential equation. If the highest order derivative that occurs is the first, the equation is called a first-order differential equation. Some first-order differential equations may be solved by the method of separation of variables. In that method, by considering the derivative to be a quotient of differentials, we rewrite the equation so that each side contains only one variable and a differential is not in a denominator. Integrating both sides of the resulting equation gives the solution. This solution involves a constant of integration and is called the general solution of the differential equation. If the unknown function must satisfy the condition that it has a specific function value for a given value of the independent variable, then a particular solution may be found.

Differential equations arise when we know a relation involving the rate of change of a function. For example, if a quantity N at time t is such that it changes at a rate proportional to the amount present, then

$$\frac{dN}{dt} = kN \qquad \text{where } k \text{ is a constant}$$

The solution of this differential equation is

$$N = N_0e^{kt}$$

where N_0 is the quantity present at $t = 0$. The value of k may be determined when the value of N is known for a given value of t (other than $t = 0$). If k is positive, then N follows an exponential law of growth; if k is negative, N follows an exponential law of decay. If N represents a quantity of a radioactive element, then

$$\frac{dN}{dt} = -\lambda N \qquad \text{where } \lambda \text{ is a positive constant}$$

Thus, N follows an exponential law of decay, and hence,

$$N = N_0e^{-\lambda t}$$

The constant λ is called the decay constant. The time for one-half of the element to decay is the half-life of the element:

$$\text{half-life} = \frac{\ln 2}{\lambda} \approx \frac{0.69315}{\lambda}$$

A quantity N may follow a rate of growth given by

$$\frac{dN}{dt} = KN(M-N) \qquad \text{where } K, M \text{ are constants}$$

Solving this differential equation gives a function of the form

$$N = \frac{M}{1+be^{-ct}} \qquad \text{where } b, c \text{ are constants}$$

which is called a logistic function. Many population sizes can be described by a logistic function. In this case, M represents the limit of the size of the population. A logistic function is also used in analyzing the spread of a rumor.

Newton's law of cooling states that the temperature T of a cooling body at time t changes at a rate proportional to the difference $T - a$ where a is the ambient temperature. Thus,

$$\frac{dT}{dt} = k(T-a) \qquad \text{where } k \text{ is a constant}$$

The solution of this differential equation can be used to determine, for example, the time at which a homicide was committed.

An integral of the form

$$\int_a^\infty f(x)\,dx \qquad \int_{-\infty}^b f(x)\,dx \qquad \text{or} \qquad \int_{-\infty}^\infty f(x)\,dx$$

is called an improper integral. The first two integrals are defined as follows:

$$\int_a^\infty f(x)\,dx = \lim_{r\to\infty}\int_a^r f(x)\,dx$$

and

$$\int_{-\infty}^b f(x)\,dx = \lim_{r\to-\infty}\int_r^b f(x)\,dx$$

If $\int_a^\infty f(x)\,dx$ (or $\int_{-\infty}^b f(x)\,dx$) is a finite number, we say that the integral is convergent; otherwise, it is divergent. The improper integral $\int_{-\infty}^\infty f(x)\,dx$ is defined by

$$\int_{-\infty}^\infty f(x)\,dx = \int_{-\infty}^0 f(x)\,dx + \int_0^\infty f(x)\,dx$$

If both integrals on the right side are convergent, $\int_{-\infty}^\infty f(x)\,dx$ is said to be convergent; otherwise, it is divergent.

Review Problems

Problem numbers shown in color indicate problems suggested for use as a practice chapter test.

In Problems 1–22, determine the integrals.

1. $\int x \ln x \, dx$

2. $\int \frac{1}{\sqrt{4x^2+1}} \, dx$

3. $\int_0^2 \sqrt{4x^2+9} \, dx$

4. $\int \frac{16x}{3-4x} \, dx$

5. $\int \frac{15x-2}{(3x+1)(x-2)} \, dx$

6. $\int_e^{e^2} \frac{1}{x \ln x} \, dx$

7. $\int \frac{dx}{x(x+2)^2}$

8. $\int \frac{dx}{x^2-1}$

9. $\int \frac{dx}{x^2\sqrt{9-16x^2}}$

10. $\int x^{1/3} \ln \sqrt{x} \, dx$

11. $\int \frac{9\,dx}{x^2-9}$

12. $\int \frac{x}{\sqrt{2+5x}} \, dx$

13. $\int 49xe^{7x} \, dx$

14. $\int \frac{dx}{2+3e^{4x}}$

15. $\int \frac{dx}{2x \ln 2x}$

16. $\int \frac{dx}{x(2+x)}$

17. $\int \frac{2x}{3+2x} \, dx$

18. $\int \frac{dx}{x^2\sqrt{4x^2-9}}$

[16]19. $\int \frac{5x^2+2}{x^3+x} \, dx$

[16]20. $\int \frac{3x^3+5x^2+4x+3}{x^4+x^3+x^2} \, dx$

[17]21. $\int \frac{\ln(x+1)}{\sqrt{x+1}} \, dx$

[17]22. $\int (\ln x)^3 \, dx$

23. Find the average value of $f(x) = 3x^2 + 2x$ over the interval $[2, 4]$.

24. Find the average value of $f(t) = t^2e^t$ over the interval $[0, 1]$.

In Problems 25 and 26, solve the differential equations.

25. $y' = 3x^2y + 2xy \qquad y > 0$

26. $y' - 2xe^{x^2-y+3} = 0 \qquad y(0) = 3$

In Problems 27–30, determine the improper integrals if they exist.[18] *Indicate those that are divergent.*

27. $\int_3^\infty \frac{1}{x^3} \, dx$

28. $\int_{-\infty}^0 e^{2x} \, dx$

29. $\int_1^\infty \frac{1}{2x} \, dx$

30. $\int_{-\infty}^\infty xe^{1-x^2} \, dx$

31. **Population** The population of a city in 1985 was 100,000 and in 2000 was 120,000. Assuming exponential growth, project the population in 2015.

32. **Population** The population of a city doubles every 10 years due to exponential growth. At a certain time, the population is 40,000. Find an expression for the number of people N at time t years later. Give your answer in terms of ln 2.

33. **Radioactive** If 95% of a radioactive substance remains after 100 years, find the decay constant and, to the nearest percent, give the percentage of the original amount present after 200 years.

34. **Medicine** Suppose q is the amount of penicillin in the body at time t, and let q_0 be the amount at $t = 0$. Assume that the rate of change of q with respect to t is proportional to q and that q decreases as t increases. Then we have $dq/dt = -kq$, where $k > 0$. Solve for q. What percentage of the original amount present is there when $t = 5/k$?

35. **Biology** Two organisms are initially placed in a medium and begin to multiply. The number N of organisms that are present after t days is recorded on a graph with the horizontal axis labeled t and the vertical axis labeled N. It is observed that the points lie on a logistic curve. The number of organisms present after 6 days is 300, and beyond 10 days the number approaches a limit of 450. Find the logistic equation.

36. **College Enrollment** A college believes that enrollment follows logistic growth. Last year enrollment was 1000, and this year it is 1100. If the college can accommodate a maximum of 2000 students, what is the anticipated enrollment next year? Give your answer to the nearest hundred.

37. **Time of Murder** A coroner is called in on a murder case. He arrives at 6:00 P.M. and finds that the victim's temperature is 35°C. One hour later the body temperature is 34°C. The temperature of the room is 25°C. About what time was the murder committed? (Assume that normal body temperature is 37°C.)

38. **Annuity** Find the present value, to the nearest dollar, of a continuous annuity at an annual rate of 6% for 12 years if the payment at time t is at the annual rate of $f(t) = 10t$ dollars.

[19]39. **Hospital Discharges** For a group of hospitalized individuals, suppose the proportion that has been discharged at the end of days t is given by

$$\int_0^t f(x)\,dx$$

where $f(x) = 0.008e^{-0.01x} + 0.00004e^{-0.0002x}$. Evaluate

$$\int_0^\infty f(x)\,dx$$

[16]Problems 19 and 20 refers to Section 15.2.

[17]Problems 21 and 22 refer to Section 15.1.

[18]Problems 27–30 refer to Section 15.7.

[19]Problems 39 and 40 refer to Section 15.7.

[19]**40. Product Consumption** Suppose that $A(t)$ is the amount of a product that is consumed at time t and that A follows an exponential law of growth. If $t_1 < t_2$ and at time t_2 the amount consumed, $A(t_2)$, is double the amount consumed at time t_1, $A(t_1)$, then $t_2 - t_1$ is called a doubling period. In a discussion of exponential growth, Shonle[20] states that under exponential growth, "the amount of a product consumed during one doubling period is equal to the total used for all time up to the beginning of the doubling period in question." To justify this statement, reproduce his argument as follows. The amount of the product used up to time t_1 is given by

$$\int_{-\infty}^{t_1} A_0 e^{kt}\, dt \qquad k > 0$$

where A_0 is the amount when $t = 0$. Show that this is equal to $(A_0/k)e^{kt_1}$. Next, the amount used during the time interval from t_1 to t_2 is

$$\int_{t_1}^{t_2} A_0 e^{kt}\, dt$$

Show that this is equal to

$$\frac{A_0}{k} e^{kt_1}\left[e^{k(t_2-t_1)} - 1\right] \qquad (1)$$

If the interval $[t_1, t_2]$ is a doubling period, then

$$A_0 e^{kt_2} = 2A_0 e^{kt_1}$$

Show that this relationship implies that $e^{k(t_2-t_1)} = 2$. Substitute this value into Equation (1); your result should be the same as the total used during all time up to t_1, namely, $(A_0/k)e^{kt_1}$.

[20]J. I. Shonle, *Environmental Applications of General Physics* (Reading, MA: Addison-Wesley Publishing Company, 1975).

41. Revenue, Cost, and Profit The following table gives values of a company's marginal-revenue (MR) and marginal-cost (MC) functions:

q	0	3	6	9	12	15	18
MR	25	22	18	13	7	3	0
MC	15	14	12	10	7	4	2

The company's fixed cost is 25. Assume that profit is a maximum when MR = MC and that this occurs when $q = 12$. Moreover, assume that the output of the company is chosen to maximize the profit. Use the trapezoidal rule and Simpson's rule for each of the following parts.

(a) Estimate the total revenue by using as many data values as possible.

(b) Estimate the total cost by using as few data values as possible.

(c) Determine how the maximum profit is related to the area enclosed by the line $q = 0$ and the MR and MC curves, and use this relation to estimate the maximum profit as accurately as possible.

Mathematical SNAPSHOT

Dieting

Today there is great concern about diet and weight loss. Some people want to lose weight in order to "look good." Others lose weight for physical fitness or for health reasons. In fact, some lose weight because of peer pressure. Advertisements for weight control programs frequently appear on television and in newspapers and magazines. In many bookstores, entire sections are devoted to diet and weight control.

Suppose you want to determine a mathematical model of the weight of a person on a restricted caloric diet.[21] A person's weight depends both on the daily rate of energy intake, say C calories per day, and on the daily rate of energy consumption, which is typically between 15 and 20 calories per day for each pound of body weight. Consumption depends on age, sex, metabolic rate, and so on. For an average value of 17.5 calories per pound per day, a person weighing w pounds expends $17.5w$ calories per day. If $C = 17.5w$, then his or her weight remains constant; otherwise weight gain or loss occurs according to whether C is greater or less than $17.5w$.

How fast will weight gain or loss occur? The most plausible physiological assumption is that dw/dt is proportional to the net excess (or deficit) in the number of calories per day, $C - 17.5w$. That is,

$$\frac{dw}{dt} = K(C - 17.5w)$$

where K is a constant. The left side of the equation has units of pounds per day, and $C - 17.5w$ has units of calories per day. Hence the units of K are pounds per calorie. Therefore, you need to know how many pounds each excess or deficit calorie puts on or takes off. The commonly used dietetic conversion factor is that 3500 calories is equivalent to one pound. Thus $K = 1/3500$ lb per calorie.

Now, the differential equation modeling weight gain or loss is

$$\frac{dw}{dt} = \frac{1}{3500}(C - 17.5w) \qquad \textbf{(2)}$$

If C is constant, the equation is separable and its solution is

$$w(t) = \frac{C}{17.5} + \left(w_0 - \frac{C}{17.5}\right)e^{-0.005t} \qquad \textbf{(3)}$$

where w_0 is the initial weight and t is in days. In the long run, note that the equilibrium weight (that is, the weight as $t \to \infty$) is $w_{eq} = C/17.5$.

For example, if someone initially weighing 180 lb adopts a diet of 2500 calories per day, then we have $w_{eq} = 2500/17.5 \approx 143$ lb and the weight function is

$$\begin{aligned} w(t) &\approx 143 + (180 - 143)e^{-0.005t} \\ &= 143 + 37e^{-0.005t} \end{aligned}$$

[21]Adapted from A. C. Segal, "A Linear Diet Model," *College Mathematics Journal* 18, no. 1 (1987), 44–45. By permission of the Mathematical Association of America.

Figure 15.6 shows the graph of $w(t)$. Notice how long it takes to even get close to the equilibrium weight of 143 lb. The half-life for the process is $(\ln 2)/0.005 \approx 138.6$ days, about 20 weeks. (It would take about 584 days, or 83 weeks, to get to 145 lb.) This may be why so many dieters give up in frustration.

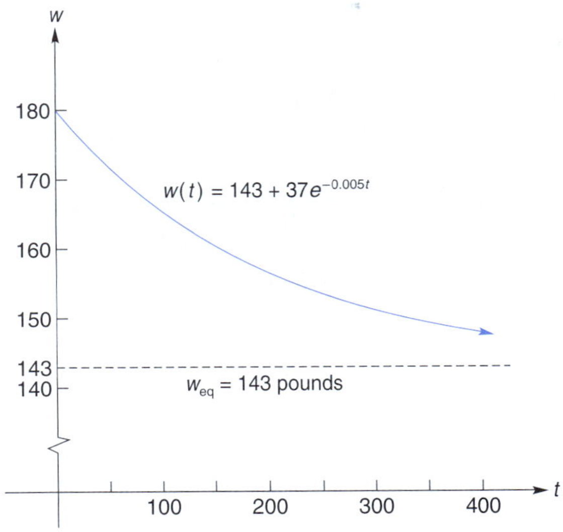

FIGURE 15.6 Weight as a function of time.

Exercises

1. If a person weighing 200 lb adopts a 2000-calorie-per-day diet, determine, to the nearest pound, the equilibrium weight w_{eq}. To the nearest day, after how many days will the person reach a weight of 175 lb? Obtain the answer either algebraically or using a graphing calculator.

2. Show that the solution of Equation (2) is given by Equation (3).

3. The weight of a person on a restricted caloric diet at time t is given by $w(t)$. [See Equation (3).] The difference between this weight and the equilibrium weight w_{eq} is $w(t) - w_{\text{eq}}$. Suppose it takes d days for the person to lose half of the weight difference. Then

$$w(t+d) = w(t) - \tfrac{1}{2}[w(t) - w_{\text{eq}}]$$

By solving this equation for d, show that $d = \dfrac{\ln 2}{0.005}$.

4. Ideally, weight loss goals should be set in consultation with a physician. In general, however, one's ideal weight is related to one's height by the body mass index (BMI), which equals weight in kilograms divided by height in meters squared. The optimal BMI range is 18.5 to 24.9.

   How many pounds would a 5′6″-tall, 180-pound woman need to lose to be within the ideal BMI range? (Be mindful of units as you calculate the answer.) To the nearest day, how long would it take for her to lose this much weight on a 2200-calorie-per-day diet?

   Further information on weight and dieting can be found at
   www.consumer.gov/weightloss/setgoals.htm.

5. What are the pros and cons of a "crash" diet, one based on drastic changes in eating habits to achieve rapid weight loss?

CHAPTER

16

Continuous Random Variables

Suppose you are designing a cellular phone network for a large urban area. Ideally, the system would always have enough capacity to meet demand. However, you know that demand fluctuates. Some periods of increased demand can be foreseen, such as holidays when many people call their families. But other times are not predictable, such as after an earthquake or some other natural disaster, when many people may be calling emergency services or checking in with friends and relatives. Building and operating a phone system with enough capacity to handle any sudden rise in demand, no matter how great, would be hugely expensive. How do you strike a balance between the goal of serving customers and the need to limit costs?

A sensible approach would be to design a system capable of handling the load of phone traffic under normally busy conditions, and to accept the fact that on rare occasions, heavy traffic will lead to overloads. You cannot always predict when overloads will occur, since disasters such as earthquakes are unforeseen occurrences. But some good *probabilistic* predictions of future traffic volume would suffice. You could build a system that would meet demand 99.4% of the time, for example. The remaining 0.6% of the time, customers would simply have to put up with intermittent delays in service.

A probabilistic description of traffic on a phone network is an example of a probability density function. Such functions are the focus of this chapter. They have a wide range of applications—not only calculating how often a system will be overloaded, for example, but also calculating its average load. This allows prediction of such things as average power consumption and average volume of system maintenance activity.

OBJECTIVE *To introduce continuous random variables; to discuss density functions, including uniform and exponential distributions; to discuss cumulative distribution functions; and to compute the mean, variance, and standard deviation for a continuous random variable.*

16.1 Continuous Random Variables

Density Functions

In Chapter 9, the random variables that we considered were primarily discrete. Now we shall concern ourselves with ***continuous* random variables.** A random variable is continuous if it can assume any value in some interval or intervals. A continuous random variable usually represents data that are measured, such as heights, weights, distances, and periods of time.

For example, the number of hours of life of a calculator battery is a continuous random variable X. If the maximum possible life is 1000 hours, then X can assume any value in the interval $[0, 1000]$. In a practical sense, the likelihood that X will assume a single specified value, such as 764.1238, is extremely remote. It is more meaningful to consider the likelihood of X lying within an *interval*, such as that between 764 and 765. (Thus, $764 < X < 765$.) In general, *with a continuous random variable, our concern is the likelihood that it falls within an interval and not that it assumes a particular value.*

As another example, consider an experiment in which a number X is randomly selected from the interval $[0, 2]$. Then X is a continuous random variable. What is the probability that X lies in the interval $[0, 1]$? Because we can loosely think of $[0, 1]$ as being "half" the interval $[0, 2]$, a reasonable (and correct) answer is $\frac{1}{2}$. Similarly, if we think of the interval $[0, \frac{1}{2}]$ as being one-fourth of $[0, 2]$, then $P(0 \leq X \leq \frac{1}{2}) = \frac{1}{4}$. Actually, each one of these probabilities is simply the length of the given interval divided by the length of $[0, 2]$. For example,

$$P\left(0 \leq X \leq \frac{1}{2}\right) = \frac{\text{length of } [0, \frac{1}{2}]}{\text{length of } [0, 2]} = \frac{\frac{1}{2}}{2} = \frac{1}{4}$$

Let us now consider a similar experiment in which X denotes a number chosen at random from the interval $[0, 1]$. As you might expect, the probability that X will assume a value in any given interval within $[0, 1]$ is equal to the length of the given interval divided by the length of $[0, 1]$. Because $[0, 1]$ has length 1, we can simply say that the probability of X falling in an interval is the length of the interval. For example,

$$P(0.2 \leq X \leq 0.5) = 0.5 - 0.2 = 0.3$$

and $P(0.2 \leq X \leq 0.2001) = 0.0001$. Clearly, as the length of an interval approaches 0, the probability that X assumes a value in that interval approaches 0. Keeping this in mind, we can think of a single number such as 0.2 as the limiting case of an interval as the length of the interval approaches 0. (Think of $[0.2, 0.2 + x]$ as $x \to 0$.) Thus, $P(X = 0.2) = 0$. In general, *the probability that a continuous random variable X assumes a particular value is* 0. As a result, **the probability that X lies in some interval is not affected by whether or not either of the endpoints of the interval is included or excluded.** For example,

$$\begin{aligned} P(X \leq 0.4) &= P(X < 0.4) + P(X = 0.4) \\ &= P(X < 0.4) + 0 \\ &= P(X < 0.4) \end{aligned}$$

Similarly, $P(0.2 \leq X \leq 0.5) = P(0.2 < X < 0.5)$.

We can geometrically represent the probabilities associated with a continuous random variable X. This is done by means of the graph of a function $y = f(x)$ such that the area under this graph (and above the x-axis) between the lines $x = a$ and $x = b$ represents the probability that X assumes a value between a and b. (See Figure 16.1.) Since this area is given by the definite integral $\int_a^b f(x)\,dx$, we have

$$P(a \le X \le b) = \int_a^b f(x)\,dx$$

We call the function f the *probability density function* for X (or simply the *density function* for X) and say that it defines the *distribution of* X. Because probabilities are always nonnegative, it is always true that $f(x) \ge 0$. Also, because the event $-\infty < X < \infty$ must occur, the total area under the density function curve must be 1. That is, $\int_{-\infty}^{\infty} f(x)\,dx = 1$. In summary, we have the following definition.

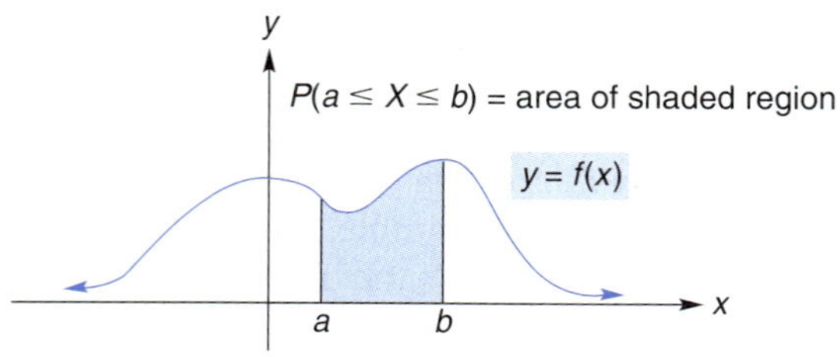

FIGURE 16.1 Probability density function.

Definition

If X is a continuous random variable, then a function $y = f(x)$ is called a ***(probability) density function*** for X if and only if it has the following properties:

1. $f(x) \ge 0$
2. $\int_{-\infty}^{\infty} f(x)\,dx = 1$
3. $P(a \le X \le b) = \int_a^b f(x)\,dx$

To illustrate a density function, we return to the previous experiment in which a number X is chosen at random from the interval $[0, 1]$. Recall that

$$P(a \le X \le b) = \text{length of } [a, b] = b - a \qquad \textbf{(1)}$$

where a and b are in $[0, 1]$. We shall show that the function

$$f(x) = \begin{cases} 1 & \text{if } 0 \le x \le 1 \\ 0 & \text{otherwise} \end{cases} \qquad \textbf{(2)}$$

whose graph appears in Figure 16.2(a), is a density function for X. To do this, we must verify that $f(x)$ satisfies the three conditions stated in the definition of a density function. First, $f(x)$ is either 0 or 1, so $f(x) \ge 0$. Next, since $f(x) = 0$

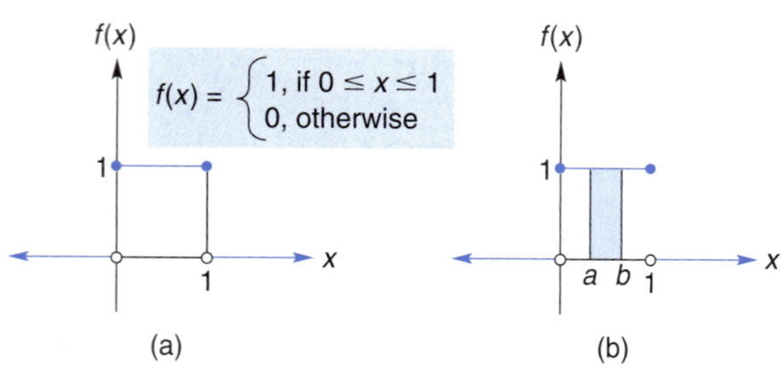

FIGURE 16.2 Probability density function.

for x outside $[0, 1]$,

$$\int_{-\infty}^{\infty} f(x)\,dx = \int_0^1 1\,dx = x\Big|_0^1 = 1$$

Finally, to verify that $P(a \le X \le b) = \int_a^b f(x)\,dx$, we compute the area under the graph between $x = a$ and $x = b$ [Figure 16.2(b)]. We have

$$\int_a^b f(x)\,dx = \int_a^b 1\,dx = x\Big|_a^b = b - a$$

which, as stated in Equation (1), is $P(a \le X \le b)$.

The function in Equation (2) is called the **uniform density function** over $[0, 1]$, and X is said to have a **uniform distribution.** The word *uniform* is meaningful in the sense that the graph of the density function is horizontal, or "flat," over $[0, 1]$. As a result, X is just as likely to assume a value in one interval within $[0, 1]$ as in another of equal length. A more general uniform distribution is given in Example 1.

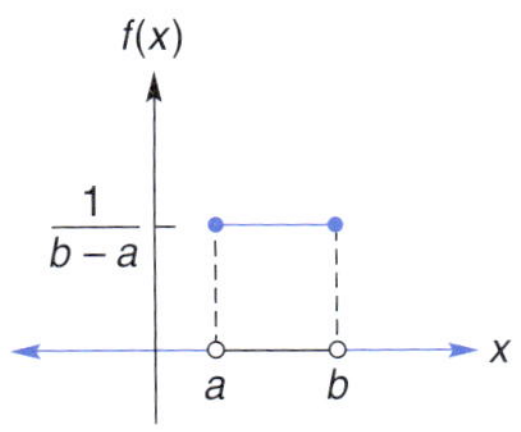

FIGURE 16.3 Uniform density function over $[a, b]$.

EXAMPLE 1 Uniform Density Function

The uniform density function over $[a, b]$ for the random variable X is given by

$$f(x) = \begin{cases} \dfrac{1}{b-a} & \text{if } a \le x \le b \\ 0 & \text{otherwise} \end{cases}$$

(See Figure 16.3.) Note that over $[a, b]$, the region under the graph is a rectangle with height $1/(b-a)$ and width $b-a$. Thus, its area is given by $[1/(b-a)][b-a] = 1$ so $\int_{-\infty}^{\infty} f(x)\,dx = 1$, as must be the case for a density function. If $[c, d]$ is any interval within $[a, b]$, then

$$P(c \le X \le d) = \int_c^d f(x)\,dx = \int_c^d \frac{1}{b-a}\,dx$$

$$= \frac{x}{b-a}\Big|_c^d = \frac{d-c}{b-a}$$

For example, suppose X is uniformly distributed over the interval $[1, 4]$ and we need to find $P(2 < X < 3)$. Then $a = 1$, $b = 4$, $c = 2$, and $d = 3$. Therefore,

$$P(2 < X < 3) = \frac{3-2}{4-1} = \frac{1}{3}$$

▶ **Now work Problem 3(a)–(g).**

APPLICATIONS IN PRACTICE 1

Density Function

Suppose the time (in minutes) passengers must wait for an airplane is uniformly distributed with density function $f(x) = \frac{1}{60}$, where $0 \le x \le 60$, and $f(x) = 0$ elsewhere. What is the probability that a passenger must wait between 25 and 45 minutes?

EXAMPLE 2 Density Function

The density function for a random variable X is given by

$$f(x) = \begin{cases} kx & \text{if } 0 \le x \le 2 \\ & \text{otherwise} \end{cases}$$

where k is a constant.

a. *Find k.*

Solution: Since $\int_{-\infty}^{\infty} f(x)\,dx$ must be 1 and $f(x) = 0$ outside $[0, 2]$, we have

$$\int_{-\infty}^{\infty} f(x)\,dx = \int_0^2 kx\,dx = \frac{kx^2}{2}\Big|_0^2 = 2k = 1$$

Thus, $k = \frac{1}{2}$, so $f(x) = \frac{1}{2}x$ on $[0, 2]$.

b. *Find* $P(\frac{1}{2} < X < 1)$.

Solution:

$$P\left(\frac{1}{2} < X < 1\right) = \int_{1/2}^{1} \frac{1}{2}x\,dx = \frac{x^2}{4}\bigg|_{1/2}^{1} = \frac{1}{4} - \frac{1}{16} = \frac{3}{16}$$

c. *Find* $P(X < 1)$.

Solution: Since $f(x) = 0$ for $x < 0$, we need only compute the area under the density function between 0 and 1. Thus,

$$P(x < 1) = \int_0^1 \frac{1}{2}x\,dx = \frac{x^2}{4}\bigg|_0^1 = \frac{1}{4}$$

▸ **Now work Problem 9(a)–(d), (g), (h).**

APPLICATIONS IN PRACTICE 2

Exponential Density Function

The life expectancy (in years) of an automobile's brake pads is distributed exponentially with $k = \frac{1}{10}$. If the brake pads' warranty lasts 5 years, what is the probability that the brake pads will break down after the warranty period?

EXAMPLE 3 Exponential Density Function

The ***exponential density function*** *is defined by*

$$f(x) = \begin{cases} ke^{-kx} & \text{if } x \geq 0 \\ 0 & \text{if } x < 0 \end{cases}$$

where k is a positive constant, called a ***parameter,*** *whose value depends on the experiment under consideration. If X is a random variable with this density function, then X is said to have an* ***exponential distribution.*** *Let* $k = 1$. *Then* $f(x) = e^{-x}$ *for* $x \geq 0$, *and* $f(x) = 0$ *for* $x < 0$ *(Figure 16.4).*

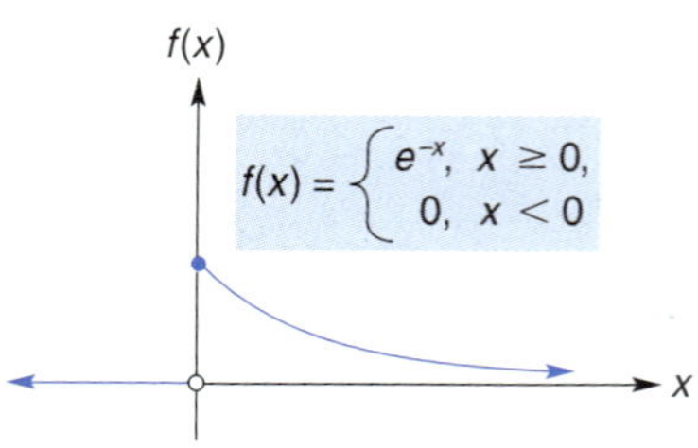

FIGURE 16.4 Exponential density function.

a. *Find* $P(2 < X < 3)$.

Solution:

$$P(2 < X < 3) = \int_2^3 e^{-x}\,dx = -e^{-x}\big|_2^3$$
$$= -e^{-3} - (-e^{-2}) = e^{-2} - e^{-3} \approx 0.086$$

b. *Find* $P(X > 4)$.

Solution:

$$P(X > 4) = \int_4^{\infty} e^{-x}\,dx = \lim_{r\to\infty}\int_4^r e^{-x}\,dx$$
$$= \lim_{r\to\infty} -e^{-x}\bigg|_4^r = \lim_{r\to\infty}(-e^{-r} + e^{-4})$$
$$= \lim_{r\to\infty}\left(-\frac{1}{e^r} + e^{-4}\right) = 0 + e^{-4}$$
$$\approx 0.018$$

Alternatively, we can avoid an improper integral because

$$P(X > 4) = 1 - P(X \leq 4) = 1 - \int_0^4 e^{-x}\,dx$$

▸ **Now work Problem 7(a)–(c), (e).**

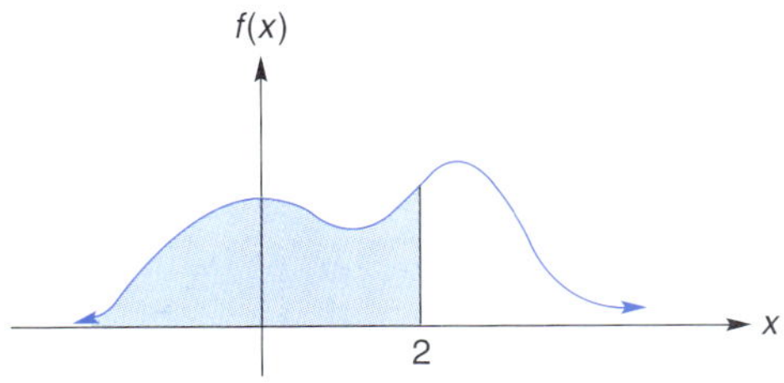

FIGURE 16.5 $F(2) = P(X \leq 2) =$ area of shaded region.

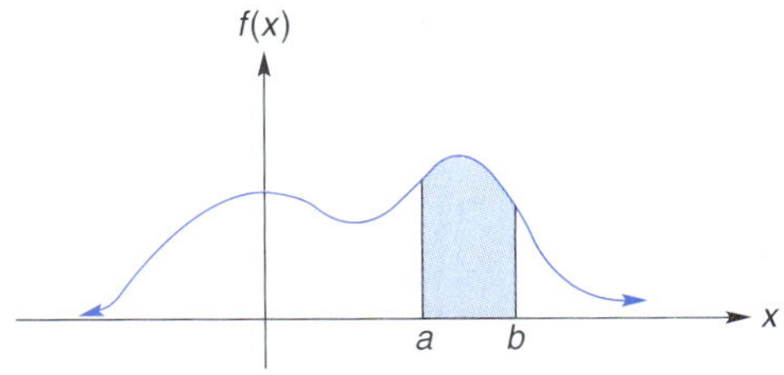

FIGURE 16.6 $P(a < X < b)$.

The **cumulative distribution function** F for the continuous random variable X with density function f is defined by

$$F(x) = P(X \leq x) = \int_{-\infty}^{x} f(t)\, dt$$

For example, $F(2)$ represents the entire area under the density curve that is to the left of the line $x = 2$ (Figure 16.5). Under certain conditions of continuity, it can be shown that

$$F'(x) = f(x)$$

That is, the derivative of the cumulative distribution function is the density function. Thus, F is an antiderivative of f, and by the Fundamental Theorem of Integral Calculus,

$$P(a < X < b) = \int_{a}^{b} f(x)\, dx = F(b) - F(a) \qquad \textbf{(3)}$$

This means that the area under the density curve between a and b (Figure 16.6) is simply the area to the left of b minus the area to the left of a.

EXAMPLE 4 **Finding and Applying the Cumulative Distribution Function**

Suppose X is a random variable with density function given by

$$f(x) = \begin{cases} \frac{1}{2}x & \text{if } 0 \leq x \leq 2 \\ 0 & \text{otherwise} \end{cases}$$

as shown in Figure 16.7.

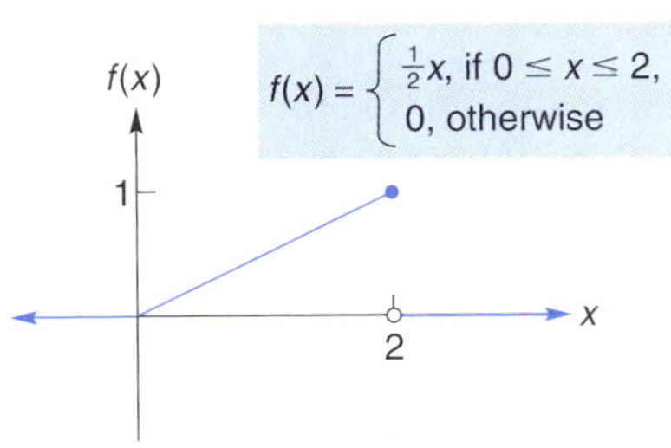

FIGURE 16.7 Density function for Example 4.

a. *Find and sketch the cumulative distribution function.*

Solution: Because $f(x) = 0$ if $x < 0$, the area under the density curve to the left of $x = 0$ is 0. Hence, $F(x) = 0$ if $x < 0$. If $0 \leq x \leq 2$, then

$$F(x) = \int_{-\infty}^{x} f(t)\, dt = \int_{0}^{x} \frac{1}{2}t\, dt = \left.\frac{t^2}{4}\right|_{0}^{x} = \frac{x^2}{4}$$

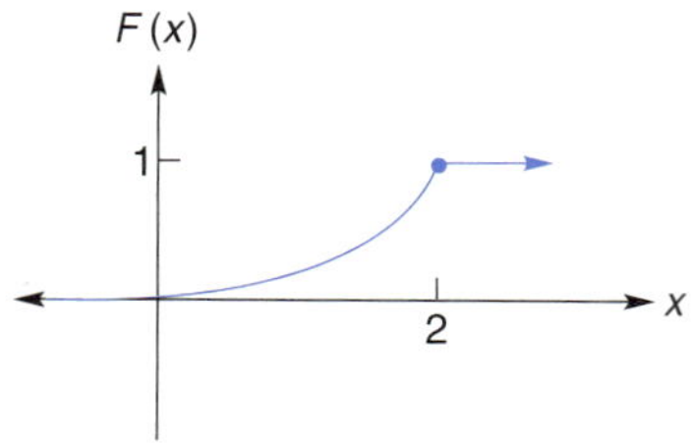

FIGURE 16.8 Cumulative distribution function for Example 4.

Since f is a density function and $f(x) = 0$ for $x < 0$ and also for $x > 2$, the area under the density curve from $x = 0$ to $x = 2$ is 1. Thus, if $x > 2$, the area to the left of x is 1, so $F(x) = 1$. Hence, the cumulative distribution function is

$$F(x) = \begin{cases} 0 & \text{if } x < 0 \\ \dfrac{x^2}{4} & \text{if } 0 \le x \le 2 \\ 1 & \text{if } x > 2 \end{cases}$$

which is shown in Figure 16.8.

b. *Find $P(X < 1)$ and $P(1 < X < 1.1)$.*

Solution: Using the results of part (a), we have

$$P(X < 1) = F(1) = \frac{1^2}{4} = \frac{1}{4}$$

From Equation (3),

$$P(1 < X < 1.1) = F(1.1) - F(1) = \frac{1.1^2}{4} - \frac{1}{4} = 0.0525$$

▶ **Now work Problem 1.**

Mean, Variance, and Standard Deviation

For a random variable X with density function f, the **mean** μ [or **expectation** $E(x)$] is given by

$$\mu = E(X) = \int_{-\infty}^{\infty} x f(x)\, dx$$

and can be thought of as the average value of X in the long run. The **variance** σ^2 [or $\mathrm{Var}(X)$] is given by

$$\sigma^2 = \mathrm{Var}(X) = \int_{-\infty}^{\infty} (x - \mu)^2 f(x)\, dx$$

You may have noticed that these formulas are similar to the corresponding ones in Chapter 9 for a discrete random variable. It can be shown that an alternative formula for the variance is

$$\sigma^2 = \mathrm{Var}(X) = \int_{-\infty}^{\infty} x^2 f(x)\, dx - \mu^2$$

The **standard deviation** is

$$\sigma = \sqrt{\mathrm{Var}(X)}$$

For example, it can be shown that if X is exponentially distributed (see Example 3), $\mu = 1/k$ and $\sigma = 1/k$. As with a discrete random variable, the standard deviation of a continuous random variable X is small if X is likely to assume values close to the mean but unlikely to assume values far from the mean. The standard deviation is large if the reverse is true.

APPLICATIONS IN PRACTICE 3

Finding the Mean and Standard Deviation

The life expectancy (in years) of patients after they have contracted a certain disease is exponentially distributed with $k = 0.2$. Use the information in the paragraph preceding Example 5 to find the mean life expectancy and the standard deviation.

EXAMPLE 5 Finding the Mean and Standard Deviation

If X is a random variable with density function given by

$$f(x) = \begin{cases} \frac{1}{2}x & \text{if } 0 \le x \le 2 \\ 0 & \text{otherwise} \end{cases}$$

find its mean and standard deviation.

Solution: The mean is simply given by

$$\mu = \int_{-\infty}^{\infty} x f(x)\,dx = \int_0^2 x \cdot \frac{1}{2}x\,dx = \left.\frac{x^3}{6}\right|_0^2 = \frac{4}{3}$$

By the alternative formula for variance, we have

$$\sigma^2 = \int_{-\infty}^{\infty} x^2 f(x)\,dx - \mu^2 = \int_0^2 x^2 \cdot \frac{1}{2}x\,dx - \left(\frac{4}{3}\right)^2$$

$$= \left.\frac{x^4}{8}\right|_0^2 - \frac{16}{9} = 2 - \frac{16}{9} = \frac{2}{9}$$

Thus, the standard deviation is

$$\sigma = \sqrt{\frac{2}{9}} = \frac{\sqrt{2}}{3}$$

▶ **Now work Problem 5.**

We conclude this section by emphasizing that a density function for a continuous random variable must not be confused with a probability distribution function for a discrete random variable. Evaluating such a probability distribution function at a *point* gives a probability. But evaluating a density function at a point does not. Instead, the *area* under the density function curve over an *interval* is interpreted as a probability. That is, probabilities associated with a continuous random variable are given by integrals.

Exercise 16.1

*1. Suppose X is a continuous random variable with density function given by

$$f(x) = \begin{cases} \frac{1}{6}(x+1) & \text{if } 1 < x < 3 \\ 0 & \text{otherwise} \end{cases}$$

(a) Find $P(1 < X < 2)$
(b) Find $P(X < 2.5)$
(c) Find $P(X \ge \frac{3}{2})$
(d) Find c such that $P(X < c) = \frac{1}{2}$. Give your answer in radical form.

2. Suppose X is a continuous random variable with density function given by

$$f(x) = \begin{cases} \dfrac{1000}{x^2} & \text{if } x > 1000 \\ 0 & \text{otherwise} \end{cases}$$

(a) Find $P(3000 < X < 4000)$
(b) Find $P(X > 2000)$

*3. Suppose X is a continuous random variable that is uniformly distributed on [2, 6].

(a) What is the formula of the density function for X? Sketch its graph.
(b) Find $P(2 < X < 3)$
(c) Find $P(0 < X < 1)$
(d) Find $P(X \le 3.5)$
(e) Find $P(X > 3)$
(f) Find $P(X = 2)$
(g) Find $P(X < 7)$
(h) Find μ
(i) Find σ
(j) Find the cumulative distribution function F and sketch its graph. Use F to find $P(X < 3)$ and $P(2 < X < 5)$.

4. Suppose X is a continuous random variable that is uniformly distributed on [0, 5].

(a) What is the formula of the density function for X? Sketch its graph.
(b) Find $P(1 < X < 3)$
(c) Find $P(4.5 \le X < 5)$
(d) Find $P(X = 4)$
(e) Find $P(X > 2)$

(f) Find $P(X < 5)$ **(g)** Find $P(X > 5)$
(h) Find μ **(i)** Find σ
(j) Find the cumulative distribution function F and sketch its graph. Use F to find $P(1 < X < 3.5)$.

***5.** Suppose X is uniformly distributed on $[a, b]$.
(a) What is the density function for X?
(b) Find μ **(c)** Find σ^2 and σ

6. Suppose X is a continuous random variable with density function given by

$$f(x) = \begin{cases} k & \text{if } a \le x \le b \\ 0 & \text{otherwise} \end{cases}$$

(a) Show that $k = \dfrac{1}{b-a}$ and thus X is uniformly distributed.
(b) Find the cumulative distribution function F.

***7.** Suppose the random variable X is exponentially distributed with $k = 3$.
(a) Find $P(2 < X < 5)$ **(b)** Find $P(X < 3)$
(c) Find $P(X > 5)$
(d) Find $P(\mu - \sigma < X < \mu + \sigma)$
(e) Show that the area under the density function is 1.
(f) Find the cumulative distribution function F

8. Suppose the random variable X is exponentially distributed with $k = 0.5$.
(a) Find $P(X > 4)$ **(b)** Find $P(0.5 < X < 2.6)$
(c) Find $P(X < 5)$ **(d)** Find $P(X = 4)$
(e) Find c such that $P(0 < X < c) = \frac{1}{2}$.

***9.** The density function for a random variable X is given by

$$f(x) = \begin{cases} kx & \text{if } 0 \le x \le 4 \\ 0 & \text{otherwise} \end{cases}$$

(a) Find k **(b)** Find $P(2 < X < 3)$
(c) Find $P(X > 2.5)$ **(d)** Find $P(X > 0)$
(e) Find μ **(f)** Find σ
(g) Find c such that $P(X < c) = \frac{1}{2}$
(h) Find $P(3 < X < 5)$

10. The density function for a random variable X is given by

$$f(x) = \begin{cases} \frac{1}{2}x + k & \text{if } 2 \le x \le 4 \\ 0 & \text{otherwise} \end{cases}$$

(a) Find k **(b)** Find $P(X \ge 2.5)$
(c) Find μ **(d)** Find $P(2 < X < \mu)$

11. Waiting Time At a bus stop, the time X (in minutes) that a randomly arriving person must wait for a bus is uniformly distributed with density function $f(x) = \frac{1}{10}$, where $0 \le x \le 10$ and $f(x) = 0$ otherwise. What is the probability that such a person must wait at most seven minutes? What is the average time that a person must wait?

12. Soft-drink Dispensing An automatic soft-drink dispenser at a fast-food restaurant dispenses X ounces of cola in a 12-ounce drink. If X is uniformly distributed over $[11.92, 12.08]$, what is the probability that less than 12 ounces will be dispensed? What is the probability that exactly 12 ounces will be dispensed? What is the average amount dispensed?

13. Emergency Room Arrivals At a particular hospital, the length of time X (in hours) between successive arrivals at the emergency room is exponentially distributed with $k = 3$. What is the probability that more than one hour passes without an arrival?

14. Electronic Component Life The length of life X (in years) of a computer component has an exponential distribution with $k = \frac{2}{5}$. What is the probability that such a component will fail within three years of use? What is the probability that it will last more than five years?

OBJECTIVE *To discuss the normal distribution, standard units, and the table of areas under the standard normal curve (Appendix D).*

16.2 The Normal Distribution

Quite often, measured data in nature—such as heights of individuals in a population—are represented by a random variable whose density function may be approximated by the bell-shaped curve in Figure 16.9. The curve extends indefinitely to the right and left and never touches the x-axis. This curve, called

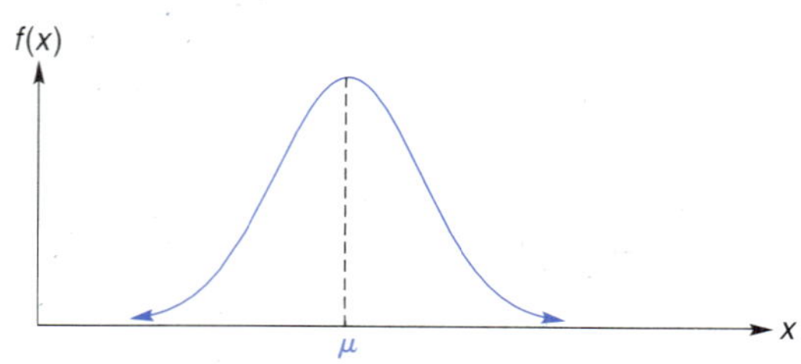

FIGURE 16.9 Normal curve.

the **normal curve,** is the graph of the most important of all density functions, the *normal density function*.

Definition

A continuous random variable X is a ***normal random variable*** or has a ***normal*** (or Gaussian)[1] ***distribution*** if its density function is given by

$$f(x) = \frac{1}{\sigma\sqrt{2\pi}} e^{-(1/2)[(x-\mu)/\sigma]^2} \qquad -\infty < x < \infty$$

The function f is called the ***normal density function.*** The parameters μ and σ are the mean and standard deviation of X, respectively.

Observe in Figure 16.9 that $f(x) \to 0$ as $x \to \pm\infty$. That is, the normal curve has the x-axis as a horizontal asymptote. Also note that the normal curve is symmetric about the vertical line $x = \mu$. That is, the height of a point on the curve d units to the right of $x = \mu$ is the same as the height of the point on the curve that is d units to the left of $x = \mu$. Because of this symmetry and the fact that the area under the normal curve is 1, the area to the right (or left) of the mean must be $\frac{1}{2}$.

Each choice of values for μ and σ determines a different normal curve. The value of μ determines where the curve is "centered," and σ determines how "spread out" the curve is. The smaller the value of σ, the less spread out is the area near μ. For example, Figure 16.10 shows normal curves C_1, C_2, and C_3, where C_1 has mean μ_1 and standard deviation σ_1, C_2 has mean μ_2, etc. Here C_1 and C_2 have the same mean but different standard deviations: $\sigma_1 > \sigma_2$. C_1 and C_3 have the same standard deviation but different means: $\mu_1 < \mu_3$. Curves C_2 and C_3 have different means and different standard deviations.

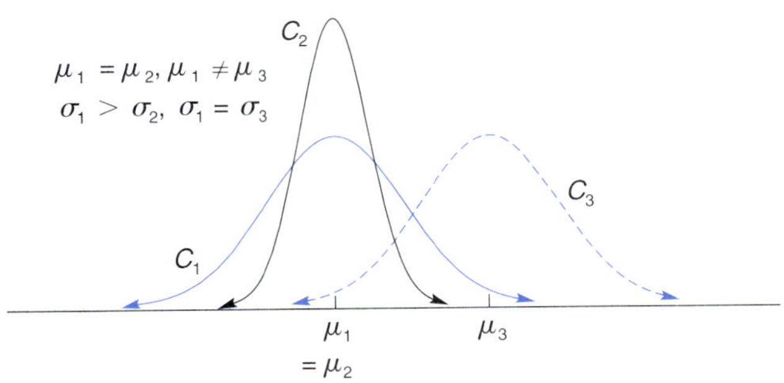

FIGURE 16.10 Normal curves.

The standard deviation plays a significant role in describing probabilities associated with a normal random variable X. More precisely, the probability that X will lie within one standard deviation of the mean is approximately 0.68:

$$P(\mu - \sigma < X < \mu + \sigma) = 0.68$$

In other words, approximately 68% of the area under a normal curve is within one standard deviation of the mean (Figure 16.11). Between $\mu \pm 2\sigma$ is about 95% of the area, and between $\mu \pm 3\sigma$ is about 99.7%:

$$P(\mu - 2\sigma < X < \mu + 2\sigma) = 0.95$$

$$P(\mu - 3\sigma < X < \mu + 3\sigma) = 0.997$$

[1]This distribution is named after the German mathematician Carl Friedrich Gauss (1777–1855).

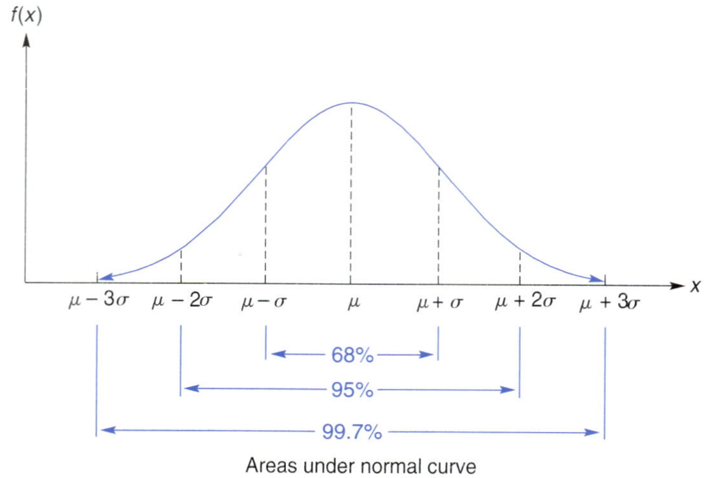

FIGURE 16.11 Probability and number of standard deviations from μ.

You are encouraged to become familiar with the percentages in Figure 16.11.

Thus, it is highly likely that X will lie within three standard deviations of the mean.

EXAMPLE 1 Analysis of Test Scores

Let X be a random variable whose values are the scores obtained on a nationwide test given to high school seniors. Suppose, for modeling purposes, that X is normally distributed with mean 600 and standard deviation 90. Then the probability that X lies within $2\sigma = 2(90) = 180$ points of 600 is 0.95. In other words, 95% of the scores lie between 420 and 780. Similarly, 99.7% of the scores are within $3\sigma = 3(90) = 270$ points of 600—that is, between 330 and 870.

▸ **Now work Problem 17.**

If Z is a normally distributed random variable with $\mu = 0$ and $\sigma = 1$, we obtain the normal curve of Figure 16.12, called the **standard normal curve.**

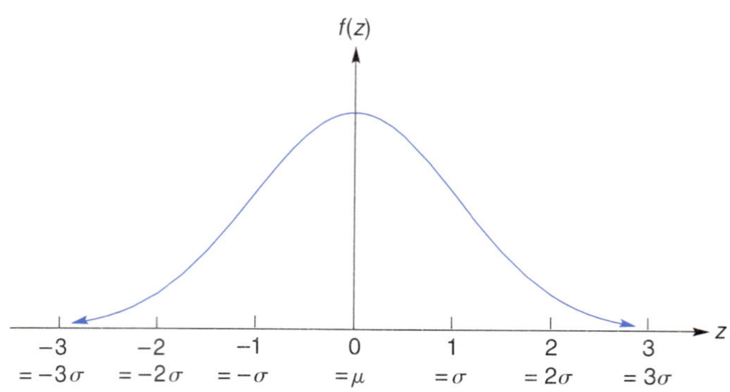

FIGURE 16.12 Standard normal curve: $\mu = 0, \sigma = 1$.

Definition

A continuous random variable Z is a ***standard normal random variable*** (or has a ***standard normal distribution***) if its density function is given by

$$f(z) = \frac{1}{\sqrt{2\pi}} e^{-z^2/2}$$

The function f is called the ***standard normal density function.*** The variable Z has mean 0 and standard deviation 1.

Because a standard normal random variable Z has mean 0 and standard deviation 1, its values are in units of standard deviations from the mean, which are called **standard units.** For example, if $0 < Z < 2.54$, then Z lies within 2.54 standard deviations to the right of 0, the mean. That is, $0 < Z < 2.54\sigma$. To find the probability $P(0 < Z < 2.54)$, we have

$$P(0 < Z < 2.54) = \int_0^{2.54} \frac{e^{-z^2/2}}{\sqrt{2\pi}}\,dz$$

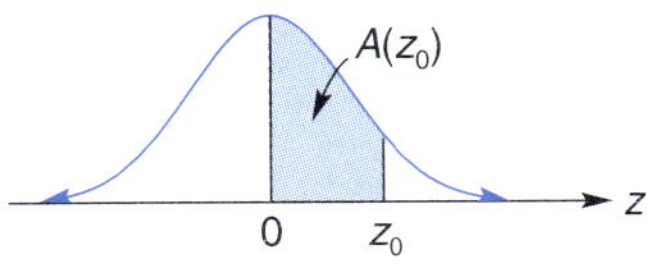

FIGURE 16.13 $A(z_0) = P(0 < Z < z_0)$.

The integral on the right cannot be evaluated by elementary methods. However, values for integrals of this kind have been computed and put in tabular form.

One such table is given in Appendix D. The table there gives the area under a standard normal curve between $z = 0$ and $z = z_0$, where $z_0 \geq 0$. This area is shaded in Figure 16.13 and is denoted by $A(z_0)$. In the left-hand columns of the table are z-values to the nearest tenth. The numbers across the top are the hundredths' values. For example, the entry in the row for 2.5 and column under 0.04 corresponds to $z = 2.54$ and is 0.4945. Thus, the area under a standard normal curve between $z = 0$ and $z = 2.54$ is 0.4945:

$$P(0 < Z < 2.54) = A(2.54) = 0.4945$$

Similarly, you should verify that $A(2) = 0.4772$ and $A(0.33) = 0.1293$

Using symmetry, we compute an area to the left of $z = 0$ by computing the corresponding area to the right of $z = 0$. For example,

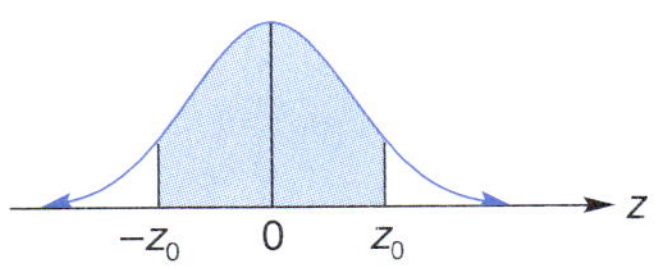

FIGURE 16.14 $P(-z_0 < Z < 0) = P(0 < Z < z_0)$.

$$P(-z_0 < Z < 0) = P(0 < Z < z_0) = A(z_0)$$

as shown in Figure 16.14. Hence, $P(-2.54 < Z < 0) = A(2.54) = 0.4945$.

When computing probabilities for a standard normal variable, you may have to add or subtract areas. A useful aid for doing this properly is a rough sketch of a standard normal curve in which you have shaded the entire area that you want to find, as Example 2 shows.

EXAMPLE 2 **Probabilities for Standard Normal Variable *Z***

a. *Find* $P(Z > 1.5)$.

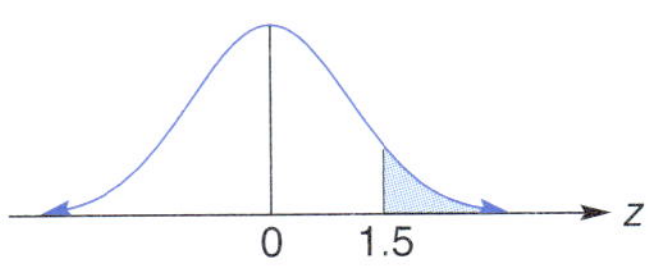

FIGURE 16.15 $P(Z > 1.5)$.

Solution: This probability is the area to the right of $z = 1.5$ (Figure 16.15). That area is equal to the difference between the total area to the right of $z=0$, which is 0.5, and the area between $z=0$ and $z=1.5$, which is $A(1.5)$. Thus,

$$\begin{aligned} P(Z > 1.5) &= 0.5 - A(1.5) \\ &= 0.5 - 0.4332 = 0.0668 \qquad \text{(from Appendix D)} \end{aligned}$$

b. *Find* $P(0.5 < Z < 2)$.

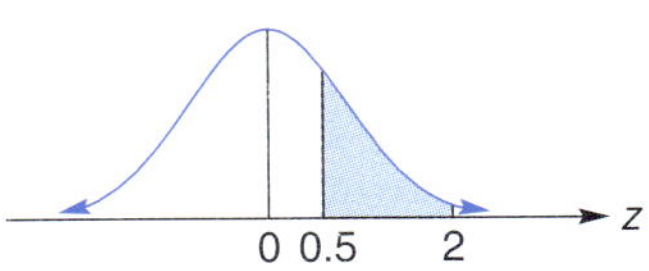

FIGURE 16.16 $P(0.5 < Z < 2)$.

Solution: This probability is the area between $z=0.5$ and $z=2$ (Figure 16.16). That area is the difference of two areas. It is the area between $z=0$ and $z=2$, or $A(2)$, minus the area between $z=0$ and $z=0.5$, or $A(0.5)$. Thus,

$$\begin{aligned} P(0.5 < Z < 2) &= A(2) - A(0.5) \\ &= 0.4772 - 0.1915 = 0.2857 \end{aligned}$$

c. *Find* $P(Z \leq 2)$.

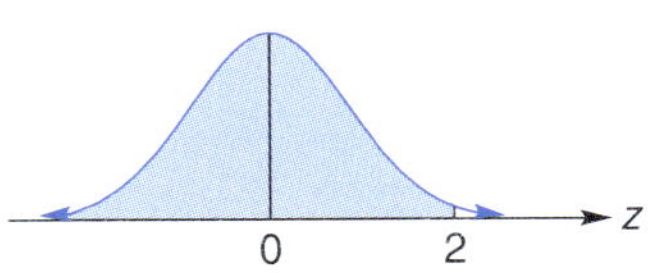

FIGURE 16.17 $P(Z \leq 2)$.

Solution: This probability is the area to the left of $z = 2$ (Figure 16.17). That area is equal to the sum of the area to the left of $z = 0$, which is 0.5, and the area between $z = 0$ and $z = 2$, or $A(2)$. Thus,

$$P(Z \leq 2) = 0.5 + A(2)$$
$$= 0.5 + 0.4772 = 0.9772$$

▶ **Now work Problem 1.**

EXAMPLE 3 **Probabilities for Standard Normal Variable Z**

a. *Find* $P(-2 < Z < -0.5)$.

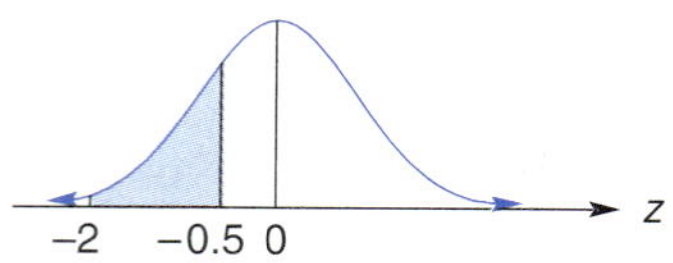

FIGURE 16.18 $P(-2 < Z < -0.5)$.

Solution: This probability is the area between $z = -2$ and $z = -0.5$ (Figure 16.18). By symmetry, that area is equal to the area between $z = 0.5$ and $z = 2$, which was computed in Example 2(b). We have

$$P(-2 < Z < -0.5) = P(0.5 < Z < 2)$$
$$= A(2) - A(0.5) = 0.2857$$

b. *Find* z_0 *such that* $P(-z_0 < Z < z_0) = 0.9642$.

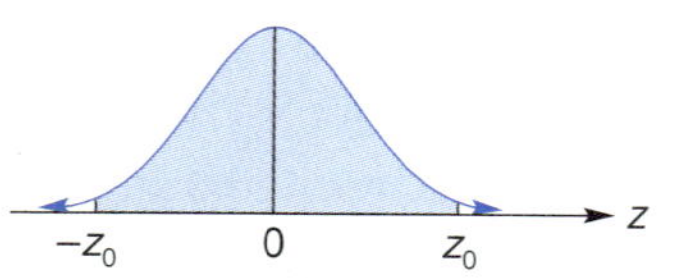

FIGURE 16.19 $P(-z_0 < Z < z_0) = 0.9642$.

Solution: Figure 16.19 shows the corresponding area. Because the total area is 0.9642, by symmetry the area between $z = 0$ and $z = z_0$ is $\frac{1}{2}(0.9642) = 0.4821$, which is $A(z_0)$. Looking at the body of the table in Appendix D, we see that 0.4821 corresponds to a Z-value of 2.1. Thus, $z_0 = 2.1$.

▶ **Now work Problem 3.**

Transforming to a Standard Normal Variable Z

If X is normally distributed with mean μ and standard deviation σ, you might think that a table of areas is needed for each pair of values of μ and σ. Fortunately, this is not the case. Appendix D is still used. But you must first express a given area as an equivalent area under a standard normal curve. This involves transforming X into a standard variable Z (with mean 0 and standard deviation 1) by using the following change-of-variable formula:

$$Z = \frac{X - \mu}{\sigma} \quad \textbf{(1)}$$

Here we convert a normal variable to a standard normal variable.

On the right side, subtracting μ from X gives the distance from μ to X. Dividing by σ expresses this distance in terms of units of standard deviation. Thus, Z is the number of standard deviations that X is from μ. That is, Equation (1) converts units of X into standard units (Z-values). For example, if $X = \mu$, then using Equation (1) gives $Z = 0$. Hence, μ is zero standard deviations from μ.

Suppose X is normally distributed with $\mu = 4$ and $\sigma = 2$. Then to find—for example—$P(0 < X < 6)$, we first use Equation (1) to convert the X-values 0 and 6 to Z-values (standard units):

$$z_1 = \frac{x_1 - \mu}{\sigma} = \frac{0 - 4}{2} = -2$$
$$z_2 = \frac{x_2 - \mu}{\sigma} = \frac{6 - 4}{2} = 1$$

It can be shown that

$$P(0 < X < 6) = P(-2 < Z < 1)$$

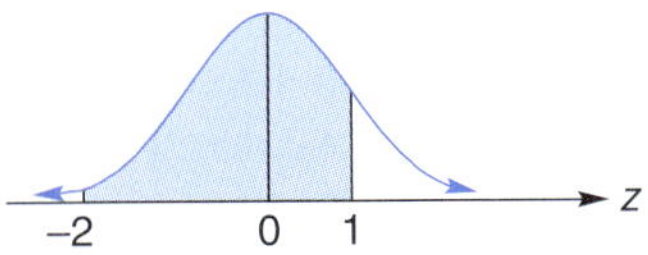

FIGURE 16.20 $P(-2 < Z < 1)$.

This means that the area under a normal curve with $\mu = 4$ and $\sigma = 2$ between $x = 0$ and $x = 6$ is equal to the area under a standard normal curve between $z = -2$ and $z = 1$ (Figure 16.20). This area is the sum of the area A_1 between $z = -2$ and $z = 0$ and the area A_2 between $z = 0$ and $z = 1$. Using symmetry for A_1, we have

$$\begin{aligned} P(-2 < Z < 1) &= A_1 + A_2 = A(2) + A(1) \\ &= 0.4772 + 0.3413 = 0.8185 \end{aligned}$$

EXAMPLE 4 **Employees' Salaries**

The weekly salaries of 5000 employees of a large corporation are assumed to be normally distributed with mean \$640 and standard deviation \$56. How many employees earn less than \$570 per week?

Solution: Converting to standard units, we have

$$\begin{aligned} P(X < 570) &= P\left(Z < \frac{570 - 640}{56}\right) \\ &= P(Z < -1.25) \end{aligned}$$

This probability is the area shown in Figure 16.21(a). By symmetry, that area is equal to the area in Figure 16.21(b) that corresponds to $P(Z > 1.25)$. This area is the difference between the total area to the right of $z = 0$, which is 0.5, and the area between $z = 0$ and $z = 1.25$, which is $A(1.25)$. Thus,

$$\begin{aligned} P(X < 570) &= P(Z < -1.25) = P(Z > 1.25) \\ &= 0.5 - A(1.25) = 0.5 - 0.3944 \\ &= 0.1056 \end{aligned}$$

That is, 10.56% of the employees have salaries less than \$570. This corresponds to $0.1056(5000) = 528$ employees.

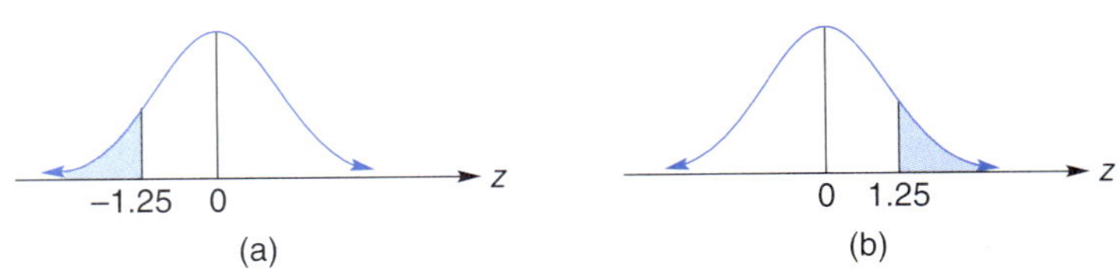

FIGURE 16.21 Diagram for Example 4.

▸ **Now work Problem 9.**

Exercise 16.2

*1. If Z is a standard normal random variable, find each of the following probabilities:

(a) $P(0 < Z < 1.7)$ **(b)** $P(0.43 < Z < 2.89)$
(c) $P(Z > -1.23)$ **(d)** $P(Z \leq 2.91)$
(e) $P(-2.51 < Z \leq 1.3)$ **(f)** $P(Z > 0.03)$

2. If Z is a standard normal random variable, find each of the following:

(a) $P(-1.96 < Z < 1.96)$ **(b)** $P(-2.11 < Z < -1.35)$
(c) $P(Z < -1.05)$ **(d)** $P(Z > 3\sigma)$
(e) $P(|Z| > 2)$ **(f)** $P(|Z| < \frac{1}{2})$

In Problems 3–8, find z_0 such that the given statement is true. Assume that Z is a standard normal random variable.

*3. $P(Z < z_0) = 0.5517$ **4.** $P(Z < z_0) = 0.0668$
5. $P(Z > z_0) = 0.8599$ **6.** $P(Z > z_0) = 0.4960$
7. $P(-z_0 < Z < z_0) = 0.2662$
8. $P(|Z| > z_0) = 0.3174$

*9. If X is normally distributed with $\mu = 16$ and $\sigma = 4$, find each of the following probabilities:
(a) $P(X < 27)$ (b) $P(X < 10)$
(c) $P(10.8 < X < 12.4)$

10. If X is normally distributed with $\mu = 200$ and $\sigma = 40$, find each of the following probabilities:

(a) $P(X > 150)$
(b) $P(210 < X < 250)$

11. If X is normally distributed with $\mu = -3$ and $\sigma = 2$, find $P(X > -2)$.

12. If X is normally distributed with $\mu = 0$ and $\sigma = 1.5$, find $P(X < 3)$.

13. If X is normally distributed with $\mu = 25$ and $\sigma^2 = 9$, find $P(15 < X \leq 29)$.

14. If X is normally distributed with $\mu = 8$ and $\sigma = 1$, find $P(X > \mu - \sigma)$.

15. If X is normally distributed such that $\mu = 40$ and $P(X > 54) = 0.0401$, find σ.

16. If X is normally distributed with $\mu = 16$ and $\sigma = 2.25$, find x_0 such that the probability that X is between x_0 and 16 is 0.4641.

*17. **Test Scores** The scores on a national achievement test are normally distributed with mean 500 and standard deviation 100. What percentage of those who took the test had a score between 300 and 700?

18. Test Scores In a test given to a large group of people, the scores were normally distributed with mean 70 and standard deviation 10. What is the least whole-number score that a person could get and yet score in about the top 15%?

19. Adult Heights The heights (in inches) of adults in a large population are normally distributed with $\mu = 68$ and $\sigma = 3$. What percentage of the group is under 6 feet tall?

20. Income The yearly income for a group of 10,000 professional people is normally distributed with $\mu = \$60{,}000$ and $\sigma = \$5000$.

(a) What is the probability that a person from this group has a yearly income less than \$46,000?
(b) How many of these people have yearly incomes over \$75,000?

21. IQ The IQs of a large population of children are normally distributed with mean 100.4 and standard deviation 11.6.

(a) What percentage of the children have IQs greater than 125?
(b) About 90% of the children have IQs greater than what value?

22. Suppose X is a random variable with $\mu = 10$ and $\sigma = 2$. If $P(4 < X < 16) = 0.25$, can X be normally distributed?

OBJECTIVE *To show the technique of estimating the binomial distribution by using the normal distribution.*

16.3 The Normal Approximation to the Binomial Distribution

We conclude this chapter by bringing together the notions of a discrete random variable and a continuous random variable. Recall from Chapter 9 that if X is a binomial random variable (which is discrete), and if the probability of success on any trial is p, then for n independent trials, the probability of x successes is given by

$$P(X = x) = {}_nC_x p^x q^{n-x}$$

where $q = 1 - p$. You would no doubt agree that calculating probabilities for a binomial random variable can be quite tedious when the number of trials is large. For example, just imagine trying to compute ${}_{100}C_{40}(0.3)^{40}(0.7)^{60}$. To handle expressions like this, we can approximate a binomial distribution by a normal distribution and then use a table of areas.

To show how this is done, let us take a simple example. Figure 16.22 gives a probability histogram for a binomial experiment with $n = 10$ and $p = 0.5$. The rectangles centered at $x = 0$ and $x = 10$ are not shown because their heights are very close to 0. Superimposed on the histogram is a normal curve, which approximates it. The approximation would be even better if n were larger. That is, as n gets larger, the width of each unit interval appears to get smaller, and the outline of the histogram tends to take on the appearance of a smooth curve. In fact, *it is not unusual to think of a density curve as the limiting case*

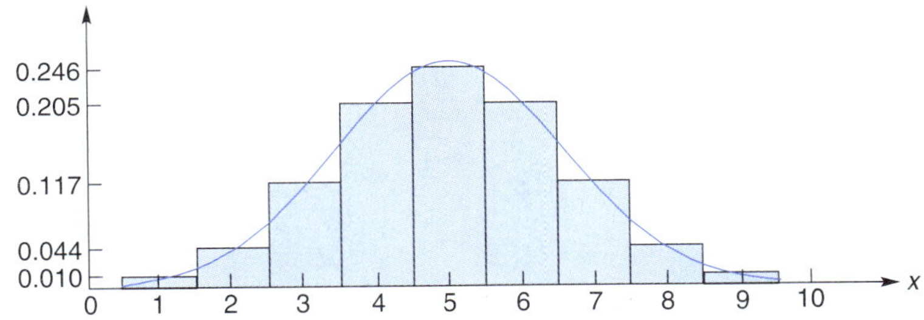

FIGURE 16.22 Normal approximation to binomial distribution.

of a probability histogram. In spite of the fact that in our case n is only 10, the approximation shown does not seem too bad. The question that now arises is "Which normal distribution approximates the binomial distribution?" Since the mean and standard deviation are measures of central tendency and dispersion of a random variable, we choose the approximating normal distribution to have the same mean and standard deviation as that of the binomial distribution. For this choice, we can estimate the areas of rectangles in the histogram (that is, the binomial probabilities) by finding the corresponding area under the normal curve. In summary, we have the following:

If X is a binomial random variable and n is sufficiently large, then the distribution of X can be approximated by a normal random variable whose mean and standard deviation are the same as for X, which are np and $\sqrt{npq}$, respectively.

Perhaps a word of explanation is appropriate concerning the phrase "n is sufficiently large." Generally speaking, a normal approximation to a binomial distribution is not good if n is small and p is near 0 or 1, because much of the area in the binomial histogram would be concentrated at one end of the distribution (that is, at 0 or n). Thus, the distribution would not be fairly symmetric, and a normal curve would not "fit" well. A general rule that you can follow is that the normal approximation to the binomial distribution is reasonable if np and nq are at least 5. This is the case in our example: $np = 10(0.5) = 5$ and $nq = 10(0.5) = 5$.

Let us now use the normal approximation to estimate a binomial probability for $n = 10$ and $p = 0.5$. If X denotes the number of successes, then its mean is

$$np = 10(0.5) = 5$$

and its standard deviation is

$$\sqrt{npq} = \sqrt{10(0.5)(0.5)} = \sqrt{2.5} \approx 1.58$$

The probability function for X is given by

$$f(x) = {}_{10}C_x(0.5)^x(0.5)^{10-x}$$

We approximate this distribution by the normal distribution with $\mu = 5$ and $\sigma = \sqrt{2.5}$.

Suppose we estimate the probability that there are between 4 and 7 successes, inclusive, which is given by

$$P(4 \leq X \leq 7) = P(X = 4) + P(X = 5) + P(X = 6) + P(X = 7)$$

$$= \sum_{x=4}^{7} {}_{10}C_x(0.5)^x(0.5)^{10-x}$$

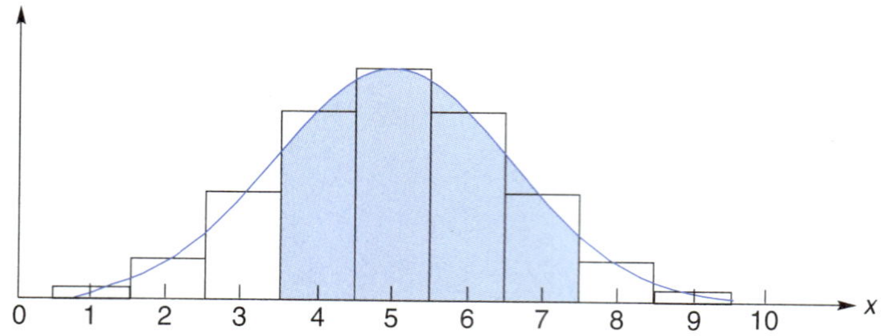

FIGURE 16.23 Normal approximation to $P(4 \le X \le 7)$.

This probability is the sum of the areas of the *rectangles* for $X = 4, 5, 6$, and 7 in Figure 16.23. Under the normal curve, we have shaded the corresponding area that we shall compute as an approximation to this probability. Note that the shading extends not from 4 to 7, but from $4 - \frac{1}{2}$ to $7 + \frac{1}{2}$, that is, from 3.5 to 7.5. This "continuity correction" of 0.5 on each end of the interval allows most of the area in the appropriate rectangles to be included in the approximation, and *such a correction must always be made*. The phrase *continuity correction* is used because X is treated as though it were a continuous random variable. We now convert the X-values 3.5 and 7.5 to Z-values:

$$z_1 = \frac{3.5 - 5}{\sqrt{2.5}} \approx -0.95$$

$$z_2 = \frac{7.5 - 5}{\sqrt{2.5}} \approx 1.58$$

Thus,

$$P(4 \le X \le 7) \approx P(-0.95 \le Z \le 1.58)$$

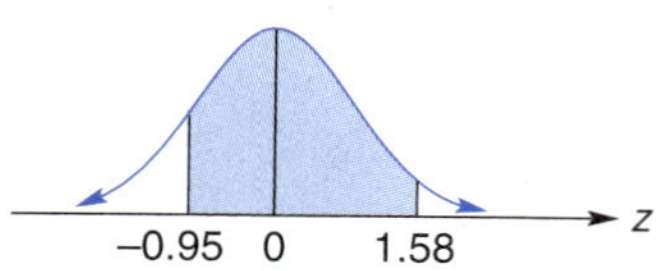

FIGURE 16.24 $P(-0.95 \le Z \le 1.58)$.

which corresponds to the area under a standard normal curve between $z = -0.95$ and $z = 1.58$ (Figure 16.24). This area is the sum of the area between $z = -0.95$ and $z = 0$, which, by symmetry, is $A(0.95)$, and the area between $z = 0$ and $z = 1.58$, which is $A(1.58)$. Hence,

$$\begin{aligned} P(4 \le X \le 7) &\approx P(-0.95 \le Z \le 1.58) \\ &= A(0.95) + A(1.58) \\ &= 0.3289 + 0.4429 = 0.7718 \end{aligned}$$

This result is close to the true value, 0.7734 (to four decimal places).

APPLICATIONS IN PRACTICE 1

Normal Approximation to a Binomial Distribution

On a game show, the grand prize is hidden behind one of four doors. Assume that the probability of selecting the grand prize is $p = \frac{1}{4}$. There were 20 winners among the last 60 contestants. Suppose that X is the number of contestants that win the grand prize, and X is binomial with $n = 60$. Approximate $P(X = 20)$ by using the normal approximation.

EXAMPLE 1 **Normal Approximation to a Binomial Distribution**

Suppose X is a binomial random variable with $n = 100$ and $p = 0.3$. Estimate $P(X = 40)$ by using the normal approximation.

Solution: We have

$$P(X = 40) = {}_{100}C_{40}(0.3)^{40}(0.7)^{60}$$

which was mentioned at the beginning of this section. We use a normal distribution with

$$\mu = np = 100(0.3) = 30$$

and

$$\sigma = \sqrt{npq} = \sqrt{100(0.3)(0.7)} = \sqrt{21} \approx 4.58$$

Do not ignore the continuity correction.

Converting the corrected X-values 39.5 and 40.5 to Z-values gives

$$z_1 = \frac{39.5 - 30}{\sqrt{21}} \approx 2.07$$

$$z_2 = \frac{40.5 - 30}{\sqrt{21}} \approx 2.29$$

Therefore,

$$P(X = 40) \approx P(2.07 \leq Z \leq 2.29)$$

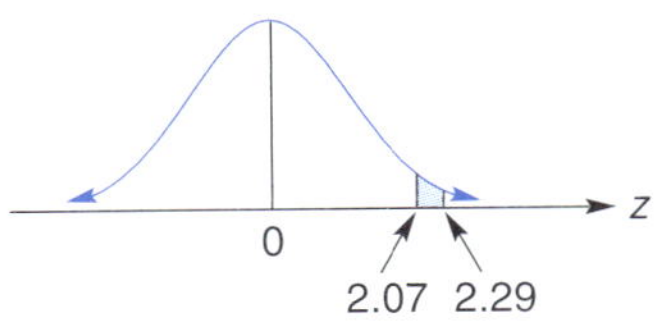

FIGURE 16.25 $P(2.07 \leq Z \leq 2.29)$.

This probability is the area under a standard normal curve between $z = 2.07$ and $z = 2.29$ (Figure 16.25). That area is the difference of the area between $z = 0$ and $z = 2.29$, which is $A(2.29)$, and the area between $z = 0$ and $z = 2.07$, which is $A(2.07)$. Thus,

$$\begin{aligned} P(X = 40) &\approx P(2.07 \leq Z \leq 2.29) \\ &= A(2.29) - A(2.07) \\ &= 0.4890 - 0.4808 = 0.0082 \qquad \text{(from Appendix D)} \end{aligned}$$

▸ **Now work Problem 3.**

EXAMPLE 2 **Quality Control**

In a quality-control experiment, a sample of 500 items is taken from an assembly line. Customarily, 8% of the items produced are defective. What is the probability that more than 50 defective items appear in the sample?

Solution: If X is the number of defective items in the sample, then we shall consider X to be binomial with $n = 500$ and $p = 0.08$. To find $P(X \geq 51)$, we use the normal approximation to the binomial distribution with

$$\mu = np = 500(0.08) = 40$$

and

$$\sigma = \sqrt{npq} = \sqrt{500(0.08)(0.92)} = \sqrt{36.8} \approx 6.07$$

Converting the corrected value 50.5 to a Z-value gives

$$z = \frac{50.5 - 40}{\sqrt{36.8}} \approx 1.73$$

Thus,

$$P(X \geq 51) \approx P(Z \geq 1.73)$$

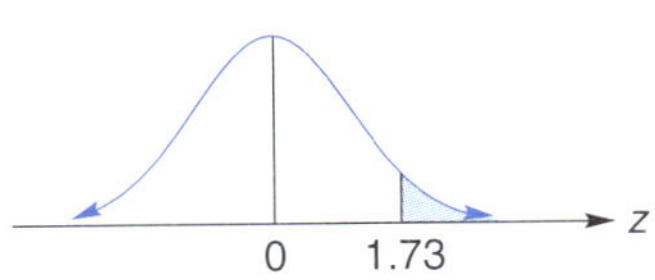

FIGURE 16.26 $P(Z \geq 1.73)$.

This probability is the area under a standard normal curve to the right of $z = 1.73$ (Figure 16.26). That area is the difference of the area to the right of $z = 0$, which is 0.5, and the area between $z = 0$ and $z = 1.73$, which is $A(1.73)$. Hence,

$$\begin{aligned} P(X \geq 51) &\approx P(Z \geq 1.73) \\ &= 0.5 - A(1.73) = 0.5 - 0.4582 = 0.0418 \end{aligned}$$

▸ **Now work Problem 7.**

Exercise 16.3

In Problems 1–4, X is a binomial random variable with the given values of n and p. Calculate the indicated probabilities by using the normal approximation.

1. $n = 150, p = 0.4$; $P(X \leq 52), P(X \geq 74)$
2. $n = 50, p = 0.3$; $P(X = 19), P(X \leq 18)$

*3. $n = 200, p = 0.6$; $P(X = 125), P(110 \leq X \leq 135)$

4. $n = 25, p = 0.25$; $P(X \geq 7)$
5. **Die Tossing** Suppose a fair die is tossed 300 times. What is the probability that a 5 turns up between 45 and 60 times, inclusive?
6. **Coin Tossing** For a biased coin, $P(H) = 0.4$ and $P(T) = 0.6$. If the coin is tossed 200 times, what is the probability of getting between 90 and 100 heads, inclusive?

*7. **Truck Breakdown** A delivery service has a fleet of 60 trucks. At any given time, the probability of a truck being out of use due to factors such as breakdowns and maintenance is 0.1. What is the probability that 7 or more trucks are out of service at any time?

8. **Quality Control** In a manufacturing plant, a sample of 100 items is taken from the assembly line. For each item in the sample, the probability of being defective is 0.06. What is the probability that there are 3 or more defective items in the sample?
9. **True-False Exam** In a true-false exam with 25 questions, what is the probability of getting at least 13 correct answers by just guessing on all the questions? If there are 100 questions instead of 25, what is the probability of getting at least 60 correct answers by just guessing?
10. **Multiple-Choice Exam** In a multiple-choice test with 50 questions, each question has four answers, only one of which is correct. If a student guesses on the last 20 questions, what is the probability of getting at least half of them correct?
11. **Poker** In a poker game, the probability of being dealt a hand consisting of three cards of one suit and two cards of another suit (in any order) is about 0.1. In 100 dealt hands, what is the probability that 16 or more of them will be as just described?
12. **Taste Test** A major cola company sponsors a national taste test, in which subjects sample its cola as well as the best-selling brand. Neither cola is identified by brand. The subjects are then asked to choose the cola that tastes better. If each of the 35 subjects in a supermarket actually have no preference and arbitrarily chose one of the colas, what is the probability that 25 or more of them choose the cola from the sponsoring company?

16.4 Review

Important Terms and Symbols

Section 16.1 continuous random variable; density function; uniform density function; uniform distribution; exponential density function; exponential distribution; cumulative distribution function; mean, μ; variance, σ^2; standard deviation, σ

Section 16.2 normal random variable; normal distribution; normal density function; standard normal curve; standard normal random variable; standard normal distribution; standard normal density function; standard units

Section 16.3 continuity correction

Summary

A continuous random variable X can assume any value in an interval or intervals. A density function for X is a function that has the following properties:

1. $f(x) \geq 0$
2. $\displaystyle\int_{-\infty}^{\infty} f(x)\,dx = 1$
3. $P(a \leq X \leq b) = \displaystyle\int_a^b f(x)\,dx$

Property 3 means that the area under the graph of f and above the x-axis from $x = a$ to $x = b$ is $P(a \leq X \leq b)$. The probability that X assumes a particular value is 0.

The continuous random variable X has a uniform distribution over $[a, b]$ if its density function is given by

$$f(x) = \begin{cases} \dfrac{1}{b-a} & \text{if } a \leq x \leq b \\ 0 & \text{otherwise} \end{cases}$$

X has an exponential density function f if

$$f(x) = \begin{cases} ke^{-kx} & \text{if } x \geq 0 \\ 0 & \text{if } x < 0 \end{cases}$$

where k is a positive constant.

The cumulative distribution function F for the continuous random variable X with density function f is given by

$$F(x) = P(X \leq x) = \int_{-\infty}^{x} f(t)\,dt$$

Geometrically, $F(x)$ represents the area under the density curve to the left of x. By using F, we are able to find $P(a \leq x \leq b)$:

$$P(a \leq x \leq b) = F(b) - F(a)$$

The mean μ of X [or expectation $E(X)$] is given by

$$\mu = E(X) = \int_{-\infty}^{\infty} xf(x)\,dx$$

the variance is given by

$$\sigma^2 = \text{Var}(X) = \int_{-\infty}^{\infty} (x-\mu)^2 f(x)\,dx$$
$$= \int_{-\infty}^{\infty} x^2 f(x)\,dx - \mu^2$$

The standard deviation is given by

$$\sigma = \sqrt{\text{Var}(X)}$$

The graph of the normal density function

$$f(x) = \frac{1}{\sigma\sqrt{2\pi}} e^{-(1/2)[(x-\mu)/\sigma]^2}$$

is called a normal curve and is bell shaped. If X has a normal distribution, then the probability that X lies within one standard deviation of the mean μ is (approximately) 0.68; within two standard deviations, the probability is 0.95; and within three standard deviations, it is 0.997. If Z is a normal random variable with $\mu = 0$ and $\sigma = 1$, then Z is called a standard normal random variable. The probability $P(0 < Z < z_0)$ is the area under the graph of the standard normal curve from $z = 0$ to $z = z_0$ and is denoted $A(z_0)$. Values of $A(z_0)$ appear in Appendix D.

If X is normally distributed with mean μ and standard deviation σ, then X may be transformed into a standard normal random variable by the change-of-variable formula

$$Z = \frac{X-\mu}{\sigma}$$

With this formula, probabilities for X may be found by using areas under the standard normal curve.

If X is a binomial random variable and the number n of independent trials is large, then the distribution of X may be approximated by using a normal random variable with mean np and standard deviation $\sqrt{npq}$, where p is the probability of success on any trial and $q = 1 - p$. It is important that continuity corrections be considered when one estimates binomial probabilities by a normal random variable.

Review Problems

Problem numbers shown in color indicate problems suggested for use as a practice chapter test.

1. Suppose X is a continuous random variable with density function given by

$$f(x) = \begin{cases} \frac{1}{3} + kx^2 & \text{if } 0 \leq x \leq 1 \\ 0 & \text{otherwise} \end{cases}$$

(a) Find k
(b) Find $P(\frac{1}{2} < X < \frac{3}{4})$
(c) Find $P(X \geq \frac{1}{2})$
(d) Find the cumulative distribution function.

2. Suppose X is exponentially distributed with $k = \frac{1}{4}$. Find $P(X > 1)$.

3. Suppose X is a random variable with density function given by

$$f(x) = \begin{cases} \frac{2}{25}x & \text{if } 0 \leq x \leq 5 \\ 0 & \text{otherwise} \end{cases}$$

(a) Find μ **(b)** Find σ

4. Let X be uniformly distributed over the interval [2, 6]. Find $P(X < 5)$.

Let X be normally distributed with mean 20 and standard deviation 4. In Problems 5–10, determine the given probabilities.

5. $P(X > 22)$ **6.** $P(X < 21)$ **7.** $P(14 < X < 18)$
8. $P(X > 10)$ **9.** $P(X < 16)$ **10.** $P(23 < X < 33)$

In Problems 11 and 12, X is a binomial random variable with $n = 100$ and $p = 0.35$. Find the given probabilities by using the normal approximation.

11. $P(25 \leq X \leq 47)$ **12.** $P(X = 48)$

13. Heights of Individuals The heights (in inches) of individuals in a certain group are normally distributed with mean 68 and standard deviation 2. Find the probability that an individual from this group is taller than 6 ft.

14. Coin Tossing If a fair coin is tossed 500 times, use the normal approximation to the binomial distribution to estimate the probability that a head comes up at least 215 times.

Mathematical SNAPSHOT

Cumulative Distribution from Data

What Section 16.3 said about histograms of discrete random variables is even more directly true for continuous random variables: The probability density curve can be thought of as the limiting case of a probability histogram. Indeed, this fact is often used to explain the idea of a probability density function.

With continuous variables, a histogram divides the range of possible values into a series of intervals, called bins. Above each bin is a bar whose height indicates how much of data set lies in that bin. Figure 16.27 illustrates this. The rightmost bin is the interval from 8 to 10. Because one-fifth of the data values lie in that bin, the bar covers one-fifth of the area of all the bars put together.

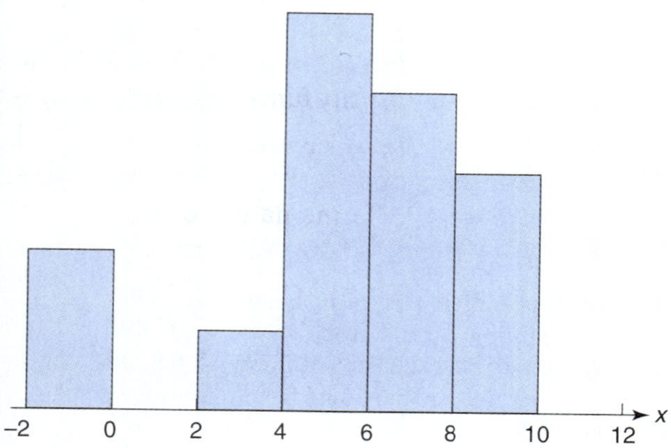

FIGURE 16.27

A probability density curve is the limit of a histogram's outline as the data set gets very large and the bin size gets very small. Figure 16.28 illustrates a larger data set and smaller bin size.

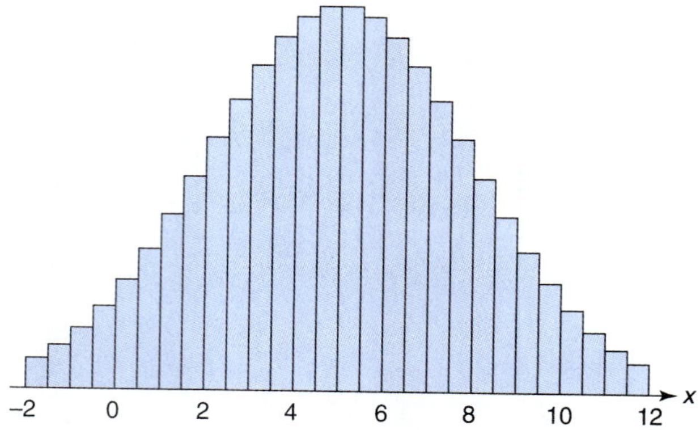

FIGURE 16.28

Unfortunately, dozens of data values are normally needed before a histogram's contours begin to smooth out. As a practical matter, we might want to "cheat" and get an idea of the shape of the probability density function using fewer data values—and without having to draw a histogram.

Here is a way to do that: First use the data values to plot points that reveal the *cumulative distribution* curve, and then use this curve to infer the shape of the probability density curve. The steps are as follows:

Step 1: Determine n, the sample size.

Step 2: Arrange the data values in order, least to greatest. The smallest data value is $v(1)$, the next-smallest is $v(2)$, and so on.

Step 3: On a set of coordinate axes, plot n points with coordinates $\left(v(i), \dfrac{i - \frac{1}{2}}{n}\right)$ for i ranging from 1 to n.

Step 4: Determine what cumulative distribution function the plot suggests.

Step 5: Find the probability density function as the derivative of the cumulative distribution function.

In Step 3, we are plotting each data value $v(i)$ against the experimental probability, based on the data, that a new value would be less than $v(i)$. This probability is calculated by taking the number of values below $v(i)$, namely $i - 1$; adding a term of $\frac{1}{2}$ to split the ith data point in two and count half of it as "below $v(i)$"; and dividing the result by n, the total number of values.

Let us see how this works with data values generated using a known probability density function. On a graphics calculator, we use the *rand* command or its equivalent to generate 15 values (Step 1), using the uniform density function that equals 1 on the interval from 0 to 1 and equals 0 elsewhere. One such run produces the following values (ordered from lowest to highest and rounded to three decimal places): 0.043, 0.074, 0.093, 0.198, 0.293, 0.311, 0.399, 0.400, 0.409, 0.566, 0.654, 0.665, 0.760, 0.919, 0.967 (Step 2). When the corresponding probabilities are plotted (Step 3), the result is as shown in Figure 16.29.

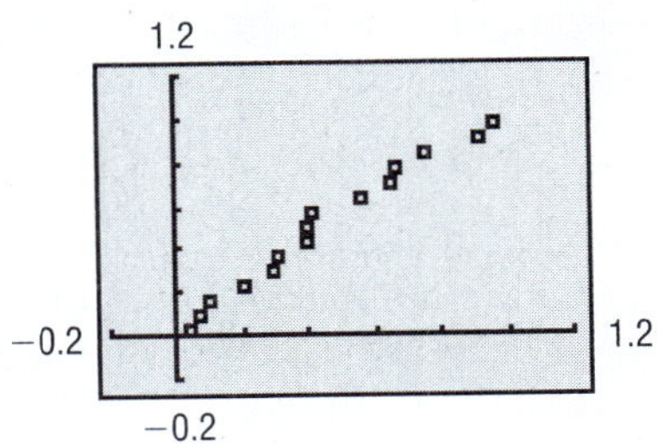

FIGURE 16.29

Even if we did not know how the points had been generated, we could still notice that all data values are between 0 and 1 and that the plotted points lie fairly close to the line $y = x$. This suggests (Step 4) the cumulative distribution function

$$F(x) = \begin{cases} 0 & x < 0 \\ x & 0 \le x \le 1 \\ 1 & x > 1 \end{cases}$$

The derivative of this function with respect to x (Step 5) is

$$P(x) = \begin{cases} 0 & x < 0 \\ 1 & 0 < x < 1 \\ 0 & x > 1 \end{cases}$$

which is exactly the function that was used to generate the data values.[2] The method works.

A word of caution, however. By nature, cumulative distribution functions are increasing from left to right and therefore all have a broadly similar shape. The plot in Figure 16.29, for example, might instead be interpreted as reflecting the cumulative distribution function

$$F_2(x) = \begin{cases} 0 & x < 0 \\ 1 - e^{-2x} & 0 \le x \end{cases}$$

The fit would not be quite so good as for $F(x)$, but it would not look out of the question either. So the method described here for identifying a possible cumulative distribution function and its derivative, the probability density function, is best supplemented by other information, such as knowledge of the process producing the data.

Exercises

1. Generate your own run of 15 values using a uniform density function, and plot the results to obtain a picture of the cumulative distribution function. How well do your results fit with the known distribution function?

2. Repeat Problem 1 for a normal density function. (On the TI-83 Plus, use the *randNorm(* command.) In Step 4, perform a logistic regression. Although the cumulative distribution function for a normal density function is *not* actually a logistic function, it has a very similar shape, so that a logistic function can be used as an approximation. Then, if your calculator has the capability, plot the derivative of the logistic function. Otherwise, note where the logistic function's slope is greatest and what happens to the slope at the ends of the curve. How does the behavior of the derivative compare with the behavior of the normal density function used to generate the data values?

3. Visit a Web site with a table of data values and see if you can determine the nature of the probability density function. You might try the near-real-time earthquake list at the U.S. Geological Survey (wwwneic.cr.usgs.gov), or one of the many data sets published by the U.S. Census (www.census.org). Why do you think the data are distributed as they are?

[2]Notice that $P(i)$ is undefined for $x = 0$ and $x = 1$. This does not matter.

CHAPTER

17

Multivariable Calculus

We know from Chapter 13 how to maximize a company's profit when both revenue and cost are written as functions of a single quantity, namely the number of units produced. But of course the production level is itself determined by other factors—and, in general, no single variable can represent them.

The amount of oil pumped from an oil field each week, for example, depends on both the number of pumps and the number of hours that the pumps are operated. The number of pumps in the field will depend on the amount of capital originally available to build the pumps as well as the size and shape of the field. The number of hours that the pumps can be operated depends on the labor available to run and maintain the pumps. In addition, the amount of oil that the owner will be willing to have pumped from the oil field will depend on the current demand for oil—which is related to the price of the oil.

Maximizing the weekly profit from an oil field will require a balance between the number of pumps and the amount of time each pump can be operated. The maximum profit will not be achieved by building more pumps than can be operated or by running a few pumps full-time.

This is an example of the general problem of maximizing profit when production depends on several factors. The solution involves an analysis of the production function, which relates production output to resources allocated for production. Because, in general, several variables are needed to describe the resource allocation, the most profitable allocation cannot be found by differentiation with respect to a single variable, as in preceding chapters. The more advanced techniques necessary to do the job will be covered in this chapter.

OBJECTIVE *To discuss functions of several variables and to compute function values. To discuss three-dimensional coordinates and sketch simple surfaces.*

17.1 Functions of Several Variables

Suppose a manufacturer produces two products, X and Y. Then the total cost depends on the levels of production of *both* X and Y. Table 17.1 is a schedule that indicates total cost at various levels. For example, when 5 units of X and 6 units of Y are produced, the total cost c is 17. In this situation, it seems natural to associate the number 17 with the *ordered pair* (5, 6):

$$(5, 6) \mapsto 17$$

TABLE 17.1

No. of Units of X Produced, x	No. of Units of Y Produced, y	Total Cost of Production, c
5	6	17
5	7	19
6	6	18
6	7	20

The first element of the ordered pair, 5, represents the number of units of X produced, while the second element, 6, represents the number of units of Y produced. Corresponding to the other production situations, we have

$$(5, 7) \mapsto 19$$

$$(6, 6) \mapsto 18$$

and

$$(6, 7) \mapsto 20$$

This correspondence can be considered an input-output relation where the inputs are ordered pairs. With each input, we associate exactly one output. Thus, the correspondence defines a function f whose domain consists of (5, 6), (5, 7), (6, 6), (6, 7) and whose range consists of 17, 19, 18, 20. In function notation,

$$f(5, 6) = 17 \qquad f(5, 7) = 19$$

$$f(6, 6) = 18 \qquad f(6, 7) = 20$$

We say that the total-cost schedule can be described by $c = f(x, y)$, a function of the two independent variables x and y. The letter c is the dependent variable.

Turning to another function of two variables, we see that the equation

$$z = \frac{2}{x^2 + y^2}$$

defines z as a function of x and y:

$$z = f(x, y) = \frac{2}{x^2 + y^2}$$

The domain of f is all ordered pairs of real numbers (x, y) for which the equation has meaning when the first and second elements of (x, y) are substituted for x and y, respectively, in the equation. Thus, the domain of f is all ordered pairs except (0, 0). For example, to find $f(2, 3)$, we substitute $x = 2$ and $y = 3$ into the expression $2/(x^2 + y^2)$. We obtain $f(2, 3) = 2/(2^2 + 3^2) = 2/13$.

APPLICATIONS IN PRACTICE 1

Functions of Two Variables

The cost per day for manufacturing both 12-ounce and 20-ounce coffee mugs is given by $c = 160 + 2x + 3y$, where x is the number of 12-ounce mugs and y is the number of 20-ounce mugs. What is the cost per day of manufacturing

1. 500 12-ounce and 700 20-ounce mugs?
2. 1000 12-ounce and 750 20-ounce mugs?

EXAMPLE 1 Functions of Two Variables

a. $f(x, y) = \dfrac{x+3}{y-2}$ is a function of two variables. Because the denominator is zero when $y = 2$, the domain of f is all (x, y) such that $y \neq 2$. Some function values are

$$f(0, 3) = \frac{0+3}{3-2} = 3$$

$$f(3, 0) = \frac{3+3}{0-2} = -3$$

Note that $f(0, 3) \neq f(3, 0)$.

b. $h(x, y) = 4x$ defines h as a function of x and y. The domain is all ordered pairs of real numbers. Some function values are

$$h(2, 5) = 4(2) = 8$$

$$h(2, 6) = 4(2) = 8$$

Note that the function values are independent of the choice of y.

c. If $z^2 = x^2 + y^2$ and $x = 3$ and $y = 4$, then $z^2 = 3^2 + 4^2 = 25$. Consequently, $z = \pm 5$. Thus, with the ordered pair (3, 4), we *cannot* associate exactly one output number. Hence $z^2 = x^2 + y^2$ does not define z as a function of x and y.

▸ **Now work Problem 1.**

EXAMPLE 2 Temperature-Humidity Index

On hot and humid days, many people tend to feel uncomfortable. The degree of discomfort is numerically given by the temperature-humidity index, THI, *which is a function of two variables,* t_d *and* t_w:

$$\text{THI} = f(t_d, t_w) = 15 + 0.4(t_d + t_w)$$

where t_d *is the dry-bulb temperature (in degrees Fahrenheit) and* t_w *is the wet-bulb temperature (in degrees Fahrenheit) of the air. Evaluate the* THI *when* $t_d = 90$ *and* $t_w = 80$.

Solution: We want to find $f(90, 80)$:

$$f(90, 80) = 15 + 0.4(90 + 80) = 15 + 68 = 83$$

When the THI is greater than 75, most people are uncomfortable. In fact, the THI was once called the "discomfort index." Many electric utilities closely follow this index so that they can anticipate the demand that air-conditioning places on their systems.

▸ **Now work Problem 3.**

If $y = f(x)$ is a function of one variable, the domain of f can be geometrically represented by points on the real-number line. The function itself can be represented by its graph in a coordinate plane, sometimes called a two-dimensional coordinate system. However, for a function of two variables, $z = f(x, y)$, the domain (consisting of ordered pairs of real numbers) can be geometrically represented by a *region* in the plane. The function itself can be

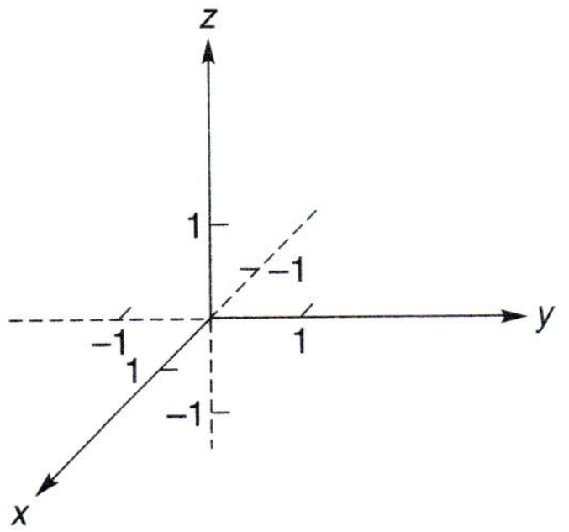

FIGURE 17.1 Three-dimensional rectangular coordinate system.

geometrically represented in a ***three*-dimensional rectangular coordinate system.** Such a system is formed when three mutually perpendicular real-number lines in space intersect at the origin of each line, as in Figure 17.1. The three number lines are called the x-, y-, and z-axes, and their point of intersection is called the origin of the system. The arrows indicate the positive directions of the axes, and the negative portions of the axes are shown as dashed lines.

To each point P in space, we can assign a unique ordered triple of numbers, called the *coordinates* of P. To do this [see Figure 17.2(a)], from P, we construct a line perpendicular to the x, y-plane—that is, the plane determined by the x- and y-axes. Let Q be the point where the line intersects this plane. From Q, we construct perpendiculars to the x- and y-axes. These lines intersect the x- and y-axes at x_0 and y_0, respectively. From P, a perpendicular to the z-axis is constructed that intersects the axis at z_0. Thus, we assign to P the ordered triple (x_0, y_0, z_0). It should also be evident that with each ordered triple of numbers we can assign a unique point in space. Owing to this one-to-one correspondence between points in space and ordered triples, an ordered triple may be called a point. In Figure 17.2(b), points (2, 0, 0), (2, 3, 0), and (2, 3, 4) are shown. Note that the origin corresponds to (0, 0, 0). Typically, the negative portions of the axes are not shown unless required.

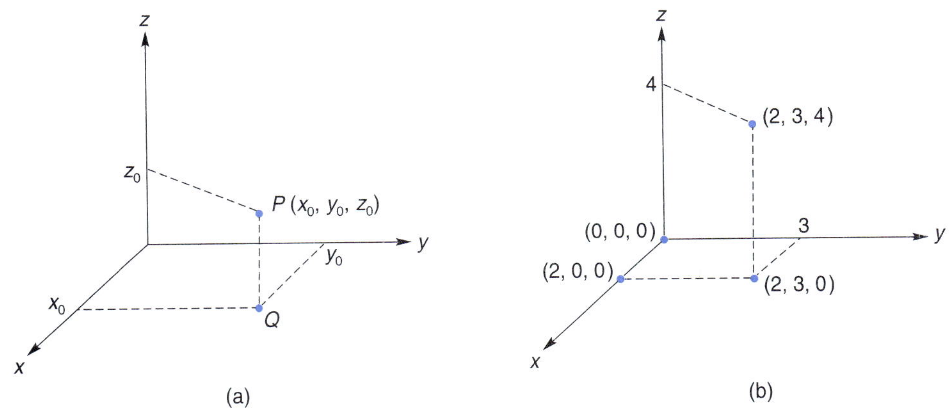

FIGURE 17.2 Points in space.

We can represent a function of two variables, $z = f(x, y)$, geometrically. To each ordered pair (x, y) in the domain of f, we assign the point $(x, y, f(x, y))$. The set of all such points is called the *graph* of f. Such a graph appears in Figure 17.3. You can consider $z = f(x, y)$ as representing a surface in space.[1]

Chapter 10 discussed the continuity of a function of one variable. If $y = f(x)$ is continuous at $x = x_0$, then points near x_0 will have their function values near $f(x_0)$. Extending this concept to a function of two variables, we say that the function $z = f(x, y)$ is continuous at (x_0, y_0) when points near (x_0, y_0) have their function values near $f(x_0, y_0)$. Loosely interpreting this concept, and without delving into it in great depth, we can say that a function of two variables is continuous on its domain (that is, continuous at each point in its domain) if its graph is an "unbroken surface."

Until now, we have considered only functions of either one or two variables. In general, a **function of n variables** is a function whose domain consists of ordered n-tuples $(x_1, x_2, \ldots, x_n)$. For example, $f(x, y, z) = 2x + 3y + 4z$ is a

[1] We shall freely use the term *surface* in the intuitive sense.

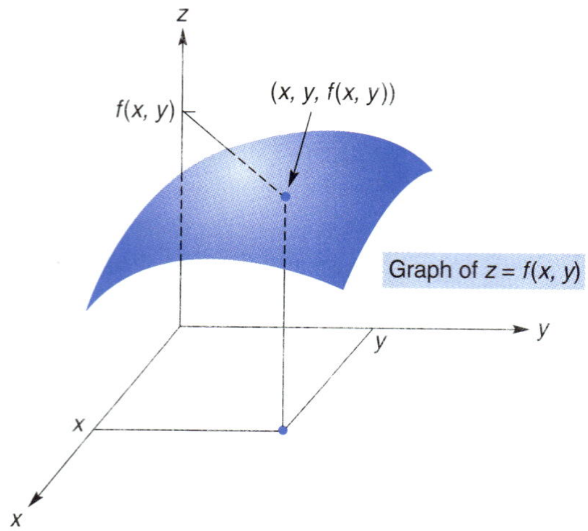

FIGURE 17.3 Graph of a function of two variables.

function of three variables with a domain consisting of all ordered triples. The function $g(x_1, x_2, x_3, x_4) = x_1x_2x_3x_4$ is a function of four variables with a domain consisting of all ordered 4-tuples. Although functions of several variables are extremely important and useful, we cannot visualize the graphs of functions of more than two variables.

We now give a brief discussion of sketching surfaces in space. We begin with planes that are parallel to a coordinate plane. By a "coordinate plane" we mean a plane containing two coordinate axes. For example, the plane determined by the x- and y-axes is the ***x,y*-plane.** Similarly, we speak of the ***x,z*-plane** and the ***y,z*-plane.** The coordinate planes divide space into eight parts, called *octants*. In particular, the part containing all points (x, y, z) such that x, y, and z are positive is called the **first octant.**

Names are not assigned to the remaining seven octants.

Suppose S is a plane that is parallel to the x,y-plane and passes through the point $(0, 0, 5)$. [See Figure 17.4(a).] Then the point (x, y, z) will lie on S if and only if $z = 5$; that is, x and y can be any real numbers, but z must equal 5. For this reason, we say that $z = 5$ is an equation of S. Similarly, an equation of the plane parallel to the x,z-plane and passing through the point $(0, 2, 0)$ is $y = 2$ [Figure 17.4(b)]. The equation $x = 3$ is an equation of the plane passing through $(3, 0, 0)$ and parallel to the y,z-plane [Figure 17.4(c)]. Now let us look at planes in general.

In space, the graph of an equation of the form

$$Ax + By + Cz + D = 0$$

where D is a constant and A, B, and C are constants that are not all zero, is a

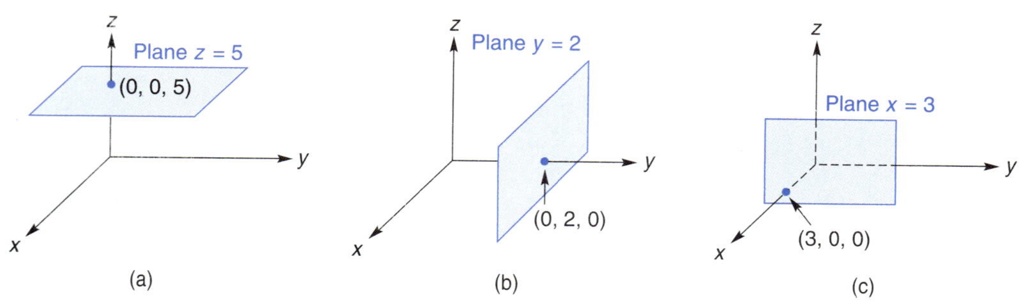

FIGURE 17.4 Planes parallel to coordinate planes.

plane. Since three distinct points (not lying on the same line) determine a plane, a convenient way to sketch a plane is to first determine the points, if any, where the plane intersects the x-, y-, and z-axes. These points are called *intercepts*.

EXAMPLE 3 Graphing a Plane

Sketch the plane $2x + 3y + z = 6$.

Solution: The plane intersects the x-axis when $y = 0$ and $z = 0$. Thus, $2x = 6$, which gives $x = 3$. Similarly, if $x = z = 0$, then $y = 2$; if $x = y = 0$, then $z = 6$. Therefore, the intercepts are $(3, 0, 0)$, $(0, 2, 0)$, and $(0, 0, 6)$. After these points are plotted, a plane is passed through them. The portion of the plane in the first octant is shown in Figure 17.5(a); however, you should realize that the plane extends indefinitely into space.

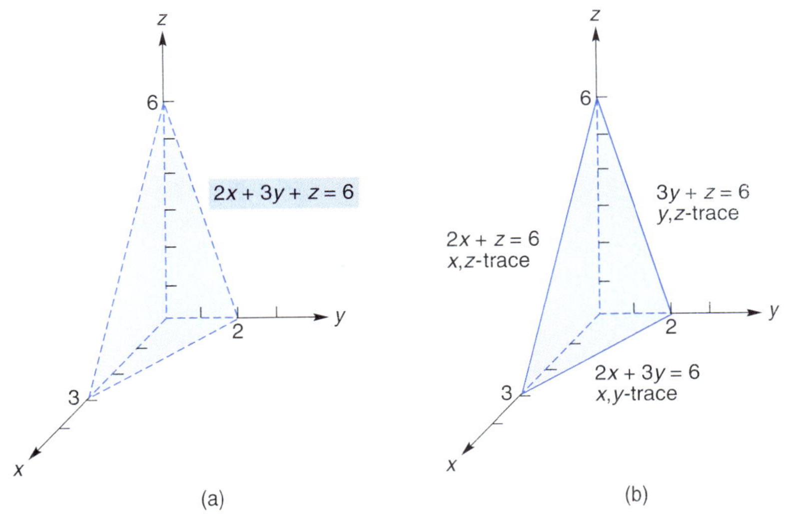

FIGURE 17.5 The plane $2x + 3y + z = 6$ and its traces.

▶ **Now work Problem 19.**

A surface can be sketched with the aid of its **traces.** These are the intersections of the surface with the coordinate planes. To illustrate, for the plane $2x + 3y + z = 6$ in Example 3, the trace in the x,y-plane is obtained by setting $z = 0$. This gives $2x + 3y = 6$, which is an equation of a *line* in the x,y-plane. Similarly, setting $x = 0$ gives the trace in the y,z-plane: the line $3y + z = 6$. The x,z-trace is the line $2x + z = 6$. [See Figure 17.5(b).]

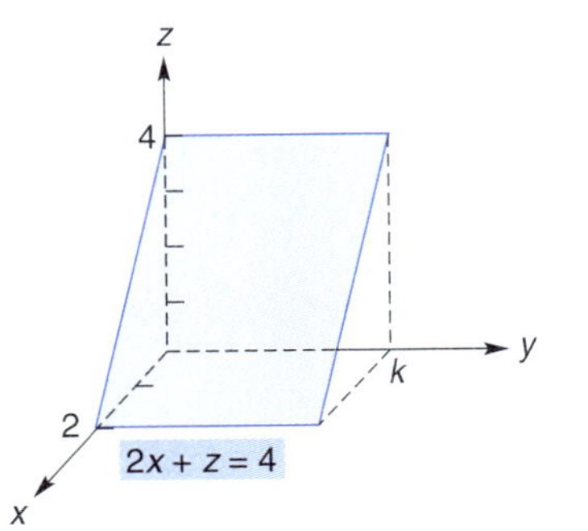

FIGURE 17.6 The plane $2x + z = 4$.

Note that this equation places no restriction on y.

EXAMPLE 4 Sketching a Surface

Sketch the surface $2x + z = 4$.

Solution: This equation has the form of a plane. The x- and z-intercepts are $(2, 0, 0)$ and $(0, 0, 4)$, and there is no y-intercept, since x and z cannot both be zero. Setting $y = 0$ gives the x,z-trace $2x + z = 4$, which is a line in the x,z-plane. In fact, the intersection of the surface with *any* plane $y = k$ is also $2x + z = 4$. Hence, the plane appears as in Figure 17.6.

▶ **Now work Problem 21.**

Our final examples deal with surfaces that are not planes, but whose graphs can be easily obtained.

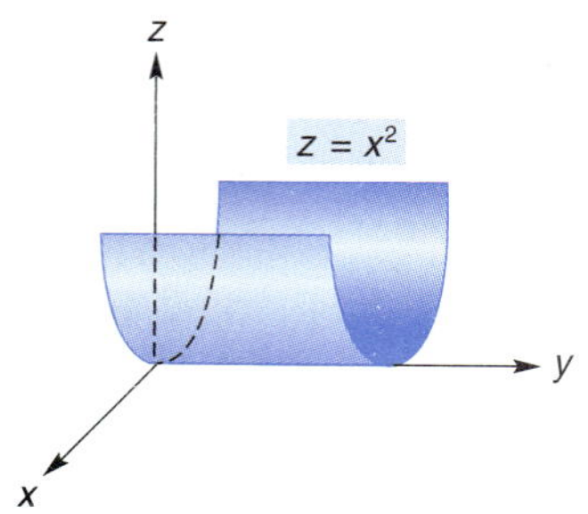

FIGURE 17.7 The surface $z = x^2$.

EXAMPLE 5 Sketching a Surface

Sketch the surface $z = x^2$.

Solution: The x,z-trace is the curve $z = x^2$, which is parabola. In fact, for *any* fixed value of y, we get $z = x^2$. Thus, the graph appears as in Figure 17.7.

▶ Now work Problem 25.

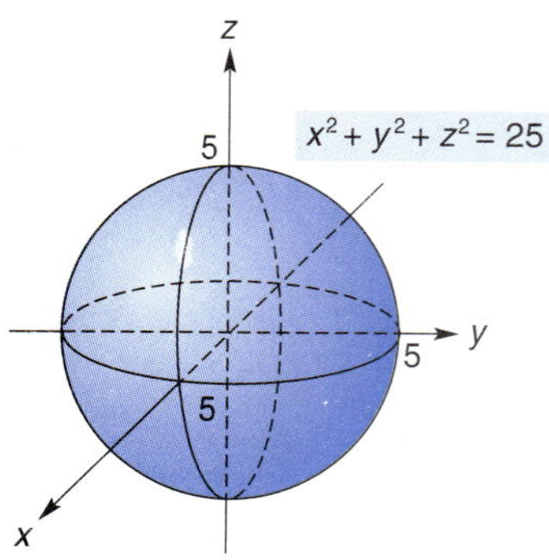

FIGURE 17.8 The surface $x^2 + y^2 + z^2 = 25$.

EXAMPLE 6 Sketching a Surface

Sketch the surface $x^2 + y^2 + z^2 = 25$.

Solution: Setting $z=0$ gives the x,y-trace $x^2+y^2=25$, which is a circle of radius 5. Similarly, the y,z- and x,z-traces are the circles $y^2 + z^2=25$ and $x^2+z^2=25$, respectively. Note also that since $x^2+y^2=25-z^2$, the intersection of the surface with the plane $z=k$, where $-5 \leq k \leq 5$, is a circle. For example, if $z=3$, the intersection is the circle $x^2 + y^2=16$. If $z = 4$, the intersection is $x^2+y^2=9$. That is, cross sections of the surface that are parallel to the x,y-plane are circles. The surface appears in Figure 17.8; it is a sphere.

▶ Now work Problem 27.

Exercise 17.1

In Problems 1–12, determine the indicated function values for the given functions.

*1. $f(x, y) = 4x - y^2 + 3;\quad f(1, 2)$

2. $f(x, y) = 3x^2y - 4y;\quad f(2, -1)$

*3. $g(x, y, z) = e^x(2y + 3z);\quad g(0, -4, 2)$

4. $g(x, y, z) = x^2yz + xy^2z + xyz^2;\quad g(3, 1, -2)$

5. $h(r, s, t, u) = \dfrac{rs}{t^2 - u^2};\quad h(-3, 3, 5, 4)$

6. $h(r, s, t, u) = \ln(ru);\quad h(1, 5, 3, 1)$

7. $g(p_A, p_B) = 2p_A(p_A^2 - 5);\quad g(4, 8)$

8. $g(p_A, p_B) = p_A\sqrt{p_B} + 10;\quad g(8, 4)$

9. $F(x, y, z) = 3;\quad F(2, 0, -1)$

10. $F(x, y, z) = \dfrac{x}{yz},\ F(0, 0, 3)$

11. $f(x, y) = e^{x+y};\quad f(x_0 + h, y_0)$

12. $f(x, y) = x^2y - 3y^3;\quad f(r + t, r)$

13. **Ecology** A method of ecological sampling to determine animal populations in a given area involves first marking all the animals obtained in a sample of R animals from the area and then releasing them so that they can mix with unmarked animals. At a later date a second sample is taken of M animals and the number of these that are marked, S, is noted. Based on R, M, and S, an estimate of the total population of animals in the sample area is given by

$$N = f(R, M, S) = \frac{RM}{S}$$

Find $f(400, 400, 80)$. This method is called the *mark-and-recapture procedure.*[2]

14. **Genetics** Under certain conditions, if two brown-eyed parents have exactly k children, the probability that there will be exactly r blue-eyed children is given by

$$P(r, k) = \frac{k!\left(\frac{1}{4}\right)^r\left(\frac{3}{4}\right)^{k-r}}{r!(k-r)!} \qquad r = 0, 1, 2, \ldots, k$$

Find the probability that, out of a total of four children, exactly three will be blue-eyed.

[2]E. P. Odum, *Ecology* (New York: Holt, Rinehart and Winston, 1966).

In Problems 15–18, find equations of the planes that satisfy the given conditions.

15. Parallel to the x,z-plane and also passes through the point $(0, -4, 0)$.

16. Parallel to the y,z-plane and also passes through the point $(-2, 0, 0)$.

17. Parallel to the x,y-plane and also passes through the point $(2, 7, 6)$.

18. Parallel to the y,z-plane and also passes through the point $(-4, -2, 7)$.

In Problems 19–28, sketch the given surfaces.

*19. $x + y + z = 1$

20. $2x + y + 2z = 6$

*21. $3x + 6y + 2z = 12$

22. $x + 2y + 3z = 4$

23. $x + 2y = 2$

24. $y = 3z + 2$

*25. $z = 4 - x^2$

26. $y = z^2$

*27. $x^2 + y^2 + z^2 = 1$

28. $3x^2 + 2y^2 = 1$

OBJECTIVE *To compute partial derivatives.*

17.2 Partial Derivatives

Figure 17.9 shows the surface $z = f(x, y)$ and a plane that is parallel to the x,z-plane and that passes through the point $(x_0, y_0, f(x_0, y_0))$ on the surface. An equation of this plane is $y = y_0$. Hence, any point on the curve that is the intersection of the surface with the plane must have the form $(x, y_0, f(x, y_0))$. Thus, the curve can be described by the equation $z = f(x, y_0)$. Since y_0 is constant, $z = f(x, y_0)$ can be considered a function of one variable, x. When the derivative of this function is evaluated at x_0, it gives the slope of the tangent line to this curve at the point $(x_0, y_0, f(x_0, y_0))$. (See Figure 17.9.) This slope is called the *partial derivative of f with respect to x* at (x_0, y_0) and is denoted $f_x(x_0, y_0)$. In terms of limits,

$$f_x(x_0, y_0) = \lim_{h \to 0} \frac{f(x_0 + h, y_0) - f(x_0, y_0)}{h} \tag{1}$$

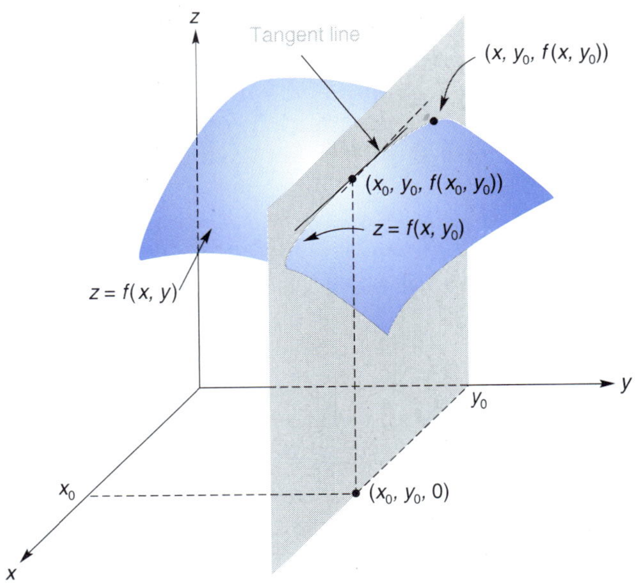

FIGURE 17.9 Geometric interpretation of $f_x(x_0, y_0)$.

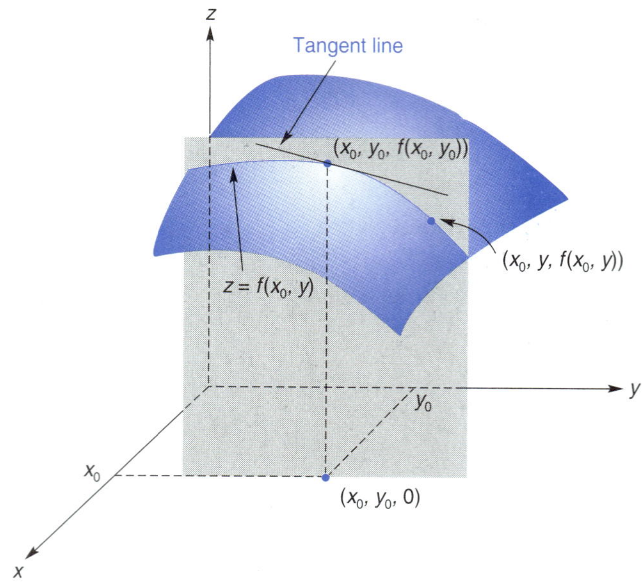

FIGURE 17.10 Geometric interpretation of $f_y(x_0, y_0)$.

On the other hand, in Figure 17.10, the plane $x = x_0$ is parallel to the y,z-plane and cuts the surface $z = f(x, y)$ in a curve given by $z = f(x_0, y)$, a function of y. When the derivative of this function is evaluated at y_0, it gives the slope of the tangent line to this curve at the point $(x_0, y_0, f(x_0, y_0))$. This slope is called the *partial derivative of f with respect to y* at (x_0, y_0) and is denoted $f_y(x_0, y_0)$. In terms of limits,

$$f_y(x_0, y_0) = \lim_{h \to 0} \frac{f(x_0, y_0 + h) - f(x_0, y_0)}{h} \tag{2}$$

This gives us a geometric interpretation of a partial derivative.

We say that $f_x(x_0, y_0)$ is the slope of the tangent line to the graph of f at $(x_0, y_0, f(x_0, y_0))$ *in the x-direction;* similarly, $f_y(x_0, y_0)$ is the slope of the tangent line *in the y-direction.*

For generality, by replacing x_0 and y_0 in Equations (1) and (2) by x and y, respectively, we get the following definition.

Definition

If $z = f(x, y)$, the ***partial derivative of f with respect to x***, denoted f_x, is the function given by

$$f_x(x, y) = \lim_{h \to 0} \frac{f(x + h, y) - f(x, y)}{h}$$

provided that the limit exists.

The ***partial derivative of f with respect to y***, denoted f_y, is the function given by

$$f_y(x, y) = \lim_{h \to 0} \frac{f(x, y + h) - f(x, y)}{h}$$

provided that the limit exists.

By analyzing the foregoing definition, we can state the following procedure to find f_x and f_y:

This gives us a mechanical way to find partial derivatives.

Procedure to Find $f_x(x, y)$ and $f_y(x, y)$

To find f_x, treat y as a constant, and differentiate f with respect to x in the usual way.

To find f_y, treat x as a constant, and differentiate f with respect to y in the usual way.

EXAMPLE 1 Finding Partial Derivatives

If $f(x, y) = xy^2 + x^2y$, find $f_x(x, y)$ and $f_y(x, y)$. Also, find $f_x(3, 4)$ and $f_y(3, 4)$.

Solution: To find $f_x(x, y)$, we treat y as a constant and differentiate f with respect to x:

$$f_x(x, y) = (1)y^2 + (2x)y = y^2 + 2xy$$

To find $f_y(x, y)$, we treat x as a constant and differentiate with respect to y:

$$f_y(x, y) = x(2y) + x^2(1) = 2xy + x^2$$

Note that $f_x(x, y)$ and $f_y(x, y)$ are each functions of the two variables x and y. To find $f_x(3, 4)$, we evaluate $f_x(x, y)$ when $x = 3$ and $y = 4$:

$$f_x(3, 4) = 4^2 + 2(3)(4) = 40$$

Similarly,

$$f_y(3, 4) = 2(3)(4) + 3^2 = 33$$

▶ **Now work Problem 1.**

Notations for partial derivatives of $z = f(x, y)$ are in Table 17.2. Table 17.3 gives notations for partial derivatives evaluated at (x_0, y_0). Note that the symbol ∂ (not d) is used to denote a partial derivative. The symbol $\partial z/\partial x$ is read "the partial derivative of z with respect to x."

TABLE 17.2

Partial Derivative of f (or z) with Respect to x	Partial Derivative of f (or z) with Respect to y
$f_x(x, y)$	$f_y(x, y)$
$\frac{\partial}{\partial x}[f(x, y)]$	$\frac{\partial}{\partial y}[f(x, y)]$
$\frac{\partial z}{\partial x}$	$\frac{\partial z}{\partial y}$

TABLE 17.3

Partial Derivative of f (or z) with Respect to x Evaluated at (x_0, y_0)	Partial Derivative of f (or z) with Respect to y Evaluated at (x_0, y_0)
$f_x(x_0, y_0)$	$f_y(x_0, y_0)$
$\left.\frac{\partial z}{\partial x}\right\|_{(x_0, y_0)}$	$\left.\frac{\partial z}{\partial y}\right\|_{(x_0, y_0)}$
$\left.\frac{\partial z}{\partial x}\right\|_{\substack{x=x_0\\y=y_0}}$	$\left.\frac{\partial z}{\partial y}\right\|_{\substack{x=x_0\\y=y_0}}$

EXAMPLE 2 Finding Partial Derivatives

a. *If* $z = 3x^3y^3 - 9x^2y + xy^2 + 4y$, *find* $\frac{\partial z}{\partial x}, \frac{\partial z}{\partial y}, \left.\frac{\partial z}{\partial x}\right|_{(1,0)}$ *and* $\left.\frac{\partial z}{\partial y}\right|_{(1,0)}$.

Solution: To find $\partial z/\partial x$, we differentiate z with respect to x while treating y as a constant:

$$\begin{aligned}\frac{\partial z}{\partial x} &= 3(3x^2)y^3 - 9(2x)y + (1)y^2 + 0\\ &= 9x^2y^3 - 18xy + y^2\end{aligned}$$

Evaluating the latter equation at $(1, 0)$, we obtain

$$\left.\frac{\partial z}{\partial x}\right|_{(1,0)} = 9(1)^2(0)^3 - 18(1)(0) + 0^2 = 0$$

To find $\partial z/\partial y$, we differentiate z with respect to y while treating x as a constant:

$$\begin{aligned}\frac{\partial z}{\partial y} &= 3x^3(3y^2) - 9x^2(1) + x(2y) + 4(1)\\ &= 9x^3y^2 - 9x^2 + 2xy + 4\end{aligned}$$

Thus,

$$\left.\frac{\partial z}{\partial y}\right|_{(1,0)} = 9(1)^3(0)^2 - 9(1)^2 + 2(1)(0) + 4 = -5$$

b. *If* $w = x^2e^{2x+3y}$, *find* $\partial w/\partial x$ *and* $\partial w/\partial y$.

Solution: To find $\partial w/\partial x$, we treat y as a constant and differentiate with respect to x. Since x^2e^{2x+3y} is a product of two functions, each involving x, we use the product rule:

$$\begin{aligned}\frac{\partial w}{\partial x} &= x^2\frac{\partial}{\partial x}(e^{2x+3y}) + e^{2x+3y}\frac{\partial}{\partial x}(x^2)\\ &= x^2(2e^{2x+3y}) + e^{2x+3y}(2x)\\ &= 2x(x+1)e^{2x+3y}\end{aligned}$$

To find $\partial w/\partial y$, we treat x as a constant and differentiate with respect to y:

$$\frac{\partial w}{\partial y} = x^2\frac{\partial}{\partial y}(e^{2x+3y}) = 3x^2e^{2x+3y}$$

▶ **Now work Problem 27.**

We have seen that, for a function of two variables, two partial derivatives can be considered. Actually, the concept of partial derivatives can be extended to functions of more than two variables. For example, with $w = f(x, y, z)$ we have three partial derivatives:

the partial with respect to x, denoted $f_x(x, y, z)$, $\partial w/\partial x$, etc.;

the partial with respect to y, denoted $f_y(x, y, z)$, $\partial w/\partial y$, etc.;

and

the partial with respect to z, denoted $f_z(x, y, z)$, $\partial w/\partial z$, etc.

To determine $\partial w/\partial x$, treat y and z as constants, and differentiate w with respect to x. For $\partial w/\partial y$, treat x and z as constants, and differentiate with respect to y. For $\partial w/\partial z$, treat x and y as constants, and differentiate with respect to z. With a function of n variables, we have n partial derivatives, which are determined in the obvious way.

EXAMPLE 3 Partial Derivatives of a Function of Three Variables

If $f(x, y, z) = x^2 + y^2z + z^3$, *find* $f_x(x, y, z)$, $f_y(x, y, z)$, *and* $f_z(x, y, z)$.

Solution: To find $f_x(x, y, z)$, we treat y and z as constants and differentiate f with respect to x:

$$f_x(x, y, z) = 2x$$

Treating x and z as constants and differentiating with respect to y, we have

$$f_y(x, y, z) = 2yz$$

Treating x and y as constants and differentiating with respect to z, we have

$$f_z(x, y, z) = y^2 + 3z^2$$

▶ **Now work Problem 23.**

EXAMPLE 4 Partial Derivatives of a Function of Four Variables

If $p = g(r, s, t, u) = \dfrac{rsu}{rt^2 + s^2t}$, *find* $\dfrac{\partial p}{\partial s}$, $\dfrac{\partial p}{\partial t}$, *and* $\left.\dfrac{\partial p}{\partial t}\right|_{(0,1,1,1)}$.

Solution: To find $\partial p/\partial s$, first note that p is a quotient of two functions, each involving the variable s. Thus, we use the quotient rule and treat r, t, and u as constants:

$$\frac{\partial p}{\partial s} = \frac{(rt^2 + s^2t)\dfrac{\partial}{\partial s}(rsu) - rsu\dfrac{\partial}{\partial s}(rt^2 + s^2t)}{(rt^2 + s^2t)^2}$$

$$= \frac{(rt^2 + s^2t)(ru) - (rsu)(2st)}{(rt^2 + s^2t)^2}$$

Simplifying gives

$$\frac{\partial p}{\partial s} = \frac{ru(rt - s^2)}{t(rt + s^2)^2} \qquad \text{(a factor of } t \text{ cancels)}$$

To find $\partial p/\partial t$, we can first write p as

$$p = rsu(rt^2 + s^2t)^{-1}$$

Next, we use the power rule and treat r, s, and u as constants:

$$\frac{\partial p}{\partial t} = rsu(-1)(rt^2 + s^2t)^{-2}\frac{\partial}{\partial t}(rt^2 + s^2t)$$

$$= -rsu(rt^2 + s^2t)^{-2}(2rt + s^2)$$

so that

$$\frac{\partial p}{\partial s} = -\frac{rsu(2rt + s^2)}{(rt^2 + s^2t)^2}$$

Letting $r = 0, s = 1, t = 1$, and $u = 1$ gives

$$\left.\frac{\partial p}{\partial t}\right|_{(0,1,1,1)} = -\frac{0(1)(1)[2(0)(1) + (1)^2]}{[0(1)^2 + (1)^2(1)]^2} = 0$$

▶ **Now work Problem 31.**

Exercise 17.2

In Problems 1–26, a function of two or more variables is given. Find the partial derivative of the function with respect to each of the variables.

*1. $f(x, y) = 4x^2 + 3y^2 - 7$

2. $f(x, y) = 2x^2 + 3xy$

3. $f(x, y) = 2y + 1$

4. $f(x, y) = e$

5. $g(x, y) = 3x^4y + 2xy^2 - 5xy + 8x - 9y$

6. $g(x, y) = (x+1)^2 + (y-3)^3 + 5xy^3 - 2$

7. $g(p, q) = \sqrt{pq}$

8. $g(w, z) = \sqrt[3]{w^2 + z^2}$

9. $h(s, t) = \dfrac{s^2 + 4}{t - 3}$

10. $h(u, v) = \dfrac{8uv^2}{u^2 + v^2}$

11. $u(q_1, q_2) = \frac{3}{4}\ln q_1 + \frac{1}{4}\ln q_2$

12. $Q(l, k) = 2l^{0.38}k^{1.79} - 3l^{1.03} + 2k^{0.13}$

13. $h(x, y) = \dfrac{x^2 + 3xy + y^2}{\sqrt{x^2 + y^2}}$

14. $h(x, y) = \dfrac{\sqrt{x + 9}}{x^2y + y^2x}$

15. $z = e^{5xy}$

16. $z = (x^2 + y)e^{3x+4y}$

17. $z = 5x\ln(x^2 + y)$

18. $z = \ln(5x^3y^2 + 2y^4)^4$

19. $f(r, s) = \sqrt{r + 2s}\,(r^3 - 2rs + s^2)$

20. $f(r, s) = \sqrt{rs}\, e^{2+r}$

21. $f(r, s) = e^{3-r}\ln(7 - s)$

22. $f(r, s) = (5r^2 + 3s^3)(2r - 5s)$

*23. $g(x, y, z) = 3x^2y + 2xy^2z + 3z^3$

24. $g(x, y, z) = 2xy^2z^6 - 4x^2y^3z^2 + 3xyz$

25. $g(r, s, t) = e^{s+t}(r^2 + 7s^3)$

26. $g(r, s, t, u) = rs\ln(2t + 5u)$

In Problems 27–34, evaluate the given partial derivatives.

*27. $f(x, y) = x^3y + 7x^2y^2$; $f_x(1, -2)$

28. $z = \sqrt{5x^2 + 3xy + 2y}$; $\left.\dfrac{\partial z}{\partial x}\right|_{\substack{x=0\\y=2}}$

29. $g(x, y, z) = e^x\sqrt{y + 2z}$; $g_z(0, 6, 4)$

30. $g(x, y, z) = \dfrac{3x^2y^2 + 2xy + x - y}{xy - yz + xz}$, $g_y(1, 1, 5)$

*31. $h(r, s, t, u) = (s^2 + tu)\ln(2r + 7st)$; $h_s(1, 0, 0, 1)$

32. $h(r, s, t, u) = \dfrac{7r + 3s^2u^2}{s}$; $h_t(4, 3, 2, 1)$

33. $f(r, s, t) = rst(r^2 + s^3 + t^4)$; $f_s(1, -1, 2)$

34. $z = \dfrac{x^2 + y^2}{e^{x^2+y^2}}$; $\left.\dfrac{\partial z}{\partial x}\right|_{\substack{x=0\\y=0}}$, $\left.\dfrac{\partial z}{\partial y}\right|_{\substack{x=1\\y=1}}$

35. If $z = xe^{x-y} - ye^{y-x}$, show that

$$\frac{\partial z}{\partial x} + \frac{\partial z}{\partial y} = e^{x-y} - e^{y-x}$$

36. **Stock Prices of a Dividend Cycle** In a discussion of stock prices of a dividend cycle, Palmon and Yaari[3] consider the function f given by

$$u = f(t, r, z) = \frac{(1+r)^{1-z}\ln(1+r)}{(1+r)^{1-z} - t}$$

where u is the instantaneous rate of ask-price appreciation, r is an annual opportunity rate of return, z is the fraction of a dividend cycle over which a share of stock is held by a midcycle seller, and t is the effective rate of capital gains tax. They claim that

$$\frac{\partial u}{\partial z} = \frac{t(1+r)^{1-z}\ln^2(1+r)}{[(1+r)^{1-z} - t]^2}$$

Verify this.

37. **Money Demand** In a discussion of inventory theory of money demand, Swanson[4] considers the function

$$F(b, C, T, i) = \frac{bT}{C} + \frac{iC}{2}$$

and determines that $\dfrac{\partial F}{\partial C} = -\dfrac{bT}{C^2} + \dfrac{i}{2}$. Verify this partial derivative.

[3] D. Palmon and U. Yaari, "Taxation of Capital Gains and the Behavior of Stock Prices over the Dividend Cycle," *American Economist* 27, no. 1 (1983), 13–22.

[4] P. E. Swanson, "Integer Constraints on the Inventory Theory of Money Demand," *Quarterly Journal of Business and Economics* 23, no. 1 (1984), 32–37.

38. Interest Rate Deregulation In an article on interest rate deregulation, Christofi and Agapos[5] arrive at the equation

$$r_L = r + D\frac{\partial r}{\partial D} + \frac{dC}{dD} \qquad (3)$$

where r is the deposit rate paid by commercial banks, r_L is the rate earned by commercial banks, C is the administrative cost of transforming deposits into return-earning assets, and D is the savings deposit level. Christofi and Agapos state that

$$r_L = r\left[\frac{1+\eta}{\eta}\right] + \frac{dC}{dD} \qquad (4)$$

where $\eta = \dfrac{r/D}{\partial r/\partial D}$ is the deposit elasticity with respect to the deposit rate. Express Equation (3) in terms of η to verify Equation (4).

39. Advertising and Profitability In an analysis of advertising and profitability, Swales[6] considers a function f given by

$$R = f(r, a, n) = \frac{r}{1 + a\left(\dfrac{n-1}{2}\right)}$$

where R is the adjusted rate of profit, r is the accounting rate of profit, a is a measure of advertising expenditures, and n is the number of years that advertising fully depreciates. In the analysis, Swales determines $\partial R/\partial n$. Find this partial derivative.

[5] A. Christofi and A. Agapos, "Interest Rate Deregulation: An Empirical Justification," *Review of Business and Economic Research* 20 (1984), 39–49.

[6] J. K. Swales, "Advertising as an Intangible Asset: Profitability and Entry Barriers: A Comment on Reekie and Bhoyrub," *Applied Economics* 17, no. 4 (1985), 603–17.

17.3 Applications of Partial Derivatives

OBJECTIVE *To develop the notions of partial marginal cost, marginal productivity, and competitive and complementary products.*

From Section 17.2, we know that if $z = f(x, y)$, then $\partial z/\partial x$ and $\partial z/\partial y$ can be geometrically interpreted as giving the slopes of the tangent lines to the surface $z = f(x, y)$ in the x- and y-directions, respectively. There are other interpretations: Because $\partial z/\partial x$ is the derivative of z with respect to x when y is held fixed, and because a derivative is a rate of change, we have

Here we have "rate of change" interpretations of partial derivatives.

$\dfrac{\partial z}{\partial x}$ is the rate of change of z with respect to x when y is held fixed.

Similarly,

$\dfrac{\partial z}{\partial y}$ is the rate of change of z with respect to y when x is held fixed.

We shall now look at some applications in which the "rate of change" notion of a partial derivative is very useful.

Suppose a manufacturer produces x units of product X and y units of product Y. Then the total cost c of these units is a function of x and y and is called a **joint-cost function.** If such a function is $c = f(x, y)$, then $\partial c/\partial x$ is called the **(partial) marginal cost with respect to** x and is the rate of change of c with respect to x when y is held fixed. Similarly, $\partial c/\partial y$ is the **(partial) marginal cost with respect to** y and is the rate of change of c with respect to y when x is held fixed.

For example, if c is expressed in dollars and $\partial c/\partial y = 2$, then the cost of producing an extra unit of Y when the level of production of X is fixed is approximately two dollars.

If a manufacturer produces n products, the joint-cost function is a function of n variables, and there are n (partial) marginal-cost functions.

EXAMPLE 1 **Marginal Costs**

A company manufactures two types of skis, the Lightning and the Alpine models. Suppose the joint-cost function for producing x pairs of the Lightning model and y pairs of the Alpine model per week is

$$c = f(x, y) = 0.07x^2 + 75x + 85y + 6000$$

where c is expressed in dollars. Determine the marginal costs $\partial c/\partial x$ *and* $\partial c/\partial y$ *when* $x = 100$ *and* $y = 50$*, and interpret the results.*

Solution: The marginal costs are

$$\frac{\partial c}{\partial x} = 0.14x + 75 \quad \text{and} \quad \frac{\partial c}{\partial y} = 85$$

Thus,

$$\left.\frac{\partial c}{\partial x}\right|_{(100,50)} = 0.14(100) + 75 = 89 \tag{1}$$

and

$$\left.\frac{\partial c}{\partial y}\right|_{(100,50)} = 85 \tag{2}$$

Equation (1) means that increasing the output of the Lightning model from 100 to 101, while maintaining production of the Alpine model at 50, increases costs by approximately \$89. Equation (2) means that increasing the output of the Alpine model from 50 to 51 and holding production of the Lightning model at 100 will increase costs by approximately \$85. In fact, since $\partial c/\partial y$ is a constant function, the marginal cost with respect to y is \$85 at all levels of production.

▶ **Now work Problem 1.**

EXAMPLE 2 **Loss of Body Heat**

On a cold day, a person may feel colder when the wind is blowing than when the wind is calm because the rate of heat loss is a function of both temperature and wind speed. The equation

$$H = (10.45 + 10\sqrt{w} - w)(33 - t)$$

indicates the rate of heat loss H (in kilocalories per square meter per hour) when the air temperature is t (in degrees Celsius) and the wind speed is w (in meters per second). For $H = 2000$*, exposed flesh will freeze in one minute.*[7]

a. *Evaluate H when* $t = 0$ *and* $w = 4$.

Solution: When $t = 0$ and $w = 4$,

$$H = (10.45 + 10\sqrt{4} - 4)(33 - 0) = 872.85$$

[7] G. E. Folk, Jr., *Textbook of Environmental Physiology,* 2d ed. (Philadelphia: Lea & Febiger, 1974).

b. *Evaluate $\partial H/\partial w$ and $\partial H/\partial t$ when $t = 0$ and $w = 4$, and interpret the results.*

Solution:

$$\frac{\partial H}{\partial w} = \left(\frac{5}{\sqrt{w}} - 1\right)(33 - t), \quad \left.\frac{\partial H}{\partial w}\right|_{\substack{t=0\\w=4}} = 49.5$$

$$\frac{\partial H}{\partial t} = (10.45 + 10\sqrt{w} - w)(-1), \quad \left.\frac{\partial H}{\partial t}\right|_{\substack{t=0\\w=4}} = -26.45$$

These equations mean that when $t = 0$ and $w = 4$, increasing w by a small amount while keeping t fixed will make H increase approximately 49.5 times as much as w increases. Increasing t by a small amount while keeping w fixed will make H *decrease* approximately 26.45 times as much as t increases.

c. *When $t = 0$ and $w = 4$, which has a greater effect on H: a change in wind speed of 1 m/s or a change in temperature of 1°C?*

Solution: Since the partial derivative of H with respect to w is greater in magnitude than the partial with respect to t when $t = 0$ and $w = 4$, a change in wind speed of 1 m/s has a greater effect on H.

▸ **Now work Problem 13.**

The output of a product depends on many factors of production. Among these may be labor, capital, land, machinery, and so on. For simplicity, let us suppose that output depends only on labor and capital. If the function $P = f(l, k)$ gives the output P when the producer uses l units of labor and k units of capital, then this function is called a **production function.** We define the **marginal productivity with respect to** l to be $\partial P/\partial l$. This is the rate of change of P with respect to l when k is held fixed. Likewise, the **marginal productivity with respect to** k is $\partial P/\partial k$ and is the rate of change of P with respect to k when l is held fixed.

EXAMPLE 3 Marginal Productivity

A manufacturer of a popular toy has determined that the production function is $P = \sqrt{lk}$, where l is the number of labor-hours per week and k is the capital (expressed in hundreds of dollars per week) required for a weekly production of P gross of the toy. (One gross is 144 units.) Determine the marginal productivity functions, and evaluate them when $l = 400$ and $k = 16$. Interpret the results.

Solution: Since $P = (lk)^{1/2}$,

$$\frac{\partial P}{\partial l} = \frac{1}{2}(lk)^{-1/2}k = \frac{k}{2\sqrt{lk}}$$

and

$$\frac{\partial P}{\partial k} = \frac{1}{2}(lk)^{-1/2}l = \frac{l}{2\sqrt{lk}}$$

Evaluating these equations when $l = 400$ and $k = 16$, we obtain

$$\left.\frac{\partial P}{\partial l}\right|_{\substack{l=400\\k=16}} = \frac{16}{2\sqrt{400(16)}} = \frac{1}{10}$$

and

$$\left.\frac{\partial P}{\partial k}\right|_{\substack{l=400\\k=16}} = \frac{400}{2\sqrt{400(16)}} = \frac{5}{2}$$

Thus, if $l = 400$ and $k = 16$, increasing l to 401 and holding k at 16 will increase output by approximately $\frac{1}{10}$ gross. But if k is increased to 17 while l is held at 400, the output increases by approximately $\frac{5}{2}$ gross.

▶ **Now work Problem 5.**

Competitive and Complementary Products

Sometimes two products may be related such that changes in the price of one of them affect the demand for the other. A typical example is that of butter and margarine. If such a relationship exists between products A and B, then the demand for each product is dependent on the prices of both. Suppose q_A and q_B are the quantities demanded for A and B, respectively, and p_A and p_B are their respective prices. Then both q_A and q_B are functions of p_A and p_B:

$$q_A = f(p_A, p_B) \quad \text{demand function for A}$$

$$q_B = g(p_A, p_B) \quad \text{demand function for B}$$

We can find four partial derivatives:

$$\frac{\partial q_A}{\partial p_A} \quad \textit{the marginal demand for } \text{A} \textit{ with respect to } p_A$$

$$\frac{\partial q_A}{\partial p_B} \quad \textit{the marginal demand for } \text{A} \textit{ with respect to } p_B$$

$$\frac{\partial q_B}{\partial p_A} \quad \textit{the marginal demand for } \text{B} \textit{ with respect to } p_A$$

$$\frac{\partial q_B}{\partial p_B} \quad \textit{the marginal demand for } \text{B} \textit{ with respect to } p_B$$

Under typical conditions, if the price of B is fixed and the price of A increases, then the quantity of A demanded will decrease. Thus, $\partial q_A/\partial p_A < 0$. Similarly, $\partial q_B/\partial p_B < 0$. However, $\partial q_A/\partial p_B$ and $\partial q_B/\partial p_A$ may be either positive or negative. If

$$\frac{\partial q_A}{\partial p_B} > 0 \quad \text{and} \quad \frac{\partial q_B}{\partial p_A} > 0$$

then A and B are said to be **competitive products** or **substitutes.** In this situation, an increase in the price of B causes an increase in the demand for A, if it is assumed that the price of A does not change. Similarly, an increase in the price of A causes an increase in the demand for B when the price of B is held fixed. Butter and margarine are examples of substitutes.

Proceeding to a different situation, we say that if

$$\frac{\partial q_A}{\partial p_B} < 0 \quad \text{and} \quad \frac{\partial q_B}{\partial p_A} < 0$$

then A and B are **complementary products.** In this case, an increase in the price of B causes a decrease in the demand for A if the price of A does not change. Similarly, an increase in the price of A causes a decrease in the demand for B when the price of B is held fixed. For example, cameras and film are complementary products. An increase in the price of film will make picture-taking more expensive. Hence, the demand for cameras will decrease.

EXAMPLE 4 **Determining Whether Products Are Competitive or Complementary**

The demand functions for products A and B are each a function of the prices of A and B and are given by

$$q_A = \frac{50\sqrt[3]{p_B}}{\sqrt{p_A}} \quad \textit{and} \quad q_B = \frac{75p_A}{\sqrt[3]{p_B^2}}$$

respectively. Find the four marginal-demand functions, and also determine whether A and B are competitive products, complementary products, or neither.

Solution: Writing $q_A = 50p_A^{-1/2}p_B^{1/3}$ and $q_B = 75p_Ap_B^{-2/3}$, we have

$$\frac{\partial q_A}{\partial p_A} = 50\left(-\frac{1}{2}\right)p_A^{-3/2}p_B^{1/3} = -25p_A^{-3/2}p_B^{1/3}$$

$$\frac{\partial q_A}{\partial p_B} = 50p_A^{-1/2}\left(\frac{1}{3}\right)p_B^{-2/3} = \frac{50}{3}p_A^{-1/2}p_B^{-2/3}$$

$$\frac{\partial q_B}{\partial p_A} = 75(1)p_B^{-2/3} = 75p_B^{-2/3}$$

$$\frac{\partial q_B}{\partial p_B} = 75p_A\left(-\frac{2}{3}\right)p_B^{-5/3} = -50p_Ap_B^{-5/3}$$

Since p_A and p_B represent prices, they are both positive. Hence, $\partial q_A/\partial p_B > 0$ and $\partial q_B/\partial p_A > 0$. We conclude that A and B are competitive products.

▸ **Now work Problem 19.**

Exercise 17.3

For the joint-cost functions in Problems 1–3, find the indicated marginal cost at the given production level.

*1. $c = 7x + 0.3y^2 + 2y + 900$; $\frac{\partial c}{\partial y}$, $x = 20$, $y = 30$

2. $c = x\sqrt{x+y} + 5000$; $\frac{\partial c}{\partial x}$, $x = 40$, $y = 60$

3. $c = 0.03(x+y)^3 - 0.6(x+y)^2 + 9.5(x+y) + 7700$; $\frac{\partial c}{\partial x}$, $x = 50$, $y = 80$

For the production functions in Problems 4 and 5, find the marginal productivity functions $\partial P/\partial k$ and $\partial P/\partial l$.

4. $P = 15lk - 3l^2 + 5k^2 + 500$

*5. $P = 1.582l^{0.192}k^{0.764}$

6. **Cobb-Douglas Production Function** In economics, a Cobb-Douglas production function is a production function of the form $P = Al^{\alpha}k^{\beta}$, where A, α, and β are constants and $\alpha + \beta = 1$. For such a function, show that

(a) $\partial P/\partial l = \alpha P/l$

(b) $\partial P/\partial k = \beta P/k$

(c) $l\frac{\partial P}{\partial l} + k\frac{\partial P}{\partial k} = P$. This means that summing the products of the marginal productivity of each factor and the amount of that factor results in the total product P.

In Problems 7–9, q_A and q_B are demand functions for products A and B, respectively. In each case, find $\partial q_A/\partial p_A$, $\partial q_A/\partial p_B$, $\partial q_B/\partial p_A$, $\partial q_B/\partial p_B$ and determine whether A and B are competitive, complementary, or neither.

7. $q_A = 1000 - 50p_A + 2p_B$; $q_B = 500 + 4p_A - 20p_B$

8. $q_A = 20 - p_A - 2p_B$; $q_B = 50 - 2p_A - 3p_B$

9. $q_A = \frac{100}{p_A\sqrt{p_B}}$; $q_B = \frac{500}{p_B\sqrt[3]{p_A}}$

10. **Canadian Manufacturing** The production function for the Canadian manufacturing industries for 1927 is estimated by[8] $P = 33.0l^{0.46}k^{0.52}$, where P is product, l is

[8] P. Daly and P. Douglas, "The Production Function for Canadian Manufactures," *Journal of the American Statistical Association* 38 (1943), 178–86.

labor, and k is capital. Find the marginal productivities for labor and capital, and evaluate when $l = 1$ and $k = 1$.

11. Dairy Farming An estimate of the production function for dairy farming in Iowa (1939) is given by[9]

$$P = A^{0.27} B^{0.01} C^{0.01} D^{0.23} E^{0.09} F^{0.27}$$

where P is product, A is land, B is labor, C is improvements, D is liquid assets, E is working assets, and F is cash operating expenses. Find the marginal productivities for labor and improvements.

12. Production Function Suppose a production function is given by $P = \dfrac{kl}{k+l}$

(a) Determine the marginal productivity functions.

(b) Show that when $k = l$, the marginal productivities are equal.

*__13. MBA Compensation__ In a study of success among graduates with a master of business administration (MBA) degree, it was estimated that for staff managers (which include accountants, analysts, etc.), current annual compensation (in dollars) was given by

$$z = 43{,}960 + 4480x + 3492y$$

where x and y are the number of years of work experience before and after receiving the MBA degree, respectively.[10] Find $\partial z/\partial x$ and interpret your result.

14. Status A person's general status S_g is believed to be a function of status attributable to education, S_e, and status attributable to income, S_i, where S_g, S_e, and S_i are represented numerically. If

$$S_g = 7\sqrt[3]{S_e}\sqrt{S_i}$$

determine $\partial S_g/\partial S_e$ and $\partial S_g/\partial S_i$ when $S_e = 125$ and $S_i = 100$, and interpret your results.[11]

15. Reading Ease Sometimes we want to evaluate the degree of readability of a piece of writing. Rudolf Flesch[12] developed a function of two variables that will do this, namely,

$$R = f(w, s) = 206.835 - (1.015w + 0.846s)$$

where R is called the *reading ease score,* w is the average number of words per sentence in 100-word samples, and s is the average number of syllables in such samples. Flesch says that an article for which $R = 0$ is "practically unreadable," but one with $R = 100$ is "easy for any literate person." (a) Find $\partial R/\partial w$ and $\partial R/\partial s$. (b) Which is "easier" to read: an article for which $w = w_0$ and $s = s_0$, or one for which $w = w_0 + 1$ and $s = s_0$?

16. Model for Voice The study of frequency of vibrations of a taut wire is useful in considering such things as an individual's voice. Suppose

$$\omega = \frac{1}{bL}\sqrt{\frac{\tau}{\pi\rho}}$$

where ω (a Greek letter read "omega") is frequency, b is diameter, L is length, ρ (a Greek letter read "rho") is density, and τ (a Greek letter read "tau") is tension.[13] Find $\partial\omega/\partial b$, $\partial\omega/\partial L$, $\partial\omega/\partial\rho$, and $\partial\omega/\partial\tau$.

17. Traffic Flow Consider the following traffic-flow situation. On a highway where two lanes of traffic flow in the same direction, there is a maintenance vehicle blocking the left lane. (See Figure 17.11.) Two vehicles (*lead* and *following*) are in the right lane with a gap between them. The *subject* vehicle can choose either to fill or not to fill the gap. That decision may be based not only on the distance x shown in the diagram, but also on other factors (such as the velocity of the *following* vehicle). A *gap index* g has been used in analyzing such a decision.[14,15] The greater the g-value, the greater is the propensity for the *subject* vehicle to fill the gap. Suppose

$$g = \frac{x}{V_F} - \left(0.75 + \frac{V_F - V_S}{19.2}\right)$$

where x (in feet) is as before, V_F is the velocity of the *following* vehicle (in feet per second), and V_S is the velocity of the *subject* vehicle (in feet per second). From the diagram, it seems reasonable that if both V_F and V_S are fixed and x increases, then g should increase. Show that this is true by applying calculus to the function g. Assume that x, V_F, and V_S are positive.

---

[9]G. Tintner and O. H. Brownlee, "Production Functions Derived from Farm Records," *American Journal of Agricultural Economics* 26 (1944), 566–71.

[10]Adapted from A. G. Weinstein and V. Srinivasen, "Predicting Managerial Success of Master of Business Administration (M.B.A.) Graduates," *Journal of Applied Psychology* 59, no. 2 (1974), 207–12.

[11]Adapted from R. K. Leik and B. F. Meeker, *Mathematical Sociology* (Englewood Cliffs, NJ: Prentice-Hall, 1975).

[12]R. Flesch, *The Art of Readable Writing* (New York: Harper & Row Publishers, 1949).

[13]R. M. Thrall, J. A. Mortimer, K. R. Rebman, and R. F. Baum, eds., *Some Mathematical Models in Biology,* rev. ed., Report No. 40241-R-7. Prepared at University of Michigan, 1967.

[14]P. M. Hurst, K. Perchonok, and E. L. Seguin, "Vehicle Kinematics and Gap Acceptance," *Journal of Applied Psychology* 52, no. 4 (1968), 321–24.

[15]K. Perchonok and P. M. Hurst, "Effect of Lane-Closure Signals upon Driver Decision Making and Traffic Flow," *Journal of Applied Psychology* 52, no. 5 (1968), 410–13.

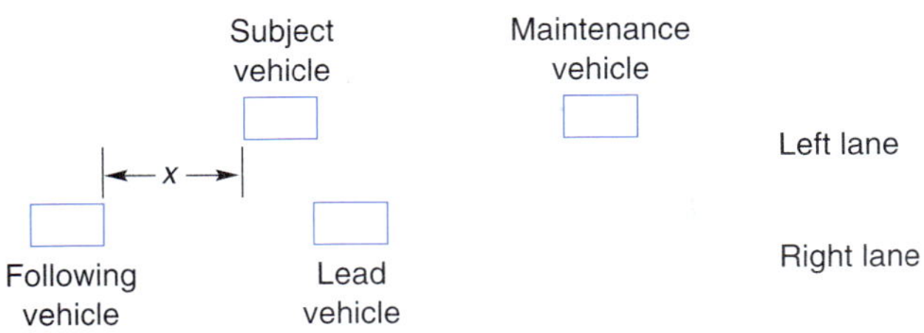

FIGURE 17.11 Diagram for Problem 17.

18. Demand Suppose the demand equations for related products A and B are

$$q_A = e^{-(p_A+p_B)} \quad \text{and} \quad q_B = \frac{16}{p_A^2 p_B^2}$$

where q_A and q_B are the number of units of A and B demanded when the unit prices (in thousands of dollars) are p_A and p_B, respectively.

(a) Classify A and B as competitive, complementary, or neither.

(b) If the unit prices of A and B are \$1000 and \$2000, respectively, estimate the change in the demand for A when the price of B is decreased by \$20 and the price of A is held constant.

***19. Demand** The demand equations for related products A and B are given by

$$q_A = \frac{30\sqrt{p_B}}{p_A^{2/3}} \quad \text{and} \quad q_B = \frac{50p_A}{p_B^{1/3}}$$

where q_A and q_B are the quantities of A and B demanded and p_A and p_B are the corresponding prices (in dollars) per unit.

(a) Find the values of the two marginal demands for product A when $p_A = 8$ and $p_B = 64$.

(b) If p_B were reduced to 60 from 64, with p_A fixed at 8, use part (a) to estimate the corresponding change in demand for product A.

20. Joint-Cost Function A manufacturer's joint-cost function for producing q_A units of product A and q_B units of product B is given by

$$c = \frac{q_A^2\left(q_B^3 + q_A\right)^{1/2}}{17} + q_A q_B^{1/3} + 600$$

where c is in dollars.

(a) Find the marginal-cost functions with respect to q_A and q_B.

(b) Evaluate the marginal-cost function with respect to q_A when $q_A = 17$ and $q_B = 8$. Round your answer to two decimal places.

(c) Use your answer to part (b) to estimate the change in cost if production of product A is decreased from 17 to 16 units, while production of product B is held constant at 8 units.

21. Elections For the congressional elections of 1974, the Republican percentage, R, of the Republican–Democratic vote in a district is given (approximately) by[16]

$$\begin{aligned} R &= f(E_r, E_d, I_r, I_d, N) \\ &= 15.4725 + 2.5945E_r - 0.0804E_r^2 - 2.3648E_d \\ &\quad + 0.0687E_d^2 + 2.1914I_r - 0.0912I_r^2 \\ &\quad - 0.8096I_d + 0.0081I_d^2 - 0.0277E_rI_r \\ &\quad + 0.0493E_dI_d + 0.8579N - 0.0061N^2 \end{aligned}$$

Here E_r and E_d are the campaign expenditures (in units of \$10,000) by Republicans and Democrats, respectively; I_r and I_d are the number of terms served in Congress, *plus* one, for the Republican and Democratic candidates, respectively, and N is the percentage of the two-party presidential vote that Richard Nixon received in the district for 1968. The variable N gives a measure of Republican strength in the district.

(a) In the Federal Election Campaign Act of 1974, Congress set a limit of \$188,000 on campaign expenditures. By analyzing $\partial R/\partial E_r$, would you have advised a Republican candidate who served nine terms in Congress to spend \$188,000 on his or her campaign?

(b) Find the percentage above which the Nixon vote had a negative effect on R; that is, find N when $\partial R/\partial N < 0$. Give your answer to the nearest percent.

22. Sales After a new product has been launched onto the market, its sales volume (in thousands of units) is given by

$$S = \frac{AT + 450}{\sqrt{A + T^2}}$$

where T is the time (in months) since the product was first introduced and A is the amount (in *hundreds* of dollars) spent each month on advertising.

(a) Verify that the partial derivative of sales volume with respect to time is given by

$$\frac{\partial S}{\partial T} = \frac{A^2 - 450T}{(A + T^2)^{3/2}}$$

(b) Use the result in part (a) to predict the number of months that will elapse before the sales volume begins to decrease if the amount allocated to advertising is held fixed at \$9000 per month.

[16] J. Silberman and G. Yochum, "The Role of Money in Determining Election Outcomes," *Social Science Quarterly* 58, no. 4 (1978), 671–82.

Let f be a demand function for product A and $q_A = f(p_A, p_B)$, where q_A is the quantity of A demanded when the price per unit of A is p_A and the price per unit of product B is p_B. The partial elasticity of demand for A with respect to p_A, denoted η_{p_A}, is defined as $\eta_{p_A} = (p_A/q_A)(\partial q_A/\partial p_A)$. The partial elasticity of demand for A with respect to p_B, denoted η_{p_B}, is defined as $\eta_{p_B} = (p_B/q_A)(\partial q_A/\partial p_B)$. Loosely speaking, η_{p_A} is the ratio of a percentage change in the quantity of A demanded to a percentage change in the price of A when the price of B is fixed. Similarly, η_{p_B} can be loosely interpreted as the ratio of a percentage change in the quantity of A demanded to a percentage change in the price of B when the price of A is fixed. In Problems 23–25, find η_{p_A} and η_{p_B} for the given values of p_A and p_B.

23. $q_A = 1000 - 50p_A + 2p_B$; $p_A = 2$, $p_B = 10$

24. $q_A = 40 - 2p_A - 3p_B$; $p_A = 3$, $p_B = 2$

25. $q_A = 100/(p_A\sqrt{p_B})$; $p_A = 1$, $p_B = 4$

OBJECTIVE *To find partial derivatives of a function defined implicitly.*

17.4 Implicit Partial Differentiation[17]

An equation in x, y, and z does not necessarily define z as a function of x and y. For example, in the equation

$$z^2 - x^2 - y^2 = 0 \tag{1}$$

if $x = 1$ and $y = 1$, then $z^2 - 1 - 1 = 0$, so $z = \pm\sqrt{2}$. Thus, Equation (1) does not define z as a function of x and y. However, solving Equation (1) for z gives

$$z = \sqrt{x^2 + y^2} \quad \text{or} \quad z = -\sqrt{x^2 + y^2}$$

each of which defines z as a function of x and y. Although Equation (1) does not explicitly express z as a function of x and y, it can be thought of as expressing z *implicitly* as one of two different functions of x and y. Note that the equation $z^2 - x^2 - y^2 = 0$ has the form $F(x, y, z) = 0$, where F is a function of three variables. Any equation of the form $F(x, y, z) = 0$ can be thought of as expressing z implicitly as one of a set of possible functions of x and y. Moreover, we can find $\partial z/\partial x$ and $\partial z/\partial y$ directly from the form $F(x, y, z) = 0$.

To find $\partial z/\partial x$ for

$$z^2 - x^2 - y^2 = 0$$

we first differentiate both sides of Equation (1) with respect to x while treating z as a function of x and y and treating y as a constant:

$$\frac{\partial}{\partial x}(z^2 - x^2 - y^2) = \frac{\partial}{\partial x}(0)$$

$$\frac{\partial}{\partial x}(z^2) - \frac{\partial}{\partial x}(x^2) - \frac{\partial}{\partial x}(y^2) = 0$$

$$2z\frac{\partial z}{\partial x} - 2x - 0 = 0$$

Because y is treated as a constant, $\dfrac{\partial y}{\partial x} = 0$.

Solving for $\partial z/\partial x$, we obtain

$$2z\frac{\partial z}{\partial x} = 2x$$

$$\frac{\partial z}{\partial x} = \frac{x}{z}$$

[17] This section may be omitted without loss of continuity.

To find $\partial z/\partial y$, we differentiate both sides of Equation (17.4) with respect to y while treating z as a function of x and y and treating x as a constant:

$$\frac{\partial}{\partial y}(z^2 - x^2 - y^2) = \frac{\partial}{\partial y}(0)$$

$$2z\frac{\partial z}{\partial y} - 0 - 2y = 0 \qquad \left(\frac{\partial x}{\partial y} = 0\right)$$

$$2z\frac{\partial z}{\partial y} = 2y$$

Hence,

$$\frac{\partial z}{\partial y} = \frac{y}{z}$$

The method we used to find $\partial z/\partial x$ and $\partial z/\partial y$ is called *implicit partial differentiation.*

EXAMPLE 1 **Implicit Partial Differentiation**

If $\dfrac{xz^2}{x+y} + y^2 = 0$, *evaluate* $\dfrac{\partial z}{\partial x}$ *when* $x = -1$, $y = 2$, *and* $z = 2$.

Solution: We treat z as a function of x and y and differentiate both sides of the equation with respect to x:

$$\frac{\partial}{\partial x}\left(\frac{xz^2}{x+y}\right) + \frac{\partial}{\partial x}(y^2) = \frac{\partial}{\partial x}(0)$$

Using the quotient rule for the first term on the left, we have

$$\frac{(x+y)\dfrac{\partial}{\partial x}(xz^2) - xz^2\dfrac{\partial}{\partial x}(x+y)}{(x+y)^2} + 0 = 0$$

Using the product rule for $\dfrac{\partial}{\partial x}(xz^2)$ gives

$$\frac{(x+y)\left[x\left(2z\dfrac{\partial z}{\partial x}\right) + z^2(1)\right] - xz^2(1)}{(x+y)^2} = 0$$

Solving for $\partial z/\partial x$, we obtain

$$2xz(x+y)\frac{\partial z}{\partial x} + z^2(x+y) - xz^2 = 0$$

$$\frac{\partial z}{\partial x} = \frac{xz^2 - z^2(x+y)}{2xz(x+y)} = -\frac{yz}{2x(x+y)} \qquad z \neq 0$$

Thus,

$$\left.\frac{\partial z}{\partial x}\right|_{(-1,2,2)} = 2$$

▶ **Now work Problem 1.**

EXAMPLE 2 Implicit Partial Differentiation

If $se^{r^2+u^2} = u\ln(t^2+1)$, determine $\partial t/\partial u$.

Solution: We consider t as a function of r, s, and u. By differentiating both sides with respect to u while treating r and s as constants, we get

$$\frac{\partial}{\partial u}(se^{r^2+u^2}) = \frac{\partial}{\partial u}[u\ln(t^2+1)]$$

$$2sue^{r^2+u^2} = u\frac{\partial}{\partial u}[\ln(t^2+1)] + \ln(t^2+1)\frac{\partial}{\partial u}(u) \qquad \text{(product rule)}$$

$$2sue^{r^2+u^2} = u\frac{2t}{t^2+1}\frac{\partial t}{\partial u} + \ln(t^2+1)$$

Therefore,

$$\frac{\partial t}{\partial u} = \frac{(t^2+1)[2sue^{r^2+u^2} - \ln(t^2+1)]}{2ut}$$

▶ **Now work Problem 13.**

Exercise 17.4

In Problems 1–11, find the indicated partial derivatives by the method of implicit partial differentiation.

*1. $2x^2 + 3y^2 + 5z^2 = 900$; $\partial z/\partial x$

2. $z^2 - 5x^2 + y^2 = 0$; $\partial z/\partial x$

3. $2z^3 - x^2 - 4y^2 = 0$; $\partial z/\partial y$

4. $3x^2 + y^2 + 2z^3 = 9$; $\partial z/\partial y$

5. $x^2 - 2y - z^2 + x^2yz^2 = 20$; $\partial z/\partial x$

6. $z^3 - xz - y = 0$; $\partial z/\partial x$

7. $e^x + e^y + e^z = 10$; $\partial z/\partial y$

8. $xyz + 3y^3x^2 - \ln z^3 = 0$; $\partial z/\partial y$

9. $\ln(z) + 9z - xy = 1$; $\partial z/\partial x$

10. $\ln x + \ln y - \ln z = e^y$; $\partial z/\partial x$

11. $(z^2 + 6xy)\sqrt{x^3+5} = 2$; $\partial z/\partial y$

In Problems 12–20, evaluate the indicated partial derivatives for the given values of the variables.

12. $xz + xyz - 5 = 0$; $\partial z/\partial x, x = 1, y = 4, z = 1$

*13. $xz^2 + yz - 7x^4 = 0$; $\partial z/\partial x, x = 0, y = 0, z = 0$

14. $e^{zx} = xyz$; $\partial z/\partial y, x = 1, y = -e^{-1}, z = -1$

15. $e^{yz} = -xyz$; $\partial z/\partial x, x = -e^2/2, y = 1, z = 2$

16. $\sqrt{xz + y^2} - xy = 0$; $\partial z/\partial y, x = 2, y = 2, z = 6$

17. $\ln z = 4x + y$; $\partial z/\partial x, x = 5, y = -20, z = 1$

18. $\dfrac{rs}{s^2+t^2} = t$; $\partial r/\partial t, r = 0, s = 1, t = 0$

19. $\dfrac{s^2+t^2}{rs} = 10$; $\partial t/\partial r, r = 1, s = 2, t = 4$

20. $\ln(x + y + z) + xyz = ze^{x+y+z}$; $\partial z/\partial x, x = 0, y = 1, z = 0$

21. **Joint-Cost Function** A joint-cost function is defined implicitly by the equation

$$c + \sqrt{c} = 12 + q_A\sqrt{9 + q_B^2}$$

where c denotes the total cost (in dollars) for producing q_A units of product A and q_B units of product B.

(a) If $q_A = 6$ and $q_B = 4$, find the corresponding value of c.

(b) Determine the marginal costs with respect to q_A and q_B when $q_A = 6$ and $q_B = 4$.

OBJECTIVE *To compute higher-order partial derivatives.*

17.5 Higher-Order Partial Derivatives

If $z = f(x, y)$, then not only is z a function of x and y, but also, f_x and f_y are each functions of x and y. Hence, we may differentiate f_x and f_y to obtain **second-order partial derivatives** of f. Symbolically,

$$f_{xx} \text{ means } (f_x)_x \qquad f_{xy} \text{ means } (f_x)_y$$

$$f_{yx} \text{ means } (f_y)_x \qquad f_{yy} \text{ means } (f_y)_y$$

In terms of ∂-notation,

$$\frac{\partial^2 z}{\partial x^2} \text{ means } \frac{\partial}{\partial x}\left[\frac{\partial z}{\partial x}\right] \qquad \frac{\partial^2 z}{\partial y \partial x} \text{ means } \frac{\partial}{\partial y}\left[\frac{\partial z}{\partial x}\right]$$

$$\frac{\partial^2 z}{\partial x \partial y} \text{ means } \frac{\partial}{\partial x}\left[\frac{\partial z}{\partial y}\right] \qquad \frac{\partial^2 z}{\partial y^2} \text{ means } \frac{\partial}{\partial y}\left[\frac{\partial z}{\partial y}\right]$$

Note that to find f_{xy}, we first differentiate f with respect to x. For $\partial^2 z/\partial x \partial y$, we first differentiate with respect to y.

We can extend our notation beyond second-order partial derivatives. For example, f_{xxy} (or $\partial^3 z/\partial y\, \partial x^2$) is a third-order partial derivative of f, namely, the partial derivative of f_{xx} (or $\partial^2 z/\partial x^2$) with respect to y. A generalization regarding higher-order partial derivatives to functions of more than two variables should be obvious.

EXAMPLE 1 **Second-Order Partial Derivatives**

Find the four second-order partial derivatives of $f(x, y) = x^2y + x^2y^2$.

Solution: Since

$$f_x(x, y) = 2xy + 2xy^2$$

we have

$$f_{xx}(x, y) = \frac{\partial}{\partial x}(2xy + 2xy^2) = 2y + 2y^2$$

and

$$f_{xy}(x, y) = \frac{\partial}{\partial y}(2xy + 2xy^2) = 2x + 4xy$$

Also, since

$$f_y(x, y) = x^2 + 2x^2y$$

we have

$$f_{yy}(x, y) = \frac{\partial}{\partial y}(x^2 + 2x^2y) = 2x^2$$

and

$$f_{yx}(x, y) = \frac{\partial}{\partial x}(x^2 + 2x^2y) = 2x + 4xy$$

▶ **Now work Problem 1.**

The derivatives f_{xy} and f_{yx} are called **mixed partial derivatives.** Observe in Example 1 that $f_{xy}(x, y) = f_{yx}(x, y)$. Under suitable conditions, mixed partial derivatives of a function are equal; that is, the order of differentiation is of no concern. You may assume that this is the case for all the functions that we consider.

EXAMPLE 2 Mixed Partial Derivative

Find the value of $\left.\dfrac{\partial^3 w}{\partial z\,\partial y\,\partial x}\right|_{(1,2,3)}$ *if* $w = (2x + 3y + 4z)^3$.

Solution:

$$\begin{aligned}\frac{\partial w}{\partial x} &= 3(2x + 3y + 4z)^2\frac{\partial}{\partial x}(2x + 3y + 4z)\\ &= 6(2x + 3y + 4z)^2\\ \frac{\partial^2 w}{\partial y\,\partial x} &= 6\cdot 2(2x + 3y + 4z)\frac{\partial}{\partial y}(2x + 3y + 4z)\\ &= 36(2x + 3y + 4z)\\ \frac{\partial^3 w}{\partial z\,\partial y\,\partial x} &= 36\cdot 4 = 144\end{aligned}$$

Thus,

$$\left.\frac{\partial^3 w}{\partial z\,\partial y\,\partial x}\right|_{(1,2,3)} = 144$$

▶ **Now work Problem 3.**

EXAMPLE 3 Second-Order Partial Derivative of Implicit Function[18]

Determine $\dfrac{\partial^2 z}{\partial x^2}$ *if* $z^2 = xy$.

Solution: By implicit differentiation, we first determine $\partial z/\partial x$:

$$\begin{aligned}\frac{\partial}{\partial x}(z^2) &= \frac{\partial}{\partial x}(xy)\\ 2z\frac{\partial z}{\partial x} &= y\\ \frac{\partial z}{\partial x} &= \frac{y}{2z} \qquad z \neq 0\end{aligned}$$

Differentiating both sides with respect to x, we obtain

$$\begin{aligned}\frac{\partial}{\partial x}\left[\frac{\partial z}{\partial x}\right] &= \frac{\partial}{\partial x}\left[\frac{1}{2}yz^{-1}\right]\\ \frac{\partial^2 z}{\partial x^2} &= -\frac{1}{2}yz^{-2}\frac{\partial z}{\partial x}\end{aligned}$$

Substituting $y/(2z)$ for $\partial z/\partial x$, we have

$$\frac{\partial^2 z}{\partial x^2} = -\frac{1}{2}yz^{-2}\left(\frac{y}{2z}\right) = -\frac{y^2}{4z^3} \qquad z \neq 0$$

▶ **Now work Problem 23.**

[18]Omit this example if Section 17.4 was not covered.

Exercise 17.5

In Problems 1–10, find the indicated partial derivatives.

*1. $f(x, y) = 4x^2y$; $f_x(x, y)$, $f_{xy}(x, y)$

2. $f(x, y) = 2x^3y^2 + 6x^2y^3 - 3xy$; $f_x(x, y)$, $f_{xx}(x, y)$

*3. $f(x, y) = 7x^2 + 3y$; $f_y(x, y)$, $f_{yy}(x, y)$, $f_{yyx}(x, y)$

4. $f(x, y) = (x^2+xy+y^2)(x^2+xy+1)$; $f_x(x, y)$, $f_{xy}(x, y)$

5. $f(x, y) = 9e^{2xy}$; $f_y(x, y)$, $f_{yx}(x, y)$, $f_{yxy}(x, y)$

6. $f(x, y) = \ln(x^2 + y^2) + 2$; $f_x(x, y)$, $f_{xx}(x, y)$, $f_{xy}(x, y)$

7. $f(x, y) = (x + y)^2(xy)$; $f_x(x, y)$, $f_y(x, y)$, $f_{xx}(x, y)$, $f_{yy}(x, y)$

8. $f(x, y, z) = xy^2z^3$; $f_x(x, y, z)$, $f_{xz}(x, y, z)$, $f_{xy}(x, y, z)$

9. $z = e^{\sqrt{x^2+y^2}}$; $\dfrac{\partial z}{\partial y}, \dfrac{\partial^2 z}{\partial y^2}$

10. $z = \dfrac{\ln(x^2 + 5)}{y}$; $\dfrac{\partial z}{\partial x}, \dfrac{\partial^2 z}{\partial y\, \partial x}$

In Problems 11–16, find the indicated value.

11. If $f(x, y, z) = 7$, find $f_{yxz}(4, 3, -2)$

12. If $f(x, y, z) = z^2(3x^2 - 4xy^3)$, find $f_{xyz}(1, 2, 3)$

13. If $f(l, k) = 5l^3k^6 - lk^7$, find $f_{kkl}(8, 1)$

14. If $f(x, y) = 3x^3y^2 + xy - x^2y^2$, find $f_{xxy}(5, 1)$ and $f_{xyx}(5, 1)$

15. If $f(x, y) = y^2e^x + \ln(xy)$, find $f_{xyy}(1, 1)$

16. If $f(x, y) = x^3 - 6xy^2 + x^2 - y^3$, find $f_{xy}(1, -1)$

17. **Cost Function** Suppose the cost c of producing q_A units of product A and q_B units of product B is given by

$$c = \left(3q_A^2 + q_B^3 + 4\right)^{1/3}$$

and the coupled demand functions for the products are given by

$$q_A = 10 - p_A + p_B^2$$

and

$$q_B = 20 + p_A - 11p_B$$

Find the value of

$$\frac{\partial^2 c}{\partial q_A\, \partial q_B}$$

when $p_A = 25$ and $p_B = 4$.

18. For $f(x, y) = x^4y^4 + 3x^3y^2 - 7x + 4$, show that

$$f_{xyx}(x, y) = f_{xxy}(x, y).$$

19. For $f(x, y) = 8x^3 + 2x^2y^2 + 5y^4$, show that

$$f_{xy}(x, y) = f_{yx}(x, y).$$

20. For $f(x, y) = xe^{y/x}$, show that

$$xf_{xx}(x, y) + yf_{xy}(x, y) = 0.$$

21. For $z = \ln(x^2 + y^2)$, show that $\dfrac{\partial^2 z}{\partial x^2} + \dfrac{\partial^2 z}{\partial y^2} = 0$.

[19]22. If $3z^2 - 2x^3 - 4y^4 = 0$, find $\dfrac{\partial^2 z}{\partial x^2}$.

[19]*23. If $z^2 - 3x^2 + y^2 = 0$, find $\dfrac{\partial^2 z}{\partial y^2}$.

[19]24. If $2z^2 = x^2 + 2xy + xz$, find $\dfrac{\partial^2 z}{\partial x\, \partial y}$.

[19]Omit Problems 22–24 if Section 17.4 was not covered.

OBJECTIVE *To show how to find the partial derivative of a function of functions by using the chain rule.*

17.6 Chain Rule[20]

Suppose a manufacturer of two related products A and B has a joint-cost function given by

$$c = f(q_A, q_B),$$

where c is the total cost of producing quantities q_A and q_B of A and B, respectively. Furthermore, suppose the demand functions for the products are

$$q_A = g(p_A, p_B) \quad \text{and} \quad q_B = h(p_A, p_B),$$

where p_A and p_B are the prices per unit of A and B, respectively. Since c is a function of q_A and q_B, and since both q_A and q_B are themselves functions of p_A and p_B, c can be viewed as a function of p_A and p_B. (Appropriately, the variables q_A and q_B are called *intermediate variables* of c.) Consequently,

[20]This section may be omitted without loss of continuity.

we should be able to determine $\partial c/\partial p_A$, the rate of change of total cost with respect to the price of A. One way to do this is to substitute the expressions $g(p_A, p_B)$ and $h(p_A, p_B)$ for q_A and q_B, respectively, into $c = f(q_A, q_B)$. Then c is a function of p_A and p_B, and we can differentiate c with respect to p_A directly. This approach has some drawbacks—especially when f, g, or h is given by a complicated expression. Another way to approach the problem would be to use the chain rule (actually *a* chain rule), which we now state without proof.

Chain Rule

Let $z = f(x, y)$, where both x and y are functions of r and s given by $x = x(r, s)$ and $y = y(r, s)$. If f, x, and y have continuous partial derivatives, then z is a function of r and s, and

$$\frac{\partial z}{\partial r} = \frac{\partial z}{\partial x}\frac{\partial x}{\partial r} + \frac{\partial z}{\partial y}\frac{\partial y}{\partial r}$$

and

$$\frac{\partial z}{\partial s} = \frac{\partial z}{\partial x}\frac{\partial x}{\partial s} + \frac{\partial z}{\partial y}\frac{\partial y}{\partial s}$$

Note that in the chain rule, the number of intermediate variables of z (two) is the same as the number of terms that compose each of $\partial z/\partial r$ and $\partial z/\partial s$.

Returning to the original situation concerning the manufacturer, we see that if f, q_A, and q_B have continuous partial derivatives, then, by the chain rule,

$$\frac{\partial c}{\partial p_A} = \frac{\partial c}{\partial q_A}\frac{\partial q_A}{\partial p_A} + \frac{\partial c}{\partial q_B}\frac{\partial q_B}{\partial p_A}$$

EXAMPLE 1 **Rate of Change of Cost**

For a manufacturer of cameras and film, the total cost c of producing q_C cameras and q_F units of film is given by

$$c = 30q_C + 0.015q_Cq_F + q_F + 900$$

The demand functions for the cameras and film are given by

$$q_C = \frac{9000}{p_C\sqrt{p_F}} \quad \text{and} \quad q_F = 2000 - p_C - 400p_F$$

where p_C is the price per camera and p_F is the price per unit of film. Find the rate of change of total cost with respect to the price of the camera when $p_C = 50$ and $p_F = 2$.

Solution: We must first determine $\partial c/\partial p_C$. By the chain rule,

$$\frac{\partial c}{\partial p_C} = \frac{\partial c}{\partial q_C}\frac{\partial q_C}{\partial p_C} + \frac{\partial c}{\partial q_F}\frac{\partial q_F}{\partial p_C}$$

$$= (30 + 0.015q_F)\left[\frac{-9000}{p_C^2\sqrt{p_F}}\right] + (0.015q_C + 1)(-1)$$

When $p_C = 50$ and $p_F = 2$, then $q_C = 90\sqrt{2}$ and $q_F = 1150$. Substituting these values into $\partial c/\partial p_C$ and simplifying, we have

$$\left.\frac{\partial c}{\partial p_C}\right|_{\substack{p_C = 50 \\ p_F = 2}} \approx -123.2$$

▸ **Now work Problem 1.**

The chain rule can be extended. For example, suppose $z = f(v, w, x, y)$ and v, w, x, and y are all functions of r, s, and t. Then, if certain conditions of continuity are assumed, z is a function of r, s, and t, and we have

$$\frac{\partial z}{\partial r} = \frac{\partial z}{\partial v}\frac{\partial v}{\partial r} + \frac{\partial z}{\partial w}\frac{\partial w}{\partial r} + \frac{\partial z}{\partial x}\frac{\partial x}{\partial r} + \frac{\partial z}{\partial y}\frac{\partial y}{\partial r}$$

$$\frac{\partial z}{\partial s} = \frac{\partial z}{\partial v}\frac{\partial v}{\partial s} + \frac{\partial z}{\partial w}\frac{\partial w}{\partial s} + \frac{\partial z}{\partial x}\frac{\partial x}{\partial s} + \frac{\partial z}{\partial y}\frac{\partial y}{\partial s}$$

and

$$\frac{\partial z}{\partial t} = \frac{\partial z}{\partial v}\frac{\partial v}{\partial t} + \frac{\partial z}{\partial w}\frac{\partial w}{\partial t} + \frac{\partial z}{\partial x}\frac{\partial x}{\partial t} + \frac{\partial z}{\partial y}\frac{\partial y}{\partial t}$$

Observe that the number of intermediate variables of z (four) is the same as the number of terms that form each of $\partial z/\partial r$, $\partial z/\partial s$, and $\partial z/\partial t$.

Now consider the situation where $z = f(x, y)$ such that $x = x(t)$ and $y = y(t)$. Then

$$\frac{dz}{dt} = \frac{\partial z}{\partial x}\frac{dx}{dt} + \frac{\partial z}{\partial y}\frac{dy}{dt}$$

Use the partial derivative symbols and the ordinary derivative symbols appropriately.

Here we use the symbol dz/dt rather than $\partial z/\partial t$, since z can be considered a function of *one* variable t. Likewise, the symbols dx/dt and dy/dt are used rather than $\partial x/\partial t$ and $\partial y/\partial t$. As is typical, the number of terms that compose dz/dt equals the number of intermediate variables of z. Other situations would be treated in a similar way.

EXAMPLE 2 **Chain Rule**

a. *If* $w = f(x, y, z) = 3x^2y + xyz - 4y^2z^3$, *where*

$$x = 2r - 3s \qquad y = 6r + s \qquad z = r - s$$

determine $\partial w/\partial r$ *and* $\partial w/\partial s$.

Solution: Since x, y, and z are functions of r and s, then, by the chain rule,

$$\begin{aligned}\frac{\partial w}{\partial r} &= \frac{\partial w}{\partial x}\frac{\partial x}{\partial r} + \frac{\partial w}{\partial y}\frac{\partial y}{\partial r} + \frac{\partial w}{\partial z}\frac{\partial z}{\partial r}\\ &= (6xy + yz)(2) + (3x^2 + xz - 8yz^3)(6) + (xy - 12y^2z^2)(1)\\ &= x(18x + 13y + 6z) + 2yz(1 - 24z^2 - 6yz)\end{aligned}$$

Also,

$$\begin{aligned}\frac{\partial w}{\partial s} &= \frac{\partial w}{\partial x}\frac{\partial x}{\partial s} + \frac{\partial w}{\partial y}\frac{\partial y}{\partial s} + \frac{\partial w}{\partial z}\frac{\partial z}{\partial s}\\ &= (6xy + yz)(-3) + (3x^2 + xz - 8yz^3)(1) + (xy - 12y^2z^2)(-1)\\ &= x(3x - 19y + z) - yz(3 + 8z^2 - 12yz)\end{aligned}$$

b. *If* $z = \dfrac{x + e^y}{y}$, *where* $x = rs + se^{rt}$ *and* $y = 9 + rt$, *evaluate* $\partial z/\partial s$ *when* $r = -2$, $s = 5$, *and* $t = 4$.

Solution: Since x and y are functions of r, s, and t (note that we can write $y = 9 + rt + 0 \cdot s$), by the chain rule,

$$\begin{aligned}\frac{\partial z}{\partial s} &= \frac{\partial z}{\partial x}\frac{\partial x}{\partial s} + \frac{\partial z}{\partial y}\frac{\partial y}{\partial s}\\ &= \left(\frac{1}{y}\right)(r + e^{rt}) + \frac{\partial z}{\partial y} \cdot (0) = \frac{r + e^{rt}}{y}\end{aligned}$$

If $r = -2$, $s = 5$, and $t = 4$, then $y = 1$. Thus,

$$\left.\frac{\partial z}{\partial s}\right|_{\substack{r=-2\\ s=5\\ t=4}} = \frac{-2+e^{-8}}{1} = -2+e^{-8}$$

▶ **Now work Problem 13.**

EXAMPLE 3 Chain Rule

a. *Determine $\partial y/\partial r$ if $y = x^2\ln(x^4+6)$ and $x = (r+3s)^6$.*

Solution: By the chain rule,

$$\begin{aligned}\frac{\partial y}{\partial r} &= \frac{dy}{dx}\frac{\partial x}{\partial r}\\ &= \left[x^2\cdot\frac{4x^3}{x^4+6} + 2x\cdot\ln(x^4+6)\right][6(r+3s)^5]\\ &= 12x(r+3s)^5\left[\frac{2x^4}{x^4+6} + \ln(x^4+6)\right]\end{aligned}$$

b. *Given that $z = e^{xy}$, $x = r - 4s$, and $y = r - s$, find $\partial z/\partial r$ in terms of r and s.*

Solution:

$$\begin{aligned}\frac{\partial z}{\partial r} &= \frac{\partial z}{\partial x}\frac{\partial x}{\partial r} + \frac{\partial z}{\partial y}\frac{\partial y}{\partial r}\\ &= (ye^{xy})(1) + (xe^{xy})(1)\\ &= (x+y)e^{xy}\end{aligned}$$

Since $x = r - 4s$ and $y = r - s$,

$$\begin{aligned}\frac{\partial z}{\partial r} &= [(r-4s)+(r-s)]e^{(r-4s)(r-s)}\\ &= (2r-5s)e^{r^2-5rs+4s^2}\end{aligned}$$

▶ **Now work Problem 15.**

Exercise 17.6

In Problems 1–12, find the indicated derivatives by using the chain rule.

*1. $z = 5x + 3y$, $x = 2r + 3s$, $y = r - 2s$; $\partial z/\partial r$, $\partial z/\partial s$

2. $z = x^2 + 5xy + 7y^2$, $x = r^2 - 2rs$, $y = r - 5s^2$; $\partial z/\partial r$, $\partial z/\partial s$

3. $z = e^{x+y}$, $x = t^2 + 3$, $y = \sqrt{t^3}$; dz/dt

4. $z = \sqrt{8x + y}$, $x = t^2 + 3t + 4$, $y = t^3 + 4$; dz/dt

5. $w = x^2z^2 + xyz + yz^2$, $x = 5t$, $y = 2t + 3$, $z = 6 - t$; dw/dt

6. $w = \ln(x^2 + y^2 + z^2)$, $x = 2 - 3t$, $y = t^2 + 3$, $z = 4 - t$; dw/dt

7. $z = (x^2 + xy^2)^3$, $x = r + s + t$, $y = 2r - 3s + 8t$; $\partial z/\partial t$

8. $z = \sqrt{x^2 + y^2}$, $x = r^2 + s - t$, $y = r - s + t$; $\partial z/\partial r$

9. $w = x^2 + xyz + y^3z^2$, $x = r - s^2$, $y = rs$, $z = 2r - 5s$; $\partial w/\partial s$

10. $w = e^{xyz}$, $x = r^2s^3$, $y = \ln(r - s)$, $z = \sqrt{rs^2}$; $\partial w/\partial r$

11. $y = x^2 - 7x + 5$, $x = 19rs + 2s^2t^2$; $\partial y/\partial r$

12. $y = 4 - x^2$, $x = 2r + 3s - 4t$; $\partial y/\partial t$

*__13.__ If $z = (4x + 3y)^3$, where $x = r^2s$ and $y = r - 2s$, evaluate $\partial z/\partial r$ when $r = 0$ and $s = 1$.

14. If $z = \sqrt{5x + 2y}$, where $x = 4t + 7$ and $y = t^2 - 3t + 9$, evaluate dz/dt when $t = 1$.

*__15.__ If $w = e^{2x+3y}(x^2 + 4z^2)$, where $x = rs$, $y = 2s - 3r$, and $z = r + s$, evaluate $\partial w/\partial s$ when $r = 1$ and $s = 0$.

16. If $y = x/(x - 5)$, where $x = 2t^2 - 3rs - r^2t$, evaluate $\partial y/\partial t$ when $r = 0$, $s = 2$, and $t = -1$.

17. Cost Function Suppose the cost c of producing q_A units of product A and q_B units of product B is given by

$$c = \left(3q_A^2 + q_B^3 + 4\right)^{1/3}$$

and the coupled demand functions for the products are given by

$$q_A = 10 - p_A + p_B^2$$

and

$$q_B = 20 + p_A - 11p_B$$

Use a chain rule to evaluate $\dfrac{\partial c}{\partial p_A}$ and $\dfrac{\partial c}{\partial p_B}$ when $p_A = 25$ and $p_B = 4$.

18. Suppose $w = f(x, y)$, where $x = g(t)$ and $y = h(t)$.

(a) State a chain rule that gives dw/dt.

(b) Suppose $h(t) = t$, so that $w = f(x, t)$ where $x = g(t)$. Use part (a) to find dw/dt and simplify your answer.

19. (a) Suppose w is a function of x and y, where both x and y are functions of s and t. State a chain rule that expresses $\partial w/\partial t$ in terms of derivatives of these functions.

(b) Let $w = 2x^2\ln|3x - 5y|$, where $x = s\sqrt{t^2 + 2}$ and $y = t - 3e^{2-s}$. Use part (a) to evaluate $\partial w/\partial t$ when $s = 1$ and $t = 0$.

20. Production Function In considering a production function $P = f(l, k)$, where l is labor input and k is capital input, Fon, Boulier, and Goldfarb[21] assume that $l = Lg(h)$, where L is the number of workers, h is the number of hours per day per worker, and $g(h)$ is a labor effectiveness function. In maximizing profit p given by

$$p = aP - whL$$

where a is the price per unit of output and w is the hourly wage per worker, Fon, Boulier, and Goldfarb determine $\partial p/\partial L$ and $\partial p/\partial h$. Assume that k is independent of L and h, and determine these partial derivatives.

[21] V. Fon, B. L. Boulier, and R. S. Goldfarb, "The Firm's Demand for Daily Hours of Work: Some Implications," *Atlantic Economic Journal* 13, no. 1 (1985), 36–42.

OBJECTIVE *To discuss relative maxima and relative minima, to find critical points, and to apply the second-derivative test for a function of two variables.*

17.7 Maxima and Minima for Functions of Two Variables

We now extend the notion of relative maxima and minima (or relative extrema) to functions of two variables.

Definition

A function $z = f(x, y)$ is said to have a ***relative maximum*** at the point (x_0, y_0)—that is, when $x = x_0$ and $y = y_0$—if, for all points (x, y) in the plane that are sufficiently close to (x_0, y_0), we have

$$f(x_0, y_0) \geq f(x, y) \quad (1)$$

For a ***relative minimum,*** we replace $\geq$ by $\leq$ in Equation (1).

To say that $z = f(x, y)$ has a relative maximum at (x_0, y_0) means, geometrically, that the point (x_0, y_0, z_0) on the graph of f is higher than (or is as high as) all other points on the surface that are "near" (x_0, y_0, z_0). In Figure 17.12(a), f has a relative maximum at (x_1, y_1). Similarly, the function f in Figure 17.12(b) has a relative minimum when $x = y = 0$, which corresponds to a *low* point on the surface.

Recall that in locating extrema for a function $y = f(x)$ of one variable, we examine those values of x in the domain of f for which $f'(x) = 0$ or $f'(x)$ does not exist. For functions of two (or more) variables, a similar procedure is followed. However, for the functions that concern us, extrema will not occur

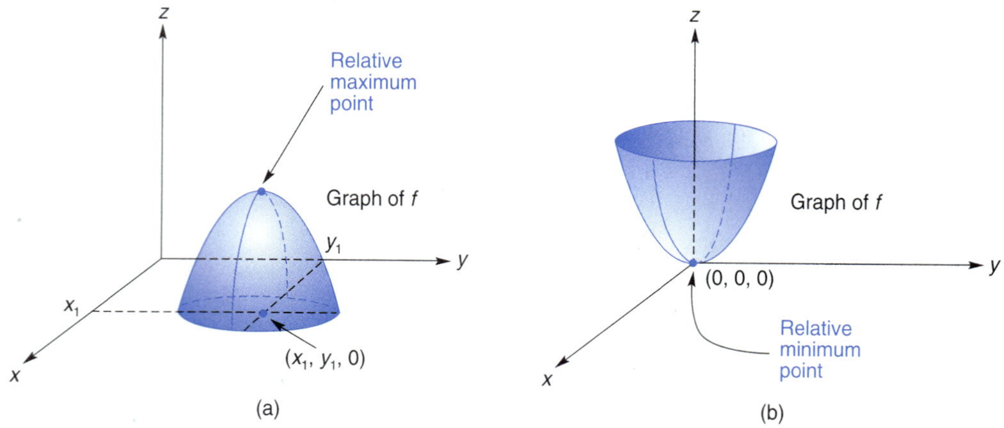

FIGURE 17.12 Relative extrema.

where a derivative does not exist, and such situations will be excluded from consideration.

Suppose $z = f(x, y)$ has a relative maximum at (x_0, y_0), as indicated in Figure 17.13(a). Then the curve where the plane $y = y_0$ intersects the surface must have a relative maximum when $x = x_0$. Hence, the slope of the tangent line to the surface in the x-direction must be 0 at (x_0, y_0). Equivalently, $f_x(x, y) = 0$ at (x_0, y_0). Similarly, on the curve where the plane $x = x_0$ intersects the surface [Figure 17.13(b)], there must be a relative maximum when $y = y_0$. Thus, in the

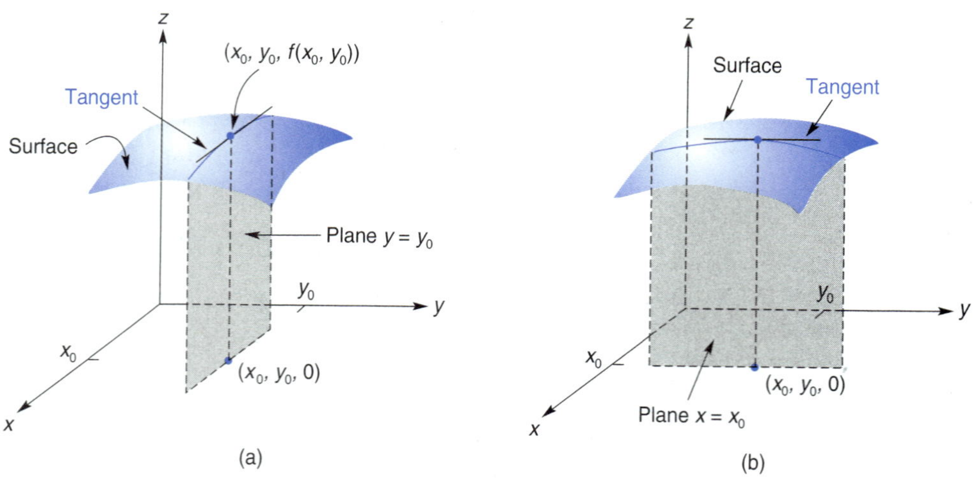

FIGURE 17.13 At relative extremum, $f_x(x, y) = 0$ and $f_y(x, y) = 0$.

y-direction, the slope of the tangent to the surface must be 0 at (x_0, y_0). Equivalently, $f_y(x, y) = 0$ at (x_0, y_0). Since a similar discussion applies to a relative minimum, we can combine these results as follows:

Rule 1

If $z = f(x, y)$ has a relative maximum or minimum at (x_0, y_0), and if both f_x and f_y are defined for all points close to (x_0, y_0), it is necessary that (x_0, y_0) be a solution of the system

$$\begin{cases} f_x(x, y) = 0 \\ f_y(x, y) = 0 \end{cases}$$

CAUTION

Rule 1 does not imply that there must be an extremum at a critical point. Just as in the case of functions of one variable, a critical point can give rise to a relative maximum, a relative minimum, or neither. A critical point is only a *candidate* for a relative extremum.

A point (x_0, y_0) for which $f_x(x, y) = f_y(x, y) = 0$ is called a **critical point** of f. Thus, from Rule 1, we infer that, to locate relative extrema for a function, we should examine its critical points.

Two additional comments are in order: First, Rule 1, as well as the notion of a critical point, can be extended to functions of more than two variables. For example, to locate possible extrema for $w = f(x, y, z)$, we would examine those points for which $w_x = w_y = w_z = 0$. Second, for a function whose domain is restricted, a thorough examination for absolute extrema would include a consideration of boundary points.

EXAMPLE 1 Finding Critical Points

Find the critical points of the following functions.

a. $f(x, y) = 2x^2 + y^2 - 2xy + 5x - 3y + 1.$

Solution: Since $f_x(x, y) = 4x - 2y + 5$ and $f_y(x, y) = 2y - 2x - 3$, we solve the system

$$\begin{cases} 4x - 2y + 5 = 0 \\ -2x + 2y - 3 = 0 \end{cases}$$

This gives $x = -1$ and $y = \frac{1}{2}$. Thus, $(-1, \frac{1}{2})$ is the only critical point.

b. $f(l, k) = l^3 + k^3 - lk.$

Solution:

$$\begin{cases} f_l(l, k) = 3l^2 - k = 0 & (2) \\ f_k(l, k) = 3k^2 - l = 0 & (3) \end{cases}$$

From Equation (2), $k = 3l^2$. Substituting for k in Equation (3) gives

$$0 = 27l^4 - l = l(27l^3 - 1)$$

Hence, $l = 0$ or $l = \frac{1}{3}$. If $l = 0$, then $k = 0$; if $l = \frac{1}{3}$, then $k = \frac{1}{3}$. The critical points are therefore $(0, 0)$ and $(\frac{1}{3}, \frac{1}{3})$.

c. $f(x, y, z) = 2x^2 + xy + y^2 + 100 - z(x + y - 100).$

Solution: Solving the system

$$\begin{cases} f_x(x, y, z) = 4x + y - z = 0 \\ f_y(x, y, z) = x + 2y - z = 0 \\ f_z(x, y, z) = -x - y + 100 = 0 \end{cases}$$

gives the critical point (25, 75, 175), as you may verify.

▸ **Now work Problem 1.**

EXAMPLE 2 Finding Critical Points

Find the critical points of

$$f(x, y) = x^2 - 4x + 2y^2 + 4y + 7.$$

Solution: We have $f_x(x, y) = 2x - 4$ and $f_y(x, y) = 4y + 4$. The system

$$\begin{cases} 2x - 4 = 0 \\ 4y + 4 = 0 \end{cases}$$

gives the critical point $(2, -1)$. Observe that we can write the given function as

$$\begin{aligned} f(x, y) &= x^2 - 4x + 4 + 2(y^2 + 2y + 1) + 1 \\ &= (x - 2)^2 + 2(y + 1)^2 + 1 \end{aligned}$$

and $f(2, -1) = 1$. Clearly, if $(x, y) \neq (2, -1)$, then $f(x, y) > 1$. Hence, a relative minimum occurs at $(2, -1)$. Moreover, there is an *absolute minimum* at $(2, -1)$, since $f(x, y) > f(2, -1)$ for *all* $(x, y) \neq (2, -1)$.

▸ **Now work Problem 3.**

Although in Example 2 we were able to show that the critical point gave rise to a relative extremum, in many cases this is not so easy to do. There is, however, a second-derivative test that gives conditions under which a critical point will be a relative maximum or minimum. We state it now, omitting the proof.

Rule 2 Second-Derivative Test for Functions of Two Variables

Suppose $z = f(x, y)$ has continuous partial derivatives f_{xx}, f_{yy}, and f_{xy} at all points (x, y) near the critical point (x_0, y_0). Let D be the function defined by

$$D(x, y) = f_{xx}(x, y) f_{yy}(x, y) - [f_{xy}(x, y)]^2$$

Then

1. if $D(x_0, y_0) > 0$ and $f_{xx}(x_0, y_0) < 0$, f has a relative maximum at (x_0, y_0);
2. if $D(x_0, y_0) > 0$ and $f_{xx}(x_0, y_0) > 0$, f has a relative minimum at (x_0, y_0);
3. if $D(x_0, y_0) < 0$, f has neither a relative maximum nor a relative minimum at (x_0, y_0);
4. if $D(x_0, y_0) = 0$, no conclusion about an extremum at (x_0, y_0) can be drawn, and further analysis is required.

EXAMPLE 3 Applying the Second-Derivative Test

Examine $f(x, y) = x^3 + y^3 - xy$ for relative maxima or minima by using the second-derivative test.

Solution: First we find critical points:

$$f_x(x, y) = 3x^2 - y \qquad f_y(x, y) = 3y^2 - x$$

In the same manner as in Example 1(b), solving $f_x(x, y) = f_y(x, y) = 0$ gives the critical points $(0, 0)$ and $(\frac{1}{3}, \frac{1}{3})$.
Now,

$$f_{xx}(x, y) = 6x \qquad f_{yy}(x, y) = 6y \qquad f_{xy}(x, y) = -1$$

Thus,

$$D(x, y) = (6x)(6y) - (-1)^2 = 36xy - 1$$

Since $D(0, 0) = 36(0)(0) - 1 = -1 < 0$, there is no relative extremum at $(0, 0)$. Also, since $D(\frac{1}{3}, \frac{1}{3}) = 36(\frac{1}{3})(\frac{1}{3}) - 1 = 3 > 0$ and $f_{xx}(\frac{1}{3}, \frac{1}{3}) = 6(\frac{1}{3}) = 2 > 0$, there is a relative minimum at $(\frac{1}{3}, \frac{1}{3})$. At this point, the value of the function is

$$f\left(\frac{1}{3}, \frac{1}{3}\right) = \left(\frac{1}{3}\right)^3 + \left(\frac{1}{3}\right)^3 - \left(\frac{1}{3}\right)\left(\frac{1}{3}\right) = -\frac{1}{27}$$

▶ **Now work Problem 7.**

EXAMPLE 4 **A Saddle Point**

Examine $f(x, y) = y^2 - x^2$ for relative extrema.

Solution: Solving

$$f_x(x, y) = -2x = 0 \quad \text{and} \quad f_y(x, y) = 2y = 0$$

we get the critical point $(0, 0)$. Now we apply the second-derivative test. At $(0, 0)$, and indeed at any point,

$$f_{xx}(x, y) = -2 \qquad f_{yy}(x, y) = 2 \qquad f_{xy}(x, y) = 0$$

Because $D(0, 0) = (-2)(2) - (0)^2 = -4 < 0$, no relative extremum exists at $(0, 0)$. A sketch of $z = f(x, y) = y^2 - x^2$ appears in Figure 17.14. Note that, for the surface curve cut by the plane $y = 0$, there is a *maximum* at $(0, 0)$; but for the surface curve cut by the plane $x = 0$, there is a *minimum* at $(0, 0)$. Thus, on the *surface*, no relative extremum can exist at the origin, although $(0, 0)$ is a critical point. Around the origin the curve is saddle shaped, and $(0, 0)$ is called a *saddle point* of f.

The surface in Figure 17.14 is called a hyperbolic paraboloid.

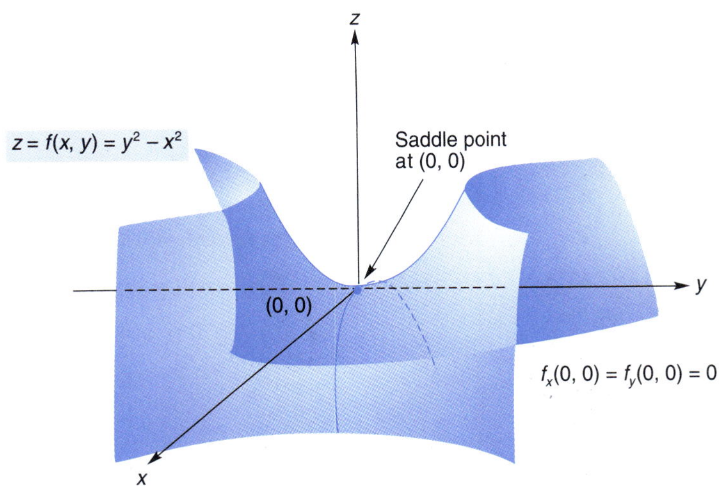

FIGURE 17.14 Saddle point.

▶ **Now work Problem 11.**

EXAMPLE 5 **Finding Relative Extrema**

Examine $f(x, y) = x^4 + (x - y)^4$ for relative extrema.

Solution: If we set

$$f_x(x, y) = 4x^3 + 4(x - y)^3 = 0 \quad (4)$$

and

$$f_y(x, y) = -4(x - y)^3 = 0 \tag{5}$$

then, from Equation (5), we have $x - y = 0$, or $x = y$. Substituting into Equation (4) gives $4x^3 = 0$, or $x = 0$. Thus, $x = y = 0$, and (0, 0) is the only critical point. At (0, 0),

$$f_{xx}(x, y) = 12x^2 + 12(x - y)^2 = 0$$

$$f_{yy}(x, y) = 12(x - y)^2 = 0$$

and

$$f_{xy}(x, y) = -12(x - y)^2 = 0$$

Hence, $D(0, 0) = 0$, and the second-derivative test gives no information. However, for all $(x, y) \neq (0, 0)$, we have $f(x, y) > 0$, whereas $f(0, 0) = 0$. Therefore, at (0, 0) the graph of f has a low point, and we conclude that f has a relative (and absolute) minimum at (0, 0).

▶ **Now work Problem 13.**

Applications

In many situations involving functions of two variables, and especially in their applications, the nature of the given problem is an indicator of whether a critical point is in fact a relative (or absolute) maximum or a relative (or absolute) minimum. In such cases, the second-derivative test is not needed. Often, in mathematical studies of applied problems, the appropriate second-order conditions are assumed to hold.

EXAMPLE 6 Maximizing Output

Let P be a production function given by

$$P = f(l, k) = 0.54l^2 - 0.02l^3 + 1.89k^2 - 0.09k^3$$

where l and k are the amounts of labor and capital, respectively, and P is the quantity of output produced. Find the values of l and k that maximize P.

Solution: To find the critical points, we solve the system $P_l = 0$ and $P_k = 0$;

$$P_l = 1.08l - 0.06l^2 \qquad P_k = 3.78k - 0.27k^2$$

$$= 0.06l(18 - l) = 0 \qquad = 0.27k(14 - k) = 0$$

$$l = 0, l = 18 \qquad k = 0, k = 14$$

There are four critical points: (0, 0), (0, 14), (18, 0), and (18, 14).

Now we apply the second-derivative test to each critical point. We have

$$P_{ll} = 1.08 - 0.12l \qquad P_{kk} = 3.78 - 0.54k \qquad P_{lk} = 0$$

Thus,

$$D(l, k) = P_{ll}P_{kk} - [P_{lk}]^2$$
$$= (1.08 - 0.12l)(3.78 - 0.54k)$$

At (0, 0),

$$D(0, 0) = 1.08(3.78) > 0$$

Since $D(0, 0) > 0$ and $P_{ll} = 1.08 > 0$, there is a relative minimum at (0, 0). At (0, 14),

$$D(0, 14) = 1.08(-3.78) < 0$$

Because $D(0, 14) < 0$, there is no relative extremum at (0, 14). At (18, 0),

$$D(18, 0) = (-1.08)(3.78) < 0$$

Since $D(18, 0) < 0$, there is no relative extremum at (18, 0). At (18, 14),

$$D(18, 14) = (-1.08)(-3.78) > 0$$

Because $D(18, 14) > 0$ and $P_{ll} = -1.08 < 0$, there is a relative maximum at (18, 14). Hence, the maximum output is obtained when $l = 18$ and $k = 14$.

▶ **Now work Problem 21.**

EXAMPLE 7 **Profit Maximization**

A candy company produces two types of candy, A *and* B, *for which the average costs of production are constant at* \$2 *and* \$3 *per pound, respectively. The quantities* q_A, q_B *(in pounds) of* A *and* B *that can be sold each week are given by the joint-demand functions*

$$q_A = 400(p_B - p_A)$$

and

$$q_B = 400(9 + p_A - 2p_B)$$

where p_A *and* p_B *are the selling prices (in dollars per pound) of* A *and* B, *respectively. Determine the selling prices that will maximize the company's profit* P.

Solution: The total profit is given by

$$P = \begin{pmatrix} \text{profit} \\ \text{per pound} \\ \text{of A} \end{pmatrix} \begin{pmatrix} \text{pounds} \\ \text{of A} \\ \text{sold} \end{pmatrix} + \begin{pmatrix} \text{profit} \\ \text{per pound} \\ \text{of B} \end{pmatrix} \begin{pmatrix} \text{pounds} \\ \text{of B} \\ \text{sold} \end{pmatrix}$$

For A and B, the profits per pound are $p_A - 2$ and $p_B - 3$, respectively. Thus,

$$\begin{aligned} P &= (p_A - 2)q_A + (p_B - 3)q_B \\ &= (p_A - 2)[400(p_B - p_A)] + (p_B - 3)[400(9 + p_A - 2p_B)] \end{aligned}$$

Notice that P is expressed as a function of two variables, p_A and p_B. To maximize P, we set its partial derivatives equal to 0:

$$\begin{aligned} \frac{\partial P}{\partial p_A} &= (p_A - 2)[400(-1)] + [400(p_B - p_A)](1) + (p_B - 3)[400(1)] \\ &= 0 \qquad \text{(product rule)} \end{aligned}$$

$$\begin{aligned} \frac{\partial P}{\partial p_B} &= (p_A - 2)[400(1)] + (p_B - 3)[400(-2)] + 400(9 + p_A - 2p_B)](1) \\ &= 0 \qquad \text{(product rule)} \end{aligned}$$

Simplifying the preceding two equations gives

$$\begin{cases} -2p_A + 2p_B - 1 = 0 \\ 2p_A - 4p_B + 13 = 0 \end{cases}$$

whose solution is $p_A = 5.5$ and $p_B = 6$. Moreover, we find that

$$\frac{\partial^2 P}{\partial p_A^2} = -800 \qquad \frac{\partial^2 P}{\partial p_B^2} = -1600 \qquad \frac{\partial^2 P}{\partial p_B \partial p_A} = 800$$

Therefore,

$$D(5.5, 6) = (-800)(-1600) - (800)^2 > 0$$

Since $\partial^2 P/\partial p_A^2 < 0$, we indeed have a maximum, and the company should sell candy A at \$5.50 per pound and B at \$6.00 per pound.

▶ **Now work Problem 23.**

EXAMPLE 8 **Profit Maximization for a Monopolist**[22]

Suppose a monopolist is practicing price discrimination by selling the same product in two separate markets at different prices. Let q_A be the number of units sold in market A, where the demand function is $p_A = f(q_A)$, and let q_B be the number of units sold in market B, where the demand function is $p_B = g(q_B)$. Then the revenue functions for the two markets are

$$r_A = q_A f(q_A) \quad \text{and} \quad r_B = q_B g(q_B)$$

Assume that all units are produced at one plant, and let the cost function for producing $q(= q_A + q_B)$ units be $c = c(q)$. Keep in mind that r_A is a function of q_A and r_B is a function of q_B. The monopolist's profit P is

$$P = r_A + r_B - c$$

To maximize P with respect to outputs q_A and q_B, we set its partial derivatives equal to zero. To begin with,

$$\begin{aligned}\frac{\partial P}{\partial q_A} &= \frac{dr_A}{dq_A} + 0 - \frac{\partial c}{\partial q_A} \\ &= \frac{dr_A}{dq_A} - \frac{dc}{dq}\frac{\partial q}{\partial q_A} = 0 \qquad \text{(chain rule)}\end{aligned}$$

Because

$$\frac{\partial q}{\partial q_A} = \frac{\partial}{\partial q_A}(q_A + q_B) = 1$$

we have

$$\frac{\partial P}{\partial q_A} = \frac{dr_A}{dq_A} - \frac{dc}{dq} = 0 \qquad (6)$$

Similarly,

$$\frac{\partial P}{\partial q_B} = \frac{dr_B}{dq_B} - \frac{dc}{dq} = 0 \qquad (7)$$

From Equations (6) and (7), we get

$$\frac{dr_A}{dq_A} = \frac{dc}{dq} = \frac{dr_B}{dq_B}$$

But dr_A/dq_A and dr_B/dq_B are marginal revenues, and dc/dq is marginal cost. Hence, to maximize profit, it is necessary to charge prices (and distribute

[22]Omit this example if Section 17.6 was not covered.

output) so that the marginal revenues in both markets will be the same and, loosely speaking, will also be equal to the cost of the last unit produced in the plant.

▸ **Now work Problem 25.**

Exercise 17.7

In Problems 1–6, find the critical points of the functions.

*1. $f(x, y) = x^2 + y^2 - 5x + 4y + xy$

2. $f(x, y) = x^2 + 4y^2 - 6x + 16y$

*3. $f(x, y) = 2x^3 + y^3 - 3x^2 + 1.5y^2 - 12x - 90y$

4. $f(x, y) = xy - x + y$

5. $f(x, y, z) = 2x^2 + xy + y^2 + 100 - z(x + y - 200)$

6. $f(x, y, z, w) = x^2 + y^2 + z^2 - w(x - y + 2z - 6)$

In Problems 7–20, find the critical points of the functions. For each critical point, determine, by the second-derivative test, whether it corresponds to a relative maximum, to a relative minimum, or to neither, or whether the test gives no information.

*7. $f(x, y) = x^2 + 3y^2 + 4x - 9y + 3$

8. $f(x, y) = -2x^2 + 8x - 3y^2 + 24y + 7$

9. $f(x, y) = y - y^2 - 3x - 6x^2$

10. $f(x, y) = x^2 + y^2 + xy - 9x + 1$

*11. $f(x, y) = x^2 + 3xy + y^2 + x + 3$

12. $f(x, y) = \frac{x^3}{3} + y^2 - 2x + 2y - 2xy$

*13. $f(x, y) = \frac{1}{3}(x^3 + 8y^3) - 2(x^2 + y^2) + 1$

14. $f(x, y) = x^2 + y^2 - xy + x^3$

15. $f(l, k) = 2lk - l^2 + 264k - 10l - 2k^2$

16. $f(l, k) = l^2 + k^2 - 2lk$

17. $f(p, q) = pq - \frac{1}{p} - \frac{1}{q}$

18. $f(x, y) = (x - 3)(y - 3)(x + y - 3)$

19. $f(x, y) = (y^2 - 4)(e^x - 1)$

20. $f(x, y) = \ln(xy) + 2x^2 - xy - 6x$

*21. **Maximizing Output** Suppose

$$P = f(l, k) = 1.08l^2 - 0.03l^3 + 1.68k^2 - 0.08k^3$$

is a production function for a firm. Find the quantities of inputs l and k that maximize output P.

22. **Maximizing Output** In a certain automated manufacturing process, machines M and N are utilized for m and n hours, respectively. If daily output Q is a function of m and n, namely,

$$Q = 4.5m + 5n - 0.5m^2 - n^2 - 0.25mn$$

find the values of m and n that maximize Q.

In Problems 23–35, unless otherwise indicated, the variables p_A and p_B denote selling prices of products A and B, respectively. Similarly, q_A and q_B denote quantities of A and B that are produced and sold during some time period. In all cases, the variables employed will be assumed to be units of output, input, money, etc.

*23. **Profit** A candy company produces two varieties of candy, A and B, for which the constant average costs of production are 60 and 70 (cents per lb), respectively. The demand functions for A and B are given by

$$q_A = 5(p_B - p_A) \quad \text{and} \quad q_B = 500 + 5(p_A - 2p_B)$$

Find the selling prices p_A and p_B that maximize the company's profit.

24. **Profit** Repeat Problem 23 if the constant costs of production of A and B are a and b (cents per lb), respectively.

*25. **Price Discrimination** Suppose a monopolist is practicing price discrimination in the sale of a product by charging different prices in two separate markets. In market A the demand function is

$$p_A = 100 - q_A$$

and in B it is

$$p_B = 84 - q_B$$

where q_A and q_B are the quantities sold per week in A and B, and p_A and p_B are the respective prices per unit. If the monopolist's cost function is

$$c = 600 + 4(q_A + q_B)$$

how much should be sold in each market to maximize profit? What selling prices give this maximum profit? Find the maximum profit.

26. **Profit** A monopolist sells two competitive products, A and B, for which the demand functions are

$$q_A = 3 - p_A + 2p_B \quad \text{and} \quad q_B = 5 + 5p_A - 2p_B$$

If the constant average cost of producing a unit of A is 3 and a unit of B is 2, how many units of A and B should be sold to maximize the monopolist's profit?

27. **Profit** For products A and B, the joint-cost function for a manufacturer is

$$c = 1.5q_A^2 + 4.5q_B^2$$

and the demand functions are $p_A = 36 - q_A^2$ and $p_B = 30 - q_B^2$. Find the level of production that maximizes profit.

28. **Profit** For a monopolist's products A and B, the joint-cost function is $c = 2(q_A + q_B + q_A q_B)$, and the demand functions are $q_A = 20 - 2p_A$ and $q_B = 10 - p_B$. Find the values of p_A and p_B that maximize profit. What are the quantities of A and B that correspond to these prices? What is the total profit?

29. **Cost** An open-top rectangular box is to have a volume of 6 ft^3. The cost per square foot of materials is \$3 for the bottom, \$1 for the front and back, and \$0.50 for the other two sides. Find the dimensions of the box so that the cost of materials is minimized. (See Figure 17.15.)

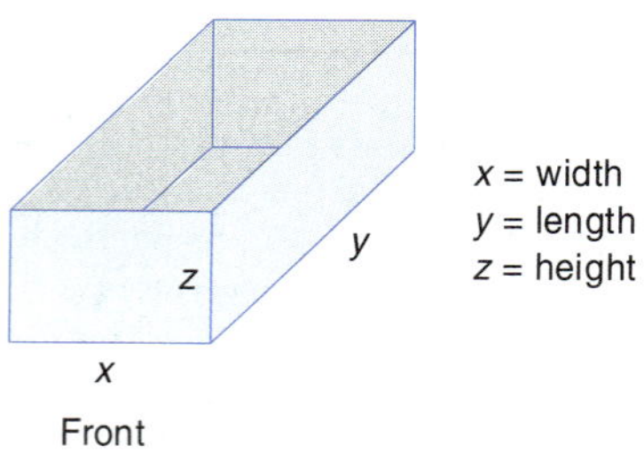

FIGURE 17.15 Diagram for Problem 29.

30. **Collusion** Suppose A and B are the only two firms in the market selling the same product. (We say that they are *duopolists*.) The industry demand function for the product is

$$p = 92 - q_A - q_B$$

where q_A and q_B denote the output produced and sold by A and B, respectively. For A, the cost function is $c_A = 10q_A$; for B, it is $c_B = 0.5q_B^2$. Suppose the firms decide to enter into an agreement on output and price control by jointly acting as a monopoly. In this case, we say they enter into *collusion*. Show that the profit function for the monopoly is given by

$$P = pq_A - c_A + pq_B - c_B$$

Express P as a function of q_A and q_B, and determine how output should be allocated so as to maximize the profit of the monopoly.

31. Suppose $f(x, y) = -2x^2 + 5y^2 + 7$, where x and y must satisfy the equation $3x - 2y = 7$. Find the relative extrema of f, subject to the given condition on x and y, by first solving the second equation for y. Substitute the result for y in the given equation. Thus, f is expressed as a function of one variable, for which extrema may be found in the usual way.

32. Repeat Problem 31 if $f(x, y) = x^2 + 4y^2 + 6$, subject to the condition that $2x - 8y = 20$.

33. Suppose the joint-cost function

$$c = q_A^2 + 3q_B^2 + 2q_A q_B + aq_A + bq_B + d$$

has a relative minimum value of 15 when $q_A = 3$ and $q_B = 1$. Determine the values of the constants a, b, and d.

34. Suppose that the production function $q = f(k, l)$ has a relative maximum value when $k = 35$ and $l = 30$. Suppose also that all the second derivatives of f exist at the point (35, 30) and that $f_{kk}(35, 30) = 0$.

(a) Determine whether
   i. $f_{kl}(35, 30)$ is a positive number,
   ii. $f_{kl}(35, 30)$ is a negative number, or
   iii. $f_{kl}(35, 30)$ is zero, or
   iv. whether it is impossible to reach one of these conclusions.

(b) Determine whether
   i. $f_{ll}(35, 30)$ must be positive,
   ii. $f_{ll}(35, 30)$ must be negative, or
   iii. $f_{ll}(35, 30)$ must be zero, or
   iv. whether it is impossible to reach one of these conclusions.

35. **Profit from Competitive Products** A monopolist sells two competitive products, A and B, for which the demand equations are

$$p_A = 35 - 2q_A^2 + q_B$$

and

$$p_B = 20 - q_B + q_A$$

The joint-cost function is

$$c = -8 - 2q_A^3 + 3q_A q_B + 30q_A + 12q_B + \frac{1}{2}q_A^2$$

(a) How many units of A and B should be sold to obtain a relative maximum profit for the monopolist? Use the second-derivative test to justify your answer.

(b) Determine the selling prices required to realize the relative maximum profit. Also, find this relative maximum profit.

36. **Profit and Advertising** A retailer has determined that the number of TV sets he can sell per week is

$$\frac{5x}{2+x} + \frac{2y}{5+y}$$

where x and y represent his weekly expenditures (in dollars) on newspaper and radio advertising, respectively. The profit is \$250 per sale, less the cost of advertising, so the weekly profit is given by the formula

$$P = 250\left[\frac{5x}{2+x} + \frac{2y}{5+y}\right] - x - y$$

Find the values of x and y for which the profit is a relative maximum. Use the second-derivative test to verify that your answer corresponds to a relative maximum profit.

37. Profit from Tomato Crop The revenue (in dollars per square meter of ground) obtained from the sale of a crop of tomatoes grown in an artificially heated greenhouse is given by

$$r = 5T(1 - e^{-x})$$

where T is the temperature (in °C) maintained in the greenhouse and x is the amount of fertilizer applied per square meter. The cost of fertilizer is $20x$ dollars per square meter, and the cost of heating is given by $0.1T^2$ dollars per square meter.

(a) Find an expression, in terms of T and x, for the profit per square meter obtained from the sale of the crop of tomatoes.

(b) Verify that the pairs

$$(T, x) = (20, \ln 5) \quad \text{and} \quad (T, x) = \left(5, \ln \frac{5}{4}\right)$$

are critical points of the profit function in part (a). [*Note:* You need not derive the pairs.]

(c) The points in part (b) are the only critical points of the profit function in part (a). Use the second-derivative test to determine whether either of these points corresponds to a relative maximum profit per square meter.

OBJECTIVE *To find critical points for a function, subject to constraints, by applying the method of Lagrange multipliers.*

17.8 Lagrange Multipliers

We shall now find relative maxima and minima for a function on which certain *constraints* are imposed. Such a situation could arise if a manufacturer wished to minimize a joint-cost function and yet obtain a particular production level.

Suppose we want to find the relative extrema of

$$w = x^2 + y^2 + z^2 \tag{1}$$

subject to the constraint that x, y, and z must satisfy

$$x - y + 2z = 6 \tag{2}$$

We can transform w, which is a function of three variables, into a function of two variables such that the new function reflects constraint (2). Solving Equation (2) for x, we get

$$x = y - 2z + 6 \tag{3}$$

which, when substituted for x in Equation (1), gives

$$w = (y - 2z + 6)^2 + y^2 + z^2 \tag{4}$$

Since w is now expressed as a function of two variables, to find relative extrema we follow the usual procedure of setting the partial derivatives of w equal to 0:

$$\frac{\partial w}{\partial y} = 2(y - 2z + 6) + 2y = 4y - 4z + 12 = 0 \tag{5}$$

$$\frac{\partial w}{\partial z} = -4(y - 2z + 6) + 2z = -4y + 10z - 24 = 0 \tag{6}$$

Solving Equations (5) and (6) simultaneously gives $y = -1$ and $z = 2$. Substituting into Equation (3), we get $x = 1$. Hence, the only critical point of Equation (1) subject to the constraint represented by Equation (2) is $(1, -1, 2)$. By using the second-derivative test on Equation (4) when $y = -1$ and $z = 2$, we have

$$\frac{\partial^2 w}{\partial y^2} = 4 \qquad \frac{\partial^2 w}{\partial z^2} = 10 \qquad \frac{\partial^2 w}{\partial z \partial y} = -4$$

$$D(-1, 2) = 4(10) - (-4)^2 = 24 > 0$$

Thus w, subject to the constraint, has a relative minimum at $(1, -1, 2)$.

This solution was found by using the constraint to express one of the variables in the original function in terms of the other variables. Often this is not practical, but there is another technique, called the method of **Lagrange multipliers,**[23] that avoids this step and yet allows us to obtain critical points.

The method is as follows. Suppose we have a function $f(x, y, z)$ subject to the constraint $g(x, y, z) = 0$. We construct a new function F of *four* variables defined by the following (where λ is a Greek letter read "lambda"):

$$F(x, y, z, \lambda) = f(x, y, z) - \lambda g(x, y, z)$$

It can be shown that if (x_0, y_0, z_0) is a critical point of f, subject to the constraint $g(x, y, z) = 0$, there exists a value of λ, say, λ_0, such that $(x_0, y_0, z_0, \lambda_0)$ is a critical point of F. The number λ_0 is called a **Lagrange multiplier.** Also, if $(x_0, y_0, z_0, \lambda_0)$ is a critical point of F, then (x_0, y_0, z_0) is a critical point of f, subject to the constraint. Thus, to find critical points of f, subject to $g(x, y, z) = 0$, we instead find critical points of F. These are obtained by solving the simultaneous equations

$$\begin{cases} F_x(x, y, z, \lambda) = 0 \\ F_y(x, y, z, \lambda) = 0 \\ F_z(x, y, z, \lambda) = 0 \\ F_\lambda(x, y, z, \lambda) = 0 \end{cases}$$

At times, ingenuity must be used to solve the equations. Once we obtain a critical point $(x_0, y_0, z_0, \lambda_0)$ of F, we can conclude that (x_0, y_0, z_0) is a critical point of f, subject to the constraint $g(x, y, z) = 0$. Although f and g are functions of three variables, the method of Lagrange multipliers can be extended to n variables.

Let us illustrate the method of Lagrange multipliers for the original situation, namely,

$$f(x, y, z) = x^2 + y^2 + z^2 \qquad \text{subject to} \qquad x - y + 2z = 6$$

First, we write the constraint as $g(x, y, z) = x - y + 2z - 6 = 0$. Second, we form the function

$$\begin{aligned} F(x, y, z, \lambda) &= f(x, y, z) - \lambda g(x, y, z) \\ &= x^2 + y^2 + z^2 - \lambda(x - y + 2z - 6) \end{aligned}$$

Next, we set each partial derivative of F equal to 0. For convenience, we shall write $F_x(x, y, z, \lambda)$ as F_x, and so on:

$$\begin{cases} F_x = 2x - \lambda = 0 & (7) \\ F_y = 2y + \lambda = 0 & (8) \\ F_z = 2z - 2\lambda = 0 & (9) \\ F_\lambda = -x + y - 2z + 6 = 0 & (10) \end{cases}$$

From Equations (7)–(9), we see immediately that

$$x = \frac{\lambda}{2} \qquad y = -\frac{\lambda}{2} \qquad z = \lambda \tag{11}$$

[23]After the French mathematician Joseph-Louis Lagrange (1736–1813).

Substituting these values into Equation (10), we obtain

$$-\frac{\lambda}{2} - \frac{\lambda}{2} - 2\lambda + 6 = 0$$

$$-3\lambda + 6 = 0$$

$$\lambda = 2$$

Thus, from Equation (11),

$$x = 1 \qquad y = -1 \qquad z = 2$$

Hence, the only critical point of f, subject to the constraint, is $(1, -1, 2)$, at which there may exist a relative maximum, a relative minimum, or neither of these. The method of Lagrange multipliers does not directly indicate which of these possibilities occur, although from our previous work, we saw that it is indeed a relative minimum. In applied problems, the nature of the problem itself may give a clue as to how a critical point is to be regarded. Often the existence of either a relative minimum or a relative maximum is assumed, and a critical point is treated accordingly. Actually, sufficient second-order conditions for relative extrema are available, but we shall not consider them.

EXAMPLE 1 Method of Lagrange Multipliers

Find the critical points for $z = f(x, y) = 3x - y + 6$, subject to the constraint $x^2 + y^2 = 4$.

Solution: We write the constraint as $g(x, y) = x^2 + y^2 - 4 = 0$ and construct the function

$$F(x, y, \lambda) = f(x, y) - \lambda g(x, y) = 3x - y + 6 - \lambda(x^2 + y^2 - 4)$$

Setting $F_x = F_y = F_\lambda = 0$, we have

$$\begin{cases} 3 - 2x\lambda = 0 & (12) \\ -1 - 2y\lambda = 0 & (13) \\ -x^2 - y^2 + 4 = 0 & (14) \end{cases}$$

From Equations (12) and (13), we can express x and y in terms of λ. Then we shall substitute for x and y in Equation (14) and solve for λ. Knowing λ, we can find x and y. To begin, from Equations (12) and (13), we have

$$x = \frac{3}{2\lambda} \quad \text{and} \quad y = -\frac{1}{2\lambda}$$

Substituting into Equation (14), we obtain

$$-\frac{9}{4\lambda^2} - \frac{1}{4\lambda^2} + 4 = 0$$

$$-\frac{10}{4\lambda^2} + 4 = 0$$

$$\lambda = \pm\frac{\sqrt{10}}{4}$$

With these λ-values, we can find x and y. If $\lambda = \sqrt{10}/4$, then

$$x = \frac{3}{2\left(\frac{\sqrt{10}}{4}\right)} = \frac{3\sqrt{10}}{5} \qquad y = -\frac{1}{2\left(\frac{\sqrt{10}}{4}\right)} = -\frac{\sqrt{10}}{5}$$

Similarly, if $\lambda = -\sqrt{10}/4$,

$$x = -\frac{3\sqrt{10}}{5} \qquad y = \frac{\sqrt{10}}{5}$$

Thus, the critical points of f, subject to the constraint, are $(3\sqrt{10}/5, -\sqrt{10}/5)$ and $(-3\sqrt{10}/5, \sqrt{10}/5)$. Note that the values of λ do not appear in the answer; they are simply a means to obtain the solution.

▶ **Now work Problem 1.**

EXAMPLE 2 **Method of Lagrange Multipliers**

Find critical points for $f(x, y, z) = xyz$, *where* $xyz \neq 0$, *subject to the constraint* $x + 2y + 3z = 36$.

Solution: We have

$$F(x, y, z, \lambda) = xyz - \lambda(x + 2y + 3z - 36)$$

Setting $F_x = F_y = F_z = F_\lambda = 0$ gives, respectively,

$$\begin{cases} yz - \lambda = 0 \\ xz - 2\lambda = 0 \\ xy - 3\lambda = 0 \\ -x - 2y - 3z + 36 = 0 \end{cases}$$

Because we cannot directly express x, y, and z in terms of λ only, we cannot follow the procedure in Example 1. However, observe that we can express the products yz, xz, and xy as multiples of λ. This suggests that, by looking at quotients of equations, we can obtain a relation between two variables that does not involve λ. (The λ's will cancel.) Proceeding to do this, we write the foregoing system as

$$\begin{cases} yz = \lambda & (15) \\ xz = 2\lambda & (16) \\ xy = 3\lambda & (17) \\ x + 2y + 3z - 36 = 0 & (18) \end{cases}$$

Dividing each side of Equation (15) by the corresponding side of Equation (16), we get

$$\frac{yz}{xz} = \frac{\lambda}{2\lambda} \quad \text{so} \quad y = \frac{x}{2}$$

This division is valid, since $xyz \neq 0$. Similarly, from Equations (15) and (17), we get

$$\frac{yz}{xy} = \frac{\lambda}{3\lambda} \quad \text{so} \quad z = \frac{x}{3}$$

Now that we have y and z expressed in terms of x only, we can substitute into Equation (18) and solve for x:

$$x + 2\left(\frac{x}{2}\right) + 3\left(\frac{x}{3}\right) - 36 = 0$$

$$x = 12$$

Thus, $y = 6$ and $z = 4$. Hence, (12, 6, 4) is the only critical point satisfying the given conditions. Note that in this situation, we found the critical point without having to find the value for λ.

▶ **Now work Problem 7.**

EXAMPLE 3 **Minimizing Costs**

Suppose a firm has an order for 200 units of its product and wishes to distribute its manufacture between two of its plants, plant 1 and plant 2. Let q_1 and q_2 denote the outputs of plants 1 and 2, respectively, and suppose the total-cost function is given by $c = f(q_1, q_2) = 2q_1^2 + q_1q_2 + q_2^2 + 200$. How should the output be distributed in order to minimize costs?

Solution: We minimize $c = f(q_1, q_2)$, given the constraint $q_1 + q_2 = 200$. We have

$$F(q_1, q_2, \lambda) = 2q_1^2 + q_1q_2 + q_2^2 + 200 - \lambda(q_1 + q_2 - 200)$$

$$\begin{cases} \dfrac{\partial F}{\partial q_1} = 4q_1 + q_2 - \lambda = 0 & (19) \\ \dfrac{\partial F}{\partial q_2} = q_1 + 2q_2 - \lambda = 0 & (20) \\ \dfrac{\partial F}{\partial \lambda} = -q_1 - q_2 + 200 = 0 & (21) \end{cases}$$

We can eliminate λ from Equations (19) and (20) and obtain a relation between q_1 and q_2. Then, solving this equation for q_2 in terms of q_1 and substituting into Equation (21), we can find q_1. We begin by subtracting Equation (20) from Equation (19), which gives

$$3q_1 - q_2 = 0 \quad \text{so} \quad q_2 = 3q_1$$

Substituting into Equation (21), we have

$$\begin{aligned} -q_1 - 3q_1 + 200 &= 0 \\ -4q_1 &= -200 \\ q_1 &= 50 \end{aligned}$$

Thus, $q_2 = 150$. Accordingly, plant 1 should produce 50 units and plant 2 should produce 150 units in order to minimize costs.

▶ **Now work Problem 13.**

An interesting observation can be made concerning Example 3. From Equation (19), $\lambda = 4q_1 + q_2 = \partial c/\partial q_1$, the marginal cost of plant 1. From Equation (20), $\lambda = q_1 + 2q_2 = \partial c/\partial q_2$, the marginal cost of plant 2. Hence, $\partial c/\partial q_1 = \partial c/\partial q_2$, and we conclude that, to minimize cost, it is necessary that the marginal costs of each plant be equal to each other.

EXAMPLE 4 **Least-Cost Input Combination**

Suppose a firm must produce a given quantity P_0 of output in the cheapest possible manner. If there are two input factors l and k, and their prices per unit are fixed at p_l and p_k, respectively, discuss the economic significance of combining input to achieve least cost. That is, describe the least-cost input combination.

Solution: Let $P = f(l, k)$ be the production function. Then we must minimize the cost function

$$c = lp_l + kp_k$$

subject to

$$P_0 = f(l, k)$$

We construct

$$F(l, k, \lambda) = lp_l + kp_k - \lambda[f(l, k) - P_0]$$

We have

$$\begin{cases} \dfrac{\partial F}{\partial l} = p_l - \lambda\dfrac{\partial}{\partial l}[f(l, k)] = 0 & (22) \\ \dfrac{\partial F}{\partial k} = p_k - \lambda\dfrac{\partial}{\partial k}[f(l, k)] = 0 & (23) \\ \dfrac{\partial F}{\partial \lambda} = -f(l, k) + P_0 = 0 \end{cases}$$

From Equations (22) and (23),

$$\lambda = \frac{p_l}{\dfrac{\partial}{\partial l}[f(l, k)]} = \frac{p_k}{\dfrac{\partial}{\partial k}[f(l, k)]} \qquad (24)$$

Hence,

$$\frac{p_l}{p_k} = \frac{\dfrac{\partial}{\partial l}[f(l, k)]}{\dfrac{\partial}{\partial k}[f(l, k)]}$$

We conclude that when the least-cost combination of factors is used, the ratio of the marginal productivities of the input factors must be equal to the ratio of their corresponding unit prices.

▶ **Now work Problem 15.**

Multiple Constraints

The method of Lagrange multipliers is by no means restricted to problems involving a single constraint. For example, suppose $f(x, y, z, w)$ were subject to constraints $g_1(x, y, z, w) = 0$ and $g_2(x, y, z, w) = 0$. Then there would be two lambdas, λ_1 and λ_2 (one corresponding to each constraint), and we would construct the function $F = f - \lambda_1 g_1 - \lambda_2 g_2$. We would then solve the system

$$F_x = F_y = F_z = F_w = F_{\lambda_1} = F_{\lambda_2} = 0$$

EXAMPLE 5 **Method of Lagrange Multipliers with Two Constraints**

Find critical points for $f(x, y, z) = xy + yz$, subject to the constraints $x^2 + y^2 = 8$ and $yz = 8$.

Solution: Set

$$F(x, y, z, \lambda_1, \lambda_2) = xy + yz - \lambda_1(x^2 + y^2 - 8) - \lambda_2(yz - 8)$$

Then

$$\begin{cases} F_x = y - 2x\lambda_1 = 0 \\ F_y = x + z - 2y\lambda_1 - z\lambda_2 = 0 \\ F_z = y - y\lambda_2 = 0 \\ F_{\lambda_1} = -x^2 - y^2 + 8 = 0 \\ F_{\lambda_2} = -yz + 8 = 0 \end{cases}$$

You would probably agree that this appears to be a challenging system to solve. Thus, ingenuity will come into play. Here is one sequence of operations that will allow us to find the critical points. We can write the system as

$$\begin{cases} \dfrac{y}{2x} = \lambda_1 & (25) \\ x + z - 2y\lambda_1 - z\lambda_2 = 0 & (26) \\ \lambda_2 = 1 & (27) \\ x^2 + y^2 = 8 & (28) \\ z = \dfrac{8}{y} & (29) \end{cases}$$

Substituting $\lambda_2 = 1$ from Equation (27) into Equation (26) and simplifying gives the equation $x - 2y\lambda_1 = 0$, so

$$\lambda_1 = \frac{x}{2y}$$

Substituting into Equation (25) gives

$$\frac{y}{2x} = \frac{x}{2y}$$

$$y^2 = x^2 \qquad (30)$$

Substituting into Equation (28) gives $x^2 + x^2 = 8$, from which it follows that $x = \pm 2$. If $x = 2$, then, from Equation (30), we have $y = \pm 2$. Similarly, if $x = -2$, then $y = \pm 2$. Thus, if $x = 2$ and $y = 2$, then, from Equation (29), we obtain $z = 4$. Continuing in this manner, we obtain four critical points:

$$(2, 2, 4) \quad (2, -2, -4) \quad (-2, 2, 4) \quad (-2, -2, -4)$$

▶ Now work Problem 9.

Exercise 17.8

In Problems 1–12 find, by the method of Lagrange multipliers, the critical points of the functions, subject to the given constraints.

*1. $f(x, y) = x^2 + 4y^2 + 6;\quad 2x - 8y = 20$

2. $f(x, y) = -2x^2 + 5y^2 + 7;\quad 3x - 2y = 7$

3. $f(x, y, z) = x^2 + y^2 + z^2;\quad 2x + y - z = 9$

4. $f(x, y, z) = x + y + z;\quad xyz = 27$

5. $f(x, y, z) = 2x^2 + xy + y^2 + z;\quad x + 2y + 4z = 3$

6. $f(x, y, z) = xyz^2;\quad x - y + z = 20\ (xyz^2 \neq 0)$

*7. $f(x, y, z) = xyz;\quad x + 2y + 3z = 18\ (xyz \neq 0)$

8. $f(x, y, z) = x^2 + y^2 + z^2;\quad x + y + z = 3$

*9. $f(x, y, z) = x^2 + 2y - z^2;\quad 2x - y = 0,\ y + z = 0$

10. $f(x, y, z) = x^2 + y^2 + z^2;\quad x + y + z = 4,\ x - y + z = 4$

11. $f(x, y, z) = xyz$; $x + y + z = 12$, $x + y - z = 0$ $(xyz \neq 0)$

12. $f(x, y, z, w) = 3x^2 + y^2 + 2z^2 - 5w^2$; $x + 6y + 3z + 2w = 4$

***13. Production Allocation** To fill an order for 100 units of its product, a firm wishes to distribute production between its two plants, plant 1 and plant 2. The total-cost function is given by

$$c = f(q_1, q_2) = 0.1q_1^2 + 7q_1 + 15q_2 + 1000$$

where q_1 and q_2 are the numbers of units produced at plants 1 and 2, respectively. How should the output be distributed in order to minimize costs? (You may assume that the critical point obtained does correspond to the minimum cost.)

14. Production Allocation Repeat Problem 13 if the cost function is

$$c = 3q_1^2 + q_1q_2 + 2q_2^2$$

and a total of 200 units are to be produced.

***15. Maximizing Output** The production function for a firm is

$$f(l, k) = 12l + 20k - l^2 - 2k^2$$

The cost to the firm of l and k is 4 and 8 per unit, respectively. If the firm wants the total cost of input to be 88, find the greatest output possible, subject to this budget constraint. (You may assume that the critical point obtained does correspond to the maximum output.)

16. Maximizing Output Repeat Problem 15, given that

$$f(l, k) = 60l + 30k - 2l^2 - 3k^2$$

and the budget constraint is $2l + 3k = 30$.

17. Advertising Budget A computer company has a monthly advertising budget of \$60,000. Its marketing department estimates that if x dollars are spent each month on advertising in newspapers and y dollars per month on television advertising, then the monthly sales will be given by $S = 90x^{1/4}y^{3/4}$ dollars. If the profit is 10% of sales, less the advertising cost, determine how to allocate the advertising budget in order to maximize the monthly profit. (You may assume that the critical point obtained does correspond to the maximum profit.)

18. Maximizing Production When l units of labor and k units of capital are invested, a manufacturer's total production q is given by the Cobb-Douglas production function $q = 6l^{2/5}k^{3/5}$. Each unit of labor costs \$25 and each unit of capital costs \$69. If exactly \$25,875 is to be spent on production, determine the numbers of units of labor and capital that should be invested to maximize production. (You may assume that the maximum occurs at the critical point obtained.)

19. Political Advertising Newspaper advertisements for political parties always have some negative effects. The recently elected party assumed that the three most important election issues X, Y, and Z, had to be mentioned in each ad, with space x, y, and z units, respectively, allotted to each. The combined bad effect of this coverage was estimated by the party's backroom boys as

$$B(x, y, z) = x^2 + y^2 + 2z^2$$

Aesthetics dictated that the total space for X and Y together must be 20, and realism suggested that the total space allotted to Y and Z together must also be 20 units. What values of x, y, and z in each ad would produce the lowest negative effect? (You may assume that any critical point obtained provides the minimum effect.)

20. Maximizing Profit Suppose a manufacturer's production function is given by

$$16q = 65 - 4(l - 4)^2 - 2(k - 5)^2$$

and the cost to the manufacturer is \$8 per unit of labor and \$16 per unit of capital, so that the total cost (in dollars) is $8l + 16k$. The selling price of the product is \$64 per unit.

(a) Express the profit as a function of l and k. Give your answer in expanded form.

(b) Find all critical points of the profit function obtained in part (a). Apply the second-derivative test at each critical point. If the profit is a relative maximum at a critical point, compute the corresponding relative maximum profit.

(c) The profit may be considered a function of l, k, and q (that is, $P = 64q - 8l - 16k$), subject to the constraint

$$16q = 65 - 4(l - 4)^2 - 2(k - 5)^2$$

Use the method of Lagrange multipliers to find all critical points of $P = 64q - 8l - 16k$, subject to the constraint.

Problems 21–24 refer to the following definition. A utility function is a function that attaches a measure to the satisfaction or utility a consumer gets from the consumption of products per unit of time. Suppose $U = f(x, y)$ is such a function, where x and y are the amounts of two products, X and Y. The marginal utility of X is $\partial U/\partial x$ and approximately represents the change in total utility resulting from a one-unit change in consumption of product X per unit of time. We define the marginal utility of Y in similar fashion. If the prices of X and Y are p_X and p_Y, respectively, and the consumer has an income or budget of I to spend, then the budget constraint is

$$xp_X + yp_Y = I$$

In Problems 21–23, find the quantities of each product that the consumer should buy, subject to the budget, that will allow maximum satisfaction. That is, in Problems 21 and 22, find values of x and y that maximize $U = f(x, y)$, subject to $xp_X + yp_Y = I$. Perform a similar procedure for Problem 23. Assume that such a maximum exists.

21. $U = x^3y^3$; $p_x = 2,\ p_y = 3,\ I = 48\ (x^3y^3 \neq 0)$

22. $U = 40x - 5x^2 + 4y - 2y^2$; $p_x = 2,\ p_y = 3,\ I = 10$

23. $U = f(x, y, z) = xyz$; $p_x = 2$, $p_y = 1,\ p_z = 4,\ I = 60\ (xyz \neq 0)$

24. Let $U = f(x, y)$ be a utility function subject to the budget constraint $xp_x + yp_y = I$, where p_x, p_y, and I are constants. Show that, to maximize satisfaction, it is necessary that

$$\lambda = \frac{f_x(x, y)}{p_x} = \frac{f_y(x, y)}{p_y}$$

where $f_x(x, y)$ and $f_y(x, y)$ are the marginal utilities of X and Y, respectively. Show that $f_x(x, y)/p_x$ is the marginal utility of one dollar's worth of X. Hence, maximum satisfaction is obtained when the consumer allocates the budget so that the marginal utility of a dollar's worth of X is equal to the marginal utility per dollar's worth of Y. Performing the same procedure as that for $U = f(x, y)$, verify that this is true for $U = f(x, y, z, w)$, subject to the corresponding budget equation. In each case, λ is called the *marginal utility of income.*

OBJECTIVE *To develop the method of least squares and introduce index numbers.*

17.9 Lines of Regression[24]

To study the influence of advertising on sales, a firm compiled the data in Table 17.4. The variable x denotes advertising expenditures in hundreds of dollars, and the variable y denotes the resulting sales revenue in thousands of dollars. If each pair (x, y) of data is plotted, the result is called a **scatter diagram** [Figure 17.16(a)].

TABLE 17.4

Expenditures, x	2	3	4.5	5.5	7
Revenue, y	3	6	8	10	11

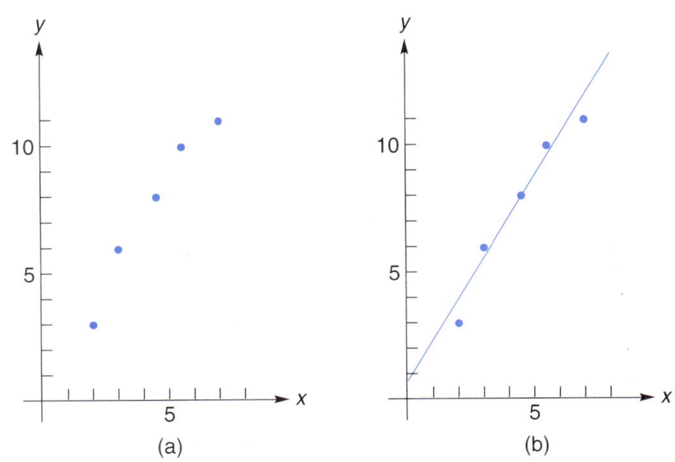

FIGURE 17.16 Scatter diagram and straight-line approximation to data points.

From an observation of the distribution of the points, it is reasonable to assume that a relationship exists between x and y and that it is approximately linear. On this basis, we may fit "by eye" a straight line that approximates the given data [Figure 17.16(b)] and, from this line, predict a value of y for a given value of x. The line seems consistent with the trend of the data, although other lines could be drawn as well. Unfortunately, determining a line "by eye" is

[24]This section may be omitted without loss of continuity.

not very objective. We want to apply criteria in specifying what we shall call a line of "best fit." A frequently used technique is called the **method of least squares.**

To apply the method of least squares to the data in Table 17.4, we first assume that x and y are approximately linearly related and that we can fit a straight line

$$\hat{y} = \hat{a} + \hat{b}x \tag{1}$$

that approximates the given points by a suitable objective choice of the constants $\hat{a}$ and $\hat{b}$ (read "a hat" and "b hat," respectively). For a given value of x in Equation (1), $\hat{y}$ is the corresponding predicted value of y, and $(x, \hat{y})$ will be on the line. Our aim is that $\hat{y}$ be near y.

When $x = 2$, the observed value of y is 3. Our predicted value of y is obtained by substituting $x = 2$ in Equation (1), which yields $\hat{y} = \hat{a} + 2\hat{b}$. The error of estimation, or vertical deviation of the point (2, 3) from the line, is $\hat{y} - y$, or

$$\hat{a} + 2\hat{b} - 3$$

This vertical deviation is indicated (although exaggerated for clarity) in Figure 17.17. Similarly, the vertical deviation of (3, 6) from the line is $\hat{a} + 3\hat{b} - 6$, as is also illustrated. To avoid possible difficulties associated with positive and negative deviations, we shall consider the squares of the deviations and shall form the sum S of all such squares for the given data:

$$S = (\hat{a} + 2\hat{b} - 3)^2 + (\hat{a} + 3\hat{b} - 6)^2 + (\hat{a} + 4.5\hat{b} - 8)^2 + (\hat{a} + 5.5\hat{b} - 10)^2 + (\hat{a} + 7\hat{b} - 11)^2$$

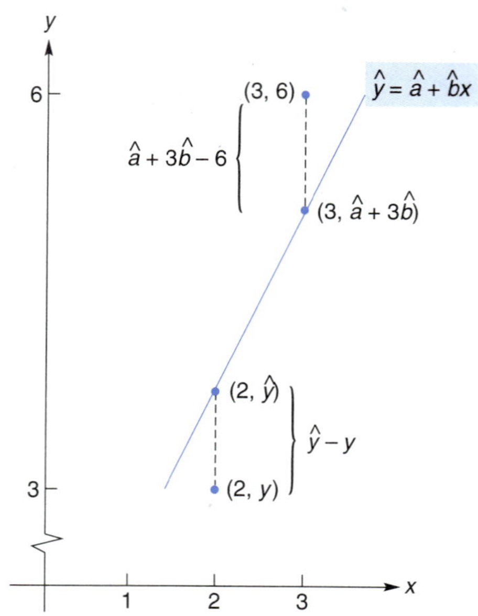

FIGURE 17.17 Vertical deviation of data points from straight-line approximation.

The method of least squares requires that we choose as the line of "best fit" the one obtained by selecting $\hat{a}$ and $\hat{b}$ so as to minimize S. We can minimize S with respect to $\hat{a}$ and $\hat{b}$ by solving the system

$$\begin{cases} \dfrac{\partial S}{\partial \hat{a}} = 0 \\ \dfrac{\partial S}{\partial \hat{b}} = 0 \end{cases}$$

We have

$$\begin{aligned} \frac{\partial S}{\partial \hat{a}} &= 2(\hat{a} + 2\hat{b} - 3) + 2(\hat{a} + 3\hat{b} - 6) + 2(\hat{a} + 4.5\hat{b} - 8) \\ &\quad + 2(\hat{a} + 5.5\hat{b} - 10) + 2(\hat{a} + 7\hat{b} - 11) = 0 \\ \frac{\partial S}{\partial \hat{b}} &= 4(\hat{a} + 2\hat{b} - 3) + 6(\hat{a} + 3\hat{b} - 6) + 9(\hat{a} + 4.5\hat{b} - 8) \\ &\quad + 11(\hat{a} + 5.5\hat{b} - 10) + 14(\hat{a} + 7\hat{b} - 11) = 0 \end{aligned}$$

which, when simplified, give

$$\begin{cases} 5\hat{a} + 22\hat{b} = 38 \\ 44\hat{a} + 225\hat{b} = 384 \end{cases}$$

Solving for $\hat{a}$ and $\hat{b}$, we obtain

$$\hat{a} = \frac{102}{157} \approx 0.65 \qquad \hat{b} = \frac{248}{157} \approx 1.58$$

It can be shown that these values of $\hat{a}$ and $\hat{b}$ lead to a minimum value of S. Hence, in the sense of least squares, the line of best fit $\hat{y} = \hat{a} + \hat{b}x$ is

$$\hat{y} = 0.65 + 1.58x \tag{2}$$

This is, in fact, the line indicated in Figure 17.16(b). It is called the **least squares line of y on x** or the **linear regression line of y on x.** The constants $\hat{a}$ and $\hat{b}$ are called **linear regression coefficients.** With Equation (2), we would predict that when $x = 5$, the corresponding value of y is $\hat{y} = 0.65 + 1.58(5) = 8.55$.

More generally, suppose we are given the following n pairs of observations:

$$(x_1, y_1), (x_2, y_2), \ldots, (x_n, y_n)$$

If we assume that x and y are approximately linearly related and that we can fit a straight line

$$\hat{y} = \hat{a} + \hat{b}x$$

that approximates the data, the sum of the squares of the errors $\hat{y} - y$ is

$$S = (\hat{a} + \hat{b}x_1 - y_1)^2 + (\hat{a} + \hat{b}x_2 - y_2)^2 + \cdots + (\hat{a} + \hat{b}x_n - y_n)^2$$

Since S must be minimized with respect to $\hat{a}$ and $\hat{b}$,

$$\begin{cases} \dfrac{\partial S}{\partial \hat{a}} = 2(\hat{a} + \hat{b}x_1 - y_1) + 2(\hat{a} + \hat{b}x_2 - y_2) + \cdots + 2(\hat{a} + \hat{b}x_n - y_n) = 0 \\ \dfrac{\partial S}{\partial \hat{b}} = 2x_1(\hat{a} + \hat{b}x_1 - y_1) + 2x_2(\hat{a} + \hat{b}x_2 - y_2) + \cdots + 2x_n(\hat{a} + \hat{b}x_n - y_n) = 0 \end{cases}$$

Dividing both equations by 2 and using sigma notation, we have

$$\begin{cases} \hat{a}n + \hat{b}\sum_{i=1}^{n} x_i - \sum_{i=1}^{n} y_i = 0 \\ \hat{a}\sum_{i=1}^{n} x_i + \hat{b}\sum_{i=1}^{n} x_i^2 - \sum_{i=1}^{n} x_i y_i = 0 \end{cases}$$

Equivalently, we have the system of so-called *normal equations:*

$$\begin{cases} \sum_{i=1}^{n} y_i = \hat{a}n + \hat{b}\sum_{i=1}^{n} x_i & (3) \\ \sum_{i=1}^{n} x_i y_i = \hat{a}\sum_{i=1}^{n} x_i + \hat{b}\sum_{i=1}^{n} x_i^2 & (4) \end{cases}$$

To solve for $\hat{b}$, we first multiply Equation (3) by $\sum_{i=1}^{n} x_i$ and Equation (4) by n:

$$\begin{cases} \left(\sum_{i=1}^{n} x_i\right)\left(\sum_{i=1}^{n} y_i\right) = \hat{a}n\sum_{i=1}^{n} x_i + \hat{b}\left(\sum_{i=1}^{n} x_i\right)^2 & (5) \\ n\sum_{i=1}^{n} x_i y_i = \hat{a}n\sum_{i=1}^{n} x_i + \hat{b}n\sum_{i=1}^{n} x_i^2 & (6) \end{cases}$$

Subtracting Equation (5) from Equation (6), we obtain

$$\begin{aligned} n\sum_{i=1}^{n} x_i y_i - \left(\sum_{i=1}^{n} x_i\right)\left(\sum_{i=1}^{n} y_i\right) &= \hat{b}n\sum_{i=1}^{n} x_i^2 - \hat{b}\left(\sum_{i=1}^{n} x_i\right)^2 \\ &= \hat{b}\left[n\sum_{i=1}^{n} x_i^2 - \left(\sum_{i=1}^{n} x_i\right)^2\right] \end{aligned}$$

Thus,

$$\hat{b} = \frac{n\sum_{i=1}^{n} x_i y_i - \left(\sum_{i=1}^{n} x_i\right)\left(\sum_{i=1}^{n} y_i\right)}{n\sum_{i=1}^{n} x_i^2 - \left(\sum_{i=1}^{n} x_i\right)^2} \tag{7}$$

Solving Equations (3) and (4) for $\hat{a}$ gives

$$\hat{a} = \frac{\left(\sum_{i=1}^{n} x_i^2\right)\left(\sum_{i=1}^{n} y_i\right) - \left(\sum_{i=1}^{n} x_i\right)\left(\sum_{i=1}^{n} x_i y_i\right)}{n\sum_{i=1}^{n} x_i^2 - \left(\sum_{i=1}^{n} x_i\right)^2} \tag{8}$$

It can be shown that these values of $\hat{a}$ and $\hat{b}$ minimize S.

Computing the linear regression coefficients $\hat{a}$ and $\hat{b}$ by the formulas of Equations (7) and (8) gives the linear regression line of y on x, namely, $\hat{y} = \hat{a} + \hat{b}x$, which can be used to estimate y for a given value of x.

In the next example, as well as in the exercises, you will encounter **index numbers.** They are used to relate a variable in one period of time to the same variable in another period, called the *base period*. An index number is a *relative* number that describes data that are changing over time. Such data are referred to as *time series*.

For example, consider the time-series data of total production of widgets in the United States for 1993–97, indicated in Table 17.5. If we choose 1994 as the base year and assign the index number 100 to it, then the other index numbers are obtained by dividing each year's production by the 1994 production of 900 and multiplying the result by 100. We can, for example, interpret the index 106 for 1997 as meaning that production for that year was 106% of the production in 1994.

TABLE 17.5

Year	Production (in thousands)	Index [1994 = 100]
1993	828	92
1994	900	100
1995	936	104
1996	891	99
1997	954	106

In time-series analysis, index numbers are obviously of great advantage if the data involve numbers of great magnitude. But regardless of the magnitude of the data, index numbers simplify the task of comparing changes in data over periods of time.

EXAMPLE 1 **Determining a Linear-Regression Line**

By means of the linear-regression line, use the following table to represent the trend for the index of total U.S. government revenue from 1995 *to* 2000 (1995 = 100).

Year	1995	1996	1997	1998	1999	2000
Index	100	107	117	127	135	150

Source: Economic Report of the President, 2001, U.S. Government Printing Office, Washington, DC, 2001.

Solution: We shall let x denote time and y denote the index and treat y as a linear function of x. Also, we shall designate 1995 by $x = 1$, 1996 by $x = 2$, and so on. There are $n=6$ pairs of measurements. To determine the linear-regression coefficients by using Equations (7) and (8), we first perform the arithmetic:

Year	x_i	y_i	$x_i y_i$	x_i^2
1995	1	100	100	1
1996	2	107	214	4
1997	3	117	351	9
1998	4	127	508	16
1999	5	135	675	25
2000	6	150	900	36
Total	21	736	2748	91
	$=\sum_{i=1}^{6} x_i$	$=\sum_{i=1}^{6} y_i$	$=\sum_{i=1}^{6} x_i y_i$	$=\sum_{i=1}^{6} x_i^2$

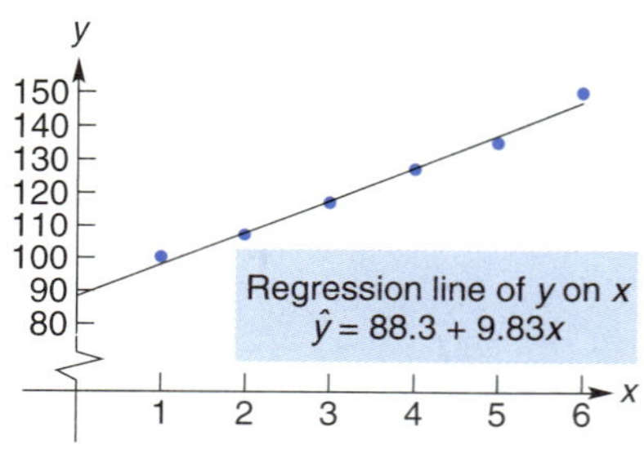

FIGURE 17.18 Linear-regression line for budget deficit.

Hence, by Equation (8),

$$\hat{a} = \frac{91(736) - 21(2748)}{6(91) - (21)^2} \approx 88.3$$

and by Equation (7),

$$\hat{b} = \frac{6(2748) - 21(736)}{6(91) - (21)^2} \approx 9.83$$

Thus, the regression line of y on x is

$$\hat{y} = 88.3 + 9.83x$$

whose graph, as well as a scatter diagram, appears in Figure 17.18.

▶ **Now work Problem 1.**

TECHNOLOGY

The TI-83 Plus has a utility that computes the equation of the least squares line for a set of data. We shall illustrate by giving the procedure for the six data points (x_i, y_i) of Example 1. After pressing STAT and ENTER, we enter all the x-values and then the y-values. (See Figure 17.19.) Next, we press STAT and move to CALC. Finally, pressing 8 and ENTER gives the result shown in Figure 17.20. (The number $r \approx 0.99448$ is called the *coefficient of correlation* and is a measure of the degree to which the given data are linearly related.)

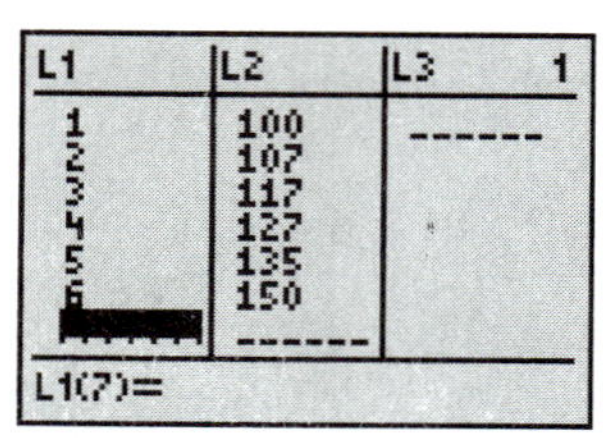

FIGURE 17.19 Data of Example 1.

FIGURE 17.20 Equation of least squares line.

Exercise 17.9

In this exercise set, use a graphing calculator if your instructor permits you to do so.

In Problems 1–4, find an equation of the least squares linear-regression line of y on x for the given data, and sketch both the line and the data. Predict the value of y corresponding to $x = 3.5$.

*1.

x	1	2	3	4	5	6
y	1.5	2.3	2.6	3.7	4.0	4.5

2.

x	1	2	3	4	5	6	7
y	1	1.8	2	4	4.5	7	9

3.

x	2	3	4.5	5.5	7
y	3	5	8	10	11

4.

x	2	3	4	5	6	7
y	2.4	2.9	3.3	3.8	4.3	4.9

5. **Demand** A firm finds that when the price of its product is p dollars per unit, the number of units sold is q, as indicated in the following table:

Price, p	10	30	40	50	60	70
Demand, q	70	68	63	50	46	32

Find an equation of the regression line of q on p.

6. **Water and Crop Yield** On a farm, an agronomist finds that the amount of water applied (in inches) and the corresponding yield of a certain crop (in tons per acre) are as given in the following table:

Water, x	8	16	24	32
Yield, y	5.2	5.7	6.3	6.7

Find an equation of the regression line of y on x. Predict y when $x = 20$.

7. **Virus** A rabbit was injected with a virus, and x hours after the injection the temperature y (in degrees Fahrenheit) of the rabbit was measured.[25] The data are given in the following table:

Elapsed Time, x	24	32	48	56
Temperature, y	102.8	104.5	106.5	107.0

Find an equation of the regression line of y on x, and estimate the rabbit's temperature 40 hours after the injection.

[25] R. R. Sokal and F. J. Rohlf, *Introduction to Biostatistics* (San Francisco: W. H. Freeman & Company, Publishers, 1973).

8. **Psychology** In a psychological experiment, four persons were subjected to a stimulus. Both before and after the stimulus, the systolic blood pressure (in millimeters of mercury) of each subject was measured. The data are given in the following table:

	Blood Pressure			
Before Stimulus, x	130	132	137	142
After Stimulus, y	138	139	144	148

Find an equation of the regression line of y on x, where x and y are as defined in the table.

For the time series in Problems 9 and 10, fit a linear-regression line by least squares; that is, find an equation of the linear-regression line of y on x. In each case, let the first year in the table correspond to $x = 1$.

9.

Production of Product A, 1993–1997 (in thousands of units)

Year	Production
1993	10
1994	15
1995	16
1996	18
1997	21

10. **Industrial Production** In the following table, let 1975 correspond to $x = 1$, 1977 correspond to $x = 3$, and so on:

Index of Industrial Production—Electrical Machinery (1997 = 100)

Year	Index
1975	77
1977	100
1979	126
1981	134

Source: Economic Report of the President, 1988, U.S. Government Printing Office, Washington, DC, 1988.

11. Computer Shipments

(a) Find an equation of the least squares line of y on x for the following data (refer to 1994 as year $x = 1$, and so on):

Overseas Shipments of Computers by Acme Computer Co. (in thousands)

Year	Quantity
1994	35
1995	31
1996	26
1997	24
1998	26

(b) For the data in part (a), refer to 1994 as year $x = -2$, 1995 as year $x = -1$, 1996 as year $x = 0$ and so on. Then $\sum_{i=1}^{5} x_i = 0$. Fit a least squares line and observe how the calculation is simplified.

12. Medical Care For the following time series, find an equation of the linear-regression line that best fits the data (refer to 1983 as year $x = -2$, 1984 as year $x = -1$, and so on):

Consumer Price Index—Medical Care, 1983–1987 (1967 = 100)

Year	Index
1983	357
1984	380
1985	403
1986	434
1987	462

Source: Economic Report of the President, 1988, U.S. Government Printing Office, Washington, DC, 1988.

OBJECTIVE *To develop some properties of homogeneous functions, including Euler's theorem.*

17.10 A Comment on Homogeneous Functions[26]

Many of the functions that are useful in economic analysis share the property of being homogeneous.

Definition

A function $z = f(x, y)$ is said to be ***homogeneous of degree n*** (n being a constant) if, for ***all*** positive real values of λ,

$$f(\lambda x, \lambda y) = \lambda^n f(x, y)$$

In words, if both x and y are multiplied by the same positive real number, then the resulting function value is a power of the number times the function value $f(x, y)$. For example, if

$$f(x, y) = x^3 - 2xy^2$$

then

$$f(\lambda x, \lambda y) = (\lambda x)^3 - 2(\lambda x)(\lambda y)^2 = \lambda^3 x^3 - 2\lambda^3 xy^2$$
$$= \lambda^3(x^3 - 2xy^2) = \lambda^3 f(x, y)$$

Thus, f is homogeneous of degree three.

[26]This section contains material from Section 17.6 and may be omitted without loss of continuity.

An important homogeneous function in economics is the Cobb-Douglas production function:

$$P = f(l, k) = Al^{\alpha}k^{1-\alpha} \qquad (\alpha \text{ and } A \text{ constants})$$

We have

$$\begin{aligned} f(\lambda l, \lambda k) &= A(\lambda l)^{\alpha}(\lambda k)^{1-\alpha} = A\lambda^{\alpha}l^{\alpha}\lambda^{1-\alpha}k^{1-\alpha} \\ &= \lambda Al^{\alpha}k^{1-\alpha} = \lambda f(l, k) \end{aligned}$$

Therefore, f is homogeneous of degree one. For example, $f(l, k) = 2l^{0.3}k^{0.7}$ is a homogeneous function of degree one.

Homogeneous production functions of degree one have an interesting property. If f is such a function, then

$$f(\lambda l, \lambda k) = \lambda f(l, k)$$

For example, if all inputs are doubled, then

$$f(2l, 2k) = 2f(l, k)$$

and output is doubled. Similarly, if all inputs are tripled, output is tripled, etc. In short, the same proportional change in each input factor of production results in the same proportional change in output.

By considering the partial derivatives of a homogeneous function, we can obtain an important result. Let $f(l, k)$ be a homogeneous production function of degree n. Then we have the identity

$$f(\lambda l, \lambda k) = \lambda^n f(l, k) \tag{1}$$

Consider the left side of Equation (1). If we set $r = \lambda l$ and $s = \lambda k$, then Equation (1) becomes

$$f(r, s) = \lambda^n f(l, k) \tag{2}$$

Now, for each side, we shall take the partial with respect to λ. For the left side, $f(r, s)$, we have, by the chain rule,

$$\begin{aligned} \frac{\partial}{\partial\lambda}[f(r, s)] &= \frac{\partial}{\partial r}[f(r, s)]\frac{\partial r}{\partial\lambda} + \frac{\partial}{\partial s}[f(r, s)]\frac{\partial s}{\partial\lambda} \\ &= \frac{\partial}{\partial r}[f(r, s)]l + \frac{\partial}{\partial s}[f(r, s)]k \end{aligned} \tag{3}$$

For the right side of Equation (2),

$$\frac{\partial}{\partial\lambda}[\lambda^n f(l, k)] = n\lambda^{n-1}f(l, k) \tag{4}$$

Using Equations (3) and (4), we set $\frac{\partial}{\partial\lambda}[f(r, s)]$ equal to $\frac{\partial}{\partial\lambda}[\lambda^n f(l, k)]$:

$$l\frac{\partial}{\partial r}[f(r, s)] + k\frac{\partial}{\partial s}[f(r, s)] = n\lambda^{n-1}f(l, k)$$

In particular, if $\lambda = 1$, then $r = l$ and $s = k$, so $f(r, s) = f(l, k)$. Thus, $\frac{\partial}{\partial r}[f(r, s)] = \frac{\partial}{\partial l}[f(l, k)]$ and $\frac{\partial}{\partial s}[f(r, s)] = \frac{\partial}{\partial k}[f(l, k)]$. Hence, we have what is called *Euler's theorem* for homogeneous functions:

$$l\frac{\partial}{\partial l}[f(l, k)] + k\frac{\partial}{\partial k}[f(l, k)] = nf(l, k) \tag{5}$$

Now, if f is homogeneous of degree one, such as the Cobb-Douglas function, then $n = 1$, and Equation (5) becomes

$$l\frac{\partial}{\partial l}[f(l,k)] + k\frac{\partial}{\partial k}[f(l,k)] = f(l,k)$$

We conclude that if we multiply the marginal productivity of each input by the quantity of the input, the sum is equal to the total product.

OBJECTIVE *To compute double and triple integrals.*

17.11 Multiple Integrals

Recall that the definite integral of a function of one variable is concerned with integration over an *interval*. There are also definite integrals of functions of two variables, called (definite) **double integrals.** These involve integration over a *region* in the plane.

For example, the symbol

$$\int_0^2 \int_3^4 xy\,dx\,dy \qquad \text{equivalently} \qquad \int_0^2 \left[\int_3^4 xy\,dx\right] dy$$

is the double integral of $f(x, y) = xy$ over a region determined by the limits of integration. That region consists of all points (x, y) in the x, y-plane such that $3 \le x \le 4$ and $0 \le y \le 2$. (See Figure 17.21.)

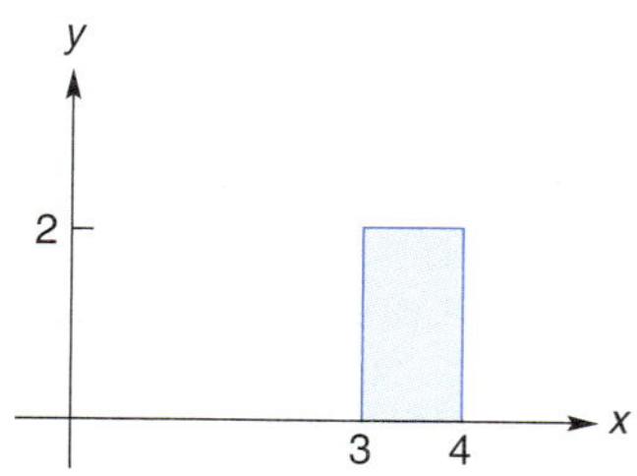

FIGURE 17.21 Region over which $\int_0^2 \int_3^4 xy\,dx\,dy$ is evaluated.

Essentially, a double integral is a limit of a sum of the form $\sum f(x, y)\Delta x \Delta y$, where, in this example, the points (x, y) are in the shaded region. A geometric interpretation of a double integral will be given later.

To evaluate

$$\int_0^2 \int_3^4 xy\,dx\,dy \qquad \text{equivalently} \qquad \int_0^2 \left[\int_3^4 xy\,dx\right] dy$$

we use successive integrations starting with the innermost integral. First, we evaluate

$$\int_3^4 xy\,dx$$

by treating y as a constant and integrating with respect to x between the limits 3 and 4:

$$\int_3^4 xy\,dx = \left.\frac{x^2 y}{2}\right|_3^4$$

Substituting the limits for the variable x, we have

$$\frac{4^2 \cdot y}{2} - \frac{3^2 \cdot y}{2} = \frac{16y}{2} - \frac{9y}{2} = \frac{7}{2}y$$

Now we integrate this result with respect to y between the limits 0 and 2:

$$\int_0^2 \frac{7}{2}y\,dy = \left.\frac{7y^2}{4}\right|_0^2 = \frac{7 \cdot 2^2}{4} - 0 = 7$$

Thus,

$$\int_0^2 \int_3^4 xy\,dx\,dy = 7$$

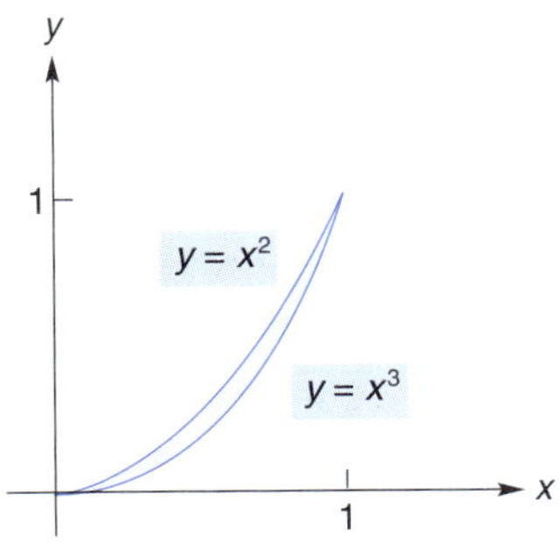

FIGURE 17.22 Region over which $\int_0^1 \int_{x^3}^{x^2} (x^3 - xy)\, dy\, dx$ is evaluated.

Now consider the double integral

$$\int_0^1 \int_{x^3}^{x^2} (x^3 - xy)\, dy\, dx = \int_0^1 \left[\int_{x^3}^{x^2} (x^3 - xy)\, dy \right] dx$$

Here we integrate first with respect to y and then with respect to x. The region over which the integration takes places is all points (x, y) such that $x^3 \le y \le x^2$ and $0 \le x \le 1$. (See Figure 17.22.) This double integral is evaluated by first treating x as a constant and integrating $x^3 - xy$ with respect to y between x^3 and x^2, and then integrating the result with respect to x between 0 and 1:

$$\begin{aligned}
&\int_0^1 \int_{x^3}^{x^2} (x^3 - xy)\, dy\, dx \\
&\quad = \int_0^1 \left[\int_{x^3}^{x^2} (x^3 - xy)\, dy \right] dx = \int_0^1 \left(x^3 y - \frac{xy^2}{2} \right) \Bigg|_{x^3}^{x^2} dx \\
&\quad = \int_0^1 \left(\left[x^3(x^2) - \frac{x(x^2)^2}{2} \right] - \left[x^3(x^3) - \frac{x(x^3)^2}{2} \right] \right) dx \\
&\quad = \int_0^1 \left(x^5 - \frac{x^5}{2} - x^6 + \frac{x^7}{2} \right) dx = \int_0^1 \left(\frac{x^5}{2} - x^6 + \frac{x^7}{2} \right) dx \\
&\quad = \left(\frac{x^6}{12} - \frac{x^7}{7} + \frac{x^8}{16} \right) \Bigg|_0^1 = \left(\frac{1}{12} - \frac{1}{7} + \frac{1}{16} \right) - 0 = \frac{1}{336}
\end{aligned}$$

EXAMPLE 1 **Evaluating a Double Integral**

Find $\int_{-1}^1 \int_0^{1-x} (2x + 1)\, dy\, dx.$

Solution: Here we first integrate with respect to y and then integrate the result with respect to x:

$$\begin{aligned}
&\int_{-1}^1 \int_0^{1-x} (2x + 1)\, dy\, dx \\
&\quad = \int_{-1}^1 \left[\int_0^{1-x} (2x + 1)\, dy \right] dx \\
&\quad = \int_{-1}^1 (2xy + y) \Bigg|_0^{1-x} dx = \int_{-1}^1 \{[2x(1 - x) + (1 - x)] - 0\}\, dx \\
&\quad = \int_{-1}^1 (-2x^2 + x + 1)\, dx = \left(-\frac{2x^3}{3} + \frac{x^2}{2} + x \right) \Bigg|_{-1}^1 \\
&\quad = \left(-\frac{2}{3} + \frac{1}{2} + 1 \right) - \left(\frac{2}{3} + \frac{1}{2} - 1 \right) = \frac{2}{3}
\end{aligned}$$

▶ **Now work Problem 9.**

EXAMPLE 2 **Evaluating a Double Integral**

Find $\int_1^{\ln 2} \int_{e^y}^2 dx\, dy$.

Solution: Here we first integrate with respect to x and then integrate the result with respect to y:

$$\int_1^{\ln 2} \int_{e^y}^2 dx\, dy = \int_1^{\ln 2} \left[\int_{e^y}^2 dx \right] dy = \int_1^{\ln 2} x \Big|_{e^y}^2 dy$$

$$= \int_1^{\ln 2} (2 - e^y)\, dy = (2y - e^y) \Big|_1^{\ln 2}$$

$$= (2 \ln 2 - 2) - (2 - e) = 2 \ln 2 - 4 + e$$

$$= \ln 4 - 4 + e$$

▶ **Now work Problem 13.**

A double integral can be interpreted in terms of the volume of a region between the x,y-plane and a surface $z = f(x, y)$ if $z \geq 0$. Figure 17.23 shows a region whose volume we shall consider. The element of volume for this region is a vertical column with height approximately $z = f(x, y)$ and base area $\Delta y \Delta x$. Thus, its volume is approximately $f(x, y)\Delta y \Delta x$. The volume of the entire region can be found by summing the volumes of all such elements for $a \leq x \leq b$ and $c \leq y \leq d$ via a double integral:

$$\text{volume} = \int_a^b \int_c^d f(x, y)\, dy\, dx$$

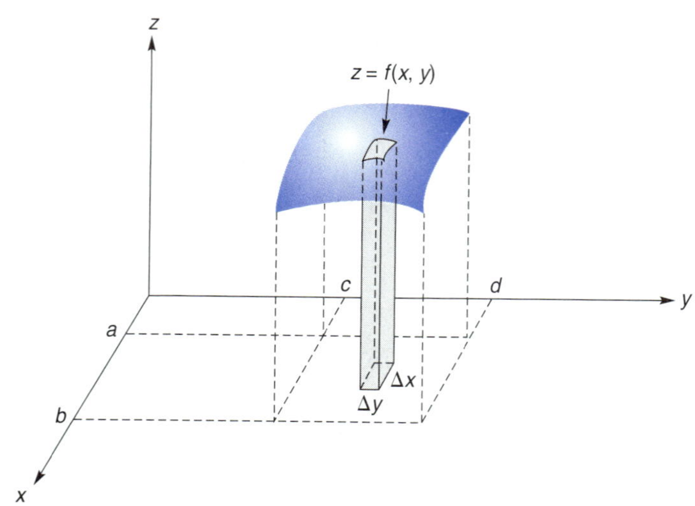

FIGURE 17.23 Interpreting $\int_a^b \int_c^d f(x, y)\, dy\, dx$ in terms of volume, where $f(x, y) \geq 0$.

Triple integrals are handled by successively evaluating three integrals, as the next example shows.

EXAMPLE 3 Evaluating a Triple Integral

Find $\int_0^1 \int_0^x \int_0^{x-y} x\,dz\,dy\,dx.$

Solution:

$$\int_0^1 \int_0^x \int_0^{x-y} x\,dz\,dy\,dx = \int_0^1 \int_0^x \left[\int_0^{x-y} x\,dz\right] dy\,dx$$

$$= \int_0^1 \int_0^x (xz)\Big|_0^{x-y} dy\,dx = \int_0^1 \int_0^x [x(x-y) - 0]\,dy\,dx$$

$$= \int_0^1 \int_0^x (x^2 - xy)\,dy\,dx = \int_0^1 \left[\int_0^x (x^2 - xy)\,dy\right] dx$$

$$= \int_0^1 \left(x^2 y - \frac{xy^2}{2}\right)\Big|_0^x dx = \int_0^1 \left[\left(x^3 - \frac{x^3}{2}\right) - 0\right] dx$$

$$= \int_0^1 \frac{x^3}{2} dx = \frac{x^4}{8}\Big|_0^1 = \frac{1}{8}$$

▸ **Now work Problem 21.**

Exercise 17.11

In Problems 1–22, evaluate the multiple integrals.

1. $\int_0^3 \int_0^4 x\,dy\,dx$

2. $\int_1^4 \int_0^3 y\,dy\,dx$

3. $\int_0^1 \int_0^1 xy\,dx\,dy$

4. $\int_0^2 \int_0^3 x^2\,dy\,dx$

5. $\int_1^3 \int_1^2 (x^2 - y)\,dx\,dy$

6. $\int_{-1}^2 \int_1^4 (x^2 - 2xy)\,dy\,dx$

7. $\int_0^1 \int_0^2 (x + y)\,dy\,dx$

8. $\int_0^3 \int_0^x (x^2 + y^2)\,dy\,dx$

***9.** $\int_1^4 \int_0^{5x} y\,dy\,dx$

10. $\int_1^2 \int_0^{x-1} 2y\,dy\,dx$

11. $\int_0^1 \int_{3x}^{x^2} 14x^2 y\,dy\,dx$

12. $\int_0^2 \int_0^{x^2} xy\,dy\,dx$

***13.** $\int_0^2 \int_0^{\sqrt{4-y^2}} x\,dx\,dy$

14. $\int_0^1 \int_{y^2}^y y\,dx\,dy$

15. $\int_{-1}^1 \int_x^{1-x} 3(x + y)\,dy\,dx$

16. $\int_0^3 \int_{y^2}^{3y} 5x\,dx\,dy$

17. $\int_0^1 \int_0^y e^{x+y}\,dx\,dy$

18. $\int_2^3 \int_0^2 e^{x-y}\,dx\,dy$

19. $\int_{-1}^0 \int_{-1}^2 \int_1^2 6xy^2z^3\,dx\,dy\,dz$

20. $\int_0^1 \int_0^x \int_0^{x-y} x\,dz\,dy\,dx$

***21.** $\int_0^1 \int_{x^2}^x \int_0^{xy} dz\,dy\,dx$

22. $\int_1^e \int_{\ln x}^x \int_0^y dz\,dy\,dx$

23. Statistics In the study of statistics, a joint density function $z = f(x, y)$ defined on a region in the x,y-plane is represented by a surface in space. The probability that

$$a \le x \le b \quad \text{and} \quad c \le y \le d$$

is given by

$$P(a \le x \le b, c \le y \le d) = \int_c^d \int_a^b f(x, y)\,dx\,dy$$

and is represented by the volume between the graph of f and the rectangular region given by

$$a \le x \le b \quad \text{and} \quad c \le y \le d$$

If $f(x, y) = e^{-(x+y)}$ is a joint density function, where $x \ge 0$ and $y \ge 0$, find

$$P(0 \le x \le 2, 1 \le y \le 2)$$

and give your answer in terms of e.

24. Statistics In Problem 23, let $f(x, y) = 12e^{-4x-3y}$ for $x, y \ge 0$. Find

$$P(3 \le x \le 4, 2 \le y \le 6)$$

and give your answer in terms of e.

25. Statistics In Problem 23, let $f(x, y) = x/8$, where $0 \le x \le 2$ and $0 \le y \le 4$. Find $P(x \ge 1, y \ge 2)$.

26. Statistics In Problem 23, let f be the uniform density function $f(x, y) = 1/8$ defined over the rectangle $0 \le x \le 4, 0 \le y \le 2$. Determine the probability that $0 \le x \le 1$ and $0 \le y \le 1$.

17.12 Review

Important Terms and Symbols

Section 17.1	three-dimensional coordinate system $f(x_1, x_2, \ldots, x_n)$ function of n variables x,y-plane x,z-plane y,z-plane octant traces
Section 17.2	partial derivative $\frac{\partial z}{\partial x}$ $f_x(x, y)$ $\left.\frac{\partial z}{\partial x}\right\vert_{(x_0, y_0)}$ $f_x(x_0, y_0)$
Section 17.3	joint-cost function production function marginal productivity competitive products complementary products
Section 17.4	implicit partial differentiation
Section 17.5	$\frac{\partial^2 z}{\partial y \partial x}$ $\frac{\partial^2 z}{\partial x \partial y}$ $\frac{\partial^2 z}{\partial x^2}$ $\frac{\partial^2 z}{\partial y^2}$ f_{xy} f_{yx} f_{xx} f_{yy}
Section 17.6	chain rule intermediate variable
Section 17.7	relative maxima and minima critical point second-derivative test for functions of two variables
Section 17.8	Lagrange multipliers
Section 17.9	scatter diagram method of least squares linear regression line of y on x index numbers
Section 17.10	homogeneous function of degree n
Section 17.11	double integral triple integral

Summary

We can extend the concept of a function of one variable to functions of several variables. The inputs for functions of n variables are n-tuples. Generally, the graph of a function of two variables is a surface in a three-dimensional coordinate system. Functions of more than two variables cannot be represented geometrically.

For a function of n variables, we can consider n partial derivatives. For example, if $w = f(x, y, z)$, we have the partial derivatives of f with respect to x, with respect to y, and with respect to z, denoted f_x, f_y, and f_z, or $\partial f/\partial x$, $\partial f/\partial y$, and $\partial f/\partial z$, respectively. To find $f_x(x, y, z)$, we treat y and z as constants and differentiate f with respect to x in the usual way. The other partial derivatives are found in a similar fashion. We can interpret $f_x(x, y, z)$ as the approximate change in w that results from a one-unit change in x when y and z are held fixed. There are similar interpretations for the other partial derivatives. A function of several variables may be defined implicitly. In this case, its partial derivatives are found by implicit partial differentiation.

Functions of several variables occur frequently in business and economic analysis, as well as in other areas of study. If a manufacturer produces x units of product X and y units of product Y, then the total cost c of these units is a function of x and y and is called a joint-cost function. The partial derivatives $\partial c/\partial x$ and $\partial c/\partial y$ are called the marginal costs with respect to x and y, respectively. We can interpret, for example, $\partial c/\partial x$ as the approximate cost of producing an extra unit of X while the level of production of Y is held fixed.

If l units of labor and k units of capital are used to produce P units of a product, then the function $P = f(l, k)$ is called a production function. The partial derivatives of P are called marginal productivity functions.

Suppose two products, A and B, are such that the quantity demanded of each is dependent on the prices of both. If q_A and q_B are the quantities of A and B demanded when the prices of A and B are p_A and p_B, respectively, then q_A and q_B are each functions of p_A and p_B. When $\partial q_A/\partial p_B > 0$ and $\partial q_B/\partial p_A > 0$, then A and B are called competitive products (or substitutes). When $\partial q_A/\partial p_B < 0$ and $\partial q_B/\partial p_A < 0$, then A and B are called complementary products.

If $z = f(x, y)$, where $x = x(r, s)$ and $y = y(r, s)$, then z can be considered as a function of r and s. To find, for example, $\partial z/\partial r$, a chain rule may be used:

$$\frac{\partial z}{\partial r} = \frac{\partial z}{\partial x}\frac{\partial x}{\partial r} + \frac{\partial z}{\partial y}\frac{\partial y}{\partial r}$$

A partial derivative of a function of n variables is itself a function of n variables. By successively taking partial derivatives of partial derivatives, we obtain higher-order partial derivatives. For example, if f is a function of x and y, then f_{xy} denotes the partial derivative of f_x with respect to y; f_{xy} is called the second-partial derivative of f, first with respect to x and then with respect to y.

If the function $f(x, y)$ has a relative extremum at (x_0, y_0), then (x_0, y_0) must be a solution of the system

$$f_x(x, y) = 0 \quad f_y(x, y) = 0$$

Any solution of this system is called a critical point of f. Thus, critical points are the candidates at which a relative extremum may occur. The second-derivative test for functions of two variables gives conditions under which a critical point corresponds to a relative maximum or a relative minimum.

The test states that if (x_0, y_0) is a critical point of f and

$$D(x, y) = f_{xx}(x, y)f_{yy}(x, y) - [f_{xy}(x, y)]^2$$

then

1. if $D(x_0, y_0) > 0$ and $f_{xx}(x_0, y_0) < 0$, f has a relative maximum at (x_0, y_0);
2. if $D(x_0, y_0) > 0$ and $f_{xx}(x_0, y_0) > 0$, f has a relative minimum at (x_0, y_0);
3. if $D(x_0, y_0) < 0$, f has neither a relative maximum nor a relative minimum at (x_0, y_0);
4. if $D(x_0, y_0) = 0$, no conclusion about an extremum at (x_0, y_0) can be drawn, and further analysis is required.

To find critical points of a function of several variables, subject to a constraint, we may use the method of Lagrange multipliers. For example, to find the critical points of $f(x, y, z)$, subject to the constraint $g(x, y, z) = 0$, we first form the function

$$F(x, y, z, \lambda) = f(x, y, z) - \lambda g(x, y, z)$$

By solving the system

$$F_x = 0 \quad F_y = 0 \quad F_z = 0 \quad F_\lambda = 0$$

we obtain the critical points of F. If $(x_0, y_0, z_0, \lambda_0)$ is such a critical point, then (x_0, y_0, z_0) is a critical point of f, subject to the constraint. It is important to write the constraint in the form $g(x, y, z) = 0$. For example, if the constraint is $2x + 3y - z = 4$, then $g(x, y, z) = 2x + 3y - z - 4$ [or $g(x, y, z) = 4 - 2x - 3y + z$]. If $f(x, y, z)$ is subject to two constraints, $g_1(x, y, z) = 0$ and $g_2(x, y, z) = 0$, then we would form the function $F = f - \lambda_1 g_1 - \lambda_2 g_2$ and solve the system

$$F_x = 0 \quad F_y = 0 \quad F_z = 0 \quad F_{\lambda_1} = 0 \quad F_{\lambda_2} = 0$$

Sometimes two variables, say, x and y, may be related in such a way that the relationship is approximately linear. When the data points (x_i, y_i), where $i = 1, 2, 3, \ldots, n$, are given to us, we can fit a straight line that approximates them. Such a line is the linear-regression line (or least squares line) of y on x and is given by

$$\hat{y} = \hat{a} + \hat{b}x$$

where

$$\hat{a} = \frac{\left(\sum_{i=1}^{n} x_i^2\right)\left(\sum_{i=1}^{n} y_i\right) - \left(\sum_{i=1}^{n} x_i\right)\left(\sum_{i=1}^{n} x_i y_i\right)}{n\sum_{i=1}^{n} x_i^2 - \left(\sum_{i=1}^{n} x_i\right)^2}$$

and

$$\hat{b} = \frac{n\sum_{i=1}^{n} x_i y_i - \left(\sum_{i=1}^{n} x_i\right)\left(\sum_{i=1}^{n} y_i\right)}{n\sum_{i=1}^{n} x_i^2 - \left(\sum_{i=1}^{n} x_i\right)^2}$$

The $\hat{y}$-values can be used to predict y-values for given values of x.

When working with functions of several variables, we can consider their multiple integrals. These are determined by successive integration. For example, the double integral

$$\int_1^2 \int_0^y (x + y)\, dx\, dy$$

is determined by first treating y as a constant and integrating $x + y$ with respect to x. After evaluating between the limits 0 and y, we integrate that result with respect to y from $y = 1$ to $y = 2$. Thus,

$$\int_1^2 \int_0^y (x + y)\, dx\, dy = \int_1^2 \left[\int_0^y (x + y)\, dx\right] dy$$

Triple integrals involve functions of three variables and are also evaluated by successive integration.

Review Problems

Problem numbers shown in color indicate problems suggested for use as a practice chapter test.

In Problems 1–4, sketch the given surfaces.

1. $2x + 3y + z = 9$

2. $z = x$

3. $z = y^2$

4. $x^2 + z^2 = 1$

In Problems 5–16, find the indicated partial derivatives.

5. $f(x, y) = 4x^2 + 6xy + y^2 - 1$; $\quad f_x(x, y), f_y(x, y)$

6. $P = l^3 + k^3 - lk$; $\quad \partial P/\partial l, \partial P/\partial k$

7. $z = \dfrac{x}{x + y}$; $\quad \dfrac{\partial z}{\partial x}, \dfrac{\partial z}{\partial y}$

8. $f(p_A, p_B) = 2(p_A - 20) + 3(p_B - 30)$; $\quad f_{p_A}(p_A, p_B)$

9. $f(x, y) = \ln\sqrt{x^2 + y^2}$; $\quad \dfrac{\partial}{\partial y}[f(x, y)]$

10. $w = \dfrac{x}{\sqrt{x^2 + y^2}}$; $\quad \dfrac{\partial w}{\partial y}$

11. $w = e^{x^2yz}$; $\quad w_{xy}(x, y, z)$

12. $f(x, y) = xy\ln(xy)$; $\quad f_{xy}(x, y)$

13. $f(x, y, z) = (x + y)(y + z^2)$; $\frac{\partial^2}{\partial z^2}[f(x, y, z)]$

14. $z = (x^2 - y)(y^2 - 2xy)$; $\partial^2 z/\partial y^2$

15. $w = e^{x+y+z} \ln xyz$; $\partial w/\partial y, \partial^2 w/\partial z\,\partial x$

16. $P = 100l^{0.11}k^{0.89}$; $\partial^2 P/\partial k \partial l$

17. If $f(x, y, z) = \frac{x + y}{xz}$, find $f_{xyz}(2, 7, 4)$.

18. If $f(x, y, z) = (6x + 1)e^{y^2 \ln(z+1)}$, find $f_{xyz}(0, 1, 0)$.

[27]19. If $w = x^2 + 2xy + 3y^2$, $x = e^r$, and $y = \ln(r + s)$, find $\partial w/\partial r$ and $\partial w/\partial s$.

[27]20. If $z = \ln(x/y) + e^y - xy$, $x = r^2s^2$, and $y = r + s$, find $\partial z/\partial s$.

[27]21. If $x^2 + 2xy - 2z^2 + xz + 2 = 0$, find $\partial z/\partial x$.

[27]22. If $z^2 + \ln(yz) + \ln z + x + z = 0$, find $\partial z/\partial y$.

23. **Production Function** If a manufacturer's production function is defined by $P = 20l^{0.7}k^{0.3}$, determine the marginal productivity functions.

24. **Joint-Cost Function** A manufacturer's cost for producing x units of product X and y units of product Y is given by

$$c = 3x + 0.05xy + 9y + 500$$

determine the (partial) marginal cost with respect to x when $x = 50$ and $y = 100$.

25. **Competitive/Complementary Products** If $q_A = 200 - 3p_A + p_B$ and $q_B = 50 - 5p_B + p_A$, where q_A and q_B are the number of units demanded of products A and B, respectively, and p_A and p_B are their respective prices per unit, determine whether A and B are competitive or complementary products.

26. **Innovation** For industry, the following model describes the rate α (a Greek letter read "alpha") at which an innovation substitutes for an established process:[28]

$$\alpha = Z + 0.530P - 0.027S$$

Here, Z is a constant that depends on the particular industry, P is an index of profitability of the innovation, and S is an index of the extent of the investment necessary to make use of the innovation. Find $\partial\alpha/\partial P$ and $\partial\alpha/\partial S$.

27. Examine $f(x, y) = x^2 + 2y^2 - 2xy - 4y + 3$ for relative extrema.

28. Examine $f(w, z) = 2w^3 + 2z^3 - 6wz + 7$ for relative extrema.

[27]Problems 19–22 refer to Section 17.4 or Section 17.6.

[28]A. P. Hurter, Jr., A. H. Rubenstein, et al., "Market Penetration by New Innovations: The Technological Literature," *Technological Forecasting and Social Change* 11 (1978), 197–221.

29. **Minimizing Material** An open-top rectangular cardboard box is to have a volume of 32 cubic feet. Find the dimensions of the box so that the amount of cardboard used is minimized.

30. The function

$$f(x, y) = ax^2 + by^2 + cxy - 10x - 20y$$

has a critical point at $(x, y) = (1, 2)$, and the second-derivative test is inconclusive at this point. Determine the values of the constants a, b, and c.

31. **Maximizing Profit** A dairy produces two types of cheese, A and B, at constant average costs of 50 cents and 60 cents per pound, respectively. When the selling price per pound of A is p_A cents and of B is p_B cents, the demands (in pounds) for A and B, are, respectively,

$$q_A = 250(p_B - p_A)$$

and

$$q_B = 32{,}000 + 250(p_A - 2p_B)$$

Find the selling prices that yield a relative maximum profit. Verify that the profit has a relative maximum at these prices.

32. Find all critical points of $f(x, y, z) = x^2yz$, subject to the condition that

$$x + 3y - 2z - 80 = 0 \ (xyz \neq 0)$$

33. Find all critical points of $f(x, y, z) = x^2 + y^2 + z^2$, subject to the constraint $3x + 2y + z = 14$.

34. **Surviving Infection** In an experiment,[29] a group of fish was injected with living bacteria. Of those fish maintained at 28°C, the percentage p that survived the infection t hours after the injection is given in the following table:

t	8	10	18	20	48
p	82	79	78	78	64

Find the linear-regression line of p on t.

[29]J. B. Covert and W. W. Reynolds, "Survival Value of Fever in Fish," *Nature* 267, no. 5606 (1977), 43–45.

35. **Equipment Expenditures** Find the least squares linear-regression line of y on x for the data given in the following table (refer to year 1993 as year $x = 1$, etc.):

Equipment Expenditures of Allied Computer Company, 1993–1998 (in millions of dollars)

Year	Expenditures
1993	15
1994	22
1995	21
1996	26
1997	27
1998	34

In Problems 36–39, evaluate the double integrals.

36. $\int_1^2 \int_0^y x^2 y^2 \, dx \, dy$

37. $\int_0^4 \int_{y/2}^2 xy \, dx \, dy$

38. $\int_1^4 \int_{x^2}^{2x} y \, dy \, dx$

39. $\int_0^1 \int_{\sqrt{x}}^{x^2} 7(x^2 + 2xy - 3y^2) \, dy \, dx$

Mathematical SNAPSHOT

Data Analysis to Model Cooling[30]

In Chapter 15 you worked with Newton's law of cooling, which can be used to describe the temperature of a cooling body with respect to time. Here you will determine that relationship in an empirical way via data analysis. This will illustrate how mathematical models are designed in many real-world situations.

TABLE 17.6

Time t	Temperature T	Time t	Temperature T
0 min	124°F	128 min	64°F
5	118	144	62
10	114	178	59
16	109	208	55
20	106	244	51
35	97	299	50
50	89	331	49
65	82	391	47
85	74	Overnight	45

Suppose that you want to create a mathematical model of the cooling of hot tea after it is placed in a refrigerator. To do this you place a pitcher containing hot tea and a thermometer in the refrigerator and periodically read and record the temperature of the tea. Table 17.6 gives the collected data, where T is the Fahrenheit temperature t minutes after the tea is placed in the refrigerator. Initially, that is, at $t = 0$, the temperature is 124°F; when $t = 391$, then $T = 47$°F. After being in the refrigerator overnight, the temperature is 45°F. Figure 17.24 gives a graph of the data points (t, T) for $t = 0$ to $t = 391$.

The pattern of these points strongly suggests that they nearly lie on the graph of a decreasing exponential function, such as the one shown in Figure 17.24. In particular, because the overnight temperature is 45°F,

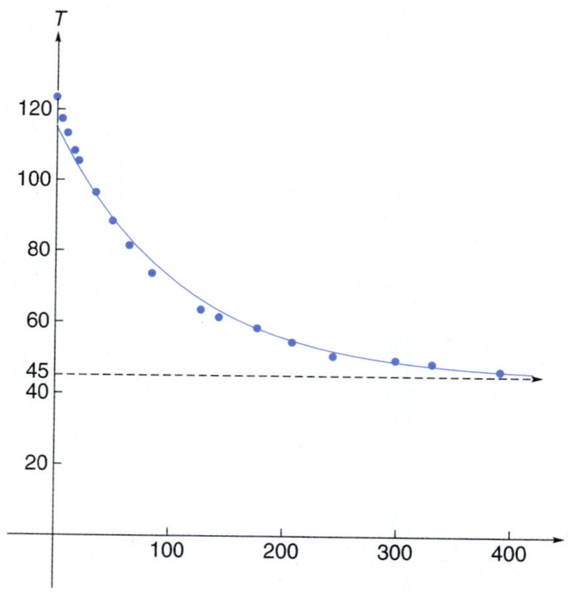

FIGURE 17.24 Data points and exponential approximation.

this exponential function should have $T = 45$ as a horizontal asymptote. Such a function has the form

$$\hat{T} = Ce^{at} + 45 \qquad (1)$$

where $\hat{T}$ gives the predicted temperature at time t, and C and a are constants with $a < 0$. (Note that since $a < 0$, then as $t \to \infty$, you have $Ce^{at} \to 0$, so $Ce^{at} + 45 \to 45$.)

[30]Adapted from Gloria Barrett, Dot Doyle, and Dan Teague, "Using Data Analysis in Precalculus to Model Cooling," *Mathematics Teacher* 81, no. 8 (November 1988), 680–84. By permission of the National Council of Teachers of Mathematics.

Now the problem is to find the values of C and a so that the curve given by Equation (1) best fits the data. By writing Equation (1) as

$$\hat{T} - 45 = Ce^{at}$$

and then taking the natural logarithm of each side, you obtain a linear form:

$$\ln(\hat{T} - 45) = \ln(Ce^{at})$$

$$\ln(\hat{T} - 45) = \ln C + \ln e^{at}$$

$$\ln(\hat{T} - 45) = \ln C + at \quad (2)$$

Letting $\hat{T}_l = \ln(\hat{T} - 45)$, Equation (2) becomes

$$\hat{T}_l = at + \ln C \quad (3)$$

Because a and $\ln C$ are constants, Equation (3) is a linear equation in $\hat{T}_l$ and t. This means that for the original data, if you plot the points $(t, \ln(T - 45))$, they should nearly lie on a straight line. These points are shown in Figure 17.25, where T_l represents $\ln(T - 45)$. Thus for the line given by Equation (3) that predicts T_l, you can assume that it is the linear regression line of T_l on t. That is, a and $\ln C$ are linear regression coefficients. By using the formulas for these coefficients and a calculator, you can determine that

$$a = \frac{17\left(\sum_{i=1}^{17} t_i T_{l_i}\right) - \left(\sum_{i=1}^{17} t_i\right)\left(\sum_{i=1}^{17} T_{l_i}\right)}{17\left(\sum_{i=1}^{17} t_i^2\right) - \left(\sum_{i=1}^{17} t_i\right)^2} \approx -0.00921$$

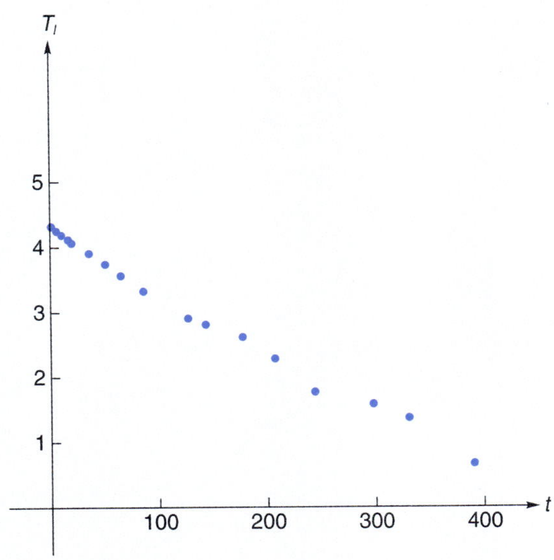

FIGURE 17.25 The points (t, T_l), where $T_l = \ln(T - 45)$, nearly lie on a straight line.

and

$$\ln C = \frac{\left(\sum_{i=1}^{17} t_i^2\right)\left(\sum_{i=1}^{17} T_{l_i}\right) - \left(\sum_{i=1}^{17} t_i\right)\left(\sum_{i=1}^{17} t_i T_{l_i}\right)}{17\left(\sum_{i=1}^{17} t_i^2\right) - \left(\sum_{i=1}^{17} t_i\right)^2}$$

$$\approx 4.260074$$

Since $\ln C \approx 4.260074$, then $C \approx e^{4.260074} \approx 70.82$. Thus from Equation (1),

$$\hat{T} = 70.82e^{-0.00921t} + 45$$

which is a model that predicts the temperature of the cooling tea. The graph of this function is the curve shown in Figure 17.24.

Exercises

1. Plot the following data points on an x,y-coordinate plane:

x	0	1	4	7	10
y	15	12	9	7	6

Suppose that these points nearly lie on the graph of a decreasing exponential function with horizontal asymptote $y = 5$. Use the technique discussed in this mathematical snapshot to determine the function.

2. Suppose that observed data follow a relation given by $y = C/x^r$, where $x, y, C > 0$. By taking the natural logarithm of each side, show that $\ln x$ and $\ln y$ are linearly related. Thus, the points $(\ln x, \ln y)$ lie on a straight line.

3. Use Newton's law of cooling (see Section 15.6) and the data points (0, 124) and (128, 64) to determine the temperature T of the tea discussed in the snapshot at time t. Assume that the ambient temperature is 45°F.

4. Try obtaining the final regression equation in the snapshot using the regression capability of a graphing calculator. First use linear regression. How does your result compare with the one in the snapshot? Then try skipping the linear-form transformation and perform an exponential regression. What difficulty do you encounter, if any? How could it be overcome?

APPENDIX A

Concepts for Calculus

We occasionally hear of an airplane running off the end of the runway during takeoff or landing. The usual problem is mechanical trouble with the aircraft, not the length of the runway itself. Runway designers know how long a runway needs to be in order for a given type of plane to land or take off, and airport controllers restrict the use of their runways to planes for which the runways are suitable.

How is the necessary runway length for a given type of airplane determined? Trial and error is not a feasible method, since human lives, as well as millions of dollars of equipment, are involved. We need a safer, more cost efficient solution—a mathematical model.

In this problem, the key variables are the minimum takeoff speed and the acceleration potential of the plane. The runway needs to be long enough for the plane, at normal acceleration, to reach the speed at which it can safely lift off.

Translating acceleration and takeoff speed into a distance, however, poses a problem. We know how to calculate the distance traveled by a vehicle moving at constant speed—distance equals the product of speed and travel time. But the speed of an aircraft taking off is constantly changing. Which speed do we use? And how do we calculate the travel time, which in this case is the time needed to reach takeoff velocity? Since the time necessary to reach the takeoff speed is dependent upon the acceleration, it must be derived from the acceleration and takeoff speed, rather than set arbitrarily.

The mathematical tool that does the job here is called *integration*—one of the cornerstone concepts of calculus. Among other things, calculus enables us to answer questions about the velocity of an object in motion given information about its acceleration, and vice versa. This appendix is an introduction to the basic concepts of calculus. It will prepare you for more advanced work.

OBJECTIVE *To gain information about the relationship between two quantities from a graph.*

A.1 Interpreting Graphs

Line Graphs

A graph is a picture of the relationship between two variable quantities. Figure A.1 shows a graph describing the relationship between the amount of insecticide sprayed on a field and the field's crop yield.

A *line graph* like this is created by first plotting data points, where each point represents a known pair of values for the two variables, and then connecting adjacent points with line segments.

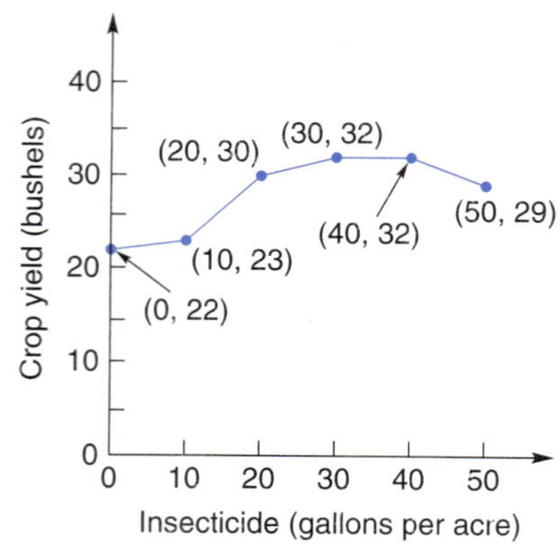

FIGURE A.1 Insecticide and crop yield.

EXAMPLE 1 **Predicting Crop Yield**

From the graph in Figure A.1, find the crop yield for each of the following insecticide amounts.

a. *No insecticide* **b.** 20 *gallons per acre* **c.** 13 *gallons per acre*

Solution:

a. When no insecticide is used, the number of gallons per acre is zero. The data point directly above 0 on the horizontal axis lies at 22 on the vertical axis. The crop yield is 22 bushels per acre.

b. The data point above 20 on the horizontal axis lies on the horizontal line through 30 on the vertical axis. Thus, when 20 gallons of insecticide per acre are used, the crop yield is 30 bushels per acre.

c. No plotted data point corresponds to 13 gallons of insecticide per acre. But the relationship between the amount of insecticide used and the crop yield is probably fairly smooth between 10 and 20 gallons, as represented by the line segment above that interval. On the segment, the point directly above 13 gallons per acre lines up with 25 bushels. So for 13 gallons, the crop yield is approximately 25 bushels per acre.

The answer found in part (c) of Example 1 is only approximate, since the actual relationship between insecticide usage and crop yield between 10 and 20 gallons of insecticide per acre is not likely to be exactly linear. But it is probably fairly smooth, and so the line segment provides a reasonable approximation to the actual relationship. The same thing applies to the segments between 0 and 10 gallons, 20 and 30 gallons, and so on.

The line segments in a line graph also show, at a glance, how the variables are changing with respect to one another in different parts of the graph.

EXAMPLE 2 **Age and Height**

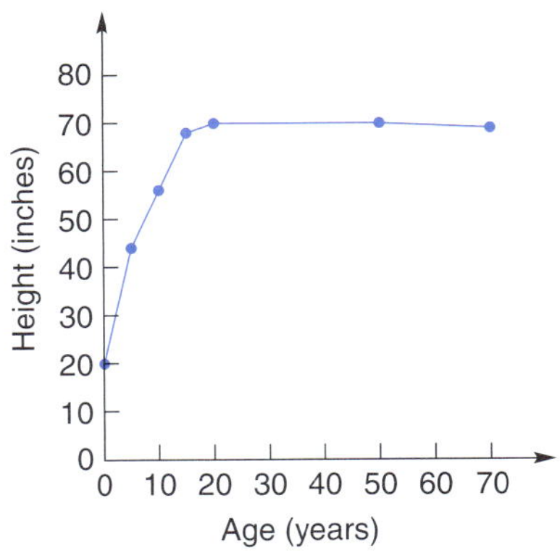

FIGURE A.2 Age and height.

Figure A.2 shows one person's height as a function of age.

a. *Over what age range did the person grow the fastest?*

Solution: A line segment slanting upward from left to right has a positive slope, which means an increase of height over time. The steepest upward-slanting segment, representing the fastest growth, is between ages 0 and 5.

b. *Over what age range did the person's height remain constant?*

Solution: The segment between ages 20 and 50 is horizontal; that is, it has slope zero, which means zero change in height over time. This makes sense: we know that a normal adult's height is fairly constant through middle age.

c. *What does the downward slant of the last line segment on the right mean?*

Solution: The downward slant represents a decrease in height over time, but a very gradual one, since the segment is not very steep.

d. *At what age was the person's height the least? the greatest?*

Solution: The lowest point on the graph represents the person's minimum height; this corresponds with age zero, or the time of birth. The highest point on the graph represents the person's maximum height; the person reaches the maximum height at age 20 and maintains that height through age 50.

The questions in Examples 1 and 2 ask for some of the most basic information we hope to obtain from graphs. First, we want to be able to determine what value of one quantity corresponds to a specific value of another quantity, as we did in Example 1 when we determined the crop yield corresponding to each amount of insecticide used per acre. The graph helped us to predict crop yield for levels of insecticide use where no exact measurement was available. We will learn more about using data and graphs to make predictions in the next section.

Second, sometimes we are interested in how a change in one quantity affects the other quantity. For example, does an increase in the amount of insecticide from 10 gallons per acre to 15 gallons per acre mean an increase in crop yield, a decrease in crop yield, or no change in crop yield? Here the slant of the graph gives the answer. At places where a graph goes up from left to right, the quantity represented on the vertical axis increases as the quantity represented on the horizontal axis increases; we describe this relationship as increasing. At places where a graph goes down from left to right, the quantity on the vertical axis decreases as the quantity represented on the horizontal axis increases; we describe this relationship as decreasing. At places where a graph is a horizontal line segment, the quantity on the vertical axis does not change as the quantity on the horizontal axis increases; we describe this relationship as constant.

The terms *increasing, decreasing,* and *constant* describe the behavior of the variable on the vertical axis as the variable on the horizontal axis increases.

Sometimes we also need to know how quickly one quantity changes with respect to changes in the other quantity. This is related to the steepness of the graph: Steeper graphs represent more rapid change. We saw this in part (a) of Example 2, where the steepest part of the graph represented the most rapid growth. We will discuss this aspect of graphs in more depth in Sections A.3 and A.4.

Other important pieces of information that we can often determine from a graph are the greatest and least values of the quantities represented. The highest point on the graph, if there is one, corresponds to the maximum value of the quantity represented on the vertical axis, whereas the lowest point, if there is one, corresponds to the minimum value. As we saw in Example 2, the maximum or minimum may occur at more than one point.

EXAMPLE 3 **Price of Stock**

The stock market in general had a rough year in 2000, but technology stocks took the greatest hit by far. Weeks of uncertainty over the outcome of the 2000 presidential election may have been one reason why technology stocks spiraled downward at the close of the year.

Figure A.3 shows the price of an Internet technology company's stock.

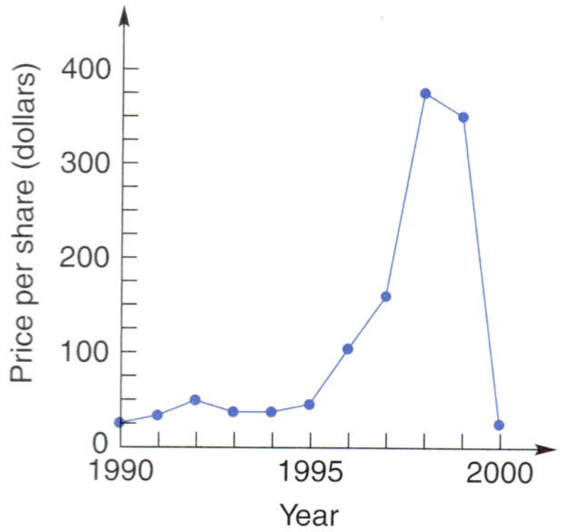

FIGURE A.3 Price of stock.

a. *What was the price per share in* 1997?

Solution: The point on the graph corresponding to the year 1997 appears to be on the same horizontal line as 150 on the vertical axis. Thus, in 1997 the price was about $150 per share.

b. *During what time period(s) was the price per share increasing?*

Solution: The line segments are going up from left to right between 1990 and 1992 and again between 1994 and 1998.

c. *During what time period was the price of stock increasing the fastest?*

Solution: The steepest upward-slanting line segment corresponds to the period between 1997 and 1998, so this was the period of fastest growth in the price.

d. *What was the highest price of the company's stock during the years represented, and in what year did it occur? What about the lowest price?*

Solution: The highest point on the graph is for the year 1998 and corresponds to the highest price, which is about $375 per share. The lowest points on the graph appear to correspond to a price of $25 per share, for the years 1990 and 2000.

Smooth Graphs

Line graphs showing the relationship between two variables usually have a few exact points and are only approximate for the other points. Generally, if more and more pairs of data values are plotted, the line graphs begin to resemble smooth curves. A compact way to express a smooth-curve relationship between two variables is through the use of mathematical formulas, or equations. In

Example 4, we examine one relationship represented by a smooth curve, and we find an equation describing the relationship.

EXAMPLE 4 **Bank Balance**

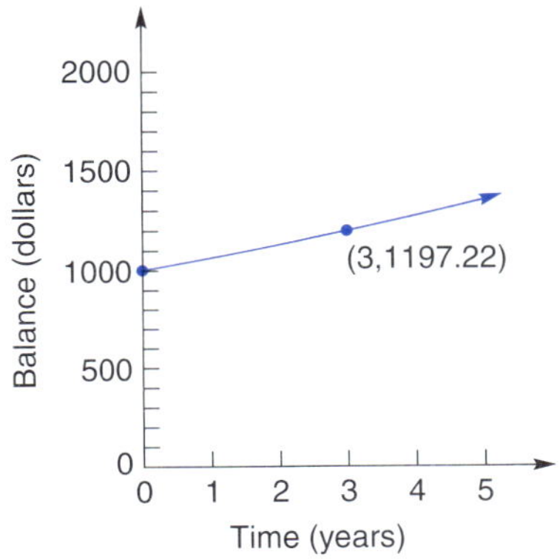

FIGURE A.4 Bank balance over time.

Figure A.4 shows the balance, over time, in a bank account where interest is compounded continuously and no deposits or withdrawals are made after the initial deposit. The steadily increasing values are due to the accumulation of interest.

The graph represents a function of the form $S = Pe^{rt}$. What quantities do P and r represent? What are their values here?

Solution: $P = S$ when $t = 0$, so P represents the initial balance in the account. The value of P is given by the place where the graph meets the vertical axis, so $P = \$1000$. The quantity r represents the interest rate; the higher r is, the faster the account grows. To find the value of r here, we first solve $S = Pe^{rt}$ or $S = 1000e^{rt}$ for r.

$$S = 1000e^{rt}$$
$$\frac{S}{1000} = e^{rt}$$
$$\ln \frac{S}{1000} = rt$$
$$r = \frac{1}{t} \ln \frac{S}{1000}$$

To calculate the value of r, we must pick a nonzero value of t and read the corresponding value of S from the graph. When t is 3, S is 1197.22. So $r = \frac{1}{3} \ln \frac{1197.22}{1000} = 0.06$. The annual interest rate for the account is 0.06, or 6%.

The value found for r is the nominal rate, or annual percentage rate. Remember that this is generally less than the effective rate, which here is $e^{0.06} - 1 \approx 0.0618 = 6.18\%$ (see Sections 5.1 and 10.3).

Example 4 raises some important questions about how much information we can gain from graphs and data. For example, how did we know that the relationship between the quantities could be represented by a function of the form $S = Pe^{rt}$? And why would we even want to find such an equation? We explore questions such as these in the next section, and we return to the discussion of steepness of a graph in Section A.3.

Exercise A.1

For Problems 1–4 use the data given in the tables to create line graphs.

1.

x	0	3	4	8	10
y	0	5	7	10	10

2.

x	0	2	3	5	8
y	0	8	4	10	6

3.

x	0	2	3	7	10
y	4	1	9	5	3

4.

x	0	4	6	7	10
y	1	4	8	7	2

5. Use the following data, which shows the population of a city in each of the given years, to create a line graph.

Year	1996	1997	1998	1999	2000
Population	5120	5342	5510	5750	6002

6. Use the following data, which shows the price per gallon of gasoline on each of the given days, to create a line graph.

Day	Monday	Tuesday	Wednesday	Thursday	Friday
Price ($)	1.55	1.62	1.62	1.58	1.67

For Problems 7–16, use the line graph in Figure A.5, which shows the total number of freshmen who enrolled in calculus during the given years at a college.

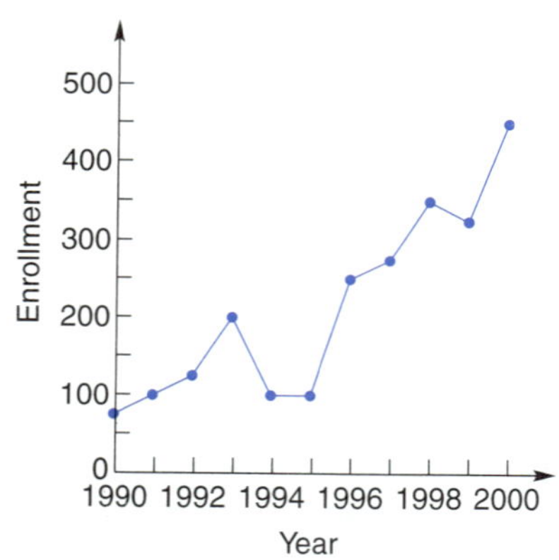

FIGURE A.5 Line graph for Problems 7–16.

7. How many freshmen enrolled in calculus during 1990?

8. How many freshmen enrolled in calculus during 1997?

9. During what time periods was the enrollment increasing? Is the slope of the line graph positive, negative, or zero during these periods?

10. During what time periods was the enrollment decreasing? Is the slope of the line graph positive, negative, or zero during these periods?

11. During what time periods was the enrollment constant? Is the slope of the line graph positive, negative, or zero during these periods?

12. During what time period did the enrollment increase most rapidly?

13. During what time period did the enrollment decrease most rapidly?

14. What was the largest enrollment during the years represented and in what year did it occur?

15. What was the lowest enrollment during the years represented and in what year did it occur?

16. During what two-year time period did the change in enrollment per year remain the same?

17. **(a)** The following information describes a stock's performance during a given week. Use the information to create a possible line graph.

- **i.** The highest price of the stock during the week was on Monday.
- **ii.** The stock's value decreased from Monday to Tuesday and from Tuesday to Wednesday. The value decreased most rapidly from Tuesday to Wednesday.
- **iii.** The value of the stock increased from Wednesday to Thursday and its value did not change from Thursday to Friday.

(b) When was the stock's value the lowest?

18. **(a)** The following information describes the number of goals scored each season for the first five seasons of a professional hockey player's career. Use the information to create a possible line graph.

- **i.** The lowest number of goals scored was during the player's first season.
- **ii.** The number of goals scored increased from the first to the second season, remained constant from the second to the third season, and increased from the third to the fourth season. The increase from the third to the fourth season was larger than the increase from the first to the second season.
- **iii.** The number of goals scored decreased from the fourth to the fifth season.

(b) Which season did the player score the most goals?

19. In the following table, d represents the distance traveled in miles after t hours of driving on a trip.

t (hours)	0	1	3	5	6	8
d (miles)	0	55	115	265	265	325

(a) Use the data to create a line graph.
(b) How far had the drivers traveled after 2 hours?
(c) Over what time interval did the drivers take a break? What is the slope of the line during this time interval?
(d) Over what time interval did the drivers travel the fastest? How does the slope of the line graph during this time interval compare to the slope of the line graph during the remaining intervals?

20. In the following table, f represents the average rate of flow per month, in cubic feet per second, of the Roaring Fork river in Colorado, t months after April.

t	0	1	2	3
f	2000	3500	2500	800

(a) Use the data to create a line graph.
(b) What was the average rate of flow during April?
(c) Between which two months did the average rate of flow increase? Is the slope of the line between these two months positive or negative?
(d) Between which two months did the average rate of flow decrease most rapidly? Is the slope of the line between these two months positive or negative?
(e) Explain why it is reasonable for the average rate of flow to change from increasing to decreasing over the months shown in the table.

21. In the following table, f represents the total number of fish held in an angler's live well t hours after a bass-fishing tournament began. Assume that $t = 0$ represents 6 A.M. and that each angler reported the total number of fish in their well every two hours.

t	0	2	4	6	8	10	12
f	0	8	14	20	20	22	26

(a) Use the data points given in the table to create a line graph.
(b) Use the graph from part (a) to determine the time interval when the greatest number of fish were caught.
(c) Over what time interval did the angler not catch any fish? What is the slope of the line during this time interval?
(d) What can be said about the number of fish caught per hour between 8 A.M. and 12 P.M.?

22. In Figure A.6, S represents the balance in dollars t years after an initial deposit was made in a bank account, where interest is compounded continuously and no deposits and withdrawals are made after the initial deposit.

(a) How much money was initially deposited in the account?
(b) As indicated in the graph, after 10 years the balance in the account is \$149.18. Use this information to find the annual interest rate r.
(c) Use the graph to estimate how long it will take for the investment to double.
(d) Use the information found in parts (a) and (b) to find the equation of the graph.
(e) Use the equation, found in part (d), to determine the amount of money in the account after 5 years.

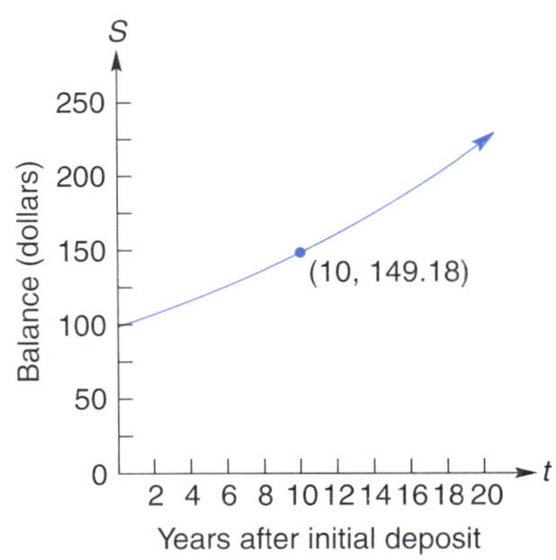

FIGURE A.6 Graph for Problem 22.

OBJECTIVE *To recognize mathematical relationships between two quantities and express those relationships with equations.*

A.2 Modeling

In the previous section, we studied the relationships between two quantities by asking certain questions about graphs. In the final example, we used information found in a graph with the knowledge that the quantities involved satisfied the equation $S = Pe^{rt}$ to write an equation for the graph. Such an equation is a mathematical **model** of the relationship between the quantities represented in the graph. In this case, the model contains sufficient information to reproduce the original data and the graph, but has the advantage of being more compact. The model also has the potential to help us predict values of the quantities that are not represented in the data or on the graph.

EXAMPLE 1 **Common Models**

Formulas are mathematical models, since they explicitly state the relationships between two or more quantities. The formula $A = \pi r^2$ from geometry describes the relationship between the length r of the radius of a circle and the area A of the circle. Given a meaningful value for either quantity, you can use the formula to find the corresponding value of the other quantity.

Mathematical modeling is the process of creating a mathematical model for a real-world situation. Modeling usually includes representing the quantities involved by appropriate variables and using given information to write statements (normally equations, but sometimes inequalities) relating the variables. The model is sometimes the end result, but more often the model is established to answer some other question about the quantities. We will study the process of mathematical modeling through a few examples.

EXAMPLE 2 **The Elements of a Model**

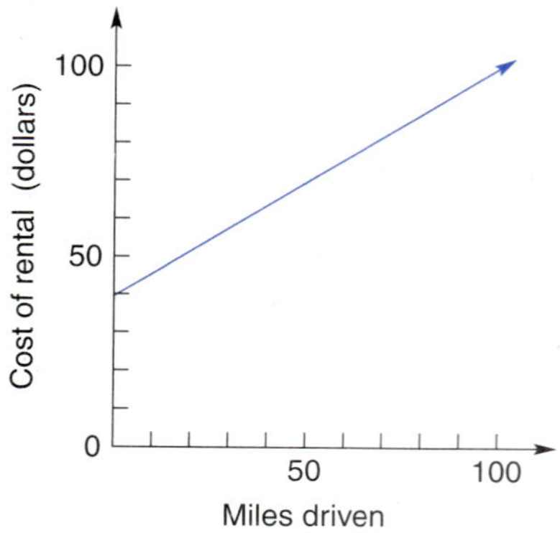

FIGURE A.7 Miles versus cost.

The graph in Figure A.7 represents the relationship between the number of miles driven and the cost of renting a 24-foot moving truck for 24 hours.

Find a mathematical model for the relationship between the number of miles driven and the cost of the rental truck. Use the model to calculate the rental cost if you drive the truck 500 miles.

Solution: The quantities are the number of miles driven and the cost of the rental. Whenever possible, we try to use variables that are meaningful in the situation. Let

$$n = \text{number of miles driven, and}$$
$$C = \text{cost of rental}$$

The next step is to write an equation relating these two variables. Here we must ask ourselves, What type of equation should this be? In Example 4 of Section A.1, we wrote an exponential equation to model the information in the graph, but is that appropriate here? Look at the graph. The graph is a line. Thus the relationship between the two variables is linear, not exponential, and should be modeled using the equation of a line $y = mx + b$, where $y = C$ and $x = n$, so $C = mn + b$.

Then we need to know the values of m and b. From the graph, we see that the point (0, 40) is on the line. Thus, $b = 40$. This \$40 is the fixed cost of renting the truck; you pay this amount regardless of the number of miles driven. The graph also appears to pass through the point (100, 100). We use the slope formula to find the value of m:

$$m = \frac{y_2 - y_1}{x_2 - x_1} = \frac{100 - 40}{100 - 0} = \frac{60}{100} = 0.60$$

This is the variable cost associated with renting the truck; you pay \$0.60 for each mile you drive the truck. Note that this quantity can be stated as a rate: \$0.60 per mile.

Thus, the mathematical model representing the relationship between these two variable is

$$C = 0.60n + 40$$

To find the rental cost for driving the truck 500 miles, let $n = 500$ in this equation and simplify.

$$C = 0.60(500) + 40 = 300 + 40 = 340$$

Thus, if you drive the truck 500 miles, the cost of the rental will be \$340.

Often the nature of the problem will put constraints on the values of the variables. Negative values, such as negative dollar amounts or negative values for time, don't make sense in some situations. Other times, a value can be too large in the positive direction. You should always question whether the values for the variable make sense.

In Example 2, the value of 500 for n is a bit large but not unreasonable. Assuming that you can average 60 miles per hour, it would take $8\frac{1}{3}$ hours to drive the truck 500 miles, a reasonable driving time in one 24-hour period. However, the cost of the rental is outrageous. The problem is that this pricing scheme is for short-term rentals where the truck will be driven relatively few miles. Long-term rental price plans would probably be more cost effective if you plan to drive the truck 500 miles.

EXAMPLE 3 Height of an Object

A ball is thrown upward from the edge of a cliff 200 feet above a river. After 1 second, the ball is 248 feet above the river, and after 3 seconds, the ball is again 248 feet above the river. Find a mathematical model expressing the relationship between the time since the ball was released and the height of the ball. How many seconds after the ball is thrown will it hit the river? Figure A.8 shows a graph of the relationship.

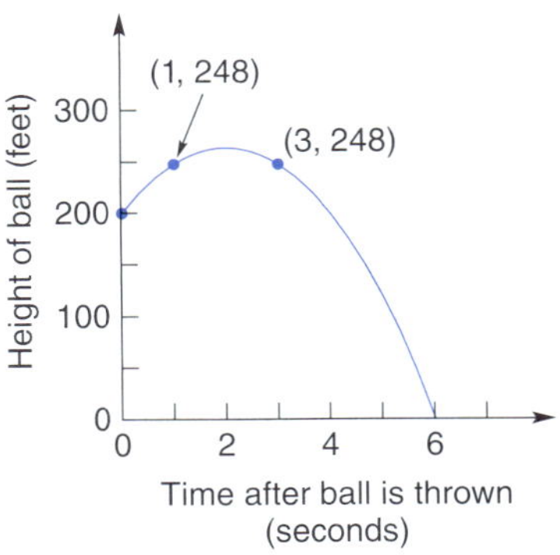

FIGURE A.8 Height versus time.

Solution: Let h be the height of the ball (in feet) and t the time after the ball is thrown (in seconds). The graph is parabolic, so the model is of the form $h = at^2 + bt + c$. We know three points on the graph (0, 200), (1, 248), and (3, 248). We need to use these to find the values of a, b, and c. Since $h = 200$ when $t = 0$, we have $c = 200$. Further, since $h = 248$ when $t = 1$ and when $t = 3$, we get the system of equations

$$\begin{cases} 248 = a + b + 200 \\ 248 = 9a + 3b + 200 \end{cases}$$

which simplifies to

$$\begin{cases} a + b = 48 \\ 9a + 3b = 48 \end{cases}$$

Multiplying the first equation by -3 and adding it to the second equation, we obtain

$$\begin{array}{rcr} -3a - 3b &=& -144 \\ 9a + 3b &=& 48 \\ \hline 6a &=& -96 \end{array}$$

Thus, $a = -16$. Substituting -16 into the equation $a + b = 48$ for a, we find that $b = 64$. Therefore, $h = -16t^2 + 64t + 200$ is the model we are seeking.

When the ball hits the river, $h = 0$, so to find the time when the ball hits the river, let $h = 0$ in the model and solve for t. Using the quadratic formula, we find that the equation $0 = -16t^2 + 64t + 200$ has two solutions: $t \approx -2$ and $t \approx 6$. We discard the negative value and conclude that the ball hits the river about 6 seconds after being thrown.

In Example 3, we are given three points on the graph explicitly, which is exactly what we need to write a quadratic model. However, in many cases, we will have far more data points than we actually need to write the appropriate equation, and the relationship may not be exactly linear or quadratic, meaning that not all the points may lie directly on the graph of the preferred model.

EXAMPLE 4 Population Change

Table A.1 gives the population of a small town at ten-year intervals, from 1950 *to* 2000. *Find a mathematical model that expresses the relationship between the population and the year. Use the model to predict the population of the town in the year* 2005.

TABLE A.1

Year	1950	1960	1970	1980	1990	2000
Population	3498	2872	2352	1928	1582	1295

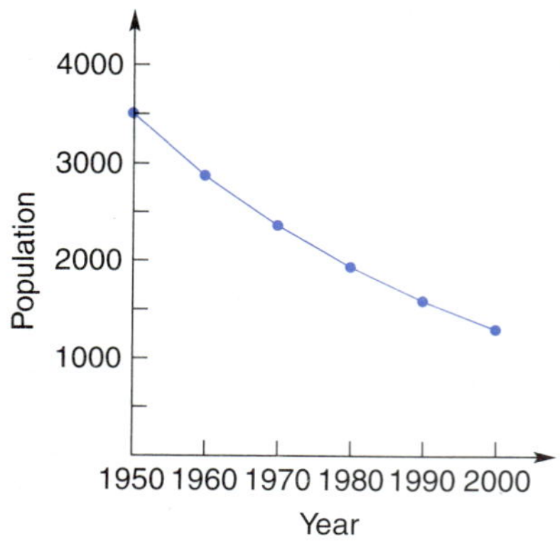

FIGURE A.9 Population versus time.

Solution: From the graph of the data, shown in Figure A.9, it may be difficult to determine whether the relationship between the year and the population is roughly linear. If the relationship is roughly linear, the changes from one decade to the next should all be approximately the same. Table A.2 lists these changes.

TABLE A.2 Population change from previous decade

Year	1960	1970	1980	1990	2000
Population change	−626	−520	−424	−346	−287

The changes listed in the table do not indicate a linear relationship. Is the relationship exponential? If it is, the percent changes from one decade to the next should all be roughly the same. Table A.3 lists the percent changes.

TABLE A.3 Percent population change from previous decade

Year	1960	1970	1980	1990	2000
Percent change	−17.9	−18.1	−18.0	−17.9	−18.1

The percent changes are all approximately the same, so an exponential equation of the form $S = Pe^{rt}$ should provide a good model of the relationship between the quantities. For convenience, we will let $t = 0$ correspond to the year 1950, $t = 10$ correspond to 1960, and so on. Thus, since $3498 = Pe^{r \cdot 0}$, we have that $P = 3498$. To find the value of r, we must use one other point on the graph. Since the points do not fit an exponential model exactly, different points will yield slightly different values of r. We choose to use the values $t = 50$ and $S = 1295$ (year 2000).

$$1295 = 3498e^{50r}$$
$$\frac{1295}{3498} = e^{50r}$$
$$\ln\frac{1295}{3498} = 50r$$
$$r = \frac{\ln\frac{1295}{3498}}{50} \approx -0.01987$$

The model describing the town's population is $S = 3498e^{-0.01987t}$. To predict the town's population in 2005, let $t = 55$ in this equation and simplify.

$$S = 3498e^{-0.01987 \cdot 55} \approx 1173$$

If the population of the town continues to decrease at the same rate, there will be about 1173 people living in the town in 2005.

In Example 4, we compared changes in the population of the town from decade to decade in order to choose the correct model. We will study more about change in the next two sections.

Exercise A.2

In Problems 1–8, use the information given to write equations which model each situation. Make certain to define variables when necessary.

1. The area of a trapezoid is one-half the product of its height and the sum of its bases.
2. The cost of producing x units of a product is determined by a fixed cost of \$1000 plus a variable cost of \$40 per unit.
3. A line passes through $(1, 2)$ and $(-3, 4)$.
4. A parabola passes through $(2, 6)$ with vertex $(-1, 2)$.
5. A \$15,525 car depreciates at a rate of 9% each year.
6. The pH level of a substance is the opposite of the common logarithm of the concentration of hydrogen ions, $[H^+]$, in the substance.
7. The probability of the union of two mutually exclusive events is the sum of the probabilities of those events.
8. A parabola passes through $(-1, 8)$, $(0, 2)$, and $(4, 18)$.

In Problems 9–16, graph the data. Determine what kind of model (linear, quadratic, exponential, or logarithmic) fits all the data points; then find the equation of the graph which passes through each data point.

9.

x	−3	−1	0	2	4
y	−1	3	5	9	13

10.

x	−3	−2	1	3	5
y	5	3	−3	−7	−11

11.

x	−1	0	1	3	4
y	$\frac{1}{3}$	1	3	27	81

12.

x	−1	0	1	2	3
y	$-\frac{1}{2}$	−2	−8	−32	−128

13.

x	−2	0	1	2	4
y	21	5	3	5	21

14.

x	−4	−3	−2	0	2
y	−5	−2	−1	−5	−17

15.

x	$\frac{1}{2}$	1	2	8	32
y	−1	0	1	3	5

16.

x	$\frac{1}{3}$	1	3	27	81
y	−1	0	1	3	4

In Problems 17–22, graph the data and determine what kind of model (linear, quadratic, exponential, or logarithmic) fits the data best.

17. (0, 0), (0.5, 3.5), (1, 5.1), (1.5, 6.7), (2, 1.3), (3, 11.3), (6.5, 34.1), (7, 23), (7.5, 25), (8.2, 26.6).

18. (0.01, −4.61), (0.1, −2.30), (0.2, 7), (0.25, −1.40), (0.5, −0.70), (0.8, −0.24), (1, 0), (3, 1.10), (4, −3.32), (4.2, 1.44), (7, 1.92), (7.3, 1.98), (10, 2.31).

19. (−8, 67), (−7.5, 59.3), (−3, 12), (−1.1, 4.2), (1, 4), (1.5, −2), (2.3, 8), (3.1, 12.8), (5, 27.3), (5.1, 29.1), (6, 38.6), (7, 20.1).

20. The following data shows the enrollment of a college for the given years.

Year	Enrollment	Year	Enrollment
1990	30,017	1996	35,945
1991	30,697	1997	36,550
1992	32,011	1998	37,993
1993	33,210	1999	39,001
1994	34,100	2000	39,992
1995	34,775		

21. The following data shows the number of milligrams of a radioactive substance, A, left after t days of decay.

t	A (mg)	t	A (mg)
0	10,000	6	450
1	6400	7	280
2	3400	8	168
3	2200	9	102
4	1300	10	60
5	778		

22. The following data shows the profit, P (in dollars), from the sale of x units of a product.

x	P (dollars)	x	P (dollars)
150	3330.18	1700	13,514.19
342	7200.91	1783	12,582.07
566	13,750.67	2007	5000.35
886	14,100.38	2300	4500.02
1300	15,400.05	2450	889.15

23. A small company which produces fly-fishing reels has a fixed cost of \$100 per month plus a variable cost of \$40 per reel.

(a) Find an equation to model the cost, C, of producing r reels per month.
(b) How much will it cost to produce 10 reels per month?
(c) How many reels can the company produce per month if the company can only afford costs of \$1000 per month.

24. A ball which was thrown upward from 5 feet off the ground was recorded to have reached a height of 49 feet 1 second after it was thrown, and a height of 61 feet 2 seconds after it was thrown. The height of the ball, h, after t seconds can be modeled by the standard form of a quadratic equation, $h = at^2 + bt + c$.

(a) Write an equation which models the height of the ball at time t.
(b) At what time will the ball reach its maximum height, and what is this height?

25. A model rocket was shot upward from the ground and was recorded to have reached a height of 64 feet after 1 second and 96 feet after 3 seconds. The height of the rocket, h, after t seconds can be modeled by the standard form of a quadratic equation, $h = at^2 + bt + c$.

(a) Write an equation which models the height of the rocket at time t.
(b) At what time will the rocket reach its maximum height, and what is that height?

26. Each March the NCAA basketball tournament begins. The purpose of the tournament is to determine the best college basketball team in the nation. A total of 64 teams are invited to play in this tournament. A team is out of the tournament once it loses.

(a) How many teams remain after the first round?
(b) How many rounds are played?
(c) Create a data table which shows the number of teams, n, that remain at the end of each round, r. Make certain to include data which indicates the number of teams which enter the tournament.

(d) Graph the data from part (c) and connect it with a smooth curve.

(e) Find an equation which models the curve in part (d).

27. An initial population of 2000 bacteria increases at a rate of 200% per day.

(a) Create a data table showing the number of bacteria present at the end of each day for the first five days. Choose appropriate variables to represent the independent and dependent variables.

(b) Find an equation which models the data.

(c) Determine how many bacteria are present at the end of 10 days.

28. Newton's law of cooling states that the surface temperature of an object cools at a rate proportional to the difference between the temperature of the object and that of the surrounding environment (ambient temperature).

(a) Using this law it can be shown that the temperature, T, of an object at time t is given by $T(t) = A + (T_0 - A)e^{-kt}$, where A is the ambient temperature, T_0 is the initial temperature of the object and k is the cooling constant. A corpse, when discovered by forensic scientists, had a body temperature of 85°F. After 2 hours the body had cooled to 74°F in an ambient temperature of 68°F. Use this information to find the cooling constant k.

(b) It can also be shown that the time of death, t_d, is given by $t_d = -\frac{1}{k}\ln\left(\frac{98.6 - A}{T_0 - A}\right)$. Use this formula, along with the information from part (a), to find how long the person had been dead before the scientists discovered the body.

(c) Interpret the meaning of the negative sign in the formula given in part (b).

29. **(a)** Identify the points which are least likely to fit the model found in Problem 17.

(b) Disregarding those points found in part (a), find an equation that seems to fit the remainder of the data.

(c) Use the equation from part (b) to predict the y-value when $x = 14$.

30. **(a)** Identify the points which are least likely to fit the model found in Problem 22.

(b) Disregarding those points found in part (a), find an equation that seems to fit the remainder of the data.

(c) Use the equation from part (b) to predict the profit when 3000 units are produced.

(d) Interpret your result from part (c).

OBJECTIVE *To understand average rate of change and its relationship to slope and steepness of a graph.*

A.3 Average Rate of Change

The Rate of Change

In mathematics, we work with the relationships among various quantities, which we represent with variables. Variables are, by definition, quantities that can change in value. A basic question that we often ask about related quantities is how a change in one quantity affects the other(s). More specifically, we can ask the question, How quickly does the value of y change as x changes? The answer to this question is what we call the **rate of change** of y with respect to x.

Graphically, we can ask an equivalent question: How steep is the graph of $y = f(x)$? Thus, the rate of change of y with respect to x can be thought of as the steepness of the graph of $y = f(x)$. When the value of y changes by the same amount each time the value of x is increased by 1 unit, the rate of change of y with respect to x is constant and the relationship between the variables is linear. In this case, the rate of change is the slope of the line.

EXAMPLE 1 **Eliminating a Federal Program: Linear Approach**

Suppose that during the 2001 session, the U.S. Congress votes to eliminate a federal program by 2006 by gradually reducing the program's budget to zero during the five years. If the program's 2001 budget is \$12.4 million and the budget is to be reduced by the same amount each year, then the budget will need to be reduced \$2.48 million per year during the five years. Since the annual reduction in the program's budget remains \$2.48 million from year to year, the relationship between the year t and the program's budget b (in millions of dollars) is linear and can be represented by the equation

$$b = 12.4 - 2.48(t - 2001), \quad 2001 \le t \le 2006.$$

The graph of this equation, shown in Figure A.10, is a line whose slope is −2.48; that is, the rate of change of the program's budget between 2001 and 2006 is −2.48 million dollars per year.

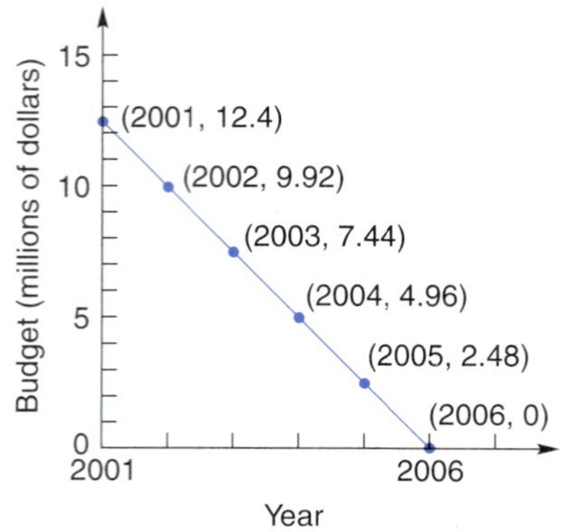

FIGURE A.10 Year versus budget.

However, as we saw in Section A.2, not all relationships between two variables are linear. Linear relationships are characterized by a constant rate of change. When the rate of change varies, some other type of relationship may exist. In these cases, the rate of change will not be so easily described.

EXAMPLE 2 **Eliminating a Federal Program: Exponential Approach**

Suppose that the federal program from Example 1 is to be eliminated by reducing the budget by 50% in each of the years 2002–2005 and then to zero in 2006. Since the budget is reduced by the same percentage in each of the first four years, the relationship between the year and the budget is exponential during these years. Table A.4 lists the change in each year's budget from the previous year's budget. Figure A.11 shows a graph of this information.

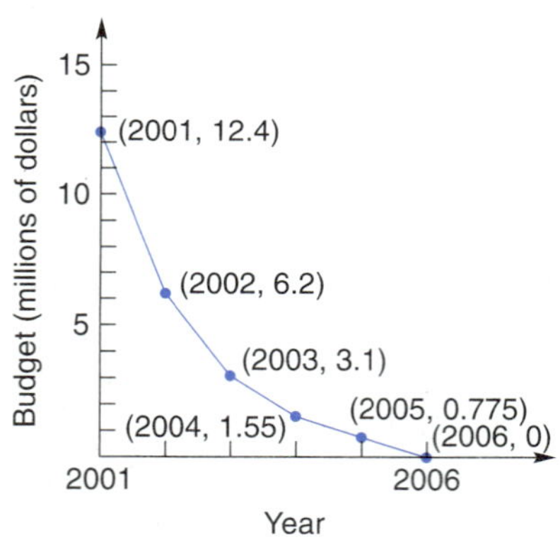

FIGURE A.11 Year versus budget.

TABLE A.4 Budget change (millions of dollars)

Year	2002	2003	2004	2005	2006
Budget change	−6.2	−3.1	−1.55	−0.775	−0.775

We cannot express this change in the budget in terms of a fixed dollar amount per year, since the change varies from year to year.

Even though the average rate of change for the budget in Examples 1 and 2 is the same, the amount spent on the program in years 2002–2005 is vastly different. Using the method of Example 1, $24.8 million will be spent during the remaining years of the program, but only $11.625 million will be spent if we use the method of Example 2.

Average Rate of Change

Our first step toward describing the rate of change in a nonlinear relationship is the *average rate of change*. For relationships such as those in Examples 1 and 2, where one quantity is time, the average rate of change is the change in the other quantity during the time period (value at the end − value at the beginning) divided by the length of time. So for both Examples 1 and 2, the average rate of change of the program's budget from 2001 to 2006 is $\frac{-12.4}{5} = -2.48$ million

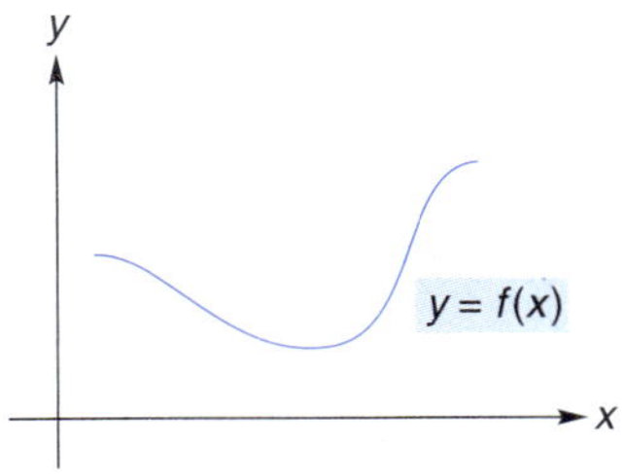

FIGURE A.12 $y = f(x)$.

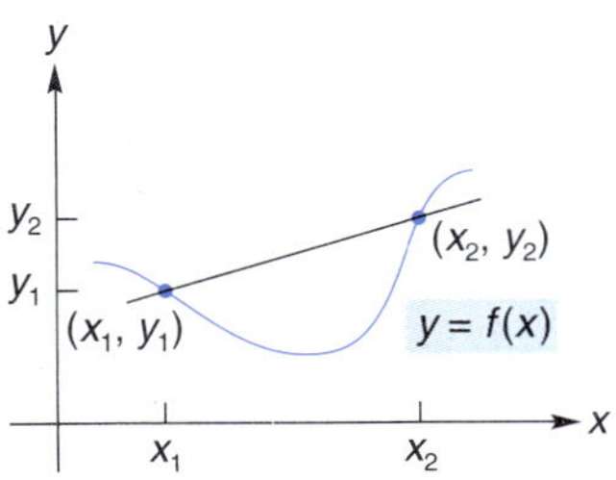

FIGURE A.13 Secant line passing through the points (x_1, y_1) and (x_2, y_2).

dollars per year, since the budget decreases from \$12.4 million in 2001 to \$0 in 2006 and the length of the time period is 5 years.

It is important to note that the average rate of change in this example is calculated for the time period 2001–2006. If we consider only the period from 2001 to 2002, then the average rate of change for the method of Example 2 would be −6.2 million dollars per year, but the average rate of change for the method of Example 1 would still be −2.48 million dollars per year.

It is not necessary that one of the variables represent time in order to find an average rate of change. We now discuss the average rate of change in more generality.

Suppose that the relationship between the two variable quantities in question can be represented as a function $y = f(x)$ as shown in Figure A.12. For any two distinct values of x in the domain of f, say x_1 and x_2, there are corresponding points, (x_1, y_1) and (x_2, y_2), on the graph of f. Through these two points, there is a line, called a **secant line** of the graph (Figure A.13). We define the average rate of change of the function f from x_1 to x_2 to be the slope of the secant line passing through the points (x_1, y_1) and (x_2, y_2).

Average Rate of Change

Let (x_1, y_1) and (x_2, y_2) be distinct points on the graph of $y = f(x)$. The **average rate of change** of f from x_1 to x_2 is $\dfrac{y_2 - y_1}{x_2 - x_1}$.

Thus, through this concept of the average rate of change, we can, in a sense, measure the steepness of a graph over an interval in terms of the slope of a line, which is already very familiar to us.

It is again important to point out that the average rate of change of f is calculated using two points on the graph of f. In general, different pairs of points will result in different average rates of change when f is nonlinear. This is a distinct difference from the linear case, where the preceding formula yields the same result regardless of the points used.

EXAMPLE 3 Average Rate of Change

Sketch the graph of $f(x) = x - 4x^2$. Then sketch the secant line for each interval of x-values, and find each corresponding average rate of change of f.

a. $[-2, -1]$ **b.** $[-0.5, 1.5]$

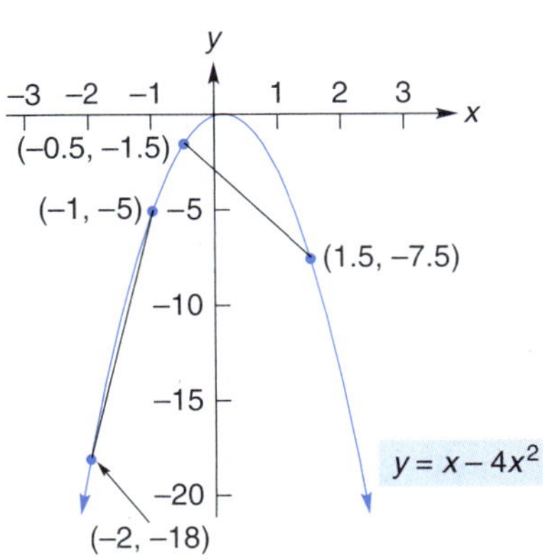

FIGURE A.14 $f(x) = x - 4x^2$.

Solution: The graph with the secant lines corresponding to the intervals is shown in Figure A.14. We find the average rates of change by calculating the slopes of the lines.

a. $$\text{Average rate} = \frac{y_2 - y_1}{x_2 - x_1} = \frac{-5 - (-18)}{-1 - (-2)} = \frac{13}{1} = 13$$

b. $$\text{Average rate} = \frac{y_2 - y_1}{x_2 - x_1} = \frac{-7.5 - (-1.5)}{1.5 - (-0.5)} = \frac{-6}{2} = -3$$

In both parts of the preceding example, there are places on the interval from x_1 to x_2 where the curve slants more steeply than the secant line, and there are places where the curve slants less steeply than the secant line. But the net effect of all the rises and falls, over the interval x_1 to x_2, is the same as the steady rise of the line. This, intuitively, is what we mean by saying that the slope of the line is the average rate of change of the function.

It is not necessary to graph a function in order to calculate an average rate of change.

EXAMPLE 4 **Average Driving Speed**

Suppose that Karl is driving to work. His distance from home is a function of his time on the road. Fifteen minutes after he starts he is 7 miles from home, and 20 minutes after he starts he is 11 miles from home. Find his average rate of change in position—his average driving speed, in other words—for the interval between 15 and 20 minutes.

Solution: Since $11 - 7 = 4$ and $20 - 15 = 5$, Karl traveled 4 miles in 5 minutes. His average rate is $\dfrac{4\text{ miles}}{5\text{ minutes}}$, or 0.8 mile per minute. (This is equivalent to 48 miles per hour.)

EXAMPLE 5 **A Stock Portfolio**

Table A.5 shows the value of Elizabeth's stock portfolio for several days in July.

TABLE A.5 Value of stocks (thousands of $)

Date	1	3	7	10	14	17	21	24	28	31
Value	14.2	14.6	15.2	14.7	13.9	14.1	13.7	14.3	14.7	14.5

Find the average rate of change of the value for each of the following intervals.

a. *July 3 to 7* **b.** *July 10 to 28* **c.** *July 7 to 21*

Solution:

a. $\text{Average rate} = \dfrac{15{,}200 - 14{,}600}{7 - 3} = \dfrac{600}{4} = \$150 \text{ per day}$

b. $\text{Average rate} = \dfrac{14{,}700 - 14{,}700}{28 - 10} = \dfrac{0}{18} = \$0 \text{ per day}$

c. $\text{Average rate} = \dfrac{13{,}700 - 15{,}200}{21 - 7} = \dfrac{-1500}{14} \approx \$107.14 \text{ per day}$

Now suppose that the graph of a function f includes the distinct points $(x_0, f(x_0))$ and $(x_1, f(x_1))$. We let $h = x_1 - x_0$, so that x_1 becomes $x_0 + h$ and the average rate of change of f on the interval $[x_0, x_1]$, or $[x_0, x_0 + h]$, becomes $\dfrac{f(x_0 + h) - f(x_0)}{h}$. This expression, which can often be simplified, plays an important role in the next section, where we continue our study of rates of change.

EXAMPLE 6 **Average Rate of Change Over an Interval $[x_0, x_0 + h]$.**

Let $g(x) = x^2$. Find the average rate of change of g for each interval of x-values.

a. $[2, 2 + h]$ **b.** $[x_0, x_0 + h]$

Solution:

a.
$$\text{Average rate} = \frac{g(2+h) - g(2)}{h} = \frac{(2+h)^2 - 2^2}{h}$$
$$= \frac{(4 + 4h + h^2) - 4}{h} = \frac{4h + h^2}{h} = 4 + h$$

b. $$\text{Average rate} = \frac{g(x_0+h)-g(x_0)}{h} = \frac{(x_0+h)^2 - x_0^2}{h}$$
$$= \frac{(x_0^2 + 2x_0h + h^2) - x_0^2}{h} = \frac{2x_0h + h^2}{h} = 2x_0 + h$$

In Section A.4, we will move from the average rate of change, which is measured over an interval, to the instantaneous rate of change, which is measured at a point.

Exercise A.3

In Problems 1–4, calculate the average rate of change between each pair of consecutive points, and use the results to determine if the relationships between the variables are linear or nonlinear over the given domains.

1.

x	0	1	2	3	4
y	4	1	−2	−5	−8

2.

x	0	1	2	3	4
y	−1	1	7	17	31

3.

x	0	1	2	3	4
y	0	1	8	27	64

4.

x	0	1	2	3	4
y	5	3	1	−1	−3

5. Sketch a graph of $f(x) = 0.5x^2$. Then sketch the secant line for each interval of x-values, and find each corresponding average rate of change.

(a) $[0, 3]$ **(b)** $[-4, 1]$
(c) $[-2, 2]$ **(d)** $[2, 2.01]$

6. Sketch a graph of $f(x) = -2x^2$. Then sketch the secant line for each interval of x-values, and find each corresponding average rate of change.

(a) $[0, 2]$ **(b)** $[-2, 3]$
(c) $[-1, -2]$ **(d)** $[3, 3.01]$

7. Sketch a graph of $g(x) = x^3 + 3x$. Then sketch the secant line for each interval of x-values, and find each corresponding average rate of change.

(a) $[1, 3]$ **(b)** $[-2, 2]$
(c) $[0, 1]$ **(d)** $[0, 0.1]$

8. Sketch a graph of $g(x) = -x^3 + 2x$. Then sketch the secant line for each interval of x-values, and find each corresponding average rate of change.

(a) $[2, 3]$ **(b)** $[-3, 3]$
(c) $[-2, -1]$ **(d)** $[-0.1, 0]$

9. Carmella was 42 inches tall when she was 8 years old, and she was 60 inches tall when she was 12 years old. Find her average rate of growth between 8 and 12 years.

10. Harold weighed 60 pounds when he was 6 years old, and he weighed 225 pounds when he was 16 years old. Find his average change in weight between 6 and 16 years.

11. A plumber charges \$65 for a half-hour job and \$128 for a two-hour job. Find the average rate of change, in dollars per hour, between a half-hour and two hours.

12. An overdue book at a library costs \$2.20 if it is 1 week late and \$8.75 if it is 24 days late. Find the average rate of change, in dollars per day, between 1 week and 24 days.

13. The following table shows the high temperature for several days in March:

Date	3	5	8	9	11	14	18	21	25	30
Temp (°F)	63	57	49	52	61	70	65	73	69	79

Find the average rate of change of temperature for each interval.

(a) March 8 to 14 **(b)** March 14 to 18
(c) March 5 to 21 **(d)** March 3 to 30

14. The following table shows the average snowfall in February for several years:

Year	1990	1991	1992	1993	1994	1995	1996	1997	1998	1999	2000
Snowfall (inches)	16	15	22	28	10	33	36	21	12	8	36

Find the average rate of change of snowfall for each interval.

(a) February 1992 to February 1999
(b) February 1990 to February 2000
(c) February 1994 to February 1996
(d) February 1991 to February 1997

15. Let $g(x) = 5$. Find the average rate of change of g for each interval of x-values.

(a) $[1, 4]$ **(b)** $[-1, 2]$
(c) $[4, 4+h]$ **(d)** $[x_0, x_0+h]$

16. Let $f(x) = -2x + 4$. Find the average rate of change of f for each interval of x-values.

(a) $[2, 5]$ **(b)** $[-3, -1]$
(c) $[1, 1+h]$ **(d)** $[x_0, x_0+h]$

17. Let $f(x) = x^2 + 2x - 5$. Find the average rate of change of f for each interval of x-values.

(a) $[0, 5]$ **(b)** $[3, 8]$
(c) $[3, 3+h]$ **(d)** $[x_0, x_0+h]$

18. Let $g(x) = -x^2 - 2x$. Find the average rate of change of g for each interval of x-values.

(a) $[4, 7]$ **(b)** $[-2, 0]$
(c) $[3, 3+h]$ **(d)** $[x_0, x_0+h]$

19. Let $g(x) = -2x - 6$.

(a) Find the slope of the secant line which passes through $(x_0, g(x_0))$ and $(x_0+h, g(x_0+h))$.
(b) Repeat part (a) with $x_0 = 3$ and $h = 0.1$.
(c) Repeat part (a) with $x_0 = 3$ and $h = 0.01$.
(d) Repeat part (a) with $x_0 = 3$ and $h = 0.001$.
(e) Why does the slope of the secant line remain the same in parts (a)–(d)?

20. Let $f(x) = x^2$.

(a) Find the slope of the secant line which passes through $(x_0, f(x_0))$ and $(x_0+h, f(x_0+h))$.
(b) Repeat part (a) with $x_0 = 2$ and $h = 0.1$.
(c) Repeat part (a) with $x_0 = 2$ and $h = 0.01$.
(d) Repeat part (a) with $x_0 = 2$ and $h = 0.001$.
(e) For $x_0 = 2$, what number is the slope of the secant line through $(x_0, f(x_0))$ and $(x_0+h, f(x_0+h))$ approaching as h gets smaller?

21. Find the equation of the function $x(t)$ if the average rate of change is a constant 2 meters per second and $x = 5$ when $t = 1$.

22. Sketch the graph of a nonlinear function whose average rate of change between any two points is always negative.

23. Sketch the graph of a nonlinear function whose average rate of change between any two points is always positive.

24. Sketch the graph of a nonlinear function whose average rate of change is negative for any two points in the interval $(-\infty, 0)$ and is positive for any two points in the interval $(0, \infty)$.

25. If $C(x)$ is a function which determines the cost of producing x units of a certain product, interpret the meaning of the average rate of change over the interval $[x_0, x_0+h]$.

OBJECTIVE *To understand the meaning and mathematical definition of an instantaneous rate of change and of a derivative.*

A.4 Instantaneous Rate of Change: The Derivative

One of the weaknesses of the average rate of change studied in the previous section is its dependence on an interval of x-values; when we are interested in the steepness of a graph, we must consider its steepness over some interval, and different intervals result in different average rates of change. We would like to know the rate of change, or steepness, at a particular point on the graph, and we would like that rate of change to be well defined and unique. The number we seek is called the *instantaneous rate of change*.

EXAMPLE 1 **Driving Speed**

Suppose that you have been driving two hours on the freeway and you note that you have traveled 128 miles during that time. Then your average speed is 64 miles per hour, which is below the posted speed limit of 65 miles per hour.

The equipment used to check the speed of automobiles actually calculates the average rate of change over a very small time interval. However, for now we will consider this to be the instantaneous rate of change.

While you are doing these calculations, a police officer registers your speed at 73 miles per hour. The officer is not interested in your average speed for the past two hours but the speed you are traveling at that instant. The measurement of 73 miles per hour is your instantaneous rate of change at the time the officer registers your speed.

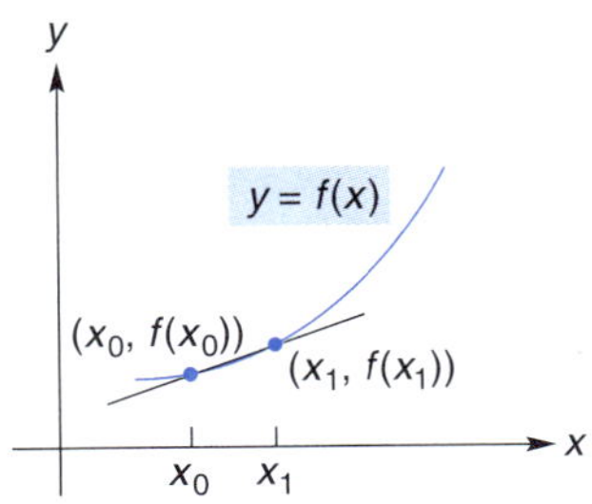

FIGURE A.15 Secant line through $(x_0, f(x_0))$ and $(x_1, f(x_1))$.

Suppose that we want to know the instantaneous rate of change of a function f at a particular x-value, x_0, or, equivalently, the steepness of the graph of f at the point $(x_0, f(x_0))$. We can use average rates of change to estimate this steepness by choosing another point $(x_1, f(x_1))$ on the graph of f, preferably a point close to $(x_0, f(x_0))$. The secant line through these two points approximates the graph of f between x_0 and x_1, and its slope can be used as an approximation of the steepness of the graph at x_0. When $(x_1, f(x_1))$ is close to $(x_0, f(x_0))$, this approximation may be quite good, as shown in Figure A.15.

Now let $h = x_1 - x_0$, so that x_1 becomes $x_0 + h$ and the average rate of change of f on the interval $[x_0, x_1]$, or $[x_0, x_0 + h]$, becomes $\dfrac{f(x_0+h) - f(x_0)}{h}$, as we saw in Section A.3. When $(x_1, f(x_1))$ is close to $(x_0, f(x_0))$, the value of h is close to zero. Generally, the closer h is to zero, the closer $(x_1, f(x_1))$ will be to $(x_0, f(x_0))$ and the better the average rate of change $\dfrac{f(x_0+h) - f(x_0)}{h}$ will approximate the instantaneous rate of change of f at x_0.

The key to the instantaneous rate of change at x_0 is to consider the average rate of change $\dfrac{f(x_0+h) - f(x_0)}{h}$ as the value of h gets closer and closer to zero. In many cases, the closer h gets to zero, the closer $\dfrac{f(x_0+h) - f(x_0)}{h}$ gets to some specific real number. If such a real number exists, it is known as the limit as h approaches zero of $\dfrac{f(x_0+h) - f(x_0)}{h}$ and is denoted by

$$f'(x_0) = \lim_{h \to 0} \frac{f(x_0+h) - f(x_0)}{h}$$

The number $f'(x_0)$ (read, "f prime of x_0") is the **instantaneous rate of change** of f at x_0.

For convenience, we will refer to the instantaneous rate of change of a graph at a particular point as the slope of the graph at that point.

EXAMPLE 2 **Slope of a Graph**

Let $f(x) = x^3$. Find the slope of the graph of f for each of the following values of x.

a. $x = 0$ **b.** $x = 4$ **c.** $x = c$

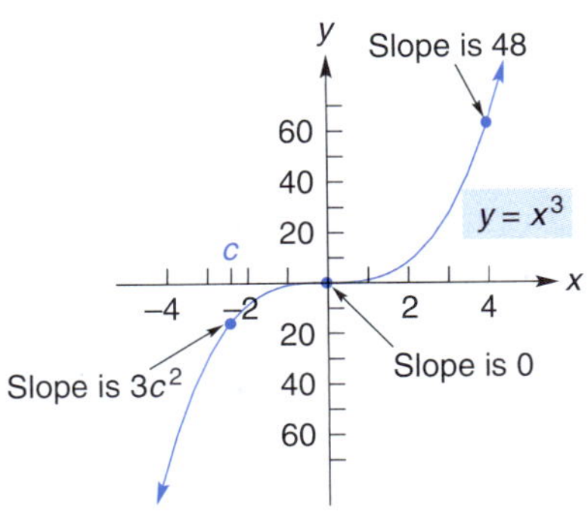

FIGURE A.16 Slope of $f(x) = x^3$ at $x = 0$, $x = 4$, and $x = c$.

Solution: Figure A.16 shows the graph and the points of interest.

a. $$\frac{f(0+h) - f(0)}{h} = \frac{h^3 - 0^3}{h} = h^2$$

As the value of h approaches 0, the value of h^2 also approaches 0; thus, $\lim_{h\to 0} h^2 = 0$. So the slope of the graph at $x = 0$ is 0. Equivalently, $f'(0) = 0$.

b. $$\frac{f(4+h)-f(4)}{h}=\frac{(4+h)^3-4^3}{h}=\frac{(64+48h+12h^2+h^3)-64}{h}$$
$$=\frac{48h+12h^2+h^3}{h}=48+12h+h^2$$

As the value of h approaches 0, the values of $12h$ and h^2 both approach 0, and, consequently, the value of $48+12h+h^2$ approaches 48; thus, $\lim_{h\to 0}(48+12h+h^2)=48$. So the slope of the graph at $x=4$ is 48 and thus $f'(4)=48$.

c. $$\frac{f(c+h)-f(c)}{h}=\frac{(c+h)^3-c^3}{h}=\frac{(c^3+3c^2h+3ch^2+h^3)-c^3}{h}$$
$$=\frac{3c^2h+3ch^2+h^3}{h}=3c^2+3ch+h^2$$

As the value of h approaches 0, the values of $3ch$ and h^2 both approach 0, and, consequently, the value of $3c^2+3ch+h^2$ approaches $3c^2$; thus, $\lim_{h\to 0}(3c^2+3ch+h^2)=3c^2$. So the slope of the graph at $x=c$ is $3c^2$, or $f'(c)=3c^2$.

If a line through a point on a graph has the same slope as the graph at that point, then the line is called a **tangent line** to the curve.

EXAMPLE 3 **Tangent Line**

Find the equation of the tangent line to the graph of $f(x)=3-x^2$ at the point where $x=1$. Graph the curve and the tangent line.

Solution: The y-coordinate of the point on the graph of f with $x=1$ is $f(1)=3-1^2=2$, so the tangent line passes through the point (1, 2). To find the slope of the tangent line, we find the slope of the graph of f at the point (1, 2).

$$\frac{f(1+h)-f(1)}{h}=\frac{[3-(1+h)^2]-(3-1^2)}{h}$$
$$=\frac{3-(1+2h+h^2)-2}{h}$$
$$=\frac{1-1-2h-h^2}{h}$$
$$=\frac{-2h-h^2}{h}$$
$$=-2-h$$

As the value of h approaches 0, the value of $-2-h$ approaches -2. So the slope of the graph of f at $x=1$ is -2. The tangent line, then, passes through (1, 2) with slope -2. Using the point-slope formula, we obtain the equation $y-2=-2(x-1)$, which is equivalent to $y=-2x+4$. Figure A.17 shows the graph of $f(x)$ and the tangent line.

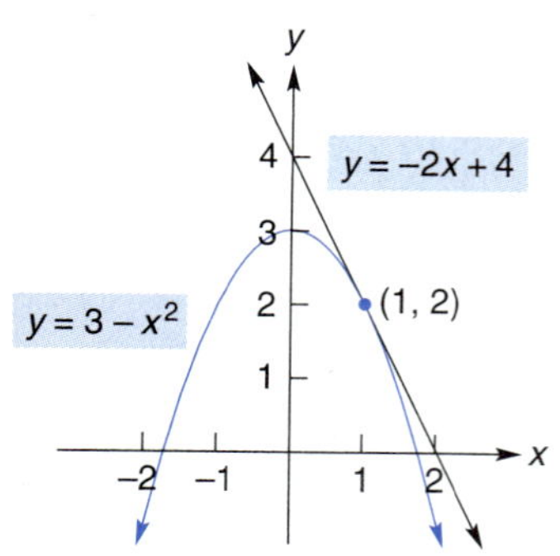

FIGURE A.17 Tangent line to the graph of $f(x)=3-x^2$ at the point (1, 2).

The Derivative

The rate of change of a function is, in general, different at each point on the function—in other words, it varies depending on the value of x. So it makes perfect sense to think of the rate of change as a function in its own right. We

When $y = f(x)$, we can write df/dx and f' as dy/dx and y', respectively.

write the rate of change of f with respect to x as $\frac{df}{dx}$, read "dee f, dee x". This function is called the **derivative** of f with respect to x. When there is no possible confusion about the variable with respect to which the derivative is found, we can use the prime symbol and write f' instead of $\frac{df}{dx}$.

Often a function given by a simple formula also has a derivative given by a simple formula.

EXAMPLE 4 Finding the Derivative

Let $f(x) = \frac{1}{x+2}$. *Find the derivative* $f'(x)$. *Assume that* $x \neq 2$.

Solution: First we write and simplify the expression $\frac{f(x+h)-f(x)}{h}$ for the function $f(x) = \frac{1}{x+2}$. Note that we use the general variable x instead of a specific value x_0 when finding $f'(x)$.

$$\frac{f(x+h)-f(x)}{h} = \frac{\frac{1}{(x+h)+2} - \frac{1}{x+2}}{h} = \frac{(x+2)-(x+h+2)}{h(x+2)(x+h+2)}$$

$$= \frac{-h}{h(x+2)(x+h+2)} = \frac{-1}{(x+2)(x+h+2)}$$

As the value of h approaches 0, the value of $x+h+2$ approaches $x+2$. Thus,

$$f'(x) = \frac{-1}{(x+2)(x+2)} = -\frac{1}{(x+2)^2}$$

The formula for the derivative can then be used to find the slope of the graph at particular points.

EXAMPLE 5 Using the Derivative to Find the Slope

Use the result from Example 4 to find the slope of the graph of $f(x) = \frac{1}{x+2}$ *for each value given.*

a. $x = 0$ **b.** $x = 6$

Solution:

a. $f'(0) = -\frac{1}{(0+2)^2} = -\frac{1}{4}$

At $x = 0$, the slope of the graph is $-\frac{1}{4}$.

b. $f'(6) = -\frac{1}{(6+2)^2} = -\frac{1}{8^2} = -\frac{1}{64}$

At $x = 6$, the slope of the graph is $-\frac{1}{64}$.

EXAMPLE 6 Finding the derivative

For $y = \sqrt{5x+11}$, *find* $\frac{dy}{dx}$ *and evaluate this derivative at* $x = 1$ *to find the slope of the curve* $y = \sqrt{5x+11}$ *at the point* $(1, 4)$.

Solution: Consider $f(x) = \sqrt{5x+11}$, and compute the slope of the curve $y = f(x)$ at the point $(x, \sqrt{5x+11})$.

$$\begin{aligned}
\frac{f(x+h)-f(x)}{h} &= \frac{\sqrt{5(x+h)+11}-\sqrt{5x+11}}{h} \\
&= \frac{(\sqrt{5(x+h)+11}-\sqrt{5x+11})(\sqrt{5(x+h)+11}+\sqrt{5x+11})}{h\cdot(\sqrt{5(x+h)+11}+\sqrt{5x+11})} \\
&= \frac{(\sqrt{5(x+h)+11})^2-(\sqrt{5x+11})^2}{h\cdot(\sqrt{5(x+h)+11}+\sqrt{5x+11})} \\
&= \frac{5(x+h)+11-(5x+11)}{h\cdot(\sqrt{5(x+h)+11}+\sqrt{5x+11})} \\
&= \frac{5h}{h(\sqrt{5(x+h)+11}+\sqrt{5x+11})} \\
&= \frac{5}{\sqrt{5(x+h)+11}+\sqrt{5x+11}}
\end{aligned}$$

As h approaches 0, we get $\dfrac{dy}{dx} = \dfrac{5}{2\sqrt{5x+11}}$. Letting $x = 1$, we find that the slope of $y = \sqrt{5x+11}$ at $(1, 4)$ is $\frac{5}{8}$.

When the function in question is linear, the derivative is a constant function.

EXAMPLE 7 **The Derivative of a Linear Function**

Let $f(x) = 7x - 11$. Find the derivative $f'(x)$.

Solution: First we write and simplify the expression $\dfrac{f(x+h)-f(x)}{h}$ for the function $f(x) = 7x - 11$.

$$\begin{aligned}
\frac{f(x+h)-f(x)}{h} &= \frac{[7(x+h)-11]-(7x-11)}{h} \\
&= \frac{7x+7h-11-7x+11}{h} \\
&= \frac{7h}{h} = 7
\end{aligned}$$

Since 7 does not vary as h approaches 0, $f'(x) = 7$. Note that this is the slope of the line $y = 7x - 11$.

As Example 7 illustrates, the derivative of a linear function is a constant function equal to the slope of the line. It also follows that, since the slope of a horizontal line is zero, the derivative of a constant function is zero.

In the study of economics, a *total-cost function,* $c = f(q)$, gives the total cost c of producing and marketing q units of a product. The *marginal cost* is the instantaneous rate of change of c, or $f(q)$ with respect to q. Thus marginal cost is $\dfrac{dc}{dq}$ and can be interpreted as the approximate cost to produce one additional unit.

EXAMPLE 8 **Marginal Cost**

Suppose the cost of producing q compact discs is given by $c = 0.001q^2 + 4q + 300$. *Find the marginal cost when 1500 compact discs are produced.*

Solution: Let $f(q) = 0.001q^2 + 4q + 300$. Then

$$\begin{aligned}\frac{f(1500+h) - f(1500)}{h} &= \frac{[0.001(1500+h)^2 + 4(1500+h) + 300] - [0.001(1500)^2 + 4(1500) + 300]}{h} \\ &= \frac{(0.001h^2 + 7h + 8550) - 8550}{h} \\ &= 0.001h + 7\end{aligned}$$

As h approaches 0, $0.001h + 7$ approaches 7. Thus, $\frac{df}{dq}(1500) = 7$, and the marginal cost is \$7 per unit.

We will study other applications of derivatives in Section A.8.

Exercise A.4

In Problems 1–4, determine whether the rate given is an average rate of change or an instantaneous rate of change.

1. An administrative assistant types 70 words per minute.
2. The advertised gas mileage for a car is 28 miles per gallon.
3. The tachometer reading of an automobile is 5000 revolutions per minute.
4. Dan lost control of his car when his speed reached 120 miles per hour.

In Problems 5–10, find the slope of the graph of each function f at the given value of x.

5. $f(x) = x^2 \quad x = -4$
6. $f(x) = \sqrt{x} \quad x = 4$
7. $f(x) = \frac{1}{x} \quad x = 3$
8. $f(x) = 2x \quad x = -1$
9. $f(x) = x^2 - x \quad x = 1$
10. $f(x) = x^3 + 8 \quad x = -2$

In Problems 11–14, find the equation of the tangent line to the graph of each function f at the given point.

11. $f(x) = x^2 - x \quad (1, 0)$
12. $f(x) = x^3 + 8 \quad (-2, 0)$
13. $f(x) = \sqrt{x} \quad (1, 1)$
14. $f(x) = \frac{1}{x+3} \quad \left(-1, \frac{1}{2}\right)$

For each function $y = f(x)$ *in Problems 15–27, find* $\frac{dy}{dx}$.

15. $y = 11$
16. $y = 5x$
17. $y = 5x + 11$
18. $y = 10x^2$
19. $y = 10x^2 + 11$
20. $y = 10x^2 + 5x + 1492$
21. $y = \frac{1}{5x + 11}$
22. $y = \frac{1492}{5x + 11}$
23. $y = \frac{3}{12 - 5x}$
24. $y = \frac{3}{12 - 5x} + 11$
25. $y = \sqrt{3x - 7}$
26. $y = \sqrt{12 - 5x}$
27. $y = \sqrt{12 - 5x} + 10x^2 + 11$
28. Suppose the cost of producing q copies of a magazine is given by $c = 1.5q + 8000$. Find the marginal cost $\frac{dc}{dq}$, and evaluate the expression when $q = 15{,}300$ copies are produced.
29. Suppose the cost of producing q rugs is given by $c = 0.05q^2 + 28q + 5000$. Find the marginal cost $\frac{dc}{dq}$, and evaluate the expression when $q = 75$ rugs are produced.
30. Suppose the cost of producing q computers is given by $c = 0.0001q^3 - 0.08q^2 + 1000q + 20{,}000$. Find the marginal cost $\frac{dc}{dq}$, and evaluate the expression when $q = 900$ computers are produced.
31. A ball is thrown straight upward at the rate of 32 feet per second. Its height, in feet, after t seconds is given by the function $h(t) = -16t^2 + 32t$. Find the instantaneous velocity $\frac{dh}{dt}$, and evaluate the expression for each value of t.
    (a) $t = 0$ seconds
    (b) $t = 2$ seconds
    (c) $t = 3$ seconds

OBJECTIVE *To develop a method for finding an approximation for the area defined by a curve.*

A.5 Area under A Curve

In Section A.4, we investigated the rate of change of a curve. We now turn to a seemingly unrelated topic: the area under a curve. Later we will see that these two topics are in fact closely connected.

Finding Exact Areas

Suppose we want to find the area of a region in the coordinate plane bounded by mathematical functions. In some cases, we can find the area using simple geometric formulas.

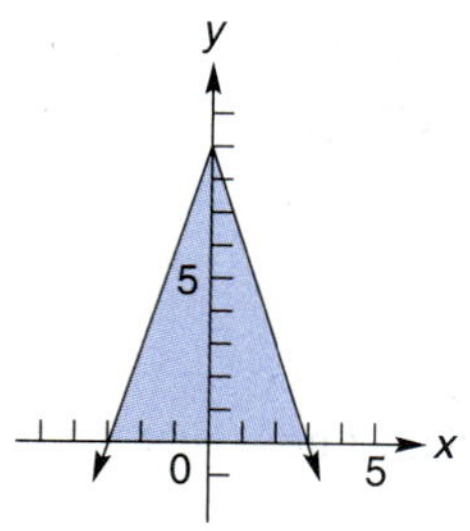

FIGURE A.18 $y = 9 - |3x|$.

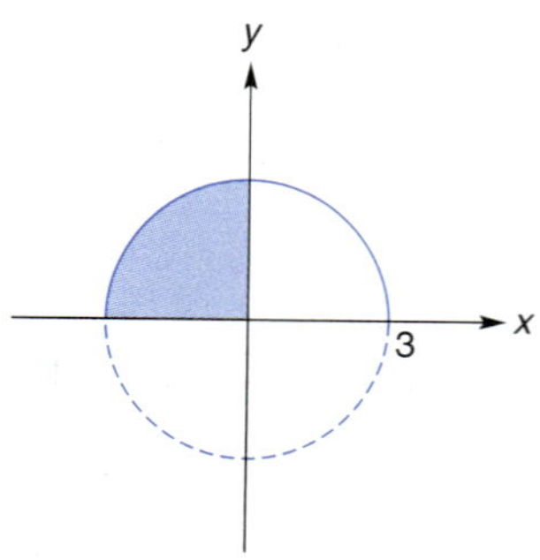

FIGURE A.19 $y = \sqrt{9 - x^2}$.

EXAMPLE 1 **Simple Areas**

Find the areas of the regions described.

a. *The region bounded by the x-axis and the graph of the function* $y = 9 - |3x|$.

Solution: The region, shown in Figure A.18, is a triangle with base $b = 6$ and height $h = 9$. The area is $\frac{1}{2}bh = \frac{1}{2}(6)(9) = 27$ square units.

b. *The region to the left of the y-axis bounded by the x-axis and the graph of the function* $y = \sqrt{9 - x^2}$.

Solution: The region, shown in Figure A.19, is one-fourth of a circle with radius $r = 3$. The area is $\frac{1}{4}\pi r^2 = \frac{1}{4}3^2\pi = \frac{9}{4}\pi$ square units.

EXAMPLE 2 **Composite Areas**

Find the area of the region above the x-axis bounded by the x-axis and the following equations:

$$y = \sqrt{9 - (x + 3)^2} \quad \textit{from } x = -6 \textit{ to } x = 0$$

$$y = 2x \quad \textit{from } x = 0 \textit{ to } x = 2$$

$$y = 4 \quad \textit{from } x = 2 \textit{ to } x = 6$$

Solution: First, graph the equations and label the regions that correspond to each equation. (See Figure A.20.)

The area of the whole region is the sum of the areas of the labelled regions.

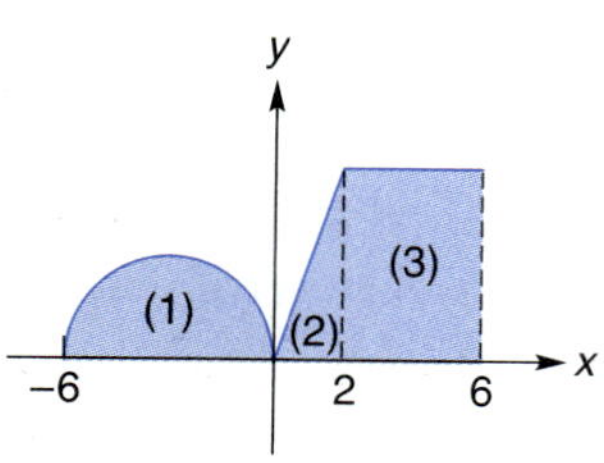

FIGURE A.20 Area for Example 2.

Region (1): This region is a semi-circle with radius $r = 3$. The area is $\frac{1}{2}\pi r^2 = \frac{1}{2}\pi(3)^2 = \frac{9}{2}\pi$ square units.

Region (2): This region is a triangle with base $b = 2$ and height $h = 4$. The area is $\frac{1}{2}bh = \frac{1}{2}(2)(4) = 4$ square units.

Region (3): This region is a rectangle with base $b = 4$ and height $h = 4$. The area is $bh = (4)(4) = 16$ square units.

The area of the entire region = area(1) + area(2) + area(3)

$$= \left(\frac{9}{2}\pi + 4 + 16\right) \text{ square units}$$

$$= \left(\frac{9}{2}\pi + 20\right) \text{ square units}$$

Finding Approximate Areas

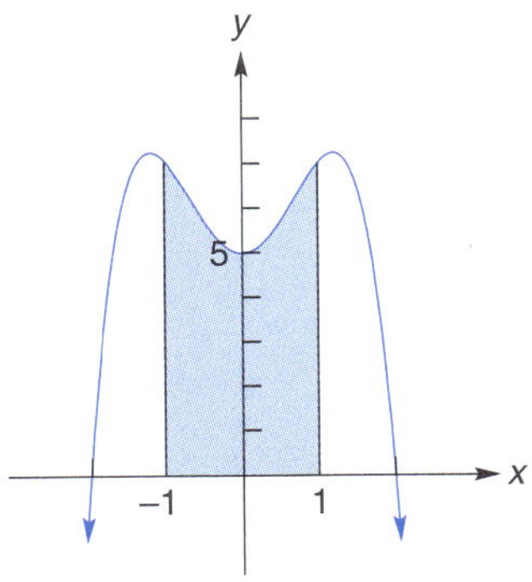

FIGURE A.21 Region bounded by x-axis and $f(x) = -x^4 + 3x^2 + 5$ from $x = -1$ to $x = 1$.

The area under some curves cannot be found using simple formulas. For instance, how can we find the area of the region shown in Figure A.21? This problem is similar to the one we faced earlier, when we wanted to find the slope of a curve at a given point.

One approach we might take is finding the *approximate* area under the curve. We can use rectangles to estimate this area. For example, to use four rectangles to estimate the region in Figure A.21, we divide the interval into four subintervals: $[-1, -0.5]$, $[-0.5, 0]$, $[0, 0.5]$, and $[0.5, 1]$. Then we draw a rectangle for each subinterval. The base of each rectangle is on the x-axis, and the top side of each rectangle intersects the graph of $f(x)$. One possible intersection point is where the function crosses the right end of the subinterval. We can use this point to calculate the height of each rectangle. Thus, we use heights of $f(-0.5) = 5.6875$, $f(0) = 5$, $f(0.5) = 5.6875$, and $f(1) = 7$. The width of each rectangle is 0.5 unit. The approximate area is therefore $0.5(5.6875+5+5.6875+7) = 0.5(23.375) = 11.6875$. The rectangles are shown in Figure A.22.

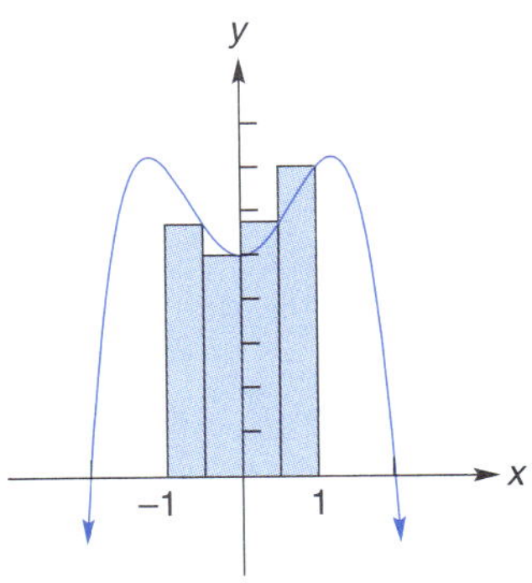

FIGURE A.22 Area approximation using rectangles.

It is not necessary to use the right endpoint of each subinterval, but calculations are usually simplest if you use one of the following conventions for the height of the rectangle in each subinterval $[a, b]$:

- Left-point rule: Use the value at the left endpoint of the subinterval. That is, use $f(a)$.
- Midpoint rule: Use the value at the midpoint of the subinterval. That is, use $f\left(\frac{a+b}{2}\right)$.
- Right-point rule: Use the value at the right endpoint of the subinterval. That is, use $f(b)$.

EXAMPLE 3 **Approximating an Area Using Rectangles**

Approximate the area of the quarter-circular region in part (b) of Example 1, bounded by the x-axis and the curve $y = \sqrt{9 - x^2}$. Compute the difference between the approximation and the exact area.

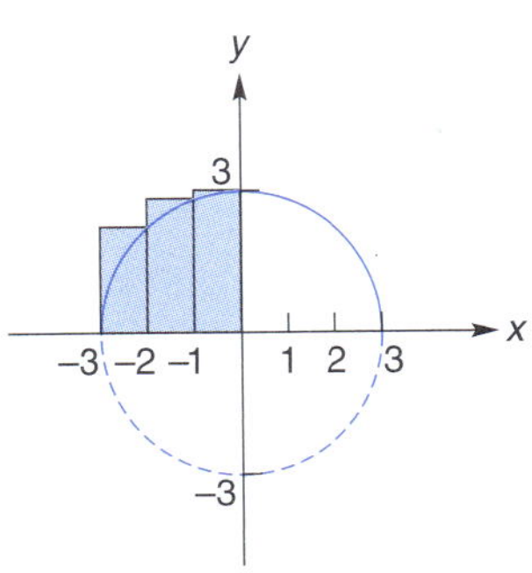

FIGURE A.23 A three-rectangle approximation.

a. *Use three equal-width rectangles, as shown in Figure A.23, to approximate the area.*

Solution:

Area of first rectangle: the base is 1 and the height is $y = \sqrt{9 - (-2)^2} \approx 2.2361$. The area is approximately $1(2.2361)$ or 2.2361.

Area of second rectangle: the base is 1 and the height is $y = \sqrt{9 - (-1)^2} \approx 2.8284$. The area is approximately $1(2.8284)$ or 2.8284.

Area of third rectangle: the base is 1 and the height is 3. The area is $1(3) = 3$.

The total approximated area is $2.2361 + 2.8284 + 3 = 8.0645$. The difference between this and the exact area:

$$\left|\frac{9}{4}\pi - 8.0645\right| \approx |7.0686 - 8.0645| \approx 0.9959$$

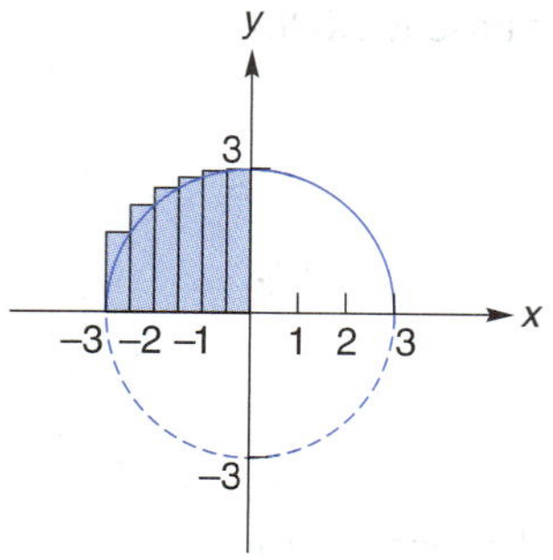

FIGURE A.24 A six-rectangle approximation.

b. *Use six equal-width rectangles, as shown in Figure A.24, to approximate the area.*

Solution:

Area of the first rectangle: the base is $\frac{1}{2}$ and the height is $\sqrt{9-\left(-\frac{5}{2}\right)^2} \approx 1.6583$. The area is $\frac{1}{2}(1.6583) \approx 0.8292$.

Area of the second rectangle: the base is $\frac{1}{2}$ and the height is $\sqrt{9-(-2)^2} \approx 2.2361$. The area is $\frac{1}{2}(2.2361) \approx 1.1181$.

Area of the third rectangle: the base is $\frac{1}{2}$ and the height is $\sqrt{9-\left(-\frac{3}{2}\right)^2} \approx 2.5981$. The area is $\frac{1}{2}(2.5981) \approx 1.2991$.

Area of the fourth rectangle: the base is $\frac{1}{2}$ and the height is $\sqrt{9-(-1)^2} \approx 2.8284$. The area is $\frac{1}{2}(2.8284) = 1.4142$.

Area of the fifth rectangle: the base is $\frac{1}{2}$ and the height is $\sqrt{9-\left(-\frac{1}{2}\right)^2} \approx 2.9580$. The area is $\frac{1}{2}(2.9580) = 1.4790$.

Area of the sixth rectangle: the base is $\frac{1}{2}$ and the height is 3. The area is $\frac{1}{2}(3) = 1.5$.

The total approximated area is $0.8292 + 1.1181 + 1.2991 + 1.4142 + 1.4790 + 1.5 = 7.6396$. The difference between this and the exact area is $|7.0686 - 7.6396| = 0.5710$.

In Example 3, note that the calculated areas of 8.0645 and 7.6396 are overestimates of the actual area of the region, since the rectangles extend above the graph of $f(x)$. By using more rectangles with smaller widths in Example 3(b), we get a better approximation to the area of the region.

Another method of estimating areas uses trapezoids instead of rectangles. For each subinterval $[a, b]$, we use a trapezoid with vertical bases of length $f(a)$ and $f(b)$, respectively, and horizontal "height" of $b-a$. The area of this trapezoid is $\frac{1}{2}(b-a)[f(a)+f(b)]$. The trapezoid rule is illustrated in Example 4.

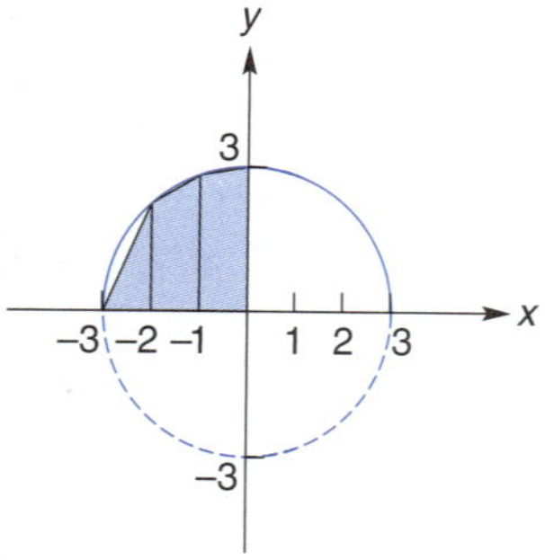

FIGURE A.25 A three-trapezoid approximation.

EXAMPLE 4 Approximating an Area Using Trapezoids

Find the sum of the areas of the three equal-height trapezoids shown in Figure A.25, where the bases of each trapezoid have one end on the x-axis and the other end on the curve $y = \sqrt{9-x^2}$.

Solution:

Area of the first trapezoid: This is really a triangle, with base 1 and height $\sqrt{9-(-2)^2} \approx 2.2361$. The area is $\frac{1}{2}(1)(2.2361) \approx 1.1181$.

Area of the second trapezoid: The height is 1 and the left base is 2.2361. The right base is $\sqrt{9-(-1)^2} \approx 2.8284$. The area is $\frac{1}{2}(1)(2.2361+2.8284) \approx 2.5323$.

Area of the third trapezoid: The height is 1 and the left base is 2.8284. The right base is 3. The area is $\frac{1}{2}(1)(2.8284 + 3) \approx 2.9142$.

The total approximated area is $1.1181 + 2.5323 + 2.9142 \approx 6.5646$.

The difference between this and the exact area is $|7.0686 - 6.5646| \approx 0.5040$.

In Examples 3 and 4, we worked with a region whose exact area is known. This illustrates that, for a given number of "slices," the trapezoid-based approximation is more accurate than the rectangle-based approximation. But the accuracy of the rectangle method can be improved by taking more slices. As the number of slices grows larger and larger, the error (represented by the collective area of the wedge-shaped regions at the top of each slice) grows smaller and smaller. This idea of "more slices" is the basis of integral calculus. In Section A.7, we will create enough slices to find *exact* areas.

It should be noted that for some regions it is not practical to find the exact area using integral calculus. For such regions, the trapezoid- and rectangle-based methods remain useful tools for finding approximate areas. (The trapezoid-based method is discussed further in Section 14.9.)

Exercise A.5

In Problems 1 and 2, find the area of the region between each graph and the x-axis.

1.

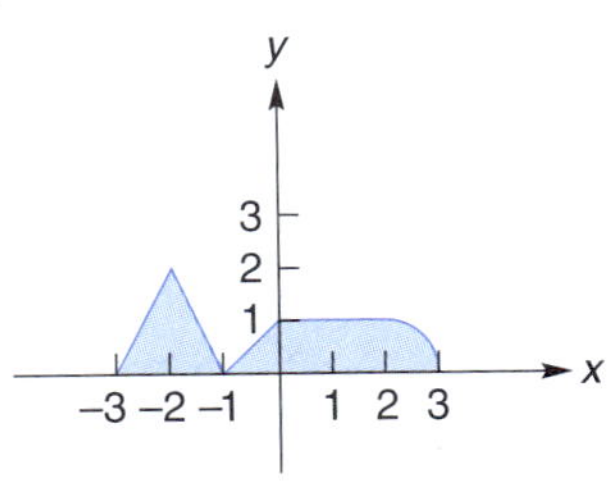

FIGURE A.26 Diagram for Problem 1

2.

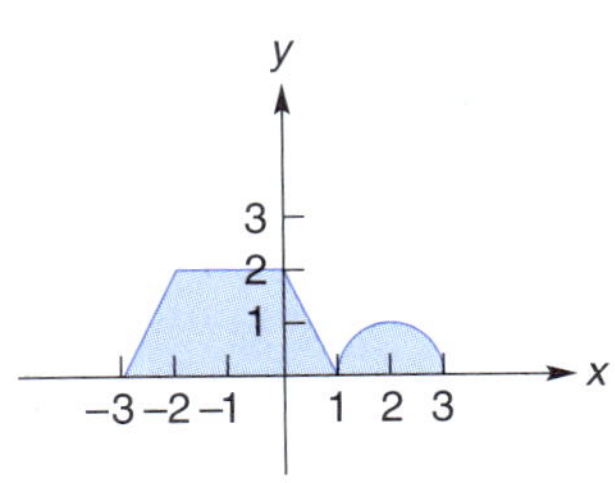

FIGURE A.27 Diagram for Problem 2

In Problems 3–6, find the area of the shaded region.

3.

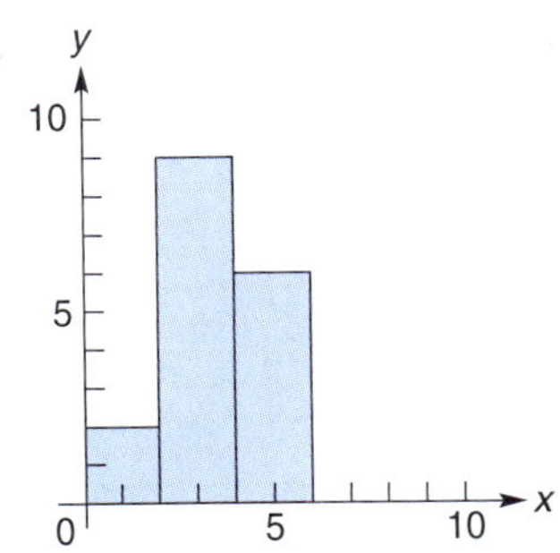

FIGURE A.28 Diagram for Problem 3

4.

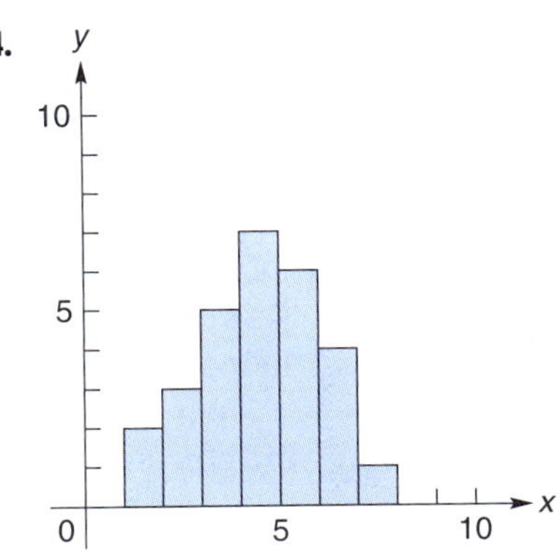

FIGURE A.29 Diagram for Problem 4

5.

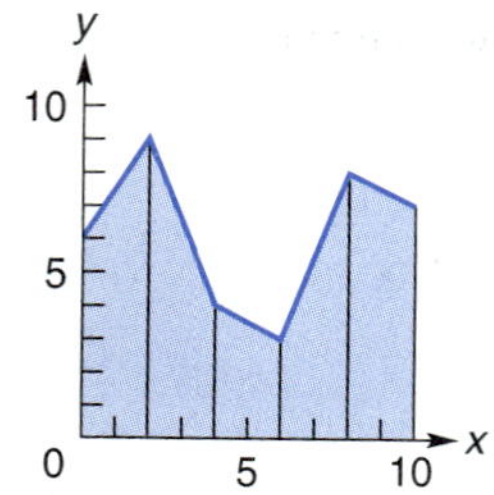

FIGURE A.30 Diagram for Problem 5

6.

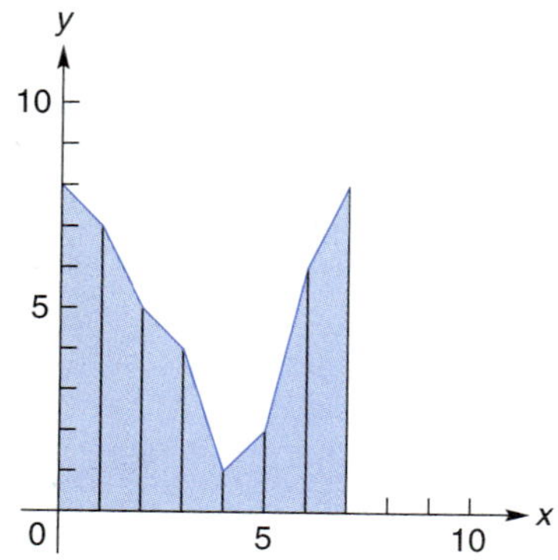

FIGURE A.31 Diagram for Problem 6

In Problems 7–10, (a) find the area of the region. Then estimate the area of the region using (b) four rectangles at equally spaced subintervals, (c) eight rectangles at equally spaced subintervals, and (d) four trapezoids at equally spaced subintervals. Use the right-point rule for the rectangle estimations. (e) Which estimation gives the best approximation to the actual area you found in part (a)?

7. The region bounded by the graph of $y = 3 + |x - 3|$ and the x-axis on the interval $[0, 8]$.

8. The region bounded by the graph of $y = \sqrt{36 - x^2}$ and the x-axis, to the left of the y-axis.

9. The region bounded by the graph of $y = \sqrt{16 - x^2}$ and the x-axis, to the right of the y-axis.

10. The region bounded by the graph of $y = \sqrt{64 - x^2}$ and the x-axis.

In Problems 11–14, estimate the area of the region using (a) four rectangles at equally spaced subintervals, and (b) four trapezoids at equally spaced subintervals. Use the right-point rule for the rectangle estimations. (c) Which estimation gives the best approximation to the actual area?

11. The region between the graph of $f(x) = x^2$ and the x-axis for x-values in the interval $[1, 5]$. The actual area is $41\frac{1}{3}$ square units.

12. The region between the graph of $f(x) = x^3$ and the x-axis for x-values in the interval $[0, 4]$. The actual area is 64 square units.

13. The region between the graph of $h(x) = x^2 + x$ and the x-axis for x-values in the interval $[2, 6]$. The actual area is $85\frac{1}{3}$ square units.

14. The region between the graph of $h(x) = x^2 + x + 3$ and the x-axis for x-values in the interval $[-2, 2]$. The actual area is $17\frac{1}{3}$ square units.

15. Graph the functions $f(x) = -x^2 + 3x + 4$ and $g(x) = -x^2 + 3x$ on the same axes. Compare the areas bounded by each curve and the x-axis on the x-interval $[0, 3]$. What do they have in common? How do they differ?

●● OBJECTIVE *To state sums in sigma notation and evaluate such sums.*

A.6 Sigma Notation

In Section A.5, we used rectangles to estimate the area under a curve, over some interval. We divided the interval into subintervals, which we used as bases for the rectangles. We used the value of the curve at each subinterval for the height of each rectangle. We then found the areas of these rectangles, and summed them to approximate the area under the curve.

When many terms are being added together, it is sometimes both possible and convenient to write the sum using a special notation. For example, suppose we are adding four terms, each of which has the form $3k$, where k is replaced, in succession, by 5, 6, 7, and 8. In **sigma notation**, we can write this sum as $\sum_{k=5}^{8} 3k$. The $\sum$ is the Greek capital letter sigma, from which we get the letter S. It stands for "sum," and the expression $\sum_{k=5}^{8} 3k$ stands for the sum of all numbers $3k$

where k ranges from 5 to 8 in steps of 1. (The steps in sigma notation are always 1 in size.)

$$\sum_{k=5}^{8} 3k = 3(5) + 3(6) + 3(7) + 3(8) = 15 + 18 + 21 + 24 = 78$$

This notation is sometimes called *summation notation,* and 5 and 8 are called the **limits of summation.** Summation notation can only be used when all the terms can be represented by a single variable expression—in this case, $3k$.

EXAMPLE 1 Evaluating Sums

Evaluate the given sums.

a. $\sum_{n=3}^{7}(5n-2)$

Solution:

$$\sum_{n=3}^{7}(5n-2) = [5(3)-2] + [5(4)-2] + [5(5)-2] + [5(6)-2] + [5(7)-2]$$
$$= 13 + 18 + 23 + 28 + 33$$
$$= 115$$

b. $\sum_{j=0}^{6}(j^2+1)$

Solution:

$$\sum_{j=0}^{6}(j^2+1) = (0^2+1) + (1^2+1) + (2^2+1) + (3^2+1) + (4^2+1)$$
$$+ (5^2+1) + (6^2+1)$$
$$= 1 + 2 + 5 + 10 + 17 + 26 + 37$$
$$= 98$$

EXAMPLE 2 Writing a Sum Using Sigma Notation

Express the sum $14 + 16 + 18 + 20 + 22 + \cdots + 100$ *in sigma notation.*

Solution: There are many ways to express this sum in sigma notation. One method is to notice that the values being added are $2n$, for $n = 7$ to 50. The sum can thus be represented as

$$\sum_{n=7}^{50} 2n$$

Another method is to notice that the values being added are $2k + 12$, for $k = 1$ to 44. The sum can thus be represented as

$$\sum_{k=1}^{44}(2k+12)$$

Since sigma notation is used to represent the addition of terms, we can utilize the properties of addition when performing operations on sums written in sigma notation. By applying these properties, we can create a list of properties and formulas for sigma notation.

By the Distributive Property of Addition,

$$ca_1 + ca_2 + \cdots + ca_n = c(a_1 + a_2 + \cdots + a_n)$$

So, in sigma notation,

$$\sum_{k=m}^{n} ca_k = c\sum_{k=m}^{n} a_k \tag{1}$$

Note that c must be constant with respect to k for Equation (1) to be used.

By the Commutative Property of Addition,

$$a_1 + b_1 + a_2 + b_2 + \cdots + a_n + b_n = a_1 + a_2 + \cdots + a_n + b_1 + b_2 + \cdots + b_n$$

So, in sigma notation,

$$\sum_{k=m}^{n} (a_k + b_k) = \sum_{k=m}^{n} a_k + \sum_{k=m}^{n} b_k \tag{2}$$

Sometimes we want to change the limits of summation. We can do this, as long as we keep the step size at 1.

$$\sum_{k=m}^{n} a_k = \sum_{k=p}^{n+p-m} a_{k+m-p} \tag{3}$$

Sigma notation sums can also be "chained together" when the upper limit of one summation is one less than the lower limit of the other summation:

$$\sum_{k=m}^{p-1} a_k + \sum_{k=p}^{n} a_k = \sum_{k=m}^{n} a_k \tag{4}$$

In addition to these four basic rules, there are some other rules worth noting. The first two follow, respectively, from Equations (1) and (2).

$$\sum_{k=1}^{n} c = cn \tag{5}$$

$$\sum_{k=m}^{n} (a_k - b_k) = \sum_{k=m}^{n} a_k - \sum_{k=m}^{n} b_k \tag{6}$$

Establishing the next three formulas requires a proof method called mathematical induction, which we will not demonstrate here.

$$\sum_{k=1}^{n} k = \frac{n(n+1)}{2} \tag{7}$$

$$\sum_{k=1}^{n} k^2 = \frac{n(n+1)(2n+1)}{6} \tag{8}$$

$$\sum_{k=1}^{n} k^3 = \frac{n^2(n+1)^2}{4} \tag{9}$$

EXAMPLE 3 **Applying the Properties of Sigma Notation**

Evaluate the given sums.

a. $\sum_{j=30}^{100} 4$ **b.** $\sum_{k=1}^{100} (5k+3)$ **c.** $\sum_{k=1}^{200} 9k^2$

Solution:

a.
$$\sum_{j=30}^{100} 4 = \sum_{j=1}^{71} 4 \quad \text{[by Equation (3)]}$$
$$= 4\left(\sum_{j=1}^{71} 1\right) \quad \text{[by Equation (1)]}$$
$$= 4 \cdot 71 \quad \text{[by Equation (5)]}$$
$$= 284$$

b.
$$\sum_{k=1}^{100} (5k+3) = \sum_{k=1}^{100} 5k + \sum_{k=1}^{100} 3 \quad \text{[by Equation (2)]}$$
$$= 5\left(\sum_{k=1}^{100} k\right) + 3\left(\sum_{k=1}^{100} 1\right) \quad \text{[by Equation (1)]}$$
$$= 5\left(\frac{100 \cdot 101}{2}\right) + 3(100) \quad \text{[by Equations (7) and (5)]}$$
$$= 25{,}250 + 300$$
$$= 25{,}550$$

c.
$$\sum_{k=1}^{200} 9k^2 = 9\sum_{k=1}^{200} k^2 \quad \text{[by Equation (1)]}$$
$$= 9\left(\frac{200 \cdot 201 \cdot 401}{6}\right) \quad \text{[by Equation (8)]}$$
$$= 24{,}180{,}300$$

The symbol Δ is the Greek letter *delta*, which mathematicians often use to represent the change in a quantity.

Now we have a tool for representing and evaluating the sum of many terms, such as the rectangles we summed in Section A.5 to find the area of a region under a curve. The subintervals used to create such a sum need not have equal length. However, we will use equally spaced subintervals to keep the calculations simple. We will use the expression Δx to denote the width of each subinterval. As you can imagine, using smaller subintervals will result in a more accurate estimate for the area of the region. Of course, small subintervals can also result in very tedious calculations. We can use the properties and formulas of sigma notation to greatly simplify these calculations.

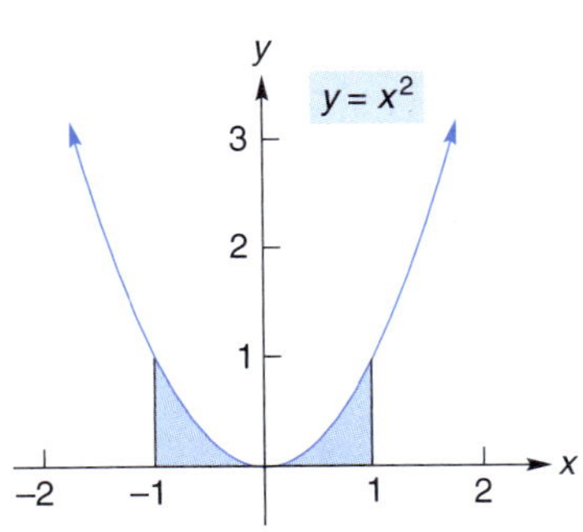

FIGURE A.32 A bounded region.

EXAMPLE 4 **Approximating an Area Using Sigma Notation**

Approximate the area of the region bounded by $f(x) = x^2$ and the x-axis on the interval $[-1, 1]$. (See Figure A.32.) Use the midpoint rule and $\Delta x = 0.01$.

Solution: Since the length of the interval is 2, the number of subintervals is $N = \frac{2}{\Delta x} = \frac{2}{0.01} = 200$. The subintervals are $[-1, -0.99]$, $[-0.99, -0.98]$, and so on. Using the midpoint rule, we must evaluate $f(x)$ at $x = -0.995, -0.985$, and so on. Therefore, $x_k = 0.01k - 1.005$. The sum is

$$\begin{aligned}\text{Area} &\approx \Delta x \cdot \sum_{k=1}^{N} f(x_k) = 0.01 \sum_{k=1}^{200} f(0.01k - 1.005)\\ &= 0.01 \sum_{k=1}^{200} (0.01k - 1.005)^2\\ &= 0.01 \sum_{k=1}^{200} (1.010025 - 0.0201k + 0.0001k^2)\\ &= 0.01 \left\{ 200(1.010025) - 0.0201 \left[\frac{200(201)}{2}\right] + 0.0001 \left[\frac{200(201)(401)}{6}\right] \right\}\\ &= 0.01(202.005 - 404.01 + 268.67)\\ &= 0.01(66.665)\\ &= 0.66665\end{aligned}$$

Later, you will be able to calculate the exact area of this region by evaluating a definite integral. Our approximation is close to the exact value for the area, which is $\frac{2}{3}$.

Exercise A.6

In Problems 1 and 2, give the lower limit of summation, the upper limit of summation, and the index of summation for each expression.

1. $\sum_{t=12}^{17} (8t^2 - 5t + 3)$ **2.** $\sum_{m=3}^{450} (8m - 4)$

In Problems 3–6 evaluate the given sums.

3. $\sum_{i=1}^{7} 6i$ **4.** $\sum_{p=0}^{4} 10p$

5. $\sum_{k=3}^{9} (10k + 16)$ **6.** $\sum_{n=10}^{14} (3n - 5)$

In Problems 7–12 express the given sums in sigma notation.

7. $36 + 37 + 38 + 39 + \cdots + 60$

8. $1 + 4 + 9 + 16 + 25$

9. $5^3 + 5^4 + 5^5 + 5^6 + 5^7 + 5^8$

10. $11 + 15 + 19 + 23 + \cdots + 71$

11. $2 + 4 + 8 + 16 + 32 + 64 + 128 + 256$

12. $10 + 100 + 1000 + \cdots + 100{,}000{,}000$

In Problems 13–26 evaluate the given sums.

13. $\sum_{k=1}^{52} 10$ **14.** $\sum_{k=35}^{135} 2$

15. $\sum_{k=1}^{n} \left(5 \cdot \frac{1}{n}\right)$ **16.** $\sum_{k=1}^{200} (k - 100)$

17. $\sum_{k=51}^{100} 10k$ **18.** $\sum_{k=1}^{n} \frac{n}{n+1} k$

19. $\sum_{k=1}^{20} (5k^2 + 3k)$ **20.** $\sum_{k=1}^{100} \frac{3k^2 - 200k}{101}$

21. $\sum_{k=51}^{100} k^2$ **22.** $\sum_{k=1}^{50} (k + 50)^2$

23. $\sum_{k=1}^{10} \left\{ \left[4 - \left(\frac{2k}{10}\right)^2\right] \left(\frac{2}{10}\right) \right\}$

24. $\sum_{k=1}^{100} \left\{ \left[4 - \left(\frac{2}{100}k\right)^2\right] \left(\frac{2}{100}\right) \right\}$

25. $\sum_{k=1}^{n}\left\{\left[4-\left(\frac{2}{n}\cdot k\right)^2\right]\frac{2}{n}\right\}$

26. $\sum_{k=1}^{n}\frac{k^2}{(n+1)(2n+1)}$

In Problems 27 and 28 approximate the area of the region bounded by the given curve and the x-axis over the given interval. Use the midpoint rule and the interval size given.

27. $f(x) = -x^2 + 3x$ over the interval $[0, 3]$ $\Delta x = 0.05$

28. $g(x) = x^2 + x$ over the interval $[0, 2]$ $\Delta x = 0.01$

OBJECTIVE *To use definite integrals to represent the limit of a sum over a given interval and to evaluate integrals.*

A.7 Areas and Integrals

In Section A.5 we saw that the area under a curve described by a function $f(x)$ can be approximated using rectangles or trapezoids to model the shape of sections of the curve. For instance, a sequence of rectangles with a constant width but varying heights can be used to model the region. We can then approximate the area under the curve by summing the areas of the rectangles. We can improve the approximation by increasing the number of rectangles used to model the area. When we do this, we are decreasing the width of each rectangle. In this section we discuss a technique for finding the area by allowing the number of rectangles to increase without bound.

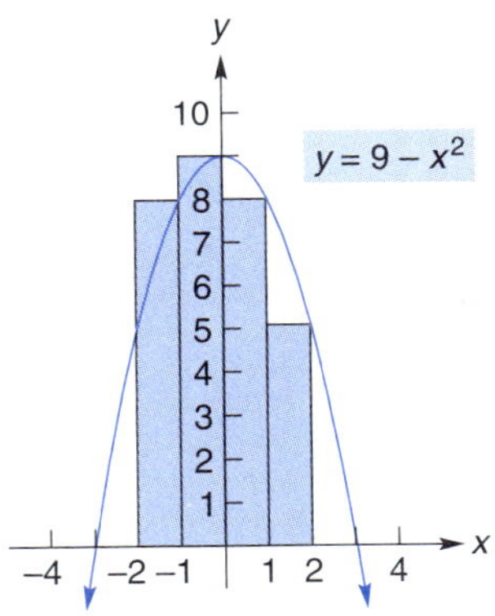

FIGURE A.33 Approximation by rectangles with $\Delta x = 1.0$.

Finding Area from A Sequence of Sums

Figure A.33 shows an approximation of the area under $f(x) = -x^2 + 9$ from $x = -2$ to $x = 2$ using rectangles. In this case, we divided the interval $[-2, 2]$, whose width is 4 units, into a sequence of 4 equal-sized intervals. We use Δx to represent the width of each subinterval which is the width of each rectangle used to model the area. In this case, the width of each rectangle is

$$\Delta x = \frac{\text{interval length}}{\text{number of rectangles}} = \frac{4}{4} = 1.$$

We can approximate the area under the curve between $x = -2$ and $x = 2$ by summing the areas of each of these rectangles.

$$\sum_{i=1}^{4}[-(-2+i(1))^2+9](1) = 8+9+8+5 = 30$$

We can improve our approximation by increasing the number of rectangles used to model the area. If we double the number of rectangles, the width of each rectangle becomes $\Delta x = \frac{4}{8} = 0.5$. (See Figure A.34.)

When we sum the areas of each of these rectangles, we get

$$\sum_{i=1}^{8}[-(-2+i(0.5))^2+9](0.5)$$
$$= 3.375 + 4 + 4.375 + 4.5 + 4.375 + 4 + 3.375 + 2.5$$
$$= 30.5$$

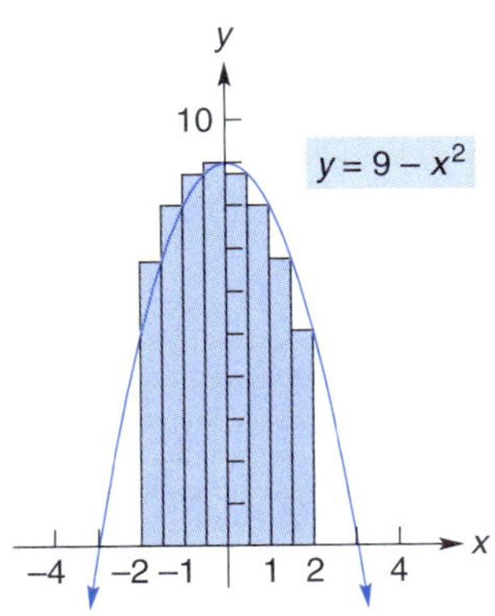

FIGURE A.34 Approximation by rectangles with $\Delta x = 0.5$.

By increasing the number of rectangles over the interval $[-2, 2]$, we can get better and better approximations. In Example 1, we show how to approximate this area using n rectangles.

EXAMPLE 1 **The Limit of an Area Approximation**

Approximate the area under the curve $f(x) = -x^2 + 9$ over the interval $[-2, 2]$ using n rectangles.

Solution: Using the properties and formulas of sigma notation introduced in Section A.6, we get

$$\begin{aligned}
&\sum_{i=1}^{n}\left[-\left(-2+i\left(\frac{4}{n}\right)\right)^2+9\right]\left(\frac{4}{n}\right)\\
&=\sum_{i=1}^{n}\left(-4+\frac{16i}{n}-\frac{16i^2}{n^2}+9\right)\left(\frac{4}{n}\right)\\
&=\sum_{i=1}^{n}\left(\frac{20}{n}+\frac{64i}{n^2}-\frac{64i^2}{n^3}\right)\\
&=\frac{20}{n}\sum_{i=1}^{n}1+\frac{64}{n^2}\sum_{i=1}^{n}i-\frac{64}{n^3}\sum_{i=1}^{n}i^2\\
&=\frac{20}{n}(n)+\frac{64}{n^2}\cdot\frac{n(n+1)}{2}-\frac{64}{n^3}\cdot\frac{n(n+1)(2n+1)}{6}\\
&=20+\frac{32(n+1)}{n}-\frac{32(n+1)(2n+1)}{3n^2}
\end{aligned}$$

The exact area of the region is the limit of this approximation as the number of rectangles increases without bound. As n gets larger and larger, $\frac{n+1}{n}$ approaches 1. Similarly, as n gets larger and larger, $\frac{(n+1)(2n+1)}{n^2}$ approaches 2. Thus,

$$\begin{aligned}
\lim_{n\to\infty}\left(20+\frac{32(n+1)}{n}-\frac{32(n+1)(2n+1)}{3n^2}\right)&=20+32\cdot 1-\frac{32}{3}\cdot 2\\
&=20+32-\frac{64}{3}\\
&=30\frac{2}{3}
\end{aligned}$$

Definite Integrals

Fortunately, a method has been developed for calculating the limit of this type of sum in an easier way. This method is called the *definite integral*. It uses some new notation. Representing the sum from Example 1 using definite integral notation, we have

$$\lim_{n\to\infty}\sum_{i=1}^{n}\left[-\left(-2+i\left(\frac{4}{n}\right)\right)^2+9\right]\left(\frac{4}{n}\right)=\int_{-2}^{2}(-x^2+9)\,dx$$

Just as the $\sum$ tells you to find a sum, the integral sign, $\int$, which looks like an elongated S, tells you to find the limit of a sum, which we interpret as an area. We read $\int_a^b f(x)\,dx$ as "the integral of f-of-x from a to b." It means find the area under $f(x)$ over the interval $[a, b]$. This is called a **definite integral,** because the **limits of integration** a and b are specified.

EXAMPLE 2 **Area Using Integral Notation**

Express the area under the line $f(x) = x + 5$ from $x = 0$ to $x = 10$ using integral notation.

Solution:

$$\int_0^{10} (x + 5)\, dx$$

EXAMPLE 3 **Graphing a Region Defined by Integral Notation**

Shade the region defined by $\int_1^3 (-x^2 + 3x + 4)\, dx$.

Solution: The integral represents the area under the graph of $f(x) = -x^2 + 3x + 4$ between $x = 1$ and $x = 3$. Graph $f(x)$ and shade the region bounded by $f(x)$, $x = 1$, $x = 3$, and $y = 0$. (See Figure A.35.)

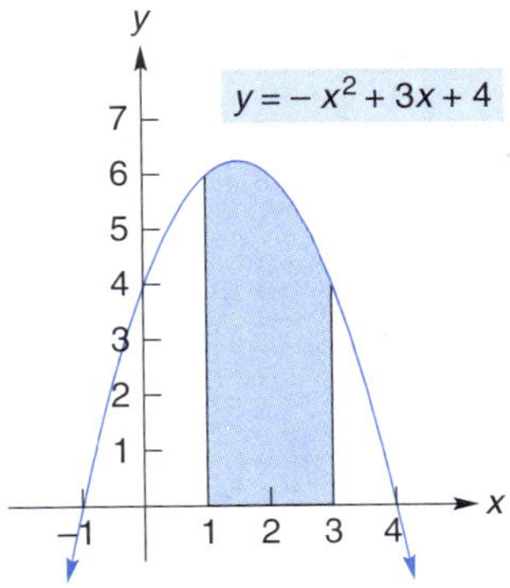

FIGURE A.35 The region defined by $\int_1^3 (-x^2 + 3x + 4)\, dx$.

Finding areas is only one of many applications of definite integrals. Definite integrals can be used to represent any situation in which you want to approximate a sum of values which can be modeled by a continuous function over a given interval.

For instance, suppose you want to find the profit of a company for the month of September and you are given a function $P(t) = 20{,}000\frac{t - 15}{15}$ that models the profit as a function of the day of September. You can represent the monthly profit using integral notation.

$$\int_0^{30} P(t)\, dt = \int_0^{30} 20{,}000\frac{t - 15}{15}\, dt.$$

Figure A.36 shows the region described by this definite integral.

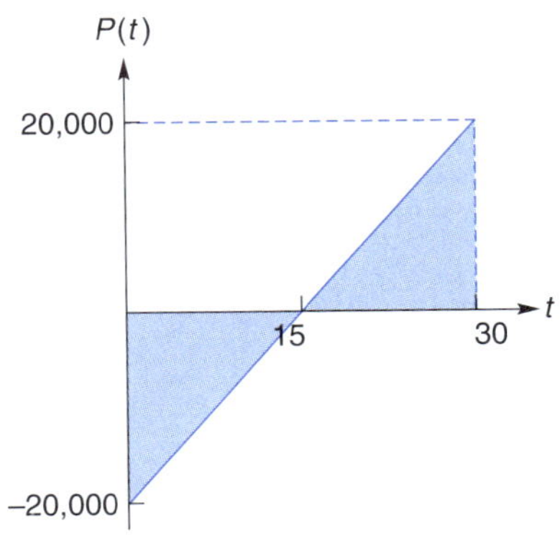

FIGURE A.36 The region defined by $\int_0^{30} 20{,}000\frac{t - 15}{15}\, dt$.

Prior to this, we have only considered functions over intervals where their graphs lie above the x-axis. That is, the function values have not been negative over any part of the interval. In this case, we are using a definite integral to represent the monthly profit. The profit is sometimes positive, indicating a gain, and sometimes negative, indicating a loss. When shading the region represented by the integral, we shade the region between the curve and the x-axis. When the curve is above the x-axis, we shade below the curve; when the curve is below the x-axis, we shade above the curve.

If we were to approximate this definite integral by summing areas of rectangles, the "area" of each rectangle would be the product of its "height" and its width, with the "height" being defined by a value of the function. If the value of the function were negative, the "area" would also be negative. Thus, our sum would contain positive and negative numbers. Using the symmetry of the graph in Figure A.36, you may be able to see that the value of this definite integral is 0, because the area above the x-axis is the same as the area below the x-axis.

EXAMPLE 4 **Graphing Area Defined by Integral Notation**

a. *Shade the region defined by $\int_0^5 (2x - 5)\, dx$ and determine whether it represents a positive number, a negative number, or 0.*

Solution: The integral represents the area of the region between $f(x) = 2x - 5$ and the x-axis on the interval $[0, 5]$. Graph $f(x) = 2x - 5$. Shade the regions between $f(x)$ and the x-axis from $x = 0$ to $x = 5$. (See Figure A.37.)

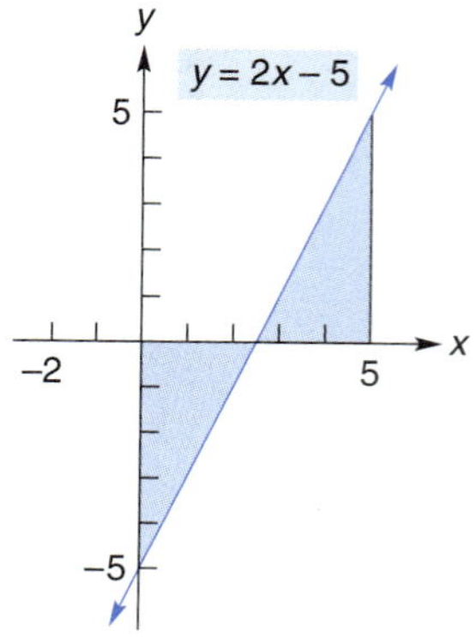

FIGURE A.37 The region defined by $\int_0^5 (2x-5)\,dx$.

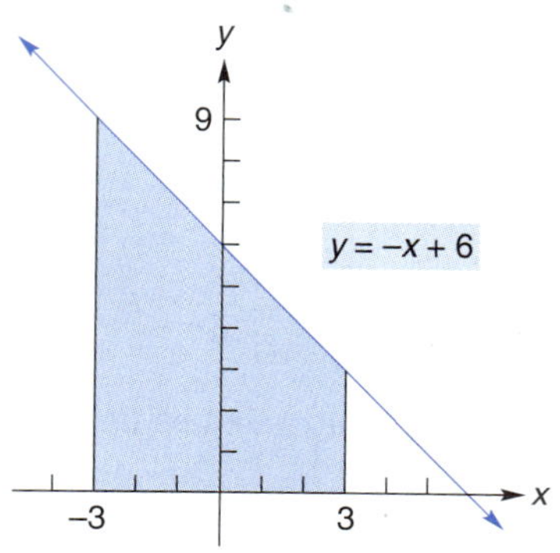

FIGURE A.38 The region defined by $\int_{-3}^{3} (-x+6)\,dx$.

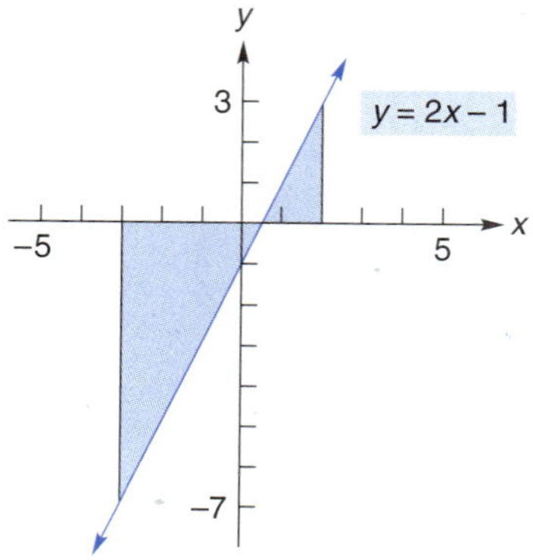

FIGURE A.39 The region defined by $\int_{-3}^{2} (2x-1)\,dx$.

Since the area below the x-axis is the same as the area above the x-axis, $\int_0^5 (2x-5)\,dx = 0$.

b. *Shade the region defined by* $\int_{-3}^{3}(-x+6)\,dx$ *and determine whether it represents a positive number, a negative number, or* 0.

Solution: The integral represents the area of the region between $f(x) = -x+6$ and the x-axis on the interval $[-3, 3]$. Graph $f(x) = -x+6$. Shade the region between $f(x)$ and the x-axis from $x = -3$ to $x = 3$. (See Figure A.38.)

Since the region lies entirely above the x-axis, $\int_{-3}^{3}(-x+6)\,dx$ represents a positive number.

c. *Shade the region defined by* $\int_{-3}^{2}(2x-1)\,dx$ *and determine whether it represents a positive number, a negative number, or* 0.

Solution: The integral represents the area of the region between $f(x) = 2x-1$ and the x-axis on the interval $[-3, 2]$. Graph $f(x) = 2x-1$. Shade the regions between $f(x)$ and the x-axis from $x = -3$ to $x = 2$. (See Figure A.39).

Since the region below the x-axis is larger than the region above the x-axis, $\int_{-3}^{2}(2x-1)\,dx$ represents a negative number.

Evaluating Definite Integrals

Now that we have the idea, let us look at how we can use definite integrals to find areas. We will start with a simple example.

We begin by seeing if we can find a function $F(x)$ that yields the area under the graph of $f(x) = x$ from 0 to b $(b > 0)$.

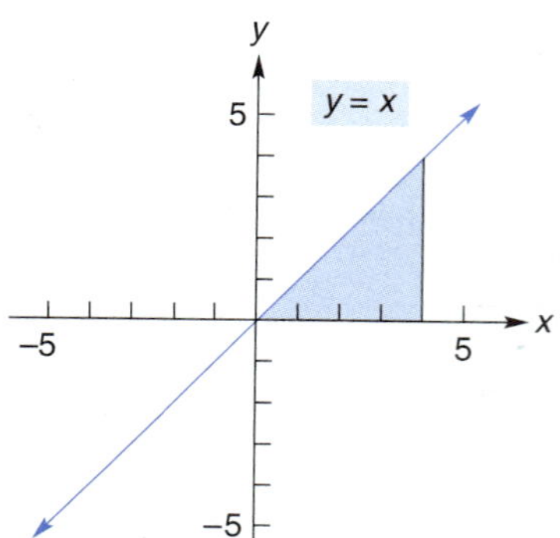

FIGURE A.40 The region defined by $\int_0^4 f(x)\,dx$.

EXAMPLE 5 Evaluating a Definite Integral

Evaluate $\displaystyle\int_0^4 f(x)\,dx$ *where* $f(x) = x$.

Solution: Since f is a linear function through the origin, the region under the graph of f is a right triangle. (See Figure A.40.) In this case the legs of the right triangle are of equal length. We can write a function $F(x)$ that gives the area of an isosceles right triangle $F(x) = \frac{1}{2}x^2$, where x is the length of a leg.

To find the area of the triangle in Figure A.40, we evaluate F at $x = 4$.

$$\int_0^4 f(x)\,dx = F(4) = \frac{1}{2}\cdot 4^2 = \frac{1}{2}\cdot 16 = 8$$

We can use $F(x)$ to calculate the area under the graph of $f(x) = x$ over any interval $[0, b]$ where $b > 0$. Now we can use this idea to find the area under the graph of $f(x) = x$ over the interval $[1, 4]$.

EXAMPLE 6 Evaluating a Definite Integral

Evaluate $\int_1^4 f(x)\,dx$ *where* $f(x) = x$.

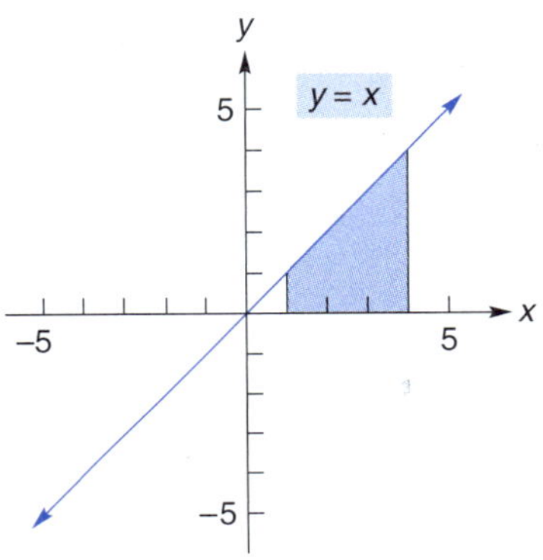

FIGURE A.41 The region under $f(x)$ from $x = 1$ to $x = 4$.

Solution: Here we want the area of the region under $f(x)$ from $x = 1$ to $x = 4$. Since $F(x) = \frac{1}{2}x^2$ gives the area under the graph from 0 to b, we can easily calculate the area from 0 to 1, $F(1)$, and subtract this area from the area from 0 to 4, $F(4)$. (See Figure A.41.)

$$\int_1^4 f(x)\,dx = F(4) - F(1) = \frac{1}{2}\cdot 4^2 - \frac{1}{2}\cdot 1^2 = \frac{16}{2} - \frac{1}{2} = \frac{15}{2}$$

Finally, let us see how the area changes as the interval changes from $[0, 4]$ to $[0, 4 + h]$. From Figure A.42, we can see that the area is increased by the area of a trapezoid with height h and bases $f(4)$ and $f(4 + h)$. This area can be expressed as

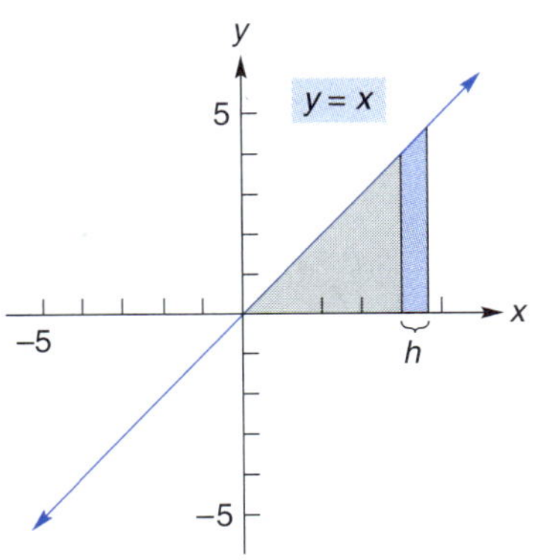

FIGURE A.42 The region under $f(x)$ from $x = 1$ to $x = 4 + h$.

$$F(4+h) = \int_0^{4+h} f(x)\,dx = F(4) + \frac{f(4) + f(4+h)}{2}\cdot h$$

For the function $f(x) = x$, we found a function $F(x)$ that gives the area under $f(x)$ over an interval $[0, b]$. We have also shown how to approximate $F(4 + h)$ using $F(x)$ and $f(x)$. Now let us see if we can generalize these ideas. We begin by assuming that there is such a function $F(x)$ for each continuous function $f(x)$. For small h, we can approximate $F(x + h)$ using $F(x)$ and the area of a trapezoid.

$$F(x+h) = F(x) + \frac{f(x) + f(x+h)}{2}\cdot h$$

Now consider the derivative of $F(x)$.

$$\begin{aligned} F'(x) &= \lim_{h\to 0} \frac{F(x+h) - F(x)}{h} \\ &= \lim_{h\to 0} \frac{\frac{f(x) + f(x+h)}{2}\cdot h}{h} \\ &= \lim_{h\to 0} \frac{f(x) + f(x+h)}{2} \\ &= f(x) \end{aligned}$$

Since the derivative of $F(x)$ is $f(x)$, we say F is an **antiderivative** of f. We cannot call it *the* antiderivative, because, as we will see, other antiderivatives are possible.

> An antiderivative of $f(x)$ is a function $F(x)$ whose derivative is $f(x)$:
> $$F'(x) = f(x)$$

The existence of more than one antiderivative is related to the fact that $x = 0$ was the left-hand boundary of the region under consideration when we found the function $F(x)$ for the area under $f(x)$. The left-hand boundary could have been any value of x. In Example 7 we find a function $G(x)$ modeling the area between the graph of $f(x) = x$ and the x-axis when $x = -5$ is the left-hand boundary. $G(x)$ is also an antiderivative of $f(x)$.

EXAMPLE 7 **Another Antiderivative**

The function $G(x) = \frac{1}{2}x^2 - \frac{25}{2}$ is also an antiderivative of $f(x)$. (You should verify this by showing that $G'(x) = f(x)$.)

$$\int_1^4 f(x)\,dx = G(4) - G(1) = \left(\frac{1}{2}\cdot 4^2 - \frac{25}{2}\right) - \left(\frac{1}{2}\cdot 1^2 - \frac{25}{2}\right)$$
$$= \frac{16}{2} - \frac{25}{2} - \frac{1}{2} + \frac{25}{2} = \frac{15}{2}$$

Note that $-\frac{25}{2}$ is the "area" between the graph of f and the x-axis from $x = -5$ to $x = 0$.

Using $G(x)$ to find the are under $f(x)$ from $x = 1$ to $x = 4$ can be visualized as in Figure A.43.

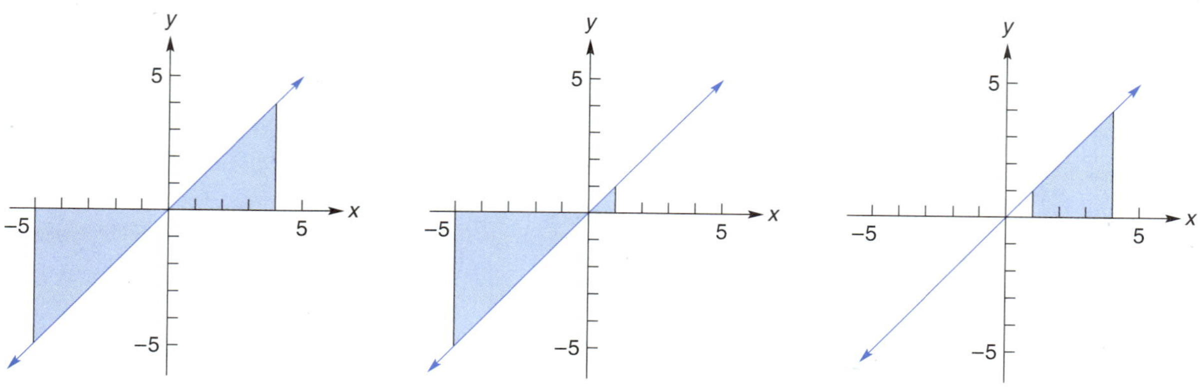

FIGURE A.43 $f(x) = x$.

Recall that in Example 6, when we evaluated $\int_1^4 f(x)\,dx$ using $F(x)$, the result was the same.

In Section A.4, we found that the derivative of a linear function $f(x) = mx + b$ is a constant function, $f'(x) = m$. Thus $G(x) = mx + b$ is an antiderivative of the constant function $g(x) = m$.

EXAMPLE 8 **Evaluating a Definite Integral**

Evaluate $\int_2^5 g(x)\,dx$ *where* $g(x) = 3$*, and interpret the result.*

Solution: $G(x) = 3x + b$ is an antiderivative of $g(x)$. Since b can be any real number, we choose $b = 0$.

$$\int_2^5 3\,dx = G(5) - G(2) = 3(5) - 3(2) = 15 - 6 = 9$$

The area between the x-axis and the function $g(x) = 3$ over the interval $[2, 5]$ is 9. (See Figure A.44.)

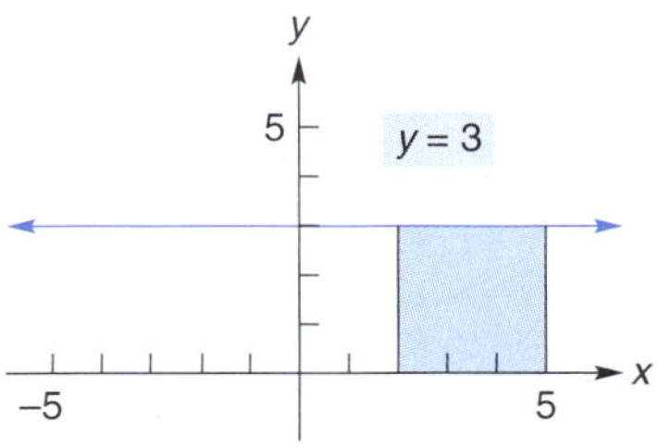

FIGURE A.44 $\int_2^5 3\,dx$.

EXAMPLE 9 **Evaluating a Definite Integral**

$F(x) = x^3 + 3x^2 + 2x$ *is an antiderivative of* $f(x) = 3x^2 + 6x + 2$*. Evaluate* $\int_2^3 f(x)\,dx$ *and interpret the result.*

Solution:

$$\begin{aligned}\int_2^3 f(x)\,dx &= F(3) - F(2)\\ &= (3^3 + 3\cdot 3^2 + 2\cdot 3) - (2^3 + 3\cdot 2^2 + 2\cdot 2)\\ &= 27 + 27 + 6 - (8 + 12 + 4)\\ &= 36\end{aligned}$$

The area between the x-axis and the function f over the interval $[2, 3]$ is 36. (See Figure A.45.)

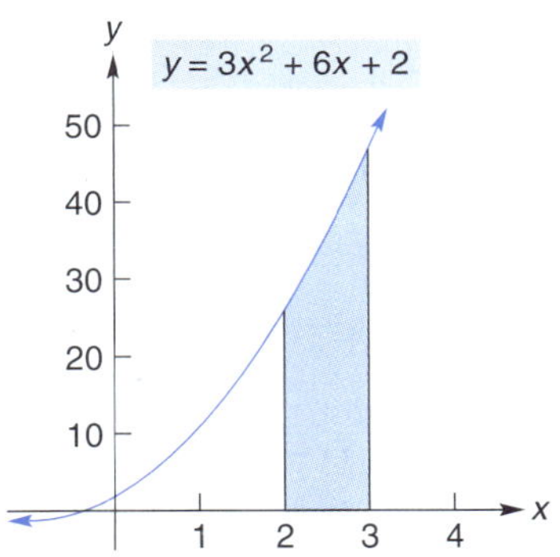

FIGURE A.45 $\int_2^3 (3x^2 + 6x + 2)\,dx$.

Now let us revisit Example 1 using integrals.

EXAMPLE 10 **Finding Area by Integration**

$F(x) = -\frac{1}{3}x^3 + 9x$ *is an antiderivative of* $f(x) = -x^2 + 9$*. Evaluate* $\int_{-2}^2 f(x)\,dx$.

Solution:

$$\begin{aligned}\int_{-2}^2 f(x)\,dx = F(2) - F(-2) &= \left[-\frac{1}{3}(2)^3 + 9(2)\right] - \left[-\frac{1}{3}(-2)^3 + 9(-2)\right]\\ &= \left[-\frac{8}{3} + 18\right] - \left[-\frac{-8}{3} - 18\right] = -\frac{16}{3} + 36\\ &= 30\frac{2}{3}\end{aligned}$$

Note that the result is the same as in Example 1.

Exercise A.7

In Problems 1–2, approximate the area under the curve over the given interval using n rectangles. Then find the limit as n increases without bound.

1. $f(x) = x + 4 \quad [-1, 3]$

2. $f(x) = -x^2 + 4 \quad [-2, 2]$

In Problems 3–8, express the area under the curve f(x) on the given interval using integral notation.

3. $f(x) = 8 \quad [4, 9]$

4. $f(x) = -4x + 5 \quad [-3, 1]$

5. $f(x) = x^2 + x + 2 \quad [-2, 5]$

6. $f(x) = -x^2 + x + 3 \quad [0, 2]$

7. $f(x) = 3^x + 5 \quad [-5, 5]$

8. $f(x) = \dfrac{1}{x} \quad [1, 5]$

In Problems 9–14, graph the region defined by the integral and determine whether it represents a positive number, a negative number, or 0.

9. $\displaystyle\int_0^6 4\,dx$

10. $\displaystyle\int_{-2}^4 \left(\frac{1}{2}x + 3\right) dx$

11. $\displaystyle\int_{-2}^4 (x^2 - 3)\,dx$

12. $\displaystyle\int_{-1}^1 -x^2\,dx$

13. $\displaystyle\int_0^3 2^x\,dx$

14. $\displaystyle\int_0^6 (x - 3)\,dx$

In Problems 15–18, (a) find $\int_0^b f(x)\,dx$ and (b) evaluate $\int_0^{10} f(x)\,dx$ for the given f(x).

15. $f(x) = 7$

16. $f(x) = 3x$

17. $f(x) = 3x + 2$

18. $f(x) = 3x + 4$

19. $F(x) = 2x$ is an antiderivative of $f(x) = 2$.

(a) Evaluate $\int_{-3}^4 f(x)\,dx$.

(b) Find another antiderivative for $f(x)$ and name it $G(x)$.

(c) Evaluate $\int_{-3}^4 f(x)\,dx$ using $G(x)$.

20. $F(x) = -x^2 + 3x$ is a antiderivative of $f(x) = -2x + 3$.

(a) Evaluate $\int_{-2}^1 f(x)\,dx$.

(b) Find another antiderivative for $f(x)$ and name it $G(x)$.

(c) Evaluate $\int_{-2}^1 f(x)\,dx$ using $G(x)$.

In Problems 21–24, evaluate the integral.

21. $\displaystyle\int_0^6 5\,dx$

22. $\displaystyle\int_{-4}^2 5\,dx$

23. $\displaystyle\int_3^{10} -2\,dx$

24. $\displaystyle\int_{-10}^{-3} -2\,dx$

In Problems 25–28, use the given antiderivative to evaluate the integral.

25. $F(x) = x^3 - \dfrac{1}{2}x^2 + x$ is an antiderivative of $f(x) = 3x^2 - x + 1$. Evaluate $\int_0^3 f(x)\,dx$.

26. $F(x) = -\dfrac{1}{3}x^3 + 2x^2 + 5x$ is an antiderivative of $f(x) = -x^2 + 4x + 5$. Evaluate $\int_{-2}^2 f(x)\,dx$.

27. $F(x) = e^x$ is an antiderivative of $f(x) = e^x$. Evaluate $\int_0^3 f(x)\,dx$.

28. $F(x) = \sqrt{x}$ is an antiderivative of $f(x) = \dfrac{1}{2\sqrt{x}}$. Evaluate $\int_1^{16} f(x)\,dx$.

● ● ● **OBJECTIVE** *Identify the meanings of derivatives and integrals in applications and apply derivatives and integrals.*

A.8 Applications of Derivatives and Integrals

In Section A.4, we introduced the derivative as the instantaneous rate of change of a function. In Section A.7, we introduced the integral as the area under a curve. At first glance, these may seem like vastly different entities, but integrals and derivatives are actually closely related. The integral of a function $f(x)$ is found using an antiderivative of $f(x)$; that is, a function $F(x)$ whose derivative is $f(x)$. Thus, if we integrate $f(x)$ and then find the derivative of the result, we end up with $f(x)$ again. On the other hand, if we find the derivative $f'(x)$ of the function $f(x)$, then $f(x)$ is an antiderivative of $f'(x)$. So if we find the derivative of $f(x)$ and then integrate the result, we would end up with $f(x)$ again, except perhaps for its constant term. In these ways, finding derivatives and finding integrals are inverse processes. In this section, we further explore the relationship between derivatives and integrals, as well as the meaning of each and how to determine when to use them in applications.

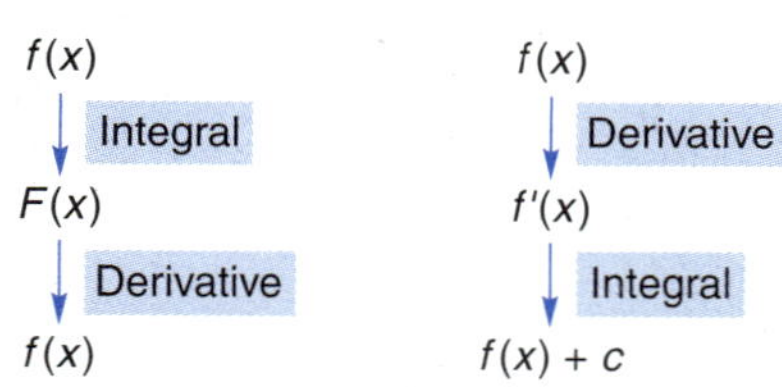

Applications Involving Derivatives

First, we consider the meaning of the derivative of a function.

EXAMPLE 1 **Meaning of a Derivative**

In a new housing development, the price of a house increases as the square footage increases. This relationship can be modeled by the function $f(x) = 80x + 20{,}000$, *where* $f(x)$ *is the price of an x-square-foot house. The derivative of* $f(x)$ *is* $f'(x) = 80$. *What does the derivative represent?*

Solution: In this case, the derivative gives the rate of change of the price of a house with respect to the square footage. Since the derivative of $f(x)$ is $f'(x) = 80$, then the price increases \$80 for each one-square-foot increase in the size of the house.

We will now consider an application similar to one that we saw earlier.

EXAMPLE 2 **Reducing a Federal Program**

Suppose that during the 2001 *session, the U.S. Congress votes to reduce the budget of a federal program from the* 2001 *budget of* \$16.8 *million to a budget of* \$4.3 *million in* 2006. *A linear model for this budget reduction can be represented by the equation*

$$f(x) = 19.3 - 2.5x \quad 1 \le x \le 6$$

where x represents the number of years after 2000 *(i.e.,* $x = 1$ *corresponds to* 2001 *and* $x = 6$ *corresponds to* 2006*).*

An exponential model for the same budget reduction can be represented by the equation

$$g(x) = 22.06e^{-0.2723x} \quad 1 \le x \le 6$$

where again, x represents the number of years after 2000. *The derivatives of the functions for the models are*

$$f'(x) = -2.5 \quad \text{and} \quad g'(x) = -6.006938e^{-0.2723x}$$

a. *Find* $f(3)$, $g(3)$, *and explain the meaning of these values.*

b. *What do the derivatives of these models represent?*

c. *Find* $f'(2)$, $g'(2)$, *and explain the meaning of these values.*

d. *Find* $f'(5)$, $g'(5)$, *and explain the meaning of these values.*

Solution:

a.

$$f(3) = 19.3 - 2.5(3) = 19.3 - 7.5 = 11.8$$

$$g(3) = 22.06e^{-0.2723(3)} = 22.06e^{-0.8169} \approx 9.746$$

Since $x = 1$ corresponds to 2001, $x = 3$ corresponds to 2003.

$f(3) = 11.8$ indicates that under the linear reduction model, the budget for the program in 2003 will be \$11.8 million.

$g(3) \approx 9.746$ indicates that under the exponential reduction model, the budget for the program in 2003 will be \$9.746 million.

b. The derivatives give the rate of change of the budget with respect to time. That is, they give the amount by which the budget for the program is increasing or decreasing at a particular time. Since the budget amounts are given in millions of dollars and the time is given in years, the units for the derivative are millions of dollars per year.

c. $f'(2) = -2.5$

$g'(2) = -6.006938e^{-0.2723(2)} \approx -3.4845$

$x = 2$ corresponds to 2002

$f'(2) = -2.5$ indicates that under the linear reduction model, the program's budget is decreasing at the rate of \$2.5 million per year in 2002. $g'(2) \approx -3.4845$ indicates that under the exponential reduction model, the program's budget is decreasing at the rate of about \$3.4845 million per year in 2002.

d. $f'(5) = -2.5$

$g'(5) = -6.006938e^{-0.2723(5)} \approx -1.5394$

$x = 5$ corresponds to 2005

$f'(5) = -2.5$ indicates that under the linear reduction model, the program's budget is decreasing at the rate of \$2.5 million per year in 2005. $g'(5) \approx -1.5394$ indicates that under the exponential reduction model, the program's budget is decreasing at the rate of about \$1.5394 million per year in 2005.

Notice in Example 2 that in 2002, the budget for the program is decreasing much more rapidly for the exponential model than it is for the linear model. In 2005, the situation is reversed: The budget is decreasing more rapidly for the linear model than it is for the exponential model. Later in this section, we will use integrals to investigate how these differences can affect the total expenditures on the program during the budget reduction period.

We can also model the pricing structure of a business. Such a model can be fairly complex since, in many situations, it is advantageous to produce or buy a greater quantity of an item, rather than a small quantity. This is known as *the economy of scale*—producing or buying more can decrease the average cost or price per unit.

EXAMPLE 3 Cost of Consulting

A computer consulting company sets up databases and forms for various other companies. The amount of time needed to prepare a database and basic forms is considerable. However, once a consultant has set up the database and prepared the basic forms, the amount of time needed to create custom forms is much less due to the consultant's familiarity with the company and its products. However, this benefit is limited. If a client wants more than 30 custom forms, an additional consultant will be needed, thus increasing the cost. The pricing for a database and forms can be modeled by $f(x) = 0.2x^3 - 18x^2 + 1040x + 10{,}320$ *where x is the number of custom forms. The derivative of* $f(x)$ *is given by* $f'(x) = 0.6x^2 - 36x + 1040$.

Use the derivative of $f(x)$ *to find and interpret* $f'(1)$, $f'(10)$, *and* $f'(25)$.

Solution:

$$f'(1) = 0.6(1)^2 - 36(1) + 1040 = 1004.6$$

$$f'(10) = 0.6(10)^2 - 36(10) + 1040 = 740$$

$$f'(25) = 0.6(25)^2 - 36(25) + 1040 = 515$$

Here, the derivative is the rate of change of the cost as the number of custom forms increases (i.e., the approximate cost for an additional custom form).

$f'(1) = 1004.6$ indicates that when one custom form is ordered, the cost for each additional custom form is approximately \$1004.60.

$f'(10) = 740$ indicates that when ten custom forms are ordered, the cost for each additional custom form is approximately \$740.

$f'(25) = 515$ indicates that when 25 custom forms are ordered, the cost for each additional custom form is approximately \$515.

Note that in Example 3, $f'(30) = 500$ and $f'(35) = 515$. This reflects the increased cost per custom form when more than 30 custom forms are required.

Applications Involving Integrals

Thus far, we have looked at applications involving derivatives. Now we consider applications involving integrals. We begin with an example that can easily be solved using multiplication. We are not suggesting that this type of problem be solved using integrals. We are only using it to illustrate the meaning of an integral.

EXAMPLE 4 Meaning of an Integral

A caterer has informed you that the cost for additional guests at a formal banquet is \$75 per guest. This can be thought of as a function $f(x) = 75$, where x is the number of additional guests. What is the meaning of $\int_0^{100} f(x)\,dx = \int_0^{100} 75\,dx$?

Solution: In this case, the integral gives the total cost for 100 additional guests.

Since the situation of Example 4 was described using a constant function, it could easily be solved with multiplication. In Example 5, we return to the problems of modeling a budget. Here we use integrals to project the total budget over 5 years using two different models, neither of which is constant.

EXAMPLE 5 Reducing a Federal Program

Suppose that during the 2001 session, the U.S. Congress votes to reduce the budget of a federal program from the 2001 budget of \$16.8 million to a budget of \$4.3 million in 2006. A linear model for this budget reduction can be represented by the equation

$$f(x) = 19.3 - 2.5x \quad 1 \le x \le 6$$

where x represents the number of years after 2000 (i.e., $x = 1$ corresponds to 2001 and $x = 6$ corresponds to 2006). An exponential model for the same budget reduction can be represented by the equation

$$g(x) = 22.06e^{-0.2723x} \quad 1 \le x \le 6$$

where, again, x represents the number of years after 2000. Antiderivatives of the functions for the models are

$$F(x) = 19.3x - 1.25x^2 \quad and \quad G(x) = -81.0136e^{-0.2723x}$$

a. *Use $F(x)$ to evaluate $\int_1^6 f(x)\,dx$.*

b. *Use $G(x)$ to evaluate $\int_1^6 g(x)\,dx$.*

c. *What do the integrals in parts (a) and (b) represent?*

d. *Use a graph to illustrate* $\int_1^6 f(x)\,dx - \int_1^6 g(x)\,dx$ *and interpret the meaning of this difference of integrals.*

Solution:

a. Since $F(x) = 19.3x - 1.25x^2$ is an antiderivative of f,

$$\begin{aligned}\int_1^6 f(x)\,dx &= F(6) - F(1)\\ &= [19.3(6) - 1.25(6)^2] - [19.3(1) - 1.25(1)^2]\\ &= (115.8 - 45) - (19.3 - 1.25)\\ &= 52.75\end{aligned}$$

b. Since $G(x) = -81.0136e^{-0.2723x}$ is an antiderivative of g,

$$\begin{aligned}\int_1^6 g(x)\,dx &= G(6) - G(1)\\ &= -81.0136e^{-0.2723(6)} - (-81.0136e^{-0.2723(1)})\\ &\approx 45.8893\end{aligned}$$

c. For $1 \le x \le 6$, $f(x)$ and $g(x)$ give the amount budgeted for a federal program depending on whether it is being reduced under a linear or exponential model.

$\int_1^6 f(x)\,dx$ gives the total amount budgeted for the program during the reduction by the linear model. Since $\int_1^6 f(x)\,dx = 52.75$, using the linear reduction model the federal program will have expenditures of \$52.75 million over the period from 2001 to 2006.

$\int_1^6 g(x)\,dx$ gives the total amount budgeted for the program during the reduction by the exponential model. Since $\int_1^6 g(x)\,dx \approx 45.8893$, using the exponential reduction model the federal program will have expenditures of \$45.8893 million over the period from 2001 to 2006.

d. $\int_1^6 f(x)\,dx$ represents the area under the graph of $f(x)$ from $x = 1$ to $x = 6$, while $\int_1^6 g(x)\,dx$ represents the area under the graph of $g(x)$ from $x = 1$ to $x = 6$. Thus it seems reasonable that the difference between the two definite integrals gives the area *between* the two curves. In fact, this is what the difference of the integrals represents. This area is shaded in Figure A.46.

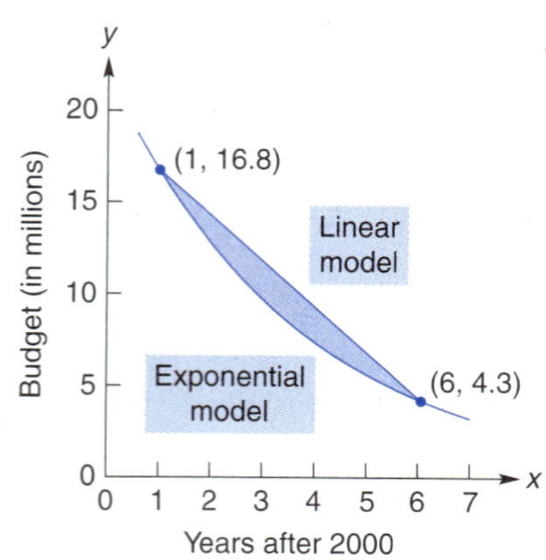

FIGURE A.46 Difference between models.

$\int_1^6 f(x)\,dx - \int_1^6 g(x)\,dx$ gives the difference in total expenditures for the two models, so $\int_1^6 f(x)\,dx - \int_1^6 g(x)\,dx \approx 52.75 - 45.8893 = 6.8607$. The total expenditures would be about \$6.86 million less from 2001 to 2006 if the budget of the program were reduced using the exponential model.

In Example 2 we found that in 2002 the budget decreased more rapidly with the exponential model that for the linear model. One result of this is that the total expenditures over the five-year period are less for the exponential model, as illustrated in Figure A.46. While it may seem obvious that the exponential model should be used to reduce the budget of the program, other considerations often affect how budgets are reduced. It may be more important to reduce services gradually, so the linear reduction model may be more realistic.

EXAMPLE 6 **Time to Completion**

A company has determined that after a one-week training period, it takes new employees an average of 960 minutes (2 days) to produce one unit. The amount of time needed to produce each additional unit decreases by 20 minutes per unit until the employee has produced a total of 36 units. At this time an employee takes an average of 240 minutes ($\frac{1}{2}$ day) to produce each unit and is considered to be "up to speed." In order to know how long of a probation period to have for new employees, it is necessary to find out how long it should take the average employee to get up to speed.

The amount of time that it takes an employee to produce unit number x can be modeled by $f(x) = 960 - 20(x - 1)$ or $f(x) = 980 - 20x$. An antiderivative of $f(x)$ is $F'(x) = 980x - 10x^2$. Use this antiderivative to determine a reasonable probation period for new employees, based on the length of time an average new employee takes to produce 36 units.

Solution: The total amount of time required for an average new employee to produce a total of 36 units is

$$\begin{aligned}\int_0^{36} f(x)\,dx &= F(36) - F(0)\\ &= [980(36) - 10(36)^2] - [980(0) - 10(0)^2]\\ &= (35{,}280 - 12,960) - 0\\ &= 22{,}320\end{aligned}$$

Thus, after 22,320 minutes, a new employee should be up to speed. Using 4.3 weeks in a month and a 40-hour workweek, this comes out to about 2.16 months. A probation period of 3 months is reasonable, since this allows time for the training period as well as allowing for some months not containing a full 4.3 workweeks.

Application Involving Derivatives and Integrals

Let us look at a motion application of the derivative and integral. The velocity of an object in motion is its instantaneous rate of change of distance with respect to time. The velocity of an object over an interval can be modeled by a function $v(t)$ over that interval. We can find the derivative of $v(t)$, $v'(t)$, to find the instantaneous rate of change of velocity with respect to time, otherwise known as acceleration. Since velocity is the instantaneous rate of change of distance with respect to time, acceleration is the instantaneous rate of change of the rate of change of distance with respect to time.

For example, suppose we have a function $v(t) = 50$, which models the instantaneous rate of change of distance of an automobile with respect to time. Distance is measured in miles and time is measured in hours. Therefore, the rate of change is 50 miles per hour, and the derivative of $v(t)$ has the units of miles per hour per hour. Here, $v'(t) = 0$, which indicates that there is no acceleration. This agrees with $v(t) = 50$, indicating a constant speed of 50 miles per hour.

Now we apply integration to this motion application. If we integrate $v(t)$ with respect to time over the interval, we find the total distance traveled during that interval. If you graph the velocity function, the area under the graph is the total distance traveled.

The distance traveled by an object over an interval of time can be modeled by a function $d(t)$. We can find the derivative of $d(t)$, $d'(t)$, to find the instantaneous rate of change of distance with respect to time t, otherwise known as velocity. If we integrate $d(t)$ over the interval, we find the total distance-time for that interval. For example, if we consider the distance in miles traveled by a taxi over some number of hours, this integral would give the mile-hours driven. This may not seem like a particularly useful quantity. However, the driver's pay, or the maintenance on the taxi (changing tires, oil), might be based on the mile-hours driven.

EXAMPLE 7 Motion Application

Your velocity while driving on a 2-hour trip can be modeled by $v(t) = -60t^2 + 120t$ over the time interval $[0, 2]$, where t is the number of hours since you started driving. The acceleration at time t, or the derivative of $v(t)$, is $v'(t) = -120t + 120$. The distance traveled over the 2-hour interval of $[0, 2]$ is the integral of $v(t)$, or $\int_0^2 (-60t^2 + 120t)\,dt$. An antiderivative of $v(t)$ is $G(t) = -20t^3 + 60t^2$.

a. *Find the instantaneous acceleration at $\frac{1}{2}$ hour, 1 hour, and $\frac{3}{2}$ hours.*

b. *Find the total distance traveled in the 2-hour interval.*

Solution:

a. The instantaneous acceleration is represented by the derivative of the velocity function. As given previously, it is $v'(t) = -120t + 120$.
At $\frac{1}{2}$ hour:

$$\begin{aligned} v'\left(\frac{1}{2}\right) &= -120\left(\frac{1}{2}\right) + 120 \\ &= -60 + 120 \\ &= 60 \end{aligned}$$

At 1 hour:

$$\begin{aligned} v'(1) &= -120(1) + 120 \\ &= -120 + 120 \\ &= 0 \end{aligned}$$

At $\frac{3}{2}$ hours:

$$v'\left(\frac{3}{2}\right) = -120\left(\frac{3}{2}\right) + 120$$
$$= -180 + 120$$
$$= -60$$

The acceleration at $\frac{1}{2}$ hour, 1 hour, and $\frac{3}{2}$ hours is $60\frac{\text{mi}}{\text{h}^2}$, $0\frac{\text{mi}}{\text{h}^2}$, and $-60\frac{\text{mi}}{\text{h}^2}$, respectively.

b. The distance traveled is found by evaluating the definite integral of the velocity function on the interval $[0, 2]$. We use $G(t) = -20t^3 + 60t^2$ given previously as an antiderivative. Over the 2-hour interval $[0, 2]$, the distance traveled is

$$G(2) - G(0) = [-20(2)^3 + 60(2)^2] - [-20(0)^3 + 60(0)^2]$$
$$= (-160 + 240) - (0 + 0)$$
$$= 80$$

The total distance traveled over the 2-hour interval is 80 miles.

Example 7 shows a context in which both the derivative and the integral of a given function are meaningful and useful. Notice how the units reflect which operation has been performed on the function. When $g(x)$ has units of miles per hour, $g'(x)$ has units of miles per hour per hour, while $\int_a^b g(x)\,dx$ has units of miles. Thinking about the units can be helpful when deciding whether to take a derivative or to integrate to solve a problem.

Exercise A.8

For Problems 1–8, determine whether you should use the derivative of the function, the integral of the function, or the function itself to solve the problem described.

1. The function $C(t) = 100e^{0.03t}$, $1960 \le t \le 2000$, describes the average annual cost of maintaining an automobile for a given year. You want to know the cumulative cost of maintaining an automobile from 1990 to 2000.

2. The function $C(t) = 100e^{0.03t}$, $1960 \le t \le 2000$, describes the average annual cost of maintaining an automobile for a given year. You want to know the average cost of maintaining an automobile in 1999.

3. The function $h(t) = -16t^2 + 128t + 3$ describes the height of a ball that has been thrown into the air. You want to know when it will hit the ground.

4. The function $h(t) = -16t^2 + 128t + 3$ describes the height of a ball that has been thrown into the air. You want to know its velocity when it hits the ground.

5. The function $v(t) = 4t - t^2$, $0 \le t \le 4$, describes the velocity of a particle at time t. You want to know the acceleration of the object when $t = 2$.

6. The function $v(t) = 4t - t^2$, $0 \le t \le 4$, describes the velocity of a particle at time t. You want to know how far the object has traveled when $t = 2$.

7. The function $v(t) = 4t - t^2$, $0 \le t \le 4$, describes the velocity of a particle at time t. You want to know how fast it is moving when $t = 2$.

8. The function $C(x) = 200 + 14x - 0.01x^2$ describes the daily production cost when x units are produced in one day. You want to know the cost of producing one more unit when $x = 17$.

9. The function $C(x) = 20 + 0.60x$ represents the cost of renting a small truck for 24 hours, where x is the number of miles the truck is driven.
   (a) Find and interpret $C(50)$.
   (b) Find and interpret $C'(50)$.

10. The function $W(S) = 300 + 0.05S$ represents a salesperson's weekly wages, where S is her weekly sales in dollars.
    (a) Find and interpret $W(4000)$.
    (b) Find and interpret $W'(4000)$.

11. A company has a fairly simple annual bonus program that rewards years of service with the company. Each year an employee receives a bonus of \$300 for each year that he or she has been with the company.

(a) Write a function $b(t)$ that expresses the annual bonus b that an employee receives with respect to years of service t.
(b) Find and interpret $b'(t)$.
(c) Interpret $\int_0^{10} b(t)\,dt$.

12. The function $E(t) = 1070e^{0.05t}$ approximates the average annual energy costs for a 2500-square-foot home in year t where $t = 0$ corresponds to 1970.

(a) Find and interpret $E(30)$.
(b) Find and interpret $E'(30)$, given that $E'(t) = 53.5e^{0.05t}$.
(c) Find and interpret $\int_0^{30} E(t)\,dt$, given that $F(t) = 21{,}400e^{0.05t}$ is an antiderivative of $E(t)$.

13. The function $r(t) = 10\ln t$ approximates the number of books Xul read in a given year, where $t = 1$ corresponds to 1991.

(a) About how many books did Xul read in 2000?
(b) Find and interpret $r'(5)$, given that $r'(t) = \dfrac{10}{t}$.
(c) Find and interpret $\int_1^{10} r(t)\,dt$, given that $R(t) = 10t(-1 + \ln t)$ is an antiderivative of $r(t)$.

14. A company with 40,000 employees has been suffering huge financial losses. The CEO has proposed to cut the workforce by about 50% over three months, and his analysts have developed two scenarios with models for implementing the cuts. The models are

$$W_l(t) = 40{,}000 - 6500t \quad \text{and} \quad W_e(t) = 40{,}000e^{-0.2231t}$$

where each represents the size of the workforce t months after the plan is implemented, $0 \le t \le 3$.

(a) Find and interpret $W_l(3)$ and $W_e(3)$.
(b) Find and interpret $W_l'(2)$ and $W_e'(2)$, given that $W_l'(t) = -6500$ and $W_e'(t) = -8924e^{-0.2231t}$.
(c) On average, each employee's salary and benefits cost the company \$4200 per month. Interpret the numbers

$$\$4200\left[\int_0^3 W_l(t)\,dt\right] = \$381{,}150{,}000 \text{ and}$$

$$\$4200\left[\int_0^3 W_e(t)\,dt\right] \approx \$367{,}426{,}091$$

(d) Which model results in the smaller workforce at the end of the three months?
(e) With which model is the size of the workforce decreasing more rapidly after two months?
(f) Which model saves the company more in salary and benefits during the three months?

15. Suppose the U.S. Congress votes in 2001 to create a new federal program that will be phased in over 10 years with an anticipated budget of \$16 billion in the year 2011. The functions $b_l(t) = 1.6t$ and $b_e(t) = 6.67\ln(1 + t)$ are proposed models for the program's budget, where $t = 0$ corresponds to 2001, $0 \le t \le 10$.

(a) Find and interpret $b_l(5)$ and $b_e(5)$.
(b) Find and interpret $b_l'(5)$ and $b_e'(5)$, given that $b_l'(t) = 1.6$ and $b_e'(t) = \dfrac{6.67}{1+t}$.
(c) Find and interpret, $\int_0^5 b_l(t)\,dt$ and $\int_0^5 b_e(t)\,dt$, given that $B_l(t) = 0.8t^2$ and $B_e(t) = 6.67(1 + t)[-1 + \ln(1 + t)]$ are antiderivatives of b_l and b_e, respectively.
(d) Which model yields the bigger budget in 2006?
(e) With which model is the budget increasing more quickly in 2006?
(f) Which model yields the greater cumulative cost for the years 2001 through 2006?

16. Suppose that the Space Shuttle takes off with speed that is given by $v(t) = \dfrac{1}{3600}t^2$, where the speed is measured in miles per second, and t represents time in seconds after liftoff. The function $v'(t) = \dfrac{1}{1800}t$ is the derivative of v, and $V(t) = \dfrac{1}{10{,}800}t^3$ is an antiderivative of v.

(a) How fast is the shuttle moving 10 seconds after liftoff?
(b) How fast is the speed of the shuttle increasing 10 seconds after liftoff?
(c) How far has the shuttle traveled in the first 10 seconds after liftoff?

17. Suppose that a car starts out at a speed that is given by $v(t) = \dfrac{7}{3600}t$, where the speed is measured in miles per second and t represents time in seconds after starting. The function $V(t) = \dfrac{7}{7200}t^2$ is an antiderivative of v.

(a) How fast is the car moving after 6 seconds?
(b) How fast is the speed of the car increasing after 6 seconds?
(c) How far does the car travel in the first 6 seconds?

18. Suppose that a company is beginning a new 10-week marketing campaign that it hopes will generate $n(t) = t^3$ new long-term customers during the week t weeks after the campaign begins, $0 \le t \le 10$. The function $n'(t) = 3t^2$ is the derivative of n, and $N(t) = \dfrac{1}{4}t^4$ is an antiderivative of n.

(a) How many new long-term customers should the campaign generate in the eighth week?
(b) How many long-term customers should the campaign generate during the 10 weeks?

A.9 Review

Important Terms and Symbols

Section A.2 model mathematical modeling

Section A.3 rate of change secant line average rate of change

Section A.4 $\frac{df}{dx}$ $f'(x_0)$ instantaneous rate of change tangent line derivative

Section A.6 $\sum$, sigma notation limits of summation

Section A.7 $\int$ definite integral limits of integration antiderivative

Summary

A graph is a picture of the relationship between two variables. A *line graph* is created by first plotting data points, where each point represents a known pair of values for the two variables, and then connecting adjacent points with line segments. The line segments in a line graph show how the variables are changing with respect to one another in different parts of the graph.

Smooth-curve graphs representing relationships between real-world variables are often created using mathematical formulas. **Mathematical modeling** is the process of creating a mathematical description, or **model,** of a real-world situation.

For a straight-line graph, the slope of the line represents the **rate of change** of the variable on the vertical axis with respect to the variable on the horizontal axis. For a given interval on a smooth-curve graph, the **secant line** is the line intersecting the curve at points whose x-coordinates are the end values of the interval. The **average rate of change** over the interval is the slope of the secant line.

If one end of an interval is held fixed and the other end is moved so that the interval shrinks to zero, then the secant line moves, approaching the **tangent line.** The **instantaneous rate of change** is the slope of the tangent line. The function $f'(x)$, which gives the slope of the tangent line through $(x, f(x))$ for any x, is called the **derivative** with respect to x of the function $f(x)$.

Sigma notation is used to abbreviate sums, and certain rules apply to it. The most basic of these are:

$$\sum_{k=m}^{n} ca_k = c\sum_{k=m}^{n} a_k \tag{1}$$

$$\sum_{k=m}^{n} (a_k + b_k) = \sum_{k=m}^{n} a_k + \sum_{k=m}^{n} b_k \tag{2}$$

$$\sum_{k=m}^{n} a_k = \sum_{k=p}^{n+p-m} a_{k+m-p} \tag{3}$$

and

$$\sum_{k=m}^{p-1} a_k + \sum_{k=p}^{n} a_k = \sum_{k=m}^{n} a_k \tag{4}$$

The **definite integral** of the function $f(x)$ over the interval bounded by the **limits of integration** a and b is written $\int_a^b f(x)\,dx$. It represents the area that lies above the x-axis and below the curve $f(x)$, minus the area that lies below the x-axis and above the curve $f(x)$. If function $F(x)$ is such that $F'(x) = f(x)$, then F is an **antiderivative** of f.

If follows from the definitions that

$$\int_a^b f(x)\,dx = F(b) - F(a)$$

Review Problems

1. Use the following data, which shows the amount of money spent on groceries per month for the given months, to create a line graph.

Month	October	November	December	January	February
Amount ($)	125	98	175	150	150

2. Using the line graph from Problem 1, during what time period did the amount spent per month on groceries increase most rapidly.

3. Using the line graph from Problem 1, during what time period did the amount spent per month on groceries decrease most rapidly.

4. The following information describes the wind speed recorded on top of Mount Washington during a given week. Use the information to create a possible line graph.

(a) The highest wind speed was recorded on Wednesday and the lowest wind speed was recorded on Friday.
(b) The wind speed decreased more rapidly from Wednesday to Thursday than from Thursday to Friday.
(c) The wind speed increased at the same rate from Monday to Tuesday as it did from Tuesday to Wednesday.

In Problems 5 and 6, use the given information to write equations which model each situation. Make certain to define variables when necessary.

5. The cost of producing x units of a product is determined by a fixed cost of \$550 plus a variable cost of \$22.50 per unit.

6. The value, v, of a \$22,000 car, after t years, if it depreciates at a rate of 8.25% a year.

7. The following data shows the height, h, of a model rocket t seconds after being shot upward from the ground.

***t* (seconds)**	1	2	3	4	5
***h* (feet)**	84	136	156	144	100

(a) Graph the data and determine what type of model (linear, quadratic, exponential, or logarithmic) fits the data.
(b) Find the equation of the graph that passes through each data point.
(c) Find the height of the rocket after 6 seconds.
(d) How long after being shot upward does the rocket return to the ground?

8. The following data shows the population of a country, P (in millions), during the given years.

Year	*P* (millions)	Year	*P* (millions)
1900	1.2	1960	5.4
1910	1.8	1970	7.1
1920	2.2	1980	12.8
1930	3.1	1990	17.9
1940	3.9	2000	24.2
1950	5.3		

(a) Graph the data and determine what type of model (linear, quadratic, exponential, or logarithmic) fits, the data best. Let t represent the number of years after 1900.
(b) Identify the points which are least likely to fit the model in part (a).
(c) Disregarding those points found in part (b), find an equation which seems to fit the remainder of the data.
(d) Use the equation in part (c) to predict the population in the year 2010.

9. Calculate the average rates of change between each pair of consecutive points in the tables, and use the results to determine if the relationships between the variables are linear or nonlinear over the given domains.

(a)

x	0	1	2	3	4
y	4	1	−2	−5	−8

(b)

x	0	1	2	3	4
y	0	1	4	9	16

10. Sketch a graph of $f(x) = -x^2 - 1$. Then sketch the secant line for each interval of x-values, and find each corresponding average rate of change.
(a) $[1, 2]$
(b) $[-1, 1]$
(c) $[-2, -1]$
(d) $[2, 2.01]$

11. The following table shows the percentage of a document which is downloaded from the internet and the time remaining to complete the downloading:

Time	5 min 2 sec	3 min 24 sec	2 min 10 sec	0 min 20 sec
Size	1% of 300K	42% of 300K	71% of 300K	99% of 300K

(a) What is the original size of the document?
(b) Use the table to create a new table which shows the size of the document, S, in kilobytes which has been downloaded and the time remaining, t, in seconds, to complete the downloading.
(c) Find the average rate of change between each pair of consecutive points.
(d) Interpret the meaning of the negative sign in each average rate of change found in part (**c**).
(e) During which time interval is the document being downloaded the fastest?

12. Let $g(x) = 2x^2 + x$. Find the average rate of change of g for each interval of x-values.
    (a) $[3, 10]$
    (b) $[-4, 0]$
    (c) $[5, 5 + h]$
    (d) $[x_0, x_0 + h]$
13. Find the equation of the function $x(t)$ if the average rate of change is a constant 4 meters per second and $x = 3$ when $t = 1$.
14. Find the slope of the graph of $f(x) = x^2 + x$ at $x = -1$.
15. Find the equation of the tangent line to the graph of $f(x) = \sqrt{x + 2}$ at the point $(2, 2)$.

In Problems 16 and 17, find the derivative $\frac{dy}{dx}$ of each function $y = f(x)$.

16. $y = x^3 + 2x$.
17. $y = \frac{1}{3 - x}$.

In Problems 18 and 19, find the area of the shaded region.

18.

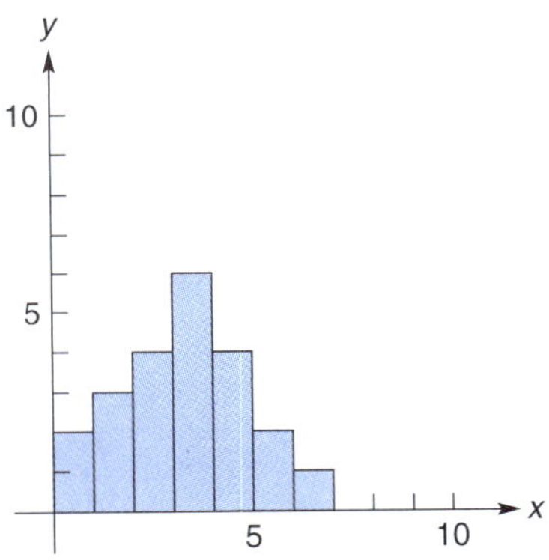

FIGURE A.47 Diagram for Problem 18.

19.

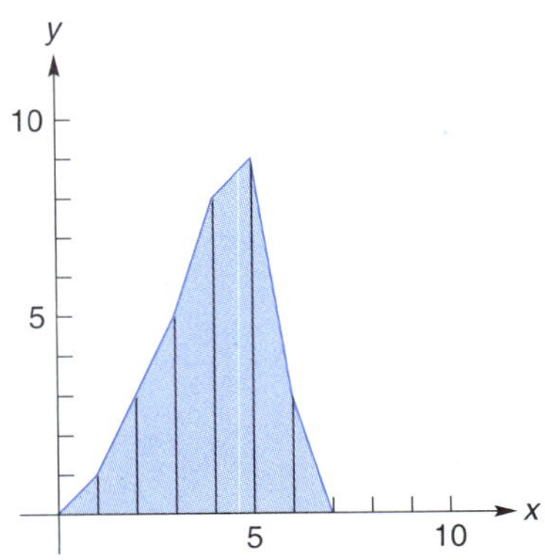

FIGURE A.48 Diagram for Problem 19.

In Problems 20 and 21, (a) find the area of each region. Then estimate the area of each region using (b) three rectangles at equally spaced subintervals, (c) six rectangles at equally spaced subintervals, and (d) three trapezoids at equally spaced subintervals. Use the right-point rule for the rectangle estimations. (e) Which estimation gives the best approximation to the actual area you found in part (a)?

20. The region bounded by the graph of $y = 1 + |x - 1|$ and the x-axis on the interval $[0, 6]$.
21. The region bounded by the graph of $y = \sqrt{9 - x^2}$ and the x-axis to the right of the y-axis.

In Problems 22 and 23, estimate the area of each region using (a) five rectangles at equally spaced subintervals, and (b) five trapezoids at equally spaced subintervals. Use the right-point rule for the rectangle estimations. (c) Which estimate gives the best approximation to the actual area?

22. The region between the graph of $f(x) = -x^2 - x + 7$ and the x-axis for x values in the interval $[-3, 2]$. The actual area is $25\frac{5}{6}$ square units.
23. The region between the graph of $f(x) = -\frac{x^3}{5} + 6x$ and the x-axis for x values in the interval $[0, 5]$. The actual area is $43\frac{3}{4}$ square units.

In Problems 24–27, evaluate the sum.

24. $\sum_{i=1}^{9} (6i + 4)$
25. $\sum_{n=8}^{16} (5n - 3)$
26. $\sum_{k=1}^{n} (2k - 1)$
27. $\sum_{k=1}^{n} \frac{k^3}{n(n+1)^2}$

In Problems 28 and 29, approximate the areas of the regions bounded by the given curves and the x-axis over the given intervals. Use the interval size given.

28. $f(x) = x^2 - 4x + 4$ over the interval $[0, 3]$ $\quad \Delta x = 0.05$
29. $f(x) = x^2 - 4x + 4$ over the interval $[0, 3]$ $\quad \Delta x = 0.01$

In Problems 30–33, express the area under the curve $f(x)$ for the given interval using integral notation.

30. $f(x) = -2x + 7 \quad [-4, 2]$
31. $f(x) = -x^2 - x + 2 \quad [-1, 1]$
32. $f(x) = 2^x + 5 \quad [-5, 5]$
33. $f(x) = 4 \log x \quad [1, 100]$

In Problems 34–37, graph the region defined by the integral.

34. $\int_0^3 5\,dx$
35. $\int_1^4 (2x + 3)\,dx$
36. $\int_{-4}^4 (x^2 - 6)\,dx$
37. $\int_{-3}^2 \left[\left(\frac{1}{2}\right)^x - 3\right] dx$

In Problems 38 and 39, (a) find $\int_0^b f(x)\,dx$ and (b) evaluate $\int_0^{10} f(x)\,dx$ for the given f(x).

38. $f(x) = 2$
39. $f(x) = x + 3$

In Problems 40–41, evaluate the integral.

40. $\int_2^7 3\,dx$
41. $\int_{80}^{100} 5000\,dx$

In Problems 42 and 43, use the given antiderivative to evaluate the integral.

42. $F(x) = 3x^2 - 5x$ is an antiderivative of $f(x) = 6x - 5$. Evaluate $\int_{-2}^{5} f(x)\,dx$.

43. $F(x) = 2x^3 - x^2 - 3x$ is an antiderivative of $f(x) = 6x^2 - 2x - 3$. Evaluate $\int_0^3 f(x)\,dx$.

44. A grandmother gives her grandson a cash birthday gift each year. The amount of the gift is \$20 for each year of his age.

(a) Write a function $b(t)$ that expresses the amount of the birthday gift b that the grandson receives with respect to his age t.
(b) Find and interpret $b'(t)$.
(c) Interpret $\int_0^{18} b(t)\,dt$.

45. The function $E(t) = 830e^{0.03t}$ approximates the average annual energy costs for a 1900 square-foot home in year t, where $t = 0$ corresponds to 1970.

(a) Find and interpret $E(31)$.
(b) Find and interpret $E'(31)$, given that $E'(t) = 24.9e^{0.03t}$.
(c) Find and interpret $\int_0^{31} E(t)\,dt$, given that $F(t) = 27{,}667e^{0.03t}$ is an antiderivative of $E(t)$.

46. Suppose the U.S. Congress votes in 2001 to reduce the budget of a federal program from \$18 billion in 2001 to \$12 billion in 2011. The functions $b_l(t) = 18 - 0.6t$ and $b_e(t) = 18e^{-0.0405t}$ are proposed models for the program's budget, where $t = 0$ corresponds to 2001, $0 \le t \le 10$.

(a) Find and interpret $b_l(5)$ and $b_e(5)$.
(b) Find and interpret $b_l'(5)$ and $b_e'(5)$, given that $b_l'(t) = -0.6$ and $b_e'(t) = -0.729e^{-0.0405t}$.
(c) Find and interpret $\int_0^5 b_l(t)\,dt$ and $\int_0^5 b_e(t)\,dt$, given that $B_l(t) = 18t - 0.3t^2$ and $B_e(t) = -444e^{-0.0405t}$ are antiderivatives of b_l and b_e, respectively.
(d) Which model yields the bigger budget in 2006?
(e) With which model is the budget decreasing more quickly in 2006?
(f) Which model yields the greater cumulative cost for the years 2001 through 2006?

47. Suppose that the Space Shuttle takes off with speed that is given by $v(t) = \frac{1}{2400}t^2$, where the speed is measured in miles per second and t represents time in seconds after liftoff. The function $v'(t) = \frac{1}{1200}t$ is the derivative of v, and $V(t) = \frac{1}{7200}t^3$ is an antiderivative of v.

(a) How fast is the shuttle moving 6 seconds after liftoff?
(b) How fast is the speed of the shuttle increasing 6 seconds after liftoff?
(c) How far has the shuttle traveled in the first 6 seconds after liftoff?

APPENDIX B

Compound Interest Tables

$r = 0.005$

n	$(1+r)^n$	$(1+r)^{-n}$	$a_{\overline{n}\rvert r}$	$s_{\overline{n}\rvert r}$
1	1.005000	0.995025	0.995025	1.000000
2	1.010025	0.990075	1.985099	2.005000
3	1.015075	0.985149	2.970248	3.015025
4	1.020151	0.980248	3.950496	4.030100
5	1.025251	0.975371	4.925866	5.050251
6	1.030378	0.970518	5.896384	6.075502
7	1.035529	0.965690	6.862074	7.105879
8	1.040707	0.960885	7.822959	8.141409
9	1.045911	0.956105	8.779064	9.182116
10	1.051140	0.951348	9.730412	10.228026
11	1.056396	0.946615	10.677027	11.279167
12	1.061678	0.941905	11.618932	12.335562
13	1.066986	0.937219	12.556151	13.397240
14	1.072321	0.932556	13.488708	14.464226
15	1.077683	0.927917	14.416625	15.536548
16	1.083071	0.923300	15.339925	16.614230
17	1.088487	0.918707	16.258632	17.697301
18	1.093929	0.914136	17.172768	18.785788
19	1.099399	0.909588	18.082356	19.879717
20	1.104896	0.905063	18.987419	20.979115
21	1.110420	0.900560	19.887979	22.084011
22	1.115972	0.896080	20.784059	23.194431
23	1.121552	0.891622	21.675681	24.310403
24	1.127160	0.887186	22.562866	25.431955
25	1.132796	0.882772	23.445638	26.559115
26	1.138460	0.878380	24.324018	27.691911
27	1.144152	0.874010	25.198028	28.830370
28	1.149873	0.869662	26.067689	29.974522
29	1.155622	0.865335	26.933024	31.124395
30	1.161400	0.861030	27.794054	32.280017
31	1.167207	0.856746	28.650800	33.441417
32	1.173043	0.852484	29.503284	34.608624
33	1.178908	0.848242	30.351526	35.781667
34	1.184803	0.844022	31.195548	36.960575
35	1.190727	0.839823	32.035371	38.145378
36	1.196681	0.835645	32.871016	39.336105
37	1.202664	0.831487	33.702504	40.532785
38	1.208677	0.827351	34.529854	41.735449
39	1.214721	0.823235	35.353089	42.944127
40	1.220794	0.819139	36.172228	44.158847
41	1.226898	0.815064	36.987291	45.379642
42	1.233033	0.811009	37.798300	46.606540
43	1.239198	0.806974	38.605274	47.839572
44	1.245394	0.802959	39.408232	49.078770
45	1.251621	0.798964	40.207196	50.324164
46	1.257879	0.794989	41.002185	51.575785
47	1.264168	0.791034	41.793219	52.833664
48	1.270489	0.787098	42.580318	54.097832
49	1.276842	0.783182	43.363500	55.368321
50	1.283226	0.779286	44.142786	56.645163

$r = 0.0075$

n	$(1+r)^n$	$(1+r)^{-n}$	$a_{\overline{n}\vert r}$	$s_{\overline{n}\vert r}$
1	1.007500	0.992556	0.992556	1.000000
2	1.015056	0.985167	1.977723	2.007500
3	1.022669	0.977833	2.955556	3.022556
4	1.030339	0.970554	3.926110	4.045225
5	1.038067	0.963329	4.889440	5.075565
6	1.045852	0.956158	5.845598	6.113631
7	1.053696	0.949040	6.794638	7.159484
8	1.061599	0.941975	7.736613	8.213180
9	1.069561	0.934963	8.671576	9.274779
10	1.077583	0.928003	9.599580	10.344339
11	1.085664	0.921095	10.520675	11.421922
12	1.093807	0.914238	11.434913	12.507586
13	1.102010	0.907432	12.342345	13.601393
14	1.110276	0.900677	13.243022	14.703404
15	1.118603	0.893973	14.136995	15.813679
16	1.126992	0.887318	15.024313	16.932282
17	1.135445	0.880712	15.905025	18.059274
18	1.143960	0.874156	16.779181	19.194718
19	1.152540	0.867649	17.646830	20.338679
20	1.161184	0.861190	18.508020	21.491219
21	1.169893	0.854779	19.362799	22.652403
22	1.178667	0.848416	20.211215	23.822296
23	1.187507	0.842100	21.053315	25.000963
24	1.196414	0.835831	21.889146	26.188471
25	1.205387	0.829609	22.718755	27.384884
26	1.214427	0.823434	23.542189	28.590271
27	1.223535	0.817304	24.359493	29.804698
28	1.232712	0.811220	25.170713	31.028233
29	1.241957	0.805181	25.975893	32.260945
30	1.251272	0.799187	26.775080	33.502902
31	1.260656	0.793238	27.568318	34.754174
32	1.270111	0.787333	28.355650	36.014830
33	1.279637	0.781472	29.137122	37.284941
34	1.289234	0.775654	29.912776	38.564578
35	1.298904	0.769880	30.682656	39.853813
36	1.308645	0.764149	31.446805	41.152716
37	1.318460	0.758461	32.205266	42.461361
38	1.328349	0.752814	32.958080	43.779822
39	1.338311	0.747210	33.705290	45.108170
40	1.348349	0.741648	34.446938	46.446482
41	1.358461	0.736127	35.183065	47.794830
42	1.368650	0.730647	35.913713	49.153291
43	1.378915	0.725208	36.638921	50.521941
44	1.389256	0.719810	37.358730	51.900856
45	1.399676	0.714451	38.073181	53.290112
46	1.410173	0.709133	38.782314	54.689788
47	1.420750	0.703854	39.486168	56.099961
48	1.431405	0.698614	40.184782	57.520711
49	1.442141	0.693414	40.878195	58.952116
50	1.452957	0.688252	41.566447	60.394257

$r = 0.01$

n	$(1+r)^n$	$(1+r)^{-n}$	$a_{\overline{n}\rvert r}$	$s_{\overline{n}\rvert r}$
1	1.010000	0.990099	0.990099	1.000000
2	1.020100	0.980296	1.970395	2.010000
3	1.030301	0.970590	2.940985	3.030100
4	1.040604	0.960980	3.901966	4.060401
5	1.051010	0.951466	4.853431	5.101005
6	1.061520	0.942045	5.795476	6.152015
7	1.072135	0.932718	6.728195	7.213535
8	1.082857	0.923483	7.651678	8.285671
9	1.093685	0.914340	8.566018	9.368527
10	1.104622	0.905287	9.471305	10.462213
11	1.115668	0.896324	10.367628	11.566835
12	1.126825	0.887449	11.255077	12.682503
13	1.138093	0.878663	12.133740	13.809328
14	1.149474	0.869963	13.003703	14.947421
15	1.160969	0.861349	13.865053	16.096896
16	1.172579	0.852821	14.717874	17.257864
17	1.184304	0.844377	15.562251	18.430443
18	1.196147	0.836017	16.398269	19.614748
19	1.208109	0.827740	17.226008	20.810895
20	1.220190	0.819544	18.045553	22.019004
21	1.232392	0.811430	18.856983	23.239194
22	1.244716	0.803396	19.660379	24.471586
23	1.257163	0.795442	20.455821	25.716302
24	1.269735	0.787566	21.243387	26.973465
25	1.282432	0.779768	22.023156	28.243200
26	1.295256	0.772048	22.795204	29.525631
27	1.308209	0.764404	23.559608	30.820888
28	1.321291	0.756836	24.316443	32.129097
29	1.334504	0.749342	25.065785	33.450388
30	1.347849	0.741923	25.807708	34.784892
31	1.361327	0.734577	26.542285	36.132740
32	1.374941	0.727304	27.269589	37.494068
33	1.388690	0.720103	27.989693	38.869009
34	1.402577	0.712973	28.702666	40.257699
35	1.416603	0.705914	29.408580	41.660276
36	1.430769	0.698925	30.107505	43.076878
37	1.445076	0.692005	30.799510	44.507647
38	1.459527	0.685153	31.484663	45.952724
39	1.474123	0.678370	32.163033	47.412251
40	1.488864	0.671653	32.834686	48.886373
41	1.503752	0.665003	33.499689	50.375237
42	1.518790	0.658419	34.158108	51.878989
43	1.533978	0.651900	34.810008	53.397779
44	1.549318	0.645445	35.455454	54.931757
45	1.564811	0.639055	36.094508	56.481075
46	1.580459	0.632728	36.727236	58.045885
47	1.596263	0.626463	37.353699	59.626344
48	1.612226	0.620260	37.973959	61.222608
49	1.628348	0.614119	38.588079	62.834834
50	1.644632	0.608039	39.196118	64.463182

$r = 0.0125$

n	$(1+r)^n$	$(1+r)^{-n}$	$a_{\overline{n}\rvert r}$	$s_{\overline{n}\rvert r}$
1	1.012500	0.987654	0.987654	1.000000
2	1.025156	0.975461	1.963115	2.012500
3	1.037971	0.963418	2.926534	3.037656
4	1.050945	0.951524	3.878058	4.075627
5	1.064082	0.939777	4.817835	5.126572
6	1.077383	0.928175	5.746010	6.190654
7	1.090850	0.916716	6.662726	7.268038
8	1.104486	0.905398	7.568124	8.358888
9	1.118292	0.894221	8.462345	9.463374
10	1.132271	0.883181	9.345526	10.581666
11	1.146424	0.872277	10.217803	11.713937
12	1.160755	0.861509	11.079312	12.860361
13	1.175264	0.850873	11.930185	14.021116
14	1.189955	0.840368	12.770553	15.196380
15	1.204829	0.829993	13.600546	16.386335
16	1.219890	0.819746	14.420292	17.591164
17	1.235138	0.809626	15.229918	18.811053
18	1.250577	0.799631	16.029549	20.046192
19	1.266210	0.789759	16.819308	21.296769
20	1.282037	0.780009	17.599316	22.562979
21	1.298063	0.770379	18.369695	23.845016
22	1.314288	0.760868	19.130563	25.143078
23	1.330717	0.751475	19.882037	26.457367
24	1.347351	0.742197	20.624235	27.788084
25	1.364193	0.733034	21.357269	29.135435
26	1.381245	0.723984	22.081253	30.499628
27	1.398511	0.715046	22.796299	31.880873
28	1.415992	0.706219	23.502518	33.279384
29	1.433692	0.697500	24.200018	34.695377
30	1.451613	0.688889	24.888906	36.129069
31	1.469759	0.680384	25.569290	37.580682
32	1.488131	0.671984	26.241274	39.050441
33	1.506732	0.663688	26.904962	40.538571
34	1.525566	0.655494	27.560456	42.045303
35	1.544636	0.647402	28.207858	43.570870
36	1.563944	0.639409	28.847267	45.115505
37	1.583493	0.631515	29.478783	46.679449
38	1.603287	0.623719	30.102501	48.262942
39	1.623328	0.616019	30.718520	49.866229
40	1.643619	0.608413	31.326933	51.489557
41	1.664165	0.600902	31.927835	53.133177
42	1.684967	0.593484	32.521319	54.797341
43	1.706029	0.586157	33.107475	56.482308
44	1.727354	0.578920	33.686395	58.188337
45	1.748946	0.571773	34.258168	59.915691
46	1.770808	0.564714	34.822882	61.664637
47	1.792943	0.557742	35.380624	63.435445
48	1.815355	0.550856	35.931481	65.228388
49	1.838047	0.544056	36.475537	67.043743
50	1.861022	0.537339	37.012876	68.881790

$r = 0.015$

n	$(1+r)^n$	$(1+r)^{-n}$	$a_{\overline{n}\rceil r}$	$s_{\overline{n}\rceil r}$
1	1.015000	0.985222	0.985222	1.000000
2	1.030225	0.970662	1.955883	2.015000
3	1.045678	0.956317	2.912200	3.045225
4	1.061364	0.942184	3.854385	4.090903
5	1.077284	0.928260	4.782645	5.152267
6	1.093443	0.914542	5.697187	6.229551
7	1.109845	0.901027	6.598214	7.322994
8	1.126493	0.887711	7.485925	8.432839
9	1.143390	0.874592	8.360517	9.559332
10	1.160541	0.861667	9.222185	10.702722
11	1.177949	0.848933	10.071118	11.863262
12	1.195618	0.836387	10.907505	13.041211
13	1.213552	0.824027	11.731532	14.236830
14	1.231756	0.811849	12.543382	15.450382
15	1.250232	0.799852	13.343233	16.682138
16	1.268986	0.788031	14.131264	17.932370
17	1.288020	0.776385	14.907649	19.201355
18	1.307341	0.764912	15.672561	20.489376
19	1.326951	0.753607	16.426168	21.796716
20	1.346855	0.742470	17.168639	23.123667
21	1.367058	0.731498	17.900137	24.470522
22	1.387564	0.720688	18.620824	25.837580
23	1.408377	0.710037	19.330861	27.225144
24	1.429503	0.699544	20.030405	28.633521
25	1.450945	0.689206	20.719611	30.063024
26	1.472710	0.679021	21.398632	31.513969
27	1.494800	0.668986	22.067617	32.986678
28	1.517222	0.659099	22.726717	34.481479
29	1.539981	0.649359	23.376076	35.998701
30	1.563080	0.639762	24.015838	37.538681
31	1.586526	0.630308	24.646146	39.101762
32	1.610324	0.620993	25.267139	40.688288
33	1.634479	0.611816	25.878954	42.298612
34	1.658996	0.602774	26.481728	43.933092
35	1.683881	0.593866	27.075595	45.592088
36	1.709140	0.585090	27.660684	47.275969
37	1.734777	0.576443	28.237127	48.985109
38	1.760798	0.567924	28.805052	50.719885
39	1.787210	0.559531	29.364583	52.480684
40	1.814018	0.551262	29.915845	54.267894
41	1.841229	0.543116	30.458961	56.081912
42	1.868847	0.535089	30.994050	57.923141
43	1.896880	0.527182	31.521232	59.791988
44	1.925333	0.519391	32.040622	61.688868
45	1.954213	0.511715	32.552337	63.614201
46	1.983526	0.504153	33.056490	65.568414
47	2.013279	0.496702	33.553192	67.551940
48	2.043478	0.489362	34.042554	69.565219
49	2.074130	0.482130	34.524683	71.608698
50	2.105242	0.475005	34.999688	73.682828

$r = 0.02$

n	$(1+r)^n$	$(1+r)^{-n}$	$a_{\overline{n}\vert r}$	$s_{\overline{n}\vert r}$
1	1.020000	0.980392	0.980392	1.000000
2	1.040400	0.961169	1.941561	2.020000
3	1.061208	0.942322	2.883883	3.060400
4	1.082432	0.923845	3.807729	4.121608
5	1.104081	0.905731	4.713460	5.204040
6	1.126162	0.887971	5.601431	6.308121
7	1.148686	0.870560	6.471991	7.434283
8	1.171659	0.853490	7.325481	8.582969
9	1.195093	0.836755	8.162237	9.754628
10	1.218994	0.820348	8.982585	10.949721
11	1.243374	0.804263	9.786848	12.168715
12	1.268242	0.788493	10.575341	13.412090
13	1.293607	0.773033	11.348374	14.680332
14	1.319479	0.757875	12.106249	15.973938
15	1.345868	0.743015	12.849264	17.293417
16	1.372786	0.728446	13.577709	18.639285
17	1.400241	0.714163	14.291872	20.012071
18	1.428246	0.700159	14.992031	21.412312
19	1.456811	0.686431	15.678462	22.840559
20	1.485947	0.672971	16.351433	24.297370
21	1.515666	0.659776	17.011209	25.783317
22	1.545980	0.646839	17.658048	27.298984
23	1.576899	0.634156	18.292204	28.844963
24	1.608437	0.621721	18.913926	30.421862
25	1.640606	0.609531	19.523456	32.030300
26	1.673418	0.597579	20.121036	33.670906
27	1.706886	0.585862	20.706898	35.344324
28	1.741024	0.574375	21.281272	37.051210
29	1.775845	0.563112	21.844385	38.792235
30	1.811362	0.552071	22.396456	40.568079
31	1.847589	0.541246	22.937702	42.379441
32	1.884541	0.530633	23.468335	44.227030
33	1.922231	0.520229	23.988564	46.111570
34	1.960676	0.510028	24.498592	48.033802
35	1.999890	0.500028	24.998619	49.994478
36	2.039887	0.490223	25.488842	51.994367
37	2.080685	0.480611	25.969453	54.034255
38	2.122299	0.471187	26.440641	56.114940
39	2.164745	0.461948	26.902589	58.237238
40	2.208040	0.452890	27.355479	60.401983
41	2.252200	0.444010	27.799489	62.610023
42	2.297244	0.435304	28.234794	64.862223
43	2.343189	0.426769	28.661562	67.159468
44	2.390053	0.418401	29.079963	69.502657
45	2.437854	0.410197	29.490160	71.892710
46	2.486611	0.402154	29.892314	74.330564
47	2.536344	0.394268	30.286582	76.817176
48	2.587070	0.386538	30.673120	79.353519
49	2.638812	0.378958	31.052078	81.940590
50	2.691588	0.371528	31.423606	84.579401

$r = 0.025$

n	$(1+r)^n$	$(1+r)^{-n}$	$a_{\overline{n}\rceil r}$	$s_{\overline{n}\rceil r}$
1	1.025000	0.975610	0.975610	1.000000
2	1.050625	0.951814	1.927424	2.025000
3	1.076891	0.928599	2.856024	3.075625
4	1.103813	0.905951	3.761974	4.152516
5	1.131408	0.883854	4.645828	5.256329
6	1.159693	0.862297	5.508125	6.387737
7	1.188686	0.841265	6.349391	7.547430
8	1.218403	0.820747	7.170137	8.736116
9	1.248863	0.800728	7.970866	9.954519
10	1.280085	0.781198	8.752064	11.203382
11	1.312087	0.762145	9.514209	12.483466
12	1.344889	0.743556	10.257765	13.795553
13	1.378511	0.725420	10.983185	15.140442
14	1.412974	0.707727	11.690912	16.518953
15	1.448298	0.690466	12.381378	17.931927
16	1.484506	0.673625	13.055003	19.380225
17	1.521618	0.657195	13.712198	20.864730
18	1.559659	0.641166	14.353364	22.386349
19	1.598650	0.625528	14.978891	23.946007
20	1.638616	0.610271	15.589162	25.544658
21	1.679582	0.595386	16.184549	27.183274
22	1.721571	0.580865	16.765413	28.862856
23	1.764611	0.566697	17.332110	30.584427
24	1.808726	0.552875	17.884986	32.349038
25	1.853944	0.539391	18.424376	34.157764
26	1.900293	0.526235	18.950611	36.011708
27	1.947800	0.513400	19.464011	37.912001
28	1.996495	0.500878	19.964889	39.859801
29	2.046407	0.488661	20.453550	41.856296
30	2.097568	0.476743	20.930293	43.902703
31	2.150007	0.465115	21.395407	46.000271
32	2.203757	0.453771	21.849178	48.150278
33	2.258851	0.442703	22.291881	50.354034
34	2.315322	0.431905	22.723786	52.612885
35	2.373205	0.421371	23.145157	54.928207
36	2.432535	0.411094	23.556251	57.301413
37	2.493349	0.401067	23.957318	59.733948
38	2.555682	0.391285	24.348603	62.227297
39	2.619574	0.381741	24.730344	64.782979
40	2.685064	0.372431	25.102775	67.402554
41	2.752190	0.363347	25.466122	70.087617
42	2.820995	0.354485	25.820607	72.839808
43	2.891520	0.345839	26.166446	75.660803
44	2.963808	0.337404	26.503849	78.552323
45	3.037903	0.329174	26.833024	81.516131
46	3.113851	0.321146	27.154170	84.554034
47	3.191697	0.313313	27.467483	87.667885
48	3.271490	0.305671	27.773154	90.859582
49	3.353277	0.298216	28.071369	94.131072
50	3.437109	0.290942	28.362312	97.484349

$r = 0.03$

n	$(1+r)^n$	$(1+r)^{-n}$	$a_{\overline{n}\rceil r}$	$s_{\overline{n}\rceil r}$
1	1.030000	0.970874	0.970874	1.000000
2	1.060900	0.942596	1.913470	2.030000
3	1.092727	0.915142	2.828611	3.090900
4	1.125509	0.888487	3.717098	4.183627
5	1.159274	0.862609	4.579707	5.309136
6	1.194052	0.837484	5.417191	6.468410
7	1.229874	0.813092	6.230283	7.662462
8	1.266770	0.789409	7.019692	8.892336
9	1.304773	0.766417	7.786109	10.159106
10	1.343916	0.744094	8.530203	11.463879
11	1.384234	0.722421	9.252624	12.807796
12	1.425761	0.701380	9.954004	14.192030
13	1.468534	0.680951	10.634955	15.617790
14	1.512590	0.661118	11.296073	17.086324
15	1.557967	0.641862	11.937935	18.598914
16	1.604706	0.623167	12.561102	20.156881
17	1.652848	0.605016	13.166118	21.761588
18	1.702433	0.587395	13.753513	23.414435
19	1.753506	0.570286	14.323799	25.116868
20	1.806111	0.553676	14.877475	26.870374
21	1.860295	0.537549	15.415024	28.676486
22	1.916103	0.521893	15.936917	30.536780
23	1.973587	0.506692	16.443608	32.452884
24	2.032794	0.491934	16.935542	34.426470
25	2.093778	0.477606	17.413148	36.459264
26	2.156591	0.463695	17.876842	38.553042
27	2.221289	0.450189	18.327031	40.709634
28	2.287928	0.437077	18.764108	42.930923
29	2.356566	0.424346	19.188455	45.218850
30	2.427262	0.411987	19.600441	47.575416
31	2.500080	0.399987	20.000428	50.002678
32	2.575083	0.388337	20.388766	52.502759
33	2.652335	0.377026	20.765792	55.077841
34	2.731905	0.366045	21.131837	57.730177
35	2.813862	0.355383	21.487220	60.462082
36	2.898278	0.345032	21.832252	63.275944
37	2.985227	0.334983	22.167235	66.174223
38	3.074783	0.325226	22.492462	69.159449
39	3.167027	0.315754	22.808215	72.234233
40	3.262038	0.306557	23.114772	75.401260
41	3.359899	0.297628	23.412400	78.663298
42	3.460696	0.288959	23.701359	82.023196
43	3.564517	0.280543	23.981902	85.483892
44	3.671452	0.272372	24.254274	89.048409
45	3.781596	0.264439	24.518713	92.719861
46	3.895044	0.256737	24.775449	96.501457
47	4.011895	0.249259	25.024708	100.396501
48	4.132252	0.241999	25.266707	104.408396
49	4.256219	0.234950	25.501657	108.540648
50	4.383906	0.228107	25.729764	112.796867

$r = 0.035$

n	$(1+r)^n$	$(1+r)^{-n}$	$a_{\overline{n}\rvert r}$	$s_{\overline{n}\rvert r}$
1	1.035000	0.966184	0.966184	1.000000
2	1.071225	0.933511	1.899694	2.035000
3	1.108718	0.901943	2.801637	3.106225
4	1.147523	0.871442	3.673079	4.214943
5	1.187686	0.841973	4.515052	5.362466
6	1.229255	0.813501	5.328553	6.550152
7	1.272279	0.785991	6.114544	7.779408
8	1.316809	0.759412	6.873956	9.051687
9	1.362897	0.733731	7.607687	10.368496
10	1.410599	0.708919	8.316605	11.731393
11	1.459970	0.684946	9.001551	13.141992
12	1.511069	0.661783	9.663334	14.601962
13	1.563956	0.639404	10.302738	16.113030
14	1.618695	0.617782	10.920520	17.676986
15	1.675349	0.596891	11.517411	19.295681
16	1.733986	0.576706	12.094117	20.971030
17	1.794676	0.557204	12.651321	22.705016
18	1.857489	0.538361	13.189682	24.499691
19	1.922501	0.520156	13.709837	26.357180
20	1.989789	0.502566	14.212403	28.279682
21	2.059431	0.485571	14.697974	30.269471
22	2.131512	0.469151	15.167125	32.328902
23	2.206114	0.453286	15.620410	34.460414
24	2.283328	0.437957	16.058368	36.666528
25	2.363245	0.423147	16.481515	38.949857
26	2.445959	0.408838	16.890352	41.313102
27	2.531567	0.395012	17.285365	43.759060
28	2.620172	0.381654	17.667019	46.290627
29	2.711878	0.368748	18.035767	48.910799
30	2.806794	0.356278	18.392045	51.622677
31	2.905031	0.344230	18.736276	54.429471
32	3.006708	0.332590	19.068865	57.334502
33	3.111942	0.321343	19.390208	60.341210
34	3.220860	0.310476	19.700684	63.453152
35	3.333590	0.299977	20.000661	66.674013
36	3.450266	0.289833	20.290494	70.007603
37	3.571025	0.280032	20.570525	73.457869
38	3.696011	0.270562	20.841087	77.028895
39	3.825372	0.261413	21.102500	80.724906
40	3.959260	0.252572	21.355072	84.550278
41	4.097834	0.244031	21.599104	88.509537
42	4.241258	0.235779	21.834883	92.607371
43	4.389702	0.227806	22.062689	96.848629
44	4.543342	0.220102	22.282791	101.238331
45	4.702359	0.212659	22.495450	105.781673
46	4.866941	0.205468	22.700918	110.484031
47	5.037284	0.198520	22.899438	115.350973
48	5.213589	0.191806	23.091244	120.388257
49	5.396065	0.185320	23.276564	125.601846
50	5.584927	0.179053	23.455618	130.997910

$r = 0.04$

n	$(1+r)^n$	$(1+r)^{-n}$	$a_{\overline{n}\vert r}$	$s_{\overline{n}\vert r}$
1	1.040000	0.961538	0.961538	1.000000
2	1.081600	0.924556	1.886095	2.040000
3	1.124864	0.888996	2.775091	3.121600
4	1.169859	0.854804	3.629895	4.246464
5	1.216653	0.821927	4.451822	5.416323
6	1.265319	0.790315	5.242137	6.632975
7	1.315932	0.759918	6.002055	7.898294
8	1.368569	0.730690	6.732745	9.214226
9	1.423312	0.702587	7.435332	10.582795
10	1.480244	0.675564	8.110896	12.006107
11	1.539454	0.649581	8.760477	13.486351
12	1.601032	0.624597	9.385074	15.025805
13	1.665074	0.600574	9.985648	16.626838
14	1.731676	0.577475	10.563123	18.291911
15	1.800944	0.555265	11.118387	20.023588
16	1.872981	0.533908	11.652296	21.824531
17	1.947900	0.513373	12.165669	23.697512
18	2.025817	0.493628	12.659297	25.645413
19	2.106849	0.474642	13.133939	27.671229
20	2.191123	0.456387	13.590326	29.778079
21	2.278768	0.438834	14.029160	31.969202
22	2.369919	0.421955	14.451115	34.247970
23	2.464716	0.405726	14.856842	36.617889
24	2.563304	0.390121	15.246963	39.082604
25	2.665836	0.375117	15.622080	41.645908
26	2.772470	0.360689	15.982769	44.311745
27	2.883369	0.346817	16.329586	47.084214
28	2.998703	0.333477	16.663063	49.967583
29	3.118651	0.320651	16.983715	52.966286
30	3.243398	0.308319	17.292033	56.084938
31	3.373133	0.296460	17.588494	59.328335
32	3.508059	0.285058	17.873551	62.701469
33	3.648381	0.274094	18.147646	66.209527
34	3.794316	0.263552	18.411198	69.857909
35	3.946089	0.253415	18.664613	73.652225
36	4.103933	0.243669	18.908282	77.598314
37	4.268090	0.234297	19.142579	81.702246
38	4.438813	0.225285	19.367864	85.970336
39	4.616366	0.216621	19.584485	90.409150
40	4.801021	0.208289	19.792774	95.025516
41	4.993061	0.200278	19.993052	99.826536
42	5.192784	0.192575	20.185627	104.819598
43	5.400495	0.185168	20.370795	110.012382
44	5.616515	0.178046	20.548841	115.412877
45	5.841176	0.171198	20.720040	121.029392
46	6.074823	0.164614	20.884654	126.870568
47	6.317816	0.158283	21.042936	132.945390
48	6.570528	0.152195	21.195131	139.263206
49	6.833349	0.146341	21.341472	145.833734
50	7.106683	0.140713	21.482185	152.667084

$r = 0.05$

n	$(1+r)^n$	$(1+r)^{-n}$	$a_{\overline{n}\rvert r}$	$s_{\overline{n}\rvert r}$
1	1.050000	0.952381	0.952381	1.000000
2	1.102500	0.907029	1.859410	2.050000
3	1.157625	0.863838	2.723248	3.152500
4	1.215506	0.822702	3.545951	4.310125
5	1.276282	0.783526	4.329477	5.525631
6	1.340096	0.746215	5.075692	6.801913
7	1.407100	0.710681	5.786373	8.142008
8	1.477455	0.676839	6.463213	9.549109
9	1.551328	0.644609	7.107822	11.026564
10	1.628895	0.613913	7.721735	12.577893
11	1.710339	0.584679	8.306414	14.206787
12	1.795856	0.556837	8.863252	15.917127
13	1.885649	0.530321	9.393573	17.712983
14	1.979932	0.505068	9.898641	19.598632
15	2.078928	0.481017	10.379658	21.578564
16	2.182875	0.458112	10.837770	23.657492
17	2.292018	0.436297	11.274066	25.840366
18	2.406619	0.415521	11.689587	28.132385
19	2.526950	0.395734	12.085321	30.539004
20	2.653298	0.376889	12.462210	33.065954
21	2.785963	0.358942	12.821153	35.719252
22	2.925261	0.341850	13.163003	38.505214
23	3.071524	0.325571	13.488574	41.430475
24	3.225100	0.310068	13.798642	44.501999
25	3.386355	0.295303	14.093945	47.727099
26	3.555673	0.281241	14.375185	51.113454
27	3.733456	0.267848	14.643034	54.669126
28	3.920129	0.255094	14.898127	58.402583
29	4.116136	0.242946	15.141074	62.322712
30	4.321942	0.231377	15.372451	66.438848
31	4.538039	0.220359	15.592811	70.760790
32	4.764941	0.209866	15.802677	75.298829
33	5.003189	0.199873	16.002549	80.063771
34	5.253348	0.190355	16.192904	85.066959
35	5.516015	0.181290	16.374194	90.320307
36	5.791816	0.172657	16.546852	95.836323
37	6.081407	0.164436	16.711287	101.628139
38	6.385477	0.156605	16.867893	107.709546
39	6.704751	0.149148	17.017041	114.095023
40	7.039989	0.142046	17.159086	120.799774
41	7.391988	0.135282	17.294368	127.839763
42	7.761588	0.128840	17.423208	135.231751
43	8.149667	0.122704	17.545912	142.993339
44	8.557150	0.116861	17.662773	151.143006
45	8.985008	0.111297	17.774070	159.700156
46	9.434258	0.105997	17.880066	168.685164
47	9.905971	0.100949	17.981016	178.119422
48	10.401270	0.096142	18.077158	188.025393
49	10.921333	0.091564	18.168722	198.426663
50	11.467400	0.087204	18.255925	209.347996

$r = 0.06$

n	$(1+r)^n$	$(1+r)^{-n}$	$a_{\overline{n}\rvert r}$	$s_{\overline{n}\rvert r}$
1	1.060000	0.943396	0.943396	1.000000
2	1.123600	0.889996	1.833393	2.060000
3	1.191016	0.839619	2.673012	3.183600
4	1.262477	0.792094	3.465106	4.374616
5	1.338226	0.747258	4.212364	5.637093
6	1.418519	0.704961	4.917324	6.975319
7	1.503630	0.665057	5.582381	8.393838
8	1.593848	0.627412	6.209794	9.897468
9	1.689479	0.591898	6.801692	11.491316
10	1.790848	0.558395	7.360087	13.180795
11	1.898299	0.526788	7.886875	14.971643
12	2.012196	0.496969	8.383844	16.869941
13	2.132928	0.468839	8.852683	18.882138
14	2.260904	0.442301	9.294984	21.015066
15	2.396558	0.417265	9.712249	23.275970
16	2.540352	0.393646	10.105895	25.672528
17	2.692773	0.371364	10.477260	28.212880
18	2.854339	0.350344	10.827603	30.905653
19	3.025600	0.330513	11.158116	33.759992
20	3.207135	0.311805	11.469921	36.785591
21	3.399564	0.294155	11.764077	39.992727
22	3.603537	0.277505	12.041582	43.392290
23	3.819750	0.261797	12.303379	46.995828
24	4.048935	0.246979	12.550358	50.815577
25	4.291871	0.232999	12.783356	54.864512
26	4.549383	0.219810	13.003166	59.156383
27	4.822346	0.207368	13.210534	63.705766
28	5.111687	0.195630	13.406164	68.528112
29	5.418388	0.184557	13.590721	73.639798
30	5.743491	0.174110	13.764831	79.058186
31	6.088101	0.164255	13.929086	84.801677
32	6.453387	0.154957	14.084043	90.889778
33	6.840590	0.146186	14.230230	97.343165
34	7.251025	0.137912	14.368141	104.183755
35	7.686087	0.130105	14.498246	111.434780
36	8.147252	0.122741	14.620987	119.120867
37	8.636087	0.115793	14.736780	127.268119
38	9.154252	0.109239	14.846019	135.904206
39	9.703507	0.103056	14.949075	145.058458
40	10.285718	0.097222	15.046297	154.761966
41	10.902861	0.091719	15.138016	165.047684
42	11.557033	0.086527	15.224543	175.950545
43	12.250455	0.081630	15.306173	187.507577
44	12.985482	0.077009	15.383182	199.758032
45	13.764611	0.072650	15.455832	212.743514
46	14.590487	0.068538	15.524370	226.508125
47	15.465917	0.064658	15.589028	241.098612
48	16.393872	0.060998	15.650027	256.564529
49	17.377504	0.057546	15.707572	272.958401
50	18.420154	0.054288	15.761861	290.335905

$r = 0.07$

n	$(1+r)^n$	$(1+r)^{-n}$	$a_{\overline{n}\vert r}$	$s_{\overline{n}\vert r}$
1	1.070000	0.934579	0.934579	1.000000
2	1.144900	0.873439	1.808018	2.070000
3	1.225043	0.816298	2.624316	3.214900
4	1.310796	0.762895	3.387211	4.439943
5	1.402552	0.712986	4.100197	5.750739
6	1.500730	0.666342	4.766540	7.153291
7	1.605781	0.622750	5.389289	8.654021
8	1.718186	0.582009	5.971299	10.259803
9	1.838459	0.543934	6.515232	11.977989
10	1.967151	0.508349	7.023582	13.816448
11	2.104852	0.475093	7.498674	15.783599
12	2.252192	0.444012	7.942686	17.888451
13	2.409845	0.414964	8.357651	20.140643
14	2.578534	0.387817	8.745468	22.550488
15	2.759032	0.362446	9.107914	25.129022
16	2.952164	0.338735	9.446649	27.888054
17	3.158815	0.316574	9.763223	30.840217
18	3.379932	0.295864	10.059087	33.999033
19	3.616528	0.276508	10.335595	37.378965
20	3.869684	0.258419	10.594014	40.995492
21	4.140562	0.241513	10.835527	44.865177
22	4.430402	0.225713	11.061240	49.005739
23	4.740530	0.210947	11.272187	53.436141
24	5.072367	0.197147	11.469334	58.176671
25	5.427433	0.184249	11.653583	63.249038
26	5.807353	0.172195	11.825779	68.676470
27	6.213868	0.160930	11.986709	74.483823
28	6.648838	0.150402	12.137111	80.697691
29	7.114257	0.140563	12.277674	87.346529
30	7.612255	0.131367	12.409041	94.460786
31	8.145113	0.122773	12.531814	102.073041
32	8.715271	0.114741	12.646555	110.218154
33	9.325340	0.107235	12.753790	118.933425
34	9.978114	0.100219	12.854009	128.258765
35	10.676581	0.093663	12.947672	138.236878
36	11.423942	0.087535	13.035208	148.913460
37	12.223618	0.081809	13.117017	160.337402
38	13.079271	0.076457	13.193473	172.561020
39	13.994820	0.071455	13.264928	185.640292
40	14.974458	0.066780	13.331709	199.635112
41	16.022670	0.062412	13.394120	214.609570
42	17.144257	0.058329	13.452449	230.632240
43	18.344355	0.054513	13.506962	247.776496
44	19.628460	0.050946	13.557908	266.120851
45	21.002452	0.047613	13.605522	285.749311
46	22.472623	0.044499	13.650020	306.751763
47	24.045707	0.041587	13.691608	329.224386
48	25.728907	0.038867	13.730474	353.270093
49	27.529930	0.036324	13.766799	378.999000
50	29.457025	0.033948	13.800746	406.528929

$r = 0.08$

n	$(1+r)^n$	$(1+r)^{-n}$	$a_{\overline{n}\rvert r}$	$s_{\overline{n}\rvert r}$
1	1.080000	0.925926	0.925926	1.000000
2	1.166400	0.857339	1.783265	2.080000
3	1.259712	0.793832	2.577097	3.246400
4	1.360489	0.735030	3.312127	4.506112
5	1.469328	0.680583	3.992710	5.866601
6	1.586874	0.630170	4.622880	7.335929
7	1.713824	0.583490	5.206370	8.922803
8	1.850930	0.540269	5.746639	10.636628
9	1.999005	0.500249	6.246888	12.487558
10	2.158925	0.463193	6.710081	14.486562
11	2.331639	0.428883	7.138964	16.645487
12	2.518170	0.397114	7.536078	18.977126
13	2.719624	0.367698	7.903776	21.495297
14	2.937194	0.340461	8.244237	24.214920
15	3.172169	0.315242	8.559479	27.152114
16	3.425943	0.291890	8.851369	30.324283
17	3.700018	0.270269	9.121638	33.750226
18	3.996019	0.250249	9.371887	37.450244
19	4.315701	0.231712	9.603599	41.446263
20	4.660957	0.214548	9.818147	45.761964
21	5.033834	0.198656	10.016803	50.422921
22	5.436540	0.183941	10.200744	55.456755
23	5.871464	0.170315	10.371059	60.893296
24	6.341181	0.157699	10.528758	66.764759
25	6.848475	0.146018	10.674776	73.105940
26	7.396353	0.135202	10.809978	79.954415
27	7.988061	0.125187	10.935165	87.350768
28	8.627106	0.115914	11.051078	95.338830
29	9.317275	0.107328	11.158406	103.965936
30	10.062657	0.099377	11.257783	113.283211
31	10.867669	0.092016	11.349799	123.345868
32	11.737083	0.085200	11.434999	134.213537
33	12.676050	0.078889	11.513888	145.950620
34	13.690134	0.073045	11.586934	158.626670
35	14.785344	0.067635	11.654568	172.316804
36	15.968172	0.062625	11.717193	187.102148
37	17.245626	0.057986	11.775179	203.070320
38	18.625276	0.053690	11.828869	220.315945
39	20.115298	0.049713	11.878582	238.941221
40	21.724521	0.046031	11.924613	259.056519
41	23.462483	0.042621	11.967235	280.781040
42	25.339482	0.039464	12.006699	304.243523
43	27.366640	0.036541	12.043240	329.583005
44	29.555972	0.033834	12.077074	356.949646
45	31.920449	0.031328	12.108402	386.505617
46	34.474085	0.029007	12.137409	418.426067
47	37.232012	0.026859	12.164267	452.900152
48	40.210573	0.024869	12.189136	490.132164
49	43.427419	0.023027	12.212163	530.342737
50	46.901613	0.021321	12.233485	573.770156

Table of Selected Integrals

Rational Forms Containing $(a + bu)$

1. $\displaystyle\int u^n\,du = \frac{u^{n+1}}{n+1} + C, \quad n \neq -1$

2. $\displaystyle\int \frac{du}{a+bu} = \frac{1}{b}\ln|a+bu| + C$

3. $\displaystyle\int \frac{u\,du}{a+bu} = \frac{u}{b} - \frac{a}{b^2}\ln|a+bu| + C$

4. $\displaystyle\int \frac{u^2\,du}{a+bu} = \frac{u^2}{2b} - \frac{au}{b^2} + \frac{a^2}{b^3}\ln|a+bu| + C$

5. $\displaystyle\int \frac{du}{u(a+bu)} = \frac{1}{a}\ln\left|\frac{u}{a+bu}\right| + C$

6. $\displaystyle\int \frac{du}{u^2(a+bu)} = -\frac{1}{au} + \frac{b}{a^2}\ln\left|\frac{a+bu}{u}\right| + C$

7. $\displaystyle\int \frac{u\,du}{(a+bu)^2} = \frac{1}{b^2}\left(\ln|a+bu| + \frac{a}{a+bu}\right) + C$

8. $\displaystyle\int \frac{u^2\,du}{(a+bu)^2} = \frac{u}{b^2} - \frac{a^2}{b^3(a+bu)} - \frac{2a}{b^3}\ln|a+bu| + C$

9. $\displaystyle\int \frac{du}{u(a+bu)^2} = \frac{1}{a(a+bu)} + \frac{1}{a^2}\ln\left|\frac{u}{a+bu}\right| + C$

10. $\displaystyle\int \frac{du}{u^2(a+bu)^2} = -\frac{a+2bu}{a^2u(a+bu)} + \frac{2b}{a^3}\ln\left|\frac{a+bu}{u}\right| + C$

11. $\displaystyle\int \frac{du}{(a+bu)(c+ku)} = \frac{1}{bc-ak}\ln\left|\frac{a+bu}{c+ku}\right| + C$

12. $\displaystyle\int \frac{u\,du}{(a+bu)(c+ku)} = \frac{1}{bc-ak}\left[\frac{c}{k}\ln|c+ku| - \frac{a}{b}\ln|a+bu|\right] + C$

Forms Containing $\sqrt{a + bu}$

13. $\displaystyle\int u\sqrt{a+bu}\,du = \frac{2(3bu-2a)(a+bu)^{3/2}}{15b^2} + C$

14. $\displaystyle\int u^2\sqrt{a+bu}\,du = \frac{2(8a^2-12abu+15b^2u^2)(a+bu)^{3/2}}{105b^3} + C$

15. $\displaystyle\int \frac{u\,du}{\sqrt{a+bu}} = \frac{2(bu-2a)\sqrt{a+bu}}{3b^2} + C$

16. $\displaystyle\int \frac{u^2\,du}{\sqrt{a+bu}} = \frac{2(3b^2u^2-4abu+8a^2)\sqrt{a+bu}}{15b^3} + C$

17. $\displaystyle\int \frac{du}{u\sqrt{a+bu}} = \frac{1}{\sqrt{a}} \ln\left|\frac{\sqrt{a+bu}-\sqrt{a}}{\sqrt{a+bu}+\sqrt{a}}\right| + C, \quad a > 0$

18. $\displaystyle\int \frac{\sqrt{a+bu}\,du}{u} = 2\sqrt{a+bu} + a\int \frac{du}{u\sqrt{a+bu}}$

Forms Containing $\sqrt{a^2 - u^2}$

19. $\displaystyle\int \frac{du}{(a^2-u^2)^{3/2}} = \frac{u}{a^2\sqrt{a^2-u^2}} + C$

20. $\displaystyle\int \frac{du}{u\sqrt{a^2-u^2}} = -\frac{1}{a}\ln\left|\frac{a+\sqrt{a^2-u^2}}{u}\right| + C$

21. $\displaystyle\int \frac{du}{u^2\sqrt{a^2-u^2}} = -\frac{\sqrt{a^2-u^2}}{a^2u} + C$

22. $\displaystyle\int \frac{\sqrt{a^2-u^2}\,du}{u} = \sqrt{a^2-u^2} - a\ln\left|\frac{a+\sqrt{a^2-u^2}}{u}\right| + C, \quad a > 0$

Forms Containing $\sqrt{u^2 \pm a^2}$

23. $\displaystyle\int \sqrt{u^2 \pm a^2}\,du = \frac{1}{2}\left(u\sqrt{u^2\pm a^2} \pm a^2\ln\left|u+\sqrt{u^2\pm a^2}\right|\right) + C$

24. $\displaystyle\int u^2\sqrt{u^2\pm a^2}\,du = \frac{u}{8}(2u^2 \pm a^2)\sqrt{u^2\pm a^2} - \frac{a^4}{8}\ln\left|u+\sqrt{u^2\pm a^2}\right| + C$

25. $\displaystyle\int \frac{\sqrt{u^2+a^2}\,du}{u} = \sqrt{u^2+a^2} - a\ln\left|\frac{a+\sqrt{u^2+a^2}}{u}\right| + C$

26. $\displaystyle\int \frac{\sqrt{u^2\pm a^2}\,du}{u^2} = -\frac{\sqrt{u^2\pm a^2}}{u} + \ln\left|u+\sqrt{u^2\pm a^2}\right| + C$

27. $\displaystyle\int \frac{du}{\sqrt{u^2\pm a^2}} = \ln\left|u+\sqrt{u^2\pm a^2}\right| + C$

28. $\displaystyle\int \frac{du}{u\sqrt{u^2+a^2}} = \frac{1}{a}\ln\left|\frac{\sqrt{u^2+a^2}-a}{u}\right| + C$

29. $\displaystyle\int \frac{u^2du}{\sqrt{u^2\pm a^2}} = \frac{1}{2}\left(u\sqrt{u^2\pm a^2} \mp a^2\ln\left|u+\sqrt{u^2\pm a^2}\right|\right) + C$

30. $\displaystyle\int \frac{du}{u^2\sqrt{u^2\pm a^2}} = -\frac{\pm\sqrt{u^2\pm a^2}}{a^2u} + C$

31. $\displaystyle\int (u^2\pm a^2)^{3/2}du = \frac{u}{8}(2u^2 \pm 5a^2)\sqrt{u^2\pm a^2} + \frac{3a^4}{8}\ln\left|u+\sqrt{u^2\pm a^2}\right| + C$

32. $\displaystyle\int \frac{du}{(u^2\pm a^2)^{3/2}} = \frac{\pm u}{a^2\sqrt{u^2\pm a^2}} + C$

33. $\displaystyle\int \frac{u^2du}{(u^2\pm a^2)^{3/2}} = \frac{-u}{\sqrt{u^2\pm a^2}} + \ln\left|u+\sqrt{u^2\pm a^2}\right| + C$

Rational Forms Containing $a^2 - u^2$ and $u^2 - a^2$

34. $\displaystyle\int \frac{du}{a^2-u^2} = \frac{1}{2a}\ln\left|\frac{a+u}{a-u}\right| + C$

35. $\displaystyle\int \frac{du}{u^2-a^2} = \frac{1}{2a}\ln\left|\frac{u-a}{u+a}\right| + C$

Exponential and Logarithmic Forms

36. $\displaystyle\int e^u\,du = e^u + C$

37. $\displaystyle\int a^u\,du = \frac{a^u}{\ln a} + C, \quad a > 0, a \neq 1$

38. $\displaystyle\int ue^{au}\,du = \frac{e^{au}}{a^2}(au - 1) + C$

39. $\displaystyle\int u^n e^{au}\,du = \frac{u^n e^{au}}{a} - \frac{n}{a}\int u^{n-1}e^{au}\,du$

40. $\displaystyle\int \frac{e^{au}\,du}{u^n} = -\frac{e^{au}}{(n-1)u^{n-1}} + \frac{a}{n-1}\int \frac{e^{au}\,du}{u^{n-1}}$

41. $\displaystyle\int \ln u\,du = u\ln u - u + C$

42. $\displaystyle\int u^n \ln u\,du = \frac{u^{n+1}\ln u}{n+1} - \frac{u^{n+1}}{(n+1)^2} + C, \quad n \neq -1$

43. $\displaystyle\int u^n \ln^m u\,du = \frac{u^{n+1}}{n+1}\ln^m u - \frac{m}{n+1}\int u^n \ln^{m-1} u\,du, \quad m, n \neq -1$

44. $\displaystyle\int \frac{du}{u\ln u} = \ln|\ln u| + C$

45. $\displaystyle\int \frac{du}{a + be^{cu}} = \frac{1}{ac}(cu - \ln|a + be^{cu}|) + C$

Miscellaneous Forms

46. $\displaystyle\int \sqrt{\frac{a+u}{b+u}}\,du = \sqrt{(a+u)(b+u)} + (a-b)\ln(\sqrt{a+u} + \sqrt{b+u}) + C$

47. $\displaystyle\int \frac{du}{\sqrt{(a+u)(b+u)}} = \ln\left|\frac{a+b}{2} + u + \sqrt{(a+u)(b+u)}\right| + C$

48. $$\int \sqrt{a + bu + cu^2}\,du = \frac{2cu+b}{4c}\sqrt{a+bu+cu^2} - \frac{b^2 - 4ac}{8c^{3/2}}\ln\left|2cu + b + 2\sqrt{c}\sqrt{a+bu+cu^2}\right| + C,$$

$c > 0$

Areas Under the Standard Normal Curve

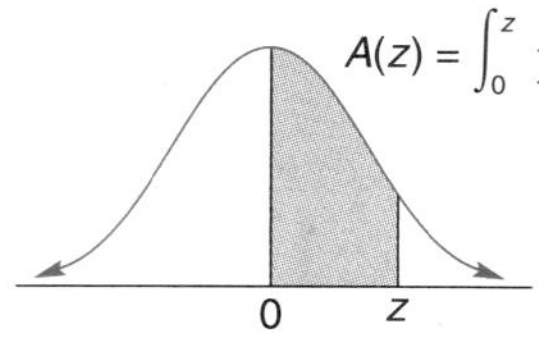

$$A(z) = \int_0^z \frac{1}{\sqrt{2\pi}} e^{-x^2/2}\, dx$$

$$A(-z) = A(z)$$

z	.00	.01	.02	.03	.04	.05	.06	.07	.08	.09
0.0	.0000	.0040	.0080	.0120	.0160	.0199	.0239	.0279	.0319	.0359
0.1	.0398	.0438	.0478	.0517	.0557	.0596	.0636	.0675	.0714	.0753
0.2	.0793	.0832	.0871	.0910	.0948	.0987	.1026	.1064	.1103	.1141
0.3	.1179	.1217	.1255	.1293	.1331	.1368	.1406	.1443	.1480	.1517
0.4	.1554	.1591	.1628	.1664	.1700	.1736	.1772	.1808	.1844	.1879
0.5	.1915	.1950	.1985	.2019	.2054	.2088	.2123	.2157	.2190	.2224
0.6	.2257	.2291	.2324	.2357	.2389	.2422	.2454	.2486	.2517	.2549
0.7	.2580	.2611	.2642	.2673	.2704	.2734	.2764	.2794	.2823	.2852
0.8	.2881	.2910	.2939	.2967	.2995	.3023	.3051	.3078	.3106	.3133
0.9	.3159	.3186	.3212	.3238	.3264	.3289	.3315	.3340	.3365	.3389
1.0	.3413	.3438	.3461	.3485	.3508	.3531	.3554	.3577	.3599	.3621
1.1	.3643	.3665	.3686	.3708	.3729	.3749	.3770	.3790	.3810	.3830
1.2	.3849	.3869	.3888	.3907	.3925	.3944	.3962	.3980	.3997	.4015
1.3	.4032	.4049	.4066	.4082	.4099	.4115	.4131	.4147	.4162	.4177
1.4	.4192	.4207	.4222	.4236	.4251	.4265	.4279	.4292	.4306	.4319
1.5	.4332	.4345	.4357	.4370	.4382	.4394	.4406	.4418	.4429	.4441
1.6	.4452	.4463	.4474	.4484	.4495	.4505	.4515	.4525	.4535	.4545
1.7	.4554	.4564	.4573	.4582	.4591	.4599	.4608	.4616	.4625	.4633
1.8	.4641	.4649	.4656	.4664	.4671	.4678	.4686	.4693	.4699	.4706
1.9	.4713	.4719	.4726	.4732	.4738	.4744	.4750	.4756	.4761	.4767
2.0	.4772	.4778	.4783	.4788	.4793	.4798	.4803	.4808	.4812	.4817
2.1	.4821	.4826	.4830	.4834	.4838	.4842	.4846	.4850	.4854	.4857
2.2	.4861	.4864	.4868	.4871	.4875	.4878	.4881	.4884	.4887	.4890
2.3	.4893	.4896	.4898	.4901	.4904	.4906	.4909	.4911	.4913	.4916
2.4	.4918	.4920	.4922	.4925	.4927	.4929	.4931	.4932	.4934	.4936
2.5	.4938	.4940	.4941	.4943	.4945	.4946	.4948	.4949	.4951	.4952
2.6	.4953	.4955	.4956	.4957	.4959	.4960	.4961	.4962	.4963	.4964
2.7	.4965	.4966	.4967	.4968	.4969	.4970	.4971	.4972	.4973	.4974
2.8	.4974	.4975	.4976	.4977	.4977	.4978	.4979	.4979	.4980	.4981
2.9	.4981	.4982	.4982	.4983	.4984	.4984	.4985	.4985	.4986	.4986
3.0	.4987	.4987	.4987	.4988	.4988	.4989	.4989	.4989	.4990	.4990
3.1	.4990	.4991	.4991	.4991	.4992	.4992	.4992	.4992	.4993	.4993
3.2	.4993	.4993	.4994	.4994	.4994	.4994	.4994	.4995	.4995	.4995
3.3	.4995	.4995	.4995	.4996	.4996	.4996	.4996	.4996	.4996	.4997
3.4	.4997	.4997	.4997	.4997	.4997	.4997	.4997	.4997	.4997	.4998
3.5	.4998	.4998	.4998	.4998	.4998	.4998	.4998	.4998	.4998	.4998

ANSWERS TO ODD-NUMBERED PROBLEMS

EXERCISE 0.1 (page 3)

1. True **3.** False; the natural numbers are 1, 2, 3, and so on.
5. True **7.** False; $\sqrt{25} = 5$, a positive integer
9. True **11.** True

EXERCISE 0.2 (page 9)

1. False **3.** False **5.** False **7.** True **9.** False
11. Distributive **13.** Associative **15.** Commutative
17. Definition of subtraction **19.** Distributive
29. -4 **31.** 5 **33.** 8 **35.** -18 **37.** 24
39. 9 **41.** $-7x$ **43.** $6 + y$ **45.** $\frac{1}{3}$ **47.** -8
49. -8 **51.** 0 **53.** $20 + 4x$ **55.** 0 **57.** 5
59. $-\frac{3}{2x}$ **61.** $\frac{3ab}{c}$ **63.** $\frac{6y}{x}$ **65.** $\frac{10}{xy}$ **67.** $\frac{7}{6}$
69. 2 **71.** $\frac{17}{12}$ **73.** $\frac{6y}{x}$ **75.** $-\frac{25}{14}$ **77.** 0
79. 0

EXERCISE 0.3 (page 15)

1. $2^5 (= 32)$ **3.** w^{12} **5.** $\frac{x^8}{y^{17}}$ **7.** $\frac{a^{21}}{b^{20}}$ **9.** $8x^6y^9$
11. x^4 **13.** x^{14} **15.** 5 **17.** -2 **19.** $\frac{1}{2}$
21. 7 **23.** 27 **25.** $\frac{1}{4}$ **27.** $\frac{1}{16}$ **29.** $4\sqrt{2}$
31. $x\sqrt[3]{2}$ **33.** $4x^2$ **35.** $4\sqrt{2} - 15\sqrt{3} + 4\sqrt[3]{2}$
37. $3z^2$ **39.** $\frac{9t^2}{4}$ **41.** $\frac{x^3}{y^2z^2}$ **43.** $\frac{5}{m^9}$ **45.** $\frac{1}{9t^2}$
47. $5^{1/5}x^{2/5}$ **49.** $x^{1/2} - y^{1/2}$ **51.** $\frac{x^{9/4}z^{3/4}}{y^{1/2}}$
53. $\sqrt[5]{(8x - y)^4}$ **55.** $\frac{1}{\sqrt[5]{x^4}}$ **57.** $\frac{3}{\sqrt[5]{w^3}} - \frac{1}{\sqrt[5]{27w^3}}$
59. $\frac{6\sqrt{5}}{5}$ **61.** $\frac{2\sqrt{2x}}{x}$ **63.** $\frac{\sqrt[3]{9x^2}}{3x}$ **65.** 4
67. $\frac{\sqrt[20]{16a^{10}b^{15}}}{ab}$ **69.** $\frac{2x^6}{y^3}$ **71.** 9 **73.** $\frac{64y^6x^{1/2}}{x^2}$
75. xyz **77.** $\frac{1}{3}$ **79.** $\frac{4y^4}{x^2}$ **81.** $x^2y^{5/2}$ **83.** $\frac{a^5c^{14}}{b^{24}}$
85. x^8 **87.** $-\frac{4}{s^5}$ **89.** $\frac{4x^4z^4}{9y^4}$

EXERCISE 0.4 (page 20)

1. $11x - 2y - 3$ **3.** $6t^2 - 2s^2 + 6$
5. $\sqrt{a} + 5\sqrt{3b} - \sqrt{c}$
7. $6x^2 - 9xy - 2z + \sqrt{2} - 4$
9. $\sqrt{2y} - \sqrt{3z}$ **11.** $-15x + 15y - 27$
13. $x^2 + 9y^2 + xy$ **15.** $6x^2 + 96$
17. $-40x^3 + 10x^2 - 20x - 50$ **19.** $x^2 + 9x + 20$
21. $w^2 - 3w - 10$ **23.** $10x^2 + 19x + 6$
25. $x^2 + 6x + 9$ **27.** $x^2 - 10x + 25$
29. $3x + 10\sqrt{3x} + 25$ **31.** $4s^2 - 1$
33. $x^3 + 4x^2 - 3x - 12$
35. $3x^4 + 2x^3 - 13x^2 - 8x + 4$ **37.** $5x^3 + 5x^2 + 6x$
39. $3x^2 + 2y^2 + 5xy + 2x - 8$
41. $8a^3 + 36a^2 + 54a + 27$
43. $8x^3 - 36x^2 + 54x - 27$ **45.** $z - 18$
47. $3x^3 + 2x - \frac{1}{2x^2}$ **49.** $x + \frac{-1}{x + 3}$
51. $3x^2 - 8x + 17 + \frac{-37}{x + 2}$
53. $x^2 - 2x + 4 - \frac{8}{x + 2}$ **55.** $x - 2 + \frac{7}{3x + 2}$

EXERCISE 0.5 (page 23)

1. $2(3x + 2)$ **3.** $5x(2y + z)$
5. $4bc(2a^3 - 3ab^2d + b^3cd^2)$ **7.** $(z + 7)(z - 7)$
9. $(p + 3)(p + 1)$ **11.** $(4x + 3)(4x - 3)$
13. $(a + 3)(a + 2)$ **15.** $(x + 3)^2$
17. $5(x + 3)(x + 2)$ **19.** $3(x + 1)(x - 1)$
21. $(6y + 1)(y + 2)$ **23.** $2s(3s + 4)(2s - 1)$
25. $u^{2/3}v^2(u + 2v^{1/2})(u - 2v^{1/2})$ **27.** $2x(x + 3)(x - 2)$
29. $4(2x + 1)^2$ **31.** $x(xy - 7)^2$
33. $(x + 2)(x - 2)^2$ **35.** $(y + 4)^2(y + 1)(y - 1)$
37. $(a + 3)(a^2 - 3a + 9)$
39. $(x + 1)(x^2 - x + 1)(x - 1)(x^2 + x + 1)$
41. $2(x + 3)^2(x + 1)(x - 1)$ **43.** $P(1 + r)^2$
45. $(x^2 + 4)(x + 2)(x - 2)$
47. $(y^4 + 1)(y^2 + 1)(y + 1)(y - 1)$
49. $(X^2 + 3)(X + 1)(X - 1)$ **51.** $y(x + 1)^2(x - 1)^2$

EXERCISE 0.6 (page 29)

1. $\frac{a + 3}{a}$ **3.** $\frac{x - 5}{x + 5}$ **5.** $\frac{3x + 2}{x + 2}$
7. $-\frac{y^2}{(y - 3)(y + 2)}$ **9.** $\frac{3 - 2x}{2x + 3}$
11. $\frac{2(x + 4)}{(x - 4)(x + 2)}$ **13.** $\frac{X}{2}$ **15.** $\frac{n}{3}$ **17.** $\frac{2}{3}$
19. $-27x^2$ **21.** 1 **23.** $\frac{2x^2}{x - 1}$ **25.** 1

27. $-\frac{(2x+3)(1+x)}{x+4}$ **29.** $x+2$ **31.** $\frac{7}{3t}$

33. $\frac{1}{1-p^2}$ **35.** $\frac{2x^2+3x+12}{(2x-1)(x+3)}$

37. $\frac{2(x+2)}{(x-3)(x+1)(x+3)}$ **39.** $\frac{35-8x}{(x-1)(x+5)}$

41. $\frac{x^2+2x+1}{x^2}$ **43.** $\frac{x}{1-xy}$ **45.** $\frac{4x+1}{3x}$

47. $\frac{(x+2)(6x-1)}{2x^2(x+3)}$ **49.** $\frac{3(\sqrt[3]{x}-\sqrt[3]{x+h})}{\sqrt[3]{x+h}\,\sqrt[3]{x}}$

51. $2-\sqrt{3}$ **53.** $-\frac{\sqrt{6}+2\sqrt{3}}{3}$ **55.** $-4-2\sqrt{6}$

57. $\frac{x-\sqrt{5}}{x^2-5}$ **59.** $4\sqrt{2}-5\sqrt{3}+14$

EXERCISE 0.7 (page 38)

1. 0 **3.** $\frac{10}{3}$ **5.** -2

7. Adding 5 to both sides; equivalence guaranteed
9. Raising both sides to the third power; equivalence *not* guaranteed
11. Dividing both sides by x; equivalence *not* guaranteed
13. Multiplying both sides by $x-1$; equivalence *not* guaranteed
15. Multiplying both sides by $(x-5)/x$; equivalence *not* guaranteed

17. $\frac{5}{2}$ **19.** 0 **21.** 1 **23.** $\frac{12}{5}$ **25.** -1

27. 2 **29.** $\frac{10}{3}$ **31.** 126 **33.** 15 **35.** $-\frac{26}{9}$

37. $-\frac{37}{18}$ **39.** $\frac{60}{17}$ **41.** $\frac{14}{3}$ **43.** 3 **45.** $\frac{25}{52}$

47. $\frac{1}{5}$ **49.** $\varnothing$ **51.** $\frac{8}{3}$ **53.** 2 **55.** 0 **57.** $\frac{7}{2}$

59. $\frac{1}{8}$ **61.** 3 **63.** $\frac{5}{13}$ **65.** $\varnothing$ **67.** 11

69. $\frac{68}{3}$ **71.** $-\frac{10}{9}$ **73.** 2 **75.** 7 **77.** $\frac{49}{36}$

79. $-\frac{9}{4}$ **81.** $r=\frac{I}{Pt}$ **83.** $q=\frac{p+1}{8}$

85. $r=\frac{S-P}{Pt}$ **87.** $a_1=\frac{2S-na_n}{n}$

89. $t=\frac{r-d}{rd}$ **91.** $n=\frac{2mI}{rB}-1$ **93.** 170 m

95. $c=x+0.0825x=1.0825x$ **97.** 3 years

99. 31 hours **101.** 20 **103.** $t=\frac{d}{r-c}$; $c=r-\frac{d}{t}$

105. $\approx$ 84 ft **107.** $-\frac{1}{2}$ **109.** 0

EXERCISE 0.8 (page 46)

1. 2 **3.** 3, 5 **5.** 3, -1 **7.** 4, 9 **9.** ± 2

11. 0, 8 **13.** $\frac{1}{2}$ **15.** $1, \frac{2}{3}$ **17.** 5, -2 **19.** $0, \frac{3}{2}$

21. 0, 1, -4 **23.** 0, ± 8 **25.** $0, \frac{1}{2}, -\frac{4}{3}$ **27.** 3, ± 2

29. 3, 4 **31.** 4, -6 **33.** $\frac{3}{2}$ **35.** $\frac{7\pm\sqrt{37}}{2}$

37. No real roots **39.** $\frac{-5\pm\sqrt{57}}{8}$ **41.** 40, -25

43. $\frac{-2\pm\sqrt{14}}{2}$ **45.** $\pm\sqrt{3}, \pm\sqrt{2}$ **47.** $2, -\frac{1}{2}$

49. $\pm\frac{\sqrt{5}}{5}, \pm\frac{1}{2}$ **51.** 3, 0 **53.** $\frac{15}{7}, \frac{11}{5}$ **55.** $\frac{3}{2}, -1$

57. 6, -2 **59.** $\frac{1}{2}$ **61.** 5, -2 **63.** $\frac{-9\pm\sqrt{41}}{4}$

65. -2 **67.** 6 **69.** 4, 8 **71.** 2 **73.** 0, 4
75. 1 **77.** $\approx$ 64.15, 3.35 **79.** 6 inches by 8 inches
83. 1 year and 10 years **85. (a)** 8 s; **(b)** 5.4 s or 2.6 s
87. 1.5, 0.75 **89.** No real root **91.** 1.999, 0.963

MATHEMATICAL SNAPSHOT—CHAPTER 0 (page 48)

1. The results agree. **3.** The results agree.

EXERCISE 1.1 (page 55)

1. 120 **3.** 48 of A, 80 of B **5.** $5\frac{1}{3}$ **7.** 1 m

9. $\approx$ 13,077 tons **11.** \$4000 at 6%, \$16,000 at $7\frac{1}{2}$%

13. \$4.25 **15.** 4% **17.** 80 **19.** \$8000
21. 1209 cartridges must be sold to approximately break even. **23.** \$116.25 **25.** 40 **27.** 46,000
29. Either \$440 or \$460 **31.** \$100 **33.** 42
35. 80 ft by 140 ft **37.** 9 cm long, 4 cm wide
39. \$112,000 **41.** 60 **43.** Either 125 units of A and 100 units of B, or 150 units of A and 125 units of B.

APPLICATIONS IN PRACTICE 1.2

1. 5375
2. $150-x_4\geq 0$; $3x_4-210\geq 0$; $x_4+60\geq 0$; $x_4\geq 0$

EXERCISE 1.2 (page 64)

1. $(4,\infty)$

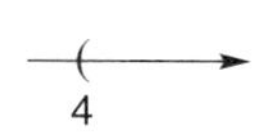

3. $(-\infty, 4]$

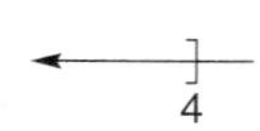

5. $\left(-\infty, -\frac{1}{2}\right]$

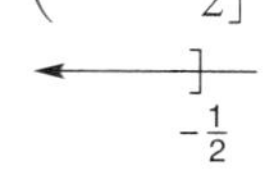

7. $\left(-\infty, \frac{2}{7}\right)$

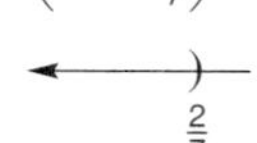

9. $(0,\infty)$

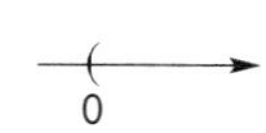

11. $\left[-\frac{7}{5}, \infty\right)$

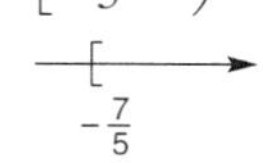

13. $\left(-\frac{2}{7}, \infty\right)$

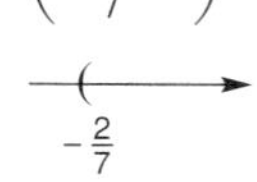

15. $\varnothing$

17. $\left(-\infty, \frac{\sqrt{3}-2}{2}\right)$

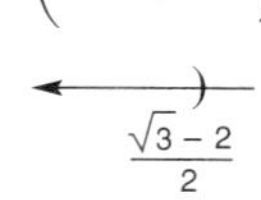

19. $(-\infty, 48)$

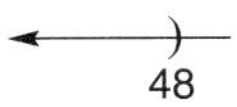

21. $(-\infty, -5]$

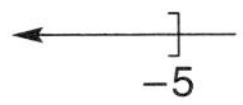

23. $(-\infty, \infty)$

25. $\left(\frac{17}{9}, \infty\right)$

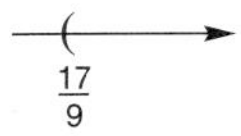

27. $(-12, \infty)$

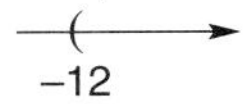

29. $(0, \infty)$

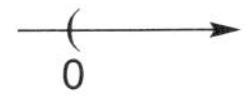

31. $(-\infty, 0)$

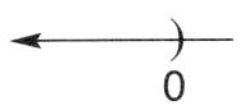

33. $(-\infty, -2]$

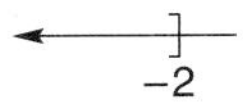

35. $444{,}000 < S < 636{,}000$ **37.** $x < 70$ degrees

EXERCISE 1.3 (page 67)

1. 120,001 **3.** 17,000 **5.** 60,000 **7.** \$25,714.29
9. 1000 **11.** $t > 36.7$ **13.** At least \$67,400

APPLICATIONS IN PRACTICE 1.4

1. $|w - 22 \text{ oz}| \le 0.3 \text{ oz}$

EXERCISE 1.4 (page 71)

1. 13 **3.** 6 **5.** 7 **7.** $-4 < x < 4$
9. $\sqrt{5} - 2$ **11. (a)** $|x - 7| < 3$; **(b)** $|x - 2| < 3$;
(c) $|x - 7| \le 5$; **(d)** $|x - 7| = 4$; **(e)** $|x + 4| < 2$;
(f) $|x| < 3$; **(g)** $|x| > 6$; **(h)** $|x - 6| > 4$;
(i) $|x - 105| < 3$; **(j)** $|x - 850| < 100$
13. $|p_1 - p_2| \le 9$ **15.** ± 7 **17.** ± 35 **19.** $13, -3$
21. $\frac{2}{5}$ **23.** $\frac{1}{2}, 3$ **25.** $(-4, 4)$
27. $(-\infty, -8) \cup (8, \infty)$ **29.** $(-14, -4)$
31. $(-\infty, 0) \cup (1, \infty)$ **33.** $\left[\frac{1}{2}, \frac{3}{4}\right]$
35. $(-\infty, 0] \cup \left[\frac{16}{3}, \infty\right)$ **37.** $|d - 17.2| \le 0.3$ m
39. $(-\infty, \mu - h\sigma) \cup (\mu + h\sigma, \infty)$

REVIEW PROBLEMS—CHAPTER 1 (page 73)

1. $(-\infty, 0]$ **3.** $\left(\frac{2}{3}, \infty\right)$ **5.** $\varnothing$ **7.** $\left(-\infty, \frac{5}{2}\right]$
9. $(-\infty, \infty)$ **11.** $-2, 5$ **13.** $\left(-\frac{1}{5}, 1\right)$
15. $\left(-\infty, -\frac{1}{2}\right] \cup \left[\frac{7}{2}, \infty\right)$ **17.** 542 **19.** 6000
21. $c < \$212{,}814$

MATHEMATICAL SNAPSHOT—CHAPTER 1 (page 74)

1. 1 hour **3.** 1 hour **5.** 600; 310

APPLICATIONS IN PRACTICE 2.1

1. (a) $a(r) = \pi r^2$; **(b)** all real numbers; **(c)** $r \ge 0$
2. (a) $t(r) = \frac{300}{r}$; **(b)** all real numbers except 0;
(c) $r > 0$; **(d)** $t(x) = \frac{300}{x}; t\left(\frac{x}{2}\right) = \frac{600}{x}; t\left(\frac{x}{4}\right) = \frac{1200}{x}$;
(e) The time is scaled by a factor c; $t\left(\frac{x}{c}\right) = \frac{300c}{x}$
3. (a) 300 pizzas; **(b)** \$21.00 per pizza;
(c) \$16.00 per pizza

EXERCISE 2.1 (page 83)

1. $f \ne g$ **3.** $h \ne k$ **5.** All real numbers except 0
7. All real numbers ≥ 3 **9.** All real numbers
11. All real numbers except $-\frac{7}{2}$
13. All real numbers except 2
15. All real numbers except 4 and $-\frac{1}{2}$
17. $1, 7, -7$ **19.** $-62, 2 - u^2, 2 - u^4$
21. $2, 4v^2 + 2v, x^4 - x^2$
23. $4, 0, x^2 + 2xh + h^2 + 2x + 2h + 1$
25. $-\frac{2}{27}, \frac{3x - 7}{9x^2 + 2}, \frac{x + h - 7}{x^2 + 2xh + h^2 + 2}$
27. $0, 256, \frac{1}{16}$ **29. (a)** $4x + 4h - 5$; **(b)** 4
31. (a) $x^2 + 2hx + h^2 + 2x + 2h$; **(b)** $2x + h + 2$
33. (a) $2 - 4x - 4h - 3x^2 - 6hx - 3h^2$;
(b) $-4 - 6x - 3h$ **35. (a)** $\frac{1}{x + h}$; **(b)** $-\frac{1}{x(x + h)}$
37. 5 **39.** y is a function of x; x is a function of y.
41. y is a function of x; x is not a function of y.
43. Yes **45.** $V = f(t) = 20{,}000 + 800t$
47. Yes; P; q **49.** 402.72 pounds per week; 935.52 pounds per week; amount supplied increases as the price increases **51. (a)** 4; **(b)** $8\sqrt[3]{2}$;
(c) $f(2I_0) = 2\sqrt[3]{2}\, f(I_0)$; doubling the intensity increases the response by a factor of $2\sqrt[3]{2}$
53. (a) 3000, 2900, 2300, 2000; 12, 10;
(b) 10, 12, 17, 20; 3000, 2300 **55. (a)** -5.13; **(b)** 2.64;
(c) -17.43 **57. (a)** 11.33; **(b)** 50.62; **(c)** 2.29

APPLICATIONS IN PRACTICE 2.2

1. (a) $p(n) = \$125$; **(b)** The premiums do not change;
(c) constant function
2. (a) quadratic function; **(b)** 2; **(c)** 3
3. $c(n) = \begin{cases} 3.50n & \text{if } n \le 5 \\ 3.00n & \text{if } 5 < n \le 10 \\ 2.75n & \text{if } n > 1 \end{cases}$ **4.** $7! = 5040$

EXERCISE 2.2 (page 88)

1. Yes **3.** No **5.** Yes **7.** No
9. All real numbers **11.** All real numbers
13. (a) 3; **(b)** 7 **15. (a)** 4; **(b)** -3 **17.** 8, 8, 8
19. $1, -1, 0, -1$ **21.** 7, 2, 2, 2 **23.** 720 **25.** 2
27. 5 **29.** $c(i) = \$4.50$; constant function
31. (a) $C = 850 + 3q$; **(b)** 250
33. $c(n) = \begin{cases} 9.50n & \text{if } n < 12, \\ 8.75n & \text{if } n \ge 12 \end{cases}$ **35.** $\frac{9}{64}$
37. (a) All T such that $30 \le T \le 39$; **(b)** $4, \frac{17}{4}, \frac{33}{4}$
39. (a) 237,077.34; **(b)** -434.97; **(c)** 52.19

41. **(a)** 2.21; **(b)** 9.98; **(c)** −14.52

APPLICATIONS IN PRACTICE 2.3

1. $c(s(x)) = c(x + 3) = 2(x + 3) = 2x + 6$
2. Let the length of a side be represented by the function $l(x) = x + 3$ and the area of a square with sides of length x be represented by $a(x) = x^2$. Then $g(x) = (x + 3)^2 = [l(x)]^2 = a(l(x))$.

EXERCISE 2.3 (page 94)

1. **(a)** $2x + 8$; **(b)** 8; **(c)** −2; **(d)** $x^2 + 8x + 15$; **(e)** 3; **(f)** $\frac{x+3}{x+5}$; **(g)** $x + 8$; **(h)** 11; **(i)** $x + 8$; **(j)** 11

3. **(a)** $2x^2 - x + 1$; **(b)** $x + 1$; **(c)** $\frac{1}{2}$; **(d)** $x^4 - x^3 + x^2 - x$; **(e)** $\frac{x^2+1}{x^2-x}$; **(f)** $\frac{5}{3}$; **(g)** $x^4 - 2x^3 + x^2 + 1$; **(h)** $x^4 + x^2$; **(i)** 90

5. 6; −32 **7.** $\frac{4}{(t-1)^2} + \frac{14}{t-1} + 1$; $\frac{2}{t^2 + 7t}$

9. $\frac{1}{v+3}$; $\sqrt{\frac{2w^2+3}{w^2+1}}$ **11.** $f(x) = x^5, g(x) = 4x - 3$

13. $f(x) = \frac{1}{x}$, $g(x) = x^2 - 2$

15. $f(x) = \sqrt[4]{x}, g(x) = \frac{x^2-1}{x+3}$

17. **(a)** $r(x) = 9.75x$; **(b)** $e(x) = 4.25x + 4500$; **(c)** $(r - e)(x) = 5.5x - 4500$

19. $400m - 10m^2$; the total revenue received when the total output of m employees is sold

21. **(a)** 14.05; **(b)** 1169.64 **23.** **(a)** 345.03; **(b)** −1.94

EXERCISE 2.4 (page 97)

1. $f^{-1}(x) = \frac{x}{3} - \frac{7}{3}$ **3.** $F^{-1}(x) = 2x + 14$

5. $r(A) = \sqrt{\frac{A}{\pi}}$ **7.** $f(x) = 5x + 12$ is one-to-one

9. $h(x) = (5x + 12)^2$, for $x \geq -\frac{5}{12}$, is one-to-one

11. $x = \frac{\sqrt{17}}{3} + \frac{2}{3}$

APPLICATIONS IN PRACTICE 2.5

1. $y = -600x + 7250$; x-intercept $\left(12\frac{1}{12}, 0\right)$; y-intercept $(0, 7250)$
2. $y = 24.95$; horizontal line; no x-intercept; y-intercept $(0, 24.95)$

3.

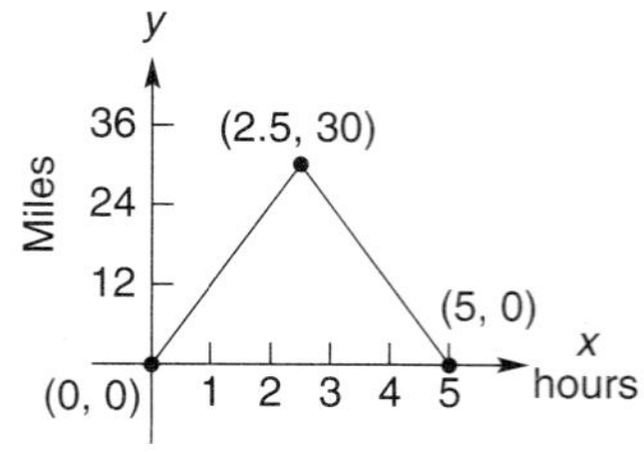

4.

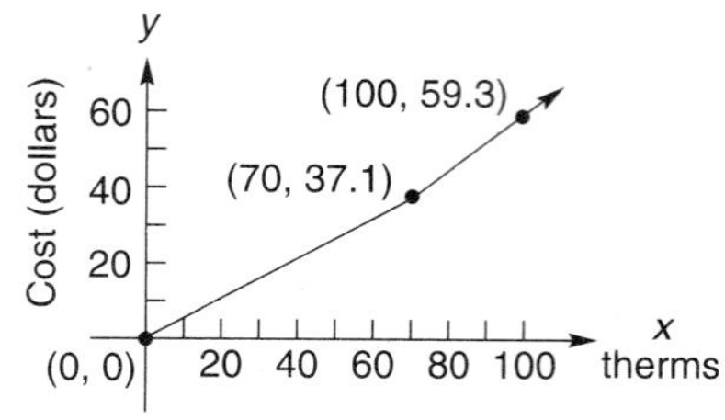

EXERCISE 2.5 (page 106)

1.

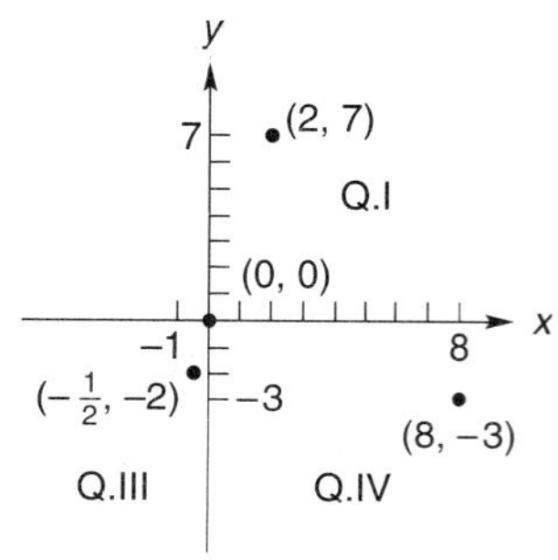

3. **(a)** 1, 2, 3, 0; **(b)** all real numbers; **(c)** all real numbers; **(d)** −2 **5.** **(a)** 0, −1, −1; **(b)** all real numbers; **(c)** all nonpositive real numbers; **(d)** 0

7. (0, 0); function; one-to-one; all real numbers; all real numbers

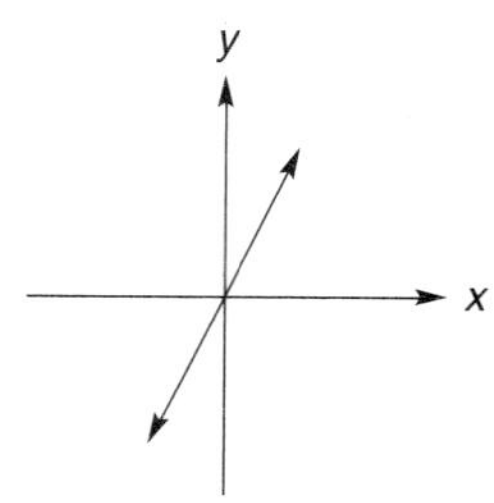

9. $(0, -5), \left(\frac{5}{3}, 0\right)$; function; one-to-one; all real numbers; all real numbers

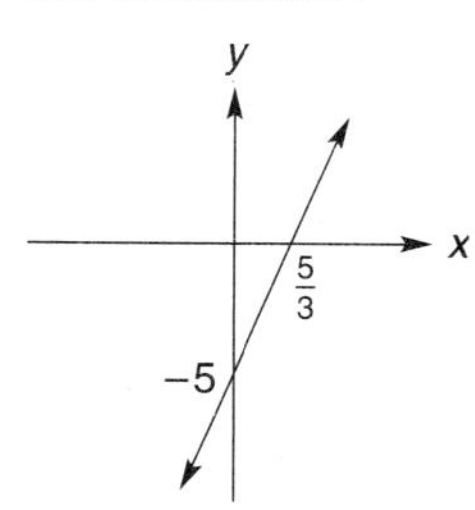

11. $(0, 0)$; function; not one-to-one; all real numbers; all nonnegative real numbers

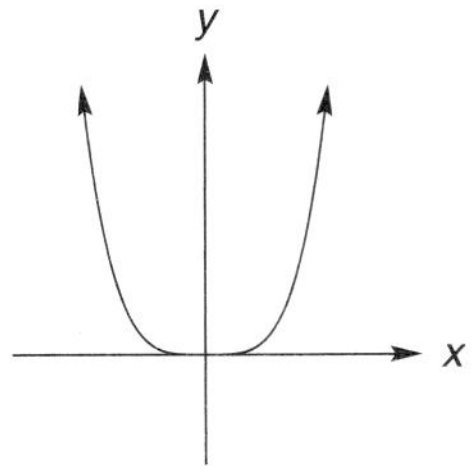

13. Every point on y-axis; not a function of x

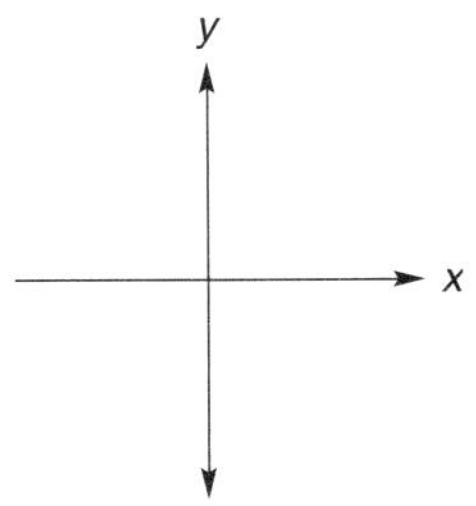

15. $(0, 0)$; function; one-to-one; all real numbers; all real numbers

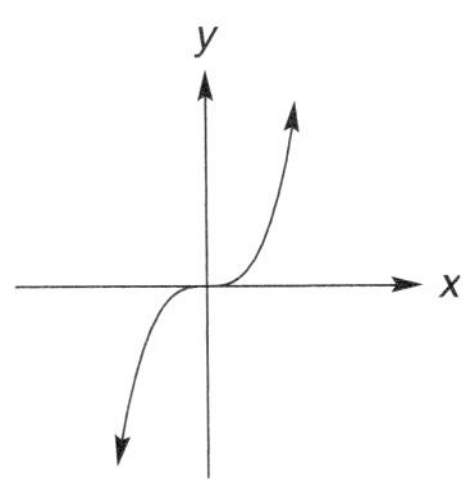

17. $(0, 0)$; not a function of x

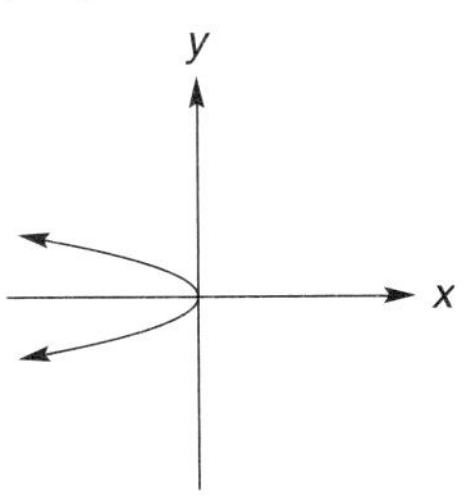

19. $(0, 2)$, $(1, 0)$; function; one-to-one; all real numbers; all real numbers

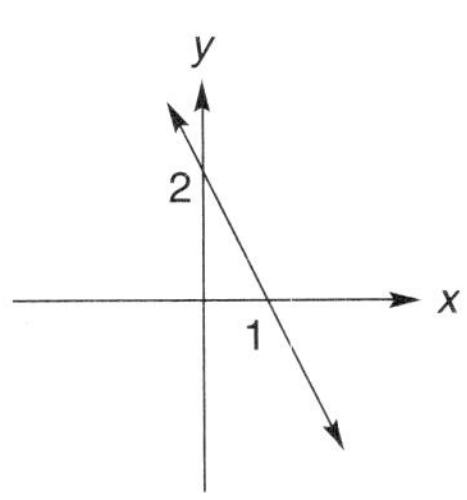

21. All real numbers; all real numbers ≤ 4; $(0, 4)$, $(2, 0)$, $(-2, 0)$

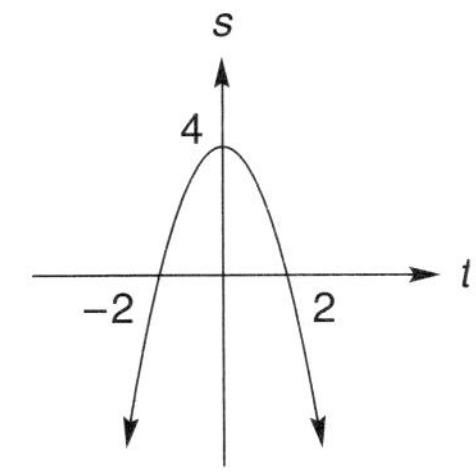

23. All real numbers; 3; $(0, 3)$

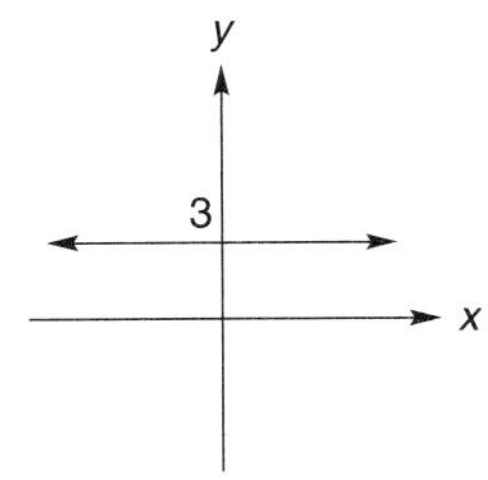

25. All real numbers; all real numbers ≥ -3; $(0, 1)$, $(2 \pm \sqrt{3}, 0)$

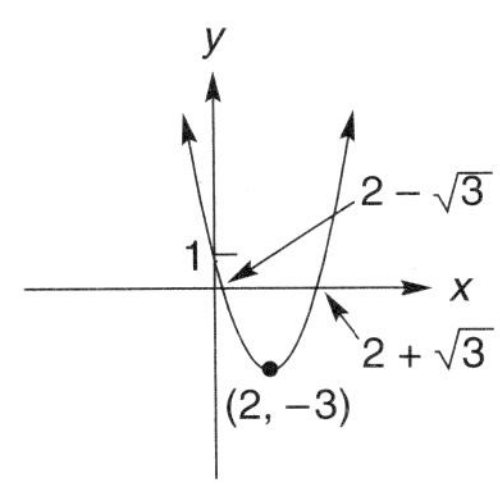

27. All real numbers; all real numbers; $(0, 0)$

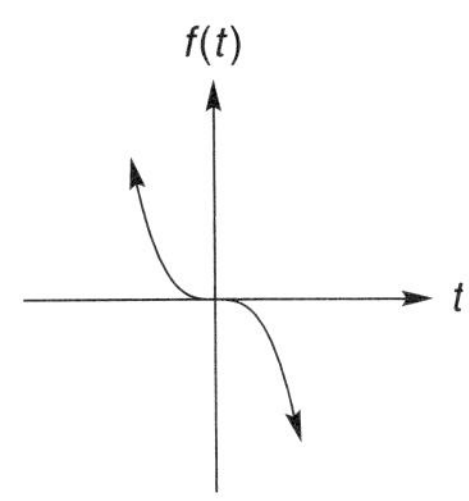

29. All real numbers ≥ 5; all nonnegative real numbers; $(5, 0)$

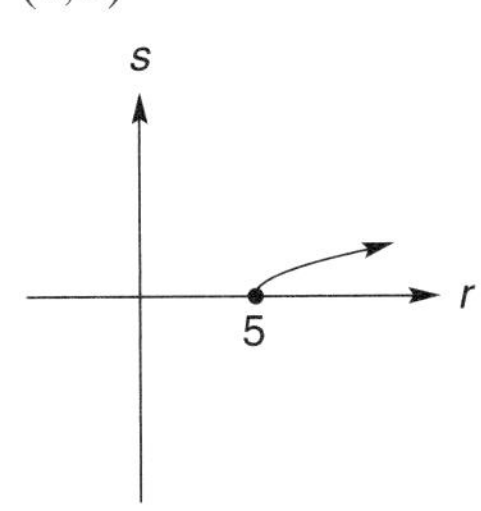

31. All real numbers; all nonnegative real numbers; $(0, 1), \left(\frac{1}{2}, 0\right)$

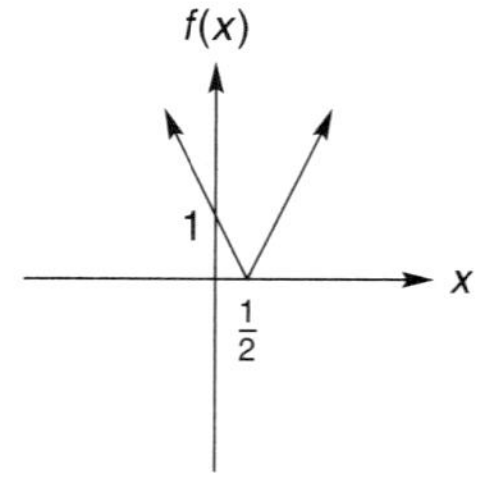

33. All nonzero real numbers; all positive real numbers; no intercepts

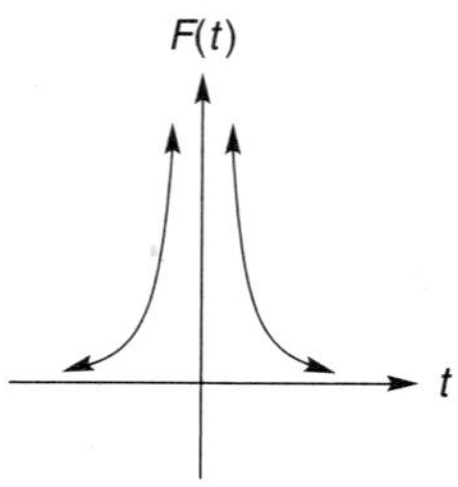

35. All nonnegative real numbers; all real numbers $1 \le c < 8$

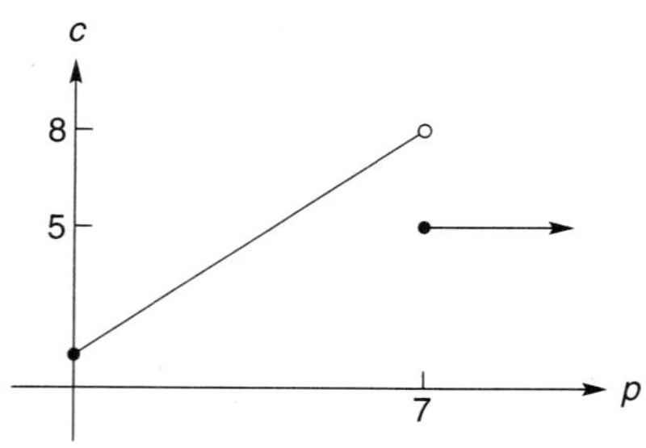

37. All real numbers; all nonnegative real numbers

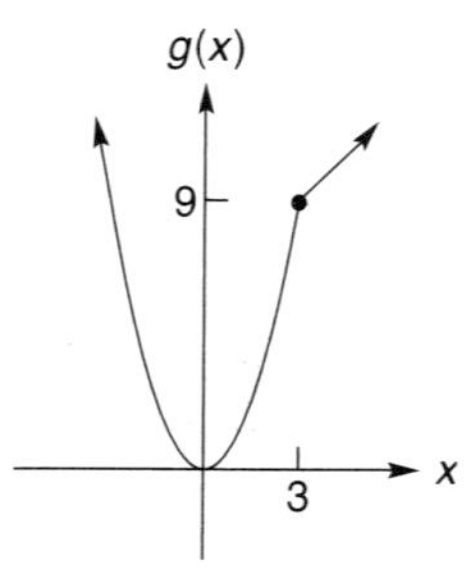

39. (a), (b), (d)

41. $y = 1800 - 175x, \left(10\frac{2}{7}, 0\right), (0, 1800)$

43. As price increases, quantity increases; p is a function of q.

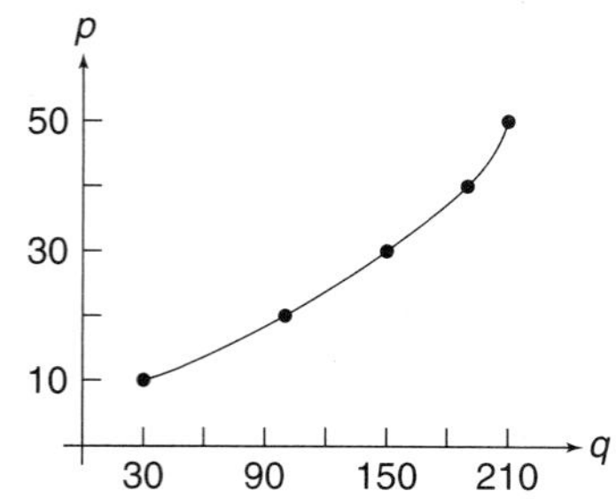

45.

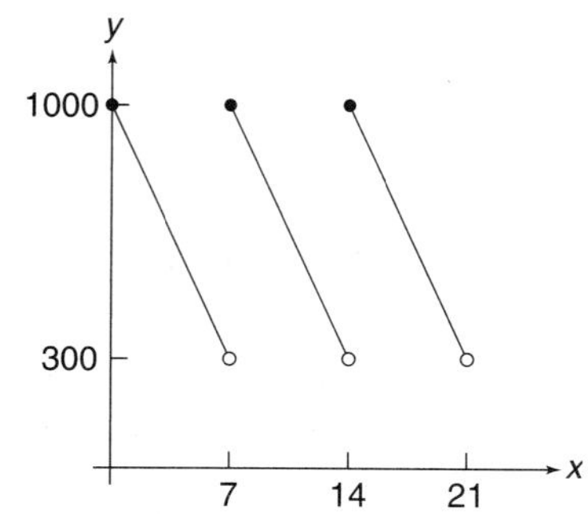

47. 0.39 **49.** $-0.61, -0.04$ **51.** -1.12
53. $-0.57, 0, 0.76, 2.31$ **55. (a)** 19.60; **(b)** -10.86
57. (a) 5; **(b)** 4 **59. (a)** 28; **(b)** $[-\infty, 28]$;
(c) $-4.02, 0.60$ **61. (a)** 34.21; **(b)** 18.68;
(c) $[18.68, 34.21]$; **(d)** no intercept

EXERCISE 2.6 (page 114)

1. $(0, 0)$; sym. about origin
3. $(\pm 2, 0), (0, 8)$; sym. about y-axis
5. $\left(\pm\frac{5}{4}, 0\right)$; sym. about x-axis, y-axis, origin
7. $(-2, 0)$; sym. about x-axis **9.** Sym. about x-axis
11. $(-21, 0), (0, -7), (0, 3)$
13. $(0, 0)$; sym. about origin **15.** $\left(0, \frac{3}{8}\right)$
17. $(3, 0), (0, \pm 3)$; sym. about x-axis

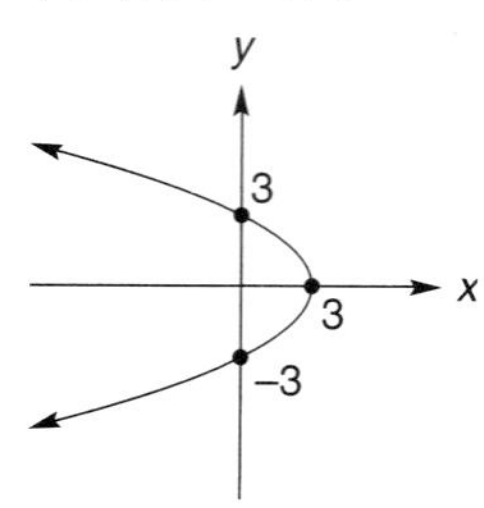

19. $(\pm 2, 0), (0, 0)$; sym. about origin

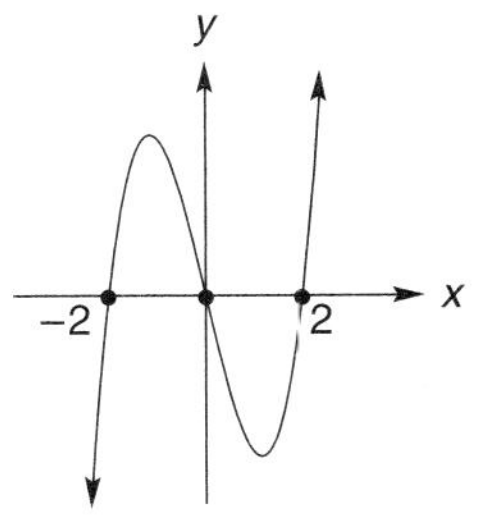

21. $(0, 0)$; sym. about x-axis, y-axis, origin

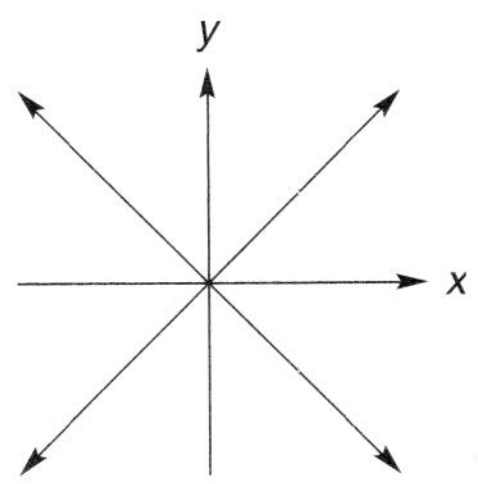

23. $(\pm 2, 0), (0, \pm 4)$; sym. about x-axis, y-axis, origin

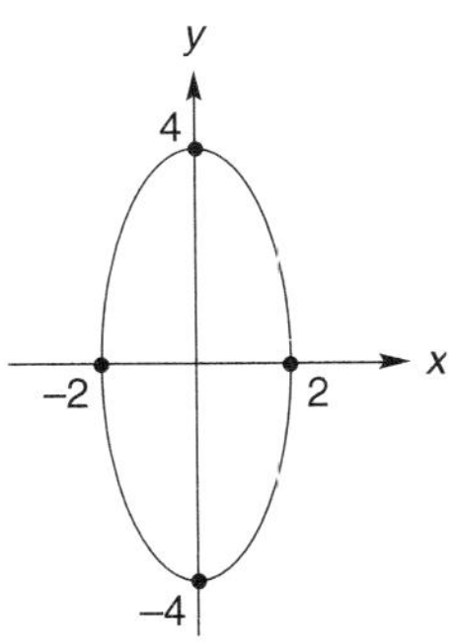

25. **(a)** $(\pm 1.18, 0), (0, 2)$; **(b)** 2; **(c)** $(-\infty, 2]$

27.

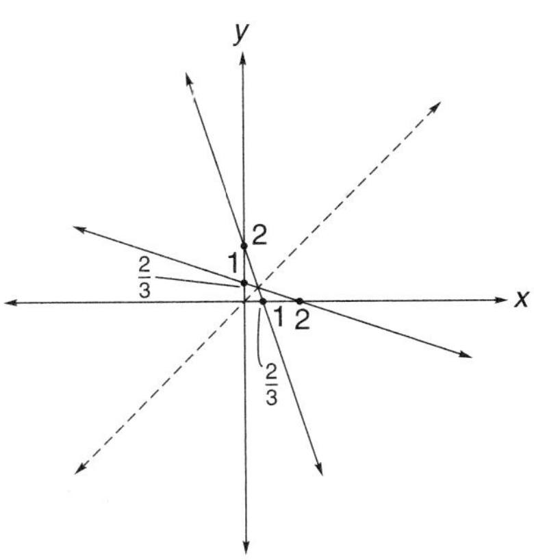

EXERCISE 2.7 (page 116)

1.

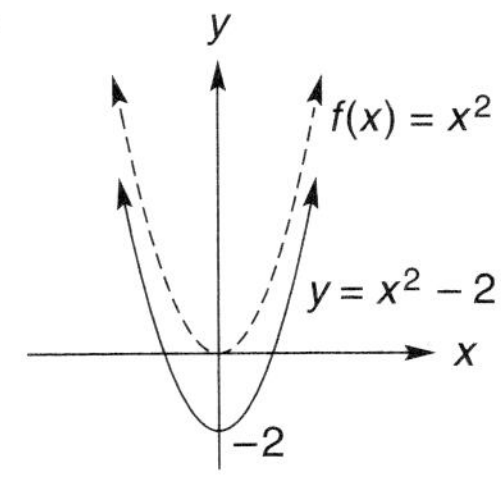

3.

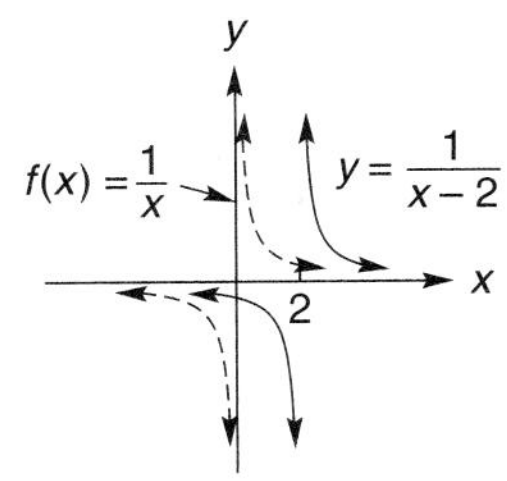

5.

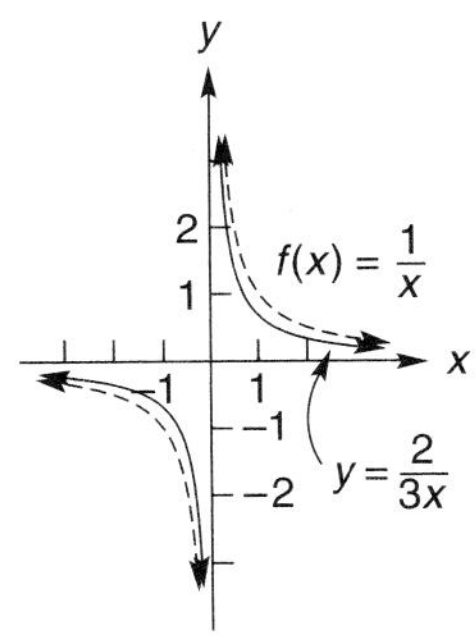

7.

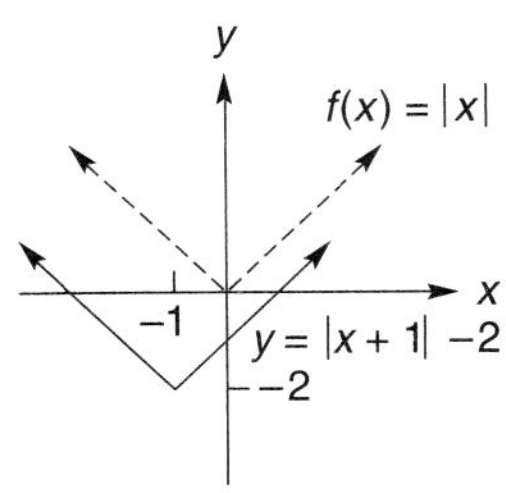

9.

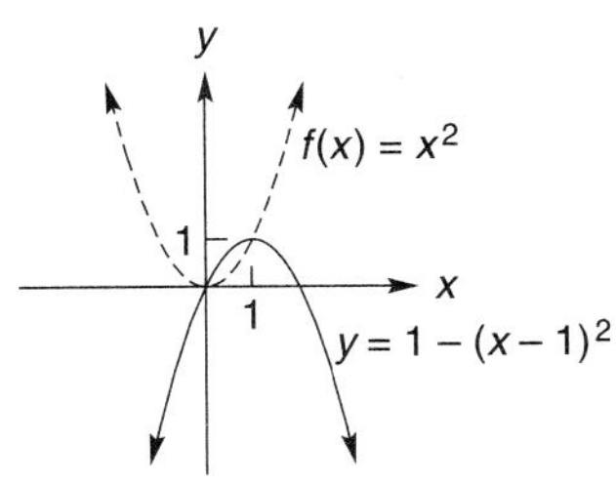

11.

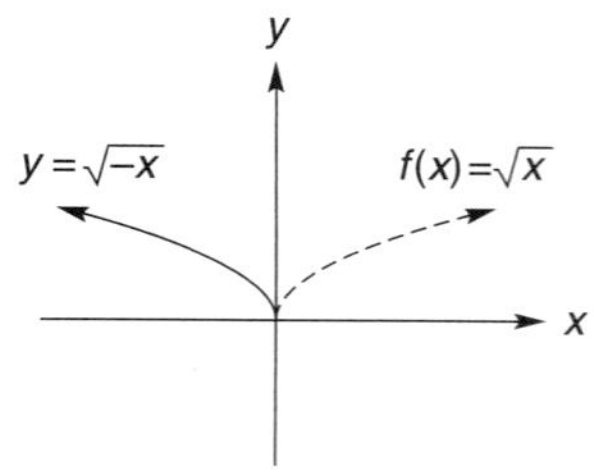

13. Translate 2 units to the right and 1 unit upward
15. Reflect about the y-axis and translate 5 units downward

REVIEW PROBLEMS—CHAPTER 2 (page 118)

1. All real numbers except 1 and 5 **3.** All real numbers
5. All nonnegative real numbers except 1
7. $7, 46, 62, 3t^2 - 4t + 7$ **9.** $0, 2, \sqrt{t}, \sqrt{x^3 - 1}$
11. $\frac{3}{5}, 0, \frac{\sqrt{x+4}}{x}, \frac{\sqrt{u}}{u-4}$ **13.** 20, −3, −3, undefined
15. **(a)** $3 - 7x - 7h$; **(b)** -7
17. **(a)** $4x^2 + 8hx + 4h^2 + 2x + 2h - 5$; **(b)** $8x + 4h + 2$ **19.** **(a)** $5x + 2$; **(b)** 22; **(c)** $x - 4$; **(d)** $6x^2 + 7x - 3$; **(e)** 10; **(f)** $\frac{3x-1}{2x+3}$; **(g)** $6x + 8$; **(h)** 38; **(i)** $6x + 1$ **21.** $\frac{1}{x-1}, \frac{1}{x} - 1 = \frac{1-x}{x}$
23. $\sqrt{x^3 + 2}$, $(x + 2)^{3/2}$
25. $(0, 0), (\pm\sqrt{3}, 0)$; sym. about origin
27. $(0, 9), (\pm 3, 0)$; sym. about y-axis

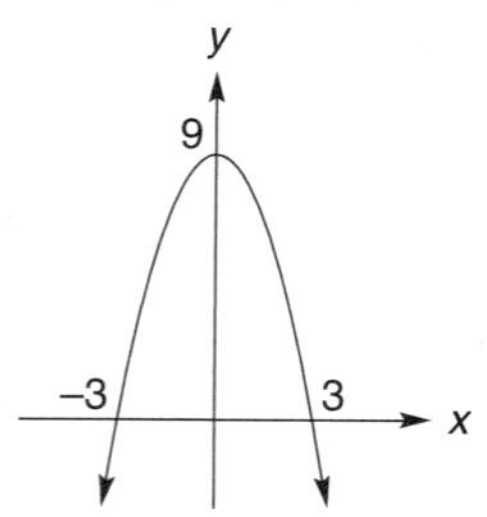

29. $(0, 2), (-4, 0)$; all $u \geq -4$; all real numbers ≥ 0

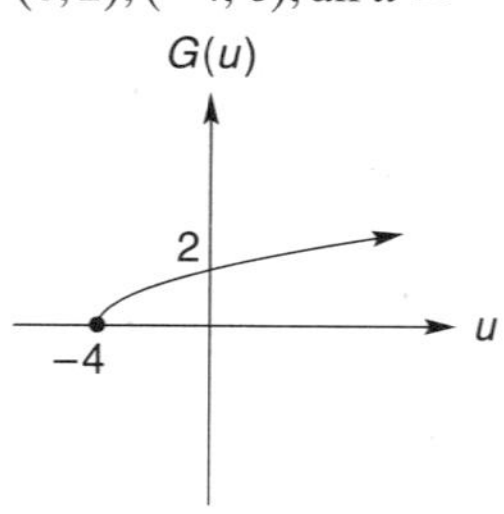

31. $\left(0, \frac{1}{2}\right)$; all $t \neq 4$; all positive real numbers

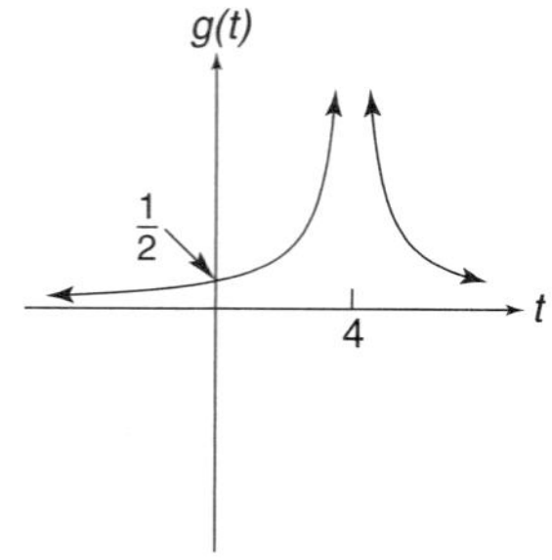

33. All real numbers; all real numbers ≥ 1

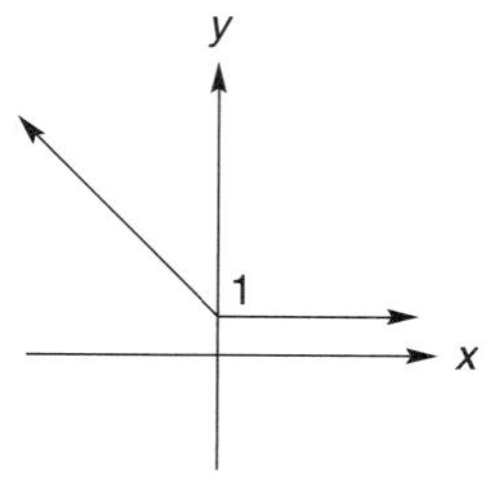

35.

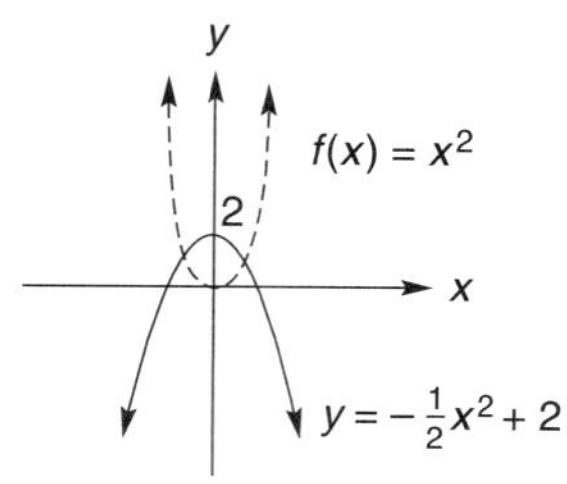

37. (a), (c) **39.** −0.67, 0.34, 1.73
41. −1.50, −0.88, −0.11, 1.09, 1.40
43. **(a)** $(-\infty, \infty)$; **(b)** $(1.92, 0), (0, 7)$
45. **(a)** None; **(b)** 1, 3

MATHEMATICAL SNAPSHOT—CHAPTER 2 (page 120)

1. \$28,321 **3.** \$87,507.90 **5.** Answers may vary

APPLICATIONS IN PRACTICE 3.1

1. −2000; the car depreciated \$2000 per year
2. $S = 14T + 8$ **3.** $F = \frac{9}{5}C + 32$
4. slope $= \frac{125}{3}$; y-intercept $= \frac{125}{3}$
5. $9C - 5F + 160 = 0$
6.

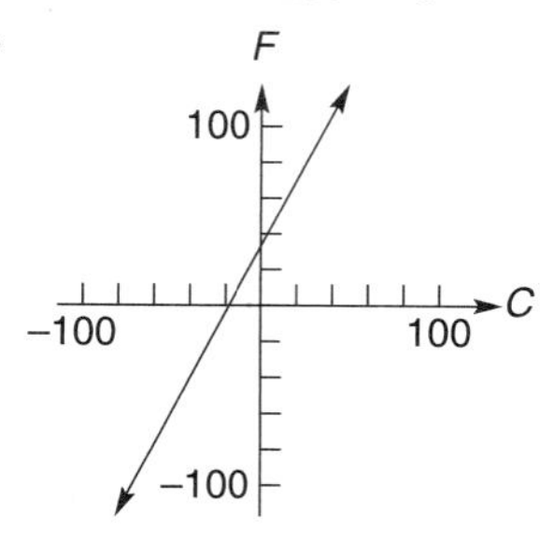

7. The slope of $\overline{AB}$ is 0; the slope of $\overline{BC}$ is 7; the slope of $\overline{CA}$ is 1. None of the slopes are negative reciprocals of each other, so the triangle does not have a right angle. The points do not define a right triangle.

EXERCISE 3.1 (page 131)

1. 3 **3.** $-\frac{1}{2}$ **5.** Undefined **7.** 0
9. $7x + y - 2 = 0$ **11.** $x + 4y - 18 = 0$
13. $3x - 7y + 25 = 0$ **15.** $8x - 5y - 29 = 0$
17. $2x - y + 4 = 0$ **19.** $x + 2y + 6 = 0$
21. $y + 3 = 0$ **23.** $x - 2 = 0$ **25.** 4; −6
27. $-\frac{1}{2}; \frac{3}{2}$ **29.** Slope undefined; no y-intercept
31. 3; 0 **33.** 0; 3
35. $2x + 3y - 5 = 0$; $y = -\frac{2}{3}x + \frac{5}{3}$
37. $4x + 9y - 5 = 0$; $y = -\frac{4}{9}x + \frac{5}{9}$
39. $3x - 2y + 24 = 0$; $y = \frac{3}{2}x + 12$
41. Parallel **43.** Parallel **45.** Neither
47. Perpendicular **49.** Perpendicular
51. $y = 4x + 14$ **53.** $y = 1$ **55.** $y = -\frac{1}{3}x + 5$
57. $x = 5$ **59.** $y = -\frac{2}{3}x - \frac{29}{3}$ **61.** (5, −4)
63. −2; the stock price dropped an average of $2 per year
65. $y = 28{,}000x - 100{,}000$ **67.** $-t + d - 1874 = 0$
71. $C = 59.82T + 769.58$ **73.** $R = 50{,}000T + 80{,}000$
75. The slope is 3.4.

APPLICATIONS IN PRACTICE 3.2

1. x = number of skis produced; y = number of boots produced; $8x + 14y = 1000$
2. $p = -\frac{3}{8}q + 1025$
3. Answers may vary, but two possible points are (0, 60) and (2, 140).

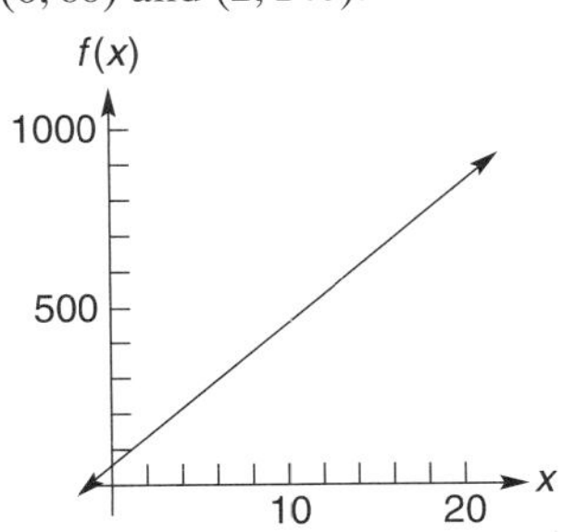

4. $f(t) = 2.3t + 32.2$ **5.** $f(x) = 70x + 150$

EXERCISE 3.2 (page 137)

1. −4; 0

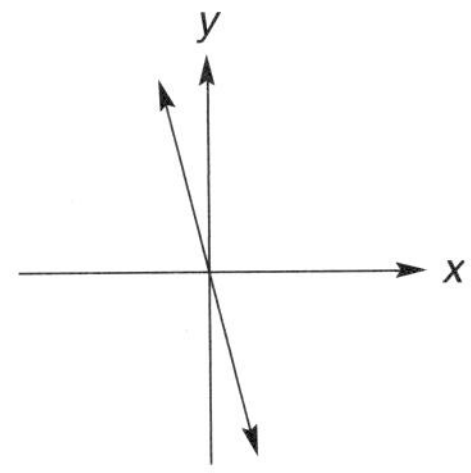

3. 5; −7

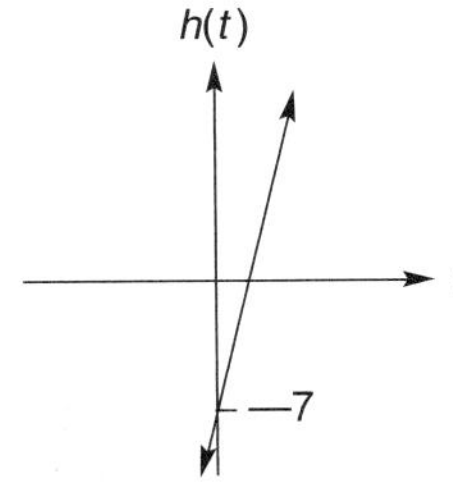

5. $-\frac{1}{7}; \frac{2}{7}$

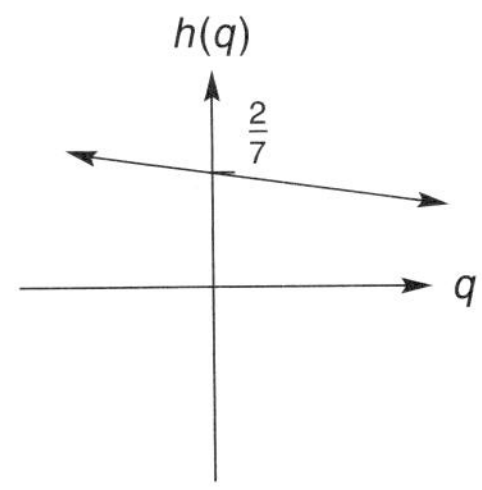

7. $f(x) = 4x$ **9.** $f(x) = -2x + 4$
11. $f(x) = -\frac{1}{2}x + \frac{15}{4}$ **13.** $f(x) = x + 1$
15. $p = -\frac{2}{5}q + 28.75$; $13.95 **17.** $p = \frac{1}{4}q + 190$
19. $c = 3q + 10$; $115 **21.** $f(x) = 0.125x + 4.15$
23. $v = -800t + 8000$; slope = −800

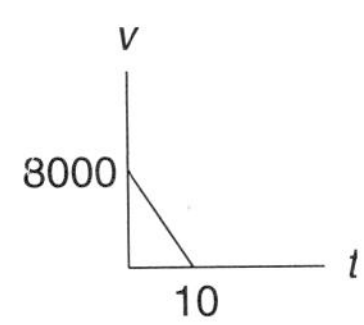

25. $f(x) = 45{,}000x + 735{,}000$ **27.** $f(x) = 64x + 95$
29. $x + 10y = 100$ **31.** **(a)** $y = \frac{5}{11}x + \frac{600}{11}$; **(b)** 12
33. **(a)** $p = 0.059t + 0.025$; **(b)** 0.556
35. **(a)** $t = \frac{1}{4}c + 37$; **(b)** add 37 to the number of chirps in 15 seconds

APPLICATIONS IN PRACTICE 3.3

1. Vertex: (1, 400); y-intercept: (0, 399); x-intercepts: (−19, 0), (21, 0)

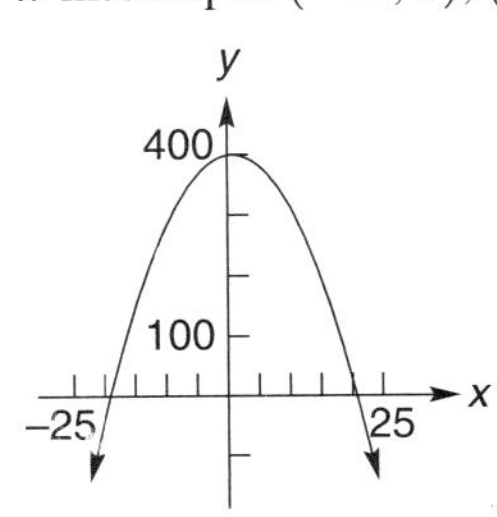

2. Vertex: $(1, 24)$; y-intercept: $(0, 8)$; x-intercepts: $\left(1 + \frac{\sqrt{6}}{2}, 0\right), \left(1 - \frac{\sqrt{6}}{2}, 0\right)$

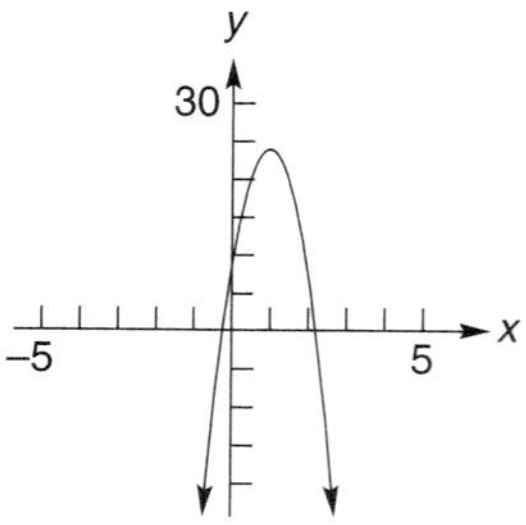

3. 1000 units; \$3000 maximum revenue

EXERCISE 3.3 (page 146)

1. Quadratic **3.** Not Quadratic **5.** Quadratic
7. Quadratic **9. (a)** $(1, 11)$; **(b)** highest
11. (a) -6; **(b)** $-3, 2$; **(c)** $\left(-\frac{1}{2}, -\frac{25}{4}\right)$

13. Vertex: $(3, -4)$; intercepts: $(1, 0), (5, 0), (0, 5)$; range: all $y \geq -4$

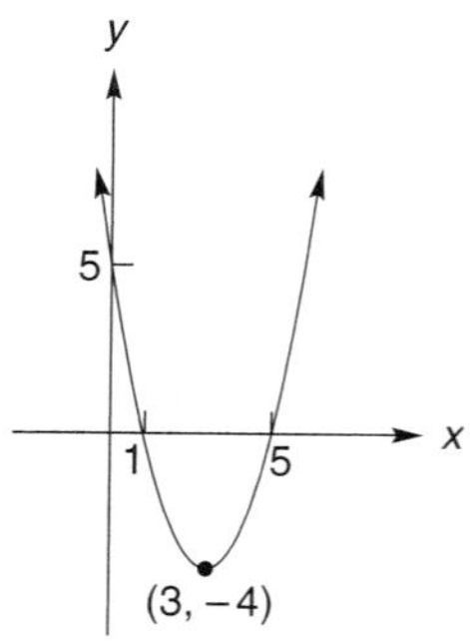

15. Vertex: $\left(-\frac{3}{2}, \frac{9}{2}\right)$; intercepts: $(0, 0), (-3, 0)$; range: all $y \leq \frac{9}{2}$

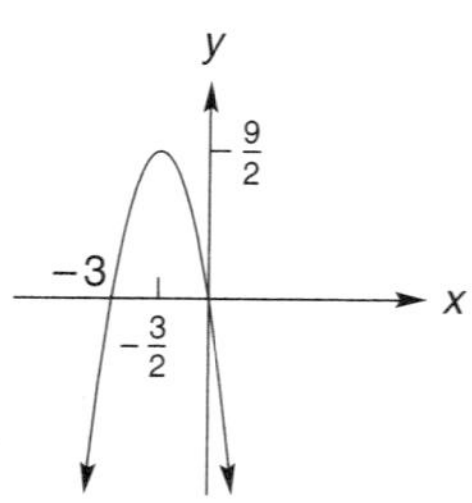

17. Vertex: $(-1, 0)$; intercepts: $(-1, 0), (0, 1)$; range: all $s \geq 0$

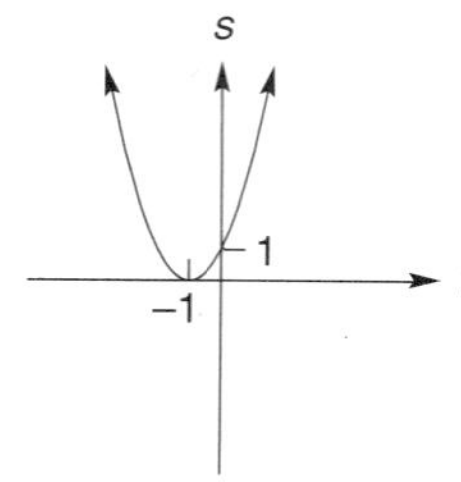

19. Vertex: $(2, -1)$; intercept: $(0, -9)$; range: all $y \leq -1$

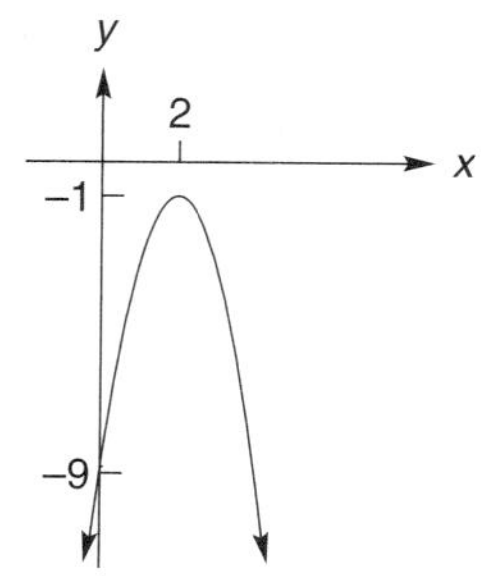

21. Vertex: $(4, -2)$; intercepts: $(4 + \sqrt{2}, 0), (4 - \sqrt{2}, 0), (0, 14)$; range: all $t \geq -2$

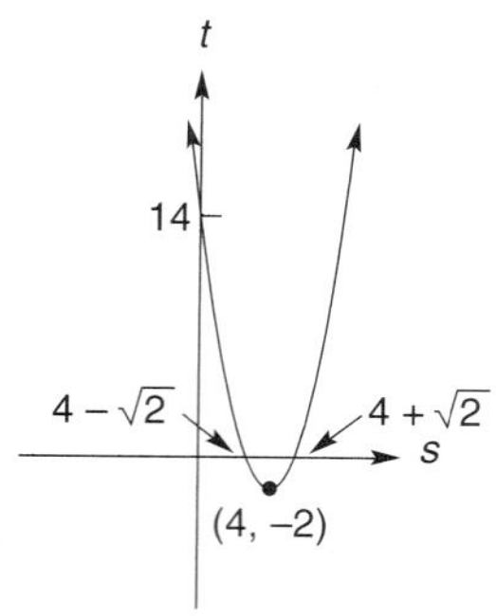

23. Minimum; $\frac{808}{49}$ **25.** Maximum; -10

27. 200 units; \$120,000 maximum revenue
29. 200 units; \$240,000 maximum revenue
31. Vertex: $(9, 225)$; y-intercept: $(0, 144)$; x-intercepts: $(-6, 0), (24, 0)$

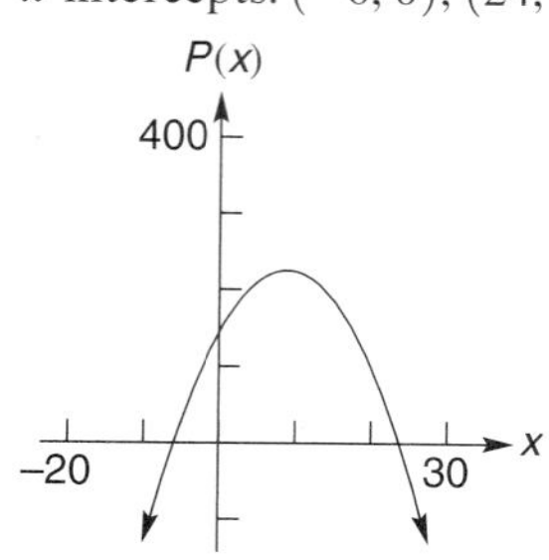

33. 70 grams **35.** ≈ 134.86 ft; ≈ 2.7 sec

37. Vertex: $\left(\frac{5}{2}, 116\right)$; y-intercept: $(0, 16)$,

x-intercepts: $\left(\frac{5+\sqrt{29}}{2}, 0\right), \left(\frac{5-\sqrt{29}}{2}, 0\right)$

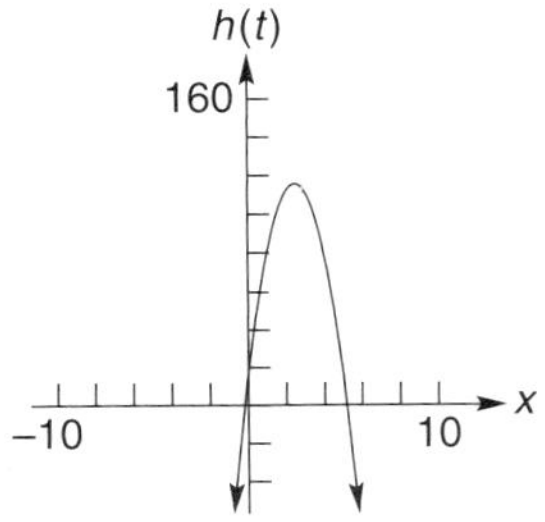

39. 50 ft × 100 ft **41.** $(1.11, 2.88)$
43. **(a)** 0; **(b)** 1; **(c)** 2 **45.** 4.89

APPLICATIONS IN PRACTICE 3.4

1. \$120,000 at 9% and \$80,000 at 8%
2. 500 of species A and 1000 of species B
3. Infinitely many solutions of the form $A = \frac{20{,}000}{3} - \frac{4}{3}r$, $B = r$ where $0 \le r \le 5000$
4. $\frac{1}{6}$ lb of A; $\frac{1}{3}$ lb of B; $\frac{1}{2}$ lb of C

EXERCISE 3.4 (page 157)

1. $x = -1, y = 1$ **3.** $x = 3, y = -1$
5. $u = 6, v = -1$ **7.** $x = -3, y = 2$
9. No solution **11.** $x = 12, y = -12$
13. $p = \frac{3}{2} - 3r, q = r$; r is any real number
15. $x = \frac{1}{2}, y = \frac{1}{2}, z = \frac{1}{4}$ **17.** $x = 2, y = -1, z = 4$
19. $x = 1 + 2r, y = 3 - r, z = r$; r is any real number
21. $x = -\frac{1}{3}r, y = \frac{5}{3}r, z = r$; r is any real number
23. $x = \frac{3}{2} - r + \frac{1}{2}s, y = r, z = s$; r and s are any real numbers
25. 420 gal of 20% solution, 280 gal of 30% solution
27. 0.5 lb of cotton; 0.25 lb of polyester; 0.25 lb of nylon
29. ≈ 285 mi/h (speed of airplane in still air), ≈ 23.2 mi/h (speed of wind)
31. 240 units of early American, 200 units of Contemporary
33. 800 calculators at Exton plant, 700 at Whyton plant
35. 4% on first \$100,000, 6% on remainder
37. 60 units of Argon I, 40 units of Argon II
39. 100 chairs, 100 rockers, 200 chaise lounges
41. 10 semiskilled workers, 5 skilled workers, 55 shipping clerks **45.** $x = 3, y = 2$ **47.** $x = 8.3, y = 14.0$

EXERCISE 3.5 (page 161)

1. $x = 4, y = -12; x = -1, y = 3$
3. $p = -3, q = -4; p = 2, q = 1$
5. $x = 0, y = 0; x = 1, y = 1.$
7. $x = 4, y = 8; x = -1, y = 3$
9. $p = 0, q = 0; p = 1, q = 1$
11. $x = \sqrt{17}, y = 2$; $x = -\sqrt{17}, y = 2; x = \sqrt{14}, y = -1; x = -\sqrt{14}, y = -1$ **13.** $x \approx 13.53, y \approx 19.06$
15. At $(10, 8.1)$ and $(-10, 7.9)$ **17.** Three
19. $x = -1.3, y = 5.1$ **21.** $x = 1.76$ **23.** $x = -1.46$

EXERCISE 3.6 (page 169)

1. The equilibrium point is $(100, 7)$.

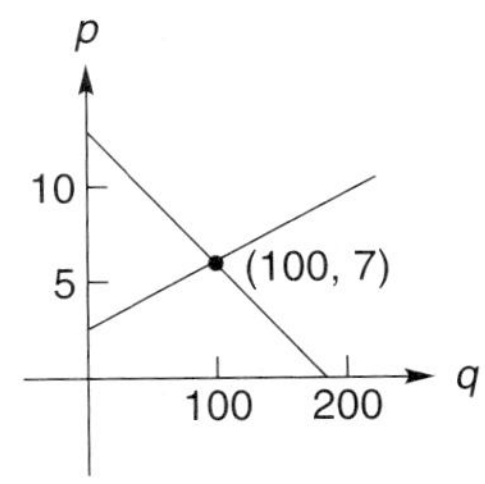

3. $(5, 212.50)$ **5.** $(9, 38)$ **7.** $(15, 5)$
9.

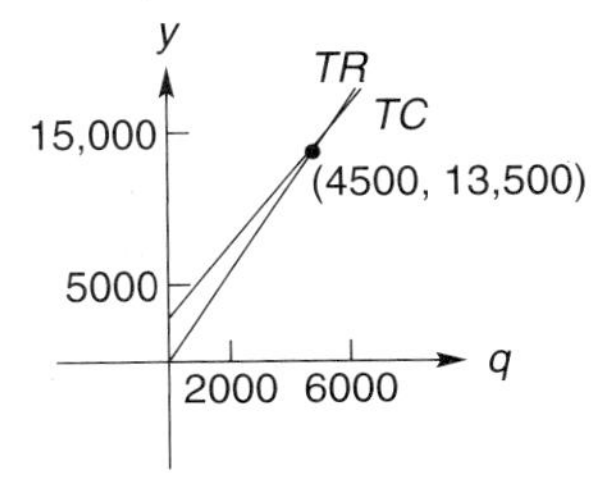

11. Cannot break even at any level of production
13. Cannot break even at any level of production
15. **(a)** \$12; **(b)** \$12.18
17. 5840 units; 840 units; 1840 units **19.** \$4
21. Total cost always exceeds total revenue—no break-even point. **23.** Decreases by \$0.70
25. $p_A = 8; p_B = 10$ **27.** 2.4 and 11.3

REVIEW PROBLEMS—CHAPTER 3 (page 171)

1. 9 **3.** $y = -x + 1; x + y - 1 = 0$
5. $y = \frac{1}{2}x - 1; x - 2y - 2 = 0$ **7.** $y = 4; y - 4 = 0$
9. $y = \frac{2}{5}x - 3; 2x - 5y - 15 = 0$ **11.** Perpendicular
13. Neither **15.** Parallel **17.** $y = \frac{3}{2}x - 2; \frac{3}{2}$
19. $y = \frac{4}{3}; 0$
21. $-5; (0, 17)$

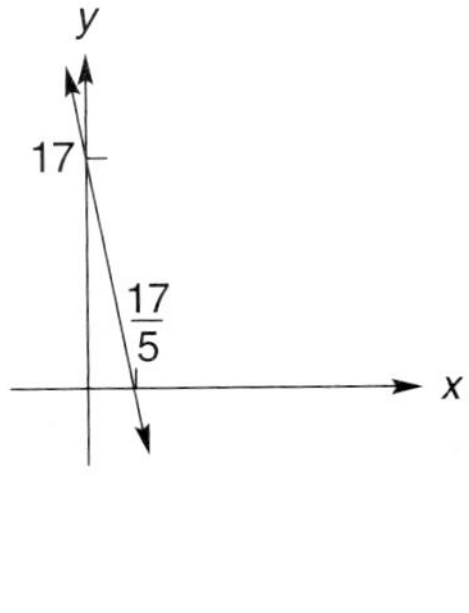

23. (3, 0), (−3, 0), (0, 9); (0, 9)

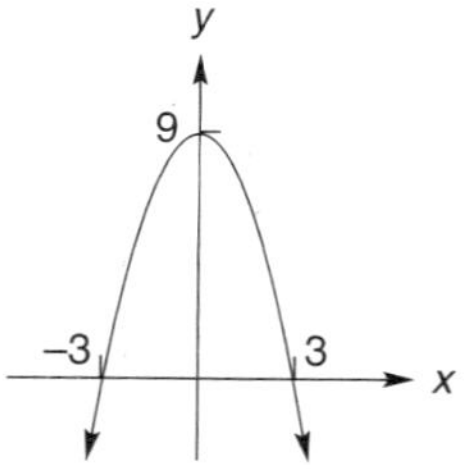

25. (5, 0), (−1, 0), (0, −5); (2, −9)

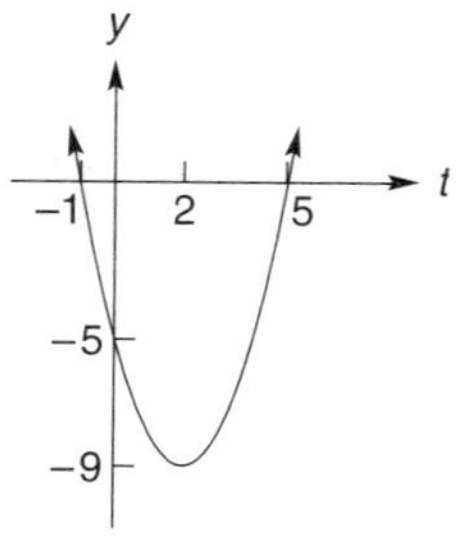

27. 3; (0, 0)

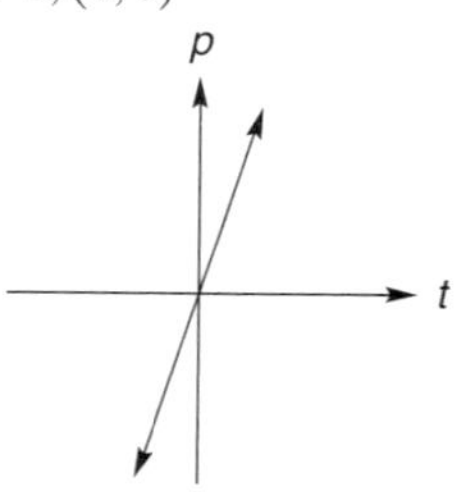

29. (0, −3); (−1, −2)

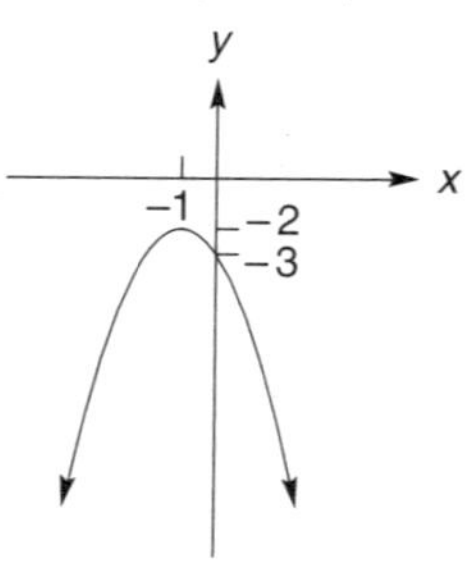

31. $x = \frac{17}{7}, y = -\frac{8}{7}$ **33.** $x = 2, y = -\frac{9}{5}$

35. $x = 8, y = 4$ **37.** $x = 0, y = 1, z = 0$

39. $x = -3, y = -4$; $x = 2, y = 1$

41. $x = -2 - 2r, y = 7 + r, z = r$; r is any real number

43. $x = r, y = r, z = 0$; r is any real number

45. $a - 4b = -7$; 13 **47.** $f(x) = -\frac{4}{3}x + \frac{19}{3}$

49. 50 units; \$5000 **51.** 6 **53.** 1250 units; \$20,000

55. 2.36 tons per square km **57.** $x = 7.29, y = -0.78$

59. $x = 0.75, y = 1.43$

MATHEMATICAL SNAPSHOT—CHAPTER 3 (page 174)

1. Advantage I is the best plan for airtimes from 85 to $153\frac{1}{3}$ minutes. Advantage II is the best plan for airtimes from $153\frac{1}{3}$ to $233\frac{1}{3}$ minutes.

3. If the initial guess is on the horizontal portion of both graphs, the calculator may not be able to find the intersection point.

APPLICATIONS IN PRACTICE 4.1

1. The shape of the graphs are the same. The value of A scales the ordinate of any point by A.

2.

Year	Multiplicative Increase	Expression
0	1	1.1^0
1	1.1	1.1^1
2	1.21	1.1^2
3	1.33	1.1^3
4	1.46	1.1^4

1.1; The investment increases by 10% every year. $(1 + 1(0.1) = 1 + 0.1 = 1.1)$

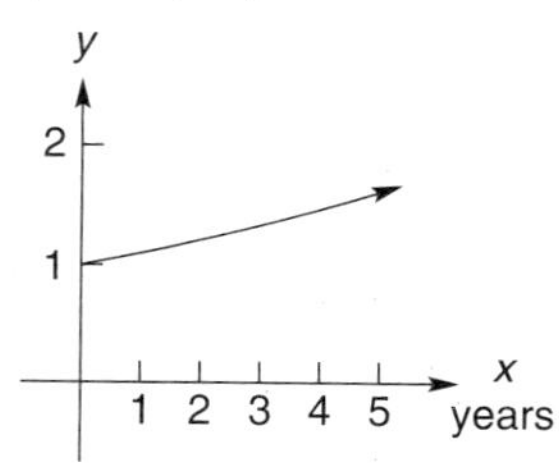

Between 7 and 8 years.

3.

Year	Multiplicative Decrease	Expression
0	1	0.85^0
1	0.85	0.85^1
2	0.72	0.85^2
3	0.61	0.85^3

0.85; The car depreciates by 15% every year. $(1 - 1(0.15) = 1 - 0.15 = 0.85)$

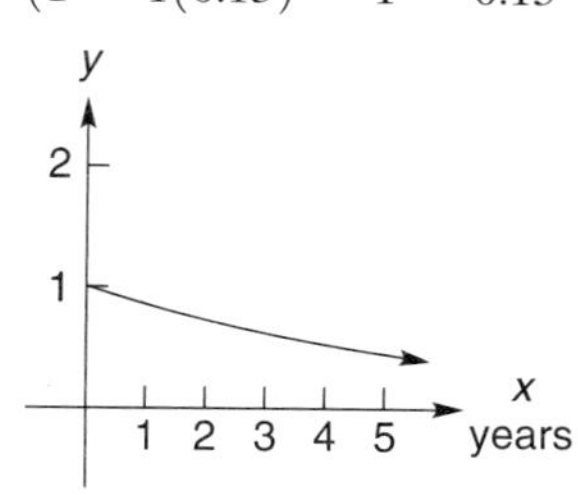

Between 4 and 5 years.

4. $y = 1.08^{t-3}$; Shift the graph 3 units to the right.

5. \$3684.87; \$1684.87 **6.** 117 employees

7.

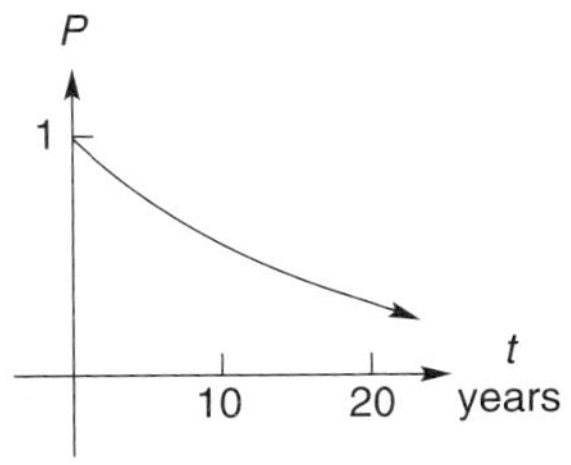

33.

Year	Multiplicative Increase	Expression
0	1	1.3^0
1	1.3	1.3^1
2	1.69	1.3^2
3	2.20	1.3^3

1.3; The recycling increases by 30% every year. $(1 + 1(0.3) = 1 + 0.3 = 1.3)$

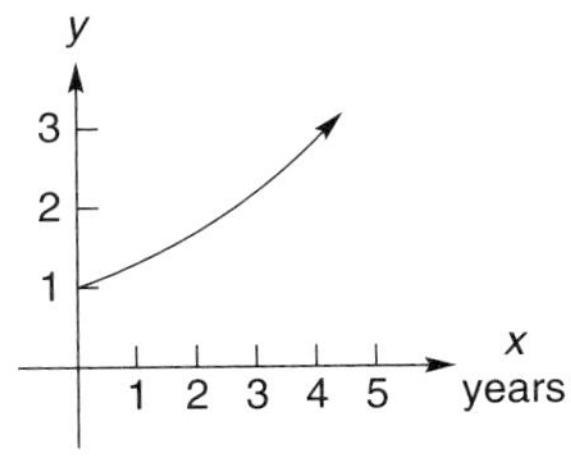

Between 4 and 5 years.

EXERCISE 4.1 (page 188)

1.

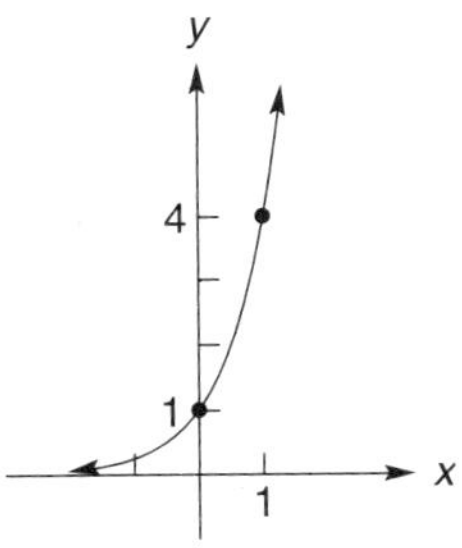

3.

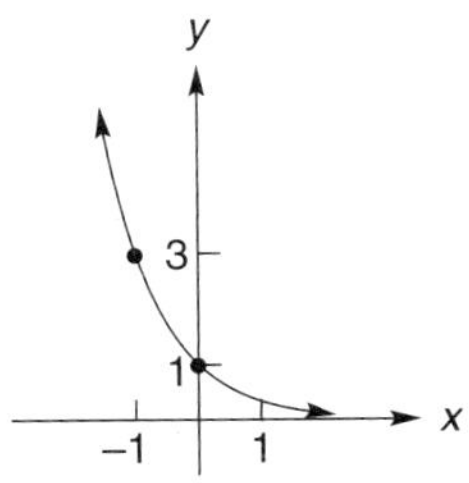

5.

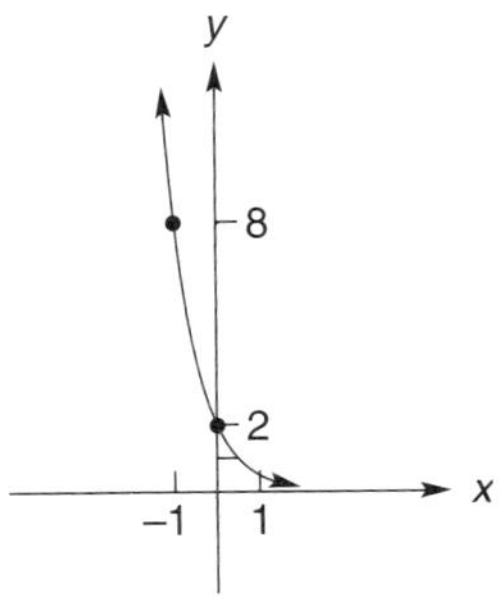

7.

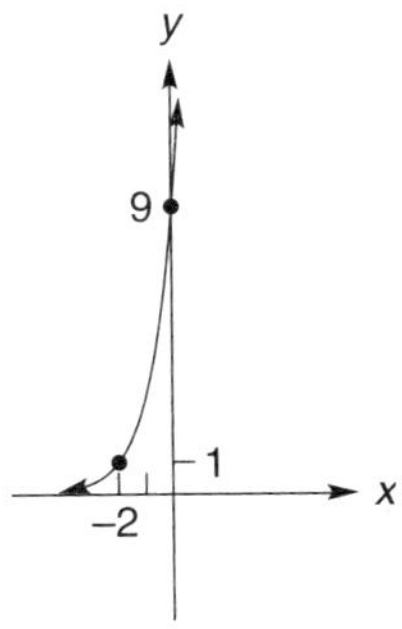

9.

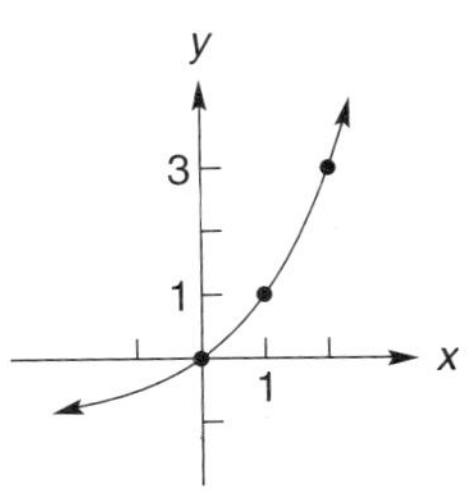

11.

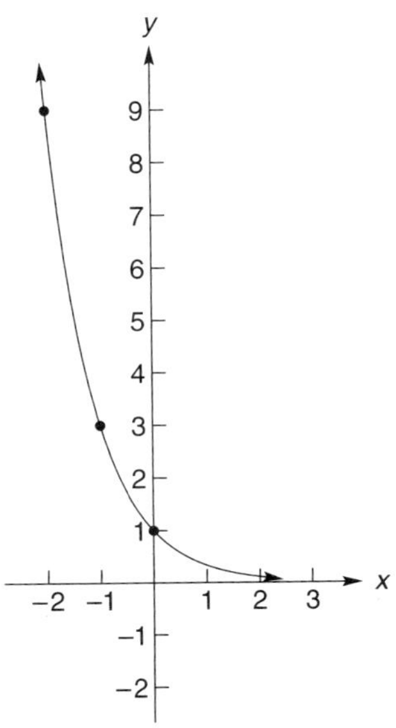

13. B 15. 138,750

17. $\frac{1}{2}$

19. **(a)** \$6014.52; **(b)** \$2014.52

21. **(a)** \$1964.76; **(b)** \$1264.76

23. **(a)** \$11,983.37; **(b)** \$8983.37

25. **(a)** \$6256.36; **(b)** \$1256.36

27. **(a)** \$9649.69; **(b)** \$1649.69 29. \$10,446.15

31. **(a)** $N = 400(1.05)^t$; **(b)** 420; **(c)** 486

35. 334, 485 37. 4.4817 39. 0.4966

41.

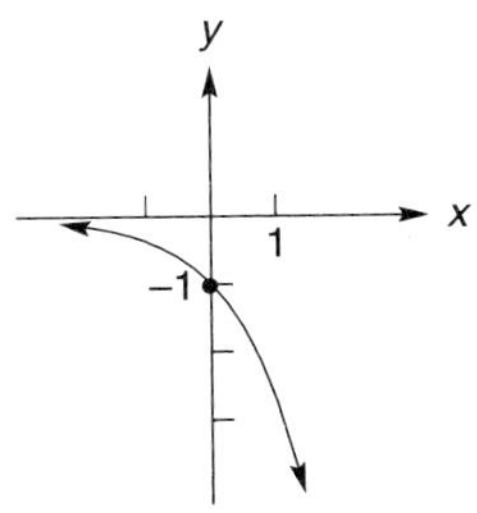

43. 0.2240

45. $(e^k)^t$, where $b = e^k$

47. **(a)** 12; **(b)** 8.8; **(c)** 3.1; **(d)** 22 hours

49. 32 years

51. 0.1465

55. 3.17

57. 4.2 min

59. 17

APPLICATIONS IN PRACTICE 4.2

1. $t = \log_2 16$; $t =$ the number of times the bacteria have doubled. 2. $\frac{I}{I_0} = 10^{8.3}$

3.

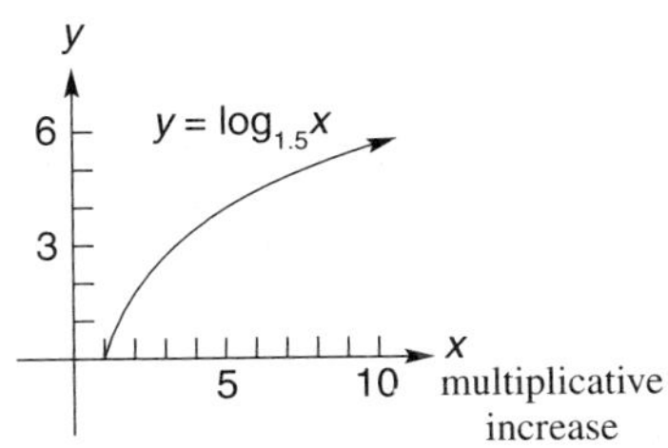

4.

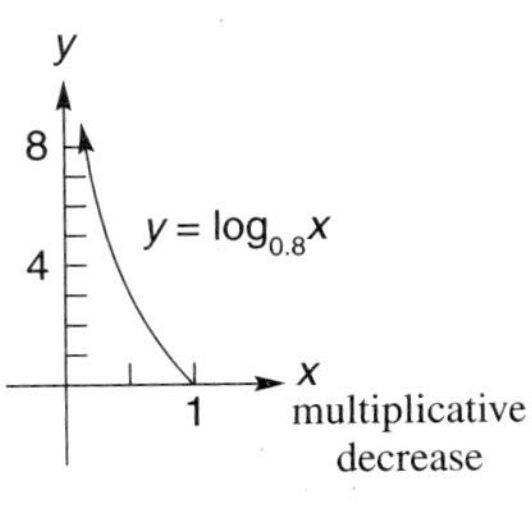

5. Approximately 13.9%

6. Approximately 9.2%

EXERCISE 4.2 (page 195)

1. $\log 10{,}000 = 4$ 3. $2^6 = 64$ 5. $\ln 20.0855 = 3$

7. $e^{1.09861} = 3$

9.

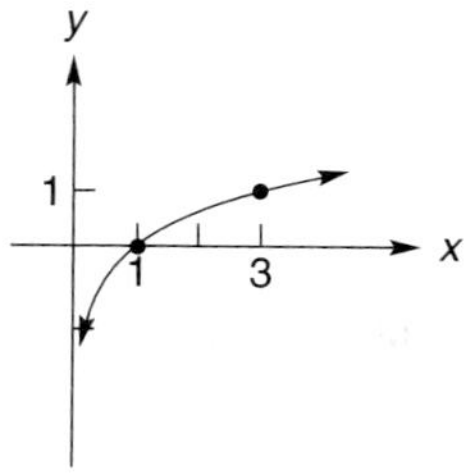

11.

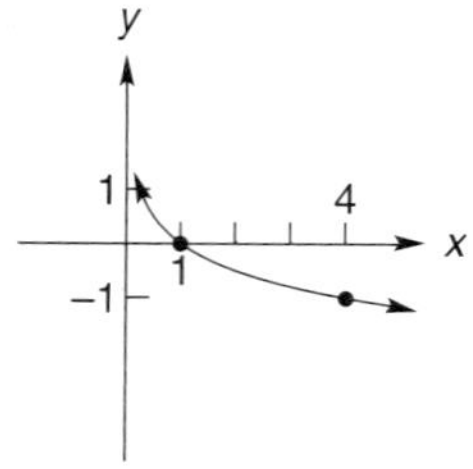

13.

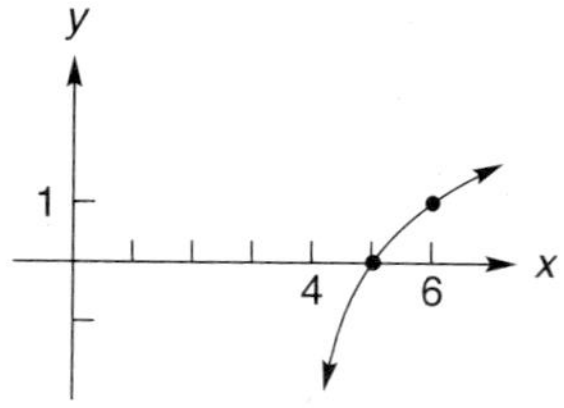

15.

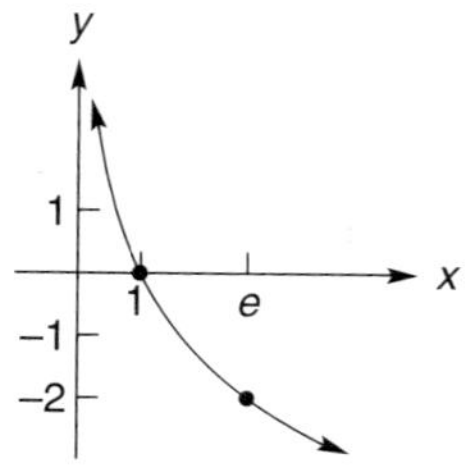

17. 2 **19.** 3 **21.** 1 **23.** −2 **25.** 0 **27.** −3 **29.** 81 **31.** 125 **33.** $\frac{1}{10}$ **35.** e^{-3} **37.** 2 **39.** 6 **41.** $\frac{1}{27}$ **43.** 2

45. $\frac{5}{3}$ **47.** 4 **49.** $\frac{\ln 2}{3}$ **51.** $\frac{5 + \ln 3}{2}$

53. 1.60944 **55.** 2.00013 **57.** $y = \log_{1.10} x$

59. 3 **61. (a)** $2N_0$;
(b) k is the time it takes for the population to double.
63. 36.1 minutes **65.** $z = y^{3/2}$
67. (a) $(0, 1)$; **(b)** $[-0.37, \infty)$ **69.** 1.10
71. 1.41, 3.06

APPLICATIONS IN PRACTICE 4.3

1. $\log(900{,}000) - \log(9000) = \log\left(\frac{900{,}000}{9000}\right) = \log(100) = 2$

2. $\log(10{,}000) = \log(10^4) = 4$

EXERCISE 4.3 (page 203)

1. $b + c$ **3.** $a - b$ **5.** $3a - b$ **7.** $2(a + b)$

9. $\frac{b}{a}$ **11.** 48 **13.** −4 **15.** 5.01 **17.** −2

19. 2 **21.** $\ln x + 2\ln(x + 1)$
23. $2\ln x - 3\ln(x + 1)$ **25.** $3[\ln x - \ln(x + 1)]$
27. $\ln x - \ln(x + 1) - \ln(x + 2)$

29. $\frac{1}{2}\ln x - 2\ln(x + 1) - 3\ln(x + 2)$

31. $\frac{2}{5}\ln x - \frac{1}{5}\ln(x + 1) - \ln(x + 2)$

33. $\log 24$ **35.** $\log_2 \frac{2x}{x + 1}$ **37.** $\log[7^9(23)^5]$

39. $\log[100(1.05)^{10}]$ **41.** $\frac{81}{64}$ **43.** 1 **45.** $\frac{5}{2}$

47. ± 2 **49.** $\frac{\ln(x + 6)}{\ln 10}$ **51.** $\frac{\ln(x^2 + 1)}{\ln 3}$

53. $y = \ln \frac{z}{7}$ **59.** $\log x = \frac{\ln x}{\ln 10}$

61. $\ln \pi$

APPLICATIONS IN PRACTICE 4.4

1. 18 **2.** Day 20 **3.** The other earthquake is 67.5 times as intense as a zero-level earthquake.

EXERCISE 4.4 (page 209)

1. 3 **3.** 2.75 **5.** −3 **7.** 2 **9.** 0.083
11. 1.099 **13.** 0.028 **15.** 5.140 **17.** −0.073
19. 2.322 **21.** 3.183 **23.** 0.483 **25.** 2.496
27. 1003 **29.** 2.222 **31.** 3.082 **33.** 3 **35.** 0.5
37. $S = 12.4A^{0.26}$ **41.** 20.5

43. $p = \frac{\log(80 - q)}{\log 2}$; 4.32 **49.** 3.33

REVIEW PROBLEMS—CHAPTER 4 (page 211)

1. $\log_3 243 = 5$ **3.** $16^{1/4} = 2$ **5.** $\ln 54.598 = 4$

7. 3 **9.** −4 **11.** −2 **13.** 4 **15.** $\frac{1}{100}$

17. −1 **19.** $3(a + 1)$ **21.** $\log \frac{7^3}{5^2}$ **23.** $\ln \frac{x^2 y}{z^3}$

25. $\log_2 \frac{x^{9/2}}{(x + 1)^3(x + 2)^4}$ **27.** $2\ln x + \ln y - 3\ln z$

29. $\frac{1}{3}(\ln x + \ln y + \ln z)$ **31.** $\frac{1}{2}(\ln y - \ln z) - \ln x$

33. $\frac{\ln(x + 5)}{\ln 3}$ **35.** 1.8295 **37.** $2y + \frac{1}{2}x$

39. $2x$ **41.** $y = e^{x^2+2}$

43.

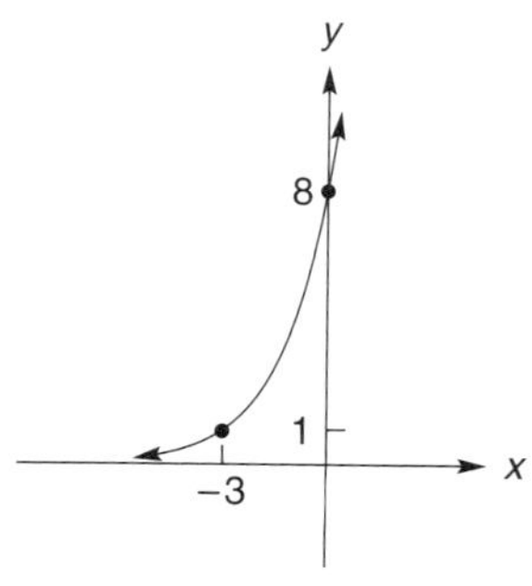

45. 5
47. 1
49. 10
51. 2^e
53. 0.880
55. −3.222
57. −1.596
59. (a) \$3829.04;
(b) \$1229.04
61. 14%

63. (a) $P = 8000(1.02)^t$; **(b)** 8323
65. (a) 10 mg; **(b)** 4.4; **(c)** 0.2; **(d)** 1.7; **(e)** 5.6
67. (a) 6; **(b)** 28 **71.** $(-\infty, 0.37]$ **73.** 2.53

75.

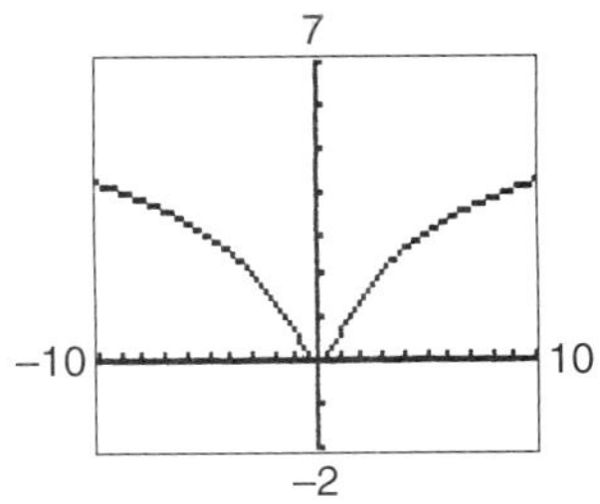

MATHEMATICAL SNAPSHOT—CHAPTER 4 (page 214)

1. (a) $P = \dfrac{T(e^{kI} - 1)}{1 - e^{-dkI}}$; **(b)** $d = \dfrac{1}{kI} \ln\left[\dfrac{P}{P - T(e^{kI} - 1)}\right]$

3. (a) 156; **(b)** 65

APPLICATIONS IN PRACTICE 5.1

1. 4.9% **2.** 7 years, 16 days **3.** 7.7208%
4. 11.25% compounded quarterly has the better effective rate of interest. The $10,000 investment is slightly better over 20 years.

EXERCISE 5.1 (page 221)

1. (a) $11,105.58; **(b)** $5105.58 **3.** 3.023%
5. 4.081% **7. (a)** 10%; **(b)** 10.25%; **(c)** 10.381%;
(d) 10.471%; **(e)** 10.516% **9.** 8.08% **11.** 9.0 years
13. $10,282.95 **15.** $38,503.23
17. (a) 18%; **(b)** $19.56% **19.** $3198.54
21. 8% compounded annually **23. (a)** 5.47%;
(b) 5.39% **25.** 11.61% **27.** 6.29%

EXERCISE 5.2 (page 226)

1. $2261.34 **3.** $1751.83 **5.** $5118.10
7. $4862.31 **9.** $6838.95 **11.** $11,381.89
13. $14,091.10 **15.** $1238.58 **17.** $3244.63
19. (a) $515.62; **(b)** profitable **21.** Savings account
23. $226.25 **25.** 9.55%

APPLICATIONS IN PRACTICE 5.3

1. 48 ft, 36 ft, 27 ft, $20\frac{1}{4}$ ft, $15\frac{3}{16}$ ft
2. $500(1.5), 500(1.5)^2, 500(1.5)^3, 500(1.5)^4, 500(1.5)^5, 500(1.5)^6$ or 750, 1125, 1688, 2531, 3797, 5695 **3.** 35.72 m
4. $176,994.65 **5.** 6.20% **6.** $101,925; $121,925
7. $723.03 **8.** $13,962.01 **9.** $45,502.06
10. $48,095.67

EXERCISE 5.3 (page 236)

1. 64, 32, 16, 8, 4 **3.** 100, 102, 104.04 **5.** $\dfrac{21{,}044}{16{,}807}$
7. 1.11111 **9.** 18.664613 **11.** 8.213180
13. $2050.10 **15.** $29,984.06 **17.** $8001.24
19. $90,231.01 **21.** $204,977.46 **23.** $24,594.36
25. $1937.14 **27.** $458.40
29. (a) $3048.85; **(b)** $648.85 **31.** $3474.12
33. $1725 **35.** 102.91305 **37.** 55,360.30
39. $131.34 **41.** $1,872,984.02
43. $205,073; $142,146

EXERCISE 5.4 (page 242)

1. $69.33 **3.** $502.84
5. (a) $221.43; **(b)** $25; **(c)** $196.43

7.

Period	Prin. Outs. at Beginning	Interest for Period	Pmt. at End	Prin. Repaid at End
1	5000.00	350.00	1476.14	1126.14
2	3873.86	271.17	1476.14	1204.97
3	2668.89	186.82	1476.14	1289.32
4	1379.57	96.57	1476.14	1379.57
Total		904.56	5904.56	5000.00

9.

Period	Prin. Outs. at Beginning	Interest for Period	Pmt. at End	Prin. Repaid at End
1	900.00	22.50	193.72	171.22
2	728.78	18.22	193.72	175.50
3	553.28	13.83	193.72	179.89
4	373.39	9.33	193.72	184.39
5	189.00	4.73	193.73	189.00
Total		68.61	968.61	900.00

11. 11 **13.** $1415.56
15. (a) $2089.69; **(b)** $1878.33; **(c)** $211.36; **(d)** $381,907
17. 23 **19.** $113,302.45 **21.** $38.64

REVIEW PROBLEMS—CHAPTER 5 (page 244)

1. $\dfrac{665}{81}$ **3.** 8.5% compounded annually
5. $586.60 **7. (a)** $1997.13; **(b)** $3325.37
9. $936.85 **11.** $886.98 **13.** $314.00
15.

Period	Prin. Outs. at Beginning	Interest for Period	Pmt. at End	Prin. Repaid at End
1	15,000.00	112.50	3067.84	2955.34
2	12,044.66	90.33	3067.84	2977.51
3	9067.15	68.00	3067.84	2999.84
4	6067.31	45.50	3067.84	3022.34
5	3044.97	22.84	3067.81	3044.97
Total		339.17	15,339.17	15,000.00

17. $1279.36

MATHEMATICAL SNAPSHOT—CHAPTER 5 (page 246)

1. $15,597.85 **3.** When investors expect a drop in interest rates, long-term investments become more attractive relative to short-term ones.

APPLICATIONS IN PRACTICE 6.1

1. 3×2 or 2×3 **2.** $\begin{bmatrix} 1 & 2 & 4 & 8 & 16 \\ 1 & 2 & 4 & 8 & 16 \\ 1 & 2 & 4 & 8 & 16 \end{bmatrix}$

EXERCISE 6.1 (page 254)

1. (a) $2 \times 3, 3 \times 3, 3 \times 2, 2 \times 2, 4 \times 4, 1 \times 2, 3 \times 1, 3 \times 3, 1 \times 1$; **(b) B, D, E, H, J**; **(c) H, J** upper triangular; **D, J** lower triangular; **(d) F, J**; **(e) G, J**
3. 6 **5.** 4 **7.** 0 **9.** 7, 2, 1, 0

11. $\begin{bmatrix} 6 & 8 & 10 & 12 \\ 10 & 12 & 14 & 16 \\ 14 & 16 & 18 & 20 \end{bmatrix}$ **13.** 120 entries, 1, 0, 1, 0

15. (a) $\begin{bmatrix} 0 & 0 & 0 & 0 \\ 0 & 0 & 0 & 0 \\ 0 & 0 & 0 & 0 \\ 0 & 0 & 0 & 0 \end{bmatrix}$; **(b)** $\begin{bmatrix} 0 & 0 & 0 & 0 & 0 & 0 \\ 0 & 0 & 0 & 0 & 0 & 0 \\ 0 & 0 & 0 & 0 & 0 & 0 \\ 0 & 0 & 0 & 0 & 0 & 0 \\ 0 & 0 & 0 & 0 & 0 & 0 \\ 0 & 0 & 0 & 0 & 0 & 0 \end{bmatrix}$

17. $\begin{bmatrix} 6 & 2 \\ -3 & 4 \end{bmatrix}$ **19.** $\begin{bmatrix} 1 & 3 & -4 \\ 3 & 2 & 5 \\ 7 & -2 & 0 \\ 3 & 0 & 1 \end{bmatrix}$

21. (a) A and **C**; **(b)** all of them

25. $x = 6, y = \frac{2}{3}, z = \frac{7}{2}$ **27.** $x = 0, y = 0$

29. (a) 7; **(b)** 3; **(c)** February; **(d)** deluxe blue; **(e)** February; **(f)** February; **(g)** 38 **31.** -2001

33. $\begin{bmatrix} 3 & 1 & 1 \\ 1 & 7 & 4 \\ 4 & 3 & 1 \\ 2 & 6 & 2 \end{bmatrix}$

APPLICATIONS IN PRACTICE 6.2

1. $\begin{bmatrix} 230 & 220 \\ 190 & 255 \end{bmatrix}$ **2.** $x_1 = 670, x_2 = 835, x_3 = 1405$

EXERCISE 6.2 (page 261)

1. $\begin{bmatrix} 4 & -3 & 1 \\ -2 & 10 & 5 \\ 10 & 5 & 3 \end{bmatrix}$ **3.** $\begin{bmatrix} -5 & 5 \\ -9 & 5 \\ 5 & 9 \end{bmatrix}$ **5.** $\begin{bmatrix} -9 & -7 & 11 \end{bmatrix}$

7. Not defined **9.** $\begin{bmatrix} -12 & 36 & -42 & -6 \\ -42 & -6 & -36 & 12 \end{bmatrix}$

11. $\begin{bmatrix} 3 & -5 & 6 \\ -2 & 8 & 0 \\ 5 & 10 & 15 \end{bmatrix}$ **13.** $\begin{bmatrix} 6 & 5 \\ -2 & 3 \end{bmatrix}$ **15. O**

17. $\begin{bmatrix} 28 & 22 \\ -2 & 6 \end{bmatrix}$ **19.** Not defined **21.** $\begin{bmatrix} -22 & -15 \\ -11 & 9 \end{bmatrix}$

23. $\begin{bmatrix} 21 & \frac{29}{2} \\ \frac{19}{2} & -\frac{15}{2} \end{bmatrix}$ **29.** $\begin{bmatrix} 4 & 2 & 20 \\ 7 & -3 & 2 \end{bmatrix}$ **31.** $\begin{bmatrix} -1 & 5 \\ 6 & -8 \end{bmatrix}$

33. Impossible **35.** $x = \frac{90}{29}, y = -\frac{24}{29}$

37. $x = 6, y = \frac{4}{3}$ **39.** $x = -6, y = -14, z = 1$

41. $\begin{bmatrix} 35 & 65 \\ 75 & 55 \\ 25 & 15 \end{bmatrix}$ **43.** 1.1 **45.** $\begin{bmatrix} 15 & -4 & 26 \\ 4 & 7 & 30 \end{bmatrix}$

47. $\begin{bmatrix} -10 & 22 & 12 \\ 24 & 36 & -44 \end{bmatrix}$

APPLICATIONS IN PRACTICE 6.3

1. \$5780 **2.** \$22,843.75 **3.** $\begin{bmatrix} 1 & \frac{8}{5} \\ 1 & \frac{1}{3} \end{bmatrix} \begin{bmatrix} y \\ x \end{bmatrix} = \begin{bmatrix} \frac{8}{5} \\ \frac{5}{3} \end{bmatrix}$

EXERCISE 6.3 (page 273)

1. -12 **3.** 19 **5.** 1 **7.** 2×2; 4
9. 3×5; 15 **11.** 2×1; 2 **13.** 3×3; 9

15. 3×1; 3 **17.** $\begin{bmatrix} 1 & 0 & 0 & 0 \\ 0 & 1 & 0 & 0 \\ 0 & 0 & 1 & 0 \\ 0 & 0 & 0 & 1 \end{bmatrix}$ **19.** $\begin{bmatrix} 12 & -12 \\ 10 & 6 \end{bmatrix}$

21. $\begin{bmatrix} 23 \\ 50 \end{bmatrix}$ **23.** $\begin{bmatrix} 1 & -4 & 2 \\ 2 & 2 & 4 \\ -3 & -2 & 3 \end{bmatrix}$ **25.** $\begin{bmatrix} -6 & 16 & 10 & -6 \end{bmatrix}$

27. $\begin{bmatrix} 4 & 6 & -4 & 6 \\ 6 & 9 & -6 & 9 \\ -8 & -12 & 8 & -12 \\ 2 & 3 & -2 & 3 \end{bmatrix}$ **29.** $\begin{bmatrix} 78 & 84 \\ -21 & -12 \end{bmatrix}$

31. $\begin{bmatrix} -5 & -8 \\ -5 & -20 \end{bmatrix}$ **33.** $\begin{bmatrix} z \\ y \\ x \end{bmatrix}$ **35.** $\begin{bmatrix} 2x_1 + x_2 + 3x_3 \\ 4x_1 + 9x_2 + 7x_3 \end{bmatrix}$

37. $\begin{bmatrix} 0 & 0 & 0 \\ 0 & -1 & 1 \\ 1 & 2 & 0 \end{bmatrix}$ **39.** $\begin{bmatrix} -1 & -20 \\ -2 & 23 \end{bmatrix}$ **41.** $\begin{bmatrix} \frac{7}{3} & 0 & 0 \\ 0 & \frac{7}{3} & 0 \\ 0 & 0 & \frac{7}{3} \end{bmatrix}$

43. $\begin{bmatrix} -1 & 5 \\ 2 & 17 \\ 1 & 31 \end{bmatrix}$ **45.** Impossible **47.** $\begin{bmatrix} 0 & 0 & -4 \\ 2 & -1 & -2 \\ 0 & 0 & 8 \end{bmatrix}$

49. $\begin{bmatrix} 3 & -1 \\ -2 & 2 \end{bmatrix}$ **51.** $\begin{bmatrix} 0 & 3 & 0 \\ -1 & -1 & 2 \end{bmatrix}$

53. $\begin{bmatrix} 2 & 0 & 0 \\ 0 & 2 & 0 \\ 0 & 0 & 2 \end{bmatrix}$ **55.** $\begin{bmatrix} 1 & -1 & 0 \\ 0 & 1 & 1 \end{bmatrix}$ **57.** $\begin{bmatrix} 6 & -7 \\ -7 & 9 \end{bmatrix}$

59. $\begin{bmatrix} 3 & 1 \\ 2 & -9 \end{bmatrix} \begin{bmatrix} x \\ y \end{bmatrix} = \begin{bmatrix} 6 \\ 5 \end{bmatrix}$

61. $\begin{bmatrix} 4 & -1 & 3 \\ 3 & 0 & -1 \\ 0 & 3 & 2 \end{bmatrix} \begin{bmatrix} r \\ s \\ t \end{bmatrix} = \begin{bmatrix} 9 \\ 7 \\ 15 \end{bmatrix}$ **63.** \$2075 **65.** \$828,950

67. (a) \$180,000, \$520,000, \$400,000, \$270,000, \$380,000, \$640,000; **(b)** \$390,000, \$100,000, \$800,000; **(c)** \$2,390,000; **(d)** $\frac{110}{239}, \frac{129}{239}$ **71.** $\begin{bmatrix} 72.82 & -9.8 \\ 51.32 & -36.32 \end{bmatrix}$

73. $\begin{bmatrix} 15.606 & 64.08 \\ -739.428 & 373.056 \end{bmatrix}$

APPLICATIONS IN PRACTICE 6.4

1. 5 blocks of A, 2 blocks of B, and 1 block of C
2. 3 of X; 4 of Y; 2 of Z **3.** A = 3D; B = 1000 − 2D; C = 500 − D; D = any amount between 0 and 500

EXERCISE 6.4 (page 285)

1. Not reduced **3.** Reduced **5.** Not reduced

7. $\begin{bmatrix} 1 & 0 \\ 0 & 1 \end{bmatrix}$ **9.** $\begin{bmatrix} 1 & 2 & 3 \\ 0 & 0 & 0 \\ 0 & 0 & 0 \end{bmatrix}$ **11.** $\begin{bmatrix} 1 & 0 & 0 & 0 \\ 0 & 1 & 0 & 0 \\ 0 & 0 & 1 & 0 \\ 0 & 0 & 0 & 1 \end{bmatrix}$

13. $x = 5, y = 2$ **15.** No solution
17. $x = -\frac{2}{3}r + \frac{5}{3}, y = -\frac{1}{6}r + \frac{7}{6}, z = r$, where r is any real number **19.** No solution
21. $x = -3, y = 2, z = 0$ **23.** $x = 2, y = -5, z = -1$
25. $x_1 = 0, x_2 = -r, x_3 = -r, x_4 = -r, x_5 = r$, where r is any real number **27.** Federal, $72,000; state, $24,000
29. A, 2000; B, 4000; C, 5000 **31. (a)** 3 of X, 4 of Z; 2 of X, 1 of Y, 5 of Z; 1 of X, 2 of Y, 6 of Z; 3 of Y, 7 of Z; **(b)** 3 of X, 4 of Z; **(c)** 3 of X, 4 of Z; 3 of Y, 7 of Z
33. (a) Let s, d, g represent the numbers of units S, D, G respectively. The six combinations are given by:

s	5	4	3	2	1	0
d	8	7	6	5	4	3
g	0	1	2	3	4	5

(b) The combination $s = 0$, $d = 3, g = 5$

APPLICATIONS IN PRACTICE 6.5

1. Infinitely many solutions: $x + \frac{1}{2}z = 0, y + \frac{1}{2}z = 0$; in parametric form: $x = -\frac{1}{2}r, y = -\frac{1}{2}r, z = r$, where r is any real number

EXERCISE 6.5 (page 290)

1. $w = -1 + 7r, x = 2 - 5r, y = 4 - 7r, z = r$, (where r is any real number)
3. $w = -s, x = -3r - 4s + 2, y = r, z = s$ (where r and s are any real numbers)
5. $w = -2r + s - 2, x = -r + 4, y = r, z = s$ (where r and s are any real numbers)
7. $x_1 = -2r + s - 2t + 1, x_2 = -r - 2s + t + 4, x_3 = r, x_4 = s, x_5 = t$ (where r, s, and t are any real numbers)
9. Infinitely many **11.** Trivial solution
13. Infinitely many **15.** $x = 0, y = 0$
17. $x = -\frac{6}{5}r, y = \frac{8}{15}r, z = r$ **19.** $x = 0, y = 0$
21. $x = r, y = -2r, z = r$
23. $w = -2r, x = -3r, y = r, z = r$

APPLICATIONS IN PRACTICE 6.6

1. Yes **2.** MEET AT NOON FRIDAY
3. $\mathbf{E}^{-1} = \begin{bmatrix} \frac{2}{3} & -\frac{1}{6} & -\frac{1}{3} \\ -\frac{1}{3} & \frac{5}{6} & -\frac{1}{3} \\ -\frac{1}{3} & -\frac{1}{6} & \frac{2}{3} \end{bmatrix}$; **F** is not invertible.
4. A: 5000 shares; B: 2500 shares; C:2500 shares

EXERCISE 6.6 (page 299)

1. $\begin{bmatrix} -1 & 1 \\ 7 & -6 \end{bmatrix}$ **3.** Not invertible **5.** $\begin{bmatrix} 1 & 0 & 0 \\ 0 & -\frac{1}{3} & 0 \\ 0 & 0 & \frac{1}{4} \end{bmatrix}$

7. Not invertible **9.** Not invertible (not a square matrix)

11. $\begin{bmatrix} 1 & -1 & 0 \\ 0 & 1 & -1 \\ 0 & 0 & 1 \end{bmatrix}$ **13.** $\begin{bmatrix} 1 & 0 & 2 \\ 0 & 1 & 0 \\ 3 & 0 & 7 \end{bmatrix}$

15. $\begin{bmatrix} 1 & -\frac{2}{3} & \frac{5}{3} \\ -1 & \frac{4}{3} & -\frac{10}{3} \\ -1 & 1 & -2 \end{bmatrix}$ **17.** $\begin{bmatrix} \frac{11}{3} & -3 & \frac{1}{3} \\ -\frac{7}{3} & 3 & -\frac{2}{3} \\ \frac{2}{3} & -1 & \frac{1}{3} \end{bmatrix}$

19. $x_1 = 10, x_2 = 20$ **21.** $x = 17, y = -20$
23. $x = 1, y = 3$ **25.** $x = -3r + 1, y = r$
27. $x = 0, y = 1, z = 2$ **29.** $x = 1, y = \frac{1}{2}, z = \frac{1}{2}$
31. No solution **33.** $w = 1, x = 3, y = -2, z = 7$
35. $\begin{bmatrix} -\frac{2}{3} & -\frac{1}{3} \\ \frac{1}{3} & -\frac{1}{3} \end{bmatrix}$ **37. (a)** 40 of model A, 60 of model B; **(b)** 45 of model A, 50 of model B **39. (b)** $\begin{bmatrix} 4 & 6 \\ 7 & 10 \end{bmatrix}$
41. Yes **43.** D: 5000 shares; E:1000 shares; F:4000 shares
45. (a) $\begin{bmatrix} 2.05 & 1.28 \\ 0.73 & 1.71 \end{bmatrix}$; **(b)** $\begin{bmatrix} \frac{84}{41} & \frac{105}{82} \\ \frac{30}{41} & \frac{70}{41} \end{bmatrix}$
47. $\begin{bmatrix} 1.80 & 1.10 & -0.46 \\ 0.35 & 1.31 & -0.17 \\ 0.44 & 0.42 & 0.59 \end{bmatrix}$
49. $w = 14.44, x = 0.03, y = -0.80, z = 10.33$

EXERCISE 6.7 (page 304)

1. $\begin{bmatrix} 1290 \\ 1425 \end{bmatrix}$; 1405 **3. (a)** $\begin{bmatrix} 134.29 \\ 162.25 \\ 234.35 \end{bmatrix}$; **(b)** $\begin{bmatrix} 68.59 \\ 84.50 \\ 108.69 \end{bmatrix}$

5. $\begin{bmatrix} 1301 \\ 1215 \\ 1188 \end{bmatrix}$ **7.** $\begin{bmatrix} 1073 \\ 1016 \\ 952 \end{bmatrix}$

REVIEW PROBLEMS—CHAPTER 6 (page 306)

1. $\begin{bmatrix} 3 & 8 \\ -16 & -10 \end{bmatrix}$ **3.** $\begin{bmatrix} 1 & 42 & 5 \\ 2 & -18 & -7 \\ 1 & 0 & -2 \end{bmatrix}$ **5.** $\begin{bmatrix} -1 & -2 \\ 5 & 22 \end{bmatrix}$

7. $\begin{bmatrix} 6 \\ 32 \end{bmatrix}$ **9.** $\begin{bmatrix} -1 & -2 \\ 2 & 1 \end{bmatrix}$ **11.** $\begin{bmatrix} 2 & 0 \\ 0 & 9 \end{bmatrix}$

13. $x = 3, y = 21$ **15.** $\begin{bmatrix} 1 & 0 \\ 0 & 1 \end{bmatrix}$ **17.** $\begin{bmatrix} 1 & 2 & 0 \\ 0 & 0 & 1 \\ 0 & 0 & 0 \end{bmatrix}$

19. $x = 0, y = 0$ **21.** No solution **23.** $\begin{bmatrix} -\frac{3}{2} & \frac{5}{6} \\ \frac{1}{2} & -\frac{1}{6} \end{bmatrix}$

25. No inverse exists. **27.** $x = 0, y = 1, z = 0$
29. $\mathbf{A}^2 = \mathbf{I}_3, \mathbf{A}^{-1} = \mathbf{A}, \mathbf{A}^{2000} = \mathbf{I}_3$

31. (a) Let x, y, z represent the weekly doses of capsules of brands I, II, III, respectively. The combinations are given by:

	x	y	z
combination 1	4	9	0
combination 2	3	6	1
combination 3	2	3	2
combination 4	1	0	3

(b) Combination 4: $x = 1, y = 0, z = 3$

33. $\begin{bmatrix} 215 & 87 \\ 89 & 141 \end{bmatrix}$ **35.** $\begin{bmatrix} 168 \\ 171.6 \end{bmatrix}$

MATHEMATICAL SNAPSHOT—CHAPTER 6 (page 308)

1. \$151.40 **3.** It is not possible; different combinations of lengths of stays can cost the same.

APPLICATIONS IN PRACTICE 7.1

1. $2x + 1.5y > 0.9x + 0.7y + 50$, $y > -1.375x + 62.5$; sketch the dashed line $y = -1.375x + 62.5$ and shade the half plane above the line. In order to produce a profit, the number of magnets of types A and B produced and sold must be an ordered pair in the shaded region.
2. $x \geq 0$, $y \geq 0$, $x + y \geq 50$, $x \geq 2y$; The region consists of points on or above the x-axis and on or to the right of the y-axis. In addition, the points must be on or above the line $x + y = 50$ and on or below the line $x = 2y$.

EXERCISE 7.1 (page 315)

1.
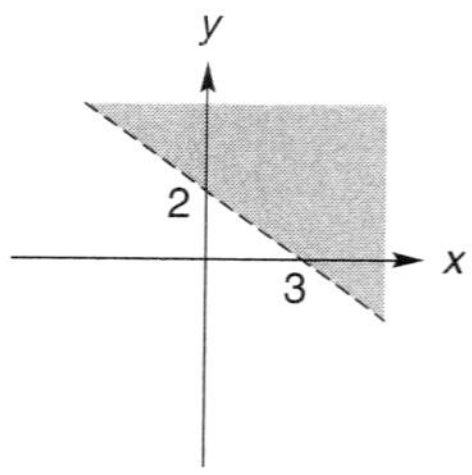

3.
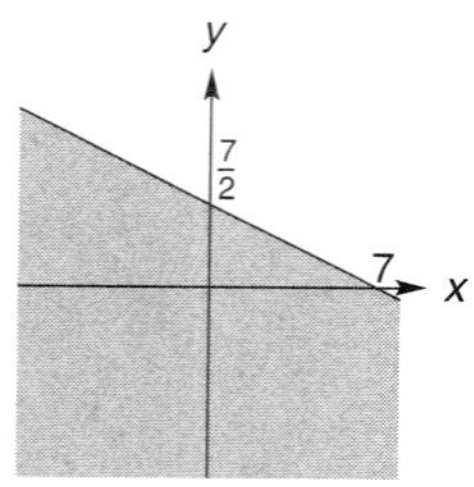

5.
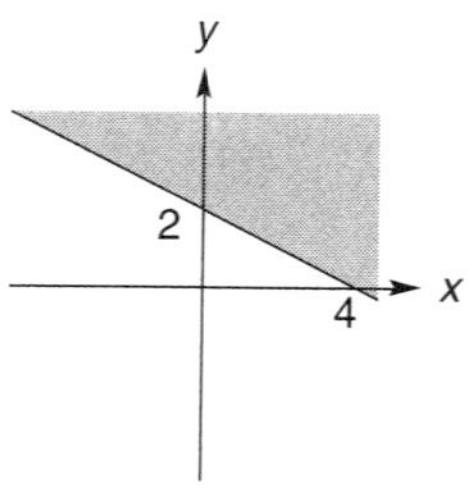

7.
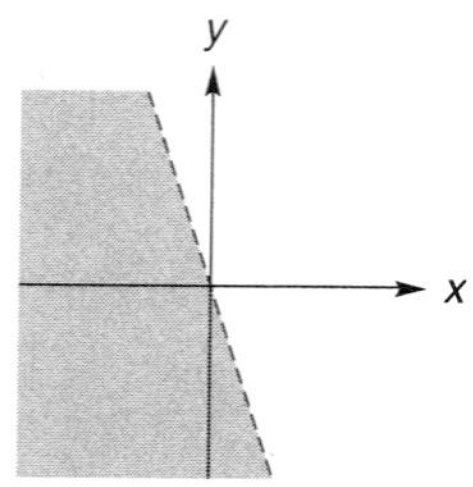

9.
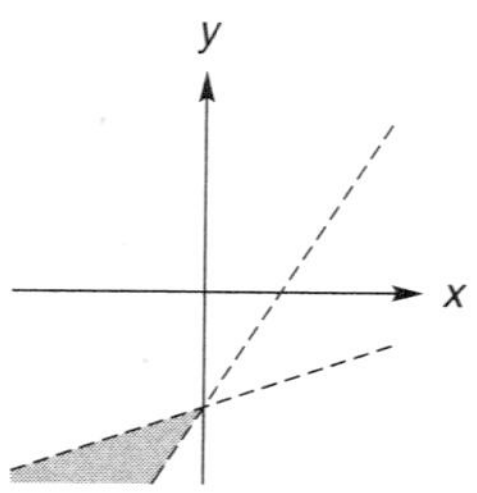

11.
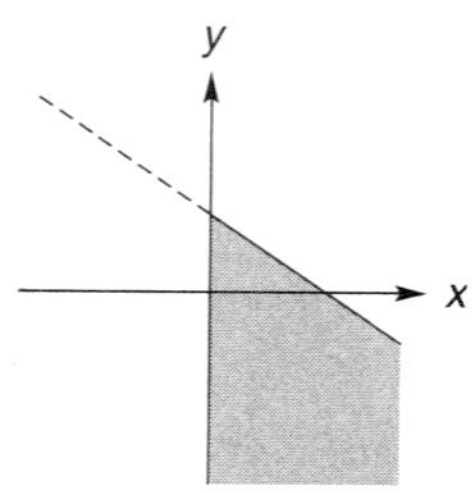

13.
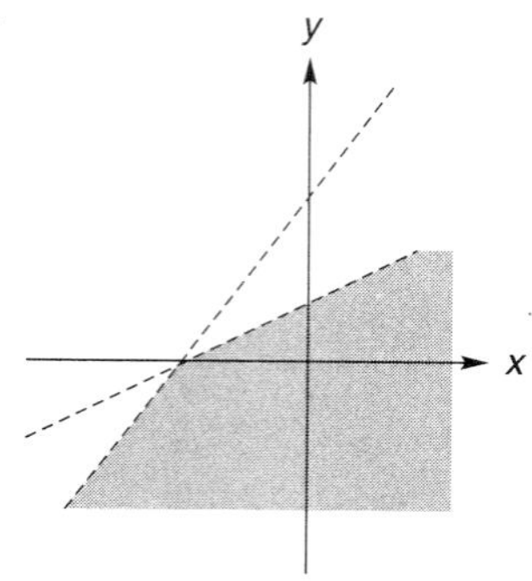

15.
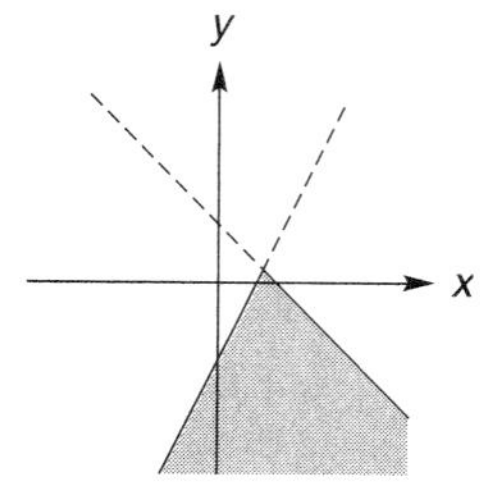

17.
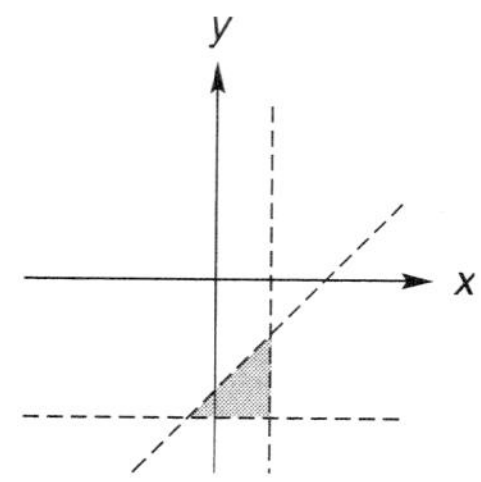

19.
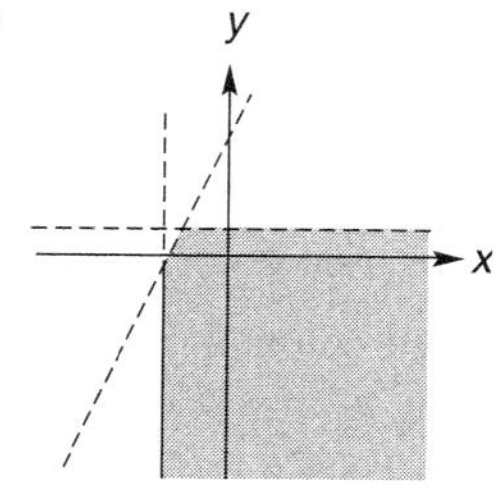

21.
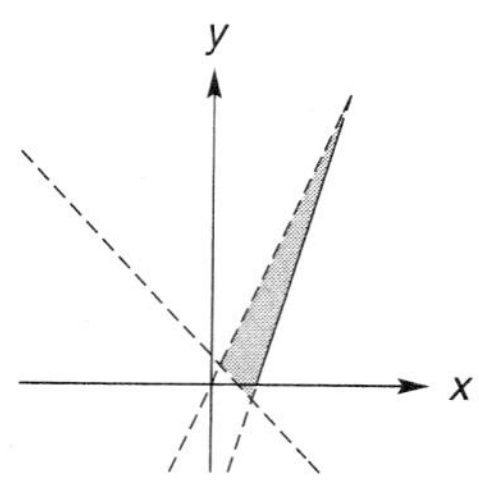

23.
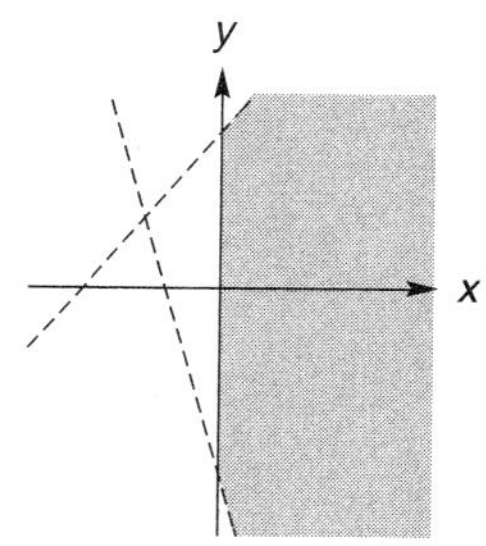

25.
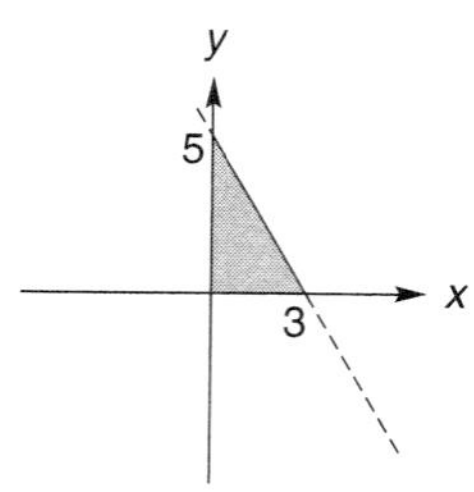

27.
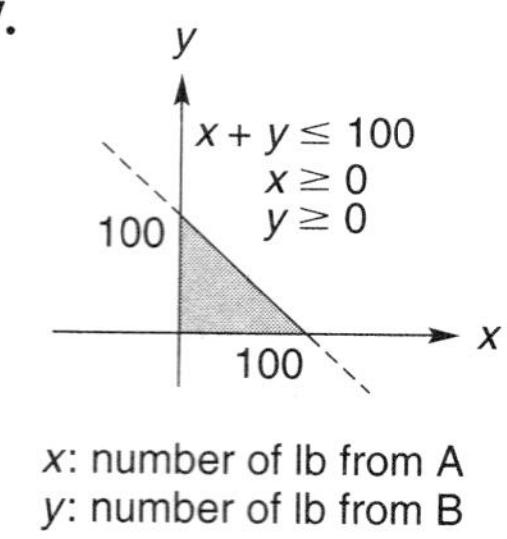

29. $x \geq 0, y \geq 0, 3x + 2y \leq 240, 0.5x + y \leq 80$

EXERCISE 7.2 (page 324)

1. $P = 75$ when $x = 15, y = 0$
3. $Z = -10$ when $x = 2, y = 3$
5. No optimum solution (empty feasible region)
7. $Z = 3$ when $x = 0, y = 1$
9. $C = 2.4$ when $x = \frac{3}{5}, y = \frac{6}{5}$
11. No optimum solution (unbounded)
13. 10 trucks, 20 spinning tops; \$110
15. 4 units of food A, 4 units of food B; \$8
17. 10 tons of ore I, 10 tons of ore II; \$1100
19. 6 chambers of type A and 10 chambers of type B
21. (c) $x = y = 75$
23. $Z = 15.54$ when $x = 2.56, y = 6.74$

25. $Z = -75.98$ when $x = 9.48$, $y = 16.67$

APPLICATIONS IN PRACTICE 7.3

1. Ship $10t + 15$ TV sets from C to A, $-10t + 30$ TV sets from C to B, $-10t + 10$ TV sets from D to A, and $10t$ TV sets from D to B, for $0 \le t \le 1$; minimum cost $780

EXERCISE 7.3 (page 328)

1. $Z = 33$ when $x = (1 - t)(2) + 5t = 2 + 3t$, $y = (1 - t)(3) + 2t = 3 - t$, and $0 \le t \le 1$
3. $Z = 72$ when $x = (1 - t)(3) + 4t = 3 + t$, $y = (1 - t)(2) + 0t = 2 - 2t$, and $0 \le t \le 1$

APPLICATIONS IN PRACTICE 7.4

1. 0 gadgets of Type 1, 72 gadgets of Type 2, 12 gadgets of Type 3; maximum profit of $20,400

EXERCISE 7.4 (page 341)

1. $Z = 8$ when $x_1 = 0, x_2 = 4$
3. $Z = 2$ when $x_1 = 0, x_2 = 1$
5. $Z = 28$ when $x_1 = 3, x_2 = 2$
7. $Z = 20$ when $x_1 = 0, x_2 = 5, x_3 = 0$
9. $Z = 2$ when $x_1 = 1, x_2 = 0, x_3 = 0$
11. $Z = \frac{16}{3}$ when $x_1 = \frac{2}{3}, x_2 = \frac{14}{3}$
13. $W = 13$ when $x_1 = 1, x_2 = 0, x_3 = 3$
15. $Z = 600$ when $x_1 = 4, x_2 = 1, x_3 = 4, x_4 = 0$
17. 0 from A, 2400 from B; $1200
19. 0 chairs, 300 rockers, 100 chaise lounges; $10,800

APPLICATIONS IN PRACTICE 7.5

1. $35 - 7t$ of device 1, $6t$ of device 2, 0 of device 3, for $0 \le t \le 1$

EXERCISE 7.5 (page 348)

1. Yes; for the tableau, x_2 is the entering variable and the quotients $\frac{6}{2}$ and $\frac{3}{1}$ tie for being the smallest.
3. No optimum solution (unbounded)
5. $Z = 12$ when $x_1 = 4 + t, x_2 = t$, and $0 \le t \le 1$
7. No optimum solution (unbounded)
9. $Z = 13$ when $x_1 = \frac{3}{2} - \frac{3}{2}t, x_2 = 6t, x_3 = 4 - 3t$, and $0 \le t \le 1$
11. $15,200. If x_1, x_2, x_3 denote the number of chairs, rockers, and chaise lounges produced, respectively, then
$x_1 = 100 - 100t$,
$x_2 = 100 + 150t$,
$x_3 = 200 - 50t$, and
$0 \le t \le 1$

APPLICATIONS IN PRACTICE 7.6

1. Plant I: 500 standard, 700 deluxe; plant II: 500 standard, 100 deluxe; $89,500 maximum profit

EXERCISE 7.6 (page 359)

1. $Z = 7$ when $x_1 = 1, x_2 = 5$
3. $Z = 4$ when $x_1 = 1, x_2 = 2, x_3 = 0$
5. $Z = 28$ when $x_1 = 8, x_2 = 2, x_3 = 0$
7. $Z = -17$ when $x_1 = 3, x_2 = 2$
9. No optimum solution (empty feasible region)
11. $Z = 2$ when $x_1 = 6, x_2 = 10$
13. 255 Standard bookcases, 0 Executive bookcases
15. 30% in A, 0% in AA, 70% in AAA; 6.6%

EXERCISE 7.7 (page 364)

1. $Z = 54$ when $x_1 = 2, x_2 = 8$
3. $Z = 216$ when $x_1 = 18, x_2 = 0, x_3 = 0$
5. $Z = 4$ when $x_1 = 0, x_2 = 0, x_3 = 4$
7. $Z = 0$ when $x_1 = 3, x_2 = 0, x_3 = 1$
9. $Z = 28$ when $x_1 = 3, x_2 = 0, x_3 = 5$
11. Install device A on kilns producing 700,000 barrels annually, and device B on kilns producing 2,600,000 barrels annually **13.** To Exton, 5 from A and 10 from B; to Whyton, 15 from A; $380 **15.** **(a)** Column 3: 1, 3, 3; column 4: 0, 4, 8; **(b)** $x_1 = 10, x_2 = 0, x_3 = 20, x_4 = 0$; **(c)** 90 in.

APPLICATIONS IN PRACTICE 7.8

1. Minimize $W = 60{,}000y_1 + 2000y_2 + 120y_3$ subject to
$300y_1 + 20y_2 + 3y_3 \ge 300$
$220y_1 + 40y_2 + y_3 \ge 200$
$180y_1 + 20y_2 + 2y_3 \ge 200$
and $y_1, y_2, y_3 \ge 0$
2. Maximize $W = 98y_1 + 80y_2$ subject to
$20y_1 + 8y_2 \le 6$
$6y_1 + 16y_2 \le 2$
and $y_1, y_2 \ge 0$
3. 5 device 1, 0 device 2, 15 device 3

EXERCISE 7.8 (page 375)

1. Minimize $W = 5y_1 + 3y_2$ subject to
$y_1 - y_2 \ge 1$
$y_1 + y_2 \ge 2$
$y_1, y_2 \ge 0$
3. Maximize $W = 8y_1 + 2y_2$ subject to
$y_1 - y_2 \le 1$
$y_1 + 2y_2 \le 8$
$y_1 + y_2 \le 5$
$y_1, y_2 \ge 0$
5. Minimize $W = 13y_1 - 3y_2 - 11y_3$ subject to
$-y_1 + y_2 - y_3 \ge 1$
$2y_1 - y_2 - y_3 \ge -1$
$y_1, y_2, y_3 \ge 0$
7. Maximize $W = -3y_1 + 3y_2$ subject to
$-y_1 + y_2 \le 4$
$y_1 - y_2 \le 4$
$y_1 + y_2 \le 6$
$y_1, y_2 \ge 0$
9. $Z = 11$ when $x_1 = 0, x_2 = \frac{1}{2}, x_3 = \frac{3}{2}$
11. $Z = 26$ when $x_1 = 6, x_2 = 1$
13. $Z = 14$ when $x_1 = 1, x_2 = 2$

15. \$250 on newspaper advertising, \$1400 on radio advertising; \$1650.
17. 20 shipping clerk apprentices, 40 shipping clerks, 90 semiskilled workers, 0 skilled workers; \$1200

REVIEW PROBLEMS—CHAPTER 7 (page 377)

1.

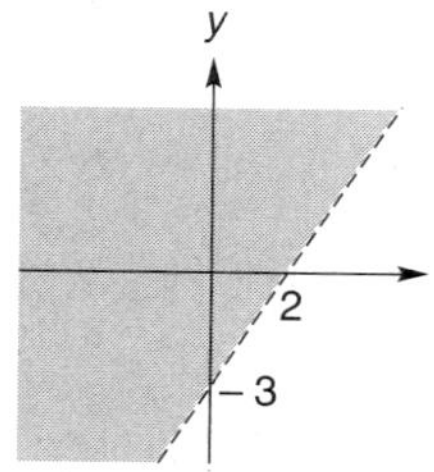

3.

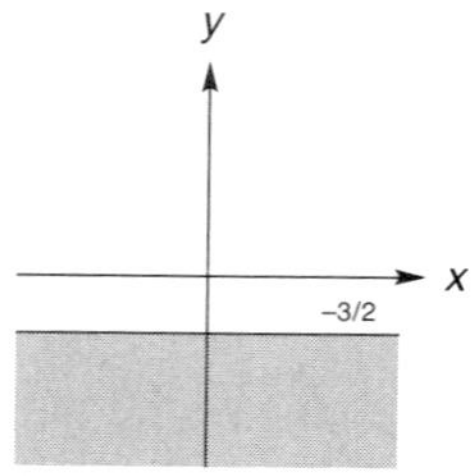

5.

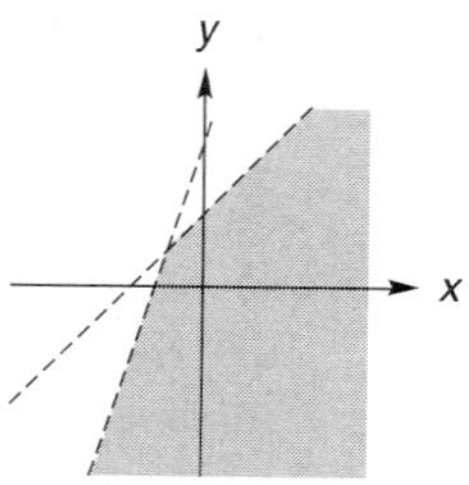

7.

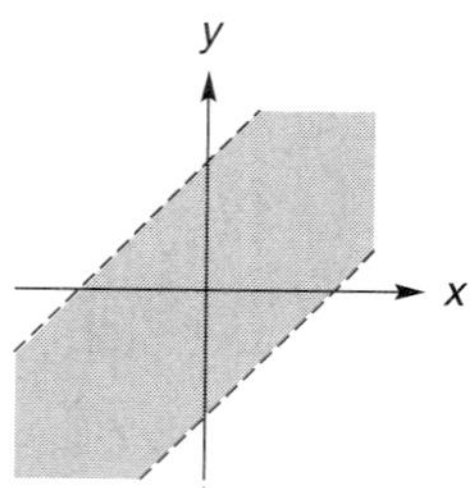

9.

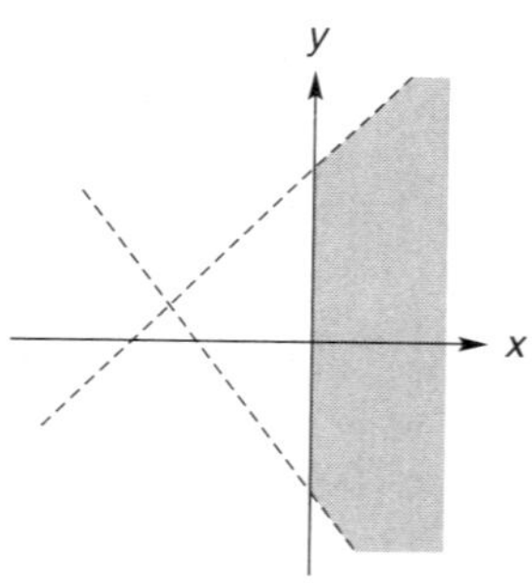

11. $Z = 3$ when $x = 3, y = 0$
13. $Z = -2$ when $x = 0, y = 2$
15. No optimum solution (empty feasible region)
17. $Z = 36$ when $x = 2 + 2t, y = 3 - 3t$, and $0 \leq t \leq 1$
19. $Z = 32$ when $x_1 = 8, x_2 = 0$
21. $Z = \frac{5}{3}$ when $x_1 = 0, x_2 = 0, x_3 = 2$
23. $Z = 24$ when $x_1 = 0, x_2 = 12$
25. $Z = \frac{7}{2}$ when $x_1 = \frac{5}{4}, x_2 = 0, x_3 = \frac{9}{4}$
27. No optimum solution (unbounded)
29. $Z = 70$ when $x_1 = 35, x_2 = 0, x_3 = 0$
31. 0 units of X, 6 units of Y, 14 units of Z; \$398
33. 500,000 gal from A to D, 100,000 gal from A to C, 400,000 gal from B to C; \$19,000
35. 10 kg of food A only
37. $Z = 129.83$ when $x = 9.38, y = 1.63$

MATHEMATICAL SNAPSHOT—CHAPTER 7 (page 380)

1. 2 minutes of radiation **3.** Answers may vary.

EXERCISE 8.1 (page 338)

1.

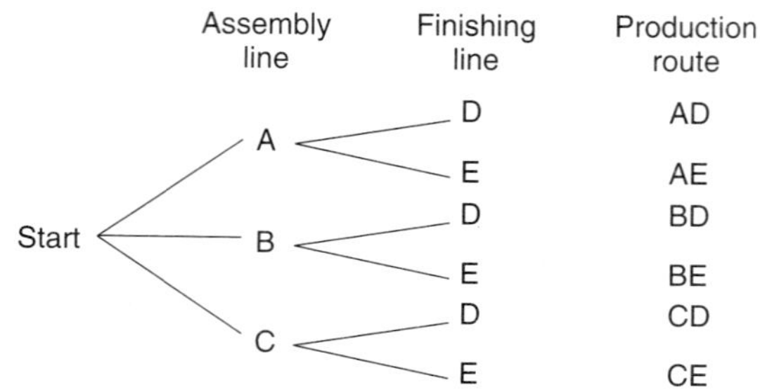

6 possible production routes

3.

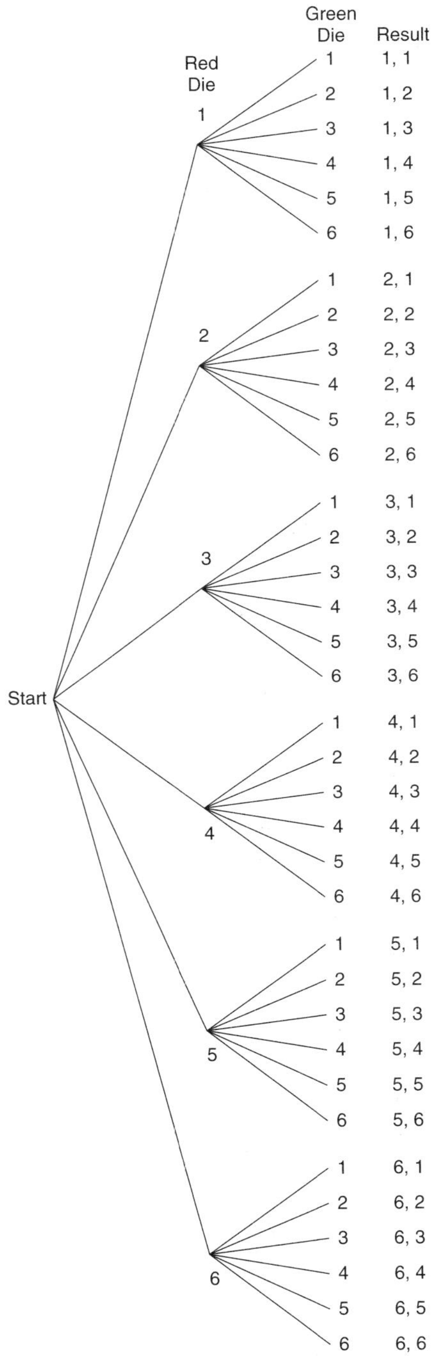

36 possible results

5. 20 **7.** 96 **9.** 1024 **11.** 20 **13.** 720
15. 720 **17.** 1000; error message is displayed
19. 6 **21.** 336 **23.** 216 **25.** 1320 **27.** 360
29. 720 **31.** 2520; 5040 **33.** 624 **35.** 24
37. **(a)** 11,880; **(b)** 19,008 **39.** 48 **41.** 2880

EXERCISE 8.2 (page 400)

1. 15 **3.** 1 **5.** 18 **9.** 2380 **11.** 715
13. $\dfrac{74!}{10! \cdot 64!}$ **15.** 56 **17.** 1680 **19.** 35
21. 720 **23.** 1680 **25.** 252 **27.** 756,756
29. **(a)** 90; **(b)** 330 **31.** 17,325 **33.** **(a)** 1; **(b)** 1; **(c)** 18 **35.** 3744 **37.** 5,250,960

APPLICATIONS IN PRACTICE 8.3

1. 10,586,800

EXERCISE 8.3 (page 410)

1. {9D, 9H, 9C, 9S}
3. {1H, 1T, 2H, 2T, 3H, 3T, 4H, 4T, 5H, 5T, 6H, 6T}
5. {64, 69, 60, 61, 46, 49, 40, 41, 96, 94, 90, 91, 06, 04, 09, 01, 16, 14, 19, 10}
7. **(a)** {RR, RW, RB, WR, WW, WB, BR, BW, BB}; **(b)** {RW, RB, WR, WB, BR, BW}
9. Sample space consists of ordered sets of six elements and each element is H or T; 64.
11. Sample space consists of ordered pairs where first element indicates card drawn and second element indicates number on die; 312.
13. Sample space consists of combinations of 52 cards taken 13 at a time; ${}_{52}C_{13}$.
15. {1, 3, 5, 7, 9} **17.** {7, 9} **19.** {1, 2, 4, 6, 8, 10}
21. S **23.** E_1 and E_4, E_2 and E_3, E_3 and E_4
25. E and H, G and H, H and I
27. **(a)** {HHH, HHT, HTH, HTT, THH, THT, TTH, TTT}; **(b)** {HHH, HHT, HTH, HTT, THH, THT, TTH}; **(c)** {HHT, HTH, HTT, THH, THT, TTH, TTT}; **(d)** S; **(e)** {HHT, HTH, HTT, THH, THT, TTH}; **(f)** $\varnothing$; **(g)** {HHH, TTT}
29. **(a)** {ABC, ACB, BAC, BCA, CAB, CBA}; **(b)** {ABC, ACB}; **(c)** {BAC, BCA, CAB, CBA}

EXERCISE 8.4 (page 421)

1. 600 **3.** **(a)** 0.8; **(b)** 0.4 **5.** No
7. **(a)** $\frac{5}{36}$; **(b)** $\frac{1}{12}$; **(c)** $\frac{1}{4}$; **(d)** $\frac{1}{36}$; **(e)** $\frac{1}{2}$; **(f)** $\frac{1}{2}$; **(g)** $\frac{5}{6}$
9. **(a)** $\frac{1}{52}$; **(b)** $\frac{1}{4}$; **(c)** $\frac{1}{13}$; **(d)** $\frac{1}{2}$; **(e)** $\frac{1}{2}$; **(f)** $\frac{1}{52}$; **(g)** $\frac{4}{13}$; **(h)** $\frac{1}{26}$; **(i)** 0
11. **(a)** $\frac{1}{624}$; **(b)** $\frac{4}{624} = \frac{1}{156}$; **(c)** $\frac{8}{624} = \frac{1}{78}$; **(d)** $\frac{39}{624} = \frac{1}{16}$
13. **(a)** $\frac{4 \cdot 3 \cdot 2}{132{,}600} = \frac{1}{5525}$; **(b)** $\frac{6084}{132{,}600} = \frac{39}{850}$
15. **(a)** $\frac{1}{8}$; **(b)** $\frac{3}{8}$; **(c)** $\frac{1}{8}$; **(d)** $\frac{7}{8}$ **17.** **(a)** $\frac{4}{5}$; **(b)** $\frac{1}{5}$
19. **(a)** 0.1; **(b)** 0.35; **(c)** 0.7; **(d)** 0.95; **(e)** 0.1, 0.35, 0.7, 0.95 **21.** $\frac{1}{10}$
23. **(a)** $\frac{1}{2^{10}} = \frac{1}{1024}$; **(b)** $\frac{11}{1024}$ **25.** $\dfrac{13 \cdot {}_4C_3 \cdot 12 \cdot {}_4C_2}{{}_{52}C_5}$
27. **(a)** $\frac{6545}{161{,}700} \approx 0.040$; **(b)** $\frac{4140}{161{,}700} \approx 0.026$

29. $\frac{1}{9}$ **31.** **(a)** 0.51; **(b)** 0.44; **(c)** 0.03 **33.** 4:1

35. 3:7 **37.** $\frac{7}{12}$ **39.** $\frac{2}{7}$ **41.** 3:1

EXERCISE 8.5 (page 434)

1. **(a)** $\frac{1}{5}$; **(b)** $\frac{4}{5}$; **(c)** $\frac{1}{4}$; **(d)** $\frac{1}{2}$; **(e)** 0 **3.** 1 **5.** 0.43

7. **(a)** $\frac{1}{2}$; **(b)** $\frac{2}{3}$ **9.** **(a)** $\frac{3}{5}$; **(b)** $\frac{2}{5}$; **(c)** $\frac{1}{2}$; **(d)** $\frac{2}{9}$

11. **(a)** $\frac{5}{8}$; **(b)** $\frac{35}{58}$; **(c)** $\frac{11}{39}$; **(d)** $\frac{8}{25}$; **(e)** $\frac{10}{47}$; **(f)** $\frac{25}{86}$

13. **(a)** $\frac{1}{2}$; **(b)** $\frac{4}{9}$ **15.** $\frac{2}{3}$ **17.** **(a)** $\frac{1}{2}$; **(b)** $\frac{1}{4}$ **19.** $\frac{2}{3}$

21. $\frac{1}{11}$ **23.** $\frac{1}{6}$ **25.** $\frac{2}{3}$ **27.** $\frac{1}{13}$ **29.** $\frac{40}{51}$

31. $\frac{8}{16{,}575}$ **33.** $\frac{11}{850}$ **35.** $\frac{2}{17}$ **37.** **(a)** $\frac{47}{100}$; **(b)** $\frac{27}{47}$

39. **(a)** $\frac{3}{4}$; **(b)** $\frac{3}{5}$ **41.** $\frac{9}{20}$ **43.** $\frac{1}{4}$ **45.** $\frac{1}{25}$

47. 0.049 **49.** **(a)** 0.06; **(b)** 0.155 **51.** $\frac{4}{31}$

EXERCISE 8.6 (page 445)

1. **(a)** $\frac{1}{4}$; **(b)** $\frac{5}{6}$; **(c)** $\frac{1}{3}$; **(d)** $\frac{2}{3}$; **(e)** $\frac{1}{12}$; **(f)** $\frac{1}{2}$; **(g)** $\frac{1}{3}$

3. $\frac{5}{6}$ **5.** Independent **7.** Independent

9. Dependent **11.** Dependent

13. **(a)** Independent; **(b)** dependent; **(c)** dependent; **(d)** no **15.** Dependent **17.** $\frac{1}{18}$ **19.** $\frac{1}{25}$

21. $\frac{3}{676}$ **23.** **(a)** $\frac{3}{10}$; **(b)** $\frac{1}{40}$; **(c)** $\frac{1}{10}$

25. **(a)** $\frac{2}{5}$; **(b)** $\frac{1}{5}$; **(c)** $\frac{7}{15}$; **(d)** $\frac{13}{15}$; **(e)** $\frac{2}{15}$

27. **(a)** $\frac{2}{15}$; **(b)** $\frac{4}{15}$ **29.** $\frac{139}{361}$ **31.** $\frac{3}{200}$

33. **(a)** $\frac{1}{1728}$; **(b)** $\frac{3}{8}$ **35.** **(a)** $\frac{15}{1024}$; **(b)** $\frac{1}{64}$; **(c)** $\frac{53}{512}$

37. 0.0106

EXERCISE 8.7 (page 454)

1. $P(E \mid D) = \frac{1}{4}, P(F \mid D') = \frac{4}{7}$ **3.** $\frac{175}{386} \approx 0.453.$

5. **(a)** $\frac{258}{937} \approx 0.275$; **(b)** $\frac{14}{3021} \approx 0.005$ **7.** $\frac{5}{8}$ **9.** $\frac{6}{31}$

11. $\frac{81}{89} \approx 0.910$ **13.** $\frac{27}{49} \approx 55.1\%$ **15.** $\frac{3}{4}$

17. $\frac{24}{29} \approx 0.828$ **19.** $\frac{4}{5}$ **21.** $\frac{14}{15} \approx 0.933$

23. **(a)** $\frac{41}{200} = 0.205$; **(b)** $\frac{24}{41} \approx 0.585$; **(c)** $\frac{23}{200} = 0.115$

25. **(a)** 0.18; **(b)** 0.23; **(c)** 0.59; **(d)** high quality

27. $\frac{7}{9} \approx 0.78$

REVIEW PROBLEMS—CHAPTER 8 (page 460)

1. 336 **3.** 36 **5.** 608,400 **7.** 32 **9.** 210
11. 462 **13.** **(a)** 2024; **(b)** 253 **15.** 34,650
17. 560 **19.** **(a)** $\{1, 2, 3, 4, 5, 6, 7\}$; **(b)** $\{4, 5, 6\}$; **(c)** $\{4, 5, 6, 7, 8\}$; **(d)** $\varnothing$; **(e)** $\{4, 5, 6, 7, 8\}$; **(f)** no
21. **(a)** $\{R_1R_2R_3, R_1R_2G_3, R_1G_2R_3, R_1G_2G_3, G_1R_2R_3, G_1R_2G_3, G_1G_2R_3, G_1G_2G_3\}$; **(b)** $\{R_1R_2G_3, R_1G_2R_3, G_1R_2R_3\}$; **(c)** $\{R_1R_2R_3, G_1G_2G_3\}$

23. 0.2 **25.** $\frac{45}{512}$ **27.** **(a)** $\frac{4}{25}$; **(b)** $\frac{2}{15}$

29. **(a)** $\frac{1}{4}$; **(b)** $\frac{1}{4}$ **31.** 3:5 **33.** $\frac{6}{7}$ **35.** $\frac{10}{13}$

37. 0.42 **39.** **(a)** $\frac{2}{11}$; **(b)** $\frac{1}{18}$ **41.** $\frac{1}{4}$

43. **(a)** $\frac{1}{3}$; **(b)** independent **45.** Dependent

47. **(a)** 0.0081; **(b)** 0.2646; **(c)** 0.3483

49. $\frac{22}{45}$ **51.** $\frac{1}{4}$ **53.** **(a)** 0.014; **(b)** $\frac{4}{7} \approx 0.57$

MATHEMATICAL SNAPSHOT—CHAPTER 8 (page 464)

1. ≈ 0.645

EXERCISE 9.1 (page 473)

1. $\mu = 1.7$; $\text{Var}(X) = 1.01$; $\sigma \approx 1.00$

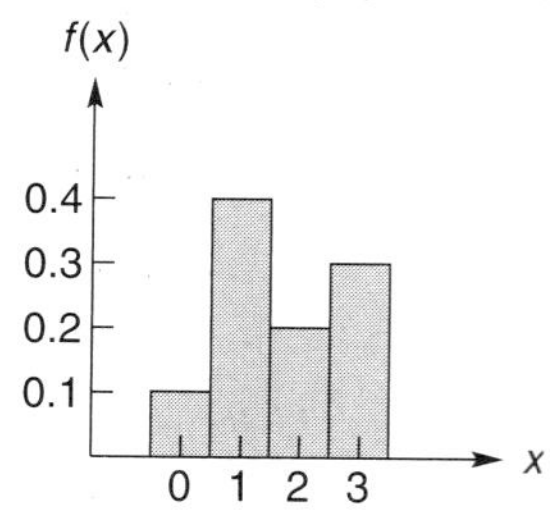

3. $\mu = \frac{9}{4} = 2.25$; $\text{Var}(X) = \frac{11}{16} = 0.6875$; $\sigma \approx 0.83$

5. **(a)** 0.1; **(b)** 5.8; **(c)** 1.56

7. $E(X) = \frac{3}{2} = 1.5$; $\sigma^2 = \frac{3}{4} = 0.75$; $\sigma \approx 0.87$

9. $E(X) = \frac{6}{5} = 1.2$; $\sigma^2 = \frac{9}{25} = 0.36$; $\sigma = \frac{3}{5} = 0.6$

11. $f(0) = \frac{1}{10}, f(1) = \frac{3}{5}, f(2) = \frac{3}{10}$

13. **(a)** −\$0.15 (a loss); **(b)** −\$0.30 (a loss) **15.** \$101.43
17. \$3.00 **19.** \$410 **21.** Loss of \$0.25; \$1

APPLICATIONS IN PRACTICE 9.2

1.

x	$P(x)$
0	$\frac{2401}{10{,}000}$
1	$\frac{4116}{10{,}000}$
2	$\frac{2646}{10{,}000}$
3	$\frac{756}{10{,}000}$
4	$\frac{81}{10{,}000}$

EXERCISE 9.2 (page 479)

1. $f(0) = \frac{9}{16}, f(1) = \frac{3}{8}, f(2) = \frac{1}{16}; \mu = \frac{1}{2}; \sigma = \frac{\sqrt{6}}{4}$

3. $f(0) = \frac{1}{27}, f(1) = \frac{2}{9}, f(2) = \frac{4}{9}, f(3) = \frac{8}{27}; \mu = 2;$ $\sigma = \frac{\sqrt{6}}{3}$ **5.** 0.001536 **7.** $\frac{96}{625} = 0.1536$ **9.** $\frac{3}{16}$

11. $\frac{165}{2048} \approx 0.081$ **13.** $\frac{1225}{3456} \approx 0.3545$ **15.** 0.002

17. (a) $\frac{9}{64}$; **(b)** $\frac{5}{32}$ **19.** $\frac{2048}{3125} \approx 0.655$ **21.** 0.7599

23. $\frac{13}{16}$ **25.** $\frac{2187}{8192} \approx 0.267$

EXERCISE 9.3 (page 489)

1. No **3.** No **5.** Yes **7.** $a = \frac{1}{3}, b = \frac{3}{4}$

9. $a = 0.3, b = 0.6, c = 0.1$ **11.** Yes **13.** No

15. $\mathbf{X}_1 = \begin{bmatrix} \frac{11}{12} & \frac{1}{12} \end{bmatrix}, \mathbf{X}_2 = \begin{bmatrix} \frac{25}{36} & \frac{11}{36} \end{bmatrix}, \mathbf{X}_3 = \begin{bmatrix} \frac{83}{108} & \frac{25}{108} \end{bmatrix}$

17. $\mathbf{X}_1 = [0.42 \quad 0.58], \mathbf{X}_2 = [0.416 \quad 0.584],$ $\mathbf{X}_3 = [0.4168 \quad 0.5832]$

19. $\mathbf{X}_1 = [0.26 \quad 0.28 \quad 0.46], \mathbf{X}_2 = [0.164 \quad 0.302 \quad 0.534],$ $\mathbf{X}_3 = [0.1766 \quad 0.3138 \quad 0.5096]$

21. (a) $\mathbf{T}^2 = \begin{bmatrix} \frac{5}{8} & \frac{3}{8} \\ \frac{3}{8} & \frac{5}{8} \end{bmatrix}, \mathbf{T}^3 = \begin{bmatrix} \frac{7}{16} & \frac{9}{16} \\ \frac{9}{16} & \frac{7}{16} \end{bmatrix}$; **(b)** $\frac{3}{8}$; **(c)** $\frac{9}{16}$

23. (a) $\mathbf{T}^2 = \begin{bmatrix} 0.50 & 0.40 & 0.10 \\ 0.23 & 0.69 & 0.08 \\ 0.27 & 0.54 & 0.19 \end{bmatrix},$

$\mathbf{T}^3 = \begin{bmatrix} 0.230 & 0.690 & 0.080 \\ 0.369 & 0.530 & 0.101 \\ 0.327 & 0.543 & 0.130 \end{bmatrix}$; **(b)** 0.40; **(c)** 0.369

25. $\begin{bmatrix} \frac{4}{7} & \frac{3}{7} \end{bmatrix}$ **27.** $\begin{bmatrix} \frac{3}{7} & \frac{4}{7} \end{bmatrix}$ **29.** $[0.5 \quad 0.25 \quad 0.25]$

31. (a) $\begin{array}{c} \\ \text{Flu} \\ \text{No Flu} \end{array} \begin{array}{c} \text{Flu} \quad \text{No Flu} \\ \begin{bmatrix} 0.1 & 0.9 \\ 0.2 & 0.8 \end{bmatrix} \end{array}$; **(b)** 37, 36

33. (a) $\begin{array}{c} \\ \text{A} \\ \text{B} \end{array} \begin{array}{c} \text{A} \quad \text{B} \\ \begin{bmatrix} 0.9 & 0.1 \\ 0.2 & 0.8 \end{bmatrix} \end{array}$; **(b)** 0.781

35. (a) $\begin{array}{c} \\ \text{D} \\ \text{R} \\ \text{O} \end{array} \begin{array}{c} \text{D} \quad \text{R} \quad \text{O} \\ \begin{bmatrix} 0.8 & 0.1 & 0.1 \\ 0.1 & 0.8 & 0.1 \\ 0.3 & 0.2 & 0.5 \end{bmatrix} \end{array}$; **(b)** 0.19; **(c)** 40%

37. (a) $\begin{array}{c} \\ \text{A} \\ \text{Compet.} \end{array} \begin{array}{c} \text{A} \quad \text{Compet.} \\ \begin{bmatrix} 0.8 & 0.2 \\ 0.3 & 0.7 \end{bmatrix} \end{array}$; **(b)** 65%; **(c)** 60%

39. (a) $\begin{array}{c} \\ 1 \\ 2 \end{array} \begin{array}{c} 1 \quad 2 \\ \begin{bmatrix} \frac{5}{7} & \frac{2}{7} \\ \frac{3}{7} & \frac{4}{7} \end{bmatrix} \end{array}$;

(b) 59.18% in compartment 1, 40.82% in compartment 2;
(c) 60% in compartment 1, 40% in compartment 2

41. (a) $\begin{bmatrix} \frac{2}{3} & \frac{1}{3} \end{bmatrix}$; **(b)** $33\frac{1}{3}\%$

REVIEW PROBLEMS—CHAPTER 9 (page 492)

1. $\mu = 1.5, \text{Var}(X) = 0.65, \sigma \approx 0.81$

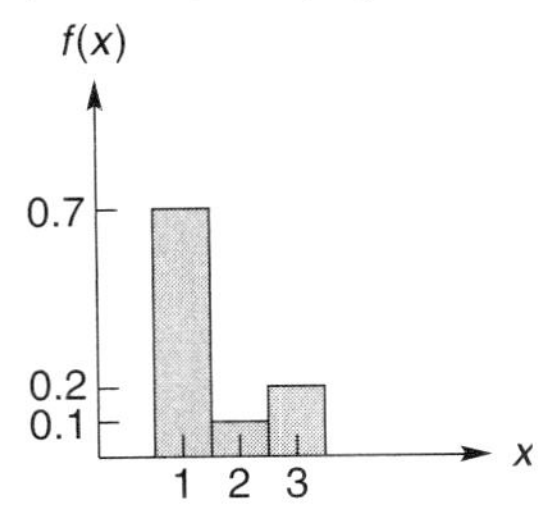

3. (a) $f(1) = \frac{1}{12}, f(2) = f(3) = f(4) = f(5)$ $= f(6) = \frac{1}{6}, f(7) = \frac{1}{12}$; **(b)** 4 **5.** $-\$0.10$ (a loss)

7. (a) \$176; **(b)** \$704,000

9. $f(0) = 0.522, f(1) = 0.368, f(2) = 0.098,$ $f(3) = 0.011, f(4) = 0.0005, \mu = 0.6; \sigma \approx 0.71$

11. $\frac{1}{64}$ **13.** $\frac{8}{81}$ **15.** $\frac{11}{27}$

17. $a = 0.3, b = 0.2, c = 0.5$

19. $\mathbf{X}_1 = [0.10 \quad 0.15 \quad 0.75], \mathbf{X}_2 = [0.130 \quad 0.155 \quad 0.715],$ $\mathbf{X}_3 = [0.1310 \quad 0.1595 \quad 0.7095]$

21. (a) $\mathbf{T}^2 = \begin{bmatrix} \frac{19}{49} & \frac{30}{49} \\ \frac{15}{49} & \frac{34}{49} \end{bmatrix}, \mathbf{T}^3 = \begin{bmatrix} \frac{109}{343} & \frac{234}{343} \\ \frac{117}{343} & \frac{226}{343} \end{bmatrix}$; **(b)** $\frac{30}{49}$;

(c) $\frac{117}{343}$ **23.** $\begin{bmatrix} \frac{1}{2} & \frac{1}{2} \end{bmatrix}$

25. (a) 76%; **(b)** 74.4% Japanese, 25.6% non-Japanese;

(c) 75% Japanese, 25% non-Japanese

MATHEMATICAL SNAPSHOT—CHAPTER 9 (page 494)

1. 7

3. Against Always Defect: $\begin{bmatrix} 0 & 1 & 0 & 0 \\ 0 & 0.1 & 0 & 0.9 \\ 0 & 1 & 0 & 0 \\ 0 & 0.1 & 0 & 0.9 \end{bmatrix}$;

Against Always Cooperate: $\begin{bmatrix} 1 & 0 & 0 & 0 \\ 0.1 & 0 & 0.9 & 0 \\ 1 & 0 & 0 & 0 \\ 0.1 & 0 & 0.9 & 0 \end{bmatrix}$;

Against regular Tit-for-tat: $\begin{bmatrix} 1 & 0 & 0 & 0 \\ 0.1 & 1 & 0.9 & 0 \\ 0 & 1 & 0 & 0 \\ 0 & 0.1 & 0 & 0.9 \end{bmatrix}$

APPLICATIONS IN PRACTICE 10.1

1. The limit as $x \to a$ does not exist if a is an integer, but it exists if a is any other value.

2. $\frac{4}{3}\pi$ cc **3.** 3616 **4.** 20 **5.** 2

EXERCISE 10.1 (page 505)

1. (a) 1; **(b)** 0; **(c)** 1 **3. (a)** 1; **(b)** does not exist; **(c)** 3 **5.** $f(-0.9) = -3.7, f(-0.99) = -3.97, f(-0.999) = -3.997, f(-1.001) = -4.003, f(-1.01) = -4.03, f(-1.1) = -4.3; -4$

7. $f(-0.1) \approx 0.9516, f(-0.01) \approx 0.9950, f(-0.001) \approx 0.9995, f(0.001) \approx 1.0005, f(0.01) \approx 1.0050, f(0.1) \approx 1.0517; 1$

9. 16 **11.** 20 **13.** -1 **15.** $-\frac{5}{2}$ **17.** 0 **19.** 5

21. -2 **23.** 3 **25.** 0 **27.** $\frac{1}{6}$ **29.** $-\frac{1}{5}$ **31.** $\frac{11}{9}$

33. 4 **35.** $2x$ **37.** -1 **39.** $2x$ **41.** $3x^2 - 8x$

43. $\frac{1}{4}$ **45. (a)** 1; **(b)** 0 **47.** 11.00 **49.** -7.00

51. Does not exist

APPLICATIONS IN PRACTICE 10.2

1. $\lim_{x\to\infty} p(x) = 0$. The graph starts out high and quickly goes down toward zero. Accordingly, consumers are willing to purchase large quantities of the product at prices close to 0.

2. $\lim_{x\to\infty} y(x) = 500$. The greatest yearly sales they can expect with unlimited advertising is $500,000.

3. $\lim_{x\to\infty} C(x) = \infty$. This means that the cost continues to increase without bound as more units are made.

4. The limit does not exist; $250.

EXERCISE 10.2 (page 514)

1. (a) 2; **(b)** 3; **(c)** does not exist; **(d)** $-\infty$; **(e)** ∞; **(f)** ∞; **(g)** ∞; **(h)** 0; **(i)** 1; **(j)** 1; **(k)** 1 **3.** 1

5. $-\infty$ **7.** $-\infty$ **9.** ∞ **11.** 0

13. Does not exist **15.** 0 **17.** ∞ **19.** 0 **21.** 1

23. 0 **25.** ∞ **27.** 0 **29.** $-\frac{2}{5}$ **31.** $-\infty$

33. $\frac{2}{5}$ **35.** $-\infty$ **37.** $\frac{11}{5}$ **39.** $-\frac{1}{2}$ **41.** ∞

43. ∞ **45.** ∞ **47.** Does not exist **49.** $-\infty$

51. 0 **53.** 1

55. (a) 1; **(b)** 2; **(c)** does not exist; **(d)** 1; **(e)** 2

57. (a) 0; **(b)** 0; **(c)** 0; **(d)** $-\infty$; **(e)** $-\infty$

59.

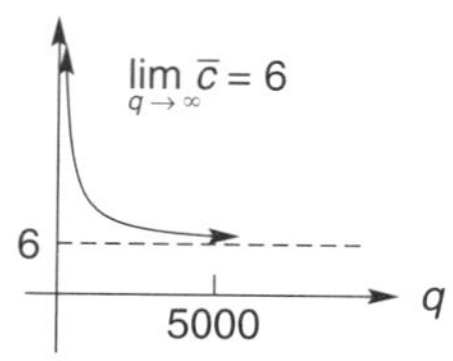

61. 20,000 **63.** 20

65. 1, 0.5, 0.525, 0.631, 0.912, 0.986, 0.998; conclude limit is 1

67. 0 **69. (a)** 11; **(b)** 9; **(c)** does not exist

EXERCISE 10.3 (page 519)

1. $5819.97; $1819.97 **3.** $1456.87 **5.** 4.08%

7. 3.05% **9.** $109.42 **11.** $778,800.78

13. (a) $39,066; **(b)** $13,671 **15.** $4.88%

17. $1264 **19.** 16 years

21. (a) $1072.51; **(b)** $1093.30; **(c)** $1072.18

23. (a) $9458.51; **(b)** This strategy is better by $26.90.

EXERCISE 10.4 (page 526)

7. Continuous at -2 and 0 **9.** Discontinuous at ± 4

11. Continuous at 2 and 0 **13.** f is a polynomial function

15. f is a rational function and the denominator is never zero.

17. None **19.** $x = -4$ **21.** None **23.** $x = -5, 3$

25. $x = 0, \pm 1$ **27.** None **29.** $x = 0$ **31.** None

33. $x = 2$

35. Discontinuities at $t = 1, 2, 3, 4$

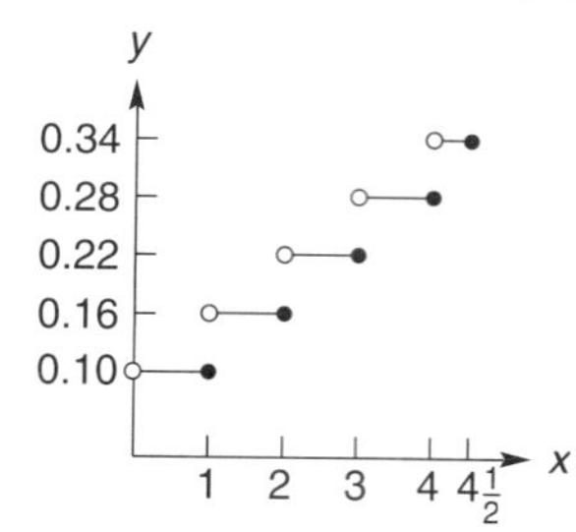

37. Yes, no, no

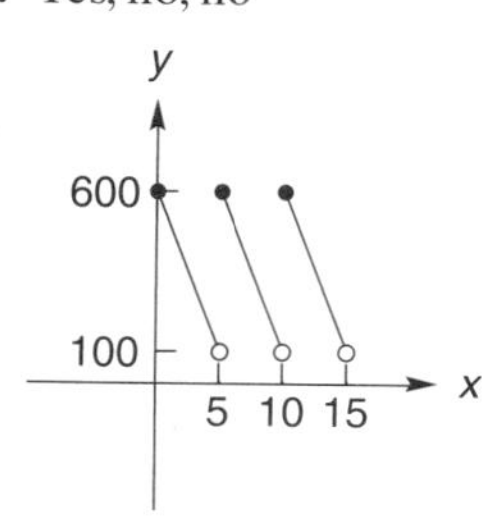

APPLICATIONS IN PRACTICE 10.5

1. $0 < x < 4$

EXERCISE 10.5 (page 530)

1. $(-\infty, -1), (4, \infty)$ **3.** $[3, 5]$ **5.** $\left(-\frac{7}{2}, -2\right)$
7. No solution **9.** $(-\infty, -6], [-2, 3]$
11. $(-\infty, -4), (0, 5)$ **13.** $[0, \infty)$ **15.** $(-3, 0), (1, \infty)$
17. $(-\infty, -3), (0, 3)$ **19.** $(1, \infty)$
21. $(-\infty, -5), [-2, 1), [3, \infty)$ **23.** $(-5, -1)$
25. $(-\infty, -1 - \sqrt{3}], [-1 + \sqrt{3}, \infty)$
27. Between 50 and 65 inclusive **29.** 17 in. by 17 in.
31. $(-\infty, -7.72]$ **33.** $(-\infty, -0.5), (0.667, \infty)$

REVIEW PROBLEMS—CHAPTER 10 (page 532)

1. -5 **3.** 2 **5.** x **7.** $-\frac{8}{3}$ **9.** 0 **11.** $\frac{5}{9}$
13. Does not exist **15.** -1 **17.** $\frac{1}{9}$ **19.** $-\infty$
21. ∞ **23.** $-\infty$ **25.** 1 **27.** $-\infty$ **29.** 8
31. 23 **33.** **(a)** \$5034.38; **(b)** \$1241.46 **35.** 6.18%
37. $\frac{\ln 2}{0.065}$
41. Continuous everywhere; f is a polynomial function.
43. $x = -3$ **45.** None **47.** $x = -4, 1$
49. $x = -2$ **51.** $(-\infty, -6), (2, \infty)$
53. $[2, \infty), x = 0$ **55.** $(-\infty, -5), (-1, 1)$
57. $(-\infty, -4), [-3, 0], (2, \infty)$ **59.** 1.00
61. 0 **63.** $[2.00, \infty)$

MATHEMATICAL SNAPSHOT—CHAPTER 10 (page 534)

1. 17%
3. An exponential model assumes a fixed repayment rate.

APPLICATIONS IN PRACTICE 11.1

1. $\frac{dH}{dt} = 40 - 32t$

EXERCISE 11.1 (page 544)

1. (a)

x-value of Q	3	2.5	2.2	2.1	2.01	2.001
m_{PQ}	19	15.25	13.24	12.61	12.0601	12.0060

(b) We estimate that $m_{\tan} = 12$.
3. 1 **5.** 3 **7.** -4 **9.** 0 **11.** $2x + 4$
13. $4q + 5$ **15.** $-\frac{6}{x^2}$ **17.** $\frac{1}{2\sqrt{x+2}}$ **19.** -4
21. 0 **23.** $y = x + 4$ **25.** $y = -3x - 7$
27. $y = -3x + 9$ **29.** $\dfrac{r}{r_L - r - \dfrac{dC}{dD}}$
31. $-3.000, 13.445$ **33.** $-5.120, 0.038$

35. For the x-values of the points where the tangent to the graph of f is horizontal, the corresponding values of $f'(x)$ are 0. This is expected because the slope of a horizontal line is zero and the derivative gives the slope of the tangent line.

APPLICATIONS IN PRACTICE 11.2

1. $50 - 0.6q$

EXERCISE 11.2 (page 551)

1. 0 **3.** $6x^5$ **5.** $80x^{79}$ **7.** $18x$ **9.** $20w^4$
11. $\frac{8}{3}x^3$ **13.** $\frac{7}{25}t^6$ **15.** 1 **17.** $8x - 2$
19. $4p^3 - 9p^2$ **21.** $-8x^7 + 5x^4$
23. $-39x^2 + 28x - 2$ **25.** $-8x^3$ **27.** $-\frac{4}{3}x^3$
29. $16x^3 + 3x^2 - 9x + 8$ **31.** $\frac{6}{5}x^3 + 7x^2$ **33.** $\frac{7}{2}x^{5/2}$
35. $\frac{3}{4}x^{-1/4} + \frac{10}{3}x^{2/3}$ **37.** $\frac{11}{2}x^{-1/2}$ or $\frac{11}{2\sqrt{x}}$ **39.** $2r^{-2/3}$
41. $-4x^{-5}$ **43.** $-3x^{-4} - 5x^{-6} + 12x^{-7}$
45. $-x^{-2}$ or $-\frac{1}{x^2}$ **47.** $-40x^{-6}$ **49.** $-4x^{-4}$
51. $-\frac{1}{2}t^{-2}$ **53.** $\frac{1}{7} - 7x^{-2}$ **55.** $-3x^{-2/3} - 2x^{-7/5}$
57. $-\frac{1}{5}x^{-6/5}$ **59.** $-x^{-3/2}$ **61.** $\frac{5}{2}x^{3/2}$
63. $9x^2 - 20x + 7$ **65.** $45x^4$
67. $\frac{1}{3}x^{-2/3} - \frac{10}{3}x^{-5/3} = \frac{1}{3}x^{-5/3}(x - 10)$ **69.** $8q + \frac{4}{q^2}$
71. $2x + 4$ **73.** 1 **75.** $4, 16, -14$ **77.** $0, 0, 0$
79. $y = 13x + 2$ **81.** $y = -4x + 6$
83. $y = x + 3$ **85.** $(0, 0), \left(\frac{5}{3}, \frac{125}{54}\right)$ **87.** $(3, -3)$
89. 0 **91.** The tangent line is $y = 9x - 16$.

APPLICATIONS IN PRACTICE 11.3

1. 2.5 units **2.** $\frac{dy}{dt} = 16 - 32t; \left.\frac{dy}{dt}\right|_{t=0.5} = 0$ feet/s
When $t = 0.5$ the object reaches its maximum height.
3. 1.2 and 120%

EXERCISE 11.3 (page 561)

1.

Δt	1	0.5	0.2	0.1	0.01	0.001
$\Delta s/\Delta t$	9	8	7.4	7.2	7.02	7.002

We estimate the velocity when $t = 1$ to be 7.0000 m/s.
With differentiation the velocity is 7 m/s.
3. **(a)** 4 m; **(b)** 5.5 m/s; **(c)** 5 m/s
5. **(a)** 8 m; **(b)** 6.1208 m/s; **(c)** 6 m/s
7. **(a)** 2 m; **(b)** 10.261 m/s; **(c)** 9 m/s
9. $\frac{dy}{dx} = \frac{25}{2}x^{3/2}$; 337.50 **11.** 0.27
13. $dc/dq = 10$; 10 **15.** $dc/dq = 0.6q + 2$; 3.8
17. $dc/dq = 2q + 50$; 80, 82, 84
19. $dc/dq = 0.02q + 5$; 6, 7

21. $dc/dq = 0.00006q^2 - 0.02q + 6$; 4.6, 11
23. $dr/dq = 0.8$; 0.8, 0.8, 0.8
25. $dr/dq = 250 + 90q - 3q^2$; 625, 850, 625
27. $dc/dq = 6.750 - 0.000656q$; 3.47
29. $P = 5{,}000{,}000R^{-0.93}$; $dP/dR = -4{,}650{,}000R^{-1.93}$
31. **(a)** -7.5; **(b)** 4.5
33. **(a)** 1; **(b)** $\frac{1}{x+4}$; **(c)** 1; **(d)** $\frac{1}{9} \approx 0.111$; **(e)** 11.1%
35. **(a)** $6x$; **(b)** $\frac{6x}{3x^2+7}$; **(c)** 12; **(d)** $\frac{12}{19} \approx 0.632$; **(e)** 63.2%
37. **(a)** $-3x^2$; **(b)** $-\frac{3x^2}{8-x^3}$; **(c)** -3; **(d)** $-\frac{3}{7} \approx -0.429$; **(e)** -42.9%
39. 3.2; 21.3%
41. **(a)** $dr/dq = 30 - 0.6q$; **(b)** $\frac{4}{45} \approx 0.089$; **(c)** 9%
43. $\frac{0.432}{t}$ **45.** \$3125 **47.** \$5.07/unit

APPLICATIONS IN PRACTICE 11.5

1. $\frac{dR}{dx} = 6.25 - 6x$ **2.** $T'(x) = 2x - x^2$; $T'(1) = 1$

EXERCISE 11.5 (page 573)

1. $(4x+1)(6) + (6x+3)(4) = 48x + 18 = 6(8x+3)$
3. $(5-3t)(3t^2-4t) + (t^3-2t^2)(-3) = -12t^3 + 33t^2 - 20t$
5. $(3r^2-4)(2r-5) + (r^2-5r+1)(6r) = 12r^3 - 45r^2 - 2r + 20$
7. $8x^3 - 10x$
9. $(x^2+3x-2)(4x-1) + (2x^2-x-3)(2x+3) = 8x^3 + 15x^2 - 20x - 7$
11. $(8w^2+2w-3)(15w^2) + (5w^3+2)(16w+2) = 200w^4 + 40w^3 - 45w^2 + 32w + 4$
13. $(x^2-1)(9x^2-6) + (3x^3-6x+5)(2x) - 4(8x+2) = 15x^4 - 27x^2 - 22x - 2$
15. $\frac{3}{2}\left[(5p^{\frac{1}{2}}-2)(3) + (3p-1)\left(5\cdot\frac{1}{2}p^{-\frac{1}{2}}\right)\right] = \frac{3}{4}\left(45p^{\frac{1}{2}} - 12 - 5p^{-\frac{1}{2}}\right)$
17. 0 **19.** $18x^2 + 94x + 31$
21. $\frac{(x-1)(5)-(5x)(1)}{(x-1)^2} = -\frac{5}{(x-1)^2}$
23. $-\frac{9}{x^7}$ **25.** $\frac{(x-1)(1)-(x+2)(1)}{(x-1)^2} = -\frac{3}{(x-1)^2}$
27. $\frac{(z^2-4)(-2)-(6-2z)(2z)}{(z^2-4)^2} = \frac{2(z^2-6z+4)}{(z^2-4)^2}$
29. $\frac{(x^2-5x)(16x-2)-(8x^2-2x+1)(2x-5)}{(x^2-5x)^2} = \frac{-38x^2-2x+5}{(x^2-5x)^2}$
31. $\frac{(2x^2-3x+2)(2x-4)-(x^2-4x+3)(4x-3)}{(2x^2-3x+2)^2} = \frac{5x^2-8x+1}{(2x^2-3x+2)^2}$
33. $-\frac{100x^{99}}{(x^{100}+7)^2}$ **35.** $4v^3 + \frac{8}{v^2}$
37. $\frac{15x^2-2x+1}{3x^{4/3}}$ **39.** $\frac{4}{(x-8)^2} + \frac{2}{(3x+1)^2}$
41. $\frac{[(x+2)(x-4)](1)-(x-5)(2x-2)}{[(x+2)(x-4)]^2} = \frac{-(x^2-10x+18)}{[(x+2)(x-4)]^2}$
43. $\frac{[(t^2-1)(t^3+7)](2t+3)-(t^2+3t)(5t^4-3t^2+14t)}{[(t^2-1)(t^3+7)]^2} = \frac{-3t^6-12t^5+t^4+6t^3-21t^2-14t-21}{[(t^2-1)(t^3+7)]^2}$
45. $3 - \frac{2x^3+3x^2-12x+4}{[x(x-1)(x-2)]^2}$ **47.** $-\frac{2a}{(a+x)^2}$
49. -6 **51.** $y = -\frac{3}{2}x + \frac{15}{2}$ **53.** $y = 16x + 24$
55. 1.5 **57.** 1 m, -1.5 m/s **59.** $\frac{dr}{dq} = 25 - 0.04q$
61. $\frac{dr}{dq} = \frac{216}{(q+2)^2} - 3$ **63.** $\frac{dC}{dI} = 0.672$
65. $\frac{7}{6}; \frac{1}{6}$ **67.** 0.615; 0.385 **69.** **(a)** 0.32; **(b)** 0.026
71. $\frac{dc}{dq} = \frac{5q(q+6)}{(q+3)^2}$ **73.** $\frac{9}{10}$ **75.** $\frac{0.7355}{(1+0.02744x)^2}$
77. $-\frac{1}{120}$

APPLICATIONS IN PRACTICE 11.6

1. $288t$

EXERCISE 11.6 (page 581)

1. $(2u-2)(2x-1) = 4x^3 - 6x^2 - 2x + 2$
3. $\left(-\frac{2}{w^3}\right)(-1) = \frac{2}{(2-x)^3}$ **5.** -2 **7.** 0
9. $18(3x+2)^5$ **11.** $30x^2(3+2x^3)^4$
13. $200(3x^2-16x+1)(x^3-8x^2+x)^{99}$
15. $-6x(x^2-2)^{-4}$
17. $-10(4x-3)(2x^2-3x-1)^{-13/3}$
19. $\frac{1}{2}(10x-1)(5x^2-x)^{-1/2}$ **21.** $\frac{1}{2}(2x-1)^{-3/4}$
23. $\frac{12}{5}x^2(x^3+1)^{-3/5}$ **25.** $-6(4x-1)(2x^2-x+1)^{-2}$
27. $-2(2x-3)(x^2-3x)^{-3}$ **29.** $-8(8x-1)^{-3/2}$
31. $\frac{7}{3}(7x)^{-2/3} + \sqrt[3]{7}$
33. $(x^2)[5(x-4)^4(1)] + (x-4)^5(2x) = x(x-4)^4(7x-8)$
35. $4x^2\left[\frac{1}{2}(5x+1)^{-\frac{1}{2}}(5)\right] + (\sqrt{5x+1})(8x) = 10x^2(5x+1)^{-\frac{1}{2}} + 8x\sqrt{5x+1}$
37. $(x^2+2x-1)^3(5) + (5x)[3(x^2+2x-1)^2(2x+2)] = 5(x^2+2x-1)^2(7x^2+8x-1)$

39. $(8x-1)^3[4(2x+1)^3(2)]+(2x+1)^4[3(8x-1)^2(8)]$
$=16(8x-1)^2(2x+1)^3(7x+1)$

41. $10\left(\frac{x-7}{x+4}\right)^9\left[\frac{(x+4)(1)-(x-7)(1)}{(x+4)^2}\right]$
$=\frac{110(x-7)^9}{(x+4)^{11}}$

43. $\frac{1}{2}\left(\frac{x-2}{x+3}\right)^{-1/2}\left[\frac{(x+3)(1)-(x-2)(1)}{(x+3)^2}\right]$
$=\frac{5}{2(x+3)^2}\left(\frac{x-2}{x+3}\right)^{-1/2}$

45. $\frac{(x^2+4)^3(2)-(2x-5)[3(x^2+4)^2(2x)]}{(x^2+4)^6}$
$=\frac{-2(5x^2-15x-4)}{(x^2+4)^4}$

47. $\frac{(3x-1)^3[40(8x-1)^4]-(8x-1)^5[9(3x-1)^2]}{(3x-1)^6}$
$=\frac{(8x-1)^4(48x-31)}{(3x-1)^4}$

49. $6\{(5x^2+2)[2x^3(x^4+5)^{-1/2}]+(x^4+5)^{1/2}(10x)\}$
$=12x(x^4+5)^{-1/2}(10x^4+2x^2+25)$

51. $8+\frac{5}{(t+4)^2}-(8t-7)=15-8t+\frac{5}{(t+4)^2}$

53. $\frac{(x^2-7)^4[(2x+1)(2)(3x-5)(3)+(3x-5)^2(2)]-(2x+1)(3x-5)^2[4(x^2-7)^3(2x)]}{(x^2-7)^8}$

55. 0 **57.** 0 **59.** $y=4x-11$

61. $y=-\frac{1}{6}x+\frac{5}{3}$ **63.** 96% **65.** 20 **67.** ≈ 13.99

69. **(a)** $-\frac{q}{\sqrt{q^2+20}}$; **(b)** $-\frac{q}{100\sqrt{q^2+20}-q^2-20}$;
(c) $100-\frac{q^2}{\sqrt{q^2+20}}-\sqrt{q^2+20}$

71. -481.5 **73.** $\frac{dc}{dq}=\frac{5q(q^2+6)}{(q^2+3)^{3/2}}$ **75.** $48\pi(10)^{-19}$

77. **(a)** $-0.001416x^3+0.01356x^2+1.696x-34.9$, -136.188; **(b)** -0.008 **79.** -4 **81.** 40

83. 86,111.37

REVIEW PROBLEMS—CHAPTER 11 (page 585)

1. $-2x$ **3.** $\frac{\sqrt{3}}{2\sqrt{x}}$ **5.** 0

7. $28x^3-18x^2+10x=2x(14x^2-9x+5)$

9. $4s^3+4s=4s(s^2+1)$ **11.** $\frac{2x}{5}$

13. $(x^2+6x)(3x^2-12x)+(x^3-6x^2+4)(2x+6)$
$=5x^4-108x^2+8x+24$

15. $100(2x^2+4x)^{99}(4x+4)$
$=400(x+1)(2x^2+4x)^{99}$

17. $-\frac{6}{(2x+1)^2}$

19. $(8+2x)(4)(x^2+1)^3(2x)+(x^2+1)^4(2)$
$=2(x^2+1)^3(9x^2+32x+1)$

21. $\frac{(z^2+4)(2z)-(z^2-1)(2z)}{(z^2+4)^2}=\frac{10z}{(z^2+4)^2}$

23. $\frac{4}{3}(4x-1)^{-2/3}$

25. $-\frac{1}{2}(1-x)^{-3/2}(-1)=\frac{1}{2}(1-x)^{-3/2}$

27. $(x-6)^4[3(x+5)^2]+(x+5)^3[4(x-6)^3]$
$=(x-6)^3(x+5)^2(7x+2)$

29. $\frac{(x+6)(5)-(5x-4)(1)}{(x+6)^2}=\frac{34}{(x+6)^2}$

31. $2\left(-\frac{3}{8}\right)x^{-11/8}+\left(-\frac{3}{8}\right)(2x)^{-11/8}(2)$
$=-\frac{3}{4}(1+2^{-11/8})x^{-11/8}$

33. $\frac{\sqrt{x^2+5}(2x)-(x^2+6)(1/2)(x^2+5)^{-1/2}(2x)}{x^2+5}$
$=\frac{x(x^2+4)}{(x^2+5)^{3/2}}$

35. $\left(\frac{3}{5}\right)(x^3+6x^2+9)^{-2/5}(3x^2+12x)$
$=\frac{9}{5}x(x+4)(x^3+6x^2+9)^{-2/5}$

37. $7(1-2z)$ **39.** $y=-4x+3$

41. $y=\frac{1}{12}x+\frac{4}{3}$ **43.** $\frac{5}{7}\approx 0.714$; 71.4%

45. $dr/dq=20-0.2q$ **47.** 0.569, 0.431

49. $dr/dq=450-q$

51. $dc/dq=0.125+0.00878q$; 0.7396

53. 84 eggs/mm **55.** **(a)** $\frac{4}{3}$; **(b)** $\frac{1}{24}$ **57.** 8π ft³/ft

59. $4q-\frac{10{,}000}{q^2}$ **61.** **(a)** 240; **(b)** $\frac{1}{100}$;
(c) no, since $dr/dm<300$ when $m=80$ **63.** 0.305

65. -0.32

MATHEMATICAL SNAPSHOT—CHAPTER 11 (page 588)

1. The slope is greater—above 0.9. More is spent; less is saved.
3. Spend $705, save $295 **5.** Answers may vary.

APPLICATIONS IN PRACTICE 12.1

1. $\frac{dq}{dp}=\frac{12p}{3p^2+4}$ **2.** $\frac{dR}{dI}=\frac{1}{I\ln 10}$

EXERCISE 12.1 (page 595)

1. $\frac{4}{x}$ **3.** $\frac{3}{3x-7}$ **5.** $\frac{2}{x}$ **7.** $-\frac{2x}{1-x^2}$

9. $\frac{6p^2+3}{2p^3+3p}=\frac{3(2p^2+1)}{p(2p^2+3)}$

11. $t\left(\frac{1}{t}\right)+(\ln t)=1+\ln t$

13. $\frac{2x^3}{2x+5}+3x^2\ln(2x+5)$ **15.** $\frac{8}{(\ln 3)(8x-1)}$

17. $2x\left[1+\frac{1}{(\ln 2)(x^2+4)}\right]$

19. $\frac{z\left(\frac{1}{z}\right)-(\ln z)(1)}{z^2}=\frac{1-\ln z}{z^2}$

21. $\dfrac{(\ln x)(2x) - (x^2 - 1)\left(\dfrac{1}{x}\right)}{(\ln x)^2} = \dfrac{2x^2 \ln(x) - x^2 + 1}{x \ln^2 x}$

23. $\dfrac{3(2x + 4)}{x^2 + 4x + 5} = \dfrac{6(x + 2)}{x^2 + 4x + 5}$ **25.** $\dfrac{9x}{1 + x^2}$

27. $\dfrac{2}{1 - l^2}$ **29.** $\dfrac{x}{1 - x^4}$ **31.** $\dfrac{4x}{x^2 + 2} + \dfrac{3x^2 + 1}{x^3 + x - 1}$

33. $\dfrac{5}{x} + \dfrac{5}{2x + 1}$ **35.** $\dfrac{2(x^2 + 1)}{2x + 1} + 2x \ln(2x + 1)$

37. $\dfrac{3(1 + \ln^2 x)}{x}$ **39.** $\dfrac{4 \ln^3(ax)}{x}$

41. $\dfrac{x}{2(x - 1)} + \ln \sqrt{x - 1}$ **43.** $\dfrac{3}{2x\sqrt{4 + 3 \ln x}}$

45. $y = 4x - 12$ **47.** $\dfrac{\ln(3) - 1}{\ln^2 3}$ **49.** $\dfrac{25}{7}$

51. $\dfrac{dq}{dp} = \dfrac{20}{2p + 1}$ **53.** $\dfrac{6a}{(T - a^2 + aT)(a - T)}$

57. 1.36

APPLICATIONS IN PRACTICE 12.2

1. $\dfrac{dT}{dt} = Cke^{kt}$

EXERCISE 12.2 (page 600)

1. $5e^x$ **3.** $2xe^{x^2+4}$ **5.** $-5e^{9-5x}$

7. $(6r + 4)e^{3r^2+4r+4} = 2(3r + 2)e^{3r^2+4r+4}$

9. $x(e^x) + e^x(1) = e^x(x + 1)$ **11.** $2xe^{-x^2}(1 - x^2)$

13. $\dfrac{e^x - e^{-x}}{3}$ **15.** $(6x)4^{3x^2} \ln 4$ **17.** $\dfrac{2e^{2w}(w - 1)}{w^3}$

19. $\dfrac{e^{1+\sqrt{x}}}{2\sqrt{x}}$ **21.** $5x^4 - 5^x \ln 5$ **23.** $\dfrac{2e^x}{(e^x + 1)^2}$

25. 1 **27.** $(1 + \ln x)e^{x \ln x}$ **29.** $-e$

31. $y - e^{-2} = e^{-2}(x + 2)$ or $y = e^{-2}x + 3e^{-2}$

33. $dp/dq = -0.015e^{-0.001q}, -0.015e^{-0.5}$

35. $dc/dq = 10e^{q/700}; 10e^{0.5}; 10e$ **37.** -5

39. e **41.** $100e^{-2}$ **47.** $-b(10^{A-bM}) \ln 10$

51. 0.0036 **53.** 0.68

EXERCISE 12.3 (page 607)

1. -3, elastic **3.** -1, unit elasticity

5. $-\dfrac{53}{52}$, elastic **7.** $-\left(\dfrac{150}{e} - 1\right)$, elastic

9. -1, unit elasticity **11.** $-\dfrac{9}{32}$, inelastic

13. $-\dfrac{1}{2}$, inelastic

15. $|\eta| = \dfrac{10}{3}$ when $p = 10$, $|\eta| = \dfrac{3}{10}$ when $p = 3$, $|\eta| = 1$ when $p = 6.50$ **17.** -1.2, 0.6% decrease

23. (b) $\eta = -2.5$, elastic; **(c)** 1 unit; **(d)** increase, since demand is elastic

25. (a) $\eta = -\dfrac{207}{15} = -13.8$, elastic; **(b)** 27.6%; **(c)** increase, since demand is elastic

27. $\eta = -1.6$; $\dfrac{dr}{dq} = 30$

29. Maximum at $q = 5$; minimum at $q = 95$

APPLICATIONS IN PRACTICE 12.4

1. $\dfrac{dP}{dt} = 0.5(P - P^2)$

2. $\dfrac{dV}{dt} = 4\pi r^2 \dfrac{dr}{dt}$ and $\left.\dfrac{dV}{dt}\right|_{r=12} = 2880\pi$ in^3/minute

3. The top of the ladder is sliding down at a rate of $\dfrac{9}{4}$ feet/second.

EXERCISE 12.4 (page 613)

1. $-\dfrac{x}{4y}$ **3.** $\dfrac{7}{3y^2}$ **5.** $-\dfrac{\sqrt{y}}{\sqrt{x}}$ **7.** $-\dfrac{y^{1/4}}{x^{1/4}}$ **9.** $-\dfrac{y}{x}$

11. $\dfrac{11 - y}{x - 1}$ **13.** $\dfrac{4y - 2x^2}{y^2 - 4x}$ **15.** $\dfrac{4y^{3/4}}{2y^{1/4} + 1}$

17. $\dfrac{1 - 6xy^3}{1 + 9x^2y^2}$ **19.** $\dfrac{xe^y - y}{x(\ln x - xe^y)}$ **21.** $-\dfrac{e^y}{xe^y + 1}$

23. $6e^{3x}(1 + e^{3x})(x + y) - 1$ **25.** $-\dfrac{3}{5}$

27. $0; -\dfrac{4x_0}{9y_0}$ **29.** $y = -\dfrac{3}{4}x + \dfrac{5}{4}$ **31.** $\dfrac{dq}{dp} = -\dfrac{1}{2q}$

33. $\dfrac{dq}{dp} = -\dfrac{(q + 5)^3}{40}$ **35.** $-\lambda I$ **37.** $-\dfrac{f}{\lambda}$ **39.** $\dfrac{3}{8}$

EXERCISE 12.5 (page 617)

1. $(x + 1)^2(x - 2)(x^2 + 3)\left[\dfrac{2}{x + 1} + \dfrac{1}{x - 2} + \dfrac{2x}{x^2 + 3}\right]$

3. $(3x^3 - 1)^2(2x + 5)^3\left[\dfrac{18x^2}{3x^3 - 1} + \dfrac{6}{2x + 5}\right]$

5. $\dfrac{\sqrt{x + 1}\sqrt{x^2 - 2}\sqrt{x + 4}}{2} \cdot \left[\dfrac{1}{x + 1} + \dfrac{2x}{x^2 - 2} + \dfrac{1}{x + 4}\right]$

7. $\dfrac{\sqrt{1 - x^2}}{1 - 2x}\left[\dfrac{x}{x^2 - 1} + \dfrac{2}{1 - 2x}\right]$

9. $\dfrac{(2x^2 + 2)^2}{(x + 1)^2(3x + 2)}\left[\dfrac{4x}{x^2 + 1} - \dfrac{2}{x + 1} - \dfrac{3}{3x + 2}\right]$

11. $\dfrac{1}{2}\sqrt{\dfrac{(x + 3)(x - 2)}{2x - 1}}\left[\dfrac{1}{x + 3} + \dfrac{1}{x - 2} - \dfrac{2}{2x - 1}\right]$

13. $x^{2x+1}\left(\dfrac{2x + 1}{x} + 2 \ln x\right)$ **15.** $\dfrac{x^{1/x}(1 - \ln x)}{x^2}$

17. $2(3x + 1)^{2x}\left[\dfrac{3x}{3x + 1} + \ln(3x + 1)\right]$

19. $4e^x x^{3x}(4 + 3 \ln x)$ **21.** 12 **23.** $y = 96x + 36$

25. $y = 6ex - 3e$ **27.** $\dfrac{1}{3e^{1.3}}$

APPLICATIONS IN PRACTICE 12.6

1. 43 and 1958

EXERCISE 12.6 (page 622)

1. 0.25410 **3.** 1.32472 **5.** −0.68233 **7.** 0.33767
9. 1.90785 **11.** 4.141 **13.** −4.99 and 1.94
15. 13.33 **17.** 2.880 **19.** 3.45

APPLICATIONS IN PRACTICE 12.7

1. $\frac{d^2h}{dt^2} = -32$ feet/sec^2 (*Note*: Negative values indicate the downward direction.)
2. $c''(3) = 14$ dollars/unit^2

EXERCISE 12.7 (page 626)

1. 24 **3.** 0 **5.** e^x **7.** $3 + 2\ln x$ **9.** $-\frac{10}{p^6}$
11. $-\frac{1}{4(9-r)^{3/2}}$ **13.** $\frac{8}{(2x+3)^3}$ **15.** $\frac{4}{(x-1)^3}$
17. $-\left[\frac{1}{x^2} + \frac{1}{(x+6)^2}\right]$ **19.** $e^z(z^2 + 4z + 2)$
21. 32 **23.** $-\frac{1}{y^3}$ **25.** $-\frac{4}{y^3}$ **27.** $\frac{1}{8x^{3/2}}$
29. $\frac{2(y-1)}{(1+x)^2}$ **31.** $\frac{y}{(1-y)^3}$ **33.** $-\frac{16}{125}$
35. $300(5x-3)^2$ **37.** 0.6 **39.** ±1
41. −4.99 and 1.94

REVIEW PROBLEMS—CHAPTER 12 (page 628)

1. $3e^x + 0 + e^{x^2}(2x) + (e^2)x^{e^2-1} = 3e^x + 2xe^{x^2} + e^2x^{e^2-1}$
3. $\frac{1}{r^2+5r}(2r+5) = \frac{2r+5}{r(r+5)}$
5. $e^{x^2+4x+5}(2x+4) = 2(x+2)e^{x^2+4x+5}$
7. $e^x(2x) + (x^2+2)e^x = e^x(x^2+2x+2)$
9. $\frac{\sqrt{(x-6)(x+5)(9-x)}}{2}\left[\frac{1}{x-6} + \frac{1}{x+5} + \frac{1}{x-9}\right]$
11. $\frac{e^x\left(\frac{1}{x}\right) - (\ln x)(e^x)}{e^{2x}} = \frac{1 - x\ln x}{xe^x}$
13. $\frac{2}{q+1} + \frac{3}{q+2}$ **15.** $-7(\ln 10)10^{2-7x}$
17. $\frac{4e^{2x+1}(2x-1)}{x^2}$ **19.** $\frac{16}{(8x+5)\ln 2}$
21. $\frac{1+2l+3l^2}{1+l+l^2+l^3}$ **23.** $(x+1)^{x+1}[1 + \ln(x+1)]$
25. $\frac{1}{t} + \frac{1}{2}\cdot\frac{1}{4-t^2}\cdot(-2t) = \frac{1}{t} - \frac{t}{4-t^2}$
27. $y\left[\frac{3}{2}\left(\frac{1}{x^2+2}\right)(2x) + \frac{4}{9}\left(\frac{1}{x^2+9}\right)(2x) - \frac{4}{11}\left(\frac{1}{x^3+6x}\right)(3x^2+6)\right]$
$= y\left[\frac{3x}{x^2+2} + \frac{8x}{9(x^2+9)} - \frac{12(x^2+2)}{11(x^3+6x)}\right]$,
where y is as given in the problem
29. $(x^x)^x(x + 2x\ln x)$ **31.** 4 **33.** −2
35. $y = 6x + 6(1 - \ln 2)$ or $y = 6x + 6 - \ln 64$

37. $(0, 4\ln 2)$ **39.** 18 **41.** 2 **43.** $-\frac{y}{x+y}$
45. $\frac{xy^2 - y}{2x - x^2y}$ **47.** $\frac{4}{9}$
49. $\frac{dy}{dx} = \frac{y+1}{y}$; $\frac{d^2y}{dx^2} = -\frac{y+1}{y^3}$
51. $f'(t) = 0.008e^{-0.01t} + 0.00004e^{-0.0002t}$ **53.** 0.90
55. $\eta = -1$, unit elasticity
57. $\eta = -0.5$, demand is inelastic
59. $-\frac{9}{16}$, $\approx \frac{3}{8}$% increase **61.** 1.7693

MATHEMATICAL SNAPSHOT—CHAPTER 12 (page 630)

1. Figure 12.11 shows that the population reaches its final size in about 45 days.
3. The tangent line will not coincide exactly with the curve in the first place. Smaller time steps could reduce the error.

APPLICATIONS IN PRACTICE 13.1

1. There is a relative maximum when $q = 2$, and a relative minimum when $q = 5$.
2. The drug is at its greatest concentration 2 hours after injection.

EXERCISE 13.1 (page 642)

1. Dec. on $(-\infty, -1)$ and $(3, \infty)$; inc. on $(-1, 3)$; rel. min. $(-1, -1)$; rel. max. $(3, 4)$
3. Dec. on $(-\infty, -2)$ and $(0, 2)$; inc. on $(-2, 0)$ and $(2, \infty)$; rel. min. $(-2, 1)$ and $(2, 1)$; no rel. max
5. Inc. on $(-\infty, -2)$ and $(1, \infty)$; dec. on $(-2, 1)$; rel max. when $x = -2$; rel. min. when $x = 1$
7. Dec. on $(-\infty, -1)$; inc. on $(-1, 3)$ and $(3, \infty)$; rel. min. when $x = -1$
9. Inc. on $(-\infty, 0)$ and $(0, \infty)$; no rel. min. or max
11. Inc. on $\left(-\infty, \frac{1}{2}\right)$; dec. on $\left(\frac{1}{2}, \infty\right)$; rel. max. when $x = \frac{1}{2}$
13. Dec. on $(-\infty, -5)$ and $(1, \infty)$; inc. on $(-5, 1)$; rel. min. when $x = -5$; rel. max. when $x = 1$
15. Dec. on $(-\infty, -1)$ and $(0, 1)$; inc. on $(-1, 0)$ and $(1, \infty)$; rel. max. when $x = 0$; rel. min. when $x = \pm 1$
17. Inc. on $(-\infty, 1)$ and $(3, \infty)$; dec. on $(1, 3)$; rel. max. when $x = 1$; rel. min. when $x = 3$
19. Inc. on $\left(-\infty, -\frac{2}{3}\right)$ and $\left(\frac{5}{2}, \infty\right)$; dec. on $\left(-\frac{2}{3}, \frac{5}{2}\right)$; rel. max. when $x = -\frac{2}{3}$; rel. min. when $x = \frac{5}{2}$
21. Inc. on $(-\infty, 5 - \sqrt{3})$ and $(5 + \sqrt{3}, \infty)$; dec. on $(5 - \sqrt{3}, 5 + \sqrt{3})$; rel. max. when $x = 5 - \sqrt{3}$; rel. min. when $x = 5 + \sqrt{3}$
23. Inc. on $(-\infty, -1)$ and $(1, \infty)$; dec. on $(-1, 0)$ and $(0, 1)$; rel. max. when $x = -1$; rel. min. when $x = 1$
25. Dec. on $(-\infty, -4)$ and $(0, \infty)$; inc. on $(-4, 0)$; rel. min. when $x = -4$; rel. max. when $x = 0$
27. Inc. on $(-\infty, -\sqrt{2})$ and $(0, \sqrt{2})$; dec. on $(-\sqrt{2}, 0)$ and $(\sqrt{2}, \infty)$; rel. max. when $x = \pm\sqrt{2}$; rel. min. when $x = 0$

29. Inc. on $(-\infty, -1)$, $(-1, 0)$, and $(0, \infty)$; never dec.; no rel. extremum

31. Dec. on $(-\infty, 1)$ and $(1, \infty)$; no rel. extremum

33. Dec. on $(0, \infty)$; no rel. extremum

35. Dec. on $(-\infty, 0)$ and $(4, \infty)$; inc. on $(0, 2)$ and $(2, 4)$; rel. min. when $x = 0$; rel. max. when $x = 4$

37. Inc. on $(-\infty, -3)$ and $(-1, \infty)$; dec. on $(-3, -2)$ and $(-2, -1)$; rel. max. when $x = -3$; rel. min. when $x = -1$

39. Dec. on $\left(-\infty, \frac{-2-\sqrt{29}}{5}\right)$ and $\left(\frac{-2+\sqrt{29}}{5}, \infty\right)$; inc. on $\left(\frac{-2-\sqrt{29}}{5}, \frac{-2+\sqrt{29}}{5}\right)$; rel. min. when $x = \frac{-2-\sqrt{29}}{5}$; rel. max. when $x = \frac{-2+\sqrt{29}}{5}$

41. Inc. on $(-\infty, -2)$, $\left(-2, \frac{11}{5}\right)$, and $(5, \infty)$; dec. on $\left(\frac{11}{5}, 5\right)$; rel. max. when $x = \frac{11}{5}$; rel. min. when $x = 5$

43. Inc. on $(-\infty, 0)$, $\left(0, \frac{18}{7}\right)$, and $(6, \infty)$; dec. on $\left(\frac{18}{7}, 6\right)$; rel. max. when $x = \frac{18}{7}$; rel. min. when $x = 6$

45. Dec. on $(-\infty, \infty)$; no rel. extremum.

47. Dec. on $\left(0, \frac{3\sqrt{2}}{2}\right)$; inc. on $\left(\frac{3\sqrt{2}}{2}, \infty\right)$; rel. min. when $x = \frac{3\sqrt{2}}{2}$

49. Dec. on $(-\infty, 0)$; inc. on $(0, \infty)$; rel. min. when $x = 0$

51. Dec. on $(0, 1)$; inc. on $(1, \infty)$; rel. min. when $x = 1$

53. Dec. on $(-\infty, 3)$; inc. on $(3, \infty)$; rel. min. when $x = 3$; intercepts: $(7, 0)$, $(-1, 0)$, $(0, -7)$

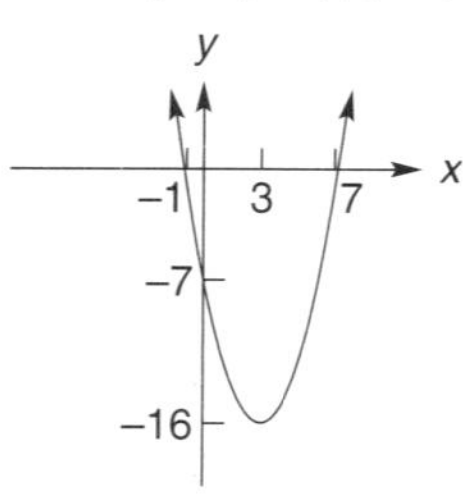

55. Dec. on $(-\infty, -1)$ and $(1, \infty)$; inc. on $(-1, 1)$; rel. min. when $x = -1$; rel. max. when $x = 1$; sym. about origin; intercepts: $(\pm\sqrt{3}, 0)$, $(0, 0)$

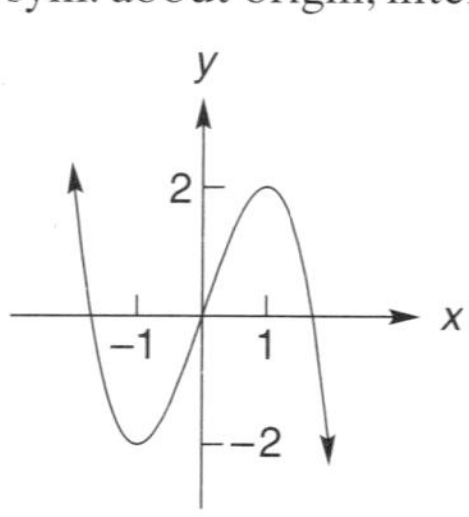

57. Inc. on $(-\infty, 1)$ and $(2, \infty)$; dec. on $(1, 2)$; rel. max. when $x = 1$; rel. min. when $x = 2$; intercept: $(0, 0)$

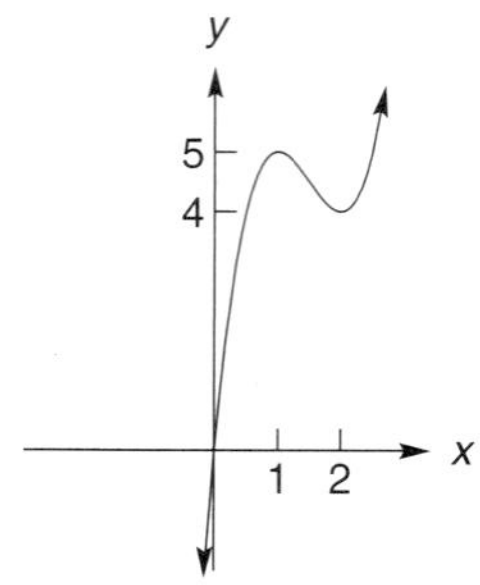

59. Inc. on $(-2, -1)$ and $(0, \infty)$; dec. on $(-\infty, -2)$ and $(-1, 0)$; rel. max. when $x = -1$; rel. min. when $x = -2, 0$; intercepts: $(0, 0)$, $(-2, 0)$

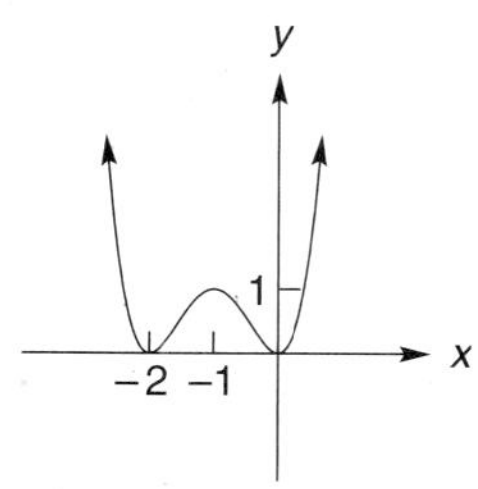

61. Dec. on $(-\infty, -2)$ and $\left(-\frac{1}{2}, 1\right)$; inc. on $\left(-2, -\frac{1}{2}\right)$ and $(1, \infty)$; rel. min. when $x = -2, 1$; rel. max. when $x = -\frac{1}{2}$; intercepts: $(1, 0)$, $(-2, 0)$, $(0, 4)$

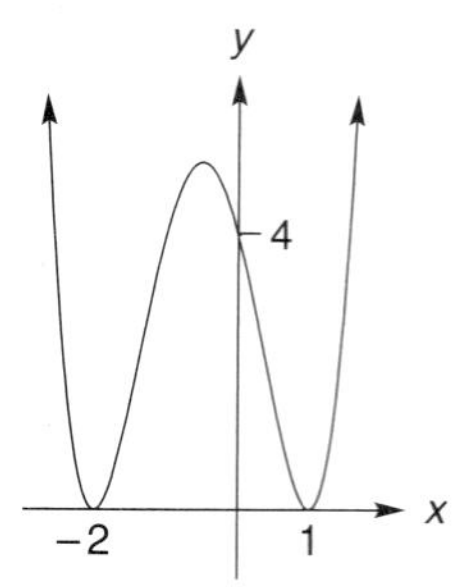

63. Dec. on $(1, \infty)$; inc. on $(0, 1)$; rel. max. when $x = 1$; intercepts: $(0, 0)$, $(4, 0)$

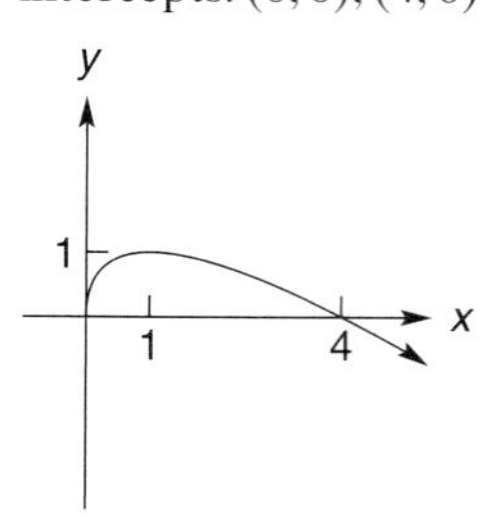

65.

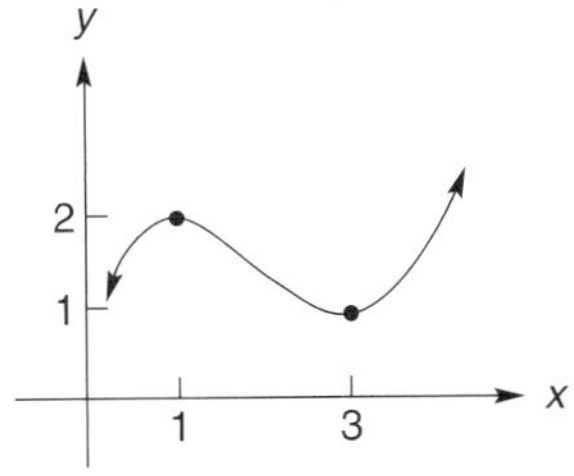

69. Never

71. 40 **75. (a)** 25,300; **(b)** 4; **(c)** 17,200
77. Rel. min.: $(-4.10, -2.21)$
79. Rel. max.: $(2.74, 3.74)$; rel. min.: $(-2.74, -3.74)$
81. Rel. min.: 0, 1.50, 2.00; rel. max.: 0.57, 1.77
83. (a) $f'(x) = 4 - 6x - 3x^2$;
(c) Dec.: $(-\infty, -2.53)$, $(0.53, \infty)$; inc.: $(-2.53, 0.53)$

EXERCISE 13.2 (page 646)

1. Maximum: $f(3) = 6$; minimum: $f(1) = 2$
3. Maximum: $f(-1) = \frac{19}{6}$; minimum: $f(0) = 1$
5. Maximum: $f(3) = 84$; minimum: $f(1) = -8$
7. Maximum: $f(-2) = 56$; minimum: $f(-1) = -2$
9. Maximum: $f(\sqrt{2}) = 4$; minimum $f(2) = -16$
11. Maximum: $f(0) = f(3) = 2$;
minimum: $f\left(\frac{3\sqrt{2}}{2}\right) = -\frac{73}{4}$
13. Maximum: $f(3) \approx 2.08$; minimum: $f(0) = 0$
15. (a) $-3.22, -0.78$; **(b)** 2.75; **(c)** 9; **(d)** 14,283

EXERCISE 13.3 (page 652)

1. Conc. up $(-\infty, 0)$, $\left(\frac{3}{2}, \infty\right)$; conc. down $\left(0, \frac{3}{2}\right)$;
inf. pt. when $x = 0, \frac{3}{2}$
3. Conc. up $(-\infty, 1)$, $(1, 7)$; conc. down $(7, \infty)$;
inf. pt. when $x = 7$
5. Conc. up $(-\infty, -\sqrt{2})$, $(\sqrt{2}, \infty)$; conc. down $(-\sqrt{2}, \sqrt{2})$;
no inf. pt.
7. Conc. down $(-\infty, \infty)$
9. Conc. down $(-\infty, -1)$; conc. up $(-1, \infty)$;
inf. pt. when $x = -1$
11. Conc. down $\left(-\infty, \frac{5}{6}\right)$; conc. up $\left(\frac{5}{6}, \infty\right)$;
inf. pt. when $x = \frac{5}{6}$
13. Conc. up $(-\infty, -1)$, $(1, \infty)$; conc. down $(-1, 1)$;
inf. pt. when $x = \pm 1$
15. Conc. up $(-\infty, 0)$; conc. down $(0, \infty)$;
inf. pt. when $x = 0$
17. Conc. up $\left(-\infty, -\frac{7}{2}\right)$, $\left(\frac{1}{3}, \infty\right)$; conc. down $\left(-\frac{7}{2}, \frac{1}{3}\right)$;
inf. pt. when $x = -\frac{7}{2}, \frac{1}{3}$
19. Conc. down $(-\infty, 0)$, $\left(\frac{3-\sqrt{5}}{2}, \frac{3+\sqrt{5}}{2}\right)$;
conc. up $\left(0, \frac{3-\sqrt{5}}{2}\right)$, $\left(\frac{3+\sqrt{5}}{2}, \infty\right)$;
inf. pt. when $x = 0, \frac{3 \pm \sqrt{5}}{2}$
21. Conc. up $(-\infty, -\sqrt{5})$, $(-\sqrt{2}, \sqrt{2})$, $(\sqrt{5}, \infty)$;
conc. down $(-\sqrt{5}, -\sqrt{2})$, $(\sqrt{2}, \sqrt{5})$;
inf. pt. when $x = \pm\sqrt{5}, \pm\sqrt{2}$
23. Conc. down $(-\infty, 1)$; conc. up $(1, \infty)$
25. Conc. down. $(-\infty, -1/\sqrt{3})$, $(1/\sqrt{3}, \infty)$;
conc. up $(-1/\sqrt{3}, 1/\sqrt{3})$; inf. pt. when $x = \pm 1/\sqrt{3}$
27. Conc. down. $(-\infty, -3)$, $\left(-3, \frac{2}{7}\right)$; conc. up $\left(\frac{2}{7}, \infty\right)$;
inf. pt. when $x = \frac{2}{7}$
29. Conc. up. $(-\infty, \infty)$
31. Conc. down $(-\infty, -2)$; conc. up $(-2, \infty)$;
inf. pt. when $x = -2$
33. Conc. down $(0, e^{3/2})$; conc. up $(e^{3/2}, \infty)$;
inf. pt. when $x = e^{3/2}$
35. Int. $(-2, 0)$, $(3, 0)$, $(0, -6)$; dec. $\left(-\infty, \frac{1}{2}\right)$;
inc. $\left(\frac{1}{2}, \infty\right)$; rel. min. when $x = \frac{1}{2}$; conc. up $(-\infty, \infty)$

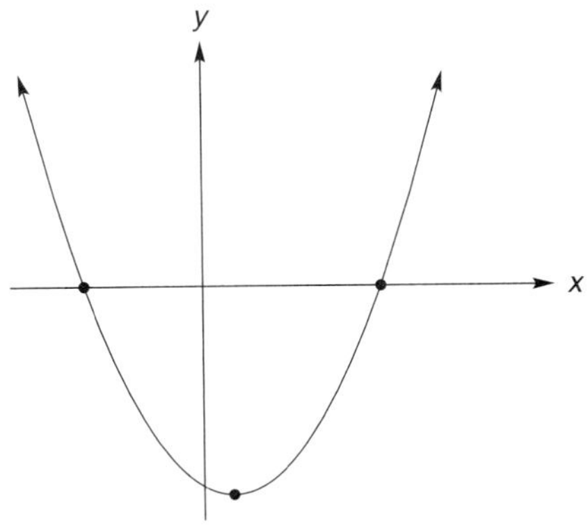

37. Int. $(0, 0)$, $(4, 0)$; inc. $(-\infty, 2)$; dec. $(2, \infty)$;
rel. max. when $x = 2$; conc. down $(-\infty, \infty)$

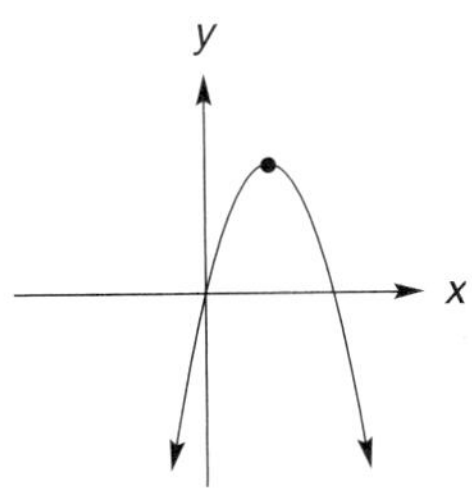

39. Int. $(0, -19)$; inc. $(-\infty, 2)$, $(4, \infty)$; dec. $(2, 4)$; rel. max. when $x = 2$; rel. min. when $x = 4$; conc. down $(-\infty, 3)$; conc. up $(3, \infty)$; inf. pt. when $x = 3$

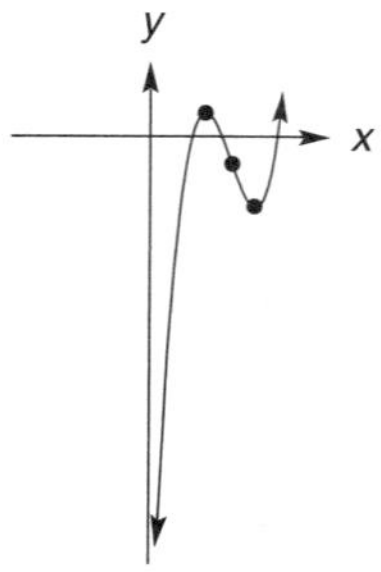

41. Int. $(0, 0)$, $(\pm 2\sqrt{3}, 0)$; inc. $(-\infty, -2)$, $(2, \infty)$; dec. $(-2, 2)$; rel. max. when $x = -2$; rel. min. when $x = 2$; conc. down $(-\infty, 0)$; conc. up $(0, \infty)$; inf. pt. when $x = 0$; sym. about origin

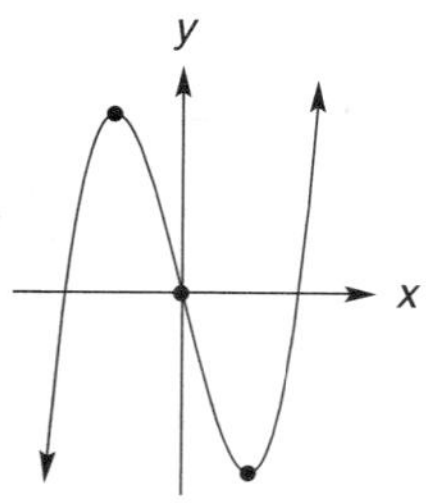

43. Int. $(0, -3)$; inc. $(-\infty, 1)$, $(1, \infty)$; no rel. max. or min.; conc. down $(-\infty, 1)$; conc. up $(1, \infty)$; inf. pt. when $x = 1$

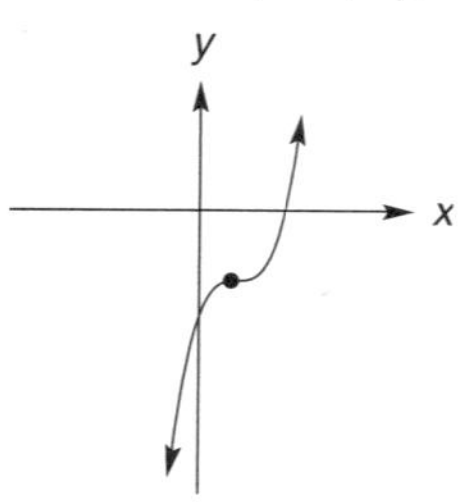

45. Int. $(0, 0)$, $(4/3, 0)$; inc. $(-\infty, 0)$, $(0, 1)$; dec. $(1, \infty)$; rel. max. when $x = 1$; conc. up $(0, 2/3)$; conc. down $(-\infty, 0)$, $(2/3, \infty)$; inf. pt. when $x = 0, x = 2/3$

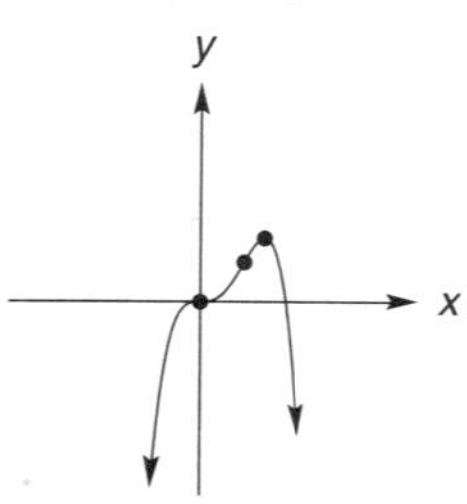

47. Int. $(0, -2)$; dec. $(-\infty, -2)$, $(2, \infty)$; inc. $(-2, 2)$; rel. min. when $x = -2$; rel. max. when $x = 2$; conc. up $(-\infty, 0)$; conc. down $(0, \infty)$; inf. pt. when $x = 0$

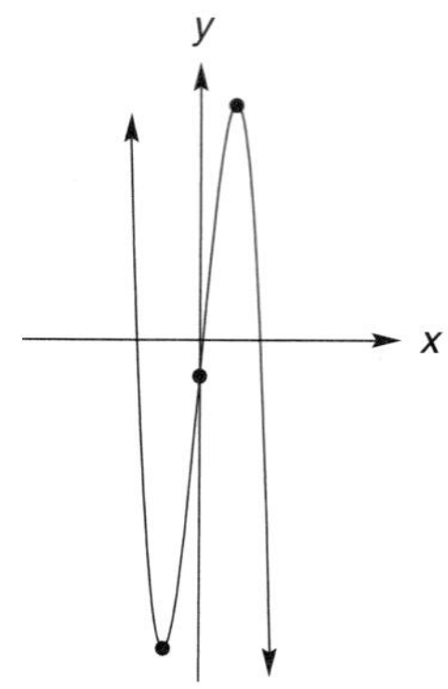

49. Int. $(0, -6)$; inc. $(-\infty, 2)$, $(2, \infty)$; conc. down $(-\infty, 2)$; conc. up $(2, \infty)$; inf. pt. when $x = 2$

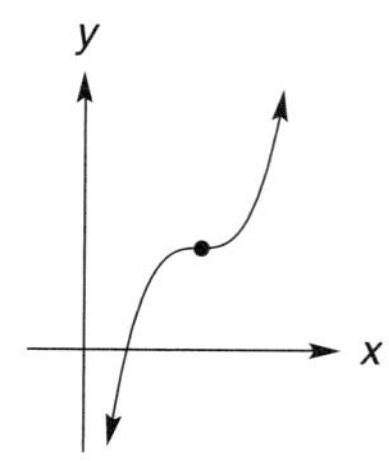

51. Int. $(0, 0)$, $(\pm \sqrt[4]{5}, 0)$; dec. $(-\infty, -1)$, $(1, \infty)$; inc. $(-1, 1)$; rel. min. when $x = -1$; rel. max. when $x = 1$; conc. up $(-\infty, 0)$; conc. down $(0, \infty)$; inf. pt. when $x = 0$; sym. about origin.

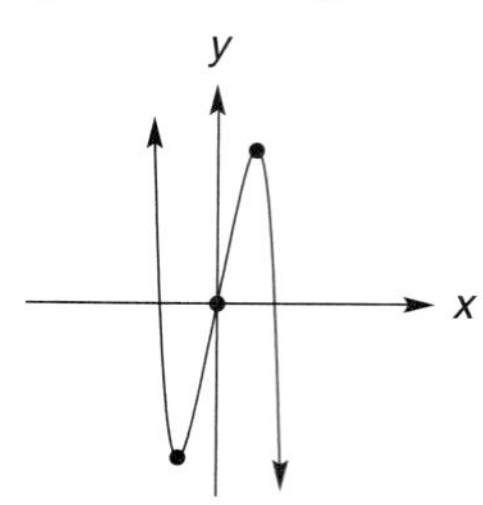

53. Int. $(0, 1)$, $(1, 0)$; dec. $(-\infty, 0)$, $(0, 1)$; inc. $(1, \infty)$; rel. min. when $x = 1$; conc. up $(-\infty, 0)$, $(2/3, \infty)$; conc. down $(0, 2/3)$; inf. pt. when $x = 0, x = 2/3$

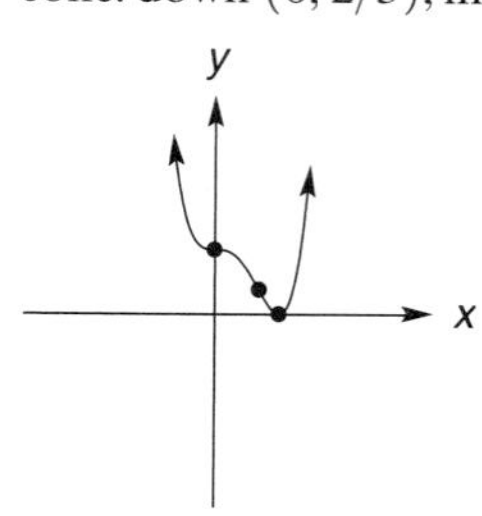

55. Int. $(0, 0)$, $(\pm 2, 0)$; inc. $(-\infty, -\sqrt{2})$, $(0, \sqrt{2})$; dec. $(-\sqrt{2}, 0)$, $(\sqrt{2}, \infty)$; rel. max. when $x = \pm\sqrt{2}$; rel. min. when $x = 0$; conc. down $(-\infty, -\sqrt{2/3})$, $(\sqrt{2/3}, \infty)$; conc. up $(-\sqrt{2/3}, \sqrt{2/3})$; inf. pt. when $x = \pm\sqrt{2/3}$; sym. about y-axis

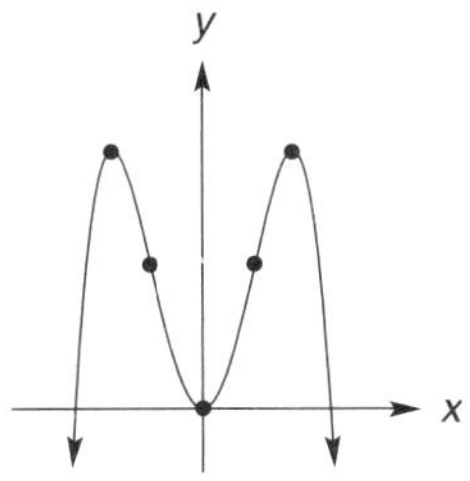

57. Int. $(0, 0)$, $(8, 0)$; dec. $(-\infty, 0)$, $(0, 2)$; inc. $(2, \infty)$; rel. min. when $x = 2$; conc. up $(-\infty, -4)$, $(0, \infty)$; conc. down $(-4, 0)$; inf. pt. when $x = -4$, $x = 0$

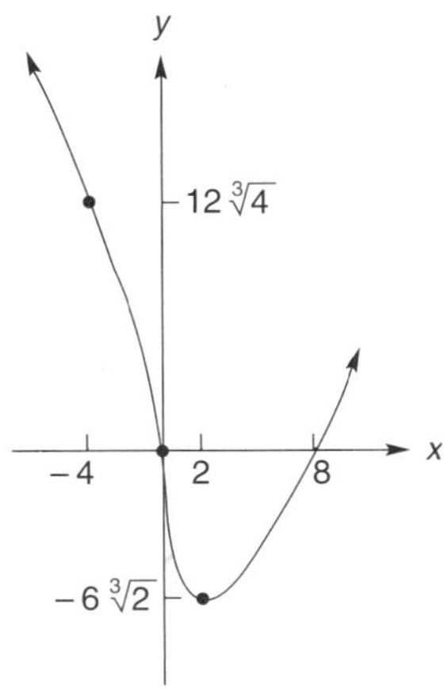

59. Int. $(0, 0)$, $(-4, 0)$; dec. $(-\infty, -1)$; inc. $(-1, 0)$, $(0, \infty)$; rel. min. when $x = -1$; conc. up $(-\infty, 0)$, $(2, \infty)$; conc. down $(0, 2)$; inf. pt. when $x = 0$, $x = 2$

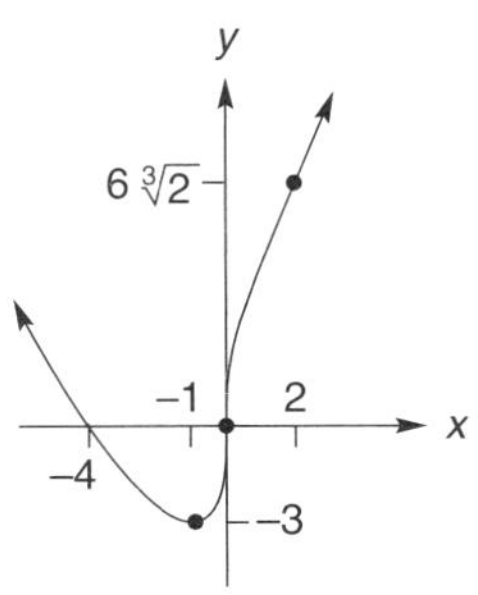

61. Int. $(0, 0)$, $\left(-\frac{27}{8}, 0\right)$; inc. $(-\infty, -1)$, $(0, \infty)$; dec. $(-1, 0)$; rel. min. when $x = 0$; rel. max. when $x = -1$; conc. down $(-\infty, 0)$, $(0, \infty)$

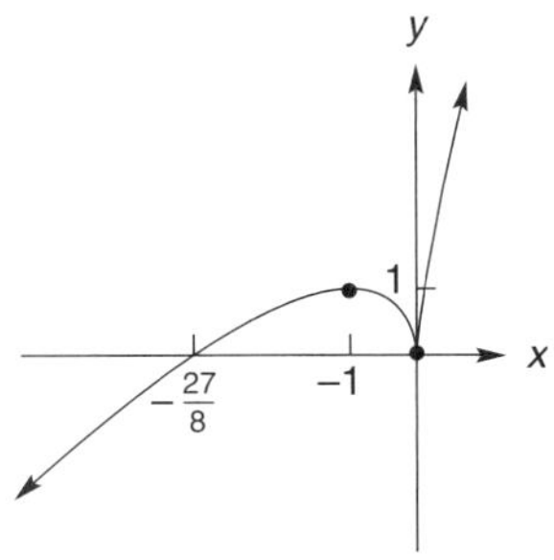

63.

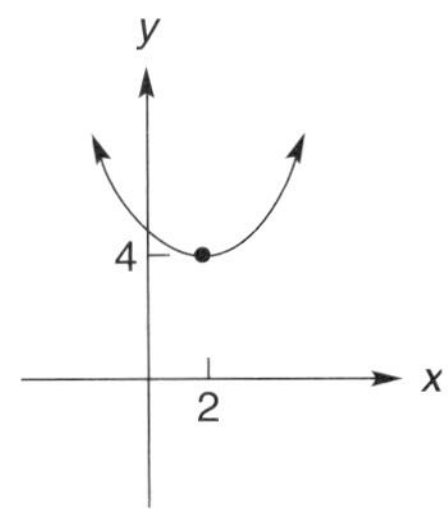

65.

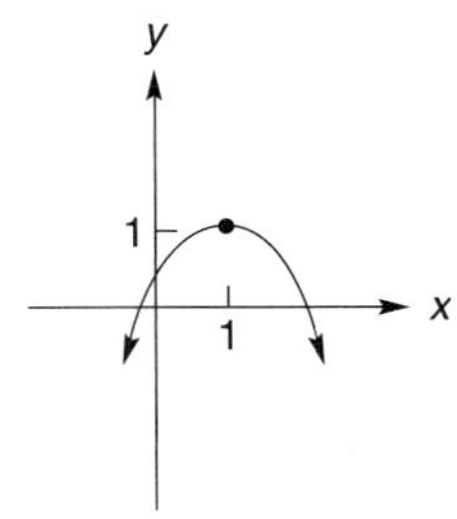

69.

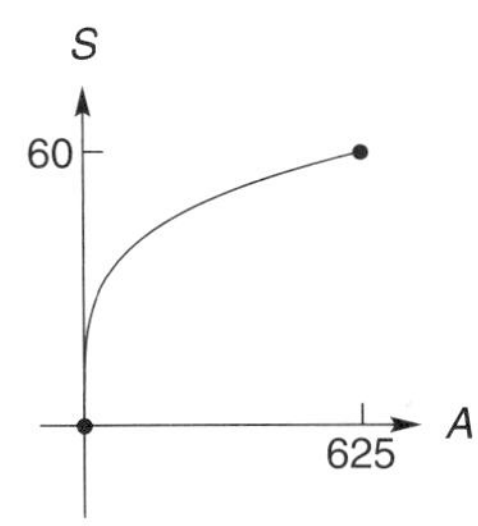

73. (b)

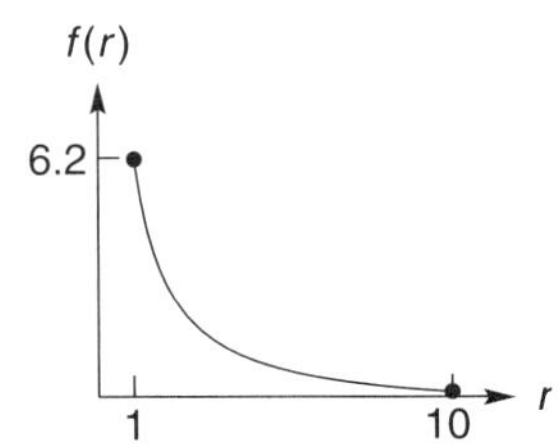

(c) 0.26

75. Two **77.** Above tangent line; concave up
79. $-2.61, -0.26$

EXERCISE 13.4 (page 656)

1. Rel. min. when $x = \frac{5}{2}$; abs. min.

3. Rel. max. when $x = \frac{1}{4}$; abs. max.

5. Rel. max. when $x = -5$; rel. min. when $x = 1$
7. Rel. min. when $x = 0$; rel. max. when $x = 2$
9. Test fails, when $x = 0$ there is a rel. min. by first-deriv. test

11. Rel. max. when $x = -\frac{1}{3}$; rel. min. when $x = \frac{1}{3}$

13. Rel. min. when $x = -5, -2$; rel. max. when $x = -\frac{7}{2}$

EXERCISE 13.5 (page 664)

1. $y = 1, x = 1$ **3.** $y = \frac{1}{2}, x = -\frac{3}{2}$

5. $y = 0, x = 0$ **7.** $y = 0, x = 1, x = -1$

9. None **11.** $y = 2, x = 2, x = -3$

13. $y = 2, x = -\sqrt{5}, x = \sqrt{5}$ **15.** $y = 7, x = 6$

17. $x = 0, x = -1$ **19.** $y = \frac{1}{4}, x = -\frac{1}{2}$

21. $y = \frac{1}{2}, x = -\frac{4}{3}$ **23.** $y = 4$

25. Dec. $(-\infty, 0), (0, \infty)$; conc. down $(-\infty, 0)$; conc. up $(0, \infty)$; sym. about origin; asymptotes $x = 0, y = 0$

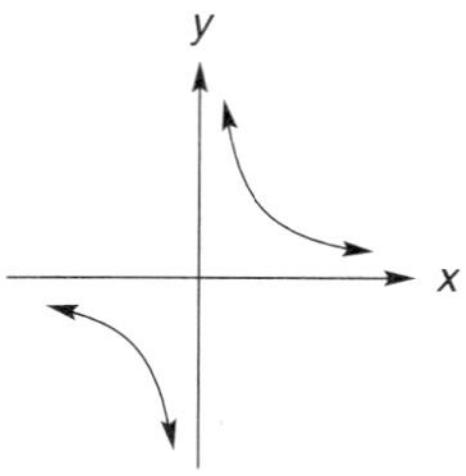

27. Int. $(0, 0)$; inc. $(-\infty, -1), (-1, \infty)$; conc. up $(-\infty, -1)$; conc. down $(-1, \infty)$; asymptotes $x = -1, y = 1$

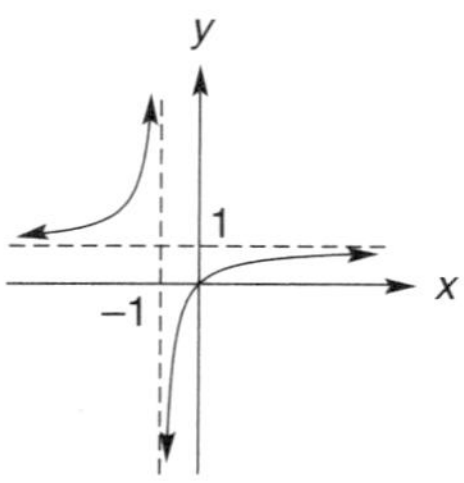

29. Dec. $(-\infty, -1), (0, 1)$; inc. $(-1, 0), (1, \infty)$; rel. min. when $x = \pm 1$; conc. up $(-\infty, 0), (0, \infty)$; sym. about y-axis; asymptote $x = 0$

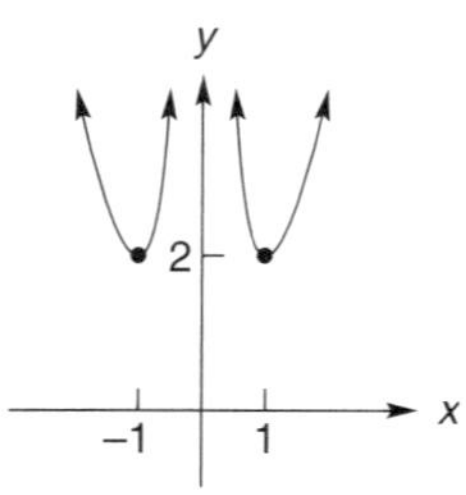

31. Int. $(0, -1)$; inc. $(-\infty, -1), (-1, 0)$; dec. $(0, 1), (1, \infty)$; rel. max. when $x = 0$; conc. up $(-\infty, -1), (1, \infty)$; conc. down $(-1, 1)$; asymptotes $x = 1, x = -1, y = 0$; sym. about y-axis

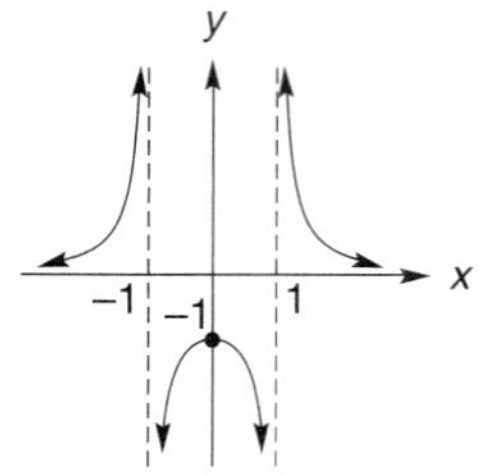

33. Int. $(-1, 0), (0, 1)$; inc. $(-\infty, 1), (1, \infty)$; conc. up $(-\infty, 1)$; conc. down $(1, \infty)$; asymptotes $x = 1, y = -1$

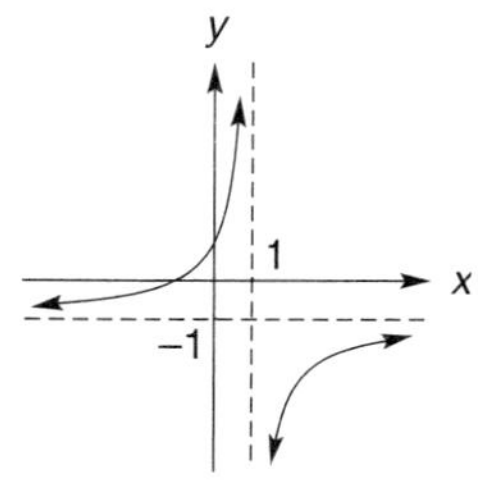

35. Int. $(0, 0)$; inc. $\left(-\infty, -\frac{8}{7}\right), (0, \infty)$; dec. $\left(-\frac{8}{7}, -\frac{4}{7}\right)$, $\left(-\frac{4}{7}, 0\right)$; rel. max. when $x = -\frac{8}{7}$; rel. min. when $x = 0$; conc. down $\left(-\infty, -\frac{4}{7}\right)$; conc. up $\left(-\frac{4}{7}, \infty\right)$; asymptote $x = -\frac{4}{7}$

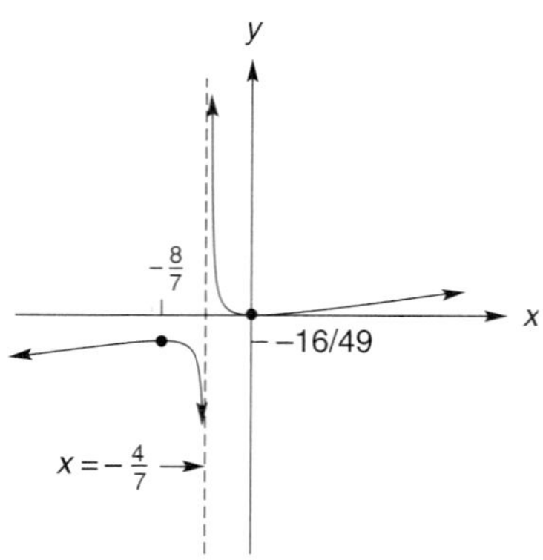

37. Int. $\left(0, -\frac{9}{8}\right)$; inc. $\left(-\infty, -\frac{2}{3}\right)$, $\left(-\frac{2}{3}, \frac{1}{3}\right)$; dec. $\left(\frac{1}{3}, \frac{4}{3}\right)$, $\left(\frac{4}{3}, \infty\right)$; rel. max. when $x = \frac{1}{3}$; conc. up $\left(-\infty, -\frac{2}{3}\right)$, $\left(\frac{4}{3}, \infty\right)$; conc. down $\left(-\frac{2}{3}, \frac{4}{3}\right)$; asymptotes $y = 0$, $x = -\frac{2}{3}$, $x = \frac{4}{3}$

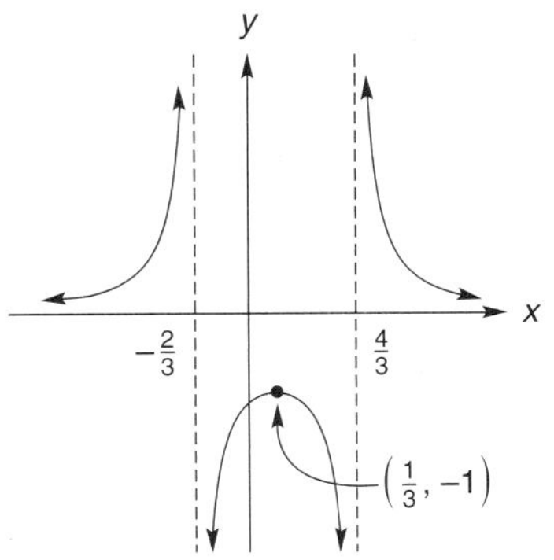

39. Int. $\left(\frac{3}{2}, 0\right)$, $\left(0, -\frac{1}{27}\right)$; dec. $\left(-\infty, -\frac{3}{2}\right)$, $\left(\frac{9}{2}, \infty\right)$; inc. $\left(-\frac{3}{2}, \frac{9}{2}\right)$; rel. min. when $x = -\frac{3}{2}$; conc. down $\left(-\infty, -\frac{9}{2}\right)$; conc. up $\left(-\frac{9}{2}, \frac{9}{2}\right)$, $\left(\frac{9}{2}, \infty\right)$; inf. pt. when $x = -\frac{9}{2}$; asymptotes $x = \frac{9}{2}$, $y = 0$

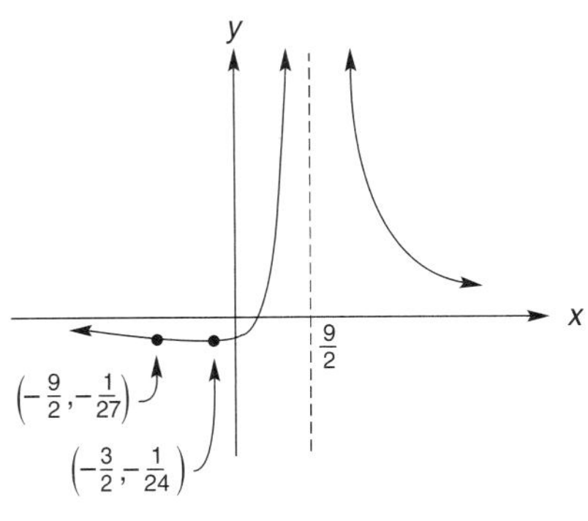

41. Int. $(-1, 0)$, $(1, 0)$; inc. $(-\sqrt{3}, 0)$, $(0, \sqrt{3})$; dec. $(-\infty, -\sqrt{3})$, $(\sqrt{3}, \infty)$; rel. max. when $x = \sqrt{3}$; rel. min. when $x = -\sqrt{3}$; conc. down $(-\infty, -\sqrt{6})$, $(0, \sqrt{6})$; conc. up $(-\sqrt{6}, 0)$, $(\sqrt{6}, \infty)$; inf. pt. when $x = \pm\sqrt{6}$; asymptotes $x = 0$, $y = 0$; sym. about origin

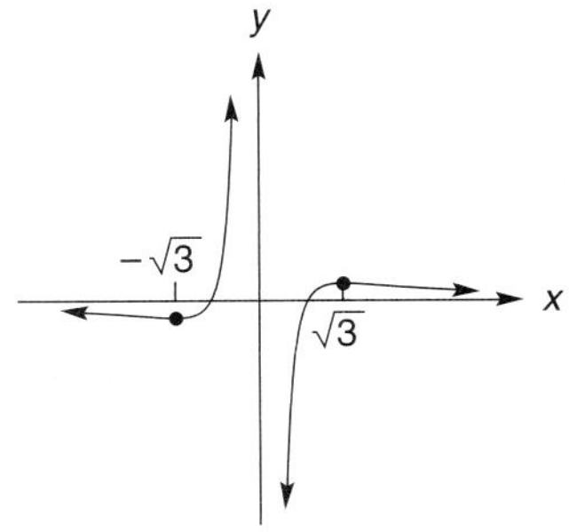

43. Int. $(0, 1)$; inc. $(-\infty, -2)$, $(0, \infty)$; dec. $(-2, -1)$, $(-1, 0)$; rel. max. when $x = -2$; rel. min when $x = 0$; conc. down $(-\infty, -1)$; conc. up $(-1, \infty)$; asymptote $x = -1$

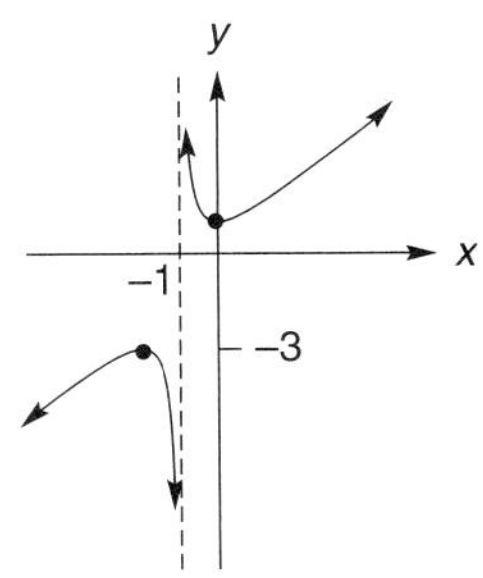

45. Int. $(0, 5)$; dec. $\left(-\infty, -\frac{1}{3}\right)$, $\left(-\frac{1}{3}, \frac{1}{3}\right)$; inc. $\left(\frac{1}{3}, 1\right)$, $(1, \infty)$; rel. min. when $x = \frac{1}{3}$; conc. down $\left(-\infty, -\frac{1}{3}\right)$, $(1, \infty)$; conc. up $\left(-\frac{1}{3}, 1\right)$; asymptotes $x = -\frac{1}{3}$, $x = 1$, $y = -1$

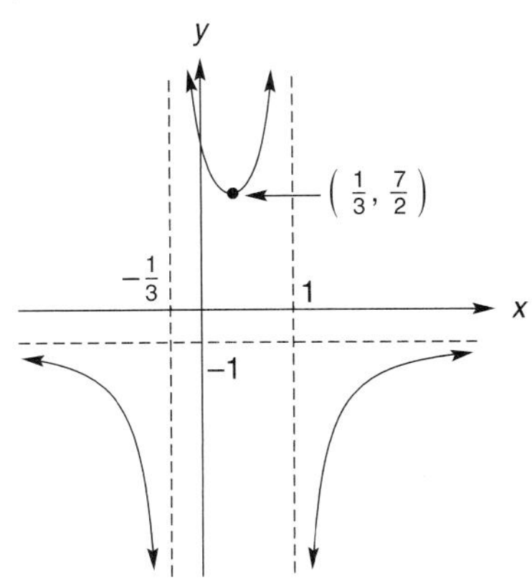

47.

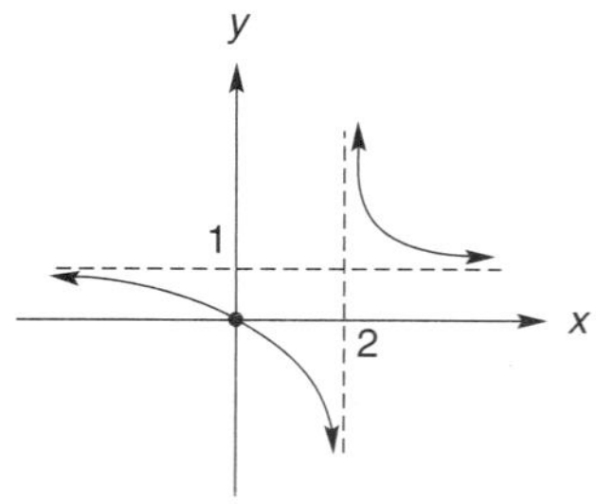

49.

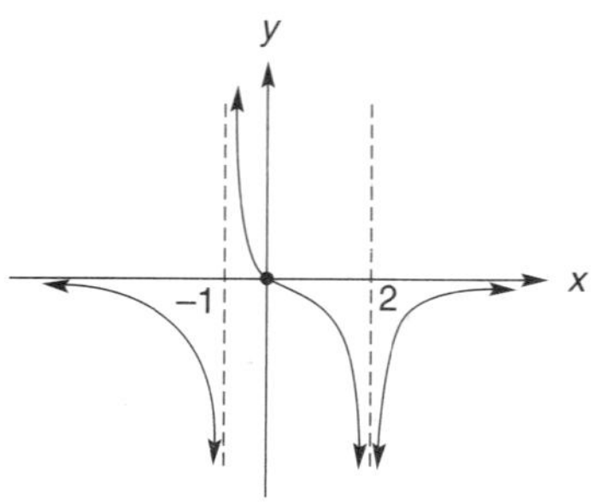

55. $x \approx \pm 2.45$, $x \approx 0.67$, $y = 2$ **57.** $y \approx 0.48$

EXERCISE 13.6 (page 674)

1. 41 and 41 **3.** 300 ft by 250 ft **5.** 100 units
7. \$15 **9. (a)** 110 grams; **(b)** $51\frac{9}{11}$ grams
11. 525 units; price = \$51; profit = \$10,525 **13.** \$22
15. 120 units; \$86,000 **17.** 625 units; \$4
19. \$17; \$86,700 **21.** 4 ft by 4 ft by 2 ft
23. 2 in.; 128 in^3.
27. 130 units, $p = \$340$, $P = \$36{,}980$; 125 units, $p = \$350$, $P = \$34{,}175$ **29.** 250 per lot (4 lots) **31.** 35
33. 60 mi/h **35.** 7; \$1000
37. $5 - \sqrt{3}$ tons; $5 - \sqrt{3}$ tons **41.** 10 cases; \$50.55

REVIEW PROBLEMS—CHAPTER 13 (page 680)

1. $y = 3, x = 4, x = -4$ **3.** $y = \frac{5}{9}, x = -\frac{2}{3}$
5. $x = 0$ **7.** $x = -\frac{15}{8}, -1$
9. Inc. $(-1, 7)$; dec. on $(-\infty, -1)$ and $(7, \infty)$
11. Dec. on $(-\infty, -\sqrt{6})$, $(0, \sqrt{3})$, $(\sqrt{3}, \sqrt{6})$; inc. on $(-\sqrt{6}, -\sqrt{3})$, $(-\sqrt{3}, 0)$, $(\sqrt{6}, \infty)$
13. Conc. up on $(-\infty, 0)$ and $\left(\frac{1}{2}, \infty\right)$; conc. down on $\left(0, \frac{1}{2}\right)$
15. Conc. down on $\left(-\infty, \frac{1}{2}\right)$; conc. up on $\left(\frac{1}{2}, \infty\right)$
17. Conc. up on $\left(-\infty, -\frac{5}{4}\right)$, $\left(-\frac{1}{4}, \infty\right)$; conc. down on $\left(-\frac{5}{4}, -\frac{1}{4}\right)$
19. Rel. max. at $x = 1$; rel. min. at $x = 2$
21. Rel. min. at $x = -1$
23. Rel. max. at $x = -\frac{2}{5}$; rel. min. at $x = 0$
25. At $x = 3$ **27.** At $x = 1$ **29.** At $x = 2 \pm \sqrt{2}$
31. Maximum: $f(2) = 16$; minimum: $f(1) = -1$
33. Maximum: $f(0) = 0$; minimum: $f\left(-\frac{6}{5}\right) = -\frac{1}{120}$
35. (a) f has no relative extrema; **(b)** f is conc. down on $(1, 3)$; inf. pts.: $(1, 2e^{-1})$, $(3, 10e^{-3})$
37. Int. $(-4, 0)$, $(6, 0)$, $(0, -24)$; inc. $(1, \infty)$; dec. $(-\infty, 1)$; rel. min. when $x = 1$; conc. up $(-\infty, \infty)$

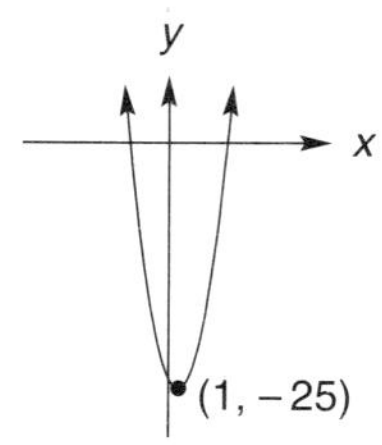

39. Int. $(0, 20)$; inc. $(-\infty, -2)$, $(2, \infty)$; dec. $(-2, 2)$; rel. max. when $x = -2$; rel. min. when $x = 2$; conc. up $(0, \infty)$; conc. down $(-\infty, 0)$; inf. pt. when $x = 0$

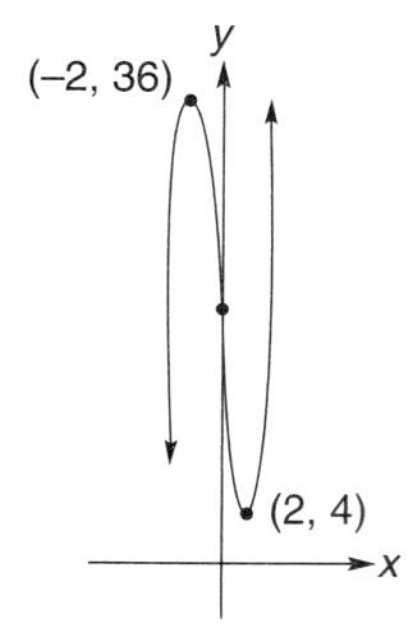

41. Int. $(0, 0)$; inc. $(-\infty, \infty)$; conc. down $(-\infty, 0)$; conc. up $(0, \infty)$; inf. pt. when $x = 0$; sym. about origin

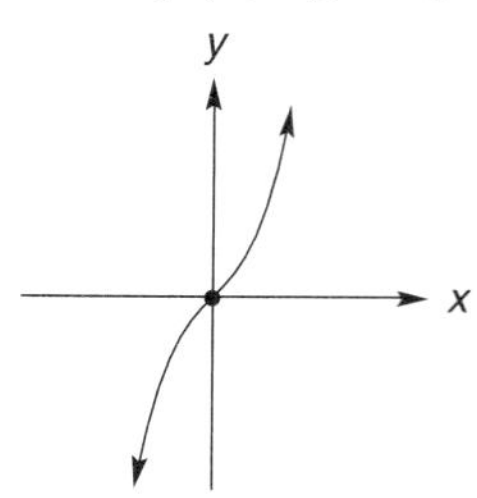

43. Int. $(-5, 0)$; inc. $(-10, 0)$; dec. $(-\infty, -10)$, $(0, \infty)$; rel. min. when $x = -10$; conc. up $(-15, 0)$, $(0, \infty)$; conc. down $(-\infty, -15)$; inf. pt. when $x = -15$; horiz. asym. $y = 0$; vert. asym. $x = 0$

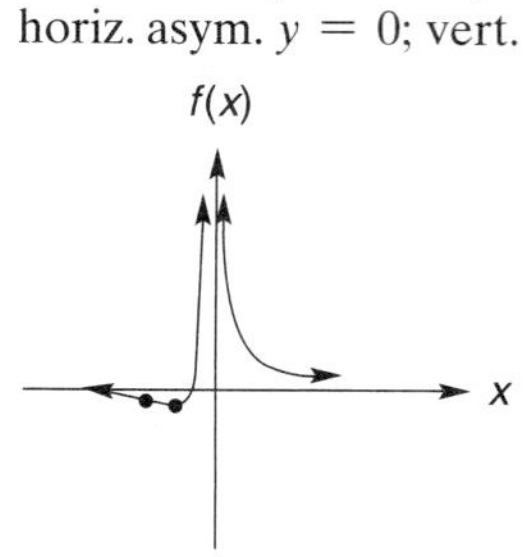

45. Int. $(0, 0)$; inc. $\left(-\infty, -\frac{1}{4}\right)$; dec. $\left(-\frac{1}{4}, \frac{1}{2}\right), \left(\frac{1}{2}, \infty\right)$; rel. max. when $x = -\frac{1}{4}$; conc. up $\left(-\infty, -\frac{1}{2}\right), \left(\frac{1}{2}, \infty\right)$; conc. down $\left(-\frac{1}{2}, \frac{1}{2}\right)$; inf. pt. when $x = -\frac{1}{2}$; horiz. asym. $y = 0$; vert. asym. $x = \frac{1}{2}$

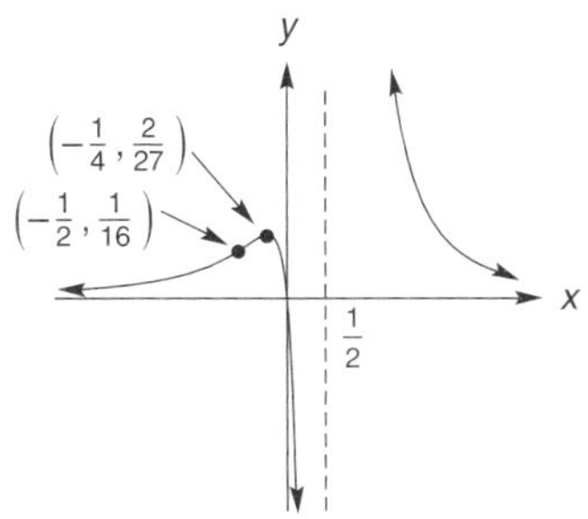

47. Int. $(0, 1)$; inc. $(0, \infty)$; dec. $(-\infty, 0)$; rel. min. when $x = 0$; conc. up $(-\infty, \infty)$; sym. about y-axis

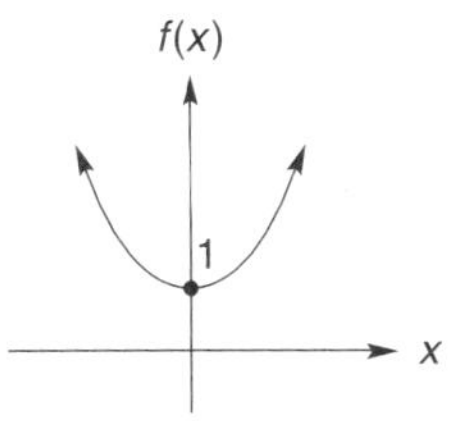

49. **(a)** False; **(b)** false; **(c)** true; **(d)** false; **(e)** false
51. $q > 2$
57. Rel. max. $(-1.32, 12.28)$; rel. min. $(0.44, 1.29)$
59. $x \approx -0.60$ **61.** 20 **63.** 200 **65.** \$2800
67. 100 ft by 200 ft **69.** **(a)** 200 stands at \$120 per stand; **(b)** 300 stands

MATHEMATICAL SNAPSHOT—CHAPTER 13 (page 684)

1. The data for 1998–2000 fall into the same pattern as the 1959–1969 data

EXERCISE 14.1 (page 691)

1. $3\,dx$ **3.** $\frac{2x^3}{\sqrt{x^4 - 9}}\,dx$ **5.** $-\frac{2}{x^3}\,dx$
7. $\frac{2x}{x^2 + 7}\,dx$ **9.** $3e^{2x^3 + 3}(12x^2 + 4x + 3)\,dx$
11. $\Delta y = -0.14$, $dy = -0.14$
13. $\Delta y = -2.5$, $dy = -2.75$
15. $\Delta y \approx 0.049$, $dy = 0.050$ **17.** **(a)** -1; **(b)** 2.9
19. 9.95 **21.** $4\frac{1}{32}$ **23.** -0.03 **25.** 1.01 **27.** $\frac{1}{2}$
29. $\frac{1}{6p(p^2 + 5)^2}$ **31.** $-p^2$ **33.** $\frac{1}{36}$ **35.** $-\frac{4}{5}$
37. 44; 41.8 **39.** 2.04 **41.** 0.7
43. $(1.69 \times 10^{-11})\pi$ cm^3 **45.** **(c)** 42 units

APPLICATIONS IN PRACTICE 14.2

1. $\int 28.3\,dq = 28.3q + C$
2. $\int 0.12t^2\,dt = 0.04t^3 + C$
3. $\int -\frac{480}{t^3}\,dt = \frac{240}{t^2} + C$
4. $\int (500 + 300\sqrt{t})dt = 500t + 200t^{3/2} + C$
5. $S(t) = 0.7t^3 - 32.7t^2 + 491.6t + C$

EXERCISE 14.2 (page 698)

1. $7x + C$ **3.** $\frac{x^9}{9} + C$ **5.** $-\frac{5}{6x^6} + C$
7. $-\frac{2}{9x^9} + C$ **9.** $-\frac{5}{6y^{6/5}} + C$ **11.** $4t + \frac{t^2}{2} + C$
13. $\frac{y^6}{6} - \frac{5y^2}{2} + C$ **15.** $t^3 - 2t^2 + 5t + C$
17. $(7 + e)x + C$ **19.** $\frac{x^2}{14} - \frac{3x^5}{20} + C$
21. $6e^x + C$ **23.** $\frac{x^{9.3}}{9.3} - \frac{9x^7}{7} - \frac{1}{x^3} - \frac{1}{2x^2} + C$
25. $-\frac{4x^{3/2}}{9} + C$ **27.** $\frac{x^{3/4}}{3} + C$
29. $\frac{x^4}{12} + \frac{3}{2x^2} + C$ **31.** $\frac{w^3}{2} + \frac{2}{3w} + C$
33. $\frac{1}{7}(z^2 - 5z) + C$ **35.** $\frac{u^{e+1}}{e + 1} + e^u + C$
37. $\frac{4x^{3/2}}{3} - \frac{12x^{5/4}}{5} + C$
39. $-\frac{3x^{5/3}}{25} - 7x^{1/2} + 3x^2 + C$
41. $\frac{x^4}{4} - x^3 + \frac{5x^2}{2} - 15x + C$ **43.** $\frac{2x^{5/2}}{5} + 2x^{3/2} + C$
45. $\frac{4u^3}{3} + 2u^2 + u + C$ **47.** $\frac{2v^3}{3} + 3v + \frac{1}{2v^4} + C$
49. $\frac{z^3}{6} + \frac{5z^2}{2} + C$ **51.** $x + e^x + C$
53. No, $F(x) - G(x)$ might be a nonzero constant
55. $\frac{1}{\sqrt{x^2 + 1}} + C$

APPLICATIONS IN PRACTICE 14.3

1. $N(t) = 800t + 200e^t + 6317.37$
2. $y(t) = 14t^3 + 12t^2 + 11t + 3$

EXERCISE 14.3 (page 703)

1. $y = \frac{3x^2}{2} - 4x + 1$ **3.** 18
5. $y = -\frac{x^4}{4} + \frac{2x^3}{3} + x + \frac{19}{12}$
7. $y = \frac{x^4}{12} + x^2 - 5x + 13$ **9.** $p = 0.7$
11. $p = 275 - 0.5q - 0.1q^2$ **13.** $c = 1.35q + 200$

15. \$7715 **17.** $G = -\frac{P^2}{50} + 2P + 20$

21. \$80 ($dc/dq = 27.50$ when $q = 50$ is not relevant to problem)

APPLICATIONS IN PRACTICE 14.4

1. $T(t) = 10e^{-0.5t} + C$ **2.** $35 \ln|t+1| + C$

EXERCISE 14.4 (page 710)

1. $\frac{(x+5)^8}{8} + C$ **3.** $\frac{(x^2+3)^6}{6} + C$

5. $\frac{3}{5}(y^3 + 3y^2 + 1)^{5/3} + C$ **7.** $-\frac{5(3x-1)^{-2}}{6} + C$

9. $\frac{1}{3}(2x-1)^{3/2} + C$ **11.** $\frac{(7x-6)^5}{35} + C$

13. $\frac{(5u^2-9)^{15}}{150} + C$ **15.** $\frac{3}{5}(27 + x^5)^{4/3} + C$

17. $e^{3x} + C$ **19.** $e^{t^2+t} + C$ **21.** $\frac{1}{14}e^{7x^2} + C$

23. $-3e^{-2x} + C$ **25.** $\ln|x+5| + C$

27. $\ln|x^3 + x^4| + C$ **29.** $-\frac{3}{4}(z^2 - 6)^{-4} + C$

31. $4 \ln|x| + C$ **33.** $\frac{1}{3}\ln|s^3 + 5| + C$

35. $-\frac{8}{3}\ln|5 - 3x| + C$

37. $\frac{2}{15}(5x)^{3/2} + C = \frac{2\sqrt{5}}{3}x^{3/2} + C$

39. $\sqrt{x^2 - 4} + C$ **41.** $\frac{1}{2}e^{y^4+1} + C$

43. $-\frac{1}{6}e^{-2v^3+1} + C$ **45.** $-\frac{1}{5}e^{-5x} + 2e^x + C$

47. $-\frac{1}{24}(3 - 3x^2 - 6x)^4 + C$ **49.** $\frac{1}{3}\ln|x^3 + 6x| + C$

51. $2 \ln|3 - 2s + 4s^2| + C$ **53.** $\frac{1}{4}\ln(2x^2 + 1) + C$

55. $\frac{1}{27}(x^3 - x^6)^{-9} + C$ **57.** $\frac{1}{4}(x^4 + x^2)^2 + C$

59. $\frac{1}{2}(4 - 9x - 3x^2)^{-4} + C$ **61.** $\frac{1}{6}e^{4x^3+3x^2-4} + C$

63. $-\frac{1}{25}(8 - 5x^2)^{5/2} + C$

65. $\frac{(2x)^{3/2}}{3} - \sqrt{2x} + C = \frac{2\sqrt{2}}{3}x^{3/2} - \sqrt{2}x^{1/2} + C$

67. $\frac{x^5}{5} + \frac{2x^3}{3} + x + C$

69. $\frac{1}{2}\ln(x^2+1) - \frac{1}{6(x^6+1)} + C$

71. $\frac{1}{3}\ln|3x - 5| + \frac{1}{27}(x^3 - x^6)^{-9} + C$

73. $\frac{2}{9}(3x+1)^{3/2} - \frac{1}{2}\ln(x^2+3)$ **75.** $2e^{\sqrt{x}} + C$

77. $-\frac{1}{4}e^{-x} + \frac{1}{4}e^x + C$ **79.** $\frac{1}{4}\ln^2(x^2 + 2x) + C$

81. $y = -\frac{1}{6}(3 - 2x)^3 + \frac{11}{2}$

83. $y = -\ln|x| = \ln|1/x|$ **85.** $160e^{0.05t} + 190$

87. $\frac{Rr^2}{4K} + B_1 \ln|r| + B_2$

EXERCISE 14.5 (page 715)

1. $\frac{x^5}{5} + \frac{4}{3}x^3 - 2x + C$

3. $\frac{1}{3}(2x^3 + 4x + 1)^{3/2} + C$

5. $-6\sqrt{4 - 5x} + C$ **7.** $\frac{4^{7x}}{7 \ln 4} + C$

9. $7x^2 - 4e^{(1/4)x^2} + C$

11. $x^2 - 3x + \frac{2}{3}\ln|3x - 1| + C$

13. $\frac{5}{14}\ln(7e^{2x} + 4) + C$ **15.** $-\frac{1}{7}e^{7/x} + C$

17. $x^2 + 4\ln|x^2 - 4| + C$ **19.** $\frac{2}{9}(\sqrt{x} + 2)^3 + C$

21. $3(x^{1/3} + 2)^5 + C$ **23.** $\frac{1}{2}(\ln^2 x) + C$

25. $\frac{1}{3}\ln^3(r+1) + C$ **27.** $\frac{3^{\ln x}}{\ln 3} + C$

29. $e^{(x^2+3)/2} + C$ **31.** $8 \ln|\ln(x+3)| + C$

33. $\frac{x^2}{2} + x + \ln|x^2 - 3| + C$

35. $\ln^{3/2}[(x^2+1)^2] + C$

37. $\frac{\sqrt{x^4 - 4x}}{2} - (\ln 7)x + C$

39. $x^2 - 8x - 6\ln|x| - \frac{2}{x^2} + C$

41. $x + \ln|x - 1| + C$ **43.** $\sqrt{e^{x^2} + 2} + C$

45. $-\frac{(e^{-x} + 6)^3}{3} + C$ **47.** $\frac{1}{5}(x^2 + e)^{5/2} + C$

49. $\frac{1}{36\sqrt{2}}[(8x)^{3/2} + 3]^{3/2} + C$

51. $-\frac{2}{3}e^{-\sqrt{s^3}} + C$ **53.** $\frac{x^2}{2} + 2x + C$

55. $\frac{\ln^2 x}{2} + x + C$ **57.** $p = -\frac{200}{q(q+2)}$

59. $c = 20 \ln|(q+5)/5| + 2000$

61. $C = 2(\sqrt{I} + 1)$ **63.** $C = \frac{3}{4}I - \frac{1}{3}\sqrt{I} + \frac{71}{12}$

65. **(a)** \$150 per unit; **(b)** \$15,000; **(c)** \$15,300

67. $2500 - 800\sqrt{5} \approx$ \$711 per acre **69.** $I = 3$

EXERCISE 14.6 (page 720)

1. 35 **3.** 0 **5.** 25 **7.** $-\frac{3}{16}$ **9.** $\frac{5}{6}$

11. $\sum_{k=1}^{19} k$ **13.** $\sum_{k=1}^{4}(2k - 1)$ **15.** $\sum_{k=1}^{10} k^2$

17. 101,475 **19.** 84 **21.** 273 **23.** 8; \$850

APPLICATIONS IN PRACTICE 14.7

1. \$5975

EXERCISE 14.7 (page 728)

1. $\frac{2}{3}$ square unit **3.** $\frac{15}{32}$ square unit

5. $S_n = \frac{1}{n}\left[4\left(\frac{1}{n}\right) + 4\left(\frac{2}{n}\right) + \cdots + 4\left(\frac{n}{n}\right)\right] = \frac{2(n+1)}{n}$

7. (a) $S_n = \frac{n+1}{2n} + 1$; **(b)** $\frac{3}{2}$ **9.** $\frac{1}{2}$ square unit

11. $\frac{1}{3}$ square unit **13.** $\frac{16}{3}$ square units **15.** 20

17. -18 **19.** $\frac{5}{6}$ **21.** 0 **23.** $\frac{11}{4}$

25. 4.3 square units **27.** 2.4 **29.** -25.5

APPLICATIONS IN PRACTICE 14.8

1. \$32,830 **2.** \$28,750

EXERCISE 14.8 (page 735)

1. 14 **3.** $\frac{15}{2}$ **5.** -20 **7.** $\frac{7}{3}$ **9.** $\frac{15}{2}$

11. $\frac{4}{3}$ **13.** 0 **15.** $\frac{5}{3}$ **17.** $\frac{32}{3}$ **19.** $-\frac{1}{6}$

21. $4 \ln 8$ **23.** e^5 **25.** $\frac{1}{3}(e^8 - 1)$ **27.** $\frac{3}{4}$

29. $\frac{38}{9}$ **31.** $\frac{15}{28}$ **33.** $\frac{1}{2}\ln 3$ **35.** $\frac{1}{2}\left(e + \frac{1}{e} - 2\right)$

37. $3 - \frac{2}{e} + \frac{1}{e^2}$ **39.** $\frac{e^3}{2}(e^{12} - 1)$ **41.** $6 + \ln 19$

43. $\frac{47}{12}$ **45.** $6 - 3e$ **47.** 7 **49.** 0 **51.** $\alpha^{5/2}T$

53. $\int_a^b -Ax^{-B}dx$ **55.** \$8639 **57.** 1,973,333

59. \$220 **61.** \$2000 **63.** 696; 492 **65.** $2Ri$

69. 0.05 **71.** 3.52 **73.** 55.39

APPLICATIONS IN PRACTICE 14.9

1. 76.90 feet **2.** 5.77 grams

EXERCISE 14.9 (page 743)

1. 413 **3.** 0.340; $\frac{1}{3} \approx 0.333$ **5.** ≈ 0.767; 0.750

7. 0.883 **9.** 2,430,733 **11.** 3.0 square units **13.** $\frac{8}{3}$

15. 0.771 **17.** $\frac{35}{6}$ km^2

EXERCISE 14.10 (page 747)

In Problems ***1–33****, answers are assumed to be expressed in square units.*

1. 8 **3.** $\frac{19}{2}$ **5.** 8 **7.** $\frac{19}{3}$ **9.** 9 **11.** $\frac{50}{3}$

13. $\frac{125}{6}$ **15.** 8 **17.** $\frac{32}{3}$ **19.** 1 **21.** 18

23. $\frac{26}{3}$ **25.** $\frac{3}{2}\sqrt[3]{2}$ **27.** $e^2 - 1$

29. $\frac{3}{2} + 2\ln 2 = \frac{3}{2} + \ln 4$ **31.** 68 **33.** 2

35. 19 square units **37. (a)** $\frac{1}{16}$; **(b)** $\frac{3}{4}$; **(c)** $\frac{7}{16}$

39. (a) $\ln\frac{5}{3}$; **(b)** $\ln 4 - 1$; **(c)** $2 - \ln 3$

41. 1.89 square units **43.** 11.41 square units

EXERCISE 14.11 (page 754)

1. Area $= \int_{-2}^{3}[(x+6) - x^2]dx$

3. Area $= \int_0^3 [2x - (x^2 - x)]dx + \int_3^4 [(x^2 - x) - 2x]dx$

5. Area $= \int_0^1 [(y+1) - \sqrt{1-y}]dy$

7. Area $= \int_1^2 [(7 - 2x^2) - (x^2 - 5)]dx$

In Problems ***9–33****, answers are assumed to be expressed in square units.*

9. $\frac{4}{3}$ **11.** $\frac{16}{3}$ **13.** $8\sqrt{6}$ **15.** 40 **17.** $\frac{125}{6}$

19. $\frac{9}{2}$ **21.** $\frac{125}{12}$ **23.** $\frac{32}{81}$ **25.** $\frac{44}{3}$

27. $\frac{4}{3}(5\sqrt{5} - 2\sqrt{2})$ **29.** $\frac{1}{2}$ **31.** $\frac{255}{32} - 4\ln 2$

33. 12 **35.** $\frac{20}{63}$ **37.** $\frac{3}{2m^3}$ square units **39.** $2^{4/3}$

41. 4.76 square units **43.** 6.17 square units

EXERCISE 14.12 (page 758)

1. CS = 25.6, PS = 38.4

3. CS = $50 \ln 2 - 25$, PS = 1.25

5. CS = 800, PS = 1000 **7.** \$426.67 **9.** \$254,000

11. CS $\approx$ 1197, PS $\approx$ 477

REVIEW PROBLEMS—CHAPTER 14 (page 761)

1. $\frac{x^4}{4} + x^2 - 7x + C$ **3.** $\frac{256}{3}$

5. $-3(x+5)^{-2} + C$ **7.** $2\ln|x^3 - 6x + 1| + C$

9. $\frac{11\sqrt[3]{11}}{4} - 4$ **11.** $\frac{y^4}{4} + \frac{2y^3}{3} + \frac{y^2}{2} + C$

13. $\frac{4z^{3/4}}{3} - \frac{6z^{5/6}}{5} + C$ **15.** $\frac{1}{3}\ln\frac{57}{5}$

17. $\frac{2}{27}(3x^3 + 2)^{3/2} + C$ **19.** $\frac{1}{2}(e^{2y} + e^{-2y}) + C$

21. $\ln|x| - \frac{2}{x} + C$ **23.** 111 **25.** $\frac{7}{3}$

27. $4 - 3\sqrt[3]{2}$ **29.** $\frac{3}{t} - \frac{2}{\sqrt{t}} + C$ **31.** $\frac{3}{2} - 5\ln 2$

33. $4(x^{3/2} + 1)^{3/2} + C$ **35.** 1 **37.** $\frac{(1 + e^{3x})^3}{9} + C$

39. $\frac{2\sqrt{10^{3x}}}{\ln 10} + C$ **41.** $y = \frac{1}{2}e^{2x} + 3x - 1$

In Problems 43–57, answers are assumed to be expressed in square units.

43. $\frac{4}{3}$ **45.** $\frac{16}{3}$ **47.** $\frac{125}{6}$ **49.** $6 + \ln 3$ **51.** $\frac{2}{3}$

53. 36 **55.** $\frac{125}{3}$ **57.** $e - 1$

59. $p = 100 - \sqrt{2q}$ **61.** \$1900 **63.** 0.5507

65. 15 square units **67.** CS $= 166\frac{2}{3}$, PS $= 53\frac{1}{3}$

73. 24.71 square units **75.** CS ≈ 1148, PS ≈ 251

MATHEMATICAL SNAPSHOT—CHAPTER 14 (page 764)

1. (a) 225; **(b)** 125

3. (a) \$2,002,500; **(b)** 18,000; **(c)** \$111.25

APPLICATIONS IN PRACTICE 15.1

1. $S(t) = -40te^{0.1t} + 400e^{0.1t} + 4600$

2. $P(t) = 0.025t^2 - 0.05t^2 \ln t + 0.05t^2(\ln t)^2 + C$

EXERCISE 15.1 (page 770)

1. $\frac{2}{3}x(x+5)^{3/2} - \frac{4}{15}(x+5)^{5/2} + C$

3. $-e^{-x}(x+1) + C$ **5.** $\frac{y^4}{4}\left[\ln(y) - \frac{1}{4}\right] + C$

7. $x[\ln(4x) - 1] + C$

9. $10x(x+1)^{3/2} - 4(x+1)^{5/2} + C$
$= 2(x+1)^{3/2}(3x-2) + C$

11. $-\frac{x}{10(5x+2)^2} - \frac{1}{50(5x+2)} + C$

13. $-\frac{1}{x}(1 + \ln x) + C$ **15.** $e^2(3e^2 - 1)$

17. $\frac{1}{2}(1 - e^{-1})$, parts not needed

19. $2(9\sqrt{3} - 10\sqrt{2})$

21. $2x(x-1)\ln(x-1) - x^2 + C$

23. $e^x(x^2 - 2x + 2) + C$

25. $\frac{x^3}{3} + 2e^{-x}(x+1) - \frac{e^{-2x}}{2} + C$

27. $\frac{e^{x^2}}{2}(x^2 - 1) + C$

29. $\frac{2^{2x-1}}{\ln 2} + \frac{2^{x+1}x}{\ln 2} - \frac{2^{x+1}}{\ln^2 2} + \frac{x^3}{3} + C$

31. $2e^3 + 1$ square units **33.** $\frac{298}{15}$ square units

37. $\int f^{-1}(x)dx = xf^{-1}(x) - F(f^{-1}(x)) + C$

APPLICATIONS IN PRACTICE 15.2

1. $r(q) = \frac{5}{2}\ln\left|\frac{3(q+1)^3}{q+3}\right|$

2. $V(t) = 150t^2 - 900\ln(t^2 + 6) + C$

EXERCISE 15.2 (page 777)

1. $\frac{12}{x+6} - \frac{2}{x+1}$ **3.** $1 + \frac{2}{x+2} - \frac{8}{x+4}$

5. $\frac{1}{x+2} + \frac{2}{(x+2)^2}$ **7.** $\frac{3}{x} - \frac{2x}{x^2+1}$

9. $2\ln|x| + 3\ln|x-1| + C = \ln|x^2(x-1)^3| + C$

11. $-3\ln|x+1| + 4\ln|x-2| + C$
$= \ln\left|\frac{(x-2)^4}{(x+1)^3}\right| + C$

13. $\frac{1}{4}\left[\frac{3x^2}{2} + 2\ln|x-1| - 2\ln|x+1|\right] + C$
$= \frac{1}{4}\left(\frac{3x^2}{2} + \ln\left[\frac{x-1}{x+1}\right]^2\right) + C$

15. $\ln|x| + 2\ln|x-4| - 3\ln|x+3| + C$
$= \ln\left|\frac{x(x-4)^2}{(x+3)^3}\right| + C$

17. $\ln|x^6 + 2x^4 - x^2 - 2| + C$, partial fractions not required

19. $\frac{4}{x-2} - 5\ln|x-1| + 7\ln|x-2| + C$
$= \frac{4}{x-2} + \ln\left|\frac{(x-2)^7}{(x-1)^5}\right| + C$

21. $4\ln|x| - \ln(x^2+4) + C = \ln\left[\frac{x^4}{x^2+4}\right] + C$

23. $-\frac{1}{2}\ln(x^2+1) - \frac{2}{x-3} + C$

25. $5\ln(x^2+1) + 2\ln(x^2+2) + C$
$= \ln[(x^2+1)^5(x^2+2)^2] + C$

27. $\frac{3}{2}\ln(x^2+1) + \frac{1}{x^2+1} + C$

29. $18\ln(4) - 10\ln(5) - 8\ln(3)$

31. $11 + 24\ln\frac{2}{3}$ square units

EXERCISE 15.3 (page 784)

1. $\frac{x}{9\sqrt{9-x^2}} + C$ **3.** $-\frac{\sqrt{16x^2+3}}{3x} + C$

5. $\frac{1}{6}\ln\left|\frac{x}{6+7x}\right| + C$ **7.** $\frac{1}{3}\ln\left|\frac{\sqrt{x^2+9}-3}{x}\right| + C$

9. $\frac{1}{2}\left[\frac{4}{5}\ln|4+5x| - \frac{2}{3}\ln|2+3x|\right] + C$

11. $\frac{1}{8}(2x - \ln[4 + 3e^{2x}]) + C$

13. $7\left[\frac{1}{5(5+2x)} + \frac{1}{25}\ln\left|\frac{x}{5+2x}\right|\right] + C$

15. $1 + \ln\frac{4}{9}$

17. $\frac{1}{2}(x\sqrt{x^2-3} - 3\ln|x + \sqrt{x^2-3}|) + C$

19. $\frac{1}{144}$ **21.** $e^x(x^2 - 2x + 2) + C$

23. $2\left(-\frac{\sqrt{4x^2+1}}{2x} + \ln|2x + \sqrt{4x^2+1}|\right) + C$

25. $\frac{1}{9}\left(\ln|1+3x| + \frac{1}{1+3x}\right) + C$

27. $\frac{1}{\sqrt{5}}\left(\frac{1}{2\sqrt{7}}\ln\left|\frac{\sqrt{7}+\sqrt{5}x}{\sqrt{7}-\sqrt{5}x}\right|\right) + C$

29. $\frac{4}{81}\left[\frac{(3x)^6 \ln(3x)}{6} - \frac{(3x)^6}{36}\right] + C$
$= x^6[6\ln(3x) - 1] + C$

31. $4(9x - 2)(1 + 3x)^{3/2} + C$

33. $\frac{1}{2}\ln|2x + \sqrt{4x^2 - 13}| + C$

35. $-\frac{\sqrt{9 - 4x^2}}{9x} + C$

37. $\frac{1}{2\pi}(4\sqrt{x} - \ln|\pi + 7e^{4\sqrt{x}}|) + C$

39. $\frac{1}{2}\ln(x^2 + 1) + C$ **41.** $(2x^2 + 1)^{3/2} + C$

43. $\ln\left|\frac{x - 3}{x - 2}\right| + C$ **45.** $\frac{x^4}{4}\left[\ln(x) - \frac{1}{4}\right] + C$

47. $e^{2x}(2x - 1) + C$

49. $x(\ln x)^2 - 2x\ln(x) + 2x + C$

51. $\frac{2}{3}(9\sqrt{3} - 10\sqrt{2})$ **53.** $2(2\sqrt{2} - \sqrt{7})$

55. $\frac{7}{2}\ln(2) - \frac{3}{4}$ **57.** $\ln\left|\frac{q_n(1 - q_0)}{q_0(1 - q_n)}\right|$

59. **(a)** \$37,599; **(b)** \$4924 **61.** **(a)** \$5481; **(b)** \$535

EXERCISE 15.4 (page 787)

1. $\frac{7}{3}$ **3.** -1 **5.** 0 **7.** 13 **9.** \$11,050

11. \$3155

APPLICATIONS IN PRACTICE 15.5

1. $I = I_0 e^{-0.0085x}$

EXERCISE 15.5 (page 793)

1. $y = -\frac{1}{x^2 + C}$ **3.** $y = (x^2 + 1)^{3/2} + C$

5. $y = Ce^x, C > 0$ **7.** $y = Cx, C > 0$

9. $y = \sqrt[3]{3x - 2}$ **11.** $y = \ln\frac{x^3 + 3}{3}$

13. $y = \frac{4x^2 + 3}{2(x^2 + 1)}$ **15.** $y = \sqrt{\left(\frac{3x^2}{2} + \frac{3}{2}\right)^2 - 1}$

17. $y = \ln\left(\frac{1}{2}\sqrt{x^2 + 3}\right)$ **19.** $c = (q + 1)e^{1/(q+1)}$

21. 120 weeks

23. $N = 40{,}000e^{0.018t}$; $N = 40{,}000(1.2)^{t/10}$; 57,600

25. $2e^{1.08124}$ billion **27.** 0.01204; 57.57 sec

29. 2900 years **31.** $N = N_0 e^{k(t - t_0)}, t \geq t_0$

33. 12.6 units **35.** $A = 400(1 - e^{-t/2})$, 157 g/m^2

37. **(a)** $V = 21{,}000e^{(2\ln 0.9)t}$; **(b)** June 2002

EXERCISE 15.6 (page 801)

1. 58,800 **3.** 500 **5.** 1990 **7.** **(b)** 375

9. 3:21 A.M **11.** \$62,500

13. $N = M - (M - N_0)e^{-kt}$

APPLICATIONS IN PRACTICE 15.7

1. 20 ml

EXERCISE 15.7 (page 805)

1. $\frac{1}{3}$ **3.** Div **5.** $\frac{1}{e}$ **7.** Div **9.** $-\frac{1}{2}$ **11.** 0

13. **(a)** 800; **(b)** $\frac{2}{3}$ **15.** 4,000,000 **17.** $\frac{1}{3}$ square unit

19. 20,000 increase

REVIEW PROBLEMS—CHAPTER 15 (page 808)

1. $\frac{x^2}{4}[2\ln(x) - 1] + C$ **3.** $5 + \frac{9}{4}\ln 3$

5. $\ln|3x + 1| + 4\ln|x - 2| + C$

7. $\frac{1}{2(x + 2)} + \frac{1}{4}\ln\left|\frac{x}{x + 2}\right| + C$

9. $-\frac{\sqrt{9 - 16x^2}}{9x} + C$ **11.** $\frac{3}{2}\ln\left|\frac{x - 3}{x + 3}\right| + C$

13. $e^{7x}(7x - 1) + C$ **15.** $\frac{1}{2}\ln|\ln 2x| + C$

17. $x - \frac{3}{2}\ln|3 + 2x| + C$

19. $2\ln|x| + \frac{3}{2}\ln(x^2 + 1) + C$

21. $2\sqrt{x + 1}[\ln(x + 1) - 2] + C$ **23.** 34

25. $y = Ce^{x^3 + x^2}, C > 0$ **27.** $\frac{1}{18}$ **29.** Div

31. 144,000 **33.** 0.0005; 90%

35. $N = \frac{450}{1 + 224e^{-1.02t}}$ **37.** 4:16 P.M **39.** 1

41. **(a)** 207, 208; **(b)** 157, 165; **(c)** 41, 41

MATHEMATICAL SNAPSHOT—CHAPTER 15 (page 810)

1. 114; 69 **5.** Answers may vary

APPLICATIONS IN PRACTICE 16.1

1. $\frac{1}{3}$ **2.** 0.607

3. Mean 5 years, standard deviation 5 years

EXERCISE 16.1 (page 819)

1. **(a)** $\frac{5}{12}$; **(b)** $\frac{11}{16} = 0.6875$; **(c)** $\frac{13}{16} = 0.8125$;
(d) $-1 + \sqrt{10}$

3. **(a)** $f(x) = \begin{cases} \frac{1}{4}, & \text{if } 2 \leq x \leq 6 \\ 0, & \text{otherwise} \end{cases}$

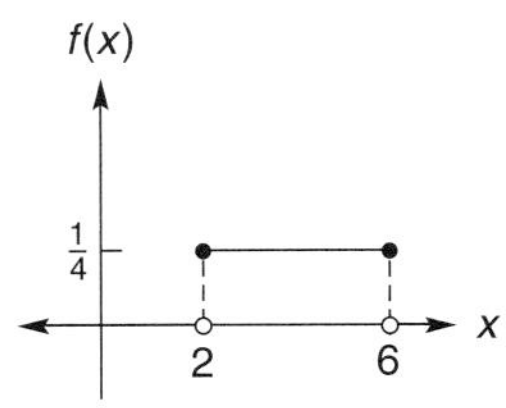

(b) $\frac{1}{4}$; **(c)** 0; **(d)** $\frac{3}{8}$; **(e)** $\frac{3}{4}$; **(f)** 0; **(g)** 1; **(h)** 4; **(i)** $\frac{2}{\sqrt{3}}$;

(j) $F(x) = \begin{cases} 0, & \text{if } x < 2 \\ \dfrac{x-2}{4}, & \text{if } 2 \le x \le 6 \\ 1, & \text{if } x > 6 \end{cases}$

$P(X < 3) = \frac{1}{4}, P(2 < X < 5) = \frac{3}{4}$

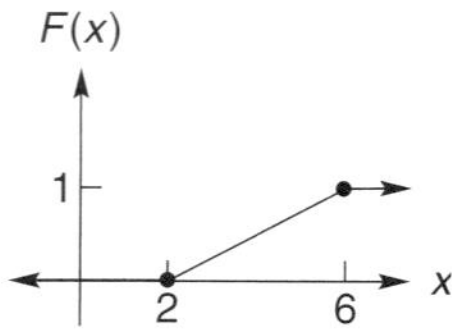

5. (a) $f(x) = \begin{cases} \dfrac{1}{b-a}, & \text{if } a \le x \le b \\ 0, & \text{otherwise} \end{cases}$

(b) $\frac{a+b}{2}$; **(c)** $\sigma^2 = \frac{(b-a)^2}{12}, \sigma = \frac{b-a}{\sqrt{12}}$

7. (a) $-e^{-15} + e^{-6} \approx 0.0248$; **(b)** $-e^{-9} + 1 \approx 1$; **(c)** $e^{-15} \approx 0.000$; **(d)** $-e^{-2} + 1 \approx 0.865$;

(f) $F(x) = \begin{cases} 0, & \text{if } x < 0 \\ 1 - e^{-3x}, & \text{if } x \ge 0 \end{cases}$

9. (a) $\frac{1}{8}$; **(b)** $\frac{5}{16}$; **(c)** $\frac{39}{64} \approx 0.609$; **(d)** 1; **(e)** $\frac{8}{3}$; **(f)** $\frac{2\sqrt{2}}{3}$; **(g)** $2\sqrt{2}$; **(h)** $\frac{7}{16}$ **11.** $\frac{7}{10}$; 5 min **13.** $e^{-3} \approx 0.050$

EXERCISE 16.2 (page 825)

1. (a) 0.4554; **(b)** 0.3317; **(c)** 0.8907; **(d)** 0.9982; **(e)** 0.8972; **(f)** 0.4880 **3.** 0.13 **5.** −1.08 **7.** 0.34 **9. (a)** 0.9970; **(b)** 0.0668; **(c)** 0.0873 **11.** 0.3085 **13.** 0.9078 **15.** 8 **17.** 95% **19.** 90.82% **21. (a)** 1.7%; **(b)** 85.6

APPLICATIONS IN PRACTICE 16.3

1. 0.0396

EXERCISE 16.3 (page 830)

1. 0.1056; 0.0122 **3.** 0.0430; 0.9232 **5.** 0.7507 **7.** 0.4129 **9.** 0.5; 0.0287 **11.** 0.0336

REVIEW PROBLEMS—CHAPTER 16 (page 831)

1. (a) 2; **(b)** $\frac{9}{32}$; **(c)** $\frac{3}{4}$;

(d) $F(x) = \begin{cases} 0, & \text{if } x < 0 \\ \dfrac{x}{3} + \dfrac{2x^3}{3}, & \text{if } 0 \le x \le 1 \\ 1, & \text{if } x > 1 \end{cases}$

3. (a) $\frac{10}{3}$; **(b)** $\sqrt{\frac{25}{18}} \approx 1.18$ **5.** 0.3085 **7.** 0.2417 **9.** 0.1587 **11.** 0.9817 **13.** 0.0228

MATHEMATICAL SNAPSHOT—CHAPTER 16 (page 832)

1. The result should correspond to the known distribution function. **3.** Answers may vary

APPLICATIONS IN PRACTICE 17.1

1. (a) \$3260; **(b)** \$4410

EXERCISE 17.1 (page 840)

1. 3 **3.** −2 **5.** −1 **7.** 88 **9.** 3 **11.** $e^{x_0+h+y_0}$ **13.** 2000 **15.** $y = -4$ **17.** $z = 6$

19.

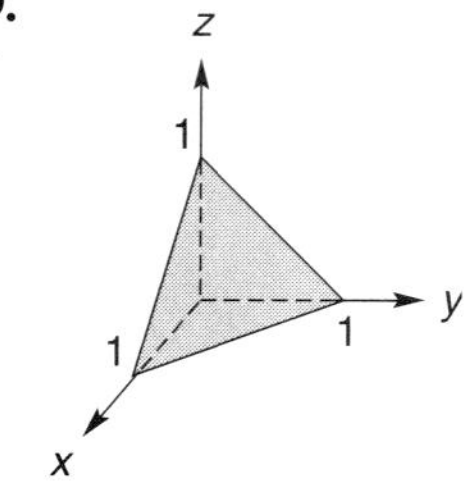

21.

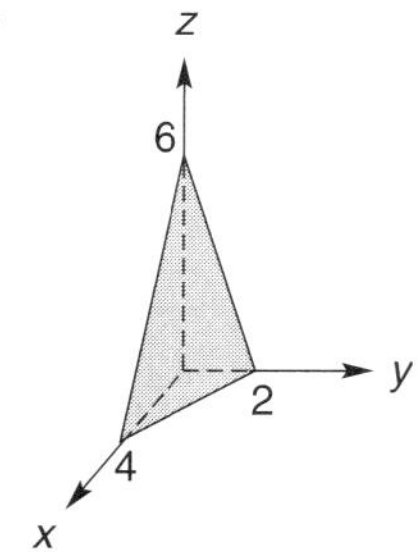

23.

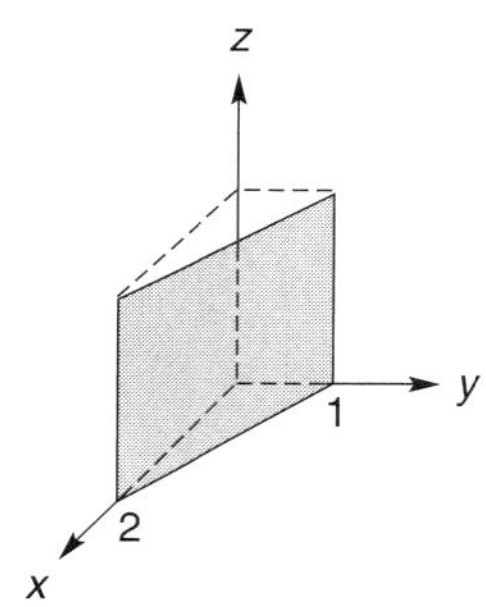

25.

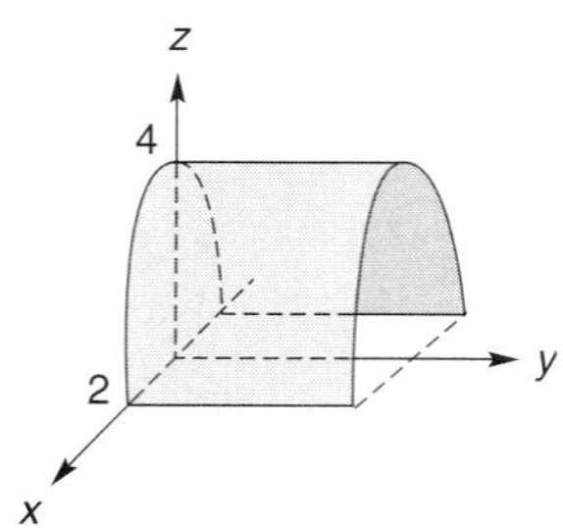

27.

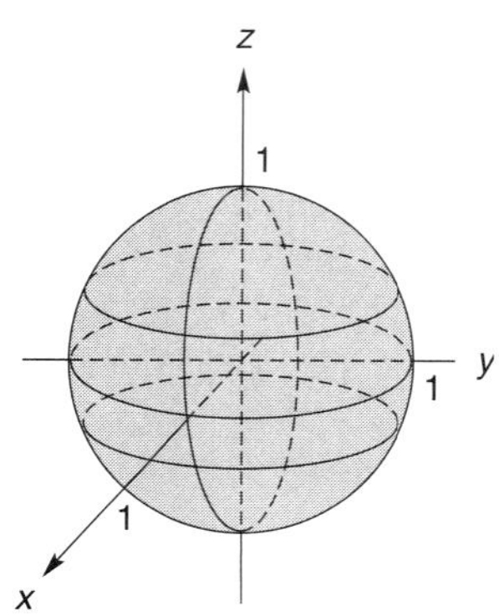

EXERCISE 17.2 (page 846)

1. $f_x(x, y) = 8x; f_y(x, y) = 6y$
3. $f_x(x, y) = 0; f_y(x, y) = 2$
5. $g_x(x, y) = 12x^3y + 2y^2 - 5y + 8;$ $g_y(x, y) = 3x^4 + 4xy - 5x - 9$
7. $g_p(p, q) = \frac{q}{2\sqrt{pq}}; g_q(p, q) = \frac{p}{2\sqrt{pq}}$
9. $h_s(s, t) = \frac{2s}{t-3}; h_t(s, t) = -\frac{s^2+4}{(t-3)^2}$
11. $u_{q_1}(q_1, q_2) = \frac{3}{4q_1}; u_{q_2}(q_1, q_2) = \frac{1}{4q_2}$
13. $h_x(x, y) = (x^3 + xy^2 + 3y^3)(x^2 + y^2)^{-3/2};$ $h_y(x, y) = (3x^3 + x^2y + y^3)(x^2 + y^2)^{-3/2}$
15. $\frac{\partial z}{\partial x} = 5ye^{5xy}; \frac{\partial z}{\partial y} = 5xe^{5xy}$
17. $\frac{\partial z}{\partial x} = 5\left[\frac{2x^2}{x^2+y} + \ln(x^2 + y)\right]; \frac{\partial z}{\partial y} = \frac{5x}{x^2+y}$
19. $f_r(r, s) = \sqrt{r + 2s}\,(3r^2 - 2s) + \frac{r^3 - 2rs + s^2}{2\sqrt{r+2s}};$ $f_s(r, s) = 2(s - r)\sqrt{r + 2s} + \frac{r^3 - 2rs + s^2}{\sqrt{r+2s}}$
21. $f_r(r, s) = -e^{3-r}\ln(7 - s); f_s(r, s) = \frac{e^{3-r}}{s-7}$
23. $g_x(x, y, z) = 6xy + 2y^2z; g_y(x, y, z) = 3x^2 + 4xyz;$ $g_z(x, y, z) = 2xy^2 + 9z^2$
25. $g_r(r, s, t) = 2re^{s+t};$ $g_s(r, s, t) = (7s^3 + 21s^2 + r^2)e^{s+t};$ $g_t(r, s, t) = e^{s+t}(r^2 + 7s^3)$

27. 50 29. $\frac{1}{\sqrt{14}}$ 31. 0 33. 26

39. $-\frac{ra}{2\left[1 + a\frac{n-1}{2}\right]^2}$

EXERCISE 17.3 (page 851)

1. 20 3. 1374.5
5. $\frac{\partial P}{\partial k} = 1.208648l^{0.192}k^{-0.236}; \frac{\partial P}{\partial l} = 0.303744l^{-0.808}k^{0.764}$
7. $\frac{\partial q_A}{\partial p_A} = -50; \frac{\partial q_A}{\partial p_B} = 2; \frac{\partial q_B}{\partial p_A} = 4; \frac{\partial q_B}{\partial p_B} = -20;$ competitive
9. $\frac{\partial q_A}{\partial p_A} = -\frac{100}{p_A^2 p_B^{1/2}}; \frac{\partial q_A}{\partial p_B} = -\frac{50}{p_A p_B^{3/2}};$ $\frac{\partial q_B}{\partial p_A} = -\frac{500}{3p_B p_A^{4/3}}; \frac{\partial q_B}{\partial p_B} = -\frac{500}{p_B^2 p_A^{1/3}};$ complementary
11. $\frac{\partial P}{\partial B} = 0.01A^{0.27}B^{-0.99}C^{0.01}D^{0.23}E^{0.09}F^{0.27};$ $\frac{\partial P}{\partial C} = 0.01A^{0.27}B^{0.01}C^{-0.99}D^{0.23}E^{0.09}F^{0.27}$
13. 4480; if a staff manager with an M.B.A. degree had an extra year of work experience before the degree, the manager would receive \$4480 per year in extra compensation.
15. (a) -1.015; -0.846; (b) One for which $w = w_0$ and $s = s_0$.
17. $\frac{\partial g}{\partial x} = \frac{1}{V_F} > 0$ for $V_F > 0$. Thus if x increases and V_F and V_s are fixed, then g increases.
19. (a) When $p_A = 8$ and $p_B = 64$, $\frac{\partial q_A}{\partial p_A} = -5$ and $\frac{\partial q_A}{\partial p_B} = \frac{15}{32}$; (b) Demand for A decreases by approximately $\frac{15}{8}$ units.
21. (a) No; (b) 70% 23. $\eta_{p_A} = -\frac{5}{46}, \eta_{p_B} = \frac{1}{46}$
25. $\eta_{p_A} = -1, \eta_{p_B} = -\frac{1}{2}$

EXERCISE 17.4 (page 856)

1. $-\frac{2x}{5z}$ 3. $\frac{4y}{3z^2}$ 5. $\frac{x(yz^2 + 1)}{z(1 - x^2y)}$ 7. $-e^{y-z}$
9. $\frac{yz}{1 + 9z}$ 11. $-\frac{3x}{z}$
13. $\frac{\partial z}{\partial x}(2xz + y) = 28x^3 - z^2$ So $\frac{\partial z}{\partial x}$ cannot be determined for $x = y = z = 0$ 15. $-\frac{4}{e^2}$ 17. 4
19. $\frac{5}{2}$ 21. (a) 36; (b) With respect to q_A, $\frac{60}{13}$; with respect to q_B, $\frac{288}{65}$

EXERCISE 17.5 (page 859)

1. $8xy$; $8x$ 3. 3; 0; 0
5. $18xe^{2xy}$; $18e^{2xy}(2xy + 1)$; $72x(1 + xy)e^{2xy}$
7. $3x^2y + 4xy^2 + y^3$; $3xy^2 + 4x^2y + x^3$; $6xy + 4y^2$; $6xy + 4x^2$
9. $\frac{zy}{\sqrt{x^2 + y^2}}; \frac{z}{(x^2 + y^2)^{\frac{3}{2}}}[x^2 + y^2\sqrt{x^2 + y^2}]$ 11. 0
13. 28,758 15. $2e$ 17. $-\frac{1}{8}$ 23. $-\frac{y^2 + z^2}{z^3} = -\frac{3x^2}{z^3}$

EXERCISE 17.6 (page 862)

1. $\frac{\partial z}{\partial r} = 13; \frac{\partial z}{\partial s} = 9$ 3. $\left[2t + \frac{3\sqrt{t}}{2}\right]e^{x+y}$
5. $5(2xz^2 + yz) + 2(xz + z^2) - (2x^2z + xy + 2yz)$
7. $3(x^2 + xy^2)^2(2x + y^2 + 16xy)$
9. $-2s(2x + yz) + r(xz + 3y^2z^2) - 5(xy + 2y^3z)$
11. $19s(2x - 7)$ 13. 324 15. $\frac{40}{e^9}$
17. When $p_A = 25$ and $p_B = 4$, $\frac{\partial c}{\partial p_A} = -\frac{1}{4}$ and $\frac{\partial c}{\partial p_B} = \frac{5}{4}$
19. (a) $\frac{\partial w}{\partial t} = \frac{\partial w}{\partial x}\frac{\partial x}{\partial t} + \frac{\partial w}{\partial y}\frac{\partial y}{\partial t}$; (b) $-\frac{20}{3\sqrt{2} + 15e}$

EXERCISE 17.7 (page 871)

1. $\left(\frac{14}{3}, -\frac{13}{3}\right)$ **3.** $(2, 5), (2, -6), (-1, 5), (-1, -6)$

5. $(50, 150, 350)$ **7.** $\left(-2, \frac{3}{2}\right)$, rel. min.

9. $\left(-\frac{1}{4}, \frac{1}{2}\right)$, rel. max.

11. $\left(\frac{2}{5}, -\frac{3}{5}\right)$; $D = -5 < 0$ no relative extremum

13. $(0, 0)$, rel. max.; $\left(4, \frac{1}{2}\right)$, rel. min.; $\left(0, \frac{1}{2}\right)$, $(4, 0)$, neither

15. $(122, 127)$, rel. max. **17.** $(-1, -1)$, rel. min.

19. $(0, -2)$, $(0, 2)$, neither **21.** $l = 24, k = 14$

23. $p_A = 80, p_B = 85$

25. $q_A = 48, q_B = 40, p_A = 52, p_B = 44$, profit $= 3304$

27. $q_A = 3, q_B = 2$ **29.** 1 ft by 2 ft by 3 ft

31. $\left(\frac{105}{37}, \frac{28}{37}\right)$, rel. min. **33.** $a = -8, b = -12, d = 33$

35. **(a)** 2 units of A and 3 units B;
(b) Selling price for A is 30 and selling price for B is 19. Relative maximum profit is 25.

37. **(a)** $P = 5T(1 - e^{-x}) - 20x - 0.1T^2$;
(c) Relative maximum at $(20, \ln 5)$; no relative extremum at $\left(5, \ln\frac{5}{4}\right)$

EXERCISE 17.8 (page 879)

1. $(2, -2)$ **3.** $\left(3, \frac{3}{2}, -\frac{3}{2}\right)$ **5.** $\left(0, \frac{1}{4}, \frac{5}{8}\right)$

7. $(6, 3, 2)$ **9.** $\left(\frac{2}{3}, \frac{4}{3}, -\frac{4}{3}\right)$ **11.** $(3, 3, 6)$

13. Plant 1, 40 units; plant 2, 60 units

15. 74 units (when $l = 8, k = 7$)

17. \$15,000 on newspaper advertising and \$45,000 on TV advertising

19. $x = 5, y = 15, z = 5$

21. $x = 12, y = 8$ **23.** $x = 10, y = 20, z = 5$

EXERCISE 17.9 (page 887)

1. $\hat{y} = 0.98 + 0.61x$; 3.12 **3.** $\hat{y} = 0.057 + 1.67x$; 5.90

5. $\hat{q} = 82.6 - 0.641p$ **7.** $\hat{y} = 100 + 0.13x$; 105.2

9. $\hat{y} = 8.5 + 2.5x$

11. **(a)** $\hat{y} = 35.9 - 2.5x$; **(b)** $\hat{y} = 28.4 - 2.5x$

EXERCISE 17.11 (page 893)

1. 18 **3.** $\frac{1}{4}$ **5.** $\frac{2}{3}$ **7.** 3 **9.** $\frac{525}{2}$ **11.** $-\frac{58}{5}$

13. $\frac{8}{3}$ **15.** -1 **17.** $\frac{e^2}{2} - e + \frac{1}{2}$ **19.** $-\frac{27}{4}$

21. $\frac{1}{24}$ **23.** $e^{-4} - e^{-2} - e^{-3} + e^{-1}$ **25.** $\frac{3}{8}$

REVIEW PROBLEMS—CHAPTER 17 (page 895)

1.

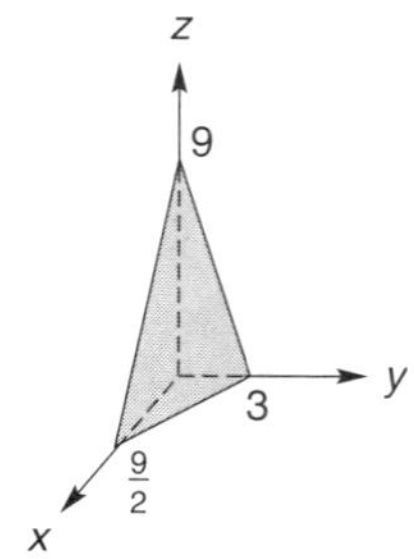

3.

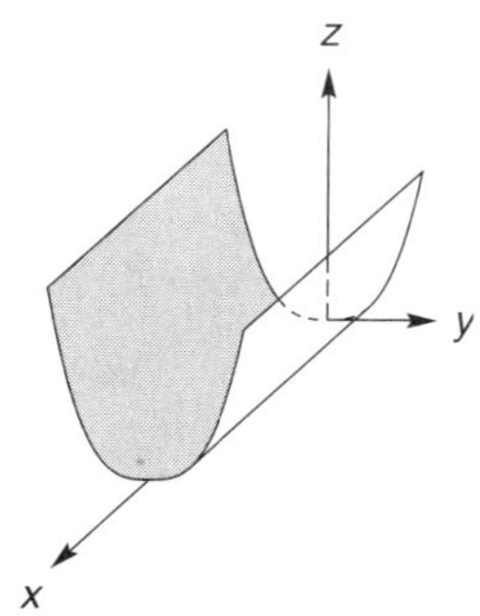

5. $8x + 6y$; $6x + 2y$ **7.** $\frac{y}{(x+y)^2}$; $-\frac{x}{(x+y)^2}$

9. $\frac{y}{x^2 + y^2}$ **11.** $2xze^{x^2yz}(1 + x^2yz)$ **13.** $2(x + y)$

15. $e^{x+y+z}\left[\ln xyz + \frac{1}{y}\right]$; $e^{x+y+z}\left[\ln xyz + \frac{1}{x} + \frac{1}{z}\right]$

17. $\frac{1}{64}$ **19.** $2(x + y)e^r + 2\left(\frac{x + 3y}{r + s}\right)$; $2\left(\frac{x + 3y}{r + s}\right)$

21. $\frac{2x + 2y + z}{4z - x}$ **23.** $\frac{\partial P}{\partial l} = 14l^{-0.3}k^{0.3}$; $\frac{\partial P}{\partial k} = 6l^{0.7}k^{-0.7}$

25. Competitive **27.** $(2, 2)$, rel. min.

29. 4 ft by 4 ft by 2 ft

31. A, 89 cents per pound; B, 94 cents per pound

33. $(3, 2, 1)$ **35.** $\hat{y} = 12.67 + 3.29x$

37. 8 **39.** $\frac{1}{30}$

MATHEMATICAL SNAPSHOT—CHAPTER 17 (page 898)

1. $y = 9.50e^{-0.22399x} + 5$ **3.** $T = 79e^{-0.01113t} + 45$

EXERCISE A.1 (page 905)

1.

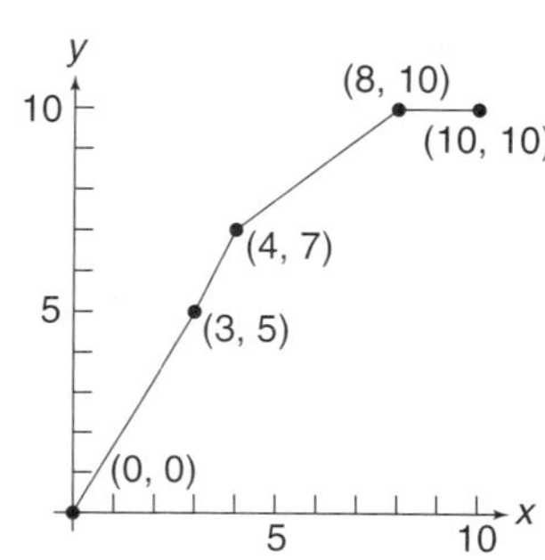

3.

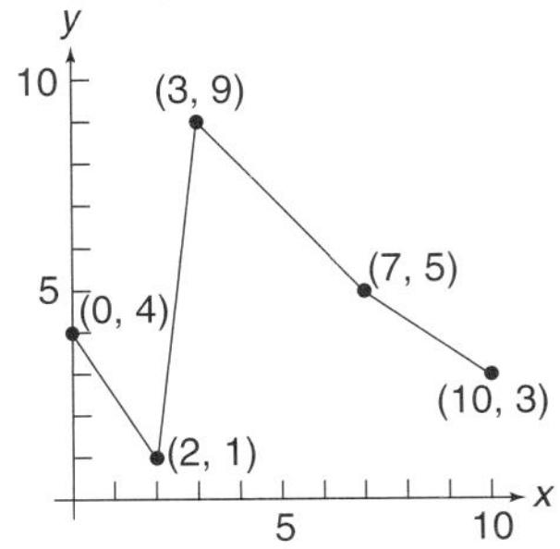

5.

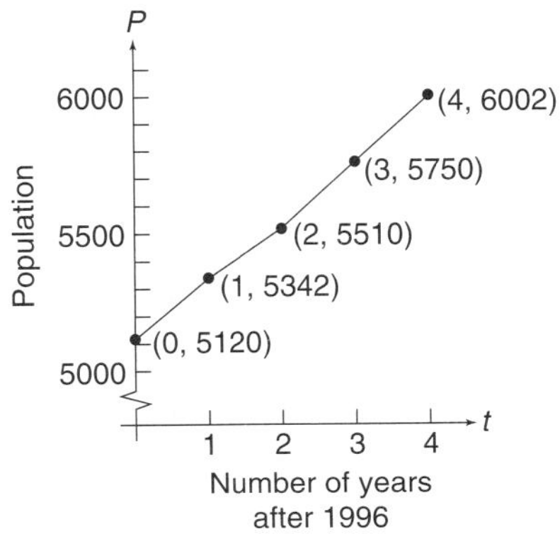

7. 75 **9.** Between 1990 and 1993, 1995 and 1998, 1999 and 2000; positive **11.** Between 1994 and 1995; zero **13.** Between 1993 and 1994 **15.** 75 students; 1990 **17. (a)** Possible graph: ; **(b)** Wednesday

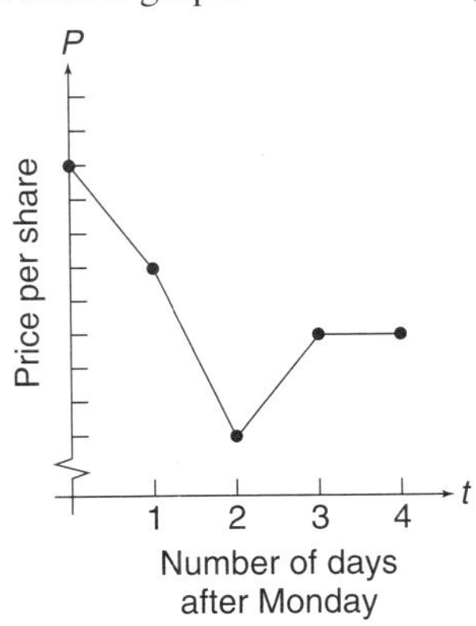

19. (a) ;

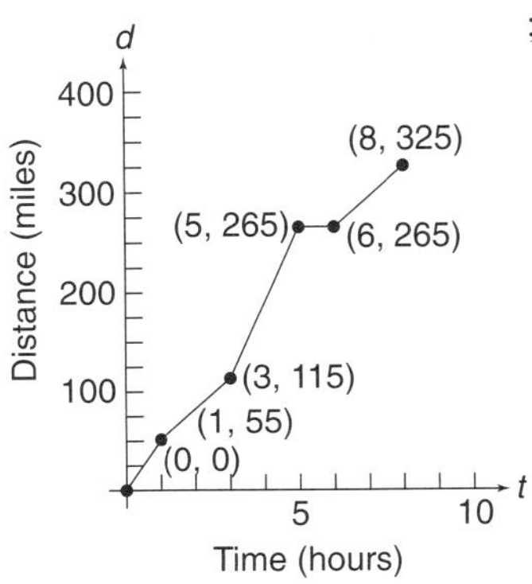

(b) approximately 85 mi; **(c)** between 5 and 6 h; 0;

(d) between 3 and 5 h; The slope of the line graph during this time interval is greater than the slope of the line graph during the remaining intervals.

21. (a) ;

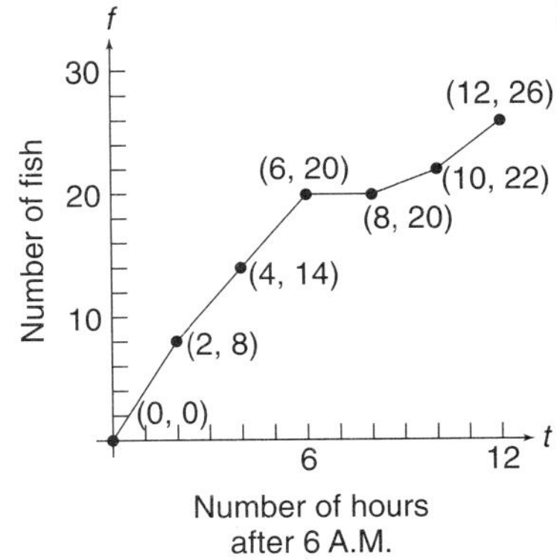

(b) between 6:00 A.M. and 8:00 A.M.; **(c)** between 12:00 P.M. and 2:00 P.M.; 0; **(d)** the number of fish caught per hour remained constant.

EXERCISE A.2 (page 911)

1. $A = \frac{1}{2}h(b_1 + b_2)$ **3.** $y = -\frac{1}{2}x + \frac{5}{2}$

5. $y = 15{,}525(0.91)^x$ **7.** $P(E \cup F) = P(E) + P(F)$

9. ; linear; $y = 2x + 5$

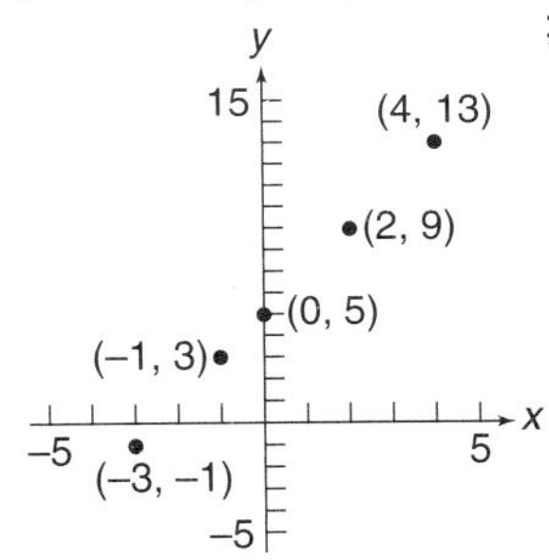

11. ; exponential; $y = 3^x$

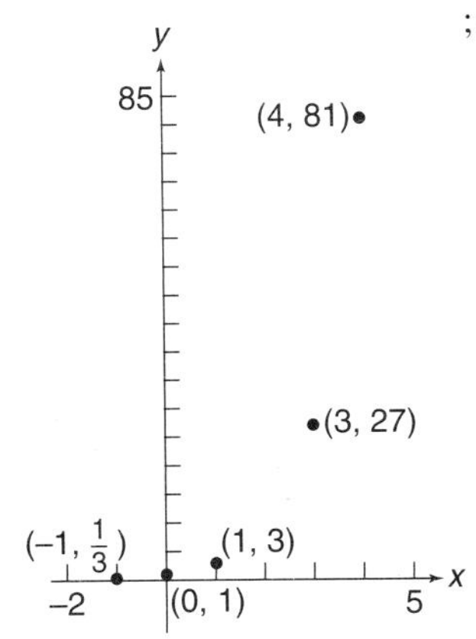

13.

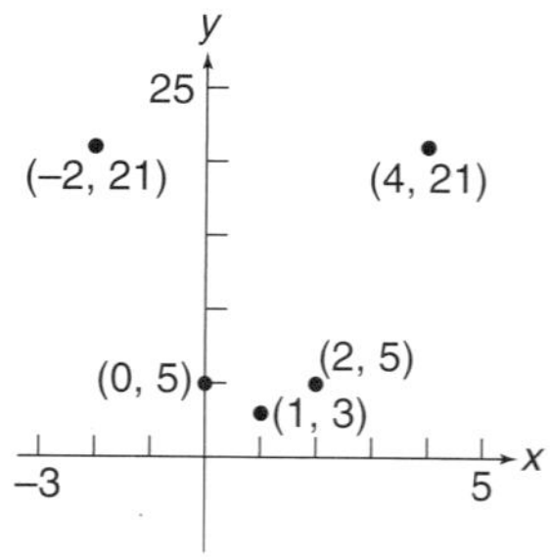

; quadratic; $f(x) = 2(x - 1)^2 + 3$ or $f(x) = 2x^2 - 4x + 5$

15.

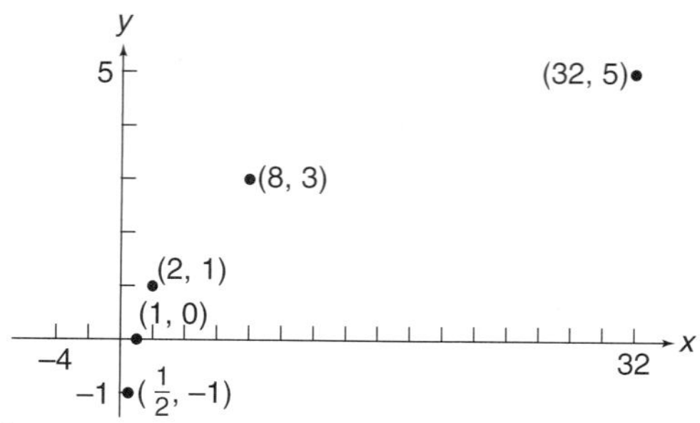

; logarithmic; $y = \log_2 x$

17.

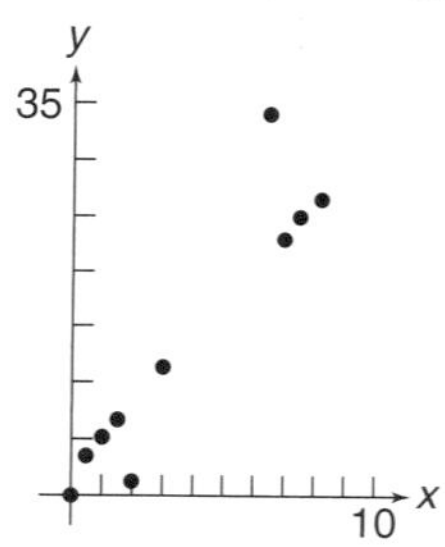

; linear

19.

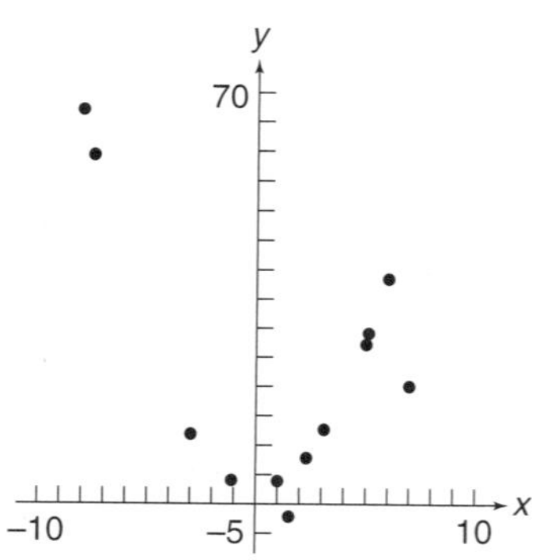

; quadratic

21.

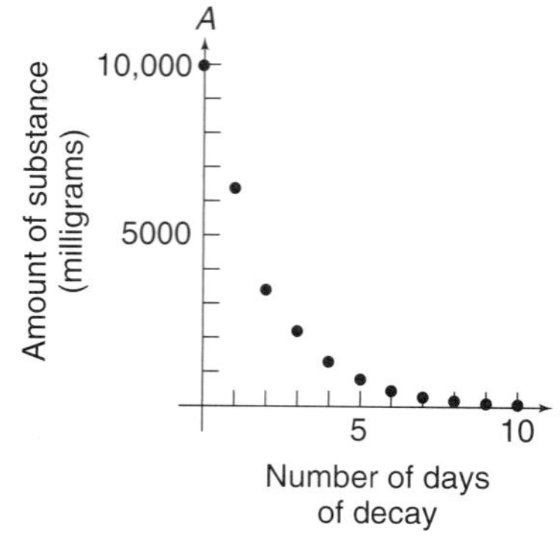

; exponential

23. **(a)** $C = 100 + 40r$; **(b)** \$500; **(c)** 22 reels
25. **(a)** $h = -16t^2 + 80t$; **(b)** 2.5 sec, 100 ft
27. **(a)**

t	0	1	2	3	4	5
N	2000	6000	18,000	54,000	162,000	486,000

;
(b) $N = 2000(3)^t$; **(c)** 118,098,000 bacteria
29. **(a)** (2, 1.3) and (6.5, 34.1); **(b)** Answers may vary, but should be close to $y = 3.1x + 1.5$;
(c) Answers may vary, but should be close to $y = 44.9$.

EXERCISE A.3 (page 917)

1. −3, −3, −3, −3; linear **3.** 1, 7, 19, 37; nonlinear
5.

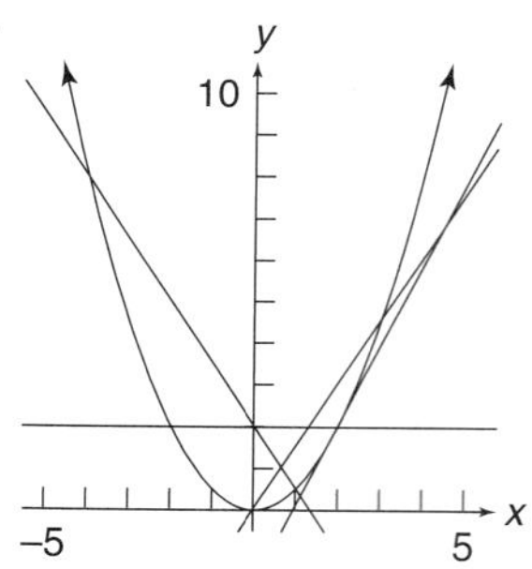

; **(a)** 1.5; **(b)** −1.5; **(c)** 0; **(d)** 2.005

7.

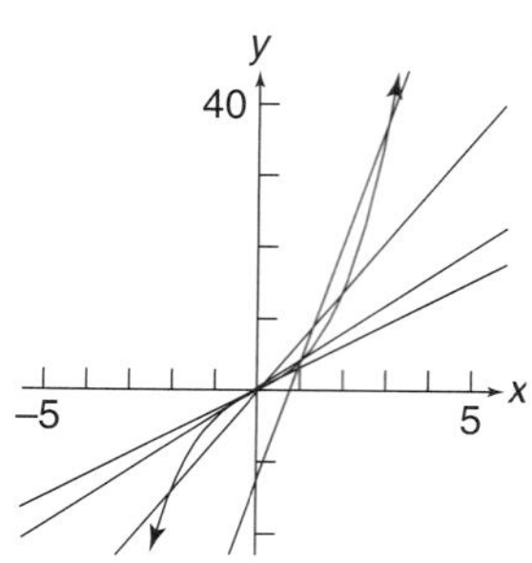

; **(a)** 16; **(b)** 7; **(c)** 4; **(d)** 3.01

9. 4.5 in. per yr **11.** \$42 per h
13. **(a)** 3.5 degrees per day; **(b)** −1.25 degrees per day; **(c)** 1 degree per day; **(d)** ≈ 0.59 degree per day
15. **(a)** 0; **(b)** 0; **(c)** 0; **(d)** 0
17. **(a)** 7; **(b)** 13; **(c)** $h + 8$; **(d)** $2x_0 + h + 2$
19. **(a)** −2; **(b)** −2; **(c)** −2; **(d)** −2;
(e) Since $g(x)$ is linear, the average rate of change between any two points is constant **21.** $x(t) = 2t + 3$

23. Possible graph:

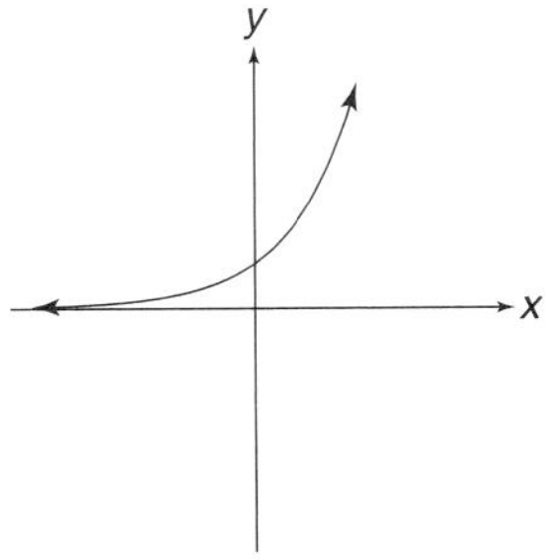

25. Average cost per unit over the interval

EXERCISE A.4 (page 923)

1. Average **3.** Instantaneous **5.** -8 **7.** $-\frac{1}{9}$ **9.** 1 **11.** $y = x - 1$ **13.** $y = \frac{1}{2}x + \frac{1}{2}$ **15.** 0 **17.** 5 **19.** $20x$ **21.** $-\frac{5}{(5x + 11)^2}$ **23.** $\frac{15}{(12 - 5x)^2}$ **25.** $\frac{3}{2\sqrt{3x + 7}}$ **27.** $-\frac{5}{2\sqrt{12 - 5x}} + 20x$ **29.** $\frac{dc}{dq} = 0.1q + 28$; \$35.50 per rug **31.** $\frac{dh}{dt} = -32t + 32$; **(a)** 32 ft/sec; **(b)** -32 ft/sec; **(c)** -64 ft/sec

EXERCISE A.5 (page 927)

*In Problems **1–13**, answers are assumed to be expressed in square units.*

1. $4\frac{1}{2} + \frac{\pi}{4}$ **3.** 34 **5.** 61 **7.** **(a)** 41; **(b)** 44; **(c)** 42; **(d)** 42; **(e)** parts (c) and (d) **9.** **(a)** ≈ 12.57; **(b)** ≈ 9.98; **(c)** ≈ 11.36; **(d)** ≈ 11.98; **(e)** part (d) **11.** **(a)** 54; **(b)** 42; **(c)** trapezoid **13.** **(a)** 104; **(b)** 86; **(c)** trapezoid **15.**

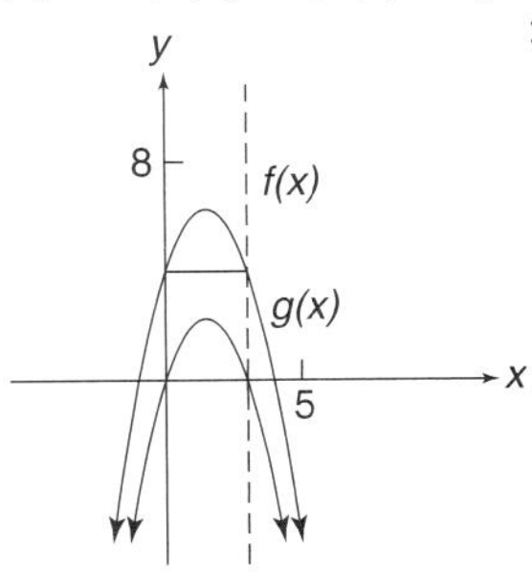

; The area under $f(x)$ can be divided into 2 sections (see graph). The top section is equivalent to the area under $g(x)$, so they have that area in common. The bottom section is a rectangle that the area under $g(x)$ does not include

EXERCISE A.6 (page 932)

1. 12, 17, t **3.** 168 **5.** 532 **7.** $\sum_{i=36}^{60} i$ **9.** $\sum_{j=3}^{8} 5^j$ **11.** $\sum_{i=1}^{8} 2^i$ **13.** 520 **15.** 5 **17.** 37,750 **19.** 14,980 **21.** 295,425 **23.** $4\frac{23}{25}$ **25.** $8 - \frac{4(n+1)(2n+1)}{3n^2}$ **27.** 4.500625 square units

EXERCISE A.7 (page 940)

1. $8\left(\frac{n+1}{n}\right) + 12; 20$ **3.** $\int_4^9 8\,dx$ **5.** $\int_{-2}^{5} (x^2 + x + 2)\,dx$ **7.** $\int_{-5}^{5} (3^x + 5)\,dx$ **9.**

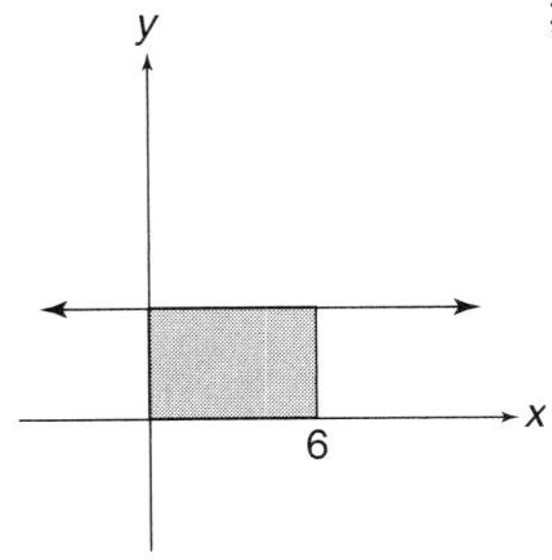

; positive

11.

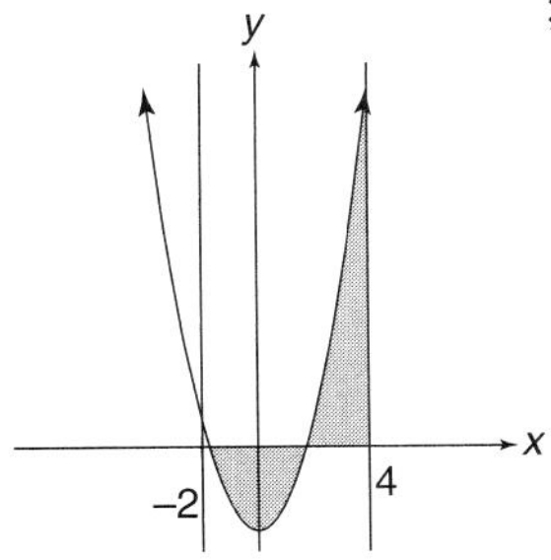

; positive

13.

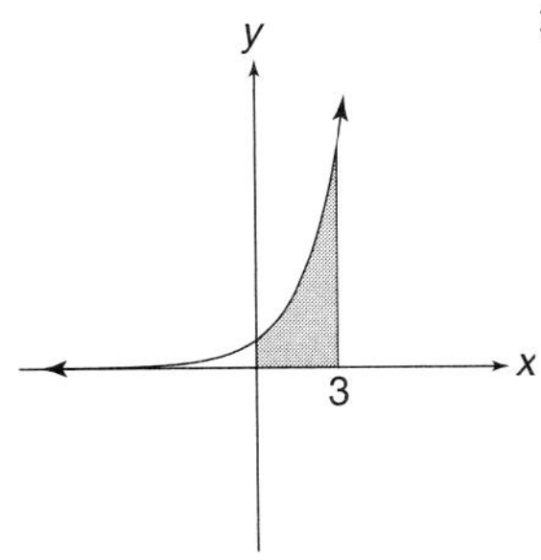

; positive

15. **(a)** $7b$; **(b)** 70 **17.** **(a)** $\frac{3}{2}b^2 + 2b$; **(b)** 170 **19.** **(a)** 14; **(b)** $G(x) = 2x + b$, where b can be any real number; **(c)** 14 **21.** 30 **23.** -14 **25.** $25\frac{1}{2}$ **27.** $e^3 - 1 \approx 19.09$

EXERCISE A.8 (page 947)

1. Integral **3.** Function itself **5.** Derivative **7.** Function itself **9.** **(a)** 50; The cost of the rental is \$50.00 when you drive the truck 50 miles; **(b)** 0.60; When you have driven the truck 50 miles, the cost is increasing at the rate of \$0.60 per mile **11.** **(a)** $b(t) = 300t$; **(b)** $b'(t) = 300$; The employee's bonus increases at the rate of \$300 per year; **(c)** The integral $\int_0^{10} b(t)dt$ approximates the sum of an employee's annual bonuses during the first ten years with the company **13.** **(a)** 23; **(b)** 2; In 1995, the number of books that Xul reads annually was increasing at the rate of about 2 books per year;

(c) 140.26; Between 1991 and 2000, Xul read about 140 books **15. (a)** 8; 11.95; In 2006, the program's budget would be \$8 billion with model b_l and \$11.95 billion with b_e; **(b)** 1.6; 1.11; In 2006, the program's budget is increasing at the rate of \$1.6 billion per year with model b_l and about \$1.11 billion per year with b_e; **(c)** 20; 38.36; In the first five years of the program, the cumulative budget would be about \$20 billion with model b_l and about \$38.36 billion with b_e; **(d)** b_e; **(e)** b_l; **(f)** b_e **17. (a)** 0.012 mi/sec; **(b)** 0.002 mi/sec^2; **(c)** 0.035 mile

REVIEW PROBLEMS (page 949)

1.

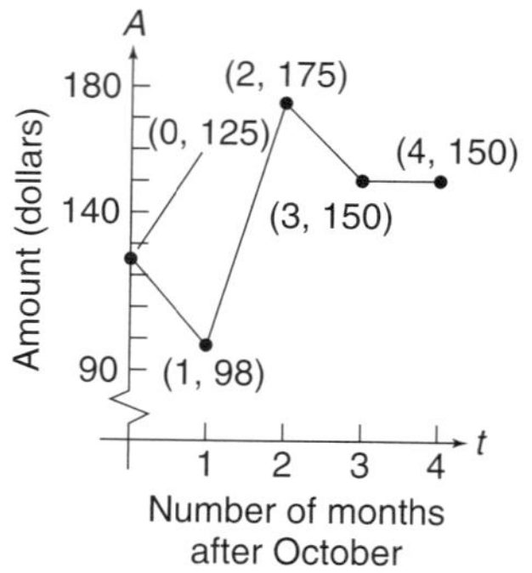

3. between October and November
5. $C = 550 + 22.50x$
7. (a)

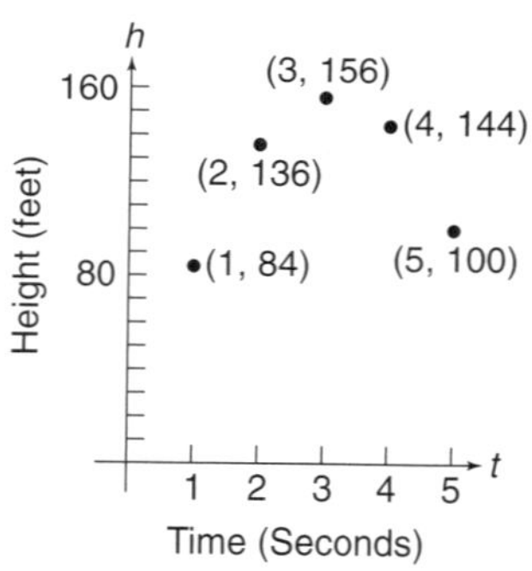

, quadratic;

(b) $h = -16t^2 + 100t$; **(c)** 24 ft; **(d)** 6.25 s
9. (a) $-3, -3, -3, -3$, linear; **(b)** 1, 3, 5, 7, nonlinear
11. (a) 300 kilobytes;
(b)

time	302 sec	204 sec	130 sec	20 sec
size	3 K	126 K	213 K	297 K

;

(c) ≈ -1.26 kilobytes per second, ≈ -1.18 kilobytes per second, ≈ -0.76 kilobytes per second;
(d) The negative sign indicates that as the amount of time left decreases, the amount of the document which has been downloaded increases; **(e)** 302 seconds to 204 seconds

13. $x(t) = 4t - 1$ **15.** $y = \frac{1}{4}x + \frac{3}{2}$

17. $\frac{dy}{dx} = \frac{1}{(3 - x)^2}$ **19.** 29 square units

21. (a) ≈ 7.07 square units; **(b)** ≈ 5.06 square units; **(c)** ≈ 6.14 square units; **(d)** ≈ 6.56 square units; **(e)** part (d)

23. (a) 45 square units; **(b)** 42.5 square units; **(c)** neither is better **25.** 513 **27.** $\frac{n}{4}$

29. 2.999975 square units

31. $\int_{-1}^{1} (-x^2 - x + 2)\, dx$ **33.** $\int_{1}^{100} 4 \log x\, dx$

35.

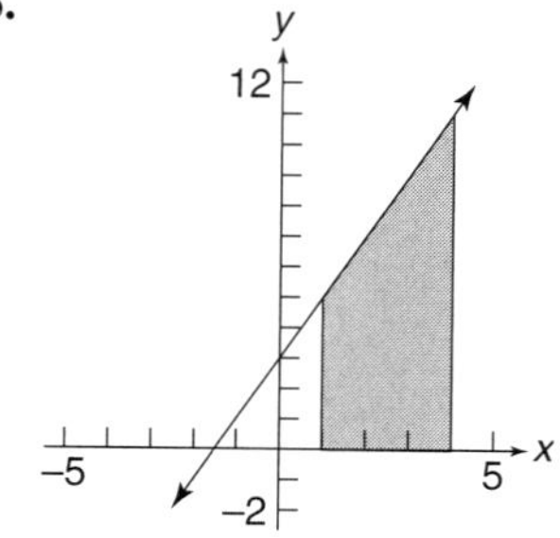

37.

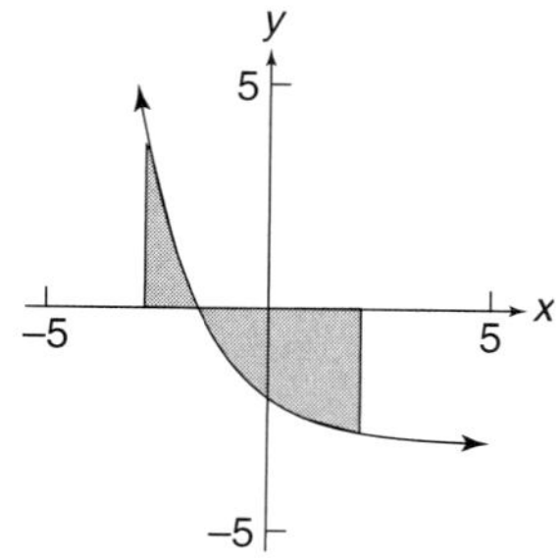

39. (a) $\frac{1}{2}b^2 + 3b$; **(b)** 80 **41.** 100,000 **43.** 36

45. (a) 2103.64; The energy costs for a 1900 square-foot home were about \$2103.64 in 2001; **(b)** 63.11; In 2001, the energy costs for a 1900 square-foot home were increasing at a rate of about \$63.11 per year; **(c)** 42,455.27; The cumulative energy costs for a 1900 square-foot home between 1970 and 2001 were about \$42,455.27
47. (a) 0.015 mi/sec; **(b)** 0.005 mi/sec^2; **(c)** 0.03 mi

INDEX

F

G

Q

R

S

T

PHOTO CREDITS

Chapter 0, page 48: Adrian Mueller/Fabrik Studios/Aurora & Quanta Productions Inc
Chapter 1, page 74: German Meneses Photography
Chapter 2, page 120: David Young Wolf/PhotoEdit
Chapter 3, page 174: German Meneses Photography
Chapter 4, page 214: Bill Aron/PhotoEdit
Chapter 7, page 380: The Image Works
Chapter 9, page 494: Bonnie Kamin/PhotoEdit
Chapter 10, page 534: German Meneses Photography
Chapter 11, page 588: The Image Works
Chapter 12, page 630: Picture Desk, Inc./Kobal Collection
Chapter 13, page 684: Ralf-Finn Hestoft/CORBIS–NY
Chapter 15, page 810: Cameron/CORBIS–NY
Chapter 17, page 898: Richard Megna/Fundamental Photographs
COVER, AP/Wide World Photos

Business Relations

$$\text{Interest} = (\text{principal})(\text{rate})(\text{time})$$

$$\text{Total cost} = \text{variable cost} + \text{fixed cost}$$

$$\text{Average cost per unit} = \frac{\text{total cost}}{\text{quantity}}$$

$$\text{Total revenue} = (\text{price per unit})(\text{number of units sold})$$

$$\text{Profit} = \text{total revenue} - \text{total cost}$$

Ordinary Annuity Formulas

$$A = R\frac{1-(1+r)^{-n}}{r} = Ra_{\overline{n}|r} \qquad \text{(present value)}$$

$$S = R\frac{(1+r)^{n}-1}{r} = Rs_{\overline{n}|r} \qquad \text{(future value)}$$

Graphs of Elementary Functions

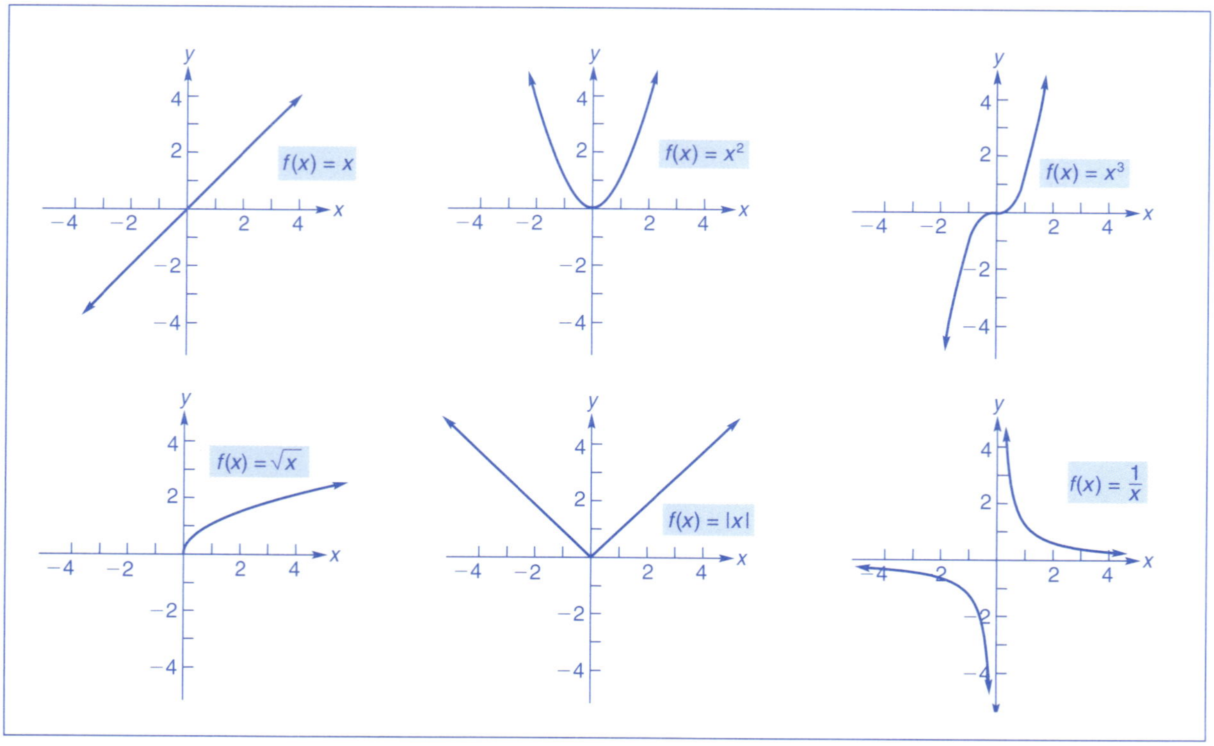

Differentiation Formulas

$$\frac{d}{dx}(c) = 0$$

$$\frac{d}{dx}(x^n) = nx^{n-1}$$

$$\frac{d}{dx}[cf(x)] = cf'(x)$$

$$\frac{d}{dx}[f(x) \pm g(x)] = f'(x) \pm g'(x)$$

$$\frac{d}{dx}[f(x)g(x)] = f(x)g'(x) + g(x)f'(x)$$

$$\frac{d}{dx}\left[\frac{f(x)}{g(x)}\right] = \frac{g(x)f'(x) - f(x)g'(x)}{[g(x)]^2}$$

$$\frac{dy}{dx} = \frac{dy}{du} \cdot \frac{du}{dx}$$

$$\frac{d}{dx}(u^n) = nu^{n-1}\frac{du}{dx}$$

$$\frac{d}{dx}(\ln u) = \frac{1}{u}\frac{du}{dx}$$

$$\frac{d}{dx}(e^u) = e^u\frac{du}{dx}$$

$$\frac{d}{dx}(\log_b u) = \frac{1}{(\ln b)u} \cdot \frac{du}{dx}$$

$$\frac{d}{dx}(a^u) = a^u(\ln a)\frac{du}{dx}$$

Integration Formulas

We assume that u is a differentiable function of x.

$$\int k\,dx = kx + C$$

$$\int x^n\,dx = \frac{x^{n+1}}{n+1} + C, \qquad n \neq -1$$

$$\int e^x\,dx = e^x + C,$$

$$\int kf(x)\,dx = k\int f(x)\,dx$$

$$\int [f(x) \pm g(x)]\,dx = \int f(x)\,dx \pm \int g(x)\,dx$$

$$\int u^n du = \frac{u^{n+1}}{n+1} + C, \qquad n \neq -1$$

$$\int e^u du = e^u + C$$

$$\int \frac{1}{u}\,du = \ln|u| + C, \qquad u \neq 0$$